International Conference on Modeling of Casting, Welding and Advanced Solidification Processes 2012

(MCWASP XIII)

IOP Conference Series: Materials Science and Engineering
Volume 33

Schladming, Austria
17 - 22 June 2012

ISBN: 978-1-63439-844-2
ISSN: 1757-8981

Printed from e-media with permission by:

Curran Associates, Inc.
57 Morehouse Lane
Red Hook, NY 12571

proceedings.com

Some format issues inherent in the e-media version may also appear in this print version.

Printed by Curran Associates, Inc. (2015)

For permission requests, please contact the Institute of Physics
at the address below.

Institute of Physics
Dirac House, Temple Back
Bristol BS1 6BE UK

Phone: 44 1 17 929 7481
Fax: 44 1 17 920 0979

techtracking@iop.org

Additional copies of this publication are available from:

Curran Associates, Inc.
57 Morehouse Lane
Red Hook, NY 12571 USA
Phone: 845-758-0400
Fax: 845-758-2634
Email: curran@proceedings.com
Web: www.proceedings.com

Table of contents

Volume 33

MCWASP XIII: International Conference on Modeling of Casting, Welding and Advanced Solidification Processes
17–22 June 2012, Schladming, Austria

Accepted papers received: 4 April 2012
Published online: 3 July 2012

Preface

Papers

Ingot and Shape Casting

Continuous Casting and Direct Chill Casting

Electromagnetic Coupling

Thermomechanics

Prediction of Defects

Meso/Macroscale Modeling of Structure and Segregation

Numerical Methods

Preface

Due to fast-paced development in computer technologies during the last three decades, computer-based process modeling has become an important tool for the improvement of existing process technologies and the development of new, innovative technologies. With the help of numerical process simulations, complex and costly experimental trials can now be reduced to a minimum. For metallurgical processes in particular, computer simulations are of outstanding importance, as the flow and solidification of molten alloys or the formation of microstructure and defects can hardly be observed experimentally. Corresponding computer simulations allow us inside views into the key process phenomena and so offer great potential for optimization.

In 1980 the conference series 'Modeling of Casting, Welding and Advanced Solidification Processes (MCWASP)' was started up, and has now been continued by holding the 13th international conference on 'Modeling of Casting, Welding and Advanced Solidification Processes', MCWASP XIII, in Schladming, Austria, from June 17–22 2012. Around 200 scientists from industry and academia, coming from 20 countries around the globe attended 78 oral and 50 poster presentations on different aspects of solidification-related modeling topics.

Besides process-related sessions such as (i) Ingot and Shape Casting, (ii) Continuous Casting and Direct Chill Casting, (iii) Directional Solidification and Zone Melting, (iv) Welding, and (v) Centrifugal Casting, a larger focus was put on (vi) Experimental Investigation and In-Situ Observations. In recent years, this topic has been significantly strengthened as advanced synchrotron technologies allow fantastic in-situ observations of phenomena happening inside small metallic samples. These observations will definitely serve as a benchmark for the modeling community. Further macroscopic aspects of advanced solidification science were tackled in the sessions (vii) Electromagnetic Coupling, (viii) Thermomechanics, (ix) Thermodynamics and Solidification Paths, and (x) Prediction of Defects. As microstructure prediction is one of the key disciplines of solidification modeling, sessions such as (x) Meso/Macroscale Modeling of Structure and Segregation, (xi) Formation of Macrosegregation, and (xii) Structure Formation at Microscale were added. Finally some new modeling ideas not being presented in the aforementioned sessions were presented in a small additional session named (xiii) Numerical Methods.

In addition to the new findings obtained by using advanced in-situ observation techniques, significant progress has been made on modeling the formation of microstructures, both at micro and meso/macroscale. Here, three-dimensional simulations of complex situations, e.g. polyphase solidification or grain growth in the presence of melt convection can now be performed quite effectively. However, the community is still challenged when a combined treatment of different phenomena is necessary. Efforts on the numerical description of solid skeleton deformation in combination with (interdendritic) melt flow have just begun. Furthermore, the combination of models for microstructure prediction at the microscale with process models at the macroscale (including melt convection and equiaxed grain motion) is still unsatisfactory. This is also true for combining process models with complex thermodynamics (e.g. for alloys consisting of many solute elements). The dependency of the results on the used numerical scheme, or the fact that the numerical grid might not be fine enough, especially in 3D calculations, are also topics which need our future attention. In conclusion, it can be stated that although the community has gained celebrated success over the last decades, we are still challenged by the complexity of the physical phenomena occurring during the solidification of melts.

Leoben, June 2012

Andreas Ludwig
Conference Chairman
Department of Metallurgy
University of Leoben
Franz-Josef-Strasse 18
8700 Leoben
Austria

Peer review statement

All papers published in this volume of *IOP Conference Series: Materials Science and Engineering* have been peer reviewed through processes administered by the proceedings Editors. Reviews were conducted by expert referees to the professional and scientific standards expected of a proceedings journal published by IOP Publishing.

A Method to set process parameters of local squeeze in HPDC

I Ohnaka[1], JD Zhu[2], A Sugiyama[1] and F Kinoshita[3]

[1]ieSol Co./Osaka University, Toyonaka-shi, Higashitoyonaka-cho 1-32-12, Osaka 560-0003, Japan
[2]Multi-Flow Software CO.LTD, Canada
[3]QUALICA Inc. , Suita-shi, Japan

E-mail:ohnaka@ams.eng.osaka-u.ac.jp

Abstract. In HPDC processes, pressurization of the mushy regions via local squeeze pins and plunger during the intensification stage is very important to decrease porosity defects. In order to better understand the process and to be able to set the process parameters more properly, we developed a simulation code to solve the flow field of mushy regions and porosity formation based on the following main assumptions: 1) Local squeeze pins and plunger tip can be treated as pressure boundaries, while the squeeze pins are treated as moving boundaries. 2)The intensification and local squeeze pressure can propagate inside regions where the solid fraction is less than critical values. 3) In the pressurized regions, the cast metal is treated as a Newtonian fluid and both the solid and liquid flow together when the solid fraction is low, and the D'Arcy flow only exists when the solid fraction goes higher. 3) Only the liquid flows in the unpressurized regions, following the D'Arcy's law. 4) Porosities grow in elements when the pressure decreases below critical values depending on the element condition. The simulation was applied to HPDC castings and showed that the pressure-duration was similar to the measured one though the simulated pressure drop in the pressurized region was much smaller than the measured one. The pressure drop, pressure-duration and pin-penetration depth were discussed and the simulation seems to be helpful to determine local squeeze parameters. Future challenges are also discussed.

1. Introduction

In high pressure die-casting(HPDC), pressurization of the mushy region is very important to decrease the porosity defects, because many gas bubbles are entrained during mold filling and risers to feed solidification contraction are not available. In particular, local squeeze, namely pressurization with local squeeze pins is often applied on thick-wall regions of the casting prone to porosity defects[1]. Because better propagation of applied pressure decreases porosity more significantly, several works have been reported on pressure propagation and the effect of pressure on the quality of castings[2-8]. However, very few can be found on porosity simulation in HPDC[9-11], especially the simulation of effect of local squeeze pins [12,13].

This paper presents a practical simulation method of the pressure propagation and porosity formation to properly set process parameters of local squeeze.

2. Simulation method

Because exact simulation of pressure propagation in the HPDC is very difficult, following assumptions and equations have been used in the simulation:

1) Boundaries between casting and the plunger tip or local squeeze pins are treated as pressure boundaries(figure 1).
2) The intensification and local squeeze pressures propagate inside regions where the solid fraction is less than a critical value, f_{gate} for intensification and f_{sqz} for the local squeeze, respectively. Namely, the pressures from the plunger tip continue to propagate into the casting as long as the gate does not "freeze off" by reaching the critical value f_{gate} across the cross-section. The regions where external pressure can and cannot propagate are called pressurized and unpressurized regions, respectively.

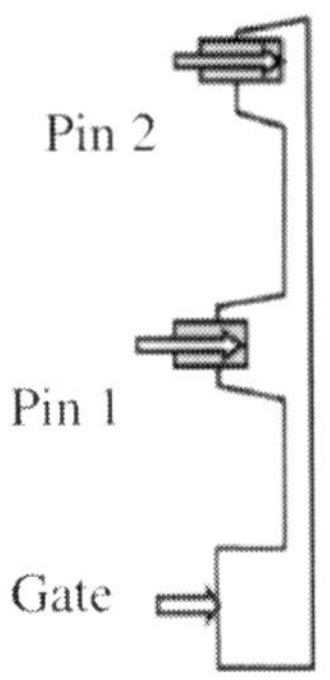

Figure 1. Pressure boundary

3) The pressurized regions are assumed as a Newtonian fluid, where both the solid and liquid flow together in the regions where solid fraction is less than g_{SC} (see equation 5) and only the liquid flows in the regions where solid fraction is higher than g_{SC}, following the D'Arcy law. The following discrete equations developed with the Direct Finite Difference Method[14-16] have been used, while the gravity and convection terms are neglected. Note that these equations are used only for flow elements without porosities. The elements with porosities are treated as pressure boundary.

Mass balance (for elements);

$$(\rho_S - \rho_L)_i \frac{\partial V_{Si}}{\partial t} = \sum (\rho S U)_{ij} \tag{1}$$

Momentum balance (for staggered elements);

$$\bar{\rho} V_i \frac{\mathbf{u}_i^{t+\Delta t} - \mathbf{u}_i^t}{\Delta t} = S(P_i - P_j) + \sum (\mu \alpha S \frac{\mathbf{u}_j - \mathbf{u}_i}{d_{ij}}) - \frac{(\alpha - 1)\mu_0 f_L}{2} (\frac{V_i}{K_i} + \frac{V_j}{K_j}) \mathbf{u}_i \tag{2}$$

where ρ : density[kg/m^3], μ : effective viscosity[$Pa \cdot s$], μ_0 :viscosity of pure liquid[$Pa \cdot s$],
Δt: time step[s], d: distance between nodal elements[m], **u**: velocity[m/s], S: surface area[m^2] ,
U:velocity on element surface[m/s], V: volume[m^3], K: permeability[m^2], f_L : liquid fraction[-],
α : 1 for the low-fraction-solid regions and 0 for high-solid-fraction regions.
Subscript i,j : element number, s : solid, L: liquid

4) In the unpressurized regions, the liquid flow follows the D'Arcy's law.

$$U_{ij} = \frac{1}{\mu(\frac{d_i}{K_i} + \frac{d_j}{K_j})} \{P_j - P_i + \rho_L g (h_j - h_i)\} \tag{3}$$

The permeability, K, is calculated wih equztion 4 that is a simplified Ergan's equation[17].

$$K = \frac{(1 - f_S)^3 d_s^2}{100 f_S^2} \tag{4}$$

where g : gravitational acceleration, f_S : solid fraction, P : pressure, h: melt head, d_S : grain diameter of solid phase.

5) The effective viscosity in the pressurized region is calculated with Mori-Ototake's equation[18], assuming equiaxed solidification structure.

$$\mu = \mu_0 [1 + \frac{0.5 d_S S_a}{1/f_S - 1/f_{SC}}] \quad \text{for} \quad f_S < f_{SC} \tag{5}$$

where S_a :specific surface area of grain, f_{SC} :critical fraction solid. Although S_a varies with time, it is assumed as a constant here (see table 1). It was found that the pressure drop is not sensitive to S_a via several case studies.

6) The temperature field is solved with the thermal energy balance equation described in Ref[14-16], neglecting the convection term. The latent heat of fusion is considered by the temperature recovery method.
7) The volume change of gas entrapped during mold filling follows the equation of state for ideal gas:

$$PV = m_0 RT \tag{6}$$

where P : pressure of the gas[Pa], V : volume of the gas[m^3], R : gas constant[J/kgK], T : temperature[K], m_0 : mass of the gas[kg].

8) The penetration depth of squeeze pins is calculated from the volume shrink during solidification of their territories bounded by the critical solid fraction. The compression of porosities is also considered. The pin penetration terminates when the solid fraction ahead of the pins, say 5mm from the pin tip, becomes greater than a critical value.
9) The thermal properties of squeeze pins are the same with those of the casting and there is no thermal resistance between the casting and the squeeze pins, while the region where the pins exist is treated as a solid region. Solidification simulation for the case where a pin was inserted by 8mm showed that the fraction solid ahead of the pin tip rapidly increased at the very early stage, but soon the solidification rate decreased approaching rather similar value to the case where no pin exists.

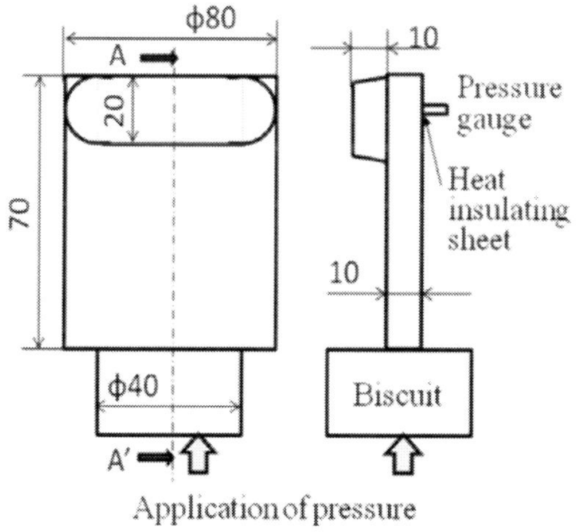

Figure 2. Casting specimen used by Iwata et.al[4]

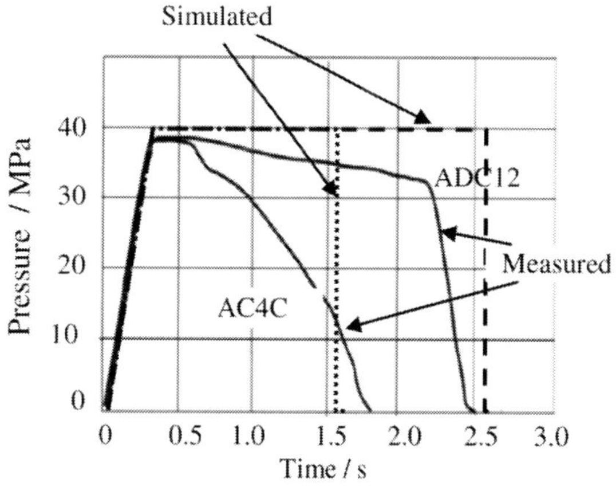

Figure 3. Comparison of measured and simulated pressure change

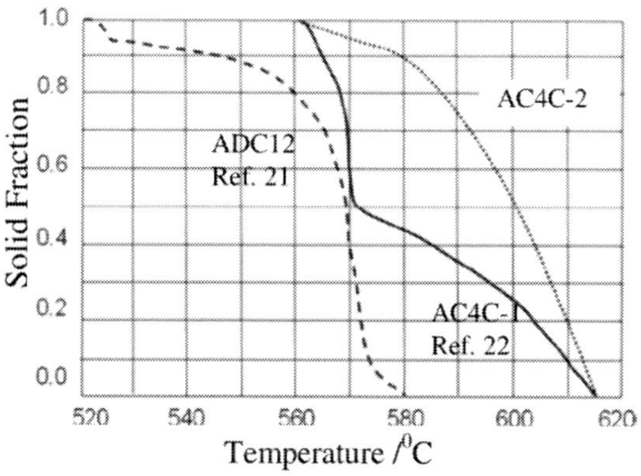

Figure 4. Fraction solid and temperature relationship used in the simulation

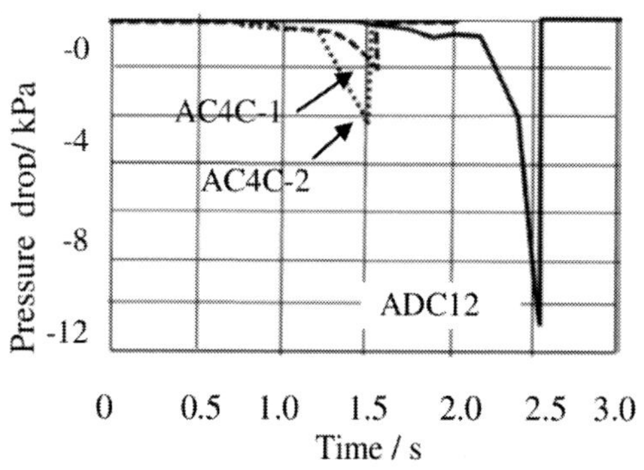

Figure 5. Simulated pressure drops at the measuring point

Table 1. Physical properties used in the simulation

Physical parameters	AC4C	ADC12
Density [kg/m^3]	2400	2480
Specific Heat[kJ/(kgK)]	1.19	1.19
Thermal cond.[W/mK]	100	100
Latent heat [kJ/kg]	425	495
Liquidus Temp.[°C]	615	580
Solidus Temp.[°C]	562	520
Ejection Temp.[[°C]	620	600
Shrinkage coef. [%]	6	4
f_{gate}, f_{sqz}	0.9	0.99
d_S in Eq.4 and 5 [μ m]	20	20
μ_0 S_a in equation 5	200	200
f_{SC} in equation 5 [-]	0.45	0.45

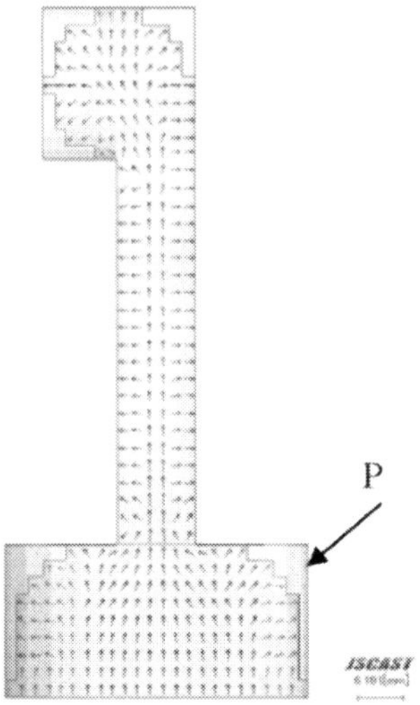

Figure 6. Simulated flow pattern at 1.2s for AC4C alloy (Cross-section AA' in figure 2)

This is because the pin temperature rapidly increased and kept due to its low thermal conductivity (SKD61) compared to that of the casting. Therefore, the assumption is reasonable for this work.

10) Porosity is calculated from the flow field assuming pores start to grow at elements when the pressure there becomes lower than a critical value. Once the critical pressure is reached, the pressure in the elements is set to zero and the net-loss of melt volume in these elements is calculated and used as the increase of porosity volume (porosity growth).

3. Simulation of pressure propagation

The measured data of Iwata et al.[4] were selected to evaluate the simulation accuracy of high pressure propagation, because the casting dimensions are clearly described as in figure 2. They measured the

Figure 3 compares the measured and simulated pressure changes with using solid fraction - temperature relationship shown in figure 4. In the simulations the gate pressure was set as follows; 0MPa at 0s, 40MPas at 0.3s, 40MPa between 0.3 and 3s, and 0MPa at 3.1s. Other data used in the simulation can be found in table 1.

Figure 5 shows simulated pressure drops (pressure at the measuring point - pressure at the biscuit [40MPa]) and figure 6 is the simulated flow field for the AC4C casting at 1.2s, showing a reasonable flow pattern.

Although the simulated pressure drop is much smaller than the measured one, the pressure-duration is rather similar between the simulation and measurement as shown in figure 3. Further, the pressure drop strongly depends on the solid fraction - temperature relationship of casting alloys as shown in figure 3 and figure 5, though the ADC12 casting showed a larger drop unlike the experimental result.

The reasons for the difference in the pressure drop may be as follows:

1) Boundary condition

Actually the boundary at the plunger tip is not a uniform pressure boundary. This is because the tip movement is controlled by the surrounding of the biscuit (P in figure 6) when the surrounding solidifies. If its strength is high enough, the force of the tip is balanced with the reaction force of the surrounding, resulting in pressure drop in the central part with lower strength. Solving such problem requires a structural analysis with considering the thermo-mechanical behaviour of the solidified part and the contact of the casting with the mold.

2) Basic equation and viscosity
The basic equation (equation 2) considers only the shear resistance which strongly depends on the flow velocity. Because the velocity of the flow is usually low, resultant pressure drop is also low. Although a higher effective viscosity causes a higher pressure drop, even a viscosity 1000 times larger than coal-tar did not cause such high pressure drops as measured.
3) Measurement error
Although the pressure sensor was set at the boundary between the casting and mold with a heat-insulating-sheet insert, the solidification near the sensor may decrease the pressure on the sensor when the solidified region has a strength that varies with alloy composition. The displacement of the sensor due to the mold expansion may also contributes to the pressure drop.

Currently it is not clear yet which is the main reason caused the difference in the pressure drops. However, the estimation of the pressure-duration may be more important than that of the pressure drop, because the former affects the casting quality more significantly [4].

The reason why the pressure-duration affects the porosity defects may be explained as follows. From assumption 6, equation 7 can be used to calculate the gas volume change if the temperature change is small;

$$V = P_0 V_0 / P \tag{7}$$

where P_0 is the initial gas pressure, namely, the pressure at the end of mold filling, and is usually much lower than the intensification pressure, P, for example $P_0 \approx$ 1MPa and $P \approx$ 40MPa. Therefore the entrapped gas becomes very small in a very short period. When the pressure decreases, however, the gas in the porosities expands. With a longer pressure-duration, the progress of solidification during this period will be larger and the gas will expand less. This may be supported by the fact that a kind of extrusion of cast metal occurred when a squeeze pin was pulled out too early(figure 7), while the effect of pressure decrease induced by the pulling-out cannot be excluded.

Another hypothesis for the effect of pressure-duration might be gas absorption into the liquid phase during the period. However this effect may be small, because the melt surface of the gas bubbles may be covered with an oxide film preventing the gas diffusion, and the pressure-duration was in the order of a few seconds in this particular experiment.

Based on the fact that complete blocking of the pressure propagation requires a certain thickness of solidified shell, the pressure-duration in simulation was determined by the time when the solid fraction at 1.7mm from the mold, where the sensor is attached reached the critical solid fraction. Sensitivity was also checked by changing the distance from 1.7mm to 0.8mm, and it was found that the pressure-duration becomes 20% shorter. As can be seen in figure 3, both simulated and measured results show a same trend, the pressure-duration of AC4C is shorter than ADC12. This is mainly due to the smaller latent heat and higher solidus temperature of AC4C as compared with those of ADC12(table 1). Because usual castings produced by the HPDC have many thin and thick parts unlike the casting shown in figure 2, the pressure-duration may be determined mainly by the solidification of thin parts and the starting time to push the pins. Therefore, the proposed method using the critical solid fraction, f_{gate} and f_{sqz} may be useful as shown below.

4. Local squeeze casting

Simulation was carried out on a local squeeze casting of ADC12 where two squeeze pins were set on the thick parts as shown in figures 8 and 10. Although the gas entrapment during mold filling was simulated (the method is reported in [11]), it's effect didn't appear in the porosity simulation. This is because the entrapped gas was compressed to very small volume during the intensification stage. It is reported that the local squeeze drastically decreased the porosity defects as shown in figure 8[1]. In the actual casting, the gate pressure of 52MPa was applied until 3s and squeeze pin pressure of 195MPa from 1s to various pulling-out times. The pin-diameter was 25mm(total pin cross-sectional area was 980mm^2).

IOP Conf. Series: Materials Science and Engineering **33** (2012) 012001 doi:10.1088/1757-899X/33/1/012001

Figure 9 shows the simulation result for the case where the local squeeze is not applied, showing many porosities formed in the product. When the local squeeze is applied from 3.5s to the end of the solidification, most of the defects are eliminated(figure 10) just as the observation(figure 8). The squeeze-pin-penetration depth was about 14 mm as shown in figure 11(when the solidification shrinkage coefficient is 5%, it increases to 17.8mm). It is greatly affected by the solidification at the head of the pins, because the pin movement is terminated when the solid fraction becomes greater than the critical value.

Figure 7 Section through the center of squeeze pins in figure 10, showing kind of extrusion of the metal (Pull-out time: 12s after mold filling [1]).

Figure 8 Section through the center of squeeze pins in Fig.10, showing good quality (squeeze starting time: 3s , pulling-out time:20s)

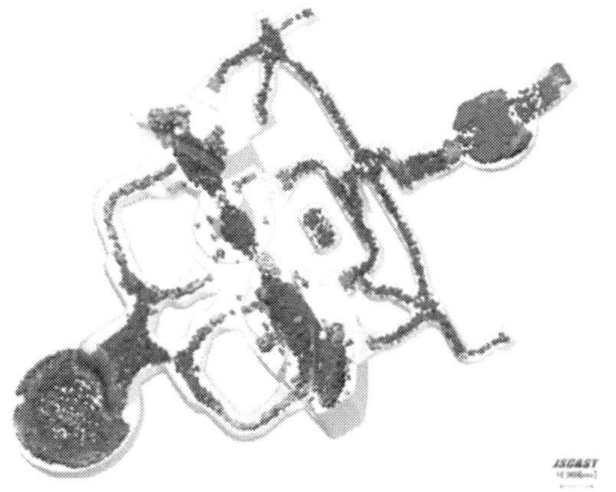

Figure 9 Simulated porosity without local squeeze

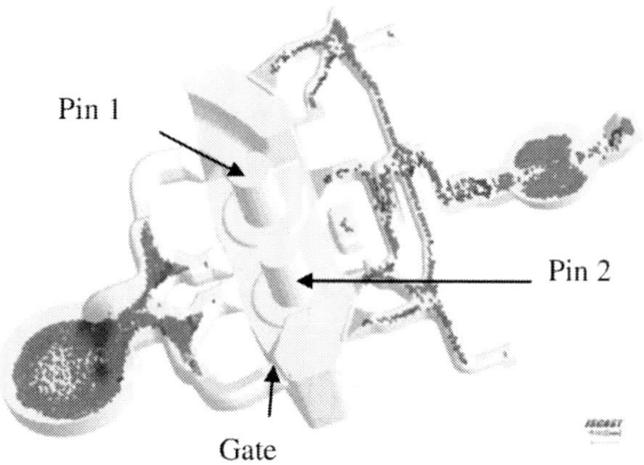

Figure 10 Simulated porosity with local squeeze

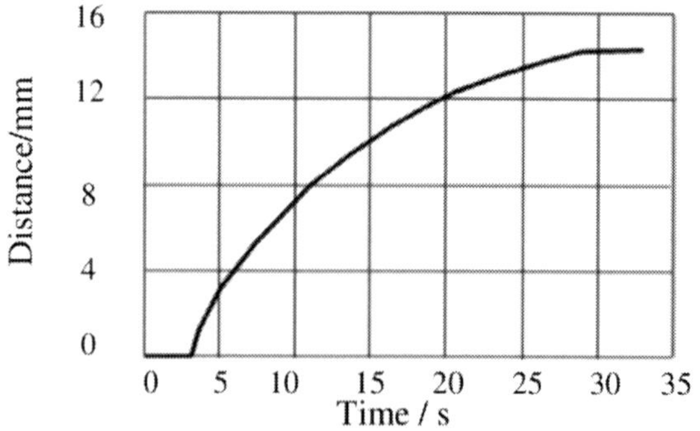

Figure 11 Simulated travel curve of the squeeze pins

The pin-penetration depth measured for the casting ranged from 20 to 27 mm depending on the squeeze starting time[1], for example, 27mm for 2s, 22mm for 5s. If the penetration depth is calculated from the volume of solidification shrinkage of the product ($475 \times 10^3 mm^3$), it is about $19 \times 10^3 mm^3$, hence the penetration depth is 19mm. However, it should be less than 19mm as the simulation shows, because solidification proceeds before starting the squeeze in the product. The large value of 27mm that was obtained when the squeeze starting time was 2 seconds after the mold filling may be due to the pushing back of the plunger tip, because the intensification time was 3 seconds and the squeeze pressure was higher than the plunger tip pressure.

However, the value of over 20mm for the starting time after 4 seconds cannot be explained only by the solidification shrinkage. The reasons for the larger penetration depth may be as follows;

1) Mold expansion

Usually the mold temperature increases until 10 or 20 second after the mold filling depending on the mass of casting and mold cooling conditions. Therefore, the mold should expand after the mold filling, resulting in increase of casting volume and hence the penetration depth. This also could affect the porosity defects.

2) Compression of porosities

If it took time to compress the porosities for some reasons, the porosities could remain even after the intensification stage and increase the penetration depth.

Although we cannot reject the latter, the former might be more possible.

5. Concluding remarks

A simulation method for setting process parameters after mold filling in HPDC has been developed and the comparison of simulation with observations showed the followings: 1) The pressure-duration was similar to the measured one and determined by solidification at the measuring point. 2) In the pressurized region, no significant pressure drop was obtained from the simulation. 3) The pressure propagation varies with the solid fraction-temperature relationship, though simulated pressure drops were much smaller than measured ones. 4) Why the longer pressure-duration causes smaller porosity defects may be explained by the suppression of the expansion of the compressed porosity gas due to the progress of solidification during the pressure-duration.

Although the proposed simulation method should use proper critical solid fractions to determine the pressurized regions and termination of squeeze pins, and cannot consider the mold expansion, it can roughly estimate the interaction between the plunger and squeeze pin, time variation of the pin penetration depth and degree of porosities. Therefore, it seems the simulation is helpful to set the process parameters.

However, there are still many challenges to improve the simulation accuracy such as visco-plastic analysis as a contact problem between the casting and mold, which enable the simulation without using the critical solid fractions. The consideration of the pin material properties and thermal resistance between the pin and casting are also required.

References

[1] Zhu J, Yokoyama H, Ohnaka I, Murakami T, and Cockcroft S, Prediction of Casting Defects in High-pressure Die Casting Processes by Using JSCAST, 2009 *Modeling of Casting, Welding and Advanced Solidification Processes XI 1,* (Warrendale, TMS), 361-368

[2] Nishi N, Sasaki H, Hirahara T and Takahashi Y , 1988 *Imono* **60** 777-783

[3] Iwamoto N and I.Kuboki 1994 *Trans. Japan Die Casting Conference,* JD94-11 p 94

[4] Iwata Y,Sugiyama Y, Iwahori H and Awano Y, 2000 *J.Japan Foundry Engineering* **72** 263-267

[5] Ikeda S, Matsumoto Y, Murakami M, Xiong S, and Hu B 2008 Relation btween Casting Conditions and the Molten Metal Pressure in a Die Mold, *Trans. 2008 Japan Die Casting Congress,* JD08-18, 119-123

[6] Kato E, Nishiyama N, Nomura H, Asai K and Tanigawa S, 2002 *J. JFES* **74** 370-376

[7] Tanigawa S, Asai K, Yang Y, Nomura H and Kato E, 2003 *J.JFES* **75** 525-531

[8] Ueki T,Tashiro M,Tajima A,Kato H,Yageta K and Kambe H, 2010 Measurement of melt and die behavior in die-casting process, *Trans. 2010 Japan Die Casting Congress*, JD10-25, 157-162
[9] Kubo K 1997 Method for Numerically Predicting Casting Defects, NADCA, T97-015
[10] Kubo K, Asao H, and Deki N, 2008 Progress in Computer Simulation of Die Casting, *Japan Die Casting Congress*, JD08-30 197-203
[11] Kimatsuka A, Ohnaka I, Zhu J, Sugiyama A and Kuroki Y,Mold Filling Simulation for Predicting Gas Porosity *2006 Modeling of Casting, Welding and Advanced Solidification Processes XI I*, (Warrendale, TMS), 603-610
[12] Li S, Sannakanishi S, Tsumori M, Akase M and Anzai K, 2003 Method for Numerical Prediction of Porosity Defects in Squeeze Casting, Trans. NADCA, T03-14
[13] Li S, Mine K,Sannakanishi S and Anzai K 2006 Qunatitative Simulation of Shrinkage Porosity by Pressure in Squeeze Casting, *Japan Die Casting Congress*, JD06-30, 199-204
[14] Ohnaka I, 1979 Classification of Numerical Methods for Transient Heat Transfer Problems and Improved Inner Nodal Point Method, *ISIJ*, **65** 1737-1746
[15] Zhu J and Ohnaka I, 1996 Three Dimensional Computer Simulation of Mold Filling of Casting by Direct Finite Difference Method", *J. of JFS*, **68** 668-673
[16] J.D.Zhu and I.Ohnaka, 1995 Treatment of Free Surface Boundary Condition for the Simulation of Mold Filling, *Modeling of Casting, Welding and Advanced Solidification Processes VI???*, (Warrendale, TMS) 359-372
[17] Ergan S, Fluid flow through packed columns, Chem.Eng.Prog., 1952 **48** 89-93
[18] Mori Y and Ototake N, On the Viscosity of Suspensions, J. Chemical Engineering of Japan, 1956 **20** 488-494
[19] Ohnaka I, Zhu J, Sugiyama A and Kinoshita F, 2011 Proc. 11th Asian Foundry Congress, Foundry Inst. Chinese Mechanical Eng. Soc. Guanzhou China, 152-157
[20] Sugiyama Y, Iwahori H, Yonekura K and Oguchi Y, 1994 *IMONO* **66** 412-417
[21] Iimai H, Kameyama Y, Suzuki Y, Nagayama Y, Kato E and Nomura H, 2001 Denso Technical Review, **6**(2) 100-106
[22] Sim J, Jang Y, Moon J, Kim J, Min K and H C, 2009 *ISIJ International* , **49** 1700-1709

IOP Conf. Series: Materials Science and Engineering **33** (2012) 012002 doi:10.1088/1757-899X/33/1/012002

Simulation of microstructure and mechanical properties of aluminum components during casting and heat treatment

M Schneider[1] and W Schaefer[1], E Sjölander[2], S Seiffeddine[2] and IL Svensson[2]

[1]MAGMA Gießereitechnologie GmbH, Aachen, Germany
[2]Jönköping University, School of Engineering, Materials and Manufacturing – Casting, Sweden

Email: `M.Schneider@magmasoft.de`

Abstract. The combination of alloy composition, metallurgy and local solidification conditions all lead to the development of locally varying microstructure and defects during the production of aluminum alloy castings. In turn, the local microstructure and defects play a critical role in determining the local mechanical properties. In this paper, the integrated modeling and simulation of the development of microstructure in aluminum alloy components from casting through heat treatment will be described. The predicted microstructure at the end of this process chain is used to predict the distribution of mechanical properties in the casting. This information can be used as input for e.g. durability calculations during the design phase of a new component. It can also be used to optimize the casting or heat treatment process to obtain the required mechanical properties using the most cost-effect manufacturing process. The use of the integrated process simulation as well as a comparison with measurements is illustrated using a prototypical cylinder head casting as an example.

1. Introduction

The combination of the alloy composition, the casting conditions (e.g. cooling rates), and post solidification heat treatment lead to a wide range of possible tensile properties in cast Al-Si alloys. The modeling of the distribution of these properties in a cast and heat treated component can be an important tool in selecting the alloy and processing conditions to meet the requirements for a given cast component [3].

Models for the simulation of the microstructure and tensile strength distribution in an as-cast component based on the alloy and casting conditions are available (see e.g. [2]). Although equations describing the variation in the tensile properties in the heat treated condition for certain alloys have been investigated (see [3] for a complete review), simulation models for these properties in the heat treated condition are not currently available. The aim of this paper is to describe a through process model for the prediction of the tensile properties in cast Al-Si components, based on the relevant microstructural parameters, during both casting and heat treatment as illustrated in Fig. 1. The focus in the work described here has been primarily on solution treatment and artificial aging.

In the approach described here, the local cooling conditions during the casting process (Fig. 1, point 1) are considered in determining the amount and distribution of Cu or Mg rich phases, which form at the end of the solidification interval in the relevant alloys. During solution treatment, the dissolution of the Cu/Mg rich phases is modeled to predict the distribution of these hardening relevant elements in the matrix for the subsequent processing steps (Fig. 1, point 2). A simplified model for the reduction in the levels of the hardening relevant elements for slower quench rates is used (Fig. 1, point 3). The remaining dissolved Cu/Mg is used to form precipitates in a controlled manner during artificial aging and achieved the desired hardening. The model estimates the rate of growth of these precipitates in order to account for their effect on the strengthening during aging (Fig. 1, point 4).

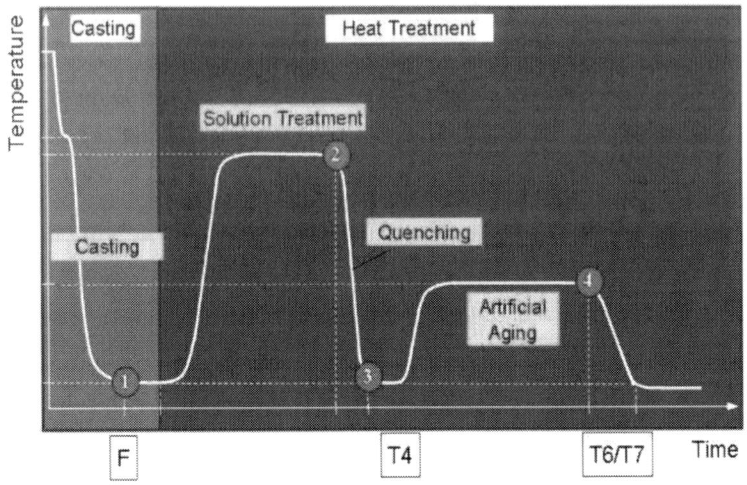

Figure 1. Temperature history during casting and heat treatment.

In the following sections, the modeling of each of the steps shown in Fig. 1 is described including their integration in a macroscopic simulation. The application of the model to a demonstrator aluminum component is then illustrated, and preliminary comparisons with measurement are discussed.

2. Prediction of As-Cast Microstructure

The prediction of the local distribution of Cu/Mg rich phases and defects (e.g. porosity) during solidification is necessary, as the as-cast structure is the starting point for heat treating. The simulation of the solidification and development of the relevant microstructural phase fractions has been carried out using a commercial casting process simulation software [1]. The solidification model, which has been described in detail elsewhere [2], accounts for the alloy composition, thermodynamic phase equilibrium, and growth kinetics to determine the local solidification path and microstructures as well as the related latent heat release for each element in the enmeshed casting at each time step. The locally different cooling conditions leads to locally varying phase fractions, also for the precipitation hardening relevant Cu and Mg rich phases. Microstructural parameters like the grain size and the secondary dendrite arm spacing (SDAS) are also predicted.

3. Modeling Microstructure and Property Development during Heat Treatment

To model a T6/T7 heat treatment with the process steps solution treatment, quenching and artificial aging, it is necessary to consider the diffusion-controlled solid phase transformations that occur during each step. The goal of the heat treatment is to dissolve as-cast phases and to replace them with precipitates in the desired number and size, which inhibit the movement of dislocations through the matrix and result in improved mechanical properties.

3.1. Solution Treatment

The solution treatment model presented here (and described in detail elsewhere [3]) can be used for hypoeutectic AlSiMg, AlSiCu and AlSiMgCu alloys, meaning a Si content between 6% and 12% and further additions of Mg or Cu or both. During solidification, the π-$Al_8Mg_3FeSi_6$ phase is the main phase containing Mg forming in AlSiMg alloys, while the Al_2Cu is the main phase containing Cu forming in AlSiCu and AlSiCuMg alloys having high Cu concentrations. The goal of the solution treatment is to dissolve the Mg and Cu rich phases and create as many vacancies in the crystal lattice as possible. A higher vacancy density in the matrix leads to faster diffusion. To model this process [4], the diffusion of Cu/Mg according to Fick's Law is calculated within a unit cell. The unit cell size is determined based on the distance between Mg or Cu rich particles. The initial concentration profile is based on back-diffusion calculated during the solidification simulation. The concentration at the edge of the unit cell at the particle-matrix interface is calculated considering thermodynamic equilibrium [5][6] with an appropriate database [7]. The diffusion of Cu in the Al-Si eutectic phase is an order of magnitude larger than diffusion in the primary α-aluminum phase [8]. In this case, it is adequate to model only the latter process, as it is rate-determining. Simulations carried using these assumptions show good agreement with the measured Cu distribution for varying solution treatment times for various SDAS [4]. Fig. 2 shows the measured and simulated Mg concentration along a slice through a dendrite arm for an Al-8Si-0.32Mg alloy as a function of solution treatment time for a SDAS of 28 µm.

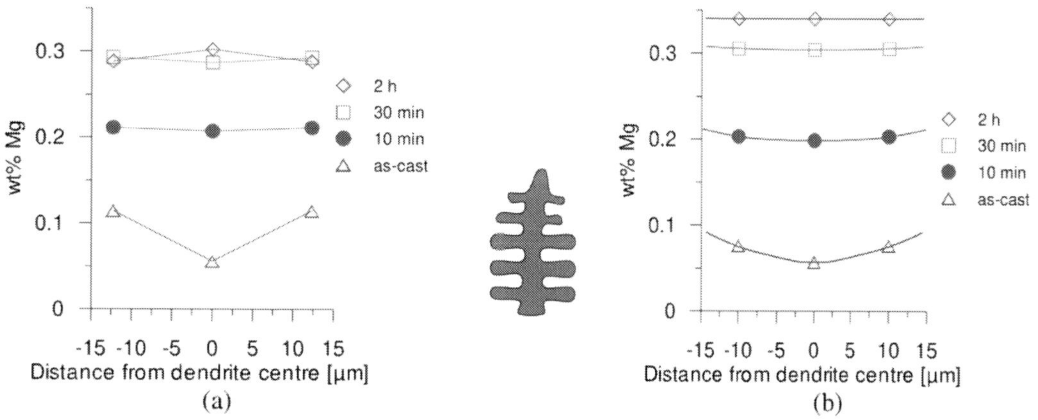

Figure 2. Measured (a) and simulated (b) Mg distribution across a dendrite arm in an Al-8Si-0.32Mg alloy for a SDAS of 28 µm for varying solution treatment times at 530°C [4].

3.2. Quenching

During quenching, the objective is to achieve a supersaturation in both the Mg/Cu concentration as well as vacancies in the lattice. That means that it is necessary to achieve a fast cooling rate to avoid kinetic processes where undesirable precipitates form or changes in the lattice structure occur. This cannot always be achieved, for example due to low rates of heat extraction in thicker wall sections or due to the fact that the casting would tear due to the generation of residual stresses. In this case, precipitates form which do not have the right size or distribution to obstruct movement of dislocations along the crystal planes. However, the growth of these precipitates also consumes some of the dissolved Mg/Cu, which is then no longer available to build and grow nuclei during aging. Summarizing measurements carried out by a number of authors shows a dependency in the maximum value of the achieved yield strength during aging on the quenching rate, as shown schematically in Fig. 3 (see [9] for a summary). The exact form of this relationship is alloy dependent.

As a first approach, the data from these quenching experiments has been used to estimate the amount of Mg/Cu "consumed" during quenching for a given cooling rate. Further experimental work and detailed modeling are required to improve the prediction of these amounts, especially at low cooling rates.

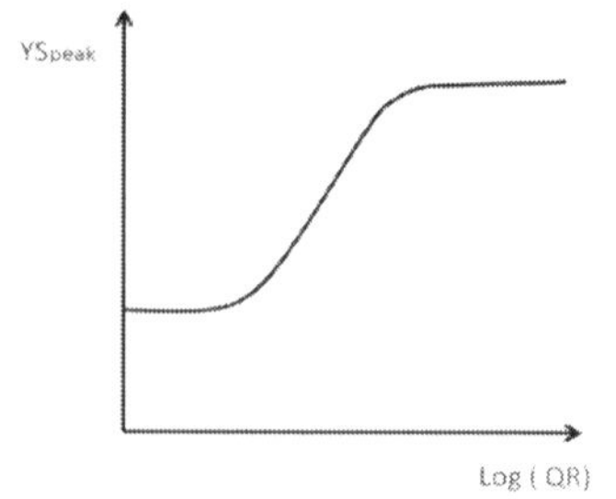

Figure 3. Qualitative description of the variation of the maximum achieved yield strength (YS_{peak}) during aging as a function of quench rate (QR).

3.3. Artificial Aging and Properties

During artificial aging, precipitates with a desired number and size should be created in a controlled process. The supersaturated vacancy density leads to an acceleration of this process. The artificial ageing model (described in detail in reference [10]) has been developed for hypoeutectic AlSiMg alloys. The model can handle Mg concentrations between 0.2 wt% and 0.6 wt% and ageing temperatures between 150°C and 210°C. Similar to solution treatment, the modeling of this process [10] considers diffusion of dissolved Mg through the crystal matrix to form and grow spherical precipitates with radius r_p and concentration C_p (see Fig. 4). The concentration at the interface between precipitate and matrix is calculated based on thermodynamic equilibrium. The size of the unit cell r_c is prescribed by the nuclei density, and the amount of Mg available to form precipitates is dependent on the previous process steps. Symmetry is assumed for the concentration at the edge of the unit cell. When all of the Mg in the matrix has been consumed, the precipitates coarsen, meaning that the phase fraction remains the same, the precipitate density decreases and the radius continues to grow.

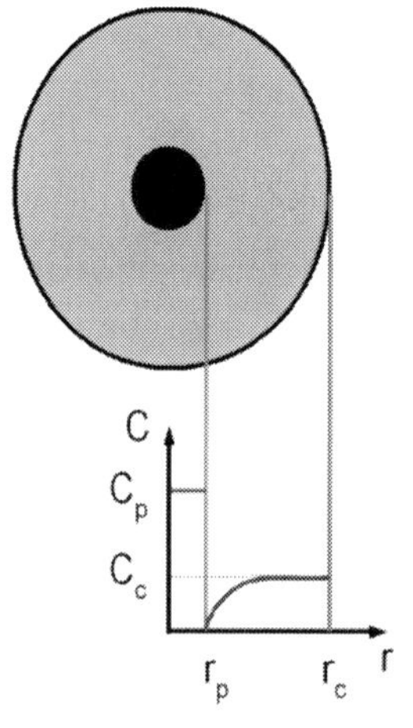

Figure 4. Schematic showing the geometry and the concentration distribution for the diffusion process during artificial aging after [10].

The prediction of the mechanical properties is based on the size and volume fraction of the precipitates, the average distance between them, and the Mg concentration in the crystal matrix [10][11][12]. For the calculation of the yield strength, the intrinsic strength of the aluminum, the increase in the strength through dissolved Mg and the contributions through precipitates being either sheared or looped by dislocations are considered. Fig. 5 shows the measured and simulated changes in the yield strength for AlSiMg alloys with varying Mg concentrations. Normally, these curves show a steep increase, a plateau and then a slow decrease in the yield strength with aging time. The agreement between model and measurement is good.

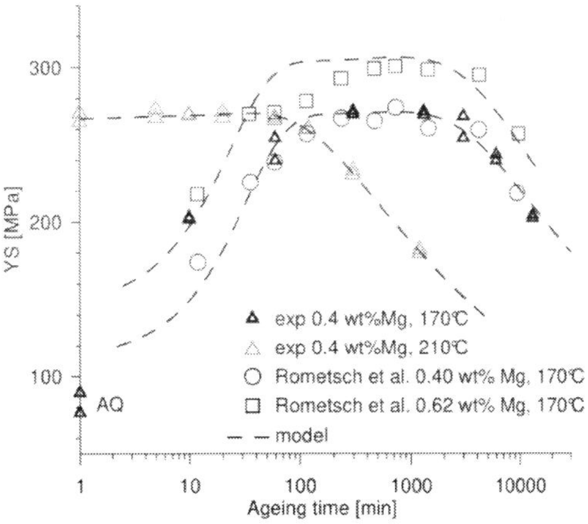

Figure 5. Comparison of measured and simulated changes in yield strength with aging time for AlSiMg alloys with varying Mg contents [10].

4. Application to a Cylinder Head Demonstrator

A demonstrator cylinder head was investigated using the modeling approach described here within the European research project NADIA. The geometry was chosen to have characteristics representative of a real cylinder head (see Fig. 6), but allow a simple extraction of samples for various measurements. The cylinder head was cast using a primary AlSi7MgCu0.5 alloy with varying cooling conditions on the fire deck and with varying porosity levels (controlled through the melt hydrogen level). The heads were then subject to a T6 or T7 heat treatment. Temperatures were measured using thermocouples in the casting during molding filling, solidification and cooling, and during heat treatment. Residual stresses were measured after casting and heat treatment, the microstructure at several locations on the fire deck was investigated, and samples taken from the head were used to determine the yield strength and other properties at various locations. The measured temperatures were used to fine tune the boundary conditions for the simulation for each process step, including quenching in both air and in water.

Figure 6. Demonstrator cylinder head (left) and casting with gating and risers (right). The casting is produced in a core package with a steel bottom plate including a cooled copper bar.

The casting was bottom poured with a filling time of around 15 sec. The simulated solidification time varied from around 40 s on the fire deck to around 450 s close to the riser. The riser itself is completely solidified after approx. 700 s. Fig. 7 shows the predicted SDAS distribution as an important characteristic of the microstructure. The copper cooling bar below the fire deck leads to shorter solidification times and to a corresponding smaller SDAS in that region. A comparison of the SDAS with measurements showed reasonable agreement as shown in Fig. 7.

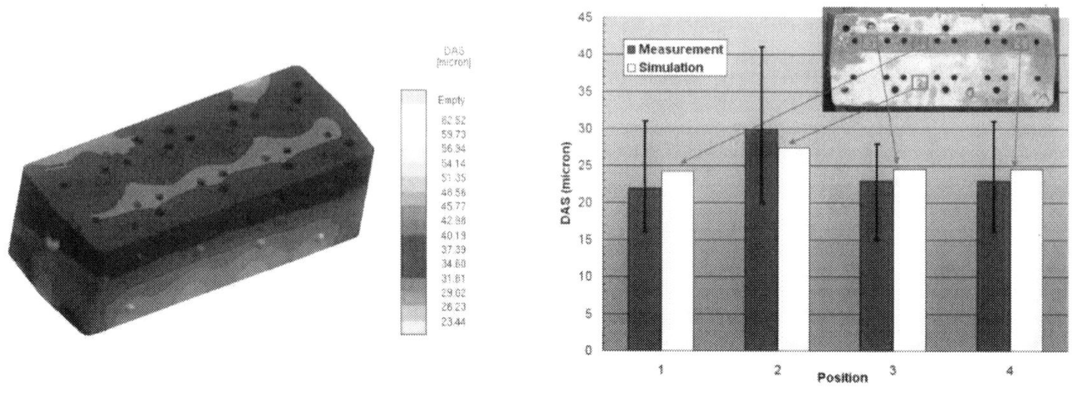

Figure 7. Predicted SDAS on the casting fire deck (left) and comparison with measured SDAS (right).

Fig. 8 shows the predicted as-cast Mg_2Si phase fraction, with variations from an average value of around 0.4% due to locally varying solidification conditions throughout the casting. The variations are due to locally differing cooling rates, and the average value is in good agreement with calculations based on the CALPHAD method [5][6][7].

Figure 8. Predicted as-cast Mg_2Si phase fraction.

The distance between Mg rich particles is needed to model the solution treatment. For cast components this distance is believed to be close to SDAS [4]. SDAS was thereby used to obtain Fig. 9 which shows the predicted Mg concentration distribution in the cylinder head after solution treatment at 530 °C for 240 minutes for an AlSi7MgCu0.5 alloy. The concentration is in average in agreement with measurements for a similar alloy in the literature [13]. Also shown in Fig. 9 is the predicted cooling rate for quenching in a 60 °C water bath. In the outer sections of the cylinder head, the quenching rate lies around 50 °C/s while the thicker inner sections have experience quenching rate of around 10 °C/s. The maximum value is relatively high and will typically not be reached in industrial production, where a number of heads in a wire basket are quenched simultaneously.

The yield strength model for AlSiMg alloys was used for the AlSi7MgCu0.5 alloy as both alloys have β'' as the main strengthening precipitate [14]. Fig. 10 shows the calculated yield strength distribution after artificial aging for 140 min at 210 °C. Measurements in the thinner outer walls showed a yield strength of approx. 260 MPa, in good agreement with the predictions. However, the simulated yield strength in the thicker inner sections is about 25 MPa lower than that measured.

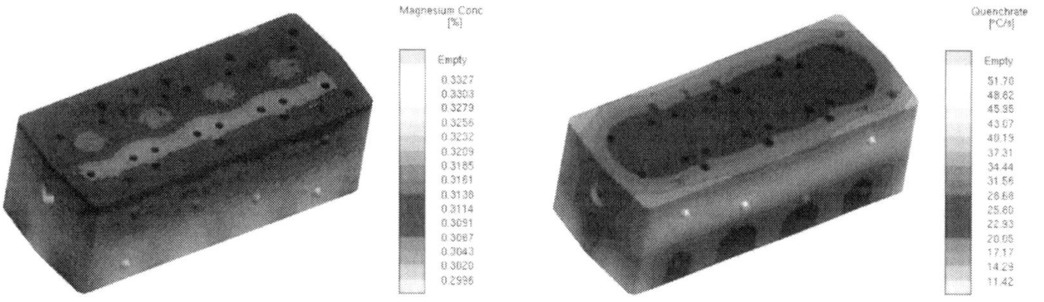

Figure 9. Predicted Mg concentration after solution treatment (left) and local quenching rate for a water quench (right).

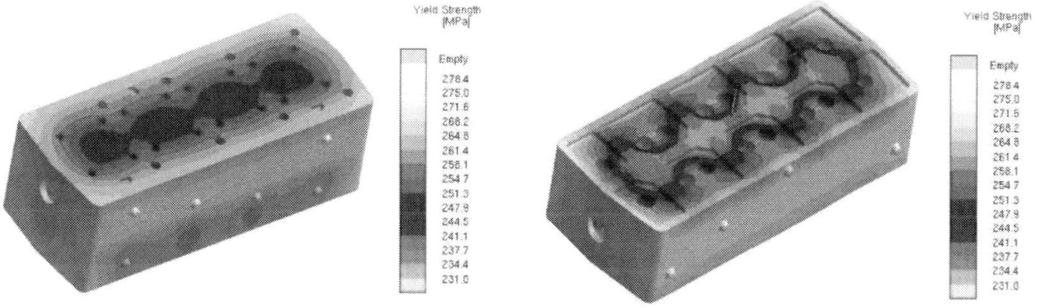

Figure 10. Calculated yield strength distribution after artificial aging for 140 min at 210 °C (fire deck view at left, a section through the casting at right).

In Fig. 11, the maximum yield strength which was achieved locally during aging and the corresponding time required to reach this strength are shown. For this casting, the maximum strength is achieved at about the same time (approx. 1 hour) and the strength does not decrease significantly in the further 80 min of aging. The simulated distribution of the yield strength clearly shows the importance of quenching in obtaining the casting's properties. The measurements also show a smaller variation in the yield strength across the casting as predicted by the simulations.

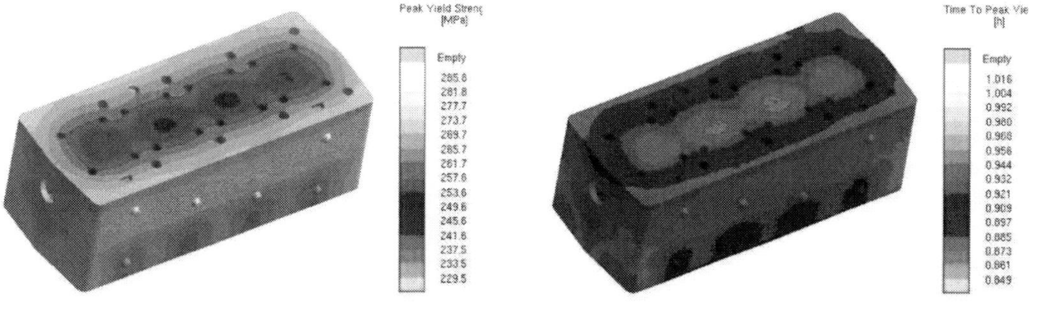

Figure 11. Maximum value of the yield strength achieved during aging (left) and the time to reach the maximum (right) for aging at 210 °C.

5. Conclusions and Outlook

In this paper an overview of a methodology for the integrated simulation of casting and heat treatment of aluminum castings has been presented, in order to prediction both the microstructure and mechanical properties in the heat treated casting. The model has been applied to a demonstrator cylinder head casting and shows its application for industrially relevant components. The objective of these activities is to provide both the foundryman as well as the casting designer with tools to predict the local properties in a castings. With this information, an optimization of both the component design as well as the casting and heat treatment processes on the computer is possible.

Without question, further work is required to extend and confirm the applicability of the model. For example, the predicted mechanical properties are not dependent on defects such as porosity. Further, the model needs to be extended to handle alloys where both Mg and Cu are present in significant amounts. The quenching process has the most significant influence on the predicted properties, and both further experiments and model development are required to better account for the relevant phenomena.

Despite these limitations, it is possible simulate the complete manufacturing process and to transfer the predicted microstructural parameters and properties to other simulation tools e.g. for the simulation of thermo-mechanical fatigue. This makes it possible to investigate the influence of both the manufacturing process as well as the casting geometry on the fatigue properties using simulation before a prototype is produced, saving both time and costs.

Acknowledgments

The authors would like to acknowledge the financial support through the EU project Nadia – "New Automotive components Designed for and manufactured by Intelligent processing of light Alloys" (NMP-2004-SME 3.4.4.5, contract n.026563-2).

References

[1] MAGMA5 Version 5.1 User Handbook, 2011
[2] Wessén M, Svensson IL, Seifeddine S, Olsson J, and Schäfer W 2006 in *Modeling of Casting, Welding and Advanced Solidification Processes XI* eds. CA Gandin and M Bellet, (TMS: Warrendale, PA)
[3] Sjölander E 2011 *Heat treatment of Al-Si-Cu-Mg casting alloys* PhD Thesis (Jönköping University: Sweden)
[4] Sjölander E and Seifeddine S 2011 Optimisation of solution treatment of cast Al-7Si-03Mg and Al-8Si-3Cu-0.5Mg alloys *Met Trans A (submitted)*
[5] Saunders N 2004 in *Proceedings of 9th International Conference on Aluminium Alloys* eds JF Nie, AJ Morton and BC Muddle (Materials Australasia Ltd: Victoria)
[6] Anderson JO, Helander T, Höglund L, Shi PF, and Sundman B, 2002 *CALPHAD* **26** 273-312
[7] TTAL5 – Al-based Alloys Database Version 5.1 2006 (ThermoTech Ltd: Guildford, UK)
[8] Zhang D, Peng J and Liu T 2006 *Mater. Sci. Eng. A* **425** 78-82
[9] Sjölander E and Seifeddine S 2010 *J. Mat. Proc. Tech.* **210** 1249-1259
[10] Sjölander E, Seifeddine S and Svensson IL 2011 *Int. J. Cast Met. Res.* **24** 338-346
[11] Rometsch PA and Schaffer GB 2002 *Mater. Sci. A* **25** 424–434
[12] Shercliff HR and Ashby MF 1990 *Acta Metal. Mater.* **38** 1789–1802
[13] Wang QG and Davidson CJ 2001 *J. Mat. Sci.* **36** 739–750
[14] Li Y, Brusethaug S and Olsen A 2006 *Scripta Mater.* **52** 99-103

IOP Conf. Series: Materials Science and Engineering 33 (2012) 012003 doi:10.1088/1757-899X/33/1/012003

The application of integrated computational material engineering (ICME) in metal castings simulation

Jianzheng Guo[1], Weisheng Cao[2], and Mark Samonds[1]
[1]ESI US R&D, Columbia, MD, USA
[2]CompuTherm LLC, Madison, WI 53719 USA

E-mail: jianzhengguo@gmail.com

Abstract. Integrated Computational Materials Engineering (ICME) is emerging as a methodology for developing advanced materials, manufacturing processing, and engineering components in a faster and more cost effective way. For casting processes, ICME involves many physical phenomena such as thermodynamics, heat transfer, fluid flow, stress, defect formation, microstructure evaluation, and thermophysical and mechanical properties. In this paper, the integration of thermodynamic calculations, thermophysical and mechanical property predictions, and the prediction of microstructure and defects during solidification and heat treatment will be presented. Such integration is helpful to understand the effects of alloy chemistry and processing conditions, and their relationship to microstructure, defect formation, and the final mechanical properties from solidification to heat treatment. Eventually the alloy chemistry and processing parameters can be optimized with the help of the integrated computational modelling.

1. Solidification model

The purpose of casting solidification micro-modelling is to predict the microstructure of a casting, such as grain size. Understanding the casting solidification process and microstructure formation will greatly facilitate casting design and quality control. Computer modeling also provides the basis for computer aided manufacturing and product lifecycle management by being able to predict mechanical properties, quality, and useful life [1-5]. Solidification models also provide the necessary input information for heat treatment simulations.

There are several ways to simulate the microstructure formation during alloy solidification, such as deterministic methods or stochastic methods. For this deterministic method, the density of grains that have nucleated in the bulk liquid at a given moment during solidification is a deterministic function (e.g. a function of under-cooling). Modeling of solidification processes and microstructural features has benefited from the introduction of averaged conservation equations and the coupling of these equations with microscopic models of solidification. Rappaz et al.[6-7] proposed a model using averaging methods to predict the growth of equiaxed grains under isothermal conditions. The nucleation is based on the model proposed by Thevoz and co-workers [8]. At a given under-cooling, the grain density n is calculated by the integral of the nucleation site distribution from zero under-cooling to the current under-cooling.

The density of grain nuclei can be calculated by:

$$n(\Delta T(t)) = \frac{n_{\max}}{\sqrt{2\pi}\cdot \Delta T_{\delta}} \int_{0}^{\Delta T(t)} \exp\left(-\frac{(\Delta T(t) - \Delta T_{n})^{2}}{2\Delta T_{\sigma}^{2}}\right) d(\Delta T(t)) \quad (1)$$

where n_{max} is the maximum grain nuclei density, ΔT_δ the standard deviation under-cooling, ΔT_n the average under-cooling. Reference [1] gives details about the model.

The secondary dendrite arm spacing is simplified by:

$$\lambda_2 = 5.5(Mt_f)^{1/3} \tag{2}$$

The averaged microstructure, such as grain size and secondary dendrite arm spacing, can be calculated based on the chemistry and cooling conditions. In the same time, the eutectic phase information, such as volume fraction and eutectic laminar spacing, can be predicted if there is eutectic phase formed according to the alloy chemistry and cooling conditions.

2. Thermophysical properties calculation

For any casting or heat treatment modeling, thermophysical properties of the alloys are one of the critically required inputs. An extensive database for the calculation of thermophysical properties has been developed [9]. These properties include density, specific heat, enthalpy, latent heat, electrical conductivity, thermal conductivity, liquid viscosity, Young's modulus, and Poisson's ratio. As an example, A357 alloy is studied in this paper. There is a large amount of eutectic at the end of solidification, in the form of silicon based on thermodynamic calculation. There are several possible intermetallic phases, such as Al_3Ti, $Al_8FeMg_3Si_6$, and Mg_2Si. The main intermetallic phase is Mg_2Si, which is observed experimentally as well. The intermetallic phase, Mg_2Si, will be assumed as the main precipitation during the heat treatment process in this paper here. Based on the thermodynamic calculation of the solidification path, the thermophysical properties mentioned above can be calculated. Figure 1 shows the calculated density and thermal conductivity for this alloy during solidification.

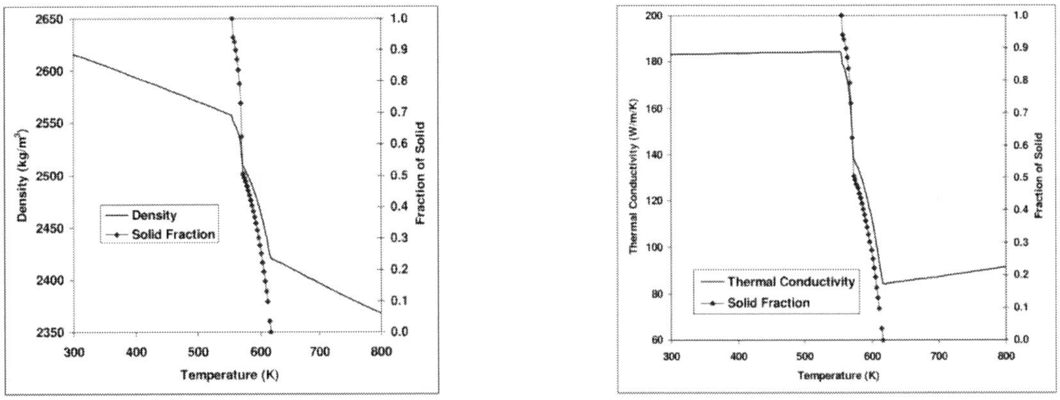

Figure 1 Calculated thermal-physical properties, (left) density and (right) thermal conductivity

3. Heat treatment modelling

For A357 alloys, the most common heat treatment process is T6, which consists of a solution heat treatment, water or oil quenching, and aging. To take advantage of the precipitation hardening, it is critical to produce a supersaturated solid solute by solution heat treatment at a temperature slightly below the eutectic temperature in order to dissolve the hardening elements, such as Mg and Si into the alpha phase to create a nearly homogenous solid solution. At the same time, the eutectic silicon will go through the process of spheroidizing and coarsening too. The particle size, shape, and spacing all affect the final properties. However in the current model, some assumptions are made, such as a) solution heat treatment time is long enough so that the equilibrium state is reached at the solution temperature; b) quench is perfect, hence the casting is supersaturated at room temperature after quench; and c) the effect of spheroidization and coarsening of eutectic silicon on properties is neglected. Only precipitation by Mg_2Si is studied extensively on hardening in this paper.

The precipitation can be simulated by a kinetic model [10]. In general, a system contains a matrix phase and a number of precipitate phases. The precipitation behavior of different precipitate phases can be described by different kinetic models, such as the Kampmann/Wagner Numerical (KWN) [11] model, and the Fast-Acting model [12]. Both models can be used to simulate the co-precipitation of phases with various morphologies, with concurrent processes of nucleation, growth and coarsening. With the KWN model, the particle size distributions (PSD) of various precipitate phases can be obtained in addition to the temporal evolution of the average size and volume fraction as obtained from the Fast-Acting model.

In this paper, the KWN model is implemented in a numerical framework [11]. The following is a brief introduction to the KWN model along with its sub-models for nucleation, growth and coarsening. More detailed descriptions can be found in Refs. [10, 11, 13, 14].

First, at each simulation step, the number of new particles is calculated using classical nucleation theory. The transient nucleation rate is given by

$$J = N_v N \beta^* \exp(-\frac{\Delta G^*}{kT}) \exp(-\frac{\tau}{t}) \tag{3}$$

where $\Delta G^* = \frac{4\pi}{3}(R^*)^2 \sigma$, R^* is the radius of the critical nucleus and $R^* = -\frac{2\sigma}{\Delta G_v}$, ΔG_v is the chemical driving force for nucleation and σ is the interfacial energy of the matrix and particle interface. The pre-exponential terms in equation (3) are N_v, the nucleation site density, Z, the Zeldovich factor, and β^*, the atomic attachment rate. t is the time, τ the incubation time for nucleation, k the Boltzmann constant and T the temperature. A multi-component version of the parameters has been proposed by Kozeschnik et al. [15] and implemented here. The nucleation site density and nucleation driving force are modified to deal with heterogeneous nucleation.

Next the growth of existing particles for each size class is assumed to be diffusion-controlled, where the Gibbs-Thomson size effect is also taken into account. The growth model for multi-component alloys proposed by Morral and Purdy [16] is adopted and modified to handle the growth/dissolution of various precipitate phases with different morphologies. The motion rate of the curved interface, e.g., the interface of a spherical or lens-like precipitate, and the edge of a plate-like precipitate, is given by,

$$v = \frac{dR}{dt} = \frac{K}{R}(\frac{1}{R^*} - \frac{1}{R}) \tag{4}$$

With

$$K = \frac{2\sigma V_m}{(\Delta C^{\alpha\beta}][M]^{-1}[\Delta C^{\alpha\beta})} \tag{5}$$

where R is the radius of the interface, R^* is the radius of the critical nucleus, σ is the interfacial energy between matrix (α) and precipitate phase (β), V_m is the molar volume of the precipitate phase, $(\Delta C^{\alpha\beta}]$ and $[\Delta C^{\alpha\beta})$ are the row and column vector of the solute concentration difference between α and β, and $[M]$ is the chemical mobility matrix. The motion rate of the planar interface, e.g. thickening of the plate, is described by

$$v = \frac{dh}{dt} = \frac{K'}{h} \tag{6}$$

with $K' = \dfrac{\Delta G_m^*}{(\Delta C^{\alpha\beta}][M]^{-1}[\Delta C^{\alpha\beta})}$, where h is the half thickness of the plate, and ΔG_m^* is the transformation driving force.

Finally, the PSD and the volume fraction of the particles are updated, and the matrix composition is recalculated based on mass conservation at the end of each simulation step. The updated values are then used in the next time step.

$$C_i^* = (1 - \sum_P^{n_P} \phi_P) C_i^\alpha + \sum_P^{n_P} \phi_P C_i^P \tag{7}$$

where C_i^* is the initial mole fraction of component i, ϕ_P is the mole fraction of precipitate phase P, n_P is the number of precipitate phases, C_i^α is the mole fraction of component i in the matrix phase α, and C_i^P is the mole fraction of component i in precipitate phase P.

It should be noted that coarsening in the KWN model arises naturally and becomes dominant when the supersaturation is sufficiently low. In the Fast-Acting model, however, the coarsening is considered explicitly. Accordingly, a model based on the LSW theory [17, 18] is employed, which characterizes the curvature-driven particle growth. The multi-component version can be written as,

$$\overline{R}^3 - \overline{R}_0^3 = \frac{4}{9} Kt \tag{8}$$

where $\overline{R}$ is the average particle radius at time t, $\overline{R}_0$ is the average particle radius at the onset of coarsening, and K is the kinetic parameter defined in equation (5). Its accuracy and efficiency are ensured by controlling the adaptive step size with a fifth-order Runge-Kutta scheme. At each simulation step, the same nucleation and growth models as given in equations (3), (4) and (6), are used for the estimation of the average size changes. The thermodynamic equilibrium information, the driving force, and the diffusivity data needed for both KWN and Fast-Acting simulations are obtained by directly coupling with the automatic thermodynamic calculation. Therefore, no initial values are required for the thermodynamic calculation at every simulation step.

4. Mechanical properties prediction

Solid solution strengthening and precipitation hardening contribute to the strength of aluminum alloys. The precipitation hardening is mainly from aging heat treatment. In this paper, the eutectic silicon spheroidizing and coarsening are not modeled. The total yield strength is assumed to be the sum of as-cast strength and the precipitation strength [19].

$$\sigma = \sigma_c + \sigma_p \tag{9}$$

Where σ is the total yield strength, σ_c is the as-cast yield strength, and σ_p is due to precipitation hardening.

Based on the microstructure information calculated from solidification micromodeling, the strength of cast aluminum can be predicted which includes solid solution strengthening. For A357 alloys, where there is a large amount of eutectic phase, the mechanical properties are determined by the combination of the properties from both the primary dendrite phase and the eutectic phase.

First solid solution strengthening is calculated, and then the effect of secondary dendrite arm spacing on strength of the primary dendrite phase, and the eutectic laminar spacing on the strength of the eutectic phase are added. The solid solution strengthening of one phase is described by the following equation:

$$\sigma_o = \sum k_j C_j^{2/3} \tag{10}$$

A Hall-Petch type equation is used to describe the strength variation due to change in secondary dendrite arm spacing for the primary dendrite phase and the eutectic laminar spacing for the eutectic phase. The as-cast yield strength can be written:

$$\sigma_c = f_p(\sigma_o + \frac{k_p}{\lambda_2^{1/2}}) + f_e \frac{k_e}{\lambda^{1/2}} \tag{11}$$

where f_p and f_e are volume fractions of primary and eutectic phases respectively. σ_0 , k_p , and k_e are solid solution stress and the Hall-Petch coefficients for the primary and eutectic phase, λ_2 and λ are the secondary dendrite arm spacing and eutectic laminar spacing.

As for precipitation, the hardening can be calculated by:

$$\sigma_p = \frac{M}{bR}(2\beta G b^2)^{-1/2} \sqrt{\frac{3f}{2\pi}} \bar{F}^{3/2} \tag{12}$$

Where M is the Taylor factor, G is the shear modulus of the Al matrix, β is a constant close to 0.5, b is the magnitude of the burgers vector, R is mean particle size, f is the volume fraction of the precipitates, $\bar{F} = \frac{\sum N_i F}{\sum N_i}$ is mean obstacle strength

$$F_i = \begin{cases} 2\beta G b^2 \frac{R_i}{R_c} & \text{if } R_i \le R_c (weak) \\ 2\beta G b^2 & \text{if } R_i > R_c (strong) \end{cases} \tag{13}$$

N_i and N_c are the number density of the size class R_i and the critical precipitation particle size

5. Predictions and validations

For validation, an experiment of an A357 sand casting with chills on some locations is performed. Figure 2 illustrates the experimental modeling setup. There are four cast iron chills used. With such a setup, cooling conditions varied dramatically for different locations. After quenching, the castings are under the process of aging at different temperatures and times. The comparison of the modeling and experiment includes two stages, one just after casting and the other after aging heat treatment. The microstructure and mechanical properties are compared.

5.1 *As-cast microstructure and mechanical properties*

The solidification process is simulated based on the micromodel presented above. As mentioned earlier, because of the use of chills, the solidification times are very different across the casting. For the chilled section, the solidification time is as short as 29 seconds. For a location which is not chilled, but with similar wall thickness, the solidification time is 242 seconds, which is an order of magnitude longer. Because of the different solidification rates, the microstructure is not uniform in the casting. For the primary phase, it is assumed that only equiaxed dendrites are formed. For a faster cooling, the dendrite grain size is smaller.

Experimentally, the secondary arm spaces are measured. For comparison, figure 3 shows the measured and predicted secondary arm spacing at two locations, one on the chilled section and the other one on an unchilled section. The exact values can be found in Table 1. It shows that the difference between measured and predicted values is very small. Based on equation (2), the secondary arm spacing is proportional to the one third power of the solidification time. This can be verified by the values of the secondary arm spacing and solidification time at those two locations numerically as well. The solidification time varies by a factor of 9 which gives roughly a three times difference in secondary arm spacing.

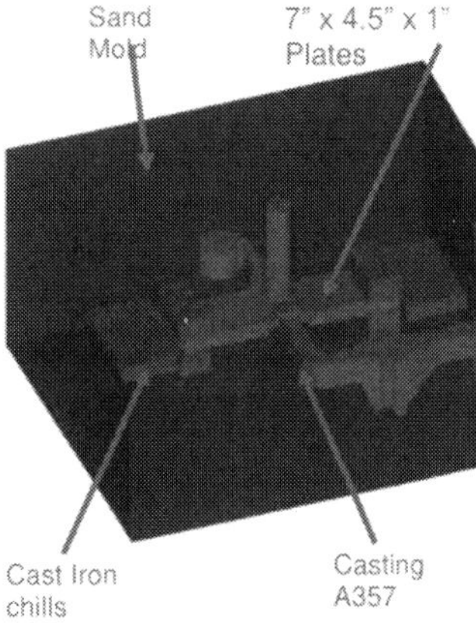

Figure 2 Setup for modelling

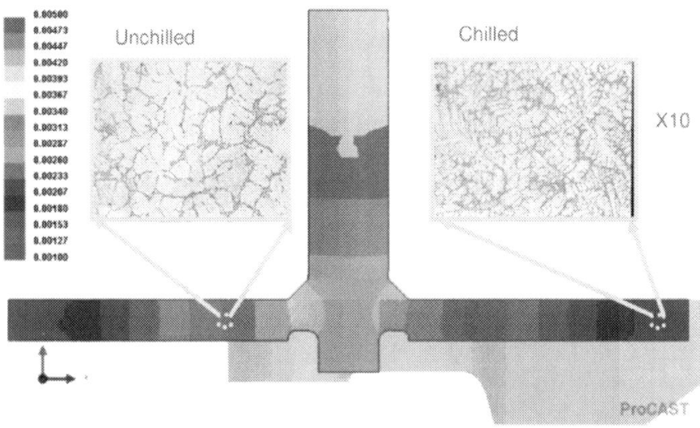

Figure 3 Comparison of secondary dendrite arm spacing between experiment and prediction

Table 1 Secondary dendrite arm spacing comparison between experiment and simulation

SDAS	Measurement(mm)	Prediction(mm)	Error(%)
chilled	0.0335	0.03556	6.0
unchilled	0.0876	0.0838	4.3

With the results of the microstructures, the as-cast room temperature yield strength can be calculated from equation (11). The room temperature as cast yield strength is shown in figure 4. With the finer microstructure in the chilled section, because of the fasting cooling, the yield strength is higher. But the difference is relatively small compared to that of SDAS and solidification time. The yield strength difference between the chilled section, 83.8MPa, and the unchilled section, 74.8MPa, is less than 10MPa, around 10%. Even though most in service castings are heat treated, the prediction of as-cast microstructure and mechanical properties is still important. First, the as-cast microstrucutre and properties affect the final structure and properties. More importantly, as-cast properties are critical for stress analysis during the casting process.

5.2 *Microstructure and mechanical properties after heat treatment*

The ultimate goal of the simulation is to predict the final part microstructure and mechanical properties. In this case, it is the final properties after heat treatment. Since we assume the solution treatment and quench are ideal and the aging is under isothermal conditions, the microstructure of the precipitates is the same across the part. Since the as-cast yield strength is calculated from the solidification micromodeling and the microstructure during aging is predicted by the kinetic model, the final mechanical properties, yield strength in this case, can be calculated based on equation (9).

Figure 4 shows the yield strength change during the aging at the temperature of 150°C. The yield strength increases from the as-cast strength with time until at around 13 hours it reaches the peak. The maximum yield strength can be as high as 300MPa. With further aging, the strength starts to decrease, which is over aging. The change of the yield strength with time can be explained by the precipitate change. Before reaching the peak, the precipitates are in the process of nucleation and growth stages. The precipitate volume fraction increases more rapidly than the size. Hence the yield strength increases according to equation (12). With too long aging (i.e. over aging), when the precipitates reach the coarsening stage, the volume fraction of the precipitates reaches to a constant value, but the size of the precipitates are still increasing. Based on equation (12), the strength will decrease.

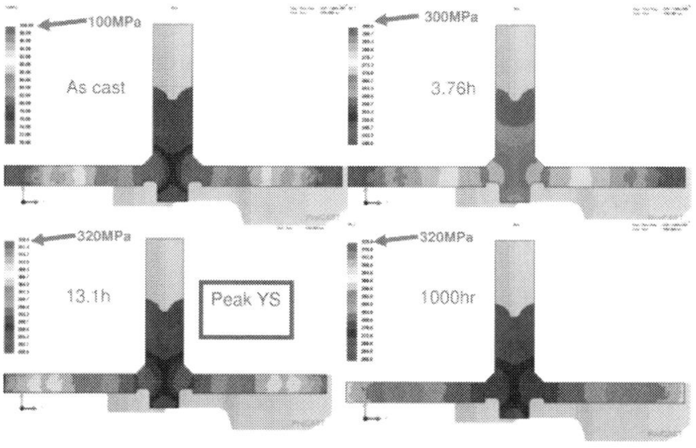

Figure 4 Yield Strength evolutions with Aging temperature of 150°C

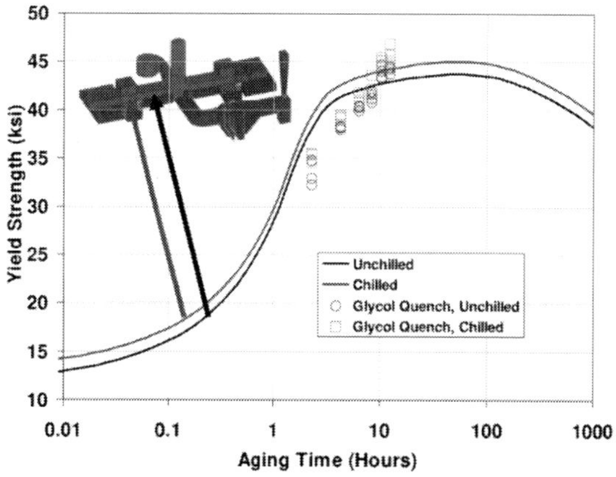

Figure 5 Comparisons between experiment and simulation with Aging at 310F

If the aging temperature continues to increase to 230°C or even 250°C, it takes an even shorter time to reach the peak strength. The peak times are 136 and 41 seconds respectively for aging temperatures of 230°C and 250°C. We can conclude that it is difficult to control the aging process if the aging temperature is too high in order not to reach the over aging stage.

Table 3 summaries the aging temperature effect on the optimal yield strength and the time needed to reach the optimal on the chilled section of the casting. The aging temperature does not affect much the optimal yield strength when it changes from 150°C to 250°C. But the time needed to reach the optimal point is fully controlled by the aging temperature. Increasing the aging temperature will accelerate the precipitation process. For example, it only needs 41 seconds to reach the peak when the aging temperature is 250°C. It can be too fast to accurately control the process if the aging temperature is too high. Figure 5 shows the comparison of current prediction and experiment results. The experimental results are from two locations in the part, one at a chilled section and the other at an unchilled section. The predictions match the experiments well. Further improvement of the model can improve the prediction accuracy. This includes the modeling of solution treatment, quenching, and the eutectic silicon spheroidizing and coarsening during aging.

Table 2 Aging temperature effect on optimal yield strength and the time to reach the optimal

Aging Temperature (°C)	Optimal YS (MPa)	Time to reach the optimal (s)
150	306	47160
190	310	3132
230	300	144
250	290	36

6 Conclusions

The microstructure and the final yield strength are predicted and compared with experimental results thanks to the integration of thermodynamic calculations, thermophysical and mechanical property predictions, and the prediction of microstructure and defects during solidification and heat treatment. Good agreement is obtained. Further improvement of the model is needed. It is necessary to fully couple the solidification and the whole heat treatment process, such as solution treatment and quenching.

References

[1] Rappaz M and Boettinger W 1999 *Acta Mater.* Vol. **47** 3205-3219
[2] Boettinger W, Coriell S, Greer A, Karma A, Kurz W, Rappaz M, and Trivedi R 2000 *Acta Mater.* **48**, 43-70
[3] Guo, J and Samonds M 2007 *Journal of Materials Engineering and Performance.* **16** pp 680-684
[4] Guo, J and Samonds M 2005 *Materials Science and Technolog.* **3** pp 41-50
[5] Kermanpur A, Mehrara M, Varahram N,and Davami P 2008 *Materials Science and Technology.* v **24** pp 100-106
[6] Rappaz M and Thevoz P 1987 *Acta Metall.* **35** pp 1487-1497
[7] Rappaz M and Thevoz P 1987 *Acta Metall.* **35** pp 2929-2933
[8] Thevoz P, Desbiolles J, and Rappaz M 1989 *Metall. Trans. A* **20** pp 311–322
[9] Guo J and Samonds M 2007 *J. Phase Equilibrium and Diffusion.* Volume **28** pp 58-63
[10] Cao W, Chen S, Zhang F, Wu K, Yang Y, Chang Y, Schmid-Fetzer R, and Oates W 2009 *Calphad.* **33** pp 328-342
[11] Kampmann R and Wagner R 1984 *Decomposition of alloys: the early stages,* ed. P. Haasen, et al. (Oxford: Pergamon Press) pp 91-103
[12] Langer J and Schwartz A 1980 *Physical Review A.* **21**(3) pp 948-958
[13] Robson J 2004 *Materials Science and Technology.* **20** pp 441-448
[14] Robson J, Jones M, and Prangnell P 2003 *Acta Materialia.* **51**(5) pp 1453-1468
[15] Kozeschnik E, Holzer I, and Sonderegger B 2007 *J of Phase Equilibria and Diffusion.* **28**(1) pp 64-71
[16] Morral J and Purdy G 1994 *Scripta Metallurgica Et Materialia.* **30**(7) pp 905-908
[17] Lifshitz I and Slyozov V 1961 *J. Phys. Chem. Solids.* **19** p 35
[18] Wagner C 1961 *Z. Elektrochem.* **65** p 581
[19] Guo J, Cao W, and Rolle C 2011 *Materials Science Forum.* ***690*** *pp 315-318*
[20] Rometsch P and Schaffer G 2002 *Materials Science and Engineering A.* **325** pp 424-434

Improving the directional solidification of complex geometries through taper addition

C A Monroe[1,3], D A Gorsky[1,2], R K Huff[1], and R V Grandhi[2]

[1] Caterpillar Inc., 14009 Old Galena Road, Mossville, IL 61552, USA
[2] Wright State University, 3640 Colonel Glenn Hwy, Dayton, OH 45435, USA
[3] University of Alabama at Birmingham, 1075 13th Street South, Birmingham, AL 35294-4440, USA

Email: camonroe@uab.com

Abstract. The Niyama criterion identifies centerline shrinkage in simulations of directionally solidifying alloys. Adding taper to the casting surface is an effective strategy to eliminate regions of critical Niyama criterion and improve the soundness in castings. However for complex geometries, there may not be an intuitive taper solution. The Niyama criterion will validate these complex tapers but not guide the design of that taper. When evaluated near the end of solidification, a thermal criterion such as the cooling rate on the surface of the casting can serve as an identification of efficient areas to taper. The cooling rate is used to calculate an effective depth (ED) which is representative of the half thickness of the underlying section. The spatial gradient of the effective depth (EDG) can be correlated to the Niyama criterion at the casting's centerline to predict the magnitude of existing taper. The EDG can then be used as a constraint for a minimal taper optimization. The EDG was calculated on two flat plates, one with and without taper, to show that this is a viable methodology for taper identification. The method is also performed on a real axle housing with know dispersed porosity to demonstrate its usefulness on three-dimensional castings.

1. Introduction

During the design of a component and its casting process, many changes to geometry and rigging add time and cost to production. Optimization can be used to reduce the lead time to develop a finished product, for example by automating the exploration of casting design and rigging to reduce casting defects such as shrinkage porosity. Optimization techniques including topology or shape/size optimization and design of experiments have been used in the past to change riser size and connection dimensions[1,2]. Little work, however, has been done to discuss the optimization of a casting's geometric boundaries to improve its soundness and directional solidification. The benefit of such an optimization would be that the design of a part can be evaluated for castability before it is sent to a foundry for further rigging.

The current method is focused on the improvement of casting geometry and not casting process variables such as orientation, rigging, flow, chills, pouring temperature, etc. Under this restriction, the main tool available to the designer to improve castability is the geometry itself i.e. tapering. By effectively using a taper, small isolated hotspots and poor directional solidification can be eliminated to leave large unfed hotspots which drive a more consistent rigging practice for the part. The Niyama criterion has been developed to evaluate the directional solidification of geometries.

In directionally solidifying alloys, the temperature based Niyama criterion correlates well with regions of micro-shrinkage[3]. This micro-shrinkage is caused by a low pressure drop across the mushy zone leading to a small driving force for flow to restore the volume deficit due to solidification contraction. The Niyama criterion is directly proportional to the pressure drop in the mushy zone for constant cooling rate and thermal gradient conditions[4]. When the Niyama criterion is lower than the determined critical value, then micro-shrinkage is likely to occur. Two thermal conditions contribute to a low Niyama criterion: low thermal gradients and high cooling rates. When the thermal gradient is lower at the same cooling rate, the mushy zone length is longer with a further distance for fluid to travel to feed the solidification shrinkage. When the cooling rate is higher for the same thermal gradient, the fluid has less time to travel to feed the solidification shrinkage. The Niyama criterion is calculated as follows:

$$Ny = G/\sqrt{\dot{T}} \tag{1}$$

where G is the thermal gradient and $\dot{T}$ is the cooling rate. In casting simulation, the Niyama criterion is calculated and stored at an evaluation temperature everywhere in the geometry. Using Equation 1, regions of low Niyama criterion can be identified and can typically occur at the centerline of casting geometries. One recommendation to improve the soundness of geometries is to add taper[5]. Taper addition will increase the thermal gradient at the centerline and reduce the region of weakly fed mushy zone length. The struggles of creating a good casting design hinges on the ability to taper sections and preprocess the design to "build-in" directional solidification and create easily accessible hotspots for risers. In complex geometries, the current method of taper addition requires experience and a trial and error method to be successful.

The method proposed is an aid to apply flat plate feeding distance rules to complex geometries. To accomplish this, the depth from the surface of the complex geometry to the centerline of the section can be determined by the surface cooling rate. This effective depth is a length that is similar to the feeding modulus of the section but represented on the surface of the casting rather than at the centerline[6]. The surface gradient of the effective depth is directly a measure of the taper of the geometry. The concept of effective depth and effective depth gradient can be used by a designer on complex geometries to check for ill-tapered regions and create taper optimization design spaces within those regions for subsequent optimization to improve the Niyama criterion of the section. Finally, an optimization approach which locally adds the minimum taper to create a successful centerline thermal gradient can be achieved.

2. Method and Development of an Effective Depth

To relate the centerline Niyama criterion to the surface cooling rate, a one-dimensional analysis of solidification in a finite thickness casting with mold is used. The following boundary conditions are assumed in this analysis: a symmetric boundary condition on the free end of the casting so that the model represents a half-thickness of the casting, the casting and mold are "merged" at the casting surface so no heat is lost at their interface, and there is a convective boundary condition between the mold and air as visualized in Figure 1.

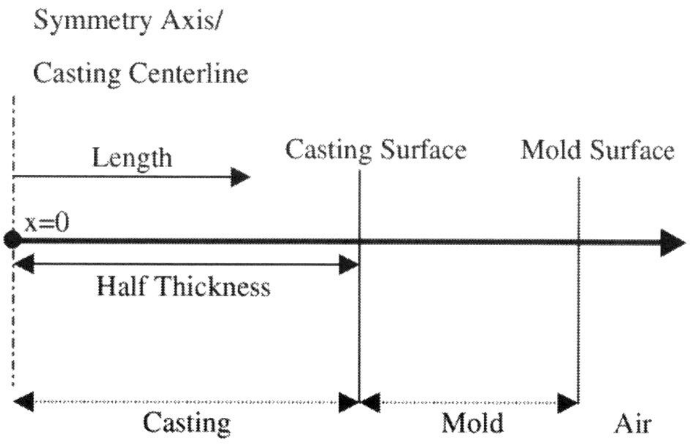

Figure 1. 1-dimensional casting with mold and boundary conditions

A one-dimensional transient heat transfer equation was used to calculate the temperature of the casting and mold from initial conditions to the evaluation temperature of the Niyama criterion. The chosen evaluation temperature for this work was selected by a dimensionless temperature evaluated as

$$\theta_{cr} = \frac{T_{eval} - T_{mold}}{T_{initial} - T_{mold}} = 0.91257, \qquad (2)$$

where T_{eval} is the chosen evaluation temperature, T_{mold} is the initial mold temperature, $T_{initial}$ is the initial casting temperature. The typical temperature used to evaluate the Niyama criterion is 10% of the liquidus and solidus temperature range above the solidus temperature[5]. As mentioned in the reference, the functional correspondence of the Niyama criterion at different evaluation temperatures remains the same. The constant material properties and initial conditions of the one-dimensional transient heat transfer equation are given in Table 1. Constant material properties were used to establish the trends of the thermal gradient and cooling rate for finite thickness castings. The constant properties are characteristic of a chemically bonded silica sand mold and low alloy liquid steel.

Table 1. Constant Material Properties and Initial Conditions.

Casting	Initial Temperature [℃]	1600
	Density [kg m^{-3}]	7022
	Thermal Conductivity [W m^{-1} K^{-1}]	40
	Specific Heat [J kg^{-1} K^{-1}]	790
Mold	Initial Temperature [℃]	20
	Density [kg m^{-3}]	1600
	Thermal Conductivity [W m^{-1} K^{-1}]	0.865
	Specific Heat [J kg^{-1} K^{-1}]	1170
Air	Initial Temperature [℃]	20
	Heat Transfer Coefficient [W m^{-2} K^{-1}]	100

Figure 2(a) shows the responses of the one-dimensional heat transfer calculation at the evaluation temperature for three casting half-thicknesses. The centerline of the casting is at zero length and only results in the casting are shown. First, consider the variation of the cooling rate, thermal gradient, and Niyama criterion along the casting length for the 0.5 inch or 12.7 mm half thickness indicated by (ii). The Niyama criterion and the thermal gradient increase in magnitude from the center of the casting to the surface. The cooling rate decreases in magnitude slowly from the center of the casting to the surface. As the casting surface is approached, the magnitude of the cooling rate decreases quickly with respect to distance from the casting surface. As the casting section increases, the cooling rate decreases as shown in the results of (i) – (iii). The Niyama criterion and the thermal gradient are not plotted for the remaining half thicknesses because they follow similar trends as the half inch half-thickness.

The magnitudes of the thermal gradient and cooling rate at the casting surface were found to vary non-linearly with respect to the casting half thickness. The regressions of the non-linear relationships of surface cooling rate ($\dot{T}_s$) and surface thermal gradient (G_s) with respect to the casting half-thickness are linear and are found in Figure 2(b). The surface cooling rate function is found to be more sensitive to the casting thicknesses than thermal gradient function. This sensitivity along with the high correlation between the regression to the calculated surface responses makes the surface cooling rate a good basis for the definition of the effective depth.

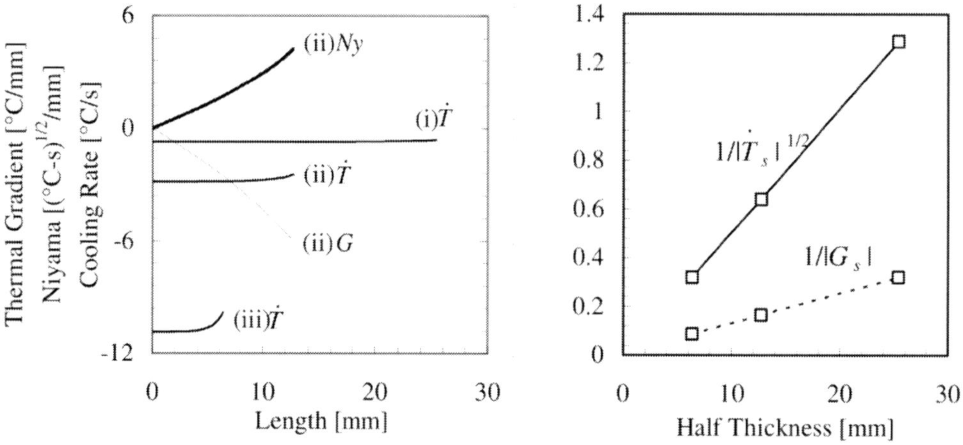

a) **b)**

Figure 2. (a) Cooling rate, thermal gradient, and Niyama criterion results at the evaluation temperature for a one-dimensional heat transfer simulation of three half thicknesses: (i) 25.4 mm (1 in), (ii) 12.7 mm (0.5 in), and (iii) 6.35 mm (0.25 in) **(b)** Linear regressions of the relationships between the surface cooling rate and surface thermal gradient at the evaluation temperature to the casting half thickness.

The effective depth (ED) is a calculated approximation of the half-thickness of a casting based on the surface cooling rate at a particular location at the Niyama criterion evaluation temperature. The slope of the surface cooling rate curve from Figure 2(b) determines the value of the constant, in mm-(℃/s)$^{1/2}$, that converts the surface cooling rate to the effective distance as shown in the following equation

$$ED = 19.746/\sqrt{\dot{T}_s}\,. \tag{3}$$

Other needed constants could be determined for changes in material properties or even for temperature dependent material properties and interfacial heat transfer coefficients. In this study, the intent is to evaluate the geometry so the constant properties and the relationship in Equation (3) is sufficient to determine the utility of this approach. Once the effective depth is determined from the surface cooling rate, then the level of taper on the surface can be established. The gradient of the effective depth (EDG) is calculated by,

$$EDG = \left|\nabla ED\right|_{surface} \tag{4}$$

where ∇ED is the spatial gradient of ED along the surface of the geometry. EDG shows the locations on the surface of the casting that require more material to promote feeding and directional solidification to reduce the micro-shrinkage in the area below it. Similar to specifying a critical Niyama criterion value, a critical EDG can be determined by a correlation to the Niyama criterion value. Once correlated, the EDG represents the critical Niyama criterion profile that lies below the casting surface.

The application of this ED and EDG criterion is straightforward on complex geometries. Using the constant properties and initial conditions specified above, the surface cooling rate can be calculated at the Niyama criterion evaluation temperature. Applying Equation (3), the surface cooling rate can be converted to the effective depth and from this value the gradient on the surface is calculated and the effective depth gradient is found. With these new criteria, an established taper minimum may be evaluated on complex geometries using casting simulation software such as MAGMAsoft user API routines to aid design decisions[8]. The API routines in MAGMAsoft allow a user to create new criteria functions in C code using any of the simulation results.

3. Results

A flat plate with a riser was created with a riser zone length (RZL) and end zone length (EZL) that do not overlap in the plate's longest dimension. This geometry ensures that the plate during solidification would form a region of low values of Niyama criterion; ideally to have the casting form regions of Niyama criterion with magnitudes below 0.7. The dimensions of the flat plate and riser are shown in Figure 3(a). A second plate with a taper area determined by the EDG method was also made to eliminate the regions of low Niyama criterion, and the taper dimensions are shown in Figure 3(b). By evaluating the effective depth, the most effective locations to add taper can be determined. On a flat plate this solution is well defined but for a complex geometry this is not clear because the solidification paths of a casting are not always intuitive.

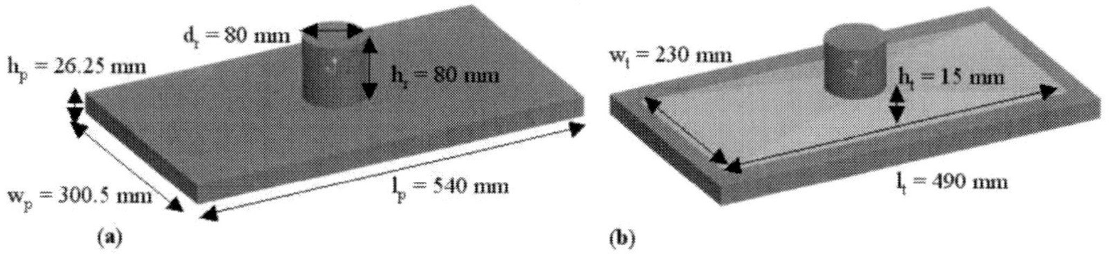

Figure 3. (a) Plate Dimensions and **(b)** Taper Dimensions

The result of the surface cooling rate and effective depth for the top quarter of the flat plate is shown in Figure 4. The edge of the plate has a faster cooling rate than the center or near the riser. This faster cooling rate makes the edge of the flat plate appear thin to the ED criterion. Conversely, the slow cooling rates near the riser on the flat plate make the ED appear thick. Only in the broad constant cooling rate region of the flat plate is the true thickness of the plate represented. This constant value of cooling rate at ~2.8 ℃/s corresponds to an ED of 11.8 mm which is in reasonable agreement with the actual half thickness of plate considering the distance from the surface to the center of the control volume. This region of constant cooling rate and ED indicates that this region has zero taper which prompts the need for material addition.

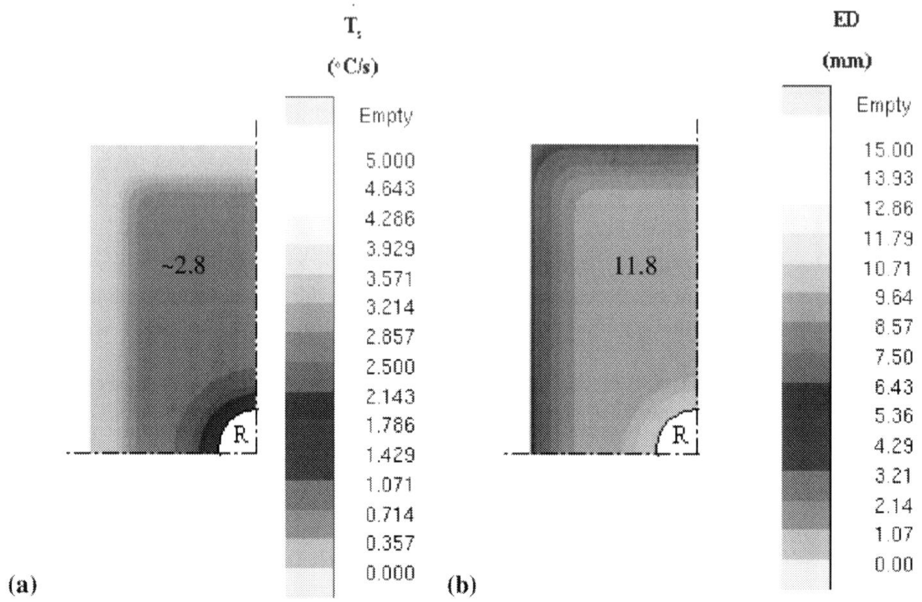

Figure 4. Flat plate **(a)** surface cooling rate and **(b)** effective depth

Figures 5(a) and (b) show the critical (<0.1) Niyama criterion region on the centerline and EDG on the surface of the flat plate, respectively. The EDG is calculated from the result shown in Figure 4(b) has been scaled between 0 and 0.03 to represent locations with less than 3% taper. In Figure 5, the large region without sufficient taper for directional solidification can be seen and correlates well with the extent of critical Niyama criterion shown. The taper is added over this region to eliminate the Niyama indications on the centerline. Figure 5(c) and (d) show the result from the taper addition described in Figure 3(b) of the critical (<0.1) Niyama criterion region and EDG on the surface, respectively. In this case, the critical region of Niyama is eliminated and the taper is above 0.03 everywhere in the EDG result. The taper result shows that the method is consistent with earlier work on feeding distance[7]. The utility of the method is evident on a complex part, where geometry is not uniform. In this case, the region that should be tapered and the approach to adding taper to the surface for promoting feeding are not intuitive. However, by using the EDG, one can identify improperly fed regions and determine appropriate locations and directions of taper to most effectively utilize risers and feed at all sections of a casting.

To further explore the utility of the ED and EDG method, a real axle housing example with dispersed porosity in an unrisered state is shown. The solidification of the axle housing was determined to be poor for castability due to the presence of a many isolated small hotspots. The riser requirement for feeding all of these hotspots sufficiently without manipulating the axle housing's geometry and to guarantee a sound casting would produce a low casting yield and makes the part expensive and requires a longer time to produce. This methodology is being applied to the axle housing to explore potential design changes in the geometry to create a more straightforward casting on which to apply risers. The geometry of the axle housing is shown in Figure 6. A general description of this geometry is a tube with holes and clevis connection points. There are broad regions of the geometry which are uniform in wall thickness contributing to potential unsound areas.

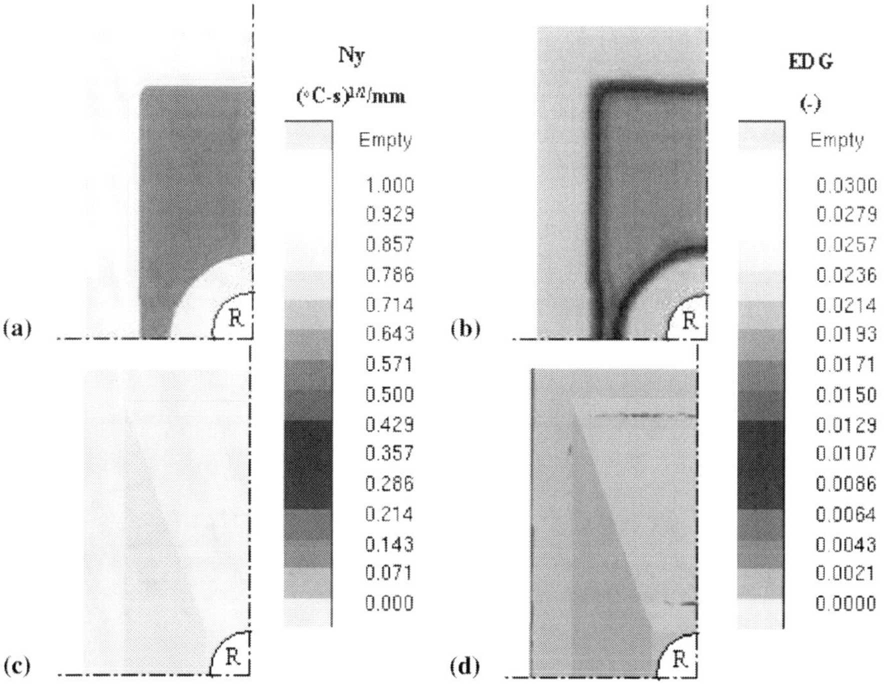

Figure 5. Flat plate **(a)** Critical Niyama criterion in the casting (Ny<0.1) and **(b)** EDG on the casting surface; Taper Plate **(c)** Critical Niyama criterion in the casting (Ny<0.1) and **(d)** EDG on the casting surface

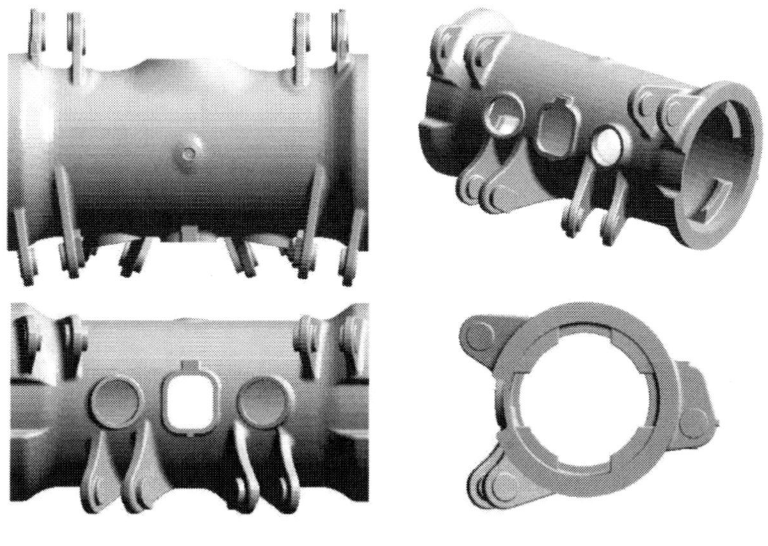

Figure 6. Axle Housing Geometry

Using the method in consideration, the cooling rate, Niyama criterion, ED, and EDG are calculated and shown in Figure 7. Since the Niyama criterion profile of the axle housing (Figure 7c) is so pervasive throughout its geometry, analyzing this criterion by itself would conclude little about how to effectively eliminate the critical Niyama criterion regions. However, Figure 7d shows the uniform walls with a resultant low magnitude taper from the EDG criterion that gives direction to a taper addition. These regions should be tapered from the thick areas to the thin sections, as dictated from the EDG figure rather than by geometric thicknesses, to provide the best geometry for directional solidification. The low taper EDG areas also correlate well to the Niyama criterion pattern at the casting centerline which supports the correlation found in the flat plate example. Once sufficient taper is added to the geometry, these deficiencies will be removed and the geometry castability will be improved.

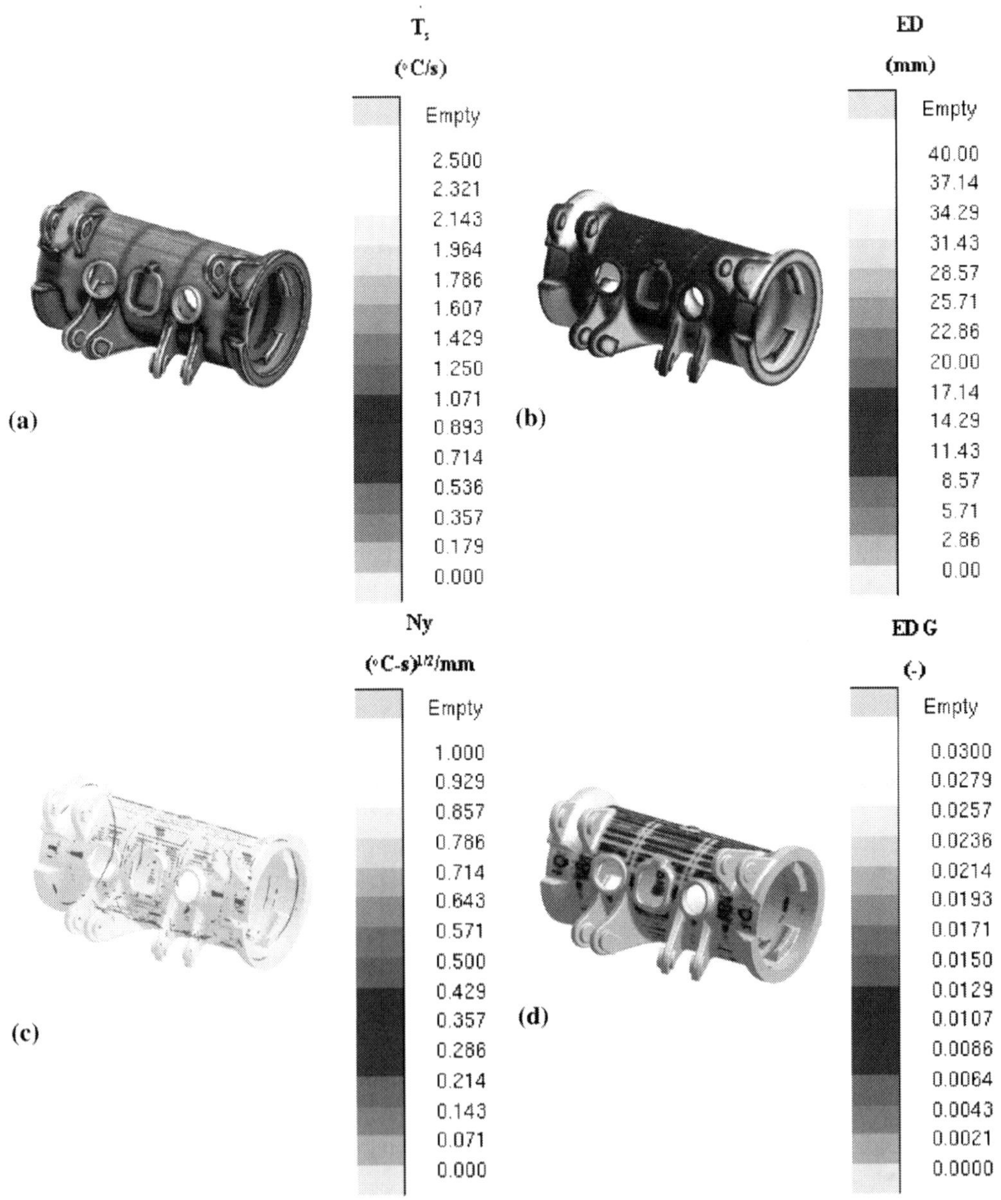

Figure 7. Axle Housing **(a)** surface cooling rate **(b)** effective depth **(c)** (Ny<0.1) and **(d)** EDG

4. Summary

Niyama criterion based taper optimizations of the surface geometry of a casting can be performed if a design space is identified that affects critical Niyama criterion regions. The effective depth gradient calculation based on the casting surface cooling rates at the Niyama criterion evaluation temperature can be used to identify regions of a three-dimensional casting that needs taper. The effective depth correlates to the half thickness of casting sections. The effective depth gradient based on the effective depth is a reasonable basis for identifying poorly tapered regions of the casting geometry. This effective depth gradient can serve as a constraint on an optimization problem to determine the minimum taper addition for improved Niyama criterion results. From this identified taper, the cost/benefit tradeoff from adding material and potential machining versus improved soundness can be determined.

Acknowledgements with Disclaimer

Research under American Foundry Society (AFS) Sub-Contract # AFS-ALCS2010-06 was sponsored by US Army ARDEC - Benet Laboratories and managed by AFS on behalf of the US Army Contracting Command Joint Munitions & Lethality contracting center and was accomplished under cooperative agreement number W15QKN-11-2-0001. The views and conclusions contained in this document are those of the authors and should not be interpreted as representing the official policies, either expressed or implied, of AFS, US Army ARDEC- Benet Laboratories or the U.S. Government. US Army ARDEC-Benet Labs, the U.S. Government or AFS is authorized to reproduce and distribute reprints for government purposes notwithstanding any copyright notation heron.

References

[1] Tavakoli R and Davami P 2009 *Structural and Multidisciplinary Optimization* **39** pp 519-530
[2] Morthland T, Byrne P, Tortorelli D, and Dantzig J 1995 *Metall Mater Trans B* **26**, pp 871-885
[3] Niyama E, Uchida T, Morikawa M, and Saito S 1982 *AFS Cast Met. Res. J.* **7**, pp 52-63
[4] Carlson K D and Beckermann C 2009 *Metall. Mater. Trans.A* **40A**, pp 163-175
[5] Carlson K D, Ou S, Hardin R, and Beckermann C 2002 *Metall. Mater. Trans. B* **33B** pp 731-740
[6] Wlodawer, R 1966 *Directional Solidification of Steel Castings* Pergamon Press
[7] Ou S, Carlson K D, Hardin R, and Beckermann C 2002 *Metall. Mater. Trans. B* **33B** pp 741-755
[8] MAGMASOFT v4.4, MAGMA GmbH, Aachen, Germany

IOP Conf. Series: Materials Science and Engineering **33** (2012) 012005 doi:10.1088/1757-899X/33/1/012005

Modeling and Optimizing Ti-6Al-4V Ingot Production

R Shuster, C Reilly, D M Maijer and S L Cockcroft

Department of Materials Engineering, University of British Columbia, Vancouver, BC, Canada V6T 1Z4

Email: carl.reilly@ubc.ca

Abstract. Control of chemistry and shrinkage void in the final stages of the consolidation processes employed to produce Ti alloy ingots is critical from the standpoint of productivity as there is a direct correlation to the amount of material that must be removed prior to further downstream processing. The application of power to the top surface during this stage allows the raising of the depth of shrinkage voids; however, it can also cause excessive evaporation of volatile elements within the alloy. The balancing of these two factors represents a classic optimization problem. A mathematical model describing the final stage of a commercial consolidation process has been developed to assist in the optimization of the process. The model solves the coupled thermal-fluid flow problem including solute conservation and evaporation. Experimental measurements consisting of the sump depth, pool profile marking, and local composition analysis have been used to validate the model predictions under various casting conditions.

1. Introduction

Titanium and its alloys possess a number of characteristics that make them suitable for a broad range of applications including structural, chemical and medical. Commercial extraction processes produce a porous metal in the solid state, referred to as sponge that requires consolidation prior to further downstream processing of the metal.

One of the commercial consolidation processes used for titanium is the Electron Beam Cold Hearth Re-melting (EBCHR), the subject of this paper. A schematic of an EBCHR furnace is shown in Figure 1. EBCHR furnaces operate under a vacuum in the range of 0.5 Pa and use high-powered electron beam guns to provide heat. The process uses a series of water-cooled copper hearths to first melt and then refine the metal prior to casting. The molten metal is held within a solid shell that forms against the copper producing an autogenous lining, which is necessary as titanium is highly reactive in the liquid state. In the casting process, the molten liquid flows from the refining hearth via the pour extension into the ingot mold, which is also water-cooled copper. To start the process a "starter block" is placed within the mold. Once the metal level reaches a specified height within the mold the starter block drops downward and the ingot is withdrawn at a rate that coincides with the flow of molten metal entering the mold. Electron beams scan the surface of the melting hearth, refining hearth and mold to keep the material molten and to control the flow. A variety of hearth and mold configurations are used commercially to produce ingot for various applications.

One of the advantages of the EBCHR process is that the refining hearth is spatially and temporally decoupled from the casting process. The temporal decoupling allows the residence time of liquid within the refining hearth to be set independent of the solidification time within the mold allowing Low-Density Inclusions to be removed (long liquid metal residence times). The spatial decoupling enables High-Density Inclusions to settle to the bottom of the refining hearth prior to entering the mold. Thus the process is suitable for handling both virgin sponge and revert feedstock. Consequently, the EBCHR process has emerged as a key technology in the production of premium and rotating grade material for aerospace applications.

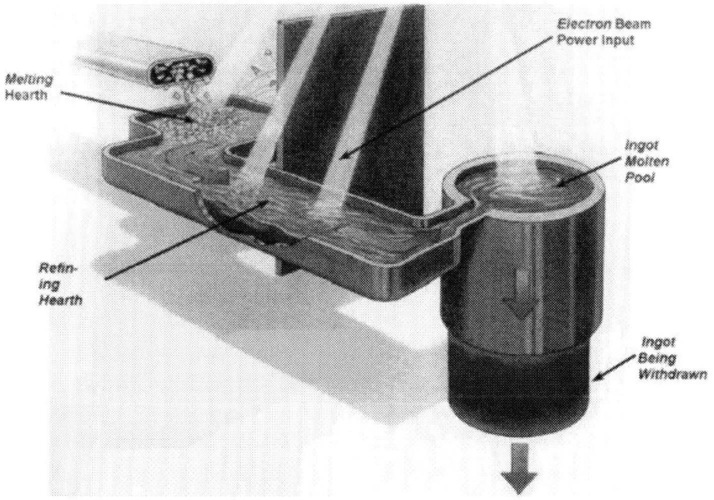

Figure 1 – Schematic of the EBCHM casting process

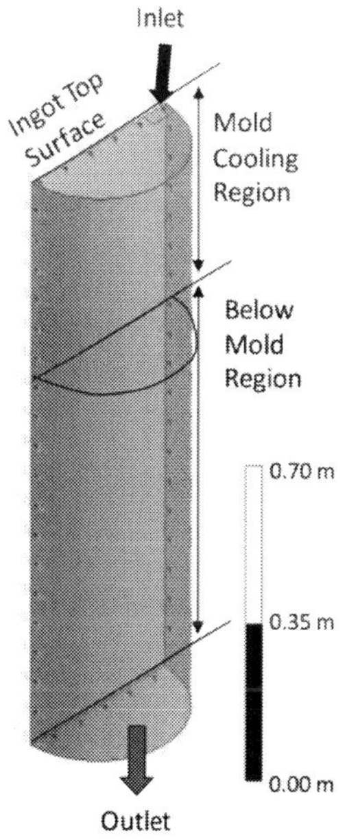

Figure 2 – Analysis domain schematic

Unfortunately, while the features described above are important from a defect removal standpoint, the large surface area in the refining hearth exposed to vacuum also promotes significant loss of volatile alloy constituents, such as aluminum and chromium. Likewise, the free surface of the ingot, which is also kept hot with an electron gun, adds to this problem. Both these losses can be effectively compensated for during steady state casting, by an appropriate increase in the volatile alloy content of the charge material. At the end of the casting process, however; liquid metal flow to the mold is stopped and depending on the hot-topping practice alloy chemistry control can become a challenge.

One approach is to power off. In the absence of additional heat input, the top surface freezes quickly limiting volatile alloy loss. Unfortunately, this approach encapsulates liquid below the surface leading to large shrinkage voids. In the case of commercially pure (CP) titanium, the ingot can be further processed (deformed) without concern as the voids can be effectively welded and removed during rolling or forging. For Ti-6Al-4V, which is an α/β alloy, this is not the case as the inside surfaces of the void are enriched in Al, which is an α-phase stabilizer, and therefore the end of the ingot must be cropped to remove all void prior to further processing, representing a loss in productivity. This leads to an optimization problem between void formation (reduced by hot-topping) and volatile loss of an alloy constituent (increased by hot-topping).

Process modeling has the potential to allow an optimal solution to be achieved without the need for numerous iterations involving costly and time-consuming trials. In the case of the EBCHR process, thus far the refining hearth has been the focus of the majority of process model development aimed at understanding and predicting inclusion removal. A three-dimensional model to predict the depth of liquid metal in the cold hearth has previously been developed [1]. Additional work has involved the application of the Langmuir equation to predicted volatile constituent losses during electron beam melting and refining [2-5]. J. Bellot *et al.* [6] developed a three-dimensional thermophysical model of the refining hearth to predict aluminum volatilization and inclusion removal. In terms of the casting process associated with the EBCHR process, Shyy *et al.* [7-8] have conducted two mathematically based studies in which they examined the effect of casting speed and gravity on pool depth and shape in EB cast Ti-6Al-4V round (electrode) ingot during steady state operation. The work focused on the numerical aspects of the problem; relatively simple process boundary conditions were adopted and there was no direct comparison between the predicted pool depth and measurements from an actual ingot.

Given the limited amount of past model development and the need to optimize the last stages of the casting process in terms of the balance between void formation and alloy loss there is a need for the development a casting process model that describes the thermal-fluid-compositional aspects of the process.

2. Mathematical Model

A mathematical model of the EBCHR casting process associated with casting Ti-6Al-4V alloy was developed in ANSYS CFX V12.0. The model is based upon the governing equations for the conservation of mass, momentum and energy [9] and includes the important phenomena such as thermal and compositional buoyancy, evaporation of aluminum, an industrial EB power boundary condition and resistance to flow in the mushy zone (Darcy flow). The calculation domain used for the 3-D analysis is shown in Figure 2, and encompasses a ½ section of the top portion of a round ingot assuming a vertical symmetry plane bisecting the metal inlet. The energy, momentum and mass input to the top surface are all assumed symmetric. A Eularian frame of reference is used for the analysis in which the domain is held fixed relative to the mold and the cast material is advected through it. The length of the domain was selected to be approximately double the steady state casting sump depth. The mesh consists of two types of elements: eight-noded isoparametric brick elements and six-noded isoparametric prism elements.

Buoyancy was implemented by adding a source term in the momentum equation (Equation 1) where ρ_{adj} is the temperature and composition dependent density and ρ_{ref} is the reference density (density at Solidus temperature).

$$S_{M,buoy} = (\rho_{adj} - \rho_{ref})g \tag{1}$$

Flow attenuation in the mushy zone is modeled by including a momentum source term using Darcy's Law (Equation 2); where μ is the viscosity, u is the velocity and K is the permeability as defined by the Carman-Kozeny equation (Equation 3), where f_s is the fraction solid, s_0 is the solid/liquid interface area per unit volume of solid, and k is the Kozeny constant. The fraction solid is interpolated from published data [10].

$$S_{M,Darcy} = -\frac{\mu}{K}u \tag{2}$$

$$K = \frac{(1-f_s)}{ks_0^2 f_s^2} \tag{3}$$

2.1. Material Properties

The material properties of Ti-6Al-4V used were based on the data published by Mills [10]. It proved necessary to extrapolate this data to be consistent with the liquidus (1625 °C) and solidus (1595 °C) temperatures established by Shy [7-8] and scale this data within the S/L phase change range using fraction solid evolution data from Pandat (product of *Computherm LLC*). The density below the solidus temperature is held constant at 4198 $kg \cdot m^{-3}$ to maintain mass conservation in the solid zone. Above the solidus temperature, the density is defined to be both temperature and aluminum concentration dependent. The temperature dependent density is taken from published data [10] and the variation due to concentration is determined based upon a rule of mixtures.

The release of latent heat during solidification and the β to α phase transformation are included in the model using the effective specific heat method where the specific heat (C_p) is modified based on the evolution of Fraction Liquid (f_L) and Fraction of β phase (f_β) (Equation 4).

$$C_{peff} = C_p - Ls_{/L}\frac{df_L}{dT} - L\alpha_{/\beta}\frac{df_\beta}{dT} \quad (4)$$

2.2. Boundary Conditions

In the EBCHR process for round ingot production, liquid titanium from the cold hearth enters a water-cooled copper mold at the inlet, Figure 1. The the surfaces of the computation domain can be divided into five separate boundary conditions regions: 1) the liquid inlet located on the top surface of the domain/ingot; 2) the remainder of the top surface, which is subject to EB heating and radiation losses; 3) the side surface of the domain/ingot within the mold; 4) the side surface of the domain/ingot below the mold; and 5) the bottom face of the domain or outlet.

The liquid inlet is incorporated into the model as an 'inlet face' located on the top surface as shown in Figure 2. The inlet face is assumed to be a small rectangle (slightly curved on the face shared with the ingot side) with an area of 58 cm^2. It is likely the pour stream retains some horizontal momentum as the vertical distance between the pour lip and the metal surface in the mold is only a few millimeters (also evident in images taken of the top surface). To include the horizontal momentum of the pour stream, the direction of the inlet flow steam is set to 45° from vertical. Due to the commercial sensitivity of the data, the flow rate and velocities of liquid Ti-6Al-4V through the inlet are not disclosed. The inlet metal temperature is assumed to be at 2121.8 K (~130 K superheat).

The top surface of the domain/ingot, apart from the inlet area described above, is treated as a free-slip wall in the model where the velocity of the liquid metal is only suppressed normal to the surface. Strictly speaking, this is a free surface and may in practice fluctuate vertically by a small amount associated with variability in the melt rate and the addition of feedstock upstream.

Heat is input to the top surface of the domain/ingot via one or more electron beams. At the metal surface, the kinetic energy of the electron beam is converted to thermal energy, with losses due to backscattering of electrons [11]. The beam or beams are programmed to scan a prescribed pattern in which the beam spot pauses at a series of N points on the surface of the ingot for a prescribed time (dwell time). The sum of the dwell time for all points in the pattern yields the time it takes for the pattern to repeat.

Denoting the time for a complete pattern as t_N and the dwell time at the ith impingement point as, t_i the power fraction for the ith point can be calculated as:

$$r_i = \frac{t_i}{t_N} \quad (5)$$

The electron beam is assumed to impart thermal energy in the form of a Gaussian distribution. Thus, the heat flux, q_i ($W{\cdot}m^{-2}$) from the ith point in the beam pattern on the ingot top surface is given in Equation 6, where P_{EB} (W) is the gun power, f_{abs} is the fraction of the electron beam power converted to thermal energy, s is the standard deviation of the Gaussian distribution, and L_i is the distance from point i. To describe the heat flux at any point on the ingot surface (x, y), L_i is replaced with the following equation, which includes the ith impingement point location (x_i, y_i).

$$q_i = f_{abs}P_{EB}r_i\frac{1}{S\sqrt{2\pi}}exp\left[-\frac{L_i^2}{2S^2}\right] \quad (6)$$

$$L_i = \sqrt{(x-x_i)^2 + (y-y_i)^2} \quad (7)$$

Additionally, the absorption factor, f_{abs} for Ti-6Al-4V has been set equal to 0.7. The total power of the electron beam (P_{EB}), assuming no losses, has been set to 600 kW.

For each location on the ingot top surface, the time-average heat flux q_{EB} ($W \cdot m^{-2}$) is calculated by summing the contributions of the N points in the repeating pattern using the following expression:

$$Q_{EB} = f_{abs} P_{EB} \frac{1}{S\sqrt{2\pi}} \sum_{i=1}^{N} \frac{t_i}{t_N} exp\left[-\frac{(x-x_i)^2+(y-y_i)^2}{2S^2}\right] \quad (8)$$

In industrial practice, the electron beam pattern is developed by overlapping a series of paths with standard shapes such as ellipses, circles, or straight lines. The specific pattern used by each facility is developed and adapted through trial-and-error and is highly proprietary.

The heat loss caused by aluminum evaporation is calculated via Equation 9; where $\dot{m}$ is the mass flux of evaporated aluminum and L_{evap} is the latent heat of vaporization. At this stage, segregation during solidification has not been included in the model.

$$q_{evap} = \dot{m} L_{evap} \quad (9)$$

Due to the vacuum environment, conduction and advection heat transfer from the top surface are neglected. Radiation is the primary mechanism for heat loss and is described by Equation 10 where ε is the emissivity of the top surface, σ is the Stefan-Boltzmann constant, T is top surface temperature and T_∞ is wall temperature of the water-cooled EB furnace chamber (Assumed to be 10 °C).

$$q_{rad} = \varepsilon\sigma(T^4 - T_\infty^4) \quad (10)$$

The mass flux $\dot{m}$ of aluminum evaporation into the vacuum is calculated by the Langmuir equation (Equation 11) where γ_i is the activity coefficient of the specified element i, M_i is Molar mass, P^0_i is the vapor pressure of element i at absolute temperature T and X_i is the mole fraction.

$$\dot{m} = \frac{\gamma_i M_i P_i^0 X_i}{\sqrt{2\pi M_i R T}} \quad (11)$$

The side surface of the ingot is in direct contact with the copper mold. A heat transfer coefficient h is used to model the heat extraction to the mold wall where T is the temperature of sample and T_{mold} is the temperature of the mold. At high ingot surface temperatures, a high heat transfer coefficient is assumed to account for liquid metal contact with the model. As the surface of the ingot solidifies, a gap forms between the ingot and mold, due to volumetric contraction. The heat transfer coefficient was ramped linearly with temperature to a value consistent with radiation heat loss at the ramp-end temperature. The parameters for this boundary condition were obtained by fitting the pool profile to experimental results. Free-slip wall and zero mass flux boundary conditions are applied on the side surface. For the bottom surface, it is assumed to be radiation heat loss only as conduction and advection are be negligible due to the vacuum environment. Here, the mold far field temperature is assumed to be 10 °C.

$$q = h(T - T_{mold}) \quad (12)$$

3. Experimental Methodology

In order to validate the mathematical model three 17" diameter Ti-6Al-4V ingots were cast with off-spec feedstock. In order to produce a variety of results for validation, electron beam power and hot top duration were varied while the remaining production parameters were held constant. Experiments 1 to 3 are respectively; baseline EB power and baseline hot top time, baseline EB power and increased hot top time and reduced EB power and baseline hot top time.

Two strategies were used to characterize the depth and shape of the sump. The first method involved the addition of copper turnings into the liquid melt pool. Copper is soluble in liquid titanium, allowing chemical sampling upon solidification to reveal the shape of the sump at the time of addition. The second method used was the addition of tungsten markers into the liquid melt pool. Due to the limited solubility and high density relative to titanium, the tungsten markers sink to the bottom of the sump and become trapped in the mushy zone. Upon solidification the ingots were sectioned (sawn and machined) and etched using a solution of 0.5% HF and 5% lactic acid.

IOP Conf. Series: Materials Science and Engineering **33** (2012) 012005 doi:10.1088/1757-899X/33/1/012005

Once the melt pool had been identified a chemical sampling plan was developed. An example of the sampling plan can be seen in Figure 3. A series of locations were selected down the vertical centerline at 0.1m (4") intervals (Locations 1-15 in Figure 3). In addition one sample per slice was selected near the edge of the liquid pool (locations 16-22 in Figure 3) and finally two samples were chosen in areas, which were solid at the time of hot topping (locations 23 and 24 in Figure 3). At each sample location a core was taken and sent for chemical analysis.

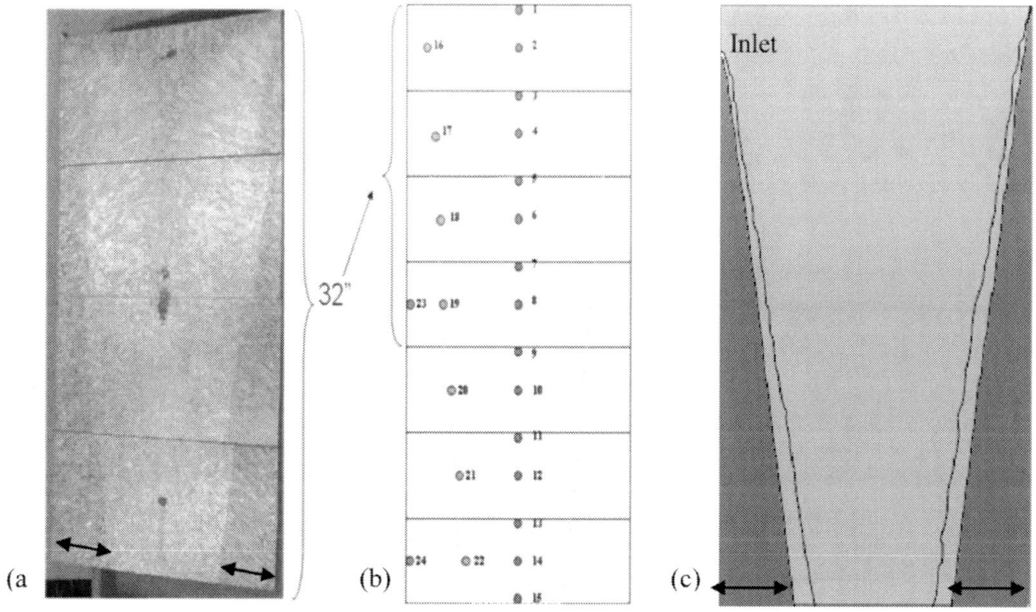

Figure 3 - (a) example experimental results, (b) chemical sampling strategy and (c) model prediction from Experiment 2, showing the top 0.81 m (32 inches) of the ingot. The arrows define the location of the sump width measurements. The lines on the right-hand-side image delineate the solidus and liquidus isotherms.

4. Results and Discussion

The experimental results shown in Figure 3 (a) clearly shows the location of the liquid pool at the end of steady state casting, which correlates well with that predicated by the model shown in Figure 3 (c). The sump profile can be seen to be nearly symmetric about the centre line of the ingot, apart from near the top due to the effect of the metal inlet (top left of image), which displaces the liquidus isotherms downward. In Experiment 2 (Figure 3) the distance between the outside of the ingot and the apparent liquid pool boundary was measured at 0.81 m (32 inches) from the top of the ingot. The results were that the solid shell was approximately 10.3 cm thick on the pour side and 10.8 cm on the opposite side. These compare with 10.8 and 11.2 cm predicted by the model at the same locations. The tungsten markers were not located in any of the slices taken of the ingot. Other additions made to the ingots are also visible (darkened areas near the centreline) but were part of another project and are not considered in this work.

Figure 4 (a) shows that the temperature at the end of steady state casting ranges from approximately 2200 to 850 K and that velocities are in the range of 0 to 9 mm s^{-1}. Figure 4 (b) through d) show the liquidus profiles at different times from the beginning of the hot-top, for the three experimental conditions. The location of the last liquid (final void depth) can be seen to be significantly higher for the experiment with the longer hot top duration (Figure 4 (c), Experiment 2) compared with the base-case conditions. It can also be seen that reducing the EB power during hot topping had a negligible impact on the location of the last liquid (Figure 4 (d), Experiment 3) compared to the base-case condition.

IOP Conf. Series: Materials Science and Engineering **33** (2012) 012005 doi:10.1088/1757-899X/33/1/012005

The compositional analysis can be seen in Figure 5. Two regimes of behavior can be seen: the first, is a region of approximately constant alloy chemistry representing steady state casting as indicated by the arrow in the figure; and the second, a region near the top of the ingot representing the evolution of chemistry within the sump during hot-topping. *Note: in the first regime there is some difference in alloy chemistry between the experiments due to use of off-spec feedstock material, which is addressed in the model by adjusting the initial alloy composition.*

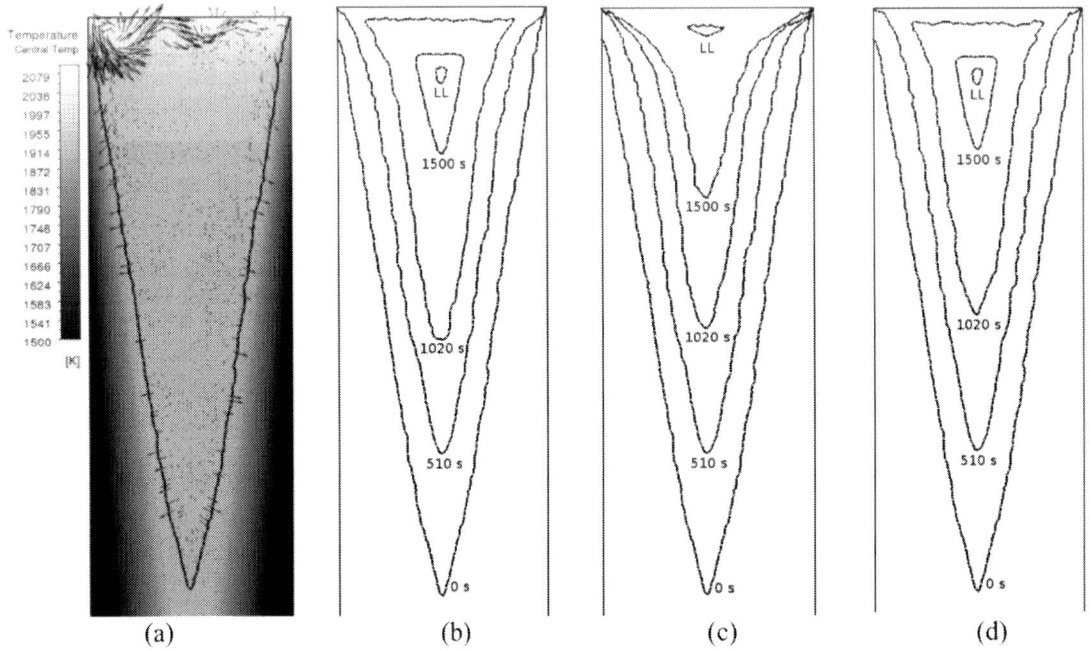

Figure 4 - Model results; (a) temperature contour plot with velocity vectors at the end of steady state casting, including liquidus line. Images (b) through (d) depict the liquidus line location at the times indicated (*0 represents the beginning of the hot top*) for experimental conditions 1 through 3, respectively. LL delineates the last liquid to solidify.

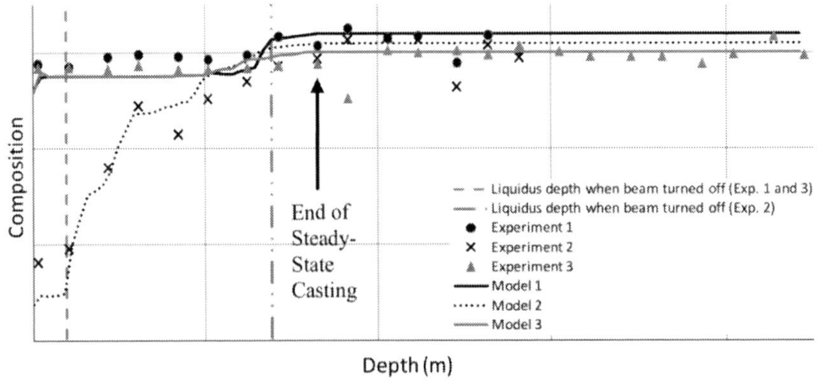

Figure 5 – Compositional analysis results with respect to aluminum for the three experiments measured along the ingot centerline.

Figure 5 shows the variation in composition along the ingot centerline with depth. The plot shows good correlation between the experimental and modeling results. It can be seen that there is a significant difference in the concentration profiles between the standard hot top time (Experiments 1 and 3) and the extended hot top case (Experiment 2). In the standard hot top time castings (Experiments 1 and 3) there is a relatively small but uniform drop in alloy content seen in the material that solidified after the end of steady state casting. In contrast there is a more significant decrease in alloy content after the end of steady state casting in the extended hot top casting (Experiment 2).

5. Conclusions

A computational thermal-fluid-compositional model of the EBCHR casting process has been developed for alloy Ti-6Al-4V. The model has been validated against an experimental sump profile and chemical sampling obtained from a commercially cast off-spec 17 inch diameter ingot. The experimental and modelling results show good correlation. The results show a pronounced difference in concentration profile between the standard and extended hot top cases due to the greater time for evaporation to occur and gradual reduction in sump depth with time. A negligible difference was seen with reduction in hot top power.

The sump depth has been shown to be greatly affected by the hot top time, whereas the EB power was seen to have negligible effect on the sump depth under the range of conditions examined.

References

[1]. Tripp, D.W., A. Mitchell, and M. Haydon. in *Proc. EBMR-State of the Art*. 1986. Reno.
[2]. Akhonin, S., et al., 2003 Mathematical modeling of aluminum evaporation during electron-beam cold-hearth melting of Ti-6Al-4V ingots. *Metallurgical and Materials Transactions B*. **34**(4): p. 447-454.
[3]. Ivanchenko, V., O. Ivasishin, and S. Semiatin, 2003 Evaluation of evaporation losses during electron-beam melting of Ti-Al-V alloys. *Metallurgical and Materials Transactions B*. **34**(6): p. 911-915.
[4]. Powell, A., et al., 1997 Analysis of multicomponent evaporation in electron beam melting and refining of titanium alloys. *Metallurgical and Materials Transactions B*. **28**(6): p. 1227-1239.
[5]. Semiatin, S., V. Ivanchenko, and O. Ivasishin, 2004 Diffusion models for evaporation losses during electron-beam melting of alpha/beta-titanium alloys. *Metallurgical and Materials Transactions B*. **35**(2): p. 235-245.
[6]. Bellot, J., D. Ablitzer, and E. Hess, 2000 Aluminum volatilization and inclusion removal in the electron beam cold hearth melting of Ti alloys. *Metallurgical and Materials Transactions B*. **31**(4): p. 845-854.
[7]. Shyy, W., et al., 1992 Modeling of turbulent transport and solidification during continous ingot casting. *International journal of heat and mass transfer*. **35**(5): p. 1229-1245.
[8]. Shyy, W., et al., 1993 Effect of turbulent heat transfer on continuous ingot solidification. *Journal of Engineering Materials and Technology, Transactions of the ASME*. **115**(Compendex): p. 8-16.
[9]. CFX, A., *ANSYS CFX V12.0 User manual*. 2011.
[10]. Mills, K.C., *Recommended values of thermophysical properties for selected commercial alloys*. 2002, Cambridge: Woodhead publishing Ltd.
[11]. Shiller, S., U. Heisig, and S. Panzer, *Electron Beam Technology*. 1982, New York: John Wiley & Sons.

IOP Conf. Series: Materials Science and Engineering **33** (2012) 012006 doi:10.1088/1757-899X/33/1/012006

Though process numerical simulation of a heavy hydro turbine blade casting

J W Kang, T J Wang, T Y Huang and B C Liu

Key Laboratory for Advanced Materials Processing Technology, Ministry of Education, Department of Mechanical Engineering, Tsinghua University, Beijing 100084, China

Email: kangjw@tsinghua.edu.cn, huangty@tsinghua.edu.cn, liubc@tsinghua.edu.cn

Abstract. Heavy Hydro turbine castings, made of martensitic stainless steel, of spatially twisted shape, are susceptible to deformation during both casting and heat treatment processes. During production, uneven heating and cooling, phase transformation, stress and stain are responsible for the deformation of castings. However, the current research most focus on casting process or heat treatment process or both but separately, martensitic phase transformation is usually neglected in casting process. In this paper, a coupled thermo-martensitic phase transformation-stress model is established. And it is implemented by secondary development of ABAQUS, which deals the finite element model change and incorporates multi thermal and mechanical boundaries during casting and heat treatment processes as well, such as the contact pair between casting and mold during casting process. The stress analysis of a heavy hydro blade casting was performed by this system. The stress, displacement and phase are obtained. The effect of martensitic phase transformation on the stress and deformation is discussed. The simulated deformation is compared with measured results, and their agreement is improved by the through process simulation.

1. Introduction

Deformation is a typical problem for large and thin sectioned castings during casting and heat treatment processes, for example, hydro turbine blade, which is of strong requirement in the hydro power plant construction in China, such as Three Gorges Project and other projects on the upper stream of the Yangtze River. The authors have performed research work in the numerical simulation of hydro turbine blade castings during casting and heat treatment process separately [1-4]. The hydro turbine blade casting is made of martensitic chromium-nickel steel, undergoes martensitic phase transformation during casting and as well as heat treatment. But, there is no report about the full coupling of heat transfer, phase transformation and stress evolution, and complicated thermal and mechanical boundaries between casting and sand mold during casting process. Martensitic phase transformation is neglected during casting process. Martensitic phase transformation during heat treatment process was considered with heat transfer and stress analysis for blade casting [4]. The results showed that martensitic phase transformation has significant influence of deformation. Therefore, it is necessary to consider phase transformation during casting process and perform through process analysis of casting and heat treatment of castings. And the initial stress and dimension of casting will contribute to stress and deformation during heat treatment; however, it is neglected in the previous work. That is another factor of this paper to do the through process analysis to realize the integration of casting and heat treatment process. Li performed the through process analysis of aluminum wheel casting from casting, heat treatment and machining to predict the final properties of the casting [5]. However, the heat treatment and phase transformation of aluminum alloy is different from that in steel castings. In this paper, a mathematical model of heat transfer-phase transformation-stress is established for the martensitic steel, and it is realized by the secondary development of ABAQUS. Then it is used to perform the analysis of a heavy hydro turbine blade casting during casting and heat treatment processes.

2. Constitution equations

During casting and heat treatment processes of castings, there is heating and cooling process accompanied with phase transformation and stress evolution. These three phenomena are closely related, as illustrated in Figure 1. The heating and cooling results in phase transformation and latent heat released during phase transformation contributes to temperature field. Because of uneven heating and cooling, stress evolves. For this Cr13%-Ni4% stainless steel, martensite transforms into austenite during heating and holding above solution temperature, while austenite transforms into martensite during normalizing process. The different specific volumes of phases also resulted in stress. And there is plasticity induced by the transformation from austenite to martensite.

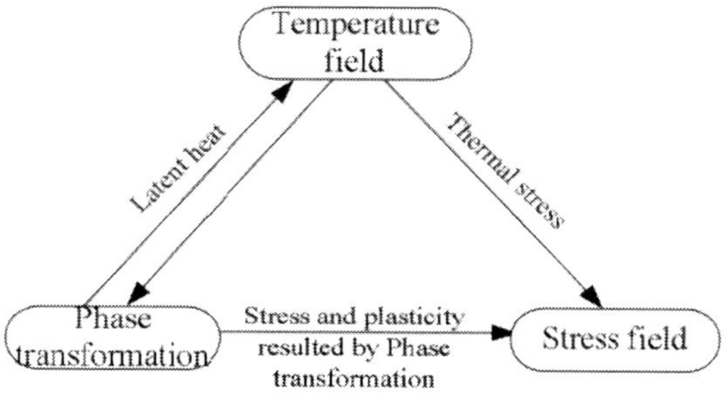

Figure 1. Schematic diagram of temperature, phase transformation and stress analysis.

The fraction of martensite during phase transformation from austenite to martensite, a kind of non diffusion controlled, is denoted by Koistinen-Marburger equation as (1)

$$\xi_{\mathrm{M}} = 1 - \exp(c(T - M_{\mathrm{s}})) \quad (1)$$

Where, M_s is the start temperature, c is the kinetic coefficient.

During heat treating, the total strain increment $d\varepsilon_{ij}$ is composed of elastic $\mathrm{d}\varepsilon_{ij}^{\mathrm{cp}}$, plastic $\mathrm{d}\varepsilon_{ij}^{\mathrm{th}}$, thermal $\mathrm{d}\varepsilon_{ij}^{\mathrm{th}}$, phase transformation $\mathrm{d}\varepsilon_{ij}^{\mathrm{tr}}$ which is resulted by the variation of specific volume of new and old phases and plasticity $\mathrm{d}\varepsilon_{ij}^{\mathrm{tp}}$ which is induced by phase transformation even though no yield occurs, as follows,

$$\mathrm{d}\{\varepsilon\} = \mathrm{d}\{\varepsilon^{\mathrm{e}}\} + \mathrm{d}\{\varepsilon^{\mathrm{cp}}\} + \mathrm{d}\{\varepsilon^{\mathrm{th}}\} + \mathrm{d}\{\varepsilon^{\mathrm{tr}}\} + \mathrm{d}\{\varepsilon^{\mathrm{tp}}\} \quad (2)$$

The elastic strain is determined by Hooke's law, as follows,

$$\mathrm{d}\{\varepsilon^{e}\} = [\overline{D^{e}}]^{-1}(T, \xi_{\mathrm{M}})\mathrm{d}\{\sigma\} \quad (3)$$

where $[\overline{D^{e}}]^{-1}(T, \xi_{\mathrm{M}})$ is the elastic modulus matrix, which are functions of temperature T and volume fraction of martensite phase ξ_{M} , $\mathrm{d}\{\sigma\}$ is the stress increment tensor.

The plastic strain is as follows,

$$d\{\varepsilon^{cp}\} = d\bar{\varepsilon}^{cp}\frac{\partial\bar{\sigma}}{\partial\{\sigma\}} \quad (4)$$

Where, $d\bar{\varepsilon}^{cp}$ is the equivalent plastic strain increment, $\bar{\sigma}$ is the equivalent stress.

The thermal strain is as follows,

$$d\{\varepsilon^{th}\} = \bar{\alpha}(T,\xi_M)dT \quad (5)$$

Where $\bar{\alpha}(T,\xi_M)$ is the thermal expansion coefficient calculated by mixture rule, a function of temperature and microstructure.

The strain resulted by phase transformation is

$$d\{\varepsilon^{tr}\} = \beta_M(T)d\xi_M \quad (6)$$

where $\beta_M(T)$ is the expansion coefficient of martensite phase, a function of temperature T.

The plasticity of phase transformation is determined by Greenwood-Johnson model,

$$d\{\varepsilon^{tp}\} = 3K\{s\}(1-\xi_M)d\xi_M \quad (7)$$

Where K is the coefficient of martensitic phase transformation, $\{s\}$ is the stress deviation.

During tempering process, only thermal stress is considered for there is no phase transformation.

Substitute equations (3)~(7) into (2), and by the stress-strain relationship get the following equation,

$$d\{\sigma\} = [\bar{D}^e]\left[d\{\varepsilon\} - \bar{\alpha}(T,\xi_M)dT - (\beta_M(T) + 3K\{s\}(1-\xi_M))d\xi_M - \frac{\partial\bar{\sigma}}{\partial\{\sigma\}}d\bar{\varepsilon}^{cp}\right] \quad (8)$$

Considering the Von Mises yield criterion with the isotropic hardening,

$$f(\bar{\sigma},\bar{\varepsilon}^{cp}) = \bar{\sigma} - H(\int d\bar{\varepsilon}^{cp}) = 0 \quad (9)$$

Where, H is the hardening coefficient.

As the plastic loading, stress is exerted on yield surface, i.e., $df(\bar{\sigma},\bar{\varepsilon}^{cp}) = 0$, then,

$$\left\{\frac{\partial\bar{\sigma}}{\partial\{\sigma\}}\right\}^T d\{\sigma\} - H' d\bar{\varepsilon}^{cp} = 0 \quad (10)$$

Substitute equation (8) into (10),

$$d\bar{\varepsilon}^{cp} = \frac{\left\{\frac{\partial\bar{\sigma}}{\partial\{\sigma\}}\right\}^T [\bar{D}^e]}{H' + \left\{\frac{\partial\bar{\sigma}}{\partial\{\sigma\}}\right\}^T [\bar{D}^e]\frac{\partial\bar{\sigma}}{\partial\{\sigma\}}}\left[d\{\varepsilon\} - \bar{\alpha}(T,\xi_M)dT - (\beta_M(T) + 3K\{s\}(1-\xi_M))d\xi_M\right] \quad (11)$$

Substitute equation (11) into (8),

$$d\{\sigma\} = \left[[\bar{D}^e] - \frac{[\bar{D}^e]\frac{\partial\bar{\sigma}}{\partial\{\sigma\}}\left\{\frac{\partial\bar{\sigma}}{\partial\{\sigma\}}\right\}^T [\bar{D}^e]}{H' + \left\{\frac{\partial\bar{\sigma}}{\partial\{\sigma\}}\right\}^T [\bar{D}^e]\frac{\partial\bar{\sigma}}{\partial\{\sigma\}}}\right]\left[d\{\varepsilon\} - \bar{\alpha}(T,\xi_M)dT - (\beta_M(T) + 3K\{s\}(1-\xi_M))d\xi_M\right] \quad (12)$$

Then, the stress- strain relationship with consideration of martensitic phase transformation is

$$\mathrm{d}\{\sigma\}=[\bar{D}^{ep}]\left[\mathrm{d}\{\varepsilon\}-\alpha(T,\xi_{\mathrm{M}})\mathrm{d}T-(\beta_{\mathrm{M}}(T)+3K\{s\}(1-\xi_{\mathrm{M}}))\mathrm{d}\xi_{\mathrm{M}}\right] \tag{13}$$

where, $[\bar{D}^{ep}]$ is the equivalent elasto-plastic matrix, which is related to temperature, fraction of martensite and stress level, is denoted as

$$[\bar{D}^{ep}]=\left[[\bar{D}^{e}]-\frac{[\bar{D}^{e}]\dfrac{\partial\bar{\sigma}}{\partial\{\sigma\}}\left\{\dfrac{\partial\bar{\sigma}}{\partial\{\sigma\}}\right\}^{T}[\bar{D}^{e}]}{H'+\left\{\dfrac{\partial\bar{\sigma}}{\partial\{\sigma\}}\right\}^{T}[\bar{D}^{e}]\dfrac{\partial\bar{\sigma}}{\partial\{\sigma\}}}\right] \tag{14}$$

The above stress-strain model is established by the secondary development the commercial code ABAQUS, and a simulation platform is established for the through process analysis for castings coupling the casting/mold thermal and boundary conditions and the shape variation during the processes. The flowchart of the platform is as shown in Figure 2. During the through process analysis, finite element models and boundary conditions have to be updated. During casting process, both casting and sand mold are considered, and their thermal and mechanical boundaries are disposed. During shakeout, only the casting with riser is considered. And casting without riser is considered and the constraint is by fixture during heat treatment.

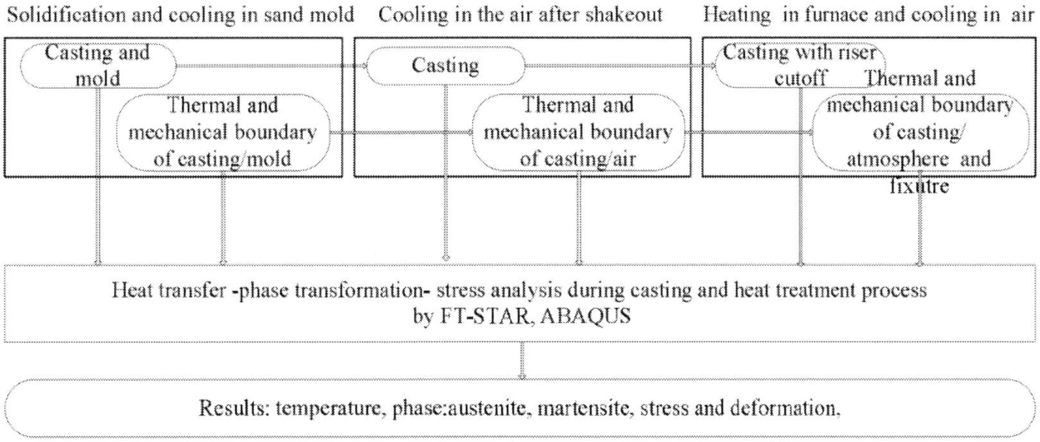

Figure 2. Coupled thermal-phase transformation-stress for casting and heat treatment process.

3. Numerical simulation

The hydro turbine blade casting for Three Gorges Project is analyzed by the above system. It exhibits x shape, with net weight 16.5t and dimension 4687mm×4268mm×1753mm. It undergoes casting in sand mold and shakeout and then cutoff of risers and pads, heat treatment including normalizing and twice tempering. During casting process the casting and sand mold is considered by contact element. During normalizing process the casting is cooled by forced air flow. The blade is placed upside down during heat treating with respect to the position in casting process. The area in contact with the fixture is fixed, serving as constraints. The finite element models are shown in Figure 3.

Here M_s is 276℃, c=0.014 and K= 8.74×10^{-5} MPa^{-1} , all are determined by experiments[6].The mechanical properties of this martensitic stainless steel such as elastic modulus, yield stress, hardening coefficient and thermal expansion for austenite and martensite are shown in Figure 4 as were determined by experiments [6]. Mixture rule is used to calculate the mechanical properties as both phases appear during phase transformation.

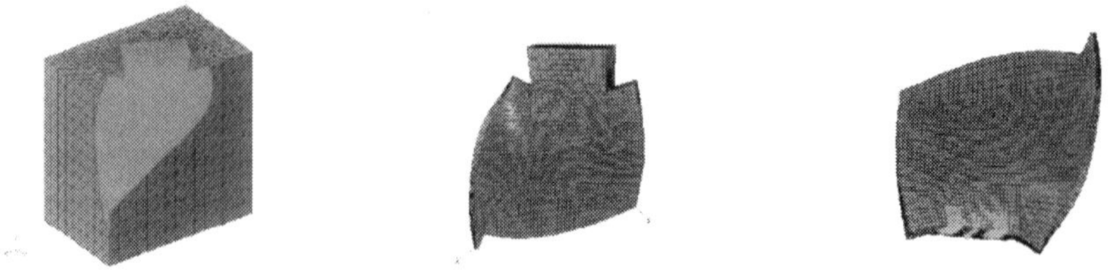

a) b) c)

Figure 3. The finite element model of the hydro turbine blade and mold. a) casting and mold during casting process, b)casting, c) casting with riser cutoff for heat treatment process.

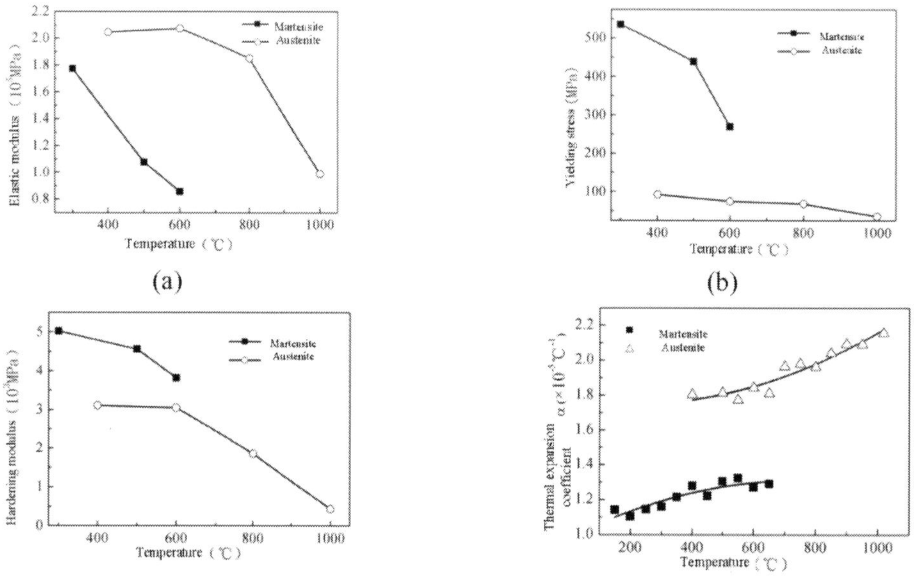

(a) (b)

(c) (d)

Figure. 4 Thermal mechanical properties of the Ni-Cr martensitic stainless steel. (a) Elastic modulus, (b) Yielding stress, (c) Hardening coefficient, (d) Thermal expansion coefficient.

3.1. Results and analysis

The temperature fields, phase transformation and stress and displacement after 50-hour cooling in sand mold are shown in Figure 5. It can be seen that martensitic transformation just begins at the two corners and the maximum displacement occurs at the same corners, and the bottom edge is of bigger stress state.

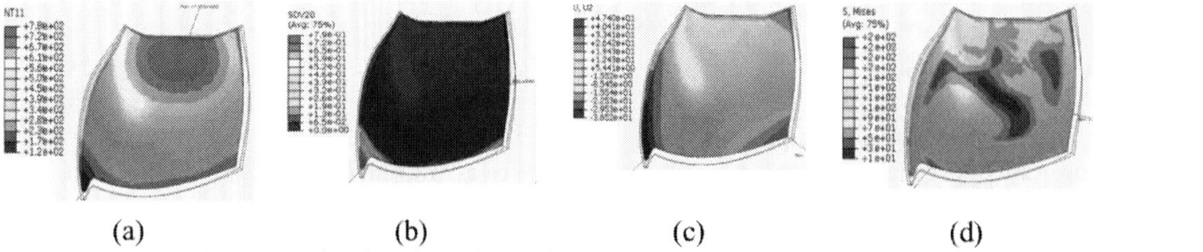

(a) (b) (c) (d)

Figure 5. Simulated results as the casting cools in sand mold 50 hours.
(a) temperature (b) martensite fraction (c) y displacement (d) stress

Temperature, phase transformation, stress and displacement of corner A and B (as shown in Figure 3b) are shown in Figure 6. It can be seen that the cooling of the turbine blade during normalizing and tempering is faster than that of casting. During casting process there is also martensitic phase transformation as it cools below 276℃, so the as cast microstructure is martensite. During normalizing martensite changes back into austenite during heating and austenite is kept until it is cooled below 276℃. After normalizing, martensite is kept during the twice tempering with no phase transformation occurrence.

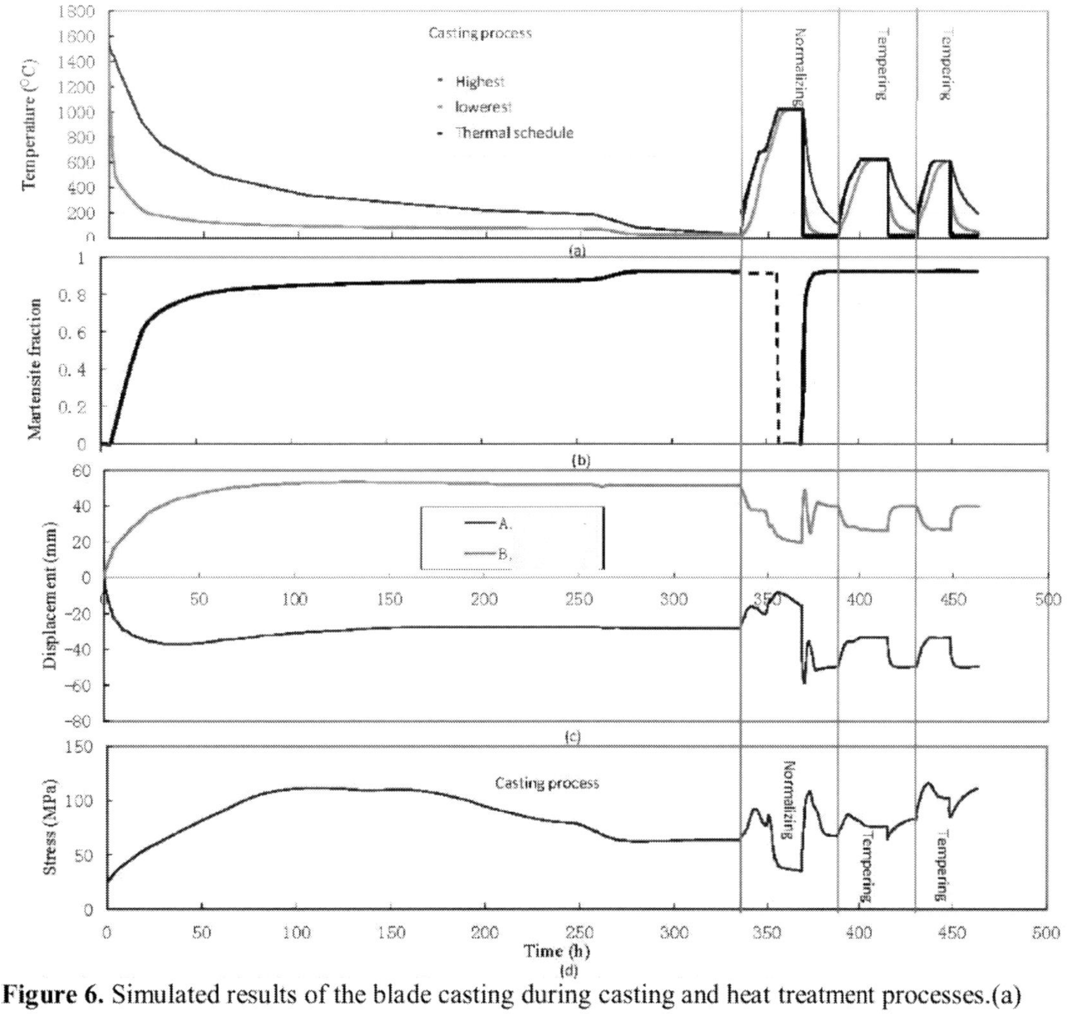

Figure 6. Simulated results of the blade casting during casting and heat treatment processes.(a) Cooling and heating curves of corner A (b) Martensite fraction (c) Displacement variation with time of corners A and B (d) Stress

The maximum deformation happens at the two corners A and B which are of the biggest curvature and thinnest. The displacement directions of corners A and B are opposite because of their initial bending in opposite direction; generally, the final deformation tends to make the blade flatter. Corner A deforms 28mm during casting process and 21.8mm further during heat treatment. The displacement of corner B in y direction is 51.7mm during casting. During heat treatment there seems no big deformation in total, the total variation is 11.7mm mainly contributed by normalizing process. The final deformation of corner A is -49.8mm while corner B 40mm. No regard of heating or cooling for normalizing and tempering, there is deformation, but their deformation is basically in opposite way, so in tempering there is no deformation contribution. However, it is special during normalizing cooling because of fast cooling and martensitic phase transformation which causes deformation recovery, so there is some net deformation.

The average equivalent stress shows a peak during casting process, reaching 110MPa, and the residual stress during casting process is only 60MPa.During heat treating there is big variation of residual stress, especially in normalizing, but the final stress is around 120MPa.

3.2. Effect of phase transformation on stress and displacement

The effect of martensitic phase transformation on stress and development are shown in Figures 7 and 8, respectively. It can be seen that martensitic phase transformation has great significant on stress and displacement. In casting process, martensitic phase transformation reacts deformation, makes it keep a certain level or deform in opposite way, i.e., a certain kind of recovery, reduces the deformation of both corner A and B around 10mm. Meanwhile, martensitic phase transformation decreases the stress of the blade. Martensitic phase transformation results in expansion, which counteracts the contraction during cooling, and phase transformation resulted plasticity releases stress. Therefore, stress and deformation of the blade is affected by both uneven heating or cooling and martensitic phase transformation at the end of casting and normalizing processes. To control deformation, inverse deformation can be given referring to the simulated deformation results. The actual inverse deformation of corner A is 45mm and corner B 30mm, which are close to the simulated y displacement 49.8mm and 40mm.

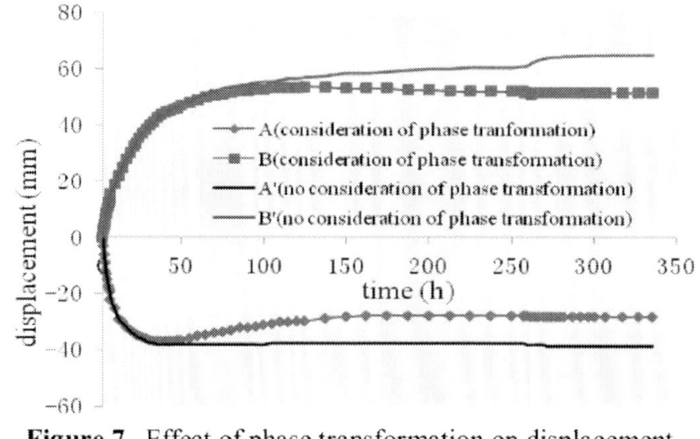

Figure 7. Effect of phase transformation on displacement.

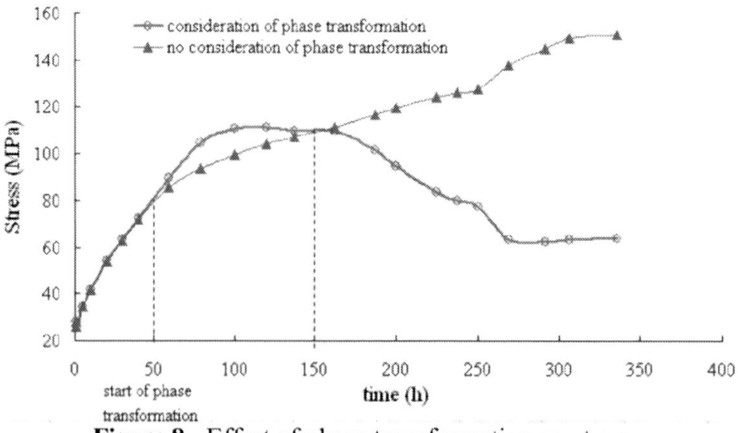

Figure 8. Effect of phase transformation on stress.

3.3. Comparison of measured and simulated displacement results

The simulated deformation results are compared to the measured one by laser measurement device, as shown in Figure 9, it can be seen their tendency is in agreement. The measured distance is actually not exactly the value in y direction, which is the reason resulting in the discrepancy.

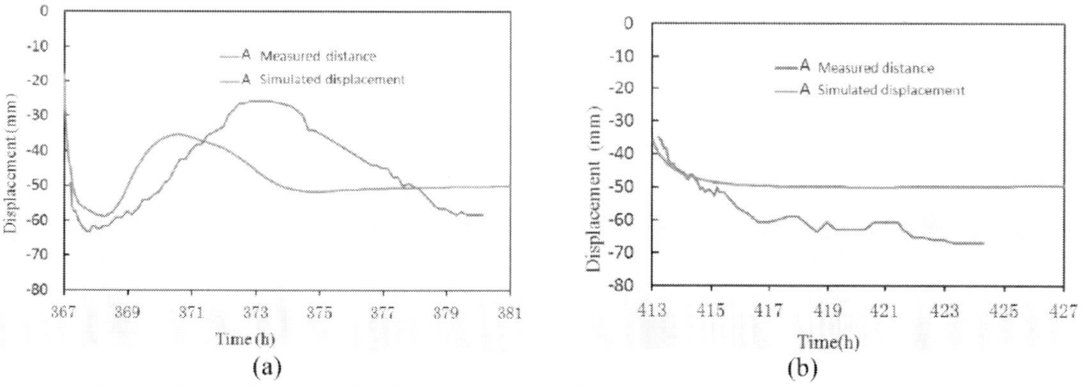

Figure 9. Comparison of the simulated displacement of corner A with measured result. (a) Cooling in Normalizing, (b) Tempering.

4. Conclusion

The coupling of heat transfer, phase transformation and stress analysis model is established for martensitic steel and the analysis platform is constructed by secondary development of ABQUS. This platform is applied into the numerical simulation of the heat treating process of the Three Gorges Project hydro turbine blade casting. The main deformation occurs at the corners A and B. Corner A deforms 28mm and 21.8mm further during heat treatment. While, corner B deforms 51.7mm during casting process, while it recovers some to 40mm during heat treatment. During heat treatment, the deformation mainly occurs during normalizing, during tempering there is no significant deformation. During casting and normalizing martensitic phase transformation greatly affects the stress and deformation of the casting.

Acknowledgement

The project is funded by Major National Sci-Tech Project of China No 2011ZX04014-052.

References

[1] Zhang J F, Kang J W, Liu B C, Wu Y, Zhang J S, Rong Z C and Zhang C C 2008 *International Journal Of Cast Metals Research* **21** 304-307
[2] Zhang J F, Kang J W, Wu Y, Zhang J S, Rong Z C, Zhang C C and Liu B C 2008 *Supplemental Proceedings: Volume 3: General Paper Selections* (The Minerals, Metals & Materials Society) p67-72
[3] Kang J W, Liu X G, Duan Y B, Huang T Y and Liu B C 2007 *Proc. of 5th Decennial International Conference On Solidification Processing* (Sheffield: University of Sheffield) p554-557
[4] Kang J W, Zhang R, Huang T Y and Liu B C 2010 *TMS Annual Meeting* (The Minerals, Metals & Materials Society) 553-561
[5] Li P, Maijer D M, Lindley T C, Lee P D 2007 *Materials Science and Engineering* **A 15** 20-30
[6] Zhang R, Kang J W and Huang T Y 2010 *Journal of Tsinghua University(Science and Technology)* **50** 1183-1187,1192

IOP Conf. Series: Materials Science and Engineering **33** (2012) 012007 doi:10.1088/1757-899X/33/1/012007

Predictions of misruns using three-phase coupled mold-filling and solidification simulations in low pressure turbine (LPT) blades

S. Jana, O. Kättlitz, F. Hediger, J.Jakumeit[a] and J. Aguilar
Access e. V., Intzestr. 5, 52072 Aachen, Germany
[a] j.jakumeit@access.rwth-aachen.de

Abstract. New alloy developments such as γ-TiAl aim at weight reduction and improvement of performance capabilities of aircraft engines. A drawback of TiAl is its low fluidity, which easily leads to misruns during the casting process.

In this work a three-phase mold filling and solidification simulation methodology has been established and validated against casting trials. It uses the finite-volume method and arbitrary polyhedral control volumes to solve the governing equations. A High-Resolution Interface-Capturing (HRIC) scheme has been established as state-of-the-art for modeling multiphase flows with sharp interfaces using the so called Volume-of-Fluid (VOF) model. This multiphase model has been extended to casting processes to predict velocity, pressure and temperature distribution for all three phases, namely the gas, melt and solidified phase.

Since LPT blades for aircraft engine applications are of widely differing in geometry with less than 1 mm thickness at the trailing edges, the effect of surface tension and wetting angle is dominant in these wall-bounded flows during the filling stage. These effects can only be calculated correctly if the interface between the phases is sharp and mesh quality is high. For the later automatically generated body-fitted polyhedral meshes with thin prism layers are used. In addition resistance of the dendrite network to melt flow must be adequately modeled. Here, an additional source term in the momentum equation based on Kozeny-Carman relation for permeability estimation is used.

A detailed analysis of filling and solidification is presented to study the performance of the simulation method.

1. Introduction

Misruns are severe casting defects, occurring when the liquid metal does not fill the entire mould cavity. Possible reasons for misruns are low fluidity, lower pouring temperature and inadequate mould design. Hence, $\gamma-$TiAl requires special processing techniques and in this regard Access has developed a production route based on centrifugal investment casting [1]. Traditional empirical correlations or experimental techniques are not adequate in describing complex thermo-fluid dynamic phenomena that occur during these kinds of casting processes, which necessitates modern computational fluid dynamic-based approaches.

In the casting process designed by Access, the γ-TiAl alloy of desired specification is melted by ISM (induction skull melting) under the influence of magnetic fields. The liquid metal is transported through a funnel into the preheated shell mold, which can be rotated up to 450 rpm. The centrifugal casting technique helps to maintain very high purity standards with good homogeneity, but makes the melt flow highly complex due to the combined effect of Lorenz forces, gravitational, surface tension, Coriolis and centrifugal forces acting on the melt. The entire casting process is carried out at around 0.1 mbar and the mould filling time is less than 1.5 seconds.

The multiphysical nature of the problem, involving the presence of gas, melt and solid in extremely thin-walled geometries, makes numerical modeling complicated. To the knowledge of the authors only very few and todate no successful attempt at predicting misruns in such a kind of centrifugal casting processes have been reported in the literature. Sung and Kim [2] attempted to model γ−TiAl turbo-charger wheel castings, but were unable to reproduce misruns in simulation as observed in real casting. Though Wang et. al. [3] were successful in modeling gas entrapments during tilt casting process, they did not explicitly show misruns due to solidifying melt.

This work was carried out using STAR-Cast [4,5] which is currently being developed by Access and CD-Adapco. Polyhedral meshes with thin prism layers have been applied, offering several advantages over tetrahedral and hexahedral meshes [6]. In the first section, a brief description of the numerical model is presented. Next, a simple geometry is studied to evaluate mesh size and time step size necessary for these kinds of calculations. In the last section a gravity cast LPT blade is simulated and the numerical results are compared with the casting trials.

2. Physical background and numerical methodology

The problem can be described by transport equations based on continuum behavior using a one phase mixture approach [12]. The equations describing mass, momentum, energy and phase conservation in a volume V, bounded by the surface S are given below (for details see [7]).

2.1. Governing equations

$$\frac{\partial}{\partial t}\int_V \rho \,\mathrm{d}V + \oint_S \rho\mathbf{v}\cdot\mathrm{d}\mathbf{s} = 0 \quad (1)$$

$$\frac{\partial}{\partial t}\int_V \rho\mathbf{v}\,\mathrm{d}V + \oint_S \rho\mathbf{v}\mathbf{v}\cdot\mathrm{d}\mathbf{s} = \oint_S \mathbf{T}\cdot\mathrm{d}\mathbf{s} + \int_V \rho\,\mathbf{f}_b\,\mathrm{d}V \quad (2)$$

$$\frac{\partial}{\partial t}\int_V \rho h\,\mathrm{d}V + \oint_S \rho h\mathbf{v}\cdot\mathrm{d}\mathbf{s} = \oint_S \mathbf{q}\cdot\mathrm{d}\mathbf{s} + \int_V \mathbf{T} : \mathrm{grad}\,\mathbf{v}\,\mathrm{d}V \quad (3)$$

$$\frac{\partial}{\partial t}\int_V C_i\,\mathrm{d}V + \oint_S C_i\mathbf{v}\cdot\mathrm{d}\mathbf{s} = \int_V s_{C_i}\,\mathrm{d}V \quad (4)$$

where t is the time, ρ is the density, $\mathbf{v}$ is the velocity vector, $\mathbf{T}$ is the Cauchy stress tensor, $\mathbf{f}_b$ is the body force, h is the thermal enthalpy, $\mathbf{q}$ is the heat flux vector, while C_i is the volume concentration of the phase i, s_{C_i} is the phase source term due to phase change (e.g. melting-solidification).

Source terms: Assuming that the mushy zone acts like a porous media, the fluid velocity resistance in the mushy zone can be approximated as a pressure drop [8]:

$$-\frac{\partial p}{\partial x_i} = \frac{\mu}{K}\mathbf{v} + \frac{C_E}{\sqrt{K}}\rho|\mathbf{v}|\mathbf{v} \quad (5)$$

where C_E is also known as Ergun's coefficient and K is the permeability. Permeability can be deduced from the Kozney-Carman's equation [9] as $K(f_s,\lambda_2) = \frac{(1-f_s)^3}{f_s^2}\frac{\lambda_2^2}{180}$, where f_s is the volume fraction solid, and λ_2 is the secondary dendrite-arm spacing.

The normal force due to surface tension is treated using the continuum surface force (CSF) model proposed by [10], which defines a volumetric source in the momentum equation that is expressed as:

$$s_\sigma = -\sigma\nabla \cdot \left(\frac{\nabla C_{n_l}}{|\nabla C_{n_l}|}\right)\nabla C_{n_l} \tag{6}$$

Local Planar Resistance or Porous Baffle: To simulate the escape of gas through the shell moulds and from other walls, we monitor pressure along non-wetted wall surfaces and compute the penetration velocity based on a pressure difference between the inside and outside wall surfaces and a resistance coefficient given by

$$p_w - p_{amb} = \rho_{scn}(a + bu_n)u_n \tag{7}$$

where a and b are the resistance coefficients, obtained from experiments, p_w is the pressure on the inside wall, ρ_{amb} is the specified reference pressure for the model (assumed to be the pressure outside the mould), ρ_{scn} is the density of the escaping fluid and u_n is the penetration velocity.

Volume of Fluid: The current approach for mould filling is based on the formulation proposed by [11]. In VOF, in addition to the governing equations an additional advection equation of void fraction is solved (as in equation 4). The convective fluxes are calculated using HRIC (high-resolution interface capturing scheme) which preserve the accuracy and sharpening of the interface.

2.2. Finite volume based simulation methodology
The governing equations were discretized based on a fully conservative finite-volume method using collocated, non-orthogonal, body-fitted grids and solved using SIMPLE algorithm. The transient term is discretized based on implicit Euler segregated concept.

3. Basic tests to establish modeling framework
3.1. Case definition
To obtain an idea of the mesh and time step size required, tests were carried out in the simplified geometry shown in Figure 1 (left Figure). These studies are described in detail in the PhD thesis by S. Jana and only the essential points are summarized. Isothermal mold filling simulations were carried out by applying a linearly increasing velocity up to 1 m/s (in 0.1 seconds) at the inlet. Studies revealed that in order to obtain a sharp melt-gas interface within 1 (max. 2) cells, a mesh of at least 8 cells across the thin-walled geometry (see Figure 1-a) is necessary. Correspondingly, a time step size of at least 1.0×10^{-4} seconds (recommended value is 5.0×10^{-5} seconds) is needed. A higher time step will induce smearing of the melt-gas front and the key features of mould filling will be missed. Studies related to the effect of heat transfer due to the fast incoming melt also revealed a similar mesh and time step size. When switching off the surface tension model the melt flows easily into the thinner parts of the geometry (see Figure 1-b). Similar situations can be observed when using a lower surface tension coefficient and a lower wetting angle. Using a surface tension coefficient of $\sigma = 1.2\ N/m$ and a wetting angle of 160° (typical values observed for γ-TiAl) shows that the melt has difficulties to flow into the thinner edged regions (see Figure 1-c).

4. Validation using gravity cast LPT blade
4.1. Problem description
To test the methodology presented above the simulation was carried out on a gravity-cast LPT blade having a trailing edge thickness of 0.7 mm. The schematic diagram of the geometry is

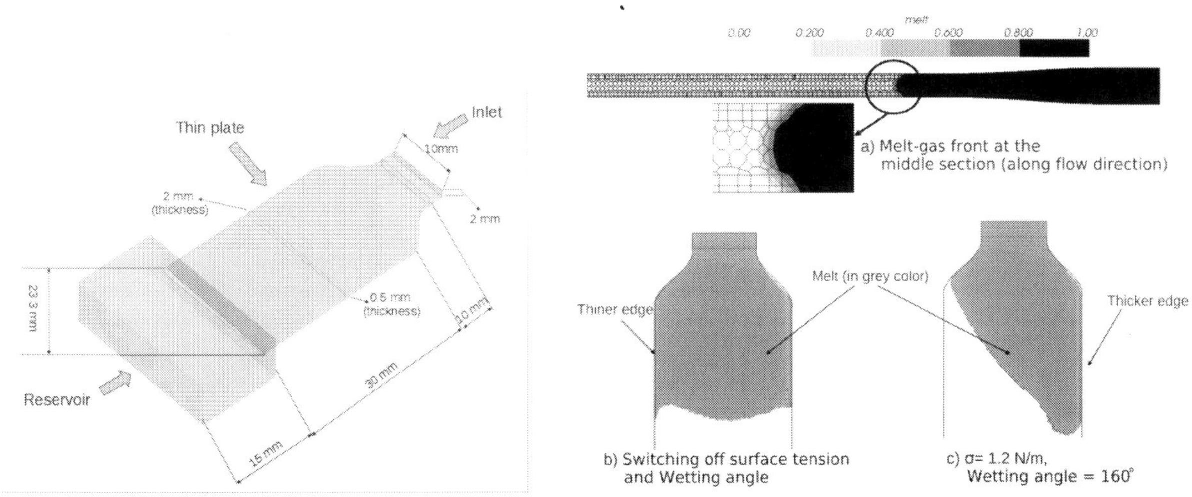

Figure 1. Left figure shows schematic diagram with dimensions of the thin plate geometry used for the basic tests. Right figure has three subfigures: figure a) shows the mesh necessary to capture the melt-gas front, figures b) and c) shows the effect of surface tension forces and wetting angle (results are shown after 0.075 secs of simulation)

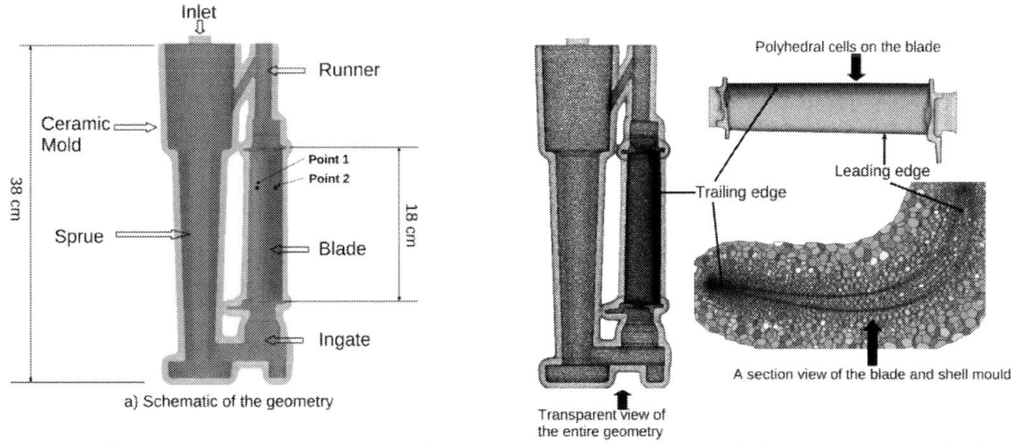

Figure 2. Figure a) shows schematic diagram with dimensions of the gravity-cast LPT blade. Figure b) shows the mesh details used in the current study.

shown in Figure 2. The cast part consists of sprue, ingate system, LPT blade and runner system. The blade is positioned vertically with the hub facing up. The objective of this simulation study is to reproduce and analyze the misruns observed at three different positions of the blade a) in the trailing edge b) in the hub c) in the leading edge (see Figure 5).

Mesh generation A ceramic shell mold with mean thickness 8 mm was generated using a reference coarse mesh. A finer surface mesh control was specified with the surface of the blade between 0.1 mm and 1.5 mm. Two prism layers of size 0.25mm on both the sides of the shell mold interfaces were specified. At the center of the sprue and at the bottom of the geometry a finer mesh was created to capture all essential details of the incoming melt and its splashing on impact. The volume mesh generated consisted of 1.4 million polyhedral cells, with the blade

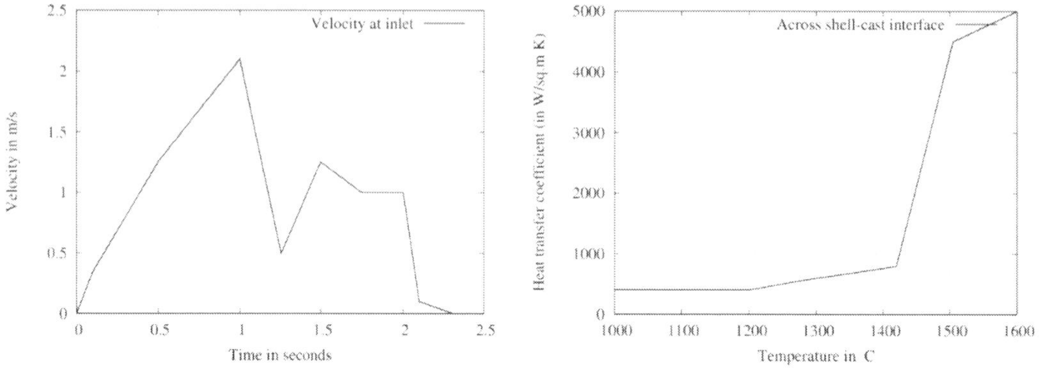

Figure 3. Figure a) shows the inlet velocity profile (over an inlet diameter of 26 mm). Figure b) shows the heat transfer coefficient used across shell-mould and metal.

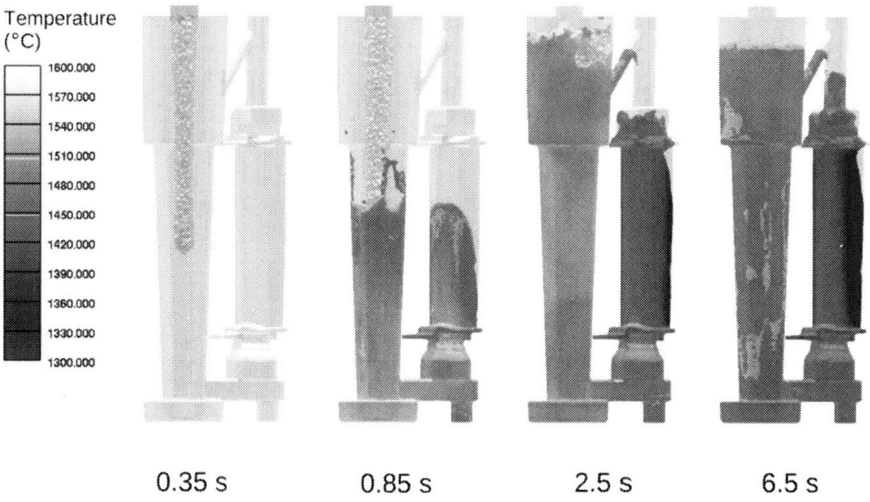

Figure 4. Figure shows the sequence of fully-coupled mould filling and solidification simulations at various times (liquidus temperature (green line) = 1505°C, solidus temperature (cyan line) = 1448°C)

consisting of around 0.4 million polyhedral cells. The details of the mesh is shown in Figure 2.

Boundary conditions and material properties

- *Inlet conditions* To obtain the inlet flow conditions, isothermal pouring of the melt from the ISM crucible to the funnel was performed. The mass flux obtained from these simulations was specified as a velocity inlet over a circular diameter of 26 mm. The melt inflow time was approx. 2.3 seconds (see Figure 3-a). The first initial melt was assumed to be 1570°C (at $t = 0$), linearly increasing to 1590°C (at $t = 0.5$ s), then remaining constant up to 2.3 seconds.

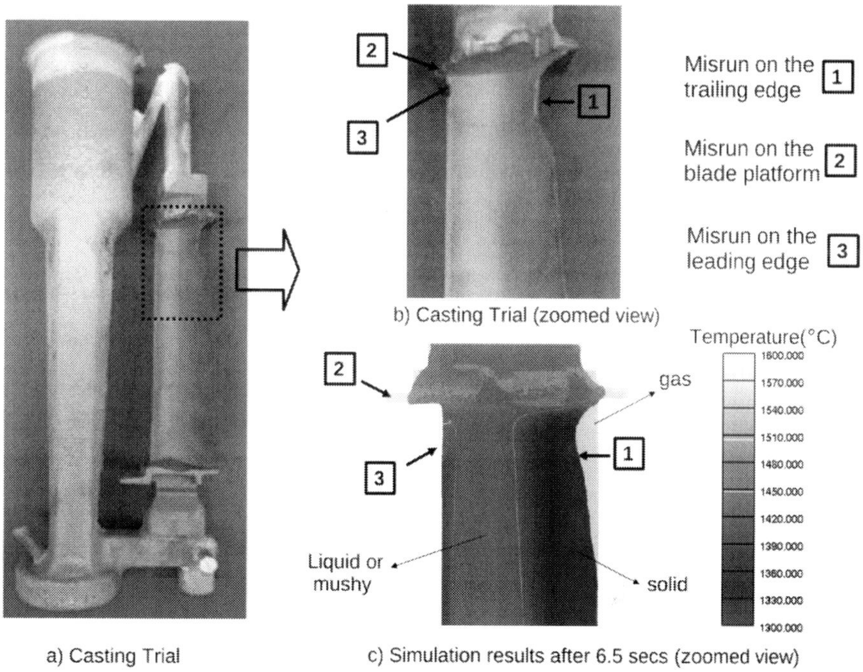

Figure 5. Figures shows a comparison of casting trials with simulation predictions. Figure a) shows the final casting of the gravity-cast LPT blade setup. Figure b) shows the magnified view of the region of misruns c) Simulation results after 6.5 seconds. (liquidus temperature (green line) = 1505°C, solidus temperature(cyan line) = 1448°C)

- *Shell mold interface* An average global heat transfer coefficient (HTC) was applied, which was calibrated for such flow and heat transfer conditions. It was assumed that HTC values decrease as the melt loses momentum and due to air gap formation (see Figure 3-b). The shell was assumed porous and the gas was allowed to escape easily by a lower value of $a = 10000$ (see equation 7).
- *Other surfaces* The outer surface of the shell mold and surfaces open to the environment had a low HTC value (100 W/m^2K) to mimic cooling by surrounding air. Free surface radiation was specified. Gas was allowed to escape easily from the surfaces open to the environment

Initial boundary conditions and material properties
The shell mold and the air inside the mold was assumed to be at 1040 $°C$. The material had the specification GE-48-2-2 (with typical composition in wt% Ti 59.4, Al 33, Nb 4.87, Cr 2.64 Fe 0.052), and thermophysical properties were obtained from the STAR-Cast material database[4].

4.2. Simulation Procedure
After the geometry is meshed, the initial/boundary conditions are applied. Under-relaxation factor(URF) for temperature was 1.0, for pressure = 0.1 and for momentum equations = 0.5). The Kozeney-Carman-based permeability term also needs a strong under-relaxation factor (= 0.01). Otherwise, oscillations in the convergence of momentum equations may occur. The maximum global residual was specified to be 10^{-5}. During the simulation a variable time step size was used. When the melt enters the ingate and the blade region a time step size of $\Delta t = 1.0 \times 10^{-4}$ seconds and at all other times a higher value of $\Delta t = 2.5 \times 10^{-4}$ seconds was used. In total 6.5 seconds of process time was simulated, which was carried out on 60 processors and took approximately 36 hours.

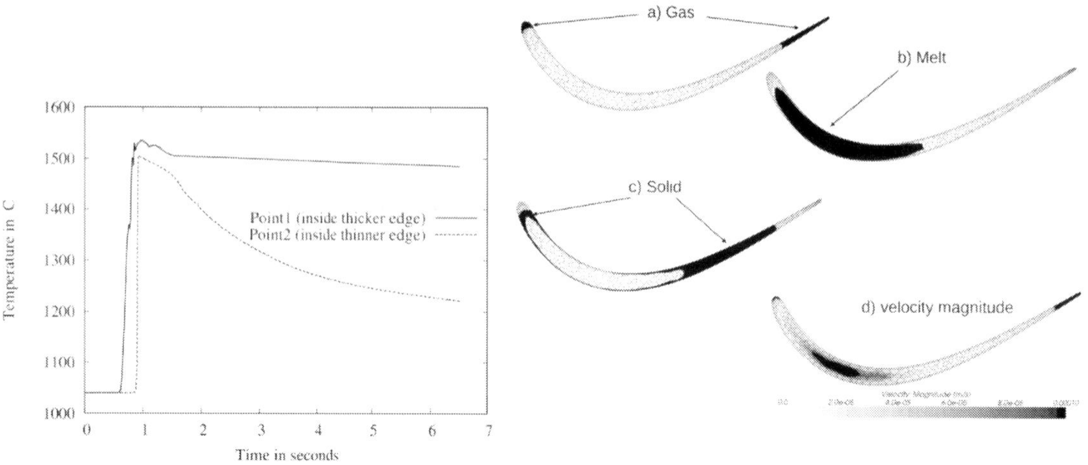

Figure 6. The left figure shows temperature history at two different points measured at a distance 17 cm from the bottom (Point 1 and Point 2 in the figure are described in figure 2. Figure a), b) and c) gas, solid and melt fractions at a cross-section across from below the hub of the blade (after 2.5 s of process time). Figure d) shows the velocity magnitude.

4.3. Results and discussion

The melt assumes momentum due to gravity and falls through the sprue. The melt makes an impact at the bottom of the sprue, and then takes a U-turn into the ingate system. Here, a lot of initial splashing is observed. The melt then fills up the shroud and travels upwards. Inside the blade the melt front assumes a parabolic shape (see figure 4 mold filling at 0.85 seconds). It can be clearly seen that the melt does not completely fill thinner regions of the blade and the surface of the melt in these regions is below liquidus temperature. The melt travels further upwards but has insufficient momentum to fill the uppermost parts of the blade completely. At 2.5 seconds (see in figure 4), a section of the melt near the trailing edge is already under solidus temperature. Even though there is a significant amount of melt height in the sprue, the melt is unable to push itself further into the thin trailing edge. At 6.5 seconds the simulation gives major misruns, which was also found in the casting trial. The misruns are (see figure 5):

- Along the topmost trailing edge (just below the hub). The shape of the misrun is very much comparable to the casting trials
- Thinner edges on the hub
- Finally near the top corner of the leading edge.

A major reason for misruns is that the melt in the thinner regions of the blade cools faster than the thicker regions, which can be seen in Figure [6]. Figure 6 a), b) and c) shows the presence of gas, liquid and solid phases at a cross-section of the blade. It can be seen that (in Figure 6-d)) velocities exists only in the gas/melt phases and that zero velocity in the solid phase.

5. Conclusions and Future Work

A three-phase mold filling and solidification methodology has been presented and a comparison to the casting trials showed excellent agreement with the shape and size of the misruns. The key features are:

- Polyhedral meshes with thin prism layers are quite efficient in predicting misruns in LPT blade-like geometries. The mesh must be fine enough (8 cells across the blade) and a low time step size of 1.0×10^{-4} s is necessary to prevent smearing of the melt-gas front.
- A sharp melt-gas interface is necessary to calculate surface tension and wetting angle effects correctly, which are the basis for a correct misrun prediction.

- A global average heat transfer coefficient was specified between the metal and the shell mold and calibrated by experiments.
- The thinner regions of the blade cool faster than the thicker regions, which results in the observed misruns.

The results demonstrate that the developed methodology is capable to predict misruns in thin geometries for gravity casting. The methodology is currently extended to centrifugal investment casting of LPT blades.

Acknowledgments
Authors would like to thank Mr. Robert Guntlin, Managing Director, Access e.V. for enthusiasm, encouragement and support during this work and to Prof. Milovan Perić and Prof. Samir Muzaferija (CD-adapco) for providing useful suggestions. Finally authors are grateful to OEM for using their LPT blade geometry. Financial support from BMWi in the frame of LUFO IV program and DFG (194/JA 853/2-1) projects are gratefully acknowledgment.

References
[1] Aguilar J, Schievenbusch A and Kättlitz O 2011 *Intermetallics* **19** 757
[2] Sung S and Kim Y 2007 *Intermetallics* **15** 465
[3] Wang H, Djambazov G and Pericleous K A 2011 *China Foundry* **8** 274
[4] www.star-cast.com
[5] Jakumeit J, Laqua R, Peric M and Schreck E 2006 *Proceedings of Modeling of Casting, Welding and Advanced Solidification Processes* **XI** 63
[6] Peric M 2004 *ERCOFTAC Bulletin* **62**
[7] Demirdzic I and Muzaferija S 1995 *Com. Met. in App. Mech. and Engg.* **125** 235
[8] Sabau A S and Viswanath S 2002 *Metall. Mater. Trans. B* **33B** 131
[9] Poirier D R, Yeum K and Mapples A L 1987 *Metall. Mater. Trans. A* **18** 1979
[10] Brackbill J U, Kothe D B and Zemach C 1992 *Journal of Computational Physics* **100** 335
[11] Muzaferija S and Peric M 1999, *Computation of free surface flow using interface-tracking and interface-capturing methods* chapter 2, ed Mahrenholtz and M. Markiewicz (WIP Press, Southampton) p 59
[12] Jakumeit J, Laqua R, Hecht U, Goodheart K and Peric M 2007 Coupled mould filling and solidification simulation applied to centrifugal casting of TiAl parts *Proceedings of the 5th Decennial International Conference on Solidification Processing SP07* (Sheffield), pp. 292

Computer simulation applied to jewellery casting: challenges, results and future possibilities

Dario Tiberto* **and Ulrich E. Klotz**
FEM Research Institute for Precious Metals & Metal Chemistry, Katharinenstrasse 17, 73525 Schwäbisch Gmünd, Germany

* Corresponding author. Metallurgy Department, FEM, Germany. Tel: +49 7171 1006 714; *e-mail*: tiberto@fem-online.de

Abstract. Computer simulation has been successfully applied in the past to several industrial processes (such as lost foam and die casting) by larger foundries and direct automotive suppliers, while for the jewelry sector it is a procedure which is not widespread, and which has been tested mainly in the context of research projects.
On the basis of a recently concluded EU project, the authors here present the simulation of investment casting, using two different softwares: one for the filling step (Flow-3D®), the other one for the solidification (PoligonSoft®). A work on material characterization was conducted to obtain the necessary physical parameters for the investment (used for the mold) and for the gold alloys (through thermal analysis).
A series of 18k and 14k gold alloys were cast in standard set-ups to have a series of benchmark trials with embedded thermocouples for temperature measurement, in order to compare and validate the software output in terms of the cooling curves for definite test parts.
Results obtained with the simulation included the reduction of micro-porosity through an optimization of the feeding channels for a controlled solidification of the metal: examples of the predicted porosity in the cast parts (with metallographic comparison) will be shown.
Considerations on the feasibility of applying the casting simulation in the jewelry sector will be reached, underlining the importance of the software parametrization necessary to obtain reliable results, and the discrepancies found with the experimental comparison.
In addition an overview on further possibilities of application for the CFD in jewellery casting, such as the modeling of the centrifugal and tilting processes, will be presented.

1 Introduction

CFD simulation is being commonly and successfully used in several industrial processes (such as lost foam and die casting) by larger foundries and direct automotive suppliers (such as BMW Leichtmetallgießerei [9-10]) while for the jewellery sector it is a procedure which is not widespread, and which has been tested mainly in the context of research projects for a few years now.
Publications on casting simulation for jewellery applications date in fact back to 2005, [2-4,8,11] and focused on the behaviour of the metal flow and of the solidification in the static casting process, in which the filling of the mould occurs by gravity / applied pressure, and dealt with benchmark casting trials monitored via thermocouples or sensors and the comparison with the simulation output; these first steps used as test parts relatively thick objects, such as a coin or a ball ring, and compared the filling and solidification time with the experimentally measured ones (using as test materials gold and silver alloys).
The work conducted by Fischer-Bühner was initially focused on sterling silver [4], and was conducted with the MAGMASOFT® software: it covered several aspect of the casting process, from the turbulence during filling to the solidification of the parts and the influence of the feed sprue positioning. In a second phase of his research [1], the attention was shifted on gold alloys and the use

of the PoligonSoft ®, again to investigate the solidification and the formation of porosity , providing a satisfactory experimental comparison with thermocouple measurement and metallography of the cast parts.

While the work from J. Fischer-Bühner was mainly focused on the thermal aspects and cooling times of the parts, the work carried out by Actis-Grande [5-6] et al. analysed in detail the filling of the mould, starting with simple test parts, and then moving on to more complex parts such as filigree objects. The results of the software obtained with the software Flow-3D® (in terms of filling time) were validated with a high frequency (1000 Hz) acquisition system, necessary because of the high speed of the metal (the total filling is in fact complete in less than 0.5-0.7 seconds). The software proved reliable in predicting the lack of filling in the thin sections of the filigree when the mould temperature was too low. Additionally, a comparison between the three softwares Flow-3D, ProCAST, and MAGMASOFT was carried out analysing their behaviour on the same test tree [7].

With this work (and following the steps of the previous research) the authors had the objective of extending the horizons on the CFD possibilities in jewellery investment casting, testing the softwares with different alloys, in order to find out the eventual limitations and discrepancies with the experimental data, and to investigate further possibilities (different processes, like the centrifugal or tilting ones)

2 Simulations

2.1 Alloy and investment properties

Three different measuring facilities have been applied for determination of the thermophysical properties of the selected gold alloys: for the the thermal expansion a high temperature dilatometer (Netzsch DIL 402 C), for the investigation of the melting range, the heat of solidification and the specific heat capacity a Differential Scanning Calorimeter (Netzsch DSC 404 C Pegasus), for the measurement of the thermal diffusivity a Laser Flash Apparatus (Netzsch LFA 427)

By using the equation [$\lambda(T) = \rho \cdot c_p(T) \cdot a(T)$] that links the density ρ, the specific heat c_p , and the thermal diffusivity a, the temperature-dependent thermal conductivities for all selected alloys were calculated (see Figure 1).

Because the heat transfer through the investment material during the casting process has a large influence on the cooling behaviour of the cast part, the thermal conductivity of the gypsum bonded investment "Silk", selected as standard material, was determined.

For this purpose the comparative method was applied: using this method the thermal conductivity of the material to be investigated is measured relatively to the well-known conductivity of a certified reference material.

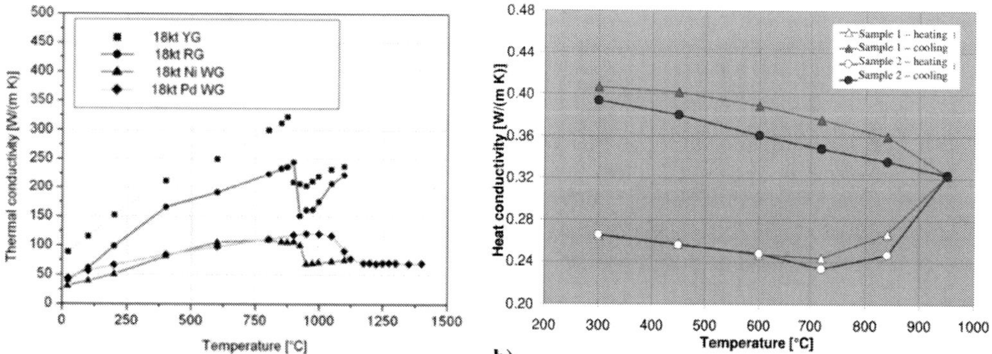

b)

a)

Figure 1 – Thermal conductivities a) 4 selected 18kt gold alloys b) 2 samples of the investment material Silk

Figure 2 - Wax tree with 16 thermocouples

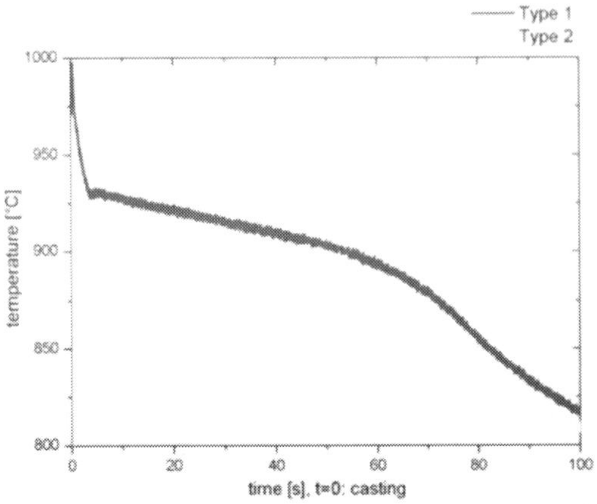

Figure 3 - Different types of acquired cooling profiles

2.2 *Experimental casting trials*

In order to investigate non-equilibrium cooling conditions as benchmarking for the simulation, casting trials with thermocouple measurements were realized by using of 4 different 18k gold alloys (yellow, red, Ni-white and Pd-white) and application of standard casting patterns. Thereby solidification time profiles for various locations could be acquired during each experiment.

Furthermore comprehensive metallographic investigations were performed on the as-cast test pattern regarding the determination of porosity levels and appropriate locations.

Figure 2 shows one of the typical wax trees with 16 integrated thermocouples at different positions, where a ball ring has been used as a standard pattern.

2.3 *Filling and cooling simulations*

The simulations were carried out using two different softwares: Flow3D 9.4® and Poligon 12.1.

Flow3D was used to obtain a temperature profile of the casting at the end of the filling step, and then such profile was transferred to Poligon for the cooling step.

Flow-3D uses the Finite Difference Model, a method which discretizes the spatial domain into small cells to form a volume mesh (or grid), and the software has its own internal multi-block meshing system, which allows to increase the detail on specified areas of the geometry (see Figure 4b), and it adopts then a "control volume" approach to solve the Navier-Stokes equations: for the examples here presented a grid with 550.000 cells was adopted, with a cell size between 0.0005 and 0.009mm; Poligon uses instead the Finite Element Method and requires the grid to be fitted on the geometries (both cast and mould), then to be processed externally, in order to obtain a tetrahedral solid mesh of the model that can be used for the computational step (see Figure 4c): the model here adopted has 100.000 nodes and 580.000 elements, with sizes that range between 0.5 and 1mm. The benchmarking was initially performed on the tree used for the casting of the 18k simple alloys, and the model for such tree was assembled in the 3D-CAD Rhino 4.0 (see Figure 4a): for the casting experiments the same 4 selected 18k gold alloys were used which had been investigated concerning their thermophysical properties in detail before (yellow gold, red gold, white gold and Pd white gold).

After the casting trials all the measured cooling curves were compared with the software output with reference to the particular position at the tree and / or the ball ring (Figure 5 shows an example).

2.4 *Results obtained with CFD*

The ability of the CFD to predict the flowing of a fluid in a cavity was exploited to optimize the sprue system of a cast part, to reduce turbulence during filling and most of all to insure a proper temperature distribution and a directional solidification, in order to reduce the formation of shrinkage porosity.

IOP Conf. Series: Materials Science and Engineering **33** (2012) 012008 doi:10.1088/1757-899X/33/1/012008

The filling simulation was used to test different sprue solutions on the parts: an example can be seen in Figure 7a and Figure 7b, which show the evaluation of two different kind of feeding possibilities, and the first one (Figure 7a), though asymmetrical, seems to provide a more uniform temperature distribution in the part at the end of the filling. Subsequently the solidification simulation was applied to identify in advance which geometries could lead to potential formation of porosity, and which places were more likely to be critical, and used to optimize the sprue system consequently (see Figure 9); Figure 8 shows the matching between the software output and the metallography performed on the part.

The heating up of the investment right after the casting was also calculated for two different mould temperatures (500°C - Figure 10a, and 600°C - Figure 10b), because a very hot mould will facilitate the formation of porosity due to the chemical reaction between the investment and the molten metal, but this effect cannot be taken into account by the software, and its influence must be therefore evaluated according to the operator's experience. This can be generalized to the entire approach that one should have with the simulation in order to apply it in a feasible and efficient way: as a tool able to point out certain problems (wrong feeding system, or mould temperatures for example) which need to be solved according to the operator's experience. In this sense, the CFD can be used as a method to optimize certain aspects of the process, without having the expectation of finding out the exact machine parameters to obtain a porosity free and perfectly filled casting.

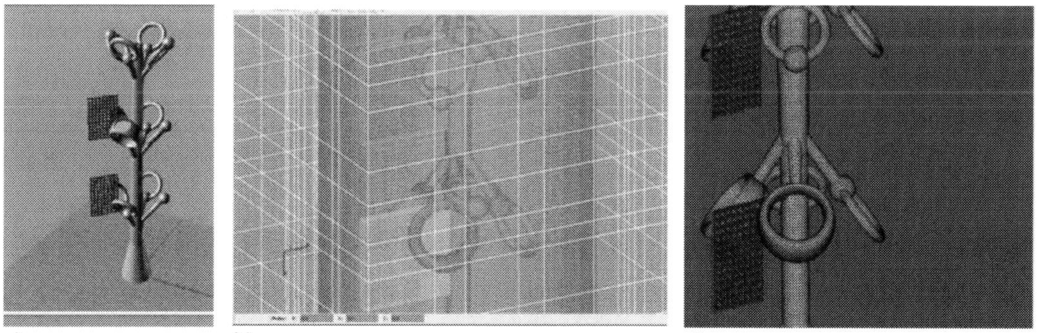

a) b) c)

Figure 4 – a) 3D Model of the tree / b) Flow3D meshing / c) Detail of the solid meshing of the model for the Poligon software

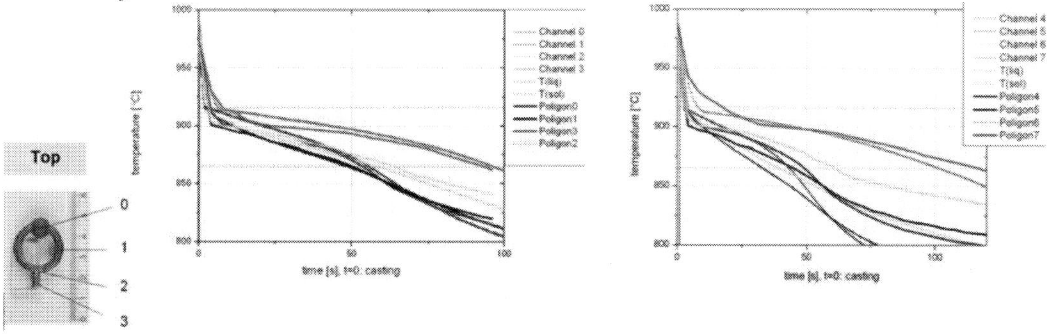

a) b)

Figure 5 - Comparison between simulation and experimental data a) test part at top of the tree / b) test part at the bottom of the tree

IOP Conf. Series: Materials Science and Engineering **33** (2012) 012008 doi:10.1088/1757-899X/33/1/012008

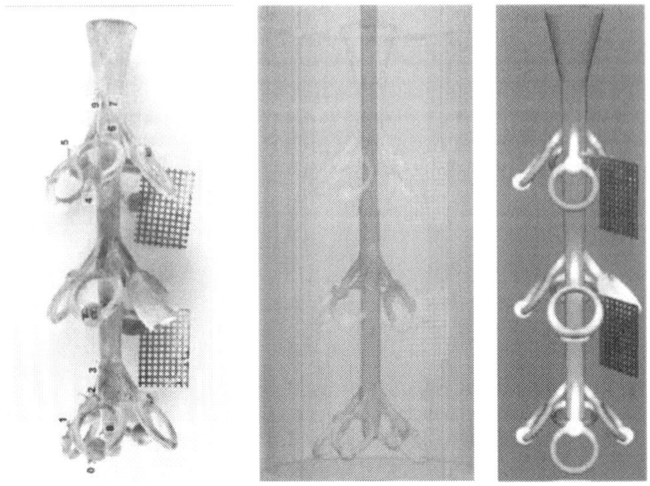

a) b) c)

Figure 6 – a) Casting tree with positions of temperature measurement / b) Filling simulation / c) Solidification profile

a)

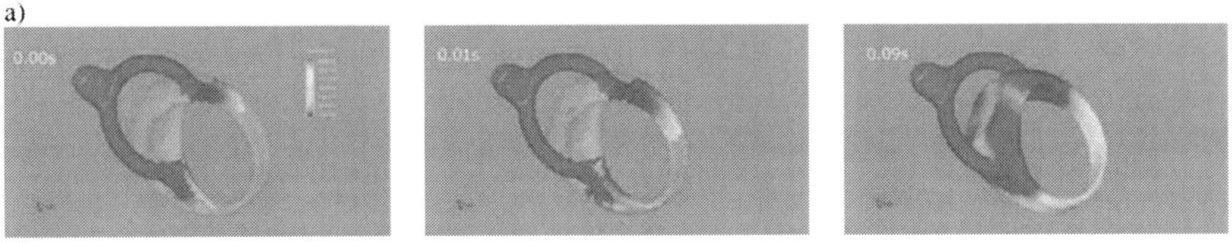

b)

Figure 7 - Example of filling simulation of a ring a) with feeding sprue on the side, and b) with symmetrical feeding system

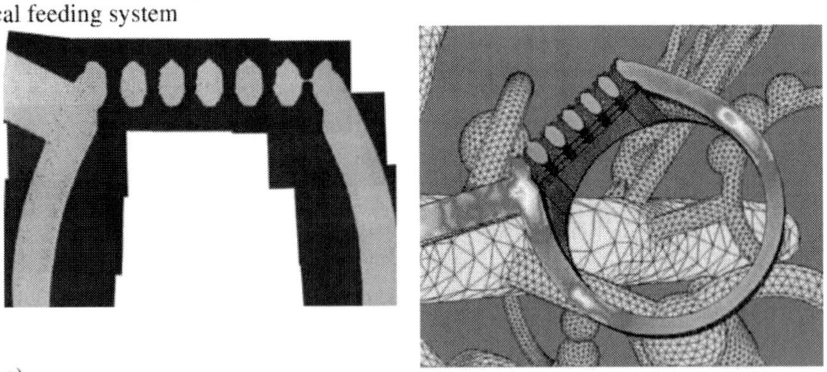

a) b)

Figure 8 - Example of b) predicted porosity and a) matching metallography on the cast part.

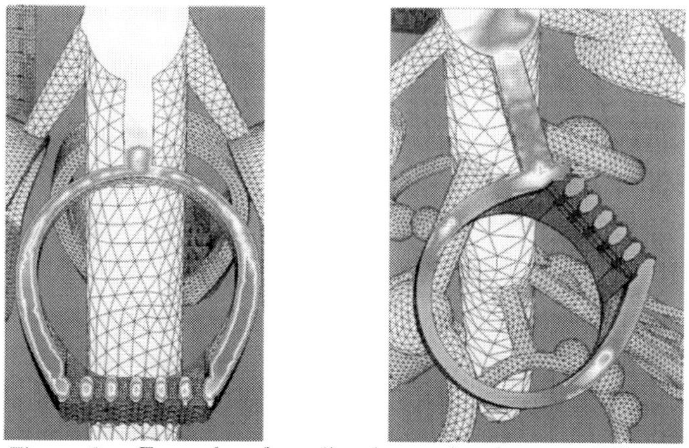

Figure 9 - Example of predicted porosity for two different sprues. Red and yellow colour indicate highest porosity levels.

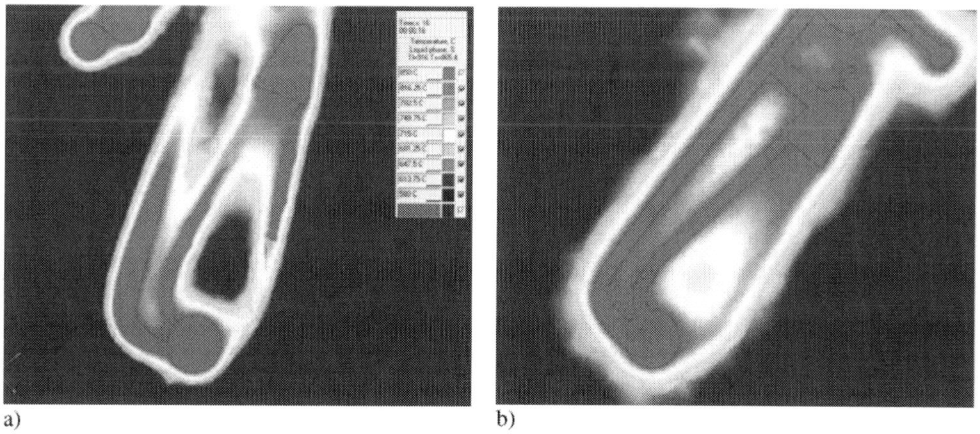

a) b)

Figure 10 - Example of investment heating with a flask temperature of a) 500°C and b) 600°C

2.5 Matching and refining of the results, and the importance of parametrization

The various temperature curves which were acquired during the casting experiments allowed to compare 1) different position on the trees (at the top and at the bottom) and 2) different type of cooling profiles: this allowed to evaluate the response of the software and its limitations.

Two different cooling profiles were observed (shown in Figure 3): a first one with a short plateau followed by a constant cooling rate, and a second one with a long plateau followed by fast cooling.

The software offers a good matching both with the first type (as already shown in Figure 5) and also with the second one, but in this case a particular attention was needed to tune-up the material parameters and the internal fit parameters of the software to get a proper response from the calculation, and match the long plateau registered by the thermocouples (Figure 13a, shows the mismatch before the optimization, and Figure 13b shows the optimized output).

The parameters that needed to be optimized are related to the heat transfer: in the Poligon software one can define the most significant regions by selecting element faces of the solid mesh (both for the mold and for the casting), and then assign one or more heat transfer coefficients to each area: by varying the values assigned to each region the cooling of the cast part can be controlled, and the output fine tuned.

The basic regions defined in this case were: the surface of the cast part and the inner surface of the mould (to take into account the heat exchange between the hot metal and the investment), the outer surface of the mould (to simulate the cooling in air or protective atmosphere such as Ar) and the top of the casting (where the riser is exposed, and a significant heat flow takes place); the different areas can be seen in Figure 12.

It was then observed that the software (after the proper optimization) offers a good matching with the measured cooling curves on test parts positioned at the tip of the tree, while for the pieces which are placed near the bottom, close to the riser (where the heat flow towards the outside is more consistent) the temperature mismatch between reality and simulation is more pronounced, and a proper tuning could not be reached (see Figure 5b): the software tends in fact to overestimate the solidification time in this part of the casting, and the experimentally measured curves are steeper than the output obtained from the simulation.

The adjustment and refinement of the results through the software parameterization is (most of the times) possible when a series of experiments with thermocouples and metallographical investigations have been run in advance, and can offer a basis for benchmarking. In "normal" cases (to approach an industrial everyday situation) the background work of benchmarking tests is not always available, and thinking of optimizing the software for every different alloy and investment (which has a great influence on the heat transfer and therefore on the cooling of the metal) is obviously not possible, both for time and costs reasons.

To this it must be noted that the investment casting procedure is influenced by several factors: performance of the refractory material (which can vary from batch to batch), chemical reaction between the hot metal and the mould, relevant influence of the furnace during the burn-out of the flask (e.g. two flasks prepared in two different ovens can have a significantly different performance).

All these aspects cannot be taken into account in the simulation, and therefore it is better to know how to employ it in a meaningful way: for example being able to evaluate the effect of the mould temperature (as in the example shown in Figure 10) and the possible influence on the investment reaction that could lead to gas porosity.

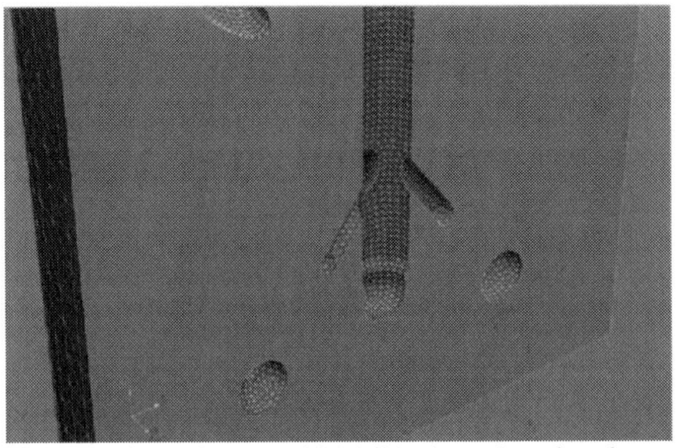

Figure 11 - Inner surface of the mould (visible through a virtual cross-section)

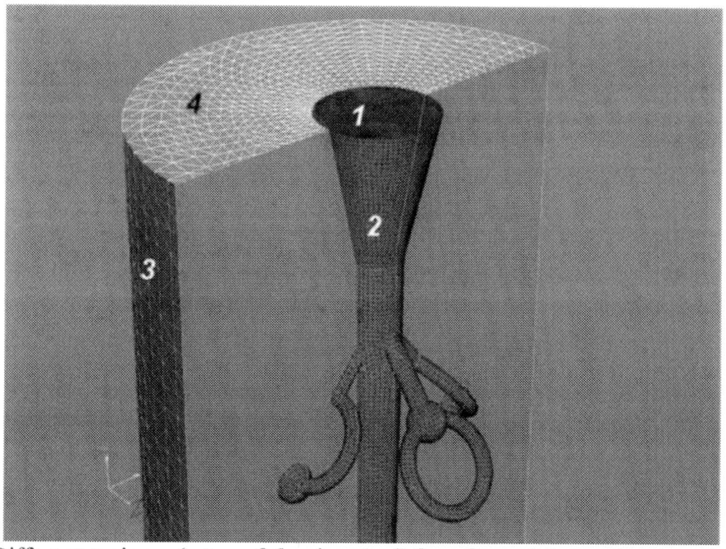

Figure 12 – Different regions: 1- top of the riser (red) 2-surface of the casting (purple) 3- external surface of the mould (blue) and 4- top of the mould

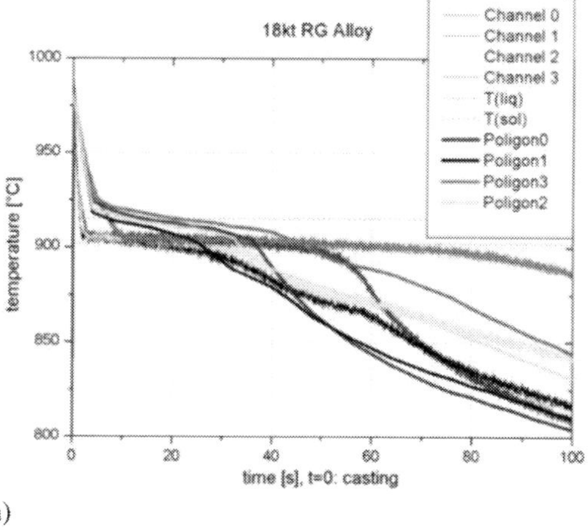

a)

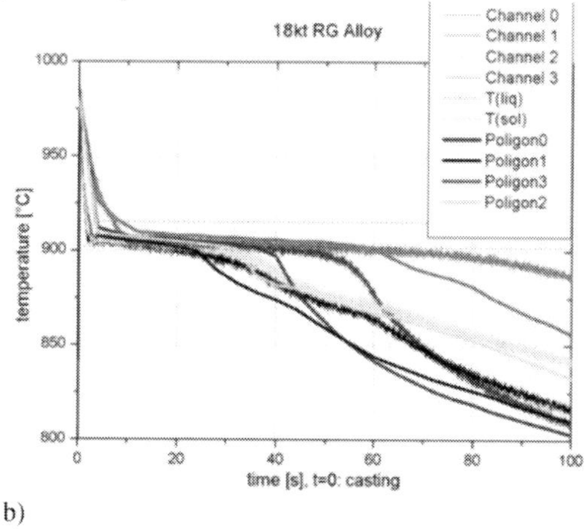

b)

Figure 13 – Comparison between experiment and software output a) before and b) after the parameters tune-up)

2.6 Further applications currently in development

Further applications, which are currently being tested and will be applied in a future research project, are the centrifugal and the tilting casting processes. The CFD can be applied to the centrifugal process to establish how the centrifugation influences the filling of the parts according to their positioning, or to the type of adopted main sprue (Figure 14a).

The simulation can be also applied to the tilting process, and first tests have been focused on the rotation angle and the speed of rotation, and how these parameters condition the flowing of the metal out of the crucible (see Figure 14b and Figure 15). Different crucible geometries could also be tested to choose the best one for the pouring of the metal. For instance such tests could be used during the machine development to achieve a proper design of the tilt mechanism and the machine geometry.

This case shows the potential of the CFD to predict the effect of the turbulences during the pouring of the melt in tilt casting.

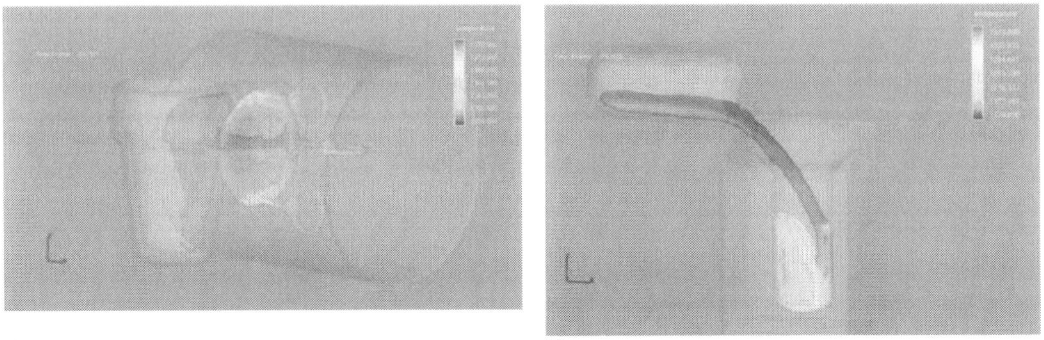

a) b)

Figure 14 - Example of a) centrifugal casting / b) tilt casting

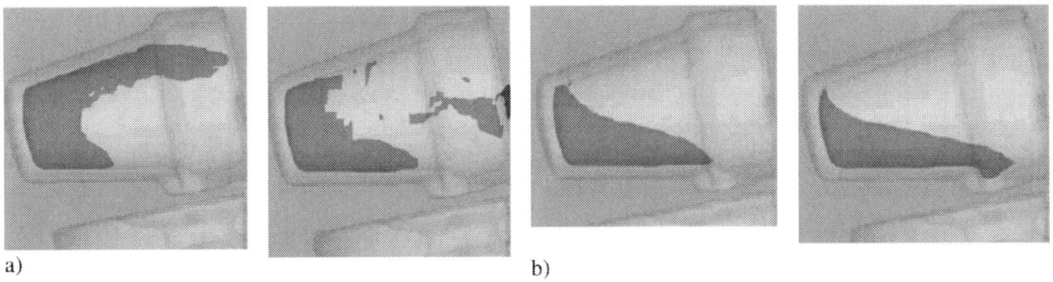

a) b)

Figure 15 - Example of two different rotation speeds in a tilt casting machine, with corresponding turbulence of the liquid metal in the crucible

3 Summary and Outlook

The work here presented demonstrates the applicability of the CFD simulation in the jewellery sector to a wide range of processes (static, centrifugal and tilt casting), exploiting the different possibilities offered by the softwares commercially available: mold filling, solidification and prediction of porosity, investment heat-up.

The quality of the results depends of course on the work previously conducted on the material characterization (investment and alloy properties) and then on the software tune-up and parametrization: good results have been obtained in terms of porosity prediction and sprue optimization.

Variations can be although found in the matching of the output when a wide range of conditions is taken into account (different alloys, investments, position on the casting tree, etc), and in the everyday industrial reality the simulation could be disadvantaged when compared to an experimental trial (in terms of "quick" problem solving); it remains a solution to be exploited in the design and planning phase, where its potential of "virtual laboratory" would assume more weight to test different solutions.

Further research will be focused on deepening the possibilities of the CFD for the centrifugal and tilting processes (applied for example to precious metals as Pt), also as a valuable help to design and engineer the casting machines and the crucible geometries.

Acknowledgements

The European Commission is acknowledged for financial support of this study under the contract no. 222179-INTOGOLD. The authors thank the partners of the INTOGOLD project for the good co-operation and the provision of materials. Special thanks are to the members of the metallurgy department of FEM, and to Mr. Marco Actis Grande (Politecnico di Torino, Italy) and Mr. Jörg Fischer-Bühner (Legor, Italy) for their precious co-operation.

References

[1] Fischer-Bühner J 2007 Advances in the prevention of investment casting defects assisted by computer simulation *The Santa Fe Symposium on Jewelry Manufacturing Technology* 149-72

[2] Actis-Grande M 2005 Computer simulation of the investment casting process: experimental validation *TCN CAE Conference* 63-73

[3] Actis-Grande M 2006 Computer simulation of the investment casting process: experimental validation *4th International Jewelry Symposium*

[4] Fischer-Bühner J 2006 Computer simulation of jewelry investment casting *The Santa Fe Symposium on Jewelry Manufacturing Technology* 193-216

[5] Actis-Grande M, Porta L and Tiberto D 2007 Computer simulation of the investment casting process: widening of the filling step *The Santa Fe Symposium on Jewelry Manufacturing Technology* 1-18

[6] Actis-Grande M and Wannarumon S 2009 Numerical simulation of Investment casting of Gold Jewelry: Experiments and validations *World Academy of Science, Engineering and Technology* **55** 82-6

[7] Actis-Grande M and Wannarumon S 2009 Comparison of Computer Fluid Dynamic software programs applied to jewelry investment casting process *World Academy of Science, Engineering and Technology* **55** 88-95

[8] Wright J 2005 Computer simulation and jewellery production *The Santa Fe Symposium on Jewelry Manufacturing Technology* 521-33

[9] Lang H 2010 Use of modern flow simulation tools for process optimization *Flow-3D User Meeting*

[10] Lang H 2011 Improvement of the simulation through validation of boundary conditions and inlet sizes *Flow-3D User Meeting*

[11] Alonso A and Franco L 2005 Computer optimization of vertical gating system designs *Foundry Management and Technology* **133 (10)** 22-4

The improvement of aluminium casting process control by application of the new CRIMSON process

X Dai[1*], M Jolly[1], B Zeng[1]

[1]School of Mechanical Engineering, University of Birmingham, Edgbaston, Birmingham, B15 2TT, UK

[*]Corresponding author: xiaojundai@yahoo.com

Abstract. All The traditional foundry usually not only uses batch melting where the aluminium alloys are melted and held in a furnace for long time, but also uses the gravity filling method in both Sand Casting Process (SCP) and Investment Casting Process (ICP). In the gravity filling operation, the turbulent behaviour of the liquid metal causes substantial entrainment of the surface oxide films which are subsequently trapped into the liquid and generate micro cracks and casting defects. In this paper a new CRIMSON process is introduced which features instead of gravity filling method, using the single shot up-casting method to realize the rapid melting and rapid filling mould operations which reduce the contact time between the melt and environment thus reducing the possibility of defect generation. Another advantage of the new process is the drastic reduction of energy consumption due to shortened melting and filling time. Two types of casting samples from SCP and ICP were compared with the new process. The commercial software was used to simulate the filling and solidification processes of the casting samples. The results show that the new process has a more improved behaviour during filling a mould and solidification than the two conventional casting processes.

1. Introduction

Foundry practice has been driven by the requirement of improving product quality and minimisation of the production costs. Both casting production process and energy efficiency play critical roles in the foundry to gratify these requirements. Besides the pressure of rising energy cost and the restriction of stern environment protection legislation it is certified that under the right socioeconomic conditions efficiency optimisation of an industrial process can be an important step toward increased industrial sustainability [1]. Selection of the appropriate casting process, facility and minimisation of the energy consumption will be the vital factors for the foundry to become competitive against rivals in a tough market.

In traditional foundries using SCP and ICP, the capacity of melt furnace for aluminium alloy usually ranges from dozens of kilograms to several tonnes. The liquid metal is held at about 700 °C in a holding furnace before it is transferred to a ladle and poured into a casting (shell)mould at a pouring station. It can take long time, or even dozens of hours, for the liquid metal in a batch to be finished and any residual metal is poured off to be re-used or scrapped for re-melting or refining in a secondary processing plant [2].

When the liquid metal reacts with hydrogen, oxygen and water in the surrounding atmosphere quality problems can happen. Once the melted aluminium alloy is exposed to the air an oxide surface layer is created. For the period of filling a sand or shell mould the turbulent filling behaviour of the liquid metal due to gravity is easy to make the oxide films on the surface of the liquid cracked and trapped into the liquid. Also the long exposing time of the liquid surface with the surrounding air during melting, transferring and filling will augment the thickness of the oxide film on the liquid surface and the level of hydrogen absorption. All of these will bring about layers of cracked oxide films, porosity and shrinkage which damage the integration of the micro-structure of the alloy, leading to degraded mechanical properties of the final product [3, 4].

A new CRIMSON (Constrained Rapid Induction Melting Single Shot Up-Casting) process has been co-invented by the researchers and engineers form both University of Birmingham and a local company, N-Tec LTD. The aims in this new process are to improve the casting quality and reduce the energy consumption within light-metal casting industry. The tactic of the new process is to melt just enough amount of alloy in a closed crucible of an induction furnace and to use the counter-gravity filling method to fill a single (shell) mould, thus assure smooth liquid alloy flow behaviour and in the same time avoid unnecessary energy consumption. The new process features rapid melting, quick bottom filling and minimising holding time hence reduces the opportunity of hydrogen absorption and generation of surface oxide film [5].

In this paper, two cases have been investigated: one is the "Tensile bar" where the SCP was applied; another one is the "Filter housing" where the ICP was applied. Using the commercial CFD software to simulate the processes of filling and solidification in sand mould or shell mould, both cases have been compared with the new process. The simulation results for filling the sand (shell) moulds of three different casting processes were compared to see how the new process could avoid the turbulent behaviour of liquid metal flow during filling sand (shell) moulds. The calculation and analysis of energy consumption were completed to see what the difference is between the current melting processes and the new facility. As a result, the quality issue and the potential energy saving for the new process can be found. This comparison is only one of a number being carried out under the support of an EPRSC project where four conventional casting processes will be benchmarked for their scrap rates and energy usage.

2. Runner system design and simulation tools

2.1. Comparison of gravity SCP, ICP and CRIMSON process

A runner system using gravity SCP for tensile bar is shown in Figure 1(a) where the system consists of basin, down sprue, runner, ingate, riser, feeder and six tensile bars. A runner system of sand mould for tensile bar using CRIMSON process is presented in Figure 1(b) where the system consists of runner, ingate, riser, feeder and six tensile bars. A runner system of shell mould for filter housing using CRIMSON process is shown in Figure 1(c) where the system consists of runner, ingate, feeder and four filter housing parts. (For the commercial reason, the runner system of current gravity ICP from a local company could not be presented here, However, its filling process, yield and energy consumption can be demonstrated in the "Results and discussion" section of this paper)

2.2. Simulation software and simulation conditions

In order to simulate the liquid metal flow in these three runner systems, the commercial CFD simulation software were used. A 'velocity magnitude' method was used to predict the flow behaviour of liquid metal during filling a mould. The simulation was implemented using a Workstation with 16.00GB RAM and eight 2.66GHz CPUs.

For the simulation of the three runner systems, Finite Difference Method (FDM) was used to generate the mesh which includes about 170,000 control volumes (cells) for Figure 1(b) and 17,667,900 control volumes (cells) for Figure 1(c). The filling flow rate of 0.25 $L.s^{-1}$ [6] and a pressure of 9 kPa were used for these two runner systems. For the runner system of gravity sand casting Figure 1(a), the mesh has about 240,000 control volumes (cells). Same main conditions for the simulation of these three runner systems are: pouring temperature is 700 °C and the around atmosphere pressure is 1 atm (1.013×10^5 kPa). Pouring height for the gravity filling method is 50 cm.

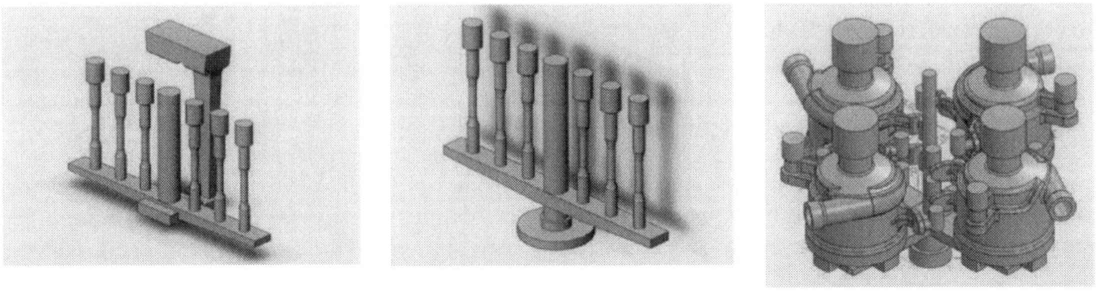

(a) Gravity sand casting (b) CRIMSON process for tensile bar (c) CRIMSON process for filter housing

Figure 1. Different runner systems.

3. Test equipment for measuring the energy consumption

In this paper, the experiment and analysis of energy consumption are based on the following assumptions:

- The system assessed is at continuous steady state which includes fuel flow, air flow rate, melting rate, flue gas parameters and thermal conduction through furnace wall;
- The fuel and combustion products behave as ideal gas mixtures;
- The environment temperature and pressure are taken as standard 25°C and 1 atm respectively;
- The electric energy consumption is only applied to the part of electricity resistance heating or induction heating in furnaces, not applied to the motors and control devices which are neglected for the convenient calculation and simplicity;
- The natural gas composition is considered as pure propane due to the small amount of N_2, CO_2, H_2S and H_2O included;
- The lost metal during drossing is neglected for the calculation simplicity.

3.1. CRIMSON facility

The layout and main components of the new facility are described in Figure 2 where its main functions and features are:

- High power Induction furnace (275 KW): it is used to quickly heat and melt the metal to the required pouring temperature. Usually each time, a billet of the required size and calculated amount of metal is put in, also the composition of the billet should be consistent with the casting component that will be poured and produced;
- Up-caster: when the crucible with the melted metal inside is ready, it is moved and clamped in the right position in Up-caster and a mould is located on the top of pouring position, a piston in the Up-caster will raise and push the melted metal in the crucible into the mould;
- Computer-controlled operation table of the Up-caster: the movement of the piston in Up-caster is automatically controlled by the pre-programmed computer program;
- Mould transfer stop: after pouring, cooling down and solidification, the mould can be moved to the transfer stop, waiting for lifting and cleaning;

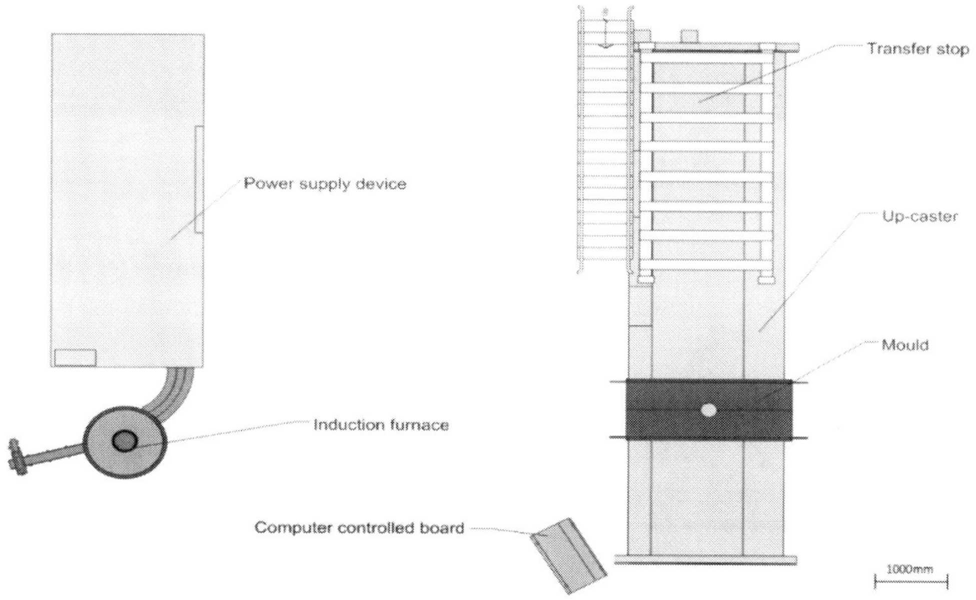

Figure 2. Schematic plan of the CRIMSON facility

3.2. *Cosworth facility*

Grainger & Worrall (G&W) LTD is using a melting furnace with a capacity of 4 tonne (Figure 3) where the gas is used to preheat and melt aluminium ingot and the electricity is used to adjust the melted liquid alloy to a required overheating temperature. The holding time for the furnace is up to 4-5 days. For aluminium alloy the overheating temperature is 760 °C and the pouring temperature is 700 °C. This furnace is used for gravity SCP.

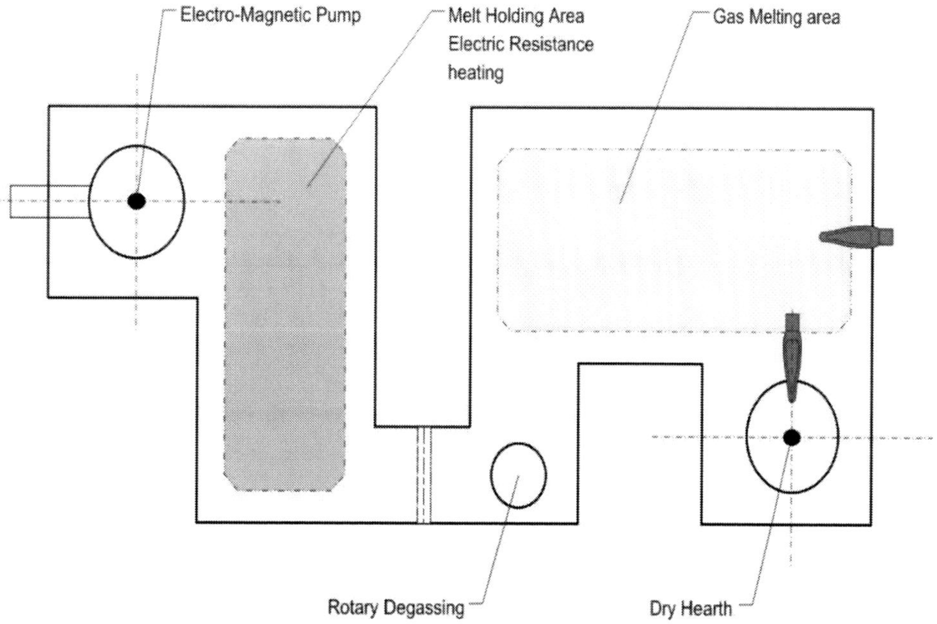

Figure 3. Schematic of the aluminium melting furnace in G&W LTD

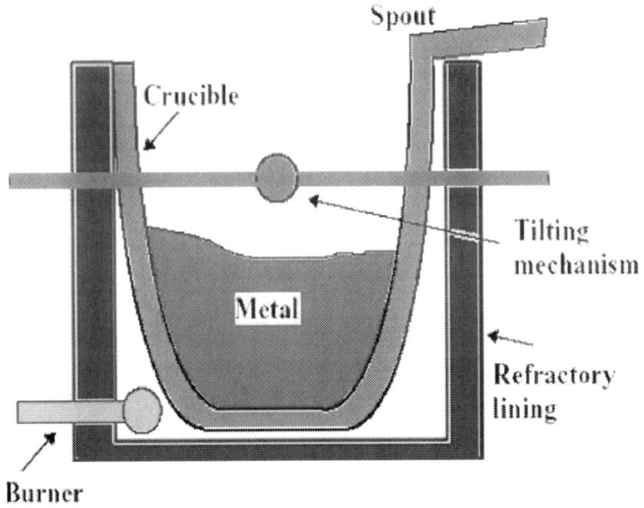

Figure 4. A typical crucible furnace

3.3. Crucible furnace

Aeromet LTD. is currently using the conventional crucible furnaces, each with a capacity of 400 kg. The natural gas is used to heat and melt aluminium alloys. The furnace has an average 2.5 hours' melting period and 1.5 hours' holding period. The overheating temperature of aluminium alloy is 780 °C. The normal pouring temperature is 700 °C. The typical crucible furnace is shown in Figure 4. This furnace is used for gravity ICP.

4. Results and discussion

4.1. Simulation

For the simulation of runner system of tensile bar (sand mould) using CRIMSON process, the filling time is 6.08 seconds. The simulation took about 30 minutes. The velocity magnitude of the liquid metal during filling is depicted in Figure 5 where the filling velocity of liquid metal can be observed using the velocity scale.

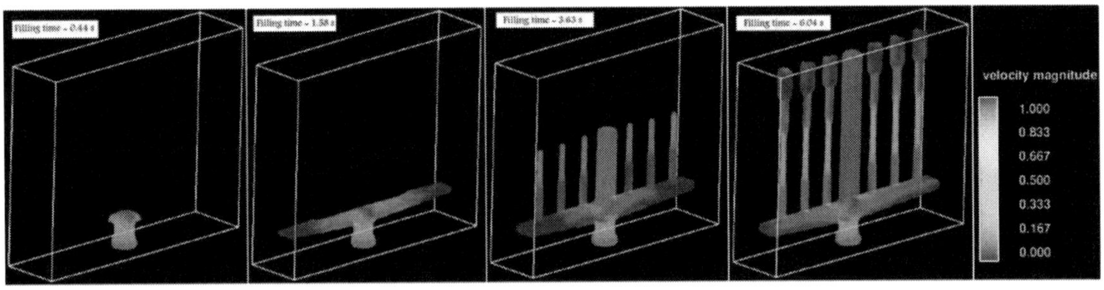

Figure 5. Numerical simulation of runner system of tensile bar using CRIMSON process.

For the simulation of runner system of tensile bar (sand mould) using gravity SCP, the filling time is 3.67 seconds. The simulation took about 20 minutes. The velocity magnitude of the liquid metal during filling a mould is described in Figure 6 where the velocity of liquid flow can be judged using the velocity scale.

For the simulation of runner system of filter housing (shell mould) using CRIMSON process, the filling time is 8.96 seconds. The simulation took about 6hours. The velocity magnitude of the liquid metal during filling a mould is described in Figure 7 where the velocity of liquid flow can be judged using the velocity scale.

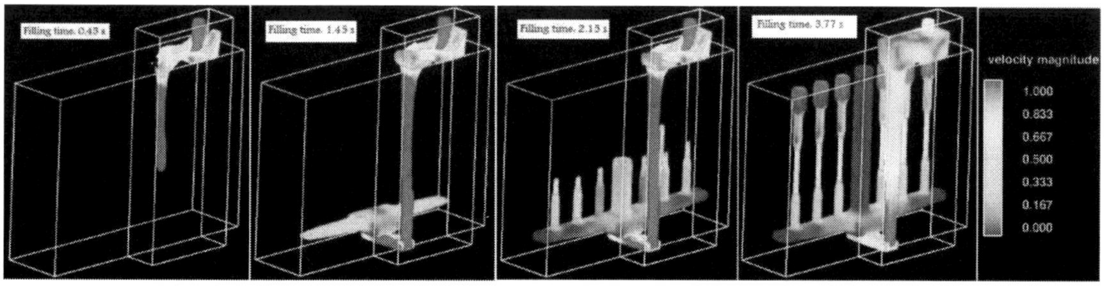

Figure 6. Numerical simulation of runner system of tensile bar using gravity SCP.

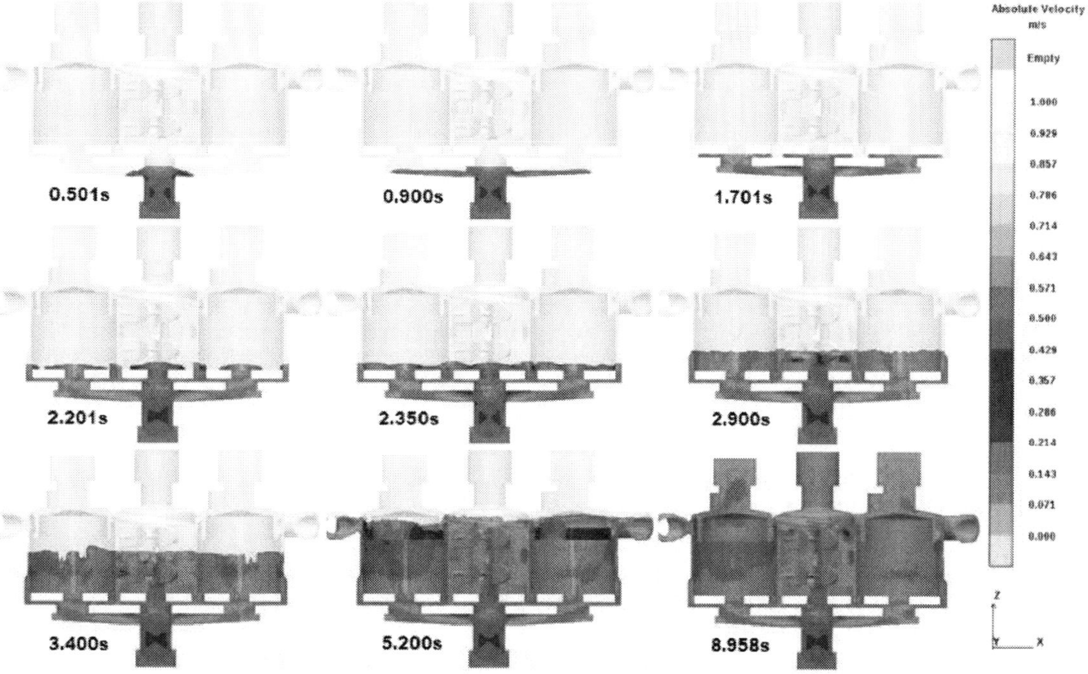

Figure 7. Numerical simulation of runner system of filter housing using CRIMSON process

From Figure 5 and Figure 7, it was found that the maximum velocities of liquid metal flow during filling are 0.4 $m.s^{-1}$ and 0.45 $m.s^{-1}$ respectively. Both the maximum velocities are less than 0.5 $m.s^{-1}$. The quiet flow behaviour of liquid metal in these two runner systems during filling the sand/shell moulds was considered appropriate for avoiding the generation of trapped oxide films, porosity and other casting defects [8]. In addition the counter-gravity filling method in CRIMSON process decreases the exposure time to the air which will reduce the opportunity of generating oxide films.

From Figure 6 and the filling process of the current gravity ICP from a local company, the maximum velocities of liquid metal flow in downsprue and in runner during filling are more than 1.0 m.s^{-1}, respectively. This means that the violent and turbulent follow flow behaviour will easily crack the oxide films on the liquid and make them trapped into the liquid. Although the filter existed in the runner system can leach the coarse trapped oxide films from the liquid which depends on the size of the holes and effectiveness of the filter, the fine oxide films will still pass through the filter and remain in the liquid. After solidification, these remaining fine oxide films will generate defects such as porosity, shrinkage etc [3, 4]. In the mean time, the turbulent flow behaviour of liquid metal in the downsprue and runner will readily make the air or hydrogen entrapped into the liquid where the porosity or bubble will be formed which will damage the mechanical properties of the casting [9]. In addition the long transferring time from furnace to ladle and the usage of the traditional gravity filling method make the melted liquid metal exposed to the air for long time, which increases the opportunity of generating oxide films on the surface of the liquid metal.

According to the abovementioned discussion, it is assumed that the counter-casting filling process that the new process used can considerably reduce the opportunity of oxide film generation on the surface of the liquid metal and the potential time for hydrogen absorption. As a result, the quality of the casting can be assured consequently. In comparison with the new process, the filling process of the conventional gravity SCP and ICP have the violent flow behaviour which will easily make the oxide films on the liquid metal and the air entrapped into the liquid which will generate different type of casting defects and damage the mechanical properties of casting. It should be pointed out here that all these results from the computer simulations will be validated by the actual experiments such as filling mould, strength test and SEM observation of next stage.

4.2. Energy efficiency

The energy consumption of the three types of melting processes was measured [7, 10]. The results and the calculated energy efficiency [7, 10] were demonstrated in Table 1.

Table 1. Energy consumption and energy efficiency of the three different melting facilities.

	Measured Energy consumption	Energy efficiency	Normal energy efficiency of furnace	Melting and holding time (hours)
Aeromet (Gas)	8.65 MJ.kg^{-1} [10] (2401.88 kWh.tonne^{-1})	13.86% [10]	Gas: 7~19%[10]	Melting 2.5 [10] Holding 1.5
G&W (Gas+Ele)	27.86 MJ.kg^{-1} (Gas: 17.78 MJ.kg^{-1} Ele: 10.08 MJ.kg^{-1})[7]	Gas: 5.65% Ele: 1.70% (*0.85%)[7]	Gas: 7~19% Ele:59~76% (*29.5~38%)[10]	Melting + holding 96~124 [7]
CRIMSON Induction furnace	1.98 MJ.kg^{-1}[7,10]	57.82%[7,10] (*28.91%)[10]	Ele:59~76%[7,10] (*29.5~38%)[10]	2 minutes for melting 4kg A354 3 minutes for melting 6kg A356

(*the value is in consideration of conversion efficiency of 50% in a natural gas-fired power plant)

As shown in Table 1, in consideration of conversion efficiency of 50% in a natural gas-fired power plant, the thermal efficiency of the melt furnace at G&W for electricity is 0.85 % [7]which is far lower than the energy efficiency of the new process (28.91%[10]). It means that there is lot of energy loss for the current melting process at G&W due to the long holding time. Therefore, it is suggested that if the current long melting and holding process at G&W could be replaced by the new process (induction heating) the thermal efficiency in using electricity will be increased up to 28%. It is estimated that

34.00 GJ.tonne^{-1} (9.44 MWh.tonne^{-1}) [7] could be saved for producing every tonne of aluminium casting alloys when using the new process.

Similar to the Aeromet company, the thermal efficiency of the crucible furnace using natural gas is 13.86% [10] which is lower than the thermal efficiency of the new process (28.91%) [10]. If the current melting and holding process at Aeromet could be replaced by the new process, the thermal efficiency will increase up to 15.05 %. When melting the same amount of aluminium alloy, it is estimated that 4.69 GJ.tonne^{-1} (1.30 MWh.tonne^{-1}) [10] could be saved.

5. Conclusions and future works

From the computer simulation results, the maximum flow velocity of the new CRIMSON process during filling a mould is less than 0.5 m.s^{-1}. The total flow behaviour is quiet and stable which potentially reduce the opportunity of generation of oxide films and other casting defects. In comparison with the new process, the conventional gravity SCP and ICP have turbulent flow behaviour during filling a sand/shell mould where the maximum velocities of liquid flow in the downsprue and runner are more than 1.0 m.s^{-1}. The violent flow behaviour will not only easily make the oxide films on the liquid metal cracked and entrapped into the liquid but also make the air trapped into the liquid. These entrapped oxide films and air will probably generate different types of casting defects such as porosity, shrinkage and bubble etc which will damage the mechanical properties of the final casting. To validate the computer simulation results, the experiments on comparing the mechanical properties of the tensile bar using different filling processes and related SEM observation on micro structure will be implemented in next stage.

The investigation on energy consumption and energy efficiency of three processes has showed that the new process is a novel method for reducing energy consumption. If the traditional melting process at G&W and Aeromet could be replaced by the new process, the estimated energy savings for the former could be of the order of 34 GJ.tonne^{-1} (9.44 MWh.tonne^{-1}) and for the later could be the order of 4.7 GJ.tonne^{-1} (1.30 MWh.tonne^{-1}).

The other issues of the energy efficiency for the foundry will be considered too where not only the melting process is included, other relevant processes should be considered.

Acknowledgement

This research project is sponsored by the Engineering and Physical Sciences Research Council (EPSRC) of the UK under the grant of EP/G060096/1. Many thanks to the University of Birmingham, N-Tec LTD. Aeromet Ltd and Grainger & Worrall Ltd for providing the experiment equipment and data.

References

[1] BSC Incorporated 2005 Advanced Melting Technologies: Energy Saving Concepts and Opportunities for the Metal Casting Industry, *Report of U.S. Department of Energy*
[2] Casting method saves energy 2009 **25** *Prof Eng Magaz* http://www.profeng.com/archive/2009/2206/22060076.htm
[3] Dai X, Yang X, Campbell J and Wood J 2003 *Mater Sci Eng A* **A354** 315-325
[4] Dai X, Yang X, Campbell J and Wood J 2004 *Mater Sci Technol* **20**(4) 505-513
[5] Jolly M *2010 TMS Annual Meeting & Exhibition* Seattle WA
[6] Gebelin J C, Lovis M and Jolly M *TMS2006*
[7] Dai X and Jolly M 2012 *Appl Energy* **89** issue 1 111-116
[8] Campbell J, 1991, *Casting*, (Butterworth-Heinemann, Oxford)
[9] Divandari M 2001 Mechanisms of bubble damage in castings *PhD Thesis* Uni. of Birmingham
[10] Dai X, Jolly M and Zeng B *TMS2012 Symposium: Energy Technologies and Carbon Dioxide Management* Orlando Florida USA

Determination of the metal/die interfacial heat transfer coefficient of high pressure die cast B390 alloy

Yongyou Cao[1,2], **Zhipeng Guo**[1,2] **and Shoumei Xiong**[1,2]
[1] Department of Mechanical Engineering, Tsinghua University, Beijing 100084, China
[2] State Key Laboratory of Automobile Safety and Energy, Tsinghua University, Beijing 100084, China

E-mail: smxiong@tsinghua.edu.cn

Abstract. High-pressure die cast B390 alloy was prepared on a 350 ton cold chamber die casting machine. The metal/die interfacial heat transfer coefficient of the alloy was investigated. Considering the filling process, a "finger"-shaped casting was designed for the experiments. This casting consisted of five plates with different thicknesses (0.05 inch or 1.27 mm to 0.25 inch or 6.35 mm) as well as individual ingates and overflows. Experiments under various operation conditions were conducted, and temperatures were measured at various specific locations inside the die. Based on the results, the interfacial heat transfer coefficient and heat flux were determined by solving the inverse heat transfer problem. The influence of the mold-filling sequence, sensor locations, as well as processing parameters including the casting pressure, die temperature, and fast/slow shot speeds on the heat transfer coefficient were discussed.

1. Introduction

High-pressure die casting (HPDC) is one of the most economical casting processes for mass producing net-shaped parts. Due to the excellent properties of die castings, an increasing number of die casting products are currently used in the automotive, aerospace, medical, electronic, and other industries. Techniques such as computer-assisted design and engineering, which have rapidly developed in recent years, are applied in the modeling and simulation of the filling and solidification processes in HPDC. The use of these techniques significantly optimizes processes and saves costs. However, computer-based techniques are only beneficial when the material properties as well as the boundary and initial conditions used as inputs are correct [1]. The interfacial heat transfer coefficient (IHTC), characterizing the thermal resistance between the metal and the mold, is believed to be the one of most important parameters during the solidification process for computer simulations [2-3]. Numerous studies [1-5] on determining the IHTC under various casting conditions have been conducted.

In the current paper, a die casting experiment was conducted using a "finger"-shaped casting of B390 alloy. The metal/die IHTC was determined according to the temperature readings obtained at different locations inside the die by solving the inverse heat transfer problems. The influence of the mold filling sequence, sensor locations, as well as processing parameters including the casting pressure, die temperature, and fast/slow shot speeds on the heat transfer coefficient were discussed.

2. Experiments

2.1. Finger-shaped casting

IOP Conf. Series: Materials Science and Engineering **33** (2012) 012010 doi:10.1088/1757-899X/33/1/012010

A specially designed casting, namely, a finger-shaped casting, was used in the current study. As shown in Figure 1, this casting has five plates with different thicknesses from T1 (0.05 inch or 1.27 mm) to T5 (0.25 inch or 6.35 mm), with an interval of 0.05 inch (or 1.27 mm). The biscuit of the casting was designed with a diameter of 60 mm and a thickness of about 20 mm. Each plate is 203 mm long (along the metal filling direction) and 19 mm wide. At the interface between the casting and die of each plate, one-dimensional heat transfer was assumed.

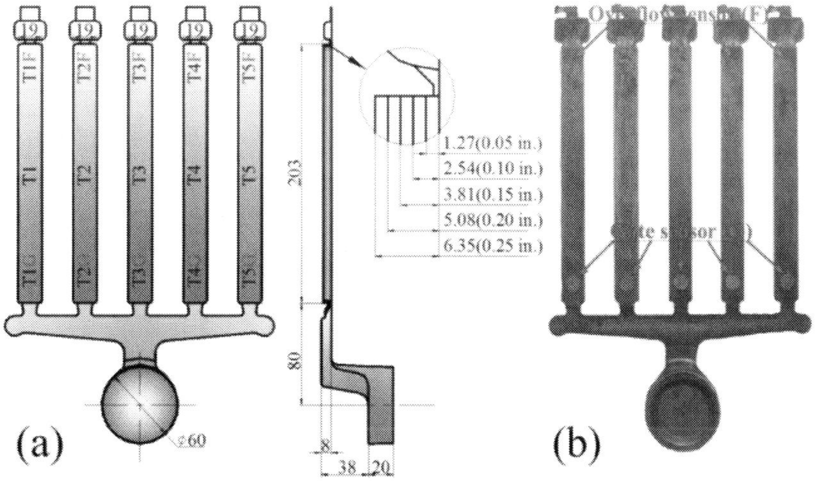

Figure 1. Finger-shaped casting: (a) configuration, and (b) actual casting showing the location of sensors.

Commercial aluminum (Al) alloy B390 (Al-17Si-4Cu) was poured into a TOYO 350 ton cold chamber high-pressure die casting machine. The chemical compositions of the alloy and the die material H13 steel is given in Table 1.

Table 1. Chemical composition of B390 alloy (Al-17Si-4Cu) and H13 steel.

Element wt.%	Si	Cu	Mg	Fe	Zn	Mn	Ni	Sn	Al
B390	16.54	4.63	0.58	0.68	0.82	0.14	0.055	0.060	Bal.
Element wt.%	C	Mn	Si	S	P	Cr	Mo	V	Fe
H13	0.396	0.36	0.94	<0.005	<0.025	5.05	1.25	0.82	Bal.

2.2. Die configuration and sensor installation

To gain a sufficiently rapid response time to follow the HPDC process and accurately measure the temperatures inside the die, a special temperature sensor unit (TSU) was designed. As shown in Figure 2, at each distance (1, 3, and 6 mm) from the front wall of the TSU, two thermocouples were adjusted. The thermocouples were grounded sheathed K-type thermocouples with 0.5 mm outside diameter and 0.045 mm wire diameter. These thermocouples were inserted into 1.1 mm diameter holes and vacuum nicro-brazed to the TSU body of H13 steel. To investigate the influence of filling, two TSUs were located at both ends of each finger plate close to the gate (G) and overflow (F), as illustrated in Figure 1(b).

Figure 3 displays the 10 TSUs containing 60 thermocouples embedded inside the stationary die until the front wall approached the cavity surface. The cooling as well as heating lines were present in the die at the distance of 25 mm from the parting face in order to control the mold temperature during HPDC. Real-time temperature data were then recorded using a data acquisition system manufactured by the Integrated Measurement Corporation (Berlin Germany) with a sampling rate of 500 Hz.

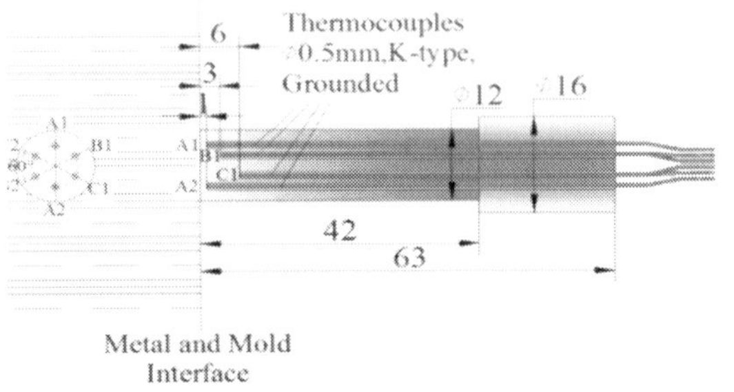

Figure 2. Configuration of the temperature sensor unit.

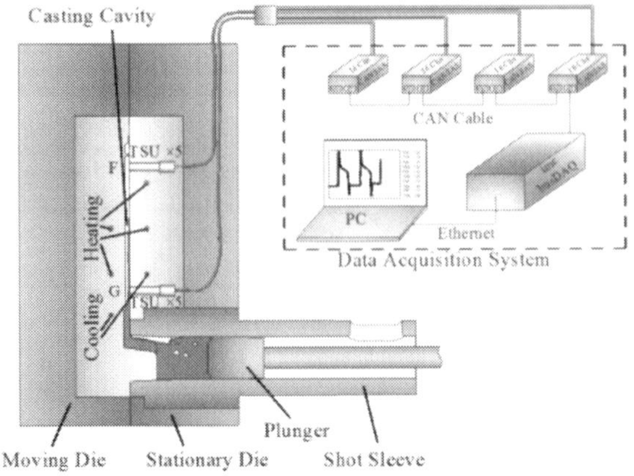

Figure 3. Graphical installations of temperature sensor units (TSUs) and data acquisition system in the cold-chamber die caster.

Table 2 lists the thermal properties of B390 alloy and H13 steel. The processing parameters, including the casting pressure (P), die temperature (T_d), fast shot speed (v_H), slow shot speed (v_L), and pouring temperature (T_P) were varied in each set of experiment. Others were fixed at the same condition shown in Table 3. Nearly 240 shots were performed; the first 20 shots were conducted to preheat the dies to thermal equilibrium.

Table 2. Thermal properties of related materials.

Thermal properties	B390	H13
Thermal conductance λ (W·m^{-1}·°C^{-1})	134	31.2 – 0.013 T[a]
Specific heat C (J·kg^{-1}·°C^{-1})	963	478 – 0.219 T
Density ρ (kg·m^{-3})	2730	7730 – 0.24 T
Solidus temperature T_S (°C)	505	1471
Liquidus temperature T_L (°C)	650	1404
Latent heat L_s (J·kg^{-1})	389000	209350

[a] T stands for temperature (°C).

Table 3. Values of related processing parameters.

Processing parameters	Basic condition	Variable
Pouring temperature T_P (°C)	730	700 and 760
Die temperature T_d (°C)	50	150
Casting pressure P (MPa)	69.98	101.09, 83.81, 67.39, and Min[a]
Slow shot speed v_L (m·s^{-1})	0.2	0.1, 0.3, 0.4, and 0.8
Fast shot speed v_H (m·s^{-1})	1.75	1.25, 1.5, 2, and 2.5

[a] Min stands for no intensification pressure.

3. Results and discussion

3.1. Heat transfer estimation

Figure 4 shows a sample of the measured temperature profiles of the 30 cycles at the T4G position under the basic condition. The temperature profiles of the same position illustrate cyclic characteristics in shape while the die casting phases were sequentially performed. Given that these cycles were performed under the same operation condition, the results showed that the process was quite reproducible. The temperature curves measured at the same distance, such as A1 and A2 or C1 and C2 almost overlapped. The maximum difference between the two temperature profiles measured at the same distance never exceeded 4 °C, indicating that the heat transfer process can be reasonably assumed to be one dimensional.

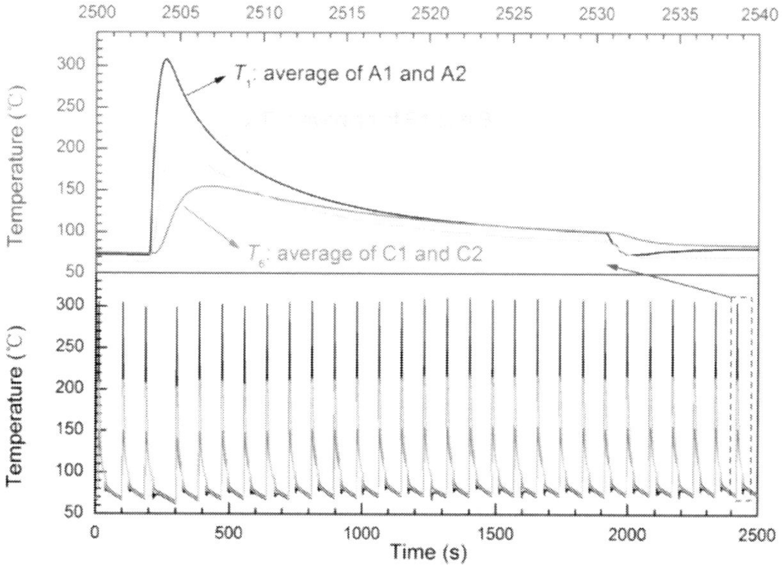

Figure 4. Sequential temperatures of temperature sensor unit at the T4G position for 30 cycles under the basic condition.

The IHTC cannot be directly calculated using the measured die temperatures because the interfacial heat flux density (IHFD) cannot be directly measured during casting and solidification. However, the computer program used an inverse method [4-6] based on the principle of Beck [7-8]. The average of measured temperatures at 1 and 6 mm from the cavity surface (T_1 and T_6) were used to evaluate IHFD, IHTC, and the die surface temperature. The inversely calculated temperature at 3 mm from the cavity surface, namely T_{3c}, was compared to the average temperature at B1 and B2 measured at the same distance to validate the inverse modeling. Once the shot was performed, the IHTC abruptly increased until reaching the peak value, maintained its value at a higher level, and then sharply decreased.

An analysis using the inverse method at the T4G position with respect to the last cycle of the measured temperatures (T_1 and T_6) was subsequently performed, as shown in Figure 4. The curves of IHFD (q) and IHTC (h), as well as the die surface temperature, casting center, and surface temperatures designated by T_{ds}, T_{cc}, and T_{cs} were then obtained, as shown in Figure 5. There was a very good fit between the measured (T_{3m}) and calculated (T_{3c}) temperatures at 3 mm, indicating that the inverse estimation results were quite reliable.

The casting surface temperature (T_{cs}) abruptly dropped after the shot was performed, only taking 72 ms to drop below the liquidus temperature. This result indicated that the molten alloy immediately lost its superheat after it made contact with the die cavity surface, and after a fast heat transfer at the metal/die interface. This phenomenon can be attributed to the prompt rise in the die surface temperature (T_{ds}). The casting surface temperature then continuously decreased, indicating a much smaller cooling rate, as also evidenced in Figure 5 from the smaller slope of the curve.

Corresponding to the rapid decrease in the casting surface temperature, the IHTC abruptly increased immediately after the shot was performed. This abrupt increase was also associated with the rapid increase in the IHFD until the peak value of 5.97×10^6 W·m^{-2} was reached. The IHTC kept growing until reaching 24.08×10^3 W·m^{-2}·K^{-1} when the IHTC started to fluctuate, rising and falling rapidly. The abrupt decrease in the IHTC was due to the fact that the close contact previously achieved between the casting and the die deteriorated. This deterioration was probably caused by the lack of the required pressure transferred from inside as the solidification process proceeded. An analysis of the IHFD curve revealed that after the peak value was reached, the heat flux exponentially decayed until the value was at a much lower level.

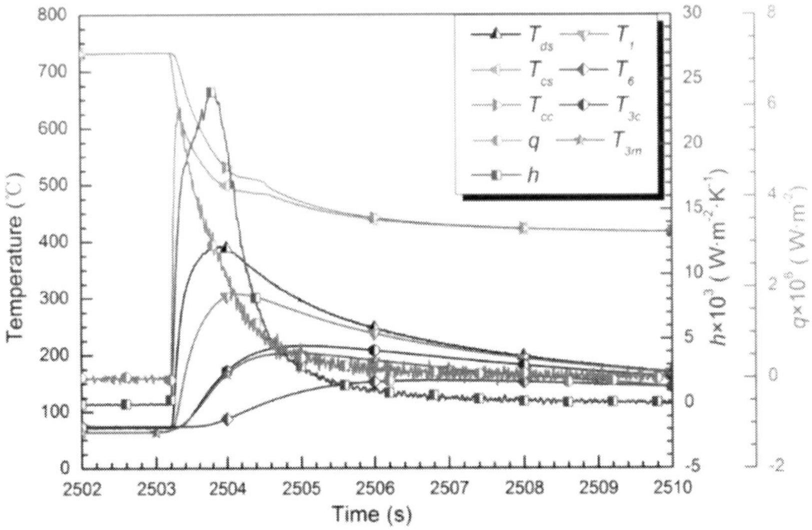

Figure 5. Typical results at the T4G position during the 30th cycle under the basic condition. T_1, T_6, and T_{3m} denote the average measured temperatures at 1, 6, and 3 mm from the cavity surface, respectively. T_{3c} is the inversely calculated temperature at 3 mm from the cavity surface. T_{ds} is the die surface temperature. T_{cs} is the casting surface temperature. T_{cc} is the casting center temperature. q is the interfacial heat flux density. h is the interfacial heat transfer coefficient.

3.2. Influence of interfacial heat transfer

3.2.1. *Casting geometry and sensor location.* Figure 6 details the related heat transfer curves of IHTC and IHFD for all positions during the 30th cycle under the basic condition, except for the T4F position due to data recording failure. The peak values of IHFD and IHTC are listed in Table 4. Similar trends in the profiles of both IHTC and IHFD were found. However, upon careful considerations of the different sensor locations and thicknesses, obvious differences between the corresponding interfacial heat transfer behaviors of the IHTC and IHFD curves (mainly, the shape and peak) were observed.

First, the peak values of IHTC and IHFD varied with the different positions and casting thicknesses. The IHTC peak of T3G near the gate reached the peak value of 64.29×10^3 W·m^{-2}·K^{-1}, the second peak value was 42.82×10^3 W·m^{-2}·K^{-1} of T5G, and the lowest peak was 15.28×10^3 W·m^{-2}·K^{-1} of T1G near the gate of the thinnest plate T1. Similar patterns existed in the IHFD.

Second, the high retention time of the IHTC varied with the different casting thicknesses. Figure 6 clearly shows that this time gradually extends with increased casting thickness.

Finally, the mold filling sequence of each plate highly depended on the shape of the runner system. The different horizontal distance from each plate to the vertical runner and different size of each ingate could directly influence the interfacial heat transfer behavior. The interfacial heat transfer behavior of T1F and T3G could be good examples for well explaining this point. By comparing the heat transfer profiles of the four positions in two plates (T1 and T3) to others, different trends and special features could be found. For example, due to the ingate nearest the vertical runner in T3G, the headmost filling during the fast shot phase caused the overheating of the alloy impacting the surface of the die cavity. This phenomenon created the distinguishing feature of the filling, leading to the maximum IHTC. As for the thinnest plate T1, given the farthest distance to the vertical runner and the smallest ingate, the molten alloy abruptly sprayed into the cavity with the simulated velocity of around 110 m·s^{-1}

compared with the average ingate velocity of 50 m·s^{-1} during the fast shot phase. Then, the metal jet first hit the sensor surface of the die cavity near the overflow (T1F) instead of that near the gate. Consequently, the IHFD near the overflow (T1F) reached its peak value 7.92 × 10^6 W·m^{-2} higher and 76 ms earlier than that near the gate (T1G), even compared with other positions. T1 plate was the thinnest (only 0.05 inch or 1.27 mm); hence, the quickly occurring solidification led to the slimmest IHTC profile. Helenius et al. [9] have similarly proposed that the peak value of the IHTC is greatly dependent on the status of the contact between the molten alloy and the die surface where the thermocouples are installed. A high IHTC peak value is observed when the melt alloy directly hits the location below which the thermocouples are adjusted during their initial contact.

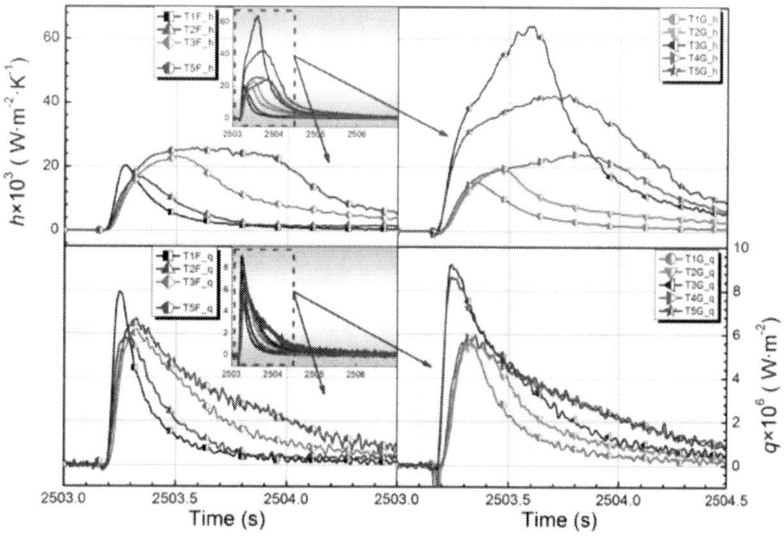

Figure 6. Profiles of IHTC and IHFD at all positions during the 30th cycle under the basic condition.

However, if the effect of the mold filling is ignored, the general trend of the IHTC peaks near the gate is higher than that near the overflow. The high retention time of the IHTC gradually extends with increased casting thickness.

Table 4. Values of q_{max} and h_{max} in Figure 6.

Sensor locations		q_{max} values		h_{max} values	
		Time[b] (s)	$q \times 10^6$ (W·m^{-2})	Time[b] (s)	$h \times 10^3$ (W·m^{-2}·K^{-1})
Near overflow (F)	T1F	2503.248	7.92	2503.272	20.37
	T2F	2503.276	5.83	2503.340	16.81
	T3F	2503.328	6.21	2503.508	23.19
	T4F[a]	-	-	-	-
	T5F	2503.324	6.72	2503.592	25.97
Near gate (G)	T1G	2503.324	5.69	2503.356	15.28
	T2G	2503.328	5.61	2503.444	19.60
	T3G	2503.240	9.20	2503.604	64.29
	T4G	2503.348	5.97	2503.824	24.08
	T5G	2503.244	8.61	2503.768	42.82

[a] Data at the T4G position failed.
[b] Time of prompt rise point of q and h was about 2503.170 s, as shown in Figure 6.

Figure 7 shows the distribution of the IHTC and IHFD peaks at all positions of the finger-shaped casting for 30 cycles under the basic condition, except the T4F position. Considering the particular filling feature for plate T3, wherein the average of IHTC peaks was about 60 × 10^3 W·m^{-2}·K^{-1}, the heat transfer behavior was ignored in the following discussion. As for the positions near the gate, the peak range of the IHTC was from 15 × 10^3 W·m^{-2}·K^{-1} to 42 × 10^3 W·m^{-2}·K^{-1}. A thicker casting corresponded to a higher average of IHTC peaks. The average of IHTC peaks at the T5G position of the thickest plate was about 42 × 10^3 W·m^{-2}·K^{-1}, whereas that at T1G was about 15 × 10^3 W·m^{-2}·K^{-1}. As for the positions near the overflow, a similar pattern was found, but the range of the averages was from 16 × 10^3 W·m^{-2}·K^{-1} to 26×10^3 W·m^{-2}·K^{-1}, which was lower than that near the gate. The fastest ingate velocity of the thinnest plate T1 severely vibrates the IHTC peaks at the T1F position. This vibration is greater than that on the peaks near the gate.

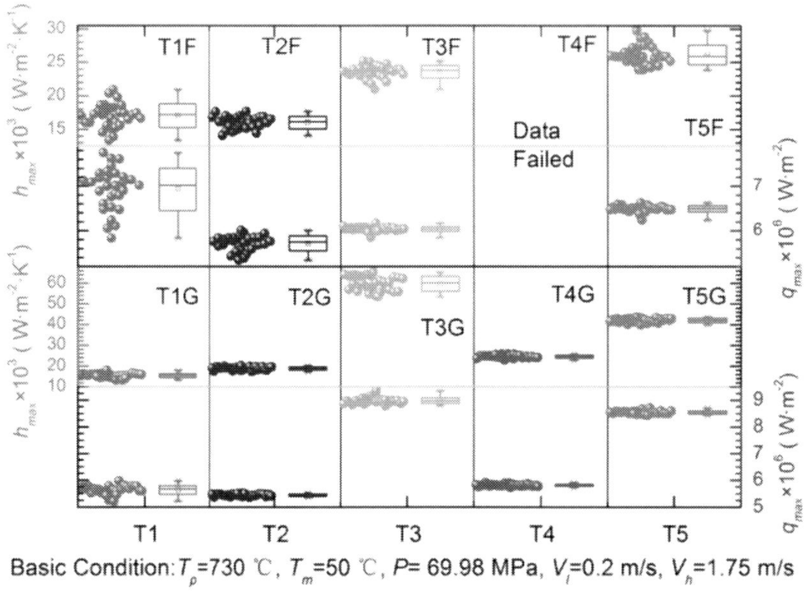

Figure 7. All peak values of IHTC and IHFD at all positions for 30 cycles under the basic condition. T1 to T5 are the five plates from the thinnest to the thickest, respectively.

3.2.2. Initial die surface temperature. After a great deal of data analysis of the finger-shaped casting of B390, the following findings were obtained.

(1) The casting pressure had nearly no effect on the IHTC peak value. With increased intensification pressure, the distribution of IHTC peaks was more discrete because of the molten alloy more sharply impacting the die surface.

(2) The peak value of IHTC near the gate increased with increased slow shot speed, especially for the plate T5.

(3) The peak value of IHTC near the gate mildly decreased with increased fast shot speed.

During HPDC process, the die temperature changes as the casting is sequentially performed and the initial die surface temperature (the die surface temperature before injection) on other hand could characterize the variation of the heat transfer between the metal and the die to some degree. After the thermal equilibrium inside the die was achieved, the initial die surface temperatures at different locations changed in a similar manner: the initial die surface temperature gradually grew as the castings were sequentially performed. Considering previous research on metal/die interfacial heat transfer behavior using the "step" shape casting for AM50 and ADC12 alloys, Guo et al.[5-6, 10] have found that the influence of processing parameters on the IHTC was mainly on the peak value. The shape of the IHTC profile was not significantly affected. The initial die surface temperature (T_{IDS}) had the dominant influence on the IHTC peak value (h_{max}) out of all the processing parameters. By correlation analysis, the following relationship was found to have the best fit:

$$h_{max} = \exp(B_h) \times T_{IDS}^{2+A_h} \tag{1}$$

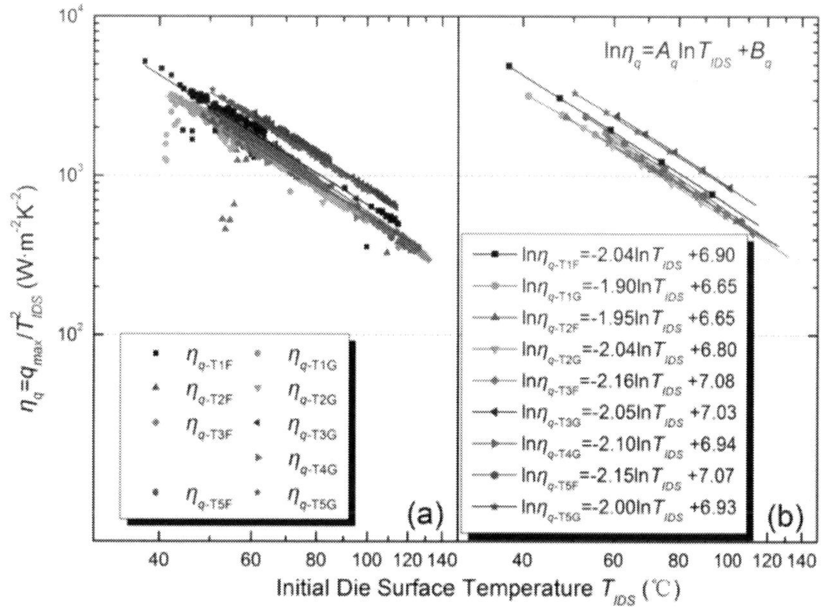

Figure 8. IHFD peaks as a function of the initial die surface temperature under all conditions.

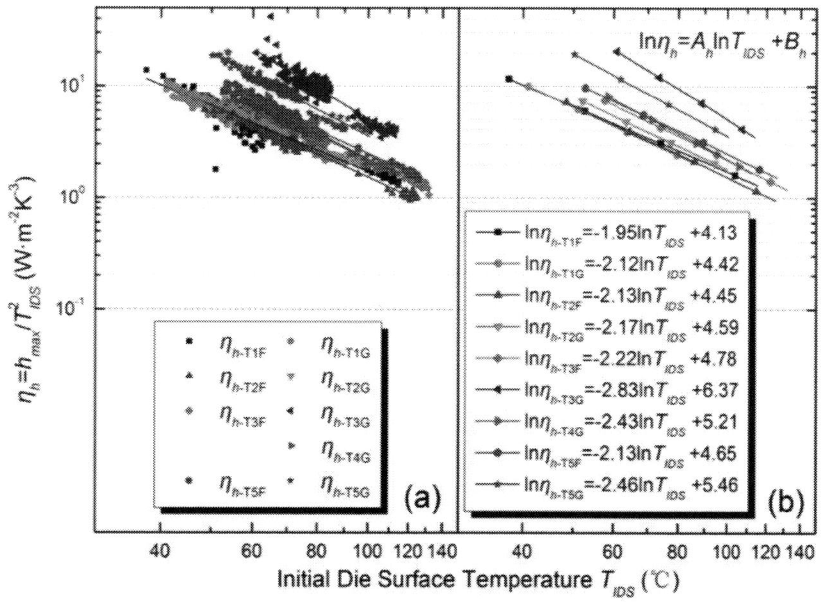

Figure 9. IHTC peaks as a function of the initial die surface temperature under all conditions.

Figures 8 and 9 illustrate the IHFD and IHTC peaks at 9 positions of 5 plates for about 240 cycles under all conditions as a set of functions of the initial die surface temperature. One of the prominent characteristics was that all data followed a negative slope versus the initial die surface temperature.

IOP Conf. Series: Materials Science and Engineering **33** (2012) 012010 doi:10.1088/1757-899X/33/1/012010

The other parameters did not show such a large influence, including the casing pressure, the slow and fast shot speeds, as well as the pouring temperature. The peak values of the IHFD and IHTC were dominated by the initial die surface temperature, and changed according to:

$$\ln\left(\frac{q_{\max}}{T_{IDS}^2}\right) = \ln \eta_q = A_q \ln T_{IDS} + B_q \tag{2}$$

$$\ln\left(\frac{h_{\max}}{T_{IDS}^2}\right) = \ln \eta_h = A_h \ln T_{IDS} + B_h \tag{3}$$

where A_q, B_q, A_h, and B_q are the coefficients of fitting.

Equations (2) and (3) show that the peak values of IHFD and IHTC decrease with increased initial die surface temperature.

The coefficients of the fitting the peak value of IHFD and IHTC varied with the sensor location.

By considering the IHTC as a function of the temperature gap at the interface, the effect of the initial die surface temperature can be easily understood. When the influence of the initial die surface temperature on the IHTC is considered, the explanation becomes easier. However, the possibility of the initial temperature of the die influencing the temperature of the liquid metal at the interface is not without merit, and should therefore influence the surface tension. As a consequence, the micro-contact conditions may change, resulting in the modifications of the heat resistance and interfacial heat transfer behavior. According to the relationship between the peak value of IHTC and the initial die surface temperature, the proper peak value of IHTC could be used for solidification simulation, as well as the function of temperature or solidification fraction considering the casting thickness.

4. Summary and conclusions

A detailed method for measuring the heat transfer of B390 alloy during HPDC has been established using a finger-shaped casting. The IHFD and IHTC have been successfully determined based on the inverse method. Based on the results, the following conclusions are drawn.

(1) The IHTC quickly increases right after the die casting shot until reaching the peak, and then slowly decreases. The peak range of the IHTC between the B390 alloy and H13 steel is from 15×10^3 W·m^{-2}·K^{-1} to 80×10^3 W·m^{-2}·K^{-1}.

(2) The mold-filling sequence of each plate directly influences the interfacial heat transfer behavior. The general value of the IHTC peak near the gate (from 15×10^3 W·m^{-2}·K^{-1} to 42×10^{-3} W·m^{-2}·K^{-1}) is higher than that near the overflow (ranging from 16×10^3 W·m^{-2}·K^{-1} to 26×10^3 W·m^{-2}·K^{-1}). The high retention time of the IHTC gradually extends with increased casting thickness.

(3) The initial die surface temperature has the most dominant influence on the IHTC peak value among all the processing parameters. With increased initial die surface temperature, the IHTC peak decreases. The coefficients of fitting the IHFD and IHTC peaks vary with the sensor location.

Acknowledgments

The current research was funded by the Ministry of Science and Technology (MOST) of China under contract nos. 2011ZX04001-71, 2011BAE21B00, 2011ZX04014-52, and 2010DFA72760. The die casting experiments were conducted with the aid of engineers from TOYO Machinery Co., Ltd.

References

[1] Hamasaiid A, Dour G, Dargusch M, Loulou T, Davidson C and Savage G 2006 Heat transfer at the casting/die interface in high pressure die casting - Experimental results and contribution to modeling *Modeling of Casting, Welding and Advanced Solidification Processes - XI (Opio, France, 28 May-2 June 2006) (TMS)* ed C A Gandin and M Bellet pp 1205-1216

[2] Pehlke R D 1964 Unidirectional analysis of heat transfer during continuous casting *Met. Eng. Q.* **4** 42
[3] Pehlke R D and Berry J T 2002 Heat transfer at the mold/metal interface in permanent mold casting of light alloys *Proc. from the 2nd Int. Aluminum Casting Technology Symp. (Columbus, USA, 7-9 October) (ASM International)* Ed M Tiryakioğlu and J Campbell pp 177-184
[4] Lau F, Lee W B, Xiong S M and Liu B C 1998 Study of the interfacial heat transfer between an iron casting and a metallic mould *J. Mater. Process. Technol.* **79** 25
[5] Guo Z P, Xiong S M, Liu B C Li M and Allison J 2008 Effect of process parameters, casting thickness, and alloys on the interfacial heat-transfer coefficient in the high-pressure die-casting process *Metall. Mater. Trans. A* **39A** 2896
[6] Guo Z P 2009 Study on metal – die interfacial heat transfer behaviour during high pressure die casting process *Doctor of Engineering* (Beijing: Tsinghua University)
[7] Beck J V, Blackwell B and Haji-Sheikh A 1996 Comparison of some inverse heat conduction methods using experimental data *Int. J. Heat Mass Tran.* **39** 3649
[8] Helenius R, Lohne O, Arnberg L and Laukli H I 2005 The heat transfer during filling of a high-pressure die-casting shot sleeve. *Mater. Sci. Eng.* **413-414A** 52
[9] Guo Z P, Xiong S M, Liu B C Li M and Allison J 2009 Understanding of the influence of process parameters on the heat transfer behavior at the metal/die interface in high pressure die casting process. *Sci. in China Ser. E-Tech. Sci.* **52** 172

Experiment and simulation study on particulate distribution in A356/SiCp suction casting

F Sun, H Zhao, Y Zhao, P Dong and F Chen

School of Mechanical and Automotive Engineering, South China University of Technology, Guangzhou, China

E-mail: hdzhao@scut.edu.cn

Abstract. In this investigation, an A356/10%100μm SiCp upward suction casting with a cylinder cavity of 40 mm diameter and 400 mm height was conducted with stirring technique. The SiCp contents in the casting different sections were quantitatively measured with micro digital image analysis system, and then were compared and analyzed. To simulate the casting mold filling, the composite mixture was simulated based on the Euler method while the particulate flow was calculated with the Lagrangian method for predicting the SiCp distribution. The flow of the mixture and particulate phases was analyzed and discussed. The simulated particulate distribution was compared and validated with the experimental results. Both results have shown that the particulate contents decreased along the upward mold filling direction, and that the particulates were prone to be trapped near the mold wall at the filling beginning part due to local lower temperature and higher viscosity and they distributed towards the center as the filling distance increases.

1. Introduction

Al-based matrix composites exhibit many advantages over the monolithic Al alloys including high specific strength, stiffness, elastic modulus, and enhanced wear resistances [1]. Casting process is a widely used method to make the composite components with high efficiency and low cost. For the particulate reinforce castings, the particulate distribution and segregation have significant effect on the casting mechanical properties and friction performance. The particulate segregation may take place during three periods: stirring, mold filling, and solidification periods. Hashim [2] and Naher [3] investigated composite slurry flow and particulate distribution in crucible during composite melt stirring preparation, and then pointed out the main factors about the particulate distribution. As for the particulate distribution during solidification, Zhang [4] observed that more particulates concentrated in the center of an Al7%Si-10%B_4C die casting, and concluded that grains formed from the casting surface pushed the particulates to the centre; Cetin [5] studied the effect of solidification rate on SiC particle spatial distribution in A356 alloy composites, and concluded that clustering of particles was most pronounced at distance ranging about 150 to 500μm. However the investigations on the particle segregation during composite casting mold filling are very limited and desirable.

In this study, an A356-based composite upwards suction casting with reinforced SiC particulates was conducted. The SiC contents in the casting different parts were quantitatively measured, and then analyzed. Based on the Euler and Lagangrian methods, the composite casting mold filling was simulated, and the particulate distribution in the casting was compared and validated with the experimental measurement.

2. Experimental works

2.1. Experiment procedures

Figure 1 shows the experimental system to fabricate the composite casting, which included a graphite crucible, an electric heater, and a H13 cylinder mold of 40mm diameter and 400 mm height connected with an air pump. In the experiment, a weighted quantity of A356 alloy used as matrix materials was melted in the crucible to 700℃. The weighted SiC particulates with an average diameter of 100 μm for 10% nominal reinforced fraction were washed in chemical reagent ethanol with an ultrasonic cleaner, and then were heated at 70℃ for 12 hours to remove moisture. Before adding the particulates into the molten alloy, they were pre-treated by heating at 700℃ for 4 hours. The pretreatment may improve the wettability of the particulates and the matrix liquid due to formed SiO_2 [6] on the particulate surfaces and can decrease their temperature differences. The particulates were added into the liquid alloy from the top of the crucible and mixed with a mechanical stirrer at a speed of 300rpm for 20min. Following reheated to 700℃, the mixed slurry were degassed and fluxed.

In the suction casting process, the die was preheated to 300℃ in a heat treatment furnace and immersed into the slurry, and the air pump was turned on and a measured 0.075MPa vacuum pressure was formed in the die cavity to suck the slurry flowing upward.

The obtained composite casting was of 350mm height less than the die height since the die bottom part was immersed into the slurry. The casting was cut into two halve along the longitude direction. As shown in Figure 1, three cross-sections of a half casting at different heights named S1, S2 and S3 were selected for the assessment of the particulate contents and distribution. The sections polished with sand papers were scanned with a Leica microscope to get digital photos.

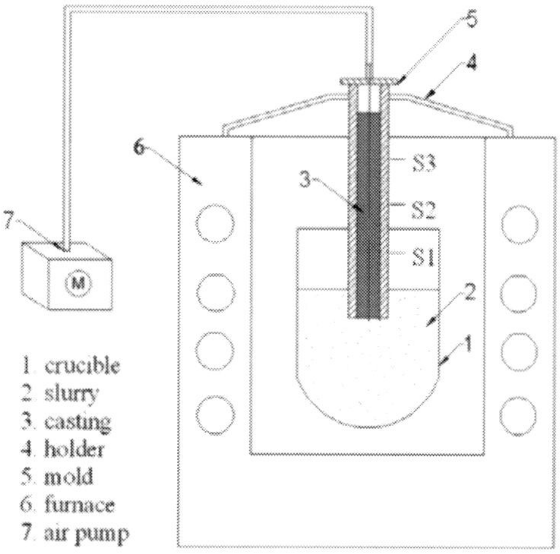

Figure 1. Experimental system for the composite suction casting.

2.2. Results and analysis

Figure 2 displays the macro photographs of the three sections. It was seen that in the S3 section the particulates distributed uniformly in the whole section whereas in the other two sections the particulates were richer near the mold wall compared to the central region.

Figure 2. Macro photographs of the selected sections, (a) S1, (b) S2 and (c) S3.

Figure 3 shows the metallographs of the middle and boundary locations of the section S1 and S3. It was clear that the particulates with irregular shapes had good wettability with the matrix. The literatures have reported that there usually existed porosities in composite castings due to gas entrapped during stirring process and clustered particulates [7]. However, as shown in Figure 3, the porosities in the experimental casting were very few and almost all particulates distributed individually in the matrix. This indicated the validity of the fabrication method for the composite slurry.

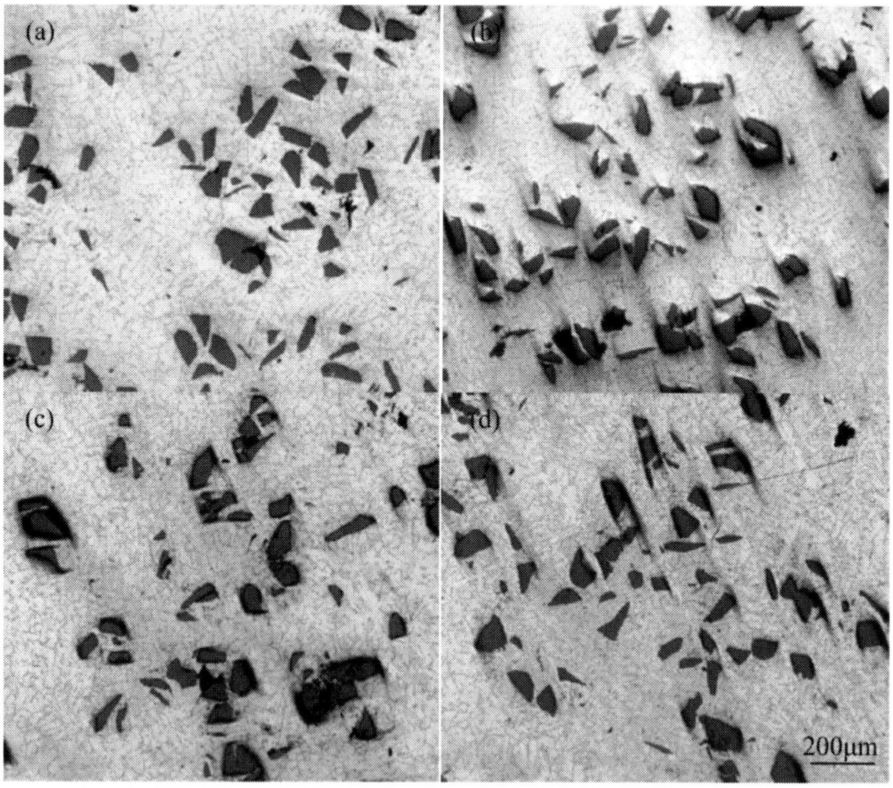

Figure 3. Metallographs of the locations in the sections (a) R/2 in S1, (b) R in S1, (c) R/2 in S3 and (d) R in S3 in the casting

The metallurgical photos, like the photos shown in Figure 3, at the central, 10mm and 20mm locations in the radius direction of each section were taken with Leica Materials Workstation software attached to the microscope. For each location, over five photos were made from the different micro-fields. Then these photos were quantitatively analyzed with Image-Pro Plus software for assessment of the particulate fraction at each location. The average measured fraction at each location is list in the Table 1. The average fraction of the all measured values was 5.7% less than the nominal particulate fraction of 10% since some particulates were lost during the stirring, degassing and fluxing processes.

The fractions in the section S1, S2, and S3 showed that the particulates in the casting were decreasing as the filling distance increased. In the section S1 and S2, the particulate fractions went up along the radius direction, while in the section S3 more particulates segregated in the central part.

Table 1. The measured particulate fractions at the section different locations

Section	r=0	r=10mm	r=20mm
S1	0.039	0.067	0.117
S2	0.028	0.061	0.104
S3	0.044	0.022	0.027

3. Numerical simulation

3.1. Mathematical Model

During the mold filling of the experiment composite casting, three phases including air, particulate and molten matrix alloy flowed. The particulate phase was dispersed and the air one was a continuous fluid. According to computational fluid dynamics [8], the matrix alloy could also be regarded as a continuum phase because the particulate fraction was less than 10%.

In the simulation of the casting mold filling, assuming that the liquid metal density, heat capacity and thermal conductivity were constant, the VOF method of the Euler approach was applied to solve the air and mixture alloy flow by solving continuity, momentum energy and volume equations as following [9]:

Continuity equation
$$\nabla \cdot \vec{u} = 0 \tag{1}$$

Momentum equation
$$\frac{\partial \vec{u}}{\partial t} + (\vec{u} \cdot \nabla)\vec{u} = \vec{f} - \frac{1}{\rho}\nabla P + \nu \nabla^2 \vec{u} \tag{2}$$

Energy equation
$$\frac{\partial T}{\partial t} + \vec{u} \cdot \nabla T = \frac{\lambda}{\rho c_p}\nabla^2 T + \frac{S_T}{\rho} \tag{3}$$

Volume fraction equation
$$\frac{\partial f_V}{\partial t} + \vec{u} \cdot \nabla f_V = 0 \tag{4}$$

where $\vec{u}$ is flow velocity, ν is dynamic viscosity, ρ is density, c is heat capacity, λ is thermal conductivity, t is time, P is static pressure, T is temperature, $\vec{f}$ is mass force, S_T is source term, f_V is the function of volume fraction.

On the basis of the flow calculation of the mixture alloy, the particulate velocity in the casting mold filling was predicted by using the Lagrangian approach which is proper to describe movement and trajectory of a discrete phase. The particulate velocity in a direction was determined by

$$\frac{\mathrm{d}u_p}{\mathrm{d}t} = F_D(u - u_p) + \frac{g_x(\rho_p - \rho)}{\rho_p} + F_x \tag{5}$$

where F_x is an additional acceleration (force/unit particulate mass) term, $F_D(u - u_p)$ is the drag force per unit particulate mass and

$$F_D = \frac{18\mu}{\rho_p d_p^2} \frac{C_D Re}{24} \tag{6}$$

where u is the mixture alloy velocity, u_p is the particulate velocity, μ is the viscosity of the mixture fluid, ρ is the mixture alloy density, ρ_p is the density of the particulate, d_p is the particulate diameter, Re is the relative Reynolds number, C_D is the drag coefficient [10].

The mixture alloy viscosity was treated as a temperature-dependent variable while its density, specific heat and thermal conductivity were dependent on the phase fractions. They were calculated by the below equations [8,11]

$$\frac{\mu}{\mu_{fT}} = 1 + 2.5C_V;\ \rho = \rho_f(1 - C_V) + \rho_p C_V;$$

$$c = c_p C_M + c_f(1 - C_M); \tag{7}$$

$$\lambda = \lambda_f \left[\frac{2\lambda_f + \lambda_p - 2C_V(\lambda_f - \lambda_p)}{2\lambda_f + \lambda_p + C_V(\lambda_f - \lambda_p)} \right]$$

where C_M is the particulate mass fraction, C_V is the particulate volume fraction, μ_{fT} is the fluid viscosity at different temperature, c_p is the particulate heat capacity, c_f is the fluid heat capacity, λ_p is the particulate thermal conductivity, λ_f is the fluid thermal conductivity.

3.2. Simulation and Results

The model simulating the mold filling and particulate movement was solved with the scheme embodied in the FLUENT code. The FLUENT allows users to incorporate user-defined physics characteristics and calculation equations into its main program. Due to the casting cylindrical geometry, a 2D grid system with 16,512 mesh cells including the mold and mold cavity was employed. Since the casting particulates were of irregular shape as shown in Figure 3, the shape coefficient of the particulates was set to 0.6. The physical data used in simulation are mentioned in Table 2. In the transport calculation, the flow field of the continuous phases was solved first, and then the particulate's movement was determined according to Equation 5 and then their distribution and the mixture alloy momentum were updated.

Table 2. Thermal physical data used in simulation

	Density (kg/m^3)	Heat capacity [J/(kg·℃)]	Thermal conductivity [W/(m·℃)]	Viscosity * (10^{-3} Pa·s)	
Fluid	ρ_f=2500	c_f=963	λ_f=110	T=700 ℃	1.28
Particulate	ρ_p=3200	c_p=1400	λ_p=126	T=665 ℃	2
Mold	ρ_m=7630	c_m=502	λ_m=30	T=620 ℃	320

* Partial viscosity data used in simulation.

The flow velocity distribution at different filling time is shown in Figure 4, in which the arrows point to the locations of the interface between the air and mixture alloy fluids. There was a long and narrow layer near the boundary wall in the mixture alloy at the beginning filling part of the mold, in which the velocities were very smaller than those in the central part. The reason for that is the alloy near the wall had the lower temperature and higher viscosity due to quick heat transfer between the mixture alloy and H13 mold. Comparing Figure 4a, 4b and 4c, the layer became longer as the filling proceeding.

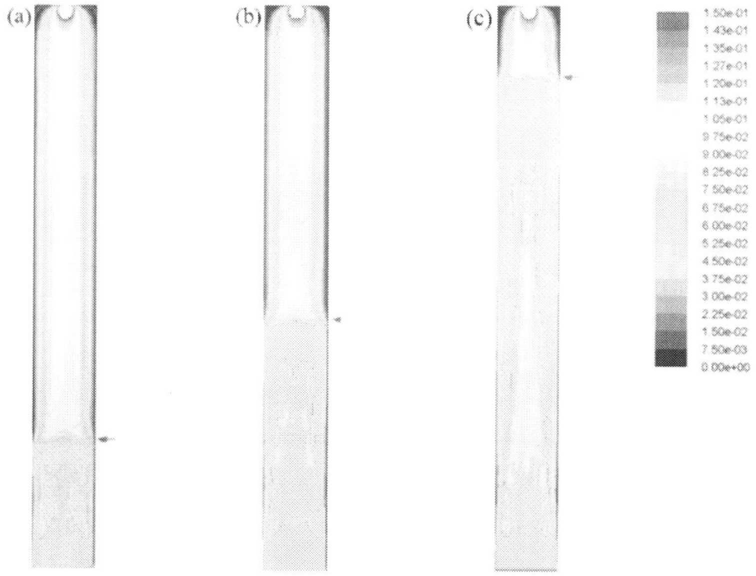

Figure 4. The flow velocity distribution at different filling time, (a) 1.07s, (b) 2.15s and (c) 4.30s.

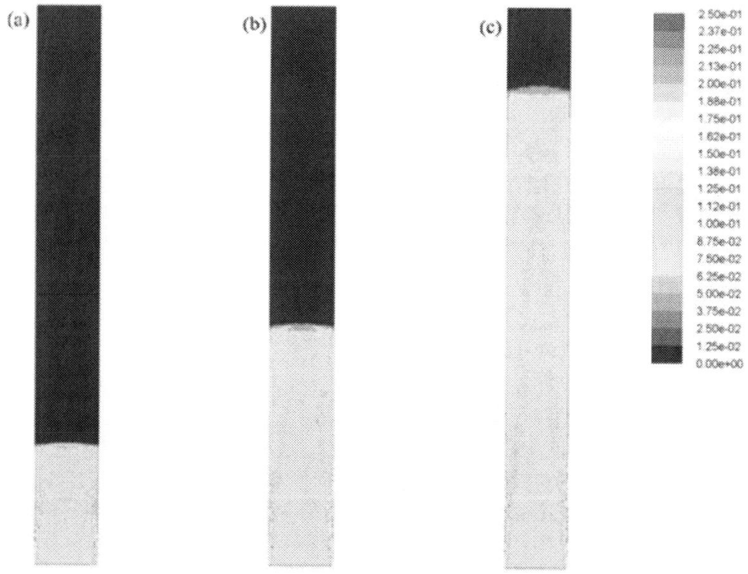

Figure 5. The particulate distribution at different filling time, (a) 1.07s, (b) 2.15s and (c) 4.30s.

Figure 5 shows the particulate distribution at different filling time. As can be seen in Figure 5a, at the beginning of the filling, more particulates segregated near the wall above the inlet and a particulate rich layer formed. With the filling continuing, this layer became thicker as given in Figure 5b. But in the later filling process the particulates tended to move to the central part of the mold cavity so that more particulates distributed in the central part of the casting upper part.

4. Validation and discussion

The simulated and experimental results of the final particulate fraction at the three sections are plotted in Figure 6 for the comparison purpose. In order to analyze the experimental and simulated results conveniently, ξ (the ratio of the actual particulate volume fraction to the largest fraction) was used to represent the relative fraction.

As can be seen in Figure 6a, both two results showed that within the section S1 there were more particulates near the boundary wall than in the central part. When the mixture alloy entered the die cavity after the casting filling started, the temperature drop of the alloy at the vicinity of the die wall was higher due to energy lost through the wall surface. As a result, the alloy there had lower temperature and higher viscosity. Since drag force is proportional to viscosity (Equation 6), the particulates were prone to be entrapped near the boundary wall. Therefore, more particulates were observed at the vicinity of the wall.

From Figure 6c and Table 1, it was known that the average particulate fractions decreased along the upward mold filling direction. This may result from the relative skip velocity between the mixture alloy and particulates as well as the particulate settling. The skip velocity is depended on the drag force to the discrete phase from the continuum fluid (the first term in the right hand of Equation 5). The settling effect due to the gravity was considered by the second term in the right hand of Equation 5.

In the Figure 6c, both two results showed that as for the section S3 the particulates were in uniform distribution and a little more particulates were found in the central part. This was different from the simulation results of Zr_2O_3 particulate distribution in a Pb20%Sn casting [12], in which more particulates segregated near the mold wall.

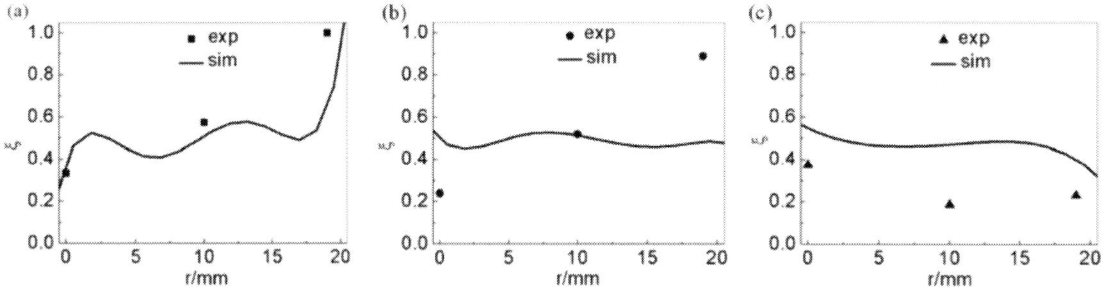

Figure 6. Comparison of simulated and experimental results of the particulate distribution in the selected sections, (a) S1, (b) S2 and (c) S3.

The above comparison showed that the experimental and simulated results of the particulate distribution in the section S1 and S3 were in good agreement. However, in the section S2 the simulation results indicated the uniform particulate distribution whereas the experiment analysis showed that the particulates segregated at the vicinity of the mold wall (please see Figure 6b). Comparing the particulate distribution in the section S1 and S2 in Figure 2 and 6 demonstrated that the particulate rich layer occurred near the mold wall at the casting lower half part. The simulation results successfully predicted occurrence of this particulate segregation layer as shown in Figure 5 but underestimated the layer height. This underestimation was possibly caused by neglecting boundary slip in the simulation setup since the mold inside surface was well machined and was of very low surface roughness. Otherwise the mixture alloy should have higher velocity along the mold wall [13,14] possibly resulting in a higher particulate segregation layer near the wall. Another factor for the simulation disagreement was that the particulate interaction was not considered in the simulation.

5. Conclusion

(1) The A356/SiCp upward suction casting was conducted, and the particulate contents in the casting different sections were quantitatively measured and analyzed. The simulation of the mixture alloy and particulate flow during the casting mold filling was carried out based on the Euler and Lagrangian methods, respectively.

(2) Both the experimental and simulation results have shown that as the filling distance increased, the particulate fractions decreased because of the drag force and settling effect, and that the particulates tended to move to the central part resulting in the uniform particulate distribution in the casting upper part.

(3) The results indicated that at the filling beginning part there was the particulate rich layer at the vicinity of the mold wall due to local lower temperature and higher viscosity. Unfortunately the simulation underestimated the height of this layer, which was possibly caused by neglecting slip of the mixture alloy on the wall in the simulation.

Acknowledgement

The research is supported by the projects of National Natural Science Foundation of China (50975093) and National Science & Technology Pillar Program by Science and Technology Ministry of China (2011BAE21B02).

References

[1] Miracle D B 2005 *Compos. Sci. Technol.* **65** 2526
[2] Hashim J, Looney L, Hashmi M S J 2002 *J. Mater. Process. Technol.* **123** 258
[3] Naher S, Brabazon D, Looney L 2007 *Compos. Part* A. **38** 719
[4] Zhang Z, Chen X G, Charette A 2007 *J. Mater. Sci.* **42** 7354
[5] Cetin A, Kalkanli A 2008 *J. Mater. Process. Technol.* **205** 1
[6] Bindumadhavan P N, Chia T K, Chandrasekaran M, Wah H K, Lam O L N and Prabhakar 2001 *Mater. Sci. Eng.* A **315** 217
[7] Prabu S B, Karunamoorthy L, Kathiresan S and Mohan B 2006 *J. Mater. Process. Technol.* **171** 268
[8] Saleh J M 2002 *Fluid Flow Handbook* (New York: McGraw-Hill)
[9] Cebeci T, Shao J P, Kafyeke F and Laurendeau E 2005 *Computational Fluid Dynamics for Engineers* (California: Horizons Publishing Inc., Long Beach)
[10] Hadier A, Levenspiel O 1989 *Powder Technol.* **58** 63
[11] Braszczynski J, Zyska A 2000 *Mater. Sci. Eng.* **A 278** 195
[12] Mat M D, Aldas K 2006 *Appl. Math. Comput.* **177** 300
[13] Mirbagheri S M H, Shrinparvar M and Ashory H 2004 *Mater. Des.* **25** 655
[14] Zhao H D, Wang F, Li Y Y, Xia W 2009 *J. Mater. Process. Technol.* **209** 4537

IOP Conf. Series: Materials Science and Engineering **33** (2012) 012012 doi:10.1088/1757-899X/33/1/012012

A numerical benchmark test for continuous casting of steel

B Šarler[1,3], R Vertnik[1,2] and K Mramor[3]

[1] Laboratory for Multiphase Processes, University of Nova Gorica, Vipavska 13, SI-5000 Nova Gorica, Slovenia

[2] Štore Steel d.o.o., Technical Development, Železarska 3, SI-3220 Štore, Slovenia

[3] COBIK Center of Excellence for Biosensorics, Automation and Process Control, Velika pot 22, SI-5250 Solkan, Slovenia

E-mail: bozidar.sarler@ung.si, robert.vertnik@store-steel.si, katarina.mramor@cobik.si

Abstract. There is a continuously developing need for benchmarking of solidification simulations - from the theoretical as well as from the applied points of view. The benchmarking is usually done in two parts. The verification part confirms the proper numerical solution (Are we solving the equations correctly?) and the validation part (Are we solving the right equations?) confirms the proper response of the simulations regarding the experimental evidence. The history of related benchmarking shows differences of the results between different numerical methods, and differences in comparison with the experiments when solving even quite simple solidification situations. The present benchmark test proposes the verification benchmark for continuous casting of steel. Since the simulations of the temperatures, velocities, pressures and concentrations in the continuous casting represent a multiscale and multiphysics problem of high complexity (far away from a closed form solution), the verification can be done only by comparing the results of different numerical methods for solving the same governing equations. This has been done in the context of continuous casting by breaking the considered two dimensional problem into several sub problems by increasing complexity of geometry (straight in vertical direction, curved), boundary conditions (linear, non-linear), material properties (only Fe and Fe - C), microscopic considerations (Lever rule and Scheil rule). The governing equations complexity first involves only convective - diffusive heat transport with a predetermined velocity field, and assumed laminar and turbulent velocity calculations afterwards, modelled by a low Reynolds number turbulence model. Further complications involve the presence of the electromagnetic forces. The paper represents guidelines for the presentation of the numerical method, discretisation and results, as well as some of the results, obtained by the commercial finite volume based code and our in-house meshless method based laboratory code.

1. Introduction

Continuous casting [1] is the most common process in the production of steel. The process starts by pouring the molten metal into to the water cooled mould, where cooling intensity is high enough to solidify steel around the inner surface of the mould and generates the solid shell with molten metal in the centre of strand. After several minutes, the strand is pulled into the secondary cooling system, which contains water spray systems with much smaller cooling intensity as in the water channels in the mould, and rollers, which support and guide the strand up to the end of the casting machine. The quality of the casted product (round or square billets, blooms or slabs) depends mainly on the process parameters in the mould region, where complex physical phenomena occur. The liquid metal, poured with a high velocity from the submerged entry nozzle (SEM) into the mould, produces turbulent flow with several re-circulating zones. Large heat fluxes, extracted from the mould, are a consequence of very high flow rates of the cooling water in the mould channels. It is impossible to measure temperature and velocity field inside the strand due to the very high temperatures of steel and inaccessibility of the region during the process. The numerical models [2] help us to better understand the casting behaviour and to further improve and optimize the process, particularly in such experimentally sophisticated situations. Various numerical methods have already been used to simulate the described problem, such as Finite Volume Method (FVM) [3-7], Finite Element Method (FEM) [8], and more advanced meshless method, like the Local Radial Basis Function Collocation Method (LRBFCM) [9-11]. Only very recently, different computational models (different governing equations, combined with different numerical methods) have been compared to assess the typical flow situations in continuous casting process [12] and their relation to model experiments. The solidification simulations are usually tested on a sequence of benchmarks that follow NAFEMS- type heat conduction simulations [13], convective diffusive cases with phase change [14], natural convection cases [15], solid-liquid specific cases with freezing of pure substance [16,17] and solidification of binary alloy [18,19]. However, until now, there is a lack of benchmark tests that would enable verification of numerical methods for solving transport phenomena in continuous casting of steel in an ordered way. There are no systematic well documented benchmark tests currently existing even for such a simple continuous casting models like the slice model [20]. In this paper, a simple benchmark test for the numerical solution of the continuous casting of steel is proposed. The main aim of the test is to compare the results of different numerical methods of the same physical model.

The presented physical model considers turbulent fluid flow with electromagnetic forces, solidification, and macrosegregation. The solution procedure is based on explicit time discretization and fractional step method [21] for velocity-pressure coupling. Turbulent flow is modelled by the two-equation eddy-viscosity model with the low-Re corrections. The numerical results are shown for the case without electromagnetic forces and macrosegregation, and with simplified microsegregation model. The meshless LRBFCM is used for the spatial discretization. The comparison is made with the results obtained by the commercial software Fluent, which is based on the FVM. The objective of this exercise is to test the ability of different numerical methods and algorithms to agree on a solution and to produce a reference numerical solution of the given problem.

2. Governing equations

The system of equations, describing the heat and fluid flow with solidification in the continuous casting of steel, is derived based on the mixture continuum formulation by Bennon and Incropera [22] and Reynolds time-averaging approach for modelling incompressible turbulent flow [23]. The time-averaged transport equations for mass, energy, momentum and solute conservation are

$$\nabla \cdot \mathbf{u} = 0, \tag{1}$$

$$\rho \frac{\partial \mathbf{u}}{\partial t} + \rho \nabla \cdot (\mathbf{u}\mathbf{u}) = -\nabla P + \nabla \cdot \left[2(\mu + \mu_t)\mathbf{S} \right] - \frac{2}{3}\rho \nabla k - \mu_L \frac{K_0 \left(1 - f_L^2\right)}{f_L^3} (\mathbf{u} - \mathbf{u}_S) + \mathbf{F}_b + \mathbf{F}_m, \tag{2}$$

$$\rho \frac{\partial h}{\partial t} + \rho \nabla \cdot (\mathbf{u}h) = \nabla \cdot (\lambda \nabla T) + \rho \nabla \cdot \left[f_S (h_L - h_S)(\mathbf{u} - \mathbf{u}_S) \right] + \nabla \cdot \left(f_L \frac{\mu_t}{\sigma_T} \nabla h_L \right), \tag{3}$$

$$\rho\frac{\partial C}{\partial t}+\rho\nabla\cdot(\mathbf{u}C)=\nabla\cdot\left(f_S D_S \nabla C_S+f_L D_L \nabla C_L\right)+\rho\nabla\cdot\left[(\mathbf{u}-\mathbf{u}_S)(C_L-C)\right]+\nabla\cdot\left(\frac{f_L\cdot\mu_t}{\sigma_C}\nabla C_L\right) \quad (4)$$

with $\mathbf{u}$, P, h, T and C for velocity, pressure, enthalpy, temperature and solute concentration, respectively, and ρ, μ, μ_t, f, λ, D are standing for density, molecular dynamic viscosity, turbulent dynamic viscosity, volume phase fraction, thermal conductivity, and diffusivity of the material. In above equations, subscripts L and S denote liquid and solid phase. $\mathbf{S}$ stands for the strain-rate tensor. The fourth term in equation (2) represents the Darcy term, where K_0 is the morphology constant of the porous medium and $\mathbf{u}_S$ is velocity of the solid phase. The fifth term in equation (2) represents the buoyancy forces, defined as

$$\mathbf{F}_b=\rho\left[\beta_T\mathbf{g}(T-T_{ref})+\beta_C\mathbf{g}(C_L-C_{ref})\right] \quad (5)$$

where β_T, β_C, T_{ref} and C_{ref} stand for thermal expansion coefficient, solute expansion coefficient, reference temperature and reference concentration, respectively. The turbulent dynamic viscosity is defined as

$$\mu_t=\rho c_\mu f_\mu\frac{k^2}{\varepsilon} \quad (6)$$

where k and ε are kinetic energy and dissipation rate, respectively. They are calculated by the following transport equations

$$\rho\frac{\partial k}{\partial t}+\rho\nabla\cdot(\mathbf{u}k)=\nabla\cdot\left[\left(\mu_L+\frac{\mu_t}{\sigma_k}\right)\nabla k\right]+P_k-\rho\varepsilon+\rho D+\mu_L\frac{K_0\left(1-f_L^2\right)}{f_L^3}k, \quad (7)$$

$$\rho\frac{\partial \varepsilon}{\partial t}+\rho\nabla\cdot(\mathbf{u}\varepsilon)=\nabla\cdot\left[\left(\mu_L+\frac{\mu_t}{\sigma_\varepsilon}\right)\nabla \varepsilon\right]+\rho(c_{1\varepsilon}f_1-c_{2\varepsilon}f_2\varepsilon)\frac{\varepsilon}{k}+\rho E+\mu_L\frac{K_0\left(1-f_L^2\right)}{f_L^3}\varepsilon. \quad (8)$$

In equations (3,4,6-8) c_μ, f_μ, $c_{1\varepsilon}$, f_1, $c_{2\varepsilon}$, f_2, σ_T, σ_C, σ_k and σ_ε are the closure coefficients, while D and E are additional source terms of the low-Reynolds turbulent model [23]. The last term on the left side of equation (2) represents the electromagnetic force acting on the conductive material moving through the induced magnetic field, defined as

$$\mathbf{F}_m=\mathbf{J}\times\mathbf{B} \quad (9)$$

where $\mathbf{J}$ and $\mathbf{B}$ are the induced current density and the magnetic flux density. The current density is calculated by Maxwell equations [5,6] under assumption that the induced magnetic field is negligible in comparison with external applied magnetic field

$$\nabla\cdot\mathbf{J}=0, \quad (10)$$

$$\mathbf{J}=\sigma(\mathbf{E}+\mathbf{u}\times\mathbf{B}), \quad (11)$$

$$\mathbf{E}=-\nabla\phi \quad (12)$$

where σ is electrical conductivity, $\mathbf{E}$ is electric field density and ϕ is electrical potential, calculated from the following equation

$$\nabla^2\phi=\nabla\cdot(\mathbf{u}\times\mathbf{B}). \quad (13)$$

The constitutive temperature-enthalpy relationship h_S and h_L in equation (3) are defined as

$$h_S = \int_{T_{ref}}^{T} c_{pS} dT \,, \tag{14}$$

$$h_L = \int_{T_{ref}}^{T_S} c_{pS} dT + \int_{T_S}^{T} c_{pL} dT + h_m = h_S(T) + \int_{T_S}^{T} (c_{pL} - c_{pS}) dT + h_m \,, \tag{15}$$

respectively. The liquid fraction f_L, concentration of solute in liquid phase C_L, partition ratio k_p, and liquidus temperature T_L are determined based on the level rule microsegregation model, i.e.

$$f_L = 1 - \frac{1}{1-k_p}\frac{T-T_L}{T-T_m}, C_L = \frac{C}{1+f_S(k_p-1)}, k_p = \frac{C_S}{C_L}, T_L = T_m + (T_e - T_m)\frac{C}{C_e} \qquad (16,17,18,19)$$

where T_m, T_e and C_e are melting temperature, eutectic temperature and solute concentration in eutectic alloy, respectively. They are obtained from the phase diagram for the iron-carbon system.

3. Solution procedure

Our solution procedure is based on the novel meshless technique that follows the following steps. We seek the solution of the equations for the velocity, pressure, temperature, concentration, turbulent kinetic energy, dissipation and current density at time $t+t_0$ by assuming known fields $\mathbf{u}$, p, T, C, k, ε and $\mathbf{J}$ at time t_0 and known boundary conditions at time $t > t_0$. The coupled set of mass conservation equation (1) and momentum conservation equations (2) are solved by the fractional step method [21], where the continuity of the mass (1) is considered by constructing the pressure Poisson's equation. The governing equations are discretized by using the explicit time discretization. This leads to the following algorithm, summarized through the following 8 steps:

1) The intermediate velocity components are calculated without the pressure gradient

$$\mathbf{u}^* = \mathbf{u}^n + \frac{\Delta t}{\rho}\left[-\rho\nabla\cdot(\mathbf{u}\mathbf{u}) + \nabla\cdot\left[2(\mu+\mu_t)\mathbf{S}\right] - \frac{2}{3}\rho\nabla k - \mu_L\frac{K_0(1-f_L^2)}{f_L^3}(\mathbf{u}-\mathbf{u}_S) + \mathbf{F}_b + \mathbf{F}_m\right]^n . \tag{20}$$

2) The pressure Poisson equation is treated

$$\nabla^2 P^{n+1} = \frac{\rho}{\Delta t}\nabla\cdot\mathbf{u}^* \tag{21}$$

by solving the related sparse matrix [11].

3) The velocity components are corrected by the pressure gradient

$$\mathbf{u}^{n+1} = \mathbf{u}^* - \frac{\Delta t}{\rho}\cdot\nabla P^{n+1} . \tag{22}$$

4) After the solution of the velocity field, the transport equations for energy is solved

$$h^{n+1} = h^n + \frac{\Delta t}{\rho}\left[-\rho\nabla\cdot(\mathbf{u}h) + \nabla\cdot(\lambda\nabla T) + \rho\nabla\cdot\left[f_S(h_L-h_S)(\mathbf{u}-\mathbf{u}_S)\right] + \nabla\cdot\left(f_L\frac{\mu_t}{\sigma_T}\nabla h_L\right)\right]^n . \tag{23}$$

5) The temperature field is calculated from the enthalpy field, calculated in previous step, using the inverse of the constitutive temperature-enthalpy relationships (14,15), i.e. $T = T(h)$.

6) Species concentration, turbulent kinetic energy and dissipation rate fields are calculated. They are discretized by the same way as the transport equation (23).

7) The current density field is updated by the following procedure. First, the electrical potential is obtained by solving the Poisson equation (13). After that, the electric field intensity is calculated from the equation (12). Finally, the current density is obtained from the equation (11).
8) The turbulent viscosity (6), solute concentrations (17,18) and liquidus temperature (19), are updated for each node. The solution is set ready for the next time step.

All derivatives of the transport equations are calculated by the LRBFCM [9,11,13,24], where the collocation is made locally on overlapped sub-domains. On each sub-domain, the scalar function Φ (standing for temperature, species concentration, velocity components, pressure, turbulent kinetic energy, turbulent dissipation rate, magnetic flux components, and current density components) is represented over a set of (in general) non-equally spaced nodes $\mathbf{p}_n$; $n = 1,2,...,N$ in the following way

$$\Phi(\mathbf{p}) \approx \sum_{k=1}^{K} \psi_k(\mathbf{p})\alpha_k, \quad \psi_k(\mathbf{p}) = \left[r_k^2 + c^2\right]^{1/2} \qquad (24,25)$$

where ψ_k stands for the multi-quadric radial basis shape functions [25], α_k for the coefficients of the shape functions, and K represents the number of the shape functions. In equation (25) c represents the shape parameter and r_k the radial distance between two points in the sub-domain. The detailed procedure of calculating the derivatives with the LRBFCM can be found in [11,13].

4. Benchmark description

4.1. Geometry

The proposed benchmark represents a simplified 2D model of the continuous casting process. The geometry of the simplified casting machine is shown in figure (1). A billet with a square cross-section is assumed. The computational domain represents half of the longitudinal section of the billet with length 1.8 m, shown in figure (2). These simplifications are used mainly for straightforward verification of the developed numerical model, and in order to reduce the CPU time of the simulations. The dimensions of the casting geometry are billet width $a_b = 0.14\text{m}$, diameter of the SEN $d_2 = 0.035\text{m}$, mould height $l_m = 0.8\text{m}$ and height of the coils $l_{EMS} = 0.1\text{m}$, respectively.

4.2. Material properties

Generally, the material properties of steel are temperature dependent. However, for benchmark purpose, constant but different values are used for each phase. They are given in table (1).

Table 1. Material properties.

ρ (kg m-3)	c_p (J kg^{-1}K^{-1})	λ (W m^{-2}K^{-1})	μ (kg/m s)	T_L (K)	T_S (K)	h_m (J/kg)	K_0 (m^{-2})	σ (Ω^{-1}m^{-1})
7200	700	30	0.006	1760	1680	250000	1.6×10^{8}	7.14×10^{5}
D_S (m^{2}s^{-1})	D_L (m^{2}s^{-1})	β_T (K^{-1})	β_C (/)					
1.6×10^{-11}	1.0×10^{-8}	1.0×10^{-4}	4.0×10^{-3}					

4.3. Boundary conditions

The boundary conditions for the velocity, temperature, turbulent kinetic energy, dissipation rate, species concentration, and current density are set as follows:

SEN outlet: Constant values are prescribed for all variables $u_{inlet,x}=0$, $u_{inlet,y}=a_b u_{c,y}/d_2$, $k_{inlet}=1.5\left(I_t u_{inlet,y}\right)^2$, $\varepsilon_{inlet}=\dfrac{c_\mu^{0.75}k_{inlet}^{1.5}}{0.07d_2}$, $T_{inlet}=T_c$, $C_{inlet}=C_c$ and $\dfrac{\partial J_y}{\partial y}=0$, where $u_{c,y}$ is the casting velocity in the casting direction, and I_t is the turbulent intensity calculated from the following relation

$$I_t=0.16\mathrm{Re}^{-1/8},\ \mathrm{Re}=\rho u_{inlet,y}d_2/\mu. \qquad (26,27)$$

Normal gradient of pressure and current density are set to zero, i.e. $\dfrac{\partial P}{\partial y}=0$ and $\dfrac{\partial J_y}{\partial y}=0$.

Billet end: The pressure outlet is prescribed, where the following boundary conditions are used

$$P=0,\frac{\partial u_x}{\partial y}=0,\frac{\partial u_y}{\partial y}=0,\frac{\partial k}{\partial y}=0,\frac{\partial \varepsilon}{\partial y}=0,\frac{\partial T}{\partial y}=0,\frac{\partial C}{\partial y}=0,\frac{\partial J_y}{\partial y}=0. \qquad (28)$$

Top surface (meniscus): At the meniscus, the symmetry boundary conditions are used (free surface flow). The normal derivative of all variables is set to zero, except the vertical velocity u_y and the current density J_y which are set to zero.

Moving walls: The walls with the solidified steel move with casting velocity along the casting direction. At the walls, where the liquid phase exists, the no-slip boundary conditions for the velocity are set. In the mould, the Robin boundary condition is used, with the surface heat transfer coefficient equal to 2000 W/(m^{2}K). Bellow the mould, at the secondary cooling system, the heat transfer coefficient is equal to 800 W/(m^{2}K). The normal derivative of the solute concentration is set to zero. The current density J_x is set to zero.

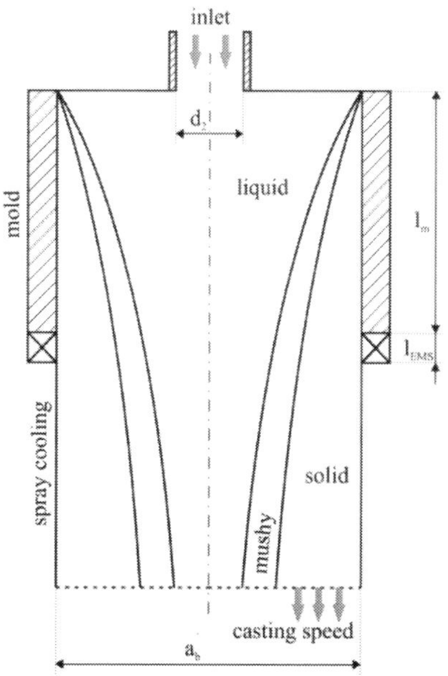

Figure 1. Benchmark geometry with typical dimensions.

Figure 2. Computational geometry.

5. Numerical results

In this conference paper, the preliminary numerical results are obtained without electromagnetic forces in momentum equation and with the following simplified microsegregation model

$f_L = (T - T_L)/(T - T_m)$. Species concentration is assumed to be constant and equal for each phase, i.e. $C_L = C_C = C$, which at the present excludes the species transport equation from the solution procedure. The low-Re turbulence model by Abe-Kondoh-Nagano (AKN) [26] is selected, where the normal distance in closure coefficients is calculated up to the wall. The simulation was performed on the node arrangement with 61616 nodes. The steady-state solution shown in figures is approached by a false transient calculation using a fixed time-step of 10^{-4} s. The results are compared with the commercial CFD package Fluent, with the 147408 quadrilateral cells, AKN turbulence model, SIMPLE velocity-pressure coupling, and second-order spatial discretization. The following process parameters were used: reference Case 1: $u_{cast} = 1.75$ m/min and $T_{cast} = 1800$ K, enhanced casting speed Case 2: $u_{cast} = 1.75$ m/min and $T_{cast} = 1800$ K, and reduced casting temperature Case 3: $u_{cast} = 1.75$ m/min and $T_{cast} = 1780$ K.

Figure (3) represents the absolute velocity field in the billet for Case 1. The comparison between LRBFCM and FVM shows very similar flow streaming and re-circulating regions. The temperature field of Case 1 in the range $T_S \le T \le T_{cast}$ is shown in figure (4) with the black line standing for liquidus isotherm. The difference in the position of the liquidus temperature at the symmetry line is about 0.1 m, which is expected due to the complex physical phenomena of the modelled process. Figure (5) shows a comparison between centreline temperature for Cases 1-2-3, calculated with FVM and LRBFCM. Figure (6) shows a comparison between surface temperatures for Cases 1-2-3, calculated with FVM and LRBFCM. Figure (7) shows a comparison between horizontal component of the velocity for Cases 1-2-3, calculated with FVM and LRBFCM. Figure (8) shows a comparison between vertical component of the velocity for Cases 1-2-3, calculated with FVM and LRBFCM. The figures (5-8) show excellent agreement between the calculations with both methods and similar response to the change of process parameters, despite different numerical methods used.

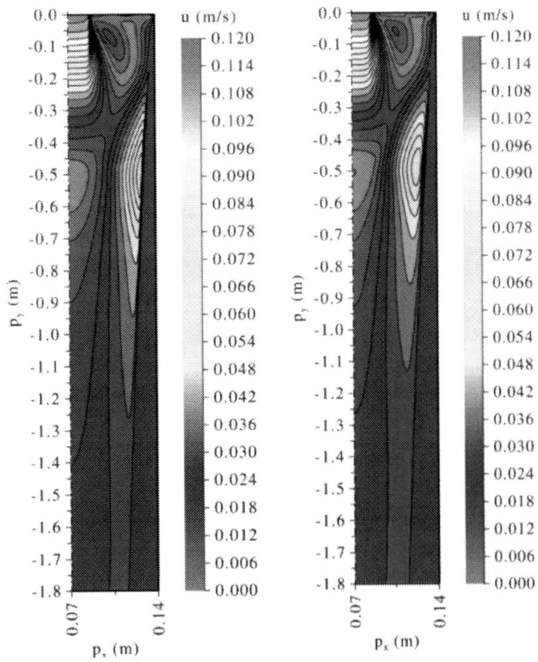

Figure 3. Absolute velocity field. Left: LRBFCM, right: FVM.

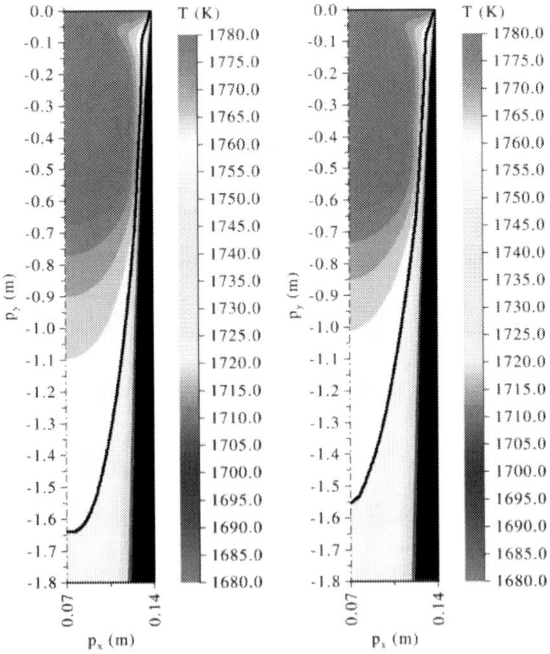

Figure 4. Temperature field. Left: LRBFCM, right: FVM. Black line marks liquidus isotherm.

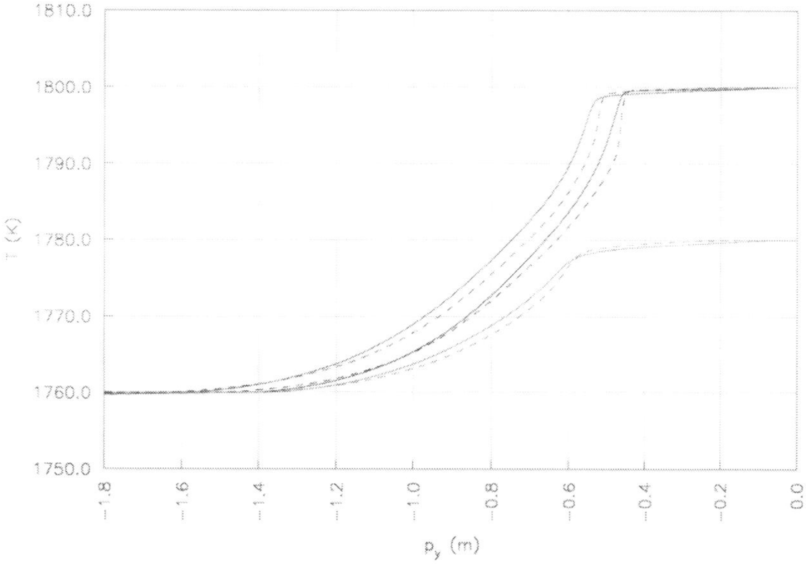

Figure 5. Centreline temperatures for Cases 1-2-3. Solid line LRBFCM, dashed line FVM. Case1: black, Case 2: blue, Case 3: red. Casting direction from right to left.

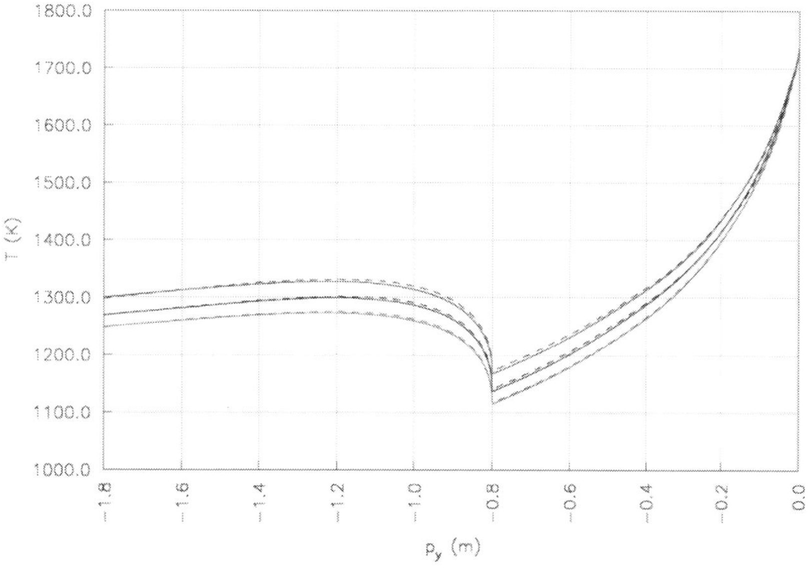

Figure 6. Surface temperatures for Cases 1-2-3. Solid line LRBFCM, dashed line FVM. Case1: black, Case 2: blue, Case 3: red. Casting direction from right to left.

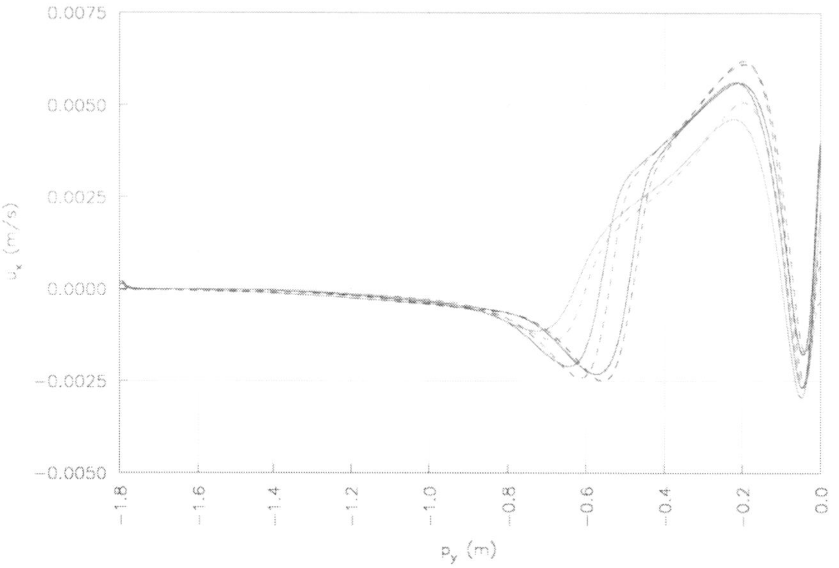

Figure 7. Horizontal velocity component u_x at $p_x = 0.105\,\text{m}$ for Cases 1-2-3. Solid line LRBFCM, dashed line FVM. Case1: black, Case 2: blue, Case 3: red. Casting direction from right to left.

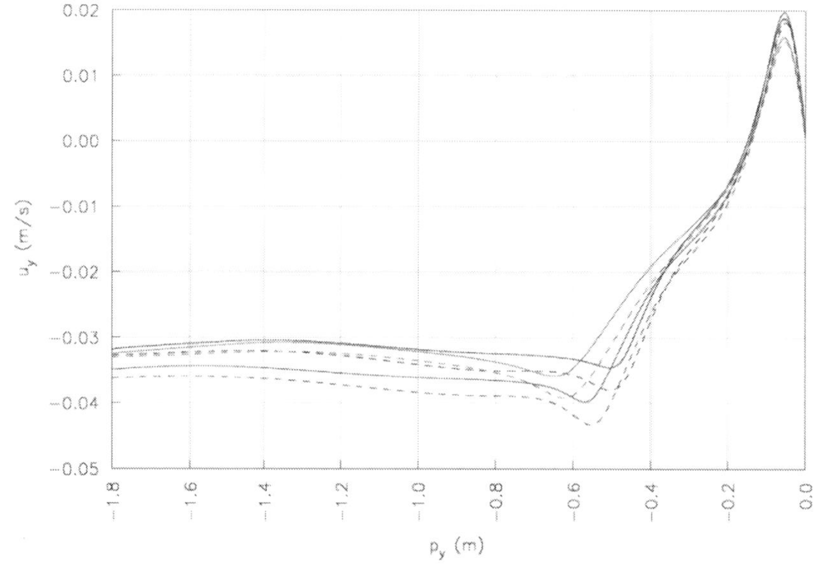

Figure 8. Vertical velocity component u_y at $p_x = 0.105\,\text{m}$ for Cases 1-2-3. Solid line LRBFCM, dashed line FVM. Case1: black, Case 2: blue, Case 3: red. Casting direction from right to left.

6. Conclusions

This paper represents a numerical benchmark proposed for the continuous casting of steel. The geometry and material properties of the continuous casting process are simplified in order to enable straightforward performance comparisons with various numerical methods. In this paper, our in-house developed meshless numerical method, based on the local radial basis function collocation, is used to get the steady state solution of the turbulent fluid flow with heat transfer and solidification. The results were obtained without electromagnetic forces in the momentum equation and with a simplified microsegregation model with constant and equal concentration of species in each phase. The numerical calculations based on the FVM, were performed with the CFD package Fluent (version 13), and compared with the results obtained by the meshless method. Very good agreement was achieved. The benchmark will be further upgraded and adapted, based on the feedback from the colleges in the field of the numerical modelling of casting processes. A web page of the benchmark results with systematically tabulated results of different numerical methods will be established, similar to [18].

Acknowledgements
The comparison exercise is proposed within Slovenian state funded research projects J2-4120, L2-3651 and P2-0379 and co-funded by Štore Steel company. The Centre of Excellence for Biosensors, Instrumentation and Process Control is an operation financed by the European Union, European Regional Development Fund and Republic of Slovenia, Ministry of Higher Education, Science and Technology. The financial support is kindly acknowledged.

References
[1] Irwing W R 1993 *Continuous Casting of Steel*, (London: The Institute of Materials)
[2] Thomas B G 2002 Modeling of the Continuous Casting of Steel - Past, Present Future, www.brimacombecourse.org/pdf/2001_Thomas.pdf
[3] Zhao B, Thomas B G, Vanka S P and O'Malley R J 2005 *Metall. Mater. Trans. B* **36** 801-23
[4] Reza Aboutalebi M, Hasan M and Guthrie R I L 1993 *Metall. Mater. Trans. B* **26** 731-44
[5] Kim D S, Kim W S and Cho K H 2000 *ISIJ Int.* **40** 670-76
[6] Kang K G, Ryou H S and Hur N K 2005 *Numer. Heat Tr. A* **48** 461-81
[7] Kubo N, Ishi T, Kubota J and Ikagawa T 2004 *ISIJ Int.* **44** 556-64
[8] Moon C H and Hwang S M 2003 *Int. J. Numer. Meth. Eng.* **57** 315-39
[9] Vertnik R, Založnik M and Šarler B 2006 *Eng. Anal. Bound. Elem.* **30** 847-55
[10] Vertnik R, Šarler B and Senčič B 2009 Simulation of turbulent flow and heat transfer in continuous casting of billets by a meshless method. *3rd International Conference on Simulation and Modelling of Metallurgical Processes in Steelmaking-STEELSIM2009 (Leoben, Austria, 8-10 September 2009)* ed A Ludwig (Leoben, Austria: ASMET) pp 155-61
[11] Vertnik R 2010 *Heat and fluid flow simulation of the continuous casting of steel by a meshless method-dissertation* (Nova Gorica, SLO: University of Nova Gorica, XVII) p 225, http://www.p-ng.si/~vanesa/doktorati/karakterizacija/8Vertnik.pdf
[12] Chaudhary R, Ji C, Vanka S P, and Thomas B G 2011 *Metall. Mater. Trans. B* **42** 987-1007
[13] Šarler B and Vertnik R 2006 *Comput. Math. Appl.* **51** 1269-82
[14] Vertnik R and Šarler B 2006 *Int. J. Numer. Method H.* **16** 617-640
[15] Kosec G and Šarler B 2009 *Int. J. Numer. Method H.* **18** 868-82
[16] Kosec G and Šarler B 2009 *Comput. Model. Eng. Sci.* **47** 191-216
[17] Gobin D and Le Quéré P 2000 *Comput. Assist. Mech. Eng. Sci.* **7** 289-306
[18] Bellet M, Combeau H, Fautrelle Y, Gobin D, Rady M, Arquis E, Budenkova O, Dussoubs B, Duterrail Y, Kumar A, Gandin C A, Goyeau B, Mosbah S and Založnik M 2009 *Int. J. Therm. Sci.* **48** 2013-16
[19] Kosec G, Založnik M, Šarler B and Combeau H 2011 *Comput. Mater. Con.* **22** 169-59
[20] Šarler B, Vernik R, Šaletić S and Cesar J 2005 *Berg- Huettenmaenn. Monatsh.* **150** 300-6
[21] Chorin A J 1967 *J. Comput. Phys.* **2** 12-26
[22] Bennon W D and Incropera F P 1988 *Numer. Heat Tr. A-Appl.* **13** 277-96
[23] Wilcox, D C 1993 *Turbulence modeling for CFD* (California: DCW Industries, Inc.)
[24] Vertnik R and Šarler B 2009 *Comput. Model. Eng. Sci.* **44** 66-95
[25] Buhmann M D 2003 *Radial Basis Function: Theory and Implementations* (Cambridge: Cambridge University Press)
[26] Abe K, Kondoh T and Nagano Y 1994 *Int. J. Heat Mass Tran.* **37** 139-51

IOP Conf. Series: Materials Science and Engineering **33** (2012) 012013 doi:10.1088/1757-899X/33/1/012013

Industrial application of a numerical model to simulate lubrication, mould oscillation, solidification and defect formation during continuous casting

Pavel E Ramirez Lopez[1], Ulf Sjöström[1], Thomas Jonsson[1], Peter D Lee[2], Kenneth C Mills[3], Marko Petäjäjärvi[4] and Jarno Pirinen[5].

[1] Casting and Flow Simulation Group, Swerea MEFOS,
Aronstorpsvägen 1, SE 974 37 Luleå, SWEDEN
[2] School of Materials, The University of Manchester,
Rutherford Appleton Laboratory, Didcot, Oxon, OX11 OFA, UNITED KINGDOM
[3] Department of Materials, Imperial College London,
Prince Consort Road, SW7 2AZ, London, UNITED KINGDOM
[4] Research Engineer, TRC / Melt Metallurgy,
Outokumpu Tornio Works, FIN-95490 Tornio, FINLAND
[5] Development Engineer, Continuous Casting,
Rautaruukki Oyj, Tervaraitti 4 B 26, 90100 Oulu, FINLAND

E-mail: prl@mefos.se

Abstract. In recent years, the addition of the slag phase to numerical models of the Continuous Casting (CC) process has opened the door to a whole new range of predictions. These include the estimation of slag infiltration and powder consumption (lubrication), heat transfer and cooling through the cooper mould (solidification) and investigating the effect of operational parameters such as mould oscillation and powder composition on surface quality / defect formation. This work presents 2D and 3D CC models capable of describing the dynamic behaviour of the liquid/solid slag in both the shell mould-gap and bed as well as its effects on heat extraction and shell formation. The present paper also illustrates the application of the model to a variety of casters and the challenges faced during its implementation. The model attained good agreement on the prediction of mould temperatures and shell thicknesses as well as slag film formation and heat flux variations during the casting sequence. The effect of different oscillation strategies (sinusoidal and non-sinusoidal) was explored in order to enhance powder consumption and surface quality. Furthermore, the modelling approach allows one to predict the conditions leading to irregular shell growth and uneven lubrication; these are responsible for defects such as, stickers, cracking and, in the worst case scenario, to breakouts. Possible mechanisms for defect formation are presented together with strategies to enhance process stability and productivity of the CC machine.

1. Introduction

The behaviour of the casting powder and the way it influences heat transfer and solidification are amongst the most complex, but important, phenomena occurring in Continuous Casting. This behaviour includes a variety of thermo-chemical reactions, such as phase transformations, fluid flow and heat transfer mechanisms. For instance, practically all the heat transfer modes (convection, conduction and radiation) and combustion occur within the mould during casting. Firstly, convection currents and heat dissipation appear in the metal and the liquid slag pool in response to density changes during phase transformation and cooling. Secondly, conduction takes place through the shell, solid slag film, sintered bed and copper mould during solidification. Finally, radiation accounts for ~10-15% of the total heat extracted from the melt[1]. The variety of phases and heat transfer mechanisms occurring at the meniscus and slag bed are depicted in Figure 1.

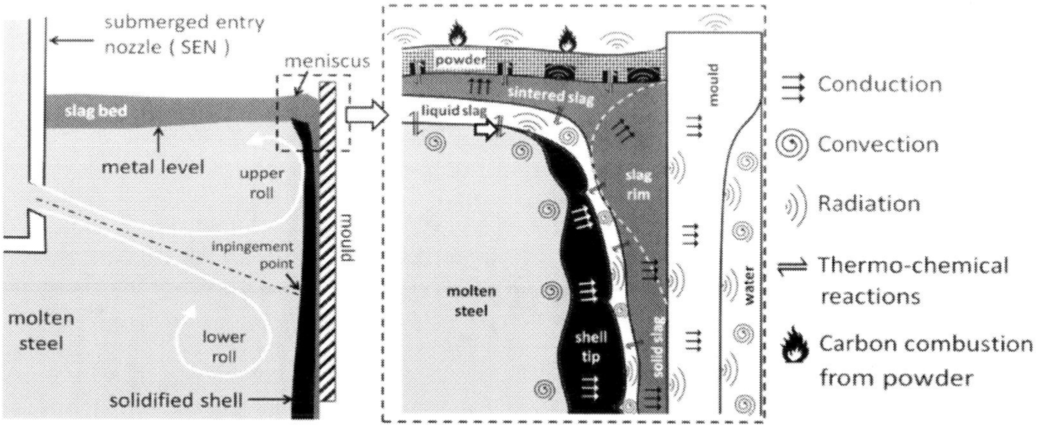

Figure 1: Heat transfer mechanisms at the meniscus.

The dynamic behaviour of the slag needs to be included in the simulations to provide a more realistic representation of the process. However, it is often treated separately because of its complexity and the computational effort required to solve it. In most of the previous modelling studies the incorporation of slag into the calculations is avoided and limited to the use of measured heat fluxes (from thermal monitoring systems) as boundary conditions which define the heat extraction rate to the mould [2, 3]. To a lesser extent, the slag-film has been represented through an equivalent thermal resistance system which gives a better idea of the effect of slag properties on heat transfer and solidification, but is usually based on arbitrary definitions of the film thickness[4-6]. Thus, these models overlook the dynamic behaviour of the slag, which includes the effect of powder composition changes during the heat, the transient influence of mould oscillation on powder consumption and the role of slag infiltration on defects such as oscillation marks, stickers, entrapment, etc. Only the investigations reported by Meng et al. [7] and the current authors [8] have addressed the modelling of the transient slag-film behaviour and its impact on infiltration.

The current research addresses the problem of adding slag to the top of the steel in the mould, and its implications on numerical simulations. This includes the creation of a model that couples the metal-slag flow, heat transfer, mould oscillation and solidification within the mould. Previous publications by the authors have described the theory behind this model[9] as well as its validation with industrial observations[10]. Furthermore, the transient approach used in the simulations allows one to predict defects such as oscillation marks [8] and to deduct a unified mechanism for their formation [11]. In this manuscript the challenges faced during the industrial application of the model are detailed and specific cases arising from its implementation and some selected results are presented.

2. Fully-integrated multi-phase approach

The modelling methodology recently introduced by the authors[9, 10] couples the flow dynamics in metal, slag and gas phases (if present) to the heat flux and solidification during mould oscillation. The fluid model is based on the Navier-Stokes equations for compressible viscous flow coupled with the Volume of Fluid (VOF) method. This allows one to predict the slag/metal interface and makes no assumptions regarding the shape of the meniscus (i.e. avoiding a pre-fixed Bikerman profile or separating the slag from the metal domain). The VOF scheme predicts the interface steel/slag by solving a single set of continuity and momentum equations for a phase mixture (otherwise, the scheme is reduced to a typical single-phase Navier-Stokes approach) [12]. Since steel is the primary phase, the continuity equation for the volume fraction is given by:

$$\frac{\partial}{\partial t}(\alpha_{steel}\rho_{steel}) + \nabla\cdot(\alpha_{steel}\rho_{steel}\vec{v}) = \sum(\dot{m}_{slag\to steel} - \dot{m}_{steel\to slag}) \tag{1}$$

Where (α) is the phase fraction, (ρ) is the density, ($\vec{v}$) is the overall velocity vector and $\dot{m}$ is the mass transfer between the phases. The effective density (ρ_{mix}) and the effective viscosity (μ_{mix}) for the mixture phase are given as:

$$\rho_{mix} = \alpha_{steel}\rho_{steel} + (1-\alpha_{slag})\rho_{slag} \tag{2}$$

$$\mu_{mix} = \alpha_{steel}\mu_{steel} + (1-\alpha_{slag})\mu_{slag} \tag{3}$$

Thus, the momentum equation becomes:

$$\frac{\partial}{\partial t}(\rho_{mix}\vec{v}) + \nabla\cdot(\rho_{mix}\vec{v}\vec{v}) = -\nabla P + \nabla\cdot\left[\mu_{mix}(\nabla\vec{v} + \nabla\left(-\frac{2}{3}\nabla\cdot\vec{v}I\right)\right] + \rho_{mix}\vec{g} - S_s + S_\gamma \tag{4}$$

Where (P) is the pressure, ($\vec{g}$) is the gravitational force vector and the last two terms are the momentum sinks due to solidification and interfacial tension. (S_γ); the latter results from the interfacial tension between steel and slag and is calculated through the CSF model proposed by Brackbill et al. [13]:

$$S_{\gamma_{slag-metal}} = \gamma_{slag-metal}\frac{\rho_{mix}\kappa\nabla\alpha_{slag}}{\frac{1}{2}(\rho_{steel} + \rho_{slag})} \tag{5}$$

Where ($\gamma_{slag-metal}$) is the slag-metal interfacial tension as calculated by Chung and Cramb[14] as a function of the powder and metal compositions, and (κ) is the local surface curvature as defined by Brackbill et al. [13]. The approach for solving heat transfer in the liquid phases is comparable to the multiphase approach, with the continuity and momentum equations sharing the energy Equation [15]:

$$\frac{\partial}{\partial t}(\rho_{mix}E) + \nabla\cdot(\vec{v}(\rho_{mix}E + P)) = \nabla\cdot(K_{eff}\nabla T + (\tau_{eff}\cdot\vec{v})) \tag{6}$$

where (E) is the enthalpy and (K_{eff}) denotes the effective thermal conductivity for the phase mixture. An improvement from prior work [9, 10] includes a correction to K_{eff} after each iteration to account for the blurring effects (gradient at the interface) inherent to VOF [12].

The additional steps consist of locating the slag-metal interface and updating the averaged value of K_{eff} for $\alpha_{steel} \le 0.5$ to the appropriate value of K_{slag} as a function of temperature. Although radiation conduction in the slag (K_{rad}) has not been explicitly included in the model, its effects on heat transfer have been added by enhancing K_{slag} through: ($K_{slag} = K_{cond} + K_{rad}$), where: K_{rad} is ~3-5% of the slag conductivity (K_{cond}) as demonstrated by Wang et al. [16]. Unfortunately, this method does not fully represent the effects of crystallization in the slag film, which are time and composition dependent. However, Mills et al. [1, 17] concluded that radiation accounts for ~15% of the total heat transfer in the mould, which makes ($K_{slag} = K_{cond} \cdot 0.05$) a reasonable assumption. Finally, the solid wide and narrow faces are included in the analysis by solving the heat conduction equation for solids:

$$\frac{\partial}{\partial t}\left(\rho_{copper} h\right) = \nabla \cdot \left(K_{copper} \nabla T\right) \tag{7}$$

Where (h) is the sensible enthalpy and (K_{copper}) is the thermal conductivity of the copper mould (~400 W/m-K). The solidification routine is based on the lever rule as detailed elsewhere [18]. Additionally, an enthalpy-porosity technique that treats the mushy zone as a porous medium is used to include the velocity sink due to solidification. This sink term (S_s) is introduced in Equation (4) and depends on the solid fraction at each cell through:

$$S_s = \frac{(1 - f_l)^2}{\left(0.001 + f_l^{\,2}\right)} A_{mush}\left(\vec{v} - v_{pull}\right) \tag{8}$$

where $A_{mush} = 1 \times 10^8$, v_{pull} is the pulling velocity and f_l is the liquid fraction of steel. Turbulence effects are added through the k-ε model which has proven fast and robust in prior works [19, 20] and is described elsewhere[15]. Finally, solidification effects on turbulence are included by adding a sink term to the k (turbulent kinetic energy) and ε (turbulent energy dissipation rate) expressions[18].

3. Model Particularities

Industrial application of the model led to a variety of changes in order to account for the conditions on the casting floor; these have often been overlooked in more "fundamental" studies, but are extremely important to the process. These include specific details of the mould geometry (e.g. mould taper, coating, etc), powder feeding procedures, oscillation strategies, as well as preferred practices from operators. These particularities are described in the following sections. A summary of the standard boundary conditions to run the model and solution procedures has been presented elsewhere [9].

3.1. Interfacial thermal resistance and mould coating

The heat transfer between shell and mould (i.e. across the slag film) is extremely important since excessive heat removal leads to thicker shells that crack when the slab is withdrawn. In contrast, insufficient heat removal can lead to thin shells that cause breakouts and mould level changes. Heat transfer in the slag-film is a complex process involving lattice conduction and radiation and is also affected by both the thickness and degree of crystallisation of the slag [16, 21]. Since the crystalline phase has a higher density than the glassy phase within the film, crystallisation is accompanied by shrinkage. This produces rugosity and loss of contact with the mould, creating an additional thermal resistance denominated interfacial thermal resistance (r_{int}) as noted by Spitzer et al. [21] (Figure 2). Previous studies have linked r_{int} to the total film resistance and solved algebraically the heat transfer as a system of electrical resistances as presented in Figure 4 [22, 23]. The present model avoids this by solving the actual infiltration of liquid-slag into the gap and using its thermal properties to differentiate between the phases.

These properties include the variation of thermal conductivity and C_p as a function of temperature as shown in Figure 3. Consequently, the thermal resistances in the slag-film (r_{film}) are reduced to:

$$r_{\text{int}} + \overbrace{\left(\frac{d_{cry}}{K_{cry}}\right)_{solid-slag} + \left(\frac{d_{glassy}}{K_{glassy}}\right)_{solid-slag} + \left(\frac{d_{liquid}}{K_{liquid}}\right)_{liquid-slag}}^{r_{film}} = r_{\text{int}} + \frac{d_{film}}{K_{eff}} \tag{9}$$

Figure 2: Surface rugosity on slag sample quenched against a copper mould. After Holzhauser et al.[24)]

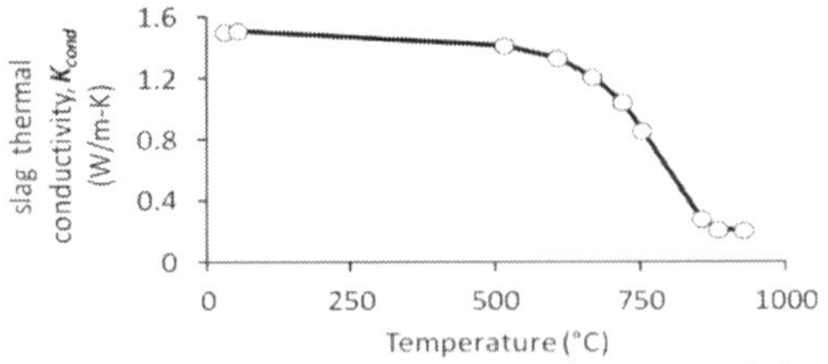

Figure 3: Example of slag thermal conductivity properties as a function of temperature (implemented in the model).

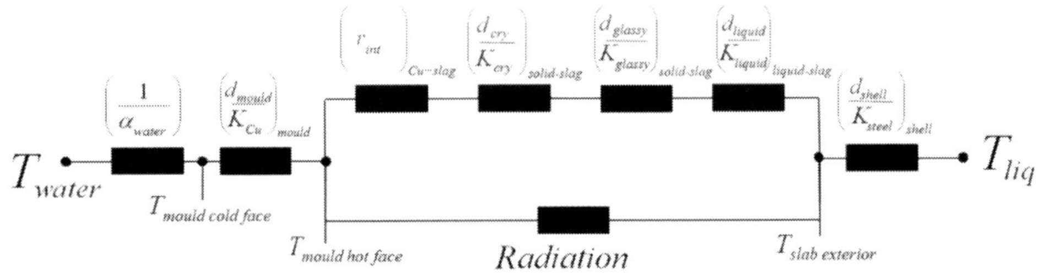

Figure 4: Thermal resistances between liquid steel and cooling water on CC.

Where d_{film} is the thickness of the slag film (a result of the calculations) and K_{eff} is the corrected thermal conductivity for the slag when $\alpha_{steel} \leq 0.5$ or the thermal conductivity of steel if $\alpha_{steel} > 0.5$. The magnitude of r_{int} is critical on the simulations and depends on the slag basicity as described on prior work [10)]:

$$r_{\text{int}} = 9 \times 10^{-5} (\text{CaO/SiO}_2)^{4.19} \tag{10}$$

Modelling of the Nickel or Chromium coating that is often applied to the mould hot face was done according to the specifications of each caster. For some cases, the thickness is constant along the mould length, whilst for others the thickness varies from top to bottom in a similar way as the taper. Both cases were implemented as an additional resistance ($r_{coating}$) in the gap, (Equation 11).

$$r_{gap} = \frac{d_{film}}{K_{eff}} + r_{\text{int}} + \overbrace{\frac{d_{coating}}{K_{coating}}}^{r_{coating}} \tag{11}$$

3.2. Powder viscosity

The behaviour of slag after infiltration depends on its break temperature (T_{br}), crystallisation rate and viscosity (η) as it travels downwards and freezes (solidifies) against the mould. This behaviour was included in the simulations through an exponential increase of viscosity below T_{br} as introduced by Meng et al. [25)]. Nevertheless, this approach is not sufficient to account for the events occurring on the slag bed. Therefore, additional viscosity properties were used to model the slag evolution from room temperature to the melting point (Figure 5). A natural result of this implementation is the prediction of a slag rim and powdered-sintered layers in the bed without the aid of any special function to describe their profile [8)].

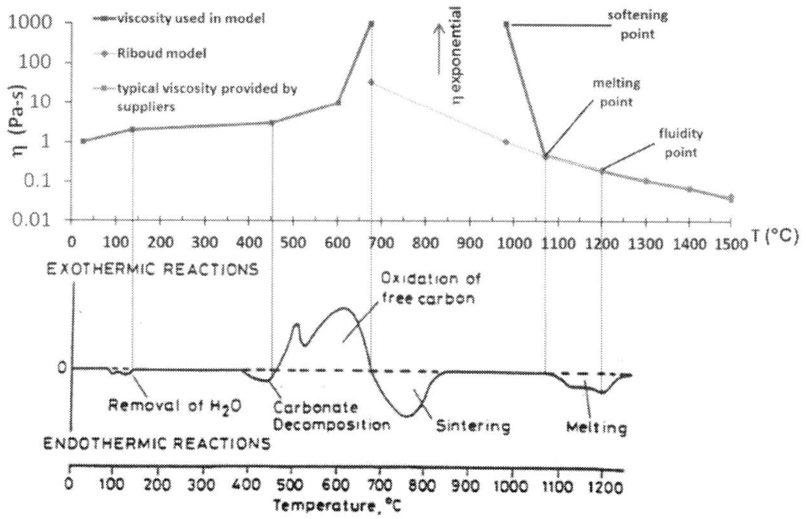

Figure 5: Slag viscosity used in the simulations *vs* evolution of slag during heating. Lower portion of the diagram after Mills [17].

3.3. Powder feeding and bed temperature

A pressure inlet condition was applied to the top of the mould to simulate the effects of a system open to the atmosphere. The slag is supplied at this boundary (as in the real process) by adding powder to maintain the desired bed thickness. Here, critical differences in casting practices have deep implications on modelling. For some casters, the casting powder is added until it fills completely the mould cavity, which creates a thick powder bed. In contrast, other operational practices maintain only a powder layer of several mm (30-60 mm) above the metal level. In modelling terms, the thin powder bed case represents the use of a third phase (air on top of the bed), and the supply of a certain volume fraction of powder through this boundary. The thick bed case requires only a 2-phase model where only slag is supplied at the boundary. In both cases, the temperature at the top of the bed is a critical boundary condition which can be measured using pyrometers or an infrared camera.

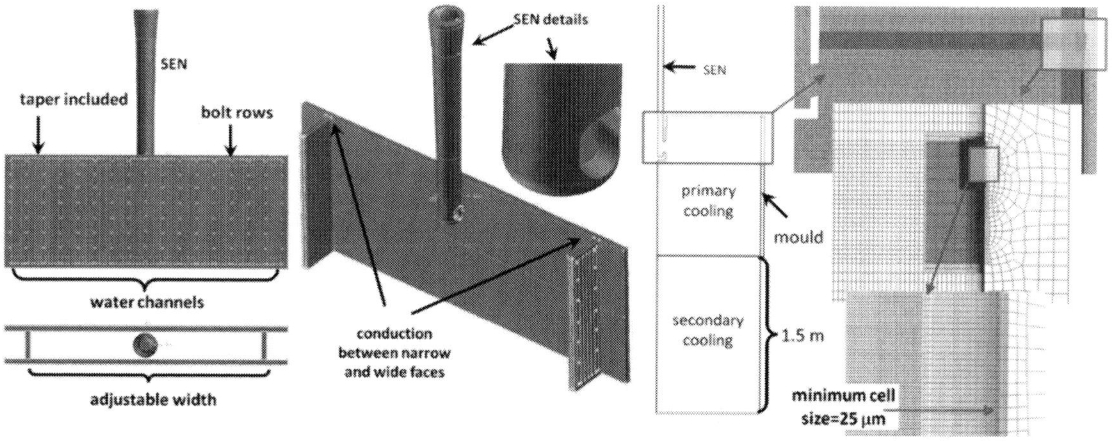

Figure 6: Example of CAD models and computational grid used on the simulations

3.4. Geometry

In order to provide a realistic representation of the caster, full CAD models based on the exact dimensions of the SEN and mould were created. These CAD's were later used to produce detailed 2D and 3D grids from the real components including water channels layout, bolt systems, fillets and SEN design features. Finally, the machine radius (for curved moulds) and taper (for straight moulds) are also included in the simulations (Figure 6).

4. Industrial Application: Selected results

After validation, the modelling methodology has been applied to various conventional and thin-slab caster configurations. Selected results for both a 3D, ¼ model and some 2D models are presented next.

4.1. Metal flow

Metal flow predictions shown in Figure 7 describe the formation of a standing wave resulting from a particular SEN design which causes most of the metal leaving the ports to flow horizontally. This creates a strong discharging jet which breaks into a small upper roll and a larger lower roll. The predictions were compared with nail bed measurements (for a similar configuration) and a good agreement on standing wave height and extension (Figure 8) was found.

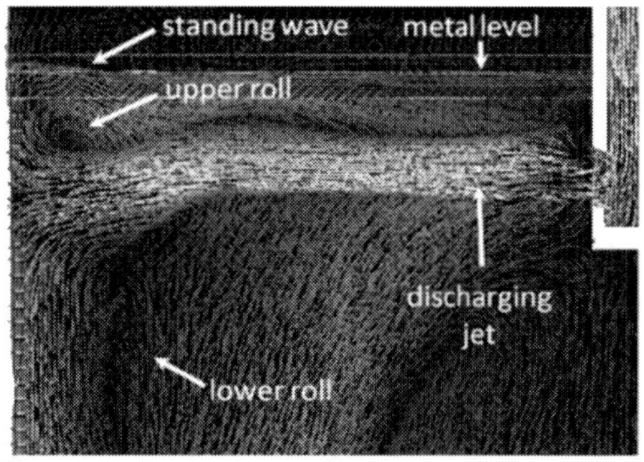

Figure 7: Jet angle predictions on 3D vs 2D

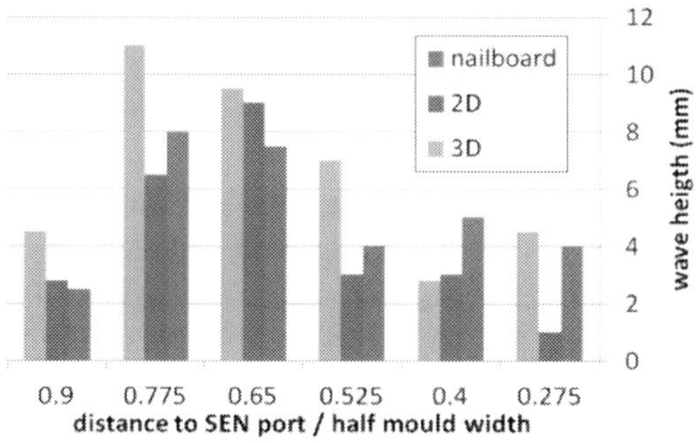

Figure 8: Standing wave predictions on 3D vs 2D

Three dimensional simulations are time consuming, but must be performed to verify the applicability of 2D models to a flow pattern that is inherently 3-dimensional (e.g. CC mould). Discharging jet angles and predicted metal levels are compared, for identical configurations, in 2D and 3D in Figure 9. The maximum standing wave height is less pronounced for the 2D model than for the 3D case (11 mm *vs* 9 mm) but presents a better agreement with the nail board measurements. Moreover, the differences in jet axis were only 1.5 degrees, which demonstrates the viability of using 2D calculations to study caster performance (*e.g* oscillation settings, steel grades and powder composition) (Figure 10). Nevertheless, 2D calculations cannot describe the entire shape of the wave, which is required to address issues such as corner cracking, irregular slag infiltration, longitudinal cracking, etc.

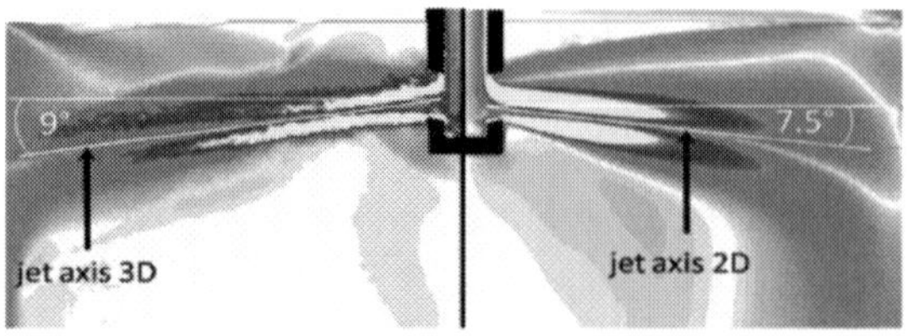

Figure 9: Jet angle predictions on 3D *vs* 2D

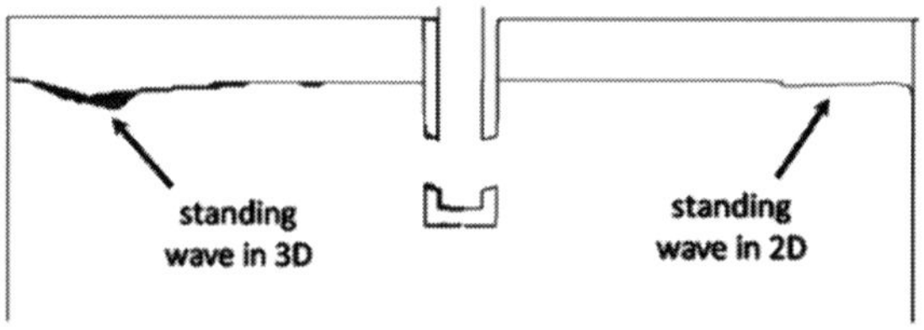

Figure 10: Standing wave predictions on 3D *vs* 2D

4.2. Heat transfer

Mould temperature calculations for a different slab caster are presented in Figure 11. These predictions for the shell and mould show a typical profile with the highest temperatures occurring between 40 to 50 millimetres below the metal level followed by a gradual decrease along the casting direction. The simulations clearly show temperature fluctuations, due to mould oscillation, which result later in internal irregularities in shell growth as seen in the casting practice [26]. This result departs radically from previous modelling works where smooth parabolic shells were predicted [6, 27]. Moreover, increases of temperature in the mould cooling water (ΔT) agreed well with predictions (Figure 12).

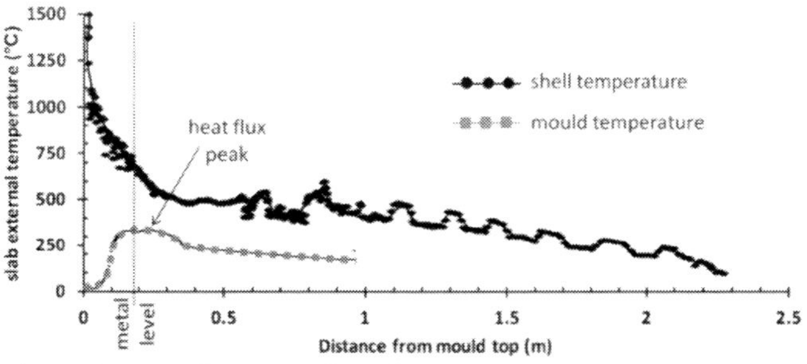

Figure 11: Predicted shell and mould temperatures measured from the mould top.

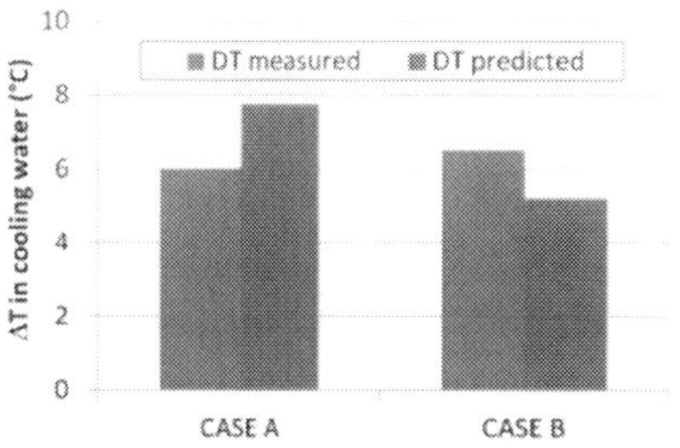

Figure 12: Predicted ΔT in cooling water vs measurements

Furthermore, different oscillation modes were implemented in the model according to particular caster settings. Heat fluxes for sinusoidal modes were presented at MCWASP XII[8] including validation by comparing predictions with a lab-based mould simulator introduced by Badri et al.[28, 29]. An improved model version includes non-sinusoidal modes as presented in Figure 13, which shows the predicted heat flux evolution for 2 cases under identical casting conditions but different oscillation modes.

The variation between maximum and minimum heat-fluxes over the cycle varies considerably with oscillation mode; the variations for the non-sinusoidal mode are more abrupt, reaching up to 1.5 MW/m^2-K, c.f. the sinusoidal mode with variations of ~0.5 MW/m^2-K. A crucial difference between modes is the time spent on the high heat flux region. For the non-sinusoidal mode, the peaks tend to be extremely fast which leaves most of the cycle in the low heat-flux area. In contrast, the sinusoidal mode remains longer in the high heat flux region. This strong variability of heat transfer produces irregular cycles of shell growth and thinning during solidification, as described in the next section.

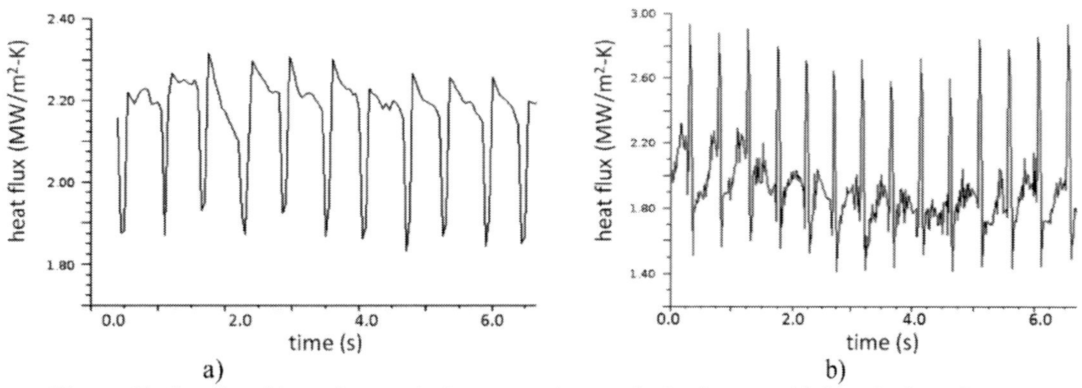

a) b)

Figure 13: Predicted heat flux variations over 6 seconds for 2 runs with identical casting conditions but different oscillation modes, a) Sinusoidal b) Non-sinusoidal

4.3. Solidification and powder consumption

The shell thickness (d_{shell}) is easily measured by tracking points along the solidification front and the liquid slag film thicknesses (d_{liquid}). Prior work has demonstrated the model's validity to predict trends observed in the industrial practice for sinusoidal modes under a variety of oscillation settings[9, 10]. This motivated the authors to compare powder consumption predictions for the 2 oscillation modes shown in Section 4.2 (Figure 13) on Figures 14 and 15.

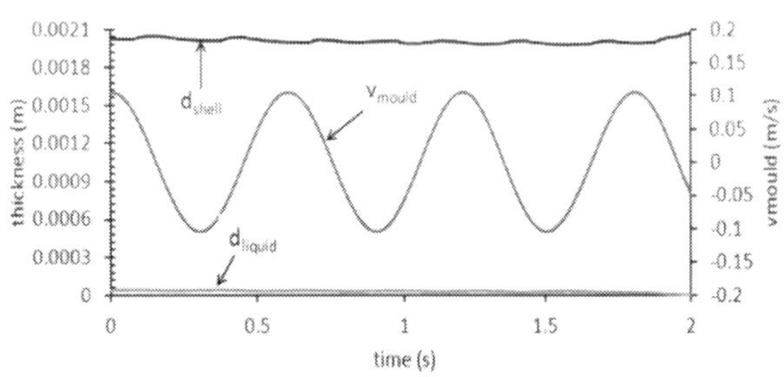

Figure 14: Shell and liquid film thicknesses for sinusoidal oscillation mode

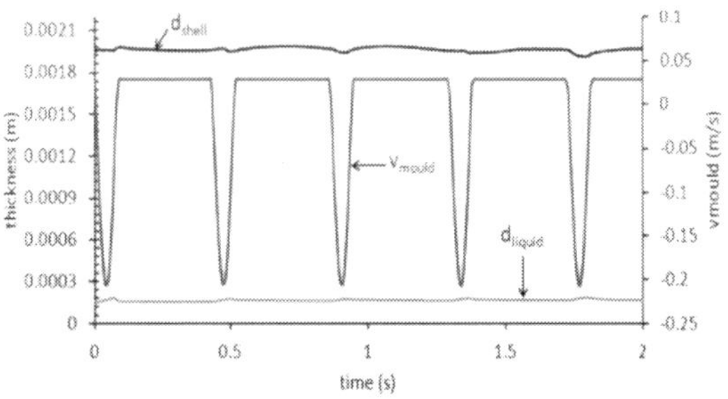

Figure 15: Shell and liquid film thicknesses for non-sinusoidal oscillation mode (α=0.65)

The non-sinusoidal mode appears to be more beneficial in terms of a smoother shell by avoiding the long term heat flux variations and replacing them with a fast t_n that affects the growth only during steep changes in mould velocity (v_{mould}), (Figure 15). However, the non-sinusoidal mode also produces a slightly thinner and weaker shell which may be more prone to cracking and may also result in mould level variations. Nevertheless, the major benefit of non-sinusoidal oscillation is to increase the powder consumption as reflected by the liquid slag-film thickness (d_{liquid}), which goes from 200 μm for the non-sinusoidal case to less than 100 μm for the sinusoidal mode. These predictions are in line with industrial observations [30, 31] including the decrease in overall heat flux and mould temperature for non-sinusoidal oscillation as noted by Suzuki [30]. Practical application of these results includes the calculation of mould friction and overall powder consumption to produce improved oscillation strategies for specific SEN-mould combinations. These strategies are beyond the scope of this manuscript.

5. Summary and conclusions

A numerical model of the CC process, (which explicitly incorporates the presence of slag) has been successfully implemented into the industrial practice. The model includes specific information about the casters modelled (including mould-SEN geometry, powder and steel compositions, oscillation settings) and operator's practices which are very relevant to the process but are often overlooked in numerical simulations. Addition of these practical constraints to the model bring it closer to reality and make possible the prediction of phenomena which are still difficult to observe in the plant (such as heat flux evolution during the oscillation cycle and changes in shell internal quality). A series of conclusions were drawn from the direct industrial implementation of the model. These include:

1. Addition of the slag bed and powder properties to simulations is necessary for an accurate representation of the meniscus. The fact that the model is able to predict infiltration as a direct result of the calculations (without a predefined slag-film or meniscus curvature) is enough reason to add slag into simulations. This makes possible not only the calculation of the slag film but also predicting variations on its thickness for different oscillation modes.
2. The implementation of geometrically realistic models for the mould and SEN, plus the calculation of the slag-metal surface provides a good representation of the flow pattern in the mould as reflected by the metal level (standing wave) predictions.
3. A precise implementation of operators' practices such as powder feeding relates for the first time the heat transfer on the bed (including calculation of the liquid pool) to the amount of slag

consumption and transient gap shape that makes possible to predict changes on heat flux and shell thickness during an oscillation cycle for different oscillation modes.

4. Implementation of different oscillation modes makes possible to explain phenomena observed on the casting floor, such as, the decrease in overall temperature, internal quality enhancement and increase in powder consumption for non-sinusoidal vs sinusoidal modes.

References

[1] S. S. Ozawa, M. M. Susa, T. T. Goto and R. Endo: *ISIJ International*, **46** (2006), 413.
[2] B. G. Thomas: *Metall. Mater. Trans. B*, **33** (2002), 795-812.
[3] A. C. Mapelli, A. W. Nicodemi and A. A. Marcandalli: *Ironmak. Steelmak.*, **30** (2003), 265-272.
[4] R. Saraswat, A. B. Fox, K. C. Mills, P. D. Lee and B. Deo: *Scand. J. Metall.*, **33** (2004), 85-91.
[5] J. W. Cho, T. Emi, H. Shibata and M. Suzuki: *ISIJ International*, **38** (1998), 834.
[6] Y. Meng and B. G. Thomas: *Metall. Mater. Trans. B*, **34** (2003), 685-705.
[7] X. Meng and B. G. Thomas: *Metall. Mater. Trans. B*, **34B** (2003), 707-25.
[8] P. E. Ramirez-Lopez, P. D. Lee and K. C. Mills: *Modeling of Casting, Welding and Advanced Solidification Processes XII*, **12** (2009), 61-68.Vancouver, CA.
[9] P. E. Ramirez Lopez, P. D. Lee and K. C. Mills: *ISIJ International*, **50** (2010), 425-434.
[10] P. E. Ramirez-Lopez, P. D. Lee, K. C. Mills and B. Santillana: *ISIJ International*, **50** (2010), 1797-1804.
[11] P. E. Ramirez Lopez, K. C. Mills, P. D. Lee and B. Santillana: *Metall. Mater. Trans. B*, **43B** (2012), 109-122.
[12] J. L. Liow, M. Rudman and P. Liovic: *ISIJ International*, **41** (2001), 225-233.
[13] J. U. Brackbill, D. B. Kothe and C. Zemach: *J. Comput. Phys.*, **100** (1992), 335-54.
[14] Y. Chung and A. W. Cramb: *Metall. Mater. Trans. B*, **31B** (2000), 957.
[15] B. E. Launder and D. B. Spalding: Lectures in Mathematical Models of Turbulence, *Cambridge University Press*, (1972).
[16] W. Wang, K. Gu, L. Zhou, F. Ma, I. Sohn, D.-J. Min, H. Mastsuura and F. Tsukihashi: *ISIJ International*, **51** (2011), 1838-1845.
[17] K. C. Mills: Mould Powders for Continuous Casting, *Imperial College London*, (2003), 211.
[18] ANSYS Fluent 6.3.26 - User's Guide: ANSYS Inc. (1995-2007),
[19] Q. Yuan, S. Sivaramakrishnan, S. P. Vanka and B. G. Thomas: *Metallurgical and Materials Transactions; B; Process Metallurgy and Materials Processing Science*, **35B** (2004), 967-982.
[20] P. E. Ramirez-Lopez, R. D. Morales, R. Sanchez-Perez, L. G. Demedices and P. O. Davila: *Metall. Mater. Trans. B*, **36** (2005), 787-800.
[21] K. Spitzer, K. Schwerdtfeger and J. Holzhauser: *Steel research*, **70** (1999), 430-436.
[22] B. B. G. Thomas and B. B. G. Ya Meng: *Metall. Mater. Trans. B*, **34B** (2003), 707-25.
[23] K. Sorimachi, T. Yamauchi and A. Yamauchi: *Ironmak. Steelmak.*, **29** (2002), 203-208.
[24] J. Holzhauser, K. Spitzer and K. Schwerdtfeger: *Steel Research*, **70** (1999), 252-258.
[25] B. Bourouga and J. Gilles: *Int J Mater Form*, **3** (2010), 77-88.
[26] B. Santillana, B. G. Thomas, G. Botman and E. Dekker: *7th European Continuous Casting Conference*, (2011), Dusseldorf, Germany.
[27] M. Hasan and S. H. Seyedein: *Can. Metall. Q.*, **37** (1998), 213-228.
[28] A. Badri, T. T. Natarajan, C. C. Snyder, K. D. Powers, F. J. Mannion and A. W. Cramb: *Metall. Mater. Trans. B*, **36B** (2005), 373-83.
[29] A. Badri, T. T. Natarajan, C. C. Snyder, K. D. Powers, F. J. Mannion and A. W. Cramb: *Metall. Mater. Trans. B*, **36B** (2005), 355.
[30] M. Suzuki, S. Miyahara, T. Kitagawa, S. Uchida, T. Mori and K. Okimoto: *Tetsu-to-Hagane*, **78** (1992), 113-120.
[31] T. Araki and M. Ikeda: *Canadian Metallurgical Quarterly*,, **38** (1999), 295-300.

IOP Conf. Series: Materials Science and Engineering **33** (2012) 012014 doi:10.1088/1757-899X/33/1/012014

Modeling of the flow-solidification interaction in thin slab casting

A Vakhrushev[1], M Wu[1,2], A Ludwig[2], Y Tang[3], G Hackl[3], G Nitzl[4]
[1]Christian-Doppler Lab for Adv. Process Simulation of Solidification & Melting,
[2]Chair of Simulation & Modelling of Metallurgical Processes,
University of Leoben, Franz-Josef-Str. 18, 8700 Leoben, Austria
[3]RHI AG, Technology Center, Magnesitstrasse 2, 8700 Leoben, Austria
[4]RHI AG, Wienerbergstrasse 9, 1100 Vienna, Austria

E-mail: alexander.vakhrushev@unileoben.ac.at

Abstract. A key issue for modelling the thin slab casting (TSC) is to consider the evolution of the solid shell, which strongly interacts with the turbulent flow and in the meantime is subject to continuous deformation due to the funnel shape (curvature) of the mould. Here an enthalpy-based mixture solidification model with consideration of turbulent flow [Prescott and Incropera, *ASME HTD*, vol. 280, 1994, pp. 59] is employed, and further enhanced to include the deforming solid shell. The solid velocity in the fully-solidified strand shell and partially-solidified mushy zone is estimated by solving the Laplace's equation. Primary goals of this work are to examine the sensitivity of the modelling result to different model implementation schemes, and to explore the importance of the deforming and moving solid shell in the solidification. Therefore, a 2D benchmark, to mimic the solidification and deformation behaviour of the thin slab casting, is firstly simulated and evaluated. An example of 3D TSC is also presented. Due to the limitation of the current computation resources additional numerical techniques like parallel computing and mesh adaptation are necessarily applied to ensure the calculation accuracy for the full-3D TSC.

1. Introduction

The thin slab casting (TSC), for its advantages of integration of the casting-rolling production chain, energy saving, high productivity and near net shape, is likely to replace the conventional slab casting for producing flat/strip products [1-2]. However, the frequently reported problems like the sensibility to breakout and edge/surface cracks have challenged the metallurgists to consider modeling tools to optimize and control the TSC parameters [3-5]. The key issue for modelling TSC is to consider the evolution of the solid shell, which interacts strongly with the turbulent flow and in the meantime is subject to continuous deformation due to the funnel shape (curvature) of the mould.

An enthalpy-based mixture solidification model with consideration of turbulent flow [6-9] was introduced by the current authors to model conventional continuous casting [10-11], where the motion of the solid shell was assumed to be everywhere parallel and constant. According to Voller et al., the treatment of the motion of the solid phase has a dramatic influence on the convection of the latent heat, hence on the shape of the evolving mushy zone [12-15]. Obviously, the assumption of parallel-constant solid velocity does not apply to TSC, where the strand shell is subject to continuous deformation due to the funnel shape (curvature) of the mould. Therefore, goals of this work are (1) to extend the previous model by considering the deforming solid shell; (2) to examine the sensitivity of the modelling result to different model implementation schemes; (3) and to explore the importance of the deforming and moving solid shell in the solidification. Modelling results on a 2D benchmark and a full-3D TSC are presented.

2. Model description

Solidification and turbulence flow

An enthalpy-based mixture solidification model [12-14] is applied. This mixture combines liquid ℓ-phase and solid s-phase, which are quantified by their volume fractions, f_ℓ and f_s. The morphology of the solid phase is usually dendritic, but here we consider the dendritic solid phase as a part of the mixture continuum. The mixture continuum changes continuously from a pure liquid region, through the mushy zone (two phase region), to the complete solid region. The evolution of the solid phase is determined by the temperature according to a $f_s - T$ relation (e.g. Gulliver-Scheil),

$$f_s = \begin{cases} 0 & T > T_{\text{liquidus}} \\ 1 - \left((T_f - T)/(T_f - T_{\text{liquidus}})\right)^{\frac{1}{k_p - 1}} & T_{\text{liquidus}} \geq T > T_{\text{Eutectic}} \\ 1 & T_{\text{Eutectic}} \leq T. \end{cases} \quad (1)$$

Only one set of Navier-Stokes equation, which applies only to the domain of the bulk melt and mushy zone, is solved in the Eulerian frame of reference.

$$\nabla \cdot \vec{u} = 0, \quad (2)$$

$$\rho \frac{\partial \vec{u}}{\partial t} + \rho \nabla \cdot (\vec{u} \otimes \vec{u}) = -\nabla p + \nabla(\mu_{\text{eff}} \nabla \cdot \vec{u}) + \vec{S}_{\text{mon}}, \quad (3)$$

$$\text{where } \vec{u} = \begin{cases} \vec{u}_\ell & \text{bulk melt region} \\ f_\ell \vec{u}_\ell + f_s \vec{u}_s & \text{mushy zone} \\ \vec{u}_s & \text{solid region.} \end{cases} \quad (4)$$

Here $\vec{u}_s$, the solid velocity, is estimated by solving Laplace's equations (see below). The momentum sink due to the drag of the solid dendrites in the mushy zone is modeled by the Blake-Kozeny law:

$$\vec{S}_{\text{mon}} = -\frac{\mu_\ell}{K} \cdot (\vec{u} - \vec{u}_s) \quad (5)$$

The permeability, *K*, is modeled as function of the primary dendrite arm spacing λ_1 [16]:

$$K = \frac{f_\ell^3}{f_s^2} \cdot 6 \cdot 10^{-4} \cdot \lambda_1^2. \quad (6)$$

The energy equation applies to the entire domain,

$$\rho \frac{\partial h}{\partial t} + \rho \nabla \cdot (\vec{u} h) = \nabla \cdot \lambda_{\text{eff}} \nabla T + S_e. \quad (7)$$

Here h is the sensible enthalpy of the solid $h_s = h_{\text{ref}} + \int_{T_{\text{ref}}}^{T} c_p dT$. At a given temperature the liquid phase is assumed to have an enthalpy of $h_\ell = h_s + L$. Release of latent heat by solidification, *L*, is treated in the source term of the energy equation,

$$S_e = \rho L\, \partial f_s / \partial t + \rho L \nabla \cdot (f_s \vec{u}_s). \quad (8)$$

A low Reynolds number k - ε model was introduced by Prescott and Incropera [6-9] to handle the turbulence during solidification. In current studies a realizable k - ε model was employed providing improved performance for flows involving boundary layers under strong pressure gradients and strong streamline curvature. The governing equations for the turbulence are

$$\frac{\partial(\rho k)}{\partial t}+\nabla\cdot(\rho\vec{u}k)=\nabla\cdot\left(\left(\mu_\ell+\frac{\mu_t}{\mathrm{Pr}_{t,k}}\right)\nabla k\right)+G-\rho\varepsilon-\frac{\mu_\ell}{K}\cdot k,\tag{9}$$

$$\frac{\partial(\rho\varepsilon)}{\partial t}+\nabla\cdot(\rho\vec{u}\varepsilon)=\nabla\cdot\left(\left(\mu_\ell+\frac{\mu_t}{\mathrm{Pr}_{t,\varepsilon}}\right)\nabla\varepsilon\right)+\rho C_{1\varepsilon}S\varepsilon-C_{2\varepsilon}\rho\frac{\varepsilon^2}{k+\sqrt{\nu\varepsilon}}.\tag{10}$$

The turbulent Prandtl numbers for k : $\mathrm{Pr}_{t,k}$=1.0, and for ε : $\mathrm{Pr}_{t,\varepsilon}$=1.2; G is the shear production of turbulence kinetic energy; $S=\sqrt{2S_{ij}S_{ij}}$, $S_{ij}=0.5(\partial u_j/\partial x_i+\partial u_i/\partial x_j)$; $\nu=S\cdot k/\varepsilon$. A simple approach is used to modify the turbulence kinetic energy in the mushy zone. It is assumed that within a coherent mushy zone turbulence is dampened by shear which is linearly correlated with the reduction of the mush permeability. The influence of turbulence on the momentum and energy transports are considered by the effective viscosity, $\mu_{\mathrm{eff}}=\mu_\ell+\mu_t$, and the effective thermal conductivity, $\lambda_{\mathrm{eff}}=\lambda_\ell+\lambda_t$, where $\mu_t=\rho C_\mu k^2/\varepsilon$, $\lambda_t=f_\ell\mu_t c_{p,\ell}/\mathrm{Pr}_{t,h}$, C_μ is a function of velocity gradient and ensures positivity of normal stresses; $\mathrm{Pr}_{t,h}$ is the turbulent Prandtl number for energy equation (0.85).

Above governing equations of the mixture solidification model were implemented in an OpenFOAM® CFD software package [17-18].

Solid velocity

A linear elasticity model [19] is further simplified to estimate the solid velocity. If we assume that in the solid domain the elastostatics condition applies and the body force is ignorable, the governing equation obtained is called Navier-Cauchy equation or elastostatic equation:

$$(\lambda+\mu)\,\nabla(\nabla\cdot\vec{\delta})+\mu\,\nabla\cdot\nabla\vec{\delta}=0,\tag{11}$$

where $\vec{\delta}$ is the displacement vector. So called Lamé parameters λ, μ are

$$\lambda=\frac{E\nu}{(1+\nu)(1-2\nu)},\quad \mu=\frac{E}{2(1+\nu)},\tag{12}$$

where E represents Young's modulus and ν is Poisson's ratio. If the solid shell is incompressible and its deformation is at small strains ($\nu=0.5$), then a volume conservation condition is fulfilled:

$$\nabla\cdot\vec{\delta}=0,\tag{13}$$

and the first term of Eq. (11) is forced to zero as well:

$$\nabla\cdot\nabla\vec{\delta}=0.\tag{14}$$

Transforming Eq.(13) and (14) from Lagragian frame into Eulerian frame by considering $\vec{u}_s=\partial\vec{\delta}/\partial t$, we obtain volume-conserved Laplace's equations:

$$\begin{cases}\nabla\cdot\nabla\vec{u}_s=0,\\ \nabla\cdot\vec{u}_s=0.\end{cases}\tag{15}$$

In 2D case, these volume-conserved Laplace's equations can be solved with a $\omega-\psi$ method (Method I) [20]. A stream function ψ and a curl of the solid velocity $\omega=\nabla\times\vec{u}_s$ are defined by:

$$u_s^x = \frac{\partial \psi}{\partial y}, \quad u_s^y = -\frac{\partial \psi}{\partial x}, \quad \omega = \frac{\partial u_s^y}{\partial x} - \frac{\partial u_s^x}{\partial y}, \tag{16}$$

with $\vec{u}_s = (u_s^x, u_s^y)^T$. Therefore, corresponding system of Eq. (15) can be written in form of:

$$\begin{cases} \dfrac{\partial^2 \psi}{\partial x^2} + \dfrac{\partial^2 \psi}{\partial y^2} = -\omega, \\ \dfrac{\partial^2 \omega}{\partial x^2} + \dfrac{\partial^2 \omega}{\partial y^2} = 0. \end{cases} \tag{17}$$

This $\omega - \psi$ method provides accurate solution, but it applies only to the 2D case. An alternative and approximation method (Method II), applicable to both 2D and 3D, is to solve the one-phase Navier-Stokes equation with an 'infinite solid viscosity'. In the current work the approximation method is to be justified by comparison with the $\omega - \psi$ method on the 2D base.

3. Benchmark (2D) configuration

A 2D benchmark is defined, as shown in Figure 1. The melt with nominal composition of Fe-0.34wt.%C fills continuously through the inlet into the domain with constant temperature (1850 K). The casting section is gradually reduced to mimic the solidification and shell deformation in TSC. Other material properties being used refer to [11]. Solid velocity is calculated with the configuration of Figure 1(a). The whole domain is filled with the solid which is extruded downwards with the constant speed $\vec{u}_{\text{pull}}$=0.07 m/s, being set at the outlet. Free slip condition is applied at the walls and non-rotational condition ($\nabla \times \vec{u}_s = 0$) is used at the inlet. Right boundary represents symmetry plane. Flow-solidification simulation is configured in Figure 1(b). A mass balance between the inlet and the outlet is fulfilled: $\rho_\ell \vec{u}_{\text{in}} A_{\text{in}} = \rho_s \vec{u}_{\text{pull}} A_{\text{out}}$, where A_{in} and A_{out} are the inlet and outlet surface areas. At the walls free-slip condition is assumed.

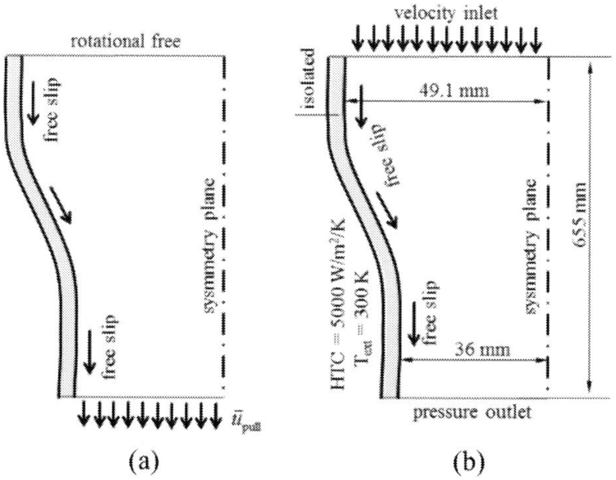

Figure 1. Configuration of a 2D benchmark (a) for solid velocity calculation and (b) for the solidification–flow calculation. The geometry in vertical direction is scaled by 1/8 (the same hereafter).

4. Results and discussions

In the presented studies solid velocities calculation is decoupled from the simulation of the melt flow. They are initially estimated by solving (15) with an assumption that the whole domain is filled with the solid and are used later on for the solidification modeling in the mush region only. Thereby liquid core in the center can accelerate or retard for balancing the total mass flow rate. The calculated solid velocity, $\vec{u}_s$, is shown in Figure 2 (a)-(b). Solid phase enters the domain in parallel to the straight wall. In the section-reduction region the solid is extruded and its velocity is gradually increased. The surface profile is forced to move along the curved wall. Comparison of the calculation results by two different methods (I and II) is made in Figure 2(b)-(d). The maximum error caused by method II, solving a simplified Navier-Stokes equation with an 'infinite solid viscosity', is 0.8%, falling in the engineering tolerance. This solid velocity will only be used by the flow-solidification model, Eq.(5) and (8), in the region where solid phase exists.

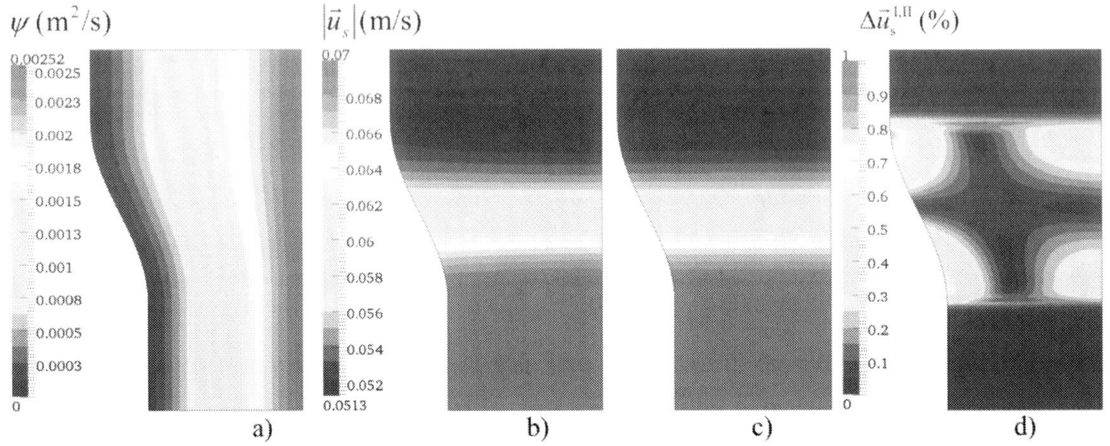

Figure 2. Calculated solid velocity: (a) stream function ψ, (b) $|\vec{u}_s|$ obtained with $\omega-\psi$ Method I, (c) $|\vec{u}_s|$ obtained with 'infinite solid viscosity' Method II and (d) velocity difference between the two methods ($\Delta\vec{u}_s^{I,II} = \left|\vec{u}_s^{I} - \vec{u}_s^{II}\right| / \left|\vec{u}_s^{I}\right| \cdot 100$).

Table 1. Parameter study of the flow-solidification model

	Flow regime	Treatment of latent heat (Eq.(8))	$f_s^{integral}$ (vol.%)*
Case I	laminar	S_e	9.38
Case II	laminar	ignoring $\rho L \nabla \cdot (f_s \vec{u}_s)$ in S_e	15.34
Case III	turbulent	S_e	8.81
Case IV	turbulent	ignoring $\rho L \nabla \cdot (f_s \vec{u}_s)$ in S_e	14.04

* $f_s^{integral}$: total solid phase (vol.%) in the whole calculation domain at the steady state.

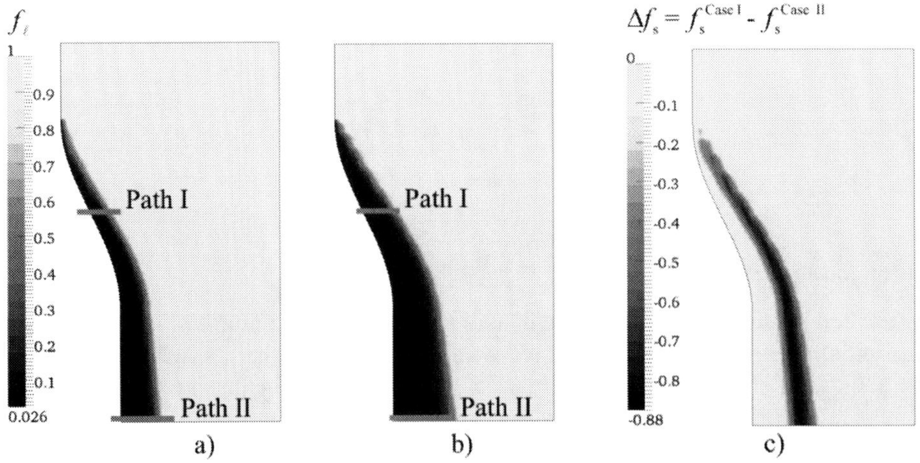

Figure 3. Predicted steady state solidification with a model considering only laminar flow: f_ℓ distribution for (a) Case I and (b) Case II; (c) difference in f_s between Case I and Case II.

In order to investigate different model assumptions, e.g. the influence of solid velocity and turbulence, on the solid shell formation by solidification, 4 simulation cases are defined (Table 1). For the boundary conditions refer to Figure 1(b). The predicted solid shell formation for the Case I and II (only laminar flow is considered) at the steady state is shown in Figure 3. Obviously the treatment of the advection of latent in the energy equation is extremely important. Ignorance of the advection term, $\rho L\nabla\cdot\left(f_s\vec{u}_s\right)$, in Case II will to a great extent overestimate the solid shell thickness. More precise analyses of the solid phase distributions along Path I and II, marked in Figure 3(a) and (b), are made in Figure 4.

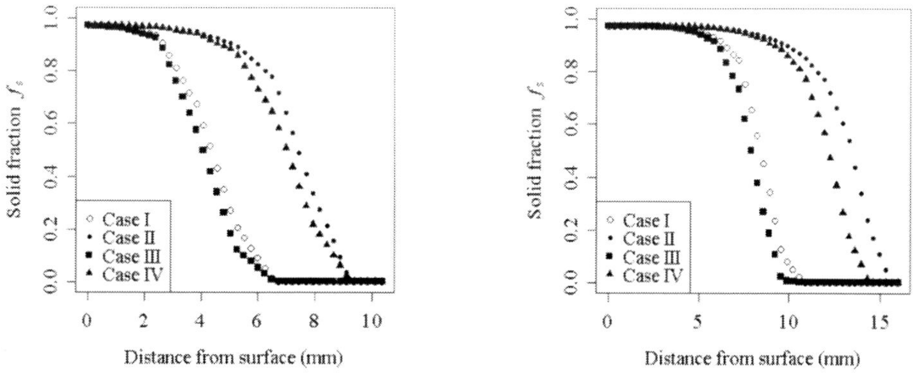

a) b)

Figure 4. Solid volume fraction distributions of different simulation cases along (a) Path I and (b) Path II are compared.

Similar calculations were carried out for the turbulent flow regime, but are not displayed here because the global phase distribution shows similar pattern to the Case I and II. Instead the influence of the turbulence on the solid shell formation is analysed (Figure 5). Comparison between Cases III and I shows that the presence of turbulence hinders the solid shell formation. Ignorance of the advection term, $\rho L\nabla\cdot\left(f_s\vec{u}_s\right)$, will also overestimate the solid shell thickness. More precise analyses of the solid phase distributions alone two Path I and II, marked in Figure 3(a) and (b), for the simulation Cases III and IV are also made in Figure 4.

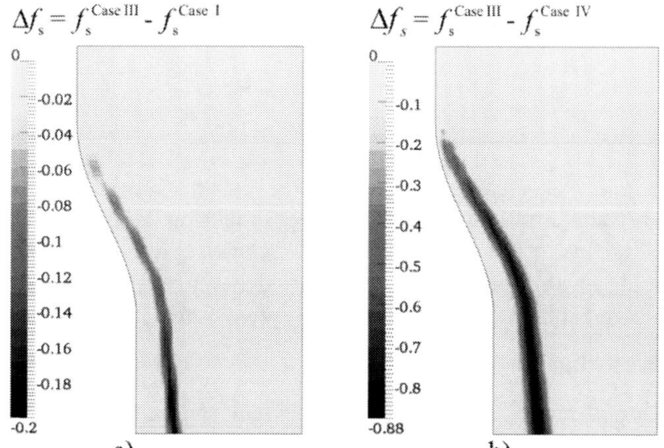

a) b)

Figure 5. Influence of turbulence on the solid shell formation, i.e. the difference in f_s distribution (a) between Cases III and I, (b) between Cases III and IV.

Additionally a mesh and time step dependency of the numerical solution was examined. A low latent heat relaxation factor (0.05) [11] along with a relatively large number of iterations (50 per time-step) allowed using a relatively large time steps without problem of divergence. It was shown that the increase in time step did not influence the final steady state solution. To improve the accuracy a necessity to use separate refinement regions for the temperature and solid fraction fields was approved previously by authors [21]. Thereby based on the error analysis of the energy equation and the resolution criterion of the Gulliver-Scheil correlation (Eq.(1)), consecutive mesh refinements were made (Figure 6). Eventually mesh independent results were obtained.

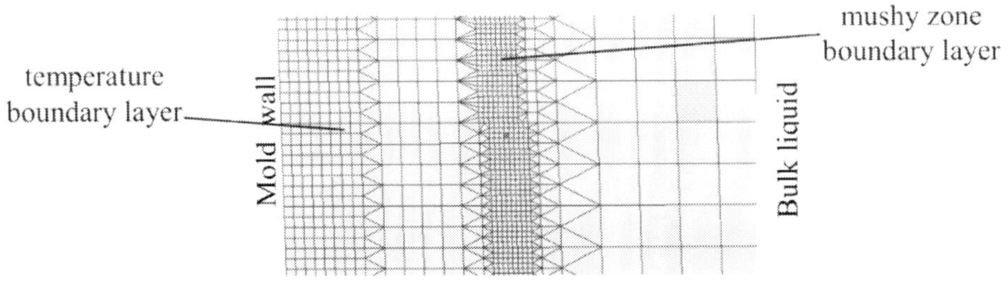

Figure 6. Mesh refinement to track the temperature boundary layer and solidification front

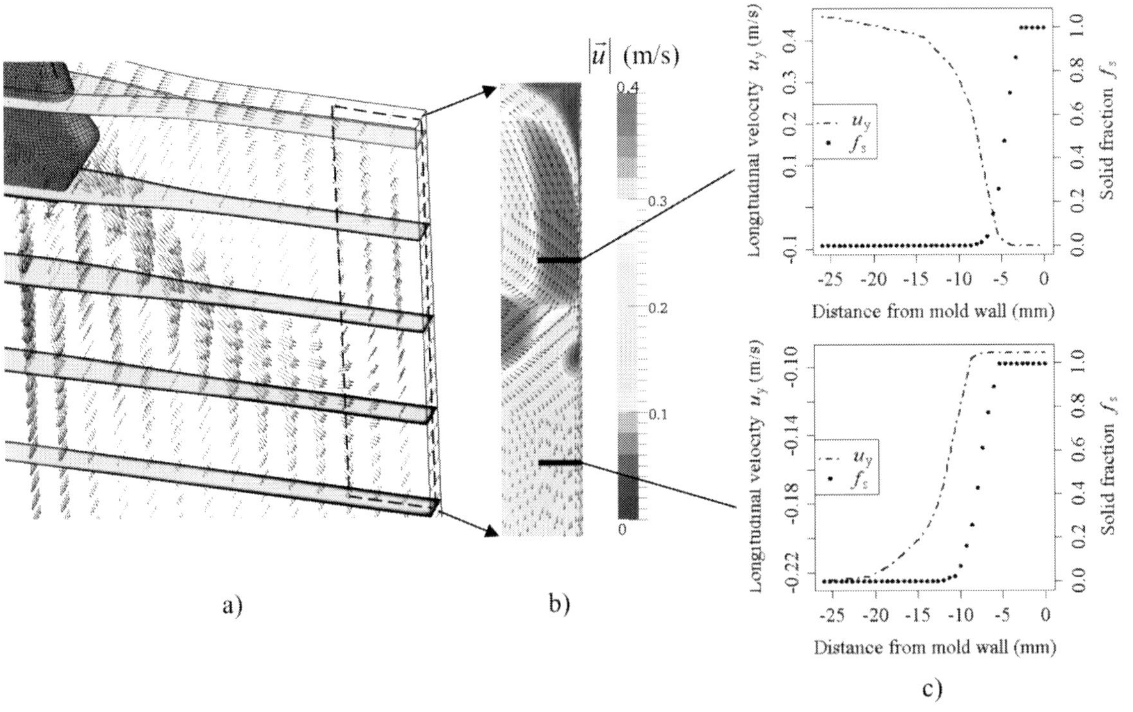

Figure 7. Quasi steady state simulation result of an engineering TSC. a) 3D distribution of the velocity vector field and evolution of the solid shell (dark region in the 5 cross sections); b) zoomed velocity field in the central plane near the narrow face; c) detailed velocity (u_y component) profile and solid volume fraction along two paths cross the mushy zone.

5. Example of thin slab casting

Based on the aforementioned model a simulation of the real engineering TSC (width 1726 mm and thickness 72 mm) was performed, and the calculation result is shown in Figure 7. The calculation domain includes the submerge entry nozzle and entire mold region and part of water cooled strand (till 2000 mm from meniscus). To ensure the calculation accuracy numerical techniques like parallel computing and mesh adaptation are necessarily applied. More than 1 million computational cells are used to resolve the interdendritic flow in the mushy zone (Figure 7(c)). In [17,18] Laplace's equation was solved for solid velocities with a constant vertical component assumption restricting longitudinal deformations. The new approach doesn't have such a limitation and helps to mimic the mush better. Detailed analysis of the 3D simulation result and evaluation will be presented in later publication.

6. Summary

Combination of flow-solidification calculation with the deforming solid shell, oriented to the application in thin slab castings, is suggested. The numerical model is implemented in an OpenFOAM® CFD open source software package. As no additional structure mechanics software is used, the motion of the solid shell is estimated on the base of a volume-conserved Laplace's equations. Validity of the proposed approach is verified in a 2D benchmark. Preliminary modelling result of a 3D thin slab casting is also presented to demonstrate the functionality of the numerical model. Further verifications, especially against experiments, are desired.

Acknowledgment

The financial support by RHI AG, the Austrian Federal Ministry of Economy, Family and Youth and the National Foundation for Research, Technology and Development is gratefully acknowledged. The authors acknowledge the fruitful discussions with Prof. Brian G. Thomas, University of Illinois at Urbana-Champaign, USA.

References

[1] Yin RY 2009 *J. Iron & Steel Res.* **16** (Supplement 1) 1.
[2] Birat JP and Bobadilla M 2006 *Proc. McWASP XI* (Eds: Gandin CA and Bellet M, TMS Publications) 33.
[3] Tang Y, Krobath M, Nitzl G, Eglsaeer C and Morales R 2009 *J. Iron & Steel Res.* **16** (Supplement 1) 173.
[4] Thomas BG 2001 *Brimacombe Lecture, 59th Electric Furnace Conf.* (Iron & Steel Soc.) 3.
[5] Tian X, Zou F, Li B and He J 2010 *Metall. Mater. Trans.* **41B** 112.
[6] Prescott PJ and Incropera FP 1994 *Transp. Phen. in Mater. Proc. & Manuf.- ASME HTD* **280** 59.
[7] Prescott PJ and Incropera FP 1994 *J. Heat Transfer* **116** 735.
[8] Prescott PJ, Incropera FP and Gaskell DR 1994 *Trans. ASME* **116** 742.
[9] Prescott PJ and Incropera FP 1995 *Trans. ASME* **117** 716.
[10] Pfeiler C, Thomas BG, Wu M, Ludwig A and Kharicha A 2008 *Steel Res. Int.* **79** 599.
[11] Wu M, Vakhrushev A, Nummer G, Pfeiler C, Kharicha A and Ludwig A 2010 *Open Transport Phenomena J. – Bentham Open* **2** 16.
[12] Voller VR and Prakash C 1987 *Int. J. Heat Mass Transfer* **30** 1709.
[13] Voller VR, Brent AD and Prakash C 1989 *Int. J. Heat Mass Transfer* **32** 1719.
[14] Voller VR, Brent AD and Prakash C 1990 *Appl. Math. Modeling* **14** 320.
[15] Chakraborty PR and Dutta P 2011 *Metall. Mater. Trans.* **42B** in press (DOI: 10.1007/s11663-011-9585-3).
[16] Gu JP and Beckermann C 1999 *Metall. Mater. Trans.* **33A** 1357.
[17] Vakhrushev A, Ludwig A, Wu M, Tang Y, Nitzl G and Hackl G 2010 *Proc. OSCIC'10.* (Munich, Nov. 4-5).
[18] Vakhrushev A, Ludwig A, Wu M, Tang Y, Nitzl G and Hackl G 2011 *Proc. ECCC* (Düsseldorf, June 27 – July 01).
[19] Slaughter WS 2002 *The Linearized Theory of Elasticity* (Boston: Birkhäuser).
[20] Roache PJ 1998 *Fundamentals of Computational Fluid Dynamics* (New Mexico: Hermosa Publishers)
[21] Vakhrushev A, Ludwig A, Wu M, Tang Y, Nitzl G and Hackl G 2011 *Proc. OSCIC'11.* (Paris Chantilly, Nov. 3-4).

IOP Conf. Series: Materials Science and Engineering **33** (2012) 012015 doi:10.1088/1757-899X/33/1/012015

Simulation of transient fluid flow in mold region during steel continuous casting

R Liu[1], B G Thomas[1], and J Sengupta[2]

[1] Department of Mechanical Science and Engineering, University of Illinois at Urbana-Champaign, 1206 West Green Street, Urbana, IL, 60801, USA
[2] ArcelorMittal Global R&D Hamilton, ArcelorMittal Dofasco Inc., 1390 Burlington Street East, Hamilton, ON, L8N 3J5 Canada

E-mail: bgthomas@illinois.edu, and joydeep.sengupta@arcelormittal.com

Abstract. A system of models has been developed to study transient flow during continuous casting and applied to simulate an event of multiple stopper-rod movements. It includes four sub-models to incorporate different aspects in this transient event. A three-dimensional (3-D) porous-flow model of the nozzle wall calculates the rate argon gas flow into the liquid steel, and the initial mean bubble size is estimated. Transient CFD models simulate multiphase flow of steel and gas bubbles in the Submerged Entry Nozzle (SEN) and mold and have been validated with experimental data from both nail dipping and Sub-meniscus Velocity Control (SVC) measurements. To obtain the transient inlet boundary conditions for the simulation, two semi-empirical models, a stopper-rod-position based model and a metal-level-based model, predict the liquid steel flow rate through the SEN based on recorded plant data. Finally the model system was applied to study the effects of stopper rod movements on SEN/mold flow patterns. Meniscus level fluctuations were calculated using a simple pressure method and compared well with plant measurements. Insights were gained from the simulation results to explain the cause of meniscus level fluctuations and the formation of sliver defects during stopper rod movements.

1. Introduction

Continuous casting is the predominant way of steel-producing. During this process, liquid steel flows from the tundish bottom, through the upper tundish nozzle (UTN), and submerged entry nozzle (SEN), and jets from the bifurcated ports near the SEN bottom into the water-cooled mold where a solid shell of steel solidifies to contain the liquid pool. Argon gas is injected through the porous refractory walls of the UTN in order to prevent nozzle clogging and reoxidation. It forms gas bubbles, which greatly affect the flow pattern in the mold and also help to remove some inclusions by floating them to the liquid flux layer. Flow rate is controlled by vertical motion of a stopper rod into the top of the UTN. The liquid steel jets from the nozzle carry nonmetallic inclusions from upstream into the liquid steel pool. Mold powder melts into liquid flux on the top surface of the liquid steel pool, which helps to prevent oxidation, as shown in Figure 1 [1]. Liquid slag may be entrained into the liquid steel pool via various mechanisms, as summarized by Hibbeler and Thomas [2]. Some of the inclusion particles, slag droplets and argon bubbles which touch the solidifying shell may become entrapped into the solidifying shell, and become defects in the final product. The flow of steel in the liquid pool has a huge impact on the transport and entrapment of inclusions, but genuine transient events have received little study.

In the current work, a transient event with multiple stopper rod movements at ArcelorMittal Dofasco's No.1 continuous caster is investigated via numerical simulation. The recorded histories of the stopper rod position and meniscus level in Figure 2 [3] show sudden movements of the stopper rod and intensive fluctuations of the meniscus level, accompanying a nearly constant casting speed. Sliver images and locations were obtained and analyzed from the downstream coil samples, and a defect was identified to be caused by the multiple stopper rod movements.

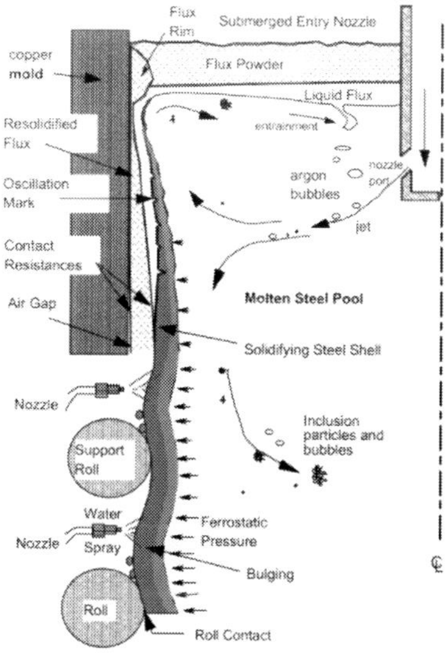

Figure 1. Complex transport processes in mold region [1]

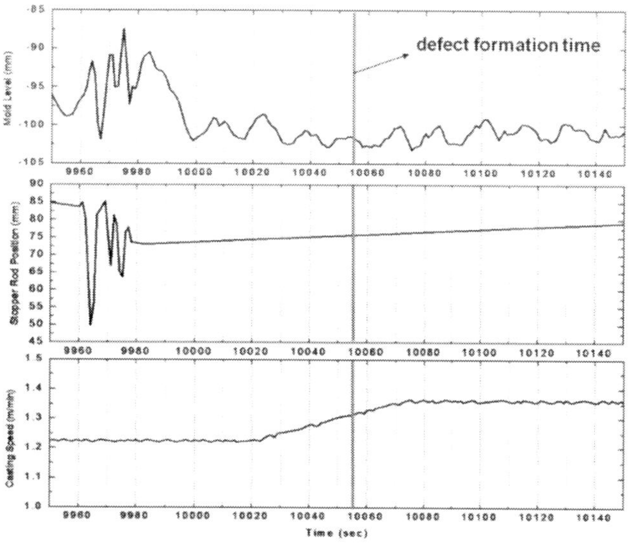

Figure 2. Recorded stopper rod position, mold level and casting speed [3]

Four sub-models were developed to incorporate the different aspects of this complex problem. Each sub-model has been validated independently using plant/lab measurements. Then, the complete model system is applied to simulate the transient multiphase flow phenomena during this process, and the mechanism for the defect formation during this process is discussed.

2. Model Description and Validation

The complete model system for this transient simulation consists of 1) a porous-flow model to study gas flow through heated UTN refractory as well as initial mean bubble size at UTN inner surface, 2) a stopper-position-based model, 3) a metal-level-based model to predict liquid steel flow rate inside SEN during stopper rod movements which is required as boundary condition for the two-phase flow simulations, and 4) an Eulerian-Mixture CFD model to simulate argon-steel two-phase flow in the nozzle and mold region, and a pressure-based post-processing method to estimate meniscus level.

2.1. Porous-flow model of argon gas in nozzle wall

Argon gas is injected into the lower part of the UTN, through grooves and slits to distribute it around the nozzle refractory. Applied back pressure forces the argon to diffuse through the porous refractory walls to form bubbles when it exits from the tiny pore openings, (active sites) at the UTN inner surface, as shown in the schematic in Figure 3. This injected argon gas significantly influences liquid steel flow in the nozzle and mold region. Two key parameters required to simulate argon-steel two phase flows are the gas flow rate and the initial bubble size injected into the steel. In this work, a new porous-flow model was developed, combined with other models, validated and applied to determine these two parameters for the subsequent transient flow simulations.

The "cold" argon flow rate is measured at standard temperature and pressure (STP) before entering the nozzle. This flow rate is much smaller than the "hot" argon flow rate, due to its great thermal expansion upon entering the liquid steel. To determine this flow rate, a heat transfer analysis was first performed to obtain the temperature distribution across the UTN wall.

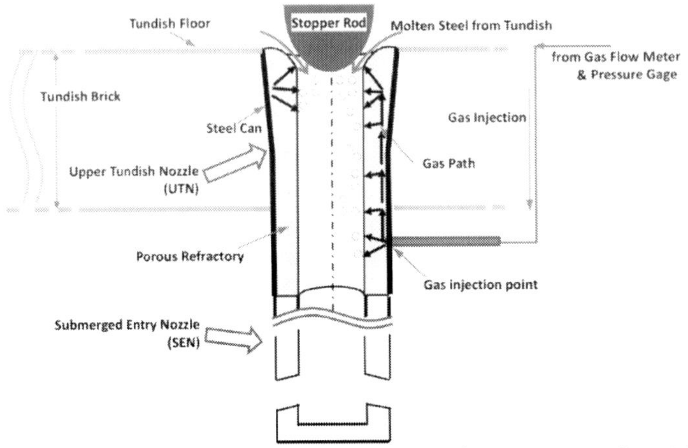

Figure 3. Schematic of gas bubble formation locations from UTN surface (slits not shown)

Then, Equation (1) is solved for the pressure distribution. It is derived from the continuity equation, Darcy's law, and the ideal gas law, to incorporate the effects of pressure and temperature on the gas volume [4]. The effect of gas viscosity change with temperature on refractory permeability K_D is considered using the viscosity-temperature correlation by Dawe [5].

$$\nabla\cdot\left(K_D\nabla p\right)=-\frac{RT}{p}\left[\nabla\left(\frac{p}{RT}\right)\cdot\left(K_D\nabla p\right)\right] \quad (1)$$

Next, the gas velocity distribution exiting the UTN surface is computed from Darcy's law via user-defined functions (UDF) with commercial CFD package FLUENT [6], distributed by ANSYS, Inc. The initial mean bubble size is then estimated, based on this velocity distribution and the corresponding number of active sites from the empirical correlation by Lee [7], using the semi-analytical two-stage model for predicting bubble size in downward cross-flow from Bai [8]. Details on the steps to estimate initial mean bubble size can also be found in [4].

To validate the porous-flow model, a static test was performed using a real UTN from ArcelorMittal Dofasco's No. 1 CC, which was cut in half, sealed at the cut surface, and placed in a water tank. Air was injected through the input hose connected to the outer wall and the locations of bubbles emerging from the inner wall are photographed in Figure 4a). The locations are compared with the preferential bubble locations predicted by the porous flow model, as indicated in Figure 4b) via the gas velocity distribution. Regions of higher gas velocity generate more and larger bubbles [8].

Note that the UTN in this experiment was cut in half along a plane perpendicular to the half-nozzle domain used in the simulation. Because of the good wetted condition between UTN refractory and water, the hydraulic pressure inside the nozzle, and capillary effects, water can enter some of the pores at the UTN inner surface and block the pore exits entirely. Thus, a threshold exists to prevent bubbles from exiting regions of low back pressure. During the experiment, back pressure was increased slowly, so the gas flow rate increased gradually from zero. The porous-flow model predicts a very non-uniform distribution of air velocity over the UTN inner surface, in Figure 4(b). Local pressure at the lowest-velocity locations was likely insufficient to overcome the threshold. It is important to note that the regions where no bubbles are observed in Figure 4(a) match well with the regions of low predicted gas velocity in Figure 4(b). Similarly, the high-velocity regions match with regions where large bubbles were observed. Thus a reasonable match was found between the simulation and experimental observations.

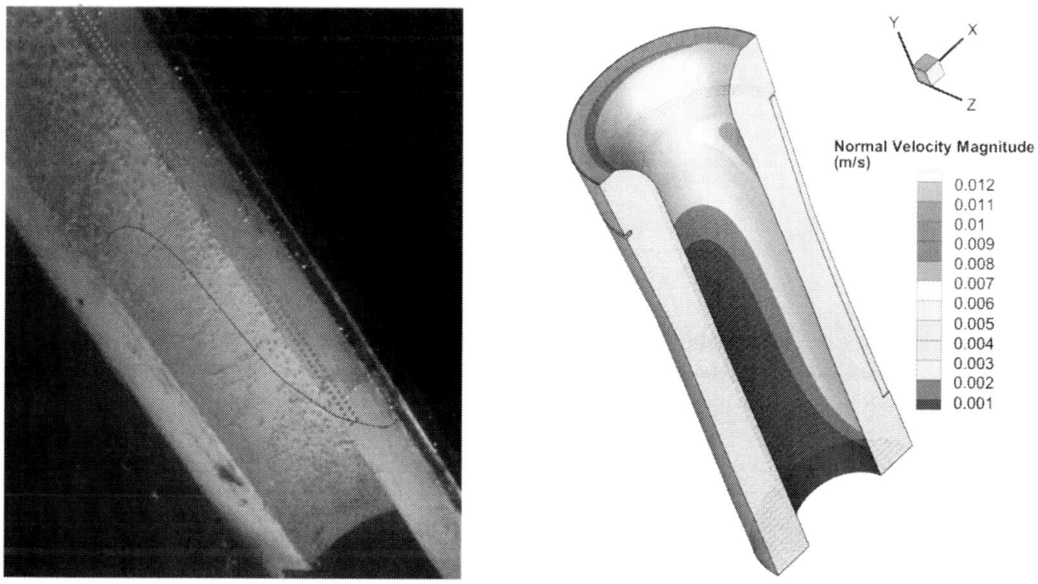

(a) Observed preferential bubbling locations (b) Predicted gas velocity distribution

Figure 4. Comparison between predicted and observed bubbling sites at UTN inner surface

2.2. Stopper-position-based model of SEN steel flow rate

During steady-state continuous casting, the liquid steel flow rate in the SEN equals the throughput at mold exit. During a transient event, however, steel flow rate in the SEN varies with time, as indicated by the rapid fluctuations of the average meniscus level. Two different models were developed in this work to predict the liquid steel flow rate into the SEN.

Firstly, a semi-analytical model, given in Equation (2), is derived from Bernoulli's equation to predict flow rate based on the measured stopper rod opening position and other parameters [3]. It is validated with plant measurements, as shown in Figure 5.

$$Q_{SEN} = A_{SEN}\left(\frac{2g\left(f_{tundish}h_{tundish} - h_{sen\_sub} + L_{SEN}\right)}{1+0.5\left(\frac{A_{SEN}}{C_2 h_{SRO}^2}\right)^2 + \left(\frac{A_{SEN}}{C_2 h_{SRO}^2} - 1\right)^2 + C_1\frac{L_{SEN}}{D_{SEN}} + C_3}\right)^{0.5} \qquad (2)$$

In this equation, A_{SEN} is the SEN inner bore cross-section area; h_{sen_sub} is the submergence depth of SEN, $h_{tundish}$ is the total height of the tundish; $f_{tundish}$ is the tundish weight fraction; L_{SEN} is the total length of SEN; D_{SEN} is the SEN inner bore diameter; h_{SRO} is the stopper rod opening. The three parameters in the equation, C_1, C_2 and C_3, are adjustable coefficients that represent different pressure head losses: C_1 is for friction, C_2 is for the stopper rod gap and C_3 is for clogging. The influence of C_2 and C_3 on the predicted relation between stopper rod position and flow rate is shown in Figure 5(a) and 5(b) . The effect of argon gas injection on the pressure head loss at the stopper rod gap is accounted for in C_2. In this work, C_2 was calibrated using the plant trial data in Figure 5(a), which had the same gas volume fraction as in the current transient study (10% gas). The effect of friction factor on this relation is negligible.

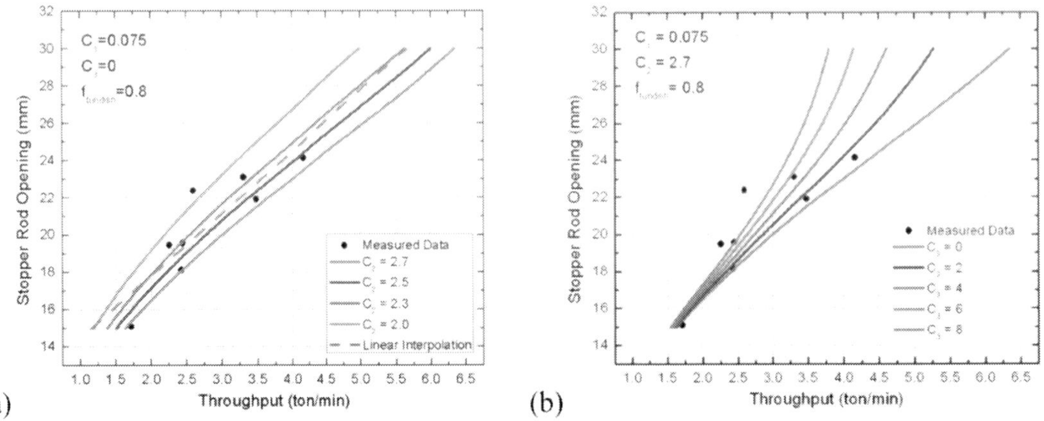

(a) (b)

Figure 5. Effect of C_2 and C_3 on stopper-position-based model predictions of flow rate [3]

2.3. Metal-level-based model of SEN steel flow rate

A metal-level-based model was derived to predict flow rate based on an overall mass conservation of the system. Knowing the casting speed, the time variations of the liquid steel flow rate in the SEN are reflected in variations of the average position of the meniscus level. Using the recorded meniscus level and casting speed in the plant, the SEN flow rate can be approximated by Equation (3) [3].

$$Q_{SEN} = \frac{h_m(i+1) - h_m(i-1)}{2\Delta t}\left(W \cdot T - \frac{\pi d_{SEN,outer}^2}{4}\right) + V_{cast}(i) \cdot W \cdot T \tag{3}$$

In this equation, the index i represents the current time; Δt is the time step size, $V_{cast}(i)$ is the casting speed at time i; W is the mold width, T is the mold thickness; h_m is the measured meniscus level at time i; and $d_{sen,outer}$ is the SEN outer bore diameter. The measured meniscus location, midway across the mold, is assumed to be representative of the average liquid level in the mold.

2.4. Multiphase fluid flow model

Argon-steel two-phase flow in the nozzle and mold was simulated with a 3-D Eulerian-Mixture model, which solves for the mass and momentum conservation equations of the argon-steel mixture phase. The continuity equation of the mixture is shown in equation (4) and (5), and the momentum conservation equation is in equation (6).

$$\frac{\partial \rho_m}{\partial t} + \nabla \cdot (\rho_m \mathbf{u}_m) = 0 \tag{4}$$

$$\text{where } \mathbf{u}_m = \frac{\alpha_s \rho_s \mathbf{u}_s + \alpha_a \rho_a \mathbf{u}_a}{\rho_m}, \text{ and } \rho_m = \alpha_s \rho_s + \alpha_a \rho_a \tag{5}$$

$$\frac{\partial(\rho_m \mathbf{u}_m)}{\partial t} + \nabla \cdot (\rho_m \mathbf{u}_m \mathbf{u}_m) = -\nabla p + \nabla \cdot \left[\mu_m \left(\nabla \mathbf{u}_m + (\nabla \mathbf{u}_m)^T\right)\right] + \rho_m \mathbf{g} + \nabla \cdot (\alpha_a \rho_a \mathbf{u}_{dr,a} \mathbf{u}_{dr,a}) \tag{6}$$

The variables α_a and α_s represent the volumetric fractions of argon phase and liquid steel phase respectively, which are found by solving equation (7), and knowing that α_a and α_s sum to 1.

$$\frac{\partial(\alpha_a \rho_a)}{\partial t} + \nabla \cdot (\alpha_a \rho_a \mathbf{u}_m) = -\nabla \cdot (\alpha_a \rho_a \mathbf{u}_{dr,a}) \tag{7}$$

Drift velocity $\mathbf{u}_{dr,a}$ on the right hand side of equation (6) and equation (7) is defined in equation (8).

$$\mathbf{u}_{dr,a} = \mathbf{u}_{as} - \frac{\alpha_a \rho_a}{\rho_m} \mathbf{u}_{as}, \text{ where the relative velocity is } \mathbf{u}_{as} = \mathbf{u}_a - \mathbf{u}_s \tag{8}$$

The mixture model is then closed using an algebraic slip formulation for the relative velocity $\mathbf{u}_{as}$ assuming that local equilibrium between phases is reached over a short spatial length, as shown in equation (9) [6], where the drag function f_{drag} is taken from Schiller and Naumann [9], and d_a is the argon bubble diameter. The standard k-ε model was applied to model turbulence in the mixture phase.

$$\mathbf{u}_{as} = \frac{(\rho_a - \rho_m) d_a^2}{18 \mu_s f_{drag}} \left(\mathbf{g} - (\mathbf{u}_m \cdot \nabla) \mathbf{u}_m - \frac{\partial \mathbf{u}_m}{\partial t} \right) \tag{9}$$

The computational domain includes the nozzle and the liquid pool in the mold region, with the solidification front interface as the domain boundary. No-slip wall boundary condition is adopted at mold top surface, as the sintered slag layer serves as a solid wall. Mass and momentum sinks [10] are imposed at the layer of computational cells next to the shell boundary, to account for the liquid steel crossing the boundary due to solidification. Similar mass and momentum sinks are applied to quantify the argon gas escaping from the top surface. This model has been validated with extensive plant measurements using both the simple but powerful nail-dipping method and measurements using a Sub-meniscus Velocity Control (SVC) device. These two different measurement methods validate each other, based on their agreement in plant trials such as shown in Figure 6 [11].

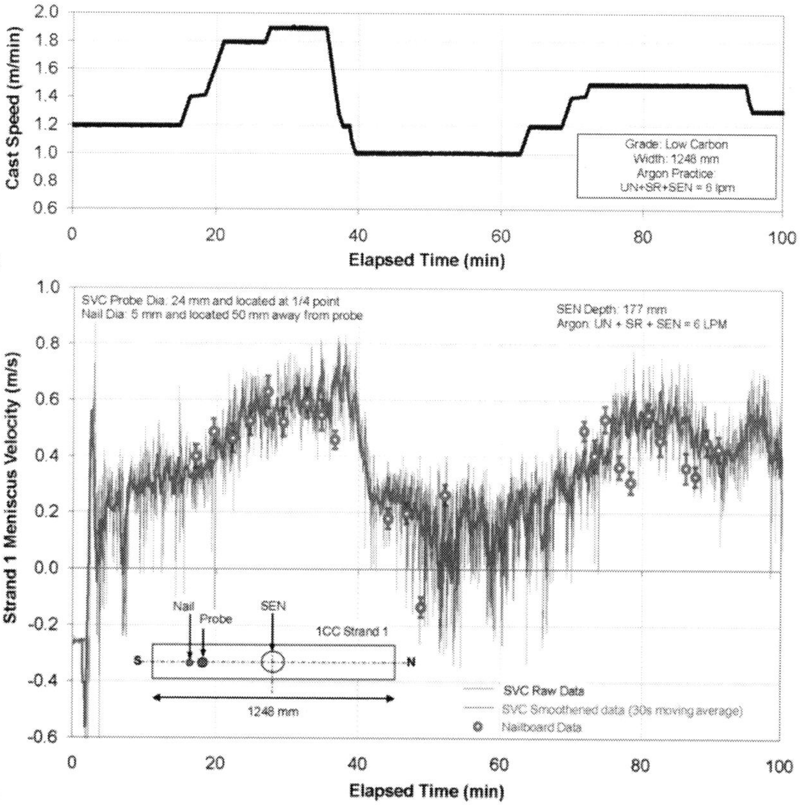

(a)

(b)

Figure 6. Plant trial comparing two surface velocity measurement methods [11]
(a) Casting speed history; (b) Meniscus velocity history from SVC and nail dipping tests.

One quarter of the SEN and mold were chosen as the computational domain for the validation cases [3]. Figure 7 shows the geometry and mesh of ~0.23 million hexahedral structured cells. The Dofasco No. 1 caster was 225mm thick with a bifurcated SEN with 15-deg downward ports, described elsewhere [3]. Three cases were simulated with Argon flow rate fixed at 6 SLPM, width of 983mm, submergence depth of 186mm, 2.5m long domain, and casting speeds of 1.5, 1.7 and 1.9 m/min. The calculated surface velocities are compared with results of the plant nail dipping tests in Figure 8. A reasonable match is obtained, which tends to validate the model. These results also confirm that increasing gas volume fraction (by decreasing casting speed), tends to change the double-roll flow pattern into a complex flow pattern, as expected from the water model results. Velocities near the SEN are directed towards the narrow face, opposing the classic surface flow towards the SEN. This reverse flow near the SEN is caused by liquid dragged upward there by the rising bubbles.

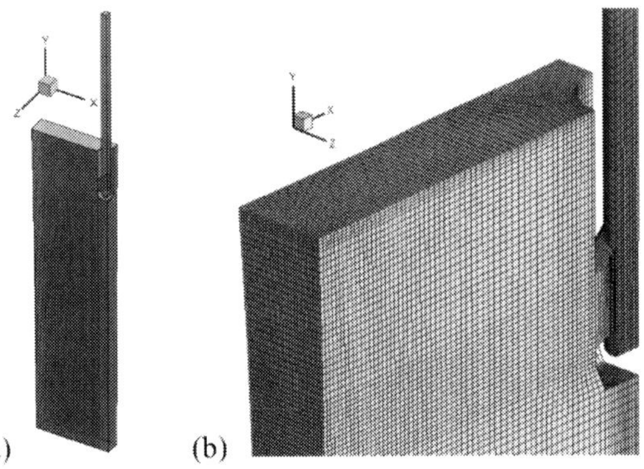

Figure 7. Computational domain geometry and mesh [3]

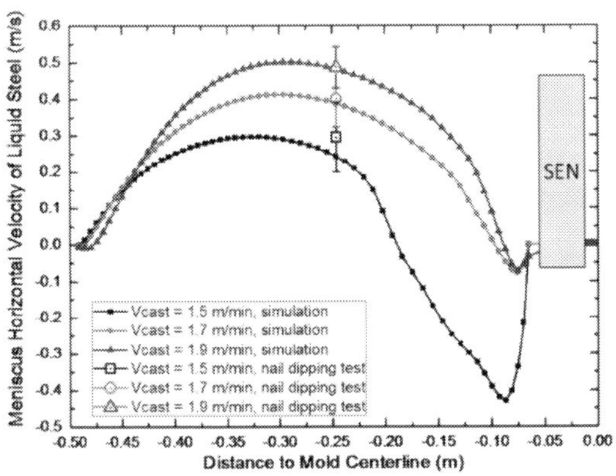

Figure 8. Comparison between predicted and measured surface velocities [3]

3. Results and Discussions

The system of models is applied to simulate the transient event described previously. The two semi-empirical models, stopper-position-based model and metal-level-based model, are used to generate the history of SEN inlet velocity boundary condition; then the porous-flow model was utilized to calculate the argon flow rate entering the UTN in hot condition, and bubble size was estimated; the transient two-phase flow simulation was run with Eulerian-Mixture model. Finally, the computed pressure at meniscus was converted to predict mold level, and compared with the measurements. The mesh consists of 0.8 million hexahedral cells in total, and time step size was 0.01sec. No-slip wall is adopted as the boundary condition at mold top surface, as the sintered slag layer serves as a solid wall.

3.1. Predicted SEN inlet liquid steel flow rate

The flow rate of the liquid steel in the SEN predicted by the stopper-position-based model and the metal-level-based model are compared in Figure 9. Note that translating the metal-level-based results back in time by 1.2sec (dashed line in Figure 9) makes the two predicted curves roughly match. This time interval represents the time delay for the steel to flow from the stopper into the mold. The SEN flow rate from stopper-position-based model is adopted in the subsequent simulation. A major flow rate drop occurred around 9965sec, and the corresponding flow pattern changes between 9955 and 9971.7sec are investigated in the following section.

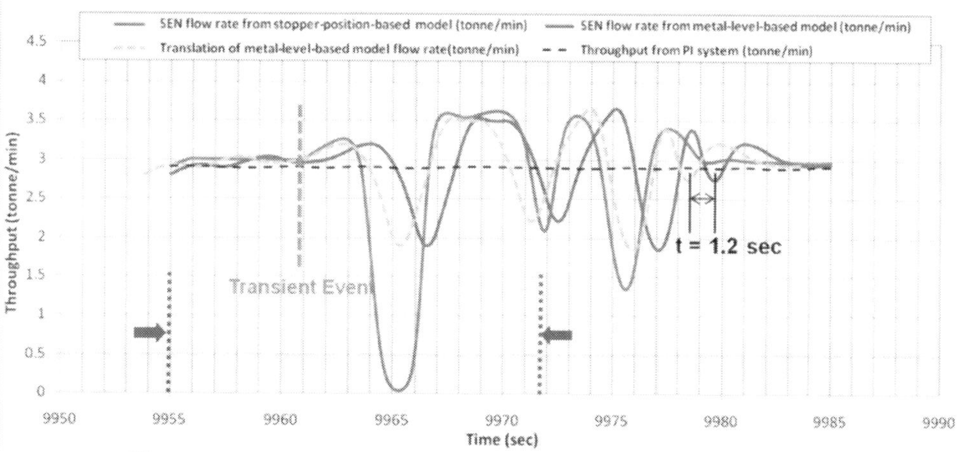

Figure 9. Predicted SEN flow rates from stopper-position-based and metal-level-based model

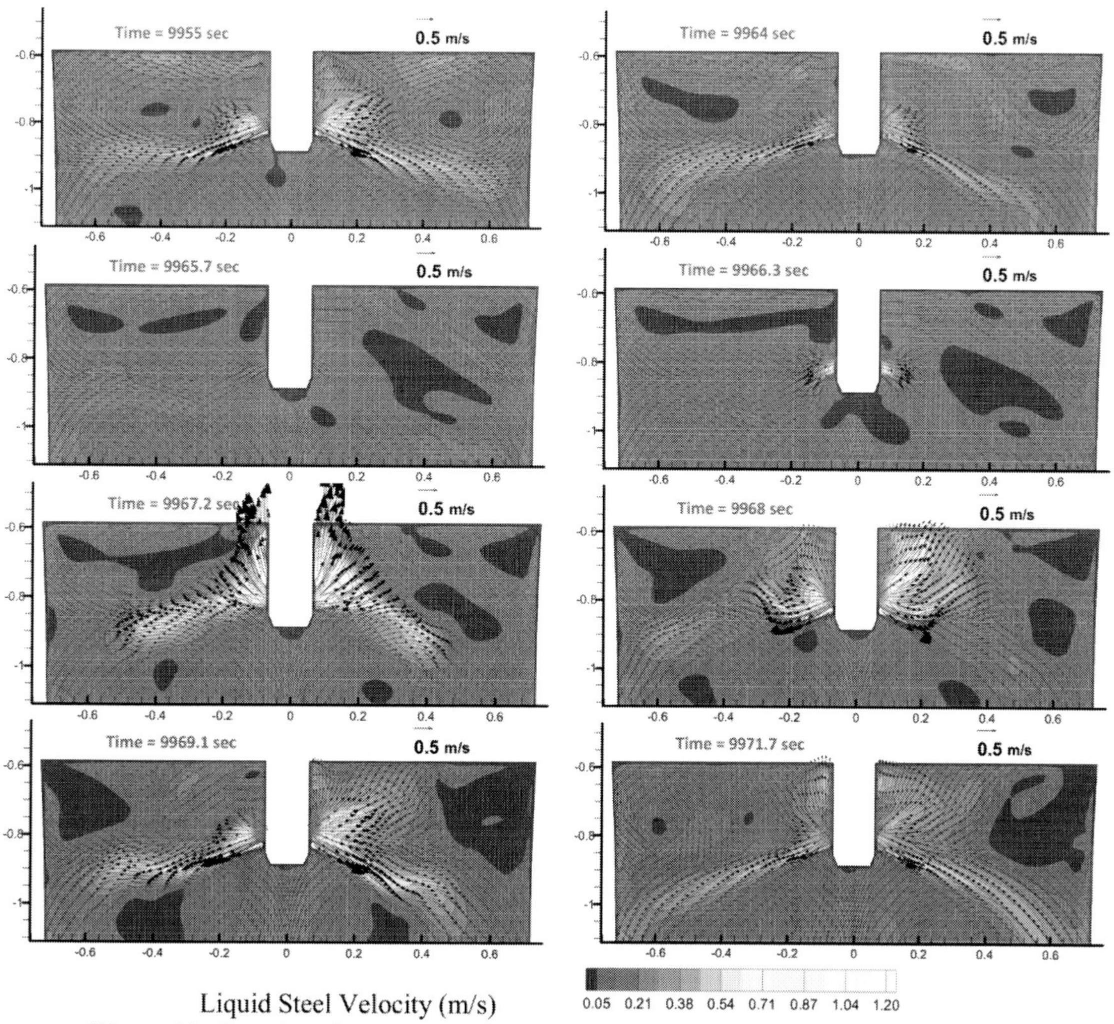

Liquid Steel Velocity (m/s)

Figure 10. Transient flow patterns during stopper rod movement (axis distances in m)

3.2. Flow pattern evolution during the transient event

Evolution of the flow pattern in the mold simulated during the 16.3sec time interval (indicated with vertical dotted blue lines in Figure 9) of multiple stopper-rod movements is shown in Figure 10. Each frame is plotted at the center plane between broad mold faces. The initial quasi-steady state flow field (at 9955sec) is observed to be a symmetric double-roll flow pattern. As inlet liquid steel flow rate decreases, (eg. at 9964sec), the strength of the jets decreases. Jet strength continues to decrease (9965.7sec) and then starts to recover (at 9966.3sec). Then, (at 9967.2sec), a strong burst of liquid steel shoots up towards the meniscus near the SEN, and significant disturbance of the meniscus is observed, which likely causes liquid slag droplets to get entrained into the liquid pool. This phenomenon is probably caused by the strong buoyancy force from a large amount of rising argon gas. Between time 9968 and 9969.2sec, the upward liquid stream towards meniscus becomes less intensive, and liquid steel jets towards the narrow faces begin to develop, and wobbling of the jets is observed. Finally, at time 9971.7sec, the jet swinging disappears, and the steady flow pattern is re-established.

3.3. Comparison of Predicted and Measured Mold Level

The flow pattern changes caused by the stopper rod movements also affect the mold level profile and cause fluctuations at the top surface, which can be detrimental to steel quality. In this simulation, a flat wall is imposed at the top surface, so to predict mold level, a simple pressure calculation is used:

$$\Delta h = \frac{p - p_0}{\rho_L g} \tag{10}$$

where Δh is the mold level deviation, p is the pressure at top surface, and p_0 is the pressure corresponding to the reference mold level used in determining Δh. ρ_L is the liquid density, and g is the gravitational acceleration. Displacement of the liquid slag layer is neglected in this equation, because the entire layer was judged to be thin enough to simply rise and fall with the steel surface profile variations.

The predicted mold level during the simulated 16.3sec interval with stopper rod movements agrees very well with the measured unfiltered mold level data, at the midway point between SEN and narrow face along the centerline, as shown in Figure 11. Note that there is significant asymmetry between the left and right sides of the mold, owing to chaotic turbulence.

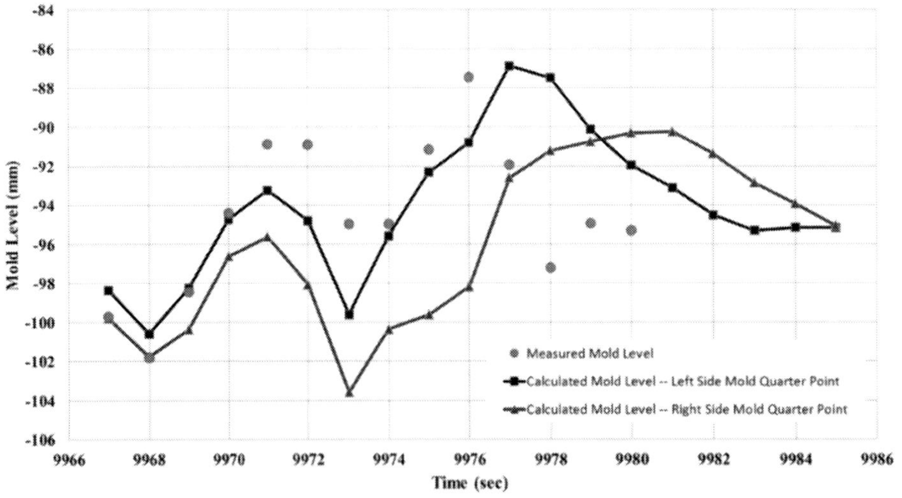

Figure 11. Comparison of calculated and measured mold level

4. Conclusions

A complete system of models is presented in this work to study transient argon-steel two-phase flow in the SEN and mold region during continuous casting process. Each submodel has been validated independently and the following conclusions can be drawn:

1. Gas flow through the hot porous refractory of the nozzle walls is nonuniform and greatly affects the gas flow rate injected into the liquid steel. The porous-flow model presented here is an important tool to determine this flow distribution and to estimate the initial mean bubble size in the nozzle, which are two crucial parameters in liquid-gas two phase flow simulations.
2. Methods such as the stopper-position-based model and the porous-flow model are required to predict the time-dependent flow rates of steel and argon in the nozzle, to provide accurate inlet conditions for transient simulations.
3. Simulated flow during a transient event suggests that significant disturbances of the meniscus occur during stopper rod movements, which may be responsible for slag entrapment leading to formation of sliver defects in the final product.
4. The model system is validated by the match between predicted and measured mold level at the plant.

Acknowledgements
The authors would like to thank the members of the Continuous Casting Consortium in University of Illinois at Urbana-Champaign for their support of our research work, and also to B. Forman at ArcelorMittal Global R&D in East Chicago for his help with the static test. ArcelorMittal Global R&D in Maizières, France is acknowledged for providing the SVC devices for trials. The authors also want to show their gratitude towards S. Chung, M. Trinh and D. Currey at ArcelorMittal Dofasco Inc. in Canada for their kind help with the plant trials.

References

[1] Meng Y and Thomas B G 2003 *Metall. Mater. Trans. B* **34B(5)** 685-705
[2] Hibbeler L and Thomas B G 2010 *Proc. AISTech 2010 Steelmaking Conf.* (Pittsburgh)
[3] Liu R, Sengupta J, Thomas B G *et al* 2011 *Proc. AISTech 2011 Conf.* (Indianapolis) p1619-1631
[4] Liu R, Thomas B G and Sengupta J 2011 *CCC Annual Meeting* (Urbana-Champaign)
[5] Dawe R and Smith E 1969 *Science* **163** 675-676.
[6] FLUENT ANSYS Inc. 2007 *FLUENT 6.3-Manual* (Lebanon, NH)
[7] Lee G, Thomas B G and Kim S 2010 *Met. Mater. Int.* **16(3)** 501-506
[8] Bai H and Thomas B G 2001 *Metal Mater. Trans. B.* **32B(6)** 1143-1159
[9] Schiller L and Naumann Z 1935 *Z. Ver. Deutsch. Ing.* **77** 318
[10] Yuan Q, Thomas B G and Vanka S P 2004 *Metal. Mater. Trans. B.* **35B(4)** 685-702
[11] Liu R, Sengupta J, Thomas B G *et al* 2011 *Proc. TMS 2011 Conf.* (San Diego) p 51-58

A dynamic control water distribution model of steel in continuous casting

Jian-Xun Fu, Weng-Sing Hwang*, De-Chang Tsai, MingHsiu Tsai, Chien-Hsun Wang
Department of Materials Science and Engineering & Research Center for Energy Technology and Strategy , National Cheng Kung University, Tainan, Taiwan

*E-mail: wshwang@mail.ncku.edu.tw

Abstract. After investigation in many continuous casting shop of steel, a dynamic water distribution model is proposed for flexible control on secondary cooling in continuous casting. In this model, the water cooling intensity is determined by the model casting speed instead of the real casting speed. When the casting speed is steady, the model casting speed is equal to the real casting speed. When the real casting speed is changing, the model casting speed according to calculating algorithm to adjust and approaches to the real one, but there is a time delay between them, so it can avoid the slab surface temperature fluctuated due to casting speed changes. The secondary cooling can be dynamically controlled by monitoring the model casting speed. The compare of the simulation results and the measured results reveals that the temperature field and thickness of slab shell in simulations agree very well with the real production situations.

1. Introduction

Secondary cooling of slab in continuous casting is closely related to the quality of casting slab and production yield. In a continuous casting, the water cooling intensity needed in each zone are computed by the control program of secondary cooling according to the steel grade and processing parameters, such as casting speed, superheat, water cooling intensity. The super or less secondary cooling water may lead to the cracking [1,2], bulging[3-5] and broadening [6-10] of slab. Secondary cooling water plays a determined role in the production of free-defect slabs.

Static control model, linear water cooling intensity control and manual adjusting of water according to the casting speed can not meet the requirements for the production of high-quality slab.[11,12] In the dynamic control model of water distribution, the steel grade, the casting speed, the slab surface temperature and the casting state are considered.[13-15] The computer calculates the slab temperature every period of time according to the control model of secondary cooling, and then derives the needed cooling water volum for each zone of slab after comparing the calculated temperature with the preset one. In the continuous casting, the static control model and dynamic control model take different treatments when the casting speed changes. In the static control model, the water cooling intensity linear change with the real casting speed while in the dynamic control model, the water-cooling intensity changes gradually with the slab surface temperature. It is a fine dynamic control model, with which the cooling water and slab temperature can be precisely controlled. The static control model can meet the processing requirements only if the casting speed keeps unchanged. Otherwise, when the casting speed changes, the slab surface temperature will fluctuate and impact the quality of slab.

The dynamic control model includes unsteady-state control model, temperature feedback control model, and computer online simulation control model.[2,16-19] In the unsteady-state control model, the thickness of slab shell is the one-dimensional function of time. This control model can solve the problem of casting speed change, but the controlling accuracy needs to be improved. In the temperature feedback control model, thermometric equipments are needed, and the accurate temperature is hard to obtain under the extreme condition of high temperature. As to the computer online simulation control model, a great amount of calculation is needed, thus real-time online control can not be promised.

In this work, a novel dynamic water distribution model, i.e. flexible control water distribution model is proposed after investigation in many continuous casting shop, and the principle of new model is different from above three dynamic models. In order to check the effectiveness of this dynamic control model, the continuous casting at steady casting speed and under variable processing conditions was numerically simulated and experimental verification.

2. Mathematical Model and Principle

2.1 Mathematical Model

The solidification model is developed according to the metallurgical criteria for slab and target surface temperature controlling principle. The 2D heat transfer mathematical equation is shown as follows:

$$\rho C_p \frac{\partial T}{\partial t} = \frac{\partial}{\partial x}\left(K\frac{\partial T}{\partial x}\right) + \frac{\partial}{\partial y}\left(K\frac{\partial T}{\partial y}\right) + S_0 \quad (1)$$

Where ρ, C_p, T, t , K , x, y and S_0 represent density, specific heat capacity, temperature, time, conductivity, coordinate of the length, width direction, and source term, respectively.
In the equation (1)

$$S_0 = \rho L_f \frac{\partial f_s}{\partial t} \quad (2)$$

where f_s, L_f represent solid fraction , latent heat ($J.kg^{-1}$), respectively. It is a linear relationship between fraction solid and temperature.

2.2 Principle of dynamic Control Water Distribution Model

In this model, the water cooling intensity needed for the secondary cooling is determined by the model casting speed instead of the real casting speed. The model casting speed is a virtual concept. To obtain the model casting speed, we consider the slab as a great number of individual elements along the slab movement. Each elements has different casting speeds at specific moment. When the casting speeds is at a steady state, the model casting speed is equal to the real casting speed. When the real casting speed changed, the model casting speed will approach to the real one according to the dynamic calculating algorithm. The secondary cooling can be dynamically controlled by monitoring the model casting speed and adjust water cooling intensity. The trend of the water cooling intensity and slab surface temperature is in accordance with the model casting speed, but a time delay from the real casting speed.

In the process of continuous casting, the discontinuously change of casting speed will lead to the discontinuously change of residence time of slab in each element, so the water cooling intensity and the slab surface temperature is severe changes.

In order to let each element of slab keep similar solidifying state, the water cooling intensity for secondary cooling will not change discontinuously in real casting process. Therefore, the flexible control model of water for secondary cooling is reasonable.

When the casting speed changes mightily and discontinuously, but the model casting speed in flexible control model and the water cooling intensity changes gradually and continuously. In the flexible control model, the water cooling intensity in each element of slab changes differently according to the influence of casting speed change. When the casting speed changes, the heat state and the first zone slab change first correspondingly. In order to maintain the heat balance, the water cooling intensity at the first zone should be adjusted first, then at the series of following zones, one by one. The change value of water cooling intensity also decreases in the serial number of zones. Therefore, the flexible control model is able to guarantee the surface quality of slab. The flexible control water distribution model has four advantages:

(1) The experience of the static water has been used to a largest degree;

(2) It does not rely on temperature monitoring instruments;

(3) In the control process, there is a few heat transfer calculation which reduces the computational time;

(4) It is particularly applicable to non-steady process control.

It is noted that the calculation should be of sufficient accuracy in the developing of model. In addition, appropriate assumption of heat transfer is needed and possible.

3. Results and discussion

As an example of SPHC low carbon steel, the flexible control water distribution model was analyzed basing on reliability. The parameters in simulation are taken from the process parameter of the #4 continuous caster (VAR caster) of Xinyu Iron and Steel Co., Ltd. of China. The composition of SPHC steel is show in Table 1.

liquidus temperature(T_l): 1528.9 ℃; solidus temperature (T_s): 1493.1 ℃.

Table 1. The composition of SPHC steel

C, %	Si, %	Mn, %	P, %	S, %	Al, %
0.05	0.05	0.20	0.02	0.012	0.03

3.1 Analysis the temperature fluctuations at constant casting speed

The surface temperature and solidification condition of low carbon steel slab at the casting speed of 0.80~1.25 m/min was analyzed using the calculating algorithm of the flexible control model in offline. The water cooling intensity was assessed using the original caster, and the target temperature of the SPHC steel was derived by reversing the setting of the water cooling intensity. The casting speed and the corresponding water cooling intensity were inputted into the program of heat transfer to calculate the slab surface temperature, and the obtained results are shown in Figure 1.

It can be seen from Figure 1 that when the casting speed is in the range of 0.80~1.00 m/min, the slab surface temperature fluctuates obviously and the slab solidified quickly. Especially, when the casting speed is 0.80 m/min and 0.85 m/min, the slab surface temperature at straightening zone is lower than 800℃, the slab surface temperature at the final stage of solidification is lower than 750℃.

In these two cases, the temperature is below the low temperature brittleness temperature, cracks are easily to appear in the slab, and the slab temperature do harm to the internal and external qualities of the slab. Therefore, the cooling water volume corresponding to the casting speed of 0.80~1.00 m/min is unsuitable to the continuous casting of the low carbon steel slab. If slab are cast at the speed of 0.80~1.00 m/min, the cooling water volume has to be re-optimized to make sure the slab in the secondary cooling zone and to promise the good slab quality.

When the casting speed is in the range of 1.05~1.20 m/min in Figure 2, the surface temperature of each slab is relatively uniform, and it increase slightly with the increase of casting speed. The slab surface temperature distributes in 880℃~960℃, which is beyond the temperature range of the low temperature brittleness temperature of low carbon steel, and there are few fluctuations with the surface temperature of each slab.

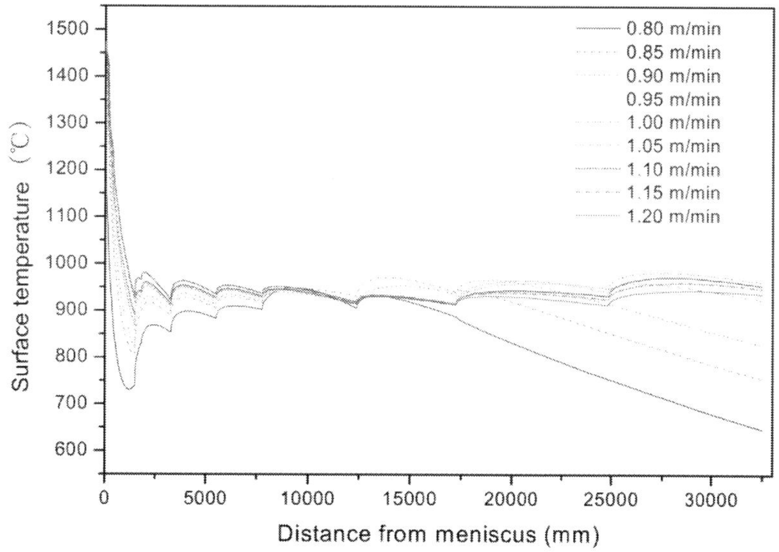

Figure 1. The slab surface temperature at different constant speeds.

It illustrates that the water cooling intensity when casting speed in the range of 1.05~1.20 m/min, not only satisfies the metallurgy criterion, but also meets the requirements of steady temperature. Accordingly, the slab surface temperature at different position of the secondary cooling zone can be matched with the cooling water volume. Then the target temperature of the SPHC steel can be derived by averaging the surface temperature at the same position.

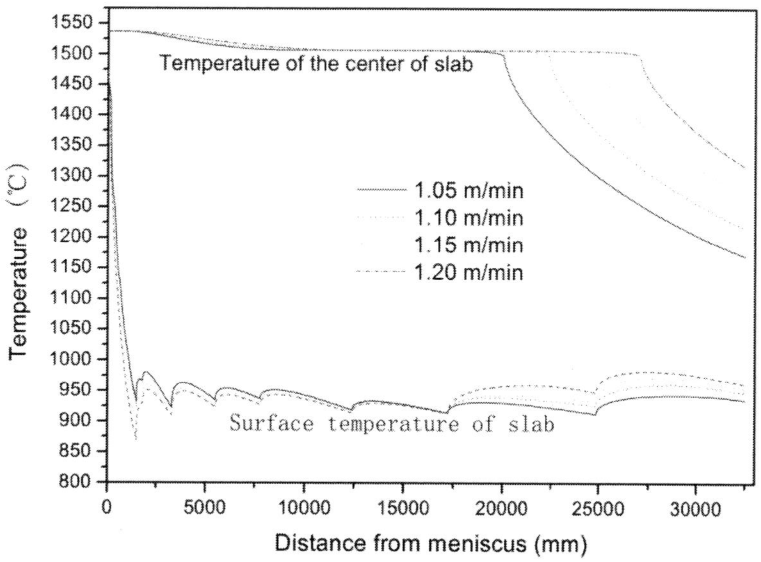

Figure 2. The center temperature of slab at different casting speed

In this way, huge costs can be saved comparing to the thermal simulation method to calculate the target temperature. The slab target temperature is the control criteria of the cooling water volume in the secondary cooling zone.[15]

3.2 Analysis the temperature fluctuations when casting speed change

The temperature fluctuations are analyzed when casting speed change with two kinds of situation. With the decrease of casting speed, the surface temperature on each zone of slab rise firstly due to the cooling water decrease immediately. Finally, the drop approaching to equilibrium values. The superheat of molten steel increase, the surface temperature of slab at the first zone is influenced most obviously with the casting speed, and the subsequent zone is influenced only slightly. At the second cooling zones, the surface temperature of slab is about 800℃~1000℃, which is out of the temperature range of the low temperature brittleness, satisfying the metallurgical criteria, and avoids the crack and bulging of slab.

With the increase of casting speed, the surface temperatures in the secondary cooling zone first drop due to the cooling water increased immediately, and then rise approaching to equilibrium values. The equilibrium surface temperature of slab is close to that before the casting speed changed. It should be noted if the casting speed has to be increased and the acceleration rate is more than 0.4m/min, the slab surface temperature will be great fluctuations , and no quality assurance.

For example, when the casting speed changes in the range of 0.8~1.2 m/min discontinuously, the model casting speed of each zone is also in the same range and changes slowly. Because the change of the casting speed first influence the residence time of slab in the 1st zone, the model casting speed changes obviously in the 1st zone, and it reduces step by step in the following zones. The cooling water volume of secondary cooling is proportional to the model casting speed, thus it changes in the same tendency with the model casing speed.

When the casting speed changes, the model casting speed of flexible control model and the slab surface temperature are shown in Figure 3 and Figure 4. Figure 3 is the relationship between surface temperature at different zones and casting times, the model casting speed was calculated by the algorithm of flexible control model.

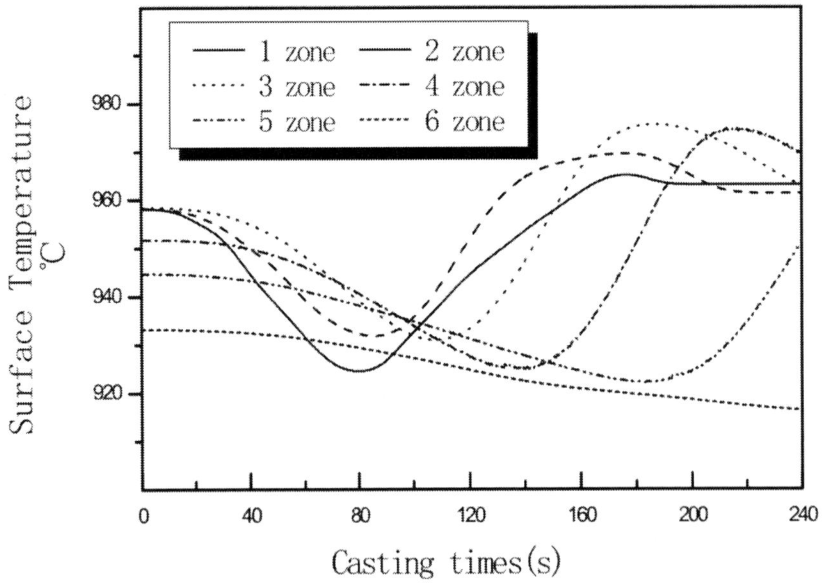

Figure 3. The slab surface temperature fluctuations with casting times

3.3 The solidification process and it's experimental verification

Figure 4 is shows the casting speed and the model casting speed at different zones fluctuations with casting time. At the beginning when the casting speed change, the model casting speed lags behind the casting speed change. This can led to the model speed maintain relatively stable, so the water cooling intensity is relatively stable.

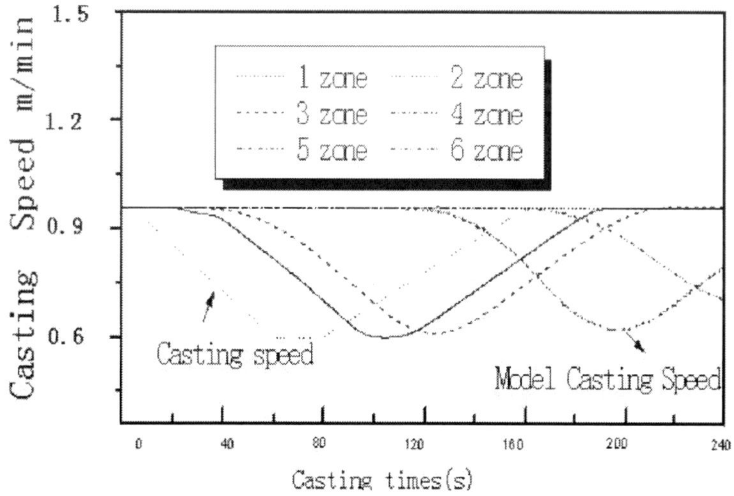

Figure 4. The model casting speed and casting speed fluctuations with casting time

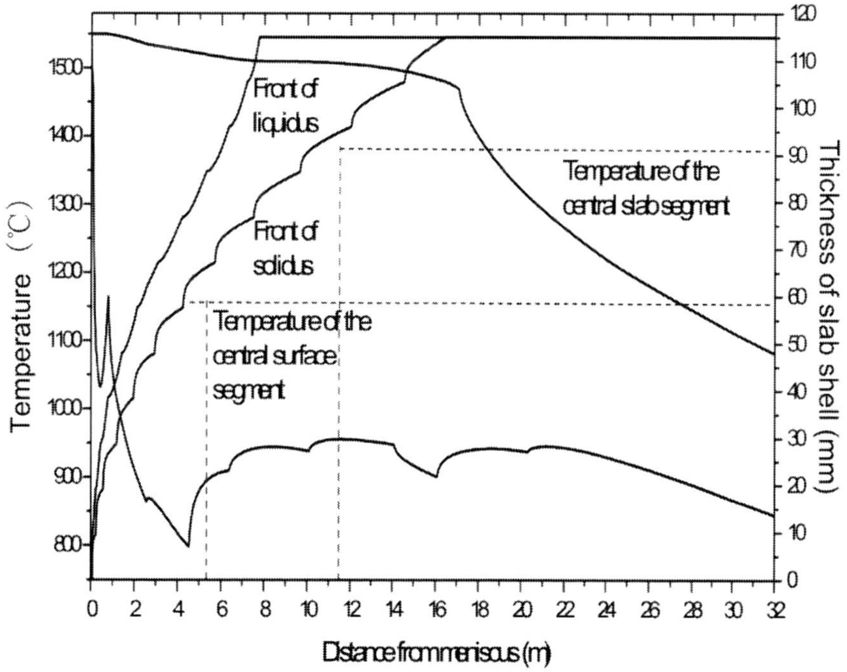

Figure 5. Analysis of the solidification of ultra low carbon steel.

The solidification process numerical simulations were made to analyze of SPHC steel. The process and the processing parameters are same as the casters in Xinyu Iron & Steel Co., Ltd. The casting speed is 0.90 m/min, the specific water cooling is 1.00 l/kg, the solid ratio is defined where *fs*=

0.85,[22] and the pouring temperature is 1542 ℃. The solidification state of SPHC steel is shown in Figure 5.

To verify the obtained simulation results of fully solidification position, The nailing experiment was carried out in #4 caster in Xinyu Iron & Steel Co., Ltd. The nailing experiment was used to detect the thickness of the solidified shell. When low melting temperature sulfide is shot into the slab with the nail, the sulfide will spread rapidly at the melting zone. Therefore it can find the solidified interface. According to the method of acid etching can be measured the thickness of the solidified shell.

The fully solidification position of slab determined from the simulation agrees very well with the shot nail test position at 17.3m from meniscus. The corresponding measured field of temperature is also agreed with the simulation results. The results of the shot nail test are shown in Figure 6.

In the real production process, there is an optimal relationship among the casting speed, the casting temperature, and specific water cooling, which constrains the slab surface temperature in the satisfaction range, and it is the guidelines to set the cooling water volume. The flexible control system is able to adjust the water cooling intensity at each zones of slab in real-time according to the steel grade, the superheat, the casting speed, and the slab surface temperature at different time, and consequently to make the slab be uniformly cooled.

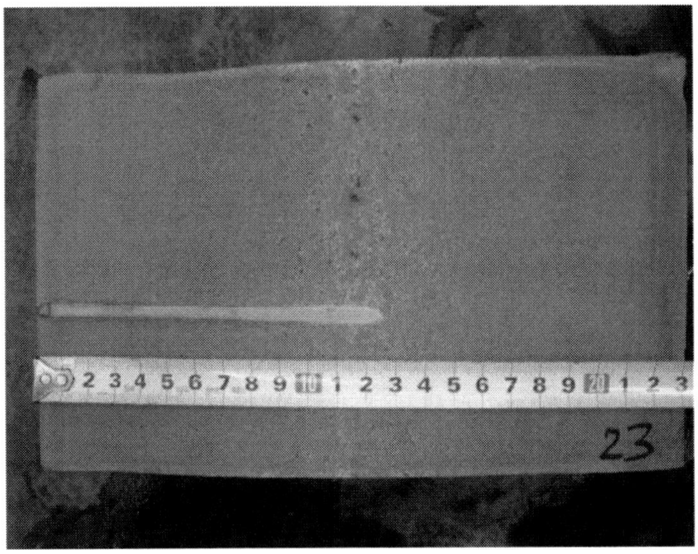

Figure 6. Photo of the nailing experiment

4. Conclusions

A novel flexible and dynamic control model is proposed for water cooling intensity control in the secondary cooling zone. In this model, the water cooling intensity is adjusted by the model casting speed instead of the real casting speed. When the casting speed is steady, the model casting speed is equal to the real casting speed. When the real casting speed is changing, the model casting speed according to calculating algorithm to adjust and approaches to the real one, so it can avoid the slab surface temperature fluctuated due to casting speed changes. The secondary cooling can be dynamically controlled by monitoring the model casting speed.

Casting speed has a dramatic effect on the heat transfer and solidification of slab in continuous casting. When the casting speed increases 0.05 m/min, the slab surface temperature rises about 7~8 ℃, and the center temperature rises about 30~35℃ in the terminal of slab.

Solidification process numerical simulations and experimental verification were carried out to check the model. It is revealed that the calculated surface temperature and thickness of slab shell agree very well with the data from real continuous casting, which illustrates the effectiveness of the model.

Acknowledgements

The authors would like to thank the National Science Council of Taiwan for financially supporting this research (NSC 100-2221-E-006-091-MY3), & "Towards the Top-Tier University and Elite Research Center Development Plan" promoted by the Ministry of Education of Taiwan (D100-23003).

References

[1] Brimacombe J K and Sorimachi K 1977 *Metallurgical Transactions* **3** 489
[2] Louhenkilpi Seepo1993 *Metallurgical Transactions* **3** 685
[3] Yoshii A and Kihara S 1986 *Trans ISIJ* **26** 891
[4] Maeno S, Wada K, Itoh Y and Nagano Y 1983 *Trans ISIJ* **24** 83
[5] Siyasiya C, van Rooyen G T and Stumpf W E J 2005 *South Afr Inst Mining Metall* **105** 473
[6] Fu J X, Li J S and Zhang H 2010 *Journal of Iron and Steel Research International* **8** 20
[7] Fu J X, Li J S and Zhang H 2010 *Acta Metallurgica Sinica* **46** 91
[8] Fu J X, Li J S and Zhang H 2011 *Science China Technological Sciences* **54** 1228
[9] Fu J X, Li J S and Zhang H 2010 *Int. Journal of Minerals, Metallurgy and Materials* **17** 723
[10] Fu J X, Hwang W X and Li J S 2011 *Steel Research Int.* **82** 5
[11] Xia G and Schiefermüller A 2010 *Steel Research Int.* **81** 652
[12] Euntai Kim 2001 *IEEE Trans on Systems* **31** 107
[13] Camisani F R, Craig I K, and Pistorius P C 2000 *ISIJ International* **40** 469
[14] Camisani F R, Craig I K and Pistorius P C 1998 *ISIJ International* **38** 147
[15] Spitzen K H, Harste K and Weber B 1992 *ISU International* **32** 848
[16] Moon C H and Hwang C M 2003 *Ironmaking and Steelmaking* **30** 48
[17] Royzman S E 1996 *Ironmaking and Steelmaking* **23** 46
[18] Liu W H, Xie Z and Ji Z P 2008 *Journal of Iron and Steel Research International* **15** 14
[19] Bouchard D, Hamel F G and Serge F 2001 *ISIJ International* **41** 1465
[20] Dalhuijsen A J and Segal A 1986 *Int. J. Numer. Methods Eng.* **23** 1807
[21] Lally B, Biegler L and Henein H 1990 *Metallurgical Transactions* **21B** 761
[22] Kobayashi S, Nagamichi T and Gunji K 1988 *Trans. Iron Steel Inst. Japan* **28** 543

Modelling of heat transfer and solidification processes in horizontal twin-roll casting of magnesium AZ31

A. Miehe and U. Gross

Institute of Thermal Engineering, TU Bergakademie Freiberg, Gustav-Zeuner-Str. 7, 09599 Freiberg, Germany

E-mail: anja.miehe@iwtt.tu-freiberg.de

Abstract. Magnesium is a desirable material for lightweight construction due to its low density and high strength-to-weight ratio. The disadvantage of low cold formability may be overcome for sheet production by combining casting and forming within horizontal twin-roll casting (TRC). During this process, the molten alloy is fed between two counter-rotating rolls water cooled from the inside. In this paper, numerical simulations using the commercial CFD software Star-CCM+ are presented. The code provides the model based on an incompressible, steady-state solver for laminar flow. To describe the solidification, an exponential increase in viscosity for decreasing temperature is employed in addition to the fraction-of-solid approach based on a linear equation for the release of latent heat. The counter-rotating rolls are exposed to a constant heat transfer coefficient and specify the characteristic velocity of the case. Using the full domain for the simulation of the TRC process provides asymmetric velocity profiles resulting in asymmetric temperature profiles and thus temperature differences between the two rolls. Symmetric results should not be enforced by using only half domain as the asymmetric profiles evoke recirculation zones differing in size which might influence the solidification rates.

1. Introduction

With a density of 1 740 kg/m^3 at 20 °C, magnesium is 36 % lighter per unit volume than aluminium (2 700 kg/m^3), and 78 % lighter than iron (7 800 kg/m^3) [1]. A Young's modulus of about 45 GPa or the shear modulus of 17 GPa imply a high strength-to-weight ratio in mechanical properties. Consequently, magnesium seems to be the choice for light weight production. The main disadvantage of magnesium alloys is the low cold formability. During production, magnesium is therefore best formed at temperatures above 225 °C, for example directly after the casting process. A technology to combine the two production steps namely casting and forming is given by twin-roll casting. The first mentioning of twin-roll casting dates back to the middle of the nineteenth century, naming an invention of Sir Henry Bessemer in 1865 [2] for the manufacturing of steel strip. The scheme of the model geometry according to the pilot plant of the Insitute of Metal Forming at the Freiberg University of Mining and Technology and the MgF Magnesium Flachprodukte GmbH Freiberg [3] is show in figure 1. During twin-roll casting, from the left side the molten metal is fed between two counter-rotating rolls being water cooled from the inside. At the surface of the two rolls, the metal starts to solidify. The thin solidified layers grow until they meet in the so called *final point of solidification*, which has to lay before the minimum roll distance, called *roll gap*, is reached. At the kissing point, casting is finished and the rolling process starts. The setback indicates the distance between the roll gap and the end of the die. According to [3], the pilot plant can produce magnesium alloy AZ31 sheet with a thickness of 4 − 7 mm and a width up to 700 mm.

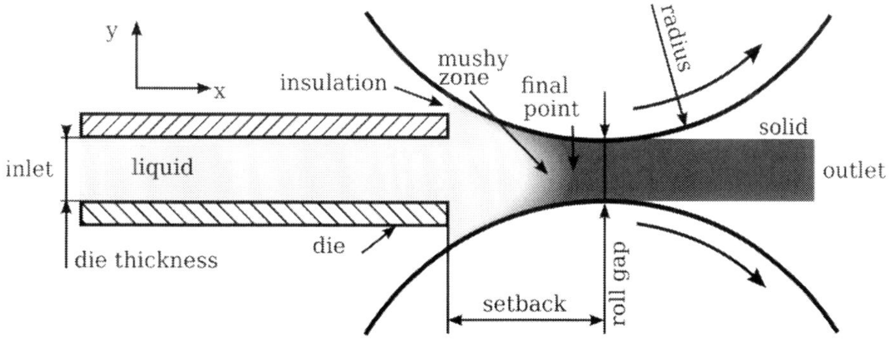

Figure 1. Not-true to scale simulation domain: the roll gap, the minimal distance between the two rolls, is as large as the die thickness.

The first attempt to simulate twin-roll casting tends back to 1981 [4], modelling the heat transfer and the fluid flow of vertical twin-roll casting of pure aluminium. The symmetry of geometry and boundary conditions are used and only one half of the domain is modelled. In a similar manner, the vertical twin-roll casting of a tin-lead alloy follows also using boundary conditions to couple the fluid and the solid region at the interface[5]. The largest amount of twin-roll casting simulations is done for steel [6–14], typically referring to vertical casting. It is common for most of the researchers to model only half domain. Going away from steel, twin-roll casting simulations are done for aluminium or tin-lead alloys by some researchers [15–18]. Presenting simulations of horizontal twin-roll casting of magnesium alloys, there are only a few papers [19–25]. Unfortunately, despite quite low Reynolds numbers some researchers using the magnesium alloy AZ31 [22, 23, 25] employed a turbulence model. Furthermore, an open field of research is the validation of the simulation results with experiments. The models in the cited papers vary from steady-state, viscosity and apparent heat capacity model to d'Arcy and the enthalpy approach. Besides the finite volume method, the finite element method is often used in combination with a deformation analysis; besides the VoF method, interface front tracking is also employed.

2. Model Description and Assumption

2.1. Physical Assumptions

The cases presented here are based on simplifying assumptions.

- The geometry is reduced to two dimensions due to the large aspect ration of about 150, but not relating to a possible symmetry with respect to the centerline between the rolls.
- The fluid is Newtonian and the material properties of the magnesium alloy AZ31 are constant and equal for the liquid and the solid state.
- The release of latent heat is coupled with a linear fraction-of-solid approach.
- There is a no-slip condition of the magnesium to the rolls.
- The heat transfer coefficient at the rolls is constant.
- Heat losses by radiation are negligible.
- There is no entrainment of gas.

2.2. Governing Equations

The balance equations to mathematically describe the twin-roll casting process are presented in the following section based on literature [26, 27].

$$\nabla \cdot \mathbf{u} = 0 \tag{1}$$

$$\nabla \cdot (\rho \mathbf{u}\mathbf{u}) = -\nabla p + \nabla \cdot \left(\mu \left(\nabla \mathbf{u} + (\nabla \mathbf{u})^T\right)\right) + \rho \mathbf{g} \tag{2}$$

$$\nabla \cdot (\rho \mathbf{u} H) = \nabla \cdot (k \nabla T) \tag{3}$$

$$\text{with } H(T,t) = \int_{T_{ref}}^{T} c_p(T)\mathrm{d}T + (1 - f_s)\Delta H_s$$

A very simple approach to model the fraction-of-solid between solidus and liquidus is the employed linear transition [28].

$$f_s = \frac{T_{\text{liq}} - T}{T_{\text{liq}} - T_{\text{sol}}} \quad \forall T,\ T_{\text{sol}} \leq T \leq T_{\text{liq}} \tag{4}$$

2.3. Boundary Conditions and Material Properties

The thermophysical properties of magnesium alloy AZ31 are presented in table 1. The boundary conditions according to the domain shown in figure 1 are displayed in table 2.

Table 1. Material date of the magnesium alloy AZ31 used in the simulation [29]

property	unit	value	property	unit	value
liquidus temperature	°C	630	solidus temperature	°C	575
latent heat	kJ/kg	377	specific heat	J/(kg K)	1 050
thermal conductivity	W/(m K)	80	density	kg/m^3	1 750
dynamic viscosity	kg/(m s)	$T > 630$°C 0.001,		$T < 575$°C 1 000	
		else $\eta = 0.001 \cdot \exp\left[-13.82 \cdot (T - T_{\text{liq}})/(T_{\text{liq}} - T_{\text{sol}})\right]$			

Table 2. Boundary conditions as indicated in figure 1, $T_u = 30$ °C

boundary	thermal condition		kinematic condition	
inlet	superheat	700 °C	constant pressure	$p = 0$ Pa
die	superheat	700 °C	no-slip wall	–
insulation	convection	$\alpha = 20$ W/(m K), T_u	no-slip wall	–
rolls	convection	$\alpha = 10\,000$ W/(m K), T_u	constant rotation	$v = 0.815$ rpm
solid surface	convection	$\alpha = 20$ W/(m K), T_u	constant velocity	$v = 2.15$ m/min
outlet	zero gradient	–	constant velocity	$v = 2.15$ m/min

The domain dimensions are normalized with the die thickness d. The length of the die is about $8\,d$, the setback is about $15\,d$ and the length of the solid sheet about $3\,d$. The sheet is as thick as the die and the radius of the rolls is about $85\,d$. In order to obtain reproducible results, the symmetric geometry is changed to a slightly asymmetric one by displacing the inlet/die region by $\pm 0.1\,d$.

3. Results and Discussion

Simulating only half domain or without any inflow region will produce the result presented in figure 2. A redirection of the flow occurs before the beginning of solidification. The flow aligns according to the characteristic rotational velocity. The results show symmetry with respect to the centerline. This kind of symmetry is also enforced when using an inflow region but only one half of the domain together with a symmetry boundary condition.

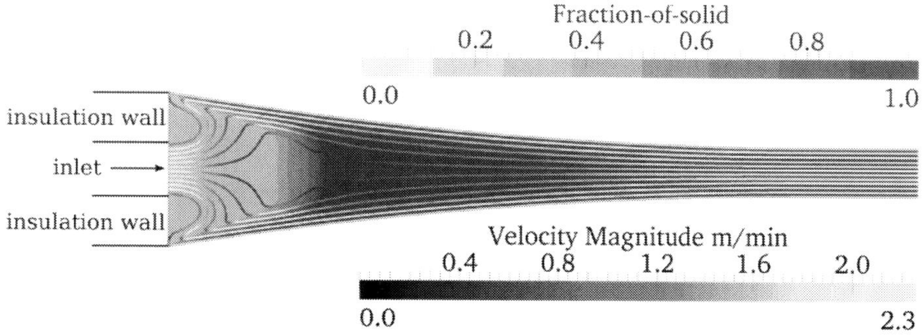

Figure 2. Fraction-of-solid in grey shades and streamlines coloured by the velocity magnitude for the symmetric case without an inflow region

Figure 3 shows the effect of a slight asymmetry with the inflow region moved downwards by $-0.1\,d$ compared to the possible symmetric case. This is done to achieve repeatable solutions as otherwise the stream will tend either to the upper or the lower roll by accident. Fraction-of-solid and streamlines coloured by the velocity magnitude are presented. The grid employed contains 50 cells over the die thickness resulting in about 87 000 cells in total. The results presented here are obtained with a quad mesh using 4 wall-parallel layers at the roll boundaries. It can be seen that the flow tends to the closer roll and that two recirculation zones differing in size form. A redirection of flow occurs before the beginning of solidification from the main stream in the middle of the domain to the rotating rolls advecting the solidified shells. Tests with the setup at hand have shown that the velocity profile is maintained symmetric up to a Reynolds number of 150 calculated with the die thickness as the characteristic length and the rotational speed as the characteristic velocity. The results presented here showing asymmetric solutions are simulated with a Reynolds number of about 300.

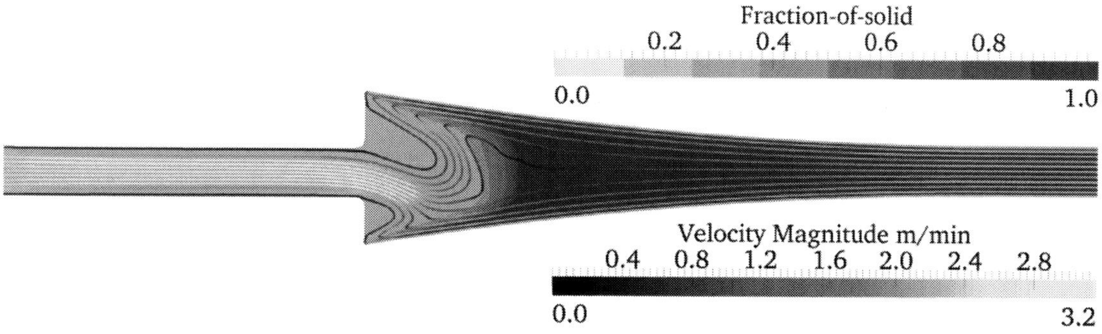

Figure 3. Fraction-of-solid in grey shades and streamlines coloured by the velocity magnitude for the asymmetric case of displaced inlet/die region by $-0.1\,d$ (downwards)

The tendency of the flow to approach the nearest roll evokes temperature differences between the two roll surfaces as shown in figure 4. The differences in absolute temperature emphasize different cooling conditions which might evoke differing solidification rates and thus inhomogeneous material properties over the sheet thickness. The further away from the mushy zone, the more symmetric the temperature profile is and therefore the smaller the temperature differences. Similar results, that is vice versa, are obtained using an asymmetric domain of displaced inlet/die region by $+0.1\,d$ (upwards). The temperature difference is respectively.

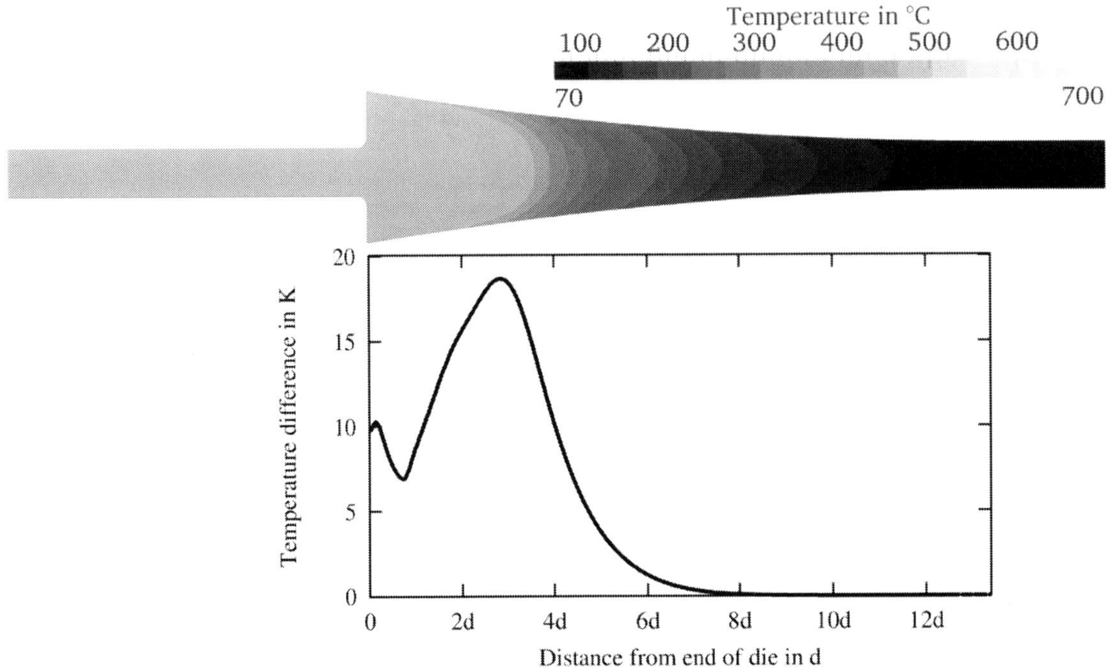

Figure 4. Temperature distribution within the solution domain and display of temperature differences between the roll surfaces shown in the diagram. The alignment of the two figures is done according to the examined roll boundaries. The diagram has the length of these boundaries.

An important parameter during the solidification is the solidification rate which determines the temperature change during time within the solidification zone. The result can be used as a measure for the expected grain sizes which in turn is a measure for the material quality. The achieved solidification rates are presented in figure 5. As the simulation at hand is steady-state,

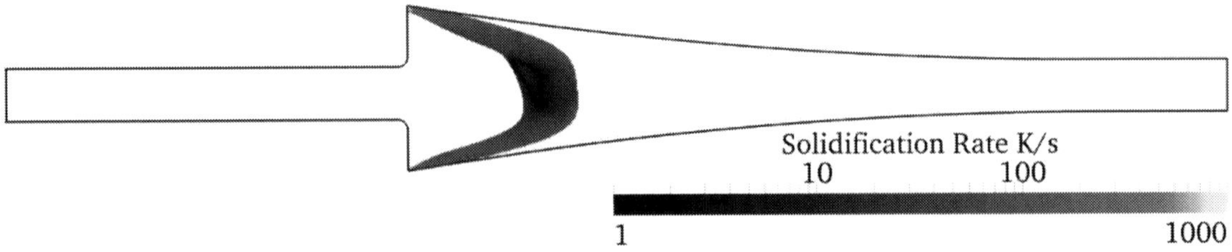

Figure 5. Calculated solidification rates for twin-roll casting of the magnesium alloy AZ31

the solidification rate is calculated using the following equation which is only evaluated within the solidification zone.

$$v_{sol} = \frac{\partial T}{\partial t} = \frac{\partial T}{\partial \mathbf{x}} \cdot \frac{\partial \mathbf{x}}{\partial t} \tag{5}$$

Figure 5 shows that the solidification rate reaches a maximum with the first roll contact. Higher values occur near to the rolls whereas the solidification rate is decreasing towards the centerline. A closer look to the solidification region with maximum rates indicates that the mesh resolution in the contact corner of the gas gap and the roll in not yet sufficient. The solidification rate of 1 000 K/s is thus more a rough estimation.

To verify the grid independence, tests were done starting with 10 cells over the die thickness, continuing with 20, 50, 100 and 200 cells. Figure 6 shows the comparison of the temperature and the velocity magnitude taken along a line across the solution domain in y-direction twice the die thickness behind the end of the die. It can be seen that a reasonable grid independence is reached for 50 cells. The benefit in accuracy for 100 or 200 cells is small compared to the computational cost.

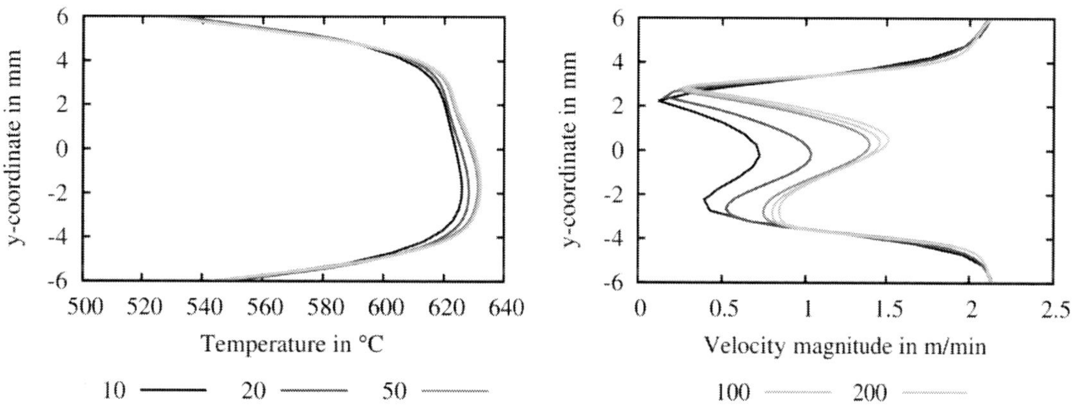

Figure 6. Comparison of grid independence according to number of cells in die thickness: profiles taken at $2\,d$ distance behind the end of the die

4. Summary and Conclusions

In this paper, the numerical investigation of horizontal twin-roll casting of the magnesium alloy AZ31 is shown. Constant thermophysical properties with the exception of a variation in viscosity and a linear fraction-of-solid approach for the release in latent heat are employed. The results show that the use of a pressure inlet combined with an inflow region of a certain length and the full domain is essential to obtain reasonable results. Employing only half the solution domain or omitting the inflow length enforce symmetry which is a priori incorrect and highly depends on the characteristic velocity of the system. The asymmetric velocity profiles evoke differences in the temperatures along the roll surfaces. Asymmetric recirculation zones additionally influence the solidification rates. These differences in solidification rates have been presented together with grid independence calculations employing 10, 20, 50, 100 and 200 cells within the die thickness. The results show, that except for the determination of the solidification rate close to the first roll contact, the mesh containing 50 cells within the die thickness provides reasonable results at reasonable computational cost. The impact of the gas entrainment and the meniscus forming from the end of the die to the rolls also acting on the recirculations is not taken into account.

The next steps within the project are to integrate the Scheil model for the determination of the fraction-of-solid and to use temperature dependent material properties. Furthermore, the Cozeny-Karman model will be used to incorporate the effect of solidification in the momentum equation instead of the viscosity approach. Influences of parameters like superheat, shape and temperature of the die, casting speed, setback and roll gap and material will be investigated.

Acknowledgments
This study is supported by the European Social Fund (ESF) within a project of a group of young researchers: Micro-crystalline magnesium material for sheet products with high-quality properties.

References

[1] Kammer D C 2000 *Magnesium Taschenbuch* (Aluminium Verlag Marketing & Kommunikation GmbH)

[2] Bessemer H 1865 Improvement in the manufacture of iron and steel

[3] Kawalla R, Oswald M, Schmidt C, Ullmann M, Vogt H P and Cuong N 2008 *Magnesium technology 2008: Proceedings of the symposium sponsored by the Magnesium Committee of the Light Metals Division of The Minerals, Metals & Materials Society (TMS).* ed Pekguleryuz M (Warrendale, Pa.: TMS, Minerals, Metals & Materials Society) pp 177–182

[4] Miyazawa K and Szekely J 1981 *Metallurgical transactions. A, Physical metallurgy and materials science* **12 A** 1047–1057

[5] Saitoh T, Hojo H, Yaguchi H and Kang C G 1989 *Metallurgical and Materials Transactions B* **20** 381–390

[6] Hwang J, Lin H, Hwang W and Hu C 1995 *ISIJ International* **35** 170–177

[7] Hwang J, Lin H, Jang J, Hwang W and Hu C 1996 *ISIJ International* **36** 690–699

[8] Chang J and Weng C I 1997 *International Journal for Numerical Methods in Engineering* **40** 493–509

[9] Chang J and Weng C I 1998 *International Journal of Heat and Mass Transfer* **41** 475–487

[10] Guthrie R and Tavares R 1998 *Applied Mathematical Modelling* **22** 851–872

[11] Gupta M and Sahai Y 2000 *ISIJ International* **40** 144–152

[12] Santos C, JA S J and Garcia A 2000 *Journal of Materials Processing Technology* **102** 33–39

[13] Sellger R and Ernenputsch L 2006 *Proceedings of the COMSOL Users Conference Frankfurt*

[14] Dai L, Yao S, Hao H, Wang T, Zheng X and Jin J 2006 *Modeling of Casting, Welding and Advanced Solidification Processes - XI* vol 2 pp 831–836

[15] Bradbury P J and Hunt J D 1995 *Modeling of Casting, Welding and Advanced Solidification Processes* pp 739–746

[16] Hwang S M and Kang Y H 1995 *Journal of Engineering for Industry* **117** 304–315

[17] Sarioglu K and Thevoz P 2003 *Light Metals 2003: Proceedings* (*TMS Annual Meetings* no 132) (San Diego: TMS Aluminium Committee) pp 747–751

[18] Volkova O, Sahebkar Moghaddam B, Lachmann S and Scheller P R 2007 *Metall* **61** 314–318

[19] Ju D, Zhao H, Hu X, Ohori K and Tougo M 2005 *Thermal flow simulation on twin roll casting process for thin strip production of magnesium alloy* (*Materials Science Forum* vol 488-489)

[20] Hu X D and Ju D Y 2006 *Transactions of Nonferrous Metals Society of China* **16** 586–590

[21] Hu X, Ju D and Zhao H 2007 *Thermal flow simulation and concave type slot nozzle design for twin roll casting of magnesium alloy AZ31* (*Materials Science Forum* vol 539-543)

[22] Zeng J 2007 *Einfluss der Strömung und Erstarrungsbedingungen auf die Homogenität von kontinuierlich gegossenem Magnesium Dünnband* Thesis Rheinisch-Westfälische Technische Hochschule Aachen

[23] Zeng J, Koitzsch R, Pfeifer H and Friedrich B 2009 *Journal of Materials Processing Technology* **209** 2321–2328

[24] He Y and Essadiqi E 2010 *TMS Annual Meeting* pp 545–552

[25] Zhao H, Li P and He L 2011 *Journal of Materials Processing Technology* **211** 1197–1202

[26] Ferziger J H and Perić M 2002 *Numerische Strömungsmechanik* (Springer-Verlag Berlin Heidelberg)

[27] Bennon W D and Incropera F 1987 *International Journal of Heat and Mass Transfer* **30** 2161–2170

[28] Swaminathan C R and Voller V R 1992 *Metallurgical Transactions B* **23B** 651–664

[29] Gale W and Totemeier T 2004 *Smithells Metals Reference Book (8th Edition)* (Elsevier)

Three dimensions thermal-mechanical model of the billet in continuous casting petal-like mould

Jing Li, Tongmin Wang*, Li Wu, Zhiqiang Cao, Tingju Li
School of Materials Sciences and Engineering, Dalian University of Technology, 116024, People's Republic of China
E-mail: *tmwang@dlut.edu.cn

Abstract. Petal-like mould is a novel mould which has been applied to the steel industry in recent years. The behavior of the petal-like billet in continuous casting mould plays an important role in designing mould. It is hard to be in situ measured during continuous casting, however, can be worked out by the way of numerical simulation. But the research about the model of the billet in petal-like mould is very little. A 3D finite-element model has been built to simulate the thermal and stress fields of the molten steel in petal-like mould in this paper. The dynamic thermal boundary condition and the effect of ferrostatic pressure have been considered in the model. The temperature and stress in the billet have been predicted by this model.

Keywords: Continuous casting, Numerical simulation, Mould, Steel billet, Shrinkage, Solidification

1. Introduction

Continuous casting now accounts for over ninety percentage of the steel production in the world. The qualities of billets in continuous casting largely depend on the solidification and shrinkage process of molten steel in mould. This process involves many important problems among which cracks are the most vital one. Hence, mould is one of the most important equipments in a continuous casting machine and it is always expected to optimize the taper of the mould in order to reduce the gaps forming. This relies on accurate predictions of temperature, deformation and stress distributions during the solidification process. However, the solidification of molten steel occurs at a very high temperature. It seems hardly appropriate to acquire the shrinking curve of the billets through experimental determination in view of the economical and technical reasons. To help understand the process, finite element models are widely used to predict the temperature and stress, improve design of continuous casting mould, and eliminate casting defects.

Since as early as 1970s, much effort has been made in developing a heat transfer model for continuous casting.[1,2] Based on these investigations, some researchers built the stress field mathematical model, and made an analysis by the coupled thermo-mechanical finite elements.[3-5] Thomas[6] has built a 2D slice model for billet, the thermal-mechanical behavior of billets was then simulated with three types of mould configuration, which produces hot corner, cold corner and equal

* Tongmin Wang: School of Materials Science and Engineering, Dalian University of Technology, Dalian 116085, China Phone: +086-411-84706790 Fax: +086-411-84706790

surface temperature around the perimeter, respectively. In his work, optimal taper profiles were predicted according to the shrinking distortion of the billets. A 2D thermo-elasto-plastic finite element model has been built by Han and Lee[7] to analyse the thermal-mechanical behavior of the continuous casting strand. This model considered the effect of microsegregation of solute elements using the thermal-mechanical model of mushy zone and δ/γ phases. With a mathematical billet thermal and solidification model, Samarasekera[8] has designed new mould taper for steels of high and low carbon grades in high speed continuous casting.

Petal-like mould is a novel mould which has been applied to the steel industry in recent years. The petal-like inner cavity mould provides for the billet to be shaped, and hence for cooling to be regulated, during travel of the billet through the shaping portion of the mould.[9] But analysis for the solidification process of the molten steel in petal-like mould is involved little. In our previous work,[10-12] a 2D slice model has been established to study the solidifying and shrinking behavior of the billets in a petal-like inner cavity mould. Although reliable predictions can be obtained through the analysis with the models, it is still questionable to accurately reproduce the 3D heat transfer and mechanical state in the mould.[13] In this paper, a 3D finite element model was made to calculate the thermal and stress fields in the billets in a petal-like inner cavity mould. This work may be regarded as a follow up on our previous study devoted to better understand the solidification and shrinking process of molten steel in the mould. A dynamic-loaded boundary condition was used to replace the movement of the billet in mould. As a non-negligible factor influencing the shell distortion of the billet during solidification, the ferrostatic pressure was also considered in the model. Temperature distribution, shell thickness, stress evolution and shrinking distortion for three kinds of billets were studied by this model.

2. Simulation model

2.1. Assumptions

Since continuous casting is a very complex process, the following assumptions were made to simplify the model:

- Neglecting the heat transfer between the meniscus and atmosphere;
- It is assumed that the material of the billet follows the Von Mises yield criterion;
- No flow effect of the liquid steel in the billet was considered.

2.2. Governing equations

Solidification with a phase change of the molten billet during continuous casting is an unsteady-state process. So, the 3D unsteady-state heat conduction equation was solved first. The heat governing equation was given in Eq. 1:

$$\rho(T)c(T)\frac{\partial T}{\partial t} = \frac{\partial}{\partial x}\left(k(T)\frac{\partial T}{\partial x}\right) + \frac{\partial}{\partial y}\left(k(T)\frac{\partial T}{\partial y}\right) + \frac{\partial}{\partial z}\left(k(T)\frac{\partial T}{\partial z}\right) + \dot{q} \tag{1}$$

Where T is the temperature (K); ρ (T) is the density of the billet (kg m^{-3}), which varies with T; t is the time (s); x, y, z is the three components in Cartesian coordinate (m); k(T) is the thermal conductivity (W m^{-1} K^{-1}); c(T) is the specific heat (J kg^{-1} K^{-1}); and .q is the latent heat source (W m^{-3}).

The mechanical behaviour of the billet during solidification includes elastic, inelastic (plastic and creep), and thermal strain. The creep strain was ignored in the strain-stress model. Then the increment total strain can be described by the following equation:[14]

$$\{d\varepsilon\} = \{d\varepsilon\}_e + \{d\varepsilon\}_p + \{d\varepsilon\}_T \tag{2}$$

Where $\{d\varepsilon\}_e$, $\{d\varepsilon\}_p$, $\{d\varepsilon\}_T$ are the elastic, plastic, and thermal strain increment, respectively. So, it depends on the temperature results obtained from the thermal field calculation.

2.3. Petal size determination

The petal-like inner cavity in the upper part accompanying with an excessive taper employed to change the size of the petal in the casting direction, and then the mould inner cavity becomes the traditional round shape close to the exit. The petal geometry in the upper part of the mould can accommodate the shrinkage of liquid steel as small casting risers, and thus obtain a rather round billet after shrinkage. When the convex billet formed in the upper part of the mould moved downwards into the round part of the mould, the billet will suffer a slight shaping from mould wall. Therefore the solidified shell can fully contact with mould wall, which leads to a more uniform temperature and stress distribution. As a result, the petal-like mould can avoid some quality problems such as crack, breakout etc. during continuous casting. Our previous work already validated that the mould with six petals was the better one, so only the 3D mould with six petals was analyzed in present work.

A 3D round model was built to calculate the total shrinkage of the billet from the meniscus to the exit of mould. The results of the total shrinkage (maximum shrinkage) for three steels were shown in table 1. The total shrinkage was used to determine the original size of petal as shown in Fig. 3. Detailed scheme was described in our previous paper.[10] This enabled the starting of the thermal and stress simulation by which the taper can be worked out.

Table 1 The results of total shrinkage of the round billet

Type of steel	37Mn5V	St45eq	X42
Shrinkage(mm)	1.195	1.134	1.887

Since the model is symmetrical, only one of the six petals was calculated to save the computer time. Fig. 1 shows the solid model, the finite element mesh model and the one-sixth cross section of the solid model. The length of the billet is 700mm and the radius of the cross section is 103mm. Because of the temperature varies fast closed to the billet surface, the mesh was refined gradually from center to surface.

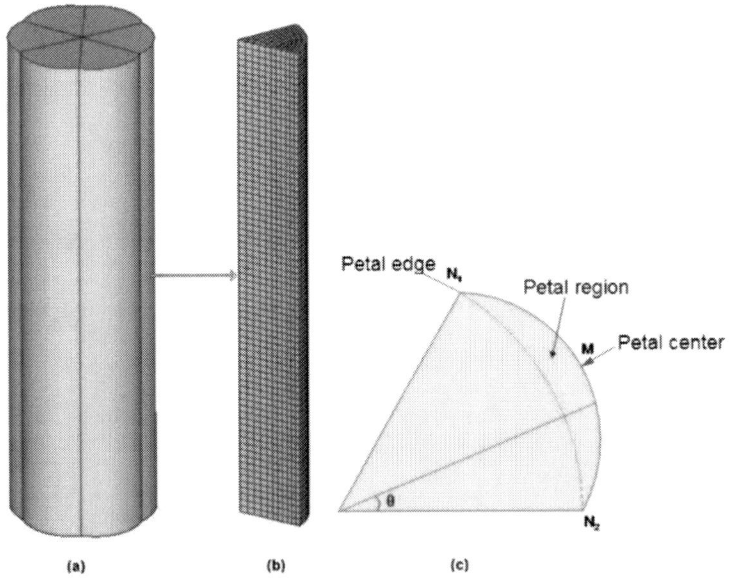

Figure 1. (a) Solid model; (b) One-sixth 3D finite-element mesh model; (c) One-sixth cross section of the solid model

2.4. Simulation parameters and boundary conditions

As the temperature of molten steel decreases continuously with phase changes during continuous casting, it is necessary to use varying material parameters. The varying parameters include density, heat conductivity, specific heat, thermal expansivity, elastic modulus and yield stress.[15]

The pouring temperature of molten steel was chosen as reference temperature. In this work, three kinds of billets with different compositions were simulated. Information of the three steels was given in table 2.

Table 2 The operating condition of the three different steels

Type of steel		37Mn5V	St45eq	X42
Main Composition(%)	C	0.36	0.185	0.125
	Si	0.35	0.25	0.25
	Mn	1.35	0.55	1
Pouring Temperature(°C)		1513	1531	1542
Liquidus Temperature(°C)		1492	1510	1517
Solidus Temperature(°C)		1427	1465	1473

The pouring temperature was defined as the initial temperature in the thermal field simulation. According to Savage-Pritchard law,[16] the heat flux in the mould interface can be described as a time-dependent function. In the present model, the heat flux function is given as:[9]

$$q = -(3.071 - 0.361\sqrt{\frac{z}{v}}) \tag{3}$$

Where z is the distance below meniscus, and v is continuous casting speed.

A coordinate-dependent thermal boundary condition was dynamically loaded to simulate the relative movement of the billet within the mould. The billet moves downwards in the mould, so the heat flux is loaded upwards along the billet surface. As shown in Fig. 2, at the initial time (t=0), heat flux began to be loaded from the meniscus (z=0). At t=nΔt (Δt is the time step), heat flux was loaded at the surface nΔz (Δz=Δt•v) away from the meniscus. That is to say, at each time step, the heat flux at each location of billet surface was loaded with the calculation of Equation (7). In this work, the continuous casting speed was set at 2 m/min.

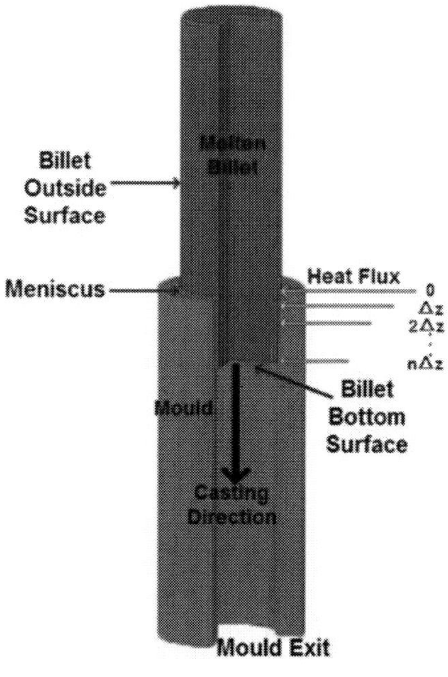

Figure 2. Schematic diagram of dynamic loading thermal boundary condition

For stress analysis, the displacement constraint at z direction was loaded at the bottom face of the billet, while the outside surface of the billet was free. The shell of the billet was pressed towards the outward direction by the ferrostatic pressure of molten steel. In general, the ferrostatic pressure was directly imposed on the internal surface of billet shell, namely the solidus line Ts, as shown in Fig. 3. Besides, there exist equiaxed crystals and dendritic crystals in mushy zone. The ferrostatic pressure would not be transferred to the surface of billet shell when it was imposed on the equiaxed crystals. The dendritic crystals usually form close to the surface of billet shell owing to the severe temperature gradient there. So when ferrostatic pressure was imposed on the dendritic crystals, it would be finally transferred to the surface of billet shell. In view of the above analysis, it is feasible to just impose the ferrostatic pressure on the internal surface of billet shell. In simulation, an eight-node element with six faces was used. In a case that the temperatures of the six faces of an element are all above (in mushy zone or liquid zone) or all below (solid zone) the solidus, the ferrostatic pressure should not be loaded; Otherwise, the case means that the current element just locates at the solidus line, the ferrostatic pressure should be loaded.

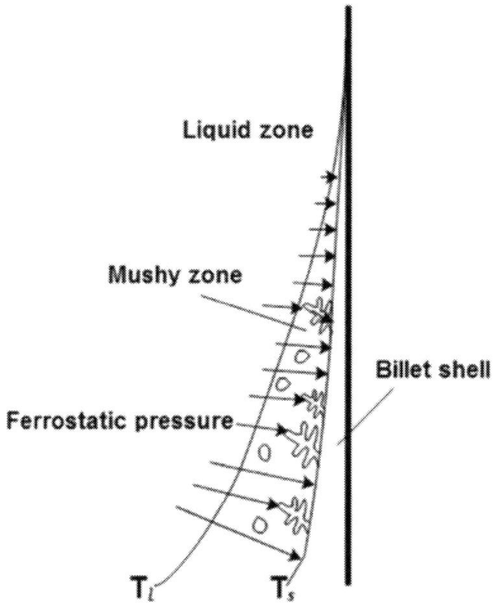

Figure 3. Schematic of how ferrostatic pressure was loaded at the internal boundary

3. Results and discussion

Fig. 4 shows the calculated thickness of the billet shell for the three steels, which were compared with the measurements of Thomas[17], show good agreement.

Fig. 5 shows the surface temperature distribution along the continuous casting direction. Near the meniscus, the liquid steel suddenly contacts the chilled mould geting intense cooling. Therefore a significant temperature decrease was found near the meniscus. The intense cooling zone is not big because solidified shell began to form rapidly. Accompanying the formation of the solidified shell, the latent heat is released, which makes the temperature decrease slowly (20mm below meniscus). With the increasing of shell thickness, the solidification front moved towards interior of billet and no latent heat directly affected the temperature of billet surface. Moreover, the solidified shell can still contact the mould due to the ferrostatic pressure. Therefore, the temperature of billet surface falls rapidly in the range of 250~400 mm. Over the next 400 mm the temperature drop decreases slowly because a large and static air gap has formed in this area that reduces the thermal transfer significantly.

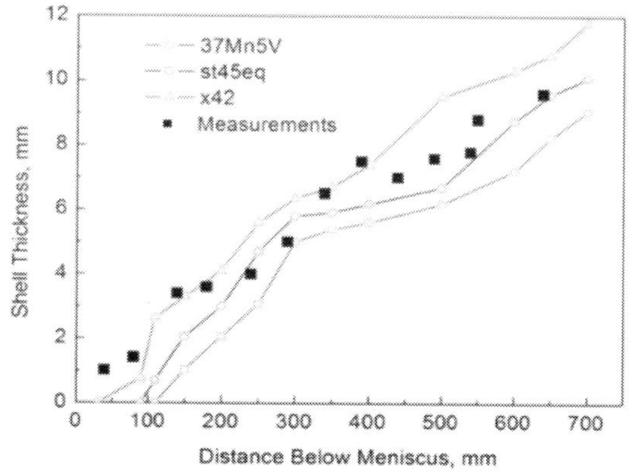

Figure 4. Thickness predicted of the billet shell for these three steels, and comparison with experimental measurements of Thomas[20].

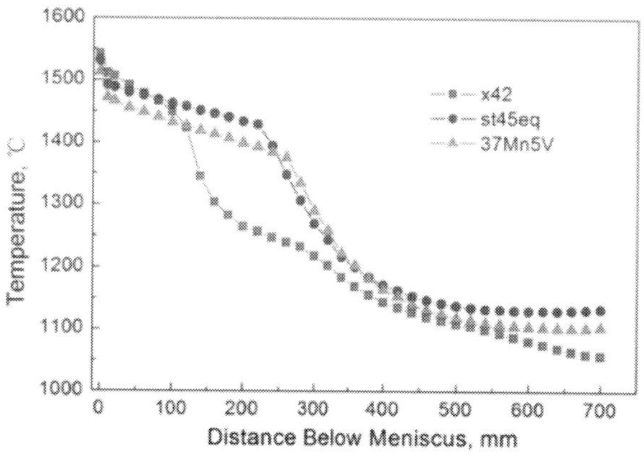

Figure 5. Temperature predicted on the shell surface of the billet in the mould for three different steels

Radial displacement of the billet is shown in Fig. 6. Near 200mm and 500mm below the meniscus, the surface of steels 37Mn5V and st45eq move outward because of the slower temperature decrease as compared with the neighbouring parts. The supporting function of the mould was not considered, so the positive radial displacement can be seen near 200mm. The radial displacement of steel 37Mn5V with considering ferrostatic pressure was compared with that without considering ferrostatic pressure as shown in Fig. 7. The former is obviously smaller than the latter because the ferrostatic pressure acts on the inner surface of shell against the shrinkage. The ferrostatic pressure has no significant effect on the shrinkage near the meniscus (liquid zone) and the exit of mould (strong shell). Fig. 8 shows the distrubition of the radial displacement in a petal at the mould exit. Radial displacement of the petal center is larger than that of the petal edge for all the three steels.

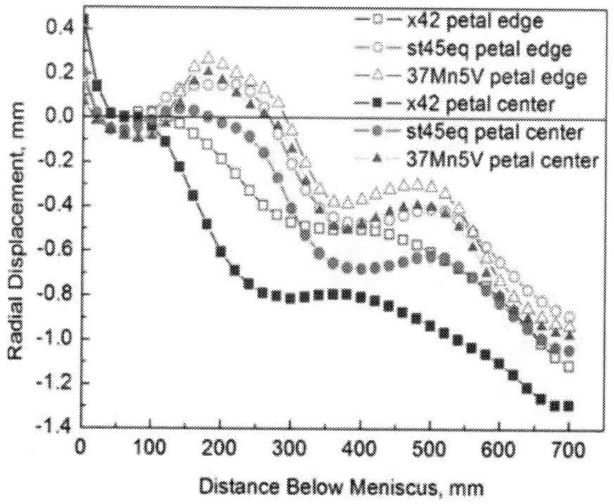

Figure 6. Radial displacement variation with distance below meniscus

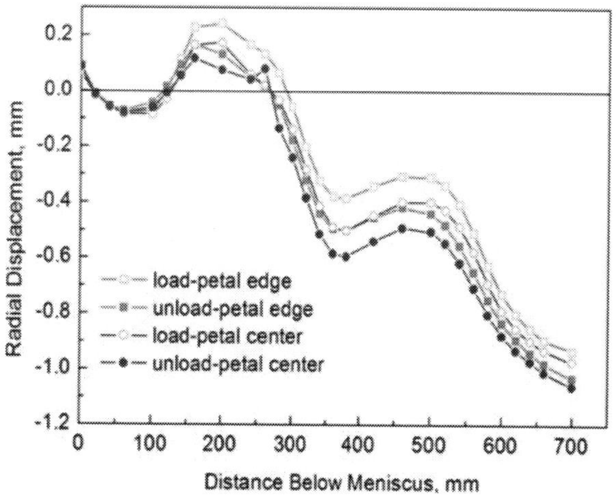

Figure 7. Comparing the radial displacement of 37Mn5V with and without loading ferrostatic pressure

The distribution of the 1st principal stress and temperature from billet center to the surface at mould exit is given in Fig. 9. The result is consistent with Boley and Weiner18 analytical solution. The stress rises at about 85mm from the billet center, which takes place in the mushy zone. The tensile stress gets the maximum value at 97mm from the billet center for steel st45eq and 37Mn5V, and at 92mm for steel x42. Then stress decreases rapidly which occurs in solid area, and tensile stress transforms into compressive stress. Owing to the subsurface shrinking faster than surface, surface is under compression stress, which accordingly causes the tensile stress in subsurface near the solidification front that may lead to subsurface cracks.19 At the surface of the billet, compressive stress gets the maximum value. So it is not easy to have a surface cracks at mould exit.

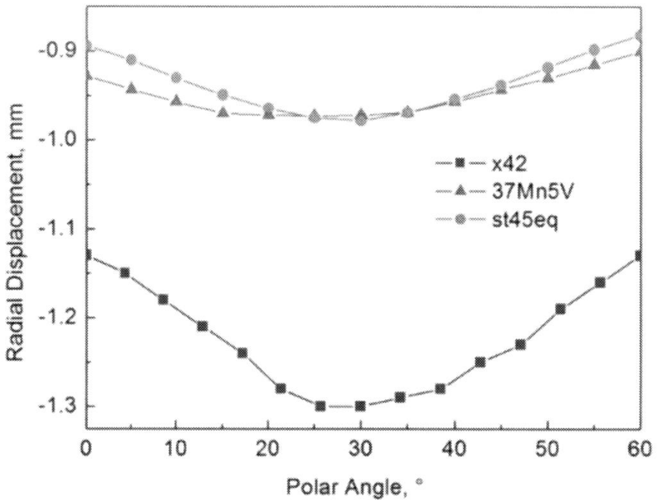

Figure 8. Radial displacement distrubition in a petal for the three steels

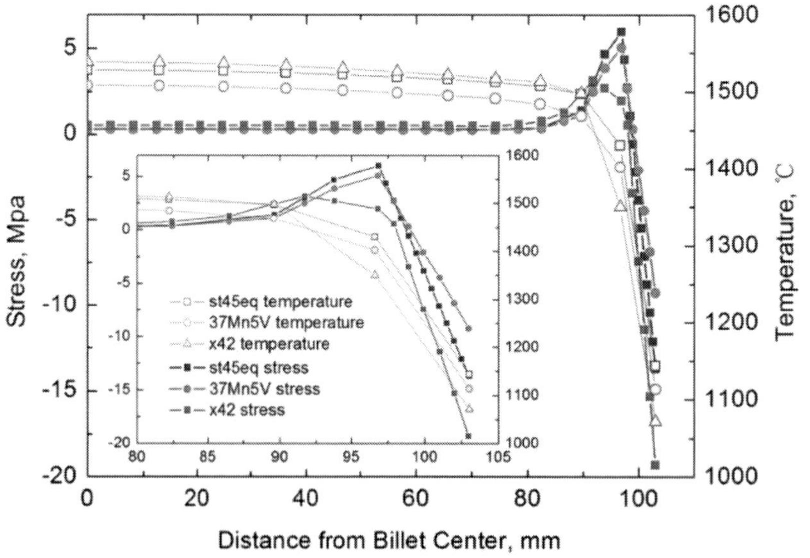

Figure 9. Stress and temperature predicted from billet center to the surface at mould exit

4. Conclusions

A transient, three-dimensional, finite-element model has been proposed to simulate the temperature, stress and shrinkage of the billet in mould during continuous casting. The effects of ferrostatic pressure are considered in this model. The tapers for petal center and edge have been designed respectively according to the shrinkages of the petal-like mould, and the following conclusions can be attained:

- The temperature decreases fastest at 250mm to 400mm below meniscus for steels st45eq and 37Mn5V, and at 130mm to 200mm below meniscus for steel x42.
- The radial displacement of the petal center is larger than the petal edge for the three steel grades.
- The radial displacement of the billet with considering ferrostatic pressure is smaller than that without considering ferrostatic pressure. The difference value has a maximum (0.15mm) at 490mm below meniscus.
- There exists tensile stress in mushy area and compression stress in solid phase area; moreover, tensile stress reach maximum at almost solidification front; compression stress reach maximum at billet shell surface.

Acknowledgements

The authors thank the National Natural Science Foundation of China (Nos. 50601003, 50971032, 51071035), the Fundamental Research Funds for the Central Universities, and Liaoning BaiQianWan Talents Program.

References

[1] Mizikar E A 1967 J. *Trans. Met. Soc. AIME* **239**, 1747-1760.
[2] Irving W R, Perkins A 1977 J. *Ironmaking Steelmaking* **4**, (5), 292-299.
[3] Thomas B G, Samarasekera L V, Brimacombe J K 1987 J. *Metall. Mater. Trans. B* **18**B, 131-147.
[4] Grill A, Brimacombe J K, Weiberg F 1976 J. *Ironmaking Steelmaking* **3**, (1), 38-47.
[5] Koric S, Hibbeler L C, Thomas B G 2009 J. *Int. J. Numer. Meth. Engng* **78**, 1-31.
[6] Li C S, Thomas B G 2003 C. *2003 ISSTech Steelmaking Conference, on 'Ideal Taper prediction for Billet Casting,' Indianapolis, USA, April* ISS-AIME, 685-700
[7] Han H N, Lee J E, Yeo T J 1999 C. *ISIJ Int.* **39**, (5), 445-454.
[8] Chow C, Samarasekera I V, Walker B N 2007 J. *Ironmaking Steelmaking* **34**, (1), 76-82.
[9] Kawa F, Stilli A 1994 C. *'Continuous casting mold for steel', United States, 5360053* **11**, 1.
[10] T. M. Wang, S. W. Cai, J. Li, J. J. Xu, Z. N. Chen, J. Zhu, Z. Q. Cao, T. J. Li 2010 J. *China foundry* **7**, (1), 61-67.
[11] Wang T M, Cai S W, Xu J, Du Y Y, Zhu J, Xu J J, Li T J 2010 J. *Ironmaking Steelmaking* **37**, (5), 341-346.
[12] Cai S W, Wang T M, Xu J J, Li J, Cao Z Q, Li T J 2011 J. *Mater. Res. Innov* **15**, (1), 29-35.
[13] Risso J M, Huespe A E, Cardona A 2006 J. *Int.J.Numer.Meth.Engng* **65**, 1355–1377.
[14] Kozlowski P F, Thomas B G, Azzi J A, Wang H 1992 J. *Metallurgical Transactions A* **23**A, 903-918.
[15] Kelly J E, Michalek K R, Connor T G, Thomas B G, Dantzig J A 1988 J. *Metallurgical Transactions A* **19**A, 1589-2601
[16] Savage J, Pritchard W H 1954 J. *Journal of Iron and Steel Institute* **178**, 269-277.
[17] Li C S, Thomas B G 2004 J. *Metall. Mater. Trans. B* **35**B, 1151-1172.
[18] Weiner J H, Boley B A 1963 J. J. Mech. Phys. Solids 11, 145-54

Effect of inlet geometry on macrosegregation during the direct chill casting of 7050 alloy billets: experiments and computer modelling

L Zhang[1,2], D G Eskin[3], A Miroux[1,2], T Subroto[1,2] and L Katgerman[1]
[1] Delft University of Technology, Department of Materials Science and Engineering, Mekelweg 2, 2628CD Delft, The Netherlands
[2] Materials innovation institute, Mekelweg 2, Delft 2628CD, The Netherlands
[3] Brunel University, BCAST, Uxbridge UB8 3PH, UK

Email: Liang.Zhang@tudelft.nl

Abstract. Controlling macrosegregation is one of the major challenges in direct-chill (DC) casting of aluminium alloys. In this paper, the effect of the inlet geometry (which influences the melt distribution) on macrosegregation during the DC casting of 7050 alloy billets was studied experimentally and by using 2D computer modelling. The ALSIM model was used to determine the temperature and flow patterns during DC casting. The results from the computer simulations show that the sump profiles and flow patterns in the billet are strongly influenced by the melt flow distribution determined by the inlet geometry. These observations were correlated to the actual macrosegregation patterns found in the as-cast billets produced by having two different inlet geometries. The macrosegregation analysis presented here may assist in determining the critical parameters to consider for improving the casting of 7XXX aluminium alloys.

1. Introduction

Macrosegregation is one of major and irreversible defects in direct chill (DC) casting of aluminum alloys. Usually, the variation of the concentration of alloying elements throughout the cross section of the billet results in thermal and mechanical properties variation, which impair the quality of final products [1, 2].

The cause of macrosegregation is the relative movement of segregated liquid and solid during solidification [2, 3]. In DC casting billet of aluminum alloy, there are two main driving forces of this relative movement. The first one is convection flow in the liquid and slurry zones, such as thermo-solutal convection (due to temperature and concentration gradients) and forced melt flow (e.g. pouring, stirring). The other is shrinkage-driven flow in the mushy zone. The later is due to the pressure difference over the solidifying layer of the mushy zone. Here the mushy and slurry zones are the two parts of the transition region during solidification corresponding to the different stages of solidification in DC casting of aluminum alloy. The limit between the two zones is the coherency isotherm. The condition of coherency can be defined as the moment when solid grains begin to interact with each other [5]. The slurry zone is the region between liquidus and coherency isotherm, where solid grains float freely. The region between coherency isotherm and solidus is called the mushy zone, where the macroscopic movement of solid grains is fully restricted.

Due to these driving forces, the extent of segregation in a real DC casting depends not only on the thermo-physical properties of the alloy but also on the actual casting parameters influencing the solidification process (e.g. casting speed, cooling rate and melt feeding). A proper correlation between casting parameters and the extent of segregation will help in determining the best casting practice, whereby a more uniform composition in the as-cast material can be achieved. Although invention and commercial use of DC casting could be dated back to the 1930s, and a number of studies both experimental and numerical [6-9] were dedicated to the effects of process parameters on the formation of macrosegregation, relatively little is known about macrosegregation under different melt flow feeding schemes.

The aim of this paper is to study the effect of inlet geometry on macrosegregation during the DC casting of 7050 alloy billets. Different melt feeding designs (melt distributor) were analyzed, combined with computer simulation of the solidification patterns, to obtain a full understanding of the evolution of macrosegregation during DC casting.

2. Experimental procedure and computer simulation

Two round AA7050 billets produced with different melt feeding schemes were selected in this research from industrial DC casting billets cast at Tata Steel Europe, The Netherlands. The casting temperature was around 680 °C. Both billets were grain refined by same amount of an Al5Ti1B master alloy. The diameter of cast billets is 315 mm. The melt was poured to the DC casting mold (without hot top) from the launder either through a semi-horizontal feeding distributor or a vertical feeding distributor, as shown in Fig.1 (a) and (b), respectively. In the semi-horizontal feeding scheme, the melt was diverted to four branches through a cross-shaped splitter, while in vertical feeding scheme, all melt was poured directly in the center of the mold. The detail casting information at the steady state casting and the average alloy composition is shown in Table 1.

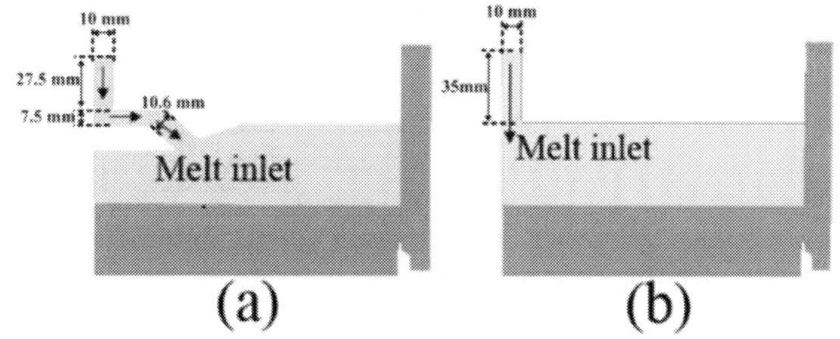

Figure 1. Geometry of melt feeding distributors used in experiments: **a,** semi-horizontal melt feeding scheme; **b,** vertical melt feeding scheme. Half of the billet is shown. Centerline is on the left.

Table 1. The composition and casting parameters of selected billets

Alloy	Composition, wt%					Melt feeding scheme	Casting speed, mm/min	Casting temperature, °C	Cooling water flow rate, l/min
	Cu	Mg	Zn	Zr	Ti				
AA7050	2.2	2.1	6.15	0.13	0.03	Semi-horizontal	50	680	40
						Vertical	50	680	35

Slices of the cast billets were sawn at positions corresponding to the steady state stage of casting. The composition across the billet diameter was measured by a spark spectrum analyzer (Spectro, Kleve, Germany). The distance between each measured point is about 1 cm. After the composition measurement, the long striped sample was examined from the surface to the center in an optical microscope after being cut, ground, polished, and electro-oxidized at 20 VDC in a 3% HBF_4 water solution. Grain size was measured using the linear intercept method and the statistical analysis of the results was performed.

In order to analyze the sump profile and flow pattern in transition region under different melt feeding schemes, ALSIM6 was used. The simulated geometry consisted of the mold, water jet, bottom block and the casting moving domain (i.e. Arbitrary Eulerian Lagrangian description domain). In this 2D computer simulation, the heat, fluid flow, mechanical properties, casting and solidification modules were included in the ALSIM6 modelling set-up. The ALSIM program also calculates the heat transfer along the billet wall from the cooling water flow rate. The air gap formed between the mold and billet, which reduces the heat transfer, is also taken into account for heat transfer calculation. In average the heat transfer is found higher for the semi-horizontal feeding than for the vertical feeding geometry. The Navier–Stokes equations are solved in the model by considering a Darcy solidification term. Low Rayleigh number (LRN) turbulent energy–pseudoturbulent dissipation (κ-ε) model was chosen to solve the turbulence problem in this simulation. The velocity of the solid phase is equal to the casting speed throughout the transition region. A more detailed description of the software setup and models involved can be found elsewhere [10, 11]. The thermo-physical parameters for AA7050 alloy and mechanical database used in this simulation are given in Refs. [12, 13].

3. Results

In the semi-horizontal feeding billet, due to the inhomogeneous melt inlet caused by four inlets and potential implications for the macrosegregation, the composition was measured in three directions (at 45 degree angle from each other), The segregation patterns were similar, which attests for the negligible effect of this feeding scheme on the symmetry of the segregation. So in this paper, we only present the composition measurements from one diameter. The deviation of Cu, Mg and Zn concentrations from the average composition across the billet diameter are shown in Fig. 2(a). A negative central segregation for these elements can be observed. In the vertical feeding billet, as shown in Fig. 2(b), the maximum negative segregation for Cu, Mg and Zn appears 40-50 mm away from the center instead of strong centerline negative segregation. And also the extent of maximum segregation is getting lager as compared with the semi-horizontal feeding billet at the same casting speed. The Cu deviation almost reached 10 rel.%. At the chilled surface, a strong positive segregation can be observed in both semi-horizontal and vertical feeding billets. We can also see that the extent of deviation of alloying elements in both melt feeding schemes closely follows the magnitude of the partition coefficient. Such as partition coefficient for Cu is 0.17, which is far away from 1, results in a stronger segregation as compared with Mg, Zn, which partition coefficients are 0.43 and 0.45, respectively.

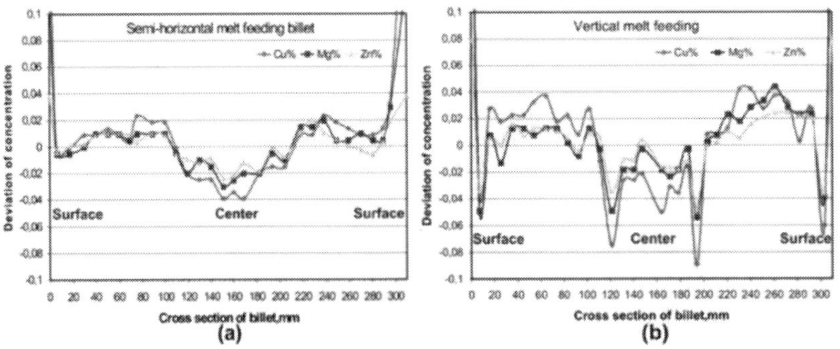

Figure 2. Deviation of alloying elements across billets diameter: **a,** semi-horizontal melt feeding billet; **b,** vertical melt feeding billet.

The grain size variation across the billet diameter is shown in Fig. 3. In the semi-horizontal feeding billet, the grains are coarser towards the center, while the minimum grain size was observed near the surface. The distribution of grain structure in the vertical melt feeding billet is totally different as compared to that obtained upon semi-horizontal feeding. The minimum grain size was found in the center of billets. In addition, in the vertical melt feeding billet, a layer of coarser grains can be found around 40 mm away from the center, as illustrated in Fig. 4. The thickness of that layer is approximately 1 mm. The exact positions are shown in Fig. 3 with dashed lines. It should be mentioned that the presence of coarse grains near the center was not taken into account during measurement of grain size in Fig. 3 because this layer is too thin as compared with the whole billet diameter.

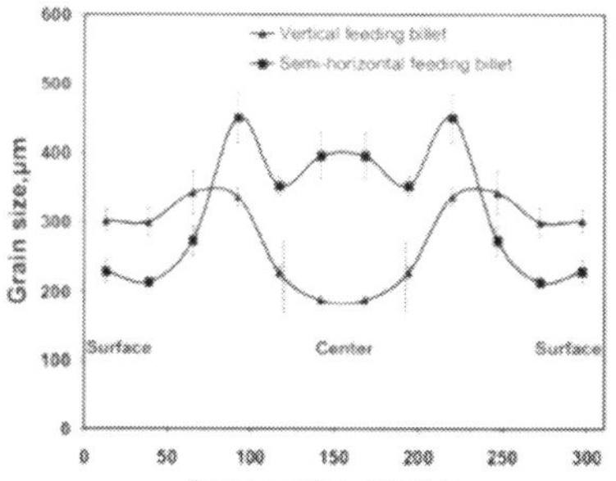

Figure 3. Grain size across billets diameter: **a,** semi-horizontal melt feeding billet; **b,** vertical melt feeding billet.

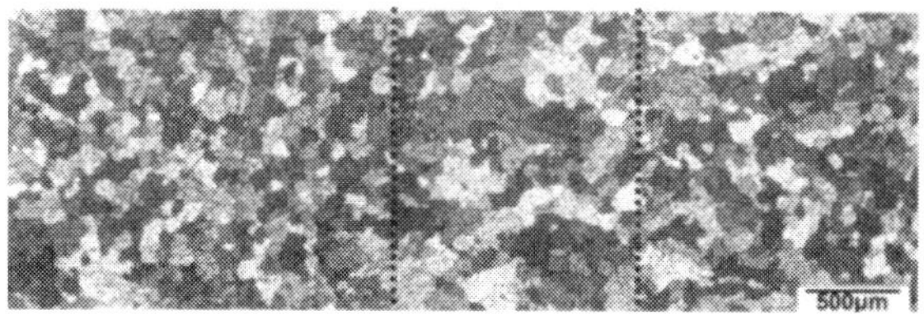

Figure 4. A typical layer of coarse grains near the center (40 mm away from the center) found in the billet with the vertical melt feeding. Coarse grains layer is shown in the dashed line area, see also in Fig. 3. Note that the length of Fig.4 is only about 4 mm.

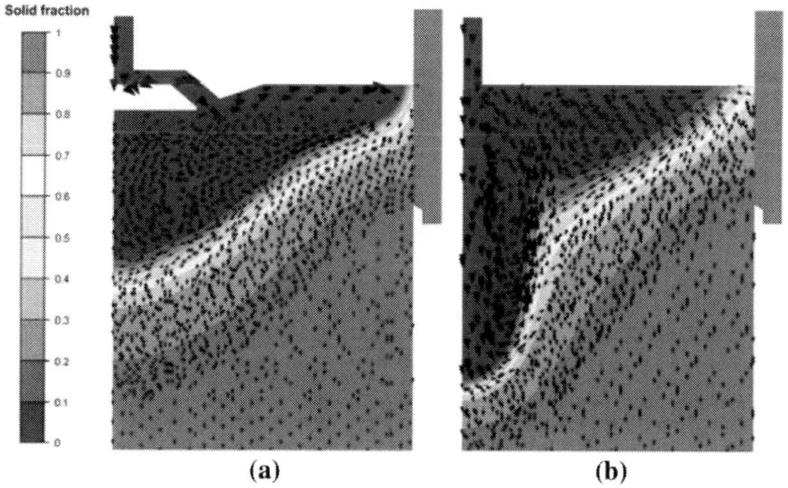

Figure 5. Simulated sump profiles and flow patterns during DC casting: **a,** semi-horizontal melt feeding billet, maximal magnitude of velocity around melt inlet area is 2.4 mm/s; **b,** vertical melt feeding billet, maximal magnitude of velocity around melt inlet area is 30 mm/s. Centerline is on the left.

Sump profiles and flow patterns in the billet were simulated for both melt feeding schemes, as shown in Fig. 5. In the billet cast with semi-horizontal feeding (Fig. 5(a)), the slope of solidification front is relative gentle. The sump depth in the center (from the melt level in the mold to the solidus) is about 12 cm. The movement of solid and liquid phases (velocity) in the liquid and in the transition region is most intense near the surface, while a moderate upstream flow can be observed in the center. In the billet cast with vertical feeding, vertical melt flow inlet in the center causes a cliff-shaped sump, as shown in Fig. 5(b). The slope increases dramatically at the position 40 mm away from the center. The solidification front drops vertically and the solidification rate at this point is almost zero. The sump depth in the center is about 17 cm. The flow is more pronounced and directed downwards in the center and upwards off-center instead of being intense near the surface as in the case of semi-horizontal feeding. More details about sump profiles and flow patterns in these simulations will be discussed in the next section.

4. Discussion

4.1. Relative movement of melt flow in slurry zone

It is well known that the convection flow in the liquid and slurry zone results in the redistribution of alloying element in the DC casting billet. The velocity patterns of this convection flow are shown in Figs. 5(a) and (b). Fig. 6 gives the melt velocity in the slurry zone across the billet from the center to the surface. The solid fraction at coherency is taken as 0.5 for our alloy, because both billets in this research are grain refining. The flow velocity in the slurry zone shown in Fig. 6 is taken around the 0.4 solid fraction isoline.

As shown in Fig. 5(a) and Fig. 6, in semi-horizontal feeding billet, the penetration of melt flow into the slurry zone is more pronounced near the surface area. This means that the effect of washing out of solute-rich liquid from the slurry zone in this area is larger than in the quarter and center of the billet. This solute-rich liquid moves then towards the quarter and central areas along with the convection flow, which results in the lack of alloying elements near the surface area and the positive segregation in the quarter position.

In the vertical feeding billet, the flow pattern changed totally because of different melt inlet (Fig. 5(b)). There are two convection flows as illustrated in Fig. 7. One is near the center area, counter-clockwise. Another is at the quarter and surface of the billet, in the clockwise direction. Two flows collide at the position 40-50 mm away from the center. As we can see from Fig. 6, the most pronounced washing out of solute-enrich liquid from the slurry zone happens around the cliff-shaped solidification front (40 mm away from the center). This strong upstream flow in the slurry zone might wash out the enriched solute either to the center or to the surface area, which causes negative segregation in this area, as shown in Fig. 2(b). It should be mentioned here that it has been reported that strong forced unidirectional flow in the center may cause a consistent change of the concentration along the billet axis, which can be detected in the last stages of casting [14]. However, in this research, we only selected one slice for each billet in the middle of steady state casting and did not take this effect into account.

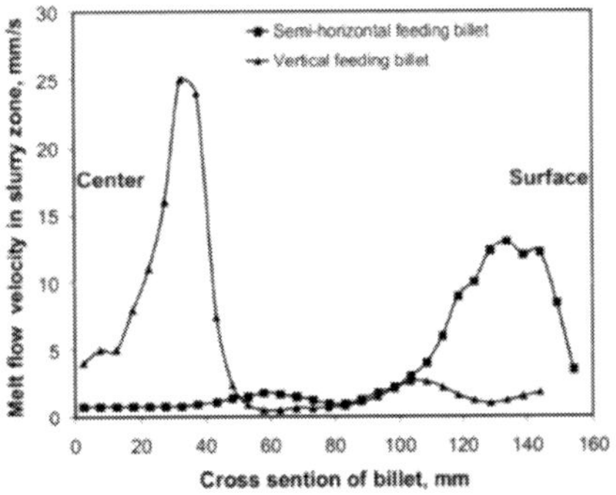

Figure 6. Computer-simulated magnitude of melt flow velocity in the slurry zone in both semi-horizontal and vertical feeding billet.

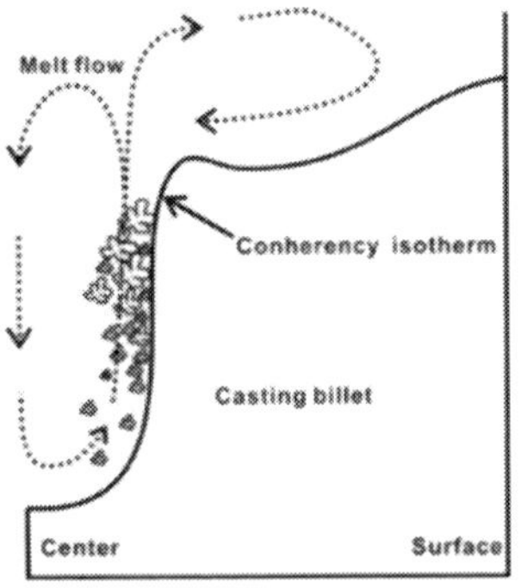

Figure 7. Schematic view of relative movement of melt and grains along cliff-shaped sump in vertical melt feeding billet

4.2. Relative movement of solid grains in slurry zone

The movement of solid grains along with convection flow is one of the mechanisms of macrosegregation. The 'floating' grains, which form at earlier stage of solidification, are usually depleted of the main alloying elements in Al alloys such as Cu, Mg, Zn. These grains float with convection flow and usually settle in the center of DC casting billet, causing centerline negative segregation [15].

In this research, we could not find any typical 'floating' grains in both the semi-horizontal and vertical billets. However, a layer of coarse grains found in the vertical melt feeding billet (Fig. 4) is really interesting and might be related to near-center negative segregation (40 mm away from the center) in Fig. 2(b) if we interpret this phenomenon as follows.

Fig. 7 is the schematic view of cliff-shaped sump in the vertical melt feeding billet. A cliff-shaped sump leads to an almost vertical solidification front approximately 40 mm away from the center. The grains, which are formed in the slurry zone around this vertical solidification front or formed earlier at other position then arrived in this area with melt flow, will stack and grow at the same position of almost vertical solidification front because of strong upstream melt flow and gravity. It also worth to notice that in Fig. 5(b), there is a collision area of clockwise and counter-clockwise flows, as mentioned before. In this area there is almost no horizontal velocity component. In this case, solid grains cannot be displaced from that cliff-shaped front due to stack of grains, but solute enriched melt can. As a result, the strong upstream flow will continue to take the solute enriched liquid away from this slurry zone but leave the solute depleted grains. This might enhance negative segregation in this cliff shaped sump area (40 mm away from the center).

4.3. Solidification shrinkage-driven flow in mushy zone

As an important mechanism of macrosegregation, the solidification shrinkage-driven flow in the mushy zone usually determines the inverse (negative) centerline segregation in DC casting billets of aluminum alloys. This shrinkage flow is caused by the pressure difference over the solidifying layer of the mushy zone, which results in the movement of the solute-rich liquid to the deeper part of mushy zone in the direction perpendicular to the solidification front.

A simple analytical model was suggested by Du and Eskin in order to analyze the effect of the mushy zone geometry on the degree of shrinkage-induced macrosegregation in DC casting aluminium alloy [16]. In this research, we used this model to make a simple analysis for segregation of Cu (Mg and Zn follow the same trend) and the effect of melt inlet on shrinkage-induced macrosegregation.

The main idea of this model is that the centerline negative segregation is caused by the horizontal component of the shrinkage-driven flow from the center to the surface of the billet. By taking solidification shrinkage and solidification front inclination into account, the total amount of the transferred solute at different positions during the solidification period (L_m/V_{cast}) is as follows [16]

$$L_h = \int_0^{L_M/V_{cast}} C_l f_l V_{cast} \beta (\sin 2\alpha)/2dt \qquad (1)$$

Where V_{cast} is the casting speed; β is shrinkage ratio (0.1 for an aluminum alloy); α is the local slope of the coherency isotherm; L_m is the vertical thickness of the mushy zone; C_l is liquid phase concentration; and f_l is the volume fraction of liquid phase. This equation can be reduced for Cu (partition coefficient k_{Cu} is taken as 0.171) to [16]

$$L_h = 0.78 C_0 L_M \beta (\sin 2\alpha)/2 \qquad (2)$$

The derivative of Eq. (2) dL_h/dr (r is the radial distance from the billet center) is the net efflux and is a measure of the macrosegregation caused by solidification shrinkage. The detail derivation of these equations can be found in Ref [16].

Fig. 8(a) is the distribution of transferred solute L_h from the center to the periphery of the semi-horizontal feeding billet. This curve represents the balance between the amount of solute coming to one point and the amount of the solute coming out of this point. When the slope is positive, the in-coming solute is less than out-coming solute, which causes negative segregation at this position by shrinkage driven flow. On the contrary, the negative slope results in positive solute income at this position. Fig. 8(b) is the relative segregation of Cu ($-(dL_h/dr)/C_{cu}$), obtained by the derivative of the polynomial fit shown in Fig. 8(a) with respect to radial distance from the center of semi-horizontal billet (with the negative sign). It is clearly shown that shrinkage driven flow causes a negative segregation in the center of the billet until the position 60 mm away from the center, then a positive segregation occurs. Near the surface of the billet, the segregation changes to negative again.

If we compare the calculated shrinkage-induced segregation in Fig. 8(b) with real Cu segregation distribution in the semi-horizontal feeding billet (Fig. 2(a)), a good agreement on the distribution of Cu segregation can be found. It should be noted that the segregation curve in Fig. 8(b) reflects only the calculated segregation caused by shrinkage-driven flow. As discussed before, the movement of solute-enrich liquid in the slurry zone from the surface to the center contributed to a centerline positive segregation. Taking into this account, negative 6.5 rel.% deviation of Cu in the center caused by the shrinkage flow (Fig. 8(b)) might be compensated by the effect of convection flow in the slurry zone. As a result, the negative centerline segregation in the real casting billet reduces to about 4 rel.%, as shown in Fig. 2(a).

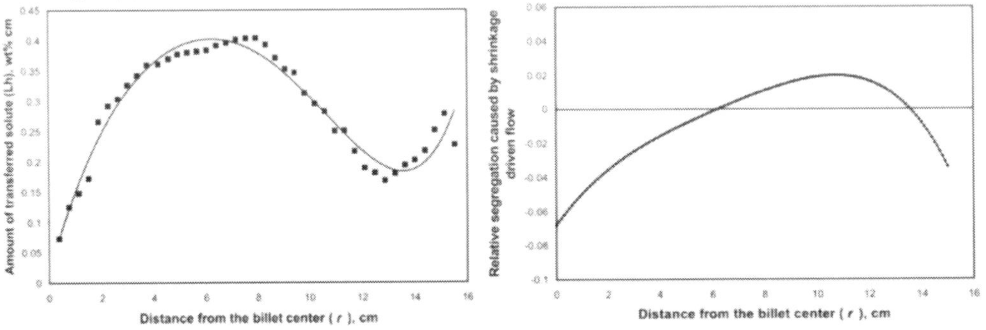

(a) (b)

Figure 8. Effect of shrinkage driven flow on macrosegregation in the semi-horizontal melt feeding billet: **a,** amount of transferred solute from the center to the surface and its polynomial fit; **b,** relative segregation caused by shrinkage driven flow, obtained by the derivative of the polynomial fit

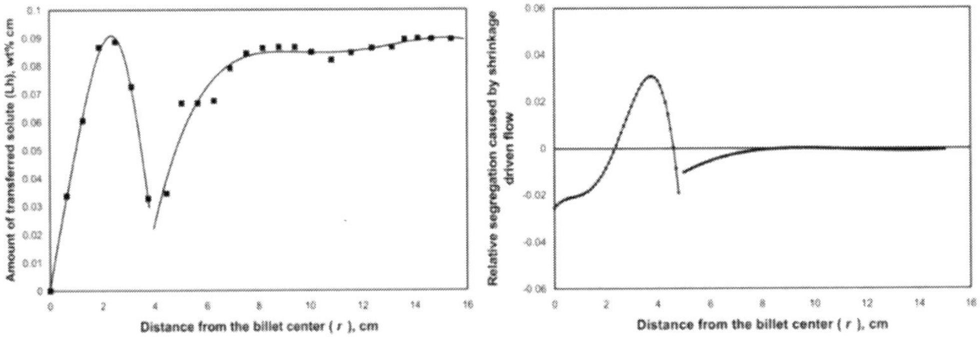

(a) (b)

Figure 9. Effect of shrinkage driven flow on macrosegregation in vertical melt feeding billet: **a,** amount of transferred solute from the center to the surface and its polynomial fits; **b,** relative segregation caused by shrinkage driven flow, obtained by the derivative of the polynomial fit

Same analysis has also been done for the vertical melt feeding billet. Fig. 9(a) is the distribution of transferred solute L_h from the center to the periphery of the vertical feeding billet. Due to relatively complex distribution of L_{h}, we divided the distribution in Fig. 9(a) into two parts and made the polynomial fits separately in order to increase fitting accuracy. By the derivative of L_{h} distribution in Fig. 9(a), we can also obtain the relative Cu segregation from the center to the surface of the vertical feeding billet, as shown in Fig. 9(b). The agreement between shrinkage-induced segregation in Fig. 9(b) and real segregation distribution in Fig. 2(b) is not as good as in semi-horizontal feeding billet. But we still can see that there is an obvious negative segregation at the position 40-50 mm away from the center, which shows the shrinkage driven flow might also contribute to the strong near-center negative segregation in addition with the convention flow and "floating" grains as discussed above.

The shrinkage driven flow also causes a positive segregation at the position 25-40 mm away from the center, as shown in Fig. 9(b). Although the positive segregation in this area does not appear in real segregation curve in Fig. 2(b) (though a slight increase in the concentration can be observed), it becomes logical when the melt flow in slurry zone is considered. This positive segregation area caused by shrinkage driven flow coincides with the position where a strong upstream flow is. Therefore the liquid near the coherency isotherm could be already depleted by this upstream flow and might not be rich enough to allow the shrinkage driven flow to create such a positive segregation. In addition, the redistribution of solute to the surface area in combination with a flat region of almost no shrinkage-induced segregation (from quarter to the surface of the billet), leads to a positive segregation in this region, as shown in Fig. 2(b).

As for the strong positive segregation at chilled billet surface we observed in both billets, similar results have been reported elsewhere [6, 15, 17]. It is a result of shrinkage driven flow toward to the chilled surface, where there is no out-coming flow. Thus, the solute will accumulate at the chilled surface and usually lead to a negative segregation in the sub-surface region, as shown near the surface area in Fig. 2 (a) and (b).

5. Conclusions

The melt inlet feeding scheme influences the macrosegregation in DC casting AA7050 billets by changing the flow patterns and sump profiles during casting. The solute-rich liquid washed out from the slurry zone by convection flow usually leads to the lack of alloying element in this area and an accumulation of solute downstream of this convection flow. The shrinkage-driven flow in the mushy zone is the dominant mechanism to determine the centerline negative segregation in a DC casting AA7050 billet. However, the final macrosegregation in casting billets is a clear result of a combination effect of convection flow in slurry zone and shrinkage-driven flow in the mushy zone. The relative movement of solid grains with melt flow also might play a role on macrosegregation in some local areas.

Acknowledgements

This research was carried out under project number MA.10210 in the framework of the Research Program of the Materials innovation institute M2i (www.m2i.nl). The author would like to express their gratitude to Dr. Démian Ruvalcaba and Mr. Jacob van Oord (Tata Steel Research, Development & Technology, The Netherlands) for their support and inputs. L. Zhang would like to thank China Scholarship Council and Prof. J.S. Zhang for providing the scholarship for the PhD study at TU Delft.

References

[1] Nadella R, Eskin D G, Du Q, Katgerman L 2008 *Progr. Mater. Sci.* **53** 421
[2] Beckermann C 2008 *ASM handbook:casting* **15** 348
[3] Flemings M C 2000 *ISIJ International* **40** 833
[4] Eskin D G and Katgerman L 2010 *Mater. Sci. Forum* **630** 193
[5] Arnberg L, Bäckerud L and Chai G 1996 *Solidification characteristics of aluminum alloys* **3** (Des Plaines: American Foundrymen's Society)
[6] Eskin D G, Zuidema J, Savran V I and Katgerman L 2004 *Mater. Sci. Eng. A* **384** 232
[7] Gariepy B and Caron Y 1991 *Light metals* (Warrendale, TMS, Rooy Elwin L editor) 961
[8] Ferreira I L, Garcia A and Nestler B 2004 *Scripta Metall.* **50** 407
[9] Venneker B C H and Katgerman L 2002 J. Light. Met. **2** 149
[10] Mortensen D 1999 *Metall. Mater. Trans. B* **30B** 119.
[11] Mortensen D, Fjær H G, Lindholm D, Rudshaug M and Sørheim E A 2011 *Mater. Sci. Forum* **693** 187
[12] Ludwig O, Drezet J-M, Martin C L and Suery M 2005 *Metall.Mater.Trans.*A **36A** 1525
[13] Lalpoor M, Eskin D G, Ruvalcaba D, Fjær H G, TenCate A, Ontijt N and Katgerman L 2011 *Mater. Sci. Eng. A* **528** 2831
[14] Eskin D G, Jafari A and Katgerman L 2011 *Mater.Sci. Technol.* **27** 890
[15] Eskin D G, Nadella R and Katgerman L 2008 *Acta Mater.* **56** 1358
[16] Eskin D G, Du Q and Katgerman L 2006 *Scripta Metall.* **55** 715
[17] Nadella R, Eskin D G and Katgerman L 2008 *Metall.Mater.Trans.*A **39A** 450

Simulation study on continuous casting process of Al/Al bimetal round billet under multi-electromagnetic

Li Wu, Tongmin Wang*, Ying Fu, Jing Li, Zhiqiang Cao, Tingju Li
School of Materials Science and Engineering, Dalian University of Technology, Dalian 116024, People's Republic of China
*E-mail: tmwang@dlut.edu.cn (T. Wang)

Abstract. In order to investigate the magnetic force distribution, flow field distribution and temperature field distribution during continuous casting process of Al/Al bimetal round billet under multi-electromagnetic fields, a comprehensive mathematical model has been developed. The model is based on a combination of the commercial finite element package ANSYS and the commercial finite volume package FLUENT, with the former for calculation of the multi-electromagnetic fields and the latter for calculation of the magnetic driven fluid flow, heat transfer, temperature field and solidification. There was a good agreement between the calculate results and the experiment results. Comparison of the calculated results of the continuous casting process without electromagnetic with that of the continuous casting process under multi-electromagnetic fields indicated that velocity patterns, temperature profiles and the sump depth are modified remarkably by the application of multi-electromagnetic field during the continuous casting process.

***Keywords*:** Numerical simulation, Continuous casting, multi-electromagnetic fields, Bimetal round billet, Aluminium alloy

1. Introduction

As a new type of material, clad material has become one of research focuses in material field. Continuous casting is an ideal method for producing clad metal because it realizes liquid state contact between two metals directly, therefore the problem of low bonding strength can be solved radically [1,4]. Recently, alternative magnetic field and pulsed magnetic field which can significantly refine solidified structure and properties of the billet has gained wide application in materials processing industry due to the advantages of high energy density, non-contact and controllable [5,10]. Moreover, many scholars have studied the effect of magnetic field by numerical methods [11,12].

Among aluminium alloys, 3003 has excellent corrosion resisting property but low strength, and 4004 has high strength but its corrosion resisting property is poor. The 3003/4004 clad aluminium can combine their advantages and therefore be widely applied in many fields, especially in various heat exchangers [13]. In this paper a comprehensive mathematic model was developed to describe the interaction of the multiple physics fields(multi-electromagnetic fields, fluid flow, heat transfer and solidification) during the continuous casting process of Al/Al bimetal round billet under multi-electromagnetic fields. The effects of electromagnetic field on melt flow, temperature field and solidification were analyzed.

* Corresponding author. Tel.: +86-411-84706790; fax: +86-411-84706790.

2. The mathematical model

2.1. Governing equation for electromagnetic field

In order to gain the distribution of the electromagnetic field in the billets, the magnetic flux density B and the electric strength E must be obtained by solving Maxwell's equations, which can be written as:

$$\nabla \times E = -\frac{\partial B}{\partial t};\nabla \times B = \frac{\varepsilon}{\mu}\frac{\partial E}{\partial t} + J;\nabla \cdot E = \frac{\rho e}{\varepsilon};\nabla \cdot B = 0 \tag{1}$$

where ε is the permittivity and μ is the permeability; J is the conduction current and should be solved with Ohm's law:

$$J = \sigma(E + U \times B) \tag{2}$$

2.2. Governing equations for flow field and temperature field

The governing equations describing the flow in the molten pool are:

Continuity equation:

$$\frac{\partial \rho}{\partial t} + \nabla \cdot (\rho U) = 0 \tag{3}$$

Momentum equation:

$$\frac{\partial(\rho U)}{\partial t} + \nabla \cdot (\rho UU) = \rho g - \nabla P + \nabla \cdot (\mu_{eff} \nabla U) + J \times B \tag{4}$$

Energy equation:

$$\frac{\partial(\rho T)}{\partial t} + \nabla \cdot (\rho UT) = \nabla \cdot (\frac{k}{c_p} \nabla T) + S \tag{5}$$

where ρ is the density of liquid melt; u_i and u_j are time average velocity; p is the pressure; μ_{eff} is the effective viscosity coefficient; g is gravitational acceleration; C_p is specific heat; k is the thermal conductivity; S is the thermal source and includes Joule heat and latent heat of solidification.

2.3. Solidification modeling

For the solidification process of a multi-component component alloy, the whole region is divided into three regions: liquid, solid and mushy region. Moreover, the latent heat is released in the mushy region during the solidification process. In order to account for the release of latent heat during solidification in the energy equation, a fixed grid enthalpy method advocated by many authors are used in this work. In this technique, the total enthalpy, H, is decomposed into the sensible heat, $h = \int_{T_{\text{ref}}}^{T} c_p dT$, and nodal latent heat, $\Delta H = f_L \Delta H_f$, in the energy equation as

$$H = \int_{T_{\text{ref}}}^{T} c_p dT + f_L \Delta H_f \; ; f_L = \begin{cases} 1 & T > T_L \\ \dfrac{T - T_S}{T_L - T_S} & T_S \le T \le T_L \\ 0 & T < T_S \end{cases} \tag{6}$$

where ΔH_f is the latent heat of fusion, T_{ref} is the reference temperature, f_L is the liquid fraction, and T_L and T_S are liquidus and solidus temperatures respectively.

Moreover, for modeling the molten metal flowing in the mushy region, a Darcy porosity approach is adopted, in which, the mushy region is assumed to behave as a porous medium and fluid flow is governed by Darcy's law as follow:

$$u_i = -\frac{\mu}{K_{\text{P}}}\left(\frac{\partial \text{P}}{\partial x_i} - \rho g_i\right) \tag{7}$$

where K_p is the permeability, and $K_{\text{P}} = \dfrac{f_L^3}{C(1 - f_L)^2}$.

2.4. Assumptions and simplified mathematic model

In this research, electromagnetic field, fluid flow, heat transfer and solidification must be coupledly modeled in order to describe accurately the interaction of the multi-physics fields during this process. But in order to simplify the question, some assumptions are given as follows:
(1) Medium is uncompressible Newtonian fluid;
(2) Ignore the interaction between alloys and the effect of fluctuations in the flow field of liquid;
(3) The alloy liquid in mould is treated as homogeneous medium, and ignore the additional magnetic field produced by the flow of aluminium liquid in the mould;
(4) Thickness of air gap and solid fraction is proportional in the mould.

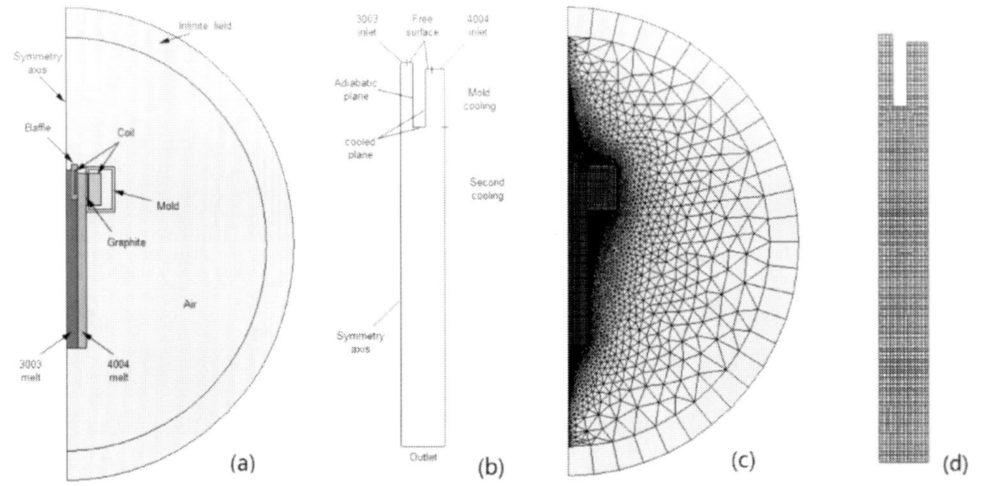

Figure 1. Geometry model in ANSYS (a) and FLUENT (b), and mesh in ANSYS (c) and FLUENT (d).

3. Numerical procedures and physical parameters used in the simulation

In order to model the casting process, coupled modeling of electromagnetic field and others fields must be carried out. Moreover, this interaction is single direction, that is, electromagnetic field has effect on others fields but no effect by contraries. So the governing equations of electromagnetic field and others fields were solved by using the commercial software ANSYS and FLUENT, respectively. The electromagnetic force and Joule heat is added to the conservation equation of momentum and energy as the source term during solving flow field. The geometry models and the mesh which are used in ANSYS and FLUENT are shown in figure 1, where the coils inside the mold and the baffle are single phase coils, which have 240 turns and 40 turns, respectively. Moreover, their dimensions are given in table 1.

Table 1. The dimensions of geometry model used based on ANSYS and FLUENT

Geometry model used based on ANSYS

3003	4004	Mold	Baffle	Coil inside the mold
0.07×0.6m	0.03×0.59m	0.1×0.15m	0.02×0.12m	0.045×0.11m
Coil inside the baffle	Graphite	Air	Infinite field	
0.01×0.08m	0.02×0.11m	Radius=0.7m	Radius=0.8m	

Geometry model used based on FLUENT

3003 inlet	4004 inlet	Mold cooling	Second cooling	Others
0.01 m	0.02 m	0.13 m	0.46 m	Shown as in figure 1

In the paper, three different casting processes have been modeled and the electricity and casting parameters are shown in table.2. 3003 and 4004 aluminium alloy were used as the experiment materials. Physical properties of the materials are given in table.3. There is no electromagnetic field supplied during process 1. During process 2, only sine current is applied in the coil inside the mold and no electromagnetic field is supplied in the baffle. During process 3, sine current is applied in the coil inside the mold and pulse current is supplied in the coil inside the baffle. The waveform of the pulse current is shown in figure 2.

Table 2. Electricity and casting parameters in different processes

Electricity parameters	Peak	Frequency	Pulse duration
Sine current	20 A×240 turns	2500 Hz	-
Pulse current	200 A×40 turns	5 Hz	2.8 ms

Casting parameters	Pouring temperature (3003-4004)	Casting speed	Current in the coil inside the mold	Current in the coil inside the baffle
Process 1	987 K-927 K	85 mm/min	-	-
Process 2	987 K-927 K	85 mm/min	Sine current	-
Process 3	987 K-927 K	85 mm/min	Sine current	Pulse current

Table 3. Physical properties of materials

Region	3003	4004	Mold	Baffle	Coil	Graphite	Air
Density(kg/m^3)	2750	2710	-	-	-	-	-
Viscosity(Pa·s)	0.0013	0.0013	-	-	-	-	-
Specific heat(J/kg·K)	893	864	-	-	-	-	-
Thermal conductivity(W/(m·K))	193	155	-	-	-	-	-
Latent heat of fusion(kJ/kg)	386	481.5	-	-	-	-	-
Solidus temperature(K)	916	830	-	-	-	-	-
Liquidus temperature(K)	927	846.5	-	-	-	-	-
Relative permeability	1	1	1	1	1	1	1
Electric resistivity • 10^{-8} (Ω·m)	4.2	4.31	29.2	73	1.68	1420	-

4. Results of the simulation

4.1. Results of electromagnetic field

Figure 3 shows the comparison of magnetic flux density at the inner surface of baffle the calculated and the measured results under different current intensity. As seen from Figure3, the magnetic flux density increases with the increasing of current intensity. And the numerical results match well with the experimental measurements. The maximum difference between measured and simulation values is 9.5%.

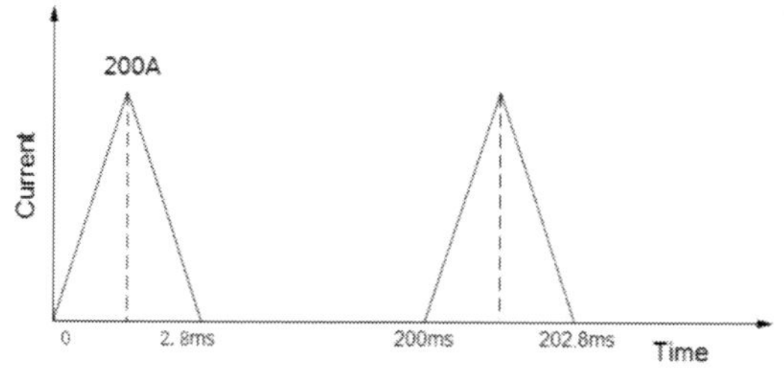

Figure 2. Variation of the pulsed electricity in the coil inside the baffle.

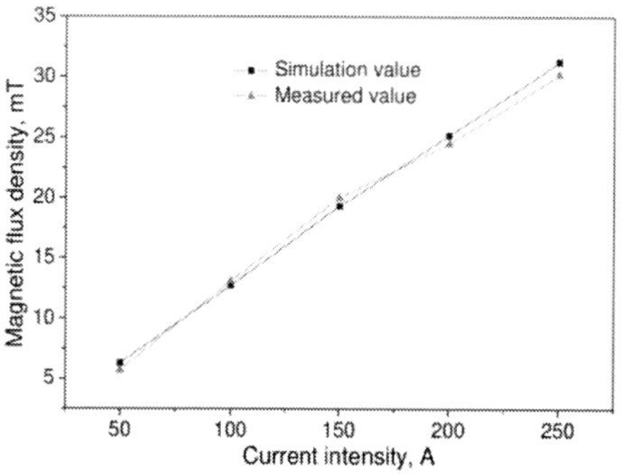

Figure 3. Comparison of magnetic flux density at the inner surface of baffle between the calculated and the measured.

Figure 4 shows the calculated results for magnetic flux density, electromagnetic force and Joule heat of the electromagnetic field in process 2 (i.e. only sine current flowing in the coil inside the mold).

Figure 4 (b, d and f) give the distributions of magnetic flux density, electromagnetic force and Joule heat along line 1, respectively. The location of line 1 is shown in figure 4 (a) and it is 40 mm above the baffle bottom. The magnetic flux density, electromagnetic force and Joule heat decrease distinctly with the decrease of distance from the centreline of billet because of the skin effect of the alternative electromagnetic field. The maximum values of them emerge at the centre zone of the mold, that is, closed to the centreline of the coil inside the mold. Moreover, the right hand parts in figure 4 (a and c) are magnetic flux density vectors and electromagnetic force vectors, respectively. The direction of magnetic flux density is approximately parallel to the centreline of the billet within the electromagnetic penetration depth and the direction of electromagnetic force is always perpendicular

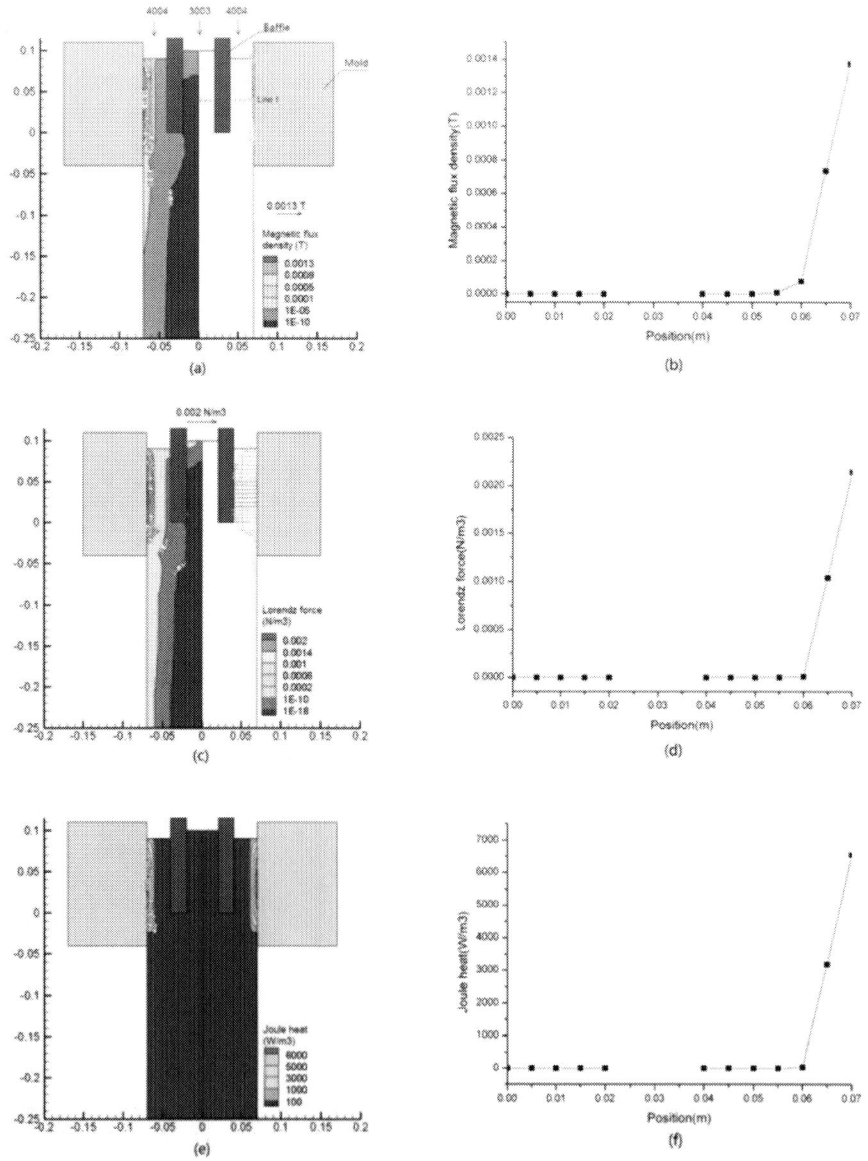

Figure 4. The calculated results of the electromagnetic field in process 2 (also in process 3 during the pulse diapauses): (a) magnetic flux density contours and vectors; (c) electromagnetic force contours and vectors; (e) Joule heat contours; (b) distribution of magnetic flux density along line 1; (d) distribution of electromagnetic force along line 1; (f) distribution of Joule heat along line 1.

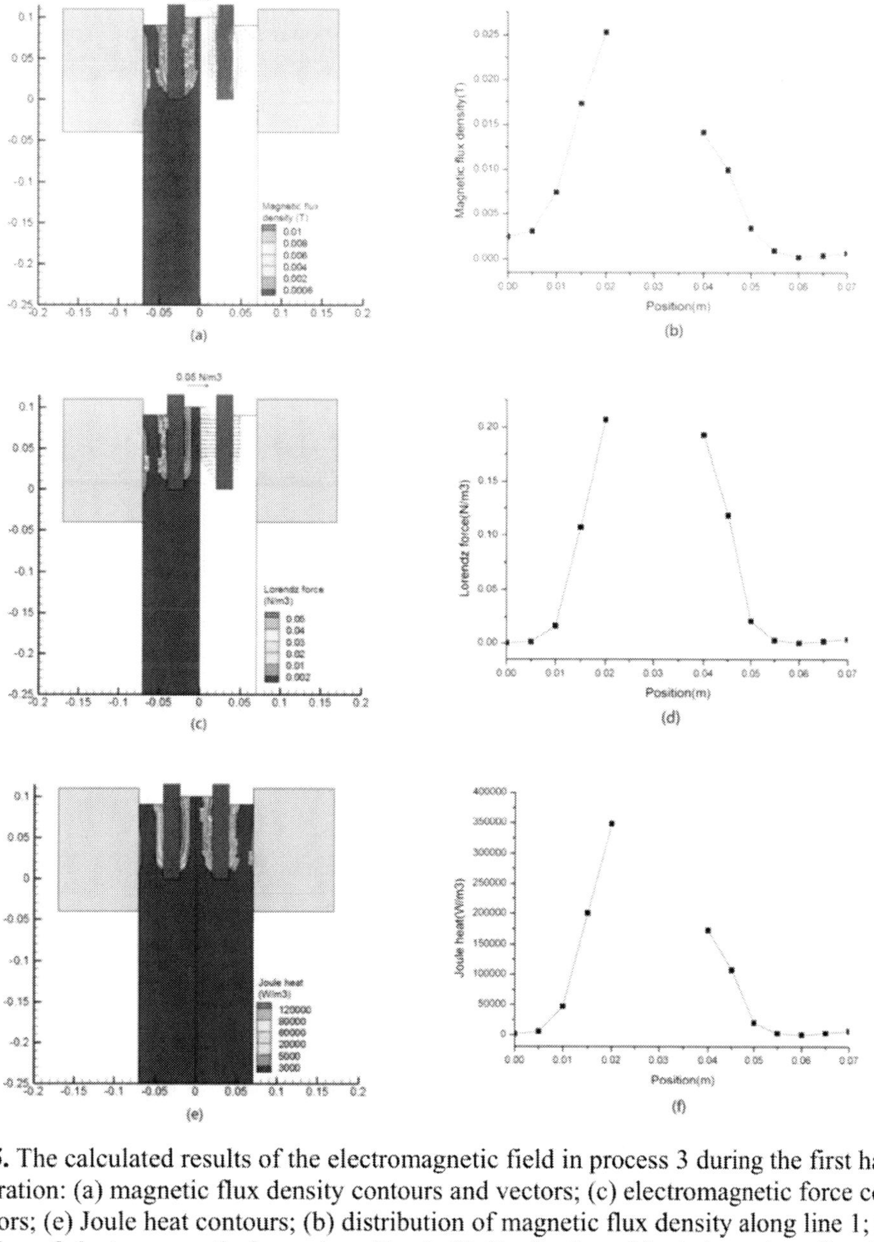

Figure 5. The calculated results of the electromagnetic field in process 3 during the first half of pulse duration: (a) magnetic flux density contours and vectors; (c) electromagnetic force contours and vectors; (e) Joule heat contours; (b) distribution of magnetic flux density along line 1; (d) distribution of electromagnetic force along line 1; (f) distribution of Joule heat along line 1.

to the direction of magnetic flux density. The electromagnetic volume force can produce magnetic pressure on the surface of 4004 aluminium alloy and 4004 melt is soft contact with the mold, which results in better surface quality of the billet. The magnetic force also generates convection slightly in 4004 melt as shown in figure 7 (b).

When sine current is flowing in the coil inside the mold and pulse current is flowing in the coil inside the baffle at the same time (i.e. process 3), the electromagnetic field distribution changes periodically with three periods as a cycle. Figure 5 and figure 6 show the calculated results of the electromagnetic field in process 3 during the first half of pulse duration and the second half of pulse duration, respectively. The electromagnetic field distribution during the pulse diapause is the same as in process 2. As we can see, the effect of magnetic force on 4004 melt near the mold is the same as in process 2. Moreover, pulse magnetic field generates concussive force in 3003 melt and also produces magnetic pressure on the surface of 4004 melt near the baffle, which is transient but powerful. Concussive effect results in vigorous forced convection as shown in figure 7 (c). And magnetic pressure also makes 4004 melt soft contact with the baffle, which results in better interface quality of 3003 and 4004.

Moreover, Joule heat generated in these processes is large and plays an important role in the distribution of temperature field and liquid fraction as shown in figure 9 and figure 11.

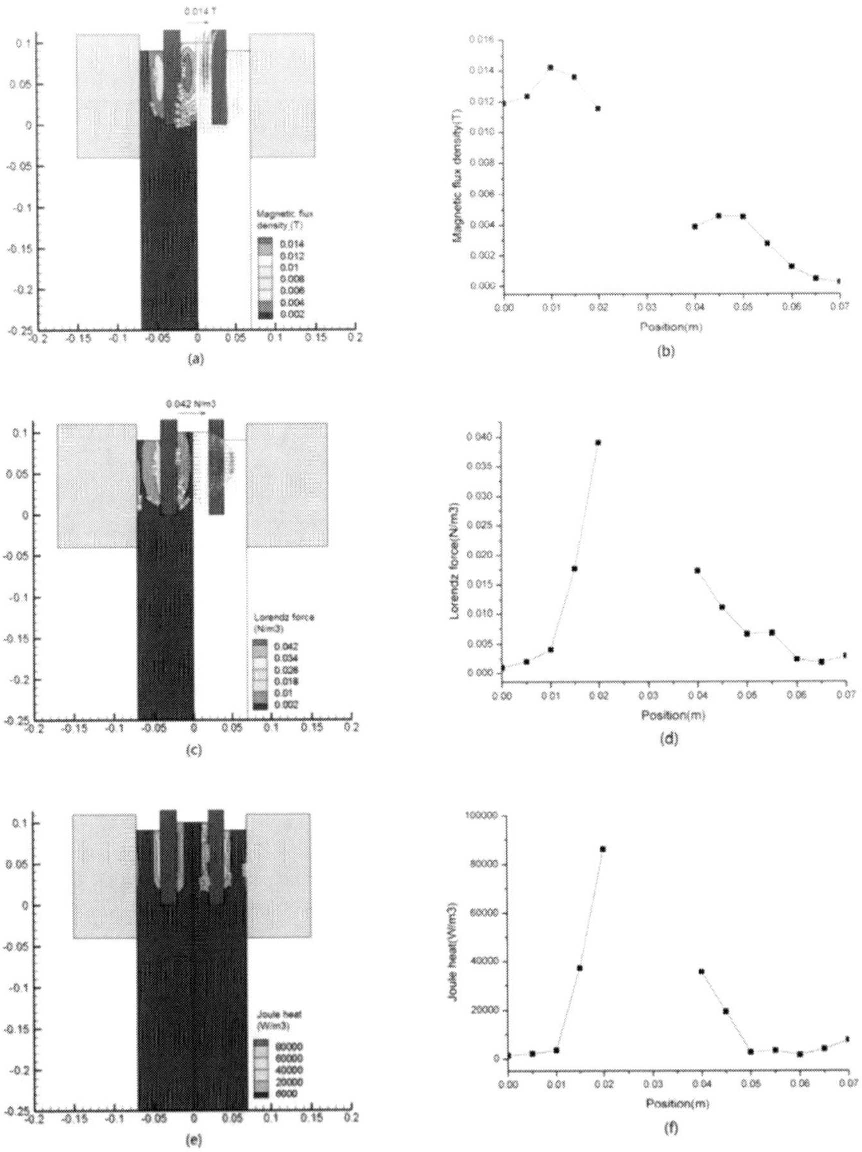

Figure 6. The calculated results of the electromagnetic field in process 3 during the second half of pulse duration: (a) magnetic flux density contours and vectors; (c) electromagnetic force contours and vectors; (e) Joule heat contours; (b) distribution of magnetic flux density along line 1; (d) distribution of electromagnetic force along line 1; (f) distribution of Joule heat along line 1.

4.2. Results of velocity, temperature and liquid fraction

Figure 7, figure 8 and figure 9 show the calculated results of velocity contours and vectors, temperature contours and liquid fraction contours in different processes, respectively. Figure 10, figure 11 and figure 12 give the distributions of velocity, temperature and liquid fraction along line 1 and line 2 (its location is 2.5 mm below the baffle bottom as shown in figure 7 (a)).

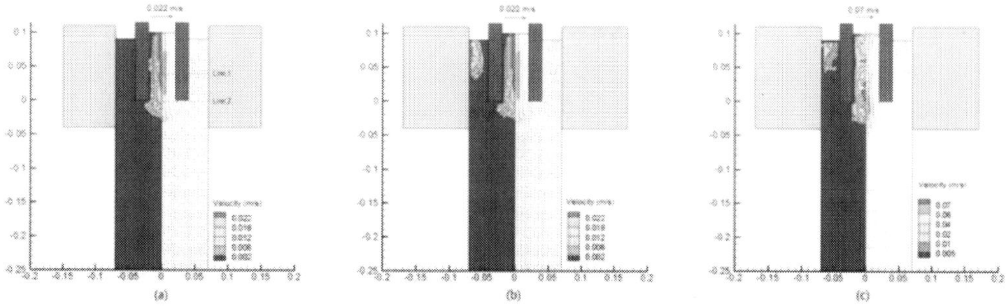

Figure 7. The calculated results of velocity contours and vectors during the three different processes: (a) process 1; (b) process 2; (c) process 3.

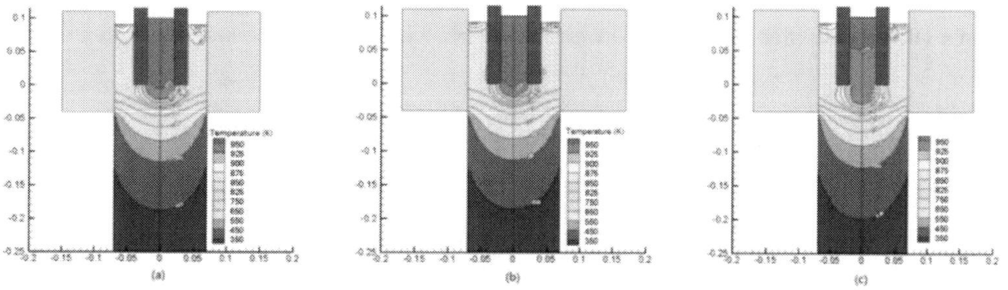

Figure 8. The calculated results of temperature during the three different processes: (a) process 1; (b) process 2; (c) process 3.

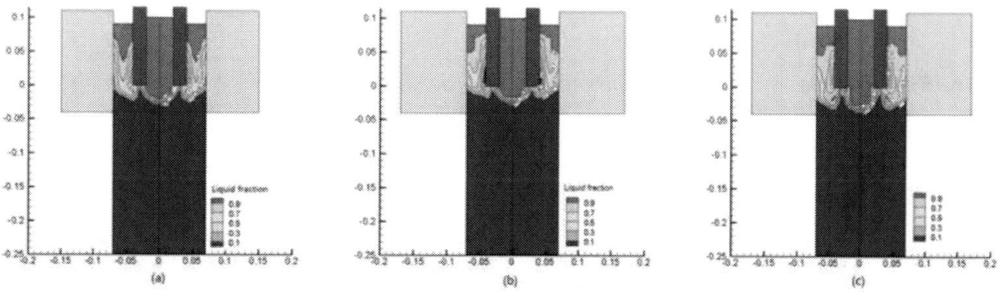

Figure 9. The calculated results of liquid fraction during the three different processes: (a) process 1; (b) process 2; (c) process 3.

During process 1, as we can see in figure 7 (a), figure 8 (a) and figure 9 (a), the melt is poured into the liquid pool from two inlets and then flows downwards along the symmetry axis to reach the solidification front. 4004 melt outside the baffle solidifies first, while 3003 melt inside the baffle is hardly cooled until it is contacted with the shell of 4004. Therefore, the shell of 4004 aluminium alloy remelts at the original contact location of 3003 and 4004. During process 2, magnetic force results in slight convection of 4004 melt as seen in figure 7 (b), and cooling effect changes because of the generation of Joule heat and melt convection as we can see in figure 8 (b) and figure 9 (b). The cooling rate of 4004 melt decreases near the mold and increases near the baffle, which results in right moving of the sump and thickening the shell of 4004 aluminium alloy near the baffle. During process 3, because the generation of Joule heat and the rush on solidification front by the vigorous melt convection inside the baffle, the sump of 3003 melt increases obviously as seen in figure 9 (c). The magnetic pressure and generation of Joule heat outside the baffle result in decrease of the cooling rate of 4004 melt near the baffle and the thickness of shell near the baffle decreases.

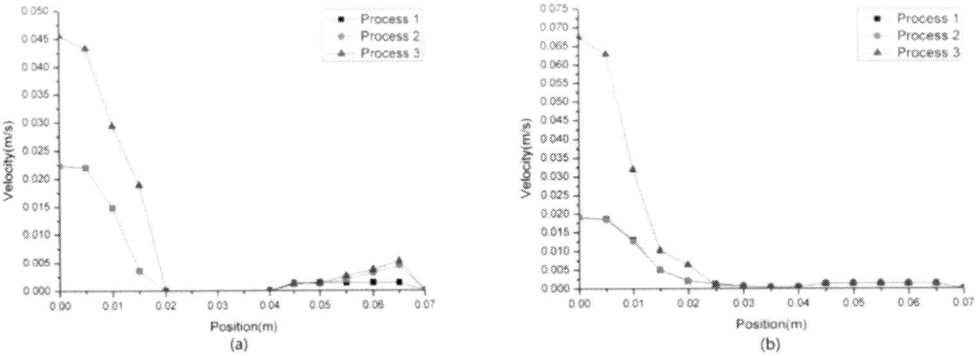

Figure 10. Velocity distributions in different processes along line 1 (a) and line 2 (b).

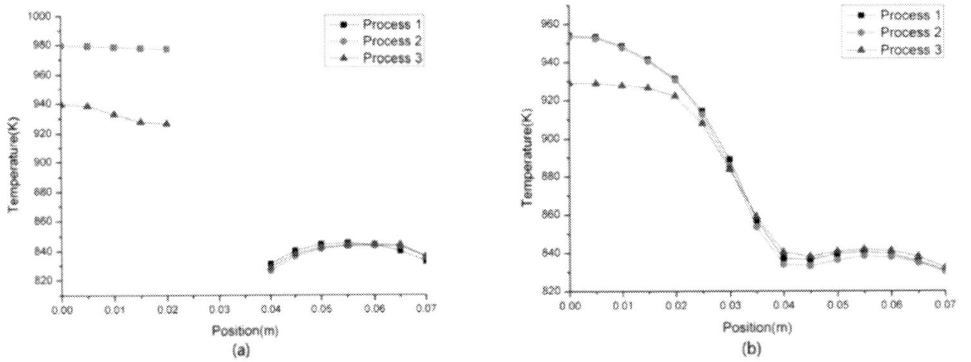

Figure 11. Temperature distributions in different processes along line 1 (a) and line 2 (b).

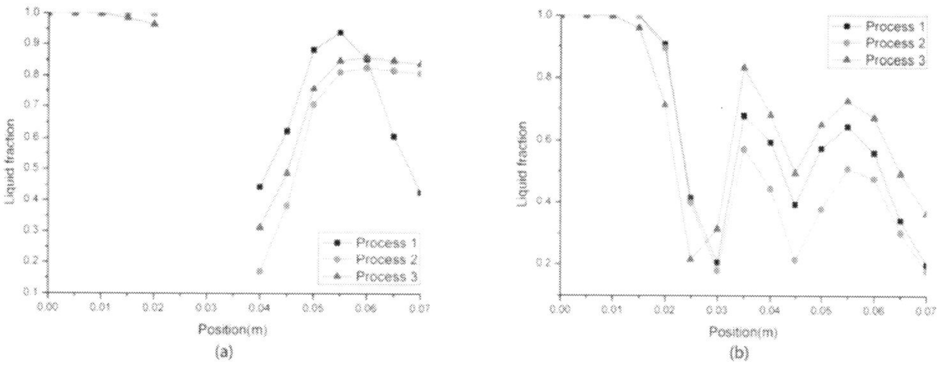

Figure 12. Distributions of liquid fraction in different processes .along line 1 (a) and line 2 (b).

5. Conclusions

1. A comprehensive mathematic model was performed by means of ANSYS and FLUENT to describe the interaction of the multiple physics fields during the continuous casting process of Al/Al bimetal round billet under multi-electromagnetic fields.

2. Alternative electromagnetic field produced magnetic pressure on the surface of outer melt and the outer melt was soft contact with the mold. Moreover, pulse electromagnetic field generated concussive magnetic force in the inner melt.

3. The magnetic force generated by pulse electromagnetic field is transient but powerful and the melt convection forced in the inner melt is more vigorously than that in the outer melt.

4. The melt convection and generation of Joule heat changed the cooling effect, which results in the change of distribution of temperature field, thickness of the shell and feature of the sump. The sump of outer melt became shallower and moved to the outside.

Acknowledgements

The authors thank the National Natural Science Foundation of China (Nos. 50601003, 50971032, and 51071035), the Fundamental Research Funds for the Central Universities, and Liaoning BaiQianWan Talents Program.

References

[1] Marukovich E I, Dozmarov V V and Voytenko I G 1998 *Foundry. Prod. Metall.* **1** 24
[2] Marukovich E I, Branovitsky A M and Na Y S 2006 *Mater. Des.* **27** 1016
[3] Xie J X, Sun D Q and Wu C J 2000 *J. Mater. Eng.* **4** 38
[4] Li Y Y, Fei J and Chen W P 2002 *J. Cent. South. Univ. Tech.* **9** 229
[5] Jin W Z, Li T J and Yin G M 2007 *Sci. Technol. Adv. Mater.* **8** 1
[6] Zhang X L, Li T J, Teng H T, Xie S S and Jin J Z 2008 *Mater. Sci. Eng.* A **475** 194
[7] Cao Z Q, Jia F, Zhang X G, Hao H and Jin J Z 2002 *Mater. Sci. Eng.* A **327** 133
[8] Gao Y L, Li Q S, Gong Y Y and Zhai Q J 2007 *Mater. Lett.* **61** 4011
[9] Li Q S, Song C J, Li H B and Zhai Q J 2007 *Maert. Sci. Eng.* A **466** 101
[10] Zi B T, Ba Q X and Cui J Z 2000 *Scr. Mater.* **43** 377
[11] Zhang H T, Nagaumi H, Zuo Y B and Cui J Z 2007 *Mater. Sci. Eng.* A **448** 189
[12] Ma X P, Yang Y S and Wang B 2009 *INT. J. HEAT. MASS. TRAN.* **52** 5285
[13] Miller W S, Zhuang L and Bottema J 2000 *Mater. Sci. Eng.* A **280** 37

A multiscale slice model for continuous casting of steel

B Šarler[1,4], R Vertnik[1,2], A Z Lorbiecka[1], I Vušanović[3] and B Senčič[2,4]

[1] Laboratory for Multiphase Processes, University of Nova Gorica, Vipavska 13, SI-5000, Nova Gorica, Slovenia.
[2] Štore Steel d.o.o, Technical Developement, Železarska 3, SI-3220 Štore, Slovenia.
[3] Faculty of Mechanical Engineering, University of Montenegro, George Washington st., 81000 Podgorica, Montenegro.
[4] Centre of Excellence for Biosensorics, Automation and Process Control, Velika pot 22, Solkan, Slovenia.

E-mail: bozidar.sarler@ung.si, robert.vertnik@store-steel.si, alorbiecka@ung.si, bojan.sencic@store-steel.si

Abstract. A simple Lagrange-an traveling slice model has been applied for the prediction of the relations between process parameters, macrosegregation and solidification grain structure formation (equiaxed to columnar and columnar to equiaxed transition) during the continuous casting process of steel billets. The main advantage of the slice model is its very fast calculation time in comparison with the complete 3D heat and fluid flow model which might need calculation time, measured in days. The slice models thus allows for fast optimisation and even for on-line simulation. The heat and species transfer models are based on the mixture continuum assumptions with Lever solidification rule and enhanced thermal and solutal diffusivities for heuristic accounting of fluid flow effects. The grain structure evolution model is based on the Gaussian nucleation rule, and KGT growth model, coupled to the macroscopic heat and species transfer models. The heat and species transfer models are solved by the meshless technique by using local collocation with radial basis functions. The grain structure evolution model is solved by the point automata technique, a novel meshless variant of the cellular automata method. A comparison of the results with the experimental data for steel grade 51CrV4 is shown in terms of macrosegregation and grain structure across the billet. Simulations and comparisons have been carried out for nominal casting conditions, reduced casting temperature, and reduced casting speed. The model predicts surprisingly well the qualitative features of the macrosegregation and grain structure patterns. Possible refinements of the model with respect to other physical mechanisms are discussed.

1. Introduction

Continuous casting [1] is currently the most common process for production of steel. The process can be accompanied by a range of quality problems such as shape defects, cracking, inclusions, non-uniform microsegregation, macrosegregation, porosities, etc. Respectively, there is a very big demand for related process modeling, which is used in the design of casters, proper setting of process parameters, and on-line control of the casters. The models [2] range from very simple, usually used in the on-line mode in the regulation of the plant, to very sophisticated multiscale and multiphysics approaches that run off-line for several days and tackle the specific aspects of the process like melt – casting powder interaction, inclusions dynamics, oscillation mark formation, etc. The focus in this paper is the so called slice model. This model neglects all interactions in the direction of the casting, and takes into account only the interactions perpendicular to the casting in form of a traveling slice. Respectively, a quasi-three dimensional picture of the strand status can be obtained much faster than when solving a complete 3D configuration. This model [3,4] performs well for heat transfer, but of course could not predict any realistic [5] turbulent fluid flow effects. Several important solidification [6] parameters, such as the shell thickness along the casting direction, the metallurgical length and the temperature of the strand surface can be calculated satisfactory accurate. Recently, the slice model has been upgraded with the solution of the grain structure [7] and macrosegregation [8] for simple Fe-C steel. The macrosegregation has been taken into account by assuming the enhanced thermal diffusivity of the liquid phase. Both, the simple grain structure and the simple macrosegregation model predict surprisingly well the qualitative features of the strand. It is the purpose of this paper to upgrade the simple slice model for estimating the grain structure and the macrosegregation in the strand to multicomponent steel. We are aware that the quantitative picture, obtained from such a simple model, might be too crude, however, on the other hand, such a simple model can be used in the on-line mode and give the plant operators a quick insight into the quality changes and trends in the strand as a function of the process parameter changes.

2. Slice model assumptions

A schematics of the continuously cast strand is shown in figure 1 with shaded traveling slices. The temperature and the concentration field of the slice can be computed from known time dependent boundary conditions. The slice time is on the other hand associated with the position in the strand. The Cartesian coordinate z measures the length from the meniscus. The strand geometry is assumed straightened along this coordinate, which is a geometrical simplification of the real curved casting process. In all slice models the mass, heat, momentum, and solute transfers in the casting direction are neglected. The z coordinate can thus be considered parabolic, while the x and y coordinates are elliptic. In this way all fields at a given z coordinate depend only on the slice history, including its cooling intensity as a function of time. The slices form at the z_{start} longitudinal coordinate of casting and travel in the direction of the $\mathbf{i}_z$ base vector with the casting speed $\mathbf{V}_{cast}$, (see figure 1). For calculating the cooling intensity of the slice as a function of time, a connection between the z coordinate of the casting machine and the slice history t is needed, which is in general:

$$z(t)=\int_{t_{start}}^{t} V_{cast}(t')\cdot dt'+z_{start}, V_{cast}(t)=\mathbf{V}_{cast}(t)\cdot\mathbf{i}_z \qquad (1)$$

where t_{start} represents the initial time of a slice. In the case when the casting speed and other process parameters are steady, we obtain the following simple connection between the z coordinate of the casting machine and the slice history $t(z)=t_{start}+(z-z_{start})/V_{cast}$.

3. Macroscopic transport model

3.1. *Governing equations*

A well-known continuum mixture model [9] is applied for solving the temperature and concentration of multicomponent steel in this study. It is assumed, consistently with the slice model postulates, that the macroscopic transport occurs only in travelling *x–y* cross sectional slice, moving from the top (horizontal position of slice) to the exit of casting rig (vertical position of slice) as seen on figure 1. The coordinate system is fixed with the slice, while the temperature and the boundary conditions are assumed as time dependent, based on constant casting speed and position of slice from the top to the bottom.

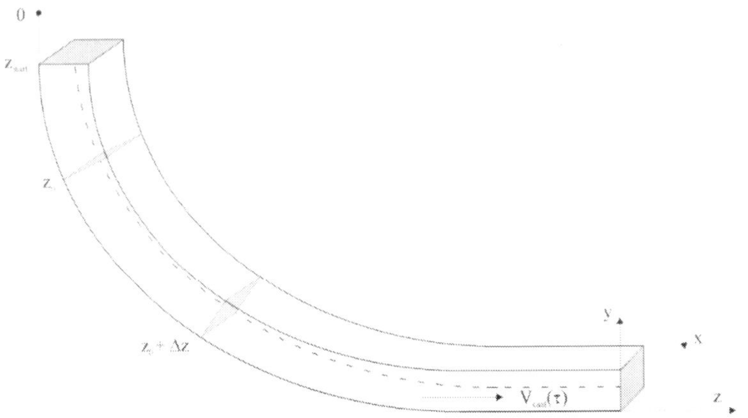

Figure 1. Slice traveling schematics in the strand.

The boundary temperatures of the travelling slice were obtained from the Robin type energy balance equations for slice boundaries, defined by the known heat transfer coefficients, and the temperature of the cooling fluid. Let us assume steel with N alloying elements. The related governing equation for heat and N equations for species transport are:

$$\frac{\partial}{\partial t}(\rho h)=\nabla\cdot(k\nabla T),\ \frac{\partial}{\partial t}(\rho C_n)=\nabla\cdot(\rho f_L D_{Ln}\nabla C_n)+\nabla\cdot(\rho f_L D_{Ln}\nabla(C_{Ln}-C_n));\ n=1,\ldots,N \quad (2,3)$$

where T stands for temperature, and the mixture quantities of steel (density $\rho = g_S\rho_S + g_L\rho_L$, thermal conductivity $k = g_S k_S + g_L k_L$, mixture enthalpy $k = f_S h_S + f_L h_L$, and mixture concentration $C_n = f_S C_{Sn} + f_L C_{Ln}$) were defined by using appropriate solid and liquid properties, weighted with either mass (f_S, f_L) or volume (g_S, g_L) fractions of solid S and liquid L phase.

3.2. Microsegregation model

In order to calculate the local temperature and liquid concentration of alloying elements, which is the driving force for macrosegregation, a supplementary microsegregation model is necessary. The supplementary microsegregation model calculates temperature T, solid C_S and liquid C_L concentrations, and mass fractions of solid f_S and liquid phase f_L as well. Mixture enthalpy and concentrations from the macro model (2,3) were used as input parameters for the micro model and they vary in general, while the solidification proceeds. In this study, simple Lever rule has been used. The governing equations for the supplementary microsegregation model are as follows:

- Mixture and phase enthalpies:

$$h = f_L h_L + f_S h_S; h_L = c_L T + (c_S - c_L)T_{sol} + h_f; h_S = c_S T, \quad (4)$$

- Mixture concentration, lever rule concentration balance, liquidus temperature

$$C_n = f_L C_{Ln} + (1-f_L)C_{Sn},\ C_{Sn} = k_n C_{Ln},\ T = T_M + \sum_{n=1}^{N} m_{Ln} C_{Ln}, \quad (5)$$

where T_{sol}, and h_f are solidus temperature and latent heat of phase change, k_n and m_{Ln} are the partition coefficients and the liquidus slopes of the alloying components (n=C, Cr, Ni). The influence of the turbulent fluid flow on macrosegregation is taken into account through the artificially increased liquid diffusivity (D_L=$50D_{nom}$).

IOP Conf. Series: Materials Science and Engineering **33** (2012) 012021 doi:10.1088/1757-899X/33/1/012021

3.3. Geometry, material properties, boundary and initial conditions

In this study, a quadratic shape of traveling slice is used with dimensions $L_x = L_y = 0.14$ m. The physical properties are $\rho_{S(L)}$=7870(6990) kg/m^3, $c_{S(L)}$=824.9(1395.8) J/kgK, $k_{S(L)}$=25(39.9) W/mK, and $h_f = 2.711\text{x}10^5$ J/kg. The diffusion coefficient in the solid is neglected in this study, since it is more than one order of magnitude lower compared to the liquid one. The initial conditions for traveling slice in this study are determined by the inlet superheat $\Delta T_{in} = 20$ K and initial uniform concentration of alloying Carbon, Chromium and Nickel respectively: $C_C = 0.51$ wt%, $C_{Cr} = 1.091$ wt%, C_{Ni}=0.105wt% (main alloying components of steel 51CrV4). The heat flow from the slice to the cooling liquid is determined through the heat transfer coefficients: $h_{mold} = 2000$ W/m^{2}K for the mold section until $z = 0.8$ m, and $h_{spray} = 800$ W/m^{2}K for the spray cooling section until the end of casting machine ($z = 10.8$ m). The reference cooling water temperature in Robin boundary conditions is set to $T_{ref} = 30$ °C. The casting speed is $V_{cast} = 1.75$ m/min. The explicit time discretization step length in this study is set to $\Delta t = 0.1$ s for temperature and concentration calculations.

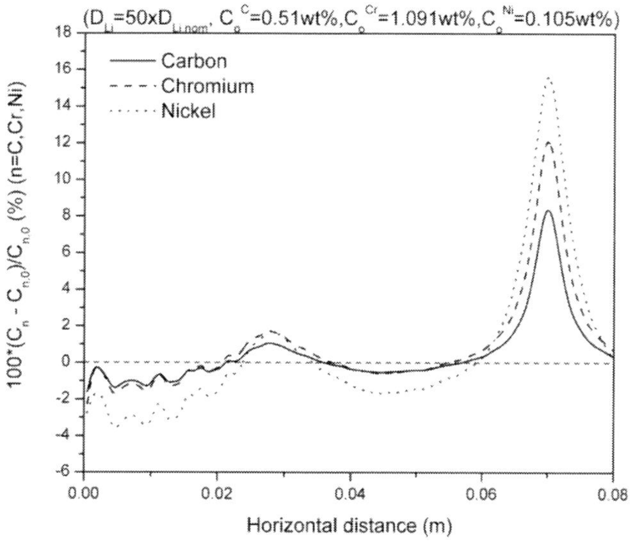

Figure 2. Calculated macrosegregation of C, Cr and Ni through the center (at 0.07 m) of the billet, parallel with the side.

The denominated horizontal cross section profiles of alloying carbon, chromium and nickel of the Fe-0.51wt%C-1.091wt%Cr-0.105wt%Ni steel, depicted in figure 2, were calculated by assuming the simple Lever rule for calculation of local temperature, phase fractions, and compositions. For all three elements there is a negative segregation near the surface, followed by a possitive segregation arround 3 cm from surface and then negative, and finally peak positive centreline segregation. The peak at approximately 0.03 m appears in the model due to the change of heat flux on the boundary from mould to the spray cooling region. This behaviour is in line with the experimental observations. However, equations (2-3) with artificially increased liquid diffusivity of course represent a quite rough approximation to the real and complex fluid flow pattern and deformation of the mushy zone. Increased diffusion on the macro scale reflects on the micro level as accelerated freezing in the mushy zone, and lower solid composition.

4. Grain structure model

Most of the grain growth models focus on the binary alloys [10, 11] and very little attention is focused to multicomponent alloys [12]. In this section a general theoretical model is discussed to describe the CET/ECT growth transformation under the solidification of a multicomponent alloy.

4.1. Nucleation model

In the present study we adopted continuous nucleation model in which two different continuous nucleation modes were considered: at the surface (index s) and in the bulk (index b). The increase of grain density dn which corresponds to an undercooling increase $d(\Delta T)$ is described as follows

$$\left(\frac{dn}{d(\Delta T)}\right)_{\varsigma} = \frac{n_{\max,\varsigma}}{\sqrt{2\pi}\Delta T_{\sigma,\varsigma}} \exp\left[-\frac{\Delta T - \Delta T_{\max,\varsigma}}{\Delta T_{\sigma,\varsigma}\sqrt{2}}\right]^2 \tag{6}$$

where: $\Delta T_{\max,s}$, $\Delta T_{\max,b}$, $\Delta T_{\sigma,s}$, $\Delta T_{\sigma,b}$, $n_{\max,s}$ and $n_{\max,s}$ represents the mean nucleation undercooling at the surface, the mean nucleation undercooling in the bulk, the standard deviation at the surface, the standard deviation for the bulk and the maximum density of nuclei that can form in the melt for the surface and bulk, respectively. The local undercooling $\Delta T = T_{liq}(C_1, C_2, \ldots C_N) - T$ is calculated from the $T_{liq}(C_1, C_2, \ldots C_N)$ and T from the macroscopic model, defined above.

4.2. Grain growth model

The following model was used as the model of the growth kinetics. The dendrite tip satisfies the Ivantsov solution, and the supersaturation Ω_n is given by

$$\Omega_n = \mathrm{Iv}(\mathrm{Pe}_n);\ \mathrm{Pe}_n = \frac{VR}{2\mathrm{D}_{\mathrm{Ln}}};\ \mathrm{Iv}(\mathrm{Pe}_n) = \mathrm{Pe}_n \exp(\mathrm{Pe}_n)\mathrm{E}_1(\mathrm{Pe}_n) \tag{7,8,9}$$

where R, D_{Ln}, Pe_n, E_1 are the dendritic tip radius, the diffusion coefficient of the solute in the liquid, Péclet number and the exponential integral function. Langer and Muller-Krumbhaar [13] have shown that the dendrite tip grows closely to a state called marginally stable. The approximate relationship can be written $R = \lambda_s$, with λ_s is the critical wavelength of the solid-liquid interface at the limit of stability. For multicomponent alloys the wavelength of the marginally stable front is given by the equation [10]

$$\omega^2\Gamma = \sum_{n=1}^{N} m_{vn} G_{c_n} \zeta_{c_n} - \mathrm{G};\ \omega = \frac{2\pi}{R_n};\ m_{vn} = \mathrm{m}_n\left\{1 + \frac{k_n - k_{vn}\left[1 - \ln(k_{vn}/k_n)\right]}{1 - k_n}\right\},$$

$$G_{c_n} = \frac{-V(1-k_{vi})\mathrm{c}_{0n}}{\left[1-(1-k_{vn})\mathrm{Iv}(\mathrm{Pe})\right]D_{Ln}};\ \zeta_{c_n} = 1 - \frac{2k_{vn}}{\left[1+(2\pi/\mathrm{Pe})^2\right]^{0.5} - 1 + 2k_{vn}} \tag{10,11,12,13,14}$$

where ω, m_{vn}, k_{vn}, c_{0n}, D_{Ln}, Γ, G_{c_n}, G, ζ_{c_n}, V are the wave number, the velocity dependent liquidus slope of component n, velocity dependent partition coefficient of component n, concentration of component n in the liquid at the dendrite tip, diffusion coefficient of solute in the liquid of component n, Gibbs-Thomson parameter, the concentration gradient of species n in the liquid at the dendritic tip, mean temperature gradient at the interface, stability parameter and the interface velocity, respectively. By using the equation (10) the tip radius can be obtained

$$R_n = 2\pi\Gamma^{0.5}\left(\sum_{n=1}^{N} m_{vn} G_{c_n} \zeta_{c_n} - \mathrm{G}\right)^{-0.5} \tag{15}$$

Several models have been presented to describe the dependence of partition coefficient on velocity. This have been discussed by Boettinger et.al [14]. The Aziz [15] solution for non-equilibrium partition coefficient is used $k_{vn} = (k_n + a_0 V / D_{Ln})/(1 + a_0 V / D_{Ln})$ where k_n and a_0 are the equilibrium partition coefficient and the characteristic length scale in order of the atomic distance estimated to be between 0.5 and 5 nm. The tip undercooling can be expressed by the sum of the following undercoolings $\Delta T^* = \Delta T_c + \Delta T_R$ where ΔT_c and ΔT_R are the constitutional and curvature undercooling [16,17] respectively.

$$\Delta T_c = \sum_{n=1}^{N} m_n c_{0n} \left(1 - \frac{m_{vn} / m_n}{1-(1-k_{vn})\mathrm{Iv}(\mathrm{Pe})} \right); \; \Delta T_K = \frac{2\Gamma}{R_n} \qquad (16,17)$$

5. Solution procedures

The solution procedure is as follows. For each slice, the temperature field is solved first, followed by the solution of the concentration fields. Afterwards, the grain structure model is solved, based on the temperature and the concentration fields. The macroscopic fields are assumed to influence the nucleation and the growth of the grains. There is no feedback assumed to take place between the grain growth model and the macroscopic models. The meshless Local Radial Basis Function Collocation Method (LRBFCM) [18] is applied to solve the macroscopic equations (2,3). This method is able to accurately and efficiently solve involved solidification problems such as the turbulent flow [19] with solidification and the macrosegregation as a consequence of the thermosolutal flow [20]. The grain structure model is solved by the Point Automata (PA) technique [16]. This method is similar to the cellular automata method. However, the behaviour is generated by the pre-defined rules based on the local relationship between the nearest neighbouring random points that cover the domain instead of the neighbouring cells. Each random node seed grows with respect to the neighbourhood configuration which is associated with the position of the neighbouring points which fall into a circle [16] with radius R. It means that each point can contain different number and position of the neighbours, which gives various possibilities of neighbourhoods. The detailed description of this novel method can be found in [11, 16]. The KGT model is used as the model of the growth kinetics. Equation (10) is solved numerically by using the false position method. By simultaneously solving equations (10) together with (11), (12), (13), (14) and (15) the variation of Péclet number with the growth velocity V can be calculated. Finally, the relationship between the tip undercooling (16), (17) and the growth velocity can be obtained. The undercooling ΔT is interpolated from the macro model for each random node separately in order to find the growth velocity.

Table 1. Nucleation parameters

quantity		value	unit
mean surface nucleation undercooling	$\Delta T_{\max,s}$	0.6	K
mean bulk nucleation undercooling	$\Delta T_{\max,b}$	12	K
standard deviation for the surface area	$\Delta T_{\sigma,s}$	0.2	K
standard deviation for the bulk area	$\Delta T_{\sigma,b}$	0.5	K

The slice is covered by uniform rectangular 140 x 140 macroscopic node arrangement and approximately 20 random PA points are associated with each macroscopic node. The nucleation, macrosegregation, and ECT/CET model parameters are presented in table 1 and table 2.

6. Validation

The boundary conditions and material properties in the present paper have been made simple in order to put focus on the principle structure of the model and to allow for recalculation of the results. The realistic boundary conditions of the billet caster are described in details in [3,7]. The heat transport in the mould takes into account the heat transport mechanisms through the casting powder, across the air-gap (if it exists), to the mould surface, in the mould, and to the mould cooling water. The heat transport mechanisms in the secondary cooling zone take into account the effects of the casting velocity, strand surface temperature, spray nozzle type, spray water flow, temperature and pressure, radiation and cooling through the rolls contact. The mentioned basic heat transfer mechanisms are modified with regard to running water and rolls stagnant water at relevant positions. The proper response of the thermal model was validated through infrared thermography [22]. A sensitivity study of the parameters of the grain structure model for Fe-C steel was performed in [7] regarding the macroscopic model, and in [16] regarding the microscopic model. The proper qualitative response of the macrosegregation model was represented in [8] for Fe-C steel.

Table 2. The physical parameters used in calculations [21]

quantity		value	unit
liquidus temperature	T_{liq}	1755.0	K
initial concentration of carbon	c_{0C}	0.510	wt.%
initial concentration of nickel	c_{0Ni}	0.105	wt.%
initial concentration of chromium	c_{0Cr}	1.091	wt.%
partition coefficient for carbon	k_C	0.34	-
partition coefficient for nickel	k_{Ni}	0.9	-
partition coefficient for chromium	k_{Cr}	0.860	-
slope of the liquidus line with respect to the C concentration	m_C	-62.3	K/ wt.%
slope of the liquidus line with respect to the Ni concentration	m_{Ni}	-3.2	K/ wt.%
slope of the liquidus line with respect to the Cr concentration	m_{Cr}	-1.8	K/ wt.%
diffusion coefficient in liquid for C	D_{LC}	$2\text{x}10^{-9}$	m^2s^{-1}
diffusion coefficient in liquid for Ni	D_{LNi}	$4.92\text{x}10^{-9}$	m^2s^{-1}
diffusion coefficient in liquid for Cr	D_{LCr}	$2.67\text{x}10^{-9}$	m^2s^{-1}
Gibbs-Thomson coefficient	Γ	$1.9\text{ x}10^{-7}$	Km
length scale for solute trapping	a_0	$5\text{x}10^{-9}$	m

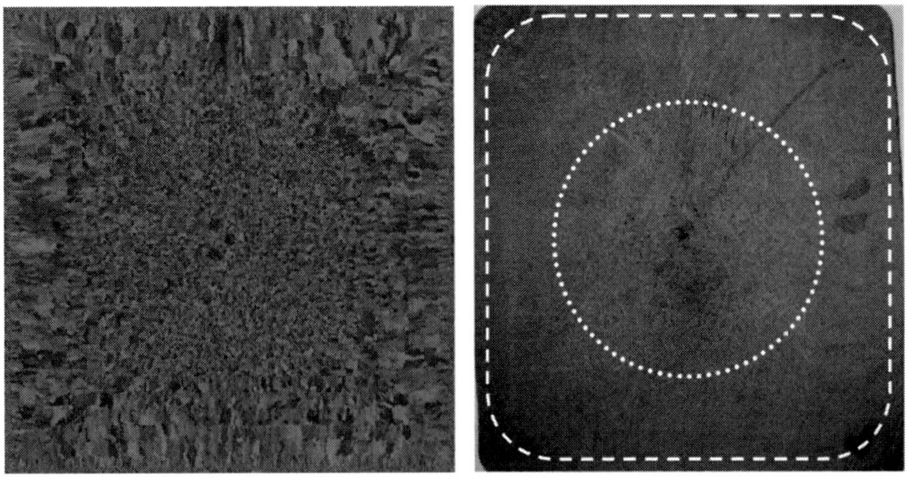

Figure 3. Left: simulated grain structure, and right: Baumann print of the 51CrV4 steel billet.

Figure 3 shows reasonable agreement between calculated (left) and measured (right) grain structure. The primary austenite grain size was between 5 and 7, measured according to standard ASTM E 112.

7. Conclusions

A new, simple slice model for estimating macrosegregation and grain structure in continuously cast strand is represented in the present paper. The model has been demonstrated for billet with 140 mm side and Fe-C-Cr-Ni alloy. A reasonable agreement with the expected macrosegregation and experimental grain structure has been achieved. The model can be used for higher component steel grades as well. The main feature of the model is its simplicity (of course on the expense of accuracy) and related efficiency, which allows the model to be used also in the on-line mode in the perspective. The influence of the fluid flow on macrosegregation is taken into account through artificially increased liquid diffusion coefficient, which mimics the advection of solute enriched liquid steel from the mushy region to the liquid sump in the slice center. The developed model will be used in order to graphically show the on-line status of the strand as well as to provide the data for the plant automation [3]. The coupling, harmonization and further experimental verification of the represented modules with respect to other steel grades, as well as bloom and slab formats is the subject of our future investigations.

Acknowledgement: The present work was supported from the Slovenian Research Agency through grant L2-3651, Slovenian - Montenegro bilateral project SI-CG4-2012/13 and Štore Steel company. The Centre of Excellence for Biosensors, Instrumentation and Process Control is an operation financed by the European Union, European Regional Development Fund and Republic of Slovenia, Ministry of Higher Education, Science and Technology. The financial support is kindly acknowledged.

Reference

[1] Irwing W R 1993 *Continuous Casting of Steel* (London: The Institute of Materials)
[2] Thomas B G 2002 Modeling of the Continuous Casting of Steel - Past, Present Future, www.brimacombecourse.org/pdf/2001_Thomas.pdf
[3] Šarler B, Vernik R, Šaletić S and Cesar J 2005 *Berg- Huettenmaenn. Monatsh.* **150** 300
[4] Louhenkilpi S 2003 *Materials Science Forum* **414-415** 445.
[5] Chaudhary R C, Ji C, Vanka S P and Thomas B G 2011 *Metall. Trans.* **42B** 987
[6] Dantzig J A and Rappaz M 2009 *Solidification* (Lausanne: EPFL Press).
[7] Lorbiecka A Z, Vertnik R, Gjerkeš H, Manojlović G, Senčič B, Cesar J, and Šarler B 2009 *Computers, Materials & Continua* **8** 195
[8] Vušanović I, Vertnik R and Šarler B 2011 *IOP: Conf. Series: Mat. Sci. Eng.* **27** 012056
[9] Bennon W D and Incropera F P 1987 *International Journal of Heat and Mass Transfer* **30** 2161
[10] Kurz W and Fisher D J 1981 *Acta metall.* **29** 11
[11] Lorbiecka A Z and Šarler B 2011 *Cellular Automata: Innovative Modelling for Science and Engineering* ed Alejandro (Rijeka) InTech p 197
[12] Lin X, Li Y, Wang M, Feng L, Chen J and Huang W 2003 *Science in China* **46** 475
[13] Langer J S and Muller-Krumbhaar H 1985 J. Cryst. Growth **72** 578
[14] Boettinger W J, Shechtman D, Schaefer R J and Biancaniello F S 1984 Metall. Trans. **15A** 55
[15] Aziz M J 1982 J. Appl. Phys. **53** 1158
[16] Lorbiecka A Z and Šarler B 2009 3rd International Conference of Simulation and Modelling of Metallurgical Processes in Steelmaking, ed A Ludwig (Leoben, Austria) p 192
[17] Lorbiecka A Z, Šarler B., 2010 Computers, Materials & Continua **18** 69
[18] Šarler B and Vertnik R 2006 *Computers and Mathematics with Applications* **51** 1269
[19] Vertnik R and Šarler B 2011 *Adv. Appl. Math. Mech.* **3** 259
[20] Kosec G, Založnik M, Šarler B and Combeau H 2011 *Computers, Materials & Continua* **22** 156
[21] Vannier I, Combeau H and Lesoult G 1993 Materials Science and Engineering A173 317
[22] Gjerkeš H, Šarler B and Vertnik R 2009 3rd International Conference of Simulation and Modelling of Metallurgical Processes in Steelmaking, ed A Ludwig (Leoben, Austria) p 206

IOP Conf. Series: Materials Science and Engineering **33** (2012) 012022 doi:10.1088/1757-899X/33/1/012022

Comprehensive CFD simulation of the ladle cycle toward lining service life optimization

Mansour Al-Harbi[1]; Abdullah Al-Nefai[2]; Sami Al-Jarallah[2]; Abdullah Al-Dossary[2]; Yannam Jagan[2].
[1]SABIC Technology Center-Jubail, P.O.Box 11669, Jubail Industrial City 31961
[2]SABIC Affiliate, P.O.Box 36536, Jubail Industrial City 31961

E-mail: harbimn@sabic.com

Abstract. The objective of this study is to optimize ladle cycle to enhance the lining service life. This optimization would be achieved by a better understanding of the lining thermal conditions during the varies stages of the ladle cycle by a mathematical simulation of the process. A 2D axisymmetric mathematical model was developed to predict the ladle lining and steel shell temperature during the full ladle cycle. The estimated steel shell temperature during drying and preheating process of a newly lined ladle was in good agreement with the literature and the plant expectation. This study recommend to make some changes in the current ladle cycle practice to reduce the abnormal lining temperature variation. Also recommend to enhance the preheating burners for sufficient lining drying/preheating. This then would improve the lining service life and reduce the refractory consumptions rate.

1. Introduction
The objective of this study is to optimize ladle cycle to enhance the lining service life. As first step towards optimisation of ladle treatment work, mathematical modelling of heat transfer in the ladle has been undertaken. A numerical model considering heat transfer is developed which can be used for prediction of the ladle lining temperature field during ladle cycle process. The model is based on first principle equations.

Refractory-lined ladles are used in steel mill for the sake of transporting molten steel from steelmaking furnace to a continuous casting machine. Ladle lining will absorb a significant amount of heat when it comes into contact with molten steel. Overall process efficiency is clearly affected by the temperature of ladle lining before molten steel is poured into the ladle. If lining temperature is far lower than that of molten steel following phenomena may occur: molten steel solidifies at lining surface, thermal shock appears inside the lining, and the molten steel temperature becomes insufficient for casting. Although the last problem can be surmounted by raising the degree of superheating of molten steel, this inevitably results in additional energy consumption. Preheating ladle lining prior to tapping is therefore necessary.

2. Mathematical model
A 2D axisymmetric mathematical model was developed to predict the ladle lining and steel shell temperature during operation. The newly lined ladle start with a drying followed by a

prolong preheating before it gets into the operation cycle. The ladle goes through the different operation conditions in sequence during its life cycle as sketch in figure 1.

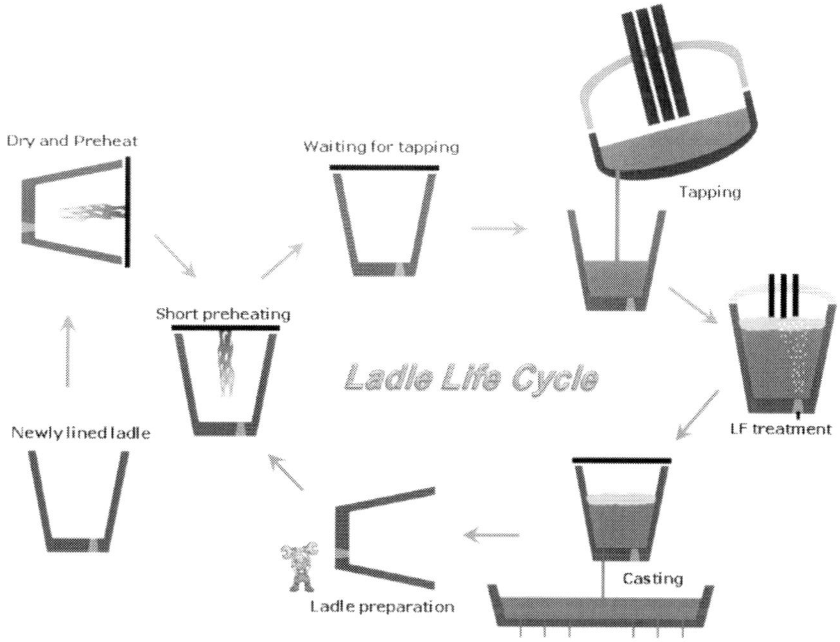

Figure 1. A sketch demonstrate the ladle life cycle.

2.1. Mathematical formulation

A 2D axisymmetric mathematical model of the ladle cycle was developed. The simulation was based on the following assumptions:

- The model start with a completely new lining at room temperature (30 °C).
- A transient solution approach was applied in this simulation.
- The simulation for each stage was done separately. However, the end lining temperature results at any stage was taken as an initial conditions for the following stage as per the ladle cycle sequence in figure 1.
- A central porous plug position was assumed to allow the use of 2D axisymmetric model.
- Tapping stage was ignored because it happens very fast. Therefore, after "waiting for tapping" stage the ladle was filled with the molten steel at 1600 °C to start the LF treatment stage.
- To avoid given an estimated constant heat loss rate from the ladle shell to atmosphere, the air space within 2 meter surrounding the ladle was counted in the model. This assumption was made where the air temperature above this distance from the ladle shell is almost at room temperature, however, the heat radiation was accounted in this model.

2.2. Transport equations

Based on the above assumptions, the following transport equations were solved.

- Mass conservation equation for each single phase and dissolved elements of interest.

$$\frac{\partial \rho}{\partial t} + \nabla \cdot (\rho \vec{\upsilon}) = S_m \qquad (1)$$

where $\vec{\upsilon}$ is velocity. Equation (1) is the general form of the mass conservation equation, where S_m is the source mass term.

- Momentum equation

$$\frac{\partial}{\partial t}(\rho \vec{\upsilon}) + \nabla \cdot (\rho \vec{\upsilon} \vec{\upsilon}) = -\nabla p + \nabla \cdot (\bar{\bar{\tau}}) + \rho \vec{g} + \vec{F} \tag{2}$$

where p is the static pressure, $\bar{\bar{\tau}}$ is the stress tensor described below, and $\rho \vec{g}$ and $\vec{F}$ are the gravitational and external body forces respectively.

$$\bar{\bar{\tau}} = \mu \left[\left(\nabla \vec{\upsilon} + \nabla \vec{\upsilon}^{T} \right) - \frac{2}{3} \nabla \cdot \vec{\upsilon} I \right] \tag{3}$$

where μ is the molecular viscosity, I is the unit tensor, and the second term on the right hand side is the effect of volume dilation.

- Energy conservation equation

$$\frac{\partial}{\partial t}(\rho E) + \nabla \cdot (\vec{\upsilon}(\rho E + p)) = \nabla \cdot (k_{eff} \nabla T - \sum_j h_j j_j + (\bar{\bar{\tau}}_{eff} \cdot \vec{\upsilon})) + S_h \tag{4}$$

where k_{eff} is the effective conductivity, h_j and j_j are the sensible enthalpy and the diffusion flux of species j respectively, and S_h is the volumetric heat source e.g. heat of chemical reaction.
The turbulent flow was solved using a ***k-ε*** model [1], the model constants $C_{1\varepsilon}$, $C_{2\varepsilon}$, $C_{3\varepsilon}$, σ_k , and σ_ε are shown in table (1). These constants have been estimated to fit a water/air experimental measurements, and it's used widely in the field of steelmaking process simulation [2,3].

Table-1. The ***K-ε*** turbulent model constants [1].

$C_{1\varepsilon}$	$C_{2\varepsilon}$	$C_{3\varepsilon}$	C_μ	σ_k	σ_ε
1.44	1.92	1.3	0.09	1.0	1.3

2.3. Solution approach
The ladle is going through a varies operational conditions during its cycle. Therefore, a different simulation technique is required for each stage. For example whenever there is a drying/preheating exercise the species module must be applied to predict the combustion process, also whenever the molten steel present the multiphase module must be applied to predict the flow behaviour. Therefore, the model was designed to give this capability to switch between varies conditions and mathematical modules using the same grid. This approach would limit the starting temperature assumption to the first drying stage, which was a room temperature. Thereafter, the final lining temperature results at any stage would be used as an initial temperature for the next stage and so on. So this approach clearly reduce the assumptions given by the user. Figure 2 demonstrate the model geometry with some features, for example the gas burner can be switch ON/OFF depending on the running stage.

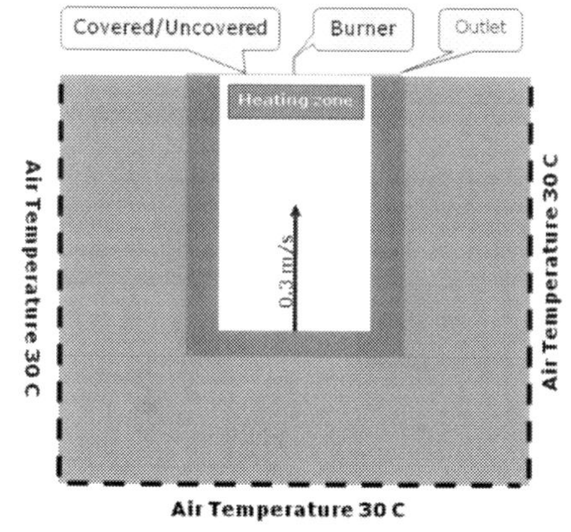

Figure 2. The model geometry parameters.

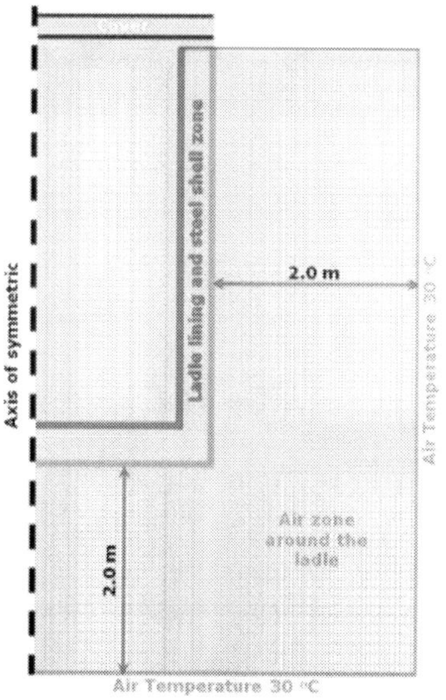

Figure 3. The generated model grid

2.3.1. Grid
During the ongoing progress of this project the grid sensitivity were examine using different grid size. The optimum grid which provide a good results accuracy with the less computational time is shown in figure 3. The number of cells in this grid is about 41,500 cells.

2.3.2. Gas burner
In this simulation the species model is turn "ON" to simulate the natural gas combustion to predict the generated heat, gas temperature and flow rate coming out of the burner. The gas temperature measured by thermocouple attached to the dryer/preheater roof was used to validate the predicted results and adjust the combustion model parameters.

2.3.3. Ladle cover
This cover can be removed according to the ladle cycle.

2.3.4. Heating zone:
This heating source was used to replace the arcing energy during the LF treatment. This heating source was given a fixed temperature based on several test run of the simulation model. This condition provides a good heating rate to the molten steel and results in a good numerical stability of the solution. .

2.3.5. Stirring:
The gas stirring porous plug was replaced by a given molten steel velocity at the location[4].

2.3.6. Drain hole
This outlet located on the bottom wall is set open only during casting stage.

3. Results and Discussion

Figure 4 illustrate ladle lining temperature after the completion of drying and preheating stages which runs in total for about 20 hours. During the ladle preheating operation the bottom wall expected to soak more heat because it takes the hottest gases impact in addition to flame radiations.

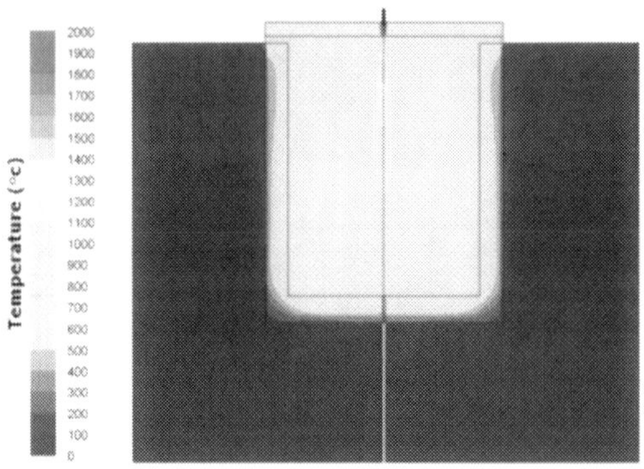

Figure 4. Temperature distribution at the end of preheating stage.

In this paper the model prediction results were presented at a pre-selected points and lines along the ladle lining as described in figure 5.

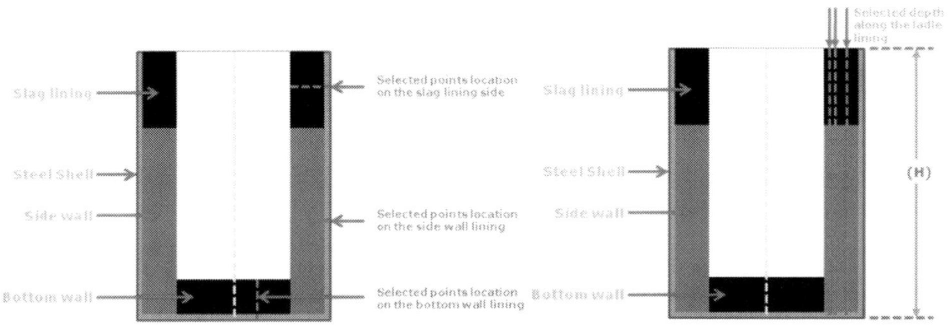

Figure 5. Selected points location (left) for temperature monitor during the drying and preheating stages. Selected lines location (right) along the ladle height for temperature monitor during the cycles stages (right)

Figure 6 suggest that the bottom side is receiving higher heat flow which results a higher wall temperature. Moreover, the steel shell temperature goes up to 300°C at the ladle bottom which is expected. The sharp fluctuating of the hot face temperature is due facing the very dynamic flames and gas movement. This fluctuating was vanish shortly through the lining wall which can be seen by the temperature results at 20mm through the wall.

Generally the ladle lining temperature was increased gradually over the heating stage where the inner ladle surface temperature reached up to 1000 °C at the bottom side and exceed 900 °C at the slag side which is in good agreement with the literature [5,6]. Also the steel shell temperature reached up to 300 °C at the bottom side and less than 250 °C from the side wall which is match with the plant expectation. Furthermore, it can be seen that the ladle lining tip (see figure 4) which exposed to the outflow gases is heated up more than the slag side, which is match with the plant expectation. This can be justified by that the gases first hit the bottom side and pass through the side wall which would consume its heating energy, therefore, by the time it reaches to the slag side wall it would be relatively cooled down. However, all the gases must leave through the small available space between the burner cover and the ladle tip. This would drive to increase the flow velocity at the outlet to satisfy the mass balance. Therefore, the heat transfer rate would be improved due to this increase in velocity, which caused the tip higher temperature discussed above.

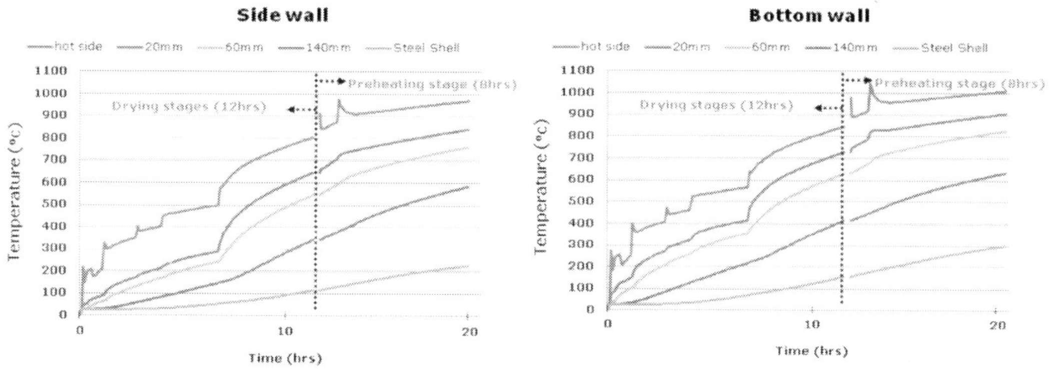

Figure 6. Temperature development during the drying and preheating operation at a selected points located on the side and bottom walls.

Once the ladle dried and preheated it gets into the ladle cycle given in figure 1. First the preheated ladle wait for tapping under cover, this waiting time might reduce the lining temperature. The lining and steel shell predicted temperature after waiting for 10 and 30 minutes respectively are illustrated in figure 7.

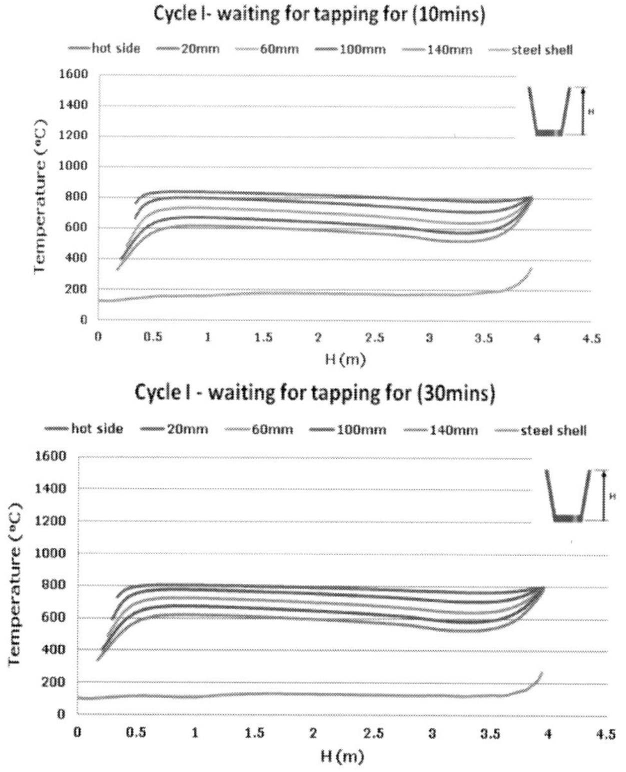

Figure 7. Predicted temperature along the ladle height after 10 & 30 minutes of waiting for tapping under cover.

It can be concluded that during the waiting time the covered ladle refractory lining manage to keep its heat perfectly. However, the drop of the steel shell temperature can be observed during the longer waiting time comparing the two figures while the refractory lining temperature remain almost the same. This can be justified due to the insulation layer attached to the steel shell which reduce refractory lining heat losses to the steel shell. In this simulation the results after waiting covered for 10 minutes were used as an initial lining temperature for LF treatment.

In LF treatment stage the lining hot side temperature lifted up dramatically once it gets into contact with the molten steel at the beginning of LF treatment is illustrated in figure 8, however, the heat generated by the molten steel propagate deeper inside lining by the end of LF treatment.

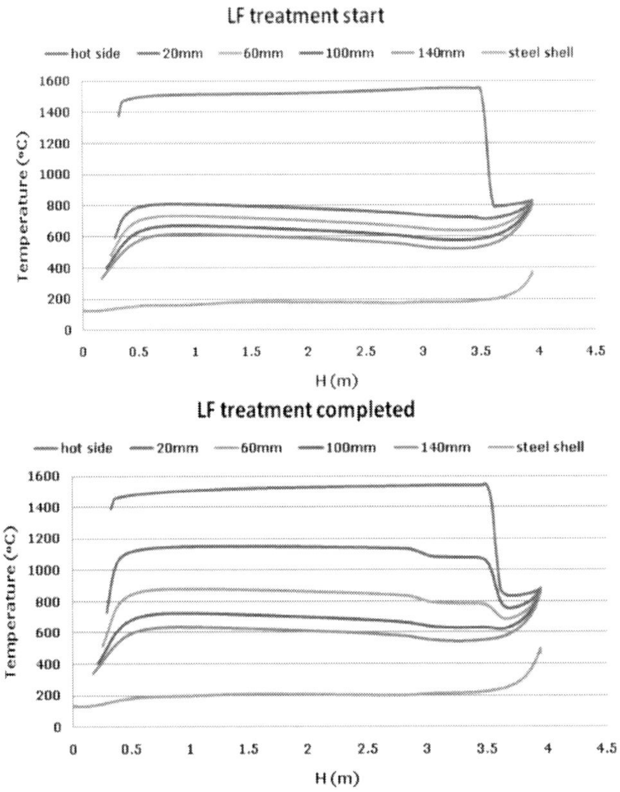

Figure 8. Predicted temperature along the ladle height at the beginning and the end of LF treatment.

It can be concluded from figure 8 above that the influence of the heat generated during the LF treatment is limited to the depth below 100mm through the lining. Although the steel shell was raised up to 200 °C comparing to it condition after waiting of 10 min prior tapping. This steel shell temperature is in good agreement with the plant expectation for a newly lining ladle.

In this simulation the full ladle was waiting for 10 minutes prior casting stage. The lining predicted temperature at the beginning and middle of casting stage are illustrated in figure 9.

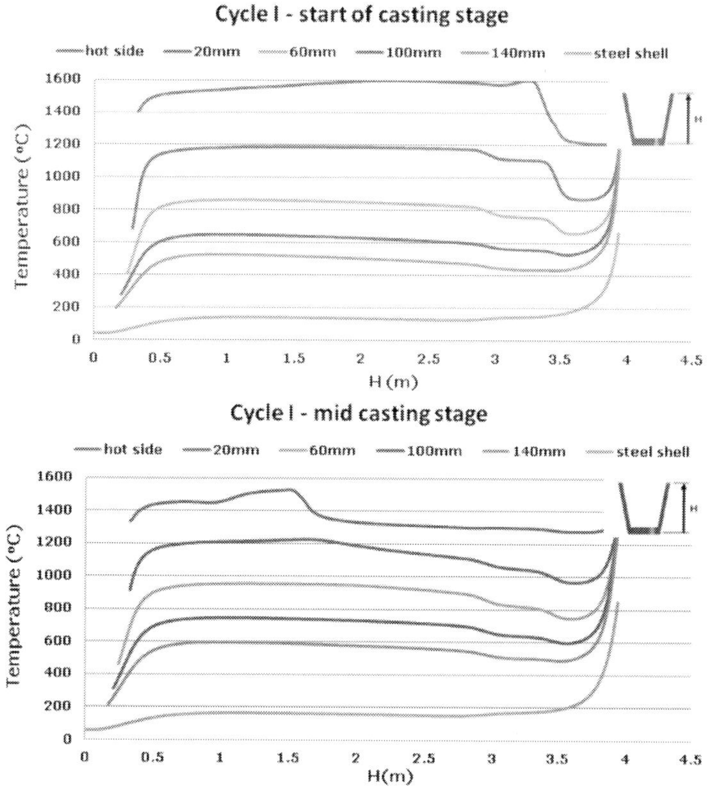

Figure 9. Predicted temperature along the ladle height at the beginning and mid of casting process.

During casting process the figure above shows that the temperature of the hot side released from molten metal contact drops below 1400 °C. On the other hand during this relatively long casting process time the lining temperature below the hot side gets normalized. During the ladle preparation period after the casting process the model didn't show any dramatically drop in temperature as illustrated in the comparison between figure 10 and 11.

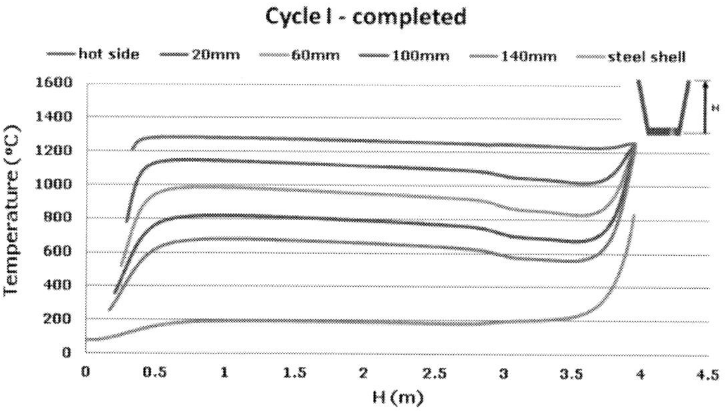

Figure 10. Predicted temperature along the ladle height at the end of cycle I.

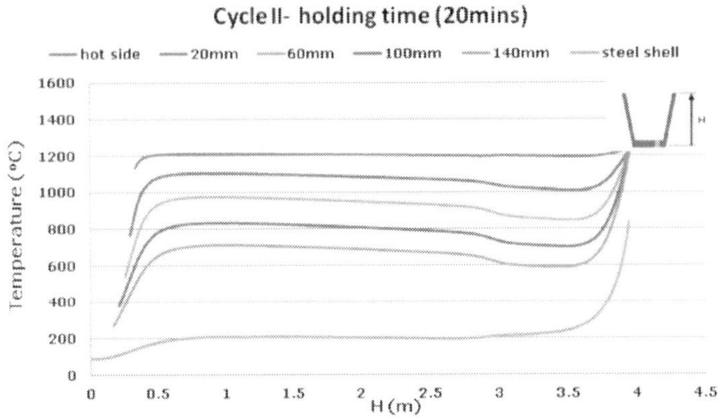

Figure 11. Predicted temperature along the ladle height at the end of preparation stage.

In the ladle cycle given in this paper ladle preparation stage is followed by a short preheating practice. Unfortunately in this model applying the combustion model using the ladle preparation stage lining temperature results as an initial conditions leads to unrealistic steel shell temperature drop, therefore, the short preheating stage was replaced by waiting under cover. Figure 10 illustrate the lining temperature after waiting under cover for 20 minutes.

4. Conclusion

- The estimated steel shell temperature during drying and preheating process of a newly lined ladle was in good agreement with the literature and plant estimation.
- The current drying and preheating procedure is capable to raise the lining hot side temperature above 1000°C as recommended by the refractory material supplier.
- The lining material is capable to store and maintain the massive heat received during LF treatment and casting process for long time as long the ladle kept under cover.
- During the ladle cycle most likely the first time the molten steel hit the newly lined ladle would give the biggest temperature raise during its service life. Therefore, the initial drying and preheating practice is of great influence on the lining service life.
- In this present work the capability of the model to predict the lining temperature history during the ladle cycle was demonstrated. This strength would be utilized to enhance the ladle lining service life and improve the cycle procedure.

References

[1] Launder B and Spalding D 1974 *Computer Meth. in Appl. Mech. and Eng.* **3** 269
[2] Castillejos E A, Salcudean M E, and Brimacombe J K 1989 *Meta. Trans. B* **20** 603
[3] Xia J L, Ahokainen T, and Holappa L 2001 *Scand. J. of Meta.* **30** 69
[4] Al-Harbi M N, Atkinson H V, and Gao S 2007 *AIST2007(CD).*
[5] CHEN J and WANG P 1992 *China Steel Technical Report* **6** 117
[6] MUCCIARDI F and GRANDILLO A 1987 Iron and Steel Engineer **12** 24

IOP Conf. Series: Materials Science and Engineering **33** (2012) 012023 doi:10.1088/1757-899X/33/1/012023

Modeling and Numerical Optimization of Withdrawal Rate in Directional Solidification Process

V Monastyrskiy
Head of CAE division, Metallurgical department,
FSUE GTERPC "SALUT", Budionny av.16, Moscow, 108115, RF

E-mail: saprlp@salut.ru

Abstract. Ni-based superalloys are currently used as materials for single-crystal (SC) or directionally solidified (DS) turbine blades produced in Bridgman or Liquid Metal Cooling (LMC) furnaces. Valid selection of the process parameters providing proper quality of the casting depends on casting geometry and furnace design. Process parameters should vary with the solidified cross section of the casting. Withdrawal rate is the most important parameter which affects solidification process more severely than the heater temperature. The technique providing for a valid withdrawal rate selection is based on the critical solidification rate calculation from experimental or numerical modeling data. A range of withdrawal rates providing for the required type and quality of the macrostructure is an output of the technique application. Full-scale and computing experiments were carried out to validate the optimization technique. The performance capability of the optimization technique is demonstrated in terms of the in situ composite solidification process. As a result of the solidification process, a real turbine blade was obtained at the desired temperature gradient and solidification rate supporting conditions of the planar solidification front.

1. Introduction

Directional solidification is widely used to produce castings with single crystal and columnar grains structure. In the process of directional solidification, a mould with molten metal overheated above the liquidus temperature is withdrawn from the high-temperature furnace to the cooling zone. Under these conditions, the temperature gradient in the melt and the solid phase growth rate depend upon the withdrawal rate and the casting geometry. The withdrawal rate and the heater temperature are the only parameters which can be controlled during directional solidification.

Influence of the heating and cooling zone temperatures on solidification conditions is a permanent topic of many investigations since the first experiments on directional solidification [1-3]. Inverse modeling of such process parameters as the thermal gradient in Bridgman and LMC processes is discussed in [4]. A recent review of the different methods of numerical optimization and inverse modeling is given by Ransing and Sood [5]. Optimization of the time dependent heater temperature control in Bridgman process is discussed in [6].

The principles of the valid withdrawal rate selection providing for the desired macro- and microstructure with maximum process output are discussed in the present paper.

2. Directional Solidification Technique and Optimization Criterion

In LMC (Liquid Metal Cooling) the mould with solidifying alloy is transferred from the area of the high-temperature heater to the cooling zone. To get a high value for the temperature gradient in the mushy (two-phase) zone of casting, the heating and cooling zones are separated by the baffle. A tin or aluminum melt bath is used as a cooling medium.

The optimization procedure was applied to the directional solidification of the VKLS-20 [7] eutectic superalloy. The stability of planar solidification front was used as a prerequisite quality criterion. Therefore, the composite structure observed in an in-situ cast specimen was an evidence of planar and stable solidification front.

The composite growth stability criterion can be written as:

$$G/W \geq (G\ W)_{crit} \tag{1}$$

where G is the temperature gradient in the mushy zone; W is the growth rate; $(G\ W)_{crit}$ is the critical value of the gradient-to-rate ratio, defining the boundary of stability of the planar solidification front for the given alloy.

Table 1. VKLS-20 composition (wt%)

C	Cr	Al	Mo	W	Co	Nb	V	Ni
0.4	4.3	6.2	0.75	12	9	4.2	0.7	bal

3. Temperature Gradient

In steady-state conditions of directional solidification in the constant cross-section casting, the temperature gradient G at the solidification front and maximum temperature gradient in the casting G_{max} can be defined from the analytical solution of the stationary heat model [3]. When the mushy zone is located far from the ends of the casting, those values of interest can be expressed as follows:

$$G = (k_h P\ \lambda_m S)(T_h - T_E) \tag{2}$$

$$G_{max} = \frac{k_c P\ \lambda_m S}{1 + \ k_c\ k_h}(T_h - T_c) \tag{3}$$

where λ_m is the thermal conductivity of the metal, k_h is the coefficient of heat transfer from the surface of the heater to the metal through the ceramic shell; k_c is the heat transfer coefficient in the cooling zone; P, S are the perimeter and cross-section area of the casting; T_h, T_c, T_E are the heater, cooler and eutectic temperature, respectively.

Depending on the position of the front towards the heating and cooling zones, a heat flow through the mould wall can be directed in or out of the casting. Accordingly, the solidification front can have planar, concave upward or downward shape.

The expressions (2),(3) are related to the process conditions when the solidification front is located near the baffle, while the heat flow through the mould walls near the solidification front is a negligible value, and the shape of the front has a small curvature.

For the temperatures T_h, T_c, T_E, presenting practical interest, the maximum point of the temperature gradient G_{max} never is in the mushy zone.

In order to increase the thermal gradient G at the growth front, one needs to increase the heater temperature and to intensify the heat transfer through the mould wall. An increase of the heater temperature must be compensated by the increase of the cooling intensity. It is necessary so as to keep dynamic equilibrium in the heater-casting-cooler system providing an optimal location of the mushy zone against the baffle.

Intensification of the heat transfer into the cooler results only in the increase of the maximum thermal gradient observed in the casting, but G remains unaltered due to the growth front position [3] change. As it follows from (2), in ideal case, there is no direct relation between the thermal gradient G and cooling medium conditions.

The expression (2) is a derived relation basing on the fundamental proposition which is valid not only for the casting with constant cross-section: the increase of the thermal gradient G necessarily requires the increase of the power of the heat source providing the heat flow through the solidification front.

In the steady-state approximation, which is valid at low solidification rates with planar front, the power of the heat source depends only on the heater temperature T_h. Non-stationary members of the heat balance equation making allowance for the contribution of the thermal capacity and latent heat liberation are negligible.

4. Withdrawal Rate Optimization Technique

In accordance with (1), to provide the solidification of an in-situ composite casting it is necessary to support a sufficiently high thermal gradient, while the solidification rate being reasonably low. Theoretically, a sufficiently low constant solidification rate satisfying condition (1) can be always assured. However, during directional solidification of the complex shape castings G and W values change according to the casting shape, furnace design and mould position even if the heater temperature and withdrawal rate remain constant. A constant withdrawal rate V which enables in-situ composite forming along the complex shaped casting, is usually much less than the average withdrawal rate which can be achieved at the optimal varying of the withdrawal rate during the solidification process. Low constant withdrawal rates result in forming of coarse microstructure, deterioration of mechanical properties and unreasonably low process yield.

The technique of the valid withdrawal rate selection is based on the assumptions that within the framework of the stationary heat model, the relationship between W and V can be expressed as

$$W(x_S)\ V(x_S) = \alpha(x_S)\,, \tag{4}$$

where $\alpha(x_S)$ is the response function depending on casting geometry and heat conditions of the solidification process in a given casting unit; x_S is the coordinate of the growth front.

The function $\alpha(x_S)$ is independent of V and, therefore, can be determined in the trial solidification process with any suitable $V(x_S)$, for example, with $V = const$. Using (1) and (4), it is not difficult to obtain an expression for the critical withdrawal rate V_{crit}:

$$V_{crit}(x_S) = \frac{A_{exp}(x_S)}{(G\ W)_{crit}}\, V_{exp}(x_S)\,, \tag{5}$$

where $A_{exp}(x_S)$ is the $G\ W$ ratio, measured from the trial solidification process as the function of growth front position, $V_{exp}(x_S)$ is the withdrawal rate in the experiment.

The critical rate V_{crit} is a maximum withdrawal rate at the solidification for the given cross-section of a casting under the stable planar growth front conditions.

The condition of the planar front stability rewritten via the parameters of the technological process is in the form of

$$V_{opt}(x_S) \le V_{crit}(x_S)\,, \tag{6}$$

The solidification time for the composite zone of required length is determined in the following way

$$\tau = \int_{x_{S0}}^{x_S} \frac{dx_S}{\alpha(x_S)V(x_S)}, \tag{7}$$

The equation (7) allows us to convert the chosen law of the mould replacement $V_{opt}(x_S)$ to the function $V_{opt}(\tau)$ which is more suitable for practical use.

Initial $A_{exp}(x_S)$ data can be determined from experiment or from numerical simulation.

5. Numerical Simulation

The figure 1 presents a 3D-model of the casting imitating a real turbine blade, and a 3D-model of the DS unit. The change of the velocity W_{exp} of equilibrium liquidus temperature isotherm at a constant withdrawal rate $V_{exp} = 12$ mm/h is shown in the figure 2. The data were computed with the help of the POLIGON [8] software package, for the superalloy VKLS-20 [7] under the following process conditions: $T_h = 1700^\circ C, T_c = 700^\circ C$. For better understanding how the presented data correspond with the geometry of the casting, the casting profile is shown in the same figure.

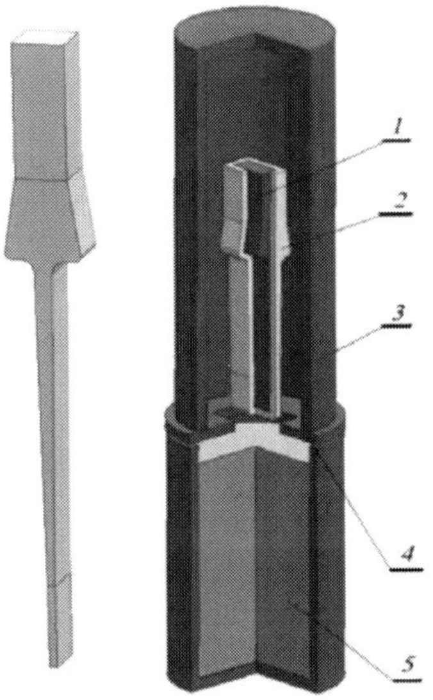

Figure 1. 3D-models of the turbine blade imitation and laboratory directional solidification unit.
1 – solidified superalloy; 2 – ceramic mold; 3 – heater; 4 – baffle; 5 – liquid metal cooling bath.

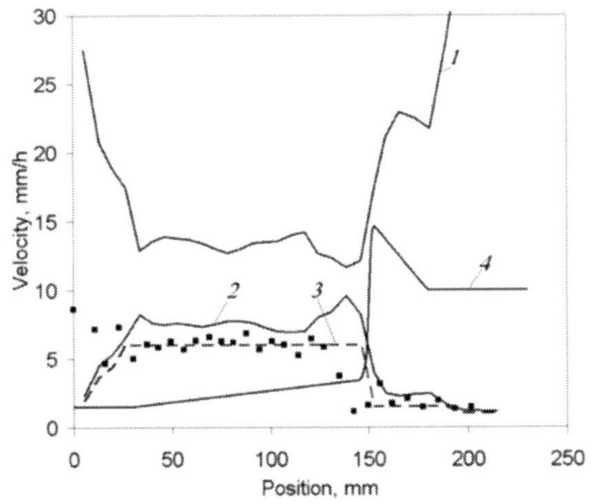

Figure 2. Profiles of computed parameters during solidification process with the constant withdrawal rate $V_{exp} = 12$ mm/h and with variable withdrawal rate $V_{opt}(x_S)$. 1 – W_{exp}; 2 – $V_{crit}(x_S)$; 3 – $V_{opt}(x_S)$; 4 – casting longitudinal section. W_{opt} is marked as solid squares.

The curve $V_{crit}(x_S)$ bounds a space of acceptable withdrawal rate. Any directional solidification process, presented by the curve $V(x_S)$ lying under the curve $V_{crit}(x_S)$, fulfils the conditions for the planar front stability (1). The solidification rate curve W_{opt}, calculated for one of the possible $V_{opt}(x_S)$ is shown in figure 2. The critical ratio G/W was assumed to be 150 Kh/sm^2.

IOP Conf. Series: Materials Science and Engineering **33** (2012) 012023 doi:10.1088/1757-899X/33/1/012023

The simulation results for the casting directionally solidified, with a constant withdrawal rate and with variable withdrawal rate $V_{opt}(x_S)$ from the figure 2 (see curve 3), are presented in the figure 3 where regions of $G/W < (G\,W)_{crit}$ are marked red. As it is shown in the figure, variable withdrawal rate fulfilling the condition (6) at the same time meets the condition (1) along the full length of the casting. The left-hand map shows the simulation result before optimization, the right-hand plot demonstrates the result after optimization. The red regions at the bottom and the top of the casting appeared due to inability to control the solidification front velocity at a given heater temperature. The dark region at the transition area from the thin to the thick part of the casting is connected with the jump of the cross-section area of the casting.

The described optimization technique operates with the values averaged over cross-section. At casting parts where the function of the geometry factor $1\,S\,dS\,dx$ is changing too fast, the local values of the thermal gradient and solidification rate can sharply differ from average values. $S(x)$ is the cross-section area function. In actual practice, the fast change of the geometry factor almost never can be compensated by process parameters varying.

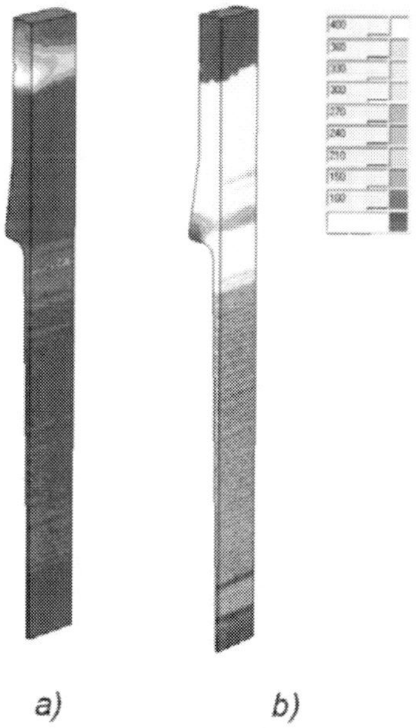

Figure 3. The G/W maps before (left) and after (right) optimization. The regions with G/W values which are lower then $(G\,W)_{crit}$ are marked red. The dimension of G/W is K·h/sm^2.

6. Experimental validation

To test the optimization technique, castings with smoothed transition areas from thin to thick part and a real turbine blade were solidified in an LMC unit. In the first stage of the experiment, the castings were solidified with thermocouples, uniformly positioned along the vertical axis of the casting. To measure the temperature, W-Re thermocouples (W-6 pct Re–W-20 pct Re) were used. The wires were encased in alumina ceramic core of rectangular cross-section. The ceramic core has 6 longitudinal holes, each of the holes was occupied by a thermocouple. An outline of the rigging used in the experiments is shown in the figure 4.

The thermocouple data recorded during the experiment, were used to calculate the thermal gradient G in the mushy zone and the solidification rate W. The method of calculation is described in [9]. Thermal gradient values measured for three castings solidified at different withdrawal rates, varying in the range of 4-8 mm/h, are shown in the figure 5. Calculated $G\,W$ curve presented in the figure 6 demonstrates the values below the threshold of the plane front stability for the thick part of the casting. While for the thin part of the castings, the withdrawal rate can be substantially larger 8 mm/h. Clearly, with the variable withdrawal rate plotted via the $V_{crit}(x_S)$ curve it is possible to achieve a significant increase in the unit output.

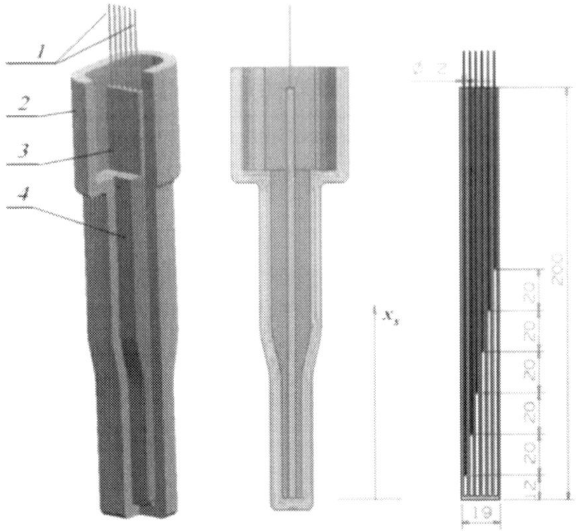

Figure 4. An outline of the rigging used in the experiments. *1* – thermocouples; *2*- ceramic mould; *3* – ceramic core; *4* – metal.

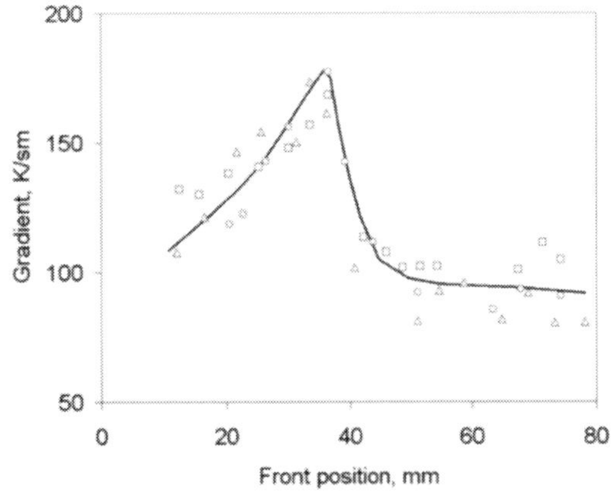

Figure 5. Thermal gradient at the isothermal surfaces of solidus measured during directional solidification with different withdrawal rates.

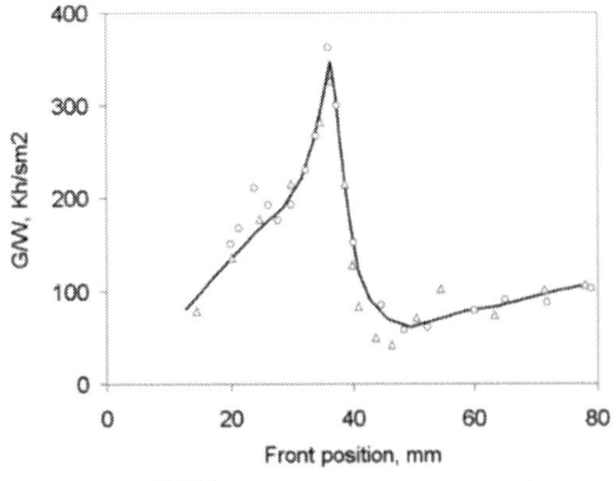

Figure 6. G/W calculated from experimental data for directional solidification with the withdrawal rate of 8 mm/h.

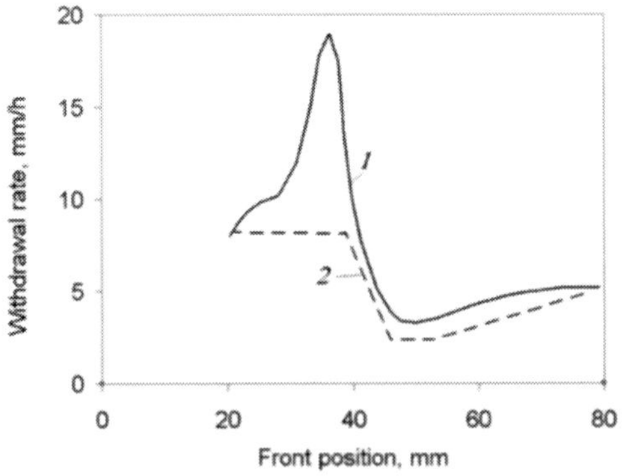

Figure 7. Calculated curve of the critical withdrawal rate (*1*) and withdrawal rate curve, chosen for experimental trial.

Critical withdrawal rate curve calculated from these experimental data, shows that the constant withdrawal rate fulfilling the condition (1), cannot exceed the value of 2.5 mm/h (figure 7). Variable withdrawal rate can be changed depending on the solidification front location, from 2.5 mm/h in the thick part of the casting up to 18 mm/h for the thin part.

The figure 7 shows variable withdrawal rate $V_{opt}(x_S)$ realized in the second stage of the experiment. Several samples were cast with $V_{opt}(x_S)$, and the microstructure of the samples was examined. Typical microstructure of the samples cast at the constant ($V = 8\,\text{mm/h}$) and variable withdrawal rates $V_{opt}(x_S)$ are shown in figure 8. The figure demonstrates in-situ composite structure $\gamma\ \gamma' - MeC$ in the part of the casting where $G\ W$ is large. Dendrites of the γ-phase denote a loss of growth front stability in the thick part at the uniform withdrawal rate 8 mm/h.

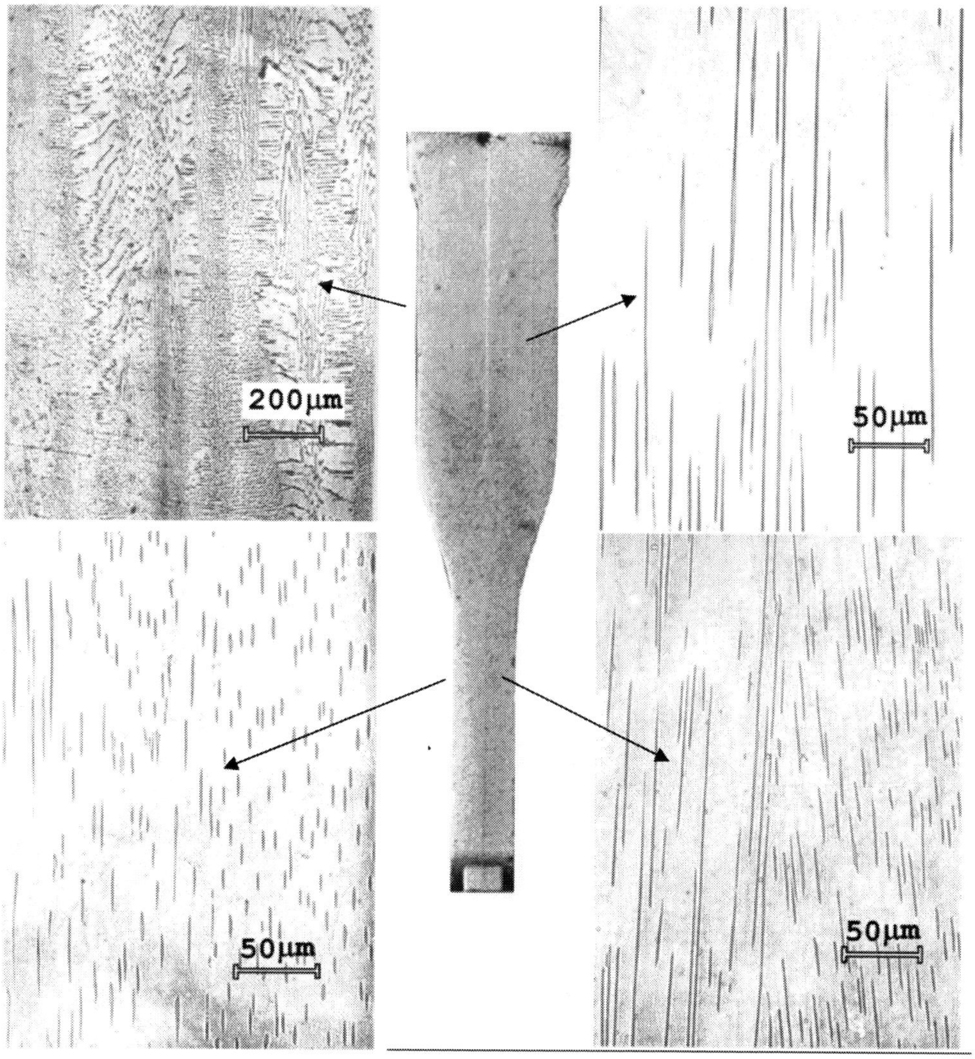

Figure 8. Microstructure of the casting before (left) and after (right) optimization.

7. Conclusions

A presented optimization technique determines the range of the withdrawal rate values meeting the specified quality casting criterion. The thermal gradient to solidification rate ratio is used in the present work as one of possible quality criteria. All features of the solidification process including casting geometry, cast unit design and cooling method (Bridgman or LMC) easily can be taken into account by the only one trial casting with thermocouples. In a difficult case of industrial directional solidification process simulation, software can be used to obtain the thermal conditions in the mushy zone and to calculate the critical withdrawal rate. The technique of the critical withdrawal rate calculation for cast samples and a real turbine blade were successfully executed in laboratory and industrial conditions.

References

[1] Giamei A F and Erickson J S 1976 *Superalloys 1976* ed. D R Muzyka, W H Couts et al (Warrendale, PA: TMS) pp 405-24

[2] An S U, Larionov V, Monastyrski V, Monastyrskaia E, Grafas I, Oh J M, Lim O D, Kim S H, Lee J H and Seo D Y 2000 *Superalloys 2000* ed. T M Pollock, R D Kissinger et al (Warrendale, PA: TMS) pp 247-55

[3] Monastyrskii V P 2004 Achievement of High Thermal Gradient at Directional Solidification of Superalloy Single Crystal Castings *Fizika i Khimiya Obrabotki Materialov (Physics and Chemistry of Materials Treatment)* **6** 77

[4] Monastyrskii V P and Logunov A V 1984 Inverse Heat Conduction Problem for Directional Solidification *Physics and Chemistry of Materials Treatment* **1** 28

[5] Ransing R S and Sood M P 2005 Review of Optimization Methods for Casting Simulation *Shape Casting: The John Campell Symposium* ed M Tiryakioglu and P N Crepeau (Warrendale, PA: TMS) pp 327-36

[6] Jakumeit J, Emmerich M and Hediger F 2006 *Proc. MCWASP-XI* ed C A Gandin and M Bellet (Warrendale, PA: TMS) pp 1019-26

[7] *Mashinostroenie. Enciklopedia (Machine-Building. Encyclopedia)* vol II-3 ed Fridlander I N et al, (Moscow: Mashinostroenie, 2001) pp 546-52

[8] POLIGON is a trademark of CSoft Moscow RF www.poligonsoft.ru.

[9] Monastyskii V P and Zeigarnik Y A 2002 Inverse Heat Conduction Problem for Analysis of Directional Solidification Experimental Data *Heat Transfer 2002: Proc. 12-th Int. Heat Transfer Conference (Grenoble, Fr, 18-23 August 2002*) (Elsevier SAS) pp 243-48

IOP Conf. Series: Materials Science and Engineering **33** (2012) 012024 doi:10.1088/1757-899X/33/1/012024

Thermal simulation of thermoelectric material by zone-melting technique

Yi-Ru Chen, Weng-Sing Hwang*, Huey-Lin Hsieh, Tsai-Kun Huang**, Jenn-Dong Hwang***, and Min-Hsiung Hung**

Department of Materials Science and Engineering, National Cheng Kung University, Tainan 70101, Taiwan

**New materials Research and Development Department, China Steel Corporation, Kaohsiung 81233, Taiwan

***Thermal Management Materials & device Lab, Division of Metallic Materials Research, Material & Chemical Research Laboratory/ITRI, Hsinchu 31040, Taiwan

E-mail: wshwang@mail.ncku.edu.tw

Abstract. The thermoelectric conversion efficiency of Bi-Te alloys is significantly affected by its microstructure. The columnar crystal structure can improve the thermoelectric conversion efficiency by raising the electrical conductivity σ and reduce the thermal conductivity κ, because of $ZT=\alpha^2\sigma T/\kappa$ [1]. Zone-melting process is a method can directionally solidify and purify the ingot by a moving heater along the solid ingot. In this study, we setup a zone-melting model and use computer simulation techniques to simulate the heat transfer field and microstructure of Bi-Te alloys. We compared the vertical temperature gradients and the horizontal temperature gradients with different simulation parameters. In view of the fact that operation of the process is not easy, simulation system has not yet been verified accurately. The simulation results will be verified by the experiment. Furthermore, the verified numerical model will be used to investigate the optimal processing parameters.

1. Introduction

In steel industries and other high temperature processes generate large amounts of waste heat. In response to the development of green energy, we use thermoelectric effect to transfer temperature differences to electric voltage for recovering waste heat. Materials under consideration for thermoelectric device applications include Bi-Te alloy, Pb-Te alloy, and Ge-Si alloy. Bismuth telluride based alloys have widely used near room temperature since its figure of merit ZT is near 1, where $ZT=\alpha^2\sigma T/\kappa$ is the characteristic performance of thermoelectric materials. The parameters of α, σ, T, and κ are Seebeck coefficient, electrical conductivity, Kelvin temperature, and thermal conductivity [1]. In presently researches, bismuth telluride based alloys were prepared by smelting and powder metallurgy. Smelting includes Brigdeman method, Czochralski method, and zone-melting method while powder metallurgy includes hot pressing and hot extrusion [2]. In this study, we use zone-melting method, i.e. directional solidification method, to prepare n-type and p-type Bi-Te alloys in order to obtain columnar structure. Growth of columnar structure can improve the magnitude of ZT by decreasing thermal conductivity and increasing electric conductivity.

We used tetrahedron mesh grid to establish the model by finite element method to simulate the temperature field of heat conduction and thermal radiation in zone-melting process. In addition, we coupled Cellular Automaton (CA) model with finite element method to become the CAFE model, which was required for the solidification process, nucleation and growth process. CAFE model can successfully predict the transition from columnar grain to equiaxed grain. And then we discussed the result of numerical simulation to get optimized parameters, such as casting size, process temperature, zone length, growth rate, and temperature gradient. In this process, it is necessary to be noted that horizontal temperature gradient must be low at the interface, whereas vertical temperature gradient must be high. In the other hand, controlling growth rate was important. It should be slow enough to let columnar structure grow. But there was considerable macro-segregation in composition along ingot because growth rate was slow (0.1mm/min). So growth rate is one of the points we discussed.

In order to get a similar columnar microstructure, process parameters must be strictly controlled. Let the axial temperature gradient is high, the radial temperature gradient is low, and reduce the segregation generated. Speed and temperatures of heater and cooler will affect the length of molten zone and vertical temperature gradient of the solid-liquid interface. By increasing the heat input, a higher value of speed can be adopted, because the interfacial temperature gradient is also increased [3]. Having faster cooling rate can suppress Se segregation, which often formed along the grain boundary of n-type alloy [4]. However, moving too fast will not have enough time to grow into big grain of columnar crystal. By slowing down the speed, the Seebeck coefficient can increase, the total thermal conductivity and lattice thermal conductivity can decrease, and then ZT increase [5]. If the speed is too slow (0.6cm/hr) would cause the segregation of Te [6]. For p-type alloy, a homogeneous Bi-Sb alloy was treated for two minutes just above its melting temperature, causing segregation of Bi-rich phase at the grain boundaries, thus forming local inhomogeneities [7]. From the literature we can see the temperature gradient and the speed ratio (G / V) will affect the growth of crystal. For Bi_2Te_3 crystal, the crystal characteristic faces of (0 1 5), (1 0 10), and (1 1 0) are clearly developed in the solidified ingot. Radial temperature gradient determines the shape of the solid-liquid interface. The major factor which governs the shape of interface is the ratio of the heat lost by conduction from the interface along the solid, to that lost laterally through the sides of the quartz. The rate of growth of the crystal depends on the rates of loss of the latent heat of solidification from the interface. The interface convexity and deformation decrease with the increasing heating power and zone length. Previous studies were more likely to adjust the temperature gradient 25 ℃ / cm at the interface. Molten zone cannot be too long or too short. Especially, as the molten zone increase, zone necking near the upper interface increase [8]. With a higher superheating, the axisymmetric solution is not stable under normal gravity. To operate a stable axisymmetric mode, a lower superheating and a shorter zone are necessary [9].

2. Experimental procedure

High purity (99.99%) Bi, Te, Se, and Sb granules were washed with 10% nitric acid, acetone, and distilled water to remove surface oxide layers. Appropriate amounts of these elements were weighted to make n-type composition $Bi_2Te_{2.7}Se_{0.3}$ doped with 0.05wt% BiI_3 and p-type composition $Bi_{0.5}Sb_{1.5}Te_3$ containing 3wt% excess Te. The inside wall of the quartz tube was carbon-coasted by the acetone cracking. Bi, Te, Se, and Sb in vacuumed quartz tube were melted in rocking furnace to ensure composition homogeneity. The alloys were grown by zone-melting method with heating temperature 720℃ and kept this temperature for an hour at the bottom of quartz. After the bottom of the alloys melted, the quartz started to move up with growth rates ranging from 1.0 to 1.5 cm/hr. For comparing the differences of temperature fields in different sample sizes, we used 17ø, 24ø, and 28ø zone-melting equipment, showed as Figure 1.

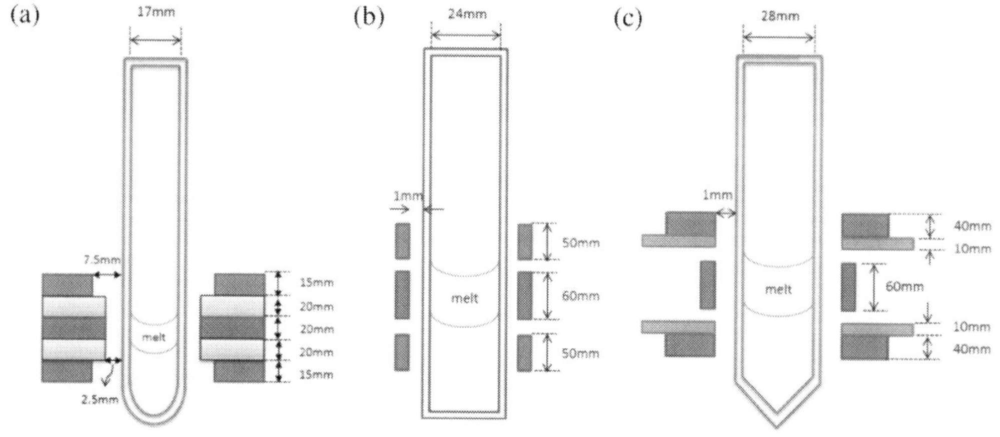

Figure 1. (a) 17ø, (b) 24ø, and (c) 28ø zone-melting equipment.

3. Mathematical formulation

3.1. Physical model and basic assumptions

Schematic diagram adopted is shown in Figure 1. Thermoelectric materials melt with 17mm, 24mm, and 28mm diameters was encapsulated by quartz. The heater and ambient temperature of each segment are given. For Figure 1(a), the length of the heater was 20mm. Heater was sandwiched in between two fire-resistant materials, which were both 20mm long. Outside of the fire-resistant materials were coolers. The interval between quartz tube and heater was 2.5mm, and the interval between quartz tube and coolers was 7.5mm. For Figure 1(b), the lengths of heater and coolers were 60mm and 50mm, respectively. The interstices between quartz tube and both heater and coolers was 1mm. For Figure 1(c), there were two coolers, 40mm, supported by 10mm thick aluminum plates. The heater was 60mm long.

To simplify the problem, the following assumption were introduced：

- The system was in a pseudo-steady state and is axisymmetric.
- The melt/quartz interface was non-deformable.
- Thermo-physical properties of the melt were assumed to be constant.
- Gravity was constant.

3.2. Mathematical model [9,10,11,12]

With respect to calculating the temperature distribution during solidification, this study adopted Fourier's fundamental laws of heat transfer and extended it to a three-dimensional model. The resulting governing equation is given by Eq. 1:

- $$\rho C_p \frac{\partial T}{\partial t} = \frac{\partial}{\partial x}\left[\lambda \frac{\partial T}{\partial x}\right] + \frac{\partial}{\partial y}\left[\lambda \frac{\partial T}{\partial y}\right] + \frac{\partial}{\partial z}\left[\lambda \frac{\partial T}{\partial z}\right] + \rho \Delta H \frac{\partial f_s}{\partial t} \tag{1}$$

where ρ is the density, λ is the thermal conductivity, C_p is the heat capacity under constant pressure, H is the latent heat and f_s is the solid fraction of the material used.

Heat transfer from the system to the ambient is by both radiation and convection according to the energy balance along the quartz surface:

- $$-k_a(\boldsymbol{n} \cdot \nabla T) = h(T - T_a) + \varepsilon_a \sigma (T^4 - T_a{}^4) \tag{2}$$

where h is the heat transfer coefficient, ε_a the thermal emissivity of the quartz, σ is the Stefan-Boltzmann constant, and T_a is the effective ambient temperature.

There was heat exchange between the surrounding (furnace) and the quartz, dictated by the effective ambient temperature $T_a(z,t)$ specified along the zone length. For simplicity, the effective temperature is assumed to be a Gaussian distribution. The effective ambient temperature distribution due to the furnace is given as follow:

- $$T_a(z,t) = (T_p - T_{a\infty}) exp\left\{-\left[\frac{(z-z_p)}{a}\right]^2\right\} + T_{a\infty} \tag{3}$$

where T_p and $T_{a\infty}$ are the peak and background temperatures, respectively. The parameter a is related to the width of the distribution, and z_p is the position of T_p, shown as Figure 2.

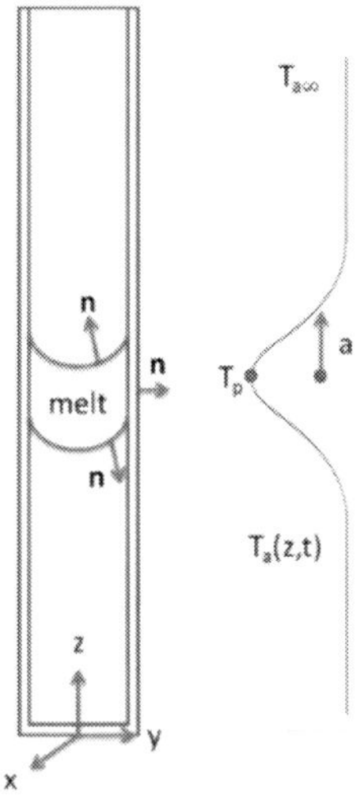

Figure 2. Effective ambient temperature distribution for computation.

Numerical simulation models of microstructure can be widely classified into deterministic and probability methods. Phase transformation dynamics of deterministic model is based on the formulas of solution to deal with some of the volume element. Length can be used to represent the characteristics of microstructure such as crystal size or phase space. Probability model guides elements to simulate randomly. Rappaz combined the advantages of deterministic and probability models to form cellular automaton (CA) method. CA method is based on heterogeneous nucleation and a continuous nucleation model which explains the relationship between nucleation density and undercooling, showed as Figure 3.

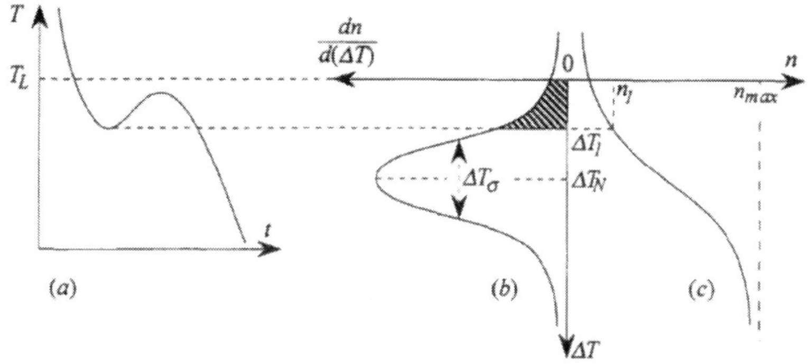

Figure 3. The relationship between nucleation density and undercooling.

The heterogeneous nucleation on the surface of the casting and in the bulk liquid were described by two different Gaussian distribution curves, showed as Figure 4.

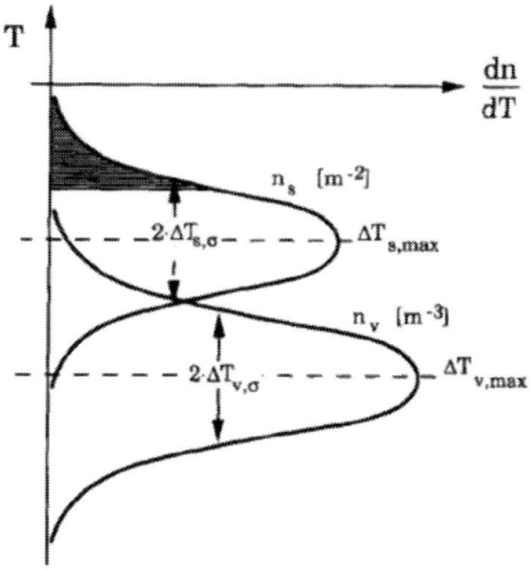

Figure 4. The heterogeneous nucleation on the surface of the casting and in the bulk liquid.

The relationship between nucleation density and undercooling is given by Eq. 4 and Eq. 5:

- $n(\Delta T) = \int_0^{\Delta T} \frac{dn}{d(\Delta T')}\left[1 - f_s(\Delta T')\right]d(\Delta T')$ (4)
- $\frac{dn}{d(\Delta T')} = \frac{n_{max}}{\sqrt{2\pi}(\Delta T_\sigma)} exp\left[-\frac{1}{2}\left(\frac{\Delta T' - \Delta T_n}{\Delta T_\sigma}\right)^2\right]$ (5)

where $n(\Delta T)$ is density of grain, $f_s(\Delta T')$ is solid fraction, n_{max} is the maximum density of nuclei, ΔT_n means nucleation undercooling , and $\frac{dn}{d(\Delta T')}$ is found according to the Gaussian distribution.

When the undercooling reaches nucleation conditions, specific CA cells start to nucleate. In a time step δt, temperature of a component reduces δT, then the undercooling increases $\delta(\Delta T)$. At this time, new grain density of nucleation on the liquid metal surface showed as below:

- $\delta n_s = n_s[\Delta T - \delta(\Delta T)] - n_s(\Delta T) = \int_{\Delta T}^{\Delta T+\delta(\Delta T)} \frac{dn_s}{d(\Delta T')} d(\Delta T')$ (6)

The Kurz-Giovanola-Trivedi(KGT) model is applied to compute the growth rate of grains, the growth rate can be expressed as Eq.7:

- $\upsilon = a_2 \Delta T^2 + a_3 \Delta T^3$ (7)

where a_2 and a_3 are growth kinetic parameters dependent upon the composition of the alloy.

4. Results and discussion

In this research, let columnar structure grow with ingot sizes, growth rates and heating time controlled. The parameters in the simulation system are listed in Table 1.

Table 1. Physical properties and some input parameters.[12]

	Emissivity	Temperature	Conductivity	Density	Specific heat	Liquidus and solidus
quartz	0.93	30℃	---	---	---	---
heater	0.80	850℃	---	---	---	---
cooler	0.56	30℃	---	---	---	---
ingot	---	---	1.4W/m•K	783Kg/ m^3	194.4KJ/Kg•K	568℃-530℃

4.1. Effects of ingot size

We used 17ø, 24ø, and 28ø zone-melting equipment to compare the differences of temperature fields. Temperature gradients at the interface of solidification were showed in Table 2.

Table 2. Varied temperature gradients with different casting sizes. (Unit: ℃/mm)

	17 ø	24 ø	28 ø
Vertical temperature gradient	8.0	5.3	4.6
Horizontal temperature gradient	1.6	2.3	2.3

Vertical temperature gradient got smaller as the size become larger, while horizontal temperature gradient was raised slightly, as Figure 5. Neither of these two changes was conducive to the growth of columnar crystals. Therefore, in order to increase production by increased the size of ingot, other process parameters should be changed for enhancing the vertical temperature gradient, like growth rate and heating time.

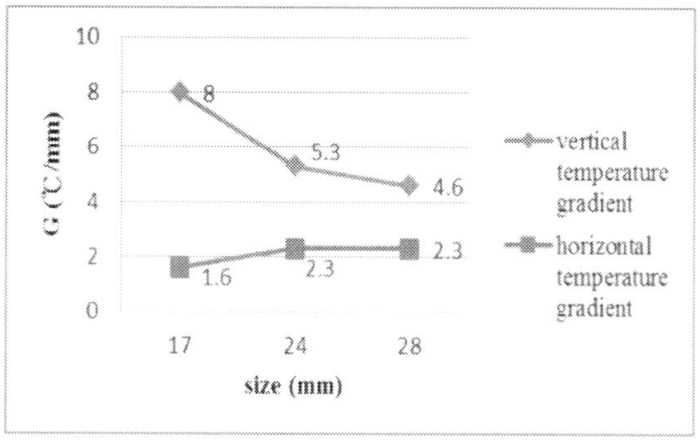

Figure 5. Varied temperature gradients with different casting sizes.

4.2. Effects of growth rate for temperature gradients

There was Table 3 showed below, describing vertical temperature gradients and horizontal temperature gradients in slow growth rate (1cm/hr) and high growth rate (1.5cm/hr).

Table 3. Varied temperature gradients with different casting sizes and growth rates. (Unit: ℃/mm)

	Growth rate	1cm/hr	1.5cm/hr
17 ø	Vertical temperature gradient	8.0	2.6
	Horizontal temperature gradient	1.6	1.5
24 ø	Vertical temperature gradient	5.3	2.9
	Horizontal temperature gradient	2.3	1.8
28 ø	Vertical temperature gradient	4.6	3.4
	Horizontal temperature gradient	2.3	2.1

Discussed separately, varied vertical temperature gradients with different casting sizes and growth rates compared in Figure 6, and varied horizontal temperature gradients with different casting sizes and growth rates compared in Figure 7. It can be seen from Figure 6 that vertical temperature gradients became smaller when growth rate was higher, especially for 17 ø, dropped from 8.0℃/mm to 2.6℃/mm. It was a bad effect for the growth of columnar crystal, so slow growth rate should be taken. However, the change of horizontal temperature gradient was insignificant.

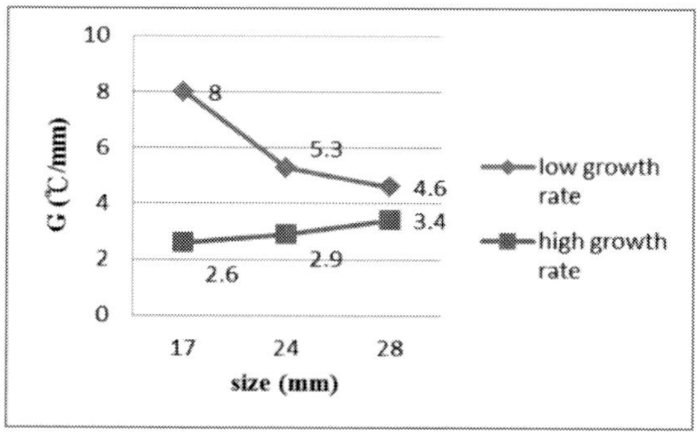

Figure 6. Varied vertical temperature gradients with different casting sizes and growth rates.

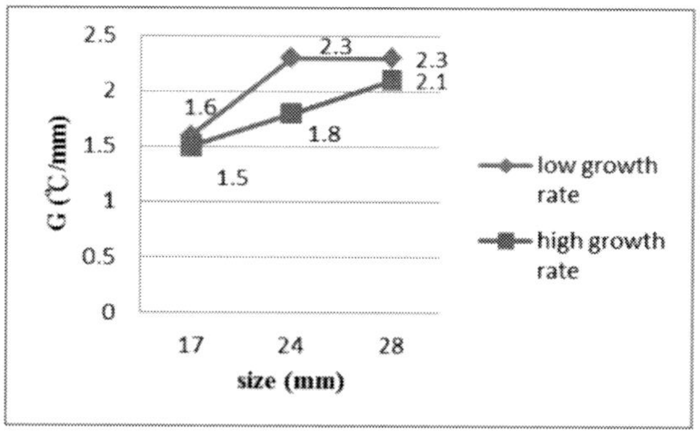

Figure 7. Varied horizontal temperature gradients with different casting sizes and growth rates.

4.3. Effects of growth rate for molten zone length

In addition to affecting the temperature gradient, growth rate impacted the length and stability of the molten zone significantly, showed as Table 4. Because of the different specification between 17ø, 24ø, and 28ø zone-melting equipment, molten zone length could not be compared with different ingot size directly. But it can be observed that zone length decreased at high growth rate, and it became unstable and asymmetric. In simulation system, the decreases of zone length (△L) were 7mm and 11mm, i.e. about 10~16% decreasing. In a previous paper, asymmetry of axial heat loss resulted in vertical heterogeneous. Bi_2Se_3 generated in the middle more than both ends of ingot [13].

Table 4. Varied zone length with different casting sizes and growth rates. (Unit: mm)

	Growth rate: 1cm/hr	Growth rate: 1.5cm/hr	△L
17 ø	60	49	11
24 ø	64	57	7
28 ø	62	51	11

4.4. Effects of heating time

Before heater and cooler started to move, the bottom of ingot was heated for one or two hours. The result of simulation in different heating time is displayed in Table 5. Heating in two hours made vertical temperature gradient increase for 17ø ingot. It was good for growing columnar structure. But in the other hand, long heating time made vertical temperature gradient decrease for large size ingot, 24ø and 28ø. And the change of horizontal temperature gradient was insignificant.

Table 5. Varied temperature gradients with different heating times. (Unit: ℃/mm).

	Heating time	1hr	2hr
17 ø	Vertical temperature gradient	7.2	8.0
	Horizontal temperature gradient	1.6	1.6
24 ø	Vertical temperature gradient	5.6	5.3
	Horizontal temperature gradient	2.5	2.3
28 ø	Vertical temperature gradient	5.7	4.6
	Horizontal temperature gradient	2.0	2.3

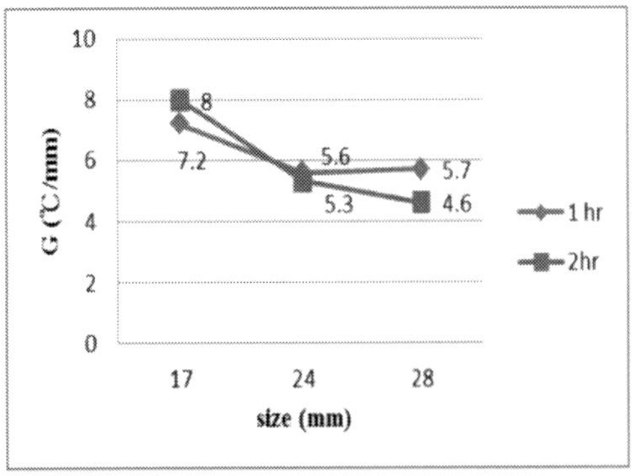

Figure 8. Varied vertical temperature gradients with different heating times.

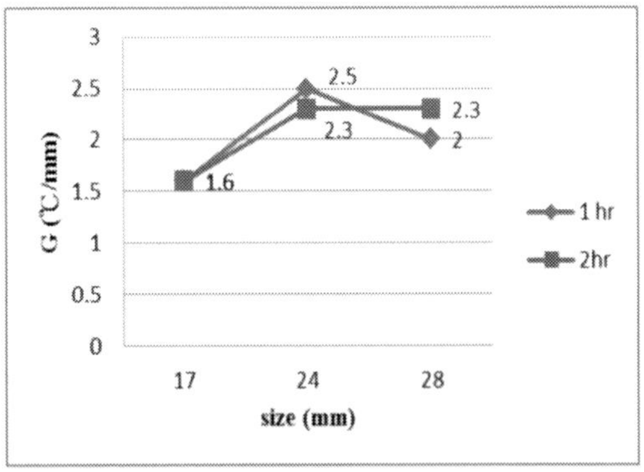

Figure 9. Varied horizontal temperature gradients with different heating times.

4.5. Microstructure
The simulated microstructure morphology of longitudinal section and cross section of 17ø ingot are showed as figure 10(a) and 11(a). We compared simulation result with experiment result in figure 10(b) and 11(b) roughly. It can be regarded as similarity between simulation and experiment. There is a certain degree of reliability of this simulation system.

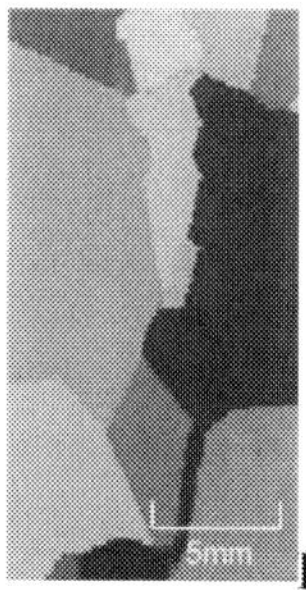

Figure10(a)

Figure10(b)

Figure 10. The (a) simulation result and the (b) experiment result of longitudinal section of 17ø ingot.

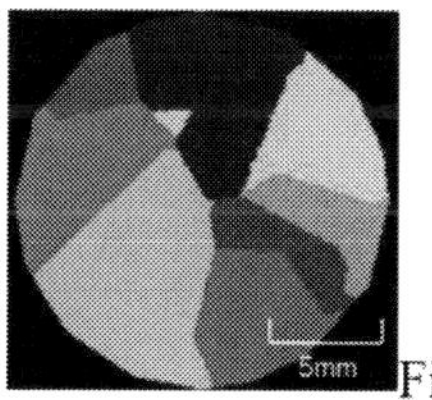

Figure11(a)

Figure11(b)

Figure 11. The (a) simulation result and the (b) experiment result of cross section of 17ø ingot.

5. Conclusion
Vertical temperature gradient got smaller as the size become larger, while horizontal temperature gradient was raised slightly. When growth rate was higher, vertical temperature gradients dropped, especially for 17 ø, from 8.0℃/mm to 2.6℃/mm. Molten zone length decreased at high growth rate, and it became unstable and asymmetric. Long heating time made vertical temperature gradient decrease for large size ingot, 24ø and 28ø, and the change of horizontal temperature gradient was insignificant. To sum up, we can get better columnar crystal with slow growth rate (1cm/hr) and short heating time for large ingot.

Acknowledgment
This study is supported by China Steel Corporation, ITRI and the Bureau of Energy, Ministry of Economic Affairs (101-D0204-2) in Taiwan, which is gratefully acknowledged. A part of the present work was also supported by the Research Center for Energy Technology and Strategy, National Cheng Kung University in Taiwan.

References
[1] Feng S K, Li S M, Luo Q Y and Fu H Z 2011 *Advanced Materials Research* **197-198** 1109-1112
[2] Ainsworth L 1956 *Proceedings of the Physical Society* B **69** 606-612

[3] Spim J A, Bernadou M J S and Garcia A 2000 *J. Alloys and Compounds* **298** 299-305
[4] Kim T S and Chun B S 2007 *J. Alloys and Compounds* **437** 225-230
[5] Yim W M and Fitzke E V 1968 *Solid State Science* 556-560.
[6] Heon P H, Young J O, Dow B H and Eui P Y 2002 *Original*
[7] Ceresara S, Codecasa M and Passaretti F 2011 *J. Electric Materials* **40** 557-560
[8] Lan C W 2003 *J. Crystal Growth* **208** 597-612
[9] Lan C W and Liang M C 2000 *J. Crystal Growth* **208** 327-340
[10] Lan C W and Yang D T 1998 *International J. Heat and Mass Transfer* **41** 4351-4373
[11] Li M W, Hu W R, and Chen S X 2004 *International J. Heat and Mass Transfer* **47** 2941-2947
[12] Perezgrande I, Rivas D and Depablo V 2002 *J.Cryst. Growth* **246** 37-54
[13] Lahalle C 1996 *J. Phys. and Chem. of Solids* **57** 1713-1717

IOP Conf. Series: Materials Science and Engineering **33** (2012) 012025 doi:10.1088/1757-899X/33/1/012025

3D finite element simulation of TIG weld pool

X Kong[1,2,3], O Asserin[1], S Gounand[1], P Gilles[2], JM Bergheau[3] and M Medale[4]
[1]CEA, DEN, DANS, DM2S, F-91191 Gif-sur-Yvette, France
[2]AREVA NP, Paris La Défense, France
[3]LTDS, Ecole Nationale d'Ingénieurs de Saint-Etienne, France
[4]Ecole Polytechnique Universitaire de Marseille, Marseille, France
E-mail: xiaofei.kong@enise.fr, olivier.asserin@cea.fr

Abstract. The aim of this paper is to propose a three-dimensional weld pool model for the moving gas tungsten arc welding (GTAW) process, in order to understand the main factors that limit the weld quality and improve the productivity, especially with respect to the welding speed. Simulation is a very powerful tool to help in understanding the physical phenomena in the weld process. A 3D finite element model of heat and fluid flow in weld pool considering free surface of the pool and traveling speed has been developed for the GTAW process. Cast3M software is used to compute all the governing equations. The free surface of the weld pool is calculated by minimizing the total surface energy. The combined effects of surface tension gradient, buoyancy force, arc pressure, arc drag force to drive the fluid flow is included in our model. The deformation of the weld pool surface and the welding speed affect fluid flow, heat flow and thus temperature gradients and molten pool dimensions. Welding trials study is presented to compare our numerical results with macrograph of the molten pool.

1. Introduction

Gas tungsten arc welding (GTAW) is a widely used technique for joining steel in nuclear industry where the weld quality must be ensured. An increased productivity of welding would have a very important economic impact for the industrial sector. Although GTAW is already a very efficient process, it can be improved, especially with respect to the welding speed. The aim of this paper is to propose a three-dimensional weld pool model for the moving GTAW process, in order to understand the main factors that limit the weld process and improve the productivity. Simulation is very helpful in understanding the physical phenomena in the weld pool and optimizing the process. In the recent years, many research dealing with the simulation of welding processes have been carried out and great progress has been made. Tanaka [1] took into account tungsten cathode, arc plasma and anode in a 2D axisymmetric unified model, but the deformation of the free surface of weld pool isn't taken into account. Debroy and al. [2] developed a 3D model with a moving heat source but the Marangoni coefficient is assumed constant. Lu and al. [3] developed a coupled model in 2D axisymmetric with the interface deformation, and the weld pool shape calculated agrees better with the experiment than the model without the free surface. Wu and al. [4] studied both front and back side deformations of fully penetrated GTAW weld pool surfaces, they gave a detailed derivation of the partial differential equation for predicting both surface deformation. Kim and Na [5] presented a 3D quasisteady heat and fluid flow analysis for the moving heat source of the gas metal arc welding

(GMAW) process with the free surface, and Hu and al. [6] modelled the transient weld pool dynamics under the periodic impingement of filler droplets. However, there is no impingement of droplets in GTAW which is very different from GMAW. Most of the present mathematical model are realised with the method of finite volume or finite difference, nowadays there is a increasing tendency to apply finite element methods. Among the reasons can be ability of the finite element method to handle complex coupled thermomechanical mechanisms with various and complex boundary conditions.

In this paper, a 3D finite element model considering free surface, moving source and temperature dependence of thermophysical properties for the GTAW is performed. The free surface profile is calculated by minimizing the total energy of the surface, the arc pressure, the arc drag force and the heat flux are given by a 2D axisymmetric arc plasma unified model [7]. The continuity equation, momentum equation and energy conservation equation are solved together by the Cast3M software [8], until a steady solution is found. The calculated GTAW weld bead geometry shows favourable agreement with experimental results.

2. Mathematical model

The 3D model of the welding pool with taking into account the traveling speed and the free surface is shown below. The assumptions are the following:

- a stationary solution is sought;
- the liquid metal is considered a Newtonian fluid;
- it is assumed that the flow of liquid metal is not turbulent;
- it is assumed that the Boussinesq approximation can be used;
- the speed of the power source v_s is assumed constant and the equations are solved in a reference frame moving with the source;
- in the fluid phase, the equations of conservation of mass, momentum and energy are solved, only the energy equation remains in the solid phase.

The unknowns of the problem are:
- $\mathbf{v}$ velocity in the fluid phase in the reference frame linked to the source ($\mathbf{v} = \mathbf{v_s}$ in the solid phase);
- p, the pressure in the fluid phase;
- h and T, the enthalpy and the temperature.

2.1. Continuity equation

In the weld pool region, the continuity equation for an incompressible fluid holds:

$$\nabla.\mathbf{v} = 0 \tag{1}$$

where v is the velocity of the fluid in the weld pool.

2.2. Momentum equation

The flow of the liquid metal in the weld pool conforms to the following equations:

$$\rho(\nabla\mathbf{v}).(\mathbf{v} - \mathbf{v_s}) = -\nabla p + \nabla.\mu(\nabla\mathbf{v} + \nabla^T\mathbf{v}) + \mathbf{f_{Bou}} + \mathbf{f_{Ext}} \tag{2}$$

$$\rho = \rho_0 + \frac{\partial\rho}{\partial T}(T - T_{ref}) \tag{3}$$

where $\mathbf{v_s}$ is the velocity of the power source, p is the hydrostatic pressure, μ is the viscosity, the buoyancy force: $\mathbf{f_{Bou}} = \rho_0\mathbf{g}\beta(T - T_{ref})\mathbf{u_z}$ where ρ_0 is the mass density at the temperature

$T_{ref} = 1730K$ and u_z is a vertical unit vector, the force of extinction speed: $\mathbf{f_{Ext}} = -AF_s(T)\mathbf{v}$ avec $A = 10^{12} >> 1$ where F_s is the solid fraction [9].
The boundary conditions are :

$$\mathbf{v} = \mathbf{v_s} \ at \ \partial\Omega_{s/l} \tag{4}$$

$$\mathbf{v}.\mathbf{n} = 0 \ at \ \partial\Omega_f \tag{5}$$

$$(\mu(\nabla\mathbf{v} + \nabla^T\mathbf{v}).\mathbf{n}).\mathbf{t} = \mathbf{f_{Marangoni}}.\mathbf{t} + \mathbf{f_{drag}}.\mathbf{t} \ at \ \partial\Omega_f \tag{6}$$

where $\partial\Omega_{s/l}$ represent the liquid/solid interface and $\partial\Omega_f$ is the free surface, $\mathbf{n}$ and $\mathbf{t}$ are the normal and the tangential direction to surface, $\mathbf{f_{drag}}$ is the arc drag force which will be given in the next subsection as a kind of surface source, $\mathbf{f_{Marangoni}} = \frac{\partial\gamma}{\partial T}\nabla T$, $\frac{\partial\gamma}{\partial T}$ is given by

$$\frac{\partial\gamma}{\partial T} = -A_r - R\Gamma_s ln(1 + k_1 a_s exp(-\Delta H_0/RT)) - \frac{k_1 a_s \Gamma_s \Delta H_0 exp(-\Delta H_0/RT)}{(1 + k_1 a_s exp(-\Delta H_0/RT))T} \tag{7}$$

where γ_m is the surface tension at melting temperature, A_r is the opposite of $d\gamma/dT$ for a pure FeS alloys, R is the gas constant, Γ_s is the surface excess at saturation, a_s is the activity of sulfur, ΔH_0 is the standard heat of adsorption, and k_1 is a constant related to entropy of segregation [10].

2.3. *Energy conservation equation*

The energy conservation can be written as follows :

$$\rho c_p(\nabla h).(\mathbf{v} - \mathbf{v_s}) = \nabla.\lambda\nabla T \tag{8}$$

where c_p is the specific heat, h is the enthalpy, λ is the thermal conductivity.
The boundary conditions are :

$$h = h_0 \ on \ \partial\Omega_{in} \tag{9}$$

$$\lambda\nabla T.n = 0 \ on \ \partial\Omega_{out} \tag{10}$$

$$\lambda\nabla T.n = q_r + q_c + q_s \ on \ \partial\Omega_{surf} \tag{11}$$

where $\partial\Omega_{in}$ and $\partial\Omega_{out}$ are the surface entering and leaving of the calculated domain, $\partial\Omega_{surf}$ represent the other surfaces of the workpiece. The convective heat loss is $q_c = -h_c(T - T_\infty)$, and the radiative loss is $q_r = -\epsilon\sigma(T^4 - T_\infty^4)$, q_s is the heat flux which will be given in the next subsection.

2.4. *The free surface*

The surface depression will form a shape which minimizes the total energy. The total energy to be minimized includes the surface energy due to the change in area of the pool surface, the potential energy in the gravitational field and the the work performed by the arc pressure displacing the pool surface [4].

$$\frac{\gamma}{R} = \rho_0\mathbf{g}\varphi + P_a + \lambda \tag{12}$$

R and φ are the radius curvature and the elevation, φ is the elevation, γ represent the surface tension which is a variable in function of temperature. λ is the Lagrange multiplier which can be determined from keeping constant volume of the workpiece. The surface tension depends on temperature, for FeS alloys, following the model presented by Sahoo et al [10] γ is :

$$\gamma = \gamma_m - A_r(T - T_f) - RT\Gamma_s ln(1 + k_1 a_s exp(-\Delta H_0/RT)) \tag{13}$$

P_a is the arc pressure distribution at the free surface which will be given in the next subsection.

2.5. Treatment of latent heat

In order to treat phase change in the weld pool liquid/solid interface, the method of V.R. Voller [9], [11] is used. The total enthalpy H is represented as the integral of heat capacity with respect to temperature and can be written as follows:

$$H(T) = \int_{T_{ref}}^{T} \rho c_p(T)dT \tag{14}$$

where $\rho c_p(T)$ is the volumetric heat capacity which depends on temperature, and where T_{ref} is a reference temperature. In the industrial application, it is assumed that the phase change takes place over a finite interval $[T_s, T_l]$, the enthalpy function can be given by:

$$H(t) = \begin{cases} \int_{T_{ref}}^{T} \rho c_s(T)dT, \ for\ T \leq T_s, \\ \int_{T_{ref}}^{T_s} \rho c_s(T)dT + \int_{T_s}^{T} \rho \frac{\partial L}{\partial T}dT, \ for\ T_s < T \leq T_l, \\ \int_{T_{ref}}^{T_s} \rho c_s(T)dT + \rho L + \int_{T_l}^{T} \rho c_l(T)dT, \ for\ T > T_l, \end{cases} \tag{15}$$

where T_s and T_l are the solidus and liquidus temperatures, and $\rho c_s(T)$ and $\rho c_l(T)$ are the temperature dependent volumetric heat capacities, L is latent heat.

2.6. Thermophysical properties

The temperature dependance of thermophysical properties is considered in this paper instead of supposing constant values. The thermophysical properties in solid and liquid $(c_p, h, \rho, \lambda, \mu, \beta)$ for the material 316L steel are taken from Kim [12], for $\frac{\partial \gamma}{\partial T}$ and γ, the formula of Sahoo [10] is used. Other thermophysical properties and parameters used in this paper are summarized in Table 1.

2.7. Determination of surface sources

The study focusses on the weld pool behaviour, classical Gaussian distribution for the heat flux and arc pressure could have been used. However, this would have affected the simulation predictive capabilities, because the parameters of Gaussian distribution are not related to the welding operating parameters. So an hybrid 2D-3D approach is performed: the boundary conditions used in the three-dimensional weld pool model (heat flux, arc pressure and arc drag force) are calculated with a 2D axisymmetric arc plasma model [7]. The present GTAW experiment are carried out under following conditions: 316L steel workpiece with dimensions $200 \times 50 \times 6\ mm^3$, the welding current is 150 A, the arc voltage is 10 V, the arc length is 2 mm, the welding speed is 15 cm/min, the shielding gas is argon with flow rate 20 l/min, the tungsten electrode is of diameter 3.2 mm with tip angle 30°. Then these welding conditions are used in the 2D axisymmetric arc plasma model, figures 1, 2 and 3 show the 2D-computed heat flux, arc pressure and arc drag force which will be used in the 3D finite element weld pool model.

3. Numerical model

The finite element method is used to discretize the equations of the problem. Quadratic and linear finite elements are used for the variable **v** and for the variables p, h and T. The method of weighted residuals is applied to seek the cancellation of integral terms associated with our problem. The coupling between the problem of fluid flow and heat transfer is taken into account with a suitable algorithm for solving the uncoupled finite element formulation (Table 2). The system of equations to be solved is non-linear, it is found that the thermal non-linearities are resolved fairly with the enthalpy formulation used, but the non-linearities associated with the

Table 1. Thermophysical properties of 316 stainless steel and welding conditions

Nomenclature	Symbol	Value (unit)
Sulfur activity	a_s	$0.03\ wt.\%$
Negative of $\frac{d\gamma}{dT}$ for pure metal	A_γ	$4.3 \times 10^{-4}\ N.m^{-1}K^{-1}$
Gas constant	R	$8314.3\ J.kg^{-1}.mol^{-1}.K^{-1}$
Surface tension at melting temperature	γ_m	$1.27\ N.m^{-1}$
Constant related to entropy of segregation	k_1	0.00318
Surface excess at saturation	Γ_s	$1.3 \times 10^{-8}\ kg.mol.m^{-2}$
Convection coefficient	h_c	$80\ W.m^2.K^{-4}$
Radiation emissivity	ϵ	0.4
Latent heat of fusion	L	$2.47 \times 10^{-5}\ J.kg^{-1}$
Solidus temperature	T_s	$1670\ K$
Liquidus temperature	T_l	$1730\ K$
Ambient temperature	T_∞	$293\ K$
Welding voltage	U	$10\ V$
Welding current	I	$150\ A$
Welding speed	v_s	$2.10^{-3}\ m.s^{-1}$
Maximum arc pressure	P_{max}	$407\ Pa$
Thickness of workpiece	H_w	$6.10^{-3}\ m$
Width of workpiece	W_w	$5.10^{-2}\ m$
Length of workpiece	L_w	$2.10^{-1}\ m$

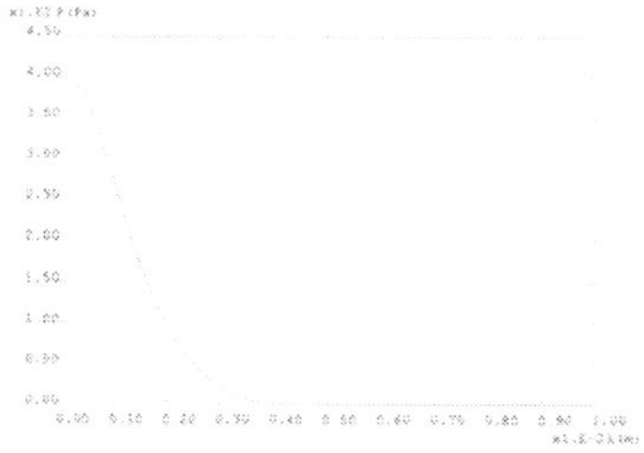

Figure 1. Radial evolution of arc pressure [7].

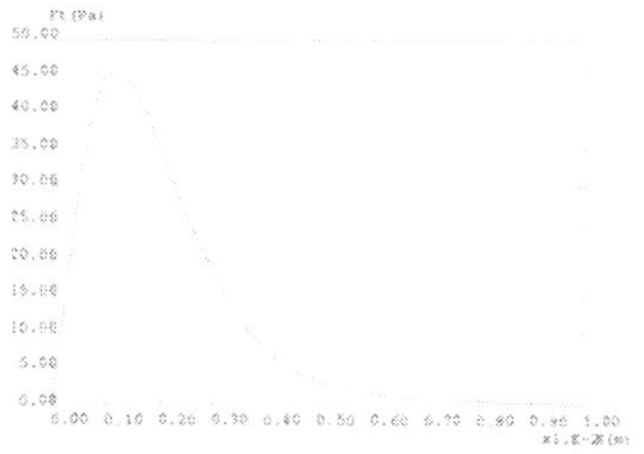

Figure 2. Radial evolution of arc drag force [7].

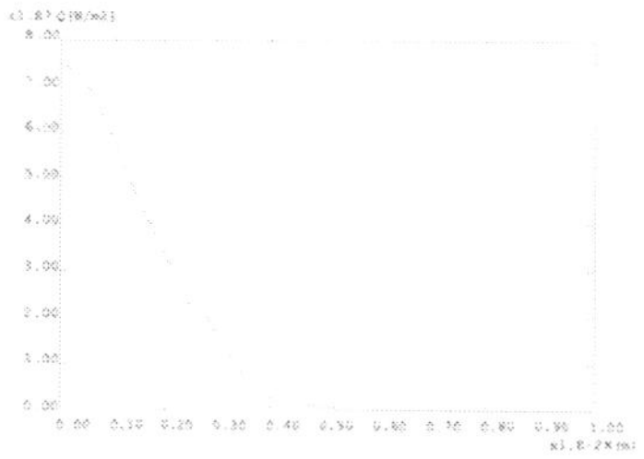

Figure 3. Radial evolution of heat flux [7].

fluid are generally more difficult to treat, for example the non-linearities due to the convective term and the phase change. The strategy is as follows: at the beginning of the calculation, the driving force $\mathbf{f_{Marangoni}}$ is initialized to zero, the thermal problem is solved. When there is a convergence on the unknowns of the non-linear problem, the intensity of the driving force f_α is increased with a coefficient α. And for improving the convergence, the convection terms are upwinded using a Streamline-Diffusion method.

Table 2. Algorithm for solving the non-linear problem

Initial condition: $(\mathbf{v}, p, h, T, \alpha)^0$ at $i = 0$
Repeat
$i = i + 1$
Determination of the fluid mesh
Calculation of the increments $\delta\mathbf{v}, \delta p$ and update $(\mathbf{v}, p)^i = (\mathbf{v}, p)^{i-1} + (\delta\mathbf{v}, \delta p)$
Calculation of the free surface profile, update $\varphi^i = \varphi^{i-1} + \delta\varphi$ and mesh movement
Calculation of the increment δh , update $h^i = h^{i-1} + \delta h$ and calculation $T^i = T(h^i)$
when $\delta_{inc} = (\delta\mathbf{v}, \delta p, \delta h) < \delta_{conv}$, $\alpha = min(\alpha * f_\alpha, 1)$
until $\delta_{inc} < \delta_{conv}$ and $\alpha = 1$

4. Results and discussion

Figure 4 show the temperature and velocity field with heat transfer and fluid flow model taking into account the free surface, as shown in this figure, the base metal ahead of the heat source has not yet melted. The weld top surface of the region directly under the heat source is strongly deformed under the arc pressure, the molten metal is pushed to the rear part of the weld pool by combined effect of all the driving forces. As the monitoring location moves away from the heat source, the weld pool surface shows less depression because of reduction of arc pressure. Furthermore, the accumulation of the liquid metal in the rear part is clearly shown.

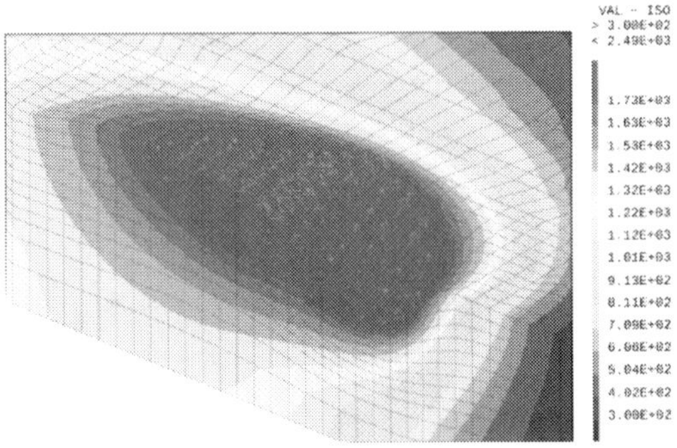

Figure 4. Calculated weld pool with considering free surface.

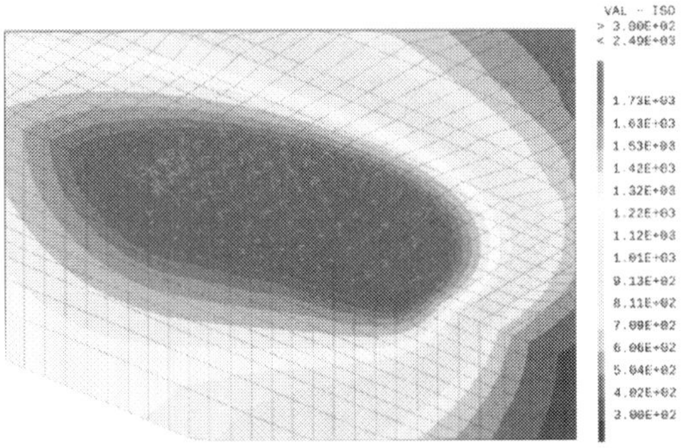

Figure 5. Calculated weld pool without considering free surface.

4.1. Weld temperature and fluid velocity field

In order to understand the influence of the free surface on the weld temperature and fluid velocity, two simulations are performed using the experimental welding parameters: one takes into account the free surface in our model (figure 4) and the other one doesn't (figure 5). Table 3 summarize the calculated weld pool characteristics, the maximal temperature in the case without free surface is 220 K higher than the one with free surface. When the free surface is taken into account, the fluid velocity is increased by the depression of free surface, and it will bring more heat energy to the rear part of weld pool, increase the volume of the weld pool, reduce the maximal temperature and finally change the isothermals.

The figures 6 and 7 show the temperature field of these two cases at the bottom surface of workpiece. In front of the heat source, the difference in the isothermals is quite small, on the other hand, the isothermals are very different behind the heat source. The isothermals are elongated for the case with free surface than the one with fixed surface, and the temperature difference exceeds 100 K in a huge area. Therefore, it is important to consider the free surface in the simulation of GTAW process.

Table 3. Weld pool characteristics in the models with and without free surface

Model	T_{max} (K)	u_{max} $(m.s^{-1})$	v_{max} $(m.s^{-1})$	w_{max} $(m.s^{-1})$	Width (mm)	Depth (mm)	Volume (mm^3)
With free surface	2486	0.31	0.32	0.15	3.60	2.36	70.6
Without free surface	2706	0.18	0.23	0.06	3.60	1.50	51.7

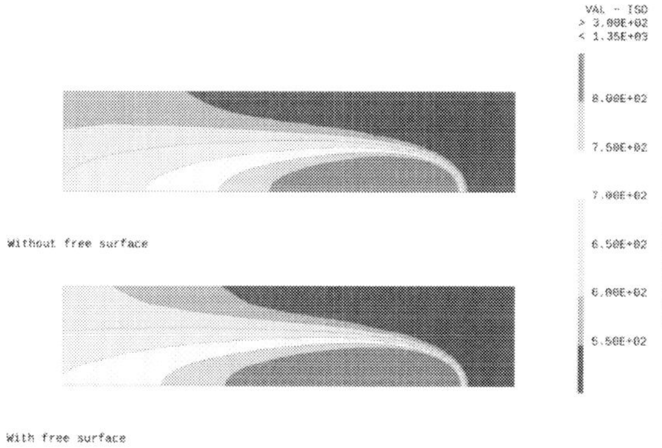

Figure 6. Calculated temperature field at the bottom surface of workpiece for the case without and with considering free surface.

Figure 7. Difference of calculated temperature field at the bottom surface of workpiece with and without considering free surface.

4.2. Weld bead geometry

The crosssection macrograph of welded joint is shown in figure 8, which also indicate the calculated weld bead shape by the models with and without free surface. As shown in this figure, the width of weld pool with a flat surface is wider than experiment and the depth is smaller, on the other hand, the shape of the weld and the penetration are better predicted when we take account into the free surface. As shown in figure 4 and figure 5, fluid flow and heat are taken away from the center of weld pool to the edge, so the deepest position of the weld pool is not just below the moving heat source, but after it because of the convection. When the surface is free, the front part of weld pool is strongly depressed by the arc pressure, so the fluid velocity at the surface is bigger and more fluid flow and heat are taken to the back part of weld pool which lead to a deeper penetration. For this experiment, the heat transfer and fluid flow model with free surface is better to predict the weld bead geometry.

5. Conclusion

A three-dimensional model of heat transfer and fluid flow has been presented in this paper, which can calculate the temperature and velocity field, predict the weld pool shape and size.

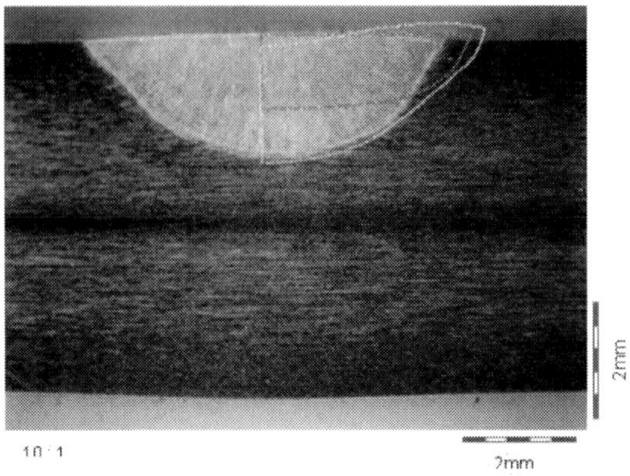

Figure 8. Experimental (green line) and calculated weld pool shape (yellow line model with free surface and red line model without free surface.

The following conclusions can be given:

(1) The model solves the continuity, momentum conservation, and energy conservation equations with taking account into the free surface and phase change, and a steady solution for the temperature, velocity field and the free surface of weld pool can be found.

(2) Taking account into the free surface in the simulation of weld pool has a great influence on the fluid velocity, the isothermals and the dimensions of the weld bead.

(3) Experimental results show that there is a good agreement between the simulated profile of weld bead by our 2D-3D approach model and the experimental one.

6. Reference

[1] Tanaka M, Terasaki H, Ushio M and Lowke J 2002 *Metallurgical and Materials Transactions* A **33** 2043–2052
[2] Mundra K and Debroy T 1996 *Numerical Heat Transfer* **29** 115
[3] Lu F, Tang X, Yu H and Yao S 2006 *Journal of Applied Physics* **35** 458
[4] Wu C, Chen J and Zhang Y 2007 *Computational Materials Science* **39** 635–642
[5] Kim J and Na S 1994 *ASME J. Eng. Ind* **116** 78–85
[6] Hu J, Guo H and Tsai H 2008 *International Journal of Heat and Mass Transfer* **51** 2537
[7] Brochard M 2008 *A unified model of gas tungsten arc welding including electrode, arc plasma and molten pool* Ph.D. thesis Commissariat à l'Energie Atomique, France
[8] Cast3M 2008 http ://www-cast3m.cea.fr
[9] Brent A, Voller V and Reid K 1988 *Numerical Heat Transfer* **13** 297–318
[10] Sahoo P, Debroy T and McNallan M 1988 *Metallurgy and Materials Transactions* B **19** 483–491
[11] Nedjar B 2002 *Computers and Structures* **80** 9–21
[12] Kim C 1975 *Thermophysical properties of stainless steels* Master's thesis Argonne national laboratory

IOP Conf. Series: Materials Science and Engineering 33 (2012) 012026 doi:10.1088/1757-899X/33/1/012026

A combined enthalpy / front tracking method for modelling melting and solidification in laser welding

G Duggan[1], WU Mirihanage[2], M Tong[1], DJ Browne[1]
[1] School of Mechanical and Materials Engineering, University College Dublin, Belfield, Dublin 4, Ireland
[2] Department of Physics, Norwegian University of Science and Technology, Trondheim, Norway.

E-mail: gregory.duggan@ucd.ie

Abstract. The authors present an integrated meso-scale 2D numerical model for the simulation of laser spot welding of a Fe-Cr-Ni steel. The melting of the parent materials due to the applied heating power is an important phenomenon, leading to the formation of the weld pool and the subsequent conditions from which solidification proceeds. This model deals with the dynamic formation of the weld pool whereby melting may be occurring at a given location while solidification has already commenced elsewhere throughout the weld pool. Considering both melting and possible simultaneous solidification in this manner ensures a more accurate simulation of temperature distribution. A source based enthalpy method is employed throughout the calculation domain in order to integrate the melting model with the UCD front tracking model for alloy solidification. Melting is tracked via interpolation of the liquidus isotherm, while solidification is treated via both the tracking of the advancing columnar dendritic front, and the nucleation and growth of equiaxed dendrites using a volume-averaging formulation. Heterogeneous nucleation is assumed to take place on TiN grain refiner particles at a grain refiner density of 1000 particles per mm^2. A mechanical blocking criterion is used to define dendrite coherency, and the columnar-to-equiaxed transition within the weld pool is predicted.

1. Introduction

The solid-to-liquid phase transformation, i.e. the melting of the parent material in the initial stages of welding, leads to the formation of the weld pool. The geometry and temperature distribution within the weld pool, once the heat input has stopped, have a significant influence on the subsequent microstructural development (e.g. columnar-to-equiaxed transition - CET) as solidification proceeds. A key concern then is the accurate modelling of the weld pool formation as this provides the starting point from which solidification commences. This paper focuses primarily on the development of a dynamic phase transformation scheme within the weld pool. To this end, fluid flow within the weld pool is not considered at this time but it is acknowledged that this will be an important step to be taken in future work to allow the results to be compared to experiments.

The welding process may be split into 3 stages. Initially there is pure melting of the parent material as a result of the heat input. This is followed directly by a period whereby both melting and solidification may occur throughout the weld pool simultaneously, as is discussed in further detail in section 4. Lastly there is pure solidification within the weld pool when the melting front has ceased to advance into the parent material. The first and last stages will be outlined in sections 2 and 3 followed by the combined melting and solidification phase in section 4.

2. Melting

The field of research on melting is a relatively new one when compared to research on solidification. There is a tendency to try to adapt knowledge directly from the much more extensively developed research on solidification. However it has been shown in experiments that there exist asymmetries between melting and solidification. To the authors knowledge, there has yet to be a melting model developed based on the kinetics of the process although there is research being carried out in the area [1-3]. As this is the case, a widely used form of the enthalpy scheme, developed by Swaminathan and Voller [4], is implemented to model the initial melting stage of the welding process.

2.1. *Enthalpy Scheme*

Using an enthalpy scheme allows the calculation domain to be considered as a continuum, avoiding the necessity to explicitly address the discontinuity associated with phase change at the solid/liquid interface. The apparent specific heat capacity method is used to model the melting phase transformation that occurs during the heating of the work-piece via the laser heat input. The laser heat input is modelled as a Gaussian distribution.

3. Solidification

In order to model solidification the energy equation, shown in the discretised form, in equation 1, is solved explicitly over a fixed grid to model heat transfer and provide the temperature profile across the domain. A source term, shown in equation 2, is included to formulate the release of latent heat. A truncated Scheil equation is used to model the change in solid fraction due to the growth of grains and the thickening of the mushy zones.

$$\rho C_p \frac{\Delta x \Delta y}{\Delta t}(\Delta T) = -\sum (\vec{q}.\vec{n})^{n+1}\,\Delta a \; + \; E \tag{1}$$

$$E = \rho \Delta x \Delta y\, L\,(\; V_m (g_s^n - g_s^{n-1}) + g_s (V_m^n - V_m^{n-1})\;) \tag{2}$$

where ρ is density, C_p is the specific heat, Δx and Δy are the grid spacing, t is time, T is temperature, Δa is the area the flux q is acting upon i.e. Δx or Δy in the 2D case. E is the source term associated with the release of latent heat, V_m is the captured volume/ dendritic mushy volume in the current control volume (CV) while g_s is the solid fraction within this mushy volume, the subscripts m and s stand for mushy zone and solid phase respectively. The superscripts n, $n+1$ and $n-1$, refer to the current time step, the next time step and the previous time step respectively. The 1st term at right hand side of equation 2 may be considered to be the latent heat release due to the thickening of the mushy zone (E_t) while the 2rd term is the latent heat release due to the advancement (E_a) of solidification front.

3.1. *Columnar Front Tracking*

The front in question is defined at a mesoscopic level; it encompasses the growing dendrite tips. The front is defined by massless markers at the points where the interface intersects the underlying fixed orthogonal grid. These markers are advanced into the liquid melt, as solidification proceeds, based on the local level of undercooling. The Burden and Hunt model [5] is used to calculate the dendrite tip growth velocities (alternative growth laws may be easily incorporated). The details of this front tracking scheme are fully outlined elsewhere [6,7].

3.2. *Equiaxed Grain Growth*

The model for equiaxed grain growth as fully outlined elsewhere [8] is based on a volume averaging approach. Key to the model is the nucleation of equiaxed grains from inoculants within the undercooled liquid ahead of the columnar front. These inoculants are assumed to follow a log-normal size distribution having a uniform density distribution. As in the columnar front tracking case, the equiaxed envelope is assumed to grow via a dendrite tip growth kinetics model based on the level of undercooling in the liquid. Growth of the equiaxed envelopes as well as thickening of the inter-dendritic mushy zones is incorporated into the overall latent heat calculation for both columnar and equiaxed growth.

3.3. Inoculant particles

The development of effective grain refiners for steel alloys, in order to improve mechanical properties, has been recognized as an area of research with great potential for cost savings. New applications call for ever increasing levels of control over the mechanical properties of steel.

Research into the use of grain refiners in cast steels has been undertaken sparingly in comparison to the extensively studied field of grain refiners in aluminium alloys. Some of the earliest research pertaining to steels was carried out by Bramfitt [9] in 1970, who evaluated the effectiveness of various powders as inoculant particles in steel based on the level of undercooling required to initiate solidification. TiN, TiC and SiC were found to be the most effective inoculant particles. In recent years this area is becoming the focus of intensive research. TiN grain refiners are the most commonly studied for a variety of steels [10-13].

In welding an equiaxed grain structure is desirable as opposed to the typical coarse grained, low-ductility, fully columnar grain structure found in many welds. However, fully columnar structure is typical of most welds due to the very high thermal gradients and the speed of the solidification front. By including inoculant particles within the weld material, heterogeneous nucleation may be promoted in the undercooled liquid ahead of the columnar front, thus a CET may be initiated.

Work carried out by Koseki *et al* [14], on the weld solidification of Fe-Cr-Ti alloys in the presence of TiN grain refiners, showed that CET may be initiated with the correct level of grain refiner density present in the weld material. The nucleation and growth of equiaxed grains was shown to result from the combined effects of the heterogeneous nucleation sites provided by the TiN inoculant particles as well as the increased undercooling due to the soluble titanium in the weld material. The present work, although not considering the effects of soluble titanium in the weld material, uses a similar grain refiner density and size distribution of TiN particles as those used in Koseki *et al* [14] and Park [15].

4. Integration of melting and solidification routine

Once the heating power of welding is switched off, the free surface of liquid metal will begin to cool quickly due to radiative/ convective heat loss to the surroundings. This results in solidification commencing at the outermost parts of the weld pool. While this is occurring, heat transferred from superheated liquid to the solid phase is still proceeding resulting in further melting. This means melting and solidification of the parent material co-exist for some finite period once the heating power is switched off as shown in figure 1.

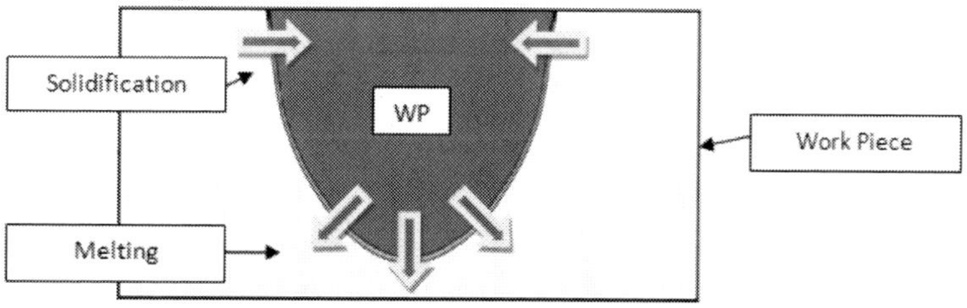

Figure 1. Melting and solidification occurring simultaneously within the weld pool.

The phase transformation model has to be able to track both melting and solidification simultaneously. The original front-tracking model for solidification as outlined in section 3, has been updated in order to consider melting. To facilitate the application of a single set of energy equations to the calculation domain to treat both melting and solidification simultaneously, a source term based enthalpy method is employed to replace the apparent specific heat capacity method previously outlined in section 2.

The energy equation for the front tracking formulation, shown by equation 1, is still valid in the source term based enthalpy method. The source term E is split into 2 components E_a and E_t. In the case of melting, E_t becomes the absorption of latent heat due to the change of volume fraction of solid phase. The contribution of latent heat due to advancement of solidification front E_a does not exist in melting. Therefore, E is calculated in equation 3 by appropriately selecting the source terms E_a and E_t as shown in equations 4 and 5, based on whether melting or solidification is occurring within a given CV.

$$E = E_t + E_a \tag{3}$$

$$E_t = \begin{cases} \rho\Delta x\Delta y\, LV_{cv}(g_s^n - g_s^{n-1}) & ,melting \\ \rho\Delta x\Delta y\, LV_m(g_s^n - g_s^{n-1}) & ,solidification \end{cases} \tag{4}$$

$$E_a = \begin{cases} 0 & ,melting \\ \rho\Delta x\Delta y\, L\, g_s(V_m^n - V_m^{n-1}) & ,solidification \end{cases} \tag{5}$$

where V_{cv} is the volume of a CV. The existing numerical scheme of assigning and advancing computational markers [6,7] during the solidification process cannot be applied in the case of melting. Therefore a new numerical scheme for marker assignment and advancement to track the boundary of the weld pool with a Lagrangian grid of computational markers, has been developed as follows.

Once the heat input is stopped, the initial front is calculated by assuming it is given by the liquidus isotherm (T_L). The temperatures along the walls of every CV in the simulation domain are scanned using bilinear interpolation. Once T_L is encountered, a computational marker is assigned. This scanning procedure is performed along both the horizontal and vertical directions. Using linear interpolation along the correct sequence of markers, a proper front is defined as illustrated in figure 2. This procedure is only implemented once, at the moment the heating input ceases.

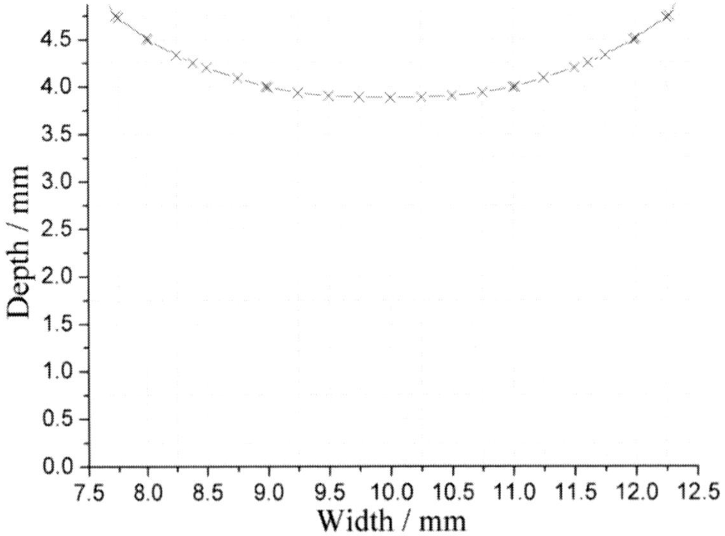

Figure 2. Assigning initial markers on grid lines (vertical gridlines 2 x Δx for clarity).

Once the front is defined, during the following process of the simulation, the temperature of each marker can be calculated by interpolating temperature values from the surrounding node temperatures.

Markers are designated as "Enthalpy Markers" if their temperature is greater than or equal to T_L, while makers with temperatures below T_L i.e. undercooled markers, are designated "Front Tracking Markers".

4.1. Updating temperature field

Two types of CV designations are used in order to ensure the correct source terms are used when updating the temperature field. A CV will be designate as an "Enthalpy CV" if it is intersected by an "Enthalpy Marker" where the next marker downstream of this is also an "Enthalpy Marker" as illustrated in figure 3.

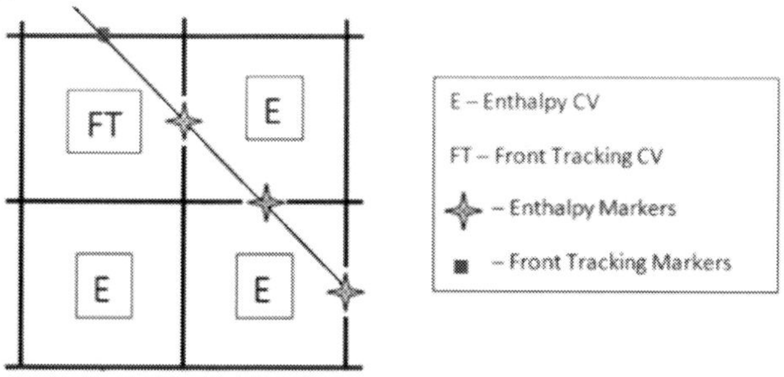

Figure 3. Assigning of a CV types based on the type of markers intersecting it.

A CV containing "Front Tracking Markers" will be designated a "Front Tracking CV", while CV's which do not contain markers are by default set to "Enthalpy CV". The mushy volume fraction i.e. the volume fraction captured by the front in a CV, is calculated based on the location of intersecting markers for CV's designated "Front Tracking CV" only. Details of the way to calculate the volume fraction of mushy zone can be found in the work of Browne and Hunt [6]. This mushy volume fraction is incorporated in the source terms as outlined in equation 4.

Once the temperature field has been updated the marker temperatures are updated as well as the marker types and CV types accordingly. The model then updates the time and repeats the process again until the specified end time.

4.2. Advancing markers

The Burden and Hunt dendrite tip growth kinetics model [5] is used to advance markers appropriately based on level of marker undercooling (T_L - T_{marker}). Only markers that are designated "Front Tracking Markers" are advanced in this manner. A marker will advance (assuming it is undercooled) into the liquid at the local normal to the front. Once markers have been advanced, the position of new markers are redefined following the procedures outlined in [6,7]. The temperatures of the new markers are redefined and the markers are designated Enthalpy / Front Tracking appropriately. The CV types are also updated at this time

The new locations of Enthalpy markers are also calculated at this stage by interpolating the new position of the T_L isotherm. If melting is occurring at a marker location, it will move outwards to the new grid/ T_L intersection point. Enthalpy markers will proceed in this manner as the melting begins to slow until a point where the marker becomes undercooled as its type will be switched to "Front Tracking Marker" and it will proceed to grow into the liquid as described previously.

5. Columnar to Equiaxed Transition

Columnar and equiaxed volume fractions are calculated separately in each CV. In order to predict CET, three blocking criteria are implemented [8]:

IOP Conf. Series: Materials Science and Engineering **33** (2012) 012026 doi:10.1088/1757-899X/33/1/012026

1. If the average equiaxed dendrite diameter is less than the primary columnar inter-dendrite arm spacing the columnar front will continue to grow. The primary columnar inter-dendrite arm spacing is calculated according to the empirical relation outlined in [16].
2. If the average equiaxed dendrite diameter is greater than the primary columnar inter-dendrite arm spacing, but no coherent equiaxed dendrite network has formed, then equiaxed dendrites are expected to grow simultaneously ahead of the columnar front
3. A coherent equiaxed dendrite network has formed; this will act as a barrier to physically block the columnar front, i.e. CET will occur.

6. Details of case study

The sample size is 20 mm by 5mm. The heat flux from the laser is assumed to follow a Gaussian distribution. The power of the laser is 1404 W, the radius is 2.5 mm and the heating duration is 0.80 s. Heat loss due to convection and radiation is assumed through the boundaries of the sample. An emissivity value of 0.5 is assumed and a heat transfer coefficient of 8 W/m^{-2} K [17] is assumed between the air and the sample. The air temperature is assumed to be 300 K. The alloy used is Fe-15.9 wt% Cr -14.1 wt% Ni; thermophysical properties used are listed in table 1. The grain refiner density was typically taken to be 1000 particles per mm^2 [14], while this value was varied in order to assess the effectiveness of the grain refiners in promoting CET.

Table 1. Thermophysical properties used in the case study

Property	Value	Unit	Source
Eutectic temperature (T_e)	1679	K	[18]
Liquidus temperature (T_L)	1710	K	[18]
Specific heat (C_p)	780	J/kgK	[13]
Density (ρ)	7250	kg/m^3	[14]
Thermal Conductivity (K)	35	W/mK	[14]
Latet heat of Fusion (L)	1.854×10^5	J/kg	[14]
Diffusivity of Cr in liquid (D_l)	1.8×10^{-9}	m^2/s	[13]
Solid/liquid interfacial energy (γ)	0.299	J/m^2	[18]
Volumetric entropy of fusion (ΔS_v)	7.71×10^5	J/m^3K	[18]
Gibbs Thomson coefficient (Γ)	3.88×10^{-7}	mK	[18]
Liquidus Slope (m_L)	-2.9	K/wt%	[18]
Partition coefficient (α)	0.86	-	[18]
Burden and Hunt growth coefficient (C)	0.8×10^{-5}	m/K^2s	[18]

7. Results and Discussion

7.1. *Melting*

For the initial melting stage, the laser heat input lasts for a duration of 0.8s. During this period the weld pool develops to a width of 4.8mm and a depth of 1.12mm. As is apparent in figure 4, the high rate of heating results in a very narrow (yellow) mushy zone between the T_s and T_L isotherms. The low thermal conductivity of the alloy (relative to a typical value in laser spot welding of an aluminium alloy [19]) means the weld pool retains heat very well and it is slow to dissipate through the free surface and the base material. This results in a wide shallow weld pool.

The initial position of the front is shown in both figures 5 and 6 via the solid blue line, while the advanced position of the front at two subsequent time steps are depicted via the red dashed lines. Melting at the base of the weld pool is evident while solidification proceeds along the upper sides. There is a significant amount of continued melting for a further 0.06s from when the laser is switched off. The model accurately reproduces this important stage of the process whereby the weld pool is allowed to fully develop before solidification proceeds at all points.

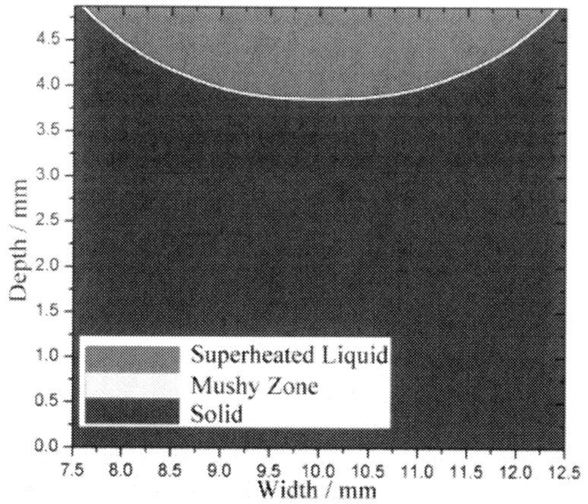

Figure 4. Weld pool at 0.8 s, heat input just turned off. Narrow yellow, mushy region apparent.

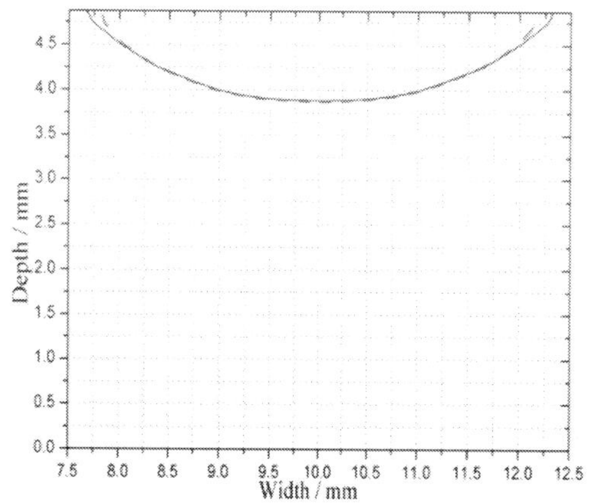

Figure 5. Advancement of melting front at base of weld pool, while solidification proceeds at the sides at 0.82 s.

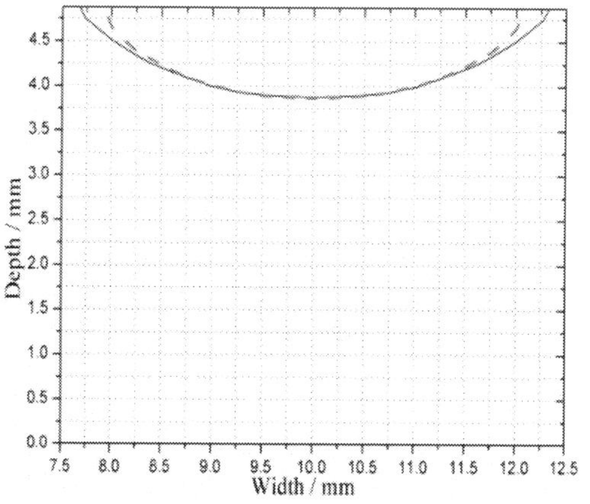

Figure 6. Further development of the weld pool as melting persists at the base of the weld pool at 0.84 s.

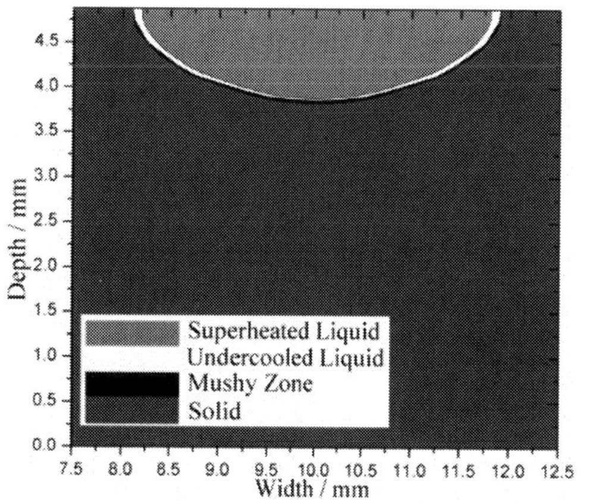

Figure 7. Undercooled liquid region (white) apparent ahead of the columnar front at 0.86 s.

7.2. Solidification

The progression of the columnar front is plotted in figures 7-9. The undercooled liquid region is illustrated via the white area ahead of the columnar front in figure 7. As can be seen this undercooled region is initially quite narrow, this is due to the low conductivity of the alloy which maintains a high thermal gradient ahead of the columnar front. While this is the case, any equiaxed grains that do nucleate ahead of the columnar front are quickly overgrown as per criterion number 2 in section 5. These equiaxed grains are given little opportunity to grow and reach coherency in order to block the columnar growth. The slow dissipation of heat within the weld pool maintains high thermal gradients until at a later stage, the thermal gradients have fallen enough to allow a significant undercooled liquid region to develop. Figures 8 and 9 show the contours of solid fraction within the weld pool, blue (1.0) being fully solid. The position of the columnar front (white dashed line) illustrates the level of equiaxed growth in the undercooled region ahead of the columnar front. The conditions within the weld pool are now such that the equiaxed grains can grow to the point of coherency and block the columnar front. CET occurs at approximately 0.95s. The remaining liquid within the weld pool from this point on solidifies with an equiaxed microstructure. The location of the columnar front at the point of CET is shown in figure 9 via the white dashed line.

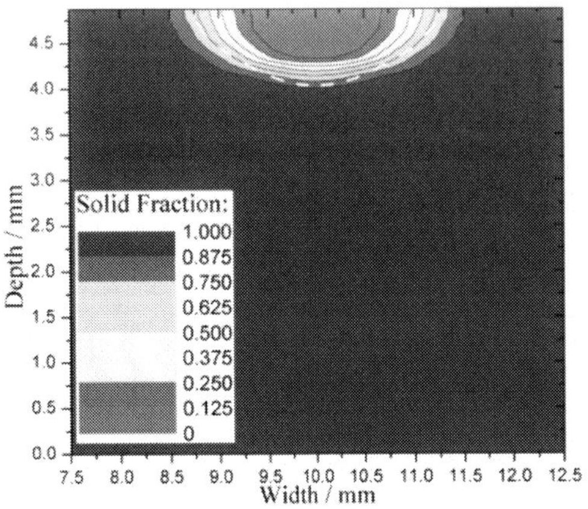

Figure 8. Equiaxed growth apparent ahead of columnar front at 0.92 s.

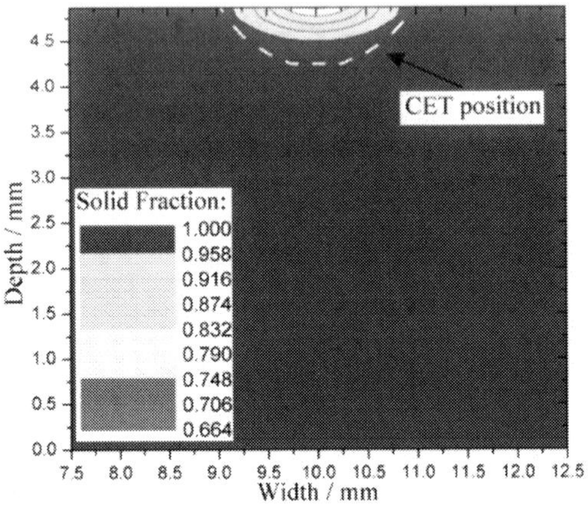

Figure 9. Undercooled liquid at 0.96 s, while the position of CET is indicated in the solid (note change in legend).

These results agree well with those observed in [14] where CET was seen to occur during the welding of a similar alloy with the same grain refiner density. It is also evident that reducing the grain refiner density results in CET occurring later in the process.

This model directly predicts the microstructure of the weld and the presence of CET without having to rely on plots of the thermal gradient (G) against the local solidification velocity (v). Typically G/v plots have been used in conjunction with the Hunt Diagram [20] in order to infer the type of microstructure present e.g. [21]. The model presented here provides a more direct representation of how the microstructure changes throughout the weld pool during solidification, based on the thermodynamics and kinetics of the process.

At this stage these numerical results have not been validated against experimental laser spot welding results as a fluid flow model will need to be incorporated to generate meaningful results as regards the solidification kinetics, to compare to experiments.

8. Conclusion

An integrated melting and solidification model of 2D laser spot welding is presented. The front tracking method has previously been successfully extended to 3D [22] with good results. The front tracked in this case consists of triangular surfaces as opposed to line segments. To carry out the extension to 3D in this case would require considerable programming effort but primarily a considerable amount of geometrical calculations would be required. The grid of surfaces defining the front would need to be continuously refined / coarsened in order to ensure a level of accuracy that would capture the important features of the weld pool. This paper focuses on developing the phase transformation model and thus the extension to 3D is left to future work.

An enthalpy scheme is used to model the initial melting stages of the process, while a newly adapted UCD front tracking method is used to track subsequent melting and solidification. Equiaxed nucleation upon TiN grain refiner particles is considered and their subsequent growth to coherency is tracked in order to predict CET. A mechanical blocking criterion is employed to predict CET, which is seen to occur in the late stages of the process, when the thermal gradient has reduced enough to allow a significant undercooled liquid region to develop and promote the growth of equiaxed grains.

Acknowledgments
This research work is supported by the European Commission as part of the FP7 programme, as the project, Modelling of Interface Evolution in Advanced Welding; contract number no. NMP3-SL-2009-229108.

References

[1] Buchmann M and Rettenmayr M, 2008 *Journal of Crystal Growth* **310(21)** 4623-27
[2] Rettenmayr M, 2009 *International Materials Reviews* **54** 1-17
[3] Benielli D, Bergeon N, Jamotchian H, Billia B and Voge PH, 2002 *Physical Review* **65 E** 051604
[4] Swaminathan CR and Voller VR, 1993 *International Journal of Numerical Methods for Heat & Fluid Flow* **3(3)** 233-44
[5] Burden MH and Hunt JD, 1974 *Journal of Crystal Growth* **22** 109-16
[6] Browne DJ and Hunt JD, 2004 *Numerical Heat Transfer* B **45** 395-419
[7] McFadden S and Browne DJ, 2009 *Applied Mathematical Modelling* **33(3)** 1397-1416
[8] Mirihanage WU and Browne DJ, 2009 *Computational Materials Science* **46 (4)** 777-84
[9] Bramfitt BL, 1970 *Metallurgical Transactions* **1** 1987-1995
[10] Tuttle R, 2010 *International Journal of Metalcasting/summer 10* 17-25
[11] Poole WJ and Weinberg F, 1998 *Metallurgical and Materials Transactions* **29A** 885-861
[12] Wang M, Cheng G, Qiu S, Zhao P and Gan Y, 2010, *International Journal of Minerals, Metallurgy and Materials* **17(3)** 276-281
[13] Löser W and Herlach DM, 1992 *Metallurgical Transactions* **23A** 1585-1591
[14] Koseki T, Inoue H, Fukuda Y and Nogami A, 2003 *Science and Technology of Advanced Materials* **4** 183-195
[15] Park JH, 2011 *CALPHAD: Computer Coupling of Phase Diagrams and Thermochemistry* **35** 455-462
[16] El-Bealy M and Thomas BG, 1996 *Metallurgical and Materials Transactions* **27B** 689-693
[17] Mohammad Hossenini H, Mirzaei M, Azari S, 2006 *Modelling of Casting, Welding and Advanced Solidification Processes* **XI** 903
[18] Koseki T and Flemings MC, 1995 *Metallurgical and Materials Transactions* **26A** 2991-2999
[19] Duggan G, Tong M and Browne DJ, 2011 *IOP Conference Series : Materials Science and Engineering* **27** 012077
[20] Hunt JD, 1984 *Materials Science Engineering* **65** 75-83
[21] Villafuerte JC, Pardo E and Kerr HW, 1990 *Metallurgical Transactions* **21A** 2009-2019
[22] Seredynski, M, 2009 Micro-macroscopic model of binary mixture solidification, PhD thesis, Warsaw University of Technology (in Polish)

IOP Conf. Series: Materials Science and Engineering 33 (2012) 012027 doi:10.1088/1757-899X/33/1/012027

Effects of electrode bevel angle on argon arc properties and weld shape

W C Dong, S P Lu[1], D Z Li and Y Y Li
Shenyang National Laboratory for Materials Science, Institute of Metal Research, Chinese Academy of Sciences, 72 Wenhua Road, Shenyang 110016, China

E-mail: shplu@imr.ac.cn

Abstract. A numerical modeling of coupled welding arc with weld pool is established using FLUENT software for moving shielded GTA welding to systematically investigate the effects of electrode bevel angle on the argon arc properties as well as the weld shape on SUS304 stainless steel. The calculated results show that the argon arc is constricted and the peak values of heat flux and shear stress on the weld pool decrease with increasing electrode bevel angle, while the radial distribution of heat flux and shear stress varying slightly. The weld shape is controlled by the pool flow patterns driving by the surface tension, gas shear stress, electromagnetic force and buoyancy. The Marangoni convection induced by surface tension plays an important role on weld shapes. All the weld shapes are wide and shallow with low weld metal oxygen content, while the narrow and deep weld shapes form under high weld metal oxygen content, which is related with the oxygen concentration in the shielding gas. The weld depth/width (D/W) ratio increases with increasing electrode bevel angle for high weld metal oxygen content and is not sensitive to the electrode bevel angle under low weld metal oxygen content. The calculated results for the weld shape, weld size and weld D/W ratio agree well with the experimental ones.

1. Introduction

Since weld geometry and thermal cycles are significantly influenced by thermal and fluid flow conditions, the quantitative understanding of the heat transfer and fluid flow in the weld pool is important for engineers and scientists. Besides, the accurate prediction of the temperature and velocity fields is a prerequisite for phase transformations modeling taking place in the welding process [1, 2]. However, it is difficult to measure the various physical phenomena to the welding arc and weld pool by experiment because of the high temperature of arc and the small size of the weld pool. To achieve this objective, many mathematical models for the welding arc and the weld pool were established in recent years [3-7]. The importance of convective heat transfer in weld pool has been recognized and the effect of different driving forces such as surface tension and electromagnetic force on the weld shape has been investigated. Zacharia et al. [4, 5], Wang et al. [6] and Dong et al. [7] developed a 3D mathematical model to simulate the convection and heat transfer of a GTA weld pool by considering the surface tension as a function of temperature and sulfur or oxygen concentration. The simulated results showed that Marangoni convection induced by surface tension has an important effect on the

[1] To whom any correspondence should be addressed.

fluid flow and heat transfer, and is considered to be one of the principal factors for the increased penetration. Moreover, authors found that the gas shear stress from welding arc is another dominating driving force affecting the liquid pool flow pattern, except for the surface tension in the argon arc [8]. The influences of active element oxygen and welding parameters on the fluid flow and weld shape are investigated systematically using a coupled model of welding arc with weld pool [9].

The electrode bevel angle is an importance welding parameter. The simulated results by Lee et al. [10] showed that the increase of the electrode bevel angle results in the decrease of the shear stress on the molten pool surface and the increase of the weld depth under 30 ppm sulfur content in weld pool. In this study, the effects of electrode bevel angle on argon arc properties and weld shape are investigated. In order to reduce the computer calculation time, a quasi-steady state welding model is established using FLUENT software. In addition, the predicted weld shape, pool size and weld depth/width ratio of SUS304 stainless steel for both high and low weld metal oxygen content is compared with the experimental ones.

2. Mathematical model

The arc plasma gas is pure argon, assumed to be 2D axisymmetric in a steady-state condition under local thermodynamic equilibrium (LTE). Due to symmetry along vertical plane passing through the welding direction, only half of the workpiece is taken as the simulated domain in the 3D weld pool model. The deformation of the pool surface due to the arc jet impingement is not considered. Results from Hong et al. [11, 12] show that the predict level of turbulence in the AISI 304 stainless steel weld pools is not intense. Our previous study [9] also showed that the calculated weld width, depth and D/W under laminar and turbulence agrees well with the experimental one. In order to reduce the calculation time, the laminar flow is assumed for the weld pool and the arc column. The generalized steady-state equation which expresses the conservation of a flow variable Φ in the finite volume approach is [13]:

$$\nabla \cdot (\rho \vec{v} \Phi) = \nabla \cdot (\Gamma_{\Phi} \nabla \Phi) + S_{\Phi} \tag{1}$$

where Γ_{Φ} the diffusion coefficient, S_{Φ} the source term, and Φ represents the scalar flow variable that must be solved in various conservation equations. With the appropriate choice of the transport coefficient and the source term, equation (1) is used to express the conservation of mass, energy and momentum, as well as the electric and the magnetic potential.

The detailed formulae and explanations are given in Ref. 14 for welding arc model. In practice, the heat source moves with a certain constant velocity. Thus, when viewed in a fixed coordinate system (x', y, z), the welding problem is unsteady. The solution of the conservation of mass, momentum and energy in this coordinate system requires a large number of grids for accurate representation of the moving, time-dependent position of the heat source and the spatial variation of the heat flux. Furthermore, very small time steps are needed to ensure the accuracy and stability of the solution [8, 9]. In the present study, a coordinate system which moves with the heat source (x, y, z) is adopted, and the problem becomes quasi-steady a short time after the start of welding. Under such conditions, the heat source and the molten metal under the heat source are fixed in space, and the material enters and leaves the computational domain at the welding speed. The material inlet velocity is taken as equal in magnitude to the welding speed but opposite in direction. The additional source term $\partial(\rho u_0 \vec{v})/\partial x$ for the effect of moving arc on the weld pool convection is implemented in the momentum equation and the energy equation for weld pool model, where u_0 is the welding speed.

In order to trace the weld pool liquid/solid interface, i.e. the phase change, the total enthalpy H is represented by a sum of sensible heat h and latent heat content ΔH, i.e. $H=h+\Delta H$. The sensible heat h is expressed as $h=\int C_p dT$, where C_p is the specific heat and T is the temperature. The latent heat content ΔH is given as $\Delta H=f_L L$, where L is the latent heat of fusion. The liquid fraction f_L is assumed to vary linerly with temperature in the mushy zone $f_L=(T-T_S)/(T_L-T_S)$, where T_L and T_S are the liquidus and the solidus temperature, respectively.

A schematic plot showing the boundary conditions for the welding arc and the workpiece is given in figure 1. For the welding arc model, the most critical boundary condition is the electrical potential at the cathode surface AB. The current density J_z is given by

$$J_z = I/\left(2\pi r_h^2\right)\cdot\exp\left(-br\right) \tag{2}$$

$$b = 2\pi\int_0^{r_c} rJ_z dr \tag{3}$$

where r_h presents cathode spot radius and r_c presents the cutoff radius of cathode. At the cathode surface AB, the temperature is assumed to be 3500 K. Thus r_c is assumed to be 3 mm, and r_h is assumed to be 0.5 mm [3]. Moreover, a special treatment is given to the cathode and anode surfaces for considering the non-LTE condition in the boundary layer. The boundary conditions for the other variables are given in Ref. 14. For the weld pool model, the heat flux, current density and plasma drag force on the pool surface are calculated by the welding arc model. The temperature coefficient of the surface tension for an alloy containing oxygen can be obtained from a semi-empirical treatment given by Sahoo et al. [15]. At all surfaces, the vector potential components flux are set to zero. The bottom surface for the electric potential is electrically isolated and hence the current density at this boundary is zero. For the other surfaces, the electric potential is zero.

The thermophysical properties of the pure argon are taken from the tabulated data of Boulos et al. [16]. The radiation loss term for argon in the energy conservation equation is taken from the work of Evans et al. [17]. Calculations have been performed for an SUS304 stainless steel. The thermodynamic and transport properties of this alloy were derived from Ref. 9.

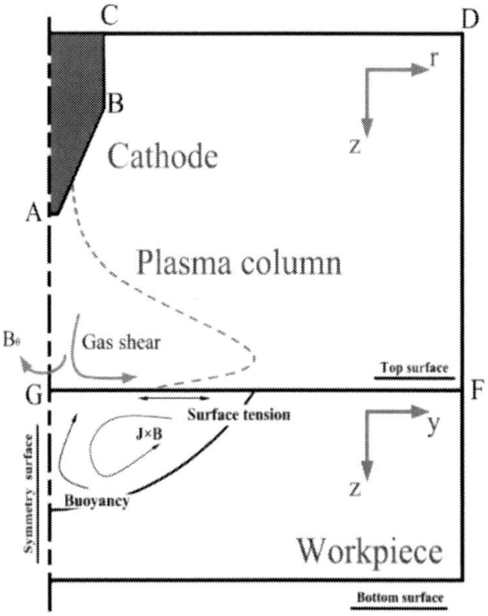

Figure 1. Schematic plot of weld cross-section showing boundary conditions used in calculation.

A non-uniform 100×30×30 grid of points system was utilized for the computational domain of 100 mm×25 mm×10 mm and the origin of the calculation is located at the center of the plate, with finer grid sizes applied near the weld pool region to improve the computation accuracy because of the higher temperature and the velocity gradients. In the arc region, these grid points are refined on the axis of the discharge and near the electrodes surface. The dimensions of the smallest grids are the thickness of the anode fall region (0.1 mm×0.1 mm). The system of partial differential equations which constitutes the model is discretized and solved iteratively by the segregated solver of the CFD commercial FLUENT© code [18]. The source terms in momentum and energy equations and the extra scalar equations needed for the electromagnetic variables are added by writing UDF (User Defined Functions) programme. The standard SIMPLE algorithm is used for the pressure-velocity coupling and a second order interpolation scheme is employed in all the equations of the model.

3. Experimental procedure

SUS304 stainless steel with the dimension of 100 mm×50 mm×10 mm was used as the substrate with the average composition of 0.06%C, 0.44%Si, 0.96%Mn, 8.19%Ni, 18.22%Cr, 0.027%P, 0.002%S and 0.002%O and the rest of Fe. Before welding, the surface of the plate was ground in order to remove all scale and was degreased by acetone. The 2% thoriated tungsten electrode was used in the experiments by direct current electrode negative (DCEN) polarity power source with a mechanized system, in which the test piece moves at a set speed 2mm/s. The welding current is 160 A, electrode gap is 2 mm, and electrode bevel angles are 15°, 30°, 45°, 60°, 90° and 120°, respectively. The pure Ar shielding and Ar-0.3%O_2 shielding were used to obtain the low and high oxygen concentration in weld pool. After welding, the weld beads were sectioned and specimens for weld shape observation were prepared using the standard metallographic techniques and etched by an HCl+Cu_2SO_4 solution to reveal the weld shape and dimensions. The cross-sections of the weld beads were photographed using an optical microscope. The weld metal oxygen content was analyzed using an oxygen/nitrogen analyzer.

4. Results and discussion

4.1. Effects of electrode bevel angle on argon arc properties

Figure 2 shows the temperature field of argon arc under 200 A welding current, 5 mm arc length and 60° electrode bevel angle. The typical bell shape of the arc periphery expressed by the isotherm of 21000 K and 10000 K is clearly observed. In figure 3, the temperature profiles of 200 A welding argon arc with 10 mm arc length and 60° electrode bevel angle are compared with the experimental results [19]. It can be seen that the agreement between the calculated and the measured temperature over 11000 K is very good. However, below 11000 K, the calculated values decrease sharply. This behaviour is probably due to the LTE assumption used in the present study. It is known that the outer fringes of the arc are not in LTE. Therefore, it can be concluded that the calculated temperatures below 11000 K are not reliable.

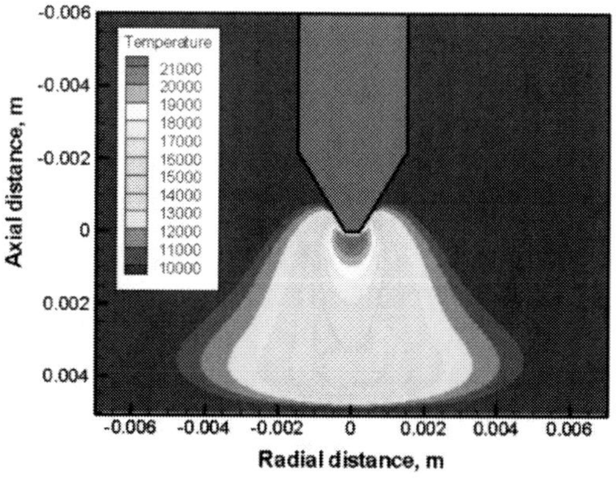

Figure 2. Temperature field of argon arc.

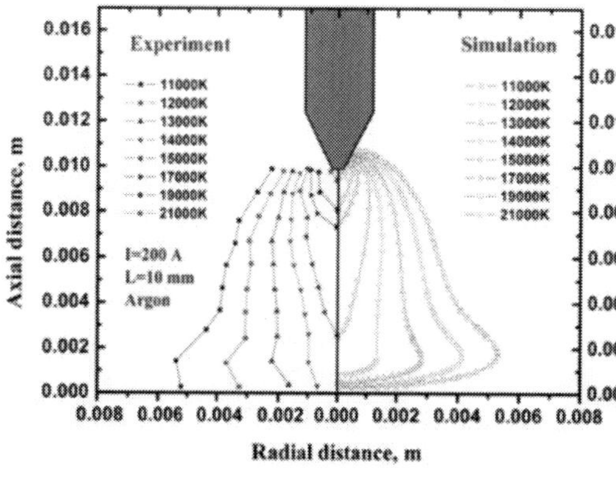

Figure 3. Comparison between experimental and calculated temperature contours in the welding arc for argon arc.

Figure 4 shows the comparison of argon arc temperature contours for 30°, 60°, 90°, 120° electrode bevel angles. As the electrode bevel angle becomes blunter, the diameter of the argon arc decreases which means that the arc is constricted. Glickstein [20] has observed by experiment that the arc becomes more constricted as the conical tip angle of the tungsten electrode increases. The calculated result agrees well with the experimental one.

Among the various arc properties, the heat flux to the workpiece and the shear stress are important items which effects on the heat transfer and fluid flow in the weld pool directly. Figure 5 shows the radial variation of the anode heat flux and the shear stress with different electrode bevel angles at 160 A welding current and 2 mm arc length, respectively. The maximum of heat flux at the anode surface decreases with the increasing electrode bevel angle, and its distribution radius changes slightly. As the radial distance increases, the shear stress increases initially and then decreases. The shear stress increases by one time as the electrode bevel angle changes from 120° to 15°. The higher the shear stress on pool surface is, the stronger the outward convection on liquid pool is. And the heat flux is easily transferred from the pool center to the edge making the weld pool wide and shallow.

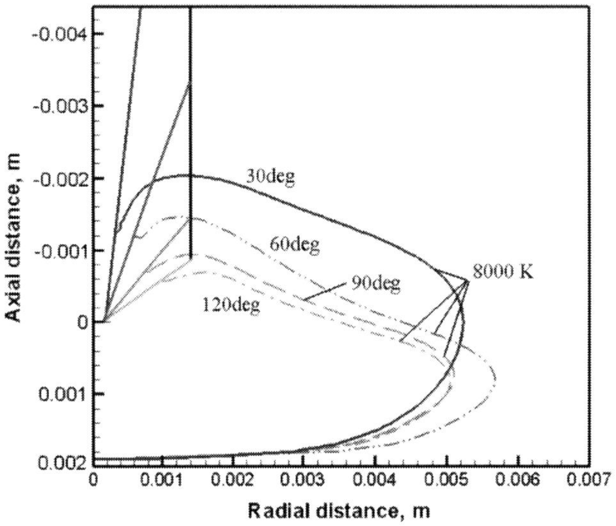

Figure 4. Argon arc temperature contours for different electrode bevel angles.

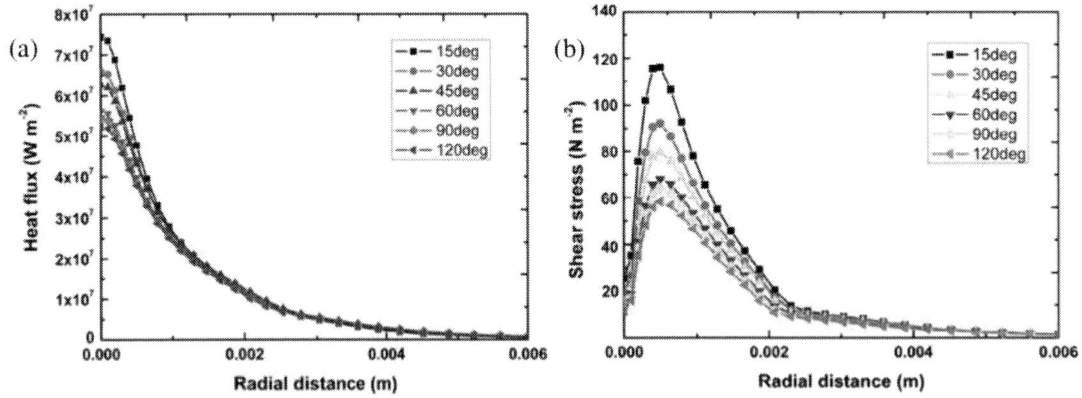

Figure 5. Effect of electrode bevel angle on (a) heat flux (b) shear stress.

4.2. Effects of electrode bevel angle on weld shape

The coupled welding arc with weld pool model was used to calculated the heat transfer and fluid flow in the weld pool under different electrode bevel angles for both high (130 ppm) and low (15 ppm) weld metal oxygen content. The corresponding welding parameters are as follows: welding current 160 A, arc length 2 mm and welding speed 2 mm/s.

Figure 6 shows the computed 3D temperature and velocity fields for a weld pool under 60° electrode bevel angle and 130 ppm weld metal oxygen content. The typical elongated pool shape due to the arc movement can be obtained. As expected, the temperature gradients are greater in front of the moving heat source than behind the heat source. For clarity, the velocity fields in only two sections of the computational domain are shown. It can be seen that the liquid metal moves from the periphery with relatively lower temperature to the center with relatively higher temperature. This is due to the liquid metal with a high concentration of surface active elements, which results in a positive temperature coefficient of surface tension [7].

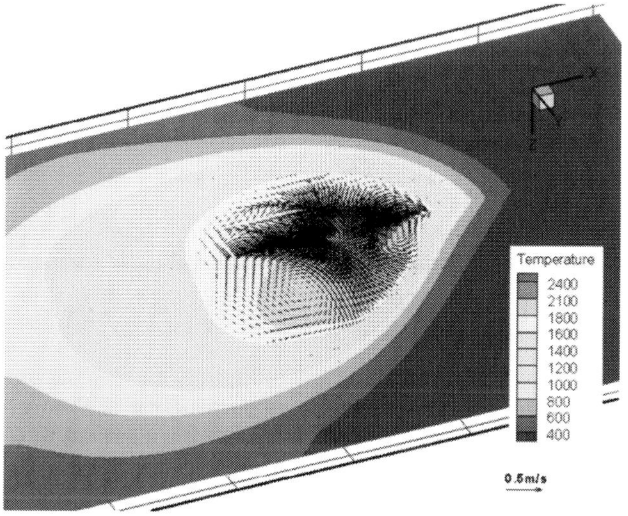

Figure 6. Calculated weld pool temperature and velocity field.

The size of the weld pool in the experiment and the calculation for different electrode angles under low and high weld metal oxygen content is listed in table 1. The comparison of weld shapes between the experiment and the simulation for 30° and 120° electrode bevel angles is shown in figures 7 and 8, respectively. The size and shape of the weld pool in the experiment and the calculation agrees well. The weld shapes are wide and shallow under low weld metal oxygen content as shown in figures 7(a) and 8(a). However, for high weld metal oxygen content, deep and narrow weld shapes are formed as shown in figures 7(b) and 8(b). In the author's previous study [7], it has been found that the direction of Marangoni convection induced by surface tension would be changed from outward to inward with the weld metal oxygen content over a critical value, and hence the weld shape changes from wide and shallow to deep and narrow, and the welding efficiency is increased. Consequently, the weld shape is determined by the weld metal oxygen content.

Table 1. The size of the weld pool between experiment and calculation for different electrode bevel angles under low and high weld metal oxygen content.

Electrode bevel angle (deg)	Low weld metal oxygen content (15ppm)				High weld metal oxygen content (130ppm)			
	Depth(mm)		Width(mm)		Depth(mm)		Width(mm)	
	Exp.	Sim.	Exp.	Sim.	Exp.	Sim.	Exp.	Sim.
15	2.12	1.90	7.53	8.50	3.50	3.70	7.59	7.40
30	2.12	2.00	8.36	8.45	3.65	3.80	7.32	7.30
45	2.07	2.10	8.39	8.45	3.63	3.85	7.06	7.20
60	2.21	2.10	7.97	8.30	3.67	3.90	7.03	7.15
90	2.07	2.25	8.19	8.25	3.83	4.00	6.82	6.90
120	2.23	2.20	6.97	8.20	3.54	4.10	6.15	6.80

Figure 9 shows the predicted and the measured weld D/W ratio under different electrode bevel angles for both high and low weld metal oxygen content. The predicted result is in good agreement with the measured one. When the electrode bevel angle is increased, the shear stress on the molten pool surface decreases. The strength of outward convection induced by shear stress is weakened. Thus, the weld width is decreased and the temperature gradient on the molten pool surface increases. Under low weld metal oxygen content, the outward Marangoni convection is strengthened. Through the interactive reaction of two convections, the weld width slightly decreases and the weld D/W ratio is not sensitive to the electrode bevel angle. When the weld metal oxygen content is high (130 ppm), the inward Marangoni convection is strengthened with increasing electrode bevel angle because of the increase of temperature gradient on the molten pool surface. Therefore, the weld width decreases, and the weld depth and the weld D/W ratio increases as the electrode bevel angle increases.

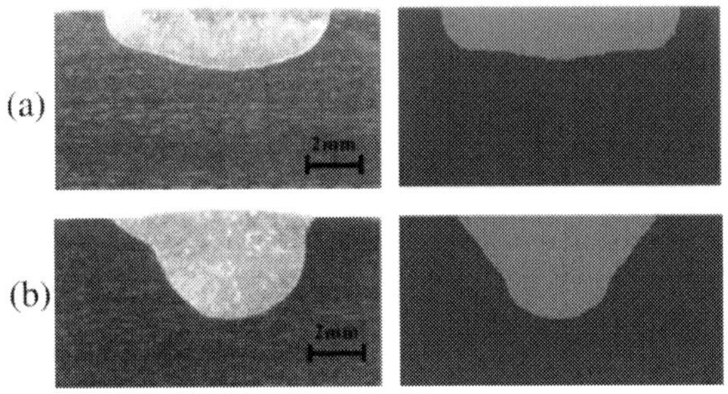

Figure 7. Comparison between experiment and calculation for weld shape (a) 15 ppm (b) 130 ppm under 30° electrode angle.

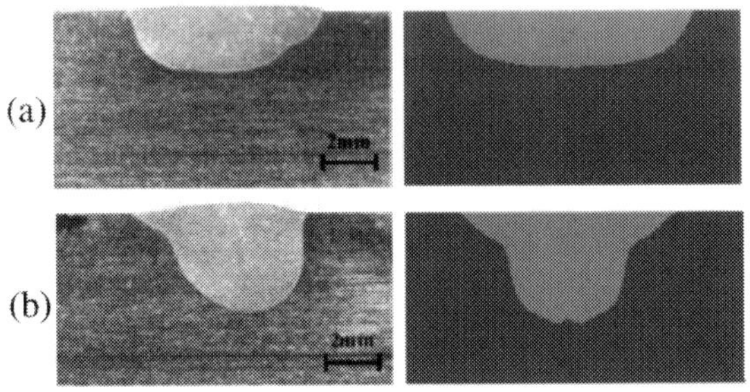

Figure 8. Comparison between experiment and calculation for weld shape (a) 15 ppm (b) 130 ppm under 120° electrode angle.

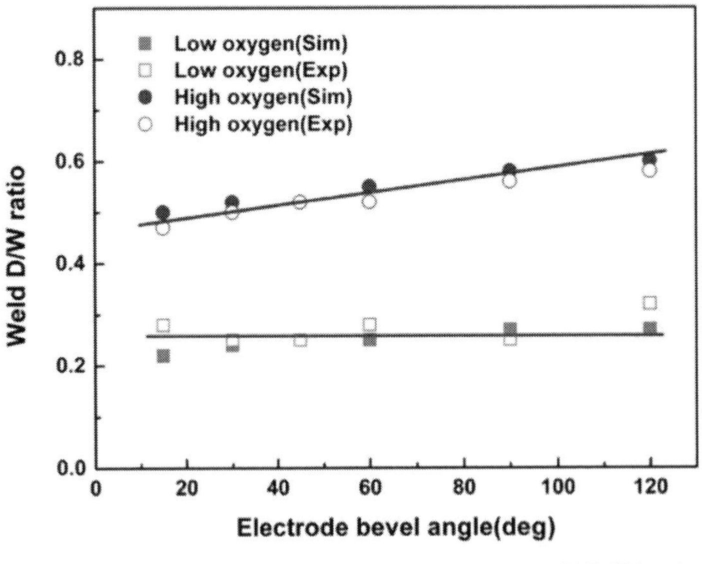

Figure 9. Comparison between experimental and calculated weld D/W ratio for different electrode angle under low (15 ppm) and high (130 ppm) weld metal oxygen content.

5. Conclusions

The effects of electrode bevel angle on welding arc properties and weld shape during GTA welding on SUS304 stainless steel were carried out using a coupled welding arc with weld pool model. Under the analogous conditions, experiments were performed to verify the simulated results. The following conclusions could be summarized:

(1) The calculated arc temperature contours agree well with the experimental results. The argon arc becomes more constricted as the electrode bevel angle increases. The peak values of heat flux and shear stress decrease and the radial distribution changes slightly with increasing electrode bevel angle.
(2) The weld shape depends on the weld metal oxygen content. Under low weld metal oxygen content, the temperature gradient on the molten pool surface increases due to the weaker shear stress with increasing electrode bevel angle, which will enhance the outward Marangoni convection. Through the interactive action of two convections, the weld width slightly increases and the weld D/W ratio is not sensitive to the electrode bevel angle. Under high weld metal oxygen content, the increase of weld penetration and the decrease of weld width cause the weld D/W ratio to continuously increase with increasing electrode bevel angle. The predicted weld shape, weld size and weld D/W ratio agrees well with the experimental ones.

Acknowledgments
The authors are grateful for the financial support from the National Science Foundation of China (Grant No. 51104142) and the Jiangsu Key Laboratory of Advanced Metallic Materials (Grant No. AMM200901).

References

[1] Kou S 2003 *Welding Metallurgy, 2nd ed* (New Jersey: Wiley-Interscience)
[2] David S A, Babu S S and Vitek J M 2003 *JOM* **55** 14
[3] Freton P, Gonzalez and Gleizes J J 2000 *J. Phys. D: Appl. Phys.* **32** 263
[4] Zacharia T, David S A, Vitek J M and DebRoy T 1989 *Weld. J.* **68** 499s
[5] Zacharia T, David S A, Vitek J M and DebRoy T 1989 *Weld. J.* **68** 510s
[6] Wang Y, Shi Q and Tsai H L 2001 *Metall. Mater. Trans. B* **32** 145
[7] Dong W C, Lu S P, Li D Z and Li Y Y 2008 *Acta Metall. Sin.* **44** 249
[8] Dong W C, Lu S P, Li D Z and Li Y Y 2009 *Int. J. Modern Phys. B* **23** 1597
[9] Lu S P, Dong W C, Li D Z and Li Y Y 2009 *Sci. Technol. Weld. Join.* **14** 509
[10] Lee S Y and Na S J 1997 *Weld. J.* **76** 484s
[11] Hong K, Weckman D C, Strong A B and Zheng W 2003 *Sci. Technol. Weld. Join.* **8** 313
[12] Hong K, Weckman D C, Strong A B and Zheng W 2002 *Sci. Technol. Weld. Join.* **7** 125
[13] Patankar S V 1980 *Numerical Heat Transfer and Fluid Flow* (New York: McGraw-Hill)
[14] Lu S P, Dong W C, Li D Z and Li Y Y 2009 *Compt. Mater. Sci.* **45** 327
[15] Sahoo P, DebRoy T and McNallan M J 1988 *Metall. Trans. B* **19B** 483
[16] Boulos M I, Fauchais P and Pfender E 1994 *Thermal Plasmas-Fundamentals and Applications, Vol. 1* (New York: Plenum)
[17] Evans D L, Tankin R S 1967 *Phys. Fluids* **10** 137
[18] Fluent Inc. 2005 *FLUENT User's Manual* (Lebanon: NH)
[19] Hsu K C and Pfender E 1983 *J. Appl. Phys.* **54** 3818
[20] Glickstein S S 1999 *Weld. J.* **78** 343s

IOP Conf. Series: Materials Science and Engineering **33** (2012) 012028 doi:10.1088/1757-899X/33/1/012028

Design of a braze alloy for fast epitaxial brazing of superalloys

S Piegert[1], **B Laux**[2] **and J Rösler**[3]

[1] Siemens AG, Sector Energy, Huttenstraße 12, 10553 Berlin, D

[2] MTU Maintenance GmbH, Münchner Straße 31, 30855 Langenhagen, D

[3] Institute of Materials, Technical University Braunschweig, Langer Kamp 6, 38106 Braunschweig, D

E-mail: sebastian.piegert@siemens.com

Abstract. For the repair of directionally solidified turbine components made of nickel-based superalloys, a new high-temperature brazing method has been developed. Utilising heterogeneous nucleation on the crack surface, the microstructure of the base material can be reproduced, i.e. single crystallinity can be maintained. In contrast to commonly used eutectic braze alloys, such as nickel-boron or nickel-silicon systems, the process is not diffusion controlled but works with a consolute binary base system. The currently applied epitaxial brazing methods rely on isothermal solidification diffusing the melting point depressants into the base material until their concentration is reduced so that the liquid braze solidifies. Contrary, the identified Ni-Mn consolute system enables a temperature driven epitaxial solidification resulting in substantially reduced process duration. The development of the braze alloys was assisted using the CALPHAD software Thermo-Calc. The solidification behaviour was estimated by kinetic calculations with realistic boundary conditions. Finally, the complete system, including braze alloy as well as substrate material, was modelled by means of DICTRA. Subsequently, the thermodynamic properties of the braze alloys were experimentally analysed by DSC measurements. For brazing experiments 300 μm wide parallel gaps were used. Complete epitaxial solidification, i.e. the absence of high-angle grain boundaries, could be achieved within brazing times being up to two orders of magnitude shorter compared to diffusion brazing processes. Theoretically and experimentally evaluated process windows reveal similar shapes. However, a distinct shift has to be stated which can be ascribed to the limited accuracy of the underlying thermodynamic databases.

1. Introduction

Advanced Ni-based superalloys such as DS and SX alloys are the materials of choice for the most demanding applications within modern gas turbines. Metal temperatures of 1000 °C are often exceeded and the components undergo severe thermomechanical loading. Hence, it is obvious that components exposed to those conditions require refurbishment to extend their life. Nowadays, two distinct braze repair processes are available:

- conventional brazing [1; 2] and
- diffusion brazing [3; 4; 5].

Conventional brazing can be conducted in economically feasible times. However, the mechanical properties are rather limited [6]. Contrary, diffusion brazing or transient-liquid phase bonding is a high strength repair technique, but, the time required for isothermal epitaxial solidification of a 300 μm wide gap can be estimated being at least as long as 50 h with B containing braze alloys [4; 6; 7; 8; 9; 10].

The aim of this study is to develop a non-diffusion based epitaxial brazing process where solidification is triggered by cooling, i.e. the process shall combine the economically short brazing times of conventional brazing and the mechanical integrity of diffusion brazing. The process relies on temperature driven epitaxial single-phase solidification of the braze joint via a planar or cellular/dendrite-like moving solid-liquid interface from the joint interfaces towards its centre as in DS and SX casting processes [11]. A solidification behaviour as aimed for was observed in rudiments and described by Hoppe [12]. However, the standard Ni-based braze alloys are eutectic systems with B and/or Si as melting point depressants (MPD). Hence, temperature driven solidification leads to a final eutectic solidification reaction disrupting the epitaxial structure of the braze joints.

As a consequence new braze alloys were developed. A substantial role played the initial definition of a process window by means of Thermo-Calc and DICTRA modelling due to the absence of experimental thermodynamic data of the according alloy systems. Subsequently, the approach of fast epitaxial temperature driven solidification was experimentally evaluated and process boundaries characterised.

2. Alloy development strategy

Other than the to date developed epitaxial brazing processes, the new brazing technique shall be applicable for wide gaps, i.e $> 250\,\mu$m, and to be carried out in economical feasible process times. Hence, the most important step is to replace the diffusion based isothermal solidification approach by a temperature driven solidification.

The basic idea of the new epitaxial brazing process relies on the following assumptions: i) solidification via cooling, ii) heterogeneous epitaxial nucleation of the solidifying liquid on the liquid-substrate interface, and iii) single-phase solidification. All three items are necessary conditions. Only if all three conditions occur, the process will result in an epitaxially solidified braze joint.

Basically, the new brazing process is carried out as a conventional brazing process, i. e. heating till T_B, a short hold time at T_B, and subsequent cooling until the joint is completely solidified. The major difference with respect to conventional brazing processes is that final eutectic solidification reactions shall not occur.

Based on the fundamental approach, numerous requirements in regard to the potential braze alloys need to be fulfilled. Basically, a combination of solidification requirements, chemical and mechanical integrity as well as service conditions needs to be addressed. Hence, the following major braze alloy requirements can be identified:

- heterogeneous epitaxial nucleation on the substrate alloys,
- single-phase solidification,
- chemical integrity,
- mechanical integrity, and
- service condition resistance.

The list demonstrates the multidimensionality of the demands. However, only the most important requirements can be evaluated here and are briefly introduced whereby the emphasis is put on the first two items and the service boundary conditions.

Adequate bonding, i.e. heterogeneous nucleation and subsequent epitaxial growth can only be provided if the braze alloy properly wets the substrate [13; 14; 15]. Usually the MPDs are rather reactive and fulfil the role of a "fluxing agent" by dissolving the thin oxide layers on the superalloys. Elements such as B, Si, P, Ti and Mn are reported to be applicable [16; 17].

An identical crystal structure and a comparable lattice parameter are the prerequisites for epitaxial growth. Moreover, nucleation of solid within the liquid braze joint has to be avoided. Otherwise stray grains would occur. A rather high undercooling for homogeneous nucleation of the braze alloy would be of advantage with respect to the stability of the process. According to literature, pure Ni can be undercooled up to 290 K [13]. However, in technical systems the obtainable amount of undercooling rather depends on the number and size of heterogeneous nuclei available [13; 14].

Besides, the single-phase solidification of the braze alloy during cooling is of substantial importance. Precipitation of a second phase from the liquid or directly on the solid-liquid interface would interfere the epitaxial solidification. Hence, a high partition coefficient $k_0 \rightarrow 1$ combined with an adequate distance from the maximum solubility of the MPD in the matrix element is favourable [13; 14]. It becomes obvious that $c_0 \leq k_0 \cdot c_{\max}^{\mathrm{MPD}}$ must hold, where $c_{\max}^{\mathrm{MPD}}$ is the maximum solubility of MPD in the matrix element and c_0 is the nominal composition of the MPD. Disadvantageous interactions with the components of the substrate due to dissolution effects should also be avoided.

Rather obvious limitations, such as T_{sol} and T_{liq} of the braze alloy, can be deduced from the service conditions and the thermodynamic properties of the substrate materials. It is thus beneficial to keep $T_{\mathrm{liq}}^{\mathrm{braze}}$ between 1150 and 1250 °C and $T_{\mathrm{sol}}^{\mathrm{braze}}$ above 1100 °C. Even at T_{B}, which generally lies at least 10 K above $T_{\mathrm{liq}}^{\mathrm{braze}}$, incipient melting of the DS or SX substrate alloys shall be avoided so that T_{B} must not exceed 1260 °C [18].

Considering the above discussed restrictions regrading the shape of the phase diagrams, nucleation and service conditions, only a nickel-based braze alloy can fulfil the requirements. Combining Ni-MPD characteristics such as k_0 and $x_{\max\,\mathrm{sol}}$, the Ni-Mn system seems to be promising as the phase diagram emphasises (fig. 1(a)).

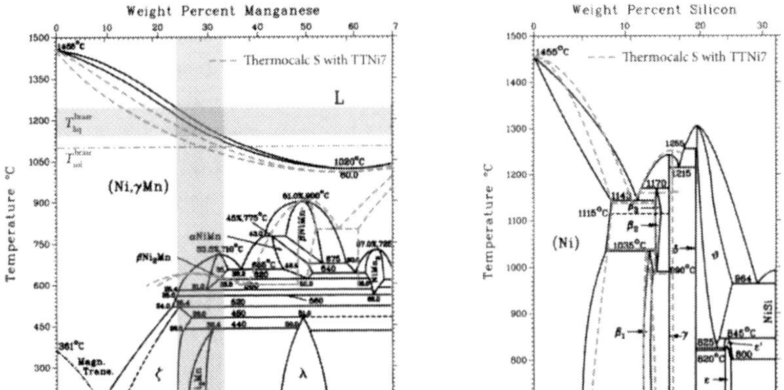

(a) Ni-Mn binary phase diagram taken from [19]

(b) Ni-Si binary phase diagram taken from [19]

Figure 1. Potential MPD combination including the according limits of solidus and liquidus temperatures (1150 °C $\leq T_{\mathrm{liq}}^{\mathrm{braze}}$ 1250 °C and $T_{\mathrm{sol}}^{\mathrm{braze}} \geq$ 1100 °C). The binaries as provided by calculations are labelled via the dashed lines.

The system shown is positive aceotropic with full solubility of the MPD in the matrix element Ni below the solidus line. This implies that the crystal structure is retained throughout the whole composition range at high temperatures. It can also be recognised that a rather narrow melting interval is present so that k_0 is rather high.

However, the efficiency of the proposed MPD (Mn) is limited. A high amount of it needs to be added to achieve the desired melting interval as discussed above. The grey shaded areas in fig. 1(a) assign those requirements. Therefore, alloying of a secondary MPD might be meaningful. The secondary MPD needs to fulfil similar requirements as discussed for the primary MPD. To reduce the overall amount of MPD within the braze alloy a higher degree of efficiency is essential. Hence, Si, Ge, Ga or even Ti are potential secondary MPDs [20; 21; 22].

Based on a patent which describes brazing of steels and nickel alloys using Ni-Mn-Si braze alloys without the formation of detrimental phases on very narrow gaps (50 μm) it was decided to implement Si as secondary MPD [23]. The binary Ni-Si phase diagram is displayed in fig. 1(b).

3. Computational definition of a process window

3.1. Computational setup

Thermodynamic characterisation of the Ni-Mn-Si ternary system was carried out by the CALPHAD software Thermo-Calc S applying the thermodynamic database TTNi 7 [24]. The simulation of the solidification behaviour of the braze alloys themselves and the substrate-braze joint system was realised by means of DICTRA with the MobNi1 database.

Preliminary calculations, as shown in fig. 3(b), revealed that the solidification of the desired Ni-Mn-Si alloys could potentially terminate in the eutectic reaction

$$L \rightarrow \gamma + Mn_6Si_7Ni_{16} \ . \tag{1}$$

The reaction can be reflected by a model as given in fig. 2(a) starting with 100 % liquid and assigning the solid phases as "passive" to both sides of the liquid. As soon as the solvus temperature of the respective phase is reached it becomes active and grows into the liquid. At any time and temperature there is thermodynamic equilibrium at the solid-liquid interfaces. The width of the simulation domain reflects the half-width of a standard braze joint as described in sec. 4.1.

To model the solidification of a liquid braze joint, the Ni-based superalloy was attached to the right of the liquid as illustrated in fig. 2(b). Only half of the gap width was included in the model due to reasons of symmetry. At the centre line of the joint $Mn_6Si_7Ni_{16}$ and γ' were attached as passive phases being an indication of a final eutectic reaction if their status changed to active throughout the simulation. The γ'-phase was added due to the fact that Si rich alloys tend to form a γ-γ'-eutectic.

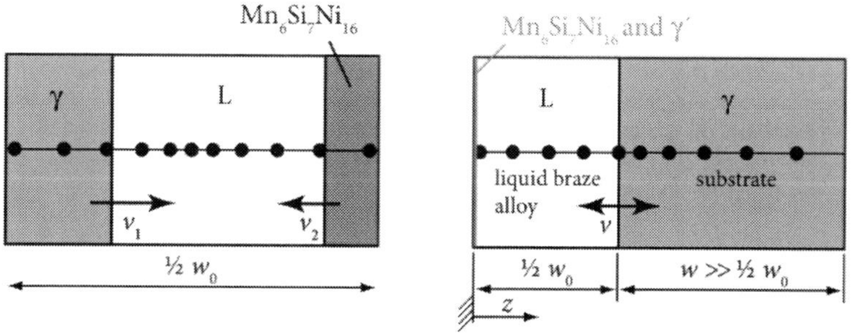

(a) Single cell model with liquid as active phase and γ as well as $Mn_6Si_7Ni_{16}$ attached to either sides of liquid as inactive phases.

(b) Geometry of the DICTRA model as utilised for the calculation of fast epitaxial brazing.

Figure 2. Overview of the boundary conditions and the model of the eutectic solidification of Ni-Mn-Si braze alloys.

IOP Conf. Series: Materials Science and Engineering **33** (2012) 012028 doi:10.1088/1757-899X/33/1/012028

The advantage of simulating only the eutectic reaction of the braze material itself is the limited calculation domain combined with only three components present (see fig. 2(a)). This lowers the calculation effort significantly compared to the model shown in fig. 2(b). However, it does not reflect the dissolution of the substrate alloy which might lead to inaccuracy.

3.2. Results

First thermodynamic calculations of the ternary Ni-Mn-Si system are shown in fig. 3 in the form of quasibinary sections. All diagrams show that the addition of Si as secondary MPD does not lead to a significant widening of the solidus-liquidus interval, i. e. k_0 is large. Furthermore, the boundary conditions, such as the required solidus and liquidus temperatures of the braze alloy, as proposed above, can be achieved. Both MPDs still show a high solubility in the Ni-matrix although they tend to form the stoechiometric $Mn_6Si_7Ni_{16}$-phase which is a cubic $Fm\bar{3}m$ crystallographic structure of the $Mn_{23}Th_6$ prototype with $a = 11.06$ Å [25]. In fig. 3 $Mn_6Si_7Ni_{16}$ is labelled as `G_PHASE`.

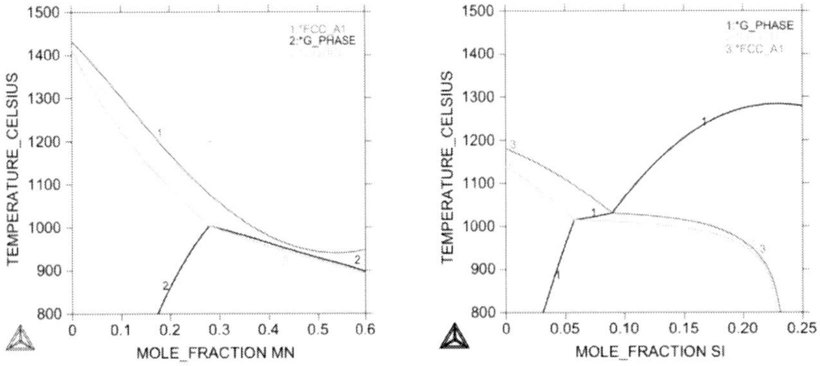

(a) $x_{\mathrm{Si}} = 5$ at.-% (b) $x_{\mathrm{Mn}} = 25$ at.-%

Figure 3. Quasibinary phase diagrams of the Ni-Mn-Si system with varying Mn and Si contents.

In order to quantify under which conditions $Mn_6Si_7Ni_{16}$ is formed in the Ni-Mn-Si space, solidification was calculated by means of the model shown in fig. 2(a) with a meaningful cooling rate of 0.01 K/s (compare to tab. 1). As an example the results of a Ni - 15 Mn - 6 Si (all values in at.-%) alloy are plotted in fig. 4(a). Various cooling rates are compared to equilibrium and Scheil solidification. It can be seen that the cooling rate of 0.01 K/s is just low enough to avoid the formation of $Mn_6Si_7Ni_{16}$.

To obtain a process window of possible braze alloys the boundary conditions were merged on the liquidus surface of the Ni-Mn-Si system in fig. 4(b). Basically, three distinct boundary conditions span the composition space: i) $T_{\mathrm{liq}} = 1250\,°\mathrm{C}$, ii) $T_{\mathrm{sol}} = 1100\,°\mathrm{C}$, and iii) single-phase solidification at a cooling rate of 0.01 K/s. The first two conditions can easily be obtained from isothermal sections at the distinctive temperatures. The third condition was derived according to the procedure described above identifying compositions which sacrifice single-phase solidification under the given cooling rate.

The brazing processes of the alloys given in tab. 1 were simulated using the model shown in fig. 2(b). The result for braze alloy Lot 3 is plotted in fig. 5. The hold time at brazing temperature T_{B} amounted to 30 min. The plot shows that Mn piles up to high concentrations in the gap centre leading to a thin layer of $Mn_6Si_7Ni_{16}$. Hence, according to the calculation solidification of braze alloy Lot 3 terminates in a final eutectic reaction under the given boundary conditions (see tab. 1).

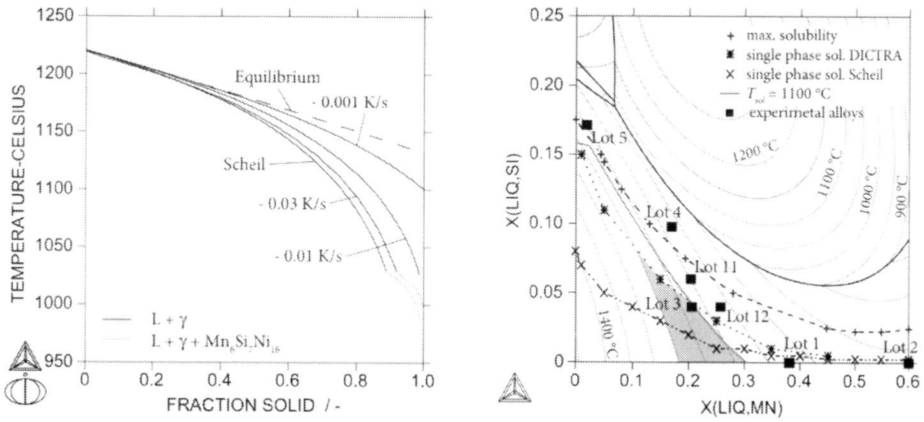

(a) Diagram of the fraction of solid formed at particular temperatures depending on the applied cooling rate.

(b) Liquidus surface of the Ni-Mn-Si ternary system with superposed information from thermodynamic equilibrium, Scheil and diffusion calculations ($\dot{T} = -0.01\,\mathrm{K/s}$). The bold black lines assign monovariants. The remaining window for a braze alloy according to the requirements is highlighted in grey.

Figure 4. Illustration of the solidification behaviour of the ternary braze alloy Ni-15 Mn-6 Si (all values given in at.-%) (left) and the liquidus surface of the Ni-Mn-Si ternary system with overlaid information (right).

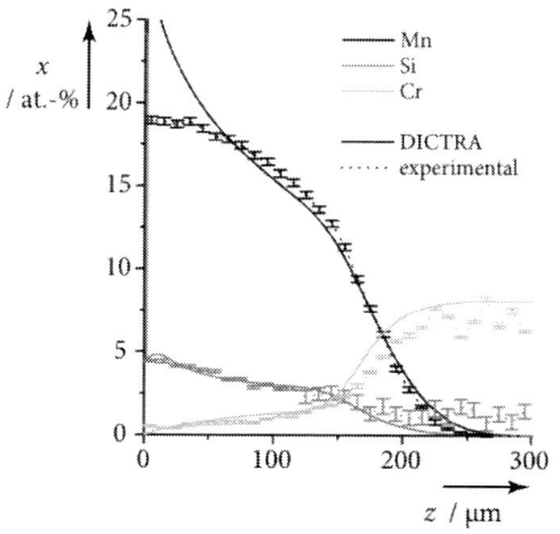

Figure 5. Theoretically and experimentally obtained element distributions after the standard heat treatment of braze alloy Lot 3 (see tab. 1); $z = 0$ assigns the gap centre and $z = 150\,\mu$m the initial gap widths of the joint.

The Si concentration shows a distinctive hump at the initial solid-liquid interface and moderately increases towards $z = 0$. However, a final drop can be stated due to the eutectic reaction. Cr is an element present in the substrate alloy. Its presence in the gap centre can mainly be attributed to the dissolution of the substrate during widening (compare to fig. 9).

4. Experimental validation

4.1. Experimental methods

For experimental validation of the process window various braze alloys were produced in an arc furnace under argon atmosphere. The solvus temperatures and the melting intervals of the braze alloys were measured on 50 to 150 mg weighing bulk samples in a NETSCH DSC 404C at heating and cooling rates of 10 K/min.

Brazing trials were carried out on 2 mm thick sheet material of the single crystalline superalloy René N 5. In the sheet material a 300 μm wide gap was introduced perpendicular to $\langle 001 \rangle$ whereby the braze alloy was placed as a slice of bulk material on top of the gap. The key data of the brazing cyles for the respective alloys are listed in tab. 1. The brazing temperature T_B is held for 30 min followed by linear cooling down to T_F within a time t_{BF}. The according cooling rate is given in the last column of tab. 1. T_F lies well below the solius of the braze alloy. After reaching T_F the furnace was switched off.

Table 1. Braze alloys for experimental validation of the fast epitaxial brazing concept including the braze heat treatment boundaries.

	components			DSC		Thermo-Calc		brazing cycle			
alloy	Ni	Mn	Si	T_{sol}	T_{liq}	T_{sol}^{TC}	T_{liq}^{TC}	T_B	T_F	t_{BF}	$\dot{T} \cdot 10^{-3}$
	/ at.-%			/ °C		/ °C		/ °C		/ min	/ K/s
Lot 1	bal.	38.2		1114	1150	1057	1075	1160	1000	300	-8.9
Lot 2	bal.	60.0		1022	1022	1010	1010	1040	925	300	-6.4
Lot 3	bal.	20.6	4.0	1176	1250	1104	1176	1260	1000	300	-14.4
Lot 4	bal.	17.0	9.8	1031	1184	1048	1121	1150	1000	60	-41.7
Lot 5	bal.	1.9	17.1	1124	1238	1101	1195	1260	1000	300	-14.4
Lot 11	bal.	20.4	6.0	1173	1248	1062	1148	1260	1000	300	-14.4
Lot 12	bal.	25.7	4.0	1139	1197	1051	1118	1210	1000	300	-11.7

Subsequent to brazing all specimens were etched with V2A pickle and investigated by means of light microscopy. V2A etchant turned out to be feasible to visualise grain boundaries as well to remove the final deformation layer after polishing for proper EBSD analyses.

For more detailed microstructural imaging the SE and BSE detectors of a LEO 1550 scanning electron microscope were utilised. Element distributions were measured by means of EDX applying an acceleration voltage of 20 keV. In order to verify the crystal structure of the braze joints, EBSD measurements were carried out with an HKL Nordlys detector. Misorientation maps were developed where the crystal orientation within the joint is related to that of the adjacent substrate. Grain boundaries were assigned to a distinct misorientation of 15 °. However, 8 ° boundaries were monitored as well since this is substantial for the mechanical properties.

4.2. Results

An overview of the investigated braze alloys can be gained from fig. 4(b) and tab. 1. The alloys can be subdivied into binary Ni-Mn systems and ternary Ni-Mn-Si with moderate and high amounts of the secondary MPD Si. This classification is reflected in fig. 6.

The binary alloys show a very narrow melting interval (Lot 1) and the aceotropic composition (Lot 2) only a melting point (fig. 6(a)). The solidus and liquidus temperatures given in tab. 1 coincide well with the experimental binary phase diagram shown in fig. 1(a). According to the cooling curves single-phase solidification should occur. Wider peaks can be observed for the brazes moderately alloyed with Si. However, the cooling curves verify single-phase solidification showing only one peak. This pattern changes entirely for the higher Si containing alloys. The melting interval increases to above 100 K as tab. 1 reveals. Braze alloys Lot 4 and 5 shown in fig. 6(c) solidify under formation of at least two phases as deduceable from the cooling curves.

The results of brazing trials on 300 μm wide braze joints according to the parameters listed in tab. 1 are summarised for Lot 2, Lot 4 and Lot 5 in fig. 7. The solidification of Lot 2 results in

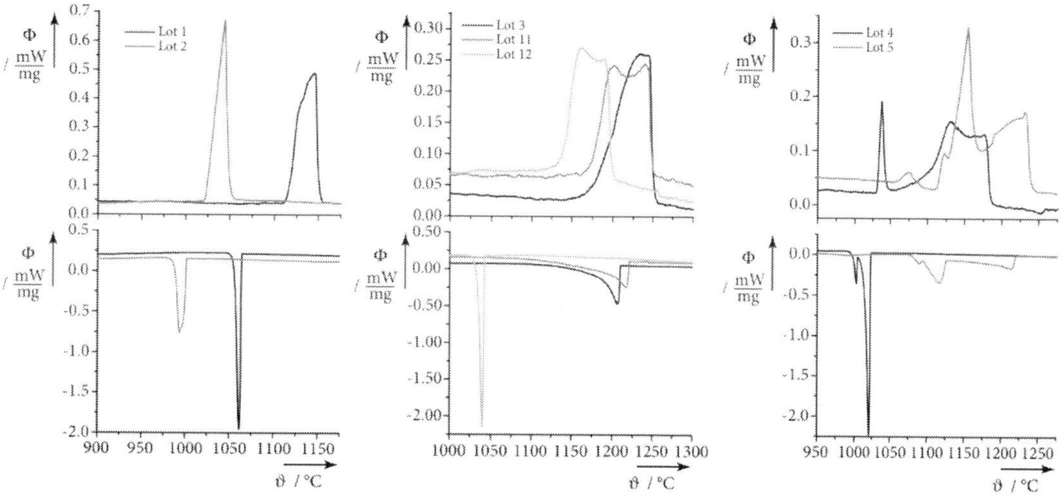

(a) Binary Ni-Mn alloys.
(b) Ternary Ni-Mn-Si alloys with $x_{Si} \leq 6$ at.-%.
(c) Ternary Ni-Mn-Si alloys with $x_{Si} \geq 6$ at.-%.

Figure 6. Heating (top of each plot) and cooling (bottom of each plot) curves of DSC measurements of the experimental braze alloys as listed in tab. 1.

a needle-like microstructure with large stray grains and defects (fig. 7(a)). Brazing with Lot 4 terminates in a eutectic reaction in the joint centre as given in eq. 1 (fig. 7(b)). However, the rest of the joint is epitaxially bonded to the substrate. A similar behaviour was found to occur in braze joints brazed with Lot 5 as shown in fig. 7(c). Here, solidification ends up with the formation of bulk γ' which appears as dark phase in the gap centre. This was confirmed via EDX and EBSD measurements.

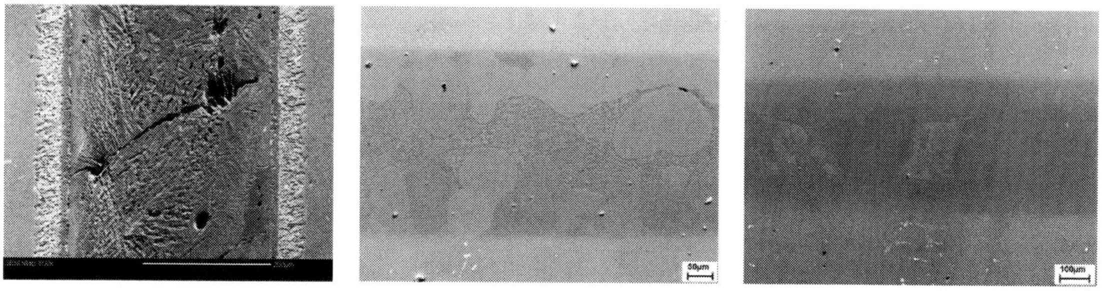

(a) SE micrograph of a joint brazed with Lot 2.
(b) SE micrographs of a joint brazed with Lot 4.
(c) SE micrograph of a joint brazed with Lot 5

Figure 7. Micrographs of various brazed joints. The brazing parameters of each alloy can be obtained from tab. 1.

Figure 8 depicts the results of the EBSD investigation of a standard gap brazed with Lot 3 applying the parameters as listed in tab. 1. The left micrograph on the left shows the SE image of the investigated area. In fig. 8(b) the misorientation map is displayed. It can be realised that neither subgrains nor stray grains are present. Discrete changes in crystal orientation, as typical for grain boundaries, are absent, only slight gradual changes by a maximum of 6 ° (green colour) can be observed.

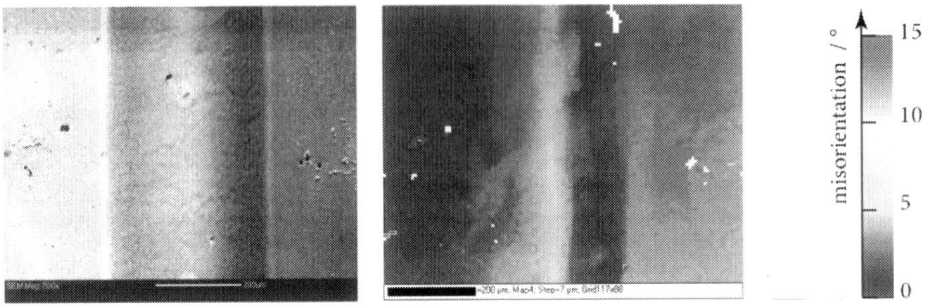

(a) SEM micrograph of the brazed joint.

(b) EBSD map of misorientation distribution with respect to the substrate orientation.

(c) colour scheme for fig. 8(b)

Figure 8. EBSD analysis of a gap brazed with Lot 3 according to tab. 1.

Similar results as shown in fig. 8 could be obtained from braze trials with braze alloys Lot 1, Lot 11, and Lot 12. It is noteworthy that Lot 11 is a limiting case in so far as wide regions are solidified in a single-phase way. However, few micrometer wide eutectic lines occurred within the joint centre under standard brazing conditions as listed in tab. 1.

5. Discussion

5.1. Experimental and theoretical results

A comparison between experimental and theoretically obtained melting intervals reveals that both T_{liq} and T_{sol} are generally underestimated (compare values in tab. 1). However, the melting interval, i.e. $T_{\text{liq}} - T_{\text{sol}}$, decreases in the same range. Obviously, there are discrepancies between the TTNi7 database and the experimental phase diagrams as plotted in fig. 1. The dashed lines indicate the calculated binaries. Within the Ni-Mn system melting is predicted at temperatures being too low.

Moreover, the kinetic calculations reveal some discrepancies in regard to the experimentally observed behaviour of the systems. Generally, simulations overestimate the occurrence of the final eutectic reaction due to a massive pileup of the Mn concentration towards the centre of the gap as fig. 5 shows. EDX measurements, in comparison, do not indicate this particular phenomenon. The reasons for this behaviour can be found in the overestimated melting interval for the Ni-Mn system, i.e. the calculated k_0 is too small, and in the lower T_{B} used for the calculations. T_{B} had to be adjusted to achieve comparable dissolution width of the base material in calculations and experiments.

5.2. Proof of concept

Five out of seven experimental alloys fulfil the requirements regarding T_{sol} and T_{liq} (compare tabs. 1 and 2). Multi-phase solidification was predicted for braze alloys Lot 4 and 5 by means of DSC analyses (fig. 6(c)) as well as by thermodynamic calculations as shown in fig. 4(b). Finally, temperature driven epitaxial brazing was achieved with four of the selected experimental braze alloys. An overview of the respective results is given in tab. 2 revealing that single-phase solidification in DSC measurements is sine qua non but not necessarily a sufficient condition. The interaction with the base material needs to be taken into account as well.

In regard to brazing quality, the best results were achieved with braze alloys showing a moderate Si content, i.e. with Lot 3 and Lot 12 as fig. 4(b) indicates. As mentioned above, braze alloy Lot 11 is a limiting case showing sparsely distributed indications of a centre line eutectic.

Table 2. Summary of the results of the standard brazing cycles of tab. 1 in combination with the fulfilment of the requirements with respect to T_{sol} and T_{liq} as well as single-phase solidification.

alloy	T_{sol}	T_{liq}	single-phase DSC	trial
Lot 1	✓	✓	✓	✓
Lot 2			✓	
Lot 3	✓	✓	✓	✓
Lot 4		✓		
Lot 5	✓	✓		
Lot 11	✓	✓	✓	✓
Lot 12	✓	✓	✓	✓

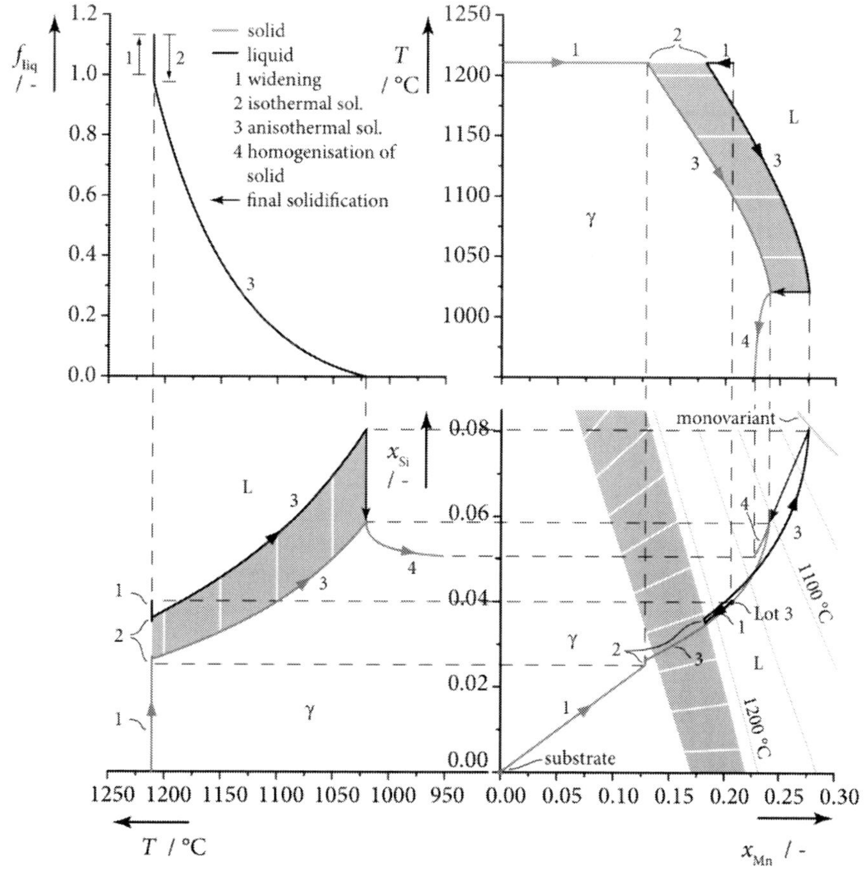

Figure 9. Temperature driven single-phase solidification of a 200 μm wide with Lot 3 brazed joint simulated by means of DICTRA. The solidification path is shown from various perspectives illustrating the relation between solid (grey bold lines) and liquid (black bold lines) at the solid-liquid interface. Within the $\gamma + L$ fields the tie lines (white lines) are shown as DICTRA assumes thermodynamic equilibrium between solid and liquid phase across the interface.

5.3. Mechanisms of fast epitaxial brazing

Thermodynamic and kinetic calculations are a sufficient tool to understand the mechanisms during temperature driven solidification in the systems at hand since an in-situ observation of the processes is not possible. An overview is provided by fig. 9 indicating an isothermal projection of the Ni-Mn-Si- ternary at T_B (bottom right), the two projections of T vs. Si and Mn content (lower left and upper right), as well as the amount of liquid f_{liq} (upper left).

According to fig. 9 the following four distinct phases can be distinguished: 1) widening of joint/dissolution of substrate, 2) isothermal solidification at T_B within the 30 min hold time, 3) anisothermal (temperature driven) solidification and 4) homogenisation of braze joint after complete solidification. The particular stages of the process can also be found in conventional and diffusion brazing processes. However, the combination of those steps while maintaining single-phase epitaxial solidification is unique.

6. Conclusions

In summary, three major findings could be accomplished within the scope of the present study:

- a basic concept was developed - MPDs and a window of possible braze alloy compositions were defined theoretically;
- the underlying mechanisms were evaluated and in principle understood, i. e. solute pileup and the role of k_0, heterogeneous nucleation, advancement of the solid-liquid interface etc., and
- a proof of concept was successfully conducted.

Besides, it was shown that the combination of experimental methods and simulation generates a kind of symbioses leading to improved understanding of newly developed material systems at very early stages of investigation.

Even though experimental and theoretical findings show some divergence, it was possible to generate first experimental alloys. The final and experimentally validated process window is indicated in fig. 10. Experimental and theoretically obtained process windows show an overlap demonstrating the feasibility of the theoretical approach.

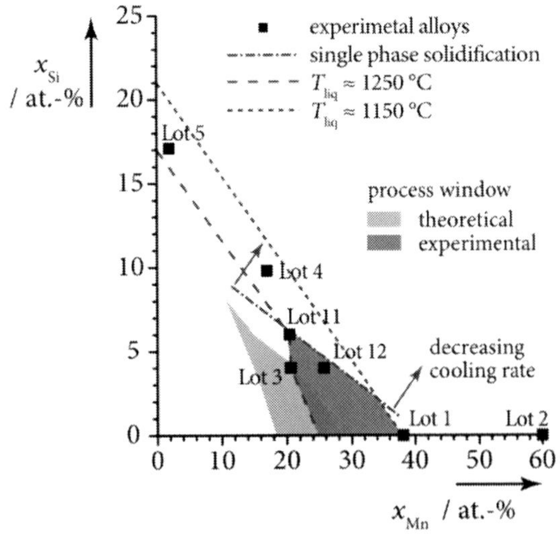

Figure 10. Comparison of the process windows as obtained from theoretical predictions by means of Themo-Calc and DICTRA calculations (light-grey area, compare to fig. 4(b)) and experimental results (dark-grey area). The isotherms are combined results of DSC measurements as carried out within the scope of the work at hand (see tab. 1) and published binary phase diagrams (see figs. 1(a) and 1(b)) [19]. The boundary line for single-phase solidification is estimated from the brazing results of the standard brazing cycles in tab. 1. A reduced cooling rate would increase the size of the process window which is indicated by the arrows (compare to fig. 4(a)).

References

[1] Demo W A and Ferrigno S J 1992 *Advanced Materials and Processes* 43 – 45
[2] Ellison K A, Lowden P and Liburdi J 1994 *Journal of Engineering for Gas Turbines and Power* **116** 237 – 242

[3] Duvall D S, Owczarski W A and Paulonis D F 1974 *Welding Journal* 203 – 214

[4] Schnell A, Stankowski A and de Marcos E 2006 *ASME Turbo Expo 2006: Power for Land, Sea and Air* Proceedings of GT2006 (American Society of Mechanical Engineers) pp 949 – 961

[5] Sinclair C W, Purdy G R and Morral J E 2000 *Metallurgical and Materials Transactions A* **31** 1187 – 1192

[6] Broomfield R W 2000 *Parsons 2000 Advanced Materials for 21st Century Turbines and Power Plant: Proceedings of the 5th International Charles Parsons Turbine Conference* ed Strang A The Institute of Materials (London, UK: IOM Communications) pp 741 – 754

[7] Heinz P 2008 *Diffusionslöten von einkristallinen Nickelbasis Superlegierungen mit binären Nickel-Germanium-Lotwerkstoffen* Dissertation Universität Erlangen-Nürnberg

[8] Idowu O A, Richards N L and Chaturvedi M C 2005 *Materials Science and Engineering* **A 397** 98 – 112

[9] Zhongci R, Shuncai W and Yunrong Z 1996 *Scripta Materialia* **34** 163 – 168

[10] Zhou Y, Gale W F and North T H 1995 *International Materials Review* **40** 181 – 196

[11] Donachie M J and Donachie S J 2003 *Superalloys: A Technical Guide* 2nd ed (Materials Park, Ohio: ASM International)

[12] Hoppe B 2003 *Verhalten gelöteter Nickelbasis-Superlegierungen unter thermomechanischer Ermüdungsbelastung* Dissertation Technische Universität Carolo-Wilhelmina zu Braunschweig Braunschweig, Germany

[13] Gottstein G 2007 *Physikalische Grundlagen der Materialkunde* 3rd ed (Springer-Verlag)

[14] Kurz W and Fisher D J 1998 *Fundamentals of Solidification* 4th ed (Switzerland: Trans Tech Publications Inc)

[15] Lugscheider E, Schmoor H and Eritt U 1995 *Development of Filler Metalls* (*DVS-Berichte* vol 166) (Düsseldorf: DVS-Verlag GmbH)

[16] Lugscheider E and Humm S 2002 *Advanced Engineering Materials* **4** 138 – 142

[17] Meier A, Gabriel V, Chidambaram P R and Edwards G R 1995 *Materials and Manufacturing Processes* **10** 625 – 641

[18] Bürgel R 2006 *Handbuch Hochtemperatur-Werkstofftechnik* 3rd ed (Wiesbaden, Germany: Vieweg & Sohn Verlag / GWV Fachverlage GmbH)

[19] Massalski T B (ed) 1986 *Binary Alloy Phase Diagrams* vol 2 (Metals Park, Ohio: Scott, W. W.)

[20] Heinz P, Volek A, Singer R F, Dinkel M, Pyczak F, Göken M, Ott M, Affeldt E and Voßberg A 2008 *Defect and Diffusion Forum* **273 – 276** 294 – 299

[21] Laux B, Piegert S and Rösler J 2009 *Metallurgical and Materials Transactions A* **40** 138 – 149

[22] Laux B, Piegert S and Rösler J 2010 *Journal of Engineering for Gas Turbines and Power* **132** 032101 URL `http://link.aip.org/link/?GTP/132/032101/1`

[23] Shirai M, Sasabe K and (Assignee) T W I S P 2004 Nickel based filler metal for brazing United States Patent Application Publication 2004/0009090 A1

[24] Lukas H L, Fries S G and Sundman B 2007 *Computational Thermodynamics* (Cambridge, UK: Cambridge University Press)

[25] Villars P 1997 *Pearsons's Handbook Desk Edition: Crystallographic Data for Intermetallic Phases* vol 2 (Materials Park, Ohio: ASM International)

An integrated framework for multi-scale multi-physics numerical modelling of interface evolution in welding

M Tong[1,9], J Liu[2], Yu Xie[2], H B Dong[2], R L Davidchack[2], J Dantzig[3], D Ceresoli[4], N Marzari[4,10], A Cocks[4], C Zhao[5], I Richardson[5], A Kidess[5], C Kleijn[5], L Hoglund[6], S W Wen[7], R Barnett[8] and D J Browne[1]

[1]University College Dublin, Ireland
[2]University of Leicester, UK
[3]Ecole Polytechnique Federale de Lausanne, Switzerland
[4]University of Oxford, UK
[5]Delft University of Technology, the Netherlands
[6]Royal Institute of Technology, Sweden
[7]Tata Steel, R, D&T, UK
[8]TWI Ltd, UK.

E-mail:mingming.tong@ucd.ie

Abstract. The project *Modelling of Interface evolution in advanced Welding* (MIntWeld) is a 4-year international research project funded by the European Commission under their FP7 programme. Its main target is to develop a numerical toolbox which can be used to predict the evolution of interfaces during welding. There are various interfaces involving multiple phenomena and different spatial scales, which can be simulated using corresponding numerical modelling methods respectively. The modelling methods include quantum dynamics, molecular dynamics, phase field, phase field crystal, computational fluid dynamics, phase transformation and heat transfer, thermodynamics, continuum mechanics and life and defects prediction. Although each modelling method is based on different physical theories and involves different scales, they are not isolated. Therefore, this project aims to design a common framework which couples each model with the upstream and/or downstream model at the relevant neighbouring length scales. The data exchange framework which underpins the coupling of the models is described, and typical examples addressing the solution to the challenges faced, such as those of data interpolation between one discretisation of the computational domain and another, are discussed. Initial successes from the model-linking efforts of the authors are also presented.

[9]To whom any correspondence should be addressed.
[10]Currently working at Ecole Polytechnique Federale de Lausanne, Switzerland

1. Introduction

As a major technique of joining materials, welding has attracted intensive research wordwide. In the process of welding, several interfaces are simultaneously present and moving, spanning many different spatial scales and involving multiple physical phenomena. The properties of the interfaces have dominant effects on the macroscopic performance (e.g. mechanical properties) of the as-welded work piece. Multi-scale multi-physical numerical modelling is essential for accurately predicting the evolution of interfaces during the process of welding. Although a variety of related numerical models have been developed separately at different scales, such as the models on atomic scale [1,2], mesoscale (from cm to tens of μm) [3,4] and macroscale [5,6], they have not been successfully integrated [7].

The project *Modelling of Interface evolution in advanced Welding* (MIntWeld) is a 4-year international research project funded by the European Commission under their FP7 programme. In the project, for the purpose of modelling interface evolution in welding, a variety of numerical models on different scales are being developed (as shown in table 1) in order to study the relevant physical phenomena, which span a wide range of spatial scales (from atomic to macro) and a wide range of time scales. A data exchange framework (DEF) has been developed in order to link the respective single-scale models together. This model linking strategy aims to make the respective numerical models benefit the models on their neighbouring spatial scales and, in particular, to build a multiscale modelling toolbox that will enable smooth technology transfer from MIntWeld researchers to industry. Scientifically, this integrated multi-scale modelling work is trying to numerically predict the formation of the weld pool, the advance of the solidification interface, evolution of microstructure, element segregation at the grain boundary and their dependence on welding process parameters. Moreover, the failure of welds during solidification of the weld pool (hot cracking) and in service (cold cracking) may be predicted based on the simulations of interface evolution.

Table 1. Research groups and their numerical models involved in MIntWeld.

Model	Spatial scale	Responsible institute *superscripts as per author affiliation list*
Ab-initio	Quantum dynamics	Oxford[4]
Molecular dynamics (MD)	Atomic	ULEIC[2]
Phase field (PF) and phase field crystal (PFC)	Micrometer	EPFL[3]
Phase field (PF)	Micrometer	ULEIC[2]
Front tracking (FT)	Mesoscale (millimeter)	UCD[1]
CFD	Macroscale (centimeter or larger)	TUD[5]
Thermodynamics	Continuum	KTH[6]
Life prediction	Continuum	Oxford[4]
Elastoplastic mechanics (based on finite element software)	Macroscale	Tata Steel[7]
Heat treatment and in service performance of welds	Mesoscale and macroscale	TWI[8]

This paper addresses the computational aspect of the model linking activity (work package 2) in the MIntWeld project. There are two separate work packages (work packages 4 and 5) especially dedicated to the validation of numerical models. In WP4, synchrotron experiments and TEM and 3D atom-probe techniques are employed in related experiments in order to validate the numerical models on nano and micro scales. In WP5, in the meantime, welding trials and characterization are being conducted in laboratory and pilot plants in order to perform full scale validation. Because the validation work is very comprehensive, involving contributions from 7 experimental partners, details are beyond the scope of this contribution, but will be published later and separately elsewhere when results become available.

2. Data exchange framework (DEF)

The numerical models listed in Table 1 are not isolated from each other but are being linked together, in order to appropriately formulate the multi-physics and predict the evolution of interfaces involved in the process of welding. A data exchange framework has been designed to facilitate such linking. The first step is the analysis of the potential data flow involved.

2.1. *Data flow among numerical models of MIntWeld*

The data transfer between different MIntWeld numerical models is shown schematically in figure 1. In this figure, the arrows indicate the direction of data flow and the text briefly clarifies the details of the data to be transferred. WP3 and WP4 refer to the modelling work-packages of the MIntWeld project. Firstly, it can be seen that there are some one-way data flows. For example, the ab-initio model of Oxford will transfer data to the downstream molecular dynamics (MD) model of ULEIC; however, the MD model does not need to transfer data back to the upstream ab-initio model. Another example of one-way data transfer is that from the front-tracking (FT) model of UCD and the CFD model of TUD to the finite element modelling work of Tata Steel. Because Tata Steel is the end user of the FT and CFD models, there is no feedback of data from it. Secondly, besides the one-way data flows, there are two-way data flows, in which two numerical models mutually exchange data. For example, the FT model of UCD and the CFD model of TUD mutually exchange data.

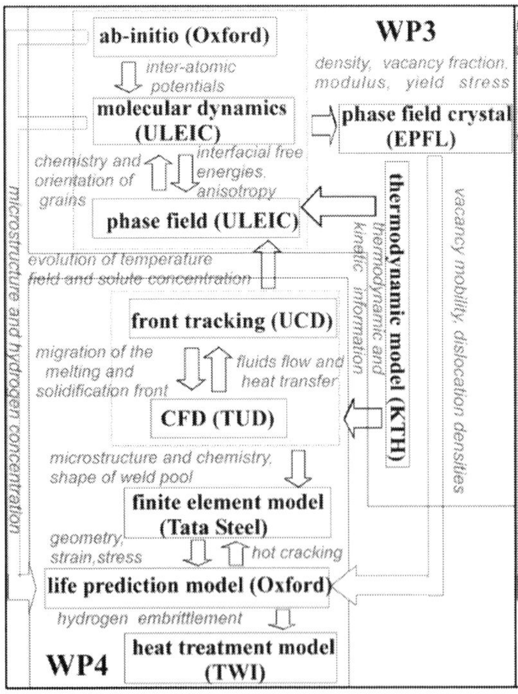

Figure 1. Data flow among MIntWeld numerical models

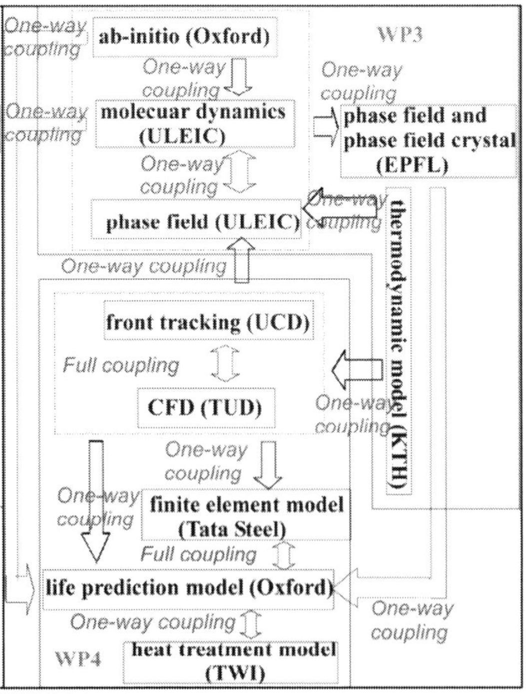

Figure 2. Design of the MIntWeld data exchange framework (DEF).

IOP Conf. Series: Materials Science and Engineering **33** (2012) 012029 doi:10.1088/1757-899X/33/1/012029

2.2. Design of DEF

Based on the analysis of data flow illustrated in figure 1, a data exchange framework (DEF) has been designed, which is illustrated in figure 2. Here, it can be seen that most of the linking of neighbouring models can be done through one-way coupling. In such one-way couplings, the data flow is either one-way or two-way, but the inputs to the beneficiary model are not affected by the outputs from it. Full coupling has to be employed, in order to link models where outputs affect inputs. They include the linking of the FT model (UCD) with CFD model (TUD) and the linking of the life prediction model (Oxford) with the finite element model (Tata Steel). The reason for the use of full coupling is that the data flow is not only two-way but also has to be updated frequently according to the outputs of the beneficiary models. In order to facilitate the full coupling of mutual models, their relevant source codes have to work together within the same software package, which means the mutual computer programs have to be integrated. In total, there are 12 one-way couplings and 2 full-couplings. It should be highlighted that, when addressing model linking, only the models that are working on neighbouring spatial scales are directly linked as designated in the DEF of figure 2. In the next section of the paper, details are presented of progress made on examples of important model linking.

3. Practical linking of MIntWeld models

The MIntWeld project was progressing to its mid-term when this paper was composed. Accompanied by development of respective MIntWeld models by corresponding MIntWeld teams, some MIntWeld models are now being linked together in practice as designed in DEF.

3.1. Linking the ab-initio model with the molecular dynamics (MD) model

With one-way coupling, the ab-initio model of Oxford has one-way communication with the MD model of ULEIC. The reason for this linking between ab-initio and empirical force fields is that whereas ab-initio simulations can be very accurate as they provide a quantum-mechanical description of the electrons without *a priori* assumptions on the nature of the chemical bonding between individual atoms, they are extremely computational intensive. On a modern supercomputer, the length and time scales accessible to ab-initio simulations are a few hundreds atoms and tens of picoseconds (1 ps = 10^{-12} s). Larger system sizes (1,000,000 atoms) and time scales (few ns, 1 ns = 10^{-9} s) are now routinely modelled using classical MD simulations with empirical force fields.

In order to construct new empirical force fields from ab-initio trajectories, Oxford chose the *force matching method* which consists of adjusting the force field parameters to reproduce as closely as possible the forces acting on every atom in a trajectory calculated by ab-initio methods. The force fields are validated by simulating solid-solid, solid-liquid interfaces and grain boundaries that show local atomic arrangements that have not been included in the ab-initio training set. The data formats listed in Table 2 illustrate the data exchange protocol between the mutual models.

Table 2. Data formats used in the linking of ab-initio model with MD model

Data	Description	File formats
EAM force field	Tabulated functions that define the atomic interactions between atoms	setfl (ASCII, can be used both in LAMMPS and DL_POLY)
Atomic trajectories	For each MD snapshot: - cell parameters - list of atoms - cohesive energy - stress tensor For each atom: - positions and forces	potfit trajectory format (ASCII) LAMMPS dump format (ASCII)

Thermodynamic data	Cohesive and mixing energy Equation of state Elastic constants Diffusion coefficients	ASCII text and tables OpenOffice spreadsheets+charts
Structural and correlation functions	Structure factor Radial distribution function Layer-resolved atomic density	ASCII text and tables OpenOffice spreadsheets+charts

Oxford and ULEIC have agreed to use the Embedded-Atom-Model (EAM) force field for classical simulations. So far, Oxford has determined the coefficients of an EAM inter-atomic potential function with its ab-inito modelling and has provided the results to the ULEIC group, which has been using these coefficients to run simulations based on their MD model.

3.2. *Linking of the MD model with phase field (PF) models*

One of the key tasks in the MIntWeld project is the prediction of microstructures formed during solidification in the weld pool. The principal microstructural pattern is dendrite, a tree-like shape, and its morphology is strongly dependent on the material properties, especially the anisotropy in liquid-solid interfacial energy. Although PF model is very good at modelling the growth of dendrites during solidification, its input parameter of interfacial energy and its anisotropy have to be fed by other modelling work. At the atomic level, MD model has been extensively used to study the properties of interfaces. In this task, MIntWeld researchers combine PF modelling to compute dendritic growth morphologies with MD modelling to determine the essential input parameters. The linking of MD and PF is a form of one-way coupling. The MD modelling results thus provide the interfacial free energy and its anisotropy as input parameters required by the PF method.

3.2.1 *Calculation of anisotropy of solid/liquid interface energy with the MD model*

For a macroscopically rough interface, the interfacial free energy and its anisotropy can be expressed in terms of an expansion in orthogonal cubic harmonics of Fehlner and Vosko as:

$$\gamma(n)/\gamma_0 = 1+\varepsilon_1\left(\sum_{i=1}^{3} n_i^4 - \frac{3}{5}\right)+\varepsilon_2\left(3\sum_{i=1}^{3} n_i^4 + 66 n_1^2 n_2^2 n_3^2 - \frac{17}{7}\right) \quad (1)$$

where $\{n_1, n_2, n_3\}$ are the Cartesian components of the unit vector $\boldsymbol{n}$ normal to the interface, γ_0 is the orientationally averaged interfacial free energy and ε_1 and ε_2 are expansion coefficients that quantify the anisotropy. The MD simulation results of the interface energy and its anisotropy, for pure iron, are itemized in Table 3.

Potential: Structure:	ABCH* BCC	Pair* BCC	MH(SA)[2*] BCC	This work BCC
γ_0 (mJ/m^2)	206(10)	221(14)	175(11)	163.4
ε_1 (%)	1.6±0.12	1.3±0.18	3.3±0.18	2.88
ε_2 (%)	−0.04±0.28	0.26±0.28	0.24±0.32	−0.0612
γ_{100} (mJ/m^2)	207	222	178	165
γ_{110} (mJ/m^2)	206	220	174	163
γ_{111} (mJ/m^2)	205	221	174	162
$\frac{\gamma_{100}-\gamma_{110}}{2\gamma_0}$ (%)	0.4±0.4	0.5±0.5	1.0±0.6	0.7
$\frac{\gamma_{100}-\gamma_{111}}{2\gamma_0}$ (%)	0.5±0.4	0.4±0.6	1.0±0.2	1.0

Table 3. Comparison of values of the interface free energy of pure Fe obtained by various potentials

*D.Y.Sun, M.Asta, Phys.Rev.B, 69, 174103 (2004)

IOP Conf. Series: Materials Science and Engineering **33** (2012) 012029 doi:10.1088/1757-899X/33/1/012029

Besides the interface energy and its anisotropy of solid/liquid interface, other properties such as kinetic coefficient of the solid/liquid interface, melting point and diffusion coefficient were also calculated with the MD simulation.

3.2.2 Employing the inputs received from simulation results of the MD model in the PF models

The phase-field method has emerged as the primary computational means to compute dendritic growth morphology. This method has proven very effective for computing microstructure evolution, but the results depend on the material properties that are used as input data. As noted earlier, the interfacial energy is one of the most important of these data. The anisotropy of interfacial energy, obtained from the MD simulation results of ULEIC as shown in Table 3, is being used as input in the phase-field simulations at ULEIC and EPFL.

EPFL has been using an adaptive finite element based PF model to investigate the importance of the interfacial energy anisotropy on the growth of dendrites. The effect of the anisotropy coefficients is illustrated in figure 3, which shows dendrite morphologies obtained with various combinations of the coefficients. In all cases, the coordinate axes correspond to the <100> crystal direction, and all of the dendrites are shown in the same spatial orientation. One can clearly see that there is a separate region in parameter space where primary <100> dendrites are obtained, another where primary <110> dendrites are obtained, and an intermediate region where mixed structures appear. The boundaries between the regions are similar to those computed by Haxhimali [8] in pure materials. For further details about these calculations, please see Friedli, et al. [9].

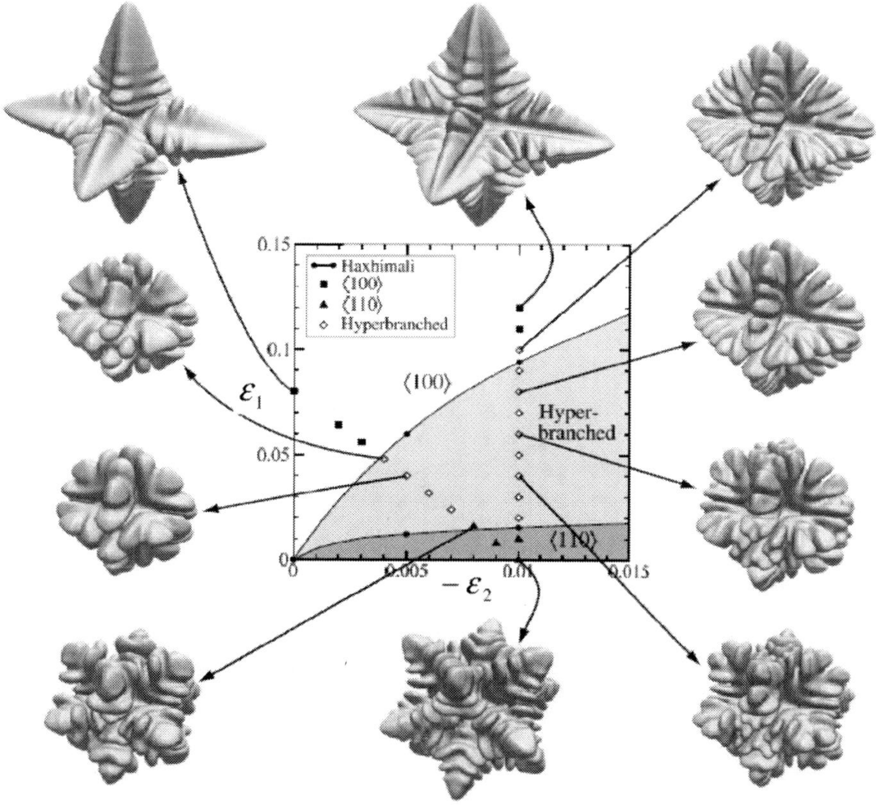

Figure 3. Computed alloy dendrites (EPFL) for various values of the anisotropy parameters.

The MD calculations, performed at ULEIC and described in the preceding section, provided the following values: $\varepsilon_1 = 0.0288$, $\varepsilon_2 = -0.000612$. Note that these values are very small, and very close to the origin in figure 3. This presents something of a challenge for accurate phase-field modeling, because there is always some grid anisotropy present in such calculations. When the actual anisotropy is very small, as in this case, the calculations must be done very carefully to eliminate the grid anisotropy. To that end, a series of calculations was performed in which the grid anisotropy was measured, so that it could be compensated for. Note that the grid is Cartesian, so the grid anisotropy reflects the underlying four-fold symmetry. The grid anisotropy varies with grid spacing, and it was found that reliable calculations could be performed with grid spacing $dx = 0.75W_0$, where W_0 is the width of the diffuse interface in the PF calculations, where the measured value $\varepsilon_1(grid) = -0.014$. Note also that the magnitude of the grid anisotropy is nearly half of the actual anisotropy predicted from the MD simulations, emphasizing the importance of the compensation calculation. Looking at figure 3, one can see that without the grid anisotropy correction, it is easy to erroneously cross the boundary where <100> dendrites are formed, entering the mixed morphology region.

An example dendrite morphology computed using these anisotropy parameters is shown in figure 4. This calculation corresponds to a dendrite grown into an initially uniform temperature melt at dimensionless undercooling ($c_p\Delta T/L_f$) equal to 0.4. An initial solid seed was placed in the corner of a cube computational domain, with the coordinate axes parallel to the following crystallographic directions:

$$\hat{\mathbf{x}} \parallel \langle 100 \rangle \;\; ; \;\; \hat{\mathbf{y}} \parallel \langle 01\bar{1} \rangle \;\; ; \;\; \hat{\mathbf{z}} \parallel \langle 011 \rangle$$

Orientation of the coordinate axes in this way permits a check of the grid anisotropy correction, by comparing the length of the <100> arms in each crystallographic direction.

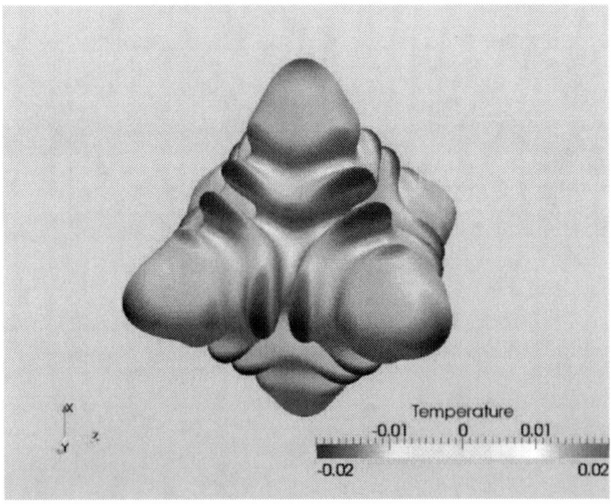

Figure 4. Computed Fe dendrite (EPFL) using parameters obtained from MD simulations performed at ULEIC. The temperature range is scaled such that zero corresponds to the equilibrium freezing temperature. Deviations from that value are associated with the local mean curvature of the interface. The primary arms are parallel to <100>.

3.3. *Linking the front tracking (FT) model with the CFD model*

During either the melting or solidification stage of the welding process, the interface that separates the liquid metal and solid work piece migrates towards either the solid phase or the liquid phase as a result of phase transformation. In the CFD model of TUD the weld pool geometry is not defined a-priori but determined at run-time during a simulation. The FT model of UCD [10,11], which is being developed, is able to predict both the melting and solidification processes of welding, thereby providing the geometry of the weld pool to the CFD model. The full coupling between the FT and CFD models, as designed in the DEF of figure 2, will significantly advance the modelling of fluid flow and melting and solidification during the process of welding.

In the full coupling of both models, the key concern is the real time inter-model data transfer. In order to facilitate such data transfer, the respective source codes of UCD and TUD will be integrated into a single software platform. Because UCD and TUD are using two different sets of computational grids in their respective numerical models, a scheme of data interpolation is key to the linking of the models, by facilitating the smooth real time data transfer of mainly thermal history and evolution of the geometry of weld pool. To some extent, the data interpolation dominates the feasibility and performance of the model linking in this case. UCD and TUD have agreed to split the linking into two steps. In the first step of the implementation of model linking, these two teams have been executing a one-way linking, in which major efforts have been made to develop a suitable scheme for data interpolation. The full coupling of mutual models will be done once the one-way linking has been successfully performed.

3.3.1. Numerical scheme for data interpolation

As schematically shown in figure 5, TUD is using a non-structured Eulerian grid to discretize the governing equations of its CFD model [12] and UCD is using a structured Lagrangian grid to resolve the interface between liquid phase and solid phase and run its FT model. In order to transfer data between the CFD model of TUD and the FT model of UCD, UCD has developed a novel scheme for data interpolation. It is called UCD-flexible-interpolation (UCDFI) scheme in the context of this paper. The basic idea of UCDFI scheme is to interpolate data with random data stencils, which makes this scheme very flexible.

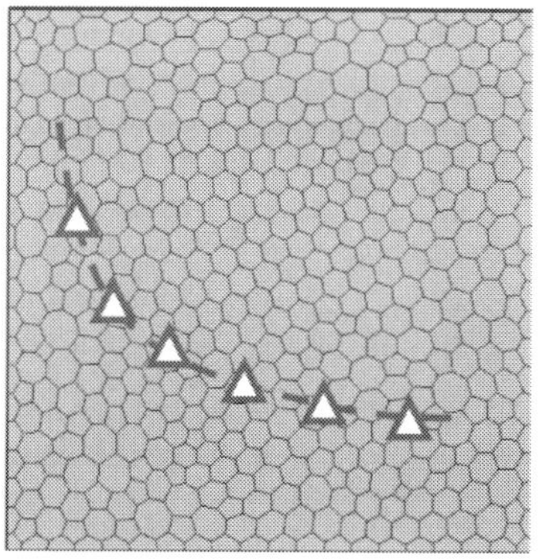

Figure 5. Position of CFD cells s (blue polygons) and front tracking markers (red triangles)

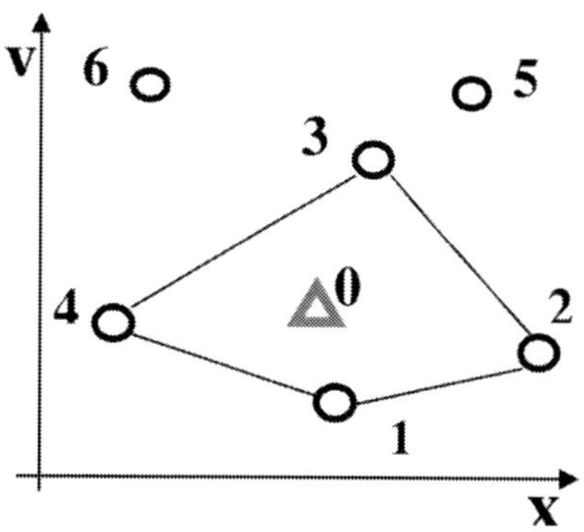

Figure 6. Position of the front tracking marker (red triangles) and CFD cells (black circles) in global coordinates

Figure 6 shows the basic configuration of UCDFI scheme. Point #0 is a computational marker used in FT model, where the value of any parameter of interest is to be calculated using interpolation. Points #1~ 6 are computational cells of the fluid flow model of TUD. Points #1~4 enclose point 0 with a four-node nonorthogonal quadrilateral in the global coordinate system, in which the position of points can be expressed as (x_i, y_i), i=0, 1, 2, 3, 4. The parameters of interest of points #1~4 are used as stencils of the data interpolation. Based on isoparametric transformation of the position of points #0~4 and a bi-linear interpolation in the natural coordinate, the data interpolation in the global coordinate can be determined as:

$$T_o = \sum_{i=1}^{4} N_i(\varepsilon_o, \eta_o) T_i \tag{2}$$

where T_0 is the parameter of interest (e.g. temperature) to be calculated, T_i is the corresponding parameter of data stencils (i=1~4), N_i is the coefficient of interpolation at CFD node number i, $\varepsilon_{0,}\eta_0$ is the position of Point #0 in the natural coordinate.

3.3.2. Results of the data interpolation

The TUD team used its CFD model to simulate the evolution of temperature field during the process of casting as shown in figure 7. This case study corresponds to the one presented by Browne and Hunt [13], in which liquid Al–2wt% Cu alloy (initially at $700^{\circ}C$) solidifies in a cavity mould due to cooling. The color scales in figure 7 represent a snap shot of the temperature field of the alloy at the time of 60 second since the start of cooling. UCD received these simulation results of temperature field from TUD and applied UCDFI scheme to interpolate the values of temperature on to front tracking markers, in order to formulate the advancement of solidification front with the FT model. The advancement of the solidification front (with front tracking method) in this case had no feedback to the temperature field (with the release of latent heat). This is a one-way coupling of the FT model of UCD with the CFD model of TUD. It was used to test the data interpolation scheme.

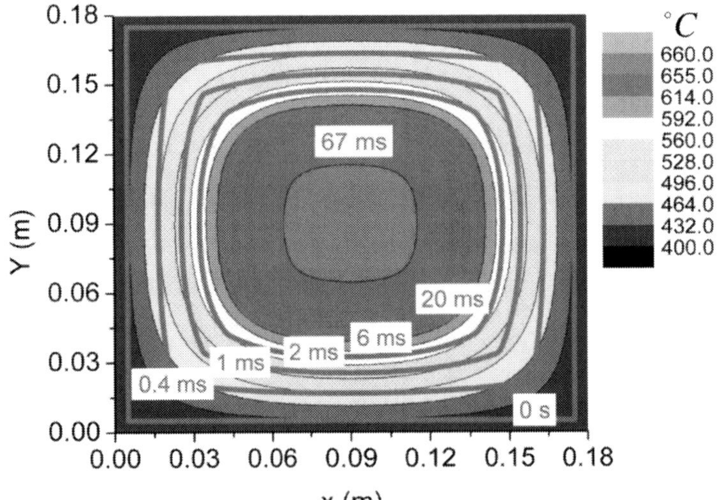

Figure 7. Temperature field (colour scales), at the time of 60 second since the start of cooling, predicted by TUD with its CFD model and the position of columnar solidification front (grey curves)

The temperature field, as represented by colour scales in figure 7, is used as the input of FT simulation. The grey curves denote the position of the solidification front at different times, as predicted by the FT model with the input of temperature and UCDFI scheme. It can be seen that, as time elapses, the solidification front advances inwards. Because the temperature around the corners of the mould is relative low, the corners of the initially rectangular solidification front advance relatively quickly. Accordingly, the overall figure of the solidification front gradually becomes smooth. At the time of 67 ms, the shape of solidification front evolves to be identical to the shape of contour of temperature.

It has to be highlighted that this is a one-way coupling. The FT model is using only the temperature field at a given time (color scales in figure 7) as its input, which represents a snapshot of the CFD simulation results at the time of 60 second since the start of cooling. Because the temperature near the wall of the mould is very low, the solidification front propagates very quickly. In future simulations that integrate the FT model and the CFD model with full coupling, the solidification front cannot, of course, exist in the blue and green zones of figure 7 in such circumstances, because they are already completely solid. Therefore, the advancement of the solidification front will not be this fast. The only purpose of using this case study here is simply to test the UCDFI scheme and its compatibility with FT model and CFD model.

3.4. Linking the thermodynamic model of KTH with other numerical models of MIntWeld

Thermodynamic data are a key type of input for many MIntWeld numerical models. The work of KTH, in MIntWeld, is trying to feed the beneficiary teams with these key inputs.

KTH is developing an advanced hybrid method, by incorporating data interpolation into directly interfacing the respective simulation software of MIntWeld teams with a package that provides the thermodynamic data in real time. To illustrate the method, assume that 2 parameters are varied during the simulation. For example, these parameters could be composition variables in a ternary alloying system at isothermal temperature. An initial calculation of the equilibrium state for a certain point, marked with a cross in figure 8(a), would require 3 equilibrium calculations (figure 8(b)). Parameters that fall within the shaded triangle would then not require any more equilibrium calculations but can be obtained from interpolation (figure 8(c)). When the varied parameters falls outside this space (figure 8(d)), the cost of extending the interpolation space is low as only one more equilibrium calculation is needed (figure 8(e)). By this operation the interpolation space has been extended and other values can be obtained from interpolation in the shaded areas (figure 8(f)).

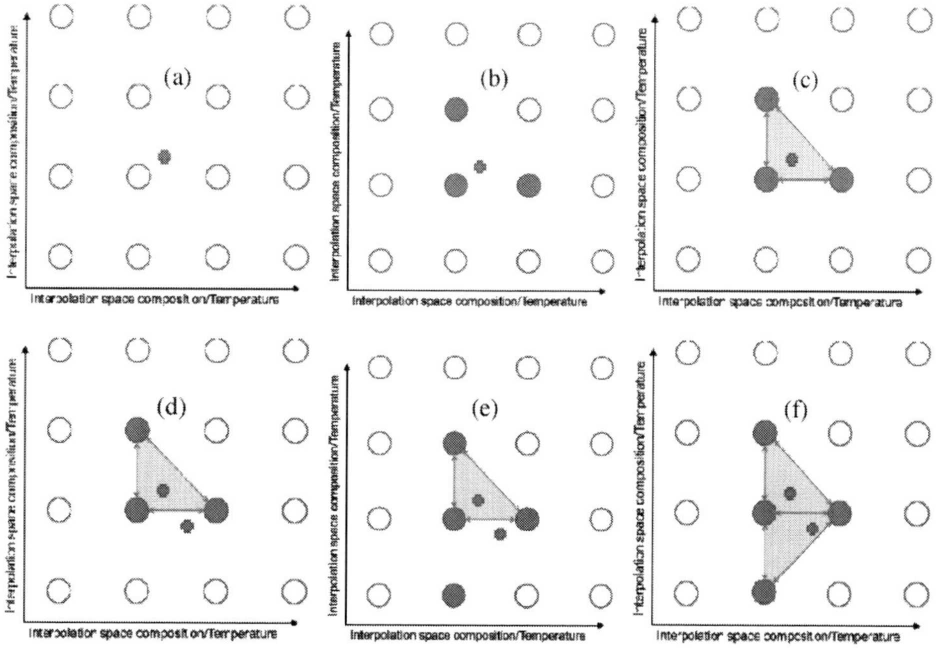

Figure 8. Schematics of the hybrid numerical model for the calculation of thermodynamics data combining real time calculation and data interpolation

The method is extendable up to any number of parameters that are varied giving a multidimensional interpolation space. This type of approach using the Thermo-Calc programming interface TQ, has shown that numerical performance can be improved by about two orders of magnitude as compared to calculating the values every time, while maintaining acceptable accuracy.

3.5 Linking life prediction model with other numerical models of MIntWeld

Oxford is developing computational finite element (FE) models for hot cracking and hydrogen embrittlement, designed to link to upstream and downstream models, as shown in figure 1 and 2.

- Hydrogen embrittlement --

The focus of this activity is modelling hydrogen embrittlement of welded components under in-service conditions typical of marine applications. It was decided that the focus will be on welds between dissimilar materials (a steel and nickel based alloy). There is controversy in the scientific literature about the mechanisms for crack growth in this (and other) material systems – with proposed mechanisms based on hydrogen either affecting the cleavage strength of interfaces or the plastic properties of the material. The models will allow for each of these possibilities. They require as input, cohesive properties of interfaces and plastic properties of the surrounding matrix and how these are influenced by hydrogen content. Activities at TWI and ULEIC will generate information about the plastic properties in different regions of a weld adjacent to the interface between the parent steel and Ni alloy weld and how these are influenced by hydrogen content. This information will be used to develop constitutive laws for the different constituents. Experimental studies at ULEIC/TWI will provide information about crack path that will guide the modelling effort. First Principle and Molecular Dynamics modelling at Oxford/EPFL/ULEIC will provide information about the work of fracture across interfaces and how this is influenced by hydrogen concentration. This behaviour will be captured in the form of a cohesive zone model that will be used in the FE studies. The first principle and molecular dynamics simulations might also be able to provide information about the diffusivity of H and the strength of hydrogen traps in the material.

- Hot Cracking --

The focus of this activity will be on similar welds, both commercial and model systems. Oxford will focus on modelling a new test configuration being developed by Tata steel. Hot cracking occurs during the latter stages of solidification in the weld pool as two solidification fronts merge. In order to model this situation Oxford will get information about microstructure, chemistry and defect development from the models being developed by UCD. This information will inform the development of a cohesive zone model (represent the behaviour at the mesoscale) which will be incorporated in a macroscopic finite element analysis.

Although the linking of the life prediction model of the Oxford team has not yet been implemented practically, the fruitful collaboration between Oxford and the other associated groups has significantly contributed to the design of an advanced DEF.

4. Summary and conclusion

In conventional research projects on welding, individual research teams work on their respective parts of the overall project separately. In the MIntWeld project, however, the research teams are not merely developing their own numerical models but also trying to link the numerical models of different teams together. The linking of different numerical models significantly contributes to resolving the multi-physics and multi-scale problem of interface evolution during the process of welding.

A data exchange framework has been designed in order to guide the linking of MIntWeld models. Under this framework, different numerical models on neighbouring spatial scales are linked together. The linking can be implemented by either one-way coupling or full coupling. As illustrated by the progress of practical linking of numerical models on neighbouring spatial scales so far, respective numerical models have significantly benefited from the linking. For example, the PF model obtained its input of interfacial energy and its anisotropy from the MD model, and the MD model obtained its input of inter-atomic potential from the ab-initio model. The successful and smooth flow of data among these three different numerical models of different teams and on different spatial scales clearly shows the potential for model linking. The approach taken concentrates on linking each length scale with its neighbour; information from very small scale phenomena will percolate through to the process scale of welding, and information will also return in the opposite direction.

As with other multi-scale modelling research, in the DEF of MIntWeld, errors may propagate between the submodels on neighbouring scales. So far, since the respective submodels are still under development, it is difficult to quantitatively predict such error propagation. However, the research partners have been working very carefully when developing their own submodels, in order to make individual computational errors as small as possible.

The smooth practical transfer of data between different numerical models, as envisaged and realised in the MIntWeld data exchange framework is expected to lead to a unique, coherent multi-scale physics description of weld behaviour, from which new understanding and control over material behaviour is anticipated. For example, with this multi-scale modelling, the numerical simulation can be used to directly predict the geometrical evolution of the weld pool, the thermal and spatial evolution of the heat affected zone, columnar-equiaxed grain structure transition in the weld pool, the morphology and segregation of grain boundaries (of either columnar or equiaxed dendrites) and cracking in laser or arc welding process, as a function of the process parameters and materials properties used as inputs.

Acknowledgments

This research work is supported by the European Commission as part of the FP7 programme, as the project, Modelling of Interface Evolution in Advanced Welding; contract number no. NMP3-SL-2009-229108.

References

[1] Jiao Z, Song C, Lin T and He P 2011 *Computational Materials Science* **50** 3585-89
[2] Song J and Srolovitz D 2009 *Journal of the Mechanics and Physics of Solids* **57** 776–787
[3] Zhan X, Wei Y and Dong Z 2008 *Journal of Materials Processing Technology* **208** 1–8
[4] Tan W, Bailey N and Shin Y 2011 Computational Materials Science **50** 2573–85
[5] J. Zhou, H.L. Tsai 2008 *International Journal of Heat and Mass Transfer* **51** 4353–66
[6] G. Xu, J. Hub, H.L. Tsai 2009 *International Journal of Heat and Mass Transfer* **52** 1709–24
[7] Grong O 1997 *Metallurgical modelling of welding, 2nd Rev edition* (Maney Publishing)
[8] Haxhimali T, Phase-field simultion study of dendritic crystal growth morphologies for cubic and hexagonal symmetries, *PhD Thesis, Northeastern University,* 2006
[9] Friedli J, Dantzig J, Di Napoli P, and Rappaz M, Phase-field modeling of the dendrite orientation in Al-Zn, *submitted to Modelling of Casting, Welding and Advanced Solidification Processes, XIII 2012,17-22 June, Austria*
[10] Duggan G, Tong M, Browne DJ 2011 An integrated meso-scale numerical model of melting and solidification in laser welding, *IOP Conf. Series: Materials Science and Engineering* **27** paper 012077
[11] Duggan G, Mirihanage W U, Tong M, Browne D J 2011 A Combined Enthalpy / Front Tracking Method for modelling Melting and Solidification in Laser Welding *submitted to Modelling of Casting, Welding and Advanced Solidification Processes, XIII 2012,17-22 June, Austria*
[12] Saldi Z, Kidess A, Kenjeres S and Kleijn C 2010 Marangoni driven free surface microflows in liquid metal pools *Proceedings of the 2nd European Conference on Microfluidics - Microfluidics 2010* (Toulouse, France, 8-10 December, 2010)
[13] Browne D and Hunt J 2004 *Numerical Heat Transfer, Part B: Fundamentals* **45** 395-419

IOP Conf. Series: Materials Science and Engineering **33** (2012) 012030 doi:10.1088/1757-899X/33/1/012030

Modelling of horizontal centrifugal casting of work roll

Zhian Xu[1], Nannan Song[3], Rob Val Tol[2], Yikun Luan[3], Dianzhong Li[3]

[1] RWP Belgium, Technologiepark 935, 9052 Zwijnaarde, Belgium
[2] SIRRIS, Technologiepark 935, 9052 Zwijnaarde, Belgium
[3] Shenyang National Laboratory for Materials Science, Institute of Metal Research, Chinese Academy of Sciences, 72 Wenhua Road, Shenyang, 110016, China

E-mail: zhian.xu@experto.be

Abstract. A numerical model to simulate horizontal centrifugal roll castings is presented in this paper. In order to simulate the flow fluid and solidification of horizontal centrifugal roll casting correctly, the model uses a body fitted mesh technique to represent the geometry. This new method maps a plate layer mesh to a circular mesh. The smooth body fitted mesh method gives more accurate calculation results for cylindrical geometries. A velocity depending on the angular velocity and inner radius of the mould is set up as a velocity boundary condition. The fluid flow coupled with heat transfer and solidification in a rapidly rotating roll is simulated. A gravity free falling method is applied as a pouring condition. A moveable pouring system is also used in the simulations. High speed steel is used to produce the work roll. Two different gating positions and a moveable gating system are simulated in this paper. Results show that the position of pouring system has a significant influence on the temperature distribution. The temperature distribution at a fixed central pouring system is more favourable than the distribution from a side pouring system. A moving gating system method is a better way to obtain a uniform temperature field in centrifugal casting and offers an alternative for existing techniques.

1. Introduction

Work rolls in contact with the metal surface to be rolled deform the metal into the plastic regime during the rolling process. Thus, the quality of work rolls has significant influence on the surface property, dimensional precision and rolling productivity of rolled steel products. Therefore, the outer layer of work roll requires a high wear ability and red hardness. The multilayered multiple metal work roll, using high speed steel (HSS) as the outer layer, catches the rolling companies' eyes for its better quality. Horizontal centrifugal casting technique, as shown in figure 1, is one of the most efficient processes for producing pipes, tubes, cylinder sleeves, rolls and so on [1-2]. For HSS work rolls, the widely used producing method is the centrifugal casting process. This technique generates centrifugal force by rotating a cylindrical mould and to press the molten metal on the mould wall to form the desired shape [3-4]. The filling process is difficult to investigate using conventional methods [5]. Using computer aided engineering to design the process could lead to money savings, shorter trial and error cycles, reducing energy consumption and pollution. To realize these goals, mathematic models for filling and solidification of centrifugal casting process are implemented in Experto-ViewCast software. The centrifugal casting process of high speed steel is simulated by using the this software.

The main concern of this paper is how to use the new method to simulate the mould filling of HSS work rolls. The solidification details will be discussed on the another paper.

In this paper the effects of different pouring parameters on the quality of HSS work roll are investigated. Steady gating systems are investigated and a moving gating system is designed to gain a more uniform temperature field for the outer layer[6].

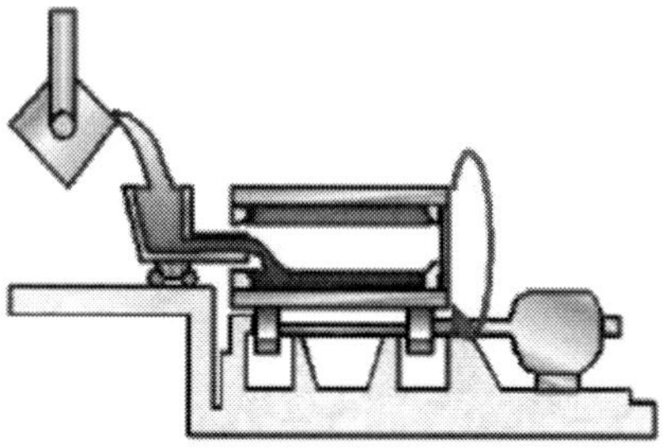

Figure 1. Schematic diagram of horizontal centrifugal casting

2. Mathematical model

The mathematic models based on the force analysis of the liquid under centrifugal casting process are built up. The fluid field calculation based on the N-S equation as following:
Navier-Stokes equation:

$$\begin{aligned}
\rho\left(\frac{\partial u}{\partial t}+u\frac{\partial u}{\partial x}+v\frac{\partial u}{\partial y}+w\frac{\partial u}{\partial z}\right)&=-\frac{\partial p}{\partial x}+\eta\left(\frac{\partial^2 u}{\partial x^2}+\frac{\partial^2 u}{\partial y^2}+\frac{\partial^2 u}{\partial z^2}\right)+\rho g_x\\
\rho\left(\frac{\partial v}{\partial t}+u\frac{\partial v}{\partial x}+v\frac{\partial v}{\partial y}+w\frac{\partial v}{\partial z}\right)&=-\frac{\partial p}{\partial y}+\eta\left(\frac{\partial^2 v}{\partial x^2}+\frac{\partial^2 v}{\partial y^2}+\frac{\partial^2 v}{\partial z^2}\right)+\rho g_y\\
\rho\left(\frac{\partial w}{\partial t}+u\frac{\partial w}{\partial x}+v\frac{\partial w}{\partial y}+w\frac{\partial w}{\partial z}\right)&=-\frac{\partial p}{\partial z}+\eta\left(\frac{\partial^2 w}{\partial x^2}+\frac{\partial^2 w}{\partial y^2}+\frac{\partial^2 w}{\partial z^2}\right)+\rho g_z
\end{aligned} \tag{1}$$

The mass conservation equation:

$$\frac{\partial u}{\partial x}+\frac{\partial v}{\partial y}+\frac{\partial w}{\partial z}=0 \tag{2}$$

VOF function is used to track the free surface:

$$\frac{\partial F}{\partial t}+u\frac{\partial F}{\partial t}+v\frac{\partial F}{\partial y}+w\frac{\partial F}{\partial z}=0 \tag{3}$$

Energy conservation equation:

$$\frac{\partial T}{\partial t}+u\frac{\partial T}{\partial x}+v\frac{\partial T}{\partial y}+w\frac{\partial T}{\partial z}=\frac{\lambda}{\rho c}\left(\frac{\partial^2 T}{\partial x^2}+\frac{\partial^2 T}{\partial y^2}+\frac{\partial^2 T}{\partial z^2}\right)+S \tag{4}$$

In the inner free surface and outside of die, the radiation effect must be considered in both flow calculation and solidified metal. The boundary conditions convection and radiation are given by formula (5).

$$-\lambda\cdot\frac{\partial T}{\partial x}=h\cdot(T_{w1}-T_{sur1})+\varepsilon\cdot\sigma\cdot({T_{w1}}^4-{T_{sur1}}^4) \tag{5}$$

Where, g_x, g_y and g_z are the components of the gravity acceleration in x, y and z direction, and u, v and w are velocity components. F is the volume fraction used to track the liquid-air interface with a VOF (Volume of Fluid) method[7]. This method can also be used at a condition of a high turbulence flow. wl and $surl$ subscripts represent free surface and surrounding separately.

ρ : density (kg/m3).
t : time (s)
p : pressure (Pa)
η : the dynamic viscosity (Pa.s)
ε : the emissivity
σ : Stefan-Boltzmann's constant ($Wm^{-2}K^{-4}$)
c : specific heat capacity (W/K)
T: temperature (K)
λ : thermal conductivity (W/m/K)
S : heat source (J/kg)
h : convection heat transfer coefficient ($Wm^{-2}K^{-1}$)

3. Meshing

The geometric models of the outer layer with different gating systems are shown in figure 2. Figure 2(a) is the model of the steady gating system. Figure 2(b) is the model of moving gating system.

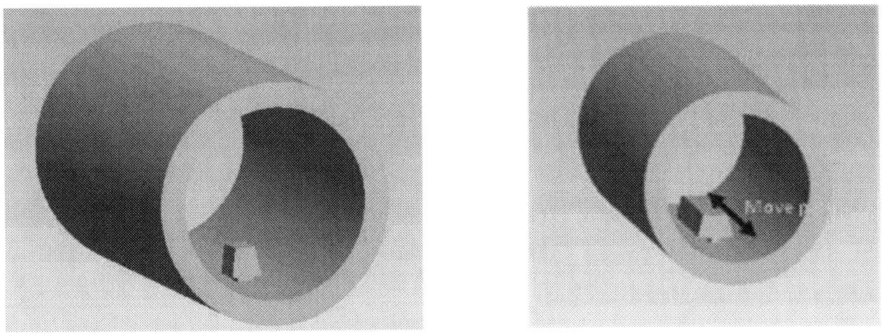

(a)Model of steady gating system at 1/3 of the mould (b) model of moving gating system

Figure 2. Geometry models of the work layer of the roll with different gating systems

A conventional finite volume method is used. Meshes based on rectangular coordinates produce 'stair step' grids as shown in figure 3. This type of grids affects the velocity distribution of the flow field close to the circular boundary. No matter how fine the grids are, the mesh pattern at the boundary of the circular geometry keeps a 'stair-step' like shape [8].

In order to overcome the disadvantage of 'stair step' grids, a body fitted mesh technique is used to represent the geometry. It maps a plate layer mesh as displayed at figure 4(a) to a circular mesh figure 4(b). For a mesh based on cylindrical coordinates the grid cells at the boundary are aligned with the circular geometry. An improved program to simulate horizontal centrifugal casting is developed based on this method. For the meshing of a cylindrical mould, The meshing angle is started from 0 ° and ended at 360 °.

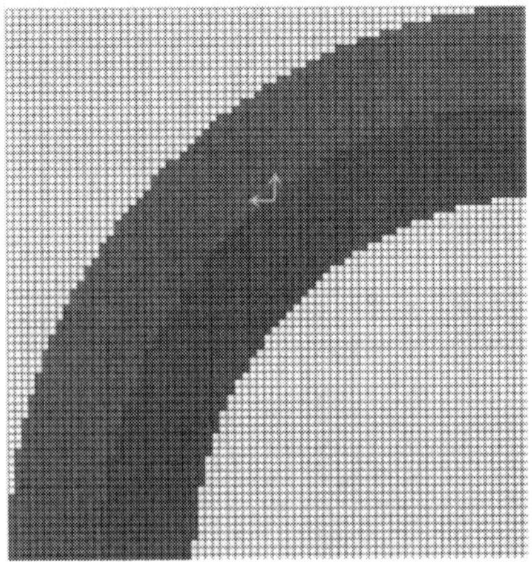

Figure 3. A 'stair step' like mesh

A role with process of dimension of Ø180×430mm roll is meshed using both meshing methods, rectilinear and cylindrical. The comparison of the both meshes demonstrates that the conventional finite difference method produces 'stair step' grids even in the case of very fine meshes as shown in figure 3. This 'stair step' pattern has a significant influence on the velocity distribution of the flow field at the circular boundary. However, with the body fitted mesh technique, even with a coarse mesh

, the grids at the boundary are aligned with the circular geometry in figure 4. This way more accurate simulation results were achieved with significantly less number of cells. It should be noted that this mapped mesh as illustrated at figure 4 can be easily to extend a finer mesh. It is necessary to use a fine mesh to do the solidification calculation.

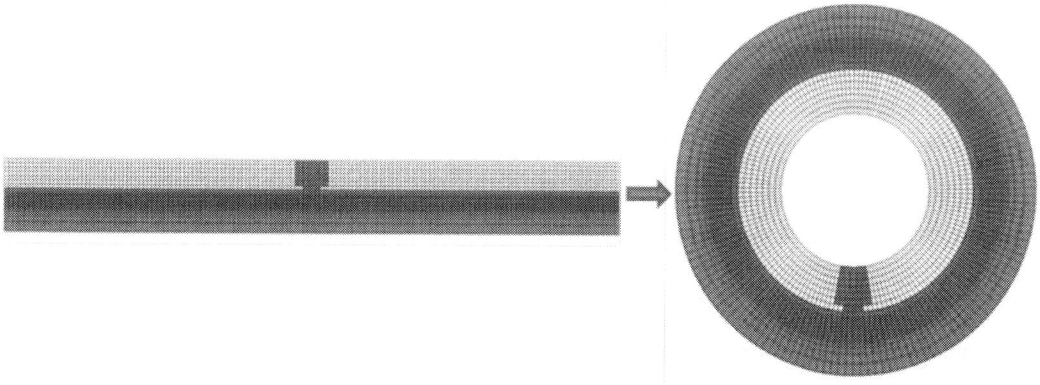

(a) Left: A plate layer mesh (b) Right: mesh mapped to a cylindrical system

Figure 4. A body fitted mesh. Red indicates metal. The pink is defined as mould material, and the yellow area as air.

4. Inlet condition

The detailed model of the moving gating system is shown in figure 5. The pouring cup can move in both sides. For the moving gating system, the velocity of the moving gate is designed to gain the most uniform filling result. The moving speed of the gate has a relationship with the length of the mould which is given by the formula 5.

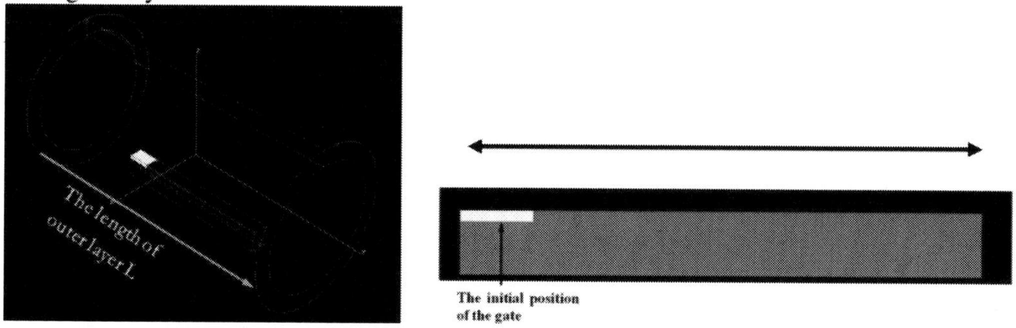

(a) the position of the moving cup (b) the moving distance of the gate in the mould

Figure 5. The detail of moving gating system. (a) the position of the moving cup (b) the moving distance of the gate in the mould

$$v_g = \frac{2n_g L(abc)}{3V_l t} \tag{5}$$

Where:

v_g : moving velocity of the gate ($m \cdot s^{-1}$)

t : moving time of the gate(t)

n_g : moving frequency of the gate

a : the length of pouring gate (m)

L : the length of the outer layer (m)
V_l : the volume of the liquid needed(m^3)
b : the width of pouring gate(m)
c : the thickness of pouring gate(m)

5. Boundary and initial conditions

5.1. Boundary conditions

Boundary conditions are needed to be specified at the inside and the outside of the mould. A detailed description of the boundary conditions used for this simulation is given as follows.

In order to realize the real rapidly-rotating mould, a constant velocity is given to the inner side of mould. The boundary velocity is calculated using the following formula:

$$V = r \times \omega \tag{6}$$

Where, V is the tangential velocity of the rotating mould, r is the mould inner radius and ω is the angular velocity of the mould. A no slip boundary condition is applied to the metal elements which are close to the mould wall. The centrifugal force is assumed to be high enough to prevent the formation of gaps between the inner wall of the mould and the solidifying metal cylinder [9].

Heat transfer by radiation, diffusion and convection between metal and air is taken into account. A full slip velocity condition is applied at the metal-air interface.

5.2. Initial conditions

The initial mould temperature is 200 °C, which was obtained from experimental data [10] for all cases. The rotation speed of the mould is 600rpm. High speed steel is used in producing the outer layer of the work roll [11]. The pouring temperature of the liquid is 1420 °C. A gravity free falling method is applied as an initial pouring condition. The metal stream is perpendicular to the mould. There is no axial direction on initial velocity component. Three different pouring positions have been simulated: two steady gating systems with different gating position and one moving gating system. For the steady gating system, one is located at one third of the length of the mould and the other one is set at the middle of the mould. The filling time is 30 s. Initially, the liquid has no axial velocity. In this case the velocity of the moving gating system is 0.1 m·s^{-1}.

6. Simulation results

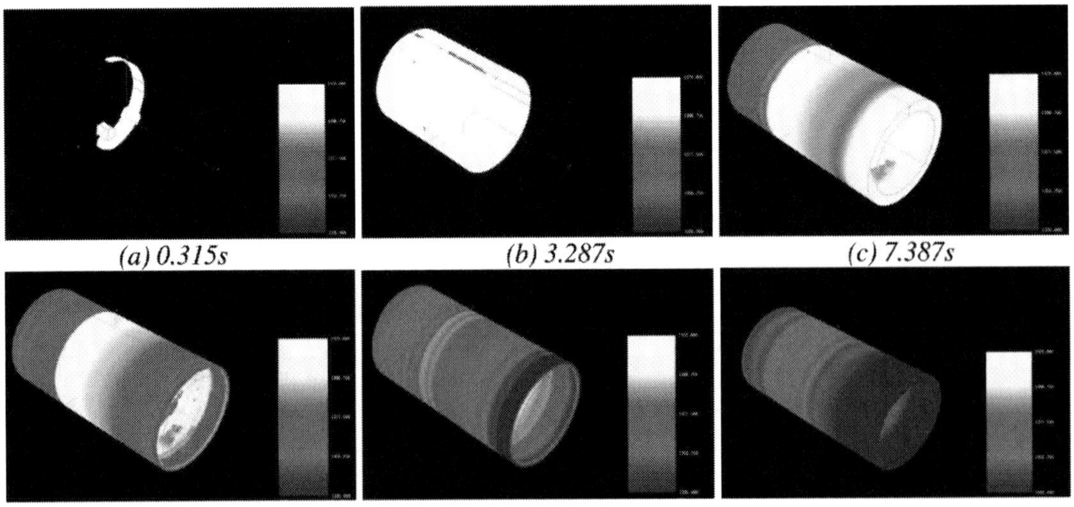

(a) 0.315s *(b) 3.287s* *(c) 7.387s*
(d) 9.437s *(e) 13.177s* *(f) 30.001s*

Figure 6. Temperature distribution of gating system at the 1/3 of the at different time steps

Based on the above conditions, simulations of different gating systems are carried out. The aim is to simulate the temperature field of the horizontal centrifugal casting process and give suitable filling parameters for the foundry.

Figure 6 shows the flow pattern and temperature distribution at different time steps in case of the gating system located at the 1/3 of the mould. Color indicates temperature (scale 1335-1420°C). From figure 5(b) it can be seen that when the gating system is at 1/3 of the mould, the liquid first reaches the side which is close to the gating system. This leads to a non-uniform temperature distribution. At the end of the filling process, the temperature of the side which is far away from gating system reaches the solidus temperature and has the lowest temperature in the entire roll. This gating system causes the formation of 'circular isothermal field', which only happens in centrifugal casting.

Figure 7 gives the flow pattern and temperature distribution of the central steady gating system at different time steps. It can be seen that when the gating system is put at the middle of the mould the temperature distribution is more uniform than that at the 1/3 of the mould. Figure 7(b) shows that the liquid flows to both sides of the mould evenly. This is different from that of figure 6(b) in which the liquid reaches one side first, then reaches to the other side. At the beginning, it has a uniform temperature distribution along the whole roll as shown figure 7(b).

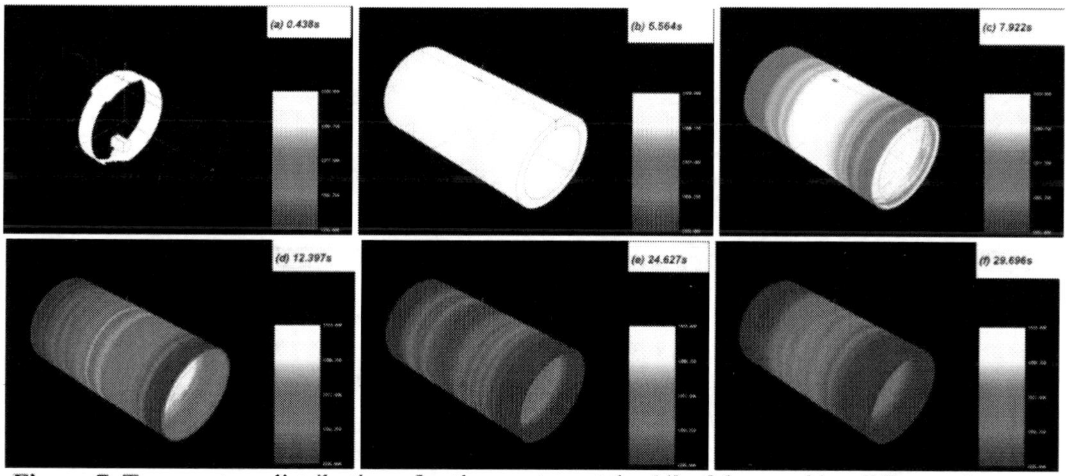

Figure 7. Temperature distribution of gating system at the 1/2 of the mould at different time steps.

The difference between these temperature distributions indicates that the position of the gating system has a significant effect on the flow and temperature fields, thus the overall heat transfer process. However, both temperature profiles are unfavourable, because of large temperature gradients between warm and colder areas. The large temperature difference between the roll shoulder and the roll body might cause casting defects, as shown in figure 8.

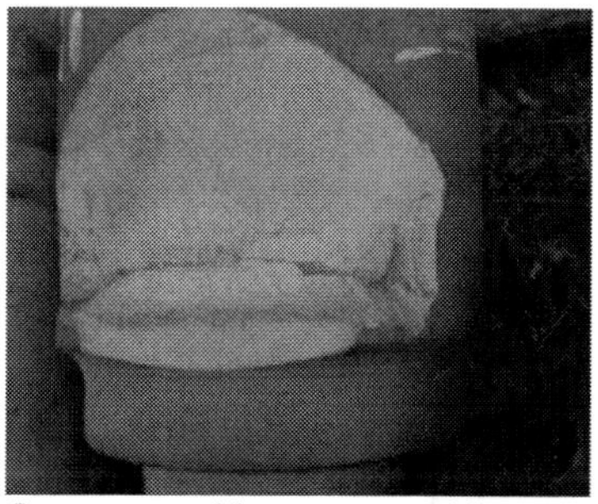

Figure 8. Sapling defects caused by a non-uniform temperature difference

Based on the simulation results shown above, a moving gating system is designed. The simulated results of mould filling sequence by using the moving gating are illustrated at figure 9. Figure 9(e) shows the temperature field of outer layer when gating system moves to one side of the mold. At that time the temperature of the whole roll body is not uniform. But when the gating system moves back to the original place, the temperature distribution becomes more uniform figure 9(i). With this kind of moving circle, it creates a uniform temperature field during mould filling.

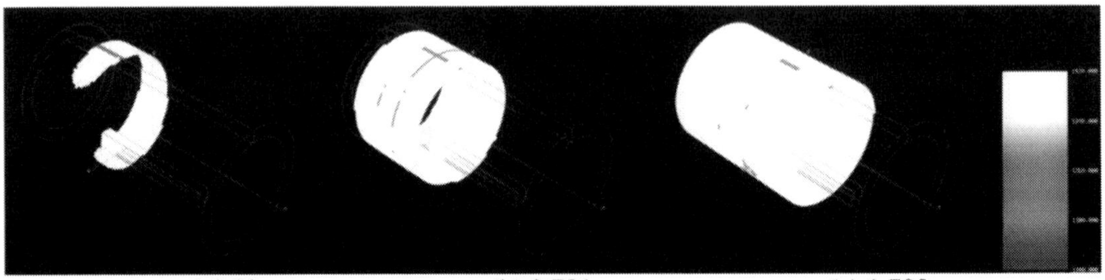

a) 0.20s (b) 0.701s (c) 1.703s

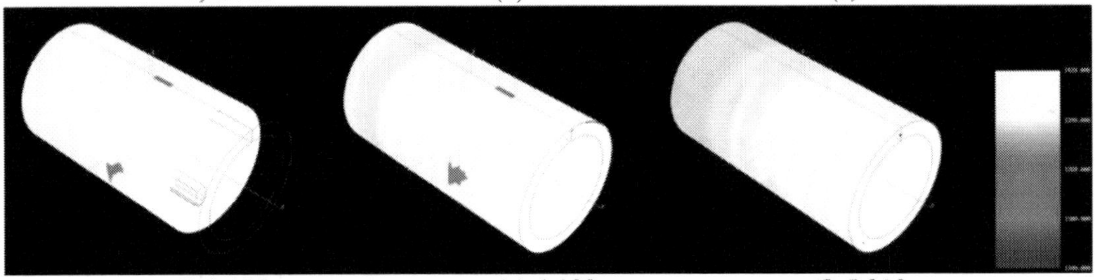

d) 3.106s (e) 4.108s (f) 5.210s

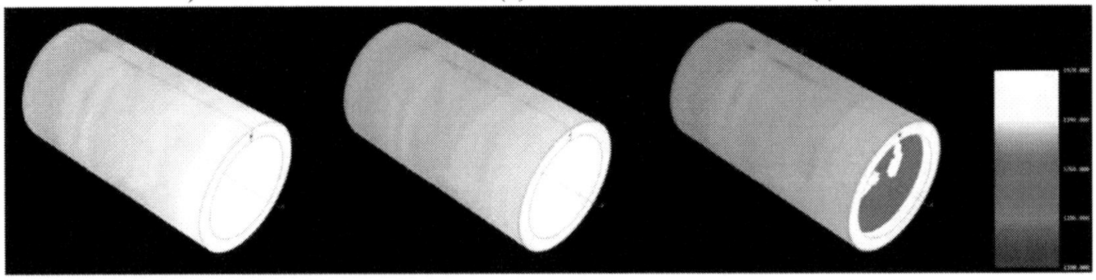

g) 6.713s (h) 8.116s (i) 9.001s

Figure 9. Mould filling and Temperature distribution of the moving gating system at different time steps

Compared the simulation results of figure 9 with the steady gating systems as shown in figure 7 and figure 8, the moving gating system creates a more uniform temperature field than the steady gating system. The temperature field between the roll shoulder and the roll body is quite small for the moving gating system. No 'circular isothermal field' formed. This provides good temperature field for the following intermediate and inner core filling and solidification.

The moving gating system is more suitable for large dimension work rolls [12], because for a small dimension work roll the moving velocity and pouring velocity are difficult to control. For a large dimension work roll, pouring from both sides of the roll or design an inner filling ladle to obtain a uniform temperature field for horizontal centrifugal casting are known techniques [13]. The moving gating system offers an alternative for these technologies.

7. Conclusions

(1) The results show that the mathematical model developed can simulate the flow field and distribution of the temperature for horizontal centrifugal casting. The body-fitted mesh which uses fewer grids is better than the conventional finite-difference 'stair-step' mesh for roll casting simulation.
(2)The simulation results show that a moving gating system results in a more uniform temperature field than a steady gating system and offers an alternative for other techniques to create a more uniform temperature field in horizontal centrifugal castings.
(3)The implemented model provides a tool for engineers to improve the centrifugal roll casting technique with stationary or moving gating systems.

Acknowledgments

The authors would like to thank the financial support from the CAS/SAFEA International Partnership Program for Creative Research Teams.

References

[1] Ju D Y and Inoue T 1992 Analysis of solidification and viscoplastic stresses incorporating a moving boundary: an application to simulation of the centrifugal casting process *Journal of Thermal Stresses* **15** 109-128
[2] Fu H G Xing J D and Xiao Q 2008 A study of segregation mechanism in centrifugal cast high speed steel rolls *Materials Science and Engineering A* **479** 253–260
[3] Wang C Y and Gao J W 2000 Modeling the solidification of functionally graded materials by centrifugal casting *Materials Science and Engineering A* **292** 207–215
[4] Kim K T 2007 Fabrication and characterization of BSCCO-2212 tube prepared by centrifugal casting *Physica C* 460–463
[5] Li C Y and Wu S P 2006 Model experiment of mold filling process in vertical centrifugal casting *Journal of Materials Processing Technology* **176** 268–272
[6] Song N N Xu Z A Kang X H Li D and Z Houbaert Y 2009 Mathematic Simulation of Centrifugal Casting of Work Roll *3rd International Conference on Simulation and Modelling of Metallurgical Processes in Steelmaking* Leoben Austria
[7] Nichols B D Hirt C Wand Hotchkiss R S 1980 SOLAVOF: A solution algorithm for transient fluid flow with multiple free boundaries *Tech. Rep. LA-8355* LOS ALAMOS
[8] Hajipour A Kermanpur A and Mahmoudi S 2008 Numerical simulation of metal flow and solidification in the multi-cavity casting moulds of automotive components *Journal of materials processing technology* **206** 62-68
[9] Ryadno A *et al* 1983 Mathematical simulation of centrifugal casting of pipes, *HEAT TRANSFER-Soviet Research* **15** No.1 134-140
[10] Robillard L Vasseur P and Zhao M 1998 Mixed convection in a low rotation horizontal cylinder *Int. Comm.Heat Mass Transfer* **25** No. 7 1031-1040
[11] Song N N *et al* 2009 Computer Simulation of Core Filling Process of Cast High Speed Steel Rolls *China Foundry* **6**(4) 314-318
[12] Xie J X Jiang S C Liu H B and Chen X H 2002 The production of centrifugal casting of nodular cast iron big roll (in Chinese) *CFHI Technology* 60-61
[13] Bi X Q 2000 The improving of the pouring system for horizontal centrifugal casting (in Chinese) *Mechenical Engineer* **5** 42

IOP Conf. Series: Materials Science and Engineering **33** (2012) 012031 doi:10.1088/1757-899X/33/1/012031

Numerical simulation of centrifugal casting of pipes

E Kaschnitz[1]
Österreichisches Gießerei-Institut, Parkstraße 21, 8700 Leoben, Austria

E-mail: erhard.kaschnitz@ogi.at

Abstract. A numerical simulation model for the horizontal centrifugal pipe casting process was developed with the commercial simulation package Flow3D. It considers - additionally to mass, energy and momentum conservation equations and free surface tracking - the fast radial and slower horizontal movement of the mold. The iron inflow is not steady state but time dependent. Of special importance is the friction between the liquid and the mold in connection with the viscosity and turbulence of the iron. Experiments with the mold at controlled revolution speeds were carried out using a high-speed camera. From these experiments friction coefficients for the description of the interaction between mold and melt were obtained. With the simulation model, the influence of typical process parameters (e.g. melts inflow, mold movement, melt temperature, cooling media) on the wall thickness of the pipes can be studied. The comparison to results of pipes from production shows a good agreement between simulation and reality.

1. Introduction

Ductile iron water pipes and piles for deep foundation engineering are produced using the "de Lavaud" horizontal centrifugal pipe casting technique. In this process, the liquid iron is filled through an open channel into a fast rotating hot-work tool-steel mold that is slightly tilted. The tool-steel is water-cooled at the outside. In order to distribute the metal, the spinning mold including its cooling system is shifted axially in a controlled movement.

The numerical simulation of this process is challenging, no description of its numerical simulation was found in literature. Some papers describe the less complex vertical centrifugal casting process [1-6]. Besides taking into account the mass, energy and momentum conservation equations and the free surface tracking, the following phenomena have to be considered for the horizontal centrifugal casting process of pipes: (i) relatively fast radial movement of the mold; (ii) relatively slow but time dependent movement of the mold in axial direction; (iii) time dependent inflow of the iron along the channel; (iv) friction between the liquid and the mold; (v) turbulence and viscosity of the iron. The situation is hindered by an unfavorable relation between the relatively thin wall thickness of the tube (in the millimeter range) and its overall length (several meters).

After overcoming instability difficulties, it was possible to develop a valid model with Flow3D. In this work, the basic principles of the solution method are discussed; examples of numerical simulations and practical experiments are presented.

[1] To whom any correspondence should be addressed.

2. Simulation model

The geometry of the model consists of a mold from tool-steel and a closing core from cold-box bonded silica sand, both parts are spinning and moving (figure 1). An open feeding channel is extended into the mold, this channel is fixed. The outer side of the mold is water-cooled (significant heat flow), the inner side contains air (almost no heat loss).

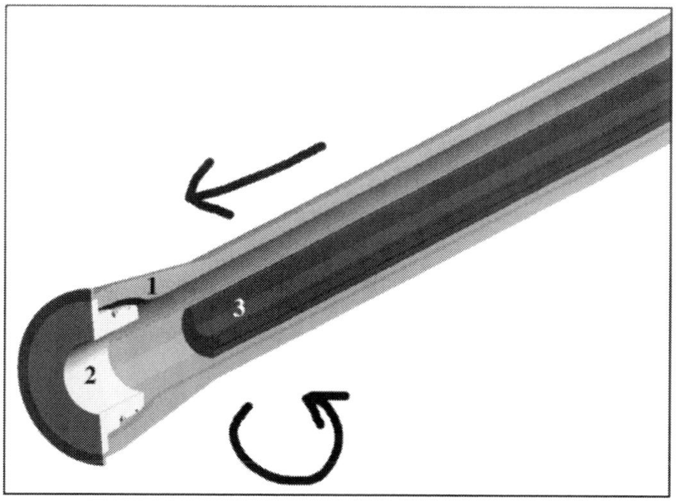

Figure 1. Vertical cut through the geometry of the model (1, mold; 2, sand core; 3, open feeding channel). The arrows show the direction of the movement of mold and core.

The mold and the core (but not the open channel) are cylindrical. As the melt flow follows the inner surface of the mold most of the time, a cylindrical coordinate system was chosen. In comparison to a regular Cartesian coordinate system, most of the unfavorable flow between neighboring cells under an angle of 45 degree is avoided.

For faster and better computations, the (cylindrical) mesh is aligned to the mold that is slightly tilted. This means the gravity vector gets the same tilt in the axial direction and splits in two components. Using the cylindrical mesh, the so-called "Non-inertial-reference-frame"-model of Flow3D has to be introduced to be able to define the gravity vector appropriately [7].

The relatively slow but time dependent movement of the mold is described by the "General-moving-obstacles"-model of Flow3D. This model shifts the assigned obstacles (mold and core) according to the pre-defined motion. Additional source terms according to the loss of mass, momentum and energy are added to the conservation equations at mesh cells in the vicinity of the shifted obstacles to compensate for these cells now covered by the obstacle [7].

This approach is not suitable for the relatively fast spinning of the mold and core, because the time steps become increasingly smaller with increasing spin velocity. Meaningful simulations would take months or years. For the rotational movement of mold and core, the so-called "Ospin"-model of the software package Flow3D was used. This model does not spin the obstacle itself, but transfers an appropriate amount of momentum as shear force from the obstacle in contact with the melt (inner surface of the mold and outer surface of the core) to the melt itself. The simulation model is comparable to a merry-go-round, where the motionless person outside pushes the child in the seat for rotational movement.

The free surface and solidification behavior of the fluid was modeled with Flow3D in a standard way. But Flow3D does no momentum and mass conservation calculations for already solidified cells. For this reason, as soon as the melt in a cell is solidified, the rigid cell is moved artificially by the predefined rotational and horizontal movement. The rotation speed leads only to a limited centrifugal pressure in the melt (the wall-thickness is rather small): the solidification behavior is not assumed to be influenced by the rotation.

The simulated pipes have a wall thickness of approximately five millimeter and are several meter long. This means that even big meshes with long computation times (up to three weeks) are relatively coarse. For that reason, special care has to be taken to choose appropriate values for viscosity and the turbulence model as well as the amount of shear stress that is transferred from mold to melt. A Renormalization-Group method is used for the transport equations for turbulent energy and its dissipation (k-epsilon model). A surface roughness factor provides additional shear stress between wall and the first cell layer of metal. These parameters are slightly dependent on the relatively rough mesh. This dependency should vanish for better meshes, but these are not to compute in reasonable time-spans nowadays. In order to approximate the fluid flow with correct parameters, practical experiments were carried out.

3. Experiments

The experimental setup consists of an electrical engine driving a horizontal cylindrical mold at precisely controlled revolution speed [8]. The control provides either constant speed or an angular velocity ramp at constant acceleration. The mold used for the experiments was a 300 mm piece of an original 5 meter hot-work tool-steel mold with its original inner surface. The sample mold has two lids at the front and at the backside; the front side lid is transparent to be able to watch the fluid inside (figure 2).

Figure 2. Experimental setup for the spinning experiments (1, mold; 2, engine; 3 gear; 4, high-speed camera).

Two different fluids were used for the experiments: colored water at room temperature and Rose's alloy (Bi, 50 wt-%; Pb, 25 wt-%; Sn, 25 wt-%) at approximately 100°C. The behavior of the fluid at different revolution speeds and different acceleration ramps was filmed from the front by a high-speed digital camera at 600 frames per second.

Figure 3 shows the shape of the free surface of the metal at a certain revolution speed. As the contrast of high-speed movies is poor in principle (and even worse in a printed paper), the interface position in the figure is marked by red arrows. The metal undergoes three phases during acceleration: (i) it forms a wedge shaped surface along the cylinder; (ii) the free surface decays and a curtain forms; (iii) when the angular velocity is high enough, the looping condition is satisfied and suddenly the curtain vanishes and the metal turns stable along the inner surface of the mold. This behavior depends mainly on viscosity of the melt and the surface condition of the mold. Figure 4 shows two simulations with different surface roughness, in one case the melt is unstable and forms a curtain, the other case shows a stable spinning of the melt.

In a second experimental setup, the engine and gear system was turned by an angle of 90 degree to use a part of the mold as a turntable. With that setup, liquid cast iron could be used instead of Rose's alloy. Using the gained knowledge from the experiments, meaningful input parameters were found for the simulation of the specific industrial pipes.

Figure 3. Shape of the free surface of the Rose's alloy at a selected revolution speed.

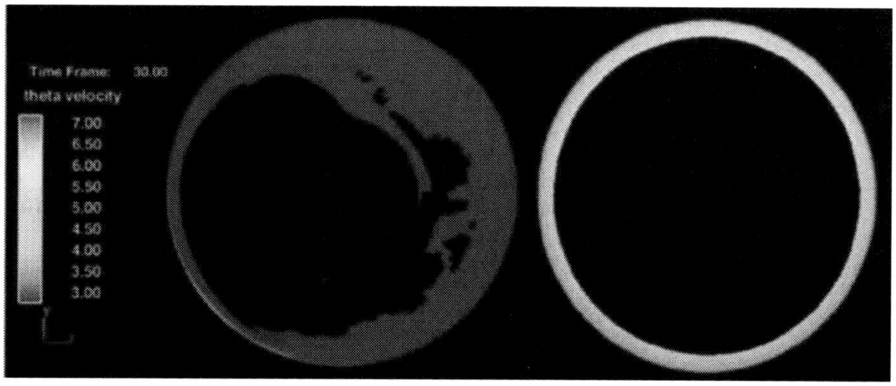

Figure 4. The simulation with low surface roughness (left) shows an unstable melt curtain, the simulation with an appropriate surface roughness (right) shows a stable melt following the looping condition.

4. Results

Figure 5 shows an example of the temperature distribution during filling of a pipe in a front view with mold and core faded out.

The wall-thickness of the produced pipes should be as uniform as possible. Therefore, the most interesting simulation result is the predicted wall-thickness of a pipe along its overall length. From each individual simulation with specific production parameters, the position of each fluid cell, its size and fill fraction was read out from Flow3D in order to compute the wall-thickness along the length. Depending on production parameters, the shape of the pipe can be quite different. As an example, figure 6 shows the wall-thickness of several pipes as a function of length.

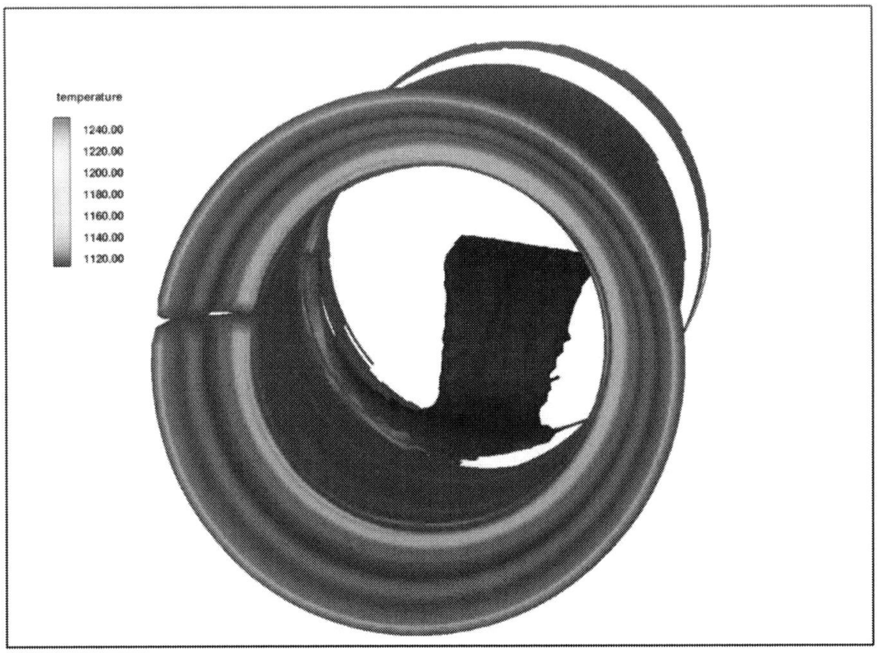

Figure 5. Simulated temperature distribution during filling of a pipe in a front view. The hot inflowing melt (in red color) is distributed along the inner tube-shaped steel mold due its rotation.

5. Conclusions

Following conclusions can be drawn:

- The De-Lavaud process can be simulated with Flow3D, taking advantage of the manifold modeling capabilities of this software
- Long calculation times (15 to 20 days) per pipe are common
- In comparison to reality, the influence of the main process parameters can be reproduced in the simulation
- Computed wall thickness matches within the relatively coarse grid size with measurements of standard pipes
- In simulation, each single process parameter can be varied and its influence on wall thickness can be studied
- Combinations of process parameters and hard (or not) to change parameter sets were simulated
- The impact of important parameters was identified.

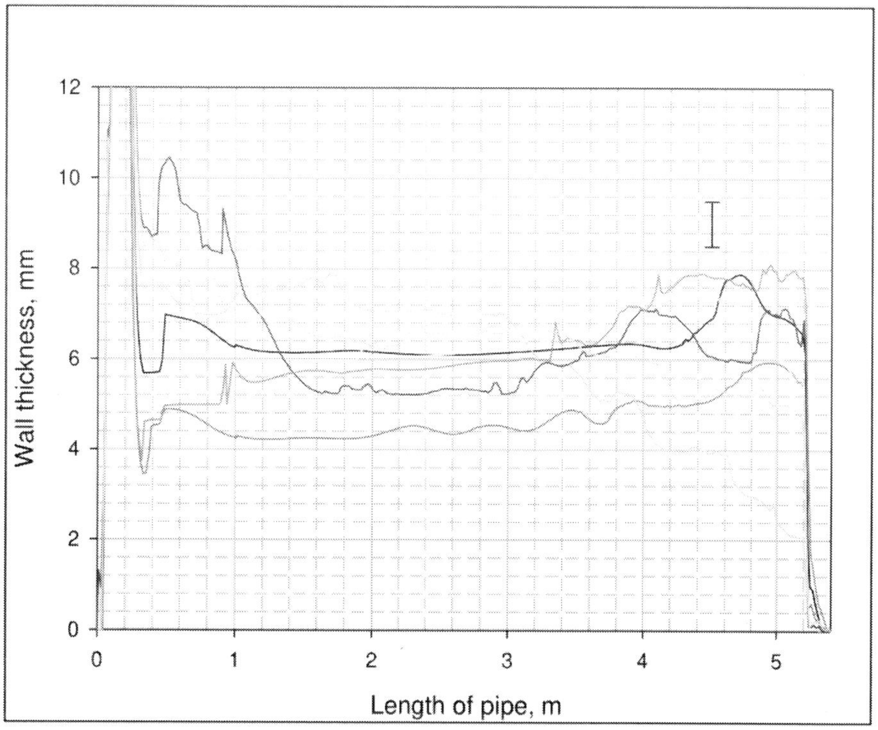

Figure 6. Examples for several simulated wall-thicknesses along the length of pipes using different process parameter sets.

References

[1] Wu S P, Liu D R, Guo J J, Li C Y, Su Y Q, Fu H Z 2006 *Mat. Sci. Eng. A-Struct* **426** 240-9
[2] Fu P X, Kang X H, Ma Y C, Liu K, Li D Z 2008 *Intermetallics* **16** 130-8
[3] Gao J W, Wang C Y 2000 *Mat. Sci. Eng. A-Struct* **292** 207-15
[4] Wu S P, Li C Y, Guo J J, Su Y Q, Lei X Q, Fu H Z T 2006 *Nonferr. Metal. Soc.* **16** 1035-40
[5] Shevchenko D M, McBride D, Humphreys N J, Croft T N, Whitey P, Green N R, Cross M 2009 Centrifugal casting of complex geometries: computational modelling and validation experiments *Modeling of Casting, Welding and Advanced Solidification Processes - XII* eds S L Cockroft and D M Maijer (Warrendale, PA: The Minerals, Metals & Materials Society (TMS)) pp 77-84
[6] McBride D, Humphreys N J, Croft T N, Whitey P, Green N R, Cross M 2009 Complex free surface flows for mould filling using centrifugal casting *Modeling of Casting, Welding and Advanced Solidification Processes - XII* eds S L Cockroft and D M Maijer (Warrendale, PA: The Minerals, Metals & Materials Society (TMS)) pp 459-66
[7] 2008 *Flow3D Version 9.3 – User Manual* (Santa Fe, NM: Flow Science Inc.)
[8] Lackner R 2011 *Diploma thesis* (Leoben, Austria: Montanuniversität Leoben)

IOP Conf. Series: Materials Science and Engineering **33** (2012) 012032 doi:10.1088/1757-899X/33/1/012032

Shallow water model for horizontal centrifugal casting

J Boháček[1], A Kharicha[2], A Ludwig[2] and M Wu[2]

[1]Materials Center Leoben Forschung GmbH, A-8700 Leoben, Roseggerstr. 12, Austria
[2]Chair for Simulation and Modelling of Metallurgical Processes, Dept. Metallurgy, University of Leoben, 8700 Leoben, Franz-Josef-Str. 18/III, Austria

E-mail: jan.bohacek@mcl.at

Abstract. A numerical model was proposed to simulate the solidification process of an outer shell of work roll made by the horizontal centrifugal casting technique. Shallow water model was adopted to solve the 2D average flow dynamics of melt spreading and the average temperature distribution inside the centrifugal casting mould by considering the centrifugal force, Coriolis force, viscous force due to zero velocity on the mould wall, gravity, and energy transport by the flow. Additionally, a 1D sub-model was implemented to consider the heat transfer in the radial direction from the solidifying shell to the mould. The solidification front was tracked by fulfilling the Stefan condition. Radiative and convective heat losses were included from both, the free liquid surface and the outer wall of the mould. Several cases were simulated with the following assumed initial conditions: constant height of the liquid metal (10, 20, and 30 mm), uniform temperature of the free liquid surface (1755 K). The simulation results have shown that while the solidification front remained rather flat, the free surface was disturbed by waves. The amplitude of waves increased with the liquid height. Free surface waves diminished as the solidification proceeded.

1. Introduction

The horizontal spin (or centrifugal) casting process (HSC) has several advantages over the conventional gravity casting and many other casting processes. Due to a high degree of metallurgical purity, minimum of shrinkage cavities, and homogeneous microstructure with very fine crystals, the centrifugally cast products generally show superior mechanical properties such as the rupture strength, the rupture strain, and the Young modulus [1]. Two physical phenomena contribute to these outstanding features. The first one is the high centrifugal force. The solidification usually starts immediately at the moment when the liquid metal gets in first contact with the mould wall, and cloud of fine crystals nucleate near the mould region. A certain portion of them is brought back into bulk by the high centrifugal force and these free crystals serve as new nucleation sites, continue to grow in the bulk, hence fine structure is achieved throughout. Moreover, the high centrifugal force supresses the possible occurrence of shrinkage, the air gap between the mould and the solidified shell, and enhances the solidification rate. The second phenomenon is the inherent vibration, which has also a positive impact on the mechanical properties by promoting a higher solidification rate and turbulent flow.

The impact of the centrifugal force, which is determined by the product of radius and the square of angular frequency, is twofold. A too low angular frequency can cause so-called raining i.e. metal droplets can fall down from the upper part of the inside surface of the casting product [2]. On the other hand, excessive speeds usually accompanied by strong vibrations can lead to longitudinal cracks caused by the hoop stress in the initially solidified shell. Furthermore, increasing the angular frequency promotes segregation. Therefore, the centrifugal force, together with many other process parameters like the pouring temperature, pouring rate and the mould coating etc., need the optimal control.

Quite amount of experimental studies on the centrifugal effect were performed. For example, Chirita and co-workers [3] found that the centrifugal force acts on the casting process by three separate mechanisms: the centrifugal pressure, the inherent vibration, and the fluid dynamics. First, the centrifugal pressure as one of the promoting mechanisms is naturally responsible for the elimination of the shrinkage (the gap between the mould and the shell) and the shrinkage porosity. The quality improvement could be also theoretically attributed to the shift of the phase diagram, which commonly happens in the squeeze casting. In the HSC this was however not proven. Next, the inherent vibration caused by quasi-static imbalances in the rotating system can enhance the melt mixing, which naturally leads to a higher heat transfer rate. The presence of the vibration promotes a transition from the lamellar to the fibrous morphology of the casting microstructure. The vibration was also proved to increase the eutectic volume fraction. However, a critical value of centrifugal acceleration was also found at which the grains tend to coarsen. They also reported some drop in the volume fraction and size of pores, but again a critical value of centrifugal acceleration most likely exists. Finally, the fluid dynamics plays an important role especially during the filling, when it washes newly solidified crystals out from the wall back into the melt. Here, these fragments stand for new nucleation sites and promote solidification of equiaxed crystals. The fluid dynamics also enhances mixing and solidification rate, consequently.

Only few numerical studies have been devoted to the horizontal spin casting. In the work of [4], an effort for a comprehensive description of the flow dynamics of the melt inside the horizontally rotating mould was done using the VOF model (STAR-CD). Two phases were considered, the liquid metal and the surrounding air. The time step was notably large (~0.01 s) which implies a very rough calculation. Nevertheless, the results were found to be in quite good agreement with experimental data. Zagórski and co-workers [5] used the similar VOF model to study the initial stage of mould filling during the vertical centrifugal casting of metal matrix composite reinforced with SiC. Additionally, a discrete phase model was used to track SiC particles. The problem was solved as 2D axisymmetric with swirl component of velocity and all calculations were terminated at 1 s of real time. A simple solidification model for planar front, coupled with equations for force balance on dispersed ceramic particles, was used by Raju and co-workers [6] to describe the centrifugal casting process of metal matrix composite. The flow dynamics was ignored. Segregation of particles was extensively studied with respect to heat transfer coefficient between the mould and the casting. Interesting work was done by Drenchev and co-workers [7] on macrostructure formation during centrifugal casting of particle reinforced metal matrix composites. System of PDEs was solved for particle movement, heat conduction inside the mould and casting, and the solidification considering a linear temperature profile in the mushy zone.

The aim of the current paper is to develop a mathematical model capable of simulating the main features of the flow dynamics inside a horizontally rotating mould coupled with heat transfer and solidification. We target at the entire HSC process i.e. the casting of the pure shell and the intermediate shell with the filling system included. The High Speed Steel (HSS) is used for the casting of the pure (or outer) shell, while the intermediate layer is made of the grey cast iron. The HSC duration is approximately 30 min. Due to the nature of multi-physics spanning different time scales (flow dynamics vs. solidification rate) the numerical model has to be simple and robust. In this paper, the model was tested without the filling system. For this reason, a certain height of the liquid metal (10, 20, and 30 mm) with a constant temperature of 1755 K was evenly imposed everywhere inside the mould in the beginning of each simulation.

2. Numerical model

The present work is based on a so-called shallow water model. The fundamental model theory can be found in literatures [8]-[10]. The shallow water equations were originally derived in one dimensional form as so-called Saint-Venant equations [11]. This model was implemented by the authors in the commercial CFD package ANSYS-FLUENT. The calculation domain comprises 4 layers (Figure 1), representing the mould, the solid metal, the liquid melt and air, respectively. More details on the model settings can also be found elsewhere in [12]-[13].

2.1. Flow modelling

Flow is only necessarily solved in the layers of liquid melt and air. The solid layer is assumed to be in perfect contact with the mould, rotating with a pre-described velocity. The air flow is not of significant importance, since momentum transfer with the liquid melt is ignorable. The flow in the liquid melt layer is considered to be laminar. Therefore, a parabolic velocity profile is imposed with zero shear stress on the free surface. Due to a high angular frequency Ω (~70 rads^{-1}) of the system the liquid melt is mainly rotating with the mould; thus, the model can be defined in the rotating frame of reference, and fictitious forces (centrifugal and Coriolis forces) are included in the momentum equations as source terms. The continuity equation is defined as:

$$\frac{\partial(h\rho)}{\partial t} + \nabla \cdot \left(h\rho\vec{u}\right) = S_{m,l\to s}, \tag{1}$$

where h denotes the height of the liquid layer and $S_{m,l\to s}$ is the mass transfer from liquid to solid due to solidification. The momentum equations for x and y coordinates are defined as:

$$\begin{aligned} &\frac{D}{Dt}\left(h\rho\vec{u}\right) + \rho h\Omega^2\left(R_{in} - h - \delta\right)\nabla(h+\delta) + \rho h\left|\vec{\Omega}\right|\left(2u_y\nabla\delta + \frac{5}{2}u_y\nabla h + \frac{5}{4}h\nabla u_y\right) = \\ &-h\nabla p + h\mu\nabla^2\vec{u} - \rho h\vec{g}\left[\left(0, \sin\left(\left|\vec{\Omega}\right|t + \frac{y}{R_{in}}\right)\right) + \cos\left(\left|\vec{\Omega}\right|t + \frac{y}{R_{in}}\right)\nabla(h+\delta)\right] - 3\mu\frac{\vec{u}}{h} + F_{l\to s} \end{aligned}, \tag{2}$$

The terms on the left-hand side of equation (2) represent inertia forces, which includes centrifugal and Coriolis forces. On the right-hand side of equation (2), the first term is the force induced due to the gradient of the static pressure. The second term handles the momentum losses resulting from stress tensor but only in tangential and axial direction. The last three terms represent the gravity force, the viscous force in the radial direction due to the no-slip BC on the mould wall, and the momentum transfer from liquid to solid due to solidification, respectively.

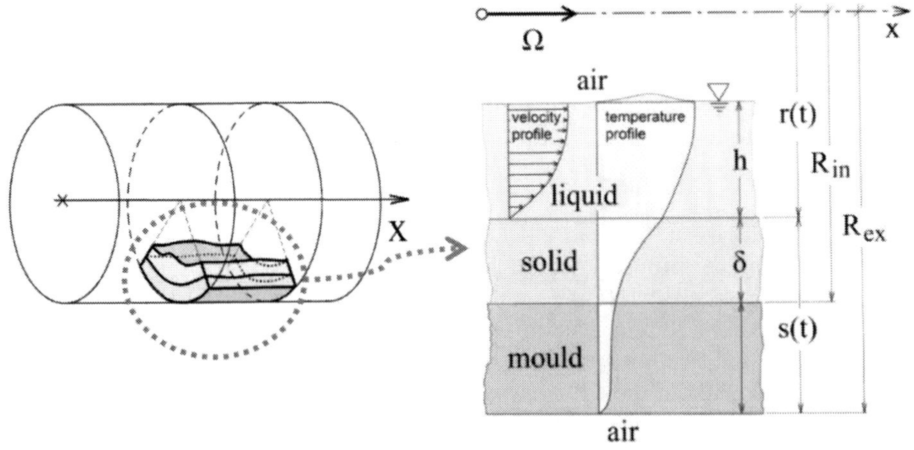

Figure 1. Schematic of a part of the horizontal centrifugal casting section.

2.2. *Mould deformation and vibration*

As discussed above, vibrations can strongly influence solidification and final quality of the product, consequently. Measurement of vibrations during the real HSC is a standard procedure and vibrations are always present. There are several possible origins of vibrations. One is due to large mould dimensions that the mould is in static imbalance because of the poor roundness of cylindrical surfaces. This becomes even worse during the filling, when the melt is unevenly spread inside the mould. Another reason is that temperature gradients through the thickness of the mould cause thermo-mechanical stresses and deformations. The present model treats vibrations as a time dependent perturbation of gravity as proposed in [14]. Corresponding acceleration can be defined as:

$$\vec{a_v} = (0, \quad \xi\cos(\omega t + \beta), \quad \xi\sin(\omega t + \beta))^T, \tag{3}$$

where ξ is the amplitude of acceleration due to vibrations, ω is the angular frequency associated to vibrations, and β is the phase. The phase β specifies the position where the oscillation begins. In the model β is set constant ($\pi/2$). The angular frequency ω is set to the double of the angular frequency Ω assuming an elliptical cross section of the mould. Besides, axial deformation of the mould due to temperature gradients is also taken into account using the following sine-like axis profile:

$$\tan\theta = A\frac{\pi}{\lambda}\cdot\sin\left(\frac{\pi x}{\lambda}\right), \tag{4}$$

where A is the amplitude of bending, x is the axial coordinate, θ is the angle between the x-axis and the deformed mould centreline, and λ is the distance between two coaxial rollers. The presence of axis bending influences the magnitudes and the orientation of all forces (centrifugal, Coriolis and gravity).

2.3. *Heat transfer and solidification*

The assumption of a fully developed laminar flow of the liquid melt naturally dictates a parabolic velocity profile in the radial direction. Concerning the temperature evolution, the Prandtl number Pr is responsible for the delay between the fully developed velocity and temperature profiles. In this case, the thermal diffusion is stronger than viscous forces ($Pr<1$); therefore, the flow can be treated as a thermally fully developed flow. According to Oosthuizen [15], temperature profile of thermally fully developed flow resembles the velocity profile. Therefore, the temperature profile is assumed to be parabolic as well. 2D average temperature field is obtained by solving the energy transport equation:

$$\frac{D}{Dt}(h\rho T) = \nabla\cdot(k\nabla T) + S_{T,l\to s}, \tag{5}$$

where k is the thermal conductivity and $S_{T,l\to s}$ is the heat source due to the heat conducted to the mould. Solidification is governed by coupling of the Stefan condition for the moving liquid-solid interface with the heat conduction to the mould. Since heat fluxes can be expected mainly in radial direction, only a 1D heat conduction equation is solved. It was derived in cylindrical coordinates for dimensionless space variable x^* defined by:

$$x^* = \frac{x - r(t)}{R_{ex} - r(t)}, \tag{6}$$

where x and $r(t)$ are distances of actual position and the moving interface from the mould axis, respectively. R_{ex} is the outer radius of the mould. Then, the heat conduction equation in the radial direction has the following form:

$$s(t)^2 \frac{\partial T}{\partial t} = \alpha \frac{\partial^2 T}{\partial x^{*2}} + \left(\frac{\alpha s(t)}{x^* s(t) + r(t)} + s(t)(x^* - 1)v \right) \frac{\partial T}{\partial x^*}, \tag{7}$$

where $s(t)=R_{ex}-r(t)$, α is thermal diffusivity. The velocity of moving interface, v, is determined by the 1D Stefan condition [16]:

$$\rho L v = \frac{k}{s(t)} \frac{\partial T}{\partial x^*}\bigg|_{x^*=0} - Q_{l\to s} + Q_{s\to a}, \tag{8}$$

where L is the latent heat. The first term on the right-hand side is the heat diffused into the solid shell. The second term signifies the heat flux from liquid to solid. The last term stands for the heat flux from liquid to air. From the physical point of view, this heat flux should be applied in equation (5); however, this naturally implies that solidification might start from the free liquid interface. It is expected that due to the density difference between liquid and solid all possible germens solidifying on the free surface will be centrifuged towards the mould wall. Therefore, the heat flux between liquid and air is applied in equation (8). Both equations (7) and (8) are one-dimensional, but are solved for each element of the 2D grid, on which the shallow water equations are solved. Heat transfer from free liquid surface to air is described by convective heat losses and radiation. Radiative heat transfer is further divided into heat redistribution inside the mould and heat losses via extremities. Losses due to the heat convection are defined as:

$$Q_{convection} = HTC_{in}(T_s - T_a), \tag{9}$$

where T_s and T_a are temperatures of the free liquid surface and air (323 K). HTC_{in} is the convective heat transfer coefficient. In the model it was set to a constant value of 100 $Wm^{-2}K^{-1}$. Radiative heat redistribution inside the mould is derived assuming the black body theory as follows:

$$Q_{rad,in} = f_x \sigma \varepsilon (T_s^4 - T_{avg}^4), \tag{10}$$

where f_x is the view factor depending on axial position (approximated using 4th degree polynomial) and σ and ε are the Stefan-Boltzmann constant and the emissivity ($\varepsilon = 0$), respectively. T_{avg} is the average temperature of free liquid interface given by:

$$T_{avg}^4 = \frac{\sum_{i=1}^{cells} f_{xi} T_s^4}{\sum_{i=1}^{cells} f_{xi}}, \tag{11}$$

Similarly, heat losses via extremities are:

$$Q_{rad,out} = (1 - f_x)\sigma\varepsilon(T_s^4 - T_a^4). \tag{12}$$

Similar relations are found for heat losses from the external mould wall. The convective HTC_{out} was set to a constant value of 150 $Wm^{-2}K^{-1}$. Temperature of air T_a is the same as in equation (9). Temperature of external mould wall is obtained from equation (7). Finally, radiative heat losses are calculated using equation (12) except that T_s is replaced with temperature of external mould wall and f_x is set to zero.

3. Results and discussion

The horizontal casting set-up is shown in Figure 2. The contour map represents the height of liquid metal after 3 s.

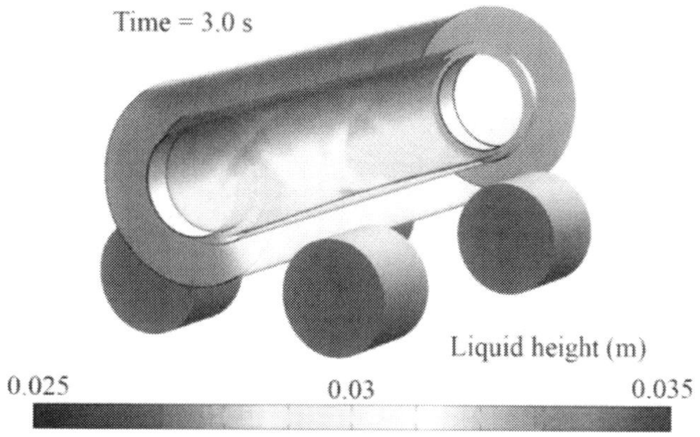

Figure 2. Liquid metal (height) distribution inside the mould.

Mould dimensions and material properties are listed in Table 1. Difference between external and internal radius of the mould gives the mould thickness (0.203 m). To keep the model simple and comprehensible all material properties are constant.

Table 1. Mould dimensions and material properties.

parameter	value (m)	property name	value
length of the mould	3.2	density	6800 kg/m^3
internal radius of the mould R_{in}	0.372	thermal conductivity	30 W/m-K
external radius of the mould R_{ex}	0.575	specific heat	750 J/kg-K
distance between two coaxial rollers λ	1.92	latent heat [a]	250 kJ/kg
axis bending A	0.002	dynamic viscosity	0.006 Pas

[a] used in Stefan condition only.

Several numerical cases were studied with varying initial liquid height h. Initial conditions can be summarized as follows: uniform distribution of liquid metal inside the mould, velocities set to zero, uniform temperature of the mould (433 K), temperature of free surface equal to the filling temperature (1755 K). Additionally, the mould is spinning at 633 rpm, which corresponds to an angular frequency Ω of 71.2 rad/s. The temperature of the liquid-solid interface is set to the liquidus temperature (1586 K).

For evaluation of results the internal mould wall was unfolded into a horizontal plane with x-axis and y-axis corresponding to axial and circumferential position, respectively. In Figure 3, the plot shows a snapshot of waves on the free surface of the melt spreading on already solidified shell. While the free surface evidently fluctuates, the liquid-solid interface remains smooth and nearly flat. The reason for that is due to the difference of characteristic times. The characteristic time of the flow can be defined as a reciprocal value of a wave frequency. It can be roughly estimated from the angular velocity Ω of the mould. It is approximately 0.1 s. The characteristic time of the solidification can be computed as the liquid height h divided by the solidification rate. Assuming the highest solidification rate occurring in the beginning (~0.01 m/s) the corresponding characteristic time is of the order of 1 s. The difference between characteristic times is thus at least one order of magnitude. In addition, the absence of the filling system and consequently the uniform temperature distribution along the internal mould wall also promotes uniform solidification.

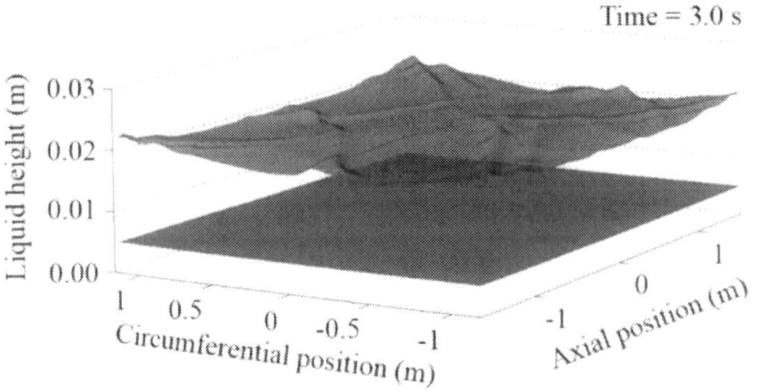

Figure 3. Waves travelling on free surface (blue) of melt with solid shell (red) beneath.

In Figure 4, transient temperature profiles are plotted in the radial direction for the case of the initial liquid height of 10 mm. The position of temperature record corresponds to the point (0, 0) in Figure 3. Dotted lines represent the corresponding parabolic temperature profiles in the liquid metal, while solid lines denote temperature profiles in both the solid metal and the mould. Only a small part of the mould close to the internal wall is depicted with coordinates ranging from 0 to 0.02 m. In the beginning, due to very high temperature gradient in the mould solidification is very fast (~0.01 m/s). However, due to heat conducted to the mould temperature gradient weakens and the solidification rate drops as time progresses. The solidification rate drops from more than 0.01 m/s below 0.001 m/s. In less than 10 s the melt is consumed and solidification is finished. Heat transfer however continues and temperature of the free surface further drops due to heat losses and heat conduction to the mould. Despite high temperatures inside the mould, a good thermal conductivity, and a perfect contact between the solid shell and the mould, temperatures of the outer part of the mould are still decreasing even after 25 s due to prevailing heat losses to air.

In Figure 5, three cases with different initial liquid heights (10, 20, and 30 mm) are compared. The data were recorded at the point (0, 0) (Figure 3). In the beginning, the heat conduction to the mould is so dominant that it leads to the same solidification rate for all three cases. Later on, curves diverge mainly because the solidification rate is also governed by the temperature gradient within the liquid metal. Markers indicate the end of solidification. It takes 85 s to solidify for the case of 30 mm of the melt height.

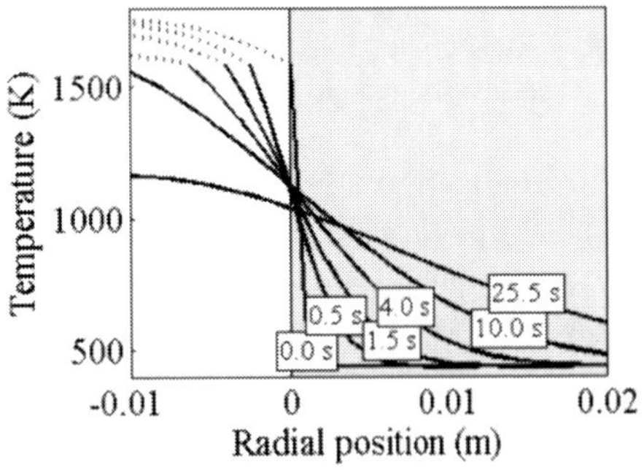

Figure 4. Typical temperature profiles in radial direction at different times; parabolic temperature profile inside the melt (dotted line) and computed temperature profile inside the solidified shell and the mould.

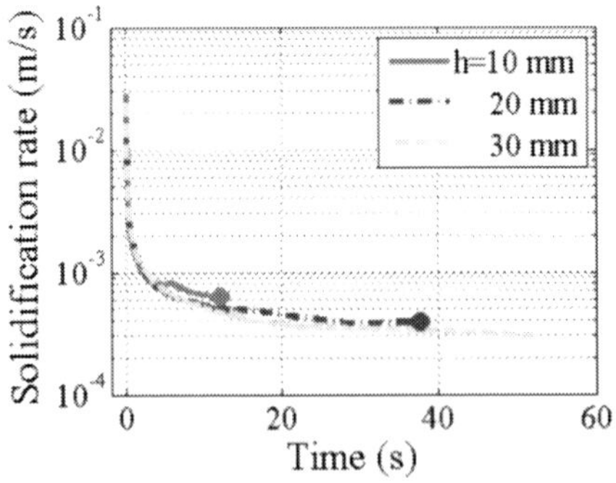

Figure 5. Solidification rate vs. time for different initial liquid heights (10, 20, and 30 mm); Markers signify end of solidification.

Solidification dynamics is naturally different from the mould centre to the extremities. Rate of solidification depends mainly on heat flux to the mould, but it is also influenced by heat flux from the liquid metal and heat losses from the free surface (Eq. 8). In Figure 6, contributions of each type of heat flux are compared in terms of temperature gradients plotted against time for the case for an initial liquid height 10 mm and without axis deformation.

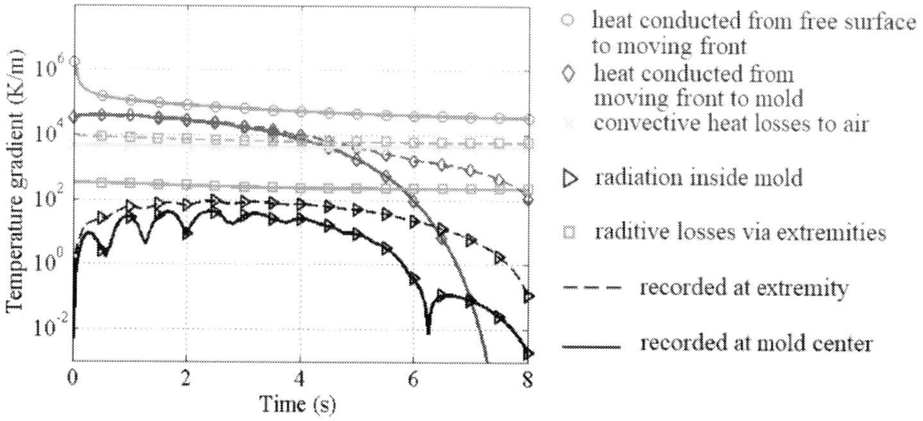

Figure 6. Temperature gradients as function of time.

Heat conduction from the moving interface to the solid (red circle) obviously dominates all other contributions during the whole process and is independent from the position in the mould. On the contrary, the radiative heat transfer inside the mould is the least dominant (black right-pointing triangle). The convective heat losses to air (green cross) are driven by the temperature of the free surface T_s. T_s almost does not depend on position in the mould because heat is mainly transported by the flow. This implies convective heat losses to air are independent of position. During the solidification, the free surface temperature T_s varies only from the initial temperature (1755 K) to the liquidus temperature (1586 K). Therefore, convective heat losses to air do not vary much in time. Radiative heat losses through extremities (magenta square) are nearly unchanged, but they decrease significantly from the mould extremity to the mould centre. Separation of curves (blue diamond) representing heat conduction in the liquid reflects a higher solidification rate near mould extremities.

4. Conclusions and future prospects

A numerical model of horizontal centrifugal casting of work roll has been developed based on the shallow water equations for multi-layer system. Flow dynamics is solved in the tangential and the axial direction considering flow variables averaged along the liquid height. The heat conduction in the radial direction was coupled with the Stefan condition for the moving liquid-solid interface. The mould deformation or more specifically, the deformation of the mould centreline was considered. Further, vibrations were introduced by perturbation of gravity.

Several simulations were conducted to study the flow dynamics of the liquid melt, wave patterns of the free surface, solidification. Results show wave propagation on the free surface. Wave amplitudes grow with increasing height of liquid metal. In the beginning of each simulation a single longitudinal wave was developed due to the counteracting of gravity and inertia forces. Later on, it was scattered into other longitudinal waves with smaller amplitudes. A lifetime of longitudinal waves was dependent on the amplitude of the axis bending. Generally, the higher the bending, the faster the transition to a developed flow regime. If there were only longitudinal (or horizontal) waves, the flow dynamics would not provide sufficient mixing in the axial direction which would most likely led to uneven microstructure along the mould length. On the other hand, in the tangential direction the mixing would be contrarily very strong. Purely annular waves would rather result in a uniform temperature field of the liquid metal. In the conducted simulations, neither longitudinal nor annular waves were observed in the developed flow regime. The waves were generally propagating in transversal direction, smoothing out temperature variations and thus, promoting a uniform solidification represented by a flat moving front. As the solidification proceeded, viscous forces became dominant and free surface fluctuations were lowered gradually. While the free surface was dynamically distorted by travelling waves, the liquid-solid interface remained smooth and nearly flat. Solidification rate was higher at extremities than in the mould centre due to higher radiative heat losses. Radiative heat transfer inside the mould was found to have only a negligible effect on solidification. Effect of vibrations was found to be negligible. The final solidified shell was however naturally influenced by the axis bending. The thickness of the shell varied along both, the tangential and the axial direction.

The current model is still under development and it is still to be extended to account for the moving mushy zone by using an extra layer with variable viscosity. Mould filling is also an important phenomenon and will be included. Further, thermal resistance due to mould coating and a possible air gap caused by shrinkage of the shell has to be included.

Acknowledgement
This work is financially supported by the Materials Center Leoben and the Eisenwerk Sulzau–Werfen.

References

[1] Chirita G, Stefanuscu I, Barbosa J, Puga H, Soares D and Silva F S 2009 On assessment of processing variables in vertical centrifugal casting technique *Int. J. of Cast Metals Research* **22** 382–9
[2] Keerthiprasad K S, Murali M S, Mukunda P G and Majumdar S 2010 Analysis of fluid flow in centrifugal casting, *Front. Mater. Sci. China* **4** 103–10
[3] Chirita G, Stefanuscu I, Soares D and Silva F S 2006 Centrifugal versus gravity casting techniques over mechanical properties, *Anales de Mecánica de la Fractura* **1** 317–22
[4] Keerthiprasad K S, Murali M S, Mukunda P G and Majumdar S 2010 Numerical simulation and cold modeling experiments on centrifugal casting, *Metall.Mater. Trans. B* **42** 144–55
[5] Zagórski R and Śleziona J 2007 Pouring mold during centrifugal casting process, *Archives of Materials Science and Engineering* **28** 441-4
[6] Raju P S S and Mehrotra S P 2000 Mathematical modelling of centrifugal casting of metal matrix composites, *Material Transactions JIM* **41** 1626–35
[7] Drenchev L, Sobczak J, Malinoc S and Sha W 2003 Numerical simulation of macrostructure formation in centrifugal casting of particle reinforced metal matrix composites. Part 1: model desription, *Modelling Simul. Mater. Sci. Eng.* **11** 635–49
[8] Pedlosky J 1987 *Geophysical Fluid Dynamics* (New York: Springer-Verlag)
[9] Knauss J A 1996 *Introduction to Physical Oceanography* (New Jersey: Prentice Hall)
[10] Dellar P J and Salmon R 2005 Shallow water equations with a complete Coriolis force and topography, *Phys. of Fluids* **17** 1-23
[11] Saint-Venant A J C 1871 Théorie du movement non permanent des eaux, avec application aux crues de rivieres et a l'introduction des marrees dans leur lit, *C. R. Acad. Sc. Paris* **73** 147-54
[12] Bohacek J, Kharicha A, Ludwig A and Wu M 2011 Liquid metal inside a horizontally rotating cylinder under vibrations, *Proc. Int. Conf. Experimental Fluid Mechanics* vol 2 (Liberec) pp 564-73
[13] Bohacek J, Kharicha A, Ludwig A and Wu M 2011 Shallow water equations for a liquid film in horizontally rotating cylinder, *Proc. Int. Conf. Computational Mechanics* (Pilsen)
[14] Martinez G, Garnier M and Durand F 1987 Stirring phenomena in centrifugal casting of pipes, *Applied Scientific Research* **44** 225–39
[15] Oosthuizen P and Naylor D 1999 Introduction to Convective Heat Transfer Analysis *(Boston: WCB/McGraw-Hill)*
[16] Dantzig J A and Rappaz M 2009 Solidification *(Lausanne: EPFL Press)*

IOP Conf. Series: Materials Science and Engineering **33** (2012) 012033 doi:10.1088/1757-899X/33/1/012033

In-situ observation of transient columnar dendrite growth in the presence of thermo-solutal convection

Wajira U. Mirihanage, Lars Arnberg, Ragnvald H. Mathiesen
Norwegian University of Science and Technology, N-7491 Trondheim, Norway.
E-mail : wajira.mirihanage@ntnu.no

Abstract. The time evolution of tip radii and velocities of growing columnar dendrites have been extracted from an image sequence measured in situ by synchrotron X-ray video microscopy during directional solidification of Al-15%Cu-9%wtSi alloy. At the scale of the primary dendrites, a highly transient growth was observed in conjunction with liquid flow caused by modest thermo-soutal convection. The nominal spatial and temporal resolutions employed in the experiment were 1.4 μm at approximately 100ms frame rate, respectively. A very prominent mutual inter-relation between the solute field and flow velocity field and the dendrite growth rate and morphology is evident in these observations. Image processing allowed extraction of the corresponding projection averaged solute field in each frame, and an estimate of the solute boundary velocity field by frame-to-frame tracking of its motion. These measurements may eventually support development of statistical dendrite kinetics models for transient states in alloy solidification processes that could be applied in microscopic scale solidification models such as phase field.

1. Introduction

The microstructure of metallic materials has a strong influence on its mechanical properties [1] and therefore, a better understanding of microstructure formation processes and their dependence on various parameters is of a major importance in materials science, particularly in the field of metallurgy. The progress and morphological evolution of solid-liquid fronts during non-equilibrium solidification in metallic alloys is still far from being fully understood; hence the effects of various interrelated physical parameters on dendrite solidification is strongly in demand for improved insight on how microstructure forms during alloy solidification process. Such interrelated parameters include growth speed, dendrite tip radius, solutal field ahead of the dendrite tip and the thermo-solutal convection adjacent to the solid-liquid interface etc [2-8].

Synchrotron X-ray radiography [9-13] has gradually been established as a promising scientific tool to observe and understand non-equilibrium solidification of metallic alloys in-situ during growth. The methods have been demonstrated over a wide range of applications, and been focused at different aspects from finer microscopic details at optimal spatio-temporal resolutions over a small field of view and also studies at lower resolutions where the field of view covers the complete sample. This contribution presents preliminary measurements from an experiment conducted combining high spatial, temporal resolutions and high-quality image contrast, providing fully time resolved microscopic information of the dendrite morphological evolution and the spatiotemporal evolution of its adjacent solute boundary in presence of natural convection.

IOP Conf. Series: Materials Science and Engineering **33** (2012) 012033 doi:10.1088/1757-899X/33/1/012033

2. Experimental

The experiment was carried out at beam line ID6 in the European Synchrotron Radiation Facility (ESRF) during a test campaign of an ultra-high-resolution Transmission X-ray Microscopy (HRTXM) [14]. It should be noted, however, that the data presented here where measured in conventional microradiography, not employing the HRTXM-mode.

An incident 14.5 keV monochromatic X-ray beam provided a photon flux at the sample of ~ 10^{13} photons/ (s·mm^2) with a field of view of approximately 0.9 X 0.7 mm^2. Image capture was performed by a Sensicam-CCD camera, with 1376 x 1040 pixels and 12 bit dynamical range, employing maximum frame rate of 10Hz. The solidification experiment was conducted with a Bridgemen furnace that used several times in the past and detailed elsewhere [11]. The directional solidification parallel with gravity was carried out with a 136±5μm thick 12 mm X 25 mm Al-15%Cu-9%wtSi alloy sample.The temperature gradient and sample pulling velocity were held constant at 43 K/mm and 20 μm/s, respectively, throughout the experiment.

3. Image processing and results

The raw data constitute a sequence of 354 images, corresponding to monitoring of the solidification dynamics for about 35s. The data has been subjected to a set of routines for image processing and analysis. As the analysis work is still going on, some of the results presented have been subjected to preliminary analysis and are not yet brought to the level of quantitative precision which we expect to reach with the complete analysis. Initially, images are passed through several filters to remove artefacts that stems from bad pixels, structures in the incident beam intensity profile, spots on the camera scintillator or dust in the visible optics path. After noise/spurious pixel removal and flat fielding, the first step towards quantitative analysis is binary masking/segmentation of each frame, employing adaptive thresholding for the gray level of each image. The final set of binary masks separate fully liquid pixels from those where the projection image is a superposition of wave transmission through any combination of solid and liquid. After masking, a Laplacian operator is used to extract the edges from the binarized masks, as shown in Figure 1.

Figure 1. (a) Contrast-enhanced image (b) Binary mask (c) Detected edges of the dendrites

The sets of edge points denoting a dendrite tip area were selected and fitted with least-square parabolas, from which the tip radius of curvature could be extracted. Simultaneously, time derivatives of the parabola locii were used to extract dendrite tip velocities, after being transformed to a set of absolute coordinates by correcting for the sample translation. The relative solute concentration levels in the liquid along the growth direction from the dendrite tip was obtained via considering the pixel gray scale values along the imaginary line drawn from the dendrite tip towards to the growth direction in this preliminary analysis.

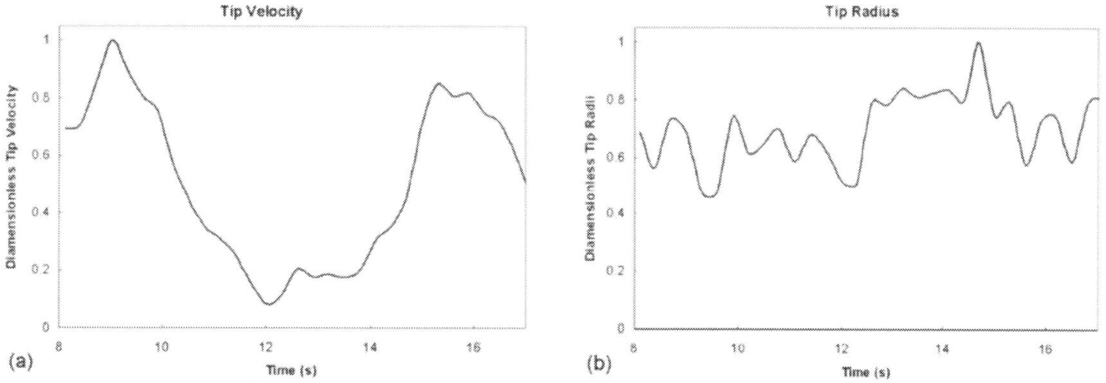

Figure 2. Measured (a) dendrite tip growth velocity and (b) radii

For the work presented here, one particular dendrite tip from the image sequence has been selected (the dendrite that shown left in Figure 1), and its tip radius and growth velocity were extracted for a selected part of the image sequence. Figurer 2 presents the time evolution of (a) dendrite tip growth velocity and (b) tip radii for the selected part of the sequence. Please note that in this work we only analyse 9 seconds interval starts from t = 8.1s of the complete image sequence and each point analysed is spaced with several non-analysed frames in between. Accordingly, the fluctuations shown in the tip radius represent real transient changes, and not variations that relate to the accuracy of the estimates.

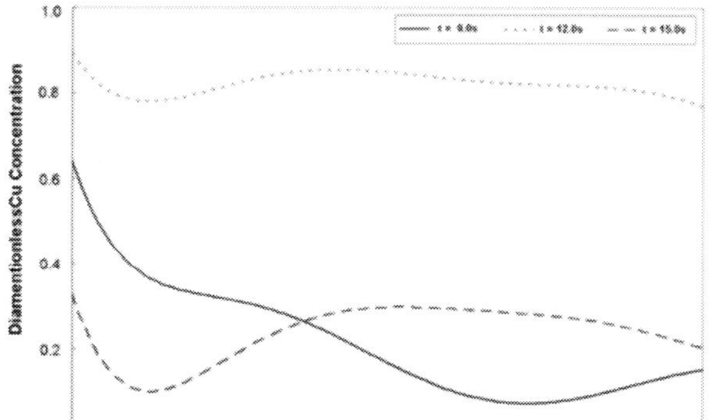

Figure 3. Dimensionless Cu concentration profiles ahead of the dendrite tip

Similarly, the evolution of the relative 2D projected Cu concentration field in front of the dendrite tip was obtained via pixel gray scale values after normalising the flat fielded and noise filtered data against the transmitted intensity through bulk liquid. The solute analysis presented here is preliminary. Full 2D analysis and subsequent 3D *xy-time* reconstruction of the liquid data along the same principles outlined previously [10] is commencing. Figure 3, show three selected Cu concentrations at times, t = 9.0s, t = 12.0s and t = 15.0s ahead of the selected dendrite tip.

4. Discussion

One of the interesting observations of this image sequence is presence of recognizable and fully temporal and 2D spatially resolved solute convection during the solidification. At the macroscopic scale, the flow rates are relatively low; however at the scale of dendrite-tip, flow rates are substantial, typically as the same magnitude as the tip velocities. In the considered part of the image sequence (as shown in Figure 4), it is possible to see wave like movement of solute flow ahead of the dendritic front. At the beginning low Cu concentration iso-contours are moving leftward and a Cu enriched zone

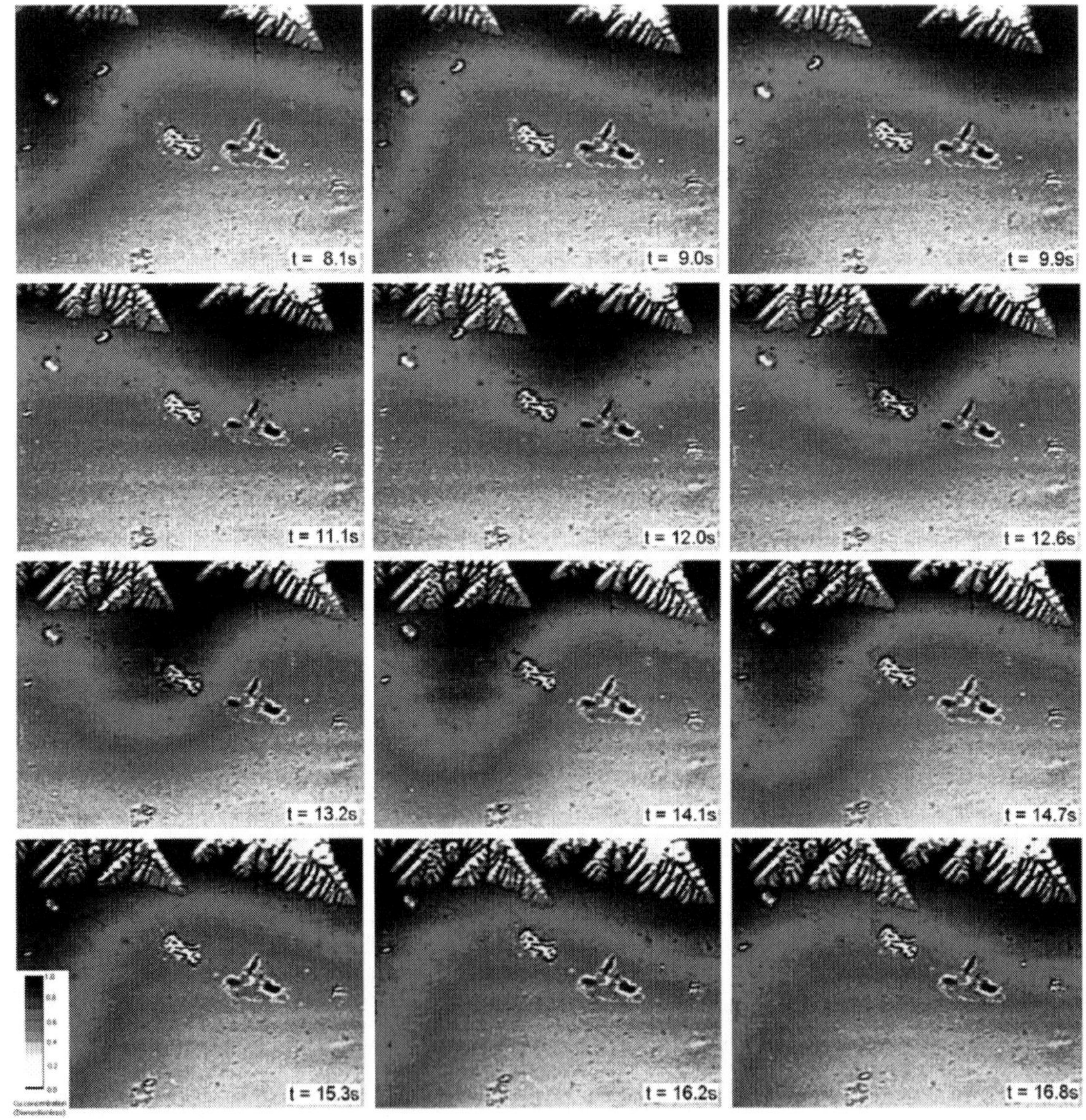

Figure 4. Visually evident thermo-solutal convective flow

in front of the dendrite tip(our selected dendrite tip) is depleted due to the flow. Then Cu concentration around the tip increases rather quickly (approximately at t=12s). Under these conditions, it is obvious that the sedimentation of Cu-rich melt from the interdendritic channels of the columnar array is the source for the settling flow of solute which causes temporary enrichment at the growth front. In periods the solute plumes ejecting from the mush practically erodes the solute undercooling at the tips. Thereafter, a second wave of Cu concentration depletion around the considered dendrite tip area begins again. Similar cycle is repeating approximately every 7 seconds in this sequence, almost in a sinusoidal manner. This repeatable wave like solute concentration fluctuation is a results of the phenomenon that combines sedimentation Cu rich liquid that concentrated between columnar channels and sideways macroscopic heat advection since the isotherms inside the cell is somewhat curved towards the sample edges. With such a small field of view relative to the full size of the cell, the thermal centre of the sample is easily located outside our field of view. In our interpretation, it looks as though, until certain Cu concentration has built-up between the columnar channels, solute diffusion and sedimentation is predominantly washed away by the macro scale thermal flow. When a certain Cu concentration has had time to build up and exceeded a certain density; it eject as a plume from the mush and feeds into this advective flow and gives rise to the sinusoidal wavy Cu-concentration pattern that transports across the columnar array. This transport patterns causes to fluctuations of Cu concentration ahead of a dendrite tip substantially as shown in Figure 3.

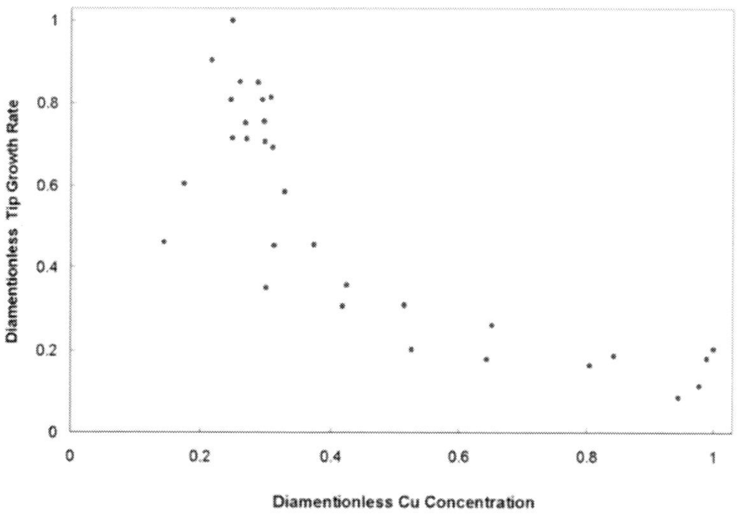

Figure 5. Tip growth rate vs. average dimensionless Cu concentration

From the preliminary analysis, it is already clear that the thermo-solutal flow induce solute field fluctuations in front of the dendrite tip, causing considerable dendrite tip growth velocity and curvature fluctuations during solidification. It is evident that the dendrite growth and morphological fluctuations correlated with fluctuations in the solute profile ahead of the dendrite tip. As a preliminary demonstration, we provide a comparison of the average dimensionless Cu concentration within 100μm of the dendrite tip against the growth rate, as shown in Figure 5. At this stage the data points only serve to show the basic trend between two parameters.

5. Summary

An image sequence of Al-15%Cu-9%wtSi alloy solidification with high temporal and spatial resolution was obtained via synchrotron video microscopy. The image sequence show considerable thermo-solutal convection induced effects on the dendrite morphology and growth rate. The spatiotemporal resolution of the image sequence will allow extraction of dendrite tip radii evolution with reasonably high accuracy, together with tip growth, solute fields and finally solute flow via image processing techniques. The results presented in this contribution are preliminary and made at a stage when measurements and analysis is still commencing.

Acknowledgements
The European Synchrotron Radiation Facility is acknowledged for granting access to the ID6 beamline for the experiment. Furthermore Drs. Anatoly Snigirev and Irina Snigireva, ESRF experimental division, are greatly acknowledged for their contributions and assistance in this work. The analysis work is carried out under the FP7 project MINTWELD.

References
[1] J. S. Langer 1992 Phys. Today **45-10** 24
[2] S.R. Coriell, M.R. Cordes, W.J. Boettinger, R.F. Sekerka 1980 J. Crystal Growth **49-1** 13
[3] R. C. Kerr, A. Woods, M. G. Worster, H. E. Huppert 1990 J . Fluid Mech. **216** 323
[4] S. H. Davis 1990 J. Fluid Mech. **212** 241
[5] M. G. Worster 1991 J . Fluid Mech. **224** 335
[6] M. G. Worster 1997 Annu. Rev. Fluid Mech. **29** 91
[7] C. Beckermann, H.J. Diepers, I. Steinbach, A. Karma, X. Tong, 1999 J. Computational Physics **154** 468
[8] M. Asta., C. Beckermann, A. Karma, W. Kurz, R. Napolitano, M. Plapp, G. Purdy, M. Rappaz , R. Trivedi, 2009 Acta Mater. **57** 941
[9] R. H. Mathiesen, L. Arnberg, F. Mo, T. Weitkamp , A. Snigirev 1999 Phys Rev Lett. **83** 5062
[10] R. H. Mathiesen, L. Arnberg 2005 Acta Mater. **53** 947
[11] R H. Mathiesen, L. Arnberg, K. Ramsoskar, T. Weitkamp, C. Rau, A. Snigirev 2002Metall Mater Trans B **33** 613
[12] T. Schenk, H. N. Thi, J. Gastaldi, G. Reinhart, V. Cristiglio, N. Mangelinck-Noël, H.Klein, J. Härtwig, B. Grushko, B. Billia, J. Baruchel 2005 J. Crystal Growth **275** 205
[13] H. Yasuda, T. Nagira, M. Yoshiya, N. Nakatsuka, A. Sugiyama, K. Uesugi, K. Umetani 2011 ISIJ Int. **51-3** 402
[14] A. Bosak, I. Snigireva, K.S. Napolskii and A. Snigirev Adv. Mater 2010 **22** 3256

X-ray tomographic microscopy analysis of the dendrite orientation transition in Al-Zn

Jonathan Friedli[1], Julie L. Fife[1,2], Paolo Di Napoli[1] and Michel Rappaz[1]

[1] Laboratoire de Simulation des Matériaux, Institut des Matériaux, Ecole Polytechnique Fédérale de Lausanne, Station 12, CH-1015 Lausanne, Switzerland
[2] Laboratory for Synchrotron Radiation, Paul Scherrer Institut, CH-5232 Villigen, Switzerland

E-mail: jonathan.friedli@a3.epfl.ch

Abstract. Recently, Gonzales and Rappaz [Met. Mat. Trans. A37:2797, 2006] showed the influence of an increasing zinc content on the growth directions of aluminum dendrites. ⟨100⟩ and ⟨110⟩ dendrites were observed below 25wt.% and above 55wt.% zinc, respectively, whereas textured seaweeds and ⟨320⟩ dendrites were observed at intermediate compositions. Considering the complexity of these structures, it is necessary to first characterize them in further details and second, to model them using the phase field method. The so-called Dendrite Orientation Transition (DOT) was thus reinvestigated in quenched Bridgman solidification samples. The combination of X-ray tomographic microscopy and electron backscattered diffraction (EBSD) analysis on a whole range of compositions, from 5 to 90wt.% Zn, allowed insights with unprecedented details about texture, growth directions and mechanisms of the aforementioned structures. We show that seaweeds rather than dendrites are found at all intermediate compositions. Their growth was confirmed to be constrained within a (100) symmetry plane. However, new findings indicate that the observed macroscopic texture does not necessarily correspond to the actual growth directions of the microstructure. Further, it seems to operate by an alternating growth direction mechanism and could be linked to the competition between the ⟨100⟩ and ⟨110⟩ characters of regular dendrites observed at the limits of the DOT. These characters, as well as 3D seaweeds, are observed in phase-field simulations of equiaxed growth and directional solidification, respectively. This study emphasizes the importance of accurate experimental data to validate numerical models and details the progress that such combinations provide for the understanding of growth mechanisms.

1. Introduction

Aluminum has a weak anisotropy of the solid-liquid interfacial energy $\gamma_{s\ell}$ (about 1%) [1] that can easily be perturbed by the addition of an alloying element with a different crystal structure, such as zinc, an hcp element. Recently, a systematic effect of this alloying element has been identified by Gonzales *et al* [2–4]. These studies showed that an increase in the zinc content in Al-Zn alloys continuously changes the dendrite growth direction from ⟨100⟩ to ⟨110⟩ in a {100} plane (Fig. 1). At intermediate compostions, *i.e.*, between 25 and 55 wt.% , Gonzales *et al* observed seaweeds at the beginning and end of this dendrite orientation transition (DOT) (areas shaded in grey in Fig. 1) and ⟨320⟩ dendrites around 50 wt.% Zn. The seaweed microstructures did not exhibit clear trunks and arms but were nevertheless textured. This continuous DOT has a direct

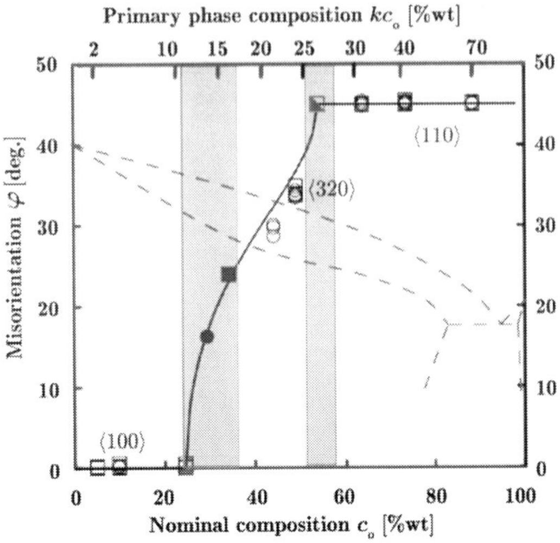

Figure 1. Angle between the ⟨100⟩ direction and the growth direction of Al-Zn dendrites as a function of the nominal concentration, c_0, of Zn. The equilibrium concentration of the first solid, kc_0, is indicated on the upper scale. Open circles and squares correspond to Bridgman and directional solidification specimens, respectively. The two shaded areas near 25 and 55 wt.% correspond to concentrations for which textured seaweed structures were observed (filled circles and squares). The high temperature portion of the Al-Zn phase diagram is superimposed in dashed lines. [2]

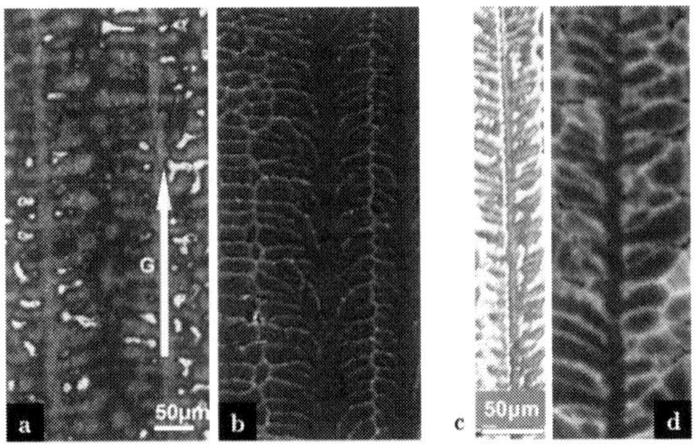

Figure 2. Bridgman solidified Al-10 wt.% Zn (a,b) and Al-50 wt.% Zn (c,d) samples. a) Metallographic section in a (100) plane with secondary arms seen at 90° from the trunk [2]. b) (100) section cutting a trunk in an X-ray tomographic microscopy dataset from a quenched sample exhibiting a much more precise view of the secondary arms. c) Reported ⟨320⟩ dendrites [2] showing a trunk and very irregular secondary arms. d) Cut on the X-ray tomographic microscopy stack with an orientation identical to c) through a trunk in a (001) plane showing arms that actually grow out of plane. Note that the magnification is identical for a) and b), and for c) and d), but it does not exactly match between (a,b) and (c,d).

influence on the texture of directionally solidified alloys [3], and thus has important implications for controlling the mechanical properties of cast alloys.

However, seaweed morphologies were, up to now, almost exclusively analysed in 2D and found in organic alloys. Furthermore, the existing studies [5–7] never quenched the growing structures and thus, solid-state diffusion during further solidification and cooling smeared the microsegregation pattern formed at an earlier stage. These major difficulties, inherent to the analysis methods used so far for Al-Zn alloys, explain the poor understanding of 3D seaweed formation and orientation selection, and therefore also the lack of a generalized prediction criteria. A study with adapted 3D methods is essential to shed light on such problems, for example by X-ray tomography observations made either *in situ* on solidifying specimens [8–10] or *ex situ* on quenched specimens [11].

Therefore, we propose a method enabling the characterization of seaweed and high multiplicity dendritic structures and extend the DOT analysis in Al-Zn alloys. Our experimental procedure encompasses Bridgman solidification and quenching for a wide composition range (from 5 to 90 wt.% Zn), followed by X-ray tomographic microscopy and electron backscattered diffraction (EBSD) measurements. This paper, which presents the experimental results, is complemented by another contribution presenting the 3D phase field simulations.

2. Experiments

2.1. Bridgman solidification

The Bridgman setup is a modified version of the vertical tubular furnace presented by Gonzales and Rappaz [2] in their DOT analysis. The modification consisted of a reduction in diameter of the copper water-cooled tube from 10 mm to 7 mm, allowing for a reduction of the specimen diameter and thus a more efficient quenching. The alloy rods were inserted in quartz or alumina tubes with 4 mm inner diameter and heated 50 °C above the liquidus temperature. The samples were then pulled through the furnace at four different velocities (1, 4, 10 and 20 mm/min) in a thermal gradient between 5 and 11 K/mm depending on the alloy and tube properties. The temperature gradient was measured by two K-type thermocouples separated by a distance of 10 mm in the alloy. After 100 mm of solidification, the specimen was dropped and quenched in water. Gonzales and Rappaz [2] already showed that the pulling velocity, in the range of their Bridgman and directional solidification experiments (*i.e.* 0.06-3 mm/s), did not influence the DOT. The four velocities in the present case were thus chosen to generate structures fine enough to be representative in the sample volume and coarse enough to be easily distinguishable from the liquid after quenching. The range of cooling rates in the Bridgman furnace varies from 0.08 to 3 K/s. While the best contrast between the primary solid and the quenched liquid is obtained at low pulling velocities, the cooling rate of the quench (400-600 K/s) leads to a fair contrast even in the most unfavorable set of experimental conditions, as confirmed in a previous study of Pompe *et al* [11]. The metallographic preparation of the quenched samples allowed a precise location of the tip of the structure and thus the location of the tomographic microscopy sample to be extracted. The good phase separation generated by the quench originates from the fact that this method gets rid of backdiffusion during cooling, which greatly simplifies both the subsequent etching procedures and the interpretations of the observed morphologies. This is clearly seen in Fig. 2 where the microstructures reported by Gonzales et al [2] obtained for two alloy compositions are compared with the present ones captured by X-ray tomographic microscopy.

2.2. X-ray tomographic microscopy

In order to characterize the complex growth morphologies in 3 dimensions, synchrotron-based X-ray tomographic microscopy experiments were performed at the TOMCAT beamline of the Swiss Light Source (Paul Scherrer Institute, Villigen, Switzerland) [12]. Since the analysis volume accessible through this technique is smaller than the Bridgman sample dimensions, each specimen had to be machined down to the appropriate size at the precise location of the tip of the quenched growth morphology. Diameters from 700 μm for highly-alloyed to 2400 μm for low-alloyed samples were selected and the specimens then scanned at energies ranging from 20 to 40 keV with pixel sizes from 0.74 to 1.85 μm. Multiple scans of the height of the sample were taken, aligned, and then merged into one stack to allow for better visualization of the scanned volume.

2.3. EBSD

The reconstructed tomographic microscopy data stacks were then searched for widely developed secondary arms in a grain reaching up to the quench solidification tip. A transverse cut was then realized with a diamond blade saw at that precise location in the column (*e.g.*, dashed line on Fig. 3b). The sections were then prepared for EBSD mapping.

The SEM micrograph on which the EBSD orientation map is based (*e.g.*, Fig. 3c) was used to retrieve the plane to which it corresponds in the tomographic microscopy dataset. As Fig. 3b shows very well, it is not always exactly perpendicular to the axis of the sample, but this deviation can be accounted for. The orientation map itself was then superimposed on the tomographic microscopy cut (Fig. 3d), the grains identified and their orientation determined. To ensure the best possible alignment, when possible, the maps spanned the whole sample. Knowing the misorientation between the EBSD coordinate system and the tomographic microscopy dataset, the orientation of the grains are then rotated from the former to the latter. All planes and directions identified in the tomographic microscopy stack could then be associated with a crystallographic orientation (*e.g.*, Fig. 3e).

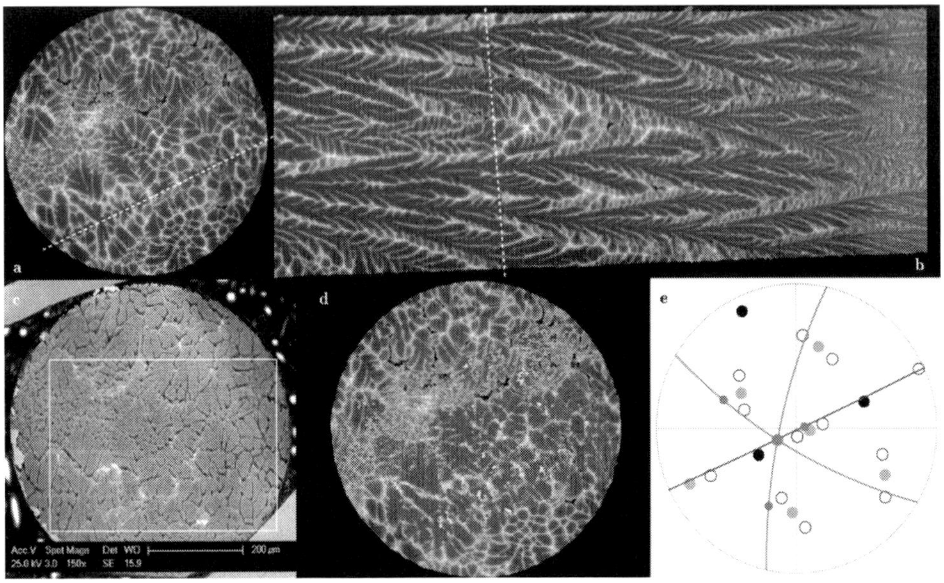

Figure 3. Transverse (a) and longitudinal section (b) of a tomogram of a quenched Al-50 wt.% Zn sample solidified at 4 mm/min. In the longitudinal section, the thermal gradient is from left to right and the quenched interface can be seen on the right of the image. The dashed lines depict the location of the longitudinal and transverse sections respectively. (c) SEM micrograph of the cut corresponding to (a) through a tomographic microscopy sample, the white rectangle indicates the location of the EBSD map. (d) Same section as (a) with the superimposed EBSD orientation map showing two grains, in green and blue. (e) Typical pole figure in the tomographic microscopy stack coordinate system. The blue and red lines correspond to planes of interest. The red and green dots represent the directions of primary trunks and secondary arms respectively, measured in the 3D reconstruction. The crystallographic orientations obtained from the EBSD map for the corresponding grain are shown by black ⟨100⟩, grey ⟨110⟩ and open ⟨320⟩ circles.

3. Results and discussion

As can already be seen in Figs. 2 and 3, the proposed method greatly improves the quality and quantity of information that can be extracted from Bridgman type experiments and enables the clear identification of microstructural features. By quenching, and thus limiting the amount of time the sample remains at high temperatures, the problems of back-diffusion, coalescence and coarsening are avoided and microstructural features are unambiguously identified. Furthermore, as back-diffusion is greatly reduced, the original zinc microsegregation pattern is kept and allows for good contrast during X-ray tomographic microscopy observations.

Two examples of the improvement in interpretation are shown in Fig. 2. In dilute Al-10 wt.% Zn, the secondary arms were believed to grow at an angle of 90° with respect to the trunk (Fig. 2a). However, Fig. 2b reveals that when cutting exactly through the trunk,

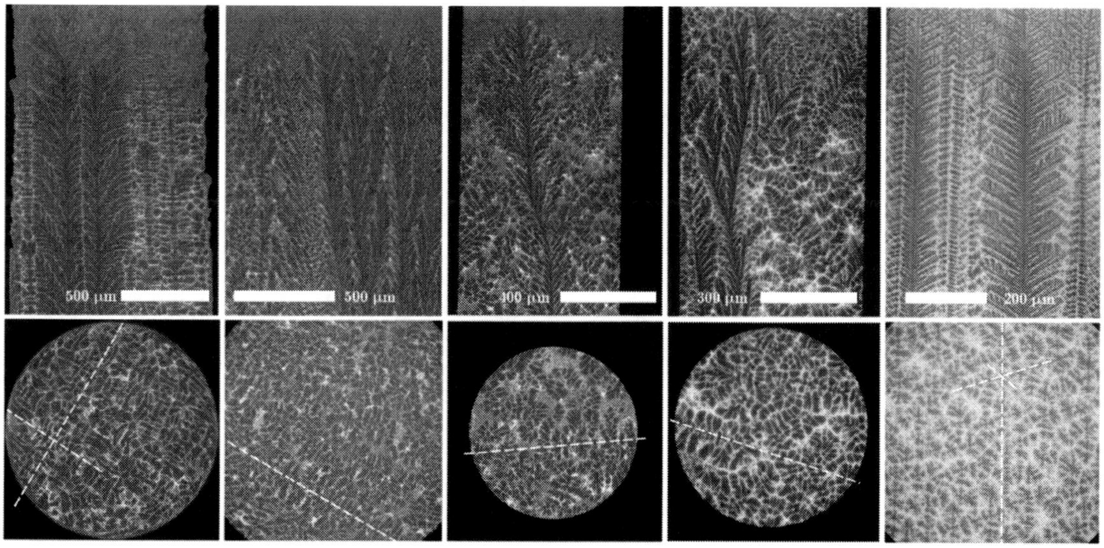

Figure 4. Longitudinal (top) and transverse (bottom) cuts through the X-ray tomography stack. From left to right: Al-20 wt.% Zn, Al-28 wt.% Zn, Al-42 wt.% Zn, Al-55 wt.% Zn solidified at 10 mm/min and Al-82 wt.% Zn solidified at 20 mm/min. The trace of the longitudinal section is shown by a dashed white line on the transverse cut. Note that whereas the length scale is the same for longitudinal and transverse sections of each alloy, the magnification varies from one alloy to another.

secondary arms start growing with a smaller angle from the trunk. Moreover, the arms have a tendency to bend and grow almost perpendicularly to the trunks, resulting in a palm-tree morphology. Further analysis of this growth behavior can be found in another paper by the same authors [13]. In the case of the high multiplicity ⟨320⟩ dendrite morphology reported by Gonzales and Rappaz [2] (Fig. 2c), the comparison with the tomographic microscopy data of Fig. 2d shows that the irregularity in the secondary arms comes from out-of-plane growth, suggesting an even more complex growth mechanism than previously expected. These salient features allow for the reexamination of the DOT in Al-Zn.

3.1. Microstructure analysis

X-ray tomographic microscopy was performed on Al-Zn alloys from 5 to 90 wt.% zinc and Fig. 4 shows longitudinal and transverse sections through the stack of samples of various compositions. First, EBSD measurements show an identical angular dependence of the dendrite behavior or seaweed texture as in the DOT for the whole range of Al-Zn alloys, and thereby confirm the results already obtained by Gonzales and Rappaz [2, 3]. However, as the last section foretold, the secondary dendrite arms in low-alloyed aluminum and the trunk and arm directions in the intermediate composition range where seaweeds are observed cannot be deduced from the DOT misorientation curve of Gonzales and Rappaz (Fig. 1).

3.2. Symmetries

The tomographic microscopy data was scouted for specific symmetry planes, appearing in the transverse cuts (Fig. 4 bottom row). Three types of symmetries have been observed in Al-Zn alloys. Between 5 and 20 wt.% Zn, trunks with ⟨100⟩ orientation and secondary arms growing in {100} planes are observed. In transverse sections those typically appear as crosses with a 90° angle (Al-20 wt.% Zn in Fig. 4 bottom row). At zinc concentrations higher than 55 wt.% , dendrites (both trunks and arms) grow along well defined ⟨110⟩ directions. Indeed, in the former studies, the transformation at the eutectic temperature (381 °C) provided some sort of a self-quench [2–4], making the structure clearly appear (*e.g.*, in Al-82 wt.% Zn Fig. 4 bottom row). This cut also shows the growth advantage, and thus the larger lateral extension, that ⟨110⟩ arms at 60° from the trunk have over the shorter ⟨110⟩ arms that grow at 90°. In the range of 25 to 55 wt.% zinc, these symmetries break down and an apparently random structure is observed. Nonetheless, a careful look through the X-ray tomographic microscopy stacks allows one to identify individual symmetry planes in each grain, such as the ones delineated in the bottom row of Fig. 4 for Al-28, 42 and 55 wt.% Zn.

The EBSD measurements show that in all intermediate compositions investigated, the symmetry plane is a (100) plane, as shown in more details in the example of Al-50 wt.% Zn (Fig. 3). This finding is in agreement with the previous studies by Gonzales *et al* since the DOT operates in a (100) plane.

3.3. Growth mechanisms

Low and high alloyed Al-Zn alloys exhibit primary dendrite trunks which clearly grow along well-defined ⟨100⟩ and ⟨110⟩ directions, respectively. At intermediate compositions, the situation is more complex since none of these alloys produce a typical dendritic structure, and thus no trunk in the classical sense is present. The three central figures of Fig. 4 (top) shows longitudinal cuts in the symmetry (100) plane of three such alloys: they each exhibit a seaweed-type structure. Instead of well-established trunks, the growth seems to occur with an alternating tip-splitting mechanism with very short arms. The structure is not random however, since certain orientations seems to be preferred in the alternating splitting process. Even if the morphology differs significantly from that of a typical dendrite, one can still identify certain features using the names of more commonly observed structures. We will call a feature a "trunk" if it is aligned within about 20° of the thermal gradient, and extends over a distance large enough to be clearly identifiable by a number of side branches. Although this definition is imprecise, the trunks in Fig. 4 are nevertheless clear. Any smaller feature which emerges from such a trunk will be called an arm.

To make a meaningful assertion about the trunks, a minimum number showing similar characteristics should be present in a cut. Thus, the sample showing (100) symmetry plane with the largest number of trunks (Al-50 wt.% Zn, Fig. 3b) was selected and examined in more detail (see Fig. 5). As shown by the solid and dashed white lines in Fig. 5a, two types of trunks were observed: those oriented close to the thermal gradient (dashed lines, *e.g.*, B) show secondary arms on both sides of the trunk in the (100) plane, whereas those further inclined (solid lines, *e.g.*, A1) exhibit arms on only one side in this plane. Note that since the secondary arms are so short and rarely straight, their direction, even taken as the average of the angle with the trunk found at the root, is subject to large errors. The five pole figures shown at the bottom of Fig. 5 correspond to the transverse section of Fig. 3d for this single grain, *i.e.*, the ⟨100⟩ (black dots), ⟨110⟩ (grey dots) and ⟨320⟩ (open circles) are the same. The orientations of trunks and arms in these pole figures are put in relation with the crystallographic orientation of the grain. In Fig. 5e, trunk B grows within a few degrees of the ⟨320⟩ direction whilst trunks A are much more regularly oriented in two directions at $\sim 14°$ with respect to ⟨320⟩ (Fig. 5b).

Once examined out of this longitudinal section, the secondary arms emerging from type A trunks are found in two other planes outlined by the two red plane traces in the corresponding stereographic projections. Fig. 5c and d illustrate this configuration for the trunks A1 and A2. The orientation of these arms in the out-of-plane sections is so ill-defined that it is not possible to identify them with any precise crystallographic direction. The in-plane secondary arms orientation on the other hand, corresponds to the direction of the trunks laying on the other side of the ⟨320⟩ direction, *i.e.*, the in-plane arms of trunk A1 have the orientation of trunk

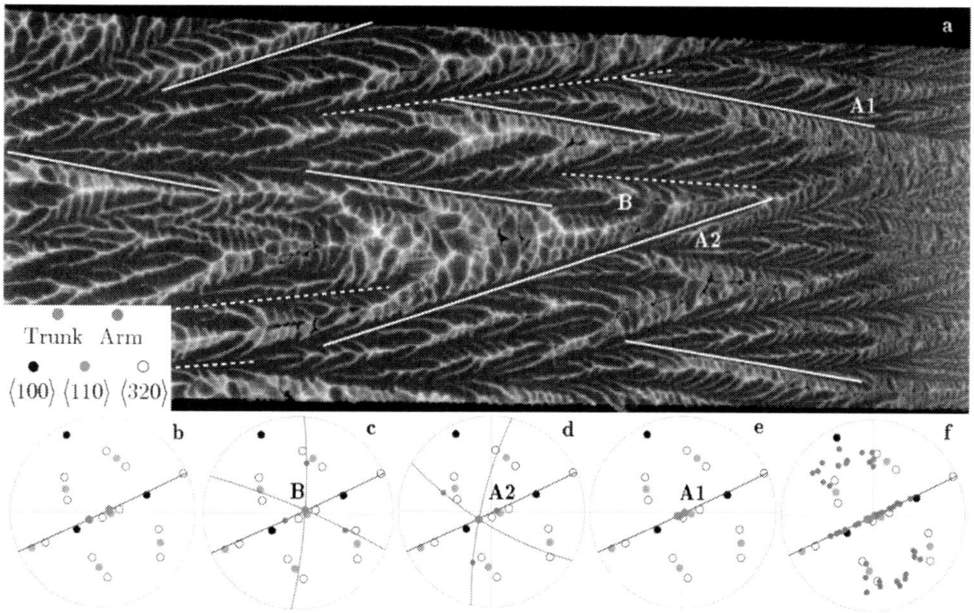

Figure 5. a) (100) symmetry plane in Al-50 wt.% Zn, showing two types of trunks. Trunks with secondary arms on only one side in the (100) plane, highlighted by solid white lines, and trunks with arms on both sides identified by dashed white lines. b)-f) are the pole figures obtained from the EBSD map in Fig. 3d with the ⟨320⟩ direction being the closest to the vertical, *i.e.*, the thermal gradient direction. The planes and directions found in the X-ray tomography stack are superimposed on the pole figures. The (100) symmetry plane is represented in blue, red and green dots represent trunks and arms, respectively. b) shows the orientation of the A-type trunks. The configuration of the trunk, secondary arms and (red) planes in which the latter lay are shown in c) and d) for trunk A1 and A2, respectively. e) shows the orientation of the B-type trunks, close to the ⟨320⟩ direction. f) shows the directions of the secondary arms associated to the B-type trunks.

A2 and vice-versa. This probably explains why over long distances, a zigzag of the trunks is observed. As such a trunk encounters a perturbation (*e.g.*, the solute field from another trunk or thermal fluctuations), it could switch sides and take the alternate direction. Another example of this phenomenon is shown for Al-42 wt.% Zn in the top row of Fig. 4.

The observation of B-type trunks shows a very different configuration. The scatter of the directions in Fig. 5f shows that the secondary arms do not seems to have any preferred growth direction. The potential resulting growth disadvantage could very well explain their short lateral extension compared to the type A configuration.

Although the phase contrast is good, the high variability of features and the feature sizes in contact with the primary solid make the segmentation very tedious and so far impossible to automate. Furthermore, this is a static post-mortem observation of a highly dynamic process and *in situ* investigations could certainly improve the understanding of this growth morphology.

The observations in Al-50 wt.% Zn show that the measurement of the texture of such a sample can be deceptive. Indeed, although the mean growth direction, *i.e.*, the texture, is ⟨320⟩, by far not all actual growth directions correspond to this orientation as evidenced by A-type trunks.

4. Conclusion

A method for investigating growth morphologies in metallic systems, featuring X-ray tomographic microscopy of quenched Bridgman samples, was presented. Focusing on the analysis of complex microstrucutres in Al-Zn alloys, the advantages of this method over other techniques was shown by the clear identification of microstructural features not observable before.

The investigations of the whole hypoeutectic range of Al-Zn alloys, firstly confirmed the texture observations made by Gonzales and Rappaz [2]. At zinc concentrations higher than 55 wt.% , dendrites grow in well defined ⟨110⟩ directions. Below 25 wt.% Zn, primary trunks grow along ⟨100⟩. Seaweed structures were identified at all compositions between 25 and 55 wt.% zinc and could, for the first time, be characterized in 3D in a metallic alloy with unprecedented details about texture, growth directions and mechanisms. It was shown that seaweeds are far from random and their growth is constrained within a (100) symmetry plane. However, new findings show that the observed macroscopic texture did not necessarily correspond to the actual growth direction. The growth seems to operate by an alternating growth direction mechanism and could be linked to the competition between ⟨100⟩ and ⟨110⟩ characters, which was simulated using phase field under equiaxed growth conditions [4,14].

Although progresses were made in the understanding of the microstructure formation mechanisms in Al-Zn alloys, some open questions remain. Advancements in the X-ray tomographic microscopy field provide unprecedented spatial and temporal resolution and allow for *in situ* observations of solidification, which could broaden the spectrum of information available for interpretation. The study of seaweed would certainly most benefit from this technological progress. Recently, TOMCAT has incorporated a laser-based heating system [15] that provides near-isothermal and directional heating capabilities in a broad range of temperatures, which enables *in situ* observations of growth phenomena in metallic and other materials systems. Finally, a companion paper in this proceedings [14] will show that phase field simulations for alloys exhibiting a complex, but weak, solid-liquid interfacial energy produce hyperbranched dendrites under equiaxed conditions (see also [4]), but much more complex structures, similar to the ones observed here, when such alloys are solidified in a thermal gradient.

Acknowledgments

The authors gratefully acknowledge the financial support of the Fonds National Suisse pour la recherche scientifique (Grant No 200020-113260 and 121598) as well as the financial and technical support of the Paul Scherrer Institute (Proposal No. 20080240, 20080846, 20090118, 20090795). They would also like to thank Sam MacDonald and André Phillion for their previous assistance during the X-ray tomography sessions at the Swiss Light Source. All electron microscopy measurements were obtained at the Centre Interdisciplinaire de Microscopie Electronique (CIME) at Ecole Polytechnique Fédérale de Lausanne (EPFL).

References

[1] Liu S, Napolitano R E and Trivedi R 2001 *Acta Materialia* **49** 4271–4276
[2] Gonzales F and Rappaz M 2006 *Metallurgical and Materials Transactions A: Physical Metallurgy and Materials Science* **37** 2797–2806
[3] Gonzales F and Rappaz M 2008 *Metallurgical and Materials Transactions A: Physical Metallurgy and Materials Science* **39** 2148–2160
[4] Haxhimali T, Karma A, Gonzales F and Rappaz M 2006 *Nature Materials* **5** 660–664
[5] Mullis A, Dragnevski K and Cochrane R 2004 *Materials Science and Engineering A* **375-377** 157–162
[6] Assadi H, Oghabi M and Herlach D 2009 *Acta Materialia* **57** 1639–1647
[7] Gonzales F 2009 *Etude des morphologies de solidification dans les alliages aluminium-zinc* Ph.D. thesis Ecole Polytechnique Fdrale de Lausanne
[8] Terzi S, Taylor J, Cho Y, Salvo L, Suéry M, Boller E and Dahle A 2010 *Acta Materialia* **58** 5370–5380
[9] Limodin N, Salvo L, Boller E, Suéry M, Felberbaum M, Gaillègue S and Madi K 2009 *Acta Materialia* **57** 2300–2310

[10] Ludwig O, Dimichiel M, Salvo L, Suéry M and Falus P 2005 *Metall Mater Trans* **36** 1515
[11] Pompe O and Rattenmayr M 1998 *Journal of Crystal Growth* **192** 300–306
[12] Stampanoni M, Groso A, Isenegger A, Mikuljan G, Chen Q, Bertrand A, Henein S, Betemps R, Frommherz U, Bhler P, Meister D, Lange M and Abela R 2006 *Progress in Biomedical Optics and Imaging - Proceedings of SPIE* **6318** 1605–7422
[13] Friedli J, Fife J, Di Napoli P and Rappaz M 2012 *Acta Materialia,* ***Submitted***
[14] Friedli J, Di Napoli P, Rappaz M and Dantzig J 2012 *Proceedings of MCWASP 2012* ed Ludwig A
[15] Fife J L, Rappaz M, Pistone M, Celcer T, Mikuljan G and Stampanoni M 2012 *Journal of Synchrotron Radiation,* ***Accepted***

IOP Conf. Series: Materials Science and Engineering **33** (2012) 012035 doi:10.1088/1757-899X/33/1/012035

In situ X-ray monitoring of convection effects on segregation freckle formation

N. Shevchenko, S. Eckert, S. Boden and G. Gerbeth

Helmholtz-Zentrum Dresden-Rossendorf, P.O. Box 510119, 01314 Dresden, Germany

E-mail: n.shevchenko@hzdr.de

Abstract. Visualizations of the solidification process under the influence of thermo-solutal convection were obtained by means of X-ray radioscopy within a Hele-Shaw cell filled with Ga-25wt%In alloy. The density-driven melt flow causes the formation of vertical segregation channels ("chimneys") in the mushy zone. The development of the chimneys and the probability of their surviving depend sensitively on the spatial and temporal properties of the flow field. The second part of this study concerns the case when the natural convection is superimposed by an electromagnetically driven flow perpendicular to the dendrites growth direction. The main effects of forced melt flow are a modification of the dendritic structure and flow-induced redistribution of the solute concentration.

1. Introduction

Thermosolutal convection has been identified as the main reason for the development of solute-rich channels in the mushy zone during the solidification of metallic alloys [1-4]. The occurrence of such segregation channels in fully-solidified castings has to be considered as a serious casting defect.

A sufficient understanding of the strongly coupled interaction between the melt flow and the solidification requires an authentic knowledge of the velocity field especially in the vicinity of the solidification front. During an upwards solidification solutal convection may arise from an unstable density stratification at the solid-liquid interface if the solute component rejected during the freezing of the primary crystals is lighter than the initial composition of the melt. In turn, the melt convection redistributes the solute concentration in the boundary layer and the mushy zone considerably leading to dramatic changes in growth rate and direction [5-8]. Further effects concern the selection of the secondary branches of growing dendrites [9]. In general, the flow impact might be different for different regions of the mushy zone. The growth rate is increased by a removal of the solute, whereas an influx of solute-rich melt causes the formation of segregation freckles by a remelting of already solidified crystals.

Many experimental investigations have been carried out using aqueous analogues as model liquids, where the convective flow pattern can be observed visually by diverse optical diagnostic techniques [10-12]. The dimensions of segregation channels occurring in such systems were shown to be comparable to those found by metallographic examinations in fully solidified metallic castings [2]. Although these similarities give point to extrapolate the findings from the transparent materials to metal alloys to a certain extent, corresponding measurements in real metallic materials would attract wide interest. Real-time observations of solidification processes in opaque metal alloys can be realised by means of the X-ray radioscopic method [13]. Only a few studies were made so far disclosing information with respect to the flow structure near the growth front. Koster et al. [14] studied natural convection in liquid gallium and gallium alloys. Qualitative images of the flow field were derived from precise measurements of local density differences from temperature gradients. However, the spatial resolution achieved in these experiments was not sufficient to perceive details of the flow structure being in the order of the typical dendrite spacing. More recent studies applied the optical flow approach to estimate the liquid flow field during the solidification of Ga-In alloys [7, 15].

The presented work is devoted to the visualization of the development of segregation channels during the bottom-up solidification of a Ga-25wt%In alloy under the influence of solutal convection. The impact of an additional forced convection is considered in the second part of this paper.

2. Experimental setup

A series of directional bottom-up solidification experiments were carried out using a Ga–25wt%In alloy ($T_{Liquidus}$=25 °C, $T_{Solidus}$=15.3 °C). The alloy made from 99.999% Ga and 99.999% In was melted and filled into a Hele-Shaw quartz cell with a parallel gap of 150 µm. The quartz cells were produced in two geometric versions: a hexagonal shape with a surface area of about 35 x 30 mm^2 (see figure 1(a)) and a square form with a surface area of about 30 x 30 mm^2 (see figure 1(b)).

The cooling of the cell during the experiment was realized by attaching a linear array of Peltier elements to the bottom part of the cell. An electric heater was mounted on the upper part of the solidification cell to heat up the alloy before the solidification experiment. In the course of the solidification the temperature gradient across the cell and the cooling rate were controlled by a simultaneous regulation of electric heater and Peltier cooler. For monitoring the temperature during the process two miniature K-type thermocouples (∅ 0.1 mm) were contacted to the face of the cell. The complete temperature field was measured in additional experiments by an infrared camera (Trotec IC 080 LV) equipped with a LCD monitor and an image sensor having a resolution of 384 x 288 pixel. Our experimental setup allowed for performing the solidification process in two different modes either by keeping a constant temperature gradient between the attached thermocouples or by adjusting a constant heating power. In the present experiment we use the constant heating power regime. The experiments within this paper were carried out at cooling rates of 0.01 K/s and heating powers of either 0.4 W or 1.7 W.

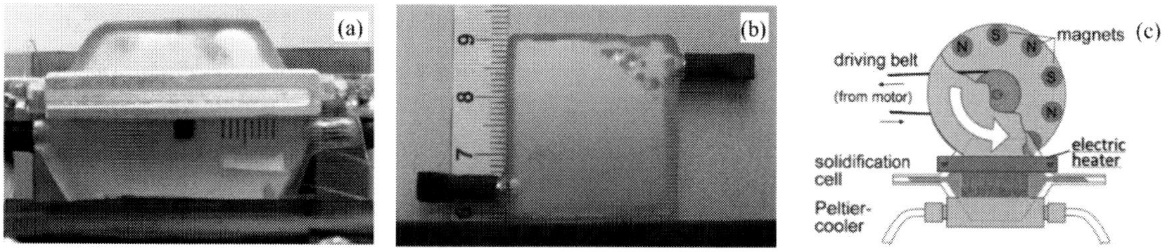

Figure 1. (a) Hele-Shaw hexagonal quartz cell; (b) Hele-Shaw square quartz cell and (c) schematic view of the solidification cell equipped with electric heater, Peltier elements and magnetic wheel.

The electromagnetically driven flow was produced by a rotating wheel with two parallel disks (∅ 52 cm) containing at their inner sides a set of permanent magnets (NdFeB) with alternating polarisation. The top part of the solidification cell was positioned in the gap between the disks. Rotation speeds of the magnetic wheel were chosen in the range of 30 – 50 revolutions per minute (rpm). Additional details of the experimental setup have been published elsewhere [7].

The solidification experiments were monitored by an X-ray radioscopic set-up delivering images with a spatial resolution of a few microns [13]. A microfocus X-ray tube (XS225D, Phoenix|X-ray) generated a horizontally aligned, divergent beam which passes through the broad face of the solidification cell and impinges an X-ray image intensifier (TH9438HX 9″, Thales). The image intensifier converted the X-rays into a two-dimensional visible light distribution which is recorded by a CCD camera (CF8/1, Kappa) with a scan rate of 50 half frames per second. For reducing the noise level during acquisition single images were integrated over a period of 440 ms or 1 s, respectively. The X-ray measurements were performed using a tube voltage of 63 kV and an electrical current of 150 μA.

At the beginning of the solidification experiment the Ga–25wt%In alloy was heated up to 10 K above the liquidus temperature. The compositional homogeneity of the molten alloy was controlled in real time by X-ray radiography. For this purpose, a reference image was recorded before starting the cooling process. The local composition of the alloy was determined from the relative brightness in the acquired X-ray images whereas estimations of the local flow field were obtained by means of the Optical Flow approach [15].

3. Results and discussion

3.1. Natural convection

The Ga–25wt%In alloy was solidified in vertical direction from the bottom of the solidification cell (see figure 1(a)). In total, 20 solidification cycles were carried out within this study. The thermal conditions of the respective experiments are presented in Table 1. The electrical heater at the upper part of the cell was operated with two different heating powers at 0.4 W and 1.7 W, respectively.

Table 1. Observation of chimneys and solidification parameters for different heating regimes for cooling rate 0.01 K/s (ΔT, temperature gradient; V, solidification rate).

Heating power (W)	Solidification cycles		$\Delta T_{horizontal}$ (K/mm)	$\Delta T_{lateral}$ (K/mm)	V (μm/s)	Ref. in Fig. 2
	Total	# CH				
0.4	10	4	0.5 – 0.9	0.1 – 0.2	8.1	a
1.7	10	10	1.3 – 1.7	0.3 – 0.5	5.2	b

CH, number of solidification cycles were at least 1 stable chimney was observed

An enhancement of the heating power leads to changes in the temperature fields (fig. 2 (a) and (b)) and increases both the vertical and the lateral temperature gradient (fig. 2(c)). It is also demonstrated in Figure 2 that the isotherms in the melt are not ideally flat and perpendicularly aligned to the vertical thermal gradient. The additional lateral temperature difference arises from the specific hexagonal geometry of the solidification cell. Because of the conical shape of the side walls there is no uniform cooling of the sample. The strength of this lateral gradient depends on several parameters such as the heat input, the applied cooling power or the distance between the heater and cooler position which has a fixed value of 10 mm in our experiments.

Figure 3 shows the development of the temperature gradients during the solidification, which continuously increase with reducing distance between the solid-liquid interface and the position of the electric heater. As might be expected, the higher power fed in by the electrical heater causes an increase of the temperature gradients. The vertical temperature gradient is responsible for changes of the undercooling region ahead of the solidification front. The occurrence of a lateral temperature gradient is supposed to produce an additional specific convection pattern driving a horizontal flow along the solidification front.

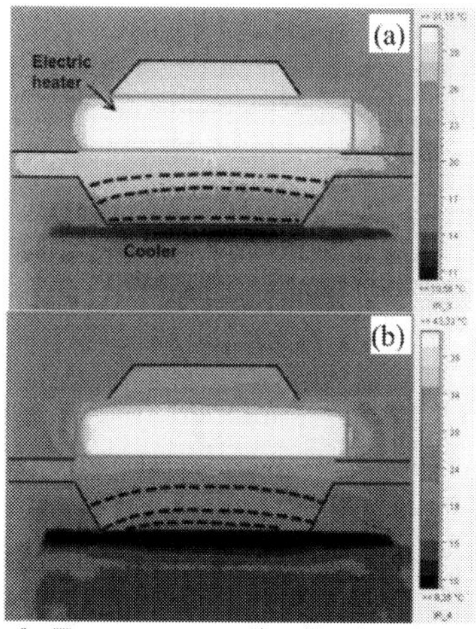

Figure 2. Temperature distribution in a solidification cell obtained in experiments with different heating power:
a) $P_H = 0.4$ W; b) $P_H = 1.7$ W.

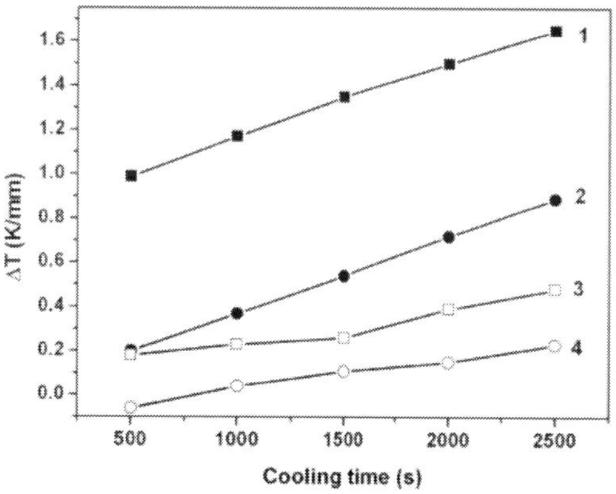

Figure 3. Temperature gradient vs. cooling time at different heating power P_H:
Vertical gradient 1 – $P_H = 1.7$ W; 2 – $P_H = 0.4$ W;
Lateral gradient 3 – $P_H = 1.7$ W; 4 – $P_H = 0.4$ W.

The X-ray visualizations revealed a sensitive dependence of the probability of the chimney formation on variations of the temperature distribution. Moreover, it was observed that fluctuating conditions of the flow and concentration field can also disturb the continued existence of early formed chimneys. Thus, the stability of the chimneys becomes an important feature of process which receives special attention in the following subsections.

3.1.1. Solidification process at low temperature gradient. Solidification experiments realised at the lower heating power (0.4 W) and the associated lower temperature gradients do not lead necessarily to the formation of sustainable segregation channels. Within this study we observed four persistent chimneys in 10 experimental runs. Figure 4 shows some snapshots of the developing microstructure recorded at early stages of the solidification process and at the end of cooling at temperature of 5 °C. The dark dendrites correspond to the primary In crystals (In-2wt%Ga) which grow from the bottom of the cell. The solutal convection shows up in form of rising plumes above the mushy zone. The development of vertical segregation channels and the dendritic network morphology is strongly affected by the initial nucleation at the bottom of the solidification cell and the appearance of considerable perturbations of the dendritic structure in the lower mushy zone. We consider in Figure 4 three different examples observed in the experiments. The relatively low temperature gradient results in an uneven growth of the dendrites and an erratic shape of the solidification front. Perturbations of the growing structures show a local accumulation of solute-rich melt which provokes the formation of distinct plumes in the liquid region. The comparison of the dendritic structures obtained in an early state and a later state of the mushy zone demonstrates the following: the formation of a stable chimney occurs primarily at positions with initial growth defects or grain boundaries, but, not every initial segregation channel evolves into a stable chimney. The latter is clearly shown in figures 4(a) and (b), whereas the other two examples exhibit strong segregation channels in the later stage of the solidification experiment. The analysis of our experiments revealed a remarkable tendency for a preferred chimney formation at the centre as shown, for instance, in the figures 4 (c) and (d). A segregation freckle appears accidentally in the central part of the cell, subsequently a stable vertical channel is formed. The occurrence of such central chimneys is promoted by the flow pattern ahead of the solidification front. We can observe a converging flow ahead of the mushy zone leading to a concentration of the rising plumes in the central part of the solidification cell (figure 4 (c)). The velocity inside the ascending plumes was determined to be in the range of 50 – 150 μm/s, whereas the assessments of the lateral flow adjacent to the solidification front yields values of about 20 - 50 μm/s. These plumes provide a suction effect in the mushy zone which causes an entrainment of solute-rich interdendritic melt. The solute pileup reduces the local undercooling and promotes a remelting of the dendritic network. Solute-rich fluid rises through these chimneys and erodes surrounding dendrites. Therefore, the form of the channel is subjected to remarkable variations. A contrary effect becomes apparent just at the solid-liquid interface. The inwards shifting of the plumes brings new In-rich liquid from the side regions towards the tips of the growing dendrites leading to an accelerated growth.

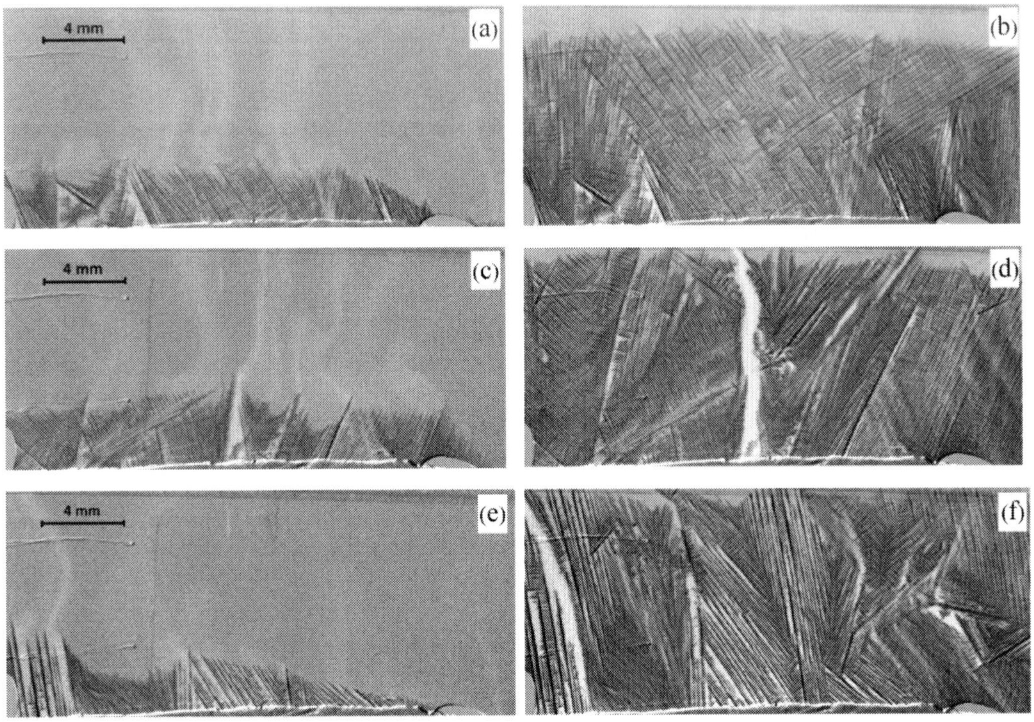

Figure 4. Dendritic structures observed at low heating power (P_H = 0.4 W) in three different experimental runs: (a,c,e) early stages of the solidification; (b,d,f) later stages of the solidification

A further example is displayed in Figure 5 containing a sequence of frames recorded from a solidification experiment. During the initial period of the solidification process the solidification front shows an uneven growth. Three incipiently dominating plumes, which are related to individual fast-growing dendrites, can be observed in the early stage (figure 5(a)). An accumulation of solute becomes visible in the mushy zone beneath the plumes. With the ascending solute-rich liquid a vortex flow is generated in the melt which in turn brings indium-rich liquid downwards. This leads to an irregular growth of the dendrites. The rising flow along the dendrite trunks obstructs the forming of side arms whereas a pronounced growth of secondary arms occurs at the opposite side due to the influx of In-enriched melt (figures 5(b-d)). Two protruding dendrites meet each other in the right half of our observation window. The large segregation channel is temporarily blocked by the left dendrite overgrowing the dendrite coming form the right-hand side (figure 5(d)). On one hand, the built-up solute leads to a remelting of the mid part of the dendrite bridging the chimney. On the other hand, stimulated by the high In-concentration entering from the right the dendrite forepart continues to grow and produces further solute. Thus, a strong chimney is maintained showing a conspicuous bending at the crossover point of both dendrites (figure 5(f)). It is interesting to note that this chimney survive while another distinct chimney on the left is overgrown starts to vanish before the end of the observation time. This image sequence clearly reveals that the development of the chimneys is strongly coupled with the advection of solute concentration by the melt flow. A strong vortex on the right hand-side stabilizes the forming segregation channel whereas an intermittent decay of the plume caused by an overturning of dendrites obstructs the further development of the chimney.

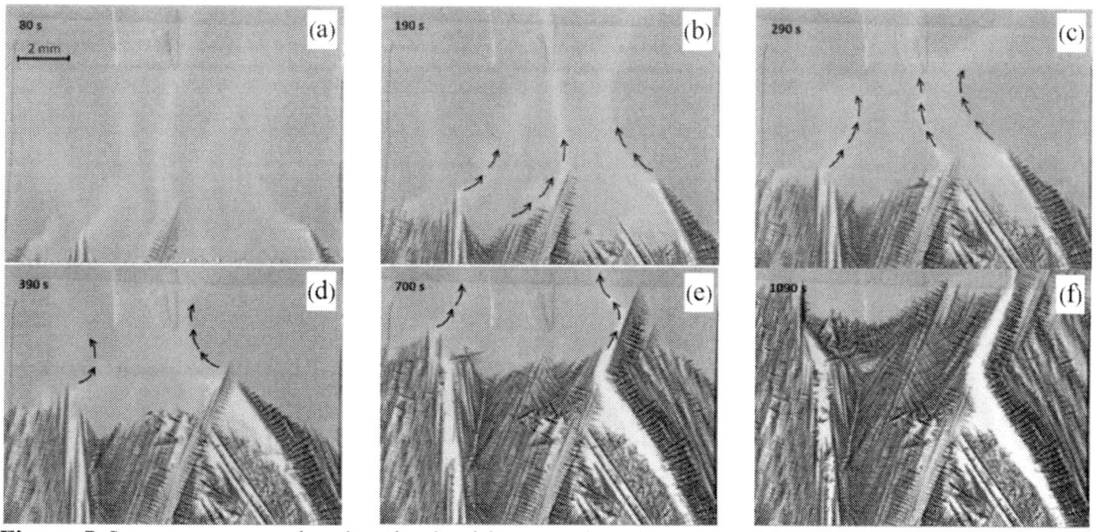

Figure 5. Image sequence showing the dendrite growth and the formation of segregation freckles at low heating power ($P_H = 0.4$ W): series of selected images recorded at different time steps.

3.1.2. Solidification process at high temperature gradient. Figure 6 shows a sequence of image frames recorded from the solidification process at the higher temperature gradient resulting from a heating power of 1.7 W. This thermal condition limits the undercooling region to a narrow zone adjacent to the solidification front. As a consequence a rather smooth and even shape of the solidification front can be observed in figure 6. In all ten solidification experiments performed under such conditions at least one stable chimney can be detected. As already noticed for the situation at the smaller heating power the highest probability for the development of a sustainable segregation channel is found at the centre of the solidification cell. Again this phenomenon is accompanied by the occurrence of a global melt flow in the liquid zone consisting of two convection rolls imposing a converging tangential flow along the solid-liquid interface. This double vortex flow spans over the entire cell volume. The velocities of the lateral flow close to the solidification front was determined to be in the range between 30 and 70 μm/s, that do not differ so much from the values determined in case of the lower heating power.

Now, the question arises concerning the driving mechanism for this global melt convection. At the beginning of this section we have already pointed out the existence of a lateral temperature gradient. However, as shown in figure 2 the coldest region appears in the centre of the solidification cell. That means, the corresponding melt flow along the solidification front has to be directed outward, which is contrary to our observations. Concurrently, lower temperatures in the centre are equivalent to higher solidification rates and a higher release of solute. Therefore, a slight density difference in lateral direction can evolve within the bulk liquid. Moreover, there is a tendency of attraction between the neighbouring plumes, which results from the entrainment of ambient fluid around a jet. The resulting lateral pressure provokes a convergence of both jet positions. Thus, the dominating central plume is further intensified by attracting and merging with neighbouring plumes. The formation of a strong and stable rising jet ahead of the mushy zone has a deep impact on the dendritic network. Figure 6 shows a solidification cycle without any significant growth defect during the initial state of the process. Nevertheless, the continuous strengthening of the central plume causes a distinct solute accumulation in the mushy zone behind. A remelting of dendrites happens and the formation of a chimney can be detected. The chimneys appear at a later stage of the solidification process and their extension is often restricted to the upper region of the mushy zone. These observations show a close resemblance to the mechanism of chimney formation due the mushy layer mode of convection [10, 17]. This mechanism differs from the case presented in the previous section where the development of the segregation channel is triggered by an initial growth defect.

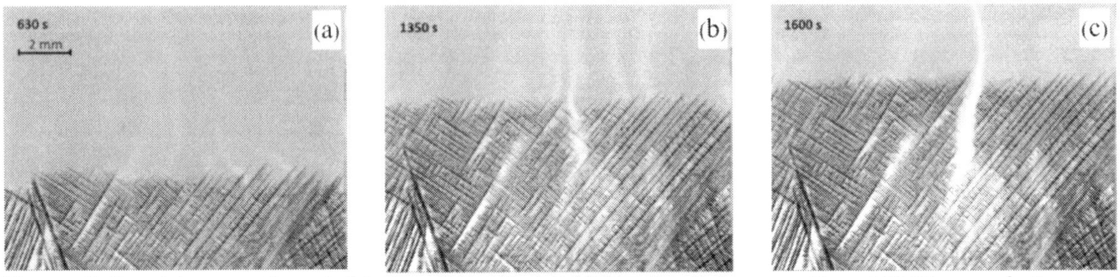

Figure 6. Dendrite growth and chimney formation at higher heating power $P_H = 1.7$ W; snapshots recorded at different time steps: (a) 630 s, (b) 1350 s, (c) 1600 s.

3.2. Electromagnetically forced convection

Further experiments have been carried out for the case when the natural convection is superimposed by an electromagnetically driven flow. The same thermal boundary conditions as for the experiments described in section 3.1 were applied. The first phase of the solidification was conducted without any external melt agitation. After a certain time when the solidification front reached the observation window, the magnetic wheel (see figure 1 (c)) was set into a counter-clockwise rotation resulting in a single vortex melt flow in clockwise direction. Adjacent to the solidification front the melt flow is almost perpendicular with respect to the growth direction of the dendrites. Figure 7 presents an experiment with a wheel rotation rate of 50 rpm and a cooling rate of 0.01 K/s. Estimations of the velocities arising from the external forcing obtained by the optical flow approach revealed values between 0.5 and 1 mm/s. At t = 0s (figure 7(a)) the magnetic wheel was put into rotation. The solutal plumes are quickly replaced by a dominating horizontal flow as shown in figure 8. As reported within a recent study [16] the main effect from the lateral inflow is the asymmetric development of the secondary dendrite arms due to the uneven solute concentration at the upstream and the downstream side of the dendrite trunks. Already existing segregation channels might be eliminated by an amplified growth of secondary arms at the downstream side of the dendrites. However, the flow presses a considerable amount of the released solute into the mushy zone. As a result new pronounced segregation channels are formed. Moreover, a higher density of chimneys can be found as compared to the situation of the pure thermosolutal convection.

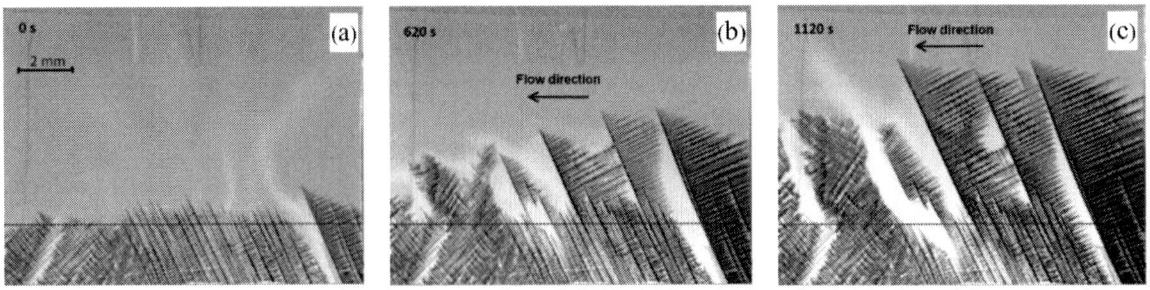

Figure 7. Dendrite growth at a heating power $P_H = 1.7$ W. (a) just before the initiating of the magnetic stirring; (b) and (c) under the influence of an electromagnetically-driven melt flow at different time steps.

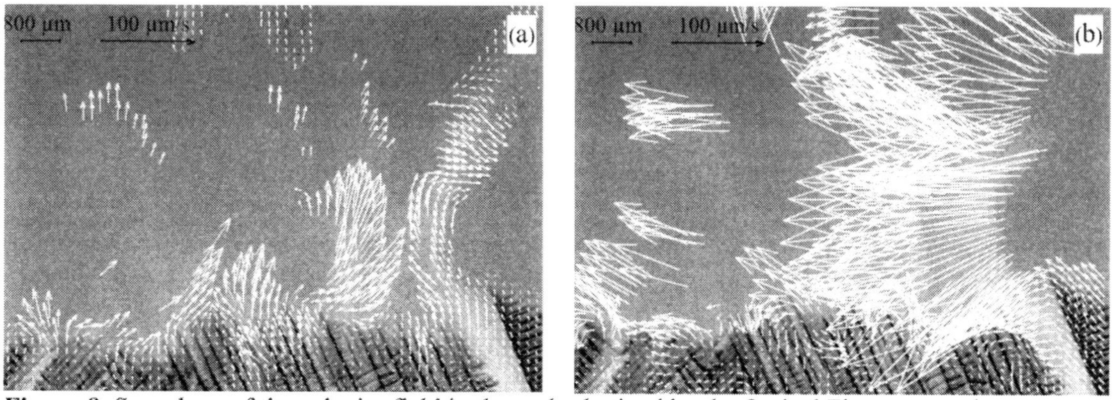

Figure 8. Snapshots of the velocity field in the melt obtained by the Optical Flow approach: (a) thermosolutal plumes; (b) forced convection driven by the magnetic wheel.

4. Conclusions

The formation of segregation channels in a Ga-25wt%In alloy was studied within this paper for the situation of thermosolutal convection and an superimposed magnetically-driven flow. The development of the chimneys is significantly affected by the melt flow. The best conditions for chimney formation were found at positions behind strong and stable plumes. Fluctuations arising for both the local and global flow pattern and the dendrite growth process may disturb or even suppress the chimney development. In many cases already existing channels become blocked or overgrown. The generation of a dominating forced flow almost perpendicular to the main growth direction leads to an enrichment of solute in the mushy zone and creates new segregation freckles.

Further work is in progress in order to classify our experimental data in respect of the known mechanisms for chimney formation under the influence of different convection modes in a mushy layer system [17].

Acknowledgements

This work was financially supported by Deutsche Forschungsgemeinschaft (DPG) in the framework of the collaborative research centre SFB 609 "Electromagnetic Flow Control in Metallurgy, Crystal Growth and Electrochemistry".

References

[1] Hellawell A, Sarazin JR, and Steube RS 1993 *Phil. Trans.: Phys. Sci. Eng.* **345** 507
[2] Sample AK and Hellawell A 1984 *Metall. Trans.* **15A** 2163
[3] Tewari SN and Shah R 1992 *Metall. Trans.* **23A** 3383
[4] Bergman MI, Fearn DR, Bloxham J and Shannon MC 1997 *Metall. Mater. Trans.* **28A** 859
[5] Murakami K, Fujiyama T, Koike A and Okamoto T 1983 *Acta metall.* **31** 1425
[6] Medina M, Du Terrail Y, Durand F and Fautrelle Y 2004 *Metall. Mater. Trans. B* **35B** 743
[7] Boden S, Eckert S and Gerbeth G 2010 *Mater. Letters* **64** 1340
[8] Willers B, Eckert S, Michel U, Haase I and Zouhar G 2005 *Mater. Sci. Eng.* **A402** 55
[9] Steinbach S and Ratke L 2007 *Metall. Mater. Trans. A* **38A** 1388
[10] Peppin SSI, Huppert MG and Worster MG 2008 *J. Fluid Mech.* **599** 465
[11] Murray BT, Wheeler AA and Glicksman ME 1995 *J. Cryst. Growth* **154** 386
[12] Zhang M and Maxworthy T 2002 *J. Fluid Mech.* **470** 247
[13] Mathiesen RH and Arnberg L 2005 *Acta Mater.* **53** 947
[14] Koster JN, Seidel T and Derebail R 1997 *J. Fluid Mech.* **343** 29
[15] Boden S, Eckert S, Willers B and Gerbeth G 2008 *Metall. Mater. Trans.* **39A** 61
[16] Shevchenko N, Boden S, Eckert S and Gerbeth G 2011 *J.Phys.: Conference Series* in press
[17] Worster MG 1992 *J. Fluid Mech.* **237** 649

IOP Conf. Series: Materials Science and Engineering **33** (2012) 012036 doi:10.1088/1757-899X/33/1/012036

Massive transformation from δ phase to γ phase in Fe–C alloys and strain induced in solidifying shell

H Yasuda[1,2], T Nagira[1], M Yoshiya[1], A Sugiyama[3], N Nakatsuka[1], M Kiire[1], M Uesugi[1], K Uesugi[4], K Umetani[4], K Kajiwara[4]

[1]Department of Adaptive Machine Systems, Osaka University, Osaka 565-0871, Japan
[2]Centre of Advanced Structural and Functional Materials Design, Osaka University, Osaka 565-0871, Japan
[3]Osaka Sangyo University, Daito, Osaka 574-8530, Japan
[4]JASRI/SPring-8, Sayocho, Hyogo 679-5198, Japan

E-mail: yasuda@ams.eng.osaka-u.ac.jp

Abstract. Time-resolved *in situ* observations using synchrotron radiation X-rays showed that γ nucleation difficulties resulted in a massive transformation from the δ phase to the γ phase. The massive transformation occurred even in high-carbon steel (0.45 mass% C) at a cooling rate of 0.33 K/s or more. A crystallographic relationship in which the close-packed (111) plane of the γ phase tended to coincide with the close-packed (110) plane of the δ phase suggested that the γ phase nucleated in a favorable plane in the δ phase. Observations of the sequence in the initial stage of solidification showed that the induced strain and strain rate in the massive transformation were larger than those expected in peritectic solidification. These observations will help us to understand deformation of solidifying shells and to build a physical model.

1. Introduction

The observation of solidification and related phenomena of interest *in situ* helps us to understand microstructural evolution and defect formation, and to build models for understanding and validating numerical simulations. In third-generation facilities such as SPring-8 [1], hard monochromatized X-rays (10–100 keV), which can penetrate metal alloys, are available for X-ray imaging. These X-rays have several advantages: (a) monochromatic light, which enhance contrast in transmission images; (b) high coherency, which achieves high spatial resolution; and (c) brilliant light, which improves time resolution. Solidification of Sn, Zn, Al alloys [2–20] and Fe alloys [21–24] have been observed.

In general, two different solidification morphologies have been known in peritectic systems [25-27]. One is the peritectic reaction, in which growth of the peritectic phase is controlled by diffusion through the peritectic phase. The other is the peritectic transformation, in which diffusion in the liquid phase at the triple junction controls growth of the peritectic phase. In Fe-C alloys, formation of γ phase (peritectic phase, fcc) is essentially considered by the peritectic reaction between δ phase (primary phase, bcc) and liquid phase, or/and the peritectic transformation at the δ/γ/L junction (refereed to as the peritectic solidification). The transformation from the δ phase to the γ phase causes volume shrinkage of about 1% and consequently results in unevenness and concave deformations of solidifying shells [28-30]. The strain / stress and deformation was analyzed by considering volume change thorough the peritectic solidification [30].

Recently, another type of solidification has been identified in Fe-C alloys [23]. The δ phase can massively transform into the γ phase. The strain and stress induced on a solidifying shell as a result of the massive transformation could differ from those induced in the peritectic solidification. It is therefore of interest to observe solidification followed by δ/γ transformation. This paper presents time-resolved X-ray imaging and Laue diffraction measurements of δ/γ transformation in Fe-C alloys. In addition, preliminary observations using a high-speed X-ray detector are also presented. Sequence of solidification and δ/γ transformation is proposed on the basis of *in situ* observations. This sequence, validated by direct observations, will help us to build a model for evaluating stress/strain induced in a solidifying shell.

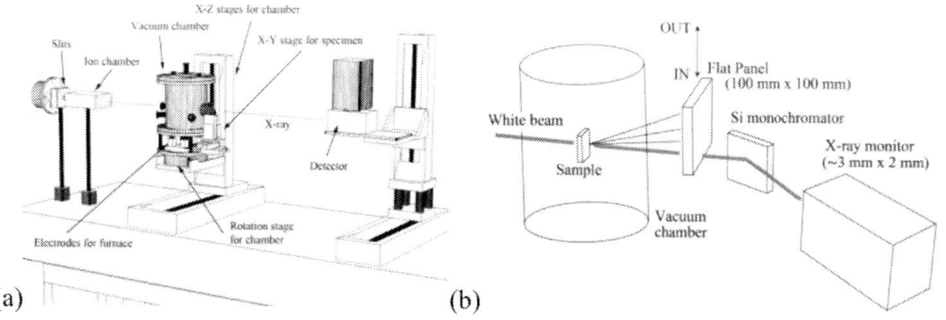

Figure 1. Setup for (a) time-resolved *in situ* observations [22] and (b) time-resolved Laue diffraction of solidification and transformation in Fe–C alloys.

2. Experiments

The observation experiments were performed at the BL20B2 beamline at SPring-8 [31]. Figure 1 shows the setup for observing solidification of Fe–C alloys. A thin specimen in a vacuum chamber and an X-ray detector are placed along the X-ray beam stream. An image detector consisting of an X-ray direct-sensing pickup tube (SATICON) [32] was used to obtain the transmission images (1024 × 1024 pixels, 10-bit resolution, pixel size: 5 × 5 μm^2). The typical frame rate was 1–4 fps and the spatial resolution was approximately 10–20 μm. An Fe–C specimen of dimensions 7× 7 × 0.1 mm^3 was set in a specimen cell was made of Al_2O_3 plates and BN plates.

Time-resolved diffraction measurement was performed at the BL28B2 beamline [31]. White X-rays were used to measure diffraction spots. As shown in Fig. 1(b), a Si monochromator was placed behind the specimen. When a flat panel (1000 × 1000 pixels, pixel size: 100 × 100 μm^2, 1 fps), which measured the diffraction spots, was removed from the X-ray beamline, transmission images were observed by a detector (CCD camera coupled with an optical lens and YAG:Ce screen). An X-ray beam with a size of 3 mm × 3 mm was used for transmission images, and X-ray beams of sizes 300 × 300 μm^2 or 50 × 50 μm^2 were used for diffraction.

Transmission images with higher frame rate as high as 250 fps at the BL20XU beamline [31]. An X-ray-imaging detector (1000 × 1000 pixels, pixel size: 5 × 5 μm^2), which consisted of a CMOS-type high-speed camera, an optical lens, and a phosphor screen, was used to observe the transmission images with high time resolution.

3. X-ray imaging of peritectic solidification

3.2. X-ray transmission images

Figure 2 shows the solidification of Fe–0.45 mass% C–0.6 mass% Mn–0.3 mass% Si alloy (referred to as Fe-0.45C alloy) at a cooling rate of 0.16 K/s and 0.83 K/s [23]. In the case of cooling at 0.167 K/s, δ dendrites grew in advance from the bottom to the top (111 s). Transformation from the δ phase to the γ phase proceeded from the roots to the tips of the δ dendrites (133 s). When the γ phase reached the tips of the δ dendrites, γ dendrites started to grow ahead of the δ dendrite tips (160 s). The γ dendrites changed their growth directions because the preferred growth direction, <100>, of the γ dendrites differed from that of the mother δ dendrites. Growth of the γ phase into the liquid phase was not clear, although the volume should increase by 18% as a result of the peritectic reaction, as calculated on the basis of the phase diagram. Since the transformation from the δ phase to the γ phase occurred in the mushy region, the observed transformation mode was categorized as a peritectic-like mode in this study.

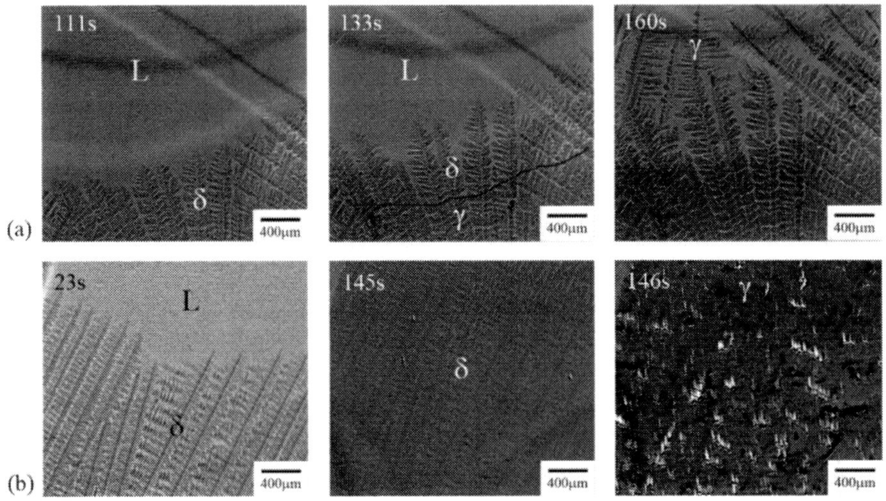

Figure 2. Peritectic transformation modes observed in Fe–0.45C alloy by X-ray imaging [23]. (a) Peritectic-like mode at a cooling rate of 0.167 K/s and (b) massive-like mode at a cooling rate of 0.83 K/s.

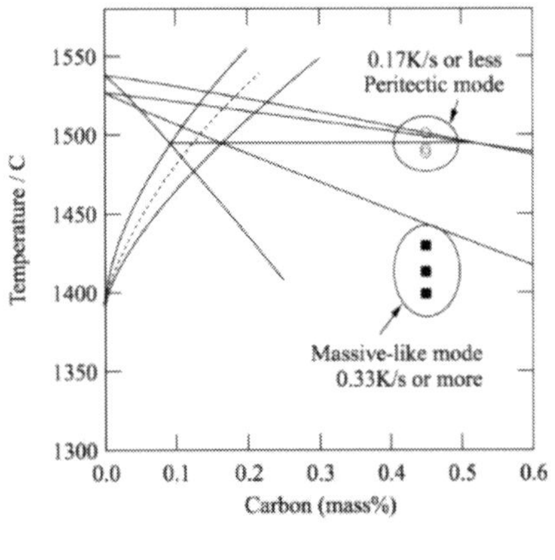

Figure 3. δ/γ transformation temperatures of Fe–0.45 mass% C–0.6 mass% Mn–0.3 mass% Si alloy measured by *in situ* observations.

Another dominant mode was observed, as show in Fig. 2(b). In the case of cooling at 0.83 K/s, δ dendrites grew and solidification was nearly completed before γ formation (23 s and 145 s). The δ phase was transformed into the γ phase in the solid state within 1 s (145 s and 146 s). The white region at 146 s corresponds to grooves caused by shrinkage in going from the δ phase to the γ phase. X-ray diffraction caused the dark regions, which were frequently observed after the δ/γ transformation. Many dark regions produced just after the transformation suggested that relatively large strain was induced in the γ phase.

The δ/γ transformation in Fig. 2(b) occurred at a temperature more than 100 K below the liquidus temperature. No solute redistribution for the δ/γ transformation was required if the δ phase was undercooled in the single γ phase region. A massive transformation or martensitic transformation could therefore occur in the transformation mode. The transformation mode that occurs in the solid δ phase was categorized as a massive-like mode.

Figure 3 shows the δ/γ transformation temperatures with the phase diagram. *In situ* observations at a cooling rate ranging from 0.17 K/s to 0.83 K/s proved that the peritectic-like mode was exceptional, even in high-carbon steel (Fe-0.45C). Since the δ phase was often cooled to temperatures 100 K below the liquidus temperature, nucleation of the γ phase seemed to be difficult.

The massive-like mode is completely different to the conventional peritectic reaction in terms of the solute redistribution and the induced strain. It is therefore important to reinvestigate the mechanism of δ/γ transformation and to know the influence of the transformation mode on formation of defects, such as deformation, cracks, and unevenness of the solidifying shell.

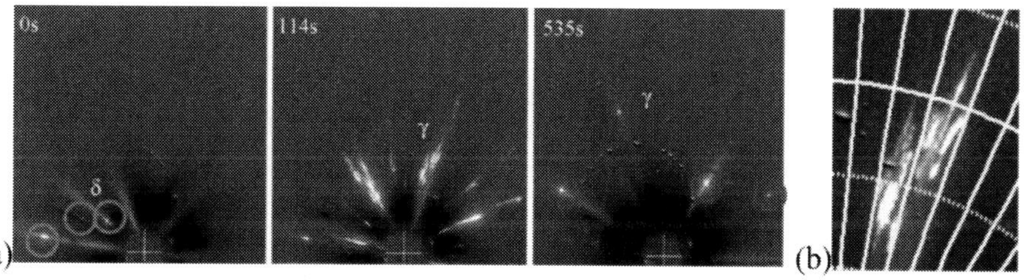

Figure 4. (a) Laue diffractions of δ dendrite (0 s), γ phase just after the transformation (114 s), and another γ grain coarsening (535 s) in Fe–0.45C alloy at a cooling rate of 0.17 K/s and (b) close-up of a diffraction spot at 114 s. The division of lines in (b) indicates 5° to the normal vector of the diffraction plane. X-ray beam size was 300 × 300 μm^2.

3.2. Laue diffraction

Figure 4(a) shows time-resolved diffraction patterns during solidification of Fe–0.45C alloy. Since the diffraction images were normalized using that of the liquid phase before cooling, the diffraction spots of the Al_2O_3 windows were eliminated. The diffraction spots indicated by red circles at 0 s were observed when δ dendrites grew in the observation region. The massive-like transformation occurred at 114 s. The spots of the δ phase disappeared, and largely distorted γ spots appeared. The distorted diffraction indicated that the γ phase produced by the massive-like transformation contained relatively large strain. The distorted spots (114 s) were replaced by non-distorted spots (535 s) during further cooling. These replacements are caused by coarsening of γ grains during cooling, and release of the strain induced by the massive-like transformation.

A close-up view of a distorted spot of the γ phase is shown in Fig. 4(b). The spot exhibited a rather complicated shape and the divergence angle of the observation area (300 × 300 μm^2) was roughly 10°. Lattice defects were therefore mainly introduced in the γ phase.

The X-ray beam was reduced to analyze the differences between the crystallographic orientations of the δ and γ phases. Figure 5(a) shows the differential diffraction patterns before and after the δ/γ transformation. The crystallographic orientations of the δ and γ phase are also shown in Fig. 5(b). The close-packed plane {110} plane of the δ phase tended to coincide with the close-packed plane {111} plane of the γ phase in most cases. However, the relationship between the directions in the close-packed planes changed in every experiment; some relationships between the crystallographic orientations of the δ and γ phases, such as the K–S relationship usually found in martensitic transformations, were not observed.

3.3. Transmission imaging with higher time resolution
Preliminary observations with higher frame rates were carried out to determine the morphology and velocity of the δ/γ interface during the massive-like transformation. Figure 6 shows the δ/γ interface passed through the observation area in the massive-like transformation of Fe-0.3C alloy. δ solidification was completed without γ phase formation. The γ phase started to grow from the right-hand top of the observation area (0.084 s). The velocity was roughly 25 mm/s. Systematic measurements have not been made, but the velocity ranged from several tens of millimeters per second to 200 mm/s. Since the dendrite growth velocity was, at most, several 100μm/s under the observation conditions, the growth of the γ phase was much faster than that of the δ dendrites.

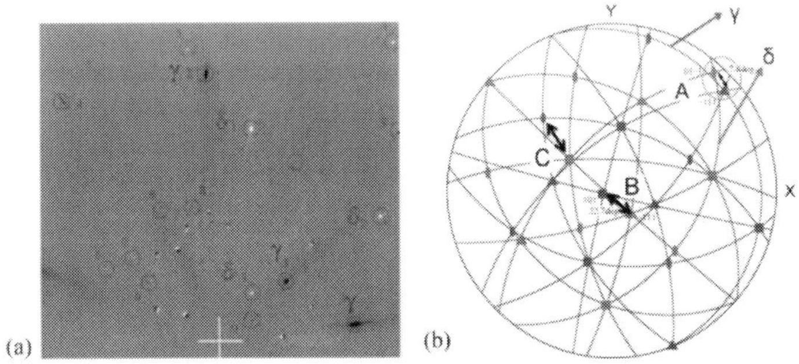

Figure 5. (a) Differential Laue pattern before and after δ/γ transformation in Fe–0.45C alloy at 0.17 K/s and (b) crystallographic orientations of δ and γ phases. The diffraction pattern was obtained by subtracting the δ diffraction pattern from the γ diffraction pattern.

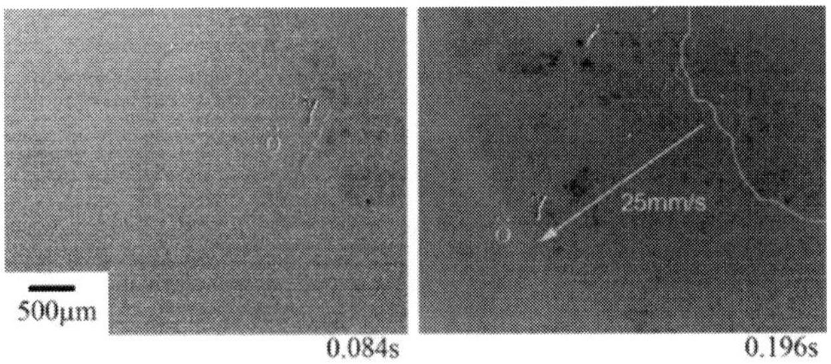

Figure 6. Motion of δ/γ interface during the δ/γ transformation in Fe-0.3C alloy. The δ/γ transformation occurred at approximately 80 K below the liquidus temperature.

4. Massive transformation and induced strain

4.1. Selection of the massive transformation
In situ observations showed that γ nucleation did not occur at the peritectic temperature and the δ phase was easily undercooled in the single γ phase region. Consequently, the massive-like transformation from the δ phase to the γ phase occurred. Selection of the massive-like transformation is therefore related to γ nucleation difficulties. The interfacial energy between the δ and γ phases is 0.7 J/m^2, and the interfacial energies between the L and δ phases and between the L and γ phases are 0.2 J/m^2 and 0.3 J/m^2, respectively [33, 34]. The relatively high interfacial energy between the δ and γ phases implied that the δ phase was not a preferred γ nucleation site.

(111) γ tended to coincide with (110) δ, even though a relationship between the directions on the close-packed planes was observed. γ nucleation events, which minimized the additional interfacial energy, resulted in this crystallographic relationship. Thus, the massive-like transformation was not martensitic.

The diffusion length is defined as D/V (D: diffusivity, V: interfacial velocity). If the diffusivity of carbon is of the order of 10^{-8} m^2/s and the interfacial velocity is 0.1 m/s, the diffusion length is estimated to be 0.1 μm. Even if carbon is redistributed at the δ/γ interface, the thickness of the carbon diffusion layer ahead of the δ/γ interface is less than 1 μm. In addition, as shown in Fig. 3, the T_0 line of δ–γ equilibrium, which defines possible undercooling for the massive transformation, is several hundred Kelvin higher than the observed transformation temperatures. The experimental results therefore showed that the massive transformation could occur.

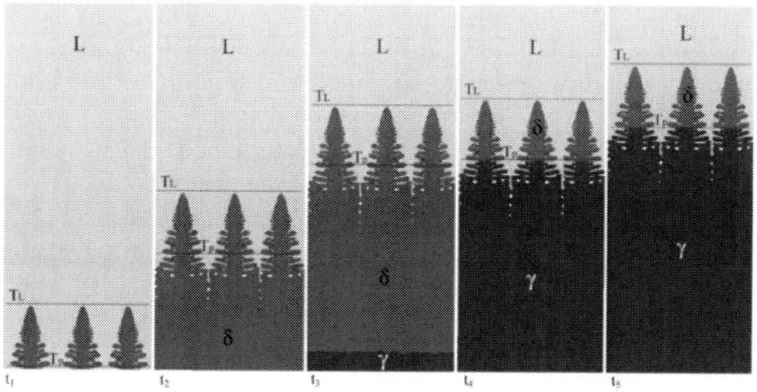

Figure 7. Development of solidifying shell and the massive transformation (Fe-0.18C). T_L and T_p are the liquidus temperature (δ) and the peritectic temperature, respectively.

4.2. Formation of solidifying shell

It is interesting to consider the solidification sequence in Fe–C alloys on the basis of our *in situ* observations. Figure 7 shows a schematic illustration of a massive transformation from a mold (bottom). Here, the solidification of Fe–0.18 mass% C is considered.

First, the peritectic solidification, in which the γ phase is produced in the mushy region of L + δ, is considered. δ dendrites grow as a primary phase. When the temperature of the δ phase in contact with the mold reaches a temperature slightly below the peritectic temperature, the γ phase nucleates and follows δ dendrite growth. According to the phase diagram (Fig. 3), the liquid phase remains at γ phase formation. The volume change due to the δ/γ transformation can be considerably compensated for by melt flow from the top to the bottom of the dendrites. Note that tensile stress may be induced in the surface region because melt flow does not completely compensate for the volume change.

Next, the massive transformation, which was observed *in situ*, is considered. The γ phase does not nucleate when the temperature of the δ phase reaches the peritectic temperature ($t = t_2$). The γ phase nucleates at $t = t_3$ and rapidly grows toward the tips of the δ dendrites. When the δ/γ interface reaches the isothermal plane of the peritectic temperature ($t = t_4$), the peritectic solidification begins. Since the massive transformation ($t_3 < t < t_4$) occurs in the solid δ phase, the volume change caused by the δ/γ transformation is not compensated for by melt flow in the mushy region. A larger strain is therefore induced in the surface region of the solidifying shell.

In addition, the growth velocity of the γ phase ($t_3 < t < t_4$) in the massive transformation is much higher than the growth velocity of the δ dendrites, and consequently larger strain rates are also induced in the surface region. Since the solid phase nearly at the solidus temperature generally exhibits viscoelastic properties, the massive transformation can cause large tensile stresses in the surface region.

In a conventional continuous casting of steel, since cooling rate at the surface region is roughly 160 K/s [35], the δ phase grows for 1s until the massive-like transformation occurred at 100K below the peritectic temperature. Thickness of solidifying shell could be in the order of mm [35]. Therefore the massive-like transformation could occur in the solidifying shell with a thickness of several mm.

It has been suggested that the volume decrease caused by δ/γ transformation induces tensile stress in the surface region and consequently causes concave bending of the solidifying shell, leading to unevenness of the shell thickness and cracks. The massive transformation observed by time-resolved *in situ* imaging gives an alternative sequence for building a model to determine the strain/stress of solidifying shells in Fe–C alloys.

5. Summary

Time-resolved *in situ* observations were performed to observe δ/γ transformations occurring during or after solidification.

(1) Difficulties in γ nucleation resulted in undercooling of the δ phase into the single γ region. Undercooling as high as 100 K was easily achieved at a cooling rate of 0.33–0.83 K/s.

(2) Massive transformation from the δ phase to the γ phase was dominant, even at Fe-0.45C alloy. Under the observation conditions, peritectic solidification, in which the γ phase is produced in the mushy region, was exceptional. It was sometimes observed at 0.167 K/s.

(3) Time-resolved Laue diffraction measurements determined the crystallographic relationship between the δ and γ phases. The close-packed (111) γ plane tended to coincide with the close-packed (110) δ plane. However, a constant relationship between the directions on (111) γ and (110) δ was not identified. Nucleation events in the δ phase resulted in the crystallographic relationship.

(4) The velocity of the δ/γ interface was preliminarily measured. The velocity ranged from roughly several tens of millimeters per second to 200 mm/s. The velocity suggested that the massive transformation was selected in the δ/γ transformation.

(5) The sequence in the initial stage of solidification is proposed on the basis of our observations. The induced strain and strain rate in the massive transformation can be larger than those expected in peritectic solidification. These observations give valuable insights and help us to understand the deformation of solidifying shells, and to build a physical model.

Acknowledgments

The synchrotron radiation experiments were performed as a long-term project (2007A0014-2009B0014) and general projects at BL20B2 and BL20XU at SPring-8 (JASRI). This study was supported by the Global COE Program (Center of Excellence for Advanced Structural and Functional Materials Design) from MEXT, a Grant-in-Aid for Scientific Research (A). The *in situ* observation of solidification for Fe alloys was also supported by the ISIJ Innovative Program for Advanced Technology (ISIJ).

References

[1] http://www.spring8.or.jp/
[2] Mathiesen R H, Arnberg L, Mo, F, Weitkamp T, Snigirev A 1999 *Phys. Rev. Lett.* **83** 5062.
[3] Mathiesen R H, Arnberg L, Ramsoskar K, Weitkamp T, Rau C, Snigirev A 2002 *Metall. Mater. Trans. B* **33** 613.
[4] Yasuda H, Ohnaka I, Kawasaki K, Sugiyama A, Ohmichi T, Iwane J, Umetani K 2004 *J. Cryst. Growth* **262** 645.
[5] Mangelinck-Noel N, Nguyen-Thi H, Reinhart G, Schenk T, Cristiglio V, Dupouy M D, Gastaldi J, Billia B, Hartwig J 2005 *J. Phys. D* **38** A28.
[6] Mathiesen R H, Arnberg L 2005 *Mater. Sci. Eng. A* **413** 283.
[7] Mathiesen R H, Arnberg K 2005 *Acta Mater.* **53** 947.
[8] Ohnaka I, Yasuda H, Sugiyama A, Ohmichi T 2005 *Trans. Ind. Inst. Metals* **58** 603.
[9] Schenk T, Thi N H, Gastaldi J, Reinhart G, Cristiglio V, Mangelinck-Noel N, Klein H, Hartwig J, Grushko B, Billia B, Baruchel J 2005 *J. Cryst. Growth* **275** 201.
[10] Li B, Brody H D, Black D R, Burdette H E, Rau C 2006 *J. Phys. D* **39** 4450.
[11] Li B, Brody H D, Black D R, Burdette H E, Rau C 2006 *Measurement Science & Technology* **17** 1883.

[12] Mathiesen R H, Arnberg L, Bleuet P, Somogyi A 2006 *Metall. Mater. Trans. A* **37A** 2515.
[13] Yasuda H, Ohnaka I, Sugiyama A, Nagira T, Tsukihara N, Kawasaki K, Umetani K 2006 *Proc. Int. Conf. on Modeling of Casting, Welding and Advanced Solidification Processes* 375.
[14] Arnberg L, Mathiesen R H 2007 *JOM*, **59** 20.
[15] Ruvalcaba D, Mathiesen R H, Eskin D G, Arnberg L, Katgerman L 2007 *Acta Mater.* **55** 4287.
[16] Yasuda H, Nagira T, Harada H, Sawai T, Sindoh H, Nishimura K 2007 *Proc. 7th Int. Conf. Zinc and Zinc alloy Coated Steel Sheet* 545.
[17] Reinhart G, A. Buffet, H. Nguyen-Thi, B. Billia, H. Jung, N. Mangelinck-Noel, N. Bergeon, T. Schenk, J. Hartwig, J. Baruchel, *Metall. Mater. Trans. A*, **39A** (2008) 865.
[18] Yasuda H, Yamamoto Y, Nakatsuka N, Nagira T, Yoshiya M, Sugiyama A, Ohnaka I, Umetani K, Uesugi K 2009 *Int. J. Cast Met. Res.*, **21** 125.
[19] Wang T M, Xu J J, Xiao T Q, Xie H L, Li J, Li T J, Cao Z Q 2010 *Phys. Rev. E*, **81** 042601.
[20] Boden S, Willers B, Eckert S, Gerbeth G 2009 *Int. J. Cast Metals Research* **22** 30.
[21] Yasuda H, Yamamoto Y, Nakatsuka N, Nagira T, Yoshiya M, Sugiyama A, Ohnaka I, Umetani K, Uesugi K 2008 *Int. J. Cast Met. Res.* **22** 15.
[22] Yasuda H, Nagira T, Yoshiya M, Nakatsuka N, Sugiyama A, Uesugi K, Umetani K 2011 *ISIJ Int.* **51** 402.
[23] Yasuda H, Nagira T, Yoshiya M, Uesugi M, Nakatsuka N, Kiire M, Sugiyama A, Uesugi K, Umetani K 2012 *Materials Science and Engineering* **27** 012084.
[24] Nagira T, Gourlay C M, Sugiyama A, Uesugi M, Kanazawa Y, Yoshiya M, Uesugi K, Umetani K, Yasuda H 2011 *Scr. Mater.* **64** 1129.
[25] Fredriksson H 1976 *Met. Sci.*, **10** 76.
[26] Hillert M, Solidification and Casting of Metals, The Metals Society, London, (1979).
[27] St. John D H 1990 *Acta Metall.* **38** 631.
[28] Singh S N, Blazek K E 1974 *J. Met.* **26** 17.
[29] Saeki T, Ohguchi S, Yamamoto T, Misumi H, Tsuneoka S 1982 *Tetsu-to-Hagane* **68** 1773.
[30] Matsumiya T, Saeki T, Tanaka J, Ariyoshi T 1982 *Tetsu-to-Hagan* **68** 1782.
[31] Goto S, Takeshita K, Suzuki Y, Ohashi H, Asano Y, Kimura H, Matsushita T, Yagi N, Isshiki N, Yamazaki H 2001 *Nucl. Instr. and Meth. A*, **467-468** 682.
[32] Umetani K, Yagi N, Suzuki Y, Ogasawara Y, Kajiya F, Matsumoto T, Tachibana H, Goto M, Yamashita T, Imai S, Kajihara Y 2000 *Proc. SPIE* **3977** 522.
[33] Van Vlack L H 1951*Trans. A.I.M.M.E.* **191** 251.
[34] Kurz W, Fisher D J 1989 Fundamentals of solidification, Trans. Tech. Pub.
[35] Hanao M, Kawamoto M, Yamanaka A 2008 *ISIJ Int.* **48** 180.

In situ, time-resolved tomography for validating models of deformation in semi-solid alloys

K M Kareh[1], P D Lee[2] and C M Gourlay[1]

[1]Department of Materials, Imperial College London, London, SW7 2AZ, UK
[2]School of Materials, The University of Manchester, M1 7HS, UK

Email: kristina.kareh05@imperial.ac.uk; peter.lee@man.ac.uk

Abstract. The development and validation of physically-based models of semi-solid alloy deformation requires experimental data on the microstructural response to load. In this work we present three-dimensional *in situ* synchrotron results of the indirect extrusion of semi-solid aluminium-copper alloys. Globular Al-15wt.% Cu specimens were deformed at solid fractions of approximately 0.7 during time-resolved x-ray tomography. Key phenomena of the response to load were observed and quantified, with a focus on particle translations and rotations, and the resulting displacement field. This example shows that crystal displacements can be quantified during semi-solid deformation of bulk specimens, allowing for datasets to be produced suitable for validating models of complex phenomena during semi-solid flow.

1. Introduction

Defects in castings are often the result of deformation of the semi-solid microstructure. Shrinkage and convection, as well as deliberate deformation in the case of semi-solid processing (SSP), affect macrosegregation, feeding and tearing behaviour. In order to predict and minimize the resulting casting defects, mush rheology must be understood, quantified and successfully modelled. An important step in developing accurate models of defect formation is *in situ* imaging and quantification of the microstructural response to load.

In situ studies on semi-solid deformation have mainly involved radiography of thin samples [1-4] with fewer studies using tomography of bulk 3D samples [5]. Recent radiography studies have shown that semi-solid metals can deform with certain defining characteristics of granular materials [1, 2]. A granular material can be defined as an assembly of discrete particles, saturated with liquid and/or gas, with particles that can be rigid or deformable [6]. Semi-solid metals with an equiaxed microstructure have been shown to exhibit Reynolds' dilatancy during shear [7, 8], a phenomenon whereby a compacted granular material expands (decreases its particle packing-density via particles pushing each other apart) in order to rearrange under load. This phenomenon has been observed and studied in a wide range of other fields and is important in the deformation of granular materials such as soils [9], and partially-molten rock [10, 11]. Additionally, a range of other deformation phenomena have been reported to occur in semi-solid alloys, including agglomeration/disagglomeration and deformation of the solid. As a result of this complexity, it is necessary to directly observe the deformation mechanisms *in situ* and link the imaged crystal-scale mechanisms to the macroscopic semi-solid deformation response, in order to develop and refine rheology models that are physically-based.

The aim of this work is to develop experimental techniques that enable the three-dimensional imaging of the crystal-scale response of semi-solid alloys to load and confirm some of the crystal-scale deformation mechanisms taking place. This will then enable the production of datasets that can be used to inform and validate models.

2. Methods

An as-cast, grain-refined Al-15wt%Cu alloy was heat treated at ~5°C above the eutectic temperature for 8 days to produce a large-scaled globular microstructure suitable for time-resolved tomography of semi-solid deformation. The resulting microstructure contained ~70 vol% α-Al with globular morphology and a mean grain size of 470 μm ± 10 μm, expressed as an equivalent sphere diameter.

In situ semi-solid deformation was conducted on a bespoke compression rig with a load cell of 1-500N ± 0.1N, a displacement range of 150 mm ± 200 nm, and a continuous ram rotation over 360° ± 0.001°, at a rotation speed of 15° per second. A 1000°C furnace mounted on the rig heated the semi-solid and controlled the temperature. The loading mode used in this work was indirect extrusion, applied by a cell constructed from boron nitride in two parts. A boron nitride cylindrical cup with ID = 10 mm, wall thickness = 1 mm, and height = 10 mm was used to contain the sample, while a ∅ = 8 mm boron nitride cylinder with a 1 mm bore at its centre was used as a hollow ram and die.

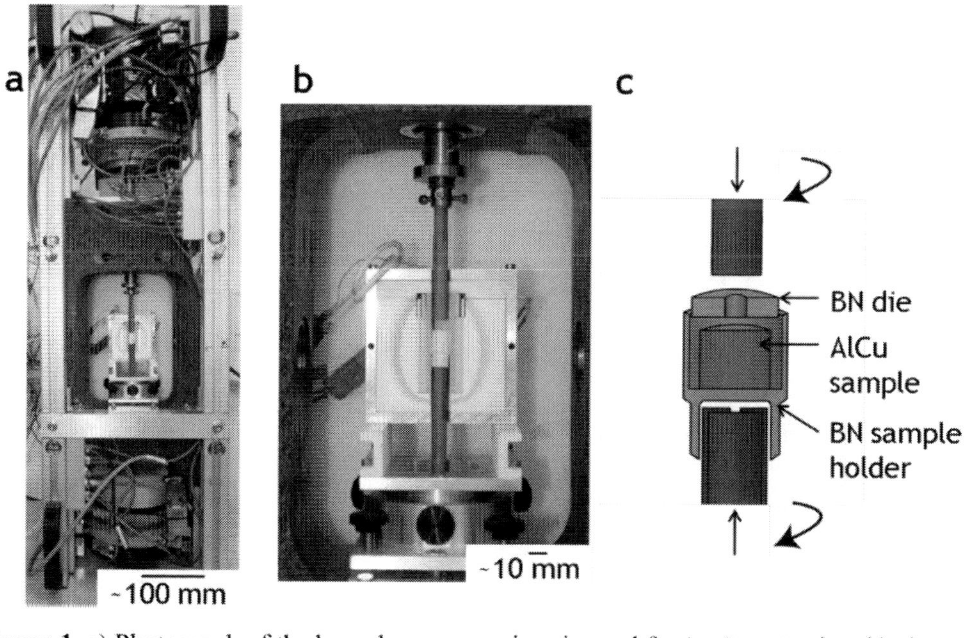

Figure 1. a) Photograph of the bespoke compression rig used for *in situ* extrusion; b) close-up of the furnace and rams; c) schematic of the indirect extrusion set up.

The rig was incorporated into beamline I12 at the Diamond Light Source (Didcot, Oxfordshire, UK) to conduct fast synchrotron X-ray tomography. A superconducting multipole wiggler was used as an X-ray source and a Si double crystal monochromator was used to produce an X-ray energy of 53 keV, the lowest possible on this beamline. For image detection, X-rays were converted to visible light using a LuAg:Ce scintillator with a thickness of 500 μm, magnified and recorded by an 800 × 600 CMOS chip on a Phantom v7.3 (Vision Research) rapid capture camera. This set-up resulted in a field of view of 9.8 × 7.3 mm with a 12.25 μm pixel size. Projected images were recorded at 60 frames per second. Each tomogram was computed from 720 projections taken at 0.5° intervals, producing one tomogram every 12 s. 24 continuous tomograms were taken during an experiment. Reconstructions were performed using reconstruction software based on the filtered back projection method [12].

In an experiment, the specimen was heated to 558°C, approximately 10°C above the eutectic temperature and corresponding to a solid fraction of 0.73, assuming equilibrium was attained during the 8-day semi-solid heat treatment. The specimen was held at this temperature for five minutes. Isothermal extrusion was then performed with a ram displacement rate of 10 μm.s^{-1} during continuous rotation and continuous image collection. This approach enabled 24 tomograms to be collected during a total displacement of 1.2 mm. After deformation, the furnace was switched off and opened, allowing the specimen to air cool.

The fully-solid material was also studied before and after deformation. Microtomography scans were performed at 85 kV and a current of 110 μA using a laboratory X-ray source (Phoenix X-ray Systems & Services, GmbH, Wunstorf, Germany), with an exposure time of 3200 ms. 720 images were captured on a detector of either 512×512 pixels or 1024×512 pixels over one revolution of the specimen. A Cu filter was placed in front of the target to minimise low-energy X-rays. Optical microscopy was performed after grinding using SiC papers of increasing grit values and polished using a 1:1 water-diluted colloidal silica solution.

Image processing was performed using Avizo 6.3 (VisageImaging, GmbH, Berlin, Germany), ImageJ (U.S. NIH, Bethesda, Maryland, USA [13]), and MATLAB 7.1 (The MathWorks Inc., Natick, MA, 2000). The image processing procedure is overviewed in Figure 2. Figure 2 a) shows an as-reconstructed slice from one tomogram, and Figures 2 b-d) show the steps to threshold the liquid, solid and pores and to separate all the globules. A non-linear diffusion filter and a median 3×3×3 filter were applied to the initial volume which was then thresholded. The volume was binarised and the grains were separated using a distance map and a watershed algorithm. The subsequent separated grains within the volume were then filtered to remove small bubbles and voxel-size artefacts.

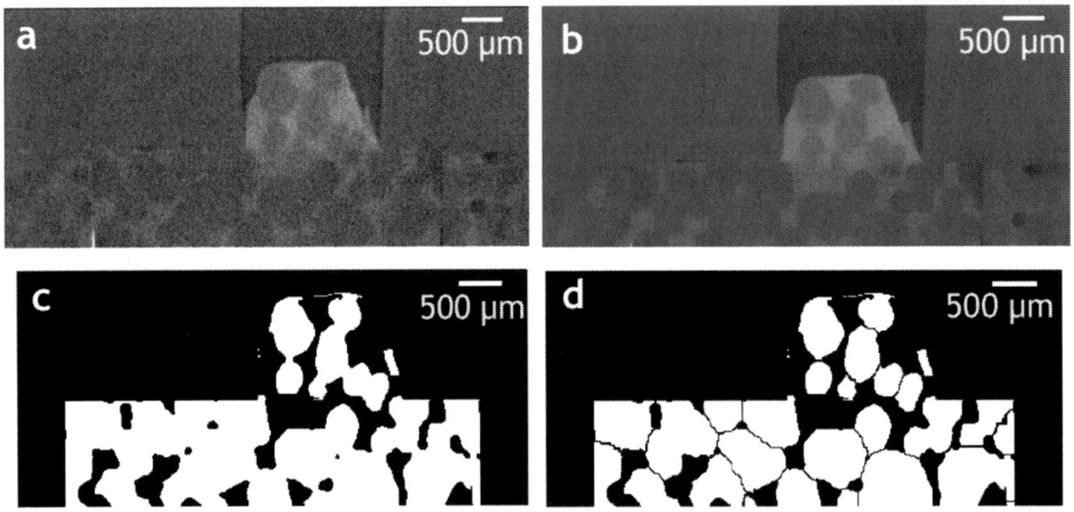

Figure 2. Image processing route as shown on an xz slice of tomogram 13, whereby a) is the original reconstructed image; b) is the image processed using a non-linear diffusion filter; c) represents the thresholded grains; d) shows the individual grains after the application of a watershed algorithm; and e) represents a rendering of the entire thresholded and watershed volume.

Digital volume correlation was performed on the dataset after a non-linear diffusion filter using DaVis DVC 8.07 (LAVision GmBH, Goettingen, Germany). The 3D displacement field was measured during semi-solid extrusion using an initial window size of 148×148×148 pixels to a final window size of 32×32×32 pixels with passes ranging from 2 to 5 and an overlap of 75% between windows.

3. Results and Discussion

The microstructure prior to partial remelting in the rig, as shown in Figure 3, is comprised of globular primary α-Al grains in a fine Al+Al_2Cu eutectic. The grains, having been held 5°C above the eutectic temperature for 8 days, contain features characteristic of extended coarsening [14]. For example, Figure 3 a) (i) shows two grains which are separated by a thin film of eutectic (liquid during heat treatment), while Figure 3 a) (ii) shows two agglomerated grains in the process of coalescing with a solid neck between them. The α-Al globules can be seen in three-dimensions in Figure 3 b), in which arrow (i) highlights an example of an agglomerate of three globules.

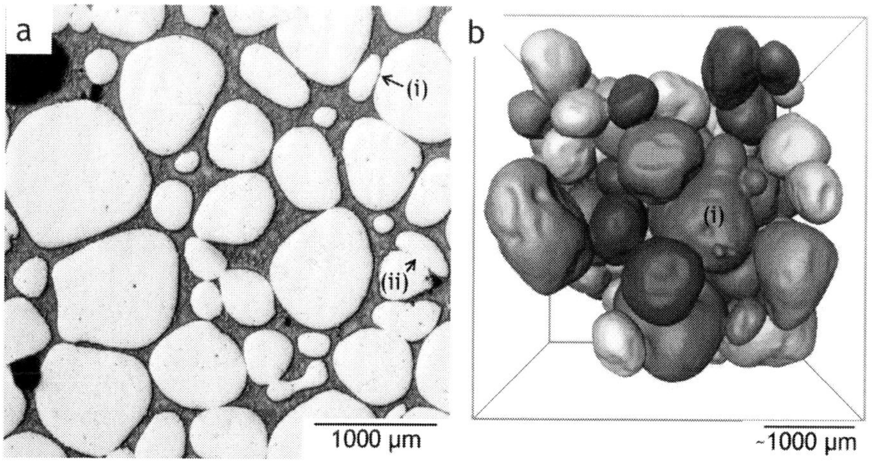

Figure 3. a) Optical micrograph showing the typical globular microstructure: (i) shows two globules separated by a liquid film, while (ii) highlights two globules with a solid neck. b) A rendering of the α-Al in 3D using a laboratory x-ray source, with (i) highlighting a 3-particle agglomerate.

Figure 4 shows three moments during semi-solid extrusion, spanning a total ram displacement of ~0.5 mm. Figure 4 a) are *xz* slices at a constant *y*-value, where the alloy can be seen to flow upwards into the bore. Figure 4 b) are the corresponding watershed and rendered 3D volumes of the solid. Note that the liquid, pores and boron nitride have been removed from Figure 4 b). Since the microstructural response to indirect extrusion can be visualised in three dimensions, the mechanisms of the two-phase flow can be studied.

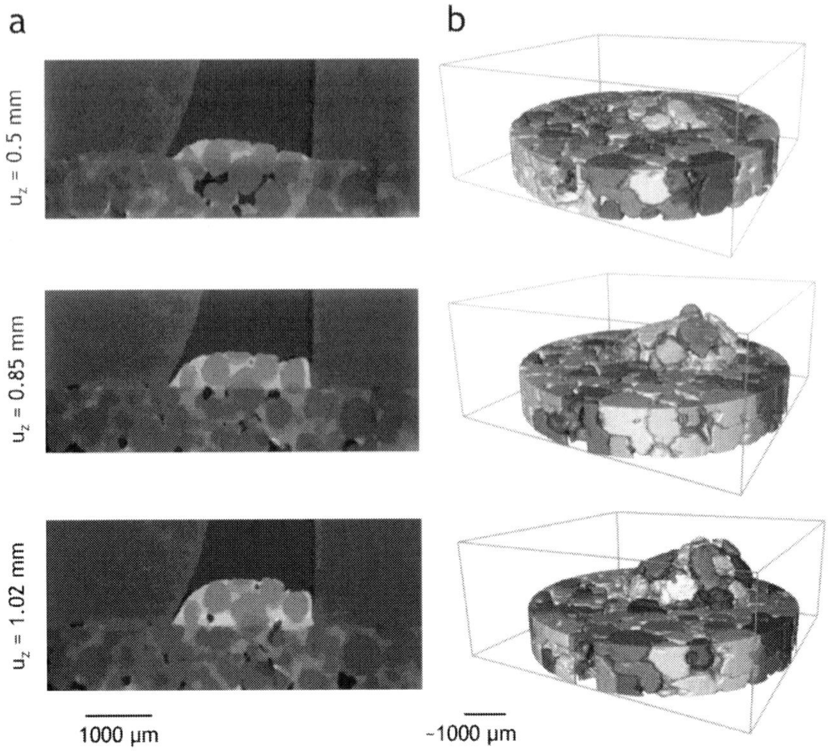

Figure 4. The microstructure at three instances during semi-solid extrusion a) *xz* slices; b) 3D rendering of the solid.

Globules at the surface of the extruded top rotate during extrusion (Figure 4), and the rotation of one globule is quantified in Figure 5. Material flow through the extrusion die is therefore by a combination of translation and rotation of the globules, coupled with the flow of interstitial liquid.

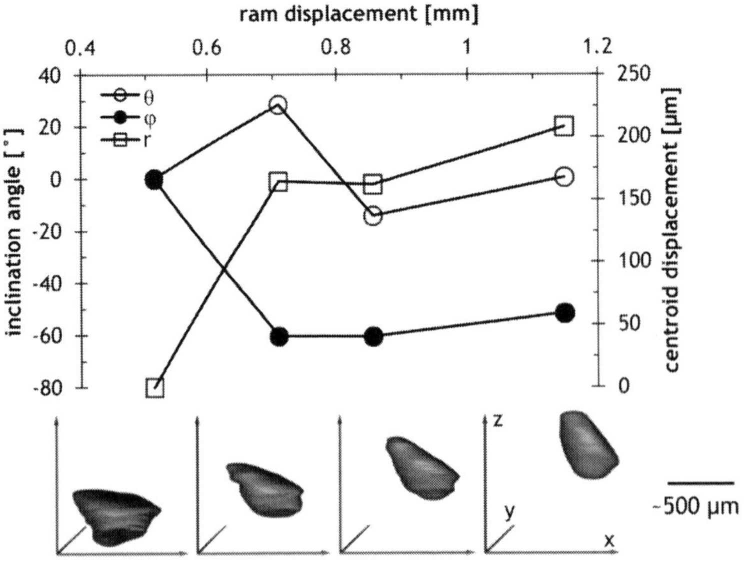

Figure 5. Rotation (θ, φ) and translation (r) of an individual grain in the extruded top.

Extrusion also caused significant macrosegregation with an increase in liquid fraction in the extruded top. Note that the liquid is Cu-enriched and appears brighter in the reconstructed slices. Figure 4 a) shows the increasing liquid fraction as semi-solid material is pushed upwards through the bore of the die. The solid-liquid separation phenomenon is examined in more detail in Figure 6.

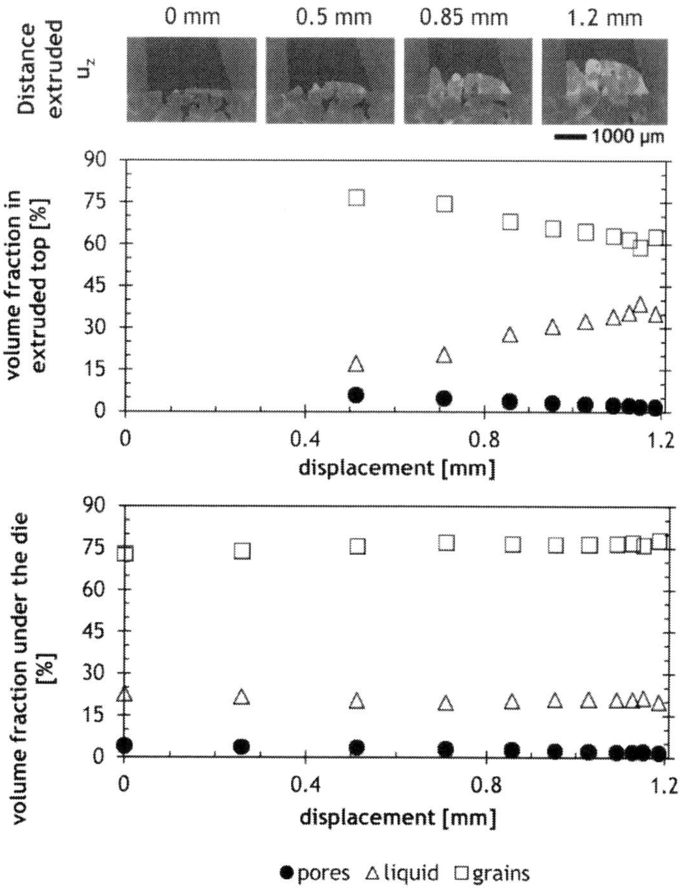

Figure 6. Evolution of the volume fractions in the field of view during extrusion: a) shows the volume fraction of solid, liquid, and pores in the extruded top. 250 µm of extruded material was chosen as a cut-off, below which an 'extruded top' does not exist; b) shows the volume fraction of solid, liquid, and porosity directly below the die.

In the extruded top there is a significant decrease of volume fraction of solid from 75% to 62%, accompanied by an increase in liquid, whose volume fraction increases from 18% to 35% (Figure 6). The fraction of porosity also decreases somewhat in the extruded top as porosity is pushed to the surface. In contrast, in the region beneath the die, the solid fraction increases from 73-78%, indicating compaction and the porosity decreases as the pores are pushed outwards and downwards, towards the bottom of the specimen. Figure 6 also demonstrates that solid-liquid segregation occurs from the beginning of extrusion. This is in line with previous bulk extrusion and backward extrusion experiments in multiple alloy systems [15-17], where low extrusion speeds lead to solid-liquid segregation at the beginning of extrusion, with the extruded volume containing a higher liquid fraction due to liquid being squeezed outwards under the ram pressure.

By running digital volume correlation on tomograms chosen such that the distance extruded between two time steps is a constant displacement increment of approximately 0.25 mm, it can be seen that the largest cumulative displacement vectors (Figure 7 a)) are located in the extruded region as well as directly underneath the bore of the die, where both the liquid and the solid are moving upwards through it. The direction of the vectors under the sides of the die shows flow outwards and towards the sides of the boron nitride cup. Since the specimen was 8 mm in diameter and the cup had an inner diameter of 12 mm, this indicates that the specimen has undergone 'upsetting', i.e. the diameter of the specimen increases via compression until it has reached the sides of the cup and becomes constrained, and extrusion simultaneously. Figure 7 b) shows displacement vectors projected onto an *xz* slice through the specimen: it can be seen that the sides of the specimen, which are not underneath the bore of the die, undergo a displacement outwards of less than 300 μm, while the region in and underneath the bore of the die undergo a displacement upwards of 800 μm to more than 1 mm. At the edges of the die, therefore, there are vectors of different direction and magnitude. It can thus be inferred that underneath the edges of the die there are regions of strain localization. Finally, the 'eddy' current seen on the right hand side in Figure 7 b) is due to the rotation of one grain in this slice. For this specimen, the large-scaled microstructure makes the images inherently discrete and affects the near-continuum assumption of DVC. An alternative way to study the displacement and strain fields is to track and plot discrete centroid translations and globule rotations (not shown here).

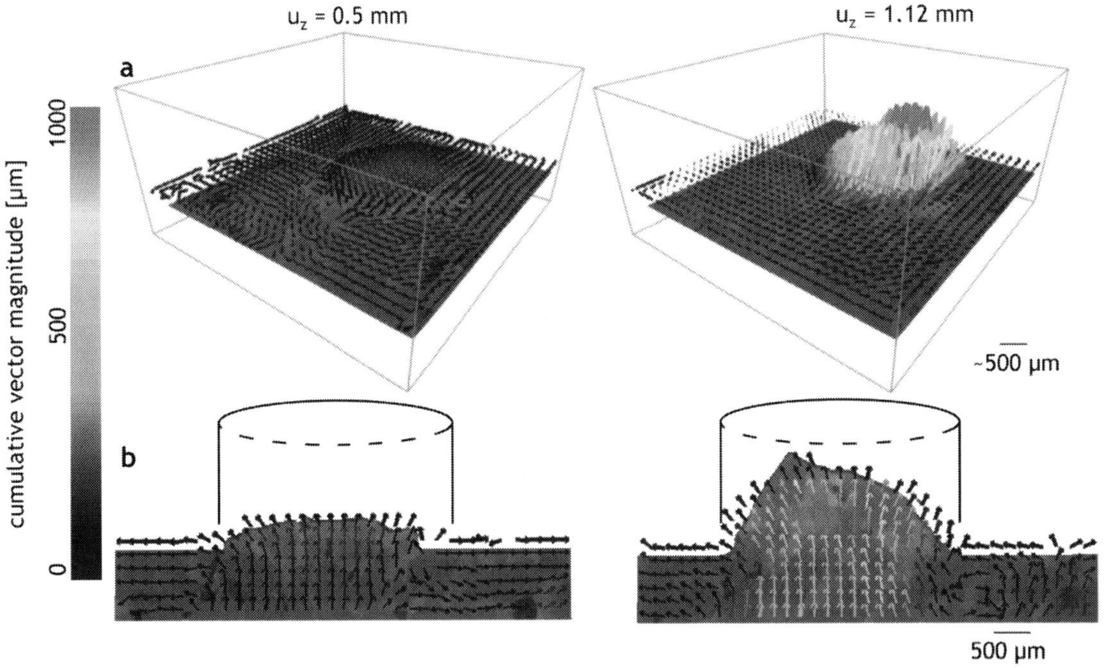

Figure 7. Digital volume correlation for a constant displacement of 0.25 mm per volume, showing the magnitude of the displacement in a) 3D, where the length of the vectors is proportional to the total displacement; and b) the magnitude of the total displacement projected on a single *xz* slice, with the bore of the die shown for clarification.

4. Conclusions

Semi-solid extrusion has been used in this work as an example of the new insights and validation data that can be generated by a hot tension-compression rig coupled with *in situ* tomography. The rig and associated furnace enable the heating and isothermal holding of a bulk specimen, as well as the rotation and ram displacement needed for time-resolved tomography imaging.

Using an adequate microstructure with respect to the detector resolution enables multiple filters to be used to process the data generated and identify grains, grain centroids and other characteristics such as grain volume and grain surface area. This is a prerequisite for any subsequent microstructure tracking which can then be performed in order to further our understanding of the interaction mechanisms between grains. Digital volume correlation has also been shown to successfully generate displacement vectors in three dimensions which can then be used to compute local strain fields and study localisation phenomena.

All the above techniques, both in terms of imaging and analysis, can therefore be used to successfully produce datasets that can validate existing models for semi-solid alloy deformation.

Acknowledgements
The authors would like to thank Dr. Peter Rockett for help in designing the rig and furnace, Richard Hamilton for help on the experiments, and Beamline I12 Diamond Light Source (EE6218) together with Drs. Thomas Connelly and Robert Atwood whom were always helpful. The authors gratefully acknowledge the support of the EPSRC for the equipment and experimental costs (EP/F001452/1 and EP/I02249X/1)) and KMK the support of a Hydro Aluminium PhD studentship.

References
[1] Nagira T, Gourlay C M, Sugiyama A, Uesugi M, Kanzawa Y, Yoshiya M, Uesugi K, Umetani K and Yasuda H 2011 *Scripta Materialia* **64** 1129-32
[2] Gourlay C M, Dahle A K, Nagira T, Nakatsuka N, Nogita K, Uesugi K and Yasuda H 2011 *Acta Mater.* **59** 4933-43
[3] Phillion A B, Hamilton R W, Fuloria D, Leung A C L, Rockett P, Connolley T and Lee P D 2011 *Acta Mater.* **59** 1436-44
[4] Zabler S, Rack A, Rueda A, Helfen L and Garcia Moreno F 2010 *Physica Status Solidi. A* **207** 718-23
[5] Terzi S, Salvo L, Suery M, Limodin N, Adrien J, Maire E, Pannier Y, Bornert M, Bernard D, Felberbaum M, Rappaz M and Bollerr E 2009 *Scripta Materialia* **61** 449-52
[6] Mitarai N and Nori F 2006 *Advances in Physics* **55** 1 - 45
[7] Gourlay C M and Dahle A K 2007 *Nature* **445** 70-3
[8] Gourlay C M, Meylan B and Dahle A K 2008 *Acta Mater.* **56** 3403-13
[9] Ristow G H 2000 *Pattern formation in granular materials* vol 164 (Berlin Heidelberg New York: Springer)
[10] Scott D R 1996 *Nature* **381** 592-5
[11] Paterson M S 2001 *Tectonophysics* **335** 51-61
[12] Titarenko V, Titarenko S, Withers P, De Carlo F and Xiao X 2010 *Journal of synchrotron radiation* **17** 689-99
[13] Rasband W S 1997-2009
[14] Limodin N, Salvo L, Suery M and DiMichiel M 2007 *Acta Mater.* **55** 3177-91
[15] Gebelin J C, Suery M and Favier D 1999 *Materials Science and Engineering A* **272** 134-44
[16] Loué W R, Suery M and Querbes J L 1993 *Processing of Semi-Solid Alloys and Composites; Cambridge, Massachusetts; USA* 10
[17] Pinsky D A, Charreyron P O and Flemings M C 1984 *Metallurgical Transactions B* **15** 173-81

IOP Conf. Series: Materials Science and Engineering **33** (2012) 012038 doi:10.1088/1757-899X/33/1/012038

Quantifying damage accumulation during the hot deformation of free-cutting steels using ultra-fast synchrotron tomography

C Puncreobutr[1], P D Lee[2,6], M Kaye[3], D Balint[3], D Farrugia[4], T Connolley[5], and J Lin[3]

[1] Department of Materials, Imperial College London, London, SW7 2AZ, UK
[2] School of Materials, The University of Manchester, Manchester, M1 7HS, UK
[3] Department of Mechanical Engineering, Imperial College London, London, SW7 2AZ, UK
[4] Tata Steel Research Development & Technology, Swinden Technology Centre, South Yorkshire, S60 3AR, UK
[5] Diamond Light Source Ltd, Harwell Science & Innovation Campus, Didcot, OX11 0DE, UK

E-mail: peter.lee@manchester.ac.uk

Abstract. Many different approaches have been proposed to simulate the nucleation and evolution of damage during the hot forming of steels. However, there is a lack of time-resolved, three dimensional quantification of the evolution of damage, a requirement for validation of the kinetic and morphological model predictions. One very significant industrial case is the hot forming of free-cutting steels (FCS), where small additions of heavy metal inclusions are added to enhance the machinability and surface quality of the steel. In this paper, we present the *in situ* ultra-fast synchrotron X-ray tomographic observations and quantification of FCS during hot deformation, including measurement of applied load. This allowed the correlation of strength to the different stages of the cracking process. The results are augmented with high spatial resolution interrupted studies. The interrupted tomographs enabled the quantification of the volume fraction, equivalent diameter, spatial distribution and orientation of inclusions and damage at various strain levels. The combination of these two studies provides a benchmark experiment for the validation of physically-based finite element models, both directly, and via constitutive equations for the time/temperature dependent effects of dislocation density, damage, strain rate and temperature.

1. Introduction

During the hot forming stages of the steel making process, the dominant mechanism of damage nucleation and evolution can vary depending on a number of parameters including strain rate, forming temperature, material composition and microstructure. Some of these dependencies are not well known, hence an improved scientific understanding of them will help steel makers control product quality. Many different approaches have been proposed to simulate the plastic deformation and damage evolution during hot forming of steels [1-3]. However, there is very limited data to validate these models. One very significant industrial case is the hot forming of free-cutting steels (FCS). In an FCS, small quantities of heavy metal elements are added to resulphurised low carbon steel to enhance

[6] Author to whom any correspondence should be addressed.

the machinability and surface quality of the product. However, the presence of inclusions which improve machinability are also known to promote damage at high temperatures by acting as void nucleation sites and stress raisers [4, 5], when conditions of triaxial strain are above a specific threshold. Prior metallurgical investigations revealed that at typical FCS hot forming temperatures (850-1100°C), a significant disparity in stress is present at MnS-matrix interfaces. This leads to debonding of these interfaces during straining and subsequently a higher damage growth rate in the vicinity of inclusion interfaces where strain, principal stress and triaxiality localisation are acting [6].

In order to quantify the inclusions and damage throughout the microstructure in three dimensions, synchrotron X-ray microtomography (XMT) has been used by several authors [7, 8]. Although the measurements were done through *in situ* tensile tests, the deformation was interrupted when the X-ray projections was captured. This allowed the material to recrystallise which severely affected the damage propagation process. Thus the time-resolved and three dimensional quantification of damage evolution is still a key requirement for the validation of the kinetic and morphological model predictions. In this paper, we present the first 3D, *in situ*, ultra-fast synchrotron X-ray tomographic observations and quantification of FCS during hot deformation. This *in situ* result, together with a high spatial resolution interrupted study, provide a good validation for damage evolution model.

2. Experimental methods

The following describes how the *in situ* and interrupted synchrotron X-ray microtomography experiments were conducted to obtain the three dimensional damage evolution results. The methods for quantifying the evolution of damage are also described.

2.1. In situ synchrotron X-ray microtomography experiment

The leaded free-cutting steel used in the *in situ* observation was cut 45.5 mm away from an as-cast billet wide-face surface. The material was machined into a uniaxial tensile test piece with 1 mm diameter and 6 mm gauge length. Synchrotron X-ray microtomography during *in situ* tensile tests were performed at Beamline I12 at the Diamond Light Source using a 53 keV monochromatic beam. An 800x600 pixel high speed CMOS camera, lens coupled to a single crystal scintillator was used to acquire radiographic images during the experiment. The spatial resolution was 12.22 μm/pixel. At the beginning of the experiment, the sample was heated to 860°C to simulate the typical hot forming surface temperature of an FCS. Then, the load was applied at a constant cross-head displacement of 5 μm/s using a bespoke rig and furnace (as shown in Figure 1a, b). Radiographs were continuously acquired at an integration time of 16 ms while the sample was continuously rotated and under load. This experimental configuration resulted in a full 3D tomographic volume every 12s and 24 3D volumes were acquired from the undeformed to final fracture stages.

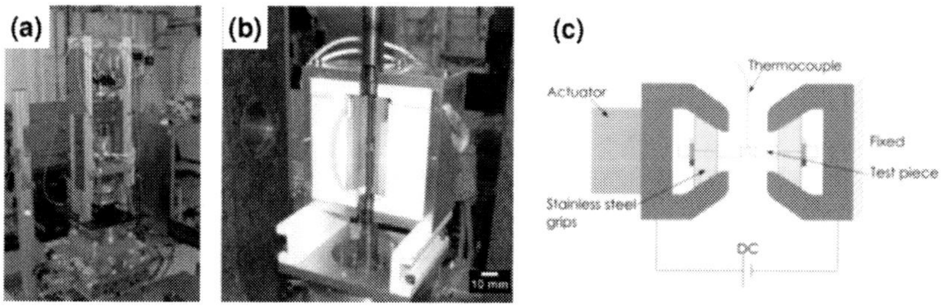

Figure 1. Experimental rig and furnace for (a, b) *in situ* observation and (c) interrupted observation.

2.2. Interrupted synchrotron X-ray microtomography experiment

The same composition of leaded free-cutting steel as used in the *in situ* observation was cut 200 mm away from the billet surface and was machined into uniaxial tensile test samples with a notch diameter of 6 mm to fit the X-ray tomographic field of view (FOV). The tensile tests were conducted using a Gleeble 3800 where the sample was clamped between two electrically conductive jaws as shown in Figure 1c. Before testing, the sample was reheated to the upper range (1200ºC) of the typical industrial hot forming operations, and then cooled and held at 1000ºC during deformation. (Note, the *in situ* was performed at the lower end of the temperature range due to furnace constraints). The displacement rate was controlled to maintain a constant strain rate of 0.1 s^{-1}. Two different levels of strain-to-failure ($\varepsilon/\varepsilon_f$) of 78% and 90% after necking were analysed in different test samples by interrupting the test before failure. In order to freeze the level of damage evolution within the microstructure, samples were quenched in water immediately after the desired level of deformation was reached.

Strained samples at strain-to-failure of 78% and 90% together with a virgin sample (no strain applied) were cut in half, with one side examined metallographically (using optical and electron microscopy) and the other using X-ray synchrotron tomography. Specimens were scanned using a monochromatic X-ray beam of 76.5 keV at beamline I12, Diamond Light Source. A 4008x2672 pixels camera with a resolution of 1.8 μm was used to obtain a high resolution of the evolution of the damage in the interrupted microstructure. A series of 1441 projections was taken over 180 degrees at an exposure time of 1 s, resulting in a full set of data required for reconstructing a 3D volume.

2.3. Image processing and quantifying methods

The radiographic projections from each scan were reconstructed using in-house reconstruction software based on the filtered back projection (FBP) method [9] in order to get a 3D volume of each sample. An orthogonal slice of the reconstructed 3D volume is shown in Figure 2a. A 3D non-linear diffusion filter was then applied to reduce the noise (Figure 2b). The data was then thresholded to obtain a binarised data of microstructural damage (inclusions and voids) at different strain levels (Figure 2c).

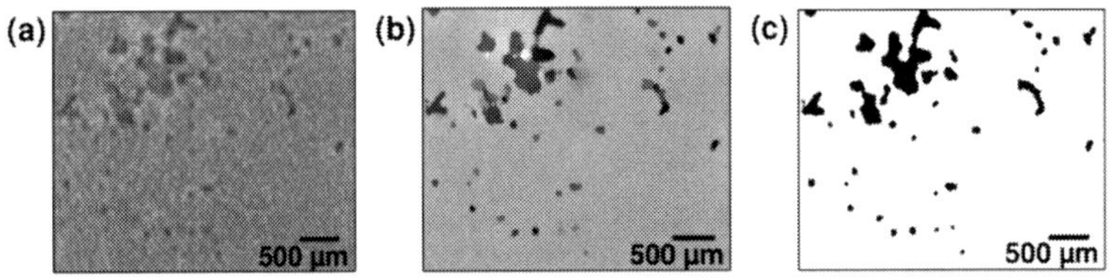

Figure 2. (a) Original orthogonal slice from interrupted observation (b) applied 3D non-linear diffusion filter (c) thresholded slice.

In the *in situ* observation results, 9 time-resolved volumes at different levels of strain were aligned using a 3D affine registration method. The volumes of size 3000x3000x2500 μm^3 were then cropped inside the gauge to ensure the cracking region was visible at all times. In the interrupted observation results, two sets of 3D volumes at each strain level were selected and quantified at different locations to represent the varied localised damage coalescence. The first volume of 900x2000x3000 μm^3 was selected at the centre of the sample gauge. The second volume of 550x1100x1550 μm^3 was chosen right next to the surface of the sample.

To quantify the microstructural damage (both *in situ* and in the interrupted studies), a marching cube algorithm [10] was used to calculate the enclosed volume and surface area of inclusions/damage. The spatial distribution measurement of inclusions/damage was also taken into account for the high-resolution interrupted study. The size and spacing distributions (denoted as ω) at different levels of strain were calculated using equation (1) where d is equivalent diameter of the inclusions and l is the nearest neighbour between inclusion centroids.

$$\omega = \left(\frac{d}{l}\right)^2 \quad (1)$$

3. Damage evolution models

Two damage evolution models were simulated using finite element modelling (FEM). The first model was an axisymmetric model of the uniaxial test piece. The second was a microscopic model, incorporating individual inclusions. Each model will be described separately. The uniaxial model used the strain rate boundary conditions as used in the interrupted tests. A quad dominated structured mesh was used, with a minimum element size of 50 μm and explicit integration. The material model is based on a temperature and strain rate dependent viscoplastic model developed by Foster, which was implemented via the Variable User MATerial, VUMAT, subroutine in Abaqus 6.10 [6]. The microstructural model used the same material model but a free default tetrahedral meshing approach was taken. The positions and volumes of individual inclusions in the FEM simulation were taken from the XMT results of the undeformed sample. The inclusions were modelled as porosity as the energy for damage nucleation was considered negligible. There is a range of stress and strain values in the necked region of the uniaxial test piece and it is believed that the displacement of nodes can be extracted and used as the nodes of the microscopic model. This global model and sub model approach permits a more realistic boundary condition for the micro model.

4. Results and discussion

The damage evolution of a free-cutting steel sample that was subjected to a constant cross-head displacement of 5 μm/s at 860°C is shown in Figure 3. Because the tensile test was done *in situ*, the stress can be measured as a function of the strain applied allowing the correlation of the strength to the different stages of damage evolution.

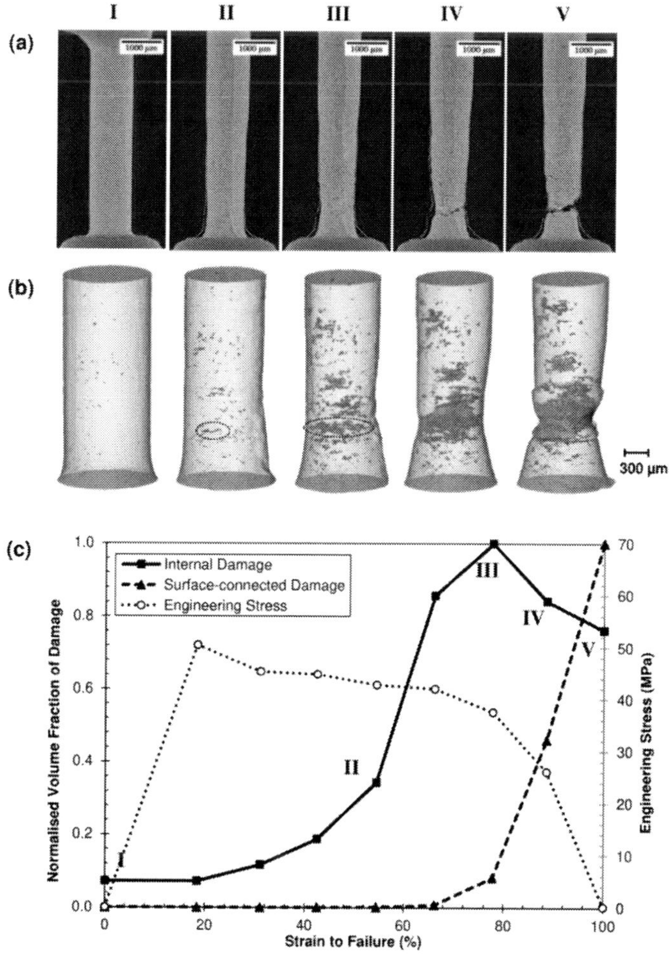

Figure 3. In situ cracking evolution of steel sample at 860°C. (a) 2-D orthogonal cross-sections; (b) 3-D reconstruction after tomography; (c) quantification of surface-connected damages and internal damages. Strains are I=0%, II=54%, III=78%, IV=89% and V=100%.

The 2D orthogonal cross section (Figure 3a) and 3D rendered (Figure 3b) images show that prior to the application of strain (frame I) there are a number of fine MnS inclusions homogenously distributed. After the initial strain was applied (frame II), the sample started necking near the bottom of the gauge (possibly because of the lower temperature closer to both ends of the gauge). Damage initiated on inclusions. Oxide scale around the sample can also be observed. Further damage and damage coalescence was observed at 78% of strain-to-failure (frame III). A large surface-connected damage region at the core of the sample (highlighted as a dashed circle), was found to develop from prior internal damage at the core. This confirms that the strain localisation and high triaxiality at the core plays an important role in damage evolution. At 89% strain, the damage is accentuated and reaches the sample surface (frame IV). The quantification result (Figure 3c) indicates that the built-up internal damage massively transforms into surface-connected damage at this stage, with a sudden drop in the engineering stress. At the maximum strain (frame V), the failure occurs mainly by localised damage coalescence near the bottom of the gauge. This is confirmed by the decrease of internal damage (blue) and rapid increase of surface-connected damage (red).

The damage evolution obtained from the *in situ* experiment was confirmed by a high-resolution interrupted experiment and simulation results as follows:

Axisymmetric finite element modelling of the uniaxial test piece was carried out to establish the local stress and strain in the interrupted tensile testing samples. The distributions of strain and stress triaxiality (σ_h/σ_{eq}) are shown in Figure 4a and Figure 4b respectively, showing a wide range of strain along the gauge length (Figure 4a). However, the difference is small in the radial direction, e.g. between the region near the centre (highlighted as region I) and the region near the surface (highlighted as region II). Different behaviour is observed in the distribution of stress triaxiality in the neck (Figure 4b). The triaxiality is much more localised in the region near the centre compared to the surface.

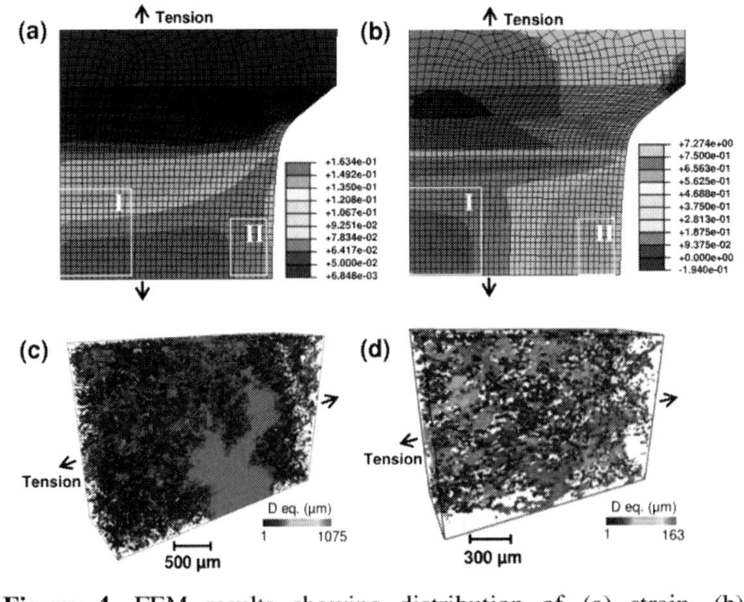

Figure 4. FEM results showing distribution of (a) strain, (b) triaxiality (σ_h/σ_{eq}) compared to damage agglomeration from interrupted XMT results at the region, (c) near the centre, I, and (d) near the surface, II.

The same regions as highlighted in Figure 4a and Figure 4b were selected from interrupted XMT results and analysed. The damage agglomeration at the strain-to-failure of 90% in the regions near the

IOP Conf. Series: Materials Science and Engineering **33** (2012) 012038 doi:10.1088/1757-899X/33/1/012038

centre (region I), and near the surface (region II), are shown in Figure 4c and Figure 4d respectively. In order to facilitate visualisation, the damage voids are labelled in color by their equivalent diameter. The results clearly show that the damage is much larger, with a higher degree of agglomeration, in the centre where the triaxiality is initially the highest. The largest coalescence of voids at the centre of sample is about 7 times larger than the largest cluster of connect voids near the surface. This indicates that the stress triaxiality plays a significant role in the evolution of damage. This agrees well with Gurson [11] that the rate of damage growth and coalescence increases exponentially with increasing triaxiality.

The evolution of damage and its coalescence at different levels of strain was also considered. The sub-volume of the same region as in I (shown in Figure 4) in the initial material and from interrupted XMT results at 78% and 90% of strain to failure was analysed. The visualisation of inclusions and damage at each strain level is shown in Figure 5a-c. The initial inclusions are labelled in red and the damage/voids are labelled in blue. This shows that the damage/voids initiated and coalesced around the inclusions by the time 78% of strain-to-failure was reached (Figure 5b). The morphology of inclusions was also observed to elongate in the tensile direction. At the higher strain-to-failure ratios ($\varepsilon/\varepsilon_f$=90%), both the extent of damage and its coalescence increased, with the bulk of the damage forming a half millimetre long crack (left hand side of Figure 5c). This agrees well with the microscopic model. The FEM model, which includes individual inclusions, shows that when the material is under strain, there is a localisation of stress around the inclusions (see the change from Figure 5d to Figure 5e). This emphasizes the crucial effect of localised stress on damage initiation. The separation distance also plays a major role in promoting stress/strain localization, where the inclusions are close to each other. This is confirmed by Rhines [12], where the diameter to separation distance ratio can affect the strain concentration enhancement. The stress localisation grows to the larger area and bridges to other surrounding inclusions when the applied strain is increased (Figure 5f).

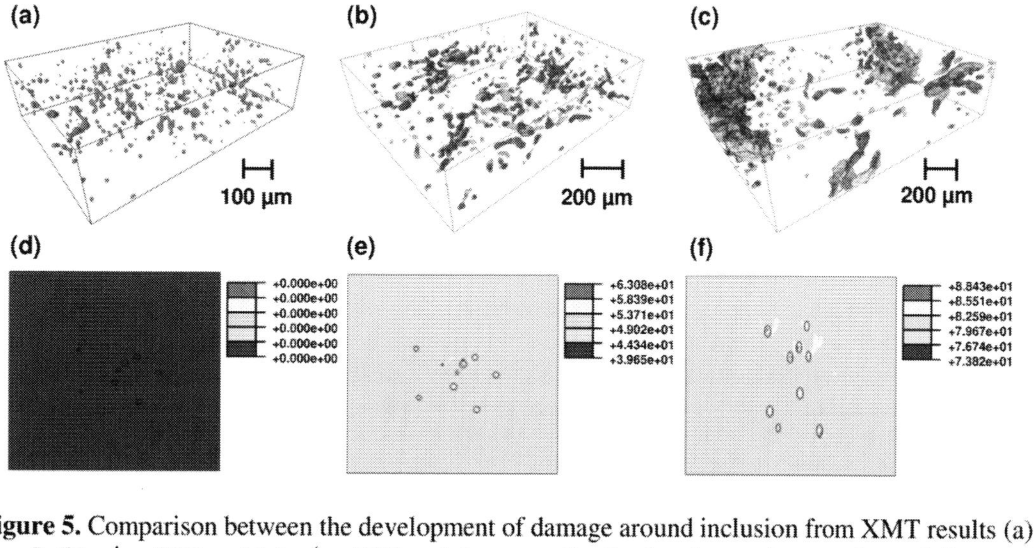

Figure 5. Comparison between the development of damage around inclusion from XMT results (a) $\varepsilon/\varepsilon_f$=0, (b) $\varepsilon/\varepsilon_f$=78% and (c) $\varepsilon/\varepsilon_f$=90% and the stress distribution from microscopic modeling (d-f).

In order to determine the effects of localised damage coalescence and separation distance in different regions of the sample, the same regions highlighted in Figure 4a-b, were analysed quantitatively. The equivalent damage diameters are plotted versus the normalised volume fractions at various strains in both the regions near the centre (Figure 6a) and near the surface of sample (Figure 6b). (Note that there is a second large peak in void fraction for the strain-to-failure of 90% due to a single large coalesced void that is not shown so that the comparison between strains can be made). Figure 6 shows that the mode diameter increases when the level of strain increases. Figure 6 quantitatively demonstrates the observations that damage initiates and grows around the inclusions as the straining progresses. Two peaks are also observed at 78% of strain-to-failure at the surface region (Figure 6b); one peak at a small diameter but with a high volume fraction, while the other is at large diameter with low volume fraction. These two peaks become a single peak at large diameter with the high volume fraction at 90% of strain-to-failure. This confirms that increasing stress localisation around the inclusions leads to damage coalescence, and coalesced damage tends to bridge with the surrounding damage when the critical localised stress is reached.

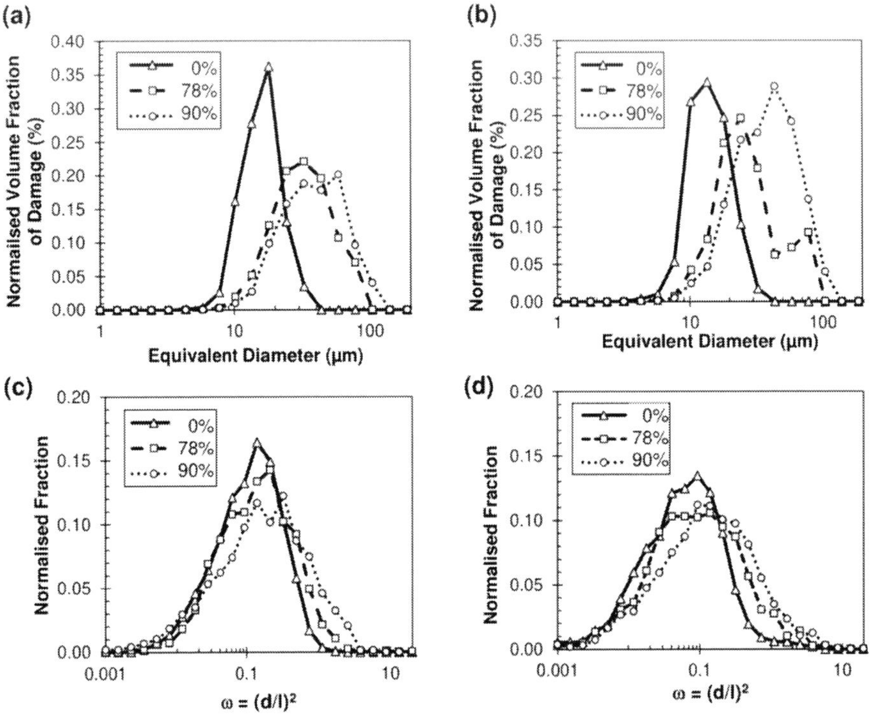

Figure 6. Comparison of volume fraction and separation distance between the damage near the centre (a, c) and the surface (b, d) of sample.

The size and spacing distribution (ω) for each inclusion and damage at different locations of the sample is plotted in Figure 6c and Figure 6d. The modes of the distributions from the location near the centre and the surface are approximately the same at 0.1. The distributions at different strain levels are also found to be roughly constant throughout deformation in both the centre and surface regions. Thus, the increase in diameter (from damage coalescence) is offset by the increase in spacing due to void growth.

5. Conclusions

The combination of both *in situ* and interrupted synchrotron X-ray microtomography is a useful tool to provide a benchmark experiment for the validation of physically-based finite element models. The relationship between stress-strain, temperature and microstructural damage is obtained. The results confirm the significant roles of strain localisation and triaxiality on damage agglomeration. The mode diameter increases as strain increases, indicating that increasing stress localisation around the inclusions leads to damage coalescence. The same behaviour was found at the regions near the centre as well as at the surface of material, but the level of damage evolution is much larger at the centre where the triaxiality is much greater. The size and spacing distribution was also measured and found to remain constant throughout deformation in both the centre and surface regions.

Acknowledgments
The authors would like to acknowledge the provision of beamtime and help from the scientists at the Diamond Light Source's I12 Beamline, especially R.C. Atwood and M. Drakopoulos, as well as R.W. Hamilton and P. Rockett for help with the design of the P2R Rig and running synchrotron experiments. The authors also would like to thanks Tata Steel, EPSRC (EP/F001452/1), M-ORS and the Thai Government Scholarship for materials and financial support.

References
[1] Horstemeyer M F and Gokhale A M 1999 A void–crack nucleation model for ductile metals *Int J Solids Struct* **36** 5029-55
[2] Hahn G T, Kanninen M F and Rosenfield A R 1972 Fracture Toughness of Materials *Annu Rev Mater Sci* **2** 381-404
[3] Foster A D, Lin J, Farrugia D C J and Dean T A 2006 Experimental investigation of the spatial variation in viscoplastic response of free-machining steels in hot rolling conditions *J Mater Process Tech* **177** 497-500
[4] Chunhui L 2001 Evolution of voids close to an inclusion in hot deformation of metals *Comp Mater Sci* **21** 360-74
[5] Farrugia D, Husain Z, Santisteban V, Llanos J, Bianchi J, Vescovo P, Pera J, Rodriguez-Ibabe J, Revilla C, Fredriksson J, Demurger J, Forestier R and Kieber B 2009 *PACROLP: the prediction and avoidance of cracking in long product hot rolling* (Luxembourg: Office for Official Publications of the European Communities)
[6] Foster A 2007 Damage modelling in free machining steels: in *Mechanical Engineering*, The University of Birmingham (PhD)
[7] Maire E, Bouaziz O, Di Michiel M and Verdu C 2008 Initiation and growth of damage in a dual-phase steel observed by X-ray microtomography *Acta Metall Mater* **56** 4954-64
[8] Weck A, Wilkinson D S and Maire E 2008 Observation of void nucleation, growth and coalescence in a model metal matrix composite using X-ray tomography *Mat Sci Eng A* **488** 435-45
[9] Titarenko V, Titarenko S, Withers P J, De Carlo F and Xiao X 2010 Improved tomographic reconstructions using adaptive time-dependent intensity normalization *J SynchroTron Radiat* **17** 689-99
[10] Lorensen W E and Cline H E 1987 Marching cubes: A high resolution 3D surface construction algorithm: in *Computer graphics and interactive techniques*, (New York, NY: ACM) pp 163-9
[11] Gurson A L 1977 Continuum Theory of Ductile Rupture by Void Nucleation and Growth: Part I-Yield Criteria and Flow Rules for Porous Ductile Media *J Eng Mater Technol* **99** 2-15
[12] Rhines W J 1961 Ductile fracture by the growth of pores: in *Mechanical Engineering*, M.I.T. (M.S.)

Real time imaging on dendrite morphology evolution during alloy solidification under electric field

Jing Zhu[1], Tongmin Wang [1*], Zongning Chen[1], Jingjing Xu[1], Honglan Xie[2], Tiqiao Xiao[2], Tingju Li[1]
1 School of Materials Sciences and Engineering, Dalian University of Technology, 116024
2 Shanghai Institute of Applied Physics, Chinese Academy of Sciences, Shanghai, 201204

E-mail: tmwang@dlut.edu.cn

Abstract. The dynamic dendrites growth of metallic alloy during the solidification can not be in-situ observed since the metallic alloy is in optically opaque. It is also not possible to in-situ study the influence of electric current on dendrites growth behaviour. By using synchrotron radiation imaging technique, the evolution of dendrite morphology under an electric current is observed. Results indicate that the direct current can significantly suppress the dendrites growth and the dendrite tip tends to be round from sharp. Increasing the current intensity, the dendrite tip splits due to the current crowding and solute pileup around the round tip. The pulsed electric current can significantly enhance the solidification rate and refine the primary and secondary dendrite arm spaces because of the thermal and mechanical shock around solid/liquid interface.

Keywords: Synchrotron radiation, Solidification, Dendritic grain, Electric field

1. Introduction

The application of electric current during metal solidification process has been tried since 1960s, and many researches have studied the influence of electric current on the solidifying metals and alloys. All the results revealed a phenomenon that the electric current can efficiently modify the behavior of the solidifying alloys [1,2]. Thus, the influence of electric current, such as direct current (DC) [3,4], alternating current (AC) [5] and electric current pulse (ECP) [6-8] on solidifying alloy were detail studied. Significant grain refinement efficiency on eutectic or noneutectic microstructure has been achieved for low melting Sn-Pb alloy, middle Al-Cu, Al-Si alloys, even for high melting steel and iron. It is well known that a series of effects, such as electromigration, electricity-to-heat, electromagnetic force induced by the electric current can impact on the solidifying alloys. Hence, a changeable kinetics or thermodynamic grain growth can be obtained.

* Corresponding author at: School of Materials Sciences and Engineering, Dalian University of Technology, Dalian 116024, RP China. Tel.: +86-411-84706790; fax: +86-411-84706790.

However, the real-time grain growth during alloys solidification is hard to observe due to the opaque and a high temperature condition of the metals and alloys. Most of experiments have been done by analyzing the final microstructure or observing a transient microstructure by quenching. The kinetics of grain growth can not be in-situ observed. Adversely, the multidate information may be missed simply through the quenching method. A cutting-edge experimental technique was recently developed and demonstrated to be successful in studies of alloys during solidification which was defined as synchrotron radiation. Researches on low melting Sn-based alloys and Al-based alloys in European synchrotron radiation (ESRF) [9,10], Japanese synchrotron radiation (Spring8) [11] and American synchrotron radiation (APS) [12] were mostly reported. Thus, synchrotron radiation is a unique method to in situ observes solidifying alloys.

Our previous researches carried out on 4W1A beamline at Beijing focused on the low melting temperature Sn-Bi and Sn-Pb alloys. The phenomena of dendrites fragmentation/floating, columnar-to-equiaxed transitions and the competitive growth of dendrites were in situ observed [13,14]. Based on these, research on solidification behavior of Sn-Bi alloy under an electric field was performed on the BL13W1 beamline at Shanghai. In present work, the evolution of dendrite morphology under an electric field and its mechanism were fully illustrated [15].

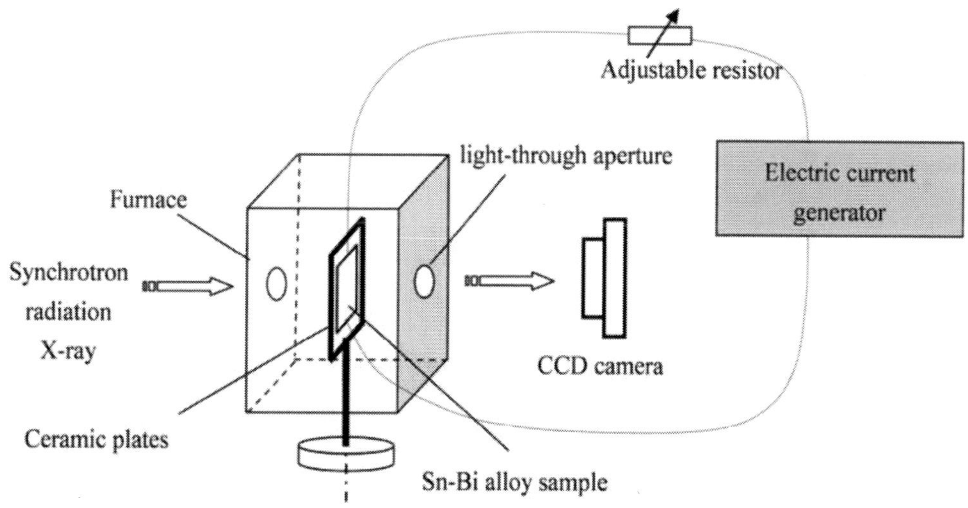

Figure 1. Schematic of synchrotron radiation imaging experiment under electric field.

2. Experimental

The experiments were carried out at the beamline BL13W1 of Shanghai Synchrotron Radiation Facility (SSRF), China. In-line X-ray phase imaging method was used in this work, as shown in Fig. 1. A fast readout-low noise CCD camera with a circular view window (Φ750 pixels) was used to collect the images. Its pixel size was down to 1.85 μm. In present experiments, the energy of synchrotron X-ray was 20 KeV and the sample-to-detector distance was 15-25 cm. The exposure time per frame was about 2-7 s.

A special furnace was designed with a window in the middle for X-ray passing through the sample. The system includes double furnaces. Each furnace consists of two parallel copper plates and a heat source, which is surrounded with a heat insulation material. The furnaces are used to melt the alloy sample and to control the temperature gradient. The temperatures of hot furnace and cold furnace can be controlled independently by two thyristor controllers. The alloy sample can move freely along the vertical direction between the two furnace rooms.

A purity of 99.99% Sn and 99.99% Bi were melted in a vacuum melting furnace for preparation of Sn-12wt.%Bi alloy. A well polished thin Sn-12wt.%Bi alloy sample with a thickness of 100 μm, which was cut into rectangular of 1.0×2.0 mm^2, and sandwiched between two 0.28-mm-thick ceramic plates (25 mm in width, 50 mm in height). The sample was fixed on the center of heating furnace and two strips of aluminum foil, as electrodes, were respectively set on the upper part and lower part of sample. In this experiment, the alloy was melted completely and then cooled homogeneously at cooling rates of 1.5 K/min. Evolution of dendrite morphology was imaged and recorded by CCD camera.

3. Results and discussion

3.1. The evolution of dendrite morphology with direct current

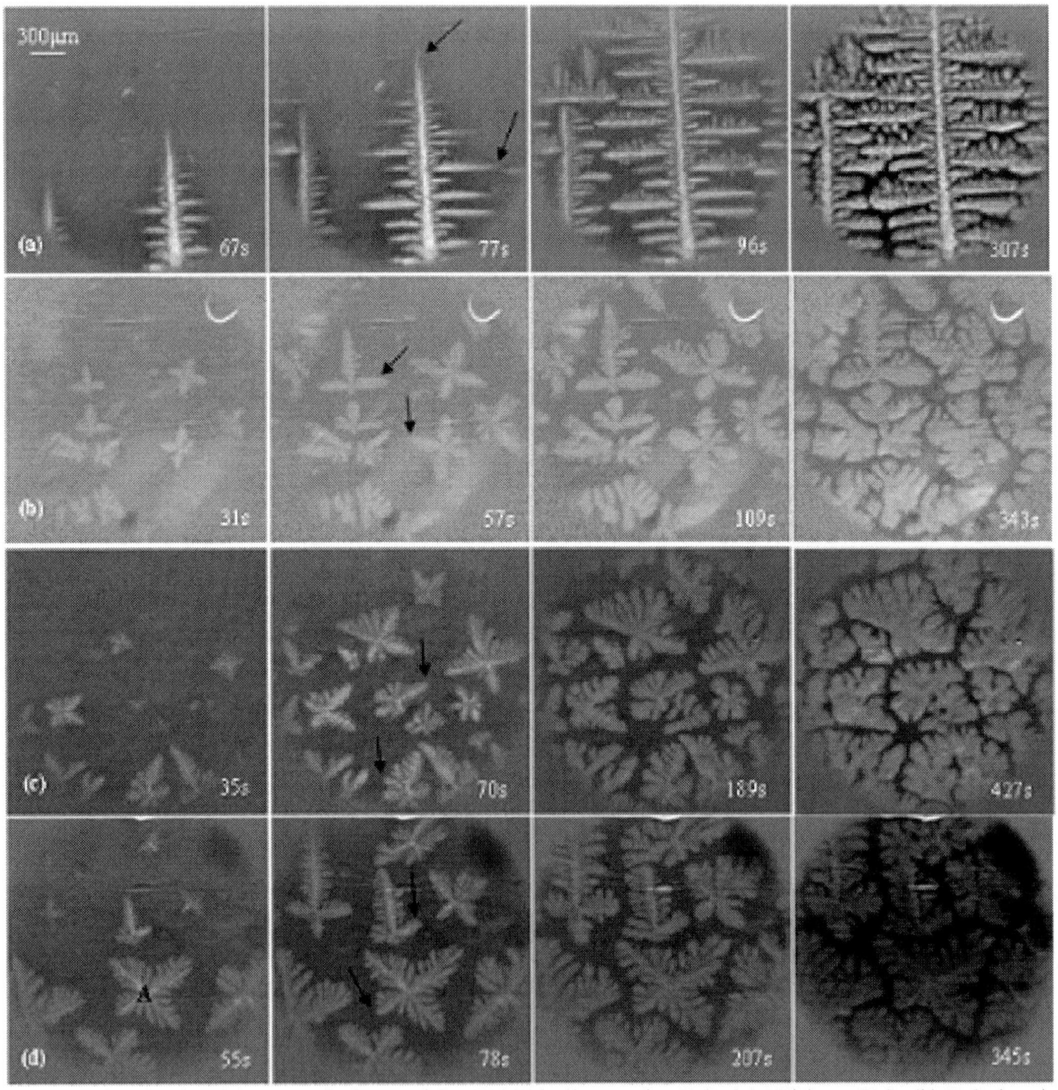

Figure 2. The dendrite morphology of Sn-12%Bi alloy under DC: (a) Without DC, (b) DC density of 7 A/cm^2, (c) DC density of 19 A/cm^2, (d) DC density of 32 A/cm^2 [15].

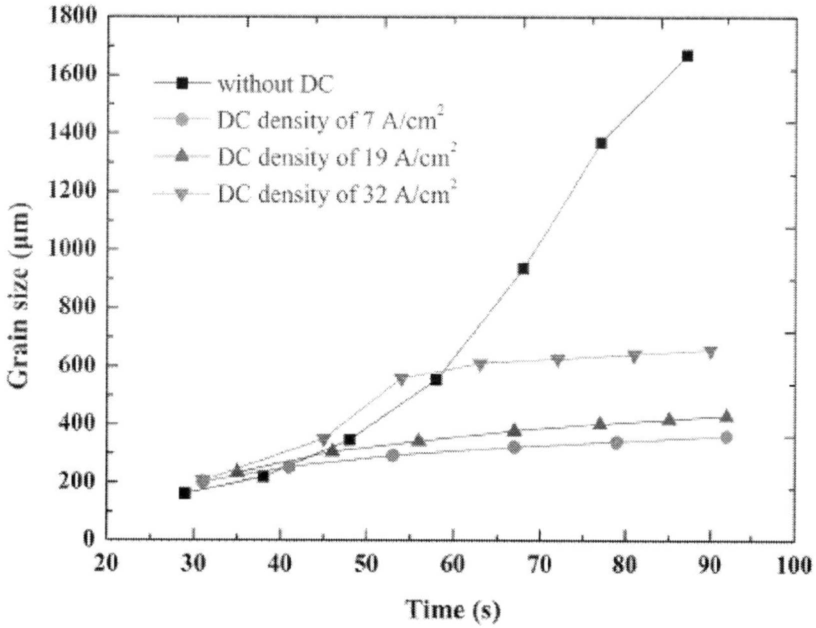

Figure 3. Variation in mean grain size with time under different DC density.

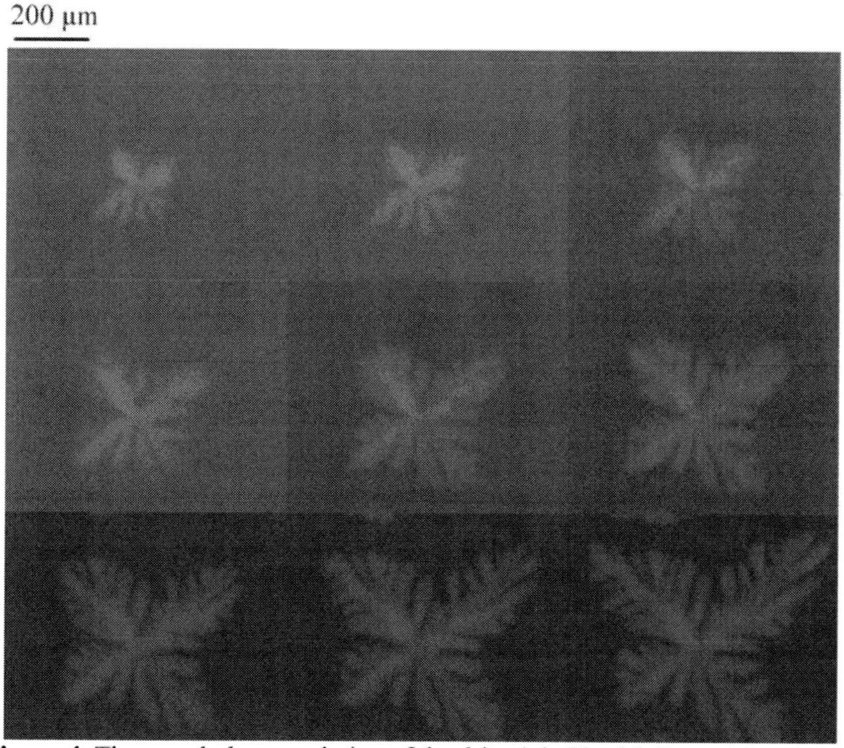

Figure 4. The morphology evolution of dendrite A in Fig. 2d (Record time is 10 s).

Fig. 2 shows the evolution of dendrite morphology with various electric current densities of 0, 7, 19, 32 A/cm^2, respectively. Without DC, the primary and secondary dendrite tips revealed a sharp shape, as shown in Fig. 2a. The columnar dendritic grains grew fast and finally reached a large size of

2 mm. With a DC density of 7 A/cm^2, however, the primary arms grew to be thick and the tips of the dendrites tended to be round or even flat, as shown in Fig. 2b. The growth of dendrites was significantly suppressed and the grains tended to be an equiaxed cellular morphology. With increasing DC density (19 A/cm^2, 32 A/cm^2), some primary arm tips tended to split and formed concave pits (Figs. 2c and 2d).

Fig. 3 shows the mean grain size together varying with time. The mean grain size was obtained by averaging all the grains within the images recorded in the experiment. Without DC, the curve presented ascending tendency during all the solidification stages which indicated the grain grew continually. The grain size finally reached a large size of around 1671 μm. The grain size turned a sudden decrement when DC was imposed. At the early stage of solidification, the grain was growing based on the variation tendency of curves. However, the curves approximated the straight line at the final stage of solidification which meant the grain almost stopped growing. It is due to the significant suppression of DC on dendrites growth. With increasing DC density from 7 A/cm^2 to 32 A/cm^2, the grain size was first decreased and then increased slightly. It might be a result of the tip-split of primary dendrite arm, or a fluctuation of solute, free energy and chemical potential, or something else due to DC field [15].

An explanation to the influence mechanism of DC on the evolution of dendrite morphology was given based on our *in situ* observations and the previous work by other researches [4,5]. Since the electrical conductivity of solid Sn-12 wt %Bi alloy is much bigger than that of liquid alloy, the electric streamlines prefer passing through solid to liquid. They easily gather in a sharp tip of dendrite to generate much more heat there due to both the Joule heating effect and the current crowding effect. The heat causes a rising temperature around the sharp dendrite tip by which the tip growth can be significantly suppressed. Accordingly, the sharp tip tends to be round or even flat in a self-adaptive way to scatter the electric streamlines for reducing Joule heat in order to keep growing [15]. At last, the dendrites present an equiaxed cellular morphology.

To better illustrate this phenomenon, Fig. 4 clearly shows the split process of the single grain A marked in Fig. 2d. When electric current density increased, the special tip-split scenario occurred to the primary arm resulting in a "branching" of the primary dendrite. With solidification proceeding, the solute is rejected gradually to the front of a growing dendrite tip during solidification. It is more difficult for the rejected solute to transversely diffuse from the tip center toward its lateral sides in case of a round or flat (higher DC density case) tip rather a sharp one. The pileup solute becomes an obstacle opposite to the tip growing by changing the constitutional undercooling, thus a concave pit is formed. The pit is expended with continuous growing of tip and the segregation of solute. Meanwhile, a tip is split into two tips. The two new tips will attract the electric streamlines again due to the current crowding effect. The Joule heat produced by electric current will slow down the growing rate of tips. Therefore, the evolution of tip split is actually determined by both the pileup solute and the Joule heat [15].

Two grains (Sn-12wt%Bi alloy) with different crystal orientations, a DC density of 19 A/cm^2 were designed using the ANASYS^{TM} software in order to analyze the imaging results above. The area in grain boundary is solid, and others are liquid, as shown in Fig. 5. The simulation results revealed that the electric streamlines favored to go through the solid grain area other than the liquid area. And the electric streamlines tended to concentrate in the dendrite tips. This may attribute to the different electrical conductivity between the solid and liquid areas. Solid area with high electrical conductivity was easy for the electric current to go through. More Joule heat induced by electrothermal effect would gather in the dendrite tips by which the grain tips were melted and the tip growth can be significantly suppressed. Accordingly, the sharp tip tended to be round or even flat in a self-adaptive way to scatter the electric streamlines for reducing Joule heat in order to keep growing.

In the solidification process, the liquid area i.e. the solid-liquid interface frontier, would be rich in Bi due to the solute redistribution effect (k_e<1). Without DC, the dendrite tips would tend to be sharp. The solute is rejected gradually to the front of a growing dendrite tip during solidification (diffusion area nearly to be 360°), a fast growth tip is thus obtained. With DC treatment, however, it is more difficult for the rejected solute to transversely diffuse from the tip center toward its lateral sides in case of a round or flat tip rather than a sharp one. The pileup solute becomes an obstacle opposite to the tip growing by changing the constitutional undercooling, thus a concave pit is formed. In summary, under DC treatment, the phenomenon of dendrite split tips changed both the grain size and morphology and the related mechanical properties of the metal can be finally influenced, better or worse. Thus, further research of this phenomenon is important for both the solidification theory and the industrial project.

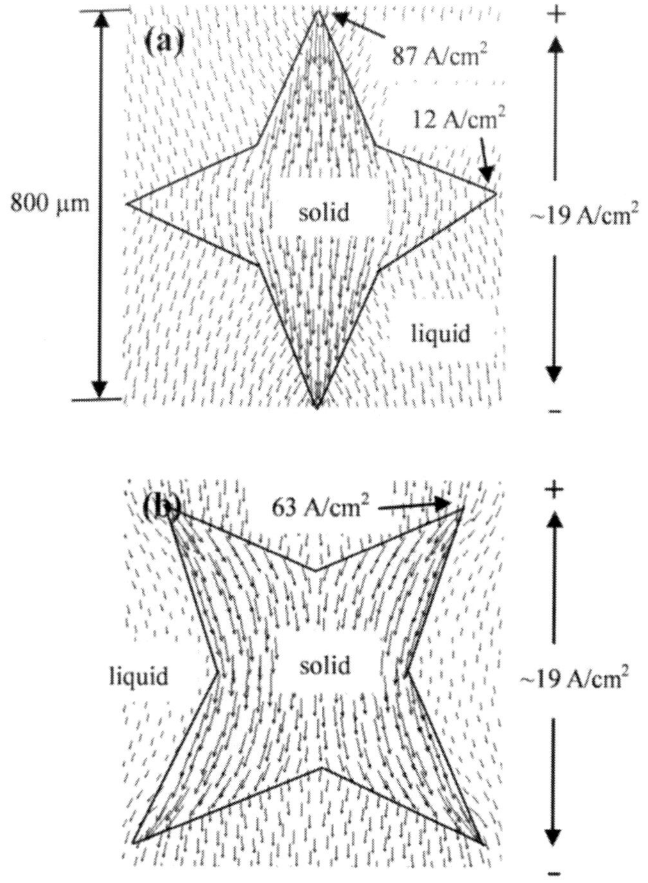

Figure 5. (a) Crystal orientation paralleling to electric current orientation, (b) Crystal orientation orienting 45 degrees to electric current orientation.

3.2. The evolution of dendrite morphology with electric current pulse

Fig. 6 shows the dendrites evolution of Sn-12wt%Bi alloy with and without ECP treatments. When ECP imposed on the solidifying alloy, the primary and secondary dendrites are obviously refined due to the electromagnetic vibration effect. The peak value of the ECP is several or even dozens times than the normal electric current. And the peaks are changed once within an electricity cycle. These would generate much more heat which can change the thermodynamics potential barrier and diffusion activation energy. At the same time, the pulse current and its induced pulse magnetic field interact with each other, and a significantly destabilizing effect is formed [16]. Hence, much more dissociated dendrites, i.e. much more nucleus, are provided for the nucleation substrates and a fine scale grain size is achieved.

It is evident that the growth rate of primary dendrite was significantly related to the ECP condition, by which the kinetics of dendrite growth was affected with notable changes. Table 1 is the dendrites growth data select measured in Fig. 6. Without ECP, the growth rate of the primary dendrite trunk was generally below 50 μm/s. In other case, when ECP was imposed, the growth rate exceeded 100 μm/s at the initial growth stage. The too-fast growth rate caused continuing release of latent heat and hence decreased the growth rate in the subsequent stage.

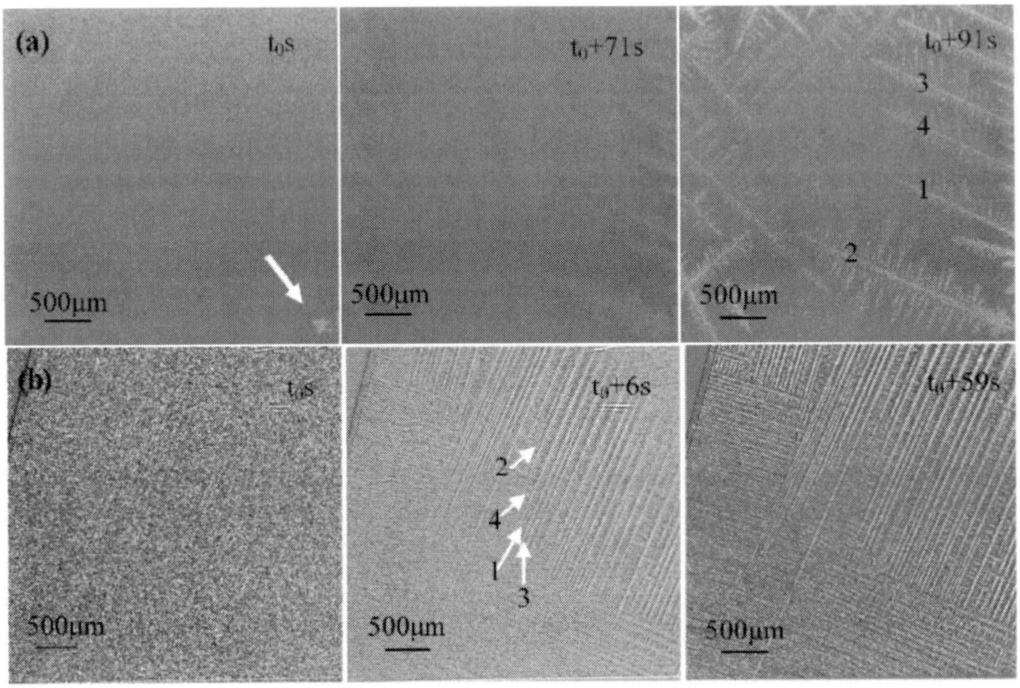

Figure 6. The dendrite morphology of Sn-12%Bi alloy under ECP: (a) Without ECP, (b) With ECP. The parameters used are [Voltage, Current, Reversing time]=[7 V, 2.7 A, 0.5s].

Table 1. The compare of the growth rate of primary dendrite tip without and with ECP.

	Dendrite 1 (μm/s)	Dendrite 2 (μm/s)	Dendrite 3 (μm/s)	Dendrite 4 (μm/s)
The dendrite growth rate without ECP	44.32	48.34	54.01	56.16
The dendrite growth rate with ECP	112.15	174.65	219.59	224.22

4. Conclusions

In situ imaging of dendrite morphology evolution under electric field has been successfully achieved at the 3rd generation SSRF of China, using in-line holography imaging method. The experimental results were analyzed and the conclusions as following:

(1) With DC, the growth of primary and secondary dendrites is significantly suppressed. This mainly attributes to electric current gathering in dendrite tip to generate Joule heat which makes the temperature of dendrite tip increasing. Thus the growth rate of the dendrites tip slows down.

(2) With DC, the dendrite tip tends to be flat or even round. With electric current density increasing, much more solute piles and can not promptly rejected as a result of the dendrite tip splits.

(3) With ECP, the primary and secondary dendrite arms are obviously refined and the solidification rate increases because of oscillating and destabilizing of pulse current.

Acknowledgements

The authors gratefully acknowledge the support of National Natural Science Foundation of China(Nos.50601003,50971032,51071035),the Fundamental Research Funds for the Central Universit -ies, and Liaoning BaiQianWan Talents Program. The authors wish to thank all the staff members of the BL13W1 beam line of SSRF.

References:

[1] Fan J H, Zhai Q J. J 2004 *Mech. Eng.* **40** 10
[2] Asai S 2000 *Sci. Technol. Adv. Mat.* **1** 191
[3] Misra A K 1985 *Met. Trans.* **16A** 1354
[4] Chang G W, Yuan J P, Wang Z D, Wu C J, Wang X H, Hu H Q 2000 *Trans. Nonferrous Met. Soc. China* **10** 611
[5] Coriell S R, McFadden G B, Wheeler A A, Hurle D T J 1989 *J. Cryst. Growth* **94** 334
[6] Nakada M, Shiohara Y, Flemings M C 1990 *ISIJ int.* **30** 27
[7] Barnak J P, Sprecher A F, Conrad H 1995 *Scripta Metall. Mater.* **32** 879
[8] Liao X L, Zhai Q J, Luo J, Chen W J, Gong Y Y 2007 *Acta Mater.* **55** 3103
[9] Mathiesen R H, Arnberg L, Mo F, Weitkamp T, Snigirev A 1999 *Phys. Rev. Lett.* **83** 5062
[10] Ruvalcaba D, Mathiesen R H, Eskin D G, Arnberg L, Katgerman L 2007 *Acta Mater.* **55** 4287
[11] Yasuda H, Ohnaka I, Kawasaki K, Sugiyama A, Ohmichia T, Iwane J, Umetan K 2004 *J. Cryst. Growth* **262** 645
[12] Li B, Brody H D, Kazimirov A 2004 *Phys Rev E* **70** 062602
[13] Wang T M, Xu J J, Huang W X, Li J, Cai S W, Li T J, Jin J Z 2009 *Modeling of Casting, Welding, and Advanced Solidification Processes – XII, Edited by Cockcroft S L, Maijer D M, TMS* 613
[14] Wang T M, Xu J J, Li J, Huang W X, Liu S C, Li T J 2010 *Sci. China E* **53** 1278
[15] WANG T M, XU J J, XIAO T Q, XIE H L, LI J, LI T J, CAO Z Q 2010 *Phys. Rev. E* **81** 042601
[16] Fang W 2005 Lanzhou University of Technology

IOP Conf. Series: Materials Science and Engineering **33** (2012) 012040 doi:10.1088/1757-899X/33/1/012040

Observation of the initiation and propagation of solidification cracks by means of *in situ* synchrotron X-ray radiography

J.W. Aveson[1], G. Reinhart[2,3], B. Billia[2,3], H. Nguyen-Thi[2,3], N. Mangelinck-Noël[2,3], T.A. Lafford[4], C.A. Vie[1], J. Baruchel[4], H.J. Stone[1]

[1]Department of Materials Science and Metallurgy, University of Cambridge, Pembroke Street, Cambridge, CB2 3QZ

[2]Aix Marseille Univ & [3]CNRS, IM2NP UMR 6242 Campus Saint-Jerome, Case 142 13397 Marseille Cedex 20, France

[4]European Synchrotron Radiation Facility, 6 Rue Jules Horowitz, 38043 Grenoble cedex 9, France

E-mail: `johnaveson@gmail.com`

Abstract. We report on the *in situ*, time resolved, observation of solidification cracking in a thin sample of an Al-15wt%Sn alloy at ESRF BM05. During the experiment, solidification cracking was seen to occur during natural cooling of the sample at a solid fraction of ~95%, between directionally agglomerated dendritic networks. Through detailed analysis, three stages of crack growth were observed: crack initiation, which typically occurs asymmetrically with the liquid film separating from one side of the dendritic network prior to full detachment; crack propagation, where the liquid film generally detaches from both sides simultaneously; and crack coalescence. We correlate our observations using scanning electron microscopy, which shows that voids between grains, and also spikes, can appear asymmetrically on either side of the crack surface.

1. Introduction

Solidification cracking, or hot tearing, is a significant problem in melt processing, for instance, in direct chill castings [1], single crystal and directionally solidified superalloy castings [2], and welding [3]. It occurs at intergranular and interdendritic boundaries during the final stage of solidification. At this stage, the remaining liquid is increasing in solute concentration and decreasing in volume fraction. In general, it is accepted that hot tearing occurs due to poor liquid feeding to compensate for the solidification shrinkage and thermal deformation. Just as with pores, hot tears can form sub-surface defects in cast products, and so are of extreme concern since they act as stress concentrators and hence are potential initiation sites for fatigue cracking.

In recent years, great advances have been made in the modelling and simulation of solidification cracking, mainly through analyses based upon the principles of granular physics. The Gourlay-Dahle experiment showed dilatancy in a mechanically deformed equiaxed mush, exemplifying the existence and importance of granular behaviour in solidification [4]. This was subsequently reinforced through *in situ* radiography experiments [5]. Vernede, Rappaz and co-workers went on to use the methodologies of granular simulations in discrete element simulations of hot tearing [6–9]. These studies show that hot tearing is a highly complex phenomenon, which involves heat flow, fluid flow under capillary action and various other factors including alloy composition and process parameters.

The potential driving forces for hot tearing are well-established: solid contraction in a thermal gradient; solidification shrinkage; and differential contraction between the mould and the solid.

Many authors have examined the hot tear susceptibility in terms of the stress and strain rate sensitivity, allowing for coarse-graining into finite element models. As a result, there are many hot tear criteria and these have been comprehensively reviewed by Eskin *et al.* [1, 10]. Guven and Hunt pointed out a distinction between *interdendritic* and *intergranular* hot tearing: interdendritic hot tearing can be induced when shrinkage driven flow is impeded; furthermore, the irregular structure and increased solute content at grain boundaries leads to increased propensity to intergranular hot tearing [11]. Thus, it is pertinent to perform *in situ* investigations of the microstructural effects on crack initiation and propagation at the dendritic length scale in order to complement the viewpoint already obtained by several authors.

Many of the recent advances in the understanding of hot tearing have been achieved by the application of synchrotron X-ray imaging. Firstly, Phillion *et al.*, performed three dimensional imaging experiments on the aluminium alloy AA5182 in order to characterise the morphology of hot tear damage [12]. More recently *in situ* ultrafast X-ray tomography and radiography experiments have shown the highly localised nature of hot tearing and cracking in isothermal tensile deformation of semi-solid Al-Cu alloys [13, 14].

Using *in situ* radiography, we have been able to observe cracking during natural cooling following the directional solidification of an Al-Sn alloy. We identify three stages in crack growth which we have termed crack initiation, crack propagation along liquid channels separating dendritic grains, and crack coalescence. In this paper, we present these stages with both evidence from radiography and supporting scanning electron microscopy.

2. Experimental methods

The Al-Sn binary system is well-suited to the study of hot tearing owing to the limited solubilities of the end members of the phase diagram and a broad freezing range, increasing the time that the alloy spends in the brittle temperature range (BTR) during which the solid fraction f_s, is between 0.85 and 0.95 [16]. An alloy of composition Al - 15 wt% Sn was prepared by melting in a sealed silica tube before being homogenised at 180°C for 72 hours. Specimens with dimension $37 \times 6 \times 0.2$ mm^3 were produced and polished in order to have a uniform thickness and to reduce spurious nucleation on scratches and other defects. Laboratory and synchrotron radiographs of the as-prepared specimens showed no evidence of porosity, hot tearing or macro segregation.

The *in situ* radiography experiment was performed at the BM05 station of the European Synchrotron Radiation Facility. The apparatus, beamline configuration and experimental protocol are well established and detailed in various publications [17, 18]. A bespoke two-zone Bridgman furnace was used for the experiments with thermal control offered by regulating the power to the heaters and an initial temperature gradient of approximately 40 K cm^{-1}. During the first stage of the experiment, cooling was achieved by reducing the temperature at the top of the specimen at a rate of 0.75 K min^{-1} whilst maintaining a constant temperature at the base of the specimen.

After the solidification was essentially complete, the power to both heaters was switched off and the sample allowed to cool naturally. During this second phase, the temperature gradient along the sample rapidly decreased towards zero as seen in Figure 2.

During solidification, radiographic images were recorded with a Fast Readout Low Noise (FReLoN) camera at a frequency of 1Hz and an exposure time of 0.3s. The incoming synchrotron

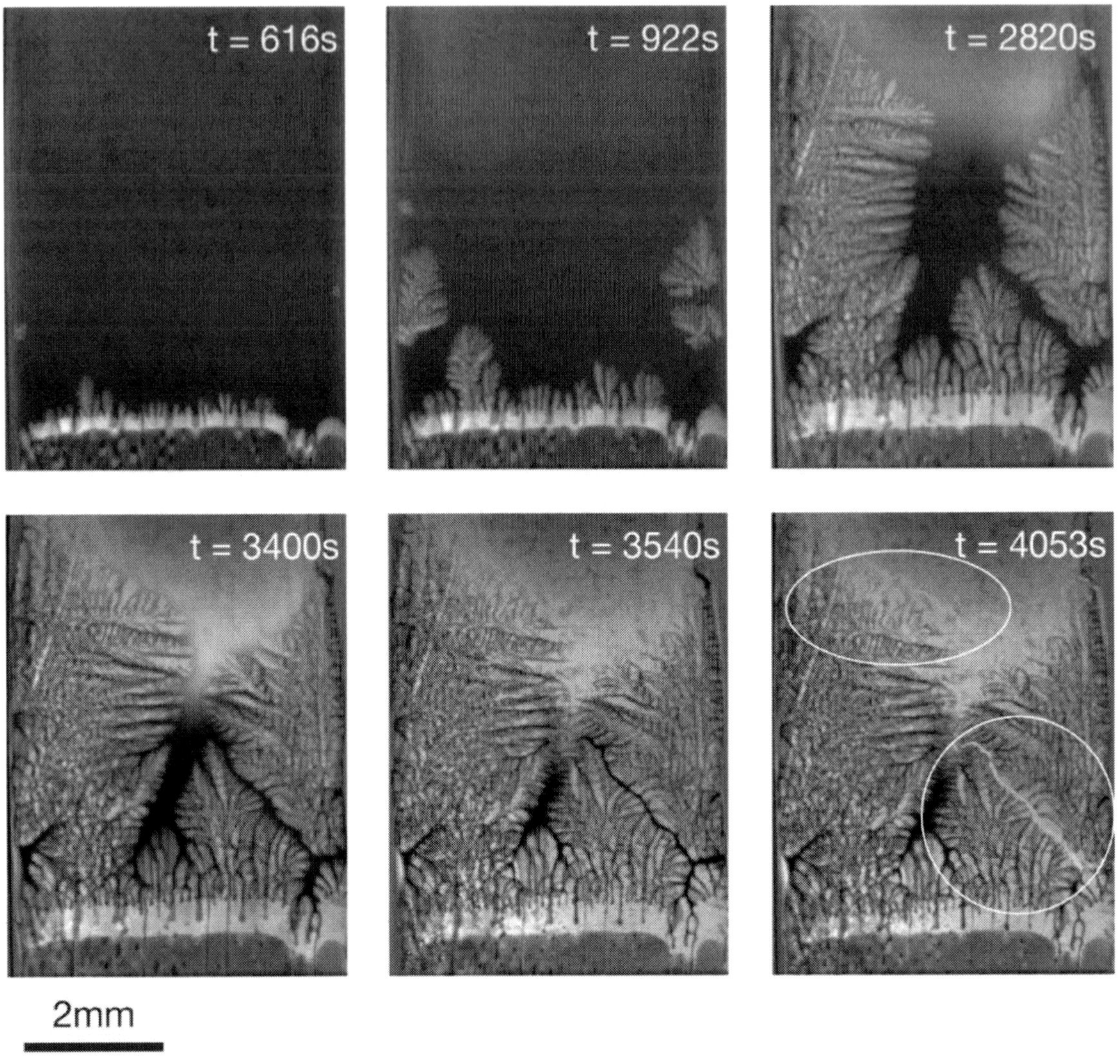

Figure 1. General overview of the solidification experiment. A mixed columnar-equiaxed microstructure is observed to emerge with hot tearing occurring at the later stages of solidification. The times shown are those elapsed from the moment at which cooling was initiated. Two cracks are encircled, and the crack under consideration is identified in the white circle at the bottom right of the last frame ($t = 4053$ s).

radiation was tuned to a photon energy of 29.2 keV using a Si (111) two crystal monochromator. By selecting an energy just above the absorption edge in the Sn spectrum, stronger contrast between solid and liquid can be obtained. The as-recorded images were processed using ImageJ [19]. All images were divided by a reference image, which showed a homogeneous liquid with a solid seed at the bottom of the field of view at the start of the experiment. This image processing eliminated spurious features not associated with the solidification [18]. In addition, image subtraction was employed with respect to a reference image to highlight microstructural changes between acquisitions. In each figure shown in the later sections, the first recorded image has been used as a subtraction reference. After the ESRF experiment, examination of

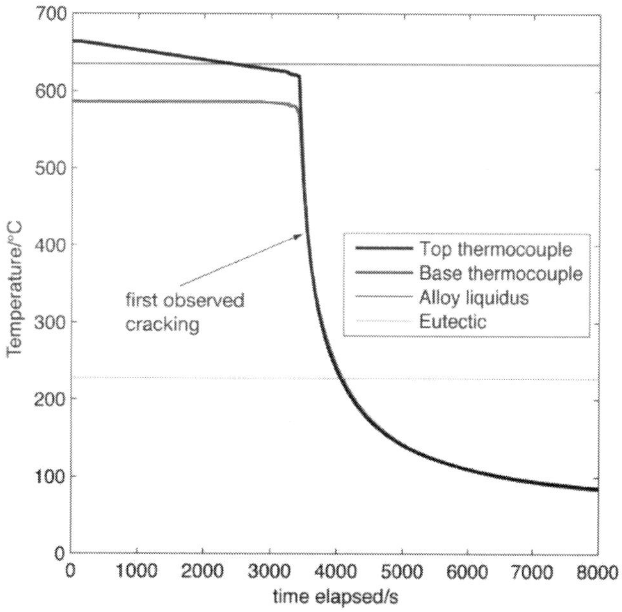

Figure 2. Thermometry data from the experiment is provided for reference with Figure 1: growth occurred predominately during controlled cooling and the hot cracking initiated midway through natural cooling.

the cracked specimen was performed using a JEOL 5500 LV SEM at 10kV.

3. Results

Figures 1 and 2 shows the whole field of view and the temperatures recorded by the two thermocouples during the experiment. In the first four images, a thermal gradient exists across the specimen with the temperature at the bottom of the field of view lower than that at the top. At $t = 616$ s, cellular and dendritic features were observed to grow out into the liquid on an otherwise smooth solid liquid interface. Two thirds of the way down the image, equiaxed grains can be seen to have nucleated and grown out to form the directionally agglomerated structure displayed at $t = 3400$ s. The top of the image appears lighter in the last image owing to shrinkage driven flow down into the specimen. This leads to the formation of a meniscus, effectively reducing the thickness which must be traversed by the beam and increasing the transmitted intensity.

Intergranular cracks were observed to form either at the boundaries between different columnar grains (bottom circle in Figure 1) or at the junction of columnar and equiaxed regions (upper circle in Figure 1). Cracks can be identified by a transition from black (liquid channel) to lightest grey (crack), encircled in the image at $t = 4053$ s. The equilibrium phase diagram suggests that cracking initiates in this case at a solid fraction of $f_s = 0.955$, which is comparable with the results of Guven and Hunt [11]. The crack observed at the lower right hand corner of the field of view is examined in more detail in the later sections. Figures 3 - 5 show sets of radiographs in which we identify three crack stages: crack initiation, crack propagation and crack coalescence.

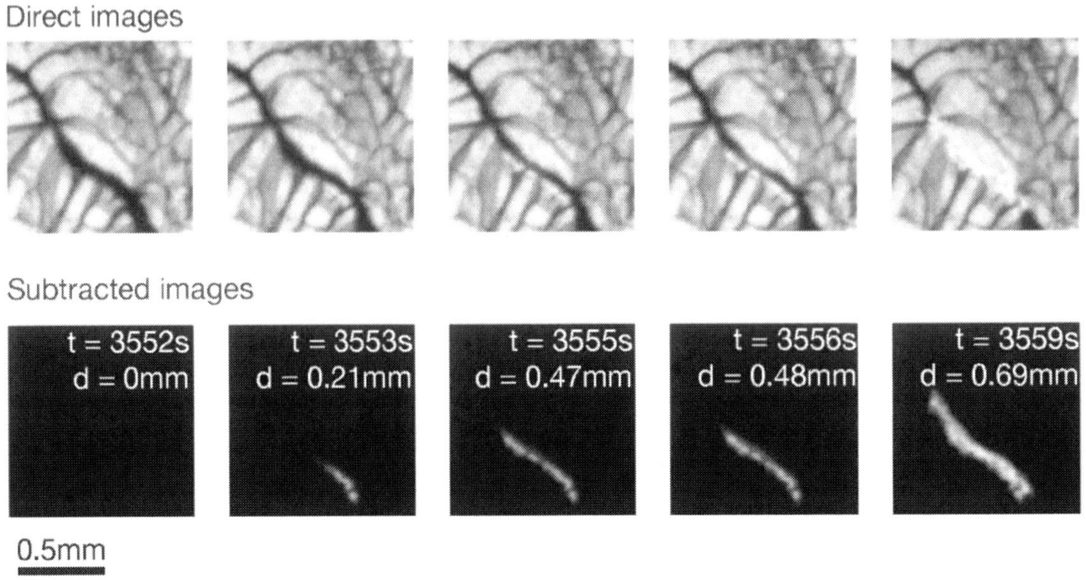

Figure 3. Asymmetric initiation of a hot tear. A localised detachment of the melt is seen at the tips of a set of dendrites of the bottom grain at $t = 3553$ s. The localised detachment extends along the lower dendritic networks before opening fully at $t = 3559$ s. Crack lengths were calculated using a simple algorithm and show that growth of the crack occurs in a stepwise manner.

3.1. Crack initiation

Figure 3 exemplifies the asymmetric detachment mechanism observed in the initiation of the intergranular crack. The first direct image at $t = 3552$ s shows the dendritic microstructure just prior to any observed cracking. In the bottom centre of the image, a set of dendrites is observed. Whilst it is visible in the direct images, the subtracted image highlights a bright white feature. This can be readily identified as the thinning and detachment of the liquid film at the dendrite tips. This detachment is due to the faster downward motion of the lower grain that results in the stretching of the liquid channel. The subsequent images clearly show the crack extending along the lower solid-liquid interface, along the dendrite tips. Careful comparison of the direct and subtracted images shows that the detachment has first occurred at the dendrite tips on the lower side of the liquid channel and extends along the channel lengthways, before widening at $t = 3559$s. Once the crack has established on one side, capillary forces extrude the remaining liquid, resulting in the full detachment of the liquid film from the opposing solid liquid interface and the oxide layer acting as a mould.

3.2. Crack propagation

Figure 4 shows the symmetric propagation of a crack towards a triple junction. It is clear from the subtracted images that there is no asymmetry in the propagation of the crack as it approaches the triple junction. Instead, the void approaches the triple junction with both solid-liquid interfaces detaching almost simultaneously. However, when it reaches the large dendrite tip at the bottom right hand corner of the image, the crack bifurcates into the liquid channel on the right with liquid first detaching asymmetrically from the large dendrite opposing its straight propagation.

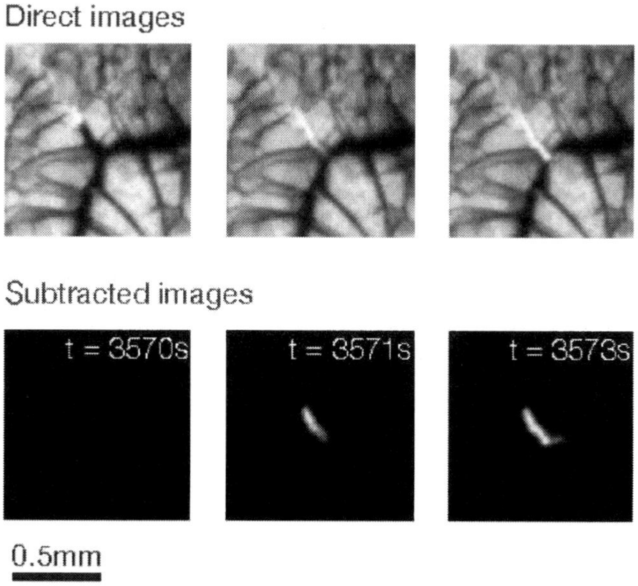

Figure 4. Symmetric crack propagation along a liquid channel down to a triple junction, where the crack propagation bifurcates asymmetrically into the channel on the right when it encounters the lower dendrite network.

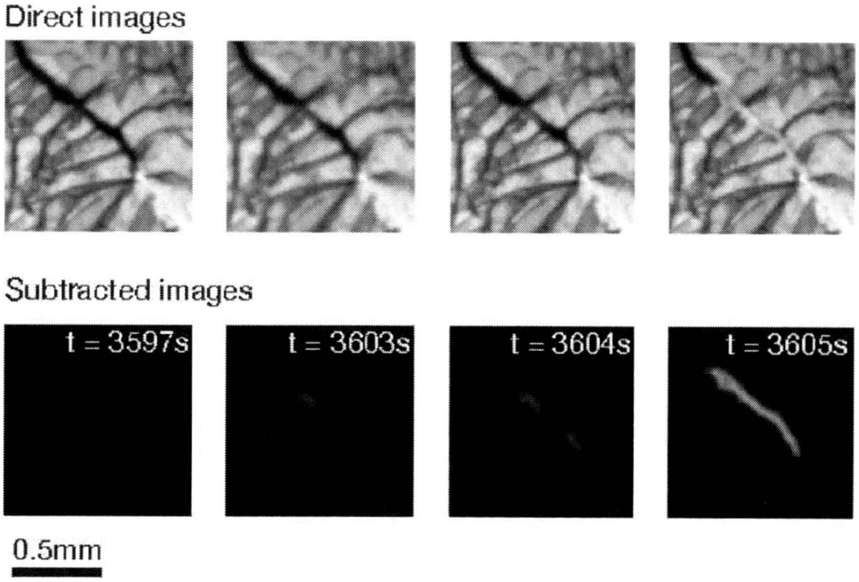

Figure 5. Initiation of a new crack and its coalescence with a pre-existing crack

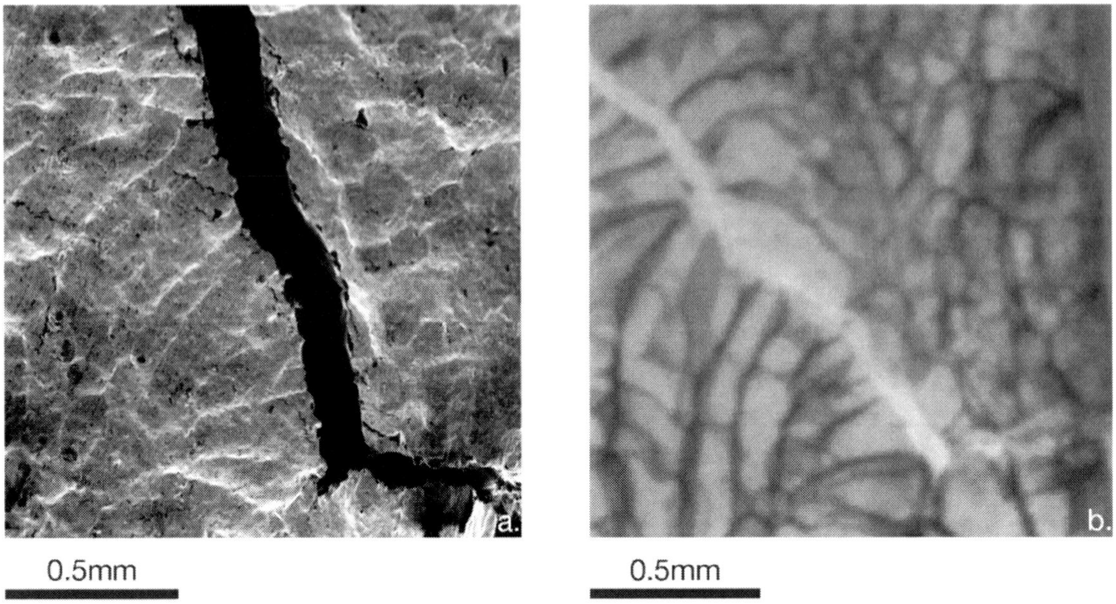

Figure 6. Scanning electron micrographs of the crack viewed in the radiography experiment. Stud-like features are shown on the side from which the liquid film detached. A detail of a radiograph from the same region: the radiograph is that shown in Figure 1 at $t = 4053$ s.

3.3. Crack coalescence

Figure 5 shows the coalescence of two cracks. In the direct image at $t = 3597$ s, a pre-existing crack is seen at the bottom right hand corner of the image. In the upper middle of the images, a second thinning event has occurred ahead of the main crack. Then, the main crack extends towards the newly nucleated detachment in a symmetric manner. The two cracks subsequently join in a time interval that is not clearly resolvable at the present image acquisition rate.

4. Discussion

In the previous section, we presented the three observed stages of hot tearing, namely crack initiation, crack propagation and crack coalescence. During the experiment, crack initiation and propagation was found to be mostly asymmetric in nature, suggesting this is the kinetically favoured process for hot tearing to proceed in this situation. This is plausible, as beyond relaxing mechanical stress induced by the moving apart of the dendritic grains on the two sides of the liquid channel, which may itself be asymmetric, asymmetric detachment also provides a reduced barrier to crack initiation. Indeed, the detachment turns a solid-liquid interface into a pair of solid surfaces which comes with an energetic penalty. Thus, since only one detachment event is required for asymmetric propagation, it provides an energetically favourable route for initial detachment. The increased capillary forces owing to the increased liquid-void surface area of the remaining liquid then provide a driving force for liquid extraction.

Previous scanning electron microscopy on metallic alloys have shown that 'spikes' form on hot tear surfaces (see e.g. Figures 7-9 in by Farup *et al.* [20]). Figure 6 shows similar features. Unlike the previous study, the features appear on one side only of the crack, corresponding to the side on which detachment occurred first.

5. Conclusions

In situ, time resolved radiography has been used to identify features in the initiation and propagation of inter granular solidification cracks, namely the occurrence of three distinct stages in crack growth: initiation, propagation into liquid channels separating dendritic grains and crack coalescence. This leads to interesting microstructural features, with similarities and contrasts to be drawn with prior studies, especially regarding asymmetric features such as bridging spikes.

Further efforts will be necessary to examine the applicability of these results. In particular, the reduced thickness of the system may have an effect on its hot tearing behaviour. Furthermore, the issue of the dynamical interaction between melt and the soft oxide layer that encases the solidifying sample should deserve specific attention. This is, of course, a recurrent, open question in aluminium alloy casting [21,22].

The asymmetric mode of propagation described deserves further investigation, both through physical modelling and further experiment. Indeed, this may have important implications on crack initiation and propagation. Fuller quantitative analysis of this data is already underway to facilitate further interpretation.

Acknowledgements

The authors are grateful for provision of beam time by the European Synchrotron Radiation Facility. X. Guichard and T. Schenk are acknowledged for their assistance in conducting the experiments. JWA acknowledges a studentship under the EPSRC/Rolls-Royce plc. Strategic Partnership.

References

[1] Eskin D and Katgerman L 2007 *Metallurgical and Materials Transactions A* **38** 1511
[2] Zhou Y, Volek A and Singer R 2006 *Journal of Materials Research* **21** 2361
[3] Easterling K 1992 *Introduction to the physical metallurgy of welding* 2nd ed (Butterworth Heinemann)
[4] Gourlay C and Dahle A 2007 *Nature* **445** 70–73
[5] Gourlay C, Dahle A, Nagira T, Nakatsuka N, Nogita K, Uesugi K and Yasuda H 2011 *Acta Materialia* **59** 4933
[6] Vernède S, Jarry P and Rappaz M 2006 *Acta Materialia* **54** 4023–4034
[7] Vernède S and Rappaz M 2007 *Acta Materialia* **55** 1703–1710
[8] Vernède S, Dantzig J and Rappaz M 2009 *Acta Materialia* **57** 1554–1569
[9] Phillion A, Vernède S, Rappaz M, Cockcroft S and Lee P 2009 *International Journal of Cast Metals Research* **22** 240–243
[10] Eskin D, Suyitno and Katgerman L 2004 *Progress in Materials Science* **49** 629–711
[11] Guven Y and Hunt J 1988 *Cast Metals* **1** 104
[12] Phillion A, Cockcroft S and Lee P 2008 *Materials Science and Engineering A* **491** 237
[13] Terzi S, Salvo L, Suéry M, Limodin N, Adrien J, Maire E, Pannier Y, Bornert M, Bernard D, Felberbaum M, Rappaz M and Boller E 2009 *Scripta Materialia* **61** 449–452
[14] Phillion A, Hamilton R, Fuloria D, Leung A, Rockett P, Connolley T and Lee P 2011 *Acta Materialia* **59** 1436–1444
[15] Vernède S 2007 *A granular model of solidification as applied to hot tearing* Ph.D. thesis Ecole Polytechnique Federale de Lausanne
[16] Dantzig J and Rappaz M 2009 *Solidification* (EPFL Press)
[17] Nguyen-Thi H, Jamgotchian H, Gastaldi J, Härtwig J, Schenk T, Klein H, Billia B, Baruchel J and Dabo Y 2003 *Journal of Physics D: Applied Physics* **36** A83–A86
[18] Buffet A, Reinhart G, Schenk T, Nguyen-Thi H, Gastaldi J, Mangelinck-Noel N, Jung H, Hartwig J, Baruchel J and Billia B 2007 *Physica Status Solidi* **204** 2721–2727
[19] Abramoff M, Magalhaes P and Ram S 2004 *Biophotonics International* **11** 36
[20] Farup I, Drezet J M and Rappaz M 2001 *Acta Materialia* **49** 1261–1269
[21] Jolly M 2005 *JOM* **57** 19–28
[22] Miresmaeili S, Campbell J, Shabestari S and Boutorabi S 2005 *Metallurgical and Materials Transactions A* **36** 2341

IOP Conf. Series: Materials Science and Engineering **33** (2012) 012041 doi:10.1088/1757-899X/33/1/012041

In-Situ Observation of Horizontal Centrifugal Casting using a High-Speed Camera

Hisao Esaka[1,3], **Kohsuke Kawai**[2,3], **Hiroshi Kaneko**[2,3] **and Kei Shinozuka**[1,3]
[1] Department of Materials Science and Engineering, National Defense Academy
[2] Undergraduate student, National Defense Academy
[3] 1-10-20 Hashirimizu, Yokosuka 239-8686 Japan

E-mail: hesaka@nda.ac.jp

Abstract. In order to understand the solidification process of horizontal centrifugal casting, experimental equipment for in-situ observation using transparent organic substance has been constructed. Succinonitrile-1 mass% water alloy was filled in the round glass cell and the glass cell was completely sealed. To observe the movement of equiaxed grains more clearly and to understand the effect of movement of free surface, a high-speed camera has been installed on the equipment. The most advantageous point of this equipment is that the camera rotates with mold, so that one can observe the same location of the glass cell. Because the recording rate could be increased up to 250 frames per second, the quality of movie was dramatically modified and this made easier and more precise to pursue the certain equiaxed grain. The amplitude of oscillation of equiaxed grain (= A_t) decreased as the solidification proceeded.

1. Introduction

Horizontal centrifugal casting is widely used for production of pipes, rolls and so on. Advantage of this process is to press molten metal against the permanent mold wall due to centrifugal force. Owing to this force, fine solidified structure and sound castings without porosity and/or non-metallic inclusion can be obtained [1]. On the other hand, there are sometimes problems of macroscopic segregation. The countermeasure to the segregation problems has not been established because the formation mechanism of macroscopic segregation is still unclear. Further, solidification process is not well understood [2-4] because the centrifugal casting itself is essentially dynamic process.

Thus, an experimental technique of in-situ observation using succinonitrile(SCN)-water alloy system for horizontal centrifugal casting process using an organic substance has been developed by the present authors [5, 6]. Here, an important result was unveiled, such as the oscillation of equiaxed grains synchronizing to rotation of mold. However, a video camera used in the previous study was normal and it could record only 30 frames per second. Therefore, the oscillation of equiaxed grains was not clearly observed.

Therefore, in this study, in order to understand the movement of equiaxed grains in horizontal centrifugal casting an in-situ observation has been made using a high speed camera.

2. Experimental Procedure

In order to make an in-situ observation of horizontal centrifugal casting, basically the same equipment as the previous study was used [5,6]. A schematic view of experimental apparatus is shown in Figure 1. A glass cell, which contained succinonitrile-water alloy, was vertically set along the rotating axis. The glass cell, 100 mm in outer diameter and 1.3 mm gap, was constructed from two pieces of round glass and some small pieces of glass with epoxy resin. A high-speed camera was equipped on the rotating shaft as well as a normal camera. The weight balance was completely changed due to equipping a heavy camera. Thus, the apparatus has been reconstructed. The high-speed camera can record 2000 frames per second at maximum. There is a memory device in the

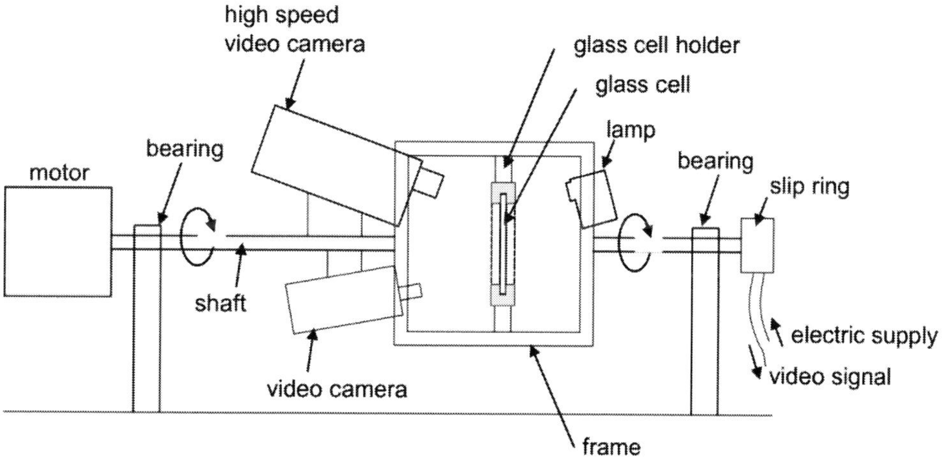

Figure 1. A schematic drawing of experimental apparatus for horizontal centrifugal casting.

high-speed camera and its capacity is approximately 10 seconds at the recording rate of 250 fps. Thus, a normal camera was necessary for monitoring the solidification of SCN alloy. In addition to this, a trigger system was newly equipped to record the solidification process with the high-speed camera. The shape and dimension of the glass cell, which contained succinonitrile-water alloy was the same as before.

Using this system, one could observe the relative movement of equiaxed grain with respect to the mold wall. The observed area was approximately 40 mm x 50 mm. Red and green laser beams were used to indicate the recording position during rotation.

Basically the solidification proceeded inwards from the outer surface, since a positive temperature gradient was imposed. In this study, rotation rate ranged from 250 to 500 rpm, which corresponded to 3.5 to 13 G at the outer surface of the glass cell. The movement of equiaxed grains was analyzed by the software, which can pursue a certain equiaxed grain.

3. Experimental Results and Discussions

The outline of solidification process of model experiment for horizontal centrifugal casting using

SCN-water alloy system is followings. This description is basically the same to the experimental conditions performed in this research.

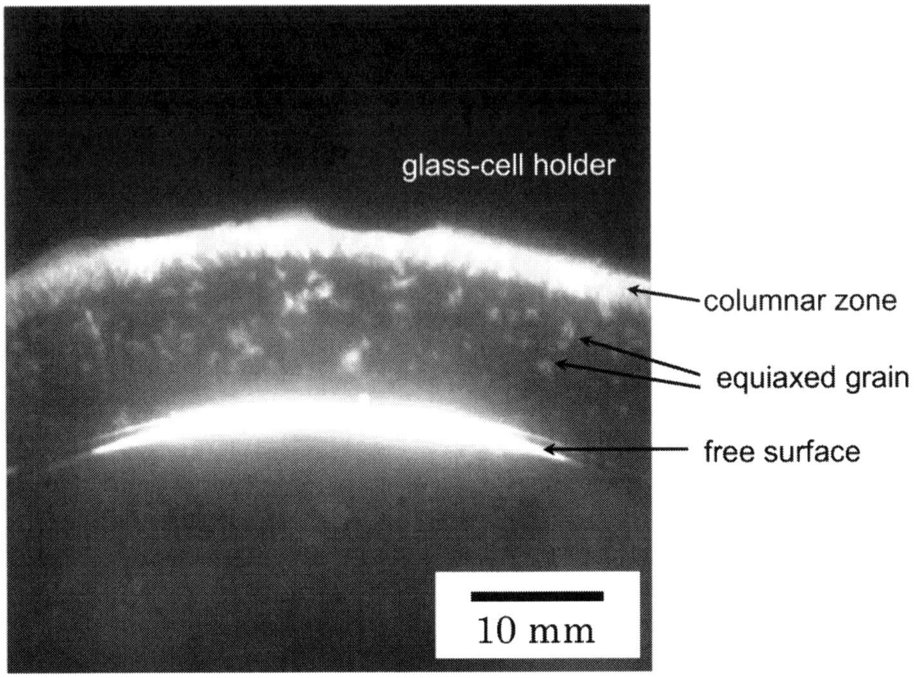

Figure 2. A snap-shot of in-situ observation for horizontal centrifugal casting at a rotation rate of 250 rpm.

At the initial stage of process, liquid phase was forced to be pressed to the outer region of glass cell. The columnar dendrites started to grow from the outer region of the glass cell. After a while, many grains (= equiaxed grains) appeared in the liquid phase. Unfortunately it is not possible to investigate the formation mechanism of equiaxed grains, such as fragmentation, because of low magnification. Some of equiaxed grains looked like typical dendrite with well-developed dendrite arms. These equiaxed grains moved right to left periodically and regularly. As solidification proceeded, the amplitude of movement of equiaxed grains gradually decreased. The amplitude of equiaxed grain increased as the position of equiaxed grain is getting closer to the free surface. When the solidification process completed, the equiaxed grains fully stopped. As far as observation performed, it could not be recognized the movement of equiaxed grain was different according to the size and morphology of equiaxed grains. However, when the size of equiaxed grains exceeded about 1 mm, the smooth movement was impeded because equiaxed grains contacted to the glass plate.

Figure 2 shows the snap-shot of in-situ observation at the rotation rate of 250 rpm. Equiaxed grains as well as columnar grain are observed in this figure. Because of the sufficient light of the lamp used, the shutter speed was 1/10000 – 1/20000 second. Therefore, equiaxed grain, which rapidly oscillated, could be clearly observed. During this observation, columnar grain did not move in the movie, on the other hand, equiaxed grains oscillated mainly left and right in this figure.

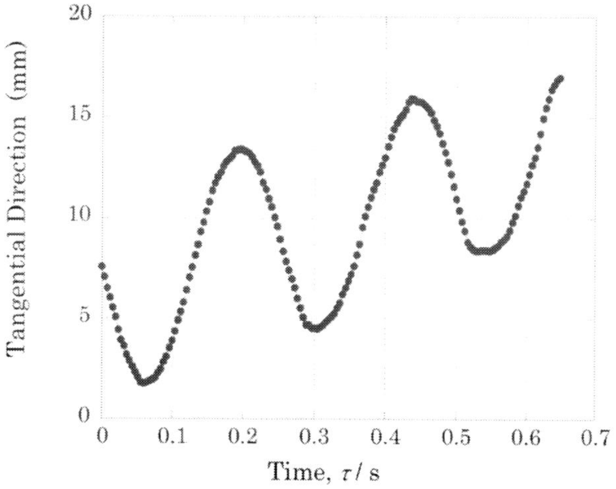

Figure 3. Movement of a certain equiaxed grain in tangential direction at a rotation rate of 250 rpm. Recording rate: 250 fps.

The movement of a certain equiaxed grain was pursued with software during about 1 second. The result is shown in Figure 3. In this case, the movement in x-direction, namely tangential direction is indicated. The previous data of analysis of movement of equiaxed grain [6] is shown in Figure 4 for comparison. This was obtained from the movie, which was recorded by a normal video camera. Since

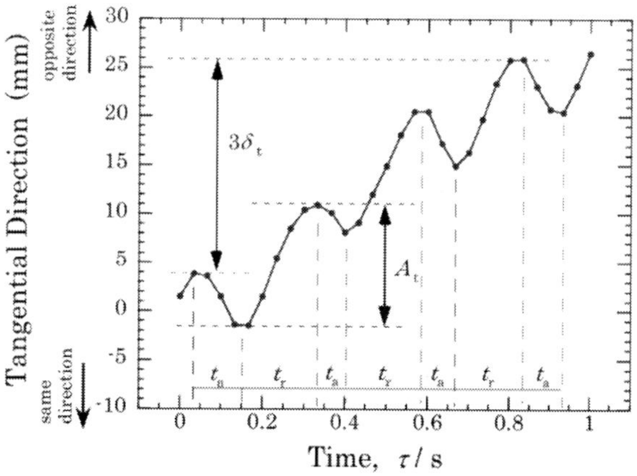

Figure 4. Movement of a certain equiaxed grain in tangential direction at a rotation rate of 250 rpm. Recording rate: 30 fps.

the detailed experimental conditions such as amount of molten alloy in the glass cell and timing for data acquisition were different, the experimental results are slightly different. However, comparing Figure 3 with Figure 4, the movement of equiaxed grain is smooth and can be precisely analyzed because of high recording rate. In case of normal video camera, recording rate is 30 frames per second. This means that the difference in angle per frame is 50° at the rotation rate of 250 rpm. On the other hand, when the recording rate is 250 fps the difference in angle per frame is only 6°.

Therefore, we can measure the movement of equiaxed grain smoothly and precisely. It could be also concluded that recording rate of 250 fps was enough for analyzing the movement of equiaxed grains at a rotation rate of up to 500 rpm.

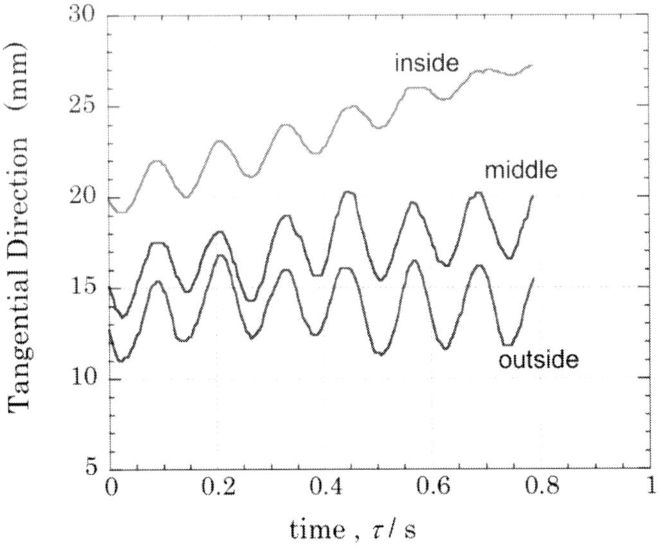

Figure 5. Movement of equiaxed grains as a function of position at the initial stage of solidification. Rotation rate: 500 rpm.

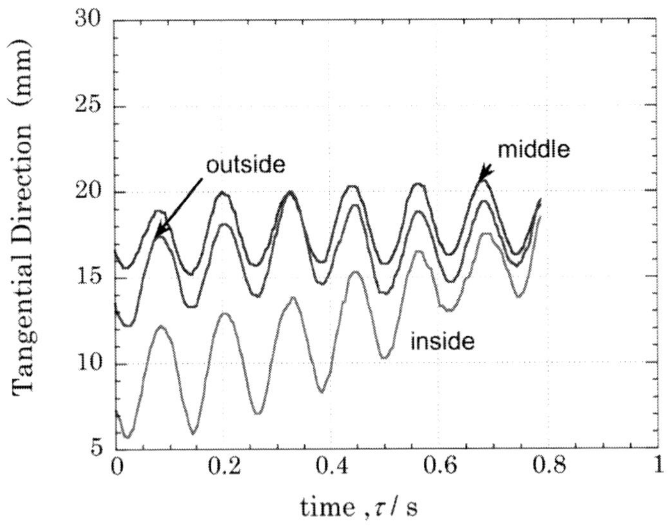

Figure 6. Movement of equiaxed grains as a function of position at the middle stage of solidification. rotation rate: 500 rpm.

Figure 5 shows the traces of equiaxed grains along tangential direction at the initial stage of solidification, when some equiaxed grains formed in the liquid phase. The position of equiaxed grain is in three levels, namely inside, middle and outside. Three sets of data are shown separately for convenience. Equiaxed grains oscillated periodically and traveled to + direction, which is anti-rotational direction. These oscillations were synchronous with the rotation. The amplitude of oscillation of inside grain seems to be smaller than that of other positions. Travelling to anti-rotational direction is quite clear for the inside equiaxed grain which was near the free surface. On the other hand, equiaxed grain near the columnar zone, which is marked 'outside', did not travel much. The fluctuation of free surface caused anti-rotational motion of liquid phase. But, the influence of free surface became small near the columnar zone.

Figure 6 shows the traces of equiaxed grains along tangential direction at the middle stage of solidification, when the almost all area of liquid phase was covered by equiaxed grains. Here, the amplitudes of oscillation were nearly the same independent of the position of equiaxed grains. Travelling to anti-rotational direction is again clear for the grain which located near the free surface.

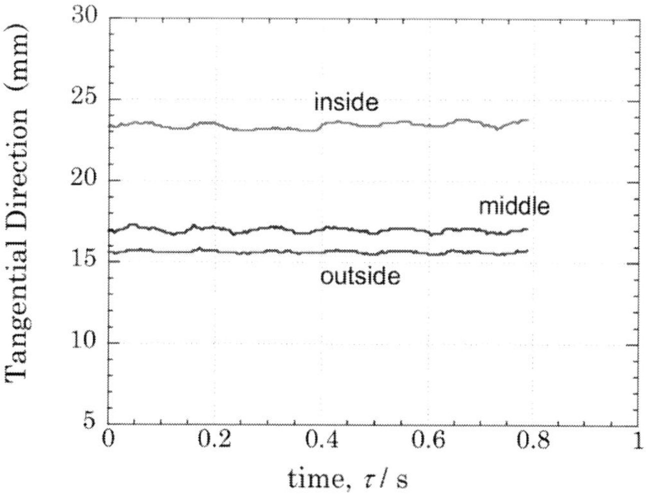

Figure 7. Movement of equiaxed grains as a function of position at the final stage of solidification. Rotation rate: 500 rpm.

Figure 7 shows the traces of equiaxed grains at the final stage of solidification. Fraction solid might be over 80% and equiaxed grains were oscillating with very small amplitude since solidification fully proceeded, apparent viscosity of the mushy zone may be high. At this stage, there is no or little difference in movement of equiaxed grain concerning to the location of equiaxed grain.

A schematic drawing for the acceleration affected by rotation and gravity is shown in Figure 8. In this figure, rotation direction is anti-clock-wise and there is one material point indicated by solid circle. Victors for acceleration are indicated every 60 degrees. Assuming that there is no inertia, the equiaxed grains advanced with respect to glass-cell in case of moving downward, namely between position A to D. On the other hand, in case of moving upward, the equiaxed grains go backward, or position D to A. Since the view point in this study was on the rotating system, the equiaxed grains were observed as oscillating in the tangential direction. Therefore, it is reasonable that the frequency of oscillation agrees with the rotation rate. To contrast this, there are little movement along radial direction. This may be due to low centrifugal force even in case of 500 rpm.

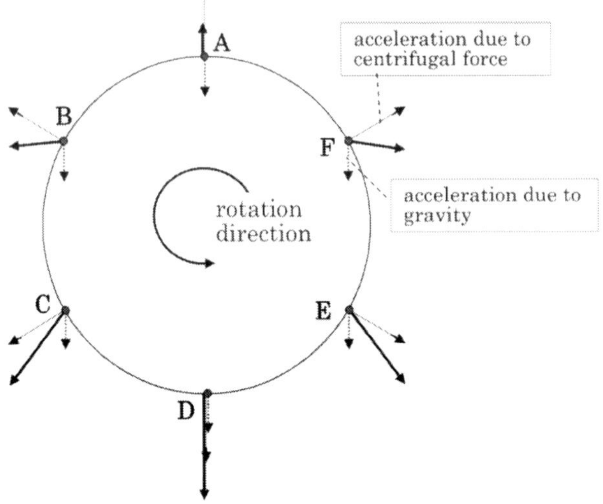

Figure 8. Change in acceleration vectors as a function of location of material point in horizontal centrifugal casting.

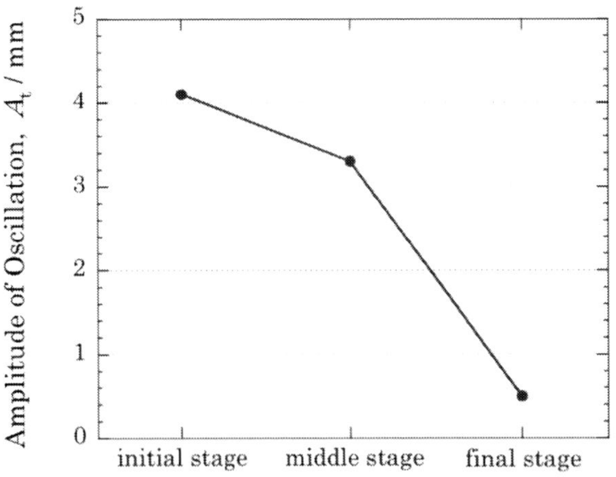

Figure 9. Change of amplitude in tangential direction of oscillation of equiaxed grain at the middle position in the liquid as a function of process of solidification. Rotation rate: 500 rpm.

The amplitude of oscillation is along tangential direction defined as shown in Figure 4, A_t, and measured at the middle position in the liquid zone. Figure 9 shows the qualitative relation between proceeding of solidification and amplitude of oscillation along tangential direction. As seen, the amplitude decreased as solidification proceeded. Although the experimental data is not enough, due to increase of apparent viscosity with proceeding of solidification the movement of equiaxed grain decreased with time and finally stopped completely.

The relations between solid fraction and apparent viscosity have been investigated [7]. Since the main object of these researches is semi-solid casting, relatively high shear force has been applied. Therefore, equiaxed grains obtained in these studies were granular or even spherical. As shown in Figure 2, the morphology of equiaxed grains observed was not spherical but dendritic. Although the morphology of equiaxed grains is different, the relationship between fraction solid and apparent viscosity is not basically different. Since the viscosity of solid-liquid mixture increases with time, it can be reasonably understood the change of amplitude with time.

4. Conclusions

The experimental apparatus for in-situ observation of horizontal centrifugal casting process using high-speed camera has been developed. Using this apparatus, following conclusions have been derived.

1) The movement of equiaxed grains could be clearly recorded using the high-speed camera.

2) Equiaxed grains oscillated with respect to the mold wall. The movement of equiaxed grains is clearly seen and can be precisely analyzed even the rotation rate was 500 rpm.

3) Amplitude of oscillation of equiaxed grains decreased as solidification proceeded, since the apparent viscosity increased with increasing fraction solid.

References

[1] A. Royer and S. Vasseur: *Metals Handbook*, Vol. 15, 9th ed., TMS(1988), 296.

[2] I. Kang and I. Ohnaka: *J. Japan Foundry Eng. Soc.*, **69**(1997), 119.

[3] I. Kang and I. Ohnaka: *J. Japan Foundry Eng. Soc.*, **69**(1997), 240.

[4] I. Kang and I. Ohnaka: *J. Japan Foundry Eng. Soc.*, **69**(1997), 815.

[5] H. Esaka, J. Hiramoto, S. Miyahara and K. Shinozuka: *Materials Science Forum*, **649**(2010), 325.

[6] H. Esaka, J. Hiramoto, S. Miyahara, K. Tokusashi, K. Shinozuka: *Tetsu-to-Hagané*, **97**(2011), 369.

[7] P. A. Joly and R. Mehrabian: *J. Mater. Sci.*, **11**(1976), 1393.

Simultaneous observation of melt flow and motion of equiaxed crystals during solidification using a dual phase Particle Image Velocimetry technique

A Kharicha, M Stefan-Kharicha, A Ludwig and M Wu
Chair for Simulation and Modelling of Metallurgical Processes, Dept. of Metallurgy, University of Leoben, A-8700 Leoben, Austria

E-mail: abdellah.kharicha@mu-leoben.at

Abstract. A dual phase Particle Image Velocimetry (PIV) technique is applied to study the flow and coupled equiaxed-columnar solidification in an ingot of NH_4Cl-H_2O solution. This technique is able to monitor simultaneously both velocities of the liquid flow and motion of equiaxed crystals. Before solidification started, a stable thermal buoyancy driven flow, laminar and 2D, was initialized in the ingot. As soon as solidification began, solutal buoyancy together with falling crystals destabilized the initial laminar flow into multiple chaotic cells, and a 3D turbulent regime was achieved. The kinetic energy for the flow was calculated. The interaction between the equiaxed grains and the melt flow was analyzed according to relative velocities. The settling velocity of an isolated equiaxed crystal was found to be 41 times smaller than the calculated one of a spherical crystal of equivalent size. The coupling between the fluid flow and the equiaxed crystal was found to be important in area of high crystal density. Chaotic and turbulent behaviors were found to be damped in regions of high equiaxed crystal density.

1. Introduction

Solidification is a multi-disciplinary field involving thermodynamics, fluid dynamics and solid mechanics, heat and mass transfer, and other disciplines [1-3]. Most challenging problems in solidification modeling are the complex interactions between physical phenomena occurring at different length scales ranging from atomic rearrangement over single crystal-melt interactions, to heat extraction, momentum and species transport at the system level.

Due to its transparency, NH_4Cl-H_2O solution was used for many experimental studies in the past [1-4], as the perfect analogue for solidification of metallic alloys.

The Particle Image Velocimetry (PIV) technique used here, for distinguishing between the two different signals, employs fluorescent tracer particles [5, 6]. The fluorescent tracers scatter the light at a wavelength different from that scattered by the equiaxed crystals. The use of two cameras is necessary. The first camera mounted with a green filter (band pass filter of 532 nm) detects only the laser light scattered by the equiaxed crystals. The second camera uses an orange filter (570 nm) through which the light reemitted by the fluorescent dye (Rhodamine B) can be recorded. This technique was successfully applied to two-phase water-bubble flows, with simultaneous separation and measurement of the different phases. Here it is applied to investigate the liquid and equiaxed crystal two phase flow.

2. Experimental Procedure

The experimental set-up is composed of: the test cell, the illumination system, the optics and the computer. The test cell (Figure 1) was cubic (10x10x1 cm^3) with three sides made of brass, and commercial glass for the front and back; the upper part was open. The cast cell was cooled down homogeneously through the three brass walls. The illumination system was an Nd-Yag double pulsed

Figure 1. Rectangular cavity used for controlled solidification experiments. The inner dimensions are 10x10x1 cm^3.

laser and the optics were 2 CCD cameras mounted in parallel and perpendicular to the test cell. The laser and the CCD cameras were piloted by the computer via the commercial software FlowManager. A hypereutectic 29.5 wt.% NH_4Cl – H_2O solution was solidified under controlled temperatures. The cooling rate was 1.2°C/min for the brass walls. The liquid solution was poured into the cell at 45°C, after 30 min the cooling was started from 45°C until 5°C. The air temperature around the test cell was 24.7°C.

3. Results

The solidification process follows different stages.The first stage (labeled by TH) involved only liquid, whereas the last four (labeled by PTH, TU, CC, and MF) involved solidification as well.

3.1 Thermal stage (TH): stable thermal buoyancy

When the walls temperature decreases below the temperature of the solution, the liquid close to the vertical walls has a tendency to sink by thermal buoyancy. However, in the experiment it took 8 min. until a stable clockwise convection pattern established. The direction of the flow was downwards near the walls and upwards at the center (Figure 2). As cooling became stronger and stronger the magnitude of downwards flow velocity slightly increased with time. It is clear that the hydrodynamic pattern of this stage was found to be very stable and symmetric. Stage TH lasted only around 1-2 minutes before evidences for solidification appeared.

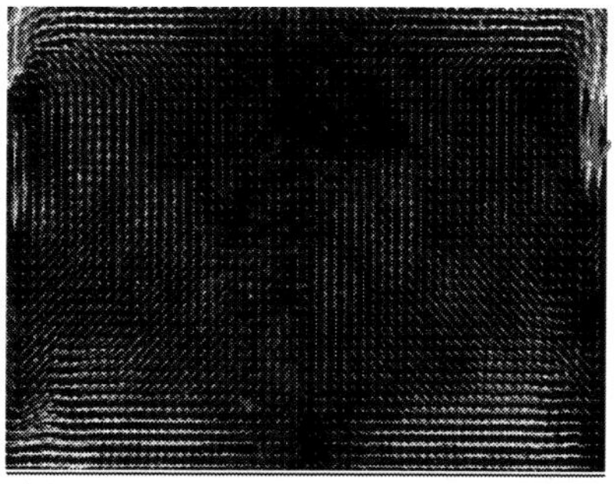

Figure 2. Stable thermal buoyancy stage at 8 min after the cooling started The velocity vectors are linearly scaled between v_{min} = 0.08 mm/s and v_{max} = 1.12 mm/s (red arrow).

3.2 Perturbation of the thermal state (PTH): solidification perturbed thermal state

Visible solidification, mainly columnar, started at about 9-10 minutes after beginning of cooling. It did not appear uniformly all over the walls surface, but rather in islands of columnar dendritic crystals (Figure 3a). In the vicinity of each columnar island the flow is immediately disturbed by the presence of one or two small eddies (Figure 3b). However, during about the first 3-4 minutes of visible solidification, the main pattern of the flow (symmetry, magnitude, main direction) is not changed. Sometimes the strong downward jet is able to detach columnar structures of up to 5 mm in size from the cooling wall. This phenomenon happens mainly at this stage, rarely in later stages. Few equiaxed crystals were also occasionally observed at the vicinity of the vertical walls (Figure 3b). This stage lasted around 3-4 minutes.

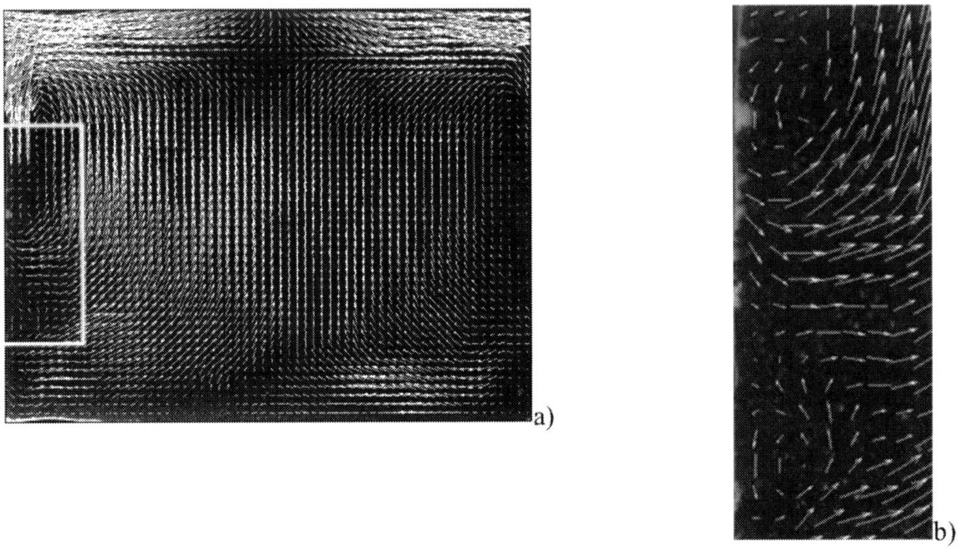

Figure 3. a) Perturbed thermal stage (PTH), at t = 9 min after switching on cooling. The velocity vectors were scaled between $v_{min} = 0.052$ mm/s and $v_{max} = 0.73$ mm/s (red arrow); b) zoom- in of yellow square.

3.3 Turbulent flow state (TU)

As the solidification proceeded, equiaxed crystals increased in number, and perturbations of the main flow increased in magnitude. A new flow regime with large number of small eddies appeared (Figure 4a-b). Now solidification covered totally the walls and filled about 5% of the picture (at the end ~30% is reached). The lack of coherence of the observed eddies, and the presence of strong velocity gradients even far from the walls, is a strong indication of the three dimensional and turbulent character of this regime. In this stage large visible equiaxed crystals appeared only near the walls. However, the movement of the small crystals could be evaluated by correlation. This stage lasted about 2 minutes.

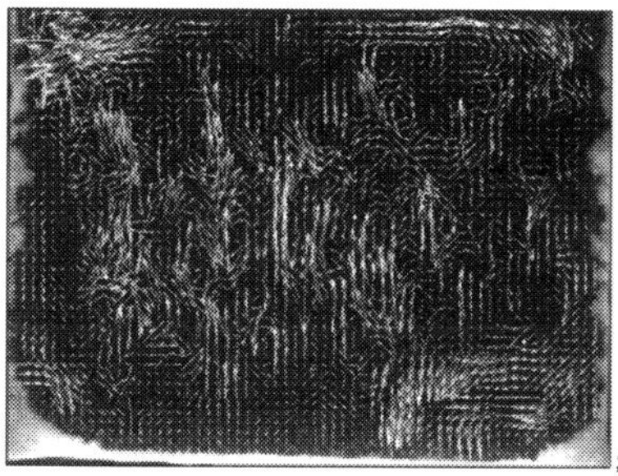

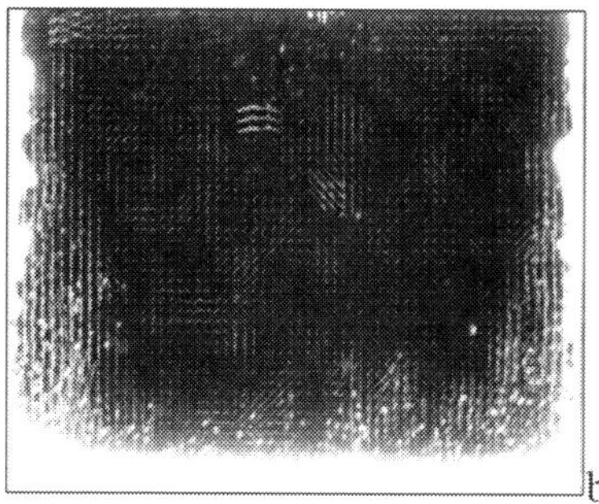

a) b)

Figure 4. Turbulent regime (TU). The liquid velocity is shown in (a) and the velocity of the equiaxed crystals is shown in b), at t = 13 min. The velocity vectors are scaled between v_{min} = 0.051 mm/s and v_{max} = 0.709 mm/s for the liquid (a) and between v_{min} = 0.252 mm/s and v_{max} = 3.52 mm/s for the equiaxed crystals (b).

3.4 Coherent chaotic flow state (CC)

Out of the turbulent flow stage a regime where the flow map is coherent but still chaotic appeared (Figure 5a). The characteristic eddy size was of about 1.5-2 cm, and thus larger than the distance between the two glass plates. The equiaxed crystallization was very strong (Figure 5b), and the leading solidification mechanism. The velocity magnitude was at its maximum. Large crystals appeared even in the bulk far from the vertical walls, and often fell, grouped in columns, or in agglomeration. Note that during this stage most of the solidification process happened. This included the initiation and growth of freckles (flow channels in the mush) where traces can be seen in Figure 6a. At the end of this stage, the solidification front had almost reached its final position. Stage CC lasted about 10 minutes.

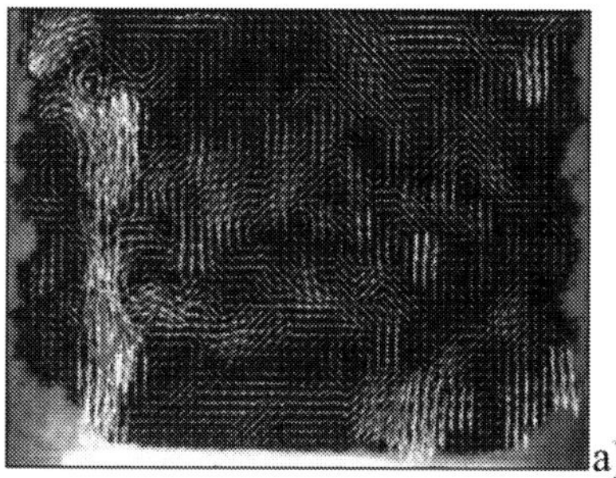

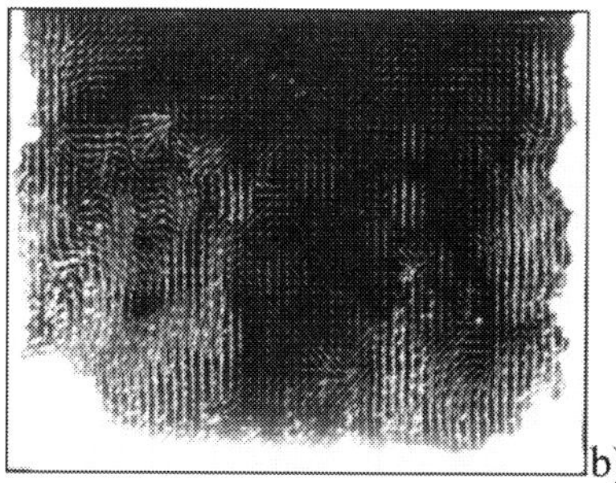

a) b)

Figure 5. Examples showing the melt velocities (a) and the equiaxed crystals velocities (b), for the so-called coherent chaotic regime (CC). The two-cameras PIV measurement was done at t = 15 min. The velocity vectors are scaled between v_{min} = 0.111 mm/s and v_{max} = 1.56 mm/s (red arrow) for the liquid (a) and between v_{min} = 0.278 mm/s and v_{max} = 3.87 mm/s for the equiaxed crystals (b).

3.5 Meandering flow state (MF): quasi-steady meandering flow occurs

The small eddies disappeared progressively and gave place to a meandering flow traveling from the top to the bottom (Figure 6a). The flow is stratified into 5 to 7 circulation layers. It is possible to follow the flow streamlines from the top to the bottom of the cell. The number density of equiaxed crystals decreased in the vicinity of the vertical solidification front, and increased in the bulk in the form of relatively large crystal column or in agglomeration (Figure 6b). But globally equiaxed falls were not as strong as in the previous CC regime. The meander flow seems to penetrate the bottom mushy zone. In this regime, the number of equiaxed crystals and the magnitude of the flow velocity were continuously decreasing in time until extinction. Finally about 30% of the cell volume is filled with solid. Stage MF lasted around 5-10 minutes until the typical meandering flow disappeared.

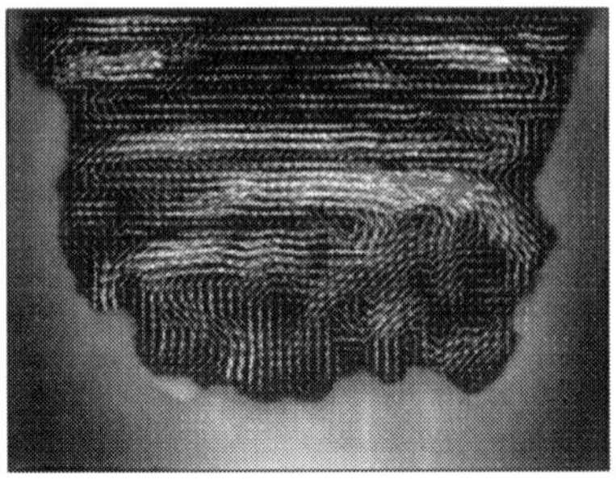
a)

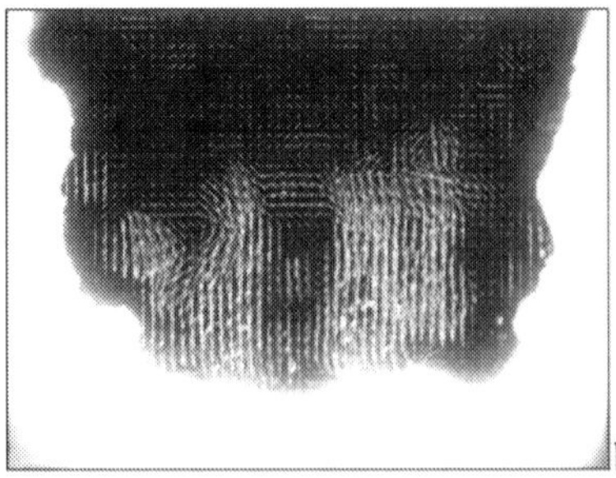
b)

Figure 6. Examples showing the melt velocities (a) and the equiaxed crystals velocities (b) for the meandering flow regime (MF). The two-cameras PIV measurement was done at t = 30 min. The velocity vectors are scaled between v_{min} = 0.067 mm/s and v_{max} = 0.466 mm/s (red arrow) for the liquid (a) and between v_{min} = 0.155 mm/s and v_{max} = 2.17 mm/s for the equiaxed crystals (b).

4 Analysis and Discussion

4.1 The evolution in time of the total kinetic energy of the liquid flow

The evolution in time of the total kinetic energy of the liquid flow calculated with the measured velocity magnitude U as

$$KE = \sum_{N} \frac{1}{2} U^2 \qquad (1)$$

is presented in Figure 7. N represents the total number of vectors all over the surface cell and it is constant. When solidification happens the vectors in the mushy zone are zero. Although the presence of some instabilities, no large change in the kinetic energy is observed before the start of solidification. Even if we would have taken the large velocity jets along the walls, which are not completely resolved, into account the kinetic energy would not change dramatically.

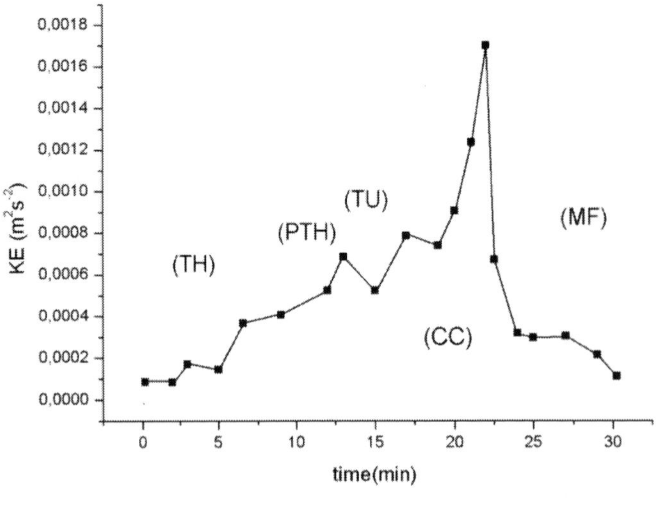

Figure 7. Evolution of the liquid kinetic energy, during the different flow regimes.

The kinetic energy increased continuously during the perturbed thermal and turbulent regimes. The maximum is reached close to the end of the coherent chaotic regime at about t = 22 minutes. Later, in

IOP Conf. Series: Materials Science and Engineering **33** (2012) 012042 doi:10.1088/1757-899X/33/1/012042

the stratified regime (MF) the kinetic energy decreases to levels similar to that of the thermal buoyancy states. This evolution of KE does not correspond to what is reported by Skudarnov [7], where the velocity decreases in magnitude all along with time. The reason for that might be the fact that in our experiment we end up in a partly solidified sample, where all dynamics fade out with time.

At around 25 minutes the lower number of equiaxed crystals promotes a steadier and slower melt flow structure (Figure 6a). However, the flow path seems to be controlled by the local curves of the dendrite tip envelop. It is probable that the meandering flow regime is created by an auto-organisation of the solutal convection in curved constrained geometry (from top to bottom). The concentration differences are small since the evolution of the dendrite tip envelop front is nearly stopped, but large enough to generate a flow with a KE similar to that of the thermal states.

4.2 Relative velocity

The relative velocity was estimated as the difference between measured crystal velocity and flow velocity. To obtain accurate crystal velocities it is necessary to choose an area where the number of crystals is large enough for the PIV correlation. The fundamental aspect is the relation between the relative velocity and the crystal size. Such a relation exists for spherical particles, and is known as Stoke's law. To check its validity for equiaxed crystals it is necessary to extract the relative velocity for isolated crystals, not in crystal agglomeration. In Figure 8 the calculated relative velocities of single crystals as function of the measured size are plotted. The maximum horizontal diameter of the crystal (d) was calculated from the pictures (1600*1186 pixels), by counting the number of bright pixels. Due to an overexposure of the pictures, necessary for a good PIV, the size of the crystals may be overestimated.

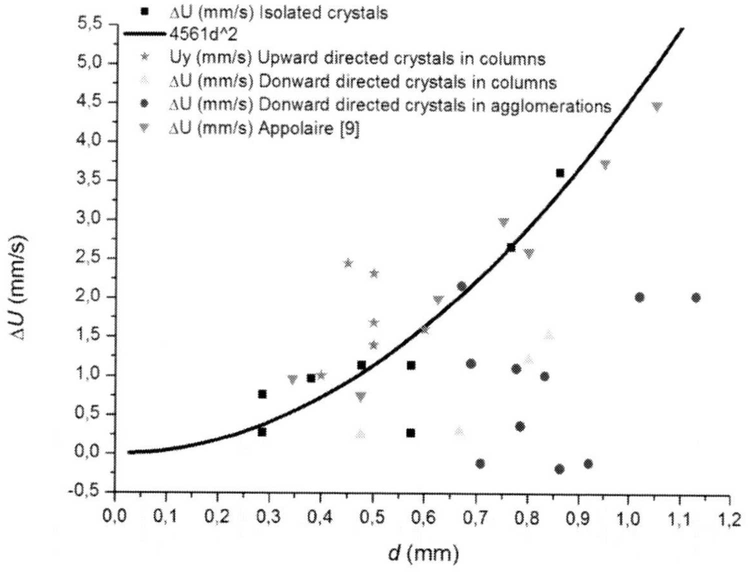

Figure 8. Relative velocity for single crystals versus the crystal size diameter (d), and comparison between rising (red) and falling (green) velocities magnitude in chimneys.

The results for isolated crystals show an increase of the relative velocity with the measured size. The magnitudes of the relative velocities found for corresponding sizes are very similar to those reported by Appolaire [8, 9]. The correlation follows a clear power tendency. Based on the Stokes's law the relative velocity of a compact sphere can be calculated by:

$$\Delta U = \frac{d^2 g}{18 C_D \mu}(\rho_A - \rho_W) = \alpha \cdot d^2 \qquad (2)$$

where ρ_A is the density of the sphere, ρ_W the density of the surrounding fluid, g the gravity constant, C_D the drag coefficient, μ the dynamic viscosity of the fluid and d is the diameter of the sphere. α gathers the materials properties in one single factor.

Using the measured sizes and velocities, the particle Reynolds number $Re_p = \rho_A \Delta U d / \mu$ is always smaller than 1, which allows the use of equation (2). Equation (2) was used to fit the experimental data of Figure 8. With the materials properties given in Table 1, the value of α can be calculated to be 188402.3. However, the fit to the measuring points from Figure 8 gives $\alpha = 4561$. Thus the computed relative velocity for spheres (0.1 m/s for d = 0.75 mm) is almost 41 times bigger than the measured ones.

Table 1: Material properties important for the discussion of the results

Property	Symbol	Value	Ref.
density of the liquid (water)	ρ_W	1078 kg·m^{-3}	Beckermann et al. [10]
density of solid ammonium chloride	ρ_A	1527.4 kg·m^{-3}	Rowe et al. [11]
dynamic viscosity of the liquid	μ	1.3 × 10^{-3} kg·m^{-1}·s^{-1}	Beckermannet al. [10]

Due to the fact that the equiaxed crystals are not spherically shaped, and to their rough fractal surface, it was expected that spheres have a larger relative velocity than the equiaxed crystals. This is the reason why numerical simulations were performed by using modified drag laws to account for the equiaxed shapes [12-15]. These drag laws, predict systematically larger drag coefficients (C_D>1) than for spheres. In our case the drag coefficient is indeed larger and equal to 41.3.

Based on the experimental correction of equation (2) the equiaxed relaxation time is estimated to be:

$$\tau_e = \alpha \frac{\rho_A - \rho_W}{\rho_A} d^2 \leq 10^{-4} s \qquad (3)$$

Compared to the measurement frequency (250 ms), it is very small. It can be assumed that the crystals reach instantaneously their terminal relative velocity. Thus, the polynomial increase of ΔU observed in Figure 8can be directly related to the growth of volume during the crystal fall.

By comparing Figure 4a and Figure 4b, we can notice that variations of the liquid velocities are smoother in areas where a large number of equiaxed crystals exist. This phenomenon is particularly visible if we compare the flow coherence in the bulk with the high equiaxed density regions. The flow in these regions is well structured in clear streamlines or in large eddies. Obviously, the presence of crystals flows, through drag interaction, damps the turbulence eddies. We can suppose that due to the extension of the area of falling crystals and to the decrease in magnitude of ΔC, a laminarization of the entire domain occurs, leading to a transition to the laminar chaotic regime (Figure 5).

5 Conclusions

A special Particle Image Velocimetry (PIV) was used to measure the velocity field during a columnar/equiaxed solidification process. The single-phase measurement technique developed for PIV has been extended to cope with liquid-equiaxed flows. It was applied to NH_4Cl solidification in a die cast cell of 10*10*1 cm^3 in which both the liquid velocity and the crystals velocity maps could be extracted. The separation between the two velocity fields could be obtained with the use of fluorescent tracer particles that emit different waves when excited with the laser wavelength. Two pictures were taken simultaneously with two different cameras; the first one recorded only the laser light through a green filter, while the second one used an orange filter to record only waves emitted by the fluorescent particles. Thus the first camera could follow only the crystals paths, while the second camera recorded the liquid velocity field.

During the cooling down different regimes were observed. Before solidification, the flow is controlled by only thermal buoyancy. As soon as the solidification starts, turbulence and chaos are generated in the liquid due to solutal convection. In the same time strong equiaxed crystal rains occur at the vicinity of the solidification fronts. At the end of the process, the flow becomes steadier and auto-organizes in a meandering and stratified structure.

In general no clear relation was found between the size and the settling velocities of the crystals. However isolated crystals were found to fall much slower than spheres of equivalent apparent sizes, in agreement to what is generally believed. This fact could be explained due to the fractal crystal shape. For equiaxed crystals with diameters smaller than 1 mm, the drag coefficient was estimated to be 41.3.

The coupling between the fluid flow and the equiaxed crystal was found to be important in area of high crystal density. In these regions, the liquid flow has shown smoother velocity variations than in the chaotic or the turbulent bulk. This observation is a proof that equiaxed crystals tend to damp a part of the turbulence. Simultaneously the velocities in the bulk flow were found to accelerate during the period of strong crystal fall near the vertical walls. Perhaps in larger systems as in industrial scale a generation of turbulence could be observed. However in the present configuration, it is not yet clear how much is the contribution of the equiaxed rains to the generation of chaos with respect to the solutal buoyancy. Spatio-temporal behavior of the liquid and crystal flow suggested the presence of instabilities and auto-organization often observed in the field of nonlinear systems and chaos. This indicates that very complex behaviors can be expected from the interaction between flow and solidification in larger systems.

Acknowledgements
The research was funded by the Austrian Science Fund (FWF): P22614-N22.

References

[1] Beckermann C and Viskanta R 1993 *Appl. Mech. Rev* **46** 1-27
[2] Beckermann C 1993 *Int. Conf. Modeling of Casting, Welding and Advanced Solidification Processes VI* 'ed T S Piwonka*et al'* (Warrendale, Pennsylvania:TMS) pp 181-92
[3] Rappaz M 1989 *Int. Mater. Rev.* **34** 93-123
[4] Khalitov D A, Longmire E K 2002 *Exp. Fluids* **32** 252-68
[5] Hassan YA, Schmidl W D and Ortiz-Villafuerte J 1998*J. of Visualization* **1** 291-301
[6] Delauré Y M C, Chan V SS and Murray D B 2003 *Exp. Thermal and Fluid Sci.* **27** 911-926
[7] Skudarnov P V, Lin C X, Wang M H, Pradeep N and Ebadian M A 2002 *Int. J. of Heat and mass Transfer* **45** 5191-200
[8] Appolaire B, Albert V, Combeau H and Lesoult G *ISIJ International* **39** 263-70
[9] Appolaire B Ph.D. INPL 1999
[10] Beckermann C and Wang C Y 1996*Metall Mater TransA* **27** 2795-84
[11] RoweRC, SheskeyPJ and QuinnME 2009 *Handbook of Pharmaceutical Excipients* 6th ed. 'ed R C Rowe *et all*' (London, England: Pharmaceutical Press)
[12] Badillo A, Ceynar D and Beckermann C 2007 *J. of Crystal Growth* **309** 197-215
[13] Badillo A, Ceynar D and Beckermann C 2007 *J. of Crystal Growth* **309** 216-24
[14] Appolaire B, Combeau H andLesoult G 2008 *Mater. Sci. Eng. A* **487** 33-45
[15] Wu M and Ludwig A 2009 *Acta Mater* **57** 5632-44

IOP Conf. Series: Materials Science and Engineering **33** (2012) 012043 doi:10.1088/1757-899X/33/1/012043

Observation of peritectic reaction in Ag-Sn alloys

Yuhi Hattori[1,3], **Hisao Esaka**[2,3] **and Kei Shinozuka**[2,3]

[1] Graduate student, National Defense Academy

[2] Department of Material Science and Engineering, National Defense Academy

[3] 1-10-20 Hashirimizu, Yokosuka 239-8686, Japan

E-mail: em50036@nda.ac.jp

Abstract. In general, peritectic reaction is known that the secondary phase grows as it surrounds the primary phase. However, relationship between primary and secondary phase is still uncertain. Therefore, to clarify this, an experimental work has been preformed using Ag-35 mass%Sn alloy. The alloy was melted in an alumina crucible and then cooled at a constant cooling rate. The specimen temperature was measured with a thermocouple. Thereafter, the specimen was dropped into a water bath at a various temperature to quench the solid/liquid interfacial morphology and solute distribution. Then the specimen was polished and was etched, if necessary. The result of the thermal history, supercooling and recalescence was observed near the peritectic temperature. The observation by scanning electron microscope (SEM) unveiled many fine protrusions (=ε-phase) on a primary ζ-phase, which is quenched at early state of peritectic reaction. In addition, concentration profile was measured by energy dispersive X-ray spectrometry (EDX) and ζ, ε and liquid phases were identified. As a result, it was confirmed that ε-phase grew along the surface of ζ-phase and it did not grow into the ζ-phase. Further, result of electron backscatter diffraction (EBSD) analysis, ε-phase was found to be single crystal within the range of this study.

1. Introduction

The solidification of carbon steel contains peritectic reaction. The peritectic reaction in carbon steel is described: L + δ-Fe → γ-Fe. This may cause hot cracking, such as surface cracking in continuous casting slab and weld metals.[1][2] In addition, the refinement of as-cast γ-Fe grains is required for producing heavy plates, pipes and bars in order to prevent cracking during rolling. However, since δ-Fe does not exist at room temperature, the relationship between δ-Fe and γ-Fe can not be investigated. Due to this reason, the research works that focused on the mechanism of peritectic reaction, especially the nucleation and growth of the secondary phase, are not so much.[3] In this study, therefore, in order to clarify the relationship between primary phase and secondary phase, observation of peritectic reaction carried out using Ag-Sn alloy.[4][5]

2. Experimental procedures

Figure 1 shows a part of the equilibrium phase diagram of Ag-Sn system.[6] Ag-Sn alloy has two peritectic reactions. In this experimental work, the alloy with low peritectic temperature was used, because of easiness for handling. One of the most advantageous points is that the phases for peritectic reaction will remain at room temperature. The alloy composition used was Ag-35 mass%Sn, which is hyper-peritectic composition. The mother alloy was prepared from 99.99 % purity Ag and Sn.

A schematic diagram of experimental apparatus is shown in Figure 2. In this experiment, peritectic reaction process was observed by using a quenching technique. Furnace temperature was measured by a K-Type thermocouple inserted from the top of the electric furnace. The Ar gas was introduced from the top of the furnace. The specimen (diameter 9 mm × height 20 mm) was placed in an alumina crucible (inner diameter 11 mm × depth 35 mm) and the temperature of specimen was measured with the K-type thermocouple (ϕ 0.1 mm) inserted into the crucible. The electric furnace was kept at 700 °C for 1 hour and was then cooled at a constant cooling rate of 3 °C/min.

First, the thermal history of specimen from 700 °C to room temperature was measured. Subsequently, at various temperatures around the peritectic temperature, the specimen was dropped into the bucket placed under the electric furnace in order to quench the solid/liquid morphology and solute distribution in the specimen. Vertical cross section of the specimen was observed by SEM (Scanning Electron Microscope) and characterized by EDS (Energy Dispersive x-ray Spectroscopy). The specimen was etched by $HCl+FeCl_3$ aqueous solution, if necessary.

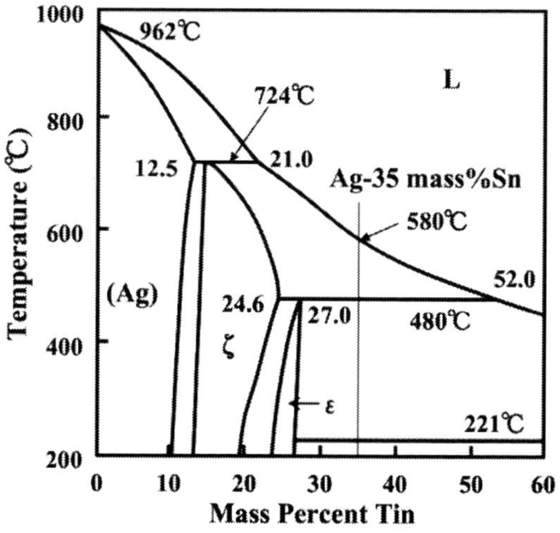

Figure 1. Equilibrium phase diagram of Ag-Sn system.[6]

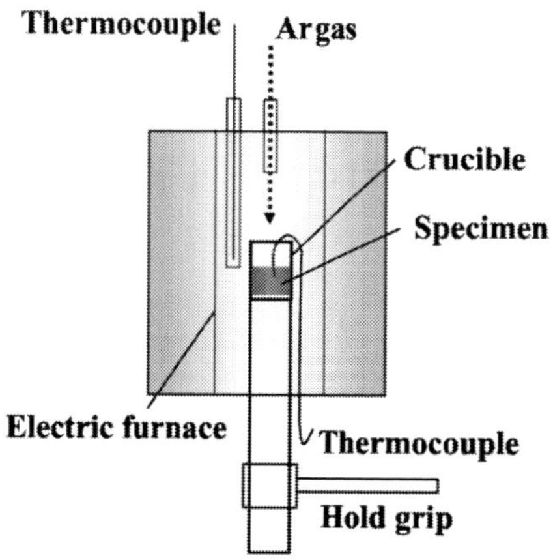

Figure 2. Schematic diagram of experimental apparatus.

3. Experimental results and discussions

3.1. Thermal history

3.1.1. Whole thermal history

Figure 3 shows the result of temperature measurement in the specimen during experiment. During cooling, cooling rate was slightly changed at liquidus temperature (T_L). Supercooling and recalescence was observed at peritectic temperature (T_P) and eutectic temperature (T_E). The observed T_L, T_P and T_E are about 578 °C, 481.1 °C and 220.4 °C, respectively and these are consistent with the values of Ag-35 mass%Sn alloy.

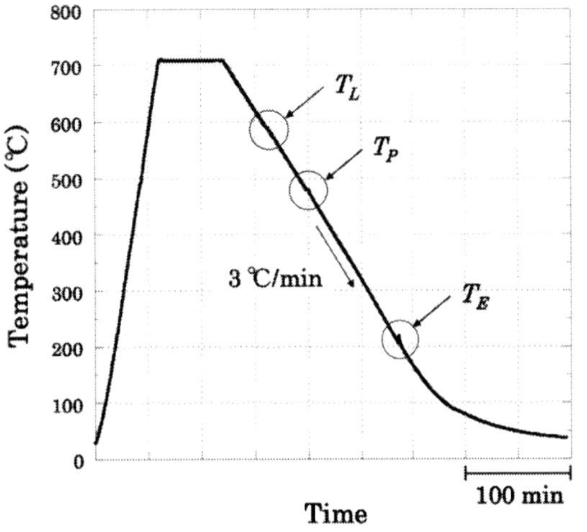

Figure 3. Thermal history obtained in this study.

3.1.2. Thermal history around peritectic temperature

Figure 4 shows the enlarged view of thermal history around peritectic temperature. Supercooling and recalescence was observed and the amount of supercooling was about 1K. This indicates that ε-phase (Ag_3Sn) nucleated with this supercooling as a second phase. In this case, the recovery of temperature was relatively slow. This may be attributed that release of latent heat per unit time was small because of the slow growth of ε-phase.

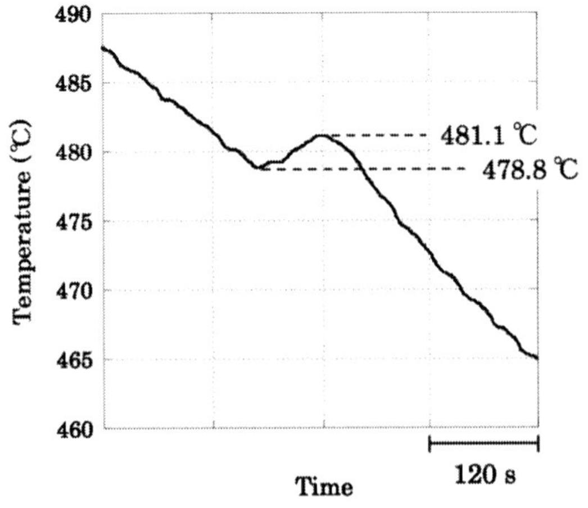

Figure 4. Thermal history around peritectic temperature.

3.2. Observation of microstructure of quenched specimen

3.2.1. Observation with low magnification

Figure 5 shows the timing of quenching and microstructures with low magnification of the specimens quenched at various point. In all specimens, there are generally coarse dendrites and they are very similar. In the specimen #1, there should be only primary ζ-phase (Ag_4Sn) when the specimen was quenched. Thus, ζ-phase crystallized in the shape of dendrites. In the specimens #2~ #6, ε-phase should have formed on the ζ-phase, though, there are little difference with the specimen #1.

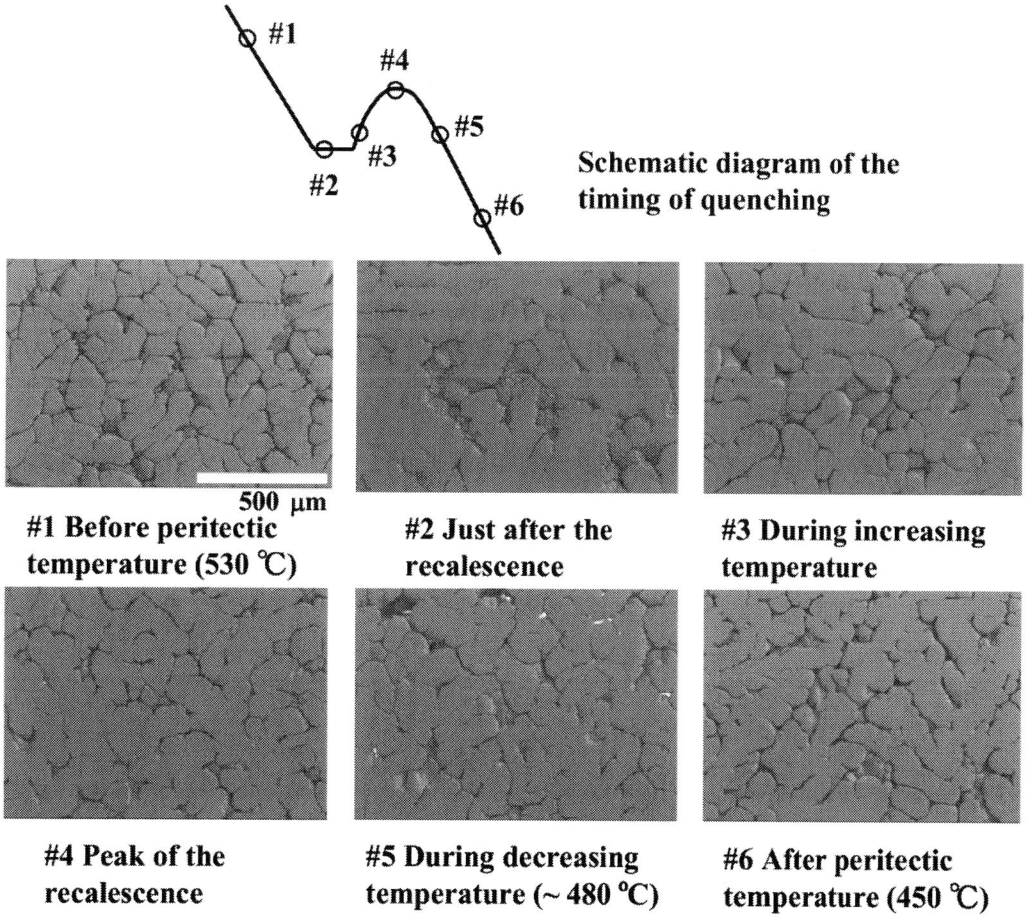

Figure 5. Solidification structure observed by SEM (Low magnification).

3.2.2. Observation with high magnification

Figure 6 shows solidification structure with high magnification observed at the same specimens as indicated in Fig. 5. Since the liquid region when the specimen was quenched was almost removed by deep-etching, the observed areas were mainly the surface of the solid phase. In the specimen #1, there are many small and angular protrusions on the surface of ζ-dendrites. These small protrusions may be ζ-phase which grew from the surface of ζ-dendrite during quenching.

In the specimen #2, ε-phase has formed from the result of thermal analysis, but the morphology is similar to #1. These small protrusions may be ζ-phase just like #1. It is considered that ζ-phase was not wholly covered by ε-phase because of slow and limited growth of ε-phase. In the specimen #3, plate-like crystals as well as very fine and angular protrusions on the surface were observed. Since this specimen was quenched during the peritectic temperature, ε-phase probably covered ζ-phase dendrite. It is thought that fine and angular crystals may be ε-phase and they grew during quenching process because ε-phase still could grow into the liquid phase when the specimen was quenched. In the specimens #4 ~ #6, it is thought that ε-phase further grew from the result of thermal history. Since the driving force of the growth of ε-phase in the liquid became small, the surface of the solid phase was found to be smooth.

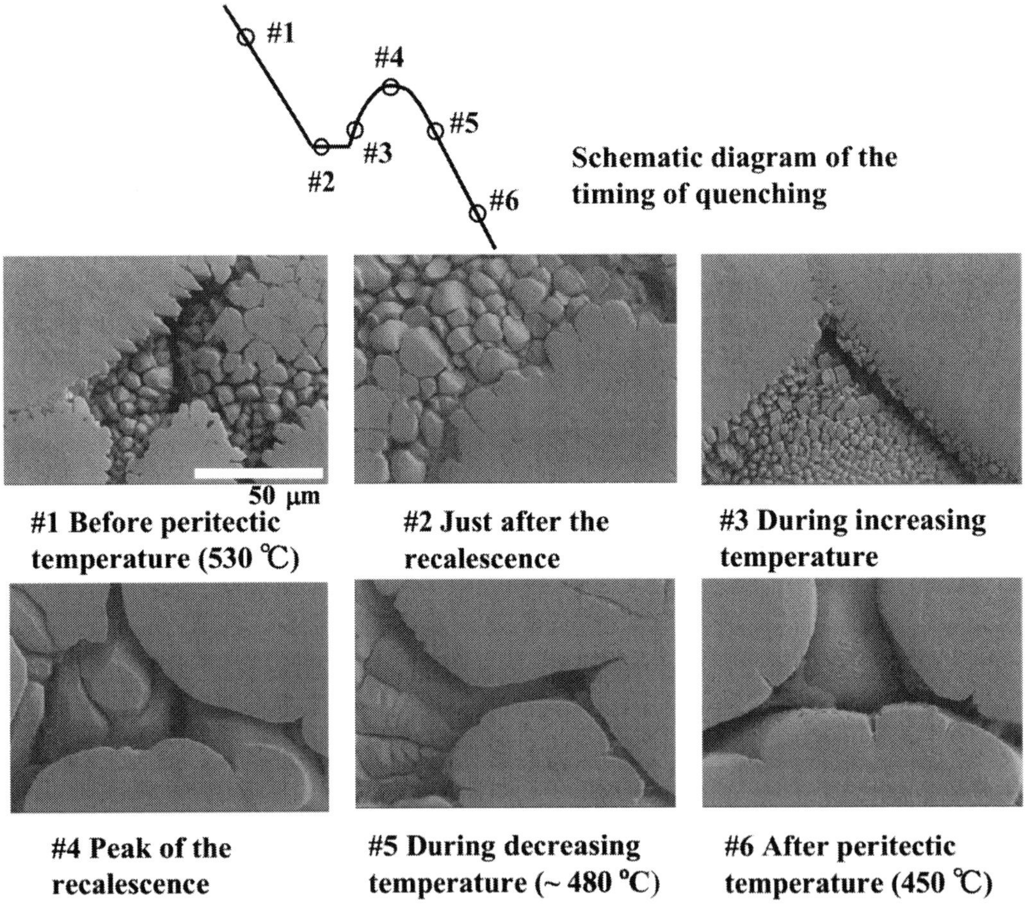

Figure 6. Solidification structure observed by SEM (High magnification).

3.3. Results of EDS analysis

The result of line spectrum analysis of specimen #2 is shown in Fig. 7. Concentration of Ag and Sn were analyzed along the line indicated in Fig. 7 a). The vertical axis is counts per second and indicates the qualitative value of Ag and Sn concentration. In the range of about 20 μm from the surface of dendrite it is found that concentration of Sn is slightly higher than the other region. This region may be ε-phase. Semi-quantitative analysis of EDS was also performed on the points indicated in Fig. 7 a). The results indicates that the average Sn content was 21.5 mass% in points1-3. On the other hand, the average Sn content was 23.7 mass% in points 4-6. The expected values of Sn concentration in ζ-phase and ε-phase from equilibrium phase diagram are 24.6 and 27.0 mass%Sn, respectively. The difference of Sn concentration in ζ-phase and ε-phase almost agrees with that of the equilibrium phase diagram, though the values are slightly different. Therefore, it can be concluded that ε-phase can be distinguished from ζ-phase from distribution of Sn. In addition, it was found that there was Sn-enriched zone and Ag-denuded zone near the ζ/ε interface.

The result of line spectrum analysis of specimen #4 is shown in Figure 8. Again, concentration of Sn in outer region of dendrite is higher than that in the central region. Here, the outer region may be ε-phase and its thickness is approximately 20 μm. There is no Sn-enriched zone near the ζ/ε interface. It may be homogenized by solid phase diffusion. Even in the specimen cooled at 3 °C/min to room temperature, the concentration of Sn in the central region of dendrite is approximately 21 %. This confirmed that ζ-phase remains in the center of dendrite at room temperature. The thickness of ε-phase was approximately 20 μm in the range of this study. This indicates that ε-phase grows in the peritectic reaction and it grows little after that.

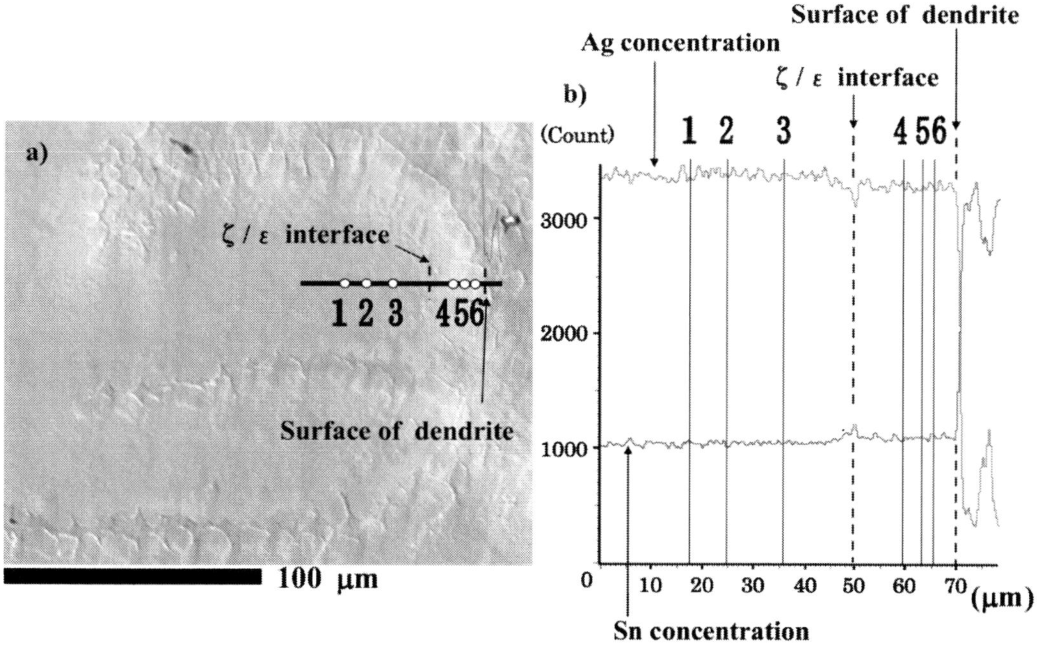

Figure 7. Result of EDS analysis along the line in specimen #2.

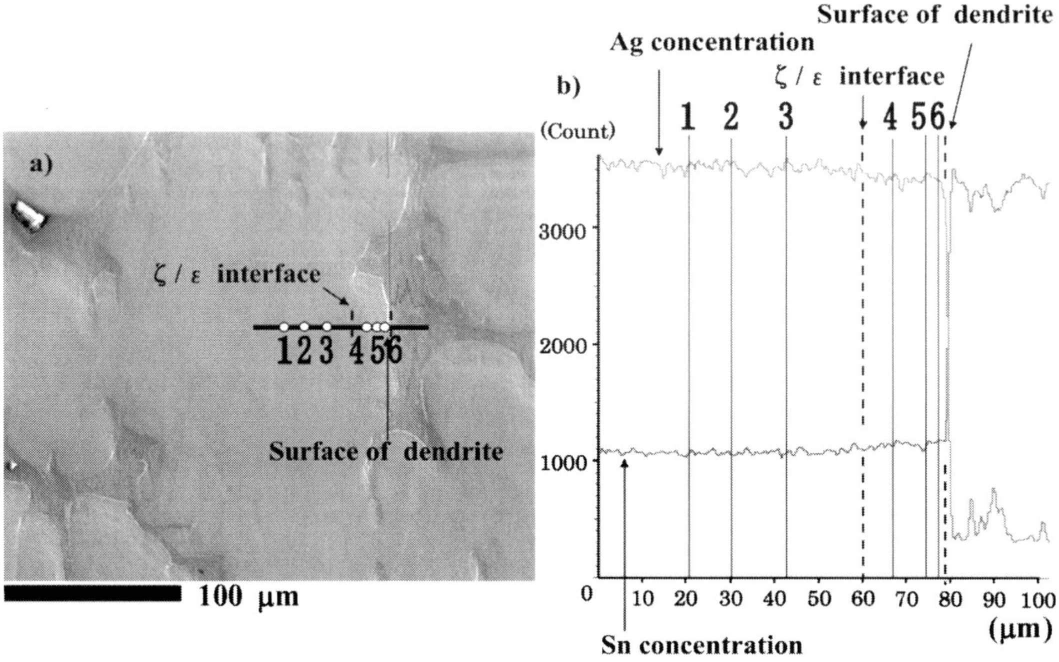

Figure 8. Result of EDS analysis along the line in specimen #4.

3.4. Results of EBSD analysis

Figure 9 shows SE image of the specimen quenched at peritectic temperature. Light gray rim, which corresponds to ε-phase, is clearly seen in this figure. Using this specimen, EBSD (electron backscatter diffraction) analysis was performed. The result is shown in Figure 10. Since crystallographic structures of ζ and ε phases are quite similar, it is difficult to crystallographically distinguish ε-phase from ζ-phase for the moment. Thus, in this study, crystallographic characterization of ε-phase which formed on the periphery of ζ-phase was investigated. Three Kikuchi patterns obtained at b)-d) are also shown in Figure 10. These patterns are exactly the same and this indicates that ε-phase which formed on ζ-phase is a single crystal.

Therefore, a process of peritectic reaction in this alloy system may be following: A ε-crystalline nucleates at a certain site on the surface of ζ-dendrite. Then it grows along the surface of ζ-dendrite and surrounds all surface of ζ-phase. Here, another nucleation of ε-phase does not take place. The growth rate of ε-phase should be slow, therefore solute distribution in the liquid at ζ/liquid interface may be fixed at ζ/ε interface. Thus, Sn-enriched zone was observed at ζ/ε interface in the specimen #2 as shown in Fig. 7. The thickness of ε-phase is thin and ε-phase dose not thicken much.

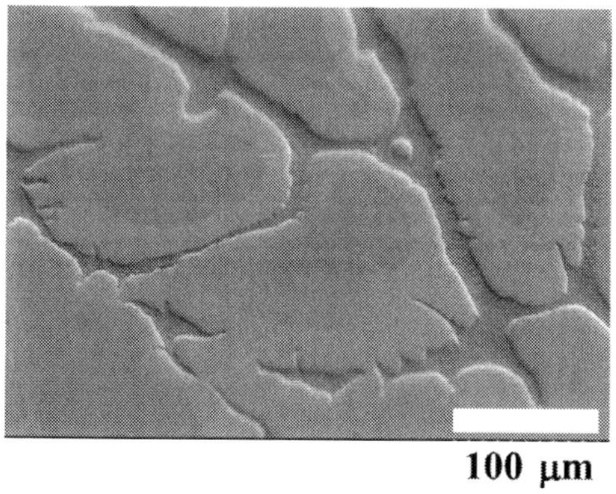

Figure 9. SE image of solidified structure in the specimen #2.

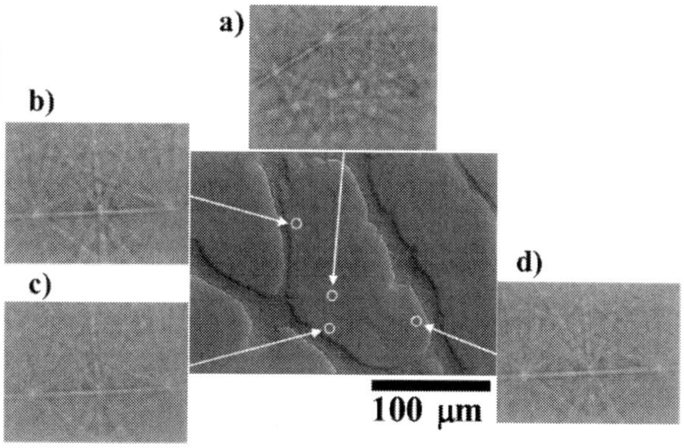

Figure 10. Results of EBSD analysis performed in the specimen #2.

4. Conclusions

In order to investigate the process of peritectic reaction, experiments which combined thermal analysis and quenching method were performed using the Ag-35 mass%Sn alloy. The following conclusions were drawn.

(1) Primary ζ-phase is intermetallics and crystallizes in the form of dendrites.

(2) ε-phase forms on the primary of ζ-phase dendrite, and does not thicken much.

(3) ε is a single crystal condition and solidified in the form of rim in the range of present experiment.

References

[1] T. Matsumiya, T. Saeki, J. Tanaka and T. Ariyoshi: *Tetsu-to-Hagané*, **68**(1982), 1782.

[2] K. Tamaki, H. Kawakami and J. Suzuki: *J. JWS*, **2**(2002), 266.

[3] T. Takahashi, K. Ohsasa and J. Tanaka: *Tetsu-to-Hagané*, **73**(1987), 99.

[4] J. A. Sartell and D. J. Mack: *J. Inst. Metals*, **93**(1964-65), 19.

[5] H. Fredriksson and T. Nylen: *Metal Sci.*, **16**(1982), 283.

[6] T. B. Massalski: *"Binary Alloy Phase Diagrams"*, TMS(1986), 71.

IOP Conf. Series: Materials Science and Engineering 33 (2012) 012044 doi:10.1088/1757-899X/33/1/012044

Correlation between microstructure and yield strength of a high-strength cold rolled enameling steel

M Jiang [a], F J Zheng [a], X F Huang [a], T Dai [a], B W Krakauer [b], M F Zhu [a]
[a]Jiangsu Key Lab for Advanced Metallic Materials, Southeast University, China
[b]AO Smith Corporate Technology Center, Milwaukee, WI, 53224 USA

Email: zhumf@seu.edu.cn

Abstract. To develop strategies for modifying the enameling steel chemical composition and processing to minimize yield strength reduction during the enamel-fire anneal (EFA), the reasons for yield strength reduction of a high strength cold rolled enameling steel were studied by combining experiments and thermodynamic calculations. The yield strength reductions were evaluated from the primary strengthening mechanism using existing empirical relationships that relate microstructural parameters to yield strength. The results show that yield strength reductions resulted primarily from an increase in grain size and a decrease in dislocation density. The EFA appeared to have a negligible effect on the solubilities of Mn, Si, and P, and the solubilities are nearly equal to the bulk concentrations. On the other hand, the elements C and N are nearly completely tied up as precipitates in the form of cementite and AlN. Thus, they have a negligible contribution to solid solution strengthening.

1. Introduction

Glass-lined thin-walled pressure vessels are predominantly used for water heater storage tanks. The minimum thicknesses used for these pressure vessels are dictated by a safety requirement that stipulates minimal permanent deformation when a tank is subject to a static pressure load. As steel has become increasingly expensive, there is a need to drive the thicknesses to lower values. The ability to use thinner steels is hampered by the enamel-fire anneal (EFA) process that requires heat treating the pressure vessels in the range of 800~900℃ [1], since the yield strength decreases during the EFA [2,3]. Thus, in order to figure out how to reduce the steel thicknesses used to produce glass-lined thin-walled pressure vessels, the reasons for yield strength reduction due to the EFA should be understood.

Some studies have been performed on the mechanical properties of the enameling steels. Takahashi *et al* [1] reported that the after-fire yield strength is lower than the before-fire strength, but the reasons for yield strength reduction were not discussed. Researchers have reported that during firing, critical grain growth may occur with a simultaneous loss of yield strength [2,4]. For the Ti-bearing enameling steels, the volume fraction and size of fine precipitates also change during firing [3,5]. These experimental studies about the relation between microstructures and yield strength are, however, not complete enough to identify possible strategies for minimizing the yield strength reduction during the enameling firing process, because they either do not report on any of the metallurgical factors responsible for yield strength reduction or they only report on a few of the possible known factors, such as grain size and precipitates.

In this work, the microstructures of a high strength cold rolled (CR) enameling steel, used for these pressure vessels, were systematically characterized by combining experiments and thermodynamic calculations. Based on the characterized microstructures, numerical estimates for the components of yield strength were determined.

2. Experimental

The material used for the present investigation is a cold rolled (CR) enameling steel sheet. The chemical composition is given in table 1. The EFA was simulated by annealing the specimens with dimensions of 200×80×1.4mm in a box type furnace. The steel was annealed at 860℃ for 5 min. The average heating rate and cooling rate were about 20℃/min and 170℃/min, respectively.

Table 1. Chemical composition of the enameling steel (wt%).

C	Mn	Si	S	P	N	Al	Cu	Cr	Mo	Nb	Ti	Fe
0.06	1.29	0.45	0.003	0.048	0.013	0.034	0.07	0.03	0.01	0.002	0.002	bal.

Tensile tests were performed on standard sheet tensile specimens at a crosshead speed of 6 mm per min by using a CMT5105 tensile machine. The microstructures were observed with a FEI-Sirion scanning electron microscope (SEM) and a TECNAI G2 20 transmission electron microscope (TEM). The average grain sizes and pearlite volume fractions of the specimens were measured by means of the Image Tool software [6]. The dislocation density was determined by the line intercept method [7-9].

The thermodynamic calculations of the equilibrium phases and element solubilities were performed using Pandat that is a phase equilibrium calculation software of multi-component alloys [10].

3. Results and discussion

3.1. Microstructure and yield strength

Figure 1 shows SEM images of the CR steel before-and-after the EFA. Before the EFA, the microstructure is ferrite and free cementite. After the EFA, grain size increases, but it is not classified as critical grain growth. A small amount of pearlite appears at the grain boundaries, and the quantity of free cementite decreases. Figure 2 shows TEM images of dislocations in the steel before and after the EFA. Before the EFA, the steel has a moderate dislocation density, and dislocation tangles are observed in some areas. After the EFA, the dislocation density decreases significantly.

The grain sizes, pearlite volume fractions, dislocation densities, and yield strength of the steel before and after the EFA were measured, and are shown in table 2. The EFA increases the grain size of the steel by about 2 μm, and the percentage change in grain size is 21%. The pearlite volume fraction increases from 0 to 2.51%. The dislocation density decreases by 65%. The EFA causes the yield strength to decrease.

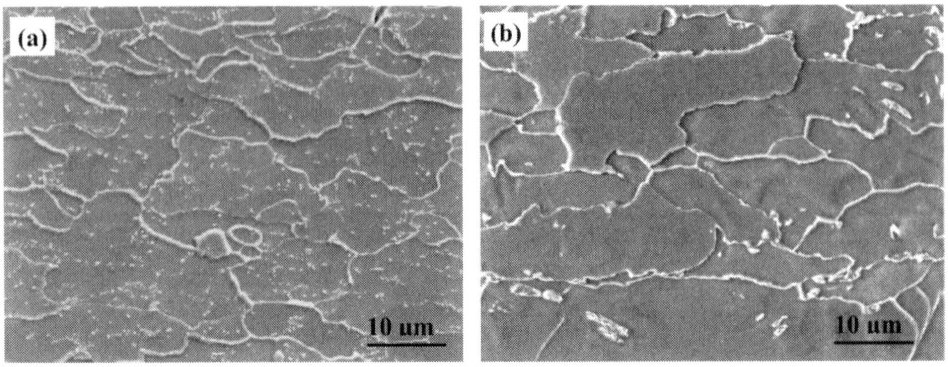

Figure 1. SEM images of the steel (a) before and (b) after the enamel-fire anneal (EFA).

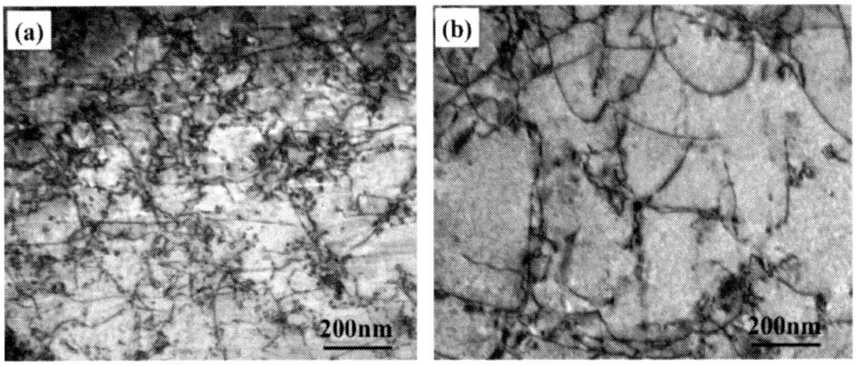

Figure 2. Dislocations in the steel (a) before and (b) after the enamel-fire anneal (EFA).

Table 2. Grain size, pearlite volume fraction, dislocation density, and yield strength of the steel.

Condition	Grain size (μm)	Pearlite volume fraction (%)	Dislocation density (10^9 cm/cm^3)	Yield Strength (MPa)
Before the EFA	8.7	0	12.2~18.2	412.61
After the EFA	10.5	2.51	4.2~6.3	351.11

3.2. Equilibrium phases and element solubilities

Figure 3 shows the mole fraction of the equilibrium phases in the CR steel as a function of temperature, obtained from the thermodynamic calculations using Pandat. According to figure 3, there are many equilibrium phases in the steel at room temperature, such as ferrite, $Mn_{23}C_6$, Fe_3P, AlN, Mo_6C, etc. However, microstructure observations show that except for ferrite and AlN, most of the equilibrium phases have not been found. On the other hand, a small amount of free cementite was observed as shown in figure 1, although almost no cementite exists at the room temperature according to the thermodynamic calculation as shown in figure 3 (b). It is known that the thermodynamic calculation is based on the equilibrium state, while the practical cases are mostly in the nonequilibrium. Thus, there must be some difference between the calculated data and experimental results.

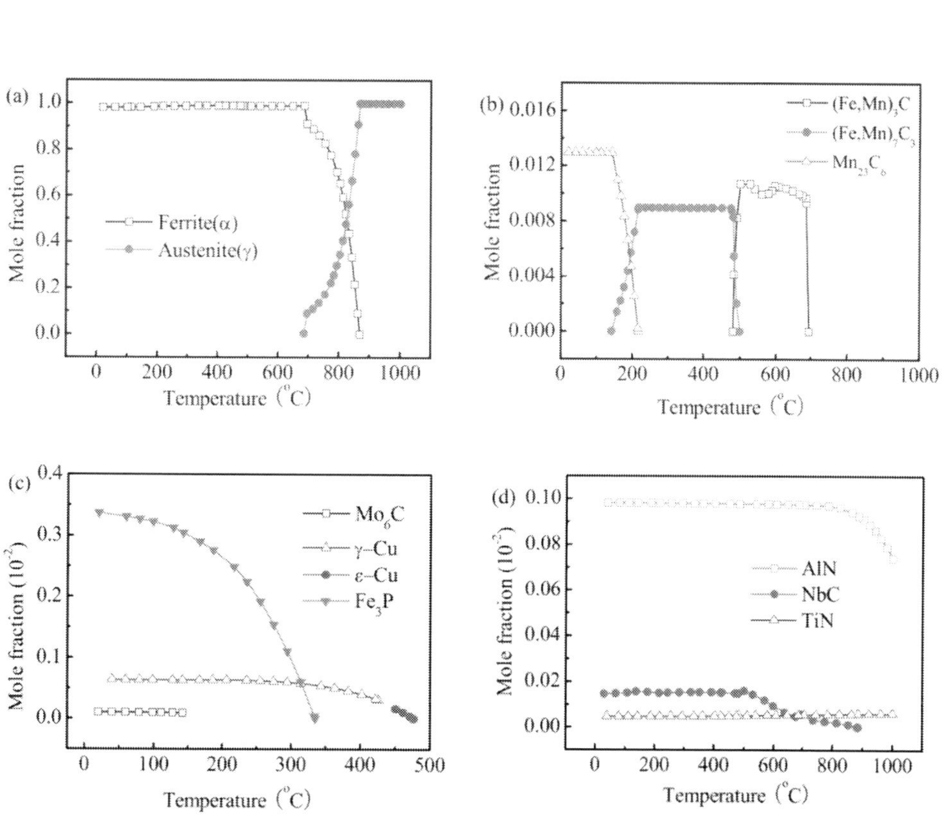

Figure 3. The equilibrium phases in the steel as a function of temperature: (a) ferrite and austenite, (b) $(Fe,Mn)_3C$, $(Fe,Mn)_7C_3$, and $Mn_{23}C_6$, (c) Mo_6C, γ-Cu, ε-Cu, and Fe_3P, (d) AlN, NbC, and TiN.

To estimate the contribution of the solid solution component to the yield strength reduction, the solubilities of the alloy elements before and after the EFA should be determined. Since it is difficult to experimentally measure the solubilities of elements, in the present work, the element solubilities of the steel before and after the EFA were estimated by thermodynamic calculations. Figure 4 shows the concentrations of Mn, P, Si, C, and N in austenite, ferrite, and precipitates of the steel at different temperatures obtained from the Pandat calculations. The elements Mn and P (see figures. 4 (a) and (b)) form precipitates when the temperature is below about 700℃ and 300℃, respectively. Pandat predicts that the precipitates are $(Mn,Fe)_7C_3$, $Mn_{23}C_6$, Fe_3P, etc. However, these phases were not observed during the microstructure analyses. Figure 4 (c) shows that the element Si is in complete solid solution. Figures. 4 (d) and (e) indicate that a significant portion of C and N is tied up as the precipitates. According to Pandat calculation and microstructure observation, the precipitates should be cementite and AlN.

Since the mechanical properties were measured at room temperature, the solubilities of Mn, P, Si, C, and N at room temperature should be obtained using the data in figure 4. However, the thermodynamic calculations of solubility shown in figure 4 are based on the equilibrium state. It is apparently not reasonable to use the equilibrium solubility of the elements before and after the EFA at the room temperature. It should be reasonable to estimate the element solubilities using the data in figure 4 in combination with the considerations of the element diffusivities and the processing conditions.

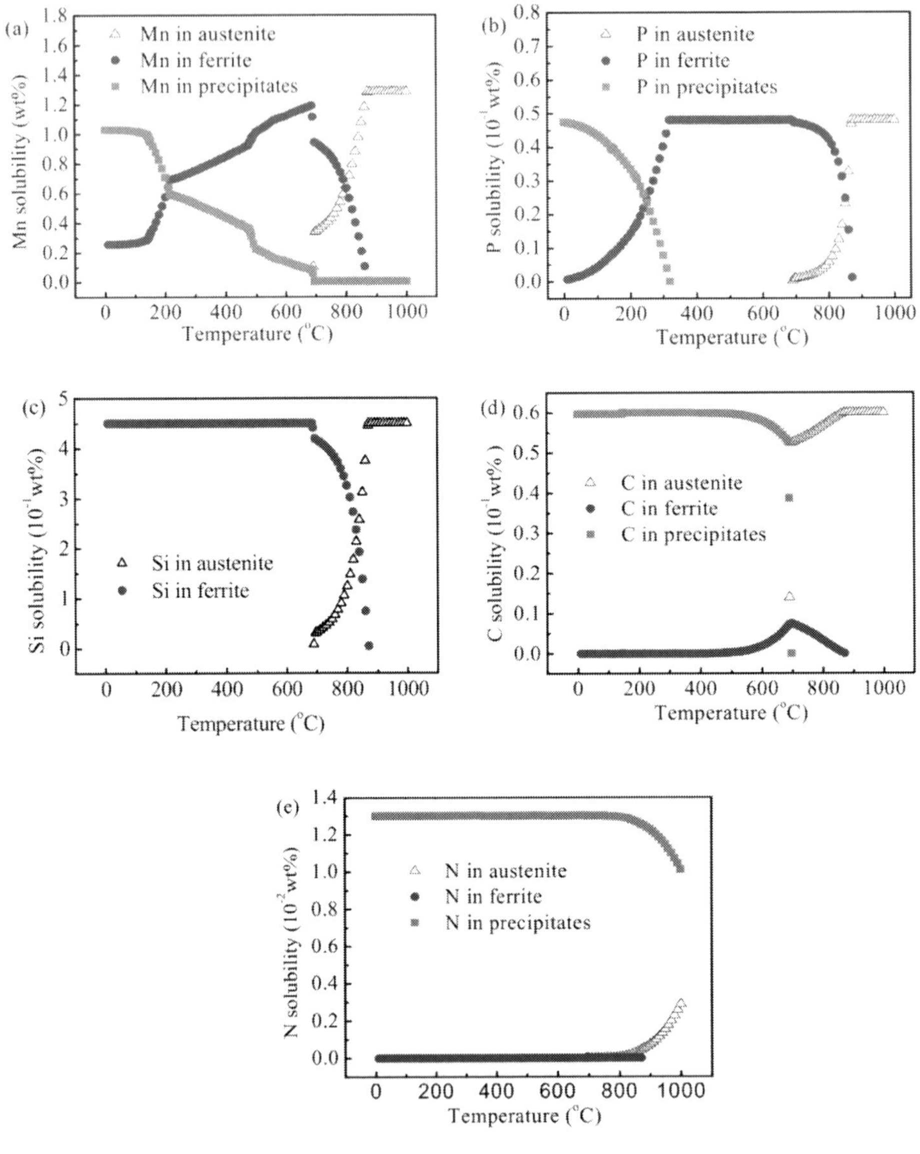

Figure 4. Pandat calculations of the solubilities of (a) Mn, (b) P, (c) Si, (d) C, and (e) N in austenite, ferrite, and precipitates of the cold rolled (CR) steel as a function of temperature.

Before the EFA, the steel was stress relief annealed at 700 ℃ for 20 h in a box anneal at the steel mill. Under those conditions, it is reasonable to assume that the solubilities of the elements dissolved in the matrix have reached equilibrium. In addition, since the diffusivities of the elements Mn, Si, and P are very low, around the magnitudes of 10^{-47}-10^{-16} m^2/s when the temperature is below 700℃, we assume negligible diffusion distances upon cooling. Accordingly, the solubilities of Mn, Si, and P were calculated at 700℃. The calculated solubilities of Mn, Si, and P are 1.2849%, 0.45%, and 0.048%, respectively. After the EFA, the steel was annealed at 860℃ for 5 min. The diffusivities of Mn, Si, and

P are in the range of 10^{-18}-10^{-15} m^2/s at 860℃. It is considered that the solubilities of those elements have reached the equilibrium during this anneal. Thus, the solubilities of Mn, Si, and P after the EFA were calculated at 860℃, and the calculated solubilities of Mn, Si, and P are 1.2849%, 0.45%, and 0.048%, respectively.

The diffusivities of the elements C and N are significantly larger than that of Mn, Si, and P, and are in the magnitudes of 10^{-12} ~ 10^{-10} m^2/s at 500℃~900℃ and 10^{-21} ~ 10^{-12} m^2/s at 25℃~500℃. We assume, therefore, that the solubilities of C and N are not changed by the EFA. The solubilities of C and N before and after the EFA were calculated over the temperature range of 25~500℃. The calculated solubilities of C and N are 1.20×10^{-22}~2.77×10^{-4} and 3.66×10^{-37}~2.14×10^{-10}. The calculated results show that the solubilities of the five elements before and after the EFA are identical. We assume, therefore, that the EFA has a negligible effect on the solubilities of Mn, Si, P, C, and N. The solubilities of Mn, Si, and P are close to the initial concentrations of those elements in both steels, whereas the solubilities of C and N are much lower than the initial concentrations. Although the elements C and N are strong solid solution strengtheners, for these enameling steels the effects of C and N will be diminished. The elements are, however, probably critical to achieve small grain sizes and to restrict grain growth during annealing [11].

3.3. Numerical estimates of the components of yield strength reduction

The components of the total yield strength, σ_y, of low carbon steels, exhibiting a ferrite-pearlite microstructure, can be represented by [12-14]:

$$\sigma_y = \sigma_0 + \sigma_s + \sigma_g + \sigma_p + \sigma_d \tag{1}$$

where σ_0 is the inherent friction stress, σ_s is the solid solution strengthening, σ_g is the grain size strengthening, σ_p is the precipitation strengthening, and σ_d is the dislocation strengthening. In general, σ_0 is constant for a given steel. In the present work, σ_0 was taken as 30 MPa [8] and assumed to not change before and after the EFA.

The contribution of solid solution strengthening, σ_s, in equation (1) was determined using the following empirical equation [15]:

$$\sigma_s = 4570[C] + 4570[N] + 37[Mn] + 83[Si] + 470[P] \tag{2}$$

where $[X_i]$ is the concentration of element X_i dissolved in the matrix, in wt.%. The values of X_i were estimated by Pandat as discussed in Section 3.2.

The grain size strengthening component in equation (1) was determined using the Hall–Petch equation [15-17]:

$$\sigma_g = k_y d^{-1/2} \tag{3}$$

where k_y is the grain boundary unlocking term, usually taken as 15.1–18.1 $Nmm^{-3/2}$ for high-angle boundaries [16]. In the current work, k_y was taken equal to 17.4 N $mm^{-3/2}$ [15,17]. The quantity d is ferrite grain diameter.

The contribution of dislocation strengthening was calculated with the equation [8,18-20]:

$$\sigma_d = a\mathrm{G}b\rho^{1/2} \tag{4}$$

where a is a constant taken as 0.38, G is the matrix shear modulus and equals to 8.3 GMPa for ferrite steels, b is the magnitude of the Burgers vector taken as 0.248 nm for ferrite [8,20], and ρ is the dislocation density.

For conventional C-Mn steels, the contribution of precipitation strengthening is usually negligible [12], and our microstructure observations corroborate this fact. Thus, the precipitation strengthening, σ_p, is not considered in the present work.

The contributions of various sources to the total yield strength of the steel and changes of the different strength factors after the EFA are given in table 3. For the before and after EFA conditions, grain size strengthening constitutes 44 to 47% of the yield strength, solid solution strengthening contributes 26 to 30%, and dislocation strengthening contributes 16 to 23%. Thus, before and after the EFA, the grain size strengthening provides the most important contributions to the yield strength for the steel. The second important factor is the solid solution strengthening, followed by the dislocation strengthening. The EFA appears to have a negligible effect on the solid solution strengthening component of the yield strength. The change in dislocation density and grain size are the primary factors responsible for yield strength reduction of the steel. The dislocation density decrease produces the primary contribution (about 70%) to the yield strength reduction.

Table 3. Before and after enamel-fire anneal (EFA) yield strength contributions (MPa).

Condition	Friction stress, σ_0	Solid solution strengthening, σ_s	Grain size strengthening, σ_g	Dislocation strengthening, σ_d	Calculated yield strength, σ_y*	Measured yield strength σ_Y
Before EFA	30	107~109	187	86~106	410~432	413
After EFA	30	107~109	170	51~62	358~371	351
Change	0	0	-17	-(35~44)	-(52~61)	-62

*Calculated yield Strength, $\sigma_y = \sigma_0 + \sigma_s + \sigma_g + \sigma_d$.

4. Conclusions

For the CR enameling steel analyzed in this study, dislocation density decrease and grain growth are the major contributors to yield strength reduction during the EFA. Dislocation density decrease is responsible for about 70% of the reduction in yield strength. The elements that contribute to the solid solution strengthening mechanism are Mn, P, and Si at the bulk concentration levels, and the EFA appears to have a negligible effect on this strengthening component of the yield strength. On the other hand, the elements C and N are nearly completely tied up as precipitates in the form of cementite and AlN. Thus, they have a negligible contribution to solid solution strengthening, but the elements are probably critical to achieve small grain sizes and to limit grain growth during the EFA.

Acknowledgements

The authors wish to thank CompuTherm Company, USA, for generously providing us with the license of Pandat software. This work was financially supported by the AO Smith Corporate Technology Center, USA.

References

[1] Takahashi I, Matsumoto Y and Aruga I 1981 *Kawasaki Steel Technical Report* **3** 57
[2] Jessman R J 1992 United States Patent, No. 5 137 584
[3] Wang D M, Zhang W S, Dong E S and Liu Z W 2008 XXI International Enamellers Congress (Shanghai) p 293
[4] Kostrubanic J M 1998 United States Patent, No. 5 782 999
[5] Sun Q S and Jin L 2007 *Journal of Iron and Steel Research* **19** 47
[6] Image Tool program developed at the University of Texas Health Science Center at San Antonio 1997 http://ddsdx.uthscsa.edu
[7] Young S K, Sung S K, Yong M C and Kyung S I 2003 *J. Nucl. Mater.* **317** 117
[8] Li Q 2003 *Mater. Sci. Eng. A* **361** 385
[9] Graça S, Colaço R, Carvalho P A and Vilar R 2008 *Mater. Lett.* **62** 3812
[10] Chen S -L, Daniel S, Zhang F, Chang Y A, Yan X -Y, Xie F -Y, Schmid-Fetzer R and Oates W A 2002 *Calphad* **26** 175
[11] Stas´ko R, Adrian H and Adrian A 2006 *Mater. Charact.* **56** 340
[12] Pradhan R R 1981 *J. Heat Treating* **2** 73
[13] Gawne D T and Lewis G M H 1986 *J. Mater. Sci.* **21** 547
[14] Misra R D K, Nathani H, Hartmann J E and Siciliano F 2005 *Mater. Sci. Eng. A* **394** 339
[15] Mao X P, Huo X D, Sun X J and Chai Y Z 2010 *J. Mater. Process. Technol.* **210** 1660
[16] Yu H, Chen Q X, Kang Y L and Sun Y 2005 *Mater. Charact.* **54** 347
[17] Yi H L, Du L X, Wang G D and Liu X H 2008 *J. Iron. Steel. Res. Int.* **15** 76
[18] Zhang Z Y, Yu H S, Wang S Q, Wang H T and Min G H 2010 *J. Mater. Sci. Technol.* **26** 151
[19] Jin Z Y and Cui Z S 2010 *J. Mater. Sci. Technol.* **26** 1063
[20] Wang R Z and Zhang H T 2007 *Acta. Metall. Sin.* **43** 1082

Using thermoelectric magnetohydrodynamics to control microstructural evolution

A Kao[1] and K Pericleous
School of Computing and Mathematical Sciences, University of Greenwich, Old Royal Naval College, Park Row, London, SE10 9LS

E-mail: a.kao@greenwich.ac.uk

Abstract. Using a fully coupled 3-dimensional numerical model, predictions of the changes to the morphology for equiaxed and columnar crystals in the presence of an external magnetic field are calculated. Significant changes include deflection of the dendrite arms, changes to the tip velocity and initiation of secondary branching. The changes to the morphology are also highly dependent on both the orientation and strength of the magnetic field. The mechanism for these changes can be identified by investigating into the Thermoelectric Magnetohydrodyhamics of the system. A wide variety of unexpected flow features are predicted in the vicinity of the solid front, including global circulations that encompass a dendrite, microcirculations around the dendrite tips, inter-dendritic flow between the secondary arms and complex helical flow along the dendrite tips. This numerical investigation provides a theoretical foundation of how the employment of external magnetic fields may be used to influence and control solidification processes.

1. Introduction

Recently attention has been directed towards the feasibility of using external magnetic fields to control the evolution of the dendrite morphology. This is of particular interest as current in-situ mechanisms generally rely solely on controlling the thermal gradients via some form of external cooling. The addition of a magnetic field acting on the liquid/solid boundary, provides an additional independent control parameter. During solidification, thermoelectric currents form due to variations of the interfacial temperature and a discontinuity in the Seebeck coefficient. In the presence of an external magnetic field these currents interact to form a Lorentz force which in the liquid phase drives fluid motion. This effect known as Thermoelectric Magnetohydrodynamics (TEMHD) causes convective transport to redistribute heat and mass altering the local free energy at the interface and modifying solidification. Changes to the dendrite morphology have been observed experimentally [1,2] and these changes have been attributed to this effect. However, only a primitive fundamental understanding of the underlying mechanism on the scale of a dendrite currently exists. This paper focuses on highlighting a variety of flow mechanisms through a fully coupled 3-dimensional numerical model providing an extension to previous work [3-6], where the focus was on simplified 3-dimensional solutions and quasi 3D-2D solutions.

[1] To whom any correspondence should be addressed.

The Seebeck effect [7] describes how an e.m.f ($\Delta\Psi$) forms when two materials with a Seebeck power (ΔS) are placed in thermal contact and that at the boundary this e.m.f varies with the surface temperature. A generalised form of Ohm's law includes a contribution from the Seebeck effect.

$$\mathbf{J} = \sigma\left(\mathbf{E} + \mathbf{u}\times\mathbf{B} - S\nabla T\right) \tag{1}$$

At the interface the potential difference can be written as

$$\Delta\Psi = \Delta ST \tag{2}$$

The volumetric enthalpy in terms of the specific heat capacity (c), liquid fraction (f) and latent heat (L), and concentration potential (V) in terms of the concentration (C) and partition co-efficient (k) are defined by

$$H = cT + fL \qquad\qquad V = \frac{C}{f(1-k)+k} \qquad (3), (4)$$

The conservation of enthalpy and solute are governed by

$$\frac{\partial H}{\partial t} = \nabla\cdot(K\nabla T) - \nabla\cdot(\mathbf{u}H) \qquad\qquad \frac{\partial C}{\partial t} = \nabla\cdot(D\nabla V) - \nabla\cdot(\mathbf{u}C) \qquad (5), (6)$$

For an idealised case the thermal conductivity $K_s = K_l$ and the Scheil assumption is used for the mass diffusivity $D_s = 0$. The interfacial temperature (T^i) is defined by the Gibbs-Thompson condition

$$T^i = T_m - \frac{\gamma(\theta,\phi)T_m}{L}\kappa - m\left(C_0 - C_l^i\right) \tag{7}$$

where the second term accounts for local free energy through surface energy anisotropy and curvature and the final term accounts for solute partitioning in binary alloys. This equation leads to a variation in surface temperature and the formation of thermoelectric currents irrespective of any external magnetic field. In the presence of an external magnetic field these currents interact to form a Lorentz force which is introduced into the full Navier-Stokes equations as well as a solidification resistance term to calculate the fluid flow.

$$\frac{\partial \rho\mathbf{u}}{\partial t} + \nabla\cdot(\rho\mathbf{u}\mathbf{u} - \mu\nabla\mathbf{u}) = -\nabla p + \mathbf{J}\times\mathbf{B} - R(f) \tag{8}$$

2. Results

2.1 Single Crystal Simulations

Consider supercooled solidification of an equiaxed dendrite in the bulk of the material. It has been demonstrated [3-6] that under these conditions thermoelectric currents will circulate between the cold tip and the hot root of the crystal. The current is localised very close to the interface with a maximum at the tip. In the presence of a magnetic field thermoelectric currents interact with it and the resulting cross product generates a Lorentz force that drives fluid motion. For the case of a field aligned along the direction of solidification a series of vortices form. A pair of vortices form at each of the tips tangential to the magnetic field, a global circulation traverses the entire dendrite passing above (and below) the tangential arms and a circulation forms around the tips parallel to the magnetic field. Each of these features can be seen in figure 1 where, in this example, the crystal is at an early stage of growth and the magnetic field is orientated in the positive *z* direction with an infinitesimally small magnitude. As the magnetic field increases, significant morphological changes can occur due to the convective transport of heat and mass. The term moderate magnetic field will be used frequently within this paper. Quantifying this statement is not a straightforward process. The numerical model is dimensionless and represents a family of solutions, where the scaling factors are determined by dimensioned material properties such as thermal diffusivity and electrical conductivity. The strength of magnetic field required to observe significant morphological changes is also dependent on many operational parameters including (and by no means limited to) the electrical properties of the material, the viscosity, the external thermal conditions, the dendrite arm spacing and the tip velocity. For example, in undercooled growth where the tip velocity is very fast, a moderate field could be in the region of 4T to 6T, while for a directionally solidified cast similar changes may occur around 1T.

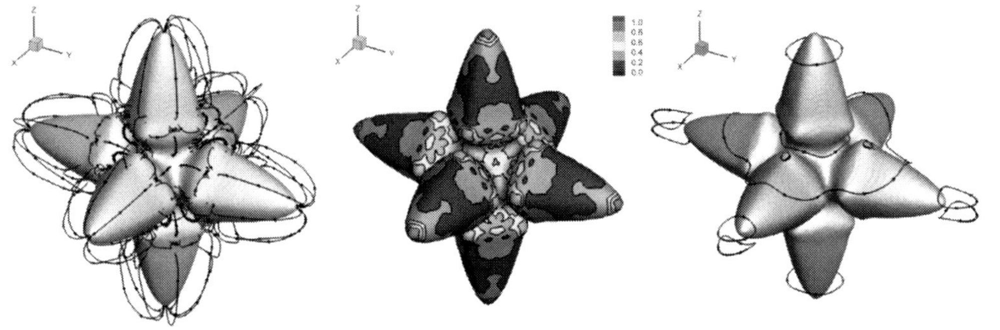

Figure 1. Early stages of dendrite growth in the presence of an infinitesimally small magnetic field. Left: Direction of **J**. Centre: Relative magnitude of **J** at the surface. Right: Direction of **u**.

Figures 2 and 3 show a comparison of the transient morphology development between a dendrite with zero magnetic field and under a moderate *z*-oriented field. Artificial noise was introduced into the heat balance at the interface to initiate secondary growth. There are two motivations for introducing such a noise; the first is numerical as the method here is derived directly from Voller [8], who uses smoothing functions at the interface when calculating the curvature. Consequently this prevents the anisotropy in the surface energy from forming secondary branches. The second is to account for physical random fluctuations analogous to Brownian motion. When a dendrite develops secondary branches, the system becomes increasingly complex as each secondary branch also contributes to the formation of thermoelectric currents and becomes intimately coupled to the Lorentz force and fluid flow. A deeper explanation of the influence of secondary branching is given in Kao *et al.* [6], however to highlight the underlying mechanisms, this noise has purposely been omitted from all cases except for figure 2. Otherwise the initiation of secondary arms is caused purely by TEMHD effects disrupting the thermal and solute boundary layers.

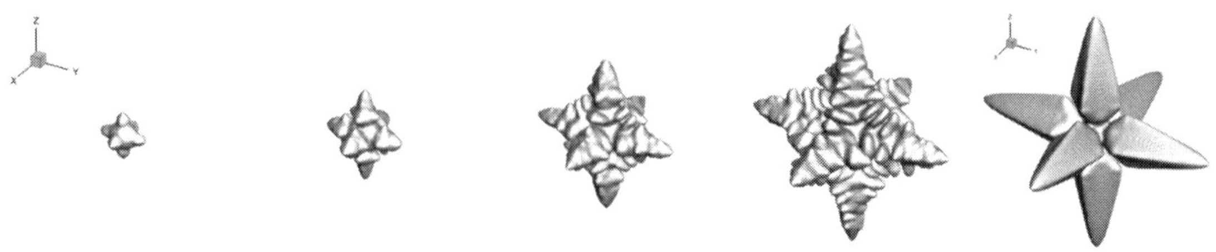

Figure 2. Transient morphology of a single dendrite growing with an artificially introduced noise and no magnetic field. Far right: A dendrite with no artificial noise.

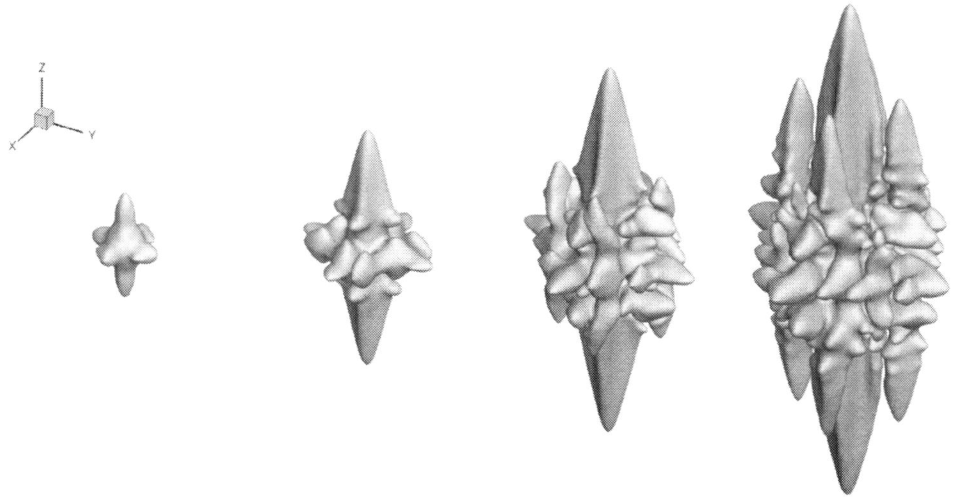

Figure 3. Transient morphology of a single dendrite growing under a moderate Bz magnetic field.

There are several notable differences between the cases with a magnetic field and without. In the magnetic field case the secondary branches are predominantly growing in a clockwise sense and this is due to the tangential tip vortices and global circulation that preferentially transport cold bulk material towards one side of the dendrite arm. The tangential tips are slower due to a deflection in the growth direction caused by the flow; the arm grows off its preferential direction as the tip vortex redistributes hot and solute rich material back towards the tip. These mechanisms where originally proposed via a quasi 3D-2D model [3-4]. However, under full 3-dimensional conditions it is also seen that the **B**-parallel tip is faster than in the zero magnetic field case. To demonstrate the mechanism for this change and also its applicability in conventional systems consider a directionally solidified dendrite with an imposed thermal gradient in the direction of solidification (figure 4). Assuming that this imposed external gradient dominates the interfacial temperature of the dendrite then the thermoelectric potential difference can be represented as

$$\Delta\Psi = \Delta STz \tag{9}$$

where the direction of solidification and thermal gradient are both in z. This creates a simplified yet qualitatively similar system of the primary z tip in figure 3. The bottom surface of the domain is considered to be a wall with the same electrical properties as the dendrite, thus **J** continues through this wall and does not circulate at the root of the dendrite. Assuming that the dendrite is no longer growing having reached equilibrium with the external thermal gradient and that this is a single crystal located somewhere in the bulk of a material then the system can be considered to be steady state and a constant pressure can be applied to the external boundaries. A physical representation of this situation is that of single crystal growth in a turbine blade after the dendrite has emerged from the grain selector. In the presence of a parallel magnetic field the Lorentz force will be formed only by the interaction with the radial component of **J**, located close to the tip of the dendrite. The force will be rotational causing a vortex around the tip to form in a similar fashion to figure 1. To conserve angular momentum of this vortex a centrifugal force will form that can only be balanced by a radial pressure gradient as both the viscous and electromagnetic damping terms will only have angular components, thus a low pressure region forms at the tip. Consequently a pressure gradient in the z direction will form, which can only be balanced by viscosity as the driving and damping terms of the Lorentz force act only in the x-y plane. The net effect is a jet directed toward the z-tip, which turns into helical flow that spirals down the dendrite. This incident flow onto the tip increases the local free energy and for transient cases increases the tip velocity as seen in figure 3. Thermal buoyancy will act against the downward velocity while mass buoyancy will be entirely dependent on the relative densities of the solute and the bulk material. However for the purpose of this exercise buoyancy effects have been neglected and the system can be considered in micro gravity conditions.

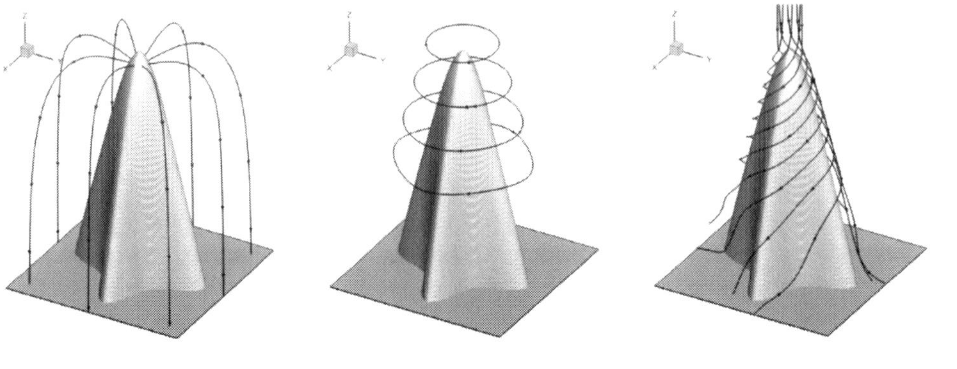

Figure 4. Single crystal growth from a surface with thermoelectric conditions analogous to directional solidification under a *Bz* magnetic field. Left: **J**. Centre: Lorentz force. Right: **u**.

As an extension to the previous example and also a more realistic interpretation of conventional directional solidification systems, consider the crystal to now be solidifying amongst many crystals in a regular cubic array. The tangential boundaries can be considered to be periodic for each of the flow variables, while the upper boundary remains at a constant pressure. Figure 5 shows the morphology and the various flow features for this case. A helical flow forms around the dendrite in a similar fashion to the previous example, however to satisfy continuity a secondary counter-flowing helical flow away from the front forms between consecutive dendrites. This flow is not driven by the Lorentz force as **J** contains only a z component and will not interact with the magnetic field, but is a consequence of viscous forces from the original helical flow. In between these two flow features a complex toroidal helical vortex forms.

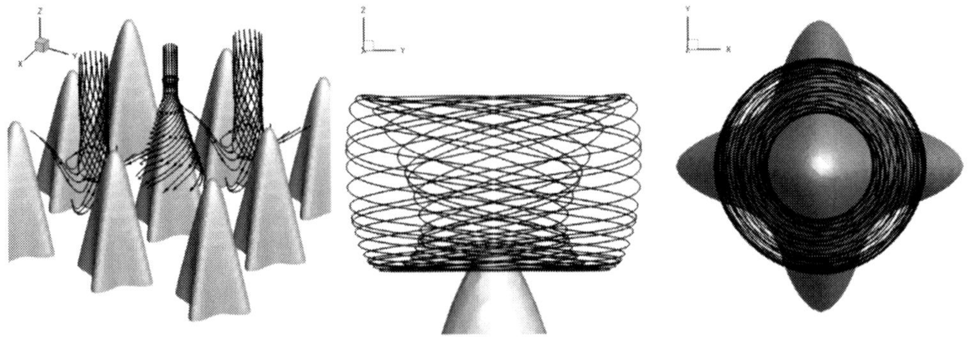

Figure 5. Growth of a regular array from a surface with thermoelectric conditions analogous to directional solidification under a z magnetic field. Left: Helical jets form at the tip and in between the primary arms. Centre: Toroidal helical flow in side view. Right: Toroidal helical flow top view.

2.2 Multiple Crystal Simulations

Returning to free-growth conditions, consider a small cluster of five nuclei solidifying from a surface. Figure 6, shows the transient growth for a z-orientated field. Many of the features of the single crystal case can also be observed in this case, including secondary branching and an extension of the primary z tip. However the interaction between neighbouring crystals provides an interesting insight into the overall morphology. Figure 7, shows the morphology looking down the z tip and a similar behaviour is observed to previous 2D work [5]; a competition to release latent heat occurs and the central dendrite becomes stunted in growth in the x-y plane, the external branches are slightly deflected from the crystallographic orientation and secondary branching is predominantly in the clockwise sense. Further examination into the fluid flow between the primary arms indicates that a relatively high velocity can exist within the inter-dendritic regime. This can be attributed to the influence of the secondary branches [8], which under these conditions will also contribute to the local thermoelectric currents driving a Lorentz force. An indication of the relative fluid flow is given in figure 8.

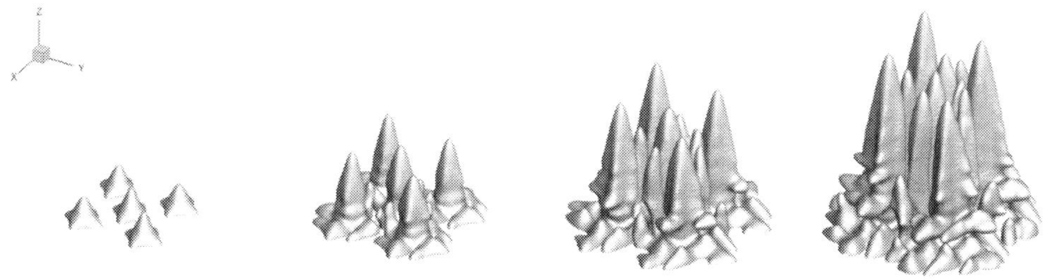

Figure 6. Transient morphology of multiple dendrites growing under a moderate Bz magnetic field.

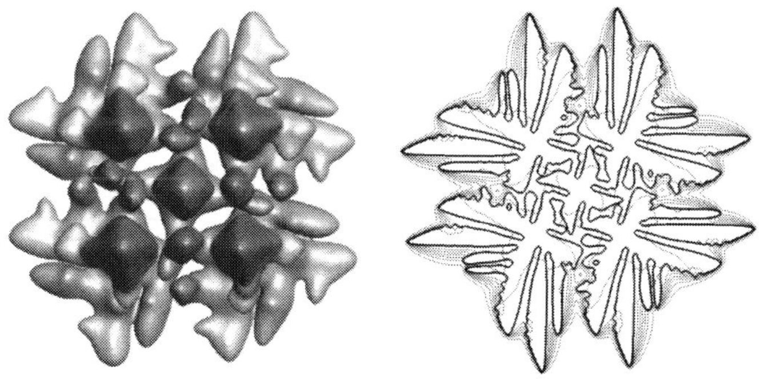

Figure 7. Comparison of morphology in the plane parallel to the magnetic field. Left: Current 3-dimensional results. Right: Quasi 3D-2D results from Kao *et al.* [5].

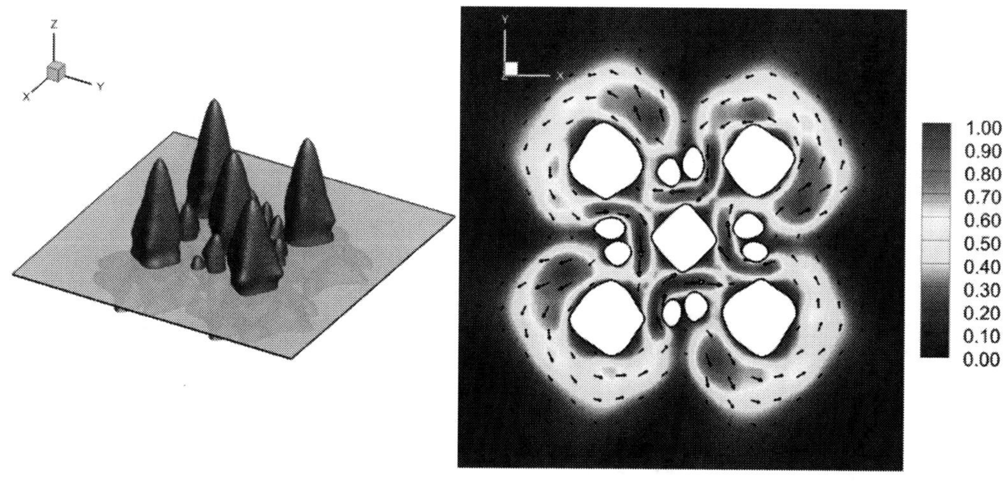

Figure 8. Fluid flow in multiple crystals under a Bz magnetic field. Left: Location of slice. Right: Relative **u**.

Under a magnetic field transverse to the direction of growth an entirely different morphology is obtained. Figure 9 shows the transient morphology of the same initial cluster of nuclei under a *y*-oriented magnetic field. The *z* tip is considerably slower than in the *z*-oriented magnetic field case, secondary branching is initiated and a complex 'interlocking' mechanism between secondary *y* branches and tertiary *z* branches is observed which is certain to influence mechanical properties. Figure 10 shows a cross section in the *x*-*y* plane revealing a relatively high velocity flow passing through the cluster of dendrites on the inter-dendritic scale. In this case the current that emanates from the primary *z* tips and passes down to the root will also interact with the magnetic field generating a force in the *x* direction. The wall where the dendrites nucleate from prevents a global circulation similar to the single crystal case from forming in the *x*-*z* plane and the flow now passes between the dendrite arms. In both the *z* and *y* magnetic field cases the secondary arm spacing is notably reduced and as a consequence of bulk flow being introduced into the inter-dendritic region allowing secondary branches to grow further into the primary arm spaces.

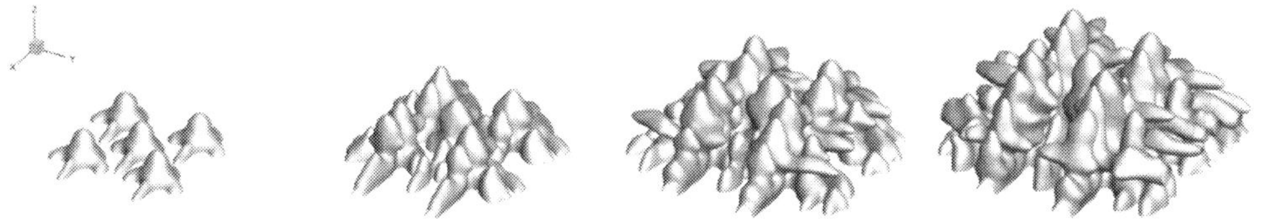

Figure 9. Transient morphology of multiple dendrites growing under a moderate B*y* magnetic field.

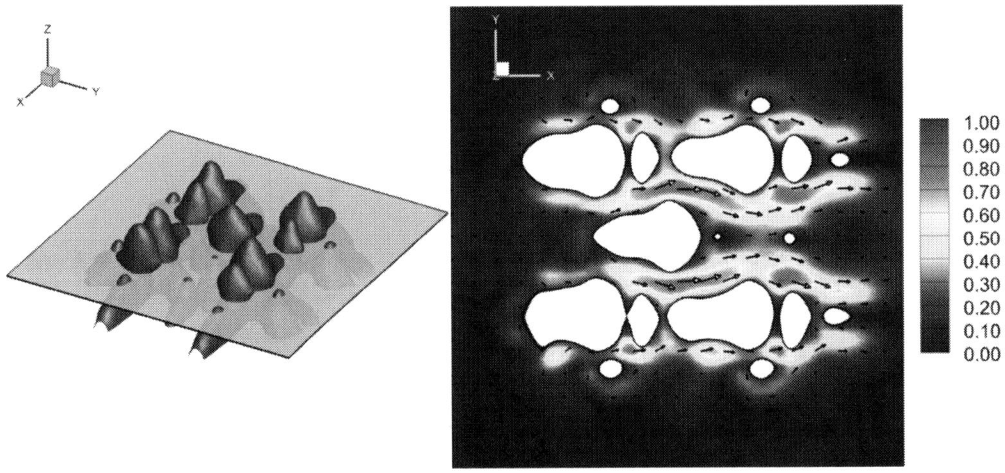

Figure 10. Fluid flow in multiple crystals under a *By* magnetic field. Left: Location of slice. Right: Relative **u**.

In many of the examples shown a common result through various mechanisms is that of a microscopic liquid jet. To give an appreciation to the magnitudes involved, it can be shown that under moderate-high magnetic field conditions a good approximation to the velocity in the plane of the Lorentz force is given by [6-7]

$$|\mathbf{u}| = \frac{\Delta S|\nabla T|}{|\mathbf{B}|} \tag{10}$$

Taking favourable extremities for solidification processes, where for undercooled growth it is possible for $\nabla T = O10^{7}$ K/m and $\Delta S = O10^{-5}$ V/K. In the presence of a 10T field gives $\mathbf{u} = O10$ m/s. However, it is anticipated that the maximum velocity will be confined to a microscopic boundary layer either within the inter-dendritic region or a circulation close to the interface and will not extend into the bulk of the material, where electromagnetic damping will dominate.

3. Conclusion

The introduction of micro fluid flow through TEMHD into solidification has been shown experimentally to have profound effects on the evolution of the microstructure. These changes are highly dependent on both the strength and orientation of the magnetic field. The mechanisms presented provide an insight into the complex nature of how these changes may occur. With a greater understanding of the underlying fundamental mechanisms it is conceivable that this effect can be used to control microstructural evolution and through it material properties.

4. References

[1] Li X, Fautrelle Y and Ren Z 2007 Influence of an axial high magnetic field on the liquid–solid transformation in Al–Cu hypoeutectic alloys and on the microstructure of the solid *Acta Materialia* **55** pp 1377-1386

[2] Moreau R, Laskar O and Tanaka M 1993 Thermoelectric magnetohydrodynamic effects on solidification of metallic alloys in the dendritic regime *Materials Science and Engineering* **173** pp 93-100

[3] Kao A, Djambazov G, Pericleous K and Voller V 2009 Thermoelectric MHD in dendritic solidification *Magnetohydrodynamics* **45** (3) pp 305-315

[4] Kao A, Pericleous K, Patel M K and Voller V 2009 Effects of magnetic fields on crystal growth *Int.J. Cast Metals Research* **22** (1-4) pp 147-150

[5] Kao A, Pericleous K, Patel M K and Voller V 2009 Thermoelectric effects on alloy solidification microstructure *Modeling of Casting, Welding, and Advanced Solidification Processes - XII (MCWASP XII). The Minerals, Metals & Materials Society, Warrendale, Pennsylvania, USA* pp 521-528

[6] Kao A and Pericleous K 2011 The effect of secondary arm growth on thermoelectric magnetohydrodynamics *Proceedings of the 8th International PAMIR Conference on Fundamental and Applied MHD, Borgo, Corsica, France* pp 837-841

[7] Shercliff J A 1979 Thermoelectric magnetohydrodynamics *J. Fluid Mechanics* **91** part 2 pp 231-251

[8] Voller V 2007 An enthalpy method for modeling dendritic growth in a binary alloy *Int J. Heat Mass Transfer* **51** pp 823-834

Simulation of a directional solidification of a binary Al-7wt%Si and a ternary alloy Al-7wt%Si-1wt%Fe under the action of a rotating magnetic field

O Budenkova[1,4], F Baltaretu[2], J Kovács[3], A Roósz[3], A Rónaföldi[3], A-M Bianchi[2] and Y. Fautrelle[1]

[1] SIMAP/EPM, Grenoble Institute of Technology, France
[2] Technical University of Civil Engineering, Bucharest, Romania
[3] HAS/MU/Materials Science Research Group, Miskolc-Egyetemváros, Hungary

E-mail: olga.budenkova@gmail.com

Abstract. We propose an approach for the solidification of metallic alloys taking into account the turbulence in the liquid phase. For modeling of the turbulent flow a k-ε model is used. This model is applied for solidification of AlSi-based alloys under the action of rotating electromagnetic field of intensity up to 65mT. Two cases have been considered, namely solidification of a binary Al-7wt%Si and a ternary alloy Al-7wt%Si-1wt%Fe. It is shown that increase of the electromagnetic field leads to the formations of various types of channels.

1. Introduction

Segregation is a very important and well-known phenomenon which appears during the solute-rich alloy solidification often resulting in deterioration of the properties of the material. Segregation may occur on various scales in the ingot. Microsegregations are linked to composition variations on the scale of the lowest solid structure, whereas macrosegregations concern the scale of the product. Convection may strongly affect the 'regular' segregation linked to the partitioning of solute at the liquid-solid interface. Convection may originate due to thermo-solutal buoyancy [1,2] but also due to other forces, e.g. electromagnetic stirring [3] or volume changes due to the shrinkage. It has been shown experimentally and numerically that fluid flow, especially electromagnetically driven flows, can produce macrosegregations [4-7] and change the structure [8]. The present paper analyses the effect of the turbulent forced convection on the segregations during solidification of alloys. The forced convection is generated by a rotating magnetic field (referred hereafter to as RMF).

The actual numerical models related to solidification of alloys involve coupled heat, momentum and solute transport phenomena in a two-phase media. The most recent models are based on volume averaging techniques [9-11]. Although fruitful, that approach has various limitations. The choice of the representative elementary volume may be questionable due to the multiplicity of the spatial scales. Indeed, the volume-averaging technique exerts an uncontrolled cut-off of the spatial scales. This may be crucial, firstly, when we deal with the meso-scale structures like the segregated channels. Secondly,

[4] To whom any correspondence should be addressed

the liquid motion outside the mushy zone can be turbulent, especially when electromagnetic stirring is used, e.g., in continuous casting of steel or aluminium processing. In that case, the spatial averaging procedure related to a phase transition may not easily be extended to account for a turbulent character of the liquid flow. A statistical phase averaging technique allows one to avoid problems related to various spatial scales. The present investigation consists in modeling directional solidification experiments performed on ground under a RMF in the frame of ESA-MICAST project [12]. Aluminium based alloys are considered, a binary Ai-7wt%Si and a ternary Al-7wt%Si-1wt%Fe. The general equations are briefly presented in Section 2 and in the Appendix. After a short description of the solidification facility in Section 3, the turbulent flow field is analyzed in Section 4. Then, the results of the modeling of the solidification are provided in Section 5 for the binary alloy and Section 6 for the ternary alloy.

2. Numerical approach

The model which is dealt with is obtained from a two-phase approach based on the ensemble averaging procedure [13,14]. One of the main advantages of that approach is the possibility to account naturally for fluctuations when the flow is turbulent as in the present paper. We consider a binary/ternary fluid mixture which is solidifying in the columnar regime. The approach is described in previous papers and a short summary is provided in Annex 1. As far as the ternary alloy is concerned, an additional transport equation is used to account for the transport and segregation of the iron. For the micro-modeling for the binary alloy we use an approach based on the dendritic envelope proposed by Rappaz and Thevoz [15] with its extension for the ternary alloy proposed by Rappaz and Boettinger in [16]. Simulations are performed in ANSYS/Fluent software and are based on Euler-Euler two-phase model which can employ either laminar or turbulent fluid flow. In order to account for a phase transition and phase interaction with a chosen approach the home-made codes have been developed and introduced in a program via user-defined functions.

3. Description of the geometry

A schematic view of the facility used in SGMU (Miskolc, Hungary) is given in figure 1. The solidification is directional with an imposed thermal gradient and cooling rate. A cylindrical crucible of internal diameter of 8mm is placed inside a stirrer, which generates a rotating magnetic field with the frequency 50 Hz with a single pole pair. For more detailed description of the set-up the reader is referred to [17].

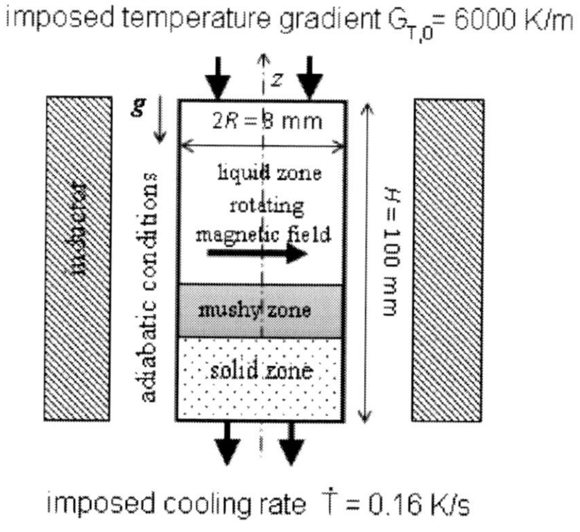

Figure 1. Sketch of the geometry and solidification conditions

4. The rotating turbulent flow

The aim of the present section is to characterize the electromagnetically-driven flow. The flow generated by the rotating magnetic field consists of an azimuthal flow with superimposed meridional recirculations. Owing to the large aspect ratio *H*/*R*, *H* and *R* being respectively the height and the radius of the sample, we use an analytical approximation of the azimuthal component F_θ of the electromagnetically forces $\mathbf{F}_{em}$ valid for an infinitely long geometry, namely

$$F_\theta = \frac{1}{2}\sigma\omega B_0^2 r\, S\,,\ S = 1 - \frac{u_\theta}{\omega R} \tag{1}$$

S being the slip, $\sigma\ (\Omega^{-1}\cdot m^{-1})$ and $\omega = 2\pi\cdot f$ respectively denoting the electrical conductivity of the melt and the angular frequency of the applied magnetic field B_0 (maximum value in tesla), r and R is a radial coordinate in a crucible and the radius of the latter, given in meter. The force expression (1) is valid since the electromagnetic skin depth $\delta = 2.5\cdot 10^{-2}$ m is much larger than the pool radius. In the present case we have assumed that the slip was close to unity since ωR is equal to 1.25 m/s and is large compared to the azimuthal velocity u_θ.

The main non-dimensional parameter which characterizes the flow regime is the magnetic Taylor number Ta_m defined as

$$Ta_m = \frac{\sigma\omega B_0^2 R^4}{2\rho v^2} \tag{2}$$

ρ and v being the density and the kinematic viscosity of the liquid. In the cases considered below, the values of the magnetic Taylor number ranges from 10^4 to $5\cdot 10^5$. The intensity of the fluid flow is characterized with a Reynolds number which is defined as $\mathrm{Re} = U_{max} R / v$, U_{max} is the maximal calculated magnitude of the velocity.

In our previous work [18] simulations for $Ta_m = 10^4$ and $Ta_m = 2.32\cdot 10^4$ ($B_0 = 13.5mT$ and $B_0 = 20mT$) were performed with an axisymmetric laminar model for the fluid flow. For these two cases the value of U_{max} was 0.11 m/s and 0.2 m/s, respectively, that corresponds to $\mathrm{Re} = 409$ and $\mathrm{Re} = 767$. Despite the moderate values of Re, the flow at $Ta_m = 10^4$ already displays instabilities in a form of vortices generated along the crucible and moving toward its ends (figure 2a). Note that the results are consistent with calculations performed in the laminar case for weaker magnetic Taylor number [6, 7]. With increase of a magnetic field, both the rotational and the translational velocity of the vortices increase (figure 2b) and not all of them could be resolved. It is known that these instabilities originate near the side wall where the rotation flow decays to zero [19] and that further increase of the external force results in various non-stable convective regimes and finally leads to the turbulent flow. In order to take into account the turbulence we used a standard *k*-ε model, referred hereafter to as k_t-ε_t to avoid any confusion (see some details in Annex). Firstly, we applied the k_t-ε_t turbulent model to the case calculated with a laminar approach for $Ta_m = 2.32\cdot 10^4$ ($B_0 = 20mT$). An instantaneous stream function obtained with a turbulent model for the mean flow velocity is shown in figure 2c. The flow exhibits a structure with two meridional vortices along the crucible. Maximal value of the mean velocity of the liquid is 10.5 cm/s and the oscillating velocity estimated via turbulent kinetic energy gives a value about 3 cm/s. Further we have performed several simulations for different values of the magnetic field in order to understand the effect of the latter on the liquid flow. It was found that increase of the electromagnetic force leads to the re-appearance of multiple vortices (figure 2d), similar to ones obtained with the laminar model. In numerical simulations with a turbulent model a transition from the two-vortices flow to a multiple vortices regime occurs at $Ta_{mc} = 1.75 \cdot 10^5$ ($B_0 = 48mT$).

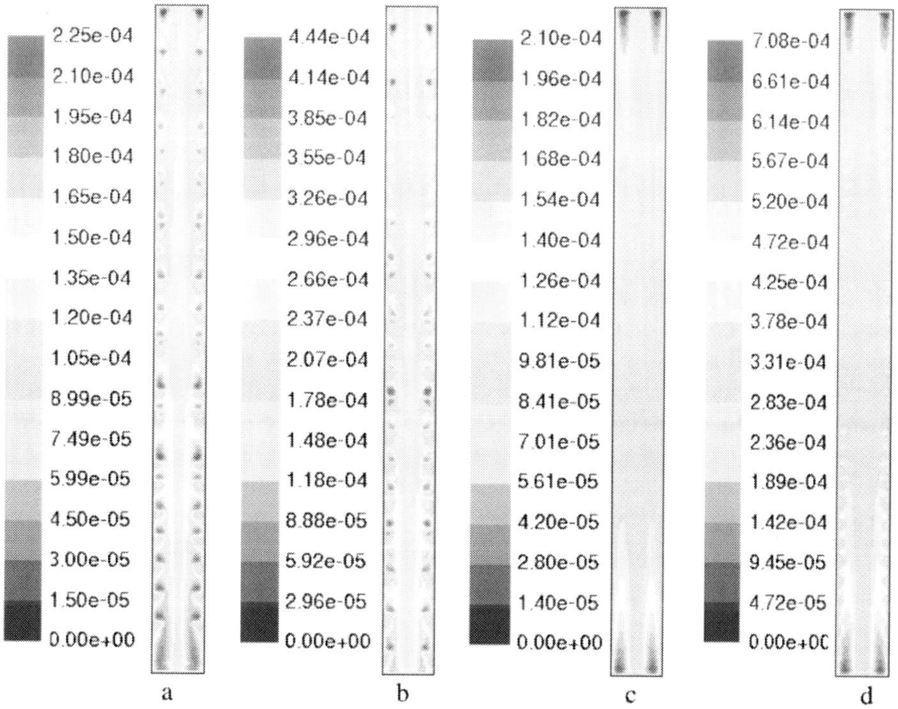

Figure 2. Instantaneous stream functions of flow of a molten Al-7%wtSi alloy in a crucible of a height $0.1m$ and a radius $4 \cdot 10^{-3} m$ under the action of various RMF.
Laminar model: a) $Ta_m = 1.0 \cdot 10^4$ ($B_0 = 13.5mT$) b) $Ta_m = 2.32 \cdot 10^4$ ($B_0 = 20mT$)
Turbulent model: c) $Ta_m = 2.32 \cdot 10^4$ ($B_0 = 20mT$) d) $Ta_m = 2.32 \cdot 10^5$ ($B_0 \approx 65mT$)

The dependence of a flow Reynolds number on the Taylor number shown in Fig. 3 confirms this observation since at the critical value $Ta_{mc} = 1.75 \cdot 10^5$ the slope of the curve changes. It is noticeable that the magnitude of the velocities may reach quite high values.

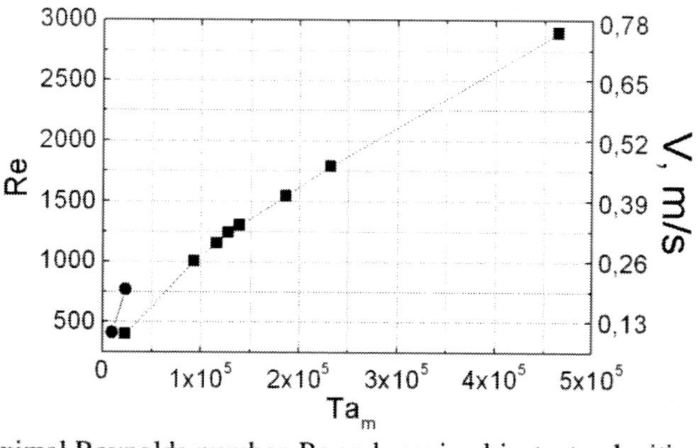

Figure 3. Maximal Reynolds number Re and maximal instant velocities as a function of the magnetic Taylor number Ta_m, computed with a laminar (circles) and a turbulent (squares) model for fluid flow

5. Solidification of a binary Al-7wt% Si alloy

In the present section the macrosegregations promoted in the Al-7wt%Si ingot by the forced flow is discussed. The thermophysical properties of the alloy can be found elsewhere [7], geometry and solidification conditions are given in figure 1. Apart from them, the model requires the primary and the secondary dendritic arm spacing as input parameters which were selected to be $\lambda_1 = 300\mu m$ and $\lambda_2 = 50\mu m$, respectively that correspond to the values measured in the experimental samples. Note that the value of λ_2 has a crucial effect on the intensity of the flow inside the mushy zone and, consequently, on the segregation.

The figures 4a and 4b present segregation pattern after 770s of solidification process for the same RMF, $B_0 = 20mT$ ($Ta_m = 2.32 \cdot 10^4$) obtained with a laminar and a turbulent model, respectively. In both cases a central channel strongly enriched with the Si with a radius $\approx 0.6mm$ is formed. Both simulations presented a periodical structure along the sample adjacent to this central channel. A characteristic time which corresponds to a spatial period of the strikes is quite long and does not seem to be linked to the oscillations of the liquid flow which are much more rapid. The strikes are much more pronounced for the calculations obtained with a laminar model. In general, it can be said that results obtained with a turbulent model repeat those obtained with a laminar model in their "smoother" form. It is well seen in figure 5 (bottom part) where the curve 2 has oscillations similar to curve 1 but with smaller amplitude. Also, the channel obtained with a turbulent model has a regular (cylindrical) form whereas in a laminar case it is disturbed. Generally, segregation patterns obtained in calculations for $B_0 = 20mT$ corresponds to experimental results described in [7] and [17].

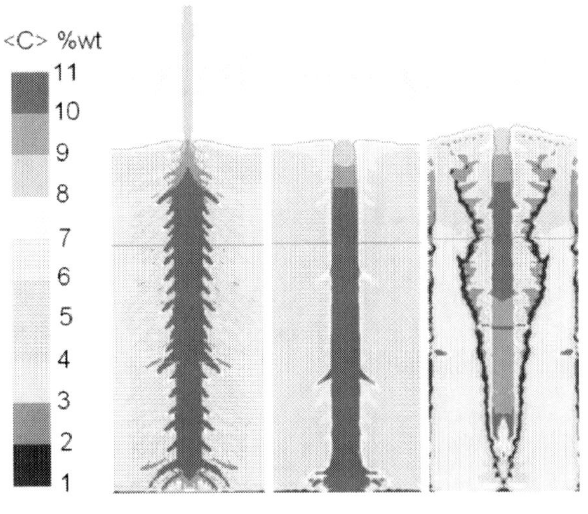

a b c

Figure 4. Segregation pattern after 770s of solidification of Al-7%wtSi sample, obtained with:
a) $B_0 = 20mT$, $Ta_m = 2.32 \cdot 10^4$, laminar model;
b) $B_0 = 20mT$, $Ta_m = 2.32 \cdot 10^4$, turbulent model;
c) $B_0 \approx 65mT$, $Ta_m = 2.32 \cdot 10^5$, turbulent model.
Radius of the sample is 4mm, length of the sample is 0.10m of which 0.025m is shown here; the lines mark the limits of the mushy zone.

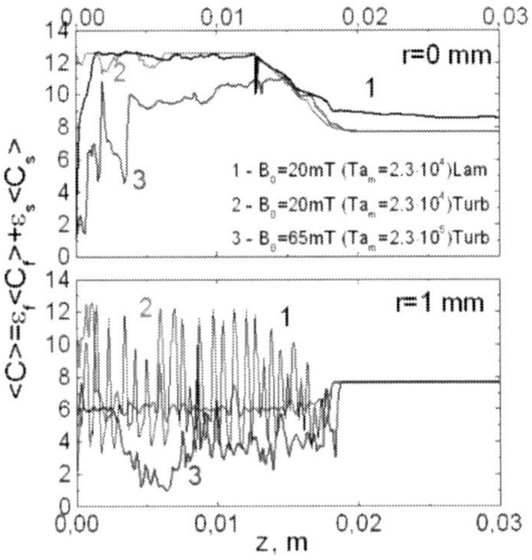

Figure 5. Longitudinal distribution of the average Si concentration along the axis of symmetry and at the distance 1mm from it after 700s of solidification of Al-7%wtSi sample. Curves 1,2,3 correspond to the cases a,b,c in figure 4

In figure 4c a segregation pattern obtained with a turbulent model after 770s of solidification of Al-7%wtSi sample under the action of a strong magnetic field, $B_0 \approx 65mT$ ($Ta_m = 2.32 \cdot 10^5$) is presented. As in both previous cases, a central channel along with a strike structure adjacent to it persists. However, for this case the strikes are modulated by a layer which has a sinusoidal form and is strongly depleted of Si. A similar layer appears near the side wall that is related to vortices concentrated there as shown in figure 2d. Calculations performed for a longer period of the solidification of AlSi sample under the action of RMF with the $B_0 \approx 65mT$ ($Ta_m = 2.32 \cdot 10^5$) have shown that after a transitional period of ~800s the lateral and the central depleted layers have started to overlap periodically as can be seen in figure 6a. This interaction between layers leads to the formation of elongated pools with the concentration of Si within 6-7%wt (i.e. close to the nominal one) separated by thin radial "dams" where concentration of Si is within 3-4%wt. This segregation pattern with double periodicity corresponds qualitatively to the experimental results observed in solidification of a sample of a ternary alloy Al-7%wtSi-1%Fe under the action of RMF with $B_0 \approx 150mT$, $f = 50Hz$, shown in figure 6b.

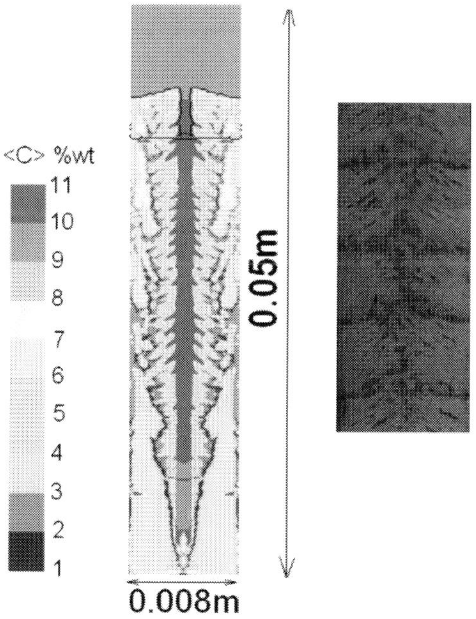

a b

Figure 6. Segregation patterns:
a) obtained numerically after 1800s of solidification of a binary Al-7%wtSi sample, with a turbulent model for the case $B_0 \approx 65mT$ ($Ta_m = 2.32 \cdot 10^5$). The radius and the length of the sample are 0.008m and 0.1m, respectively and 0.05m of the sample are shown; the lines mark the limits of the mushy zone
b) observed in a solidified sample (of the same size) of a ternary alloy Al-7%wtSi-1%Fe under the thermal gradient 8000K/m with a cooling rate 0.64K/s under the action of RMF with $B_0 \approx 150mT$, $f = 50Hz$. Darker areas extended from the centre of the sample to its sides are beta Al_5FeSi intermetallics.

It must be stressed, however, that the presence of the third component in the alloy modifies conditions for the solidification and can affect significantly the development of the mushy zone and, consequently, the solute transport through the latter. Therefore, we have started to develop a model for the solidification of a ternary alloy. The approach and results which should be considered as preliminary are presented in the next section.

6. Solidification of a ternary Al-7%wt Si-1%wt Fe alloy

In calculations the dependence of the liquidus temperature T_L on concentrations of the two solutes $C_{l,1}$, $C_{l,2}$ was taking in a form $T_L = T_m + m_1 C_{l,1} + m_2 C_{l,2}$ where $m_i = \partial T_L / \partial C_{l,i}$ is a "partial slope" of the liquidus surface. The values for m_1 and m_2 were extrapolated from the phase diagrams of binary alloys AlSi and AlFe. Interaction between the solutes has not been considered. Following Rappaz and Boettinger [16], we have used different averaged concentrations for interdendritic liquid and for solid-liquid interface. Another assumption made in [16] about independent solute fields given by the Ivantsov solution has been also employed. This assumption along with a modified condition of dendrites growth at marginal stability constitute a set of non-linear equations for definition of solutes concentrations at solid-liquid interface and dendrites' tip velocity correspondingly. Note that the model requires two segregations coefficients which were taken from the binary alloys as well. That is segregation coefficients for Si and Fe were, respectively, $k_1(Si) = 0.133$ and $k_2(Fe) = 0.029$.

Calculations were performed for the solidification conditions specified in figure 1 and for a magnetic field with low intensity $B_0 = 13.5mT$ ($Ta_m = 1.0 \cdot 10^4$), therefore, a laminar model was used for calculation of the fluid flow. In figure 7 the segregation patterns for the two solutes is presented for a relative average concentration, i.e., for the value $C_{r,i}$ defined as

$$C_{r,i} = \frac{\langle C_i \rangle - C_{i,0}}{C_{i,0}} \tag{3}$$

where i denotes either Si or Fe and $C_{i,0}$ is a nominal (initial) concentration of the solute i. This is done because qualitatively the segregation patterns are similar for both solutes since they are defined by the same fluid motions. However, quantitatively the segregation for Fe should be more pronounced because it has smaller segregation coefficient. This is consistent with the results presented in figure 6 where the relative concentration of iron in the mushy zone is lower than that of Si and the enrichment of the pure liquid in the central channel is higher.

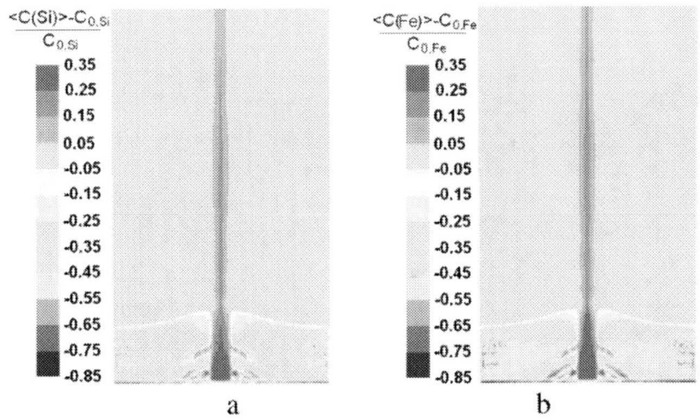

Figure 7. Distribution of $C_{r,Si}$ (a) and $C_{r,Fe}$ (b) defined by eq.3 and calculated with a laminar model after 100s of solidification of the Al-7%wtSi-1%wtFe sample, $R = 4mm$, $B_0 = 13.5mT$ ($Ta_m = 1.0 \cdot 10^4$)

According to these results a formation of the solid β-phase characterized by a cubic matrix of FeSi can be expected in the central part of the sample.

7. Conclusions

We have proposed a statistical approach to handle the solidification process with a turbulent flow in the liquid phase. That approach is globally very similar to the volume averaging process especially in the mush region. The interest of such an approach is however its natural capability to handle turbulent flows. It is shown that the turbulent flows generated by a RMF promote the formation of a large central channel with side secondary channels. It is also shown that the mean flow as well as turbulence penetrates in the upper part of the mushy zone. Accordingly the solute transport is significantly enhanced compared to the laminar case, and the formation of the secondary channels is promoted. The turbulent diffusivity in the mushy zone has not been addressed. The next step is to derive more precise expressions for the overall growth rate as well as the dispersive flux.

Acknowledgments
The authors acknowledge a financial support from MICAST-ESA-MAP AO-99-031 project and from TÃMOP-4.2.1.B-10/2/KONV-2010-0001 project in the framework of the New Hungarian Development Plan.

Appendix
We distinguish two main phases, i.e., liquid and solid. The phase averaging procedure is used to establish the equations governing the averaged quantities such that the mean velocity $\mathbf{v}_l$, temperature T, liquid and solid concentrations C_s and C_l. The velocity of the solid phase is zero. An envelope model described in [15] and [10,11] is used to parametrize the solute transfer from the dendrite to the liquid bulk. For simplicity, the physical properties of the solid and liquid zones are identical. The general equations governing the mean values of the various fields are gathered in Table 1 below.

Table 1. Constitutive equations for the purely columnar binary model

Conservation equations	Solid	Liquid
Dendrites	$\frac{\partial n}{\partial t} = n_c^0 \delta(T - T_{liq})$, $n_c^0 = \left[\frac{4}{3}\pi(\lambda_1/2)^3\right]^{-1}$	
Mass	$\frac{\partial(\rho\varepsilon_s)}{\partial t} = \Gamma_s$	$\frac{\partial(\rho\varepsilon_f)}{\partial t} + \nabla.(\rho\varepsilon_f \mathbf{v}_f) = -\Gamma_s$
Momentum for the liquid phase	$\frac{\partial(\rho\varepsilon_f \mathbf{v}_f)}{\partial t} + \nabla.(\rho\varepsilon_f \mathbf{v}_f \mathbf{v}_f) = -\varepsilon_f \nabla p + \nabla\left[\varepsilon_f(\mu_f + \mu_t)\left(\nabla\mathbf{v}_f + (\nabla\mathbf{v}_f)^T\right)\right] + \rho'\varepsilon_f \mathbf{g} - K_f \varepsilon_f \mathbf{v}_f + \varepsilon_f \mathbf{F}_{em}$ $\rho' = \rho\left[1 + \beta_T(T - T_0) + \beta_T(C_f - C_0)\right]$	
Energy	$\frac{\partial(\rho c T)}{\partial t} + \nabla.(\rho c \varepsilon_f \mathbf{v}_f T) = \nabla.(\lambda \nabla T) + \Gamma_s L$	
Solute concentration	$\frac{\partial(\varepsilon_s \rho C_s)}{\partial t} = \nabla(\rho\varepsilon_s D_s \nabla C_s) + C_s^* \Gamma_s$	$\frac{\partial(\varepsilon_f \rho C_f)}{\partial t} + \nabla.(\rho\varepsilon_f \mathbf{v}_f C_f) = \nabla(\rho\varepsilon_f (D_f + D_{ft})\nabla C_f) - C_s^* \Gamma_s$, $D_{ft} = \nu_t$
Auxiliary conservation equations		
Mass conservation in the interdendritic liquid	$\frac{\partial(\rho\varepsilon_d)}{\partial t} = \Gamma_g - \Gamma_s$, $\Gamma_g = \rho S_g \overline{w}_g$	
Solute conservation in the extradendritic liquid	$\varepsilon_f \rho_f \frac{\partial C_f}{\partial t} + \varepsilon_f \rho_f \mathbf{v}_f \nabla C_f = \nabla(\varepsilon_f \rho_f D_f \nabla C_f) + \left(\frac{\rho_f S_g D_l}{\delta_{l-d}} - \Gamma_g\right)(C_l^* - C_f)$	
Solute conservation in the interdendritic liquid	$\Gamma_s(C_l^* - C_s^*) = \varepsilon_d \rho_f \frac{\partial C_l^*}{\partial t} + \varepsilon_d \rho_f \mathbf{v}_f \nabla C_l^* + \left(\frac{\rho_f S_g D_l}{\delta_{l-d}}\right)(C_l^* - C_f)$	
Auxiliary relations		
Constraints	$\varepsilon_l + \varepsilon_d + \varepsilon_s = 1$, $\varepsilon_l + \varepsilon_d + \varepsilon_s = 1$, $\varepsilon_d + \varepsilon_s = \varepsilon_g$, $\varepsilon_d + \varepsilon_l = \varepsilon_f$	
Interfacial equilibrium for solute	$C_s^* = \begin{cases} k C_l^* & if\ \Gamma_s \geq 0 \\ C_s & if\ \Gamma_s < 0 \end{cases}$	$C_l^* = \frac{T - T_f}{m}$

Radius of the dendrite envelope	Radius of the grain (dendrite)	Area density of the grain envelope
$R_1 = \left(\frac{4}{3}\pi n\right)^{-1/3}$	$a = R_1 \varepsilon_g^{1/3}$	$S_g = 4\pi a^2 n$
Diffusion length	$\varphi_g = D_l \frac{C_l^* - C_l}{\delta_{l-d}}$ where $\delta_{l-d} = \delta\left[1 - \frac{10a^2\delta + 10a\delta^2 + 4\delta^3}{5(R_1^3 - a^3)}\right]$ with $\delta = \begin{cases} D_l / \overline{w}_g & \text{if } Pe_g > 2 \\ (R_1 - a)/2 & \text{if } Pe_g \leq 2 \end{cases}$ $Pe_g = \frac{\overline{w}_g (R_1 - a)}{D_l} >> 2$	
Grain growth kinetics	$\Gamma_g = \rho S_g \overline{w}_g$ and $\overline{w}_g = \frac{D_l m (k-1) C_l^*}{\pi^2 \Gamma} Pe_t^2$	
Turbulence model equations		
Equation for the liquid turbulent kinetic energy k_t	$\frac{\partial(\varepsilon_f k_t)}{\partial t} + \nabla.\left(\varepsilon_f \mathbf{v}_f k_t\right) = \nabla.\left((\varepsilon_f \nu_{ft} / \sigma_k)\nabla k_t\right) - \varepsilon_f (P_k - \varepsilon_t)$ P_k is the usual production term	
Equation for the turbulent dissipation rate ε_t	$\frac{\partial(\varepsilon_f \varepsilon_t)}{\partial t} + \nabla.\left(\varepsilon_f \mathbf{v}_f \varepsilon_t\right) = \nabla.\left((\varepsilon_f \nu_{ft} / \sigma_\varepsilon)\nabla \varepsilon_t\right) - \varepsilon_f \frac{\varepsilon_t}{k_t}(C_{1\varepsilon} P_k - C_{2\varepsilon}\varepsilon_t)$	

The momentum transfer coefficient K_f between the liquid and solid is expressed by means of the isotropic permeability of the fluid given by the well-known Carman-Kozeny law. The turbulent diffusivity is set to zero in the mushy zone.

References

[1] Flemings M C, Nereo G E 1967 *Trans. Met. Soc. AIME*, vol 239 p 1449
[2] Flemings M C, Mehrabian R, Nereo G E 1968 *Trans. Met. Soc. AIME* vol 242 p 41
[3] Medina M, Du Terrail Y, Durand F and Fautrelle Y 2004 *Metall. Mat. Trans.* vol 35B p 743
[4] Zaidat K, Mangelinck-Noel N, Dupouy M D, Moreau R 2005 *J Cryst. Growth* vol 275 p 501
[5] Steinbach S and Ratke L 2007 *Met. Mat. Trans. A* vol 38A p 1388
[6] Hainke M, Friedrich J and Müller G 2004 *J. Materials Science* vol 39 p 2011
[7] Noeppel A, Ciobanas A, Wang X D, et al 2010 *Met. Mat. Trans. B* vol 41 p 193
[8] Willers B, Eckert S, Nikrityuk P A, Rabiger D, Dong J, Eckert K and Gerbeth G 2008 *Met. Mat. Trans. B* 2008 vol 39B p 304
[9] Ganesan S and Poirier D 1990 *Metallurgical Trans. B* vol 21B 173
[10] Wang C and Beckermann C 1993 *Metallurgical Trans. A* vol 24A p 2787
[11] Wang C and Beckermann C 1996 *Metallurgical Trans. A* vol 27A p 2754
[12] Steinbach L R S, Muller G, Hainke M et al 2006 *Mat. Sci. Forum* vol 508 (Solidification and Gravity IV) p 131
[13] Ciobanas A and Fautrelle Y 2007 *J. Phys. D: Appl. Phys.* vol 40 p 3733
[14] Ciobanas A and Fautrelle Y 2007 J. Phys. D: Appl. Phys. vol 40 p 4310
[15] Rappaz M and Thevoz Ph 1987 *Acta Metall.* vol 35 p 1487
[16] Rappaz M and Boettinger W J 1999 *Acta Mater.* vol 47 p 3205
[17] Kovács J, Rónaföldi A, Kovács Á, Roósz A 2009 *Trans. Indian Inst. Met.* vol 62 p 461
[18] Budenkova O, Noeppel A, Kovács J, Rónaföldi A, Roósz A, Bianchi A-M, Baltaretu F, Medina M and Fautrelle Y 2010 *Mat. Sci. Forum* vol 649 (Solidification and Gravity V) p 269
[19] Saric W S 1994 *Annu. Rev. Fluid Mech.* vol 26 p 379

IOP Conf. Series: Materials Science and Engineering **33** (2012) 012047 doi:10.1088/1757-899X/33/1/012047

Adjustment and verification of macroscopic melt flow during solidification by means of various AC magnetic fields

S. Eckert[1], D. Räbiger[1], T. Vogt[1], S. Franke[2], J. Czarske[2], G. Gerbeth[1]

[1]Helmholtz-Zentrum Dresden-Rossendorf (HZDR), P.O. Box 510119, 01314 Dresden, Germany

[2]Technische Universität Dresden, Helmholtzstrasse 18, 01069 Dresden, Germany

E-mail: s.eckert@hzdr.de

Abstract. We present an experimental study concerning the solidification of AlSi alloys exposed to a pulsed rotating magnetic field. Isothermal flow measurements were carried out in order to understand the flow structures resulting from the application of time-modulated magnetic fields. These investigations revealed transient flow regimes showing distinct inertial oscillations and coherent vortex structures. An intense melt flow with periodic reversals of the flow direction at the solidification front can be created by a suitable choice of the magnetic field parameters. Such resonant states of the flow pattern have been proven to provide beneficial conditions for solidification processes. Optimised flow conditions realized in a solidifying melt result in a significant grain refinement without provoking the formation of harmful segregation freckles.

1. Introduction

Electromagnetic stirring during solidification has been proved to be a striking method for achieving a purposeful alteration of the microstructure of casting ingots, such as grain refinement or the promotion of a transition from a columnar to an equiaxed dendritic growth (CET). However, the imposition of a rotating (RMF) or a travelling magnetic field (TMF) also causes problems like the occurrence of typical segregation pattern or a deflection of the upper free surface. A permanent radial inward (RMF and downward TMF) or outward (upward TMF) flow along the solidification front is responsible for the transport of solute to the axis or the wall of the ingot resulting in typical freckle segregation pattern filled with alloy of eutectic composition [1-6]. Several studies have been devoted to overcoming the handicaps of rotary stirring with the specific goal to generate a vigorous stirring in the bulk without considerable deformations of the free surface [7-9]. It was shown recently, that the application of modulated AC magnetic fields offers considerable potential for optimizing the melt stirring [10, 11]. Especially, the secondary flow can be organized in such a way that periodic reversals of the flow direction occur adjacent to the solidification front, which has been proven as an important method to prevent flow-induced macrosegregation [12]. It became apparent that a careful adjustment of the modulation parameters is required in order to create intense secondary flows with periodic reversals of the flow direction.

Within this paper we especially consider the effect of a pulsed RMF on the liquid metal flow solidifying inside a circular cylinder, whereas the RMF is applied in form of successive rectangular pulses. Previous investigations revealed the existence of an optimal pulse length T_P, where a maximum intensity of a periodic meridional flow can be observed [11]. In that case the corresponding pulse frequency f_P relates to the eigenperiod of inertial waves in a developed regime, as given by Greenspan [13]. We present typical results of the flow structures obtained by flow measurements under isothermal conditions. Selected magnetic field configurations were applied on the directional solidification of Al – 7 wt pct Si alloys. The impact of the pulsed electromagnetic stirring was evaluated in terms of temperature measurements during solidification and examinations of the grain structure and the phase distribution inside the solidified samples.

2. Experimental setup

2.1. Isothermal flow measurements

A schematic view of the experimental setup for the isothermal flow measurements is shown in figure 1. A cylindrical vessel made of Plexiglas was used with an aspect ratio $A = H_0/2{\cdot}R_0 = 1$. The size of the inner diameter of $D = 2{\cdot}R_0$ and the height H_0 was chosen to be 67.5 mm. The cylinder is closed by rigid lids and filled with the eutectic alloy GaInSn. The experiments were performed in the magnetic induction system PERM at HZDR, wherein the fluid vessel was placed concentrically. In order to preclude flow artefacts arising from symmetry deviations of the experimental setup (vertical alignment, conformity of both the cylinder and the magnetic field axis), special care was necessary to ensure a precise positioning of the cylinder inside the magnetic system. The homogeneity of the magnetic field was checked using a 3-axis Gauss meter (Lakeshore model 560, sensor type MMZ2560-UH). Within a radius of 35 mm, which approximately corresponds to the radial dimension of the container, the variance of the magnetic field strength was found to be smaller than 3%.

The ultrasound Doppler velocimetry (UDV) was used to perform measurements of the fluid flow inside the cylinder [11]. The DOP2000 velocimeter (model 2125, Signal Processing SA, Lausanne) equipped with an 4 MHz transducer (TR0405LS, acoustic active diameter 5 mm) was applied to measure the primary azimuthal flow at two vertical positions. By attaching the ultrasonic transducer to the wall of the fluid container, the acoustic coupling between sensor and fluid was realized through the curved cylindrical surface. Moreover, a flow mapping of the meridional flow was realised by using a new measuring system operating with a linear array of 25 single transducers attached to the plane-parallel bottom wall of the flow container (see figure 1). Details of the measuring system are given in [14]. The spatial resolution in lateral direction varies from 5 mm at the sensor to approximately 7.5 mm at the lid of the fluid vessel. In axial direction a spatial resolution of about 1.4 mm was achieved. The velocity data were acquired with sampling frequencies between 3.5 and 25 Hz. The accuracy of the velocity data can be assessed to be better than 0.15 mm/s.

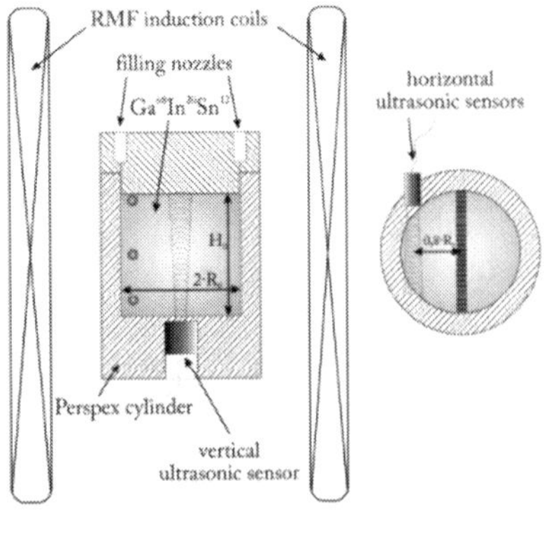

Figure 1. Schematic view of the experimental setup for the isothermal flow measurements

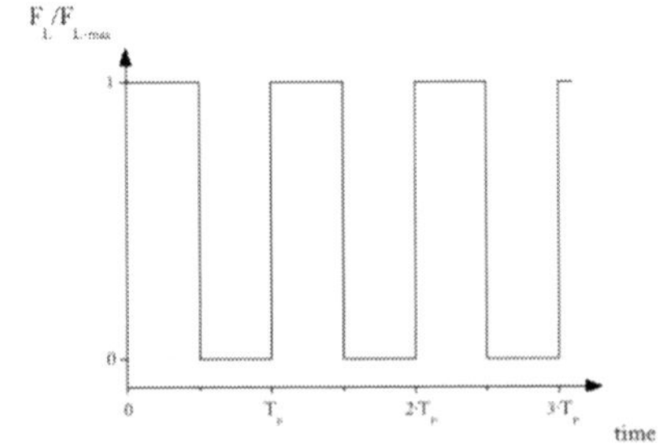

Figure 2. Modulation scheme of the driving Lorentz force F_L

2.2. Solidification experiments

Solidification experiments were performed using an Al - 7 wt pct Si alloy prepared from 99,99 pct Al and 99.999 pct Si. The raw alloy was modified with 200 ppm Sr. The set-up and the realization of the solidification experiments are briefly described here. More details can be found in [12]. The metal alloy was solidified directionally from the bottom in a double-walled cylindrical stainless steel mould. A schematic drawing of the experimental set-up is depicted in Figure 3(a). The mould has an internal diameter of 50 mm and a height of 100 mm. The filling height for each charge was 60 mm in the liquid state. The mould was positioned at a water-cooled copper chill which was kept at a constant temperature of about 20°C. In the experiments presented here the cooling of the sample started 30 seconds before the action of the electromagnetic field. The solidification experiments were performed at the COMMA (COMbined MAgnetic field) facility of HZDR. A field frequency of 50 Hz was selected being small enough to neglect any influence arising from the skin effect. The pulse duration of the RMF was arbitrarily chosen to be 4 seconds. The magnetic field strength was varied from 3.3 mT to 8.5 mT. Temperature measurements were performed using an arrangement of type-K thermocouples. For this purpose three thermocouples were installed along the ingot axis at fixed vertical positions at 1.5, 25 and 50 mm. The sampling rate for each thermocouple was 0.2 s.

A further setup is in preparation which will enable for a simultaneous monitoring of the temporary position of the growth front, the temperature distribution and the flow pattern during solidification. Such experiments are supposed to provide an elaborate data base for the validation of numerical simulations. A schematic view of the new experimental setup is depicted in figure 3(b). The mould consists of a double-walled, cylindrical container which inner cylinder is heated by means of a resistance heater. The slit between the inner and the outer cylinder is evacuated in order to minimize the radial heat flux. A free surface of the melt is avoided in this experiment. This approach guarantees defined boundary conditions at the upper position. Moreover, side effects arising from perturbations of the free surface can be neglected. A floating lid will be installed in order to balance expansion or shrinking during the heating/solidification cycles. This lid contains specific ports for the installation of ultrasonic sensors for measurements of the flow field and the detection of the position of the solid-liquid interface [15]. Different assemblies of thermocouples in the bulk alloy and inside the side walls and the bottom of the container will be used to measure the temperature field.

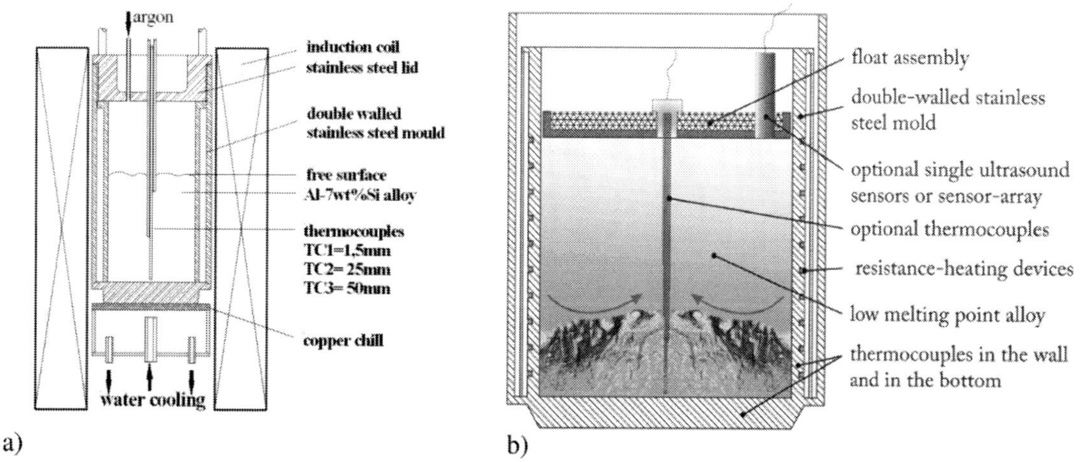

Figure 3. Schematic view of the experimental setup a) used for the AlSi7 solidification experiments and b) the new configuration allowing for simultaneous observations of the front position, the temperature field and the velocity field

3. Results and discussion

3.1. Isothermal Flow Measurements

The modulation scheme of the pulsed RMF as considered within this study is displayed in figure 2. As a consequence the fluid flow experiences a sequence of spin-up and spin-down processes. The duration of the particular pulse cycles T_P turns out to be a crucial control parameter. Measurements of both the primary and the secondary flow have been carried out for different values of T_P. Figure 4 shows time series of the mean vertical velocity and the local azimuthal velocity u_{phi} for three different pulse durations. Here, the vertical velocity u_z in figure 4(a) has been averaged across the vertical mid-plane of the cylinder as follows:

$$u = \frac{1}{H_0 R_0^2} \int_{H_0/2}^{H_0} \int_{-R_0}^{R_0} r\sqrt{u_z^2}\, dr\, dz \tag{1}$$

After a transient behaviour related to the initiation of the electromagnetic driving force the measurements reveal a distinct periodicity of both the primary and the secondary flow, whereas the cycle duration corresponds to the particular pulse cycle T_P. The rising edges of a particular RMF pulse appear as gray branch lines in the diagrams. While the flow oscillations appear to be very regular for a pulsing at T_P = 15 s (figure 4(a)), the oscillation amplitude of the secondary flow increases significantly for T_P = 20 s. More precisely, a doubling of the frequency occurs with respect to $1/T_P$ and every second peaks becomes very intense. This phenomenon, which disappears again with further increase of the pulse cycle, is similar to a resonant circuit. Another indication can be noticed in figure 4(b), where the azimuthal velocity has been recorded at two different vertical positions. These two curves exhibit a distinct phase shift for T_P = 20 s.

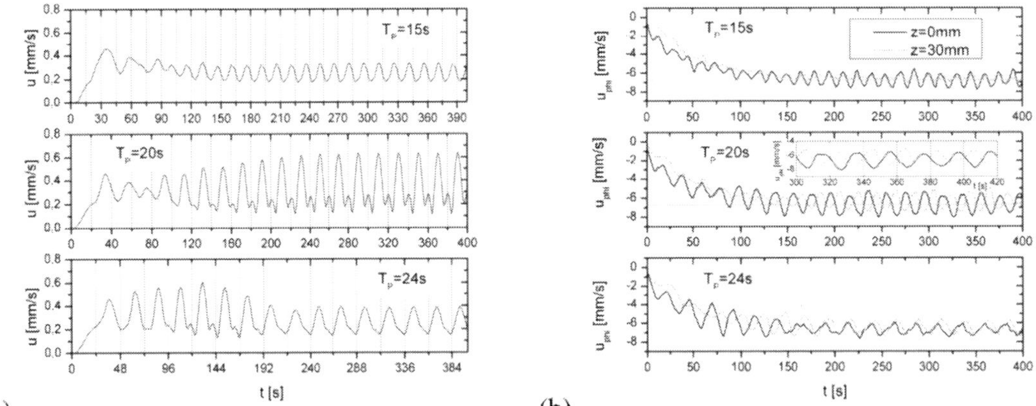

(a) (b)

Figure 4. Time-resolved measurements of (a) the vertical velocity and (b) the azimuthal velocity (B_0 = 0.46 mT)

Figure 5 shows two snapshots of the flow pattern obtained for a pulse cycle of T_P = 20 s. The typical structure of the toroidal double vortex can be detected in figure 5(a). The secondary flow in the central region is directed upwards in the bottom part and downwards in the upper part of the cylinder, respectively. The moment when this double vortex appears coincide with the large peak of the secondary flow in figure 4(a). The smaller peak of u corresponds to the snapshot shown in figure 5(b). Here, we can observe an inversion of the flow direction, a so-called "inverse" double vortex is formed at lower intensity.

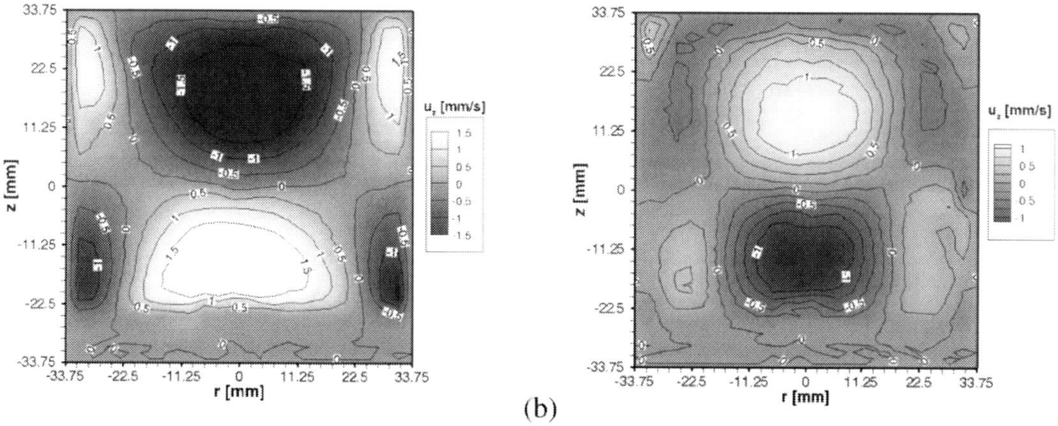

(a) (b)

Figure 5. Snapshots of the vertical velocity across the vertical mid-plane of the cylinder recorded by the US sensor array for T_P = 20 s at B_0 = 0.46 mT: (a) typical double vortex of the secondary flow; (b) "inverse" double vortex

The mean values of the vertical velocity averaged over both the total measuring time and the vertical cylinder cross section are drawn in figure 6(a) vs. the pulse duration cycle T_P for a magnetic field strength of 0.46 mT. It becomes obvious that the intensity of the secondary flow exhibits a pronounced maximum around $T_P \approx 20$ s, where distinct periodic reversals of the secondary flow have been observed. This resonance peak is shifted to lower T_P with increasing magnetic field strength as depicted in figure 6(b), in particular the maximum is found at T_P = 2.9 s for B_0 = 2.25 mT. Moreover, further maxima of the magnitude of the averaged secondary flow appear at larger pulse durations.

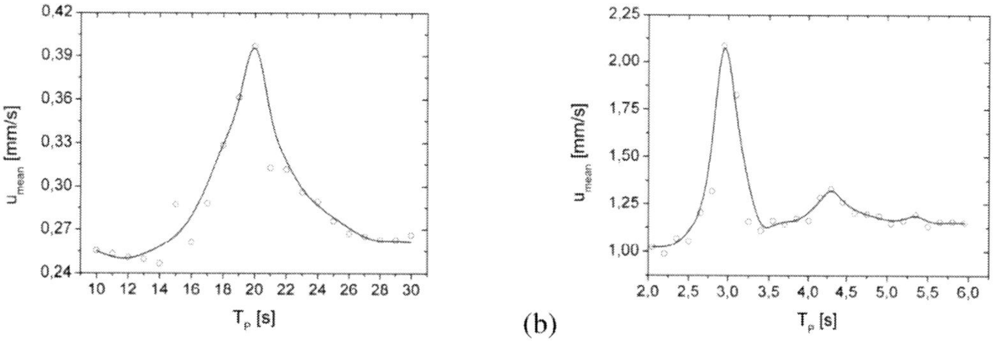

(a) (b)

Figure 6. Dependence of the intensity of the secondary flow on the duration of the pulse cycle T_P: (a) B_0 = 0.46 mT; (b) B_0 = 2.25 mT

Figure 7 shows snapshots of the vertical flow obtained for pulse durations of T_P = 4.3 s and 5.3 s, respectively, which correspond to the second and third peak in figure 6(b). It is interesting to note that new flow pattern emerge marked by an increment of vortex cells in radial direction.

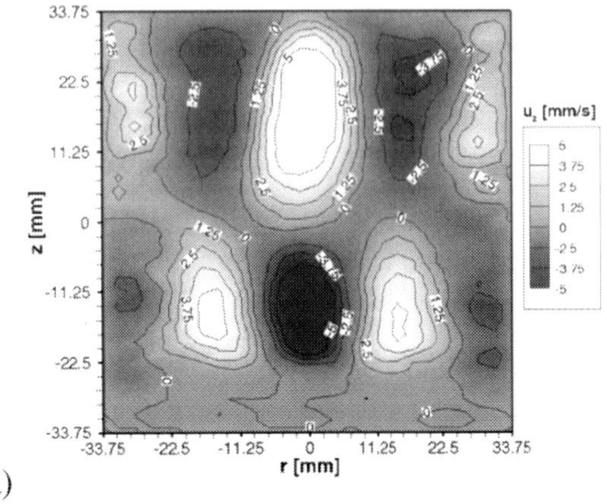

(a)

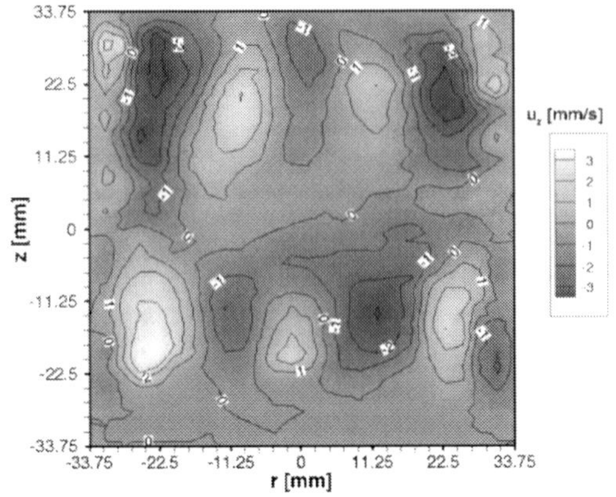

(b)

Figure 7. Snapshots of the vertical velocity across the vertical mid-plane of the cylinder recorded by the US sensor array at B_0 = 2.25 mT: (a) T_P = 4.3 s; (b) T_P = 5.3 s

3.2. Solidification experiments

The following solidification experiments constitute a first study to demonstrate the capability of pulsed magnetic fields for electromagnetic melt stirring during solidification. In figure 8 we compare different samples which were grown under the influence of a continuously applied RMF at a field strength of 8.5 mT and pulsed RMF's with magnetic fluxes of 3.3 and 8.5 mT, respectively. In both cases a pulse duration T_P of 4 s was chosen. The photographs in figure 8 (a-c) depict the solidified macrostructure obtained on the longitudinal section. A columnar zone with vertically orientated grains in the bottom region becomes visible while the magnetic field remains off-state. The initiation of the magnetic field leads to an immediate refinement of the large columnar grains. A mixture of equiaxed and columnar grains occurs whereas the columnar grains become deflected with respect to the vertical line and oriented towards the incoming flow. The solidified structure changes into a purely equiaxed one at higher positions within the samples. Only small differences between the three samples presented here become obvious with respect to the macroscopic grain structure. However, the analysis of the phase distribution reveals significant variations of the eutectic content across the longitudinal section (see figure 8 (d-f)). We observe a strong enrichment of the eutectic component for the non-modulated magnetic field in the columnar region along the axis and in the zone where the CET occurs (figure 8(d)). The application of RMF pulses reduces the macrosegregation effects significantly. The field configuration with B_0 = 3.3 mT and T_P= 4 s (see figure 8(f)) delivers an almost uniform distribution of the eutectic phase within the longitudinal cross section, whereas a solute enrichment near the axis remains visible for the configuration in figure 8(e).

The flow measurements in the previous section showed the possibility of achieving an intense melt flow adjacent to the solidification front with periodic reversals of the flow direction. The secondary flow induced by a conventional RMF along the growth front is always directed inwards and conveys the rejected solute towards the sample axis. The formation of pronounced segregation channels can obviously be prevented by the periodic flow without any preferred direction which exhibits vanishing mean values for a long-time average. Figure 8(g) illustrates the impact of the electromagnetic stirring on cooling curves recorded at a height of 25mm about the bottom. The turning-on of the permanently applied RMF cause a short temperature increase followed by a continuous decrease of the temperature in the further course of the cooling. The modulated stirring by the pulsed RMF is reflected by distinct temperature oscillations until arriving at the Liquidus temperature where a jump of the viscosity occurs. The temperature oscillations correlate with the periodic changes of the flow direction. This effect is also visible in figure 8(h) where the cooling rate is drawn for a position at the sample axis.

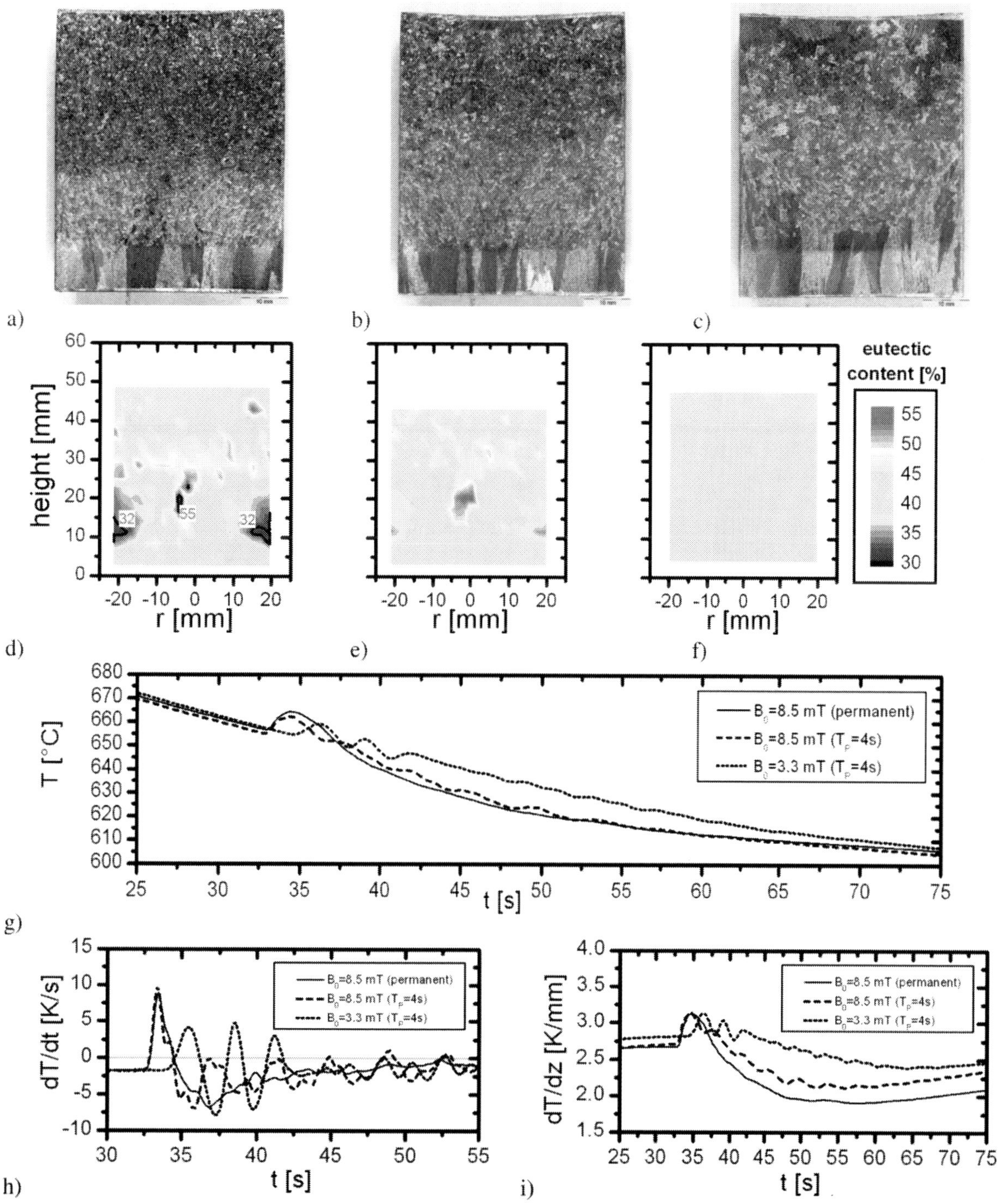

Figure 8. Comparison between (a,d) a permanently applied RMF with B_0 = 8.5 mT and two modulated RMF's with (b,e) B_0 = 8.5 mT and T_P = 4 s and (c,f) B_0 = 3.3 mT and T_P = 4 s. Pictures (a-c) present the macrostructure of the longitudinal section and (d-f) the corresponding area distribution of the eutectic phase. Figure (g) shows the cooling curves recorded on the axis at a height of 25 mm. Corresponding data of the cooling rate and the temperature gradient are presented in (h) and (i), respectively.

Remarkable fluctuations of the temperature in the upper mushy zone may promote the remelting of dendrite arms which can act as additional nuclei for grain growth.

Corresponding curves of the cooling rate and the temperature gradient are presented in the figures 8(h) and (i), respectively. The modulated stirring at a field strength of B_0=3.3 mT produces reversals of the sign of the cooling rate. It becomes obvious that the chosen pulse cycle duration of 4 seconds is not the optimum for the configuration considered here, because the oscillations become significantly damped after three periods. The inertial mode might be stimulated effectively, but the resulting rotation rate of the melt flow does not fulfill the required resonance condition for maintaining the inertial flow oscillations throughout the progressing solidification process.

4. Conclusions

The present study concerns a liquid metal flow being exposed to a pulsed RMF. The alternating power up and power down of the magnetic field generates successive spin-ups and spin-downs of the rotating fluid flow. Abrupt changes in energy injection rates promote the propagation of inertial waves through the interior of the fluid [13]. The permanent reversals of the flow direction were found at the eigenmodes of the inertial oscillations which can be excited by a suitable choice of the magnetic field parameter. Such a striking feature makes these flow regimes promising for applications in directional solidification processes. The electromagnetic control of the flow direction prevents a unidirectional transport of solute along the solidification front and, hence, avoids the formation of flow-induced segregation channels.

In this way, the electromagnetic stirring method that uses a modulated RMF offers considerable potential to enhance the stirring efficiency and to optimize the properties of castings by a well-aimed flow control during solidification. Further investigations are necessary to improve the understanding of the development of various flow regimes under the influence of a time-modulated driving force as well as the mutual impact of a progressive solidification front and the flow pattern in the melt.

Acknowledgments

The research is supported by the Deutsche Forschungsgemeinschaft (DFG) in form of the SFB 609 "Electromagnetic Flow Control in Metallurgy, Crystal Growth and Electrochemistry".

References

[1] Roplekar JK and Dantzig JA 2001 *Int. J. Cast Metals Res.* **14** 79
[2] Medina M, Du Terrail Y, Durand F and Fautrelle Y 2004 *Metall. Mater. Trans. B* **35** 743
[3] Nikrityuk PA, Eckert K and Grundmann G 2006 *Int. J. Heat Mass Transfer* **49** 1501
[4] Zimmermann G, Vitusevych VT and Sturz L 2010 *Mater. Sci. Forum* **649** 249
[5] Kovács J, Roósz A and Szőke J 2006 *Mater. Sci. Forum* **508** 263
[6] Steinbach S and Ratke L 2007 *Metall. Mater. Trans. A* **38** 1388
[7] Vives C 1993 *Metall. Trans. B* **24** 493
[8] Taniguchi S, Mataike K, Okubo M, Ando T and Ueno K 2003 *Proc. of the 4th Int. Conf. oon Electromagnetic Processing of Materials (EPM)*, Sendai, 339
[9] Spitzer K-H, Reiter G and Schwerdtfeger K 1996 *ISIJ Int.* **36** 487
[10] Kojima S, Ohnishi T, Mori T, Shiwaku K, Wakasugi I and Ohgarni M 1983 *Proc. of the 66th Steelmaking Conf.*, Atlanta, 127
[11] Eckert S, Nikrityuk PA, Räbiger D, Eckert K and Gerbeth G 2008 *Metall. Mater. Trans.* B **39** 374
[12] Willers B, Eckert S, Nikrityuk PA, Räbiger D, Dong J, Eckert K and Gerbeth G 2008 *Metall. Mater. Trans. B* **39** 304
[13] Greenspan HP 1990 The theory of rotating fluids, Breukelen Press, Brookline
[14] Franke S, Büttner L, Czarske J, Räbiger D and Eckert S 2010 *Flow Meas. Instrum.* **21** 402
[15] Eckert S, Willers B and Gerbeth G 2005 *Metall. Mater. Trans. A* **36** 267

IOP Conf. Series: Materials Science and Engineering **33** (2012) 012048 doi:10.1088/1757-899X/33/1/012048

Radial solidification of Al-Si alloys in the presence of a rotating magnetic field

V. Travnikov[1], **P.A. Nikrityuk**[2], **K. Eckert**[1], **D. Räbiger**[3], **S. Odenbach**[1], **S. Eckert**[3]

[1]Technische Universität Dresden, Institute of Fluid Mechanics, Chair of Magnetofluiddynamics, 01062 Dresden, Germany

[2]Technische Universität Bergakademie Freiberg, Reiche Zeche, Fuchsmühlenweg 9, 09596 Freiberg, Germany

[3]Helmholtz -Zentrum Dresden-Rossendorf, P.O. Box 510119, 01314 Dresden, Germany

E-mail: `Vadim.Travnikov@tu-dresden.de`

Abstract. We study the flow field during the radial solidification of an Al-7wt pct Si melt inside a cylindrical cavity with cooled side walls which is placed in a rotating magnetic field (RMF). The solidification process is simulated by means of a continuum model which is implemented in a code which solves the axisymmetric Navier-Stokes equations. We first analyze the start-up of the buoyancy-driven convection driven by the radial heat flux in absence of the RMF. We show that the initial large vortex breaks up into three smaller one. The resulting radial temperature profiles are compared with experimental data. Second, we study how the vortex structure is modified by the application of small RMF's.

1. Introduction

Rotating magnetic fields (RMF's) are widely used in solidification due to their marked influence onto the microstructure evolution, thereby stimulating grain refinement and the columnar-to-equiaxed transition [1]. The main advantage of the RMF is the possibility of a non-invasive control of the flow by means of varying different field parameters such as the magnetic induction B_0 or the frequency ω. In this work we extend previous works, devoted to the unidirectional solidification under the influence of both constant [2] and temporary modulated RMF's [3], towards the case when the solidification is driven by a radial heat flux. The flow is now caused by both the buoyancy force, arising from the radial temperature difference between cold side wall (Fig. 1a) and the hot melt in the cavity, and the Lorentz force generated by the RMF. The pure buoyancy-driven flow, appearing when $B_0 = 0$, consists of one toroidal vortex whose downflow is along the cold side wall. By contrast, for high amplitudes of the RMF, the Lorentz force dominates the buoyancy one. The resulting flow is a combination of a primary flow in azimuthal direction and a secondary flow circulating in the meridional plane analogously to the purely buoyancy-driven flow. However, the secondary flow in presence of the RMF consists of two vortices rotating in opposite direction. In presence of a free top surface, as it is the case here, they are not symmetric with respect to the horizontal mid-plane of the cylinder [4, 5]. Generally, the buoyancy- and the Lorentz-force-driven convection are in competition and the dynamics of the problem becomes more complex if both effects have to be taken into account.

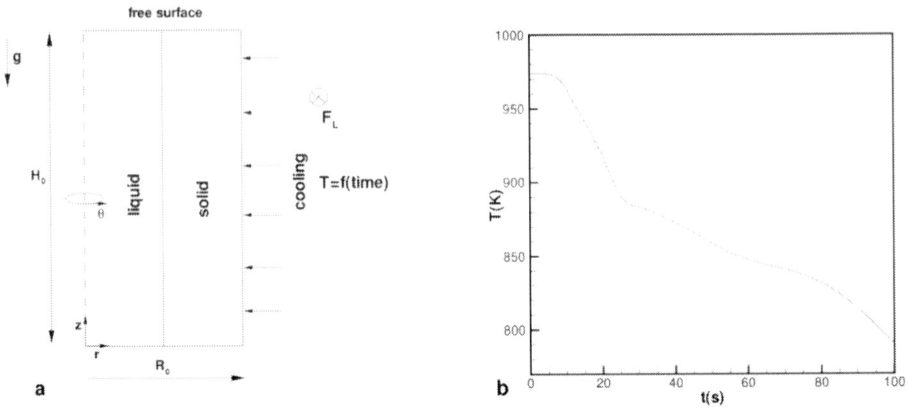

Figure 1. (a) Geometry of the problem. (b) The temperature $T_W(t)$ of the side wall as determined in the experiment by means of thermocouples.

2. Formulation of the problem

2.1. Rotating magnetic field

We consider a finite cylindrical cavity with a radius $R_0 = 0.025$ m and a height $H_0 = 0.06$ m filled with an Al-7wt pct Si alloy. Important non-dimensional parameters are the Hartmann and the Reynolds numbers, given by $Ha = B_0 R_0 \sqrt{\sigma/(2\rho\nu)}$ and $Re_\omega = \omega R_0^2/\nu$. The magnetic Taylor number representing a measure of the magnetic forcing can be derived as $Ta = Ha^2 Re_\omega = \sigma\omega B_0^2 R_0^4/(2\mu\nu)$. The magnetic field is characterized by the field induction B_0 and the frequency $\omega = 2\pi f$ ($f = 50$ Hz), whereas ρ, ν and σ stand for the material properties density, kinematic viscosity and electrical conductivity, respectively. The low-induction condition, implying that the flow does not change the magnetic field applied, is satisfied. The magnetic field induces a Lorentz force, F_L, consisting of a steady, axisymmetric and a time-independent, oscillating part of the frequency 2ω. The second part can be neglected under the conditions considered here [6]. An analytical solution can be written for the body force in a finite cylinder, which solely has an azimuthal component [7]:

$$F_L = \frac{1}{2}\sigma\omega B_0^2 r\left[1 - \frac{R_0}{r}\sum_{k=1}^{\infty}\frac{2J_1(\lambda_k \frac{r}{R_0})\cosh(\lambda_k \frac{z-\frac{1}{2}H_0}{R_0})}{(\lambda_k^2-1)J_1(\lambda_k)\cosh(\lambda_k \frac{H_0}{2R_0})}\right], \tag{1}$$

where J_1 is a Bessel function of the first kind and λ_k are the roots of $J_1'(\lambda_k) = 0$.

2.2. Solidification model and governing equations

The standard continuum model of binary metal alloy solidification [8] has been adopted for the binary alloy solidification. To simplify the problem we assume that (i) the shrinkage-driven flow is negligible ($\rho_l = \rho_s$), (ii) the flow in the liquid phase is axisymmetric and (iii) the electrical conductivities of the solid and liquid phases are the same. Based on these assumptions we have the following set of equations: The *continuity equation* has the simple form

$$\nabla \cdot \mathbf{u} = 0 \tag{2}$$

because the density is assumed to be constant. In the *Navier-Stokes equation*

$$\rho\frac{\partial \mathbf{u}}{\partial t} + \rho(\mathbf{u}\nabla)\mathbf{u} = -\nabla p + \mu\nabla^2\mathbf{u} - \frac{\mu}{K}\mathbf{u} + \rho\mathbf{g}\Big[\beta_T(T - T_{ref}) + \beta_C(C_l - C_{ref})\Big] + \mathbf{F_L}, \tag{3}$$

$\mathbf{F_L}$ is the Lorentz force and the fluid flow in the mushy zone is modeled using the Darcy's law. Here, the permeability K is calculated by using the CarmanKozeny equation

$$K = K_0 \frac{\epsilon^3}{(1-\epsilon)^2} \quad (4)$$

where K_0 is given by

$$K_0 = \frac{d^2}{180}. \quad (5)$$

d is the primary dendrite arm spacing which is set to 200 μm for the Al-7wt pct Si alloy, see references in [3]. ϵ is the volume fraction of the liquid phase.

The *energy equation* is formulated in such way that the temperature is the principal variable.

$$\rho\bar{c_p}\frac{\partial T}{\partial t} + \rho\bar{c_p}(\mathbf{u}\nabla)T = \nabla(\lambda_m \nabla T) - \rho L_m^f \frac{\partial f_l}{\partial t} \quad (6)$$

Finally, for the *species equation* holds:

$$\rho\frac{\partial C_m}{\partial t} + \rho(\mathbf{u}\nabla)C_m = \nabla(D_m \nabla C_m) + \nabla(D_m \nabla(C_l - C_m)) - \rho\mathbf{u}\nabla(C_l - C_m) \quad (7)$$

The mixture quantities, λ_m, D_m, $\bar{c_p}$ and $\bar{c_p}$ obey to the standard relations: $\lambda_m = \lambda_l\epsilon + \lambda_l(1-\epsilon)$
$\bar{c_p} = c_{pl}f_l + c_{ps}f_s$,

$D_m = D_l f_l + D_s f_s$ $C_m = C_l f_l + C_s f_s$, where f_l and f_s are the mass fraction of the liquid and solid phases, respectively. Because of the condition $\rho_l = \rho_s$, we have $\epsilon = f_l$. For the relationship between ϵ and T, the conduction-dominated solidification rule is used [3]:

$$\epsilon = \min\left[1, \max\left[0, \frac{T - T_s}{T_l - T_s}\right]\right]. \quad (8)$$

T_s and T_l are the temperatures on the solidus and liquidus lines in the phase diagram. The liquidus temperature is connected with C and liquidus line slope via $T_l = T_m + m \cdot C$. The initial temperature and mass concentration of Si are $T_{ref} = 974$K and $C_{ref} = 0.07$. Relevant material data are summarized in the Table 1.

In analogy to the experiment, the top boundary of the melt is free. This top surface can be considered as flat if the following restriction for the Froude number is fulfilled $Fr = \Omega R_0^2/(gH_0) < 0.1$ [9], where Ω is the angular velocity at the vicinity of the axis. This is the case for all B_0 investigated. Concerning the temperature boundary conditions, the top and bottom boundaries are adiabatic while the temperature of the side wall, $T_W(t)$, is a function of time, cf. Fig. 1(b). This function $T_W(t)$ was taken as boundary condition for the side wall.

3. Code validation

The solidification model described above has been implemented into the open source code of a 2D Navier Stokes solver [10], where the SIMPLE algorithm with collocated-variables arrangement is used to calculate the pressure and the velocities. For details we refer to [2]. The outer iterations were stopped if the residual of the energy equation is less than 10^{-5} and less than 10^{-10} for pressure and velocities, respectively. Different grids and time steps are tested and it was found that the grid $150CV_r \times 360CV_z$ and the time step $\Delta t = 0.1$s are a good compromise between the computation time and accuracy.

IOP Conf. Series: Materials Science and Engineering **33** (2012) 012048 doi:10.1088/1757-899X/33/1/012048

Table 1. Physical properties of an Al-7wt pct Si alloy [3].

Characteristic	Solid phase	Liquid phase
Thermal conductivity(λ_s, λ_l), $Wm^{-1}K^{-1}$	150	70
Specific heat (c_{ps}, c_{pl}), $Jkg^{-1}K^{-1}$	958	1126
Thermal expansion coefficient β_T, K^{-1}		$1.17 \cdot 10^{-4}$
Solutal expansion coefficient β_C		0.05
Latent heat L, Jkg^{-1}	387400	
Melting point T_m, K	933.5	
Eutectic temperature T_s, K	850	
Density ρ_m kgm^{-3}	2390	
Dynamic viscosity μ, Nsm^{-2}		$1.146 \cdot 10^{-3}$
Electric conductivity σ, A$V^{-1}m^{-2}$	$3.65 \cdot 10^6$	
Liquidus line slope m, K	-662.7	

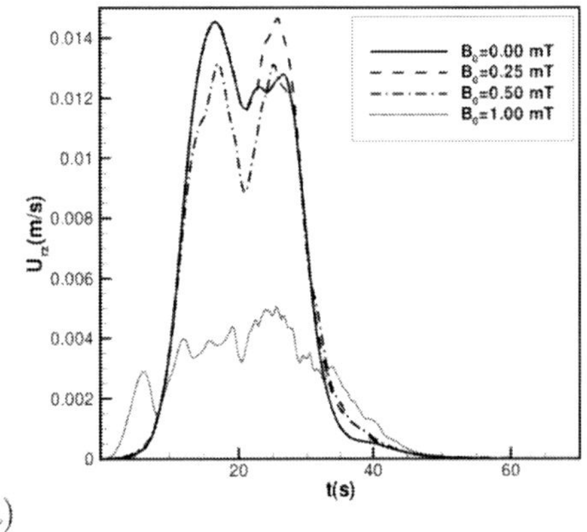

(a)

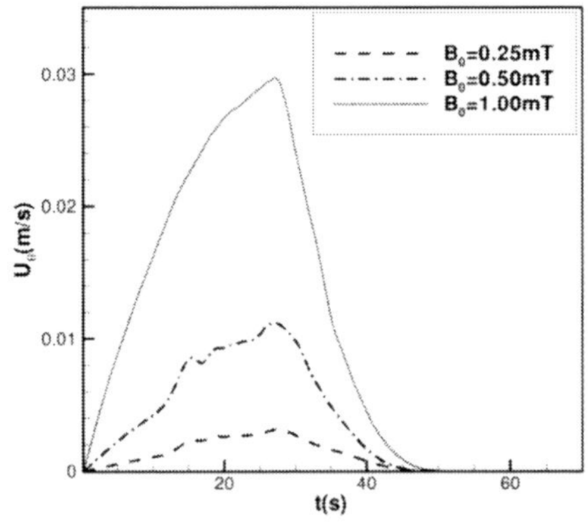

(b)

Figure 2.
The volume-averaged quantities U_{rz} (a) and U_θ (b) for different values of B_0.

4. Results

The dynamics of the solidification driven by a radial heat flux differs from that encountered during the unidirectional solidification previously studied in [2, 3]. This is due to the instantaneous onset of natural convection because the fluid close to the cooled side walls is heavier with respect to the hotter melt in the center of the cavity. This flow couples with that driven by the RMF. To characterize the resulting flow in the melt during the solidification we plot in Fig. 2 the evolution of the volume-averaged velocities U_{rz} and U_θ, given by [2]

$$U_{rz} = \frac{2}{H_0 R_0^2} \int_0^{H_0} \int_0^{R_0} r\sqrt{u_r^2 + u_z^2} dr dz, \quad U_\theta = \frac{2}{H_0 R_0^2} \int_0^{H_0} \int_0^{R_0} r\sqrt{u_\theta^2} dr dz, \tag{9}$$

with and without RMF.

The flow in the meridional (r,z)-plane, characterized by U_{rz} in Fig. 2a, generally rises with begin of the solidification, either due to the start-up of the buoyancy-driven flow or due to the spin-up of the melt, driven by the RMF. In the latter case, the Ekman pumping caused by the build-up of pressure gradients in both radial and axial direction is responsible for the flow in the (r,z)-plane. There is a distinct modification of both the structure and the amplitude of U_{rz} in dependence on the magnetic induction B_0. The purely buoyancy-driven flow, occurring when $B_0 = 0$, displays a first main maximum at $t = 16.5$ s. The azimuthal velocity component has a biggest magnitude at this time on the axis $u_z = 0.041$ m/s. On comparing with the stream lines, shown in Fig. 3a, this U_{rz} maximum is associated with the large buoyancy-driven vortex, rotating in clockwise direction, which is accompanied by a small vortex in the lower right corner. After passing this maximum, a minimum appears in Fig. 2a at $t = 21$ s which is characterized by a vertical arrangement of two smaller clockwise rotating vortices at top and bottom (Fig. 3a), a third smaller one in between, rotating in opposite direction. Out of this U_{rz} minimum, a second smaller maximum appears at $t = 26.8$ s during which the recovering of the flow structure of $t = 16.5$ s is disturbed by the division of the large vortex into two smaller ones. As consequence the magnitude of the azimuthal velocity component decreases till $u_z = 0.036$ m/s. After this maximum the progressing solidification of the melt leads to a braking of the flow until it vanishes at about $t = 35$.

A low amplitude of the RMF, e.g. $B_0 = 0.25$ mT leads in parallel with the onset of the azimuthal flow, see Fig. 2, to a re-symmetrization of the two U_{rz} maximums. This trend is present also for $B_0 = 0.5$ mT, however coupled with a reduction of the U_{rz} amplitudes. Going to higher B-fields, such as $B_0 = 1$ mT, we note that the increase in U_θ is accompanied by a considerable decrease in U_{rz}. Furthermore, the double maximum structure of the buoyancy-dominated flow disappears.

Before describing the underlying flow structure we inspect the purely buoyancy-driven case in more detail. Fig. 3b shows the isotherms in the melt for the characteristic maximums and minimum in Fig. 3a. Without convection, the isotherms would run parallel to the sidewall as shown for the late stage of solidification at $t = 35$ s when a significant braking of the convection in the mushy zone took place already. By contrast, the buoyancy-driven convection advects the isotherms in direction of the flow. As a result we obtain an ensemble of isotherms whose curvature reflects the underlying multi-vortex structure.

Fig. 5a shows the radial temperature profiles $T(r, z = H_0/2, t)$, measured by eight thermocouples as a function of time. This plot was obtained by aligning the profile $T(r, z = H_0/2, t_0 + \Delta t)$ on top of that of $T(r, z = H_0/2, t_0)$ etc. At $t = 0$ the melt is isotherm on a temperature $T = 700°C = 974\,K$. With progressing time and solidification, the volume of the cylindrical cavity becomes more and more non-isotherm and the temperature gradient in the wall-near regions increases. This behavior is correctly reflected by the numerics (Fig. 5b) which shows that the nearly isothermal volume of the cavity, i.e. the melt, shrinks with progressing time, i.e. solidification stage, and growing radial temperature gradients form inside the solid material.

We now come back to Fig. 2 to understand the underlying reasons for the changes provoked upon increasing B_0. For this reason we plot in Fig. 4 both the stream function and the isotherms for $B_0 = 0.5$ mT. We see for $t = 17$ s that the large buoyancy-driven vortex of Fig. 3a (t=16.5 s) is squeezed and sandwiched by two smaller ones which rotate in opposite direction. As a result, the U_{rz} maximum in Fig. 2 is reduced. The origin of this fact is that the upper vortex of the double vortex, driven by the Ekman pumping in the RMF-case, opposes the buoyancy-driven vortex since the first one rotates in counter-clockwise direction. The reason for the re-symmetrization of both peaks in Fig. 2a seems to be the canceling out of the third vortex in the lower right corner at $t = 25$ s for the same reasons.

IOP Conf. Series: Materials Science and Engineering **33** (2012) 012048 doi:10.1088/1757-899X/33/1/012048

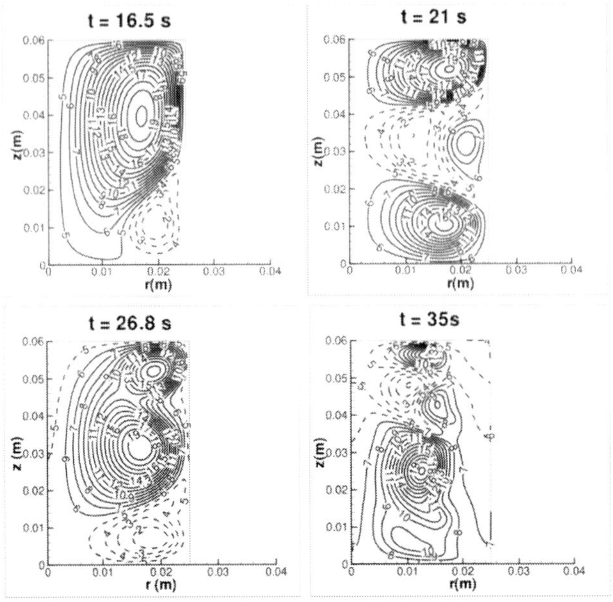

(a)

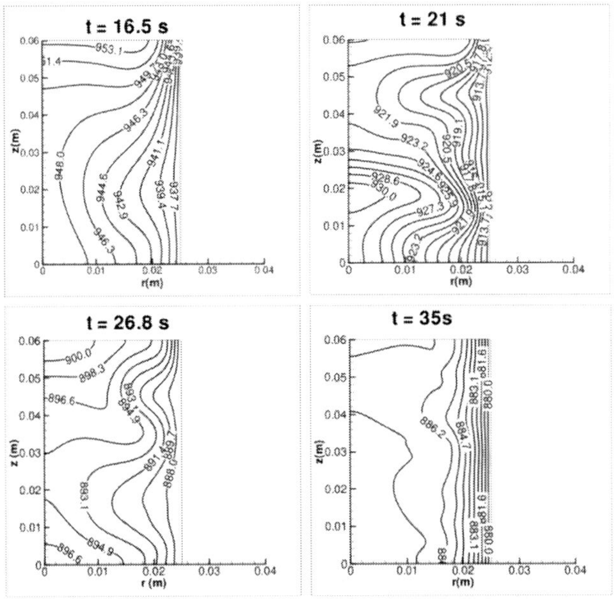

(b)

Figure 3.
Stream function for $B_0 = 0$ mT. The isolines belong to the interval $(-1.7 \cdot 10^{-3}, 7.4 \cdot 10^{-3})$ ($t = 16.5$s), $(-1.7 \cdot 10^{-3}, 5.7 \cdot 10^{-3})$ ($t = 21$s), $(-1.8 \cdot 10^{-3}, 6.7 \cdot 10^{-3})$ ($t = 26.8$s), $(-2.2 \cdot 10^{-3}, 6.1 \cdot 10^{-3})$ ($t = 35$s). (b) Temperature for $B_0 = 0$ mT.

IOP Conf. Series: Materials Science and Engineering **33** (2012) 012048 doi:10.1088/1757-899X/33/1/012048

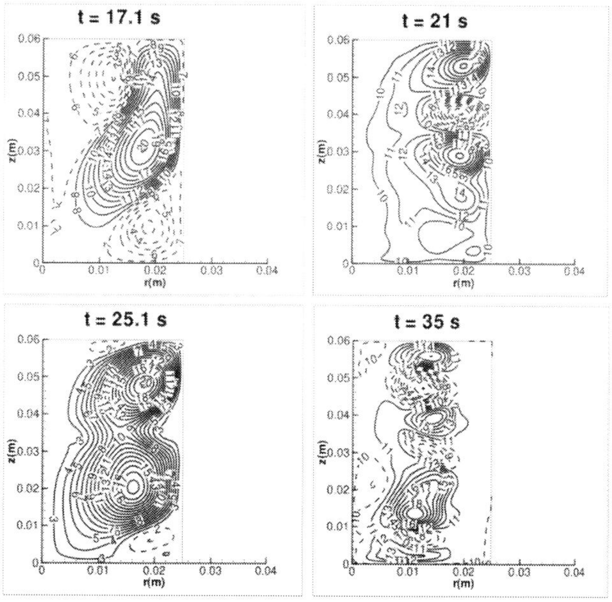

(a)

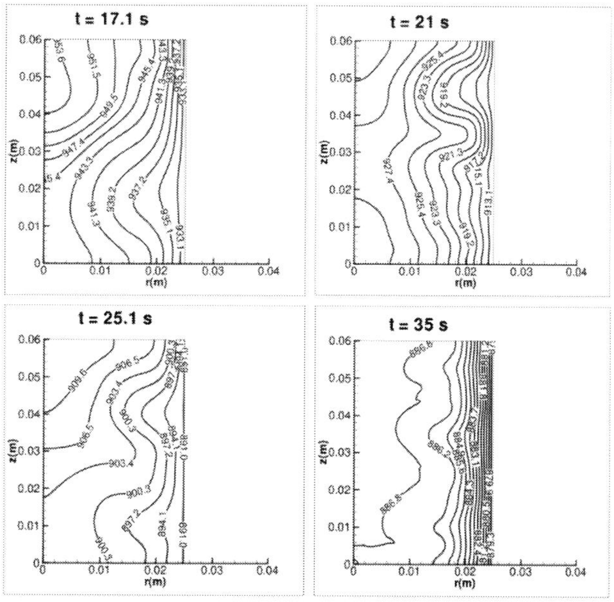

(b)

Figure 4. Stream function for $B_0 = 0.5$ mT. (a) The isolines belong to the interval $(-2.0 \cdot 10^{-3}, 6.0 \cdot 10^{-3})$ ($t = 17.1$s), $(-3.6 \cdot 10^{-3}, 4.3 \cdot 10^{-3})$ ($t = 21$s), $(-4.4 \cdot 10^{-4}, 6.6 \cdot 10^{-3})$ ($t = 25.1$s), $(-7.0 \cdot 10^{-4}, 9.0 \cdot 10^{-4})$ ($t = 35$s). (b) Temperature for $B_0 = 0.5$ mT.

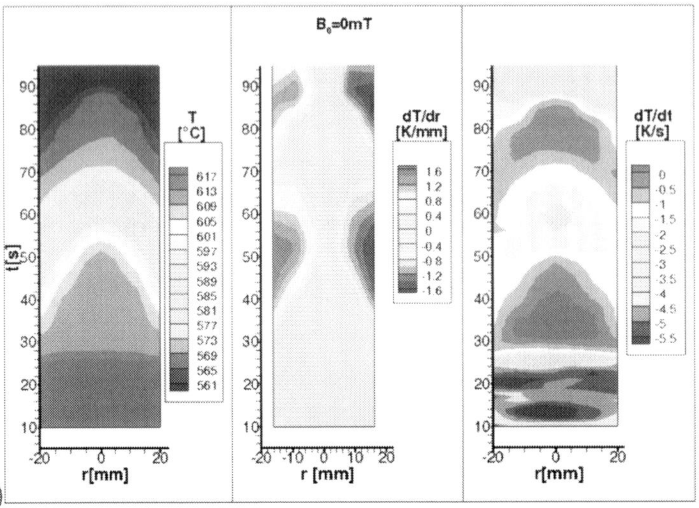

(a)

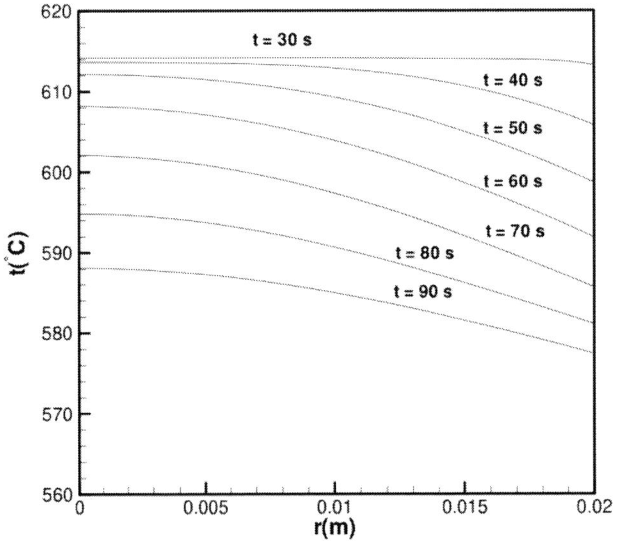

(b)

Figure 5.
(a) Experimental measurements of the temperature for $B_0 = 0$ mT. (b) Numerical calculation of the temperature shape $B_0 = 0$ mT.

5. Summary and conclusion

The flow field during the radial solidification of an Al-7wt pct Si melt was studied with and without application of an RMF. We have shown, for the case without an RMF, that the start-up of the buoyancy-driven convection leads to a large vortex which breaks up into a multi-cellular structure later on. The increase of the solid fraction with progressing solidification exerts a braking onto the flow until it finally decays completely. Upon application of weak RMF's a systematic decrease of the flow amplitudes in the meridional (r,z)-planes sets in together with a symmetrization of the double peak structure. The reason for this behavior is the competition between the upper, Ekman-driven vortex and the buoyancy-driven one. These results presented

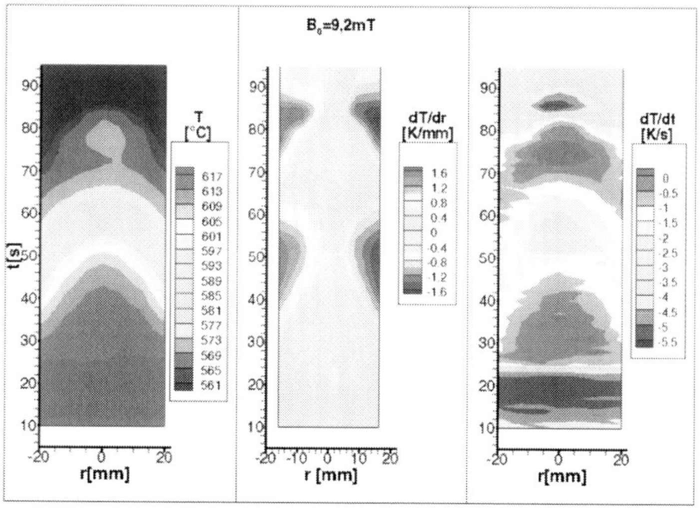

Figure 6.
Experimental measurements of the temperature for $B_0 = 9.2$ mT.

here are limited to small amplitudes of the RMF. An extension towards higher RMF requires a three-dimensional consideration of the problem.

Acknowledgments

Financial support by the Deutsche Forschungsgemeinschaft (DFG) within the Collaborative Research Center SFB 609 "Electromagnetic Flow Control in Metallurgy, Crystal Growth and Electrochemistry" is gratefully acknowledged.

References

[1] Eckert S, Willers B, Nikrityuk P A, Eckert K, Michel U, Zouhar G 2005 *Mat. Sci. Eng. A* 413-414, 211.
[2] Nikrityuk P A, Eckert K, Grundmann R 2006 *Int. J. Heat Mass Transfer* **49** 1501-1515.
[3] Willers B, Eckert S, Nikrityuk P A, Räbiger D Dong J, Eckert K, Gerbeth G 2008 *Met. Mat.Trans.* **39B** 304.
[4] Nikrityuk P A, Eckert K, Grundmann R 2004 *Magnetohydrodynamics* **40** 127.
[5] Travnikov V, Eckert K, Nikrityuk P A, Odenbach S, Vogt T, Eckert S 2011 *J. Crystal Growth.* DOI:10.1016/j.jcrysgro.2011.10.048
[6] Witkovski L M, Walker J S 1999 *Phys. Fluids* **11** 1821.
[7] Gorbachev L P , Nikitin N V, Ustinov A L 1974 *Magntenohydrodynamics* **4** 406.
[8] Bennon W D, Incropera F P 1987 *Int. J. Heat Mass Transfer*, **30**, 2161
[9] Spohn A, Mory M, Hopfinger E J 1993 *Exp. Fluids* **14** 70.
[10] Ferziger J H, Peric M 2002 *Computational Methods for Fluid Dynamics* (Berlin: Springer)

Measurement and simulation of deformation and stresses in steel casting

D Galles[1], C A Monroe[2] and C Beckermann[1]

[1]Department of Mechanical and Industrial Engineering, University of Iowa, Iowa City, IA 52242, USA
[2]Caterpillar Inc., Peoria, IL 61629, USA

E-mail: becker@engineering.uiowa.edu

Abstract. Experiments are conducted to measure displacements and forces during casting of a steel bar in a sand mold. In some experiments the bar is allowed to contract freely, while in others the bar is manually strained using embedded rods connected to a frame. Solidification and cooling of the experimental castings are simulated using a commercial code, and good agreement between measured and predicted temperatures is obtained. The deformations and stresses in the experiments are simulated using an elasto-viscoplastic finite-element model. The high temperature mechanical properties are estimated from data available in the literature. The mush is modeled using porous metal plasticity theory, where the coherency and coalescence solid fraction are taken into account. Good agreement is obtained between measured and predicted displacements and forces. The results shed considerable light on the modeling of stresses in steel casting and help in developing more accurate models for predicting hot tears and casting distortions.

1. Introduction

During cooling of a steel casting, thermal contractions result in dimensional differences between the mold and the final part. A casting is said to contract freely if there are no mechanical interactions between the mold and the contracting steel, in which case the patternmaker's shrink accurately predicts the final dimensions. However, for complex shaped castings, interactions between the mold or cores and the steel generate stresses within the casting, which cause mechanical strains and can result in gross distortions. Additionally, the buildup of stresses at high temperatures may cause hot tears, which are irreversible cracks that form in the semi-solid mushy zone [1]. To minimize these potential problems, accurate stress models that predict the deformations of a steel casting during solidification and cooling can be a valuable tool. Such models require knowledge of the mechanical properties of the steel over the entire range of temperatures and strain rates encountered during casting. The objective of this study is to predict deformations and stresses during steel casting and validate the predictions with data acquired from experiments.

Because the microstructure created during solidification can differ from that in reheated samples, obtaining experimental data in-situ during casting is preferable. However, due to the difficulty of measuring forces and displacements in steel at high temperatures, studies in which stress models of steel casting are validated by in-situ measurements are limited. Monroe and Beckermann [2] acquired force and displacement data during casting of a steel bar to validate a hot tear model. Rowan et al. [3]

used in-situ force measurements in the so-called submerged split chill contraction (SSCC) test to validate a stress model for a solidifying steel shell.

In this study, experiments involving casting of a rectangular steel bar in a sand mold are used to validate the predictions of a stress model. During the experiments, a uni-axial tensile force is applied to the ends of the bar in order to induce mechanical strains. The applied force and the displacements of the ends of the steel bar are measured in-situ as a function of time. In addition to these strained bar experiments, unrestrained bar experiments are performed where no force is applied and the bar is allowed to contract freely. The temperatures in the experiments are predicted using MAGMAsoft [4] and validated through temperature measurements. Stress simulations of the present experiments are performed using the finite-element software ABAQUS [5]. A viscoplastic constitutive model for the mechanical behavior of the steel during solidification and cooling is used that treats the mush as a compressible porous material. The temperature-dependent mechanical properties are estimated from data found in the literature. For the strained bar, the measured forces are included as a boundary condition in the stress model. Measured and predicted end displacements are compared.

2. Description of Experiments

Casting experiments were performed at the University of Northern Iowa's Metal Casting Center. The steel melt was prepared in an induction furnace. The target chemistry was ASTM A216 Grade WCB carbon steel. The molds were made from silica sand bonded with a phenolic urethane no-bake binder. A schematic of the strained bar experiment is shown in figure 1. The dimensions of the cast bar were 12 inches (305 mm) long with a 1 inch (25.4 mm) square cross section. The restrained bar setup is converted to an unrestrained bar setup by removing the restraint, load bolts, restraining bolts, coupling, and turnbuckle.

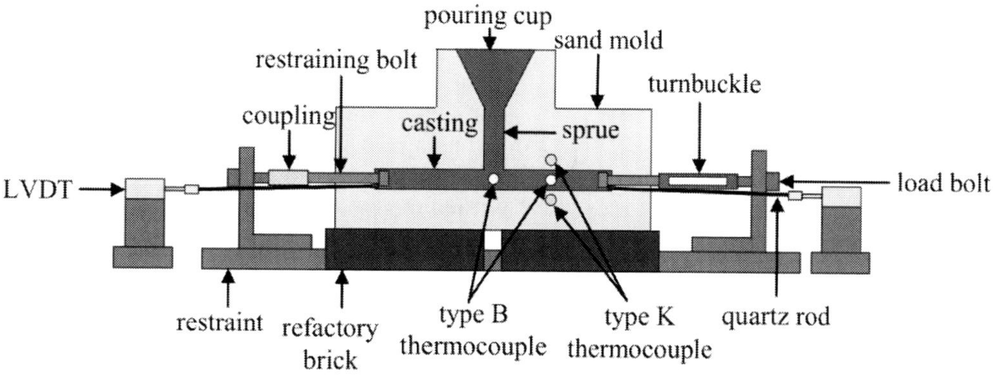

Figure 1. Experimental setup for casting of a strained steel bar.

To acquire temperature, force, and displacement data, a variety of measurement devices were used. Temperatures in the casting and mold were measured using type B and type K thermocouples, respectively. The type B thermocouples were encased in quartz tubing to protect them from the molten steel. The uni-axial force applied to the casting was measured with load bolts located on either side of the mold and anchored by a restraint frame. To transmit the force experienced by the casting to the load bolts, restraining bolts were connected to the load bolt at one end, while the other end extended into the mold cavity. To assure the restraining bolts were firmly embedded in the casting, nuts were threaded onto the ends of the restraining bolts inside the mold cavity. Once poured, the casting was allowed to partially solidify until it reached a level of coherency and could transmit stresses, at which time the turnbuckle was turned to induce stresses in the casting. The forces measured by the two load bolts were in close agreement, indicating that mechanical interactions between the casting and the mold were negligibly small. Axial displacement measurements of the ends of the cast bar were taken with LVDTs located on either side of the mold. In order to avoid damage of the LVDTs by the heat from the casting, quartz rods were used. Quartz has a very low coefficient of thermal expansion. On each side of the mold, the quartz rod was connected at one end to the LVDT, and the other end was inserted through the mold and extended 3-5 mm into the mold cavity. To ensure the quartz rod did not slip in the casting, the end of the rod inside the mold cavity was bulged into a spherical shape using an oxy-acetylene torch. The length change of the bar was calculated by adding the displacements measured by the two LVDTs.

Overall, three unrestrained and five strained bar experiments were performed. Due to space limitations, plots of all experimental results are not shown here. An example of a measured temperature evolution is shown in figure 2. The measured length changes as a function of time for one unrestrained bar experiment and for one strained bar experiment are shown in figures 4 and 5, respectively. Figure 5 also shows the measured force as a function of time in the strained bar experiment. These plots are discussed in detail below. The thermal results from all experiments were very similar throughout the entire cooling period. The displacement results showed that the evolution of the length change in the three unrestrained bar experiments was very reproducible. The length changes measured in the strained bar experiments varied depending on when and how much stress was induced in the casting by the turnbuckle.

3. Thermal Simulations

Using the experimental casting chemistry, the temperature dependent thermophysical properties and the solid fraction as a function of temperature curve were generated using IDS [6]. These properties include the density, thermal conductivity, and apparent specific heat, which takes into account the latent heat that is released during solidification and solid-state phase changes.

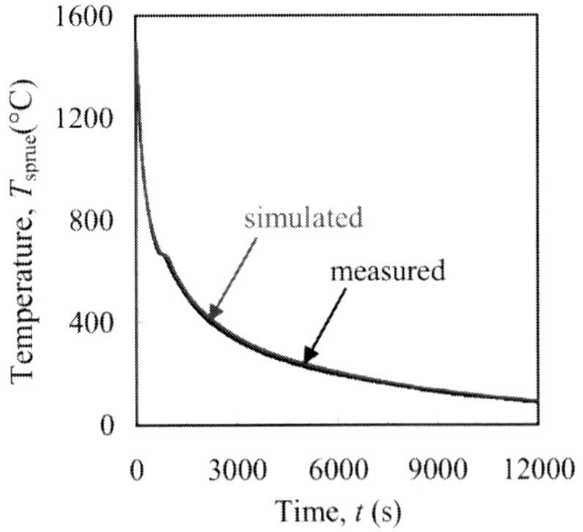

Figure 2. Comparison of measured and predicted temperatures underneath the sprue.

Using these properties along with an estimated pouring temperature and interfacial heat transfer coefficient between the steel and the mold, MAGMAsoft simulations were performed to predict the evolution of the thermal field from the pouring temperature to room temperature throughout the casting. To compare the measured temperatures from the experiments with the predicted temperatures in the MAGMAsoft model, virtual thermocouples were inserted in the model at the locations of the type B thermocouples shown in figure 1. An example of a comparison between measured and simulated temperatures for the thermocouple located below the sprue is shown in figure 2. As the molten steel initially contacts the thermocouple, its temperature increases almost instantaneously from room temperature to approximately 1500 °C. The temperature then rapidly decreases as heat is transferred from the steel to the mold and atmosphere. The temperature continues to decrease until the solid state transformation of austenite to pearlite (at approximately 675 °C). At this point, latent heat is released and the cooling rate of the steel slows to nearly zero for a brief time. This is observed in figure 2 as an inflection point in the curves at approximately 1000 seconds. After the solid state transformation, the temperature again decreases (but at a slower rate) until equilibrium with the room temperature is reached. Through a trial-and-error method, excellent agreement between measured and predicted temperatures was achieved by adjusting the pouring temperature and the interfacial heat transfer coefficient. By adjusting the pouring temperature, the time at which the measured and simulated temperatures reached the solidus temperature was matched exactly. Below the solidus temperature, the measured and predicted temperature curves were matched by decreasing the interfacial heat transfer coefficient, which was originally set to a constant value of 1000 W/m^{2}K. This decrease is realistic and reflects the formation of an air gap between the casting and the mold during cooling. For the simulated curve in figure 2, the interfacial heat transfer coefficient was decreased at temperatures below 1000 °C.

4. Stress Model

Assuming small strain theory, the total strain, ε, can be decomposed into the elastic (e), thermal (th), and viscoplastic (vp) components as $\varepsilon = \varepsilon_e + \varepsilon_{th} + \varepsilon_{vp}$. Calculations of the elastic and thermal strains were performed using standard procedures and are not discussed here.

The viscoplastic strain is determined from the flow condition. The flow condition limits the maximum stress the steel can hold by keeping the equivalent stress less than or equal to the dynamic yield stress. When the equivalent stress exceeds the dynamic yield stress, the plastic strain is increased to satisfy $\sigma_{eq} = \sigma_{dy}$. In the mushy state, the steel behaves as a compressible material and is modeled using porous media theory, where the equivalent stress is a function of both the von Mises stress and the pressure. Once the steel has completely solidified, it behaves as a regular von Mises material. The equivalent stress is calculated from

$$\sigma_{eq}^2 = A_1(g_s^*)q^2 + A_2(g_s^*)p^2 \tag{1}$$

where q is the von Mises stress and p is the pressure. The functions A_1 and A_2 are from the Cocks model and depend on the scaled solid fraction [7, 8], which is given by $g_s^* = g_s^{coal}\left[(g_s - g_s^{coh})/(g_s^{coal} - g_s^{coh})\right]$, where g_s is the solid fraction, g_s^{coh} is the coherency solid fraction, and g_s^{coal} is the coalescence solid fraction. The coherency solid fraction is the solid fraction at which the material has reached a point of coherency and can transmit stresses. The coalescence solid fraction occurs at a lower temperature and is the solid fraction at which voids begin to coalesce. The values chosen for the coherency and coalescence solid fractions for this study were 0.5 and 0.85, respectively [2, 9]. In the limit where the solid fraction is unity, equation (1) reduces to the von Mises condition, where A_1 is equal to unity and A_2 is equal to zero.

The present viscoplastic constitutive model relates the dynamic yield stress to the viscoplastic strain and strain rate and is given by

$$\sigma_{dy} = \sigma_0\left(1 + \frac{\varepsilon_{eq}}{\varepsilon_0}\right)^n\left(1 + \frac{\dot{\varepsilon}_{eq}}{\dot{\varepsilon}_0}\right)^m \tag{2}$$

where σ_0 is the initial yield stress, n is the strain hardening exponent, m is the strain rate sensitivity exponent, ε_{eq} is the equivalent plastic strain, and $\dot{\varepsilon}_{eq}$ is the equivalent strain rate. The reference shear strain, ε_0, is calculated from $\varepsilon_0 = \sigma_0 n/E$, in which E is Young's modulus. The reference strain rate, $\dot{\varepsilon}_0$, is given by $\dot{\varepsilon}_0 = A\,exp(-Q/RT)$ where A is a prefactor, Q is the activation energy, and R is the universal gas constant.

To illustrate the impact of stresses occurring in the mushy state, consider a one-dimensional tensile test. During plastic deformation, $\sigma_{eq} = \sigma_{dy}$ and stress-strain curves can be calculated directly from equations 1 and 2. For uni-axial conditions, the von Mises stress and the pressure can be written in terms of the stress, σ_{11}. Figure 3 shows the uni-axial stress as a function of the equivalent strain at a constant strain rate (10^{-4} 1/s) and

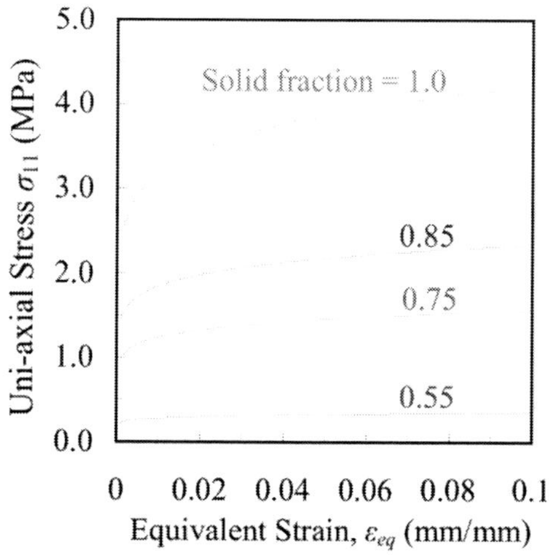

Figure 3. Stress-strain behavior in the mushy state for different solid fractions.

for different solid fraction values. At a solid fraction of 0.55, the mush is very near the coherency limit and therefore can carry only a small load. Increasing the solid fraction from 0.55 to 1.0 significantly increases the load in the mush at a given strain.

5. Mechanical Properties

For the elastic properties, a constant value of 0.3 was used for Poisson's ratio, and the temperature-dependent Young's modulus was taken from Li and Thomas [10].

For the casting experiments in this study, IDS predicted the delta-ferrite to austenite phase transformation to occur at temperatures higher than the coherency solid fraction temperature. Therefore, all viscoplastic deformation in the present experiments occurred in the austenitic phase. The unknown viscoplastic parameters in equation 2 were estimated from experimental stress-strain data for austenitic steel found in the literature. Wray [11] performed tensile tests on reheated plain carbon steel specimens at varying strain rates and carbon contents for temperatures ranging from 850°C to 1200°C. Suzuki et al. [12] performed creep tests on reheated plain carbon steel specimens with a 0.19% carbon content at varying stresses for temperatures ranging from 1200°C to 1400°C. These data were used to estimate the parameters in equation 2. Three parameters (σ_0, n, m) were estimated as linear functions of temperature, while the pre-factor in the reference strain rate, A, is carbon-content dependent [13] and was estimated using a quadratic relationship. The activation energy, Q, was estimated as a temperature-independent constant. All parameters were estimated simultaneously using a Levenberg-Marquardt least-squares minimization between predicted and experimental stresses. The resulting viscoplastic parameters are shown in table 1. They were used from the liquidus temperature down to the solid state transformation temperature of austenite to pearlite (at 675°C). Analysis of the present experimental data revealed that the viscoplastic strain in the strained bars was negligibly small below the solid state transformation temperature. Therefore, to assure no viscoplastic strain was predicted after the solid state transformation, the initial yield stress, σ_0, was set to a very high value at the onset of the solid state transformation.

Table 1. Estimated viscoplastic parameters for austenitic steel, where T is the temperature in °C and $\%C$ is the carbon content.

Parameter	Estimated Value
σ_0	$0.5729 - 1.461\text{x}10^{-5}T$ [MPa]
n	$0.2457 - 6.192\text{x}10^{-5}T$
m	$0.0492 + 9.930\text{x}10^{-5}T$
A	$24870 + 124600(\%C) + 12400(\%C)^2$ [1/s]
Q	353 [kJ/mol]

6. Stress/Displacement Simulations

Stress simulations were performed using ABAQUS, where the viscoplastic constitutive model from section 4 was implemented in a user-defined UMAT subroutine. The temperature fields as a function of time predicted by MAGMAsoft were imported into ABAQUS and mapped onto the finite element mesh. Since mechanical interactions with the mold were negligibly small, stress simulations were performed only for the steel. For the simulation of the strained bar experiments, the measured force was used as a boundary condition. This boundary condition was applied to one end of the bar, while the other end was constrained with a zero-displacement boundary condition.

An example of a comparison between the measured and predicted length changes as a function of time for an unrestrained bar experiment is shown in figure 4. A negative length change represents shrinkage of the bar as it cools. It can be seen from figure 4 that the majority of the shrinkage occurs within the first 400 s after the casting is poured. After 400 s, the solid state transformation from austenite to pearlite begins in the bar and continues until approximately 800 s. During this transformation, the bar expands, which can be seen in figure 4 as a length change increase between

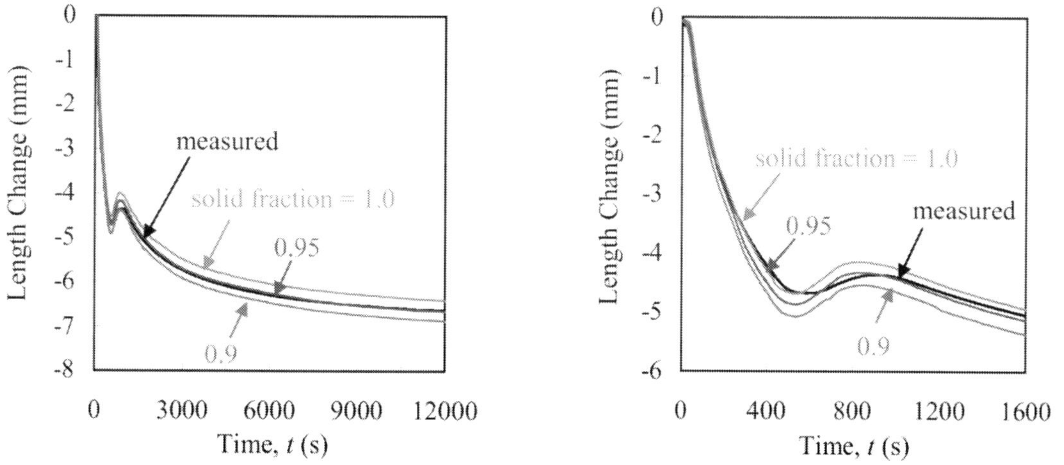

(a) large time scale (b) medium time scale

Figure 4. Comparison of measured and predicted length changes for an unrestrained bar.

400 and 800 s. After the transformation, the bar continues to shrink, albeit at a much slower rate, until it cools to room temperature. At room temperature (i.e., at ~12000 s), the measured length change of the bar is equal to about -6.7 mm, which corresponds to a total shrink of about 2.2%. Because no forces were applied in the unrestrained bar experiments, the length change in the bar is a result of thermal strains only. For the present simulations, the coefficient of thermal expansion was calculated from the density, which in turn was predicted using IDS and is presumed to be accurate. However, the temperature at which integration of the thermal strain is started in the simulations is generally unknown. A lower limit is the solidus temperature (solid fraction of unity), but it could also be higher, such that some liquid is still present in the solid matrix. Therefore, several simulations were performed in which this temperature was varied. The results in figure 4 show that the best agreement between measured and predicted length changes is obtained when integration of the thermal strain is initiated at a temperature corresponding to a solid fraction of 0.95. Finally, the slight disagreement between measured and predicted length changes seen in figure 4(b) during the solid state phase transformation can be attributed to small inaccuracies in the thermal simulation of the experiment.

Figure 5 shows the measured and predicted length changes for one of the strained bar experiments. The measured force, used in the stress simulation as a boundary condition, is also plotted as a function of time. For the first 50 s after the casting is poured, there are no forces applied, and therefore, the bar shrinks like an unrestrained bar during this period. Starting at 50 s, the turnbuckle is turned and the force builds up. As can be seen in figure 5(b), the measured force reaches a value of about 3500 N at the end of the induced straining period (at 123 s). After 123 s, the force continues to build due to thermal contraction of the restrained bar and eventually reaches a value of about 6200 N at 12000 s. This applied tensile force has a significant impact on the length changes of the bar. At about 80 s, the length of the bar begins to increase (at about 80 s). Shortly after the induced straining period, the bar length decreases again. The total measured length change of the strained bar, at 12000 s, is about -4 mm (1.3%). Therefore, the induced straining results in about 2.7 mm of viscoplastic deformation over the length of the bar. For the curve labeled "original simulation", the viscoplastic parameters listed in table 1 were used. On the large time scale of figure 5(a), the measured and predicted length changes show reasonable agreement, but on the small time scale of figure 5(b), significant differences can be observed during and after the induced straining period. This disagreement can be directly attributed to deficiencies in the calculation of the viscoplastic strain at high temperatures. Through a sensitivity study, it was determined that better agreement between the measured and predicted length changes could be achieved by adjusting the strain rate sensitivity exponent, m, in equation (2), although other

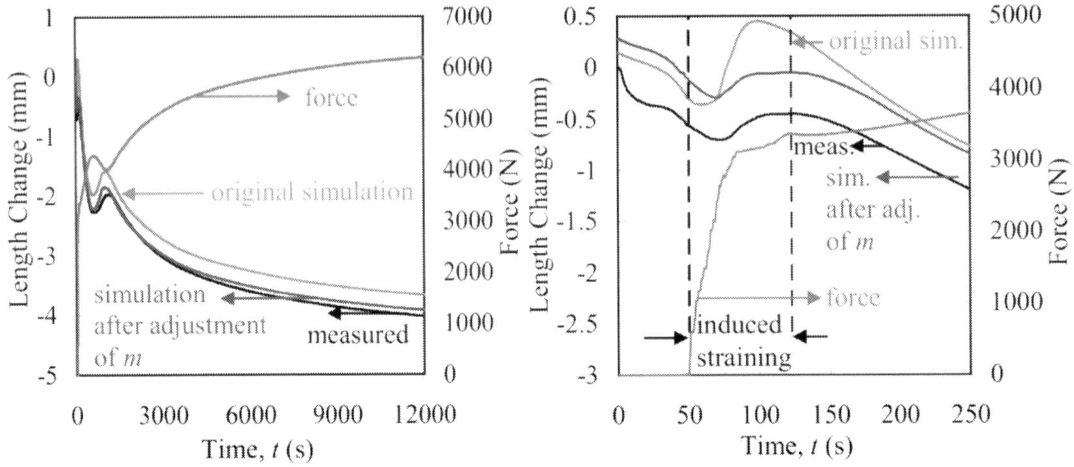

(a) large time scale (b) small time scale

Figure 5. Comparison of measured and predicted length changes for a strained bar.

viscoplastic properties could have been changed too. Figure 6 shows the original and adjusted strain rate sensitivity exponent, *m*, as a function of temperature. As given in table 1, the original *m* is a linear function of temperature. The steep decrease between 650°C and 600°C is due to the solid state transformation. The adjusted *m* is larger than the original *m* at high temperatures, decreases steeply between 1300°C and 1150°C, and is constant until 650°C. As shown in figure 5, this adjustment of *m* results in much better agreement between the measured and predicted length changes. The remaining differences are due to slight inaccuracies in the predicted thermal contraction of the bar during the first 50 s. Additional experiments are needed to corroborate the present adjustments in the strain rate sensitivity exponent. However, the comparisons in figure 5 illustrate that even minor changes in the viscoplastic properties can have a large effect on the predicted deformations. Hence, accurate predictions of steel casting distortions can only be made if the viscoplastic properties are known accurately. The deficiencies in the original simulation could also indicate that determining viscoplastic properties from tensile tests of reheated samples, as done in Wray [11] and Suzuki et al. [12], is not appropriate for use in stress models of steel casting. This is because the microstructure and microsegregation in reheated samples is likely to be different from that in just solidified steel.

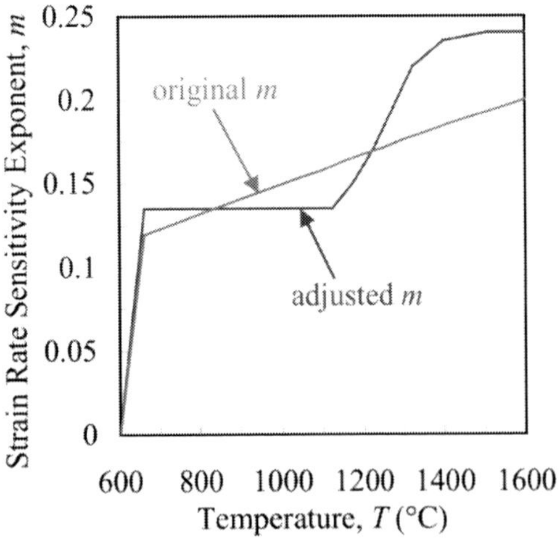

Figure 6. Original and adjusted strain rate sensitivity exponents.

The predicted equivalent plastic strain distribution in the strained bar at room temperature (12000 s) is shown in figure 7. As expected, the majority of the plastic strain occurs in the middle of the bar in the hotspot directly underneath the sprue. This is because the steel was much weaker in the middle of the bar than at the ends when the induced straining commenced. The ends cool more quickly due to the presence of the relatively cold restraining bolts. The large amounts of plastic strain predicted near the ends of the bar are not due to the induced straining but instead are a result of the compressive stresses that build up in the steel as it contracts around the restraining bolt.

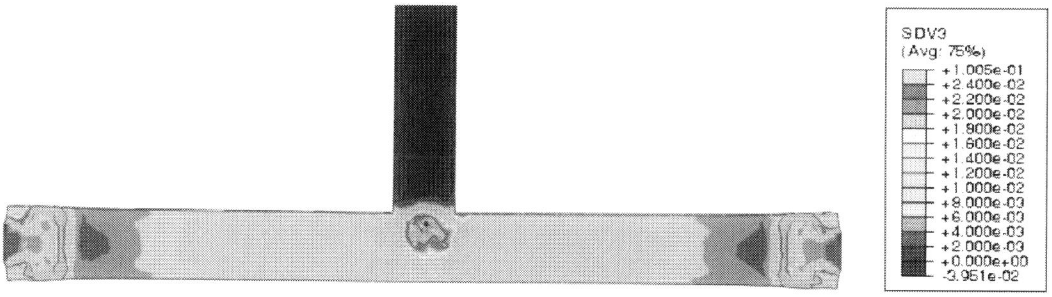

Figure 7. Room temperature equivalent plastic strain in the centre cross-section of the strained bar.

7. Conclusion

The present experiments are well suited to validate stress models of casting. They allow for the in-situ, real time measurement of displacements and forces during solidification and cooling. In addition, stresses of varying amount can be induced in the casting at pre-selected times. The present comparisons between measured and predicted length changes of the steel bar revealed a number of important issues. In particular, the use of viscoplastic properties estimated from stress-strain data for reheated steel samples appears to result in inaccurate deformation predictions. However, a relatively minor adjustment in the strain rate sensitivity exponent led to excellent agreement between measured and predicted length changes under induced straining. The results of this study are encouraging, and significant progress was made in understanding the mechanical behavior of steel during casting and predicting casting distortions and final dimensions.

References

[1] Bichler L, Elsayed A, Lee K and Ravindran C 2008 Influence of mold and pouring temeratures on hot tearing susceptibility of AZ91D magnesium alloy *Int. J. Metalcasting* **2** 45-56

[2] Monroe C and Beckermann C 2005 Development of a hot tear indicator for steel castings *Mat. Sci. and Eng. A* **413-414** 30-6

[3] Rowan M, Thomas B G, Pierer R and Bernhard C 2011 Measuring mechanical behavior of steel during solidification: modeling the SSCC test *Metal. and Mat. Trans. B* **42** 837-51

[4] MAGMAsoft [computer program] Version 4, Kackerstrasse 11, 52072 Aachen, Germany: MAGMA GmbH

[5] Abaqus/Standard [computer program] Version 6.10.2, Providence, RI: Abaqus Inc. ; 2006

[6] IDS [computer program] Version 2.0.0, Espoo, Finland: Helsinki Univ. of Technology; 1999

[7] Marin E B and McDowell D L 2005 A semi-implicit integration scheme for rate-dependent and rate-independent plasticity *Computers and Structures* **63** 30-6

[8] Cocks A C F 1989 Inelastic Deformation of porous materials *J. Mech. Phys. Solids* **37** 693–715

[9] Hardin R A and Beckermann C 2007 Effect of porosity on the stiffness of cast steel *Metal. and Mat. Trans. A* **38** 3041-8

[10] Li C and Thomas B G 2004 Thermomechanical finite-element model of shell behavior in continuous casting of steel *Metal. and Mat. Trans. B* **35B** 1151-72

[11] Wray P J 1982 Effect of carbon content on the plastic flow of plain carbon steels at elevated temperatures *Metal. Trans. A* **13A** 125-34

[12] Suzuki T, Tacke T H, Wunnenberg K and Schwerdtfeger K 1988 Creep properties of steel at continuous casting temperatures *Ironmaking and Steelmaking* **15** 90-100

[13] Wray P J 1980 *Modeling of Casting and Welding Proc.* ed H D Brody and D Apelian (Rindge, NH: The Metallurgical Society of AIME) pp 245-257

Transient and residual stresses in large castings, taking time effects into account

J Thorborg, J Klinkhammer, M Heitzer
MAGMA Gießereitechnologie GmbH, Aachen, Germany

E-mail: J.Thorborg@magmasoft.de

Abstract. Casting of large scale steel and iron parts leads to long solidification and cooling times. Solid mechanical calculations for these castings have to take the time scale of the process into account, in order to predict the transient and residual stress levels with a reasonable accuracy. This paper presents a study on the modelling of the thermo-mechanical conditions in the cast material using a unified approach to describe the constitutive behaviour. This means a classical splitting of the mechanical strain into an elastic and an inelastic contribution, where the inelastic strain is only formulated in the deviatoric space in terms of the J2 invariant. At high temperatures, creep is pronounced. Since the cooling time is long, the model includes a type of Norton's power law to integrate the significant contribution of creep to the inelastic strains. At these temperature levels, annealing effects are also dominant and hence no hardening is modelled. However, at intermediate and lower temperature levels, hardening is more pronounced and isotropic hardening is considered. Different hardening models have been studied and selected based on their ability to describe the behaviour at the different temperature levels. At the lower temperature levels, time effects decrease and the formulation reduces to a time independent formulation, like classical J2-flow theory. Several tensile and creep experiments have been made at different temperature levels to provide input data for selecting the appropriate contributions to the material model. The measurements have furthermore been used as input for extracting material data for the model. The numerical model is applied on different industrial examples to verify the agreement between measured and calculated deformations.

1. Introduction

The casting process is a complex process where many parameters influence the final quality of the considered part. To get a better understanding of the process it is useful to apply numerical modelling to analyze the transient and residual phenomena which take place. In the presented work the focus is mainly on the constitutive modelling of the mechanical system for large iron castings during solidification and cooling. As for casting in general the mechanical fields are mainly driven by the thermal gradients which are a result of the heat transfer from the part to the mold, chills and the surroundings. The casting system, the cores and the mold are furthermore a constraint to the thermal contraction and deformation in general, which can lead to stress and/or relaxation of the material. Especially at high temperature where the cast material is weak compared to the core and mold material, the thermal contraction and the constraints can lead to considerable amount of permanent deformations, which at the end of the cooling process at ambient temperature can lead to distortions and stresses. In severe cases cracks starts to form in highly loaded regions even before service loads are applied.

IOP Conf. Series: Materials Science and Engineering **33** (2012) 012050 doi:10.1088/1757-899X/33/1/012050

To get a good description of the thermo mechanical conditions it is important to select a material model which gives a proper integration of the mechanical load history. For most metal casting processes it is typically sufficient to use a classically thermo-elasto-plastic model. Based on Hooke's law for the linear elastic part and time independent plasticity for the inelastic part, in terms of the J_2 formulation in the deviatoric system. In that case time has no direct influence on the mechanical model only indirectly on the loading via the thermal gradients. Anyway, if the cooling is either very fast or very slow, time effects starts to play an important role directly. In the first case fast cooling resembles conditions comparable to quenching where the material response becomes very sensitive to the strain rate. I.e. the response becomes viscoplastic instead of 'just' plastic. This behaviour is described well by the overstress formulation where the response of the material is between the static plastic response and the elastic response. Where the static plastic response is the nonlinear response of the material if sufficient time would be available for the inelastic strain to form and develop. While the elastic response is the response if time goes to zero and no time would be available to form inelastic deformations. The actual response would be in-between these two bounds, depending on the available time during the load increment. Please see figure 1 for a simple illustration. In the second case slow cooling resembles the condition seen in e.g. larger castings where the thermal gradients are smaller and the cooling time is sufficiently long for relaxation phenomena to take place. This type of time dependency is typically well described by creep models where the inelastic strain rate is associated to the stress level in terms of e.g. a power law. This type of formulation is in contrast to the time independent model where the inelastic strain increment is associated to a stress increment and not to the stress level itself.

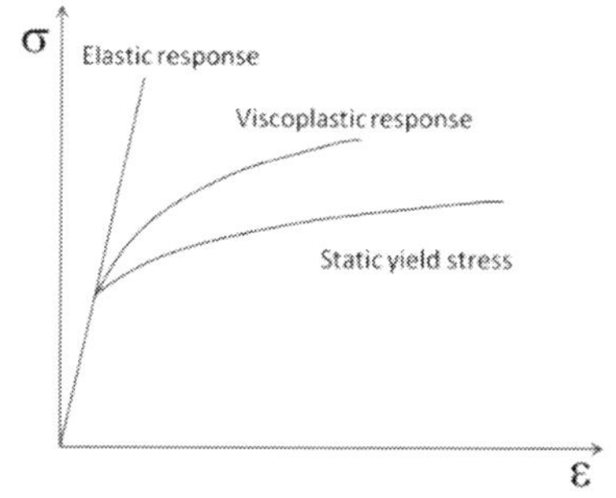

Figure 1. Viscoplastic response over stress model

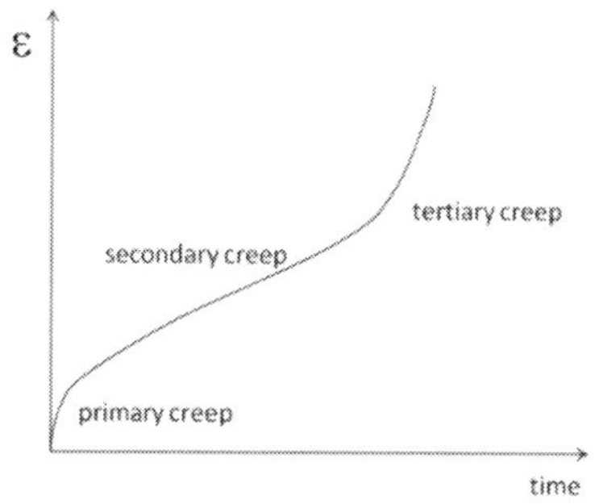

Figure 2. Creep response

As indicated above time dependency in the mechanical response of the material can be described and taken into account in different ways depending on what type of phenomena are considered. Hence, different models and approaches have been proposed in literature to take time effects into account. Since the considered work in this paper is focused on the 'slow' cooling of large iron castings the main focus is kept on the classical creep formulation where time effects are taken into account via Norton's power law, to mainly describe the nonlinear behaviour of stress relaxation at high temperature. Nevertheless, the implementation is unified in the sense that the model is generalized to include nonlinear conditions which is also taking place at lower temperatures. This is mainly achieved by selecting an appropriate hardening law, which can be adapted to different conditions at different temperature levels and by augmenting the Norton's power law, with a temperature sensitive part in terms of the Arrhenius expression where the activation energy for diffusion and temperature is taken into account.

The following section will shortly present the implemented time dependent constitutive model and the hardening law, including an overview of the numerical algorithm used to integrate the mechanical load history. This section is followed by a presentation of the experimental work, where several tensile and creep tests have been performed at different temperature and load levels, in order to have sufficient information to extract material data for the implemented model. The strategy and procedure for extracting the material data will shortly be explained. Finally, the modelling section will be followed by a validation section where calculated results will be compared to experimental results, in terms of residual stress measurements. The parts and the procedure for measuring the residual stresses, hole drilling method, will shortly be commented. The intension is to show the applicability of the unified approach to the nonlinear response of the material during the transient conditions of a slow cooling process.

2. The model - time dependent plasticity

The classical time independent plasticity formulation is based on the assumption of a yield surface which limits the elastic range and introduces a range of irreversible deformations described by the flow rule. The yield surface is generally described by a yield function, which for J2 flow theory is limited to the deviatoric stress space and hence has the following format:

$$f = \sigma_y - \sigma_e = 0 \tag{1}$$

where σ_y is the yield stress and σ_e is the effective stress, based on v.Mises' flow hypothesis, [2]. The flow rule is assumed to be associated to the yield function using the Prandtl Reuss relation:

$$\dot{\varepsilon}_{kl}^{in} = \dot{\lambda}\frac{\partial f}{\partial \sigma_{ij}} \tag{2}$$

where $\dot{\lambda}$ is a load parameter and $\partial f/\partial \sigma_{ij}$ is the outward normal to the yield surface. The assent dot does not refer to a time differentiation but to a load increment.

For Metal plasticity it is often assumed that the yield stress, σ_y follows a strain hardening function to describe the observed material behaviour. Several different hardening laws are proposed in literature depending on what type of phenomena they are supposed to describe and for what range of the mechanical load history they are applied to. However, two fundamentally different types can be found in literature. One having a continuous increase in the yield stress level as function of an increasing amount of inelastic strain, see figure 3. The second, having a saturating yield stress response as function of accumulating inelastic strain, where a steady state level of the yield stress is reached after a certain amount of inelastic strain, see figure 4.

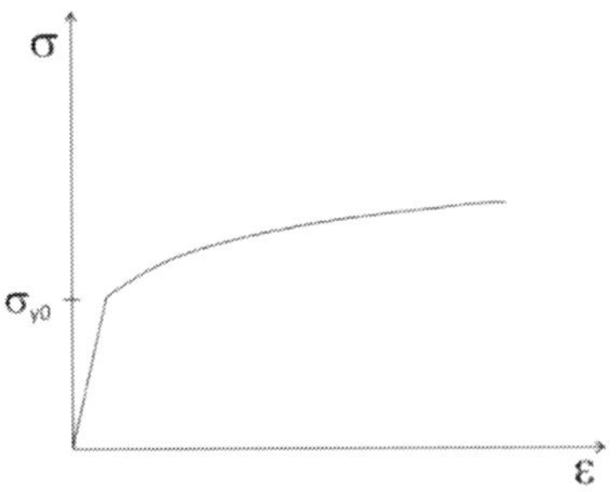

Figure 3. Power hardening law

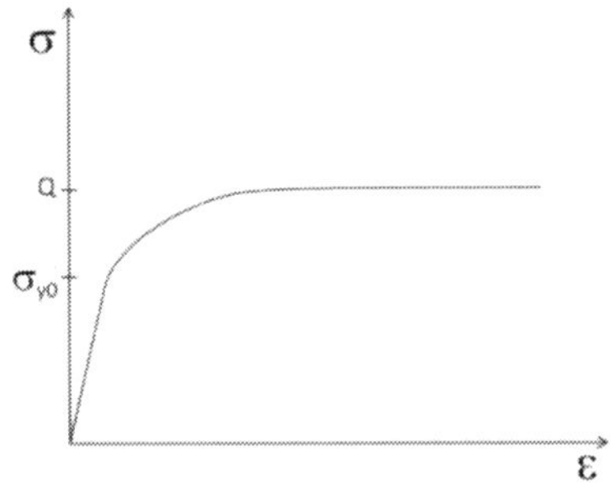

Figure 4. Saturating hardening law

A typical example of the first branch of models is a classical power law model as e.g. the Ramberg-Osgood model. Which is typically developed for lower and intermediate temperature levels.

$$\varepsilon^{in} = \alpha\frac{\sigma_{y0}}{E}\left(\frac{\sigma}{\sigma_{y0}}\right)^n \tag{3}$$

The second branch of models has a yield stress response in a closed interval given by e.g.

$$\sigma_y = \sigma_{y0} + \sigma_{sat}\left(1 - e^{-B\varepsilon^{in}}\right) \tag{4}$$

see [11], where σ_{y0} is the initial yield stress, n and B are hardening parameters and σ_{sat} is the steady state level of the yield stress. When σ_{sat} is reached no further hardening is assumed to take place and the strain rate equation will hence produce a constant strain rate for fixed load conditions as e.g. in the case of secondary creep, see later in this section. This type of hardening behaviour describes well what happens at higher temperature, where a certain range of primary creep is followed by an interval of secondary creep until the material reaches the tertiary creep range where failure occurs. Even though the power law was mainly developed to model the hardening response at lower temperature levels it can, however, be used at higher temperatures as well, by adjusting the parameters to reduce the yield stress increase as function of accumulating inelastic strain. On top of adjusting the material parameters at higher temperatures, to decrease the slope of the hardening response, it is also possible to introduce a cut off temperature level above which the effective inelastic strain rate is not used to integrate the contribution to the hardening parameter. This resembles the annealing process of the material and the response becomes ideal plastic in the sense that no further stress increase takes place and hence the strain rate becomes constant for constant load conditions. This cut off also helps to prevent artificially high amount of inelastic strain to be used as hardening parameter at lower temperature. In the current work, the power law type hardening is used on the following form

$$\varepsilon^{mech} = \frac{\sigma_y}{E}\left[\frac{1}{n}\left(\frac{\sigma}{\sigma_y}\right)^n - \frac{1}{n} + 1\right] \tag{5}$$

which has a continuous transition from Young's modulus in the elastic range to the tangent modulus in the plastic range.

Introducing strain rate sensitivity to the inelastic response of the material can be done in different ways. One way is to introduce a model based on the dynamic yield stress concept, where the yield surface is assumed to be strain rate dependent, but still having a closure of the elastic range in the case of a vanishing strain rate. This type of formulation can have the following structure, [10].

$$\sigma_{dy} = \sigma_{ref}\left(1 + \frac{\dot{\varepsilon}_e^{in}}{\dot{\varepsilon}_0}\right)^{1/m} \quad or \quad \dot{\varepsilon}_e^{in} = \dot{\varepsilon}_0 \cdot \left[\left(\frac{\sigma_{dy}}{\sigma_{ref}}\right)^m - 1\right] \; for \; \sigma_{dy} > \sigma_{ref} \tag{6}$$

This type of formulation only produces inelastic strain rates if the dynamic yield stress, σ_{dy} exceeds the reference stress level, σ_{ref}. In that sense this type of model belongs to the over stress formulation as e.g. the viscoplastic Perzyna model, where the elastic range is limited by the static yield stress level, which resembles the reference stress in equation (6). Alternatively, the time dependent mechanical conditions at high temperature are described well by assuming a power law for the effective inelastic strain rate

$$\dot{\varepsilon}_e^{in} = \dot{\varepsilon}_0 \cdot \left(\frac{\sigma_e}{g(\sigma_e)}\right)^m \tag{7}$$

where $\dot{\varepsilon}_e^{in}$ is the effective inelastic strain rate and σ_e is the effective stress. The strain hardening function $g(\sigma_e)$ accounts for the decreasing inelastic strain rate as function of the accumulating inelastic strain. It resemble an effective yield stress for the material, if the stress was measured in a tensile test with the strain rate prescribed such that $\dot{\varepsilon}_e^{in} = \dot{\varepsilon}_0$, [2]. However, in this type of constitutive law no real yield surface is assumed. Independently of the stress level the model will lead to a strain rate. At high temperature levels this model describes well the kinematic process of dislocation movements, diffusion of atoms and grain boundary sliding, [9]. But the model is also applicable at lower temperatures where the material typically has an elastic range with an inelastic response above a certain yield stress level. If the effective stress, σ_e is getting below the stress level, given by the hardening function, $g(\sigma_e)$, the inelastic strain rate will vanish. Especially at lower temperatures where the strain rate exponent is typically high, indicating low strain rate sensitivity. On the other hand if the effective stress level exceeds the stress level from the hardening function, $g(\sigma_e)$, an inelastic strain rate will immediately form and the stress level will decrease towards the stress level from $g(\sigma_e)$ and at the same time produce additional hardening of the material. This behaviour is close to classical elastic unloading below the yield stress and plastic yielding above the yield stress. In the limiting case where m goes to infinity the system directly reduces to the time independent version, where the inelastic load increment is found by the consistency condition and not by the constitutive equation (7), [2]. It should be noticed that the first model with the explicit yield limit, equation (6), will turn to the power law formulation given by equation (7) if the '+1' term is removed, hence equation (6) becomes:

$$\sigma_{dy} = \sigma_{ref}\left(\frac{\dot{\varepsilon}_e^{in}}{\dot{\varepsilon}_0}\right)^{1/m} \quad or \quad \dot{\varepsilon}_e^{in} = \dot{\varepsilon}_0 \cdot \left(\frac{\sigma_{dy}}{\sigma_{ref}}\right)^m \tag{8}$$

which resemble the constitutive law in equation (7).

Finally, the Arrhenius expression is used as a contribution to the reference strain rate as the main temperature dependent contribution to the inelastic strain rate. This contribution is assumed to follow the activation energy for diffusion, even though this is not a physically based assumption, but it has proven to be a reasonable assumption based on the experimental observations.

$$\dot{\varepsilon}_0 = Ae^{-\left(\frac{Q}{RT}\right)} \tag{9}$$

where A is not part of the Arrhenius expression but part of the power law data, extracted from the master curve, which is presented in the next section. Q is the activation energy, R the gas constant and T the absolute temperature.

To a wide extent the two models, equation (6) and (7), can be tuned to give similar response by adjusting the material date. In the present work however, the model in equation (7) has been selected due to the smooth transition between the elastic and the plastic range, via the rate sensitive term and by that being able to model creep behaviour and rate sensitivity at high and low temperature respectively.

2.1. *Numerical implementation*

To ensure a numerically stable and fast algorithm the constitutive models are implemented by using a backward Euler time discretization, [5] and [6]. This is obtained by using the well established return mapping algorithm to enforce the constitutive equation on implicit time level and by that fulfilling both the fundamental equilibrium condition and the constitutive equation on new time level. This evaluation is based on a strain driven problem where an initial elastic trial stress state is relaxed in the deviatoric system by the inelastic strain rate equation, equation (7). The inelastic strain rate is given by the assumed power law behaviour and to ensure numerical stability the reference stress in this expression is evaluated on implicit time level. To evaluate the stress conditions on new time level the effective stress function is implemented, [7]. This function resembles the scalar consistency condition used to determine the load parameter for time independent plasticity. To establish the numerical algorithm Newton Raphson iterations are used to obtain the global equilibrium, and to optimize the convergence a line search is performed after solving the system to minimize the force residual Liu and [8]. The function can be derived in the following steps, based on the decomposition of the deviatoric strain tensor

$$^{t+\Delta t}S_{ij} = {}^{t+\Delta t}S_{ij}^{tr} - 2\mu\left(\Delta e_{ij}^{cr}\right) \tag{10}$$

Insertion of the creep strain rate yields

$$^{t+\Delta t}S_{ij} = {}^{t+\Delta t}S_{ij}^{tr} - 2\mu\frac{3}{2}\Delta t\dot{\varepsilon}_0{}^{t+\Delta t}\sigma^{n\,t+\Delta t}S_{ij} \tag{11}$$

Collecting terms and taking the inner product

$$^{t+\Delta t}S_{ij}{}^{t+\Delta t}S_{ij}\left(1+3\mu\Delta t\dot{\varepsilon}_0{}^{t+\Delta t}\sigma^n\right)^2 = {}^{t+\Delta t}S_{ij}^{tr\,t+\Delta t}S_{ij}^{tr} \tag{12}$$

This can be rewritten to the following transcendental equation used to solve for the effective stress on new time level

$$f({}^{t+\Delta t}\sigma) = {}^{t+\Delta t}\sigma\left(1+3\mu\Delta t\dot{\varepsilon}_0{}^{t+\Delta t}\sigma^n\right) - {}^{t+\Delta t}\sigma^{tr} = 0 \tag{13}$$

3. Material data extraction

Material data for the inelastic strain rate model presented in the previous section is obtained by performing several tensile and creep tests at different temperature and load levels. The first step of extracting the material data is to determine the steady state inelastic strain rate for the considered experiment. At each temperature level several tests are performed at different load levels and the stress level where the transition from diffusion control to climb controlled creep is determined. This stress level is used for normalizing the experiments in a master curve where the steady state strain rate is normalized by the Arrhenius expression and plotted as function of the stress load normalized by the above mention transition stress level. An example of this type of master curve is presented below for the GJS-400 iron alloy. This master curve is based on several creep experiments, listed in the table below, table 1.

Table 1. Consider creep experiments, [13]

		Loading in percent of $R_{p0.1}$					
T [°C]	time [h]	100	90	80	70	60	50
400	7	213	192		149		
500	6	180	162	144		108	
600	5				78		56
700	4	62	56		43	37	31
800	3			22			14
900	2		27		21	15	12
1000	1						9

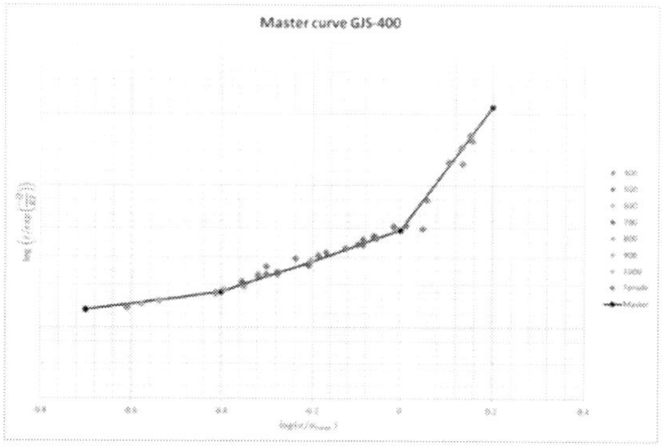

Figure 5. Master curve for GJS-400 normalized inelastic strain rate as function of normalized load level

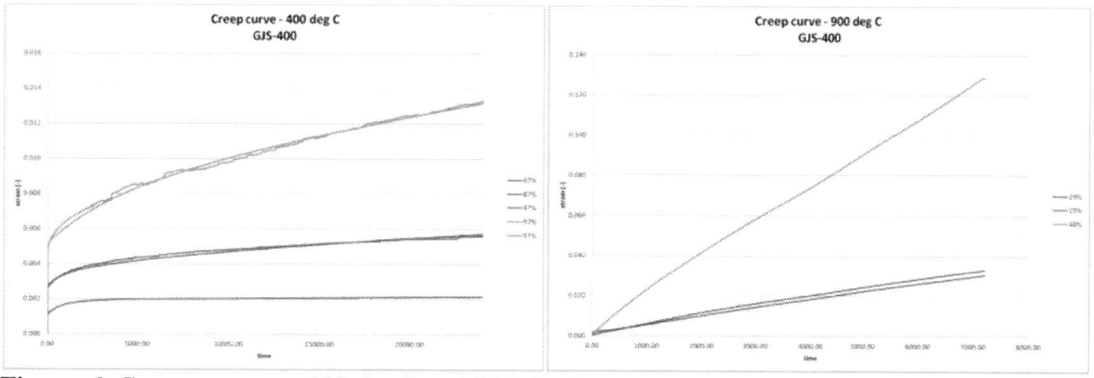

Figure 6. Creep curves at 400 degC and 900 degC, load level is given as percent of σ_{trans}

When the master curve has been plotted based on an appropriate normalization of the strain rate response with the Arrhenius expression and by that determine the temperature dependent activation energy, different regions can be identified in the mater curve. These sub linear regions are used to extract data for the power law exponent m and the strain rate constant A, which is part of $\dot{\varepsilon}_0$, in equation (9). The data are later on looked up during the simulation as function of the actual stress level with respect to the temperature dependent transition stress level. Hence, the power law constants m and A are not directly a function of the temperature, which would limit the response of the model at a certain temperature level, but they are a function of the current load level independently of the temperature level. This gives a more general behaviour of the model, where the response is not limited to a certain deformation mechanism at each temperature level.

Finally, measurements are needed to determine the classical thermo-mechanical material data, i.e. the temperature dependent E-modulus, the yield strength and the hardening parameter. To extract these data, several stress strain curves at different temperature levels have been extracted and the results have been used on a tabular form as input to the calculation. To evaluate the extraction of material data and the implementation of the material model several simple calculations on bars have been performed to reproduce the creep curves and stress strain curves.

4. Experimental validation

One examples has been used in the current work for validating the implemented model, a pump house mounting unit in cast iron GJS 400, see figure 7. The temperature region is still in the range of 600-700 °C after 3 hours. This leaves considerable amount of time for inelastic strain to develop in the material and by that relaxing the stress level in general. Different regions of tensile and compressible zones are identified and stress measurements are performed with the hole drilling method. Of course the accuracy is limited for this method and the extracted results are based on transformation of information from the strain gauge response due to stress redistribution when the material is removed by drilling/removing material in the surface. However, the method is useful to get an overview of the different tensile and compressible zones and the relative magnitude of these zones. To evaluate the influence of the strain rate formulation on the calculated mechanical fields two calculations have been performed on the part. One with classical time independent plasticity and one with the implemented strain rate formulation. Both calculations have been compared to the measured stress levels in the considered point. Some of the results are shown below in figure 8.

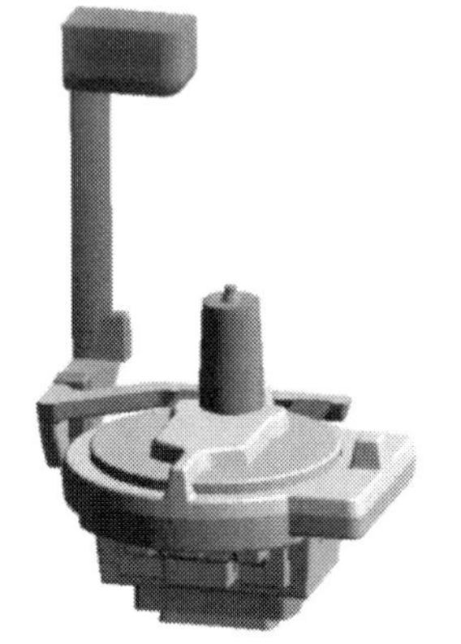

Figure 7. considered part

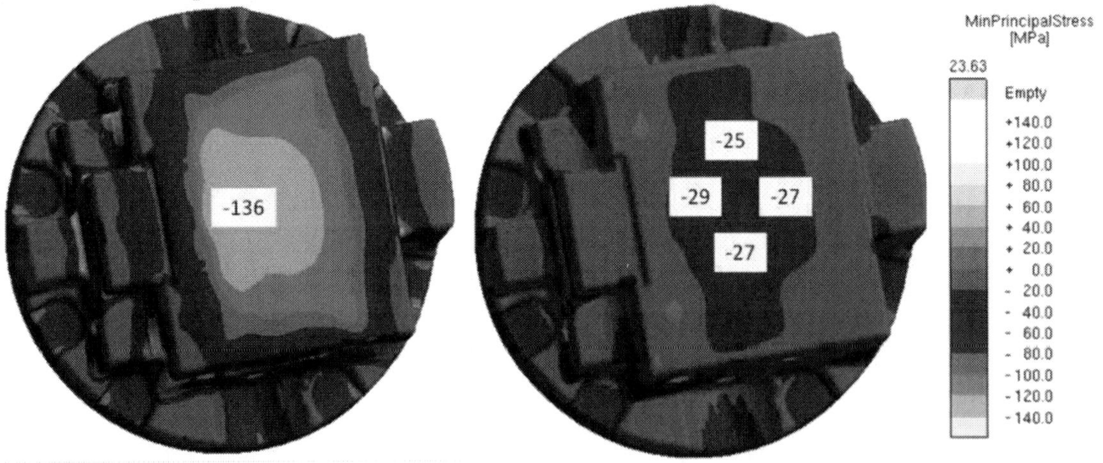

Figure 8. Lower side of the pump house mounting unit in GJS-400, Measurements -8 to -20 MPa, time independent plasticity -136 MPa, rate dependent formulation -29MPa.

It is seen from these comparisons that the strain rate sensitive formulation resemble the measurements better than the classical time independent model. Generally, the strain rate formulation predicts lower residual stress compared to the time independent plasticity formulation. Clearly two reasons can explain this observation. First of all the cooling time plays the overall and dominating reason for this, since it leaves time for relaxation. But also the fact that the rate formulation predict a more smooth description of the inelastic field compared to the more discrete time independent formulation leads generally to a more smooth stress distribution with lower peak levels in the stress tensor. This observation has also been seen in other similar calculations performed on both bigger and smaller parts not considered in this work.

Facts about the simulation: The mechanical calculation is semi-coupled to the solidification and cooling calculation. The solidifying material is treated as a strain rate sensitive solid with reduced stiffness and strength, hence the size of the calculation domain is constant during the simulation. The thermal contraction is considered to be zero above T-solidus. Contact was considered between the cast part and the cores. The number of elements is approximately 1.2 mill in the considered example and the calculation was done in approx 8 hours on a 8 core work station. The thermal load history was subdivided in 80 load steps.

5. Conclusion

Different inelastic strain rate models have been evaluated and implemented in a existing 3D cast simulation software. To keep a unified approach in the full temperature and load interval, different hardening laws have been assessed. Based on these evaluations a version of Norton's power law was selected with a contribution from the temperature dependent Arrhenius expression and hardening described by a power law with continuous tangent modulus.

Creep curves and tensile curves have been measured in order to extract material data for the implemented material model. Based on the extracted data the implementation was verified by reproducing the measured curves. Finally, the implemented model and the extracted data were used to predict residual stresses in a cast iron part. The measurements were performed with the hole drilling method and different zones of tensile and compressible regions were found and compared to the simulated results. Reasonable agreement was found between the measurements and the simulated results. The results from the time dependent formulation was in better agreement with the measured results compared to the calculated results using the time independent formulation.

References

[1] Lu Z Wang G 1996 *A simple unified theory for the cyclic deformation of metals at high temperature* (Acta Mechanica) vol 118 pp 20-28

[2] Tvergaard V 2001 *Plasticity and creep in structural materials* (Department of Mechanical Engineering, Technical University of Denmark, Lyngby)

[3] Anand L 1982 *Constitutive equations for the rate-dependent deformation of metals at elevated temperatures (J. of Eng. Materials and Technology)* vol 104 pp 12-17.

[4] Miller A 1976 *An inelastic constitutive model for monotonic, cyclic, and creep deformation: part I – Equation development and analytical procedures. (Conference on Micromechanical Modelling of Flow and Fracture)* Paper No 75-Mat-14

[5] Kleiber M 1998 *Handbook of computational solid mechanics* (New York: Springer Verlag)

[6] Simo J C 1997 *Computational inelasticity* (New York: Springer Verlag)

[7] Bathe K J 1996 *Finite Element Procedures* (New Jersey: Prentice-Hall, Inc.)

[8] Liu W K, Belytschko T. and B. Moran 2000 *Nonlinear Finite Elements for Continua and Structures* (New York: J. Wiley & Sons)

[9] Frost J and Ashby F 1982 *Deformation-Mechanism Maps: The Plasticity and Creep of Metals and Ceramics* (Pergamon Press)

[10] K.S. Chan et. al. 1984 *A survey of unified constitutive theories. Second Symposium on Nonlinear Consritutive Relations for High Temperature Applications* (NASA Lewis RC)

[11] Dunne F and Petrinic N 2005 *Introduction to Computational Plasticity*, (USA: Oxford University Press)

[12] Thorborg J. et. al. 2006 *Thermo-mechanical conditions in heat treated aluminium cast parts* (Modeling of Casting, Welding and Advanced Solidification Processes XI) eds C A Gandin and M Bellet (Warrendale, PA: TMS)

[13] AiF Forschungsvorhaben Nr. 15333 N/1 2010 *Messung und Simulation von Eigenspannungen in Gusstücken aus Gusseisen unter besonderer Berücksichtigung von Kernformstoffen*, (Düsseldorf: IFG - Institut für Giessereitechnik GmbH)

IOP Conf. Series: Materials Science and Engineering **33** (2012) 012051 doi:10.1088/1757-899X/33/1/012051

Casting and stress-strain simulations of a cast ductile iron component using microstructure based mechanical behavior

Jakob Olofsson and Ingvar L Svensson
Jönköping University, School of Engineering, Dept. Mechanical Engineering, Materials and Manufacturing – Casting, P.O. Box 1026, SE-551 11 Jönköping, Sweden

E-mail: jakob.olofsson@jth.hj.se

Abstract. The industrial demand for increased component performance with concurrent reductions in component weight, development times and verifications using physical prototypes drives the need to use the full potential of casting and Finite Element Method (FEM) simulations to correctly predict the mechanical behavior of cast components in service. The mechanical behavior of the component is determined by the casting process, and factors as component geometry and casting process parameters are known to affect solidification and microstructure formation throughout the component and cause local variations in mechanical behavior as well as residual stresses. Though residual stresses are known to be an important factor in the mechanical behavior of the component, the importance of local mechanical behavior is not well established and the material is typically considered homogeneous throughout the component. This paper deals with the influence of solidification and solid state transformation on microstructure formation and the effect of local microstructure variations on the mechanical behavior of the cast component in service. The current work aims to investigate the coupling between simulation of solidification, microstructure and local variations in mechanical behavior and stress-strain simulation. This is done by performing several simulations of a ductile iron component using a recently developed simulation strategy, *a closed chain of simulations for cast components*, able to predict and describe the local variations in not only elastic but also plastic behavior throughout the component by using microstructural parameters determined by simulations of microstructural evolution in the component during the casting process. In addition the residual stresses are considered. The results show that the FEM simulation results are significantly affected by including microstructure based mechanical behavior. When the applied load is low and the component is subjected to stress levels well below the yield strength of the material, the residual stresses highly affect the simulation results while the effect of local material behavior is low. As the applied load increases and the stress level in the component approaches and passes the yield strength, the effect of residual stresses diminishes while the effect of local mechanical behavior increases. In particular the predicted strain level is heavily affected by the use of local mechanical behavior. It is proposed that it is important to include both local mechanical behavior and residual stresses in stress-strain simulations to predict the true mechanical behavior of the component.

1. Introduction

The solidification process of cast iron is a complex interaction between physical, chemical and mechanical phenomena. Much research has been aimed at understanding the solidification of cast iron, and solidification models have been developed which are able to simulate the solidification process and predict microstructural features. By combining these solidification models with simulations of fluid flow it is possible to simulate the entire casting process of a specific component and predict the local variations in microstructure throughout the component.

The mechanical behavior of cast irons is commonly characterized by a tensile curve which shows true stress σ (Pa) versus total true strain ε_T (-). A linear elastic region described by Hooke's law ($\varepsilon_{el}=\sigma/E$, where E (Pa) is Young's modulus) and a non-linear plastic region are identified. In this work the plastic behavior is described using the Hollomon equation $\sigma = K \times \varepsilon_{pl}^{\ n}$ [1] where the strength coefficient K (Pa) and the strain hardening exponent n (-) relate true stress and true plastic strain ε_{pl} (-).

The mechanical behavior of cast irons is known to be controlled by microstructural parameters, where graphite fraction and morphology has been shown to be important contributors to the deformation behavior of the material [2-4]. In previous research models to predict Young's modulus [5, 6] and the parameters of the Hollomon equation [7, 8], i.e. the strength coefficient and the strain hardening exponent, have been developed. By combining these models with solidification models it is possible to predict both elastic and plastic mechanical behavior throughout cast iron components using casting simulation software [9].

During the solidification process of a cast iron component the solidification conditions vary throughout the component, which cause local variations in the final microstructure and mechanical behavior [6] as well as formation of residual stresses. Though the effect of residual stresses on the mechanical integrity of cast iron components has been studied [10], the effect of local variations in mechanical behavior on the stress-strain response has not been previously investigated.

The current work aims to investigate this effect by using casting process simulation and stress-strain simulations of a ductile iron heavy truck engine support. The simulations have been performed using a recently presented simulation strategy called *a closed chain of simulations for cast components* [9]. The simulation strategy combines different types of simulations, schematically illustrated in Figure 1. The results from each simulation step is the basis for the next simulation step, thus the denotation a *closed* chain of simulations. The strategy uses solidification and solid-state transformation models to predict microstructure formation and mechanical behavior on a local level throughout the component, and incorporates the variations in mechanical behavior into an FEM simulation of the mechanical behavior of the component. The effect of these variations on the behavior of the component needs to be established since it directly affects the process of designing cast iron components, which concerns the work of metallurgists as well as CAE and design engineers.

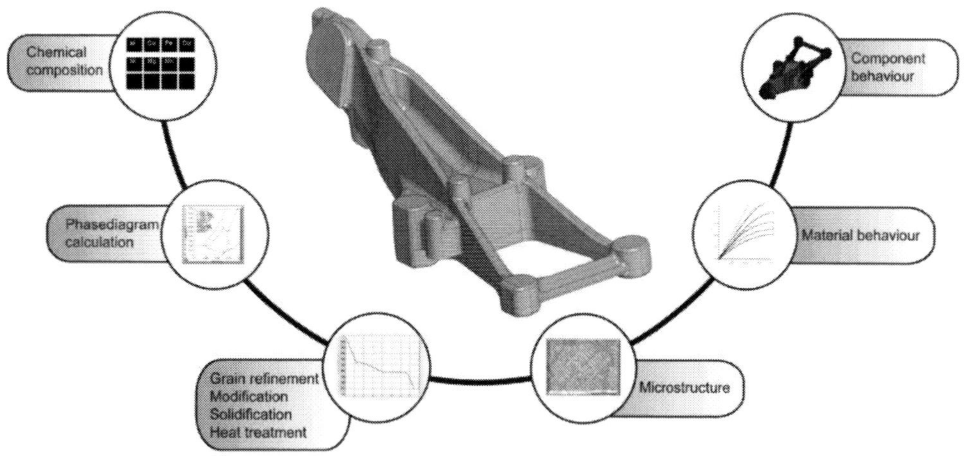

Figure 1. The closed chain of simulations for cast components [9]. The figure is reprinted with permission from Elsevier.

2. Simulation setup

2.1. Casting process simulation

The casting process simulation of the component has been performed in a development version of MAGMASoft [11] which uses solidification [12] and solid state transformation models developed by e.g. Wessén and Svensson [13] to predict microstructure evolution throughout the component, and the previously mentioned microstructural relationships [7, 8] to predict the local variations in the parameters of the Hollomon equation.

2.2. Incorporation of predicted mechanical behavior

An in-house developed software [9] has been applied to create material definitions for the FEM simulation based on the results from the casting process simulation. The materials are defined by a linear elastic region and a piecewise linear plastic region, in which a number of points along the plastic part of the stress-strain curve are specified. In this work the stress-strain curve was defined from ε_{pl}=0% to ε_{pl}=10%. This range is divided into two intervals, from 0% to 1% respectively from 1% to 10%. Each interval is described by 10 linearization points, thus in total 20 linearization points are used. These settings have in previous work been shown to be an efficient way to describe the non-linear stress-strain curve, see reference [9] where the settings are further explained.

Four different material descriptions are considered with parameters as shown in Table 1. The descriptions correspond to different ways of using data from the casting process simulation in the FEM simulation:

- LOCAL uses individual material definitions for every element of the FEM mesh, i.e. about 132 000 material definitions, to capture the local variations in mechanical behavior. This is *a closed connection* between casting process simulation and FEM simulation which uses *component specific microstructure-based mechanical behavior*, enabled by *the closed chain of simulations for cast components* [9].
- AVERAGE uses *weighted averages* of the predicted mechanical behavior parameters throughout the component to create *a very accurate homogenous description* of the material in the component. Since all parameter values throughout the FEM mesh are included in the weighted average values the variations in the parameter values are taken into account.
- MAXMIN uses the *mean of the maximum and minimum values* of the mechanical behavior parameters to create *a simple homogeneous material description*. Since only the maximum and minimum values are used to determine these mean values the variations in parameter values throughout the FEM mesh are *not* taken into account.
- TABLE uses tabulated homogeneous material data for the standard material EN-GJS-600-3, which is similar to the material used in the casting process simulation. The use of tabulated material data is a common way of performing FEM simulations of cast iron components in the industry, but do not consider any component specific effect of the casting process on the mechanical behavior of the material in the component. To characterize the plastic behavior of the material the strain hardening exponent and the strength coefficient in this work have been derived from tabulated values of yield stress σ_Y, ultimate tensile stress σ_U and ultimate elongation ε_U as $n=\ln(\sigma_U/\sigma_Y)/\ln(\varepsilon_{pl,Y}/\varepsilon_{pl,U})$ respectively $K=\sigma_Y/(\varepsilon_{pl,Y})^n$.

Table 1. Parameter values used in the different material descriptions.

Material name	E [GPa]	n [-]	K [MPa]	No. material definitions
LOCAL	168.1-175.5	0.108-0.175	707.1-1668.1	132 000
AVERAGE	170.7	0.13	1046.2	1
MAXMIN	171.8	0.14	1187.6	1
TABLE	170.0	0.25	1733.0	1

2.3. FEM simulation
A FEM mesh of the component consisting of about 132 000 second order tetrahedral elements was created in Abaqus [14] format. Four surfaces were fixed and a load was applied as shown in Figure 2. The load is linearly increased from 0 to 150 kN with a 15 kN load increase in every timestep of the simulation. This load level was chosen to achieve a maximum stress level around the estimated ultimate tensile stress for the material (600 MPa). The 0.2% offset yield stress of the material is first exceeded at an applied load of 60 kN.

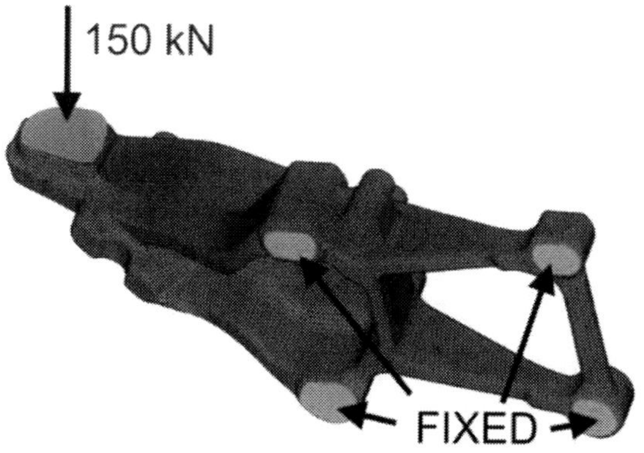

Figure 2. FEM simulation loadcase setup.

FEM simulations were performed using all four material variants, with and without including the residual stresses predicted by the casting process simulation. The simulations *without* residual stresses will be referred to using the name of the material description used in the simulation (TABLE, MAXMIN etc.) and the simulations *with* residual stresses included are referred to with the additional suffix –RS (e.g. TABLE-RS etc.). For every simulation the predicted maximum von Mises stress and maximum von Mises strain in the component at every timestep were extracted. To evaluate the accuracy of the simulations one simulation is selected as a reference, and the relative error in maximum value (*MVE*) is determined as the difference in maximum values predicted by the simulation under consideration (*C*) and the reference simulation (*R*) divided by the value predicted by the reference simulation, i.e. $MVE=(C-R)/R$. A negative value of relative error thus corresponds to an underrated maximum stress or strain value, and a positive value corresponds to an overrated value.

3. Results and discussion

3.1. Casting process simulation results
The casting process simulation shows that local variations in solidification conditions leads to different microstructures in different regions of the component. This leads to variations in Young's modulus and the parameters of the Hollomon equation throughout the component, shown in Figure 3.

3.2. Effect of local variations in mechanical behavior.
To evaluate the isolated effect of local variations in mechanical behavior the simulations without residual stresses are first considered. Comparing the results from the LOCAL simulation and the AVERAGE simulation it is found that not only does the maximum stress value change when the local variations in mechanical behavior are neglected, the location of the maximum stress may also be found in another location, see Figure 4. The variations in material behavior thus affects the stress distribution throughout the component, and the incorrect location of maximum stress or strain may be predicted when homogeneous material behavior is assumed. This effect is most significant in geometrical stress concentrations, and may lead to incorrect conclusions about the severity of stress concentrations and the mechanical performance of the component.

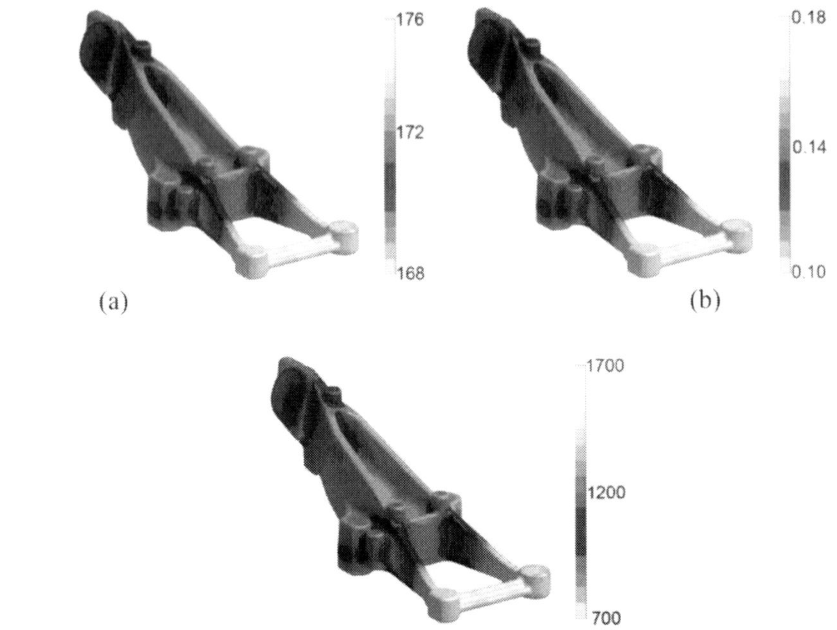

(a) (b)

(c)

Figure 3. Local variations in a) Young's modulus b) strain hardening coefficient and c) strength coefficient throughout the component predicted by the casting process simulation.

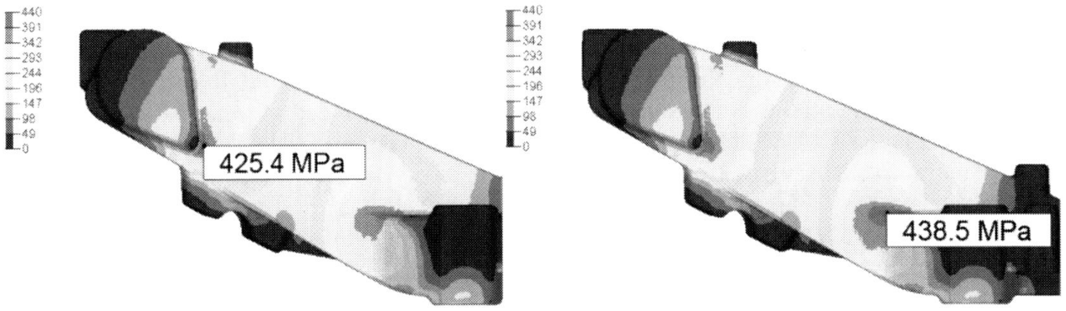

(a) **(b)**

Figure 4. Predicted maximum stress at an applied load of 75 kN by a) the LOCAL simulation and b) the AVERAGE simulation.

To numerically evaluate the influence of local material behavior the LOCAL simulation is chosen as reference simulation, and the simulations *without* residual stresses are evaluated. Figure 5 shows the obtained development of the relative error in maximum stress, Figure 5a, and maximum strain, Figure 5b, during the simulation. In the elastic region the differences in load response are only determined by the variations in Young's modulus, an effect which is small at low loads. As the load increases and approaches the yield stress the effect however increases, and it is seen that, independent of material description, the assumption of homogeneous material behavior may lead to an error of about 5% in stress level and 5-10% in strain level in the elastic region. If tabulated material data is used, an error in stress level of more than 10% is obtained. As the load level further increases and enters the plastic region the differences between the material descriptions are significant. AVERAGE gives a rather constant error in stress level of only about 2% while the error in strain level increases from 10 to 17% with increasing load. The MAXMIN gives an error of about 7% in stress level and 10-35% in strain level. In the plastic region the effect of local variations in material behavior thus become significant and must be considered to capture the stress-strain response of the component. The use of tabulated data may underrate or overrate the stress level significantly (±10-15%), and the error in strain level may vary widely (5-25%). This shows that none of the homogeneous material descriptions are able to describe the effect of local material behavior on the mechanical behavior of the component. The assumption of homogeneous material behavior leads to errors especially in predicted strain level. In order to obtain accurate FEM simulation results of a ductile iron component the local variations in mechanical behavior thus needs to be captured using solidification models and be incorporated into the FEM simulation. This motivates the use of component specific microstructure-based mechanical behavior in structural analyses of ductile iron components.

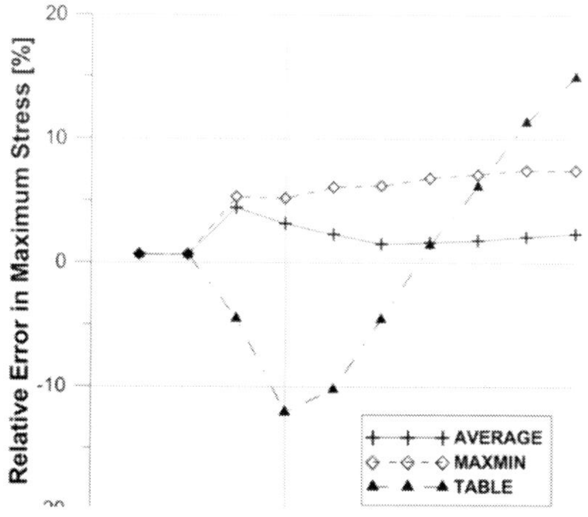

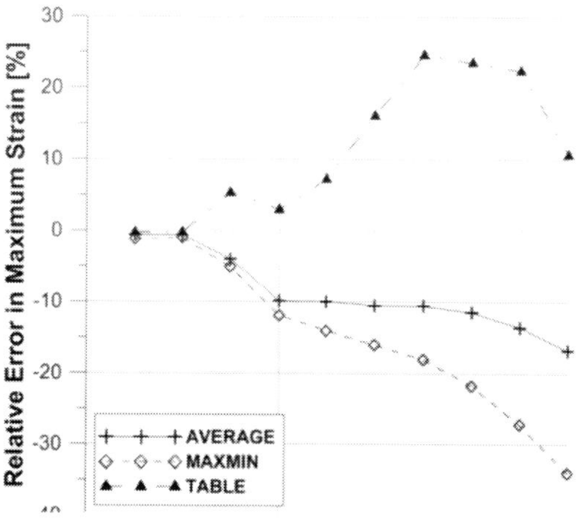

(a) (b)

Figure 5. Relative error in a) maximum stress and b) maximum strain using different material definitions. Vertical line at 60 kN indicates the beginning of plasticity, i.e. the 0.2% offset yield stress is exceeded.

3.3. *Effect of residual stresses.*

A similar evaluation is performed of the simulation results *with* consideration of residual stresses. In this case the simulation with both local mechanical behavior and residual stresses, LOCAL-RS, is selected as reference simulation, and the results from the remaining simulations *with* residual stresses are evaluated. Figure 6 shows the obtained development of relative error in maximum stress and maximum strain. Comparing Figure 5 and Figure 6 it is seen that the trends and amounts of error are very similar. The conclusions about the importance of including local variations in mechanical behavior are thus still valid when residual stresses are included.

To evaluate the relative importance of including local material behavior versus including residual stresses the results from the simulations *without* residual stresses included are re-evaluated using the LOCAL-RS simulation as reference. The obtained development of relative error in maximum stress and maximum strain is shown in Figure 7, in which the results from the simulations *without* residual stresses are shown with black curves and the results *with* residual stresses are shown with grey curves. The results show that local material behavior and residual stresses affect the mechanical behavior of the ductile iron component in different ways. When the applied load is low the contribution from the residual stresses is dominating while the effect of local material behavior is low. As the applied load increases and the stress level in the component approaches the yield stress the relative importance of residual stresses however diminishes and the selected type of material description instead determines the accuracy of the simulation results. AVERAGE leads to a low error in stress level (about 2%) but a large error in strain level (about 10-15%). MAXMIN and TABLE gives a larger error in both stress (5-10%) and strain levels (about 10-30%). LOCAL has almost no relative error in stress or strain in the plastic region, thus to correctly determine the stress and strain levels in the component subjected to plastic loads it is more important to consider local variations in material behavior than to consider residual stresses. At low loads (less than 60 kN), the residual stresses must however be included to predict the correct stress level.

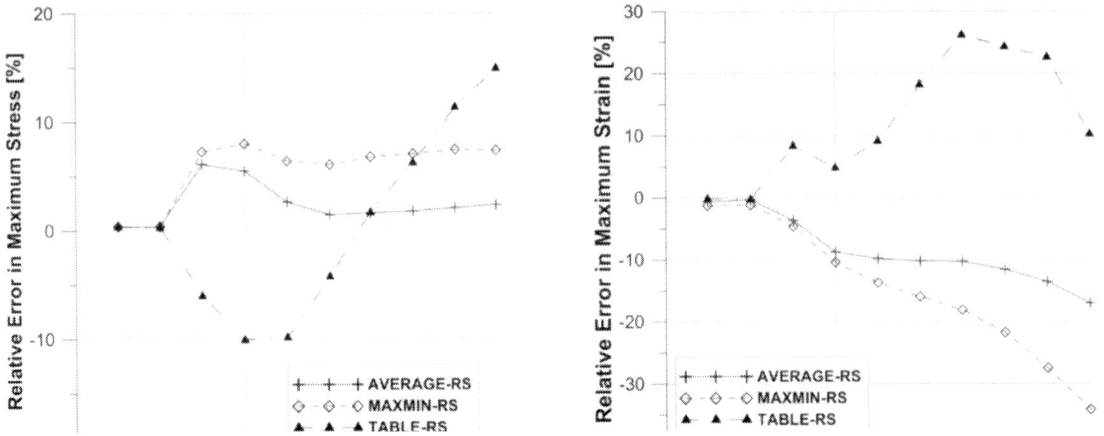

(a) (b)

Figure 6. Relative error in a) maximum stress and b) maximum strain using different material descriptions and including residual stresses.

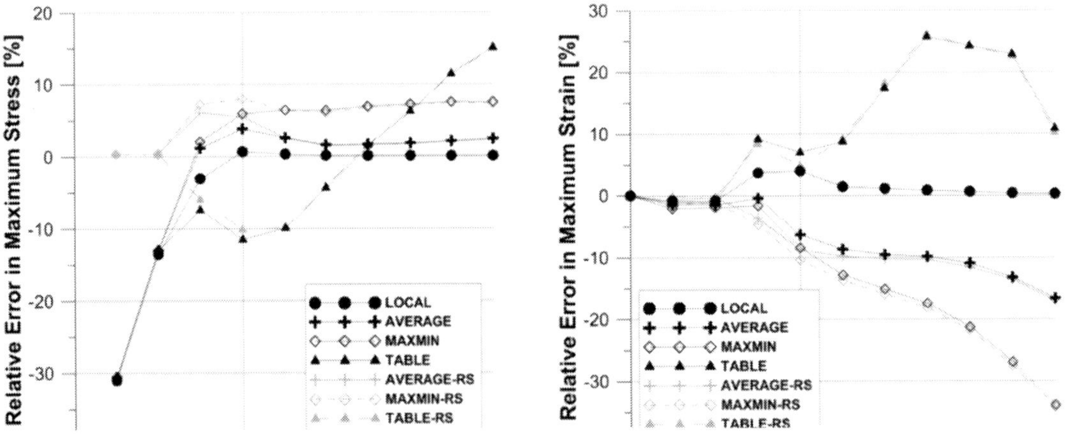

(a) (b)

Figure 7. Combined effect of using different material definitions and not including residual stresses on relative error in a) maximum stress and b) maximum strain. Results *without* residual stresses are shown with black curves and results *with* residual stresses are shown with grey curves.

4. Conclusions

- The predicted mechanical behavior of the component is significantly affected by the local variations in mechanical behavior and residual stresses caused by the local variations in solidification conditions.
- Local variations in mechanical behavior cause a stress/strain distribution throughout the component that homogeneous material descriptions are not able to describe. This effect becomes noticeable already in the elastic region.
- Residual stresses gives an important contribution to the predicted stress level when the applied load is low.
- Local variations in mechanical behavior gives a larger contribution to the predicted stress level than residual stresses when the applied load is high.
- Both local variations in mechanical behavior and residual stresses must be included to correctly determine the stress and strain level at all loads.

Acknowledgements
The Swedish Knowledge Foundation is greatly acknowledged for financing the research profile COMPCast and its subproject CCSIM in which the current work has been performed. The authors would also like to thank Scania CV AB for supplying the component geometry.

References

[1] Hollomon J H 1945 Tensile deformation *Trans. AIME* **162** 268–90
[2] Sjögren T and Svensson I L 2005 Studying elastic deformation behaviour of cast irons by acoustic emission *Int. J. Cast Met. Res.* **18** 249–56
[3] Sjögren T and Svensson I L 2007 The Effect of Graphite Fraction and Morphology on the Plastic Deformation Behavior of Cast Irons *Metall. Mater. Trans. A* **38** 840–7
[4] Sjögren T, Vomacka P and Svensson I L 2004 Comparison of mechanical properties in flake graphite and compacted graphite cast irons for piston rings *Int. J. Cast Met. Res.* **17** 65–71
[5] Sjögren T and Svensson I L 2004 Modelling the effect of graphite morphology on the modulus of elasticity in cast irons *Int. J. Cast Met. Res.* **17** 271–9
[6] Sjögren T, Wessén M, Svensson IL and Schäfer W 2006 Modeling and Simulation of Elastic Properties in Cast Compacted Graphite Iron Engine Block *Int. Conf. on Modeling of Casting, Welding and Advanced Solidification Processes XI (Opio, France)*
[7] Larsson D, Sjögren T and Svensson I L 2008 Modelling the plastic deformation behaviour of cast irons *Giessereiforschung* **60** 8–16
[8] Svensson I L and Sjögren T 2009 On modelling and simulation of mechanical properties of cast irons with different morphologies of graphite *Carl Loper Cast Iron Symposium (Madison, Wisconsin, USA)*
[9] Olofsson J and Svensson I L 2012 Incorporating predicted local mechanical behaviour of cast components into finite element simulations *Mater. Des.* **34** 494–500
[10] Gustafsson E, Hofwing M and Strömberg N 2009 Residual stresses in a stress lattice--Experiments and finite element simulations *J. Mater. Process. Technol.* **209** 4320–8.
[11] MAGMAsoft 2011 MAGMA Gmbh. http://www.magmasoft.com
[12] Fredriksson H and Svensson I L 1985 Computer Simulation of the Structure Formed During Solidification of Cast Iron *Proc. Int. Symp. on the Physical Metallurgy of Cast Iron (Stockholm, Sweden)* pp 273–84
[13] Wessén M and Svensson I 1996 Modeling of ferrite growth in nodular cast iron *Metall. Mater. Trans. A.* **27** 2209–20
[14] ABAQUS Manual v 6.9 2009 Dassault Systèmes Simulia Corp. http://www.simulia.com

IOP Conf. Series: Materials Science and Engineering **33** (2012) 012052 doi:10.1088/1757-899X/33/1/012052

A partitioned resolution for concurrent fluid flow and stress analysis during solidification: application to ingot casting

Michel Bellet, Okba Boughanmi, Grégory Fidel

MINES ParisTech - CEMEF, CNRS UMR 7635, Sophia Antipolis, France

michel.bellet@mines-paristech.fr

Abstract. In simulation of solidification processes, an open issue is the concurrent modelling of fluid flow and solid mechanics. This is critical for the prediction of cracks formed in solidified regions during the filling stage of ingot casting, or in the mould during continuous casting. The proposed approach consists of a partitioned two-step resolution strategy.

1. Introduction

When modelling casting processes, one of the critical issue is to achieve a concurrent computation of fluid flow (ingot mould filling, thermal convection, nozzle jet in continuous casting) and solid mechanics (stress-strain in solidified regions). This is of crucial importance for the prediction of surface or sub-surface cracks that may initiate in solidified regions during the filling stage of ingot casting, or during continuous casting. The current state-of-the-art consists in separating the analysis in two distinct stages: fluid flow using CFD codes and stress-strain analysis using structural codes. This induces several drawbacks regarding practical use and computational efficiency. A monolithic formulation, treating the Fluid-Solid Interaction (FSI) could be seen as a possible alternative, but is not adapted to the context of solidification, because of huge differences between liquid viscosity and solid consistency. It is then preferable to consider this FSI problem as a weak interaction problem, for which a partitioned formulation is more efficient than a monolithic one. Therefore, a two-step resolution strategy combining fluid flow and solid mechanics has been developed:

- In a first step, a solid-oriented resolution is performed, in which the liquid is modelled using an arbitrary high viscosity. From this first resolution we get the velocity field in solidified regions, and associated strains and stresses. This resolution is highly non-linear due to the elastic-viscoplastic constitutive models used for the solid material.
- A second step consists of a liquid-oriented resolution: Navier-Stokes equations, including buoyancy, and Darcy contributions in mushy zone, are considered. This resolution can be linearized and therefore is computationally less demanding than the first one.

Both resolutions are performed using a level set formulation to track the evolution of the interface air/metal. This approach is applied to the continuous simulation of the filling and cooling of a parallelepipedic ingot, showing the ability to predict stress and strain fields during the filling stage which cannot be achieved by standard casting simulation codes.

2. Two-step coupled formulation for mechanical modelling

The main limitation of a monolithic formulation is its inability to consider concurrently a low viscous liquid and an elastic-viscoplastic solid. To overcome this limitation, a 2-step resolution strategy has been developed, which consists in dividing the mechanical problem into two coupled sub-problems, which are detailed in the next sections:

- a solid-oriented resolution, from which velocities, stresses and distortions are calculated in the solidified regions;
- a liquid-oriented resolution which gives access to velocity and pressure fields in the liquid regions and in the air.

2.1. Step I: solid-oriented resolution

2.1.1. *Constitutive equations.* In the first step, the approach proposed in [1][2] is applied. A thermo-elastic-viscoplastic (TEVP) model is used for the solid state. A thermo-viscoplastic (TVP) model is used to model the behaviour of the metal both in the mushy and the liquid states. The mushy metal is considered as an homogenized non-Newtonian fluid, in which the strain-rate sensitivity is continuously increasing with the liquid fraction. At liquidus temperature, the behaviour tends to the Newtonian model, which is used in the fully liquid state. However, in Step I, the liquid is modelled using an arbitrary high value for the viscosity (typically 1 to 100 Pa s instead of nominal values around 10^{-3} Pa s). TEVP and TVP behaviour models are distinguished by a critical transition temperature T_c chosen equal to the solidus temperature.

2.1.2. *Equations solved.* The weak form of the following momentum and mass conservation equations are solved by a finite element method [1]:

$$\begin{cases} \nabla \cdot \mathbf{s} - \nabla p + \rho \mathbf{g} - \rho \dfrac{\mathrm{d}\mathbf{v}}{\mathrm{d}t} = 0 \\ \nabla \cdot \mathbf{v} - \mathrm{tr}\,\dot{\boldsymbol{\varepsilon}}^{el} - \mathrm{tr}\,\dot{\boldsymbol{\varepsilon}}^{th} = 0 \end{cases} \tag{1}$$

where $\mathbf{s}$ is the deviatoric stress tensor, p the pressure, $\mathbf{g}$ the gravity vector and $\dot{\boldsymbol{\varepsilon}}$ the strain rate tensor. When assembling the stiffness matrix and right hand side to solve the mechanical problem, tetrahedral elements are treated differently according to their average temperature: if it is lower than T_c their behaviour is TEVP, otherwise they are treated using the TVP model. In this latter case, the term $\mathrm{tr}\,\dot{\boldsymbol{\varepsilon}}^{el}$ is absent from the contribution of TVP elements in the weak form of Eq. (1b). The contributions of thermal dilatation and solidification shrinkage are included in the trace of $\dot{\boldsymbol{\varepsilon}}^{th}$, which is written as:

$$\dot{\boldsymbol{\varepsilon}}^{th} = -\frac{1}{3\langle\rho\rangle}\frac{\mathrm{d}\langle\rho\rangle}{\mathrm{d}t}\mathbf{I} \tag{2}$$

where the average density of the metal is $\langle\rho\rangle = g_s\rho_s + g_l\rho_l$, with ρ_s and ρ_l the densities in the solid state and in the liquid state, respectively. It is assumed that ρ_s and ρ_l depend on temperature, except in the solidification interval where they are constant. Under this assumption, Eq. (1) includes dilatation in the solid and in the liquid states, together with solidification shrinkage in the solidification interval. The velocity and pressure fields deduced from Step I calculation are denoted $\mathbf{v}_{\mathrm{StepI}}$ and p_{StepI}, respectively.

2.2. Step II: liquid-oriented resolution

This second step is a fluid-oriented resolution, which is supposed to provide more accurate velocities in mushy and liquid regions. For solid nodes the velocity is imposed to its calculated value in Step I.

2.2.1. *Constitutive equations.* The behaviour of the liquid in Step II is chosen to be Newtonian using the nominal dynamic viscosity in the liquid state (~ 10^{-3} Pa s). This behaviour is also applied to the liquid phase circulating in the mushy zone, but in this latter case, Darcy terms are used to model the exchange of momentum between the solid and the liquid phase.

2.2.2. *Equations solved.* The equations are issued from the spatial averaging method [6]:

$$\begin{cases} \nabla\cdot\mathbf{s}-\nabla p-\dfrac{\mu_l g_l}{\mathrm{K}}(\mathbf{v}-g_l\mathbf{v}_s)+g_l\rho_l\mathbf{g}-\rho_{l,ref}\left(\dfrac{\partial\mathbf{v}}{\partial t}+(\nabla\mathbf{v})\mathbf{v}\right)=0 \\ \nabla\cdot\mathbf{v}=-\dfrac{\rho_S-\rho_L}{\rho_L}\dfrac{\partial g_s}{\partial t}-\nabla\cdot(g_s\mathbf{v}_s) \end{cases} \tag{3}$$

In these equations, vector **v** denotes the averaged liquid velocity: $\mathbf{v}=g_l\mathbf{v}_l$. The solid velocity $\mathbf{v}_s$ is known from Step I resolution: $\mathbf{v}_s\equiv\mathbf{v}_{\mathrm{StepI}}$. The first equation is obtained after minor adaptations in order to comply with constraints associated with the fluid flow solver. The permeability K is defined by the Carman-Kozeny model, μ_l denotes the liquid dynamic viscosity and the liquid density is defined as $\rho_l=\rho_{l,ref}(1-\beta_T(T-T_{ref}))$ where β_T is the volumetric thermal expansion coefficient and the reference temperature T_{ref} is taken equal to the liquidus temperature T_L. Eq. (3b) is the global mass balance, issued from the expression of the averaged density in the mushy state: $\langle\rho\rangle=g_s\rho_S+g_l\rho_L$, with ρ_S and ρ_L the densities at solidus and liquidus temperature, respectively. The weak form of Eq. (3) is solved by a stabilized SUPG-PSPG finite element method [8], in which the velocity of solid nodes ($T < T_S$) is imposed equal to the velocity calculated in Step I resolution: $\mathbf{v}\equiv\mathbf{v}_{\mathrm{StepI}}$. The velocity and pressure fields deduced from Step II calculation are denoted $\mathbf{v}_{\mathrm{StepII}}$ and p_{StepII}, respectively.

3. Free surface modelling

3.1. Level set method

The level set method [7] is used to track the metal/air interface. The level set function α is defined as the signed distance to the interface. By convention, $\alpha < 0$ corresponds to air, $\alpha > 0$ represents the metal, and $\alpha = 0$ the interface. As the only useful information for our computation is the exact zero iso-value position, the signed distance function is truncated and approached by a sinusoidal filter, in order to limit advection calculations around the interface. The modified level set $\phi(\alpha)$ writes [9]:

$$\phi(\alpha)=-\frac{2E}{\pi}\text{ if }\alpha<-E \qquad \phi(\alpha)=\frac{2E}{\pi}\sin\left(\frac{\pi}{2E}\alpha\right)\text{if }-E<\alpha<E \qquad \phi(\alpha)=\frac{2E}{\pi}\text{ if }E<\alpha \tag{4}$$

where E is a thickness related to the local mesh size h_{msh} (E = 10 to 20 h_{msh}). In the finite element context, functions α and $\phi(\alpha)$ are P1 fields (nodal fields, linearly interpolated). Knowing the velocity field on the whole computational domain, function ϕ can be updated at each time increment, expressing the motion of the interface, while keeping its metric property (that is $\|\nabla\phi\|=1$ by the interface $\phi=0$) by means of a convective reinitialization equation described in [7] and [9]. This transport equation is stabilized with a SUPG (Streamline-Upwind-Petrov-Galerkin) method [5].

3.2. Mixed material properties

In the neighbourhood of the interface, metal and air properties have to be homogenized in order to show a smooth continuous change. The general expression of a mixing law of a generic variable ψ, is:

$$\langle\psi\rangle=\psi_m H(\phi)+\psi_a(1-H(\phi))=\psi_a+(\psi_m-\psi_a)H(\phi) \tag{5}$$

where $\langle\psi\rangle$ denotes the mixed variable, ψ_m its value in the metal, and ψ_a its value in the air. $H(\phi)$ is the mixing function, the values of which are in [0,1], and is directly deduced from the level set function ϕ. In the present study, a P0 mixing law is used. On each tetrahedron (n = 4), we have:

$$H_{P0}(\phi)=\sum_n[\phi_n]_+\Big/\sum_n|\phi_n| \tag{6}$$

4. Solution Strategy

At each time increment, the resolution algorithm is decomposed in eight successive stages:

- 1st stage. Thermal resolution. The energy equation is solved, providing the temperature distribution in the metal and in the air, and liquid metal fraction. The advection velocity in the convection term is $\mathbf{v}_{\text{StepII}} - \mathbf{v}_{msh}$ where $\mathbf{v}_{msh}$ denotes the grid velocity. The velocity field $\mathbf{v}_{\text{StepII}}$ calculated at previous time increment is used in the convection term.

- 2nd stage. Optional mechanical resolution. Step I: solid-oriented. The first folder of the momentum resolution consists of the solid-oriented resolution. This step is optional and triggered as soon as the time past since the more recent Step I resolution exceeds the value Δt_{StepI} (obtained below, stage 8). Step I resolution provides velocity and pressure fields on the whole domain: ($\mathbf{v}_{\text{StepI}}$, p_{StepI}). However, only $\mathbf{v}_{\text{StepI}}$ at nodes belonging to solid-like elements (those treated as EVP) will be used in the follow-up of the resolution scheme. The stress tensor $\boldsymbol{\sigma}$, and the generalized plastic strain $\bar{\varepsilon}$ and strain rate $\dot{\bar{\varepsilon}}$ are calculated in solid-like elements.
- 3rd stage. Mechanical resolution. Step II: liquid-oriented. The second folder of the momentum resolution consists of the fluid-oriented resolution. This resolution provides velocity and pressure fields on the whole domain: ($\mathbf{v}_{\text{StepII}}$, p_{StepII}). Note that at nodes belonging to solid-like elements, $\mathbf{v}_{\text{StepII}}$ is imposed equal to $\mathbf{v}_{\text{StepI}}$.
- 4th stage. Mesh updating. Nodes belonging to solid-like elements are considered Lagrangian. Their position is updated, provided that a Step I resolution has been performed at the current time increment. Because Δt_{StepI} is the time past since their last update, and $\mathbf{v}_{\text{StepII}} = \mathbf{v}_{\text{StepI}}$ for the considered nodes, the following updating scheme is applied: $\mathbf{x}^{new} = \mathbf{x}^{old} + \Delta t_{\text{StepI}} \mathbf{v}_{\text{StepII}}$. All other nodes are considered Eulerian and then fixed.
- 5th stage. Free surface tracking. The updating of the level set function permits interface tracking. It is achieved by the convection-reinitialization scheme mentioned in Section 3.1. The advection velocity in the convection term is $\mathbf{v}_{\text{StepII}} - \mathbf{v}_{msh}$ where $\mathbf{v}_{msh}$ denotes the grid velocity. For Eulerian nodes, $\mathbf{v}_{msh} = 0$. For Lagrangian nodes, two cases are distinguished:
    - No Step I resolution at the current increment. The Lagrangian nodes of the grid have not been updated: $\mathbf{v}_{msh} = 0$.
    - The Step I resolution has been performed at the current increment. Lagrangian nodes have been just displaced by $\Delta t_{\text{StepI}} \mathbf{v}_{\text{StepII}}$. Consequently, their grid velocity in the level set convection term is taken as: $\mathbf{v}_{msh} = (\Delta t_{\text{StepI}} / \Delta t_{\text{StepII}}) \mathbf{v}_{\text{StepII}}$.
- 6th stage. Updating mixed properties. This step consists in the mixing of material properties, according to the value of the level set function.
- 7th stage. Optional adaptive remeshing. The remeshing is guided by an error estimator of Hessian type, based on interpolation error. Details can be found in [4].
- 8th stage. Time step determination. Time step is adjusted according to the intensity of convection in air and liquid metal. The nominal time step $\Delta t \equiv \Delta t_{\text{StepII}}$ for the next increment is obtained from a CFL condition. In addition, if a Step I calculation has been performed during the current time increment, then the time step Δt_{StepI} is determined in such a way that the maximum incremental deformation in the solid be equal to a user-prescribed value.

5. Application

This academic application intends to illustrate the added value of the proposed partitioned resolution, in the case of the filling and cooling of a parallelepipedic ingot (Figure 1). Liquid steel enters through the bottom inlet at prescribed temperature T_{fill} and velocity v_{fill}. Air is ejected at the top of the cavity during filling. Two symmetry planes are defined in $z = 0$ and 2.5 mm, making the problem planar (but analyzed in 3D). Heat is extracted from the metal through the metal/mould interface, using a convection-type expression for heat flux density: $q = h_T(T - T_{ext})$ where the heat transfer coefficient and the external temperature are constant. There is no heat flux between air and mould. The metal is supposed to stick along cavity walls.

5.1. Ingot filling

Figure 2 illustrates the filling sequence. The zero-isovalue of the level set function representing the metal/air interface is shown at various times, revealing waves and air entrapment caused by impingement against mould walls in the early stages of filling. As filling proceeds, the amplitude of waves at free surface reduces until the end of filling at time 1.72 s for which 80% of the cavity is

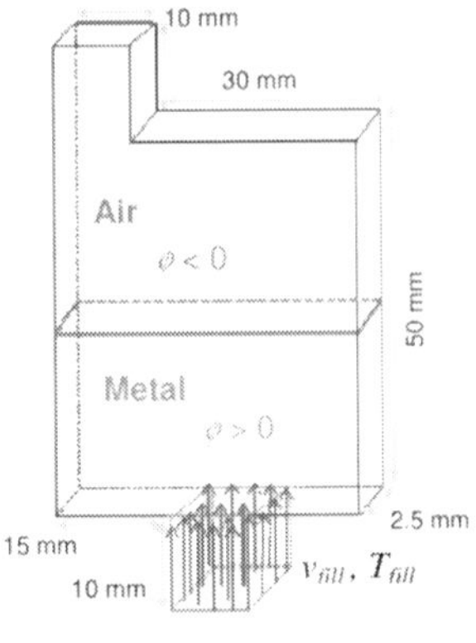

Metal properties			
ρ	$\rho_l = 7060$; $\rho_s = 7303$ kg m^{-3}	T_{fill}	1550 °C
c_p	$c_{p,l} = 717$; $c_{p,s} = 688$ J kg^{-1} K^{-1}	T_L	1478 °C
k	$k_l = 34$; $k_s = 32$ W m^{-1} K^{-1}	T_S	1425 °C
$g_l(T)$	$g_l(T) = (T-T_S)/(T_L-T_S)$	L	2.6 10^5 J kg^{-1}
β_T	7.2 10^{-5} K^{-1}	v_{fill}	0.1 m s^{-1}
h_T	10^3 W m^{-2} K^{-1}	T_{ext}	200 °C
VP	$K_{vp}(T)$, $m_{vp}(T)$; $\mu_{l,I} = 10$ Pa s; $\mu_{l,II} = 10^{-3}$ Pa s		
EVP	$K_{evp}(T)$, $m_{evp}(T)$, $E(T)$, $\sigma_y(T)$, $\nu = 0.33$		
Air properties			
c_p	1006 J kg^{-1} K^{-1}	T_0	800 °C
k	0.0242 W m^{-1} K^{-1}	β_T	3.85 10^{-4} K^{-1}
μ	$\mu_I = 1$ Pa s; $\mu_{II} = 10^{-4}$ Pa s	ρ	1.125 kg m^{-3}

Figure 1: Definition of test case. Geometry and material properties.

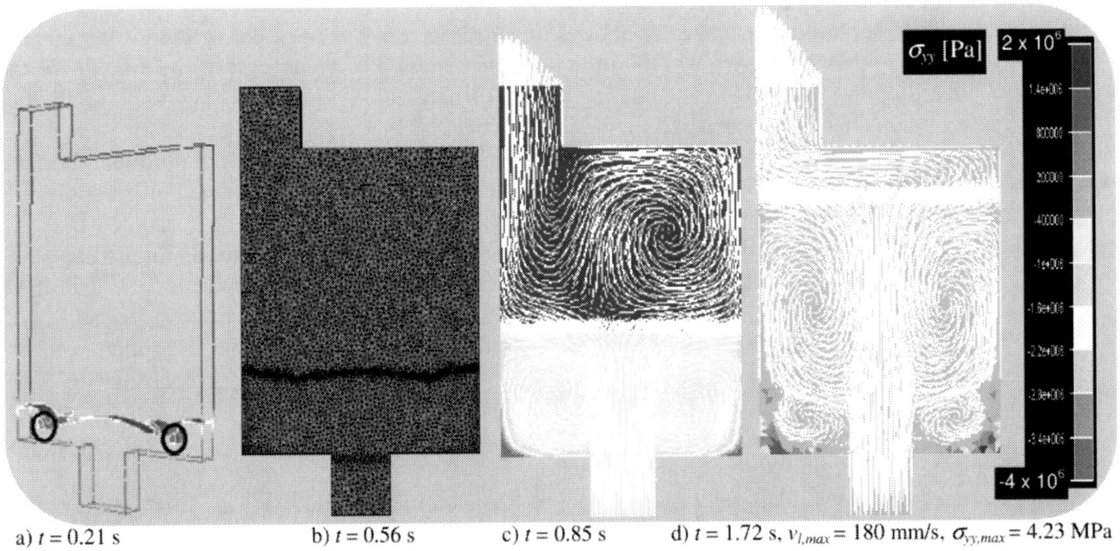

a) $t = 0.21$ s b) $t = 0.56$ s c) $t = 0.85$ s d) $t = 1.72$ s, $v_{l,max} = 180$ mm/s, $\sigma_{yy,max} = 4.23$ MPa

Figure 2: a) Free surface evolution in early stages of filling; b) adapted mesh during filling; c) fluid flow in metal and air, and solid fraction at mid-filling; d) fluid flow and vertical stress in solid shell at the end of filling.

filled. The Figure also shows mesh adaptation, fluid flow in terms of velocity vectors, together with the formation of stress in the solid shell during filling.

5.2. Ingot cooling

After filling, the ingot is cooled down under the same heat extraction conditions. At time 10 s, during solidification, the fluid flow and stresses in the solid are shown in Figure 3, using four different computational strategies. It can be seen in Figure 3a that a sole solid-oriented resolution (that is a resolution with an arbitrary high value of liquid viscosity, 10 Pa s) shows only a liquid feeding flow that compensates solidification shrinkage. Such a simulation gives access to stresses in solidified regions but is not representative of fluid flow. As shown in Figure 3b, this result can be improved by taking a lower viscosity value, 1 Pa s. However this value can be considered as the lower bound for such a formulation, and convection velocities are still too low.

In Figure 3c, the result of a sole liquid-oriented resolution is displayed. In such a resolution, the metal is considered as a non-Newtonian fluid whatever its temperature. As this formulation permits lower viscosity values, fluid flow is well modelled. The increase of the viscoplastic consistency when decreasing temperature gives rise to the freezing of solidified regions. However, stresses in these solidified regions are null. Because of the fluid-like behaviour, deviatoric stresses are proportional to strain-rates: no flow, no stress!

Finally, Figure 3d illustrates the proposed partitioned solution scheme. It can be seen that both fluid flow and stresses can be correctly computed.

Comparing cases c and d with a and b, it can be seen that the calculation of fluid flow directly affects the extension of the solidified region and the shape of the metal free surface. This difference is due to the smoo-thing of temperature gradients in the liquid pool that results from fluid flow: we change from a diffusive regime when no flow, to a convective regime when natural convection is correctly modelled.

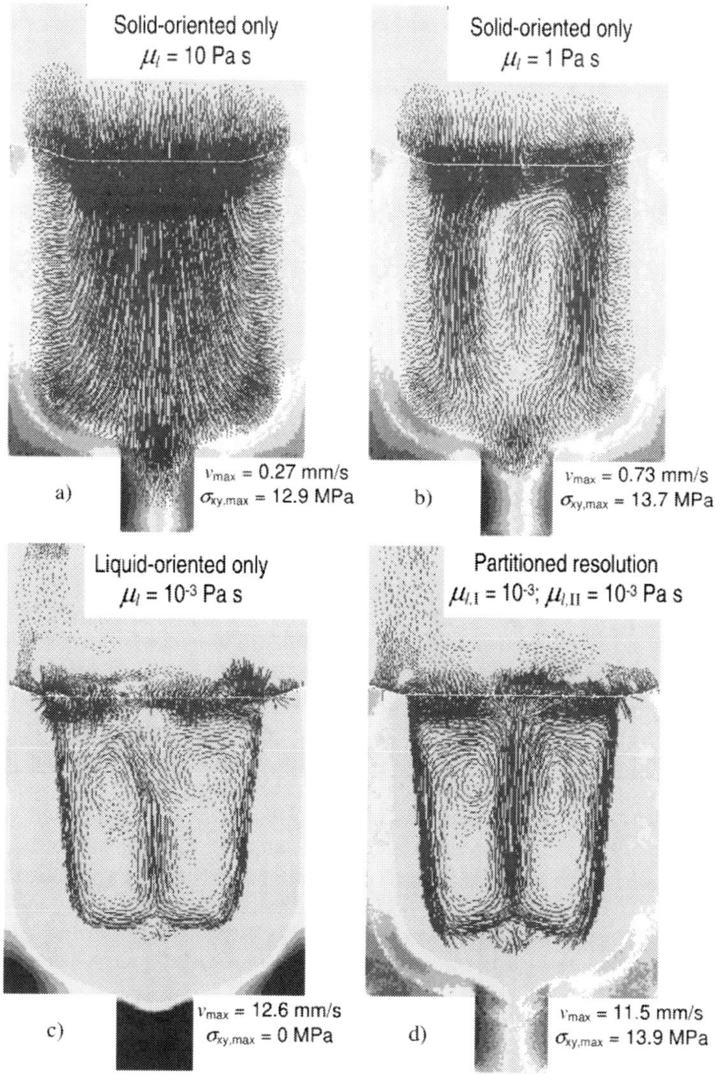

Figure 3: Four different results obtained by different formulations. Only the proposed partitioned scheme gives access to fluid flow and stresses in solid regions (shear stress represented here).

6. Conclusion

The proposed method demonstrates the feasibility of consistent and concurrent nume-rical resolution of fluid flow and solid mechanics. A future perspective is its implementa-tion in the solidification code Thercast®. This should provide an efficient tool for optimizing casting processes.

Acknowledgement

This work has been supported by the French "Agence Nationale de la Recherche" (project "Cracracks" 07-MAPR-0008).

References

[1] Bellet M and Fachinotti V D 2004 *Comp. Meth. Appl. Mech. Eng.* **193** 4355-81
[2] Bellet M, Jaouen O and Poitrault I 2005 *Int. J. Num. Meth. Heat Fluid Flow* **15** 120-142
[3] Fachinotti V D and Cardona A 2007 *Int. J. Numer. Meth. Engng.* **70** 728-755
[4] Hamide M, Massoni E and Bellet M 2008 *Int. J. Num. Meth. Engng.* **73** 624-641
[5] Hughes T J R 1987 *Int. J. Num. Meth. Fluids* **7** 1261–75
[6] Ni J and Beckermann C 1991 *Metall. Trans.* B **22** 349-361
[7] Osher S and Sethian J 1988 *J. Comput. Physics* **79** 12-69
[8] Tezduyar T E and Osawa Y 2000 *Comp. Meth. Appl. Mech. Eng.* **190** 411-430
[9] Ville L, Silva L and Coupez T 2011 *Int. J. Num. Meth. Fluids* **66** 324-344

Macroscopic modelling of semisolid deformation for considering segregation bands induced by shear deformation

S Morita[1], H Yasuda[1,2], T Nagira[1], C M Gourlay[3], M Yoshiya[1], A Sugiyama[4]

[1] Department of Adaptive Machine Systems, Osaka University, Suita, Osaka 565-0871, Japan
[2] Centre of Advanced Structural and Functional Materials Design, Osaka University Suita, Osaka 565-0871, Japan
[3] Department of Materials, Imperial College, London, SW7 2AZ, U.K.
[4] Osaka Sangyo University, Daito, Osaka 574-8530, Japan

E-mail: yasuda@ams.eng.osaka-u.ac.jp

Abstract. *In-situ* observation was carried out to observe deformation of semi-solid Fe-2mass%C steel with 65% solid and globular morphology by X-ray radiography. Deformation was predominantly controlled by the rearrangement of globules. The solid particles were pushed into each other and rearrangement caused lower solid fraction regions to form. On the basis of the observation, a macroscopic model that introduces a normal stress acting on the solid due to collisions and rearrangement is proposed. The solid particles are treated as a non-Newtonian fluid. The stiffness parameters, which characterize the flow of the solid, are introduced. Stability of semisolid to fluctuations in solid fraction during simple shear was analysed. Shear deformation can be stably localized in the semisolid with a certain solid fraction range. The model essentially reproduces band segregation formation.

1. Introduction

Partially solid alloys (semisolid) are often deformed during solidification due to solidification shrinkage and external forces. For example, significant stress operates on the semisolid in centrifugal casting, high-pressure die-casting or twin-roll casting. In addition, the semisolid is poured into a mould in semisolid processing [1-8]. It is well known that semisolids exhibit various mechanical phenomena. Thixotropic properties [9,10], agglomeration / disagglomeration [9], deformation of solid globules [11], and dilatancy during crystal rearrangement [12-14] have all been reported. Furthermore, mush rheology is closely related to casting defects such as band segregation.

Figure 1 shows an example of band segregation in a centrifugal casting of an Al alloy [15]. The positive segregation was formed concentrically. The flow in the solid-liquid two-phase region, which is induced by gravity, resulted in the band segregation [15]. Band segregation has also been observed in high-pressure die casting [16]. Segregation bands were shown to have the characteristics of the dilatant shear bands commonly found in granular materials including soils [12,16]. It was pointed out that the band segregation formed due to strain localization in the semisolid.

Recently, in-situ observation of deformation of semisolid was carried out using X-ray imaging in a synchrotron radiation facility, SPring-8 [14,17]. Direct evidence was presented for granular deformation mechanisms in both globular and equiaxed-dendritic samples, including shear-induced dilation of the solid assembly.

In this paper, in-situ observation of shear deformation in Fe-2mass%C is presented to demonstrate the deformation mechanism. Then, a macroscopic model is proposed with the aim of developing a single, relatively simple, model spanning solid fractions from suspension flow to a viscoplastic solid skeleton. The approach introduces a normal stress acting on the solid particles that depends on the solid fraction in order to account for the effects of collisions and rearrangement of the solid particles. Stability of semisolid to simple shear is then discussed to help build a model for numerical simulation.

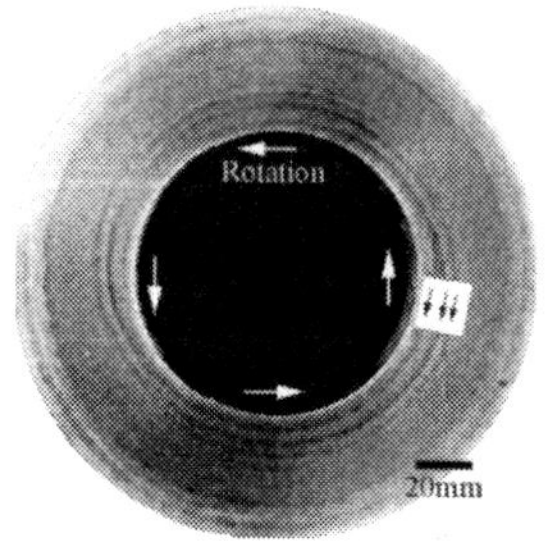

Figure 1. An exmaple of segregation band in centrifugal casting (after I. Ohnaka) [15].

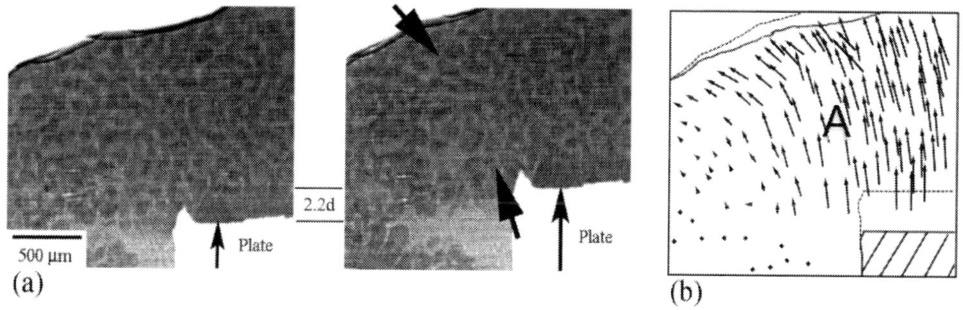

Figure 2. In-situ observation of shear deformation in Fe-2C alloy. (a) Sequence of shear deformation and (b) motion of solid particles in a period of 4.5s. 'd' is mean grain diameter.

2. In-situ observation of shear deformation

In-situ observation was performed in BL20B2 of SPring-8. Detailed procedures are described in the previous study [17]. The semisolid microstructure of high carbon steel (Fe-2C-1Mn-0.5Si in mass%) was generated by partially remelting the as-cast alloy. The average grain size in the sample was 120 μm.

Figure 2 shows the deformation of globular semisolid γ-Fe with solid fraction of 0.65 [17]. An Al_2O_3 plate moved at a rate of 40 μm/s applying direct-shear with a strain rate of 10^{-3}/s. Rearrangement of solid grains, induced by the motion of the plate, produced liquid-filled spaces between globules (black arrows in Fig. 2(a)). The motion vectors of the solid particles are shown in Fig. 2(b). The divergence of the motion vectors in region A means that the solid particles spread peripherally and the liquid flowed into the region. Thus, rearrangement resulted in the dilation of the solid packing-density. The *in-situ* observation proved that the rearrangement of solid grains causes shear-induced dilation of a sample. See [14,17] for more detail and further evidence.

3. Macroscopic model

3.1. Strain tensor of solid particles

In the model [18], solid particles are treated as a viscoplastic fluid. The in-situ observation showed that collisions and rearrangement of solid particles can dilate the liquid-filled spaces during shear [14, 17].

A hydrostatic stress, σ_s, operating on the solid particles is therefore introduced as a simple approach to simulate the dilation. The full stress tensor of the solid particles is given by

$$\begin{bmatrix} \sigma_{11}^s & \sigma_{12}^s & \sigma_{13}^s \\ \sigma_{21}^s & \sigma_{22}^s & \sigma_{23}^s \\ \sigma_{31}^s & \sigma_{32}^s & \sigma S_{33} \end{bmatrix} = \begin{bmatrix} \sigma_s + \tau_{11}^s & \tau_{12}^s & \tau_{13}^s \\ \tau_{21}^s & \sigma_s + \tau_{22}^s & \tau_{23}^s \\ \tau_{31}^s & \tau_{32}^s & \sigma_s + \tau_{33}^s \end{bmatrix}. \quad (1)$$

The deviator stress tensor of solid particles is expressed by the following equation.

$$\tau_{ij}^s = 2C_1\dot{\varepsilon}_{ij}^s \quad (2)$$

Since the solid is treated as a fluid, the expression is the same as that for the liquid phase. Note that the stiffness C_1 depends on solid fraction, as will be discussed subsequently.

σ_s is the component of the normal stress acting on the solid which depends on the degree of collisions / rearrangement of the solid and the solid fraction via another stiffness parameter, C_3. In this paper, simple shear is considered to help understand band segregation. In simple shear, Fig. 3, σ_s is simply expressed as

$$\sigma_s = -\sqrt{6}C_3\left|\dot{\varepsilon}_{12}^s\right| \quad (3)$$

As the shear strain rate increases, the frequency of collisions and rearrangement increases. Here, the normal stress is assumed to be proportional to the shear strain rate. Since σ_s is hydrostatic, it should not depend on the coordinate system. Thus, σ_s is essentially a function of the invariants of the strain rate tensor.

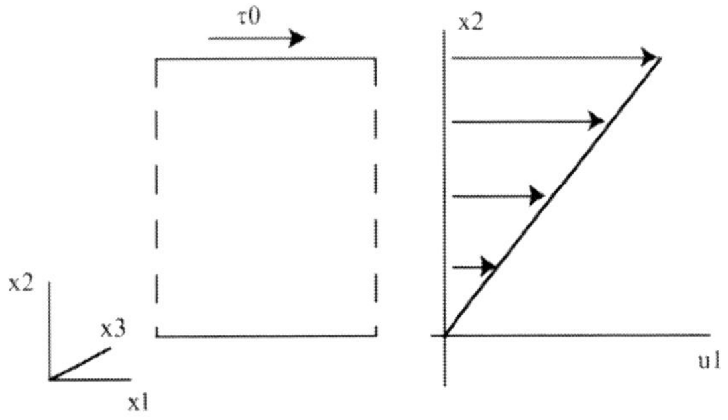

Figure 3. Simple shear for considering stability of semisolid to perturbations in solid fraction.

3.2. Governing equations

Governing equations of liquid and solid particles are expressed using the Navier-Stokes equations.

$$\rho\frac{\partial}{\partial t}\left(f_l u_i\right) + \rho\frac{\partial}{\partial x_j}\left(f_l u_i u_j\right) = -f_l\frac{\partial P}{\partial x_i} + \frac{\partial}{\partial x_j}\tau_{ij}^l - F_i^{sl}, \quad (4a)$$

$$\rho\frac{\partial}{\partial t}\left(f_s v_i\right) + \rho\frac{\partial}{\partial x_j}\left(f_s v_i v_j\right) = -f_s\frac{\partial P}{\partial x_i} - \frac{\partial \sigma_s}{\partial x_i} + \frac{\partial}{\partial x_j}\tau_{ij}^s + F_i^{sl} \quad (4b)$$

$$\frac{\partial f_l}{\partial t} + \frac{\partial}{\partial x_i}\left(f_l u_i\right) = 0 \quad (4c)$$

$$\frac{\partial f_s}{\partial t} + \frac{\partial}{\partial x_i}\left(f_s v_i\right) = 0 \quad (4d)$$

$$f_l + f_s = 1 \quad (4e)$$

The force of F^{sl}_i is the interaction between liquid and solid particles. In the case that Reynolds number of particles is sufficiently smaller than unity in the suspension regime, Stokes' law can be used as the interaction. Thus, the force per unit volume is given by

$$F_i^{sl} = n_s f_i^{sl} = -\frac{18\mu}{d_s^2} f_s \left(v_i - u_i \right) \tag{5}$$

For solid fractions in the suspension regime, as the diameter decreases, the force F^{sl}_i increases and the motion of solid particles tends to follow the liquid flow.

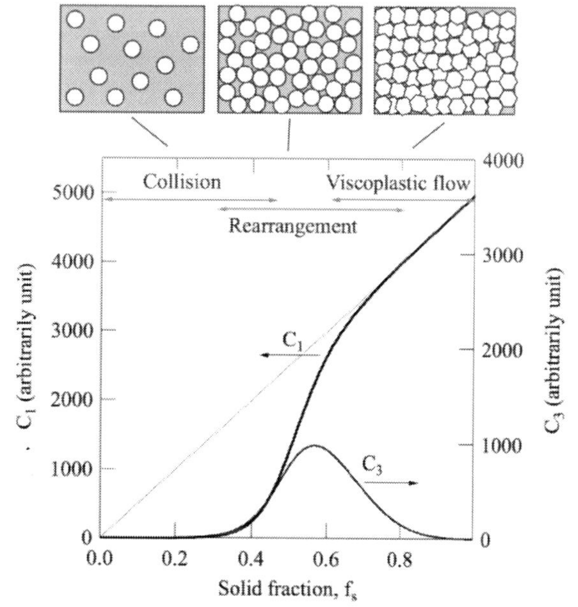

Figure 4. Stiffness parameters, C_1 and C_3, as a function of f_s.

3.3. Stiffness of solid particles

It is of interest to consider the stiffness parameters, C_1 and C_3 in Eq. (2) and (3). The stiffness parameters depend on the volume fraction, size and shape of solid particles. Here, only the influence of solid fraction on the parameters is considered.

The stiffness C_1 originates from collisions, rearrangement and/or viscoplastic deformation of the solid particles, depending on the solid fraction. Figure 4 shows the probable f_s dependence of C_1 and schematic configuration of solid particles in liquid phase. At f_s=0, C_1 and dC_1/df_s are μ (viscosity of liquid phase) and 0, respectively. C_1 starts to increase with f_s when collisions frequently occur. C_1 then becomes much larger when the solid fraction is sufficient for solid particles to be in contact and additional shear stress is required to drive the dilation that enables rearrangement. The viscoplastic flow of the solid phase then becomes dominant at higher solid fraction. In general, the flow stress of the individual solid particles can be simply expressed as $\sigma = A\dot{\varepsilon}$. The parameter A is more than 10^3 times larger than the viscosity of liquid phase. Therefore, C_1 will be in the order of $10^3\mu$ at f_s~1.

The trend of C_3 is also shown in Fig. 4. As shown in eq. (3), the value of C_3 determines the hydrostatic stress level that is generated by collisions / rearrangement of the solid particles. Since C_3 is created by solid collisions / rearrangement, C_3 is zero at f_s~0, and has a positive value at intermediate f_s when dilation occurs due to solid collisions / rearrangement. At the highest solid fractions, shear deformation is controlled by viscoplastic flow of the solid phase, and rearrangement is less dominant. Thus, C_3 is zero at f_s~1. As a result, C_3 exhibits a maximum at a certain f_s, as shown in Fig. 4.

The trends of C_1 and C_3 are qualitatively explained. Further efforts are required to evaluate C_1 and C_3 as a function of solid fraction. Even though the f_s dependences of C_1 and C_3 are not known precisely, the trends shown in Fig. 4 are expected to be valid. Next, it is of interest to study the influence of these trends on the mechanical response of semisolid to simple shear.

4. Stability of the semisolid to shear stress

4.1. Assumptions and simplified equations

Simple shear of the semisolid, Fig. 3, is considered. The following assumptions are introduced as the initial conditions. Any variables do not depend on x_1. Flow velocity of solid particles in the x_1 direction is the same as that of liquid phase.

$$u_1 = v_1 \tag{6a}$$

The flow velocities in the x_3 direction are always zero.

$$u_3 = v_3 = 0 \tag{6b}$$

Mass conservation in the x_2 direction gives the following relationship.

$$f_l u_2 + f_s v_2 = 0 \tag{6c}$$

Equations (6a)-(6c) give the simplified governing equations as follows. The Navier-Stokes equations in x_1 direction give

$$\rho \frac{\partial v_1}{\partial t} = \frac{\partial}{\partial x_2}\left(\tau_{12}^{l} + \tau_{12}^{s}\right) = \frac{\partial}{\partial x_2}\left\{\left[\left(1-f_s\right)\mu + f_s C_1\right]\frac{\partial v_1}{\partial x_2}\right\}. \tag{7a}$$

The term $\bar{\mu} = (1-f_s)\mu + f_s C_1$ is the apparent viscosity of the semisolid. The trend of $\bar{\mu}$ calculated from C_1 basically agrees with the measured viscosity of semisolid. The Navier-Stokes equations in x_2 direction give

$$\frac{\partial P}{\partial x_2} = -\rho \frac{\partial}{\partial x_2}\left(\frac{f_s}{1-f_s} v_2 v_2\right) - \frac{\partial \sigma_s}{\partial x_2} + \frac{\partial}{\partial x_2}\left(\tau_{22}^{l} + \tau_{22}^{s}\right).$$

The equation means that the pressure gradient in x_2 direction is induced to satisfy the mass conservation. Thus, the equation of motion in x_2 direction is expressed without the pressure.

$$\begin{aligned}\rho f_s \frac{\partial v_2}{\partial t} = & -\rho v_2 \frac{\partial f_s}{\partial t} - \rho \frac{\partial}{\partial x_2}\left(f_s v_2 v_2\right) + \rho f_s \frac{\partial}{\partial x_2}\left(\frac{f_s}{1-f_s} v_2 v_2\right) \\ & + f_s \frac{\partial}{\partial x_2}\left\{2\left(1-f_s\right)\mu \frac{\partial}{\partial x_2}\left(\frac{f_s}{1-f_s} v_2\right)\right\} + \left(1-f_s\right)\frac{\partial}{\partial x_2}\left[2 f_s C_1 \frac{\partial v_2}{\partial x_2}\right] \\ & - \left(1-f_s\right)\frac{\partial}{\partial x_2}\left(\frac{C_3}{\sqrt{6}}\left|\frac{\partial v_1}{\partial x_2}\right|\right) - \frac{18\mu}{d_s^2}\frac{f_s}{1-f_s} v_2\end{aligned} \tag{7b}$$

Equations (7a), (7b) and (4d) are the simplified governing equations for analysing the stability of the solid fraction (i.e. solid packing-density) to simple shear.

4.2. Stability analysis

A solution of the simplified governing equations at steady state is given by

$$f_s = f_s^0 \tag{8a}$$

$$\frac{\partial v_1}{\partial x_2} = G \quad (G>0) \tag{8b}$$

$$v_2 = 0 \tag{8c}$$

A perturbation, given by eq. (9), is introduced. γ is the relative fluctuation of the solid fraction

$$f_s = f_s^0\left[1 + \gamma \sin\left(\omega x_2\right)\right] \tag{9}$$

Equation (9) substituted into eq. (4d) gives the time derivative of the solid fraction induced by the perturbation.

$$\frac{\partial f_s}{\partial t} = f_s^0 \frac{d\gamma}{dt}\sin(\omega x_2) = -f_s \frac{\partial v_2}{\partial x_2} - v_2 \frac{\partial f_s}{\partial x_2} = 0 \tag{10a}$$

The perturbation does not directly influence the distribution of the solid fraction. On the other hand, the influence of the perturbation appears in the time derivatives of v_1 and v_2.

$$\rho \frac{\partial v_1}{\partial t} = G\mu f_s^0 \gamma\omega \left[\frac{C_1}{\mu} + f_s^0 \frac{\partial}{\partial f_s}\left(\frac{C_1}{\mu}\right) - 1 \right] \cos(\omega x_2) \tag{10b}$$

$$\rho f_s \frac{\partial v_2}{\partial t} = -\frac{1}{\sqrt{6}} G\left(1-f_s^0\right) f_s^0 \gamma\omega \left(\frac{\partial C_3}{\partial f_s}\right) \cos(\omega x_2) \tag{10c}$$

The acceleration rate of the relative fluctuation of solid fraction is obtained using eqs. (4d), (10b) and (10c).

$$\frac{\partial^2 f_s}{\partial t^2} = f_s^0 \frac{d^2\gamma}{dt^2}\sin(\omega x_2) = -\frac{\partial}{\partial t}\left[\frac{\partial}{\partial x_2}(f_s v_2)\right]$$

Thus,

$$\rho \frac{d^2\gamma}{dt^2} = K\gamma \tag{11a}$$

$$K = -\frac{1}{\sqrt{6}} \tau_0 \omega^2 \frac{\left(1-f_s^0\right)\dfrac{\partial C_3}{\partial f_s}}{\left(1-f_s^0\right)\mu + f_s^0 C_1} \tag{11b}$$

Here, K is the stability parameter and the following relationship is used.

$$G = \frac{\tau_0}{\bar{\mu}} = \frac{\tau_0}{\left(1-f_s^0\right)\mu + f_s^0 C_1}$$

Equation (11a) is the equation of motion for the fluctuation in solid fraction. If the stability parameter, K, is positive, the fluctuation steps away from the original perturbation. Namely, if K>1, the perturbation causes segregation of the solid particles. In contrast, the perturbation is reduced or oscillated if K is negative. Thus, if K<1, the semisolid is stable to simple shear, leading to the uniform deformation.

Figure 5(a) shows the value of K as a function of the solid fraction. In this estimation, C_1 and C_3 shown in Fig. 4 are used. K is negative for solid fractions less than a lower critical solid fraction, f_{ds}, as which $\partial C_3/\partial f_s = 0$. K becomes positive when $f_s > f_{ds}$. At high solid fraction, viscoplastic flow of the solid phase becomes dominant and K converges into zero. According to Fig.5a, K is almost zero at $f_s >$ 0.9 (referred to as f_{vc}).

When $f_s < f_{ds}$, the normal stress acting on the solid, σ_s, increases with increasing the solid fraction. The normal stress induced by the perturbation attenuates a non-uniform solid fraction. In addition, v_1 and v_2 are changed by the perturbation. As a result, a fluctuation of the solid fraction is not simply attenuated, but it is transient. Thus, the fluctuation cannot remain during simple shear. It is suggested that shear deformation is not likely localized and the band segregation is not formed.

When the solid fraction is between f_{ds} and f_{vs}, it is unstable to simple shear because a fluctuation of the solid fraction is amplified. The semisolid is likely to separate into the two parts with solid fractions of f_{ds} and f_{vs}. The separated two parts are stable to simple shear. The apparent viscosity of the lower solid fraction part is much smaller than that of the higher solid fraction part. Therefore, the shear deformation is stably localized in the lower solid fraction part, leading to band segregation in the solidification structure, as shown in Fig 5(b).

By introducing a solid normal stress created by collisions/rearrangement of the solid particles, the model analytically predicts that an instability essentially exists for the semisolid and reproduces the localization of shear deformation and the band segregation. Evaluation of C_1 and C_3 by experiments is required to simulate the localization of the shear deformation and the band segregation formation.

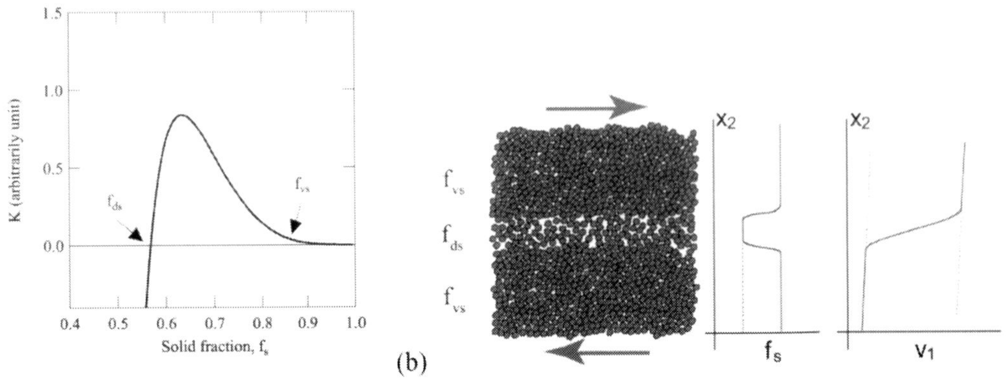

(a) (b)

Figure 5. (a) Stability parameter K and solid fraction f_s. C_1 and C_3 in Fig. 4 are used for the calculation. (b) Schematic illustration of the localization of the shear deformation.

5. Summary

In-situ experiments were carried out to observe the shear deformation of semi-solid Fe-2mass%C steel at 65% solid by X-ray radiography. The deformation was predominantly controlled by rearrangement of globular solid particles. The solid particles were pushed into each other and rearrangement caused shear-induced dilation of the solid particle assembly.

A model is proposed in which the solid particles are treated as a non-Newtonian fluid. On the basis of the in-situ observation, the normal stress operating on the solid particles is introduced in the model. The stiffness parameters C_1 and C_3, which represent the apparent viscosity and the normal stress, respectively, are qualitatively estimated.

A stability analysis for simple shear shows that the semisolid with solid fraction between f_{ds} and f_{vs} becomes unstable to shear stress and is likely to separate into a lower solid fraction part and a higher solid fraction part. Shear deformation is stably localized in the separated semisolid. The analysis suggests that band segregation is formed when the localization of the shear deformation is stabilized. Further efforts for evaluating C_1 and C_3 are required to help validate the model.

Symbols

u_i : flow velocity of liquid. i is direction
v_i : flow velocity of solid particles
f_s : fraction of solid
f_l : fraction of liquid
ε^s_{ij} : Strain rate tensor of solid particles $\dot{\varepsilon}^s_{ij}=(1/2)[(\partial v_i/\partial x_j)+(\partial v_j/\partial x_i)]$
ε^l_{ij} : Strain rate tensor of liquid particles $\dot{\varepsilon}^l_{ij}=(1/2)[(\partial v_i/\partial x_j)+(\partial v_j/\partial x_i)]$
σ^s_{ij} : Full stress tensor of solid particles, including hydrostatic stress
σ^l_{ij} : Full stress tensor of liquid, including hydrostatic pressure
τ^s_{ij} : Deviator stress tensor of solid particles
τ^l_{ij} : Deviator stress tensor of liquid
σ_s : Normal stress induced by collisions/rearrangement of solid particles
P : Pressure in liquid phase
F_i^{sl} : Interaction between liquid and solid particles
d_s : Diameter of solid particles
$\bar{\mu}$: Apparent viscosity of semisolid

IOP Conf. Series: Materials Science and Engineering **33** (2012) 012053 doi:10.1088/1757-899X/33/1/012053

Acknowledgements
This study was supported by the Global COE Program (Centre of Excellence for Advanced Structural and Functional Materials Design) from MEXT, and a Grant-in-Aid for Scientific Research (A). The synchrotron radiation experiments were performed as a long-term project (2007A0014- 2009B0014), and as general projects at BL20B2 and BL20XU beamlines at SPring-8 (JASRI). The *in situ* observation of solidification in Fe alloys was also supported by "ISIJ Innovative Program for Advanced Technology" from the Institute of Iron and Steel, Japan, and a "Royal Society Daiwa Anglo-Japanese Foundation International Exchanges Award".

References
[1] Tseng A A, Horský J, Raudenslý M, Kotrbácek P 2001 *Mater. Design* **22** 83.
[2] Kikuchi M, Kopp R 2002 *Ann. of the CIRP* **51** 653.
[3] Omar M Z, Palmiere E J, Howe A A, Atkinson H V, Kapranos P 2005 *Mater. Sci. Eng. A* **395** 53.
[4] Ramadan M, Takita M, Nomura H, El-Bagoury N 2006 *Mater. Sci. Eng. A* **430** 285.
[5] W. Püttgen, W. Bleck, G. Hirt, H. Shimahara, *Adv. Eng. Mater.* **9** (2007) 231–245.
[6] Sugiyama S, Li J, Yanagimoto J 2007 *Mater. Trans.* **48** 807.
[7] Pierret J C, Passili A, Vaneetveld G, Lecomte-Beckers J, Cezard P, Bigot R 2008 *Int. J. Mater. Form* 1011.
[8] Rassili A, Atkinson H V 2010 *Trans. Nonferrous Met. Soc. China* **20** s1048.
[9] Young K P, Riek R G, Flemings M C 1979 *Metals Technology* **6** 130.
[10] Flemings M C 1991 *Metall. Trans.* **22A** 957.
[11] Tzimas E, Zavaliangos A 1999 *Acta Mater.* **47** 517.
[12] Gourlay C M, Dahle A K 2007 *Nature* **445** 70.
[13] Gourlay C M, Meylan B, Dahle A K 2008 *Acta Mater.* **56** 3403.
[14] Gourlay C M, Dahle A K, Nagira T, Nakatsuka N, Nogita K, Uesugi K, Yasuda H 2011 *Acta Mater.* **59** 4933.
[15] Kang I, Ohnaka I 1997 *J Japan Foundry Engineering Society* **69** 240.
[16] Gourlay C M, Laukli H I, Dahle A K 2007 *Metall. Mater. Trans. A* **38** 1833.
[17] Nagira T, Gourlay C M, Sugiyama A, Uesugi M, Kanazawa Y, Yoshiya M, Uesugi K, Umetani K, Yasuda H 2011 *Scr. Mater.* **64** 1129.
[18] Yasuda H, Gourlay C M, Morita S, Nagira T, Yoshiya M, to be submitted.

IOP Conf. Series: Materials Science and Engineering **33** (2012) 012054 doi:10.1088/1757-899X/33/1/012054

Numerical tensile test on a mushy zone sample

Jean-François Zaragoci, Luisa Silva, Michel Bellet, Charles-André Gandin
MINES ParisTech CEMEF, UMR CNRS 7635, 06904 Sophia Antipolis, France

E-mail: luisa.silva@mines-paristech.fr

Abstract. Time sequences of 3D images of an Al-Cu alloy in the mushy state are obtained using in situ and real-time X-ray microtomography during a tensile test. Surface meshes of phase interfaces are built from these images with the help of a Marching Cubes algorithm. The signed distance to the surface meshes is then computed on a finite element mesh of the volume of the phases. These signed distance functions enable to track the interfaces between the phases in an implicit way. A numerical representation of the real microstructure is thus obtained, allowing to perform a numerical tensile test which is compared with the experimental tensile test. Results retrieve the general dynamic behaviour of the strain field evolution and opens promising perspectives for further interpretations of experimental results based on numerical simulations.

1. Introduction

Nowadays, numerical simulations of complex systems are mandatory in many fields. The first aspect to consider for performing analyses on such system is the creation of a realistic numerical representation of the object under study. This is the case in phase transformations taking place in the context of materials science. Several phases may coexist in a given material, delimited by interfaces. A mesh of the structure of the phases is then needed to study the heat, mass and momentum transfers taking place at phase interfaces and in the bulk phases. Results of such analyses for the entire material are not trivial because of the various phase properties and interface behaviours.

When it is not possible to consider a numerical representation of the real specimen, statistical representative elementary volumes (REV) can be used. Advantages lead in the possibility to easily change the morphology of the REV. The sample behaviour can then be studied with respect to various parameters. For instance, by using a modified Voronoi tessellation, Phillion et al. showed the effect of microstructural features on the semi-solid tensile deformation of an aluminium alloy such as solid fraction, porosity and grain size [1].

During the last decade, X-ray microtomography has considerably developed [2-4]. This imaging technique gives access to in situ three-dimensional (3D) morphology of a multiphase sample. Thus, Terzi et al. obtained a set of 3D images showing morphological evolutions in a few millimeters sample of an aluminium copper alloy during tensile testing [5]. It is currently possible to reach a spatial and time resolution of less than 1 μm and 1 s, respectively, on millimeters domain size [4].

The ambition of this paper is to describe a way to transform data coming from X-ray microtomography into a finite element (FE) representation and to use it to run numerical simulations. In section 2, the approach chosen to obtain the FE representation is explained. Section 3 presents the model equations for the numerical tensile test and the way to solve them. Finally, results on the real morphology are presented and compared to the experimental data in section 4.

2. Finite element representation

Three steps are required to transform a 3D picture into a FE representation [6]: segmentation, surface mesh computation and volume mesh computation. The segmentation is the designation of all the phases of the multiphase domain, using the grey level of the tomography images. Because images are already segmented by Terzi et al. [5], only generations of surface and volume mesh will be explained here.

2.1. Surface mesh computation

The surface meshes used to separate the different phases are a representation of the interfaces between these phases. In the case considered here, three phases are identified with the X-ray microtomography: solid, liquid and air. Two segmentations give access to two binary images, where only black and white levels are present to separate the liquid phase from the solid+gas zone and the gas from the solid+liquid zone. These two images are sufficient to define all solid/liquid, solid/gas and liquid/gas interfaces. Two surface meshes are then computed with the help of a Marching Cubes (MC) algorithm. Voxels are the constitutive element of 3D images. A 3D image can be seen as a stack of 2D slices. A marching cube is a logical cube created from eight neighbouring voxels, four from one slice and four from the neighbouring slice [7]. For each voxel, two states are possible: inside a phase or outside it. Consequently there are $2^8 = 256$ cases for a MC. Each case corresponds to a surface separating the phases. To obtain the surface, rotational symmetry is considered [8] to reduce the problem from 256 cases to 23 patterns shown in figure 1. Once the surface is computed for one MC, the algorithm moves to the next cube until the whole image is treated.

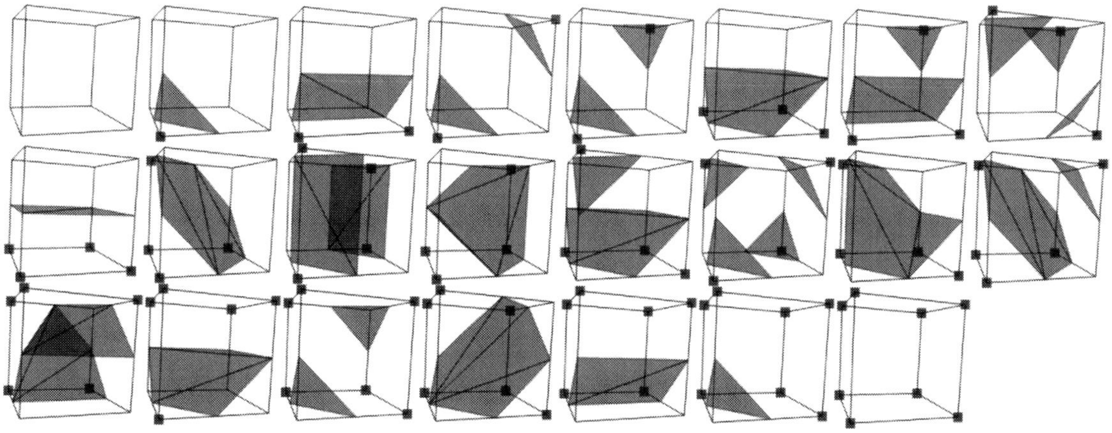

Figure 1. Illustration of the 23 patterns considered for a single marching cube and corresponding surface mesh. Vertices are assigned a dot if they are inside the liquid.

2.2. Volume mesh computation

The volume mesh computation is based on the immersed volume method (IVM) described in [9]. In our case the volume mesh used for tensile test computations must include, as for the images, three existing phases. Two level set functions are computed within this volume mesh to take into account the properties of the three phases.

A level set function is a signed distance from a surface mesh computed for all nodes of our tetrahedral volume mesh. Two level sets are computed: the liquid level set ϕ_l, representing the interface between the liquid and the solid+gas mixture, and the air level set ϕ_a, representing the interface between the air and the solid+liquid mixture. The negative sign for the liquid level set is chosen inside the liquid and inside the air for the air level set. The algorithm used to compute the distance between a triangular facet of the surface mesh and a node of the volume mesh is described in [11]. To get the signed distance, the normal to the facet of the surface mesh is used. The distance between a node and the surface mesh is the minimal distance between this node and all the facets. Because surface and volume meshes can have several millions of facets and nodes respectively, it is very time consuming to compute the distance from a given node to all the facets of the surface mesh. To decrease the CPU time cost, a hierarchical representation of the surface mesh is built at the beginning of the simulation [12].

With the IVM, each equation treated is solved on the whole computational domain [9]. Behaviour of all phases vary from Newtonian to viscoplastic, meaning that all the physical properties necessary for a simulation are defined in the whole volume mesh using a classical mixture law of the properties of each phase and the level sets defining the interfaces. In the immediate neighbourhood of an interface separating two phases, 1 and 2, a piecewise continuous mixing model is used. A mixed property α is thus defined as:

$$\alpha = H\alpha_1 + (1 - H)\alpha_2 \tag{1}$$

α_1 and α_2 are the property values for the two phases considered and H is the filling ratio of the element for phase 1, computed with the level set separating the two phases. With three phases, equation (1) is applied twice to take into account the two level sets and the properties of all the phases. In a first time, the air level set is used with air and liquid properties. Then, the solid level set, viz. the opposite of the minimum of air and liquid level sets, is used with solid properties and the results of the first mixture.

A topological mesher [10] is used with only isotropic meshes in this paper.

3. Numerical model

3.1. Equations

Simulating the morphological evolutions of a sample at this scale is a highly challenging multiphysics problem. In this work, the physical models chosen to simulate the tensile test of the aluminium copper alloy are simplified.

3.1.1. Mechanical behaviour. Because inertial and mass terms are assumed negligible, only the Stokes equations without right hand member are solved:

$$\nabla \cdot \bar{\bar{\sigma}} = \vec{0} \tag{2}$$

$$\nabla \cdot \vec{v} = 0 \tag{3}$$

$\bar{\bar{\sigma}}$ is the stress tensor and $\vec{v}$ the velocity vector of the flow. The liquid and the air are modelled as two incompressible Newtonian fluids. The stress tensor inside these phases reads:

$$\bar{\bar{\sigma}} = 2\eta\bar{\bar{\varepsilon}}(\vec{v}) - p\bar{\bar{I}} \tag{4}$$

η is the viscosity, $\bar{\bar{\varepsilon}}(\vec{v})$ the strain rate tensor and $\bar{\bar{I}}$ the identity tensor. For the solid phase, neglecting elasticity, the behaviour is viscoplastic. Stress and strain rate are related by:

$$\bar{\bar{\sigma}} = 2\eta_{eq}\bar{\bar{\varepsilon}}(\vec{v}) - p\bar{\bar{I}} = 2\left(\frac{K}{3}\dot{\varepsilon}^{m-1}\right)\bar{\bar{\varepsilon}}(\vec{v}) - p\bar{\bar{I}} \tag{5}$$

K is the consistency, m the strain rate sensitivity, $\dot{\bar{\varepsilon}}$ the equivalent strain rate and η_{eq} the apparent solid viscosity. The corresponding relation between the equivalent stress $\bar{\sigma}$ and $\dot{\bar{\varepsilon}}$ is:

$$\bar{\sigma} = K\dot{\bar{\varepsilon}}^{m} \tag{6}$$

3.1.2. Interfaces displacements. To keep the interfaces still represented by the zero isovalue of the level set functions while updating the configuration, the level sets should be moved with the velocity derived from the resolution of the Stokes equations. To this purpose, the following convection equation is solved for ϕ_l and ϕ_a :

$$\frac{\partial \phi}{\partial t} + \vec{v} \cdot \vec{\nabla}\phi = 0 \tag{7}$$

After solving equation (7), the level set functions are no longer signed distance functions. To recover this property at times t when judged necessary, a reinitialisation equation is solved [13]:

$$\frac{\partial \phi}{\partial \tau} = S(\phi)\left(1 - \left\|\vec{\nabla}\phi\right\|\right) \tag{8}$$

$S(\phi)$ is the sign of the level set whereas τ is a fictitious time step linked to the mesh size. To speed up the computation, equation (7) and equation (8) are coupled to obtain a convective reinitialisation equation [14]. Moreover, the level set functions being useful only near the interfaces, a smoothed function is rather used, through the hyperbolic tangent of the distance to each interface [15].

3.2. Finite elements

The partial derivative equations are solved with the finite element method with CIMLib, a highly parallel C++ scientific computation library described in [16], using PETSc [17] (Portable, Extensible Toolkit for Scientific Computation) for solving the linear systems and MPI (Message Passing Interface) for the parallelism.

3.2.1. Stokes equations. After mixing the consistency and the strain rate sensitivity of the three phases, Stokes equations are solved with the method introduced in [18], known as MINI-element, which satisfies the Brezzi-Babuška stability condition [19]. The linear system is solved with the conjugate residual method associated with an incomplete LU preconditioning [20]. To consider the non linear behaviour of the solid phase, the fixed point method is used: once the velocity field $\vec{v}$ is computed, a new equivalent strain rate $\dot{\bar{\varepsilon}}$ is obtained, giving rise to a new estimate for the viscosity η_{eq} to be considered for the next resolution of equation (2).

3.2.2. Convective reinitialisation equation. The standard Galerkin method leading to results with spurious oscillations well known for hyperbolic problems, a streamline upwind Petrov-Galerkin (SUPG) stabilisation method is employed [21]. The linear system is solved with the generalized minimal residual method (GMRES) [22] and an incomplete LU preconditioning.

4. Results

4.1. Finite element representation

Given its morphological complexity [5], accurate representation of the sample implies a very large number of nodes, leading to long and difficult subsequent simulations. Thus, an isotropic mesh size of 10 μm is used whereas the initial voxel size is 2.8 μm. Moreover, only a quarter of the sample is modelled. Despite these considerations, the volume mesh contains more than 3 million nodes and 16 million elements. Figure 2a shows the experimental interface between the air and the liquid+solid mixture 513 s after the beginning of the tensile test. It is attained from microtomography data with voxel representation. Figure 2b shows the same interface drawn for the FE numerical representation. The CPU time for computing the surface mesh is 14 s with one processor (RAM: 3.25 Go, 2.99 GHz Intel Core CPU). The CPU time for computing the air level set is less than 5 min with 16 processors (2.4 GHz Opteron cores linked by an Infiniband network). Excellent geometrical reconstruction of the domain boundaries with the FE mesh is demonstrated, only limited by the size of the FE mesh used.

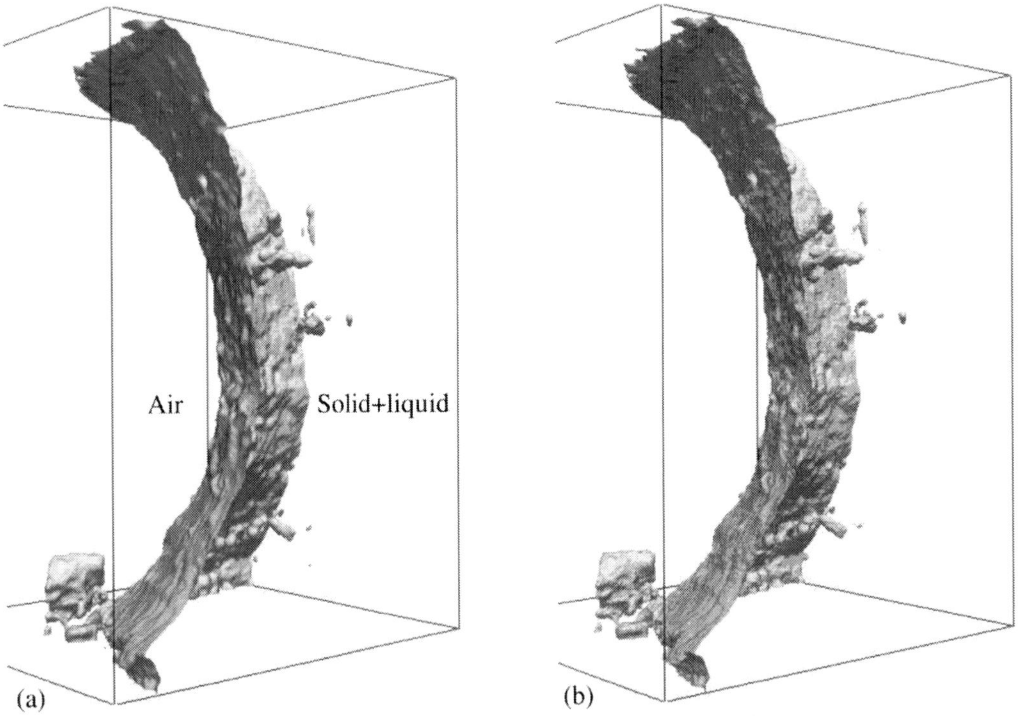

Figure 2. Metallic sample boundary between the air and the solid+liquid mixture 513 s after the beginning of the tensile test. (a) Experimental and (b) numerical boundaries. Vertical field of view: 1.43 mm.

4.2. Mechanical computation

Boundary conditions are applied for a uniaxial vertical tensile test. A vertical velocity is imposed at the bottom and at the top of the sample on all phases and vertical symmetry conditions are applied. To match the experimental conditions, the velocity at the bottom is zero and the velocity at the top is 0.1 $\mu m.s^{-1}$. For the solid, parameters are taken from a compression test campaign similar to the one made by [23], leading to m = 0.2 and K = 35.2 MPa. The apparent solid viscosity is on the order of 10^{10} Pa.s whereas viscosities to be considered are about 10^{-5} Pa.s for the air and 10^{-3} Pa.s for the liquid. With these values, the ratio between minimal and maximal viscosities is very large and no solution is found for the Stokes equations, due to a very bad conditioning of the global matrix and a limited precision. It is thus necessary to use augmented viscosities. For the liquid and the air, viscosities used in the simulations are 100 kPa.s and 1 kPa.s, respectively. These values are large by comparison with real values, but the ratio with the apparent solid viscosity is still significant and stress inside the air and the liquid are negligible. Figure 3 shows the equivalent strain rate and the equivalent stress of a 2D cross sectional cut 513 s after the beginning of the tensile test, assuming a static mechanical state. It can be seen that maximum strain rate values are found in liquid pockets. Stresses are quite heterogeneous in the solid, showing maximum values in specific regions, especially between liquid regions.

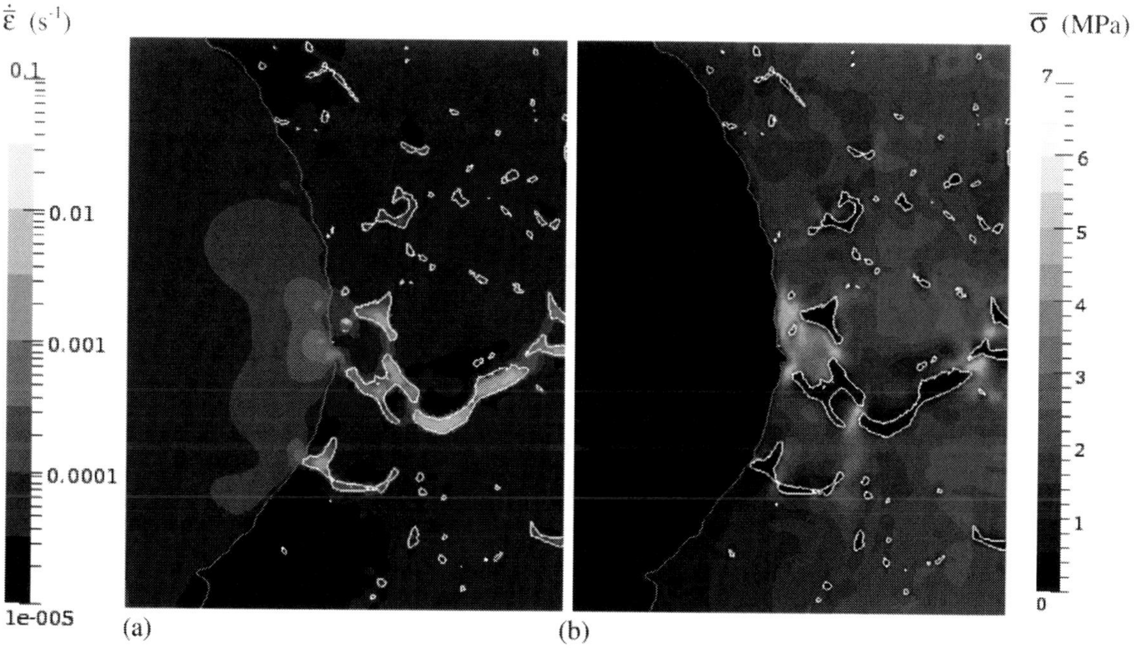

Figure 3. Mechanical results for a 2D cross sectional cut of figure 2b. (a) Equivalent strain rate and (b) equivalent stress. Grey and white contours represent the zero isovalue of the air and liquid level sets respectively, i. e. boundaries between phases.

4.3. Dynamic computation

For the experiment described in [5], a lot of X-ray images are available at incremental times. It then becomes possible to compute the associated numerical representations to perform static mechanical computations, i. e. without moving the interfaces, as in the previous section. This leads to as many calculations as recorded images with time. But the model is also developed to start with one numerical representation taken as initial condition, then solve the Stokes equations, move the interfaces and continue in a time stepping loop, until the end of the experiment. Thanks to this dynamic simulation, comparisons can be made between the evolution of experimental and numerical morphologies of the interfaces. Figure 4 illustrates such a comparison for the interface between the air and the liquid+solid mixture 1323 s after the beginning of the tensile test. To obtain this reasonable agreement with the observation using X-ray microtomography, the starting numerical representation is the one shown in figure 2, just before experimental penetration of air in the sample. The CPU time for computing this evolution is four days with 50 processors (2.4 GHz Opteron cores linked by an Infiniband network) and a time step of 13.5 s. To corroborate this qualitative comparison, the equivalent stress evolution in the solid phase is compared for the static and the dynamic computations in figure 5. Both curves show a stress decrease due to sample damage by air propagation.

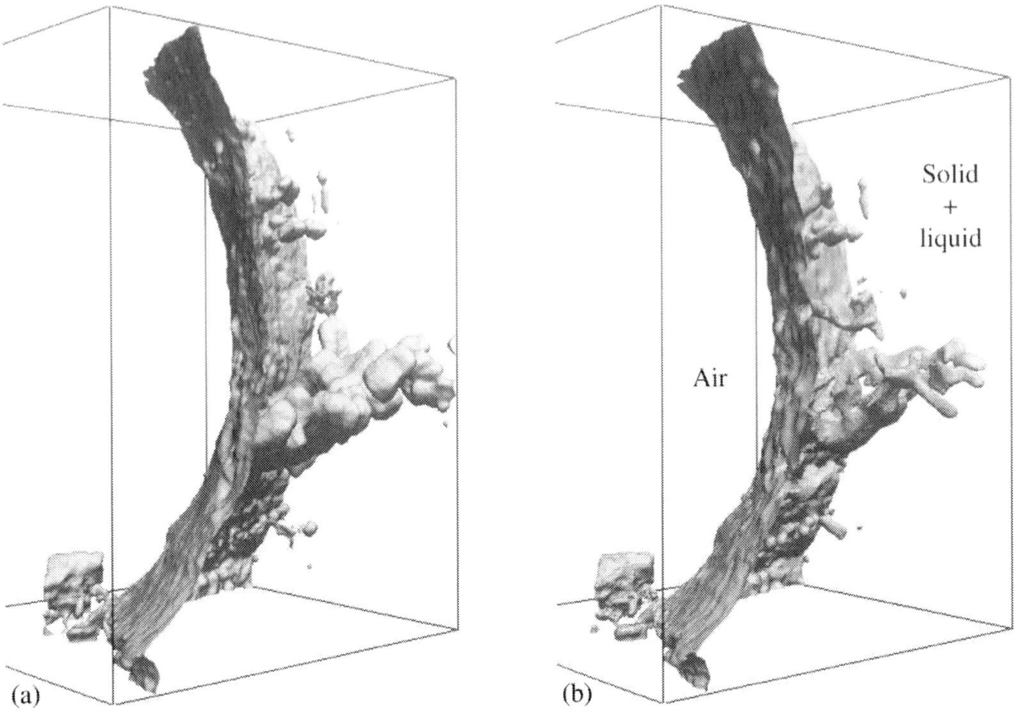

Figure 4. Boundary between the air and the solid+liquid mixture 1323 s after the beginning of the tensile test. (a) Experimental and (b) numerical boundaries. Vertical field of view: 1.43 mm.

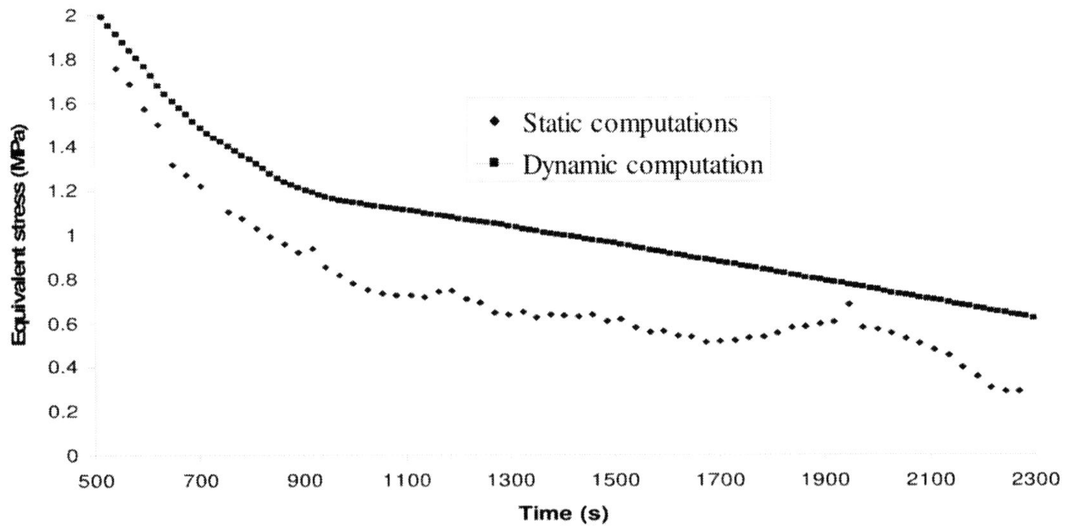

Figure 5. Equivalent stress evolution in the solid phase for static and dynamic computations.

5. Conclusion & perspectives

Transformation of X-ray microtomography data into tetrahedral 3D finite element representations has been presented in this paper, using implicit interface description with the help of level set functions. By solving conservation equations with initial and boundary conditions similar with experiments, direct comparisons between numerical and experimental results are possible. Comparisons made for a tensile test on a mushy zone sample are realistic and motivate further work.

Perspectives include: (i) more complex models taking into account solidification and additional physical phenomena such as surface tension, wetting and gas solubility, (ii) better interface representation with anisotropic meshes, (iii) larger computation to simulate the entire sample and increase space resolution. Such efforts could be conducted in parallel to more accurate experiments to gain advantage of microtomography recent progresses and development of image correlation tools to better estimate the boundary conditions to apply from experimental data.

Acknowledgements
This work was carried out in the framework of the project ANR-08-BLANC-0062 "SIMUZAL", supported by the "Agence Nationale de la Recherche", which is gratefully acknowledged. The authors would especially like to thank Bastien Mireux, Michel Suéry and Luc Salvo for providing them the experimental images and the results of the compression test campaign used in this work.

References

[1] Phillion A B, Cockcroft S L and Lee P D 2008 *Acta Mater.* **56** 4328
[2] Salvo L, Cloetens P, Maire E, Zabler S, Blandin J J, Buffière J-Y, Ludwig W, Boller E, Bellet D and Josserond C 2003 *Nucl. Instrum. Meth. B* **200** 273
[3] Baruchel J, Buffière J-Y, Cloetens P, Di Michiel M, Ferrie E, Ludwig W, Maire E and Salvo L 2006 *Scripta Mater.* **55** 41
[4] Salvo L, Suéry M, Marmottant A, Limodin N and Bernard D 2010 *C. R. Phys.* **11** 641
[5] Terzi S, Salvo L, Suéry M, Limodin N, Adrien J, Maire E, Pannier Y, Bornert M, Bernard D, Felberbaum M, Rappaz M and Boller E 2009 *Scripta Mater.* **9** 449
[6] Madi K, Forest S, Boussuge M, Gailliègue S, Lataste E, Buffière J-Y, Bernard D and Jeulin D 2007 *Comp. Mater. Sci.* **39** 224
[7] Lorensen W-E and Cline H-E 1987 *Comput. Graphics* **21** 163
[8] Nielson G M, Huang A and Sylvester S 2002 *IEEE Visualization* (Boston) 459
[9] Hachem E, Kloczko T, Digonnet H and Coupez T 2010 Stabilized finite element solution to handle complex heat and fluid flows in industrial furnace using the immersed volume method *Int. J. Numer. Meth. Fl.* online http://dx.doi.org/10.1002/fld.2498
[10] Gruau C and Coupez T 2005 *Comput. Method. Appl. M.* **157** 115
[11] Schneider P and Eberly D 2002 *Geometric Tools for Computer Graphics* (Elsevier Science Inc., New York)
[12] Bruchon J, Digonnet H and Coupez T 2009 *Int. J. Numer. Meth. Eng.* **78** 980
[13] Sussman M, Smereka P and Osher S 1994 *J. Comput. Phys.* **114** 146
[14] Ville L, Silva L and Coupez T 2011 *Int. J. Numer. Meth. Fl.* **66** 324
[15] Coupez T 2011 *J. Comput. Phys.* **230** 2391
[16] Digonnet H, Silva L and Coupez T 2007 *Proc. NUMIFORM* (Porto) **208** 269
[17] Balay S, Buschelman K, Eijkhout V, Gropp W, Kaushik D, Knepley M, Curfman McInnes L, Smith B and Zhang H *PETSc Users Manual – Revision 3* (Mathematics and Computer Science Division, Argonne National Laboratory)
[18] Arnold D, Brezzi F and Fortin M 1984 *Calcolo* **23** (4) 337
[19] Babuška I 1973 *Math. Comput.* **27** (122) 221
[20] Saad Y 2003 *Iterative Methods for Sparse Linear Systems* (SIAM)
[21] Brooks A N and Hughes T J R 1982 *Comput. Method. Appl. M.* **32** 199
[22] Saad Y and Schultz M H 1986 *SIAM J. Sci. Comput.* **7** (3) 856
[23] Fuloria D and Lee P D 2009 *Acta Mater.* **57** 5554

IOP Conf. Series: Materials Science and Engineering **33** (2012) 012055 doi:10.1088/1757-899X/33/1/012055

Simulation of distortion and residual stress in high pressure die casting – modelling and experiments

P Hofer[1], E Kaschnitz[1], P Schumacher[1,2]

[1]Österreichisches Gießerei-Institut, Parkstraße 21, 8700 Leoben, Austria
[2]Lehrstuhl für Gießereikunde, Montanuniversität Leoben, Franz-Josef-Straße 18, 8700 Leoben, Austria

E-mail: peter.hofer@ogi.at

Abstract. Two individual high-pressure die-casting geometries were developed in order to study the influence of process parameters and different alloys on the distortion behaviour of castings. These geometries were a stress lattice and a V-shaped sample tending to form residual stress due to different wall thickness respectively by a deliberate massive gating system. In the experimental castings the influence of the most important process parameters such as die temperature and die opening time and the cooling regime was examined. The time evolution of process temperatures was measured using thermal imaging. The heat transfer coefficients were adapted to the observed temperature distributions. Castings were produced from the two alloys AlSi12 and AlSi10MnMg. The distortion of the castings was measured by means of a tactile measuring device. For the alloy AlSi10MnMg thermo-physical and thermo-mechanical data were obtained using differential scanning calorimetry, laser flash technique, dilatometry and tensile testing at elevated temperatures. These data were used for modelling the material behaviour of the AlSi10MnMg alloy in the numerical model while for the alloy AlSi12(Fe) literature data were used. Process and stress simulation were conducted using the commercial FEM software ANSYS Workbench. A survey on the results of the comparison between simulation and experiment is given for both alloys.

1 Introduction

Increasing computing power of modern hardware systems opens new possibilities of simulating complex phenomena in the casting process as well as the overall simulation of the casting process. While the simulation of filling and solidification can be considered as state of the art, in numerical process simulation, the simulation of residual stresses and distortion needs to be improved in order to make more accurate predictions. This is due to a number of reasons; these include a lack of accurate input data for numerical simulation or still difficult to handle mechanical contact problems. A model approximating real casting conditions has to be chosen carefully in order to make accurate predictions. In this work trial castings with reproducible distortion behaviour were designed and produced in high-pressure die-casting (HPDC); the distortion of the final products were measured by tactile methods.

Thermo-physical and thermo-mechanical material properties were measured over a wide temperature range for the aluminium alloy AlSi10MnMg. Simulation models considering the process parameters and the material behaviour were developed.

These simulation models only contain parameters which were measured directly or were derived from measurements and are perfectly free from fit parameters without any physical meaning. The results of the simulations are compared to the distortion of the trial castings. A critical review of the model assumptions used in this work and a short preview on research tasks to further improve the experimental design and the simulation techniques are given.

2 Theory of stress prediction in casting processes

Stress simulation in casting processes is a highly complex problem consisting of several tasks. The correct prediction of temperature and solidification, as well as an accurate material model is the basis to simulate stress and distortion. These phenomena are induced by shrinkage constraints in the mould as well as the geometry of the casting part including the gating system. The thermal, mechanical and material models are not independent but can show strong interaction.

Temperature T as a function of time t is calculated by the solution of the heat transfer equation

$$\frac{\partial}{\partial t}\left(\rho c_p T\right)=\nabla\cdot\left(\lambda\nabla T\right)+\Delta h_f\left(\frac{\partial f_S}{\partial T}\right)\left(\frac{\partial T}{\partial t}\right) \tag{1}$$

where ρ is density, c_p is heat capacity, λ is thermal conductivity, Δh_f is latent heat of fusion and f_s is solid fraction [1, 2].

As thermal strains are calculated from the thermal model, every stress and distortion simulation can be only as accurate as the calculated temperature distribution. The mechanical model describes the evolution of thermal (and mechanical) strains due to thermal (and mechanical) forces. The components of the total strain $\vec{\varepsilon}^{tot}$ are described by *Eq.(2)*, the relation between thermal stress and thermal strain is given by *Eq.(3)*,

$$d\vec{\varepsilon}^{tot} = d\vec{\varepsilon}^{el} + d\vec{\varepsilon}^{(visco-)plas} + d\vec{\varepsilon}^{th} + d\vec{\varepsilon}^{E(T),v(T)}, \text{ and} \tag{2}$$

$$d\sigma = (\bar{D}^{el} - \bar{D}^{plas})\left(d\vec{\varepsilon}^{tot} - d\vec{\varepsilon}^{th}\right) - d\sigma_0, \tag{3}$$

where $d\vec{\varepsilon}^{el}$, $d\vec{\varepsilon}^{(visco-)plas}$, and $d\vec{\varepsilon}^{th}$ are elastic, (visco-)plastic, and thermal components of strain. The component $d\vec{\varepsilon}^{E(T),v(T)}$ is caused by the temperature dependence of Young's Modulus and the Poisson ratio. In *Eq.(3)* stress σ and strain are connected via the stiffness and plasticity matrix $\bar{D}^{el}, \bar{D}^{plas}$; σ_0 is a virtual component that takes into account the temperature dependency of the yield stress in the work hardening model [1].

3 Experimental

The geometry of the stress lattice and the V-shape for the trail castings were designed using a commercial numerical casting simulation tool (MAGMAsoft) in order to optimize filling, solidification and the evolution of reproducible distortion before machining the high-pressure die-casting tool. The stress lattice consists of a thick middle bar and two thin side bars which are connected by two beams perpendicular to them. The V-shape consists of a thin-walled symmetric lid and an arbitrarily massive gating system, in order to induce well measurable distortion in the lid. Because of the different cooling rates due to differences in wall thickness between certain casting areas (stress lattice) or between casting and gating (V-shape), the geometry is deformed plastically which leads to residual stress and global distortion of the casting part. Photographs of a stress lattice and a V-shaped Lid are shown in figure 1.

All experimental castings where performed using a commercial HPDC-machine (Bühler SC D/53, 530 kN closing force). In a series of experimental castings the die temperature, the closing time (the time between the injection of the melt and the ejection of the casting) of the die and the cooling regime of the sample after ejection (water quenching or natural convection) were varied systematically. Two alloys AlSi12(Fe) and AlSi10MnMg were used in order to compare results for different alloys. A list of the selected parameters is given in Table 1. The process parameters were measured and logged for the use in process modelling.

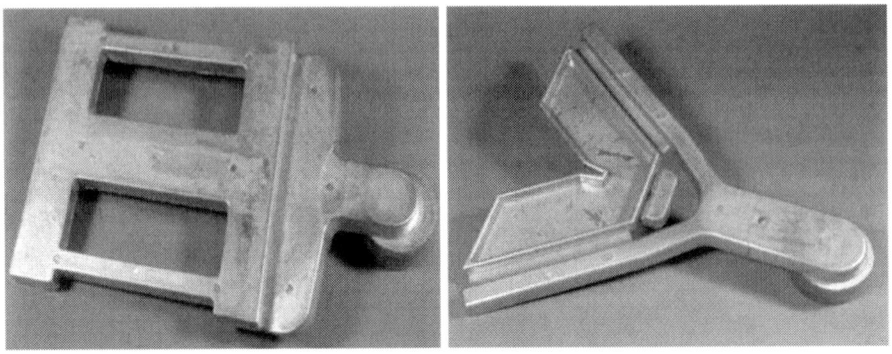

Figure 1. Stress lattice (left) and V-shaped (right) casting produced by high-pressure die-casting.

Table 1. List of varied parameters in experimental castings

		Stress lattice		V-shape	
Alloy	Die temperature [°C]	Closing time 20 sec.	Closing time 30 sec.	Closing time 5 sec.	Closing time 8 sec.
AlSi10nMg	160	x		x	x
	180	x		x	x
	200	x			
	230	x	x		
	260	x	x		
AlSi12(Fe)	140	x			
	160	x		x	x
	180	x		x	x
	200	x		x	x
	230	x	x		
	260	x			

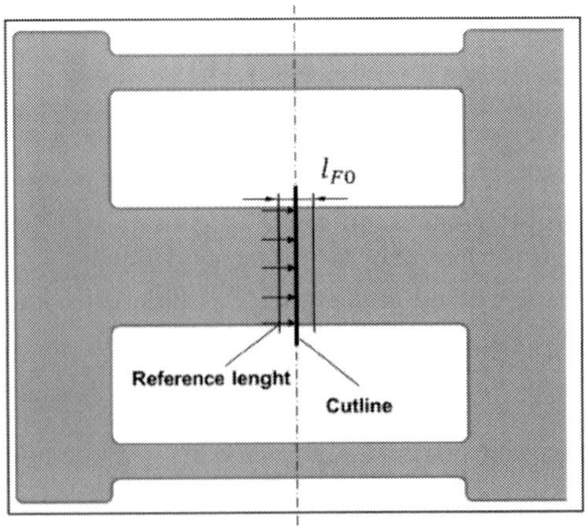

Figure 2. Evaluation of the distortion of a stress lattice by cutting of the middle bar.

Both the tool and the casting temperatures were obtained by thermal imaging using an infrared camera system (FLIR ThermaCam E45). The cooling curves of the castings (after applying a thin graphite layer in order to achieve an emmitance close to one) were measured by an infrared camera system (FLIR ThermaCam A320). The temperatures of the castings at the time of ejection were obtained by extrapolation using a polynomial fit function.

After cooling of each individual casting to ambient temperature its distortion was measured. A sketch of the stress lattice is given in figure 2. Global distortion of the stress lattice was measured by comparing the differences of a defined reference length l_{F0} before and after cutting the middle bar of the stress lattice. The residual stress was obtained by applying a simple elastic material behaviour according to Hooke's law. A similar stress lattice produced in sand casting was analysed by Fendt [3].

In figure 3 a sketch of a V-shaped lid is shown. Global distortion was evaluated by determining the distance between two selected points of the geometry. The procedure of the evaluation can be described as follows: The distance a_0 is the distance of the reference points at the die at room temperature. The distance a_1 is the distance of the same reference points at the casting at room temperature. As T_{die} is the die temperature set by the oil tempering system and T_a is room temperature, the distance of the reference points in the die at the time of ejection is given by:

$$a_{die,ej} = a_0 \cdot [1+(T_{die} - T_a) \cdot \alpha_{die}], \tag{4}$$

where α_{die} is the coefficient of thermal expansion the die material. As the casting shrinks to the die during cooling and the distance of the reference points of the casting $a_{cast,ej}$ is equal to the distance of the reference points in the die $a_{die,ej}$. As T_{cast} is the temperature of the casting at the time of ejection and α_{cast} is the coefficient of thermal expansion the casting material, the shrinkage of the casting Δl_{cast} can be calculated by:

$$\Delta l_{cast} = \left(T_{cast} - T_a\right) \cdot \alpha_{cast}. \tag{5}$$

The elongation of the distance between the reference points $\Delta a_{0\text{-}1}$ is the global distortion of the geometry caused by the residual stress and can be calculated be *Eq. 6*

$$\Delta a_{0\text{-}1} = a_1 \cdot (1+\Delta l_{cast}) - a_{die,ej}. \tag{6}$$

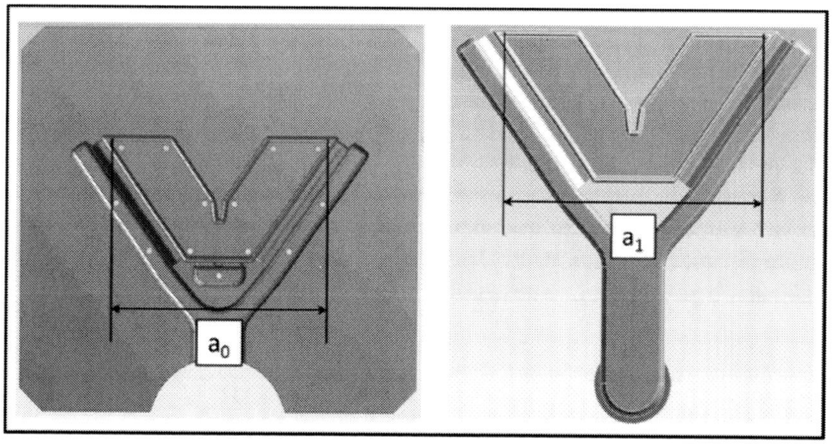

Figure 3. Evaluation of the elongation of a V-shaped bowl

Figure 4 shows a sketch of the V-shape before and after cutting the casting system. The elastic recovery of the lid is calculated by *Eq 7*:

$$\Delta a_{ela} = a_1 - a_2. \tag{7}$$

The quality of numerical simulation depends strongly on the accuracy of the input data. For the alloy AlSi12(Fe) existing literature data were judged to be sufficient. The alloy AlSi10MnMg had to be fully characterized by thermo-physical and thermo-mechanical measurements using differential-scanning-calorimetry, laser-flash technique, dilatometry and tensile testing at elevated temperatures.

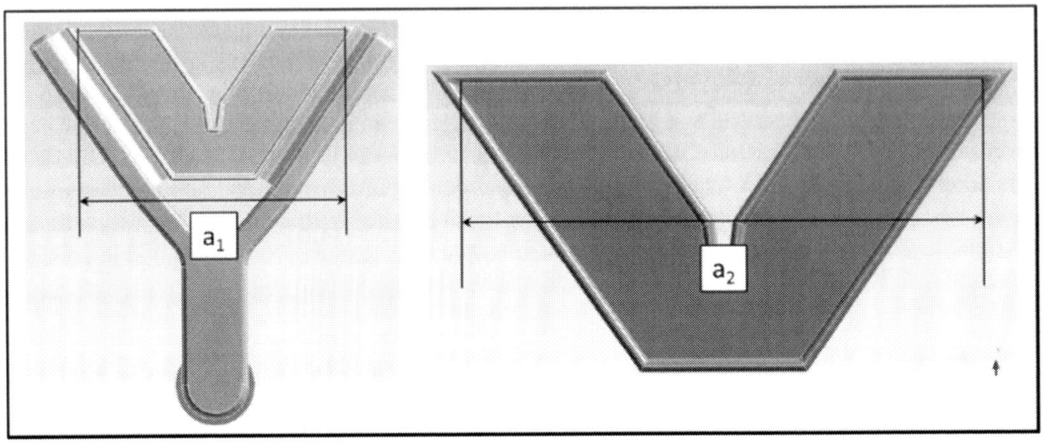

Figure 4. Evaluation of the elastic recovery of a V-shaped lid due to machining.

4 Modelling and Simulation

The transient temperature distribution, the displacement and the thermal stress were simulated with a commercial software package (ANSYS Workbench 11.0 SP1) using an FEM code. Boundary and initial conditions were taken from process data acquired during the production of the castings. The heat transfer coefficients between casting and die as well as between casting and cooling medium were obtained by adjusting the temperature simulation to the results of thermal imaging. The heat transfer coefficient between die and casting as well as the heat transfer coefficient between casting and air were considered to be temperature-dependent. The temperature loss during mould filling and the thermo-mechanical interaction between casting and mould (gap formation) had to be disregarded for the reasons of the complexity and stability of the contact modelling as well as computing power restrictions. As the friction coefficient between die and casting has been unknown the mechanical contact between die and casting was modelled as frictionless contact; there is no shear stress on the contact surfaces. The ejection forces were not taken into account for the same reasons.

The material behaviour was considered to be isotropic and homogeneous over the entire casting. For the plastic material behaviour a work hardening (multilinear-isotropic) model was used. The start of plastic flow was estimated by analysing the stress-strain curves. The plastic flow was considered to start when the stress-strain curves shows a deviation of more than one percent from linear slope. This tends to be more accurate than the yield stress ($R_{p0,2}$) which has no physical meaning. The die was modelled elastically, because plastic stress components in the die are restricted to a very thin surface layer.

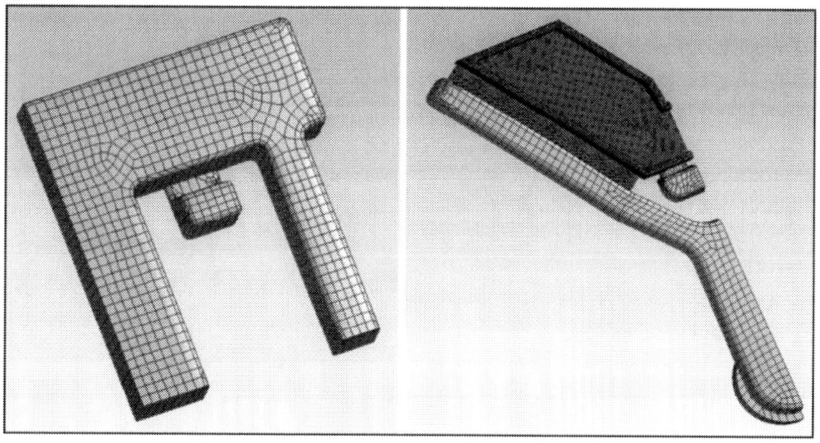

Figure 5. Geometry and enmeshment of the stress lattice (left) and the V-shape (right) for the calculation of the temperature and displacement fields.

The calculated temperature fields were applied as thermal loads in the mechanical simulation. In order to achieve finer enmeshment the stress lattice was cut along its two symmetric planes (neglecting the influence of the gating system, which is very small according to initial simulation results). The V-Shaped lid was cut along the axis of symmetry along the middle of the geometry. Figure 5 shows the meshed geometry of stress lattice and V-shape.

5 Results

Figure 6 shows the comparison between the measured and simulated strain of the stress lattice for different cooling regimes as a function of die temperature for both cast alloys. As the differences in the distortion values for different closing times (due to the low thermal conductivity of the mould material restraining the heat flow) are not significant the results in figure 6 are only shown for a closing time of 20 seconds. However the solidification of the casting is completed within 12 to 15 seconds for the whole range of different die temperatures. Experiments with incomplete solidification of the casting at the time of ejection could not be performed for technical reasons. The strain in air cooled castings shows a minimum at a die temperature of 230° C. This might be due to the relation of the average temperature level of the casting at the time of ejection and the temperature difference between the middle and the side bars. The influence of the cooling regime after ejection is high and leads to strains about twice as high for samples quenched in water than cooled by natural convection to ambient temperature. Comparing the results of AlSi10MnMg and AlSi12(Fe), the sensitivity for distortion is highly influenced by the material properties of the cast alloys. As the AlSi12(Fe)-alloy has a lower thermal expansion and a higher yield stress than the AlSi10MnMg-alloy, the measured and calculated strain values are significantly lower.

Figure 7 shows the comparison between the measured and simulated strain of the V-shaped lid for different closing times as a function of die temperature for both cast alloys. The elongation shows a slight increase according to the temperature of the die and a slight decrease according to the closing time for both alloys. The influence of the cooling regime after ejection (not shown) is smaller than for the stress lattice. The elongation values for both alloys are in the same range. Figure 8 shows the comparison of the elastic recovery of the V-shaped lid for different closing times as a function of die temperature for both cast alloys. For the alloy AlSi12(Fe) there is only little influence for both, the closing time and die temperature in the measurements. For the alloy AlSi10MnMg this influence is higher. The elastic recovery for AlSi10MnMg is higher than for AlSi12(Fe). This might be due to the higher yield stress and the lower temperature at the time of ejection.

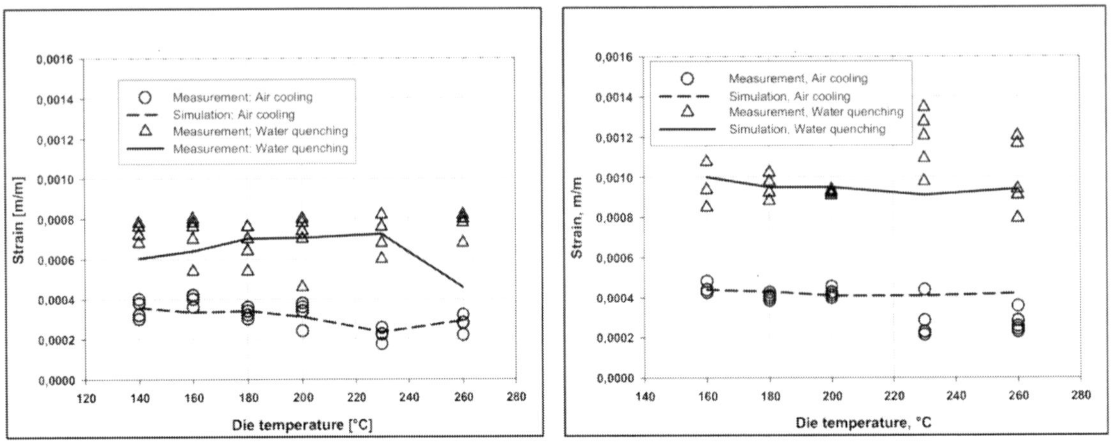

Figure 6. Comparison of measured and calculated elastic strain in the middle bar of a stress lattice cast in HPDC (20 seconds closing time) for the alloys AlSi12(Fe) (left) and AlSi10MnMg (right).

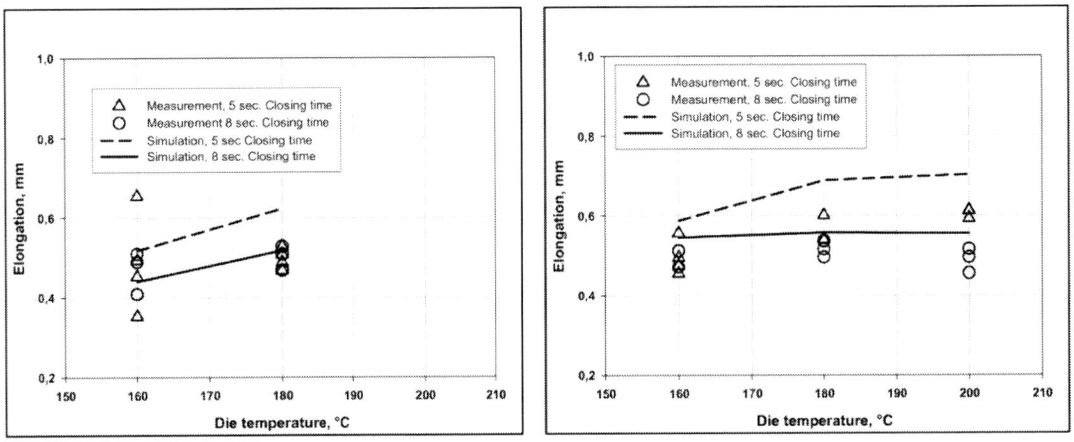

Figure 7. Comparison of measured and calculated elongation of a V-shaped lid at cast in HPDC (air cooled after ejection) for the alloys AlSi12(Fe) (left) and AlSi10MnMg (right).

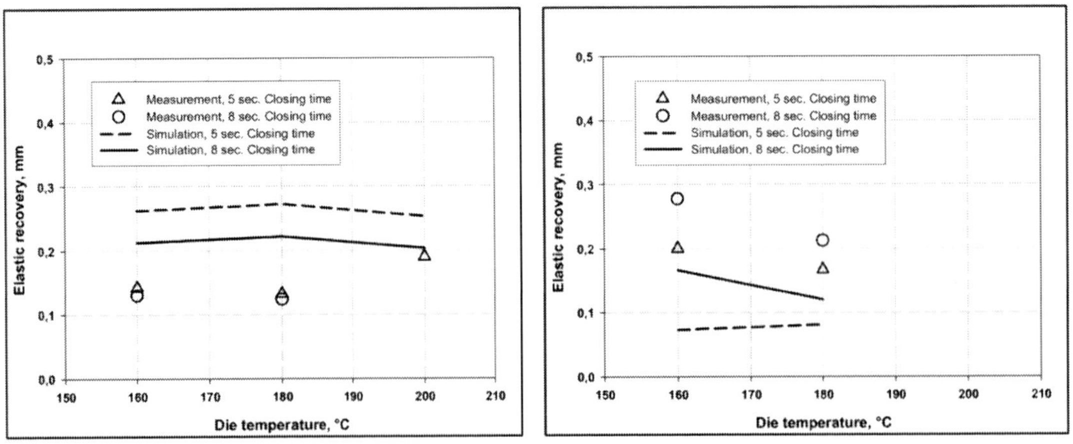

Figure 8. Comparison of measured and calculated elastic recovery of a V-shaped lid at cast in HPDC (air cooled after ejection) for the alloys AlSi12(Fe) (left) and AlSi10MnMg (right).

6 Discussion

The results of the experimental castings show that there are differences between the distortion behaviour of the stress lattice and the V-shape according to the chosen process parameters. The stress lattice is more sensitive to the cooling regime after ejection than the V-shape because of the increase of temperature difference between side and middle bars caused by water quenching. This temperature difference is the most important driving force for the distortion of the stress lattice. As cooling in the die is restricted by the relatively low thermal conductivity of the die material and because of the large thermal mass of the stress lattice, the influence of the die temperature pre-set by the oil tempering system is quite small. The V-shape is more sensitive to the die temperature and the closing time. High die temperatures and short closing times lead to higher temperatures at ejection which lead to higher deformation due to a lower yield stress. An influence of the alloy was obtained for both the stress lattice and the V-shape.

The simulation of the distortion in the stress lattice shows a good agreement between simulation and measurements for all process parameters and for both alloys. The results of the simulation of the elongation for the V-shape show also good agreement with the measurements. The agreement of the simulated and measured elastic recovery is quite poor. The reason seems to be the estimation of the start of plastic flow especially at elevated temperatures. A parameter study showed that a variation of twenty percent in the yield stress in the simulation model leads to an increase of about twenty percent in elastic recovery but only three percent in the elongation before machining. Further investigations are presently carried out to elaborate these phenomena.

Although the simulation of residual stress shows a good agreement with the results from experimental castings further research is needed. This is due to the following reasons:

- all material properties were considered to be homogenous over the entire casting
- mechanical and thermal contact has to be improved
- the mould was not considered to be deformed plastically
- the mechanical material behaviour in the initial solidification range had to be estimated due to the lack of data for that range.

The results of this work can be used for the process simulation of more complex castings. The identification of critical process parameters can help foundrymen for better planning the casting process of castings which tend to the formation of distortion.

Acknowledgement

This work was supported by the Österreichische Forschungsförderungsgesellschaft mbH (FFG),Vienna, Austria and the Steirische Wirtschaftsförderungsgesellschaft mbH (SFG), Graz, Austria.

References

[1] Ludwig A 2002 *Int. J. Thermophys.* **23** 1131-46
[2] Kaschnitz E 2005 *Gießerei-Rundschau* **52** 176-8
[3] Fendt A, *Einfluss der Wärmebehandlung auf den Eigenspannungszustand von Aluminiumgussteilen* 2000 (Technische Universität München: Munich, Germany)
[4] Hofer P, Kaschnitz E 2008 *Gießerei-Rundschau* **55** 190-3
[5] Hofer P, Kaschnitz E, Schumacher P 2010 *Gießerei-Rundschau* **57** 182-6
[6] Hofer P, Kaschnitz E, 2011 *High-Temp. High-Press.* 311-22

IOP Conf. Series: Materials Science and Engineering **33** (2012) 012056 doi:10.1088/1757-899X/33/1/012056

Modelling of shrinkage cavity defects during the wheel and belt casting process

S Dablement[1], D Mortensen[2], H Fjaer[2], M Lee[3], J Grandfield[4], G Savage[5] and V Nguyen[5]

[1] Nexans Research Centre Lens, Boulevard du Marais, BP29, 62301 Lens, France
[2] Institute for Energy Technology, P.O.Box 40, N-2027 Kjeller, Norway
[3] Nexans Research Centre Lens, 207 Sunshine Road, Tottenham, Victoria, 3012, Australia
[4] Grandfield Technology Pty Ltd, Brunswick West, Victoria, Australia
[5] CAST Cooperative Research Centre, CSIRO, Clayton, Victoria, Australia

Email: sebastien.dablement@nexans.com

Abstract. Properzi continuous casting is a wheel and belt casting process used for producing aluminium wire rod which is essential to the making of electrical cables and over head lines. One of the main concerns of Properzi process users is to ensure good quality of the final product and to avoid cast defects especially the presence of shrinkage cavity. Numerical models developed with the Alsim software, which allows an automatic calculation of gap dependent heat transfer coefficients at the metal-mould interface due to thermal deformation, are used in order to get a better understanding on the shrinkage cavity formation. Models show the effect of process parameters on the cavity defect development and provide initial guidance for users in order to avoid this kind of casting defect.

1. Introduction

Production of aluminium wire rod with a wheel and belt casting process, like the continuous Properzi process used in Nexans (figure 1), consists of pouring liquid aluminium in a mould in rotation via a tundish. The mould is composed of a grooved wheel made of copper alloy which is partially surrounded by a steel belt for containing the liquid metal. Wheel and belt are water cooled with a spray system for assuring the solidification of the liquid metal. The product thus obtained is a solid bar of trapezoidal cross section which is hot rolled in line for forming a round section wire. The aluminium wire rod is finally packaged as a coil of several tons.

As well as productivity, the main concerns of Properzi process users are to limit the rate of wasted rod wire and to assure good quality of the final product. Wheel and belt casting is a complex process where the control of each parameter and operators' know-how are fundamental for producing a good quality product. It is difficult or almost impossible to detect casting defects in line especially when the defect is at the core of the cast bar. Internal defects such as macro-segregation, porosity or cavity can be only detected after production and with destructive testing.

The main casting defect occurring is shrinkage cavity defect (figures 2 and 3) which leads to scrap an important quantity of coils for avoiding quality problem during drawing process and causes lost sales opportunity. Previous modelling studies of the wheel and belt casting process took into only account heat flow, and only two-dimension geometry are used when mechanical aspect is treated, and furthermore no one treats defect solving [1-7].

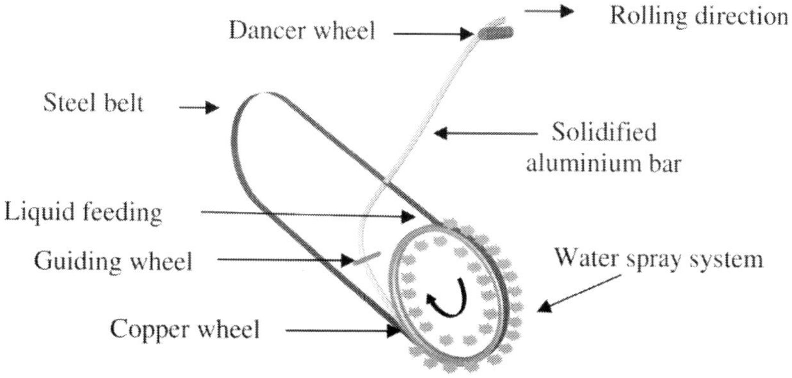

Figure 1. Schema of the Properzi process

In this study, numerical modelling has been used to better understand effect of process parameters on the shrinkage cavity formation in order to solve this issue. Alsim software has been used for developing thermo-mechanical models of the Properzi process with the advantage of calculating automatically the gap dependent heat transfer coefficients. More details are given on the development of the Alsim software in [8]. A two-dimension thermo-mechanical model has been developed for describing the case where cavity defects can be due to a too long liquid pool and a three-dimension thermo-mechanical model has been used when shrinkage cavity formation can be due to the formation of liquid metal pockets.

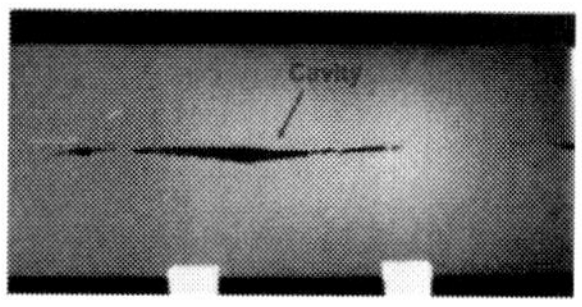

Figure 2. X-ray image of the longitudinal cross-section of a cast bar with shrinkage cavity defect

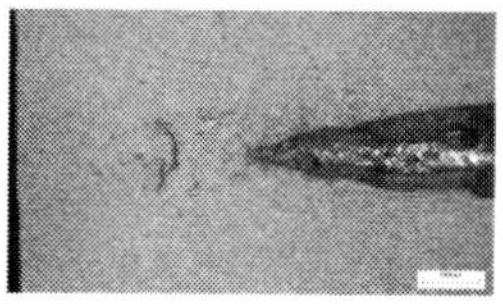

Figure 3. Metallographic image of a cast bar with shrinkage cavity defect

2. Modelling

2.1. Solved equations

In this work, the Alsim software has been used to develop the models (the 2D model includes heat and stress, the 3D model includes heat, fluid flow and stress). The Alsim software includes heat and fluid flow and stress and is based on a continuum mixture model for the solid-liquid material. A Darcy force is used in the mixture momentum equations. Thermal convection is included by the Boussinesq approximation. The turbulence is modeled by a low Reynolds number (LRN) k-e model after Launder and Sharma. The permeabilty of the mush is dependent on the morphology of the mush, it is given by the Kozeny-Carman equation.

Additional capability has been added to the software to enable full thermo-mechanical modelling of the air gap formation between the bar and the wheel and belt. The heat transfer coefficient between the bar and the wheel and belt is calculated according to the gap dimension and the air thermal conductivity i.e. a fully coupled thermo-mechanical analysis. An initial contact value is used.

For stresses and deformations, the basic equation in the stress modelling is Cauchy's equation for the momentum balance where the inertia terms are neglected:

$$\nabla \cdot \boldsymbol{\sigma} = \rho \mathbf{g} \tag{1}$$

The material is treated as an elastic-viscoplastic material for which plasticity and creep are treated in a uniform manner. The model formulation has been based on the classical small deformation theory where the total strain e, derivable from the displacement field u is subdivided into viscoplastic, elastic and thermal components.

$$\boldsymbol{\varepsilon} = \frac{1}{2}\left(\nabla \mathbf{u} + [\nabla \mathbf{u}]^{t}\right) = \boldsymbol{\varepsilon}^{vp} + \boldsymbol{\varepsilon}^{e} + \boldsymbol{\varepsilon}^{T} \tag{2}$$

The thermal strain component can be written as an integral of the (volumetric) thermal expansion coefficient β_T.

$$\boldsymbol{\varepsilon}^{T} = -\frac{1}{3}\int_{T}^{T_{sol}} \beta_T(T)dT \cdot \mathbf{I} \tag{3}$$

The constitutive equation for elasticity (Hook's law) defines how the stress depends on the elastic strain ε^{e} and the temperature dependent Young's modulus and Poisson's ratio entering the matrix D.

$$\boldsymbol{\sigma} = \mathbf{D}(T) \cdot \boldsymbol{\varepsilon}^{e} \tag{4}$$

A set of equations assuming steady state creep above a temperature T_0 and work hardening below this temperature, was applied here:

$$\bar{\sigma} = F(T) \cdot (\phi_0 + \phi)^{n(T)} \cdot \left(\dot{\bar{\varepsilon}}_p\right)^{m(T)} \tag{5}$$

$$d\phi = \begin{cases} d\bar{\varepsilon}_p & \text{when } T \le T_0 \\ 0 & \text{when } T > T_0 \end{cases} \tag{6}$$

Thermo-mechanical coupling: in the treatment of the mechanical problem in casting, the temperature development represents the major load, or the cause of stresses, due to inhomogeneous thermal contractions. The temperature field also influences the local response of this load through the temperature dependency of the elastic and viscoplastic mechanical properties. On the other hand, the resulting stresses and displacements, affect the thermal boundary conditions regarding contact zones or air gaps between ingot and mold. On the boundaries between two different materials the heat transfer coefficient α is calculated according to:

$$\alpha = Min\left(\alpha_{Contact}, \alpha_{gap}\right) \tag{7}$$

Where the gap heat transfer coefficient α_{gap} is calculated by:

$$\alpha_{gap} = \frac{1}{\dfrac{1}{\alpha_{conduction} + \alpha_{radiation}} + \dfrac{1}{\alpha_{contact}}} \tag{8}$$

Where the local displacement dependent gap value d_{gap} is used to calculate the conduction heat transfer through the gap $\alpha_{conduction}$, and the radiation part $\alpha_{radiation}$ is found from the local temperatures and emissivities on the two boundaries. The heat transfer coefficient representing the contact between the materials, $\alpha_{contact}$, was a constant quantity imposed on the calculation (dependent on lubrication, surface roughness etc). Equation (8) is a parallel coupling between two heat transfer mechanisms (conduction and radiation) which is then serial coupled with the surface heat resistance. α_{gap} approaches infinity as d_{gap} approaches zero, so $\alpha_{contact}$ is used instead of α_{gap} for small or zero gap-values (according to equation 7).

More information about solved equations is given in [9-11].

2.2. Material properties

For both models the material properties are as following:

For the aluminium cast bar the alloy AA1350 was modelled. Thermal properties were taken from the ThermProp package. The fluid properties published in [11] were applied. A thermo-elastic-viscoplastic formulation is used where the mechanical properties F(T), n(T) and m(T) functions in equation (5) were taken from [12] with adjustment. The work hardening temperature T0 in equation (6) was equal to 360°C, and the initial accumulated strain parameter φ0 in equation (5) was equal to 10E-4.

The wheel is made of silver-copper alloy and the belt is made of steel. A thermo-elastic formulation is used for the wheel and the belt. The thermal and mechanical properties of the wheel and the belt are given in table 1.

Table1. Termal and mechanical properties of the wheel and the belt

Property	**Wheel (Cu-Ag)**	**Belt (steel)**
Density (kg/m^3)	8,940	7,900
Conductivity (W/m.K)	377	48
Specific Heat (J/kg.K)	385	510
Young Modulus (MPa)	125000	200000
Poisson's ratio	0.34	0.29
Thermal Expansion Coefficient	16.5E-6	11.0E-6

3. 2D thermo-mechanical model and shrinkage cavity defect due to deep liquid pool

3.1. Shrinkage cavity defect due to deep liquid pool

Users and manufacturers of wheel and belt casting processes assume that a too deep liquid pool leads to the formation of shrinkage cavity defects.

In the case of the copper wheel and belt casting, as mentioned by R Adams in [13], when the liquid pool is too long, the gravity can pull the molten core back down out of the centre of the cast bar (figure 4). No more precision is given, but with the principle of the hydrostatic equilibrium, this phenomenon is possible only if the liquid pool goes above the tundish where the liquid is poured at the entry of the process. Another reason can be that the longer the liquid pool is the more difficult it is to feed the mushy zone.No value regarding the maximum length of the liquid pool is formally mentioned in any references but it is advisable that the bar is totally solidified before it exits the wheel and to be really sure it's better before the 180 degree angle.

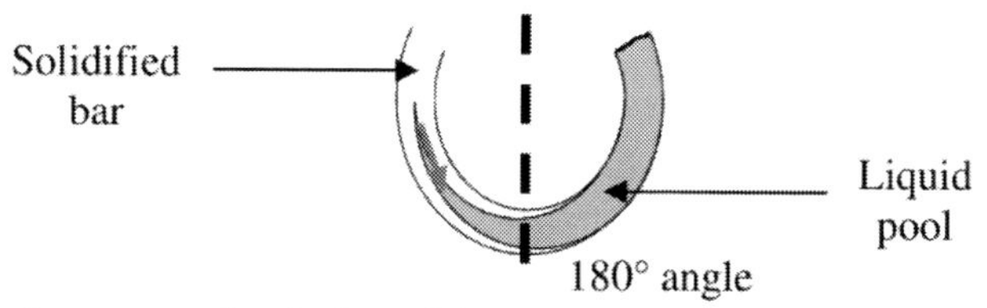

Figure 4. Figure of a deep liquid pool

3.2. 2D thermo-mechanical model

A two-dimensional thermo-mechanical model is sufficient for estimating when the aluminium bar is totally solidified according to the value of process parameters.

The geometry consists of a transversal cross-section of the wheel and belt casting process with a symmetrical axis and a meshing with a compatible interface is used (figure 5).

The initial temperature of the wheel and the belt is set at 50°C and that of the bar at the casting temperature (Tcast)

A fixed heat transfer coefficient is applied to the wheel and the belt for the water cooling with ambient temperature fixed at 30°C (cooling HTC). When the bar is in the mould, a gap dependent heat transfer coefficient is calculated according to the calculated gap dimension and the air thermal conductivity, an initial and constant contact value is used (contact HTC). Leaving the wheel (at the 210 degree angle) the bar is air cooled and a constant heat transfer coefficient of 10 W/m².K is used. A normal and tangential distributed stiffness dependent respectively on the normal and tangential displacement is applied to the bar-mould interface. A fixed distributed normal and tangential stiffness is used for the belt-wheel interface and for the outer part of the wheel which is clamped onto the rotating equipment.

There is an assumption on the total strain in the third direction. The total strain out of the plane is zero (the sum of the thermal strain, the elastic strain and the viscoplastic strain is zero, but the components are not zero)

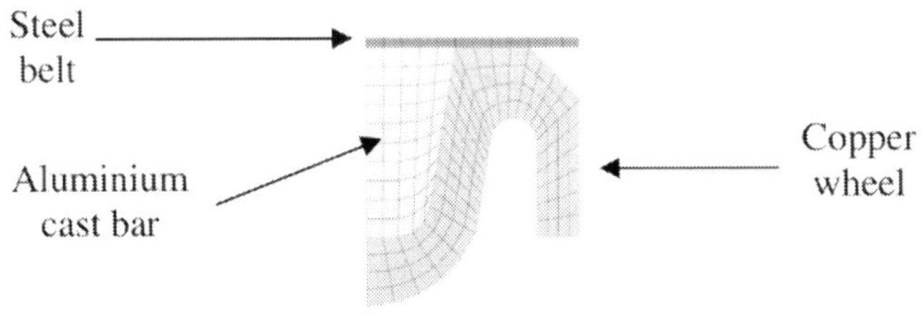

Figure 5. Geometry and mesh of the Properzi 2D model

3.3. Model results for normal conditions

Normal conditions for the Properzi process are defined with the following parameter values in the model (assumed to represent normal operating conditions of the process):

- casting temperature = 710°C
- rotation speed = 1.77 RPM
- cooling HTC = 8,000 W/m².K
- contact HTC = 4,000 W/m².K

The model predicts that the cast bar is totally solidified after a calculating time of 15 seconds i.e. at 160 degree angle, therefore with these conditions the cast bar should not have any shrinkage cavity defect. Some results of this case are presented in the following figures.

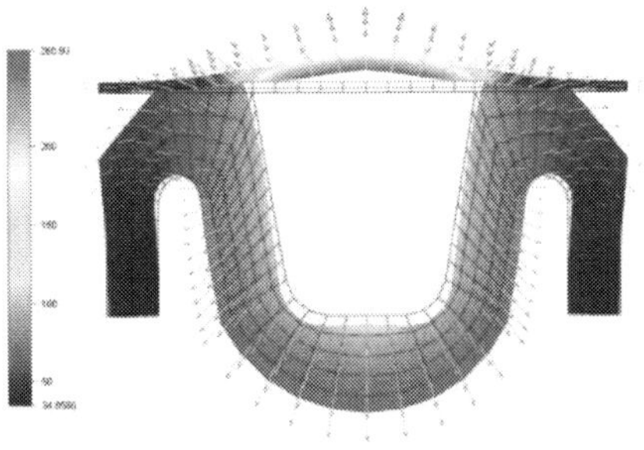

a. Mold temperature (°C) + magnified deformed geometry + undeformed mesh

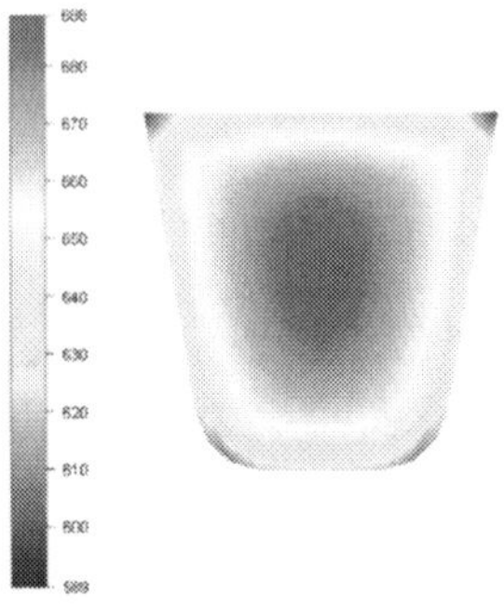

b. Temperature field of the bar (°C)

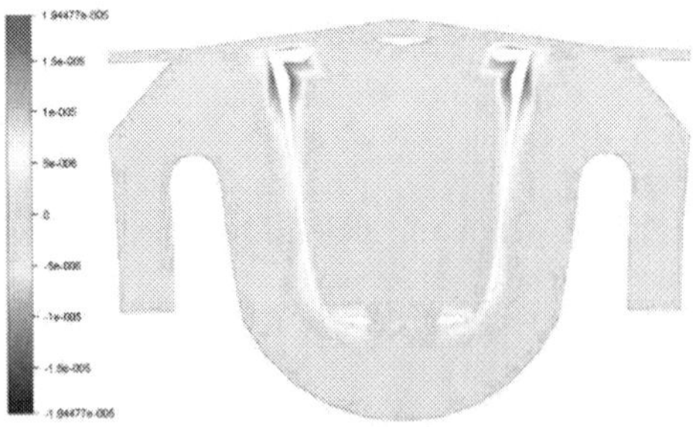

c. Air gap size (m) + magnified deformed geometry

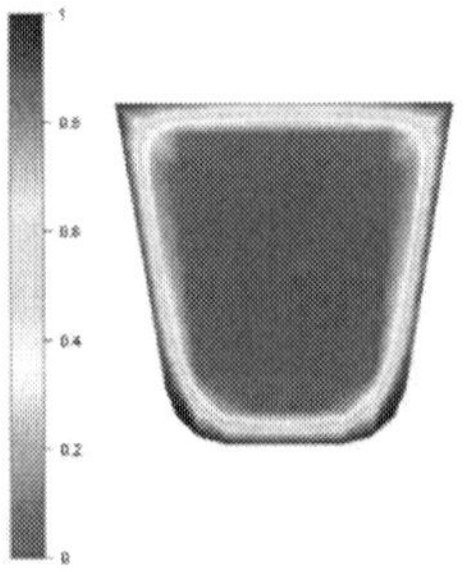

d. Solid fraction of the bar (1 = totally solidified)

Figure 6. Model results after 1.9 second calculating (20 degree angle)

3.4. Model results: effect of process parameters on the depth of the liquid pool

The model can predict the effect of each parameter on the length of the liquid pool. In the following calculations when a process parameter is studied as a variable, the other parameter values are set as in normal conditions.

As illustrated in figure 7a, the casting temperature has a poor effect on the length of the liquid pool. Indeed, even if the casting temperature is set at 780°C (i.e. 70 degrees above the casting temperature of the normal conditions) the solidification terminates at 167 degree angle.

Figure 7b shows that the rotation speed has an important effect on the length of the liquid pool and a minimal speed of 2 RPM is necessary in order that the solidification terminates after 180 degree angle.

The water cooling heat transfer coefficient starts to have an effect on the shrinkage cavity formation below a value of 2,000 W/m².K. (Figure 7c).

Figure 7d shows that the heat transfer coefficient of contact between the mould and the bar has to be below a value of 2,500W/m².K for that the solidification terminates after a 180 degree angle.

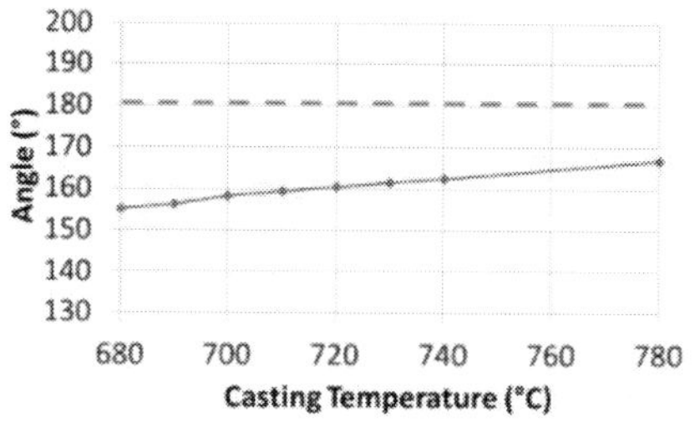

a. Casting temperature effect

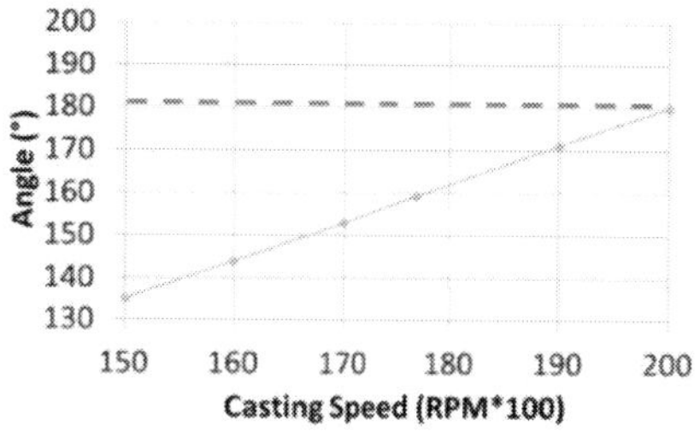

b. Rotation speed effect

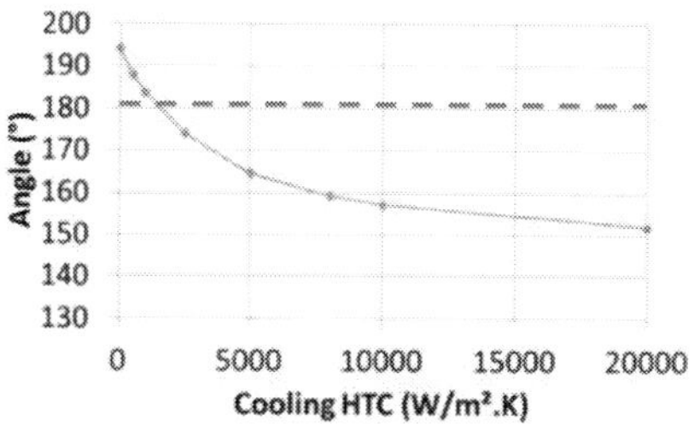

c. Cooling HTC effect

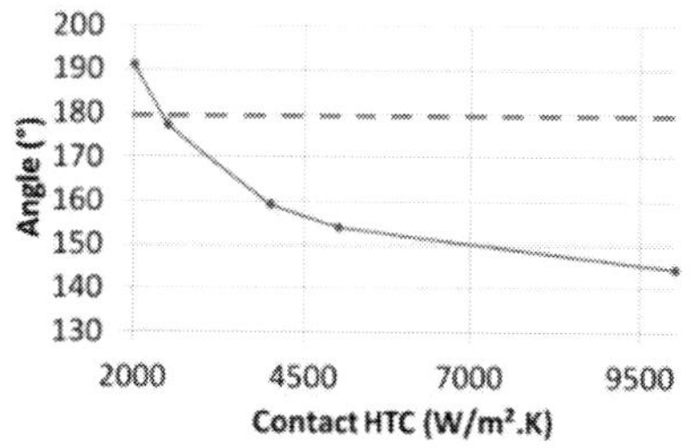

d. Contact HTC effect

Figure 7. Model results: effect of parameters on the depth of the liquid pool (in degree angle)

The two-dimension thermo-mechanical model of the Properzi process predicts some results which allow operators to get the effect of parameters on the shrinkage cavity formation into the as-cast bar. The casting temperature has almost no effect on cavity defect and this is explained by the fact that the most important part of heat to be extracted is still the latent heat of solidification even if the casting temperature is high. The rotation speed has an important effect, but in any event, it should not be the cause of the shrinkage cavity defect because it is a well-controlled process parameter and having a speed variation of 13% seems impossible. Decrease of the cooling heat transfer coefficient has an effect. Consequently, during production all phenomena which lead to decrease the water cooling have to be checked, e.g.: failure of the water spray system or scale build-up on the mould. Decrease of the contact heat transfer coefficient has an effect too. All phenomena leading to decrease heat transfer between the bar and the mould have to be checked, e.g.: bad roughness of the mould, dirt build-up on the mould or a too thick coating deposited on the mould.

4. 3D thermo-mechanical model and shrinkage cavity defect due to liquid metal pocket formation

4.1. Shrinkage cavity defect due to liquid metal pocket formation

Shrinkage cavity defect formation can occur when the heat transfer between the metal and the mould is unsteady. Indeed, in this case the liquid pool of solidification is an imperfect V-shaped sump and causes the formation of liquid metal pockets (figure 8). And as the liquid metal is surrounded by solidified metal, the feeding of metal is impossible to compensate shrinkage when liquid metal solidifies. This leads to the formation of a cavity called shrinkage cavity defect.

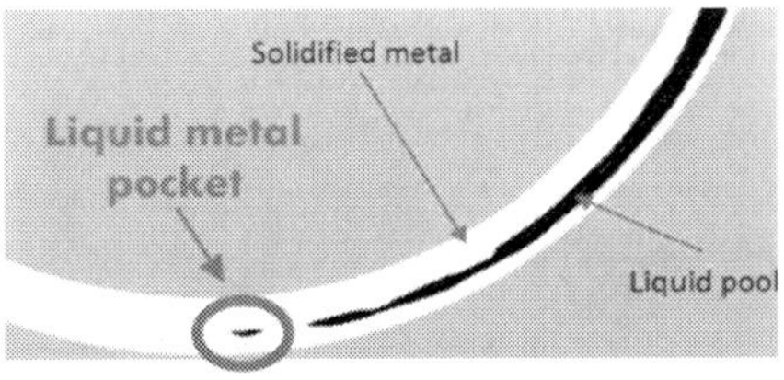

Figure 8. Figure of liquid metal pocket formation into the bar due to a non-homogeneous V-shaped sump

The two dimension thermo-mechanical model does not allow studying phenomena which could lead to have an unsteady heat transfer between the metal and the mould. A three dimension thermo-mechanical model has been used for studying cases where liquid metal pocket formation can be possible.

4.2. The 3D thermo-mechanical model

The three dimension model used here is an improved version of the model as described in a previous study [14]. In this new version the belt is totally taken into account and that allows reproducing efforts induced by the tension and the pressure wheels which ensure the good positioning of the full belt during process. This version better calculates contact between the belt and the cast but it also allows predicting the thermal and the mechanical cycles of the belt which will be useful for determining the belt life. This updated model includes the geometry of the guide wheel (which guides the bar at the exit of the wheel) and the geometry of the dancer wheel (which supports the bar and assures the speed regulation between the casting wheel and the rolling) in order to be closer to the reality and to better calculate the bar path from the casting wheel to the rolling (Figure 9).

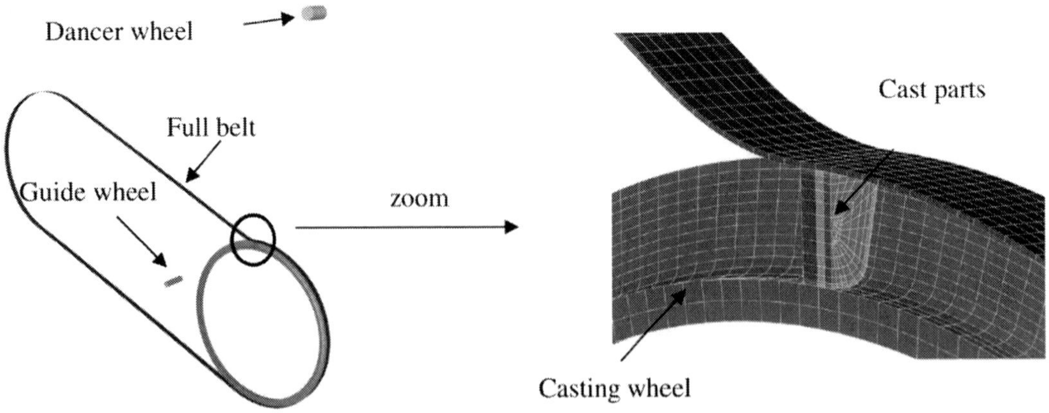

Figure 9. Geometry and mesh of the Properzi 3D model
(Cast parts: red=Euler domain, yellow=Expansion zone and green=Lagrangian domain)

The initial temperature of the wheel and the belt is set at 50°C and the inlet temperature for the bar at 710°C.

A fixed heat transfer coefficient is applied to the wheel and the belt for the water cooling with ambient temperature fixed at 30°C (cooling HTC). When the bar is in the mould, a gap dependent heat transfer coefficient is calculated according to the calculated gap dimension and the air thermal conductivity, an initial and constant contact value is used (contact HTC). Leaving the wheel (at the 210 degree angle) the bar is air cooled and a constant heat transfer coefficient of 10 W/m².K is used. A normal and tangential distributed stiffness dependent respectively on the normal and tangential displacement is applied to the bar-mould interface. A fixed distributed normal and tangential stiffness is used for the belt-wheel interface and for the outer part of the wheel which is clamped onto the rotating equipment.

The wheel, the bar and the belt are all assumed to move at the casting speed i.e. no friction or sticking is accounted for.

4.3. Model results for normal conditions

Normal conditions for the Properzi process are defined with the following parameter values in the model (assumed to represent normal operating conditions of the process):

- casting temperature = 710°C
- rotation speed = 1.77 RPM
- cooling HTC = 8,000 W/m².K
- contact HTC = 4,000 W/m².K

The model predicts that the cast bar is totally solidified at 174 degree.

4.4. Model results: effect of process parameters on the shrinkage cavity formation

Many assumptions could be made in order to try to explain the formation of liquid metal pockets which lead to the formation of shrinkage cavity defects. Several hypotheses have been tested with the 3D model in order to check if they had an impact on the heat transfer between the metal and the mould.

One of these assumptions is that a fluctuating heat transfer of contact between the liquid metal and the mould, due to a non-constant thickness of coating during process for example, could lead to the formation of liquid metal pockets. In order to test this case, all parameters are set as normal conditions except values of the contact heat transfer coefficient between the bar and the casting wheel. On the inner surface of the wheel (in contact with the bar) two boundary identifiers (Bid) have been defined in order to set a different value of the contact HTC as illustrated in figure 10.

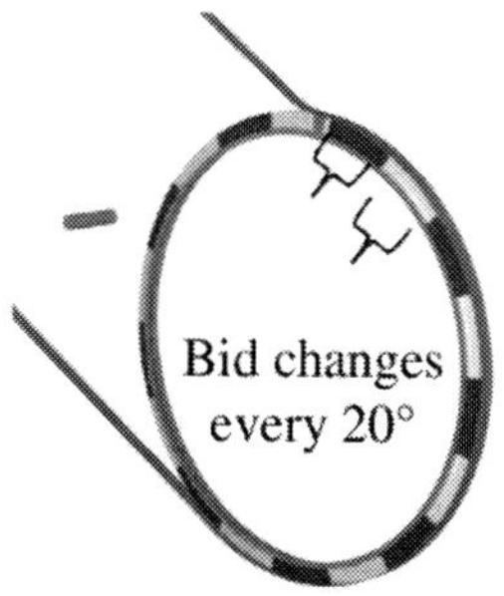

<u>Blue Bid:</u>
Constant contact HTC=4,000W/m².K

<u>Yellow Bid:</u>
Contact HTC = 2,000W/m².K when metal is liquid
Contact HTC = 4,000W/m².K when metal is solid

Figure 10. Definition of the contact HTC value between the bar and the wheel

Thus, assuming that the contact HTC value depends only on the coating deposed on the casting wheel, the model can predict the effect of a non-constant thickness of coating during process.

As illustrated in figure 11, a coating thickness of the casting wheel which changes during processing leads to the formation of liquid metal pockets which result in shrinkage cavity defects after total solidification.

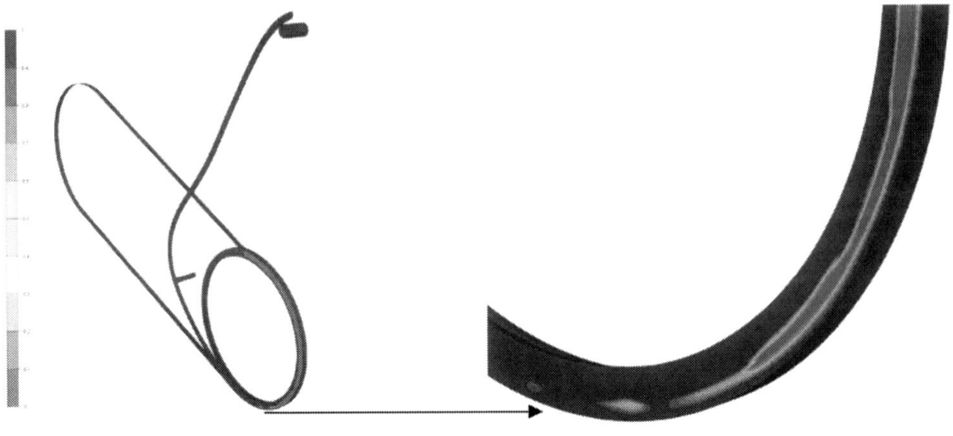

Figure 11. Liquid metal pocket formation due to non constant coating (blue is for totally solidified part)

The 3D thermo-mechanical model can predict the effect of the coating quality deposited on the mould. Depending on the nature of the coating, the model can determine the maximal variation of the coating thickness which is acceptable for avoiding the presence of shrinkage cavity defects during Properzi process.

5. Conclusions

Thanks to the Alsim software Nexans has in its possession a comprehensive model of its Properzi continuous casting process by taking into account the casting wheel, the real path of the bar and the full belt. With a full thermo-mechanical modelling of the air gap formation between the bar and the mould, the model calculates itself the heat transfer coefficient at this interface. Numerical modelling, in combination with the operators' know-how, allows Nexans to have a better understanding on the effect of process parameters on the shrinkage cavity defect formation which is impossible to detect during production. By improving the quality of the cast bar, Nexans can reduce scrap and produce better aluminium wire rod.

References

[1] Mohapatra S P, Sahoo S K, Nanda S, Palchaudhary A and Patnaik S C 2010 Thermo mechanical modelling of bar casting *Journal of Advanced Research in Mechanical Engineering* **1** 173

[2] Mohapatra S P, Sahoo S K, Nanda S, Hembram P, Palchaudhary A and Patnaik S C 2010 Numerical simulation of aluminium bar casting for wire rod production *Journal of Scientifif and Industrial Research* **69** 913

[3] Cottignies L, Duhoux V, Blais S and Duran C 2009 Heat transfer during rod casting *The minerals, metals and materials Society* 877

[4] Cheung N, Santos N S, Quaresma J M V, Dulikravich G S and Garcia A 2008 Interfacial heat transfer coefficients and solidification of an Al alloy in a rotary continuous caster *International Journal of Heat and Mass Transfer* **52** 451

[5] Lindholm D , Sørheim E, Rødscth J, Mortensen D and Heggland H 2002 Mathematical modeling of wire rod casting *The minerals, metals and materials Society* 651

[6] Liu J C, Anyalebechi P N, Scherbak M and Bachowski R 1991 Coupled thermomechanical model for continuous bar casting *The minerals, metals and materials Society* 229

[7] Davies V L and Westby O 1977 Numerical computation of the solidification of continuously cast aluminum rod *Nonferrous Wire handbook* 197

[8] Mortensen D, Fjær H, Lindholm D, Rudshaug M and Sørheim E 2011 The development of

Alsim – a modelling tool for direct chill casting, twin roll casting, wheel and belt casting and chain conveyor casting *Materials Science Forum* **693** 187

[9] Mortensen D, Henriksen B R,M'Hamdi M, Hallvard and Fjær G 2008 Coupled modelling of air-gap formation and surface exudation during extrusion ingot DC-casting *The minerals, metals and materials Society* 773

[10] Mortensen D 1999 A mathematical model of the heat and fluid flows in direct-chill casting of aluminium sheet ingots and billets *Metallurgical and Materials Transactions* **30B** 119

[11] Fjær H and Mo A 1990 ALSPEN – A mathematical model for thermal stresses in DC casting of aluminium billets *Metallurgical and Materials Transactions* **21B** 1049

[12] Haaften V M 2002 Constitutive behaviour and hot tearing during aluminium DC casting PhD-Thesis ISBN 90-9015792-1

[13] Adams R 2002 Cast bar quality basics *SCR Users Meeting*

[14] Grandfield J, Dablement S, Fjaer H, Mortensen D, Lee M, Nguyen V and Savage G 2011 3D thermo-mechanical modelling of wheel and belt continuous casting *Materials Science Forum* **693** 235

IOP Conf. Series: Materials Science and Engineering **33** (2012) 012057 doi:10.1088/1757-899X/33/1/012057

Simulating the deformation of dies in the foundry

A Chabod, Y Longa, J M Dracon, K Chailler, P Hairy, A Da Silva

C.T.I.F: Centre Technique des Industries de la Fonderie, Sèvres, France

E-mail : chabod@ctif.com, longa@ctif.com, dracon@ctif.com, chailler@ctif.com, hairy@ctif.com, dasilva@ctif.com

Abstract. Digital simulation (QuikCAST, ProCAST) is already used extensively when designing metallic dies for founding, in particular to design filling and gating systems. Simulation of the steady-state temperature cycles of dies has also been mastered. With large castings, the temperature gradient induced between the moulding surface and the rear surfaces of the die leads to deformations that may be large enough to measure, and incompatible with the required dimensional accuracy. The temperature gradient also creates thermal fatigue stresses that cause crazing of the die surface. In the study conducted by CTIF, aimed at measuring tooling deformations, various ways of measuring displacements at high temperatures (with and without contact) were investigated in order to evaluate their capabilities and limitations. An experimental device was designed - a test bench combining a metallic die having a simple geometry, in which an aluminium part could be cast, and instrumentation (temperature and displacement sensors). The deformations of the die were measured during first cycles of temperature homogenization. Concurrently, thermomechanical calculations were carried out on the same geometry using PROCAST. The calculation results are well correlated with the experimental measurements and validate the tools and the calculation methods. This thermomechanical approach makes it possible to optimize die design in the foundry and to predict high-temperature deformations as early as the design stage. Knowledge of these deformations makes it possible in turn to anticipate the geometrical and dimensional variations undergone by the castings themselves and so to improve their accuracy. The designer can act on the temperature of the die or the design of the casting, or create a die in which the expected thermal deformation is reversed so as to produce a casting having the correct dimensions. In short, thermomechanical simulation can be applied to this problem to achieve a better understanding of the phenomena and to test solutions in advance.

1. Introduction

The die design process in gravity die casting makes extensive use of simulation for the design of filling systems. Simulation of the steady-state temperature cycles of the dies has also been mastered. Shorter cycle times require temperature regulation (cooling and heating) by a heat transfer fluid (water or oil). For large castings, the thermal gradient induced between the moulding surface and the rear surfaces of the die causes deformations that may become large enough to measure and incompatible with dimensional accuracy requirements [5][6]. The goal of this study was to design an experimental test bench capable of detecting measurable levels of deformation, then use it to validate a thermomechanical simulation system.

2. Design of the test bench

2.1. Design of the device

The moulding part comprises a large diameter 400 mm and relatively thin (20mm) die made of X38CrMoVa 5.3 tool steel, designed to maximize thermomechanical deformations (Figure 1). The test bench has a table, with 4 height adjustment jacks, on which the die is placed flat, leaving access for the measurements, in particular a clear field of view for the image analysis measurements. The die is centred on a spindle sliding in a ring force-fitted in the die. Handling rings are used to manipulate the die and the table as a whole. Supports position the sensors. There are holes in the die for the thermocouples. A bracket screwed to the die acts as aiming mark for filming vertical displacements. The origin for the positioning of the sensors is the centre of the axis of the tooling.

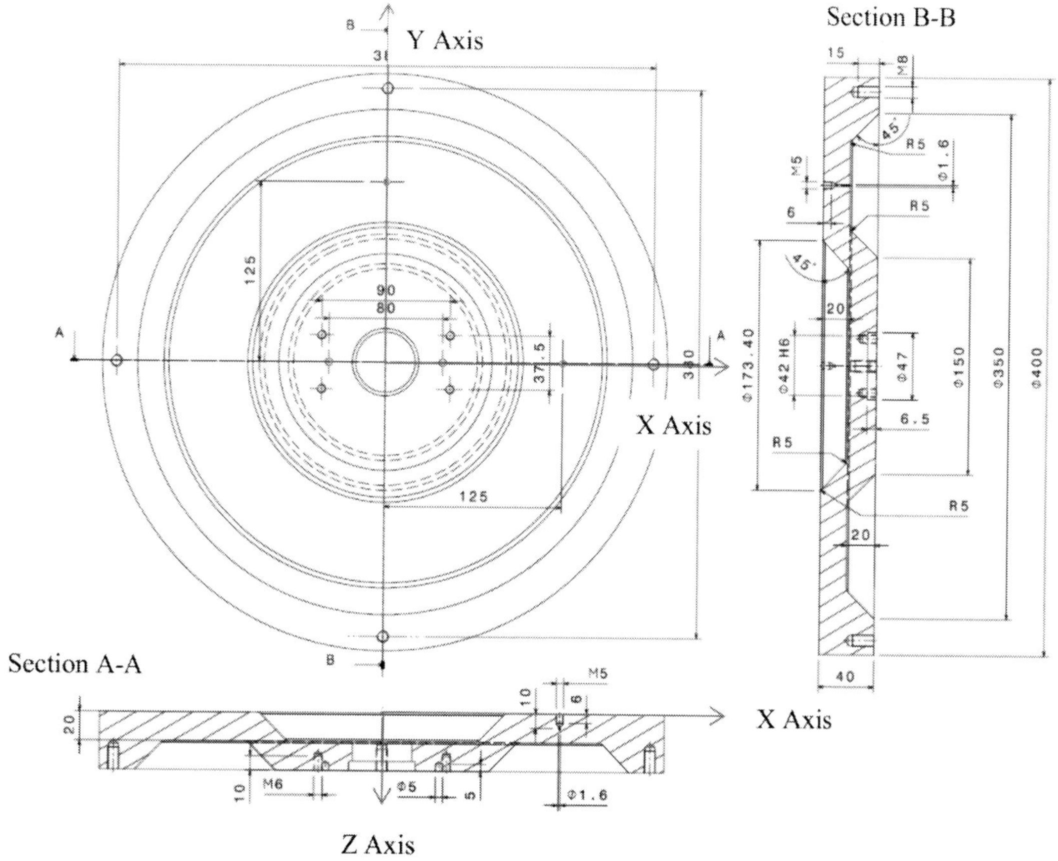

Figure 1. Design of the moulding.

Table 1. Positions of the sensors.

Sensor	X [mm]	Y [mm]	Z [mm]
LVDT Z sensor	50	0	20
LVDT R sensors	+/-200	0	20
Thermocouple	125	0	10

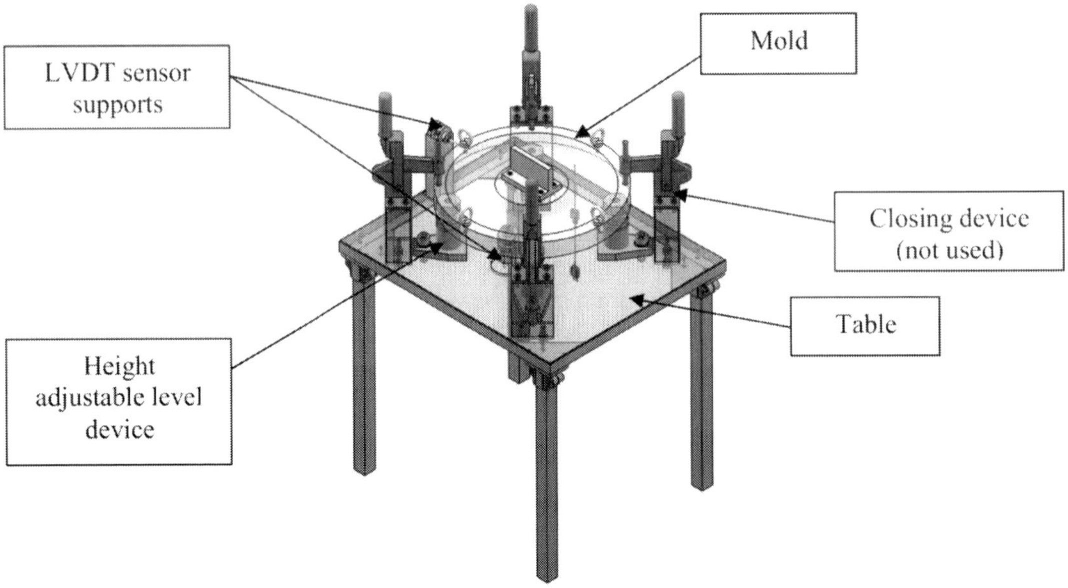

Figure 2. Experimental test bench.

The experimental bench is made of a heavy weight table, on which the mold is put on 4 supports whose level is height-adjustable (Figure 2). The level of the mold is controlled to ensure the horizontal position during pouring.

2.2. Measuring resources LVDT sensors
The measurement of displacements by LVDT inductive sensors has been described in an earlier CTIF publication [1]. These sensors are metallic, allowing use at high temperature, and their measurement precision is +/- 0.005mm (observed background noise).

2.2.1. Displacement measurement by image analysis
The technique consists in tracking the evolution of a speckle pattern on a surface by comparing successive images. Digital image correlation is widely used for displacement measurements and strain field analysis on surfaces both at ambient temperature [2] and at very high temperature [4]. We chose Matlab software with a plugin dedicated to measurement by image analysis [2]. The advantage of this technique is that acquisition is simply by digital camera (still or movie), without contact with the hot metallic die, and the digital processing can be done later. In order to validate the method, we fed precisely known input to the MATLAB analysis image software. For this, we used an image with a speckle pattern, which we stretched artificially by 10 pixels using ImageJ image processing software. We obtained a series of images simulating a tensile test with a displacement of 10 pixels and a strain of 4.1%, the size of the image being 244 pixels. The next step in the validation of the method consisted in filming tensile tests on flat steel specimens coated with a speckle paint. This step was an opportunity to employ the whole image analysis measurement system: the application of a speckle paint, filming of the tests, processing of the files, and image analysis.

The measurement results are as follows.

Table 2. Results for elongation at failure in tensile tests, by image analysis (digital movie and still cameras) and by conventional measurement per NF EN ISO 6892-1.

	Still camera	Movie camera	Test
Specimen 1	27.8%	29.9%	28.60%
Specimen 2	28%	29.3%	27.36%
Specimen 3	9.8%[a]	30.1%	28.16%

The elongation results measured by image analysis are very close to the values measured according to standard NF EN ISO 6892-1.

Although the measurement is contactless, this technique is not accurate for our equipment, because of oxidation of the measurement surfaces, convective movements of hot air perturbing the stability of the image, and, to a lesser extent, the conversion of the video files.

3. Instrumented pourings

3.1. Experimental protocol

This paragraph shows the instrumented experimental pourings performed on the tooling. The test bench includes means of measuring the radial displacement R and vertical displacement Z, using LVDT inductive sensors. A thermocouple is positioned through a drilled steel bolt, with thermal properties comparable to mold material.

Figure 3. Experimental pouring.

The operations necessary for the pouring of five AlSi7Mg0. 6 aluminium alloy castings lasted 5,000 seconds (approximately 1h20min). The spoon is gauged to control the cast part weight (mean : 3.9 kg, scatter : +/- 100 g). The casting temperature of the alloy was 700°C, and the metallic tooling was initially at the ambient temperature of 20°C. The die/metal interface was covered by a 50 μm thick DYCOTE 39 die coating.

IOP Conf. Series: Materials Science and Engineering **33** (2012) 012057 doi:10.1088/1757-899X/33/1/012057

3.2. Correlations between calculation and tests

The thermomechanical simulations with PROCAST software (Figure 4) were intended to model the displacement and temperature fields in the die in order to compare the results with the instrumented tests. The alloy used for the cast was AlSi7Mg0.6 aluminium and the die is made of XC38CrMoVa5.3 tool steel. The calculation covered a filling phase, the shaking out of the casting at t0+300s, and the beginning of the next cycle at t0+1000s. The following boundary conditions are applied. A thermal resistance is used to model die coating between mold and cast part during filling and solidification and then its value is reduced to take into account convection with air after shaking-out. The rest of the surface of the mold is subjected to convection with ambient air. Symmetry of the geometry and loads enable to model only one quarter of the bench. We can see the central axis and the height adjustable device in mechanical and thermal contact with the mold. Only 3 cycles are taken into account because of convergence difficulties due to contact between the deformed mold and the height-adjustable device.

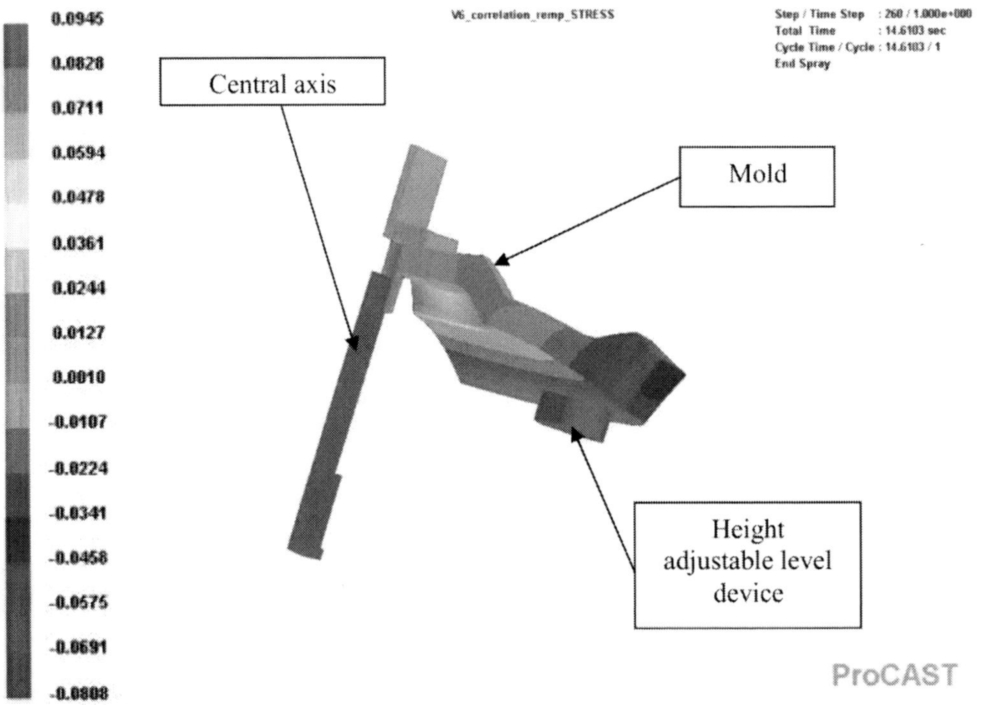

Figure 4. PROCAST thermomechanical calculations (sag in cm, deformation X50).

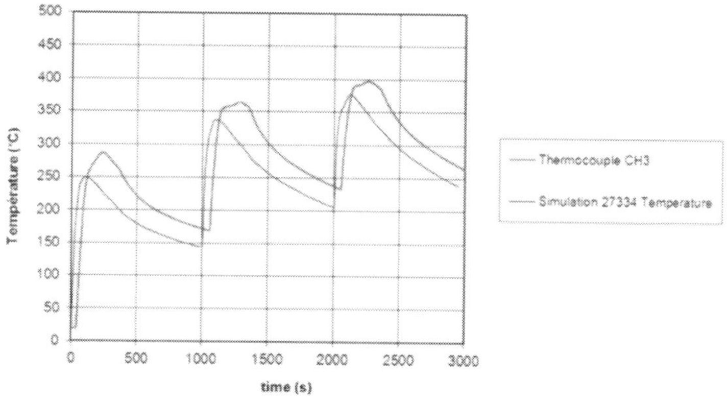

Figure 5. Comparison of the thermocouple measurements with the simulation.

The values obtained by simulation differ constantly, by approximately 25 °C, from the thermocouple measurements (Figure 5).

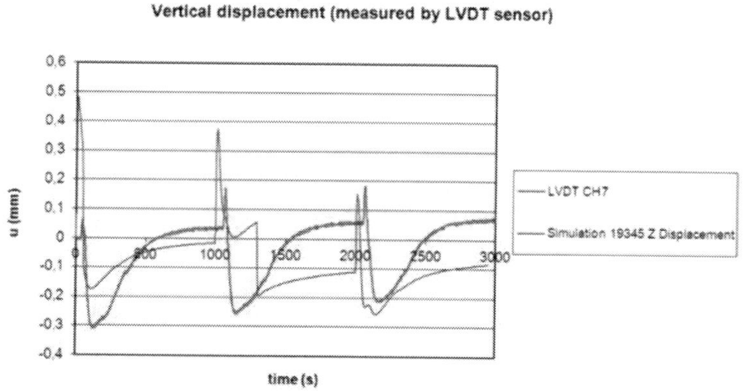

Figure 6. Comparison of the vertical displacements measured by LVDT sensors with the simulation.

The measured vertical displacements (Figure 6) are the same in shape and of the same order of magnitude in size as the measurements by LVDT sensors. We observe a first phase in which the die assumes a convex shape (sag strictly positive, cf. Figure 4), in the first instants of the cycle, where the heat input is intense, engendering swelling of the die surface. Then, the heat contributed by the alloy diffuses by conduction into the bottom part of the die, and at the same time the moulding surface cools substantially, in particular after shaking out. In the course of this second phase, the temperature is higher in the bottom part of the die, leading to a concave die shape (sag strictly negative). At the end of the cycle, the geometry of the die once again becomes convex.

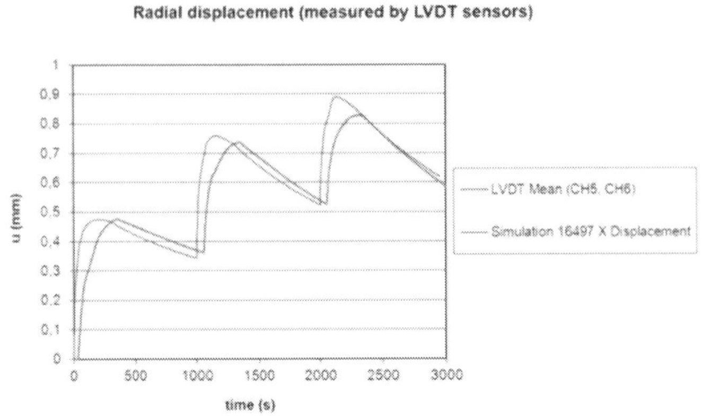

Figure 7. Radial displacements measured by LVDT sensors.

The radial displacement measurement is made using two LVDT sensors at diametrically opposite positions on the die (Figure 7). This arrangement eliminates the effects of any eccentricity of the die during the casting.

4. Conclusion

The study has led to the designing of a device that can detect deformations on a die having only one moulding part. The investigations showed that LVDT inductive sensors yield a measurement precision compatible with the amplitudes of displacement of the die. However, the measurement by image analysis was deeply perturbed by movements of hot air above the moulding surface. However, this technique has been validated at low temperature for tensile tests. The correlation between the LVDT sensor measurements and the thermomechanical simulations by PROCAST software is satisfactory.

References

[1] Journée F and Ogier G 2002 Mesure des déformations en cours de solidification *Fonderie Fondeur d'Aujourd'hui* **214** pp 10-23
[2] Eberl C 2006 Digital Image Correlation and Tracking by Christoph Eberl
[3] Hild F and Roux S 2005 Digital Image Correlation : From displacement measurement to identification of elastic properties *Strain* **42** pp 69-80
[4] Pradille C, Aime C, Ricort G, Bellet M and Chastel Y 2009 Méthode interspectre pour le suivi des déformations des aciers à chaud, 19ème Congrès Français de Mécanique Marseille
[5] Weiss K and Honsel C 2008 Stress in casting dies - simulation and practical example *Casting plant and Technology International* **02/2008** pp 28-33
[6] Rosbrook C and Kind R 1999 Casting process simulation of stresses and distortion in casting and dies *Die Casting Engineer* **September/October 1999** pp 72-76

IOP Conf. Series: Materials Science and Engineering **33** (2012) 012058 doi:10.1088/1757-899X/33/1/012058

Thermo-mechanical modeling of dendrite deformation in continuous casting of steel

J Domitner[1], J-M Drezet[2], M Wu[3,4] and A Ludwig[1,4]

[1] Christian Doppler Laboratory for Multiphase Modeling of Metallurgical Processes, University of Leoben, Franz-Josef-Straße 18, 8700 Leoben, Austria
[2] Computational Materials Laboratory, Ecole Polytechnique Fédérale de Lausanne, LSMX MX-G Ecublens, Station 12, 1015 Lausanne, Switzerland
[3] Christian Doppler Laboratory for Advanced Process Simulation of Solidification and Melting, University of Leoben, Franz-Josef-Straße 18, 8700 Leoben, Austria
[4] Chair of Simulation and Modeling of Metallurgical Processes, University of Leoben, Franz-Josef-Straße 18, 8700 Leoben, Austria

E-mail: josef.domitner@unileoben.ac.at

Abstract. In the field of modern steelmaking, continuous casting has become the major manufacturing process to handle a wide range of steel grades. An important criterion characterizing the quality of semi-finished cast products is the macrosegregation forming at the centre of these products during solidification. The deformation induced interdendritic melt flow has been identified as the key mechanism for the formation of centreline segregation. Bulging of the solidified strand shell causes deformation of the solidifying dendrites at the casting's centre. Hence, a fundamental knowledge about the solid phase motion during casting processes is crucial to examine segregation phenomena in detail. To investigate dendritic deformation particularly at the strand centre, a thermo-mechanical Finite Element (FE) simulation model is built in the commercial software package ABAQUS. The complex dendritic shape is approximated with a conical model geometry. Varying this geometry allows considering the influence of different centreline solid fractions on the dendrite deformation. A sinusoidal load profile is used to describe bulging of the solid which deforms the dendrites. Based on the strain rates obtained in the FE simulations the dendrite deformation velocity perpendicular to the casting direction is calculated. The velocity presented for different conditions is used as input parameter for computational fluid dynamics (CFD) simulations to investigate macrosegregation formation inside of a continuous casting strand using the commercial software package FLUENT.

1. Introduction

Nowadays, continuous strand casting is the main production process to manufacture a wide range of semi-finished steel products. To remain competitive, steelmaking companies must provide their continuous cast products (slabs, billets, blooms) at a desired quality on time, always keeping an eye on the actual production costs. This has become a critical factor of economic success during the past years and it will play an increasingly important role in the future. Hence, a detailed knowledge of the casting process is essential to guarantee stable operating conditions and to gain reproducible product specifications [1].

In particular the formation of macrosegregation [2-3] inside the cast products should be prevented as far as possible, since macrosegregation causes problems in further processing steps. Technical measures like mechanical softreduction (MSR) can reduce but not completely prevent this undesirable defect during casting [4-7]. Since solidification advances from the surface towards the centre, the segregated steel melt (enriched with alloying elements such as carbon, silicon, manganese, etc.) accumulates and finally solidifies at the strand centre. The accumulation is enhanced, if the shrinkage induced flow of unsegregated melt along the casting direction and therefore its mixing with the highly enriched melt is inhibited.

In slab casting, strand shell deformation (e.g. caused by bulging between adjacent guiding rolls of the casting machine) was identified as key mechanism to form positive centreline macrosegregation [8-11]. Strand surface bulging causes a periodic movement of the dendritic mushy zone inside the strand. The motion of the dendrites perpendicular to the casting direction is directly related to the deformation of the solid strand shell. However, if the dendrite trunks meet at the centre of the cast strand at the final stage of solidification, they will deform under compression.

Segregation formation is strongly influenced by relative motion between the growing dendrites and the interdendritic melt [12-14]. Hence, describing the deformation velocity of the dendrites at the strand centreline is essential to simulate macrosegregation formation. Miyazawa et al. proposed that the solid velocity component perpendicular to the casting direction decreases linearly towards the centre [8]. However, Mayer et al. and Domitner et al. used an exponentially decreasing velocity approach to consider that most deformation might occur at the dendritic tip region [11, 15]. To obtain deeper understanding about the actual deformation velocities, a thermo-mechanical Finite Element simulation model was implemented in the commercial software package ABAQUS.

2. Thermo-mechanical simulation model

2.1. Model geometry

Solidification inside of steel continuous casting strands leads to different complex-shaped dendritic morphologies. Generally, four characteristic morphology types can be distinguished: strictly oriented, mostly oriented, equiaxed and globulitic [9]. While globulitic and equiaxed dendritic solidification occurs mainly at the strand surface, oriented columnar solidification can be observed in regions between the surface and the strand centreline. Occasionally, down-sinking equiaxed crystals can be observed in the central region.

Since the current simulations are focused on investigating the deformation behaviour of columnar dendrites perpendicular to the casting direction, equiaxed crystals are not considered in the model. Besides, the complex columnar dendritic shape is approximated with a conical axisymmetric model geometry, as shown in figure 1. This simplified geometry represents a primary dendrite trunk growing perpendicular to the casting direction x and therefore parallel to the deformation direction y. The influence of secondary dendrite arms on the deformation behaviour is neglected. The model geometry has an entire length of l = 30 mm and an overall diameter of d = 1.2 mm. While d is based on a typical primary dendrite arm spacing observed at the centre region of a continuous casting strand [16-17], l is related to the local mushy zone width being estimated by numerical simulation of a 25.0 m long continuous casting strand [15].

The volume fraction of solid varies from the tip to the root of the primary dendrite. As shown in figure 1, the so-called "zero-strength" solid fraction, f_S^{zero}, corresponds to the upper boundary of the model geometry, whereas the centreline solid fraction, f_S^{cent}, corresponds to the bottom boundary representing the dendrite tip. The tip is assumed to be located at the centreline of the casting strand. In the performed simulations, dendritic deformation occurs only if $f_S < f_S^{zero}$, while the dendrite trunks are assumed to behave rigid if $f_S \geq f_S^{zero}$. The centreline solid fraction, f_S^{cent}, is calculated with

$$f_S^{cent} = f_S^{zero}\left(\frac{r^{cent}}{r^{zero}}\right)^2 . \tag{1}$$

In equation (1), r^{zero} stands for the radius of the modeled dendrite at f_S^{zero}. It is defined as

$$r^{zero} = \frac{d}{2} = \frac{l}{50}. \tag{2}$$

r^{cent} represents the dendrite's tip radius at the centre of the casting strand. In the performed simulations, five model geometries with r^{cent} = 0.1 mm, 0.2 mm, 0.3 mm, 0.4 mm and 0.5 mm are compared to investigate the influence of different f_S^{cent} on the deformation behaviour. These geometries are depicted in figure 2. The height of the conical region, h, depends on r^{cent}:

$$h = 50\left(r^{zero} - r^{cent}\right) \tag{3}$$

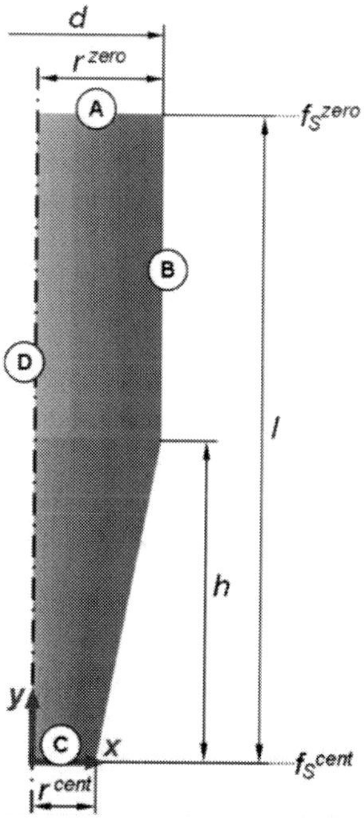

Figure 1. 2D axisymmetric model geometry used in the ABAQUS simulations. For visualization, a scaling of $x:y$ = 10:1 is applied. The letters A, B, C and D indicate the position of different boundary conditions.

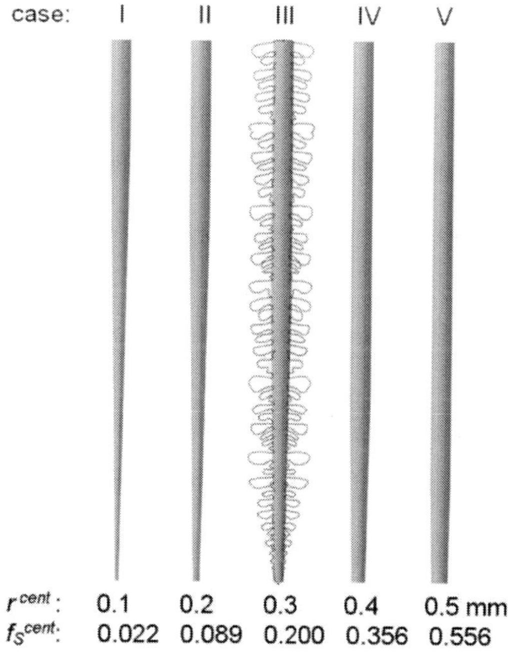

Figure 2. Different model geometries used in simulation cases I-V (actual geometry without scaling). For comparison, a dendritic contour including secondary arms is shown schematically at case III.

2.2. Boundary conditions

In figure 1, four boundaries ("A", "B", "C" and "D") are marked. At the upper boundary "A" a sinusoidal displacement is applied to deform the dendrite in y-direction. This load represents a single bulging cycle between two adjacent guiding rolls at a continuous casting plant. Figure 3 shows the time-dependent load profiles, displacement and corresponding solid velocity, $v_{S,y}^{zero}$. It is assumed that the maximum bulging amplitude decreases linearly from 0.8 mm at the exit of the continuous casting mould (strand coordinate X = 0.0 m, corresponding casting time T = 0.0 s) to 0.0 mm at the zone of strand end solidification (strand coordinate X = 22.5 m, corresponding casting time T = 1800.0 s). Hence, the bulging profile shown in figure 3 is obtained for X = 16.5 m where solidification and therefore the growing dendrites reach the strand centreline. Details about the bulging profile definition are given in [15].

Furthermore, no horizontal deformation in x-direction is allowed at the right boundary "B", whereas vertical movement in y-direction is hindered at the bottom boundary "C". To simulate a 2D axisymmetric problem, boundary "D" is treated as rotational symmetry axis.

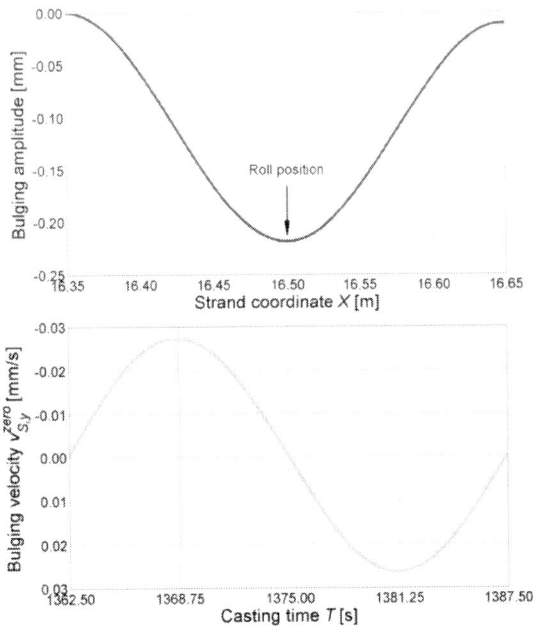

Figure 3. Load profiles used in all of the five simulation cases to deform the dendrites.

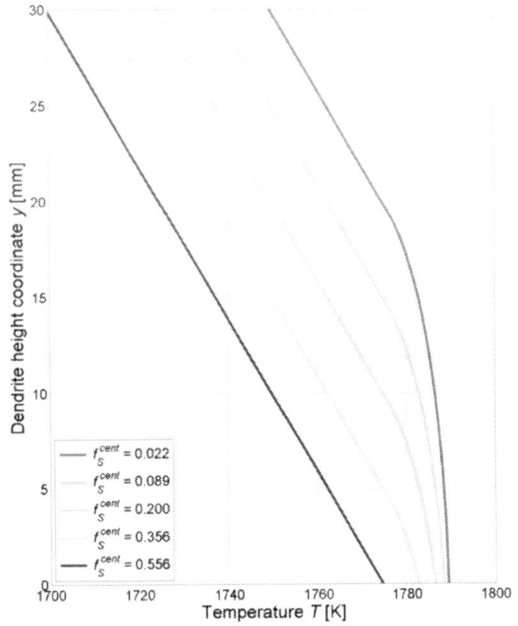

Figure 4. Solid fraction dependent temperature profiles for simulation cases I-V.

2.3. Numerical model

2.3.1. Thermal conditions

Since the mechanical properties are temperature dependent, the temperature varying only in y-direction within the dendrite is imposed in the ABAQUS simulations. Two temperature regimes are distinguished in the model. For $f_S < 0.5$ the temperature is calculated according to Scheil for a binary Fe-C-alloy, as given in equation (4). For $f_S > 0.5$ a linearly decreasing temperature is imposed using equation (5), where G is the uniform temperature gradient at the dendrite axis. Equations (4) and (5) approximate the temperature profiles which were obtained by modeling solidification inside a 25.0 m long continuous casting strand [15]. The data required to calculate the axial temperature profiles are given in table 1; the resulting temperature profiles used for simulation cases I-V are shown in figure 4.

$$T(f_S < 0.5) = T_m + C_{L,0}^{C}\, m\,(1 - f_S)^{k-1} \tag{4}$$

$$T(f_S > 0.5) = T(f_S = 0.5) - G\,(y - y(f_S = 0.5)) \tag{5}$$

Table 1. Data to calculate the temperature profiles

Melting temperature T_m	1790 K
Initial carbon concentration $C_{L,0}^{C}$	0.182 wt.%
Liquidus slope m	-116.7 K/wt.%
Distribution coefficient k	0.36
Temperature gradient G	2500 K/m

2.3.2. *Mechanical properties*

Two contributions are considered in the mechanical model for calculating the strain rate tensor $\dot{\boldsymbol{\varepsilon}}$, namely the elastic strain rate $\dot{\boldsymbol{\varepsilon}}_{el}$ and the viscoplastic strain rate $\dot{\boldsymbol{\varepsilon}}_{vp}$:

$$\dot{\boldsymbol{\varepsilon}} = \dot{\boldsymbol{\varepsilon}}_{el} + \dot{\boldsymbol{\varepsilon}}_{vp} \tag{6}$$

According to Hooke's law, $\dot{\boldsymbol{\varepsilon}}_{el}$ depends on the stress tensor $\boldsymbol{\sigma}$ and on the elasticity tensor $\mathbf{D}_{el}$, which is defined by the Young's modulus $E(T)$ and the Poisson's ratio μ:

$$\dot{\boldsymbol{\varepsilon}}_{el} = \mathbf{D}_{el}^{-1}\,\dot{\boldsymbol{\sigma}} \tag{7}$$

Considering only creep deformation at high temperatures, $\dot{\boldsymbol{\varepsilon}}_{vp}$ is expressed as

$$\dot{\boldsymbol{\varepsilon}}_{vp} = \frac{3\dot{\varepsilon}_{vp}^{eq}}{2\sigma^{eq}}\,\mathbf{s}, \tag{8}$$

where $\mathbf{s}$ stands for the deviatoric tensor and σ^{eq} is the Von Mises equivalent stress. $\dot{\varepsilon}_{vp}^{eq}$ represents the equivalent viscoplastic strain rate, which is defined as:

$$\dot{\varepsilon}_{vp}^{eq} = A\,\mathrm{e}^{-\frac{Q}{\mathrm{R}T}}\left(\sinh\left(a_\sigma \sigma^{eq}\right)\right)^n \tag{9}$$

In equation (9), R is the universal gas constant and Q is the activation energy. Further details about defining the parameters Q, A, $a_\sigma(T)$, $n(T)$, $E(T)$ and μ can be found in [10].

3. Results and discussion

3.1. Thermo-mechanical simulations

The local deformation velocity in y-direction within a given mesh element i, $v_{S,y}(i)$, is calculated with equation (10). Therein, $\dot{\varepsilon}_y(i)$ represents the y-component of the strain rate tensor obtained by the ABAQUS simulations. $\Delta y(i)$ stands for the height of the element located at the dendrite axis (x= 0 mm).

$$v_{S,y}(i) = \dot{\varepsilon}_y(i)\,\Delta y(i) \tag{10}$$

Furthermore, equation (11) is used to determine $v_{S,y}(n)$, the total deformation velocity of the mesh element n. Starting at element $i = 1$ located at the dendrite tip, the velocities $v_{S,y}(i)$ are summed up for $i = 1 \ldots n$.

$$v_{S,y}(n) = \sum_{i=1}^{n} v_{S,y}(i) \tag{11}$$

The total number of mesh elements along the dendrite axis, N, is constant for each of the five simulation cases (N= 60). According to the applied boundary conditions, $v_{S,y}(0) = 0$ at y= 0 mm and $v_{S,y}(N) = v_{S,y}^{zero}$ at y= 30 mm.

As an example, figure 5 shows the velocity profiles obtained for simulation case III at different times using equation (11). These results are focused on the first half of the bulging cycle when the casting strand is compressed in front of a guiding roll. It is assumed that only this compression causes the deformation of the dendrites at the strand centre. Due to the negligible elastic contribution at high temperature no deformation (expansion) occurs when the strand widens behind the roll. Since dendrite deformation acts against the positive y-direction, the obtained values of $v_{S,y}$ have a negative sign.

The bulging cycle retained for the simulation starts at the casting time T= 1362.5 s, which corresponds to the bulging cycle time t= 0.00 s. At this moment $v_{S,y}^{zero}$ = 0 and therefore $v_{S,y}$ = 0 too. Hence, the velocity profile is a straight horizontal line for t= 0.00 s, as depicted in figure 5 (a). However, with advancing bulging time $v_{S,y}$ increases in amplitude until maximum values are reached at $t = 6.25$ s. This results in a strongly curved velocity profile. Then, the velocity $v_{S,y}$ decreases until $v_{S,y} = 0$ at $t = 12.50$ s, which is shown again with a horizontal line in figure 5 (a). For each of the velocity profiles displayed in figure 5, $v_{S,y} = 0$ at the centerline solid fraction $f_S^{cent} = 0.20$.

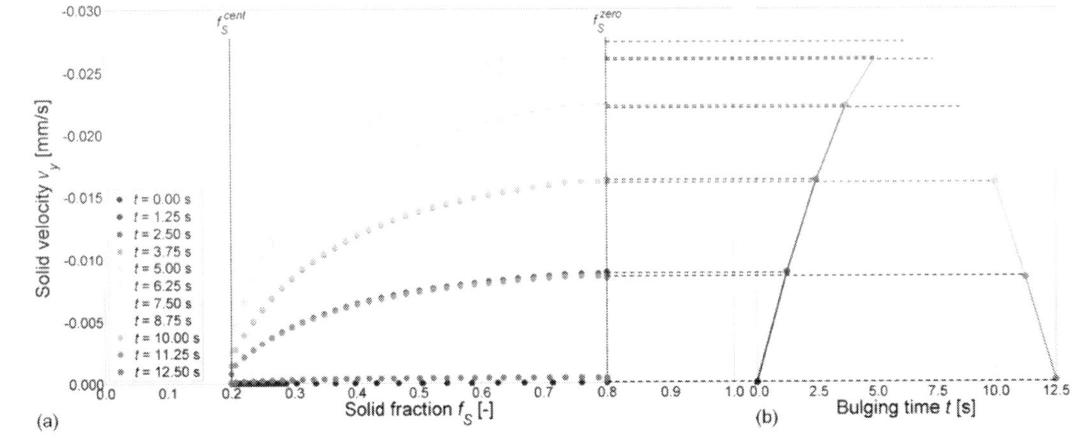

Figure 5. Profiles of solid velocity $v_{S,y}$ between f_S^{cent} and f_S^{zero} (a) and evolution of the corresponding maximum velocity at f_S^{zero} (b). The profiles are calculated for simulation case III.

Figure 6 compares the velocity profiles at $t = 6.25$ s calculated for simulation cases I-V. For each of these profiles, $v_{S,y} = 0$ at the centerline solid fraction f_S^{cent} and $v_{S,y} = v_{S,y}^{zero}$ at the zero-strength solid fraction f_S^{zero}. Since the bulging velocity profile shown in figure 3 is used for simulation cases I-V, $v_{S,y}^{zero}$ is identical for each case at a particular time t. Equations (12) and (13) are suggested to fit the calculated velocities for solid fractions between f_S^{cent} and f_S^{zero}:

$$v_{S,y} = v_{S,y}^{zero}\left(1 - f_S^{\Psi}\right) \tag{12}$$

$$\Psi = a\frac{\left(f_S^{cent} - f_S\right)}{\left(f_S^{zero} - f_S\right)^b} \tag{13}$$

In equation (13), a and b are dimensionless fit constants. With $a = 4.0$ and $b = 0.8$ the fitting curves approximate relatively well the obtained simulation data. Exemplary, the fitted velocity profiles at the maximum bulging amplitude (bulging time $t = 6.25$ s) are shown in figure 6.

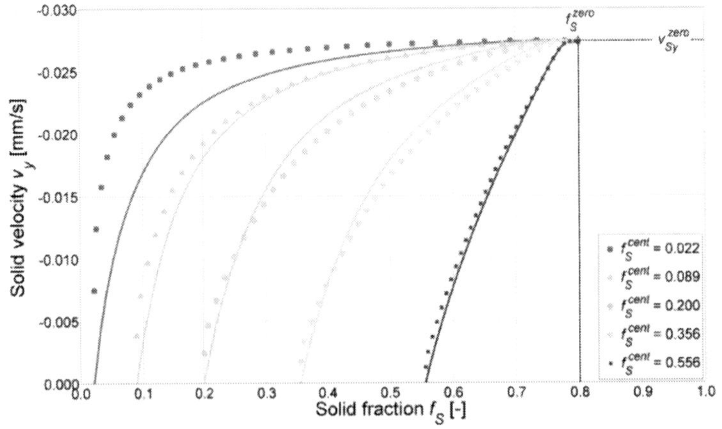

Figure 6. Velocity profiles at $t = 6.25$ s for simulation cases I-V. Each dot marks the calculated velocity $v_{S,y}$ at a certain mesh element. The continuous lines are fitting curves of $v_{S,y}$.

As shown in figure 6, equations (12) and (13) are suitable to fit the results of simulation cases II-V, but they do not fit the results of case I. Thus, these comparatively simple fit equations are not applicable to very small solid fractions. Nevertheless, this can be accepted, since previous simulations have shown that small solid fractions at the strand centre have only a minor influence on the formation of centreline macrosegregation [15].

3.2. Solidification simulations

The obtained relationship between solid fraction and deformation velocity is then used for fluid flow simulations to predict macrosegregation formation in steel continuous casting. For this reason, a two dimensional horizontal continuous casting strand is modeled. Details about the 25 m long strand geometry and about the basic model definitions are explained in [15], where the exponential approach of equation (14) is used instead of equation (12) to describe the dendrite deformation. However, the function Ψ in equation (14) is calculated with equation (13), but with $a = 50$ and $b = 0.25$.

$$v_{S,y} = v_{S,y}^{zero}\left(1 - e^{\Psi}\right) \tag{14}$$

The fundamental conservation equations used for the numerical multiphase model in [15] as well as in the current paper are described in [18-21]. Although this model basically includes three phases, the equiaxed phase is not considered in the current simulations. According to equation (15) the mixture concentration of carbon, C_M^C, is defined to quantify the macrosegregation. In this equation, ρ_S and ρ_L are the densities of the solid and of the liquid phase. C_S^C and C_L^C represent the carbon concentrations within solid and liquid, f_S and f_L are the volume fractions of both phases.

$$C_M^C = \frac{\rho_S \cdot C_S^C \cdot f_S + \rho_L \cdot C_L^C \cdot f_L}{\rho_S \cdot f_S + \rho_L \cdot f_L} \tag{15}$$

Figure 7 (b) shows the macrosegregation profiles of carbon obtained at the outlet of the simulation domain considering solid phase velocities after equations (12) (present work) and (14) (exponential approach [15]), respectively. At the outlet position, the cross-section of the continuous casting strand is totally solid ($f_S > 0.95$). Hence, it is not possible to modify the macrosegregation being observed at this position. The simulations deliver the typical macrosegregation profiles occurring in continuous casting of steel: a strongly positive segregation peak at the strand centre accompanied with a negative segregation valley beside.

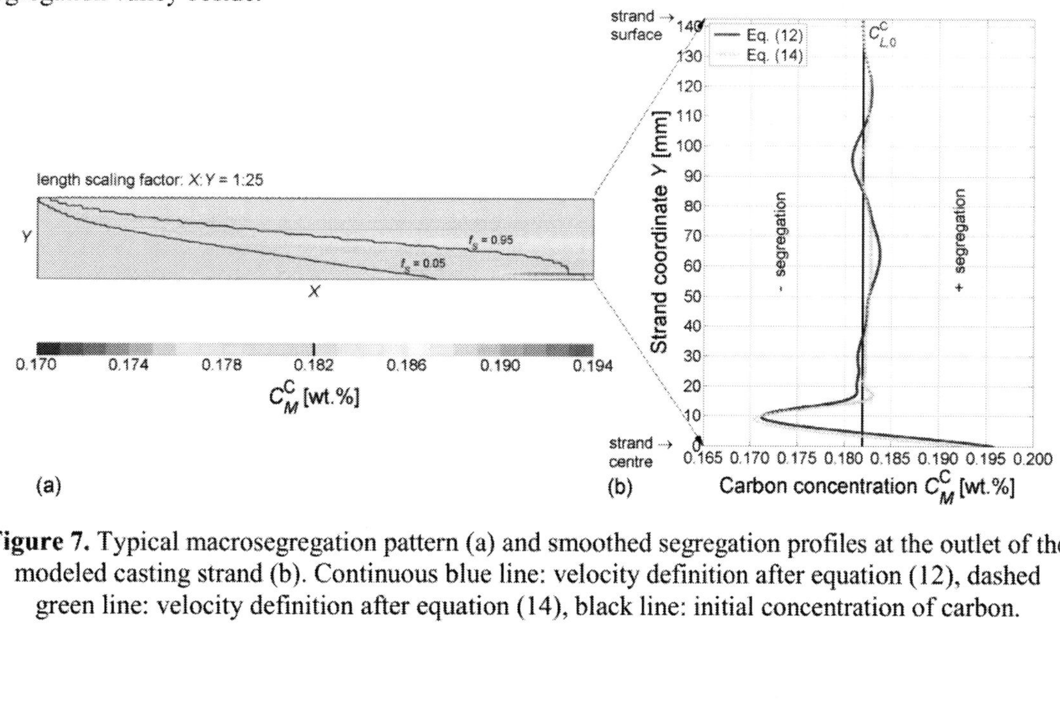

Figure 7. Typical macrosegregation pattern (a) and smoothed segregation profiles at the outlet of the modeled casting strand (b). Continuous blue line: velocity definition after equation (12), dashed green line: velocity definition after equation (14), black line: initial concentration of carbon.

4. Conclusions

In the current paper, a general possibility to combine thermo-mechanical simulations of deforming dendrites at microscale level with CFD simulations to predict centreline segregation formation in continuous casting at macroscale level is presented. In the numerical two-phase model used for CFD simulations, the solid phase velocities in both coordinate directions must be predefined. Particularly, quantifying the deformation speed of the solid dendrites at the strand centre is necessary, which is achieved with thermo-mechanical simulations. Therein, the complex dendritic morphology is approximated with a simple conical model geometry being compressed with a sinusoidal load amplitude representing strand bulging during the casting process.

The simulation results indicate that the dendritic deformation velocity is immediately influenced by the solid fraction at the centreline of the casting strand. For low centreline solid fractions, dendritic deformation is concentrated at the dendrite tips at the strand centre region. In this case a comparatively rapid decay of the solid velocity at the centre can be observed. However, with increasing centreline solid fractions the deformation zone extends along the dendrite axis. Based on these simulation results a relationship is proposed to describe the deformation velocity depending on the solid fraction. This relationship is then used in subsequent CFD-simulations to examine macrosegregation formation in strand casting. The results show a strongly positive segregation peak at the strand centre accompanied with a negative segregation valley beside, which is typical for strand casting of steel.

Due to a lack of geometry data, the complex dendritic morphology is simplified in the current thermo-mechanical simulations. Thus, the influence of side arms and the interaction of adjacent dendrites particularly at higher solid fractions are not considered. Simulations using more realistic dendritic morphologies captured with experimental methods (e.g. X-ray computed tomography or X-ray synchrotron analysis, serial sectioning techniques, etc.) would be of further interest to improve the understanding of the deformation behaviour.

References

[1] Brimacombe J K 1999 *Metall. Mater. Trans.* B **30B** 553-66
[2] Flemings M C 2000 *ISIJ Int.* **40** 833-41
[3] Lesoult G 2005 *Mater. Sci. Eng.* A **413-414** 19-29
[4] Reichert A, Tacke K-H, Harste K, Hecht M, Heilemann J, Ölmann U, Schwerdtfeger K, Barber B, Beaverstock C and Walmsley R 2002 *Casting and solidification - Strand reduction in continuous casting and its effect on product quality* (Luxembourg: European Communities)
[5] Thome R and Harste K 2004 *Steel Res. Int.* **75** 693-700
[6] Bleck W, Wang W and Bülte R 2006 *Steel Res. Int.* **77** 485-91
[7] Thome R and Harste K 2006 *ISIJ Int.* **46** 1839-44
[8] Miyazawa K and Schwerdtfeger K 1981 *Arch. Eisenhüttenwesen* **52** 415-22
[9] Stadler P, Hagen K, Hammerschmid P and Schwerdtfeger K 1982 *Stahl Eisen* **102** 451-59
[10] Kajitani T, Drezet J-M and Rappaz M 2001 *Metall. Mater. Trans.* A **32A** 1479-91
[11] Mayer F, Wu M and Ludwig A 2011 *Steel Res. Int.* **81** 660-67
[12] Beckermann C 2002 *Int. Mater. Rev.* **47** 243-61
[13] Schwerdtfeger K and Heilemann J 2006 *ISIJ Int.* **46** 70-74
[14] Eisermann H and Schwerdtfeger K 2006 *ISIJ Int.* **46** 75-81
[15] Domitner J, Wu M, Mayer F, Ludwig A, Kaufmann B, Reiter J and Schaden T 2011 *Simulations and modelling of solidification and continuous casting processes: Proc. Europ. Conf. on Continuous Casting (Düsseldorf)* S6.1-6
[16] El-Bealy M and Thomas B G 1996 *Metall. Mater. Trans.* B **27B** 689-93
[17] Cicutti C, Bilmes P and Boeri R 1997 *Acta Metall.* **37** 599-604
[18] Ludwig A and Wu M 2005 *Mater. Sci. Eng.* A **413-414** 109-14
[19] Wu M and Ludwig A 2006 *Metall. Mater. Trans.* A **37A** 1613-31
[20] Wu M and Ludwig A 2007 *Metall. Mater. Trans.* A **38A** 1465-75
[21] Wu M, Domitner J and Ludwig A 2012 *Metall. Mater. Trans.* A **43A** 945-64

IOP Conf. Series: Materials Science and Engineering 33 (2012) 012059 doi:10.1088/1757-899X/33/1/012059

Residual Thermal Stresses Simulation for Multi-crystalline Silicon Casting

Jing-Wei Huang [1], Weng-Sing Hwang [1], Chia-Hui Hwang[1] and Ya-Wen Chang [2]

[1] Department of Materials Science and Engineering, National Cheng Kung University,No.1, Daxue Rd., East Dist., Tainan City 701, Taiwan (R.O.C.)
[2] R&D Division, MOTECH Industries Inc., Tainan, Taiwan

wshwang@mail.ncku.edu.tw

Abstract. Multi-crystalline silicon is the major material used for solar cells due to its low production costs and high conversion efficiency. When melting silicon transforms from the liquid phase to the solid phase, dislocations are generated by the temperature gradient and thermal stresses. The disloactions become recombination centers for minority charge carries, decreasing conversion efficiency. Since verifying the change of stress of a silicon ingot is difficult during solidification, computer simulations are used to optimize the parameters of the multi-crystalline silicon ingot casting process. In this study, a numerical model of a silicon ingot is developed to carry out residual thermal stress simulations. The result obtained under various control conditions are dicussed.

1. Introduction

The solar photovoltaic (PV) industry has received increasing attention due to concerns over global warming. Multi-crystalline silicone (mc-Si) is the most widely used material in the casting processes of solar cells. It has a low production cost and high conversion efficiency [1, 2]. In the casting processes, the shape of the liquid-solid interface has a great impact on the quality of the obtained silicon ingot. A non-uniform shape of the liquid-solid interface induces increases in the temperature gradient and thermal stresses, thereby increasing the number of dislocations [3-10]. The dislocations become recombination centers in the ingot, decreasing conversion efficiency. In order to obtained a uniform liquid-solid interface, a heat exchange method (HEM) is used during the process. The HEM, which is significantly affected by the cooling conditions and the design of the hot-zone, leads to the unidirectional solidification of mc-Si and thus a uniform liquid-solid interface.

In casting, thermal stresses are the major type of stress in the ingot. Thermal stress results from differences in the cooling speed between different ingot parts, and causes different degrees of expansion in the ingot. Thermal stress in a silicon ingot that does not vanish when the temperature of the ingot goes from the melting point to room temperature is called residual thermal stress. When melting silicon transforms from the liquid to the solid phase, dislocations are generated by the temperature gradient and thermal stresses. Defects such as dislocations and grain boundaries may carry electrical charge, and become recombination centers for minority charge carries. Processes are thus designed to reduce thermal stresses [11-15].

The temperature of an ingot is too high to allow optical instruments to be used to investigate the condition of the ingot. It is thus difficult to verify the shape of a silicon ingot's liquid-solid interface during grain growth. An alternative is to use computer simulation, which not only substantially reduces production cost and time but can also predict the temperature gradient, heat flux, and residual thermal stresses [16, 17]. Moreover, the parameters can be adjusted to simulate various processes for comparison.

2. Mathematical method

2.1. Thermal model

To calculate the temperature distribution during crystallization, this study adopted Fourier's fundamental laws of heat transfer and extended them to a three-dimensional model. The resulting governing equation is:

$$\rho C_p \frac{\partial T}{\partial t} = \frac{\partial}{\partial x}\left[\lambda \frac{\partial T}{\partial x}\right] + \frac{\partial}{\partial y}\left[\lambda \frac{\partial T}{\partial y}\right] + \frac{\partial}{\partial z}\left[\lambda \frac{\partial T}{\partial z}\right] + \rho \Delta H \frac{\partial f_s}{\partial t}, \tag{1}$$

where ρ is the density, λ is the thermal conductivity, C_p is the heat capacity under constant pressure, H is the latent heat and f_s is the solid fraction of the material used.

Since all materials radiate energy at temperatures above absolute zero, and since radiation is a major mechanism of heat transfer at high temperatures in a furnace, the effects of radiation on the heat transfer between the specimens in the furnace during the solidification process are considered. From the surface radiations, heat flux, q, can be expressed as:

$$q = \varepsilon\sigma(T_a^4 - T_b^4), \tag{2}$$

where σ is the Stefan–Boltzmann constant, ε is the thermal emissivity of a given surface ($\varepsilon < 1$, except for a black body), and Ta and Tb are the temperatures of the specimens in the furnace, respectively.

2.2. Stress model

In this study, the stress calculation is conducted using the thermo-elasto-plastic model. The model exhibits elastic behaviour before material deformation and plastic behaviour after the yield point. In the model, Young's modulus and the yield stress change with temperature. The thermo-elasto-plastic model adopts the bilinear kinematic hardening model to simulate processes and to comply with von Mises's criterion. Figure 1 shows a diagram of thermo-elasto-plastic model [14].

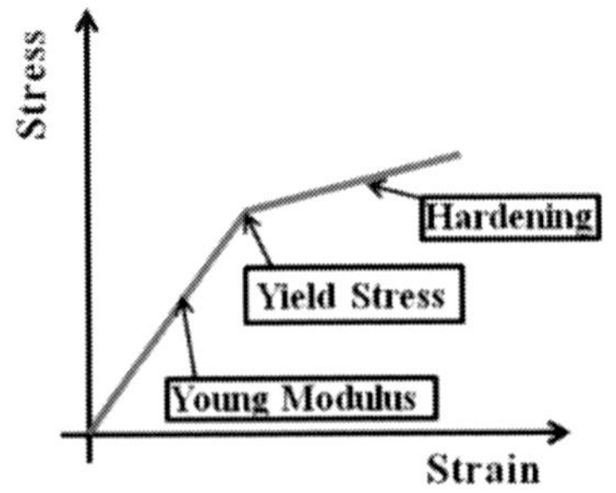

Figure 1. The diagram of thermo-elasto-plastic model.

The behaviour of silicon changes from brittle to ductile at approximately 500 °C. The model of plasticity was included in the simulation. In the thermo-elasto-plastic model, the total incremental strain components may be assumed to consist of the following components:

$$\{d\varepsilon\}=\{d\varepsilon_e\}+\{d\varepsilon_p\}+\{d\varepsilon_T\} \quad (3)$$

where $d\varepsilon_e$, $d\varepsilon_p$, $d\varepsilon_T$ are the elastic, plastic, and thermal incremental strains, respectively. The silicon parameters are shown in table 2.

Table 2. Silicon parameters [18-21].

Silicon	Room Temperature(25°C)
Young's modulus	169000MPa
Thermal Expansion	$2.568*10^{-6}$
Poisson's Ratio	0.226
Yield stress	1.9GPa
Plastic modulus	800MPa

3. Experimental method

The numerical system contains an mc-Si substrate, a crucible, a supporting plate, a heater, a cooling plate, and a furnace. Figure 2 shows the positions of these items. The system is heated from room temperature to 1450 °C, which is then held to ensure that the mc-Si completely transforms from a solid to a liquid. A heat gate at the bottom of the system is then opened, and the mc-Si starts to solidify.

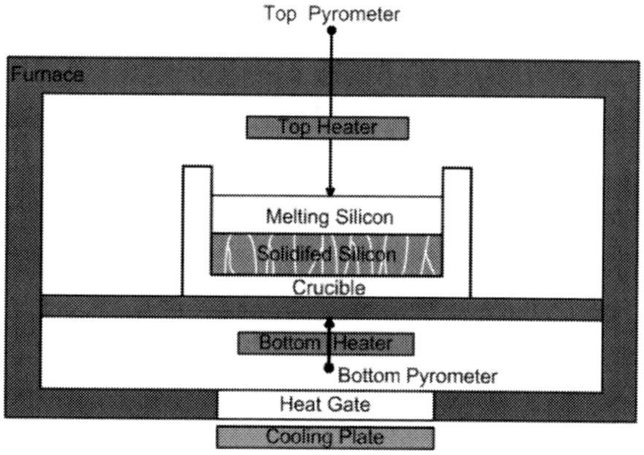

Figure 2. Schematic illustration of the furnace of silicon unidirectional growth.

The system uses three methods to transfer heat, namely thermal convection, thermal radiation, and thermal conductivity. The top heaters and bottom heaters transfer heat to the mc-Si via thermal radiation and thermal convection. When the mc-Si starts to solidify, heat is transferred through the crucible and supporting plate as a result of thermal conductivity. Finally, the system uses a cooling plate to deliver heat to the furnace exterior by thermal radiation. In this study, because the temperature of the system is extremely high, thermal radiation is a major consideration for temperature simulation using equation 2.

The quality of the mc-Si ingot is affected by the rate of crystal growth during solidification. The conversion efficiency of a solar cell is affected by the grain size of the silicon wafer, which in turn is influenced by the cooling rate required for crystallization. Thus, the control of the temperature gradient is important during solidification.

In this research, the first way to verify the initial simulation model is as following: two pyrometers are set on the top and the bottom of the furnace respectively to detect the temperature profile varies with time. The certain point of molten silicon surface will be detected by the SiO_2 rod once an hour so that numerical simulation system could be verified with information from the pyrometer. The position of pyrometer and SiO_2 rod are showed in figure 3.

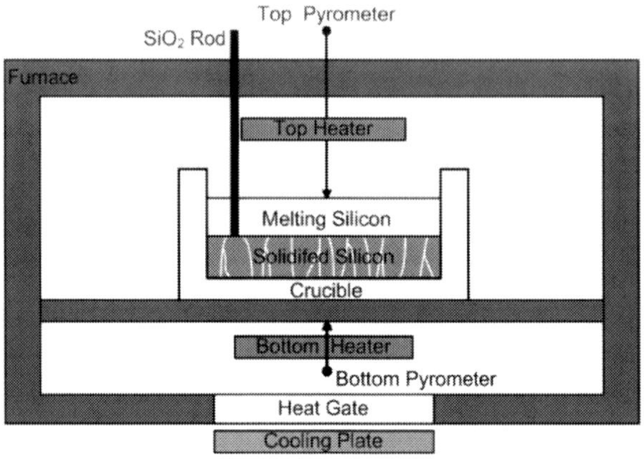

Figure 3. Schematic illustration of the way to detect solidification height and temperature profile.

In this paper, two model configurations, one with a single heat gate and one with multiple heat gates, are considered. Figures 4 and 5 show the temperature profiles of the single- and multiple-heat-gate configurations, respectively. For the single-heat-gate configuration, when the system opens the gate, there is a large heat loss initially. For the multiple-heat-gate configuration, the system opens the gate one by one to avoid large heat loss. This allows the temperature gradient to be controlled effectively during solidification.

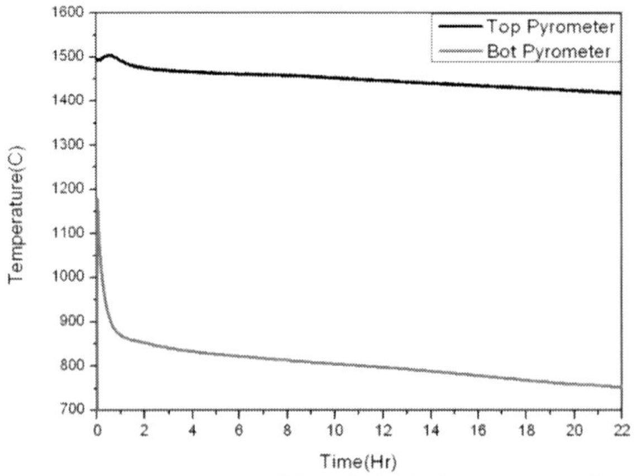

Figure 4. The temperature profiles of single heat gate configuration.

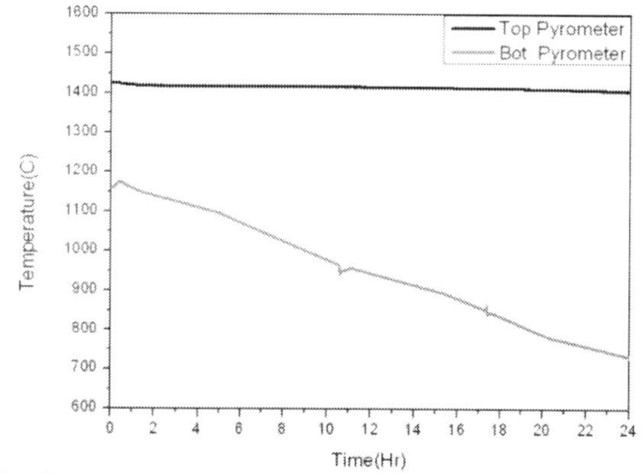

Figure 5. The temperature profiles of multiple heat gate configuration.

4. Results and Discussion

Figure 6 and 7 show the comparison of the solidification height between experiment and simulation with two configurations. In the figure, the experiment and simlution value are closed with the process time going so that simulation can be considered having the high degree of accuracy. Figure 8 shows the simulation of the solidification height between single- and multiple-heat-gate configurations. Single-heat-gate process has faster solidification speed because the large heat loss in the process beginning, and multiple-heat-gate process has smaller faster solidification speed so that stress value accumulation of multiple-heat-gate process is slower at process first.

Figures 9 and 10 show the axial distributions of stresses along the central and peripheral lines of the silicon ingot, respectively. The red line represents the multiple-heat-gate process and the black line represents the single-heat-gate process.

Along the central line of the slicon ingot, the stresses for the multiple-heat-gate process are smaller than those for the single-heat-gate process in all parts. The stresses of the multiple-heat-gate process are smaller than those of single-heat-gate process at the top of the peripheral line, but larger at the bottom of the peripheral line. With decreasing temperature gradient, the silicon ingot of the multiple-heat-gate process remains at a higher temperature than that of the ingot of the single-heat-gate process, and thus the stresses of the former are smaller. The stresses along the peripheral line of the silicon ingot are influenced by the temperature gradient and the compression of the crucible. The effect of the temperature gradient is not obvious, so the stresses in the two processes are similar. Although the mutliple-heat-gate process produces high stress values in some places, the values are smaller than those for the single-heat-gate process with a higher temperature in the furnace.

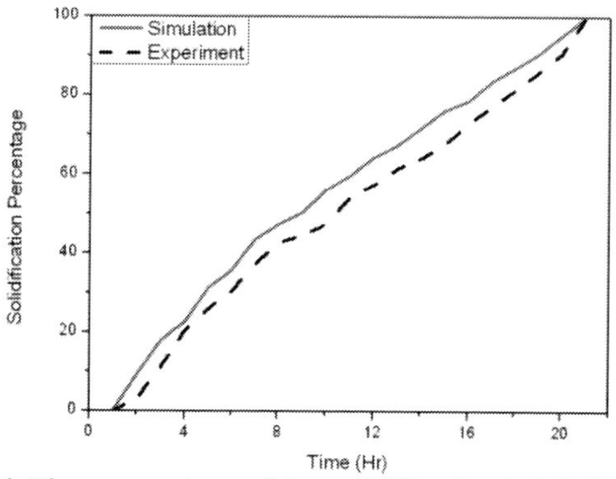

Figure 6. The comparison of the solidification height between experiment and simulation in single-heat-gate configuration.

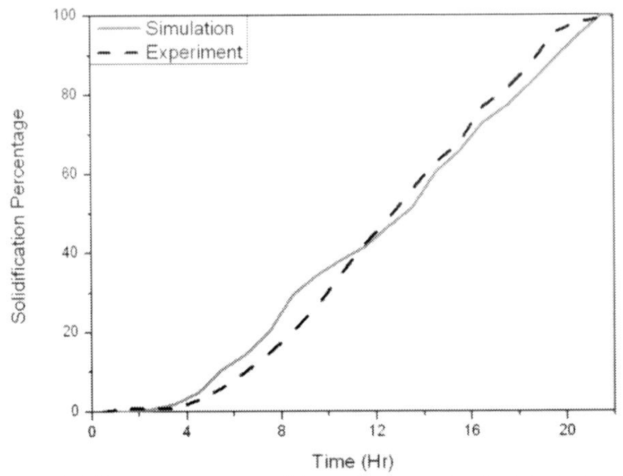

Figure 7. The comparison of the solidification height between experiment and simulation in multiple-heat-gate configuration.

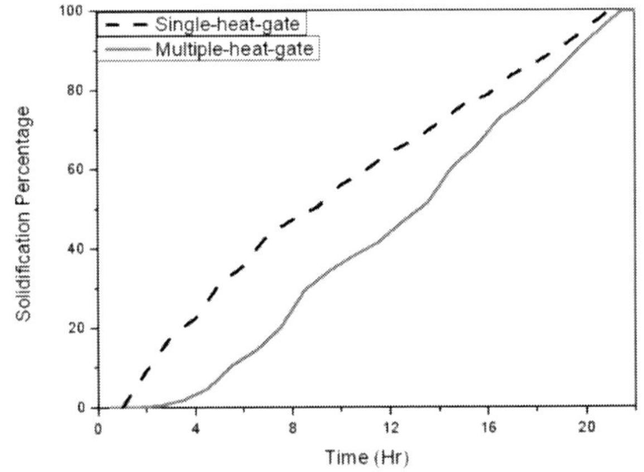

Figure 8. The simulation of the solidification height between single- and multiple-heat-gate configurations.

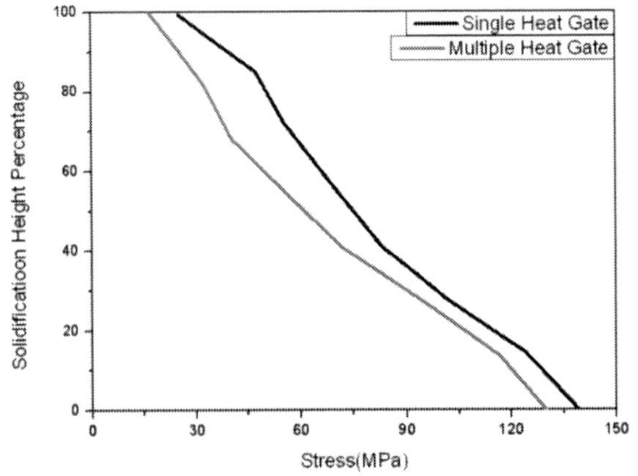

Figure 9. Axial distribution of stresses along the central line of the silicon ingot.

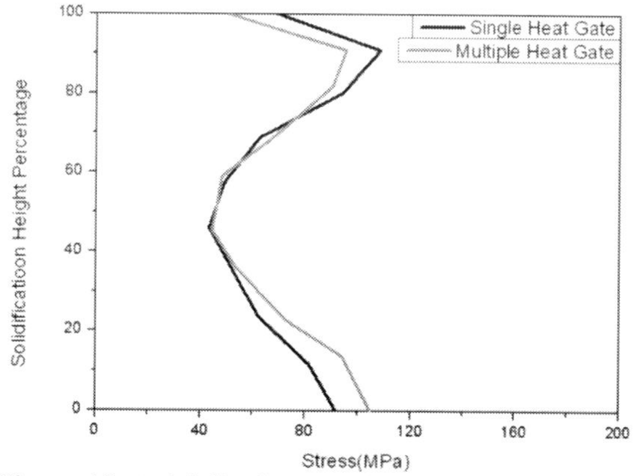

Figure 10. Axial distribution of stresses along peripheral line of the silicon ingot.

5. Conclusion

A numerical model of the residual thermal stress distribution for mc-Si casting was developed. Single- and multiple-heat-gate configurations were simulated. With the verifying the numerical simulation system, results show that the temperature gradient can be decreased and it can be determined which condition decreases stresses, resulting in increased mc-Si quality.

Acknowledgements

The authors would like to thank the Bureau of Energy, Ministry of Economic Affairs, Taiwan, for financially supporting this research under Contract no. 101-D0204-2. A part of the present work was also supported by the Research Center for Energy Technology and Strategy, National Cheng Kung University in Taiwan.

References

[1] il Kim D and Kwan Kim Y 2006 *Solar Energy Materials and Solar Cells.* **90** 1666-1672
[2] Min Kim J and Kwan Kim Y 2004 *Solar Energy Materials and Solar Cells.* **81** 217-224
[3] Koji S 1989 *Instute for Materials Research*
[4] Häßler C and Reisner E U 1999 *Solid State Phenomena* **67-68** 447-452
[5] Kvande R and Mjøs Ø 2005 *Materials Science and Engineering*: **A 413-414** 545-549
[6] M'Hamdi M, Meese E.A, Laux H and Øvrelid E.J 2006 *Materials Science Forum* **508** 597-602
[7] Kvande R and Mjøs Ø 2005 *Materials Science and Engineering*: **A 413-414** 545-549.
[8] Arafune K, Sasaki T, Wakabayashi F and Terada Y 2006 *Physica B: Condensed Matter* **376-377** 236-239
[9] Barvinschi F, Stelian C, Delannoy Y, Mangelinck N and Duffar T 2003 *Journal of Optoelectronics and Advanced Materials* **5** 293–300
[10] Moller H, Funke C, Rinio M and Scholz S 2005 *Thin Solid Films* **487(1-2)** 179-187
[11] Chen X, Nakano S and Kakimoto K 2008 *Journal of Crystal Growth* **310(19)** 4330-4335
[12] Wei J, Zhang H, Zheng L, Wang C and Zhao B 2009 *Solar Energy Materials and Solar Cells* **93(9)** 1531-1539
[13] Miyazawa H, Liu L, and Kakimoto K 2008 *Journal of Crystal Growth* **310(6)** 1142-1147
[14] Oswald M, Turek M and Bagdahn J 2010 *Int. Conf. 11th on Thermal, Mechanical and Multiphysics Simulation and Experiments in Micro-Electronics and Micro-Systems, EuroSimE 2010*

[15] Sun L, Liu' L, Zaoyang L and Wenhan Z 2011 *IEEE*
[16] Franke D, Rettelbach T, Häßler C, Koch W and Müller A 2002 *Solar Energy Materials & Solar Cells* **72** 83–92
[17] Häûler C, Stollwerck G, Koch W, Krumbe W, Müller A, Franke D and Rettelbach T 2001 *Adv. Mater* **13** 23
[18] Ying Yang T, Jyh Chen C, ChungWei L and Chi Yung C 2010 *Journal of Crystal Growth* **312(8)** 1282-1290
[19] Miyazaki N, Kuroda Y and Sakaguchi M 2000 *Journal of Crystal Growth* **218** 221-231
[20] Ono N, Kitamura K, Nakajima K and Shimanuki Y 2000 *Jpn. J. Appl. Phys.* **39** 368–371
[21] William N, Sharpe Jr, Yuan B and Vaidyanathan R 1997 *IEEE*

IOP Conf. Series: Materials Science and Engineering **33** (2012) 012060 doi:10.1088/1757-899X/33/1/012060

Neutron diffraction analysis and solidification modeling of Impulse-Atomized Al-36 wt%Ni

A. Ilbagi[1], D. Tourret[2, 3], H. Henein[1] and Ch.-A. Gandin[2, 4]
[1]Chemical and Materials Engineering, University of Alberta, Edmonton, Canada;
[2]MINES ParisTech & [4]CNRS, CEMEF UMR 7635, 06904 Sophia Antipolis, France;
[3]Deutsches Zentrum für Luft- und Raumfahrt, Institut für Materialphysik im Weltraum, 51147 Köln, Germany

E-mail: ilbagi@ualberta.ca

Abstract. Peritectic solidification reactions appear in many metallic systems such as Al-Ni alloys. However, microstructural evolution and effects of processing parameters on phase selection during peritectic reaction are not well understood. In this paper Impulse-Atomization experiments and simulations were used to study rapid solidification of Al-36 wt% Ni particles. Secondary dendrite arm spacing measurements were used to estimate the cooling rate achieved during the solidification of these particles. The weight fractions of the phases formed in different sizes of the particles were measured using neutron diffraction technique. Solidification paths were then simulated with a model previously validated for concurrent dendritic, peritectic and eutectic phase transformations in binary alloys. Model predictions are compared to the experimental results to understand the sequence of transformations that leads to the final metallurgical state of the particles.

1. Introduction

It is known that the catalytic properties of Al-Ni alloys depend on constituent phases formed during the solidification process [1]. In this paper, solidification of atomized particles of Al-36 wt%Ni alloy will be studied using neutron diffraction and Rietveld analysis. A segregation model will also be used to simulate the solidification of this alloy. The model considers finite diffusion during multiple phase transformations and growth kinetics of microstructures. It can predict the time evolution of temperature, phase fractions and average compositions of phases [2,3]. It is also coupled with thermodynamic equilibrium calculations using the CalPhaD method with the dedicated software ThermoCalc (http://www.thermocalc.se) [4,5,6]. The model has been used to simulate the solidification of droplets processed using electromagnetic levitation (EML) [3] and gas atomization (GA) [7]. However, the solidification of a droplet in EML is strongly disturbed by large stirring effects from the electromagnetic forces used for positioning and heating the sample. Also, for GA, a wide droplet size distribution is achieved and each droplet finds itself in a different local gas velocity and temperature. As a result, depending on how the droplets interact and move with the atomizing gas the evolution of chemistry and/or microstructure varies [8-10]. This makes it very difficult to identify how individual process variables can affect the resulting characteristics of individual, or group of, droplets. To minimize the effect of external forces on the solidifying droplet Impulse Atomization (IA) technique was used for this research.

IA is a single fluid atomization technique that is capable of producing spherical droplets with diameters of 100 to 1000 μm. In this technique, the liquid droplets solidify while free-falling through a stagnant inert gas. In such a system the heat transfer phenomenon between the falling droplet and the cooling gas can be readily characterized [11]. Also, disturbance from external forces in IA is minimal and the falling droplets experience levels of reduced gravity depending on the droplet size, type of the cooling gas and initial velocity [12]. IA has been used to produce variety of different alloys such as Al-Cu [13] and Al-Ni [14]. In this paper, IA technique is utilized as a tool for studying solidification of Al-36 wt%Ni.

2. Experiments

A 4-meter-high atomization tower was filled with helium and the oxygen content was reduced to less than 10 ppm. An induction furnace located at the top of the tower then started heating the 99.9% pure Al and Ni (Alfa Aesar) to make Al-36 wt%Ni melt. This composition is shown with an arrow in Al-Ni phase diagram calculated using ThermoCalc and the Public BINary thermodynamic database for alloys (PBIN) database (Figure 1). The alloy was held for 30 minutes at 100 K above its liquidus temperature. Afterwards, the liquid was pushed through orifices at the bottom of the crucible to generate liquid streams, which subsequently broke up into spherical droplets by Rayleigh instability. The droplets lose their heat as they fall through the stagnant helium atmosphere. The solidified particles were then collected in oil at the bottom of the tower. For more details of the technique see [15]. The solidified particles were subsequently washed, using toluene and ethanol, and sieved into different sizes.

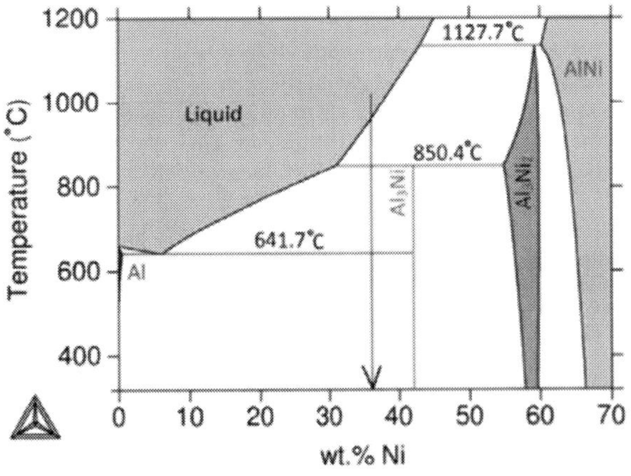

Figure 1. Al-rich side of Al-Ni phase diagram calculated using ThermoCalc and the PBIN database. Arrow represents the composition studied in this paper.

To study the microstructure, scanning electron microscopy was performed using a Zeiss Evo MA15 SEM with a 20 keV electron beam energy. EDX analysis on the SEM was done using a Bruker Silicon Drift Detector with a peak resolution of 125 eV. The variation of microstructure length scale (*i.e.* the Dendrite Arm Spacing, DAS) with droplet size was estimated using the method previously presented [3]. Back-scattered SEM images of Al-36 wt%Ni droplets of different sizes were processed. The interface between the peritectic and the eutectic structures is determined by its gray level threshold value. The DAS was then evaluated as the average spacing between intercepts of this interface with evenly spaced lines through the image. Total of 60 lines on ten different particles from each particle size were used to estimate the DAS value.

ND was used to characterize the phases formed during solidification. The experiments were conducted using a neutron beam of 1.33 Å wavelength at National Research Council of Canada - Canadian Neutron Beam Centre (NRC-CNBC) in Chalk River, ON. To obtain the weight fraction of the phases formed under different conditions, Rietveld analysis was performed using the GSAS (General Structure Analysis System) software [16].

3. Modeling

The model used in this paper was thoroughly described elsewhere [2,3,7]. It is based on a volume averaging method. While the detailed topological morphology of the microstructure is not predicted, the global evolution of the system can still be modeled. Averaged conservation equations are written for the energy, the total mass and the solute mass over a Representative Elementary Volume (REV). Assumptions are made of constant and equal densities in all phases and no mass exchange with the outer boundary of the REV (closed system). Hence, mass transfer only takes place by diffusion and phase transformation. A uniform temperature is also assumed. Heat exchange takes place at the outer boundary of the REV. The kinetics of the phase transformations are computed as a function of microstructure features such as the growth velocity of the microstructures and the secondary DAS. Such features are inputs to the model. The spherical REV represents a single atomized particle undergoing solidification.

Several microstructures successively nucleate at its center of the domain and propagate by radial growth. Each new nucleated microstructure defines a new zone (e.g. primary, peritectic, eutectic) in which several phases may coexist. Because different liquids in different zones have different compositions, they are considered as different phases. Identically, after the nucleation and growth of secondary solid phases, solid phases are split in the different zones to which they belong. The radial expansion of each new zone follows the growth kinetics of the new solid microstructure. The grain envelope, *i.e.* the boundary between the mushy zone and the bulk extradendritic liquid zone, is defined by the location of the dendrite tips and it grows according to a classical dendrite tip growth law from Kurz *et al.* [17]. For a peritectic zone, the growth kinetics is based on a simple adaptation of the dendritic theory for a solid structure facing several liquid zones [2,3]. The classic model from Jackson and Hunt [18] gives the growth kinetics of the final eutectic zone. The nucleation event of a solid phase in an undercooled liquid takes place when the temperature becomes lower than the prescribed nucleation undercooling of the solid phase. In the following no nucleation undercooling has been considered. This choice is commented and justified in the discussion section. Hence, nucleation takes place when the temperature is lower than the equilibrium temperature for the formation of the new phase. One has to take care yet to compute such a metastable equilibrium temperature with the actual interdendritic liquid composition.

Assuming a global microstructure morphology (through interface densities) and composition profiles (through diffusion lengths), the total mass and the solute mass balances are written for each phase in each zone. Within each zone, phase transformations and diffusion fluxes are assumed in a plate-like geometry in the secondary DAS. Mass exchanges are considered at the boundary of two adjacent zones, *i.e.* in the radial growth direction. The later situation actually depicts the exchanges within one – or several – given thermodynamic phase(s) belonging to two adjacent zones. Through a boundary between two zones, the average composition of a given thermodynamic phase is continuous, while at an interface between two phases, thermodynamic equilibrium is assumed. The equilibrium calculations are done with the CalPhaD-based software ThermoCalc. Hence, non-linear phase diagrams are computed together with their metastable extensions [3].

In addition to the mass balances for each phase in each zone, an average global heat balance is written for the entire domain. Stable and metastable enthalpies of phases and their partial derivatives with respect to temperature and composition are also given by thermodynamic equilibrium calculations [3]. A convective heat transfer condition is assumed at the external boundary of the droplet. In order to account for the atomization process, a time dependent heat transfer coefficient is computed with a model dedicated to the atomization process [2,13]. The Nusselt number, which characterizes heat exchange between the particle and the helium gas, is expressed using the Whitaker correlation as a function of the Reynolds number of the flow, the Prandtl number, and the viscosity of the gas. The velocity of the falling droplet appearing in the Reynolds number is integrated with respect to time from the balance between gravity and drag forces in the atomization fluid. The convective heat transfer coefficient is thus directly correlated to the velocity of the droplet [2].

The unknowns of the resulting system of ordinary differential equations (ODE) are the fractions of phases, the average compositions of phases and the temperature of the domain. For instance, in the simulations performed in this paper, involving four phases and three solidification reactions (primary dendritic, peritectic and eutectic), the whole model reduces to 19 ODEs and as many unknowns. An implicit multiple step backward method (or Gear method) is used to solve the system of equations.

4. Results and Discussions

Figure 2 shows the microstructure of solidified particles of Al-36 wt%Ni with diameters of 850 and 355 μm. Three different gray scales representing three different phases can be seen in these images. EDX analysis of the brighter phase observed in the core of dendrites showed that it contains 39 at% nickel, which is close to that of Al_3Ni_2. These dendrites are surrounded by a slightly darker gray phase. From EDX it was found that this phase contains 24 at% nickel. The morphology and the composition of this phase suggest that it is the peritectic phase, Al_3Ni. The darkest gray structure that occupies most of the remaining interdendritic region and contains more than 95 at% aluminum is the eutectic structure that formed during the last stage of the solidification [19]. Some porosity due to solidification shrinkage is observed as well, as can be seen on the top left part of Figure 2a. There is a large variation of microstructure length scale as a result of variation of particle sizes. The variation of the measured DAS with particle size is well fitted by the relation DAS[μm] = 0.0043 D[μm] + 0.3, with coefficient of determination of 0.9341, in the range [100 μm, 900 μm], where D is the particle diameter. It is clear that increasing particle size results in increasing DAS.

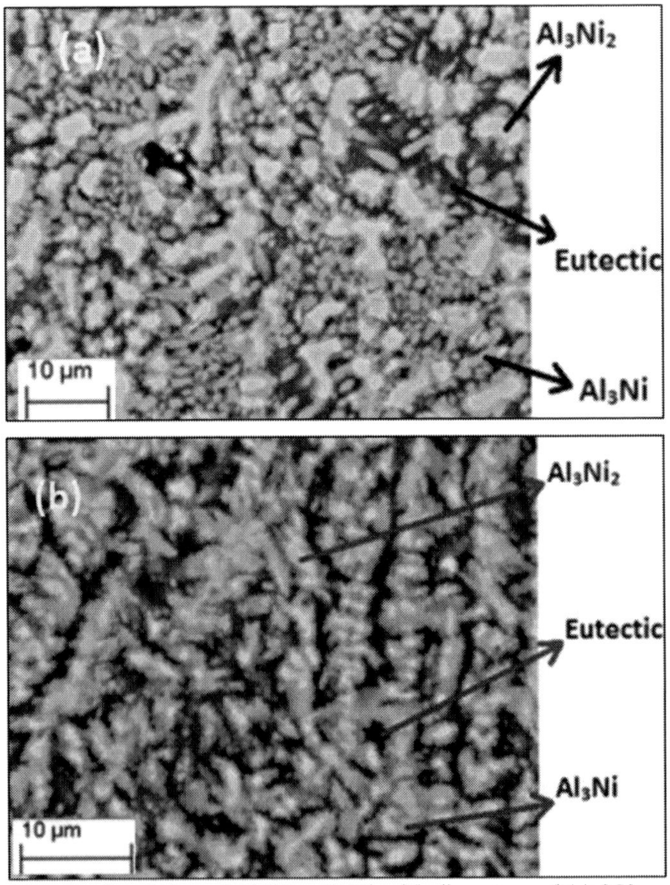

Figure 2. Microstructure of particles of Al-36 wt% Ni with diameters of (a) 850 μm and (b) 355 μm.

Profile refinement was carried out using the software GSAS on ND patterns from particles of Al-36 wt%Ni atomized in helium. Three phases (i.e. Al_3Ni_2, Al_3Ni and Al) were identified. Figure 3 shows an example of measured ND pattern of 5 grams of IA particles with diameter of 500 μm along with the calculated diffraction pattern. The top line shows the calculated diffraction pattern as obtained from GSAS software and the bottom line represents the difference between the observed and calculated diffraction pattern. The few and low oscillation peaks in this curve indicate a satisfactory agreement between measured and refined patterns.

Additionally, the simulation of the solidification of an Al-36 wt%Ni alloy particle of diameter 355 μm was achieved with the simulation parameters summarized in Table 1. Expect the diffusion coefficient in the solid phases, the simulation parameters are similar to those used in previous work [3,7]. As previously discussed [3], the values of interdiffusion coefficients in liquid Al-Ni alloys are still under debate [20-22] while for the solid phases, very few (for Al_3Ni_2 [23]) or no values for diffusion coefficients have been reported. Hence, for the liquid we use a commonly assumed appropriate value of 10^{-9} m^2s^{-1} [20]. For the solid phases, the previously used value of 5×10^{-11} m^2s^{-1} was found to give good quantitative predictions for EML and GA processed Al-Ni alloys, even though this value was a little higher than the measurements in Al_3Ni_2 by Sarkhel and Seigle [22] (3.7×10^{-12} to 3.4×10^{-11} m^2s^{-1} between 1073 K and 1273 K). Here, we presents results assuming a diffusion coefficient of 5×10^{-12} m^2s^{-1} in the solid phases, which still lies in the range experimentally assessed [22]. The results of the model will be compared with those considering the previous value of the solid diffusion coefficient.

We did not include nucleation undercooling of the solid phases in the present simulation. The possible and likely occurrence of solid phase nucleation undercooling may affect the solidification path and the final amount of constituents [3]. The wide solidification range of the considered alloy (> 300 K) gives rise to a very large range and numerous combinations of possible nucleation undercoolings. However, unlike in EML experiments, no measurements were directly performed to determine the nucleation undercoolings in the investigated droplet. Hence, as a first comparison and for the sake of simplicity, we chose to set the nucleation undercooling of solid phases to zero rather than adjusting three additional parameters in a simulations where a lot of key properties are already not well identified (e.g. diffusion coefficients). Furthermore, when high cooling rates are involved, the occurrence of high growth-related undercoolings [3,7] could have a predominant influence compared to the nucleation undercooling of solid phases. Hence, good predictions could be achieved for the GA process without accounting for nucleation undercoolings [7].

Figure 4 shows the evolution with time of (a) the temperature of the droplet with equilibrium and nucleation temperatures of the different phases and (b) the volume fractions of phases, compared with the final phase fractions according to standard predictions of the lever rule and Scheil-Gulliver model (horizontal lines). The cooling curve exhibits three recalescences, corresponding to the successive growths of the primary dendritic Al_3Ni_2, the peritectic Al_3Ni and the eutectic (Al + Al_3Ni) microstructures. Even though the nucleation undercooling is set to zero for all phases, we observe significant undercoolings before the recalescences, which corresponds to a growth-related undercooling due to the important cooling rate [3,7]. First, around t ~ 0.018 s, the growth of the primary Al_3Ni_2 dendritic structure provokes the first recalescence. After the primary recalescence, the temperature decreases and reaches a small peritectic recalescence at t ~ 0.055 s. This small temperature increase indicates that the peritectic solid already grows during the step of primary solidification, is then stopped by the primary recalescence, and finally starts again when the temperature decreases below the peritectic equilibrium temperature. Indeed, on the cooling curve, the nucleation temperature of the peritectic appears above its equilibrium temperature, thus indicating the early growth of Al_3Ni just before the primary recalescence. Due to the early formation of peritectic phase, the small amount of peritectic phase remaining to form for reaching close-to-equilibrium solidification conditions during the peritectic recalescence explains its moderate amplitude. This solidification scenario is similar to one predicted and described in details previously for gas atomized Al-Ni droplets (namely Al-35.22 wt% Ni droplets of diameter ~ 60 μm in [7]). During this peritectic recalescence, the decrease in primary Al_3Ni_2 fraction is first due to the remelting of the primary solid during the peritectic growth and to the peritectic transformation $Al_3Ni_2 \rightarrow Al_3Ni$, which lasts up to complete solidification. After the peritectic recalescence, the solidification ends when the temperature goes below the eutectic nucleation temperature (plus a small growth undercooling), as the eutectic structure grows and fills in the remaining liquid.

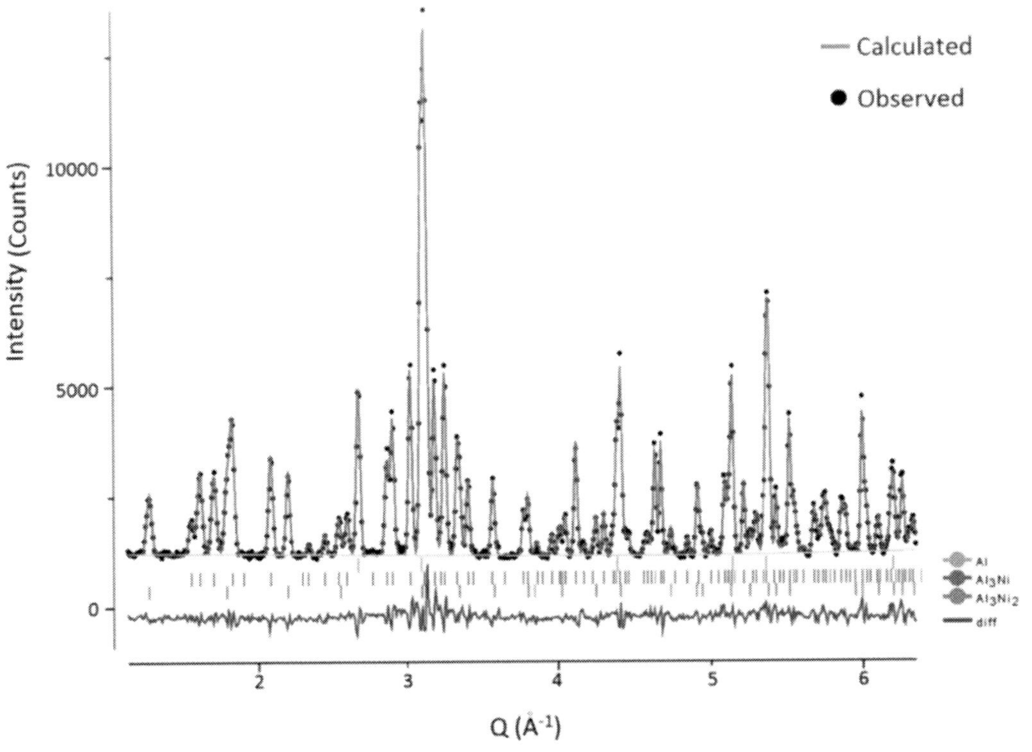

Figure 3. Profile refinement of Al-36 wt%Ni with the diameter of 500 μm using the GSAS computer code. The top line shows the calculated diffraction pattern as obtained from GSAS software and the bottom line represents the difference between the observed and calculated diffraction pattern. The small vertical bars indicate the positions of diffraction peaks of the Al (top), Al_3Ni (middle) and Al_3Ni_2 (below).

Comparison of the simulation results with those from equilibrium lever rule and Scheil-Gulliver model (Figure 4b) clearly shows that while the lever rule overestimates the amount of peritectic phase and no residual primary Al_3Ni_2, Scheil-Gulliver model reaches opposite deviations, *i.e.* underestimation of peritectic and overestimation of primary Al_3Ni_2. That is because according to Scheil-Gulliver model, no solid-state diffusion takes place, which is a requirement of peritectic transformation.

Figure 5 compares the phase fractions from simulation with those from Rietveld analysis on IA powders of different sizes. The experimental measurements show that the fraction of Al_3Ni_2 increases with increasing particle size from 355 to 850 μm. This is likely due to the fact that the increased cooling rate for smaller particle sizes results in a primary phase nucleation undercooling, which in turn reduces the amount of Al_3Ni_2. This may explain why the simulation overestimates the amount of Al_3Ni_2 at smaller particle sizes, since no nucleation undercooling was assumed in the simulations.

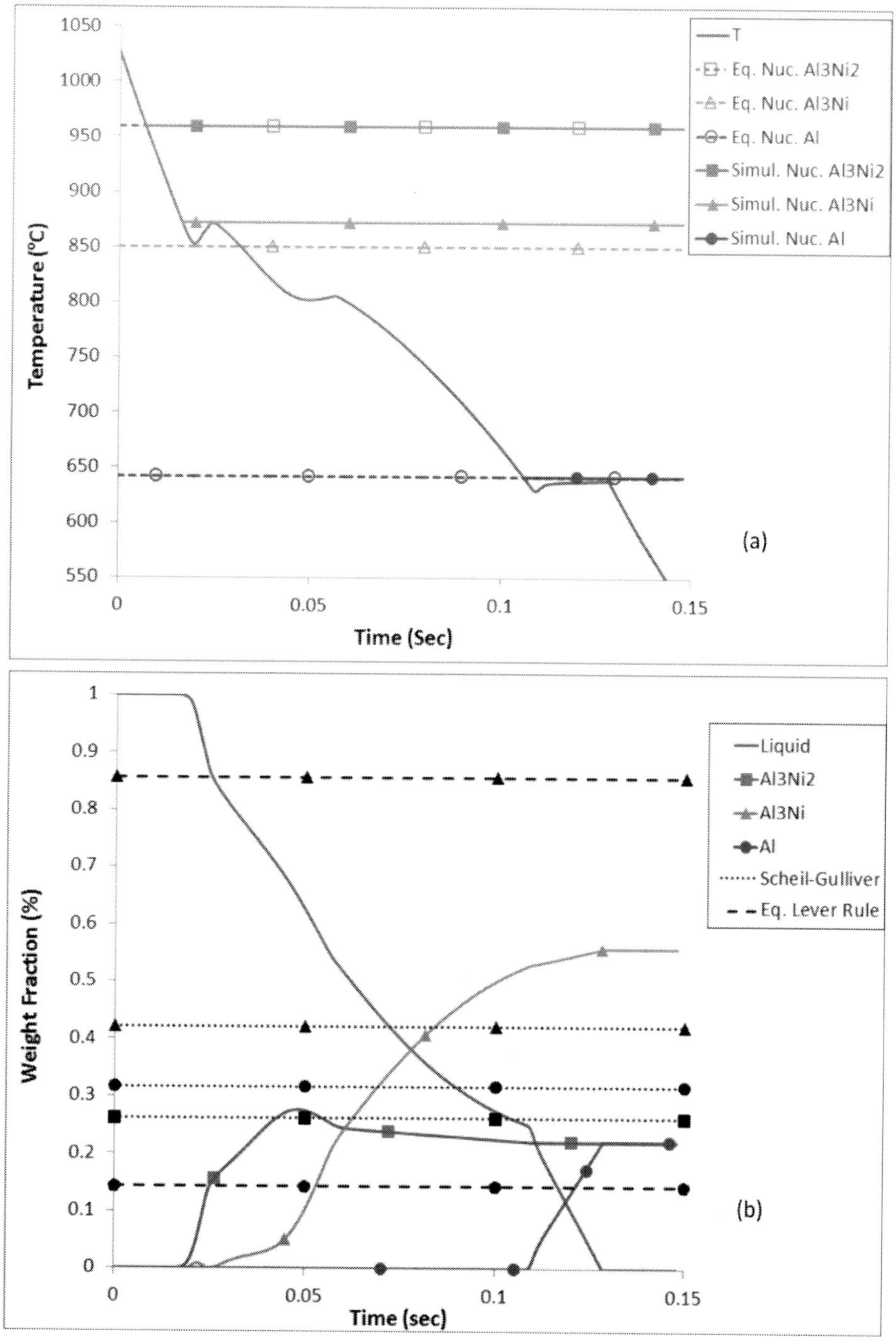

Figure 4. Time evolution of (a) the temperature and (b) the phase fractions of a solidifying IA droplet with diameter 355 μm.

Table 1. Values of thermo-physical data of the Al-36 wt%Ni alloy and numerical parameters.

Material parameter	Value	Process parameter	Value
Thermodynamic database	PBIN	External temperature [°C]	25 °C
Gibbs-Thomson coefficients [°C m]		Droplet initial velocity [m s^{-1}]	0.5
Al_3Ni_2 / liquid [3]	2×10^{-8}	Density of Helium [kg m^{-3}] [25]	$37.303\ T^{-0.9559}$
Al_3Ni / liquid [24]	3.5×10^{-8}	Viscosity of Helium[Pa s] [25]	$(4.3679\ T^{0.6702})\times 10^{-7}$
Al / liquid [24]	1.86×10^{-7}	Helium conductivity [W m^{-1} K^{-1}] [25]	$0.002778\ T^{0.7025}$
Eutectic structure contact angle		Prandtl number [Pr] [25]	0.713
		Whitaker's relation coefficients	
Al_3Ni / liquid [24]	19.22	[B][25]	2.778×10^{-3}
Al / liquid [24]	11.29	[m] [25]	0.7025
Diffusion in solid structures [m^2 s^{-1}]	$5\cdot10^{-12}$		
Diffusion in liquid [m^2 s^{-1}] [3]	10^{-9}	Particle size, D [μm]	355,500,710,850
Jackson-Hunt model series P' [18]	$0.335(f_\alpha^{eut}(1-f_\alpha^{eut}))^{1.65}$		
Density [kg m^{-3}]	4874		
Secondary dendrite arm spacing [μm]	0.0043 D[μm] + 3		
Nucleation undercooling of solids [°C]	0		
Time step [s]	1		
Initial temperature [°C]	1025		

At larger particle sizes, on the other hand, the simulation overestimates the fraction of Al_3Ni. Since no experimental diffusion coefficient in the Al_3Ni phase has yet been reported, choosing an arbitrary diffusion coefficient may result in the discrepancies observed. The diffusion coefficient in the Al_3Ni phase used here differs from the value proposed in earlier publications [3,7]. The current value of 5×10^{-12} m^{2}s^{-1} was found to give a better agreement with the measured values without influencing significantly the cooling curve. A comparison with simulation values using the previous value for the solid diffusion coefficients, 5×10^{-11} m^{2}s^{-1}, is also given in Figure 5. However, the relevance of the diffusion coefficient in the Al_3Ni phase remains limited due to the small solubility range (0.1 wt.%) introduced in the Al_3Ni phase, in order to allow diffusion in the stoichiometric phase [3].

Various thermo-physical properties of Al-Ni alloys are not well determined, which have a crucial effect on the solidification history of a particle. Furthermore, their effects highly depend on each other, and hence they cannot and should not be considered individually. If a good quantitative agreement was found with observed cooling curves and solid fractions for one set of parameters [3,7], we believe that it indicates that the entire set of parameters was well suited to the investigated process and alloy. This point stresses once again the lack of well-assessed experimental values for crucial parameters such as diffusion coefficients in each individual phase. However, the inclusion of finite solid-state diffusion phenomenon within the semi-analytical model used here, even with approximate diffusion coefficients, shows a far better agreement with measurements than the standard and widely used equilibrium models (lever rule and Gulliver-Scheil model), as can be seen in Figure 4b.

Additionally, it has been suggested that the morphology of Al_3Ni_2 and Al_3Ni may also affect the phase fractions of the solidified phases [26]. It has been shown that the extensive growth of the primary phase at lower cooling rates results in decreasing amount of available surfaces for peritectic phase nucleation. On the other hand, the morphology of Al_3Ni that leaves open channels between the liquid and the primary phase may change the phase fractions and must be taken into account in the simulation of peritectic transformation for Al-Ni alloys.

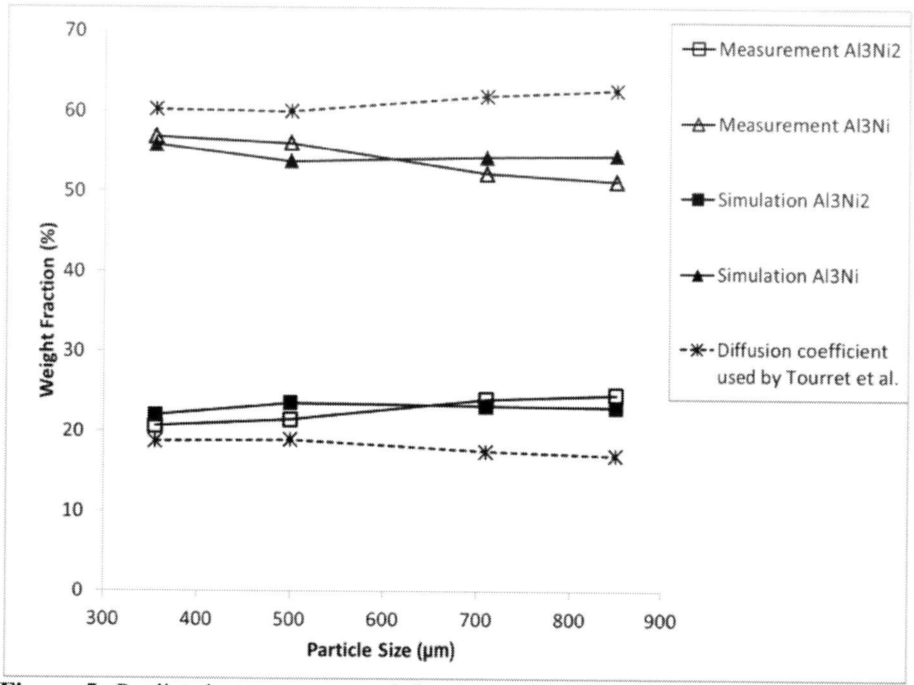

Figure 5. Predicted versus measured fractions of Al_3Ni_2, Al_3Ni and Al phases as a function of IA particle size.

5. Conclusions

Effect of particle size on phase selection and microstructural evolution during peritectic reaction in IA droplets of Al-36 wt% Ni were discussed. The fractions of phases were determined by neutron diffraction and Rietveld analysis. The amount of phases depends on the size, *i.e.* the cooling rate. A microsegregation model for the solidification of binary alloys was used that accounts for diffusion in all phases, as well as nucleation undercooling and growth kinetics of the solidifying microstructures. It considers the successive occurrence of several phase transformations, which includes one or more peritectic reactions and one eutectic reaction. The model closely predicts the measured phase fractions after the adjustment of the diffusion coefficient in solid phases. The observed discrepancies were likely due to the uncertainties on the diffusion coefficient, nucleation undercooling of different phases and the morphology of the primary and the peritectic phases.

Acknowledgements

This research was supported by Canadian Space Agency (CSA) within contract number 9F007-08-0154 and by Natural Sciences and Engineering Research Council of Canada (NSERC). In addition, the authors would like to thank National Research Council of Canada – Canadian Neutron Beam Centre, Chalk River, ON, Canada for their support in conducting the ND experiments. Funding by the European Space Agency (ESA) within the project NEQUISOL is also acknowledged.

References

[1] Devred F, Gieske A H, Adkins N, Dahlborg U, Bao C M, Calvo-Dahlborg M, Bakker J W, Nieuwenhuys B E 2009 *Appl. Catal. A* **356**(2) 154
[2] Tourret D, Gandin Ch-A, 2009 *Acta Mater.* **57** 2066
[3] Tourret D, Gandin Ch-A, Volkmann T, Herlach D M, 2011 *Acta Mater.* **59** 4665
[4] Kaufman L, Bernstein H. 1970 *Computer calculation of phase diagrams* (New York: Academic

Press)
[5] Hillert M 1998 *Phase equilibrium, phase diagrams and phase transformations* (New York: Cambridge University Press)
[6] Saunders N, Miodownik A P 1998 *CALPHAD calculation of phase diagrams: a comprehensive guide* (New York: Pergamon)
[7] Tourret D, Reinhart G, Gandin Ch-A, Iles G N, Dahlborg U, Calvo-Dahlborg M and Bao C M 2011 *Acta Mater.* **59** 6658
[8] Ziesenis J, Tillwick J, Krauss M, Uhlenwinkel V 2000 *Atomization: fundamentals and practice*, eds. Cooper K P et al. (TMS, Warrendale, PA), 61
[9] Ting J, Connor J, Ridder S 2005 *Mater. Sci. Eng. A* **390** 452
[10] Freyberg A V, Henein H, Uhlenwinkel V and Buchholz M 2003 *Metall. Mater. Trans. B* **34** 243
[11] Wiskel J B, Henein H, Maire E 2002 *Can. Metall. Q.*, **41**(1) 97
[12] Delshad Khatibi P, Ilbagi A, Beinker D and Henein H 2011 *J. Phys. Conf. Ser.*, in press, 11 pp
[13] Prasad A, Mosbah S, Henein H and Gandin Ch-A 2009 *ISIJ Int.*, **49**(7) 992
[14] Ilbagi A, Henein H and Phillion A B 2011 *J. of Mater. Sci.* **46**(19) 1855
[15] Henein H 2002 *Mater. Sci. Eng. A*, **326**(1) 92
[16] Von Dreele R B 1994 *Neutron Diffraction in Materials Science and Technology: A Comprehensive Treatment* **2B** (Wiley–VCH, New York, NY)
[17] Kurz W, Giovanola B, Trivedi R 1986 Acta Metall Mater **34** 823
[18] Jackson K A, Hunt J D 1966 T Metall Soc AIME **236** 1129
[19] Ilbagi A, Delshad Khatibi P, Swainson I P, Reinhart G and Henein H 2011 *Can. Metall. Q.* **50**(3) 295
[20] Zhang L, Du Y, Steinbach I, Chen Q, Huang B. 2010 Acta Mater **58** 3664
[21] Asta M, Morgan D, Hoyt JJ, Sadigh B, Althoff JD, de Fontaine D, et al. 1999 Phys Rev B **59** 14271
[22] Horbach J, Das SK, Griesche A, Macht M-P, Frohberg G, Meyer A. 2007 Phys Rev B **75** 174304
[23] Sarkhel AK, Seigle LL. 1982 Metall Trans A **13A** 1313
[24] Marasli N, Hunt JD. 1996 Acta Mater **44** 1085
[25] Wiskel J B, Henein H, Maire E. 2002 Can Metall Quart **41** 97
[26] Ilbagi A 2011 *Non-equilibrium containerless solidification of Al-Ni alloys* PhD Thesis, University of Alberta

IOP Conf. Series: Materials Science and Engineering 33 (2012) 012061 doi:10.1088/1757-899X/33/1/012061

Prediction of solidification path and carbide precipitation in Fe-C-V-Cr-Mo-W high speed steels

Hongwei Zhang[1], Charles-André Gandin[2,3], Jicheng He[1] and Keiji Nakajima[4]

[1] Northeastern University, Key Laboratory of Electromagnetic Processing of Materials, Shenyang, 110004, P.R. China
[2] MINES ParisTech & [3]CNRS, CEMEF UMR 7635, 06904 Sophia Antipolis, France
[4]Royal Institute of Technology, Department of Materials Science and Engineering, SE-100 44 Stockholm, Sweden

E-mail: hongweizhang@epm.neu.edu.cn

Abstract. The solidification path and precipitation of carbides in the Fe-C-V-Cr-Mo-W high speed steel system are predicted with the help of thermodynamic equilibrium calculations. The Partial Equilibrium (PE) approximation is favoured. According to experimental data for high speed steel samples, the precipitating solidification sequence of carbides, including nature, composition and amount are discussed as a function of the nominal composition of C and V. The results show that the solidification path can be reasonably predicted by the Partial Equilibrium approximation for cooling rate lower than 10 K min^{-1}. The experimental results suffer from the sensitivity limitation of the characterization methods used when the phase fraction becomes too small.

1. Introduction

The carbides in high speed steels play important roles on wear-resistant properties. In order to obtain good properties and long life-time, the nature of the carbides should be strictly controlled in the solidification structure.

In general, this is achieved through the study of solidification path using simple approximations with respect to diffusion in the solid phases. Solidification paths in a multi-component system are derived from microsegregation models. The Gulliver-Scheil (GS) approximation [1,2] can be realistic despite its omission of the diffusion of all components in solids including the interstitial elements. The Partial Equilibrium (PE) approximation [3] distinguishes the diffusion of interstitials and substitutionals as infinite and zero, respectively. The Kobayashi model [4] considers diffusion in solid but it can only be applied to dilute multi-component alloys. The Zhang et al. model [5] considers the diffusion of several components in each phase but is limited to dendritic and peritectic structures [6], no eutectic structures being yet considered as it was the case for binary alloys [7-9].

Hillert [10] analyzed the back-diffusion of components C and Cr in liquid and primary FCC phase in Fe-C-Cr alloy by using a one-dimensional numerical solution. Chen and Sundman [3] developed a Partial Eequilibrium model to predict the partial equilibrium solidification behaviour of Fe-C-Cr alloy, which is implemented in software package Thermo-Calc. They showed good agreement with the previous simulations reported by Hillert [10]. Kozeschnik et al. [11] also consider both the partial distribution of fast diffusers and the solid-solid phase transformation such as the peritectic transformation of liquid plus delta-ferrite into austenite in low-carbon steels.

In this paper, considering the back diffusion of carbon in solid, the Partial Equilibrium (PE) approximation is developed and applied to the prediction of the solidification path and carbide precipitation in the Fe-C-V-Cr-Mo-W high speed steels. The effects of C and V contents on the amounts, composition and nature of the carbides are discussed. The range of cooling rate for application is discussed through comparison between the predictions and the experimental data.

2. Experimental work

The experiments were made by Yamamoto [12]. The raw materials were melted inside an electric resistance furnace with Ar gas atmosphere. The melts were made of Fe, C, V, Cr, Mo and W in which the compositions of Cr, Mo and W were fixed to 5 wt%. The C content varies in the approximate range 1 to 4 wt% and the V content varies in the range 3 to 11 wt%. Exact alloy compositions are given in table 1. The molten alloy was poured into a cylindrical metallic mold with diameter 12 mm and length 100 mm. The solidification was studied by thermal analysis. For that purpose, 30 g rods of each alloy were remelted to a temperature 100 °C above its melting point in a furnace with SiC heater and Ar gas atmosphere, then cooled down at cooling rate of -0.17K s^{-1} (i.e. 10 K min^{-1}). The precipitation temperatures for solid phases were estimated through the thermal analysis curve. The solidification structure was observed through quenching during solidification. The phases were identified by X-ray diffraction, color etching and chemical analysis using Electron Probe Micro-Analysis (EPMA). The phase compositions were determined at the same time, i.e. close to the temperature of formation.

Table 1. Composition of alloys [12].

Alloy #	Chemical composition (wt%)				
	C	V	Cr	Mo	W
3	1.39	3.03	4.23	4.90	4.77
6	1.99	3.06	4.16	4.94	4.30
7	2.02	6.24	5.09	4.98	3.78
8	2.07	11.13	4.82	4.77	4.62
9	2.11	8.48	5.00	4.64	5.09
10	2.48	5.31	5.17	5.27	4.73
11	2.50	3.13	4.49	5.21	5.05
14	3.27	3.42	4.06	5.06	3.85

#3, #6, #11, #14 are regarded as Fe-[1-4]**C**-3V-5Cr-5Mo-5W alloys.
#6, #7, #8, #9 are regarded as Fe-2C-[3-11]**V**-5Cr-5Mo-5W alloys.
#10 was used for EPMA analyses.

3. Partial equilibrium model

Based on a global minimization technique of the Gibbs energy of a system, thermodynamic equilibrium calculation is performed providing the conditions to zero degree of freedom [13]. For a n-component system, n+2 conditions need to be set which are here the (n-1) solute compositions, the moles of the system, the temperature and pressure.

A full equilibrium calculation following the nominal composition of a system gives direct access to the lever rule (LR) approximation. The Gulliver-Scheil (GS) approximation, assuming no diffusion in solids, indicates that the new phases always precipitate from the residual liquid and the already formed solids remains frozen. The Partial Equilibrium (PE) approximation assumes no diffusion in solids for substitutionals and infinite diffusion for interstitials among which one can consider C, O, N, H and B elements in steels. The new phases still precipitate from the residual liquid as the GS model. Then the compositions and the amounts of the phases are adjusted based on equalization of the chemical potentials of the interstitials. For the present system studied, only C behaves as interstitial, all other elements being assumed as substitutional. In an imposed temperature stepping loop, with decreasing temperature, the solidification sequence then obey the following equations:

- Equal chemical potential of the interstitial component C between liquid phase, l, and a solid phase, s_1:
$$\mu_C^l = \mu_C^{s_1} \quad (1)$$
- Mass conservation of interstitial component C before and after partial equilibrium:
$$f^l\,[(w_C^l)' - (w_C^l)] / [1 - (w_C^l)'] + f^{s_1}\,[(w_C^{s_1})' - (w_C^{s_1})] / [1 - (w_C^{s_1})'] = 0 \quad (2)$$
- Unchanged u-fraction of substitutional components, such as Fe, before and after partial equilibrium in each phase:
$$(w_{Fe}^l)' = [1 - (w_C^l)'] / [1 - (w_C^l)]\,(w_{Fe}^l) \text{ and } (w_{Fe}^{s_1})' = [1 - (w_C^{s_1})'] / [1 - (w_C^{s_1})]\,(w_{Fe}^{s_1}) \quad (3)$$
- Mass conservation of substitutional components before and after partial equilibrium:
$$(f^l)' = [1 - (w_C^l)] / [1 - (w_C^l)']\,(f^l) \text{ and } (f^{s_1})' = [1 - (w_C^{s_1})] / [1 - (w_C^{s_1})']\,(f^{s_1}) \quad (4)$$

The chemical potentials of C in a phase s_1 and l are determined through phase equilibrium calculation for a given phase composition, temperature and pressure. In the above equations, (w_C^l) and $(w_C^l)'$ are the mass compositions of carbon in liquid before and after partial equilibrium, $(w_C^{s_1})$ and $(w_C^{s_1})'$ are the corresponding averaged mass composition of C in solid s_1, i.e. $(w_C^{s_1}) = \sum_k ({}^k w_C^{s_1}\,{}^k f^{s_1}) / \sum_k {}^k f^{s_1}$, where ${}^k w_C^{s_1}$ is the newly formed composition at step k in a decreasing temperature loop of a solidification sequence, f^l and $f^{s_1} = \sum_k {}^k f^{s_1}$ are the mass fraction of liquid and solid s_1 at step k, respectively.

The partial equilibrium among several phases (i.e. liquid and solid phases s_1, s_2, …, s_m) is equivalent to several partial equilibriums between two phases (e.g. liquid and solid s_1, liquid and solid s_2, … liquid and solid s_m). The mass conservation of interstitial component among all phases then follows the relation:

$$f^l\,[(w_C^l)' - (w_C^l)] / [1 - (w_C^l)'] + \sum_{j=1}^{m} \{ f^{s_j}\,[(w_C^{s_j})' - (w_C^{s_j})] / [1 - (w_C^{s_j})'] \} = 0 \quad (5)$$

The program, based on the previous work by Chen and Sundman [3], is implemented using the TQ interface for coupling with thermodynamic equilibrium calculations made by Thermo-Calc [13] and access to the thermodynamic properties of steels from the database TCFE6 [0]. The temperature step is -1 °C. The calculation finishes once the liquid fraction in mass is less than 10^{-4}.

4. Results

4.1. Solidification paths

The solidification path is predicted using the above programmed approximation. Within the composition range shown in table 1, the main precipitating carbides during solidification are MC, M_6C, M_7C_3 and M_2C. As known, the precipitation sequence of the carbides is closely related to the solute accumulated in the residual liquid. Figure 1 plots the composition evolution in residual liquid upon solidification for alloy #6. Except for the Mo and C compositions that gradually increase in the liquid, V, W and Cr increase prior to decrease. It indicates that the carbides related to V (i.e. MC), W (i.e. M_6C) and Cr (i.e. M_7C_3) successively form during the solidification sequence. The Mo-related carbides (M_2C) form later. It can also be deduced that the Mo-related carbides may be present in small amount since they form at the final stage of solidification.

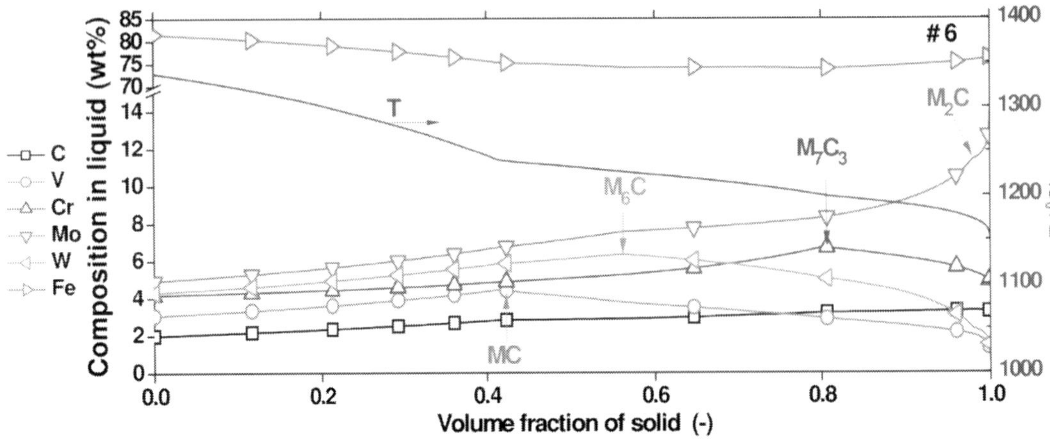

Figure 1. Variation of liquid composition for alloy #6 during the solidification sequence. Prediction assuming the PE approximation.

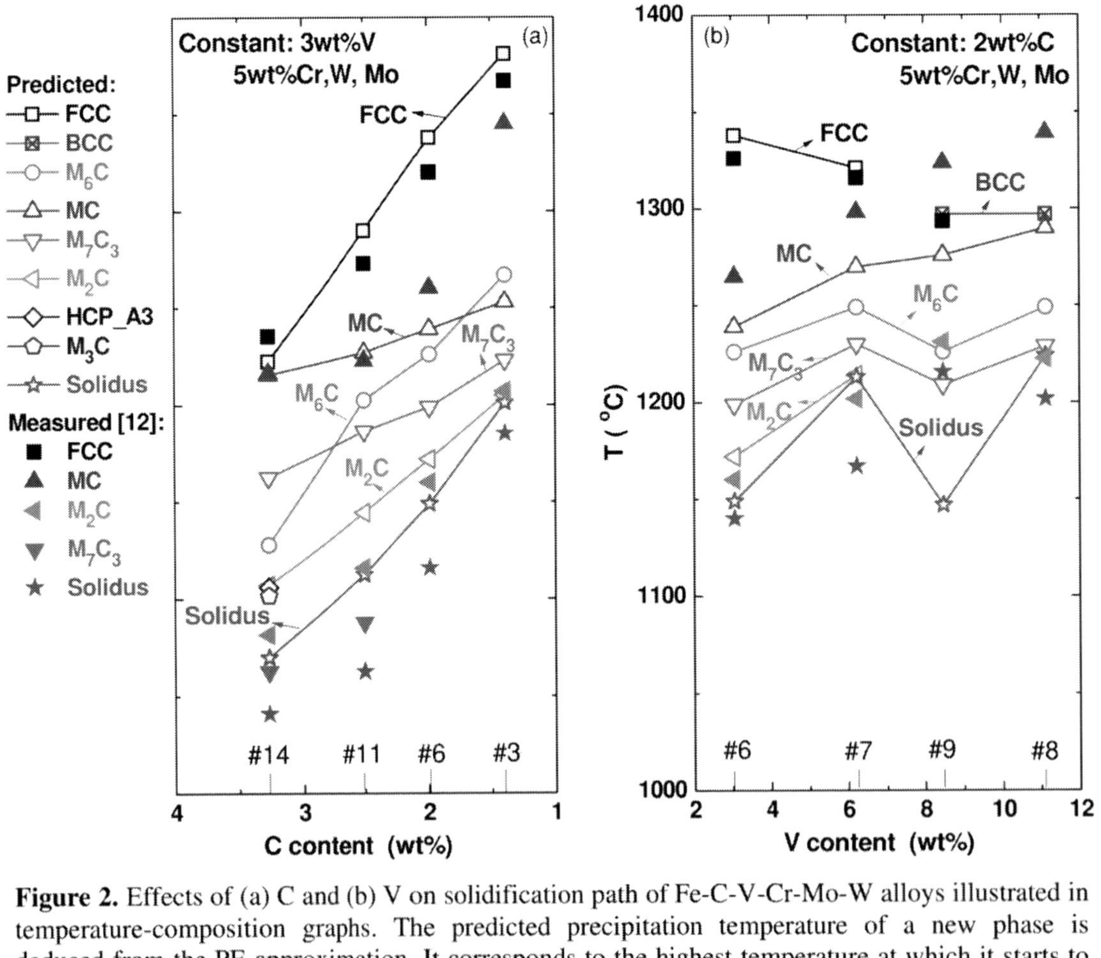

Figure 2. Effects of (a) C and (b) V on solidification path of Fe-C-V-Cr-Mo-W alloys illustrated in temperature-composition graphs. The predicted precipitation temperature of a new phase is deduced from the PE approximation. It corresponds to the highest temperature at which it starts to form. The corresponding temperature is measured from thermal analysis [12].

The carbide precipitating sequence as a function of C and V contents are illustrated in temperature - composition diagrams presented in figure 2. The curves with the open symbols denote the predicted results and the solid symbols are used for the experimental data [12]. The sequence of carbide precipitation as a function of temperature is seen on the left part of the figure 2b for alloy #6 shown in figure 1. It is found that the liquidus and solidus are predicted within a maximum 30 °C difference from the measured ones, while the formation of FCC, MC, and M_2C is retrieved by the simulation. The formation of M_6C and M_7C_3 was yet not found in the experimental analyses. It is not clear whether these carbides were not detected during the experimental analyses or the predictions are not accurate. Considering the complexity of the multicomponent systems, the comparison reached for all alloys is rather good and do provide some validation of the assessed thermodynamic data and the choice of the PE approximation.

Table 2. Compositions predicted in carbides for alloy #10 compared with EPMA data (in parentheses [12])

Carbides	Composition in u-fraction (wt%)					Note
	Fe	Cr	V	Mo	W	
MC	2.30	7.88	48.73	13.44	14.17	VC
	(2.5)	(7.5)	(56.25)	(15)	(17.5)	
M_2C	8.60	4.79	13.70	59.40	6.40	Mo_2C
	(7.5)	(15)	(11.25)	(35)	(30)	
M_7C_3	38.80	32.48	10.54	6.28	3.63	$(Fe,Cr)_7C_3$
	(57.5)	(21.25)	(5)	(11.25)	(6.25)	
M_6C	32.78	3.36	2.20	28.33	31.03	$(Fe,Mo,W)_6C$

With w_i^α the mass-fraction of a substitutional component i in phase α, and considering all substitutional components j in phase α, the u-fraction of a i in α is defined as $w_i^\alpha / \sum_j w_j^\alpha$.

4.2. Carbides composition

There are MC, M_2C, M_7C_3, M_6C, etc, carbides in the solidification structure of Fe-C-V-Cr-Mo-W alloys. As known, the composition changes in the newly formed phases during solidification. It results in a non-uniform composition distribution in the microstructure. On the other hand, the carbide compositions also changes when varying the nominal compositions of C and V. Table 2 summarizes the analyses of the carbides composition in alloy #10. As seen, the carbides MC, M_6C, M_7C_3 and M_2C contain the 6 components of the system in which one or two act as dominating component. Analyzing phase constitutions, it is known that MC is a carbide of vanadium where V dominates with u-fraction of 48.73 wt%. M_6C is a (W, Mo)- realted carbide, in which W is in u-fraction of 31 wt% and Mo is 28 wt%, respectively. M_7C_3 is a (Cr, Fe)- related carbide, where Cr and Fe are in u-fraction of 39 wt% and 32 wt%, respectively. M_2C is a carbide of molybdenum, in which Mo is in u-fraction of 59 wt%. Other carbides are also analyzed, in which M_3C is also a 6 component iron carbide with u-fraction of Fe reaching 82-83 wt%. The data in parentheses of table 2 are detected by EPMA analysis [12]. Comparison indicates that MC, M_2C, M_7C_3 contain the same 6 components and the predicted compositions for MC and M_7C_3 are almost the same as those by EPMA analysis. While the constitution of M_2C by EPMA analysis is more like the one of M_6C by prediction in which Mo and W are both the dominating components and in the same composition range. Moreover, it is indicated that Mo uniquely dominates in M_2C from the prediction. It was also shown that the experiments did not probe the existence of M_6C.

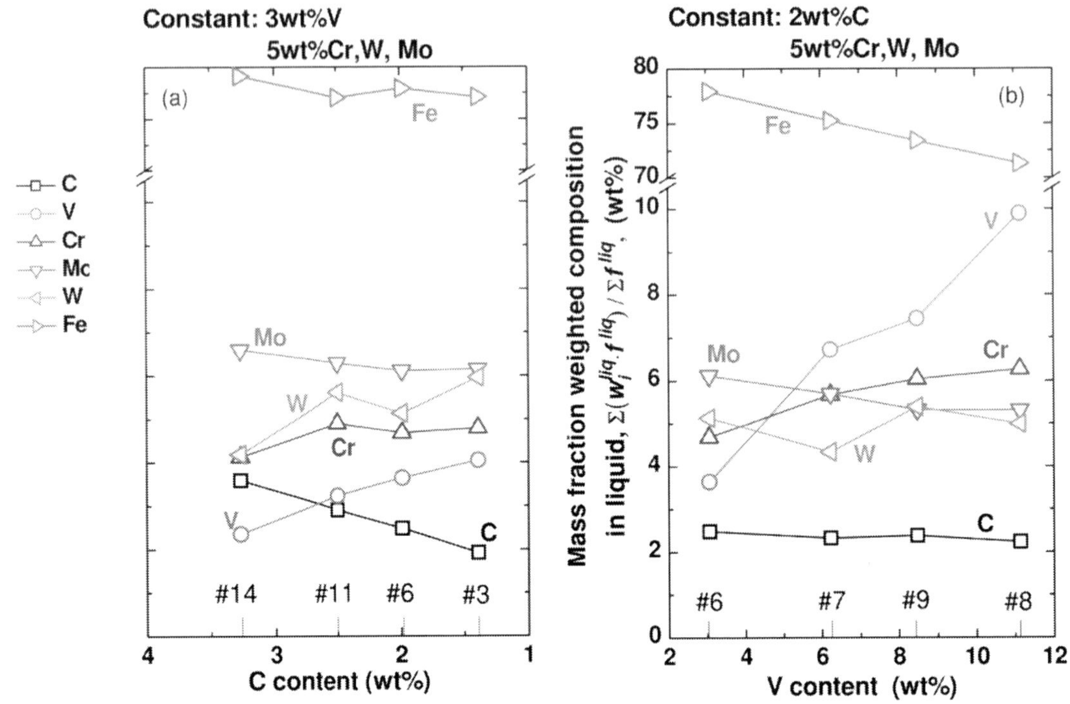

Figure 3. Effects of (a) C and (b) V on the average liquid composition during the entire solidification sequence. Prediction assuming the PE approximation.

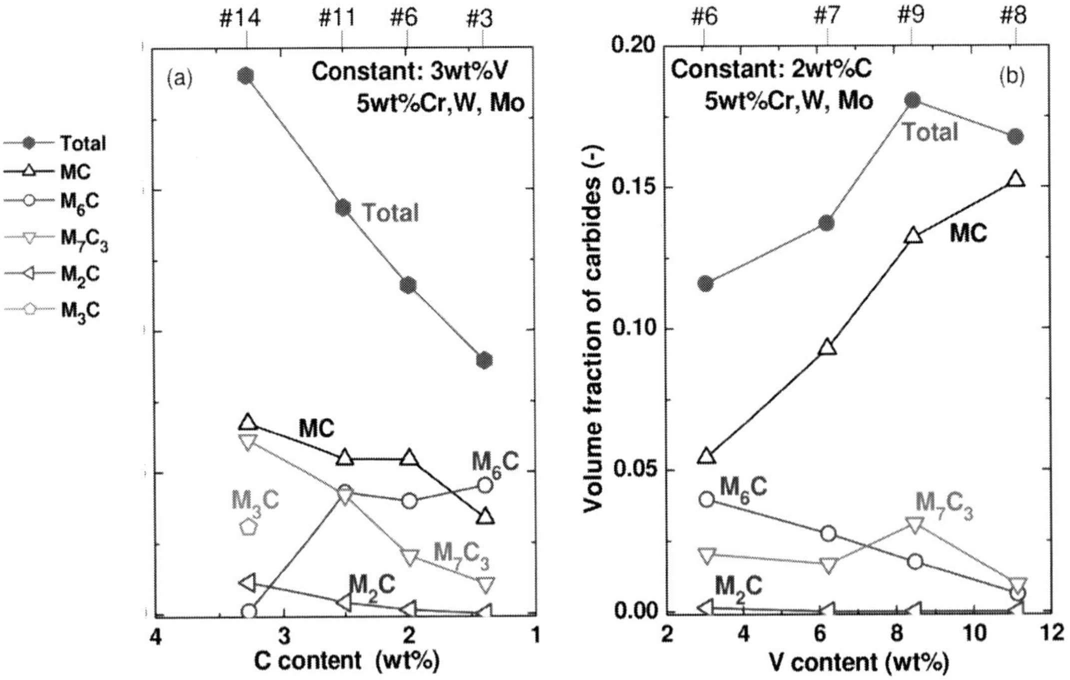

Figure 4. Effects of (a) C and (b) V on the amount and nature of the carbides formed during the entire solidification sequence. Prediction assuming the PE approximation.

5. Discussion

5.1. Effect of C

Fixing 3 wt% V and 5 wt% for Cr, W and Mo, the effects of C on the precipitation sequence of carbides is illustrated in figure 2a. With the variation of C within 1.2-3.6 wt%, the primary precipitating phase is FCC. Carbides such as MC, M_6C, M_7C_3, M_2C, and even M_3C for the highest C content, precipitate afterwards. One can compare the sequence of carbide precipitation on the same graphs for all alloys by drawing the average liquid composition during the solidification sequence. This is presented in figure 3a. It is found that with an increase of C, the compositions of Mo and C in liquid are increased gradually while the others are decreased. This indicates that with the increase of C, after the formation of the W-, V- and Cr-related carbides, the Mo-related carbides gradually form. The precipitating sequence of the carbides (figure 2a) verifies the above deductions. Following the solidification sequence, MC, M_6C and M_7C_3 carbides form firstly, afterwards M_2C form. And with the increase of C, the Mo-related carbide (M_2C) precipitates further and further away above the solidus point. These deductions are verified also from the precipitating amounts of the carbides (figure 4a) which shows that the amount of M_2C is less than 1% in volume fraction.

Analyzing the precipitating sequence, the amount as well as the composition of the carbides, one can confirm that M_6C simultaneously contains high W and Mo contents and forms in large amount at the first stage of the solidification sequence, rather than M_2C. As a result, the main carbides within the variation of C composition should be MC, M_6C and M_7C_3 instead of the measured MC, M_2C and M_7C_3 through X-ray diffraction analysis in Yamamoto's experiments [12]. Moreover, the carbides less than 1% in amount, such as M_2C in this case, can be predicted by solidification path prediction but can not be detected by X-ray diffraction due to its poor sensitivity to phase fractions less than 1%.

With the increase of C by 2wt% (from 1.3wt% to 3.3wt%, as shown in figure 4a), the volume fraction of the total carbides increases greatly by 10%. Most of the carbides increase where M_7C_3 and MC increase largely, M_6C first increases then decreases and is replaced by the other carbides such as M_3C and M_2C. The formation of these carbides closely relates to the local composition in residual liquid. As indicated in figure 3a, with the increase of C, the compositions of Fe and Mo in the residual liquid increase, which makes possible the formation of M_3C (related to Fe) and M_2C (related to Mo) but M_6C (related to both W and Mo), based on the analysis of carbide composition.

5.2. Effect of V

Keeping the composition of C as 2 wt% and varying V among 3-11wt%, as seen from figure 2b, the precipitating carbides are MC, M_6C, M_7C_3 and M_2C after the primary FCC phase when V is less than 8 wt%, or the BCC phase instead. During the formation of carbides, the accumulation or the consumption of the composition in residual liquid with the variation of V shows the same tendency as for the case of C, while the decreasing sequence is in the order of V→W→Cr→Mo. It indicates that the precipitation of carbide follows the sequence MC→M_6C→M_7C_3→M_2C. And with the increase of V, the compositions of Cr and V increase in the residual liquid but the others decrease. It indicates that the carbides vary correspondingly from Mo- and W-related carbides (M_2C, M_6C) to Cr- and V-related carbides (M_7C_3, MC and M_6C). These deductions are verified from the precipitating sequence of the carbides (figure 2b) which follows the sequence of MC, M_6C, M_7C_3 and M_2C and from the precipitating amounts of the carbides (figure 4b) which shows that the main carbides predicted are MC, M_6C and M_7C_3 within the variation of V composition. The above analysis is also deduced that it should be M_6C (with the high content of both W and Mo simultaneously) in large amount in the precipitating carbides instead of the measured M_2C through X-ray diffraction analysis in Yamamoto's experiments. With the increase of V by 8.13% (from 3 wt% to 11.13 wt% in figure 4b), the amount of the total carbides only increase by 5.5%, in which MC increases largely by 10% but the other carbides like M_7C_3 and M_6C decrease. As shown in figure 3b, the increase of V makes the compositions of Cr and V increase in the residual liquid but decreases the composition of Fe. As a result, only V related carbide MC increases while the (Cr, Fe) related carbide M_7C_3 decreases.

6. Conclusions

The precipitation sequence, the nature, the composition and the amount of carbides during the solidification of Fe-C-V-Cr-Mo-W high speed steels are discussed as a function of the nominal composition of C and V by the help of Partial Equilibrium (PE) approximation in thermodynamic calculations. It indicates that

- The main carbides formed during solidification are MC, M_6C and M_7C_3 with variations of C in the range 1.2-3.6 wt% and variations of V in the range 3-11 wt% for Fe-C-V-5wt%Cr-5wt%Mo-5wt%W alloy. Both the nature and amounts of carbides change with the C and V content:
    i. When the C content is increased, carbides change from V-, W-, and Cr-related carbides (MC, M_6C, M_7C_3) to Mo-related carbides (M_2C etc). The increase of V content tends to carbide changes from Mo- and W-related carbides (M_2C, M_6C) to the Cr- and V-related carbides (M_7C_3, MC).
    ii. The increase of C largely favours the formation of M_7C_3, M_3C and MC but M_6C carbide in amounts. With an increase of the V content, the amount of MC carbides increases largely, unlike M_7C_3 and M_6C carbides.
- Partial Equilibrium approximation, based on the thermodynamic equilibrium calculation and a reliable database, gives more reasonable prediction of the solidification path in multi-component alloy solidification process for cooling rate lower than 10 K/min, in spite of the sensitivity limitation of the characterization methods used in experiments to the phases with too small amount.

Acknowledgments
The work is supported by 111 Project [No.B07015] and Fundamental Research Funds for the Central Universities [N100409004], China, as well as by the European Space Agency [4200020277].

References

[1] Gulliver G H 1913 *J. Inst. Metals* **9** 120
[2] Scheil E 1942 *Zeitschrift für Metallkunde* **34** 70
[3] Chen Q and Sundman B 2002 *Mater. Trans.* **43** 551
[4] Kobayashi S 1988 *Trans. ISIJ* **28** 535
[5] Zhang H W, Gandin Ch-A, Ben Hamouda H, Tourret D, Nakajima K, He J C 2010 *ISIJ int.* **50** 1859
[6] Zhang H W, Gandin Ch-A, Nakajima K, He J C 2012 *MCWASP XIII proceedings volume*
[7] Tourret D, Gandin Ch-A 2009 *Acta mater.* **57** 2066
[8] Tourret D, Gandin Ch-A, Volkmann Th, Herlach DM 2011 *Acta Mater.* **59** 4665
[9] Tourret D, Reinhart R, Gandin Ch-A, Iles GN, Dahlborg U, Calvo-Dahlborg M, Bao CM 2011 *Acta Mater.* **59** 6658
[10] Hillert M, Hoglund L and Schalin M 1999 *Mater. Trans.* A **30** 1635
[11] Kozeschnik E, Rindler W and Buchmayr B 2007 *Int. J. Mater. Res.* **98** 826
[12] Yamamoto K 1999 Doctorial thesis of Kyushu University (Japan)
[13] 2008 Thermo-Calc TCCS manuals *Thermo-Calc software AB* (Stockholm, SE)
[14] Shi P 2008 TCS steels/Fe-alloys database V6.0 *Thermo-Calc Software AB* (Stockholm, SE)

IOP Conf. Series: Materials Science and Engineering 33 (2012) 012062 doi:10.1088/1757-899X/33/1/012062

Resampling technique applied to statistics of microsegregation characterization

Jacques Lacaze[1], Janin Eiken[2], Alain Hazotte[3]
1 - CIRIMAT, Université de Toulouse, BP 44362, 31030 Toulouse, France
2 - ACCESS e.V., Aachen, Germany
3 - LETAM, Université Paul Verlaine, Metz, France

E-mail: jacques.lacaze@ensiacet.fr

Abstract. Characterization of chemical heterogeneities at the dendrite scale is of practical importance for understanding phase transformation either during solidification or during subsequent solid-state treatment. Spot analysis with electron probe is definitely well-suited to investigate such heterogeneities at the micron scale that is relevant for most solidified products. However, very few has been done about the statistics of experimental solute distributions gained from such analyses when they are now more and more used for validating simulation data. There are two main sources generating discrepancies between estimated and actual solute distributions in an alloy: i) data sampling with a limited number of measurements to keep analysis within a reasonable time length; and ii) uncertainty linked to the measurement process, namely the physical noise that accompanies X-ray emission. Focusing on the first of these sources, a few 2-D composition images have been generated by phase field modelling of a Mg-Al alloy. These images were then used to obtain "true" solute distributions to which to compare coarse grid analyses as generally performed with a microanalyser. Resampling, i.e. generating several distributions by grid analyses with limited number of picked-up values, was then used to get statistics of estimates of solute distribution. The discussion of the present results deals first with estimating the average solute content and then focuses on the distribution in the primary phase.

1. Introduction

In a previous work [1], a review of the basic understanding of microsegregation build-up and its effects on cast products was made, together with a description of the most usual ways to characterize it. At present time, this characterization of solutes distribution at the scale of dendritic grains makes most often use of automatic energy (EDX) or wave-length (WDX) dispersive micro-analyzers. The procedure generally followed consists in recording N spot analyzes along a regular grid on a metallographic section of the material to be characterized. In multicomponent alloys, correlation between the contents in different alloying elements (w_i) should be analyzed to highlight the presence of minor phases [1-4] and separate primary phase from secondary phases as there can be non-monotonous evolution for some solutes during multiphase deposits [3]. Then, data are sorted and a cumulative distribution may be plotted as the fraction of points f_N giving a solute count lower (or higher) than a specified value w_i for each of the identified phases.

It is quite a usual practice in the solidification field to compare the function $w_i(f_N)$ to the so-called Scheil's model or to profiles calculated accounting for solute diffusion during solidification [5-7].

Solute distributions could as well be useful means to follow the effect of solid-state diffusion during heat treatments in the solid state that are generally designed to smooth out microsegregation [8, 9].

Very few has however been made concerning the confidence that can be put on the experimental distribution curves. Gungor [10] considered that each composition range used to sort the data could define an individual phase and applied "usual" statistics to the volume fraction estimate of each of these "virtual" phases. Actually, investigation of the statistical quality of any characterization methodology of chemical heterogeneities should best be checked on simulated images for which the distributions are known. Such studies have been performed first on very simple simulated microstructures [11-13]. Recently, more realistic images were obtained by phase field modelling on which systematic resampling was used to characterize statistics of grid analysis [14]. Since then, the proposed resampling technique has been used by Zollinger and Daloz on experimental chemical mapping of a ternary alloy [15]. The present work complements the previous study [14], it presents image generation and results from resampling, and finally discusses the best practice for microsegregation characterization.

2. Image generation

The image generation procedure is the same as previously [14], i.e. 2-D images were calculated using the phase field software Micress [16, 17] applied to the solidification of a Mg-3Al (wt.%) alloy [18]. The only difference is that the number of grain nuclei was increased from 400 to 900. As before, the grains were implemented within a calculation domain of 6000x6000 cells with a cell size Δx set at 2 μm, thus leading to an average grain size of 400 μm. Periodic conditions were assumed at the outer boundaries of the domain.

Calculations were started at 904 K for a liquidus temperature equal to 904.41 K. Temperature evolution was modelled averaged over the calculation domain by the balance between heat extraction and latent heat release. The specific heat was set to 2.8 $J.K^{-1}.cm^{-3}$ for the liquid and 1.9 $J.K^{-1}.cm^{-3}$ for the (Mg) solid solution. A constant heat extraction rate of -30 $J.cm^{-3}$ combined with a latent heat of 650 $J.cm^{-3}$ resulted in an initial decrease in temperature followed by a distinct recalescence during about 4 seconds. Driving force for growth Δg ($J.cm^{-3}.K^{-1}$) was extrapolated from a thermodynamic calculation performed for the initial composition based on Calphad data [19]:

$$\Delta g = \left[\frac{d\Delta g}{dT}\right]_{liq} \cdot \Delta T = 0.842 \cdot \{T - T_{liq}(w^0)\} \qquad (1)$$

where T [K] stands for temperature and the subscript "liq" for liquidus. The liquidus temperature T_{liq} is expressed as:

$$T_{liq}(w^l) = T_{liq}(w^0) + m_{liq} \cdot (w^l - w^0) = 904.407 - 5.215 \cdot (w^l - 3.0) \qquad (2)$$

where w^0 and w^l are the nominal and liquid aluminium contents (wt.%), and m_{liq} is the liquidus slope.

Solute partitioning at the interfaces was extrapolated from the same thermodynamic calculation by:

$$w^s = w^{s,0} + \frac{m_{liq}}{m_{sol}}(w^l - w^0) = 0.819 + \frac{-5.215}{-17.626}(w^l - 3.0) \qquad (3)$$

where w^s and $w^{s,0}$ stand for solute content in the solid in equilibrium with liquid at, respectively, w^l and the nominal content w^0. The partition coefficient between (Mg) and liquid is equal to 0.273 at the time of nucleation and increases slightly during solidification. Diffusion of Al was simulated with a constant coefficient in the liquid (10^{-5} $cm^2.s^{-1}$) and temperature dependent in the solid phase (D_{so}=12 $cm^2.s^{-1}$ and Q_s=144000 $J.mol^{-1}$) giving a diffusion coefficient of 5.6 10^{-8} $cm^2.s^{-1}$ at the liquidus temperature. In order to have a significant change in solute concentration within the final images, all simulations were stopped after a solidification time of 14 s, the solid fraction being about 0.7. The Al concentration fields were then written out to generate images.

Figure 1a shows the composition map for one grain and figure 1b the composition profile along the line A-B in figure 1a. When a nucleus has formed at the given undercooling, it grows with solute piling up away from the liquid/solid interface. This leads to the increase in the solid concentration from zero to about 100 µm in figure 1b. After some time, the growth rate decreases due to the recalescence and so does the pile-up, this corresponds to the solute decrease from 100 to 290 µm. Finally, when neighbouring solute fields in the liquid start interacting with each other, the concentration in the liquid increases and so does again the concentration in the solid, from 290 to 310 µm.

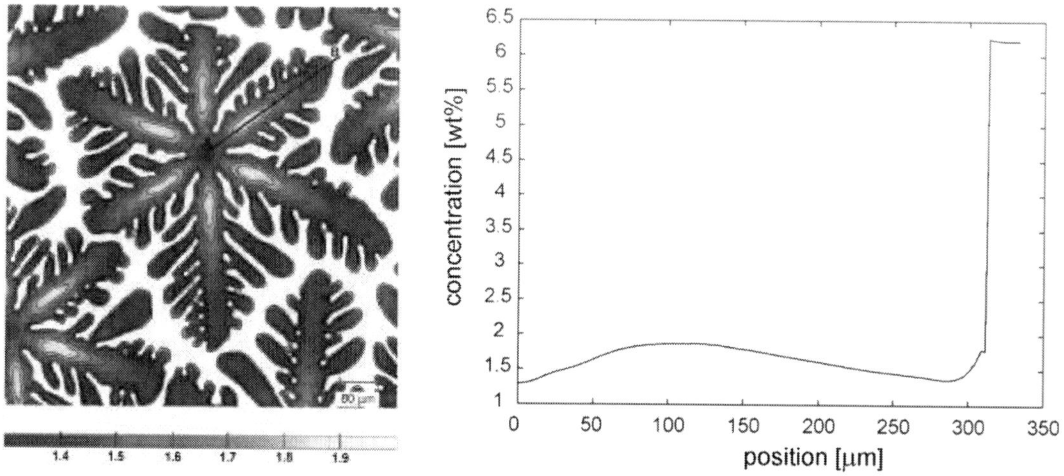

Figure 1. Mapping of the aluminium distribution in one grain (left column) and aluminium profile along the dendrite arm AB (right column).

Interfacial energies and mobilities of the solid phase were defined with hexagonal anisotropy [18] and grain orientations were set randomly. Figure 2-a shows an image for which the nuclei have been implemented with a minimal distance (320 µm) between them leading to grains with nearly equal size (the image will be referred as ES for equal-size). Figure 2b presents a zoom on a second image when the nuclei have been implemented at random (labeled RM for random in the following). A zoom of each image is presented in figure 3. In both images, the liquid in the interdendritic areas is enriched in aluminium. It could as well be considered as a eutectic with unresolved microstructure.

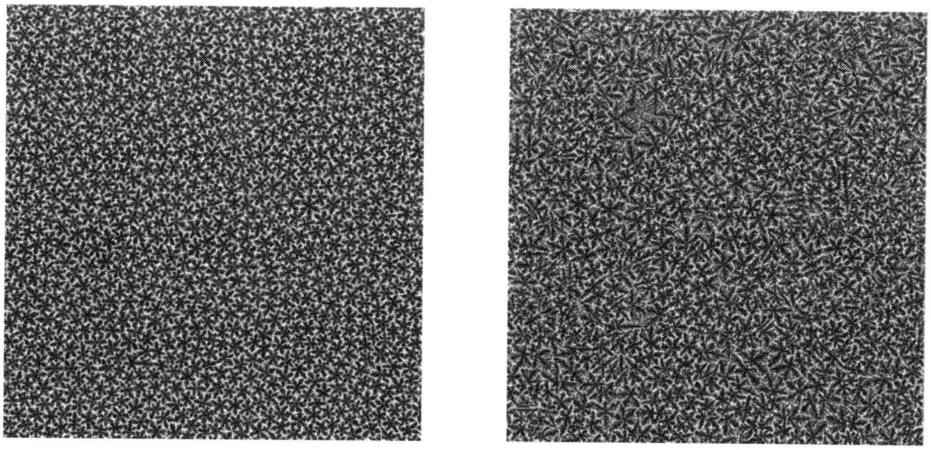

Figure 2. Images generated with 900 grains implemented: a) with geometrical constraint (left column); or b) randomly (right column). Note that the colour scale is not the same as in figure 1a. The side length is 1.2 mm.

Figure 3. Zoom of the images in figure 2. Note that the colour scale is not the same as in figure 1a. The side length is 350 μm.

The chemical composition distribution of each image (6000x6000 cells) was drawn using classes 0.1 wt.% Al in size. Figure 4a compares the distributions of the ES and RM images. They are quite similar with two main peaks corresponding to the primary and liquid phases at low and high aluminium contents respectively. The fact that the peak at low solute contents is doubled is somehow fortuitous with the side-peak at higher Al contents being related to the upper part of the recalescence stage in figure 1b.

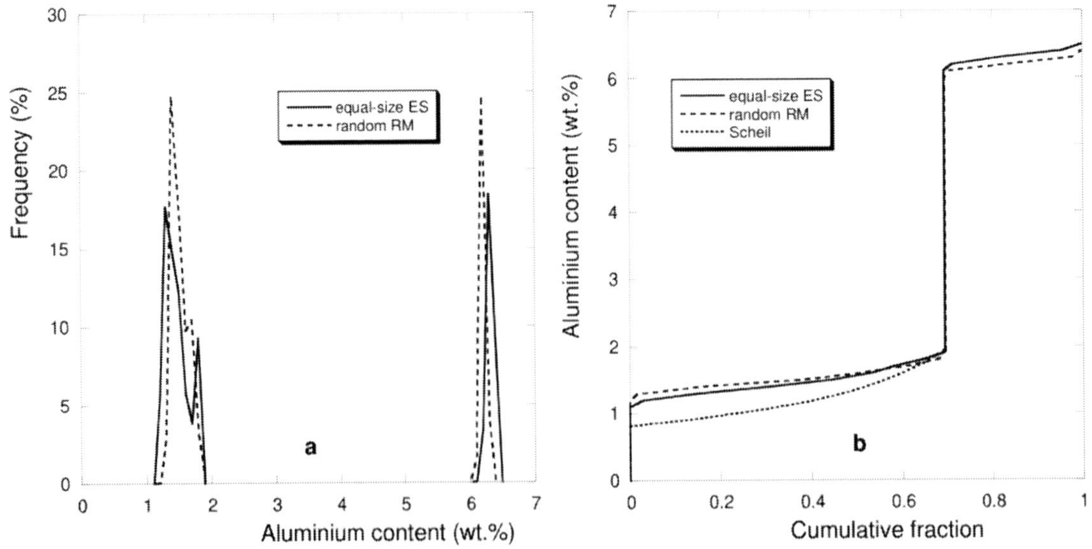

Figure 4. Distribution (a) and cumulative distribution (b) of aluminium. Solid and interrupted lines correspond to ES and RM images.
The dotted line in (b) is the profile calculated according to Scheil's model.

The corresponding cumulative distributions are plotted in figure 4b. As expected, they show very limited difference between each other in agreement with the distribution in figure 4a. The peak associated to the primary phase is slightly narrower for the RM image in figure 4a, leading to a slightly flatter curve in figure 4b when compared to that for image ES. It is worth stressing that the highest aluminium contents in the primary phase, here associated to cumulative fractions in between 0.6 and 0.7, are in fact related to the recalescence stage as described above.

It is also noted in figure 4b a negative curvature on the two curves at low solute content that is to be associated to slight differences in the extent of solid state diffusion in solid deposited. In fact, the importance of solid state diffusion is clearly demonstrated by comparing the ES and RM cumulative curves to that calculated with Scheil's model also plotted in figure 4b. As a matter of fact, the undercooling of 0.4 K before nucleation leads to a shift of the first solid from 0.819 to 0.84 wt.% Al (equations 2 and 3). This latter value is far below the minimum value observed on the ES and RM distribution curves, so that the shift of these curves to higher Al contents is due to solid state diffusion.

Solute profiles are sometimes used to evaluate partition coefficient, and this is generally done avoiding accounting for the lowest fractions in the cumulative distribution because of the negative curvature appearing there. If Scheil's model is assumed, a constant partition coefficient k can be evaluated by considering the change in solute content between two solid fractions $f_{s,1}$ and $f_{s,2}$ as:

$$k = 1 + \frac{\ln\left({}^{w_{s,1}}\!/\!{}_{w_{s,2}} \right)}{\ln\left({}^{(1-f_{s,1})}\!/\!{}_{(1-f_{s,2})} \right)} \quad (4)$$

where $w_{s,1}$ and $w_{s,2}$ are the corresponding solute contents in (Mg) primary phase. Taking $f_{s,1}$=0.05 and $f_{s,2}$=0.70 the k value obtained is 0.61 or 0.69 for ES and RM distribution respectively. These values are more than twice as large as the value used for calculating the images.

Evaluation of the average aluminium content on the whole images gives in both cases 3 wt.% Al at better than 0.005 wt.%. Using this value as the mean of the distribution, the variance of the distributions could be calculated as 4.7 and 5.03 $(wt.\%)^2$ for RM and ES images respectively. These values can be compared with the one obtained considering the material is made of two phases with average solute contents of 1.45 and 6.25 wt.% Al and volume fractions of 0.7 and 0.3 respectively. The variance for a random composition measurement is then $\{0.7.(1.45-3)^2+0.3.(6.25-3)^2\}=4.9$ wt.% Al right in between the values for RM and ES images.

Since the calculation cell used in the present work (2μm) has a dimension close to the usual size of measurement spots, these values give estimate of the intrinsic variability of the composition value that could be recorded on an actual microstructure though not accounting for physical noise. In this particular system, it is concluded that the standard deviation of the composition in one pixel, σ_0, will be about 2.17-2.24 wt.% at the very best. When averaging over N non-correlated pixel values, the 95% confidence interval of the estimated mean solute content is about $\pm 2.\sigma_0.N^{-0.5}$, i.e. ±0.44 wt.% for a grid of 10x10 points or ±0.22 wt.% for a 20x20 one. This realistic and minimum accuracy limit may still appear quite high if measurements are dedicated to the estimation of the average (nominal) concentration of a given element.

3. Resampling

In order to investigate the statistics of data recorded from single grid analysis, systematic resampling was performed by implementing series of regular grids of N=10x10 "measurement" points with a first point selected at random in the image. The grid step ΔL was the same in the x and y directions of the grid, and could be varied from one resampling series to another. In order to have significant statistics, 1000 such grids were implemented for each series, which on the whole means picking up 10^5 cell values, a number which is still small when compared to the 3.6 10^7 cells in each image.

In a first step, emphasis was put on the evaluation of the average solute content by grid analysis. Averaging this quantity as obtained through the 1000 grids of a given resampling procedure (i.e. a given grid step) showed it to give always the expected value of 3.00 wt.% at better than 0.01 wt.%. It was however noted that the standard deviation σ_{grid} on this average solute content depends on the grid spacing as already reported [14]. Figure 5 shows the evolution of σ_{grid} as function of ΔL/D, where D is the average grain size (400 μm) for both the RM and ES images. It is seen that σ_{grid} decreases rapidly for ΔL/D varying from 0.1 to 0.25, and then presents oscillations which level off slowly as the grid spacing increases, i.e. as the "measurements" get less and less correlated. At that stage, i.e. when ΔL/D is higher than 1.5-2, the curves have more or less reached the expected random value σ_0=0.22. Although the differences between ES and RM curves are not pronounced, they confirm the spatial correlations that can exist between grid spacing and microstructure size [13, 14]. In order to avoid bias due to such correlations, the grid spacing to be used should be at least twice as large as the grain size.

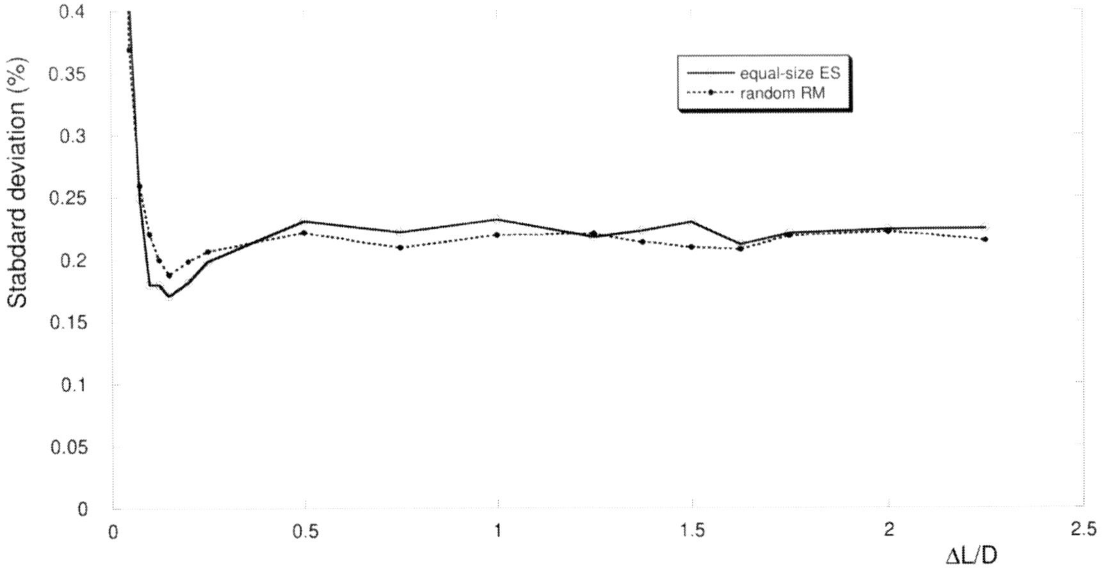

Figure 5. Evolution with grid spacing of the standard deviation of the average aluminium content estimated from 1000 grids of 100 points each.

In a second step, it appeared of interest to characterize the standard deviation for each class of solute values. Figure 6 presents the standard deviation of the number of pixel values in each class (dots) evaluated again from 1000 grids of 100 points each obtained with a grid size equal to 800 μm. It is seen that its average value for each peak is of the same order as the standard deviation for the estimate of the phase fraction V_V, either solid or liquid, namely $V_V.(1-V_V).N^{-2}$, giving 4.6% for a grid of N=100 points.

The same calculation was applied to each of the solute classes using the distribution values in figure 4a. This results in the estimates of the standard deviation shown with dots in figure 6. Such a procedure, which is exactly that suggested by Gungor [10] is seen to reproduce fairly well the standard deviation calculated by resampling.

Finally, it is worth recalling that it has been noted previously that the observed scattering in the data associated with each solute class has little effect on the cumulative distribution estimated for either the primary or liquid phase [14]. In other words, the cumulative distribution in the primary phase may be fairly well established by recording as low as 100 values along a grid with an appropriate step size. The, most of the variance in the estimate of the average solute content is related to the variance in the estimate of the liquid (or solid) fraction.

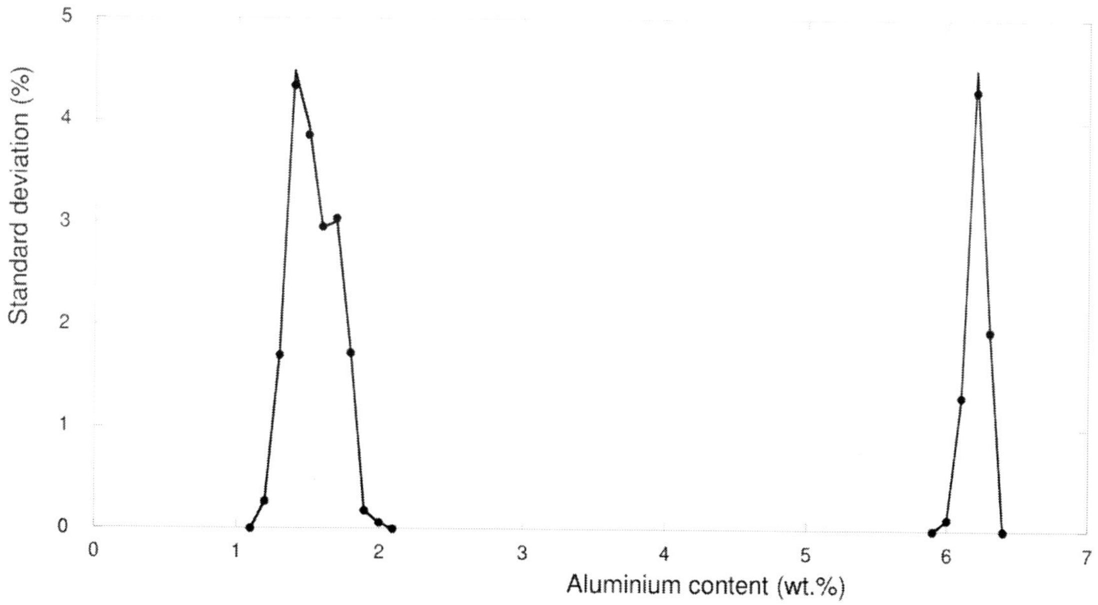

Figure 6. Standard deviation of the number of values in each solute class estimated from 1000 grids of 100 points each (solid line). Dots are the standard deviation using the distribution in figure 4a and assuming fully random distribution.

4. Discussion - conclusion

Using resampling, it has been shown that analysis along regular grids gives better results than random sampling according to the variance estimate of the measurements [13,15]. However, the variance depends on the grid spacing, and Gungor [10] was the first to stress the need to use a large grid spacing, preferably much larger than the size of the grains. Such a condition relates to the need for measurements to be independent from each other on a statistical point of view. In particular, any microstructure space periodicity may induce correlation in the measurement that could strongly bias the results. This has been discussed previously [13] and it was found that such correlations are sufficiently smoothed out when the grid spacing is 2-3 times the grain size as illustrated again in figure 5. In practice, a rough estimation of grain size or grain density can easily be obtained by quantitative optical or electron-scanning image analysis.

The classical tools for statistical analysis can be then used when measurement points are uncorrelated. For instance, the variance of estimation of the volume fraction V_V of a "phase" (liquid or eutectic areas, or regions within a given composition class) is obtained as $V_V.(1- V_V).N^{-2}$, where N is again the total number of measurement points. Applied to the present work where V_V=30%, the estimate of the liquid (eutectic) fraction from a grid of 100 points could vary in between 21 and 39 % at a 95% confidence level. For the same conditions, the estimate of the average solute content may vary in between 2.56 and 3.44 wt.% when the true value is 3 wt.%. These ranges of possible variation due to statistics appear quite large and much more measurements would be needed for accurate evaluation of such integral quantities.

In contradistinction, the results presented above show it is quite feasible to estimate solute distribution in dendritic alloys by means of a limited number of local composition measurements, typically 100 measurement points, in agreement with previous works [11, 15]. This is certainly due to compensating effects between countings made on neighbour solute classes. However, physical noise is present in real microanalysis, that can be simulated by adding a noise following a normal distribution to each composition value picked up from the image [14]. As expected, this noise also exacerbates the negative curvature of the curve at low aluminium content that is already present in the original curve as a result of the size distribution of secondary and tertiary arms [6].

2-D cuts of real structures or of 3-D simulations should certainly lead to some slight increase of the variance, for both the average solute content and solute distribution [20]. However, these effects are not expected to change significantly the conclusions of this work as the unit cells for chemical heterogeneities will remain the grains.

A last note should be made which is that, whatever the sorting method used, the present results show that cumulative distributions could at best indicate the solute segregation behavior, i.e. positive or negative segregation, but should definitely not be used for estimating partition coefficients.

References

[1] Lacaze J and Lesoult G 1991 *Proc. Nature and Properties of Semi-Solid Material* (San Diego) (Warrendale, USA: TMS) p 105
[2] Feest E A 1983 *Proc. Solididfication technology in the foundry and casthouse* (London: The Institute of Metals) p 188.
[3] Siredey N and Lacaze J 1992 *Le Vide, les Couches Minces* suppl. **261** 149
[4] Mirkovic D and Schmid-Fetzer R 2010 *Proc. Int. Conf. Magnesium alloys and their applications (Weimar)* (Weinheim, Germany: Wiley-VCH) p 1231
[5] Lacaze J and Lesoult G 1993 *Materials Science and Engineering* **A173** 119
[6] Lacaze J and Lesoult G 1995 *ISIJ International* **35** 658
[7] Yan X, Chen S, Xie F and Chang Y A 2002 *Acta materialia* **50** 2199
[8] Grosdidier T, Hazotte A and Simon A 1990 *Proc. Euromat'89* (Oberursel, Germany: DGM Informationgesellschaft mbH) p 395
[9] Hermann W, Lacaze J and Oquab D 2003 *Advanced Engineering Materials* **5** 32
[10] Gungor M N 1989 *Metall. Trans. A* **20A** 2529
[11] Yang W, Chen W, Chang K M, Mannan S and DeBarbadillo J 2000 *Metall. Mater. Trans. A* **31A** 2569
[12] Hazotte A, Lecomte J S and Lacaze J 2005 *Mater. Sci. Eng. A* **413-414** 223.
[13] Hazotte A and Lacaze J 2007 *Trans. Indian Inst. Met.* **60** 267
[14] Hazotte A, Eiken J and Lacaze J 2010 *Proc. Int. Conf. Magnesium alloys and their applications (Weimar)* (Weinheim, Germany: Wiley-VCH) p 398
[15] Zollinger J and Daloz D 2011 *Materials Characterization* **62** 1058
[16] www.micress.de
[17] Eiken J, Boettger B and Steinbach I 2006 *Physical review E* **73** 066122
[18] Eiken J 2010 *Int. J. Mater. Res.* **4** 503
[19] Schmid-Fetzer R, Janz A, Gröbner J and Ohno M 2005 *Advanced Engineering Materials* **7** 1142
[20] Lacaze J, Benigni P and Howe A 2003 *Advanced Engineering Materials* **5** 37

IOP Conf. Series: Materials Science and Engineering **33** (2012) 012063 doi:10.1088/1757-899X/33/1/012063

A multiphase segregation model for multicomponent alloys with a peritectic transformation

Hongwei Zhang[1], **Charles-André Gandin**[2,3], **Keiji Nakajima**[4] **and Jicheng He**[1]

[1] Northeastern University, Key Laboratory of Electromagnetic Processing of Materials, Shenyang, 110004, P.R. China

[2] MINES ParisTech & [3]CNRS, CEMEF UMR 7635, 06904 Sophia Antipolis, France

[4] Royal Institute of Technology, Department of Materials Science and Engineering, SE-100 44 Stockholm, Sweden

E-mail: hongweizhang@epm.neu.edu.cn

Abstract. A multiphase microsegregation model for the solidification of multicomponent alloys that experience a peritectic transformation is developed. It is based on a volume averaging method. Coupling is achieved among the conservation equations for total mass, solute mass and energy. The diffusion in extradendritic and several interdendritic liquid and solid phases, the growth kinetics of the dendritic and peritectic solidifying microstructures, as well as the velocities of the solid/liquid and solid/solid phase interfaces are considered in the model. The model is applied to the dendritic and peritectic solidification of a Fe-C-Cr alloy. Equilibrium between phases at the interface is taken into account and computed using a thermodynamic software. The occurrence of several recalescences due to the growth of the microstructure and the peritectic transformation are predicted and the solidification behaviors near recalescence are evaluated. By adjusting the parameters, the lever rule (LR), Gulliver-Scheil (GS) and partial equilibrium (PE) approximations are retrieved.

1. Introduction

Most industrial applications in metal processing deal with multicomponent alloy systems. The segregation pattern during the solidification process mainly determines the structures and properties of the final products. Usually several microstructures concurrently exist in a multicomponent alloy system such as dendritic, peritectic and eutectic ones. Better understanding and distinguishing between intra and intergranular segregations among these structures is the field of on-going research.

For the dendritic structure, Wang and Beckermann [1] developed the segregation model for equiaxed dendritic solidification considering back-diffusion in solid in a binary alloy system. Rappaz and Boettinger [2] extended the model to a multicomponent alloy with the simplification of uniform composition in interdendritic liquid region and linearized phase diagrams. Tourret and Gandin [3-5] removed these assumptions by considering non-complete mixing of solute in interdendritic liquid and coupling with thermodynamic equilibrium calculations. Their model predictions were compared with measured data with concurrent dendritic, peritectic and eutectic reactions in binary Al-Ni alloys. Zhang et al. [6] extended the model to predict a dendritic solidification in a ternary alloy system for a single dendritic solid phase.

For the peritectic structure, Kobayashi [7] proposed a multicomponent segregation model based on the mathematical solution of Brody and Fleming to predict the peritectic transformations in steels.

Thuinet and Combeau [8] developed a segregation model to predict the dendritic and peritectic structures for Fe-Ni-C alloys. In their model the distribution of solute is solved in each phase. This differs from the parabolic distribution assumed in the models proposed by Wang and Beckermann, Tourret et al. and Zhang et al. [1, 3-6]. Natsume et al. [9] also developed a simple one-dimensional free boundary model coupled with thermodynamic calculation to predict the peritectic phase transformation in a Fe-based multicomponent alloy.

The present paper presents an extension of the microsegregation model developed by Zhang et al. [6] to the peritectic transformation in multicomponent alloy systems. Direct coupling with equilibrium calculations is achieved in the Fe-C-Cr system. Detailed discussions on the recalescence and the peritectic transformation are given. Extreme cases of the model retrieving the lever rule (LR), the Gulliver-Scheil (GS) and the partial equilibrium (PE) approximations are shown.

2. Mathematical model

A multicomponent microsegregation model is proposed based on volume averaging for heat and mass exchanges in a multiphase system where all phases are fixed and have equal and constant densities, ρ [1,3-6]. Illustration of the solidifying domain is given in figure 1. The main assumptions are listed hereafter.

i- The system is one-dimensional and spherical with radius R, external surface A and volume V.

ii- Solid phases, s_1 and s_2, nucleate at the centre of the domain in an undercooled liquid, l, and propagate by radial growth. Phase s_1 develops a dendritic microstructure. Its dendrite tip position, $R^{(1)}$, borders the outer zone (0), filled with the extradendritic liquid, $l^{(0)}$. The inner mushy zone (1) is a mixture of solid phase s_1, $s_1^{(1)}$, and interdendritic liquid, $l^{(1)}$. Phase s_2 develops a peritectic microstructure. The boundary between zone (2) and zone (1) is located at position $R^{(2)}$. In zone (2), the peritectic phase $s_2^{(2)}$ grows between the interdendritic liquid $l^{(2)}$ and the primary solid phase, $s_1^{(2)}$.

iii- Mass exchanges are considered at all s/l, s/s and l/l interfaces and zone boundaries except $s_1^{(1)}/l^{(0)}$ and $s_2^{(2)}/l^{(1)}$ interfaces.

iv- The temperature is uniform over the system due to the large value of Lewis number.

v- Equilibrium is assumed at all phase interfaces.

vi- A dendrite tip growth kinetics for multicomponent alloy system is used. It is based on the marginal stability approximation [10] and the Ivantsov solution for the diffusion field outside a paraboloidal dendrite tip [11].

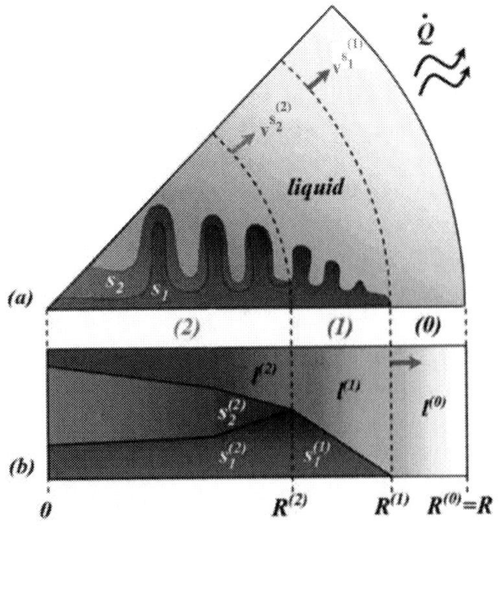

Figure 1. Schematic distribution of the dendritic phase, s_1, peritectic phase, s_2, and the liquid phase, l, in the extradendritic domain, (0), and the interdendritic domains, (1) and (2). Distinction is made between the solid phases, $s_1^{(1)}$, $s_1^{(2)}$, $s_2^{(2)}$ and the liquids of different compositions, $l^{(0)}$, $l^{(1)}$and $l^{(2)}$. The dendritic and peritectic microstructures nucleate at the centre of a spherical geometry with radius R. The boundary between mushy zone $(1)=s_1^{(1)}+l^{(1)}$ and extradendritic liquid $(0)=l^{(0)}$ is located at position $R^{(1)}$, the other boundary between mushy zone $(2)=s_1^{(2)}+s_2^{(2)}+l^{(2)}$ and mushy zone $(1)=s_1^{(1)}+l^{(1)}$ is located at position $R^{(2)}$.

2.1. *Averaged equations*

Neglecting the macroscopic diffusion fluxes, the conservation equations for the total mass and solute mass of a phase α surrounded by α/β interfaces are given by:

$$\frac{\partial g^{\alpha}}{\partial t} = \sum_{\beta(\beta\neq\alpha)} S^{\alpha/\beta}\, v^{\alpha/\beta} \tag{1}$$

$$g^{\alpha}\frac{\partial <w_i^{\alpha}>^{\alpha}}{\partial t} = \sum_{\beta(\beta\neq\alpha)} S^{\alpha/\beta}\,(w_i^{\alpha/\beta} - <w_i^{\alpha}>^{\alpha})\left(v^{\alpha/\beta} + \frac{D_i^{\alpha}}{l_i^{\alpha/\beta}}\right) \tag{2}$$

The corresponding interfacial mass and solute mass balances between phase α and β can be written:

$$v^{\alpha/\beta} + v^{\beta/\alpha} = 0 \tag{3}$$

$$(w_i^{\alpha/\beta} - w_i^{\beta/\alpha})\, v^{\alpha/\beta} + \frac{D_i^{\alpha}}{l_i^{\alpha/\beta}}(w_i^{\alpha/\beta} - <w_i^{\alpha}>^{\alpha}) + \frac{D_i^{\beta}}{l_i^{\beta/\alpha}}(w_i^{\beta/\alpha} - <w_i^{\beta}>^{\beta}) = 0 \tag{4}$$

The global energy balance during alloy solidification is given by:

$$\frac{\partial <H>}{\partial t} = \sum_{\alpha}\left(<H^{\alpha}>^{\alpha}\frac{\partial g^{\alpha}}{\partial t} + g^{\alpha}\frac{\partial <H^{\alpha}>^{\alpha}}{\partial T}\frac{\partial T}{\partial t} + g^{\alpha}\sum_{i}\frac{\partial <H^{\alpha}>^{\alpha}}{\partial <w_i^{\alpha}>^{\alpha}}\frac{\partial <w_i^{\alpha}>^{\alpha}}{\partial t}\right) = -\frac{h_{ext}\, S_{ext}}{\rho}(T-T_{ext}) \tag{5}$$

where the system exchanges heat with its surroundings maintaining at temperature T_{ext} by an apparent heat transfer coefficient, h_{ext}.

2.2. *Interfacial area concentrations and diffusion lengths*

The interfacial area concentration is defined as $S^{\alpha/\beta} = A^{\alpha/\beta}/V$. By assuming a plate-like geometry, they are expressed as $S^{l^{(1)}/l^{(0)}} = 3\, R^{(1)^2}/R^3$, $S^{s_1^{(1)}/s_1^{(2)}} = (g^{s_1^{(1)}}/g^{(1)})\,(3\, R^{(2)^2}/R^3)$ and $S^{l^{(2)}/l^{(1)}} = (g^{l^{(1)}}/g^{(1)})\,(3\, R^{(2)^2}/R^3)$ at zone boundaries and $S^{l^{(1)}/s_1^{(1)}} = 2\, g^{(1)}/\lambda_2$ and $S^{l^{(2)}/s_2^{(2)}} = S^{s_2^{(2)}/s_1^{(2)}} = g^{(2)}\,(2/\lambda_2)$ at the interfaces in zones, respectively.

The diffusion length is defined as $l_i^{\alpha/\beta} = -(w_i^{\alpha/\beta} - <w_i^{\alpha}>^{\alpha})/(\partial w_i/\partial n)|_{\alpha/\beta}$ for element i in phase α from α/β interface. By considering a parabolic composition profile, it is written as $l_i^{\alpha/\beta} = (g^{\alpha}/g^{(k)})(\lambda_2/6)$ for α phase ($\alpha = s_1^{(1)}$ and $l^{(1)}$ from $s_1^{(1)}/l^{(1)}$ interface and $s_1^{(2)}$ from $s_1^{(2)}/s_2^{(2)}$ interface as well as $l^{(2)}$ from $s_2^{(2)}/l^{(2)}$ interface) at the interfaces in zone(k) (k=1 or 2). The diffusion length for $s_2^{(2)}$ from $s_1^{(2)}/s_2^{(2)}$ and $s_2^{(2)}/l^{(2)}$ interface is $l_i^{\alpha/\beta} = (g^{s_2^{(2)}}/g^{(2)})(\lambda_2/12)$ by considering a symmetrical condition in the middle of phase $s_2^{(2)}$. In liquid, the diffusion in radial direction is also considered by assuming a quasi-stationary diffusion profile. The mathematical expressions for $l_i^{l^{(0)}/l^{(1)}}$ and $l_i^{l^{(1)}/l^{(2)}}$ are available in [3-5].

2.3. *Tip growth kinetics*

The boundary velocities for zones (1) and (2) are set equal to the tip velocities of the solid phases $s_1^{(1)}$ and $s_2^{(2)}$, respectively. The growth kinetics for dendritic growth of solid $s_1^{(1)}$ into the liquid ahead in zone (0) and peritectic growth of solid $s_2^{(2)}$ into the liquid ahead in zone (1) and (0) follows the same solution of supersaturation, equation (9), and the marginal stability limit, equation (11), with the help of the thermodynamic equilibrium at the interface, equation (6). Regarding the dendritic growth of solid $s_1^{(1)}$, it is the solution of the following equations

$$T_L(<w_i^{l^{(0)}}>^{l^{(0)}}) - T = \sum_i \frac{\partial T}{\partial w_i^{l^{(0)}/s_1^{(1)}}}(<w_i^{l^{(0)}}>^{l^{(0)}} - w_i^{l^{(0)}/s_1^{(1)}}) + \frac{2\,\Gamma^{s_1^{(1)}}}{r^{s_1^{(1)}}} \tag{6}$$

$$\Omega_i = \frac{w_i^{l^{(0)}/s_1^{(1)}} - <w_i^{l^{(0)}}>^{l^{(0)}}}{w_i^{l^{(0)}/s_1^{(1)}} - w_i^{s_1^{(1)}/l^{(0)}}} \tag{7}$$

$$k_i^{s_1^{(1)}/l^{(0)}} = \frac{w_i^{s_1^{(1)}/l^{(0)}}}{w_i^{l^{(0)}/s_1^{(1)}}} \tag{8}$$

$$\Omega_i = \mathrm{Iv}(Pe_i) = Pe_i \exp(Pe_i)\, E_1(Pe_i) \tag{9}$$

$$Pe_i = \frac{r^{s_1^{(1)}}\, v^{s_1^{(1)}/l^{(0)}}}{2\, D_i^l} \tag{10}$$

$$r^{s_1^{(1)}\,2}\, v^{s_1^{(1)}/l^{(0)}} = \frac{1}{4\pi^2} \sum_i \frac{D_i^l\, \Gamma^{s_1^{(1)}}}{(\partial T/\partial w_i^{l^{(0)}/s_1^{(1)}})\, w_i^{l^{(0)}/s_1^{(1)}}\, (k_i^{s_1^{(1)}/l^{(0)}} - 1)} \tag{11}$$

where $T_L(<w_i^{l^{(0)}}>^{l^{(0)}})$ is the temperature read on the liquidus surface in multicomponent $s_1^{(1)}$ +Liquid metastable phase diagram at composition $<w_i^{l^{(0)}}>^{l^{(0)}}$. When considering the growth of peritectic solid $s_2^{(2)}$ into the liquid ahead in zone (1) together with zone (0), the properties are substituted correspondingly. For example, the liquidus temperature in this case is $T_L(<w_i^{l^{(0+1)}}>^{l^{(0+1)}})$ instead at the averaged composition $<w_i^{l^{(0+1)}}>^{l^{(0+1)}}$. $<w_i^{l^{(0+1)}}>^{l^{(0+1)}}$ is the phase fraction- weighted composition of liquid $l^{(1)}$ and $l^{(0)}$.

2.4. *Compositions at boundary $l^{(k-1)}/l^{(k)}$ (k=1, 2) and $s_{k-1}^{(k-1)}/s_{k-1}^{(k)}$ (k=2)*

Combining equation (4) and the continuity conditions such as $w_i^{l^{(0)}/l^{(1)}} = w_i^{l^{(1)}/l^{(0)}}$ at $l^{(0)}/l^{(1)}$ boundary, the compositions at the $l^{(0)}/l^{(1)}$, $l^{(1)}/l^{(2)}$, $s_1^{(1)}/s_1^{(2)}$ boundaries can be obtained.

2.5. *Thermodynamic data and phase equilibrium*

By directly coupling with the Thermo-Calc software and using PTERN database [12] for the Fe-C-Cr system, the thermodynamic data and phase equilibrium parameters such as $<H^\alpha>^\alpha$, $\partial<H^\alpha>^\alpha/\partial T$, $\partial<H^\alpha>^\alpha/\partial<w_i^\alpha>^\alpha$, $k_i^{\alpha/\beta}$ and $\partial T/\partial w_i^{\alpha/\beta}$ can be directly accessed.

3. Algorithm

The equilibrium nucleation temperature is determined by a thermodynamic equilibrium calculation with the local alloy composition. With cooling of the system, the primary and peritectic phases nucleate gradually at the centre of the spherical domain when the local undercooling is greater than the prescribed nucleation undercoolings. The zone boundary velocity, the interfacial area concentrations and the diffusion lengths, the solid/solid and solid/interdendritic liquid interface compositions and velocities, and the zone boundary compositions are first determined. Then the total mass conservation, equation (1), the solute mass conservation, equation (2), and the global energy balance, equation (5), are solved iteratively by the Gear method until convergence. The phase fractions, the average compositions of the phases and the system temperature are obtained.

Table 1. Material parameters, initial and boundary conditions.

Parameter		Value			
Thermodynamic database		PTERN[12]			
Nominal mass composition,	$i = C$	0.6			
$w_{i,0}$ (wt %)	$i = Cr$	10			
Diffusion coefficients in liquid,	$i = C$	10^{-8}			
D_i^l ($m^2\ s^{-1}$)	$i = Cr$	10^{-9}			
		P[13]	LR	GS	PE
Diffusion coefficients in BCC phase,	$i = C$	$6{\cdot}10^{-9}$	$6{\cdot}10^{-9}$	10^{-16}	$6{\cdot}10^{-9}$
D_i^{s1} ($m^2\ s^{-1}$)	$i = Cr$	$1.7{\cdot}10^{-11}$	10^{-10}	10^{-16}	10^{-16}
Diffusion coefficients in FCC phase,	$i = C$	10^{-9}	$6{\cdot}10^{-9}$	10^{-16}	$6{\cdot}10^{-9}$
D_i^{s2} ($m^2\ s^{-1}$)	$i = Cr$	$2.7{\cdot}10^{-13}$	10^{-10}	10^{-16}	10^{-16}
Secondary dendrite arm spacing, λ_2 (m)		20×10^{-6}	6×10^{-6}	30×10^{-6}	20×10^{-6}
Gibbs-Thomson parameter for BCC/LIQ, $\Gamma^{s1}_{(1)}$ (K m)		10^{-9}		10^{-13}	
Gibbs-Thomson parameter for FCC/LIQ, $\Gamma^{s2}_{(2)}$ (K m)		$1.9{\cdot}10^{-9}$		$1.9{\cdot}10^{-13}$	
Density of the phases, ρ ($kg\ m^{-3}$)		7674			
Radius of the system, R (m)		10^{-3}			
Initial temperature (K)		1739.46			
Nucleation undercooling (K)		0			
Heat transfer coefficient, h_{ext} ($W\ m^{-2}\ K^{-1}$)		40			
Surrounding temperature, T_{ext} (K)		293			

4. Results and Discussions

The properties needed in the calculations are listed in table 1. The standard diffusion coefficients for a Fe-0.6wt%C-10wt%Cr alloy are retrieved from the MOB2 mobility database [13]. Their values in each solid phase are taken at each phase nucleation temperature and are kept constant during the simulation. All the simulations assume a constant pressure and a closed system.

4.1. Solidification path using the present model compared with LR, GS and PE approximations

The present model simulation is labelled P in table 1 and in figure 2. Adjusting the values of the diffusion coefficients and the secondary arm spacing, the solidification path predicted by the LR, GS and PE approximations can be retrieved as illustrated in figure 2.

The LR approximation expects full mixing of solutes in liquid and solids. This suggests that the values of the diffusion coefficients are the largest and the diffusion length is the thinnest compared to other approximations. To tend towards this limit, the diffusion coefficients of C and Cr in solids are all artificially increased to approach the values in liquid as retrieved from MOB2. The secondary arm spacing is set to a small value, i.e. 6 μm. As can be seen in figure 2, the thin line curves with open symbols retrieve the predictions using Thermo-Calc, i.e. using a full equilibrium approximation. The peritectic FCC phase (s_2) thus quickly and fully consumes the primary BCC phase (s_1) immediately at the peritectic point, 1726.61K. The solidification finishes with a unique FCC phase at 1672.84K.

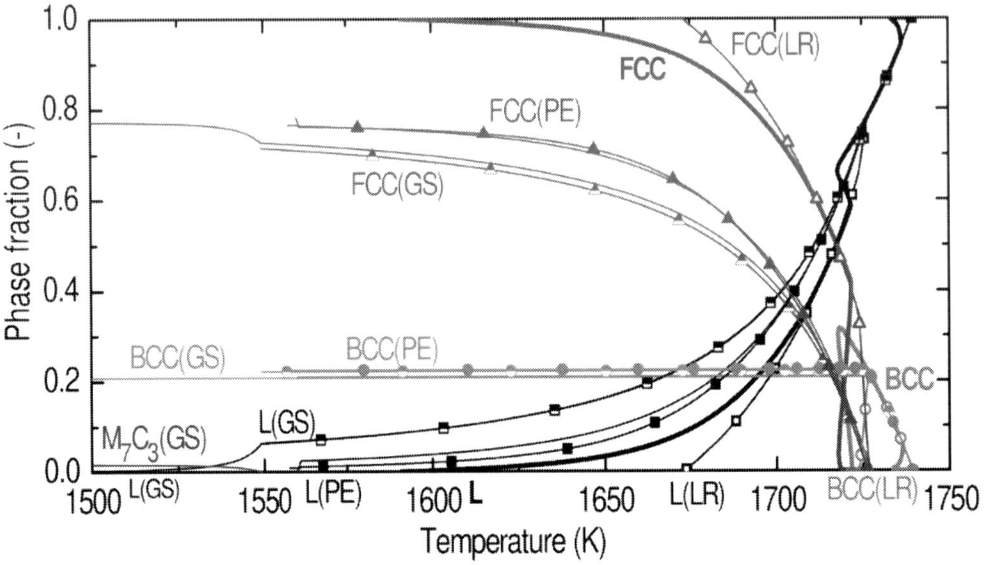

Figure 2. Solidification path predicted by the present P model. Thick lines with labels represent simulations with the parameters listed in column P of table 1. Retrieving the solidification path of LR, GS and PE approximations are denoted by the thin lines with open, half solid and solid symbols, respectively, with the parameters shown in table 1. The solidification path predicted by LR, GS and PE approximations is illustrated by the thin lines with labels but no symbols. A colour code is used to identify the (red) dendritic BCC, (blue) peritectic FCC and (black) liquid L phases. The (brown) M_7C_3 phase is also predicted with the GS and PE approximations.

The GS approximation assumes no diffusion of solutes in solids. The thickness of the diffusion layer is thus the thickest compared to other approximations. In order to approach this assumption, all diffusion coefficients of C and Cr in solid phases are set to a very small value, 10^{-16} m^2 s^{-1}. The secondary arm spacing is set to 30 μm. With such little diffusion of the solutes in solids, the BCC/FCC interface becomes fixed. As a result, the already formed BCC phase is just frozen as shown by the thin line curves with half solid symbols in figure 2. No peritectic transformation can thus form and the dendritic/peritectic interface is blocked, both the BCC and FCC phases formed being kept constant upon cooling. Moreover, the end of solidification is delayed to a low eutectic point. Here the calcualtion is just stoped at the first eutectic point, i.e. upon nucleation of M_7C_3 phase at 1549.27K.

The case of PE distinguishes the diffusivities of interstitual and substitutional solutes in solids. Here, the diffusion coefficients of C are set as LR and the ones of Cr are set as GS. The secondary arm spacing is chosen as 20 μm to provide an appropriate value of diffusion length in between the cases of LR and GS. As known, fast diffusion of C and no diffusion of Cr have no effects on the movement of BCC/FCC interface, the latter being fixed due to the absence of diffusion of one of the chemical specie. So the already formed BCC phase is also kept constant when the formation of the FCC phase takes place. This is made visible by the thin line curves with solid symbols in figure 2. There is no peritectic transformation as well. In this case, solidification finnishes at a higher nucleation temperature of M_7C_3 phase, 1560.15K.

The more general case is considered with diffusion coefficients extracted from the MOB2 database and giving a certain thickness of the diffusion length. The secondary arm spacing is set to 20 μm. The

Gibbs-Thomson parameters are in the order of 10^{-9} K m. As found in table 1, the diffusion of C is very fast, and the one of Cr is greater than PE model. As a result, the predicted solidification path (the thick lines) is in between the ones of LR and PE in figure 2. The solidification finishes at 1590.81K, i.e. before the formation of M_7C_3, and the solidification path approaches LR. The BCC phase is consumed fast by FCC phase near the peritectic point, 1726.66K. The peritectic transformation can be detected.

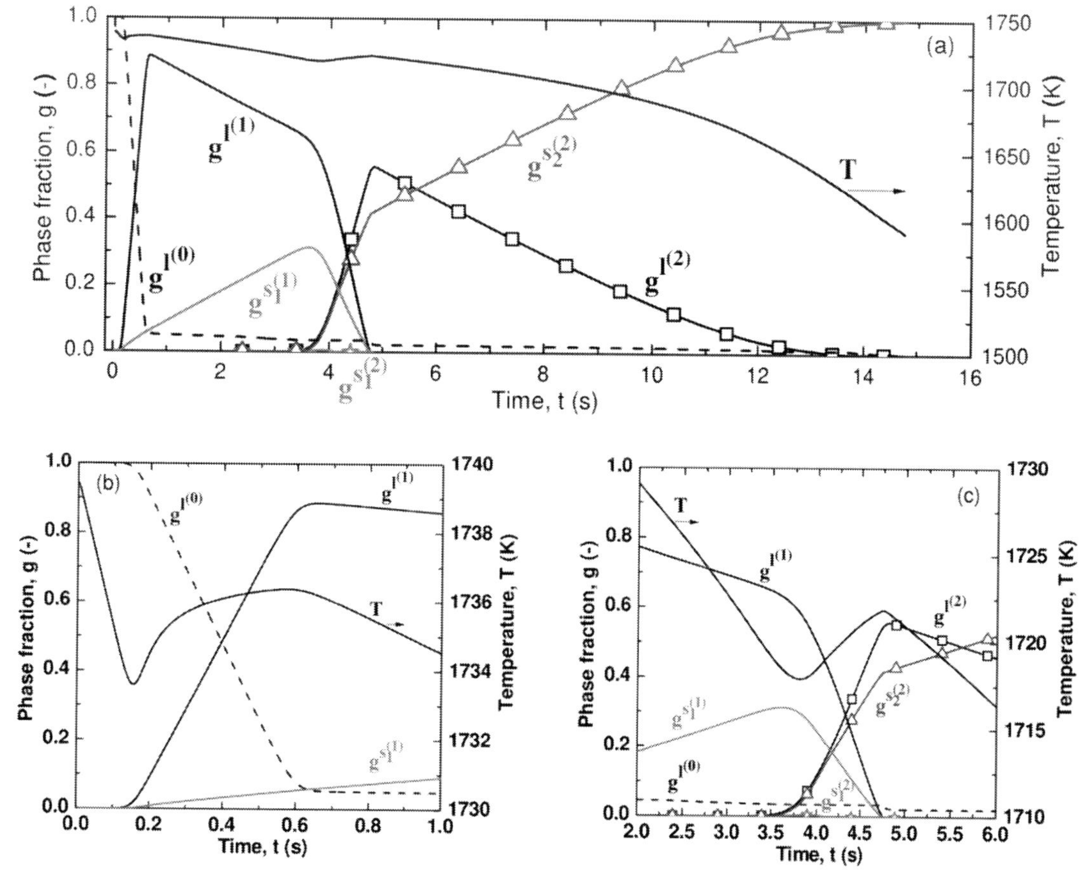

Figure 3. Time evolutions of temperature, phase fraction and zone fraction with the present model. The heat transfer coefficient h_{ext} is 40 W m^{-2} K^{-1}, other parameters adopted are listed in column P of table 1. (a) global view of the cooling curve, magnification during the recalescences predicted (b) below the liquidus temperature and (c) below the peritectic temperature.

4.2. Recalescence and evolution of the peritectic transformation

Figure 3 gives the time evolutions of the temperature and phase fraction using the parameters in column labeled P of table 1. Two recalescences can be seen on the cooling curve. They are associated with the formation of $s_1^{(1)}$=BCC primary dendritic phase and $s_2^{(2)}$=FCC peritectic phase, as shown in figure 3b and 3c, respectively. The recalescence takes place when the heat released due to the fast development of the mushy zone (zone (1) or zone (2)) is greater than the heat extracted by the surroundings. When the two contributions compensate, the maximum temperature is reached. Finally, the temperature decreases when the mushy zone is almost fully developed.

The FCC phase nucleates at about 2.4 s, as illustrated in figure 3c. This indicates the formation of zone (2). At the initial stage, FCC grows slowly in amount and BCC continues its growth into the interdendritic liquid in zone (1) up to reaching a maximum value. Then the FCC phase gradually propagates and surrounds the primary BCC phase in zone (2) upon reaction of the peritectic phase. During this period, the FCC phase grows up in large amount, replacing the liquid and the BCC phase. This is the so-called peritectic transformation. There is small amount of the BCC phase in zone (1) maintained after the consumption by the FCC phase. It is thus converted to the corresponding BCC phase in zone (2). Figure 3a and 3c show the change of the BCC fraction in zone (1) and (2). The fast increase of the mushy zone (2) makes the second recalescence. At the maximum of the recalescence, zone (1) is totally overgrown by zone (2) and disappears. The amount of zone (2) approaches its maximum at 4.8s, as well as the amounts of the interdednritic liquid inside. The peritectic transformation also finishes at the same time with the moderate Gibbs-Thomson parameters of 10^{-9} K m. Afterwards, the formation of the FCC phase from the residual liquid in zone (2) and finally in zone (0) takes place until the end of solidification.

5. Conclusion

A multiphase microsegregation model for the solidification of multicomponent alloys that experience a peritectic transformation is developed. The model includes quantitively the effects of the cooling rate and the diffusivities of solutes in both liquid and solids. The application of the model to the dendritic and peritectic solidification of a Fe-C-Cr alloy shows that

1) The present finite-diffusion microsegregation model is verified through reproducing the extreme approximations of LR, GS and PE by adjusting the diffusion coefficients and the secondary arm spacing.

2) The recalescence due to the fast development of the mushy zone and evolution of the peritectic transformation due to the solute diffusion can be properly interpreted.

3) The model shows potentials to estimate the effects of cooling rate while being coupled with a macroscopic transport model for application to the solidification of multicomponent alloys.

Acknowledgments

The work is supported by 111 Project [N B07015] and fundamental Research Funds for the Central Universities [N100409004], China.

References

[1] Wang C Y and Beckermann C 1993 *Metall. Trans.* A **24** 2787
[2] Rappaz M and Boettinger W J 1999 *Acta Mater.* **47** 3205
[3] Tourret D, Gandin Ch-A 2009 *Acta mater.* **57** 2066
[4] Tourret D, Gandin Ch-A, Volkmann Th, Herlach DM 2011 *Acta Mater.* **59** 4665
[5] Tourret D, Reinhart R, Gandin Ch-A, Iles GN, Dahlborg U, Calvo-Dahlborg M, Bao CM 2011 *Acta Mater.* **59** 6658
[6] Zhang H W, Gandin Ch-A, Ben Hamouda H, Tourret D, Nakajima K, He JC 2010 *ISIJ int.* **50** 1859
[7] Kobayashi S 1988 *Trans. ISIJ* **28** 535
[8] Thuinet L and Combeau H 2009 *Comp. Mater. Sci.* **45** 285, 294
[9] Natsume Y, Shimamoto M, Ishida H 2010 *ISIJ int.* **50** 1867
[10] Langer J S and Muller-Krumbhaar H 1977 *J. Cryst. Growth* **42** 11
[11] Dantzig J A and Rappaz M 2009 *Solidification* EPFL Press (Lausanne, CH)
[12] Shi P 2008 *Public ternary alloy solutions database V1.3* Thermo-Calc Software AB (Stockholm, SE)
[13] Helander T 1998 *MOB2-Mobility database* Thermo-Calc software AB (Stockholm, SE)

IOP Conf. Series: Materials Science and Engineering **33** (2012) 012064 doi:10.1088/1757-899X/33/1/012064

Linking up of HT-LSCM and DSC measurements to characterize phase diagrams of steels

P Presoly, R Pierer and C Bernhard
Chair of Metallurgy, Montanuniversitaet Leoben,
Franz-Josef-Straße 18, A-8700 Leoben, Austria

Email: peter.presoly@unileoben.ac.at

Abstract. The phase transformation sequence during the solidification of carbon steels strongly influences their behavior in the casting process. Therefore, most exact knowledge of the dependence of the transformation characteristics on the steel composition is of highest relevance for process and quality optimization. The influence of alloying elements like C, Mn or Si on phase transformation is well understood as far as their content is rather low. New steel grades, like high-alloyed TRIP- or TWIP-steels contain almost up to 10 wt.-% of Si and Al and 30 wt.-%Mn. The present work focuses on first results of the parallel investigation into phase transformation of Fe-Al-C alloys by means of Differential Scanning Calorimetry (DSC) and Thermo-Optical Analysis (TOA) with a High-Temperature Laser-Scanning-Confocal-Microscope (HT-LSCM). DSC is a well established method for the accurate measurement of all phase transformation temperatures accompanied by significant enthalpy changes. Due to small enthalpy changes, DSC results are limited with respect to the γ/δ-transformation. Besides dilatometry and X-ray diffraction, the optical in-situ observation of phase transformation by HT-LSCM proved to be a comprehensive method. After a short description of the methods, results for the Fe-0.4%Al-0.22%C and Fe-1.5%Al-0.22%C systems will be discussed in detail and finally compared with results from computational thermodynamics.

1. Introduction

Phase diagrams and the knowledge of phase transformations of materials represent very important information for scientists, materials engineers and process operators to understand material behavior during solidification, heat treatment and the whole further processing. Regarding steels, a special feature are the two different high-temperature phases: γ(fcc)- and δ(bcc)-iron. Figure 1 illustrates the equilibrium binary Fe-C (gray thin lines) and four ranges of different transformation sequences, which are additionally summarized with a table in this figure. Steels which solidify according to the transformation sequence of range II are of special interest, because in this case the transformation of δ-Fe to γ-Fe (i.e. the peritectic phase transition $\gamma \rightarrow L+\delta$) starts in the solid-liquid two phase region and ends in the solid. It is believed that this specific transformation sequence during solidification and subsequent cooling is responsible for an increased defect appearance (such as hot tears, surface defects, depression formation and in the worst case breakouts) in the continuous casting process [1-4]. This critical range II is often characterized by a carbon content between C_A and C_B. However, since steel is always a multi-component material it is clear that pseudo binary phase diagrams must be used to understand the transformation sequence.

Figure 1 schematically shows how alloying elements significantly influence the Fe-C phase diagram (green thick lines) and the position and temperature of the characteristic points C_A*, C_B* and C_C*. Pseudo binary Fe-C diagrams of multi-component alloys can be calculated with commercial Gibbs minimizers (CALPHAD's approach) such as ThermoCalc (TC), FactSage (FS), MTDat, PANDAT to mention just a few examples. Using such commercial software tools, the position and temperature of the characteristic points C_A*, C_B* and C_C* can be easily determined. Low alloyed steels (typical construction and engineering steel grades) and simple Cr and Ni high alloyed steels (stainless and tool-steel grades) seem to be well described in commercial thermodynamic databases.

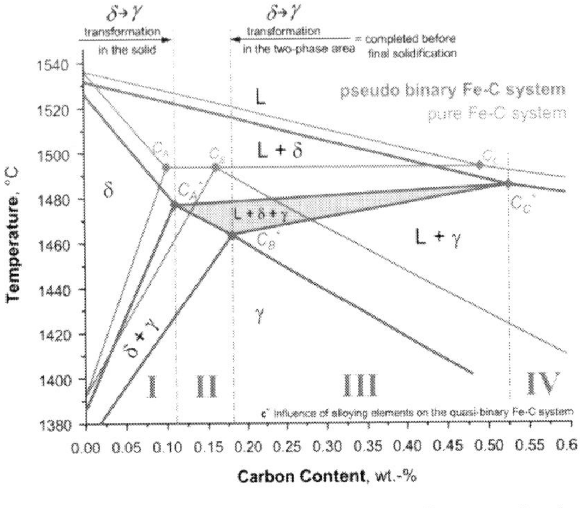

Austenit Former | Ferrite Former
Mn, Ni, Cu, P, N... | Cr, Mo, S, Ti, Al..

Different transformation characteristics in the Fe-C system

Range	Position	Phase sequence	Characteristics
I	left of C_A	L → L+δ → δ → δ+γ → γ	*primary δ-Fe solidification*
II	between C_A and C_B *= peritectic*	L → L+δ → δ+γ → γ	*peritectic transformation starts during the solidification and finish in the solid*
III	between C_B and C_C *= hyper-Peritectic*	L → L+δ → L+γ → γ	*peritectic transformation starts during the solidification finish in the L/S region*
IV	right of C_C	L → L+γ → γ	*primary γ-Fe solidification*

Figure 1. Fe-C equilibrium diagram in the high temperature range.

In contrast to the above mentioned steels, however, calculated pseudo binary phase diagrams of novel steels grades which exhibit for example higher content of Mn, Si and Al (typically TRIP and TWIP steels) show partly divergent results. This can mainly be explained by a lack of trustworthy thermodynamic data. Figure 2 shows, for example, the results of calculated pseudo binary Fe-1%Si-C (left) and Fe-1%Al-C (right) diagrams using ThermoCalc 5.0 (TC) and FactSage 6.2 (FS).

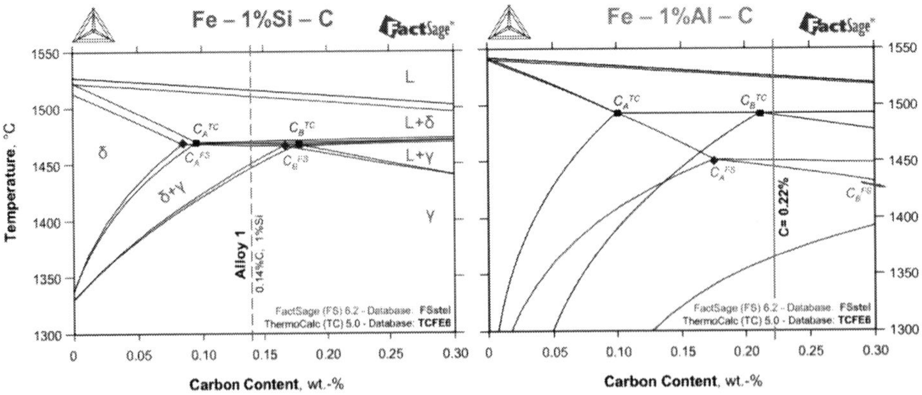

Figure 2. ThermoCalc and FactSage calculation of the pseudo binary Fe-C diagram of the system Fe-1%Si-C (left) and Fe-1%Al-C (right).

The results of the ThermoCalc and FactSage calculations show very good agreement (i.e. the difference in the transformation temperatures ΔT is smaller than 10 °C) in the case of Fe-1%Si-C alloy, but very contradictory results in the case of Fe-1%Al-C. In order to visualize these uncertainties with respect to Al in detail, Figure 3 shows the pseudo binary Fe-Al-0.22%C diagram, calculated with

ThermoCalc and FactSage. It can clearly be seen that already at rather low values of Al, both calculated phase diagrams are totally different. The above mentioned and illustrated disagreement, especially with respect to the results when Al is present, exemplarily shows the great need for experimental methods to evaluate thermodynamic calculations and to measure new systems on the one hand. On the other hand, however, sometimes only the knowledge of the transformation behavior and sequence is of interest (i.e. is a new steel grade between C_A* and C_B*). In doing so, different experimental methods can be used. The present study shows the possibilities, assets, drawbacks and limitations of two different experiments, the High-Temperature Laser-Scanning-Confocal-Microscopy (HT-LSCM) and the Differential Scanning Calorimetric (DSC) method using the example of three different alloys. These three alloys are Fe-1%Si-0.14%C, Fe-0.4%Al-0.22%C and Fe-1.5%Al-0.22%C, which are additionally illustrated in Figure 2 (left) and Figure 3.

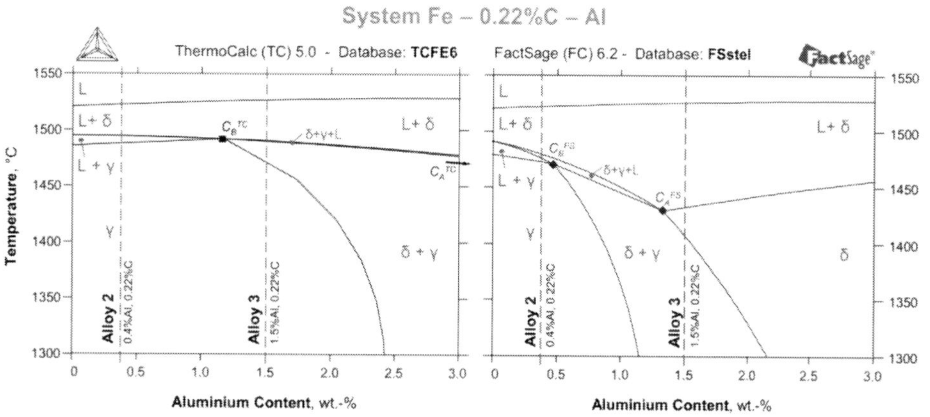

Figure 3. Test alloys and thermodynamic calculations of the pseudo binary Fe-Al system.

2. Experimental Methods

In the following section, the two applied experimental methods are described. Prior to that it should be mentioned that the production of the necessary samples was the same in both cases: The alloys were melted in an alumina crucible from high purity raw materials with a High Frequency Remelting Furnace under argon atmosphere and were centrifugally cast in a copper mould and analyzed by emission-spectroscopy.

The High-Temperature Laser-Scanning-Confocal-Microscopy is a special kind of Thermo-Optical Analysis (TOA) which enables in situ observations of surface phenomena in liquid and solid samples up to a maximum temperature of 1650°C. Since this well established method is described in detail in the relevant literature [5-8], the present study gives just a brief overview on the experimental set-up: The right hand side of Figure 4 shows an overview of the whole facility, the left hand side shows the experimental assembling together with detailed images of the sample holder in the high-temperature furnace. This infrared furnace consists of a gold-coated chamber and has the shape of a symmetric ellipse. The heating is carried out by means of a halogen lamp which is located in the bottom focal point, whereas the crucible is in the upper focal point.

Similar to other thermal analysis methods, protective gas atmosphere (argon, quality 6.0), type-S thermocouples in the sample holders and alumina crucibles were used. In the present study, the following experimental adjustments and parameters were employed. The size of the samples was 4x4x1mm (polished, but not etched). The heating rate was 500 K/min up to a temperature of 1300 °C, subsequently an isothermal holding for 5 minutes took place being followed by a slow heating up using a heating rate of 10 K/min until everything was liquid. During the whole period of the experiment, a video was produced, where a frame rate of 60 frames per second enables the observation and recording of fast processes. This video was finally analyzed in detail and the results will be illustrated later in the paper.

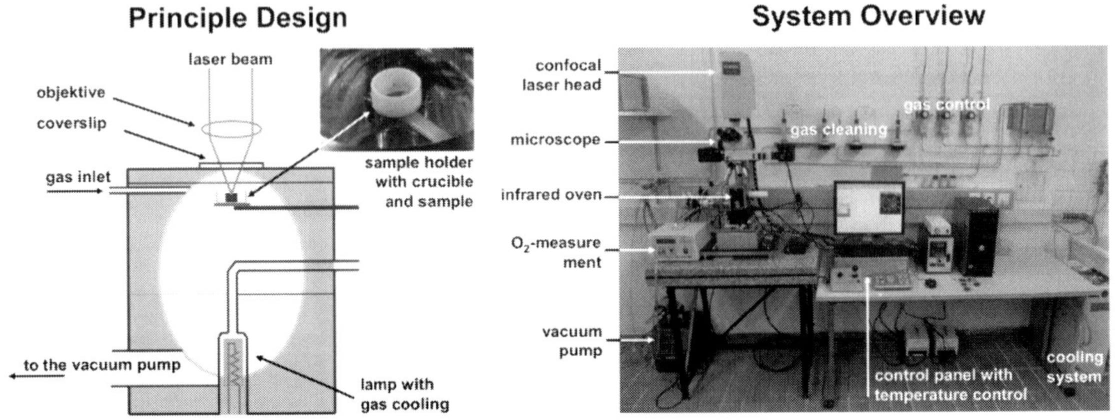

Figure 4. HT-LSCM at the Chair of Metallurgy, Montanuniversitaet Leoben.

The Differential Scanning Calorimetric measurements record all phase transitions which can be associated with an exo- or endothermic effect (=enthalpy change). The equipment used in the study was a NETZSCH DSC with a platinum DSC sensor and type-S thermocouples. The DSC consisted of a sample and reference crucibles, which are connected by a heat flow plate both exposed, ideally, to the same heating power input by convection and radiation from the cylindrical heating oven. The measurement signal is derived from the temperature difference between the sample and reference crucible expressed in terms of a heat flow. Detailed descriptions can be found for example in [9]. The experimental set-up used in the present study was calibrated by measuring the well known melting point and enthalpy of fusion of seven high purity elements (In, Sn, Al, Ag, Au, Ni and Co). All of these measurements were carried out under the same conditions in alumina crucibles with lids under protective gas atmosphere (argon, quality 6.0). In order to achieve best equilibrium conditions, grounded steel samples with a mass of 50 mg and a controlled heating rate of 10 K/min were used. In addition, the thermal resistance and the time constants of the measurement system were corrected. Finally it should be mentioned that accurate measurements of highly reactive steels with higher amounts of Al and Si requires special attention regarding the purity of the atmosphere in the DSC and their leak tightness. For this reason, different getters are used to clean the protective gas in the oven and most importantly, the samples were heated up only once in order to measure only the virginal material.

3. Results – Thermal Analysis

First of all, the results of the DSC measurement of alloy 1 (Fe-1%Si-0.14%C) will be illustrated and discussed in detail. Thereafter, the findings of alloy 2 (Fe-0.4%Al-0.22%C) and alloy 3 (Fe-1.5%Al-0.22%C) applying both experimental methods, DSC and HT-LSCM, will be analyzed in detail and finally the determined phase transformation sequence and temperatures will be compared.

The left hand side of Figure 5 shows a typical result of a DSC measurement (alloy 1: Fe-1%Si-0.14%C), whereas the illustrated curve (DSC-signal^{corr}), i.e. the measured heat flow, is adjusted by the influence of the heating rate, sample mass, crucible and the DSC-sensor configuration. In doing so, this signal correction procedure identifies the determined temperatures as equilibrium transformation temperatures. This was done by using a commercial software tool [10] and applies for all illustrated curves in the following section. Additionally, the transformation temperatures and subsequently the present phases are illustrated in the diagram. This correlation of the possibly appearing phases is not trivial and will be discussed below:

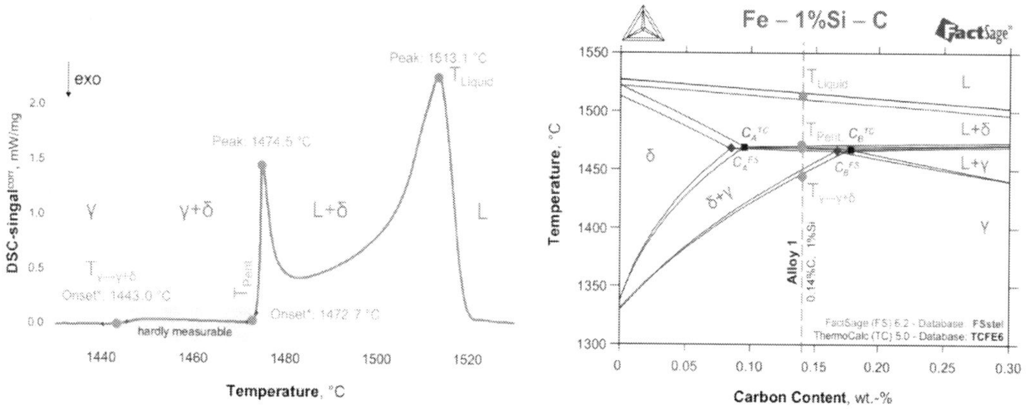

Figure 5. DSC measurements of alloy 1 (Fe-1%Si-0.14%C).

Since phase transformations are always associated with a change in enthalpy (i.e. heat will be released or consumed) and the DSC method allows the measurement of the heat flow, an occurring phase transformation causes a deviation of the horizontal baseline. The first deviation which can be seen in Figure 5 takes place at 1443 °C. However, only a very small deviation of the heat flow can be observed which points to a transformation of very small enthalpy changes. With respect to even possible phase transitions in the high temperature range (T > 1200 °C) of steels, this is typical when austenite starts to transform δ-ferrite (γ→γ+δ). Hence, the temperature of 1443 °C was identified as $T_{\gamma\rightarrow\gamma+\delta}$. The larger the temperature range of the two phase region (γ+δ), the smaller is the heat flow per unit of time. Consequently the γ→δ transformation is hardly measurable. At 1472.7 °C a very sharp peak appears which is followed by a typical melting peak of an alloy. These peaks can be associated with the peritectic phase transformation (γ→L+δ) temperature, T_{Perit} and the liquidus temperature T_{Liquid}. The sharp increase is not only due to the enthalpy change caused by the peritectic transition but also due to fact that this peritectic peak coincides with the solidus temperature. Hence, this is a clear criterion of an alloy situated between C_A and C_B (= range II). Alloys which are situated between C_B and C_C show clearly different characteristics, because the solidus temperature is lower than the peritectic temperature. Thus, the deviation from the baseline due to the melting process is followed by a sharp increase due to the peritectic transition.

Finally, the determined phase transformation temperatures are incorporated in the Fe-1%Si-C pseudo binary phase diagram, illustrated on the right hand side of Figure 5. It can be seen that the determined transformation temperatures of alloy 1 are in very good agreement (ΔT is smaller than 10 °C) with the results of the TC and FS calculation. A large number of DSC measurements in the system Fe-Si-C and Fe-Si-Mn-C showed that the standard deviation of the measurement of solid-solid transitions is $^{+}/_{-}5$ °C and of solid-liquid transitions $^{+}/_{-}2$ °C. Therefore, DSC measurements, which are very well reproducible, can be used as a benchmark.

As indicated in the introduction, the thermodynamic calculations of the ternary system Fe-Al-C are very contradictory. In order to shed light on the matter, first investigations were conducted using the HT-LSCM method. Figure 6 shows selected pictures of alloy 2 (Fe-0.4%Al-0.22%C). At 1440 °C (A2-I) shows a pure austenitic microstructure whereas a significant grain coarsening can be observed in the video. Subimage A2-II shows this grain growth in terms of a consumption of a smaller grain. Melting of the sample typically starts at the grain boundary at a temperature of 1480 °C (A2-III), which can be identified very clearly by a moving liquid phase on the surface of the sample. With further heating the melt spreads over the whole surface (A2-IV). As a result, it is not possible to determine the liquidus temperature as well as further transformations such as the peritectic. However, since the first detected phase transformation is γ→L+γ, alloy 2 is situated either between C_B and C_C (range III) or even higher than C_C (range IV). Both thermodynamic software tools predict the following transformation sequence for alloy 2: γ→L+γ→L+δ→L (= range III, see also Figure 3).

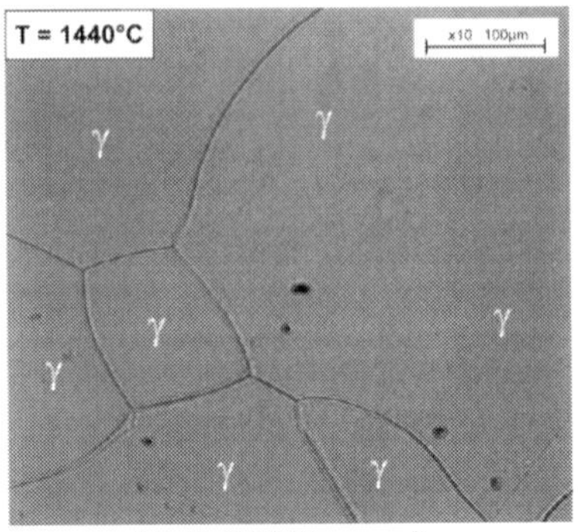

A2-I: 100% γ-grains

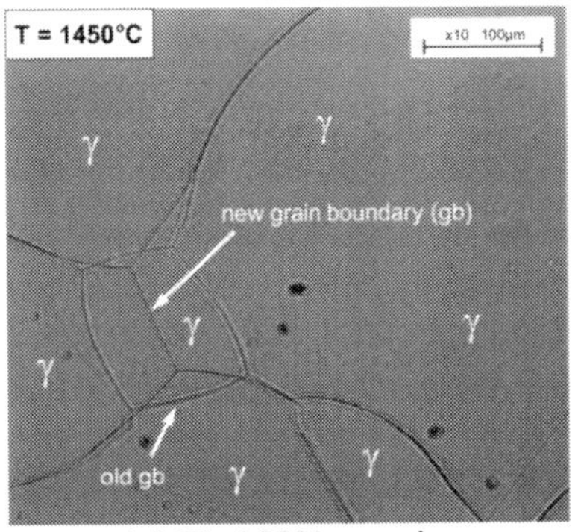

A2-II: 100% γ-grains

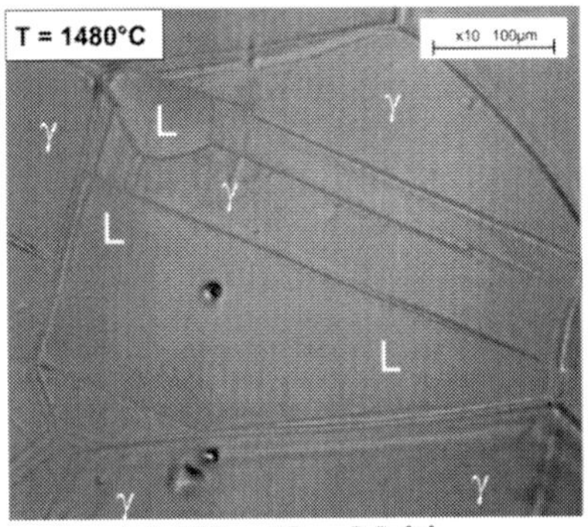

A2-III: γ+L = Melting

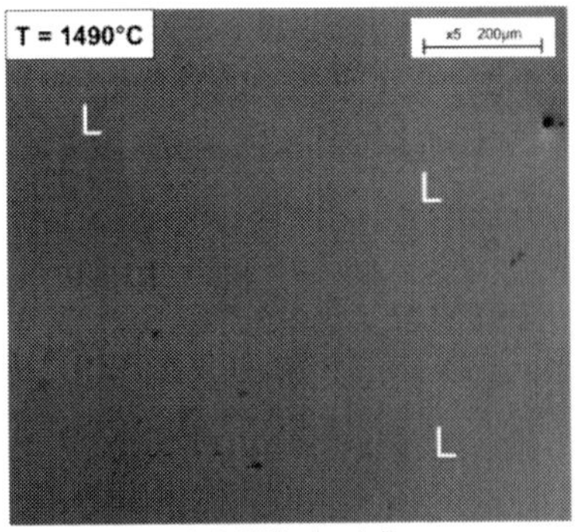

A2-IV: (γ+) L

Figure 6. HTLSCM result of alloy 2 (Fe-0.4%Al-0.22%C).

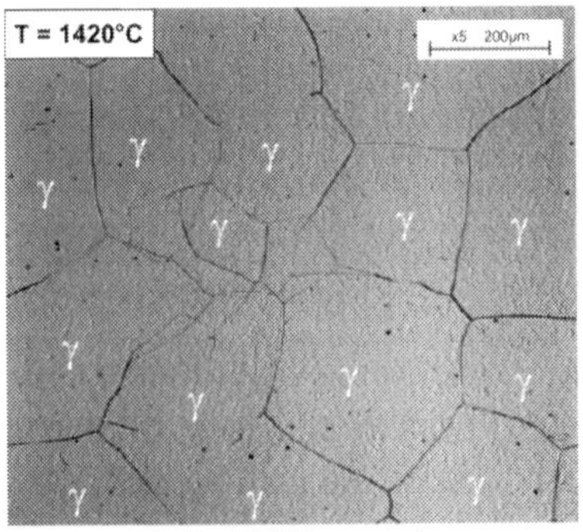

A3-I: 100% γ-grains

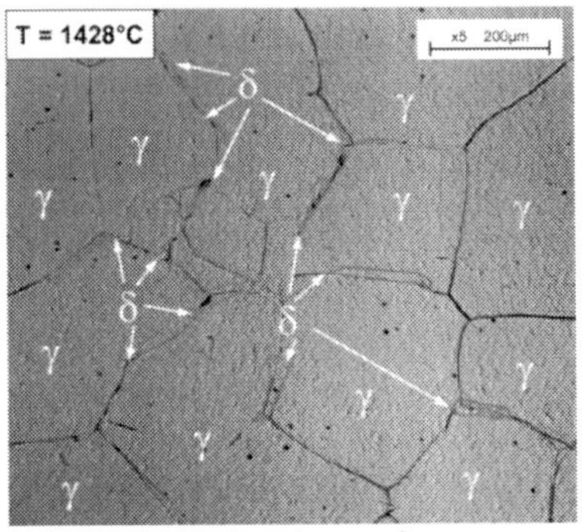

A3-II: δ-formation

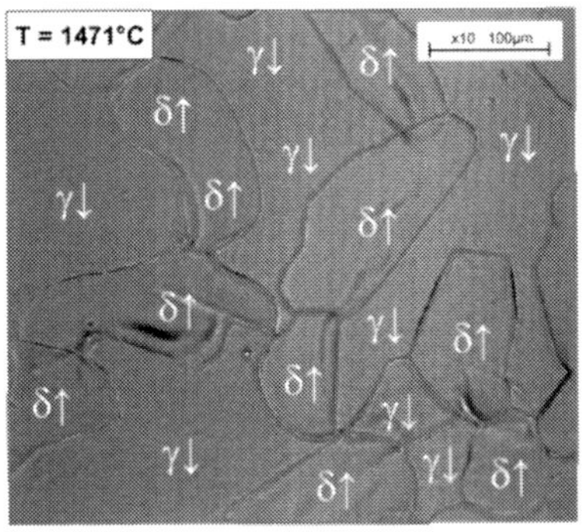

A3-III: γ+δ

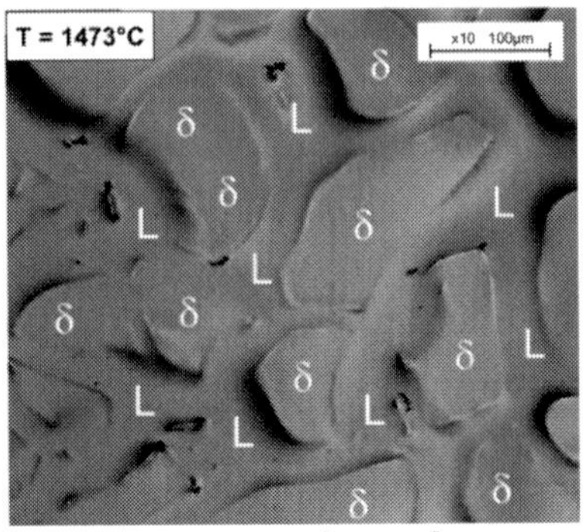

A3-IV: γ+δ→L+δ

Figure 7. HTLSCM result of alloy 3 (Fe-1.5%Al-0.22%C).

The results of the HT-LSCM investigations of alloy 3 are illustrated in Figure 7, which show a transformation sequence of γ→γ+δ→L+δ→L (= range II). In the observed high temperature range (1300-1420 °C) the microstructure consists of 100% γ-grains (A3-I) which shows also a grain growth similar to alloy 2. At 1430°C the formation of δ-ferrite starts at the γ grain boundaries and tipple points (A3-II). With further heating the phase fraction of the δ phase increases, while the γ phase decreases (A3-III). Only 2°C higher, the peritectic phase transformation occurs immediately (A3-IV). The δ phase remains, while at first only the former γ phase is liquid. With further heating the residual δ phase melts and the liquid phase spreads over the whole surface as described also for alloy 2. The experimentally determined transformation sequence can be summarized by γ→γ+δ→L+δ→L (= range II). For alloy 3 the calculated transformation sequence according to TC is γ→γ+δ→L+δ→L (range II). According to FS it is γ→γ+δ→δ→L+δ→L (range I). Hence, TC describes the results of the HT-LSCM measurements of alloy 3 better than FS.

However, it can already be pointed out that the HT-LSCM method can not be used to determine transformation temperatures precisely. Therefore, DSC measurements of alloy 2 and 3 were performed additionally. Figure 8 shows the results from these measurements with the subsequently assigned phases carried out in a similar way as described by alloy 2. Furthermore, the positions of the HT-LSCM pictures are marked in the diagrams (black arrows). The DSC trials demonstrate a transformation sequence of γ→L+γ→L+δ→L (= range III) of alloy 2 and γ→γ+δ→L+δ→L (= range II) of alloy 3. These results clearly support the findings of the HT-LSCM observations, viz. an accordance of the experimentally determined transformation sequence with the calculated one using TC.

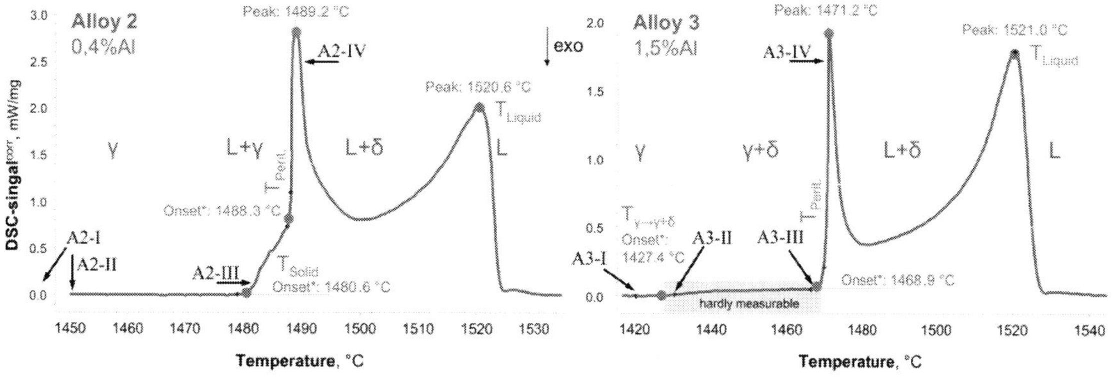

Figure 8. DSC measurements of the aluminium alloys 2 and 3 with 0.22%C.

Figure 9 represents a summarization of the thermodynamic calculations with the results from DSC measurements (red points) and the HT-LSCM method (black arrows). It is clearly noticeable that the HT-LSCM observations and the DSC measurements fit very well together and the DSC results confirm the HT-LSCM predictions on the transformation sequence. However, in order to determine transformation temperatures, the DSC method must be preferred. Comparing these values with the calculations, it can be stated that both software tools achieve a good accordance at alloy 2 (Fe-0.4%Al-0.22%C). However, considering alloy 3 (Fe-1.5%Al-0.22%C), only TC predicts the measured transformation sequence. With the exception of T_{Liquid}, the calculated transformation temperatures, using both software tools, significantly differ from the experimentally determined values.

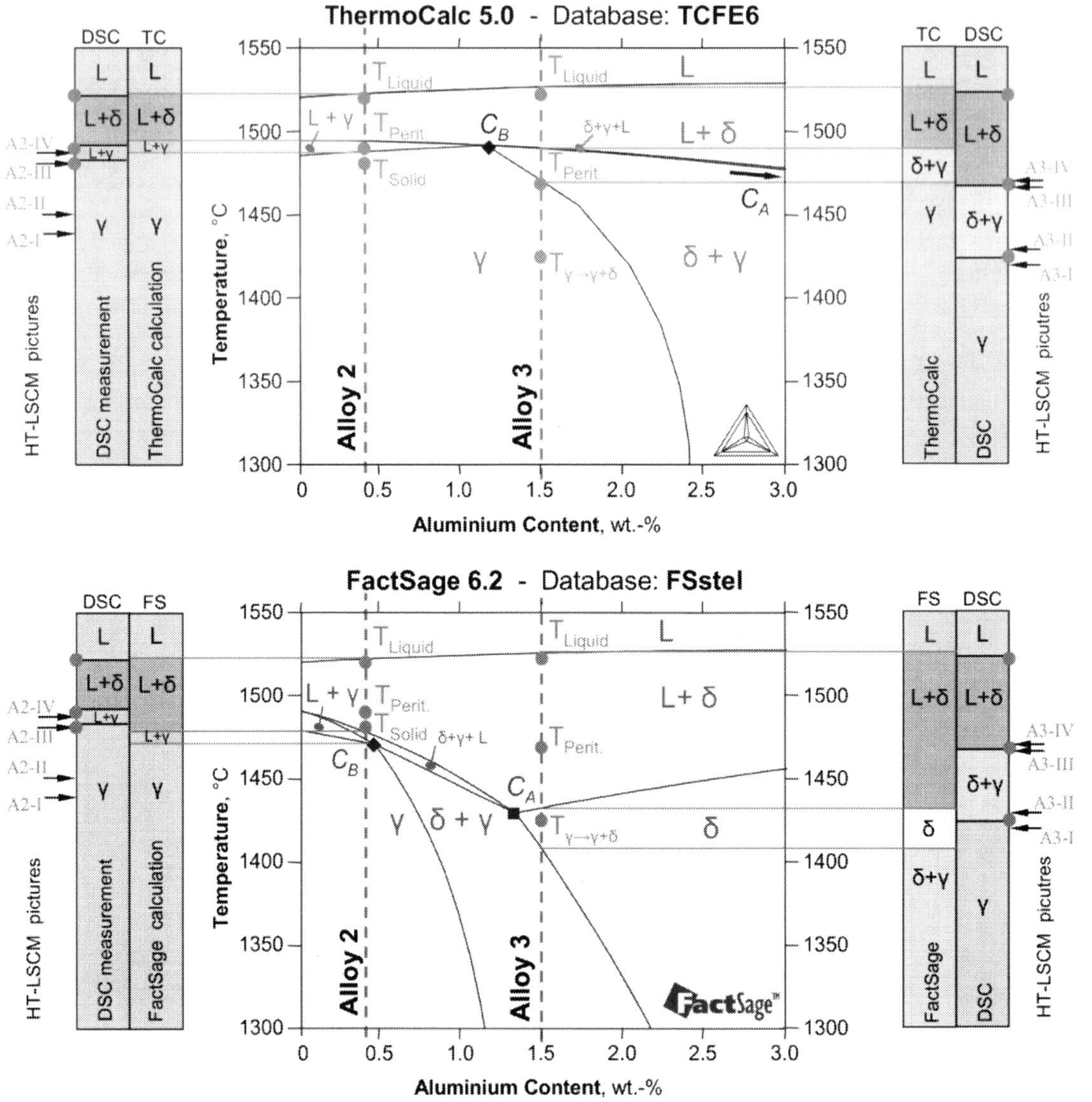

Figure 9. Compilation of the thermodynamic calculations with the measurements results.

4. Summary and Outlook

The present study has shown that there is a lack of inconsistent thermodynamic data, for example for newly developed steel grades with a higher amount of Al. As a result, experimental methods are necessary to advance these thermodynamic data in terms of phase diagrams. Therefore, the present study illustrates and discusses the high temperature phase transformations of three different alloys using two laboratory methods, the Differential Scanning Calorimetry (DSC) and the High-Temperature Laser-Scanning-Confocal-Microscope (HT-LSCM). It could be shown that DSC measurements are a perfect method to measure the solidus, peritectic and liquidus temperatures with high accuracy. Furthermore, the transformation sequence can directly be deduced from the characteristics of the measured DSC signal. HT-LSCM investigations enable a direct observation (Thermo-Optical Analysis) of the microstructure and the transformation behavior up to the melting point. Both methods complement each other perfectly:

- Measuring the $\gamma \rightarrow \delta$ transformation (generally solid-solid transformations with small enthalpy changes) by means of DSC is difficult, but works very well by applying the HT-LSCM method.
- DSC is an excellent method to determine the hyperperitectic and/or the liquidus temperature. This can not easily be realized using HT-LSCM, at least during the heating stage. However, applying the concentric solidification method [7], the HT-LSCM represents a perfect laboratory to investigate the peritectic reaction and transformation.
- Although the temperature measurements are calibrated in both cases, DSC measurements are the most suitable method to determine phase transformation temperatures with high accuracy. In contrast to determining phase transformation temperatures, the HT-LSCM is predominant when investigating grain growth, solidification, sub-cooling and time-temperature-transformations (kinetic studies).

The combination of both methods is a powerful tool and enables the determination of phase diagrams, especially in the high temperature range. This will be realized in ongoing and future research activities with a strong focus on steel in combination with alloying elements which are not well described at higher contents (e.g. Al > 0.1% and combinations of Al with Si, Mn in steels).

Acknowledgement

Financial support by the Austrian Federal Government (in particular from the Bundesministerium für Verkehr, Innovation und Technologie and the Bundesministerium für Wirtschaft und Arbeit) and the Styrian Provincial Government, represented by Österreichische Forschungsförderungsgesellschaft mbH and by Steirische Wirtschaftsförderungsgesellschaft mbH, within the research activities of the K2 Competence Centre on "Integrated Research in Materials, Processing and Product Engineering", operated by the Materials Center Leoben Forschung GmbH in the framework of the Austrian COMET Competence Centre Programme, is gratefully acknowledged.

References

[1] Grill A and Briamcombe J K 1976 *Ironmaking and Steelmaking* **2** 76
[2] Maehara Y, Yasumoto K, Tomono H, Nagamichi T and Ohmori Y 1990 *Mat. Sci. Tech.* **6** 793
[3] Murakami H, Suzuki M, Kitagawa T and Miyahara S 1992 *Tetsu-to-Hagane* **78** 105
[4] Xia G, Bernhard C, Ilie S and Fürst C 2008 Why are some peritectic steels susceptible to surface cracking formation for the continuously cast slab *6th European Conf. on Continuous Casting 2008* (Riccione, Italy: Associazione italiana di metallurgie)
[5] Yin H, Emi T and Shibata H 1998 *ISIJ International* **38** 794
[6] McDonald N J and Sridhar S 2005 *Journal of Materials Science* **40** 2411
[7] Reid M, Phelan D and Dippenaar R 2004 *ISIJ International* **44** 565
[8] Phelan D, Reid M and Dippenaar R 2006 *Met. and Mat. Trans. A* **37** 985
[9] Höhne G W, Hemminger H W and Flammersheim H J 1996 *Differential scanning calorimetry: an introduction for practitioners* (Berlin: Springer)
[10] NETZSCH DSC/DTA Corrections 2006 version 2006.01

Using orthogonal experimental design to optimize alloy composition

X M Li[1] and J J Yu
School of Materials Science and Engineering, Southeast University, Nanjing 211189, P.R.China

E-mail: xmli6@seu.edu.cn

Abstract. This paper applies orthogonal experimental design to the optimisation of alloy composition of 7000 Al alloys by studying the effects of Zn, Mg and Cu contents on the main strengthening η phase. For orthogonal experimental design, three factors and five levels are used, and total of 25 compositions are therefore obtained for examination. Phase transformation and phase diagrams were studied using computer simulation. The mechanical and physical properties of the alloys are simulated and interpreted based on microstructural analysis. It was found that with Zn content as high as 6.7wt%, Mg content between 1.9-2.2wt% and Cu between 2.2-2.6wt% the alloy will have a balanced microstructure and property output.

1. Introduction

Al-Zn-Mg-Cu aluminium alloys are widely used in aerospace and transportation industry. The properties of Al-Zn-Mg-Cu alloys (7000 series) are greatly associated with the microstructure of the material, especially stable and metastable phases. The main hardening elements Zn, Mg and Cu play an important part in the formation of the main phases such as η ($MgZn_2$) phase, S (Al_2CuMg) phase and T ($Al_2Mg_3Zn_3$) phase, among which η phase is thought to be the dominant strengthening phase [1-3]. Thus, the optimum alloy composition of 7xxx alloys can be obtained by assessing microstructural modification and property balance [4-5].

In the last two decades, the use of thermodynamic modelling via the multi-platform software has been extensively applied to complex alloys of many types, including Al-based alloys. JMatPro is designed to be used to calculate the metal material multiphase equilibria with various properties by incorporating various theoretical models and material properties database [6-8]. In this paper, we will use orthogonal experimental design to optimise alloy composition of Al-Zn-Mg-Cu alloys by studying the effects of Zn, Mg and Cu contents on the precipitation of η phase. The mechanical and physical properties of the alloys will be simulated and a preferential composition scale will be given.

2. Experimental design

In this study, the orthogonal experimental design to find out the optimum composition for balanced properties (both mechanical and physical properties) within the composition window of 7xxx series Al alloys is set up, as shown in Table 1. The discussed alloys are simulated in as-cast condition. The main alloying elements Zn, Mg and Cu are taken into consideration and the range of Zn is from 5.1-6.7%, Mg from 1.9-2.9% and Cu from 1.2-2.6% (in wt%). Three factors and five levels are used, and total of

[1] To whom any correspondence should be addressed.

25 compositions are therefore obtained in the experimental design, see in Table 2. Other minor alloying elements Fe, Si and Zr are set as 0.1wt%, while Ti and Mn as 0.05wt%.

JMatPro program developed by Sente Software is a Java based software which can be used to calculate the metal material multiphase equilibria with various properties by incorporating theoretical models and material properties database. In this study it is used to calculate the amount of η phase precipitates under different heat treatment conditions, and the proof stress as well as electrical conductivity of the alloys with various alloying element contents are also simulated and studied.

Phase diagram calculation is based on CALPHAD method (calculation of phase diagrams). Its scientific basis lies in the fact that each phase in the system can be represented by a thermodynamic model described by a mathematic equation. The system reaches equilibrium when the Gibbs energy of the entire system reaches the minimum. The equilibrium state can be a mixture of various phases or simply a stable single phase in some regions. The Gibbs energy of a phase in a multi-component system can broadly be represented by the general equation:

$$G_m = \sum_i x_i G_i^0 + RT\sum_i x_i \ln x_i + \sum_i \sum_{j>i} x_i x_j \sum_v \Omega_v (x_i - x_j)^v \tag{1}$$

where the first component is the Gibbs energy of pure components, the second indicates ideal entropy, and the third is pairwise interaction terms with interactive parameters. x_i, x_j is the mole fraction of different alloying component, T is the temperature, R is the gas constant. Ω_v is a interaction coefficient dependent on the value of v. When v is limited to a value of 0, it corresponds to the regular solution model and when values for 0 and 1 are provided, it corresponds to a sub-regular type model. In practice values do not usually rise above 2.

For thermo-physical properties such as thermal conductivity and electrical conductivity, considered in equilibrium state, the model adopts the generic form:

$$P = \sum_i x_i P_i^0 + \sum_i \sum_{j>i} x_i x_j \sum_v \Omega_{ij}^v (x_i - x_j)^v \tag{2}$$

where the first component is the physical property of pure component, and similarly the second is an indication of interaction terms with interactive parameters.

For mechanical properties, two types of strengthening mechanism namely solid solution strengthening and particle strengthening mechanism due to precipitation processes are considered in the model in the following form:

$$\sigma_y = \sigma_i + \Delta\sigma_{ss} + \Delta\sigma \tag{3}$$

where σ_i is the intrinsic strength of the matrix, $\Delta\sigma_{ss}$ is the solid solution contribution, and $\Delta\sigma_{ppt}$ indicates particle strengthening contribution. JMatPro software can do thermodynamic calculation and property modelling based on these theories and the database in the program. One can select the required interface and input the data (such as composition, temperature etc) for calculations, such as phase diagrams, strength or electrical conductivity at various conditions [6-8].

Transmission Electron Microscopy (TEM) samples were prepared by cutting a material into a thin slice (~0.3mm), subsequently punching it to a 0.3mm diameter disc and grinding to thickness around 0.15mm. The samples were electro-polished using a solution of 70% methanol and 30% nitric acid, maintained at a temperature of between –20°C and –30°C. The TEM facility was a JEOL-JEM 2010 transmission electron microscope. An accelerating voltage of 200 kV was used.

Table 1. 7xxx alloy composition changing range (wt%)

Factor	Zn content	Mg content	Cu content
1	5.1	1.9	1.2
2	5.5	2.2	1.6
3	5.9	2.5	2.0
4	6.3	2.7	2.3
5	6.7	2.9	2.6

Table 2. Alloy composition design with various Zn, Mg and Cu contents $L_{25}(3^5)$

Number	1#	2#	3#	4#	5#	6#	7#	8#	9#	10#	11#	12#	13#
Zn	5.1	5.1	5.1	5.1	5.1	5.5	5.5	5.5	5.5	5.5	5.9	5.9	5.9
Mg	1.9	2.2	2.5	2.7	2.9	1.9	2.2	2.5	2.7	2.9	1.9	2.2	2.5
Cu	1.2	1.6	2.0	2.3	2.6	1.6	2.0	2.3	2.6	1.2	2.0	2.3	2.6
Number	14#	15#	16#	17#	18#	19#	20#	21#	22#	23#	24#	25#	
Zn	5.9	5.9	6.3	6.3	6.3	6.3	6.3	6.7	6.7	6.7	6.7	6.7	
Mg	2.7	2.9	1.6	2.2	2.5	2.7	2.9	1.9	2.2	2.5	2.7	2.9	
Cu	1.2	1.6	2.3	2.6	1.2	1.6	2.0	2.6	1.2	1.6	2.0	2.3	

3. Results and discussion

3.1. Effects of alloying elements on η phase

The statistics of 25 alloys for the amount of η phase at room temperature are listed in Table 3. As explained before, for orthogonal experimental design, we consider 3 factors (Zn, Mg, Cu) and 5 levels, i.e. with different contents for each alloying element. K is average value under the same level, for instance, K_1 equals the average of 5 values of η amount under level 1. K_{max} is the maximum value out of the 5 values and K_{min} is the minimum value out of the 5 values. $R = K_{max} - K_{min}$. The values for K and R are listed in Table 4. A big R value implies a significant influence of the factor on the result. Therefore, Zn has tremendous effect on the η phase as the R of Zn is the highest. The calculated phase diagram of an Al-6.7Zn-2.3Mg-1.9Cu alloy is shown in Fig. 1. From the phase diagram it is observed that the main phases of this type of alloy include $MgZn_2$ (η phase), S phase and T phase. The η phase takes up about 10% (wt) at equilibrium state, whilst the rest of other phases are all below 2%(wt), which indicates that this phase plays an important part in determining the microstructural characteristics and properties of this alloy. Therefore, we will primarily discuss the main effects of this phase in this study. It should be noted that all figures except Figure 2 are obtained from simulation results in this work.

TEM experiment was carried out to verify the microstructure of the alloy. A TEM micrograph of an Al-Zn-Mg-Cu alloy (as cast condition), displaying the main η phase, is presented in Fig. 2. It can be seen that the fine precipitates are uniformly distributed in the matrix and the percentage of the precipitates is in general agreement with the result predicted in the calculated phase diagram. The amount of η phase at room temperature under various alloying elements is plotted in Fig. 3. From Fig. 3 it is shown that the amount of η phase can reach the maximum when Zn content is about 6.7wt%, while Mg content varies between 1.9wt% to 2.2wt% and when Cu between 2.0wt% to 2.6wt% the amount of η phase reaches a local maximum level. Previous research indicates that a high Zn/Mg ratio is beneficial to the precipitation of η phase [3, 9].

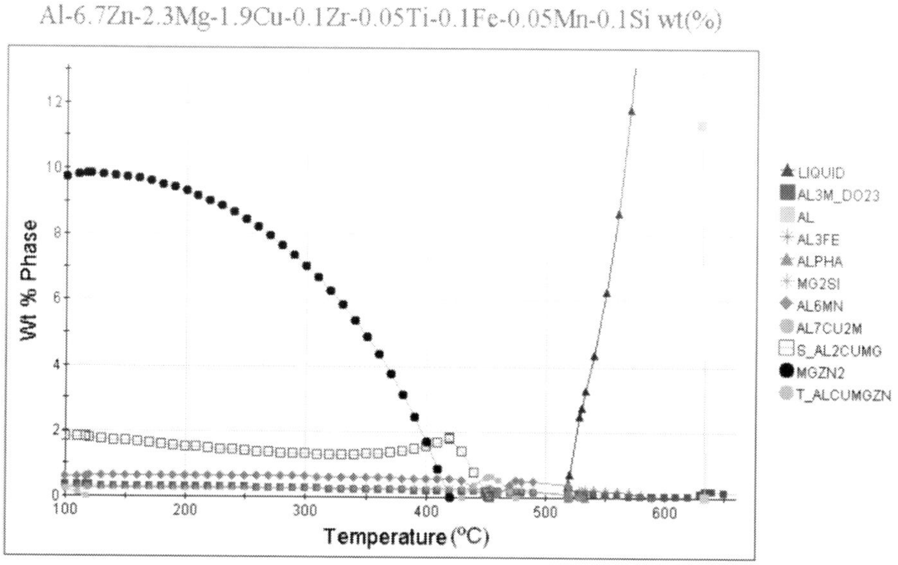

Figure 1. The calculated phase diagram of an Al-6.7Zn-2.3Mg-1.9Cu alloy.

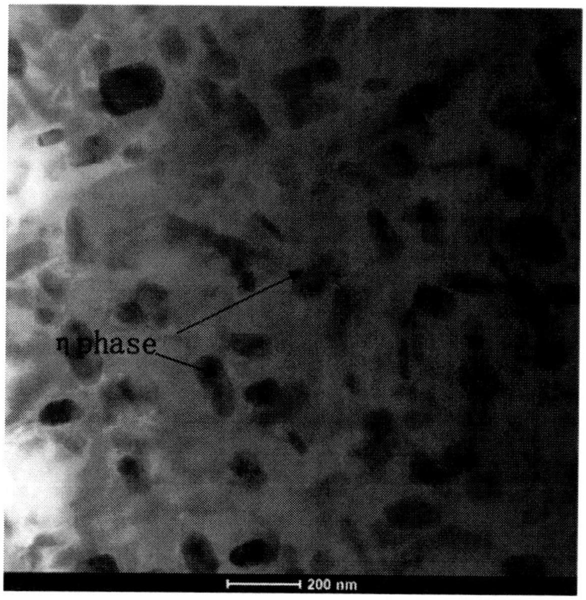

Figure 2. TEM micrograph of a cast Al-6.72Zn-2.38Mg-1.84Cu alloy.

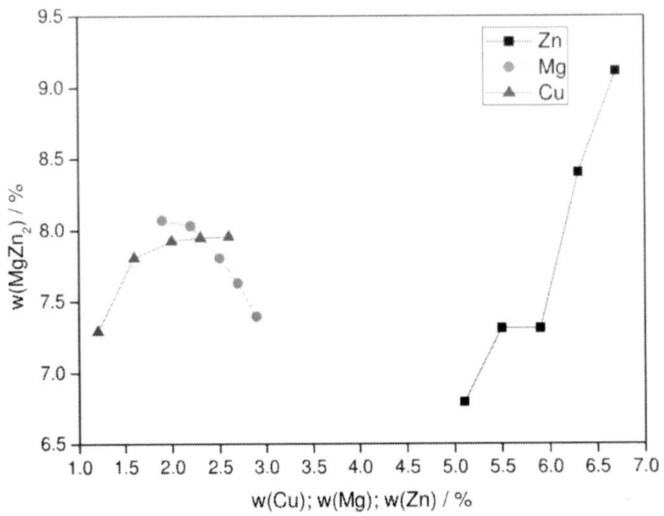

Figure 3. The amount of η phase at room temperature with various alloying elements.

The amount of η phase changing with temperature with different Zn, Mg and Cu contents is shown in Fig. 4. When Zn content increases from 5.1wt% to 6.7wt%, the amount of η precipitates at equilibrium state increases from 6.8% to 9.2%, as shown in Fig. 4(a). This is because more Zn is available in the alloy, and more η phase tends to precipitate. However, Zn content can not be added too high as other consequences may occur, for instance other coarse particles may appear, and we need to refer to phase diagram and thermo-mechanical treatment conditions [10]. The changes of Mg and Cu contents do not exhibit marked effects on the amount of η phase at equilibrium state, as shown in Fig. 4(b), (c), which implies that these two alloying elements have less significant effects than Zn on the formation of η phase. However, they play an important role in the formation of S phase and T phase which are thought to be detrimental to the toughness of the alloy.

3.2. Property simulation of Al-Zn-Mg-Cu alloys

In order to evaluate the effects of alloying elements on the mechanical properties as well as physical properties of the alloys, the proof stress and electrical conductivity of Al-Zn-Mg-Cu alloys were simulated with various Cu contents. The results are shown in Fig. 5 and Fig. 6, respectively. It is revealed that the alloys have relatively higher proof stress with the Cu content between 1.2 to 1.6wt%, whilst they have the lowest proof stress with 2.0wt% of Cu. Subsequent increase of Cu content will result in an improvement of proof stress (between 2.4-3.2wt%), but too much Cu addition (more than 2.6wt%) will cause other Cu-containing coarse particles[2-3]. The electrical conductivity generally decreases with the increase of Cu content, as shown in Fig. 6. This is because high Cu content will result in the formation of detrimental phases such as S phase and Al_7Cu_2Fe particles thus affect the mechanical properties. It is therefore estimated that the Cu content between 2.2 to 2.6wt% will have a balanced overall property output.

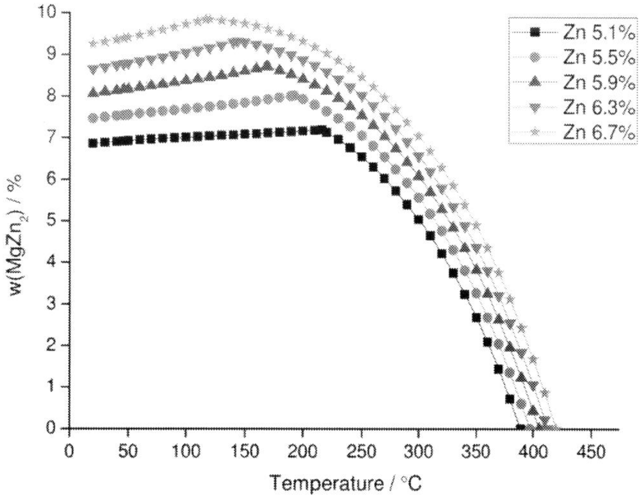
Zn 5.1%
Zn 5.5%
Zn 5.9%
Zn 6.3%
Zn 6.7%
w(MgZn2) / %
Temperature / °C

(a)

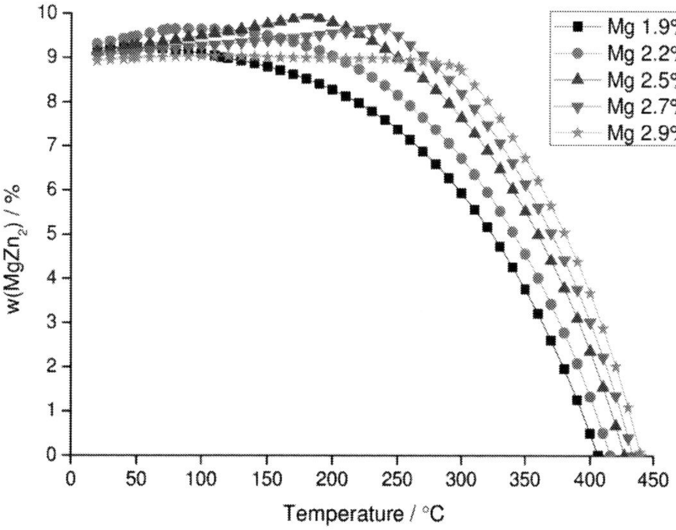
Mg 1.9%
Mg 2.2%
Mg 2.5%
Mg 2.7%
Mg 2.9%
w(MgZn2) / %
Temperature / °C

(b)

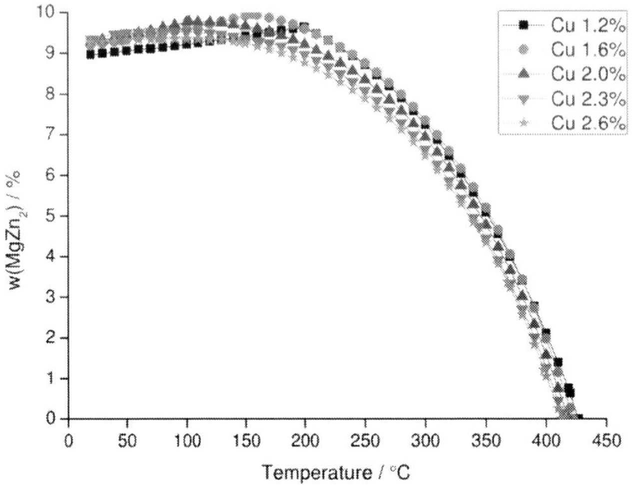

(c)

Figure 4. The amount of η phase changing with temperature at various alloying element contents. (a) 2.3wt%Mg, 1.9wt%Cu, Zn varies; (b) 6.7wt%Zn, 1.9wt%Cu, Mg varies; (c) 6.7wt%Zn, 2.3wt%Mg, Cu varies.

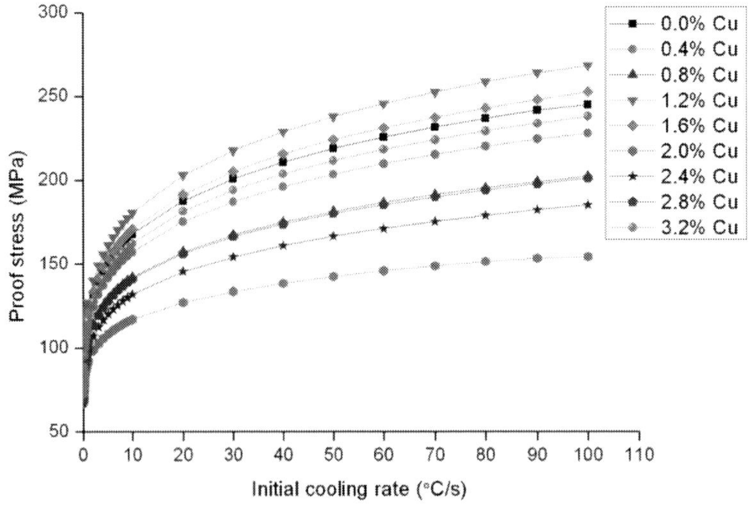

Figure 5. Proof stress of the alloys with various Cu contents between 0-3.2wt%, Zn=6.1wt%, Mg=2.3wt%.

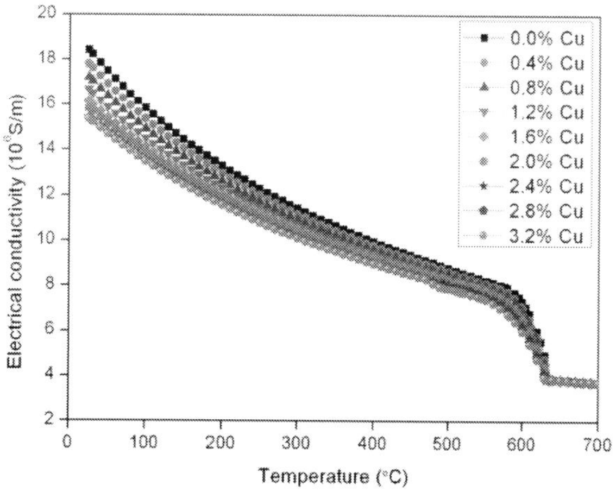

Figure 6. Electrical conductivity of the alloys with various Cu contents between 0-3.2wt%, Zn=6.1wt%, Mg=2.3wt%.

Table 3. Amount of η phase at room temperature for different alloys

Number	Zn (wt %)		Mg (wt %)		Cu (wt %)		η phase (wt %)
1#	1	5.1	1	1.9	1	1.2	6.93
2#	1	5.1	2	2.2	2	1.6	6.85
3#	1	5.1	3	2.5	3	2.0	6.77
4#	1	5.1	4	2.7	4	2.3	6.73
5#	1	5.1	5	2.9	5	2.6	6.68
6#	2	5.5	1	1.9	2	1.6	7.62
7#	2	5.5	2	2.2	3	2.0	7.54
8#	2	5.5	3	2.5	4	2.3	7.43
9#	2	5.5	4	2.7	5	2.6	7.39
10#	2	5.5	5	2.9	1	1.2	5.83
11#	3	5.9	1	1.9	3	2.0	7.75
12#	3	5.9	2	2.2	4	2.3	7.75
13#	3	5.9	3	2.5	5	2.6	7.65
14#	3	5.9	4	2.7	1	1.2	6.44
15#	3	5.9	5	2.9	2	1.6	6.96
16#	4	6.3	1	1.9	4	2.3	8.67
17#	4	6.3	2	2.2	5	2.6	8.68
18#	4	6.3	3	2.5	1	1.2	7.94
19#	4	6.3	4	2.7	2	1.6	8.37
20#	5	6.7	5	2.9	3	2.0	8.35
21#	5	6.7	1	1.9	5	2.6	9.23
22#	5	6.7	2	2.2	1	1.2	9.18
23#	5	6.7	3	2.5	2	1.6	9.08
24#	5	6.7	4	2.7	3	2.0	9.06
25#	5	6.7	5	2.9	4	2.3	9.01

Table 4. Values for K and R

	Zn (wt %)	Mg (wt %)	Cu (wt %)
K_1	6.792	8.040	7.264
K_2	7.162	8.000	7.776
K_3	7.310	7.774	7.894
K_4	8.402	7.598	7.918
K_5	9.112	7.366	7.926
R	2.320	0.674	0.662

The fact that increase of Cu content results in low electrical conductivity is difficult to explain, but earlier research suggested that the metallic additions to aluminium generally reduce its conductivity, particularly those in solid solution [11-12].

4. Conclusion

The major alloying elements of Al-Zn-Mg-Cu 7xxx series alloys were investigated using computer simulation software in terms of stable equilibrium phase, mechanical and physical properties. The orthogonal experimental design has been applied to the optimisation of alloy composition of Al-Zn-Mg-Cu alloys by studying the effects of Zn, Mg and Cu contents on the precipitation of η phase. It was found that a Zn content as high as 6.7wt%, Mg content between 1.9-2.2wt% and Cu between 2.2-2.6 wt% will have a balanced microstructure and property output. However, since this study only takes account of the major precipitating η phase without discussing the overlapping effect of other phases, hence the limitation should be addressed and this will be the topic of future studies.

Acknowledgments

The Key Laboratory for Advanced Metallic Materials of Jiangsu Province, Southeast University, China, is gratefully acknowledged for providing the usage of JMatPro 6 software. The authors would like to thank Mr Y. J. Chen from CnTech Co. Ltd for valuable technical discussions.

References

[1] Marlaud T *et al* 2010 Acta Mater. **58** 248.
[2] Li X M and Starink M J 2001 Mater Sci Tech **171** 324.
[3] Polmear I 1995 *Light Alloys–Metallurgy of the Light Metals* (St. Edmundsbury Press Ltd, UK).
[4] Park J K and Ardell A J 1988 Scr Metall., **22** 1115.
[5] Li X M and Starink M J 2011 J. Alloys Compd. **509** 471.
[6] Guo Z *et al* 2005 Mater Sci Eng A, **413-414** 465.
[7] Cao X *et al* 2004 J. Mater. Sci. **39** 2303.
[8] Saunders *et al* 2003 JOM, **55** 60.
[9] Yu J J and Li X M 2011 J. Phase Equilibria and Diffusion, **32** 350.
[10] Li X M and Starink M J 2000 Mater. Sci. Forum, **331-337** 1071.
[11] Li X M and Starink M J 2011 J. Materials Engineering and Performance, (DOI) 10.1007/s11665-011-9973-5.
[12] Mondolfo L F 1976 *Aluminium Alloys: Structure and Properties* (Butterworth & Co. Ltd, England).

Investigation of solidification in undercooled Al-rich Al-Ni alloy systems

G J Ehlen[1,2] and D M Herlach[1]

[1] Institut für Materialphysik im Weltraum, Deutsches Zentrum für Luft- und Raumfahrt (DLR), 51170 Köln, Germany
[2] Institut für Experimentalphysik IV, Ruhr-Universität, 44780 Bochum, Germany

E-mail: georg.ehlen@dlr.de

Abstract. Dendrite growth velocities as a function of undercooling on the Al-rich side of the Al-Ni system have recently been measured by electromagnetic levitation. They have shown an anomalous behaviour, which so far cannot be theoretically described. The present work investigates whether properties of the phase diagram could explain this effect. For this purpose two simplified qualitative models of sharp interface theory, one treating the forming AlNi (B2) phase as a solid-solution, one treating it as an intermetallic phase, have been developed. To treat the concentration range between 50 and 70 at% Al, sets of thermodynamical data for six different compositions have been created. Results are compared to the experimental data. In both models low undercoolings at low Ni concentrations lead to very limited velocities, because the constitutional undercooling ΔT_c overrides the kinetic undercooling ΔT_k. The experimental results for low undercoolings seem to contain an effect which cannot be described by a model of this type.

1. Introduction

As it is well known, the dendrite growth velocity V depends on its driving force, i.e. the Gibbs free energy difference ΔG between liquid and solid phase. Practically, to study the dendrite behaviour, V is measured as a function of the undercooling ΔT instead of ΔG, because ΔG is not directly accessible. Since ΔG increases with ΔT, a monotonous increase of V with ΔT is expected, as it is often observed in pure metals and solid-solution alloys [1]. A recent study on undercooled Al-rich Al-Ni intermetallic alloys using electromagnetic levitation (EML) [2], however, showed anomalous dendrite growth behaviour: instead of increasing with ΔT, V first decreases and then increases for $Al_{60}Ni_{40}$ and $Al_{65}Ni_{35}$ alloys, and decreases monotonically for $Al_{68.5}Ni_{31.5}$ and $Al_{70}Ni_{30}$ alloys, figure 1 (a). Until now, no reasonable explanations are available for such anomalous dendrite behaviour.

Several theoretical investigations have been carried out on congruently melting $Al_{50}Ni_{50}$ [1], [4] and Ni-rich Al-Ni alloys [5], [6], where a monotonous increase of V with ΔT was found. In these cases, the equilibrium freezing range ΔT_0 is zero or quite small. For the Al-rich part, however, the solidus drops much faster than the liquidus with the decrease of concentration X_{Ni}, thus resulting in a quick increase of ΔT_0 with the decrease of the Ni content (X_{Ni}). If ΔT_0 = 350 K, i.e. the maximum undercooling which is experimentally accessible, we have X_{Ni} = 0.425, figure 1 (b), which happens to be the Ni composition below which anomalous behaviour occurs. So, we wonder if there is a relationship between the anomalous dendrite behaviour and the shape of the phase diagram.

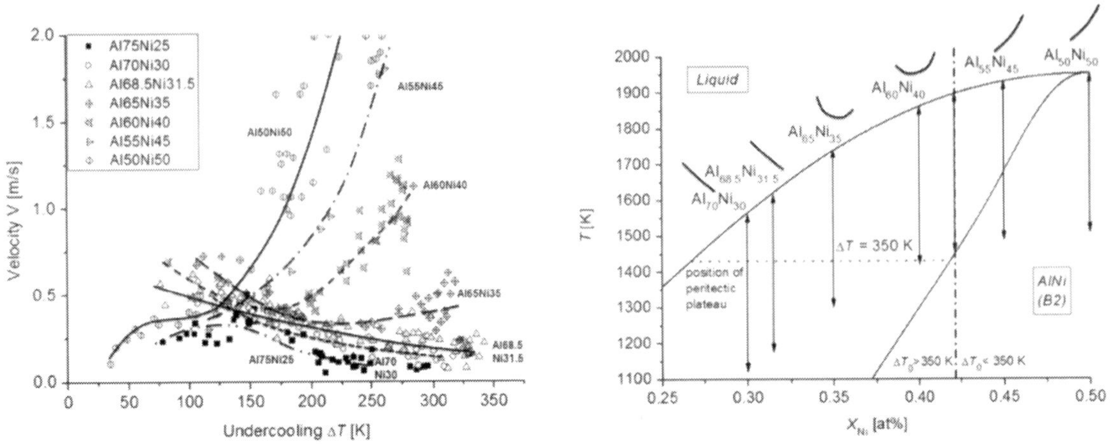

(a) (b)

Figure 1. (a) Dendrite growth velocities as a function of bath undercooling measured using a high-speed video technique on electromagnetically levitated samples of Al-rich Al-Ni alloys [2]. An anomalous dendrite growth behaviour is found for samples with Ni fraction $X_0 < 0.45$. The lines show data trends. (b) phase diagram AlNi (B2) + Liquid; for $X_{\mathrm{Ni}} \lesssim 0.425$ the equilibrium freezing range ΔT_0 exceeds the maximum undercooling $\Delta T = 350\ K$ which is experimentally accessible; schematical $\Delta T/V$ trends above the liquidus line show that this concentration marks the onset of anomalous behaviour; as nucleation of Al_3Ni_2 phase is not considered, the dotted line showing the position of the peritectic plateau is for orientation only.

The present work presents a first attempt to study the effect of the phase diagram with curved phase boundaries on the dendrite growth behaviour of undercooled Al-rich Al-Ni intermetallic alloys. What makes modelling especially difficult is a strongly curved two-phase area in a highly concentrated system, bounded by an intermetallic compound with large homogeneity range on one side and a peritectic plateau/intermetallic Al_3Ni_2 compound at the bottom. We proposed two simplified qualitative models, which do not meet all requirements of the system, but should be able to give insight into the mechanisms determining the growth velocities. Model A treats the AlNi (B2) phase as a solid-solution and applies a qualitative solute-trapping model. Model B, which should be closer to the real situation, treats AlNi (B2) phase as an intermetallic phase by using results of a combined disorder-trapping and solute-trapping model. For both models, the effect of curved phase boundaries is considered. The models are applied to different alloy compositions and comparative calculations are carried out to show their different undercooling contributions.

2. The Models

2.1. Model A: AlNi (B2) phase treated as solid-solution

The model is based on the local non-equilibrium model (LNM) by Galenko *et al.* [7], [8]. The influence of curved phase lines is described using approaches by Divenuti *et al.* [9]. We assume constant and homogeneous initial concentration X_0 (molar fraction) leading to homogeneous initial liquidus temperature $T_{L\infty} = T_L(X_0)$. The initial bath temperature is T_∞. Both heat and solute transport are functions of the total undercooling $\Delta T = T_{L\infty} - T_\infty$. When the system has reached steady state, the dendrite grows with constant velocity V and tip radius R into the melt. In this state, the liquidus temperature at the interface T_L^i and the real temperature at the interface T^i are linked to the initial values by the relations

$$T_L^i = T_{L\infty} - \Delta T_r - \Delta T_n - \Delta T_c \qquad \text{and} \qquad T^i = T_\infty + \Delta T_t \tag{1}$$

where ΔT_r represents the curvature undercooling, ΔT_n the shift of the liquidus line due to solute-trapping, ΔT_c the constitutional undercooling and ΔT_t the thermal undercooling. The difference $\Delta T_k = T_L^i - T^i$ is the kinetic undercooling, which is strongly related to the driving force for dendritic growth. ΔT_k can be written as a function of the total undercooling ΔT and the undercooling contributions as:

$$\Delta T_k = \Delta T - \Delta T_r - \Delta T_n - \Delta T_c - \Delta T_t \tag{2}$$

The general solution for the diffusion field around a shape-preserving paraboloid of revolution (which approximates the shape of a dendrite tip) with a steady state tip radius was found by Ivantsov [10] to be

$$\Omega_{t/c} = P_{t/c}\, exp(P_{t/c}) E_1(P_{t/c}) \equiv Iv(P_{t/c}) \tag{3}$$

where $\Omega_{t/c}$ is defined as the dimensionless thermal (t) or solutal (c) supersaturation, $E_1(P_{t/c})$ is the exponential integral function and $Iv(P_{t/c})$ is the Ivantsov function. $P_t = VR/2a_L$ represents the thermal, $P_c = VR/2D$ the solutal Peclet number. Here a_L is the thermal diffusivity and D the interdiffusion coefficient. Using the Ivantsov solution the dendrite tip temperature can be linked to the initial bath temperature:

$$\Delta T_t = T_Q \Omega_t = T_Q Iv(P_t) \tag{4}$$

The scaling factor $T_Q = \Delta H_f / c_{P,L}$ is the adiabatic temperature of solidification. In case of the solutal field the Ivantsov solution links the liquid concentration at the interface X_L^i to the initial bulk concentration X_0:

$$\Omega_c = \frac{X_L^i - X_0}{X_L^i(1 - k_V)} = Iv(P_c) \tag{5}$$

The scaling factor giving the dimensionless supersaturation Ω_c is the concentration difference $X_L^i(1 - k_V)$ at the solid liquid interface with k_V being the velocity dependent partitioning coefficient.

If curved phase boundaries are considered, the description of constitutional undercooling in LNM has to be modified. As the slope of the liquidus line m_V is no longer constant, but depends on concentration X_L^i at the interface and growth velocity V, no linear approximations can be used. All temperatures have to be read from the velocity dependent phase diagram. In addition, when reading temperatures from the equilibrium or velocity dependent phase diagram, the temperature shift due to the Gibbs-Thomson effect (curvature undercooling) has to be considered [9]:

$$\Delta T_r = 2\Gamma/R \tag{6}$$

where Γ is the Gibbs-Thomson coefficient and R the universal gas constant. All concentrations have to be read at the temperature $T + \Delta T_r$. This holds especially for the curvature corrected partitioning coefficients $k_e' = k_e(T + \Delta T_r)$ and $k_V' = k_V(T + \Delta T_r)$. The shift of the liquidus line due to solute-trapping becomes

$$\Delta T_n = T_{Lr}(X_0) - T_{LV}(X_0) \tag{7}$$

the constitutional undercooling

$$\Delta T_c = T_{LV}(X_0) - T_{LV}(X_L^i) \qquad V < V_D, \qquad \Delta T_c = 0 \qquad V \geq V_D \tag{8}$$

with

$$X_L^i = \frac{X_0}{1-(1-k_V')Iv(P_c)} \tag{9}$$

T_{Lr} and T_{LV} denote the liquidus lines of the curvature corrected equilibrium and kinetic phase diagrams. To calculate the kinetic phase diagram, which results from solute-trapping, we have to modify the local curvature corrected partitioning coefficient k_e':

$$k_V' = \frac{k_e'(1-(V/V_D)^2) + V/V_{DI}}{(1-(V/V_D)^2) + V/V_{DI}} \qquad V < V_D, \qquad k_V' = 1 \qquad V \geq V_D \tag{10}$$

where V_D is the diffusion velocity in the bulk melt and V_{DI} the diffusion velocity on the interface. As we do not yet have a consistent kinetic model for rigorous treatment of the concentrated solid-solution, we use an assumption, which is consistent with the LNM for dilute systems and straight phase lines to calculate the kinetic shift of the concentrations:

$$X_L^i = X_{L\,e}' \frac{1-k_e'}{1-k_V' + \ln(k_V'/k_e') + (1-k_V')^2\, V/V_D} \qquad X_S^i = X_L^i k_V' \tag{11}$$

As in model A the AlNi (B2) intermetallic phase is treated like a solid-solution with a large solubility range, the kinetic undercooling can be represented as

$$\Delta T_k = \frac{V}{\mu} \tag{12}$$

with μ being the kinetic coefficient.

To calculate the dendrite tip radius, solvability theory gives the operation point of a dendrite by taking into consideration the anisotropy of solid-liquid interfacial energy [11]. The result of solvability theory in calculating the dendrite tip radius differs from the result of the criterion of marginal stability [12] only by the anisotropy parameter. In the present case we assume that the anisotropy parameter is near by unity. Using modifications for curved phase boundaries [9] and local non-equilibrium [7], the dendrite tip radius can be expressed as

$$R = \frac{\Gamma}{\sigma^*}\left[T_Q P_t \xi_t - \frac{2X_L^i \left(\frac{dT_L}{dX}\right)_{X=X_L^i}(1-k_V')}{1-V^2/V_D^2} P_c \xi_c\right]^{-1} \tag{13}$$

with

$$\xi_t = 1 - \frac{1}{\sqrt{1+1/\sigma^* P_t^2}} \qquad \xi_c = 1 - \frac{2k_V'}{1-2k_V' - \sqrt{1+(1-V^2/V_D^2)/\sigma^* P_c^2}} \tag{14}$$

Where $\sigma^* = 1/4\pi^2$ is the stability parameter from marginal stability theory and $\xi_{t/c}$ are the thermal and solutal stability functions. The assumption of a constant stability parameter is certainly not valid for high velocities in pure systems, but in our case the dendritic growth velocities are one or two orders of magnitude smaller.

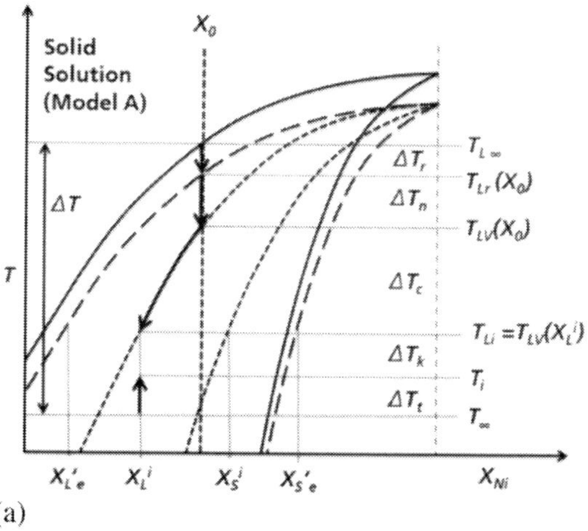

(a)

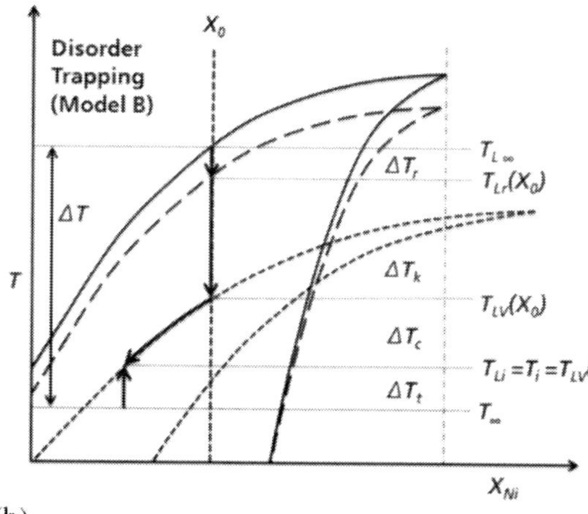

(b)

Figure 2. Definition of undercooling components and temperatures for (a) Model A; (b) Model B.

2.2. Model B: AlNi (B2) treated as intermetallic phase

In a second step we consider the AlNi (B2) to be an intermetallic phase. To keep things simple we did not implement a complete disorder-trapping model, but we qualitatively use results from Assadi *et al.* [1]. Due to disorder-trapping the liquidus and solidus lines move. With increasing velocities the peak at $X_0 = 0.5$ shifts towards higher Ni concentrations and lower temperatures, and the curves themselves are vertically compressed. This information is used to modify a phase diagram calculated with ThermoCalc considering AlNi (B2) and liquid phase only. Figure 3 shows the resulting kinetic phase diagram as a function of growth velocity.

To determine $V(\Delta T)$ most of the equations of the solid-solution model A are used, but the role of some of the undercooling terms changes. Figure 2 shows the differences. The kinetic deformation of the phase diagram due to the simultaneous disorder-trapping and solute-trapping leads to a new definition of the kinetic undercooling. In the solid-solution model A the deformation was described by the term ΔT_n. As in model B the growth velocity is directly linked to this deformation, it is redefined to be the new kinetic undercooling

$$\Delta T_k = T_{Lr}(X_0) - T_{LV}(X_0) \tag{15}$$

while

$$\Delta T_n := 0 \tag{16}$$

The constitutional undercooling is read directly from the kinetic phase diagram:

$$\Delta T_c = T_{LV}(X_0) - T_{LV}(X_L^i). \tag{17}$$

The interface temperature and the liquidus temperature at the interface are defined to be identical.

$$T^i = T_L^i = T_L(X_0) - \Delta T_r - \Delta T_k - \Delta T_c = T_\infty + \Delta T_t. \tag{18}$$

IOP Conf. Series: Materials Science and Engineering **33** (2012) 012066 doi:10.1088/1757-899X/33/1/012066

As the solute-trapping effect is already included in the kinetic phase diagram, equations 10 and 11 are no longer needed. The curvature corrected partitioning coefficient k'_V is directly read from the kinetic phase diagram as

$$k'_V = X_{SV}(T + \Delta T_r)/X_{LV}(T + \Delta T_r). \tag{19}$$

where X_{SV} and X_{LV} are the corresponding kinetic solid and liquid concentrations. Due to the nonlinear phase lines most of the terms have to be calculated iteratively.

3. Material Data

Due to the high liquidus temperatures and the aggressivity of liquid Al-Ni alloys the amount of available thermophysical data is limited. We have tried to create datasets with reasonable concentration dependencies. If such were not available, the parameters were kept constant to prevent artefacts. Table 1 shows the parameter sets used for the numerical calculations.

The choice of phase diagram is difficult, because the available diagrams differ extremely in peak temperatures, freezing ranges, partitioning coefficients and slopes. In addition, we assume the AlNi (B2) phase growing alone, so the equilibrium phase diagram from [13], which includes all phases and the peritectic reactions, cannot be used. Instead, for model A we use an equilibrium phase diagram including AlNi (B2) phase and liquid, calculated using ThermoCalc with all other phases switched off. The peak temperature of 1953 K is higher than in [13], but for the qualitative calculations the absolute temperatures are not too important. For model B we need a kinetic phase diagram. As the kinetic phase diagram by Assadi *et al.* [1] was not available in the concentration, temperature and velocity ranges needed for the calculations, we used the ThermoCalc equilibrium diagram from model A and made it velocity dependent by shifting the peak and stretching the liquidus and solidus lines, table 1. Like this we can qualitatively reproduce the properties of the kinetic phase diagram describing the influence of simultaneous disorder and solute-trapping [1].

Table 1. Material parameters for modelling of dendrite growth in different Al-rich Al-Ni alloys.

		$Al_{50}Ni_{50}$	$Al_{55}Ni_{45}$	$Al_{60}Ni_{40}$	$Al_{65}Ni_{35}$	$Al_{68.5}Ni_{31.5}$	$Al_{70}Ni_{30}$
Parameter	Symbol	Numerical value					
Heat capacity [J/(kg|mol·K)]	c_{PL}	37.9[b]	37.5[b]	36.7[b]	35.7[a]	35.0[b]	34.7[a]
Diffusion coefficient [m^2/s]	D	$1.84\cdot10^{-8}$[c]	$1.81\cdot10^{-8}$[d]	$1.76\cdot10^{-8}$[d]	$1.7\cdot10^{-8}$[d]	$1.65\cdot10^{-8}$[d]	$1.6\cdot10^{-8}$[c]
Gibbs Thomson coeff. [K·m][e]	Γ	$2.99\cdot10^{-7}$	$2.94\cdot10^{-7}$	$2.83\cdot10^{-7}$	$2.70\cdot10^{-7}$	$2.56\cdot10^{-7}$	$2.48\cdot10^{-7}$

Parameters which are used for all alloys:
Latent heat AlNi (B2): ΔH_f = 16000 J/mol [f]; Thermal diffusivity (L) $9.5\cdot10^{-6}$ m^2/s [g]; Diffusion speed in bulk liquid V_D = 40 m/s; Diffusion speed on interface V_{DI} = 20 m/s; Stability constant (from marginal stability theory) $\sigma^* = 1/4\pi^2 = 0.02533$; Kinetic growth coefficient (model A) μ = 0.025 m/(s·K) [h]

[a] computed using ThermoCalc database; [b] interpolated from [a]; [c] [14]; [d] data from [14] interpolated; [e] calculated using the negentropic model by Spaepen and Thompson [15]; [f] value for $Al_{50}Ni_{50}$ from [5]; [g] computed from electrical conductivity, [16], using Wiedemann Franz law; [h] corrected value from [17] for $Al_{50}Ni_{50}$.

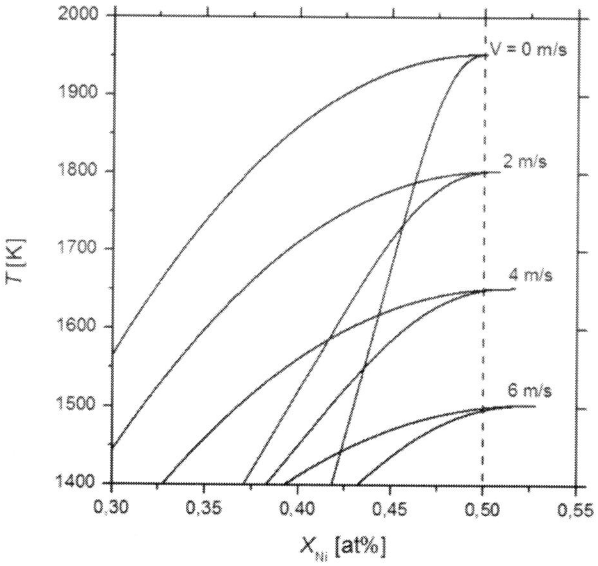

Figure 3. Kinetic phase diagram used for model B; the case $V = 0$ corresponds to the equilibrium phase diagram used for model A

4. Numerical Results and Discussion

For both models (model A: AlNi (B2) as solid-solution, model B: AlNi (B2) as intermetallic phase) the growth velocity V was calculated as a function of total bath undercooling ΔT for a number of Al-rich AlNi/Ni alloy compositions with Ni contents between 0.3 and 0.5. Figure 4 shows the results. In both models velocities decrease with decreasing Ni content, which is consistent with experimental results for high undercoolings around 300 K. While in the solid-solution case (model A) velocities depend almost linearly on ΔT, the disorder-trapping model B shows a completely different behaviour for very high undercoolings. For $\Delta T > 400$ K velocity becomes almost independent from concentration. Even if such high undercoolings may be unreachable, the calculation results show the different mechanisms how velocities are generated.

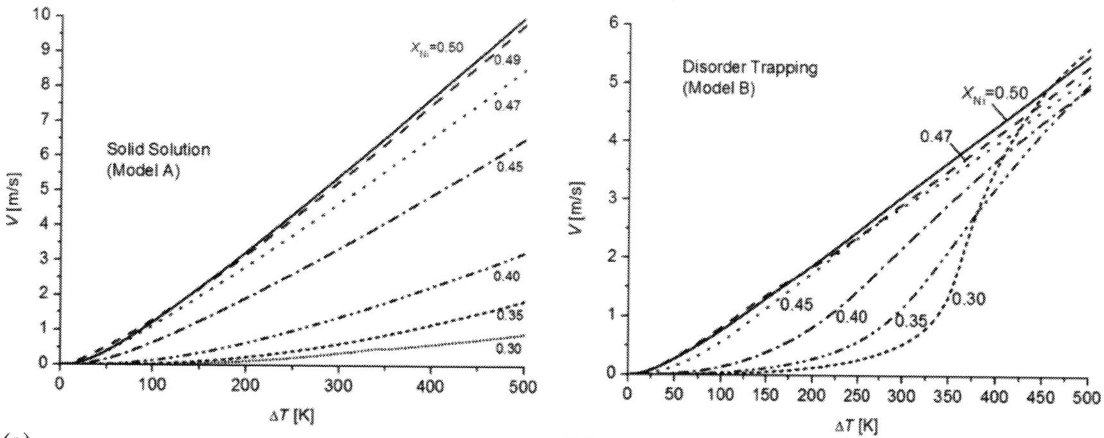

(a) (b)

Figure 4. Growth velocities of Al-rich Al-Ni alloys with Ni contents X_{Ni} between 0.3 and 0.5; (a) Model A: AlNi (B2) as solid-solution; (b) Model B: AlNi (B2) as intermetallic phase.

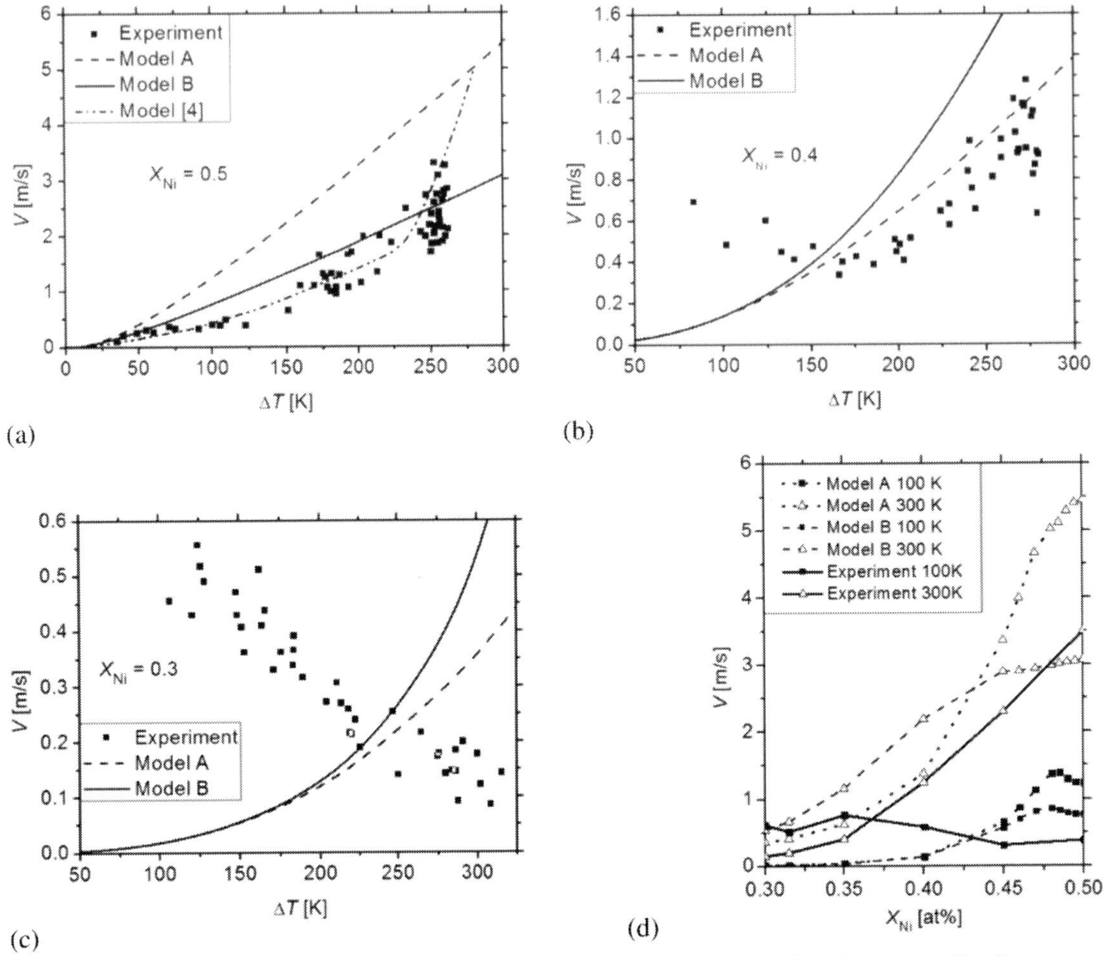

Figure 5. Comparison of experimental and calculated growth velocities V as function of bath undercooling ΔT for Al-rich Al-Ni alloys with (a) $X_{\mathrm{Ni}} = 0.5$; compared to special model for $Al_{50}Ni_{50}$ from [4]; (b) $X_{\mathrm{Ni}} = 0.4$; (c) $X_{\mathrm{Ni}} = 0.3$. (d) Concentration dependency of V for different ΔT. Results for models A and B are compared to approximate experimental values from figure 1(a).

Figure 5 compares simulations and experimental results. Figure 5 (a)-(c) shows for three examples with $X_{\mathrm{Ni}} = 0.5$, 0.4 and 0.3, that both models are able to reproduce the order of magnitude of the velocities for medium and high undercoolings. For $X_{\mathrm{Ni}} = 0.5$ and 0.4 also the trends are reproduced in this undercooling range. The computed velocities are too high, but it can be expected that the experimental data will be well described by more accurate models ([4] for $Al_{50}Ni_{50}$, cf. figure 5 (a)). For low Ni contents $X_{\mathrm{Ni}} = 0.4$ and 0.3 and low undercoolings none of the models is able to describe even qualitatively the behaviour of the velocities. The concentration dependency of the velocities, plotted for two different undercoolings in figure 5 (d), shows clearly the observed anomaly: While both models predict a monotone decrease of velocities with decreasing Ni content, and the experimental values for $\Delta T = 300$ K are consistent with this, there is a strong increase of velocities for Ni contents < 0.45 and $\Delta T = 100$ K, which is predicted by none of the models.

To understand, why in the framework of the models A and B velocities are expected to go down for decreasing Ni contents and decreasing bath undercooling, we look at the undercooling terms. Figure 6 shows the concentration dependency of the undercooling contributions for $\Delta T = 100$ K and 400 K. In both models the velocity increases with increasing kinetic undercooling ΔT_k. If we look at ΔT = 100 K, we see that in both models ΔT_k and thus V decrease with decreasing Ni content. This behaviour is a consequence of the phase diagram. Even for high velocities the constitutional undercooling can never exceed the equilibrium freezing range ΔT_0. For high Peclet numbers P_c the Ivantsov function approaches unity, i.e. ΔT_c equals ΔT_0. This means, for all compositions with ΔT_0 < 100 K, (X_{Ni} > 0.47, cf. figure 1 (b)), all velocities are principally possible. This leaves a large enough part of ΔT to be split between ΔT_t and ΔT_k and to develop quite high velocities. For alloys with low Ni content it becomes impossible to reach Iv = 1, because ΔT_0 is larger than the maximum possible ΔT of the levitation experiments. This means that the range of possible P_c and thus V is more and more limited. Already small velocities lead to high values of ΔT_c and there is not much left for ΔT_t and ΔT_k. This effect is observed similarly for both models A and B, but model B works different than model A. In this model high velocities lead to kinetic deformation of the phase diagram and to reduction of the possible ΔT_c. For small undercoolings the strong curvature of the phase line leads to ΔT_c "overwhelming" the velocities, but for large undercoolings, the velocitiy wins and suppresses ΔT_c, which explains the

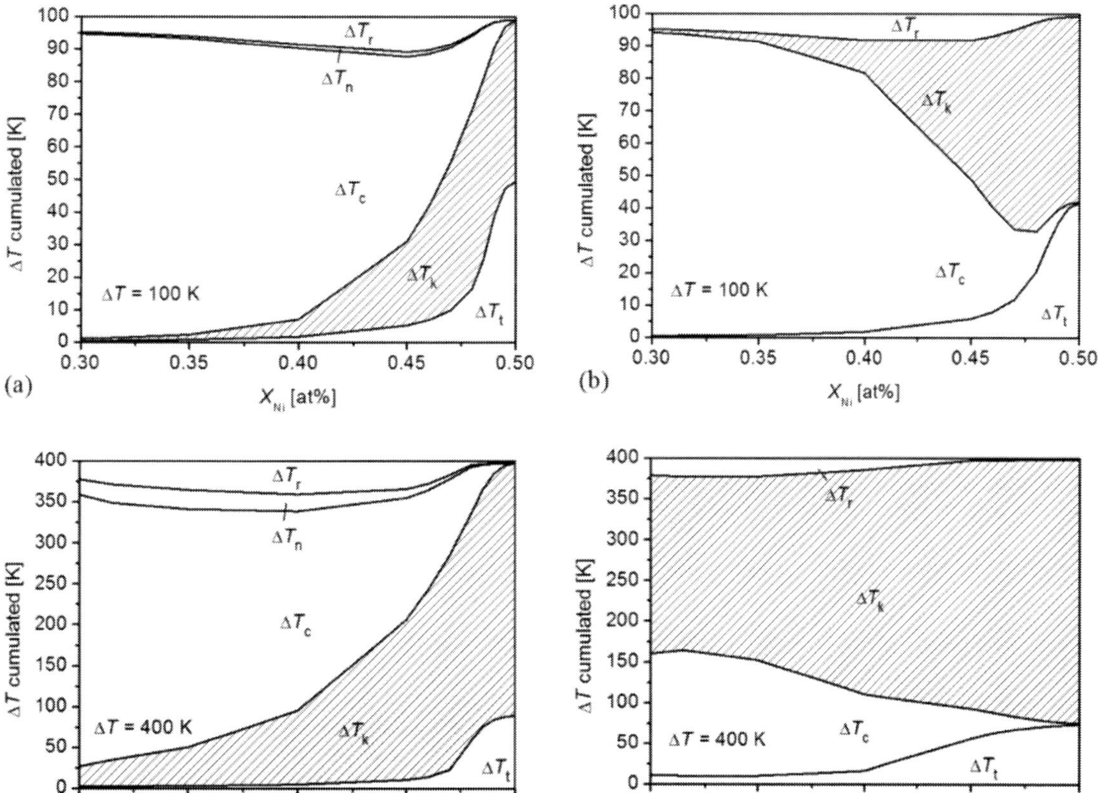

Figure 6. Concentration dependency of undercooling contributions for Al-rich Al-Ni alloys at two different bath undercoolings: (a), (b) 100 K; (c), (d) 400 K; (a), (c) Solid-solution model A; (b), (d) Disorder-trapping model B; the shaded area shows the kinetic undercooling ΔT_k, which is closely related to the growth velocity V.

quasi-linear, concentration independent behaviour of V observed in figure 4 (b). Model A does not show such an effect. For large undercoolings only the fraction of ΔT remaining for ΔT_k increases and with it the velocities V.

So in both models low undercoolings at low Ni concentrations lead to very limited velocities, because the constitutional undercooling ΔT_c overrides the kinetic undercooling ΔT_k. And this result should be independent from the quality of the models. The experimental results for low undercoolings seem to contain an effect, which cannot be described by a model of this type.

5. Conclusions and Outlook

Dendritic growth of Al-rich AlNi alloys has been described by two simplified qualitative sharp interface models, model A treating the AlNi (B2) phase as a solid-solution, model B treating it as an intermetallic phase. Both models correctly predict the order of magnitude of the velocities, including the monotonous decrease of velocities with decreasing Ni content at medium and high undercoolings, and in some cases the trends of the data. For low undercoolings and low Ni content both models fail completely. Investigation of the interaction of the undercooling contributions has shown that this is not caused by the imperfections of the models. The experimentally observed change of slope in the $V/\Delta T$ curves is due to a velocity rise in the area of medium undercoolings (ΔT = 100-150 K). For both models the properties of the phase diagram (freezing ranges and curvature) principally predict low velocities in this area and therefore cannot explain the anomalous growth behaviour.

The next steps will be to replace the rough qualitative models by more accurate thermodynamic ones. This will help to clarify which parts of the experimental data can be described by the classical theories, and which part is governed by a so far unknown effect. Another task is to find out, if the intermetallic phase Al_3Ni_2 could play a role, which could nucleate in alloys with low Ni content at high undercoolings.

Acknowledgements

The authors thank H. Assadi, J. Eiken, U. Hecht, H. Wang, V. Witusiewicz and our colleagues from DLR, especially S. Binder, P. Galenko, J. Gegner, D. Holland-Moritz, S. Klein, M. Kolbe and T. Volkmann for much helpful discussion and for providing data. The work was supported by Deutsche Forschungsgemeinschaft (DFG) under Grant No. HE 1601/26-1.

References

[1] Herlach D M 1994, *Mater. Sci. Eng.* **R12,** 177
[2] Lengsdorf R, Holland-Moritz D, and Herlach D M 2010 *Scripta Materialia* **62**, 365
[3] Assadi H, Reutzel S, Herlach D M 2006 *Acta Materialia* **54,** 2793
[4] Hartmann H, Holland-Moritz D, Galenko P K, Herlach D M 2009, *EPL*, **87**, 40007
[5] Barth M, Wei B, Herlach D M, Feuerbacher B 1994, *Mater. Sci. Eng.*, **178**, 305
[6] Assadi H, Barth M, Greer A L, Herlach D M 1998, *Acta Mater.*, **46**, 491
[7] Galenko P K, Danilov D A 1997, *Phys. Lett. A*, **235,** 271
[8] Galenko P K, Danilov D A 1999, *J. Cryst. Growth*, **197,** 992
[9] Divenuti A G and Ando T 1998 *Metall. Mat. Trans.* **29A**, 3047
[10] Ivantsov G P 1947, *Dokl. Akad. Nauk S.S.S.R.* **58,** 567
[11] Ben Amar M and Pelce P 1989, *Phys. Rev. A*, **39**, 4263
[12] Langer J S, Müller-Krumbhaar H 1978, *Acta Metall.* **26**, 1681
[13] Massalski T B ed 1990 *Binary Alloy Phase Diagrams* 2nd run (ASM International)
[14] Griesche A, Zhang B, Horbach J, Meyer A 2009, *Defect Diff. Forum* **289-292**, 705
[15] Thompson C V, Spaepen F 1983, *Acta Metall.* **31**, 2021
[16] Egry I, Brooks R, Holland-Moritz D, Novakovic R, Matsushita T, Plevachuk Y, Ricci E, Seetharaman S, Sklyarchuk V and Wunderlich R, 2010 *High Temp.-High Press.*, **38**, 343
[17] Kerrache A, Horbach J, Binder K 2008, *EPL* **81**, 58001
[18] Plevachuk Y, Egry I, Brillo J, Holland-Moritz D, Kaban I 2007, *Int. J. Met. Res.* **98**, 2

Simplified pressure model for quantitative shrinkage porosity prediction in steel castings

A V Catalina[1] and C A Monroe[1,2]

[1] Caterpillar Inc., 14009 Old Galena Road, Mossville, IL 61552, USA

[2] University of Alabama at Birmingham, 1075 13th Street South, Birmingham, AL 35294-4440, USA

Email: Catalina_Adrian_V@cat.com
Phone: +1 309 578 6314

Abstract. In order to reduce the computational effort, a simplified pressure model is proposed for predicting shrinkage-related defects in steel castings. Based on the specific conditions developing during casting solidification, the governing fluid flow equations (*i.e.*, Stokes and Darcy) have been reformulated such as the pressure field could be calculated only based on volume (density) changes occurring during the solidification process. Velocity is thus eliminated from the governing equations of fluid flow therefore simplifying the method of solution for pressure and tremendously reducing the computational effort. Model development and simulation results are presented and discussed.

1. Introduction

Metals and alloys typically undergo a negative volume change during the liquid-to-solid transformation. This volume change is responsible for a series of casting defects known as shrinkage cavities and shrinkage porosities. Because shrinkage porosity is one of the major defects encountered in foundries, with consequences ranging from high rework cost to casting rejection, development of predictive mathematical models has been an important topic of research for many years.

The number of models produced to describe the occurrence of shrinkage porosity in castings is nothing short of impressive (see for example the reviews in Reference 1 and 2). Perhaps the most simple of them is the model proposed by Niyama *et al.* [3], also known as Niyama criterion, which relates the occurrence of microporosity in steel castings to local thermal parameters (*e.g.*, cooling rate and temperature gradient). While widely used in metal casting, the Niyama criterion is, however, only a qualitative indicator. For quantitative predictions more complex models have been developed. In addition to the heat transfer and phase change problem, these models also account for the liquid flow through the interdendritic channels as well as the precipitation out of the melt and growth of gas bubbles [4, 5, 6, 7, 8].

Because the occurrence of porosity is strongly related to the evolution of local pressure of the melt in the interdendritic regions, metal pressure has to be calculated by solving the governing fluid flow equations in the fully liquid regions (Navier-Stokes) and the mushy zone (Darcy's Law) as well. This can be a very laborious task as the numerical schemes for solving the Navier-Stokes equations are complex and, for the problem at hand, the solution convergence may be difficult to obtain. In order to circumvent this problem, Pequet et al. [6] proposed a model that solves the governing equations only in the mushy zone, but establishing the appropriate boundary conditions around this zone could be by itself a laborious task and source of inaccuracies.

A simplified model for calculating the liquid alloy pressure during casting solidification is proposed in this paper. By making use of the continuity condition and taking advantage of specific conditions at the solid boundaries during the solidification process, the governing equations of flow are rewritten in a form that can be integrated for pressure without much effort. Examples of model application to shrinkage porosity prediction in steel castings will be presented. Gas precipitation out of solution and pore growth are not considered at this stage of development and therefore porosity is assumed to occur only because of insufficient feeding with liquid in the interdendritic channels.

2. Model Description

2.1. *Governing Equations*

The cooling of metal from its pouring to solidification temperature is accompanied by density changes that induce motion (flow) of the liquid metal in the volume of the casting. Under a pressure gradient, the metal flows toward the area where shrinkage is occurring until the flow is blocked either by the precipitating solid phase, by non-metallic inclusions, or by gas bubbles [9].

By assuming that at a certain moment in time a volume element of the casting is composed of a mixture of solid (s) and liquid (liq) phases as well as of some amount of porosity (p), then the volume fractions (f) of the constituents of the mixture satisfy the relationship:

$$f_s + f_{liq} + f_p = 1 \tag{1}$$

while the density of the mixture, ρ, is given by:

$$\rho = f_s\rho_s + f_{liq}\rho_{liq} + f_p\rho_p \tag{2}$$

The rate of density change and the resulting flow can be calculated from the equations of energy, mass/continuity, and liquid momentum conservation.

2.1.1. Energy Conservation. By neglecting the advection of heat due to the feeding flow, the energy conservation equation can be written as:

$$\rho \cdot C_p \cdot \partial T/\partial t = \nabla \cdot (k\nabla T) + \rho_s \cdot \Delta H_f \cdot \partial f_s/\partial t \tag{3}$$

where C_p is specific heat, T is temperature, t is time, k is thermal conductivity, ΔH_f is latent heat of fusion, and f_s is the volume fraction of solid. For the solid fraction evolution of multicomponent alloys solidifying as solid solutions the model described in Reference 8 has been employed in this work.

2.1.2. Continuity.

$$\frac{1}{\rho_{liq}}\frac{\partial \rho}{\partial t} + \nabla \cdot \vec{\mathbf{V}} = 0 \tag{4}$$

where $\vec{\mathbf{V}}$ is the superficial liquid velocity vector.

2.1.3. *Liquid Momentum Conservation.* By assuming that the flow occurs at low Reynolds number (Re << 1), then the Stokes equations can be used in the liquid regions as well as in those solidifying regions that did not reach the point of dendrites coherency (i.e., the liquid + solid mixture can flow together). Beyond the point of dendrites coherency the solid phase becomes entirely immobile and Darcy Law of flow through porous media can be applied. These equations are, respectively:

$$\nabla P = \eta \cdot \nabla^2 \vec{\mathbf{V}} + \rho_{liq} \cdot \vec{\mathbf{g}} \tag{5}$$

$$\nabla P = -\frac{\eta}{K_{mz}} \cdot \vec{\mathbf{V}} + \rho_{liq} \cdot \vec{\mathbf{g}} \tag{6}$$

where P is metal pressure, η is the dynamic viscosity of the liquid, $\vec{\mathbf{g}}$ is the gravity vector, and K_{mz} is permeability of the mushy zone that can be expressed as:

$$K_{mz} = C_\lambda \cdot \frac{\lambda_2^2}{180} \cdot \frac{(1-f_s)^3}{f_s^2} \tag{7}$$

where f_s is the solid fraction, λ_2 is the secondary dendrite arm spacing (SDAS), and C_λ is an adjustable parameter. For $C_\lambda = 1$ Equation 7 becomes identical to the expression used for K_{mz} in References 6 and 10. Other expressions for K_{mz} can also be found in the literature, such as in References 4 and 11 for instance. Because they all give quite different values for K_{mz}, the use of the adjustable parameter C_λ was chosen in Equation 7.

Equation 5 and 6 along with the continuity condition in Equation 4 allow for calculating the pressure and flow velocity fields in the solidifying casting. However, as mentioned earlier, the numerical procedure for obtaining the solution could require a high computational effort and therefore makes this type of models unattractive for practical purposes. Consequently, a simplified model is proposed in the next section. It allows one to calculate only the pressure field, but this is not necessarily a shortcoming as all the current models aiming at predicting porosity occurrence in castings are based mainly on the metal pressure and not on fluid velocity.

2.2. *Simplified Model for Metal Pressure*

The elimination of fluid velocity, $\vec{\mathbf{V}}$, from Equation 5 and 6 is at the core of the model. This can be achieved in a straightforward manner by rewriting Equation 5 and 6, respectively, as:

$$\nabla^2 P = \eta \cdot \nabla^2 (\nabla \cdot \vec{\mathbf{V}}) + \nabla \cdot (\rho_{liq} \vec{\mathbf{g}}) \tag{8}$$

$$\nabla \cdot \left(\frac{K_{mz}}{\eta} \cdot \nabla P \right) = -\nabla \cdot \vec{\mathbf{V}} + \nabla \cdot \left(\frac{\rho_{liq} K_{mz}}{\eta} \cdot \vec{\mathbf{g}} \right) \tag{9}$$

Both Equation 8 and 9 contain in their right hand side the term $\nabla \cdot \vec{V}$ which can be replaced by the density change term from the continuity equation as seen in Equation 4. Thus, by using the simplifying notation:

$$S_h = \frac{1}{\rho_{liq}} \cdot \frac{\partial \rho}{\partial t} = -\nabla \cdot \vec{\mathbf{V}} \tag{10}$$

the equations for pressure become:

- *in the liquid:* $$\nabla^2 P = -\eta \cdot \nabla^2 S_h + \nabla \cdot (\rho_{liq} \vec{\mathbf{g}}) \tag{11}$$
- *in the mushy zone:* $$\nabla \cdot \left(\frac{K_{mz}}{\eta} \cdot \nabla P \right) = S_h + \nabla \cdot \left(\frac{\rho_{liq} K_{mz}}{\eta} \cdot \vec{\mathbf{g}} \right) \tag{12}$$

In principle, Equation 11 and 12 can be solved numerically without difficulty provided that appropriate boundary conditions are available. For the flow occurring in the mushy zone or Equation 12, the boundary condition at solid boundaries is given by:

$$\nabla P = \rho_{liq} \cdot \vec{\mathbf{g}} \tag{13}$$

which can be readily derived from Equation 6 by setting the velocity equal to zero at the solid boundary (mold wall or dendrite root). Because Equation 12 was obtained by applying the divergence operator to Equation 6, it follows that the boundary condition in Equation 13 applies to Equation 12 as well. However, the condition in Equation 13 does not apply to the flow occurring in the Stokes regime in Equation 11 except for the limit case when the liquid is still. Therefore, in order to establish the solid boundary conditions for Equation 11, one has to make some simplifying yet reasonable assumptions related to the specifics of the solidification process and formation of shrinkage porosity. A first observation related to the solidification process is that it begins at the contact with the mould walls and then proceeds inwards toward the thermal center. In addition to that, shrinkage porosity can develop only within the mushy zone and not into the free liquid. Therefore, except for a relatively short early stage, it can be safely assumed that the liquid regions where flow occurs in the Stokes regime are always bounded by a layer of mushy zone where the flow is governed by Darcy's Law.

Consequently, solid boundary conditions for Stokes flow, or Equation 11, are not necessary for the problem at hand. The remaining boundary condition is that at the free surface of the melt, which usually is the atmospheric pressure, *i.e*:

$$P_{fs} = P_{atm} \tag{14}$$

where P_{fs} is the applied pressure on the free surface of the melt, such as the top surface of the downsprue or of the open risers.

3. Numerical Implementation And Computed Results

The simplified pressure model for shrinkage porosity prediction was implemented into the Finite-Volume (FV) solidification code SolCAT3D, which is an in-house developed code of Caterpillar Inc. The solidification model employed by the code is a refined version of that presented in Reference 8.

Equation 11 and 12 along with the boundary conditions shown in Equation 13 and 14 are used to calculate the pressure field for shrinkage porosity predictions. It is assumed that Stokes type flow occurs in those regions of the casting characterized by an average solid fraction $\bar{f}_s < f_{cr}$, where f_{cr} is a critical solid fraction at which the transition between the Stokes and Darcy type of flow occurs. The average solid fraction $\bar{f}_s$ is calculated as:

$$\bar{f}_s = \frac{\sum_{i=1}^{N} f_{s,i} \cdot Vol_i}{\sum_{i=1}^{N} Vol_i} \tag{15}$$

where N is the number of the first nearest neighbours also including the central FV cell, $f_{s,i}$ is the solid fraction of the cell i, and Vol_i is the volume of the cell i. Fully solidified metal cells or mould cells are excluded, *i.e.*, $Vol_i = 0$ for such cells. This averaging is used in order to ensure a smooth transition between regions characterized by Stokes' and Darcy's flow regimes, respectively.

The pressure field is computed at every time-step during solidification although the solution can be completely unrealistic before a mushy zone of $\bar{f}_s \geq f_{cr}$ develops at the contact with the mould walls. In this work, shrinkage porosity is considered to occur in those regions where metal pressure drops to negative values. The reason of choosing this criterion is because at this stage of development a model for gas porosity was not yet implemented. Similar to the approach in Reference 4, once the porosity begins developing in a certain volume element (*i.e.*, $P < P_{cr} = 0$) the pressure is no longer calculated and all the subsequent volume changes of the metal will contribute toward the development of shrinkage porosity. Because in this work flow velocity is not explicitly calculated, the pipe shrinkage is evaluated according to the procedure described by Pequet *et al.* [6].

In order to assess the capability of the proposed model, the solidification of steel castings of chemical composition shown in Table 1 was performed and is presented in this paper. Table 1 also shows the liquidus (T_{liq}) and solidus (T_{sol}) temperatures as well as the total shrinkage of the alloy in the solidification interval. T_{liq} and T_{sol} were calculated according to the model presented in Reference 8 where all the thermodynamic parameters employed by the solidification model are also presented. The density change occurring during solidification was calculated based on the relationships provided by Miettinen in Reference 12. The relationship used to calculate SDAS of the alloy was also taken from Reference 12, *i.e.*:

$$\lambda_2 = 150 \cdot 10^{-6} \cdot \dot{T}^{-0.385}, \text{[m]} \tag{16}$$

where $\dot{T}$ is the cooling rate at the liquidus temperature. The value of λ_2 was kept constant for the entire duration of solidification. The critical solid fraction, f_{cr}, at which transition from Stokes to

Darcy flow occurs was taken as f_{cr} = 0.75 while for the coefficient C_λ used to calculate the permeability of the mushy zone (see Equation 7) a value C_λ = 3 was chosen in this work.

Table 1. Chemical composition and characteristics for the steel used in this work.

C wt%	Si wt%	Mn wt%	P wt%	S wt%	Cr wt%	T_{liq} °C	T_{sol} °C	Solidification Shrinkage %
0.19	0.40	1.25	0.02	0.02	0.2	1511	1408	4.45

The castings chosen for simulations are presented in Fig.1. The casting presented in Figure 1a is an end-risered plate (25.4 x 140 x 483 mm) also used in the study of Carlson and Beckermann [13], while in Figure 1b and Figure 1c a casting used on Caterpillar's machinery is presented. The difference between Figure 1b and Figure 1c is the number of risers applied to the same casting, *i.e.*, two additional risers applied to the casting in Figure 1c. The four chills shown in Figure 1b were also considered for the configuration shown in Figure 1c, although they cannot be seen on that figure. Mould filling was not simulated and therefore the gating system was not included. The initial melt temperature was set to 1570 °C and furan-bonded sand mold was considered for all simulations.

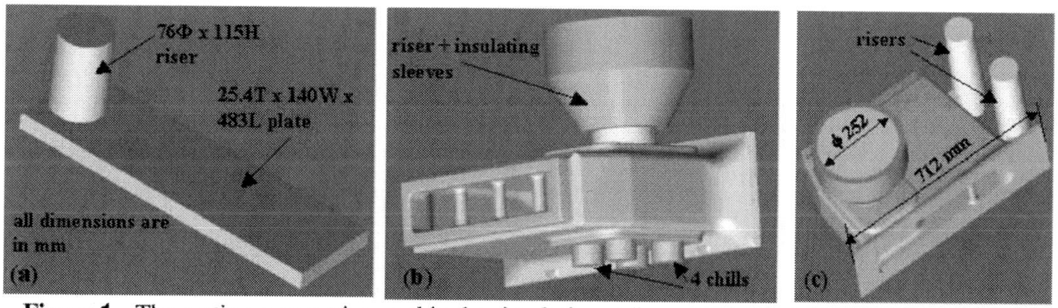

Figure 1. The casting geometries used in the simulations (dimensions are shown)

The envelope of the calculated shrinkage porosity in the plate casting is shown in Figure 2, while Figure 3 shows the calculated porosity in the mid-thickness plane of the plate. In Figure 3 the radiograph of a real plate [13] is also presented for comparison. Because the evolution of gas porosity was not simulated, a precise quantitative comparison between the radiograph in Figure 3a and simulated results of Figure 3b is not attempted at this time. However, it can be observed that the extent of the porosity regions in Figure 3a and Figure 3b are very similar. Figure 4 shows calculated pressure and solid fraction distributions in the cross-section of the plate+riser casting just before the shrinkage porosity begins to develop. It can be observed on Figure 4a that at the time when porosity begins to develop the entire liquid alloy is bounded by a solid or mushy zone layer therefore making applicable the boundary condition expressed in Equation 13. The shape exhibited by the solid

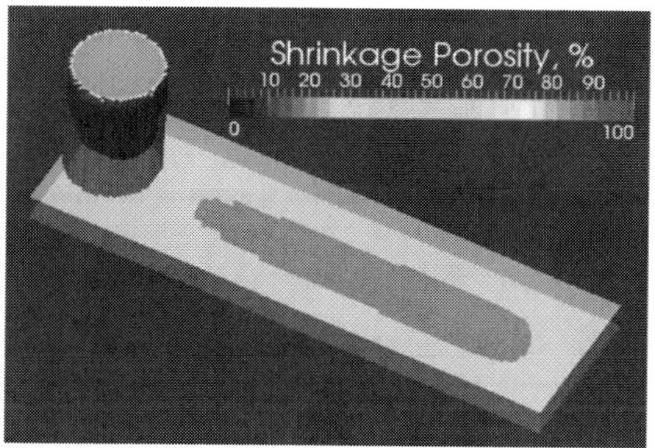

Figure 2. The envelope of calculated shrinkage porosity in the plate + riser casting

fraction contour near the top of the riser in Figure 4b is because of the piping that develops in that region. Also, the reason the pressure does not exhibit any value in that region (see Figure 4a) is because all the FV cells are either empty or fully solid.

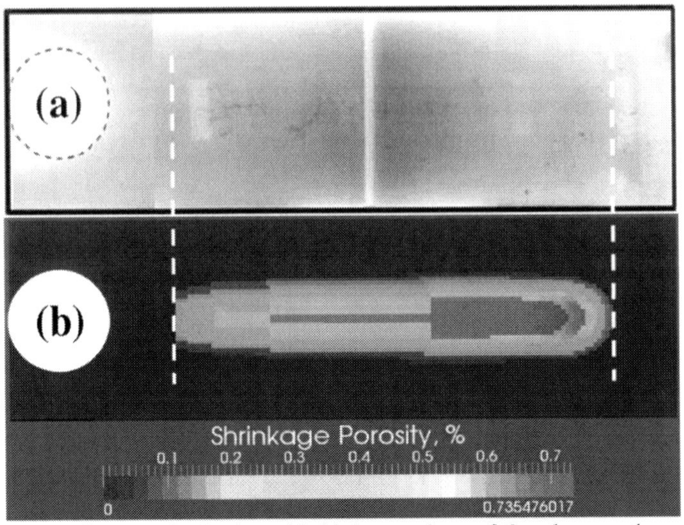

Figure 3. Porosity in the mid-thickness plane of the plate casting:
(a) - radiograph of the real plate [13];
(b) - calculated porosity

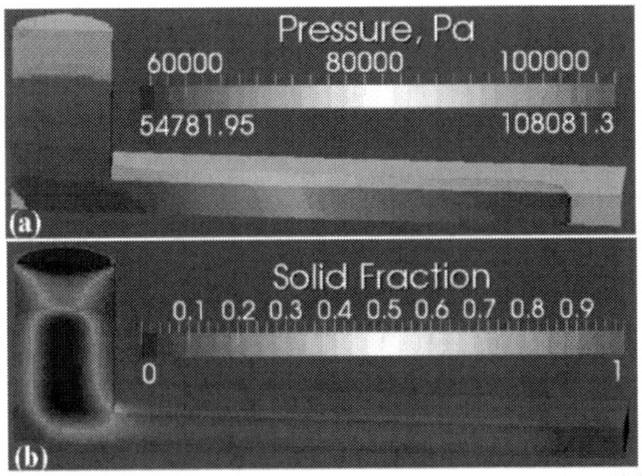

Figure 4. The calculated pressure (a) and solid fraction distribution (b) in the cross-section of the plate + riser casting just before the shrinkage porosity begins to develop

Figure 5 shows the envelope of the predicted shrinkage porosity for the casting presented in Figure 1b. Various porosity regions, away from the riser, can be observed on this figure. Perhaps the most deleterious for the casting is the porosity predicted in the hot spot marked with **A**, where it can reach a value as high as 3.5-3.7%. This is not unexpected, given the thinner walls that separates the hot spot from the riser. The region marked as **B** is also a hot spot but much smaller than that of region **A** and the predicted porosity there is therefore in the range of 0.9-1.0%.

The soundness of the casting can be much improved by using the two additional risers shown in Figure 1c. The predicted outcome for this case is presented in Figure 6 where it can be observed that the porosity of region **A** has completely disappeared. Also, the porosity region at the horizontal bottom wall of the casting decreased its size considerably. However, region **B** is still too far from the risers for a proper feeding and therefore the porosity predicted here is practically unchanged when compared to the results shown in Figure 5.

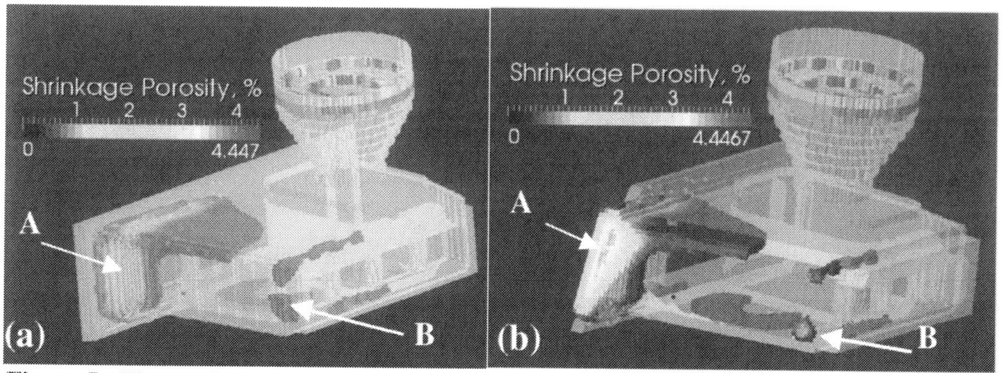

Figure 5. The predicted shrinkage porosity in the casting shown in Figure 1b; a)-the envelope of shrinkage porosity; b)-a cross section through the region containing the highest amount of porosity

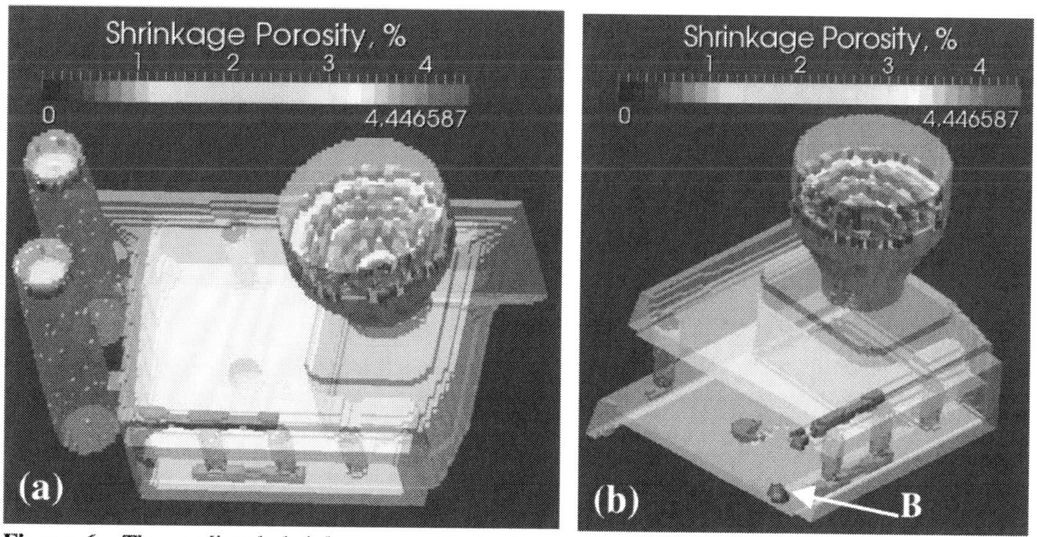

Figure 6. The predicted shrinkage porosity in the casting shown in Figure 1c; a)-the envelope of shrinkage porosity; b)-a cross-section view in a plane similar to that shown in Figure 5b.

4. Conclusions

In order to reduce the computational effort for predicting the occurrence of shrinkage porosity in castings, a simplified model for calculating the metal pressure during casting solidification has been developed and implemented in Caterpillar's casting/solidification code SolCAT3D. Based on the specific conditions developing during solidification of castings (*e.g.*, small Re number, liquid alloy bounded by a layer of mushy zone), the proposed model eliminates the velocity from the governing equations of fluid flow thus allowing for a straightforward computation of the pressure field. Examples of model application toward predicting shrinkage porosity in a simple end-risered plate as well as on a more complicated casting have been presented. While the location and size of the predicted porosity regions seem reasonable, a quantitative comparison between the predictions and amount of porosity in the real castings has not been attempted at this time. Such a comparison would be appropriate only when the gas precipitating out of the melt is also considered. Further work to include the gas effect is underway.

Acknowledgements

This work was supported by Caterpillar Inc. Responsibility for opinions and statements contained in this paper are that of the authors, not of Caterpillar Inc.

References

[1] Lee P D, Chirazi A and See D 2001 *J. Light Metals* **1** 15-30
[2] Stefanescu D M 2005 *Int. J. Cast Met. Res.* **18** (3) 129-143
[3] Niyama E, Uchida T, Morikawa M and Saito S 1982 *AFS Cast Met. Res. J.* **7** 52-63
[4] Sabau A S and Viswanathan S 2002 *Metall. Mater. Trans. B* **33B** 243-255
[5] Sung P K, Poirier D R and Felicelli S D 2002 *Modelling Simul. Mater. Sci. Eng.* **10** 551-568
[6] Pequet C, Gremaud M and Rappaz M 2002 *Metall. Mater. Trans. A* **33A** 2095-2106
[7] Carlson K D, Lin Z, Beckermann C, Mazurkevich G and Schneider M 2006 in *'Modeling of Casting, Welding and Advanced Solidification Processes-XI'* ed. Gandin C and Bellet M 627-634
[8] Catalina A V, Leon-Torres J F, Stefanescu D M and Johnson M L 2007 in *'Solidification Processing 2007'* ed. H. Johnes 699-703
[9] Stefanescu D M 2002 in *'Science and Engineering of Casting Solidification'* 1st ed, Kluwer Academic, New York 241
[10] Kubo K and Pehlke R D 1985 *Metall. Mater. Trans.* **16B** 359-366
[11] Khajeh E and Maier D M 2010 *Acta Mater.* **58** 6334-6344
[12] Miettinen J 1997 *Metall. Mater. Trans. B* **28B** 281-297
[13] Carlson K D and Beckermann C 2009 *Metall. Mater. Trans. A* **40A** 163-175

IOP Conf. Series: Materials Science and Engineering 33 (2012) 012068 doi:10.1088/1757-899X/33/1/012068

Semi-quantitative predictions of hot tearing and cold cracking in aluminum DC casting using numerical process simulator

T Subroto[1,2], A Miroux[1,2], D Mortensen[3], M M'Hamdi[4], D G Eskin[5] and L Katgerman[2]

[1] Materials innovation institute (M2i), Mekelweg 2, Delft, 2628CD, The Netherlands
[2] Dept. of Material Science and Engineering, Delft University of Technology, Mekelweg 2, Delft, 2628CD, The Netherlands
[3] Institute for Energy Technology, N-2027 Kjeller, Norway
[4] SINTEF Materials and Chemistry, P.B. 124 Blindern, N-0314 Oslo, Norway
[5] BCAST, Brunel University, Uxbridge, Middlesex, UB8 3PH, United Kingdom

E-mail: t.subroto@m2i.nl

Abstract. Cracking is one of the most critical defects that may occur during aluminum direct-chill (DC) casting. There are two types of cracking typical of DC casting: hot tearing and cold cracking. To study and predict such defects, currently we are using a process simulator, ALSIM. ALSIM is able to provide semi-quantitative predictions of hot tearing and cold cracking susceptibility. In this work, we performed benchmark tests using predictions of both types of cracks and experimental results of DC casting trials. The trials series resulted in billets with hot tearing as well as cold cracking. The model was also used to study the influence of several casting variables such as casting speed and inlet geometry with respect to the cracking susceptibility in the ingots. In this work, we found that the sump geometry was changed by the feeding scheme, which played an important role in hot tear occurrence. Moreover, increasing the casting speed also increased the hot tear and cold crack susceptibility. In addition, from the result of simulation, we also observed a phenomenon that supported the hypotheses of connection between hot tearing and cold cracking.

1. Introduction

Direct-chill (DC) casting is a common technique used to produce aluminum billets and ingots owing to its robustness and relative simplicity [1]. However due to the nature of this technique, which introduces a high temperature gradient between the ingot and cooling surface, defects can form during DC casting such as ingot distortion and cracks. The latter is considered one of the worst forms of defects because it is not only capable of reducing ingot quality, but cracks may also occur throughout the entire ingot, thus the ingot has to be fully scrapped [2]. There are two types of cracks that could occur during DC casting; hot tearing when the alloy is still in the semi solid state and cold cracking when the alloy is already in a fully solid. Cold cracking usually occurs when the alloy is in its brittle state, i.e. at a relatively low temperature [3]. This type of cracking can occur catastrophically and therefore, presents also a safety hazard.

In this work, we use process simulation software called ALSIM that is specially tuned for aluminum DC casting. ALSIM is able to predict defects that might arise during DC casting such as air gap formation, porosity and both types of cracking. The main goal of this work is to perform benchmarking between DC casting simulation in ALSIM and trial DC casting. This is to validate the crack susceptibility model with respect to different casting parameters. We performed casting trials with different casting conditions such as: casting speed and cooling rate, also changing the inlet geometry. By varying these parameters, we expect to observe the effect of fluid flow on both types of cracking susceptibility. In addition, since we have performed a simultaneous prediction of both cracks, we also expect to observe the possible connection between the two types of cracking as explained by Lalpoor et al. [4] and we will try to formulate the possible mechanism of such connection based on ALSIM calculation.

The result of this work provide a better understanding for the interpretation of cracking prediction in ALSIM and also assist the modeling efforts on the connection between hot tearing and cold cracking.

2. DC casting modeling in ALSIM

ALSIM is a finite element package with the capability to perform a coupled calculation between thermal, fluid flow and mechanical phenomena. A more detailed description regarding the ALSIM software and its physical model (thermal, fluid-flow and mechanical models) can be found elsewhere [5, 6]. In order to predict the behavior of a certain alloy during DC casting, ALSIM needs several inputs, such as the solidification path of the alloy. In addition, ALSIM needs specific information regarding the thermal, fluid and mechanical properties of the alloy, which can be obtained either using model prediction software or through experiment [4]. In the next sub-section, we provide a brief description of how ALSIM performs the heat transfer and mechanical calculations during DC casting.

2.1. Heat transfer and mechanical calculation phenomena

Heat transfer is a vital phenomenon during DC casting because it defines the solidification front, convective fluid flow, and temperature gradients which in turn governs the sump geometry and mechanical response of the alloy during the casting process. In this simulation the primary heat extraction occurs through the mold and bottom block. We assume the heat transfer coefficient of both of the cooling surface is approximately 1000 W/m^2K. However, during DC casting, often air gap is formed at the lower part of the mold, due to the thermal contraction of the ingot which reduces the heat transfer between the ingot and the mold. In ALSIM, this phenomenon is also taken into account for heat transfer calculation.

During casting the alloys undergoes phase transformation from liquid to solid which produces different mechanical behavior of the alloy in various domains of the ingot. ALSIM uses different constitutive model to represent behavior of the alloy in specific temperature ranges. At a very high temperature, where the material is fully liquid or at a low solid fraction, no mechanical calculations are performed. This is because we assumed that most of the mechanical forces are accommodated via grain translation and rotation due to the abundance of the fluid phase. At a higher solid fraction or at high temperature – fully solid state, indeed mechanical calculations are performed and the constitutive models used for these calculations are described elsewhere [7]. At the lower temperature – fully solid state, the constitutive behavior of the material is described by the extended Ludwik equation as described in [8].

The result from calculations with this constitutive behavior leads to the variables that are needed for the crack prediction calculation in ALSIM, which will be discussed in more detail in the next part.

2.2. Cracking criteria

In ALSIM, the hot tearing susceptibility is measured by taking into account both the feeding difficulties and thermal deformation effect, as these two phenomena are well known to be the main driving force for hot tearing during DC casting [9]. The criterion used is called the integrated critical strain or ICS and the formulation is as follows:

$$\Delta\varepsilon\left(w_v, w_d\right)=\begin{cases}0 & \text{for } p_l \geq p_c \\ \displaystyle\int_{t(p_l<p_c)}^{t\left(g_s=g_s^{nof}\right)}\left(w_v \cdot tr\left(\dot{\varepsilon}_s^p\right)+w_d \cdot \bar{\dot{\varepsilon}}_s^p\right)dt & \text{for } p_l < p_c\end{cases} \quad (1)$$

ICS ($\Delta\varepsilon$) measures the amount of deformation (through the plastic strain rate or $\dot{\varepsilon}_s^p$) that occurs when the mush is vulnerable to hot tearing, namely when it is in a state of feeding difficulties and the pressure in the mushy zone (p_l) is below a certain critical pressure (p_c). This calculation is finished when the solid fraction of the mush (g_s) has passed through the advanced coalescence stage ($g_s>g_s^{nof}$), because at this solid fraction, we assume that coalescence and bridging between the grains are fairly advanced, therefore no continuous liquid films are present. Thus, the alloy has obtained sufficient ductility to prevent the formation of a hot tear. The parameters w_v and w_d enable us to tune the weight of the calculation between volumetric and deviatoric strain components respectively. More details of this criterion could be found elsewhere [7].

For cold cracking, ALSIM measures the cracking susceptibility via the critical crack size or CCS. The principle idea of such criterion is as follows; if a defect (for instance: micro hot-tear, inclusion or porosity) exists within the ingot and such a defect has a dimension exceeding the CCS at a temperature of brittle fracture, then cold cracking will occur. The formulations of cold cracking criteria for different crack geometries [3] are shown in Table 1.

Table 1. Geometrical factor of different types of cracks [3]. a_c is half of CCS, K_{Ic} is plane strain fracture toughness and σ_{11} is the first principal stress.

Penny-shaped crack	Thumbnail-shaped crack
$a_c=\frac{\pi}{4}\left(\frac{K_{Ic}}{\sigma_{11}}\right)^2$	$a_c=\frac{\pi}{(2\times1.13)^2}\left(\frac{K_{Ic}}{\sigma_{11}}\right)^2$

The half of CCS (a_c) takes into account the geometry of the initial defect; whether it is a volumetric crack (penny-shaped crack) or a surface crack (thumbnail shaped crack). This criterion also takes into account the plane strain fracture toughness (K_{Ic}) from several temperature points, which are obtained experimentally [4] and the first principal stress σ_{11}, which is readily available from ALSIM calculations.

2.3. Simulation setup

In this work, we performed benchmarking between 2D DC casting simulations with trial DC casting billets. Owing to the geometric symmetry, to reduce the calculation time, we only performed the simulation on half of the geometry with the centerline of the ingot as the symmetry axis.

The simulated geometries are built according to the trial setup, which includes the melt distributor, mold, water jet, bottom block and ingot part or cast domains, i.e. the Euler, expansion and Lagrange domains (see Figure 1a). These geometries are meshed using quadrilateral elements. When the casting starts and the first part of the billet has been solidified, the bottom block will be retracted downwards and elements in the expansion zone will grow every time step vertically until a new row of elements is splitted from the expansion zone and added to the Lagrange domain below. Subsequently, the ingot shell which has just solidified is sprayed with water at room temperature and after the casting is finished, the ingot is continuously sprayed with room temperature water for several minutes to cool down.

Other than varying the casting speed and water flow rate, we also use two different inlet geometries; a semi-horizontal feeding scheme (Figure 1b) which has two flow directions – downward vertical and horizontal to the periphery of the billet and a vertical feeding scheme (Figure 1c) which has only downward vertical flow direction.

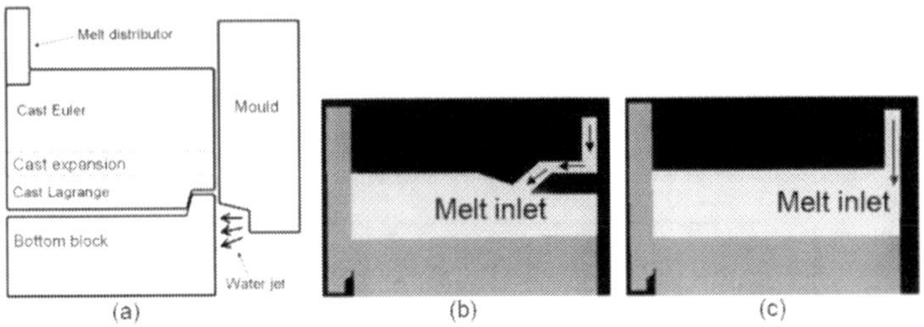

Figure 1. Generic casting setup geometry (a) and two different types of inlet flow geometry; semi-horizontal feeding scheme (b) and vertical feeding scheme (c).

The solidification path of an AA7050 alloys was calculated using JMat-Pro software. For the constitutive mechanical behavior prediction, an Al–2 wt% Cu database was used for the mushy zone regime [10], since the AA7050 data in this temperature regime is not available. For the fully solid temperature regime we used the database of an AA7050 alloy [4].

For calculation of hot tearing susceptibility, we used a critical pressure of –7000 Pa and a g_s^{nof} of 0.99. The weights for volumetric and deviatoric strain components were 1.0 and 0.0, respectively. For the cold cracking susceptibility calculation, we used our own data for plane strain fracture toughness (K_{Ic}) of the AA7050 alloy and we assumed that the initial crack geometry is penny shaped.

We performed three different casting trials with different casting conditions and the result of those castings were billets with hot tears, cold cracks and a healthy ingot without visible fracture. The details of the trial casting information are described in the next section.

3. Procedure of casting trials

The cylindrical billets were DC cast at Tata Steel Research and Development centre in IJmuiden, The Netherlands. The diameter of the ingot was 315 mm and the nominal composition of the ingots is shown in Table 2. The melt was first degassed in the furnace at 730 °C and then cast with a conventional DC casting mold as shown in Figure 1a. The casting temperature was approximately 680 °C. All of the billets were grain refined by using the same amount of Al5Ti1B for each ingot.

Table 2. Chemical composition of AA7050.

Main Alloying Elements, wt pct					
Zn	Cu	Mg	Zr	Ti	Al
6.15	2.2	2.1	0.13	0.03	balance

Table 3. Description of the casting process parameters for the AA7050 alloy.

Process parameter	Ingot 1	Ingot 2	Ingot 3
Distributor/inlet geometry	Vertical	Semi-horizontal	Semi-horizontal
Nominal casting speed (mm/min)	50	90	50
Water flow rate (l/min)	40	110	40
Cracking status	Hot tear	Cold crack	Healthy

We performed three DC casting trials with different casting parameters shown in Table 3. As a result, we have one billet with hot tear, one with cold crack and one healthy ingot that we will use as a reference. The nominal casting speed and water flow rate correspond to the steady state casting conditions after the ramping up stage.

4. Result and discussion

Figure 2 shows the cracked billets. The hot tear is shown in Figure 2a. One can see a centerline crack (between the arrow heads) that propagates as a plane along the longest axis of the ingot. This is the classical feature of a catastrophic hot tear [11], because such a crack is produced through tensile-radial mechanical deformation of the semi-solid mush due to the high temperature gradient between the center and the surface of the billet.

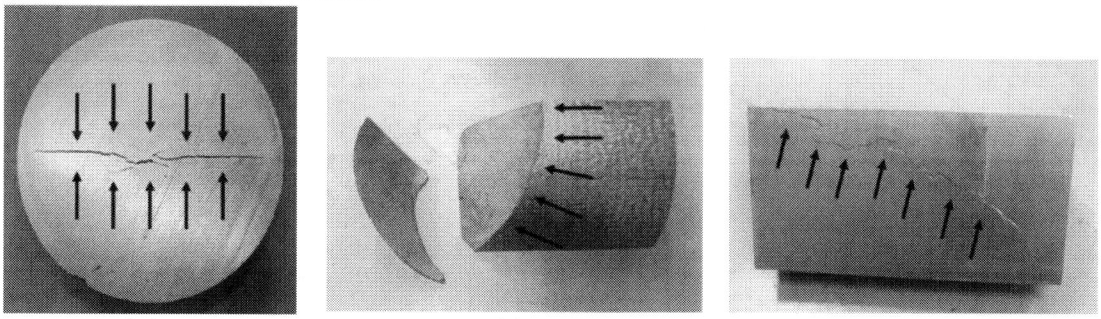

(a) (b) (c)

Figure 2. Ingot with hot tear (a) and ingot with cold crack which occurs throughout the ingot (b, c).

Figures 2b and 2c show cold cracked part of the other billet. These figures demonstrate that the cracks are propagating throughout the billet but not necessarily propagate through the centerline. For instance, in Figure 2b, the crack that splits the billet open starts from the centerline and propagates to the surface (pointed by the arrows). In Figure 2c, one can see a crack path (pointed by the arrows) of a cold crack that has not split the billet apart.

4.1. Simulation results

In Figure 3, we can see examples of ALSIM calculation from the healthy billet case. These results are taken at a simulation time of 650 seconds or when the ingot reaches a length of 500 mm (in the steady state casting condition).

Figure 3a shows a parabolic-shaped contour which roughly represents the sump shape of the ingot. In Figure 3b, one can see that areas with high tensile stresses are located in the center of the ingot and nearby the water impingement zone, and the surface part is under low tensile stress or compressive stress. Figure 3c demonstrates that the pressure is decreasing with increasing solid fraction, and reaches the lowest value in the bottom of the sump and near the mold. This phenomenon signifies the feeding difficulty which is well captured by the model. Figures 3d and 3e represents hot tearing susceptibility through ICS and cold cracking susceptibility through CCS respectively. For ICS, we can see that the highest amount is on the centerline of the ingot and the value peaked just above the bottom block, representing the startup stage of the casting. This result supports the previous findings [7, 12]. Meanwhile for cold cracking susceptibility, we can see that the center of the ingot is more prone to cold cracking than the surface, which also supports another previous work [13]. The area which has a negative CCS value (area pointed to by the arrows) has temperatures higher than 200 °C, thus no calculation is performed in this temperature regime since the state of the material at such temperatures is not favorable for cold cracking.

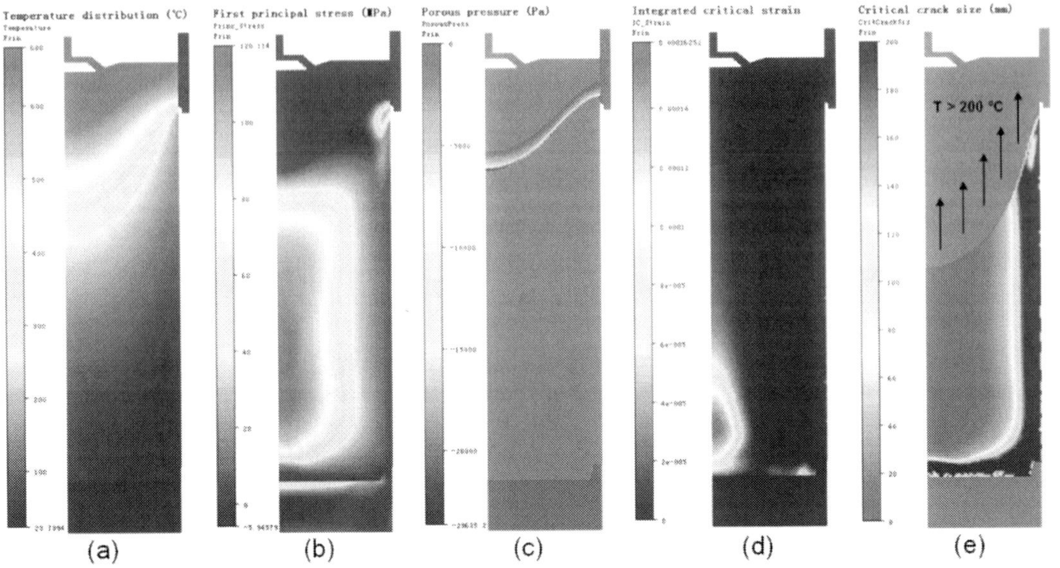

Figure 3. Example of calculation results for the healthy billet at the steady-state casting condition: temperature distribution in °C (a), first principal stress in MPa (b), porous pressure in Pa (c), integrated critical strain in arbitrary unit (d) and critical crack size in milimeter (e).

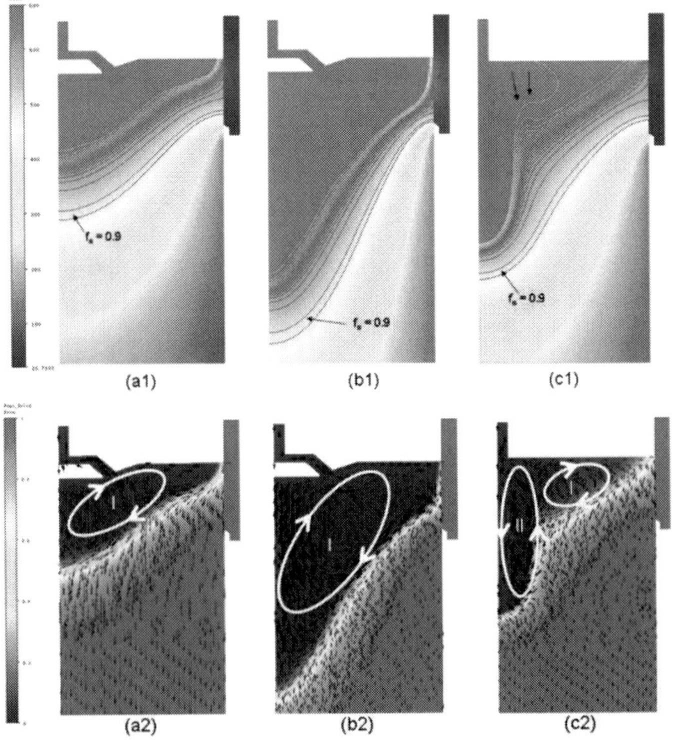

Figure 4. Comparison of temperature distribution in °C and solid fraction isolines – between 0 and 1 (a1, b1 and c1) and comparison of solid fraction and fluid flow, in the sump (a2, b2 and c2) for different cases. a – healthy, b – cold crack and c – hot tear.

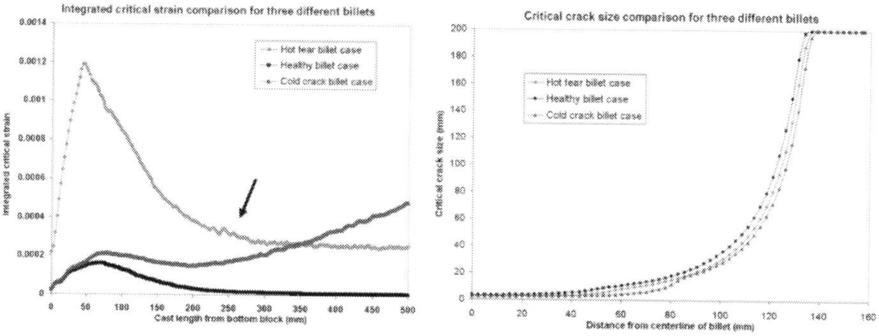

Figure 5. Examples of ICS comparison (a) and CCS comparison (b) for the three different cases.

In Figure 4 (a1, b1, c1), we can see the different sump shapes and depths for the different cases. All of these results are in the steady-state casting stage. In terms of sump depth (from metal level down to a solid fraction (f_s) of 0.9), the deepest sump occurs in the cold cracking case with a depth of 253 mm while the hot tear case has a depth of 176.5 mm and the shallowest sump corresponds to the healthy ingot with a depth of 140 mm. Although the depth varies significantly between the healthy and the cold-crack cases, the sump shape is similar and the sump depth increases linearly with casting speed. The sump geometry changes significantly when the inlet flow is changed. This is apparent when we compare the healthy and the hot tearing cases. Although the casting speed is the same, the difference in terms of sump geometry and depth is remarkable. This difference is believed to be due to the flow pattern produced by the different inlet geometries.

We also observed the flow patterns of each case at the steady-state casting condition. In the healthy case, there is only one dominating flow (pattern I within white ellipse – Figure 4 a2). This condition is also holds true for the cold-crack case (pattern I within white ellipse – Figure 4 b2). The dominating flow in both case are in the same direction, along with the natural convective flow. In the hot tearing case, however there are two dominating flows with different flow directions (pattern I is clockwise and pattern II is counter-clockwise – Figure 4 c2). These flows intersect in a region of approximately 49 mm from the ingot centerline, thus in this area there is almost no horizontal velocity component. The model also responds quite well to this phenomenon where there is a discontinuous sump gradient in this particular region (the steeper sump beside pattern II and the shallower sump beside pattern I). The variation in the flow results in a very variable sump profile especially at a lower solid fraction (denoted by the arrow heads in Figure 4 c1).

Figure 5a shows the hot tear susceptibility along the centerline of the billet from the bottom block up to 500 mm of billet height. For all of the cases, the ICS values reaches a high peak in the start-up stage, which is a common phenomenon during DC casting [7], and either vanish (healthy billet) or stabilize to a certain value (hot-tear and cold-crack cases) when entering the steady state. In the cold-crack case, after the decrease in the post startup phase, the ICS value increases but finally stabilizes at approximately 0.0006 after about 700 mm of billet length. We believe that the evolution of ICS value in such a case is due to the increase of casting speed which is almost twice of the healthy case. Although the sump in the cold crack case is the deepest among all given cases, the ICS at the startup is similar to the healthy case. The peak ICS value in the startup stage for hot tear case is approximately six times higher than for the other cases and the catastrophic hot tear was observed when the billet length reaches 250 mm (pointed by the arrow).

For cold cracking susceptibility calculation, we plotted in Figure 5b, the susceptibility at 500 mm of billet height, from the center to the surface. In general, the CCS increases dramatically after reaching the mid-radius of the ingot. We can also see that generally the cold-crack case has the lowest CCS value compared to other cases which means the highest susceptibility to cold crack is in accordance with the experimental trials.

Based on the hot tear susceptibility calculation, in the cold-crack case, after the billet reaches 700 mm of cast length, the ICS value stabilizes to a value which is approximately three times of the peak value during the start up stage. This result can be interpreted as the possibility of formation of micro (non-critical) hot tears that, due to the stress condition in the billet at high temperature and the ingot ductility [4], will not develop until the casting is finished and the billet has cooled down to its brittle state. This result supports the hypotheses about the connection between hot tearing and cold cracking [4].

5. Conclusion

From the result of this work, we draw several conclusions as follows:

1. The melt feeding scheme affects the shape of the sump by distributing hot melt and changing the temperature profile which plays an important role in the susceptibility to hot tearing in the billet.
2. Sump depth also increases with casting speed which in turn increases the thermal gradient between the centre and the surface of the billet and ultimately promotes hot tearing and cold cracking.
3. From ALSIM calculations, we obtained a supporting argument for the connection between the two cracks; cold cracks could be initiated from non-critical micro hot tears.
4. The output of the simulation is in qualitative agreement with the experimental results of trial DC casting which validates the cracking model used in ALSIM.

Acknowledgements
This research was carried out within the Materials innovation institute (www.m2i.nl) research framework, project number M42.5.09340. The authors would like to express their gratitude to Dr. Démian Ruvalcaba and Mr. Jacob van Oord (Tata Steel Research, Development & Technology, The Netherlands) for their support and inputs. Support from Modelling assisted INnovation for Aluminum DC Casting process (MINAC) community is highly appreciated.

References

[1] Granger D A 1989 *Aluminum Alloys–Contemporary Research and Applications* (London: Academic Press) pp 109-135
[2] Lalpoor M, Eskin D G and Katgerman L 2008 *Materials Science and Engineering A* **497** 186
[3] Lalpoor M, Eskin D G and Katgerman L 2010 *Metallurgical and Materials Transactions A* **Vol. 41A** 2425
[4] Lalpoor M, Eskin D G and Katgerman L 2009 *Metallurgical and Materials Transactions A* **Vol. 40A** 3304
[5] Mortensen D, Fjær H G, Lindholm D, Rudshaug M and Sørheim E A 2011 *Materials Science Forum* **693** 187
[6] Mortensen D 1999 *Metallurgical and Materials Transactions B* **Vol. 30B** 119
[7] M'Hamdi M, Mo A and Fjær H G 2006 *Metallurgical and Materials Transactions A* **Vol. 37A** 3069
[8] Fjær H G and Mo A 1990 *Metallurgical Transactions B* **Vol. 21B** 1049
[9] Eskin D G, Suyitno and Katgerman L 2004 *Progress in Materials Science* **49** 629
[10] Ludwig O, Drezet J-M, Martin C L and Suery M 2005 *Metallurgical and Materials Transactions A* **Vol. 36A** 1525
[11] Suyitno, Eskin D G, Savran V I and Katgerman L 2004 *Metallurgical and Materials Transactions A* **Vol. 35A** 3551
[12] Drezet J -M, M'Hamdi M, Benum S, Mortensen D and Fjaer H 2002 *Materials Science Forum*, **396** 59
[13] Lalpoor M, Eskin D G, Ruvalcaba D, Fjær H G, Ten Cate A, Ontijt N and Katgerman L 2011 *Materials Science and Engineering A* **528** 2831

IOP Conf. Series: Materials Science and Engineering **33** (2012) 012069 doi:10.1088/1757-899X/33/1/012069

Integrated design of castings: effect of porosity on mechanical performance

R A Hardin and C Beckermann

Department of Mechanical and Industrial Engineering, University of Iowa, Iowa City, IA 52242, USA

E-mail: becker@engineering.uiowa.edu

Abstract. Porosity can significantly reduce the strength and durability of castings in service. An integrated design approach has been developed where casting simulation is combined with mechanical performance simulations. Predictions of the porosity distribution from the casting process simulation are transferred to and used in stress and fatigue life simulations. Thus, the effect of casting quality on service performance can be evaluated. Results of a study are presented where the measured porosity distribution in cast steel specimens is transferred to an elasto-plastic finite-element stress analysis model. Methods are developed to locally reduce the mechanical properties according to the porosity present, without having to resolve individual pores. Plastic deformation is modeled using porous metal plasticity theory. The predictions are compared to tensile measurements performed on the specimens. The complex deformations and the reductions in the ductility of the specimens due to porosity are predicted well. The predicted stresses are transferred to a fatigue analysis code that takes the porosity distribution into account as well. The measured and predicted fatigue lives are also in good agreement. Finally, the results of a case study are presented that illustrate the utility of the present integrated approach in optimizing the design of a steel casting.

1. Introduction

Discontinuities in castings, like porosity, create uncertainty in a casting design's robustness. There are no standard methodologies to account for its presence in the design. Consequently, designers employ overly large safety factors to entire cast parts to ensure reliability. Contributing to the issue, the processes of designing and producing castings are usually uncoupled, except for the specification of non-destructive evaluation (NDE) requirements. Unless design engineers have test data or experience for a part, they request NDE requirements without knowing how the NDE relates to part performance. If porosity could be accurately predicted from casting simulation, and its effects on the part performance realistically modeled, engineers could develop robust designs that are tolerant of the porosity and reliable. Also, NDE could be intelligently specified to assure component performance.

In this work, porosity is measured using radiography, engineering approaches are applied to simulate the effect of porosity on stiffness, deformation, damage, fracture, and fatigue life of cast steels, and the simulation results are compared to measurements. The steels studied here (ASTM A216 Grade WCB and AISI 8630) have a combination of good ductility and strength. For example WCB steel has 248 MPa yield strength, 485 MPa tensile strength, and 22% elongation at failure as minimum tensile requirements. Failure of such ductile metals occurs on the microscale by mechanisms of void nucleation, growth and coalescence [1]. Voids can preexist as microporosity and nucleate from imperfections like second phase particles. Voids grow from increasing local stress and plastic straining. As voids nucleate and grow, the void volume fraction increases. The voids begin to interact, and the void volume fraction where interaction between voids begins is termed the critical void volume fraction f_C. As plastic strain continues increasing, local necking and coalescence occurs between voids until a connected chain of voids forms and failure occurs. The void fraction at which fracture occurs is the failure void volume fraction f_F.

Porosity has detrimental effects on the structural performance of steel castings. Literature in the areas of stiffness, strength, fatigue and component performance in the presence of porosity has been reviewed [2,3]. It is found that microporosity does not result in a measurable loss of stiffness, large stress concentrations, or stress redistribution, but it greatly affects fatigue resistance [4,5]. Microporosity reduces the ductility of metals since microvoids pre-exist before any stress is applied; the nucleation stage is bypassed. Macroporosity in metals causes gross section loss, and locally reduces the effective stiffness [6-8]. Macroporosity is not uniformly distributed throughout the entire cast part, and the casting material properties are inhomogeneous. Stress redistribution occurs in a cast part due to macropores and holes, and stress concentrations occur near them, which lead to localized plastic deformation and the development of micro-cracks causing failure. The elastic-plastic behavior of the porous materials in the 10% to 70% porosity range has a non-linear dependence on the amount of porosity [2,6-9]. Materials having porosity less than approximately 10% exhibit mechanical properties having a linear dependence on porosity amount assuming that voids do not interact [10] and by considering isolated pores [11], or a uniform distribution of pores [12]. Applying the micromechanical mechanisms of ductile failure described previously, numerous micromechanics-based models of ductile failure have been developed over the past two decades [13]. Porous metal plasticity is an often used micromechanical model of ductile failure, and it has been implemented in the commercial finite element analysis (FEA) software *ABAQUS* [14-15]. In this model, the volume fraction of porosity is a primary state variable, and the inelastic flow of the material is modeled as voids nucleate, grow and coalesce at higher strains until failure occurs.

Here steel castings with porosity were produced, made into specimens and radiographed. The porosity was quantitatively determined from the radiographs. The castings underwent fatigue and tensile testing. Using porosity data from the radiographs, finite element models of the test specimens with porosity were created and the testing simulated using elastic properties that are locally dependent on porosity and using porous metal plasticity to predict plasticity and ductile failure behavior of steel with porosity. A multi-axial strain-life model is used to predict fatigue life. Results from the simulated material behavior are compared with measurements.

2. Procedures

The effect of porosity on the elastic properties of steel was studied using cylindrical test specimens having 5 mm diameter gage section made from 8630 quenched and tempered steel. The as-cast specimen was designed to position porosity in the machined specimen gage section (details in [2]). Elastic properties were measured from stable cycle hysteresis loops during fatigue testing. The measurements were compared to simulations to determine the dependence of elastic properties on porosity that gave the best agreement. Radiographs of the cylindrical/elastic property specimens were taken from orthogonal directions to reconstruct their porosity fields for simulations [2] as shown in Figure 1. For the plasticity and ductile failure studies, WCB steel specimens were produced from 2.5 mm thick x 12.7 cm wide vertically cast, top-risered, plates of two lengths (38.1 cm and 45.7 cm). Nine plates were produced and tested. The cast plates were normalized and tempered, and machined into 1.91 cm thick tensile test coupons having a gage section width of 8.64 cm. Other dimensions were determined by the ASTM E8 tensile test standard. Radiographs were taken of the plate specimens (shown in Figure 2(a)) in which stepped gage blocks were placed. A calibration curve for radiographic grey level versus plate thickness was determined. Filtering and background corrections were performed and indications were identified and analyzed to measure the through section porosity seen in Figure 2(b). Tensile testing was performed at the University of Iowa for the sound porosity-free

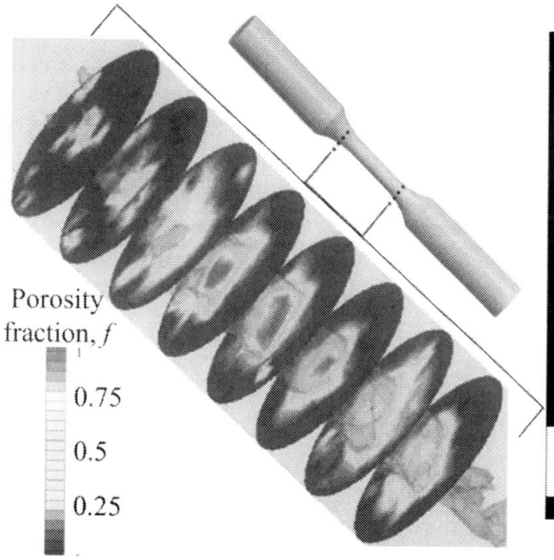

Figure 1 Porosity reconstructed in gage section of a fatigue test specimen.

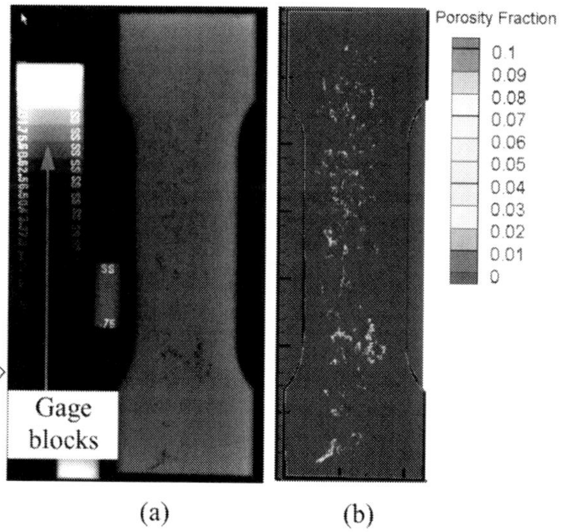

Figure 2 (a) Radiograph of plate specimen, (b) section porosity measured from radiograph.

material, and for material with porosity at SSAB North American Division, Muscatine, Iowa. After tensile testing, the porosity was observed to lie about the mid-thickness of the plates on the fracture surfaces. A finite element mesh for the tensile test plates was developed with mesh refinement at the plate mid-thickness shown in Figure 3(a). Mesh dimensions at the mid-thickness were defined so that two elements on each side about the mid-thickness of the plate were in a 2.2 mm thick region. This was the average thickness of porosity observed on the facture surfaces. The porosity was assumed to lie in this centerline region as seen in Figure 3(b). The porosity was transferred from the radiographic measurements to the FEA nodes using an average porosity about a node position to conserve porosity.

The *ABAQUS* elastic material model was used with properties depending on porosity locally as a function of porosity fraction *f* at the FEA nodes [2]. The porosity fraction is $f = V_{pore}/V_0$ where V_{pore} is the volume of porosity and V_0 is the total volume about a node in the sound metal matrix [2]. The relationship between the elastic modulus and porosity fraction used node-by-node in the FEA analysis

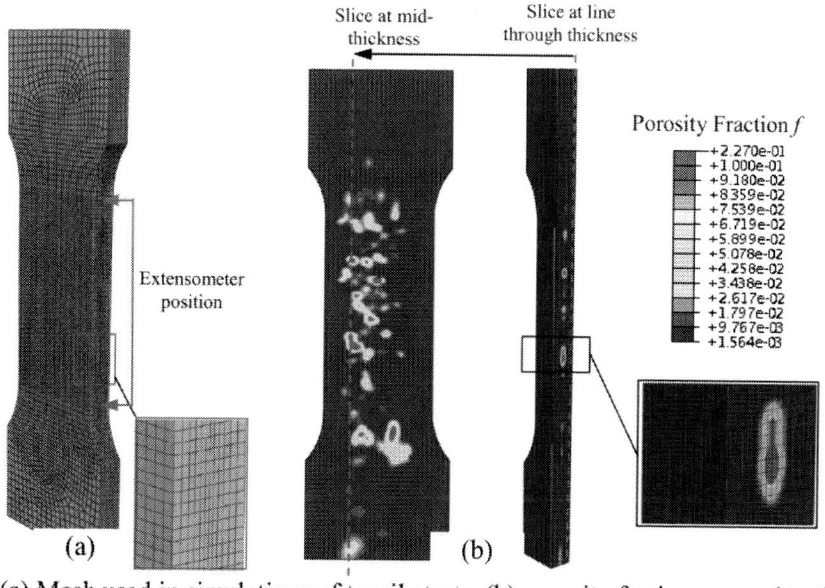

Figure 3 (a) Mesh used in simulations of tensile tests, (b) porosity faction mapped to the FEA mesh.

was $E(f) = E_0\left(1 - f/0.5\right)^{2.5}$ where E_0 was 198 GPa for WCB steel and 207 GPa for 8630 steel. The Poisson ratio ν was dependent on f using $\nu(f) = \nu_S + (f/f_\infty)(\nu_\infty - \nu_S)$ with $\nu_\infty = 0.14$, $f_\infty = 0.472$ and the Possion ratio for the sound steels was $\nu_S = 0.3$.

The porous metal plasticity material model in *ABAQUS* was used to simulate ductile plasticity and failure. Due to space limitations, only a brief description can be given here. A complete presentation of the model is found in Gurson et al. [11-14], the *ABAQUS* manual [15] and the numerical integration procedure used was developed by Aravas [16]. The model requires the user to define the hardening behavior of the metal matrix using true stress-strain data, which was determined from the tensile testing of WCB steel without porosity (sound steel). The yield condition used in the model is given by

$$\phi = \left(\frac{q}{\sigma_y}\right)^2 + 2q_1 f \cosh\left(-\frac{3}{2}\frac{q_2 p}{\sigma_y}\right) - \left(1 + q_3 f^2\right) = 0 \tag{1}$$

where f is the porosity fraction, q is the effective Von Mises Stress, p is the hydrostatic stress, σ_y is the yield stress of the sound material as a function of plastic strain, and q_1, q_2 and q_3 are material parameters. Note that when $f = 0$ (for porosity free material), the yield condition becomes $q = \sigma_y$, the Von Mises yield condition. The hydrostatic stress (pressure) p and the Von Mises Stress q are the two stress invariants $p = -(1/3)\boldsymbol{\sigma} : \mathbf{I}$ and $q = \sqrt{(3/2)\mathbf{S} : \mathbf{S}}$ where $\boldsymbol{\sigma}$ is the Cauchy stress tensor and $\mathbf{S}$ is the deviatoric stress tensor $\mathbf{S} = p\mathbf{I} + \boldsymbol{\sigma}$. The material parameters q_1, q_2 and q_3 in Eq. (1) model the interactions between voids [12]. Values used here were $q_1 = 1.5$, $q_2 = 1.0$ and $q_3 = 2.25$, where $q_3 = q_1^2$, as recommended for ductile metals [15]. The flow rule for the plastic strain rate $\dot{\varepsilon}^{pl}$ is

$$\dot{\varepsilon}^{pl} = \dot{\lambda}\frac{\partial\phi}{\partial\boldsymbol{\sigma}} = \dot{\lambda}\left(-\frac{1}{3}\frac{\partial\phi}{\partial p}\mathbf{I} + \frac{3}{2q}\frac{\partial\phi}{\partial q}\mathbf{S}\right) \tag{2}$$

where $\dot{\lambda}$ is a non-negative scalar constant of proportionality, a measure of the plastic flow rate. The plastic strain in Eq. (2) causes void growth and nucleation as it increases. The equation describing the rate of void change by growth and nucleation is $\dot{f} = (1 - f)\dot{\varepsilon}_{kk}^{pl} + A\dot{\varepsilon}_m^{pl}$ where the first term on the right hand side denotes growth of existing voids from current void fraction f and $\dot{\varepsilon}_{kk}^{pl}$, the total plastic strain rate (trace of the strain rate tensor), and the second term denotes the rate of change due to nucleation. In the nucleation term the equivalent plastic strain rate $\dot{\varepsilon}_m^{pl}$ is multiplied by a scaling coefficient $A = \frac{f_N}{s_N\sqrt{2\pi}}\exp\left[-\frac{1}{2}\left(\frac{\varepsilon_m^{pl} - \varepsilon_N}{s_N}\right)^2\right]$ and is based on the assumption that the nucleation function (A/f_N) follows a normal distribution depending on the plastic strain range about a mean value ε_N, a standard deviation s_N for a volume fraction of nucleated voids f_N. Values recommended for metals [15] were used here: $\varepsilon_N = 0.3$, $s_N = 0.1$ and $f_N = 0.04$. Coalescence and failure criteria models were used [14,15], in which the void fraction f in Eq. (1) is replaced by f^* an effective void volume fraction due to coalescence. When $f \le f_c$, f^* takes the actual void volume fraction f since coalescence has not begun. When $f > f_c$, the effective void fraction f^* increases more rapidly than f due to the coalescence with $f^* = f_c + \frac{\bar{f}_F - f_c}{f_F - f_c}(f - f_c)$. In *ABAQUS* $\bar{f}_F$ is calculated from the model material parameters from $\bar{f}_F = (q_1 + \sqrt{q_1^2 - q_3})/q_3$. When $f \ge f_F$, where f_F is the void fraction at failure the material has no load carrying capacity. In the present work, $f_c = 0.05$ and $f_F = 0.15$. Further details on the solution scheme used in *ABAQUS* are given in [14,15].

The Brown-Miller multi-axial strain life algorithm with Morrow mean stress correction and a critical plane analysis was used in the *fe-safe* software to predict crack initiation fatigue life from the FEA stress-strain output. This algorithm is recommended for ductile steels. Strain-life fatigue properties for the sound material were used; the fatigue strength coefficient σ_f', the fatigue strength exponent b, the fatigue ductility coefficient ε_f', and the fatigue ductility exponent c. These are measured for the material without discontinuities. Because porosity results in stress redistribution, and pores acting locally as notches and stress concentration sites, it was important to consider the size and geometry of the pores, and these variables relative the size of the computational grid. Therefore an adaptive subgrid technique was employed to improve the model's ability to predict the detrimental effects of porosity that a coarse computational grid cannot resolve [3]. The steps required to compute the fatigue notch factor and fatigue properties [3] can be summarized as: 1) a pore size model was used to define the maximum pore radius as function of porosity fraction f, 2) a local sub-grid model, using spherical stress concentration factors K_t, was used to correct the FEA stresses to account for un- or under-resolved local stresses at pores, and 3) a local fatigue notch factor K_f was calculated as a function of K_t, maximum pore radius and material notch sensitivity (Neuber/Peterson). The entire process was incorporated within the constraints of the commercial software *fe-safe*, and software has been developed to facilitate this process from casting simulation to performance prediction. Due to space limitations here, the reader should consult [3] for details.

3. Results

Simulation results using the porosity dependent elastic properties are given in Figure 4. All details of the simulations of the tensile tests can be found in Ref. [2]. In Figure 4(a) the non-uniform strain of a specimen due to porosity is shown. Because of the non-uniformity, the predicted strain was averaged about the circumference of the gage section at the extensometer location, and its 95% confidence interval was determined. Simulations for the 25 specimens are compared with measurements in Figure 4(b) showing good agreement for predicted versus measured strain. Note that where there is disagreement, there is also large uncertainty due to the non-uniformity in the predicted strain.

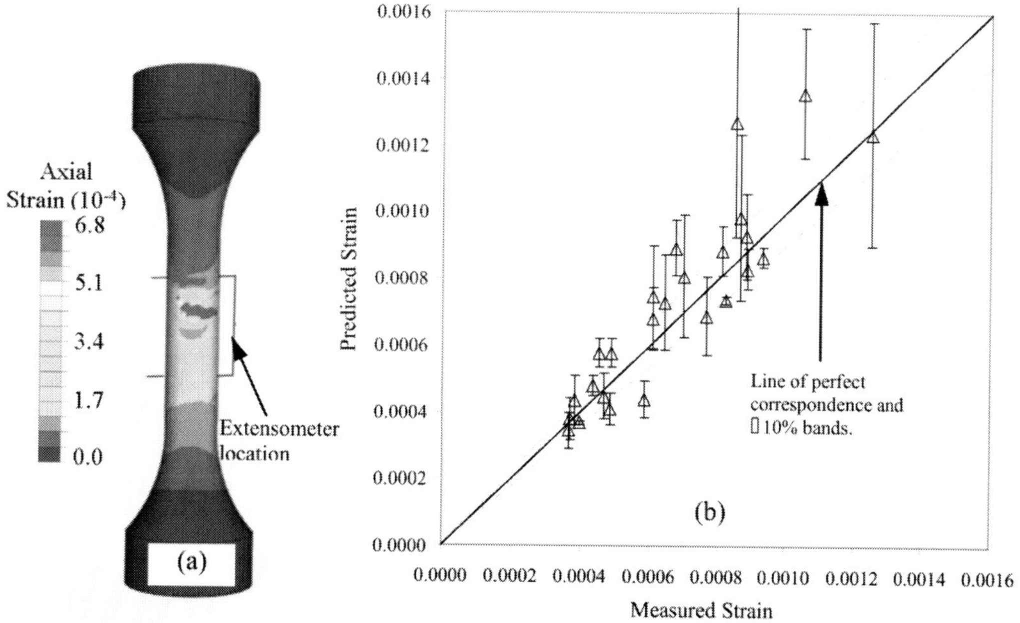

Figure 4 (a) Predicted strain on surface of test specimen, (b) measured and predicted strain from *ABAQUS* FEA simulations using E and v locally dependent on porosity fraction f.

The tensile curves of the plate specimens with porosity were simulated using the porosity fields mapped to the FEA nodes at the plates' mid-thickness. The initial porosity fraction for the sound WCB steel was set to 0.2%. This along with the porous metal plasticity model parameters ($q_1 = 1.5$, $q_2 = 1.0$,

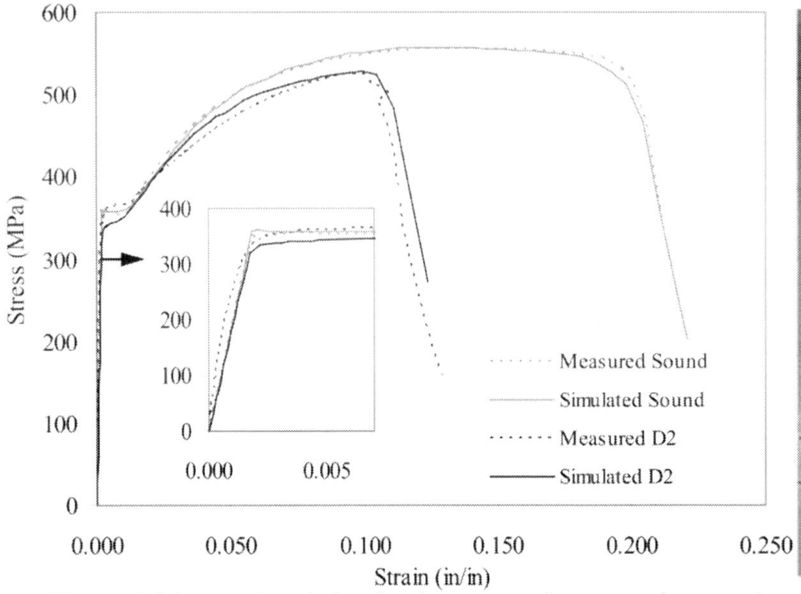

Figure 5 Measured and simulated stress-strain curves for sound material and plate D2 with porosity using ABAQUS with porous metal plasticity model.

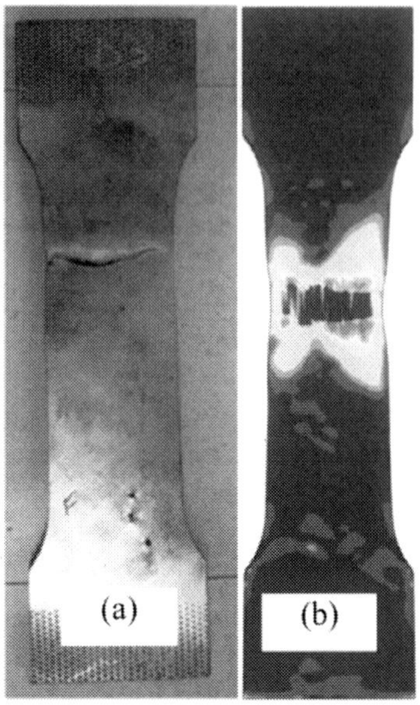

Figure 6 Plate D2: (a) after testing, (b) predicted von Mises stress at failure.

q_3= 2.25, ε_N = 0.3, s_N = 0.1, f_N = 0.04, f_c = 0.05 and f_F = 0.15) were determined by trial and error simulations, achieving good agreement between the measured and predicted tensile curves for the sound material as shown in Figure 5. The maximum elongation at failure, ultimate tensile (UT) stress, and yield stress from the measurements and predictions are within 2% for the sound steel. Plates with porosity were simulated using the model parameters, and measured nodal porosity values or 0.2% porosity if there is no measurable porosity at a node. The measured and predicted tensile curves for a plate with porosity (D2) are shown in Figure 5. The reduction in ductility is the most noticeable effect of the porosity and this is well predicted. The UT stress is well-predicted for plate D2 as well in Figure 5. In Figure 6(a) a photo of plate D2 after testing compares favorably with the predicted appearance of the final failure in Figure 6(b). In Figure 7, the measured and predicted elongation at failure results are given for all plates, and it appears there are only one or two outliers where the property is under-predicted. Note the sound data point is circled in Figure 7.

Comparisons between the measured and predicted fatigue lives for the twenty-five specimens shown in Figure 4(b) are given in Figure 8. The lowest predicted life from the *fe-safe* simulation was taken as the fatigue life. The FEA model had a node spacing of 0.25 mm. A uniform background pore size of at least 100 μm radius is used throughout these specimen (even at sound nodes), since the entire gage section was predicted to have, and observed to contain, microporosity even in radiographically sound areas. Even without assuming this, the agreement was good. However with the background microporosity in the simulations, the agreement between the measured and predicted fatigue lives was improved, as discussed in [3]. In Figure 8, the agreement appears to be within one decade, which is thought to be very good for fatigue predictions.

A case study was performed on a steering spindle casting to demonstrate how these methods of performance prediction for steel with porosity can be used in a design process. The casting is made from ASTM 148 105-85 steel and has a mass of about 8 kg in its machined condition. Boundary conditions consist of a vertical wheel load and steering forces. Porosity arising from the casting process was predicted using *MAGMAsoft* and was transferred to the FEA model using *MAGMAlink*. As seen in Figure 9(a) half-symmetry was assumed in the casting simulation, and porosity was predicted to form at two locations in the simulation; indicated as locations "A" and "B" in Figure 9(a).

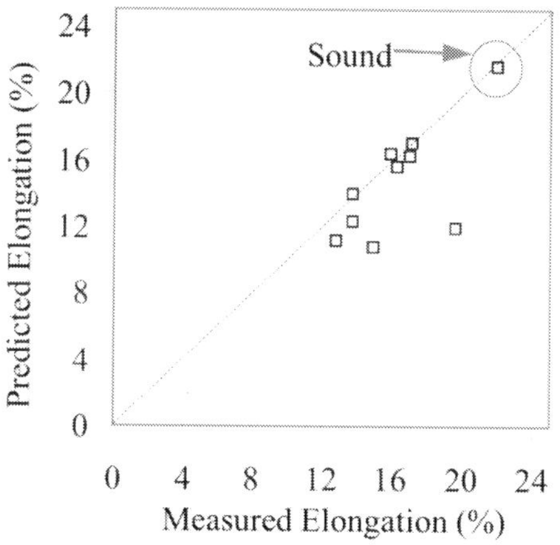

Figure 7 Predicted and measured elongation at fracture, sound WCB steel circled.

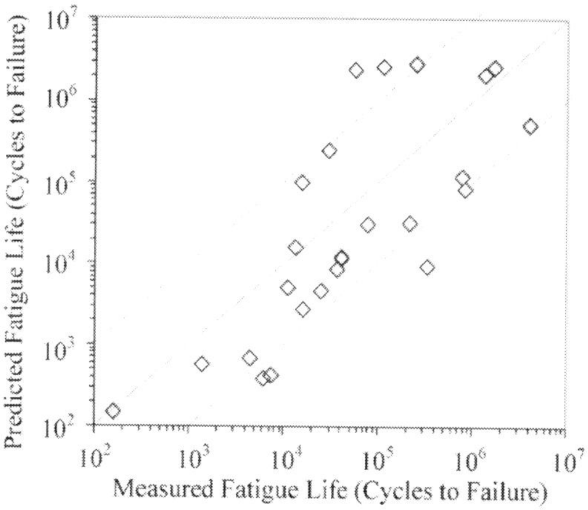

Figure 8 Measured and predicted fatigue life

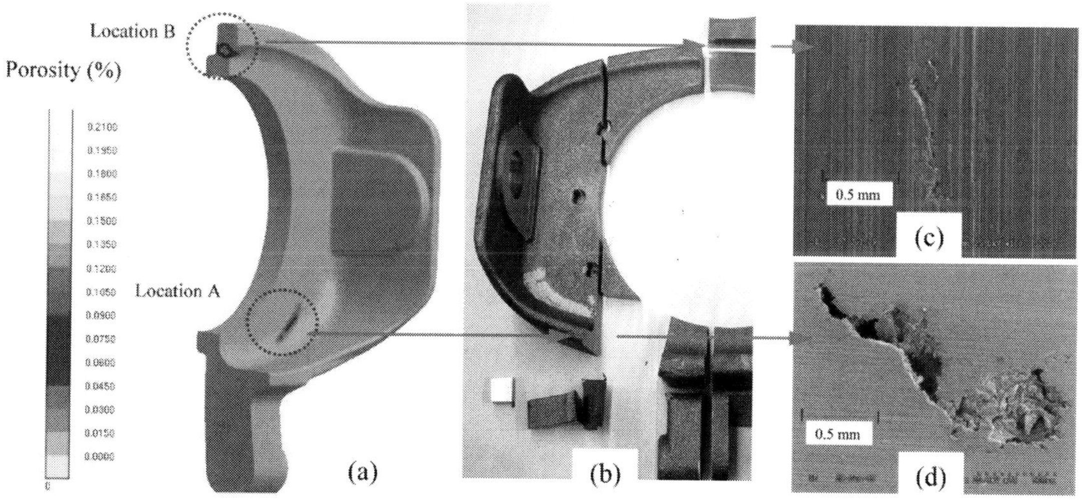

Figure 9 (a) Predicted porosity in as-cast case study part at two locations, (b) cut sections of casting with porosity, and (c) and (d) SEM images of porosity observed in sections.

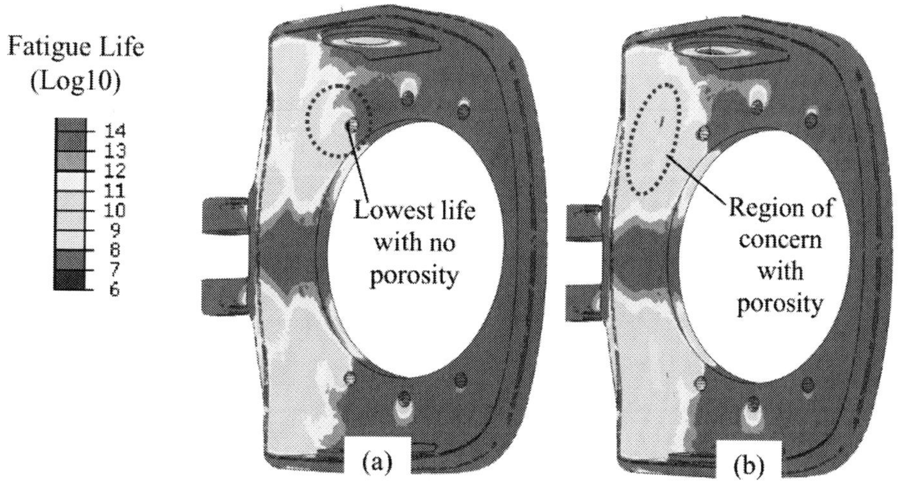

Figure 10 Predicted fatigue life (a) without and (b) with including the effect of porosity.

Parts were cast, machined and radiographed, which detected porosity at location "A". As shown in Figure 9(b), the casting was cut up in the areas of the porosity predictions, and visible porosity was found at both locations (see Figures 9(c) and (d)). The porosity field from *MAGMAsoft* was used in the FEA stress analysis using the methods of elastic-plastic analyses described earlier. The FEA results were then used in *fe-safe* to predict fatigue life for two cases: assuming no porosity and using the predicted porosity. These fatigue life predictions are given in Figures 10(a) and 10(b), respectively. Without porosity the life is predicted to be greater than 10^7 (runout condition) cycles except near a bolt hole, where the applied boundary condition is believed to cause an unrealistic stress concentration. In Figure 10(b) including the *MAGMAsoft* porosity predictions, the life is markedly lower at location "A" where porosity is predicted. This area of concern is indicated in the figure, and it was recommended that the porosity at position "A" be eliminated by modifying the casting rigging. Location "B" has a very low stresses and porosity there has no effect on fatigue life.

4. Conclusions

This work demonstrates how the mechanical and fatigue behavior of cast steel with porosity can be predicted. Radiography is used to measure the porosity so it can be included in simulations. The porous metal plasticity material model in the FEA software *ABAQUS* gave promising predictions of the effects of porosity on plasticity and ductile failure. The model requires nine parameters which were determined for WCB steel. Using these model parameters and porosity fields mapped from the radiographs, the entire tensile test curves for castings with centerline porosity were simulated. Predicted and measured tensile curves and properties were compared, and in most cases comparisons between the ultimate stress and elongation were good. Reduction in ductility in castings is perhaps the most noticeable effect of porosity and the model predicted it well. The clearest factors contributing to any disagreement in the prediction are the limitations in the porous metal plasticity model, and the assumptions and deficiencies in determining how the porosity was distributed through the thickness of the tensile plate specimens. The authors anticipate improving this work by investigating in more detail how the porosity is distributed through the plate thickness in future work. Use of the computational tools to predict discontinuities and their effects on part performance will give engineers more confidence in their casting designs and provide a path to design optimization. Also, designers can specify non-destructive testing that is more relevant to part performance. Combining casting process modeling, stress and fatigue analysis in a comprehensive and integrated approach to the design, production and non-destructive testing of castings will advance our ability to provide reliable and high performance cast components.

References

[1] Anderson T L 2005 *Fracture Mechanics: Fundamentals and Apps.* (Taylor & Francis) 219-231
[2] Hardin R and Beckermann C 2007 *Metall. Mater. Trans. A* **38A** 2992-3006
[3] Hardin R A and Beckermann C 2009 *Metall. Mater. Trans. A* **40A** 581-597
[4] Heuler p *et al* 1992 *Fatigue & Fracture of Eng. Mat. & Structures* **16** 115-136
[5] Sigl K M *et* al 2004 *Int. J. of Cast Metals Research* **17** 130-146
[6] Herakovich C T and Baxter S C 1999 *J. of Materials Science* **34** 1595-1609
[7] Roberts A P and Garboczi E J 2000 *J. Am. Ceram. Soc.* **83** 3041-3048
[8] Ramakrishnan N and Arunachalam V S 1993 *J. Am. Ceram. Soc.* **76** 2745–2752
[9] Zhang E and Wang B 2005 *Int. J. of Mech. Sci.* **47** 744-756
[10] Dewey J M 1947 *Journal of Applied Physics* **18** 578-581
[11] Gurson A L 1977 *J. of Eng. Mat. and Tech.* **99** 2–15
[12] Tvergaard V 1981 *Int. J. of Fracture Mechanics* **17** 389–407
[13] Besson J 2010 *Int. J. of Damage Mech.* **19** 3-52
[14] Needleman A and Tvergaard V 1984 *J. Mech. Phys. Solids* **32** 461–490
[15] Dassault Systèmes 2010 *ABAQUS Theory Manual V. 6.10, sec. 4.3.6 Porous Metal Plasticity*
[16] Aravas N 1987 Int. J. for Num. Methods in Eng. **24** 1395–1416

IOP Conf. Series: Materials Science and Engineering **33** (2012) 012070 doi:10.1088/1757-899X/33/1/012070

A 3D coupled hydro-mechanical granular model for the prediction of hot tearing formation

M. Sistaninia[1], A.B. Phillion[2], J.-M. Drezet[1], M. Rappaz[1]

1. Computational Materials Laboratory, Ecole Polytechnique Fédérale de Lausanne, Lausanne, Switzerland

2. Okanagan School of Engineering, University of British Columbia, Kelowna, Canada

E-mail: `meisam.sistaninia@epfl.ch`

Abstract. A new 3D coupled hydro-mechanical granular model that simulates hot tearing formation in metallic alloys is presented. The hydro-mechanical model consists of four separate 3D modules. (I) The Solidification Module (SM) is used for generating the initial solid-liquid geometry. Based on a Voronoi tessellation of randomly distributed nucleation centers, this module computes solidification within each polyhedron using a finite element based solute diffusion calculation for each element within the tessellation. (II) The Fluid Flow Module (FFM) calculates the solidification shrinkage and deformation-induced pressure drop within the intergranular liquid. (III) The Semi-solid Deformation Module (SDM) is used to simulate deformation of the granular structure via a combined finite element / discrete element method. In this module, deformation of the solid grains is modeled using an elasto-viscoplastic constitutive law. (IV) The Failure Module (FM) is used to simulate crack initiation and propagation with the fracture criterion estimated from the overpressure required to overcome the capillary forces at the liquid-gas interface. The FFM, SDM, and FM are coupled processes since solid deformation, intergranular flow, and crack initiation are deeply linked together. The granular model predictions have been validated against bulk data measured experimentally and calculated with averaging techniques.

1. Introduction

Hot tearing is a spontaneous failure of the semi-solid state that occurs in many metallic alloys and which results in an intergranular fracture profile. The thermal strains induced during casting due to both solidification shrinkage and thermal contraction of the solid skeleton can lead to the formation of hot tears. In order for simulations to successfully predict the formation of this defect during solidification, both the interaction between the solid and liquid phases as well as the evolution in microstructure that occurs during solidification must be included. It has recently been demonstrated that partially solidified alloys can exhibit the characteristics of a cohesionless granular material, including Reynolds' dilatancy, arching, jamming and stick-slip flow [1–3]. In this regard, the semi-solid behavior is quite complex in comparison with fluid flow and solid deformation. Fluid flow involves large deformations and low stresses, whereas solid deformation involves the opposite, i.e., high stresses and small deformations. Not surprisingly, modeling the semi-solid regime remains a challenging task. Most recent modeling studies of hot tearing are based on macro scale averaging methods for modeling the interaction between a deforming solid skeleton and intergranular fluid flow (e.g. [4–7]). The main weakness of such approaches is that they cannot account for the localization of straining and feeding at grain boundaries, a feature that has been clearly evidenced recently by in-situ X-ray tomography tensile tests [3]. Recently, a number of authors have begun to consider the semi-solid as a granular material to describe equiaxed-globular solidification [8–10], mechanical behavior [11–14] and liquid feeding [15–18]. Although most of the granular semi-solid deformation models were successful in modeling the mechanical behavior of semi-solid alloys before fracture initiation, they were not successful in modeling the failure stage, i.e., hot tear formation. In order to reach such a goal, the formation and percolation of voids within the intergranular liquid channels need to be considered, a feature that has not been addressed in previous models. On the other hand, previous fluid flow models [15–18] made the major assumption that the grains remain fixed. In the present study, a 3D coupled hydro-mechanical granular model has been developed to overcome the limitations.

2. Model development

The granular model is made of four separate 3D modules: (I) A Solidification Module (SM) for generating the initial solid-liquid structure [9]; (II) A Fluid Flow Module (FFM) for pressure drop calculation [18]; (III) A Semi-solid Deformation Module (SDM) for localization of straining [14]; and (IV) A Failure Module (FM) for modeling crack initiation and propagation. The FFM, SDM, and FM are coupled processes since solid deformation, intergranular flow, and crack initiation and propagation are deeply linked together and a fully hydro-mechanical coupling is used between the FFM and SDM. In the present work, the material simulated is an Al-2wt.%Cu alloy, due to its long freezing range, propensity for hot tear formation, and availability of experimental constitutive data (e.g. [7, 19, 20]).

2.1. Solidification Module

The 3D granular model of solidification known as GMS-3D [9] is an extension of the previous work by Mathier et al. [10] and Vernède et al. [21]. In this simulation, the grain geometry is derived from a Voronoi tessellation of random nucleation centres with each Voronoi region assumed to be a single grain. Each polyhedral grain derived from the Voronoi tessellation (Fig. 1) is subdivided into a set of pyramids having the nucleation center as the summit and the Voronoi facet as the base. The base of the pyramid is subdivided into triangles so as to only have tetrahedral elements that can be used either in SDM or FFM calculations (see Fig. 1(b)). Because of their regular shape, solidification within a tetrahedron can be reduced to a one-dimensional problem that is calculated using a micro-segregation model with infinite mixing in the liquid and back-diffusion in the solid. While all tetrahedra within a pyramid solidify at the same speed, since they have the same perpendicular length and hence the same Fourier number in the 1D solidification simulation, elements from different pyramids solidify at different rates. Further details of this model, together with a discussion of its limitations and domain of validity, are given in Refs. [9, 14].

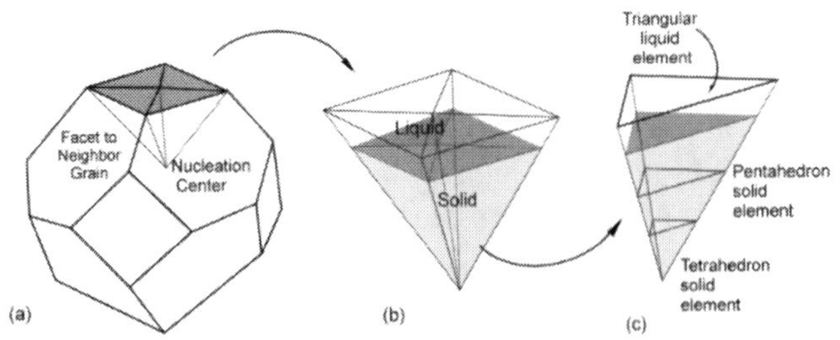

Figure 1. (a) a Voronoi region, or polyhedral final grain, (b) a small pyramidal subdivision showing both the solid portion (grey) and liquid portion (transparent), (c) a single tetrahedron decomposed into a set of solid elements

2.2. Fluid flow Module

The Navier-Stokes equations can be simplified if one considers the specific geometries of the interconnected liquid channels that remain between the solidifying polyhedral grains of the solidification module. In each triangular liquid prism separating two adjacent grain facets, it is assumed that the flow is parallel to the facet, i.e., the fluid velocity ν_ℓ has only two components, ν_x and ν_y, in a frame attached to the solid facet with the local z'-axis perpendicular to it. Under such condition, the fluid velocity vector in between the polyhedral grains is given by [18]

$$\vec{\nu_\ell} = \frac{1}{2\mu_\ell}\vec{\nabla} p_\ell [z'^2 - h^2] \tag{1}$$

where h is half-width of the liquid channel, p_ℓ the pressure, $\vec{\nu_\ell}$ the liquid velocity, μ_ℓ the dynamic viscosity of the liquid. Combined with the conservation of mass, this equation gives [18]

$$\frac{2h^3}{3\mu_\ell} \nabla^2 p_\ell = 2\beta\nu^* + (v_{sn+} - v_{sn-}) + \frac{2h}{K_\ell}\frac{\partial p_\ell}{\partial t} \tag{2}$$

where $(v_{sn+} - v_{sn-})$ is the difference between the normal (z') components of the velocity of the two neighbor grains, ν^* the solidification velocity of the solid-liquid interface, $\beta = (\rho_s/\rho_\ell - 1)$ the shrinkage factor, ρ_s, and ρ_ℓ the densities of the liquid and solid, respectively, K_ℓ the liquid bulk modulus. The left hand side of Eq. 2 is the variation of pressure required to compensate the volume change (right hand side) induced by solidification shrinkage and solid deformation (i.e., change of liquid channel width). This equation is solved using a Finite Element (FE) code written in C++. Since the flow within an element is assumed to flow parallel to the facets only, each triangular liquid channel is discretized using 3-node 2D triangular elements. Once the individual element matrices have been developed, they are assembled into a master matrix for the whole set of interconnected triangular elements (mass conservation is automatically conserved along triple lines). This global stiffness matrix is solved with a Conjugate Gradient linear iterative method using a free open access C++ template library known as IML++ [22].

2.3. Semi-solid Deformation Module

The geometry for simulating the semi-solid deformation consists of solid grains surrounded by liquid film channels. Within each grain, deformation is rather homogeneous and modeled using a continuum FE method, whilst within the entire domain, possibly discontinuous deformation is modeled using Discrete Element Method (DEM). Each solid tetrahedral element from the solidification calculation is split into three solid elements, a tetrahedron and two pentahedra, as shown in Fig. 1(c) for the FE calculation. As a result, there are approximately 120 solid elements per grain for the calculation. The output of the SM is automatically output as a finite element mesh using a C++ subroutine within GMS-3D and the numerical simulation of the SDM is performed using the commercial code Abaqus 6.9TM. The mechanical behavior of these solid elements is assumed to be elasto-viscoplastic. The Young's modulus of the solid phase is assumed to be $E_{solid} = 10$ GPa, while the Poisson's ratio was taken as $\nu_{solid} = 0.30$ [7]. The flow stress is based on the viscoplastic Ludwik's equation

$$\sigma(\varepsilon, \dot{\varepsilon}, T) = k_s(T)\varepsilon^{n(T)}\dot{\varepsilon}^{m(T)} \tag{3}$$

where σ is the stress, ε the total plastic strain experienced by the material, $\dot{\varepsilon}$ the strain rate, k_s a flow stress coefficient, n the strain hardening parameter and m the strain rate sensitivity of the material. The values for these coefficients are as follows: $m = 0.192, n = 0$, and $k_s = 40MPa.s$ [19]. In this simulation, it is assumed that the resistance of a liquid channel to separation is due only to the pressure that is exerted on the solid liquid interfaces. Furthermore, a hard contact pressure-overclosure relationship is used to prevent solid grain penetration. Shearing of the channel is modeled through a shearing parameter E_c, which is the only adjustable parameter of the model. This parameter is adjusted by comparing the mechanical response of the system to an imposed displacement with the semi-solid tensile tests available in the literature [7].

IOP Conf. Series: Materials Science and Engineering **33** (2012) 012070 doi:10.1088/1757-899X/33/1/012070

2.4. Failure Module

In order to model hot tearing, a criterion for this defect's initiation and propagation is necessary. Although multiple criteria have been suggested in the literature [see review by Eskin et al.[23]], the pressure drop due to shrinkage and deformation is a key parameter of most recently-developed hot tearing criteria. In the present study, it is assumed that there is a contact between liquid and atmosphere, which is the case of semi-solid tensile tests used to investigate hot tearing (e.g.[3]). Accordingly, the criterion can be estimated from the overpressure required to overcome capillary forces at the liquid-atmosphere interface as illustrated in Fig. 2. The Young-Laplace equation of the penetrating void (assumed cylindrical in shape) is given by:

$$p_a - p_\ell = \frac{\gamma}{R} = \frac{\gamma \cos\Theta}{h} \tag{4}$$

where p_a is the atmospheric pressure, Θ is the dihedral angle and γ the surface tension at the void-liquid interface. Considering that a thin oxide skin can have formed between the void and the liquid, the value of γ can be estimated as [24]

$$\gamma = \sigma_{\ell o} + \sigma_{og} + \tau\delta \tag{5}$$

where $\sigma_{\ell o}$ and σ_{og} are the interfacial energy between oxide skin and liquid, oxide skin and air, respectively, and δ and τ are the thickness and the tensile strength of the oxide layer. The $\gamma \cos\Theta$ value has been fixed to 3.2 Jm^{-2} in the present simulations [24]. Thus, the hot tear starts to propagate into a liquid channel connected to the atmosphere once p_ℓ is such that

$$p_\ell = p_a - \frac{3.2}{h} \tag{6}$$

Since the width of the widest channel h_{max} increases during a semi-solid tensile test, the maximal p_ℓ, i.e. $p^c_{\ell,max} = p_a - \frac{3.2}{h_{max}}$ within the widest channel also increases during tensile deformation.

2.5. Hydro-mechanical coupling

Coupling in this hydro-mechanical model occurs by carrying out each of the FFM, SDM, and FM simulations incrementally at a given fraction solid (i.e. SM output), with iterations between each increment, until bulk fracture occurs. Assuming an initial value for liquid pressure $p_\ell = p_{\ell 0}$, the simulation begins with the SDM calculation. The mechanical response of the system to the imposed displacement $\Delta u = \dot{\varepsilon} L \Delta t$, where $\dot{\varepsilon}$ is the bulk strain rate and L the dimension of the RVE is calculated and then imported into the FFM. The new thickness and the value $(v_{sn+} - v_{sn-})$ of all the channels is then calculated using the normal strain, $\varepsilon_{z'}$ of each channel. With this information, the pressure p_ℓ is calculated using the FFM. The estimated p_ℓ for the SDM calculation is then modified for the next iteration based on the new p_ℓ found in each liquid channel network. Although the pressure gradient through a liquid channel network is the driving force for the fluid flow in this network, it is negligible for the SDM calculations due to the size of the RVE and thus it can be assumed that the liquid pressure is uniform. Next, the pressure in a channel is compared to failure criterion of the FM. When a hot tear forms in a channel, it is assumed that the liquid is sucked into the rest of the liquid network and the channel becomes dry. Hence when the FM predicts a hot tear has formed in a given channel, the pressure in this channel is set to p_a for the SDM calculation. However, it does no longer participate to the overall feeding in the FFM calculation and the channel is closed by setting its displacement (local deformation) to -2h($\varepsilon_{z'} = -1$). The iterations are continued until the p_ℓ value converges, at which point the process is repeated for the next increment.

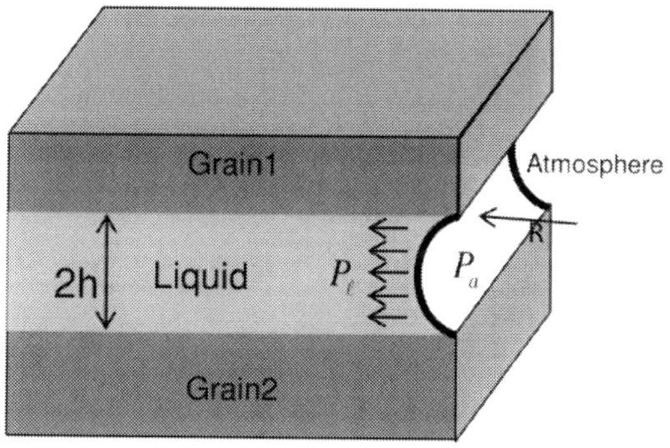

Figure 2. Schematic of two grains and a liquid channel within which a meniscus with a hemi-cylindrical shape starts forming.

3. Results and Discussion

All the results presented hereafter are shown by setting p_a to zero. In order to validate the predictions of the FFM, the permeability $\mathcal{K}$ of the mushy zone is calculated and compared to the value predicted by Carman-Kozeny's relationship [25]:

$$\mathcal{K}S_s^2 = \frac{(1-g_s)^3}{5} \tag{7}$$

where $\mathcal{K}$ is normalized by the square of the specific solid-liquid interfacial area S_s^2, and g_s is the volume fraction of solid. For the calculation, an isothermal RVE of the mushy zone with various g_s is considered with the following boundary conditions: a constant pressure p_0 on the top surface, a fixed flux q on the bottom surface while the four lateral surfaces are closed. Darcy's law is then used to calculate the permeability:

$$\mathcal{K} = \frac{q\mu_\ell}{\frac{p_1-p_0}{L}} \tag{8}$$

where p_1 is the average pressure found on the bottom side of the volume element. Figure 3 shows the pressure contours for three values of g_s (g_s=0.60, 0.90 and 0.98) where $q = 100$ μm s^{-1} and $p_0 = 0$. In this figure, although the pressure is defined only in the liquid channels, the gray color is also represented within the grains for visibility. The calculated liquid flow in each channel is represented with a plate of finite thickness proportional to the local flow normalized by the overall flow. The fluid flow is fairly well distributed in the liquid channels for the low g_s case (Fig.3(a)). However, localized feeding occurs along a few preferential paths at g_s= 0.90 (Fig. 3(b)) and even more at g_s= 0.98 (Fig. 3(c)). Also, note that the maximum local pressure drop ($p_0 - p_1$) increases with g_s; the maximum local pressure drop for g_s=0.98 is approximately 4 orders of magnitude larger than the maximum local pressure drop for g_s=0.60.

Figure 4 shows a comparison of the variation in normalised permeability, $\mathcal{K}S_s^2$, with g_s as calculated using Carman-Kozeny relation (Eq. 7) and the present FFM. The two curves are in excellent agreement, especially for $g_s > 0.4$. Please note that at low g_s, the liquid channels are wide and thus the Poiseuille laminar flow assumption is no longer valid, while the flow contribution at vortices, i.e., points where 4 grains meet, cannot be neglected anymore.

Figure 5 shows the simulated stress-strain predictions at $g_s = 0.96$ and $g_s = 0.94$ at a strain rate of $\dot{\varepsilon} = 10^{-3}s^{-1}$. The E_c value in these simulations is fixed to 2.6 MPa and all the surfaces are assumed to be closed for the FFM calculation. For the SDM calculation a symmetry boundary condition has been applied to the surfaces x=0, y=0 and z=0 of the RVE. The surface $x = L_x$ is connected to a reference node which is displaced at a fixed velocity in x-direction, while the surfaces $y = L_y$ and $z = L_z$ are free to move. The variables L_x, L_y, and L_z refer to the dimensions of the domain in the x, y, and z directions, respectively. Locations

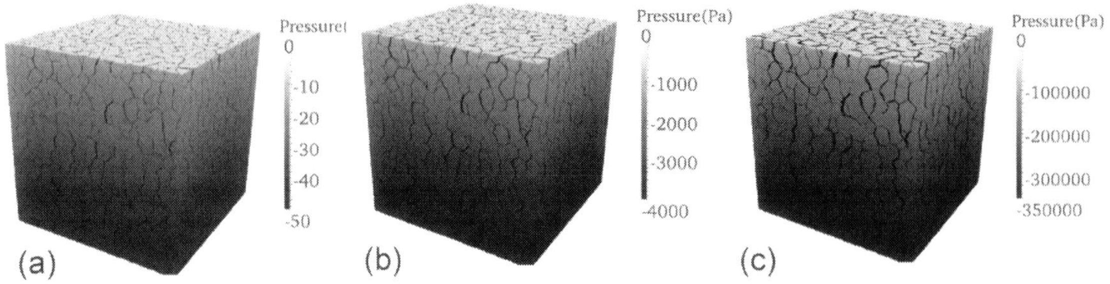

Figure 3. Fluid flow through a mushy zone volume element ($2 \times 2 \times 2$ mm^3) consisting of 8000 grains for (a) $g_s = 0.6$ (b) $g_s = 0.90$, (c) $g_s = 0.98$; the line width used to show each channel is proportional to the local flow normalized by the total flow within the volume element

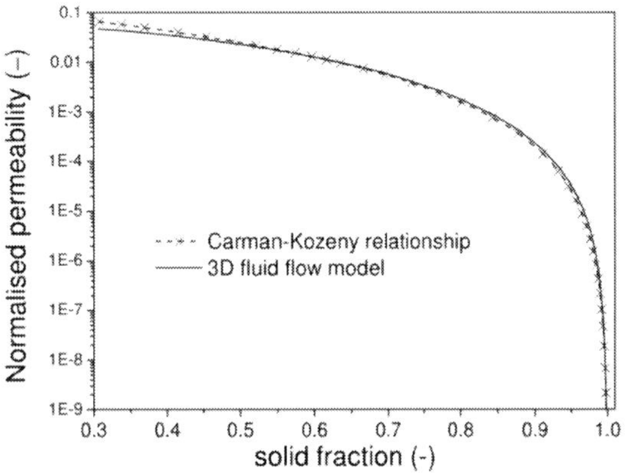

Figure 4. Normalized permeability, $\mathcal{K}S_s^2$, with g_s for a random network of grains calculated using the 3D fluid flow model and Carman-Kozeny's relationship

(a), (b), and (c) will be discussed below. As can be seen, the stress increases with increasing strain, as expected, and reaches a maximum value σ_{max} before overall failure occurs. In Figure 6, the intergranular liquid pressure drop that occurs during the tensile deformation is reported (continuous curves, output of the FFM), along with the failure pressure $p^c_{\ell,max}$ associated with the widest channel (dashed curves, output of the FM). The two curves p_ℓ and the $p^c_{\ell,max}$ cross at the strains $\varepsilon \cong 0.009$ and $\varepsilon \cong 0.012$ for $g_s = 0.96$ and $g_s = 0.94$, respectively. At this point, which corresponds to a strain one increment beyond location (a) for the curve $g_s = 0.96$, the first crack has now initiated in the widest channel connected to ambient air. As can be seen by comparing this portion of the curves in Figs. 5 and 6, the overall stress within the RVE continues to increase, but at a lower rate. With increasing strain, the crack continues to propagate within the RVE, and the rate $d\sigma/d\varepsilon$ continues to decrease until σ_{max} near (c) is reached, and the rate becomes negative.

Figure 7 shows the contour plots of the Von Mises stress in (MPa) for $g_s = 0.94$ at the three strains corresponding to (a), (b), and (c) of Figs.5 and 6. In this figure, the channels in which the Von Mises stress is not represented (white channels) correspond to those where a hot tear has formed. Finally, in Fig. 8, a comparison is made between the simulated maximum stress (stress at fracture) and the experimental results for the same Al-2wt.%Cu alloy from [7], over a range of fractions of solid between 0.8 and 0.99 at a strain rate of $10^{-3}s^{-1}$. As can be seen, this model correctly reproduces the general trends of the experimental results. However, for $g_s < 0.94$, it seems that this model overestimates the strength of this alloy.

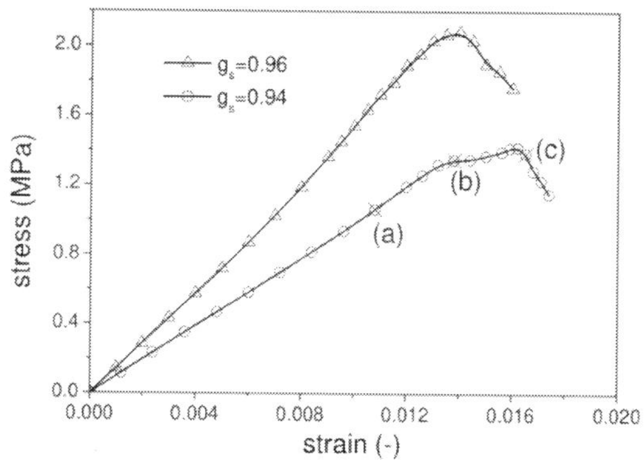

Figure 5. Calculated tensile behavior of partially solidified Al-Cu alloys Al-2wt.%Cu: stress vs. strain curves for △ $g_s = 0.96$ and ○ $g_s = 0.94$.

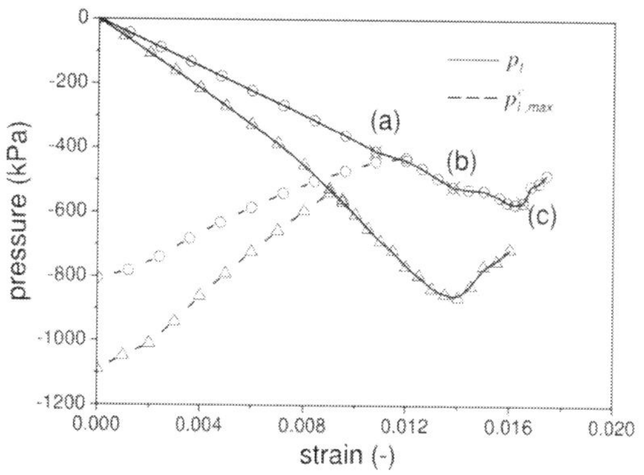

Figure 6. Liquid pressure drop (continues curves) and maximum value of the cavitation pressure (dashed curves) in the mushy zone as a function of strain corresponding to Fig.5

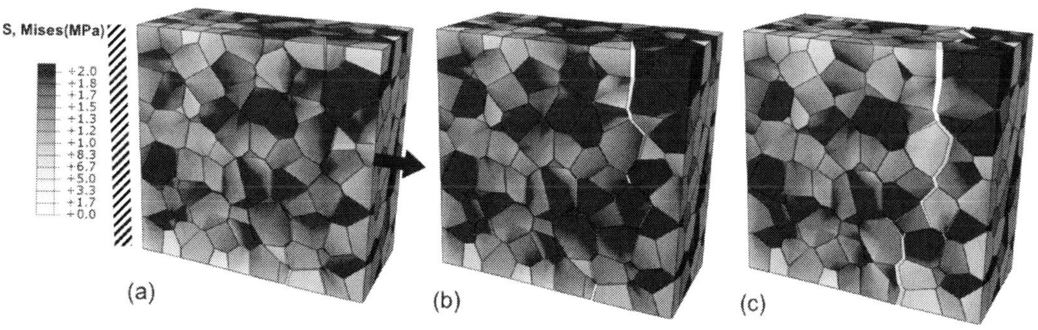

Figure 7. Contour plots of the Von Mises stress at various deformation stages and for $g_s = 0.94$. The white channels correspond to those where a hot tear has formed.

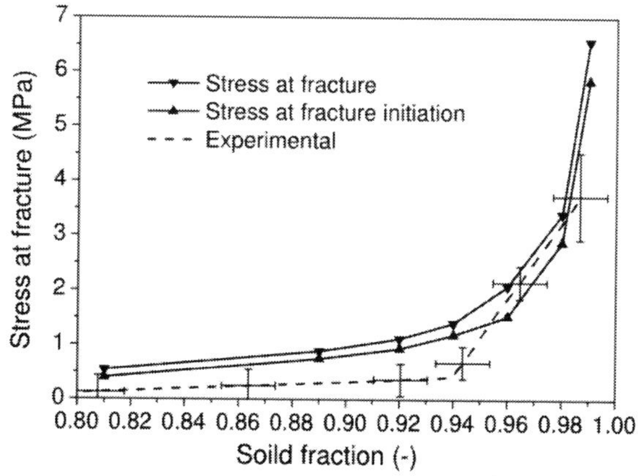

Figure 8. Comparison between the tensile strength measured in semi-solid Al-2wt.%Cu alloy [7](dashed line curve) and simulation results (continuous curves).

4. Conclusion

A 3D granular hydro-mechanical coupled model has been developed to predict hot tearing formation in solidifying alloys. This model is made up of four separate 3D modules; (I) a solidification module (II) a semi-solid deformation module (III) a fluid flow module and (IV) a failure module. This model is able to predict the overall response of the semi-solid alloys to an externally applied strain before and after fracture initiation, while accounting for the localization of strains at grain boundaries. The strength of semi-solid alloys at various g_s predicted with this model agree well with the experimental results.

Acknowledgements

The authors would like to thank the Swiss Competence Centre for Materials Science and Technology (CCMX), and partner companies within the thematic area "Multi-scale, multi-phenomena modelling of metallic systems" for funding this research.

References

[1] Gourlay C M, Dahle A K, Nagira T, Nakatsuka N, Nogita K, Uesugi K and Yasuda H 2011 *Acta Materialia* **59** 4933–4943
[2] Gourlay C M and Dahle A K 2007 *Nature* **445** 70–73
[3] Terzi S, Salvo L, Sury M, Limodin N, Adrien J, Maire E, Pannier Y, Bornert M, Bernard D, Felberbaum M, Rappaz M and Boller E 2009 *Scripta Materialia* **61** 449 – 452
[4] Bernard D, Nielsen O, Salvo L and Cloetens P 2005 *Materials Science and Engineering a-Structural Materials Properties Microstructure and Processing* **392** 112–120
[5] Monroe C and Beckermann C 2005 *Materials Science and Engineering A* **413-414** 30–36
[6] M'Hamdi M, Mo A and Martin C 2002 *Metallurgical and materials transactions a* **33** 2081–2093
[7] Ludwig O, Drezet J M, Martin C and Sury M 2005 *Metallurgical and Materials Transactions A* **36** 1525–1535
[8] Vernède S and Rappaz M 2006 *Philosophical Magazine* **86** 3779–3794
[9] Phillion A B, Desbiolles J L and Rappaz M 2009 *Modeling Of Casting, Welding, And Advanced Solidification Processes - XII* (Minerals, Metals & Materials Soc, 184 Thorn Hill Rd, Warrendale, Pa 15086-7514 Usa) pp 353–360
[10] Mathier V, Jacot A and Rappaz M 2004 *Modelling and Simulation in Materials Science and Engineering* **12** 479–490
[11] Phillion A B, Cockcroft S L and Lee P D 2008 *Acta Materialia* **56** 4328–4338
[12] Vernède S, Dantzig J A and Rappaz M 2009 *Acta Materialia* **57** 1554–1569
[13] Lahaie D J and Bouchard M 2001 *Metallurgical and Materials Transactions B* **32** 697–705
[14] Sistaninia M, Phillion A, Drezet J M and Rappaz M 2011 *Metallurgical and Materials Transactions A* **42** 239–248
[15] Phillion A, Vernede S, Rappaz M, Cockcroft S and Lee P D 2009 *International Journal of Cast Metals Research* **22** 240–243
[16] Vernède S, Jarry P and Rappaz M 2006 *Acta Materialia* **54** 4023 – 4034
[17] Dijkstra W O, Vuik C, Dammers A J and Katgerman L 2004 ed Rappaz M, Beckermann C and Trivedi R (TMS) pp 151–156
[18] Sistaninia M, Phillion A B, Drezet J M and Rappaz M 2011 *Acta Materialia* **(under review)**
[19] Fuloria D and Lee P 2009 *Acta Materialia* **57** 5554–5562
[20] Magnin B, Maenner L, Katgerman L and Engler S 1996 *Materials Science Forum* **217-222** 1209–1214
[21] Vernède S and Rappaz M 2007 *Acta Materialia* **55** 1703 – 1710
[22] Dongarra J, Lumsdaine R, Pozo R and Remington K 1994 *Proceedings of the second object oriented numerics conference* pp 214–218
[23] Eskin D G, Suyitno and Katgerman L 2004 *Progress in Materials Science* **49** 629–711
[24] Syvertsen M 2006 *Metallurgical and Materials Transactions B: Process Metallurgy and Materials Processing Science* **37** 495–504
[25] Carman P 1937 *Transactions of the Institution of Chemical Engineers* **15** 150–166

Numerical study of dendrite coherency during equiaxed solidification by the Discrete Element Method

L Yuan[1], C O'Sullivan[2], C M Gourlay[1]

[1]Department of Materials, Imperial College London, SW7 2AZ, London, UK
[2]Department of Civil and Environmental Engineering, Imperial College London, London, SW7 2AZ, UK

E-mail: lyuan@imperial.ac.uk

Abstract. Equiaxed solidification of Al alloys has been simulated by a continuum model in 2D, producing morphology variations from near-globular to highly-branched dendritic. The resulting microstructures were taken as initial samples to perform direct-shear simulations using the Discrete Element Method (DEM) and study the dendrite coherency point. Crystal rearrangement in response to direct-shear is analysed in different grain morphologies with a focus on force transmission and crystal translations and rotations. The simulations show that the coherency point decreases significantly as the morphology becomes more dendritic. Significant rotation was observed around the shear plane, leading to local dilation. The modelling results reproduce the key trends reported in prior experiments on the effect of grain size and morphology on dendrite coherency, and suggest that the coherency point depends on both the internal fraction of liquid within the crystal envelopes and also on the shape of envelopes.

1. Introduction

During equiaxed solidification, feeding difficulties start to develop when individual dendrites impinge upon their neighbors and form a network. This point is commonly termed 'dendrite coherency' and is often thought of as the transition point from mass to interdendritic feeding [1, 2]. To date, determination of the coherency point has been only via experimental measurements, either using a rheological test [1, 3, 4] or thermal analysis [1, 5]. Recent in-situ X-ray imaging of deformation in semi-solid alloys near coherency [6, 7] suggests that equiaxed alloys deform with fundamental similarities to granular materials such as water-saturated sands [8]. This suggests that the particulate Discrete Element Method (DEM), a simulation tool widely used for granular materials can be used to study the mechanical behavior of the crystal assembly near the coherency point.

Studies on a wide range of Al foundry alloys have shown that the dendrite coherency solid fraction (g_s^{coh}) is correlated with crystal morphology [1, 3, 4]. This result has begun to be interpreted by using the concept of "dendrite envelope", defined as the smooth surface connecting the outmost dendrite tips [9-12]. However, the evaluation of envelopes introduces uncertainties because of the difficulties in capturing the dendrite morphology at coherency during solidification. In this study, the formation of numerous equiaxed dendrites was simulated by a continuum model, μMatIC, in 2D [13]. The resultant microstructures at a large range of solid fractions were then imported into a commercial DEM package, PFC^{2D} [14], as initial samples to perform direct-shear, mimicking the rheological dendrite coherency test. Dendrite rearrangement in response to shear was studied at a range of solid fractions for different crystal morphologies, allowing the influence of morphology on dendrite coherency to be discussed.

2. Numerical methods

2.1. Microscale solidification model

The continuum model employed in this study solves the momentum, mass and energy transport equations, including random nucleation based on a Gaussian distribution, and phase changes based on both solute diffusion and partitioning. The detailed description of the model can be found in prior publications [15-19]. In the model, an imposed anisotropy in interfacial energy was adopted with a modified decentered square (2D) / octahedron (3D) technique [20], which ensures dendrites grow with the preferred <100> orientation. The benefit of this implementation is that it can significantly reduce the computational cost, as it allows the mesh size to be an order of magnitude coarser than other numerical models, such as phase field, which evaluate the interfacial energy in an accurate manner. The algorithm has trade-offs regarding the precision at dendrite tips. However, it maintains acceptable accuracy for the prediction of dendrite morphology and microsegregation at the microstructure level. The model has been successfully used to study solidification phenomena including columnar-to-equiaxed transition [21], the formation of porosity [15] and interdendritic convection [22].

A binary alloy, Al-7wt%Si was selected for this study, because it is commonly used in the rheological dendrite coherency test [1, 3, 4, 11, 23, 24]. The thermophysical properties are listed in Table 1, using a linearized Al-Si phase diagram. The eutectic reaction is ignored here and both the solidus and liquidus continue into the eutectic region. Four different grain sizes (*d*) from 50 to 400 μm were predicted by altering the nucleation density in the model. 225 crystals were obtained in each case with a mesh (pixel) size of 5 μm. This leads to a domain size that varies from 750 to 6000 μm.

Table 1. Physical and Elastic Properties of Al-7wt% Si used in simulations [18, 25]

Property	Unit	Value
Liquidus Temperature, T_{liq}	°C	614
Solidus Temperature, T_{sol}	°C	577
Liquidus slope, m	°C/wt%	-6.0
Partition coefficient, k	-	0.13
Solute diffusion coefficient in liquid, D_l	m^2/s	3.0×10^{-9}
Solute diffusion coefficient in solid, D_s	m^2/s	1.0×10^{-12}
Shear modulus, G	Pa	1.71×10^{10}
Poisson's Ratio, υ	-	0.384
Friction coefficient, μ	-	0.1
Effective density, ρ_e	Kg/m^3	182

2.2. Sample preparation for DEM

Figure 1 shows the procedure of preparing DEM samples. The microstructure simulated in the continuum model consist of meshes (pixels) with a local solid fraction ranging from 0 to 1 (figure 1(a)). Pixels were binarised to be either 0 or 1 by applying a threshold that keeps the overall solid fraction unchanged (figure 1(b) dark pixels). Then, each pixel with value of 1 was replaced by a disk with diameter equal to the pixel size (figure 1(c)). This 1 to 1 mapping leads to large number of disks, which requires high computational cost in DEM. An algorithm was developed to minimize the total number of disks while maintaining the exact morphology. It identifies square groupings of disks that do not intersect the crystal boundary and replace the groupings with a single large disk whose diameter equals to the side length of the square (figure 1(d)). This optimization retains the centroid and the morphology of each crystal while significantly reducing the total number of disks. Finally, the individual disks were bound with their neighbors as a rigid body to compose each individual crystal.

The final sample for DEM, generated from figure 1(a), is shown in figure 1(f). With the refinement, the number of disks in this case is reduced by 40.7% from 16806 to 9967.

Crystal envelopes were estimated from the morphology based on average interface curvature ($\bar{k}$) which is calculated by an approximate method [26]. When a concave interface is detected, the cell is counted into the envelope and assigned as solid. An example of the envelope is shown in light grey in figure 1(b). The disks comprising the envelope are shown in figure 1(e) which contains large disks and fewer disks than the original crystal (9704 in total as optimized).

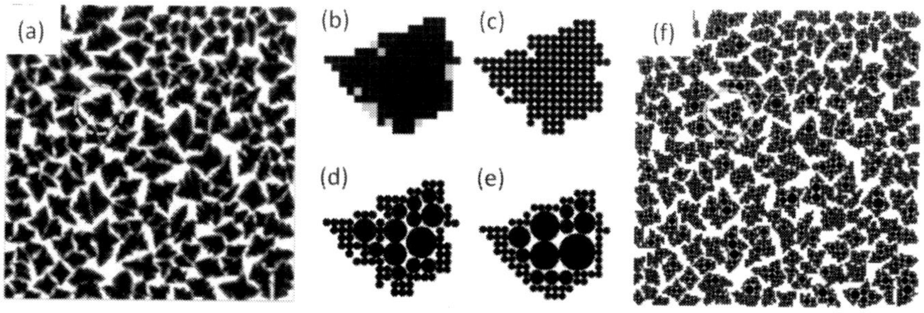

Figure 1. Preparation of DEM samples. (a) simulated microstructure with a diffuse solid-liquid interface; (b) binarised image of the crystal circled in (a), solid is black and the crystal envelope is shown in light grey; (c) mapping solid pixels to circular disks; (d) optimised disks with inner square arrays replaced with a single disk; (e) optimised disks based on the envelope; (f) Final DEM sample from the simulated microstructure in (a).

2.3. Discrete Element Method

The DEM was first introduced by Cundall and Strack for the analysis of rock and soil mechanics [27, 28]. A particle mechanics code, PFC^{2D} (Particle Flow Code in two dimensions) developed by Itasca Consulting Group, Inc., is used here to implement the DEM. It assumes that the particles are rigid bodies and forces are only transmitted through particle-particle contacts. Therefore, the deformation in the simulation arises only from the translation and rotation of crystals, and the opening and interlocking at interfaces. In PFC^{2D}, the movements and interaction forces between crystals (particles) are explicitly calculated using a finite difference method based on Newton's laws of motion.

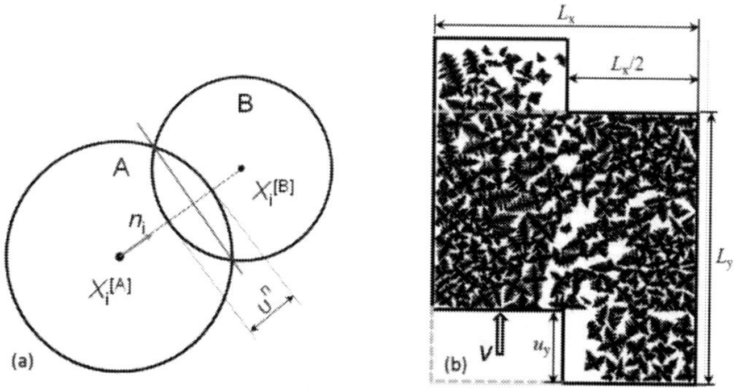

Figure 2. Setup for DEM simulations. (a) Detail of a contact between outer-disks showing the force-displacement relation; and (b) geometry of the direct-shear simulations.

Figure 2(a) shows a schematic of a contact event. The Hertz-Mindlin contact model was used in this study, which assumes a linear elastic relation between the contacting materials [29]. At the start of each time-step, contacts are updated from the known particle-particle or particle-boundary positions. The force-displacement law is then applied to each contact to calculate the contact force based on the relative motion. The normal contact force, F^n, and the increment of shear force, ΔF^s ,are given by:

$$F^n = K^n U^n \tag{1}$$
$$\Delta F^s = -K^s \Delta U^s \tag{2}$$

where U^n is the contact overlap and ΔU^s is shear displacement. K^n and K^s are the contact normal and shear stiffnesses and their values are determined from the shear modulus, G, and Poisson's ratio, ν:

$$K^n = \left(\frac{2G\sqrt{2R}}{3(1-\nu)} \right) \sqrt{U^n} \tag{3}$$

$$K^s = \left(\frac{2\left(3G^2(1-\nu)R\right)^{1/3}}{2-\nu} \right) \left|F^n\right|^{1/3} \tag{4}$$

where R is the radius of the disks. The frictional strength of the contact is given by:

$$F^s \leq \mu F^n \tag{5}$$

where μ is the friction coefficient between particles. Since little is known of the friction between crystals during the deformation of semi-solid metals, a low friction coefficient of 0.1 was used in this study since liquid lubrication is expected at contacts.

The motion of a particle is determined by the resultant force and moment vectors acting upon it. The translational and rotation motions can be calculated from:

$$F_i = m(\ddot{x}_i - g_i) \tag{6}$$
$$M_i = I_i \dot{\omega}_i \tag{7}$$

where F_i is the sum of all external forces on the particle, $\ddot{x}_i$ is the acceleration, m is the mass, g_i is the body force acceleration vector (e.g. gravity), M_i is the resultant moment, I_i is the moment of inertia and $\dot{\omega}_i$ is the angular acceleration.

To mimic the rheological dendrite coherency test, the simulated microstructures ware deformed in direct-shear by DEM, as shown in figure 2(b). Half of the domain moves upwards at a constant speed of 5×10^{-3} m/s and volume change is not permitted. The liquid in the domain is not explicitly considered but the buoyant density of crystals (the difference between the density of solid and liquid Al at the melting point) is used. The simulation stops when the shear strain (u_y/L_x) reaches 0.33. This is to ensure at least 10×10 crystals in the shear domain as recommended by ASTM D 3080-04 [30]. A more detailed study in which volume strains can arise due to shear-induced dilation is presented in reference [31].

3. Results and discussion

3.1. Mechanical response in direct shear and coherency

The microstructural and global mechanical response to direct-shear is shown in figure 3 for both crystals and their envelopes, for the samples with 50 μm mean grain size. Solid fractions varying from 65.1% to 86.8% were simulated in DEM, two of which, namely 73.2% and 76.2%, are selected to overview the simulation of crystal rearrangement and force transmission (figure 3(a)(b)(d) and (e)). Force chains are plotted as white lines between crystal centroids, in which the line thickness is proportional to the force magnitude. When shear starts, crystals are pushed upwards, creating crystal-crystal contacts. Force is transmitted from the pushing wall through contacts between crystals to the stationary walls. At the solid fraction of 73.2% (figure 3a), force chains were fleeting and continuously broken due to the rotation and translation of crystals. Figure 3b is a snapshot during deformation where there is a force chain across the domain. This force chain was then broken during the next increment of deformation. The resultant shear stress (total force on the pushing wall over the wall length) vs. shear strain is plotted in figure 3(c) in blue. It shows that the crystal assembly at 73.2% solid has little resistance to shear, with numerous low stress peaks corresponding to the increments when force chains extend across the domain. In contrast, with a small increase of solid fraction to

76.2%, the crystal assembly develops a much more spread distribution of force chains throughout the domain (figure 3(b)), and the force chains were not broken by further deformation. In particular, a strong force chain network forms from wall to wall across the shear plane, similar to previously reported contact force distribution in geomechanics DEM simulations [32]. The shear stress on the pushing wall increases dramatically after the shear strain reaches 0.1 (red curve in figure 3(c)), indicating the assembly becomes jammed. As the solid fraction increases further, the system becomes jammed at lower shear strain (figure 3). Since dendrite coherency is measured from a marked increase in resistance to shear in rheological tests, the transition between 73.2% and 76.2 % solid can be determined as the dendrite coherency point in the simulations.

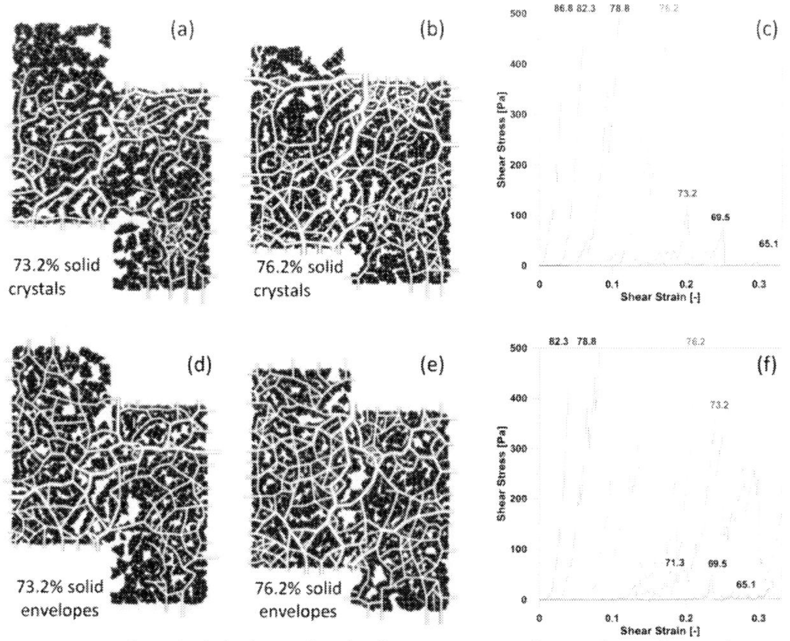

Figure 3. Microstructural and global mechanical response to direct-shear for various solid fractions during equiaxed solidification for crystals (a)-(c) and the envelopes (d)-(f) at equivalent solid fractions to (a)-(c) (d = 50 μm).. White lines are force chains with line thickness proportional to force magnitude; (c) and (f) macroscopic shear stress-strain responses on the pushing-wall. Values above each curve are solid fractions in %.

The mechanical response of crystal assemblies was compared with those of their envelopes in figure 3(d)-(f). At both 73.2% and 76.2% solid, similar force chain networks developed for both crystals and their envelopes (figure 3(d) and (e)). The shear stress-strain curves show slightly higher resistance with envelopes (figure 3(f)) than with crystals (figure 3(c)). The differences arise because the envelopes have different outer-disk arrangements than crystals (figure 1), leading to differences in contacts and force transmission. Nonetheless, the dendrite coherency point occurs in the same solid fraction range (73.2% - 76.2 %) for both crystals and envelopes. This implies that the liquid between secondary dendrite arms does not contribute to crystal-crystal contacts and has little effect on crystal movement. Thus, crystal rearrangement is essentially determined by the envelopes. From a computational viewpoint, the envelopes can significantly reduce the total disks involved (figure 2), especially when the crystal becomes dendritic (as illustrated in figure 8). For example, for a case with 199127 pixels, the DEM simulation takes 24 hours for crystals and only 5 hours for envelopes.

3.2. Effects of dendrite morphology

The mechanical response of crystal assemblies at various solid fractions and with four mean grain sizes (50 μm, 100 μm, 200 μm, and 400 μm) was simulated. The typical morphology of simulated crystals and their envelopes at the simulated coherency point is shown in figure 4. The morphology changes from near-globular to complex-dendritic as the grain size increases. The internal solid fraction at coherency (g_i^{coh}), defined as the ratio of solid fraction to the envelope fraction, is plotted against the grain size (figure 4(a)). With increasingly complex morphology, g_i^{coh} decreases approximately linearly with increasing grain size, due to the increasing liquid fraction in envelopes.

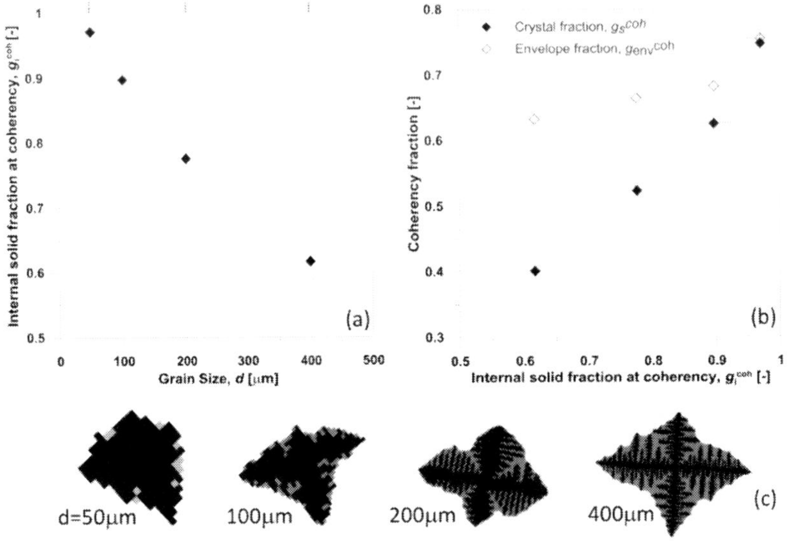

Figure 4. Influence of morphology on dendrite coherency. (a) Internal solid fraction within envelopes at coherency (g_i^{coh}) vs. grain size; (b) simulated coherency fraction vs. g_i^{coh} for both crystals and envelopes; and (c) Typical crystal morphology at coherency for each grain size. Crystals are in black, envelopes are in grey.

The solid fractions at coherency for four grain sizes are plotted against g_i^{coh} in figure 4(b). The coherency solid fraction increases from 0.4 to 0.75 as the internal solid fraction within envelopes increases and the crystal shape becomes more globular. When envelopes are taken into account, the coherency fraction for envelopes is within the much smaller range of 0.64 to 0.76. However, within this small range, the envelope fraction systematically decreases as the g_i^{coh} decreases, which suggests that the coherency point also depends on the shape of envelopes. A general trend of the envelope morphology changes can be seen in grey in figure 4(c). Note that, the simulated coherency solid fraction is higher than experimental measurement which lies between 0.15 and 0.55 [31]. This is due to the modeling of 2D crystals instead of 3D crystals in reality.

3.3. Crystal rearrangement at coherency

The initial microstructures at the simulated coherency point for four morphologies are shown in the top row of figure 5. It reveals a lighter appearance of the microstructures as the grain size increases (morphology becomes increasingly complex), due to the significant decrease of internal solid faction. The distribution of crystals in each sample is shown in the bottom row of figure 5 shortly after jamming. The greyscale shows the magnitude of accumulated rotation for each crystal. The running-average magnitude of crystal rotation across the width of the sample for four morphologies is quantified as a function of position (figure 6). Significant rotation occurs in the middle of the domain where shearing concentrates, which is also illustrated in figure 5. However, there is no general trend for the magnitude of accumulated rotation as grain size changes. This is probably because individual rotations are dependent on the random configuration of the initial distributions of crystals and most likely a larger number of crystals in the shear zone are required to explore the trend.

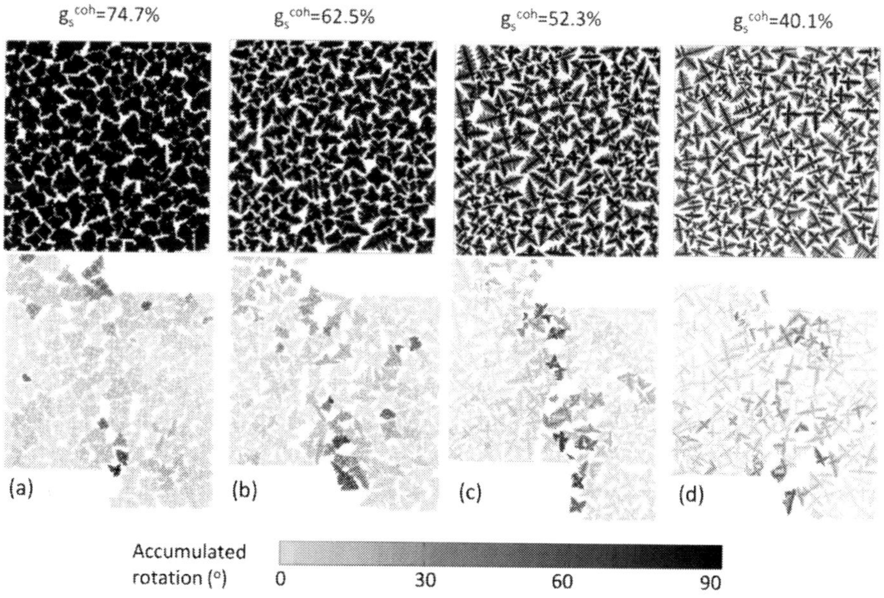

Figure 5. Microstructural response to direct-shear at the dendrite coherency point for the four morphologies of this study. Top row: simulated microstructures prior to shear; Bottom row: microstructures after jamming. The greyscale shows the magnitude of accumulated rotation.

In a crowded crystal assembly, rotating crystals need to create space between them to allow crystal motion. A dilated band in the middle of the domain can be seen in each case in figure 5 after shearing, with compaction of crystals that are far away from the band. This matches the structure of shear bands in granular media where rotation of particles is concentrated within a relatively narrow zone during shear banding process [33]. However, more crystals are required to perform DEM simulations on shear banding as natural band widths are commonly 6-20 crystals wide.

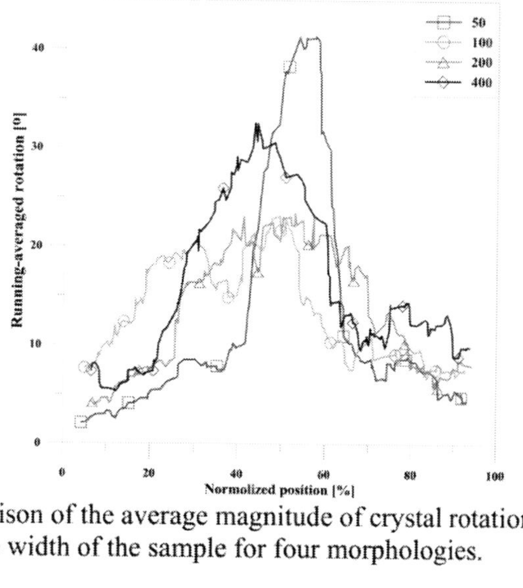

Figure 6. Comparison of the average magnitude of crystal rotation as a function of position across the width of the sample for four morphologies.

4. Conclusions

The discrete element method is used to perform direct-shear tests on crystal assemblies predicted by a 2d continuum model of equiaxed solidification. The dendrite coherency points of four morphologies, with grain sizes ranging from 50 to 400 μm, were determined by exploring the influence of dendrite morphology and solid fraction on crystal translation, rotation and force transmission. It has been found that:

1) The solid fraction at the coherency point significantly decreases as the morphology of the crystal becomes increasingly dendritic. The internal solid fraction can be used as a parameter to quantify the complex structure within the crystal envelopes.
2) The liquid between secondary dendrite arms has little contribution to crystal-crystal contacts. Crystal rearrangement is essentially determined by the envelopes.
3) The coherency point is determined by not only the internal fraction of liquid within the crystal envelopes but also on the shape of envelopes themselves.
4) Strong rotations occur in and around the shear zone, leading to a dilated band.

Acknowledgements
This project is funded by the UKs Engineering and Physical Sciences Research Council (EPSRC) grant: EP/H016848/1 and RAEng/EPSRC grant: EP/F050682/1.

References
[1] Veldman N *et al* 2001 *Metall Mater Trans A* **32** 147
[2] Dahle A K, Arnberg L 1997 *Acta Mater* **45** 547
[3] Arnberg L *et al* 1993 *Mat Sci Eng A* **173** 101
[4] Dahle A K, Arnberg L 1996 *JOM* **48** 34
[5] Malekan M, Shabestari S G 2009 *Metall Mater Trans A* **40** 3196
[6] Gourlay C M *et al* 2011 *Acta Mater* **59** 4933
[7] Nagira T *et al* 2011 *Scripta Mater* **64** 1129
[8] Zhu H P *et al* 2008 *Chem Eng Sci* **63** 5728
[9] De Groh III H *et al* 1993 *Metall Trans B* **24** 749
[10] Gourlay C M, Dahle A K 2007 *Nature* **445** 70
[11] Meylan B *et al* 2011 *Acta Mater* **59** 3091
[12] Rappaz M, Thévoz P 1987 *Acta Metall Mater* **35** 1487
[13] Lee P D *et al* 2011 http://www3.imperial.ac.uk/advancedalloys/software
[14] Itasca Consulting Group Inc 2011 http://www.itascacg.com/pfc2d/
[15] Lee P D, Hunt J D 2001 *Acta Mater* **49** 1383
[16] Wang W *et al* 2003 *Acta Mater* **51** 2971
[17] Yuan L, Lee P D 2010 *Modell Simul Mater Sci Eng* **18** 055008
[18] Lee P D *et al* 2004 *Mat Sci Eng A* **365** 57
[19] Yuan L *et al* 2009 *Int J Cast Met Res* **22** 204
[20] Gandin C A, Rappaz M 1997 *Acta Mater* **45** 2187
[21] Dong H B, Lee P D 2005 *Acta Mater* **53** 659
[22] Yuan L, Lee P D 2010 *ISIJ Int* **50** 1814
[23] Dahle A K, Arnberg L 1997 *Mat Sci Eng A* **225** 38
[24] Gourlay C M *et al* 2008 *Acta Mater* **56** 3403
[25] Nadal M, Le Poac P 2003 *J Appl Phys* **93** 2472
[26] Nastac L 1999 *Acta Mater* **47** 4253
[27] Cundall P A, Strack O D L 1979 *Géotechnique* **29** 47
[28] Hart R *et al* 1988 *Int. J. Rock Mech. Min Sci. & Geomech. Abstr.* **25** 117
[29] O'Sullivan C 2011 Particulate Discrete Element Modelling: A Geomechanics Perspective (New York: Taylor & Francis)
[30] ASTM Standard D3080 2004 (West Conshohocken, PA:ASTM International)
[31] Yuan L *et al* 2012 *Acta Mater* **60** 1334
[32] Cui L, O'Sullivan C 2006 *Géotechnique* **56** 455
[33] Bardet J P, Proubet J 1991 *Géotechnique* **41** 599

Modelling of different entrainment mechanisms and their influences on the mechanical reliability of Al-Si castings

Y Yue* **and N R Green**
School of Metallurgy and Materials, University of Birmingham, Edgbaston, Birmingham, B15 2TT, UK
E-mail: *YXY859@bham.ac.uk

Abstract. Modelling of three common entrainment mechanisms in fluid flow, namely plunging jet, return wave and rising jet, that generating defects during casting were conducted and validated by A356 alloy. Previous research highlighted surface turbulence in liquid metals could result in the fold-in of surface films into bulk liquid and leave cracks or porosity in the solidified component. An algorithm applied face normals of free surface and their interaction to capture the entrainment of surface films was used in study. The model integrated the algorithm into CFD package *FLOW-3D* and was used to track the entrainment defects formation and distribution involved in different entrainment behaviours in casting process. The defects density in local volume plotted by the model was then employed as quantitative criteria to predict the reliability of castings. Directly observation of transient flows in moulds by real-time X-ray radiography showed good correlation between real filling scenario and simulation results. Fracture strength achieved by four-point bend tests on the samples from corresponding locations with models in castings showed some extent relationship between defects density and reliability of castings quantitatively. Studies also indicated the plunging jet mechanism usually leaded to high variation of mechanical properties due to intense turbulence and entrainment.

1. Introduction

Entrainment defects such as double oxide films and entrapped bubbles occur frequently in aluminium alloy castings and are very detrimental to mechanical properties and their reproducibility [1]. The formation of entrainment defects occurs in two main ways: *a)* folding over of oxide films, and *b)* direct entrainment of gas bubbles.When liquid alloy is transferred or poured from, say, a crucible into the mould, the oxide layer formed on the free surface of the fluid by contact with the surrounding atmosphere will be easily broken or folded and then entrained into the bulk liquid by the surface turbulence (Figure 1a). The poorly bonded oxide layers, with entrapped gas, lead to the formation of cracks in the solidified castings, and provide sites for the heterogeneous nucleation of microprosity [1, 2]. Micro-bubbles can also be directly entrained into the liquid metal by flows such as plunging jets and hydraulic jumps (return waves) [3], as shown in Figure 1b. The micro-bubbles can move quickly depending on their size and density relative to that of the liquid alloy, and result in pores or bubble trail damage in casting [1].

The statistical study of fracture strengths of Al-Si alloys by Green and Campbell indicated that the mechanical strength of a casting was closely related to the density of entrainment defects. Two types of oxide films, those formed during alloy preparation (old films) and those formed during the filling process (young films) were observed. Such tangled oxide films could create discontinuities in castings and hence, reduce their reliabilities [5]. Cáceres and Selling showed that both ductility and the fracture strength of Al-Si alloys samples were mainly determined by dross or oxide films, rather than bulk porosity content,

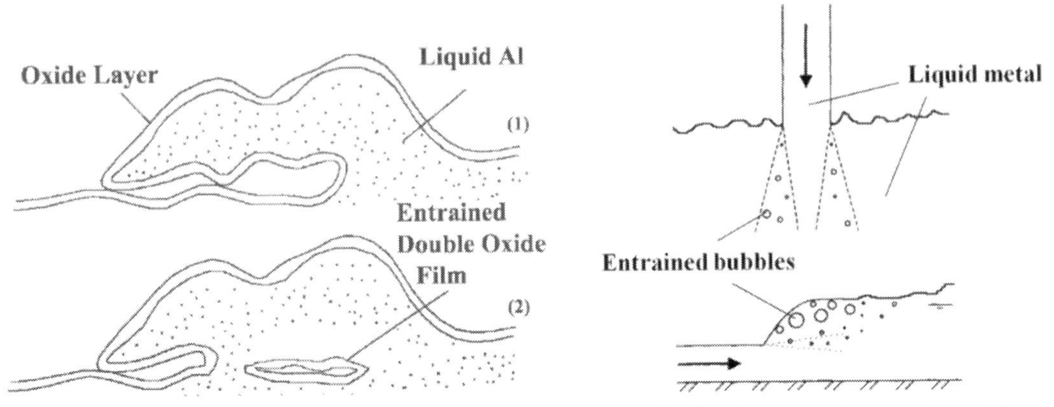

(a) Folded in double films [4] (b) Directly entrained micro-bubbles [1]

Figure 1: Two different defects formation mechanisms.

inferred from the complex morphology of oxide films on the fracture surface [6]. In regard to fatigue behaviour of Al-Si castings, Nyahumwa et al. claimed that the young oxide films or a combination of both young and old oxide films were the major crack initiators [7]. Although some studies suggested that the porosity in castings was more detrimental to fatigue properties, the importance of oxide films was still emphasised [8, 9].

Previous studies have pointed out that decreasing the surface turbulence during the filling process could increase mechanical properties and the reliability of castings [5, 10, 11]. However, little quantitative correlation has been provided to link turbulence level and reliability [12, 13], mainly due to the difficulty of the quantification of the turbulence. The influences of different entrainment mechanisms and their relative significance are not clear, either. In the present research, three very common entrainment mechanisms, namely, plunging jet (PJ), return wave (RW) and rising jet (RJ) were studied using a computer model. The formation of entrainment defects was modelled and their final distribution in a casting was predicted. Mechanical testing of bend strengths of the castings was used to validate the model.

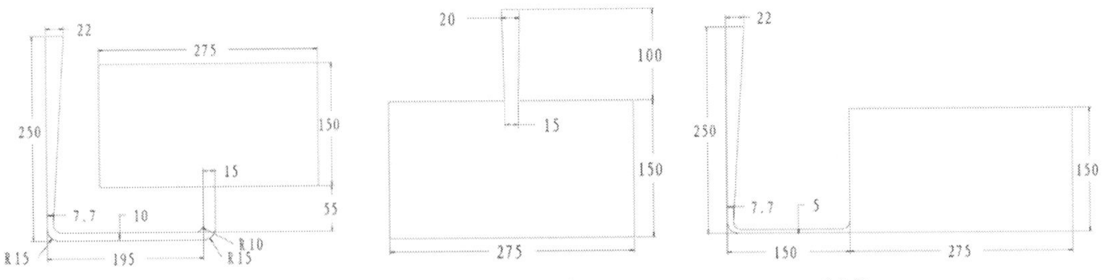

(a) Rising jet (b) Plunging jet (c) Return wave

Figure 2: The geometries of three moulds (unit: mm). The mould thickness is 15 mm.

2. Modelling method

The simulation was conducted using computational fluid dynamics (CFD) software *FLOW-3D* with a customised code to model the entrainment defects formation. The details of the customisation is presented in an unpublished work [14]. But, in brief, the customisation uses face normal of free surface to identify the relative movement of the liquid surface in each mesh element. Once the movement of adjacent free surfaces is correlated with the pre-defined entrainment forms in the code, a particle is placed in the element and the movement of the particle will be subsequently tracked in the *FLOW-3D* model. The properties of particles (density, size and interaction with the liquid) were tuned to represent the properties of double oxide films. It should be noted that the model presented here only modelled entrainment defects and did not consider the effects of thermal gradients and solidification of the castings.

Three moulds with the same cavity dimensions, but different filling systems were employed for both modelling and experiment. The geometries of the moulds are shown in Figure 2. The modelled pouring process included modelling of the rotation of a crucible and the filling of the pouring basin, followed by vertical removal of a graphite stopper to allow liquid metal to enter the downsprue. The layout of the objects and mesh settings are shown in Figure 3. The casting plate was sliced into 30 blocks to define different regions in the casings as shown in Figure 4. The number of particles in each block were exported by an in-house code to illustrate the distribution of defects, and plotted as Weibull statistics to study the relationship between predicted defects and reliability of castings.

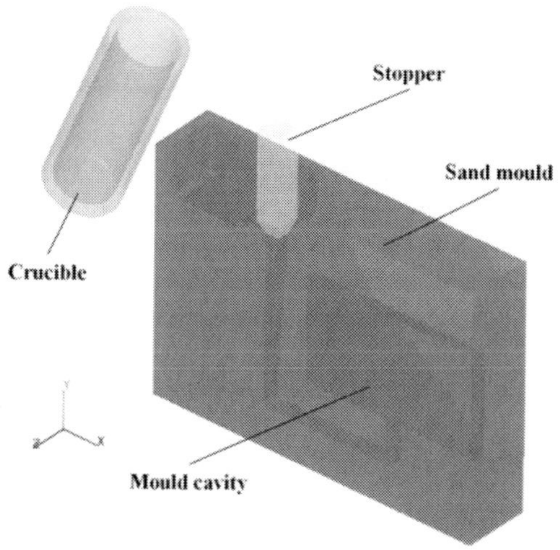

Figure 3: The schematic of objects layout.

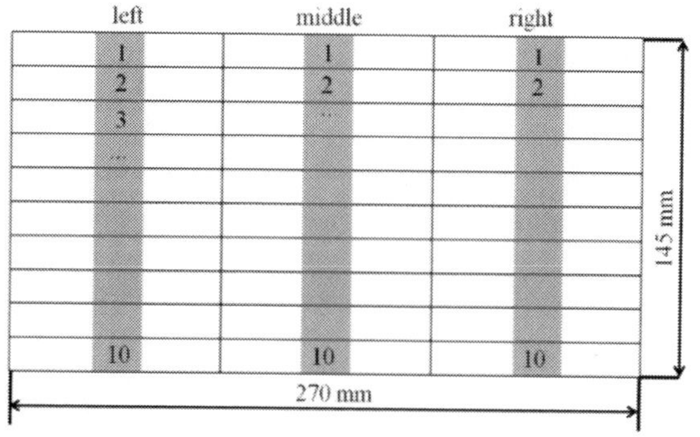

Figure 4: The layout of test bars. Gauge length is in grey.

3. Experimental

Chemically bonded sand moulds were used in the experiments. Three castings were produced for each entrainment mechanism by broadly used Al-Si-Mg alloy A356. The alloy was melted and held at 650°C for 30 min to passively degas, then rapidly heated and cast at 720 ± 5°C. Both the movement of the crucible and stopper were recorded and used as input data in the simulation. The filling process was observed by real-time X-ray radiography to study the flow structure in the mould cavity and running system, and to validate the flow model. Film X-ray was utilised to detect large defects in the castings.

The cast plates were then heat-treated to T6 condition and machined to strips with a dimension of $90\times14\times14$ mm. Four-point bend testing with inner and outer spans of 20 and 60 mm respectively were conducted on the specimens. A two-parameter Weibull distribution, as suggested by Green and Campbell, was applied to study the reliability of castings [5].

4. Results

4.1. Flow structure

Typical flow structures displayed in Figure 5 suggest that the plunging jets led to the highest entrainment level, followed by rising jet and return wave structures. In the RJ mould, since the velocity of the liquid in the mould exceeded the critical gate velocity [15], a fountain formed in the mould cavity and dropped on the main liquid body, entraining large areas of double oxide films. While the formation of entrainment defects in PJ and RW moulds was a combination of both entrained oxide films and air. The oxide films on the slower body of liquid could be rolled into the melt. Entrained air bubbles mainly concentrated in the impingement point and can be clearly viewed in the figure.

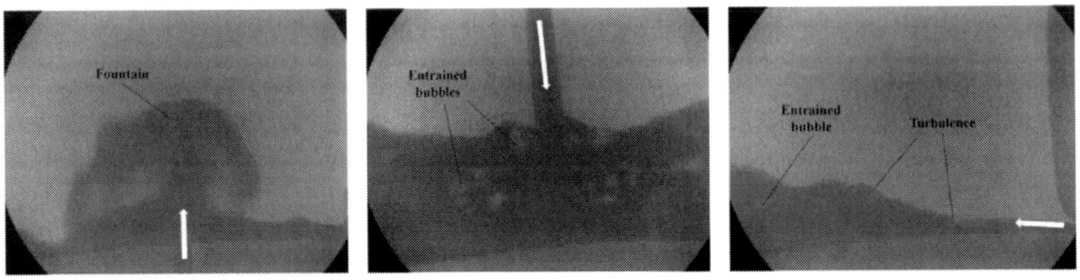

(a) Rising jet (b) Plunging jet (c) Return wave

Figure 5: Typical flow structures of three entrainment mechanisms obtained by real-time X-ray radiography. The arrows indicate the flow direction.

Comparison between the real filling scenarios and the simulation results (Figure 6) shows *FLOW-3D* can accurately simulate the free surface movement of the liquid. The swing of the fountain to the left and right in rising jet, the symmetric rotation of liquid around the plunging jet and the oscillation on the surface of the back wave were all reproduced in simulation. The velocity magnitudes of flow also plotted in modelling results to illustrate the momentum distribution in moulds. However, the formation and distribution of micro-bubbles in the liquid was not directly modelled due to the single phase nature of the model.

4.2. Defect distribution

Film X-ray observation revealed that microporosity existed in the centre of all cast plates caused by solidification shrinkage (Figure 7a). Gas pores were only found in PJ plates at the top part, near the ingate, owing to the movement of entrapped gas. Smaller defects, such as gas porosity and tangled oxide films were discovered by scanning electron microscope (SEM) examination in random selected samples. The result of RJ casting is presented here as an example of modelling result. Figure 7b quantitatively illustrates the modelled defect distribution. The warmer colour indicates the higher concentration of particles. This result shows that in the RJ mould, there were more defects in the bottom part of the plate, and more defects in the right middle of the plate comparing to the left part.

Certain trends of strength variation in different region of the cast plates were obtained in the experiments in three moulds. Based on the assumption that the lower fracture strength associated with the higher number of particle, Figure 8 compares the strengths of samples in different locations of RJ plates, with the reciprocal of the modelled defect numbers in each corresponding sample. It shows the strength of samples in the left and right parts were similar, but trends of variation from top to bottom were opposite. The simulation results show some the variation of particle counts in the plate, but the results was not fully correlated with the experiments.

4.3. Mechanical strength and reliability of castings

The cumulative plot of samples in Figure 9 shows that the bend strength of samples closely followed the Weibull distribution, and in general, samples from RJ and RW moulds had less spread in mechanical properties than those from PJ moulds. The figure also shows that besides the concentration of the strength of samples in the region of around 350-500 MPa, there is another concentration in the low strength region as highlighted in the circle.

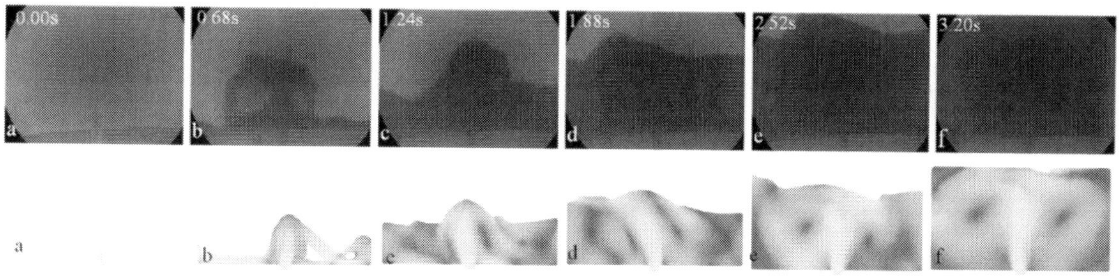

(a) Rising jet

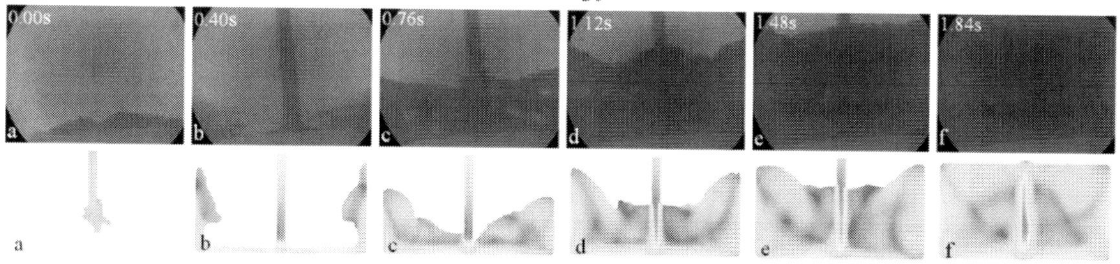

(b) Plunging jet

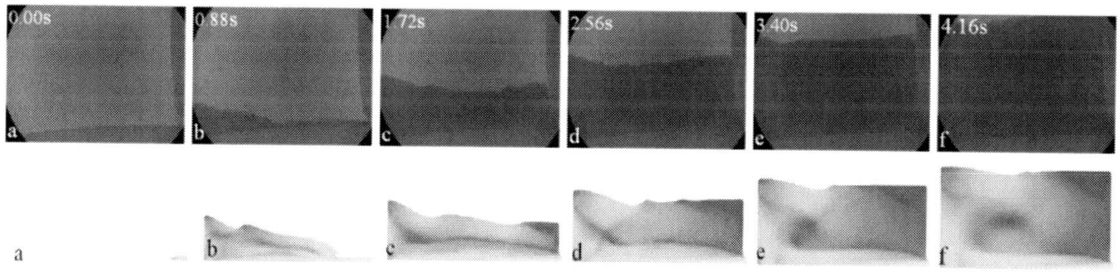

(c) Return wave

Figure 6: Filling scenario of three moulds. (top: real-time X-ray radiography; bottom: *FLOW-3D* simulation.

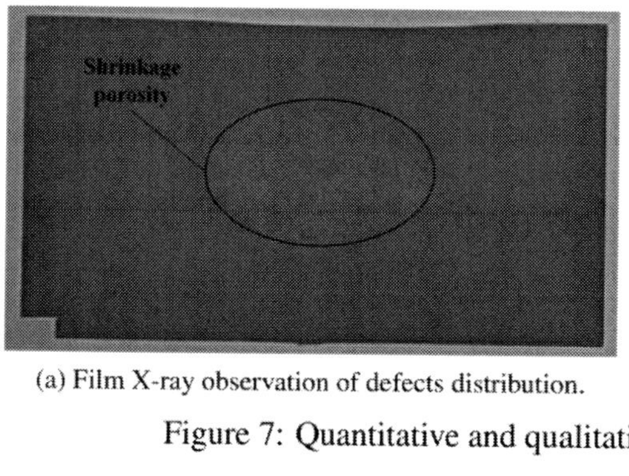

(a) Film X-ray observation of defects distribution.

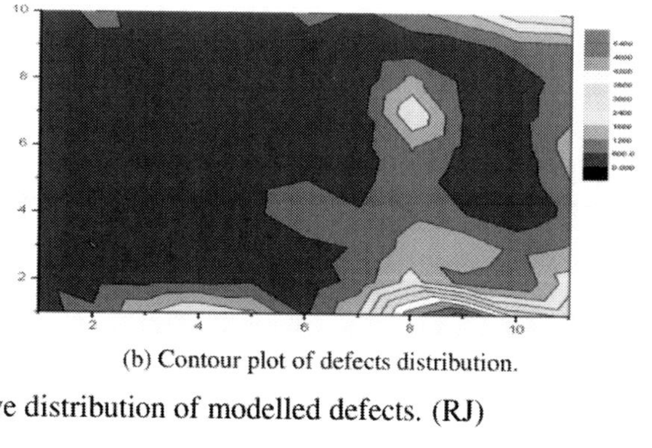

(b) Contour plot of defects distribution.

Figure 7: Quantitative and qualitative distribution of modelled defects. (RJ)

Weibull plots in Figure 11a clearly evident that samples from RW and RJ castings had higher reliability than those from PJ castings. Castings produced by RW and RJ moulds had similar reliability, resulted in similar Weibull moduli, 5.75 and 5.72 respectively. Since shrinkage porosity existed in the centre of the castings, to eliminate the effects of the shrinkage porosity on the reliability of castings, a Weibull plot of the bend strength excluding shrinkage porosity affected samples is shown in Figure 11b.

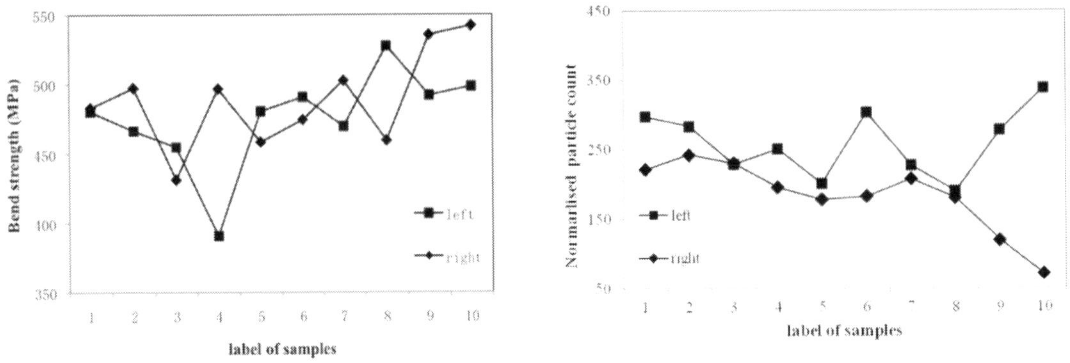

(a) Strength variation in RJ plate. (b) Normalised paticle distribution in RJ plate.

Figure 8: Defect distributions in RJ castings. (Shrinkage porosity affected data are censored)

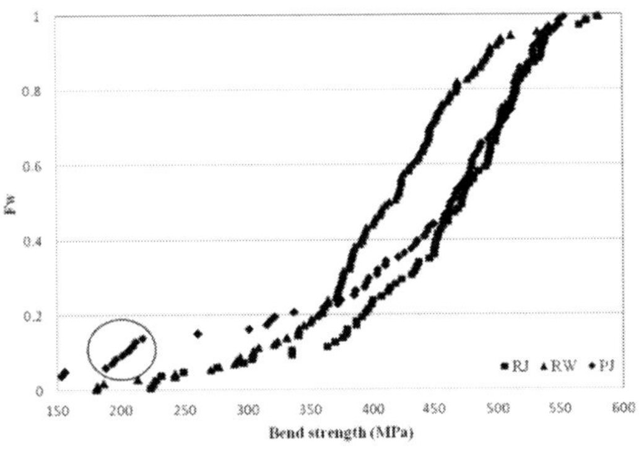

Figure 9: Cumulative plot of bend strength.

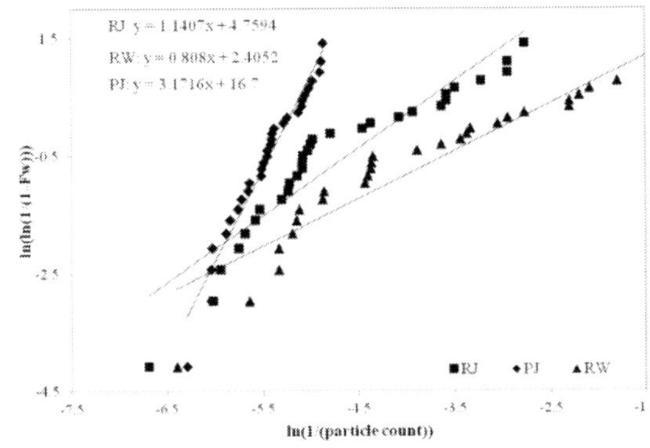

Figure 10: Weibull plot of modelling results

This shows RJ samples still had higher reliability, followed by PJ and RW samples.

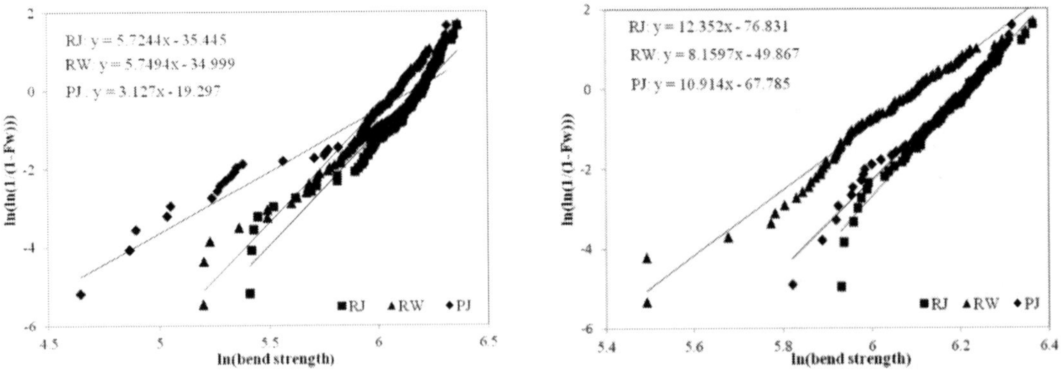

(a) All bend strength data (b) Censored bend strength data.

Figure 11: Linear Weibull plot of bend strength.

A Weibull plot of the reciprocal of the particle count in Figure 10 shows that the simulated modulus of RJ castings (1.14) was slightly higher than that of RW castings (0.81), which is agree with the experiment results in Figure 11, but both were lower than that of the PJ samples (3.17). However, considering the position parameter of the Weibull plot, a higher modulus of PJ castings indicates that their strengths have less scatter, but are constantly lower than those of others.

5. Discussion

5.1. Effects of flow structure on the entrainment level and defects distribution

Three air entrainment mechanisms of liquid with high surface turbulent flow structures all lead to surface films folded, broken and incorporated into the bulk liquid. However, different levels of entrainment are achieved by the three mechanisms. The plunging jet and return wave both entrain large volumes of gas when high velocity liquid impacts on the slower moving liquid body at the impingement points, but the flow direction determined gas movement results in different levels of gas trapped in. In PJ mould the higher velocity of liquid leads to more gas entrained and the gas pockets move downwards which may take longer time to float up. Whilst in RW mould the jet moves horizontally and gas entrained is pushed only shallowly into the melt and can easily flow to the surface. The velocity differences also lead to volume differences of entrained gas, i.e. large gas pockets in PJ mould, but micro-bubbles in RW mould. In RJ mould the collapsed jets splash onto the surface of the flow resulting in extensive surface films enfolded but less gas trapped due to the small relative velocity between the dropping splashes and the main liquid body. The movement of bubbles leaves bubble trails damage to castings, which may cause more damage in PJ and RW than in RJ moulds. In addition, the evolution of flow structures may have an impact on the entrainment level. As demonstrated in Figure 6, the chaotic turbulent level in RJ and RW moulds reduced dramatically when the moulds filled gradually, but the gas was constantly entrained by plunging jets and the momentum of liquid severely affected bubble movement. These all would have some effects on the reliabilities of castings.

The reproducible mechanical properties of samples in defined locations in each mould showed the strength variation of samples followed certain trend, while the sole difference of these castings were the flow structure in filling stage. This implied that the flow structure played an important role in the determination of the defect distribution in castings. Besides, in the two-stage casting processing (filling and solidification), defects like gas porosity can either nucleate on the pre-existing entrainment defects or form at other sites. Therefore, good running system design must consider both flow structure and thermal field distribution of liquid in the mould.

5.2. Casting reliability

The number and the distribution of defects have significant impact on the reliability of casting. Several defect types, including shrinkage porosity, gas porosity and oxide film defects were observed in all cast plates, but their different magnitudes in castings from different moulds resulted in variation in reliability. Weibull plots show that the reliability of castings reduced with increasing turbulence level, as highly turbulent flows during mould filling could create large numbers of entrainment defects and trap them in the castings. It was also found that the samples contained extensive shrinkage porosity had very low strength as shown in the circle in Figure 9, which reduced the reproducibility of castings severely. Elimination of shrinkage porosity affected data doubled the Weibull moduli of castings. But it must be mentioned that the Weibull modulus of PJ was higher than that of RW in such censored data (Figure 11b) cannot be simply understood as PJ moulds can produce more reliable castings. In fact, the censored data also removed the influence of entrainment defects which were concentrated on the middle part of the PJ mould, that left only samples with less defects being considered in the plot.

Modelling results indicated that the RJ casting had highest reliability, followed by RW castings. More particles were generated in PJ moulds and were widely distributed in the plate, resulting in a lower variation (high Weibull modulus) of particle count in different samples, but generally lower mechanical properties. The predicted reliabilities agreed with experiment, which highlighted the capability of the model to simulate entrainment events, but the inaccuracy of prediction of the defects distribution was due to some limitations of the current model.

5.3. Limitations of the current model

To simplify the simulation of entrainment defect formation and distribution, the model focused on the young oxide film formation during the filling process, which means the tracked particles only represented double oxide films defects, and defects associated with them like gas porosity. This is because *FLOW-3D* is a single phase CFD software, and hence cannot directly model air entrainment. Besides, only one type of particles can be employed in the simulation, and this limits the application of particles to predict the distribution of such varied things as double oxide films and micro-bubbles simultaneously. In addition, the reliabilities achieved by experiments revealed the combined effects of several kinds of defects, both formed during filling and solidification, whereas the model solely considered the effect of flow and it could not distinguish the different effects of double oxide films defect and defects developed by the films. It also should be noted that due to lack of evidence of the behaviour of entrainment defects in liquid, including the interactions of defects and between defects and the surrounding environment (mould wall, liquid metal), there could be some variance in simulation comparing with reality.

6. Conclusions

A model of the quantitative measurement of the formation and distribution of entrainment defects has been developed. This model was applied to assess the relative importance of three different entrainment mechanisms during mould filling, and validated through study of mechanical properties and reliability of Al-Si-Mg alloy castings. The modelled reliabilities correlate with the experimental data. The model can give general distribution map of defects, but the predicted defects distribution in local areas is not yet accurate. This is mainly owing to the limitations of the model.

Acknowledgements

The author (YY) acknowledges the financial support of the School of Metallurgy and Materials, University of Birmingham for his PhD, and useful discussions with Dr. W D Griffiths of University of Birmingham.

References

[1] Campbell J 2006 *Materials Science and Technology* **22** 127–145
[2] Liu L, Samuel A M, Samuel F H, Doty H W and Valtierra S 2003 *Journal of Materials Science* **38** 1255 – 1267
[3] Chanson H 1996 *Air bubble entrainment in free-surface turbulent shear flows* (Academic press) ISBN 0-12-168110-6
[4] Raiszadeh R and Griffiths W D 2006 *Metallurgical and Materials Transactions B* **37** 865–871
[5] Green N R and Campbell J 1994 *AFS Transactions* **114** 341–348
[6] Caceres C and Selling B 1996 *Materials Science and Engineering A* **220** 109–116
[7] Nyahumwa C, Green N R and Campbell J 1998 *AFS Transactions* 215–224
[8] Wang Q G, Apelian D and Lados D A 2001 *Journal of Light Metals* **1** 73–84
[9] Wang Q G, Crepeau P N, Davidson C J and Griffiths J R 2006 *Metallurgical and Materials Transactions B* **37** 887–895
[10] Nyahumwa C, Green N R and Campbell J 2001 *Metallurgical and Materials Transactions A* **32**
[11] Dai X, Yang X, Campbell J and Wood J 2003 *Materials Science and Engineering A* **354** 315–325
[12] Lai N, Griffiths W and Campbell J 2003 *Modeling of Casting, Welding and Advanced Solidification Processes- X*
[13] Reilly C, Green N R and Jolly M R 2009 *Metallurgical and Materials Transactions B* **40B**
[14] Reilly C, Jolly M R, Green N R and Gebelin J C 2010 Assessment of casting filling by modelling surface entrainment events using CFD *2010 TMS Annual Meeting & Exhibition*
[15] Runyoro J, Boutorabi S and Campbell J 1992 *AFS Transactions* **100** 225–234

IOP Conf. Series: Materials Science and Engineering **33** (2012) 012073 doi:10.1088/1757-899X/33/1/012073

Modelling of liquid metal flow and oxide film defects in filling of aluminium alloy castings

X Dai[1], M Jolly[1], X Yang[2], J Campbell[3]

[1]School of Mechanical Engineering, University of Birmingham, Edgbaston, Birmingham, B15 2TT, UK
[2]School of Science and Technology, Glyndŵr University, Wrexham LL11 2AW, UK
[3]Interdisciplinary Research Centre in Materials, University of Birmingham, Edgbaston, Birmingham, B15 2TT, UK

E-mail: xiaojundai@yahoo.com

Abstract. The liquid metal flow behaviours in different runner system designs have important effects on the mechanical strength of aluminium alloy castings. In this paper, a new model has been developed which is a two-dimensional program using a finite difference technique and the Marker and Cell (MAC) method to simulate the flow of liquid metal during filling a mould. In the program the Eulerian method has been used for the liquid metal flow, while the Oxide Film Entrainment Tracking Algorithm (OFET) method (a Lagrangian method) has been used to simulate the movement of the oxide film on the liquid metal surface or in the liquid metal flow. Several examples have been simulated and tested and the relevant results were obtained. These results were compared with measured bending strengths. It was found that the completed program was capable of simulating effectively the filling processes of different runner systems. The simulation results are consistent with the experiment. In addition, the program is capable of providing clearer images for predicting the distribution of the oxide film defects generated during filling a mould.

1. Introduction

The effect of liquid metal flow and surface turbulence using different designs of filling systems on the reliability of aluminium castings, and the importance of minimizing surface turbulence during filling a mould to attain reliable mechanical properties of the castings has been demonstrated by Yang et al. [1]. Their research results have shown that the proper designed running systems could maintain the ingate velocity in the range of 0.3-0.4 m/s and the final castings acquired using such runner system could obtain a greater strength. However, the use of the traditional running systems usually attains an ingate velocity well above 0.5 m/s, the critical velocity for aluminium castings [2], and the final castings have a lower fatigue strength. The results are consistent with the study conducted by Runyoro et al [3] who indicated that good quality casting could be achieved by controlling the liquid meniscus velocity below 0.5 m/s. All these studies demonstrated that there is an important relationship between the liquid aluminium flow behaviour, oxide film defect distribution, microstructures of casting defects and mechanical properties of castings for different runner systems.

Poor design of a runner system usually results in a higher priming metal flow velocity and surface turbulence so that the surface oxide films on the liquid aluminium advancing front is readily broken-up, folded-over and entrapped in the bulk liquid metal. The entrapped oxide films are frequently accompanied by different casting defects such as shrinkage pores [4], cracks and dross [5] in solidified castings.

In Al-7Si-Mg casting alloy, the casting defects associated with the oxide films may present a different morphology such as "tangled or network" [6], "layer oxide or globular oxide" [5] and "cloud or strip clustering particles" [7]. The oxide films may originate from two main sources; the stage of the melt preparation and filling. The former usually produces so-called "old" oxide films due to their formation over long time [2]. These oxide films enter a mould with careless melt preparation when using a ladle for pouring. These "old" oxide films can be entrapped and solidified in the castings to form defects in different types. However, they can be avoided by careful melt preparation. The latter, the "young" oxide films are formed in the liquid aluminium stream as it flows via the runner system into the mould in a short time [5]. As indicated by Yang et al. [1], the "young" oxide films can be broken-up and incorporated into the liquid aluminium due to surface turbulence, finally forming defects in the castings. Because of the exposure time difference of these two types of oxide films, they will present different thickness in the filling. Thiele [8] has indicated that, the thickness of oxide film measured on the liquid aluminium is 900nm at 700°C where the melt is held for 1 hour. After 5 seconds at 700°C the measured oxide film thickness on 99.5 Al is about 24 nm. When the oxidation time was increased by a factor of ten, the oxide film thickness was doubled. Therefore, the "old" and "young" oxide films can be broadly identified by their thickness.

The objective of this study is to investigate the influence of the likely oxide film defect distribution on the final mechanical strength of castings. Although the "Old" oxide films entering the mould are clearly an important source of defects in castings, their effects are in any case suppressed in this work because only differences are being assessed. The aspect of the quality of the original melt is beyond the scope of this research. Since the pouring process, and the flow of the melt through the channels that guides it into the mould cavity, last only a few seconds, the new and thin films, 'young' oxide films generated from the filling, are the concern of this investigation. Computational Fluid Dynamics (CFD) modelling was used to investigate the liquid metal flow behaviour and oxide film defect distribution in the moulds for different ingate velocities. A new numerical algorithm (2-D) – Oxide Film Entrainment Tracking (OFET) for predicting such oxide film defect distribution during the filling has been proposed. Experimental verification is undertaken by the production of Al-7Si-Mg alloy castings. Weibull statistics are employed to quantify the effects on the strength and reliability of the castings produced at two different ingate velocities, which are achieved by employing two different runner system designs. X-Ray and SEM methods were used to observe the casting defects.

2. Modelling of the oxide film entrainment

2.1. Liquid metal flow description and free surface tracking

The liquid metal flow during the filling is basically a transient free surface flow. Assuming the molten liquid metal flow to be incompressible, the governing equations with boundary conditions can be written as free surface evolution:

$$\frac{\partial F}{\partial t}+\frac{\partial (uF)}{\partial x}+\frac{\partial (vF)}{\partial y}=0 \tag{1}$$

where F is the fractional volume-of-fluid contained in a computational cell, u and v are the fluid velocity components respectively. This equation states that F moves with the liquid metal. Equation (1) has been employed in this investigation coupled with an oxide film entrainment tracking algorithm.

When equation (1) is integrated over a computational cell, the change in F in a cell reduces to fluxes of F across the cell surfaces. The computation of fluxes is described in VOF method [9], which uses a type of donor-acceptor flux approximation to avoid the smearing of the F function. The mass and momentum conservation in two-dimensional Cartesian coordinates can be expressed as:

$$\frac{\partial u}{\partial x}+\frac{\partial v}{\partial y}=0 \tag{2}$$

$$\frac{\partial u}{\partial t}+\frac{\partial (u^2)}{\partial x}+\frac{\partial (uv)}{\partial y}=-\frac{\partial p}{\partial x}+v\left(\frac{\partial^2 u}{\partial x^2}+\frac{\partial^2 u}{\partial y^2}\right)+g_x \tag{3}$$

$$\frac{\partial v}{\partial t}+\frac{\partial (uv)}{\partial x}+\frac{\partial (v^2)}{\partial y}=-\frac{\partial p}{\partial y}+v\left(\frac{\partial^2 v}{\partial x^2}+\frac{\partial^2 v}{\partial y^2}\right)+g_y \tag{4}$$

where p is the ratio of pressure to the density of the liquid metal, v is the kinematic viscosity, and (g_x and g_y) are body accelerations in both x and y directions, respectively. The pressure terms can be included by the use of the Poisson equation, which combines the continuity and momentum equations and reads:

$$\frac{\partial^2 p}{\partial x^2}+\frac{\partial^2 p}{\partial y^2}=-\frac{\partial D}{\partial t}-\frac{\partial^2 (u^2)}{\partial x^2}-2\frac{\partial^2 (uv)}{\partial x \partial y}-\frac{\partial^2 (v^2)}{\partial y^2}+\frac{1}{R_e}\left(\frac{\partial^2 u}{\partial x^2}+\frac{\partial^2 u}{\partial y^2}\right) \tag{5}$$

where D is the velocity divergence given by the left-hand side of equation (2). The reason for incorporating the Poisson equation for pressure is that there is the lack of a direct link for pressure between continuity and momentum equations. The body accelerations are neglected in equation (5) because they are not critical to the solution of the pressure Poisson equation.

The above transport equations are discretised using the finite difference method. The transient terms are discretised using forward differencing, the convective terms are discretised using the first order upwind differencing while the pressure and diffusive terms are discretised using central differencing. The resulting linerised equations are solved using the Marker and Cell (MAC) method [10].

2.2. *Oxide film model on free surface*

A kinematic approach has been utilised for modelling the movement and entrainment of the oxide films on the liquid metal surface. Because the oxide films on the free surface are usually very thin, the inertial forces of the films can be disregarded in comparison to other forces like gravity and pressure. The Lagrangian equation describing the movement of oxide films can be written as:

$$\frac{d\vec{s}}{dt}=\vec{V} \tag{6}$$

where $\vec{s}$ and $\vec{V}$ are the displacement and velocity of a tracking point on the films, respectively. The velocities of the tracking points on the films can be determined by interpolating the velocities of the surrounding liquid metal. To ensure a proper determination of the positions of the films, sufficient tracking points are introduced. At time step n, the locations and the velocities of the tracking points can be obtained from the flow calculation. The positions of these points for next time step n+ 1 can be determined by the following equations:

$$x_i^{n+1} = MIN\{x_i^n + 0.5(u_i^n + u_i^{n+1})\Delta t + CX, x_{is}^{n+1}\}$$
$$y_i^{n+1} = MIN\{y_i^n + 0.5(v_i^n + v_i^{n+1})\Delta t + CY, x_{is}^{n+1} \quad (7)$$

and where

$$CX = MAX\{x_{is}^{n+1} - (x_i^n + 0.5(u_i^n + u_i^{n+1})\Delta t), 0.0\}$$
$$CY = MAX\{y_{is}^{n+1} - (y_{is}^n + 0.5(v_i^n + v_i^{n+1})\Delta t), 0.0\} \quad (8)$$

where x_{is}^{n+1} and y_{is}^{n+1} are the points which represent the liquid metal free surface as shown in Figure 1. The MIN feature in Eq. (7) prevents the displacement of the films out of the advancing free surface while the MAX feature readjusts those points back to the surface. In Figure 1, the points with the symbol "⊗" which are joined by the dashed curve represent the positions of tracking points at time step n+1 without utilising the MIN and MAX features. The solid curves with the symbols "①" and "②" indicate the liquid metal free surfaces obtained by the VOF method at time steps n and n+1.

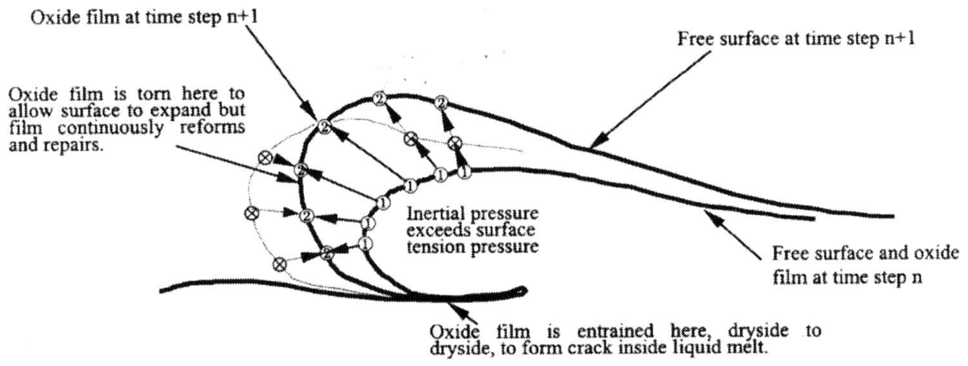

Figure 1. Tracking points on the oxide film at time steps n and n+1, the oxide film points are located at the end of arrows.

The surface segments, which consist of the tracking points, are considered as film elements. In the simulation, a surface normal that is pointing away from the dryside of the surface element is introduced. If the local inertial pressure is larger than the restraining pressure, the oxide film will be torn but the film continuously reforms and repairs to allow the free surface to expand. A new tracking point is then added to account for the newly formed oxide film which bridges the surrounding track points. After new tracking points are added, all tracking points are renumbered. If the surface normal for the oxide film elements is pointing against each other, the oxide films, dry side to dry side, are considered to be entrained into the flow [2]. In doing so, all the tracking points for the film elements have to be renumbered consecutively and a numerical algorithm to judge such fold over is proposed accordingly as described in Section 2.3.

For the movement of films already entrapped in the liquid metal, the following equation is used.

$$x_i^{n+1} = x_i^n + 0.5(u_i^n + u_i^{n+1})\Delta t$$
$$y_i^{n+1} = y_i^n + 0.5(v_i^n + v_i^{n+1})\Delta t \quad (9)$$

The treatment of films entrapped in the melt will be slightly different from treating those films on the free surface. It is assumed that once the oxide films are torn, there is no need to introduce new tracking points since no oxygen is available for the creation of more oxidised surface. This assumption neglects the real contribution that might be made by the air entrained between the double film, either trapped in macroscopic enclosed folds or bubbles, or as contained in the microscopic roughness of the oxide crystals.

2.3. Oxide film entrainment and fold over judgement - the numerical algorithm (OFET)

As shown in Figure 1, when the reflected liquid from the walls of the mould forms a backward rolling wave overlying the underneath stream, the oxide films are folded into the liquid metal, dry side to dry side. A careful observation of this phenomenon reveals that the velocities for these two overlaid opposite flowing streams are always opposite. This information can be used to judge whether or not the oxide films are entrained and folded. Suppose that the fold over has taken place as shown in Figure 2. The velocities at both the top and bottom of the computational cell are employed.

$$F_{oxide} = \mathrm{sgn}\left[\min\left(\frac{u_{i+\frac{1}{2},j+1}}{u_{i+\frac{1}{2},j}}, \frac{v_{i+\frac{1}{2},j+1}}{v_{i+\frac{1}{2},j}}\right)\right] \tag{10}$$

where $u_{i+\frac{1}{2},j}$ and $u_{i+\frac{1}{2},j+1}$ are the x-direction velocity components, and $v_{i+\frac{1}{2},j+1}$ and $v_{i+\frac{1}{2},j}$ are the y-direction velocity components at the bottom and top sides of the cell. The sgn function determines the sign of the smaller ratio between the velocity components. If F_{oxide} is negative, an oxide film entrainment or fold over is judged to occur. If F_{oxide} is positive, no oxide film fold over happens. The above OFET algorithm is expected to predict qualitatively the distribution of oxide films in the casting.

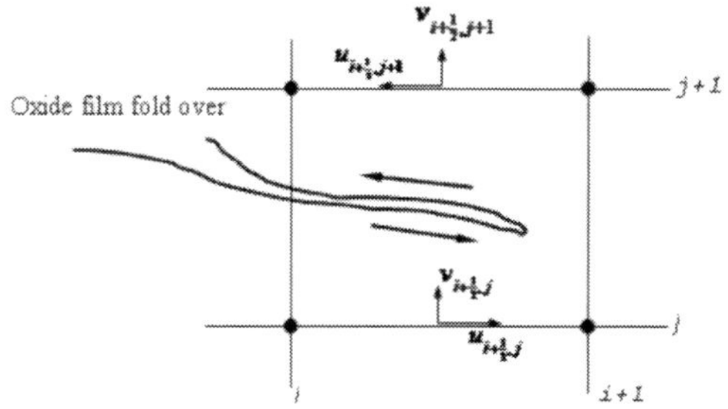

Figure 2. A sketch of how to judge the oxide film fold over

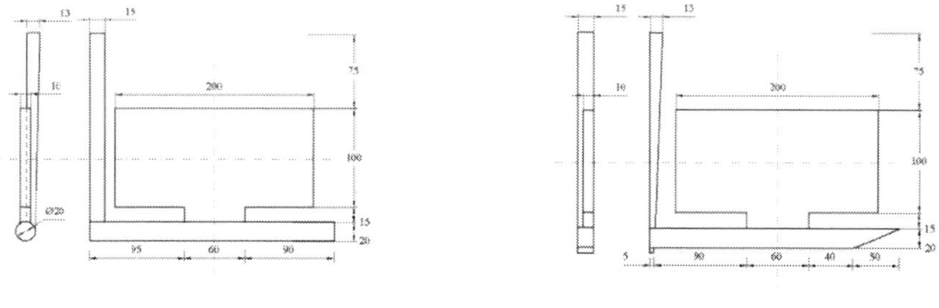

Figure 3. Runner systems (a) Vortex-flow runner system(b)Rectangular runner system

3. Experimental investigation

3.1. Runner system design

Two different runner system designs with dimensions: Vortex-flow Runner (VR) and Rectangular Runner (RR) are shown in Figure 3. Each runner system includes a pouring basin, a tapered sprue, a runner, a gate and a mould cavity. It can be seen that the only difference between the runner systems is the cross-sectional shapes of the runners. This arrangement allows a direct comparison of the liquid metal flow behaviour and the performance of the castings made using different runner systems.

3.2. Casting aluminium alloy and module materials
Commercial aluminium alloy (Al-7Si-0.4Mg) was used to make the castings. Four castings were poured, two for each runner system. All the castings were poured at a temperature of 735-740°C. After casting, all the samples were subjected to heat at 540°C for 6 hours and quench in hot water (80-90°C), then precipitation hardened at 160°C for 6 hours. Chemically bonded silica sand with AFS fineness 60 and PEPSET process was used to make the mould.

3.3. Mechanical strength test
All the castings were machined into four-point bend specimens as shown in Figure 4. The Four-point Bend Test method was employed as shown in Figure 5.

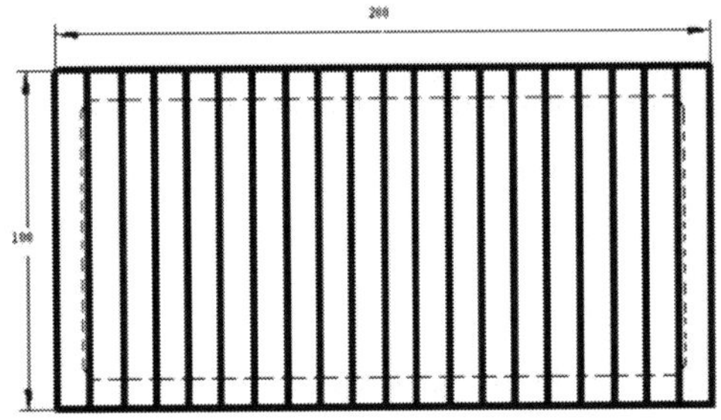

Figure 4. The sampling plan for cast plates, producing specimens for Four-Point Bend Test

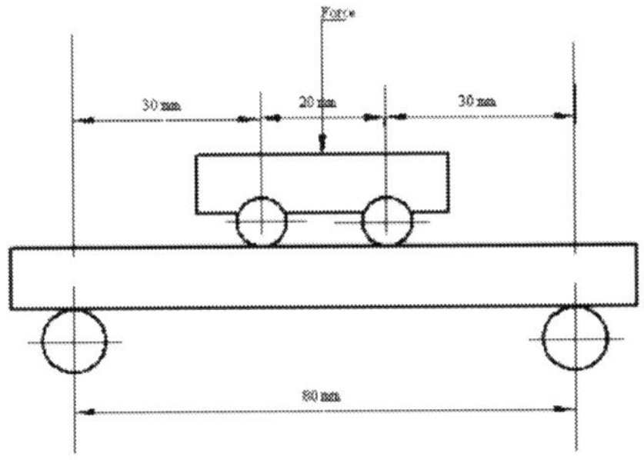

Figure 5. The sketch of Four-Point Test adopted in this work.

3.4. Observation of X-ray and SEM observation
The macroscopic defect distribution in the castings for both VR and RR runner systems was viewed by X-ray radiography. The SEM method was used to observe the micro defects on the polished fracture surface of specimen from RR runner casting

4. Results
Based on the previous work [1], the maximum flow velocity in the ingate is at the right-hand edge for both the VR and RR runner at the moment when the liquid metal front enters the mould, with the value reaching 0.4 m/s and 0.7 m/s in the cases of VR and RR runners, respectively. In the case of VR runner system, the metal velocity was maintained below 0.5 m/s for the entire filling time.

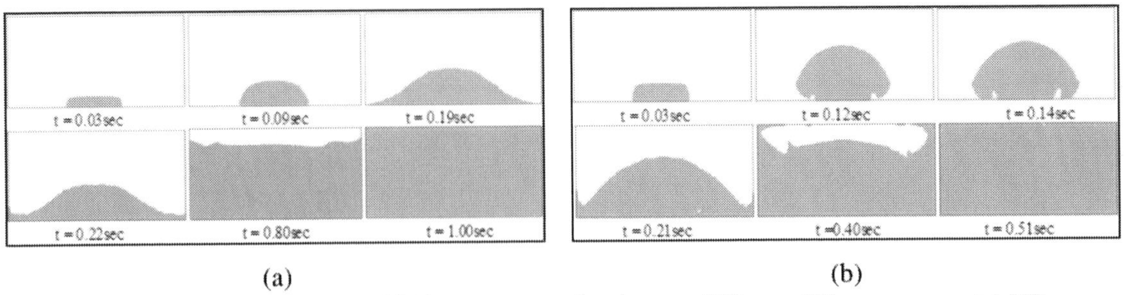

Figure 6. Numerical simulations of filling processes for the two different filling systems; (a) VR runner system; (b) RR runner system.

Numerical simulations of the filling processes for the case of two different filling geometries are shown in Figure 6a, 6b. It can be seen from Figure 6a that the liquid metal flow for the ingate velocity

0.4 m/s (equivalent to the use VR runner system) is smooth, and neither splash nor folding-over of the melt can be observed when the liquid aluminium enters the mould cavities. In contrast to the case of 0.4 m/s, the liquid metal flow in the case of ingate velocity 0.7 m/s, using RR runner system, (Figure 6b) exhibits an undesirable behaviour: the melt fountains and folds over, thus giving rise to the possibility of generating surface turbulence, which might potentially entrap oxide film defects.

Corresponding to Figure 6, the predicted distributions of entrapped oxide films in the castings using the OFET numerical algorithm are shown in Figure 7a and 7b. It can be seen from Figure 7a that the predicted susceptible oxide film defects are basically excluded from the internal zone of casting, and most of the oxide films are distributed around the edge of the casting. This implies a better mechanical strength of the castings since the bend test assesses the central areas. Figure 7b clearly shows that the entrained oxide films are concentrated into the body of the castings. The frequency plots of the UBS for specimens from Figure 4 are shown separately in Figure 8a and 8b. Figure 8c shows the corresponding Weibull plots. The lower values in the two plots represent the specimens taken from the central area of the castings. Specimens taken from the edge area of castings show higher strength values due to the "edge effect".

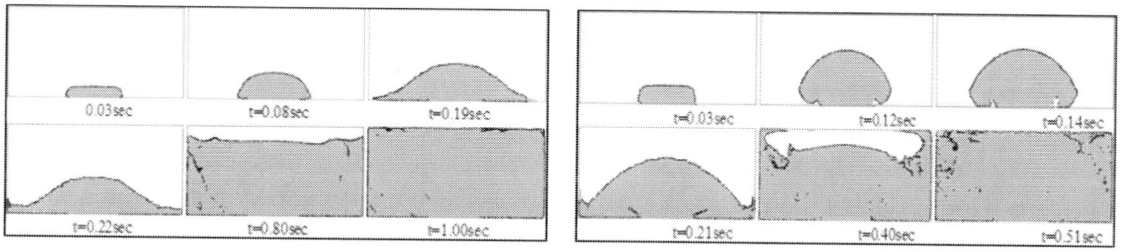

(a) (b)

Figure 7. Predicted susceptible oxide film defect distribution in the castings by using the OFET numerical algorithm. (a) VR runner system; (b) RR runner system.

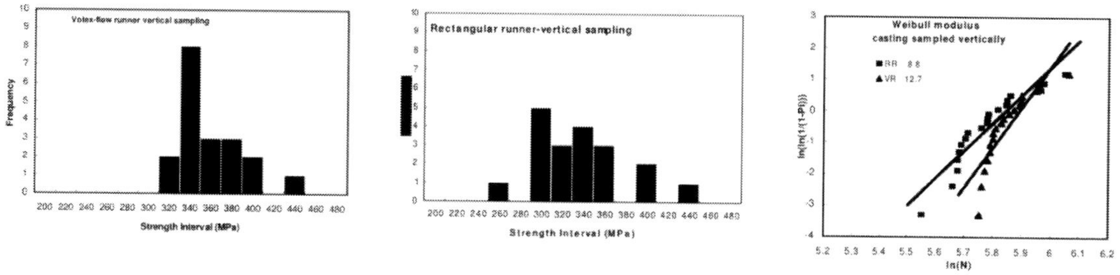

(a) (b) (c)

Figure 8. (a)UBS distributions of VR runner system; (b) UBS distributions of RR runner system; (c) Weibull plots of all the casting specimens UBS.

For vertical sampling, the Weibull moduli (the slope of the line) of the respective sets of specimens are 12.7 and 8.8. The Weibull moduli shown in Figure 8c indicate the average strength and the reliability of the castings for VR filling system is higher than that of the RR system.

Figure 9 shows the distribution of defects in the casting samples obtained using RR runner system. The shrinkage porosity distributions concentrate in the area closed to the ingate of the casting sample in the RR runner system. The radiographic image of the VR plates were featureless and are consequently not shown.

In Figure 10 SEM micrographs show the distribution of defects on the polished fracture surface of specimen from RR runner casting where Figure 10a show a tangled network of oxide films and Figure 10b is a close view of Figure 10a with gas between the double layer type oxide film.
Figure 11 shows the oxide films on the fracture surface of RR specimen using SEM method.

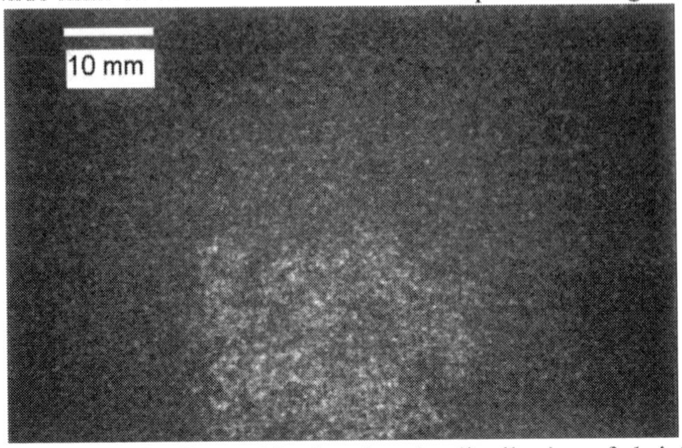

Figure 9. X-ray Radiographic showing the distribution of shrinkage porosity in the castings where is close to the ingate of cast plate (RR).

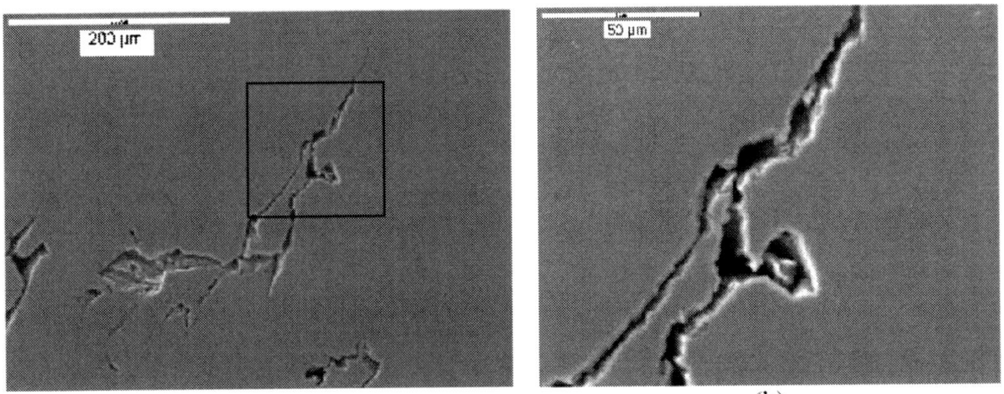

(a) (b)

Figure 10. SEM micrographs show the distribution of defects on the polished fracture surface of specimen from RR runner casting; (a) show a tangled network of oxide films; (b) is a close view of (a) with gas between the double layer type oxide film.

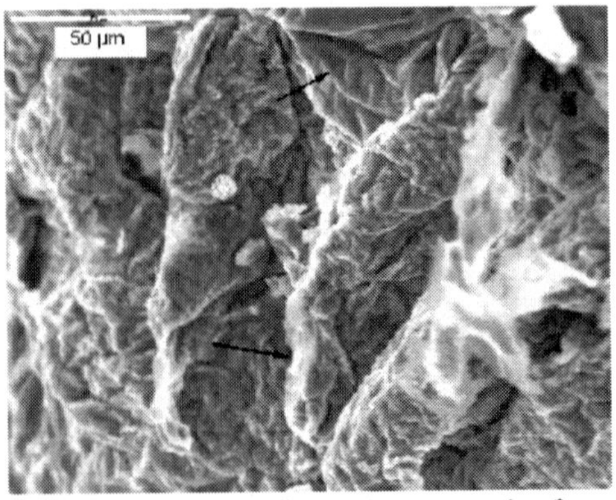

Figure 11. The oxide films shown on the fracture surface of a RR specimen using SEM method

5. Discussion

Based on the OFET simulations, the oxide film defects are occurring around the outer surface of the casting for the VR case while they are entrained into the central part of the casting for the RR case. However, the entrainment of the oxide films into the casting for the case of ingate velocity of 0.7 m/s are expected to influence the final mechanical strength of those samples in the central part of the casting.

The results for the case of ingate velocity of 0.7 m/s (the RR running system) are consistently poorer when judged by oxide film entrainment tracking simulation, strength, Weibull modulus, or by microstructure observed by using X-ray radiograph and SEM methods. This consistency strongly suggests that the castings filled with a higher ingate velocity, in particular greater than the critical velocity of 0.5 m/s, are most likely defective. Indeed, it was observed from the simulations that when the ingate velocity is greater than 0.5 m/s, the melt certainly forms fountain and folds over, thus generating the surface turbulence. This could be especially important in the brief but vulnerable period especially when the runner is still priming and the melt is just entering the mould through the ingate, so that there is freedom for the melt to jump and splash, creating additional area for the creation of new films.

The outstandingly robust result of the use of lower ingate velocity (the VR running system) demonstrated again that once the ingate velocity is less than the critical velocity of 0.5 m/s, the liquid metal fountaining, which may contribute to the generation of surface turbulence, could be avoided. This is the condition of the Weber number being smaller than unity. In order to convert the high velocity of the liquid metal at the exit of the sprue into a relatively gentle and smooth stream, advancing with a velocity less than the critical velocity for the entrainment of the surface as the metal enters the mould cavity, the filling system has to be effective in dissipating sufficient kinetic energy of the liquid metal. The use of vortex flow runner (VR) can generate a strong ordered movement of the liquid as it enters the runner via the sprue. The ordered behaviour extends along the length of the runner, assisting the reduction of ingate velocity as a result of the considerable dissipation of energy by the rapid rotation.

The above results demonstrate the vital role played by oxide films during the filling of the mould. The results confirm the expectations as indicated in the introduction, that the "young" oxide films are closely related to surface turbulence. Thus, a reduction in surface turbulence will reduce the possibilities of free surface break-up and the entrapment of surface oxide films. It seems that by optimising the runner system design and improving the flow behaviour during the filling, the fluid velocity in the ingate can be reduced to be less than the critical velocity 0.5m/s and the entrainment of "young" oxide films into the castings can be significantly reduced so that the mechanical strength and reliability of aluminium alloy castings can be enhanced.

6. Conclusions

A numerical algorithm – Oxide Film Entrainment Tracking (OFET, 2-D) for forecastingthe oxide film defects distribution in the liquid aluminium throughout the filling is proposed and incorporated into the free surface tracking code developed for simulating filling. It is found from numerical simulation that the number and distribution of the oxide film defects generated from filling are significantly affected by ingate velocities. The reduction of ingate velocity can effectively reduce casting defects, while this can be achieved by controlling the disordered behaviour of liquid metal flow and by dissipating the kinetic energy of the metal in the runner. The simulation results were validated using the methods of bending strength test and micro casting defect observation.

Acknowledgements
The authors wish to acknowledge the support of the EPSRC under contract GR14507/01. Many thanks to the assistance of Mr David Cloney, Mr John Turner, Mr Willie Logan and Ms Margaret Corrigan for assisting in the preparation of the mechanical strength test and SEM images.

References
[1] Yang X, Jolly M and Campbell J 2000 *Alum. Trans.* **2** 61-80
[2] Campbell J 1991 *Casting*. (Oxford: Butterworth-Heinemann)
[3] Runyoro J, Boutarabi S M and Campbell J 1992 *AFS. Trans.* **102** 225-234
[4] Cáceres C H and Selling B I 1996 *Mat. Sci. Engng. A* **220** 109-116
[5] Divandari M 2001 *PhD Thesis* University of Birmingham 12-16
[6] Green N R and Campbell J 1994 *AFS. Trans.* **114** 341-347
[7] Huang W L, Shu J W and Shih S T 2000 *AFS. Trans* **40** 547-560
[8] Thiele W G 1962 *Alum.* **38** 707-715, 780-786 (Translation)
[9] Hirt C W and Nichols B D 1981 *J. Comput. Phys.* **39** 201-225
[10] Harlow F H and Welch J E 1965 *Phys of Fluid.* **8-12** 2182-2189

IOP Conf. Series: Materials Science and Engineering **33** (2012) 012074 doi:10.1088/1757-899X/33/1/012074

Four-phase fully-coupled mold-filling and solidification simulation for gas porosity prediction in aluminum sand casting

J. Jakumeit[1], S. Jana[1], T. Waclawczyk[2], A. Mehdizadeh[3], A. Sadiki[3], J. Jouani[4]

1 Access e.V., RWTH-Aachen, Intzestr. 5, 52072 Aachen, Germany
2 TU Darmstadt, FNB, Dolivostr. 15, 64293 Darmstadt, Germany
3 TU Darmstadt, Institute of Energy and Power Plant Technology, Petersenstr. 30, 64287 Darmstadt, Germany
4 Foundry Institute, RWTH-Aachen, Intzestr. 5, 52072 Aachen, Germany
[a] j.jakumeit@access.rwth-aachen.de

Abstract. The impact of mold-filling and oxide film enclosure on gas porosity in A356 was investigated using a three-phase, fully-coupled, mold-filling and solidification simulation. For the prediction of gas porosity, a fourth hydrogen phase was added. At the solidification front hydrogen is rejected from the solid and accumulates in the melt. Pores nucleate if the solute gas exceeds the solubility limit. Air and melt are separated by a volume of fluid interface and special treatment of the hydrogen phase convection was necessary to limit the hydrogen to the melt. Folding of the melt surface was used as a source for oxide film entrainment. These oxide films were transported with the melt and used as nucleation sites for gas porosity formation. The influence of melt flow due to filling and oxide film distribution was analyzed using a simple 3-block test geometry. The test geometry was cast in A356 and analyzed by computer tomography to validate the porosity prediction.

1. Introduction

Porosity-related issues are a major quality problem in the casting industry and are among the most important sources of defects impacting the mechanical properties of castings. Pioneering work in porosity prediction for castings has been carried out by Fleming and co-workers [1]. A comprehensive review is provided by Lee et al. [2] and by Stefanescu [3]. Kubo and Pehlke [4] clarified the basis of empirical rules for soundness and suggested that the simultaneous occurrence of shrinkage and gas evolution is an essential mechanism in the formation of porosity defects. Lee and coworkers [5] developed a combined continuum and stochastic model of diffusion-controlled growth to simulate the formation of porosity during the solidification of aluminum alloys. Beckerman and co-workers [6, 7] extended the model to include the effect of finite-rate hydrogen diffusion on porosity formation in aluminum alloys. Their calculations showed the impact of the diffusion limited growth of pores on gas porosity. Felicelli and co-workers [8] used a continuum model of dendritic solidification coupled with diffusion of gas to predict the volume fraction of porosity caused by precipitation of hydrogen gas during solidification of aluminum plate castings.

An additional challenge is to combine the porosity model with mold-filling. During filling, the melt/gas interface is captured by the volume-of-fluid (VoF) approach. This approach makes it difficult to limit the existence of hydrogen and pores to the liquid phase, since no clearly defined interface and no nodes or faces are defined at the interface. To the knowledge of the authors only a few publications tackle this problem [9, 10].

In aluminum alloys entrainment of oxide films from the melt free surface into the melt can cause defects, since oxide films act as nucleation sites for pores [11, 12]. During a turbulent filling of the cavity the melt free surface gets folded and oxide bi-films are entrained into the melt. These initially flat bi-films can easily be filled with dissolved gas and form gas pores. In this work two new approaches for identifying the locations and quantity of oxide films entrained into the melt were developed and combined with a gas porosity model (see section 2). The models were applied to a simple test geometry and compared to experimental findings as well as a simulation featuring homogeneous distribution of nucleation sites in the melt before pouring (section 3).

2. Theory

2.1. Three phase casting simulation

The simulation of the complete casting process is based on a Finite-Volume method using control volumes (CVs) of an arbitrary polyhedral shape. The transport equations for mass, momentum, energy, and chemical species in integral form are applied to each CV, whereby the surface and volume integrals are approximated using the midpoint rule. Linear equation systems for each variable are solved using either conjugate-gradient or algebraic multi-grid iterative solvers. A generic integral form of a transport variable ϕ is used to solve mass, momentum, energy, phase fraction and hydrogen concentration conservation equations:

$$\frac{\partial}{\partial t}\int_V T_\phi \rho\phi dV + \int_S C_\phi \rho\phi\vec{v}\cdot ds = \int_S D_\phi \nabla\phi \cdot ds + \int_V S_\phi dV \tag{1}$$

where T_ϕ, C_ϕ, D_ϕ and S_ϕ are the coefficients of the transient, convective, diffusive terms and the source term, respectively. ρ is the density, $\vec{v}$ the velocity, V represents the volume of the cell and S the surface area.

For mass conservation $\phi = 1$, $T_\phi = 1$, $C_\phi = 1$, $D_\phi = 0$ and $S_\phi = 0$ applies. Pressure and velocities ($\phi = \vec{v}$, $T_\phi = 1$, $C_\phi = 1$, $D_\phi\nabla\phi = \hat{T}$ and $S_\phi = \vec{F_b}$, where $\hat{T}$ is the Cauchy stress tensor and $\vec{F_b}$ the body force, here gravity) are coupled via SIMPLE-algorithm [13, 14]. The phase change is computed using the enthalpy approach ($\phi = h$, $T_\phi = 1$, $C_\phi = 1$, $D_\phi\nabla\phi = \vec{q}$ and $S_\phi = Q_h$, where $\vec{q}$ is the heat flux and Q_h the heat source). The volume fraction of solidified liquid is determined based on temperature distribution using a tabulated fraction solid *vs* temperature curve ($fs(T)$). Latent heat L is released in proportion to the change in fraction solid ($Q_h \simeq L \cdot (\partial fs)/(\partial T)$). Flow is suppressed in regions where the volume fraction of solid phase is greater than a fixed limit. All equations are coupled using an iterative segregated solution procedure. Details of discretization and the solution method are available in [14, 15] and will not be elaborated here.

In detecting the entrainment of oxide films by folding of the melt free surface, the method used to track the motion of the free surface must be very accurate and free from artificial smearing. The volume of fluid (VOF) approach is used to tackle the problem: the entire fluid domain is considered to be filled with an effective fluid whose properties vary according to the distribution of volume fractions of melt C_m, solid C_s and gas C_g ($C_m + C_s + C_g = 1$). The transport of melt, solid and gas is computed by solving transport equations for their volume fractions (equation 1 with $\phi = C_i$, $T_\phi = 1$, $C_\phi = 1$, $D_\phi = 0$ and $S_\phi = s_{C_i}$). These equations contain only transient and convective terms (no degassing) and source term s_{C_i} for the phase change from melt to solid. In order to achieve sharpness of the interface a special discretization scheme is required. Here, an improved version of the HRIC-scheme (high-resolution interface capturing) developed by [15, 16] is used, which typically resolves the interface (i.e. the transition from volume fraction 1 to volume fraction 0) within one cell. The scheme also guarantees that computed volume fractions satisfy the boundedness criterion (i.e. the volume fraction must not become either greater than 1 or less than 0).

2.2. *Dissolved hydrogen*

The basis for the nucleation of gas pores is dissolved hydrogen in the aluminum melt. The distribution of the hydrogen is calculated by adding a dissolved gas phase to the model. In the presence of liquid, the dissolved gas C_h is modeled using the generic transport equation 1 ($\phi = C_h$, $T_\phi = 1$, $C_\phi = 1$, $D_\phi = 1$ and $S_\phi = s_{C_h}$). Transport by convection and diffusion as well as removal of gas due to growth of pores is considered. When solidification occurs, hydrogen gas becomes rejected due to lower solubility of hydrogen in the solid phase. This was taken into account in the source term s_{C_h} of equation 1, which was derived by a local balance equation for the concentration in each cell for each time step (see [17] for details).

The dissolved hydrogen should diffuse within the liquid phase but not across the melt/gas interface. In order to circumvent this difficulty, a large value on the diagonal term (Ap) of the finite volume matrix was provided for cells filled with air. Hence, diffusion of dissolved gas into the air was suppressed numerically. In this way, on very fine grids when the VOF front is captured in one single cell, the gas and the VOF front will overlap. Such numerical grid convergence is not within the scope of the current work and will be addressed in future publications.

2.3. *Oxide film entrainment detection*

For detection of the oxide film entrainment two new approaches were developed. Both can be divided into the following steps:

- detection of oxide film entrainment locations
- estimation of the oxide film density due to entrainment
- transport of the oxide films as nucleation centers for gas porosity

2.3.1. Detection of oxide film entrainment locations The two approaches differ in the first step, the detection of the locations, where oxide films are entrained into the melt.

Geometrical detection

The first approach uses geometrical analysis of the volume fraction of melt C_m and approximate conditions for detection of the interface folding. Since the VoF approach in STAR-Cast is based on the interface capturing method, the interface position is determined by the $C_m = 0.5$ iso-contour and thus the moment of the collision of the two fronts can be approximated using local vectors normal to the interfaces $\vec{n} = \nabla C_m / |\nabla C_m|$, local face velocities and volume fractions from the old/new time levels. At each time step, each cell is analyzed and marked as a location of oxide entrainment if:

- the cell and neighbor cells include interfaces
- the velocities of recognized fronts lead to a collision
- the cell contains more than 45 % melt and the quantity is growing in time

Fig. 1 schematically illustrates the detection mechanism.

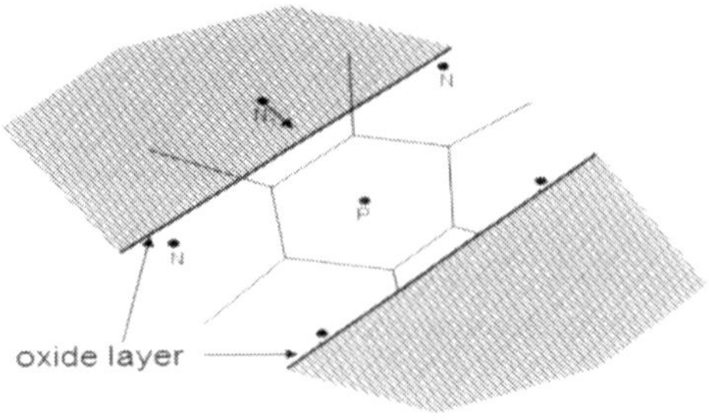

Figure 1. Principle of the geometrical oxide entrainment detection.

Weber number based detection

The second approach is based on the work of Mehdizadeh et al. [18]. Here, the dynamic Weber number is used to identify melt/gas interface, which are instable and may break up into smaller films which can be entrained into the melt. The dynamic Weber number $We = (\rho k L_t)/\sigma$ is a measure of the relation between the fluid inertial forces and the surface tension of the interface Here, ρ and σ are the density and surface tension of the melt respectively, k is the turbulent kinetic energy at the interface and L_t is the length scale based on the turbulence scale at the interface ($L_t = k^{2/3}/\epsilon$ at dissipation rate ϵ). Ranz et al. [19] found that a fluid/gas interface is mainly stabilized by surface tension. If We is larger than 1, disturbances by inertial forces can build up and lead to a break up of the interface. Therefore, cells with a melt/gas interface ($0.45 < C_m < 0.55$) and a dynamic Weber number $We > 1$ are marked as locations for oxide entrainment. A standard $k - \epsilon$ model was added to the set of transport equations to obtain the turbulent kinetic energy and dissipation rate [13].

2.3.2. Estimation of the oxide film density due to entrainment The next step was to estimate the number of small oxide films, created by the entrained interface though a break-up process. A microscopic model of this break up process would be necessary to calculate oxide film density. Such a model has not yet been developed and the calibration of its parameter by experiments will be a major challenge. As a first approach, the number of oxide film parts was assumed to be proportional to the surface area of the entrained interface. Since most polyhedral cells have a nearly round shape, the surface area of an interface through the cell can be estimated by the radius of the cell $r_c = (3 * V_c/4)^{1/3}$ (V_c = cell volume) and a factor $[4 \cdot C_m \cdot (1 - C_m)]$, which takes into account, that the surface area is largest for half-filled cells and tends to zero for completely ($C_m \to 1$) or very slightly filled cells ($C_m \to 0$). The oxide film density increase δn_{of} at a cell i marked as entrainment location is therefore given by:

$$\Delta n(i)_{of} = C_{of} \cdot [4 \cdot C_m(i) \cdot (1 - C_m(i)] \cdot 2\pi r_c(i)^2 \quad , \tag{2}$$

where C_{of} is a proportionality constant.

2.3.3. Transport of oxide films as nucleation centers for gas porosity The oxide film density n_{of}, initialized in this way was then added to the set of transport equations and traced by the solution of the Euler advection equation 1 where $T_\phi = C_\phi = 1$ (transport with the velocity field), $D_\phi = 0$ (no diffusion) and $S_\phi = \Delta n_{of}/V$. The numerical trick used for the gas species C_h was also used to keep n_{of} restricted to the melt phase. Numerical diffusion effects connected with grid size and discretization techniques are a disadvantage of this approach. The tracking of the bi-oxide films can be improved by introduction of the Lagrangian markers, which would increase the numerical effort significantly.

2.4. Gas porosity model

The final step of the gas porosity prediction is to calculate pore nucleation and growth. The gas porosity model developed by two of the authors [17] was used.

2.4.1. Pore nucleation model Three conditions need to be satisfied for the nucleation of pores:

- inclusions such as impurities or oxide films must exist and function as nucleation sites
- local liquid gas pressure must be greater than pore pressure
- liquid dissolved gas concentration must be above the saturation limit.

The advancement of the model reported here is that for condition 1, the oxide film density, calculated by the oxide entrainment models is used as nucleation site density. The pore nucleation rate is, therefore, directly proportional to the oxide film density and, in this way, depends on oxide film entrainment due to turbulent filling.

The second condition of pressure is as follows:

$$P_{gas} > P_{ambient} + P_{hydro} + P_{\sigma} + P_{mushy} \quad . \tag{3}$$

Using Sieverts law, $P_{gas} = \left[\frac{C_l}{(K_e/f)}\right]^2 P_{atm}$ can be calculated, where K_e is the equilibrium coefficient and f the activity coefficient depending on the material. Atmospheric pressure P_{atm} and hydrostatic pressure P_{hydro} result directly from the flow simulation. The pressure due to surface tension is $P_{\sigma} = 2.0\frac{\sigma}{r_p}$, where σ is the surface tension and r_p is the pore radius. The pressure drop through the dendrites of the mushy zone (P_{mushy}) is calculated using Darcy's law of flow through porous media by integration across the mushy zone from the temperature in the mushy zone up to liquidus temperature (see [17] for details).

Thirdly, for pore growth the concentration of the dissolved gas C_h must exceed the solubility limits

$$C_h > C_{lsat} \quad , or \quad C_h > C_{ssat} \quad , \tag{4}$$

C_{lsat} is the solubility limit on the liquid side and is calculated from pore pressure using Sievert's law ($C_{sat} = (K_e/f)\sqrt{P_{pore}/P_{atm}}$). This value is taken to be the concentration available at the pore surface. For the solid side solubility limit is calculated as $C_{ssat} = \kappa C_{lsat}$. Pores grow until the melt is completely solidified.

2.4.2. Pore growth After pore nucleation the pore volume may change due to changes in the surrounding temperature dT or pressure dP and due to dissolved gas, which diffuses into the pore [6, 7]. The first two effects can be calculated directly from derivatives of the ideal gas law. The growth due to diffusion is based on the mass diffusion of hydrogen into pores. The complete model is described in Jana et al.[17]

The gas porosity model was validated experimentally by Fang and Granger [20], with directional solidification of A356 in a long bar. Good agreement with experimental findings for pore volume fraction and equivalent pore diameter was found (see [17]).

3. Results

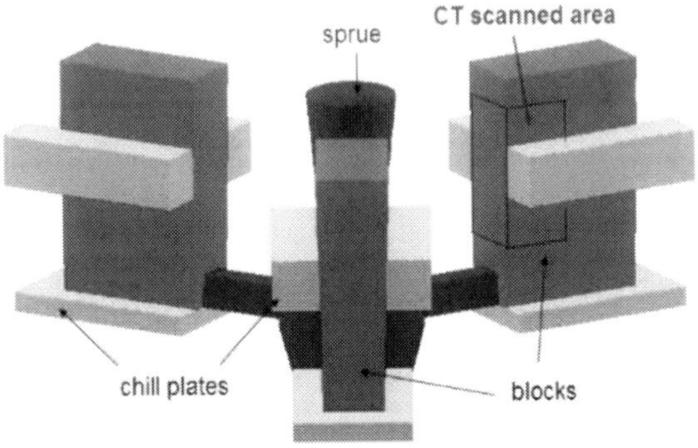

Figure 2. Simple test geometry of 3 blocks

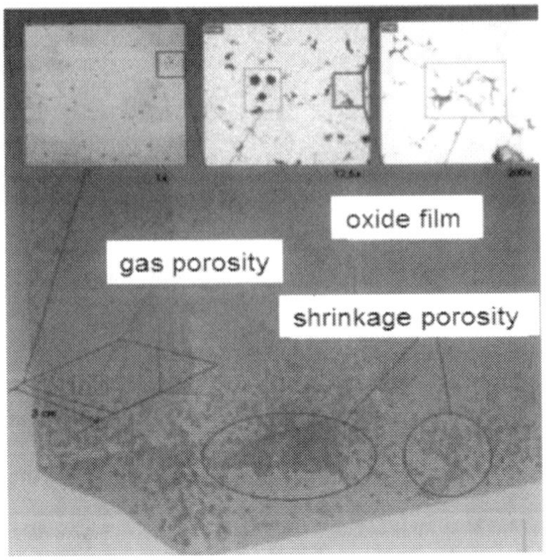

Figure 3. CT analysis of the casts blocks

The above-described oxide film entrainment and gas porosity prediction model was applied to a test geometry, consisting of three symmetrical blocks (20 x 30 x 4cm) in a Furan sand mold fed by one runner system (see Fig. 2). Chill plates are arranged to enforce porosity. Liquid $AlSi_7Mg_{0.3}$ is poured into the cavity within 18 s, and complete solidification is achieved after 6 min. Porosity in the blocks is measured using computer tomography (CT). Fig. 3 shows the porosity found in one block by CT scanning. A mixture of gas and shrinkage porosity was detected.

Three different simulations were performed to analyze the effect of the oxide film entrainment models on gas porosity prediction: one simulation for each of the two different approaches for identifying cells with oxide film entrainment, the *geometrical* and the *Weber number* approach. The third simulation does not use these models but assumes that the melt preparation leads to a uniform oxide film density of $n_{of} = 2 \cdot 10^{11} \quad 1/m^3$ (*initially impure melt*). In reality a mixture of oxide film entrainment during melt preparation and pouring will be present. Here we separate both mechanisms so as to be able to compare their impact on gas porosity distribution. For comparison, the proportionality constant of the oxide film break up model was calibrated in such a way that all three simulations show a similar average oxide film density.

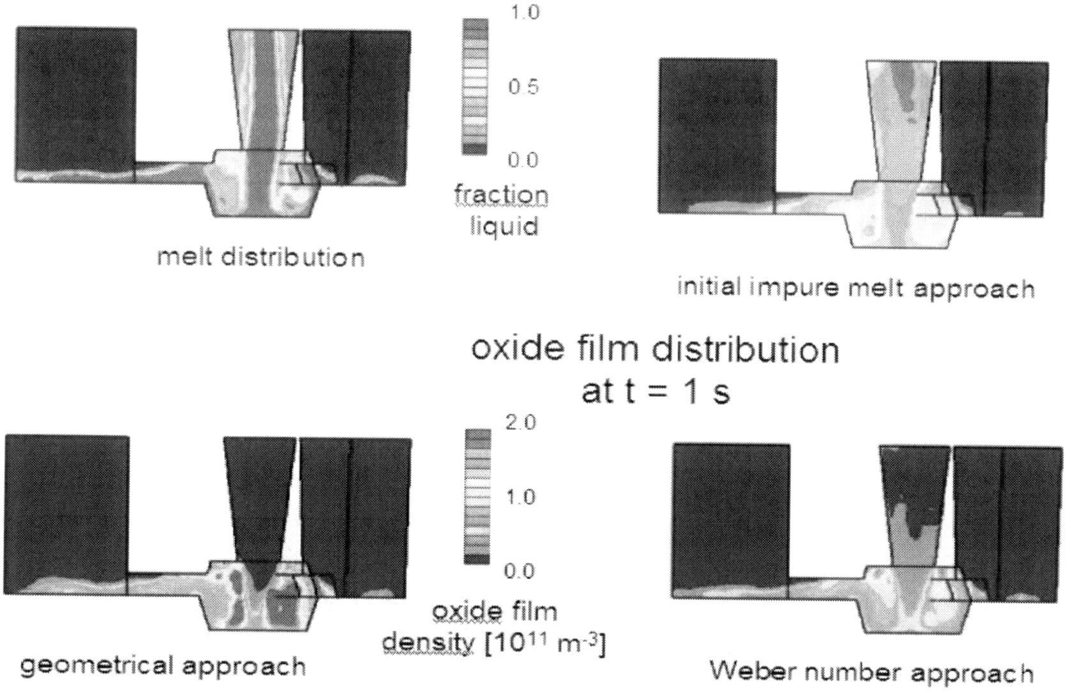

Figure 4. Oxide film distribution after 1 s calculated by the *initial impure melt* simulation (top right), the *geometrical* approach (bottom right) and *Weber number* approach. The top left plot shows the melt distribution at this time as reference.

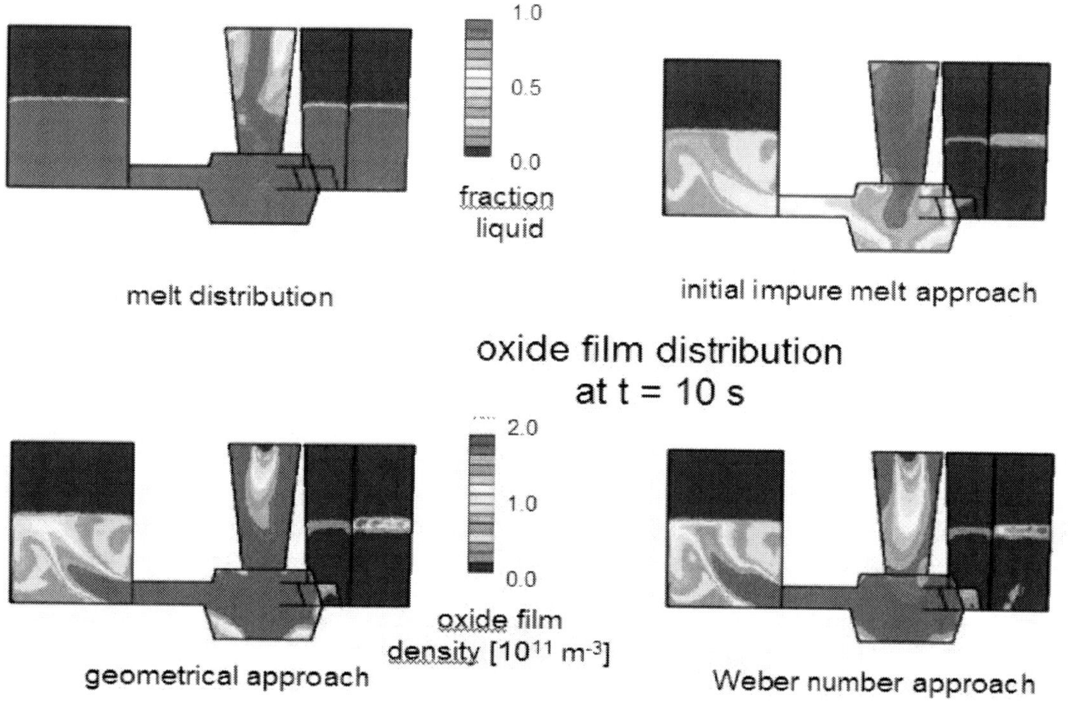

Figure 5. Oxide film distribution after 10 s (plots similar to Fig. 4).

Fig. 4 shows the oxide film distribution after 1 s of filling. The top left plot shows the melt distribution at this time as reference. For the *initial impure* simulation (top right) the oxide film distribution follows the melt flow but gets spread significantly. Apparently the numerical trick for keeping the oxide films inside the liquid domain does not function well, when a lot of splashing occurs and the VoF melt/gas interface is not sharp. The two oxide entrainment approaches show a very different distribution. Oxide films are mainly entrained when the melt hits the bottom of the cavity and a lot of splashing occurs. The free stream of melt from the inlet to the bottom does not lead to a significant amount of oxide. Both approaches show a different distribution but the main characteristics is similar.

The same plots after 10 s of filling are shown in Fig. 5. The top left plot shows, that the cavity is half-filled at this time. Oxide films are moved with the melt flow into the 3 blocks. Filling inside the blocks is smooth from bottom to top, the melt/gas interface is well-defined and the oxide films are nicely kept inside the melt. The differences between the three approaches become smaller inside the blocks. The oxide film distribution is dominated by the flow of the melt into the block.

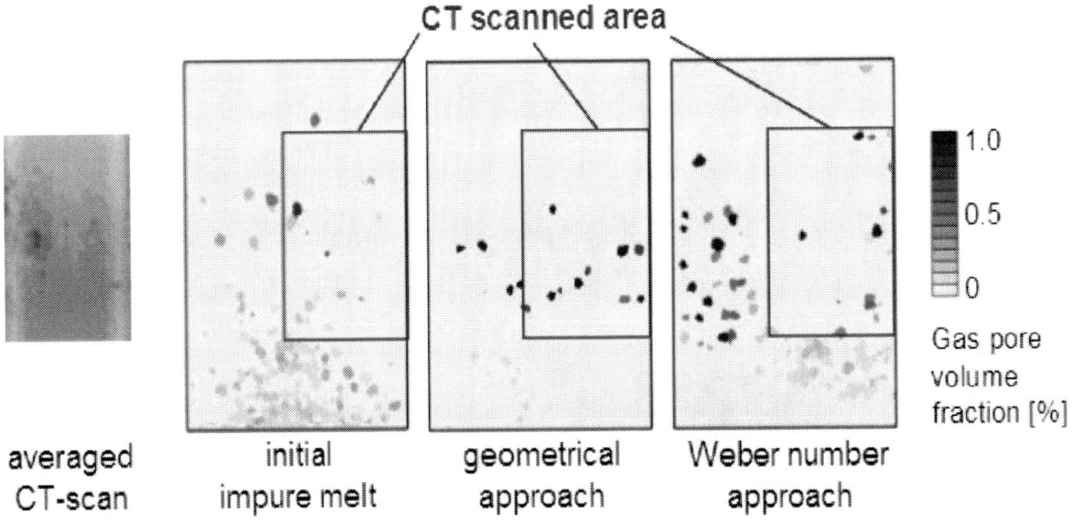

Figure 6. Gas pore distribution as a result of the three simulation approches: *initial impure melt* (second left), *geometrical* (second right) and *Weber number* (right) approach. The left plot shows the experimental result - the CT scan averaged over the thickness of the block.

Finally, Fig. 6 compares the gas pore distribution for all three simulations after 400 s, i.e. when the melt is completely solidified, with experimental findings. The amount of gas pore is in the similar range of about 0.5 % volume fraction, which is within the pore volume fraction range of the experimental findings. That all three simulations give a similar amount of gas porosity can be expected, since the oxide film break-up proportionality constant C_{of} is selected to give the same average oxide film density of $n \sim 2^{11} m^{-3}$. A more quantitative comparison is difficult since the computer simulation does not predict single pores and the estimation of the pore volume fraction from the CT picture is limited because the transition from melt to gas is not sharp.

The spatial distribution calculated by the three simulations is different. The *initial impure melt* simulation results in to pores mostly towards the bottom of the block and some of higher volume fraction below the chill plates. Using the *geometrical* and *Weber number* approach leads to a distribution with some more pronounced spots. The *geometrical* approach results in pores mostly below the side chill plates and the *Weber number* approach results in pores mainly towards the outer part of the block and along the melt flow from the block inlet. None of the simulation results show good agreement with pore distribution as shown in the CT-scan, and a conclusive distinction as to whether pores are created mostly during melt preparation or the filling process is not possible. An averaged CT-scan of the whole vertical cross-section and a better statistics on both sides, experimental and simulation, may permit such a distinction and actual research is going in this direction.

4. Conclusion

A gas porosity model was combined with a fully-coupled three-phase filling and solidification simulation to investigate the impact of the filling on the gas pore distribution in aluminum castings. Oxide bi-films, entrapped during the turbulent filling process were taken as nucleation sites for gas pore nucleation. Two models were developed to identify locations, where folding or breaking of the metal/gas interface brings oxide films from the interface into the melt. The first uses geometrical considerations and the second is based on the dynamic Weber number as a measure of the interface instability. Simulations results for both models were compared with a third simulation, which uses a melt with an initially constant oxide film density.

These three approaches were applied to a simple 3-block test geometry. The simulation result shows that the three approaches lead to different oxide film distributions during the early phase of the filling process ($t \leq 1s$). Later, the 3 blocks are filled smoothly from the bottom and the differences become smaller. Nevertheless, the three approaches lead to different gas pore distributions. Based on these preliminary results a conclusive distinction as to whether pores are created mostly during the melt preparation or the filling process, is not possible.

Acknowledgment

This work was supported by DFG (Deutsch Forschungs Gemeinschaft) (Joint Project 194 / JA 853/2-1) and CD-adapco.

References

[1] Fleming M C 1998 Solidification modeling, past and present *Modeling of Casting, Welding and Advanced Solidification Processes* ed Thomas B G and Beckermann C pp 1–13
[2] Lee P D, Chirazi A and See D 2001 *J. of Light Met.* **1** 15–30
[3] Stefanescu D M 2005 *Int. J. of Cast Metal. Res.* **18** 129–143
[4] Kubo K and Pehlke R D 2005 *Int. J. of Cast Metal. Res.* **18** 129–143
[5] Atwood R C and Lee P D 2003 *Acta Mater.* **51** 15–30
[6] Carlson K D, Lin Z, Beckermann C, Mazurkevich G and Schneider M C 2006 Modeling of porosity formation in aluminium alloys *Modeling of Casting, Welding and Advanced Solidification Processes XI* ed Gandin C A and Bellet M pp 627–634
[7] Carlson K D, Lin Z and Beckermann C 2007 *Metall. Mater. Trans. B* **38B** 541–555
[8] Felicelli S D, Wang L, Pita C M and Escobar de Obaldia E 2009 *Metall. Mat. Trans. B* **19** 265–269
[9] Muradoglu M and Tryggvason G 2008 *Journal of Computational Physics* **227** 2238–2262
[10] Schlottke J and Weigand B 2008 *Journal of Computational Physics* **227** 5215–5237
[11] Campbell J 2006 *Materials Science and Technology* **22** 127–145
[12] Reilly C, Green N and Jolly M 2009 *Metall. Mat. Trans. B* **40** 850–858
[13] Ferziger J H and M P (eds) 2002 *Computational Methods for Fluid Dynamics* (ISBN 2-540-42074-6: McGraw-Hill)
[14] Teskeredzic A, I D and S M 2002 *Num. Heat Trans., Part B* **42** 437–459
[15] Demirdzic I and Muzaferija S 1995 *Com. Met. in App. Mech. and Engg.* **125** 235–255
[16] Muzaferija S and Peric M 1999 *Computation of free surface flow using interface-tracking and interface-capturing methods* (eds. Mahrenholtz and M. Markiewicz WIP Press, Southampton) chap 2, pp 59–100
[17] Jana S, Jakumeit J and Jouani M Y 2009 Numerical predictions of gas porosity in casting processes using fully coupled mold-filling and solidification simulation. *Modeling of Casting, Welding and Advanced Solidification Processes XI*

[18] Mehdizadeh A, Jana S and Sadiki A 2011 *Progress in Comp. Fluid Dynamics* accepted for publication
[19] Ranz W E, Dreier J and William M 1964 *I and EC Fundamentals* **3** 53–60
[20] Fang Q T and Granger D A 1989 *AFS Trans.* **97** 989–1000

IOP Conf. Series: Materials Science and Engineering **33** (2012) 012075 doi:10.1088/1757-899X/33/1/012075

Simulation of the as-cast structure of Al-4.0wt.%Cu ingots with a 5-phase mixed columnar-equiaxed solidification model

M Wu[1,2], M Ahmadein[2], A Kharicha[2], A Ludwig[2], J H Li[3] and P Schumacher[3]
[1] Christian-Doppler Lab for Advanced Process Simulation of Solidification & Melting,
[2] Chair for Simulation and Modelling of Metallurgical Processes,
[3] Chair for Casting Research,
Dept. of Metallurgy, Univ. of Leoben, A-8700 Leoben, Austria

E-mail: menghuai.wu@unileoben.ac.at

Abstract. Empirical knowledge about the formation of the as-cast structure, mostly obtained before 1980s, has revealed two critical issues: one is the origin of the equiaxed crystals; one is the competing growth of the columnar and equiaxed structures, and the columnar-to-equiaxed transition (CET). Unfortunately, the application of empirical knowledge to predict and control the as-cast structure was very limited, as the flow and crystal transport were not considered. Therefore, a 5-phase mixed columnar-equiaxed solidification model was recently proposed by the current authors based on modeling the multiphase transport phenomena. The motivation of the recent work is to determine and evaluate the necessary modeling parameters, and to validate the mixed columnar-equiaxed solidification model by comparison with laboratory castings. In this regard an experimental method was recommended for in-situ determination of the nucleation parameters. Additionally, some classical experiments of the Al-Cu ingots were conducted and the as-cast structural information including distinct columnar and equiaxed zones, macrosegregation, and grain size distribution were analysed. The final simulation results exhibited good agreement with experiments in the case of high pouring temperature, whereas disagreement in the case of low pouring temperature. The reasons for the disagreement are discussed.

1. Introduction

The mechanical properties of castings (either in primary or final form) and welded joints are of great importance in the design and manufacture of engineering products. These properties can be modified by controlling the microstructure formed during the solidification. Solidification is a multi-phase / multi-scale process where the transport of heat, momentum, mass, and species interact with each other in a complex manner to provide the final morphology. Therefore, the prediction of the microstructure at the industrial scale represents a big challenge for researchers. Empirical knowledge about the formation of the as-cast structure, mostly obtained before 1980s [1]-[5], rests primarily on two critical issues: one is the origin of equiaxed crystals; one is the columnar-to-equiaxed transition (CET). The most-likely-operating nucleation mechanisms proposed were: (1) heterogeneous nucleation [6]; (2) the 'big band' theory [1]; (3) partial remelting of columnar dendrites [7]; (4) the showering down of dendrite crystals formed from the casting top surface [8]; and (5) the so-called 'separation' theory [3] which has some similarity to the 'big band' theory. Progress was also made in the understanding of

CET since the pioneer work of Hunt [9] in 1980s. A CET map, the correlation of the columnar primary dendrite tip growth velocity with the local temperature gradient at the moment of CET, was established to analyze the occurrence of CET. This CET map was later confirmed and further improved by many authors [10][15]. In the meantime, empirical correlations were proposed as indirect criteria to predict the CET for engineering castings [16]-[20].

The current authors [21]-[25] proposed a 5-phase mixed columnar-equiaxed model that treats the solidification as a multiphase transport problem which takes into account the impacts of flow and grain transport based on the above mentioned empirical knowledge. Preliminary simulations showed that the model is capable to qualitatively produce structures similar to those observed in previous experiments [1] [3] [5]. In addition, classical ingot casting experiments on Al-Cu alloy were conducted in a previous work [26] and the results were compared to simulations using the 5-phase model. The macrostructure of Al-Cu ingots could be reproduced by the model, however, the detailed structural quantities and point-to-point validation did not match the experiment. Modeling results have shown that the most sensitive parameters for the macrostructure are the nucleation parameters. Thus, the authors recommend that further evaluations and more detailed analyses of the modeling and experimental results should continue. In the current work a numerical parameter study is carried out to investigate the influence of the nucleation parameters on the as-cast structure. Previous experiments for in-situ determination of the nucleation parameters were also used to validate and improve the modeling results.

2. The numerical model

The model comprises any combination of the hydrodynamically interacting phases: equiaxed crystals, columnar crystals, and liquid melt as explained in Figure 1. It was assumed that immediately following the nucleation of equiaxed crystals in the chill zone, columnar dendrites start to grow from the casting surface. The three phases, denoted as e-, c- and ℓ-phases, are considered and quantified with their volume fractions, f_e, f_c, f_ℓ. They move with corresponding velocities, $\vec{u}_e$, $\vec{u}_c$, and $\vec{u}_\ell$. Here $\vec{u}_c$ is predefined (zero in the case of ingot casting), while $\vec{u}_e$ and $\vec{u}_\ell$ are solved numerically. The volume averaging approach was employed to formulate the conservation equations of mass, momentum, species, and energy for the three phases, as described in the Eulerian multiphase model.

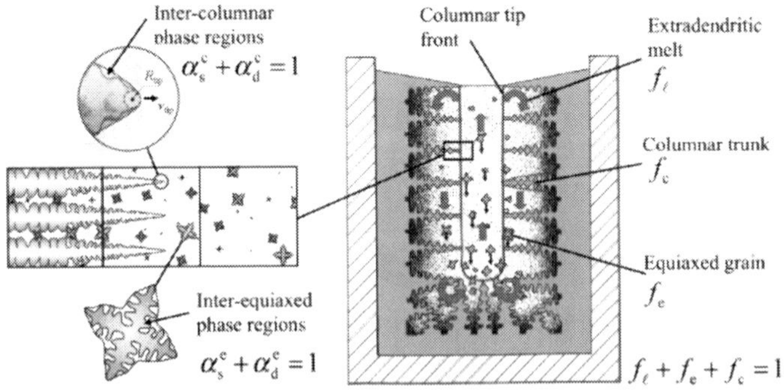

Figure 1. Schematic of the mixed columnar-equiaxed solidification in an ingot casting

A three-parameter heterogeneous nucleation law [27] is applied for the origin of equiaxed grains. These nucleation parameters can be obtained experimentally. The e-phase, the equiaxed grains attached to casting surfaces (top or side wall) are allowed to be brought away by the melt convection into the bulk region. Almost all the aforementioned nucleation mechanisms [1][3][6][8] can be taken into account except for the one due to partial remelting of columnar dendrites [7]. As the current model does not include mold filling, hence a simple idea is proposed to consider the equiaxed nuclei which have formed during mold filling, i.e. to pre-set an initial grain number density in the as-filled state, n_0.

The dendritic growth of crystals is taken into consideration as shown in Figure 1. Two distinct phase regions exist within the envelope of the crystal: the solid dendrites and interdendritic melt. It is assumed that the interdendritic melt is transported with the solid dendrites and is generally more enriched with solute element than the extradendritic melt surrounding the crystals. In this sense, a fictitious crystal boundary envelope is constructed to separate the interdendritic melt from the extradendritic melt. Therefore, five 'thermodynamic' phase regions are defined in the system: the solid dendrites and interdendritic melt in the equiaxed grain, the solid dendrites and interdendritic melt in the columnar dendrite trunk, and the extradendritic melt. They are quantified with their volume fractions, f_s^e, f_d^e, f_s^c, f_d^c, f_ℓ, and characterized by their corresponding solute concentrations, c_s^e, c_d^e, c_s^c, c_d^c, c_ℓ. Inside an equiaxed grain, volume fractions of interdendritic liquid and solid dendrites are quantified respectively with α_d^e, α_s^e, hence $f_d^e = \alpha_d^e \cdot f_e$ and $f_s^e = \alpha_s^e \cdot f_e$. Inside a columnar dendrite trunk, $f_d^c = \alpha_d^c \cdot f_c$ and $f_s^c = \alpha_s^c \cdot f_c$.

Both hard blocking [9] and soft blocking [12] mechanisms are implemented and applied to model the columnar-to-equiaxed transition (CET). The hard blocking mechanism suggests that CET might occur when the equiaxed grains ahead of the columnar dendrite tip exceeds a critical volume fraction ($f_{e,CET} = 0.49$). A recent study [28] has shown that the critical value suggested by Hunt [9] might be too large. Instead, a value of $f_{e,CET} = 0.2$ should be used. In current simulations, $f_{e,CET}$ is set to 0.49 except for the case shown in Figure 8, where $f_{e,CET} = 0.2$. The soft blocking mechanism suggests that the exhausting of the growth driving force (constitutional undercooling) due to the enrichment of solute element (rejected by the growing equiaxed grains) stops the growth of columnar primary dendrite tips. For more details regarding growth kinetics, treatment of the dendritic morphology of crystals, algorithm of tracking the columnar tip, etc. refer to previous publications [21][22]. The above solidification model is implemented in a multiphase CFD code (ANSYS Fluent 6.3.26).

3. Experiments

The nucleation parameters for the Al-4.0 wt.%Cu alloy were obtained from a previous work [29] by conducting a series of experiments based on the works of Rappaz and Gandin [30] and Rappaz [27]. The casting samples were cooled at different rates. The resulting maximum undercooling and the corresponding grain number density were experimentally determined and plotted to provide a Gaussian curve, from which the nucleation characteristic parameters were deduced (Table 1).

Classical ingot casting experiments on Al-4.0 wt.%Cu were additionally conducted. The alloy was prepared by mixing the suitable proportions of Cu and Al with commercial purity. The alloy was melted and poured into a clay-bonded graphite crucible at different pouring temperatures, namely 800, 750, and 700 °C. The initial mold temperature was about 20 °C for all samples. Ingots were cooled down in normal atmosphere. After solidification specimens were cut, ground, polished and etched for macroscopic examination. The resulting macrographs are shown in Figure 3. In addition, macrosegregation was studied for one sample (poured at 800 °C) using spectrometer analysis.

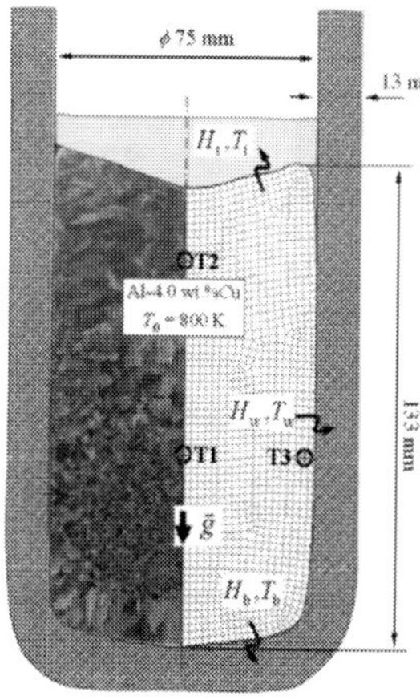

Figure 2. Ingot geometry and configuration of simulation.

Table 1. Parameters used for the simulation of the Al-4.0 wt.%Cu ingot

Thermodynamic data: $T_{liquidus}$ = 922.5 K $T_{solidus}$ = 847.15 K $T_{eutectic}$ = 821.35 K c_0 = 4.0 wt.% $c_{eutectic}$ = 0.332 wt.% Liquidus slope, m = - 260 K/wt.% Partition coefficient, k = 0.145 Latent heat = 389.32 kJ/kg	Boundary conditions: T_0 = 800, 700°C H_t = 10 W·m^{-2}·K^{-1} T_t = 290 K H_w = 120 W·m^{-2}·K^{-1} T_w = 290 K H_b = 120 W·m^{-2}·K^{-1} T_b = 290 K
Nucleation parameters: n_{min} = 10^7, 10^8 and 10^9 m^{-3} n_{max} = 1.48 x 10^{11} m^{-3} ΔT_N = 28.84 K ΔT_σ = 10.17 K n_0 = 10^7, 10^8 and 10^9 m^{-3}	Morphological parameters & hard blocking criterion: Φ^e_{env} = 0.48 Φ^e_{sph} = 0.4 λ_1 = 500 μm Φ^c_{env} = 0.80 Φ^c_{circ} = 0.5 λ_2 = 100 μm $f_{e,CET}$ = 0.49, 0.2

From a first glance at Figure 3 it seems that a lower pouring temperature favors the formation of equiaxed structure, and the equiaxed grain size is reduced with the decreasing pouring temperature. In Figure 3(a) for the ingot poured at 800 °C the region marked (A+B) contains mainly equiaxed grains. Finer equiaxed grains are located in the bottom half, 'region-A', while coarser grains are in the top, 'region-B'. It can also be observed that the equiaxed zone extended upwards to near the mold wall to form the 'region-C', where it interferes with the outer columnar structure zone. In the outer region the columnar trunks growing perpendicular to mold walls are evident, especially in the upper part of the ingot. The structure of the ingot poured at 750 °C (Figure 3 (b)) is mainly very fine equiaxed in the core 'region-A', and relatively coarse equiaxed in the exterior and upper (region-B), where it interferes with columnar structure. Further reduction of pouring temperature (Figure 3(c), poured at 700 °C) introduces even finer equiaxed grains in the core 'region-A', a band of relatively coarse equiaxed grains in the exterior 'region-B', and very coarse grains in the top 'region-C', where it interferes with some fine columnar structure. The variation in equiaxed/columnar zones and the grain size are mainly due to the different temperature gradient, the cooling rate and grain movement during solidification.

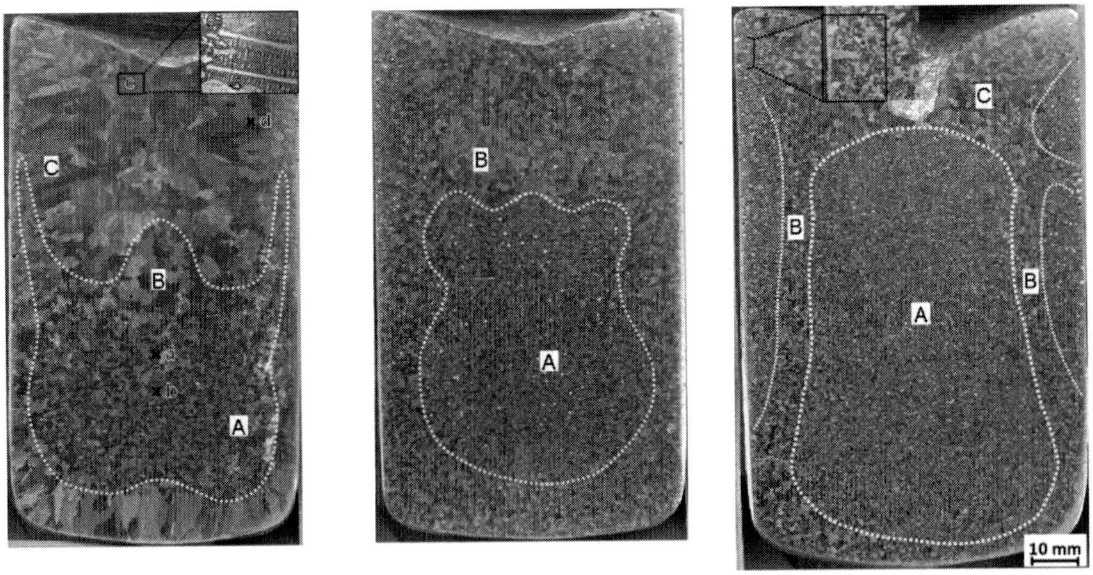

(a) (b) (c)

Figure 3. Macrostructure of the Al-4.0%Cu ingots poured at (a) 800, (b) 750, and (c) 700°C.

4. Simulation and discussion

A 2D-axisymmetric grid (Figure 2) with 1197 volume elements was constructed such that the top surface profile is similar to that of the real casting. Some parameters used for the simulation are listed in Table 1. For more details about other simulation parameters refer to [21][22]. Calculations started with a mini phase volume fraction ($f_e = f_c = 10^{-5}$), an initial grain number density of n_0 and grain diameter of 1 μm. The columnar primary dendrite tips were initialized at the side and bottom mold walls. Mold filling is not considered in the simulation. Instead, the mold was assumed to be initially filled with the molten alloy of an initial temperature, T_0, equal to the pouring temperature. Simulations were conducted for the castings with different pouring temperatures (800 and 700 °C). Adaptive time-stepping scheme was employed with a minimum time step of 10^{-4} s.

The numerically predicted macrostructure for the ingot poured at 800 °C is shown in Figure 4(a). Two zones could be easily distinguished: the equiaxed phase in the core region and the columnar phase in the exterior region. Both regions are separated by a mixed columnar-equiaxed zone. The final position of the columnar primary dendrite tips is indicated by the CET-line. The equiaxed zone is extended vertically near the sidewall with a lower volume fraction. It is additionally found that the current results based on the experimentally in-situ determined grain nucleation parameters matches better with the experiment than the previous result that was based on estimated nucleation parameters Figure 4(b) [26]. Thus the macrostructure with distinguished columnar and equiaxed zones is reproducible by the numerical model.

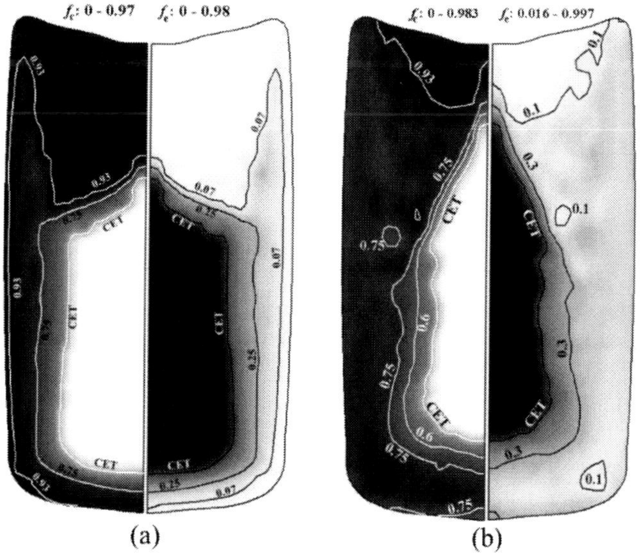

Figure 4. Predicted macrostructure of the Al-4.0wt.%Cu ingot poured at 800 °C: distribution of f_c (left-half) and f_e (right-half) phases. The CET line indicates the final position of the columnar primary dendrite tips. Simulations were performed (a) using experimentally in-situ determined nucleation parameters (Table 1); and (b) using estimated nucleation parameters $n_{max} = 10^{11}$ m^{-3}, $\Delta T_N = 3.5$ K, $\Delta T_\sigma = 0.5$ K [26].

Further structural quantities of this ingot are shown in Figure 5. The calculated grain size, d_e, shows good qualitative agreement to experiments. The grain size in the bottom half is finer than that in the top half. Coarse columnar structure (Figure 5(b)) was predicted in the upper and peripheral regions of the ingot, similar to the as-cast result in Figure 3(a). Close to the CET-line the diameter of the columnar trunk, d_c, gets smaller, and the amount of equiaxed phase becomes dominant. The calculated grain number density, n_e, as shown in Figure 5(c) agrees with the real macrograph (Figure 3(a)). The grain number density in upper region (10^7 m^{-3}) is equivalent to the minimum of each volume element allowed by the model, which virtually means that no equiaxed grains. The maximum grain number density is obtained in the lower bottom. Lowest grain number density with coarsest grain size appears at CET peak 'region-B' in Figure 3(a). A quantitative comparison of n_e, d_e, and d_c at some selected positions (a-d) as marked in Figure 3(a) is given in Table 2. The results confirmed the above discussions. The predicted volume fraction of the formed eutectic is also shown in Figure 5(d).

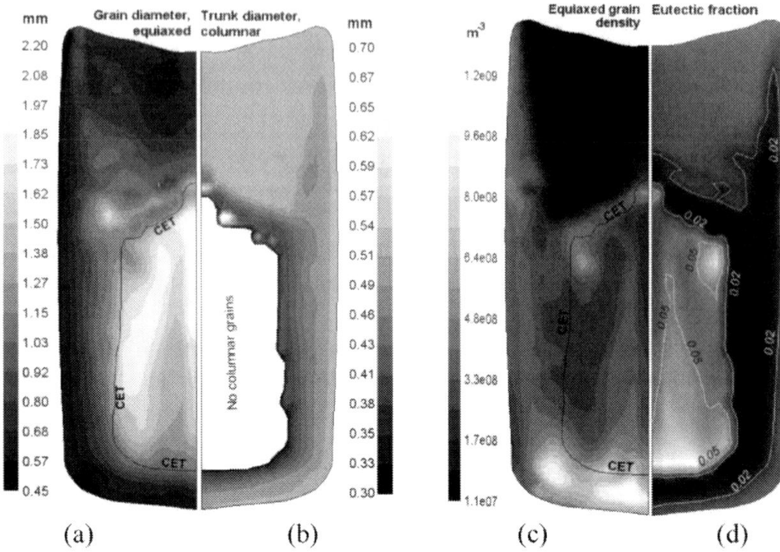

Figure 5. Additional structural quantities for the ingot poured at 800 °C: (a) d_e, (b) d_c, (c) n_e, (d) $f_{eutectic}$.

Table 2. Measured microstructural quantities *vs.* calculated.

Position	a*		b*		c*		d*	
	Exp.	Sim.	Exp.	Sim.	Exp.	Sim.	Exp.	Sim.
n_e, m^{-3}	8.5x10^8	4.6x10^8	9x10^8	5.7x10^8				
d_e, mm	1.2	1.7	1.05	1.56				
d_c, mm					0.73	0.58	0.65	0.52

*Refer to Figure 3(a) for the positions.

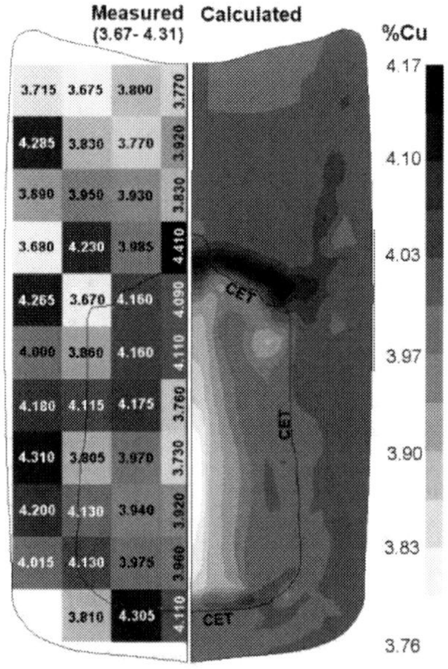

Figure 6. c_{mix} of the ingot poured at 800 °C: measurement (left) *vs.* calculation (right).

The calculated solute concentration of the solid mixture, c_{mix}, shown in Figure 6, exhibited good agreement compared to the readings of the spark analysis of the cast sample shown in Figure 6 (right). The calculated highest positive segregation is located at the top of CET and the lowest negative segregation is close to the center of CET. The top and bottom of the ingot have relatively low c_{mix}. The extremes of macrosegregation do not agree but are close to each other. The c_{mix} values at some odd locations do not agree as well. The analysis of macrosegregation formation mechanisms is out of the scope of the current paper.

The same modeling parameters for the case poured at 800 °C are applied to the case poured at 700 °C. Nevertheless, similar CET profile and similar macrostructure distribution to those shown in Figure 4(a) were obtained in contrast to the expectations. It is well established that lowering the pouring temperature increases the proportion of the equiaxed zone, i.e. promotes an earlier CET. This contradiction may be owed to the ignorance of the pouring effect. During real mold filling, nucleation of equiaxed grains starts when the molten metal hits and contacts the cold walls of the mold. In the current simulation the mold filling was ignored for the purpose of simplification and reduction of the calculation time. To compensate this ignored effect in the current simulation, the initial state of grain number density, n_0, was customized. Reasonably, n_0 is dependent on the pouring temperature and pouring method [3, 26]. Numerical parameter study was performed to investigate the influence of n_0 on the macrostructure, as shown in Figure 7. The assumption of small n_0 (10^7 m^{-3}) matches well with the case of high pouring temperature (800 °C), while a quite large n_0 (10^9 m^{-3}) should be chosen for the case poured at 700 °C, as shown in Figure 7(a) and (d). The macrostructure as predicted in Figure 7(d) agrees satisfactorily with the experimental result of Figure 3(c). One point needs mentioning here is that a minimum grain number density, n_{min} = 10^7 m^{-3} (default) is numerically preset for the incoming melt in the current nucleation model. Initialization of n_e with '$n_0 = n_{min}$' is necessary for solving the grain transport equation. This numerical parameter, n_{min}, seems to show some influence on the modeling results as shown in Figure 7(c) and (d). This influence is not yet fully understood, and it demands further investigation.

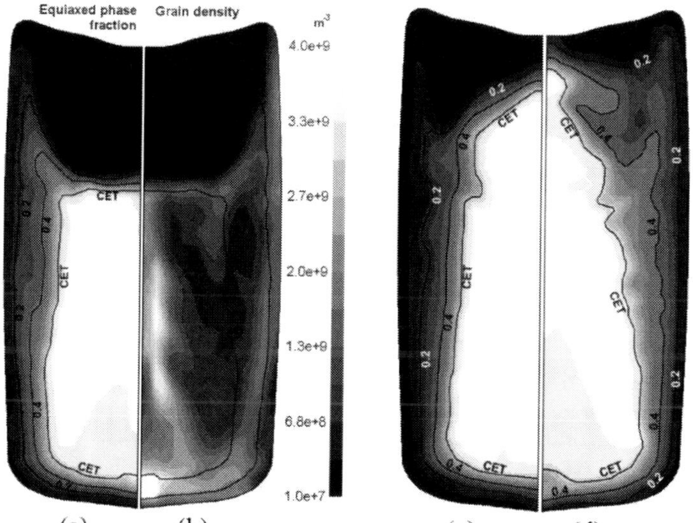

(a) (b) (c) (d)

Figure 7. Calculated macro-structural quantities for the ingot poured at 700 °C. (a) f_e and (b) n_e calculated using $n_{min} = 10^7$ and $n_0 = 10^9$, (c) f_e using $n_{min} = n_0 = 10^8$, (d) f_e using $n_{min} = n_0 = 10^9$.

The hard blocking of the columnar dendrite tip growth is also an important issue that impacts the prediction of the CET [26, 28]. Reduction of the blocking criterion, $f_{e,CET}$, from the originally suggested 0.49 [9] to the newly proposed 0.2 [28] seems to predict a larger equiaxed zone as shown in Figure 8 for the ingot poured at 700°C. Hence, lower $f_{e,CET}$ better matches with the experiment in the current ingot.

The above results and discussion highlight that simulation results of the high pouring temperature ingot exhibited better agreement with experiments in comparison with those of the lower pouring temperature ingot. The high superheat (~185°C) in the former case heats up the mold walls and reduces the opportunity of melt undercooling and nucleation during the filling of mold. Moreover, it can even re-melt the previously formed nuclei. As a result, this case matches well the model assumptions, where no filling is taken into account. On the other hands, calculated results at the lower pouring temperature (superheat ~85°C) deviate from reality, since nucleation is more likely to occur during mold filling, which is already ignored in the simulation.

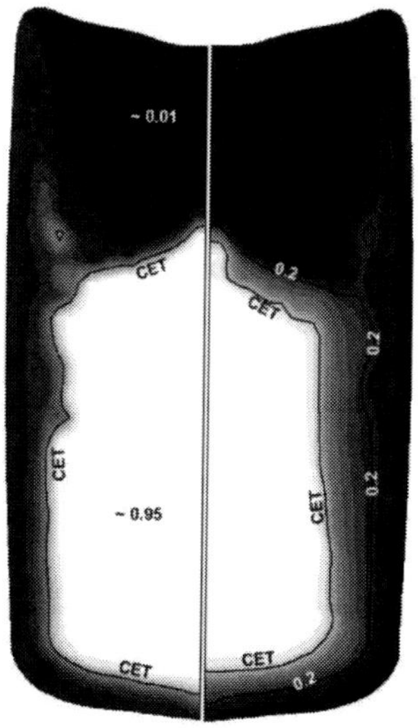

Figure 8. Influence of hard blocking criterion on CET. Predicted f_e using $f_{e,CET} = 0.2$ (left) and $f_{e,CET} = 0.49$ (right).

5. Conclusion

The as-cast structure, CET and macrosegregation in an Al-4.0 wt.% Cu ingot were successfully predicted using the 5-phase mixed columnar-equiaxed solidification model. As a further step to the previous work [26], it is found that the nucleation parameters for the equiaxed phase play a critical role, and therefore, an experimental method [29] is recommended to determine the nucleation parameters. Calculations with the in-situ determined nucleation parameters improved the modelling results to a large extent. The simulation results exhibited very good agreement with experiments at high pouring temperature. However, modeling results for the ingot poured at low pouring temperature demonstrated the importance of the effect of mold filling in the formation of the as-cast structure. Further verifications for the hard blocking criterion are also desired.

Acknowledgement

This work was financially supported by the FWF Austrian Science Fund (P23155-N24). Casting experiments were performed by Georg Nunner and Gerhard Hochleithner.

References

[1] Chalmers B 1963 *J. Austral. Inst. Met.* **8** 255
[2] Flemings M 1974 *Solidification Processing* (New York: McGraw-Hill)
[3] Ohno A 1987 *Solidification: The Separation Theory and its Practical Applications* (Berlin: Springer-Verlag)
[4] Spittle J, Dellamore G and Smith R 1967 Proc. *The Solidification of Metals* (The Iron and Steel Institute London) 318
[5] Morando R, Biloni H, Cole G and Bolling G 1970 *Metall. Trans.* **1** 1407
[6] Winegard W and Chalmers B 1954 *Trans. ASM-AIME* **46** 1214
[7] Jackson K, Hunt J, Uhlmann D, Seward III T 1966 *Trans. TMS-AIM*E **236** 149
[8] Southin R 1967 Trans. TMS-AIME **239** 220
[9] Hunt J 1984 *Mater. Sci. Eng.* **65** 75
[10] Gäumann M, Trivedi R and Kurz W 1997 *Mater. Sci. Eng. A* **226-228** 763
[11] Wang C and Beckermann C 1994 *Metall. Mater. Trans. A* **25A** 1081
[12] Martorano M, Beckermann C and Gandin Ch-A 2003 *Metall. Mater. Trans. A* **34A** 1657
[13] Martorano M and Biscuola V 2009 *Acta Mater.* **57** 607
[14] Wu M and Ludwig A 2006 *Metall. Mater. Trans. A* **37A** 1613
[15] Wu M and Ludwig A 2007 *Metall. Mater. Trans. A* **38A** 1465
[16] Gandin Ch-A. 2000 Acta Mater. **48** 2483
[17] Gandin Ch-A 2000 *ISIJ Int.* **40** 971
[18] McFadden A, Browne D and Gandin Ch-A 2009 *Metall. Mater. Trans.* A **40A** 662
[19] McFadden A and Browne D 2009 *Applied Math. Modell.* **33** 1397
[20] Siqueira C, Cheung N and Garcia A 2002 *Metall. Mater. Trans. A* **33A** 2107
[21] Wu M, Fjeld A and Ludwig A 2010 *Comp. Mater.* Sci. **50** 32
[22] Wu M, Ludwig A and Fjeld A 2010 *Comp. Mater.* Sci. **50** 43
[23] Wu M and Ludwig A 2009 *Acta Mater.* **57** 5621
[24] Wu M and Ludwig A 2009 *Acta Mater.* **57** 5632
[25] Wu M and Ludwig A 2009 *Int. J. Cast Metals Research* **22** 323
[26] Wu M Nunner G Ludwig A Li JH Schumacher P 2011 *IOP Conf. Series: Mater. Sci. Eng.* **27** 012018
[27] Rappaz M 1989 *Int. Mater. Rev.* **34** 93
[28] Martorano M and Biscuola V 2008 *Metall. Mater. Trans .A* **39A** 2885
[29] Ahmadein M Pustal B Berger R Subasic E Bührig-Polaczek A 2009 *Metall. Mater. Trans. A* **40A** 646
[30] Rappaz M and Gandin C-A 1993 *Acta Metall. Mater.*, **41** 345

IOP Conf. Series: Materials Science and Engineering **33** (2012) 012076 doi:10.1088/1757-899X/33/1/012076

Integrated modeling and heat treatment simulation of austempered ductile iron

E. Hepp, V. Hurevich, W. Schäfer
MAGMA GmbH, Kackertstr. 11, 52072 Aachen, Germany
E-mail: v.hurevich@magmasoft.de

Abstract. The integrated modeling and simulation of the casting and heat treatment processes for producing austempered ductile iron (ADI) castings is presented. The focus is on describing different models to simulate the austenitization, quenching and austempering steps during ADI heat treatment. The starting point for the heat treatment simulation is the simulated microstructure after solidification and cooling. The austenitization model considers the transformation of the initial ferrite-pearlite matrix into austenite as well as the dissolution of graphite in austenite to attain a uniform carbon distribution. The quenching model is based on measured CCT diagrams. Measurements have been carried out to obtain these diagrams for different alloys with varying Cu, Ni and Mo contents. The austempering model includes nucleation and growth kinetics of the ADI matrix. The model of ADI nucleation is based on experimental measurements made for varied Cu, Ni, Mo contents and austempering temperatures. The ADI kinetic model uses a diffusion controlled approach to model the growth. The models have been integrated in a tool for casting process simulation. Results are shown for the optimization of the heat treatment process of a planetary carrier casting.

1. Introduction

ADI materials are of special interest in automotive and mechanical engineering because they offer an advantageous combination of mechanical properties. The tensile strengths of these materials typically ranges from 800 to 1600 N/mm^2 at elongations of up to 10%. They also offer excellent fatigue strength and toughness. This combination of properties makes ADI materials comparable to heat-treated steels, but their density is 10% lower than for steel, making them suitable for light-weight applications. On the other hand, the mechanical properties react very sensitively to variations in heat treatment parameters, making an ADI casting's quality extremely process sensitive. The definition of reliable time frames for the processing steps becomes a challenging task for the complicated geometries of the heat-treated parts. What is more, the processing times largely depend on the charge size, its configuration and even on the orientation of the heat-treated parts in the furnace or the quench bath. This imposes limitations on the application of ADI materials. Consequently, numerical simulation can provide tools to master the complexity of the heat treatment process and thus enhance its predictability. Naturally, this will also contribute to the further propagation and versatile use of ADI materials in practice.

The aim of the work described here is the development of an integrated heat treatment process model that covers all processing stages of ADI castings and to establish a simulation tool for the complete process chain. The heat treatment process consists of three steps: austenitization, quenching and austempering. Existing approaches for each individual step were analyzed and those deemed most suitable were selected. The principal criteria used for the selection were robustness, availability of input data and ease of integration. These criteria are often contradictory to simulation accuracy. Hence, the verification of the models is also an important aspect. The three following sections deal with austenitization, quenching and austempering, respectively. They contain a short review of previous work, a description of the numerical approach and its verification. The integrated simulation of a planetary carrier has been carried out, with the simulation results presented in the fifth section.

All numerical investigations and model verifications were performed for a reference alloy with the composition given in Table 1. All stable and metastable phase diagrams and derived parameters like ferrite and pearlite temperatures and equilibrium carbon concentrations in different phases were obtained using calculations made with Thermocalc (TM) (1).

Table 1. Chemical composition of the reference alloy. Values are given in wt%.

C	Si	Mn	Ni	Cu
3.64	2.61	0.25	0.5	0.9

2. Austenitization

Austenitization is the first heat treatment step. Due to heating of the cast part to temperatures between 850 °C – 950 °C, the initial ferrite-pearlite phase transforms to austenite. After that, the carbon contained in the graphite particles dissolves into the newly formed austenite matrix. The temperature and duration used for austenitization are the important process control parameters in this step. Their determination depends on the chemical and ferrite-pearlite composition of the casting. The austenitization temperature defines the maximum amount of carbon that can potentially be dissolved in the austenite matrix. This is significant as this carbon stabilizes the austenite at room temperature and determines its fraction in the ADI microstructure. The duration of the process must be defined such that a diffusion-controlled homogenization of the carbon distribution in the austenite matrix is ensured. If the duration is too long, on the other hand, this may result in grain coarsening in the austenite and in a decrease in productivity and unnecessary energy consumption. A determination of the minimum possible austenitization time to achieve a complete distribution of carbon is the important result of the austenitization simulation.

In the framework of the present work, the transformation kinetics as well as carbon homogenization are considered to be determined by carbon diffusion. This "diffusion-controlled" approach represents the state-of-the-art in austenitization simulation and can be found in (2–4). The models for the transformation kinetics and carbon homogenization differ only by the initial and boundary conditions. The assumptions are the same for both of the models as described below:

- Local thermodynamic equilibrium is assumed at the phase boundaries.
- A so-called para-equilibrium (PE) approach is used. This means that an Fe-C section of the multicomponent phase diagram is considered, keeping all other elemental concentrations constant. The diffusion of Si, Mo, Ni, Mn and Cu is neglected because of their low mobility (5). The application of the PE approach to the ferrite-to-austenite transformation is justified by (6), where an investigation of ternary Fe-C-X systems using local equilibrium with (LEP) and without partitioning (LENP) has been carried out.

- Microsegregation is not taken into account, although the consideration of the influence of microsegregation could be considered similar to (7).

2.1. Problem formulation

In the references (2; 3) some simplified analytical approaches to modeling the ferrite-to-austenite transformation and subsequent carbon dissolution are presented. They are restricted to the ferrite-austenite transformation and do not consider pearlite. This leads to an overestimation of the transformation time, which is discussed in more detail below.

The austenitization begins with a pearlite-austenite transformation once the metastable temperature is reached. Because of the small characteristic diffusion length in pearlite, this transformation is considered to be controlled solely by equilibrium considerations and the release of latent heat.

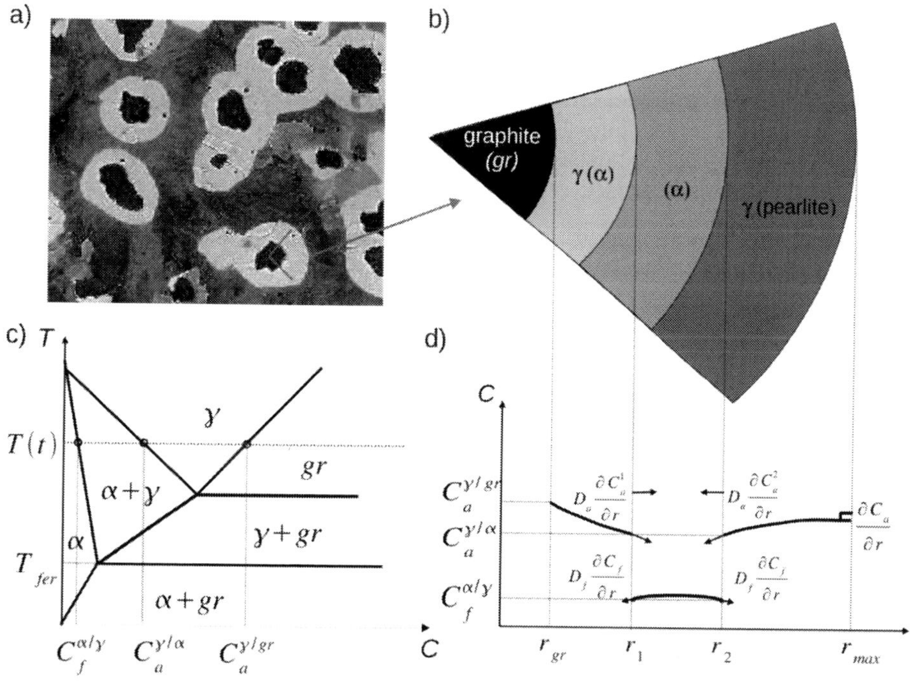

Figure 1. Diffusion-controlled ferrite-austenite transformation: a) micrograph of the initial (as-cast) microstructure; b) schematic diagram of a eutectic cell considered as the calculation domain; c) schematic showing an Fe-C section through the multicomponent phase diagram; d) schematic of the corresponding carbon concentration profiles with boundary conditions.

During subsequent heating, the ferrite-austenite transformation starts. The initial as-cast microstructure is presented in Figure 1a. The graphite spheres are surrounded by ferrite shells and are embedded in a newly transformed austenite matrix. The graphite sphere together with its diffusional surrounding area, also called a eutectic cell, form the calculation domain (see Figure 1b). Once the temperature exceeds this equilibrium temperature at the graphite-ferrite interface (see Figure 1c) an austenite layer starts to form around the graphite, denoted as $\gamma(\alpha)$. This austenite is surrounded by the ferrite shell α and the austenite, transformed from pearlite $\gamma(pearlite)$, is found on the outside of the cell. Figure 1d shows a schematic of the carbon concentration profile over the calculation domain. There are two diffusional flows into the α area. The abundance of carbon in the graphite means the diffusion through the $\gamma(\alpha)$ austenite is continuously strong. The diffusion through the $\gamma(pearlite)$ plays an important role at the beginning of the transformation and diminishes with time due to consumption of the carbon which is stored in $\gamma(pearlite)$. The diffusion problem for the calculation domain is formulated as follows:

$$\begin{aligned}
\frac{\partial C_{a1}}{\partial t} &= D_a\left[T(t)\right]\left[\frac{2}{r}\frac{\partial C_{a1}}{\partial r} + \frac{\partial^2 C_{a1}}{\partial r^2}\right], & r \in [r_{gr}, r_1] \\
\frac{\partial C_f}{\partial t} &= D_f\left[T(t)\right]\left[\frac{2}{r}\frac{\partial C_f}{\partial r} + \frac{\partial^2 C_f}{\partial r^2}\right], & r \in [r_1, r_2] \\
\frac{\partial C_{a2}}{\partial t} &= D_a\left[T(t)\right]\left[\frac{2}{r}\frac{\partial C_{a2}}{\partial r} + \frac{\partial^2 C_{a2}}{\partial r^2}\right], & r \in [r_1, r_{max}]
\end{aligned} \tag{1}$$

where C_{a1}, C_f and C_{a2} are the carbon concentrations in $\gamma(\alpha)$, α and $\gamma(pearlite)$, and D_a and D_f are the carbon diffusion coefficients in austenite and in ferrite, respectively. The system of equations (1) contains three coupled boundary problems, hence six boundary conditions as well as three initial conditions are required. The boundary and initial conditions used are listed in Table 2. Because of the assumed local thermodynamic equilibrium, all Dirichlet boundary conditions and C_f can be taken from the phase diagram (see Figure 1c). $C_{pearlite}$ and r_{max} are obtained from a previous casting process and solidification simulation performed using a commercial casting process simulation software (10).

Table 2. Initial and boundary conditions for modeling diffusion during austenitization.

zone denotation	variable	initial condition	boundary condition	coordinate
$\gamma(\alpha)$	C_{a1}	non existent at $t = 0$	Dirichlet $C_a^{\gamma/gr}$	r_{gr}
			Dirichlet $C_a^{\gamma/\alpha}$	r_1
α	C_f	C_f^0	Dirichlet $C_f^{\alpha/\gamma}$	r_1
			Dirichlet $C_f^{\alpha/\gamma}$	r_2
$\gamma(pearlite)$	C_{a2}	$C_{pearlite}$	Dirichlet $C_a^{\gamma/\alpha}$	r_2
			Neumann $\partial C_{a2}/\partial r = 0$	r_{max}

The equation system (1) contains two unknown mobile fronts r_1 and r_2. In order to make the problem well-posed, the system (1) must be supplemented by two additional mass balance equations:

$$\begin{aligned}
\frac{\partial r_1}{\partial t}\left[C_a^{\gamma/\alpha} - C_f^{\alpha/\gamma}\right] &= D_f\frac{\partial C_f}{\partial r} - D_a\frac{\partial C_{a1}}{\partial r}, \quad r = r_1 \\
\frac{\partial r_2}{\partial t}\left[C_a^{\gamma/\alpha} - C_f^{\alpha/\gamma}\right] &= D_a\frac{\partial C_{a2}}{\partial r} - D_f\frac{\partial C_f}{\partial r}, \quad r = r_2.
\end{aligned} \tag{2}$$

The system (1) and (2) is solved numerically using the finite difference method. The two fronts r_1 and r_2 move towards each other, and the transformation is completed as soon as the fronts meet.

For the subsequent carbon homogenization, there are no moving fronts. As a result, the problem can be described by a single diffusion equation with a Dirichlet boundary condition $C_a^{\gamma/gr}$ applied on r_{gr} and a Neumann boundary condition applied on r_{max}.

2.2. Results and verification

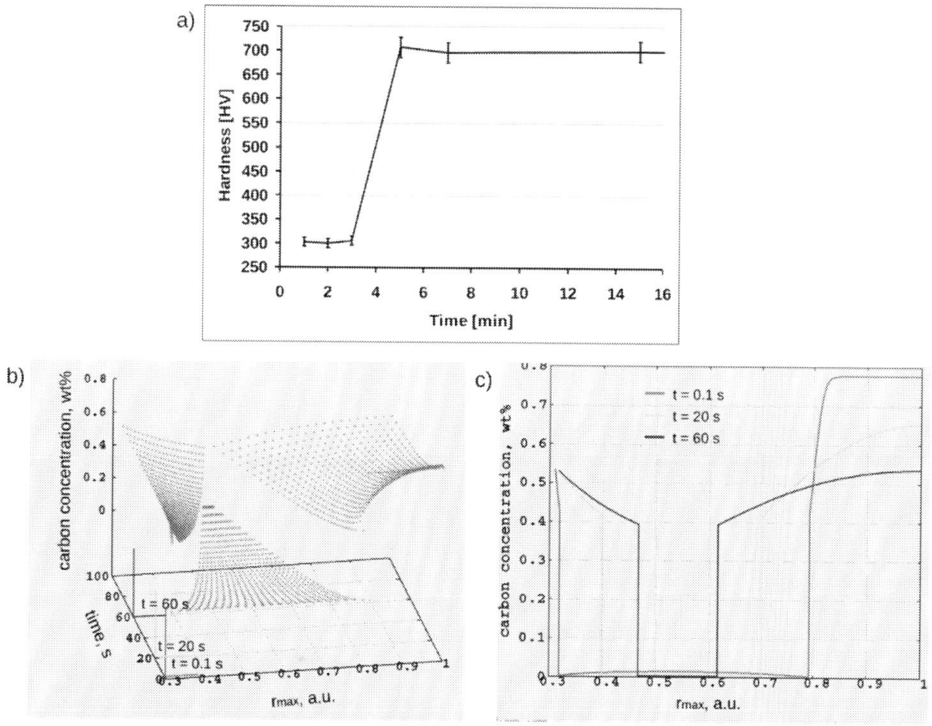

Figure 2. Ferrite-to-austenite transformation: a) measured micro-hardness as function of austenitization time for the reference alloy; b) simulated carbon concentration distribution over a eutectic cell as a function of austenitization time; c)simulated carbon concentration distribution over a eutectic cell for times of 0.1, 20 and 60 seconds.

In order to verify the austenitization model, measurements have been carried out by heating samples for different times, quickly quenching them and measuring the micro-hardness (9). These measurements are plotted in Figure 2a for the reference alloy used in this study. The sharp increase in hardness after 3 minutes indicates the ferrite-austenite phase transformation. This transformation takes about $1.5-2$ minutes. A numerical simulation has been performed for the same alloy and heating conditions. The simulated carbon concentration profile as a function of dimensionless r_{max} over time is mapped in Figure 2b. Figure 2c shows the corresponding concentration profiles for the three times indicated on the left axis of Figure 2b. The jumps in carbon concentration correspond to the mobile fronts r_1 and r_2. The calculated transformation time (about 100s) corresponds well with the measured one. Similar hardness measurements and related simulations have been carried out for another 15 alloys with varying Cu, Ni and Mo contents. The resulting simulated and measured transformation times have also shown a good correlation.

3. Quenching

Quenching represents the next step of ADI heat treatment. The aim of quenching is to obtain a strongly undercooled and carbon-saturated austenite, which is required for the following austempering transformation. The cooling must be quick enough to avoid undesirable eutectoid transformations. For this stage, the simulation primarily involves solving the heat conduction equation with appropriate boundary conditions. A possible eutectoid transformation is taken into account using an approach described in (8). A CCT diagram provides the important input data for the simulation of this transformation, specifying the areas in which the precipitation of ferrite and pearlite is still possible. The simulations performed in the present study are based on CCT diagrams measured by the Institut für Gießereitechnik (IfG) within the framework of the LEA project (9).

4. Austempering

Austempering is the last heat treatment step. During austempering, needle-shaped ADI ferrite is formed. At the same time, the remaining austenite is saturated with carbon rejected from ferrite up to around 2wt%. This carbon makes the austenite stable at room temperature. The critical parameters are the holding temperature and the austempering time. The temperature determines the maximum amount of carbon that can possibly dissolve in austenite and, consequently, the stability of the austenite at room temperature. A temperature that is too high leads to a poor carburization of the austenite. On the other hand, a temperature that is too low causes the formation of martensite. Usually, the holding temperature is in the range of 280 °C – 400 °C. The austempering time must be long enough to ensure the completion of the ADI transformation. On the other hand, an austempering time that is too long promotes the precipitation of carbides. For this process step, the aim of the simulation is to define the optimum austempering time.

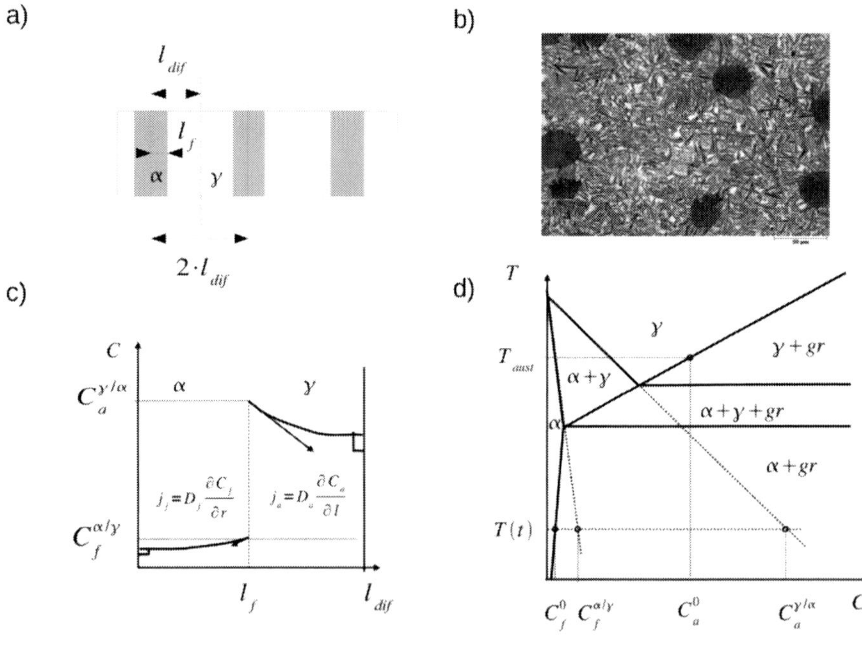

Figure 3. Approach to modeling Austempering: a) schematic diagram of the calculation domain; b) micrograph of the ADI microstructure; c) schematic of the corresponding carbon concentration profiles with boundary conditions; d) schematic showing an Fe-C section through the multicomponent phase diagram.

The austempering and austenitization models are similar: both models are based on the PE approach and use the same assumptions. The austempering model differs from the austenitization model in the geometry of the calculation domain as well as in the boundary conditions used. The calculation domain for the austempering model is shown in Figure 3a. The corresponding carbon concentration profile is indicated schematically in Figure 3c. The boundary conditions are taken from an isopleth phase diagram as shown in Figure 3d.

4.1. Ferrite nucleation

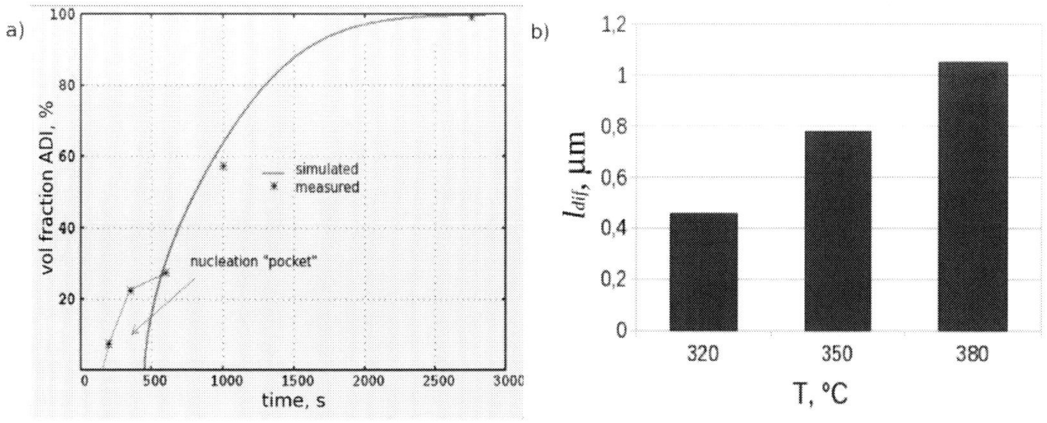

Figure 4. Nucleation of ADI ferrite: a) measured fraction ADI ferrite showing how l_{dif} can be adjusted to obtain good agreement; b) l_{dif} as a function of temperature for the reference alloy.

The nucleation of acicular ferrite determines the characteristic diffusion length and significantly influences the following transformation kinetics. Within the scope of the available models for ADI transformation kinetics, the ferrite needles, or lamellae in 3D, are assumed to be oriented parallel to each other (see Figure 3a). In this case, half of the lamellar spacing is used as the characteristic diffusion length l_{dif}. As can be seen in Figure 3b, there is no pattern in lamellae distribution, requiring a determination of an equivalent diffusion length. The scattering in ADI microstructures makes this difficult to determine through metallographic evaluation. In contrast, the current phase composition of ADI, or the transformation kinetics, could be securely measured. In the present work, the equivalent diffusion length is determined based on experimental measurements with the help of inverse analysis. The required measurements for 16 alloys have been carried out by IfG (9). The Cu, Ni and Mo contents as well as the holding temperature have been varied. The procedure for determining the equivalent diffusion length is illustrated in 4a, where measured and simulated fraction ADI ferrite results are shown. The model for the transformation kinetics is described in the next subsection. The equivalent diffusion length and the latency time are adjusted to obtain satisfactory agreement with the measurements. The measured transformation kinetics show a "pocket" at the beginning of the transformation. This "pocket" could be explained by continuous changes in the equivalent diffusion length during ferrite nucleation. The simulation uses a fixed diffusion length, and will not be able to reproduce the "pocket". Nevertheless there is a good agreement between simulated and measured kinetics in general.

Figure 4b shows the diffusion length of the reference alloy as a function of the temperature. The diffusion lengths for 320 °C and 380 °C differ by more than a factor of 2. Obviously, the temperature plays a decisive role in ferrite nucleation.

4.2. Transformation kinetics

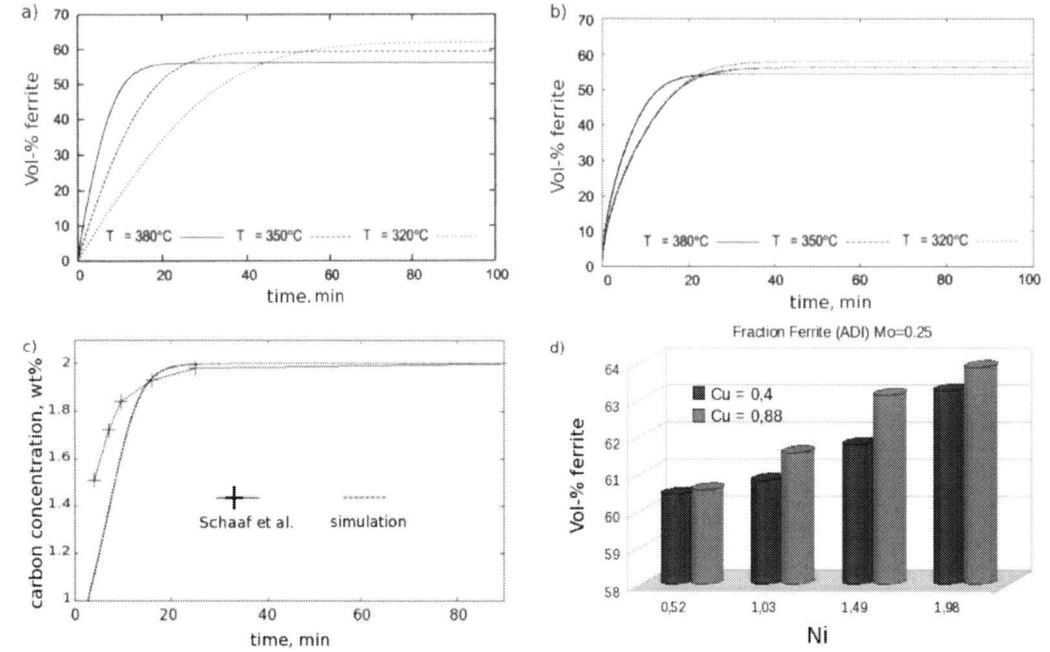

Figure 5. Transformation kinetics during austempering: a) ADI ferrite fraction as a function of time obtained by (9); b) ADI ferrite fraction as a function of time obtained by PE approach; c) comparison of predicted carbon concentration over time using the PE approach and measured by (11); d) predicted fraction of ADI ferrite as a function of Ni and Cu content using the PE approach.

In order to verify the model of the transformation kinetics, a comparison with a considerably more sophisticated reference model was drawn. The reference model was developed by the Giesserei-Institut at the RWTH Aachen (9) and considers the cross-diffusion of all components in the system. The transformation kinetics obtained by the PE model and the reference model are compared in Figure 5a and 5b, respectively. The simulations were carried out for 3 different temperatures of 320 °C, 350 °C and 380 °C. The corresponding diffusion length is taken from Figure 4b. The results for 380 °C and 350 °C show a good correlation with the reference model. But, for 320 °C the PE approach predicts a shorter ausferritization time.

Another verification is made by comparing the simulated carbon concentration profile with the average carbon concentration obtained using a neutron diffraction method (11). The experiment and simulation have been carried out for a temperature of 400 °C and the chemical composition given in (11). The measured and simulated concentrations are plotted in Figure 5c, where a good agreement can be observed. Figure 5d shows the predicted ferrite fraction in the ADI matrix as a function of Cu and Ni content. The fraction of ferrite increases with increasing Ni and Cu content, which agrees with the experimental data provided by (12).

Summarizing the above, it can be said that the PE approach provides a sufficient accuracy. This fact, combined with its fast performance, makes the PE approach ideal for the simulation of ausferritization kinetics for real castings.

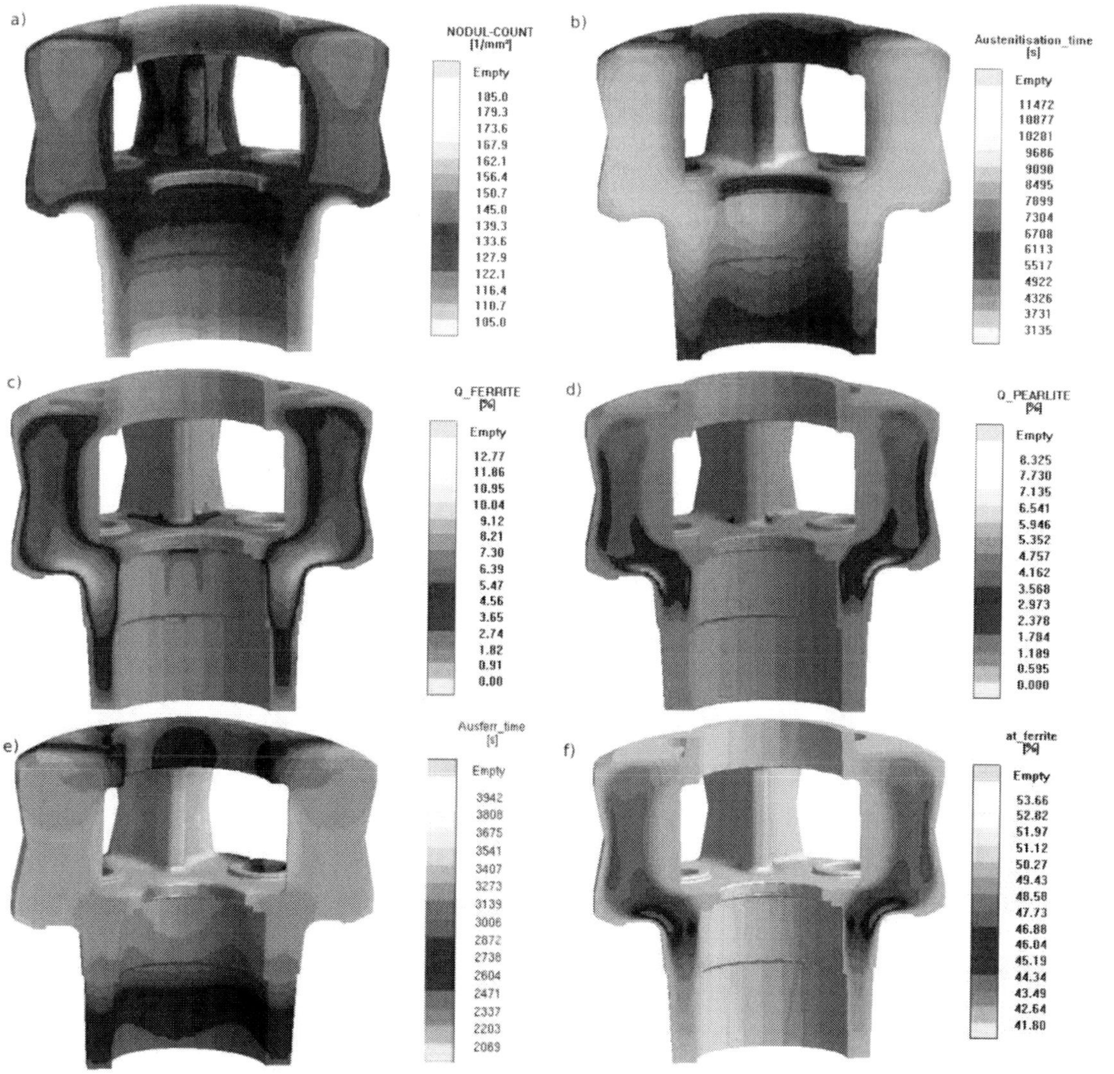

Figure 6. Simulation results for the planetary carrier casting: a) predicted nodule-count from the casting simulation; b) predicted minimum required austenitization time; c) predicted fraction of ferrite after quenching; d) predicted fraction of pearlite after quenching; e) predicted minimum required ausferritization time; f) predicted fraction ferrite in ADI.

5. Application example

Within the framework of the LEA project (9), an approximately 3400kg planetary carrier from the foundry Eickhoff in Germany was used as a demonstrator part. The heat treatment process consisted of austenitization for 6 hours at 890°C, followed by quenching and austempering for 3 hours at 360°C. For the alloy used, the ferrite and pearlite temperatures were 762°C and 734°C respectively (1). Figure 6a shows the nodule count distribution taken from the solidification and cooling simulation. It serves as the initial situation for the following austenitization and quenching predictions. The minimum time required to locally achieve a complete austenitization is shown in Figure 6b. The austenitization time correlates well with the nodule count because it is mainly controlled by carbon diffusion, and consequently by the diffusion length. The most critical area for austenitization is in the center of the casting, which has the lowest nuclei density, and takes about 3 hours. The simulation reveals a potential margin to reduce the austenitization time by nearly one-half. Figures 6c and 6d show the predicted fractions of ferrite and pearlite respectively, which are formed during quenching. The simulation indicates some ferrite and pearlite in the "thicker" wall sections, indicating that there is room for improving the quenching process. The predicted fraction of ADI ferrite is shown in Figure 6e. The lower ADI-ferrite content in the "thicker" wall sections can be explained through the lower amount of austenite due to the eutectoid transformation. Figure 6f shows the minimum required time for local austempering. Based on the simulation results, the austempering time could also be reduced down to around 1.5 hours.

6. Conclusions

A simulation tool for the integrated modeling of casting and heat treatment processing of ADI has been developed. The used models were validated based on experimental investigations and applied to an example casting. The results show that the modeling approach yields simulation models that can be effectively and accurately applied to optimize the production of ADI castings in practice.

Acknowledgements

The authors are grateful to the German Federal Ministry of Education and Research (BMBF) as well as the Forschungszentrum Jülich (PTJ) for the financial support.

References

[1] Thermocalc Version 6 User Handbook 2006

[2] Zimba J, Henwood D, Navara E and Simbi D J 1999 *Mat. Sci. and Technol.* **15** 1024

[3] Batra U, Ray S and Prabhakar S R 2005 *J of Mat. Eng. and Performance* **14(5)** 574

[4] Thomson R C, James J S and Putman D C 2005 *Mat. Sci. and Technol.* **16** 1412

[5] Fridberg J, Torndahl L E and Hillert M *et al.* 1969 *Jernkontorets annaler* **153(6)** 263

[6] Inden G and Hutchison C R 2003 *Austenite Formation and Decomposition, ISS and TMS* 65

[7] Lekakh S N and Richards V L 2010 *AFS Proceedings* 10-083 263

[8] Wessén M and Svensson I L 1994 *Metal. and Material trans.* **27(A)** 1996

[9] Schäfer W *et al.* LEA - Leichtbau mit gegossenen ADI Bauteilen, Abschlußbericht, BMBF Projekt, FKZ 03X3013, 2011

[10] MAGMA5 Version 5.1 User Handbook 2011

[11] Schaaf P, Cusenza S, Bamberger M, Amran Y, Weiss K, Meier L, Hofmann M 2009 *Int. Found. Res.* **61** 14

[12] Mi Y 1995 *Scripta Met. Mat.* **32** 1313

MCWASP XIII IOP Publishing
IOP Conf. Series: Materials Science and Engineering **33** (2012) 012077 doi:10.1088/1757-899X/33/1/012077

Direct simulation of a directional solidification experiment observed *in situ* and real-time using X-ray imaging

G Reinhart[1,3], **Ch-A Gandin**[2,3], **N Mangelinck-Noël**[1,3], **H Nguyen-Thi**[1,3], **B Billia**[1,3], **J Baruchel**[4]

[1] Aix-Marseille Univ & [3] CNRS, IM2NP UMR 6242, Campus Saint-Jérôme Case 142, 13397 Marseille Cedex 20, France
[2] MINES ParisTech & [3] CNRS, CEMEF UMR 7635, 06904 Sophia Antipolis, France
[4] ESRF, Avenue des Martyrs, BP 220, 38048 Grenoble Cedex, France

E-mail: Charles-Andre.Gandin@mines-paristech.fr

Abstract. It has been shown in the last decade that *in situ* and real-time observation of metallic alloy solidification is possible by using X-ray monitoring conducted at third generation synchrotron sources. A detailed analysis of a Bridgman experiment carried out at ESRF with an Al - 3.5 wt% Ni alloy was presented earlier [1]. This article proposes a direct simulation of the solidification of the entire sample for this experiment, in which all the dendritic grains are individually represented as they nucleate and grow in the experiment. This is possible by extracting from the radiographs a list of all the nucleated grains, including the positions and orientations of their main trunks. Simulation is performed using a two-dimensional (2D) Cellular Automaton (CA) – Finite Element (FE) model. As a result of the coupling between the CA and FE methods, consequences of the macroscopic transport of heat, liquid momentum and solute mass on the development of the dendritic grain structure are accounted for, and vice versa. The macroscopic deformation of the columnar front observed during the experiment is reproduced, as well as the columnar-to-equiaxed transition. The influence of flow patterns on macrosegregation is also discussed.

1. Introduction

Solidification of metallic alloys usually starts with the nucleation of crystals. Each crystal grows and adopts a dendritic morphology to form a single grain. During its development, each grain can be identified as an envelope which contains a mixture of a primary solid phase plus an intradendritic liquid, i.e. a mixture referred to as mushy zone. The grain morphology can be more or less elongated depending on the local history of the temperature and melt composition in its neighboring. For instance, growth in a temperature gradient may result in an elongated columnar grain structure whereas a uniform temperature would favor isotropic equiaxed grain morphology. The presence of liquid flow due to buoyancy forces also influences the grain morphology and could modify the solute transport (micro- and macro-segregation). In addition, such solute transport leads to variations of the solidification conditions and then an inhomogeneous distribution of the fraction of solid phases formed upon dendritic, peritectic and eutectic microstructures between the intra and intergranular regions.

In situ and real-time direct observation of all these dynamical phenomena is now possible by using X-ray monitoring conducted at third generation synchrotron sources [2-4]. On other hand, modeling of these observations is still a challenge, mainly because it requires models including heat and mass transfers coupled with physical phenomena taking place at various length scales. First comparisons between X-ray observations of solidification and numerical models have yet been reported [5, 6]. In the present contribution, a Bridgman experiment carried out at the European Synchrotron Radiation Facility (ESRF, Grenoble, F) with a refined Al - 3.5 wt% Ni alloy [1] is thoroughly characterized and used as a benchmark, with the goal to provide inputs for a 2D CAFE model [5]. This enables us to perform a direct simulation of the dendritic grain structure and the eutectic distribution in the entire sample accounting for fluid flow and its interaction with the growing solid microstructures. Most recent advances in coupled experimental and numerical analyses are demonstrated.

2. Experimental details

2.1. Solidification experiments at ESRF

Directional solidification experiments were performed in a Bridgman-type furnace at the ID19 imaging beamline of ESRF [7, 8]. The sample was made of an Al - 3.5 wt% Ni alloy with addition of 0.5 wt% Al-Ti-B to generate particles acting as preferred nucleation sites for new grains. Its dimensions were 40 mm in length and 6 mm in width. The thickness was reduced to 150-200 μm in order to obtain sufficient transmission of the X-ray beam. The sample was adjusted into a soft graphite crucible. Radiographs of the solidifying samples were recorded with an ESRF FReLoN (Fast Read-out Low Noise) camera in order to display the time-evolution of the microstructure. The optics was chosen to obtain a good compromise between a large field of view (15×6 mm^2) and a good spatial resolution (pixel size = 7.46 μm). X-ray beam energy was set to 13.5 keV using a double Si(111) monochromator and the acquisition frequency was about 0.3 Hz which is a compromise between satisfactory contrast and high acquisition rate.

Solidifications were performed upward, in stable conditions with respect to thermosolutal convection, with a heavy solute rejected during the solidification that increases the liquid density and a positive temperature gradient of approximately 30 $K.cm^{-1}$. For the solidification experiment that will be discussed presently, solidification started with a low pulling velocity, 1.5 $\mu m.s^{-1}$, leading to the formation a columnar dendritic structure. A permanent regime was established after approximately 4200 s. The pulling velocity was then suddenly increased to a value of 4 $\mu m.s^{-1}$ in order to activate nucleation in undercooled region above the interface and provoke a Columnar to Equiaxed Transition (CET).

2.2. Image processing and analysis

The image processing software ImageJ [9] was used in order to remove spurious artifacts on the radiographs due to the non-uniform profile of the X-ray beam, monochromator defects or surface defects on the crucible. Image processing consisted in a flat-field correction of the radiographs, and then in a division by a corrected image of the same zone recorded earlier when the alloy was fully liquid. More detail regarding this procedure can be found in [1]. The final result is an enhanced contrast and almost defect free image of the grain structure, with the Al-rich dendritic structures appearing in light grey and the Ni-rich regions such as the eutectic structure appearing in dark grey. A quantitative characterization of the solidification experiment and the final grain structure was performed in order to determine experimental parameters to be used as input, or for comparison with the numerical simulations. Because the field of view is limited to a small window, an entire image of the final grain structure was deduced by pasting partial radiographs of the sample after complete solidification (figure 1a). Segmentation of the grains was performed manually by drawing the corresponding envelopes (figure 1b). Individual grains were identified by using the particle analysis function of ImageJ. The positions of the grain nucleation centers, after sedimentation if any, were determined unambiguously from the solidification movie. The corresponding coordinates (X_i,Y_i) were defined in a reference frame placed at the bottom center of the sample as identified in figure 1b. The crystallographic orientation of the dendritic grains could not be determined during the experiment or during *post-mortem* analyses since the sample was remelted to carry out further solidification experiments. Thus, the orientation of the longest primary arm growing upward relative to the vertical growth direction, θ_i, was measured for each grain (figure 1c). Finally, the positions of the highest dendrite tip and of the eutectic front, and therefore the mushy zone height, were measured as a function of time.

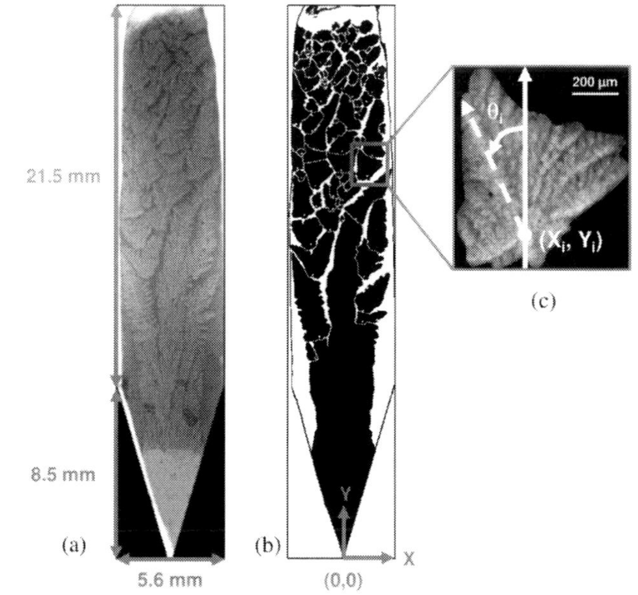

Figure 1. (a) Reconstructed image of the fully solidified sample with its characteristic dimensions used in the numerical simulations, (b) image of the grain structure after segmentation and (c) example of determination of the grain orientation and grain centre.

3. Modelling

3.1. Cellular Automaton – Finite Element modeling

A volume averaging approach is used to write conservation equations for a representative elementary volume (REV) made of a mixture of one solid phase, s, plus one liquid phase, l. No other phase being present, the sum of their volume fraction is equal to unity: $g^s + g^l = 1$. Furthermore, equal and constant densities in the phases are assumed, $\rho^s=\rho^l=\rho_0$, together with a fixed solid phase, $\mathbf{v}^s=0$. As a consequence, the average total mass conservation equation simplifies to $\nabla\cdot\langle\mathbf{v}\rangle = 0$ where the average macroscopic flow velocity reduces to $\langle\mathbf{v}\rangle = g^l \langle\mathbf{v}^l\rangle^l$, $\langle\mathbf{v}^l\rangle^l$ being the average intrinsic velocity of the liquid phase. When writing the momentum conservation over the REV, a permeability term appears that is calculated thanks to the Carman-Kozeny relationship, $[{g^l}^3 \lambda_2^2] / [180\,(1-g^l)^2]$, where a constant value of the secondary dendrite arm spacing, λ_2, is taken, deduced from measurements. The Boussinesq approximation is introduced: the liquid density, ρ^l, is kept constant in all terms of the momentum equation except for the gravity term where it is replaced by a function of the local solute composition in the liquid phase, $\langle w^l\rangle^l$, and the local temperature, T, $\rho^l = \rho_0 \left[1 - \beta_T (T - T_L) - \beta_w (\langle w^l\rangle^l - w_0)\right]$, where β_T and β_w denote the thermal and solutal expansion coefficients, respectively. References used to define the variations of the liquid density with the local solute composition in the liquid phase and the local temperature are the liquidus temperature of the alloy, T_L, and its nominal composition, w_0, respectively. The other physical parameter of the momentum equations is the dynamic viscosity of the liquid phase, μ^l. The average heat flow equation is also considered. The average enthalpy per unit mass, $\langle H\rangle = C_p\, T + g^l\, \Delta_s^l H_f$, is linked to the specific heat taken constant for the liquid and solid phases, $C_p^s = C_p^l = C_p$, and to the latent heat of fusion per unit mass, $\Delta_s^l H_f$. The thermal conductivities of the alloy in the solid, κ^s, and in the liquid, κ^l, (table 1) are taken constant. The variation is linear in the mushy zone between the eutectic temperature, T_E, and the liquidus temperature. Finally, the average conservation of the solute mass is written for a binary alloy with the average composition of solute, $\langle w\rangle$, as the main unknown. The physical parameter entering in the solute mass balance is the diffusion coefficient of the solute element in the liquid phase, D^l. A more detailed presentation of the set of equations is presented elsewhere [10] as well as in this proceedings volume [11]. The main guidelines to solve these equations in a two dimensional representation with the FE method are provided elsewhere [12, 13].

A microsegregation model is needed to convert the average composition of solute, $\langle w \rangle$, and the average enthalpy, $\langle H \rangle$, into a temperature, T, and a fraction of solid, g^s. It is based on the coupling between a CA method to track the development of the grain envelopes and a local microsegregation model based on the lever rule approximation. The FE mesh is further divided into a regular lattice of fine squares named CA cells. Each cell is defined by its center coordinates and the finite element in which it is located. Linear interpolation coefficients are computed between the FE nodes and the CA cell, thus permitting interpolation of the variables solved by the FE method onto the CA grid. Similarly, information computed onto the CA grid can be summed up and projected onto the FE mesh. Each cell is also attributed with an index that defines its state. At the beginning of a simulation starting from a superheated melt, all cells are in the liquid state. As nucleation and growth proceed, the index of cell is changed to a non-zero value. The growth of the structure in a cell is also characterized by the size of its local mushy zone. It is defined by the extension of a square, the four apices representing the dendrite trunks and arms directions. A dendrite tip kinetics defined in [14] is used, with the Gibbs-Thomson coefficient, Γ, as the main parameter. The cell is then in a mushy state, i.e. made of a mixture of solid and liquid phases. Its fractions of phases are computed by converting the projected average enthalpy and composition at the cell position into a temperature and a fraction of solid and liquid phases assuming the lever rule approximation. Once the prescribed growth temperature of the eutectic structure is reached, a simple isothermal transformation is assumed in order to transform the remaining liquid phase into a volume fraction of eutectic.

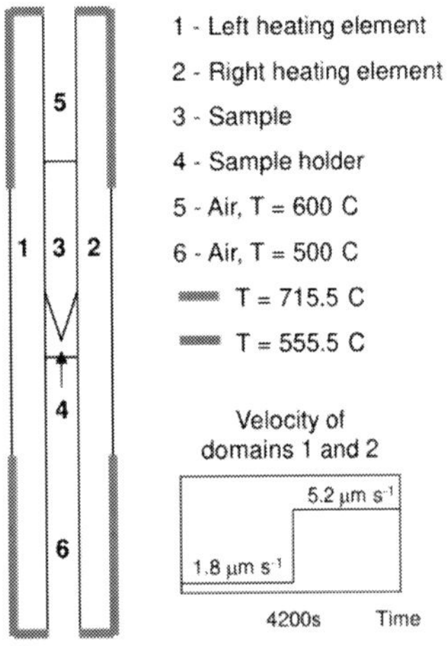

Figure 2. Simple representative geometry of the experimental device used for the numerical simulations. Domains 1 and 2 are pulled upward with a velocity profile shown in the graph, while domains 3 and 4 are fixed.

3.2. Adjustment of thermal parameters

A simple two-dimensional geometry was used (figure 2). Two heating elements, labeled 1 and 2, surround the sample and its holder, labeled 3 and 4 respectively. The thermal properties of the heating elements were defined to act as good heat conductors and the sample holder has the same thermal conductivity and heat capacity as the alloy (table 1). Cooling of the sample is provoked by translation of the heating elements with an adjusted time evolution of the velocity also shown in figure 2. Note that in the experiment, the sample and its holder are translated instead of the heating elements of the Bridgman furnace. This change is of no importance with respect to the simulation results. The size of the mushy zone is defined as the distance between the eutectic front and the highest dendrite tip position of the solidification front; it can be directly measured in the radiographs. Because the Bridgman furnace does not provide a constant imposed temperature gradient, the boundary conditions had to be adjusted at the top and bottom regions of the heating elements in order to retrieve the time evolution of the recorded mushy zone length. Their temperatures were maintained constant, equal to 555.5°C in the cold zone and 715.5°C in the hot zone, at surfaces highlighted in bold in figure 2. Heat transfer coefficients were also imposed at the bottom of the sample holder and top of the sample i.e. at boundaries in contact with the air. Values are reported in table 1. All other boundaries are simplified assuming either a perfect contact or adiabatic conditions. It is to be noticed that, due to the relative movement of the heating elements with respect to the sample and its holder, boundary conditions are not fixed in space and need to be recalculated during the time stepping algorithms to solve the average conservation equations.

Figure 3 shows the result of the heat flow adjustment on the predicted time evolution of the position of the dendritic growth front, y_d, and the eutectic growth front, y_e, as well as on the mushy zone height, y_d-y_e. They are compared with experimental measurements. Although adjustment is not perfectly achieved, the results show that the overall kinetics of the growth front is well reproduced, as well as the increase with time of the mushy zone length. The last effect is directly linked to the gradual decrease of the temperature gradient following the mushy zone development, a phenomenon that is well known in directional solidification [15].

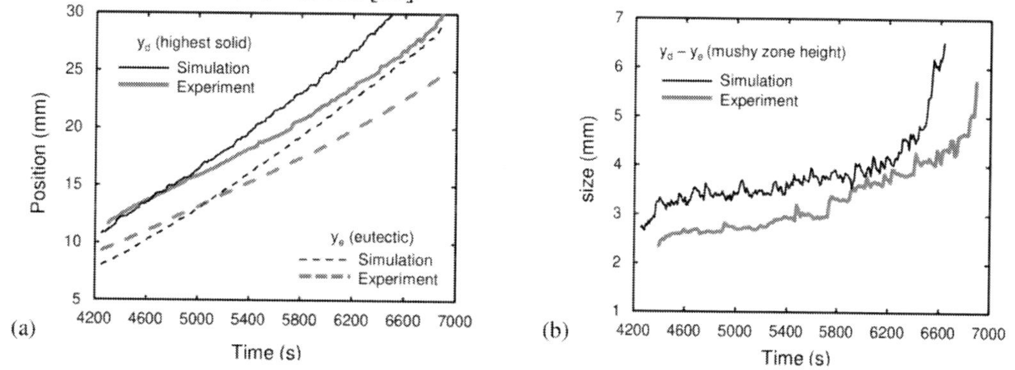

Figure 3. Measured and calculated evolutions of (a) the dendritic, y_d, and eutectic, y_e, growth fronts and (b) the mushy zone height, y_d-y_e, after increase of the pulling velocity (cf. figure 4a).

4. Results

4.1. Convective flow pattern during solidification

Even when solidification of an alloy is performed upward in stable conditions with respect to thermosolutal convection, a strong macroscopic deformation of the solidification front could be observed, initiated by the presence of a residual transverse temperature gradient and gradually amplified by solutal convection [16-18]. The occurrence of solute segregation by fluid flow is usually confirmed indirectly by observation of the solidification front (figure 4a), with a growing solid localized in the center of the sample and surrounded by eutectic after complete solidification (steepling phenomena). The convective flow pattern predicted by the numerical model is shown in figure 4c with white arrows representing the calculated fluid flow velocity field. This pattern is similar to the one previously calculated by Zhou et al. [19]. It is composed of two rolls located in the bulk liquid, on top of two other rolls located ahead of the solidifying front and penetrating into the mushy zone (figure 4d). This configuration is found to propagate upward following the progress of solidification process until the mushy zone reaches the end of the sample. Then only one roll on the left and one roll on the right remain, which disappear after complete solidification. As shown in figure 4d, the deformation of the solidification front is due to the two bottom convective rolls leading to solute accumulation on the sides of the sample. The steepling phenomenon observed experimentally is thus well reproduced and clearly visible in the simulated grain structure figure 4b. The maximum liquid velocity calculated in this region is 12 $\mu m.s^{-1}$. As expected [19] this fluid flow velocity value is larger than the actual growth rate of 6 µm.s-1 and lower than experimental value of growth velocity for which convection becomes negligible, i.e. 20 µm.s^{-1} in previous experiments performed at ESRF with the same device [20]. It is also visible in figure 4d that liquid flow is prevalent at grain boundaries, where the solid fraction is less due to solute accumulation.

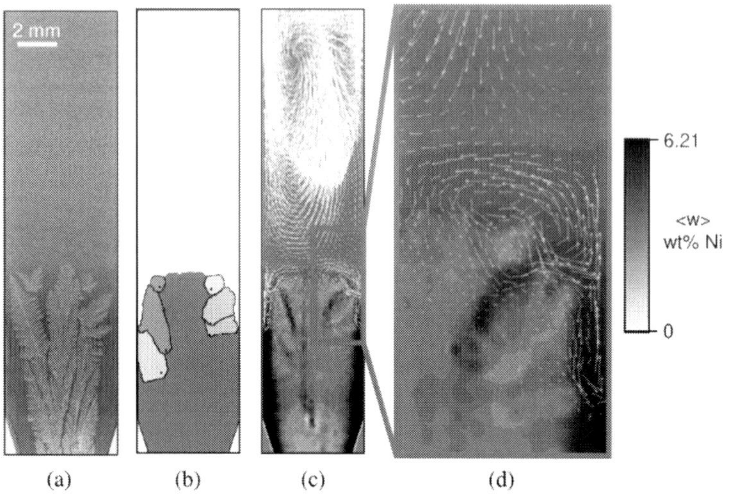

Figure 4. (a) Radiograph of the solidification front, (b) calculated grain structure and (c) composition map and fluid flow velocity field after 5200 s of pulling; the maximum velocity value in the image is approximately 135 µm s^{-1}. (d) Zoom showing the fluid flow velocity field near the solidification front in more details; the maximum velocity value in the image is approximately 12 µm s^{-1}

4.2. Eutectic fraction distribution and grain structure

The radiograph contrast is mainly linked to X-ray absorption by the chemical elements. A direct comparison can thus be made between radiography of the fully solidified sample (figure 5a) and the simulated final eutectic fraction (figure 5b) due to the fact that the eutectic fraction is also directly proportional to the local solute enrichment, i.e. to macrosegregation.

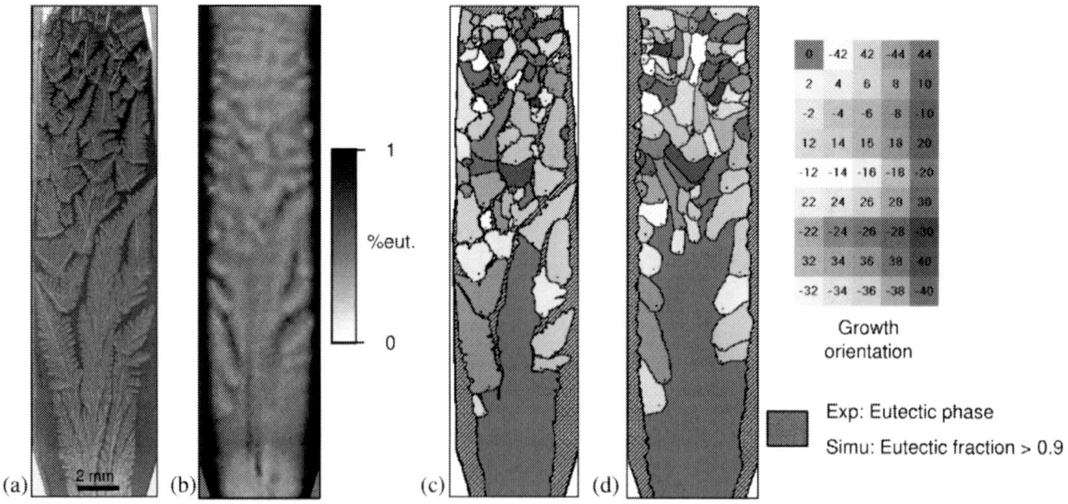

Figure 5. (a) Image of the fully solidified sample after image processing, (b) simulated final eutectic fraction, (c) experimental and (d) simulated final grain structures (color available online); the black dot in each grain indicates the position of the nucleation centre.

The distribution of inter and intragranular eutectic is partly distinguished in the simulation and similar to the experiment. Qualitatively, a higher fraction of eutectic is predicted in between the grains, as well as on the sides of the sample that correspond to the eutectic layer surrounding the grain structure observed experimentally. The last effect is made more visible in figure 5d where zones with a eutectic fraction > 0.9 are superimposed to the simulated grain structure. In the experimental images, eutectic area are clearly visible in between the grains (figure 5a and 5c), whereas this feature is not retrieved in the numerical simulation (figure 5d). This is due to the facts that (i) direct tracking of the primary eutectic structure is not included and (ii) the FE mesh size is too coarse to reproduce this observation. Finally, comparison between figures 5c and 5d also shows that the transition from a columnar to an equiaxed grain structure is remarkably well reproduced in the simulations, with the initial columnar grain blocked at the same altitude than in the experiment. Differences between experiment and simulation in grain structure in the equiaxed zone can be mainly attributed to the fact that the nucleation positions used for the simulation were chosen as the positions of the nucleation centers in the grains after sedimentation in the experiment. This choice thus leads to a different competition between the grains during their development.

5. Conclusion

A Bridgman experiment with a refined Al - 3.5 wt% Ni alloy has been thoroughly characterized by means of X-ray radiography and provides inputs for a 2D CAFE simulation. Direct simulation of the dendritic grain structure and the eutectic distribution in the entire sample accounting for fluid flow and its interaction with the growing solid has been successfully performed. Macroscopic deformation of the solidification front by convection is well reproduced, with accumulation of solute on the sides leading to a dendrite steepling phenomenon. Distribution of eutectic fraction is qualitatively retrieved and similar position for the columnar-to-equiaxed transition is obtained. The present investigation clearly reveals the interest of coupled experimental and numerical analyses in order to get a better understanding of the physical phenomena occurring during solidification processes.

Work is currently in progress to perform further analyses by comparing experimental results with numerical simulations with and without convection for getting a more precise understanding of the effect of liquid flow on macrosegregation and grain structure. The effect of nucleation undercooling will also be investigated. Simulations using 3D multiple structure tracking in CAFE modeling [11, 21] are envisaged in the future, for example to take into account the eutectic structure formation and growth.

Acknowledgements

Financial support from ESA and CNES, and technical support of the ESRF ID19 staff members is greatly acknowledged. The cellular automaton model is implemented in the 2D finite element software R2Sol developed at MINES ParisTech CEMEF.

References

[1] Reinhart G, Nguyen-Thi H, Mangelick-Noël N, Billia B, Schenk T, and Baruchel J 7-10 June 2011 *3rd Int. Conf. on Advances in Solidification Processes*, Rolduc, Netherlands.
[2] Mathiesen R H, Arnberg L, Ramsokar K, Weitkamp T, Rau C and Snigirev A 2002 *Met. Mat. Trans. B* **33B** 613–23
[3] Schenk T, Nguyen-Thi H, Gastaldi J, Reinhart G, Cristiglio V, Mangelinck-Noël N, Klein H, Härtwig J, Grushko B, Billia B and Baruchel J 2005 *J. Cryst. Growth* **275** 201–08
[4] Yasuda H, Ohnak I, Kawasaki K, Sugiyama A, Ohmichi T, Iwane J and Umetani K 2004 *J. Cryst. Growth* **262** 645–52
[5] Guillemot G, Gandin Ch-A and Bellet M 2007 *J. Cryst. Growth* **303** 58–68
[6] Delaleau P, Beckermann C, Mathiesen R H and Arnberg L 2010 *ISIJ int* **50** 1886–94
[7] Reinhart G, Mangelinck-Noël N, Nguyen-Thi H, Schenk T, Gastaldi J, Billia B, Pino P,Härtwig J and Baruchel J 2005 *Mat. Sci. Eng. A* **413-414** 384–88
[8] Nguyen-Thi H, Reinhart G, Mangelinck-Noël N, Jung H, Billia B, Schenk T, Gastaldi J, Härtwig J and Baruchel J 2007 *Met. Mat. Trans. A* **38A** 1458–64
[9] Abramoff M D, Magelhaes P J and Ram S J 2004 *Biophotonics International* **11** 36-42
[10] Gandin Ch-A 2010 Comptes Rendus Physique **11** 216-225
[11] Carozzani T, Digonnet H, Bellet M, Gandin Ch-A, MCWASP XIII (2012) this proceedings

[12] Liu W 2005 *Finite element modelling of segregation and thermomechanical phenomena in solidification processes* Ph. D. Thesis, Ecole Nationale Supérieure des Mines de Paris, Paris, F
[13] Mosbah S, Bellet M, Gandin Ch-A 2010 Metall Mater Trans **41A** 651
[14] Gandin Ch-A, Guillemot G, Appolaire B, Niane N T 2003 Mater. Sci. & Eng **A342** 44-50
[15] Dantzig J A, Rappaz M 2009 *Solidification* (Lausanne : EPFL Press) p 457
[16] Burden M H, Hebditch D J and Hunt J D 1973 *J. Cryst. Growth* **20** 121-24
[17] Nguyen Thi H, Dabo Y, Drevet B, Dupouy M D, Camel D,Billia B, Hunt J D and Chilton A 2005 *J. Cryst. Growth* **281** 654-68
[18] Bogno A, Nguyen-Thi H, Buffet A, Reinhart G, Billia B, Mangelinck-Noël N, Bergeon N, Baruchel J and Schenk T 2011 *Acta Mater.* **59** 4356-65
[19] Zhou B H, Jung H, Mangelinck-Noël N, Nguyen-Thi H, Billia B, Liu Q S and Lan W 2008 Adv. Space Res. **41** 2112-17
[20] Reinhart G 2006 *Dynamique de formation de la microstructure de solidification d'alliages métalliques: caractérisation par imagerie X synchrotron* Ph. D. Thesis, Université Paul Cézanne Aix-Marseille III, Marseille, F
[21] Carozzani T, Digonnet H, Gandin Ch-A, Modelling Simul. Mater. Sci. Eng. **20** (2012) 015010

Table 1. Values of thermo-physical data of the Al-3.5wt%Ni alloy and numerical parameters.

Parameter	Variable	Value	Unit
Nominal composition	w_0	3.5	wt%
Liquidus temperature	T_L	650	°C
Segregation coefficient	K	0.08	wt% / wt%
Liquidus slope	m_L	-3.1	wt% °C^{-1}
Eutectic temperature	T_E	641.6	°C
Gibbs-Thomson coefficient	Γ	$2.41 \cdot 10^{-7}$	°C m
Heat capacity	C_p	$2.85 \cdot 10^{-6}$	J kg^{-1} °C^{-1}
Enthalpy of fusion	$\Delta_s^l H_f$	$9.8 \cdot 10^{-8}$	J kg^{-1}
Diffusion of Ni in liquid Al	D^l	$2.2 \cdot 10^{-9}$	m^2 s^{-1}
Dynamic viscosity	μ^l	$6 \cdot 10^{-3}$	Pa s
Solutal expansion coefficient	β_w	$-8.4 \cdot 10^{-3}$	wt%$^{-1}$
Thermal expansion coefficient	β_T	$0.11 \cdot 10^{-3}$	°C^{-1}
Density	ρ_0	2450	kg m^{-3}
Thermal conductivity in the solid	κ^s	200	W m^{-1} °C^{-1}
Thermal conductivity in the liquid	κ^l	100	W m^{-1} °C^{-1}
Secondary dendrite arm spacing	λ_2	$150 \cdot 10^{-6}$	m
Primary nucleation undercooling	ΔT_n	0	°C
Location of nucleation event		Experimental input	
Cell size		$15 \cdot 10^{-6}$	m
Imposed minimum FE mesh size		$150 \cdot 10^{-6}$	m
Imposed maximum FE mesh size		$1000 \cdot 10^{-6}$	m
Objective relative error on <w>		$1 \cdot 10^{-4}$	wt%
Time step		1	s
Initial temperature		715.5	°C
Heat transfer coefficients			
Sample holder / heating elements		Perfect contact	
Sample / heating elements		Perfect contact	
Sample / sample holder		Perfect contact	
Sample / air		2000	W m^{-2} °C^{-1}
Sample holder / air		3000	W m^{-2} °C^{-1}
Heating elements / air		Adiabatic	W m^{-2} °C^{-1}

IOP Conf. Series: Materials Science and Engineering **33** (2012) 012078 doi:10.1088/1757-899X/33/1/012078

Modeling of microstructure evolution of magnesium alloy during the high pressure die casting process

Mengwu Wu[1,2] **and Shoumei Xiong**[1,2]
[1] Department of Mechanical Engineering, Tsinghua University, Beijing 100084, China
[2] State Key Laboratory of Automobile Safety and Energy, Tsinghua University, Beijing 100084, China

E-mail: smxiong@tsinghua.edu.cn

Abstract. Two important microstructure characteristics of high pressure die cast magnesium alloy are the externally solidified crystals (ESCs) and the fully divorced eutectic which form at the filling stage of the shot sleeve and at the last stage of solidification in the die cavity, respectively. Both of them have a significant influence on the mechanical properties and performance of magnesium alloy die castings. In the present paper, a numerical model based on the cellular automaton (CA) method was developed to simulate the microstructure evolution of magnesium alloy during cold-chamber high pressure die casting (HPDC) process. Modeling of dendritic growth of magnesium alloy with six-fold symmetry was achieved by defining a special neighbourhood configuration and calculating of the growth kinetics from complete solution of the transport equations. Special attention was paid to establish a nucleation model considering both of the nucleation of externally solidified crystals in the shot sleeve and the massive nucleation in the die cavity. Meanwhile, simulation of the formation of fully divorced eutectic was also taken into account in the present CA model. Validation was performed and the capability of the present model was addressed by comparing the simulated results with those obtained by experiments.

1. Introduction

Magnesium alloy is the lightest of all metals used for structural applications. With the increasing demand for light weighting, energy saving and environment protection in the automobile industry, magnesium alloy has been widely used by automobile manufacturers to replace steel, cast iron, and even aluminium alloy. Successful applications of magnesium alloy in automobile industry components include steering wheels, gear boxes, instrument panels and air intake systems, etc. [1]. The high pressure die casting (HPDC) process is a net-shape or near net-shape process with the advantages of high efficiency, considerable economic benefit and high precision of the product size, and generally, it is the dominant manufacturing process for magnesium structural applications.

The microstructure of die cast magnesium alloy has a great influence on the final performance of castings. One of the key microstructure characteristics of high pressure die cast magnesium alloy is the appearance of relatively coarse dendrites at the central region of castings. Based on experimental studies and numerical simulation of the filling sequence and solidification behaviour during the HPDC process [2-4], these coarse dendrites have been proven to originate from the nucleation and crystal growth in the melt in the shot sleeve, and they are named "Externally Solidified Crystals" or "ESCs" for short. The externally solidified crystals have a significant impact on the quality and performance of magnesium alloy die castings. For example, they obviously affect the grain size distribution, which consequently affects the mechanical properties of die castings. Meanwhile, under extremely poor conditions, for instance, the pouring temperature is too low, or the delay time of the melt in the shot sleeve is too long, cold flakes (or externally solidified products, ESP) may form and occur in the microstructure of die castings. This will severely lower the quality of the castings. Another influence of the externally solidified crystals is that they may affect the filling sequence and solidification behaviour of the melt. Based on the ESCs theory, Dahle *et al.* [5] explained the formation of defect band in the microstructure of die castings properly. Since most commercially available magnesium alloys for the HPDC process contain aluminium as the main alloying element, another microstructure characteristic of die cast magnesium alloy is the formation of fully divorced eutectic in the interdendritic liquid at the last stage of solidification in the die cavity. The constitution, morphology and distribution of the divorced eutectic in the as-cast microstructure also influence the performance of magnesium alloy die castings, especially the ductility, corrosion resistance, high temperature strength and creep resistance [6]. Accordingly, one of the key aspects of die casting technology is to investigate the effect of process parameters on the microstructure of magnesium alloy, consequently, optimize the process parameters and control the microstructure formation during the HPDC process.

In the last two decades, numerical simulation has been rapidly developed as a powerful tool for simulating and predicting the time-dependent microstructure evolution during various solidification processes. Among those deterministic and stochastic methods, the cellular automaton (CA) method has been used extensively for simulation of casting processes on both micro and macro scales with a relatively small computational load [7]. However, most of the published works using the CA method focus on the microstructure simulation of cubic metals, whereas the CA method has considerable difficulties in simulating the microstructure of magnesium alloy with hcp crystal structure. Besides, for the simulation of solidification structure of magnesium alloy during the HPDC process, the formation of externally solidified crystals and divorced eutectic ought to be considered.

In the present paper, a cellular automaton model was proposed to simulate the crystal growth of magnesium alloy. Based on the nucleation characteristics and solidification behaviour during the cold-chamber HPDC process, both of the nucleation and crystal growth in the melt in the shot sleeve and in the die cavity were taken into account. Meanwhile, the present CA model considered the formation of fully divorced eutectic at the last stage of solidification in the die cavity. The model was applied to simulate the microstructure of "cover-plate" die castings of AM60 magnesium alloy at different process parameters, and the simulated results were then compared with the experimental results.

2. Morphology and distribution of the ESCs and the divorced eutectic

During the cold-chamber HPDC process, firstly, a certain amount of melt at a certain temperature is poured into the shot sleeve. After the slow shot and fast shot phases, the melt fills the die cavity with a high velocity and pressure. Experimental and modeling studies demonstrate that when the melt is hold in the shot sleeve, especially at the slow shot phase, the melt's superheat is lost due to the impingement on the relatively cold shot sleeve wall and plunger. In this case, nucleation and crystal growth occur in the melt in the shot sleeve. At the fast shot phase, a mixture of liquid and crystals is injected into the die cavity, and during this phase, the floating crystals migrate to the central region of the die cavity due to the force of the flowing melt. Therefore, the final microstructure of die castings is not the same as commonly observed microstructure of castings with conventional casting processes. Figure 1 shows the typical microstructure at the central region of AM60 magnesium alloy die castings using OM, EBSD and SEM, respectively. It can be seen that the central region comprises a mixture of coarse dendrites and fine grains. The size of the basal fine grains is about 5~30 μm, while the equivalent grain size of the coarse dendrites is up to 100 μm. Based on the fact that the melt in the die cavity solidifies with a high cooling rate and a short solidification time, the coarse dendrites should not nucleate and grow in the melt in the die cavity. This could also help to validate the theory that the coarse dendrites in the microstructure of die castings originate from the nucleation and crystal growth in the melt in the shot sleeve.

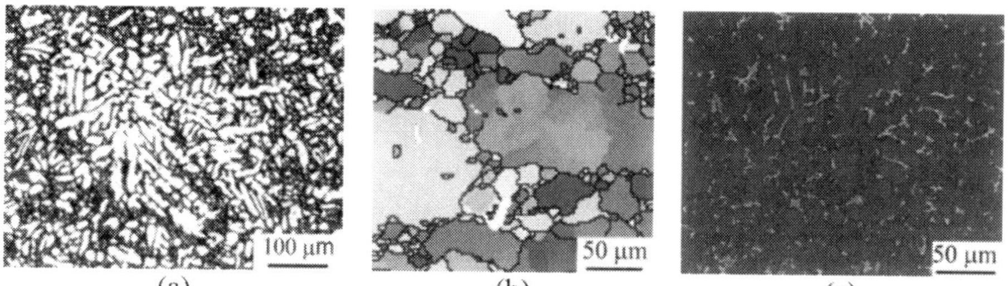

(a) (b) (c)

Figure 1. Typical microstructure at the central region of AM60 magnesium alloy die castings using OM (a), EBSD (b) and SEM (c).

Figure 2 shows the schematic diagram of a "cover-plate" die casting associated with the experimental observation and quantitative statistics of the ESCs distribution over cross section "A". A detailed description of the dimensions of the "cover-plate" die casting and the process parameters could be found in reference [8]. During the statistics of the area fraction distribution of the ESCs by using an image analysis software, those crystals which grain size was more or less equal to or larger than twice the average grain size were treated as ESCs. It can be noted from figures 2(b) and 2(c) that since the ESCs migrate to the central region of the castings during the fast shot phase of the melt into the die cavity, the percentage of the ESCs is higher at the central region than the surface layer of the castings. Meanwhile, the area fraction of the ESCs along the thickness of the die casting satisfies a Gaussian distribution, while its value at the central region may be up to 25%. Based on experimental studies [3, 4], the percentage of the ESCs is higher at the area near the gate than that of far from the gate of the castings. Besides, the percentage of the ESCs increases when decreasing the pouring temperature and the slow shot phase plunger velocity. At a lower fast shot phase plunger velocity, coarse ESCs are more easily to be observed at the cross section of the castings. In addition, the percentage of the ESCs in the die castings may also be influenced by the dimensions of the shot sleeve and the filling fraction of the melt in the shot sleeve.

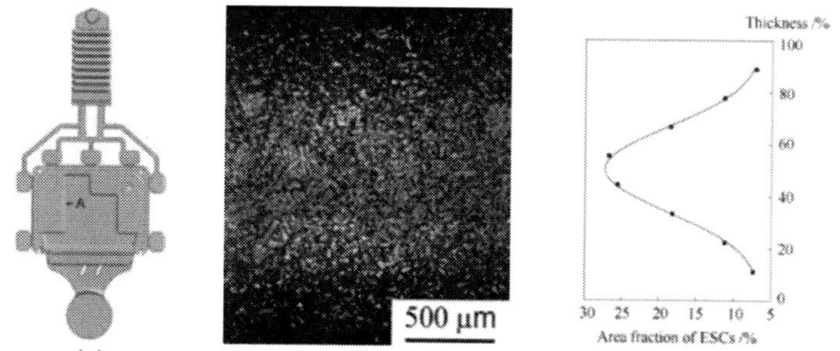

(a) (b) (c)

Figure 2. Schematic diagram of the "cover-plate" die casting (a) associated with the experimental observation (b) and quantitative statistics (c) of the ESCs distribution over cross section "A".

Since the solidification of the melt in the die cavity during the HPDC process is far away from equilibrium state, fully divorced eutectic appears in the as-cast microstructure of Mg-Al alloy as shown in figure 1(c). In other words, the eutectic α-Mg nucleates and grows attached to the primary α-Mg, while the eutectic β-$Mg_{17}Al_{12}$ nucleates and grows independently. According to the authors' previous work, islands and networks of eutectic are observed at the central region of AM60 magnesium alloy die castings, while the eutectic reveals a more dispersed and granular morphology at the surface layer.

3. Model description

Simulation of the microstructure evolution during the solidification process involves modeling of the nucleation and crystal growth at the grain-size scale and even at the atomic scale. One of the key aspects relates to the establishment of the nucleation model and also the crystal growth model suitable for the solidification process. In the present work, as for the solidification of the primary α-Mg, a nucleation model was proposed which was divided into two stages based on the nucleation characteristics and sequence during the cold-chamber HPDC process. When the melt is still in the shot sleeve, it starts to solidify in a manner resembling gravity pour permanent mould casting. In this case, the continuous nucleation model with a Gaussian distribution of nuclei is adopted [9]. When the melt enters the die cavity, enormous nucleation and crystal growth occurs in the bulk liquid metal. As for the surface layer of castings, or castings with thin thickness, the cooling rate during solidification may be as high as 1000 K/s. Such high cooling rate will induce the formation of free chill crystals. Based on experimental statistics and data fitting, the nucleation density and cooling rate satisfy a linear relationship. As for the central region of castings with thick thickness, the cooling rate during solidification is much lower due to the weakening of heat transfer and the release of latent heat. In such case, the nucleation density and cooling rate satisfy an exponential relationship.

Based on the fact that Mg has a hcp crystal structure, the dendrites of commercially available magnesium alloys grow with six-fold symmetry. In the present paper, based on the CA method, a growth model was established in which the growth kinetics was calculated from the complete solution of the transport equations. By defining a special neighbourhood configuration with the square CA cell, and using a set of capturing rules which was proposed by Beltran-Sanchez and Stefanescu for the dendritic growth of cubic crystal metals during solidification [10], modeling of dendritic growth of magnesium alloy with six-fold symmetry and different growth orientations was achieved. More details relating the nucleation model and dendritic growth model, as well as assumptions and simplifications for simulation of the solidification of the primary α-Mg during the HPDC process can be found in references [8, 11].

As for simulation of the formation of fully divorced eutectic at the last stage of solidification in the interdendritic liquid in the die cavity, firstly, a nucleation model proposed by Charbon and LeSar was adopted to simulate the eutectic nucleation [12]. Each eutectic nucleus is generated by two CA cells, one of which represents a nucleus of eutectic α-Mg, and the other represents a nucleus of eutectic β-$Mg_{17}Al_{12}$. Based on the nucleation and growth mechanism of fully divorced eutectic, the eutectic α-Mg nucleates attached to the primary α-Mg, while the eutectic β-$Mg_{17}Al_{12}$ nucleates at a certain distance from the eutectic α-Mg in the interdendritic liquid. Once the eutectic α-Mg and β-$Mg_{17}Al_{12}$ have nucleated, the cooperative and competitive growth of the two phases begins. A detailed description regarding the growth kinetics and growth algorithms of the two eutectic phases of Mg-Al alloy could be found in the authors' previous work [13].

4. Simulation results and discussion

To validate the capability of the present CA model, simulations were carried out to predict the microstructure of "cover-plate" die castings of AM60 magnesium alloy at different process parameters, and then the simulated results were compared with the experimental results. The standard process parameters during the "cover-plate" die casting experiments were as follows: pouring temperature (973 K), mould temperature (453 K), casting pressure (77 MPa), slow shot phase plunger velocity (0.15 m/s), fast shot phase plunger velocity (4.0 m/s), and filling fraction of the shot sleeve (18.3%).

4.1. Modeling of the solidification structure of primary α-Mg considering the ESCs in the shot sleeve

During simulation of the nucleation and dendritic growth of the primary α-Mg, the software Anycasting was used firstly to calculate the temperature field of the melt in the shot sleeve. After the slow shot phase, simulation of the solidification of the melt was switched from in the shot sleeve to in the die cavity, since the time of the fast shot phase was extremely short and generally at the magnitude of dozens of milliseconds. Meanwhile, in the present CA model, remelting and fragmentation of the dendrites during the fast shot phase of the melt into the die cavity was not taken into account. The temperature field of the melt in the die cavity was obtained using an inverse heat transfer model [11]. Figure 3 shows the simulated microstructure evolution of AM60 magnesium alloy at the central region of cross section "A" of the "cover-plate" die casting under the standard process parameters. The calculation domain consists of 570×430 cells with a cell size of 1.0 μm. It can be seen that the microstructure with a mixture of coarse ESCs and fine grains at the central region of the die casting was acquired via simulation. With an equivalent grain size of the ESCs up to 100 μm, the area fraction of the ESCs in figure 3(c) is about 24.5%.

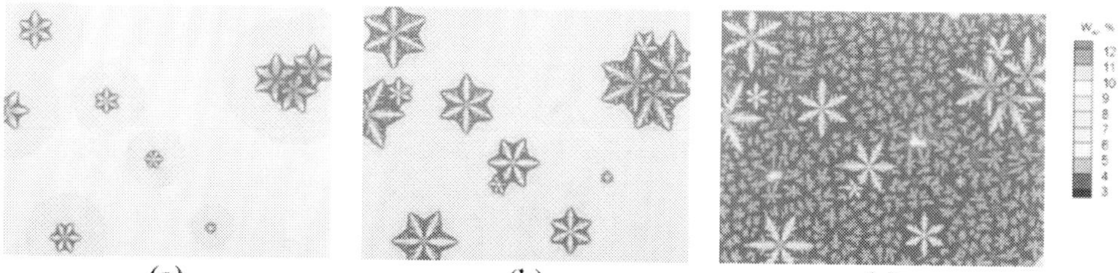

(a) (b) (c)

Figure 3. Simulated microstructure evolution of AM60 magnesium alloy at the central region of cross section "A" of the "cover-plate" die casting under the standard process parameters: the melt is still in the shot sleeve in figures 3(a) and 3(b), whereas it is in the die cavity in figure 3(c).

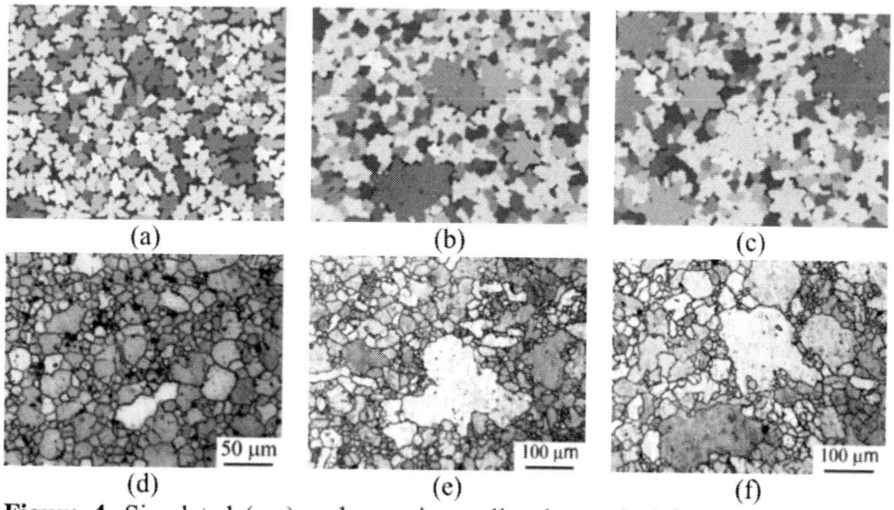

Figure 4. Simulated (a-c) and experimentally observed (d-f) microstructure of AM60 magnesium alloy at the surface layer (a, d), one-fourth thickness (b, e) and the central region (c, f) of cross section "A" of the "cover-plate" die casting under the standard process parameters.

Figure 4 shows the simulated and experimentally observed microstructure of AM60 magnesium alloy at the surface layer, one-fourth thickness and the central region of cross section "A" of the "cover-plate" die casting under the standard process parameters. The calculation domain was divided into 570×430 cells with a cell size of 0.5 μm and 1.0 μm, respectively in figure 4(a) and figures 4(b-c). It can be seen from figure 4(a) and 4(d) that the surface layer comprises uniformly fine grains. The simulated average grain size is about 14.4 μm, that's very close to the experimentally measured value 14.7 μm. Comparing figures 4(b-c) with figures 4(e-f), it also can be illustrated that the simulated results agree well with the experimental results relating the grain size and the area fraction of the ESCs.

At the one-fourth thickness and the central region of cross section "A", the simulated area fractions of the ESCs are 16.1% and 24.5%, while the experimentally measured are 16.8% and 26.7%, respectively.

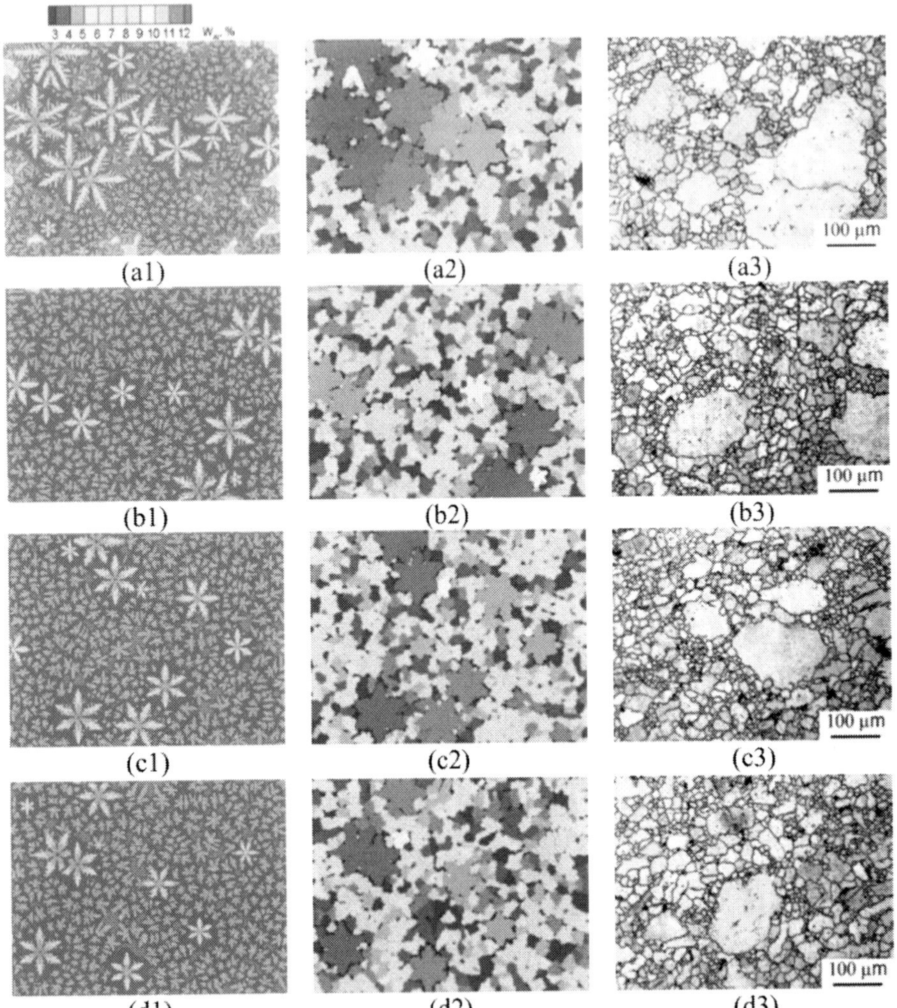

(a1) (a2) (a3)
(b1) (b2) (b3)
(c1) (c2) (c3)
(d1) (d2) (d3)

Figure 5. Simulated solute map (a1-d1), grain size distribution (a2-d2) and experimentally observed grain size distribution (a3-d3) of AM60 magnesium alloy at the central region of cross section "A" of the "cover-plate" die casting under different pouring temperatures (T_p) and slow shot phase plunger velocities (V_s): (a1-a3) T_p = 953 K, V_s = 0.15 m/s (b1-b3) T_p = 993 K, V_s = 0.15 m/s (c1-c3) T_p = 973 K, V_s = 0.3 m/s (d1-d3) T_p = 973 K, V_s = 0.4 m/s.

Simulations were further conducted to predict the microstructure of AM60 magnesium alloy at the central region of cross section "A" of the "cover-plate" die castings at different pouring temperatures (953, 973 and 993 K) and slow shot phase plunger velocities (0.15, 0.3 and 0.4 m/s). As shown in figure 5, and in conjunction with figures 4(c) and 4(f), it can be concluded that the simulated results are consistent with the experimental results. With a decrease of the pouring temperature, the percentage of the ESCs increases. Under the three different pouring temperatures of 993, 973 and 953 K, the simulated area fractions of the ESCs are 19.2%, 24.5% and 33.5%, while the experimentally measured are 19.3%, 26.7% and 32.1%, respectively. Meanwhile, the percentage of the ESCs increases when decreasing the slow shot phase plunger velocity. The reason for this is that the holding time of the melt in the shot sleeve would be longer in such case, which results in a more severely loss of the superheat of the melt then. Under the three different slow shot phase plunger velocities of 0.4, 0.3 and 0.15 m/s, the simulated and experimentally measured area fractions of the ESCs are 14.8%, 18.1%, 24.5% and 14.5%, 17.2%, 26.7%, respectively.

4.2. Modeling of the formation of fully divorced eutectic

When the temperature of the melt in the die cavity goes below the eutectic temperature, and the solute concentration in the interdendritic liquid is up to the eutectic composition, eutectic nucleation and growth begins. Figure 6 shows the simulated and experimentally observed eutectic morphology, as well as distribution at the central region of AM60 magnesium alloy die castings. The calculation domain consists of 570×430 cells with a cell size of 1.0 μm. Firstly, it can be seen from the amplified view of the rectangle region in figure 6(c) that the eutectic α-Mg indeed nucleates and grows attached to the primary α-Mg, while the eutectic β-$Mg_{17}Al_{12}$ nucleates and grows independently at the early stage of eutectic solidification. As the solidification proceeds, the two eutectic phases impinge on each other, and then the cooperative and competitive growth of the two phases begins. Comparing figure 6(c) with 6(d), it can be noted that the simulated result agrees well with the experimental result relating the morphology and distribution of the eutectic, which grows into the form of islands/networks at the grain boundary of the primary α-Mg.

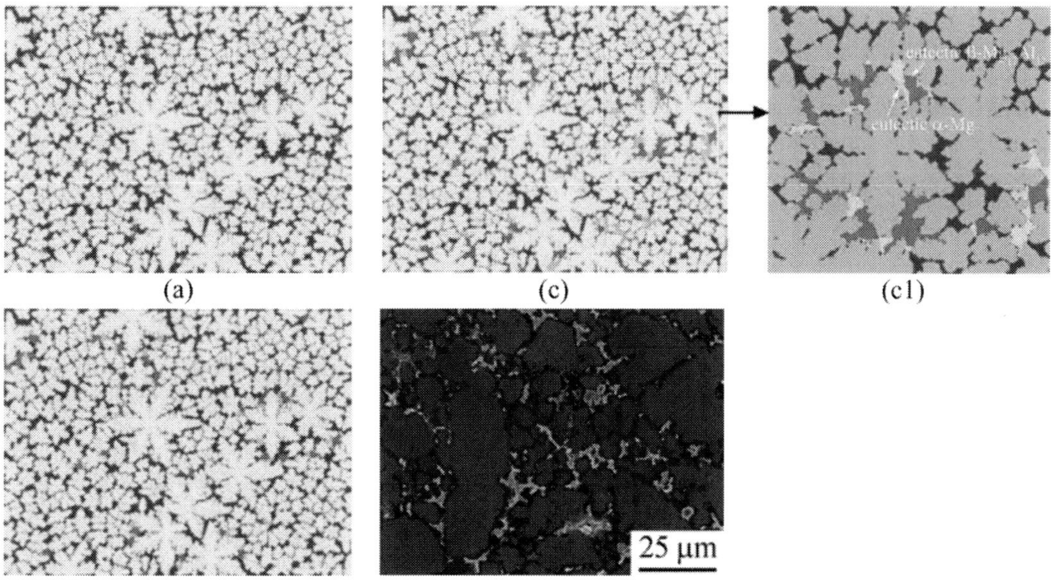

Figure 6. Simulated microstructure evolution of the eutectic (a-c) and experimentally observed eutectic morphology (d) at the central region of AM60 magnesium alloy die castings: figure 6(c1) shows the amplified view of the rectangle region in figure 6(c).

5. Conclusions

A cellular automaton model was developed to simulate the microstructure evolution of magnesium alloy during the cold-chamber HPDC process. Based on the solidification behaviour and sequence during the HPDC process, both of the nucleation and crystal growth of the primary α-Mg in the melt in the shot sleeve and in the die cavity, as well as the formation of the fully divorced eutectic at the last stage of solidification in the interdendritic liquid in the die cavity were taken into account in the present CA model. During simulation of the microstructure of a "cover-plate" die casting of AM60 magnesium alloy, it can be found that the percentage of the ESCs is higher at the central region than that at the surface layer of the die castings. With a decrease of both of the pouring temperature and slow shot phase plunger velocity, the percentage of the ESCs increases in the die castings. Meanwhile, the fully divorced eutectic grows into the form of islands/networks at the central region of the die castings. The simulated results are in accordance with the experimental results.

Acknowledgments

This work was financially supported by the Ministry of Science and Technology of China under grant Nos. 2011BAE21B00, 2011ZX04001-071, 2011ZX04014-052 and 2010DFA72760. The die casting experiments were conducted with the help of engineers from Toyo Machinery & Metal Co., Ltd.

References

[1] Friedrich H and Schumann S 2001 *J. Mater. Process. Technol.* **117** 276
[2] Helenius R, Lohne O, Arnberg L and Laukli H I 2005 *Mater. Sci. Eng.* A **413-414** 52
[3] Laukli H I, Lohne O, Sannes S, Gjestland H and Arnberg L 2003 *Int. J. Cast Met. Res.* **16** 515
[4] Wu M W and Xiong S M 2011 *Acta Metall. Sin.* **47** 528
[5] Dahle A K, Sannes S, StJohn D H and Westengen H 2001 *J. Light Met.* **1** 99
[6] Cao H and Wessén M 2004 *Metall. Mater. Trans.* A **35** 309
[7] Boettinger W J, Coriell S R, Greer A L, Karma A, Kurz W, Rappaz M and Trivedi R 2000 *Acta Mater.* **48** 43
[8] Wu M W and Xiong S M 2011 *Proc. 11th Asian Foundry Congress (Guangzhou)* vol 2 (Li R D: Shenyang/Foundry Institution of Chinese Mechanical Engineering Society) p 139
[9] Thévoz P, Desbiolles J L and Rappaz M 1989 *Metall. Trans.* A **20** 311
[10] Beltran-Sanchez L and Stefanescu D M 2002 *Int. J. Cast Met. Res.* **15** 251
[11] Wu M W and Xiong S M 2010 *Acta Metall. Sin.* **46** 1534
[12] Charbon C and LeSar R 1997 *Model. Simu. Mater. Sci. Eng.* **5** 53
[13] Xiong S M and Wu M W 2011 Experimental and modeling studies of the lamellar eutectic growth of Mg-Al alloy *Preprint* 10.1007/s11661-011-0831-8

Multiscale modeling of the solidification microstructure evolution in the presence of ultrasonic stirring

Laurentiu Nastac

THE UNIVERSITY OF ALABAMA, Department of Metallurgical and Materials Engineering, P. O. Box 870202, Tuscaloosa, AL, 35487-0202, USA

E-mail: lnastac@eng.ua.edu

Abstract. Ultrasonic treatment (UST) was studied to improve the quality of cast ingots as well as to control the solidification microstructure evolution. Ultrasonically-induced cavitation consists of the formation of small cavities (bubbles) in the molten metal followed by their growth, pulsation and collapse. These cavities are created by the tensile stresses that are produced by acoustic waves in the rarefaction phase. The cavitation threshold pressure for nucleation of the bubbles may decrease with increasing the amount of dissolved gases and especially with the amount of inclusions in the melt. A UST model was developed to predict the ultrasonic cavitation and acoustic streaming. The developed UST modeling approach is based on the numerical solution of Lilley model (that is founded on Lighthills's acoustic analogy), fluid flow, and heat transfer equations, and mesoscopic modeling of the grain structure. The UST model was applied to study the solidification of Al-based alloys) under the presence of ultrasound. It is found that the predicted ultrasonic cavitation region is relatively small, the acoustic streaming is strong and thus the created/survived bubbles/nuclei are transported into the bulk liquid quickly. The predicted grain size under UST condition is at least one order of magnitude lower than that without UST, which is in excellent agreement with the experimental data.

1. Introduction

Ultrasound technology is a powerful technological tool to affect a material. Ultrasonic vibrations can significantly affect heat and mass transfer in liquids, modify the structure and properties of solids and interfere with the interactions of solids and liquids. The primary causes are due to ultrasonic cavitation, acoustic streaming and movement of dislocations associated with propagation of ultrasound waves in media. Ultrasonically-induced cavitation consists of the formation of small cavities (bubbles) in the molten metal followed by their growth, pulsation and collapse. These cavities are created by the tensile stresses that are produced by acoustic waves in the rarefaction phase [1].

Relevant applications include structural refinements of solidifying metals with applications to continuous casting as well as to remelting processes (vacuum arc remelting (VAR), electroslag remelting (ESR), electron beam remelting (EBR) and plasma arc melting (PAM)) for Ti alloys, specialty steels and superalloys. Additional potential applications include die-casting, semi-solid metal processing, and forging of high fractions of fine spherical grains. Some critical technologies that have immediate application to manufacturing industry are shown in [1-7].

The objective of this work is to develop an ultrasonic modeling tool to assist in the scale up of the following processes: (i) An ultrasonic technology to significantly refine the solidification microstructure of castings and ingots and (ii) A nanotechnology-assisted by ultrasonic cavitation to manufacture metal-matrix-nano-composite (MMNC) castings with enhanced mechanical properties. This paper discusses the ultrasonic model development effort and simulation results for an ingot casting process.

2. Modeling approach for analysis of grain refinement with ultrasound

Fully couple energy, mass, acoustic, and compressible fluid flow models were used for this ultrasonic analysis. Several acoustic models [8] were investigated including Fows Williams and Hawkings Model, Broadband Noise Source Model (Proudman's and more recently Lilley model based on Lighthills's acoustic analogy), Jet Noise Source Model (by Goldstein and Ribner) and The Boundary Layer noise source Model. It was determined that Lilley model based on Lighthills's acoustic analogy is the most adequate for this work. Figure 1 shows the geometry of the liquid pool. Since time-domain computations were used for resolving the acoustic field, a very small time step of the order of 1e-07 s was used. The ultrasonic probe has a diameter of 20 mm, an amplitude, A = 10 microns, and a resonant frequency, f = 17.5 kHz.

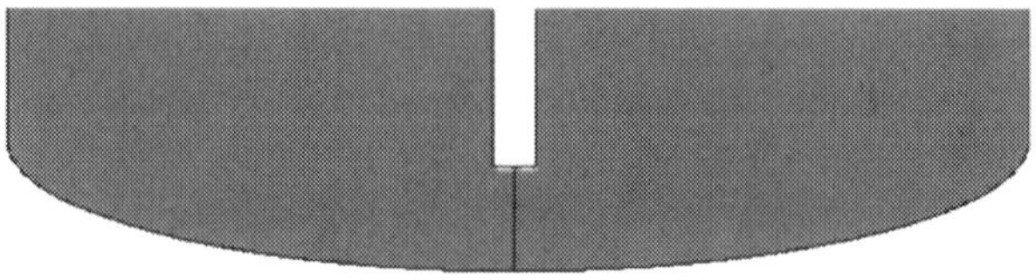

Figure 1. A356 liquid pool (the ultrasound probe is shown in the middle of the pool).

In the current multi-phase ultrasonic cavitation modeling approach, a basic two-phase cavitation model is used that consists of using the standard viscous flow equations governing the transport of phases (Eulerian multi-phase) and the k-ε turbulence model. In cavitation, the liquid-bubble mass transfer is governed by the cavity (bubble) transport equation:

$$\frac{\partial}{\partial t}(f_b \rho_b) + \nabla \cdot (f_b \rho_b \vec{V}_b) = R_G - R_C \tag{1}$$

where b subscript denotes the cavitation bubble phase, f_b is the bubble volume fraction, ρ_b is the bubble density, $\vec{V}_b$ is the bubble phase velocity, R_G and R_C are the mass transfer source terms related to the growth and collapse of the cavitation bubbles, respectively.

In equation (1), the interphase mass transfer rates per unit volume (R_G and R_C) account for the liquid and bubble phases in cavitation. They are calculated using the growth of a single bubble based on the Rayleigh-Pleasset model [9]. The model assumes no barrier for nucleation; thus, the bubble dynamics can be obtained from the general Rayleigh-Plesset equation as follows [9]:

$$R_b \frac{d^2 R_b}{dt^2} + \frac{3}{2}\left(\frac{dR_b}{dt}\right)^2 = \frac{p_b - p}{\rho_L} - \frac{2\sigma_L}{\rho_L R_b} - \frac{4\nu_L}{R_b}\frac{dR_b}{dt} \tag{2}$$

where R_b is the bubble radius, σ_L is the surface tension coefficient of the liquid phase, ρ_L is the liquid density, ν_L is the kinematic viscosity of the liquid phase, p_b is the bubble surface pressure, and p is the local far-field pressure.

3. Simulation results

3.1. Cavitation

The predictions of the flow, pressure and ultrasonic cavitation are presented in figures 2-7 for an A356 alloy at two time sequences: 8.8e-06 s and 2e-05 s. As illustrated in figure 7a, time t = 8.8e-06 s corresponds to the beginning of the cavitation event. The cavitation region will increase with time (see figure 7b, t = 2e-05s). Though the predicted ultrasonic cavitation region in figure 7 is relatively small, the acoustic streaming is strong (see figures 2-5) and thus the created/survived bubbles/nuclei are transported into the bulk liquid quickly.

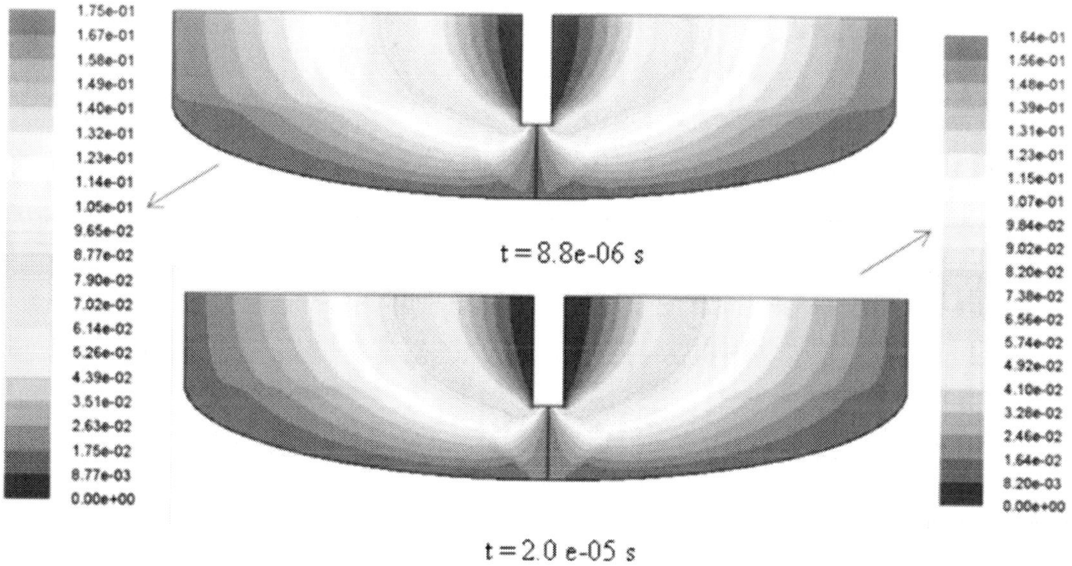

Figure 2. Flow induced by ultrasound: Flow field (streamline, in kg/s).

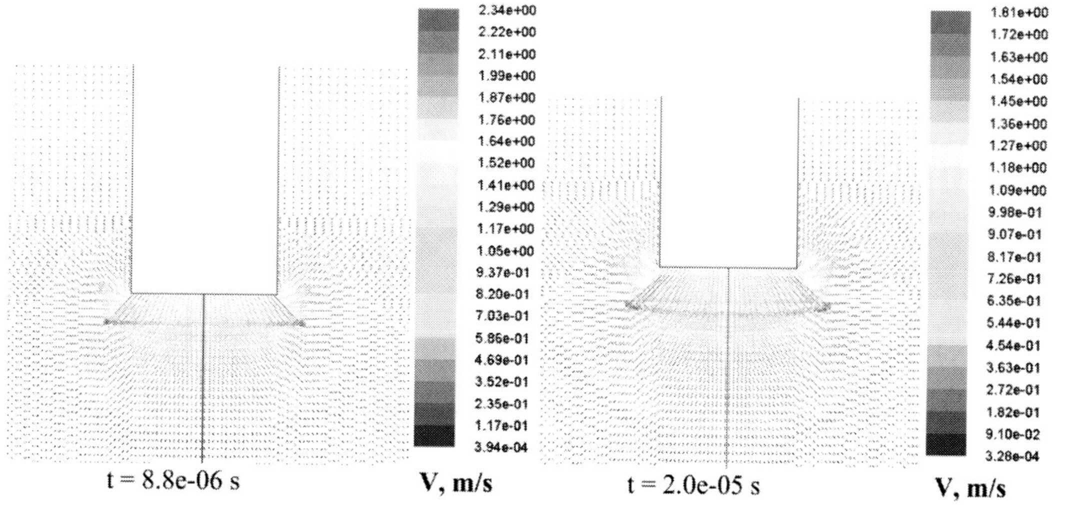

Figure 3. Flow induced by ultrasound: Flow field (velocity vector).

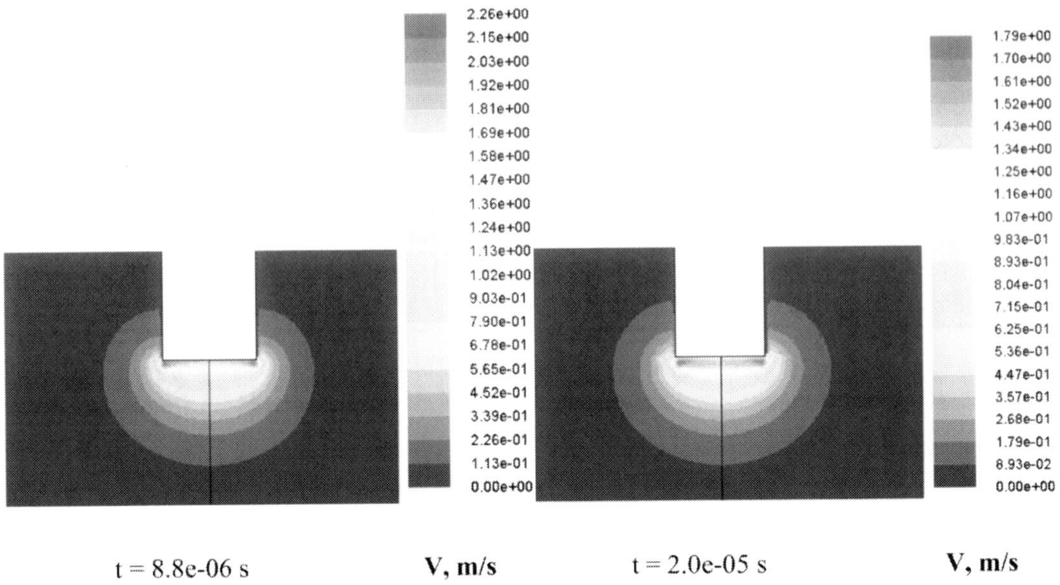

Figure 4. Flow induced by ultrasound: Flow field (velocity contour).

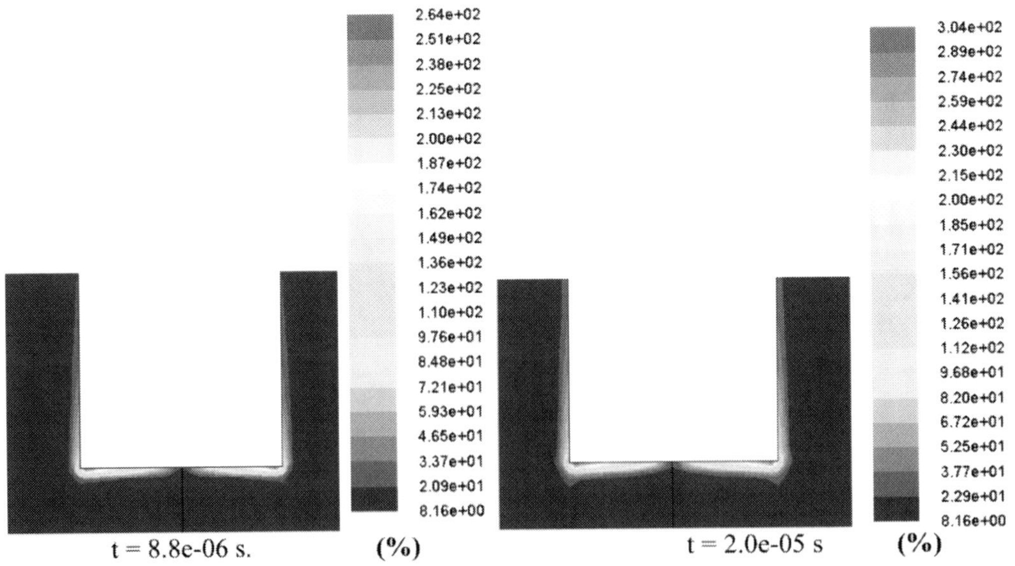

Figure 5. Flow induced by ultrasound: turbulent intensity (%).

Also, the pressure for nucleation of the bubbles (cavitation threshold pressure or P_c) in A356 alloy (see fiigure 6) is of the same order of magnitude as the experimentally determined pressures to nucleate gas bubbles in other metallic systems (see reference [1], p. 123 and reference [10]). Note also that the cavitation threshold may decrease with increasing the amount of dissolved gases (such as Hydrogen in A356) and especially with the amount of inclusions in the melt (such as Al_2O_3 in A356) (see figure 2.39 in reference [1]).

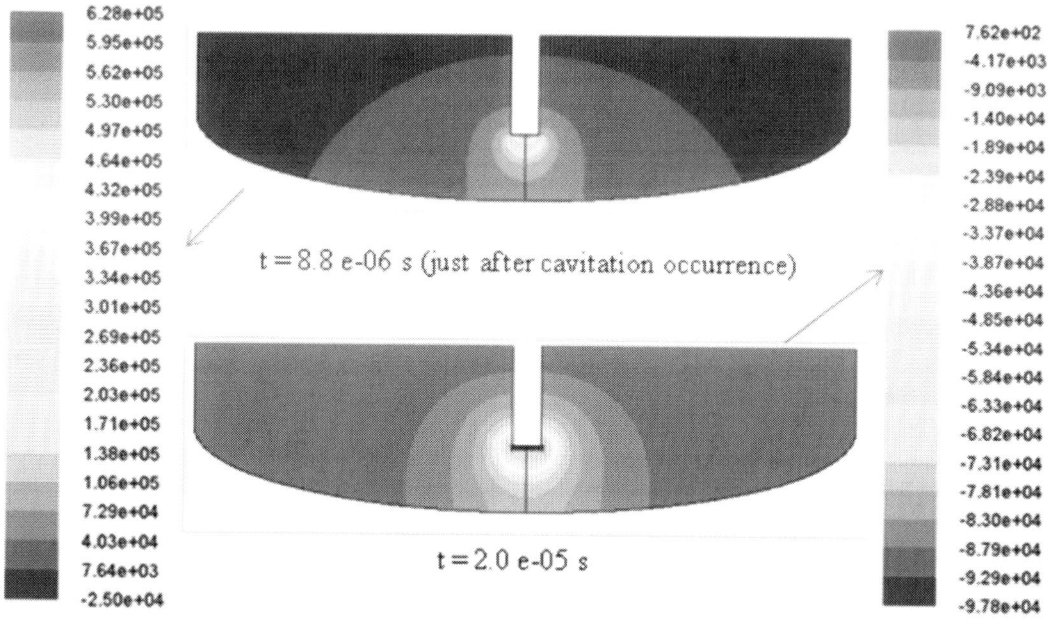

Figure 6. Flow induced by ultrasound: Pressure contour (in Pa).

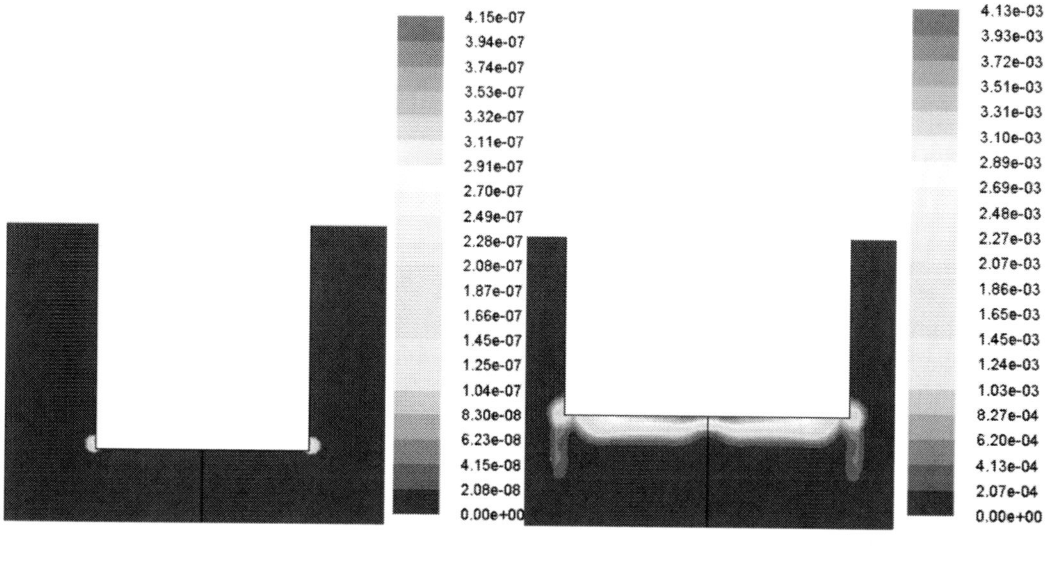

Figure 7. Cavitation induced by ultrasound. Legends show the volume fraction of cavities/bubbles/potential nuclei (cavitation region).

3.2. Prediction of the solidification structure in the presence of ultrasound

The mesoscopic model developed previously in [11, 12] is adapted to include the effect of UST on solidification structure evolution and therefore on grain refinement of alloys. The mesoscopic model includes computations of the grain size and columnar-to-equiaxed transition (CET), as well as of segregation. It also includes computations at the dendrite tip length scale (mesoscopic scale), for prediction of dendritic morphology and microsegregation patterns.

Table 1 presents the modeling data used in the microstructure modeling of alloy A356. Equation 4 in reference [13] was used to estimate the number of nuclei in the presence of UST. It was assumed in equation 4 [13] that the cavitation-induced heterogeneous nucleation is the main mechanism for grain refinement. This is also in accordance with the work reported in reference [6], that is, the cavitation-induced heterogeneous nucleation plays a more important role than the "sluggish" dendrite fragmentation process in the formation of equiaxed grains during solidification of A356 alloy castings.

The predicted solidification structure is shown in figure 8, where G_E is the equiaxed grain density and the legend illustrates the grain crystallographic orientation. The simulation of the microstructure shown in figure 8 was performed assuming a steel mold of 25 mm wall thickness. In these computations, the pouring temperature of the molten A356 alloy was 625°C and the initial temperature of the mold was 20 °C. As it can be seen from figure 8, the predicted grain size under UST condition is approximately one order of magnitude lower than that without UST. The predicted average grain size in figure 8b, which is about 200 μm, is in excellent agreement with the experimental results presented in reference [6]. This validates the mesoscopic solidification model.

Table 1. Data Used in Microstructure Modeling of Alloy A356

Solidification Kinetics Property	A356
Liquid diffusivity, D_L [m^2 s^{-1}]	3x10^{-9}
Solid diffusivity, D_S [m^2 s^{-1}]	1x10^{-12}
Liquidus slope [K wt. %$^{-1}$]	m_L = -6.5
Initial Si concentration [wt. %]	C_o = 7.0
Si partition coefficient	k = 0.14
Eutectic [wt. %]	12.6
Gibbs-Thomson coefficient, Γ [K m]	0.9 x 10^{-7}

4. Concluding remarks and future work

An approach for modeling the effects of ultrasonic vibration on the ingot structure was developed. This modeling approach was implemented into Ansys's Fluent, a commercial CFD software [14]. It was shown that the UST is a promising technology that can be used to improve the solidification structure and consequently to increase the mechanical properties of ingots and castings made of steel, Ti, Al, Mg, superalloys, etc.

Well-defined experiments on ultrasonic stirring of A356 alloys are being conducted at the University of Alabama, Solidification Laboratory, to advance the current understanding of the influence of ultrasonic cavitation and acoustic streaming on the microstructure evolution of solidifying alloys. Also, the present UST model is currently used to assist in the development and optimization of cast MMNCs. The UST modeling approach for the A356-MMNC system was introduced in [15].

The developed UST modeling approach will be used to study the globular/dendritic transition in alloys that are solidifying in the presence of ultrasonic stirring. Also, additional R&D work will be performed at the University of Alabama to further develop, refine and integrate the developed UST and microstructure models into commercial modeling and simulation software packages.

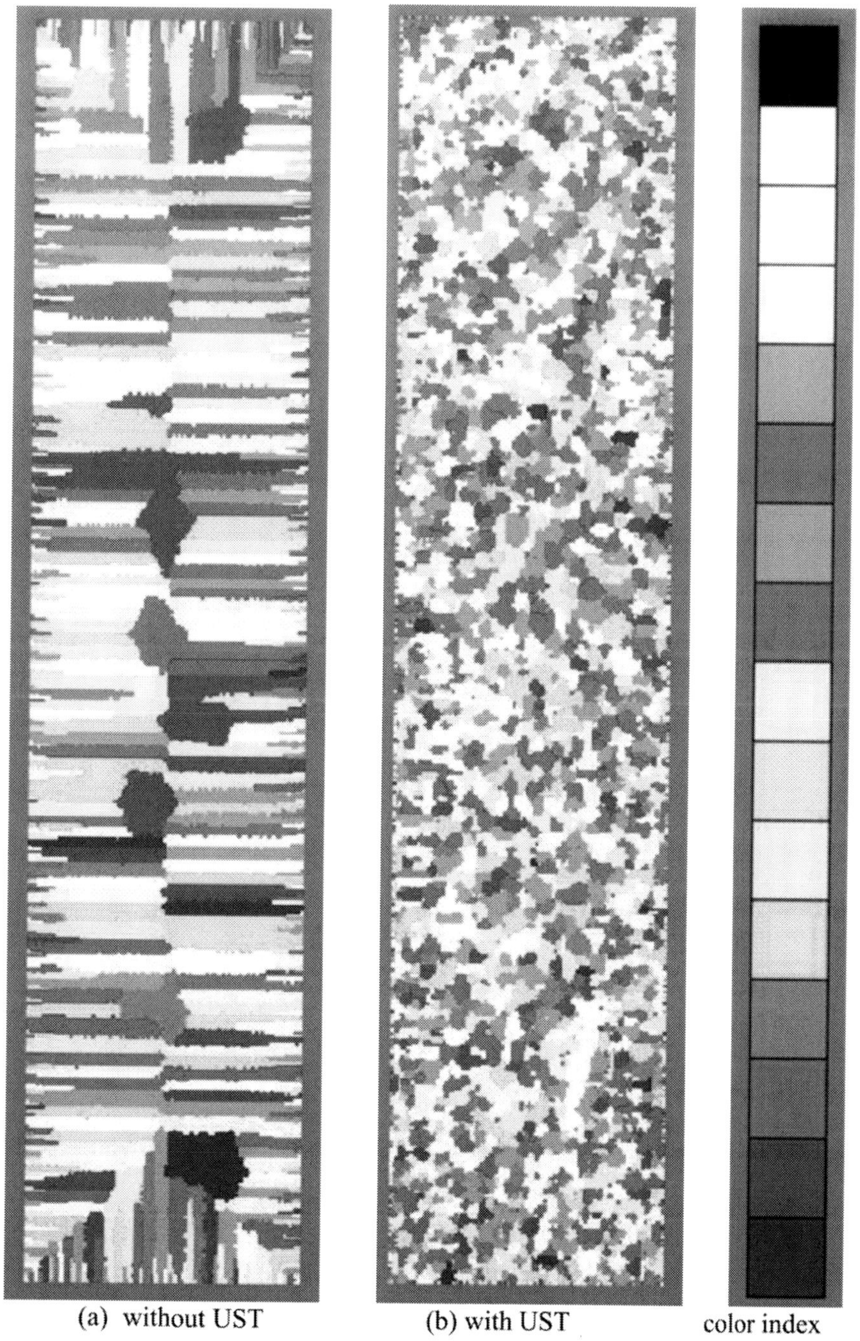

Figure 8. Ultrasonic treatment: prediction of the solidification structure for an A356 alloy (2 mm x 9 mm test geometry): (a) $G_E = 5 \times 10^7$ nuclei/m^2 and (b) $G_E = 5 \times 10^9$ nuclei/m^2.

IOP Conf. Series: Materials Science and Engineering **33** (2012) 012079 doi:10.1088/1757-899X/33/1/012079

References

[1] Abramov O V 1998 *High-Intensity Ultrasonics Theory and Industrial Applications* Gordon and Breach Science Publishers
[2] Eskin G I 1998 *Ultrasonic Treatment of Light alloy Melts* Gordon and Breach Science Publishers
[3] Rosenberg L D ed. 1969 *Sources of high-intensity ultrasound* vols. 1-2 New York: Plenum Press
[4] Campbell J 1981 *International Metals Reviews* **26** 71-108
[5] Jian X, Meek T T, and Han Q 2006 *Scripta Materialia* **54** 893-896
[6] Jian X, Xu H, Meek T T, and Han Q 2005 *Material Letters* **59** 190-193
[7] Meek T T and Han Q 2006 *Ultrasonic Processing of Materials* Final DOE Technical report, Oak Ridge National Laboratory
[8] Fluent 6.3: 2006 *User's Guide Manual* Fluent Inc.
[9] Brennen C E 1995 *Cavitation and Bubble Dynamics* Oxford University Press 51
[10] El Kaddah N H and Robertson D G C 1977 *Journal of Coloids and Interface Science* **60** 2 349-360
[11] Nastac L 2004 *Modeling and Simulation of Microstructure Evolution in Solidifying Alloys* Springer: http://www.springer.com/materials/special+types/book/978-1-4020-7831-6
[12] Nastac L 1999 *Acta Materialia* **47** 17 4253–4262
[13] Nastac L 2011 *Metallurgical and Materials Transactions B* **24** 1297-1305
[14] Fluent: http://ansys.com/Products/Simulation+Technology/Fluid+Dynamics/ANSYS+FLUENT
[15] Nastac L and Dai Y 2008 Ultrasonic Model Development and Applications to Ingot Casting Processes *Proc. 2008 International Ansys Conference (Pittsburgh, PA, August 25-28, 2008)*

IOP Conf. Series: Materials Science and Engineering **33** (2012) 012080 doi:10.1088/1757-899X/33/1/012080

Numerical simulation on austenitization of cast steel during heating process

B Su,[1] Z Q Han,[1] B C Liu,[1] Y R Zhao,[2] B Z Shen[2] and L Z Zhang[2]

[1] Key Laboratory for Advanced Materials Processing Technology (Ministry of Education), Department of Mechanical Engineering, Tsinghua University, Beijing 100084, China
[2] CITIC Heavy Industries Co., Ltd., Henan 471039, China

E-mail: zqhan@mail.tsinghua.edu.cn

Abstract. A cellular automaton model has been developed to simulate the austenitization process of ASTM A216 WCA cast steel during heating process. The dissolution of pearlite and the transformation of ferrite into austenite were simulated. The calculation domain was divided into square cells, which are characterized by certain attributes that represent the status of each cell: pearlite (P), ferrite (α), austenite (γ) or γ/α interface. The dissolution of pearlite was described by nucleation and growth of austenite. A mixed-mode model in multicomponent system was employed to calculate the growth velocity of the γ/α interface. According to Burke and Turnbull's theory, austenite grain coarsening induced by γ/γ grain boundary migration was simulated. To validate the model, dilatometric and quenching experiments were carried out. The dilatometric experiment was conducted using a Gleeble1500D with a sample 8 mm in diameter. The temperature of the sample was measured using thermocouples welded on the sample surface. In the quenching experiments, steel samples were heated to different temperatures then dropped into a water tank immediately, and the microstructure of the samples was examined to determine the fraction of the austenite. The simulated results were compared with the experimental results and the capability of the model for quantitatively predicting the microstructure evolution of the steel in heating process was assessed.

1. Introduction

Austenitization is a very important process in heat treatment of steels [1]. Austenitization usually affects the austenite grain size and the concentration homogeneity of the materials, which have a great impact on the kinetics of phase transformation in the subsequent cooling processes and the mechanical properties of steel. Therefore, it is of great significance to study the kinetics and microstructure evolution during the austenitization process of steels.

Microstructure simulation is an effective approach to optimize heat treatment processes. In recent years, several computer models, such as cellular automaton (CA) [2] and Monte Carlo [3], have been developed to simulate the microstructure evolution during austenitization process. In most of the reported models, the transformation of ferrite into austenite was governed by the diffusion of carbon. In fact, many alloying elements have a great influence on the austenitization [4-7]. Further, it is necessary to make more efforts to validate the microstructure simulation models.

In this paper, a CA model has been developed to simulate the microstructure evolution of ASTM A216 WCA cast steel during heating process. In the model, the thermodynamics and solute diffusion of the multicomponent system were taken into account by using Thermo-Calc [8] and Dictra [9] software. The dissolution of pearlite and the transformation of ferrite into austenite were simulated by using the developed model. To validate the model, dilatometric and quenching experiments were carried out using a Gleeble1500D, and metallographic examination was conducted. The simulated results were compared with the experimental results and the capability of the model for quantitatively predicting the microstructure evolution of the steel in heating process was assessed.

2. Model description

ASTM A216 WCA is a type of microalloyed medium-carbon steel. From the Fe-Fe_3C phase diagram, we know that austenitization in hypoeutectoid steels occurs in two steps: (a) P→ γ, and (b) α →γ, where P is pearlite, γ is austenite, and α is ferrite.

Nucleation of austenite grains in pearlite region takes place just above the eutectoid temperature and the transformation of pearlite into austenite is very fast since the diffusion distance for solute atoms are relatively small [3]. The transformation of ferrite into austenite occurs at a higher temperature and is completed above the (γ +α)/γ line in the phase diagram.

2.1. Dissolution of pearlite

A continuous nucleation model [10] was employed to calculate the nucleus number of austenite in the pearlite region.

$$n_\gamma(\Delta T) = f_N \exp\left(\frac{-Q_N}{k\Delta T}\right) \tag{1}$$

where ΔT is the value of overheating ($\Delta T = T - Ac_1$), $n_\gamma(\Delta T)$ is the nucleus density, f_N is the factor representing the influence of pearlite structure on the nucleation, Q_N is the activation energy of nucleation, and k is Boltzmann's constant.

Austenite grains grow into the pearlite region at a velocity $v^{\gamma/P}$ [10].

$$v^{\gamma/P}(\Delta T) = f_G \exp\left(\frac{-Q_G}{k\Delta T}\right) \tag{2}$$

where f_G is the factor representing the influence of pearlite structure on the growth rates, and Q_G is the activation energy of growth.

2.2. Transformation of ferrite into austenite

A mixed-mode model [7, 11] was employed to calculate the growth velocity of the γ /α interface, where both the finite interface mobility and the finite diffusivity of the alloying elements are taken into account.

$$v^{\gamma/\alpha} = M^{\gamma/\alpha}\Delta G_{\alpha\to\gamma} \tag{3}$$

where $v^{\gamma/\alpha}$ is the interface growth velocity, $M^{\gamma/\alpha}$ is the interface mobility, and $\Delta G_{\alpha\to\gamma}$ is the chemical driving force and can be calculated by

$$\Delta G_{\alpha\to\gamma} = \sum_{i=1}^{N} c_i^\gamma\left(\mu_i^\alpha - \mu_i^\gamma\right) \tag{4}$$

where N is the number of chemical components in the system, c_i^γ is the mole fraction of component i in γ at the interface, and μ_i^α and μ_i^γ are the chemical potentials per mole of component i in α and γ at the interface, respectively. The chemical potential was obtained using Thermo-Calc software. The interface mobility can be described by

$$M^{\gamma/\alpha} = M_0^{\gamma/\alpha} \exp\left(-\frac{E^{\gamma/\alpha}}{RT}\right) \tag{5}$$

where $M_0^{\gamma/\alpha}$ is the pre-exponential factor, $E^{\gamma/\alpha}$ is the activation energy for atom motion at the interface, and R is the gas constant.

2.3. Grain coarsening

Grain coarsening is resulted from the competitive growth of grains with the same phase. Boundary morphology is determined by how grain boundaries join together. According to Burke and Turnbull's theory, the velocity of a grain boundary segment can be expressed by [12-14]

$$v^{\gamma\gamma} = M^{\gamma\gamma} P \tag{6}$$

where $M^{\gamma\gamma}$ is the boundary mobility between austenite grains, and P is the driving force for grain boundary migration and can be calculated by

$$P = E\kappa \tag{7}$$

where E is the austenite grain boundary energy and can be calculated by the Read-Shockley equation [13, 14]

$$E = E_{\mathrm{m}}\left(\frac{\theta}{\theta_{\mathrm{m}}}\right)\left[1 - \ln\left(\frac{\theta}{\theta_{\mathrm{m}}}\right)\right] \tag{8}$$

where E_{m}=0.79J·m^{-2} and θ_{m}=15°. θ is the grain boundary misorientation and can be formulated as

$$\theta = 2\pi \frac{|S_j - S_i|}{Q} \tag{9}$$

where Q is the maximum of the crystallographic orientation number, and S_j and S_i are the crystallographic orientation numbers of grains j and i, respectively. When the grain boundary becomes a high angle boundary, θ >15°, E remains independent on the misorientation angle and E=E_{m}.

κ is the austenite grain boundary curvature and can be calculated by using the model described in reference [13].

2.4. Cellular automaton algorithm

Cellular automaton method was used in the simulation. A two-dimensional computational domain was discretized into square cells (as seen in Figure 1). Each cell has the following variables, (1) grain identifying variable, (2) cell status, which is one of P, α, γ, and γ/α interface, (3) concentration, and (4) interface migration distance. In the simulation, the effect of carbon and manganese were considered and their concentrations were obtained by solving diffusion equations. The interface migration distance was calculated based on the growth kinetics and was used in cell capturing in the CA model.

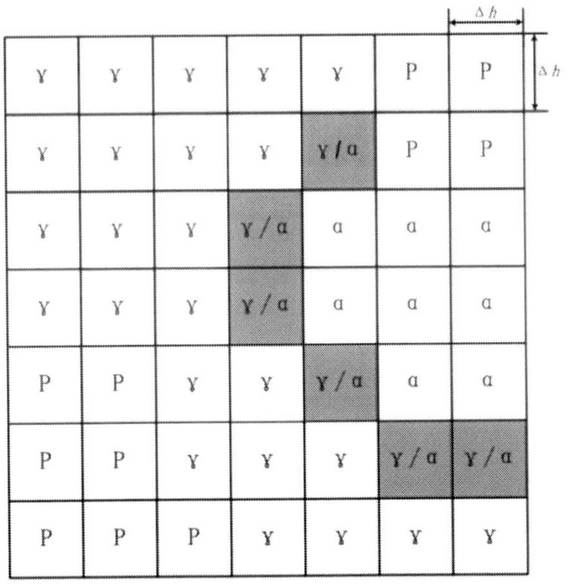

Figure 1. An illustration of the CA cells

3. Materials and experimental procedures

To validate the model, dilatometric and quenching experiments were carried out to study the variation of austenite fraction with temperature. The variation of austenite fraction results in volumetric change of the samples, so the austenite fraction can be deduced from the experimental results of the dilatometry [15, 16]. In the quenching experiment, the austenite transforms to martensite. The fraction of the austenite can be derived by measuring the fraction of the martensite in the quenched samples.

The composition of the steel is listed in Table 1. The dilatometric experiment was conducted using a Gleeble1500D with a sample 8 mm in diameter. The temperature of the sample was measured using thermocouples welded on the sample surface. The dilatometric curve for the heating rate of 1.0℃/s is illustrated in Figure 2. It can be seen that the heating process represented by the dilatometric curve can be divided in three stages. In the first stage, the dilatometric curve exhibits a linear expansion with temperature change. This is a pure thermal expansion of the steel sample without phase change. In the second stage, the slope of the dilatometric curve reduced due to the volume change caused by the dissolution of pearlite and the transformation of ferrite into austenite. This was a result of the competition between the thermal expansion and the volume change induced by austenitization. In the third stage, the sample was fully austenitized, and the curve exhibits again a linear expansion feature reflecting the thermal expansion of the austenite.

In the quenching experiments, steel samples were heated at a rate of 1.0 ℃/s to different temperatures then dropped into a water tank immediately. The microstructure of the samples was examined to determine the fraction of the austenite. Figure 2 shows the quenching temperatures on the dilatometric curve. The microstructure was characterized by using optical microscopy and different etchants. LePera's etchant was employed to identify the martensite transformed from pearlite, and 4% natal was used to distinct ferrite and martensite, and saturated aqueous picric acid was adopted to exhibit the grain boundaries of the prior austenite.

Table 1. The chemical composition of the steel (wt. %)

C	Si	Mn	S	P	Ni	Cr	Cu	Mo
0.28	0.36	0.90	0.025	0.025	0.2	0.25	0.25	0.15

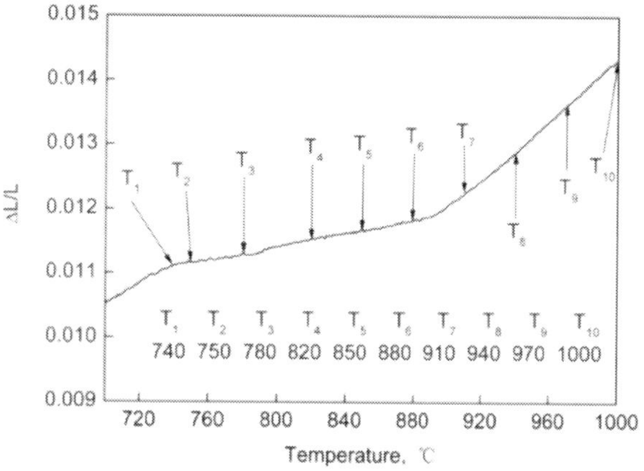

Figure 2. The dilatometric curve and the selected quenching temperatures

4. Simulation results and discussion

The initial microstructure of the samples consists of pearlite (mean interlamellar spacing is 0.63μ m) and ferrite (area fraction is 0.592 and mean grain size is 141μ m). The computational domain was divided into 150×100 rectangular cells. The cell size is 5×5μ m. The parameters used in the simulation are shown in Table 2.

Table 2. The parameters used in the simulation

Parameter	Value	Unit	References
Eutectoid temperature (Ac_1)	707	℃	Thermo-Calc software
Nucleation factor (f_N)	5.0×10^{7}	$m^{-2}\cdot s^{-1}$	-
Activation energy of nucleation (Q_N)	3.5×10^{-23}	$J\cdot mol^{-1}$	[10]
Growth factor (f_G)	5.0×10^{-5}	$m\cdot s^{-1}$	-
Activation energy of growth (Q_G)	4.1×10^{-23}	$J\cdot mol^{-1}$	[10]
Activation energy for atom motion at the γ/α interface ($E^{\gamma/\alpha}$)	147	$kJ\cdot mol^{-1}$	[2]
Pre-exponential factor ($M_0^{\gamma/\alpha}$)	0.004	$mol\cdot m\cdot J^{-1}\cdot s^{-1}$	-

The simulated microstructure evolution during heating process at the rate of 1.0℃/s is shown in Figure 3. Figure 3(a)-(c) show the transformation of pearlite into austenite, where the black area represents pearlite and the green area represents ferrite. The other color areas represent γ phase, and different colors represent γ grains with different crystallization orientations. Figure 3(d)-(f) show the transformation of ferrite into austenite. Figure 4(a) shows the as-cast microstructure of the sample. Figure 4(b)-(f) show the metallographic photos of the sample quenched at 740℃, 750℃, 880℃, 910℃ and 1000℃, respectively. The microstructure in Figure 4(a) consists of ferrite and pearlite. The microstructure in Figure 4(b) and (c) consist of ferrite, residual pearlite, and martensite. The microstructure in Figure 4(d) and (e) consist of residual ferrite and martensite. The microstructure in Figure 4(f) is martensite. All the martensite was austenite before quenching. Figure 4(f) exhibits the grain boundaries of the prior austenite. Figure 5 shows the evolution of the fraction of austenite as a function of the temperature. It can be seen that the simulated results are in good agreement with the experimental data. The pearlite dissolution is very fast and the transformation of ferrite into austenite is slower and is almost completed at 950℃.

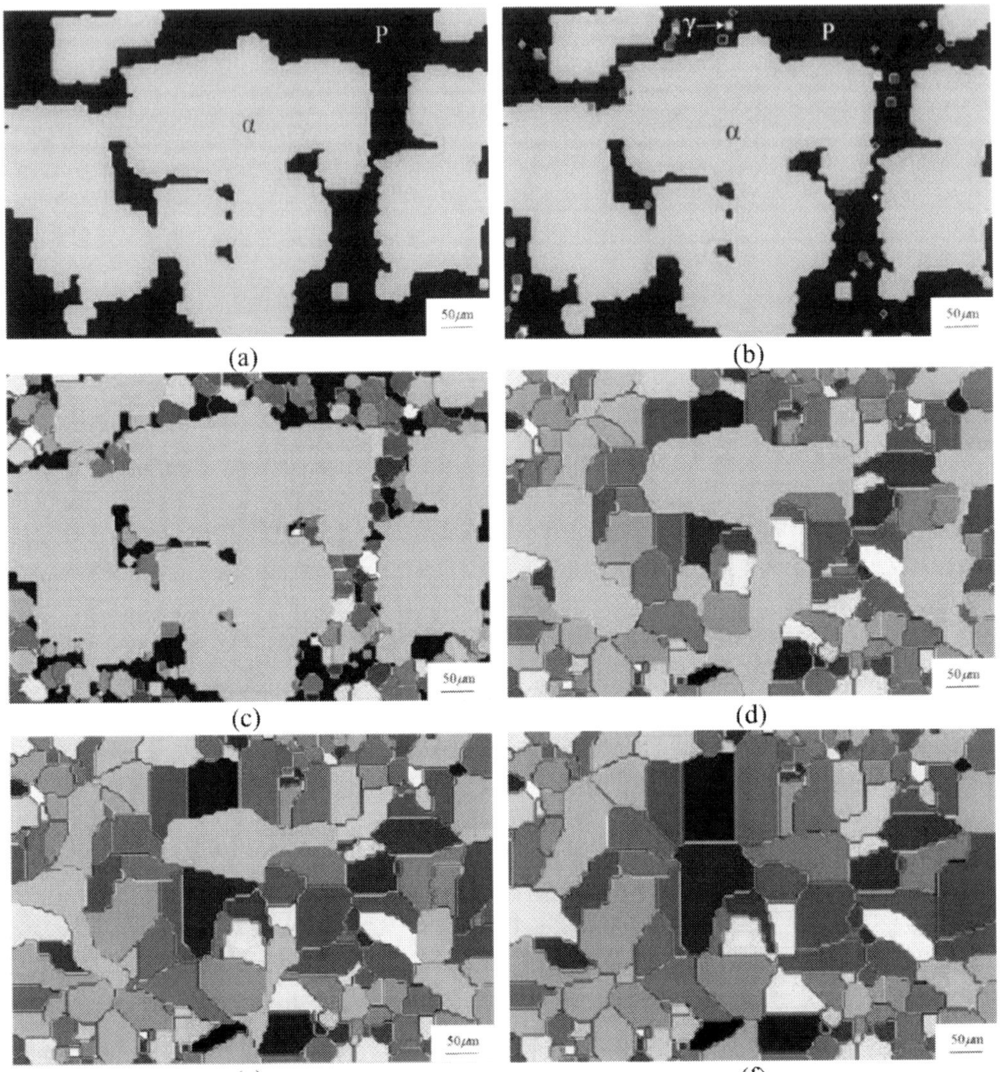

Figure 3. The simulated microstructure at different temperatures during the heating process, (a) room temperature, (b) 740℃, (c) 750℃, (d) 880℃, (e) 910℃, (f) 1000℃

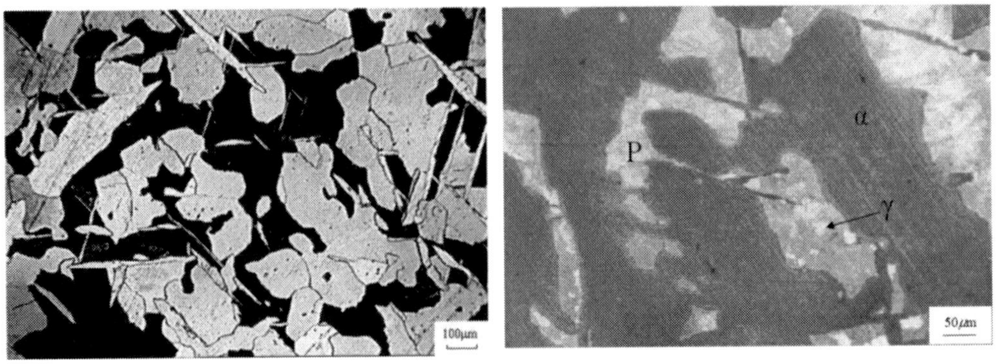

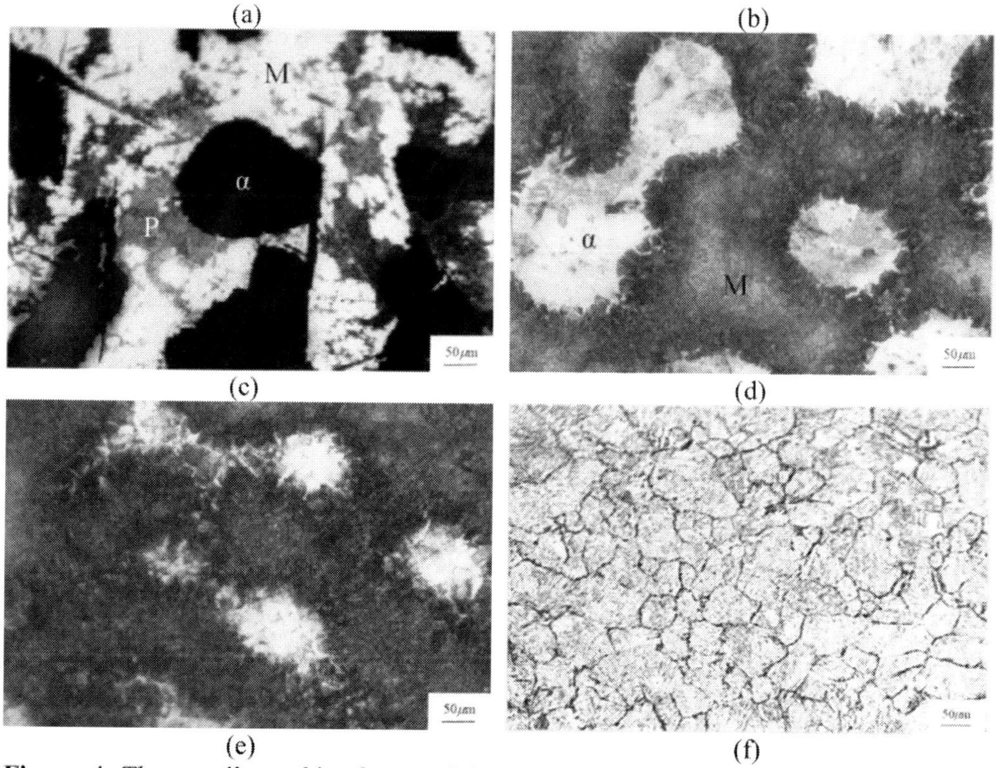

Figure 4. The metallographic photos of the samples, (a) the as-cast microstructure (4% natal), (b) the microstructure of the sample quenched at 740℃ (LePera's etchant), (c) quenched at 750℃ (LePera's etchant), (d) quenched at 880℃ (4% natal), (e) quenched at 910℃ (4% natal), and (f) quenched at 1000℃ (saturated aqueous picric acid), where M is martensite, P is pearlite, and α is ferrite

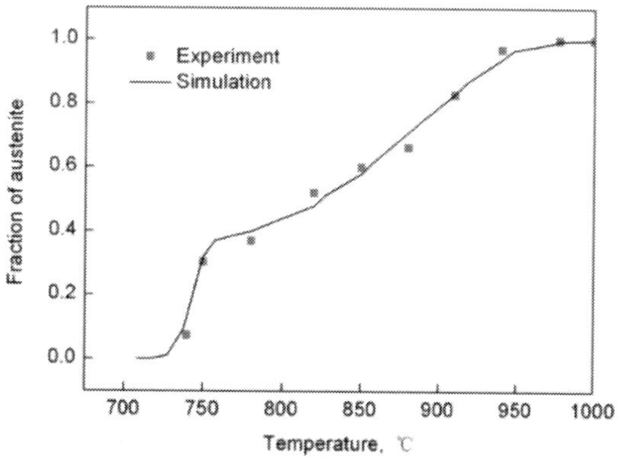

Figure 5. The fraction of austenite as function of temperature for the heating rate 1℃/s

5. Conclusion

A cellular automaton model has been developed to simulate the austenitization process of ASTM A216 WCA cast steel during heating process. The dissolution of pearlite was described by continuous nucleation and growth of austenite grains. A mixed-mode model for multicomponent system was employed to calculate the growth velocity of the γ /α interface. The austenite grain coarsening induced by γ /γ grain boundary migration was considered and the migration velocity of the grain boundaries was calculated based on Burke and Turnbull's theory. To validate the model, dilatometric and quenching experiments were carried out, and the microstructure of the quenched samples was examined and compared with the simulation results. It was shown that the simulated morphology and grain size of the austenite are in good agreement with the experimental results.

Acknowledgement
This work was funded by the National Science and Technology Major Project of China (NO.2011ZX04014-052).

References
[1] Oliveira F L G, Andrade M S and Cota A B 2007 *Materials Characterization* **58** 256
[2] Yang B J, Chuzhoy L and Johnson M L 2007 *Computational Materials Science* **41** 186
[3] Jacot A and Rappaz M 1999 *Acta Materialia* **47** 1645
[4] Odqvist J, Hillert M and Ågren J 2002 *Acta Materialia* **50** 3211
[5] Aaronson H I, Reynolds W T and Purdy G R 2004 *Metallurgical and Materials Transactions A* **35A** 1187
[6] Fazeli F and Militzer M 2005 *Metallurgical and Materials Transactions A* **36A** 1395
[7] Bos C and Sietsma J 2009 *Acta Materialia* **57** 136
[8] Andersson J, Helander T, Höglund L, Shi P F and Sundman B 2002 *Calphad* **26** 273
[9] Borgenstam A, Engstrom A, Hoghund L and Agren J 2000 *Journal of Phase Equilibria* **21** 269
[10] Caballero F G, Capdevila C and García de Andrés C 2000 *Scripta Materialia* **42** 1159
[11] Hillert M 1999 *Acta Materialia* **47** 4481
[12] Burke J E and Turnbull D 1952 *Progress in Metal Physics* **3** 220
[13] Lan Y J, Li D Z and Li Y Y 2006 *Metallurgical and Materials Transactions B* **37B** 119
[14] Li D Z, Xiao N M, Lan Y J, Zheng C W and Li Y Y 2007 *Acta Materialia* **55** 6234
[15] Dykhuizen R C, Robino C V and Knorovsky G A 1999 *Metallurgical and Materials Transactions B* **30B** 107
[16] Choi S 2003 *Materials Science and Engineering A* **363** 72

Modeling of different zones of as-cast structure of high carbon steel ingots

Zhiye Chen, Sonja Arnsfeld, Dieter Senk
Chair of Iron and Steel Metallurgy, Department of Ferrous Metallurgy, RWTH Aachen University, 52072 Aachen, Germany

E-mail: zhiye.chen@iehk.rwth-aachen.de

Abstract. Ingot casting technology has been expanded to large parts during the last decades. As the ingot sizes increase, higher quality of the as-cast semi-products is demanded, with regard to the control of casting defects like macrosegregation and structure inhomogeneity. In order to investigate macrosegregation and to estimate the as-cast structure, theoretical study and simulation work on ingot solidification are carried out at the Department of Ferrous Metallurgy of RWTH Aachen University (IEHK). A solidification model has been developed, and based on that, modeling of the structure morphology has been performed. A proper coupling of the developed solidification model with experimental results from IEHK is under investigation. This solidification model is a two-phase FVM model applied for high carbon steel with 0.6 wt.% [C]. The temperature and concentration fields of the solid and liquid phases have been calculated and these results can provide information for further prediction of the solidification structure such as CET (Columnar to Equiaxed Transition) zone and casting defects in an ingot. The structure morphology model introduces a shape factor of grains as the quantitative criterion for identification of the structure morphology. It focuses on the interaction between nuclei density and resulting macroscopic structure, and can calculate the strictly columnar zone, CET zone with mainly columnar characteristics and enclosed equiaxed crystals, and the pure equiaxed zone. The results will be presented, and the correlation of CET zone with development of macrosegregation in the inner part of an ingot will be discussed.

1. Introduction

Solidification is the most important process which influences the final quality of the as-cast semi-product. Today, the demand of heavy ingots and continuous cast products throughout the world for larger turbines and generator rotor shafts are increasing. As the size of as-cast semi-products increase, the problem of structure and segregation in large ingots and continuous cast products takes on new importance. In order to model the solidification structure, consideration of many solidification phenomena such as heat transfer, phase transformation, fluid flow, and the movement of the residual liquid and equiaxed grains is required, meanwhile the understanding of nucleation, growth kinetics, and species redistribution on interfacial scale is also demanded[1, 2].

Many modeling and simulation work, both analytically and numerically, has been done on the prediction of as-cast structure during solidification process since 1970's[3-4]. Today, the multiphase volume-average model proposed by Beckermann has become the most promising model in the field of solidification simulation[2]. However, due to the lack of information, for instance the nuclei density in the melt, nucleation undercooling, solute redistribution mechanism at the solid/liquid interface, and the local heat transfer coefficient with consideration of gap formation during solidification and the sensitivity of this model on these parameters, prediction on solidification structure from this model remains uncertain, and the application of the model for steel ingots' solidification is still very limited. Considerable works is needed to be done in order to improve this model and to validate the simulation especially for large ingot size. The present study uses the volume-averaged method to simulate the solidification process. The heat transfer, nucleation and grain growth are calculated from this solidification model and the results are used for the further calculation of the position of CET zone with mainly columnar characteristics and enclosed equiaxed crystals.

As an important feature of solidification structure, CET zone defines important properties of the as-cast products[5]. Different CET mechanisms have been proposed in the past decades[6] and the most famous two mechanism are the "mechanical blocking" mechanism proposed by Hunt[7] and the "solute interaction" mechanism proposed by Martorano and Beckermann[8]. Most of the mechanisms focus on the blocking effect of equaixed grains on columnar dendrites, either the geometrical impingement or the solutal interaction between columnar and equiaxed grains. However, the gradual transition of grains' morphology from columnar to equiaxed grains are not quantified, and it seems in those models that the grains changes from columnar into equiaxed in a sudden, which appears to be in contradiction with that observed from experimental results. The present study tries to calculate the grain morphology transition from columnar to equiaxed during solidification quantitatively, where a geometry factor is used as the index of grain elongation degree based on theory of M. Rappaz[9].

2. Mathematical Model Description

2.1 The solidification model

The solidification model in the present study is a two-phase model which describes the globular solidification for a binary Fe-C alloy. The solid and liquid phases in a representative elementary volume (REV) are treated separately and the field properties are averaged. The melt is considered to be stagnant before solidification. The melt convection and the grain movement are neglected.

2.1.1. Governing equations

The governing equations are based directly on the model of Wu and Ludwig[10]. The conservation equations which govern the mass balance in the two-phase system are

$$\frac{\partial}{\partial t}(f_l\rho_l)+\nabla\cdot(f_l\rho_l\vec{u}_l)=M_{sl} \tag{1}$$

$$\frac{\partial}{\partial t}(f_s\rho_s)+\nabla\cdot(f_s\rho_s\vec{u}_s)=M_{ls} \tag{2}$$

where ρ_l and ρ_s are density of liquid and solid steel, f_l and f_s are the volume-average volume fraction of liquid and solid phase. u_l and u_s are velocity of liquid and solid, and here $u_l=u_s=0$. M_{ls} and M_{sl} are the source terms which indicate the interfacial mass transfer due to phase transformation. The heat conservation equations are

$$\frac{\partial}{\partial t}(f_l\rho_l h_l)+\nabla\cdot(f_l\rho_l\vec{u}_l h_l)=\nabla\cdot(f_l\lambda_l\nabla\cdot T_l)+Q_{sl} \tag{3}$$

$$\frac{\partial}{\partial t}(f_s\rho_s h_s)+\nabla\cdot(f_s\rho_s\vec{u}_s h_s)=\nabla\cdot(f_s\lambda_s\nabla\cdot T_s)+Q_{ls} \tag{4}$$

where T_l and T_s are the volume-average temperature and enthalpy of the liquid and solid phase, and λ_l and λ_s are the heat conductivity coefficients. The species conservation equations are

$$\frac{\partial}{\partial t}(f_s\rho_s c_s)+\nabla\cdot(f_s\rho_s\vec{u}_s c_s)=\nabla\cdot(f_s\rho_s D_s\nabla\cdot c_s)+C_{ls} \tag{5}$$

$$\frac{\partial}{\partial t}(f_l\rho_l c_l)+\nabla\cdot(f_l\rho_l\vec{u}_l c_l)=\nabla\cdot(f_l\rho_l D_l\nabla\cdot c_l)+C_{sl} \tag{6}$$

where c_l and c_s are the volume-average composition of liquid and solid steel, and D_l and D_s are the diffusion coefficient of carbon in the liquid and solid steel. The source terms C_{ls} and C_{sl} indicate the interfacial solute redistribution due to solidification and remelting.

2.1.2. *Nucleation and growth*

The nucleation is considered as heterogeneous nucleation based on M. Rappaz's theory.

$$N = \frac{d(\Delta T)}{dt} \cdot \frac{n_{\max}}{\sqrt{2\pi} \cdot \Delta T_\sigma} \cdot e^{-\frac{1}{2}\left(\frac{\Delta T - \Delta T_N}{\Delta T_\sigma}\right)^2} \quad (7)$$

where N is the nucleation rate, ΔT is the constitutional undercooling, n_{max} is the maximum nuclei distribution, ΔT_N is the mean nucleation undercooling, and ΔT_σ is the standard deviation of the Gaussian distribution of nucleation rate. The conservation equation for nucleation is:

$$\frac{\partial n}{\partial t} + \nabla \cdot (\vec{u}_s \cdot n) = N \quad (8)$$

where n is the volume-averaged nuclei distribution. In order to define the growth of the globular/equiaxed dendrites, grains are considered to be sphere shaped and the grain growth velocity is in the radius direction, which is according to the theory of Kurz and Fisher[11]. At the liquid/solid interface, thermodynamic equilibrium is assumed. The grain growth velocity v is given by:

$$v = \frac{2 \cdot D_l}{d_e} \cdot \frac{c_l^* - c_l}{c_l^* - c_s^*} = \frac{2 \cdot D_l}{d_e(1-k)} \cdot \left(1 - \frac{c_l}{c_l^*}\right) \quad (9)$$

where d_e is the volume-averaged diameter of equiaxed grains, c_s* and c_l* are the interfacial concentrations of the solid and liquid phases. The volume-averaged mass transfer between liquid and solid phase M_{ls} and M_{sl} are defined as

$$M_{ls} = v \cdot (n \pi d_e^{\,2}) \cdot \rho_s \cdot f_l \quad (10)$$

$$M_{ls} = -M_{sl} \quad (11)$$

2.1.3. *Boundary conditions and parameters*

The solidification model is a 1-D axis symmetrical model applied on a 100 kg square-shaped steel ingot with cast iron mould (see figure 1). It is considered that the ingot is isolated at the top and the bottom, which means that this model could also be applied for a CC-billet with the width of 140 mm. The temperature of the mould inner surface is measured from a real ingot cast at IEHK RWTH Aachen University with the same geometry and the same steel grade, and then fitted as a function of time during and imported to the model as the boundary condition. Other parameter and properties used for simulation are summarized in table 1 and illustrated in **Figure 1**.

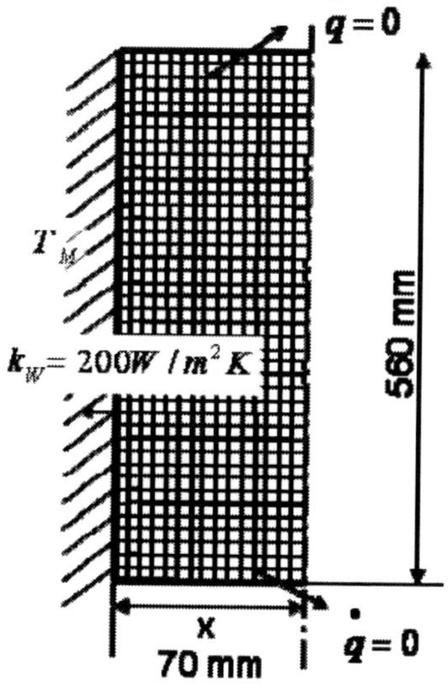

Table. 1 Assumed parameters and properties for the simulation

Initial melt temperature [°C]	1500 °C
Initial mould temperature [°C]	T_M = 80 °C
Density of liquid and solid [kg/m^3]	ρ_l = 7000, ρ_s = 7027
Thermal conductivity [W/mK]	λ_l = 33.94, λ_s = 33.94
Carbon diffusion coefficient [m^2/s]	$D_l = 2\times10^{-8}$, $D_s = 5.6\times10^{-10}$
Specific heat capacity [J/kgK]	$cp_l = cp_s$ = 808.25
Segregation coefficient of carbon	$k = 0.2+5.63\times10^{-4}T-3.98\times10^{-7}T^2$ [12]

Figure 1. Schematic diagram of assembly of half ingot and mould

2.2 The CET model

The CET model uses the analytical model for eutectic growth developed by M. Rappaz[9]. The following equations describe a nucleated grain growing under the same conditions at the impingement point of the former grains.

$$\Delta T^0(t_{est}) = \Delta T_{col} \frac{\tanh(\Delta T_g^{'}(t_{est}) - \Delta T_n^{'})}{1 + \Delta T_n^{'} \cdot \tanh(\Delta T_g^{'}(t_{est}) - \Delta T_n^{'})} \tag{12}$$

$$\Delta T^{\pi}(t_{est} - L/v_T) = \Delta T_{col} \frac{\tanh(\Delta T_g^{'}(t_{est} - L/v_T) - \Delta T_n^{'}) + \Delta T_n^{'}}{1 + \Delta T_n^{'} \cdot \tanh(\Delta T_g^{'}(t_{est} - L/v_T) - \Delta T_n^{'})} \tag{13}$$

$$\Delta T_n^{'} = \Delta T_{col} \frac{\Delta T_n}{\Delta T_{col}} \tag{14}$$

The equation (14) describes the relation between the undercooling of the nucleus to the undercooling of the columnar solidification front. A nucleus, growing ahead of the solidification front can never get a higher undercooling than the solidification front itself. At the grain center position, the undercooling is

$$\Delta T_g^{'}(t) = \frac{G \cdot v_T \cdot t}{\Delta T_{col}} \tag{15}$$

At the impingement point of two grains ($t = t_{est}$) the undercooling of the two grains must be equal:

$$\Delta T^0(t_{est}) = \Delta T^{\pi}(t_{est} - L/v_T) \tag{16}$$

$\Delta T^0(t_{est})$ needs to be numerically solved in order to calculate the shape factor S:

$$S = \frac{R_g^0(t_{est})}{R_g^0(t_{est} - L/v_T)} > 1 \tag{17}$$

Based on the temperature field model and concentration field calculated from the solidification model, parameters such as isotherm velocity v_T, temperature gradient G, undercooling of the nuclei ΔT_n, undercooling of the columnar front ΔT_{col} are defined in (18)-(21).

$$v_T = \frac{dx}{dt} \cdot \frac{f_s(i+1, j) - f_s(i, j)}{f_s(i+1, j) - f_s(i+1, j+1)} \tag{18}$$

$$\Delta T_{col} = \frac{G \cdot D_l}{v_T} \cdot + m \cdot c_0 \cdot (1 - \frac{1}{1 - (1 - \frac{1}{k}) \cdot Pe}) \tag{19}$$

$$\Delta T_n = T_s(i, j+1) - T_s(i, j) \tag{20}$$

$$G = \frac{T_l(i, j) - T_l(i, j-1)}{dx} \tag{21}$$

2.3 Numerical solution

In the present study, conservation equations are discretized based on fully implicit volume finite difference scheme, and a Matlab program has been developed to solve these conservation equations. The REV size is 3.5 mm, and the time step is 0.01 s. The error value for residual difference is 10^{-7}, and the iteration times in each time step varies from 16 to185.

The CET model simulation was carried out in several steps (**Figure 2**):

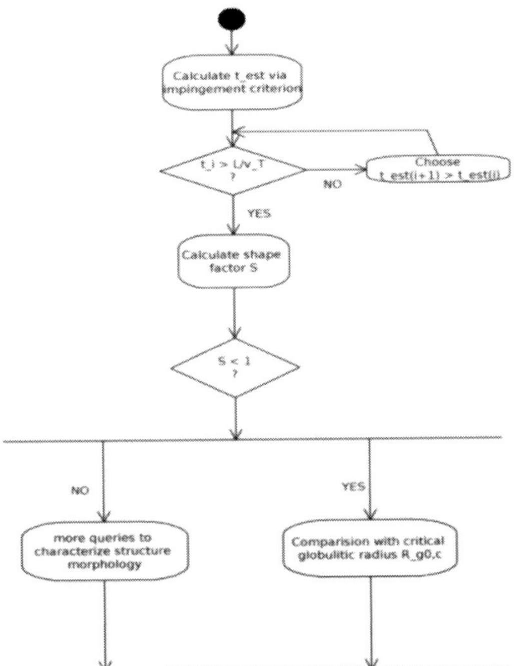

Figure 2. Program structure of the CET model

a) Calculation of the estimated time t_{est} for two particles to collide. Due to the periodic characteristics, t_{est} is numerically solved from (12) and (13).

b) Calculation of the shape factor S according to (17).

c) Comparison between t_{est} and the time t for solidification front of columnar dendrites to overgrow the growing particles. If $t_{est} < t$, CET or equiaxed grains can form, if $t_{est} >> t$, columnar structure will develop, and if $t_{est} > t$, grown particles may be enclosed by the columnar dendrites. This is defined as the static "time criterion". The shape factor S was calculated firstly without progressing of the columnar front (without time criterion). Since in reality CET and equiaxed zone can only occur when the particles have enough time to grow, S was then calculated with consideration of the growth of columnar dendrites (with time criterion).

3 Results and Discussion

3.1 Solidification sequence

The present solidification model is applied to a Fe-0.6wt.% [C] alloy. The simulated solidification time is about 600 seconds. The evolution of temperature, mass transfer rate between the liquid and solid phase, and the liquid concentration in the ingot during solidification are shown in **Figure 3**.

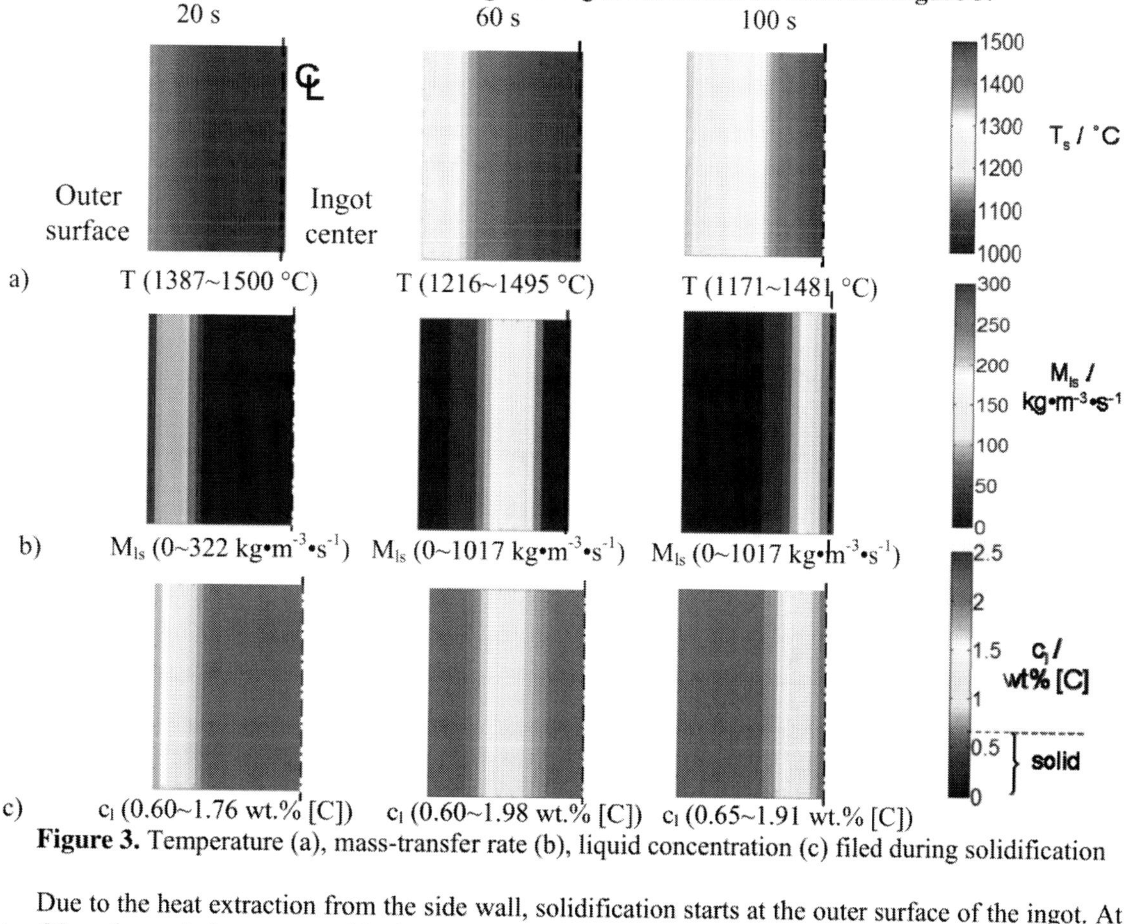

Figure 3. Temperature (a), mass-transfer rate (b), liquid concentration (c) filed during solidification

Due to the heat extraction from the side wall, solidification starts at the outer surface of the ingot. At t = 20 s, the melt in contact with the mould is cooled down to 1387 °C (a), which is far below the liquidus temperature (1475 °C) and is approaching to the solidus temperature (1360 °C), and the mass transfer rate M_{ls} of 0~322 kg/(m³s) (b) indicates the nucleation and growth of equiaxed grains. Meanwhile, the melt is enriched during grains growth from the initial concentration of 0.6 wt.% [C] up to 1.76 wt.% [C]. As the time increase to 60 s and 100 s, the melt is further cooled down and the mass transfer rated is increased to 1017 kg•m^{-3}•s^{-1}, the melt concentration is also enriched to ~1.9 wt.% [C]. This will lead to the magnitude of the thermal buoyancy effect of the melt flow and therefore influence the macrosegregation pattern.

The distribution of nuclei and the nucleation rate at different time of the ingot is illustrated in **Figure 4**. The initial nuclei distribution is set to be 1×10^{9} m^{-3}, and at t = 20 s the nuclei distribution is increased from 1×10^{9} m^{-3} up to 5×10^{9} m^{-3} in the mushy zone. It can be observed that the nucleation rate obeys the Gaussian distribution. The maximum nucleation rate at t = 20 s is ~ 8.5×10^{9} m^{-3}s^{-1}, which is larger than that of ~ 5×10^{9} m^{-3}s^{-1} at t = 100 s. At the outer surface of the ingot the temperature decreases faster than in the center, and the undercooling rate d(ΔT)/dt by both heat removal and solute enrichment is larger at the boundary, therefore the nucleation rate is larger at the boundary of an ingot.

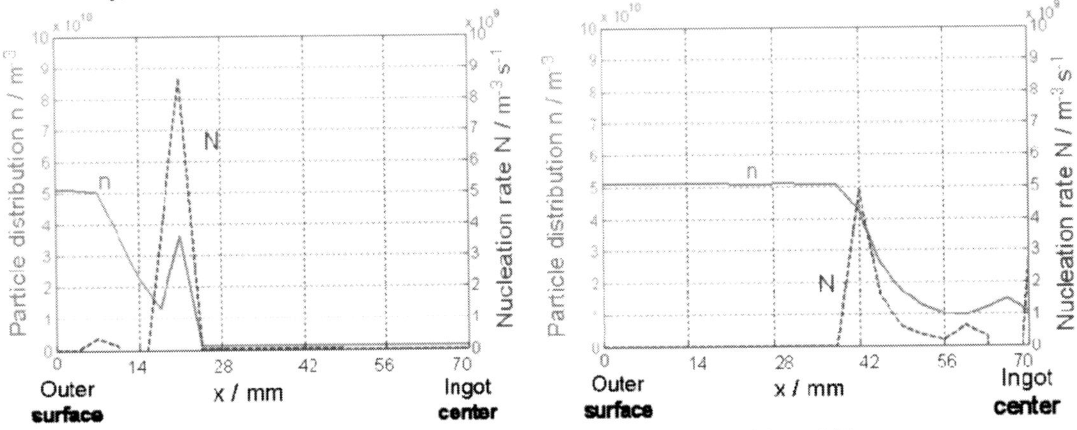

a) t = 20s b) t = 100s

Figure 4. Particle distribution and nucleation rate along the x-axis

3.2 Comparison of calculated T-curves to experiments

The measured and predicted cooling curves are compared in **Figure 5** at the ingot center, outer surface and the mould inner wall. The temperature was measured by thermocouples during pouring and solidification, and the pouring stage is not illustrated here. The predicted temperature curve at the outer surface fits to the measured curved quite well in the first half stage of solidification, and the curves at the ingot center show certain difference. At the early solidification, the predicted temperature stays at the liquidus temperature for ca. 100 s, and then it starts to decrease slowly, and the measured temperature stays at the liquidus temperature for relative long time of ca. 500 s and after that decrease dramatically. The most obvious difference stays at the very late stage of solidification (after 500 s), which might be due to the assumption that the grain growth at the solid/liquid interface is in thermodynamic equilibrium. Since at this time the liquid in the bulk melt is also enriched with solute, the interfacial solute redistribution must not stay in thermodynamic equilibrium any longer.

3.3 The CET map

As it was mentioned, solidification started from the cooled outer surface to the ingot center. It is considered that for $f_s > 0.65$, the morphology of the solidification structure would not be able to change, therefore the region with $f_s > 0.65$ is not calculated in the present study. The calculated CET map (one column represents 5 mm of the ingot width) at different time is shown in **Figure 6**, and different colors are chosen to mark different solidification structures (Time progress is 1 s in each diagram):

- Globulitic structure for $S < 1$ → green
- Equiaxed structure for $1 < S < 5$ → red
- CET for $5 < S < 10$ → orange
- Columnar structure for $S > 10$ → blue

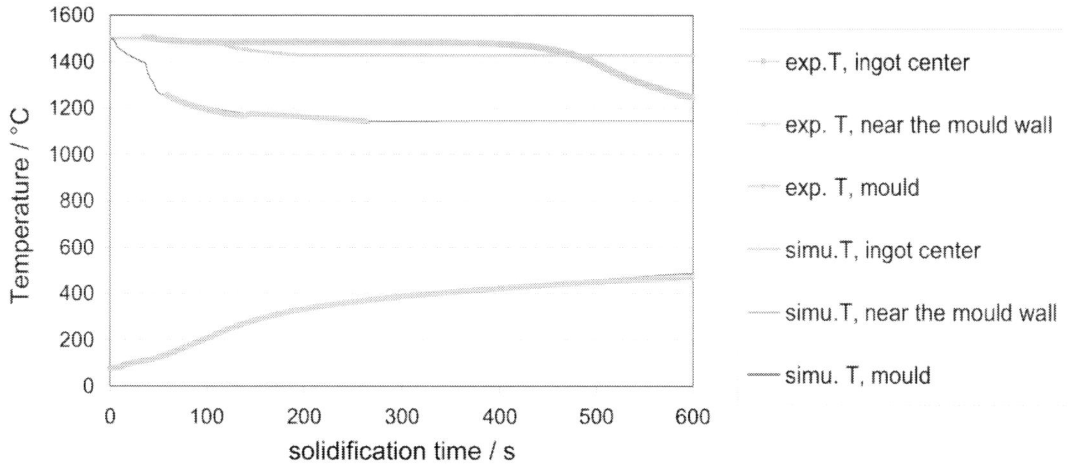

Figure 5. Comparison of measured and predicted cooling curves in the ingot and in the mould

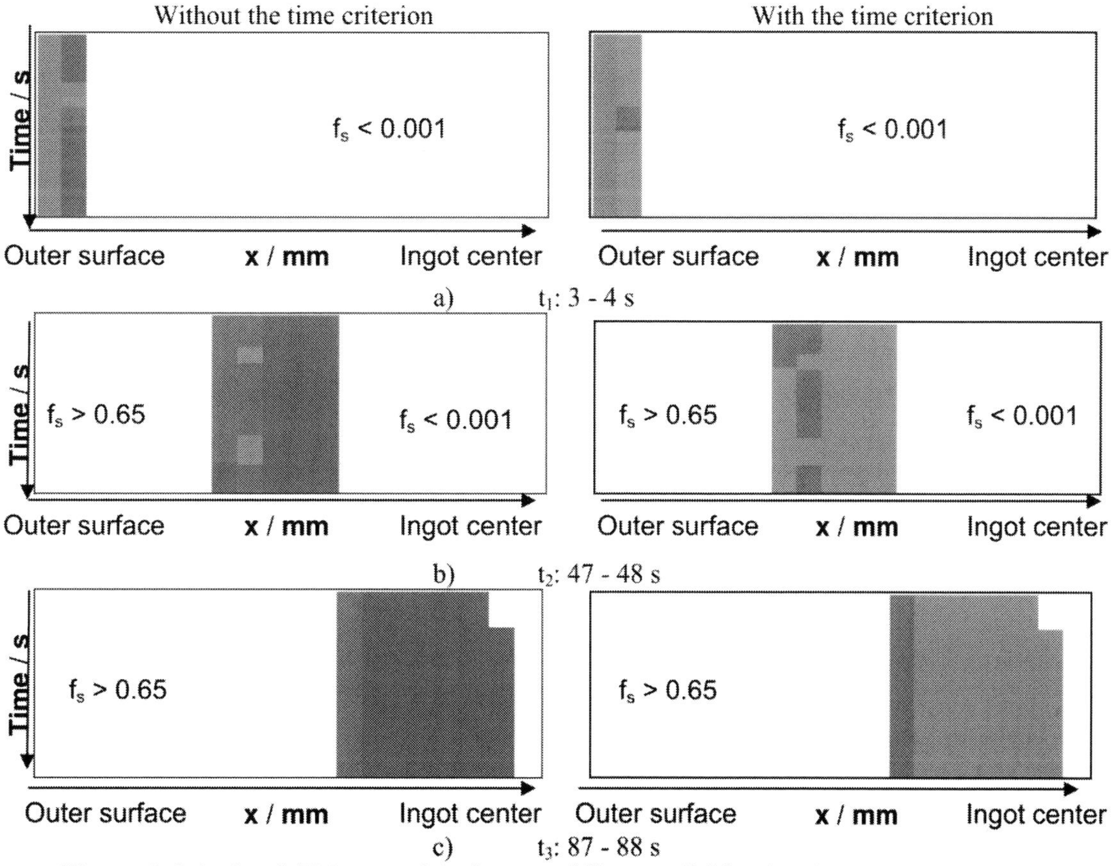

Figure 6. Calculated CET map of the ingot at different solidification time

At the beginning of solidification, globulitic chill zone with the width of 5 mm can form (a). Without the time criterion (a, left), equiaxed zone can form, and with the time criterion (a, right), the formation of the equiaxed zone is suppressed by columnar dendrites growth. In the middle of solidification (47-48 s), for the case without the time criterion, a mixed structure of CET and columnar structure is dominating right before the equiaxed zone. For the case with the time criterion, the formation of equiaxed structure is suppressed by columnar growth, and a narrow CET zone indicated by the orange stripe is formed in front of the columnar zone. The progressing of CET zone can be observed from b) and c) and the CET zone covers around 25-30 mm. The blue area on the right side of the CET zone indicates the potential columnar structure that grains are able to develop, since here $t_{est}>t$, the columnar dendrites can overgrow the growing equiaxed grains / CET structure.

Figure 7 shows the sulfur print of the as-cast ingot along the longitude direction. The chill zone and columnar zone is ca. 3 mm and 25 mm in width, respectively. The CET zone which includes the mixture of columnar and equiaxed dendrites covers a range of ca. 18 mm. Compared with figure 6, the predicted chill zone and columnar zone show nearly the same width, and the predicted CET zone is ca. 50% wider than the observed, and the center equiaxed zone can be predicted only for the case without the static time criterion. Thus a suitable time criterion which balances the development of the equiaxed grains and the growth of the columnar solidification front need to be further investigated.

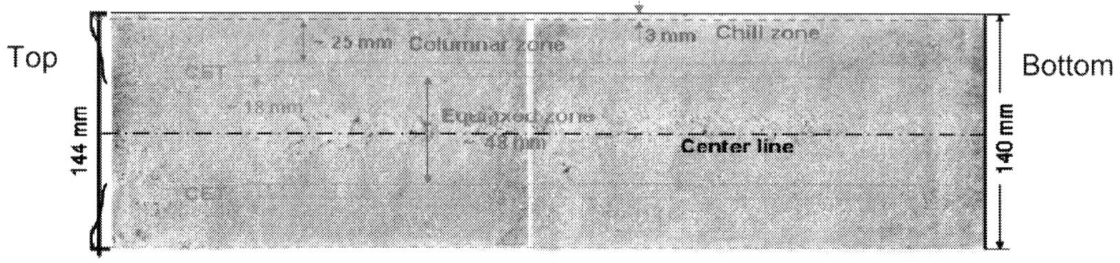

Figure 7. Sulfur print of the as-cast ingot

4 Conclusions

The multiphase/multiscale model of Ludwig has been applied to simulate the solidification of a 100 kg square shaped ingot for a Fe-0.6 wt.% [C] steel. The comparison of cooling curves between simulation and experiments shows that at the first half stage of solidification, the predicted cooling curves fit the measured curves quite well, and at the late stage of solidification (after 500 s) the curves start to depart from each other. The melt concentration during solidification can be enriched up to ca. 1.9 wt.% [C], which provides a strong thermal buoyancy force to melt convection in the mushy zone.

Based on the results from the solidification model, a CET model was developed to predict the solidification structure after the theory of M. Rappaz. The globulitic chill zone and columnar zone were predicted to be similar to the experimental results. The enclosure of CET/equiaxed structure in columnar structure was also predicted for the case with consideration of columnar dendrites growth.

Acknowledgements

Authors would like to thank the support of Promotionsstipendium from RWTH Aachen University and to acknowledge the technical assistance provided by M.Sc. Syed Bilal Hussain and B.Sc. Henning Bernhardt during numerical simulation.

References

[1] Gu J P and Beckermann C 1999 *Metall. and Mater. Trans.* A **30** 1357
[2] Ludwig A and Wu M 2002 *Metall. and Mater. Trans.* A **33** 3673
[3] Flemings M C 1966 *Trans. Metall. Soc. AIME* **236** 625
[4] Ni J and Beckermann C 1991 *Metall. Trans.* B **22** 349
[5] Wang J, Wang F, Li C and Zhang J 2010 *Steel Research Inter.* **81** 150
[6] Flood S.C and Hunt J D 1998 *ASM Handbook* vol 15 (Materials Park) p 130
[7] Hunt J D 1984 *Mater. Sci. Eng.* **65** 75
[8] Martorano M A and Beckermann C 2003 *Metall. and Mater. Trans.* A **34** 1657
[9] Rappaz M 1994 *Acta Metal. Mater.* **42** 2365
[10] Wu M and Ludwig A 2006 *Metall. and Mater. Trans.* A **37** 1613
[11]Kurz W and Fisher D J 1989 *Fundamentals of Solidification* (Trans. Tech. Publications, Switzerland) p 65
[12] Chuang Y and Schwerdtferger K 1973 *Arch. Eisenhüttenkunde* **44** 341

3-D analysis of grain selection process

Tomoka Arao[1,3], Hisao Esaka[2,3] and Kei Shinozuka[2,3]
[1] Graduate student, National Defense Academy
[2] Department of Materials Science and Engineering, National Defense Academy
[3] 1-10-20 Hashirimizu, Yokosuka 239-8686, Japan

E-mail: em50037@nda.ac.jp

Abstract. It is known that the grain selection plays an important role in the manufacturing process for turbine blades. There are some analytical or numerical models to treat the grain selection. However, the detailed mechanism of grain selection in 3-D is still uncertain. Therefore, an experimental research work using Al-Cu alloy has been carried out in order to understand the grain selection in 3-D.A mold made by Al_2O_3 was heated to 600 °C (=liquids temperature of the alloy) and was set on a water-colded copper chill plate. Molten Al-20 wt%Cu alloy was cast into the mold and unidirectional solidified ingot was prepared. The size of ingot was approximately $\phi 25\times 65^H$ mm. To obtain the thermal history, 4 thermocouples were placed in the mold. It is confirmed that the alloy solidified unidirectionally from bottom to top. Solidified structure on a longitudinal cross section was observed and unidirectional solidification up to 40 mm was ensured. EBSD analysis has been performed on horizontal cross section at an interval of ca.200 μm. These observations were carried out 7-5 mm from the bottom surface. Crystallographic orientation of primary Al phase and size of solidified grains were characterized. A large solidified grain, the crystallographic orientation of which is approximately <101> along heat flow direction, is observed near the lowest cross section. The area of <101> grain decreased as solidification proceeded. On the other hand, it is found that the area of <001> grain increased.

1. Introduction

In the process of grain selection, a grain preferred growth direction of which is close to the heat flow direction, overgrows grain that is growing more different from the heat flow direction [1-4]. It is known that the direction and size of grain plays an important role in the mechanical properties for as cast products, such as turbine blades [5]. Thus grain selection is not only important for understanding formation of solidification structure, but also fundamental for casting process. There are some analytical or numerical models to treat the grain selection [6-8]. However, the detailed mechanism of grain selection in 3-D is still uncertain. Therefore, an experimental research work has been carried out using Al-Cu alloy in order to understand the grain selection in 3-D.

2. Experimental procedures

Al-20 wt%Cu alloy was used in this study. Alloy was melted in an alumina crucible under Ar atmosphere at 750°C. After holding enough time, the molten alloy was cast into the ingot case shown in Fig.1. A mold made by Al_2O_3 was heated to 600 °C (=liquids temperature of the alloy) and was placed on a water-cooled copper chill plate. The size of ingot was approximately ϕ25×65H mm. To obtain the thermal history, 4 thermocouples (K-type, ϕ0.1 mm) were placed in the mold at intervals of 5 mm. Temperature change was recorded using a data logger at intervals of 5 ms. Solidified structure on a longitudinal cross section was observed by SEM. EBSD (electron backscatter diffraction) analysis has been performed on horizontal cross section at intervals of ca.200 μm. These were carried out 7~5 mm from the bottom surface. After polishing with SiC abrasive paper, colloidal silica was used as a final finish. Then, Ar-ion milling treatment was performed to obtain suitable surface for EBSD analysis. Crystallographic orientation of primary Al phase and size of solidified grains were characterized. HKL CHANNEL 5 was applied for EBSD analysis.

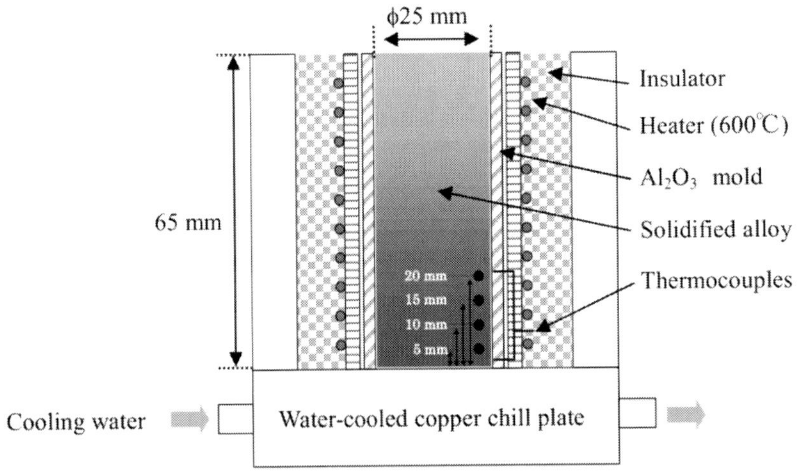

Figure 1. Experimental apparatus for unidirectional solidification.

3. Experimental results

3.1 Measurement of thermal history

Thermal history recorded is shown in Fig.2. These data were obtained from 5 mm to 20 mm from bottom surface. It was confirmed that the alloy solidified unidirectionally from bottom to top at least 20 mm.

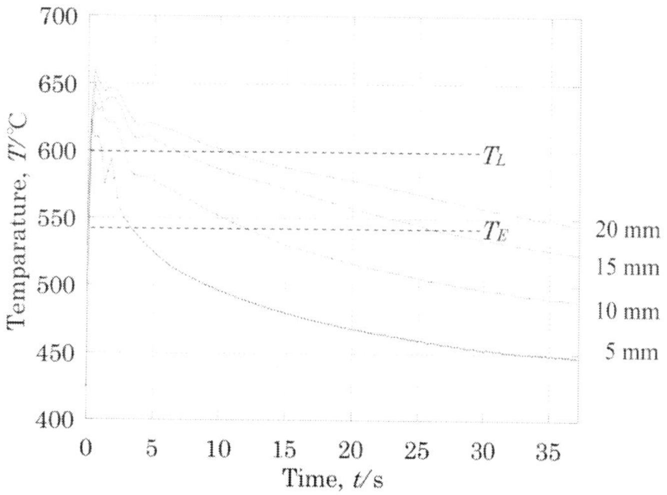

Figure 2. Cooling curves of Al-20 wt%Cu alloy obtained in this study.

3.2 Grain structure (Crystallographic orientation on longitudinal cross section)

EBSD mapping obtained from the longitudinal cross section at about 5-7 mm from the bottom is shown in Fig.3. The crystallographic orientation of primary aluminum in Z-direction are indicated. Coordinate system is also shown in this figure. Almost all solidified grains elongated along the growth direction. It is also confirmed that the grains were grown from bottom to top without influenced by the mold wall.

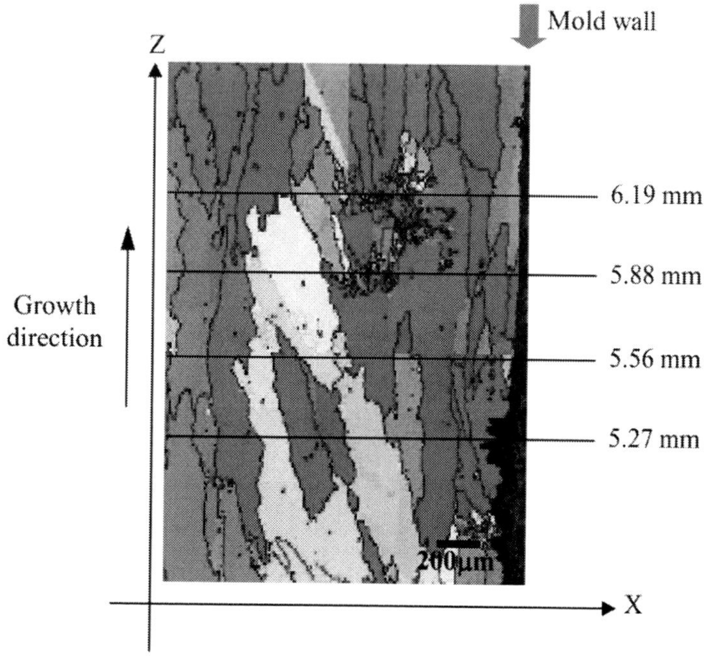

Figure 3. EBSD mapping of longitudinal cross section at 5-7 mm from the bottom.

3.3 Grain structure (Crystallographic orientation on horizontal cross section)

EBSD mappings of cross section are shown in Fig.4. These indicate crystallographic orientation in Z-direction and coordinate system is also shown. The height of cross section are: (a) 6.19 mm, (b) 5.88 mm, (c) 5.56 mm, and (d) 5.27 mm, respectively. The grain boundary was drawn by black line and was defined that the difference of orientation exceeds ten degrees. Some grains are named #1, #2, #3 and #4, as shown in Fig.4. With the growth, grain boundary between grain #1 and #2 moved gradually to +Y-direction. The grain boundary between grain #3 and #4 gradually moved to both +X-direction and +Y-direction as solidification proceeded. The grain boundary between grain #1 and #3 did not move. There are also some small grains in the upper left regions of the EBSD mapping. Actually, grain selection took place in this region as solidification proceeded. The analysis of grain selection for these grains was difficult, because the difference in height between two EBSD mappings (~200μm) was decided for relatively large grains. The reason that small grains are observed in the upper left regions of the EBSD mapping is not clear for the moment. But, solidified structure observed in the region might be affected by the mold wall.

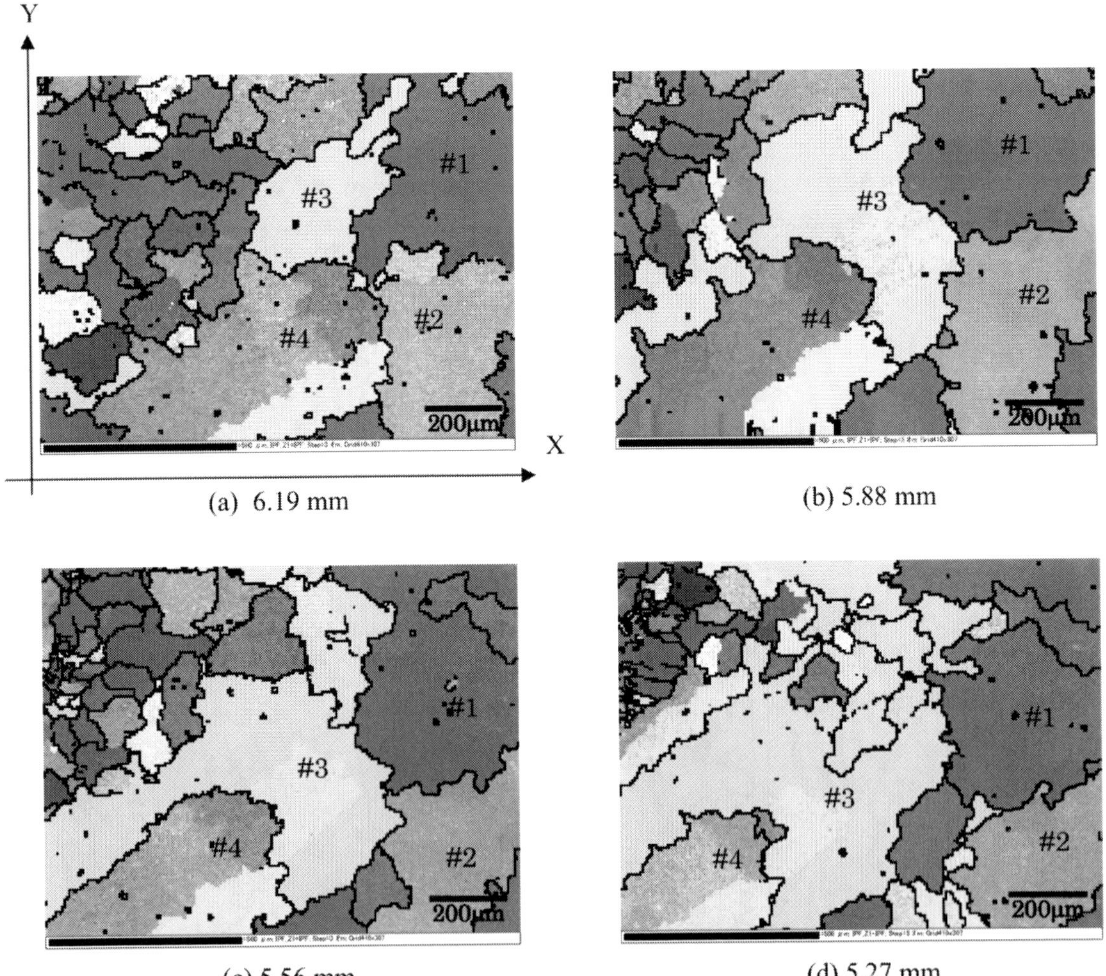

Figure 4. EBSD images of horizontal cross section at various solidification height (=distance from the bottom). (a) 6.19 mm, (b) 5.88 mm, (c) 5.56 mm, (d) 5.27 mm.

3.4 Pole figure measurement

Pole figures in Z-direction corresponding to Fig.4 are shown in Fig.5. When solidification height is 5.27 mm, it is rather random orientation in the area of measurement. As solidification proceeds, the scattering in orientation gradually decreases. As a concrete example, direction near <101> (=green) and near <111> (=blue) gradually decreases and direction near <001> (=red) seems to increases with increasing solidification height.

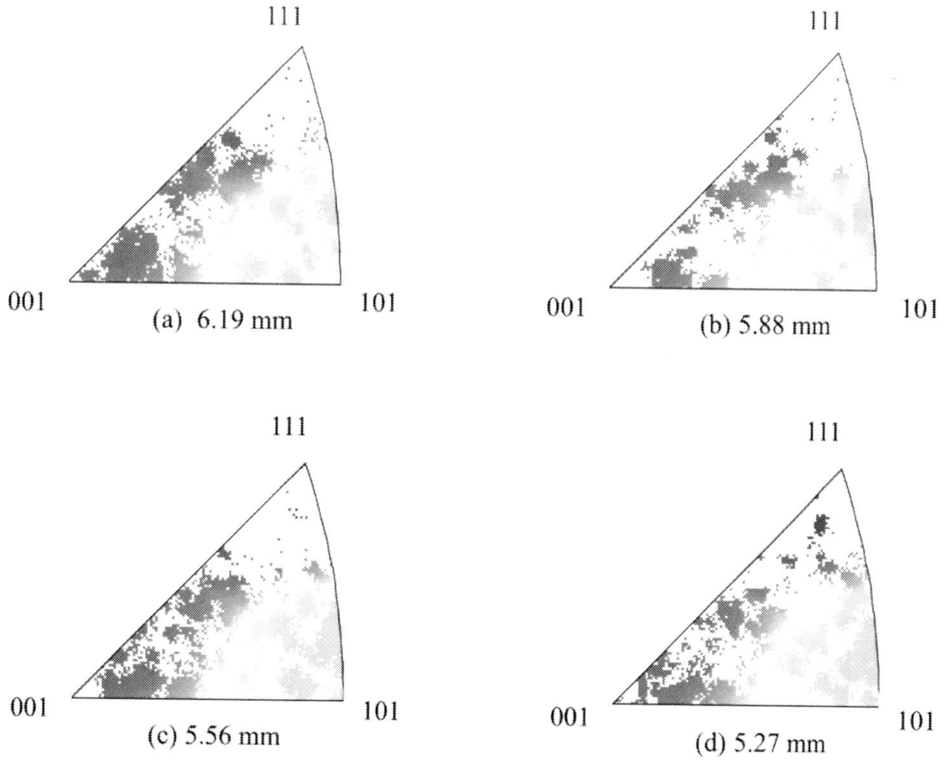

Figure 5. The pole figure images corresponding to Fig.4. (a) 6.19 mm, (b) 5.88 mm, (c) 5.56 mm, (d) 5.27 mm.

4. Discussion

4.1 Analysis using 3D-CAD

A spatial arrangement of primary dendrite trunk and coordinate system is shown in Fig.6. Zenith angle θ and rotation angle ϕ are defined in this figure. Based on the crystallographic data obtained from EBSD analysis, 3D relationship between primary dendrite trunk was analyzed using 3D-CAD.

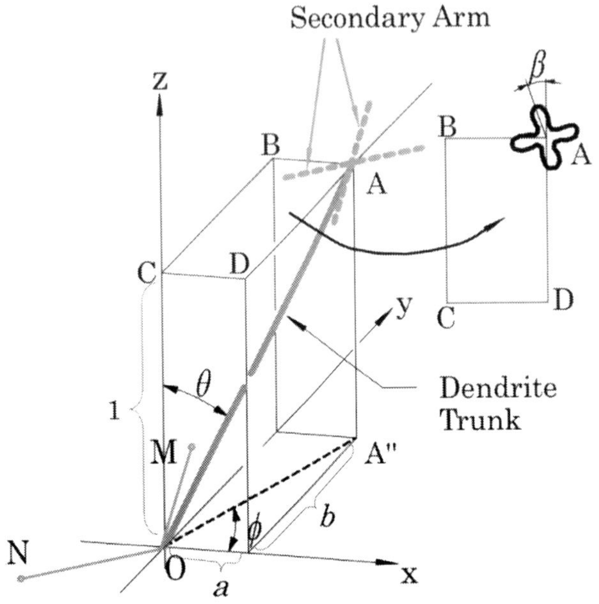

Figure 6. Definition of dendrite trunk and secondary arm plates indicating zenith angle θ and rotation angle ϕ.

4.2 Behavior of grain boundary

It is found in Fig.4 that some grain boundaries moved to certain direction. In this section, the behavior of grain boundaries will be analyzed.

(1) Grain boundary between grain #1 and #2.

As described before, grain boundary between grain #1 and #2 moved gradually to +Y-direction as solidification proceeded. The reason of this is following. According to the crystallographic data of EBSD, the directions of the primary dendrites which constitutes the grain #1 and #2 can be calculated. θ and ϕ of grain #1 and #2 are 11.4 °, 79.9 °, and 19.3 °, 2.0 °, respectively.

The grains #1 and #2 are in a row along Y-direction, as shown in Fig.4. Therefore, two dendrite trunks were drawn separating in Y- direction. The bird's eye view of this drawing is shown in Fig.7(a). As shown, the primary trunks of the grain #1 and #2 are in the position of a twist mutually. Therefore, it is not easy to understand how a grain boundary between the grain #1 and #2 forms. Thus, a simple analysis has been performed with a view in X-direction, since a motion of grain boundary in Y-direction should be discussed. The projection in X-direction is shown in Fig.7(b). Using this figure, a quasi-2D analysis can be carried out. It can be recognized from Fig.7(b) that the primary trunks of grain #1 and #2 diverge. It is known from analysis in 2-D that branching takes place and the grain boundary forms between two angles of primary trunks in case of divergence [6-8]. Therefore, the grain boundary between grain #1 and #2 may be a thick dashed line as indicated in Fig.7(b). This dashed line will move to +Y-direction with increasing Z. This indicates that a grain boundary between grain #1 and #2 moves to +Y-direction as solidification proceeds. This agrees with the experimental results described in the previous section.

(2) Grain boundary between grain #3 and #4.

As indicated before, the grain boundary between grain #3 and #4 gradually moved to both

+X-direction and +Y-direction with increasing solidification height. The reason of this is following. According to the crystallographic data of EBSD, the directions of the primary dendrites which constitutes the grain #3 and #4 can be calculated. θ and ϕ of grain #3 and #4 are 44.8 °, 74.8 °, and 32.3 °, 55.8 °, respectively. The grain #3 and #4 are in a row along both X-direction and Y-direction as shown in Fig.4. Therefore, two dendrite trunks should be drawn separating in X-direction and Y-direction. First, to understand the movement of grain boundary in X-direction the bird's eye view of this drawing separating in X-direction is shown in Fig.8(a). Again, the primary trunks of the grain #3 and #4 are in the position of a twist mutually. A simple analysis with projection in Y-direction was performed. This is shown in Fig.8(b) and it is found that dendrite trunks converge in this view. It is known from analysis in 2-D that blocking takes place in case of convergence. Thus, a grain boundary forms along the dendrite trunk which is closer to the heat flow direction than the other. The grain boundary in this case may be thick dashed line as shown in Fig.8(b). Here again, the dashed line will move to +X-direction with increasing Z and this agrees with experimental results. Secondly, to understand the behavior of grain boundary in Y-direction the bird's eye view of drawing separating in Y-direction is shown in Fig.8(c). A projection in X-direction was drawn and this is shown in Fig.8(d). Similarly, in this case branching will occur and the grain boundary may be a thick dashed line. This line moves +Y-direction with increasing Z-direction. This again agrees with the experimental observation.

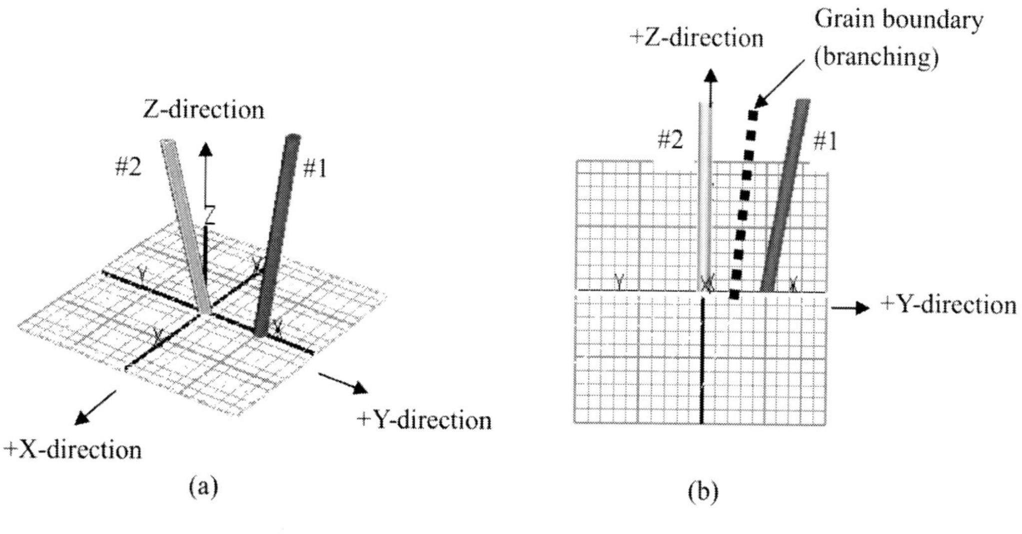

Figure 7. (a) Bird's eye view of dendrite trunks of grains #1 and #2. (b) Projection in X-direction for dendrite trunks of grains #1 and #2 and the grain boundary which may be formed between the grain #1 and #2.

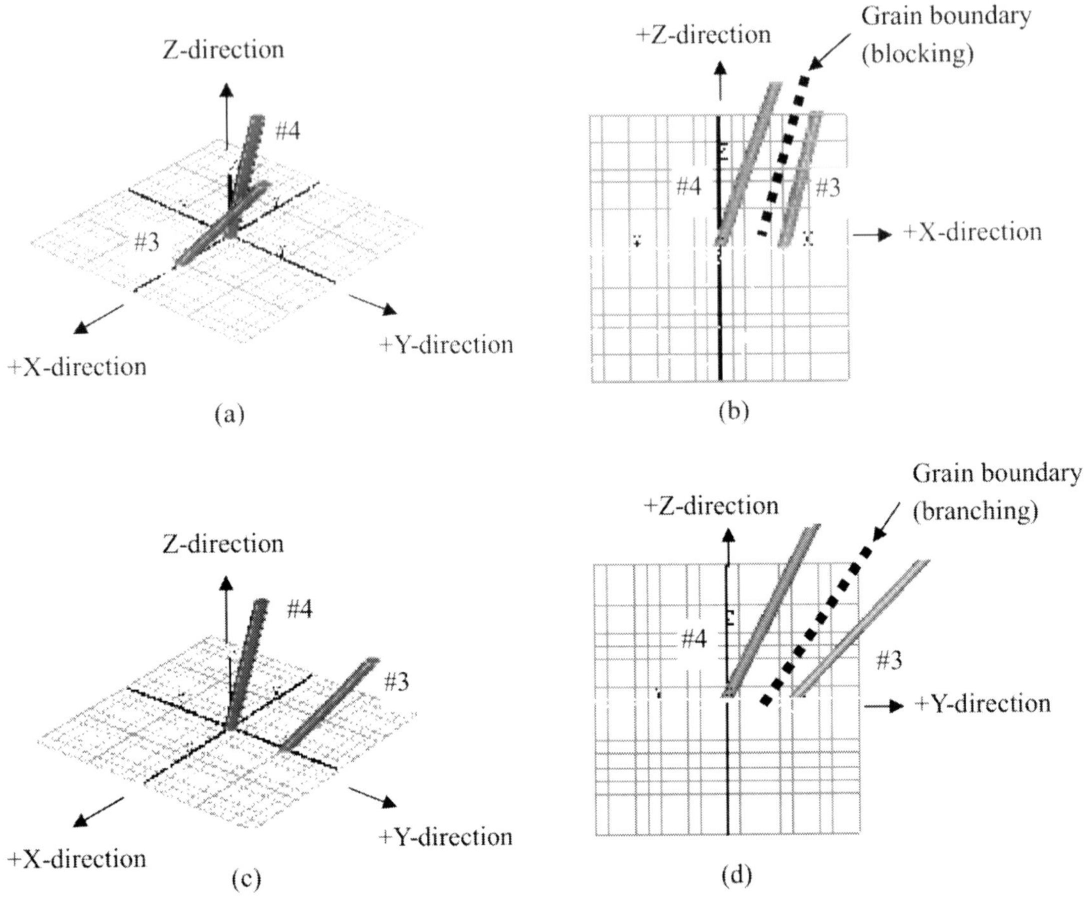

Figure 8. (a) Bird's eye view of dendrite trunks of grains #3 and #4. In this case two trunks are in row in X-direction. (b) Projection in Y-direction of Fig.8(a) and predicted grain boundary (thick-dotted line). (c) Bird's eye view of dendrite trunks of grains #3 and #4. In this case two trunks are in row in Y-direction. (d) Projection in X-direction of Fig.8(c) and predicted grain boundary (thick-dotted line).

5. Conclusions

In this study, the grain selection has been analyzed by EBSD on horizontal cross section of unidirectionally solidified Al-Cu alloy. Following conclusions have been obtained.

(1) The scattering in orientation gradually decreased as solidification proceeded.

(2) EBSD mapping on a series of cross section unveiled that a grain boundary moved towards a certain direction. The behavior of grain boundary can be explained qualitatively from the directions of dendrite trunks of both grains.

References

[1] T. F. Bower and M. C. Flemings: *Trans. Metall. AIME*, **239**(1967), 1620.
[2] M. C. Flemings: *Solidification Processing*, McGraw-Hill, New York, USA, (1974), pp. 134-176.
[3] W. Kurz and D. J. Fisher: *Fundamentals of Solidification 3rd ed.* Trans. Tech. Publications, Switzerland, (1992), pp. 63-92.
[4] D. Walton and B. Chalmers: Trans. Metall. Soc. AIME **215**(1959) 447-456.
[5] M. Durand-Charre: *The Microstructure of Superalloys*, Gordon and Breach Science Publications, Netherlands, (1997), pp. 53-66.
[6] H. Esaka, K. Shinozuka and M. Tamura: *Mater. Sci. Eng. A,* A413-414, (2005), 151-155.
[7] M. Rappaz and C. -A. Gandin: *Acta Mater.*, **41**(1993), 345-360.
[8] M. F. Zhu and C. P. Hong: *ISIJ Int.*, **42**(2002), 520

IOP Conf. Series: Materials Science and Engineering **33** (2012) 012083 doi:10.1088/1757-899X/33/1/012083

Cellular automaton modelling of ductile iron microstructure in the thin wall casting

A A Burbelko[1], D Gurgul, W Kapturkiewicz, M Górny
AGH University of Science and Technology, 23 Reymonta Str., Krakow, Poland

E-mail: abur@agh.edu.pl

Abstract. The mathematical model of the globular eutectic solidification in 2D was designed. Proposed model is based on the Cellular Automaton Finite Differences (CA-FD) calculation method. Model has been used for studies of the primary austenite and of globular eutectic grains growth during the ductile iron solidification in the thin wall casting. Model takes into account, among other things, non-uniform temperature distribution in the casting wall cross-section, kinetics of the austenite and graphite grains nucleation, and non-equilibrium nature of the interphase boundary migration.

1. Introduction

Nodular graphite cast iron, also known as ductile iron (DI), has major applications in critical engineering parts due to its mechanical properties and castability. The mechanical and physical properties of this material depend on the shape and number of the graphite grains and microstructure of the metallic matrix.

Solidification of DI was a subject of many computer modelling programs described in literature [1 -5], in which the stationary conditions of carbon diffusion in austenite is pre-assumed. Recently, a tendency for production of thin-walled castings has been observed [6-8]. In this technology, the process of the fast solidification may be very far from equilibrium and steady-state conditions [9].

The purpose of the present work is a two-dimension model development for simulation of the DI structure formation during the solidification in the condition of non steady-state temperature and diffusion fields in the thin-wall casting.

2. Model of process

The CA-FD is one of the known methods of the simulation of microstructure formation during the solidification [10, 11]. In the CA microstructure modelling the outer grain shape is the result of the simulation and does not superimposed beforehand. The model development for a one-phase microstructure evolution is a subject of the numerous researches [12 -21]. Model of the eutectic solidification of DI in the uniform temperature field and superimposed cooling rate is known [22].

Presented model is based on the CA-FD technique and will predict solidification of DI in the non-uniform temperature field during the cooling of the thin-wall casting in the sand mould. Model takes into account the continuous nucleation of austenite and graphite grains from liquid controlled by undercooling, separate non-equilibrium growth of graphite nodules and austenite dendrites at the first

[1] To whom any correspondence should be addressed.

solidification stage, and the following cooperative growth of graphite-austenite eutectic in the binary Fe-C system. A set of six cell states for microstructure modelling: three mono-phase states – "liquid", "austenite", and "graphite" – and three two-phase states were used.

At the beginning, all of the cells in the CA lattice are in the "liquid" state and have initial temperature. The analysed casting domain is in the thermal contact with a sand mould with a normal initial temperature. Nucleation and growth of the solid grains are possible when the temperature of the liquid drops below the liquidus.

2.1. Heat and mass diffusion

The numerical solution of the nonlinear Fourier equation was used for heat flow in the analyzed domains (casting and mould):

$$c\frac{\partial T}{\partial \tau} = \nabla(\lambda \nabla T) + q_T \quad (1)$$

where: T is the temperature, τ is the time, λ is the thermal conductivity, and c is the volumetric specific heat, q_T is the latent heat generation rate.

Solute diffusion in the domains of every phase was calculated in the same manner as temperature distribution, by the numerical solution of the diffusion equation with a source term at the interface:

$$\frac{\partial C}{\partial \tau} = \nabla(D\nabla C) + q_C \quad (2)$$

where D is the solute diffusion coefficient, and C is the solute concentration in this phase.

Both source functions are equal to zero outside of the interface cells. In the interface cells the value of the heat and mass sources for the finite-difference scheme are:

$$q_T = L_{\alpha\to\beta}\frac{\partial f_\beta}{\partial \tau} \quad (3)$$

$$q_C = (C_\alpha - C_\beta)\frac{\partial f_\beta}{\partial \tau} \quad (4)$$

where $L_{\alpha\to\beta}$ is the volumetric latent heat of $\alpha\to\beta$ transformation, C_α and C_β are the carbon concentrations in the vanishing and growing phases, and Δf_β is the growth of the new phase volume fraction during the time step.

For thermal and diffusion calculation in the casting region, the following boundary conditions were used: at the top and bottom of the grid – periodic boundary condition; at the left side – symmetry BC; at the right side – mould with an absence of mass flow and heat flow through the border according to Newton law:

$$q = h(T_c - T_m) \quad (5)$$

where h is the heat transfer coefficient (value of 500 $W\cdot m^{-2}K^{-1}$ was assumed), and T_c and T_m are the temperature values of the casting and mould surfaces in the contact.

The details of solution are described in [23].

2.2. Nucleation

According to [24] undercooling of the nucleation from liquid during ductile iron solidification in the thin wall reaches up to 65 K below the equilibrium temperature. Grain nucleation in industrial alloys has a heterogeneous nature. The substrates for the nucleus are randomly distributed in the bulk. Bulk distribution of differently-sized substrates also has a stochastic nature. The undercooling value of substrate activation is a function of its size. Functional relationship between the active substrate fraction and undercooling ΔT should be a feature of the probability distribution law [25]. The undercooling value of each phase should be calculated relative to the appropriate liquidus lines.

The number of active substrates in the domain V of the melt with an undercooling ΔT below the liquidus may be calculated on the basis of the cumulative distribution function $F(\Delta T)$:

$$N = N_{\max} F(\Delta T)V \tag{6}$$

where: N_{max} is the maximum specific number of substrates for nucleation.

When one substrate position doesn't have any influence on another substrate's positions, the random variable calculated as the number of substrates in any random domain V will have the Poisson statistical distribution with the mean value $\nu = N_{max}V$. For this statistical distribution the probability density function is

$$P_r(k) = e^{-\nu}\nu^k \; k! \tag{7}$$

where: e is the Euler-Mascheroni constant, and k is the estimated number of substrates.

The method of nucleation modelling for a CA lattice is known [25]. According to [25] the undercooling values randomly generated with a statistical distribution curve are attributed to randomly chosen cells. If a cell is chosen several times (i.e., if it contains more than one nucleation site), only the smaller nucleation undercooling is used. If the CA cell is too high/large in size, the calculated grain density will be underestimated. A modified version of this algorithm was used in this paper.

The probability of the lack of a nucleus in the cell ($k = 0$) according to Poisson's statistic is equal to:

$$P_r(k=0) = e^{-\nu} \tag{8}$$

The probability of one substrate ($k = 1$) in the cell is represented by the following equation:

$$P_r(k=1) = \nu e^{-\nu} \tag{9}$$

The probability of more than one substrates in one cell may be calculated as:

$$P_r(k>1) = 1 - \nu e^{-\nu}(1+\nu) \tag{10}$$

The next way of substrate placement and the undercooling of nucleation selection is proposed based on the mean number of active substrates in one cell:

$$\nu = N_{max} F(\Delta T)\upsilon \tag{11}$$

where υ is a cell's volume (or surface for 2D).

For each of the cells in the CA a random number p should be generated with an equiprobability distribution in the (0..1] range. For small domains when $\nu << 1$ this value is equal to probability of one substrate presence in the analyzed domain. Because for small cells $P_r(k=0 \text{ or } k=1) \rightarrow 1$ and $P_r(k>1) \rightarrow 0$, the condition of the precise one substrate present in the cell is the following inequality:

$$p < N_{max} F(\Delta T)\upsilon \tag{12}$$

The nucleation undercooling in this case should be estimated on the basis of the inverse function of the above-mentioned cumulative distribution curve (fractile):

$$\Delta T = F^{-1}\left[p(\upsilon N_{\max})^{-1}\right] \tag{13}$$

The solid grain will begin to grow when the undercooling exceeds the above level. The substrates are present (and nucleation is possible) only in cells with a positive ΔT value.

The method of continuous nucleation prediction used in modelling has been presented in details in the [26].

The Weibull statistical distribution was used in this paper for nucleation modelling [27]. The specific number of active substrates is given by:

$$n = N_{\max} \exp\left(-b \cdot \Delta T^{-1}\right) \tag{14}$$

where N_{max} and b is a nucleation parameters.

The undercooling of nucleation can be calculated as

$$\Delta T = -b\left\{\ln\left[p\left(\upsilon N_{\max}\right)^{-1}\right]\right\}^{-1} \tag{15}$$

where p is a random number generated with an equiprobability distribution in the (0..1] range.

The values for N_{max} and b used in the present work for the modelling of graphite and austenite grains nucleation are listed in the table 1. The solid grain will begin to grow when the undercooling exceeds the above level. The substrates are present (and nucleation is possible) only in cells with a positive ΔT value. The state of the CA cell with the active nucleus varies from "liquid" to "austenite" or "graphite". The states of adjacent liquid cells are changed to the appropriate interface. The new phase growth and volume fraction changes are only possible in the interface cells.

Table 1. Nucleation parameters

	Austenite	Graphite
b (K)	5	50
N_{max} (m^{-2})	$3 \cdot 10^8$	$1 \cdot 10^9$

2.3. Grain growth

The kinetic undercooling of the mother liquid phase is a thermodynamic driving force of the new grains' growth. Total undercooling at the solid-liquid interface, hence the difference between the equilibrium solidification temperature T_{Eq} (determined from the phase equilibrium diagram of carbon concentration obtained during simulation on the transformation front) and the real temperature T_r is equal to the sum of capillary undercooling ΔT_κ and kinetic undercooling ΔT_μ:

$$T_{Eq} - T_r = \Delta T_\kappa + \Delta T_\mu \tag{16}$$

where $\Delta T_\kappa = \Gamma\kappa$, Γ is the Gibbs-Thomson coefficient, and κ is a front curvature.

Basing on [28], it has been assumed in the computations that the interface migration rate u is a linear function of the local kinetic undercooling ΔT_μ:

$$u = \mu \Delta T_\mu \tag{17}$$

where μ is the kinetic growth coefficient.

The increment of the new phase volume fraction in the interface cells Δf over one time step $\Delta\tau$ in the square CA cells of size a was calculated using the equation proposed in [29]:

$$\Delta f = u\,\Delta\tau \cdot \left[a\left(\cos\theta + \sin\theta\right)\right]^{-1} \tag{18}$$

where θ is the angle between the X axis and the normal direction of the grain interface.

If the phase volume fraction in the interface cell increases up to 1, this cell varies they state from interface to appropriate one-phase. Additionally, this cell captures all of the adjacent ones: their states exchange to the appropriative interface.

The normal direction of the grain boundary in the interface cells was determined by the approach of the F-vector [30]. The angle θ between the growth direction (normal to the grain boundary) and the positive X-axis direction was calculated as follows:

$$\theta = \arctan\left[\sum_{i,j} y_{i,j} f_{i,j} \cdot \left(\sum_{i,j} x_{i,j} f_{i,j}\right)^{-1}\right] \tag{19}$$

where: $f_{i,j}$ is the volume fraction of the phase in the cell (i,j), and $x_{i,j}$ and $y_{i,j}$ are the relative coordinates of the adjacent cells. The summation in (19) concerning the 20 neighbouring cells gives the best results of normal direction estimation [29].

Another details of the used CA-FD model of a two-phase eutectic solidification were introduced elsewhere [31, 32].

3. Parameters for modelling

Computations of the diffusion field and microstructure were carried out on a grid of 200×1000 cells. The side of each cell was 1 μm in length. An initial uniform carbon concentration in the binary Fe-C liquid was assumed, equal to 0.04675 part of the mass fraction (eutectic equivalent S_C = 1.10).

Parameters used in the modelling are shown in the table 2.

Table 2. Thermo-physical parameters used in the modelling.

Heat conductivity ($W \cdot m^{-1} \cdot K^{-1}$):			
– liquid	λ_L	30	[16]
– austenite	λ_γ	20	[33]
– graphite	λ_{gr}	20	[16]
– mould	λ_m	0,65	
Diffusivity of carbon in (m^2/s):			
– liquid	D_L	$1.25 \cdot 10^{-9}$	[34]
– austenite	D_γ	$5 \cdot 10^{-10}$	[16]
Transition heat (J/m^3):			
– liquid – austenite	$L_{L/\gamma}$	$19.7 \cdot 10^8$	[33]
– liquid – graphite	$L_{L/gr}$	$16.2 \cdot 10^5$	
– austenite – graphite	$L_{\gamma/gr}$	$8.8 \cdot 10^5$	
Specific heat ($J \cdot m^{-3} \cdot K^{-1}$):			
– liquid	$c_{v,L}$	$5.6 \cdot 10^6$	[33]
– austenite	$c_{v,\gamma}$	$5.84 \cdot 10^6$	[16]
– graphite	$c_{v,gr}$	$1.78 \cdot 10^6$	[33]
– mould	c_m	$1.67 \cdot 10^6$	
Gibbs-Thomson coefficient for interface ($m \cdot K$):			
– austenite – liquid	$\Gamma_{\gamma/L}$	$1.9 \cdot 10^{-7}$	[33]
– graphite – liquid	$\Gamma_{gr/L}$	$7.0 \cdot 10^{-6}$	
– graphite – austenite	$\Gamma_{gr/\gamma}$	$9.45 \cdot 10^{-6}$	
Kinetic coefficient of the interface ($m \cdot s^{-1} \cdot K^{-1}$):			
– austenite – liquid	$\mu_{\gamma/L}$	10^{-3}	[35]
– graphite – liquid	$\mu_{gr/L}$	10^{-8}	
– graphite – austenite	$\mu_{gr/\gamma}$	10^{-8}	

IOP Conf. Series: Materials Science and Engineering **33** (2012) 012083 doi:10.1088/1757-899X/33/1/012083

On the long sides of this mesh the periodic boundary condition (BC) was used [36] and on the short sides of this mesh the adiabatic one was used. Initial temperature of alloy is equal to 1637 K. For the temperature field simulation the following BC were using: the periodic BC for long sides, the adiabatic one at the casting axis (left side in the figure 1) and Newton BC with heat transfer coefficient $h = 500\ \mathrm{W{\cdot}m^{-2}{\cdot}K^{-1}}$ between casting and mould (right side in figure 1).

Casting is cooled in the green sound mould. For the temperature field calculation the 5×500 "rare" mesh was used with the spatial step equal to 50 μm. Mould initial temperature was equal to $T_m = 298$ K. On the long sides of this mesh the periodic BC was used. On the short outer side the Newton BC was assumed with $h = 100\ \mathrm{W{\cdot}m^{-2}{\cdot}K^{-1}}$.

4. Results of modelling

The history of microstructure formation in the 2 mm casting is presented in figure 1.

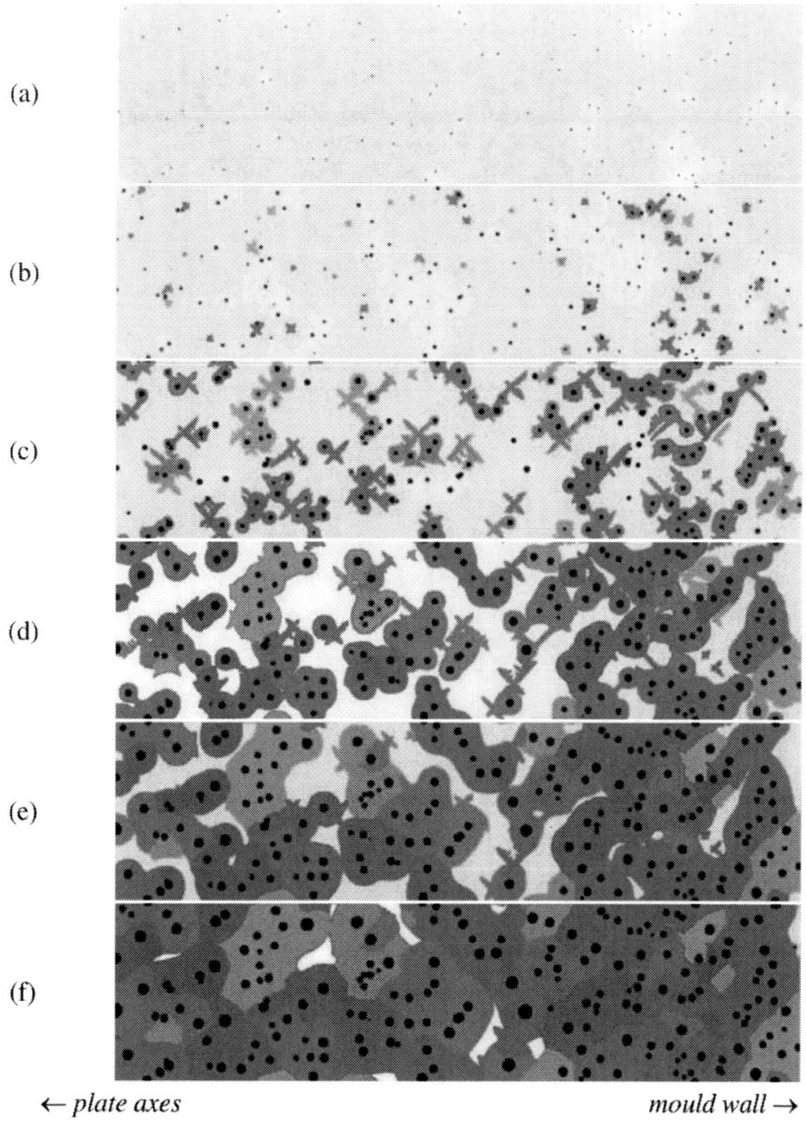

Figure 1. Microstructure of the DI (simulation results); cooling time: a) 3.8 s, b) 4.8 s, c) 6.8 s; d) 10.4 s; e) 13.4 s; f) 17.4 s.

The comparison of the virtual microstructure and microstructure of the real 2 mm casting (non-etched) is shown respectively in figures 2a and 2b. The simulated graphite nodules size and bulk distribution are similar to the experimental one.

(a)

(b)

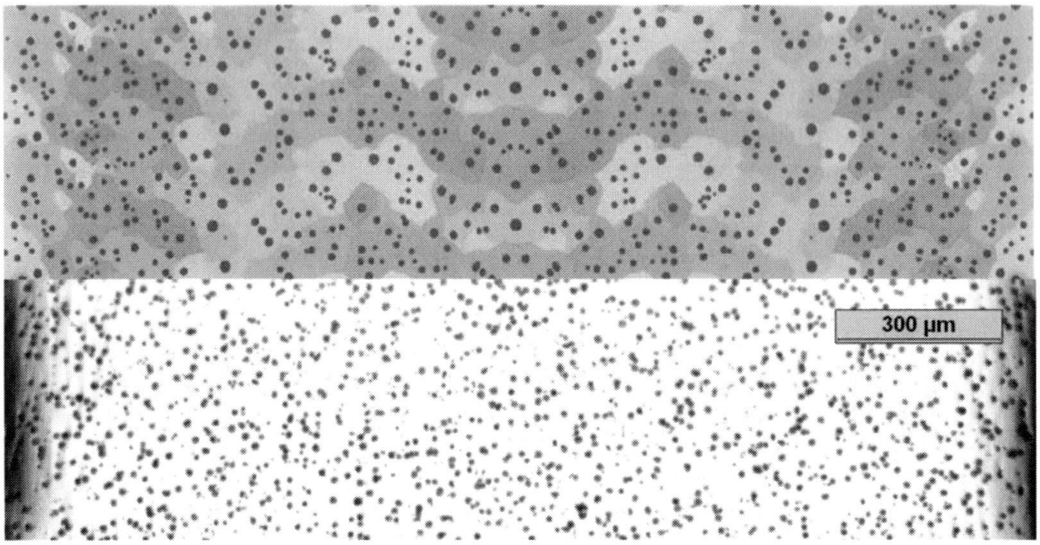

Figure 2. Final microstructure of the DI in the 2 mm wall: a) simulation for S_C = 1.1, separate austenite grains have different colour (results of modelling are repeated 4 times due to symmetry and periodic boundary conditions); b) real casting, DI with S_C = 1.1 (un-etched).

Figure 3 shows the cooling kinetics (modelling) for five simulations and the curve of the mean temperature value calculated on the base of these primary results. Primary results were obtained for the identical initial and border conditions.

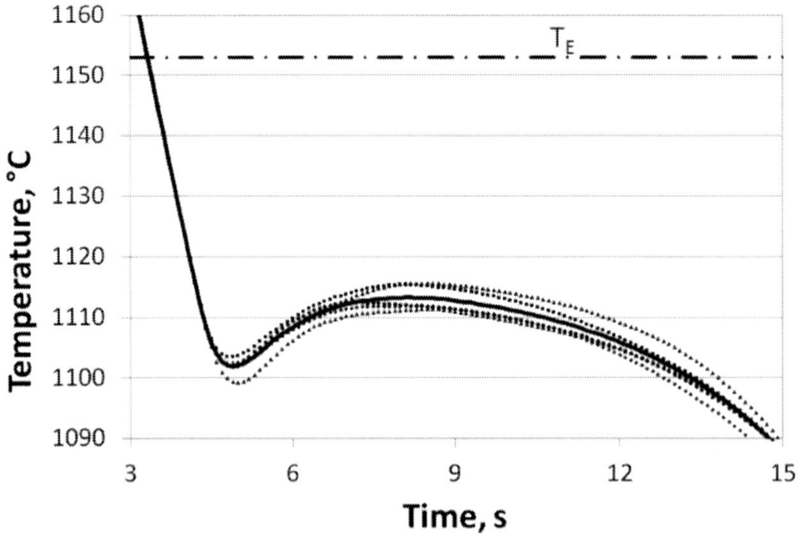

Figure 3. Five cooling curves obtained in the simulations with the identical initial and boundary conditions (points) and mean value of the temperature (solid line). Differences between presented primary cooling curves are the results of stochastic nature of the grain nucleation (random nucleation moments and places).

The observed differences between presented primary cooling curves are the results of stochastic nature of the grain nucleation (random nucleation moments and places according to description in the subsection 2.2). There is visible the big undercooling below the eutectic temperature. The range of the maximum undercooling is 50...54 K.

The kinetics of austenite and graphite growth during solidification is presented in figure 4.

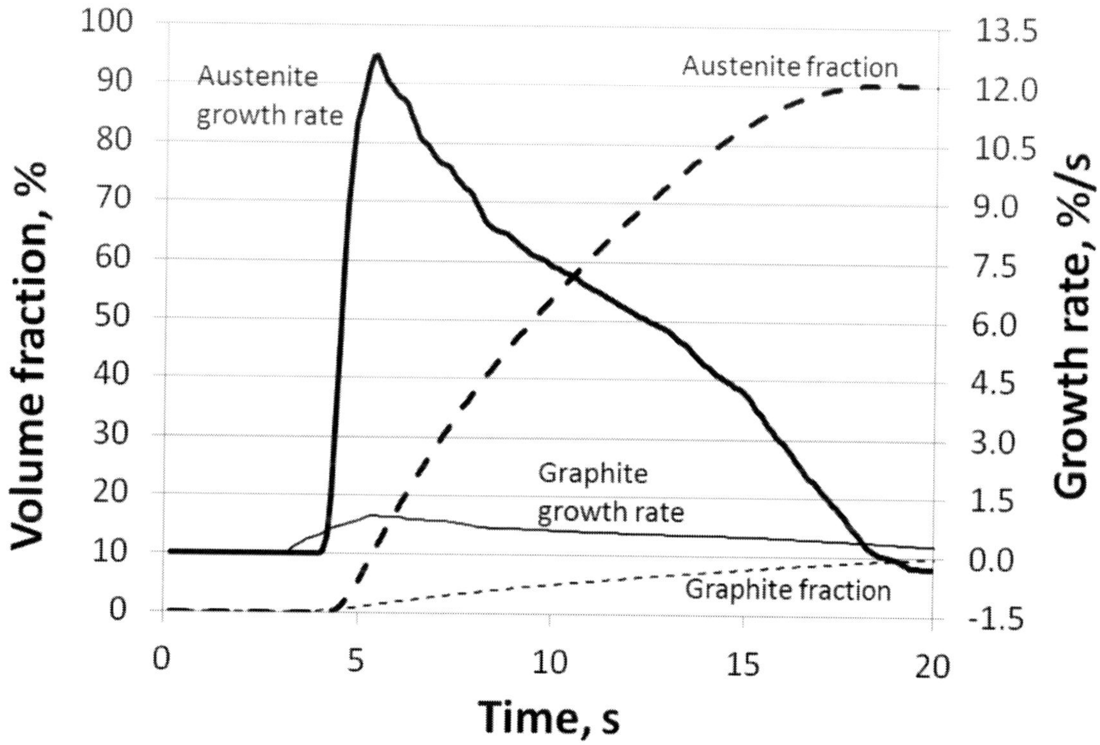

Figure 4. Kinetics of austenite and graphite growth (modelling).

5. Conclusions

The CA-FD computer model for the simulation of the DI solidification in thin wall was presented. The results of modelling were compared with the experimental microstructure. Results obtained in the 2 mm width casting wall confirm the model accuracy.

The model is well worked for thin walled ductile iron casting, for the high level undercooling below the equilibrium temperature and very fast solidification process.

Acknowledgements

This work was supported by Polish NCN project No. N N508 621140.

References

[1] Stefanescu D M, Catalina A, Guo X, Chuzhoy L, Pershing M A and Biltgen G L *1998* Prediction of room temperature mcrostructure and mechanical properties in iron castings *Modeling of Casting, Welding and Advanced Solidification Processes VIII (San Diego, CA, June 7-12, 1998)* ed B G Thomas and C Beckermann (TMS, Warrendale) pp 455–462

[2] S.M. Yoo, A. Ludwig and P.R. Sahm 1997 Numerical simulation of solidification of nodular cast iron in permanent molds *Proc. of the 4th decennial Intern. Conf. on Solidification*

Processing (Sheffield, UK, 07-10. July, 1997) ed J Beech and H Jones (Ranmoor House, Univ. of Sheffield) pp. 494–497
[3] Chang S, Shangguan D and Stefanescu D 1992 *Metal. Trans. A.* **23A** 1333
[4] Skaland T, Grong O and Grong T 1993 *Metal. Trans. A.* **24A** 2347
[5] Onsoien M, Grong O, Gundersen O and Skaland T 1999 *Metl. Mat. Trans. A.* **30A** 1053
[6] Fraś E and Górny M 2011 *Foundry Trade J. Int.* **185** 85
[7] Labrecque C and Gagne M 2003 *Int. J. of Cast Metals Res.* **16** 313
[8] Stefanescu D M, Ruxanda R E and Dix L P 2003 *Int. J. of Cast Metals Res.* **16** 319
[9] Fredriksson H, Stjerndahl J and Tinoco J 2005 *Mat. Sci. Eng. A.* **413** 363
[10] Rafii-Tabar H and Chirazi A 2002 *Physics Reports-Review Section of Physics Letters* **365** 145
[11] Lee P D, Chirazi A, Atwood R C and Wang W 2004 *Mat. Sci. Eng. A.* **365** 57
[12] Umantsev A R, Vinogradov V V and Borisov V T 1985 *Kristallografia* **30** 455 (in Russian)
[13] Rappaz M and Gandin Ch A 1993 *Acta Met. et Mater.* **41** 345
[14] Pan S and Zhu M 2010 *Acta Mater.* **58** 340
[15] Guillemot G, Gandin Ch A and Bellet M 2007 *J. of Crystal Growth* **303** 58
[16] Beltran-Sanchez L and Stefanescu D M 2004 *Metall. Mat. Trans. A.* **35** 2471
[17] Pavlyk V and Dilthey U 2004 *Modelling and Simulation in Materials Science and Engineering* **12** 33
[18] Zhu M F and Hong C P 2002 *ISIJ Int.* **42** 520
[19] Jarvis D J, Brown S G R and Spittle J A 2000 *Mat. Sci. Techn.* **16** 1420
[20] Burbelko A A, Fraś E, Kapturkiewicz W and Gurgul D 2010 *Mat. Sci. Forum* **649** 217
[21] Burbelko A A, Fraś E, Kapturkiewicz W and Olejnik E 2006 *Mat. Sci. Forum* **508** 405
[22] Zhao H L, Zhu M F and Stefanescu D M 2011 *Key Eng. Materials* **457** 324
[23] Kapturkiewicz W, Burbelko A A, Fraś E, Górny M and Gurgul D 2010 *J. of Achievments in Materials and Manufacturing Engineering* **43** 310
[24] Górny M. 2010 *ISIJ International* **50** 847
[25] Gandin Ch A and Rappaz M 1994 *Acta Metall. Mater.* **42** 2233
[26] Burbelko A, Fraś E, Gurgul D, Kapturkiewicz W and Sikora J 2011 *Key Eng. Materials* **457** 330
[27] Fraś E, Wiencek K, Burbelko A A and Górny M 2006 *Mat. Sci Forum* **508** 425
[28] Hoyt J and Asta M 2002 *Phys. Rev.* B **65** 1
[29] Burbelko A A, Kapturkiewicz W and Gurgul D 2007 Analysis of causes and means to reduce artificial anisotropy in modelling of the solidification process on cellular automaton *Proc. of the 4th decennial Intern. Conf. on Solidification Processing (Sheffield, UK, 07-10. July, 1997)* ed J Beech and H Jones (Ranmoor House, Univ. of Sheffield) pp 31–35
[30] Dilthley U and Pavlik V 1998 Numerical simulation of dendrite morphology and grain growth with modified cellular automata *Modeling of Casting, Welding and Advanced Solidification Processes VIII (San Diego, CA, June 7-12, 1998)* ed B G Thomas and C Beckermann (TMS, Warrendale) pp 589–596
[31] Burbelko A and Gurgul D 2011 *Computer Methods in Materials Science* **11** 128
[32] Gurgul D and Burbelko A A 2010 *Archives of Metallurgy and Materials* **55** 53
[33] *Tablicy fiziczeskih velichin* 1976 ed. Kikoin I.K (Moskwa Avtomizdat) (in Russian)
[34] Magnin P, Mason J T and Trivedi R 1991 *Acta Met. et Mater.* **39** 469
[35] Burbelko A 2004 *Mezomodelowanie krystalizacji metodą automatu komórkowego* (Kraków, UWND AGH) (in Polish)
[36] Chopard B and Droz M 2005 *Cellular Automata Modeling of Physical Systems* (Cambridge University Press)

IOP Conf. Series: Materials Science and Engineering **33** (2012) 012084 doi:10.1088/1757-899X/33/1/012084

A Stochastic mesoscopic model for predicting the globular grain structure and solute redistribution in cast alloys at low superheat

Laurentiu Nastac and Nagy El Kaddah

THE UNIVERSITY OF ALABAMA, Department of Metallurgical and Materials Engineering, P. O. Box 870202, Tuscaloosa, AL, 35487-0202, USA

E-mails: lnastac@eng.ua.edu and **and** Nelkaddah@eng.ua.edu

Abstract. It is well known that casting at low superheat has a strong influence on the solidification morphology and macro- and microstructures of the cast alloy. This paper describes a stochastic mesoscopic solidification model for predicting the grain structure and segregation in cast alloy at low superheat. This model was applied to predict the globular solidification morphology and size as well as solute redistribution of Al in cast Mg AZ31B alloy at superheat of 5°C produced by the Magnetic Suspension Melting (MSM) process, which is an integrated containerless induction melting and casting process. The castings produced at this low superheat have fine globular grain structure, with an average grain size of 80 μm, which is about 3 times smaller than that obtained by conventional casting techniques. The stochastic model was found to reasonably predict the observed grain structure and Al microsegregation. This makes the model a useful tool for controlling the structure of cast magnesium alloys.

1. Introduction

It is well established that the grain size, and solute segregation and precipitation of intermetallic phases within the grain have significant influence on mechanical properties of the cast alloys [1]. For typical cooling rates of sand and die castings -- in the range of 0.1-100°C/s -- the grain structure is dendritic, with segregation of the alloying elements in the matrix between the secondary dendrite arms [2, 3]. The detrimental effect of segregation on mechanical properties is generally minimized by reducing the grain size of as-cast alloy through either grain refinement and/or casting at high cooling rates [2-4].

One of the grain refinement methods is casting at low superheat. This method not only reduces the grain size, but the solidification morphology of the casting from a dendritic to a globular structure, further preventing segregation within the grains. Previous studies on casting superalloys using the Microcast X technology have shown that a superheat of approximately 10°C produces fine grains around 100 μm [5-7]. Studies on melting and casting Al-Li and Mg alloys at such low superheat using the Magnetic Suspension Melting (MSM) process have shown that the castings are homogenous, oxide free, and exhibit a fine globular grain structure [8, 9]. The average grain size for Al-Li 2090 and 8090 alloys unidirectionally-solidified using a stainless steel bottom chill mold was about 30 μm [10, 11], while that for Mg AZ31B alloy was approximately 80 µm [12, 13]. These studies have also shown that segregation of secondary and intermetallic phases did not occur within the grain, but rather only at the grain boundaries.

Recently, simulation of microstructure evolution during solidification has become a powerful tool for predicting the grain structure and segregation in cast alloys. Phase Field [14], Monte Carlo [15], Cellular Automaton (CA) [15-17], Stochastic Modeling [15-18], and Voronoi Tessellation [19] are the most common methods for simulating microstructure evolution during solidification. This paper describes a stochastic globular solidification model for cast alloys at low superheat and critically examines the model by comparing theoretically predicted against experimentally measured grain size and Al segregation in cast Mg AZ31B alloy at 5°C superheat.

2. Experimental technique

Casting of Mg AZ31B at low superheat was carried out using the MSM process, which is an integrated containerless induction melting and casting process specifically developed for reactive metals such as Mg. The experimental system is comprised of two main components: the melting compartment and the casting chamber as illustrated in figure 1. The charge is unidirectionally melted from the top to the bottom of the ingot and pours into a mold located in the casting chamber. A detailed description of the experimental system used in this study is given in reference [18].

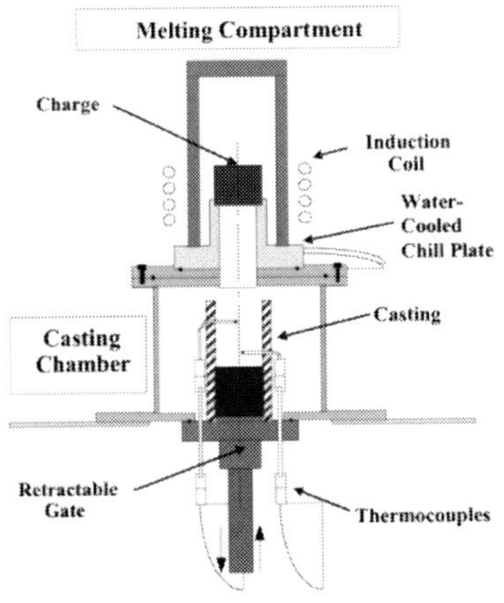

Figure 1. Schematic of the MSM system.

All experiments were conducted using an Mg AZ31B alloy with composition Mg-3.143%Al-0.66%Zn-0.377%Mn-0.051%Si. A typical experimental run involved placing short bars 76 mm in diameter and 63 mm high into the melting compartment and flushing the system with argon at a rate of 8.5 L/min. The coil was then energized, adjusting the current as necessary to maintain a constant superheat of 5°C. Upon melting the charge, the metal was poured into a bottom-chilled mold shown in figure 2. The ceramic mold was a 153 mm silica tube with 76 mm ID and 89 mm OD produced by Zircar Refractory Composites Inc. The stainless steel chill block used in this study was 76 mm in diameter and 50 mm high. A battery of thermocouples was used to measure the cooling rates at different heights in the center of the mold, and was attached to an Omega TC-08 data acquisition system.

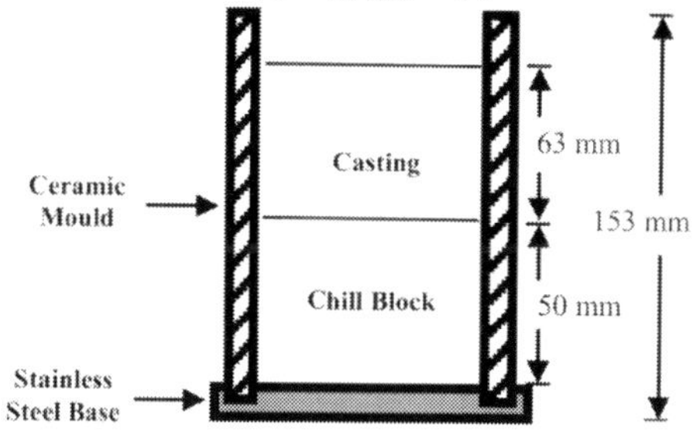

Figure 2. Sketch of the casting mold.

Characterization of the cast microstructures included optical microscopy and microprobe analysis. Optical micrographs were taken using a Nikon Epiphot 200 optical microscope, and EDS mapping of solute distribution was done using a JEOL 8600 FE SEM EPMA.

3. The stochastic globular solidification model

This section presents the formulation for a stochastic model for globular solidification. This model is based on previously developed models for dendritic solidification [15-18]. The presented globular solidification model takes into account the nucleation and growth of hexagonal crystals.

3.1. Model formulation

A stochastic mesoscopic model for grain structure evolution includes nucleation and growth kinetics, as well as the growth anisotropy and grain selection mechanisms. The geometry of the solidified material is first enmeshed in macro-volume elements for which energy, mass, and momentum transport can be calculated by a classical deterministic methods. Then, each macro-volume is subdivided into a number of cubic micro-volume elements that have a state index associated with them. In the time-step calculation, the average temperature of the specimen is calculated from an energy balance. When the average temperature is lower than the equilibrium (or nucleation) temperature, the nucleation and growth of grains begin. The structure of the stochastic micro-model consists of a regular network of microcells that resembles the macro-volume element.

The mesoscopic model is characterized by: (a) geometry of the microcell, (b) state of the microcell, (c) neighborhood configuration, and (d) transition rules that determine the state of the microcell. The geometry of microcell is chosen to be cubic in 3-D and square in 2-D calculations. Each microcell can have only two possible states: it is either liquid or solid. The ratio between the number of solid microcell s and the total number of microcells represents the fraction of solid within the macro-volume element. The neighborhood configurations are based only on the first- and second-order nearest neighbor microcell. More details about these configurations can be found in references [15, 17]. The probabilistic selection of these configurations is based upon the crystallographic anisotropy of grain growth. Solidification behavior depends to a great extent on the transition rules. In the present model, the change of state of the microcell from liquid to solid is initiated either by nucleation or by grain growth kinetics. For computational convenience, the models for the description of nucleation and grain growth kinetics are initiated based on the local solidification start time, t_L, which is calculated from the macroscopic model. The local solidification start times of microcells were obtained by interpolating the local solidification start times of macro-volume elements.

3.2. Growth anisotropy and probabilistic selection of neighborhood configuration

At the beginning of the simulation, all the microcells are liquid, and their state index is set to zero. As nucleation proceeds, some microcells become solid, and their index is changed to an integer larger than zero, *n*. The microcells in contact with the mold wall are identified with a different reference index, *m*. The index is transferred from the parent microcell to adjacent microcells, as they become solid through growth. For the case of dendritic solidification of AZ31B alloys, the integer takes into account the preferential growth of hexagonal crystals in the prism direction. The anisotropic characteristic in the prism direction certainly results in 6-fold preferred growth directions with angles between the primary trunks of dendrites of 60°. For graphical representation each integer has a color associated with it, and each microcell is a pixel on the computer screen. The crystallographic orientation of the new grains is chosen randomly among 255 orientation classes, which are the first 255 colors used for graphical representation.

In 2-D calculations, the probability, $dp(\theta)$, that a newly nucleated grain has a principal growth direction in the range $(\theta, \theta + d\theta)$ is given by:

$$dp(\theta) = \frac{3}{\pi}\, d\theta \tag{1}$$

where $3/\pi$ takes into account the six-fold symmetry of the hexagonal crystal, *i.e.*, the integral of $dp(\theta)$ from $-\pi/6$ to $\pi/6$ is equal to unity.

3.3. Nucleation

The number of grains, δN_e, that nucleate in the volume of the liquid and at the surface of the mold during one micro-time step, δt, are calculated by using nucleation site distribution, $N_e(\delta t)$. These distributions can be calculated by assuming some experimental approaches similar to those already used in deterministic models of solidification. The instantaneous nucleation model presented in [20] was used in this work to calculate the nucleation site distribution. Other nucleation models are described in references [17, 21]. Assuming no grain movement in the liquid, the grain density at any given location, *N*, can be expressed as a function of local cooling rate:

$$\delta N_e = C_0 + C_1\dot{T} + C_2\dot{T}^2 \tag{2}$$

where C_0, C_1, and C_2 are the nucleation parameters determined from experiments and $\dot{T}$ is the local cooling rate at the beginning of solidification.

The probability, dp_e, for a microcell located in the bulk of the liquid or at the surface of the mold to nucleate during the micro-time step, δt, are calculated as

$$dp_e = \frac{\delta N_e}{N^v_{\mu VE} - N^s_{\mu VE}} \tag{3}$$

where $N^v_{\mu VE}$ and $N^s_{\mu VE}$ are the number of microcells in the bulk and at the metal-mold interface, respectively. During each time-step calculation both metal-mold interface and bulk microcells are scanned and a random number, *rand*, is generated for each of them ($0 \le rand \le 1$). The nucleation of a microcell that is still liquid will occur only if $rand \le dp_e$.

3.4. Growth

It is assumed that a nucleus formed at a particular location will grow based upon a growth kinetics model and a neighborhood configuration rule previously described. As nucleation proceeds, the microcell becomes active (*n* or $m > 0$) and can grow over a distance, $d_{\mu VE}(t)$, that is given by the following equation:

$$d_{\mu VE}(t) = \int_{t_0}^{t} V(\Delta T)dt \tag{4}$$

where t_0 is the initial time, t is the actual time, and V is the growth velocity.

The initial time is defined as either the nucleation time, if growth is initiated through a nucleus, or the capturing time, if growth is initiated by capturing another microcell. At the capturing time, t_c, the initial microcell on which the growth was initiated captures the nearest neighbors based on a selected neighborhood configuration. For the case of equiaxed growth, this occurs when $d_{\mu VE}(t_c)$ is equal to the distance

$$a_\theta = a\sqrt{tan^2\theta + 1} \tag{5}$$

where a is the size of the microcell for the stochastic network and θ takes values from $-\pi/6$ to $\pi/6$. At the capturing time, the initial microcell becomes solid and its state index is transferred to the captured neighbors.

In order to reflect the probabilistic nature of grain extension, the nearest neighbors are trapped by the active microcells only when a randomly generated number, *rand*, is smaller than the capture probability, p_g, defined as follows:

$$p_g = \frac{d_{\mu VE}(t)}{a_\theta} \tag{6}$$

Further, the same procedure is used until all microcells become solid.

3.5. Growth kinetics models used for computation of $d_{\mu VE}$

The model developed by Nastac and Stefanescu [15] is applied for equiaxed solidification, where the growth velocity is described by

$$V = \mu\, \Delta T^2 \qquad \text{with} \qquad \mu = \frac{2\sigma^*}{\Gamma}\left[\frac{m_L(k-1)C_L^*}{D_L} + \frac{\rho L}{K_L}\right]^{-1} \tag{7}$$

where D_L is the liquid diffusivity, Γ is the Gibbs-Thomson coefficient, k is the partition ratio, m_L is the liquidus slope, ρ is the density, L is the latent heat of fusion, K_L is the liquid thermal conductivity, C_L^* is the liquid interface concentration, and the stability constant $\sigma^* = 1/4\pi^2$.

The melt undercooling for the system under consideration can be calculated based on the following definition/assumption:

$$\Delta T = T_L + m_L\left(C_L^* - C_o\right) - T_b \tag{8}$$

where T_L is the equilibrium liquidus temperature, $\langle C_L \rangle^L$ is the intrinsic volume-averaged liquid concentration, and T_b is the bulk temperature defined as the average temperature in the volume element. C_L^* is calculated with the segregation model described in [15, 18].

4. Results and discussions

4.1. Solidification morphology

Figures 3(a) and 3(b) show typical microstructures of cast AZ31-B alloy at superheats of 5°C and 60°C, respectively. These figures clearly show that casting at a low superheat produced globular solidification morphology while casting at high superheat is a dendritic one. It is also seen that casting at low superheat produced finer grain structures than that at high superheat.

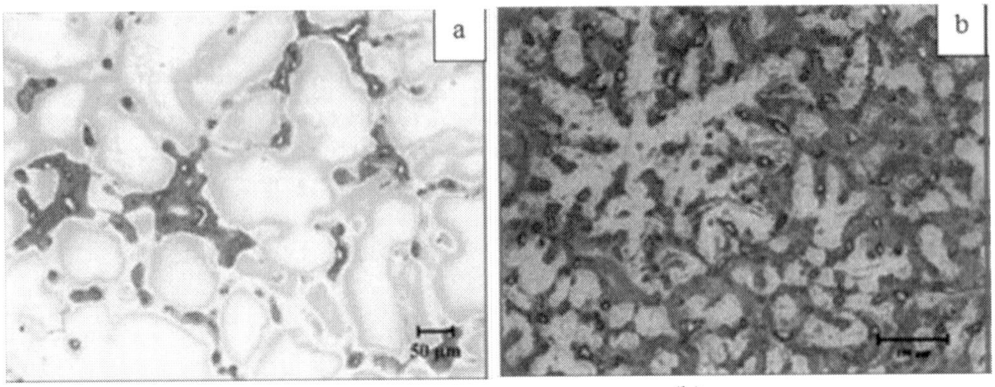

(a) (b)

Figure 3. Microstructures of MSM cast (a) and conventionally cast (b) magnesium AZ31B alloy.

The significant difference in the grain size and solidification morphology can be attributed to the combination of the high nucleation potential of the melt in the MSM process due to dendrite fragmentation during remelting of the alloy [22] and casting at low superheat. These two factors not only reduce the grain size in the casting but also act to enhance the grain-growth rate and reduce the temperature gradient at the solidification front, thus favoring the transition from dendritic to cellular solidification [16].

4.2. Stochastic mesoscopic modeling of microstructure evolution.

The simulation of the microstructure in the bottom-chill mold shown in figure 2 was carried out assuming zero heat flux at the top and side walls. In these computations, the pouring temperature of the molten AZ91B alloy was 632°C and the initial temperature of the stainless steel chill block was 20°C. Table 1 presents the data used in the microstructure modeling of Mg AZ31B alloy.

Figure 4 shows a comparison between the theoretically predicted and experimentally measured grain structure of Mg AZ31B castings. It can be seen from figure 4 that the predictions compare very well with the experimental microstructure in terms of size and morphology, thus validating the globular solidification model.

Table 1. Data Used in Microstructure Modeling of Alloy AZ31B

Solidification Kinetics Property	AZ31B
Liquid diffusivity, D_L [m^2 s^{-1}]	$3x10^{-9}$
Solid diffusivity, D_S [m^2 s^{-1}]	$1x10^{-12}$
Liquidus slope [K wt. $\%^{-1}$] [14]	$m_L = -6.6$
Initial Al concentration [wt. %]	$C_o = 3.14$
Al partition coefficient [14]	$k = 0.39$
Eutectic [wt. %]	32.3
Gibbs-Thomson coefficient, Γ [K m]	0.9×10^{-7}
Equiaxed nucleation parameters δN_e [nuclei m^{-3}]	$C_0 = 5x10^{12}$; $C_1 = 1x10^{12}$; $C_2 = 0$

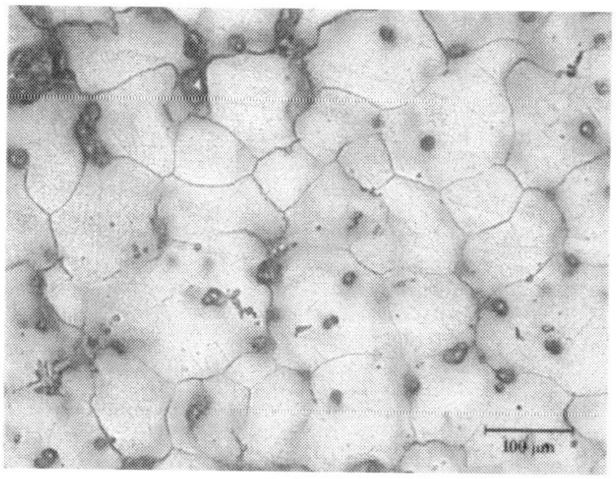

(a)

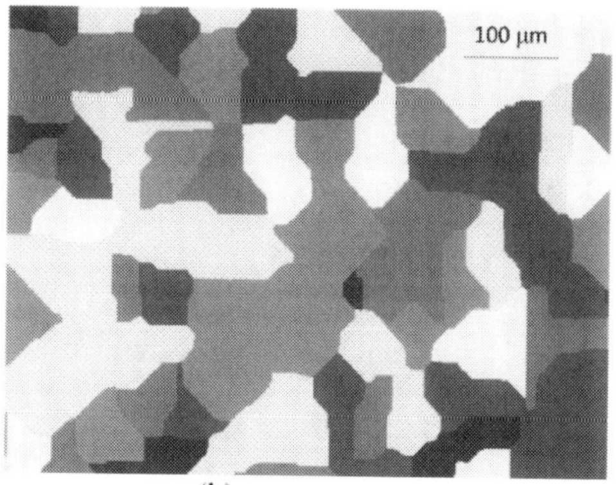

(b)

Figure 4. Comparison between (a) experimental and (b) simulated microstructures cast at 5°C superheat (solidification time, $t_s = 45s$).

4.3. Microsegregation

Figure 5a shows the predicted segregation of Al in the cast Mg AZ31B ingot at the end of solidification. This figure shows the coring phenomenon with enriched Al at the grain boundary similar to that shown in figure 3a. The measured and predicted Al segregation profile at the end of solidification within 2 globular grains is illustrated in figure 5b. The excellent agreement between the measured and predicted magnitude and profile of Al distribution further validates the segregation formulation of the model for globular solidification.

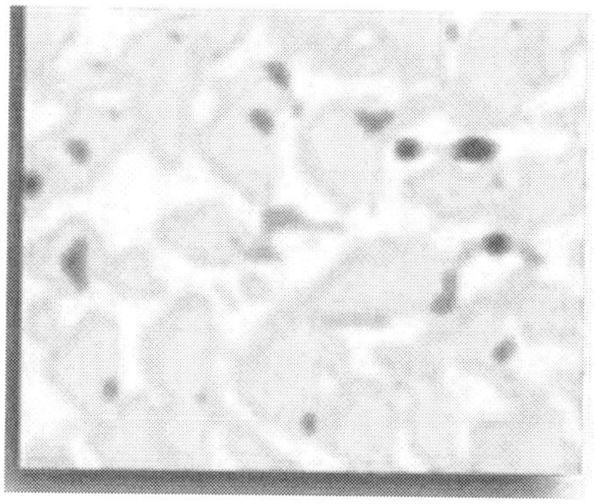

(a)

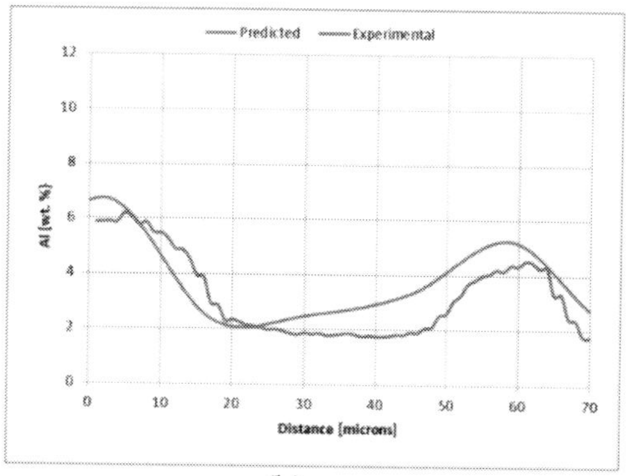

(b)

Figure 5. (a) Simulated microsegregation map of Al in cast AZ31B at 5°C superheat and (b) comparison of Al microsegregation profile within the grain against measurements.

5. Conclusions

A stochastic mesoscopic model was developed to predict the microstructure evolution in AZ31B castings. The model predictions were found to be in good agreement with the experimental measurements in terms of the microsegregation as well as of the size and morphology of the primary phase. This suggests that the developed stochastic mesoscopic model can be used to assist in controlling the microstructure of cast alloys with globular morphologies.

Acknowledgment
The authors wish to thank the National Science Foundation for funding this investigation under Grant No. CMMI-0856320.

References

[1] Stefanescu D M 1998 *Metal Casting Handbook 15* ASM International.
[2] StJohn D H 2005 *Metallurgical and Materials Trans. A* **36** 1669-79.
[3] Kurz W and Fisher D J 1986 *Fundamentals of Solidification* Trans Tech. Publications.
[4] Cao P, Qian M, and Stjohn D H 2005 *Scripta Materialia* **53** 7 841-44.
[5] Brinegar J R *et al.* 1989 *U.S. patent No. 4832112.*
[6] Ma Y *et al.* 2003 *Journal of Materials Processing Technology* **137** 35-39.
[7] L. Liu *et al.* 2005 *Materials Science and Engineering A* **394** 1-8.
[8] El-Kaddah N, Piwonka T S, and Berry J T 1991 *U.S. patent No. 5033948.*
[9] El-Kaddah N 1992 *Proc. of Int. Conf. on Magnetohydrodynamics in Process Metallurgy* Eds J. Szekely J *et al.* (TMS, 1992, p. 283).
[10] Reynolds D J, Shamsuzzoha M, and El-Kaddah N, Characterization of Al-Li Castings Produced by the Magnetic Suspension Melting Process 1997 *Proc. of the 4th Decennial International Conference on Solidification Processing* Eds. Beech J and Jones H (University of Sheffield, Sheffield, pp. 45-48).
[11] C. Mahato, M. Shamsuzzoha, and N. El-Kaddah, Solidification Morphology and Structure of Cast Al-Li 2090 Alloy at Low Superheats, 2004 *Solidification of Aluminum Alloys* Eds Chu M G *et al.* (TMS, Warrendale, OH, 2004, pp. 321-328).
[12] Rimkus N, Vihtelic L, and El-Kaddah N, Grain and Microstructure of Cast Mg AZ31-B Alloy Produced by the Magnetic Suspension Melting Process, 2010 *Proc. of Light Weight Materials for Vehicles and Components Symposium* (TMS, 2010, pp. 753-61).
[13] Rimkus N, Weaver M L, El-Kaddah N, Microstructure and Mechanical Behavior of Cast Mg AZ31B Alloy Produced by the Magnetic Suspension Melting Process *Proc. of Magnesium Technology 2011* (TMS, 2011, pp. 335-343).
[14] Warren J A and Boettinger W J 1995 *Acta Materialia*, **43** 2 689.
[15] Nastac L 2004 *Modeling and Simulation of Microstructure Evolution in Solidifying Alloys* Springer-New York: http://www.springer.com/materials/special+types/book/978-1-4020-7831-6
[16] Stefanescu D M 2002 *Science and Engineering of Casting Solidification* Springer 145-148.
[17] Nastac L, and Stefanescu D M 1997 *Modelling and Simulation in Materials Science and Engineering* **5** 391-420.
[18] Nastac L 1999 *Acta Materialia* **17** 4253-62.
[19] Vernede S and Rappaz M 2007 *Acta Materialia* **55** 1703-10.
[20] Stefanescu D M, Upadhya G, and Bandyopadhyay D, Heat Transfer-Solidification Kinetics Modeling of Solidification of Castings 1990 *Metallurgical and Materials Trans. A* **21** 997-1005.
[21] Olfield W, A Quantitative Approach to Casting Solidification: Freezing of Cast Iron, 1996 *ASM Trans.* **59** 945-59.
[22] Adams J, and El-Kaddah N, Containerless processing and characterization of high purity aluminum alloys 1993 *Proc. of the 1st Int. Conf. on Processing Materials for Properties*, (TMS, Warrendale, OH, 1993, pp. 905-08).

Prediction of Solidification and Microstructure of Inconel Alloy Using Numerical Simulation

J Roučka[1], V Kosour[2], M Kováč[3] V Krutiš[4], K Hrbacek[5]
[1] Ass.prof, BUT-FME Brno, CZ
[2] Doctor student, BUT-FME Brno, CZ
[3] Senior consultant Mecas Esi Plzeň, CZ
[4] Technical manager Mecas Esi Plzeň, CZ
[5] Prof, research manager, PBS Velká Bíteš, CZ

E-mail: roucka@fme.vutbr.cz

Abstract. Test castings of Inconel 713LC nickel alloy were prepared by pouring into shell molds in vacuum furnace. The temperature was experimentally measured in the course of solidification and cooling, and the structure was evaluated in as-cast condition. Applying thermal insulation of shells, the dependence of the structure of castings on cooling rate was monitored on test castings. The solidification pattern was solved numerically, using the ProCAST simulation program. Parameters for the simulation of as-cast structure were sought using the CAFÉ module of the ProCAST program. Calculation parameters were established with the experimentally determined temperature curves and structure.

1. Introduction

Cast blades for gas turbines and turbo-blowers and other heavy-duty components from nickel alloys are required to possess high integrity and meet the requirements for mechanical properties and, frequently, the prescribed type of microstructure. This is typically achieved by controlled solidification. The method is still based predominantly on the experience of process engineers and on conducting trials. Development of casting procedures for the above components could benefit from predicting the solidification process and the resulting microstructure by numerical simulation. In such case, computation parameters would have to be determined and verified by comparing with data for real-world castings. Finding accurate values of parameters is rather difficult.

This study describes numerical simulation of temperature field and microstructure in cast Inconel 713LC gas turbine stator blades.

Precision castings of Ni-based super alloys are mostly cast by the technology of self-supporting shell molds. The shell thickness must provide sufficient strength and it is usually 8 – 10 mm.The shells are annealed to temperatures of over 1,050 °C and by pouring their temperature ranges between 900 and 1,000 °C. The shell is usually formed by several castings located on a common gate. For reasons of economy, the aim is to have as many castings in one shell as possible and this leads to a high density of their arrangement.

As regards the heat transfer from molten metal to the mold, its accumulation and further distribution into the surrounding by pouring into self-supporting shells is quite a unique technology [1]. Whereas by metal solidification in massive molds the predominant part of heat from the solidifying casting, Q_{metal}, accumulates in the mold, Q_{mold}, the amount of heat transferred into the surrounding, Q_{surr}, is of little significance (Fig.1a). In thin-walled shells the proportion of accumulated heat is small due to their small thermal capacity, and a significant part of heat is transferred out of the mold already during the solidification (Fig. 1b). The rate of heat removal from the shell mold can be effectively controlled by applying heat insulation on the entire shell mold or on its parts. The insulation inhibits heat transfer and notably reduces the cooling rate of the metal – Fig. 1c. Insulators are often used to control solidification progress towards risers.

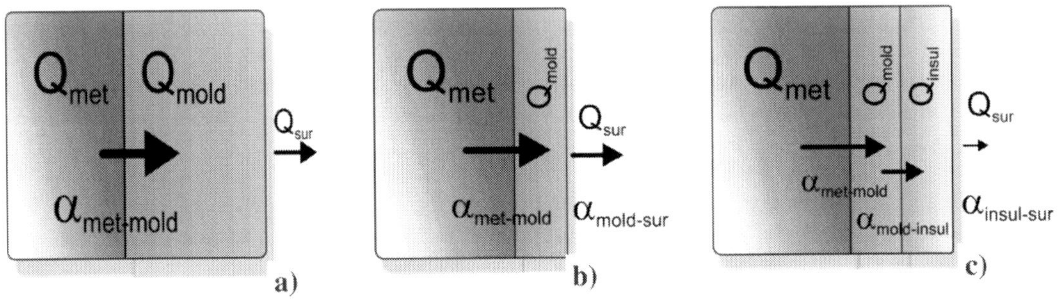

Figure 1. Distribution of heat during metal solidification:
a) massive mold, b) shell mold, c) shell mold with insulation

Because of the absence of gas atmosphere in the vacuum furnace, practically the only mechanism of heat transfer from the mold to the surrounding environment is radiation; there is no heat transfer via conduction from the mold into the neighborhood here (At the site where the shell is placed, heat is also conducted into the sand bed). This is accompanied by mutual radiation interaction with the surrounding environment; in molds of intricate shape with several castings the individual parts of the shell irradiate one another and thus affect the solidification of neighboring castings – Fig. 2. Such a mutual heat action may result in the heterogeneity of initial temperatures in various parts of the mold, changes in the progress of casting solidification, directional asymmetry of the structure, and it may affect the subsequent cooling process in the castings. When pouring into shells, local or overall insulation of shells is often applied. At points where it is applied, insulation reduces the intensity of shell cooling, in particular the radiation level, and changes the thermal effects of castings on one another.

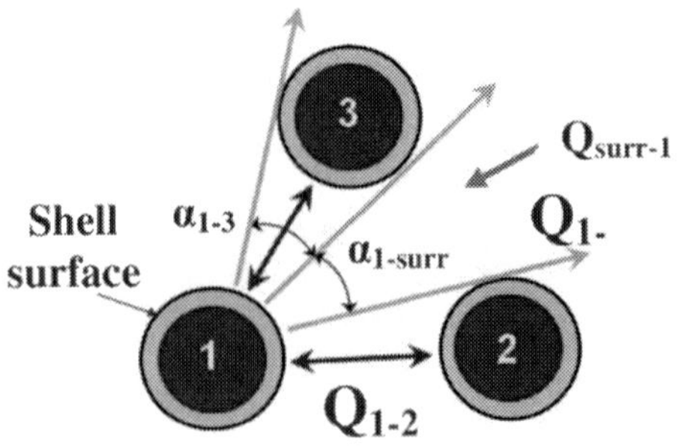

Figure 2. Radiation from outer surface of the shell. Radiation interaction between shell molds

This highly specific thermal situation in pouring into self-supporting shells must be respected in both the mold design and numerical simulation of the solidification and cooling of castings. In the case of numerical calculation it is in particular necessary to correctly characterize the conditions of heat accumulation in the shell during pouring and solidification. and, above all, the conditions of heat transfer on the outer ceramic surface, the effect of the shape configuration of the cluster, and also the effect of the mantel of the shell. Only very few verified data are available today for these calculations. In the simulations the calculation parameters were corrected in order to obtain the maximum possible agreement between calculation and reality.

2. Shell molds, melting facility, and temperature measurement

Tests were carried out while pouring in type VIM IC 5E double-chamber vacuum furnace (ALD Vacuum Technologies). Trial shell molds were produced in a standard fashion. Each frame comprised 4 vertically positioned blade patterns sharing a single gate. Two configurations heat insulation were tested. The reference configuration involved insulating the gate and runner sections of the shell mold with 13 mm-basalt fiber insulation fabric [2].

An alternative configuration included additional 6.5 mm-thick insulators on the top half of the blades.

Their purpose was to slow down solidification in the top section of the casting and induce directional solidification towards the gate.

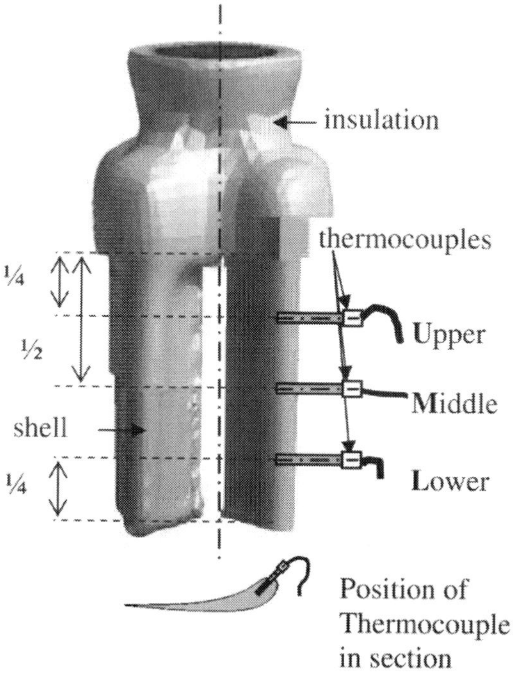

Self-supporting shells were made of molochite ceramic, 9 investment layers, shell thickness 8-9 mm. Prior to pouring, the shells were annealed to a temperature of 1,100 °C and placed in the chamber of vacuum furnace. The measurement was conducted using S type thermocouples of 0.35 mm in diameter in ceramic double-capillary tubes of dia.2.8 mm.

Temperatures were measured in blade castings at 3 height levels. They were positioned at these locations ¼, ½ and ¾ of the blade high. The shell temperature was measured too.The thermocouples deployed in metal were in protective quartz tubes.

The shell temperature was measured in the shell wall some 3 mm from the metal.– Fig. 3.

Figure 3. Shell with thermocouples

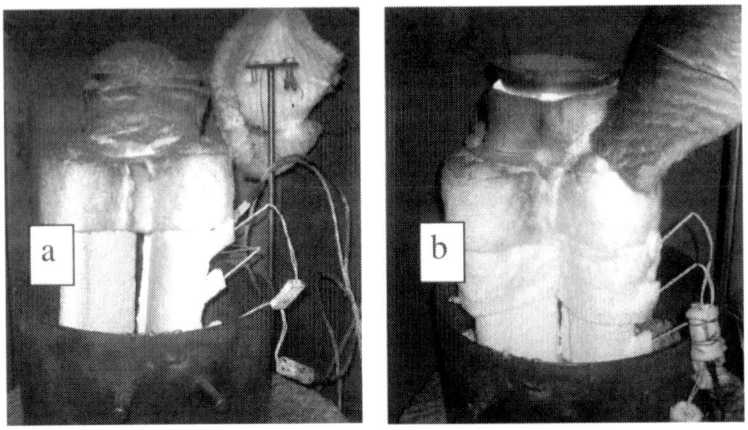

Figure 4. Shell mould with insulation: a) on the cross-shaped gate, b) on one half of the blade

The transfer of data from the vacuum furnace to the supply and recording devices is a major technical problem. The thermocouples are connected via compensation leads to an input of A/D converter ADAM 4018 (Advantech) with a sampling frequency of 10 Hz. To be protected against metal spatter and heat the converter is mounted on the lower side of casting bench. Through a bushing in the furnace wall the module ADAM 4018 is connected to the RS-232/RS-485 convector ADAM 4520 and to a measuring computer. The two modules are supplied from the PWR-242 module [3], [4]. The system must be resistant to the action of magnetic fields and utilizable in high vacuum. All the leads in the furnace are located in such a way that they cannot be damaged by the movement of inter-chamber valve, casting bench or the door (Fig. 5). The Adventech ADAM View Program was used to record the measurements [5], [6].

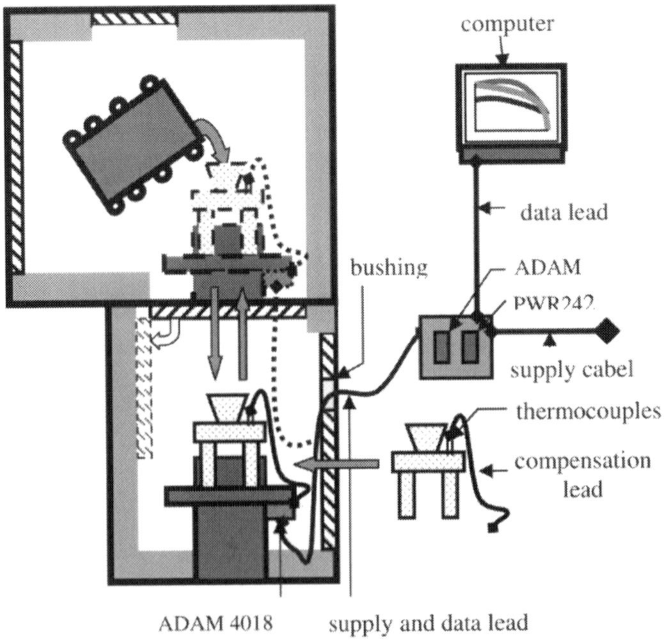

Figure 5. System of temperature measurement in vacuum furnace

IOP Conf. Series: Materials Science and Engineering **33** (2012) 012085 doi:10.1088/1757-899X/33/1/012085

3. Temperature Simulation

Figure 6 shows temperature plots for shell molds for both insulation configurations. Numerical calculation of cooling and solidification of castings requires thermal and physical characteristics to be determined for the metal and heat insulation and boundary conditions to be defined, i.e. the heat transfer between the components – Fig. 7.

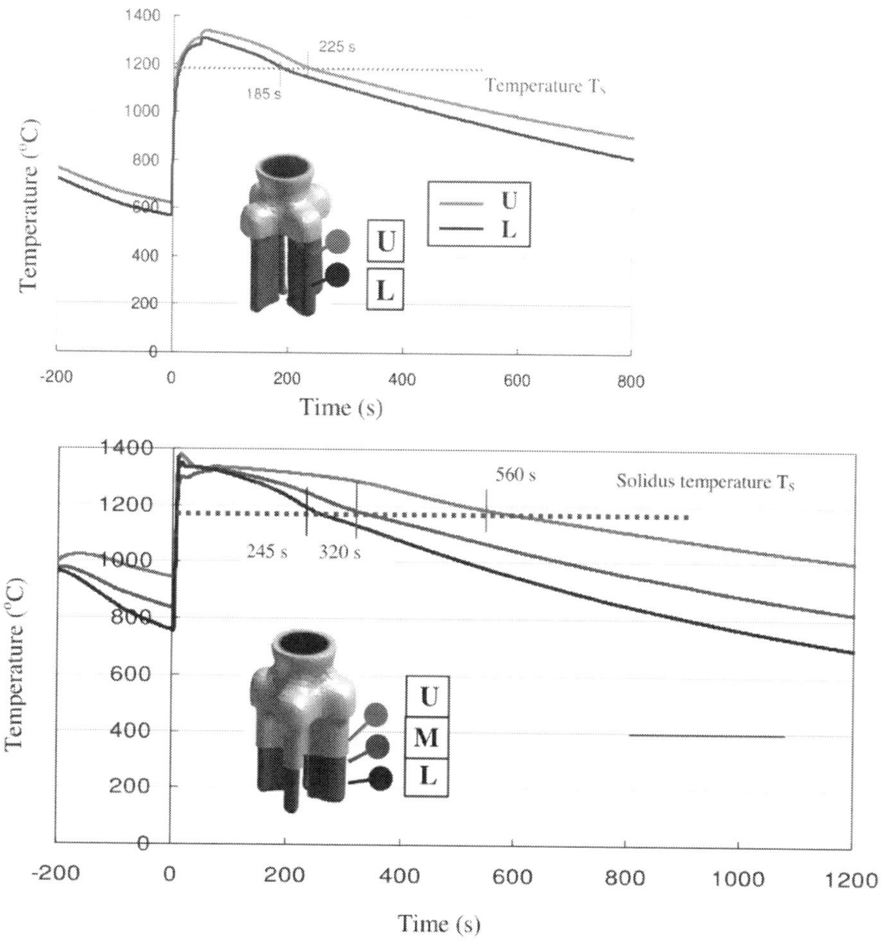

Figure 6. Measured cooling curves for shell moulds

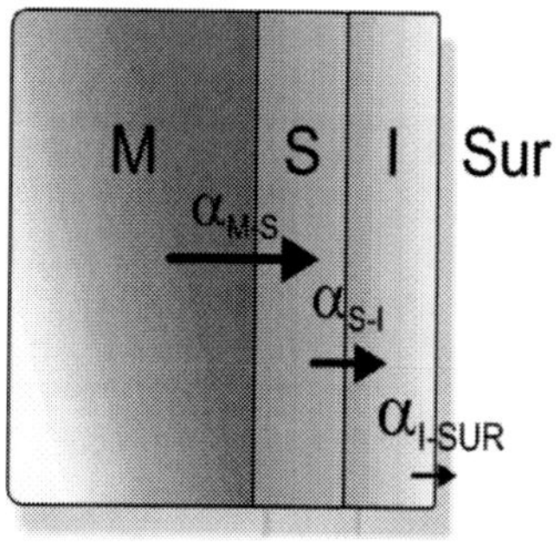

Figure 7. Heat transfer from castings to the environment.

$\alpha_{M\text{-}S}$ - heat transfer coefficient for the metal-shell mould interface

$\alpha_{S\text{-}I}$ - shell mould – insulation heat transfer coefficient

$\alpha_{I\text{-}S}$ - shell mould– environment heat transfer coefficient

The initial calculation was performed using data from the ProCast program database [7]. It showed that the experimentally measured and simulated data significantly differ and the calculation parameters will have to be validated. The validation was carried out by varying the heat transfer coefficients α for the interfaces between individual layers.

Casting-shell mould contact surfaces were described using a time-dependent coefficient with a range of values $\alpha_{met\text{-}shell}$ =1,500 – 400 W.m^{-2}.K, which took into account the changes in the character of the contact over the course of crystallization. Extraction of heat from the external surface of the shell mould or from insulators in vacuum is assumed to take place only by radiation. It is characterized by the emissivity coefficient depending on the surface temperature ε_{shell} = 0.6 – 0.8.

Figure 8 contains cooling curves which were measured and simulated after the final adjustment of calculation parameters for the mould with insulation half-covering the blades.

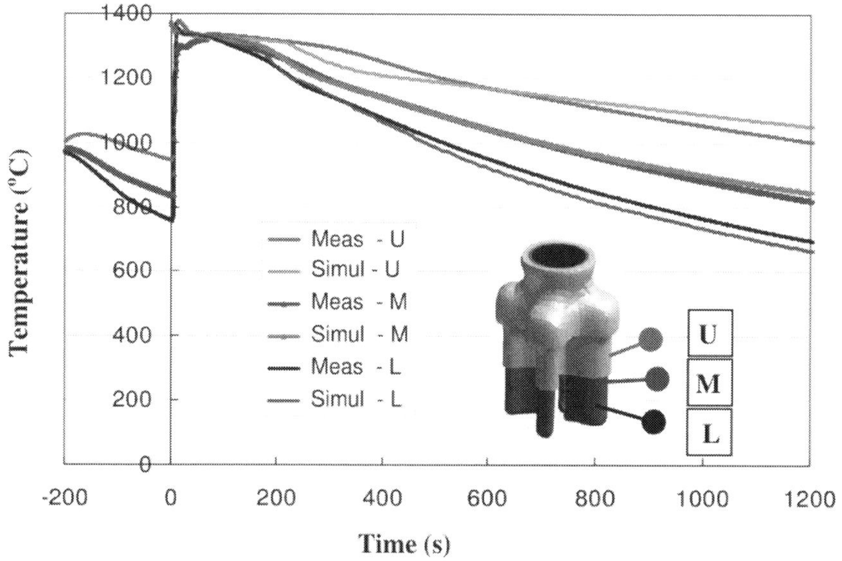

Figure 8. Comparison between real-world and simulated cooling upon adjustment

4. Microstructure of the blades castings and its simulation

The required microstructure of cast turbine blades, wheels and similar parts is specified by engineering designers typically in the form of permitted ranges of dendrites and dispersion of equiaxed grains. However, customer requirements in this field vary. The prescribed type of microstructure may be obtained by selecting proper thermal parameters for the metal and by controlling cooling rates in various sections of the mould. (There is another option in applying nucleating agents to the face of the shell mould.) Microstructures on cross-sections of these experimentally cast blades were examined using image analysis. Numbers of grains and mean, maximum and minimum grain size were measured as well.

Microstructures of the castings were simulated using CAFE module (Cellular Automatic Finite Elements Methods) [7], which is a part of the ProCAST system. The calculation is based on results of numerical simulation of temperatures, characteristics of surface and volume nucleation and parameters of kinetics of crystal growth. Parameters for some alloys are available in the database of the program. Data for other alloys must be calculated using CalcoSOFT3D, which is part of ProCAST or obtained from optimization calculations for specific alloys and types of castings. Sources of input data for calculations are very scarce and still unreliable.

The calculation requires input data on kinetics of crystal growth in dependence on the undercooling (below the equilibrium solidification temperature of the alloy) and data for surface and volume nucleation.

Nucleation is characterized by Gaussian surface and volume nucleation curves depending on the number of nucleation sites. The governing parameter is the undercooling, i.e. the temperature difference from the equilibrium transformation temperature (Fig. 9). The shape and location of the curves is given by

- undercooling, at which the maximum number of nuclei is present - dTm, DTm
- maximum number of nuclei - Gmax, Nmax
- variance – dTS, DTS

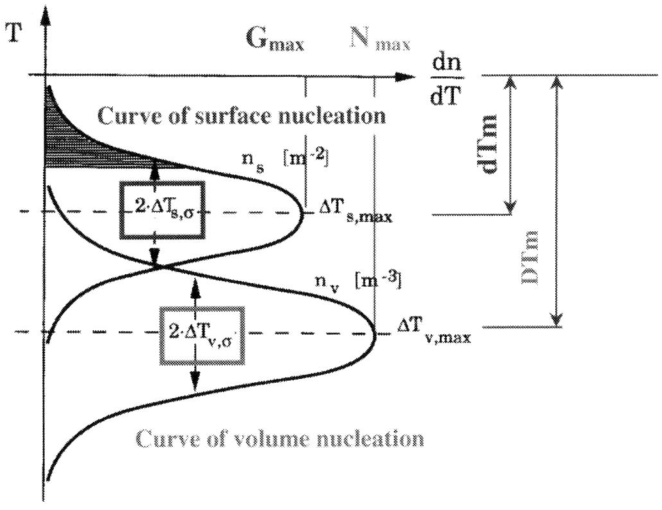

Figure 9. Surface and volume nucleation parameters

Kinetics of growth refers to the growth rate of crystals in dependence on the undercooling. Preparation can be performed with the aid of a special calculator built in the ProCast program. The calculator uses chemical composition of the alloy as input data.

The database of the program does not contain data for this particular alloy. The data was therefore sought by optimisation calculations. Various values of parameters were used to trace their impact on the resulting microstructure. It was found that the parameters have dramatic impact on the results of microstructure simulation. Figure 10 shows the actual microstructure on a cross section of a non-insulated blade one quarter length from its tip and results of several simulations with different parameters. Different parameter values lead to profoundly different microstructures. At the same time, no guidelines leading the user to the actual microstructure are available.

Figure 10. Real and simulated structures of the blade

5. Conclusion

Customers' unique and disparate requirements for precision castings from nickel-based superalloys with prescribed microstructure will necessitate the use of solidification and microstructure simulations. Simulation of cooling of shell-mould casting with its specific mechanism of heat extraction from metal is fundamentally different from simulation of casting in conventional moulds. The present study showed that results of numerical simulation can be brought to good agreement with real-world conditions. Simulation of microstructure requires further optimisation of parameters of the calculation.

References

[1] Incropera F 1990 *Fundamentals of Heat and Mass Transfer* (New York: Wiley &Sons)
[2] Zemčík L 2007 Technologie vakuového tavení žárupevných slitin *Slévárenství* **55** pp 87-91
[3] Roučka J et al 2008 *Proc. Int. Foundrymen Conf.* (Zagreb: University of Zagreb) 38-2008
[4] Roučka J et al 2008 *Proc. Metal* (Ostrava: Tanger) **183** p 89
[5] Roučka J et al 2009 Thermal Processes while Pouring into Ceramic Shells and their Numerical Simulation *Int. Journ. of Metalcasting* **61** pp 2-11
[6] Roučka J et al 2010 Solidification of Superalloys in Shell Moulds *Archives of Foundry Engineering vol* **10,** Issue 1 pp 137-146
[7] 2010 ProCAST Simulation Suite

This investigation was supported by the project of Ministery of Industry and Trade CZ, Nr. FR-TI2/104

IOP Conf. Series: Materials Science and Engineering **33** (2012) 012086 doi:10.1088/1757-899X/33/1/012086

Analysis of a numerical benchmark for columnar solidification of binary alloys

H Combeau[1*], M Bellet[2], Y Fautrelle[3], D Gobin[4], E Arquis[5], O Budenkova[3], B Dussoubs[1], Y Du Terrail[3], A Kumar[1], Ch-A Gandin[2], B Goyeau[6], S Mosbah[1], T Quatravaux[1], M Rady[5], M Založnik [1]

1 Institut Jean Lamour, CNRS – Université de Lorraine, Département SI2M, Ecole des Mines de Nancy, Parc de Saurupt CS14234, F-54042 Nancy cedex, France
2 MINES ParisTech & CNRS, CEMEF UMR 7635, 06904 Sophia Antipolis cedex, France
3 EPM-SIMAP - CNRS – INPG – Université Joseph Fourier, 38402 St Martin d'Hères cedex, France
4 FAST - CNRS – Université Pierre et Marie Curie, 91405 Orsay cedex, France
5 I2M- Dpt TREFLE - CNRS – Université de Bordeaux – Arts et Métiers ParisTech – ENSCBP, 33607 Pessac Cedex, France
6 EM2C - CNRS – Ecole Centrale Paris, 92295 Châtenay-Malabry, France

E-mail: herve.combeau@ijl.nancy-universite.fr

Abstract. During the solidification of metal alloys, chemical heterogeneities at the scale of the product develop. It is referred to as "macrosegregation". Numerical simulation tools exist in the industry. However, their predictive capabilities are not validated and are still limited. A 2D numerical benchmark is presented, based on the solidification of metallic Pb-Sn alloys. Concerning the numerical benchmark, a "minimal" common model of solidification is assumed, including columnar growth without undercooling, fixed solid, isotropic permeability of the mushy region, local thermodynamic equilibrium, lever-rule assumption for the local average composition. We focus our attention on the numerical method used to solve the average conservation equations: Finite Volume, Finite Element, Velocity-Pressure coupling treatment, scheme for convective terms, etc. At this stage of the work, we cannot exhibit a reference solution. However we draw some conclusions on the effects of the grid dependency, in particular on the location and sizes of the segregate channels. The development of both thermally and solutal driven convections in the first stage of the process (cf. low Prandtl and high Lewis numbers) and the relative independency of the convective scheme are also discussed. This presentation also have the goal to call other contributors to join this benchmark [1] in order to enrich the exercise and to reach a reference solution for this important problem in metallurgy.

1. Introduction

In the field of production of metal products, the solidification step is responsible of the formation of defects such as heterogeneities of composition (macrosegregation) and structure extending at the scale of the product [1]. The reduction of these defects is an important issue for industry because it damages the quality of the final properties of products. Numerical simulation is a valuable tool for progress in this direction. However, computer codes are not yet fully predictive. Several reasons are related to that which can be attributed on one side to the modeling aspect and on the other side to numerical aspects. The macrosegregation is caused by relative movements of solid and liquid phases such as: natural convection, flow induced by shrinkage for the liquid phase; grain motion, solid skeleton deformation for the solid phase [1]. Due to the complexity of the involved phenomena which take place over several characteristic length scales of space and time, there description by models is till now incomplete, even if only one of these phenomena is considered. The transport equations accounting for the different length scales are derived by an averaging technique [2-4] or the mixture theory [5]. By these two methods we can use a single set of equations in the whole domain (fully liquid, mushy zone and fully solid). At the beginning of the development of these models, simple assumptions concerning the transport phenomena at the microscopic scale were adopted, like the lever rule for the microsegregation. As the microsegregation is at the origin of the macrosegregation, several works have been dedicated to microsegregation models applicable for macrosegregation predictions [6-11]. Notably, macrosegregation has already been predicted with microsegregation models that were coupled with thermodynamic databases [10, 11, 13, 17]. The improvement of the description of the microstructure was also an important center of interest. By introducing the concept of the grain envelops to define a boundary between the grains and the liquid outside of the grains [7], it was possible to predict the evolution of the grain morphology. The most refined models predict the columnar to equiaxed transition [11], the grain size, the grain morphology and the macrosegregation induced by the motion of the grains and the interdendritic liquid [15]. The prediction of channel segregates was also the topic of many studies. Most of them were concerned with freckles, corresponding to a situation where the thermal gradient is oriented in the same direction as the gravity [1, 17]. Only few of them were related to other situations of solidification. Some studies were devoted to the case where the thermal gradient is perpendicular to the gravity. It was shown that the mesh size is the main parameter for channels prediction [14]. Refinement of the mesh size conduced to a change in the curvature of the channels and induced a preferential path of the flow inside the channels. Investigations on 3D effects started only recently [12, 16, 28]. Two different shapes of channel have been predicted, a first one with a tubular section whereas the second one is lamellar. Until now, the few experimental results available in the literature reported tubular channels [16]. More experimental investigations need to be done to confirm the existence of lamellar channels.

Concerning numerical simulation, difficulties are inherent to the values of the thermophysical properties in case of metals. Thermal or solutal natural convection are difficult to simulate with a good accuracy for liquid metals because of the low value of the Prandtl number (~ 0.01) on one hand and the great value of the Lewis number (~ 104). Nonlinearities in the equations like the permeability term in the momentum equation need also to be carefully accounted for in the discretization, as well as the strong coupling between the transport phenomena which necessitate developing specific algorithms [20,22].

In 2007, we decided to launch the SMACS project centered on the prediction of macrosegregation. This project involves five French laboratories: CEMEF, IJL, EM2C, EPM-SIMAP, and TREFLE. The aims of this project are one hand to compare the predictions of computer codes on base cases in natural convection and solidification (numerical benchmark); on the other hand, to qualify the quality of the prediction of the models by comparisons with experiments carried out at EPM-SIMAP [24]. For the numerical benchmark, the model has been fixed for all the partners and chosen as simple as possible with regard to the description of solidification, whereas for the second part the choice of the model was open.

This communication aims to present some of the results obtained in the benchmark part of the SMACS project. These results concern one of the two reference cases defined in the solidification part of the benchamrk and are relative to the formation of macrosegregation during the solidification of a Pb-18 wt% Sn alloy in a 2D cartesian domain.

2. Presentation of one of the two benchmark test cases in solidification

This test case consists to simulate the solidification of an alloy Pb-18 wt% Sn in a rectangular cavity of 0.1 m wide and 0.06 m high (see Figure 1).

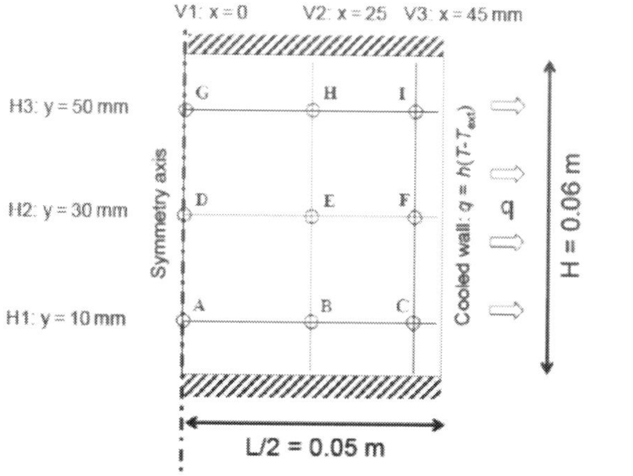

Figure 1. Description of the solidification problem (boundary conditions and locations of the horizontal and vertical lines as well as the locations of the points used for the outputs).

The top and bottom walls are assumed to be adiabatic, while the two vertical walls are cooled: a Fourier condition, the same for both sides is imposed. For the flow, a non-slip condition is imposed on all the boundaries of the domain. Due to the symmetry of geometry and boundary conditions and assuming again the symmetry of the solution, only half the field was considered. At the initial time, the alloy is assumed at rest (no velocity) and its temperature equal to the liquidus temperature corresponding to its nominal composition (285.5 ° C). The geometry of the domain, the boundary conditions and the location of the outputs are shown in Figure 1.

The model describing the solidification has been set; each partner has taken care to solve the same system of equations. The equations expressing the conservation of: energy, global mass, momentum and mass of solute were derived from the method of averaging volume. By taking the order listed above these equations are written as:

$$\rho\frac{\partial\langle h\rangle}{\partial t} + \rho c_p \nabla T\cdot\mathbf{V} - \nabla\cdot(k\nabla T) = 0 \tag{1}$$

$$\nabla\cdot\mathbf{V} = 0 \tag{2}$$

$$\rho\frac{\partial\mathbf{V}}{\partial t} + \frac{\rho}{g_l}(\nabla\mathbf{V})\mathbf{V} = \nabla\cdot(\mu_l\nabla\mathbf{V}) - g_l\nabla p - \frac{\mu_l g_l}{\mathrm{K}}\mathbf{V} + g_l\rho(1-\beta_T(T-T_0)-\beta_c(C_l-C_0))\mathbf{g} \tag{3}$$

$$\frac{\partial\langle C\rangle}{\partial t} + \nabla C_l\cdot\mathbf{V} = 0 \tag{4}$$

The definition of the variables is given in the nomenclature at the end of this document. The chosen microsegregation model corresponds to the complete thermodynamic equilibrium between liquid and solid phases (lever rule):

$$\langle C\rangle = g_l C_l + g_s C_s = \left(g_l + (1-g_l)k_0\right)\cdot C_l \tag{5}$$

$$T = T_m + m_l C_l \tag{6}$$

Full details of the case with all the thermo-physical data and definition of outputs can be downloaded on a dedicated website [25] and in ref [18].

Five computer codes have been used to simulate this test case. Four computer codes are based on the finite volume formulation (FVM) (FLUENT 6.2 (EPM-SIMAP), THETIS (TREFLE/EM2C) [2], SOLID (IJL) [20] and OpenFOAM (IJL)) and one code is based on the finite element method with unstructured triangular meshes (FEM) (R2SOL (CEMEF) [21]) (see Table 1). For FLUENT, THETIS and OpenFOAM solvers the energy conservation equation (1) is solved with the temperature as the main unknown. For R2SOL and SOLID solvers, equation (1) is solved with the mass average enthalpy as the main unknown [20]. For these two codes, the solute mass conservation equation (4) is solved with the average composition as the main unknown, following the scheme proposed by Voller [22]. In the case of EPM-SIMAP, this equation is separated in two parts, one for the liquid phase and one for the solid phase. For the discretization in space of the convective terms, a second order scheme was used in the case of TREFFLE (RST) and EPM-SIMAP (second order upwind), IJL in case of SOLID used a 1st order upwind scheme and in case of OpenFOAM used a first order upwind or a QUICK scheme, whereas CEMEF used an SUPG stabilization scheme.

Table 1. Key features of the computer codes mesh size and time step used for the results presented in this paper

Group	**Software**	**Mesh (size, average mesh size (m))**	**Time step (s)**
IJL	SOLID F.V. (upwind)	192x232 (2.6 x 10^{-4})	5 x 10^{-3}
CEMEF	R2SOL F.E. (SUPG)	46502 nodes (2.5 x 10^{-4})	5 x 10^{-3}
EPM-SIMAP	FLUENT F.V. (second order upwind)	200x240 (2,5 x 10^{-4})	5 x 10^{-3}
TREFLE	THETIS F.V. (TVD)	268x324 (1,9 x 10^{-4})	1 x 10^{-3}
IJL	OpenFOAM F.V. (upwind or QUICK)	200x240 (2,5 x 10^{-4})	5 x 10^{-3}

The meshes and time steps values used for the results presented in this paper are reported in Table 1. A sensitivity study to the mesh and time step has been carried out by each partner whose results are presented in section 4.

3. Description of the progress of the solidification

The results obtained by EPM-SIMAP are presented in this section in order to describe the evolution of solidification in the domain. The total time of solidification is about 600 seconds. From the temperature, liquid fraction and velocity fields at time 120 seconds reported in Figure 2, one can see that the isotherms are almost vertical in the region where the liquid fraction is less than 0.8 whereas there is a thermal stratification in the central part where the liquid fraction is larger. The circulation of the liquid in the domain is counterclockwise. The progressive enrichment in tin of the interdendritic liquid during solidification which decreases its density is responsible of this counterclockwise flow. This effect is dominant over the thermal effect, which explains the direction of rotation of the fluid obtained. In areas of low liquid fraction, we note the presence of channels inside which the liquid fraction is locally higher. These channels have started to form in region of high liquid fraction, early after the beginning of the solidification, after 20 seconds two channels are clearly visible. Their formation is linked to the direction of the interdendritic liquid flow with regard to the thermal gradient. A positive component of the velocity of the interdendritic liquid in the direction of the thermal gradient will induce locally a delay to the solidification which can later become a channel. One can notice that the remelting of the solid skeleton is not necessary. These channels, due to their increased permeability, are a preferential path for the liquid flow. The temperature field as can be seen in Figure 2(a) is not strongly affected by the presence of the channels.

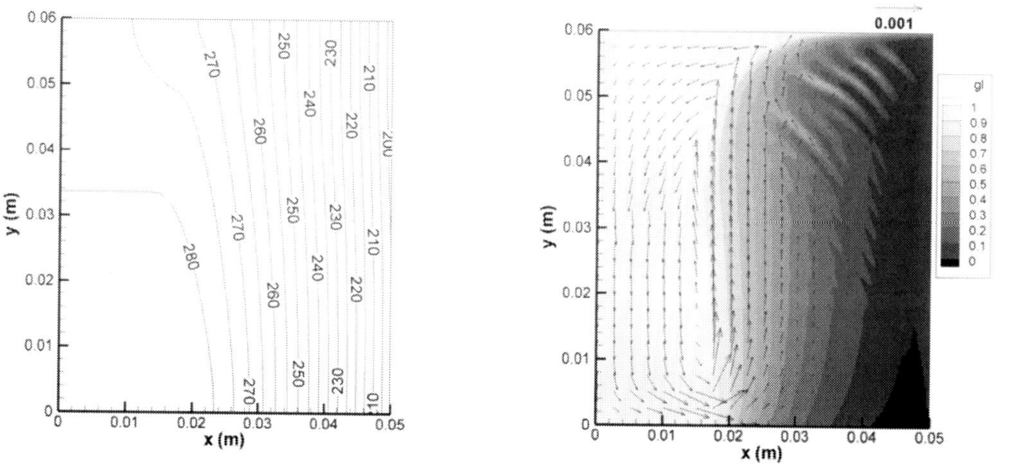

(a) (b)

Figure 2. (a) temperature field, (b) velocity field (vectors, the scale is indicated by the arrow at the top right in m s^{-1}) and liquid fraction field (grey levels) at time 120 s.

The map of the average composition in tin after the end of solidification obtained by EPM-SIMAP is shown in Figure 3(a). The upper left corner of the half cavity which is the last to solidify is positively segregated, while the bottom right region where the interdendritic liquid circulates in the reverse direction of the thermal gradient is negatively segregated. A positive segregation has also developed inside the channels associated with a negative segregation at their periphery.

4. Comparison and analysis of the results obtained by all the partners

Maps of the average concentration field at the end of solidification (t=600 s) obtained by the six codes are presented in Figure 3. The macrosegregation is in accordance with the description in the previous section and the six simulations are in qualitative agreement on the general segregation map. The principal differences are the position and the number of channel segregates. It is interesting to notice the differences between the two simulations using both OpenFOAM but with a different scheme for the discretization of the convective terms. With the upwind scheme, the channels start to form at the height of 0.04 m whereas with the QUICK scheme they start at the height of 0.045 m. More channels form in case of the upwind scheme: at the height H3 (see Figure 1 for its location), six channels are visible while only four are visible in case of the QUICK scheme. As the only difference between these two results is the discretization scheme of the convective terms, this demonstrates its strong influence on channels prediction. We can distinguish two groups of results, which coincide with the order of the convection schemes that were used. We can observe that the results from EPM-SIMAP, TREFLE and OpenFOAM QUICK, all using a second order scheme, are closer with regard to the channels. They predict the same number of channels and the height at which they formed. The same remarks can be done for the IJL and OpenFOAM upwind results. The result coming from CEMEF, the only FEM software participating to this benchmark exercise, cannot be sorted in the two previous classes. The channels start closer to the left vertical wall, and at a lower height than in the other contributions. These similarities between different contributions are re-discussed more quantitatively in a next part of this section.

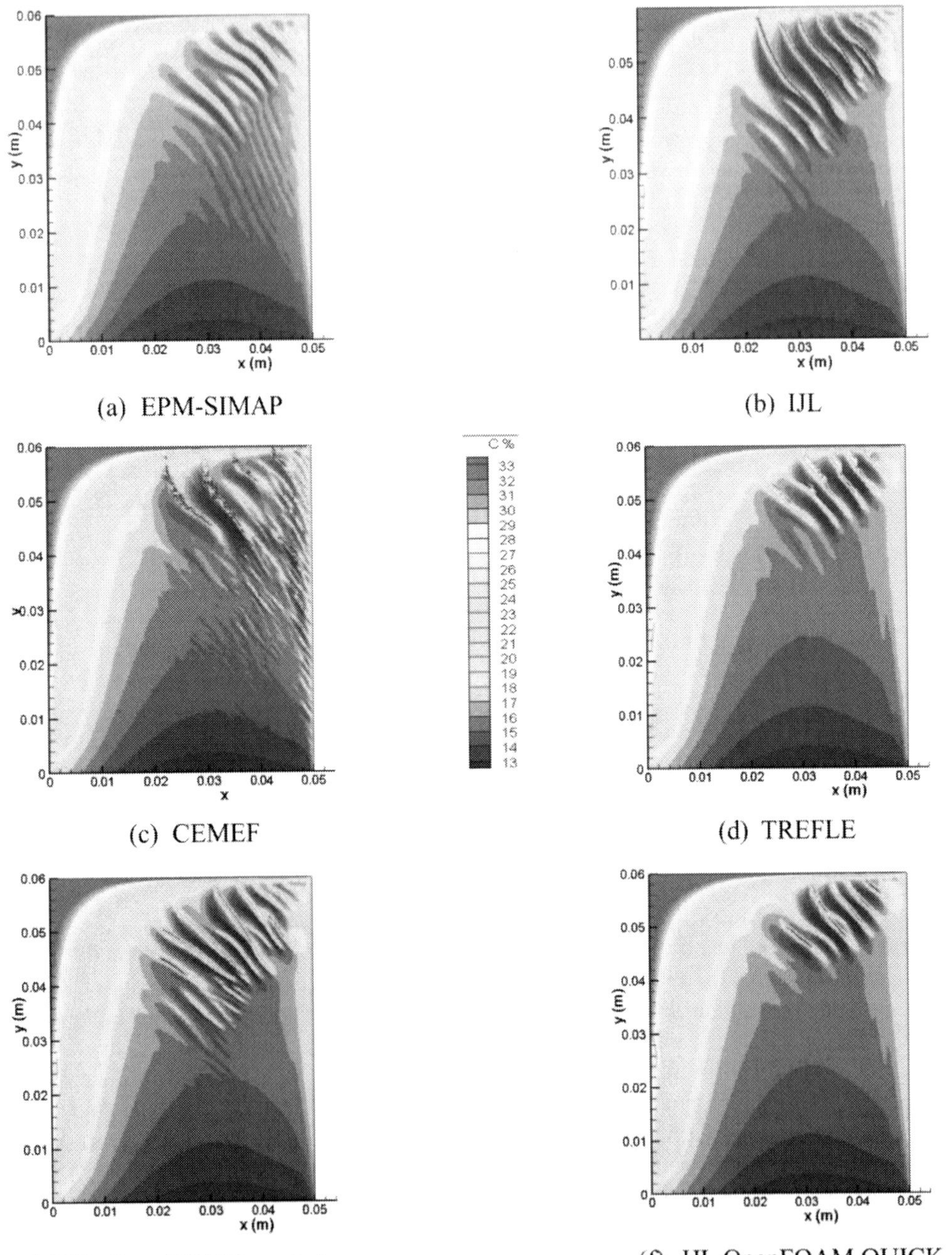

Figure 3. At the end of solidification: map of average mass concentration in Sn (wt. pct.) obtained by the different contributors. The color scale is similar for the six contributions.

In Figure 4, the horizontal profiles at height H1 and H2 (see location in Figure 1) of the final average composition in tin for all the simulations are presented. In term of macrosegregation, i.e. variations at the scale of the width of the domain, the results are similar. However, significant differences are visible in the region where the channels are located (from 0.02m to 0.05m in Figure 4(b). In this region, for all the simulations, variations of composition with stronger amplitude than for the macrosegregation can be observed. These variations correspond to the channels formed during the solidification. However, in the zoom of Figure 4(b), it can be seen that the location and the amplitude of these variations differ in function of the simulation. Notably, in the case of OpenFOAM, the results differ between a discretization of the convective terms using a first upwind scheme and a QUICK scheme, everything else being similar in the numerical scheme. It was expected that for a relatively fine mesh like those used for the results presented in this paper, such sensitivity would have been smaller. Thus, even in the case of finer meshes, these results confirm those of reference [20] in which a first comparison exercise was conducted between two computer codes for a similar situation: macrosegregation results were similar, the main differences were also observed at the level of segregated channels.

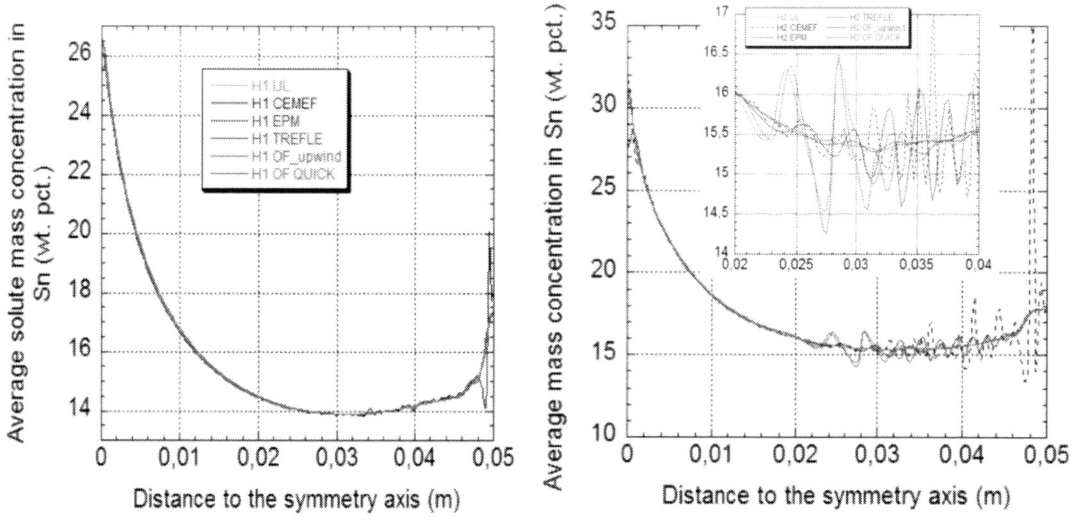

(a) (b)

Figure 4. At the end of solidification: Average mass concentration in Sn (wt. pct.): (a) along the horizontal line H1, (b) along the horizontal line H2.

In the upper part of the domain, as can be seen in Figure 3, the segregation in the channels is more severe. The horizontal profiles of concentration in Sn along the line H3 are plotted in Figure 5 for the six contributions. The height H3 is located 0.01 below the top of the cavity. As can be seen in Figure 3 this height corresponds to the mid-height of the channels. At this height, the variation of concentration across a channel is much stronger than at height H2. At the height H2, the maximum of the variation of the average concentration in Sn across a channel was of the order of 2 wt. pct, whereas this maximum is now of the order of 20 wt. pct.. The concentration varies by this magnitude on a distance of the order of 2 mm. The size of the Representative Elementary Volume (REV) used for the derivation of the volume averaged equations can be estimated by the microstructure size used to calculate the permeability. This is the only microscopic characteristic length introduced in the input data. For this benchmark case, its value is 185 μm. Thus, the average concentration in Sn varies in the order of 1.5 wt. pct over a REV located in the region of the channels, whereas this variation was only 0.15 wt. pct. at the height H2. If we look at the variation of the average concentration in Sn at the scale of the width of the domain, it is of the order of 20 wt. pct. in Sn along 0.05 m. This corresponds to a variation of concentration equal to 0.07 across a REV. To be valid, the volume averaging technique assumes that there is a separation of scales between a microscopic scale and a macroscopic scale. This implies that there is an intermediate scale corresponding to the size of the REV. At this intermediate scale, the variation of a quantity averaged in a volume of the size of the REV has to be negligible with regard to the variation at the macroscopic scale. This condition is not satisfied in the channel region, where we get a variation of 1.5 wt. pct. in Sn across a REV, which is much greater than the variation at the macroscopic scale estimated to 0.07 wt. pct. in Sn at the scale of the REV. Therefore, the assumption of scale separation is not valid any more in a channel. This can be a reason why the results differ between the different contributions for the prediction of channels.

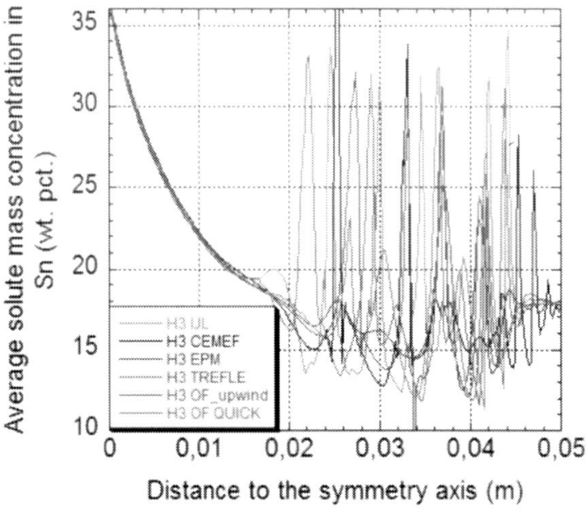

Figure 5. At the end of solidification: average mass concentration in Sn (wt. pct.) along the horizontal line H3.

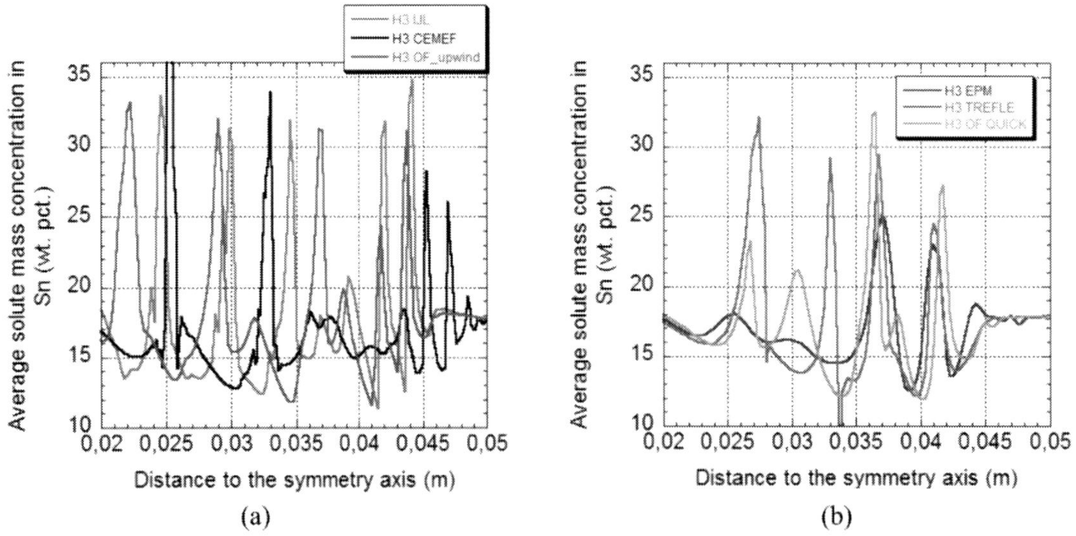

Figure 6. Zoom of the average mass concentration in Sn (wt. pct.) profiles along the horizontal line H3 at the end of solidification: (a) for IJL, CEMEF, OpenFOAM_upwind contributions, (b) for EPM, TREFLE, OpenFOAM_Quick contributions.

In Figure 6, the profiles of average concentration in Sn at the height H3 are zoomed in the region of the channels. The results are plotted on two graphs: in Figure 6.a, the results of the two finite volume codes using a first order scheme for the convective term (IJL and OpenFOAM upwind) and the finite element model (CEMEF) are plotted whereas in Figure 6.b the results of the three finite volume codes using a second order scheme for the convective terms (EPM-SIMAP, TREFLE and OpenFOAM QUICK). We have distinguished these two classes of solutions because they present similarities in terms of the amplitude of the variation of the average concentration in Sn across the channels and in terms of distance between two channels. As the global solute mass balance (equation (4)) is a purely convective equation, it is not surprising to find a high sensitivity of the order of accuracy of the interpolation scheme for the interpolation of the convective terms. It is well known [27] that a first order scheme like the upwind scheme will induce more numerical diffusion than a higher order scheme. However, even if we find similarities between the solutions produced by the same order of accuracy in space of the scheme, the results are not superimposed. For the results obtained with a finite element method (CEMEF), one can see from Figures 3 and 4.b that the channels start to form closer to the vertical right side of the domain than in the case of the other contributions.

A mesh and time step sensibility study was conducted by each contributor. While all details of these studies cannot be presented here, the most important observations are summarized. The mesh convergence observed with respect to the maximum and minimum values of the solute concentration after completed solidification is shown in Figure 7. The five codes based on the finite volume method show very similar sensibilities of the maximum value of the average concentration at the end of the solidification to the mesh density. In all cases, the order of convergence is smaller than one. For the finite element code, the maximum value seems to evolve more rapidly and is higher than for the other contributions. For all the contribution the maximum value is located in the upper left corner of the half domain which does not correspond to a channel. The sensibility of the minimum values to the mesh size is dependent of the contributor. First is has to be mentioned that the location of the minimum value is found close to the channels for all the contributors excepted for EPM-SIMAP for whose it is located in the negatively segregated region in the bottom of the cavity. As the channels are very sensitive to the mesh size and to the code, it explains why the minimum varies with the contributions.

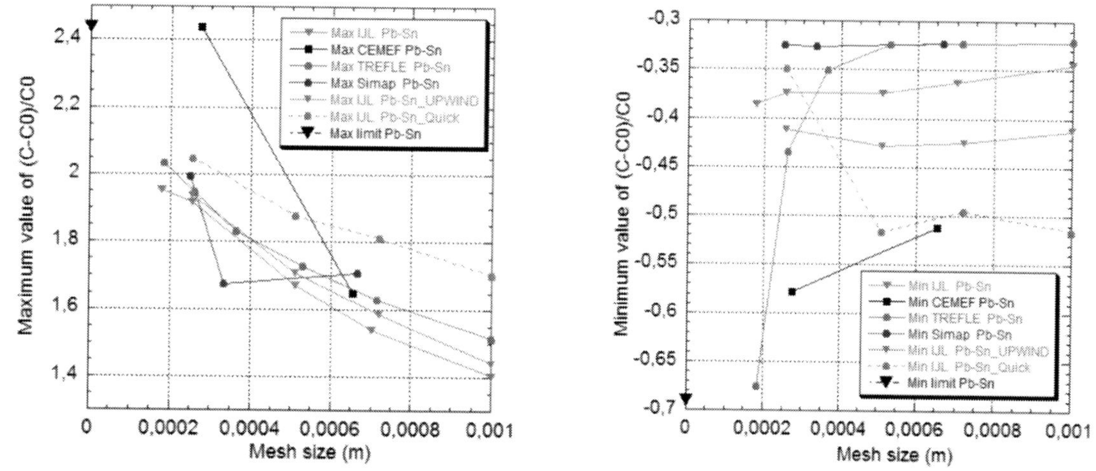

(a) (b)

Figure 7. Mesh convergence of the final relative segregation$(C-C_0)/C_0$: (a) the maximum of the field at the end of solidification, (b) the minimum of the field at the end of solidification.

The velocities of the liquid are an important element of comparison of the solutions. In Figure 8, the horizontal profiles along the horizontal line H2 of the liquid fraction and of the modulus of the intrinsic velocity of the liquid defined as: $\left\|V_l^l\right\| = \dfrac{\|V\|}{g_l}$ are presented at time 120 seconds. The shapes of the curves are similar for the six simulations. Similarly, the temperature profiles not included in this communication are superimposed. The strongest differences are observed in the channels corresponding to the local fluctuations of the liquid fraction in Figure 8(a). Similarly, for points located 0.02 m farther from the symmetry axis, differences in the modulus of the intrinsic velocity of the liquid phase are correlated with the fluctuations of the liquid fraction. Differences are also visible in the region of high liquid fraction, typically for a liquid fraction greater than 0.95.

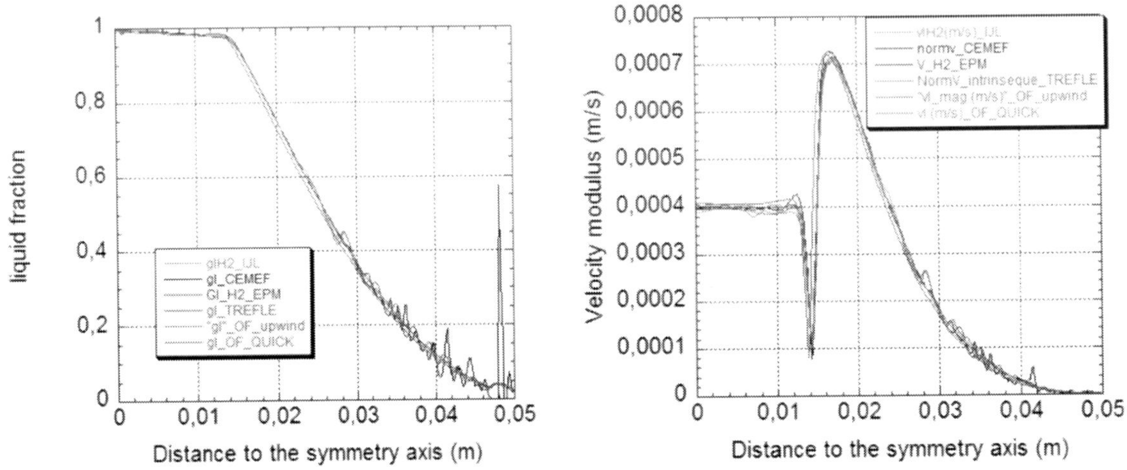

(a) (b)

Figure 8. Horizontal profiles along the line H2 (see Figure 1) at time 120 seconds, (a) liquid fraction, (b) modulus of the intrinsic velocity of the liquid phase.

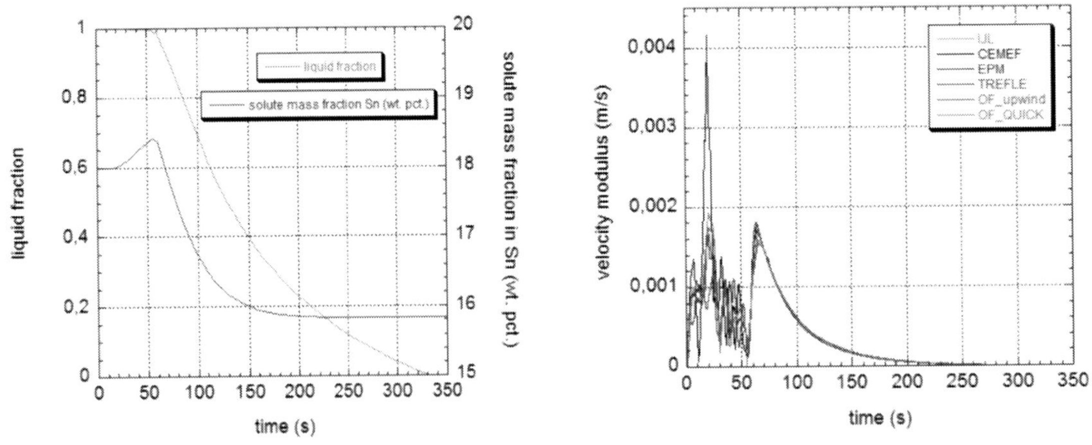

(a) (b)

Figure 9. Time evolutions at point E (see Figure 1 for its location) of (a) the solute mass fraction and liquid fraction; (b) the modulus of the intrinsic velocity of the liquid phase.

The time evolutions of the liquid fraction and of the average concentration in point E, shown in Figure 9(a), show that the macrosegregation continues to evolve even up to small liquid fractions (approximately until the liquid fraction becomes lower than 0.3). In Figure 9(b) we show the evolution of the velocity modulus in point E with time. All solutions predict overall the same evolution, after t=70s, the velocity decreases monotonically. Some notable differences in the beginning, until t=70s are visible. This initial interval corresponds to the period when liquid fractions were high. These observations suggest an explanation for the reason why the macrosegregations are similar and the channels different. As for this case, the macrosegregations develop principally at liquid fraction lower than 0.9 below which the flow is similar for all the contributions, the prediction of the macrosegregation gives similar results for all the contributions. The flow for these liquid fractions is then dominated by the three last terms of equation (3) and the convective term in this equation is negligible. So the results are less sensitive to the discretization scheme. At the reverse the channels form at high liquid fraction for which the inertia term in equation (3) stays important. Then the results are sensitive to the discretization scheme.

5. Conclusion

This communication presents the results of the comparison between different numerical procedures in the solution of an identical 'minimal' solidification model. We can conclude that even though the qualitative image of the different solutions is the same, notable differences exist in the evolutions and local behaviors. For the final segregation map in average mass concentration in Sn, the main differences are located in the channels and in their vicinity. The number of channels and the intensity of segregation inside the channels vary from one contribution to another. Even with fine meshes, it has not been possible to find a unique solution, mesh independent and similar for all the contributions. These differences find their origin in the prediction of the flow at high liquid fraction which is very sensitive to the discretization scheme. One conclusion of this study is that it seems to be unrealistic to search a reference solution when channels form during solidifcation. One common point is that channels are predicted by all contributors, which mean that at minimum the results can indicate a sensitivity to channel formation. At the reverse, concerning the macrosegregation pattern the solution is similar for all the contributions. The reason is that the macrosegregation continues to evolve until low liquid fraction (typically until the liquid fraction is greater than 0.3). The calculation of the flow in the region of low liquid fraction is less sensitive to the discretization of the convective terms because these terms become negligible. Similar conclusions have been drawn for the second solidification case investigated in this benchmark exercise for a Sn-10wt%Pb alloy [26].

This benchmark exercise is now continued by a comparison of thermal and solutal natural cases without solidification in order to check the accuracy of the codes. Moreover, a case without channels formation has been defined [23]. Two codes have been used, one based on a finite volume method with a second order scheme to discretize the convective terms and the second one based on a meshless method. Even if a unique converged solution has not been found at the moment, it seems possible by refining the mesh to get a reference solution.

Acknowledgments
This work was funded by the French National Research Agency (ANR) in the framework of the 'Programme Blanc', as part of the project SMACS (ANR-07-BLAN-190).

Nomenclature

$\langle h \rangle$	average enthalpy per unit mass, $J.\ kg^{-1}$
T	temperature, K
C_p	specific heat per unit mass, $J.kg^{-1}.K^{-1}$
$\mathbf{V}$	average velocity of the liquid, $m.s^{-1}$
k	thermal conductivity, $W.m^{-1}.K^{-1}$
g_α	volume fraction of the phase α= l or s
$\mathbf{g}$	gravity acceleration, $m.s^{-2}$
K	permeability, m^2
p	pressure (Pa)
$\langle C \rangle$	average mass solute fraction
C_l	average mass solute fraction in the liquid phase
m_l	liquidus slope (K wt. pct.$^{-1}$)
k_0	partition coefficient
Greek symboles	
β_T	thermal expansion coefficient, K^{-1}
β_c	solutal expansion coefficient, $(wt.\ pct.)^{-1}$
μ	dynamic viscosity, *Pa.s*
ρ	density, $kg.\ m^{-3}$
Subscripts and superscripts	
l	liquid phase
s	solid phase

References

[1] Beckermann C 2002 *Int. Mater. Rev.* **47** 243
[2] Ganesan S and Poirier D R 1990 *Metall. Trans.* **21B** 173
[3] Poirier D R, Nandapurkar P J, and Ganesan S 1991 *Metall. Trans.* **22B** 889
[4] Ni J and Beckermann C 1991 *Metall. Trans.* **22B** 349
[5] Bennon W D and Incropera F D 1987 *Int. J. Heat Mass Transfer* **30** 2161
[6] Combeau H, Drezet J M, Mo A and Rappaz M 1996 *Metall. Trans.* **27A** 2314
[7] Wang C Y and Beckermann C 1993 *Mater. Sci. Eng.* **171** 199
[8] Schneider M C and Beckermann C 1995 *Metall. Trans.* **26A** 2373
[9] Schneider M C, Gu J P, Beckermann C, Boettinger W J and Kattner U R 1997 *Metall. Trans.* **28A** 1517
[10] Gu J P and Beckermann C 1999 *Metall. Trans.* **30A** 1357
[11] Wu M, Fjeld A and Ludwig A 2010 *Comp. Mat. Sc.* **50** (1) 32
[12] Li J, Wu M, Hao J and Ludwig A 2012 *Comp. Mat. Sc.* **55** 407
[13] Thuinet L, Combeau H 2009 *Comp. Mat. Sc.* **45** (2) 294
[14] Kumar A, Dussoubs B, Založnik M, Combeau H 2009 *Journal of Physics D:Applied Physics* **42** (10) 105503
[15] Combeau H, Založnik M, Hans S, and Richy P E 2009 *Metall. Trans.* **40B** 289
[16] Sawada T, Oikawa K, Anzai K 2009 *MCWASP XII, TMS, Warrendale, PA, USA* 303
[17] Nakajima K, Zhang H, Oikawa K, Ohno M and Jönsson P G 2010 *ISIJ* **50** (12) 1274
[18] Bellet M and col. 2009 *Int. J. Th. Sci.* **48** (11) 2013
[19] Thétis, 2009 *http://thetis.enscbp.fr/*
[20] Ahmad N, Combeau H, Desbiolles JL, Jalanti T, Lesoult G, Rappaz J, Rappaz M and Stomp C 1998 *Metall. Trans.* **29A** 617
[21] Liu W, Xie C, Bellet M, Combeau H 2009 *Acta Metall. Sin.* **22** 233
[22] Voller V R, Brent A and Prakash C 1989 *Int. J. Heat Mass Transfer* **32** 1719
[23] Kosec G, Založnik M, Šarler B and Combeau H 2011 *Computers, Materials and Continua* **22** (2) 169
[24] Hachani L, Saadi B, Wang X D, Nouri A, Zaidat K, Belgacem Bouzida A, Ayouni-Derouiche L, Raimondi G and Fautrelle Y 2012, *Int. J. Heat and Mass Transfer* **55** 1986
[25] http://www.ijl.nancy-universite.fr/benchmark-solidification
[26] Combeau H and col. 2011 *TMS Annual Meeting* **2** 755
[27] Ferziger J H, Peric M 1997 *Springer* ISBN 3-540-59434-5
[28] De Felice V F, Tveito K O, Založnik M, Combeau H, M'Hamdi M 2012 *Modeling of Casting, Welding and Advanced Solidification Processes XIII in press*

3D CAFE simulation of a macrosegregation benchmark experiment

T. Carozzani[1], H. Digonnet[1], M. Bellet[1], Ch.-A. Gandin[1,2]
[1]MINES ParisTech & [2]CNRS, CEMEF UMR 7635, 06904 Sophia Antipolis, France

E-mail: Charles-Andre.Gandin@mines-paristech.fr

Abstract. A macrosegregation benchmark experiment is simulated using a three dimensional (3D) Cellular Automaton (CA) - Finite Element (FE) model. It consists of a Sn - 3 wt% Pb alloy solidified in a rectangular cavity. Thanks to tabulated thermodynamic properties and solidification paths with temperature and composition, the effect of natural convection and macrosegregation on cooling curves is correctly predicted. Nucleation parameters are adjusted so that the simulated grain structure correlates with the real grain structure. Although macrosegregation is well predicted, this is not the case for freckles yet observed in the solidified sample.

1. Introduction

Macrosegregation is a major problem for the casting industry. It can be due to several mechanisms involving *i*- solute transport in the liquid and in the mushy zone due to fluid flow, *ii*- grain transport because of gravity and convection, *iii*- liquid flow driven by shrinkage and deformation and *iv*- diffusion of solute in the liquid and solid phases [1].

Due to the multiple and overlapping origins of macrosegregation, quantitative validation of simulation tools is a challenging task. Experimental benchmarks have thus been developed for that purpose. A reference work is the early experiment of Hebditch and Hunt [2] in Pb - 48 wt% Sn alloy. A similar benchmark was recently developed with enhanced data acquisitions and metallurgical inspections [3, 4]. It provides detailed data on temperature evolution during solidification and measured distribution of composition and eutectic fraction in the solidified ingot. Alloy composition can also be changed and forced convection can be studied thanks to electromagnetic stirring. Few attempts have already been made to simulate the experiment (e.g. [5, 6]). However, they lack either tracking of the columnar grain structure and/or 3D considerations. Both are yet expected to have a great influence on the results if quantitative comparison is to be reached. We propose here to simulate such a benchmark experiment with a CAFE model that accounts for structure and macrosegregation in three dimensions.

2. Experimental

A Sn - 3 wt% Pb ingot is considered. Its dimensions are 10 cm width x 6 cm height x 1 cm thick. The opposite smallest surfaces, dimensions 6 cm x 1 cm, are positioned vertically in contact with heat exchangers that serve to melt or freeze the ingot by imposed temperature evolutions. The other sides of the ingot are well insulated and considered as adiabatic. The ingot is first melted by heating up to 260 °C. A 10 minutes holding time at this temperature is applied in order to homogenize the liquid while imposing a forced convection by electromagnetic stirring. Stirring is then stopped and time is reset to zero. The right-hand-side (RHS) heat exchanger is heated up to 280 °C while the left-hand-side (LHS) heat exchanger is cooled down to 240 °C. These temperatures are maintained for 1000s so that natural convection due to the temperature difference is stabilized. Finally, the RHS and LHS heat exchangers are simultaneously cooled down at -0.03 °C s^{-1} until complete solidification, so that a 40 °C difference is maintained between the LHS and RHS heat exchangers.

Temperature evolution is recorded with 50 thermocouples equally spaced by 1 cm in a regular lattice of 5 rows x 10 columns, plus 6 thermocouples in each heat exchanger. The segregation map is retrieved with X-ray imaging and quantitative chemical analyses using Inductively Coupled Plasma (ICP) spectrometry at the same 50 positions where the thermocouples are located. Experimental results presented in the next section are based on reference 4.

3. Modeling

3.1. FE method

The CAFE model has been extensively described elsewhere [7-10] and is only briefly presented hereafter. It consists of a standard FE method to solve average macroscopic equations written with the approximation of constant and equal densities for both solid and liquid phases, ρ_0, as well as a fixed solid phase.

The following set of equations is solved to compute the average enthalpy, $<H>$, the average composition, $<w>$, and the average velocity, $<\mathbf{v}>$.

Energy conservation
$$\rho_0\left(\frac{\partial <H>}{\partial t} + <\mathbf{v}>\cdot\nabla <H^l>^l\right) - \nabla\cdot(<\kappa>\nabla T) = 0 \quad (1)$$

Solute conservation
$$\frac{\partial <w>}{\partial t} + <\mathbf{v}>\cdot\nabla <w^l>^l - \nabla\cdot(D^l g^l \nabla <w^l>^l) = 0 \quad (2)$$

Momentum conservation
$$\rho_0 \frac{\partial <\mathbf{v}>}{\partial t} + \frac{\rho_0}{g^l}\nabla\cdot(<\mathbf{v}>\times<\mathbf{v}>) = \nabla\cdot(\mu\nabla <\mathbf{v}>) - g^l\nabla p + g^l \rho\, \mathbf{g} - \mathbf{M} \quad (3)$$

where $<H^l>^l$, $<\mathbf{v}>$, $<\kappa>$ and T are respectively the average liquid enthalpy, the average liquid velocity, the average thermal conductivity and the temperature, $<w^l>^l$, D^l and g^l are respectively the average liquid composition, the diffusion coefficient of the solutal element in the liquid and the volume fraction of liquid, μ, p and $\mathbf{g}$ are the dynamic viscosity, the intrinsic pressure in the liquid and the gravity vector. Diffusion in the solid is neglected here. The Boussinesq approximation is assumed in order to compute the fluid flow induced by natural convection. The density is therefore kept constant in all terms of the momentum equation except in the gravity term, where:

$$\rho = \rho_0\,(1 - \beta_T\,(T - T_0) - \beta_w\,(<w^l>^l - w^l_0)) \quad (4)$$

with β_T and β_w the thermal and solutal expansion coefficients, T_0 and w^l_0 reference temperature and composition. The volumetric friction force, $\mathbf{M}$, accounts for the interaction of the liquid with the solid phase. It is defined as $\mathbf{M} = (\mu/K)\, g^l <\mathbf{v}>$ with the local permeability given by $K = [{g^l}^3\, \lambda_2^{\,2}] / [180\,(1-g^l)^2]$. In the fully liquid region, K tends toward infinity and the classical Navier-Stokes equation is retrieved. In the solid region, K tends toward zero and the **M** term becomes dominant, leading to Darcy's relation. Details on the calculation of the permeability are given elsewhere [8]. It mainly depends on the secondary dendrite arm spacing, λ_2. The value used in the present work is directly extracted from measurements [4]. Finally, a local solidification model is used, that predicts the liquid fraction and the temperature considering an average enthalpy and an average composition, while accounting for the nucleation and growth the grain structure with some undercooling [10]. This is explained hereafter.

3.2. CA method

A regular grid of cubic cells is generated in the simulation domain, with a cell size smaller than the FE mesh. Random cells are picked where nucleation sites are allocated. A critical nucleation undercooling,

ΔT_{nucl}, is associated to each site. Its value is randomly chosen following a Gaussian distribution defined by a mean undercooling, ΔT_N, a standard deviation, ΔT_σ, and the total density n_{max} that corresponds to the integral of the distribution. If the CA cell associated with the nucleation site is still liquid when the local undercooling reaches the nucleation undercooling, then the site nucleates a new grain with a random orientation given by a set of Euler angles.
A velocity, v_p, is used to compute the growth of the half diagonals of a local octahedral shape associated with the CA cell. These diagonals corresponds to the main <100> dendrite growth directions of a cubic metal. For a cell located at the grain boundary with the free liquid, the supersaturation, Ω, is proportional to the difference between the dendrite tip composition, w^{ls}, and the average liquid composition far from the interface, $w^{l\infty}$, the latter being approximated as $\langle w \rangle$:

$$\Omega = (w^{ls} - w^{l\infty})/(w^{ls}\,(1-k)) \quad (5)$$

It is computed from a boundary layer correlation:

$$\Omega = P_{v_p} \exp(P_{v_p}) \{E_1(P_{v_p}) - E_1(P_{v_p}\,[1 + 4\,(A\,Re_{2r}^B\,Sc^C \sin(\theta/2))^{-1}])\} \quad (6)$$

with $A = 0.5773$, $B = 0.6596$, $C = 0.5249$ and where $P_{v_p} = (r\,v_p)/(2D^l)$ is the growth Péclet number defined with the dendrite tip radius r, $P_{vl} = (r\,v^l)/(2D^l)$ is the flow Péclet number proportional to the liquid velocity, v^l, $Sc = \nu/D^l$ is the Schmidt number defined with the kinematic viscosity ν, $Re_{2r} = 4P_{vl}/Sc$ is the Reynolds number, θ is the angle between the fluid flow direction and a <100> growth direction [11]. The velocity of the dendrite tip is computed with the relation given by the marginal stability criterion:

$$r^2\,v_p = (D^l/\sigma^*)\,(\Gamma/(\,m_L\,(k-1)\,w^{ls}) \quad (7)$$

with m_L the liquidus slope, k the segregation coefficient, Γ the Gibbs-Thomson coefficient and σ^* the stability constant equal to $(4\pi^2)^{-1}$ [8, 9]. The volume fraction of the propagated growth front, $g^{(i)}$, is then computed for each cell from the growing shape. $g^{(i)}$ locally accounts for the presence of structure i, here dendritic (i=1) or eutectic (i=2) for a binary Sn-Pb alloy [10]. When only the dendritic structure is present, the value $g^{(1)}=0$ corresponds to a fully liquid state. The value $g^{(1)}=1$ corresponds to a "fully mushy" state, i.e. the remaining liquid is interdendritic. Intermediate values correspond to a transition state made of mixture of a mushy zone, $g^{(1)}$, plus an extradendritic liquid, $1-g^{(1)}$. Thus the value of $g^{(1)}$ depends on nucleation events, growth kinetics, and more generally temperature evolution in the ingot in the vicinity of the FE nodes. As no growth is solved here for the eutectic structure, $g^{(2)}$ is simply set to one below the eutectic temperature, T_E.

3.2. Coupling with thermodynamic

Summation of $g^{(i)}$ over the CA cells permits to access the average information at the FE nodes. It is used by a microsegregation model for the determination of the internal phase fractions in the primary and secondary structures at the FE nodes. For that purpose, a relation between enthalpy, average composition and temperature is needed in the energy conservation [10]. The microsegregation model is based here on a tabulation of the hypoeutectic phase diagram assuming the lever rule approximation. Enthalpy of different phases α, $H^{(\alpha)}$, and internal fractions for each phase in each structure, $g^{\alpha_i^{(i)}}$, are tabulated for several average solute composition, $\langle w \rangle$ and temperature, T. Given the local composition from the solution of the solute mass conservation, $\langle w \rangle$, the average enthalpy from the energy conservation, $\langle H \rangle$, and the volume fractions of structures from the CA resolution, $g^{(i)}$, a local temperature, T, and internal phase fractions, $g^{\alpha_i^{(i)}}$, verify the relation:

$$\langle H \rangle = \sum_{\alpha_i} g^{(i)}\, g^{\alpha_i^{(i)}}(\langle w \rangle,T)\; H^{\alpha}(\langle w \rangle,T) \quad (8)$$

where $H^{\alpha}(\langle w\rangle,T))$ and $g^{\alpha_i^{(i)}}(\langle w\rangle,T)$ are computed by linear interpolations of the tabulations. Variations of heat capacities and latent heat with temperature and composition are thus implicitly taken into account, as well as possible solid phase transformations. Figure 1 illustrates the tabulation of the enthalpy and internal fraction of phases as a function of temperature in the dendritic and eutectic structures for various Pb compositions of hypoeutectic Sn-Pb alloys. As read on figure 1b, solidification takes place with a primary dendritic structure made of the BCT_A5 phase which is partially transformed into the FCC_A1 phase when the solvus temperature is reached. The later solid state phase transformation and its temperature depend on the local average composition.

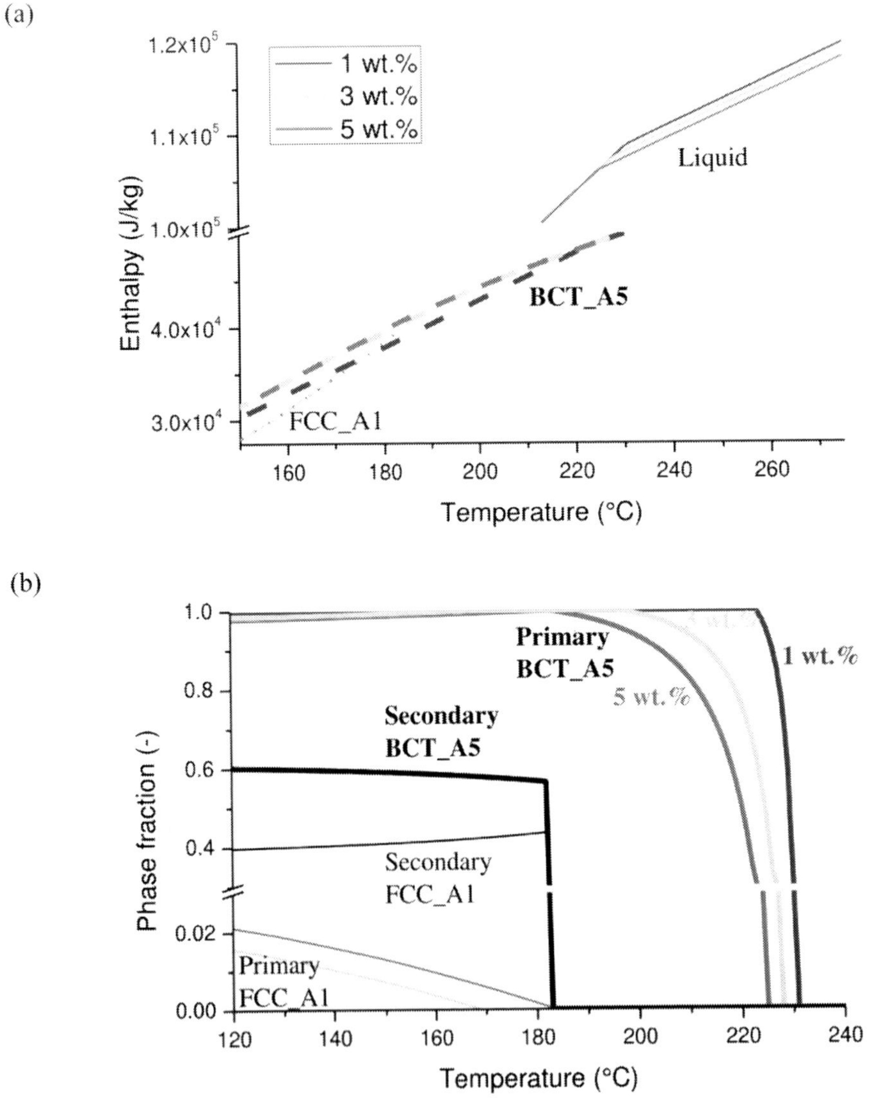

Figure 1. Tabulated (a) enthalpy and (b) internal phase fractions for (colored lines) the primary dendritic structure and (black lines) the secondary eutectic structure as a function of temperature and Pb composition. Internal phases are (thick lines) BCT_A5 and (thin lines) FCC_A1. Data extracted from the PBIN database using ThermoCalc and assuming full equilibrium.

The eutectic structure is made of a mixture of the BCT_A5 and FCC_A1 phases with internal fractions also changing with temperature. Please note that these phase names are directly extracted from the PBIN database. The present tabulation permits to account for solid state transformations, although it is not clear if they are made possible due to the limited diffusion of solute in the solid phases. This strategy of coupling yet clearly permits to be as close as possible to thermodynamic properties, thus directly using tabulated enthalpies for each phase as a function of temperature and composition as shown on figure 1a. The method that consists of an extraction of heat capacity and latent heat as a function of temperature is avoided. Material properties become dependent on alloy composition and phase fractions, the later being considerably changed due to macrosegregation. Such methodology was shown earlier to provide advantages when quantitative comparison with measurements is to be reached [12].

4. Results and discussion

The global mass conservation requires adiabatic conditions at the boundaries of the simulation domain. A zero flow velocity is also imposed on ingot's boundaries. For heat transfer, all boundaries are adiabatic except the small right and left surfaces in contact with the heat exchangers. Temperatures that are imposed on these boundaries are deduced from the heat flow in the heat exchangers and the temperature in the sample near the surface [4]. This heat flow is deduced from temperature measurements in the heat exchanger at a given height. Thermocouples positions are given elsewhere [4]. For the results presented hereafter, the selected positions are distributed in the length of the ingot, as positions 5, 35, 65 and 95 mm from LHS of the ingot surface and at mid-height, i.e. at 30 mm from the ingot base. These positions are referred to as L30, L27, L24 and L21 in the following, respectively. Temperature evolutions at mid-height on left and right sides, i.e. at 0 and 100 mm from the left side of the ingot surface deduced from the heat flow analysis in the heat exchangers at mid-height, are referred to as positions $FL3_{in}$ and $FR3_{in}$, respectively. Simulation parameters are given in table 1. Note that the liquidus slope, m_L, and the partition coefficient, k, are only used to compute the velocity of the growth front with the CA model because coupling with thermodynamic properties deduced from the PBIN database have not yet been implemented.

Figure 2 presents the simulated fluid velocity at different times, along with the computed grain structure. As expected, the interaction between the flow and the structure is very strong. Although it is not visible on figure 2, fluid flow within the mushy zone is not null. It is however much lower than in the free liquid. The convection loop seen prior to solidification, e.g. before 2400 s in figure 2, is mainly induced by the temperature difference between the LHS and RHS heat exchangers. Figure 3 permits to access the imposed time evolution of the corresponding boundaries thanks to the reported temperature history at the ingot surfaces $FL3_{in}$ and $FR3_{in}$. Starting from an initial liquid melt at around 260 °C, the LHS of the ingot is heated up and hold at around 273 °C while the RHS is cooled down and hold at around 243 °C. The 30 °C difference between the LHS and RHS deviates from the desired 40 °C difference imposed between the heat exchangers. This is due to the heat resistance between the heat exchangers and the alloy. Upon further cooling, the grain structure develops. The size of the free liquid is further decreased. At about 3300 s, while half of the sample is still fully liquid, no more superheat remains and the temperature gradient is almost zero, further dumping convection.

Predicted and experimental cooling curves are shown in figure 3, together with the predicted solid fraction at the same locations. The agreement between measured and simulated cooling curves is overall very good. During the first stages of the experiment ($t < 1600$s), the temperature gradient is well represented. No numerical oscillations are observed, which confirms that fluid convection reaches a stable state. This could not be reached in two-dimensional simulations: the fluid flow was more pronounced and no permanent regime could be achieved, leading to perturbed cooling curves with no clear temperature plateau. During solidification ($2000s < t < 4000s$), all abrupt inflexions on the cooling curves are also retrieved. They correspond to the change from a fully liquid to fully mushy state when the dendritic front overgrows the thermocouple position. An interesting slope change is retrieved shortly after time 3500 s on thermocouple L21. From the CAFE simulation, it can be interpreted as a recalescence linked with a sudden increase of the solid fraction. It is due to the nucleation of equiaxed grains in the melt taking place at this location while temperature is almost uniform leading to a fast grain growth and latent heat released sufficient to increase the temperature. This phenomenon can be observed on figure 2c where few grains have nucleated in the melt, and the columnar grain interface is destabilized due to an almost uniform temperature as well as a fully undercooled liquid.

As shown in figure 4, the simulated grain structure can also be compared with the experimental macrograph. The structure is mainly columnar and the trend for the grains to tilt upward is reproduced by the simulation. Explanation is due to the introduction of the fluid flow direction with respect to the <100> growth directions of the grains for the computation of the growth kinetic. Because the flow is going downward along the growth front, as is shown in Fig. 2b, the growth directions of the grains that are opposite to the flow can adopt a smaller undercooling. They extend faster in the undercooled liquid and lead to the present grain selection. While this correlates well with the experimental observations, the grain direction does not retrieve the strong misorientation. It is not obvious whereas this is due to the kinetic law itself, or to the growth algorithm implemented in the CA method. Further investigations should be led in this direction that would probably require more dedicated experiments. Adjustments have been made to the nucleation parameters (table 1) in order to fit the observed grain density, allowing correct prediction of the columnar-to-equiaxed (CET) transition.

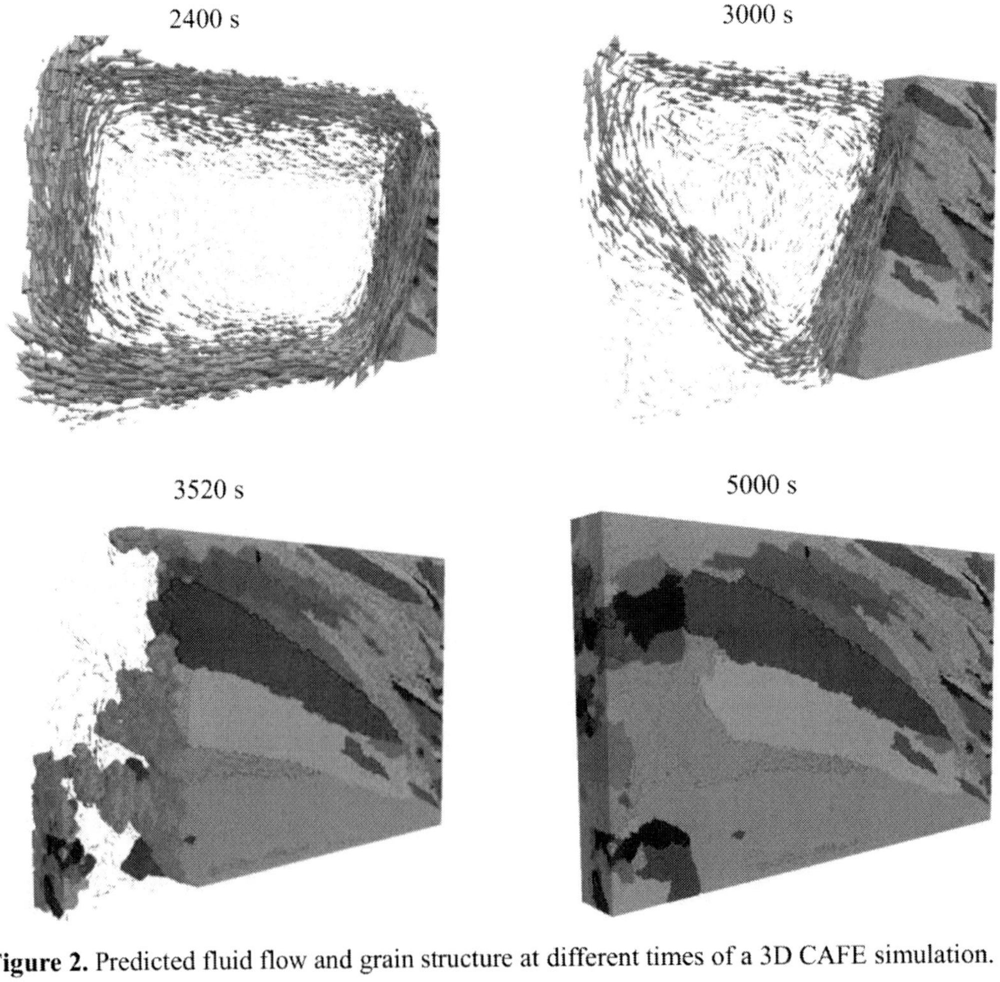

Figure 2. Predicted fluid flow and grain structure at different times of a 3D CAFE simulation.

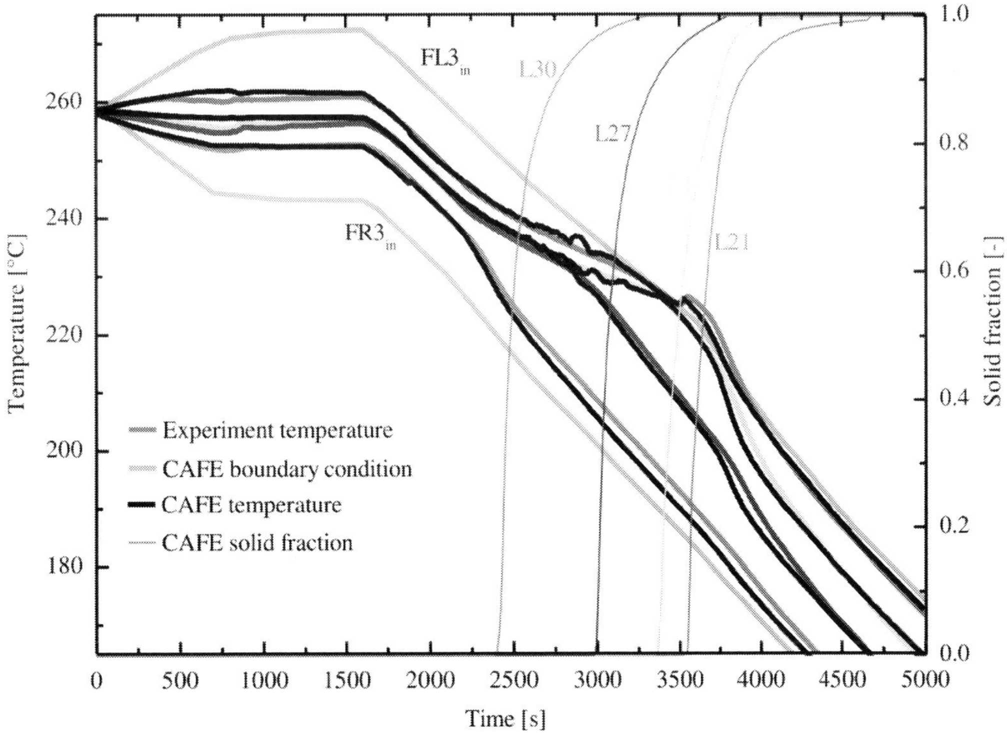

Figure 3. Experimental [4] (thick black lines) and predicted (thick colored lines) cooling curves, along with predicted solid fraction (thin lines). Curves are given for four positions L21, L24, L27 and L30, which are respectively at 5, 35, 75 and 95mm from the left side, at ingot's mid-height. Imposed temperatures at mid-height on right, $T(FR3_{in})$, and left, $T(FL3_{in})$, sides are also represented.

Figure 5 shows the predicted solute composition maps. The measured distribution is deduced from quantitative chemical analysis. The CAFE prediction is drawn at the ingot center with the same scale. Only the area delimited with the black rectangle can be compared. It corresponds to the measurements area. The global agreement in the right part of the ingot is good, as the chemical analysis would correspond to a mean value, i.e. within the thickness of the sample. In the left part of the ingot, a highly segregated zone is predicted that crosses almost all the ingot's height, whereas it is only measured at the bottom of the ingot. However, it is believed that measurements are not conducted on a sufficiently dense array. A better way to conduct comparison would then be to average the predicted quantities over a volume that is representative of the quantitative measurement.

A X-ray image of the ingot reveals segregated channels in the bottom-right part of the ingot [4]. This is not predicted by the simulation. Although the reason for this discrepancy is not yet very clear, anisotropy of the mush permeability is suspected to play a role, while it is only treated as isotropic in the present model. Freckles prediction also strongly depends on the mesh size [6]. Despite many numerical optimizations [9] and parallel computing, the current 3D simulation requires heavy resources thus limiting the minimum element size (table 1).

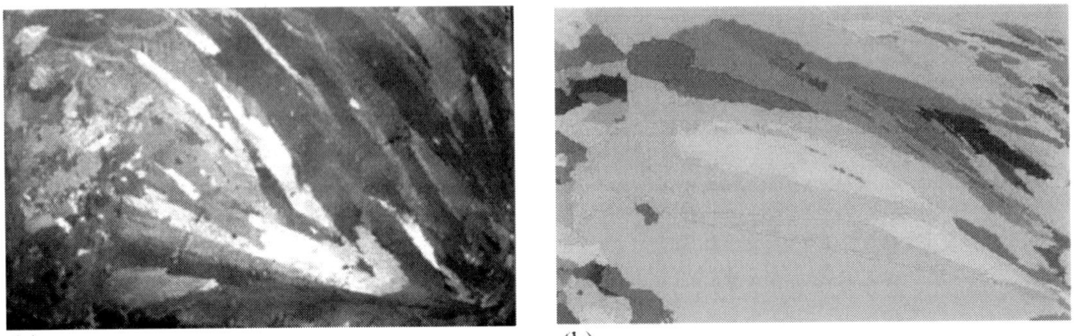

(a) (b)

Figure 4. Grains structure at ingot center (a) as observed by metallographic etching [4] and (b) as predicted by the 3D CAFE model.

5. Conclusion

A 3D CAFE model was presented and applied to macrosegregation benchmark experiment. It is able to retrieve measured cooling curves, macrosegregation and grain structure. Quantitative prediction being mainly achieved when comparing temperature evolution, it is believed that the fluid flow and growth undercooling are also well reproduced. However, prediction of segregated channel is still disappointing and may require additional resources and model improvements.

Acknowledgements

This work was funded by the Agence Nationale de la Recherche under projects SMACS and Si-X.

References

[1] Beckermann C 2002 *International Materials Reviews* **47** 243
[2] Hebditch D J and Hunt J D 1974 *Metall. Trans.* **5** 349
[3] Wang X D and Fautrelle Y 2009 *Int. J. Heat and Mass Transfer* **52** 5624
[4] Hachani L, Saadi B, Wang X D, Nouri A, Zaidat K, Belgacem Bouzida A, Ayouni-Derouiche L, Raimondi G and Fautrelle Y 2012 *Int. J. Heat and Mass Transfer* **55** 1986.
[5] Gandin Ch-A, Blaizot J, Mosbah S, Bellet M, Zimmermann G, Sturz L, Browne D J, McFadden S, Jung H, Billia B, Mangelinck N, Nguyen Thi H, Fautrelle Y and Wang X 2010 *Mater. Sci. Forum* **649** 189
[6] Combeau H, Bellet M, Fautrelle Y, Gobin D, Rady M, Arquis E, Budenkova O, Dussoubs B, Duterrail Y, Kumar A, Gandin C A, Goyeau B, Mosbah S and Zaloznik M 2011 *Model. of Casting, Welding and Adv. Solidif. Proc. XIII* (these proceedings)
[7] Gandin Ch-A, Desbiolles J L, Rappaz M and Thevoz Ph 1999 *Metall. Mater. Trans.* A **30** 3153
[8] Guillemot G, Gandin Ch-A and Combeau H 2006 *ISIJ International* **46** 880
[9] Mosbah S, Bellet M and Gandin Ch-A 2010 *Metall. Mater. Trans.* A **41** 651
[10] Carozzani T, Digonnet H and Gandin Ch-A 2012 *Modelling Simul. Mater. Sci. Eng.* **20** 015010
[11] Gandin Ch-A, Guillemot G, Appolaire B and Niane N T 2003 *Mater. Sci. Eng.* A **342** 44
[12] Tourret D, Gandin Ch-A, Volkmann T, Herlach D M 2011 *Acta Mater.* **59** 4665

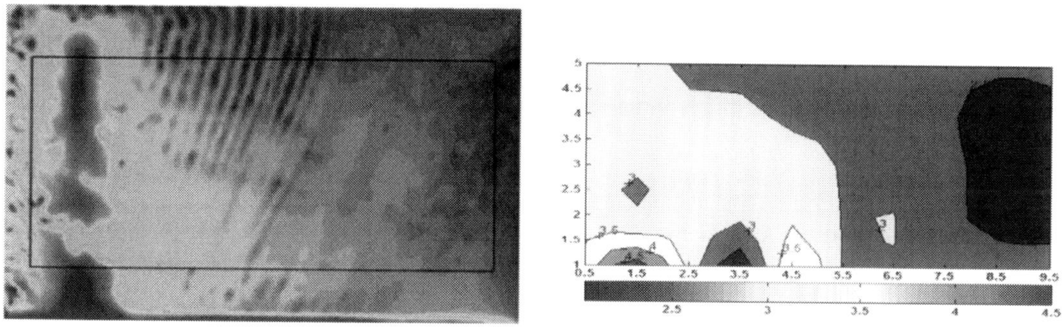

(a) (b)

Figure 5. Segregation map (a) predicted at the center of the ingot and (b) measured by chemical analysis [4]. The black rectangle in (a) corresponds to the measured area represented in (b).

Table 1. Value of simulation parameters.

Parameter	Variable	Value	Unit
Melting temperature	T_m	232.	°C
Eutectic temperature	T_e	183.	°C
Nominal composition	w_0	3.0	wt%
Eutectic composition	w_e	38.0	wt%
Segregation coefficient	k	0.0656	wt%
Liquidus slope	m_L	-1.2895	wt%
Gibbs-Thomson coefficient	Γ	$2. \cdot 10^{-7}$	°C m
Diffusion of Pb in liquid Sn	D^l	$3. \cdot 10^{-9}$	$m^2 s^{-1}$
Dynamic viscosity	μ^l	$2 \cdot 10^{-3}$	Pa s
Solutal expansion coefficient	β_w	$-5.3 \cdot 10^{-3}$	$wt\%^{-1}$
Thermal expansion coefficient	β_T	$9.5 \cdot 10^{-5}$	$°C^{-1}$
Reference temperature	T_0	228.14	°C
Reference composition	w_0^l	3.0	wt. %
Density	ρ_0	7130	$kg\ m^{-3}$
Gravity field	**g**	-9.81	$m\ s^{-2}$
Thermal conductivity in the solid	κ_s	55	$W\ m^{-1}\ °C^{-1}$
Thermal conductivity in the liquid	κ_l	33	$W\ m^{-1}\ °C^{-1}$
Secondary dendrite arm spacing	λ_2	$90 \cdot 10^{-6}$	m
Maximum nucleation density	n_{max}	10^7	m^{-3}
Average nucleation undercooling	ΔT_n	5.	°C
Nucleation undercooling standard deviation	ΔT_σ	1.	°C
CA cell size		$200 \cdot 10^{-6}$	m
FE mesh size		$780\text{-}1200 \cdot 10^{-6}$	m
Time step		0.1	s
Initial temperature		260.	°C
Initial composition		3.0	wt%
Initial velocity		0.	$m\ s^{-1}$

IOP Conf. Series: Materials Science and Engineering **33** (2012) 012088 doi:10.1088/1757-899X/33/1/012088

Three-dimensional study of macro- and mesosegregation formation in a rectangular cavity cooled from one vertical side

V F De Felice[1], K O Tveito[2], M Založnik[1], H Combeau[1], M M'Hamdi[2,3]
[1] Institut Jean Lamour, CNRS – Université de Lorraine, Département SI2M, Ecole des Mines de Nancy, Parc de Saurupt CS14234, F-54042 Nancy cedex, France
[2] Dept. of Materials Technology, NTNU, N-7491 Trondheim, Norway
[3] SINTEF Materials and Chemistry, N-0314 Oslo, Norway

E-mail: valerio.de-felice@ijl.nancy-universite.fr

Abstract. In metal alloys the solidification process leads to heterogeneities in chemical composition, which result in macroscopic (on the scale of the product) and mesoscopic (on the scale of several grains) segregation structures in the casting. Mesosegregations are a severe form of segregation in which solute concentration changes abruptly with respect to the surrounding regions and can induce structural heterogeneities in the casting. In this paper we move on from [1,2], in which channel mesosegregations and their numerical prediction were studied on a two-dimensional configuration – a small (10 × 6 cm) ingot of a Sn-10wt%Pb binary alloy. We extend the study to a slender three-dimensional cavity (10 × 6 × 1 cm) and we investigate three-dimensional effects in the formation of channel mesosegregations in a columnar mushy zone. We investigate the effects of the three-dimensionality on the structure of channels in the mushy zone and of channel mesosegregations. We present the influence of the permeability of the mushy zone on the channel structures. We show that channels can form either as lamellar or as tubular structures: at higher permeability secondary instabilities across the cavity thickness promote a transition from lamellar to tubular channels.

1. Introduction

The prediction and the control of chemical, structural and mechanical heterogeneities represent a fundamental target in metal processing. In solidification processes heterogeneities of chemical composition inevitably form. They appear as macroscopic (on the scale of the product) and mesoscopic (on the scale of several grains) segregation structures. Macro- and mesosegregation in the casting severely alter the homogeneity and the quality of the final product. Mesosegregation represents a severe form of segregation in which solute concentration varies abruptly in a certain part of the casting. These strong chemical variations can induce structural heterogeneities of the casting that impair its mechanical properties. Channel mesosegregations appear as long narrow trails of strong segregation aligned in some preferred direction. The precise mechanism of their formation is yet unexplained. In this frame, both experimental studies and numerical simulations are needed to better understand the segregation mechanisms and provide a tool to control and to minimize their effect.

The evolution of segregation depends on the entire temporal history of the solidification process, which presents a strong coupling between the momentum, mass and thermal transport. Perturbations in the mushy zone due to the liquid flow and due to the transport of solute seem to be the principal phenomena leading to instabilities that develop into channel mesosegregation structures. Accurate simulation and analysis of the fluid flow, thermal gradients and solute transport, are crucial in the investigation of channel genesis. In this work, we move from the results of [1,2], where mesosegregation for a two-dimensional rectangular cavity (10 × 6 cm) filled by a liquid Sn-10wt%Pb binary alloy was studied, and we extend the study to a three-dimensional cavity (10 × 6 × 1 cm). Some recent simulation results for this three-dimensional configuration, presented in [3], showed that mesosegregations form as lamellae, basically retaining a two-dimensional structure. Experimentally, tubular channel mesosegregations were observed on different configurations, notably in paper [4], where experimental investigations by X-ray tomography and 3D simulations were performed for a small Sn-Bi alloy casting. In the present work we suggest the existence of a transition in the mesosegregation morphology. We indicate that the influence of secondary three-dimensional flows could break up the two-dimensional lamellar structure and develop tubular channel mesosegregation structures. We study the lamellar-to-tubular transition as a function of the DAS (dendrite arm spacing) and show that at higher DAS (and thus higher permeability of the mushy zone) instabilities in the transversal direction, inducing tubular structures, are more likely to develop.

We present the problem configuration in section 2. The governing equations as well as the numerical resolution procedure are detailed in sections 3 and 4; and the three-dimensional effects in the formation of channel mesosegregations in a columnar mushy zone are presented in section 5.

2. Problem setup

The solidification of a binary Sn-10wt%Pb alloy in a three-dimensional cavity (10 × 6 × 1 cm) is simulated. The cavity is symmetrically cooled from its two vertical small faces, by an external flow at the temperature T_∞ and heat transfer coefficient α. The cooling is described by identical Fourier boundary conditions on the two external vertical walls. The top, bottom, front and back walls are thermally insulated. No-slip boundary conditions are imposed at the walls. We assumed symmetry of solution with respect to Plane S (see Figure 1) and we solved the problem on a half of the entire ingot volume. The ingot cavity is initially filled with the liquid alloy at rest; the alloy is at its nominal composition and at the liquidus temperature T_{liq}=219.14°C.

Table 1. Properties of Sn-10wt.%Pb, boundary and initial conditions

Symbol	Units	Value
c_p	J(kg°C)$^{-1}$	260
k	W(m°C)$^{-1}$	55.0
ρ_0	kg m^{-3}	7000
L	J kg^{-1}	6.1x10^4
μ_l	Pa s	1.0x10^{-3}
β_T	°C^{-1}	6.0x10^{-5}
β_C	(wt%)$^{-1}$	-5.0x10^{-3}
T_m	°C	232.0
C_e	wt%	38.1
k_p	-	0.0656
m_l	°C (wt%)$^{-1}$	-1.286
C_0	wt%	10.0
$T_0 = T_{liq}$	°C	219.14
α	Wm^{-2}°C^{-1}	400
T_∞	°C	25

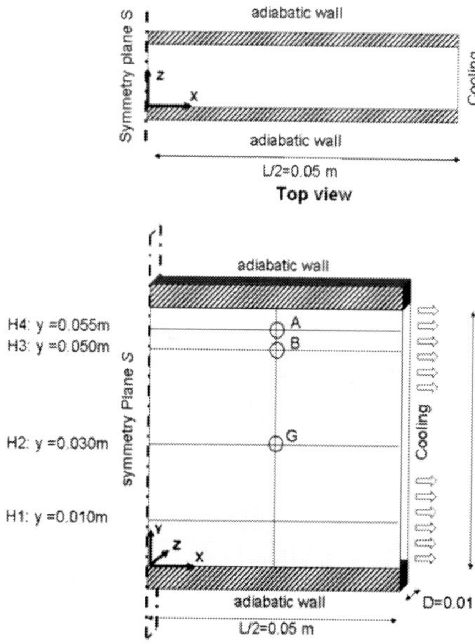

Figure 1. Problem geometry setup.

Once the solidification process is started, the solid phase grows in a columnar mushy zone that forms on the right hand side of the model geometry, moving from the cooled side toward the symmetry plane. A liquid motion is induced, driven by both thermal and solutal buoyancy.

3. Mathematical modelling

The solidification process in multi-component systems involves the presence of both liquid and solid phases as well as a mushy zone. In literature the governing equations are often derived moving from the basic two-phase flow formulation, where the equations for each phase are additively combined taking into account the interactions at the liquid-solid interface (we refer the reader to references [5,6] for further details). In this frame the balance equations (Eqs. (1–3)) for the transport of mass, heat, solute and momentum are presented in order to provide a necessary basis for the subject presented in this work. The microsegregation model (lever rule) corresponds to the assumption of a full thermodynamic equilibrium of the solid and liquid phases (Eq. (6)). The density of the liquid phase in the buoyancy term follows a linear variation with the temperature and the solute concentration (Eq. (4)). The permeability of the porous mushy zone depends on the liquid fraction and the dendrite arm spacing DAS, and follows the Kozeny-Carman law (Eq. (5)). The model is identical to the one defined in the benchmark [1] and more details about the model assumptions are given in [2].

Table 2. Model equations.

Mass and momentum: $\nabla\cdot\vec{v}=0$	$\rho_0\dfrac{\partial\vec{v}}{\partial t}+\rho_0\nabla\cdot(\vec{v}\dfrac{\vec{v}}{g_l})=-g_l\nabla p+\mu_l\nabla^2\vec{v}+g_l\tilde{\rho}\vec{g}-\dfrac{g_l\mu_l}{K}\vec{v}$	(1)
Heat: $\rho_0\dfrac{\partial h_m}{\partial t}+\rho_0\nabla\cdot(\vec{v}h_l)=\nabla\cdot(k\nabla T)$ (2)	Solute: $\dfrac{\partial\langle C\rangle}{\partial t}+\nabla\cdot(\vec{v}C_l)=0$	(3)
Density: $\tilde{\rho}=\rho_0\left[1-\beta_T(T-T_0)-\beta_C(C_l-C_0)\right]$ (4)	Permeability: $K=\dfrac{DAS^2g_l^3}{180(1-g_l)^2}$	(5)
Lever rule: $\langle C\rangle=g_sC_s+g_lC_l=C_l\left[g_l+k_p(1-g_l)\right]$; $T_{liq}=T_m+m_lC_l$; $h_m=g_sh_s+g_lh_l=c_pT+L$		(6)

4. Solution procedure

We used two different finite-volume codes to perform the computations. The first code (NTNU/SINTEF) was written at NTNU (Trondheim, Norway), in Fortran 90 and parallelised using MPI. The second code (CastFOAM) was written at IJL (Nancy, France) and is based on the OpenFOAM® platform written in C++ [9]. In the NTNU/SINTEF code, the two-phase solidification model is based on the operator splitting scheme by Založnik et al. [7] for coupling macroscopic transport and grain growth, but in the present work the model is simplified under the assumption of local thermodynamic equilibrium and stationary solid [8]. The system of equations is discretized with the finite-volume method and the SIMPLE [10] method for staggered grids is used for pressure-velocity coupling. The convective terms are discretized with the MUSCL [10] scheme, using the van Leer flux limiter [10] and a first order fully implicit Euler scheme [10] is used for the time derivative. At a given time step, the following four steps are performed successively for the NTNU/SINTEF code until the convergence test in step 5 is satisfied:

1) Resolution of the momentum and global mass balance equations:
   The SIMPLE [10] algorithm is used to treat the pressure-velocity coupling; the last known values of T and C_l are used in the calculation of the buoyancy term.
2) Resolution of the solute mass balance equation during the transport stage:
   The conservation equation of C_l^{tr} is solved according to the splitting scheme [7]:

$$\frac{\partial\left(g_l^{tr}C_l^{tr,**}\right)}{\partial t}+\nabla\cdot(\vec{v}C_l^{tr,**})=0\,,\qquad C_m^{**}=g_l^{tr}C_l^{tr,**}+g_s^{tr}C_s^{tr}$$
$$\left(g_l,g_s,C_s\right)^{tr}=\left(g_l,g_s,C_s\right)^{old}\ \text{(stationary solid)} \tag{7}$$

3) Resolution of the energy equation:

The energy equation is solved for the average enthalpy, h_m. A first order Taylor expansion is used for the temperature to obtain an implicit evaluation of the convection and conduction terms:

$$\frac{\partial h_m^{**}}{\partial t}+\nabla\cdot(\rho_0 c_p\vec{v}\tilde{T})=\nabla\cdot(k\nabla\tilde{T})$$
$$\tilde{T}=T^*+\frac{dT}{dh}\left(h_m^{**}-h_m^*\right)\,,\quad \frac{dT}{dh}=\max\left(\frac{1}{10c_p},\frac{T^*-T^{old}}{h_m^*-h_m^{old}}\right) \tag{8}$$

4) Local equilibrium:

Under the assumption of local equilibrium the liquid and solid solute concentration, liquid fraction and temperature are calculated from the new estimation of h_m and $C_m=g_s^{tr}C_s^{tr}+g_l^{tr}C_l^{tr}$. Equilibrium is defined by Eqs. (6). To solve this system, they are combined into a quadratic equation, which is solved analytically for the liquid fraction g_l:

$$a\left(g_l^*\right)^2+bg_l^*+c=0$$
$$a=(k_p-1)L\,;\quad b=\left(k_p-1\right)\left(c_pT_m-\langle h\rangle\right)-k_pL\,;\quad c=k_p\left(\langle h\rangle-c_pT_m\right)-\left(c_pmC\right) \tag{9}$$

5) Convergence test:

Steps 1-4 are repeated until the following convergence criterion is satisfied:

$$\max\left|\frac{|\vec{v}|^{**}-|\vec{v}|^{*}}{f^{norm}_{velocity}}\right|<\varepsilon\,,\ \max\left|\frac{p^{**}-p^{*}}{f^{norm}_{pressure}}\right|<\varepsilon\,,\ \max\left|\frac{C_l^{tr,**}-C_l^{tr,*}}{f^{norm}_{solute}}\right|<\varepsilon\,,\ \max\left|\frac{h_m^{**}-h_m^{*}}{f^{norm}_{enthalpy}}\right|<\varepsilon \tag{10}$$

where f^{norm} is a normalization factor for each equation. When the criterion is satisfied the time is advanced by Δt; the symbol * denotes the value of the last iteration and ** the value of the new iteration of the equation system.CastFOAM employs a different scheme for coupling transport and phase change. In each time step, the momentum and continuity equations are solved first in a step of the PISO [10] pressure-velocity coupling scheme. Then the solute conservation equation is solved in terms of the liquid concentration C_l. After that thermodynamic equilibrium (lever rule) is recovered by solving Eqs. (6). Then the heat transport equation in solved in a temperature formulation and after that the lever rule (Eqs. (6)) is applied again. This is iterated until convergence is reached and the time is advanced by Δt. The discretization is formulated on a collocated pressure-velocity arrangement, the convective terms are discretized with the QUICK [10] scheme and implicit Euler [10] time stepping is used.

5. Results and discussion

The results of simulations of solidification in the three-dimensional ingot, obtained with the NTNU/SINTEF code, are presented in this section. First, simulation results, for a test case, of the NTNU/SINTEF code are compared and validated with CastFOAM. Results in terms of macro- and meso-segregation at the end of solidification are discussed in the two cases. Then a parametric study in terms of dendrite arm spacing is performed in order to show the effects of permeability (Eq. (5)) in the mushy zone on the segregation behaviour. For the simulations a constant time step Δt=0.005 s, much smaller than the solidification, convection and diffusion characteristic times, was chosen. For the space-discretization a regular cubic mesh was chosen for the cavity: 128x128x32 volumes. The simulations end at t=600 s, after that the entire ingot is completely solidified. According to [1] and [11], the choice of these parameters provide a good spatial and time resolution, necessary for the description of the mesosegregation zone, where the shape, the length and the position of channels are very sensitive to mesh size [11].

5.1. Comparison of NTNU/SINTEF code and CastFOAM results: concentration field at the end of solidification and characterization of the mesosegregation zones.

We investigate the results at the end of solidification obtained by the two codes for Case 1 (Table 3). We first present the segregation map at the end of solidification (t=600 s) in the vertical XY symmetry plane (z=0.005 m) obtained by the two codes.

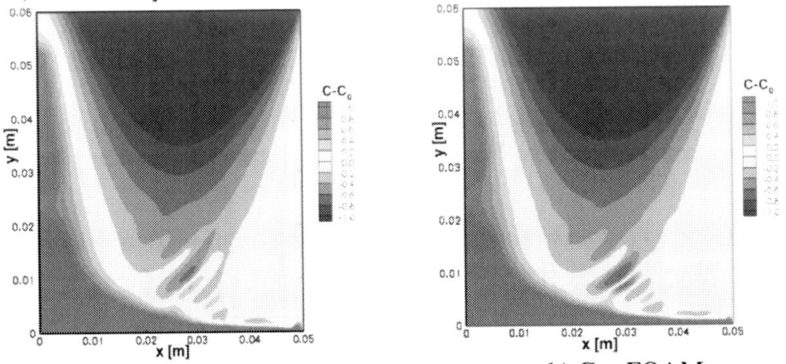

a) NTNU/SINTEF code b) CastFOAM

Figure 2. Segregation (C–C_0 in wt%Sn) at t=600 s in the vertical mid-section obtained by the NTNU/SINTEF code and by CastFOAM.

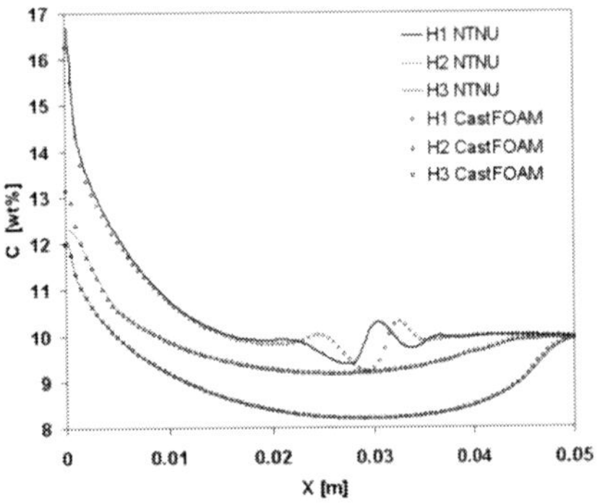

Figure 3. Average solute concentration (C in wt%Pb) profiles along the x direction in sections H1, H2 and H3 obtained by NTNU/SINTEF code (continuous lines) and CastFOAM (symbols) in the symmetry plane for Case 1.

Table 3. Dendrite arm spacing (DAS) used in the four simulations.

Case	DAS
1	65 μm
2	90 μm
3	120 μm
4	150 μm

Figure 2 shows a large negative segregation zone in the central and the upper regions of the domain, while in the bottom and the left part of the domain a large positive segregation region is observed. This is a classical result for columnar solidification: negatively segregated zones form where the liquid flow is oriented opposite to the temperature gradient and positively segregated zones, where it is oriented in the same direction. The cooling here is from the right and the natural convection creates a clockwise flow circulation. Additionally, the liquid that solidified last (on the left) has been progressively enriched in Pb content, which generates a positively segregated zone along the left vertical boundary. In bottom of the domain we observe a mesosegregated region in which narrow channels appear (aligned at about 45°). Inside them, the solute concentration is significantly higher than in the surrounding regions. The comparison between the results suggests that the global macrosegregation patterns have similar behaviour. In the meso-segregation region we observe, in both cases, the formation of the same number of channel mesosegregations with approximately the same orientation, length and location. The mesosegregation intensity prediction, however, is higher in CastFOAM, while the NTNU/SINTEF code predicts longer channels. Concentration profiles along sections H1, H2 and H3 (see Figure 1) are presented in Figure 3 and show in a more quantitative way the good agreement in macrosegregation prediction and the presence of the same number of mesosegregations, although a shift of their x position is observed. The observed small differences in mesosegregation patterns provided by the two codes are similar to the differences already observed between the software tested in the frame of the numerical benchmark on macro- and mesosegregation prediction [1]. As we will show, the structures in Case 1 are practically two-dimensional and only little variation is observed along the thickness (z direction). Comparison of results in three dimensions provided by the two codes for the remaining three cases also provided good agreement: very close segregation structures and lamellar-to-tubular channel transitions were obtained. Results obtained by the Leoben group [3] (only for Case 1) also agreed and they showed a lamellar structure of mesosegregates.

5.2. Influence of permeability on the 3D structure of the channel mesosegregates

The dendrite arm spacing (DAS) represents a crucial parameter in solidification calculations because it strongly affects the permeability (Eq. (5)) in the mushy zone. In this section a parametric study on DAS is performed and shows that as this parameter increases, the increase of permeability strongly affects the segregation evolution, leading to modifications in the macro- and mesosegregation in the ingot. Results for four values of DAS (Table 3) are presented in terms of average solute concentration at the end of solidification. In Figure 4, in order to show the three dimensional structure of the channel mesosegregates, an isosurface of the average concentration is presented and confirms that in Case 1, corresponding to the smallest DAS, the segregates are organised in form of lamellae that are close to being two-dimensional. Only a small three-dimensional effect is noticed close to the lateral walls. In Figure 4(a) we can see that for small DAS the channel mesosegregations are lamellar. In Figure 4(b) the results for the transition Case 2, corresponding to higher permeability, are presented: the two dimensional structure of the lamellae is broken and a central tubular structure appears. In Figures 4(c), and 4(d), corresponding to even higher values of DAS, the average concentration isosurfaces show the presence of a complex mesosegregate with secondary tubular structures, which branch from the central tubular channel already observed in Figure 4(b).

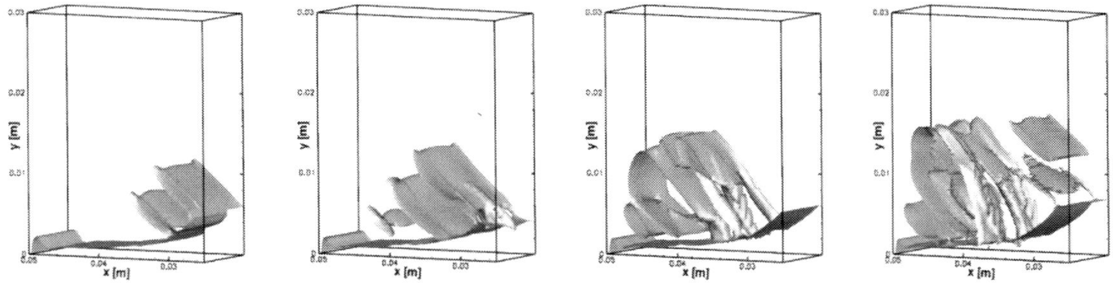

(a) Case 1 (DAS = 65μm): lamellar	(b) Case 2 (DAS = 90μm): lamellar-tubular	(c) Case 3 (DAS = 120μm): lamellar-tubular	(d) Case 4 (DAS = 150μm): branched tubular

Figure 4. Transition of the channel mesosegregation structure with increasing DAS. The segregation isosurface ($C – C_0 = 0.1$) at the end of solidification is shown.

The results of the simulations suggest a lamellar-to-tubular transition in the range of DAS [65:90μm]; from quasi two-dimensional lamellar channel mesosegregates to a fully three-dimensional structure, which increases in complexity as the DAS increases.

5.3 Lamellar to tubular transition in Case 2 (DAS = 90 μm)

In this section we analyse the destabilisation of the lamellar channel structure and the formation of the inner segregation channel for Case 2 (Figure 4(b)). In Figure 5 the formation of the channels is presented by a sequence of temporal snapshots. Three isosurfaces for the liquid fraction are shown: g_l=0.90 (light blue), g_l=0.95 (pink) and g_l=0.98 (yellow).

In Figures 5(a) and (b) we observe a first instability in the solidification front that occurs in the XY plane where the first lamella is forming. Its shape is practically two-dimensional, slight three-dimensional effects are only seen at the walls; they occur due to the delay of the front close to the boundaries. The mechanism seems to follow the same behaviour as described for the 2D case in [1,2] inducing a lamellar-type effect. At t=34s, the g_l=0.98 isosurface already presents transversal perturbations. During the time evolution (Figures 5(b)–(f)) these perturbations are damped however,and the surface becomes flat again. A delay of solidification close to the lateral boundaries is due to the recirculating flow that is more important in the inner part of the domain and promotes the cooling effect and the solidification.

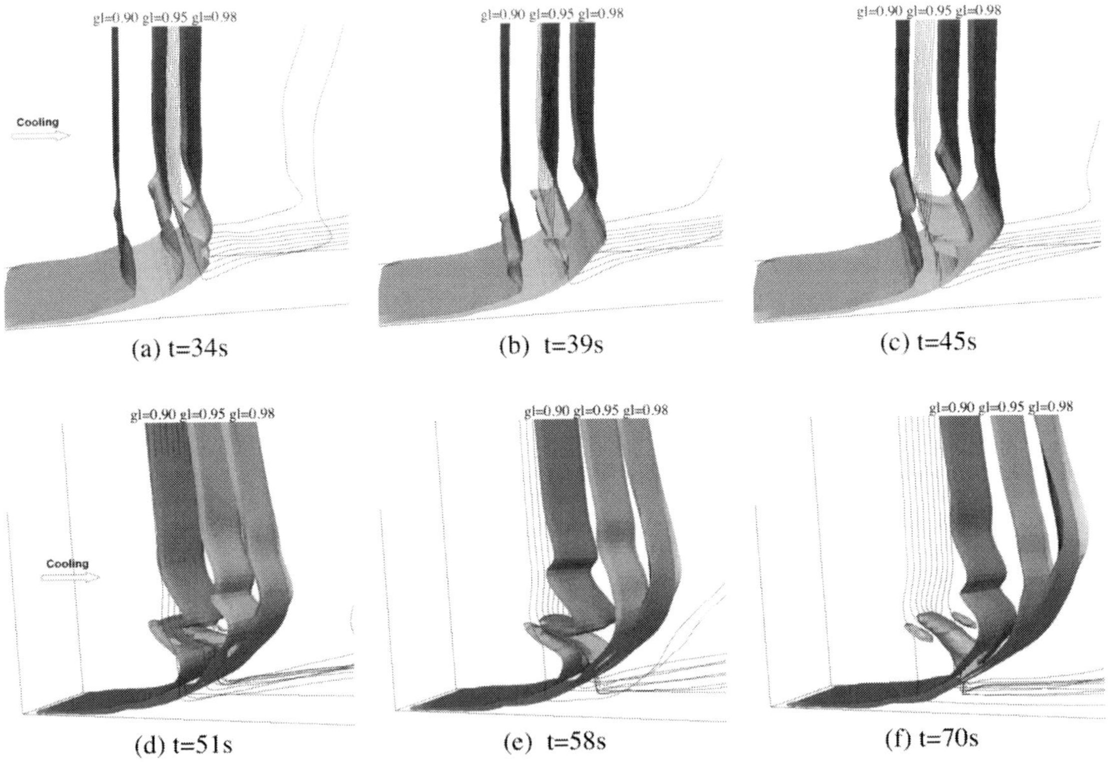

Figure 5. Transition form lamellar to tubular channels for Case 2. Isosurfaces of liquid fraction and streamlines of liquid velocity are shown. g_l=0.90 (light blue), g_l=0.95 (pink) and g_l=0.98 (yellow).

In Figure 5(c) a second instability in the isosurface g_l=0.95 occurs in the z direction. A central lamella and two lateral channels develop. This is an instability that leads to a lamellar-to-tubular transition, which progressively develops the tubular structures. In Figure 5(d), at g_l=0.95, it developed into a clear central tubular channel. At the same moment a lamellar shape is observed at g_l=0.90. We can see that streamlines now concentrate in the central part of the domain due to the higher permeability of the tubular channel, which promotes the passage of the fluid flow. The g_l=0.98 isosurface is still flat.

Between t=51s and 58s the g_l=0.98 isosurface is very quickly destabilized. In Figures 5(e) and (f) all three isosurfaces show the presence of a central tubular channel, which becomes the preferred passage for the flow – almost all the streamlines are concentrated there. As a conclusion, for the Case 2, the analysis of the mesosegregate evolution shows a regular lamellar to tubular transition – a result of a three-dimensional instability of the solidification front.

6. Conclusion

This work presents macro- and mesosegregation in a three-dimensional cavity filled by a Sn-10wt%Pb binary alloy and shows the sensibility of results with respect to the dendrite arm spacing. The structure of the mesosegregation pattern is presented in three dimensions, and major differences for four different DAS are shown at the end of solidification. The study of the evolution of the system in the DAS=90μm case (Case 2) for the first 70s suggests that after a first instability of the solidification front, which generates lamellar, quasi two-dimensional channel mesosegregation structures, a second instability of the front in the transversal direction is responsible for the development of fully three-dimensional tubular structures.

Acknowledgement

The Norwegian partners acknowledge support by Hydro, Alcoa Norway, Aleris, the Research Council of Norway and NOTUR High Performance Computing program.

References

[1] Combeau H, Bellet M, Fautrelle Y, Gobin D, Arquis E, Budenkova O, Dussourbs B, Duterrail Y, Kumar A, Mosbah S, Quatravaux T, Rady M, Gandin C.-A, Goyeau B, Založnik M 2011 *Frontiers in Solidification Science, TMS Annual Meeting 2012 (San Diego (CA), USA)* (TMS, Warrendale (PA), USA) p 755
[2] Bellet M, Combeau H, Fautrelle Y, Gobin D, Arquis E, Budenkova O, Dussourbs B, Duterrail Y, Kumar A, Mosbah S, Rady M, Gandin C.-A, Goyeau B, Založnik M 2009 *Int. J. Therm. Sci.* **48** 2013
[3] Li J, Wu M, Hao J, Ludwig A, 2012 IOP *Conf. Series: Mater. Sci. Eng.* 27 (2011) 012055.
[4] Sawada T, Oikawa K, Anzai K, 2009 *International Conference on Simulation and Modelling of Metallurgical Processes in Steelmaking (Steelsim)* (Leoben, Austria)
[5] Ni J, Beckermann C, 1991 *Metall. Trans. B* **22** 349
[6] Bennon W D, Incropera F P, 1987 *Int. J. Heat Mass Transfer* **30** 2161
[7] Založnik M, Combeau H 2010 *Comput. Mater. Sci.* **48** 1
[8] Založnik M, Kumar A, Combeau H 2010 *Comput. Mater. Sci.* **48** 11
[9] OpenFOAM®, version 1.7.1, computer program, OpenCFD Inc.
[10] Prosperetti A, Tryggvason G, Computational Methods for Multiphase flow, 2007, Cambridge University Press
[11] Kumar A, Dussoubs B, Založnik M, Combeau H 2009 *J. Phys. D: Appl. Phys.* **42** 105503

Numerical study of the impact of inoculant and grain transport on macrosegregation and microstructure formation during solidification of an Al-22%Cu alloy

K O Tveito[1], M Bedel[2], M Založnik[2], H Combeau[2], M M'Hamdi[1,3], A Kumar[2] and P Dutta[4]

[1] Dept. of Materials Technology, NTNU, N-7491 Trondheim, Norway
[2] Institut Jean Lamour, Département SI2M, CNRS – Université de Lorraine,Ecole des Mines de Nancy, Parc de Saurupt CS14234, F-54042 Nancy cedex, France
[3] SINTEF Materials and Chemistry, N-0314 Oslo, Norway
[4] Department of Mechanical Engineering, Indian Institute of Science, Bangalore, India

E-mail: knut.tveito@material.ntnu.no

Abstract. We investigate the impact of the nucleation law for nucleation on Al-Ti-B inoculant particles, of the motion of inoculant particles and of the motion of grains on the predicted macrosegregation and microstructure in a grain-refined Al-22 wt.% Cu alloy casting. We conduct the study by numerical simulations of a casting experiment in a side-cooled 76x76x254 mm sand mould. Macrosegregation and microstructure formation are studied with a volume-averaged two-phase model accounting for macroscopic heat and solute transport, melt convection, and transport of inoculant particles and equiaxed grains. On the microscopic scale it accounts for nucleation on inoculant particles with a given size distribution (and corresponding activation undercooling distribution)and for the growth of globular solid grains. The growth kinetics is described by accounting for limited solute diffusion in both liquid and solid phases and for convective effects. We show that the consideration of a size distribution of the inoculants has a strong impact on the microstructure(final grain size) prediction. The transport of inoculants significantly increases the microstructure heterogeneities and the grain motion refines the microstructure and reduces the microstructure heterogeneities.

1. Introduction

The prediction of the microstructure formation during solidification is of major importance for the industry. Micro-macro models are developed for decades now to relate the macroscopic solidification conditions to the microscopic structures obtained. Rappaz [1] gives an overview of the microscopic mechanisms of microstructure formation and the coupling between them and the macroscopic transport phenomena.

A crucial point of the microscopic modelling is nucleation. As recalled in[2], the nucleation is rarely homogeneous as it is energetically more favourable for the metal to solidify on foreign particles which are always, deliberately or not, present in the metal. Heterogeneous nucleation is not easy to calculate, as even a known distribution of refiner particles, deliberately added into the liquid metal, can be modified by the fragmentation of grains or by the agglomeration of particles. However, different nucleation models were developed, which are used in the macroscopic models.

When the experimental average grain density is known, an easy way to avoid modelling the inoculants activation is to consider an initial grain density in the melt entering the mould which matches the experimental data[3]. The grains are not generated but only redistributed by the flow. But when inoculants are taken into account, their modelling is most of the time very simplified. All the inoculants are usually supposed to be activated instantaneously at the liquidus or at a given undercooling, as justified in[4]. More rarely, an inoculant size distribution is considered trough a normal law and is discretized in classes [5]. Though there is no reason to justify that the inoculants remain fixed when the liquid moves, inoculants transport was never considered until now. Thus the goal of this study is to analyse the impact of the inoculant motion on the macrosegregation and the microstructure formation. An experiment is modelled with different hypotheses in order to see the impact of each phenomenon.

2. Model description

The model is an extension of the volume-averaged two-phase model and the solution algorithm that was described in detail in [6].Only the extensions as well as the main features of the model are presented here. The model describes and couples phenomena at two scales, which are considered to be distinct. At the macroscopic scale the transport of momentum, mass, heat, solute, grain, and inoculants are described by volume-averaged transport equations. At the microscopic scale the nucleation and the growth of globular grains are described by volume-averaged interfacial balances. The liquid flow and the movement of globular grains are described on a macroscopic level by considering the buoyancy-driven flow arising from the liquid density variations (thermo-solutal convection),the density differences between liquid and solid, and the drag forces at the grain interfaces. The Boussinesq approximation is used, and in the present work the shrinkage-induced flow is neglected. When the volume fraction of solid exceeds the imposed packing fraction, the solid phase is assumed to be stationary and the liquid interfacial drag is described with a Darcy term modelled by the Kozeny-Carman law. On a microscopic level, the model takes into account finite diffusion of solute in liquid and solid phases and assumes local thermal equilibrium and thermodynamic equilibrium at the solid-liquid interface.

Nucleation of grains is assumed to occur on grain-refiner (inoculant) particles. According to the athermal nucleation theory of Greer and co-workers [5],the critical undercooling for free growth of a grain on an inoculant particle of diameter d is given by $\Delta T_c(d)=4\Gamma/d$ where Γ is the Gibbs-Thomson constant. The number of activated particles then depends on the size distribution of the particle population, which can be represented by an exponential distribution density function: $n(d)=N_0/d_0.\exp(-d/d_0)$ [m^{-4}], where N_0 [m^{-3}]is a characteristic inoculant population density and d_0[m] is the characteristic width of the distribution. This representation holds for particles large enough. In the modelling, the size distribution is discretized into m classes of inoculants. Each class is represented by a population density N^i_{nuc} and a critical undercooling $\Delta T_c^i=4\Gamma/d^i$, depending on the average inoculant diameter of the class d^i. When the local solutal undercooling reaches the critical undercooling of the class, its inoculant density, N^i_{nuc}, is added to the grain density, N_G, and N^i_{nuc} becomes locally zero. The conservation equations for the density of each inoculant class and the grain density are Equations (1) and (2), respectively, where Φ^i represents the transfer from inoculants to grains upon nucleation. The inoculants are assumed to be transported by the liquid with its velocity and one conservation equation is solved separately for each class.

$$\frac{\partial\left(N^i_{nuc}\right)}{\partial t}+\nabla\cdot\left(N^i_{nuc}\langle\vec{v}_l\rangle^l\right)=-\Phi^i\ ,\quad \Phi^i=\begin{cases}N^i_{nuc}\delta(t) & ,\ \Delta T>\Delta T^i_c\\ 0 & ,\ \text{else}\end{cases}\ ,\quad i=1,...,m \tag{1}$$

$$\frac{\partial\left(N_G\right)}{\partial t}+\nabla\cdot\left(N_G\langle\vec{v}_s\rangle^s\right)=\sum_{i=1}^{m}\Phi^i \tag{2}$$

3. Experiment description and its modelling

The two-phase model is applied to the experiment of Kumar et al.[7], concerning horizontal solidification of a grain-refined (0.05 wt.% TiB_2) Al-22wt.%Cu alloy. The details of the experiment can be found in Ref. [7]. Here, an illustration of the experimental setup is provided in Figure 1a, showing the graphite chill and the cast cavity including its dimensions.

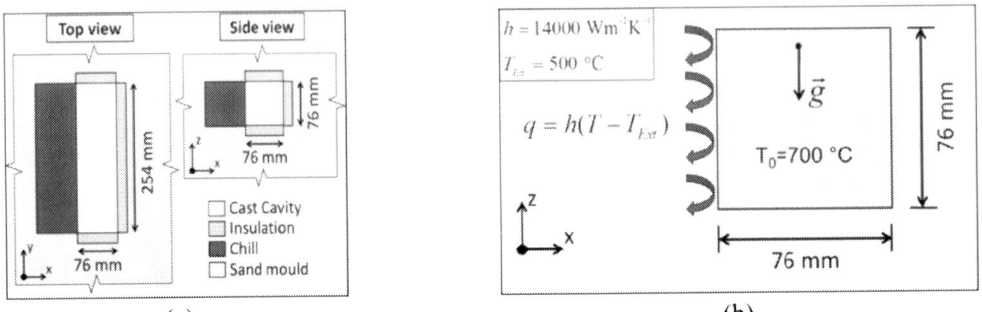

Figure 1. (a) Illustration of experiment geometry and (b) 2D model for numerical simulations.

The experiment is modelled as a2D problem in the *xz* plane as shown in Figure 1b. The cooling heat flux along the left boundary is modelled with a Fourier condition, while the other boundaries are assumed to be thermally insulated. N conditions are assigned to all boundaries and the initial temperature is 700 °C. The thermophysical data are given in Table. The constant solute diffusion coefficients in the liquid and solid phases were calculated from [8] at 599 °C, the liquidus temperature of the nominal alloy. The liquidus slope and melting point were adjusted to better fit the phase diagram for high concentrations of copper. The liquid and solid densities are constant and equal, except for the buoyancy term in the momentum equations, where the liquid density is taken from [7]. The liquid, the solid and the average ($\rho_{average}=g_l\rho_l+g_s\rho_s$) densities are shown as function of solid fraction in Figure 2, by using the lever rule model for the solidification path.

Table 1. Modified thermophysical data Al-22wt.%Cu alloy

Property	Unit	Value
Diffusion coefficient, solid	$m^2\ s^{-1}$	4.2×10^{-13}
Diffusion coefficient, liquid	$m^2\ s^{-1}$	3.8×10^{-9}
Thermal conductivity, solid	$W.m^{-1}.°C^{-1}$	120.7
Thermal conductivity, liquid	$W.m^{-1}.°C^{-1}$	57.3
Specific heat (liquid and solid)	$J.kg^{-1}.°C^{-1}$	1100
Characteristic length used in Kozeny Carman law	m	50×10^{-6}
Pseudo Al melting temperature	°C	703.86
Liquidus slope	°C wt.%$^{-1}$	-4.766
Eutectic temperature	°C	548
Partition coefficient	-	0.173
Reference liquid density	$kg\ m^{-3}$	2800.00
Solid density (buoyancy)	$kg\ m^{-3}$	2609.45
Thermal coefficient of volumetric expansion (β_T)	$°C^{-1}$	1.17×10^{-4}
Solutal coefficient of volumetric expansion (β_s)	$(wt.\%)^{-1}$	-7.3×10^{-3}
Dynamic viscosity, liquid	$N.s.m^{-2}$	1.2×10^{-3}
Packing fraction (gs)	-	0.27
Gibbs-Thomson coefficient	K.m	1.9×10^{-7}

We can notice that even when the grains move, the buoyancy effect increases with the solid fraction as it depends on the average density. The packing fraction is set to 0.27[7] and a columnar zone with stationary solid phase is assumed for the first 10 mm from the chill.

The size distribution for grain refiner particles is taken from [9], for 5kg/t of inoculants of type TiB_2(instead of the 0.5kg/t used in the experiment, in order to limit the undercooling). The distribution is modelled by an exponential function with the parameters N_0=3.56x10^{12} m^{-3} and d_0=7.1x10^{-7} m, for inoculant diameters between 1 and 10 µm. The distribution, when usually represented as a function of the particle diameter, is continuously decreasing. However, when expressed as a function of the critical undercooling, this distribution reaches a maximum at around 0.5°C, as can be seen in Figure 3. In the present study, the distribution is discretized into 15 classes: undercooling intervals are defined by increasing step from 0.1 to 0.95°C. The inoculant density distribution and the undercooling intervals used for the discretization are shown in Figure 3. Note that the exponential law was extrapolated to particles smaller than 0.1 μm to cover the whole range of undercoolings.

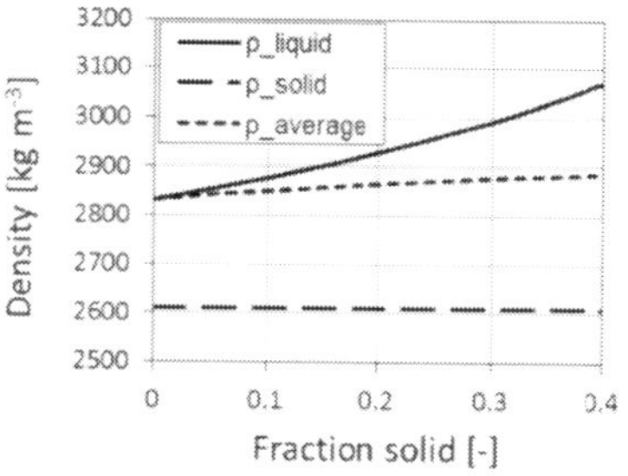

Figure 2. Densities versus the solid fraction, lever rule model.

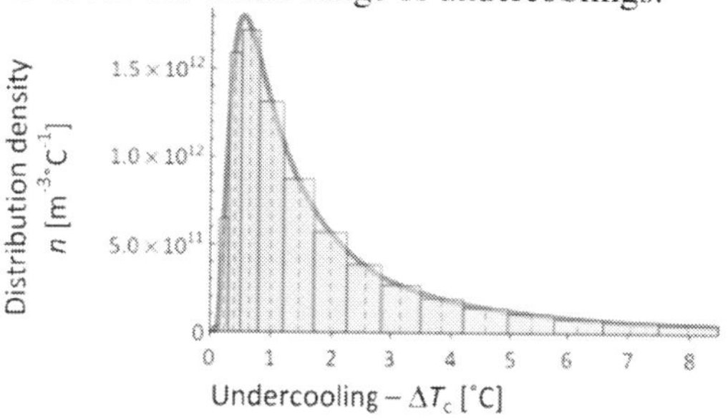

Figure 3. Distribution density function of the inoculant population (red line) and inoculant classes (blue bars). The surface of a bar corresponds to the inoculant population volume density N^i_{nuc}[m^{-3}] of a class and ΔT_c^i are indicated as dashed lines.

The transport equations are solved with a Finite Volume Method and the SIMPLE-method for staggered grid is used for pressure-velocity coupling. The convective terms are discretized with a first-order upwind scheme and for time discretization a fully implicit first-order scheme is used. For all simulations an equidistant grid of 40x40 cells and constant time step of 10^{-3}s is used.

4. Numerical results

In order to study the impact of the different phenomena on the macrosegregation formation and the microstructure, the four cases are defined with an increasing complexity. Their main characteristics are defined in Table 2. The first case is used as a reference for the macrosegregation formation: only natural convection is taken into account at the macro-scale, and all the inoculant classes are instantaneously activated at the liquidus. The inoculant density is taken as the sum of the 15 inoculant class densities defined in the previous section. Then for the Case2, the inoculant distribution is taken into account so that the local grain density depends on the reached undercooling. In Case3, the inoculants can move with the liquid and the local grain density additionally depends on inoculant transport. Finally, in Case 4, the grain motion is taken into account. In all the cases, the inoculants are uniformly distributed in the melt at t=0s.

Table 2. Cases definition

	Case 1	Case 2	Case 3	Case 4
Nucleation at the liquidus	x			
Nucleation function of the undercooling		x	x	x
Inoculant motion			x	x
Grain motion				x

4.1. Case 1: Reference case

The inoculantsandthe grains are assumed to be fixed, and the grain density is homogeneous throughout the cavity. All the inoculant classes (which correspond to an initial inoculant density of $3.135\text{x}10^{12}$ m^{-3}) are activated locally as soon as the local temperature reaches the liquidus.

The final macrosegregation pattern for Case 1 is shown in Figure 8a. The cooling from the left hand boundary introduces a horizontal temperature gradient and thermo-solutal convection results in a counter-clockwise flow pattern. Within the mushy zone the rate of segregation is described by the conservation equation for the average composition C_m:

$$\frac{\partial C_m}{\partial t} = -g_l \langle \vec{v}_l \rangle^l \cdot \nabla \langle C_l \rangle^l = -\frac{1}{m} g_l \langle \vec{v}_l \rangle^l \cdot \nabla [T - \Delta T_s] \quad , \quad T = T_m + mC_l^* \quad , \quad \Delta T_s = m\left(C_l^* - \langle C_l \rangle^l\right) \qquad \textbf{(3)}$$

where g_l is the volume fraction liquid, $\langle v_l \rangle^l$ is the intrinsic liquid velocity, m is the liquidus slope, C_l^* is the interface composition and $\langle C_l \rangle^l$ is the average liquid composition. The influence of the gradients of undercooling is dependent on the liquid Fourier number ($Fo_l = D_l t_{solidification}/R_f^2$, where R_f is the final grain radius). In the entire cavity, the liquid Fourier number is higher than 100, and according to Ref. [10], the influence of the gradients of undercooling on macrosegregation formation is then small. Within the mushy zone, negative segregation is formed in the upper part of the cavity where the liquid flows in the opposite direction of the temperature gradient, while a severe positive segregation is formed in the lower part of the cavity where the liquid flows in the same direction than the temperature gradient. This segregation pattern can be easily understood from Equation (3).

4.2. Case 2: Impact of inoculant distribution (neither inoculant nor grain motion)

By accounting for the inoculant size distribution, the competition between nucleation and growth is included in the model. As the inoculants and the grains are fixed for Case 2, the number of activated inoculant classes, and thus the final grain density, only depends on the maximum local undercooling. Generally, the solidification kinetics results in higher undercoolings reached at higher cooling rates.

The local cooling rate, and thus the undercooling, attains the highest values close to the cooled left-hand boundary where up to 12 inoculant classes are activated, resulting in a higher grain density. With increasing distance from the cooled boundary the local cooling rate and undercooling decreases, and the grain density decreases accordingly as shown in the horizontal profile in Figure 4a for Case 2.

In the present work the permeability of the mushy zone is uncoupled from the predicted grain size as the characteristic length in the Kozeny Carman law is set constant. As a result, the liquid flow is not significantly affected by the inoculant distribution and only slight differences in macrosegregation formation are seen for Case 1 compared to Case 2, and the latter case is, therefore, not presented in detail. The small differences can be attributed to the difference in undercooling and its influence on macrosegregation (see Equation (3)).

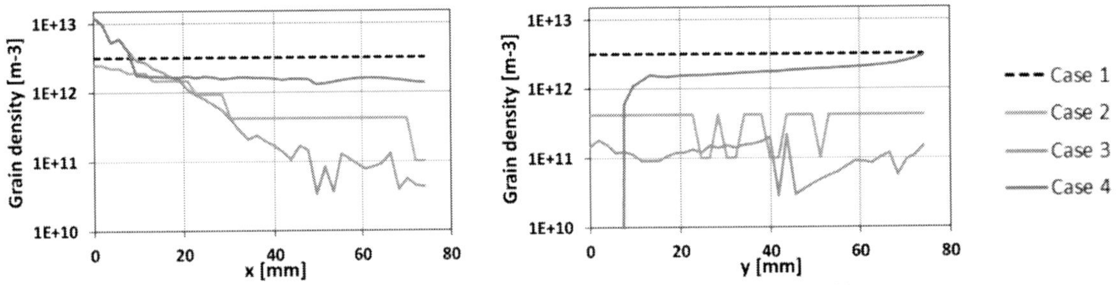

a) Horizontal profile at y=15mm b) Vertical profile at x=60mm

Figure 4. Comparison of a) horizontal and b) vertical grain density profiles for cases 1 to 4.

4.3. Case 3: Impact of inoculant motion (without grain motion)

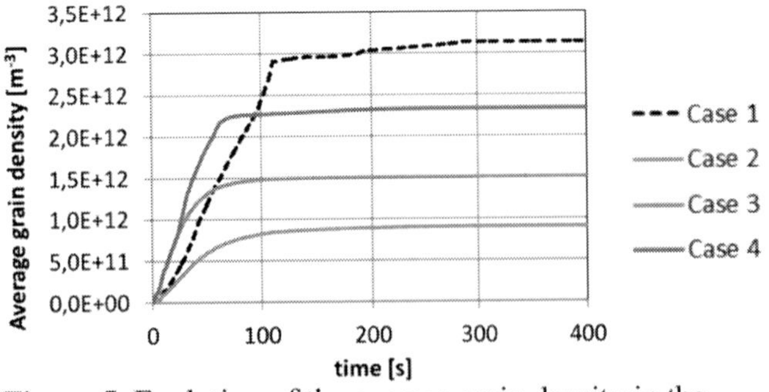

Figure 5. Evolution of the average grain density in the cavity with time.

The inoculant motion does not change the evolution of the thermal field in the cavity; therefore the differences in macrosegregation between Case 2 and Case 3are small. The microstructure, however, is strongly modified. In Figure 4a we can notice a steeper grain density gradient across the cavity, and in Figure 5we can see that globally more grains nucleate in Case 3 than in Case 2.

During the initial stages of

solidification, when the mushy zone is spreading from the chill towards the interior, the counter clockwise flow continuously brings liquid from the bulk liquid zone, carrying a high inoculant density (the inoculants in the bulk liquid were not yet activated), into the mushy zone in the top part of the cavity. If the undercooling in the mushy zone is sufficient, these additionally supplied inoculants are activated in the mushy zone. Consequently, the liquid flowing back from the mushy zone into the bulk liquid zone at the bottom of the cavity carries a low inoculant density, as can be seen in Figure 6a and c. This results in a certain inoculant segregation. On the one hand, more inoculants are supplied to the chill region, which refines its grain structure compared to Case 2, where inoculants were fixed. On the other hand, the inoculant density in the region away from the chill is diluted (Figure 6a–b), which coarsens the grain structure compared to Case 2. This explains the grain density gradient across the cavity. The inoculant segregation has however a somewhat different consequence for larger than for smaller inoculants. Large inoculants require only small undercoolings for their activation. These are easily reached across the whole cavity in Case 2 as in Case 3. The segregation of large inoculants therefore results only in a redistribution of the grain density, but does not influence the total grain population. Small inoculants require large undercoolings, which are reached only close to the chill. The segregation of small inoculants does therefore not affect the interior regions, but only increases the grain population in the chill region (Figure 6c–d). This results in an increase of the total grain population, compared to Case 2. A higher increase rate of the grain population for Case 3, seen in Figure 5 also shows that the additional grains are generated at the beginning, during the nucleation of the chill region.

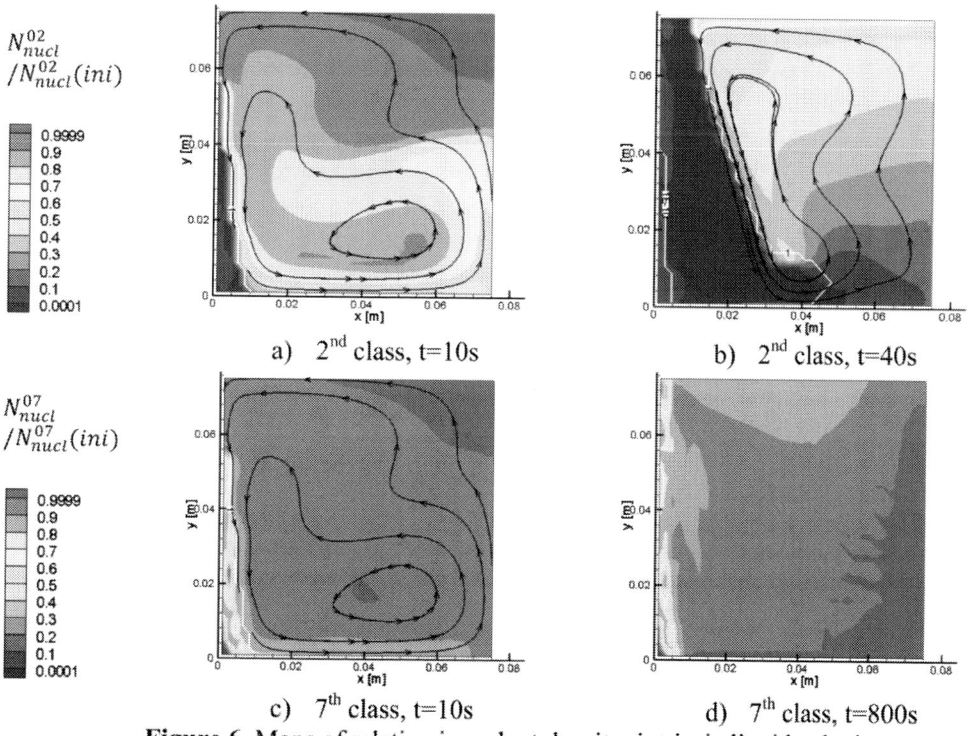

Figure 6. Maps of relative inoculant density, intrinsic liquid velocity streamlines in black and liquidus and solidus isolines in white (shown by solid fraction contour), for the 2nd and the 7th inoculant classes, at different times, for Case 3.

4.4. Case 4: Impact of grain motion

It has been seen through the first three cases that the nucleation law and the inoculant motion do not impact the macrosegregation pattern but only increase the microstructure heterogeneities. In Case 4, the grain motion is added to these two phenomena, and differences are noticed in both microstructure and macrosegregation.

Without grain motion, the flow is counter-clockwise as the richer and colder liquid is heavier. When grain motion is added, the direction of the flow could be modified as the grains, which are lighter than the liquid, tend to float upwards along the solidification front. However, the density difference between the two phases is not high enough to do so. The average density is still increasing with the solid fraction, as can be seen in Figure, so the decrease in density due to solidification is more than compensated by the liquid enrichment and cooling. Thus the main flow direction is not modified as can be seen in Figure 7, but its pattern is totally changed as a second loop appears at the bottom. This change in flow pattern induces modifications in both macrosegregation and microstructure.

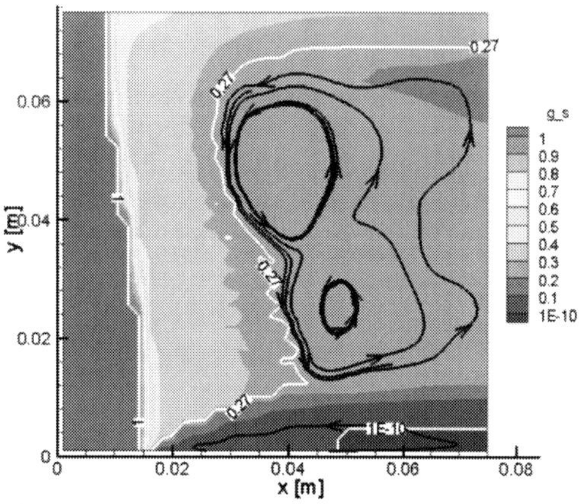

Figure 7. Map of solid fraction, liquidus, packing and solidus contours (trough solid fraction) in white and intrinsic liquid velocity streamlines in black at 120s for Case 4.

The impact of grain motion on macrosegregation can be seen by comparing the macrosegregation maps for Case 1 and Case 4 (Figure 8). The macrosegregation, which is not altered when nucleation is taken into account, is highly modified by the grain motion. The grain flotation results in accumulation of solute-lean grains towards the top of the cavity, and thus in a vertical, downward oriented concentration gradient. In Figure 7, the packing fraction isoline position is a sign of the accumulation of grains at the top. The only exception is the 10 mm thick columnar zone next to the chill, where the grains are fixed; therefore the pattern remains similar to case 3.

The microstructure is also changed by the grain motion after the columnar zone. Next to it, the newly nucleated grains are continuously swept away by the liquid. Thus the nucleation-growth competition is modified and a high undercooling is maintained. As a consequence, more grains are formed, which can be seen in Figure 5. The total number of grains increases at the same speed as in Case 3 in the columnar zone, but when the activation zone reaches the equiaxed zone, grains are formed faster. The transport of the grains leads to a more homogeneous distribution of the grain density. This is clearly seen in Figure 4a. The average grain density is higher and its distribution across the cavity is more homogeneous. The grain density also increases from the bottom to the top, as can be seen in Figure 4b, which is not surprising as the grains float.

5. Discussion and conclusion

The goal of this paper was to study the impact of the nucleation modelling and the grain motion on the microstructure and the macrosegregation formation in the case of a small casting of Al-22%Cu. We did not try to fit the experimental results, as many parameter values, as well the heat extraction rate evolution and the effective inoculant distribution are insufficiently or not at all characterized. But by studying increasingly complex modelling, tendencies are noticed and the phenomena responsible for them are identified.

The macrosegregation pattern is shown in Figure 8 for cases 1 and 4 and for the experimental data given in [7] (the values measured at the points are interpolated). We notice that the main segregation gradient is directed toward the right-bottom corner of the cavity in Case 1, and that it is almost vertical in Case 4. Qualitatively, we can compare the experimental result to these two patterns: it seems to be a combination of these two tendencies with a vertical gradient at the left side of the cavity and a gradient directed to the right-bottom corner at the end of the solidification. Thus it can be supposed that the grain motion impact is reduced at the end of solidification when only few grains remain in the mushy zone. A more precise modelling of the grain motion, taking into account the interactions between the grains, could improve this estimation.

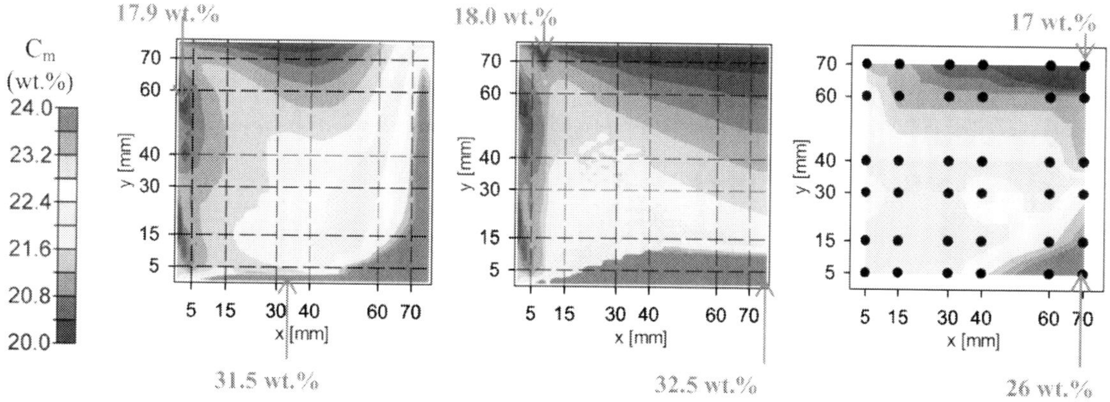

a) Case 1 b) Case 4 c) Experimental data

Figure 8. Final segregation maps for a) Case 1, b) Case 4 and c) experimental data (grid lines and points showing the measurement locations).

Inoculant motion and distribution have only small impact on the macrosegregation pattern. To check this assumption, the Case 4 was also run for a different nucleation law: inoculants were all activated at the liquidus like in Case 1. This led to the same macrosegregation pattern as for Case 4, even if the intensity of the segregation was lower.

Even though the inoculant motion has no noticeable impact on the macrosegregation, it is predominant in the microstructure formation. It is shown in this study that it increases the heterogeneities in the product (Figure 4) and that it refines the microstructure (Figure 5). This is the first study of the influence of inoculant on the microstructure and it shows that it seems unavoidable to take into account inoculant motion to correctly predict the microstructure.

This is a preliminary study and more work has to be done to properly estimate the inoculant distribution. Much higher undercoolings than what is usually considered in the nucleation models were found. Undercoolings of more than 5°C were calculated while in [5], the inoculant distribution is considered only up to 0.76°C. This would lead to think that finer particles than usually considered could act as nucleation sites. If it is not the case, agglomeration of inoculants or fragmentation of grains could be another explanation. In any case, this study points out the importance of the inoculants modelling, of their distribution as well as their motion, in microstructure prediction.

Acknowledgement
This research work is supported by Hydro, Alcoa Norway, Aleris, and the Research Council of Norway.

References

[1] Rappaz, M., *International Materials Reviews*, 1989. **34**(3): p. 93–124.
[2] Dantzig, J.A. and M. Rappaz, *Solidification*. 2009, Lausanne: EFPL Press.
[3] Reddy, A.V. and C. Beckermann, *Materials Processing in the Computer Age II*, 1995: p. 89-102.
[4] Stefanescu, D., et al., *Metallurgical and Materials Transactions A*, 1990. **21**(3): p. 997-1005.
[5] Greer, A.L., et al., *Acta Materialia*, 2000. **48**(11): p. 2823-2835.
[6] Zaloznik, M. and H. Combeau, *Computational Materials Science*, 2010. **48**(1): p. 1-10.
[7] Kumar, A., et al., *Metallurgical and Materials Transactions B*, 2011. **42**(4): p. 825-836.
[8] Du, Y., et al., *Materials Science and Engineering A*, 2003. **363**(1-2): p. 140-151.
[9] Tronche, A., *PhD Thesis*. 2000, University of Cambridge.
[10] Tveito, K.O., et al., *IOP Conference Series: Materials Science and Engineering*, 2011(in press).

IOP Conf. Series: Materials Science and Engineering **33** (2012) 012090 doi:10.1088/1757-899X/33/1/012090

Modelling of macrosegregation in steel ingots: benchmark validation and industrial application

Wensheng LI[1], **Bingzhen SHEN**[2], **Houfa SHEN**[1] **and Baicheng LIU**[1]
[1] Key Laboratory for Advanced Materials Processing Technology, MOE,
Department of Mechanical Engineering, Tsinghua University, Beijing, 100084, China
[2] CITIC Heavy Industries Co., Ltd., Luoyang, 471039, China

E-mail: shen@tsinghua.edu.cn

Abstract. The paper presents the recent progress made by the authors on modelling of macrosegregation in steel ingots. A two-phase macrosegregation model was developed that incorporates descriptions of heat transfer, melt convection, solute transport, and solid movement on the process scale with microscopic relations for grain nucleation and growth. The formation of pipe shrinkage at the ingot top is also taken into account in the model. Firstly, a recently proposed numerical benchmark test of macrosegregation was used to verify the model. Then, the model was applied to predict the macrosegregation in a benchmark industrial-scale steel ingot. The predictions were validated against experimental data from the literature. Furthermore, macrosegregation experiment of an industrial 53-t steel ingot was performed. The simulation results were compared with the measurements. It is indicated that the typical macrosegregation patterns encountered in steel ingots, including a positively segregated zone in the hot top and a negative segregation in the bottom part of the ingot, are well reproduced with the model.

1. Introduction

The production of heavy steel ingots with improved structure and chemical homogeneities is of great concern in industry [1]. Prediction of macrosegregation in industrial steel ingots is of great importance. In the last few decades numerous multiphase models have been developed to depict the solidification of steel ingots and to predict the macrosegregation. Beckermann and co-workers ([2], and references therein) proposed a multiphase model that accounts for melt convection and grain motion, which bridges the length scales between global transport phenomena and microscopic grain growth kinetics. Ludwig and co-workers [3-4] developed a series of multiphase solidification models. The most sophisticated one is a five-phase model that accounts for columnar-to-equiaxed transition, non-dendritic and dendritic crystal growth, and columnar primary dendritic tip tracking, as well as melt flow and grain motion. An overview of published multiphase models is presented in reference [5].

The application of multiphase models to industrial steel ingots is a challenging problem mainly due to the large computational resources required to accurately resolve the variety of the phenomena over the process scale. A pioneering application was performed by Combeau and co-workers [6-7], in which macrosegregation in an industrial 3.3-t steel ingot was measured and predicted. Li et al. [8] also presented simulations for macrosegregation in this benchmark steel ingot, using a two-phase multi-scale solidification model.

In this paper, recent progress made by the authors on modelling of macrosegregation in steel ingots is reported. The proposed two-phase solidification model involves the simultaneous consideration of the coupled macroscopic phenomena of heat transfer, fluid flow, solute transport, and solid movement and microscopic phenomena of grain nucleation and growth. The model was verified with a numerical benchmark test of macrosegregation. Then, the predictions of macrosegregation in a benchmark 3.3-t steel ingot were validated against experimental data from the literature. Finally, application of the model to an industrial 53-t steel ingot is provided.

2. Model description

The volume-averaged two-phase solidification model was described previously [8]. The governing equations and supplementary relations are summarized as follows.

Mass conservation equations:

$$\frac{\partial}{\partial t}(g_s\rho_s)+\nabla\cdot(g_s\rho_s\mathbf{u}_s)=M_{ls} \quad (1)$$

$$\frac{\partial}{\partial t}(g_l\rho_l)+\nabla\cdot(g_l\rho_l\mathbf{u}_l)=-M_{ls} \quad (2)$$

Momentum conservation equations:

$$\frac{\partial}{\partial t}(g_s\rho_s\mathbf{u}_s)+\nabla\cdot(g_s\rho_s\mathbf{u}_s\mathbf{u}_s)=-g_s\nabla p+\nabla\cdot(\mu_s g_s\nabla\mathbf{u}_s)+g_s\rho_s^b\mathbf{g}+K_{ls}(\mathbf{u}_l-\mathbf{u}_s) \quad (3)$$

$$\frac{\partial}{\partial t}(g_l\rho_l\mathbf{u}_l)+\nabla\cdot(g_l\rho_l\mathbf{u}_l\mathbf{u}_l)=-g_l\nabla p+\nabla\cdot(\mu_l g_l\nabla\mathbf{u}_l)+g_l\rho_l^b\mathbf{g}+K_{sl}(\mathbf{u}_s-\mathbf{u}_l) \quad (4)$$

Species conservation equations:

$$\frac{\partial}{\partial t}(g_s\rho_s C_s)+\nabla\cdot(g_s\rho_s\mathbf{u}_s C_s)=\frac{S_V\rho D_s}{\delta_s}(C_s^*-C_s)+C_s^*M_{ls} \quad (5)$$

$$\frac{\partial}{\partial t}(g_l\rho_l C_l)+\nabla\cdot(g_l\rho_l\mathbf{u}_l C_l)=\frac{S_V\rho D_l}{\delta_l}(C_l^*-C_l)-C_l^*M_{ls} \quad (6)$$

Energy conservation equations:

$$\frac{\partial}{\partial t}[(g_s\rho_s c_s+g_l\rho_l c_l)T]+\nabla\cdot[(g_s\rho_s c_s\mathbf{u}_s+g_l\rho_l c_l\mathbf{u}_l)T]=\nabla\cdot[(g_s k_s+g_l k_l)\nabla T]+M_{ls}L \quad (7)$$

Species balance at the solid-liquid interface:

$$\frac{S_V\rho D_s}{\delta_s}(C_s^*-C_s)+C_s^*M_{ls}+\frac{S_V\rho D_l}{\delta_l}(C_l^*-C_l)-C_l^*M_{ls}=0 \quad (8)$$

Grain population balance:

$$\frac{\partial n}{\partial t}+\nabla\cdot(\mathbf{u}_s n)=N \quad (9)$$

In the above equations, g is the volume fraction, ρ is the density, M_{ls} is the interfacial phase change rate, p is the pressure, μ is the viscosity, K_{ls} (= K_{sl}) is the interphase friction coefficient, C is the concentration, S_v is the interfacial area concentration, δ is the solute diffusion length, D is the mass diffusivity, c is the specific heat, k is the thermal conductivity, L is the latent heat, n is the grain density, and N is the grain nucleation rate. The subscripts "s" and "l" refer to solid and liquid, respectively, and the superscript "*" represents equilibrium at the solid-liquid interface.

The volume fractions of solid and liquid sum to unity: $g_s + g_l = 1$. The densities of both phases are assumed to be equal and constant: $\rho_s = \rho_l = \rho$, except in the buoyancy terms according to the Boussinesq approximation

$$\rho_s^b=\rho(1+\beta_{sl}) \quad (10)$$

$$\rho_l^b=\rho[1-\beta_T(T-T_{ref})-\beta_C(C_l-C_{ref})] \quad (11)$$

where β_{sl} is the solidification volume shrinkage, β_T and β_C are the thermal and solutal expansion coefficients respectively, T_{ref} and C_{ref} are reference values for the temperature and concentration.

The interfacial friction coefficient is calculated by the Gidaspow correlation [9],

$$K_{ls} = \begin{cases} \frac{3}{4} C_D \frac{g_s g_l \rho_l}{d_p} |\mathbf{u}_s - \mathbf{u}_l| g_l^{-2.65} & g_s \le 0.2 \\ 150 \frac{g_s^2 \mu_l}{g_l d_p^2} + 1.75 \frac{\rho_l g_s}{d_p} |\mathbf{u}_s - \mathbf{u}_l| & g_s > 0.2 \end{cases} \tag{12}$$

where C_D is the drag coefficient and d_p is the characterized particle diameter.

The diffusion lengths are modeled as

$$\delta_s = \frac{d_s}{10} \qquad \delta_l = \frac{d_s}{2} \tag{13}$$

where d_s is the grain diameter.

The viscosity of the solid phase is given by

$$\mu_s = \frac{(1 - g_s/g_{sc})^{-2.5 g_{sc}} - g_l}{g_s} \mu_l \tag{14}$$

where g_{sc} is the critical solid volume fraction (i.e., the packing limit).

3. Numerical procedure

The model equations presented above are similar in structure and take the general form as

$$\frac{\partial}{\partial t}(g_k \rho_k \Psi_k) + \nabla \cdot (g_k \rho_k \mathbf{u}_k \Psi_k) = \nabla \cdot (g_k \Gamma_k \nabla \Psi_k) + S_k \tag{15}$$

where Ψ_k is a typical representative variable associated with phase "k", the expression for the diffusion coefficient Γ_k and the source terms S_k can be deduced from the parent equations. The general governing conservation equation can be discretized using a finite volume method [10]. A staggered grid is employed for the discretization of the momentum equations. A power law scheme is adopted to estimate the convection-diffusion flux at the control volume faces. The transient term is treated fully implicitly.

The resolution of the velocity-pressure coupling for a multiphase flow system can be performed by the IPSA (inter-phase slip algorithm). A detailed description of this algorithm is provided in reference [11]. A variant of IPSA is proposed for the numerical implementation of the present model. The present approach could also be viewed as an extension of the well-known single-phase solution algorithm SIMPLEC (semi-implicit method for pressure-linked equations-consistent) to multiphase flows. In the derivations, the philosophy of a multiphase pressure-based solution procedure, called the mass conservation-based algorithm (MCBA) [12], is followed as well.

3.1. Pressure correction equation

To derive the pressure correction equation, the mass conservation equations of the various phases are added to yield the global mass conservation equation given by

$$\sum_k \left\{ \frac{(g_k \rho_k) - (g_k \rho_k)^{old}}{\delta t} \delta V + \Delta (g_k \rho_k \mathbf{u}_k \cdot \mathbf{S}) \right\} = 0 \tag{16}$$

where the Δ operator represents the operation

$$\Delta(\Theta) = \sum_{f=NB} \Theta_f \tag{17}$$

and **S** is the surface vector at the cell face.

The algorithm consists of two stages, the predictor stage and the corrector stage. In the predictor stage, the momentum equations can be solved based on a guessed pressure field p^*. The resulting velocity fields denoted by $\mathbf{u}_k^*$ which satisfy the momentum equations will not, in general, satisfy the mass conservation equations. Thus, a corrector stage is needed in order to obtain velocity and pressure fields that satisfy both equations. The corrections for pressure and velocity denoted by p' and $\mathbf{u}'_k$ respectively, can be introduced in the following manner

$$p = p^* + p' \qquad \mathbf{u}_k = \mathbf{u}_k^* + \mathbf{u}'_k \tag{18}$$

Hence the momentum equations solved in the predictor stage are

$$A_k \mathbf{u}_k^* = \sum\nolimits_{NB} A_{k,NB} \mathbf{u}_{k,NB}^* + \mathbf{B}_k - g_k \delta V \nabla p^* \tag{19}$$

The final solutions are

$$A_k \mathbf{u}_k = \sum\nolimits_{NB} A_{k,NB} \mathbf{u}_{k,NB} + \mathbf{B}_k - g_k \delta V \nabla p \tag{20}$$

Subtracting the above two equations from each other and rearranging yields the following equation

$$\left(A_k - \sum\nolimits_{NB} A_{k,NB}\right)\mathbf{u}'_k = \sum\nolimits_{NB} A_{k,NB}\left(\mathbf{u}'_{k,NB} - \mathbf{u}'_k\right) - g_k \delta V \nabla p' \tag{21}$$

The velocity-correction formula can be obtained by omitting the first term on the right-hand side (RHS) of equation (21)

$$\mathbf{u}'_k = -g_k D_k \nabla p' \tag{22}$$

where $D_k = \dfrac{\delta V}{A_k - \sum_{NB} A_{k,NB}}$.

By combining equations (16), (18), and (22), the final form of the pressure correction equation is obtained as

$$\sum_k \left\{\Delta(g_k \rho_k g_k D_k \nabla p' \cdot \mathbf{S})\right\} = \sum_k \left\{\frac{(g_k \rho_k) - (g_k \rho_k)^{old}}{\delta t} \delta V + \Delta(g_k \rho_k \mathbf{u}_k^* \cdot \mathbf{S})\right\} \tag{23}$$

The corrections are then applied to the velocity and pressure fields by using the following equations:

$$p = p^* + p' \qquad \mathbf{u}_k = \mathbf{u}_k^* - g_k D_k \nabla p' \tag{24}$$

3.2. Solution procedure

The discretization equations are solved iteratively in the following sequence.

(1) Guess the interfacial phase change rate M_{ls} by adopting the value of the previous time-step (zero for the first time-step).
(2) Solve the velocities $\mathbf{u}_s$ and $\mathbf{u}_l$ using the IPSA algorithm.
(3) Solve the temperature and concentration fields (T, C_s, and C_l) from the conservation equations.
(4) Check the status of each cell. If the cell is within the mushy zone, go on to the next step; otherwise, set $M_{ls} = 0$ and return to step (2).
(5) Solve the grain density n from the conservation equation.
(6) Calculate the phase change rate M_{ls} from the interfacial species balance and return to step (2).
(7) Repeat until convergence, before advancing to the next time-step.

4. Benchmark validation

The model was verified using a numerical benchmark recently proposed by Bellet et al. [13]. The test is concerned with solidification of a binary Sn-10wt%Pb alloy in a two-dimensional rectangular mould cavity of 60 mm height and 100 mm width. The mould cavity initially filled with the alloy at the liquidus temperature was cooled symmetrically on both vertical walls. Full details of the problem are available in reference [13]. Due to the symmetry, a half of the full mould domain was simulated. It is noteworthy that a fixed and rigid solid phase is assumed. The mesh size used in the present simulation was 0.5 mm and the time-step was 0.02 s. Figure 1 shows the predicted final segregation maps in the cavity. The y-axis refers to the vertical centreline of the full mould cavity. The principal characteristics of macrosegregation formed can be recognized: a positive segregation zone along the vertical axis and at the bottom, a negative segregation pocket in the upper right part of the ingot, and channel segregates in the bottom central part. It can be concluded that the present model retrieves well the macrosegregation patterns predicted in the literature [14].

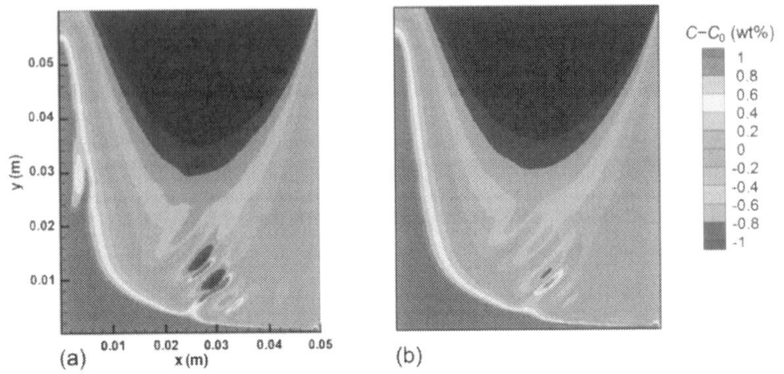

Figure 1. Final segregation pattern in Sn-10wt%Pb alloy.
(a) Literature simulation. (b) Present prediction.

Further, the model was used to predict the macrosegregation in a benchmark 3.3-t steel ingot, about 2 m in height and 0.6 m in average diameter. The movement of equiaxed crystals as well as the pipe formation at the top of the ingot was taken into account in the simulation. Equiaxed nucleation was described with a heterogeneous three-parameter Gaussian nucleation law ($\Delta T_N = 5$ K, $\Delta T_\sigma = 2$ K, and $n_{max} = 5 \times 10^{12}$ m^{-3}). The mesh size was 10 × 15 mm. The time-step was variable, in the order of 0.01 s. The predicted final macrosegregation pattern for carbon is shown in figure 2a. Two features can be observed: first, a strong positive segregation is shown at the top of the ingot; secondly, a conically shaped negative segregation zone is present at the bottom of the ingot. Measured and predicted carbon macrosegregation profiles at the vertical ingot centreline are compared in figure 2b. Although some discrepancies could be found between the results plotted, the present simulation reproduces generally well the measured segregation tendency.

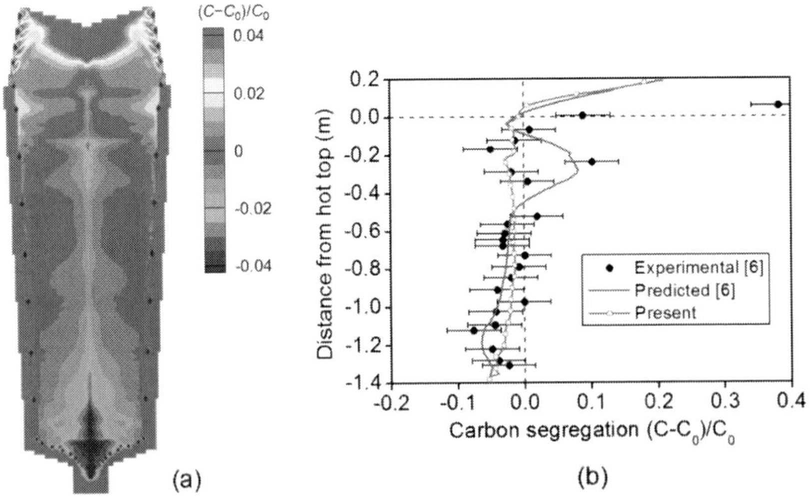

Figure 2. (a) Predicted final macrosegregation in benchmark steel ingot.
(b) Measured and predicted segregation along ingot centreline.

5. Industrial application

The model was applied to a 53-t steel ingot cast and analyzed by the steel plant of CITIC Heavy Industries Co., Ltd. The ladle composition and the dimensions of the ingot are reported in table 1 and figure 3 respectively. The liquid metal was top poured at the temperature of 1550 °C in a cast iron mould with a refractory hot top. A layer of exothermic powder was overlaid after teeming. The solidification time for such an ingot was about 15 hours. During solidification, the time evolution of the temperature was recorded by thermocouples located at different positions in the mould and in the solidifying ingot. Finally, carbon concentration was measured in a longitudinal section of the hot top.

Table 1. Ladle composition of 53-t steel ingot.

Chemical species	C	S	P	Si	Mn	Cr	Mo
Mass (wt%)	0.41	0.003	0.007	0.23	0.59	1.0	0.16

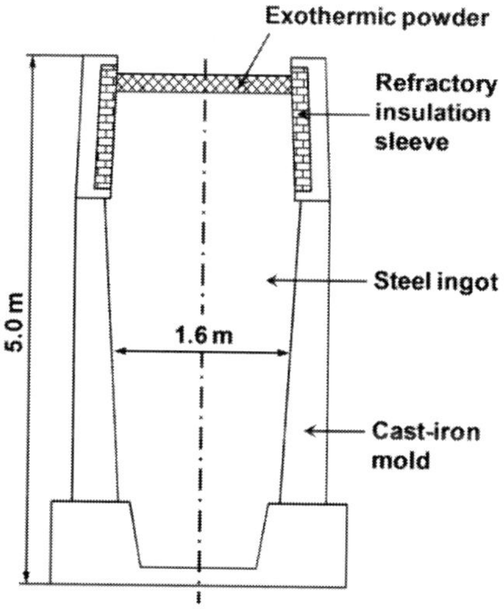

Figure 3. Schematic of 53-t steel ingot.

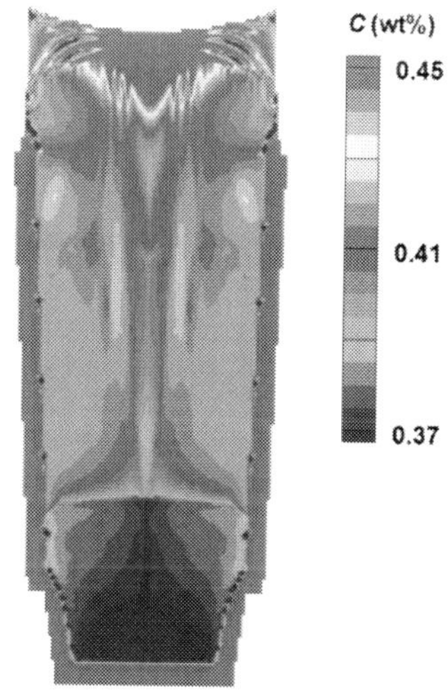

Figure 4. Predicted final macrosegregation in fully solidified ingot.

Symmetry along the centreline axis was assumed. A grid system consisting of 15,376 cells was used with mesh size of 20 mm. At the beginning of solidification, the time-step was in the order of 0.01 s. As solidification proceeds, the time-step was increased, finally up to 0.5 s. The steel alloy was simplified as a binary Fe-C alloy with a nominal composition of C_0 = 0.41 wt%C, neglecting the other chemical species. The thermophysical properties used are reported in the literature [8, 15]. The aforementioned nucleation parameters were adopted in the simulation.

Figure 4 shows the predicted final macrosegregation pattern for carbon in the fully solidified ingot. A negative segregation zone is observed at the bottom of the ingot. This zone has a characteristic cone-shape. A strong positive segregation is present in the hot top part. Further, a striking feature can be observed: the A-segregates, appeared as strong, banded channel segregations, are predicted in the top part of the ingot.

The predicted evolution of the ingot solidification is shown in figure 5. The equiaxed grains nucleated descend along columnar layer at the surface and entrain the surrounding liquid with them, inducing a downward melt flow at the mould side. The solute-lean grains are blocked at the bottom part of the ingot and lead to a negative segregation. The motion of grains has been progressively weakened as solidification proceeds (figure 5b). At a later stage of solidification (figure 5c), the thermosolutal convection dominates and establishes a counter-clockwise flow, descending at the centre and ascending at the surface of the ingot. The A-segregates develop in the hot top as a result of instabilities in the mushy zone growth that perturb the fluid flow at the scale of a few centimeters [6]. As can be seen, the thermosolutal convection is also responsible for the negative segregation at the bottom of the ingot and a positive one in the hot top part. In addition, the maximum values of liquid velocity shown in figures 5a through c are 7.78×10^{-2}, 6.36×10^{-4} and 8.90×10^{-5} m s^{-1}, respectively.

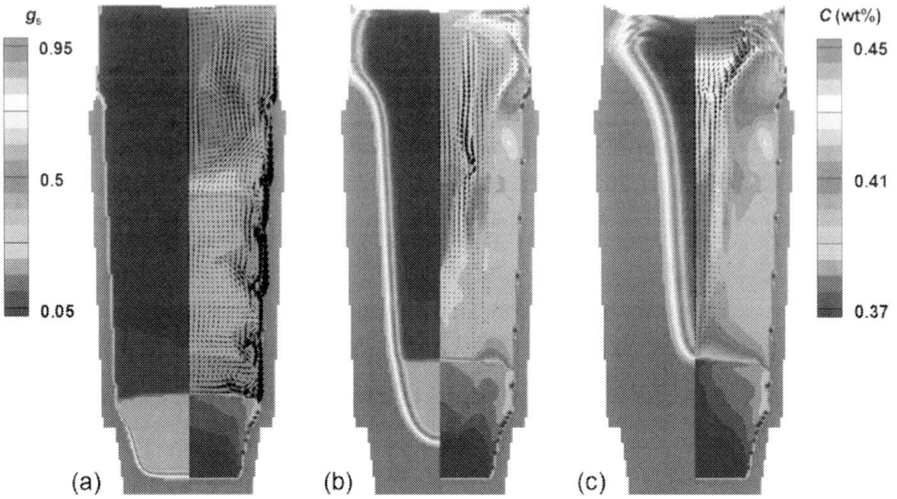

Figure 5. Predicted solidification sequence. Left: solid fraction. Right: macrosegregation and liquid velocity. (a) 0.5 h, (b) 2.5 h, and (c) 6.7 h.

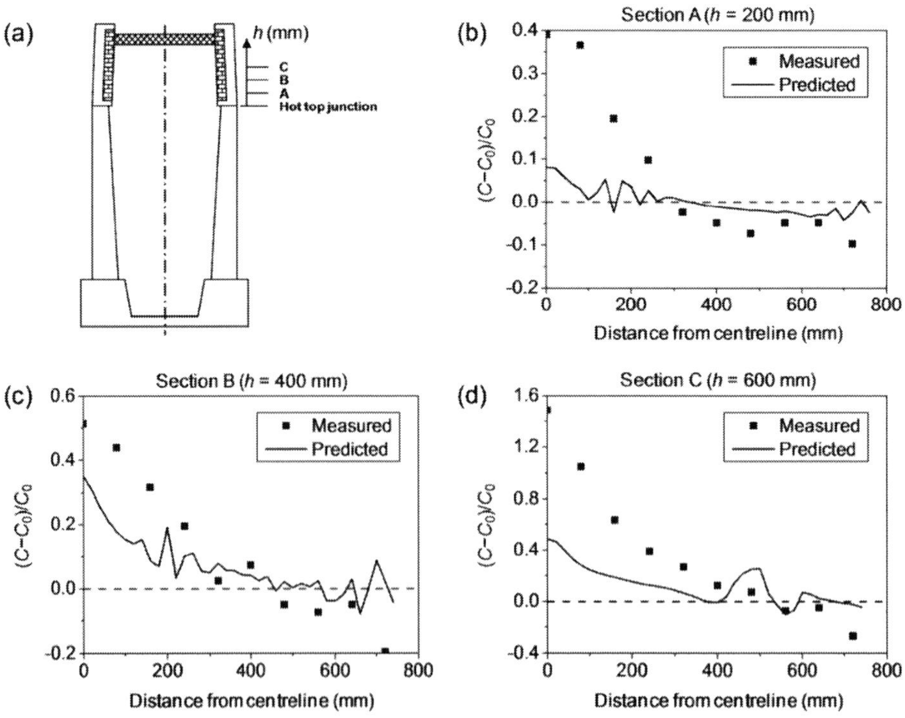

Figure 6. (a) Schematic showing different transverse sections in ingot hot-top. Measured and predicted segregation. (b) Section A. (c) Section B. (d) Section C.

Figure 6 presents the comparison of the predicted and measured macrosegregation variations along transverse sections of the hot top. From the centre to the ingot surface, the measured segregation experiences a transition from positive to negative. The predictions are generally in good agreement with the measured tendencies. However, both the positive segregation near the centre and the negative segregation near the ingot surface are underestimated in the simulations. Possible sources for the discrepancies could include uncertainties in the thermophysical data, two-dimensional simplification of geometry, or measurements errors. It should also be noted that the source of free-floating grains and the globular/dendritic morphological transition need to be determined and quantified in modelling. Furthermore, the predicted A-segregates are indicated with oscillations in the simulated segregation profiles (figures 6b though d). Nonetheless, there are not sufficient measurement points to observe the possible A-segregates. Future experiments should be designed and performed considering more details to provide a sound basis for assessing the model predictions.

6. Conclusions

A two-phase solidification model tackling melt flow, grain motion, and pipe formation has been presented. The performance of the model in the prediction of macrosegregation was evaluated on literature benchmarks of a laboratory-scale Sn-10wt%Pb ingot and an industrial 3.3-t steel ingot. The model was applied to simulate the solidification of an industrial 53-t steel ingot. Positive segregation was predicted in the hot top and conically shaped negative segregation zone was predicted in the bottom part of the ingot. The prediction of the formation of A-segregates in the top part of the ingot was made possible as well. The results show that the model is able to predict important macrosegregation patterns. Nonetheless, it is shown from comparisons with the measurements that the positive segregation in the hot top is under-predicted by the model. The simulations need to be extended to three dimensions using more accurate thermophysical properties. Further numerical work and experiments remain to be done to properly account for the microstructure and grain morphological transitions in the macroscopic solidification model.

Acknowledgments

This work was financially supported by the National Major Science and Technology Project of China (No. 2011ZX04014-052) and the National Basic Research Program of China (No. 2011CB012901).

References

[1] Lesoult G 2005 *Mater. Sci. Eng.* A **413–414** 19–29
[2] Beckermann C 2002 *Int. Mater. Rev.* **47** 243–261
[3] Wu M and Ludwig A 2009 *Acta Mater.* **57** 5621–31
[4] Wu M, Fjeld A and Ludwig A 2010 *Comput. Mater. Sci.* **50** 32–42
[5] Pardeshi R, Dutta P and Singh A K 2009 *Ind. Eng. Chem. Res.* **48** 8789–804
[6] Combeau H, Zaloznik M, Hans S and Richy P E 2009 *Metall. Mater. Trans.* B **40** 289–304
[7] Zaloznik M and Combeau H 2009 *Modeling of Casting, Welding and Advanced Solidification Processes – XII* ed Cockroft S L and Maijer D M (Warrendale, PA: TMS) pp 165–72
[8] Li W S, Shen H F and Liu B C 2011 *Int. J. Miner. Metall. Mater.* Accepted for publication
[9] Gidaspow D 1994 *Multiphase Flow and Fluidization: Continuum and Kinetic Theory Description* (New York: Academic Press)
[10] Patankar S V 1980 *Numerical Heat Transfer and Fluid Flow* (New York: Hemisphere)
[11] Karema H and Lo S 1999 *Comput. Fluids* **28** 323–60
[12] Moukalled F, Darwish M and Sekar B 2003 *J. Comput. Phys.* **190** 550–71
[13] Bellet M, Combeau H, Fautrelle Y, Gobin D, Rady M, Arquis E, Budenkova O, Dussoubs B, Duterrail Y, Kumar A, Gandin C A, Goyeau B, Mosbah A and Zaloznik M 2009 *Int. J. Therm. Sci.* **48** 2013–16
[14] Combeau H, Bellet M, Fautrelle Y, Gobin D, Arquis E, Budenkova O, Dussoubs B, Duterrail Y, Kumar A, Mosbah S, Quatravaux T, Rady M, Gandin C A, Goyeau B and Zaloznik M 2011 *TMS 2011 Annual Meeting Supplemental Proc.* vol 2 (Hoboken: John Wiley & Sons, Inc.) pp 755–62
[15] Li W S, Shen H F and Liu B C 2010 *Steel Research Int.* **81** 994–1000

IOP Conf. Series: Materials Science and Engineering **33** (2012) 012091 doi:10.1088/1757-899X/33/1/012091

Modelling macrosegregation in a 2.45 ton steel ingot

J Li[1,2], **M Wu**[1,2], **A Ludwig**[2], **A Kharicha**[2]

[1] Christian Doppler Lab for Adv. Process Simulation of Solidification and Melting,
[2] Chair of Simulation and Modelling of Metall. Processes, Univ. Leoben, Austria

E-mail: Jun.Li@unileoben.ac.at

Abstract. A three phase model for the mixed columnar-equiaxed solidification was proposed by the current authors [Wu and Ludwig 2006 *Metall. Mater. Trans.* **37A** 1613-31]. The main features of the mixed columnar-equiaxed solidification are considered: the growth of the columnar dendrite trunks from the ingot surface, the nucleation and growth of the equiaxed crystals, the sedimentation of the equiaxed crystals, the thermal and solutal buoyancy flow and its interactions with the growing crystals, the solute partitioning at the solid-liquid interface during solidification, the solute transport due to melt convection and equiaxed sedimentation, the mechanical interaction/impingement between columnar and equiaxed crystals and the columnar-to-equiaxed transition (CET). However, due to the model complexity and the limited computational capability the model has not yet applied to the large steel ingots of engineering scale. This paper is going to simulate a 2.45 ton big-end-up industry steel ingot, for which some experimental results were reported [Marburg 1926 *Iron Steel Inst.* **113** 39-176]. Here a simplified binary phase diagram for the steel (Fe-0.45 wt. %C) is considered. Comparison of the modelling results such as as-cast columnar and equiaxed zones, macrosegregation with the experimental results is made. Details about the formation sequence of the distinguished crystal zones and segregation patterns are analyzed.

1. Introduction

Macrosegregation is a very common and serious defect in large steel ingots. The typical segregation pattern in a steel ingot consists of a positive segregation in the top part, a cone-shape negative segregation in the bottom part, inverse segregation near the surface, V segregates along the centerline, and A-segregates in the columnar zone [3-4]. This inhomogeneity occurs due to a relative motion between the liquid and solid phase during solidification. It is understood that this relative motion can arise as a result of thermo-solutal convection, shrinkage-induced feeding flow, flotation and sedimentation of free moving grain, mechanical or electromagnetic stirring, flow induced by pore or gas bubble formation, deformation of the solid skeleton, and capillary force induced flow [5].

Since the first modeling attempt of Flemings and co-workers [6-7] many macrosegregation models were presented [8-11]. Gu and Beckermann [11] for the first time have applied a fully coupled, multicomponent solidification model with melt convection to a large industry-scale ingot (1.016 × 2.08 × 2.819 m^3), their simulation result showed a qualitative agreement with the positive segregation in the top region of the ingot. However, since the neglect of sedimentation of free equiaxed crystals they could not predict negative segregation at the bottom. Combeau and co-workers [12-13] presented a two phase (solid and liquid) model to study the influences of motion and morphology of equiaxed grain on a 3.3 ton big steel ingot. Some progress was made regarding to prediction of the bottom negative segregation zone, and the results have a good agreement with the experiment. Additionally, some streak-like segregates (A-segregates) were predicted.

The current authors [1, 14] for the first time developed a mixed columnar-equiaxed model which accounts directly for the nucleation and growth of the equiaxed globular grains, the growth of columnar dendrite trunks, and the effect of equiaxed grain sedimentation and melt convection. Their previous studies have successfully predicted both the cone negative segregation in the bottom region of ingot, and the columnar-to-equiaxed-transition (CET). However, due to the model complexity and the limited computational capability the previous studies were just based on the laboratorial scale of steel ingot. In this article, the mixed columnar-equiaxed is employed to study the macrosegregation in the industry-scale steel ingot on the base of the classic experiment [2].

2. Cast ingot: experiment and model description

Many industry-scale steel ingots were poured and analyzed for the segregation in the last century [2, 15]. As an example the sulphur print of a 2.45 ton big-end-up ingot (Fe-0.45 wt.%C) is shown in Figure 1(a). The nominal mixture concentration distribution ($(c_{mix}-c_0)/c_0$) is shown in Figure 1(b). As reference, this ingot is numerically studied. Configuration of this reference ingot, together with necessary boundary and initial conditions, is described in Figure 1 (c). Since the experiment was done many decades ago, due to the lack of precise process description, some process parameters and boundary conditions have to be derived on the base of assumptions. 2D axis symmetrical simulations are performed. Two cases are considered: Case I – using the mixed columnar-equiaxed three phase model; Case II – using the same model with the ignorance of the occurrence of equiaxed crystal. Table 1 lists some thermo-dynamic and physical properties.

Table1. Thermo-dynamic & physical properties[12]

Property	Units	Quantity
Melting of pure iron	K	1805.15
Liquidus slope	$K\ (wt.\%)^{-1}$	-80.45
Equilibrium partition coefficient	-	0.36
Reference density	$kg{\cdot}m^{3}$	6990
Specific heat	$J\ kg^{-1}\ K^{-1}$	500
Thermal conductivity	$W\ m^{-1}\ K^{-1}$	34.0
Latent heat	$J\ kg^{-1}$	2.71×10^{5}
Viscosity	$Kg\ m^{-1}\ s^{-1}$	4.2×10^{-3}
Thermal expansion coefficient	K^{-1}	1.07×10^{-4}
Solutal expansion coefficient	$wt.\%^{-1}$	1.4×10^{-2}
Second dendritic arm spacing	m	5×10^{-4}
Diffusion coefficient (liquid)	$m^2\ s^{-1}$	2.0×10^{-8}
Diffusion coefficient (solid)	$m^2\ s^{-1}$	1.0×10^{-9}

Details of the numerical model are described elsewhere [1, 5, 14]. A brief outline of the model and simulation settings are given below:

- A simple boundary condition is applied. Here, the value of heat transfer coefficient is taken based on the final solidification time which refers to the classical theory and experiment ones [15].
- Three phases are defined: the primary liquid phase (ℓ), the equiaxed phase (e), and the columnar phase (c). The corresponding phase fraction is f_ℓ, f_e, and f_c with $f_\ell+f_e+f_c=1$. Both the liquid and equiaxed phases are moving phases, for which the corresponding Navier-Stokes equations are solved. The columnar phase is assumed to stick to the wall, and solidifies from the wall towards the bulk melt. Thus, no momentum equation for the columnar phase is considered.
- Columnar dendrites are approximated by growing cylinders starting from the mould wall towards the casting centre. The advance of the columnar tip front was tracked during the solidification.

- Equiaxed grains are approximated as spheres. However, for the calculation of drag force the morphology of the equiaxed grains was modified as octahedral.
- A three-parameter heterogeneous nucleation law is used for the nucleation of equiaxed grains. No fragmentation and grain attachment are currently considered.
- Growth of the columnar trunk and equiaxed grain is governed by diffusion. Back diffusion is also considered.
- The permeability in the columnar mush zone is modelled according to Blake-Kozeny, while the drag law between the melt and equiaxed phase is modelled according to Wang's approach [16].
- Growth of columnar primary tips is stopped when the volume fraction of equiaxed phase in front of them reaches 0.49 (hard blocking criterion) [17].
- Packing limit for the equiaxed phase is set as $f_e + f_c \geq 0.637$ [18], and the equiaxed crystals are trapped when the trapping limit of $f_c \geq 0.2$ is reached [19].

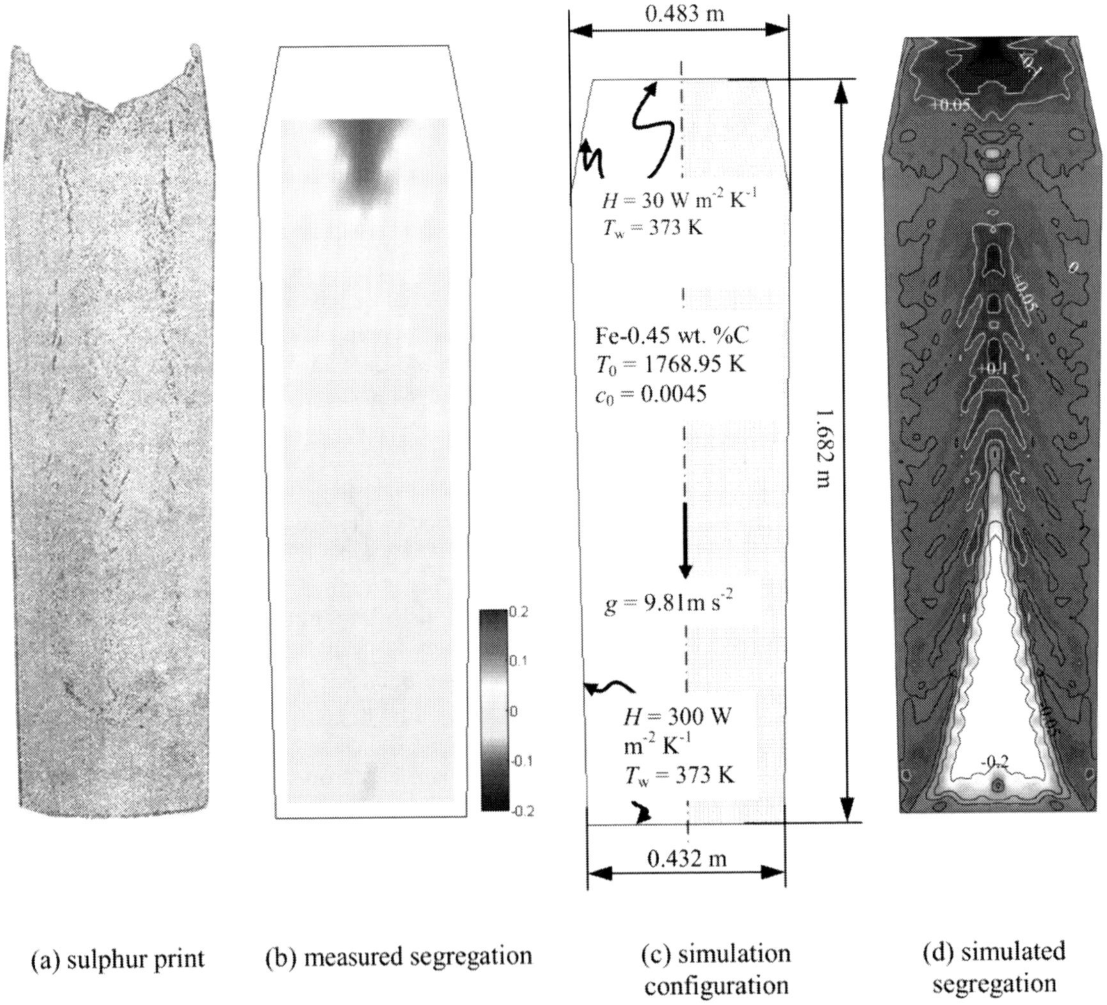

Figure 1. Configuration of a 2.45-ton industry-scale steel ingot. (a)-(b) experiment [2], (c) simulation configuration and (d) simulated macrosegregation in grey scale (black for the positive segregation and light for the negative segregation) overlapped with isolines. The macrosegregation, both experimental (b) and simulated (d), is shown for the nominal mixture concentration ($(c_{mix}-c_o)/c_o$).

IOP Conf. Series: Materials Science and Engineering **33** (2012) 012091 doi:10.1088/1757-899X/33/1/012091

3. Simulation results and discussions

3.1. Case I: mixed columnar-equiaxed solidification

In this case full three phases are considered. The nucleation parameters for the equiaxed grains are assumed: $n_{max} = 5\times10^9$ m^{-3}, $\Delta T_\sigma = 2$ K, $\Delta T_N = 5$ K. The simulated dynamic evolution of equiaxed grain volume fraction, columnar volume fraction, equiaxed grain sedimentation velocity and melt velocity at 100 s, 500 s, 1500 s, and 4000 s are shown in Figures 2(a) through (d), respectively. The final macrosegregation pattern is shown in Figure 1(d).

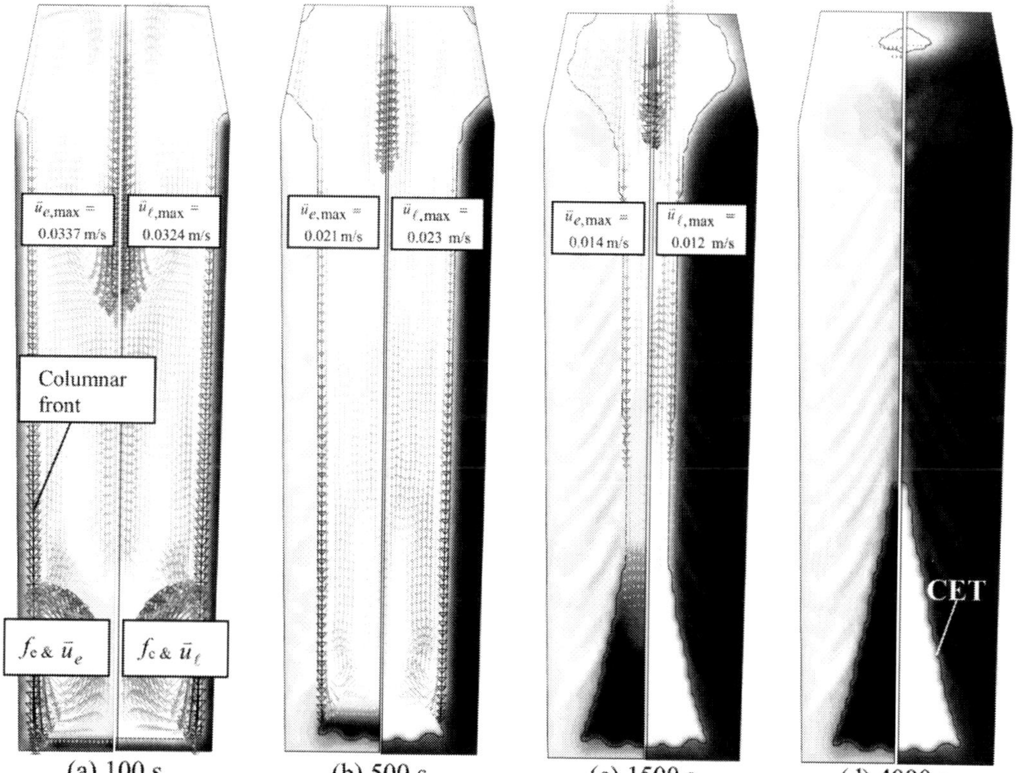

(a) 100 s (b) 500 s (c) 1500 s (d) 4000 s

Figure 2. Solidification sequence of flow fields. The volume fraction of each phase (f_e or f_c) is shown in gray scale from 0 (bright) to 1 (dark). The left half of each figure shows the evolution of equiaxed volume fraction (f_e) together with the equiaxed sedimentation velocity ($\vec{u}_e$) in black arrows. The right half of each figure shows the evolution of columnar volume fraction (f_c) together with the melt velocity ($\vec{u}_\ell$) in black arrows. The columnar dendrite tip position also marked with a black solid line.

In the early stage of solidification, the equiaxed grains nucleate and grow gradually near the mould walls and bulk melt. Since the density of equiaxed grains are bigger than the melt, the equiaxed grains sink from the upper region to the bottom region, as shown in Figure 2(a). However, one may notice that in the centre part of the ingot there are some differences. In some regions the equiaxed sedimentation velocity is upward. Because the equiaxed volume fraction in these regions is very small (i.e. the momentum is small), the equiaxed grain sedimentation velocity is easily influenced and finally dominated by the upward liquid melt flow.

The direction of the liquid melt velocity near the columnar front is contributed by three factors: the solutal buoyancy force which will lead to the upward melt flow; the thermal buoyancy force which will lead to the downward melt flow; and the equiaxed sedimentation velocity which will drag the surrounding melt flow downwards. It is clearly seen from the liquid melt flow field, which shows the downward melt flow there, that in this ingot the two downward factors are the decisive. The melt velocity direction near the columnar front is downward, however, due to the big scale of the ingot the melt velocity direction in the centre region is obscure and disorder.

As the sinking of the equiaxed grains leads to an accumulation of equiaxed grains in the bottom region of the ingot, the equiaxed grain density in the bottom region is bigger than the other regions which will increase the equiaxed solidification rate in this bottom region. As shown in Figure 2, the increasing of equiaxed volume fraction in the bottom region is faster than that in the other regions. This kind of sinking and accumulation of equiaxed grain will form a characteristic cone-shape distribution of equiaxed fraction in the bottom region.

As the columnar tip front is explicitly tracked, the simulation shows that the columnar tip fronts from both sides tend to meet in the ingot centre. However, in the lower part of the ingot the large amount of equiaxed grains stop the propagation of the columnar tip front. The final position of columnar tip indicates the columnar-to-equiaxed transition (CET) position. The CET separates the areas where only equiaxed grains appear from the areas where both columnar dendrites and equiaxed grains coexist. The CET line is predicted in this study, as seen in Figure 2(d). In addition, due to the disorder of both the sedimentation velocity and the melt convection, the sawtooth-like distribution of both the equiaxed phase and columnar phase volume fraction was found in the ingot. These sawtooth like profiles of both equiaxed phase and columnar phase volume fraction will definitely effect the final macrosegregation distribution.

Figure 1(d), presents the predicted final macrosegregation distribution. The cone-shaped negative segregation in the bottom region and a positive segregation zone in the upper region are obviously shown. The main mechanism for this cone-shaped negative segregation zone is grain sedimentation [5]. The settling grains were poor in solute and thus their pile-up results in a negative segregation at the bottom of the ingot. A further contributing factor to the strength of this negative segregation comes from the flow divergence of the residual liquid through this zone at a late solidification stage. The positive segregation in the upper region of the ingot is caused by the flow of the enriched melt in the bulk region. The solidification pattern agrees with the classical explanation of steel ingot solidification, summarized by Campbell [20].

One interesting result is detected that some streak-like patches of positive segregation were predicted near the mould wall and even in the centre region. These streak-like patches are somehow looks like the so called A-segregation [21]. Here we call these streak-like segregation as quasi-A-segregation. Most people consider that the main formation mechanism for A-segregation is: the flow disturbance in the mushy zone which will result in the flow channel and subsequently A segregation. The current authors [22-23] have studied the mechanism of channel segregation in a Sn-Pb benchmark. They denoted that the formation of channel segregation should be considered in two steps: (1) the initiation of channel, which is determined by the Rayleigh number; (2) the growth of channel which is the result of flow-solidification interactions. However, the formation mechanism of quasi-A-segregation in the current ingot must be different. In the early Sn-Pb benchmark the only driving force for the melt flow is thermal-solutal buoyancy force, whereas, in the current case the sedimentation of equiaxed grains and their inteactions with the columnar tip front and melt flow seem to play dominant role. The details about the mechanism for this kind of A-segregation are still to be investigated.

3.2. Case II: two-phase columnar solidification

Here, Case II used the same model and the same parameter as Case I. The only difference in this case is that we did not take into account the nucleation and growth of equiaxed grains. The solidification sequences and evolution of macrosegregation in two-phase columnar model are shown in Figure 3(a)-(c). The columnar tip front and volume fraction of columnar phase move from the mould walls towards the bulk melt region. For the solute-enriched interdendritic melt near the solidification front,

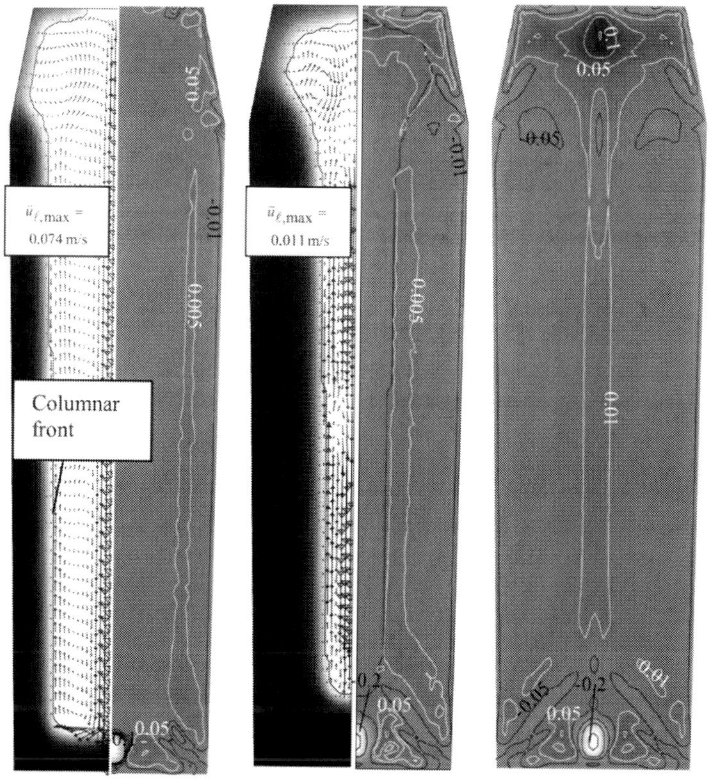

(a) 500 s (b) 1200 s (c) 3000 s

Figure 3. Solidification sequence and evolution of macrosegregation in the case of two-phase columnar solidification. In the left half of each figure the columnar volume fraction (f_c) is shown in gray scale together with the melt velocity ($\bar{u}_\ell$) in black arrows. In the right half of each figure the distribution of macrosegregation (((c_{mix}-c_o)/c_o) is shown in gray scale overlaid with macrosegregation iso-lines in solid lines. The columnar tip position was also indicated with a black solid line. Figure (c) shows the final macrosegregation distribution.

the solutal buoyancy leads to an upward flow. However, the thermal buoyancy leads to a downward flow. In this simulation case (Fe-0.45 wt. %C) the solutal buoyancy dominates over the thermal buoyancy, therefore the upward flow in the solidification front is the primary phenomena leading to the flow, as seen in Figure 3(a). The melt convection in the solidification front is upward, and this upward flow will drive the centre bulk melt to downward. Since the melt domain is large, the melt convection in the inside of bulk melt is disorder.

As seen from the final macrosegregation pattern in Figure 3(c), in the near wall region, where the solidification start from, a small degree of negative macrosegregation (-0.005) is predicted, while in the centre region, where the solidification is later on, a small degree of positive macrosegregation (0.01) is predicted. It is noted that a small region negative macrosegregation (relative macrosegregation value equal to -0.2) is found in the bottom region. In the upper region a positive macrosegregation is predicted with a relative macrosegregation value of 0.12. The reason for the negative segregation in the bottom middle region and the positive segregation in the upper middle region can be explained [5]: In the bottom middle region, since the solute enriched melt are replaced by the solute poor melt, the negative segregation tend to formation. The reason for the positive segregation in the upper middle region is that the solute poor melt is replaced by the solute enriched interdendritic melt from the solidification front.

4. Verification and discussions

The centre line segregation distributions are compared between the experiment results and the simulation results, as shown in Figure 4. The experiment shows the negative segregation in the lower part region and positive segregation in the upper region. The mixed columnar-equiaxed three-phase model also shows the negative segregation in the lower part and positive segregation in the upper part. However, the negative segregation in the lower part is predicted more seriously than the experiment result, and in the middle of upper part where the mixed three-phase model predicted the negative segregation while the experiment shows the positive segregation. The two-phase columnar solidification case predicts a negative segregation in the bottom region and a small degree negative segregation in the middle of upper middle region. However, this model predicts a small degree positive segregation in the upper part region where both the experiment and the mixed three-phase model shown the negative segregation.

The possible explanation for the overestimate of the negative segregation in the lower part by the mixed columnar-equiaxed model could be in two aspects. On one hand, the assumption of totally spherical equiaxed grain will underestimate the average concentration for each equiaxed grain which subsequently will overestimate the serious of negative segregation in the bottom equiaxed zone. On the other hand, the overestimation of equiaxed nucleation in the whole ingot will also result in the overestimation of the negative segregation.

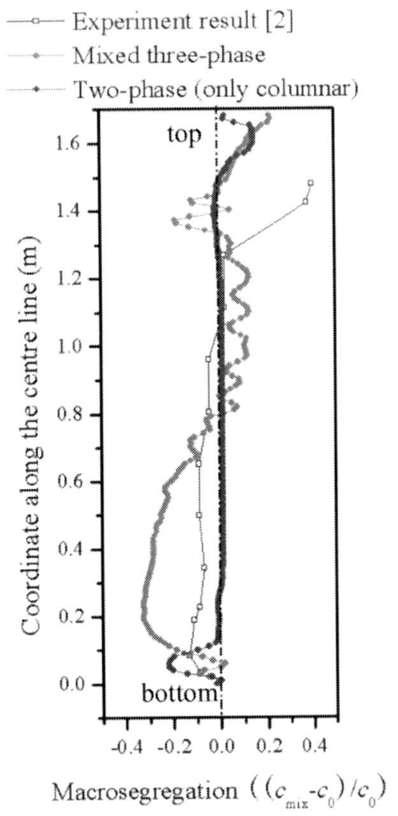

Figure 4. Comparison of macrosegregation $(((c_{mix}-c_o)/c_o)$ distribution along the centre of cast

As been observed from the lower part of the ingot, the mixed three-phase model (with the sedimentation of the equiaxed grain) overestimates the negative segregation. However, the two-phase columnar solidification case underestimates the negative segregation. Therefore, one possible conjecture is as follows: in this experiment there should contain a certain number of equiaxed grains in the lower part; this mixed three-phase model has overestimated the nucleation of equiaxed grain.

In the middle of the upper part, although the experiment shows the positive segregation, both of the simulation models predict the negative segregation. The reason for this deviation may also be in two aspects. Firstly, for the simulation case, during the final stage of solidification in the middle region of the ingot, the columnar tips will grow to the centre which will lead to the bridging and mini-ingotism in the upper part [24]. As the same mechanism as it forms in the whole steel ingot, eventually, the negative segregation will form in the middle of upper part. The second aspect is due to the ignorance of solidification shrinkage. In addition, as seen from the experiment curve, the distance between two points is in the order at least of 10 cm. It is known that in the region below the hot-topping and in the hot top part strong variations can occur. It means that the measured points did not provide sufficient resolution. People often found the negative segregation in the middle of upper part regions as well [12].

One should emphasize that the main purpose of this study is to show and verify that this mixed columnar-equiaxed model can be applied in the industry scale steel ingot, and can predict some reasonable macrosegregation patterns. We are not going to adjust some of the calculation parameters to cater for the experiment results. Since the experiments were done in many decades ago and many of the process parameters and material properties are unknown, the simulation results could only reproduce the experiment results in a qualitative manner.

5. Summary

A mixed columnar-equiaxed solidification model was applied to study the formation of macrosegregation in a 2.45 ton industry-scale steel ingot. It is verified that the experimentally reported segregation phenomena can be numerically simulated. The cone-shape negative segregation in the bottom region and a positive segregation in the upper region of the ingot were predicted. The CET line, separating the purely equiaxed zone from the coexisting columnar-equiaxed zone, was predicted. The quasi-A-segregates near the mold wall were also found. The global macrosegregation distribution agreed qualitatively with the experimental result. However, the quantitative error of the numerical simulation is still quite large due to the lack of reliable process parameters and nucleation (equiaxed) parameters. The above-mentioned simulation was based on the assumed nucleation parameters ($n_{\max} = 5\times10^9$ m^{-3}, $\Delta T_\sigma = 2$ K, $\Delta T_N = 5$ K) for the equiaxed grains. One more simulation was performed with ignorance of the occurrence of equiaxed crystal. It is interesting to found that the latter case has a better quantitative agreement with the experimental result along the casting centerline than the former case with full three phases. It implies that the assumed nucleation parameters for equiaxed crystal in the former case have overestimated the amount of equiaxed grains, or the current 2.45 ton ingot might solidify mostly with columnar morphology. This hypothesis needs further verification.

Acknowledgments

The authors acknowledge the financial support by the Austrian Federal Ministry of Economy, Family and Youth and the National Foundation for Research, Technology and Development. Jun Li would like to thank the support from the China Scholarship Council.

References

[1] Wu M and Ludwig A 2006 *Metall. Mater. Trans.* **37A** 1613-31
[2] Marburg E 1926 *J. Iron steel Inst.* **113** 39-176
[3] Hultgren A 1973 *Scand. J. Met.* **2** 217-27
[4] Flemings M C 1976 *Scand. J. Met.* **5** 1-15
[5] Wu M, Könözsy L, Ludwig A, Schützenhöfer W and Tanzer R 2008 *Steel Res. Int.* **79** 56-63
[6] Mehrabian R and Flemings M C 1970 *Metall. Trans.* **1** 455-64
[7] Fujii T, Poirier D R and Flemings M C 1979 *Metall. Trans. B* **10B** 331-39
[8] Vannier H, Combeau H, Chevrier J C and Lesoult G 1991 *in Modeling of casting, welding and advanced solidification process V, TMS* (Warrendale, PA) 189-95
[9] Ohnaka I 1986 *in state of the art of computer simulation of casting and solidification processm* editor Fredriksson H, Les éditons de physique, Les Ulis, France 211-23
[10] Schneider M C and Beckermann C 1995 *Metall. Mater. Trans. A.* **26A** 2373-87
[11] Gu J P and Beckermann C 1999 *Metall. Mater. Trans. A.* **30A** 1357-66
[12] Combeau H, Založnik M, Hans S and Richy P E 2009 *Metall. Meter. Trans. B* **40B** 289-304
[13] Založnik M and Combeau H 2010 *Int. J. Therm. Sci.* **49** 1500-09
[14] Wu M and Ludwig A 2007 *Metall. Meter. Trans. A* **38A** 1465-75
[15] Marburg E 1953 *Trans. AIME* **5** 157-72
[16] Wang C and Beckermann C 1995 *Metall. Meter. Trans. B* **26B** 111-9
[17] Hunt J D 1984 *Mater. Sci. Eng.* **65** 75-83
[18] Wu M and Ludwig A 2009 *Acta Mater.* **57** 5621-31
[19] Ludwig A and Wu A 2005 *Mater. Sci. Eng.* A **413-414** 109-14
[20] Campbell J 1991 *Castings, Butterworth Heinemann Ltd*, Oxford
[21] Moore J J and Shah N A 1983 *Inter. Metals Reviews* **28** 338-56
[22] Li J, Wu M, Hao J and Ludwig A 2012 *Comp. Mater. Sci.* **55** 407-18
[23] Li J, Wu M, Hao J, Kharicha A and Ludwig A 2012 *Comp. Mater. Sci.* **55** 419-29
[24] Moore J J *Iron & Steel Soc.* 1984 **3** 11-20

Numerical simulation of delayed pouring technique for a 360t heavy steel ingot

J Li[1], D R Liu[2], X H Kang[1] and D Z Li[1, 3]

[1] Shenyang National Laboratory for Materials Science, Institute of Metal Research, Chinese Academy of Sciences, No.72 Wen Hua Road, Shenyang 110016, China
[2] School of Materials Science and Engineering, Harbin University of Science and Technology, No.4 Lin Yuan Road, Xiang Fang District, Harbin 150040, China

E-mail: dzli@imr.ac.cn

Abstract. A continuum mathematical model for the transport phenomena in the solidification systems has been established to study the central axial macrosegregation in a 360t multi concentration pouring (MCP) steel ingot. A time explicit finite difference scheme is adopted to calculate the coupling of the temperature, concentration and velocity flow fields. The flow equations are solved by the solution algorithm for transient fluid flow (SOLA) technique. Simulations of Fe-C-Si-Mn quaternary alloy are performed. The established model has been validated by comparing with the experimental result of a 360t MCP steel ingot. The simulated carbon concentration profile along the centreline of the ingot shows a fair agreement with the measurement. The influence of the delay time for the last ladle on the macrosegregation along the centreline has been investigated. Simulation results show that the delay time for the last ladle has a significant effect on the macrosegregation, especially for the positive segregation below the hot top. A novel criterion of selecting the delay time for the last ladle has been proposed to reduce the macrosegregation. By selecting the proper delay time for the last ladle, the carbon concentration along the centreline in the ingot body can be controlled within the range of the industrial limit.

1. Introduction

Macrosegregation is a phenomenon of the deviation from the average composition over a distance larger than the dendrite arm spacing, which occurs during the solidification process of the melt [1]. It is an unavoidable phenomenon due to the different solubilities of elements in the liquid and solid phases [2]. For large ingots and castings, macrosegregation is more likely to happen due to long solidification time. Additionally, it is far more difficult to eliminate subsequently by thermo-mechanical post-treatments, and results in fateful and permanent defects in forgings [3]. Therefore, suppressing macrosegregation during the casting process has always been the focus of casting research [4-7]. In the past decades, scientists and engineers have tried many different techniques such as continuous casting [8], squeeze casting [9], electromagnetic stirring [10] and multi concentration pouring (MCP) technique to restrain the formation of the macrosegregation during the casting process. Among these techniques, the MCP technique developed in Japan in 1982 and characterized as the sequential pouring of the melt with different carbon concentrations [11], is considered to be a very promising technique due to its good maneuverability and notable effect, especially for the gigantic

[3] Author to whom any correspondence should be addressed.

mould casting ingots (300-600 t).

The MCP technique has been adopted by many heavy machinery plants around the world recently to reduce the macrosegregation in huge steel ingot. However, since the influence of the parameters on the macrosegregation is still not clear, the MCP casting parameters are generally selected according to experience, which directly results in poor application. With respect to the high experimental cost of the large steel ingot, it is more economical to study the formation of the macrosegregation during the MCP process and optimize the pouring technique using numerical simulation method. However, little information about the mathematical modeling of the MCP process has been found in the literature due to its complexity. Only Liu et al. [12] have performed the correlative simulations, with Fe-C binary alloy calculated in the model.

In this paper, a continuum mathematical model for the transport phenomena in the solidification systems has been established to investigate the formation of the macrosegregation in a 360 t MCP steel ingot. A time explicit scheme is used to solve the coupling of the temperature, concentration and velocity flow fields. Based on the explicit finite difference scheme, flow equations are solved by the SOLA technique, rather than the well-known time implicit SIMPLER algorithm. Fe-C-Si-Mn quaternary alloy is calculated in the simulation. The influence of the delay time for the last ladle on the macrosegregation in the large steel ingot has been reasonably studied and a novel criterion for selecting the delay time has been proposed.

2. Mathematical and numerical modelling

The mathematical model is referred to that proposed by Pardeshi et al. [13] and Voller et al. [14]. Here only a brief description of the model is described.

2.1. Governing equations

Mixture energy conservation equation

$$\frac{\partial[\rho H]}{\partial t}+\nabla\cdot\left(\rho c_p\bar{U}T+\rho\bar{U}\Delta H\right)=\nabla\cdot(\lambda\nabla T) \quad (1)$$

$$[\rho H]=f_s\rho h_s+(1-f_s)\rho h_l \quad (2)$$

where $\rho=\rho_l=\rho_s$, h_s is the solid phase enthalpy, h_l is the liquid phase enthalpy, H is the mixture enthalpy and λ is the heat conductivity.

Mixture solute conservation equation

$$\frac{\partial[\rho C]^k}{\partial t}+\nabla(\rho\vec{U}C_l^k)=0 \quad (3)$$

$$[\rho C]=f_s\rho_s<C_s>+(1-f_s)\rho_l C_l \quad (4)$$

where C is the mixture concentration, $\langle C_s\rangle$ and C_l are the average solid and liquid concentration respectively and the superscript k is a marker for the kth component of an m component system.

Continuity

$$\nabla\cdot(\bar{U})=0 \quad (5)$$

Momentum equations are based on the mixture theory,

V-momentum

$$\frac{\partial(\rho v)}{\partial t}+\nabla\cdot\left(\rho\bar{U}v\right)=-\frac{\partial P}{\partial x}-Kv+\nabla\cdot(\mu_l\nabla v) \quad (6)$$

W-momentum

$$\frac{\partial(\rho w)}{\partial t}+\nabla\cdot\left(\rho\bar{U}w\right)=-\frac{\partial P}{\partial z}-Kw+\nabla\cdot\left(\mu_l\nabla w\right)+\rho g\left[\beta_T\left(T-T_{ref}\right)+\sum_{j=1}^{m}\beta_c^j\left(C_l^j-C_{ref}^j\right)\right] \quad (7)$$

$$T_{ref}=\frac{\sum_{i=1}^{n}\left(T_l^i\times V_{ladle}^i\right)}{\sum_{i=1}^{n}V_{ladle}^i} \quad (8) \qquad C_{ref}=\frac{\sum_{i=1}^{n}\left(C_0^i\times V_{ladle}^i\right)}{\sum_{i=1}^{n}V_{ladle}^i} \quad (9)$$

where m=3, is the number of elements (C, Si and Mn), n is the number of ladles, T_{ref} is the average liquidus temperature of ladles, C_{ref} is the average initial concentration of ladles.
The mushy region is modelled by Carman-Kozeny relationship

$$K=\frac{K_0 f_s^2}{\left(1-f_s\right)^3} \quad (10) \qquad K_0=\frac{180\times\mu_l}{d_s^2} \quad (11)$$

2.2. *Numerical implementation*
Conservation equations are numerically solved on a staggered Cartesian mesh using a finite volume approach and uniform spaced grids. Velocity components are defined on the interfaces of the control volume and other dependent variables are defined at the center of the control volume. The upwind scheme is adopted to evaluate the contribution of the convection. Diffusion terms are discretized with standard second-order finite difference procedure. In the present study, during each macro-scale time step, an explicit-time-stepping scheme is used to solve the coupling of the temperature, concentration and momentum equations, and the continuity equations are solved by the SOLA technique. The time-step is determined by the maximum of the velocity at the previous time step. The grid size is selected to be 30 mm×30 mm.

It is worth mentioning that in this work the filling process is not taken into consideration, since the simulated ingot has a very large volume. The filling time is generally 40~50 minutes, which is negligible compared to the 70-hour solidifying time. Thus the pouring of next ladle is simply treated as putting melt above the melt of former ladles in a simultaneous and quiescent way. After pouring (actually the pouring time is zero), the calculation domain is expanded. Velocity distribution in the melt of the former ladles is inherited and set as the initial condition for the following calculation. Then, the melt from different ladles is considered as a whole and participates in the calculation of all conservation equations.

3. Validation of the established model
Figure 1 shows the schematic of a 360 t large steel ingot produced by the MCP technique. In the hot top mould, a 160 mm thick layer of fireclay and a 20 mm thick layer of insulation board were used at the circumference of the hot top. The top of the hot top was covered by a 300 mm thick layer of covering flux. The thermophysical data of the insulation materials in the hot top mold are listed in table 1. The chemical composition of the experimental steel are listed in table 2. The nominal carbon concentration of the experimental steel is 0.18 wt%. To counteract the traditional negative segregation at the lower part and positive segregation at the upper part of the ingot, the carbon concentration in ladle I and III was 0.21 wt% and 0.14 wt%, respectively. The carbon concentration in ladle II was equal to the nominal value. Additionally, the contents of Si and Mn in each ladle were uniform. The mass of the melt in each lade was 160 t (I), 160 t (II) and 40 t (III), respectively. Three ladles with different carbon concentrations were poured in succession, with uniform pouring velocity of about 8 t/min. In the experiment research, some specimens were taken from the centerline between point A and B after solidification and the carbon concentration was measured, as illustrated in figure 1. The

measured carbon concentration profile along line AB is shown in figure 2. The measured result indicates a negative segregation at the lower part and a positive segregation at the upper part of the ingot. The positive segregation at the upper part of the ingot body is severe and the maximum of the carbon concentration is 0.26 wt%, which is 0.04 wt% higher than the maximum of the industrial limit (0.16-0.22 wt%).

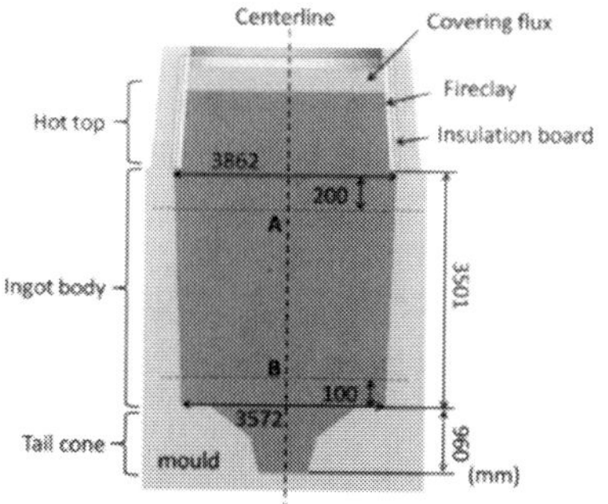

Figure 1. Schematic of vertical central plane of the 360 t steel ingot in two-dimensional.

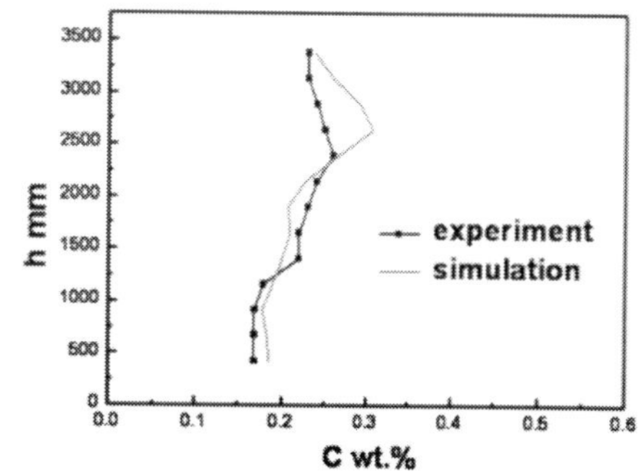

Figure 2. Comparison between experimental and simulated carbon content profile along line AB.

Table 1. Thermophysical data of the insulation materials in the hot top mold.

	ρ (Kgm^{-3})	λ (Wm^{-1}K^{-1})	c_p (KJKg^{-1}K^{-1})
Fireclay	2570	70	0.92
Insulation board	500	2.3	1.1
Covering flux	500	60	0.6

Table 2. Chemical composition of the experimental steel (wt%).

C	Si	Mn	P	S
0.16-0.22	0.15-0.30	1.2-1.5	≤0.01	≤0.015

In order to validate the established simulation model, solidification process of the 360 t MCP steel ingot with the experimental processing parameters has been simulated. Because of the axial symmetry of the steel ingot, the simulation model was restricted to the vertical central plane of the ingot and a two dimension simulation was performed. After the final ladle poured, a layer of covering flux was overlaid at the top of the ingot and then heat transfer was calculated at the covering flux/hot top interface. Before this process was finished, the top surface of the melt was treated as insulated. The experimental steel was simplified to Fe-0.18 wt%C-0.2 wt%Si-1.4 wt%Mn quaternary alloy in the simulations and the carbon concentration distribution was chosen to evaluate the macrosegregation. Thermophysical data of the Fe-C-Si-Mn alloy and the thermal boundary conditions used in simulation are referred to reference [7]. Comparison of the carbon concentration profile along the centerline between the simulation and the measured result is shown in figure 2. The simulation result presents a similar profile compared to the measurement, with only minute difference in details. It means that the established continuum mathematical model and the developed numerical scheme are reasonable and can be adopted to simulate the formation of the macrosegregation in the heavy steel ingot during the MCP processing.

4. Results and discussion

4.1. Optimization of the insulation condition in the hot top mold

Besides the conventional negative and positive segregation, the measured carbon concentration profile along the centerline of the experimental steel ingot shows another interesting phenomenon, i.e., the largest carbon content is in the ingot body rather than in the hot top as usual. This phenomenon indicates that the final solidification region is in the ingot body, which is directly caused by the poor insulation condition of the hot top. This solidification sequence would definitely induce serious positive segregation and shrinkage porosity in the ingot body. Therefore, the insulation condition of the hot top should be optimized before improving the pouring technique to reduce the macrosegregation. Better insulation materials are chosen to ensure the better insulation condition of the hot top. The thermophysical data of optimized insulation materials are listed in table 3.

Table 3. Thermophysical data of optimized insulation materials in the hot top mold.

	ρ (Kgm^{-3})	λ (Wm^{-1}K^{-1})	c_p (KJKg^{-1}K^{-1})
Fireclay	1000	0.6	1.13
Insulation board	200	0.3	1.13
Covering flux	200	0.4	1.13

Figure 3 shows the carbon concentration profile along the centerline of the 360 t MCP ingot after optimizing the insulation condition of the hot top. It clearly indicates that the point of the maximal positive segregation is shifted into the hot top, which means that a perfect sequential solidification is achieved. At the same time, the macrosegregation in the ingot body is reduced, the maximal carbon content decreases from 0.26 wt% to 0.24 wt%. Unfortunately, the 0.24 wt% carbon concentration in the ingot body is still beyond the range of the industrial limit (0.16-0.22 wt%). Namely, the carbon concentration cannot be controlled within the industrial limit for such gigantic ingot only by optimizing the insulation condition of the hot top. Therefore, pouring technique should be further optimized to reduce the macrosegregation after the insulation optimization. The following simulations are based on the insulation optimization model.

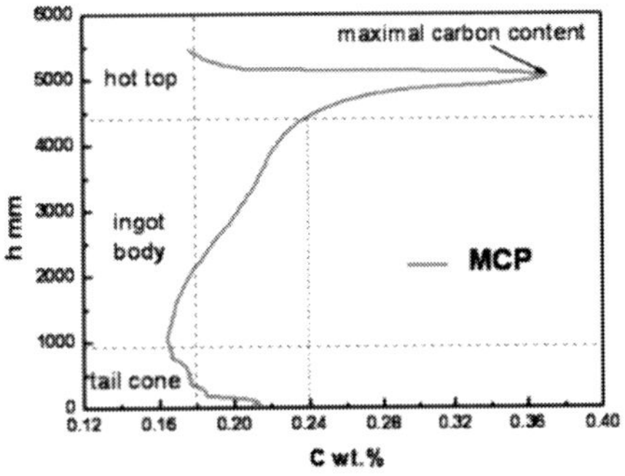

Figure 3. Carbon concentration profile along the centerline for MCP after insulation optimization.

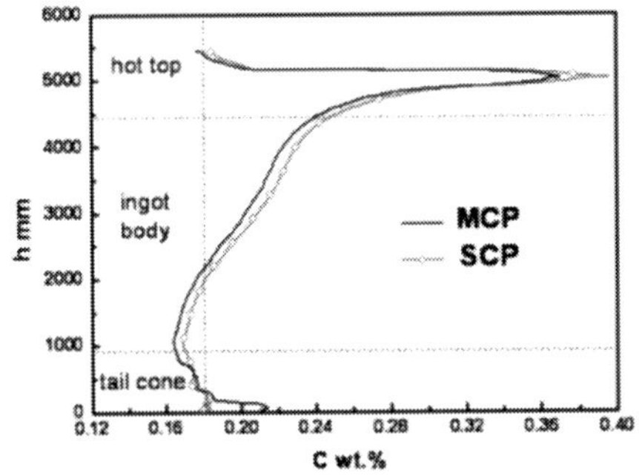

Figure 4. Comparison of carbon concentration profile along the centerline between MCP and SCP.

4.2. Comparison of MCP and single concentration pouring (SCP)

In order to clarify the effect of the pouring technique on the macrosegregation, first the solidification process of the 360 t steel ingot with the SCP technique is simulated. The initial carbon concentration in the ladle is 0.18 wt%, equaling to the nominal concentration of the experimental steel. The result is compared with that of the traditional MCP technique, in which three ladles with different carbon contents are poured in succession, as shown in figure 4. Surprisingly, there is almost no significant difference between these two curves, which means that the macrosegregation cannot be reduced efficiently by the traditional MCP technique. Figure 5 shows the solid fraction and the carbon concentration distribution in the ingot three hours after the last ladle poured in the MCP process. It can be seen that, at this moment, the solidification ratio of the whole ingot is only about 10%, while the carbon distribution is almost homogeneous in the remaining melt. The melt of different carbon concentrations from different ladles mixes quickly by natural convection before solidification. The homogenous mixing of the melt eliminates the dilute effect of the last ladle with lower carbon concentration. Thus in order to achieve the reduction of the macrosegregation, a proper interval between pouring ladles should be set to efficiently use the concentration gradient of different ladles. Generally, in order to avoid crack, there should be no interval between ladles when pouring the ingot body. So the interval between the second and the last ladle, viz., the delay time for the last ladle becomes the key point to be investigated and adjusted to control the macrosegregation. In the following discussion the multi concentration delayed pouring (MCDP) is used to describe a MCP case with a delay time for the last ladle.

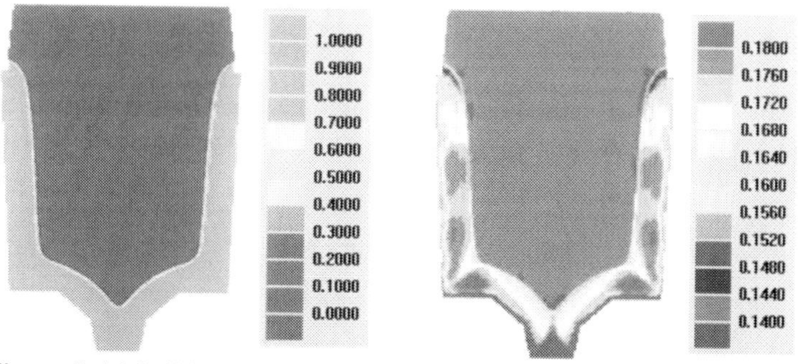

Figure 5. (a) Solid fraction and (b) carbon concentration distribution in the ingot 3 hours after the last ladle poured in the MCP process.

4.3. The criterion of selecting the delay time for the last ladle

In order to investigate the influence of the delay time for the last ladle on the macrosegregation in ingot, simulations are carried out for solidification in two MCDP cases with different delay time: (1) 6 hours and (2) 30 hours. The results are compared with the result of the MCP case (delay 0 hour), as shown in figure 6.

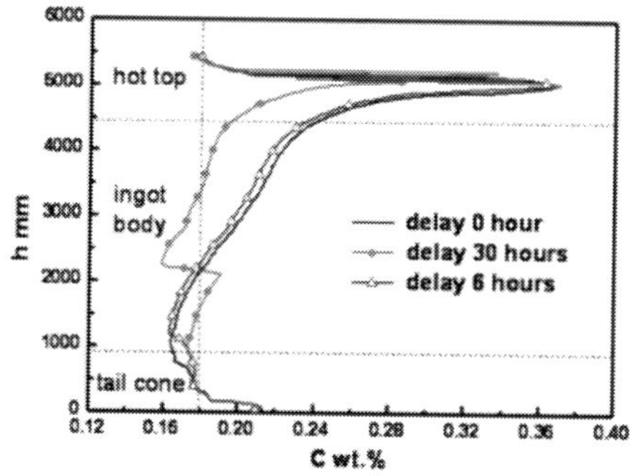

Figure 6. Comparison of carbon concentration profile along the centerline with different delay time.

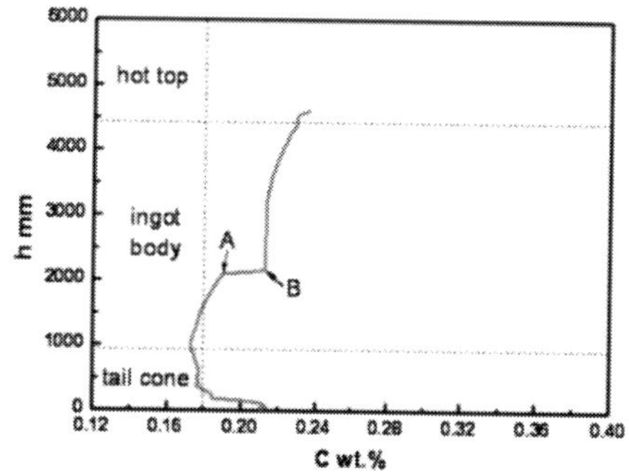

Figure 7. Carbon concentration profile along the centerline 30 hours after the first two ladles poured.

It can be seen that, when the delay time is 30 hours, viz. the last ladle is poured 30 hours after the first two ladles poured, a negative-positive-negative-positive (NPNP) style segregation formed from the bottom to the hot top along the centerline. Although the positive segregation below the hot top is reduced remarkably, this result is still not ideal due to the large concentration fluctuation and uneven concentration distribution in the ingot body. Figure 7 shows the carbon concentration profile along the centerline at the moment just before the last ladle poured, which can explain the above result clearly. The carbon content changes sharply from point A to B in figure 7, indicating the existence of the solid-liquid interface in this region at the moment. When the melt with lower carbon content in the last ladle is poured, the negative-positive segregation below point A cannot be influenced any more, as the metal here has solidified. While the upper residual melt would definitely mixes with the new poured melt to even mixed melt quickly. Then a new solidification mode is established in the upper part of the ingot, where the new negative-positive segregation will appear again after the solidification. Finally the NPNP style segregation forms in the whole ingot body. It can be concluded that a too long delay time for the last ladle will induce the NPNP style segregation. However, when the delay time is 6 hours, it can be seen that no noticeable improvement of the macrosegregation is achieved and the largest carbon concentration is still beyond the industrial limit. This is because the delay time is so short that the melt with lower carbon content mixes homogeneously with the residual carbon enriched melt before solidification, similar to the condition of the MCP described in section 4.2. It seems that the delay time of 6 hours is still not long enough to reduce the macrosegregation and a longer delay time should be considered.

From the above discussion, it can be concluded that in order to control the carbon macrosegregation in the ingot body efficiently, the delay time for the last ladle should be neither too long nor too short. To ensure the quality of the ingot, not only the carbon concentration should be controlled within the industrial limit but also the large concentration fluctuation should be avoided. So, these two facts taken together, a criterion of selecting the delay time for the last ladle is proposed. The delay time for the last ladle should be as long as possible with the precondition of preventing the appearance of the NPNP style segregation, viz., the melt with lower carbon concentration in the last ladle should be poured at the moment that the segregation along the centerline turns from negative to positive. At this moment, the melt with lower carbon content is poured to dilute the remaining higher carbon concentration melt, to efficiently control the macrosegregation in the ingot.

4.4. The selection of the delay time for the 360 t steel ingot

Firstly, a case with only the first two ladles poured is simulated, in order to eliminate the influence of the last ladle. The carbon concentration profile along the centerline after solidification is showed in figure 7. The profile presents the same pattern with the case of pouring three ladles, the macrosegregation along the centerline turns from negative to positive at the height of 1590 mm in the ingot body. In order to determine this transition time, solid fraction along the centerline is traced. It is found that after 17-hour solidification, the solidification height along the centerline just reaches 1590 mm. Consequently, the 17-hour delay time is selected for the last ladle for the 360 t steel ingot.

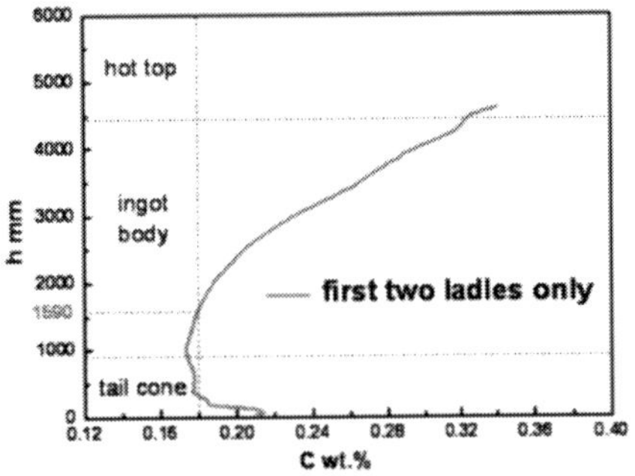

Figure 8. Carbon concentration profile along the centerline for only the first two ladles poured.

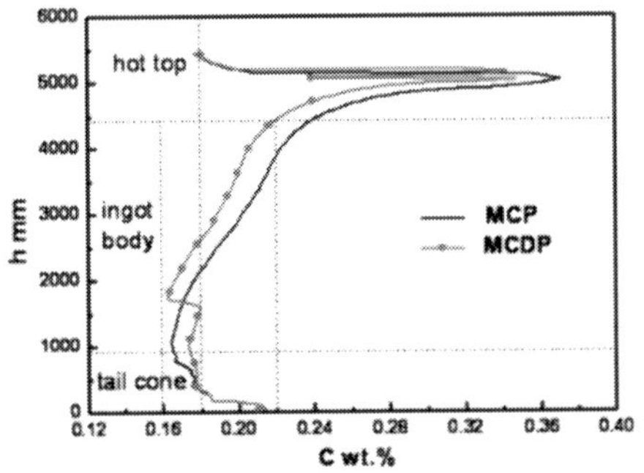

Figure 9. Comparison of carbon content profile along the centerline for MCP and MCDP (delay 17 hours).

Finally, the solidification process of the MCDP technique with 17-hour delay time is simulated. Figure 9 shows the carbon profiles along the centerline of this MCDP and the traditional MCP, which are different only on the delay time for the last ladle. It can be seen that the carbon segregation along the centerline is reduced significantly with the MCDP technique, especially for the positive segregation below the hot top. The carbon concentration of the ingot body is controlled within the range of the industrial limit (0.16-0.22 wt%). The maximal carbon content in the ingot body decreases from 0.24 wt% to 0.22 wt% and the integral area of the positive segregation along the centerline decreases significantly. Meanwhile, the NPNP style segregation fluctuation is avoided and the carbon concentration in the ingot body is more uniform. Consequently the macrosegregation in the 360 t steel ingot can be controlled efficiently by using the MCDP technique with 17-hour delay time for the last ladle.

It should be mentioned that the formation and movement of the equiaxed grains, which may influence the formation of the segregation, are not considered in the simulations. If they were considered, the area of the negative segregation at the lower part of the ingot may be enlarged in theory, due to the cloud of equiaxed grains formed at the beginning stage during the solidification. The height, where the segregation along the centerline turns from negative to positive, may be increased, and hence the optimized delay time in the MCDP technique should be prolonged accordingly. However, it cannot be quantificationally predicted that how the equiaxed grains influence the optimization of the delay time arbitrarily. In fact the simulation would become very complicated when the equiaxed grains are considered, which will be studied in the future work.

5. Conclusions

A mathematical model for the calculation of the macrosegregation in large steel ingot during the MCP process has been established and validated by comparing with the experimental result. The formation of the macrosegregation in the 360 t steel ingot has been investigated using the established model. The conclusions are summarized as follows:

1. The traditional MCP technique with no interval between ladles has no significant effect on reducing the macrosegregation in the large steel ingot. The delay pouring time for the last ladle has remarkable influence on the macrosegregation distribution, especially for the positive segregation below the hot top. In order to effectively reduce the macrosegregation, a proper delay time for the last ladle should be selected.
2. If the delay time is too long, the NPNP style segregation appears. If the delay time is too short, the positive segregation cannot be reduced noticeably. To control the macrosegregation in the large steel ingot efficiently, a criterion of selecting the delay time for the last ladle has been proposed. The last ladle should be poured at the moment when the segregation along the centerline turns from negative to positive.
3. Based on the selecting principle, the 17-hour delay time is decided for the 360 t steel ingot. With the 17-hour delayed pouring technique, the carbon concentration along the centerline in the ingot body is controlled within the industrial limit and the macrosegregation distribution is improved remarkably in the 360 t steel ingot.

Acknowledge

The authors gratefully acknowledge the financial supports from the Main Direction Program of Knowledge Innovation of Chinese Academy of Sciences (Project No.KGCX2-YW-221).

References

[1] Olsson A, West R and Fredriksson H 1986 *Scandinavian Journal of Metallurgy* **15** 104-12
[2] Maidorn C and Blind D 1985 *Nuclear Engineering and Design* **84** 285-96
[3] Lesoult G 2005 *Materials Science and Engineering* A **413-414** 19-29
[4] Gu J P and Beckermann C 1999 *Metallurgical and Materials Transactions* A **30** 1357-66
[5] Flemings M C 2000 *ISIJ International* **40** 833-41
[6] Hebditch D J and Hunt J D 1974 *Metallurgical Transactions* **5** 1557-64
[7] Combeau H, Zaloznik M, Hans S and Richy P E 2009 *Metallurgical and Materials Transactions* B **40** 289-304
[8] Choudhary S K and Ganguly S 2007 *ISIJ International* **47** 1759-66
[9] Zhong Y, Su G Y and Yang K 2003 *Journal of Materials Science & Technology* **19** 413-6
[10] Griffiths W D and McCartney D G 1996 T *Materials Science and Engineering* A **216** 47-60
[11] Tanaka Y and Sato I 2011 *Journal of Nuclear Materials* **417** 845-59
[12] Liu D R, Sang B G, Kang X H and Li D Z 2010 *International Journal of Cast Metals Research* **23** 354-63
[13] Pardeshi R, Voller V R, Singh A K and Dutta P 2008 *International Journal of Heat and Mass Transfer* **51** 3399-409
[14] Voller V R, Mouchmov A and Cross M 2004 *Applied Mathematical Modelling* **28** 79-94

IOP Conf. Series: Materials Science and Engineering **33** (2012) 012093 doi:10.1088/1757-899X/33/1/012093

Modeling of macrosegregation and solidification microstructure for Al-Si alloy under unidirectional solidification by a coupled cellular automaton – finite volume model

Hongwei Zhang[1]**, Keiji Nakajima**[2]**, Engang Wang**[1] **and Jicheng He**[1]

[1] Northeastern University, Key Laboratory of Electromagnetic Processing of Materials, Shenyang, 110004, P.R. China
[2] Royal Institute of Technology, Department of Materials Science and Engineering, SE-100 44 Stockholm, Sweden

E-mail: hongweizhang@epm.neu.edu.cn

Abstract. A coupled Cellular Automaton – Finite Volume (CA-FV) model is developed for the prediction of macrosegregation and solidification microstructure of an Al-Si binary alloy under unidirectional solidification. The enthalpy, concentration and velocities are calculated at FV scale following macroscopic fluid flow, heat and solute transfer equations. The temperature and solid fraction are updated at CA scale from both the energy balance and the nucleation, grain growth under fluid flow and entrapment procedure with the help of microsegregation approximation. The variables at FV and CA sites exchange information through topological relations. The recalescence and the inter- and intra-granular segregation during solidification are predicted. The effects of thermal/solutal buoyancy and the grain structure on molten natural convection and the segregation pattern are also analyzed. Verification of the model has been done using the previous experimental solidification morphologies by other researchers.

1. Introduction

The formation of macrosegregation is closely related to the alloy solidification process. The segregation in castings consequently leads to the formation of other defects such as hot crack, etc. As known, the solidification process of an alloy includes many phenomena, such as the macroscopic transport (i.e. fluid flow, heat and solute transfer), transformation thermodynamics (i.e. phase equilibrium, phase interface and chemical equilibrium) and solidification kinetics (solute distribution, nucleation and growth), etc. In order to control the macrosegregation and solidification microstructure for obtaining better properties of the alloy product, it is necessary to develop a reasonable model to predict the macroscopic transport and nucleation and growth procedure in actual solidification process.

Bennon and Incropera [1] firstly developed a binary macroscopic continuum transport model coupling laminar flow, heat and solute transfer process. It is then applied to predict the turbulent transport and central segregation of continuous casting billets by Aboutalebi, Hasan and Guthrie[2]. Recent years, the macroscopic transport models coupling with the nucleation and grain growth procedure is developed by Gandin and co-workers [3-6]. In the model, the heterogeneous nucleation, grain growth and solidification morphologies are predicted by microscopic models such as Cellular

Automaton (CA) method [3]. The growth kinetics for dendrite and eutectic tips in pure diffusion regime, as well as the dendrite tip growth kinetics in the presence of fluid flow [4] is described. A microsegregation model relating macro- and micro- scopic process is adopted. Through coupling the flow, heat and solute transport with CA solidification prediction and lever rule, Guillemot and Gandin [5,6] predicted the segregation profile in the solidification of Pb-Sn alloy and the channel segregation in single grain growth of Ga-In alloy. The predictions fit well with the experiments.

In the present paper, considering the influence of grain structure on the macrosegregation, the model coupling macroscopic fluid flow, heat and solute transfer and microscopic nucleation and grain growth procedure is developed. The continuum macroscopic transport equations in which the relative phase motions is clearly expressed are adopted here instead of the governing equations by volume averaging theorem in Guillemot and Gandin's model [5I. It is applied for the prediction of an Al-Si binary alloy unidirectional solidification in a square cell. The inter- and intra-granular segregation besides macrosegregation in multi-grain alloy solidification process is discussed. The effects of grain structure and thermal and solutal buoyancies on molten natural convection and the segregation pattern are also analyzed.

2. Coupled FV-CA model

A two-dimensional coupled FV-CA model for the solidification of Al-Si binary alloy is developed based on Guillemot and Gandin's idea [5]. However, the macroscopic transport equations in continuum form [1] are adopted here instead of those in average form [5]. One of its advantages is that the terms related to relative phase motions are clearly expressed in the energy and solute conservation equations. The macroscopic fluid flow, heat and solute transport is modelled by finite volume method, the microscopic nucleation and growth is described by CA method, and the local thermodynamic equilibrium is assumed to prevail and described by lever rule. The macro- and microscopic transport phenomena are coupled in a double-way.

2.1. Macroscopic transport model

The following assumptions are considered: (1) The solidification is unsteady under laminar flow with respect to a fixed coordinate system; (2) An incompressible Newtonian molten is assumed with the Boussinesq approximation for density; (3) The local thermodynamic equilibrium is considered; (4) The solidifying alloy system is viewed as a continuum comprising liquid and solid phase; (5) Physical properties such as ρ, c_p are equal and constant for liquid and solid, D_l is constant; (6) A columnar dendritic solidification is assumed where pore formation is ignored.

- Continuum transport equations

$$\frac{\partial}{\partial t}(\rho)+\frac{\partial(\rho u_i)}{\partial x_i}=0 \tag{1}$$

$$\frac{\partial}{\partial t}(\rho u_j)+\frac{\partial(\rho u_i u_j)}{\partial x_j}=-\frac{\partial p}{\partial x_i}+\frac{\partial}{\partial x_j}\left[\mu_l(\frac{\partial u_i}{\partial x_j}+\frac{\partial u_j}{\partial x_i})\right]-\frac{\mu_l}{K_p}(u_i-u_{si})+\rho B_i \tag{2}$$

$$\frac{\partial(\rho h)}{\partial t}+\frac{\partial(\rho u_j h)}{\partial x_j}=\frac{\partial}{\partial x_j}(\frac{\kappa_{eff}}{c_p}\frac{\partial h}{\partial x_j})+\frac{\partial}{\partial x_j}[\frac{\kappa_{eff}}{c_p}\frac{\partial}{\partial x_j}(h_s-h)]-\frac{\partial}{\partial x_j}[\rho(u_j-u_{sj})(h_l-h)] \tag{3}$$

$$\frac{\partial(\rho w)}{\partial t}+\frac{\partial(\rho u_j w)}{\partial x_j}=\frac{\partial}{\partial x_j}(\Gamma_{eff}\frac{\partial w}{\partial x_j})+\frac{\partial}{\partial x_j}[\Gamma_{eff}\frac{\partial}{\partial x_j}(w_l-w)]-\frac{\partial}{\partial x_j}[\rho(u_j-u_{sj})(w_l-w)] \tag{4}$$

where buoyancy force ρB_i in gravity direction is the contributions of thermal and solutal effects, $-\rho g[\beta_T(T-T_{ref})+\beta_s(w_l-w_{ref})]$. The permeability in Darcian damping term related to mushy zone is calculated as $K_p = K_0 \cdot f_l^3/(1-f_l)^2$ by Kozeny-Carman formula [7], K_0 is permeability coefficient. The efficient thermal capacity is a mixture of the ones in solid and liquid, $\kappa_{eff} = \kappa_l f_l + \kappa_s f_s$. The efficient diffusivity is $\Gamma_{eff} = f_l \rho D_l$ omitting the diffusion in solid.

- Conversion of enthalpy and concentration to temperature and solid fraction

Assuming local thermodynamic equilibrium and neglecting the curvature of solidus and liquidus lines, the temperature T and solid fraction f_s in a cell can then be deduced from the enthalpy H and concentration w by lever rule, detailed relations can be found in reference [5].

2.2. *Microscopic model for the evolution of grain structure*

- Nucleation algorithm

An instantaneous nucleation algorithm is adopted [8]. The critical nucleation undercoolings are pre-distributed at mold surface and in bulk molten, respectively. Once the local undercooling at a CA cell approaches the prescribed one during solidification, it nucleates. A crystallographic orientation α is randomly chosen among the predefined orientation classes in the range of $(-\pi/4 \sim \pi/4)$.

- Dendrite tip growth kinetics in the presence of fluid flow [4,9]

The dendrite tip growth rate v_{tip} in the presence of fluid flow is obtained by GGAN model [4], providing local undercooling ΔT, flow velocity u and angle θ between tip growth direction and flow direction with solving the following equations,

$$\Delta T = mw_0(1 - \frac{1}{1-(1-k_p)\Omega}) + \frac{2\Gamma}{r} \tag{5}$$

$$\Omega = (w^* - w_0)/[w^*(1-k_p)] \tag{6}$$

$$\Omega = Pe_v \exp(Pe_v)\left\{E_1(Pe_v) - E_1(Pe_v(1 + \frac{4}{A\mathrm{Re}_{2r}^B Sc^C \sin(\theta/2)}))\right\} \tag{7}$$

$$r^2 v_{tip} = \frac{D_l}{\sigma^*}\frac{\Gamma}{mw^*(k_p-1)} \tag{8}$$

where $Pe_v = rv_{tip}/(2D_l)$, $Pe_u = ru/(2D_l)$, $\mathrm{Re}_{2r} = 2ru/\nu = 4Pe_u/Sc$, $Sc = \nu/D_l$, constant A=0.5773, B=0.6596, C=0.5249. σ^* =0.0253. $E_1(P_v) = \int_{P_v}^{\infty} \exp(-\tau)/\tau d\tau$, is estimated by the approximation of the series [10].

With fluid flow, the growth envelope of the grain becomes an irregular quadrilateral in two-dimensional (2D) case. Thus the grain growth and entrapment is described by the decentred growth / entrapment algorithm [5, 9] in CA model.

2.3. *Coupling of macroscopic and microscopic models [5]*

The enthalpy, concentration and components of velocities are calculated by macroscopic transport equations at FV nodes under cooling conditions and interpolated to microscopic CA cells. Afterwards, 1) the temperature and solid fraction are updated from the enthalpy and concentration; 2) the nucleation is calculated by the increase of undercooling at liquid CA cell, the grain growth and entrapment are predicted by the increase of undercooling and local flow velocity at solid CA cell. As a result, the temperature and solid fraction are updated and the state of CA cell is transited. The temperature and solid fraction at CA cells are then summated to update the values at FV nodes. The process follows continuous cooling, nucleation and growth procedure until the total region is solidified.

2.4. Initial and boundary conditions for unidirectional solidification

A 2D case is assumed for the molten alloy under unidirectional solidification in a square cell. Four boundaries of the region are set as solid walls, there is no velocity at wall. The heat extract is imposed only from the left wall with a cooling rate in terms of enthalpy, $\dot{h}$=-4000 $J/kg \cdot s$, the other walls are insulated with the surroundings. There is no mass transfer at walls.

The initial conditions are set as: initial velocities $u = v = 0$, initial concentration in mass fraction w_0=0.07, initial temperature T_0=620°C, initial liquid fraction $f_l = 1$.

3. Results and Discussions

The coupled CA-FV model written in Fortran language is applied to the solidification process of Al-7wt%Si binary alloy. An ingot configuration of 0.01m×0.01m is divided into an uniform FV size 0.5mm×0.5mm and a finer CA cell size 25 μm ×25 μm. The 8 neighbours' Moore relationship is adopted in CA algorithm. The solid grains are assumed fixed. The physical properties of Al- 7wt%Si alloy and parameters used in calculations are ρ=2452.5 $kg\ m^{-3}$, c_p =1060 $J\ kg^{-1}K^{-1}$, D_l=6.45×10^{-9} $m^2\ s^{-1}$, μ_l=1.38×10^{-3} $Pa \cdot s$, k_p=0.13, κ_s = 233 − 0.110 T and $\kappa_l = 36.5 + 0.028T$ $Wm^{-1}K^{-1}$, ΔH_f =387400 $J\ kg^{-1}$, β_T =1×10^{-4} K^{-1}, β_s =-4.0×10^{-2}, m =-6.5 $K\ wt\%^{-1}$, Γ =1.96×10^{-7} $K\ m$, S_C=90.6, K_0=5.56×10^{-11} m^2, $w_{ref} = w_0$, $T_{ref} = T_0$. Nucleation density, nucleation undercooling and standard deviation at wall are n_s^*=9.6×10^3 m^{-1}, $\Delta T_{S,\max}$ =1.0K and $\Delta T_{S,\sigma}$ =0.1K, respectively. The ones in bulk melt are n_v^*=1.5×10^6 m^{-2}, $\Delta T_{v,\max}$ =6.0K and $\Delta T_{S,\sigma}$ =0.1K, respectively. These nucleation data are used for determination of the prescribed nucleation undercoolings.

3.1. Results by macro/microscopic coupling

- Cooling curves

The calculation begins at 620°C, which is above the liquidus point (614.05°C) of Al-7wt%Si alloy. The cooling curves at surface (left wall), intermediate (1/4 length to the left) position and centre are plotted in figure 1a. The temperatures on the curves of surface and intermediate position represent the averaged ones among the relating nodes. The recalescence can be seen clearly at beginning of the solidification in the curves at surface and intermediate position. It is the result of that the heat released due to grain growth in an undercooled melt is more than that extracted by the surroundings. The larger growth undercooling shown on the curve of surface is due to the larger cooling rate at left wall. The consequently larger going up of the temperature is a direct result of releasing more heat with larger increase of solid fraction. The slopes of the cooling curves become smaller afterwards. With unbalanced heat released and extracted in mushy zone, the small jumps of the temperature can also be seen during the solidification, which further decreases the cooling rate. When the temperature approaches the eutectic point (577°C), it keeps isothermal until totally solidified. Afterwards, the temperature decreases in solid state and the slopes of curves become larger again. The estimated cooling rate at left wall, intermediate and centre position during solidification is 0.347K/s, 0.320K/s and 0.309K/s, respectively.

- Fluid flow

The fluid flow in alloy solidification is driven by both thermal and solutal buoyancies, as expressed in the buoyancy formula. The reference points for temperature and concentration are set at their initial values. With decreased temperature and positive thermal expansion coefficient, the thermal buoyancy is in the direction of gravity. Furthermore, the solute Si accumulates in residual liquid as partition ratio is less than 1. The accumulation of Si concentration in liquid together with a negative solutal expansion coefficient results in that the solutal buoyancy is also in the same direction of gravity. Imposing cooling condition at left wall, the downward thermal and solutal buoyancies near

IOP Conf. Series: Materials Science and Engineering **33** (2012) 012093 doi:10.1088/1757-899X/33/1/012093

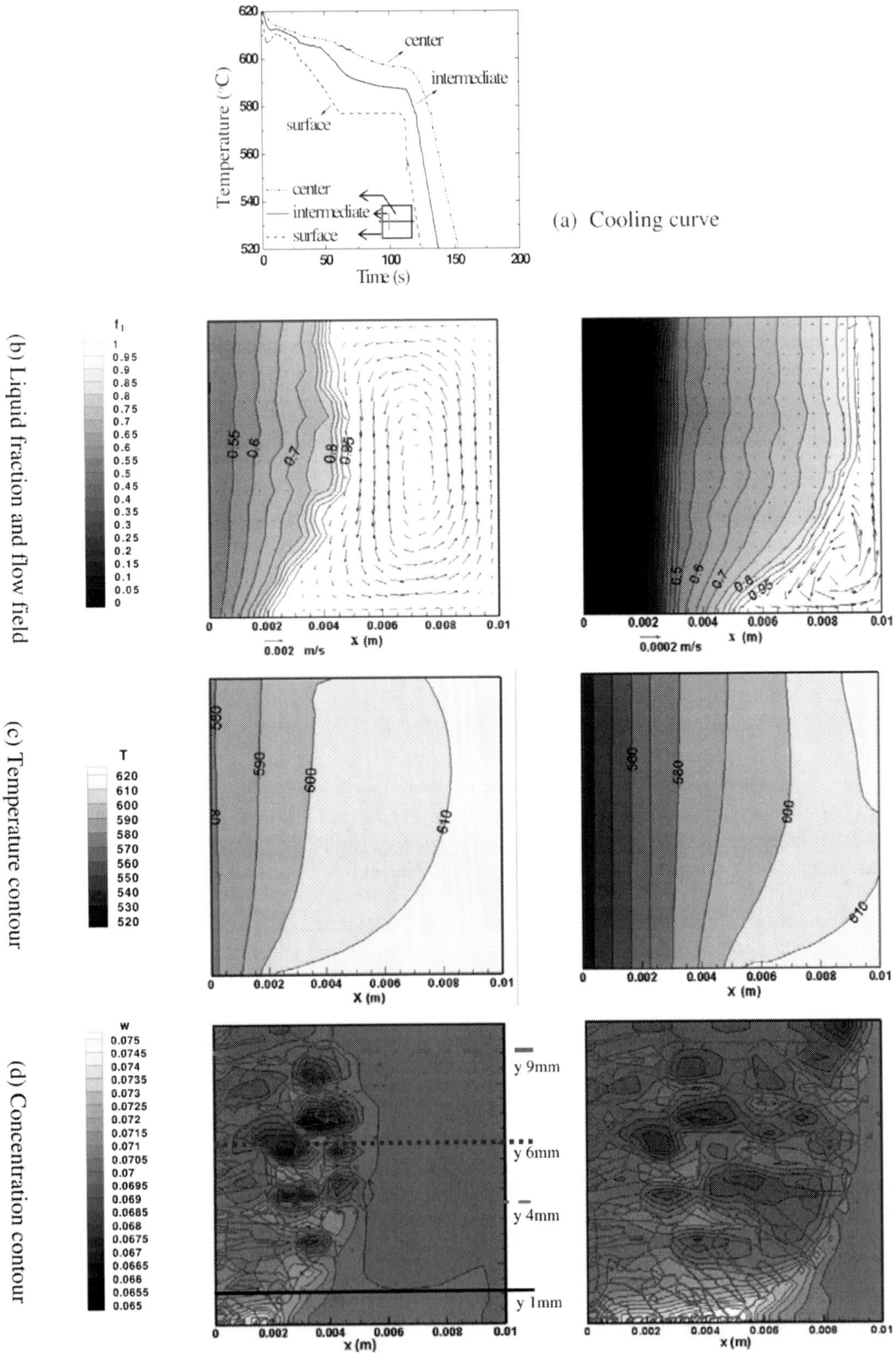

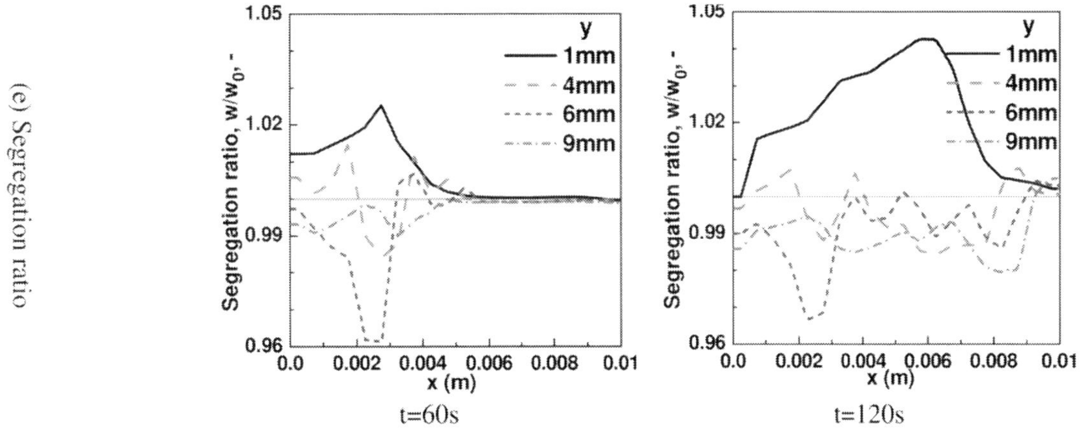

Figure 1. Calculation by CA-FV model in a 0.01m × 0.01m ingot with cooling rate $h = -4000 J/kg \cdot s$ imposing at left wall

solidification front drives the molten to flow in counter-clockwise circulation (figure 1b). This buoyancy-driven natural convection is in the magnitude of 0.002m/s for maximum velocity at 60s and damped to 0.0002 m/s at 120s.

- Temperature field and solidification morphologies

The isotherms close to the left wall are roughly vertical, as shown in figure 1c. It indicates that temperature distribution there is largely influenced by one-dimensional heat extraction (unidirectional solidification). However, the temperature distribution near solidification front is simultaneously influenced by the thermal and solutal buoyancy- driven flow. As a result, the isotherms there are shown as the unsymmetrical arc shape instead of the symmetrical one in figure 3b. The solidification is delayed at the lower part of the region compared with the isotherms without consideration of fluid flow (figure 3b).

The grain growth favouring in upstream direction can be illustrated in superimposed red-coloured grain envelopes in figure 1d. With cooling at left wall, the nucleation firstly occurs at the left wall. The nuclei near solidification front grow in the upstream direction of the counter-clockwise flow. Finally the unsymmetrical arc- shaped solidified region is formed surrounding the left wall. The grains at upper part grow faster. The coarse equiaxed grains in the solidified structure are the results of large nucleation undercooling and large cooling rate.

- Concentration field

Solute migrates following the counter-clockwise bulk molten flow, as seen in figure 1d and 2b. In the view of ingot size, the bottom part of the region is the high positive segregation zone. However, the concentration profile is much complicated by the influence of solidification and grain growth. Figure 1d overlapped the grain envelopes on concentration profile. As shown, the segregation in granular region is negative segregation, where the intragranular concentrations are much lower than the intergranular ones. It is the contributions of both the solute rejection in grain growth and solute migration following the intergranular liquid. With partition ratio less than 1 (k_p=0.13 for Si in Al-7wt%Si alloy), the redundant solute in solid grain is rejected to the intergranular liquid. However, the buoyancy-driven flow in intergranular liquid is not efficient enough to move away all the accumulated solutes or make them mixed uniformly. Thus the solutes distribute non-monotonically at inter- and intra-granular region. Figure 1e plots the segregation ratio, i.e. the ratio of local solute concentration to the initial one, along four horizontal lines at y coordinate 1, 4, 6 and 9mm, respectively, as illustrated superimposed in figure 1d. Regarding the curves labeled y=4, 6 and 9mm, the segregation ratio altering in larger and smaller values in figure 1e corresponds to the positions altering from intergranular to intragranular zone in figure 1d.

Except these, the largest value of segregation ratio on the curve labeled y=1mm denotes the position of solidification front at the bottom of the region. There the accumulation of solutes is owed to the counter-clockwise circulation in bulk molten. Although the solutes distribute almost uniformly in most part of the residual liquid with relatively strong bulk flow, the solidification front at the bottom of the region is a "dead zone" for fluid flow, as shown in figure 1b. The solutes just reside there, as shown in figure 1d and 1e at 60s and 120s, which makes it the final high concentration zone. The position where the solidification front meets with the bottom is exactly the position of maximum positive segregation, although the finally solidified part is at the right-bottom corner of the region.

- Verification of the model

Yin et al [11] made single grain solidification experiments on Ga-5wt%In alloy. Guillemot and Gandin [6] modeled the corresponding solidification process. Comparison of the calculation conditions shows that the experimental configuration is a larger rectangle with 0.048m×0.033m in vertical section. Except the difference in configuration, other conditions are similar to our simulation. In experiment, the measured cooling rate at left wall ($\dot{T}$ = -2.362 K/hr) is much higher than that at right ($\dot{T}$ = -0.328 K/hr), other walls are set as adiabatic. Similarly, in the present simulation, the molten alloy is cooled down only from left wall and other walls are set as adiabatic. On the other hand, both Ga-5wt%In and Al-7wt%Si alloy solidify in dendritic region until approaching the eutectic point. Furthermore, both of the solutal expansion coefficients for the two alloys are negative ($\beta_{s,\,\mathrm{In}}$ =-1.663x10^{-3} [11]). It indicates that the molten flow driven by solutal buoyancy has the same direction as gravity, as well as the one driven by thermal buoyancy. Both the experimental (figure 1 in [11]) and predicted results (figure 3 in [6]) for Ga-5wt%In alloy show that the grain nucleates at left wall and grows gradually to an unsymmetrical arc-shaped solidified structure surrounding the left wall. Our simulation got the similar solidification morphology. Moreover, although their work focused on single grain growth and channel segregation while our simulation is for multi-grain solidification and inter- and intra-granular segregation, we all get the negative concentration distribution in solidified structure. However, there are some differences in the results. In their experiments and prediction, the solutes accumulate to the bottom of the residual liquid with clear horizontal fringes. In our simulation, the bottom of ingot is also a high positive segregation zone, but no serious segregation fringe is shown in the residual molten region. This can be explained by multi-grain solidification and also the adoption of a square configuration for simulation. Further verification of the present model still needs to be done.

Furthermore, the effects of the thermal and solutal expansion coefficients β_T and β_s, as well as the partition ratio k_p, on the flow circulation and the segregation pattern can be summarized here. With a positive β_T, the thermal buoyancy always keeps the same direction of gravity during cooling. However, the value of β_s varies with the solute elements. With partition ratio k_p <1 and a negative β_s, as for Si in the present Al-7wt%Si alloy and In in Ga-5wt%In alloy [6,11], the accumulated solutes in liquid make the solutal buoyancy in the same downward direction of both thermal buoyancy and gravity. Thus the combined buoyancies drive the fluid to flow downwards near the solidification front. The solute migration following flow shows a positive segregation at the bottom and negative segregation at the top of the region. Contrarily, a positive β_s together with the solute accumulation in residual liquid makes the solutal buoyancy in the opposite direction of gravity. Due to the fact that the value of β_s is always 1 or 2 order higher in magnitude than that of β_T, the solutal buoyancy dominates the final flow circulation and the segregation pattern. Thus the combined upward buoyancies make the fluid flow upwards near the solidification front. It results in the negative segregation at the bottom and positive segregation at the top of the region, which is just the case in Pb-48wt%Sn alloy where $\beta_{s,\,\mathrm{Sn}}$ = 4.5x10^{-3} [5].

3.2 Comparison with the results by only macroscopic transport

Figure 2 illustrates the results by only macroscopic transport calculations (equations (1)~(4)) with updating the temperature and solid fraction by lever rule. The nucleation and grain growth are not considered. Compared with figure 1a, the cooling curves with only macroscopic transport calculation can't detect the occurrence of recalescence and the solidification time is shorter. However, from the curves, the slow down of the temperature decrease due to the release of the latent heat and the isothermal transformation at eutectic point can be seen.

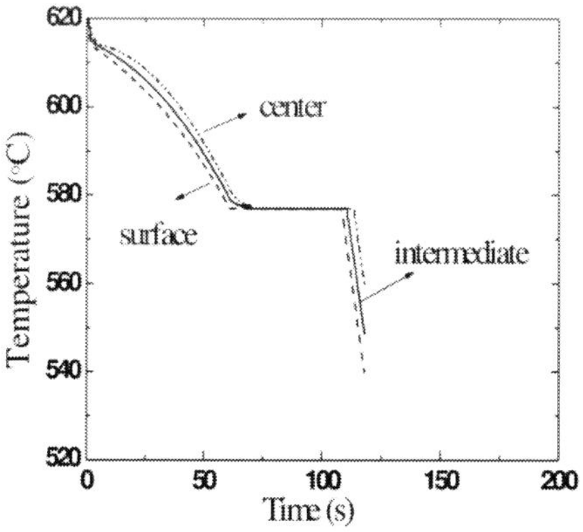

(a) Cooling curve

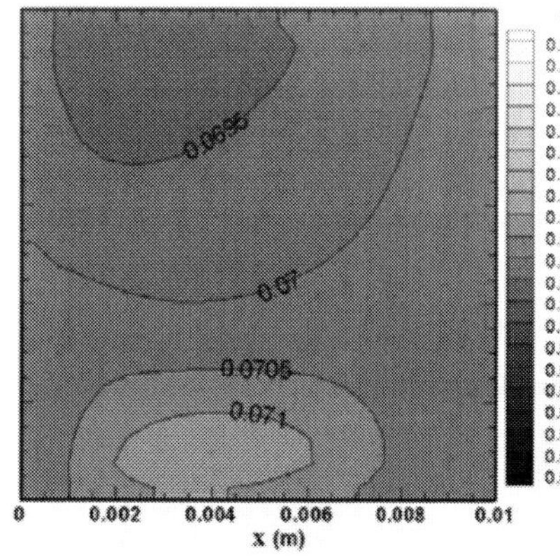

(b) Concentration contour

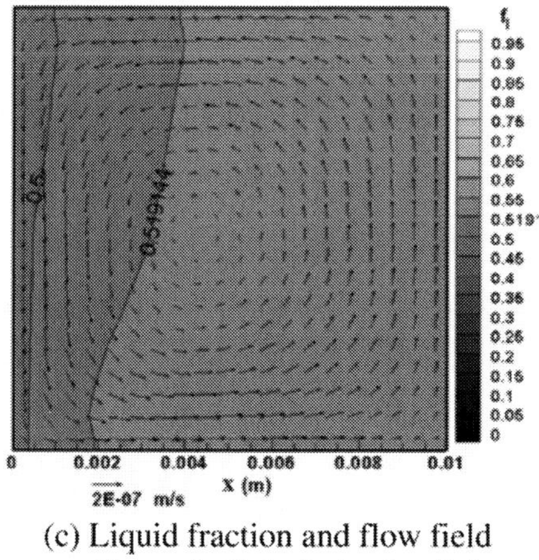

(c) Liquid fraction and flow field

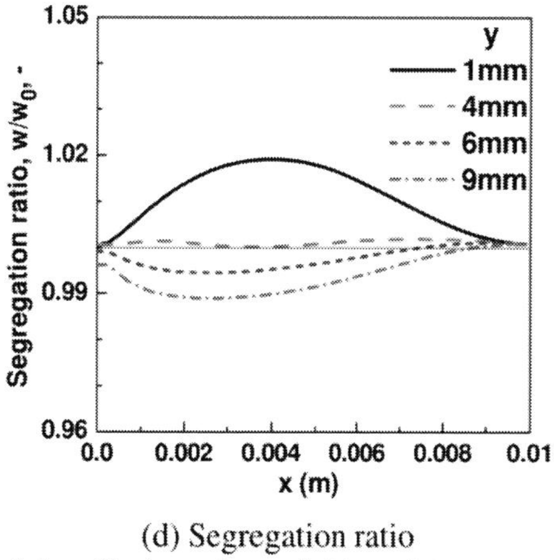

(d) Segregation ratio

Figure 2. Calculation by finite volume model at 60s in an ingot 0.01m×0.01m

The counter-clockwise circulation in liquid zone driving by thermo-solutal buoyancy can be predicted with macroscopic transport calculations. But the flow is very weak through comparison of the magnitude of flow velocities in figure 1b and 2c. The positive segregation at bottom and negative segregation at top part of the region is shown, as the result of migration with fluid flow and mushy zone solute redistribution. Comparison figure 2d with figure 1e at 60s, the prediction by only macroscopic transport can only detect the macro-scale segregation. The clear and smooth accumulation of the solutes can be seen in figure 2d. Except macro-scale segregation, the segregation in granular scale can't be predicted although it exactly affects the formation of hot crack.

3.3 Influence of fluid flow

The model without consideration of fluid flow only calculates the temperature and solute transport together.

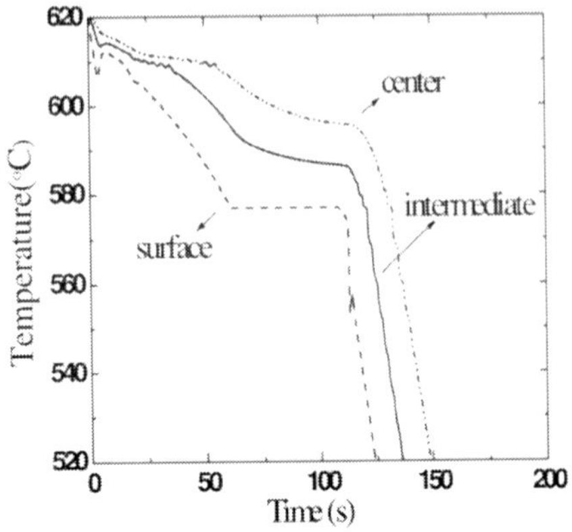

(a) Cooling curve

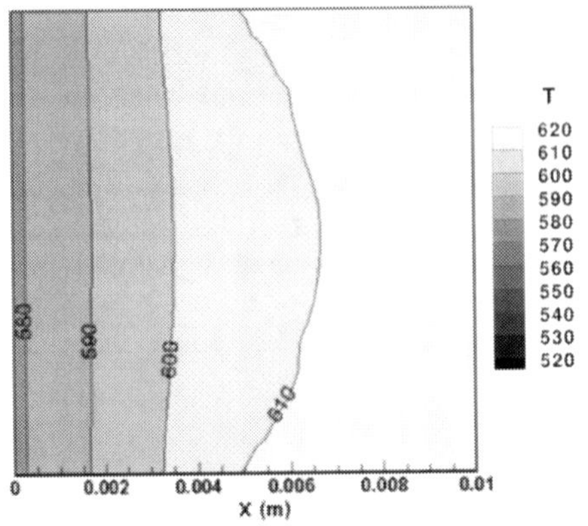

(b) Temperature contour

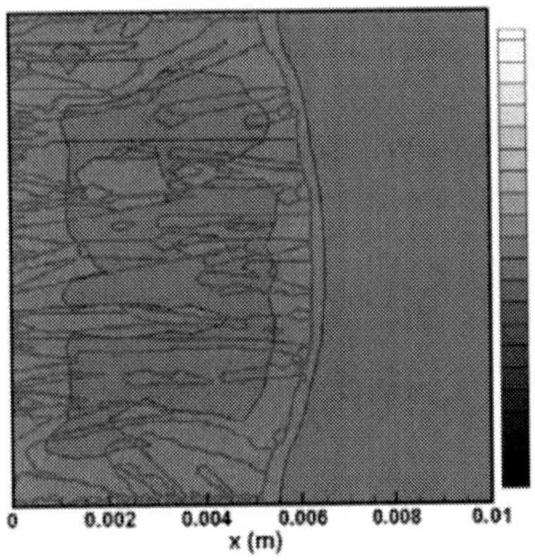

(c) Concentration contour overlapped with grain envelopes (in red color)

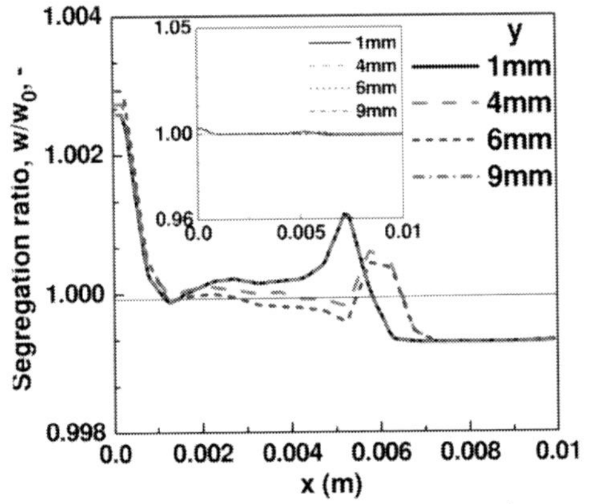

(d) Segregation ratio in cross section

Figure 3. Calculation results at 60s by CA-FV coupling model without fluid flow, imposing cooling rate $\dot{h} = -4000 J / kg \cdot s$ at left wall

Comparing figure 3 with figure 1, the transport without fluid flow relies on heat conduction and solute diffusion. The distribution similar to directional solidification is shown. With cooling from left wall, the isotherms are roughly vertical in section. The concentration distributes almost uniformly in vertical direction of the region, as illustrated in figure 3d. The final solidified zone is exactly the right part of the region. However, in the case of considering fluid flow, the final solidified zone is the right-bottom corner; furthermore, the temperature and concentration distribute differently in vertical direction of the region.

The molten flow also significantly affects the solidification morphologies. The grains without flow grow mainly as columnar ones in the opposite direction of heat flux as that in exactly unidirectional solidification process. Contrarily, the grain growth considering fluid flow reflects the conflict growth of columnar and equiaxed grains. The grain growth favours in upstream direction.

4. Conclusions

The unidirectional solidification process of an Al-7wt%Si binary alloy in a square cell with one vertical surface cooled down is simulated by the coupled CA-FV model. The roles of thermal and solutal buoyancies as well as the grain structure on the flow circulation and segregation pattern are deeply discussed in the case of that partition ratio k_p is less than 1. The results indicate that

1) The flow circulation and macrosegregation pattern was dominated by solutal but thermal buoyancy. The direction of solutal buoyancy controls the direction of convection. A negative solutal expansion coefficient results in a downward flow near the solidification front. Consequently a positive segregation at the bottom and negative segregation at the top of the region forms. Contrarily a positive value causes the opposite phenomena.
2) The multi-grain structure greatly influences the solute redistribution in granular scale. The relative concentration is positive at intergranular zone and negative at intragranular zone. Totally, the negative segregation was shown in the solidified structure.
3) The coupled CA-FV model considers the nucleation and grain growth under cooling condition and the feed back of solid fraction and temperature by nucleation and growth, as well as the upstream growth of grains. It shows potentials in prediction of the recalescence, the inter- and intra-granular segregation and the conflict growth of columnar and equiaxed grains during solidification.

Acknowledgments

The work is supported by 111 Project [B07015], 863 project [2007AA03Z518], fundamental Research Funds for the Central Universities [N100409004] and Natural Science Foundations of Liaoning Province [20072033], China.

References

[1] Bennon W D and Incropera F P 1988 *Int. J. Heat Mass Tran.* **31** 2147
[2] Aboutalebi M R, Hasan M and Guthrie R I L 1995 *Metall. Mater. Trans.* B **26** 731
[3] Rappaz M and Gandin Ch -A 1993 *Acta Metall.* **41** 345
[4] Gandin Ch -A, Guillemot G, Appolaire B and Niane N T 2003 *Mater. Sci. Eng.* A **342** 44
[5] Guillemot G, Gandin Ch -A, Combeau H, Heringer R 2006 *ISIJ Int.* **46** 880
[6] Guillemot G, Gandin Ch -A, Bellet M 2007 *J. Crystal Growth* **303** 58
[7] Asai S and Muchi I 1978 *Trans. ISIJ* **18** 90
[8] Gandin Ch-A, Desbiolles J -L, Rappaz M and Thevoz Ph 1999 *Metall. Mater. Trans.* A **30** 3153
[9] Zhang H W, Nakajima K, Xing W, Wang A L and He J C 2009 *ISIJ int.* **49** 1010
[10] Abramowitz M, Stegun I A (Eds.) 1972 *Handbook of Mathematical Functions (Tenth printing)* Dover, NY, 231
[11] Yin H B and Koster J N 1999 *J. Cryst. Growth* **205** 590

IOP Conf. Series: Materials Science and Engineering **33** (2012) 012094 doi:10.1088/1757-899X/33/1/012094

Numerical study of effect of sulphur element on mesosegregation by thermosolutal convection in Iron-Carbon-Sulphur system

D R Liu [1]**, X H Kang** [2] **and D Z Li** [2]

[1] School of Materials Science and Engineering, Harbin University of Science and Technology, No.4 Lin Yuan Road, Xiang Fang District, Harbin 150040, China

[2] Shenyang National Laboratory for Materials Science, Institute of Metal Research, Chinese Academy of Sciences, No.72 Wen Hua Road, Shenyang 110016, China

E-mail address: nacylake@126.com.

Abstract. The effects of a sulphur element on the mesosegregation formation by the multicomponent thermosolutal convection in a Fe-0.21wt pct C-X wt pct S system studied numerically. A two-dimensional solidification model that describes conservations of mass, momentum, energy and solute is presented. Simulations are performed on the Hebditch-Hunt casting. The validation of codes is carried out by the comparison with a consensus of previous numerical simulation for the Sn-5 wt pct Pb alloy. Three cases, encompassing variations of the initial concentration of the sulphur (0, 0.05, 0.1 wt pct), are studied. Simulated results show that no mesosegregates form for S=0.0 wt pct. The mesosegregates are triggered by the weak instabilities of the solidification front growth for S=0.05 wt pct. The increased fluid flow induces the local remelting, and results in the mesosegregation for S=0.1 wt pct.

1. Introduction

As a serious concern for the manufacturing process, macrosegregation during dendritic growth of alloys is directly linked to the quality of the casting and to the soundness of cast parts. The typical types of macrosegregation, which are often found in the casting, are top-positive segregation, bottom-negative segregation, and mesosegregation (V shape and A shape segregation channel) [1]. Regions with the two former patterns of segregation can be cropped off from the casting during processing. But, the mesosegregation bands cannot be removed. Furthermore, these narrow bands are characterized to have a higher segregation than that of the nearby solid regions and are positions where elements with the low melting point cluster, corresponding to detrimental properties [2]. Then, mesosegregation is considered to be a harmful kind of casting defect.

Numerical modelling of mesosegregation during the solidification of metallic alloy has received great interest, because it is useful in order to understand better its fundamental origins. Reviews on numerical study of mesosegregation are found in references [3, 4]. Key factors responsible for its formation are summarized. Flemings believed that the local melting and the resulting flow instability result in channel segregates [3]. Rady and co-workers explained that the formation of mesoscale channels is associated with thermosolutal flow interactions that follow the onset of solidification [5]. A series of research work of Založnik and Combeau presented that mesosegregation stems from the local instability of the advancement of the mushy zone front [4, 6, 7]. They stressed the local remelting does not necessarily occur to initiate a channel segregate. Using the direct macroscopic modelling technique, Guillemot and co-workers proposed that the formation of channel is instability of the growth front, which is induced by the movement and the accumulation of equiaxed grains [8].

While the majority of the attempts at modelling the formation of mesosegregation have focused on the binary system, studies that consider multicomponent solidification are needed to examine whether the binary system and the multicomponent system share the same mechanism for the mesosegregation formation.

In this research, a two-dimensional finite-difference-mesh model is presented to simulate the macrosegregation caused by multicomponent thermosolutal convection in the case of columnar dendrite solidification. The simulations are performed on the Hebditch-Hunt casting. The validation of codes is carried out by the comparison with a consensus of previous numerical simulation for Sn-5 wt pct Pb alloy in reference [7]. Macrosegregation formations are simulated during the solidification of a binary Fe-0.21wt pct C alloy and a ternary Fe-0.21C-XS (wt pct) alloy. The influences of varying the initial concentration of the sulphur on the mesosegregation are investigated. The formation mechanisms under different conditions are discussed.

2. Mathematical model

A mathematical model that describes the conservations of mass, momentum, energy and species for a fluid-solid system undergoing solidification is applied [9]. It is two-dimensional. Conservation equations are written in terms of volume-averaged quantities.

2.1. Conservation of energy

The conservation of energy can be written as follows:

$$\rho\frac{\partial[H]}{\partial t}+\rho\nabla\cdot\left(\vec{U}c_pT+\vec{U}L_{heat}\right)=\lambda\nabla\cdot(\nabla T) \tag{1}$$

$$[H]=f_sh_s+(1-f_s)h_l \tag{2}$$

where h_s is the solid phase enthalpy, h_l the liquid phase enthalpy, H the mixture enthalpy and λ the heat conductivity.

2.2. Conservation of solute

Neglecting the dispersive and macroscopic diffusion fluxes, the solute is transported through the domain by the fluid advection. The conservation of solute is written as follows:

$$\rho\frac{\partial[C]^k}{\partial t}+\rho\nabla\cdot\left(\vec{U}C_l^{\,k}\right)=0 \tag{3}$$

where $[C]^k$ is the mixture concentration.

2.3. Conservation of mass and momentum

Mass conservation $\quad \nabla\cdot\left(\vec{U}\right)=0 \quad$ (4)

The classical mixture theory leads to the following equations:

u momentum (X direction)

$$\rho\frac{\partial(u)}{\partial t}+\rho\nabla\cdot\left(\vec{U}u\right)=-\nabla P-\frac{\mu_l}{K}u+\mu_l\nabla\cdot(\nabla u) \tag{5}$$

v momentum (Y direction)

$$\rho\frac{\partial(v)}{\partial t}+\rho\nabla\cdot\left(\vec{U}v\right)=-\nabla P-\frac{\mu_l}{K}v+\mu_l\nabla\cdot(\nabla v)+\rho g\left[\beta_T\left(T-T_{ref}\right)+\sum_{k=1}^{n}\left(\beta_C\left(C_l-C_{ref}\right)\right)^k\right] \tag{6}$$

In the mushy region, the permeability K is calculated according to the Carman-Kozeny formula [9].

2.4. Calculation of solid fraction

To calculate the solid fraction, a strategy is developed based on three aspects: the coupling of the energy and species conservation equations in the mushy zone through the phase diagram; thermodynamic equilibrium at the solid-liquid interface and a Scheil microsegregation model. In order to consider the remelting, a procedure similar to the method in reference [10] is taken.

During one time step Δt, by a time-marching scheme, a quadratic equation in the solid fraction f_s^t is obtained as:

$$A + B = 0 \tag{7}$$

where

$$A = c_p T_m - [H]^t + (1 - f_s^{\,t}) L_{heat} \qquad B = c_p \sum_{k=1}^{n} \left(m_l \frac{(1 - f_s^{\,t-\Delta t}) C_l^{mid} + \beta f_s^{\,t} k C_l^{t-\Delta t}}{(f_s^{\,t} - f_s^{\,t-\Delta t}) k + (1 - f_s^{\,t}) + \beta f_s^{\,t} k} \right)^k$$

$$C_l^{mid} = \frac{[C]^t - [C]^{t-\Delta t} + (1 - f_s^{\,t-\Delta t}) C_l^{t-\Delta t}}{(1 - f_s^{\,t-\Delta t})}$$

3. Numerical validation

The validation of codes is carried out by the comparison with results in the existing literature [7]. The model alloy is Sn-5 wt pct Pb. A Hebditch-Hunt casting is hired, as shown in figure 1. Uniform grids 100×60 with grid size 0.1cm are used for all simulations. The reference [7] can be referred to for the thermophysical data and for the computational parameters.

The verification is performed by comparing present predictions of final macrosegregation map (figure 2), and values of flow velocity at 400 seconds (figure 3) with existing numerical simulation. From figure 2, the segregation values are very close. From figure 3, a little discrepancy in the flow behaviour is noted. For the present computation, the flow strength at X direction is not stronger. The insignificant differences probably result from the different numerical scheme designed to solve the conservation equations, i.e., a weak coupling between transfers of heat and species and fluid flow is used by present authors.

All in all, comparisons show a fair agreement. Therefore, the test with an acceptable outcome gives us confidence that the self-developed codes are reasonable for the following analysis.

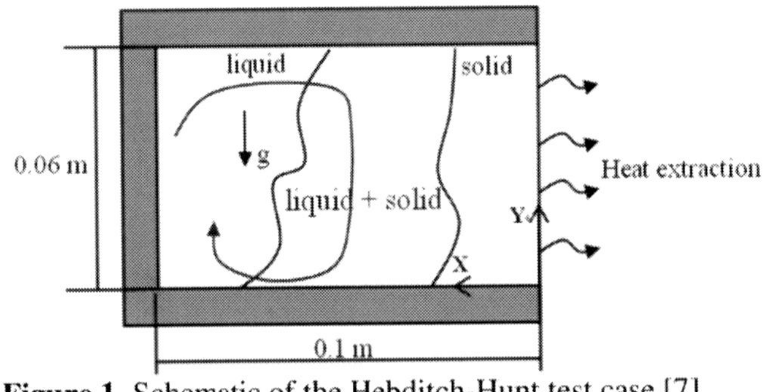

Figure 1. Schematic of the Hebditch-Hunt test case [7].

4. Results and discussion

The present mathematical model and numerical scheme are applied to simulate the macrosegregation during solidification of a binary Fe-0.21wt pct C alloy and a ternary Fe-0.21C-(0.0~0.1)S (wt pct) alloy. The two-dimensional rectangular cavity in *Section 3* is used, with the same enmeshment parameters. The solidification of alloy commences immediately at the liquidus temperature. The relevant thermophysical properties of Fe-C-S alloy are given in the reference [11]. The influences of varying the initial concentration of sulphur on the mesosegregation are investigated.

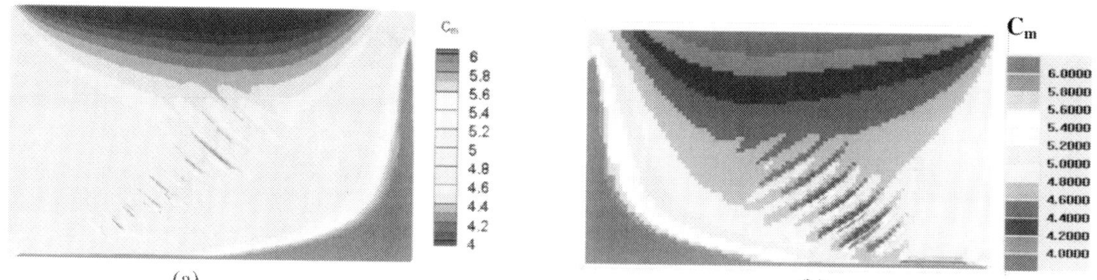

(a) (b)

Figure 2. Comparisons of final macrosegregation between (a) literature [7] and (b) present study.

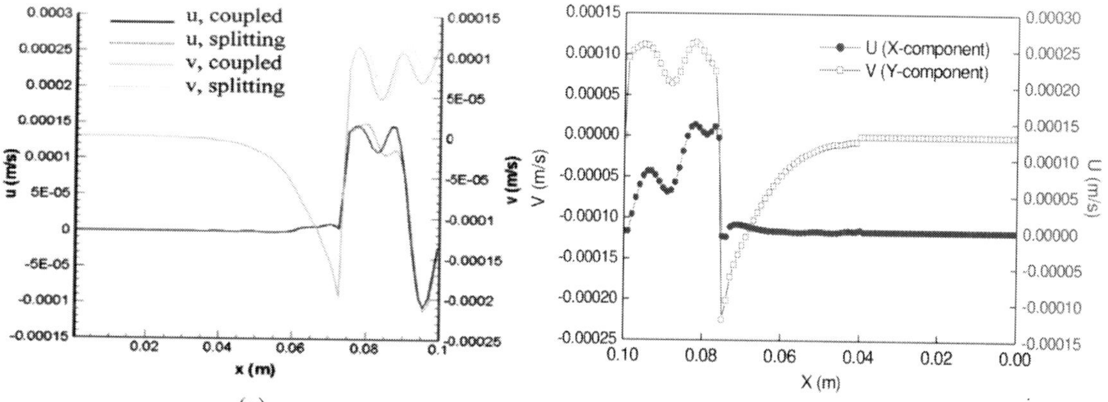

(a) (b)

Figure 3. Comparisons of values of X and Y components of $\vec{U}$ at 400s along a horizontal line (Y=15mm) between (a) literature [7] and (b) present paper.

4.1 Solidification of a ternary Fe-0.21C-X S (wt pct) alloy

The influence of the initial concentration of sulphur on the mesosegregation is shown in figure 4. The channel segregates are formed in both cases. As the initial concentration of sulphur is increased, the channels become more enriched in the solute. To reveal the mechanisms which are responsible for the observed mesosegregation, (i) a horizontal line (line-1) is selected at the vertical location Y=50mm; (ii) the solidification histories of two points (A and B) are examined. The two points locate on line-1 and are adjacent with each other in the positive and negative segregation bands; (iii) a parameter ΔT_{uc} named the local constitutional undercooling / superheat is introduced [4] as follows:

$$\Delta T_{uc} = T - T_l = T - \left[T_m + \sum_{k=1}^{n} \left(m_l C_l^{mid} \right)^k \right] \tag{8}$$

where C_l^{mid} is the liquid concentration ahead of the solidification front. During solidification course, it is a transient term calculated after the transfer stage and before the solidifying stage. When ΔT_{uc} becomes larger than zero, it indicates a departure from the equilibrium and is used to evaluate the degree of remelting at the advancing solidification front [4]; (iv) another variable (R solidification rate cm/s) is introduced and is defined as follows:

$$R = \dot{T} \times G^{-1} \tag{9}$$

where $\dot{T}$ (°C/s) and G (°C/cm) are local cooling rate and temperature gradient, respectively.

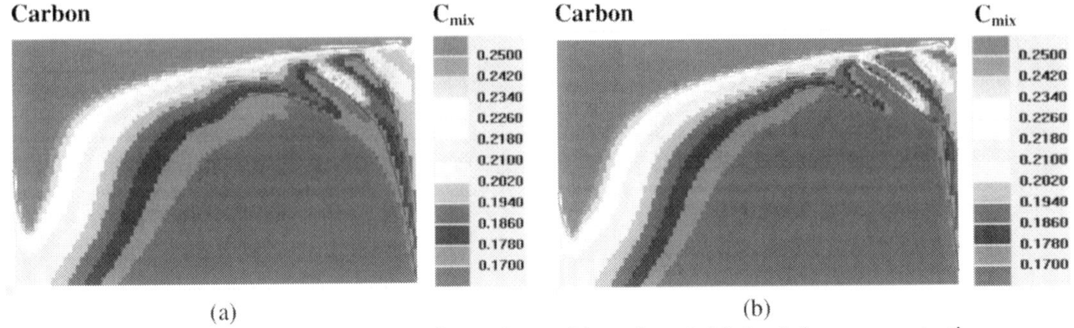

Figure 4. Final macrosegregation maps for carbon with various initial sulphur concentration, (a) S=0.05, (b) S=0.1

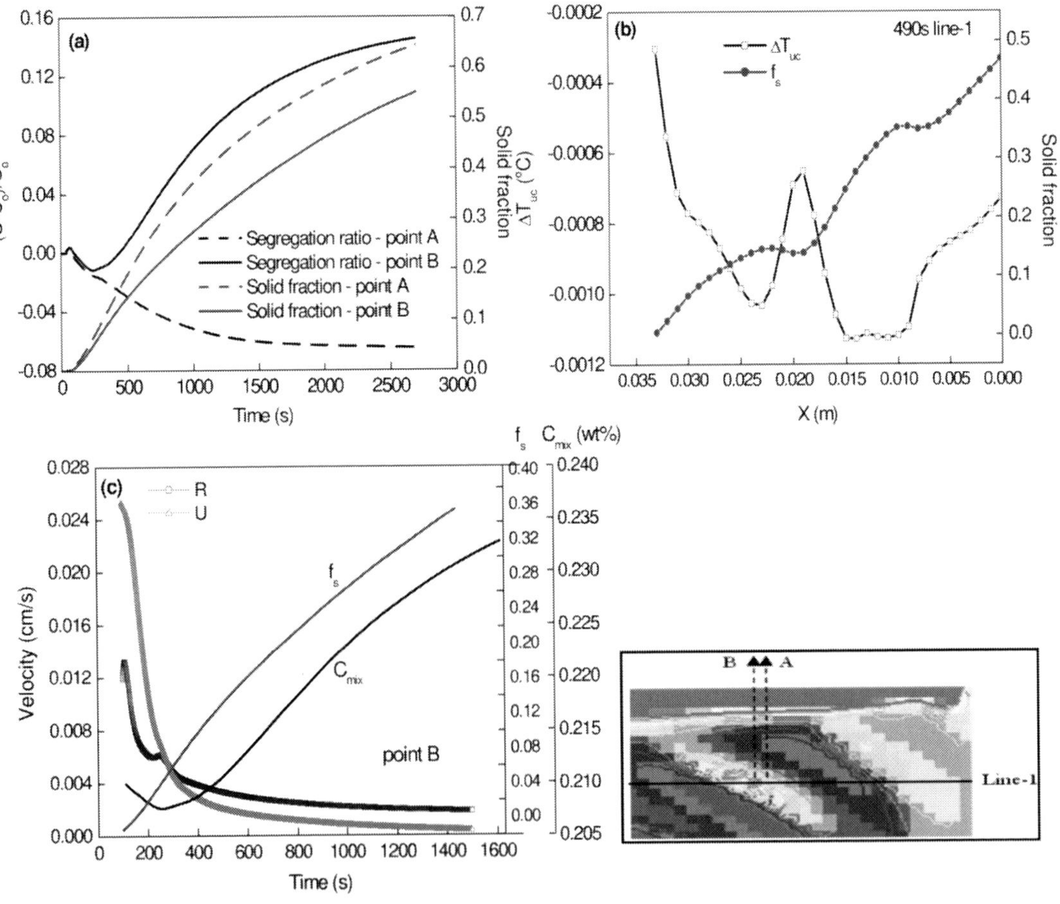

Figure 5. For S =0.05 wt pct, (a) evolution of segregation ratio and solid fraction in point A and point B; (b)variations of ΔT_{uc} and solid fraction along line-1 at 490s;(c) evolution of solidification rate, X component of flow vector, solid fraction and mixture concentration in point B.

For S=0.05wt pct, the mesosegregation formation starts at about 490 seconds when the mixture concentrations in point A and point B begin to evolve in different manners (figure 5(a)). The solid fractions, however, evolve in a similar way for the two points, i.e., increasing with time. As noted from figure 5(b), both point A and point B experience a constitutional undercooling of the order of minus 10^{-4}°C. It further proves that no local remelting shows up. These above phenomena match well with the proposed theory for mesosegregation formation in reference [4]. The local remelting is not a required condition to form channels. A very small perturbation in the isofraction of solid is enough to initiate the process of mesosegregation formation.

For S=0.1wt pct, the phenomena are different. In figure 6(a), the mesosegregation formation starts at about 450 seconds. The solid fractions do not evolve in a similar way for the two points. A small extent of local remelting occurs at the point B, and the corresponding constitutional superheat is found (figure 6(b)). The local remelting is because that the speed of flow is greater than the speed of the moving isotherm along the X direction (figure 6(c)) [3]. Both carbon and sulphur partition to the interdendritic liquid and make the liquid lighter. Therefore, the convection is enhanced not only by including the element sulphur, but also by increasing the initial concentration of sulphur according to the calculation of the buoyancy term in equation (6). Such stronger solute-rich interdendritic fluids deflect ahead of the solidification front at the top section of casting. Because the temperature of them rises by moving from the colder to the hotter, the local remelting accompanies. So, the formation of mesosegregation starts. Furthermore, as the localized melting of solid contributes greatly to the increase of the permeability, a stronger flow makes the mesosegregation more severe.

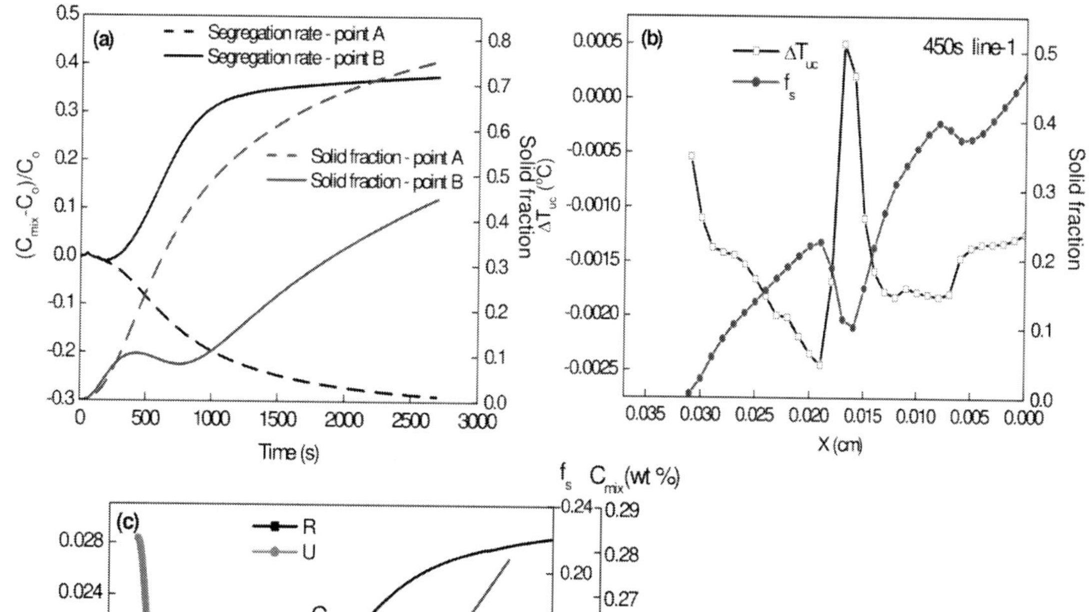

Figure 6. For S =0.1 wt pct, (a) evolution of segregation ratio and solid fraction in point A and point B; (b) variations of ΔT_{uc} and solid fraction along line-1 at 450s; (c) evolution of solidification rate, X component of flow vector, solid fraction and mixture concentration in point B.

The multicomponent thermosolutal convection pattern provides an admirable mechanism for instabilities, such as the small perturbation in the isofraction of solid and the local remelting. From figure 7 and figure 8, at the top section, a reversal flow that is perpendicular to the isofraction of solid, leaves the mushy zone and turns back downward, is the key factor in determining the two kinds of destabilization.

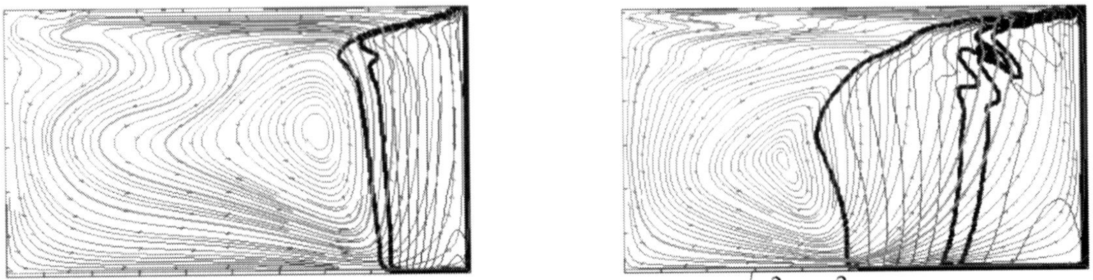

(a) $\sqrt{v^2 + u^2}\,|_{max} = 0.032\text{cm/s}$ (b) $\sqrt{v^2 + u^2}\,|_{max} = 0.0093\text{cm/s}$

Figure 7. Streamline and isofraction of solid for S=0.05 wt pct at (a) 300s and (b) 1500s

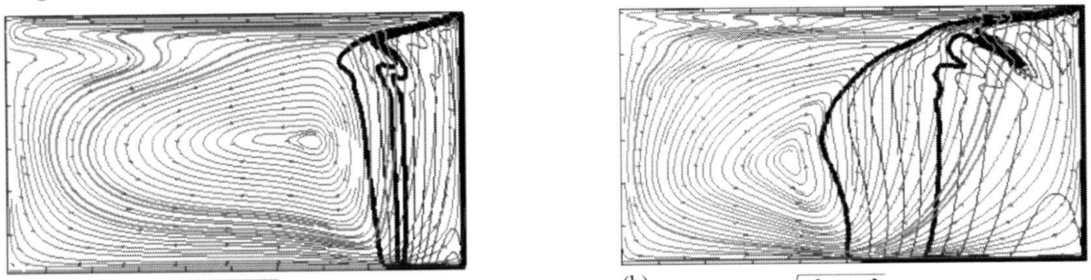

(a) $\sqrt{v^2 + u^2}\,|_{max} = 0.035\text{cm/s}$ (b) $\sqrt{v^2 + u^2}\,|_{max} = 0.0098\text{cm/s}$

Figure 8. Streamline and isofraction of solid for S=0.1 wt pct at (a) 300s and (b) 1500s.

4.2 Solidification of a binary Fe-0.21 wt pct C alloy

Figure 9 illustrates the final macrosegregation map. No mesosegregation is noted. The result depends on the evolution of the flow field as solidification proceeds. An indication of this phenomenon appears in the streamline pattern (figure 10).

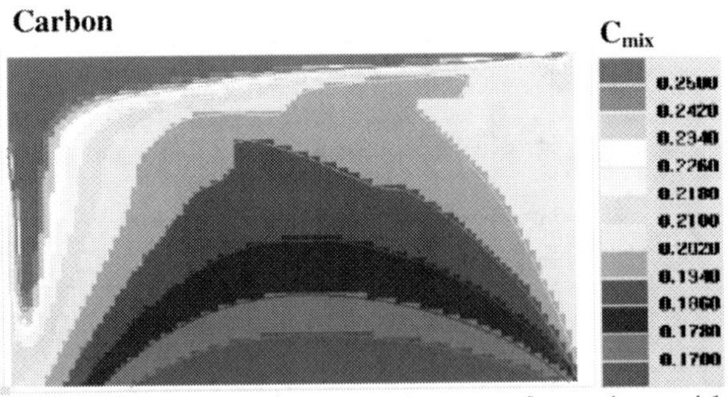

Figure 9. Final macrosegregation map for carbon with the initial sulphur concentration S=0.0.

At 300 seconds, in the top right picture, a bump on the isofraction of solid is noted, which corresponds to the position where the flow lines deviate inwards. It is primarily because of the local variations of growth rate which accompany the solutal fluctuations in the liquid. At 1500 seconds, the morphology of the solid isofraction does not change very much. As the inward deviation degree of flow is not sufficiently large, the perturbations from the interdendritic flow to the solidification front do not develop. Therefore, the mesosegregation stemming from the instabilities of solidification front or the local remelting does not exist. The overall macrosegregation pattern is simple: positive segregation in the top and central regions, and negative segregation in the bottom and right area.

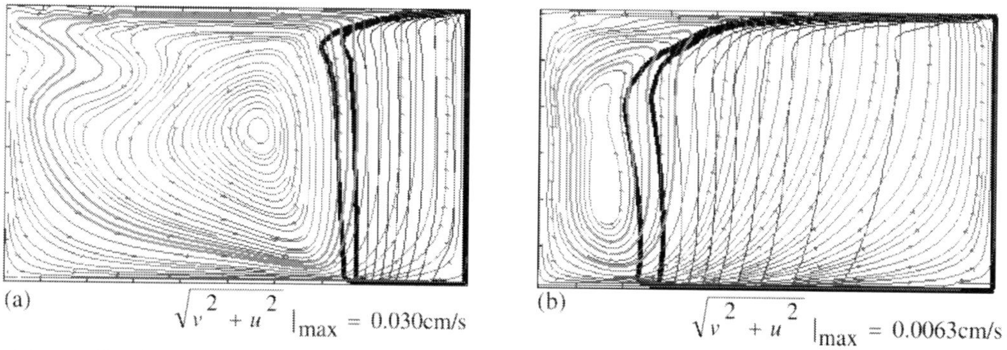

Figure 10. Streamline and isofraction of solid for S=0.0 wt pct at (a) 300s and (b) 1500s.

5. Conclusions

1. For the solidification of Fe-0.21C-0.05S (wt pct) alloy, the constitutional undercooling, instead of the constitutional superheat, occurs. Moreover, the X-component of the interdendritic liquid velocity is slower than the solidification rate. Then, the mesosegregation formation is a result of instabilities in the mushy zone growth. The local remelting is not a necessary condition.

2. By increasing the initial concentration of suphur from 0.05wt pct to 0.1wt pct, the magnitude of the multicomponent thermosolutal convection is raised. A stronger reversal flow whose X-component is faster than the solidification rate of isotherm results in the local remelting. The occurrence of the constitutional superheat corresponds to the oscillation in the isofraction of solid. So, the formation of mesosegregation is accompanied by a local remelting.

3. No mesosegregates are found for Fe-0.21C (wt pct) alloy because of the weak degree of the reversal flow from the mush to the bulk liquid. This indicates the importance of incorporating multi-elements for the macrosegregation prediction.

4. The fair agreement in the comparison with reference [7] shows the reliability of the present numerical scheme.

Acknowledge

The authors gratefully acknowledge the financial supports from the National Natural Science Foundation of China (Grant No. 51101045), from the Natural Science Foundation of HeiLongJiang Providence (Grant No. E201108) and from the Main Direction Program of Knowledge Innovation of Chinese Academy of Sciences (Grant No. KGCX2-YW-221).

References

[1] Gu J P and Beckermann C 1999 *Metall. Mater. Trans.* A**30** 1357-66
[2] Lesoult G 2005 *Mater. Sci. Eng.* A**413-414** 19-29
[3] Flemings M C 2000 *ISIJ Int.***40** 833-41
[4] Založnik M and Combeau H 2010 *Int. J. Therm. Sci.***49** 1500-09
[5] Rady M A, Satyamurty V V and Mohanty A K 1997 *Metall. Mater. Trans. B***28** 943-52
[6] Combeau H, Založnik M, Hans S and Richy P E 2009 *Metall. Mater. Trans. B***40** 289-304
[7] Založnik M, Kumar A and Combeau H 2010 *Comput. Mater. Sci.***48** 11-21
[8] Guillemot G, GandinCh- A and Combeau H 2006 *ISIJ Int.***46** 880-95
[9] Pardeshi R, Voller V R, Singh A K and Dutta P 2008 *Int. J. Heat Mass Transfer***46** 1115-37
[10] Kumar A, Dussoubs B, Založnik M and Combeau H 2009 *J. Phys. D: Appl. Phys.***42** 1-13
[11] Schneider M C and Beckermann C 1995 *Metall. Mater. Trans.* A **26** 2373-88

IOP Conf. Series: Materials Science and Engineering **33** (2012) 012095 doi:10.1088/1757-899X/33/1/012095

Multi-scale needle-network model of complex dendritic microstructure formation

Damien Tourret and Alain Karma

Department of Physics and Center for Interdisciplinary Research on Complex Systems, Northeastern University, Boston, MA 02115, USA

E-mail: d.tourret@neu.edu, a.karma@neu.edu

Abstract. We present a novel multi-scale Dendritic Needle Network (DNN) approach in order to model well-developed highly-ramified dendritic microstructures on the coarser scale of several crystal grains while retaining a faithful quantitative description of the transient dynamics of individual dendritic branches. This approach is intended to bridge the scale gap between phase-field and cellular automaton methods. The dynamics of each needle-like branch, characterized by its tip velocity V and radius ρ, is fixed by two conditions: (i) on the inner tip scale, a standard microscopic solvability condition relates ρ^2V to the strength of surface tension anisotropy, and (ii) on the outer diffusion length scale, a flux balance condition relates the product ρV^2 to a flux intensity factor extracted from a contour integral analogous to the J-integral of fracture mechanics. The method is tested for low supersaturation and reproduces the analytical solutions for both early stage and steady state growth dynamics. The results are directly compared with a quantitative phase-field simulation for an experimentally relevant supersaturation. We present as well an illustrative simulation for highly branched polycrystalline growth. This model should permit to investigate the macroscale grain evolution through the dynamics of individual primary dendrites and higher-order branches, controlled by both the intragrain history-dependent selection and the intergrain dendrite interactions.

1. Introduction

Complex dendritic microstructures arise from the subtle balance between microscopic interfacial phenomena and macroscopic thermal and solute transport [1]. To date, a gap remains between phase-field methods [2], which model primarily dendritic growth on the scale of primary and secondary branches, and coarse-grained models designed to model the dynamics of the "envelope" of an entire dendritic grain. This gap stems from the fact that existing coarse-grained models, such as mesoscopic models [3, 4] and cellular-automaton-finite-element (CA-FE) models [5, 6], do not resolve the interactions between higher order (secondary and tertiary) branches within a grain or between grains. Yet those interactions can be important both for determining the selection of the microstructure within a grain and for shaping the boundaries between grains.

We present a novel multi-scale dendritic needle network (DNN) approach that is designed to bridge the scale of individual dendrite branches with the grain scale while keeping track of the diffusive interactions between branches. This is achieved by treating each branch of the dendritic tree-like network as a thin needle crystal. The dynamics of each needle is prescribed by two conditions that jointly determine the needle tip velocity V and tip radius ρ. The first is a standard microscopic solvability condition relating ρ^2V to the strength of surface tension anisotropy. The second is a flux balance condition relating ρV^2 to the intensity of the diffusive flux at the tip on the diffusion scale. The latter is achieved with the introduction of a "flux intensity factor" (FIF), which can be calculated in 2D using a contour-integral around the needle tip that is directly analogous to the J-integral of fracture mechanics for mode III anti-plane shear. Importantly, the description of the needle dynamics remains quantitatively valid as long as the dendrite tip Péclet number is small, which is the case for many processing conditions.

In this paper, we present the theory of the DNN approach for 2D isothermal binary alloy solidification. We show the capability of this approach to reproduce quantitatively analytical solutions for both early stage and steady-state growth regimes. For an experimentally relevant supersaturation of 0.1, we compare the results of the DNN approach with a quantitative phase-field simulation of an equiaxed dendrite growth. Eventually, we illustrate the potential of the model to simulate the growth of highly branched dendritic structures and the grain interactions.

2. Dendritic Needle Network model

2.1. Needle dynamics.

The growth conditions, i.e. the tip radius ρ and velocity V, of a dendritic branch result from the competition between the large length scale diffusion flux at the dendrite tip and, at the much smaller length scale of the tip, the surface tension effects, inducing a diffusion flux along the interface towards the flatter regions behind the tip. In the model proposed here, the dendritic branches and their side branches are treated as "needle crystals" [7]. With $c(x,y)$ the solute concentration field, we define the solute supersaturation $u(x,y) = (c_l^* - c(x,y))/(c_l^* - c_s^*)$ where c_l^* (c_s^*) is the equilibrium concentration at the interface in the liquid (solid). The solid/liquid interface is assumed to be in local equilibrium with $u = d_0\kappa$ on the interface, where d_0 is the capillarity length, κ is the interface curvature, and u obeying the diffusion equation

$$\partial_t u = D\nabla^2 u \tag{1}$$

in the melt, with D the diffusion coefficient. The dynamics of a needle crystal is established by coupling solute balances of the long-range diffusion field with a solvability condition at the scale of the tip, as illustrated in figure 1.

On a large scale compared to ρ (figure 1b), the parabolic dendrite appears sharp and the surface tension effects are negligible. Close to the tip, the quasi-instantaneous diffusion can thus be assumed to follow Laplace equation $\nabla^2 u = 0$, whose solution near the tip of a sharp needle has a square root singularity of the form $|\vec{\nabla}u(\vec{r})| \sim 1/\sqrt{|\vec{r}|}$, if $|\vec{r}|$ is the distance to the tip. Hence, if x is the needle growth direction and $x = y = 0$ is the tip location (i.e. the distance to the tip is $-x$ at $y = 0$), the concentration gradient normal to the tip has the form

$$\partial_y u|_{y=0} = \mathcal{F}/\sqrt{-d_0 x} \quad \text{when} \quad x \to 0 \quad , \tag{2}$$

where $\mathcal{F}$ is a dimensionless measure of the solute flux feeding the tip region, i.e. $\int_0^{-a} \partial_y u|_{y=0}\, dx \approx 2\mathcal{F}\sqrt{a/d_0}$ if a is a short distance behind the tip. Since the region behind the tip is essentially planar on a scale large compared to ρ, mass conservation imposes that the interface velocity $dy_i(t)/dt$ is given by the flux $D\,\partial_y u$ at the interface position $y_i(t)$. Yet, for a small supersaturation $\Omega \ll 1$, the interface, which moves $\sim \Omega$ times slower than the diffusion field, may be assumed stationary at its origin and the evolution of the interface can be computed as

$$dy_i(t)/dt = D\,\partial u/\partial y|_{y=0} \quad . \tag{3}$$

Changing time t by $-x/V$ and integrating leads to the equation of the interface $y_i(x) = 2\mathcal{F}(D/V)\sqrt{-x/d_0}$, which implies that the parabolic tip radius is $\rho = 2(\mathcal{F}D/V)^2/d_0$. Then the product ρV^2 is defined as a function of $\mathcal{F}$ as

$$\rho V^2 = 2D^2\mathcal{F}^2/d_0 \tag{4}$$

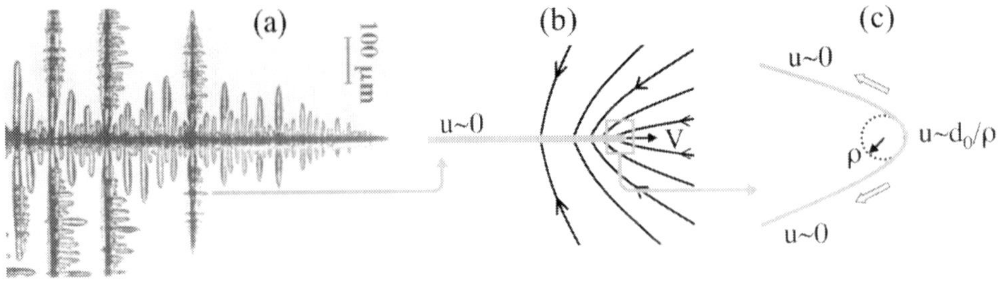

Figure 1. Schematic illustration of the "dendritic needle network" approach. The physical dendritic network, illustrated here in (a) for an ammonium chloride crystal grown from a supersaturated solution [8], is represented as a "skeleton" of thin needles. Each needle in the network is treated mathematically as a line segment as shown in (b). The diffusion equation is solved in the entire domain with the boundary condition $u = 0$ on each needle. This boundary condition follows from the fact that surface tension effects are negligible on the outer diffusion scale, so that each needle coincides with an isoconcentration line. From the solution of the diffusion field in the vicinity of each needle, one can define a "flux intensity factor" (FIF) computed from a contour integral around the tip, which fixes the product ρV^2 as described in the text. On the scale of the needle tip radius ρ shown in (c), the solid-liquid interface is not tracked explicitly in the multi-scale DDN approach. However, the dynamics of the interface on this scale is captured quantitatively through a solvability condition for the existence of a growing solution that relates the additional product $\rho^2 V$ to surface tension anisotropy, thereby fixing independently V and ρ [7]. This approach exploits the fact that, while the flux intensity factor can change in time during transient growth conditions due to inter-branch interactions, the product $\rho^2 V$ determined on the smaller scale ρ remains constant in time.

On the tip radius scale (figure 1c), capillarity effects need to be considered. The shape of the dendrite interface results from the competition between the diffusion flux and the surface tension induced flux. Since the supersaturation ahead of the curved tip is $u_{tip} \approx d_0/\rho$, the surface diffusion flux towards flatter regions with $u \approx 0$ scales as $D\nabla u \approx D\, d_0/\rho^2$. This flux must balance the solute flux at the tip, which scales as V. Hence, a condition exists which fixes the product $\rho^2 V$ as

$$\rho^2 V \;=\; 2Dd_0/\sigma \quad . \tag{5}$$

This solvability condition [9, 10, 11] has been extensively validated by phase-field simulations (e.g. [12, 13]). Thus, the needle dynamics is completely defined by the conditions (4) and (5).

2.2. Computation of the Flux Intensity Factor

The flux intensity factor (FIF), $\mathcal{F}(t)$, defined in equation (2) measures the amplitude of the diffusion field discontinuity at the tip. In 2D, this FIF may be calculated with a contour integral identical to the J-integral used in fracture mechanics to estimate the stress intensity factor (SIF) at the tip of a crack [14]. Indeed, under pure anti-plane shear loading, i.e. mode III, linear elasticity gives that the displacement field in the out-of-plane direction, w, follows $\nabla^2 w = 0$. If μ is the shear modulus, the SIF in mode III is $K_{\rm III} \equiv \lim_{x\to 0}\left\{\mu\partial_x w\sqrt{-2\pi x}\right\}$, which may be calculated exactly by a path-independent contour integral along an open contour Γ surrounding the tip [14]. Hence, if (n_x, n_y) are the coordinates of the normal $\mathbf{n}$ to the integration contour Γ, as illustated in figure 2a,

$$K_{\rm III}^2 = \mu^2 \int_\Gamma \left[\left((\partial_y w)^2 - (\partial_x w)^2\right) n_x + 2\,\partial_x w \partial_y w\, n_y\right] \mathrm{d}\Gamma \quad . \tag{6}$$

Similarly, close to the tip of a needle-like crystal, solute diffusion is very fast and the diffusion equation reduces to $\nabla^2 u = 0$ at this scale. As illustrated in figure 2, if we let $\partial_y u = -\partial_x w$ and $\partial_x u = \partial_y w$, the Laplace diffusion problem maps exactly to the mode III fracture case. Hence, the FIF defined in equation (2) is similar to $K_{\rm III}\sqrt{d_0}/(\mu\sqrt{2\pi})$ and is exactly calculated as

$$\mathcal{F}^2 = \frac{d_0}{2\pi}\int_\Gamma \left[\left((\partial_x u)^2 - (\partial_y u)^2\right)n_x - 2\,\partial_x u \partial_y u\, n_y\right] d\Gamma \quad . \tag{7}$$

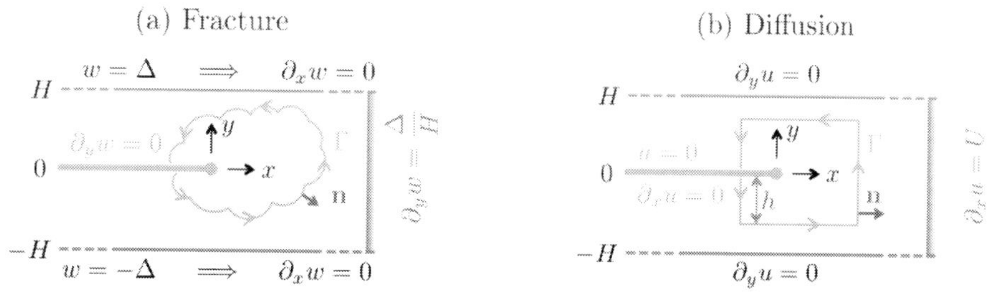

Figure 2. Similarity between (a) the linear elastic displacement at the tip of a crack under anti-plane shear loading and (b) the laplacian diffusion field at the tip of a needle crystal. In an infinite strip of width $2H$, the mode III loading (a) with no stress on surface of a semi-inifinite crack, i.e. $\partial_y w = 0$, imposed displacements $w = \pm\Delta$ at the top/bottom boundaries $y = \pm H$, resulting in $\partial_y w = \Delta/H$ at $x \to \infty$, is similar to the Laplace diffusion (b) around a needle tip where $u = 0$, with no flux at $y = \pm H$ and an imposed flux $\partial_x u = U$ at $x \to \infty$. In these cases, the stress/flux intensity factors are (a) $K_{\rm III}^2 = 2\,\mu^2\,\Delta^2/H$ and (b) $\mathcal{F}^2 = HU^2 d_0/\pi$.

2.3. Implementation

The DNN model simply consists in a diffusion equation (1) coupled with a set of discontinuities ($u = 0$) within the diffusion field, which grow according to equations (4) and (5). This is solved using an explicit finite difference scheme on a grid of uniform spacing Δx. The diffusion part is implemented on parallel graphic processing units (GPU), while the needle growth dynamics is computed without advanced parallelization. After a diffusion step, the contour integral (7) is evaluated at the tip of each needle, which then grows with its prescribed velocity, and u is set to zero on the needles. When a tip is located in between two grid points, the value of u at the point ahead of the tip is estimated through the Laplace equation $\nabla^2 u = 0$. The thickening of each needle behind the tip is integrated over time with the normal planar approximation (3). If considered, side branching events occur at a distance $N \times \rho$ behind the tip and the initial length of the side needle is set as the integrated width of the parent needle at this position.

3. Quantitative validations

3.1. Computation of the flux intensity factor

We validate the calculation of the FIF through the contour integral (7) by simulating the test case illustrated in figure 2b with a square integration contour of side $2h$ centered on the needle tip. In figure 3, we plot the relative error on $\mathcal{F}$ for different contour sizes in three different configurations. The error decreases like $\approx (h/\Delta x)^{-2.5}$. Moreover, even for the smallest contours, the error is limited to a few percents (less than 4% for $h = 2\Delta x$). Similar results were obtained with different rectangular and not-centered contours, as well when reducing the mesh size instead of increasing the size of the contour.

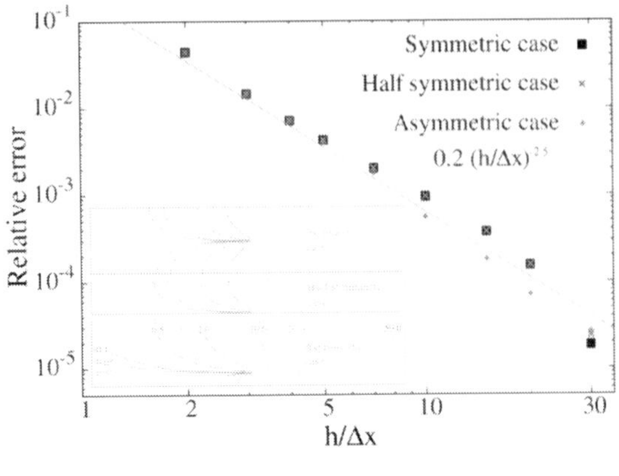

Figure 3. Relative error $\epsilon = |(\mathcal{F} - \mathcal{F}_{th.})/\mathcal{F}_{th.}|$ on the flux intensity factor $\mathcal{F}$ vs. $h/\Delta x$ for the case depicted in figure 2b, with $H = 30$, $\Delta x = 1$, $D = 1$, $d_0 = 1$ and $U = \sqrt{\pi/H} \approx 0.3236$, so that $\mathcal{F}_{th.} = 1$ (equation (7)). Insets show the tip region for the three different configurations: a needle centered in between the top and bottom boundaries; a symmetric half-domain of height H with the needle at the bottom; and a case where the needle is not vertically centered.

3.2. From early stage to steady state growth: bridging length scales

At the early stage of solidification, a crystal seed is fed by a circular region of liquid of radius $R(t) \sim \sqrt{Dt}$. For a low Ω, a cross seed grows slower than $R(t)$ and its length follows a power law $L(t) \sim t^\alpha$ with $\alpha = 3/5$ [7, 15], which was confirmed by simulations [13]. Later, once the needles catch up and grow outside the circular diffusion field, the dendritic tip radius and velocity tend to their constant steady state values ρ_s and V_s. Using the solution of the 1D diffusion problem behind the tip where the interface is essentially planar together with equations (2) to (5), one can show that the steady state operating state of the tip when $\Omega \ll 1$ is defined by [7]

$$\rho_s = \pi d_0/(\Omega^2\sigma) \quad , \tag{8}$$

$$V_s = 2D\sigma\Omega^4/(\pi^2 d_0) \quad . \tag{9}$$

In order to validate the DNN approach, we ran several tests for a growing cross-like seed at low supersaturation, e.g. for a Péclet number of $Pe_s \equiv \rho_s V_s/(2D) \approx 10^{-3}$. For small mesh sizes $\Delta x \ll \rho_s$, the early stage transient growth is reproduced with a power law exponent $\alpha \approx 3/5$. For a small integration contour ($h = 3\Delta x$), the actual computed exponent is closer to 0.627, which remains reasonably accurate. For a higher mesh size $\Delta x \sim \rho_s$, since the initial length of the needles necessary to apply the contour integral for computing $\mathcal{F}(t)$ is $L_0 = 3\Delta x$, the early stage transient is not resolved but the DNN model resolves well the evolution towards the steady state growth regime (e.g. ρ_s deviates from the definition (8) by less than 1.6% for $\Delta x = \rho_s$).

3.3. Comparison of DNN and phase-field methods

We compare the results of the DNN model to a quantitative phase-field (PF) simulation [16, 17] for a supersaturation of $\Omega = 0.1$. The PF model is described in [17] with an additional non-linear preconditioning, $\phi = \tanh(\psi/\sqrt{2})$, in order to improve its stability at higher mesh sizes [18]. A simulation of early stage equiaxed growth was performed with both models with the same domain and mesh sizes. The GPU-parallelized PF simulation was achieved in 400 minutes on a Nvidia Tesla C2050, while the partially parallelized DNN simulation took about 100 minutes The results of the simulations are compared in figure 4. Additionally, figure 4a,b show the results of a second DNN simulation with a twice higher mesh size, which lasted less than 12 minutes.

Figure 4a shows the evolution of the tip velocity and parabolic tip radius. We attribute the deviation between the DNN and the PF results ($\approx 15\%$ on ρ and $\approx 24\%$ on V) to the fact that we are approaching the limit of validity of the sharp needle approach. The DNN model is valid for low Ω when the steady state is reasonably well defined by equations (8) and (9) and the Péclet number is $\approx \Omega^2/\pi$. For $\Omega = 0.1$, the Ivantsov function, $\Omega = \sqrt{\pi Pe}\exp(Pe)\mathrm{erfc}(\sqrt{Pe})$ [20], yields that $Pe \approx 0.00364$ which sensibly differs from $\Omega^2/\pi \approx 0.00318$. Hence, this underestimation of Pe leads to the observed deviation on the tip radius, which scales like Pe^{-1} and the tip velocity,

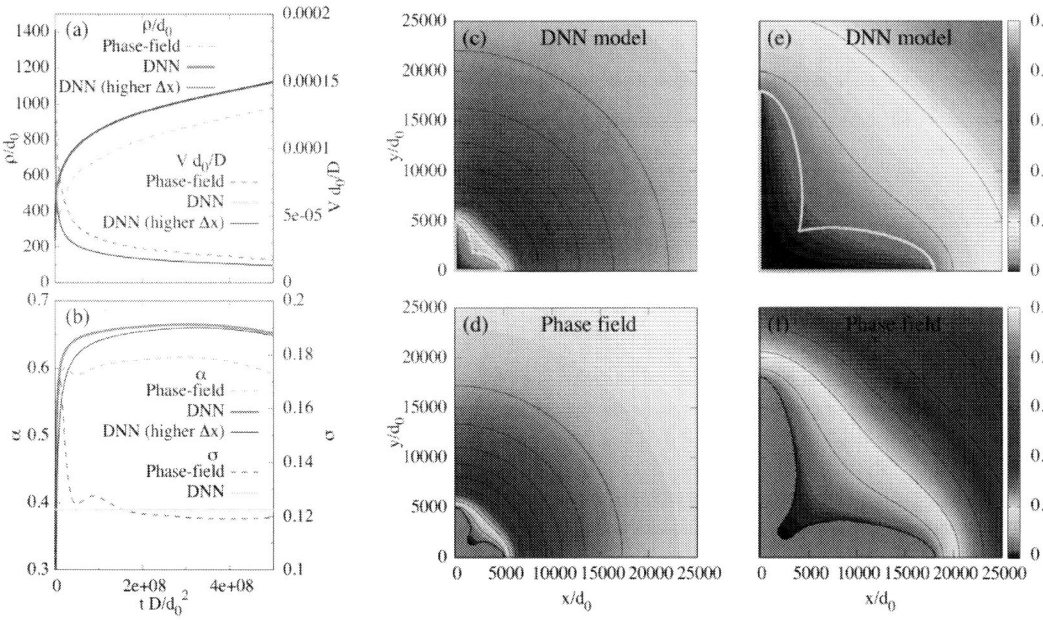

Figure 4. Results of the DNN model compared to a quantitative phase-field (PF) simulation of a free equiaxed crystal growth for an initial supersaturation $\Omega_0 = 0.1$. Figure (a) shows the time evolution of the horizontal needle tip velocity and the parabolic tip radius. For the PF simulation, it is approximated as $\rho = \rho_{tip}/(1 - 15\epsilon_4)$ [19], where ρ_{tip} is the effective tip radius from PF including a localized shape distortion near the tip [19]. Figure (b) shows the exponent of the needle length power law $L(t) \sim t^\alpha$ computed as $\alpha = d(\ln L)/d(\ln t)$ and the parabolic tip selection constant σ extracted from the PF results from the same approximation as the radius, which yields $\sigma = \sigma_{tip}(1 - 15\epsilon_4)^2$ where $\sigma_{tip} = 2Dd_0/(\rho_{tip}(t)^2 V(t))$ from the PF simulation. Figures c-f show the supersaturation $u(x, y) \equiv (c_l^* - c(x, y))/(c_l^* - c_s^*)$ (color map) with isovalues $u = 0.001$ and every 0.01 from $u = 0.01$ to $u = 0.1$ (thin black lines) when the needle length reaches $5000d_0$ (c,d) and $18000d_0$ (e,f) in the PF (d,f) and the DNN results (c,e), for which the needles are drawn as green lines and the integrated solid/liquid interface appears as a cyan line. The domain size is $\approx (41000d_0)^2$ and the mesh size $\Delta x = 40d_0$, except for the second DNN run labeled "higher Δx" (black lines in a, b) for which $\Delta x = 80d_0$. The radius (length) of the initial circular (cross) seed is $3\Delta x$. The PF simulation is performed with $W = 40d_0$, $k = 0.15$ and $\epsilon_4 = 0.02$ [17]. The DNN calculations are achieved with $h = 3\Delta x$ and $\sigma = 0.1225$.

which scales like Pe^2. With a twice higher mesh size, the DNN results show no further deviation, which indicates that a similar accuracy may be retained with even higher values of Δx.

Figure 4b shows the evolution of the needle length power law exponent, $\alpha = d(\ln L)/d(\ln t)$, to compare with the analytical value of 3/5 introduced previously [7, 13, 15]. While the PF value is very close to 0.6, the value from the DNN model is closer to 0.65. The final decrease is due to the influence of the domain boundaries on the diffusion field. We observed that α predicted by the DNN model gets closer to 3/5 as Ω decreases, which further confirms that the deviation is mostly due to Ω getting close to the limit of validity of the current approach. On the same graph, we show the value of the parabolic tip selection constant σ. Its value, which depends only on the anisotropy strength [10], is an input of the DNN model, while the PF simulation rapidly converges to a value $\sigma \approx 0.12$, as expected with a one-sided model for $\epsilon_4 = 0.02$ [10].

Figure 4c-f show the solid/liquid interface and the supersaturation field u in the liquid when the length of the needles reaches $L = 5000d_0$ (c,d) and $L = 18000d_0$ (e,f). Since the DNN model relatively underestimates the tip velocity, these lengths are reached later and the diffusion field, growing as $\sqrt{Dt}$, is expanded wider. The needles remaining sharp in the DNN model, isovalues of u appear within the solid in figure 4c,e. Even though the complex interface pattern is not resolved, the integrated DNN interface shows a reasonable agreement with the PF results.

4. First applications to highly branched structures

We tested the capability of the model to predict branched dendritic structures on a wider range of supersaturations. Several experimentally relevant behaviors were reproduced, such as the confinement of the side branches within the diffusive zone at low supersaturations. When increasing Ω, the DNN results show a competition/selection of secondary spacing together with an increase of the internal solid fraction within the grain envelope, which tends toward a wedge-shaped envelope when the velocity of the larger side branches competes with that of the main tip. These trends and grain shapes are in good agreement with experiments [21]. We present in figure 5 an illustrative simulation of the free growth of five equiaxed seeds in a domain of initial supersaturation $\Omega_0 = 0.6$. Even though the model was established for a low Péclet number, this type of analysis provides precious information about the grain dynamics and their interaction, such as the location of grains boundaries, as shown in figure 5d (even though, having the same orientation, the crystals cannot be rigorously regarded as different "grains").

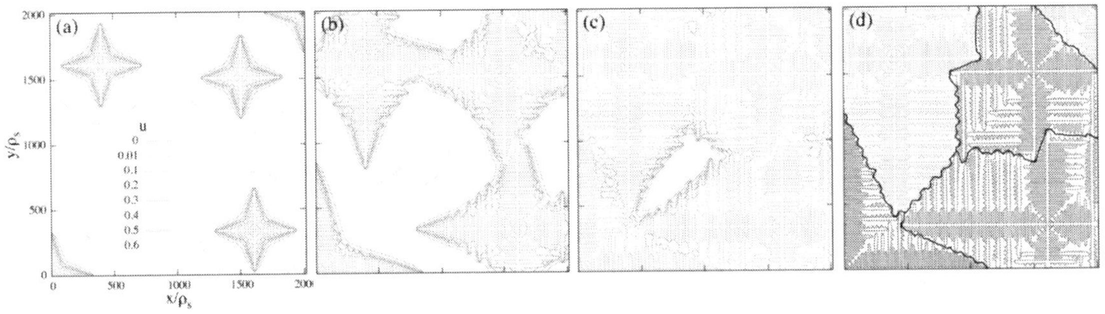

Figure 5. Growth of five equiaxed seeds in a $(2016\,\rho_s)^2$ domain of initial supersaturation $\Omega_0 = 0.6$, with $\Delta x = \rho_s$, $L_0 = 3\Delta x$, $h = 3\Delta x$. Side branching events are set to occur every $N = 20$ times the current tip radius. The evolution of the microstructure is shown at $t\,V_s/\rho_s \approx$ (a) 570, (b) 1530, (c) 2480 and (d) $t\,V_s/\rho_s > 3500$. Contour lines show the isovalues of the supersaturation u. For illustration purposes, u is set to zero on the entire width of the needles. The final microstructure is illustrated in (f) with colored grains and grain boundaries.

5. Conclusion

We have proposed a new DNN (dendritic needle network) approach to simulate the evolution of highly ramified dendritic structures. This approach is based on representing the dendritic network as an ensemble of thin needles and tracking the network evolution on the grain scale by formulating two conditions that rigorously prescribe the growth rate V of each needle in the limit where the tip radius ρ is much smaller than the solute diffusion length. The first condition relates the product ρV^2 to a flux intensity factor (FIF), which is a quantitative measure of the strength of the diffusion flux at the tip of a needle, as viewed on the outer scale of the grain. We have shown that this FIF at the tip of a needle is mathematically analogous to the stress intensity factor at the tip of a crack in fracture mechanics, and that it may be calculated using a contour integral. The second condition is a standard microscopic solvability condition that relates the product $\rho^2 V$ to the surface tension anisotropy. Jointly, those two conditions uniquely prescribe ρ and V and hence the needle dynamics. The unique scale-bridging advantage of this approach is that it describes the evolution of the entire dendritic network and inter-branch interactions on the grain scale without needing to track the solid-liquid interface dynamics on the tip scale ρ, as in a phase-field simulation. The method can therefore describe the dendritic network evolution on the grain scale while retaining a complete description of the interactions between primary and higher-order branches, in the early stage transient regime as well as in the steady state.

We have confirmed here the feasibility and the quantitative validity of the DNN approach for the isothermal solidification of a binary alloy in 2D at low supersaturations. We compared the results of the DNN approach to a benchmark phase-field simulation for an experimentally relevant supersaturation and found a reasonable agreement, with quantitative differences that stem from the assumed vanishing needle thickness in the DNN approach. Further validation of the method will involve direct comparisons of secondary branch dynamics with detailed experimental measurements [21]. The extension of this approach to non-isothermal columnar growth should provide a powerful tool to study both the intra-grain and inter-grain evolution of dendritic networks, including misoriented crystal grains interactions in directional solidification experiments [22], where high order branches play a key role in the dynamical selection of the primary spacing. A more challenging problem being pursued is the extension of the method to 3D, which requires to formulate an analog of the FIF for a thin paraboloidal needle tip region of finite thickness. Another interesting extension is the incorporation of fluid flow.

In summary, the DNN approach provides a novel computationally efficient method to model the evolution of highly ramified dendritic networks that generally form in sufficiently concentrated alloys under slow growth conditions. Further validation of the approach and its extension to non-isothermal growth and three-dimensions remain needed to make the DNN approach a useful tool for predicting microstructure evolution in industrial processes.

Acknowledgments

This research was supported by NASA under award No. NNX11AC09G.

References

[1] Dantzig JA and Rappaz M 2009 *Solidification* (Switzerland: EPFL Press)
[2] Boettinger WJ, Warren JA, Beckermann C and Karma A 2002 *Annu. Rev. Mater. Res.* **32** 163
[3] Steinbach I, Beckermann C, Kauerauf B, Li Q and Guo J 1999 *Acta Mater.* **47** 971
[4] Steinbach I, Diepers H-J and Beckermann C 2005 *J. Cryst. Growth* **275** 624
[5] Rappaz M and Gandin Ch-A 1993 *Acta Metall. Mater.* **41** 345
[6] Gandin Ch-A 2010 *C. R. Physique* **11** 216
[7] Karma A 2001 in *Branching in Nature*, Les Houches, vol 13, ed V Fleury, JF Gouyet and M Lonetti (EDP Sciences, Springer) chapter XI, p 365
[8] Chan SK, Reimer HH and Kahlweit M 1976 *J. Cryst. Growth* **32** 303
[9] Langer JS 1987 in *Chance and Matter, Lectures on the Theory of Pattern Formation*, Les Houches, Session XLVI, ed J Souletie, J Vannimenus and R Stora (Amsterdam: North Holland) p 629
[10] Barbieri A and Langer JS 1989 *Phys. Rev.* A **39** 5314
[11] Ben Amar M and Brener E 1993 *Phys. Rev. Lett.* **71** 589
[12] Provatas N, Goldenfeld N and Dantzig JA 1998 *Phys. Rev. Lett.* **80** 3308
[13] Plapp M and Karma A 2000 *Phys. Rev. Lett.* **84** 1740
[14] Rice JR 1968 *J. Appl. Mech.* **35** 379
[15] Almgren R, Dai WS and Hakim V 1993 *Phys. Rev. Lett.* **71** 3461
[16] Karma A and Rappel WJ 1998 *Phys. Rev.* E **57** 4323
[17] Karma A 2001 *Phys. Rev. Lett.* **87** 115701
[18] Glasner K 2001 *J. Comput. Phys.* **174** 695
[19] Karma A, Lee YH and Plapp M 2000 *Phys. Rev.* E **61** 3996
[20] Kurz W and Fisher DJ 1989 *Fundamentals of Solidification* (Switzerland: Trans Tech)
[21] Couder Y, Maurer J, González-Cinca R, and Hernández-Machado A 2005 *Phys. Rev.* E **71** 031602
[22] Gandin Ch-A, Eshelmann M and Trivedi R 1996 *Metall. Mater. Trans.* A **27A** 2727

IOP Conf. Series: Materials Science and Engineering **33** (2012) 012096 doi:10.1088/1757-899X/33/1/012096

Quantitative phase-field model for dendritic growth with two-sided diffusion

S Y Pan and M F Zhu

Jiangsu Key Lab for Advanced Metallic Materials, Southeast University, China

E-mail: zhumf@seu.edu.cn

Abstract. A quantitative phase-field (PF) model with an anti-trapping current (ATC) is developed to simulate the dendritic growth with two-sided diffusion. The asymptotic analysis is performed at the second-order for the PF equations coupled with nonlinear thermodynamic properties and an ATC term under the equal chemical potential condition. The PF mobility and ATC are derived based on the asymptotic analysis in the thin interface limit, and the solute drag model. Then the model is reduced to the dilute solution limit for dendrite solidification in binary alloys. The test of convergence with respect to the interface width exhibits an excellent convergent behavior of the proposed model. The performance of the model is then validated by the comparison of PF simulations to the predictions of the Gibbs-Thomson relation, and of the linearized solvability theory, for the isothermal dendritic growth of an Fe-1.5mol.%C alloy. The results demonstrate quantitative capabilities of the model that effectively suppresses the abnormal solute trapping effect when the interface is taken artificially wide. It is also found that the present model can quantitatively describe dendrite growth with various solid diffusivities, ranging from the case with one-sided diffusion to the symmetrical model.

1. Introduction

Phase-field (PF) method has emerged as a powerful tool to simulate microstructure evolution during solidification process [1]. Yet, the PF model in the sharp interface limit has encountered the efficiency difficulty to describe realistic microstructures on experimentally relevant length and time scales. Based on the thin interface limit, Karma and Rappel [2] developed a quantitative PF model to simulate dendritic growth of pure materials in a much larger system even at low undercooling.

For the case of alloy solidification, however, it is demonstrated that the PF model in the thin interface limit, with different diffusivities in the two phases, suffers from the three spurious effects of efficient chemical potential jump across the interface, interface stretching, and surface diffusion [3].

IOP Conf. Series: Materials Science and Engineering **33** (2012) 012096 doi:10.1088/1757-899X/33/1/012096

Karma *et al.* [4, 5] proposed a PF model with an anti-trapping current (ATC) to eliminate the abnormal effects in the case of one-sided dilute alloy solidification with zero solute diffusivity in the solid. The extension of the ATC scheme to the two-sided model, involving diffusion in the solid, was initially attempted by Gopinath *et al.* [6]. The ATC, however, has a singularity problem when the model is reduced to the one-sided case. Another attempt was made by Steinbach [1], and an extrapolated form of ATC was proposed. Based on the matched asymptotic analysis in the thin interface limit, Ohno and Matsuura [7] proposed a quantitative two-sided PF model with ATC for dilute alloy solidification.

This paper presents a quantitative PF model with an ATC scheme to simulate dendrite growth in binary alloys, involving asymmetrical two-sided solute diffusion. The PF mobility and ATC are derived based on the second-order asymptotic analysis in the thin interface limit, and the solute drag model [8, 9]. The asymptotic analysis is performed for the PF equations coupled with nonlinear thermodynamic properties and an ATC under the equal chemical potential condition. To validate the model against the classic theories of dendritic growth, the model is reduced to the dilute solution limit. After demonstrating the convergent behavior, the performance of the model is validated through extensive comparisons of the PF simulations to the predictions of the Gibbs-Thomson relation and of the linearized solvability theory.

2. Isotropic phase-field model with nonlinear thermodynamic properties

Based on the approach proposed by Kim [10], the PF equations coupled with nonlinear thermodynamic properties for isothermal solidification can be written as

$$M_\phi^{-1}\partial_t\phi = \mathrm{K}^2\nabla^2\phi - (8/\pi^2)\varpi f'(\phi) - g_\delta'(\phi)\Delta G_m / V_m \tag{1}$$

$$\partial_t u = \nabla\cdot(M_\delta(\phi)\nabla\mu_m / V_m + \xi a_\delta(\phi)(u_l - u_s)\partial_t\phi(\nabla\phi/|\nabla\phi|)) \tag{2}$$

where ϕ is the PF variable, M_ϕ is the PF mobility, $\nabla=\{\partial_x, \partial_y\}$ is the gradient operator for two-dimensions (2D), and ∂_t is the time derivative. $\Delta G_m = \{g^s_m(u_s)-g^l_m(u_l)-\mu_m(u_s-u_l)\}$ is the chemical driving force for isothermal solidification, $g^s_m(u_s)$ and $g^l_m(u_l)$ are the mole free energies of the solid and liquid phases, V_m is the mole volume, K is the gradient energy coefficient, ϖ is the potential height, $f(\phi)=(\pi^2/8)\phi(1-\phi)$ is a parabolic potential with double obstacles, $u=u_s(1+h_\delta(\phi))/2+u_l(1-h_\delta(\phi))/2$ is the concentration, $M_\delta(\phi)=M_l q_\delta(\phi)$ $(q_\delta(\phi)=M_s/M_l\ (1+m_\delta(\phi))/2+\ (1-m_\delta(\phi))/2)$ is the diffusion mobility at the interface, M_l and M_s are the diffusion mobility in the liquid and solid phases, and μ_m is the chemical potential, restricted by the equal chemical potential condition at the interface: $\mu_m=\mu^s_m(\partial g^s_m/\partial u_s)=g^l_m(\partial g^l_m/\partial u_l)$. The model yields $2\xi= \pi/\sqrt{2\varpi}$ and $\sigma= \pi/8\cdot\mathrm{K}\sqrt{2\varpi}$, where 2ξ is the PF interface width and σ is the interface energy. The second term on the right-hand side of equation (2) is the solute ATC term.

To perform asymptotic analysis, the dimensionless length and time scales are chosen as $d=\sigma/g_0$ and $t_0=d^2/D$, where $D=M_l g_0$ for the case $M_l>M_s$. Thus, the dimensionless transformation can be taken as follows: $\tilde{r}=r/d_0$, $\tilde{t}=t(D/d^2)$, $\tilde{\nabla}=d\nabla$, $\partial\tilde{t}=\partial t(D/d^2)$, $\tilde{v}_n=v_n(d/D)$, $\tilde{\kappa}=d\kappa$, where v_n is the interface normal velocity, and κ is the interface curvature. The dimensionless energy scale, g_0, is determined by $g_0=\max\{\Delta G_m/V_m\}$. Thus, the dimensionless phase-field model is expressed as

$$\tau_\phi\varepsilon^2\partial_{\tilde{t}}\phi = \varepsilon^2\tilde{\nabla}^2\phi - \pi^2/8\cdot f'(\phi) - a_1\cdot\varepsilon\ g'(\phi)\cdot\Delta\tilde{G} \tag{3}$$

$$\partial_{\tilde{t}}u = \tilde{\nabla}\cdot(q_\delta(\phi)\cdot\tilde{\nabla}\tilde{\mu} + \varepsilon\cdot a_\delta(\phi)\cdot(u_l - u_s)\cdot\partial_{\tilde{t}}\phi\cdot\tilde{\nabla}\phi/|\tilde{\nabla}\phi|) \tag{4}$$

where $\Delta\tilde{G}=(\Delta G_m/V_m)/g_0$, $\tilde{\mu}=(\mu_m/V_m)/g_0$, and $a_1=\pi^2/16$.

The second-order matched asymptotic analysis for the PF equations shown in equations (3) and (4) is carried out in order to obtain the PF mobility and ATC. According to the asymptotic analysis, the spurious interface effects caused by the problem of asymmetrical two-sided diffusion with an artificial interfacial width can be suppressed by solving the following three constraints: $\int_0^{+\infty}[\Lambda_\delta(\phi_0)-\Lambda_\delta(\phi_0(+\infty))]d\eta=\int_0^{-\infty}[\Lambda_\delta(\phi_0)-\Lambda_\delta(\phi_0(-\infty))]d\eta$, where $\Lambda_\delta(\phi_0)=p_\delta(\phi_0)$, $h_\delta(\phi_0)$, and $m_\delta(\phi_0)$, respectively, and $p_\delta(\phi_0)=\{h_\delta(\phi_0)-2a_\delta(\phi_0)\cdot\partial_\eta\phi_0-1-q_\delta(\phi_0(-\infty))\cdot\chi\}/q_\delta(\phi_0)$. $g_\delta(\phi)$, $h_\delta(\phi)$, and $m_\delta(\phi)$ are the interpolation functions of the chemical energy, concentration, and diffusion mobility, respectively. They are constructed as a function of the parameter, δ, by

$$g_\delta(\phi)=\frac{1+h_\delta(\phi)}{2}=\frac{1+m_\delta(\phi)}{2}=\begin{cases}0 & for \quad 0\le\phi\le\delta \quad (1-\lambda\le\eta\le+\infty)\\ (\phi-\delta)/(1-2\delta) & for \quad \delta<\phi<1-\delta \quad (-1+\lambda<\eta<1-\lambda)\\ 1 & for \quad 1-\delta\le\phi\le1 \quad (-\infty\le\eta\le-1+\lambda)\end{cases} \tag{5}$$

The function $a_\delta(\phi)$ in equation (4) required to determine the ATC, can be calculated by

$$a_\delta(\phi)=k_{at}(\phi,\delta)(2/\pi)\sqrt{\phi(1-\phi)}(M_l-M_s)/\bar{M}(\Omega),\ (\delta<\phi<1-\delta) \tag{6}$$

where $\bar{M}(\Omega)=[(1-\Omega)(M_l)^{-1}+\Omega(M_s)^{-1}]^{-1}$, $\bar{u}_0$ is the initial concentration, and $k_{at}(\phi,\delta)=[1-h_\delta^2(\phi)]/[4\phi(1-\phi)]$, $(\delta<\phi<1-\delta)$.

By applying the solvability condition of the inner expansion of the ϕ equation at the second order, the PF mobility, M_ϕ, can be expressed as

$$\pi^2/(16\xi M_\phi)=1/M_{fric}+\xi(u_s-u_l)^2/(4\bar{M}_\delta(\Omega)) \tag{7}$$

where, $k_{KF}(\delta)=[4(1-2\delta)\sqrt{(1-\delta)\delta}+\pi(1+8\delta+8\delta^2)-32\delta\arccos(\sqrt{\delta})-4(1+8\delta^2)\arcsin(\sqrt{\delta})]/[\pi(1-2\delta)^2]$, $\bar{M}_\delta(\Omega)^{-1}=k_{KF}(\delta)\bar{M}(\Omega)^{-1}$ and M_{fric} is the interface mobility caused by the solvent atoms rearranging themselves from one lattice into another lattice [8].

According to the solute drag model [8], the rate of the total Gibbs energy dissipation during the motion of the interface, v_n^2/M_{eff}, controlled by the coupled processes of the friction of the interface (solvent atoms rearrange) and the trans-interface diffusion (solute drag), can be written as

$$v_n^2/M_{eff}=v_n^2/M_{fric}+Q_{sd} \tag{8}$$

where v_n^2/M_{fric} is for overcoming the friction of the interface [9], and Q_{sd} is for overcoming the solute drag. Through the matched asymptotic analysis, we have $Q_{sd}=v_n^2\xi(u_s-u_l)^2/(4\bar{M}_\delta(\Omega))$.

According to equation (8), the effective interface mobility, M_{eff}, can be expressed as

$$1/M_{eff}=1/M_{fric}+Q_{sd}/v_n^2 \tag{9}$$

Substituting equation (9) into equation (7), the PF mobility can also be expressed as

$$M_\phi=\pi^2M_{eff}/(16\xi) \tag{10}$$

Substituting equations (9) and (10) into the solvability condition of the inner expansion of the ϕ equation at the first order, $-\Delta G_m/V_m=\pi^2v_n/(16\xi M_\phi)+\sigma\kappa$, we obtain

$$-\Delta G_m / V_m = v_n / M_{fric} + Q_{sd} / \tilde{v}_n + \sigma\kappa = v_n / M_{fric} + v_n \xi (u_s - u_l)^2 / (4\bar{M}_\delta(\Omega)) + \sigma\kappa = v_n / M_{eff} + \sigma\kappa \qquad (11)$$

Assuming M_{fric}=∞ or β_{fric}=0 for the case of solidification, where β_{fric} is the interface kinetic coefficient calculated by $\beta_{fric}=T_M/(LM_{fric})$, and L is the latent heat of pure metal, we obtain $\xi_0 = 4\bar{M}_\delta(\Omega)(u_s - u_l)^{-2}(L/T_M)\beta_{eff}$ from equation (9), where $\beta_{eff}=T_M/(LM_{eff})$ is the effective interface kinetic coefficient. Using this ξ_0, the effect of solute drag, Q_{sd}, could be maintained as a constant for different interface widths. Accordingly, the PF mobility in the thin interface limit can be expressed as

$$M_\phi^{-1} = (4/\pi^2)\xi\xi_0(u_s - u_l)^2 / \bar{M}_\delta(\Omega) \qquad (12)$$

3. Anisotropic phase-field model in the dilute binary alloy system

Taking the crystalline anisotropy of fourfold symmetry into account, the orientation dependent interface width can be expressed as $\xi(\theta) = \xi A_s(\theta)$, $2\xi(\theta) = \pi K(\theta)/\sqrt{2\varpi}$, and $\sigma(\theta) = (\pi/8)K(\theta)\sqrt{2\varpi}$, where $\theta = \arctan(\partial_y\phi/\partial_x\phi)$, $A_s(\theta) = 1 + \varepsilon_4\cos(4\theta)$, and ε_4 is the energy anisotropy strength. Thus, the PF equation for the isothermal dendritic growth in dilute binary alloys can be written as

$$M_\phi^{-1}(\theta)\partial_t\phi = \nabla\cdot K(\theta)^2\nabla\phi - \partial_x[K(\theta)K'(\theta)\partial_y\phi] + \partial_y[K(\theta)K'(\theta)\partial_x\phi] - (8/\pi^2)\varpi f'(\phi) - g_\delta'(\phi)RT_M(k_e - 1)(u_l^e - u_l)/V_m \qquad (13)$$

where $k_e=u_s^e/u_l^e$ is the equilibrium partition coefficient, u_s^e and u_l^e are the equilibrium concentration in the solid and liquid, respectively. R is the gas constant.

The orientation dependences are also introduced to the PF mobility given in equation (12) to obtain $M_\phi^{-1}(\theta) = (4/\pi^2)\xi(\theta)\xi_0(\theta)(u_s - u_l)^2/\bar{M}_\delta(\Omega) = M_\phi^{-1}A_s(\theta)^2$. For dilute binary alloys, $M_l/M_s = D_l/(k_e D_s)$. Thus, the PF mobility of dilute alloys with crystalline anisotropy can be expressed as

$$M_\phi^{-1}(\theta) = (4/\pi^2)\xi(\theta)\xi_0(\theta)RT_M(k_e - 1)^2 u_l / [V_m \bar{D}_\delta(\Omega)] \qquad (14)$$

where $\bar{D}_\delta(\Omega)^{-1} = k_{KF}(\delta)\bar{D}(\Omega)^{-1}$, $\bar{D}(\Omega) = [(1-\Omega)(D_l)^{-1} + \Omega(k_e D_s)^{-1}]^{-1}$, and $\xi_0(\theta) = \xi_0 A_s(\theta)$.

With the assumption of dilute alloys, the concentration field satisfies $u_s = k_e u_l$ and $u_l = u/p_\delta(\phi)$, where $p_\delta(\phi) = k_e(1 + h_\delta(\phi))/2 + (1 - h_\delta(\phi))/2$. The diffusion equation can thus be described as

$$\partial_t u = \nabla\cdot(q_\delta(\phi)\cdot\nabla u / p_\delta(\phi) + \xi a_\delta(\phi)(u_l - u_s)\partial_t\phi(\nabla\phi/|\nabla\phi|)) \qquad (15)$$

where $q_\delta(\phi) = k_e D_s(1+m_\delta(\phi))/2 + D_l(1-m_\delta(\phi))/2$, $a_\delta(\phi) = k_{at}(\phi,\delta)(2/\pi)\sqrt{\phi(1-\phi)}(D_l - k_e D_s)/\bar{D}(\Omega)$, D_l and D_s are the diffusion coefficient in the liquid and solid, respectively, and $k_{at}(\phi,\delta) = [1 - h_\delta^2(\phi)]/[4\phi(1-\phi)]$ ($\delta < \phi_0 < 1 - \delta$). Ω is evaluated by $\exp(-D_l/(k_e D_s))$ to satisfy the condition of $Lim_{k_e D_s/D_l \to 0}(D_l - k_e D_s)/\bar{D}(\Omega) = 1$.

4. Results and discussion

4.1. Dendrite morphology

The PF model described in Section 3 was applied to simulate dendritic growth of an Fe-C alloy. The physical parameters of Fe-C alloys were taken from reference [11]. Other simulation parameters were given as follows: $d_0=\Gamma(u_s^e-u_l^e)^{-1}|m_e|^{-1}$=0.04566 μm, β_{eff}=50 sK/m, ξ_0 = 20d_0, ε_4=0.04, δ=0.08, and the interface width was taken as 5 grid points. All the parameters used in the present work were the same as stated here, unless specified otherwise.

Figure 1 shows the concentration fields and ATC of the steady-state dendrite growth for an Fe-0.15mol.%C alloy with $\Omega_0=(u_s^e-\bar{u}_0)/(u_s^e-u_l^e)$=0.55 at the temperature of $T_0=T_M$-5K. The domain is

1600×800 grid with ξ=3.75 d_0. Different scales are used in the left and right boxes to better visualize the concentration variations in the solid and liquid, respectively. The vectors at the interface illustrate the magnitude and direction of the ATC. Note that the ATC has a maximum value at the tip. Apparently, the interface velocity reaches the highest at the tip, and the effect of abnormal solute trapping should increase with the interface velocity. Thus, the largest ATC is needed to eliminate the abnormal solute trapping at the tip.

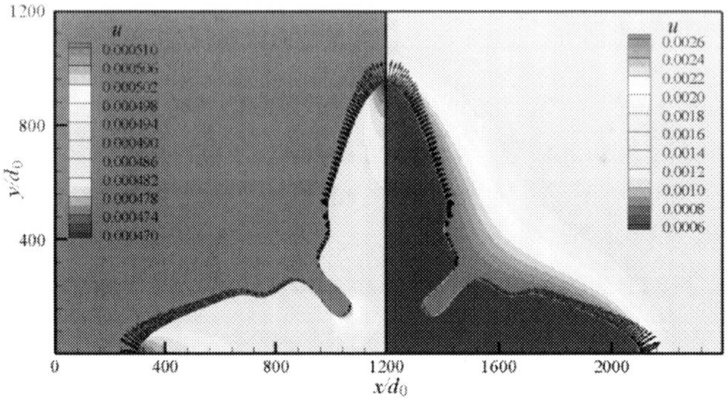

Figure 1. Simulated concentration fields of dendrite growth for an Fe-0.15mol.%C alloy at Ω_0=0.55. The left and right boxes show concentration maps in the solid and liquid, respectively. The vectors indicate the magnitude and direction of the anti-trapping current.

4.2. Computational convergent performance

The computational convergent performance of the present model was tested with respect to various interface thicknesses. In order to represent the same physical length of 1200d_0, the grid sizes used were 1600×1600, 1200×1200, 800×800, 600×600, 500×500, 400×400, 300×300, 240×240, and 200×200 for ξ=1.875d_0, 2.5d_0, 3.75d_0, 5d_0, 6d_0, 7.5d_0, 10d_0, 12.5d_0, and 15d_0, respectively. Other conditions were identical with those used for figure 1. After dendritic growth reached the steady-state, the tip velocity, v_n^*, tip radius, ρ^*, and tip composition in the liquid, u_l^*, were measured, and the average values were calculated from the measured values of several time step intervals. The partition coefficient, k_e, was computed from the tip compositions in the liquid and solid at the interface. Figure 2 shows the steady-state tip parameters, including growth velocity, radius, composition in the liquid, and the partition coefficient, as a function of interface thickness, ξ/d_0. The converged solution of tip parameters are obtained when the interface thickness, ξ/d_0, is less than about 10. It has been demonstrated that in the thin interface limit, convergence needs to satisfy the condition that the interface width must be much smaller than the radius of local curvature. Ramirez *et al.* [12] suggested that the ratio of tip radius to interface width, ρ^*/ξ, should be larger than about 10 to obtain converged results. Note that for the case of figure 2, the interface thickness ξ/d_0 = 10 is corresponding to the ratio ρ^*/ξ = 2.7376. It can be expected that for a larger tip radius, ρ^*, the converged solution could be obtained with a larger interface thickness, ξ/d_0. Consequently, the present model exhibits an excellent convergent behavior.

In the thin interface limit, the convergence is limited by two independent conditions of $\xi << \rho^*$, and $\xi << D_l/v_n^*$ [5, 10]. For the case of figure 2, D_l/v_n^* is about 95d_0. Apparently, the convergent range of the interface width in figure 2 satisfies the two limiting conditions.

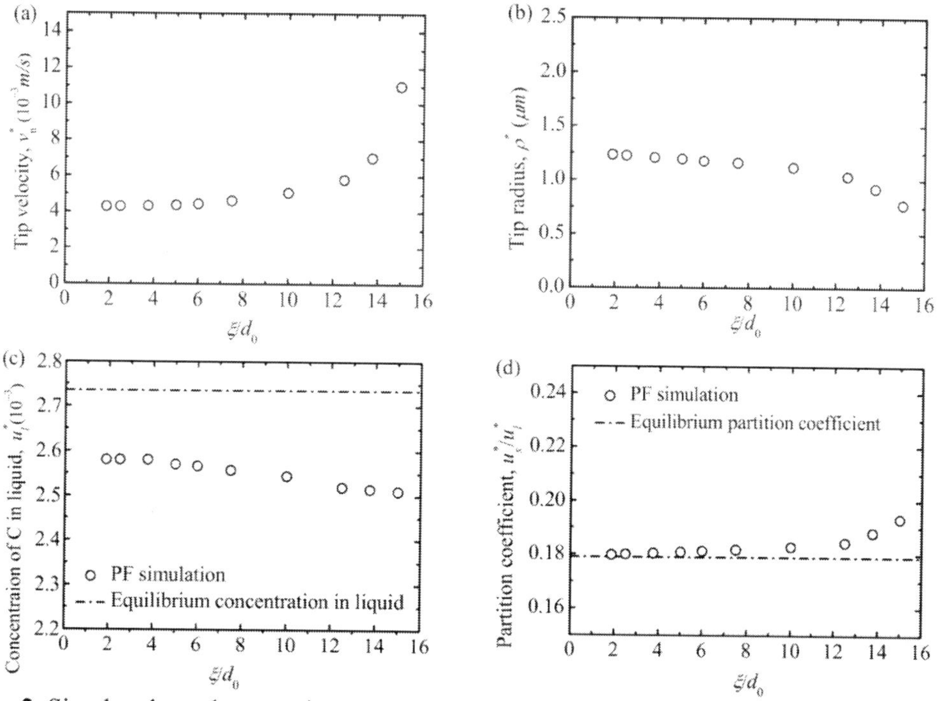

Figure 2. Simulated steady-state tip parameters as a function of interface width for an Fe-0.15mol%C alloy at Ω_0=0.55: (a) velocity, (b) radius, (c) liquid composition, and (d) partition coefficient.

4.3. Comparison with the Gibbs-Thomson relation

To validate the accuracy of the present model, the evolution of tip composition in the liquid simulated by the present PF model was compared with those calculated from the Gibbs-Thomson (GT) relation. Taking the dynamic effect and crystallographic anisotropy into account, the GT relation at the dendrite tip can be expressed as

$$u_l^* = u_l^e - \frac{V_m}{RT_M(1-k_e)} \cdot [(1+\varepsilon_4)\frac{v_n^*}{M_{eff}} + (1-15\varepsilon_4)\frac{\sigma}{\rho^*}] = u_l^e - \frac{(1+\varepsilon_4)\beta_{eff} v_n^* + (1-15\varepsilon_4)\Gamma/\rho^*}{|m_e|} \tag{16}$$

where $|m_e|=T_M^2R(1\text{-}k_e)/(LV_m)$ is the liquidus slope, and $\Gamma=T_M\sigma/L$ is the GT coefficient. In the absence of dynamic effect, i.e., β_{eff}=0, the GT relation at the dendrite tip is expressed as

$$u_l^* = u_l^e - \left[(1-15\varepsilon_4)\Gamma/\left(\rho^* |m_e|\right)\right] \tag{17}$$

Figure 3 shows the profiles of tip composition in the liquid, u_l^*, varying with time, obtained from the PF simulations, the GT relation with dynamic effect by equation (16), and the GT relation without dynamic effect by equation (17), for the two interface widths ξ=3.75d_0 and ξ=12.5d_0. The equilibrium composition is also given in the figure for comparison. The tip velocity, v_n^*, and tip radius, ρ^*, used for calculating equations (16) and (17) were obtained from the PF simulations. As shown, the profiles of PF simulation agree well with those calculated from the GT relation with dynamic effect for both interface widths, indicating that the abnormal solute trapping is indeed nearly eliminated. In addition, the data of PF simulations and the GT relation by equation (16) are lower than that obtained from the GT relation without the dynamic effect by equation (17). This difference is apparently caused by the effect of the dynamic undercooling.

The computational efficiency of the present model is acceptable. For example, the calculation times for figures 3 (a) and 3 (b) are 21050 s and 295 s, respectively, on a PC Dual-Core, CPU-i3 530.

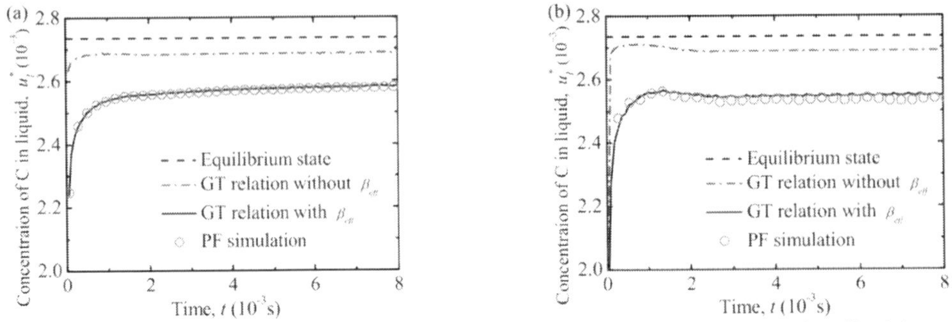

Figure 3. Comparison of the PF simulations for tip composition in the liquid to those computed from the GT relation with dynamic effect by equation (16), and the GT relation without dynamic effect by equation (17) for an Fe-0.15mol.%C alloy at Ω_0=0.55: (a) ξ=3.75d_0, and (b) ξ=12.5d_0.

4.4. Comparison with the linearized solvability theory

The second validation was performed by comparing the PF simulations with the linearized solvability theory for the effect of solid diffusivity, D_s, on the selection parameter, σ^*. According to the linearized solvability theory, the following relation is held during the steady-state dendrite growth at low undercooling [13]

$$\sigma^* = 2\sigma_0^* / (1 + k_e D_s / D_l) \tag{18}$$

where $\sigma^*=2D_l d_0/(v_n^*(\rho^*)^2)$, and $\sigma_0^*=2D_l d_0/(v_{n0}^*(\rho_0^*)^2)$ is the selection parameter for 2D non-symmetry case ($k_eD_s \neq D_l$), and for 2D symmetrical case ($k_eD_s=D_l$), respectively. For this validation, the simulations were carried out with different values of D_s by setting the realistic values of k_e and D_l as constant. The values of v^*_{n0} and ρ^*_0 used for calculating σ^*_0 in equation (18) were obtained from the PF simulations with ξ=3.75d_0. Figure 4 (a) presents the profiles of the selection parameter, σ^*, as a function of k_eD_s/D_l. The PF simulation plots were calculated by $\sigma^*=2D_l d_0/(v_n^*(\rho^*)^2)$ with the two interface widths ξ=3.75d_0 and ξ=12.5d_0. The good agreement between the PF simulation and the linearized solvability theory prediction illustrates that the present PF model can reasonably describe dendrite growth ranging from the case of $k_eD_s/D_l \approx 0$ (one-sided model) to the symmetrical case of $k_eD_s/D_l = 1$.

In the view of the linearized solvability theory [13], the selection parameter for symmetrical case, σ^*_0, is a function of interface energy anisotropy, ε_4, and the Péclet number, P_c. For two dimensions, the deduced solvability condition can be expressed by the following integral equation:

$$\int_{-\infty}^{+\infty} \frac{A(x)^{1/2}}{(1+x^2)^{3/4}} \frac{A(x)^{1/4}(1-ix)^{1/4}}{(1+x^2)^{3/8}} \exp\left(\frac{S_0(x)}{(\sigma_0^*)^{1/2}} - \frac{P_c}{2}\left(ix + \frac{x^2}{2}\right)\right) dx = 0 \tag{19}$$

where $A(x)=1-15\varepsilon_4+120\varepsilon_4 x^2(1+x^2)^{-2}$, $S_0(x) = i\int_0^x (1-is)^{1/4}(1+is)^{3/4} A(s)^{-1/2} ds$, and i is the imaginary unit. The solution of equation (19) determines σ_0^* as a function of ε_4 and P_c. The selection parameter, σ_0^*, as a function of anisotropy, ε_4, simulated by the PF model with ξ=3.75d_0 and ξ=12.5d_0 was compared with those calculated from the linearized solvability theory by equation (19) with P_c=0. For this comparison, σ_0^* of PF simulations was calculated based on equation (18) ($\sigma^*_0=\sigma^*(1+k_eD_s/D_l)/2$). The comparison is shown in figure 4 (b), and indicates that when anisotropy, ε_4, is smaller than around 0.05, the profiles obtained from the PF simulations with two interface widths agree well with the one predicted by the linearized solvability theory.

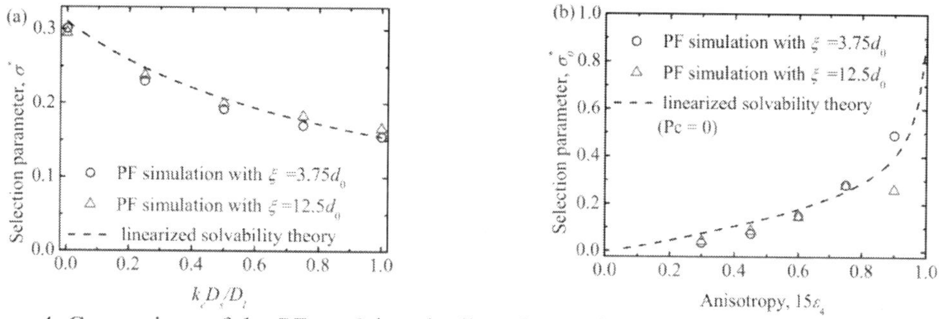

Figure 4. Comparison of the PF model to the linearized solvability theory: (a) the effect of solid diffusivity D_s on σ^*, and (b) the effect of interface energy anisotropy ε_4 on σ_0^*.

5. Conclusions

A quantitative PF model with an anti-trapping current (ATC) is developed to simulate dendritic growth involving two-sided solute diffusion. The PF mobility and ATC are derived based on the second-order asymptotic analysis and the solute drag model. Model tests are performed in terms of the convergent behavior with respect to the interface width, and the comparison of the PF simulations to the GT relation and the linearized solvability theory, for the isothermal dendritic growth of an Fe-1.5mol.%C alloy. The convergent solution is obtained when the ratio of tip radius to interface width, ρ^*/ξ, is less than 3, exhibiting an excellent convergent behavior of the proposed model. The simulated tip composition agrees well with the prediction of the GT relation involving the dynamic effect, when the interface is extended to ξ=12.5d_0. Good agreements are also found for the comparisons of the PF simulations to the predictions of the linearized solvability theory regarding the selection parameter as functions of solid diffusivity, and of interface energy anisotropy.

Acknowledgements

The work was financially supported by the AO Smith Corporate Technology Center, USA, and NSFC (Grant No. 50971042).

References

[1] Steinbach I 2009 *Modelling Simul. Mater. Sci. Eng.* **17** 073001
[2] Karma A and Rappel W J 1998 *Phys. Rev. E* **57** 4323
[3] Almgren R F 1999 *SIAM J. Appl. Math.* **59** 2086
[4] Karma A 2001 *Phys. Rev. Lett.* **87** 115701
[5] Echebarria B, Folch R, Karma A and Plapp M 2004 *Phys. Rev. E* **70** 061604
[6] Gopinath A, Armstrong R C and Brown R A 2006 *J. Cryst. Growth* **291** 272
[7] Ohno M and Matsuura K 2009 *Phys.Rev. E* **79** 031603
[8] Svoboda J, Fischer F D and Gamsjäger E 2002 *Acta Mater.* **50** 967
[9] Hillert M 1999 *Acta Mater.* **47** 4481
[10] Kim S G 2007 *Acta Mater.* **55** 4391
[11] Ohno M and Matsuura K 2010 *Acta Mater.* **58** 5749
[12] Ramirez J C, Beckermann C, Karma A and Diepers H J 2004 *Phys. Rev. E* **69** 051607
[13] Barbieri A and Langer J S 1989 *Phys. Rev. A* **39** 5314

IOP Conf. Series: Materials Science and Engineering **33** (2012) 012097 doi:10.1088/1757-899X/33/1/012097

Capillary-mediated dendritic branching

Martin E Glicksman

Allen S. Henry Chair and University Professor, Mechanical & Aerospace Engineering Dept., Florida Institute of Technology, Melbourne, Florida 32901, USA

E-mail: mglicksman@fit.edu

Abstract. We consider the Gibbs-Thomson temperature distribution as an active interface energy field with gradients and fluxes. Capillarity in conventional dendritic growth theories acts only as a passive boundary condition on the transport field that conducts latent heat into the melt to drive the solidification. Instead, we find the vector divergence of capillary-mediated interface fluxes autonomously releases or withdraws energy along a curved solid-liquid interface. Local energy conservation (a Stefan balance) shows that capillary-mediated energy slightly retards, or enhances, the local freezing rate. At special points near the tip where the capillary-mediated energy release changes sign, the surface Laplacian of the Gibbs-Thomson potential vanishes. Locally adjacent freezing rates develop opposed biases from the added or subtracted capillary energies, which 'curl', or rotate, the solid-liquid interface. A noise-free Greens function dynamic solver confirms that the rotation points induce dendritic branching. Moreover, capillary-induced rotations occur precisely at points that may be calculated exactly for different initial shapes with different interface energy anisotropies. This mechanism appears to provide a deterministic process capable of inducing branching and evolving material-specific dendritic patterns. Noise and stability play no direct roles in dendritic morphogenesis.

1. Introduction

Crystalline dendrites in metallic systems commonly develop on size scales from 10^{-6} to 10^{-2} m, and exhibit crystallographically selected primary stems with quasi-periodic side branches, and even tertiary branches. A well-branched 6-pointed dendrite of a body-centered cubic material is shown growing *in situ* in its supercooled melt in Figure 1. It is the continuous propensity for side-branching that establishes the specific patterns, length scales, and spatial distributions of chemical components within a solidifying alloy. Cast dendritic microstructures strongly affect a material's mechanical, physical, and chemical characteristics. Consequently, the growth and behavior of melt-grown dendrites remain important technical issues that are relevant to alloy casting, primary metals production, welding, and soldering. Dendrites often dominate the initial microstructures for materials produced by these industrial processes, determining their as-cast chemical segregation, grain size, and crystallographic textures.

2. Dendritic Growth Theory

Dendritic growth theory traditionally is divided into two components: 1) transport, which explains how the large-scale energetics of a $\ell \rightarrow s$ transformation operate and control dendrite speed and size; and 2) interface physics, which provides the underlying microscopic phenomena responsible for branching and directionality that characterize complex dendritic patterns in real materials. This paper explains that the two components of dendrite theory—transport and

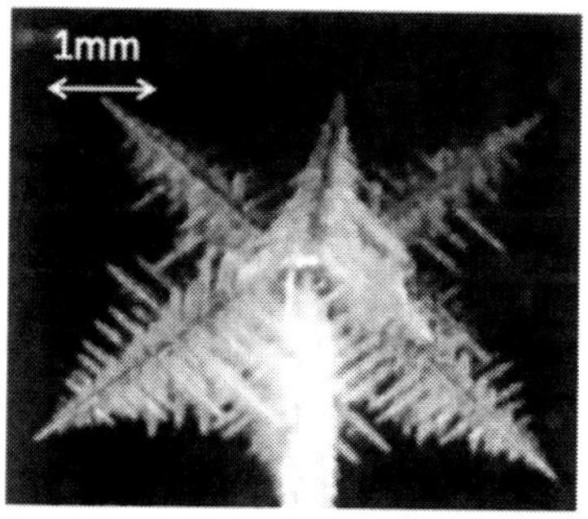

Figure 1. In situ photograph of a single free-growing dendrite of high-purity, BCC succinonitrile, CN-$(CH_2)_2$-NC. This dendrite was photographed while growing *in situ* in its transparent supercooled melt. Five of its six equivalent ⟨100⟩ cube-edge directions are visible, which are the 'easy' growth directions for the primary stems. Secondary side-branches develop along the primary stems just aft of the advancing tips, generating sheet-like arrays arranged almost periodically on orthogonal (100) cube planes. Branching continues until the residual supercooling is exhausted.

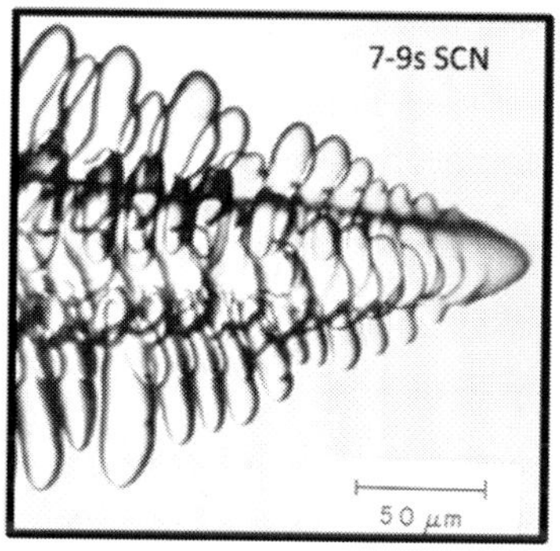

Figure 2. Higher resolution micrograph of one of the primary stems shown growing in Figure 1, fully resolving its tip structure and a growing train of nearly periodic side branches. Branches initially grow in non-orthogonal directions to the primary growth directions, ⟨100⟩, as the tips are surrounded by a transport field with strong local gradients to conduct away the latent heat of fusion. Side branches, once free of the influence of the steep transport gradients, eventually grow orthogonally in the easy growth directions dictated by cubic crystallography.

interface physics—are not actually independent aspects of the crystal growth, but are linked through an autonomous effect involving capillary-mediated energy fluxes transported along the solid-liquid interface. The interface gains or loses extra energy where capillary-mediated interface fluxes develop vector divergences that modify the local growth rates. This aspect of capillarity-mediated energy transport leads to dendritic branching and controls the evolving pattern.

2.1. Transport

Ivantsov [1], and other investigators of diffusion-controlled phase transformation [2], developed a series of well-known mathematical solutions for the thermal conduction and mass-transfer fields responsible for steady-state dendritic freezing. These analytical steady-state solutions relate a dimensionless parameter, the growth Péclet number, with the main independent variable, the supercooling or supersaturation, $\Delta\vartheta$, that provides the free energy to drive spontaneously the $\ell \rightarrow s$ phase transformation. Mathematical expressions for the steady-state transport fields surrounding branchless dendrites in both pure materials and alloys are essentially the same, differing merely in their transport coefficients that define the growth Péclet number, P, where $P \equiv \frac{VR}{2\alpha}$. The material parameter, α, appearing in the growth Péclet number represents either the thermal diffusivity, for the case of a pure (one-component) melt, or the interdiffusion coefficient, for a solidifiying alloy melt. Moreover, the energy transport solution only predicts, as a function of the melt supercooling or supersaturation, the product $V \times R$, where V is the dendrite's tip speed, and R, is its tip radius of curvature. Individual values of the two dependent variables, V and R, cannot be derived from transport theory alone.

Ivantsov analytically solved the energy equation in the melt surrounding a steadily advancing, isothermal, branchless, paraboloidal dendrite. However, finding the unique solution to the dendritic growth problem requires an independent physical condition, involving one, or both, of the dependent variables. The quest to identify the 'missing' physical condition has remained elusive, despite many attempts to identify it.

IOP Conf. Series: Materials Science and Engineering **33** (2012) 012097 doi:10.1088/1757-899X/33/1/012097

2.2. Interfacial physics

Over time, a number of creative ideas were eventually proposed for the needed interface physics [3]-[18]. Among them were the following reasonable sounding hypotheses [19]-[24]: maximum velocity; interfacial stability; shape self-consistency; minimum entropy production; marginal stability (maximum radius); selective noise amplification; microscopic solvability; trapped wave theory; and maximum entropy production. Experiments, however, showed that these hypotheses led to predictions that were inconsistent with observations. Specifically, theories failed to describe correctly dendritic branching behavior and important scaling laws relating size and growth speed. Perhaps most disappointing about these hypotheses is that none proved capable of providing even qualitative descriptions of the solidification patterns found among dendrites from materials with different crystallography and thermal properties.

3. Capillarity

A peculiar feature found among dendritic growth models is relegating capillarity to be the 'inner' boundary condition on the solid-liquid interface, to which the 'outer' (Ivantsov) transport field in the melt conforms. The Gibbs-Thomson condition used in this specific manner provides a more accurate interface condition than does the *isothermal* boundary condition, which was assumed by Ivantsov in developing his transport solutions. The use of capillarity as the boundary condition applied at the solid-liquid interface has remained unchanged for more than a half century. The reasons for limiting the Gibbs-Thomson effect as a passive boundary condition are explained next.

3.1. Gibbs-Thomson—a boundary condition

The Gibbs-Thomson effect relates the solid-liquid equilibrium temperature with the local interface curvature, at a fixed pressure. Moreover, the equilibrium temperature itself is altered imperceptibly, by only a few mK, even at a highly curved dendrite tip. The reduction of the equilibrium temperature from the bulk melting temperature, T_m, is even far less where the crystal-melt interface is flatter. This circumstance varies little with different substances, because crystals and their melts have extremely small ratios between their interfacial energy (capillarity) and their volumetric heat of fusion (transport). So, understandably, models of dendritic growth assume that capillary fields are energetically irrelevant, and seemingly inconsequential when compared with the (Ivantsov) field, which controls the energetics of the $\ell \to s$ transformation. Indeed, the Ivantsov field involves temperature differences that are orders of magnitude greater than the Gibbs-Thomson distribution. For these apparently cogent reasons, the Gibbs-Thomson distribution is used strictly as a passive boundary condition in standard models of dendritic growth. As will be shown contrary to convention, capillarity plays a subtle, and thus far unsuspected, role in the initiation and control of dendritic branching. Capillarity, in fact, provides the elusive interface physics, which, for over 50 years, remained 'hidden in plain sight'.

3.2. Gibbs-Thomson—an energy field

The Gibbs-Thomson temperature distribution is reconsidered here as a full-fledged energy field. This reassessment occurred after the author analyzed some experiments on the melting kinetics of needle-like crystals. These crystals were observed slowly melting while suspended motionless in their melt under microgravity conditions. Needle-shaped crystals initially melted self-similarly, i.e., retaining a constant geometric aspect ratio. But prior to their total extinction by melting, the crystallites altered their shape, and spheroidized [25, 26]. The onset of spontaneous spheroidization was caused by *additional heat currents generated directly from the Gibbs-Thomson temperature distribution* [27]. This finding, discovered in experiments on *melting* crystals, prompted a re-examination of the role played by the Gibbs-Thomson temperature for the reverse case of crystallite *freezing*: i.e., the role of capillarity in dendrites [27, 28].

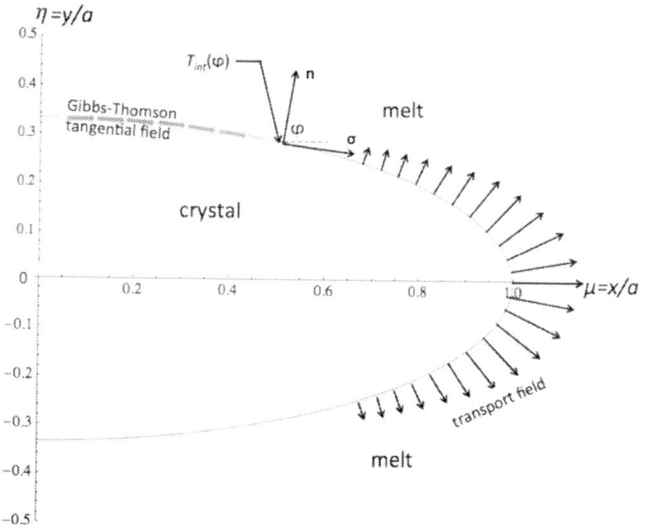

Figure 3. A 3:1 elliptically-shaped dendritic interface. The tangential Gibbs-Thomson equilibrium temperature field, $T_{int}(\varphi)$, and the external transport field act along the crystal-melt interface. Also shown are the dimensionless space coordinates, $\mu = x/a$ and $\eta = y/a$, the unit normal vector on the interface, $\vec{n}$, with its normal angle, φ, and the tangent vector, $\vec{\sigma}$, directed along the curved interface.

3.3. Tangential fields

We examine the behavior of a two-dimensional interface in the form of a semi-ellipse, with its semi-major x-axis, a, longer than its semi-minor y-axis, b. Such a slender finger-like region is suitable for analysis. See Figure 3 for the configuration of its interfacial energy fields. We scale the coordinates in units of the semi-major axis a, so that $x/a \equiv \mu$, and $y/a \equiv \eta$ define dimensionless axes for the interface. The equation of an elliptical interface transforms to

$$\eta = \frac{b}{a}\sqrt{1-\mu^2}. \tag{1}$$

The presence of an Ivantsov field suggests the transport of latent heat from the advancing interface into the surrounding supercooled melt. The Gibbs-Thomson temperature distribution, $T_{int}(\varphi)$, represents the equilibrium temperatures along the semi-ellipse, with the curved tip region, near $\mu = 1$, remaining slightly cooler than near the equator where $\mu \to 0$. The temperature along the ellipse's boundary, allowing some 4-fold anisotropy of the interfacial energy, is expressed by the Gibbs-Thomson-Herring equation [29, 30],

$$T_{int}(\varphi) = T_m - \left(\frac{\gamma_{s\ell}(\varphi) + \gamma''_{s\ell}(\varphi)}{\Delta S_f/\Omega_\ell}\right)\kappa(\varphi). \tag{2}$$

Here $\gamma_{s\ell}(\varphi) = \gamma_0(1 + \epsilon\cos(4\varphi))$, is the anisotropic crystal-melt interfacial energy density, and γ_0 is the energy modulus. The term $\gamma''_{s\ell}(\varphi)$ is the second angular derivative of the interface energy, called the 'torque'; $\Delta S_f = \Delta H_f/T_m$ is the molar entropy of fusion; Ω_ℓ is the molar volume of the melt; and $\kappa(\varphi)$ is the ellipse's curvature. Note this distribution represents *local* two-phase equilibrium, as an ellipse is not the global equilibrium configuration (Wulff shape) for 4-fold anisotropy. The small negative temperature variation, $T_{int}(\varphi) - T_m$, caused by interface curvature corresponds to the dimensionless interface potential, $\vartheta(\varphi)$,

$$\vartheta(\varphi) = \frac{T_{int}(\varphi) - T_m}{\Gamma/a} = -a[1 - 15\epsilon\cos(4\varphi)]\kappa(\varphi), \tag{3}$$

where $\Gamma/a = a^{-1}\left(\frac{\gamma_0}{\Delta S_f/\Omega_\ell}\right)$ is the capillary constant, Γ, divided by the semi-major axis a. Γ/a bears the physical unit [K], and $a\kappa(\varphi)$ is the dimensionless interface curvature. ϵ is the anisotropy strength.

The normal angle for a semi-elliptical interface is $\varphi = \frac{\pi}{2} + \arctan\left[-\frac{b}{a}\frac{\mu}{\sqrt{1-\mu^2}}\right]$, so the interface potential, $\vartheta(\mu)$ may be expressed as function of dimensionless distance, μ,

$$\vartheta(\mu) = -\frac{b}{a}\left[(1-\mu^2)\left(1-\left(\frac{b}{a}\right)^2\left(\frac{\mu^2}{-1+\mu^2}\right)\right)\right]^{-\frac{3}{2}} \quad (0 \leq \mu \leq 1). \tag{4}$$

A 2:1 ellipse is chosen for specificity to represent the solid-liquid interface, so the semi-axes in Eq.(4) assume values $a = 2$, $b = 1$. The interface potential for this ellipse is plotted in Figure 4.

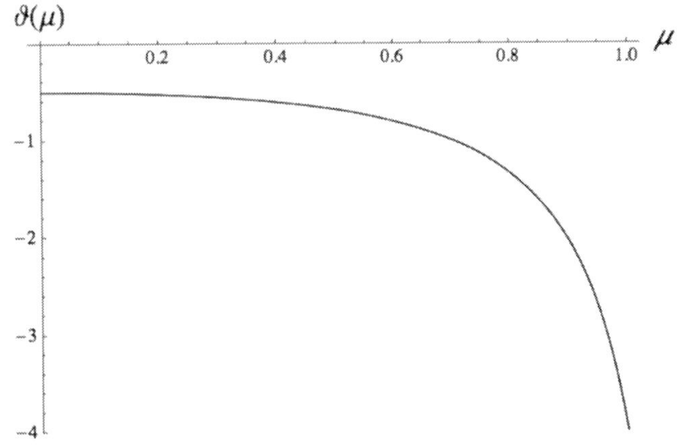

Figure 4. Gibbs-Thomson potential, $\vartheta(\mu)$, calculated for a 2:1 semi-ellipse, with a small ($\epsilon = 0.5\%$) 4-fold anisotropy. The interface potential falls initially quite slowly then decreases rapidly as the tip is approached and $\mu \to 1$. The Gibbs-Thomson gradient prompts a weak tangential thermal flux, as energy is transported from higher to lower potential along the solid-liquid interface.

The associated vector gradient of the Gibbs-Thomson field transports a tangential flux of thermal energy, $\hat{\Phi}(\mu) \cdot \vec{\sigma}$, along the interface from higher potential toward lower potential. If scaled by the interfacial conductivity, k_{int} [watts/K], this dimensionless flux equals minus the tangential derivative of the potential, $\vartheta(\mu)$, namely,

$$\vec{\nabla}_\sigma[\vartheta(\mu)] = -\hat{\Phi}(\mu) \cdot \vec{\sigma}, \tag{5}$$

where $\vec{\nabla}_\sigma[\ ]$ is the dimensionless gradient operator acting along the interface. The tangential energy flux is plotted in Figure 5 for a 2:1 semi-ellipse with a small ($\epsilon = 0.5\%$) 4-fold anisotropy of its interface energy .

3.4. Interfacial energy conservation

Although temperature variations caused by the Gibbs-Thomson effect are indeed small, any energy fluxes, irrespective of their magnitude, must still be conserved and accounted for along the interface. We assumed that the interface is sharp (zero thickness) and consequently lacks the capacity to store energy by altering its temperature, i.e., the sharp interface has zero specific heat. To evaluate the rate of local energy accumulation or dispersal from the capillary field we calculate the surface divergence of the Gibbs-Thomson gradient, or, equivalently, calculate (minus) the divergence of the tangential energy flux, namely,

$$\vec{\nabla}_\sigma \cdot \vec{\nabla}_\sigma[\vartheta(\mu)] = -\vec{\nabla}_\sigma[\hat{\mathbf{\Phi}}_\sigma(\mu)] = \dot{q}(\mu), \tag{6}$$

where the term $\dot{q}(\mu)$ denotes the dimensionless rate of capillary-mediated energy acquired or lost along the interface. The requirement for local energy conservation is satisfied by adding the capillary term, $\dot{q}(\mu)$, to the rates of latent heat production and their conduction away from the interface to the surrounding bulk phases. Conservation of energy at a freezing dendritic interface yields the following (dimensional) Stefan energy balance,

$$\left(\frac{\Delta H_f}{\Omega_\ell}\right) \vec{v}_n \cdot \vec{n} + \dot{Q} - (k_\ell \vec{\nabla} T_\ell + k_s \vec{\nabla} T_s) \cdot \vec{n} = 0. \tag{7}$$

The first term in Eq.(7) is the rate of latent heat production caused by solidification at a point, where the normal velocity of the interface is $\vec{v}_n$; the second term, $\dot{Q}$, is the dimensional rate of capillary-mediated energy release re-scaled from the tangential divergence of the Gibbs-Thomson gradients, $\dot{q}(\mu)$; and the last term is the net transport of energy conducted from the interface toward the surrounding bulk phases. The coefficients k_ℓ and k_s are the bulk thermal conductivities of the melt and crystal, respectively. These terms must balance everywhere along the solid-liquid interface. As the macroscopic transport fields change slowly compared to microscopic interfacial processes, the influence of capillarity occurs both with $\dot{Q} > 0$, for which the normal velocity is slightly *reduced* relative to its value without capillarity, and with $\dot{Q} < 0$, for which the normal velocity is slightly *increased* relative to its value without capillarity. These slight capillary-mediated biases, or modulations, of the normal interface velocity caused by capillarity are of profound importance, as they explain the occurrence of branching.

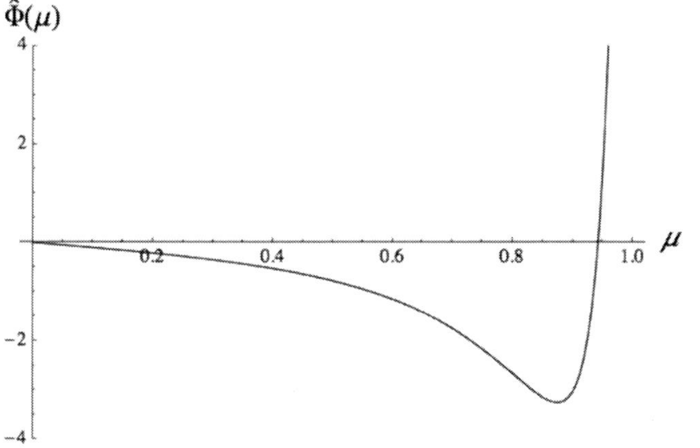

Figure 5. Dimensionless tangential flux calculated for the Gibbs-Thomson field plotted in Figure 4. Although the slope of the potential in Figure 4 appears monotone negative in the coordinate, μ, its *tangential gradient* is non-monotone, and the flux even reverses sign near the tip. This peculiarity of the Gibbs-Thomson potential gradient has important implications on the process of interface branching.

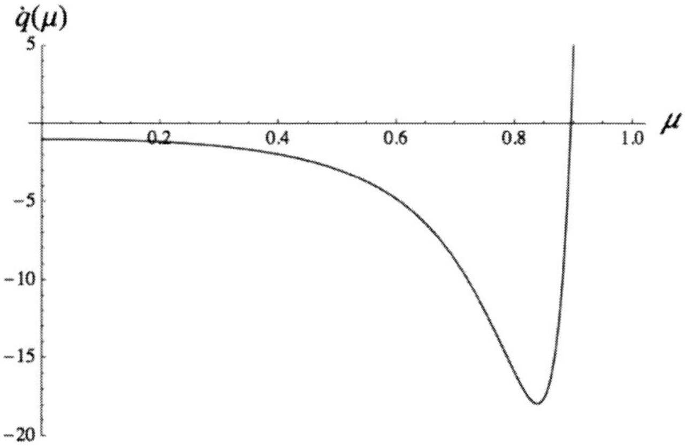

Figure 6. Divergence of the gradients of the Gibbs-Thomson potential shown in Figure 5 above versus dimensionless distance. The divergence equals the rate at which capillary-mediated energy increases, $\dot{q}(\mu) > 0$, or decreases, $\dot{q}(\mu) < 0$, at each point along the interface. Capillary-mediated energy modulates the local freezing velocity, as the total energy released at each point on the interface (latent heat plus capillary energy) is conducted away by the transport fields in the melt and the solid

IOP Conf. Series: Materials Science and Engineering **33** (2012) 012097 doi:10.1088/1757-899X/33/1/012097

4. Dendritic branching

Figure 7-left is a plot of $\dot{q}(x)$ versus the dimensional x-coordinate of the 2:1 elliptical dendrite. A root occurs at $x \approx 1.7$, where the capillary-mediated rate of energy accumulation reverses its sign. Eq.(7) suggests that where $\dot{q}(x) < 0$ (in the range $0 \leq x \leq 1.7$) the interface velocity is increases slightly by capillarity, but beyond $x \approx 1.7$, closer to the tip, $\dot{q}(x) > 0$, and the interface velocity decreases slightly. The apposing interface velocity changes lead to a 'curling' motion, or rotation, of the interface at $x \approx 1.7$. Figure 7-right shows one frame extracted from the video output from a dynamic solver evolving in time a 2:1 elliptical starting shape with the same interface energy anisotropy. The pattern developing in the video frame is adjusted to maintain a fixed area or number of pixels. The 2:1 starting ellipse, with the same area, is juxtaposed onto the video frame to allow the capillary-induced interface rotation to be viewed more easily. These data show the close correspondence between the location of a dynamically evolved rotation point, where side branch initiation eventually starts, and the root in the $\dot{q}(x)$ plot, which occurs where the surface Laplacian of the Gibbs-Thomson potential vanishes. A quantitative connection was verified for different starting shapes and interface energy anisotropies by J. Lowengrub and S. Li between the zeros of the surface Laplacian of the Gibbs-Thomson potential, the roots of the energy release rate, $\dot{q}(x)$, and the occurrence dendritic branching [31].

5. Summary

The analysis of capillary energy fields and the verification of interface rotation through dynamic data suggest that a deterministic mechanism for dendritic branching exists, requiring no additional physics to explain branching and pattern complexity. Energy released or withdrawn from the interface slightly slows, or accelerates, local freezing rates, respectively. Local freezing rates change slightly because capillary-mediated energies add to, or subtract from, the latent heat transported from the interface. Although capillary-mediated energy releases cause only miniscule enhancement, or reduction, in the local freezing rate, over time they increase the local curvature where the capillary energy rate is negative, and increase the local freezing rate. Similarly, a decrease in local curvature occurs where the capillary-mediated energy rate is positive, which reduces the local freezing rate. These small velocity and curvature changes, acting over time

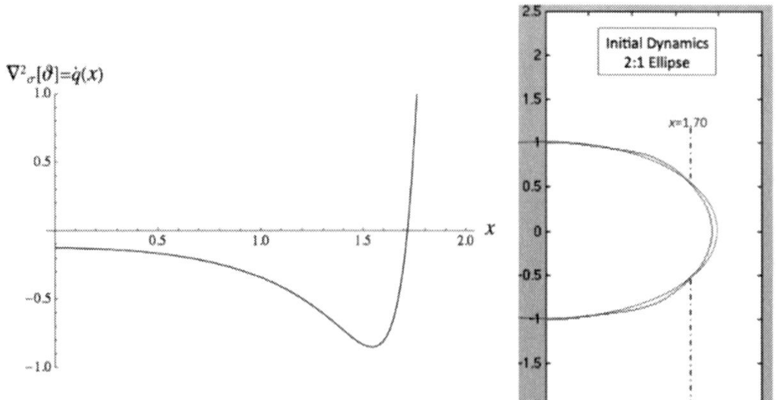

Figure 7. Check on the initial rotation points. **Left:** Capillary-mediated energy release rate, $\dot{q}$, versus distance, x/a, ($a = 1$) showing the analytically predicted rotation point where $\nabla^2_\sigma[\vartheta] = 0$. **Right:** Frame from dynamic simulation of the same starting shape and anisotropy. Initial rotation point (left panel) is confirmed by the dynamic solver at $x/a \approx 1.70$.

rotate the interface, eventually initiating side-branch formation. Mathematically, rotation points occur at locations along the interface where the surface Laplacian of the interface potential vanishes, and the capillary-mediated energy release rate reverses sign. See again Eq.(6) for this condition.

A number of starting shapes (ellipses of various eccentricities) with different 4-fold anisotropies have been tested to date against data from a precision dynamic solver. Analytically derived rotation points are confirmed dynamically. Details of these dendritic branching studies will be published elsewhere.

Acknowledgments

The author is indebted to Professor John Lowengrub, Department of Mathematics, University of California Irvine, Irvine, CA, and Professor Shuwang Li, Department of Applied Mathematics, Illinois Institute of Technology, Chicago, IL, for their enormously helpful collaboration.

References

[1] Ivantsov 1947 G P *Dokl. Akad. Nauk USSR* **58** 567
[2] Horvay G and Cahn JW 1961 *Acta Metall.* **9** 695
[3] Temkin D E 1960 *Dokl. Akad. Nauk USSR* **132** 1307
[4] Bolling G F and Tiller W A 1961 *J. Appl. Phys.* **32** 2587
[5] Holtzmann E G 1970 *J. Appl. Phys.* **41** 1460
[6] Oldfield W 1973 *J. Matl. Sci. Eng.* **11** 211
[7] Nash G E and Glicksman M E 1974 *Acta Metall.* **22**1283
[8] Langer J S and Müller-Krumbhaar H 1977 *J. Cryst. Growth,* **42** 11
[9] Langer J S and Müller-Krumbhaar H 1978 *Acta Metall.* **26** 1681
[10] Langer J S 1980 *Rev. Mod. Phys.* **52** 1
[11] Kessler D and Levine H 1986 *Phys. Rev. Lett.* **57** 3069
[12] Kessler D and Levine H 1987 *Phys. Rev. A,* **36** 2693
[13] Barbieri A and Langer J S 1987 *Phys. Rev. A,* **39** 5314
[14] Xu J J 2002 *Adv. in Mechanics and Mathematics* **1** Springer New York 213
[15] Xu J J 2007 *Intro. Dynamical Theory of Solidification Interfacial Instability* Chinese Academy Press, Beijing
[16] Chen X J Chen Y Q Xu J P and Xu J J 2008 *Frontiers of Phys. China* **3** 1
[17] Chen Y Q Tang X X and Xu J J 2009 *Chinese Phys. B* **18** 686
[18] Sekhar J A 2011 *J. Mater. Sci.*, DOI 10.1007/s10853-011-5688-0, published on-line Springer ISSN 0022-2461
[19] Glicksman M E Schaefer R J and Ayers J D 1976 *Metall. Trans. A* **7** 1747
[20] Huang S C and Glicksman M E 1981 *Acta Metall.* **29** 701
[21] Huang S C and Glicksman M E 1981 *Acta Metall.* **29** 717
[22] Esaka H and Kurz W 1985 *Z. Metall.* **76** 127
[23] Muschol M Liu D and Cummins H Z 1992 *Phys. Rev. A* **46** 1038-1050
[24] Losert W Shi B Q and Cummins H Z 1998 *Proc. Natl. Acad. Sci. USA* **95** 431
[25] Glicksman M E Lupulescu A and Koss M B 2003 *J. Thermophysics & Heat Transfer* **17** 69-76
[26] Lupulescu A Glicksman M E and Koss M B 2005 *J. Cryst. Growth* **276** 549-565
[27] Glicksman M E 2011 *J. Phys.: Conf. Ser.* **327** 012001
[28] Cheong A G and Rey A D 2002 *Phys. Rev. E* **66** 021704
[29] Herring C 1951 *Physics of powder metallurgy* Ed Kingston W E McGraw-Hill New York 143
[30] Glicksman M E 2011 *Principles of Solidification* Springer USA New York 199-210.
[31] Lowengrub J and Li S 2011 Private communication

IOP Conf. Series: Materials Science and Engineering **33** (2012) 012098 doi:10.1088/1757-899X/33/1/012098

Competitive grain growth in directional solidification investigated by phase field simulation

Junjie Li[1], Zhijun Wang[1], Yujuan Yang[2], Jincheng Wang[1]

[1] State Key Laboratory of Solidification Processing, Northwestern Polytechnical University, Xi'an 710072, P. R. China

[2] College of Mechanical Engineering, Tongji University, Shang Hai, 201804, P. R. China

Email: lijunjie@nwpu.edu.cn

Abstract. During directional solidification, the competitive dendritic growth between various oriented grains is a key factor to obtain desirable texture. In order to understand the mechanism of competitive dendritic growth, the phase field method was adopted to simulate the microstructure evolution of bicrystal samples. The simulation has well reproduced the whole competitive growth process for both diverging and converging dendrites. In converging case, besides the block of the unfavorably oriented dendrite by the favorably oriented one, the unfavorably oriented dendrite is also able to overgrow the favorable one under the condition of relatively low pulling velocity. This unusual overgrowth is dictated by the solute interaction of the converging dendrite tips. In diverging case, it was found that the grain boundary can be either inclined or parallel to the favorably oriented grain depending on the disposition of two grains.

1. Introduction

The competition of different oriented grains is a prevalent phenomenon during directional solidification of single crystal casting. The development of a sharp solidification texture depends directly on the efficiency of competitive grain growth. The widely accepted theoretical model for this competition was proposed by Walton and Chalmers [1] and then schematically summarized by Rappaz and Gandin[2,3]. It is recognized that the favorably oriented dendrite leads and blocks its unfavorably oriented neighbor due to the difference in tip undercooling.

Although these classical theory descriptions have been confirmed in many studies [2-6], some inconsistent results are also observed in recent experiments of Ni-base superalloys [7-8]. In the bicrystal volume samples with converging dendrites, Zhou et al. [7] found that the unfavorably oriented dendrite was also able to overgrow the favorably oriented one, which leads to the suppression of the well-aligned grain by the misaligned one. The orientation map of bicrystal samples with diverging dendrites obtained by Wagner et al. [8] illustrated that the relatively well-aligned grain cannot develop new dendrite arms at the grain boundary (GB), so that the GB was parallel to the misaligned grain but not inclined towards the misaligned grain.

Due to the absence of in-situ observation of competitive growth in above experiments, the detailed microstructure evolution process is still unclear. The mechanism for the unusual overgrowth manner is also difficult to be stated conclusively only from the experimental data. In this paper, the phase field method, which has experienced wide application in the field of directional solidification [9-12], is used to investigate the competitive growth during the directional solidification of 2D bicrystal sample. Some factors that may induce the unusual overgrowth are illustrated.

2. Simulation method

In order to distinguish grains with different crystalline orientations, the multiphase-field approach [13-18] was used. The phase state with $\phi_1=1$ denotes the liquid phase, whereas $\phi_i=1$ for $i\neq 1$ denotes the same solid phase but with different crystalline orientations. The sum of each phase fields coexisting in any position in the system satisfy

$$\sum_{j=1}^{n}\phi_j = 1 \tag{1}$$

Then the number of nonzero phase field variables coexisting at a given point is

$$N = \sum_{i=1}^{n} s_i \tag{2}$$

where $s_i=1$ if $\phi_i>0$ and $s_i=0$ otherwise. In the dilute solution limit the evolution equation of the phase field is given by:

$$\frac{\partial \phi_i}{\partial t} = -\frac{2}{N}\sum_{j\neq i}^{n} s_i s_j M_{ij}\left(\frac{\delta f}{\delta \phi_i} - \frac{\delta f}{\delta \phi_j} + \Delta g_{ij}\right) \tag{2}$$

where

$$\frac{\delta f}{\delta \phi_i} = \begin{cases} \sum\limits_{j\neq 1}^{n}\left(\frac{\varepsilon_{SL}^2}{2}\nabla^2\phi_j + w_{SL}\phi_j\right) & \text{for i} = 1 \\ \frac{\varepsilon_{SL}^2}{2}\nabla^2\phi_1 + w_{SL}\phi_1 + \sum\limits_{j\neq 1,i}^{n}\left(\frac{\varepsilon_{SS}^2}{2}\nabla^2\phi_j + w_{SS}\phi_j\right) & \text{for i} \neq 1 \end{cases} \tag{3}$$

$$\Delta g_{ij} = \begin{cases} 0 & \text{for } i \neq 1 \text{ and } j \neq 1 \\ 6\phi_1(1-\phi_1)\Delta S(T_m - T - mc_L) & \text{for } i = 1 \text{ and } j \neq 1 \\ -6\phi_1(1-\phi_1)\Delta S(T_m - T - mc_L) & \text{for } i \neq 1 \text{ and } j = 1 \end{cases} \tag{4}$$

The diffusion equation with anti-trapping current which compensates the effect of diffuse interface thickness is given as [17,18]:

$$\frac{\partial c}{\partial t} = \nabla\phi_1 D_L \nabla c_L + \nabla\left(\frac{\varepsilon_{SL}}{\sqrt{2w_{SL}}}(c_L - c_S)\sqrt{\phi_1(1-\phi_1)}\frac{\partial \phi_1}{\partial t}\frac{\nabla\phi_1}{|\nabla\phi_1|}\right) \tag{5}$$

Here we neglect the diffusion in solid phase which is very small and has little effect on the competitive grain growth. $c=\phi_1 c_L+(1-\phi_1)c_S$ is the mixture concentration of solid and liquid. The parameters, ε and w, with subscript SL or SS are related with the interface energy σ and interface width 2ξ of the solid-liquid interface or solid-solid interface:

$$\varepsilon_{SL} = \frac{4}{\pi}\sqrt{\xi_{SL}\sigma_{SL}}\,;\quad w_{SL} = \frac{2\sigma_{SL}}{\xi_{SL}} \tag{6}$$

$$\varepsilon_{SS} = \frac{4}{\pi}\sqrt{\xi_{SS}\sigma_{SS}}\,;\quad w_{SS} = \frac{2\sigma_{SS}}{\xi_{SS}} \tag{7}$$

The phase field mobility M_{ij} has a definite relationship with the interface mobility. For the vanishing kinetic coefficient, the phase field mobility of solid-liquid interface M_{SL} is determined as [16,17]:

$$M_{SL} = \frac{8\sigma D_L \sqrt{2w_{SL}}}{\pi \Delta S m c_L^e (1-k)\varepsilon_{SL}^3} \tag{8}$$

For the GB a simpler relationship exits between M_{SS} and GB mobility. As our focus is the dendrite tip growth but not GB motion, we have simply set M_{SS}=0.1 M_{SL}.

A binary approximation to a nickel-based superalloy was adopted, in which all the elements except Ni are taken as solute. The material properties are given as: melting point of Ni T_m=1728 K, liquidus slope m=4.47 K/wt.%, partition coefficient k=0.48, entropy of fusion ΔS=2×105 J/m^{3}K, interface energy σ=0.21 J/m^2, diffusion coefficient in liquid D_L=3×10^{-9} m^2/s and initial concentration c_0=6.5 wt.%. In order to decrease the primary dendrite spacing so as to reduce the computational size, a relatively large thermal gradient G is specified, G=60 K/mm. Our simulations were carried out at a relative low pulling velocity V=60 μm/s and a relative high pulling velocity V=1000 μm/s. Dendritic growth were obtained at both of these two velocities. No-flux boundary conditions are applied on the vertical walls. Stochastic noise obeying Gaussian distribution was added for the sidebranching dynamics. The interface width was employed as 2ζ=5Δx. A proper selection of 2ζ is related with the characteristic length scale of the microstructure, which is tip radius for dendrite growth and decreases with increased pulling velocity. Therefore a large value of 2ζ can be used at low pulling velocity. In our studies the grid spacing and time step were Δx= Δy=1 μm and Δt=6.67×10^{-5} s for V=60 μm/s, and Δx= Δy=0.25 μm and Δt=4.17×10^{-6} s for V=1000 μm/s. This discretization is a compromise between calculation time and independence of the discretization.

The temperature was specified with the gradient in the y-direction and decreased according to the cooling rate VG. In order to keep the dendrite in the calculation box, a moving frame method was used when the grid number between the dendrite tip and top boundary was less than a certain value. In each moving step the bottom region with a certain length was shifted and cut off. The values of phase field and concentration field in the shifted region were frozen in and saved. The total solidification structure can be obtained by combining all the "frozen-in" domains with the computational domain. This treatment can save simulation time and also does not affect the dynamics of dendrite tip.

3. Results and discussion

The competitive dendrite growth of two converging grains at V=60μm/s with uniform dendrite spacing of λ_1=160μm are shown in Fig. 1. The preferred crystal orientation of grain A is parallel to the thermal gradient, and the misorientation angle of grain B is 10°. The total solidification structure is divided into four parts from the bottom to the top (Figs. 1 a-d). It can be seen that the unfavorably orientated dendrites are blocked by the favorably oriented dendrite A1 at the GB in the initial stage (Fig. 1a and 1b), meanwhile dendrite A1 also dose not develop new stems toward grain B. So the GB keeps parallel to dendrite A1. This result is consistent with the prediction of classical theory model [1-3]. The unusual overgrowth observed in the experiment [7], i.e. the unfavorably oriented dendrite blocks the favorably oriented one, is also found in our simulation for a long time, see Figs. 1c and 1d. Moreover, the unusual overgrowth happens only after several unfavorably oriented dendrites have been blocked, but not for every impingement. This is also in accordance with the experimental observation [7].

It has been assumed that the solute interaction is a probable reason for the unusual overgrowth [7]. When two converging dendrites get close to each other, their concentration fields will overlap, which should affect the growth behavior of dendrite tips. Here our simulation gives a direct evidence for this assumption. Fig. 2a shows the horizontal liquid composition distributions just ahead of the favorably oriented GB dendrite (dendrite A1) when the unfavorably oriented GB dendrite (dendrite B2) gets close to it. The peak at the left side corresponds to the position ahead of dendrite A1, and the peak at the right side corresponds to the position in front of dendrite B2. It can be seen that the liquid composition in front of dendrite A1 increases with time. This is due to the severe superposition of concentration field which results in the difficulty of lateral solute diffusion. Because the driving force for dendrite growth decreases as the liquid concentration increases, the growth of dendrite A1 should slow down. This is clearly shown in Figs. 1e-1g. It can be seen that when dendrite B2 gets close to dendrite A1 (Fig. 1f), not only the growth of dendrite B1, but also the growth of dendrite A1 is retarded due to the solute field overlapping. The retarded growth of the favorably oriented GB dendrite was also observed in directional solidification experiments by analyzing the microstructure of quenched samples [7]. However, our simulation indicates that usually the lag of dendrite A1 can not be maintained. It recovers to the same height as dendrite A2 when the growth of dendrite B2 almost stops (see Fig. 1g).

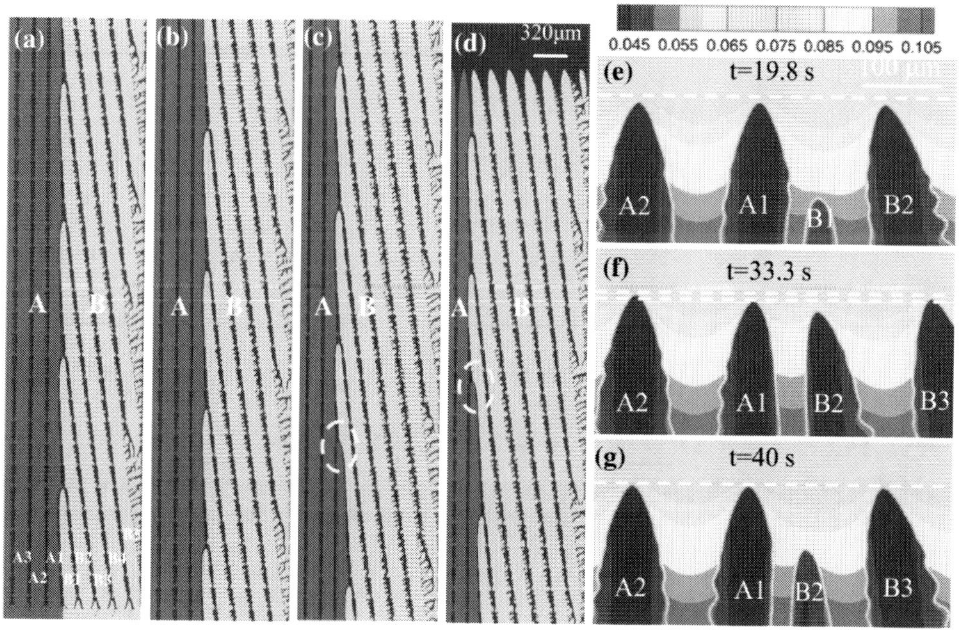

Figure 1. Competitive growth of converging dendrites at V=60μm/s with uniform spacing, λ_1=160μm, the final structures are divided into four parts (a), (b), (c) and (d) from the bottom to the top, (e)-(g) concentration field near the tip at different times.

Besides this effect, the solute field overlapping also results in the asymmetry solute field in the two sides of the favorably oriented GB dendrite (dendrite A1). Due to the severe superposition of concentration field in the right side of dendrite A1, the lateral solute gradient in this side is smaller than that in the left side. So the solute diffusion is more difficult in the right side. This makes dendrite A1 move toward the left side, i.e. the side with large spacing. This mechanism is similar to that reported by Han and Trivedi [19]. They found that in a single oriented dendrite array dendrite spacing adjusted to uniform through later motion after the elimination or creation of dendrites. The lateral displacements of dendrite tips as a function of time for the three favorably oriented dendrites are shown in Fig. 3a. It can be seen that the lateral motion of dendrite A1 increases with time. At the same time dendrite A2 also moves toward left side due to approaching of dendrite A1, but its motion is slower than dendrite A1. So the spacing between dendrite A1 and A2 decreases continuously with time. When this spacing decreases to a certain extent, the aforementioned lag of dendrite A1 due to one unfavorably oriented dendrite getting close will preserve and dendrite A1 will be blocked by the next unfavorably oriented dendrite. That is why usual overgrowth only happens after some unfavorably oriented dendrites have been blocked.

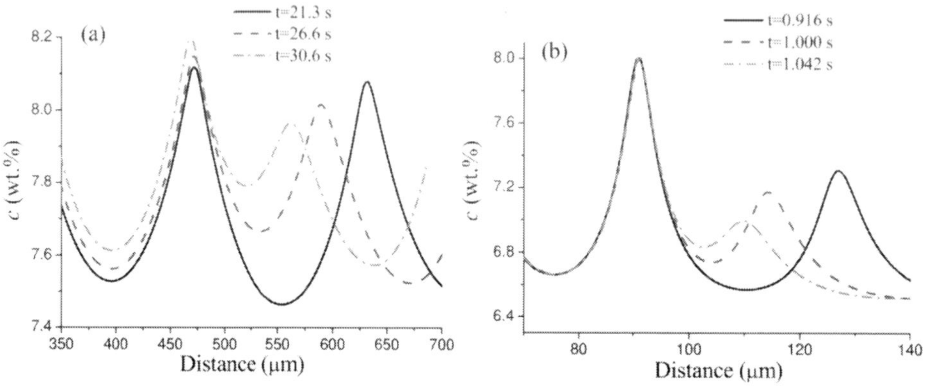

Figure 2. Liquid composition distributions along the horizontal line just ahead of the favorably oriented GB dendrite (dendrite A1) at different times, (a) the results for the condition shown in Fig. 1, (b) the results for the condition shown in Fig. 4.

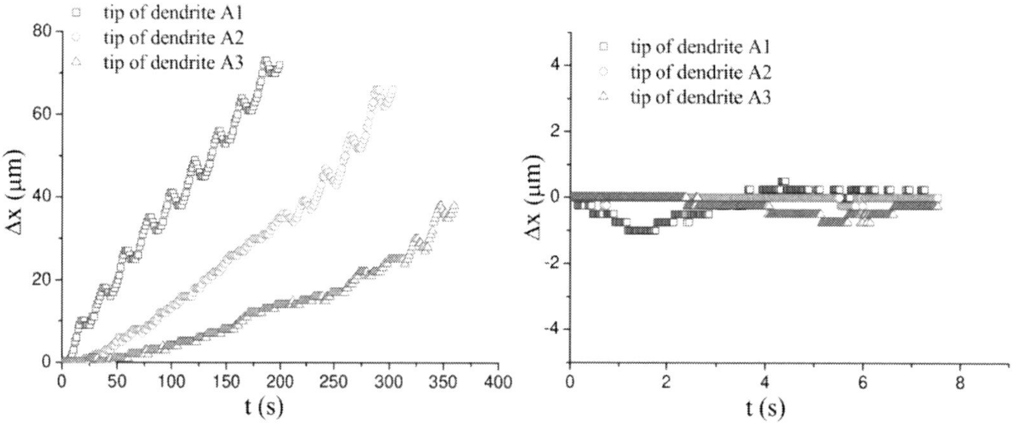

Figure 3. The lateral displacement of dendrite tips as a function of time for the three favorably oriented dendrites, (a) the results for the condition shown in Fig. 1, (b) the results for the condition shown in Fig. 4.

The competitive dendrite growth of two converging grains at relatively high pulling velocity, V=1000μm/s, is shown in Fig. 4. Because the low velocity simulation indicates the unusual overgrowth only happens when the spacing between dendrite A1 and A2 decreases to certain extent. Here we have set the dendrite spacing in grain A to be small so as to decrease the time for the occurrence of unusual overgrowth, if it would happen at this high velocity. However, the unusual overgrowth is not found in this case. When the unfavorably oriented GB dendrite (dendrite B3) gets close to the favorably oriented GB dendrite (dendrite A1), the horizontal liquid composition just ahead of dendrite A1 is almost unchanged except a small increase of the tail in the right side (see Fig. 2b). So the growth of dendrite A1 does not slow down during the approaching of dendrite B2 as shown in Figs. 4 e-g. Fig. 3b indicates that the spacing adjustment through the lateral motion of dendrite tip is also neglected at high velocity. Since the solute diffusion length at high velocity is smaller than that at low velocity, the superposition of concentration field is insignificant at high velocity. So the aforementioned effects of solute interaction on the growth behavior of dendrite tips do not work here, and the unusual overgrowth will not happen.

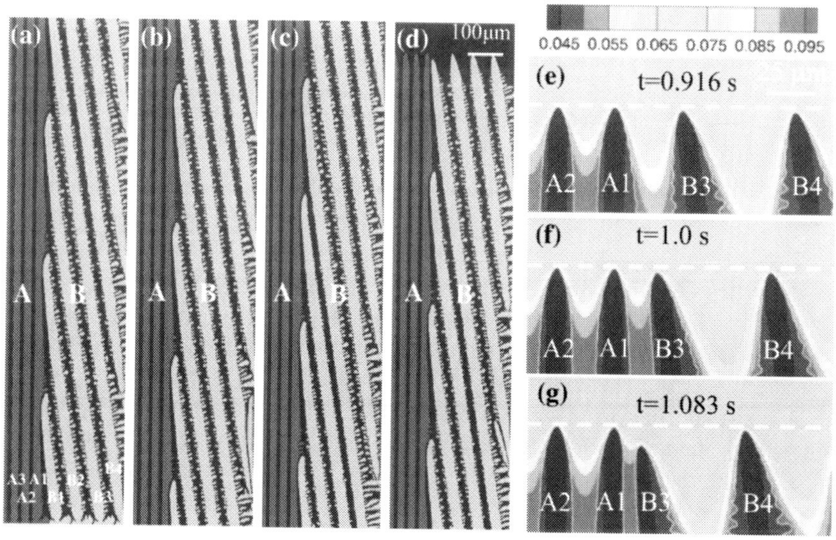

Figure 4. Competitive growth of converging dendrites at V=1000μm/s with λ_1=30μm in grain A and λ_1=60μm in grain B, the final structures are divided into four parts (a), (b), (c) and (d) from the bottom to the top, (e)-(g) concentration field near the tip at different times.

The unusual overgrowth in our simulation is caused by the solute interaction effect. Whether this effect can lead to the unusual growth is related with two characteristic lengths, i.e., the solute diffusion length l_D and the distance between the tips of favorably and unfavorably oriented dendrites l_Δ. The solute effect will be enhanced by increasing l_D or by decreasing l_Δ. The value of l_Δ increases as the thermal gradient decreases. So decreasing thermal gradient will weaken the solute effect. However, under the condition of lower thermal gradient the unusual overgrowth may also happen at a lower pulling velocity since l_D is larger. Comparisons of the characteristic lengths between simulations and experiments [4,7] are given in table 1. For the calculation of l_D in experiments the diffusion coefficient D is estimated to be 3000μm^2/s, which is the same as that in simulations. It can be seen that the value of l_Δ in our simulations is an order of magnitude smaller than that in experiments. This is due to the large thermal gradient which is approximately an order of magnitude larger than that in experiments. It can be also seen that the ratio of l_D/l_Δ for the occurrence of unusual overgrowth in experiment is smaller than that in our simulation. In the experiments of Zhou [7], the unusual overgrowth is found when l_D is approximately two times larger than l_Δ. But in our simulations the unusual overgrowth is not found when this ratio is six. This means that in our simulation the occurrence of unusual overgrowth is more difficult, and a more significant solute interaction effect by increasing diffusion length is needed to bring the unusual overgrowth. This may relate with the two dimensional nature of our simulation. In 2D case there is only one unfavorably oriented GB dendrite directly contacting with the favorably oriented GB dendrite. While the three-dimensional nature of the experimental process leads to the interaction of the favorably oriented GB dendrite with several unfavorably oriented dendrites. So the solute interaction effect should be stronger in three dimensional experiments.

Due to the two dimensional nature and unreal parameters, we cannot fully confirm that the explanation of unusual overgrowth obtained from our simulation is just the specific reason for experimental observation. But at least it is reasonable and possible. Much more works both in simulations and experiments are needed to get a thorough understanding of the unusual overgrowth.

Table 1 Comparisons of some characteristic lengths between simulations and experiments

	V (μm/s)	l_D (μm)	l_Δ (μm)	l_D/l_Δ
Experiments with unusual overgrowth[7]	16.7	179.6	80	2.25
Experiments without unusual overgrowth[4]	27	111.1	110	1.01
Simulation with unusual overgrowth	60	50	3	16.67
Simulation without unusual overgrowth	100	30	5	6

Fig. 5 shows the microstructure of two diverging grains with various dispositions. In Figs. 5a and 5b the preferred crystal orientation of grain A is parallel to the thermal gradient, and the misorientation angle of grain B is 10° and 20°, respectively. It can be seen that the gap between the diverging grains is filled by new dendrites developed from both grain A and grain B. However, due to the superiority of grain A in geometrical disposition, the development of new dendrites from grain A can lead to overgrowth of grain B, while the opposite is not true. So the GB is inclined toward grain B. This is the same as the prediction of the classical theory model [1-3]. Moreover, it can be found that the number of newly developed dendrites from grain A in Fig. 5a is larger than that in Fig. 5b after the same pulling length, which means the GB angle increases with the misorientation angle of grain B. This result is consistent with the experimental observation [7].

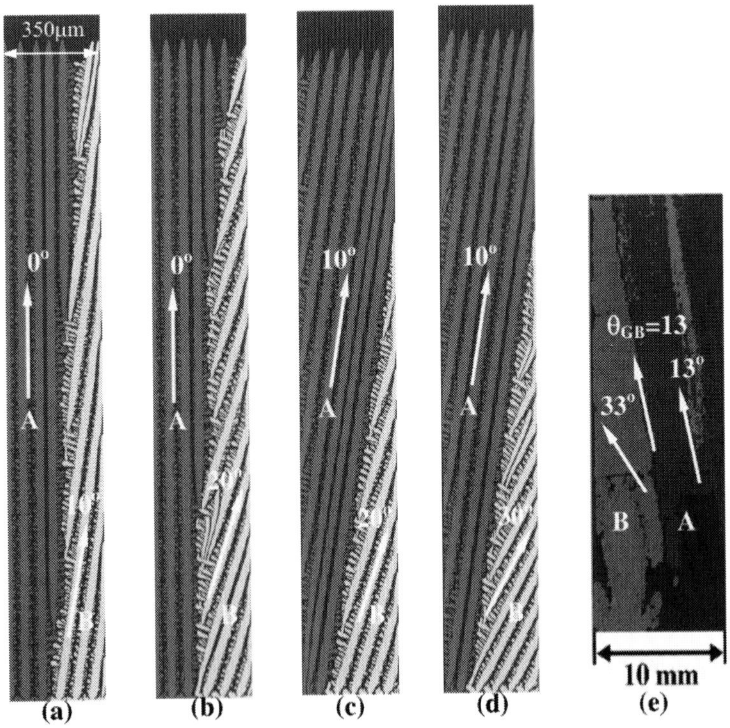

Figure 5. Competitive growth of diverging dendrites with various dispositions, (a)-(d) simulation at V=1000μm/s (d) experimental observation form reference [8].

It can be also seen from Figs. 5a and 5b that the number of newly developed dendrites from grain B is larger than that from grain A. The reason is that in the gap region the growth direction of secondary dendrite arm of the unfavorably oriented dendrite is more aligned with the heat flow direction than that of the favorably oriented dendrite. So the secondary arm in grain B is more developed and the new dendrites are easy to develop from it. When the preferred crystal orientation of grain A is also misaligned with the heat flow direction, as shown in Figs. 5c and 5d, the development of the second dendrites in grain A is so inadequate that new dendrites can not develop from grain A in the gap region. Therefore the GB is parallel to the relatively favorably oriented grain rather than inclining toward another grain as in Figs. 5a and 5b. This result is in agreement with the observation in experiments [8]. It indicates that the disposition of the two diverging grains is a key factor to determine the misorientation of GB.

4. Summary

The competitive grain growth processes for both diverging and converging dendrites have been well reproduced by our phase field simulation. Besides the usually accepted overgrowth manner, some unusual overgrowth results observed in some recent experiments are also obtained in our simulation. In converging case it is found that the overgrowth of favorably oriented dendrite by the unfavorably oriented one at the low pulling velocity is related with the effect of solute interaction, which can result in the retarded growth of dendrites at GB and the lateral motion of favorably oriented dendrite. In diverging case it is found that the development of the secondary arm and new dendrites depends on the geometrical disposition of the two grains, which determines the inclination angle of the GB.

Acknowledgements
The work was supported by National Natural Science foundation of China (Grants No. 51101124, 51001082), National Basic Research Program of China (Grants No.2011CB610401), the Fundamental Research Fund of Northwestern Polytechnical University (JC201006), Free Research Fund of State Key Laboratory of Solidification Processing (67-QP-2011).

References

[1] Walton D and Chalmers B 1959 *Trans. Metal.l Soc. AIME* **215** 447
[2] Rappaz M and Gandin C A 1993 *Acta Metal.l Mater.* **41** 345
[3] Gandin C A and Rappaz M 1994 *Acta Metall. Mater.* **42** 2233
[4] D'Souza N, Ardakani M G, Wagner A, Shollock B A and Mclean M 2002 *J. Mater. Sci.* **37** 481
[5] Zhu M F and Hong C P 2001 *ISIJ Int.* **41** 436
[6] Dong H B and Lee P D 2005 *Acta Mater.* **53** 659
[7] Zhou Y Z, Volek A and Green N R 2008 *Acta Mater.* **56** 2631
[8] Wagner A, Shollock B A and Mclean M. 2004 *Mater. Sci. Eng. A* **347**:270
[9] Steinbach I 2008 Acta Mater. **56** 4965
[10] Amoorezaei M, Gurevich S and Provatas N 2010 *Acta Mater.* **58** 6115
[11] Grurevich S, Karma A, Plapp M and Trivedi R 2010 *Phys. Rev. E* **81** 011603
[12] Wang Z J, Wang J C, Li J J, Yang G C and Zhou Y H 2011 *Phys. Rev. E* 84 041604
[13] Steinbach I, Pezzolla F, Nestler B, Seebßelberg M, Prieler R, Schmitz G J and Rezenda J L L 1996 *Physica D* 94 135
[14] Eiken J, Böttger B and Steinbach I 2006 *Phys. Rev. E* **73** 066122
[15] Kim S G, Kim W T, Suzuki T, Ode M 2004 *J. Cryst. Growth* **261** 135
[16] Kim SG 2007*Acta Mater* **55** 6513
[17] Kim SG 2007*Acta Mater* **55** 4391
[18] Steinbach I 2009 *Modelling Simul. Mater. Sci. Eng.* **17** 073001
[19] Han S H and Trivedi R 1994 *Acta Metall. Mater.* **42** 25

A new approach to multi-phase field for the solidification of alloys

P C Bollada, P K Jimack, A M Mullis

Institute of Materials Research, University of Leeds, LS2 9JT

E-mail: p.c.bollada@leeds.ac.uk

Abstract. We show that the standard approach to the modeling of multi-phase field dynamics for the solidification of alloys has three major defects and offer an alternative more successful formulation. The model contains an action made up of a free energy functional of the temperature, concentration and the phase variables. In addition there is a penalty functional for the gradients of the phase variables, which keeps all field variables continuous — the phase field method. A variation of this action is related to time derivatives of the variables in the field. At any physical point there exists up to N phases, ϕ_i, each of which represent the proportion of each phase at that point, thus implying the constraint, $\sum_{i=1}^{N} \phi_i = 1$. The standard Lagrange multiplier treatment for imposing this constraint has three major defects: non-reduction to standard single phase formulation; a dependence on the value of N; generation of unphysical additional phases. We demonstrate a multi-phase formulation that avoids these two defects and, partly as a consequence, does not generate spurious additional phases. Moreover, this aim is achieved without losing the active part that three non-zero phases should play if present at any point.

1. Introduction

In recent years the importance of phase-field simulation as a tool to understanding microstructure formation during solidification has grown significantly. The key advantage of such models is that by introducing a continuous (differentiable) phase variable, ϕ, the value of which represents the phase of the material, the need to explicitly track the solid-liquid interface is removed. Instead, the mathematically sharp interface is replaced by a diffuse interface of finite width, the motion of which may be tracked using standard techniques for partial differential equations.

Early phase-field models of solidification concentrated on single-phase systems, in which there was the liquid and only a single solid phase present. However, the phase-field concept may be extended to systems where there is more than one solid phase present, resulting in multi-phase field models. For a topical review of multi-phase field modelling in material science see [1]. In multi-phase field models the scalar variable, ϕ, is replaced by a vector, $\boldsymbol{\theta}$, where the ith element θ_i, is the amount of phase i present[1]. This extension has, though, yielded variations in the derivation techniques used to obtain the equations of motion for the interface from the starting equations, with consequent differences in the properties of the resulting models. One of the main issues to arise in multi-phase field models is that because the phases, θ_i, can act

[1] We use the notation $\theta_i, i \in [1, N]$ for the linearly dependent physical variable and $\phi_i, i \in [1, N-1]$ for the independent variables

independently, an additional condition must be applied to ensure that the sum of the phases remains everywhere constant.

2. Standard Lagrange multiplier treatment

Most phase field models of solidification (both single and multi-phase) have a common starting point, this being the definition of a free energy functional, F, of the phase variables, θ_i, concentration, c and temperature, T. The appropriate form of F for the multi-phase problem has been adapted from several sources in the literature, e.g. [2]

$$F = \int_\Omega \frac{1}{2}\sum_{i=1}^{j-1}\sum_{j=2}^{N}\Gamma^{ij}|\theta_i\nabla\theta_j - \theta_j\nabla\theta_i|^2\,\mathrm{d}^3\mathbf{x} + \int_\Omega f(\boldsymbol{\theta}, c, T)\,\mathrm{d}^3\mathbf{x} \tag{1}$$

where: Ω is an arbitrary volume; Γ^{ij} includes the gradient energy coefficients and the anisotropy between phases i and j necessary, for example, for dendritic growth; and f is the free energy density. A particularly simple example of the latter, sufficient for this paper, is given by a minor modification to the formulation of [2] (though the arguments to be presented here are independent of the precise form assumed for f):

$$f \equiv \sum_{i=1}^{j-1}\sum_{j=2}^{N} W_{ij}\theta_i^2\theta_j^2 - \sum_j m_j\theta_j^3(6\theta_j^2 - 15\theta_j + 10) + \frac{RT}{v_m}\left[c\ln(c) + (1-c)\ln(c)\right], \tag{2}$$

with the coefficients governing the concentration-dependent double-well potential extended to N phases given by

$$\begin{aligned} W_{ij} &= W_{ij}^A c + (1-c)W_{ij}^B \\ m_j &= m_j^A c + (1-c)m_j^B. \end{aligned} \tag{3}$$

Here R is the universal gas constant, v_m the molar mass (assumed constant), the constants W_{ij}^A and W_{ij}^B are entries of symmetric matrices whose values are dependent upon the double-well potential barrier between phases i and j and the constants m_i^A and m_i^B relate to the Gibbs energy of phase i, for either pure component A or B. Specifically, this formulation omits any enthalpy of mixing terms, which restricts the type of solid phases that can result to ideal binary solid solutions.

The equations governing the evolution of the phase and solute profiles can be given as

$$-\tau\frac{\partial\theta_i}{\partial t} = \frac{\delta F}{\delta\theta_i}, \quad i \in [1, N] \tag{4}$$

and

$$\frac{\partial c}{\partial t} = \nabla\cdot\left(D(\boldsymbol{\theta})c(1-c)\nabla\frac{\delta F}{\delta c}\right) \tag{5}$$

together with the constraint

$$\sum_{j=1}^{N}\theta_j = 1, \tag{6}$$

where: τ is a characteristic time equivalent to inverse mobility, which is here assumed constant; D is a function defining the local diffusivity, which is a sum of the diffusivities for each phase weighted by the amount of each phase present.

The constraint (6) implies a linear dependence of the variables indicating that the system can be represented by $N-1$ independent variables, which we denote by $\phi_i, i \in [1, N-1]$. In particular, when $N=2$ the multi-phase system is related to a single phase system with variable ϕ. This may be set to, say, $\phi = \theta_1$, but there are other equally valid alternatives.

The Lagrange multiplier method for ensuring the constraint (6) expresses (4) as

$$-\tau \frac{\partial \theta_i}{\partial t} = \frac{\delta F}{\delta \theta_i} + \Lambda, \quad i \in [1, N] \tag{7}$$

where, to guarantee $\sum_{j=1}^{N} \dot{\theta}_j = 0$, we must have

$$\Lambda = -\frac{1}{N} \sum_{j=1}^{N} \theta_j. \tag{8}$$

2.1. Non-reduction of the multiphase formulation to single phase

We now demonstrate that the standard Lagrange multiplier treatment of multi-phase field dynamics, e.g. [2], does not reduce to the equivalent single phase form. Let $F(\theta_1, \theta_2, c)$ be the free energy for an $N=2$ phase system dependent on liquid phase, θ_1 and solid phase, θ_2 and concentration, c. Then $\theta_1 + \theta_2 = 1$ and we choose a single variable ϕ so that

$$\theta_1 = \phi \tag{9}$$

and

$$\theta_2 = 1 - \phi. \tag{10}$$

The multi-phase gradient contribution for $N=2$, for example, is

$$G(\theta_1, \theta_2) = \int_\Omega \tfrac{1}{2}\Gamma^{12} |\theta_1 \nabla \theta_2 - \theta_2 \nabla \theta_1|^2 \, \mathrm{d}^3\mathbf{x} \tag{11}$$

which reduces to

$$G(\phi, 1-\phi) = \int_\Omega \tfrac{1}{2}\Gamma^{12} |\nabla \phi|^2 \, \mathrm{d}^3\mathbf{x} \tag{12}$$

The single phase equation is

$$-\tau \dot{\phi} = \frac{\delta F}{\delta \phi} \tag{13}$$

This is equivalent, in the multi-phase (two phase) variables to

$$\begin{aligned} -\tau \dot{\theta}_1 &= \frac{\delta F}{\delta \theta_1} - \frac{\delta F}{\delta \theta_2} \\ -\tau \dot{\theta}_2 &= \frac{\delta F}{\delta \theta_2} - \frac{\delta F}{\delta \theta_1} \end{aligned} \tag{14}$$

since[2]

$$\begin{aligned} \frac{\delta F}{\delta \phi} &= \frac{\partial \theta_i}{\partial \phi} \frac{\delta F}{\delta \theta_i} \\ &= \frac{\delta F}{\delta \theta_1} - \frac{\delta F}{\delta \theta_2}. \end{aligned} \tag{15}$$

[2] Note that throughout this paper repeated sufficies will imply summation, unless they appear on both sides of the equation.

Whereas in the multi-phase formulation the Lagrange multiplier gives

$$-\tau\dot{\theta}_i = \frac{\delta F}{\delta\theta_i} - \frac{1}{N}\sum_j^N \frac{\delta F}{\delta\theta_j}, \tag{16}$$

which for $N = 2$ gives

$$\begin{aligned} -\tau\dot{\theta}_1 &= \tfrac{1}{2}\frac{\delta F}{\delta\theta_1} - \tfrac{1}{2}\frac{\delta F}{\delta\theta_2} \\ -\tau\dot{\theta}_2 &= \tfrac{1}{2}\frac{\delta F}{\delta\theta_2} - \tfrac{1}{2}\frac{\delta F}{\delta\theta_1}. \end{aligned} \tag{17}$$

Thus the Lagrange multiplier approach does not reduce to the single phase formulation and in the particular case for $N = 2$ gives precisely half that of the single phase formulation.

2.2. *Spurious extra phases and N dependence*

This section shows that the formulation in [2], when used with solidification from a pure seed, say θ_2 growing into melt θ_1, leads to the unphysical formation of the other phase(s), θ_i for $i > 2$. This is illustrated in Fig. 2 (left) for $N = 3$). See also [5] p301. A new alternative model that does not exhibit spurious growth, see Fig. 2 (right), is presented in Section 3. The Lagrange multiplier formulation in [2] is

$$-\tau\dot{\theta}_i = \frac{\delta F}{\delta\theta_i} - \frac{1}{N}\sum_{j=1}^N \frac{\delta F}{\delta\theta_j} \tag{18}$$

where for isotropy (Γ^{ij} independent of $\boldsymbol{\theta}$) we have

$$\begin{aligned} \frac{\delta F}{\delta\theta_i} &= \sum_{j\neq i}^N \Gamma^{ij}\{2(\theta_i\nabla\theta_j - \theta_j\nabla\theta_i)\cdot\nabla\theta_j + (\theta_i\nabla^2\theta_j - \theta_j\nabla^2\theta_i)\theta_j\} \\ &+ \sum_{j\neq i}^N 2W_{ij}\theta_i\theta_j^2 - 30m_i\theta_i^2(1-\theta_i)^2. \end{aligned} \tag{19}$$

Consider a system of N phases but only two phases θ_1, θ_2 present in some region with no interaction with other phases. With $\theta_{i>2} = 0$ the Lagrange multiplier gives

$$\begin{aligned} -\tau\dot{\theta}_1 &= (1 - \tfrac{1}{N})\frac{\delta F}{\delta\theta_1} - \tfrac{1}{N}\frac{\delta F}{\delta\theta_2} \\ -\tau\dot{\theta}_2 &= (1 - \tfrac{1}{N})\frac{\delta F}{\delta\theta_2} - \tfrac{1}{N}\frac{\delta F}{\delta\theta_1} \\ -\tau\dot{\theta}_{i>2} &= -\tfrac{1}{N}\left(\frac{\delta F}{\delta\theta_1} + \frac{\delta F}{\delta\theta_2}\right), \end{aligned} \tag{20}$$

and clearly the growth depends on N. Moreover, it is only for $N = 2$ that a pure phase grows as single phase growth (up to a factor of two).

It should be added that with careful choice of potential, see [3], spurious growth can be mitigated. However, this only holds for $N = 3$ and the generalisation to $N > 3$ is not clear within the Lagrange multiplier formulation.

The Lagrange multiplier approach may be viewed as an $N \times N$ matrix projection, $\mathbf{P}$,

$$-\tau\dot{\boldsymbol{\theta}} = \mathbf{P}_{(N)}\frac{\delta F}{\delta \boldsymbol{\theta}}, \quad -\tau\dot{\theta}^i = P^{ij}_{(N)}\frac{\delta F}{\delta \theta^j} \tag{21}$$

where the projection equivalent to the Larange multiplier is given by

$$\mathbf{P}_{(N)} = \mathbf{I} - \mathbf{n}\mathbf{n}^T. \tag{22}$$

For example with $N = 2$, we have

$$\mathbf{n} = \frac{1}{\sqrt{2}}[1, 1]^T \tag{23}$$

is the outward normal to the line $\theta_2 = 1 - \theta_1$. If we consider the phase variables, θ_1 and θ_2 to be Cartesian coordinates, then $\mathbf{n}$ has unit length.

3. A new formulation

The essence of the new formulation is to alter the projection matrix, $\mathbf{P}$, in such a way as to eliminate the defects we have highlighted with the Lagrange multiplier approach. To this end we propose a set of properties that the matrix (metric) $\mathbf{P}$ must possess (as with the Lagrange multiplier, a new specification of $\mathbf{P}$ does not affect material properties of the model):

(i) Reduces to $n < N$ case when only n phases are present locally in a N phase system.

(ii) The projection must never be zero at any point, as this will inhibit growth from a pure phase.

(iii) The projection must be symmetric with positive or zero eigenvalues as a result of the consistency requirement between the left-hand and right-hand side components.

(iv) The metric should be degenerate and continuous: that is, it must map from dimension N to dimension $n < N$ smoothly.

(v) Triple points should be active parts of the system: this excludes the model proposed by Steinbach [4].

These criteria do not fix a unique value for $\mathbf{P}$. We now give a form for $\mathbf{P}$ which satisfies most (but not all at this stage) of the criteria.

By inspecting (14) we see that for $N = 2$ we must have

$$\mathbf{P} = \begin{bmatrix} 1 & -1 \\ -1 & 1 \end{bmatrix} \tag{24}$$

In order to generalise this result we note that we may trivially write this

$$\mathbf{P} = \frac{\theta_1\theta_2}{(1-\theta_1)(1-\theta_2)}\begin{bmatrix} 1 & -1 \\ -1 & 1 \end{bmatrix} \tag{25}$$

since $\theta_1 = 1 - \theta_2$. This suggests a more general form for $\mathbf{P}$ given by

$$\mathbf{P} = \sum_{j=2}^{N}\sum_{i=1}^{j-1}\frac{\theta_i\theta_j}{(1-\theta_i)(1-\theta_j)}(\mathbf{x}_i - \mathbf{x}_j)\otimes(\mathbf{x}_i - \mathbf{x}_j), \tag{26}$$

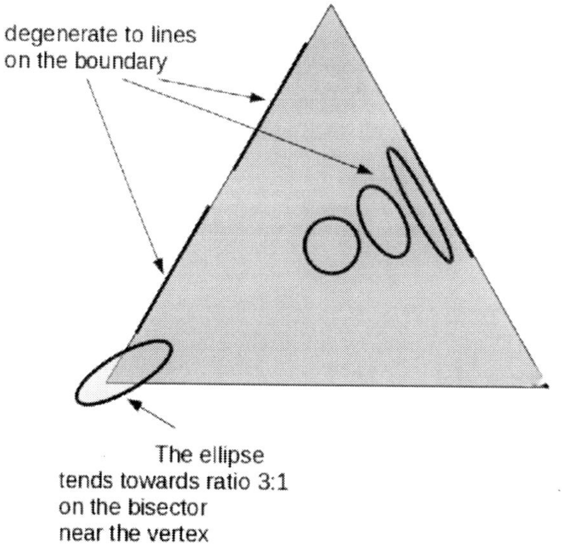

Figure 1. Eigen values and vectors displayed as ellipses as a function of position on the triangular simplex. As a point approaches the boundary the ellipse degenerates to a line, but as we approach a vertex from the centre the circle becomes elliptical.

where $\mathbf{x}_i$ are the barycentric coordinates of each vertex of the simplex representing the multiphase field, e.g. a triangle for $N = 3$, tetrahedran for $N = 4$ etc. The vertices are thus, $(\mathbf{x}_i)_j = \delta_{ij}$, so that in components, P^{kl}

$$P^{kl} = \sum_{j=2}^{N}\sum_{i=1}^{j-1} \frac{\theta_i\theta_j}{(1-\theta_i)(1-\theta_j)}(\delta_{ki} - \delta_{kj})(\delta_{li} - \delta_{lj}). \tag{27}$$

The $N = 3$ case follows as

$$\mathbf{P} = \frac{\theta_1\theta_2}{(1-\theta_1)(1-\theta_2)}\begin{bmatrix} 1 & -1 & 0 \\ -1 & 1 & 0 \\ 0 & 0 & 0 \end{bmatrix} + \frac{\theta_2\theta_3}{(1-\theta_2)(1-\theta_3)}\begin{bmatrix} 0 & 0 & 0 \\ 0 & 1 & -1 \\ 0 & -1 & 1 \end{bmatrix} + \frac{\theta_3\theta_1}{(1-\theta_3)(1-\theta_1)}\begin{bmatrix} 1 & 0 & -1 \\ 0 & 0 & 0 \\ -1 & 0 & 1 \end{bmatrix}. \tag{28}$$

where we note that if, say, $\theta_3 = 0$ we obtain the $N = 2$ case.

3.1. Some properties of $\mathbf{P}$

This section considers the new model from the perspective of eigenvectors and eigenvalues of the matrix $\mathbf{P}$ in order to see the effect on the system. The interface defined by $\theta_3 = 0$ gives the matrix

$$\mathbf{P} = \begin{bmatrix} 1 & -1 & 0 \\ -1 & 1 & 0 \\ 0 & 0 & 0 \end{bmatrix} \tag{29}$$

This has one positive eigenvalue, 2, with corresponding eigenvector $[1, -1, 0]$, which aligns with the the interface. On the other hand, in the centre of the simplex, $\boldsymbol{\theta} = [1/3, 1/3, 1/3]$, we find one eigenvalue of $3/4$ corresponding to any vector lying on the simplex. So in the centre the matrix, **P**, simply projects out the term normal to the simplex and multiplies by 3/4. On the other hand, on the interface, **P** has the effect of also projecting out the term normal to the interface. In general **P** has the double effect of projecting out the normal component and rescaling the components of the vector along the two eigenvectors. Representing **P** at any point on the simplex by an ellipse with major and minor axes of lengths and direction given by the eigenvalues and eigenvectors, we can view the action of **P** in the centre as a circle and at an interface a degenerate ellipse — a line (see Fig. 1).

Consider a typical path on an $N = 3$ simplex given by

$$\mathbf{x}(t) = [\theta_1 = t, \theta_2, \theta_3 = 1 - 2t], \quad t \in [0, \tfrac{1}{2}]. \tag{30}$$

We are interested in the property of **P** as we approach the vertex $\theta_3 = 1$ as t tends to zero. We find that **P** is an ellipse with major axis of length 3/2 pointing along the line and minor axis of length 1/2.

In contrast, for the Lagrange multiplier, **P** is represented as a circle throughout the triangle.

3.2. *Ill-defined* **P** *for a pure phase*

The model as it stands suffers from being ill-defined at any vertex, $\theta_i = 1$. This is due to a feature of our construction that, at the vertices, **P** depends on the path. For example, with $N = 3$, and $\theta_1 = 1$ we find **P** degenerates to two matrices for paths along the two adjoining edges $\theta_3 = 0$ and $\theta_2 = 0$:

$$\mathbf{P}|_{\theta_3=0} = \begin{bmatrix} 1 & -1 & 0 \\ -1 & 1 & 0 \\ 0 & 0 & 0 \end{bmatrix} \quad \mathbf{P}|_{\theta_2=0} = \begin{bmatrix} 1 & 0 & -1 \\ 0 & 0 & 0 \\ -1 & 0 & 1 \end{bmatrix}. \tag{31}$$

Thus we require, in addition to the definition (27), to define unique matrices at each vertex. Alternatives such as enforcing $\theta_i < 1$ via the initial condition, the use of a small parameter in local averaging, or modifying the potential are all problematic.

Many candidates for the value of **P** at the vertices fail, including: $\mathbf{P} = \mathbf{0}$, which inhibits growth; and $\mathbf{P} = \mathbf{I} - \mathbf{U}/N$, which introduces spurious growth. However, we found

$$\mathbf{P}_{\text{vertex}} = 2\mathbf{I} - \mathbf{U}, \text{ if } 1 - \theta_i < \delta, \text{ for any } i, \tag{32}$$

where the small parameter, $\delta \ll 1$, did not significantly alter results[3]. To discuss this we write this out for $N = 3$

$$\mathbf{P}_{\text{vertex}} \equiv \begin{bmatrix} 1 & -1 & -1 \\ -1 & 1 & -1 \\ -1 & -1 & 1 \end{bmatrix} \tag{33}$$

and assume that θ_3 and its gradients vanish. Thus,

$$\frac{\delta F}{\delta \theta_3} = 0 \tag{34}$$

Consequently the third column plays no role and $\dot{\theta}_1$ and $\dot{\theta}_2$ reduce correctly to a two phase formulation. On the other hand

$$\tau\dot{\theta}_3 = \frac{\delta F}{\delta \theta_1} + \frac{\delta F}{\delta \theta_2} \tag{35}$$

[3] For a range of 10^{-10} to 10^{-14} in δ we found the steady state growth rate was effectively unaltered in eutectic or pure phase simulation

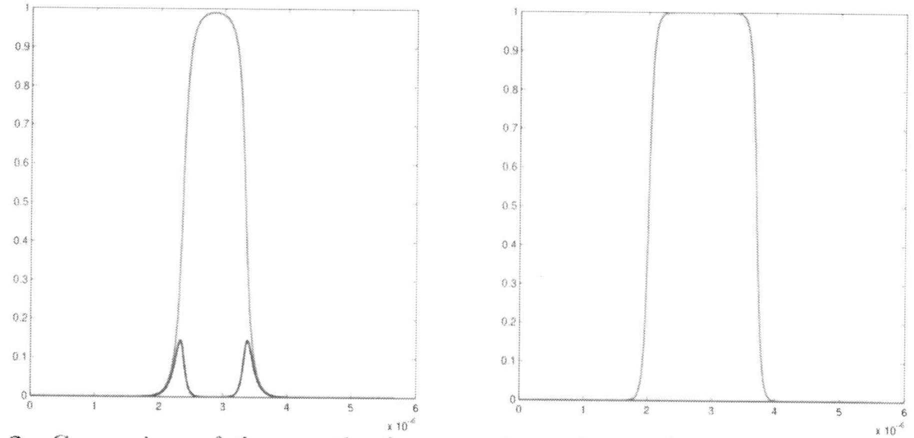

Figure 2. Comparison of the growth of a pure phase, $\theta_2 = 1$ (red continuous line), for the Lagrange multiplier model (left) and an alternative new model presented in section 3 (right). A cross section of the growing seed, θ_2 is superimposed with the cross section of phase θ_3 (blue dashed line). The liquid phase, $\theta_1 = 1 - \theta_2 - \theta_3$, is not shown. The Lagrange multiplier (left) introduces spurious growth of phase θ_3 at the growing boundary, which is completely absent using the new model (right) presented in Section 3. This also effects the speed of growth: the new model grows at a greater rate.

has a right hand side which is non-zero in general. In fact, from (19) we see that the contribution from the potential is zero leaving, for $\theta_1 = 1$:

$$\tau\dot{\theta}_3 = 2(\nabla\theta_1 \cdot \nabla\theta_1) + \nabla^2\theta_1. \tag{36}$$

Now, since $\theta_1 = 1$ we must have $\nabla\theta_1 = 0$ and $\nabla^2\theta_1 \leq 0$ implying $\tau\dot{\theta}_3 \leq 0$. Assuming negative contributions are trapped numerically (if $\theta_i < 0$ then $\theta_i = 0$) this contribution is effectively ignored.

Hence, we have shown that when θ_3 and all its gradients are vanishing, then at one of the other vertices, we find that

$$\mathbf{P}_{\text{vertex}} \equiv \begin{bmatrix} 1 & -1 & -1 \\ -1 & 1 & -1 \\ -1 & -1 & 1 \end{bmatrix} \text{ is indistinguishable from } \begin{bmatrix} 1 & -1 & 0 \\ -1 & 1 & 0 \\ 0 & 0 & 0 \end{bmatrix} \tag{37}$$

The general case, (32), easily follows.

Using this model we obtained the result given in right of Fig. 2. The model also eliminates spurious growth in eutectic solidification.

With $\mathbf{P}$ given by (32) at the vertices, otherwise (27), the final formulation is thus

$$-\tau\dot{\phi}^i = P^{ij}_{(N)}\frac{\delta F}{\delta\phi^j} \tag{38}$$

References

[1] Steinbach I 2009 *Modelling and Simulation in Materials Science and Engineering* **17** 073001+ ISSN 0965-0393 URL `http://dx.doi.org/10.1088/0965-0393/17/7/073001`

[2] Nestler B 2000 *Physica D: Nonlinear Phenomena* **138** 114–133 ISSN 01672789 URL `http://dx.doi.org/10.1016/S0167-2789(99)00184-0`

[3] Folch R and Plapp M 2005 *Physical Review E* **72** 011602+ URL `http://dx.doi.org/10.1103/PhysRevE.72.011602`

[4] Steinbach I, Pezzolla F, Nestler B, Seeszelberg M, Prieler R, Schmitz G J and Rezende J L L 1996 *Physica D* 135–147 ISSN 0167-2789 URL `http://dx.doi.org/10.1016/0167-2789(95)00298-7`

[5] Garke H, Nestler B and Stoth B, 1999 SIAM J. Appl. Math. **60**, No. 1, pp. 295-315

IOP Conf. Series: Materials Science and Engineering **33** (2012) 012100 doi:10.1088/1757-899X/33/1/012100

A modified cellular automaton method for polydimensional modelling of dendritic growth and microsegregation in multicomponent alloys

S. C. Michelic[1,2], J. M. Thuswaldner[3] and C. Bernhard[1]

[1] Montanuniversitaet Leoben, Chair of Metallurgy, Franz-Josef-Straße 18, 8700 Leoben, Austria; [2] INTECO special melting technologies, Wiener Straße 25, 8600 Bruck an der Mur, Austria; [3] Montanuniversitaet Leoben, Chair of Mathematics and Statistics, Franz-Josef-Straße 18, 8700 Leoben, Austria

E-mail: `sebastian.michelic@inteco.at`

Abstract. Numerous numerical models for simulating solidification of metals on a microscopic scale have been proposed in the past, among them are most importantly the phase-field method and models based on cellular automata. Especially the models based on cellular automata (adopting the virtual front tracking (VFT) concept) published so far are often only suitable for the consideration of one alloying element. Since industrial alloys are usually constituted of multicomponent alloys, the possibility of applying cellular automata is rather limited. With the aim of enhancing this modelling technique, a new, modified VFT model, which allows for the treatment of several alloying elements, in the low Péclet number regime is presented. The model uses the physical fundamentals of solute and heat diffusion in two dimensions as a basis for determining the solidification progress. By a new and effective approach, based on a functional extrapolation of the concentration gradient, dendritic growth in multicomponent Fe-C-Si-Mn-P-S alloys could be studied. The model shows the typical behaviour of dendritic solidification, such as parabolic tip and secondary dendrite arm formation as well as selection of preferably aligned columnar dendrites. A validation of the model is performed by the evaluation of morphological parameters and comparing them to experimentally determined values. The results for free and constrained dendritic growth effectively demonstrate the capabilities of this new model. The model is especially attractive for bridging the gap between one-dimensional microsegregation models and multidimensional morphology models with regard to modelling the complex interrelations between segregation on a multidimensional level and morphology formation.

1. Introduction

Generally speaking, the formation of microstructure during solidification is partly the basis for the properties of a metallic material. Regarding solidification-related influences on the properties, the following two factors are essential: On the one hand the topology of the microstructure itself and on the other hand the physical and chemical implications of the microstructure (in alloys most explicitly seen by solutal segregation, *i. e.*, microsegregation); both items are inevitably connected to each other. Specific and very thorough research on either phenomenon has been carried out in the past and is published in the relevant textbooks [1, 2].

In the last years special attention has been paid to providing appropriate modelling techniques especially for deepening the understanding of microstructure formation. In the literature three different approaches to microstructure modelling can be identified: Firstly, front tracking models [3–5] are used to precisely reproduce the interface between solid and liquid by a dynamically refined meshing in the region. Secondly, Phase-Field (PF) models deal with the solid-liquid interface by introducing a continuous transitional layer of finite thickness using an additional PF quantity, thus eliminating the problem of explicitly tracking the interface. Several reviews on the methodology and capabilities of the PF models are available [6–10]. Thirdly, cellular automata (CA) were either employed for mesostructure simulations during solidification (when coupled with finite element modelling, this technique is referred to as CA-FE modelling [11–13]) as well as for modelling microstructure. Early attempts were published in [14] and in [15], however having the drawback of lacking an arbitrary dendrite growth direction due to the mesh anisotropy. This drawback was resolved in [16, 17] where the authors developed a virtual front-tracking (VFT) method of the solid-liquid interface.

In any case, most of the above models were developed for simulating only binary systems (mostly Al-Cu, Al-Si or Fe-C). Despite the remarkable reproduction of dendrite morphology, columnar as well as equiaxed solidification by VFT models, the application of such models to industrial materials is limited.

Next to the question of microstructure modelling, the effect of microsegregation is of key importance on material properties. Kraft and Chang [18] give a review on more than 60 different microsegregation models that were published in the last century, a number which emphasises the importance of the topic. Virtually all models mentioned in this publication feature the study of microsegregation either by an analytical or a numerical one-dimensional approach. The models mostly differ in the topology of the calculation domain, in the number of alloying elements and phase boundaries and in their assumptions regarding the solute diffusivity in liquid and solid. However, most models have the shortcoming of neglecting physical effects (such as interface curvature); moreover, some models even do not account for thermal and solutal diffusivity. Furthermore, the interfacial growth velocity is often derived from empirical relations. Although one-dimensional models are an excellent basis for a locally focused consideration of solidification, physically correct results are questionable when the problem domain exceeds the very focused local interest of solidification.

In order to provide a possible improvement to the above mentioned shortcomings of microstructure models on the one hand and microsegregation models on the other hand a modified CA approach with the VFT scheme has been developed which considers multicomponent alloys in several dimensions. The basics of this model have been published in [19]; the present contribution will shortly summarise the model assumptions and then show some of the results to illustrate the model capabilities.

2. Short Model Outline

The developed model features a two-dimensional consideration of the microstructure formation and the microsegregation by the technique of cellular automata. In the model each cell is characterised by a number of basic quantities (where i, j and t are spatial and time indices):

- centre coordinates $\boldsymbol{x}(i, j)$
- temperature $\vartheta(i, j, t)$
- concentration $C_\tau^v(i, j, t)$ of the alloying element v in phase τ (here $\tau = L$ when liquid or $\tau = S$ when solid)
- fraction of solid $f_S(i, j, t)$
- interfacial curvature $\kappa(i, j, t)$
- growth direction $\varphi(i, j)$

According to the CA technique, the quantities in each cell at a time step t will results from the neighbouring cells' values in the current time step t and the preceding time step $t-1$. Therefore, the model procedure can be summarised as follows: Starting from a fully balanced liquid system, the chemical equilibrium is disturbed by artificially placing nuclei in the whole domain. Thus, certain cells' state is set to solid and the rejected solute is distributed to its neighbours. Due to the governing laws of solute diffusion, the concentration of neighbouring cells will then again be influenced. Additional cooling of the system will also cause thermal convection in the system. Both influences will cause the solid phase to grow (or dissolve), rejecting further solute and latent heat of fusion. Owing to the growth of the solid phase, the curvature of the nuclei will steadily change, also affecting the interfacial conditions.

The rules for determining the quantitative values of each cell are governing physical laws which are iteratively applied. These rules are already described in a published work [19] and will therefore not be repeated here. Nonetheless, it is important to mention that when modelling microstructure formation multicomponent alloys (*i. e.*, more than one alloying element in the matrix) the combination of solute diffusion and interfacial growth due to chemical (solutal) undercooling leads to an ill-posed problem due to the direct dependence of either phenomenon on each other. In the present model, the ill-posed problem is overcome by a functional extrapolation of the concentration gradient towards the solid-liquid interface (shown in Fig. 1) which yields the unknown interfacial concentration $C_{\tau}^{\upsilon,*}$. Details to this approach are derived in [19].

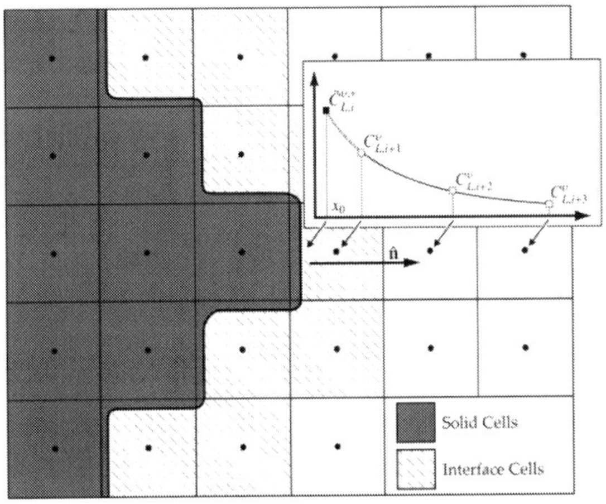

Figure 1. Illustration of the equilibrium concentration extrapolation at the interface.

2.1. Model Simplifications and Assumptions

The model uses the following simplifications and assumptions:

(i) The solid-liquid interface is not treated explicitly in the thermal field, in contrast to the treatment in the solutal field. As no heat balance is made over the interface, dendritic growth driven by thermal diffusion rather than by chemical undercooling cannot be reproduced.

(ii) Kinetic undercooling is not considered in the model which limits its application when treating rapid solidification conditions.

(iii) Cross diffusion between the alloying elements and a dependence of the partition coefficient on more elements than the superimposed binary systems is not considered, also partially lacking appropriate material data. Nonetheless, it is believed that there is a considerable influence of a multicomponent functionality of the partition coefficient, which should therefore be resolved in near future.

(iv) Fluid flow is not considered.

IOP Conf. Series: Materials Science and Engineering **33** (2012) 012100 doi:10.1088/1757-899X/33/1/012100

3. Model Results

In the following, some exemplary results of simulations with the model shall illustrate the potential of this novel approach. The material parameters used for the present study are the same as in [19]. In advance, it should be stated that fundamental comparisons of the model results with well-established analytic models—like the LGK-model [20] and the microsegregation model by Ohnaka [21]—have been carried out in advance to this study and are published in [19].

3.1. Equiaxed Solidification of a Single Nucleus

At first, an equiaxed dendrite in the system Fe-0.6%C-0.5%Si-0.5%Mn-0.005%P-0.005%S in a domain size of 100×100 µm^2 is considered. The initial conditions of the simulation are uniform liquidus temperature and a single, one-celled seed in the domain centre. The domain is globally cooled at all sides with a heat transfer coefficient of 500 W m^{-2}K^{-1}. Figure 2 shows the concentration fields for all alloying elements after a solidification time of 0.02 s.

It can be seen that typical dendritic features with branching and development of secondary dendrite arms develop, whereby the generation of secondary arms is only a product of interfacial curvature. It should however be highlighted that the secondary dendrite arm spacing which can be observed in the present example is not comparable to the final secondary dendrite arm spacing as the mechanisms of selection and coarsening are not visible at this solidification time (the model however features these mechanisms as shown later). Moreover, due to the differences in partition coefficients and diffusivity, considerable differences in the distribution of solute are obvious.

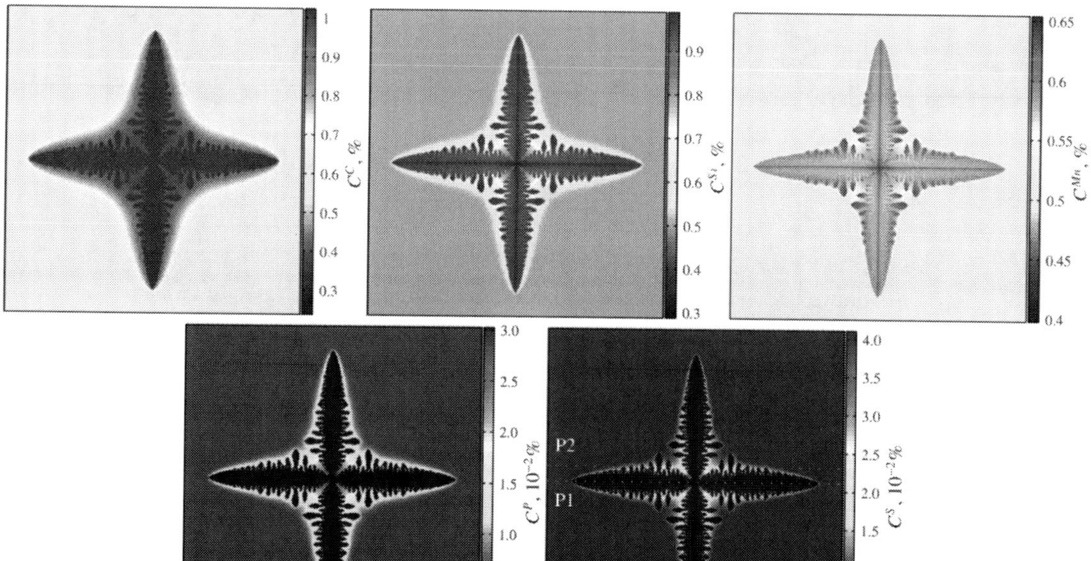

Figure 2. Concentration Fields for Free Dendritic Growth in a Multicomponent Fe-Si-Mn-P-S-alloy after a Solidification Time of 0.02 s.

3.2. Columnar Solidification

As solidification usually occurs in columnar form, a rectangular model of 300×100 µm^2 with an initially solid first row of boundary cells at one of the longer sides was taken as a basis to investigate the model's behaviour in the quaternary system Fe-0.6%-0.5%Si-0.5%Mn. The domain was cooled with a heat transfer coefficient of 500 W $m^{-2}K^{-1}$ at this particular face; adiabatic isolation was prescribed at the remaining faces.

Thus, not only the behaviour of the model regarding the formation of irregularities (finally resulting in dendrites) but also the resulting interdendritic concentration profiles can be evaluated. Figure 3 shows the C-distribution after a solidification time of 0.3 s in the total computational domain—the solidification direction is downwards. It can clearly be seen that out of the initially planar front, primary dendrites developed; the selection of dendrite arms oriented more ideally (almost opposite the heat extraction direction) is evident. The results in the lower part of the domain must be regarded carefully, since the dendrites could not grow further (no mass flow) and thus are starting growth normal to the interface. Next to the formation of primary dendrites, the development of secondary arms is also well visible. In accordance with many practical observations, this formation is more or less pronounced, depending on the available diffusion length (*i. e.*, distance to the next primary arm).

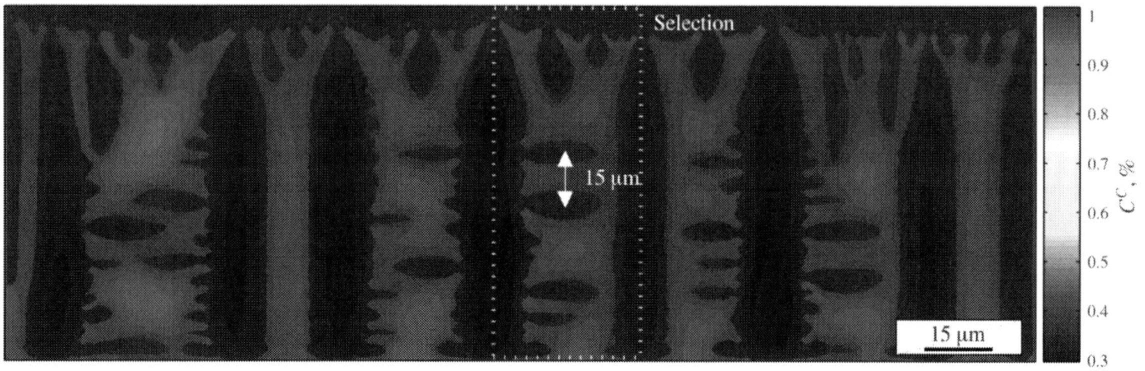

Figure 3. Simulated C-concentration field after 0.3 s.

3.2.1. Differences in Interdendritic Enrichment Figure 4a shows a highlighted selection of Fig. 3. An interdendritic point marked P1 in this figure, lying well between the tip and the trunk of the two secondary arms, is used to demonstrate the differences in interdendritic segregation depending on the distance from the dendrite trunk. It is compared to position P2 which represents the enrichment at the triple point (meeting of two secondary dendrite tips and a different primary dendrite arm).

The results in Fig. 4b show the significant changes regarding the maximum value of segregation for the two cases: Silicon, showing the strongest segregation behaviour, enriches from a starting value of 0.5 % to 3.8 % between the dendrites—at the triple point this value rises to 7.4 %. In terms of the segregation ratio C_L/C_0 this corresponds to a factor of 7.6 interdendritically and 15 at the triple point. Pertaining to the segregation behaviour of Manganese, again a considerable increase in the segregation ratio is detected: here, a factor of 2.4 versus an increase by $3.6C_0$ at the triple point. Regarding the profile of Carbon, an equalisation of the composition occurs at the triple point, leading to approximately the same segregation ratio as observed interdendritically ($C_L/C_0 = 3.3$).

Although it has to be pointed out that these absolute values are still rather qualitative (using non-constant diffusion parameters possibly changes the composition fields), it becomes clear that the present model is generally able to predict these variations of a larger scale than a conventional microsegregation model. Several solidification-related phenomena like for example macrosegregation or hot tearing involve a polydimensional view of the issue of microsegregation.

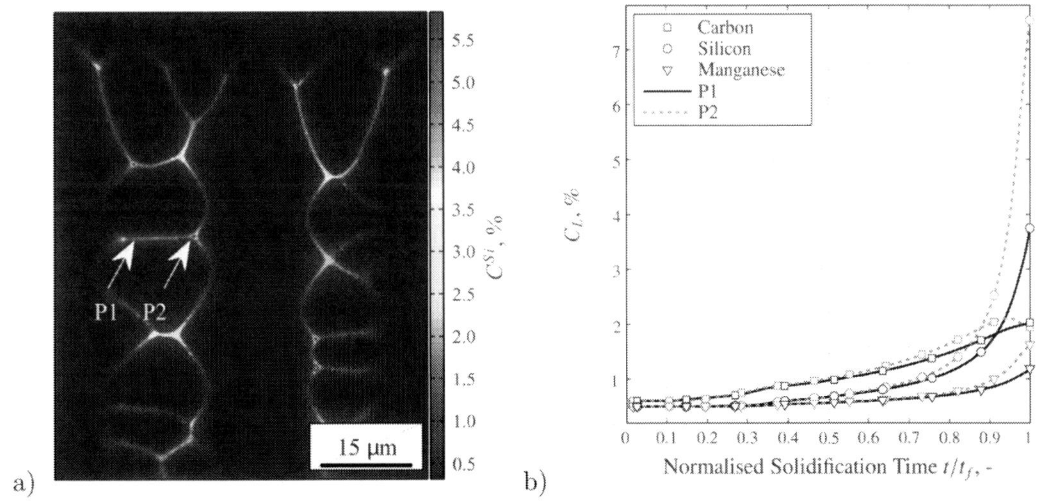

Figure 4. a: Concentration fields in columnar solidification for Si at the point to final solidification (0.6 s) and b: Illustration of the differences in interdendritic enrichment between two dendrite arms (P1) and at a triple point (P2).

3.3. Influence of Cooling Rate

The effects of changing cooling rate on a solidification structure are well known. In a second example with the same preconditions as the previous (mesh size and alloy), the effect of cooling rate on the microstructure shall be compared to values observed in reality as well as to analytical derivations for multicomponent alloys from the literature. In [22] the authors deduce a relationship between the secondary dendrite arm spacing and the solidification time. Since this empiric relation is based on observations at room temperature (*i. e.*, coarsening of dendrite arms has already taken place), evaluations of the present model are also conducted at a late stage of solidification. Furthermore, in [23] dendritic solidification in multicomponent alloys is treated from the analytical viewpoint. In this work, one finds a relationship for the secondary dendrite arm spacing as a function of local solidification time and enrichment in the liquid for multicomponent alloys. Both studies are taken as a basis for the following comparison.

Figure 5 shows the Si-distribution for a late solidification stage according to the previously mentioned conditions, however for an increased cooling with h = 1000 W $m^{-2}K^{-1}$; the formation of secondary arms is obvious. As mentioned before, not all secondary arms would actually survive the process of coarsening. Therefore, only those secondary arms which exhibit a clear tendency for growth and which will not suffer from a potential remelting or from overgrowth by a near neighbour were selected for distance measuring. In the present example, these dendrite trunks are marked with a short white dash in the figure.

The numerical results of the evaluation for different cooling rates are displayed in Fig. 6. The solidification time was determined from the simulation, *i. e.*, it represents the time when the last gaps between the dendrites close in the system. Additionally, the figure shows the prediction of the model by [22]. It has to be stated that this model was actually derived for higher solidification times (at least several seconds). However, for reasons of computation time, such simulations cannot be performed with the present model in a reasonable period of time. The figure additionally shows the results of the model by [23]. The final composition of the liquid at the base of the dendrite —as a necessary input parameter for this modell—was determined from the simulation. The consistency between the present model and the results derived from experimental observations as well as those from an analytical approach to the topic is remarkable and underlines the correctness of the assumptions in the presented model.

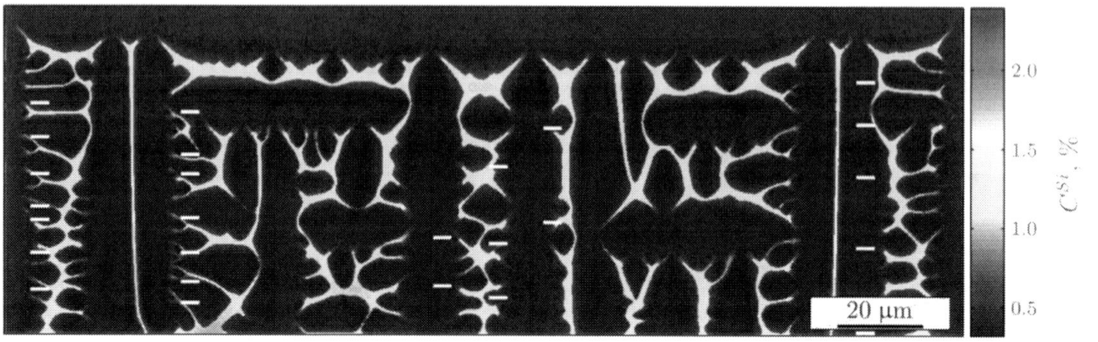

Figure 5. Si-distribution after a solidification time of 0.25 s for h = 1000 W $m^{-2}K^{-1}$. The white dashes indicate the origins of the secondary arms that were considered for the λ_2-evaluation.

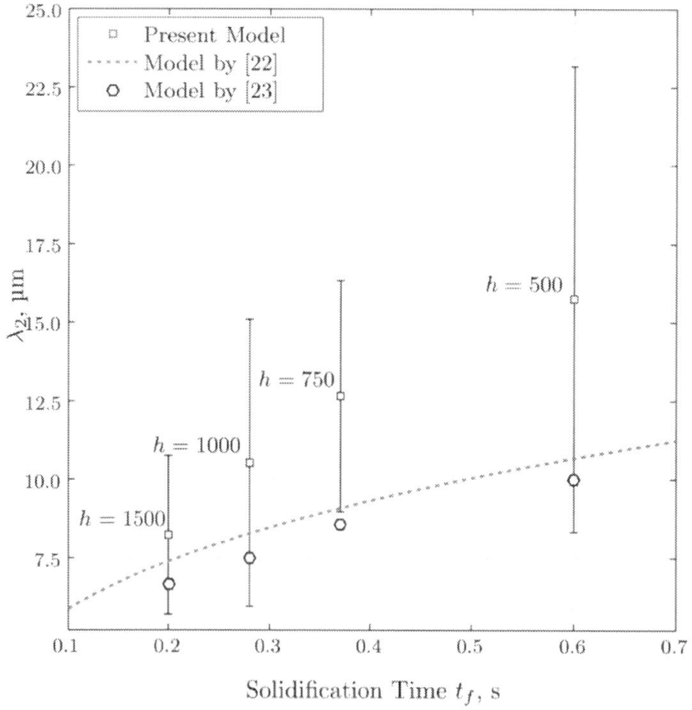

Figure 6. Comparison of λ_2-spacings calculated with the present model to those of [22] and [23]; heat transfer coefficients h in W $m^{-2}K^{-1}$.

4. Summary and Conclusions

In the present paper a recently developed model for the simulation of microstructure formation and microsegregation for multicomponent alloys has been used to study several model cases and to evaluate the capabilities of this model.

From the perspective of microstructure it could be shown that the model features unconstrained equiaxed dendritic growth, the evolution of dendritic solidification structures from a planar front and the development of secondary dendrite arms. Moreover, the influence of cooling rate on the secondary dendrite arm spacing was evaluated and compared to results from practical observations.

Next to the microstructural viewpoint, the phenomenon of microsegregation was investigated. It could be shown that by a multidimensional view on the problem of microsegregation, substantial differences regarding the microsegregation between two secondary dendrite arms and the enrichment of solute at a triple point are observed. This result becomes highly important when calculating local solidification temperatures as these differences cause a considerable decrease of the solidus temperature.

It is believed that the present model bridges the gap between phase-field models, which feature an exact resolution of dendritic features—however, at very high computational effort—and macrostructural solidification models. Thus, phenomena which are currently of high practical relevance in the field of casting of steel can in future be addressed by a quicker means than previously available. Examples of such questions are the columnar-to-equiaxed transition, the formation of hot tear segregations (which also involves the introduction of a mechanical or pseudo-mechanical model) or the study of the effects of rapid solidification conditions on the formation of segregation patterns.

In order to bring the model even closer to practise, several modification steps are planned in the future: Firstly, by obvious modification of the algorithm a second phase boundary in oder to consider the peritectic transition shall be achieved. Moreover, temperature and concentration-dependent material properties shall be employed. Later, the upgrade of the model to a three-dimensional treatment of solidification should be persued by obvious modification of the presented equations. In order to further assess the quality of the model, comparisons to phase-field models are planned.

References

[1] Kurz W and Fisher D 1998 *Fundamentals of Solidification* 4th ed (Zurich: Trans Tech Publications)
[2] Dantzig J and Rappaz M 2009 *Solidification* 1st ed Materials (Lausanne: EPFL Press)
[3] Saito Y, Goldbeck-Wood G and Müller-Krumbhaar H 1988 *Physical Review A* **38** 2148
[4] Udaykumar H, Mittal R and Shyy W 1999 *Journal of Computational Physics* **153** 535–574
[5] Nakagawa M, Natsume Y and Ohsasa K 2006 *ISIJ International* **46** 909–913
[6] Karma A and Rappel W 1996 *Physical Review E* **53** 3017–3020
[7] Boettinger W, Warren J, Beckermann C and Karma A 2002 *Annual Review on Material Research* **32** 163–164
[8] Asta M, Beckermann C, Karma A, Kurz W and Rappaz M 2009 *Acta Materialia* **57** 941–971
[9] Hecht U, Gránásy L, Pusztai T, Böttger B, Apel M, Witusiewicz V, Ratke L, De Wilde J, Froyen L, Camel D, Drevet B, Faivre G, Fries S, Legendre B and Rex S 2004 *Material Science and Engineering* **R46** 1–49
[10] Fries S, Böttger B, Eiken J and Steinbach I 2009 *International Journal of Materials Research* **100** 128–134
[11] Rappaz M and Gandin C A 1993 *Acta Metallurgica et Materialia* **41** 345–360
[12] Gandin C A and Rappaz M 1997 *Acta Materialia* **45** 2187–2195
[13] Gandin C A, Desbiolles J L, Rappaz M and Thévoz P 1999 *Metallurgical and Materials Transactions A* **30** 3153–3165
[14] Dilthey U, Pavlik V and Reichel T 1997 *Mathematical Modelling of Weld Phenomena 3* ed Cerjak H and Bhadeshia H (London: The Institute of Materials) pp 85–105
[15] Nastac L 1999 *Acta Materialia* **47** 4253–4262
[16] Beltran-Sanchez L and Stefanescu D 2003 *Metallurgical and Materials Transactions A* **34A** 367–382
[17] Beltran-Sanchez L and Stefanescu D 2004 *Metallurgical and Materials Transactions A* **35A** 2471–2485
[18] Kraft T and Chang Y 1997 *Journal of Materials* **49** 20–28
[19] Michelic S, Thuswaldner J and Bernhard C 2010 *Acta Materialia* **58** 2738–2751
[20] Lipton J, Glicksman M and Kurz W 1987 *Metallurgical and Materials Transactions A* **18** 341–345
[21] Ohnaka I 1986 *Transactions ISIJ* **26** 1045–1051
[22] Pierer R and Bernhard C 2008 *Journal of Material Science* **43** 6938–6943
[23] Rappaz M and Boettinger W J 1999 *Acta Materialia* **47** 3205–3219

IOP Conf. Series: Materials Science and Engineering **33** (2012) 012101 doi:10.1088/1757-899X/33/1/012101

Phase field simulation of multi-dendrite growth in a coupled thermal-solute-convective environment

Z. Guo[1, 4], J. Mi[2] and P.S. Grant[3]
[1] Department of Mechanical Engineering, Tsinghua University, Beijing, 100084, China
[2] Department of Engineering, University of Hull, East Yorkshire, HU6 7RX, UK
[3] Department of Materials, University of Oxford, Parks Road, Oxford, OX1 3PH, UK

E-mail: zhipeng_guo@mail.tsinghua.edu.cn

Abstract. To study the complexities in multi-dendrite growth during solidification, a fully coupled thermal-solute-convective phase field model and a parallel multigrid computing scheme have been developed for the first time for the simulation of multi-dendrite growth in a condition involving coupled heat/solute transfer, and flow convection. The interplay between fluid flow and heat/solute transfer and the resultant effects on dendrite growth, such as the effects of recirculating eddies, solute pile-up on the growth of primary and higher-order dendrite arms, and the final multi-dendrite morphology have been analysed and are discussed in detail. In particular, the development of multi-dendrites with four and six-fold symmetry in the presence of fluid flow is studied and contrasted.

1. Introduction

The dendritic microstructures of metallic alloys that are often formed in the processes of solidification are the predominant microstructural features to determine the mechanical properties of high quality castings [1-6]. Since the 1950s, intensive research has been conducted to understand the fundamentals of grain nucleation and dendritic grain evolution in different solidification conditions. Many solidification technologies have been developed to either explore the benefit of aligning dendritic microstructures in a predefined direction [7]; or to minimise the detrimental effects that dendrite growth can have on chemical and microstructural homogeneity by transforming them into equiaxed polygonal grain structures [8-9].

Since the 1990s, the development of the Phase-Field (PF) methodology and the availability of supercomputing power have made it possible to calculate and track the evolution of dendrites from immediately after nucleation to a fully developed dendritic grain, across the length scales of 10^{-9} to 10^{-4} m. Many previous studies have been focused on calculating single dendrite growth and comparing the calculated dendrite tip features (velocity, radius etc.) with classical theories [10-14]. The simulation of multi-dendrite growth with a realistic alloy and solidification condition involving coupled thermal, solute and fluid flow has not been addressed. The key challenges for realistic prediction of microstructure are:

[4] To whom any correspondence should be addressed.

(1) The development of a set of partial differential equations (PDEs) coupled with the PF variable that can describe coupled and transient phenomena during solidification involving solute transfer, heat flow and natural or forced convection [10, 15-17].
(2) The development of a robust and stable numerical scheme to solve the discretized highly non-linear PDEs, and a computationally efficient and scalable algorithm that can harness the immense computing power available from multi-core parallel computing infrastructure to calculate many dendrites (a few tens to hundreds) in a meaningful large domain (a few hundreds of micrometre across) in an acceptable time scale e.g. in $10^1 - 10^2$ computing hours rather than thousands.
In this paper, we report the development of a fully coupled PF model with a companion parallel multigrid computing scheme, which is a further advance (with the addition of forced convection) based on our very recent research of simulating multi-dendrite growth in a coupled thermal-solute environment [18]. We firstly give a brief description of the theoretical aspects of the model and then present the multi-dendrite growth for two binary alloys in a fully coupled thermal-solute-convective environment.

2. Mathematical model and numerical techniques

2.1 The phase field governing equations
The Ginzburg-Landau free energy functional is used to derive the PF equations [12]

$$F = \int dV\left(\frac{\sigma}{2}|\nabla\phi|^2 + f_{AB}(\phi, c, T)\right) \tag{1}$$

where F is the free energy, σ is the gradient energy coefficient, and f_{AB} is the free energy density function for a binary mixture respectively. With the presence of convection and assuming constant thermo-physical properties, the governing equations for the PF variable ϕ (-1 in bulk liquid and 1 in bulk solid phases), solute concentration c, and temperature T are:

$$\tau_0\frac{\partial\phi}{\partial t} = -K_\phi\frac{\delta F}{\delta\phi} \tag{2}$$

$$\frac{\partial c}{\partial t} + \vec{v}\cdot\nabla c = \nabla\cdot\left(K_c\nabla\frac{\delta F}{\delta c} - \vec{j}_{at}\right) \tag{3}$$

$$\frac{\partial T}{\partial t} + \vec{v}\cdot\nabla T = \alpha\nabla^2 T + \frac{L}{2c_p}\frac{\partial\phi}{\partial t} \tag{4}$$

where K_ϕ and K_c are constants, α is thermal diffusivity, L is latent heat, c_p is heat capacity, and $\vec{v}$ is superficial velocity. $\vec{j}_{at}$ is the so-called "anti-trapping" current developed by Karma [19]:

$$\vec{j}_a = -\frac{W_0}{\sqrt{2}}\frac{c/c_\infty}{[1+k-(1-k)\phi]}\frac{\partial\phi}{\partial t}\frac{\nabla\phi}{|\nabla\phi|} \tag{5}$$

Assuming the liquid alloy is an incompressible fluid, and by ignoring the effect of gravity, the conservation of the momentum and mass are governed by the Navier-Stokes equation and the continuity equation, respectively [20]

$$\frac{\partial\vec{v}}{\partial t} + \vec{v}\cdot\nabla\left(\frac{\vec{v}}{f_l}\right) = -f_l\nabla P + \upsilon\nabla^2\vec{v} + \vec{M}_1^d \tag{6}$$

$$\nabla\vec{v} = 0 \tag{7}$$

where $\vec{M}_1^d = -2\upsilon h_c(1-f_l)^2\vec{v}$ is a dissipative interfacial force per unit volume developed by Beckerman [21] and $h_c = 2.757$ according to an asymptotic analysis of a plane flow past the diffuse interface [21]. f_l is the local liquid fraction and $f_l = (1 - \phi)/2$, υ is the kinematic viscosity and P is the pressure.

2.2 The dimensionless governing equations

Using the similar approach detailed in [18] to replace the solute c in Eq. (3) and T in Eq. (4) with the dimensionless forms of solute $U = (2c/c_\infty/[1+k-(1-k)\phi]-1)/(1-k)$ and temperature $\theta = (T-T_M-mc_\infty)/(L/c_p)$; and furthermore scaling the length and time using the interfacial width W_0 and relaxation time τ_0, the final dimensionless governing equations are (with vanishing kinetic effects):

$$A(\psi)^2\left[\frac{1}{Le}+Mc_\infty(1+(1-k)U)\right]\frac{\partial\phi}{\partial t}=\phi(1-\phi^2)-\lambda(1-\phi^2)^2(\theta+Mc_\infty U)(1+\xi)+$$
$$\nabla\cdot\left(A(\psi)^2\nabla\phi\right)-\frac{\partial}{\partial x}\left(A(\psi)A(\psi)'\frac{\partial\phi}{\partial y}\right)+\frac{\partial}{\partial y}\left(A(\psi)A(\psi)'\frac{\partial\phi}{\partial x}\right) \tag{8}$$

$$\left(\frac{1+k}{2}-\frac{1-k}{2}\phi\right)\frac{\partial U}{\partial t}=\nabla\cdot\left(\tilde{D}\frac{1-\phi}{2}\nabla U+\frac{1}{2\sqrt{2}}\left[1+(1-k)U\right]\frac{\partial\phi}{\partial t}\frac{\nabla\phi}{|\nabla\phi|}\right)$$
$$+\frac{1}{2}\left[1+(1-k)U\right]\frac{\partial\phi}{\partial t}-\frac{1}{2}\tilde{v}\cdot\{[1+k-(1-k)\phi]\nabla U-[1+(1-k)U]\nabla\phi\} \tag{9}$$

$$\frac{\partial\theta}{\partial t}=\tilde{\alpha}\nabla^2\theta+\frac{1}{2}\frac{\partial\phi}{\partial t}-\tilde{v}\cdot\nabla\theta-\dot{q} \tag{10}$$

$$\frac{\partial\tilde{v}}{\partial t}+\tilde{v}\cdot\nabla\left(\frac{\tilde{v}}{f_l}\right)=-f_l\nabla\tilde{P}+\tilde{\upsilon}\nabla^2\tilde{v}-2\tilde{\upsilon}h_c(1-f_l)^2\tilde{v} \tag{11}$$

$$\nabla\cdot\tilde{v}=0 \tag{12}$$

where $\lambda = 15L^2/(16Hc_pT_M)$ is a scaling parameter with its reciprocal measuring the energy barrier height H. $Mc_\infty = |m|(1-k)c_\infty/(L/c_p)$, and m is the liquidus slope in the binary phase diagram (assuming the liquidus and solidus are straight lines), k is the partition coefficient, and c_∞ is the initial solute concentration in the liquid. The anisotropic effect can be simulated using $A(\psi) = (1+\varepsilon\times\cos[\omega(\psi-\psi_0)])$, and ψ is the angle between the interface normal and the x-axis; ψ_0 is the predefined orientation angle of the dendrite seeds, ε weights the magnitude of the anisotropy strength, and ω is the symmetrical or harmonic factor. To mimic the thermal fluctuations in the system, a single noise term ξ was introduced in Eq. (8) and realized using a Gaussian random number (zero mean) with specific amplitude. In Eq. (10), $\dot{q}$ is either a heat sink ($\dot{q}>0$) or heat source ($\dot{q}<0$) to mimic the cooling or heating of the entire system.

2.3 The computational domain, boundary conditions and alloys

Eqs. (8) – (12) were discretized on a two-dimensional, rectangular computational domain comprising 12288 × 6144 cells with equal grid spacing of $\Delta x = 0.8$. At the four boundaries (top, bottom, left and right), the boundary conditions for the PF variable, solute and temperature are all set to be Neumann condition with zero gradient flux; while forced convection is introduced at the top boundary in the $-y$ direction using a constant velocity of $v = v_0$, and the gradient of velocity is set to zero, i.e. $\partial v/\partial x = 0$ for both left and right boundaries, and $\partial v/\partial y = 0$ for the bottom boundary. For all boundaries, velocity in the x-axis is set to zero, i.e. $u = 0$. At the dendrite interface, no-slip boundary condition is applied.

Two alloys were used in the simulations including a four-fold Al-4wt%Cu alloy (alloy A), and a six-fold alloy B (thermo-physical properties of which are close to a impure succinonitrile alloy, but a six-fold crystal symmetry is assumed here to achieve the most difference from the Al-4wt%Cu. Besides, such assumption could also provide insights in understanding the influence of convection on six-fold dendrites, which has not been studied before). The related simulation parameters used are listed in Table 1. Simulation starts with five seeds of initial radii of $30d_0$ "inserted" inside the computing domain of an initial undercooling of θ_0 (-0.13 and -0.25 for alloys A and B, respectively). Other parameters used are $\lambda = 15$, $v_0 = -10$, $\xi = 2\times10^{-3}$ and $\dot{q} = 10^{-3}$.

Table 1. Key simulation parameters for the applied four- and six- fold alloys [10, 22].

	Alloy A	**Alloy B**
Lewis number, *Le*	10^4	10^2
Scaled liquidus slope, Mc_∞	0.035	0.10
Equilibrium partition coefficient, *k*	0.15	0.30
Anisotropy strength, ε	0.02	0.007
Prandtl number, Pr	0.002	0.2
Thermal capillary length, d_0 (m)	6.6×10^{-10}	2.81×10^{-9}

A novel parallel multigrid computing scheme based on implicit full approximation storage (FAS) multigrid algorithm [23] was developed and implemented in multi-core parallel computing architecture, details of which can be found in [18]. The significant advantage of the implicit algorithm is that time marching steps of $\sim10^4$ times larger than those employed in the explicit method can be used when solving the coupled solute and temperature equations i.e. Eq. (9) and (10), providing a $\sim10^4$ times increase in computing efficiency when tackling the multi-scale problem arising from the $10^2\sim10^4$ times difference (the Lewis number) between thermal and solute diffusivities [11-12]. Furthermore, the implicit algorithm is an inherently stable numerical scheme [24].
The parallel code was initially developed on the 640-core (1.9 TB RAM) supercomputing cluster, SAL, housed at the Oxford Supercomputing Centre using C++ language and two message passing libraries namely Message Passing Interface (MPI) [25] and Open Multi-Processing (OpenMP) [26]. The simulated cases presented in this paper were calculated using 96 computing cores (finished in 48 hours) on the supercomputing clusters (8800-cores) housed at the National Laboratory for Information Science and Technology at Tsinghua University.

3. Results and discussion

3.1 Multi-dendrite morphology of four-fold crystals

Figs. 1a and 1b show the morphologies (solute contour maps) of five dendrites of alloy A (labeled from D_1 to D_5) calculated with and without convection until $t = 12000dt$. Fig.1a and b are mirrored against each other along L_3 for convenient comparison and interpretation. P_0 denotes the initial location of the seed of dendrite D_1, and P_1 is the bottom point of L_5. Apart from L_3, which is the symmetrical line of dendrite D_1 in *y* direction, two more lines marked by L_4 and L_5 are also drawn in Figs.1a and 1b to allow the distribution of solute and temperature along those lines be extracted from the simulation for analysis.
Fig.1a shows that without convection the four primary arms of each dendrite grow maturely along the original seeded crystal orientations and the secondary dendrite arms on either side of the primary dendrite arms also grow and stretch out fully until they meet the secondary arms growing out from the adjacent primary arms, resulting in a near symmetrical secondary arm arrays at both sides of each primary arm. Taking dendrite D_1 for instance, the secondary arms grow upwards and downwards

between lines L_1 and L_2 (both approximately have an angle of 45° with respect to the x-axis), and form a symmetrical secondary dendrite arm array. Similar phenomena are shown in dendrites D_2 – D_5. However when convection is present, as shown in Fig. 1b, the near symmetrical dendrite arrays are lost, and the growth behaviour of the primary or secondary arms are directly linked to the direction of the flow. The primary arms with direction of growth oriented against the flow grow faster, while growth for those growing along the principal flow direction is reduced. The upstream primary arms are longer when convection is present and many more secondary arms develop from the upstream primary arms, arising from the interaction of the flow and its role in redistributing the solute rejected from the growing solid. Taking dendrite D_1 for example, as the flow is propagating towards the upstream primary arm tip, rejected solute at the tip is convected away and "washed" downwards, decreasing the local solute concentration as shown in Fig.1d where the solute concentration is ~1.8c_∞ with convection and ~2.0c_∞ without convection. The growth of this arm is enhanced since a lower solute concentration corresponds to a higher solidifying temperature according to Al-Cu phase diagram, resulting in a higher level of undercooling to speed up the growth of dendrites. As the solute is further washed down along the outside profile of the secondary arms, the solute concentration at those secondary arm tips is also reduced and the corresponding growth of such arms is therefore promoted. On the contrary, for arms in the downstream region, solute that is washed down from upstream accumulates locally as shown in Fig.1c where the solute concentration is ~2.2c_∞ without convection, and ~3.4c_∞ with convection. Growth of downstream primary arms is therefore reduced.

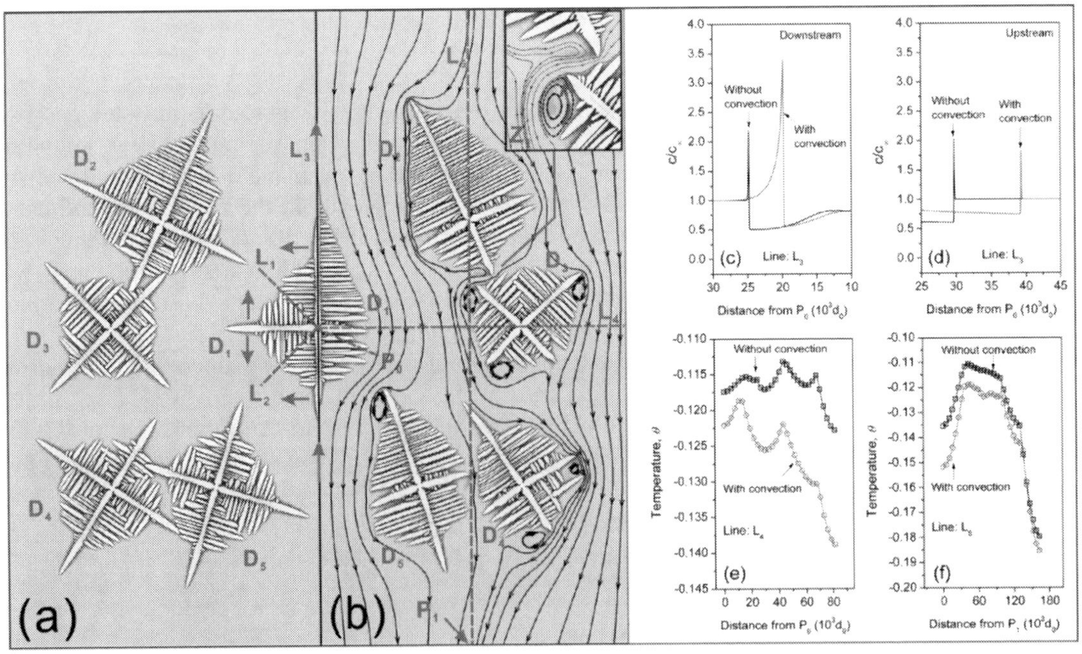

Figure 1. The morphologies of five four-fold dendrites of alloy A simulated in a coupled thermal-solute-convective system until t = 12000dt, showing contour maps of solute concentration (c/c_∞, changing from 0.38 (white colour) to 4.38 (black colour)) when convection is (a) absent, and (b) present (the two figures are mirrored along L_3). The solute distribution along line L_3 in dendrite D_1 is shown in detail in (c) and (d) for the downstream and upstream regions. Temperature profiles along lines L_4 and L_5 are also shown in (e) and (f) respectively.

Figs. 1e and 1f show that the temperature is higher at the central region of the computing domain, but lower at both top and bottom regions close to the boundary. This is because in the central region, more solid (relatively) is formed and therefore more latent heat per unit volume is released. Consequently, larger undercooling occurs at both the top and bottom regions of the computing domain, providing a greater driving force for dendrite growth, such as dendrite D_2 with longer primary arms and more secondary arms, regardless of any effect of convention.

3.2 Multi-dendrite morphology of six-fold crystals

The simulation of five six-fold dendrites of alloy B is carried out similarly as those of the alloy A and results are shown in Fig. 2 (the simulation lasts 24000 dt, two times longer than that of the four-fold case).

Fig. 2a shows that without convection each dendrite exhibits a nearly perfect six-fold symmetry structure with secondary arms stretching out from each of the six primary arms. Fig.2b shows that convection inhibits the prior symmetry and promotes the growth of both upstream primary arms and their secondary arms and again slows down downstream primary arm growth. For dendrite D_1 in Fig. 2b, the final length of the upstream primary arm is 34551d_0, ~34% longer than its mirrored counterpart in Fig.2a where convection is absent; the increase in the primary arm length is almost the same as dendrite D_1 in Fig.1 at 32%. However for alloy B, Fig. 2d shows that, with and without convection, the difference of the solute concentration at the tip of dendrite D_1 is just $0.02c_\infty$, approximately ten times less than that in Fig.1d ($0.20c_\infty$) and similarly for the downstream primary arms in Fig.2c and Fig.1c. Therefore in comparison with alloy A, solute redistribution due to convection in alloy B is comparatively weak.

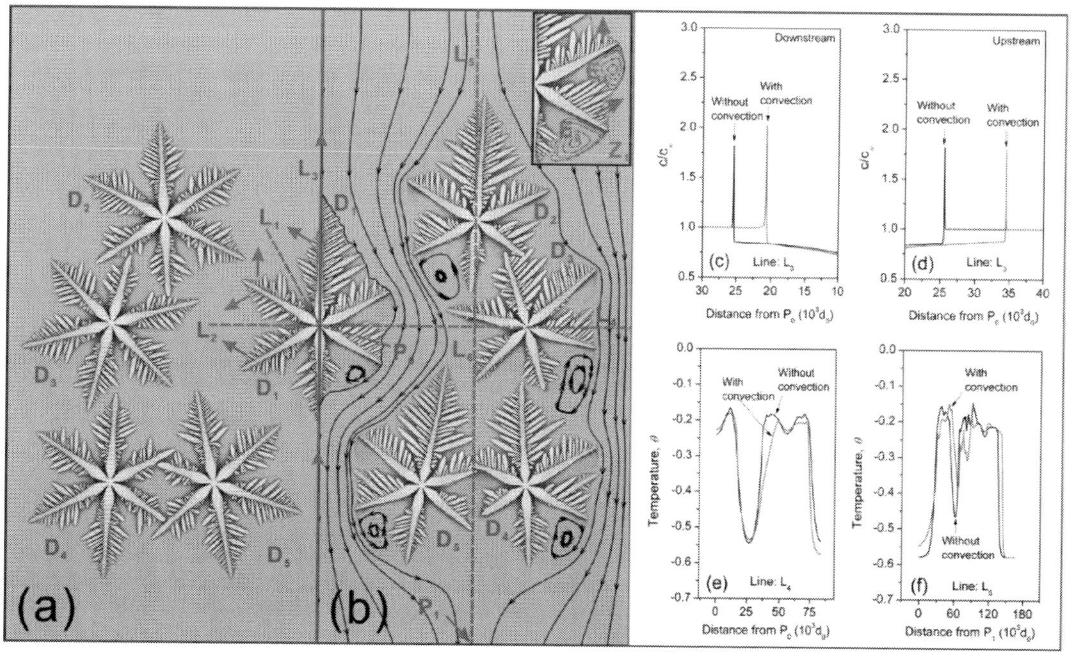

Figure 2. The morphologies of five six-fold dendrites of alloy B simulated in a coupled thermal-solute-convective system until t = 23892dt, showing contour maps of solute concentration (c/c_∞, changing from 0.38 (white colour) to 2.78 (black colour)) when convection is (a) absent and (b) present (the two figures are mirrored along L_3). The solute distribution along line L_3 is shown in (c) and (d) for the downstream and upstream regions of dendrite D_1. Temperature profiles along lines L_4 and L_5 are also shown in (e) and (f) respectively.

To clearly understand the mechanism that leads to the increase of the upstream dendrite arm length of dendrite D_1 for alloy B, the difference of upstream dendrite tip temperature i.e. $\Delta\theta_{tip} = \theta_{tip\text{-}convection} - \theta_{tip\text{-}no\ convection}$, the difference of upstream dendrite tip solute concentration i.e. $\Delta c_{tip} = c_{tip\text{-}convection} - c_{tip\text{-}no}$

$_{\text{convection}}$, and the difference of upstream dendrite tip position i.e. $\Delta d_{\text{tip}} = d_{\text{tip-convection}} - d_{\text{tip-no convection}}$ throughout solidification are extracted from the simulation and shown in Fig. 3. For dendrite D_1, Δc_{tip} increases quickly until $\sim 10\times10^3 dt$ and then remains relatively stable while $\Delta\theta_{\text{tip}}$ drops drastically and reaches -0.31 at $\sim 24\times10^3 dt$, apart from a short period of recalescence from $12\times10^3 dt$ to $16\times10^3 dt$. For dendrite D_1 of alloy A, both Δc_{tip} and $\Delta\theta_{\text{tip}}$ increase until $\sim 5\times10^3 dt$, followed by a continuous decrease in Δc_{tip}, and a decrease during $5\text{-}9\times10^3 dt$ and then a recalescence until $12\times10^3 dt$ for $\Delta\theta_{\text{tip}}$. Comparing Figs. 3a and 3b, the absolute value of $\Delta\theta_{\text{tip}}$ is always larger for alloy B than alloy A. The relative flattened Δc_{tip} shown in Fig. 3a for alloy B indicates that the growth promotion with convection in Fig. 3c can be attributed primarily to the variation of the local temperature at the dendrite tip rather than the variation of solute, as is the case for alloy A. This difference has its origin in the ~100 times higher Prandtl number for alloy B (see Table 1) and consequently heat transfer is ~100 times "slower" than that in alloy A, causing a larger local temperature difference and higher undercooling around the tip of dendrite D_1.

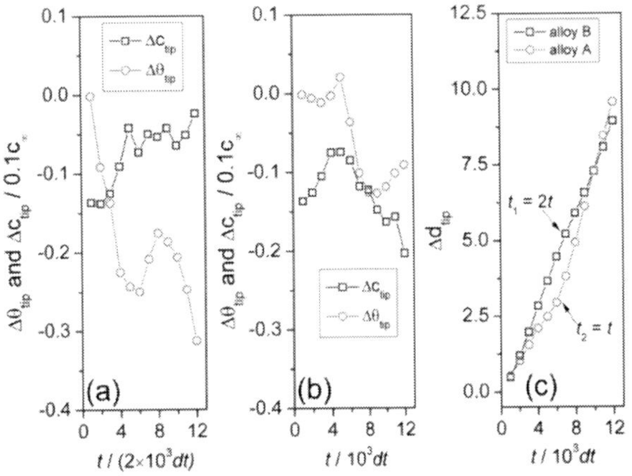

Figure 3. Temperature difference and solute concentration difference at the tip of the upstream primary arm of dendrite D_1 for (a) alloy B and (b) alloy A throughout solidification between the cases with and without convection. $\Delta\theta_{\text{tip}} = \theta_{\text{tip-convection}} - \theta_{\text{tip-no convection}}$ and $\Delta c_{\text{tip}} = c_{\text{tip-convection}} - c_{\text{tip-no convection}}$; and (c) the tip position difference of the upstream primary arm of D_1 ($\Delta d_{\text{tip}} = d_{\text{tip-convection}} - d_{\text{tip-no convection}}$) for both alloys (time is scaled with t_1 and t_2 for alloys B and A, respectively).

Recirculating eddies are created when flow passes an obstacle, such as the upper-left primary arm of dendrite D_3 as shown in the zoom-in area "Z_1" of Fig. 1b, indicated by the arrow. The recirculation in the eddy can cause solute to accumulate and inhibit growth (compare Figs.1a and 1b). Similar eddy recirculation also occurs for the six-fold dendrite morphology as shown in Fig. 2b (see zoom-in area "Z_1"). Considering dendrite D_1, eddies are created on the downstream of both the two sideways orientated primary arms (one at inclined to the flow at ~30° and one declined at ~30°), as indicated by E_1 and E_2. Growth of the secondary arms on the upper side of these primary arms is promoted (as indicated by arrows) while it is strongly inhibited on the downstream side, with barely any development of secondary arms. Because the development of the secondary arms is particularly sensitive to the flow (and associated solute and thermal effects for alloys A and B respectively) and in turn the flow is sensitive to the dendrite morphology, direct comparison between detailed morphological evolution for four and six-fold dendrite symmetry is difficult. The results here confirm the complex interplay of the alloy crystallography, thermophysical parameters and any fluid flow and underscore the importance of coupling these aspects if meaningful microstructural prediction is to be achieved.

4. Conclusions

A thermal-solute-convection phase field model for binary alloy solidification has been developed and the arising partial differential equations have been solved numerically solved using a newly developed parallel multigrid approach. Two simulations concerning multi-dendrite growth corresponding to four- and six- fold alloys have been performed. For the first time, multi-dendrite growth under realistic thermal (Le ~ 10^4) and flow (Pr ~ $10^{-1} - 10^{-3}$) convections has been simulated and the influence of both thermal and flow convections on dendrite morphology were investigated. Convection promoted the growth of upstream oriented primary arms but decelerated growth of both downstream oriented primary and side arms. The promotion of growth in the upstream direction was reduced when the dendrite growth was inclined to the primary direction of flow, and very different secondary dendrite arm patterns developed on upstream and downstream sides of these primary arms. This effect was more marked for six-fold rather than four-fold symmetry dendrites and for alloys with very high Lewis number such effect arose primarily due to solute recirculating in eddies behind primary arms; for alloys with low Lewis number, temperature differences in upstream and downstream regions were more important in growth inhibition due to their lower thermal conductivity.

Acknowledgements
The authors would like to thank the UK Royal Academy of Engineering through the Newton International Fellowship Scheme for financial support, the Oxford Supercomputer Centre and the National Laboratory for Information Science and Technology in Tsinghua University for access to supercomputing facilities and support for parallel programming.

References

[1] Asta M, Beckermann C, Karma A, Kurz W, Napolitano R, Plapp M, Purdy G, Rappaz M and Trivedi R 2009 Acta Mater **57** 941.
[2] Boettinger WJ, Warren JA, Beckermann C and Karma A 2002 Annu Rev Mater Res **32** 163.
[3] Chen L-Q 2002 Annu Rev Mater Res **32** 113.
[4] Moelans N, Blanpain B and Wollants P 2008 Calphad **32** 268.
[5] Granasy L, Pusztai T, Borzsonyi T, Warren JA and Douglas JF 2004 Nat Mater **3** 645.
[6] Haxhimali T, Karma A, Gonzales F and Rappaz M 2006 Nat Mater **5** 660.
[7] Schneider MC, Gu JP, Beckermann C, Boettinger WJ and Kattner UR 1997 Metallurgical and Materials Transactions **28A** 1517.
[8] Hitchcock M, Wang Y and Fan Z 2007 Acta Mater **55** 1589.
[9] Grant PS 2007 Metall Mater Trans **38A** 1520.
[10] Badillo A, Ceynar D and Beckermann C 2007 J Cryst Growth 309 216.
[11] Ramirez JC and Beckermann C 2005 Acta Materialia 53 1721.
[12] Ramirez JC, Beckermann C, Karma A and Diepers HJ 2004 Physical Review E 69 051607.
[13] Lu Y, Beckermann C and Ramirez JC 2005 Journal of Crystal Growth 280 320.
[14] Jeong J-H, Goldenfeld N and Dantzig JA 2001 Physical Review E 64 041602.
[15] Lan CW, Hsu CM and Liu CC 2002 J Cryst Growth **241** 379.
[16] Lan CW and Shih CJ 2004 J Cryst Growth **264** 472.
[17] Giummarra C, LaCombe JC, Koss MB, Frei JE, Lupulescu AO and Glicksman ME 2005 J Cryst Growth **274** 317.
[18] Guo Z, Mi J and Grant PS 2012 Journal of Computational Physics **231** 1781.
[19] Karma A 2001 Physical Review Letters **87** 115701.
[20] Tong X, Beckermann C, Karma A and Li Q 2001 Physical Review E **63** 061601.
[21] Beckermann C, Diepers HJ, Steinbach I, Karma A and Tong X 1999 Journal of Computational Physics **154** 468.
[22] Kurz W and Fisher DJ 1998 *Fundamentals of Solidification* (Switzerland:Trans Tech Publications Ltd).
[23] Trottenberg U, Oosterlee C and Schuller A 2001 Multigrid (London: Academic Press).
[24] Rosam J, Jimack PK and Mullis AM 2008 Acta Materialia **56** 4559.
[25] Gropp W, Lusk E and Skjellum A 1999 *Using MPI: Portable Parallel Programming with the Message-Passing Interface* (Cambridge: MIT Press).
[26] Chapman B, Jost G and van der Pas R 2007 *Using OpenMP Portable Shared Memory Parallel Programming* (Cambridge: MIT Press).

IOP Conf. Series: Materials Science and Engineering **33** (2012) 012102 doi:10.1088/1757-899X/33/1/012102

Influence of natural convection on microstructure evolution during the initial solidification transient: comparison of phase-field modeling with in situ synchrotron X-ray monitoring data

Yun Chen[1], Henri Nguyen-Thi[2,3], Dian Zhong Li[1,4], Abdoul-Aziz Bogno[2,3], Bernard Billia[2,3], Na Min Xiao[1]

[1] Shenyang National Laboratory for Materials Science, Institute of Metal Research, Chinese Academy of Sciences, Shenyang, 110016, China

[2] Aix-Marseille Univ, IM2NP, Campus scientifique de Saint-Jérôme, Case 142, 13397 Marseille Cedex 20, France

[3] CNRS, IM2NP, Campus scientifique de Saint-Jérôme, Case 142, 13397 Marseille Cedex 20, France

E-mail: dzli@imr.ac.cn

Abstract. The influence of natural convection on the evolution of the solid-liquid (s/l) interface during the initial transient of upward directional solidification was studied on Al–4 wt.% Cu alloy by coupling the two dimensional quantitative phase-field model with the Navier-Stokes equations. The simulations were compared with in situ and real-time synchrotron X-ray monitoring data. The origin of natural convection in experiment was the presence of a small unavoidable horizontal temperature gradient. Due to the stringent requirement on the phase-field interface width parameters, the simulated domain could not be chosen as large as the size of the experimental sample. As the calculated fluid flow strength would be weakened by using a smaller domain, a horizontal temperature gradient ten times larger than the estimated experimental value was applied in simulation to recover a fluid flow washing the s/l interface similarly to experiments. Direct comparison to experimental measurements demonstrated that the phase-field simulations with convection qualitatively reproduced the evolution of all the characteristic parameters measured in experiments. Based on these results, the effects of natural convection on the growth dynamics of the s/l interface during directional solidification of alloy were further clarified.

1. Introduction

Ever better understanding of microstructure evolution during solidification is crucial for improving and optimizing the casting process as well as the properties in use of materials. Depending on the applied processing parameters and the material physical parameters, a wide variety of solidified microstructures are observed in casting, welding and other solidification processes. During the past decades, numerous theoretical and numerical models were elaborated to predict characteristic features of microstructure evolution during directional solidification under pure diffusion transport of heat

[4] To whom any correspondence should be addressed.

and/or chemical species [1,2]. However, it is recognized that crystal growth from the melt is most generally dynamically interacting with fluid flow in the liquid phase [2-10]. Heat or solute transport by the flow then significantly alters the growth morphology of the s/l interface predicted by the diffusion-dominated models, such as the morphological instability and the operating state of the cell or dendrite tips. Since natural convection is usually driven by density differences in the liquid, the strength and flow pattern are controlled by the thermal and solute gradients. Today, detailed theoretical prediction of the influence of natural convection on the growth of alloys still remains a pending issue. Nevertheless, remarkable achievements have demonstrated that numerical computer simulation of crystal growth incorporating the fluid dynamics equations into the numerical calculations is a key to comprehensive studying of fluid flow effects [2,3,7-10]. However, all numerical simulations require validation using precise experimental data. Recently, taking advantage of the high synchrotron brilliance, Bogno and coworkers have characterized the s/l interface evolution by in situ and real-time X-ray imaging at European Synchrotron Radiation Facility (ESRF) during upward directional solidification of Al-Cu alloys by cooling down method [6]. In particular, these experimental results provided a series of benchmark data on microstructure evolution during initial solidification transient and a deep understanding of its dynamics, as well as on the convection effects on the long range solute distribution ahead of the solidification front. Through direct quantitative comparison of phase-field simulation with the experimental observation data, it was already gratifying that simulation described qualitatively well the evolution of the s/l interface in diffusion-dominated situation [11]. Basically due to the existence of natural convection in the experiment, discrepancies between phase-field simulation and experimental values of the characteristic parameters (solutal length, interface velocity and interface concentration in the melt) were actually found, as well as with respect to the onset of morphological instability of the s/l interface. Consequently, it is necessary to make a critical comparison of these experimental data to the predictions of phase-field simulation with fluid flow.

In this paper, we report on numerical simulations of the initial directional solidification transient on Al–4 wt.% Cu alloy that incorporate the Navier-Stokes equations into the quantitative phase-field model for binary alloy proposed by Karma [12]. The influence of fluid flow on the initial solidification transient was characterized from planar growth to the inception of the cellular structure. The simulation results enabled us to further clarify the growth dynamics of the solid microstructure in interaction with the fluid flow that occurred in Al–4 wt.% Cu directional solidification experiments.

2. Phase-field model and in situ observation data on Al-4 wt.% Cu directional solidification

2.1. The phase-field modelling in the presence of fluid flow in the melt

The quantitative two-dimensional (2D) Karma phase-field model for binary alloy solidification [12] is adopted to investigate the evolution of the s/l interface under condition of melt flow. The order parameter ψ of phase field has a constant value in solid ($\psi = 1$) and liquid ($\psi = -1$) phases, and varies smoothly across the thin diffuse interface. The governing equations of the phase-field model are following Refs [12,13]. The phase-field equation reads

$$\begin{aligned}\tau(\mathbf{n})\frac{\partial\psi}{\partial t} &= \nabla\cdot\left(W^2(\mathbf{n})\nabla\psi\right)+\partial_x\left(|\nabla\psi|^2 W(\mathbf{n})\frac{\partial W(\mathbf{n})}{\partial(\partial_x\psi)}\right)\\ &\quad+\partial_y\left(|\nabla\psi|^2 W(\mathbf{n})\frac{\partial W(\mathbf{n})}{\partial(\partial_y\psi)}\right)+\left[\psi-\lambda H\left(1-\psi^2\right)\right]\left(1-\psi^2\right)\end{aligned} \tag{1}$$

where x is the horizontal direction and y the vertical direction, $W(\mathbf{n}) = W_0\, a(\mathbf{n})$, $\tau(\mathbf{n}) = \tau_0 a^2(\mathbf{n})$, $\mathbf{n} = \nabla\psi/|\nabla\psi|$, $a(\mathbf{n}) = W_0(1-3\varepsilon_4)[1+4\varepsilon_4\frac{(\partial_x\psi)^4+(\partial_y\psi)^4}{(1-3\varepsilon_4)|\nabla\psi|^4}]$, $\lambda = a_1\xi$, $\xi = W_0/d_0$, $d_0 = \Gamma/[|m_l|C_0(1-k)]$, ε_4 is the fourfold anisotropy strength and Γ the Gibbs-Thomson coefficient, $\tau_0 = [1-(1-k)(y-V_P t)/l_T]$, $H = U +$

$(y - V_p t)/l_T$, $l_T = |m_l|C_0(1 - k)/G$. The dimensionless solute concentration U is obtained by solving the solute conservation equation with the solute antitrapping term introduced by Karma [12]. It is written including the terms describing the solute transport by convection

$$\begin{aligned}&[(1+k)-(1-k)\psi]\frac{\partial U}{\partial t}+\mathbf{u}\cdot[((1+k)-(1-k)\psi)\nabla U-(1+(1-k)U)\nabla\psi]\\&=\nabla\cdot[Dq(\psi)\nabla U+\frac{1}{\sqrt{2}}(1+(1-k)U)\frac{\partial\psi}{\partial t}\frac{\nabla\psi}{|\nabla\psi|}]+[1+(1-k)U]\frac{\partial\psi}{\partial t}\end{aligned} \tag{2}$$

where $D = a_2\lambda W_0^2/\tau_0$, $q(\psi) = [(1-\psi)+k(1+\psi)D_s/D_l]$, $C/C_0 = [1+(1-k)U][1+k-(1-k)\psi]/2$, with m_l the liquidus slope and k the solute partition coefficient. D_l and D_s are the solute diffusion coefficients in liquid and solid, respectively. In this approach, the attachment kinetics at the interface is neglected so that $a_1 = 0.8839$, $a_2 = 0.6267$.

It is assumed that the melt is incompressible

$$\frac{(1-\psi)}{2}\nabla\cdot\mathbf{u}=0 \tag{3}$$

and the fluid motion described by the Navier-Stokes equations:

$$\frac{(1-\psi)}{2}\frac{\partial\mathbf{u}}{\partial t}-\frac{1}{Re}\nabla^2\mathbf{u}+\frac{(1-\psi)}{2}(\mathbf{u}\cdot\nabla)\mathbf{u}=\frac{(1-\psi)}{2}(f_i-\nabla p). \tag{4}$$

The source term in the Navier-Stokes equations is expressed by

$$f_i=\begin{cases}-2h(1+\psi)^2(1-\psi)a_1a_2\xi^2d_0\upsilon/D_lu,\ i=x\\-2h(1+\psi)^2(1-\psi)a_1a_2\xi^2d_0\upsilon/D_lv+g\left[\alpha_T\left(T-T^*\right)+\alpha_C\left(C-C^*\right)\right]\tau_0^2/(\xi d_0),\ i=y\end{cases} \tag{5}$$

where u, v are the components of the velocity vector **u** in horizontal and vertical directions respectively. p is the pressure and $Re = W_0^2/(\upsilon\tau_0)$ with υ denotes the kinematic viscosity. The dissipative interfacial force and the buoyancy force are taken into account through the source term f_i. The constant h is equal to 2.757 as determined from an asymptotic analysis for a shear flow past a diffuse interface [3]. In the expression of the buoyancy term, the dependence of melt density on temperature and composition is written in the Boussinesq approximation [14] with g is the acceleration of gravity, α_T and α_C the thermal and solutal expansion coefficients and T^* and C^* reference values.

Table 1. Parameters used in the numerical simulations with natural convection on Al–4 wt.% Cu alloy

Kinematic viscosity, υ	5.0×10^{-3} cm^2/s
Cooling rate, R	0.3 ℃/min
Horizontal temperature gradient, G_x	56 °C/cm
Vertical temperature gradient, G_y	37 °C/cm
Thermal expansion coefficient, α_T	-1.0×10^{-4} °C^{-1}
Solute expansion coefficient, α_C	9.2×10^{-3} wt.%$^{-1}$
Reference temperature, T^*	920 ℃
Reference concentration, C^*	4 wt.%
Gravity acceleration, g	−9.8 m^2/s
Gibbs-Thomson coefficient, Γ	2.36×10^{-4} cm°C
Interface width parameter, ξ	54

2.2. Numerical implementation

The numerical simulations of Al–4 wt.% Cu directional solidification were carried out by solving equations (1) and (2) together with Navier-Stokes equations (4) and incompressibility equation (3). Heat extraction from the top boundary to the bottom was imposed by setting a constant vertical thermal gradient G_y and solidification driven by cooling both the hot and cold zones at rate R. In order to induce natural convection, a horizontal temperature gradient G_x between the two side walls was applied over the whole computational domain. The phase-field governing equations as well as the Navier-Stokes equations were solved by using the adaptive finite element method. Unless otherwise stated in Table 1, all other initial and boundary conditions as well as the physical parameters of Al–4 wt.% Cu alloy were identical to Ref. [15]. To eliminate the influence of rejected solute on the increasing of the concentration in the bulk liquid during the planar solidification, a large channel simulation domain with a size of 350×7985 (370×8451 μm²) was chosen with the dimensionless minimum length of the adaptive mesh element size dx_{min} = 0.25. The momentum and mass conservation equations were solved using the modified projection method [16] by a semi-implicit approach, in which to avoid any restriction on the time step dt, the advection term $(\mathbf{u}\cdot\nabla)\mathbf{u}$ in equation (4) was replaced by the skew-symmetric form $(\mathbf{u}\cdot\nabla)\mathbf{u} + 0.5(\nabla\cdot\mathbf{u})\mathbf{u}$, and thus the viscous step was rewritten in the form,

$$\frac{(1-\psi)}{2}\frac{\partial \mathbf{u}}{\partial t} - \frac{1}{Re}\nabla^2\mathbf{u} + \frac{(1-\psi)}{2}[(\mathbf{u}\cdot\nabla)\mathbf{u} + \frac{1}{2}(\nabla\cdot\mathbf{u})\mathbf{u}] = \frac{(1-\psi)}{2}(f_i - \nabla p) \quad . \tag{6}$$

Finally, all the boundaries of the domain were considered as walls where the flow velocity was zero, i.e. no-slip boundary condition was adopted at all these walls.

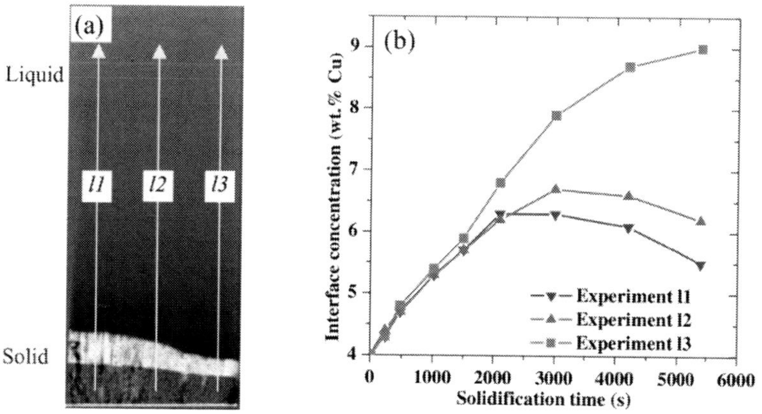

Figure 1. (a) The three vertical lines for measurement of longitudinal solute profile in both solid and liquid phases in experiment. In situ synchrotron X-ray radiograph of Al–4 wt.% Cu sample during directional solidification by cooling-down at R = 0.3 °C/min after solidification for 1480 s. Image width = 5.04 mm. (b) Variation with time of the liquid concentration at the interface measured on the three lines [6].

2.3. Experimental procedure

Synchrotron X-ray radiography was used to monitor *in situ* the evolution dynamics of the s/l interface on Al–4 wt.% Cu samples during upward directional solidification in a Bridgman furnace [6]. The samples used in this investigation were 40 mm in length, 6 mm in width and about 200 μm in thickness. The field of view was around 5×15 mm² and the experiments were carried out at the ID19 beamline of ESRF. The temperature gradient was adjusted to G = 37 °C/cm and the solidification induced by applying a constant cooling rate R = 0.3 °C/min on both hot and cold zones of the furnace. During solidification, the sample was illuminated perpendicularly by the incident monochromatic X-ray beam to investigate the interface evolution in situ and in real-time. Successive radiography images were recorded to study the interface dynamics and solute distribution. Detailed description of the experimental set-up and procedure was given in [6,17], and of the procedure for measurement of solute profiles in [18]. Since fluid flow driven by a small horizontal gradient temperature existed in experiment, the macroscopic s/l interface was not a plane perpendicular to the vertical direction, but rather showed smooth curved shape. To determine the longitudinal solute profiles in the solid and liquid phases, solute concentration was measured along three vertical lines, *l1*, *l2* and *l3* as shown in figure 1(a).

2.4. Origin of natural convection in experiment

For Al–4 wt.% Cu alloy, the solute rejected upon solidification is heavier than the solvent, and thus during upward directional solidification the temperature and solute gradients along the vertical were in stable configuration with respect to the density profile in the melt, i.e. they did not provided a driving force for hydrodynamic instability. However, in experiment the initial interface at the time solidification was started (see almost horizontal line separating the dark grey and white bands at the bottom of figure 1(a)) was generally slightly departing from planarity, even after a long thermal stabilization phase. This was due to the fact that a horizontal temperature gradient G_x close to the s/l interface was unavoidable [6]. This gradient originated first from the thermal conductivity differences between liquid, solid and crucible, and in addition from the fact that the sample was not well centered in the furnace (otherwise the s/l interface would have been symmetric with respect to the vertical centerline). At any time during planar solidification, the temperature gradient in horizontal direction can be estimated from the measured interface concentration difference between the lines *l1* and *l3* (figure 1(b)) according to the following equations

$$\Delta C = \left(C_{i3} - C_{i1}\right) = \Delta T / m_l \,,\ \Delta T = G_x \Delta x + G_y \Delta y \tag{7}$$

where C_{i1}, C_{i3} are the interface concentrations at lines *l1* and *l3*, ΔC and ΔT the differences in concentration and temperature between *l3* and *l1* at interface, and Δx and Δy the horizontal and vertical distances between these two lines at the s/l interface. Using the data at solidification time t = 1047 s, the estimated horizontal temperature gradient G_x was about 5.6 °C/cm in experiment. Unfortunately, due to the stringent requirement on the phase-field interface width parameters, the simulated domain could not be chosen as large as the size of the experimental sample. However, for a given solidification process, the flow velocity in liquid is characterized by a dimensionless Rayleigh number (Ra) which, for a fixed horizontal temperature gradient G_x and a sample of width of d, is given by $Ra_T = g\alpha_T G_x d^4/\upsilon D_T$, where D_T is the thermal diffusivity. According to Rayleigh number, the convection strength is controlled by d^4, so that the calculated fluid flow strength is weakened by using a computational domain narrower than the sample. Thus, a horizontal temperature gradient ten times larger than the experimental value was applied in numerical simulation to recover the expected fluid flow strength in terms of several calculation tests, which led to simulated front shape analogous to the s/l interface morphology observed in the experiment.

3. Results and discussion

3.1. Comparison of phase-field simulation with convection effects to in situ monitoring data

Figures 2(a-c) show the phase-field interface shape and solute distribution in liquid and solid simulated at different times using phase-field modelling with convection. The evolution of the fluid flow field in the liquid ahead of the solidification front is shown in figure 2(d-f). The upper fluid flow loop is driven by the horizontal thermal gradient, whereas the counter rotating loop in the groove is driven by the horizontal solute gradient created by bulk thermal convection and the complex shape of the solidification front. With the progress of solidification, the horizontal solute gradient is first opposing the thermal driving force as solutally-heavier melt is on the right (higher solute concentration) where it will sink. This is opposed to the thermal driving force as thermally lighter fluid is on the right. The flow strength next to the solidification front therefore keeps decreasing with time until the fluid flow along the interface gets dominated by the settling down of heavier solute-rich melt (figures 2(d, e)), which manifests by the formation of a counter rotating loop in the liquid groove on the right (figure 2(f)) accompanying the development of the cellular s/l interface. Thus, the convection driven by the horizontal solute segregation takes with time a more and more important part ahead of the cell tip front. This increasing fluid flow intensity has significant influence on the evolution of the solid-liquid interface shape, more especially in the commencement of cellular solidification, where the visible difference in cell morphology demonstrates significant effect on the growth of the cell tips. Even though the horizontal temperature gradient was arbitrarily increased ten times larger than the estimated value, the maximum velocity of flow (4.5 μm/s) is still much smaller than the flow velocity (1.1 mm/s) calculated with the estimated temperature gradient using a domain size identical to the experimental sample [11]. From figure 2(a-c), it can be seen that on the left side (cold wall), the isoconcentration lines are compressed, whereas they are stretched on the right side (hot wall, which affects locally the solute gradients and thus the growth rates. The shape of the solidification front after a long time of solidification (~2700 s) is rather different from the case controlled by purely diffusion-dominated transport. It appears as a smooth curved shape rather than a planar plane perpendicular to vertical direction, which agrees well with observations in the experiment in the early stage of solidification. The s/l interface is advancing faster on the left side than on the right side as fluid flow is descending near the left wall while it is ascending near the right wall, as shown in figure 2(b). The solute rejected in the liquid at the interface is on the whole transported away from the left wall, and accumulated at the right wall, which leads to different solute boundary layer thicknesses at the right and left sides of the s/l interface (figure 3(b)).

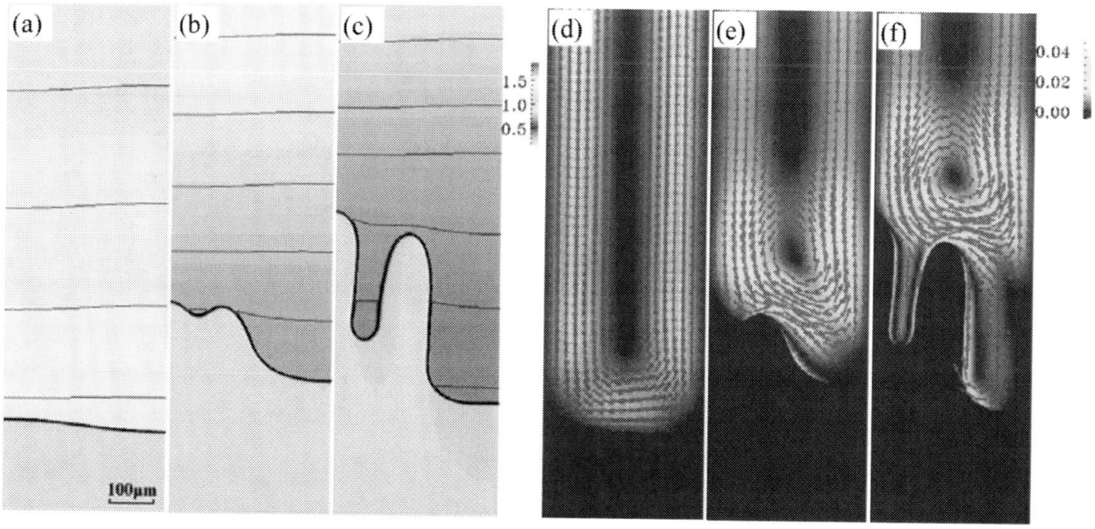

Figure 2. (a-c) Phase-field simulated solute profile (the black lines are isoconcentration lines) and s/l interface morphology after solidification of (a) 1678 s, (b) 3173 s, (c) 3675 s. Dimensionless solute concentration is scaled by C_0. (d-f) Time evolution of the simulated fluid flow field ahead of the solidification front, (d) 1678 s, (e) 3173 s, (f) 3675 s. In (d-f), fluid flow velocity is scaled by $D_l/(a_1 a_2 \xi^2 d_0) = 75.81$ μm/s. The calculated maximum fluid flow velocity is 4.5 μm/s.

In order to understand fluid flow influence on the dynamical evolution of the s/l interface morphology, the characteristic parameters of growth, namely interface concentration at the liquid side, solutal length of the solute profile and interface velocity, were calculated along two representative positions, respectively towards the left wall where fluid flow is descending similar to *l1* position in experiment and towards the right wall where fluid flow is ascending similar *l3* position in experiment. Experimental results and numerical predictions are quantitatively compared in figure 3. At the beginning of planar solidification, natural convection is weak and solidification proceeds in nearly diffusive conditions in both in simulation and experiment. Consequently, close agreement between simulations with and without consideration of fluid flow and experimental measurements is achieved. Then, descending fluid flow promotes solidification on the left side of the sample and retards it on the right side in simulation, which agrees with what is observed in experiment. The distance between the highest and the lowest levels of the smoothly curved s/l interface shape becomes larger and larger with time. This effect is more pronounced in the experiment, which shows that the fluid flow calculated in the simulation is underestimated, likely due to too strong lateral confinement effect. At later stage of planar solidification, the differences in interface concentration, solutal length and interface velocity between *l1* and *l3* become significant. Since the calculation cannot reproduce the actual fluid flow in the experiment yet, quantitative differences between the phase-field simulation and experimental observations are not surprising. However, the simulation already shows trends identical to those revealed by in situ monitoring of the experiment for the influence of fluid flow on the evolution dynamics. Namely, along *l1* the s/l interface velocity is higher than that along *l3* and the solute boundary layer is thinner than that along *l3*.

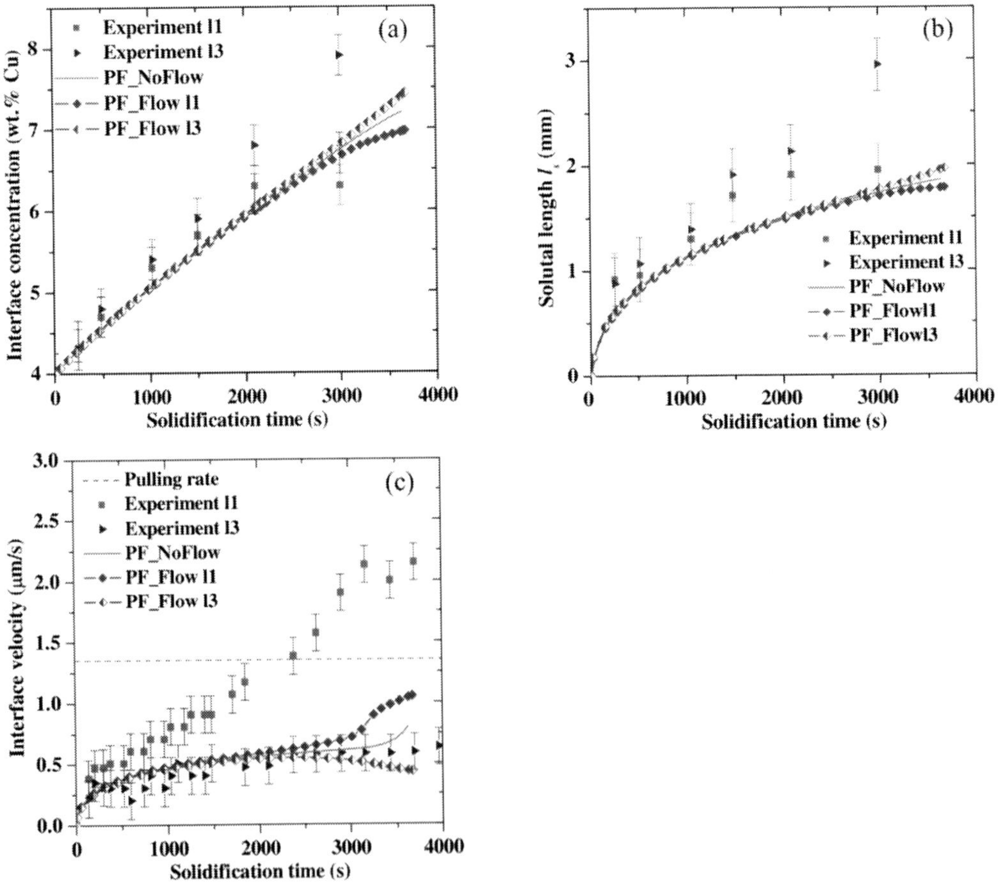

Figure 3. Comparison of phase-field predictions of the time evolution of the characteristic parameters with experimental measurements. (a) Concentration at the s/l interface on the liquid side; (b) Solutal length calculated using equation (4) in Ref. [11]; (c) Solidification front velocity.

3.2. Convection effects on the onset of morphology instability of the planar solidification front

Convection effects have been shown to be dominant on the planar to cellular and cellular to dendritic transitions [19]. The influence of natural convection on the stability of an accelerating planar interface (i.e. in transient conditions) and morphological transitions were thus examined by phase-field simulation and experiment. Despite the fact that the phase-field simulation in a narrower domain is not capable of revealing the exact microstructure observed in the experiment, the deformation behavior of the planar solidification in simulation is remarkably consistent with experimental observations as shown in figure 4(a) and (b). In simulation, the interface broke down first at *l1* after about 2800 s of solidification, a bit earlier than in the pure diffusion transport case (~3050 s), but much later than in experiment (~1480 s). Along *l3*, the interface was much more stable than that along *l1* both in simulation and experiment, which is in agreement with the expected effects of convection. The initial cell spacings measured in experiments and simulation are compared in table 2. The distance between the two cells in simulation varied from the initial wavelength of the deformed planar front 103 μm (after ~2800 s of solidification) to 165 μm after ~3675 s of solidification. This shows that the rejected solute at the left cell tip was swept into the groove between the two cells and gradually drove the right cell towards the right. The first visible wavelengths measured in experiment (figure 4(a)) after 2072 s of solidification are listed in table 2, together with the initial cell spacing in the phase-field simulation under diffusion dominated condition alone [11] (figure 4(c)). A large difference between experimental and simulation values is obtained with phase-field simulation not taking into consideration fluid flow, whereas a close agreement is achieved when natural convection was incorporated into the phase-field simulation. These results confirm that natural convection has an essential effect on the nonlinear development of morphological instability of the planar solidification front. The inhomogeneous solute

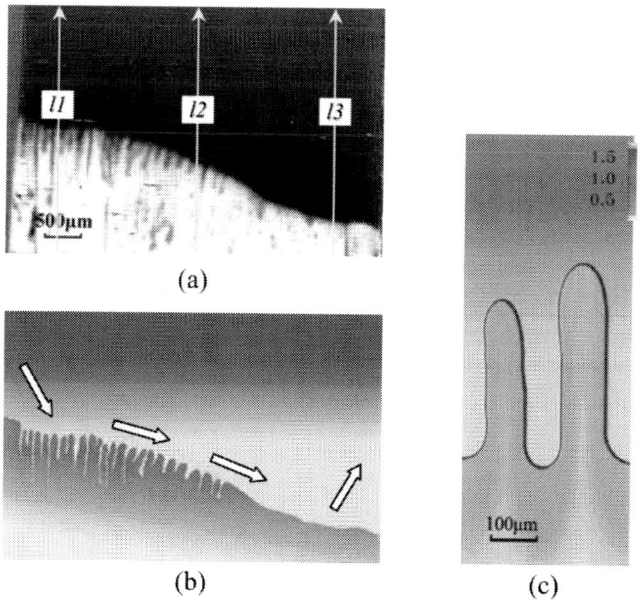

Figure 4. (a) The morphology of the solidification front observed by *in situ* and real-time synchrotron imaging after $t = 2072$ s of solidification. (b) Clear view of interface corrugations, or initial cells, and visualization of solute segregation in the melt through color change as obtained from image processing of the cellular morphology observed in radiography. The yellow arrows sketch the direction of solute transport (by fluid flow) in the melt ahead of the front. (c) Typical phase-field simulated cellular morphology under pure diffusion transport condition for computation at $t = 4508$ s [11].

distribution induced by fluid flow in the liquid just above the s/l interface leads in particular to different behaviours of the smooth interface along *l1* and *l3* with respect to morphological instability. The solute macrosegregation in the melt slowed down interface velocity at *l3* (lower concentration gradient) and promoted growth at *l1* (enhanced concentration gradient). In agreement with the Mullins-Sekerka criterion [20], the planar interface consequently lost its stability earlier at *l1* where velocity and solute gradient were higher while it remained smooth at *l3*, where velocity and solute gradient were lower. In addition, the initial wavelength of deformed planar front was reduced by the narrower solute boundary layer caused by descending flow along *l1* [11]. This is in agreement with the experiments carried out on Al-Cu alloy by Trivedi et *al.*[19], who showed that convection effects in a large diameter sample stabilized the planar interface because convection reduced the solute gradient at the interface in the liquid [19].

Table 2. The interface corrugation wavelength or cells as observed in experiment, and given by phase-field simulations for the diffusion-dominated case and for phase-field simulation taking into account natural convection.

Measured lines	Experiments (μm)	Phase-field simulation, no flow (μm)	Phase-field simulation with natural convection (μm)
l1	109	167	103
l2	163	167	—
l3	—	167	—

4. Conclusions

We have presented a quantitative comparison of phase-field simulation under the influence of natural convection with the in situ and real-time observation of the evolution of the s/l interface during directional solidification of an Al-Cu alloy by cooling down method. By applying a proper horizontal temperature gradient, the observed curve of a smooth s/l interface was achieved in the simulation. Since the simulated domain is much smaller than the experimental size the natural convection strength was weakened and the simulations were not able to exactly reproduce the experimental observations. However, the solute distribution ahead of the solidification front and its evolution in the simulation agree well with experimental measurements, as well as the solidification dynamics. Simulations confirm that fluid flow can locally change the solute gradient at the solid-liquid interface, according to the fluid flow direction versus s/l interface growth direction, and thus shifted the inception of s/l instability. In other words, analysis of the stability of the planar s/l interface should take into account not only the fluid flow intensity, but also the fluid flow patterns.

Acknowledgements

This work was supported by National Key Science & Technology Special Project, China (Grant No. 2011ZX04014-052) and National Natural Science Foundation for Young Scientists, China (Grant No. 51001096).

References

[1] Boettinger W J, Coriell S R, Greer A L, Karma A, Kurz W, Rappaz M and Trivedi R 2000 *Acta Mater.* **48** 43
[2] Asta M, Beckermann C, Karma A, Kurz W, Napolitano R, Plapp M, Purdy G, Rappaz M and Trivedi R 2009 *Acta Mater.* **57** 941
[3] Beckermann C, Diepers H J, Steinbach I, Karma A and Tong X 1999 *J. Comput. Phys.* **154** 468
[4] Bouissou P and Pelcé P 1989 *Phys. Rev. A* **40** 6673
[5] Bouissou P, Perrin B and Tabeling P 1989 *Phys. Rev. A* **40** 509
[6] Bogno A, Nguyen-Thi H, Buffet A, Reinhart G, Billia B, Mangelinck-Noël N, Bergeon N, Baruchel J and Schenk T 2011 *Acta Mater.* **59** 4356
[7] Lan C W and Shih C J 2004 *Phys. Rev. E* **69** 031601
[8] Lang Y and Peter D L 2010 *Modelling Simul. Mater. Sci. Eng.* **18** 055008
[9] Zhu M, Dai T, Lee S and Hong C 2005 *Science in China E* **48** 241
[10] Steinbach I 2009 *Acta Mater.* **57** 2640

[11] Chen Y, Bogno A-A, Xiao N M, Billia B, Kang X H, Nguyen-Thi H, Luo X H and Li D Z 2012 *Acta Mater.* **60** 199
[12] Karma A 2001 *Phys. Rev. Lett.* **87** 115701
[13] Echebarria B, Folch R, Karma A and Plapp M 2004 *Phys. Rev. E* **70** 061604
[14] Worster M G 1997 *Annu. Rev. Fluid Mech.* **29** 91
[15] Chen Y, Bogno A-A, Billia B, Kang X H, Nguyen-Thi H, Li D Z, Luo X H and Debierre J-M 2010 *ISIJ Int.* **50** 1895
[16] Guermond J-L and Quartapelle L 1998 *Int. J. Numer. Meth. Fluids* **26** 1039
[17] Billia B, Nguyen-Thi H, Mangelinck-Noel N, Bergeon N, Jung H, Reinhart G, Bogno A, Buffet A, Hartwig J, Baruchel J and Schenk T 2010 *ISIJ Int.* **50** 1929
[18] Buffet A, Thi H N, Bogno A, Schenk T, Mangelinck-Noel N, Reinhart G, Bergeon N, Billia B and Baruchel J 2010 *Mater. Sci. Forum* **649** 331
[19] Trivedi R, Miyahara H, Mazumder P, Simsek E and Tewari S N 2001 *J. Cryst. Growth* **222** 365
[20] Mullins W W and Sekerka R F 1964 *J. Appl. Phys.* **35** 444

IOP Conf. Series: Materials Science and Engineering 33 (2012) 012103 doi:10.1088/1757-899X/33/1/012103

Modelling of dendritic growth and bubble formation

W Wu[a], M F Zhu[a], D K Sun[a], T Dai[a], Q Y Han[b], D Raabe[c]

[a] Jiangsu Key Lab for Advanced Metallic Materials, Southeast University, China
[b] Department of Mechanical Engineering Technology, Purdue University, USA
[c] Max-Planck-Institut für Eisenforschung, Düsseldorf, Germany

E-mail: zhumf@seu.edu.cn

Abstract. A two-dimensional lattice Boltzmann method (LBM)-cellular automaton (CA) model is developed for the simulation of dendritic growth and bubble formation during alloy solidification. In the model, a kinetic LBM, which describes flow dynamics through the evolution of distribution functions of moving pseudo-particles, is adopted to numerically solve the gas-liquid two-phase flow based on the Shan-Chen multiphase scheme. The kinetics of dendritic growth is determined according to a local solute equilibrium approach. The present model takes into account the effect of liquid-solid phase transformation on the nucleation and growth of bubbles. The interaction mechanism between dendrites and bubbles is also embedded in the model. The wettability of a bubble on a smooth solid surface is simulated. The simulated contact angles with various interaction coefficients agree well with the data calculated from an empirical formula derived from the Young's equation. The proposed model is applied to simulate dendritic growth and bubble formation under directional solidification conditions. The simulated results are compared with those observed experimentally during solidification of a transparent organic material. The simulation results reveal some dynamic features of bubble nucleation, growth, and motion, as well as the interaction between the dendritic growth and bubble formation during solidification.

1. Introduction

The performance of casting components is primarily dependent on the solidification microstructures and defects. Porosity formed due to the precipitation of dissolved gases is one of the major casting defects that exist in castings of lightweight materials, such as aluminum and magnesium alloys. Porosity considerably deteriorates the mechanical properties of cast components, in particular their fatigue life, limiting their applications, especially under safety critical conditions. Considerable efforts have been devoted to develop models for predicting porosity in castings, including analytical solutions, criteria functions, and complex numerical models, such as Darcy's law coupled with the conservation and continuity equations, gas diffusion-controlled pore growth models and continuum-stochastic models by coupling cellular automaton (CA) and finite difference method (FDM) [1-4]. These efforts have achieved great success in predicting the volume fraction, and in modeling the coupled growth of grains and gas pores. However, the existing approaches have limited potential in describing the dynamic features associated with gas bubble formation and motion during dendritic solidification. Recent experimental observations revealed that the evolution and motion of gas bubbles in the mushy zone determine the final distribution and morphology of porosity in castings [5].

During the last two decades, the lattice Boltzmann method (LBM) has rapidly developed as a powerful tool for the numerical calculation of fluid flow. LBM is a kinetic-based approach that considers flows to be composed of a collection of pseudo-particles represented by a distribution function. The viscous fluid flow evolves automatically from the streaming and collision processes of particles that are constrained to move over a lattice. The macroscopic physical quantities can be obtained by the distribution functions of particles [6, 7]. The intrinsic features of LBM allow simulating complex fluid systems, such as multiphase and multicomponent flow phenomena under complicated boundary conditions. LBM has been coupled with phase-field (PF) and CA methods to simulate the interaction of melt convection and dendritic growth in pure materials and alloys [8-11]. However, little work has been reported on applying LBM to simulate the dendritic growth and gas bubble formation during solidification.

This paper presents a two-dimensional (2D) LBM-CA model for the simulation of the coupled growth of dendrites and gas bubbles. A two-phase flow LBM based on the Shan-Chen (SC) multiphase scheme is adopted to model the nucleation, growth, and motion of gas bubbles. The kinetics of dendritic growth is determined by a local solute equilibrium approach previously proposed by Zhu and Stefanescu (ZS) [12]. After model test through simulating the wettability of a bubble on a smooth solid surface, the simulated evolution of dendritic growth and bubble formation under directional solidification is presented and compared with those obtained experimentally.

2. Model description and governing equations

2.1. Two-phase flow LBM model

In the present work, bubble formation and motion are simulated by solving the discrete Boltzmann equations based on the SC two-component multiphase LBM scheme [13-15]. Two sets of distribution functions are introduced for the two fluid components of gas and liquid. Considering that the 2D simulation domain is placed horizontally, the buoyancy effect due to the temperature and concentration gradients under gravity is neglected. According to the Bhatnagar-Gross-Krook (BGK) approximation and taking into account the liquid-gas phase transition, the lattice Boltzmann equation (LBE) of each fluid component is expressed as

$$f_i^\sigma(\vec{x}+\vec{e}_i\Delta t, t+\Delta t) - f_i^\sigma(\vec{x},t) = -\frac{1}{\tau^\sigma}[f_i^\sigma(\vec{x},t) - f_i^{\sigma(eq)}(\vec{x},t)] + \omega_i\Delta\rho^\sigma$$
$$\sigma = 1,2, \quad i = 0,1,\ldots,8 \tag{1}$$

where $f_i^\sigma(\vec{x},t)$ is the particle distribution function (PDF), $f_i^{\sigma(eq)}(\vec{x},t)$ is the equilibrium PDF (EPDF), and τ^σ is the relaxation time of the σth component. Δt is the time step. $\Delta\rho^\sigma$ represents the density variation of the σth component caused by phase transition. $\vec{e}_i$ is the discrete velocity and ω_i is the weight coefficient.

The two-dimension nine-velocity (D2Q9) scheme [13] is employed, in which the space is discretized into a square lattice including nine discrete velocities, $\vec{e}_i$, given as

$$\vec{e}_i = c\cdot\begin{cases} (0,0) & i=0 \\ (\cos[\pi(i-1)/2], \sin[\pi(i-1)/2]) & i=1,2,3,4 \\ (\cos[\pi(i-5)/2+\pi/4], \sin[\pi(i-5)/2+\pi/4]) & i=5,6,7,8 \end{cases} \tag{2}$$

where $c=\Delta x/\Delta t$ is the lattice speed, and Δx is the lattice spacing. In the D2Q9 scheme, the weight coefficients ω_i in Eq. (1) are given as ω_0=4/9, $\omega_{1\text{-}4}$=1/9, and $\omega_{5\text{-}8}$=1/36. The EPDF $f_i^{\sigma(eq)}(\vec{x},t)$ is defined as

$$f_i^{\sigma(eq)}(\vec{x},t) = \omega_i\rho^\sigma(\vec{x},t)[1+3(\vec{e}_i\cdot\vec{u}^\sigma(\vec{x},t))/c^2 + 4.5(\vec{e}_i\cdot\vec{u}^\sigma(\vec{x},t))^2/c^4 - 1.5(\vec{u}^\sigma(\vec{x},t))^2/c^2] \tag{3}$$

The density of the σth phase, $\rho^\sigma(\vec{x},t)$, and the average flow velocity, $\vec{u}(\vec{x},t)$, are defined as

$$\rho^\sigma(\vec{x},t) = \sum_{i=0}^{8} f_i^\sigma(\vec{x},t) \tag{4}$$

$$\rho(\vec{x},t)\vec{u}(\vec{x},t) = \sum_{\sigma=1}^{2}\sum_{i=0}^{8} \vec{e}_i f_i^\sigma(\vec{x},t) \tag{5}$$

where $\rho(\vec{x},t)$ is the total density of the two components, which is obtained from

$$\rho(\vec{x},t) = \sum_{\sigma=1}^{2}\sum_{i=0}^{8} f_i^\sigma(\vec{x},t) \tag{6}$$

In equation (3), the macroscopic velocity of the σth phase, $\vec{u}^\sigma(\vec{x},t)$, is given by

$$\rho^\sigma(\vec{x},t)\vec{u}^\sigma(\vec{x},t) = \rho^\sigma(\vec{x},t)\vec{u}(\vec{x},t) + \tau^\sigma \vec{F}^\sigma(\vec{x},t) \tag{7}$$

where $\vec{F}^\sigma(\vec{x},t)$ is the total resultant force acting on the σth fluid component.

In the SC multiphase model, the fluid-fluid cohesive force acting on the σth component is defined as

$$\vec{F}_c^\sigma(\vec{x},t) = -G_c\rho^\sigma(\vec{x},t)\sum_{i=0}^{8}\omega_i\rho^{\bar{\sigma}}(\vec{x}+\vec{e}_i\Delta t,t)\vec{e}_i \tag{8}$$

where σ and $\bar{\sigma}$ denote the two different fluid components, and G_c is a parameter that controls the strength of the cohesion force. The pressure at position $\vec{x}$ and time t can be determined from the densities as

$$P(\vec{x},t) = \rho(\vec{x},t)/3 + G_c\rho^\sigma(\vec{x},t)\rho^{\bar{\sigma}}(\vec{x},t)/3 \tag{9}$$

During the phase transition from liquid to gas, the increase in gas quantity matches the decrease in liquid quantity. Thus, the source term in equation (1), $\Delta\rho^\sigma$, can be considered as $\Delta\rho^g=-\Delta\rho^l$, and evaluated as

$$\Delta\rho^g = -(\rho^g + \rho^l)\Delta\phi \tag{10}$$

where $\Delta\phi$ is the quantity of phase transition from liquid to gas, which is related to the increase in solid fraction Δf_s at the solid-liquid interface by $\Delta\phi = A\Delta f_s$, where the coefficient A is taken as 10^{-6}.

To calculate the unknown particle distribution functions at the boundary nodes, four surfaces of the domain are treated as solid walls by imposing the bounce-back rule.

2.2. The CA model of dendritic growth

As described previously, the kinetics of dendritic growth is determined according to a local solute equilibrium approach [12]. In this approach, the driving force for the evolution of the solid/liquid interface is considered to be controlled by the difference between the local interface equilibrium composition and local actual liquid composition. The interface equilibrium composition C_l^* is calculated by

$$C_l^* = C_0 + [(T_l^* - T_l^{eq}) + \Gamma K(1-15\varepsilon\cos[4(\varphi-\varphi_0)])]/m \tag{11}$$

where T_l^* is the interface temperature, T_l^{eq} is the equilibrium liquidus temperature at the initial composition C_0, ε is the degree of anisotropy of the surface tension, m is the liquidus slope, Γ is the Gibbs-Thomson coefficient, K is the curvature of the solid/liquid interface, φ is the growth angle between the normal to the interface and the x-axis, and φ_0 is the angle of the preferential growth direction with respect to the x-axis. On the other hand, the local actual liquid composition C_l is obtained by solving the following solute diffusion equation:

$$\frac{\partial C}{\partial t} = \nabla \cdot (D \nabla C) + C(1-k)\frac{\partial f_s}{\partial t} \tag{12}$$

where C is composition, D is the solute diffusion coefficient, k is the solute partition coefficient, and f_s is the solid fraction. The second term on the right-hand side denotes the amount of solute rejected at the interface. Equation (12) is solved using an explicit finite difference scheme, and the zero-flux boundary condition is implemented on the four surfaces of the calculation domain. Then, the local interface equilibrium composition C_l^* calculated by equation (11) is compared with the local actual liquid composition C_l. If the difference satisfies $\Delta C = C_l^* - C_l > 0$ in an interface cell, the solid fraction will increase. According to the solute equilibrium condition at the interface, during one time step interval, the increased solid fraction, Δf_s, of an interface cell is evaluated by

$$\Delta f_s = (C_l^* - C_l)/(C_l^*(1-k)) \tag{13}$$

If the sum of the solid fraction in an interface cell equals one, the state of this cell is assigned as solid and new interface cells are explicitly captured, while the exact solid-liquid front is implicitly scaled by the solid fraction within each interface cell. The local interface curvature K, and the growth angle φ are calculated based on the solid fraction gradient at the solid/liquid interface with the following equations:

$$K = \left[2\partial_x f_s \partial_y f_s \partial_{xy}^2 f_s - (\partial_x f_s)^2 \partial_y^2 f_s - (\partial_y f_s)^2 \partial_x^2 f_s\right] \cdot \left[(\partial_x f_s)^2 + (\partial_y f_s)^2\right]^{-3/2} \tag{14}$$

$$\varphi = \arccos\left(\partial_x f_s / \left[(\partial_x f_s)^2 + (\partial_y f_s)^2\right]^{1/2}\right) \tag{15}$$

2.3. *The coupled LBM-CA model*

The two-phase flow LBM model for gas bubble formation is coupled with the CA model for dendritic growth. The force $\vec{F}^{\sigma}(\vec{x},t)$ acting on the σth fluid component, involving the fluid-fluid cohesive force, $\vec{F}_c^{\sigma}(\vec{x},t)$, and fluid-solid adhesion force, $\vec{F}_{ads}^{\sigma}(\vec{x},t)$, is defined as

$$\vec{F}^{\sigma}(\vec{x},t) = \vec{F}_c^{\sigma}(\vec{x},t) + \vec{F}_{ads}^{\sigma}(\vec{x},t) \tag{16}$$

where the fluid-fluid cohesive force, $\vec{F}_c^{\sigma}(\vec{x},t)$, is calculated by equation (8), and fluid-solid adhesion force, $\vec{F}_{ads}^{\sigma}(\vec{x},t)$, is calculated by

$$\vec{F}_{ads}^{\sigma}(\vec{x},t) = -G_{ads}^{\sigma}\rho^{\sigma}(\vec{x},t)\sum_{i=0}^{8}\omega_i s(\vec{x}+\vec{e}_i\Delta t,t)\vec{e}_i \tag{17}$$

where $s(\vec{x}+\vec{e}_i\Delta t)$ is an indicator function that is equal to 1 or 0 for a solid or a fluid component node, respectively. G_{ads}^{σ} is the interaction strength coefficient between the solid and fluid phases.

Dendritic growth is simulated using the CA model described by equations (11)-(15). As dendrite grows, the concentration of dissolved gas in the liquid phase increases, due to the fact that gas is far less soluble in solid than in liquid. As a result, gas bubbles occur when the local gas supersaturation reaches a certain level. Then, the formed bubbles become sinks for the supersaturated gas in liquid. These phenomena are simulated by solving the LBM equations (1)-(10). Dendrites are marked as solid phase in the LBM model. The bounce-back scheme is applied both at the solid-liquid and solid-gas interfaces. The force $\vec{F}^{\sigma}(\vec{x},t)$ acting on the two fluid components is calculated by equations (16) and (17), which influences the distribution functions of the gas and the liquid. In addition, dendritic growth is prohibited at the solid-gas interface. Consequently, the interaction mechanism between dendrites and bubbles is naturally embedded in the model.

3. Results and discussion

3.1. The effect of interaction coefficient on the wettability of a gas-liquid-solid system

To test the proposed model regarding the effect of interaction coefficients on the wettability of a gas-liquid-solid system, the contact angles of a bubble on a solid surface with various interaction coefficients were simulated. According to the Young's equation, the contact angle shown in figure1 is determined by

$$\cos\theta = (\sigma_{ls} - \sigma_{gs})/\sigma_{gl} \tag{18}$$

where θ is the contact angle, σ_{gl}, σ_{gs} and σ_{ls} are the interfacial tensions of gas-liquid, gas-solid and liquid-solid, respectively. Sukop *et al.* [14] and Huang *et al.* [15] related the interfacial tensions between fluid-solid and fluid-fluid phases in the Young's equation to the cohesion factor, G_c, in equation (8), the interaction coefficients, G^{σ}_{ads}, in equation (17), and the densities of the two fluid phases, ρ^{σ}, in equation (4). Based on the approach, for a gas-liquid-solid system, the contact angle can be evaluated by

$$\cos\theta = \left(G_{ls} - G_{gs}\right)\cdot\left(G_c \frac{|\rho_g - \rho_f|}{2}\right)^{-1} \tag{19}$$

where G_{gs} and G_{ls} are the interaction coefficients between gas-solid and liquid-solid phases, and ρ_g and ρ_f are the densities of gas and liquid, respectively.

Simulations were carried out in a domain of 110×110 lu^2 (lattice unit). The relevant parameters were taken as G_c=1.0, ρ_g =0.1, ρ_f =3.0, and G_{ls}=-G_{gs}. The simulated values of contact angle are compared with the data calculated from equation (19) in figure 2. It can be seen that the contact angle decreases with the increase of G_{ls}, and the simulated results agree well with the calculations by equation (19), indicating that the present model can reasonably describe the wettability phenomenon of a gas-liquid-solid system.

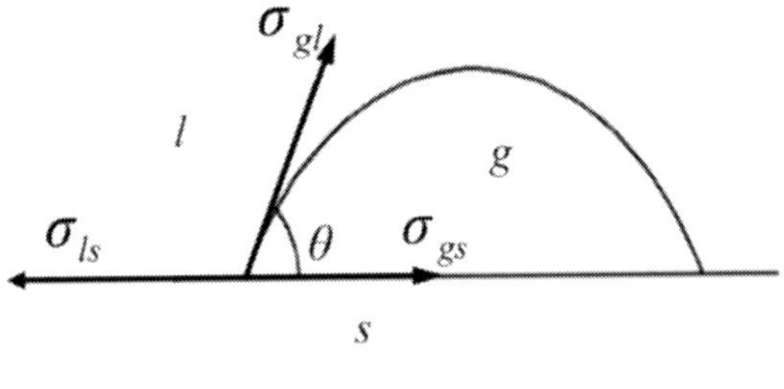

Figure 1. Schematic of contact angle.

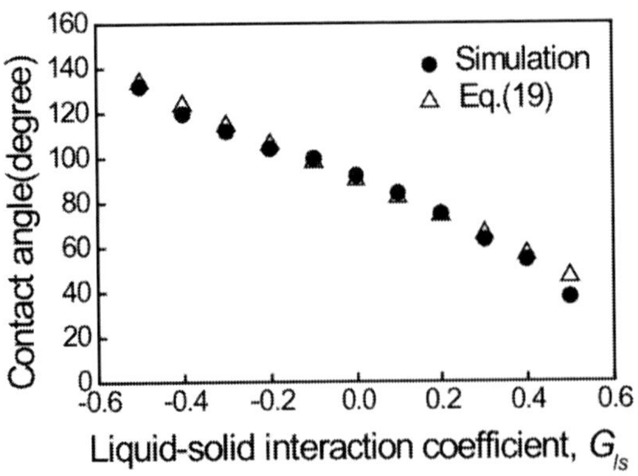

Figure 2. Contact angle as a function of liquid-solid interaction coefficient.

3.2. Dendritic growth and bubble formation under unidirectional solidification

The present model was applied to simulate the evolution of dendritic growth and bubble formation of an Al-4mass%Cu alloy under directional solidification condition. The physical properties were taken as k=0.17, m=-2.6$K/(mass\%)$, $\Gamma=2.4\times10^{-7} m\cdot K$, ε=0.02, $D_l=3\times10^{-9} m^2/s$, and $D_s=3\times10^{-13} m^2/s$. Considering that it is nearly non-wetting between gas and melt, the interaction coefficients of different phases were taken as G_c=2.0, G_{ls}=-0.9, and G_{gs}=0.9, making the contact angle close to 180°. The simulation was performed in a domain consisting of a 250×800 mesh with an uniform mesh size of 1 μm, and it was laid horizontally. Figure 3 presents the simulated evolution of columnar dendrites and gas bubbles of an Al-4mass%Cu alloy directionally solidified with a thermal gradient G=20 K/mm, initial undercooling of 3 K and a cooling rate of 4 K/s. At the beginning of the simulation, several nuclei were assigned at the wall of the left-hand side with random crystallographic orientations. The initial gas content was taken as 4.5%. As shown in figure 3, after initial competition, three dendrites with similar orientations with respect to the temperature gradient overgrew the others, together with the branching of the secondary arms. Since gas solubility is greater in liquid than in solid, as dendrites grew, the gas element was rejected from the newly formed solid into the surrounding liquid phase. With the increase of the gaseous concentration level in the liquid, the supersaturated gas element was gradually separated to form bubbles when the solid fraction reached around 14%. The tiny bubbles were mostly nucleated at the solid/liquid interface between the secondary arms (see figure 3 (a)). This is because the enriched gas concentration was highest at those locations. The formed bubbles grew through absorbing the gas elements from the nearby liquid. The growing bubbles were squeezed by the narrow secondary arm spacing and became elongated, leading to the decrease of the local bubble radius. It is known that the internal pressure of a bubble increases with the decrease of the bubble radius according the Gibbs-Thomson relation, which drives the bubbles to escape from the locations between the secondary arms to the liquid channels between dendrites as shown in figure 3 (b). Because of the effect of gas-liquid interfacial tension, the bubbles grew spherically when they were fully enveloped by the liquid. As dendritic solidification proceeded, the growth space of bubbles was restricted by the complex dendrite network. As a result, the internal pressure of the bubbles increased, forcing the bubbles to move or jump towards the regions of higher liquid fractions. When the bubbles moved and jumped in the liquid channels, the nearby bubbles might coalesce together to form a larger bubble. For example, the three bubbles inside the circle in figure 3 (c) coalesced to form one bubble marked by the arrow in figure 3 (d). The bubble moved towards to the new location marked by the arrow in figure 3 (e). Then it coalesced further with the two neighboring bubbles to form a worm-like bubble marked by the arrow in figure 3 (f). It also can be seen from figure 3 (f) that some bubbles escaped out of the mushy zone to the liquid region in front of dendrites and became spherical bubbles, while some others were entrapped in the mushy zone by the dendrite network.

The *in-situ* experiment of directional solidification was performed with cyclohexane, a transparent organic material. A glass sample tube with a rectangular cavity size of 0.2×4×150 mm was filled with liquid cyclohexane containing dissolved gases. The sample tube was placed horizontally on a copper plate. One side of the copper plate was connected to a cold stage and the other to a hot stage to establish a constant temperature gradient. The sample tube was withdrawn by a translation device with controlled velocities. Thus, directional solidification of the cyclohexane took place inside the tube along the horizontal direction. Since the thickness of sample tube was very thin, the solidification was considered approximately in two dimensions. In addition, the focus of the present work is to investigate the interaction between the dendritic growth and bubble formation in the mushy zone. Compared with the adhesion force between fluid and solid, the buoyancy force in the mushy zone in a horizontal thin sample tube should be much weaker. Consequently, the buoyancy effect under gravity during the *in-situ* experiment was also neglected. Figure 4 presents the experimental morphologies of dendrites and bubbles formed under directional solidification of cyclohexane. The temperature gradient and the withdraw rate of sample tube were about 10 K/mm and 40 $\mu m/s$, respectively. Since the rate of bubble nucleation was very fast, it was difficult to capture the moment of bubble nucleation. It was observed that small bubbles, initially nearly spherical, grew and coalesced to form the worm-like bubbles lying between the dendrite arrays as shown in figure 4 (a). In the late stage, bubbles might escape out of the mushy zone to the liquid region at the freezing front and become spherical bubbles as shown in figure 4 (b).

Considering that the simulation and experimental results were obtained from different materials and conditions, it might be difficult to make a direct quantitative comparison. Nevertheless, the simulations of the present model reproduced some phenomena observed in experiments, such as worm-like bubbles within the inter-dendritic regions, and spherical bubbles escaping out of the mushy zone.

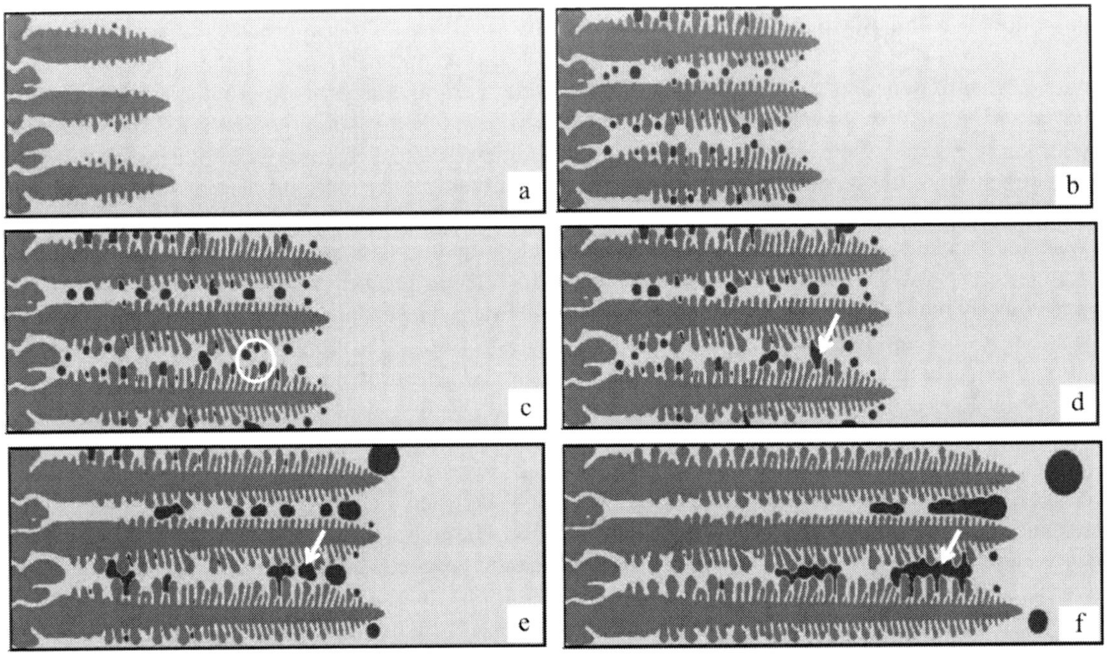

Figure 3. Simulated evolution of dendrite growth and bubble formation under directional solidification of an Al-4*mass*%Cu alloy with solid fraction of (a) 15%, (b) 28%, (c) 36%, (d) 37%, (e)43%, (f) 54% (red: solid, green: liquid, blue: gas).

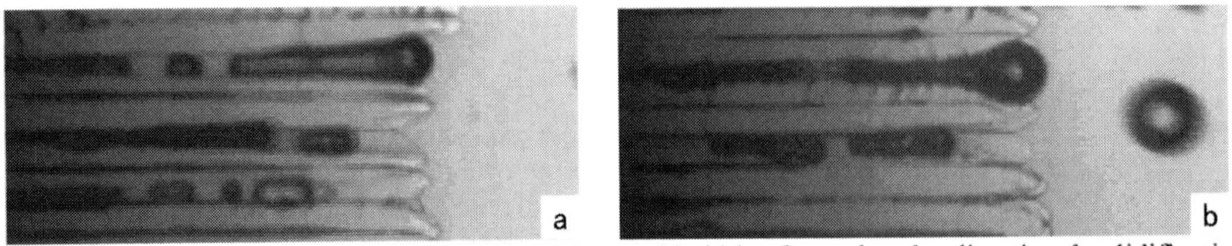

Figure 4. Experimental morphologies of dendrites and bubbles formed under directional solidification of cyclohexane. The photos were taken at (a) 21s, (b) 35s.

4. Conclusions

A two-dimensional coupled LBM-CA model is developed for the simulation of dendritic growth and bubble formation during alloy solidification. In the model, a multiphase LBM is adopted to simulate the nucleation, growth and motion of bubbles in liquid. On the other hand, the dendritic growth is simulated by a CA method where the kinetics of dendritic growth is determined using a local solute equilibrium approach. The proposed model is tested regarding the effect of interaction coefficients on the wettability of a gas-liquid-solid system. It is found that the contact angle decreases with the increase of solid-liquid interaction coefficient, and the simulated results agree well with the data calculated from an empirical formula derived based on the Young's equation. The proposed LBM-CA model is applied to simulate the evolution of dendritic growth and bubble formation during unidirectional solidification. The simulation results illustrate some phenomena, including dendritic competitive growth, the preferential nucleation locations of bubbles, and bubble growth, coalescence, deformation due to the squeezing effect imposed by the topology of the neighbouring dendrites, bubble migration in the liquid channels towards the region of higher liquid fractions, as well as the escape of bubbles out of the mushy zone to the liquid region in front of dendrites. The simulations of the present LBM-CA model provide an insight into the underlying physics of bubble nucleation, growth, and motion in the mushy zone, as well as the interaction between the dendritic growth and bubble formation during solidification.

Acknowledgements
The work was financially supported by NSFC (Grant No. 50971042). The *in-situ* experiment was performed in State key Laboratory of Solidification Processing, Northwestern Polytechnical University.

References
[1] Lee P D, Chirazi A and See D 2001 *J. Light Met.* **1** 15
[2] Stefanescu D M 2003 *Int. J. Cast Metals Res.* **18** 129
[3] Atwood R C and Lee P D 2003 *Acta Mater.* **51** 5447
[4] Dong S Y, Xiong S M and Liu B C 2004 *Mater. Sci. Technol.* **20** 23
[5] Han Q Y 2006 *Scripta Mater.* **55** 871
[6] Raabe D 2004 *Modelling Simul. Mater. Sci. Eng.* **12** R13
[7] Guo Z L, Asinari P and Zheng C G 2009 *Phys. Rev. E* **79** 026702
[8] Miller W, Succi S and Manutti D 2001 *Phys. Rev. Lett.* **86** 3578
[9] Selzer M, Jainta M and Nestler B 2009 *Phys. Status Solid B* **246(6)** 1197
[10] Sun D K, Zhu M F, Pan S Y and Raabe D 2009 *Acta Mater.* **6** 1755
[11] Yin H, Felicelli S D and Wang L 2011 *Acta Mater.* **59** 3124
[12] Zhu M F and Stefanescu D M 2007 *Acta Mater.* **55** 1741
[13] Huang H B, Thorne D T, Schaap M G and Sukop M C 2007 *Phys. Rev. E* **76** 066701
[14] Shan X W 2006 *Phys. Rev. E* 73 047701
[15] Sukop M C and Thorne D T. 2005 Lattice Boltzmann Modeling-An Introduction for Geoscientists and Engineers (2nd ed) (New York: Springer) p114

IOP Conf. Series: Materials Science and Engineering **33** (2012) 012104 doi:10.1088/1757-899X/33/1/012104

3D Phase-Field Simulation of Micropore Formation during Solidification: Morphological Analysis

H. Meidani[1], A. Jacot[1,2] and M. Rappaz[1]

[1] Computational Materials Laboratory, Institute of Materials, EPFL, Lausanne, Switzerland
[2] Calcom ESI SA, Parc Scientifique, PSE-A, Lausanne, Switzerland

E-mail: hossein.meidani@epfl.ch, michel.rappaz@epfl.ch

Abstract. A 3D multiphase-field (PhF) model has been developed in order to study the formation of a micropore constrained to grow in a solid network (i.e., pinching effect). The model accounts for the pressure difference due to capillarity between liquid and gas, the equilibrium condition at triple (solid-liquid-pore) lines, the partitioning and diffusion of dissolved gases such as hydrogen. From the predicted 3D morphology of the pore, entities such as the Interfacial Shape Distribution (ISD) are plotted and analyzed. It is shown that the mean curvature of the pore-liquid surface, and thus also the pressure inside the pore, is uniform. Despite the complex morphology of pores reconstructed using high-resolution X-ray tomography, the present PhF results suggest that a simple pinching model based on a spherical tip growing in between remaining liquid channels is a fairly good approximation.

1. Introduction

In order to better characterize and more importantly minimize the formation of micropores during the solidification of metals, a considerable effort has been dedicated to the modeling of pore formation [1]. These models, which range from very simple criteria functions [2] to sophisticated computational solutions of conservation and state equations, provide quantitative information on the effect of the processing conditions and alloy chemistry on microporosity [3–6]; including the effect of solidification shrinkage and gas segregation in the mushy zone.

Common to most advanced models of microporosity formation at the process scale is the consideration of average conservation equations. To close the problem mathematically, a constitutive model for the radius of the curvature of the pore as a function of its volume is required [5, 7–9]. If the pore is assumed to be spherical, this relationship is straightforward. However, in reality the growth of micropores is limited by the well-developed solid network. Due to numerous contacts with the solid phase, pores take a complex and highly tortuous shape. Local mean curvatures as high as $0.2\,\mu m^{-1}$ have been measured recently in aluminum alloys by X-ray tomography [10], i.e., the pressure difference between the pore and the surrounding liquid can be as high as $400\ kPa$. Since the pore is a compressible phase, this can substantially affect the pore size and fraction. For such situations, a so-called pinching model, i.e., a mathematical expression relating the radius of curvature of the pore-liquid interface, r_p, to the volume fraction of solid g_s, microstructural parameters and possibly the volume fraction of pores, g_p, is required. Different pinching models have been proposed. Initial models [8] assumed a spherical pore with a radius equal to half the dendritic arms spacing, $\lambda_2/2$. A more elaborate model of Pequet et al. [5] considers different stages: (i) the nucleated pore *escapes* its foreign substrate with a constant

radius; (ii) the pore is a growing sphere, (iii) it is pinched by the solid and its radius is given by a function of g_s and λ_2; (iv) the radius remains constant when interdendritic eutectic forms near the end of solidification. Poirier et al. [7] assumed the pores to be constrained by dendritic arms arranged in a hexagonal network, and introduced the effect of arms impingement into their model. Couturier and Rappaz [9] refined this model by assuming the pore to grow in between a regular arrangement of spherical grains (in 3D) or cylindrical dendrite arms (in 2D). The pores grow with an initial spherical shape until they come into contact with the solid phase: their radius of curvature is then assumed to be given by the largest sphere (3D) or cylinder (2D) that can fit in between the grains, respectively. In other words, they do not change their topology and are never forced to grow in between the solid.
A common point between these models is the simplified pore geometry. However, in reality, the pores have a much more complex shape. A phase-field (PhF) model, directly describing the pore morphology as a function of different parameters such as gas concentration, solid fraction and capillary forces was presented in [11], in 2D, and recently in 3D [12]. In the present contribution, the model is used to reproduce the pinching effect in 3D and study the morphology of a pore spanning over a distance larger than the dendritic arm spacing.

2. Model

A multiphase-field model is used in this study, for which a more detailed description can be found in [11, 12]. Three phase-field variables, $\phi_i(\mathbf{x}, t)$, where $i = s, l, p$ is a phase index, are introduced to show the distribution of the solid, liquid and pore, respectively, in time and space. These variables can be understood as local volume fractions and thus linked together using the constraint $\phi_s + \phi_l + \phi_g = 1$. One salient feature of the PhF method is a smooth and continuous variation of ϕ_i across any interface [13]. Spatial and temporal evolution of the phase-field variable ϕ_i is given by the following partial differential equation:

$$\dot{\phi_i} = \sum_{k \neq i} M_{ik} \begin{pmatrix} \epsilon_{ik}^2(\phi_k \nabla^2 \phi_i - \phi_i \nabla^2 \phi_k) \\ -2W_{ik}\phi_k^2(1-\phi_k)^2\phi_i(1-\phi_i)(1-2\phi_i) \\ -30\phi_i\phi_k(1-\phi_i)(1-\phi_k)\Delta G_{ik} - \Lambda \end{pmatrix} \tag{1}$$

with

$$\Delta G_{sl} = \Delta s_f \Delta T \quad \Delta G_{lp} = p_p - p_l \quad \Delta G_{sp} = 0 \tag{2}$$

and

$$W_{ik} = \frac{15\sqrt{2}\gamma_{ik}}{\delta_{ik}} \quad \epsilon_{ik}^2 = 2W_{ik}\delta_{ik}^2 \quad M_{ik} = \frac{\mu_{ik}}{\delta ik} \tag{3}$$

Here, Δs_f is the volumetric entropy of melting, ΔT is the undercooling, p_p and p_l are the pressures in the pore and liquid, respectively; ϵ_{ik}^2, M_{ik} and W_{ik} are the three PhF parameters associated with the physical parameters γ_{ik}, M_{ik} and W_{ik} describing the interfacial energy, the mobility and the interface thickness, respectively, between phases i and k [13]. The effect of the radius of curvature of the pore is taken into account through the driving force term for the liquid/pore transformation. It is expressed here as the pressure difference between the pore and liquid, which can be justified by solving the steady-state form of Eq. 1 in spherical coordinates, for a single pore surrounded by liquid. By doing so, the Laplace pressure condition is recovered:

$$p_p - p_l = 2\gamma_{lp}H \tag{4}$$

where H is the mean curvature of the pore. For a spherical pore of radius r_p, $H = r_p^{-1}$. Throughout this article, for the sake of simplicity, we will focus on aluminum alloys, where hydrogen is known to be the gas responsible of pore formation [10]. However, the model is more

general and can be applied to any alloy system in which a compressible phase forms.
The equilibrium solubility of hydrogen in the solid, c_s^H, and liquid, c_l^H, can be expressed as a function of p_p using Sievert's law, while assuming thermodynamical equilibrium at the interface:

$$c_s^H = S_s\sqrt{\frac{p_p}{p_0}} \quad c_l^H = S_l\sqrt{\frac{p_p}{p_0}} \tag{5}$$

where S_s and S_l are Sievert's constants (in $mole_H m^{-3}$) for the solid and liquid, respectively, and p_0 is the standard pressure. Taking local volumetric molar concentration of hydrogen and introducing Eq. 5 and the perfect gas law gives:

$$< c^H >= (\phi_s S_s + \phi_l S_l)\sqrt{\frac{p_p}{p_0}} + \phi_p \frac{2p_p}{RT} \tag{6}$$

The factor 2 in the last term is coming from the fact that hydrogen is in the form of H_2 in the pore.
Based on the experimental observations of Lee at al. [3, 4], finite rate hydrogen diffusion in the liquid is assumed to govern the growth kinetics of the pore. This has been taken into account in the model through a hydrogen conservation equation, neglecting any hydrogen concentration gradient in the gas phase and any hydrogen transport in the liquid due to convection:

$$\frac{\partial < c^H >}{\partial t} = \nabla \cdot (\phi_s D_s^H \nabla c_s^H + \phi_l D_l^H \nabla c_l^H) \tag{7}$$

A finite volume method with an explicit time-discretization scheme is used to numerically solve Eqs 1 and 7. At each time-step, the phase distribution is calculated by solving Eq. 1 and then, Eq. 7 is solved to calculate the concentration field. Besides, the calculation of the driving force terms of Eq. 1 requires solving the second order polynomial expression of Eq. 6 to obtain p_p from $< c^H >$. Since homogenization of the hydrogen concentration inside the pore is assumed to be infinitely fast (uniform pressure p_p), c^H is homogenized at each time-step inside the pore. For the diffusion and phase equations, no flux boundary condition is used. Therefore, the number of moles of hydrogen remains constant in the domain. Besides, due to the high computational cost of the 3D PhF calculations, the simulation domain is much smaller than the volume associated with one pore, i.e. n_p^{-1}, where n_p in the pore number density. Calculations were therefore undertaken at relatively large hydrogen contents (or supersaturation) in order to compensate for this effect.
As the model has already been presented in [11] and [12], the reader is referred to these publications for a more detailed description.

3. Analysis of pore curvature

Characterization of the pore-liquid interface in 3D requires computing its curvature in order to better understand the pinching effect. It is recalled that, for any surface patch, two principal radii of curvature, r_{p1} and r_{p2}, can be defined, from which two principal curvatures, $\kappa_1 = r_{p1}^{-1}$ and $\kappa_2 = r_{p2}^{-1}$, can be obtained. The local mean curvature, H, is then given by:

$$H = \frac{\kappa_1 + \kappa_2}{2} \tag{8}$$

For the problem at hand, only the pore-liquid interface is studied, using a similar method as in [12]. The iso-surface of the pore-liquid interface corresponding to $\phi_p = \phi_l = 0.5$ (i.e., $\phi_s = 0$) was extracted from the three-dimensional reconstruction of the results. Then, it

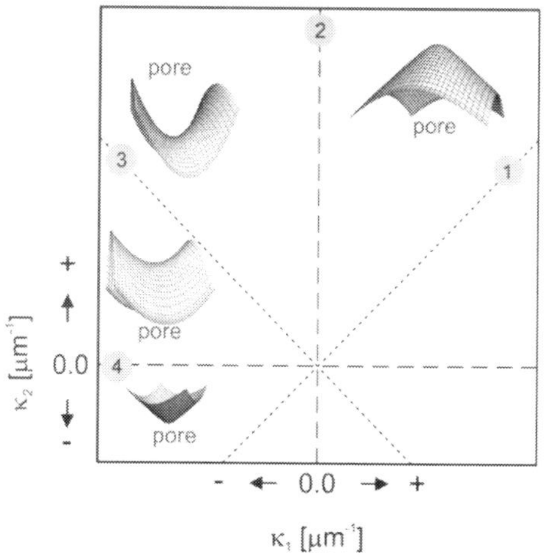

Figure 1. $\kappa_1 - \kappa_2$ space used to draw the Interfacial Shape Distribution (ISD) plot $P(\kappa_1, \kappa_2)$, representing the probability of having a surface element of the pore interface with the curvature (κ_1, κ_2). The curvature is measured with a unit surface normal pointing out of the pore.

was smoothened by triangular elements using the Avizo software (Avizo is a registered trade mark of Mercury Computer Systems, Chelmsford, Massachusetts). For each of the triangles composing this surface, which have a size close to the computational mesh, a quadratic surface was then best fitted considering a certain number (n_n) of triangular neighbors. Since curvature corresponds to first-derivatives of the unit surface normal along principal curvature directions [14], or second-derivatives of the surface positions, using nearest-neighbor triangles alone is not accurate enough. On the other hand, large values of n_n increase the calculation time and also result in a loss of accuracy by excessive smoothing of the surface. The best tradeoff between accuracy and performance was achieved with $n_n = 18$, while analyzing the curvature of a reference sphere.

For such 3D situations, a useful tool for a detailed analysis of the pore morphology based on both κ_1 and κ_2 is the Interfacial Shape Distribution (ISD) plot, first introduced by the group of Voorhees and coworkers for the analysis of coarsening of dendritic structures [15]. The ISD is a two-dimensional probability contour plot, $P(\kappa_1, \kappa_2)$, which measures the probability of having a surface patch, i.e., a triangular element in this case, with its principal curvatures falling in the range $[\kappa_1 - \Delta\kappa/2, \kappa_1 + \Delta\kappa/2]$ and $[\kappa_2 - \Delta\kappa/2, \kappa_2 + \Delta\kappa/2]$. Note that the areas of the triangular elements at which the curvature are estimated are taken into account in the statistics. The distribution $P(\kappa_1, \kappa_2)$ is then plotted in a (κ_1, κ_2) diagram in the form of isovalues. By definition, the curvatures are selected such that $\kappa_2 \geq \kappa_1$, therefore the distribution can have non-zero values only above the line $\kappa_2 = \kappa_1$, for which several regions can be distinguished in figure 1. They are separated by various lines : (1) Points on the line $\kappa_2 = \kappa_1$ correspond to pore-liquid surface elements having locally a spherical topology which can be convex ($H > 0$) or concave ($H < 0$); (2) Points along the line $\kappa_1 = 0$ ($\kappa_2 > 0$) correspond to pore elements with a convex cylindrical topology; (3) Along the line $\kappa_2 = -\kappa_1$, the pore surface has a saddle shape with a zero mean curvature; (4) Along the line $\kappa_2 = 0$ ($\kappa_1 < 0$), the pore surface element has also a cylindrical topology, but concave. In between these lines, a few morphologies have been illustrated.

4. Results and discussion

After being validated, in a similar manner as in [12], the model was used to study the morphology of a pore, pinched by a simplified solid network. Conceptually, the present porosity model can be applied to any complex geometrical situation such as a Representative Volume Element (RVE) obtained from X-ray micro-tomography [10]. However, because of the heavy computation time inherent to the PhF method and also because it is interesting to compare the present results

with analytical models of pinching [9], a simple situation is first considered here.
A simplified solid morphology, similar to the 3D model of Couturier et al. [9], was selected: it is composed of spherical solid grains arranged in the corners of a cubic network (which is referred to as *cubic RVE* throughout this article). In order to study the growth of a pore extending to distances larger than the dendritic arms spacing, a larger domain composed of two cubic RVEs was used (figure 2). An edge size of $\lambda_2 = 10\,\mu m$ was chosen, which is in fact the distance between two adjacent spherical grains. During the calculations, the radius of the solid grains is kept constant ($r_s = 5\,\mu m$), i.e. the evolution of the solid is neglected. Other parameters of the model are listed in Table 1.

Table 1. PhF model parameters

Parameter	Unit	Value
Mesh size (ΔX)	m	5.0×10^{-8}
δ_{ik}	m	1.2×10^{-8}
p_l	Pa	10^5
T	K	1000.0
M_{ls}	$m^3 J^{-1} s^{-1}$	1.0×10^{-6}
D_l	$m^2 s^{-1}$	1.0×10^{-6}
S_l	$mol_H m^{-3}$	0.7
S_s	$mol_H m^{-3}$	0.04
γ_{pl}	Jm^{-3}	0.9
γ_{ps}	Jm^{-3}	1.05
γ_{ls}	Jm^{-3}	0.15

A small spherical pore was nucleated at the center of the left octahedral space. Since the liquid is supersaturated in hydrogen, the pore grows. For a given initial hydrogen concentration c_0^H, the pore grows until equilibrium is reached, i.e., until the hydrogen content in the liquid is uniform and corresponds to the solubility limit of the final pressure p_p within the pore (Eq. 5). Initially, c_0^H was set to $7\,mol_H m^{-3}$ for which the final shape is spherical (figure 2.a). In order to obtain pores of different sizes, the hydrogen concentration was increased step-by-step, letting each time the pore to stabilize based on the new concentration. Morphologies of the pore at three instants corresponding to $c_0^H = 7, 24.3$ and $77.7\,mol_H m^{-3}$ are presented in figure 2. The different color patches in this figure indicate the local mean curvature. Areas with negative mean curvature (blue patches) are in contact with the solid and thus are characterized by a mean curvature $H_{ps} = -r_s^{-1} = -0.2\mu m^{-1}$. The main area of interest is the pore-liquid interface where the mean curvature is positive (red patches). This positive curvature is linked to the pressure difference between the pore and the liquid through the Laplace-Young equation (Eq. 4). Since p_p and p_l are homogenous, the mean curvature of the pore-liquid interface is also expected to be constant, which is the case here as shown by the uniform red color. The measured mean curvature for the pore-liquid interface shown in figure 2.a, b and c are $H = 0.505$, 0.276 and 0.282 μm^{-1}. These values are in good agreement with that obtained from the pressure difference $(p_p - p_l)$ extracted from the PhF calculations and Eq. 4 which give $H = 0.511$, 0.279 and 0.284 μm^{-1}, respectively. Each time that the pore was stabilized its radius of curvature at the pore-liquid interface, H^{-1}, was calculated using $(p_p - p_l)$ values extracted from the PhF calculations for various c_0^H values (figure 3). Considering physically the growth of a pore for a fixed value of r_s and λ_2 (and thus g_s), the maximum possible radius of a spherical pore is given by the model of Couturier, i.e., $r_p^{max} = \frac{\lambda_2\sqrt{3}}{2} - r_s$ (the conversion of r_s into g_s is straightforward up to $r_s = \lambda_2/\sqrt{2}$ and requires a few geometrical calculations above this limit, when the solid grains impinge on

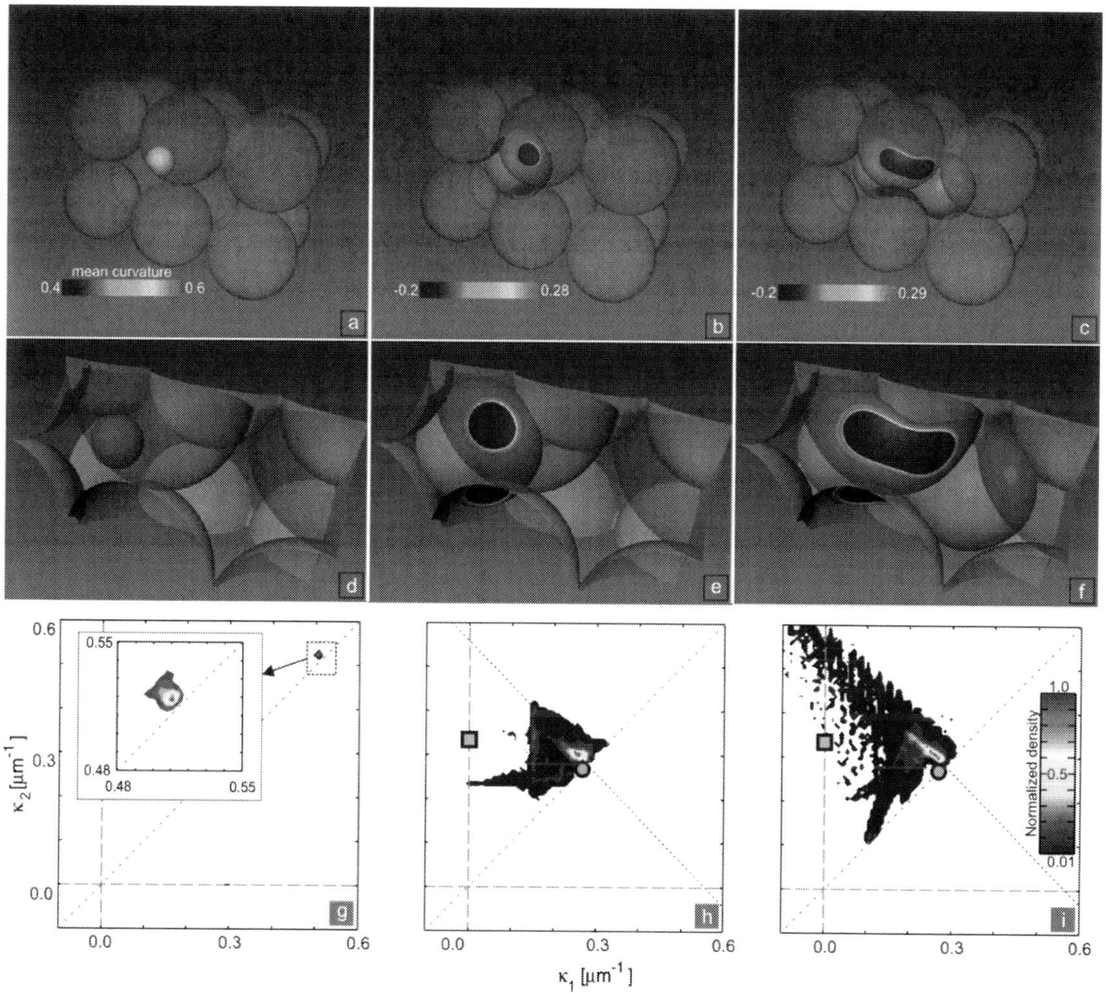

Figure 2. The morphology of pores, grown at $g_s = 0.53$ and for $c_0^H = 7$, 24.3 and $77.7 mol_H m^{-3}$ are presented in (a), (b) and (c), respectively, and with a higher magnification in (d), (e) and (f). The color of the pore shows the local mean curvature, calculated on the surface patches. The solid grains in contact with the pore are indicated by transparent spheres (cadet blue). The ISD plots corresponding to each pore are presented in (g), (h) and (i), taking into account the pore-liquid interface only. The green square and circle indicate the curvature given by the 2D and 3D model of Couturier [9], respectively.

each other [9, 12]). On the other hand, for a given value of r_s and λ_2, a minimum radius of curvature, r_p^{min} can also be defined: it corresponds to the radius of the curvature of a pore passing through the solid spheres in any face of the cubic RVE which is equal to the radius of a sphere that can be inscribed in between these solid spheres. It is thus given by $r_p^{min} = \frac{\lambda_2}{\sqrt{2}} - r_s$ up to $g_s = 0.965$. These maximum and minimum values are also presented in figure 3 with solid and dashed horizontal lines, respectively. The initial spherical pore has a radius of curvature even smaller than r_p^{min} (point **a** in figure 3). With an increase of c_0^H, it grows and comes into contact with the solid, at this point its radius is equal to r_p^{max}. Afterwards, with a further increase of c_0^H, the pore is pushed into the liquid channels with a decreasing H^{-1} (point **b** in figure 3). Continuing to increase c_0^H, the pore is pushed more into the liquid channels until it can pass through one of the 6 openings on the faces of the cubic RVE. Here, the channel in the middle of the domain has been made 1% wider than the other ones, therefore the pore can pass through it before reaching the other faces of the RVE. While passing this channel, H^{-1} falls to its minimum given by the r_p^{min} line. After this, H^{-1} starts increasing again until it reaches the following layer of solid grains ahead of it (point **c** in figure 3). As already observed in a previous contribution for 2D cases [11], H^{-1} is also bound in 3D between two limits that are dictated by the solid morphology, namely the liquid channel widths. Considering a distribution of liquid channel widths in a casting, it sounds reasonable to define a mean value $r_p^{mean} = (r_p^{min} + r_p^{max})/2$ for the pinching effect that could be used in macroscopic porosity models.

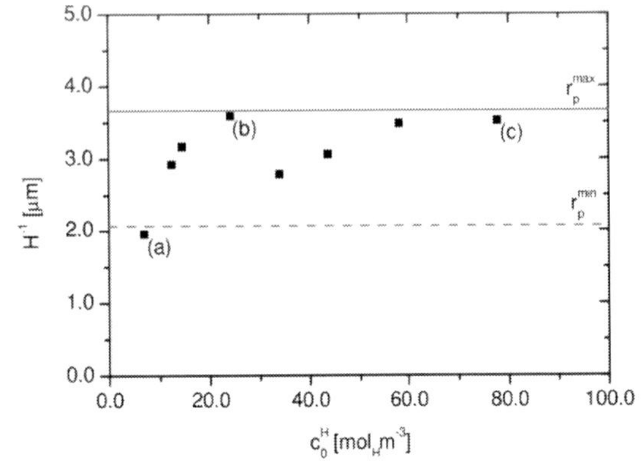

Figure 3. Radius of the curvature of the pore given by, H^{-1} for different c_0^H values calculated with the PhF model. The maximum and minimum possible values, r_p^{max} and r_p^{min}, are shown with the horizontal solid and dashed lines respectively. Points labeled (**a**), (**b**) and (**c**) correspond to 2

ISD plots corresponding to the pores presented in figure 2.a , b and c are plotted in the same figure in (g), (h) and (i), taking into account the pore interface in contact with the liquid only. It should be noted that these ISD plots also contain a noise contribution as the curvature is a second derivative of the calculated interfaces.

Figure 2.g exhibits a distribution close to the $\kappa_1 = \kappa_2$ line, which is representative of the pore spherical shape. Figure 2.h and i feature a set of points aligned along a diagonal line ($\kappa_1 + \kappa_2 = cst$), in the area where the mean curvature of the pore is positive. Extracting the pressure difference ($p_p - p_l$) from the PhF results to calculate the mean curvature of the interface gives the black dotted line shown in figure 2.h and i. As can be seen, this line matches very well the measured principal curvature values. The fact that all the surface elements of the pore-liquid interface are along this line ($\kappa_1 + \kappa_2$) $= 2H$ ensures that the pressure within the pore is indeed uniform. The maximum density of these points in figure 2.h and i furthermore corresponds to the nearly spherical *tip* of the pore located in between 4 grains, close to the faces of the cubic RVE, for which $\kappa_1 = \kappa_2$. Besides, the points in the ISD plot also extend toward the line $\kappa_1 = 0$, i.e., $\kappa_2 = 2H$, which corresponds to the cylindrical-type shape of the pore located in between 2 grains (see figure 2.b and c). The extension towards this line is more pronounced for pore **c** which is spread over two octahedral sites. We can assume that a pore extended to various liquid channels, will exhibit a distribution further closer to $\kappa_1 = 0$ line, which is in fact in agreement with X-ray microtomography observations of Felberbaum et al [10], for Al-Cu specimens having a dendritic structure.

The predictions of the 2D and 3D model of Couturier [9] are also presented in figure 2.h and i. with a green square and circle, respectively. It can be observed that the 2D model of [9] does not provide a reasonable prediction of the pore curvature. However, their 3D model gives much better estimates. In this figure, the difference between their 3D model and the PhF results is very small compared to the situations reported in [12], because the corresponding pores have

been selected close to the maximum radius r_p^{max}, as already shown in figure 3.
Finally, in any case, the spherical tip of pores growing at high solid fraction will try to follow a path of maximum liquid space, i.e., minimal pore-liquid curvature. Even in a dendritic specimen, such spaces are found near the tips of the dendrites, and not near their roots. For example, in a specimen exhibiting a columnar structure, secondary dendrite arms tips are in a configuration close to that shown in figure 2. Therefore, the pinching model with $r_p^{mean} = (r_p^{max} + r_p^{min})/2$, might actually describe better the pinching effect in a dendritic specimen.

5. Conclusion

A 3D PhF model, which is capable of describing the complex morphology of pores growing in between a pre-existing solid phase, has been developed. The model accounts for segregation and diffusion of hydrogen, capillary forces at interfaces and the Laplace-Young's condition. The results show that the presence of the solid affects the pore morphology, which then influences its internal pressure, and thus its total volume.
The effect of globular grains on the pore morphology was studied and the PhF results were compared with the analytical model of Couturier et al. [9]. Based on the comparisons, and considering the fact that there is also a distribution of grain sizes in a real casting, we propose to use a radius r_p^{mean} for predicting the curvature overpressure in macroscopic porosity calculations. In the present contribution, hydrogen diffusion in the solid is neglected due to its low solubility, but at high solid fraction, this cannot be neglected anymore [6]. On the other hand, although X-ray tomography characterization of as-solidified alloys has shed light on the curvature and complex morphology of pores, in-situ measurements during solidification should be made to further validate the present PhF results. Thanks to the advances in ultra-fast tomography techniques and the development of furnaces that can fit X-ray beam lines [16], this should be possible in the near future.

Acknowledgments

The authors would like to thank the Swiss Competence Centre for Materials Science and Technology (CCMX) and partner companies, Asulab, Constellium, Kugler Bimetal, Novelis, Rolex and Varinor, within the thematic area "Multi-scale, multi-phenomena modeling of metallic systems" for funding this research.

References

[1] Lee P D, Chirazi A and See D 2001 *Journal of Light Metals* **1** 15–30
[2] Niyama E, Uchida T, Morikawa M and Saito S 1982 *Int. Cast Met. J.* **7** 52–63
[3] Lee P D and Hunt J D 1997 *Acta Materialia* **45** 4155–4169
[4] Atwood R, Sridhar S, Zhang W and Lee P 2000 *Acta mater.* **48** 405 – 417
[5] Pequet C, Rappaz M and Gremaud M 2002 *Met. Trans.* **33 A** 2095–2106
[6] Carlson K D, Lin Z and Beckermann C 2007 *Metallurgical and Materials Transactions* **38 A** 541 – 555
[7] Poirier D, Yeum K and Maples A 1987 *Met. Trans.* **18 A** 1979–1987
[8] Kubo K and Pehlke R 1985 *Metallurgical and Materials Transactions B* **16** 359–366
[9] Couturier G and Rappaz M 2006 *Simulation of Aluminum Shape Casting Processing: From Alloy Design to Mechanical Properties* ed Wang Q, Krane M and Lee P (TMS Publ.) p 143
[10] Felberbaum M and Rappaz M 2011 *Acta Materialia* **59** 6849–6860
[11] Meidani H and Jacot A 2011 *Acta Materialia* **59** 3032–3040
[12] Meidani H, Desbiolles J L, Jacot A and Rappaz M 2012 *Acta Materialia* **(in press)**
[13] Boettinger W J, Warren J A, Beckermann C and Karma A 2003 *Annual Review of Materials Research* **32** 163–194
[14] Pressley A 2001 *Elementary Differential Geometry* (Springer-Verlag)
[15] Mendoza R, Alkemper J and Voorhees P 2003 *Metallurgical and Materials Transactions A* **34** 481–489
[16] Fife J, Rappaz M, Pistone M, Celcer T, Mikuljana C and Stampanoni M 2011 *J. Synchlotron Radiation* **(submitted)**

IOP Conf. Series: Materials Science and Engineering **33** (2012) 012105 doi:10.1088/1757-899X/33/1/012105

Numerical solution of the phase-field equation with minimized discretization error

Janin Eiken

Access e. V., Intzestraße 5, 52072 Aachen, Germany

E-mail: j.eiken@access.rwth-aachen.de

Abstract. Phase-field models are gaining importance as application-oriented tools for alloy and process optimization. In this context, numerical performance plays an essential role, but is often difficult to reconcile with accuracy. The numerical solution of the phase-field equation based on finite differences inherently implies a discretization error which may strongly bias the simulation results. A direct correlation between the accuracy of the finite-difference solution and the number of numerical cells resolving the diffuse interface profile has been observed. This presents a problem, because high numbers of interface cells are unfavorable for computational performance. To overcome the problem, an improved numerical solution of the phase-field equation is proposed, which allows obtaining highly accurate results with three or four interface cells only. The solution takes advantage of the a priori knowledge of the phase-field profile, determined by the choice of the potential function in the free energy functional. This knowledge enables an almost exact quantification and compensation of the error evoked by the discrete nature of the grid spacing and interface width. Benchmark simulations demonstrate the significant gain in both accuracy and numerical efficiency

1. Introduction

Numerical microstructure prediction has become an indispensable tool in the process of alloy and process optimization. The phase-field method is recognized as the method of choice for space-resolved microstructure simulation. It replaces the classical sharp interface description by a smooth change of a continuously defined phase-field variable over a thin transition layer. The interface motion is implicitly described by the change of the phase-field according to a differential equation which is derived from a free energy functional based on a relaxation approach:

$$\dot{\phi} = -M\frac{\delta F(\phi)}{\delta\phi} = -M\left[\frac{\partial h(\phi)}{\partial\phi}\Delta f_{\alpha\beta} + W\frac{\partial g(\phi)}{\partial\phi} - \varepsilon^2\nabla^2\phi\right] \tag{1}$$

$$with\ \ F(\phi) = \int_V \left[(1-h(\phi))\,f_\alpha + h(\phi)\,f_\beta + W\,g(\phi) + \frac{\varepsilon^2}{2}|\nabla\phi|^2\right]dV \tag{2}$$

where f_α and f_β denote the free energy densities of two phases α and β, and $\Delta f_{\alpha\beta}$ the thermodynamic driving force for transition. $h(\phi)$ is an interpolation function and $g(\phi)$ a potential function with two minima. W and ε are scaling factors for the potential and the gradient contribution. M denotes the mobility. For more details see [1]. Numerical solution of the phase-field equation based on finite differences (FD) inherently implies a discretization error which may significantly bias the simulation results [2]. This error could theoretically be minimized by a high numerical resolution of the diffuse interface area, but high numbers of interface cells are unfavorable for computational performance. To reconcile accuracy with good performance an implicit compensation of the discretization error is proposed.

2. Improved numerical solution with minimized discretization error

2.1. Direct matching of the phase-field equation to the sharp interface solution

The phase-field profile in the diffuse interface area is determined by the choice of the potential function in the free energy functional. The here proposed numerical solution is based on a parabolic potential function $g(\phi)=\phi(1-\phi)$ and the following parameters: W = 4ση-1, $\varepsilon=\sqrt{8\eta\sigma\pi^{-1}}$, M = π2μ(8η)-1 and $h'=8\pi^{-1}\sqrt{\phi(1-\phi)}$, where σ is the interfacial energy, η the interfacial width and μ the kinetic coefficient [3,4]. The nominal phase-field profile describes the travelling wave solution for a planar interface:

$$\phi_{nom}(x_n)=\frac{1}{2}-\frac{1}{2}\cos\left(\frac{\pi}{\eta}\left(x_n-x_n^0(t)\right)\right) \qquad \forall\ x_n^0(t)<x_n<x_n^0(t)+\eta \tag{3}$$

where x_n^0 is a reference position for $\phi = 0$ with changes with time due to the interface motion. This profile is defined within the interface region only. In the bulk of phase α, the phase-field variable has the constant value zero and in the bulk of phase β the value one, see figure 1. Using the first and second derivatives of the profile function (3), equation (1) can be rewritten as:

$$\dot{\phi}=-\mu\left[\left|\frac{d\phi_{nom}}{dx_n}\right|\Delta f_{\alpha\beta}-\sigma\left(\nabla^2\phi-\frac{d^2\phi_{nom}}{dx_n^2}\right)\right] \tag{4}$$

Assuming that the actual phase-field profile equals the nominal profile ($\phi(x_n) = \phi_{nom}(x_n)$) and using a geometric interpretation of the normal velocity v_n and the mean interface curvature κ in terms of the phase-field variable [5], this equation directly matches the sharp interface solution:

$$v_n=\mu\left[\Delta f_{\alpha\beta}-\sigma\kappa\right] \quad \text{using } v_n=-\left|\frac{d\phi}{dx_n}\right|^{-1}\dot{\phi} \text{ and } \kappa=\left|\frac{d\phi_{nom}}{dx_n}\right|^{-1}\left(\nabla^2\phi-\frac{d^2\phi_{nom}}{dx_n^2}\right) \tag{5}$$

2.2. Correction for discrete time stepping

Assuming differential time stepping equation (5) can be rewritten as:

$$\phi(t+dt)=\phi(t)-ds_n\frac{d\phi}{dx_n} \qquad \text{with } ds_n=\mu\,dt\left(\Delta f_{\alpha\beta}-\sigma\kappa\right) \tag{6}$$

where ds_n is the differential step by which the interface advances in normal direction. In contrast, numerical solution is based on discrete interface steps Δs_n corresponding to discrete time steps Δt. In this case, equation (13) only represents a first order approximation and continuously deforms the interface profile. This could be overcome by adding higher order terms, but a more efficient solution is to use the knowledge of the nominal profile. The new phase-field value can very precisely be evaluated by extrapolation of the profile function (3) in normal direction from x_n to $x_n - \Delta s$:

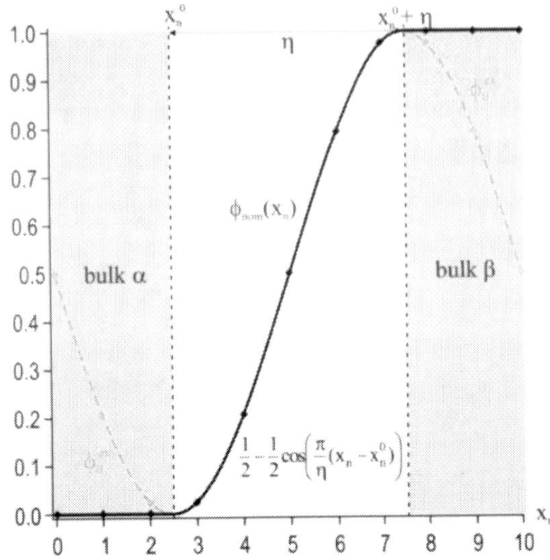

Figure 1. The nominal phase-field profile represents the stationary solution for a planar interface. In the present model, it is defined within the interfacial region only. In the bulk regions the phase-field remains constant 0 or 1.

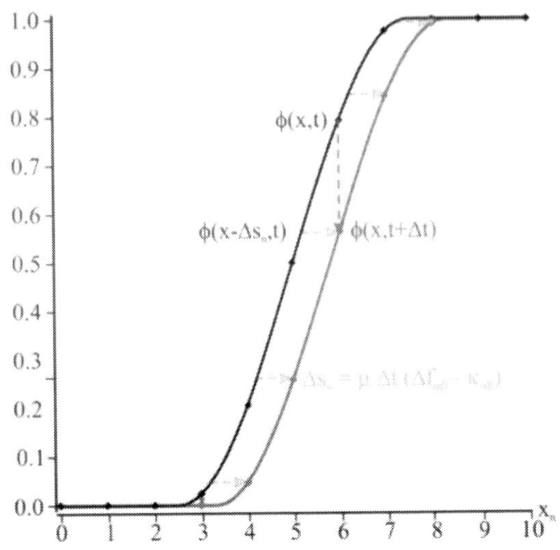

Figure 2. Motion of the diffuse interface by discrete steps. At each grid point, the profile function is solved for x_n and re-evaluated for x_n-Δs, which directly corresponds to the new phase-field value at t + Δt.

$$\phi(t+\Delta t) = \frac{1}{2} - \frac{1}{2}\cos\left(\frac{\pi}{\eta}\left(x_n - \Delta s_n - x_n^0\right)\right) = \frac{1}{2} - \frac{1}{2}\cos\left(\arccos(1-2\phi(t)) - \frac{\pi}{\eta}\Delta s_n\right) \tag{7}$$

This is illustrated in figure 2. It allows an accurate description of the interface motion, even for large steps Δs_n with the only restriction that the step has to be smaller than the grid spacing ($\Delta s_n < \Delta x$). It should be noted that, although equation (7) is bound to the here chosen profile function (3), similar discrete kinetic formulations can simply be derived for any arbitrary profile.

2.3. *Correction of differential quotients*

The matching to the sharp interface solution is only accurate under the assumption that the actual phase-field profile equals the nominal profile ($\phi(x_n) = \phi_{nom}(x_n)$). During a simulation, the actual profile may temporarily deviate from the nominal profile, e.g. due to heterogeneous driving forces. Separation of the terms in equation 3 reveals the implicit mechanism of profile readjustment:

$$\dot{\phi} = -\mu\left[\left|\frac{d\phi_{nom}}{dx_n}\right|\Delta f_{\alpha\beta} - \sigma\left(\nabla^2\phi - \frac{d^2\phi}{dx_n^2}\right) + \sigma\left(\frac{d^2\phi_{nom}}{dx_n^2} - \frac{d^2\phi}{dx_n^2}\right)\right] \tag{8}$$

Whenever the phase-field profile becomes deformed, the difference between the actual and the nominal second derivatives in interface normal direction should immediately drive the profile back to the undistorted nominal profile. However, numerical discretization based on the FD method results in an inconsistency: Due to the discretization error, first and second order derivatives of the actual profile, when being approximated by differential quotients ($D_{FD}^{I/II}$), still differ from the analytically exact derivatives ($D_A^{I/II}$) even if both profiles are equal ($\phi(x_n) = \phi_{nom}(x_n)$):

$$D_A^I = \frac{d\phi}{dx_n} = \frac{\pi}{\eta}\sqrt{\phi(x_n)(1-\phi(x_n))} \neq D_{FD}^I = \frac{\phi(x_n+\Delta x_n)-\phi(x_n-\Delta x_n)}{2\Delta x_n} \tag{9}$$

$$D_A^{II} = \frac{d^2\phi}{dx_n^2} = \frac{\pi^2}{2\eta^2}(1-2\phi(x_n)) \neq D_{FD}^{II} = \frac{\phi(x_n+\Delta x_n)+\phi(x_n-\Delta x_n)-2\phi(x_n)}{\Delta x_n^2} \tag{10}$$

This inconsistency results in artificial driving forces which deform the profile and bias the simulation result. Using the knowledge of nominal profile, this error can be quantified and compensated. Under assumption of the nominal profile ($\phi = \phi_{nom}$), correction factors $FDC^{I/II}$ can be evaluated which precisely relate the discrete differential quotients $D_{FD}{}^{I/II}$ to the analytical derivatives $D_A{}^{I/II}$:

$$D_A^I = D_{FD}^I \cdot FDC^I(\Delta x, \eta) \qquad \text{with } FDC^I(\Delta x, \eta) = \frac{\pi\Delta x/\eta}{\sin(\pi\Delta x/\eta)} \tag{11}$$

$$D_A^{II} = D_{FD}^{II} \cdot FDC^{II}(\Delta x, \eta) \qquad \text{with } FDC^{II}(\Delta x, \eta) = \frac{(\pi\Delta x/\eta)^2}{2(1-\cos(\pi\Delta x/\eta))} \tag{12}$$

In principal, such a correction could be done for arbitrary profile function. The here chosen nominal profile (3) has the advantage that the corresponding correction factors $FDC^{I/II}$ are direct functions of the numerical grid spacing Δx and the interfacial width η, but not of the phase-field variable itself. The definitions in equations (9) and (10) still presume discretization in interface normal direction ($\Delta x = \Delta x_n$). Generalized multidimensional correction factors $FDC^{I/II}$ are derived from the individual contributions $FDC_{iDim}{}^{I/II}$, evaluated for $\Delta x_{iDim} = n_{iDim}\,\Delta x$, where n_{iDim} are the components of the local interface normal vector:

$$FDC_{iDim}^I = \frac{\pi\Delta x_{iDim}/\eta}{\sin(\pi\Delta x_{iDim}/\eta)} \qquad \text{and} \qquad FDC_{iDim}^{II} = \frac{(\pi\Delta x_{iDim}/\eta)^2}{2(1-\cos(\pi x_{iDim}\Delta x/\eta))} \tag{13}$$

$$FDC^I = \sqrt{\sum_{iDim}(FDC_{iDim}^I)^2} = \frac{\pi\Delta x/\eta}{\sqrt{\sum_{iDim}\sin^2(\pi n_{iDim}\Delta x/\eta)}} \tag{14}$$

$$FDC^{II} = \sum_{iDim} FDC_{iDim}^{II} = \frac{(\pi\Delta x/\eta)^2}{2\left(\sum_{iDim}(1-\cos(\pi n_{iDim}\Delta x/\eta)\right)} \tag{15}$$

These factors can easily be evaluated and applied, independently of the other proposed corrections, everywhere in the code where differential quotients of the phase-field are evaluated based on finite differences. It should be noted that the evaluation of the interface normal vector from FD gradients ($n_{iDim} = D_{FD}{}^I{}_{iDim}/\,|D_{FD}{}^I|$) itself implies a discretization error which implicitly would have to be taken into account for an exact solution. However, this error is relative small, since the corrections for nominator and denominator ($FDC_{iDim}{}^I / FDC^I$) compensate each other in large part.

2.4. *Correction of the transition between interface and bulk region*

Compared with the hyperbolic tangent profile resulting from the classic double-well potential, the here used profile function (3) has the major advantage that it changes from zero to one over a finite interface thickness η. Therefore it has to be evaluated over a restricted interface area only, which is beneficial for numerical performance. However, special care has to be taken to treat the transition between interface and bulk regions in a numerical correct way. All interface cells with $\phi < \phi_{nom}(\Delta x)$ and $\phi > \phi_{nom}(\eta - \Delta x)$ have nearest neighbors in the bulk regions where the phase-field has been set to zero or one, respectively. This leads to an additional error in the evaluation of the differential quotients (9, 10). To overcome the problem, the phase-field values of these cells are extrapolated in normal direction into the bulk, i.e from x_n to $x_n \pm \Delta x$:

$$\phi_B^{ex(\pm)} = \phi_{nom}(x_n \pm \Delta x) = \frac{1}{2} - \frac{1}{2}\cos\left(\arccos(1 - 2\phi(t) \pm \frac{\pi}{\eta}\Delta x_n\right) \quad \forall \phi > (<) \frac{1}{2} \pm \frac{1}{2}\cos(\frac{\pi}{\eta}\Delta x) \tag{16}$$

The difference between the constant bulk value (one or zero) and the extrapolated bulk value (see figure 1) is used to evaluate the additional bulk correction terms $\Delta D_B^{I/II}$ which accurately compensate the missing contributions in the differential quotients due to nearest neighbors in the bulk regions. E.g. the correction term ΔD_B^{II} is given by:

$$\Delta D_B^{II} = \frac{\phi_B^{ex(-)}}{\Delta x^2} \quad \forall \phi < \frac{1}{2} - \frac{1}{2}\cos(\frac{\pi}{\eta}\Delta x), \qquad \Delta D_B^{II} = \frac{\phi_B^{ex(+)} - 1}{\Delta x^2} \quad \forall \phi > \frac{1}{2} + \frac{1}{2}\cos(\frac{\pi}{\eta}\Delta x) \tag{17}$$

In all other cells, the bulk correction term ΔD_B^{II} is set to zero. In a similar way, the correction term ΔD_B^{I} can be evaluated for the first derivative. An additional correction term $\Delta\phi_{IB}$ has to be added to the kinetic phase-field equation (13) to correct the extrapolated phase-field values for all interface cells leaving the interface region. In opposite direction a correction term $\Delta\phi_{BI}$ for bulk cells entering the interface region.

2.5. *The new formulation combining all corrections*

Combination of all proposed corrections allows to completely compensate the discretization error in 1D simulations with constant driving forces, as well as to significantly minimize the error in all other simulations. The new phase-field equation is based on discrete time steps and has only to be solved in interface cells ($0 > \phi > 1$) and in bulk cells with at least one neighboring interface cell:

$$\phi(t + \Delta t) = \frac{1}{2} - \frac{1}{2}\cos\left(\arccos\left(1 - 2(\phi(t) + \Delta\phi_{BI})\right) - \frac{\pi}{\eta}\Delta s_n\right) + \Delta\phi_{IB} \quad \text{with } \Delta s_n = \mu\,\Delta t\left(\Delta f_{\alpha\beta} - \sigma k_{\alpha\beta}\right) \tag{18}$$

The correction terms $\Delta\phi_{IB}$ and $\Delta\phi_{BI}$ have only to be considered for cells leaving or entering the diffuse interface area. The curvature term $\kappa_{\alpha\beta}$, which also implies profile readjustment, is given by:

$$\kappa_{\alpha\beta} = \frac{(\nabla^2\phi)_{FD} - D_A^{II}/FDC^{II} + \Delta D_B^{II}}{D_A^{I}/FDC^{I} + \Delta D_B^{I}} \quad \text{with } D_A^{I} = \frac{\pi}{\eta}\sqrt{\phi(1-\phi)} \text{ and } D_A^{II} = \frac{\pi^2}{2\eta^2}(1 - 2\phi) \tag{19}$$

The Laplace operator $\nabla^2\phi$ is solved using the standard FD method. The analytical derivatives $D_A^{I/II}$ have been brought into a consistent formulation using the corrections described in equations (9-17).
The new discrete phase-field equation (12) does not deform the interface profile provided that Δs_n is homogeneous across the interface. Inevitable profile deformations caused by inhomogeneous driving forces or numerical oscillations are precisely corrected by the penalty force inherent in the curvature term $\kappa_{\alpha\beta}$. This mechanism is also used to initialize the profile at the start of simulation or after a nu-

cleation event. Both, the improved profile conservation and the precise readjustment are essential prerequisites for accurate phase-field simulations.

3. Benchmark simulations

The new discrete phase-field formulation (18, 19) has been implemented in the MICRESS® software [6]. A series of benchmarks with well-defined analytical solutions allowed validation of a) the new corrected formulation and b) the standard formulation in comparison. Simulations were run under variation of the diffuse interface width η and the numerical grid spacing Δx. In all simulations, the time step has been chosen sufficiently small to ensure numerical stability.

3.1. 1D simulation of a planar interface moving with constant driving force

The first example describes the one-dimensional motion of a planar interface under the assumption of a constant kinetic interfacial coefficient ($\mu = 10\ mm^4J^{-1}s^{-1}$) and a constant driving force ($\Delta f_{\alpha\beta} = 0.0001\ Jmm^{-3}$). The analytic sharp-interface solution is given by $v = \mu\ \Delta f_{\alpha\beta} = 1\ \mu m\ s^{-1}$. Figure 3 shows that the accuracy of the standard formulation strongly depends on the numerical parameters. For a typical number of five interface cells and a grid spacing of 1 µm, the velocity is underestimated by almost 23%. Decrease of the spacing by an order of magnitude reduces the error to about 8%. Higher precision is only achieved by an increased number of interface cells ($\eta/\Delta x$). In contrast, the new formulation precisely reproduces the sharp-interface solution independent of the choice of the numerical parameters. This proves that the proposed corrections fully compensate the discretization error in 1D-simulations with constant forces.

3.2. 2D shrinkage in steady state

The second benchmark simulation describes the motion of a curved interface driven by capillarity forces. The velocity was evaluated after a steady state had been established. Figure 4 shows the comparison with the corresponding analytic solution $v_x = \mu\ \sigma\pi/a$. Like in the previous example, the standard formulation strongly underestimates the velocity. Again, very good agreement is found for the new corrected formulation. The deviations exceed the limit of 1% only for extremely small values of $\eta/\Delta x$ (< 3). This shows that the new formulation is also applicable in 2D and essentially minimizes the discretization error in case of curved interfaces and inhomogeneous forces

3.3. 3D simulation of a shrinking of sphere

The last example in figure 5 shows a shrinking sphere. Again, the standard formulation underestimates the kinetics of shrinkage for small numbers of interface cells, while the corrected formulation is in excellent agreement with the analytical sharp interface solution. This benchmark additionally demonstrates the applicability of the new formulation in 3D. Note that the shrinkage is well described even in the last stage, where the fraction solid reaches zero.

4. Conclusions

It has been illustrated how the a priori knowledge of the nominal phase-field profile, determined by the choice of the potential function in the free energy functional, can be used to reduce the discretisation error in numerical phase-field computations based on finite differences. This knowledge enables an almost exact quantification and compensation of the error evoked by the discrete nature of the grid spacing and time stepping. The proposed corrections have been combined in a new numerical formulation of the phase-field equation, which has been implemented in the phase-field software MICRESS®.

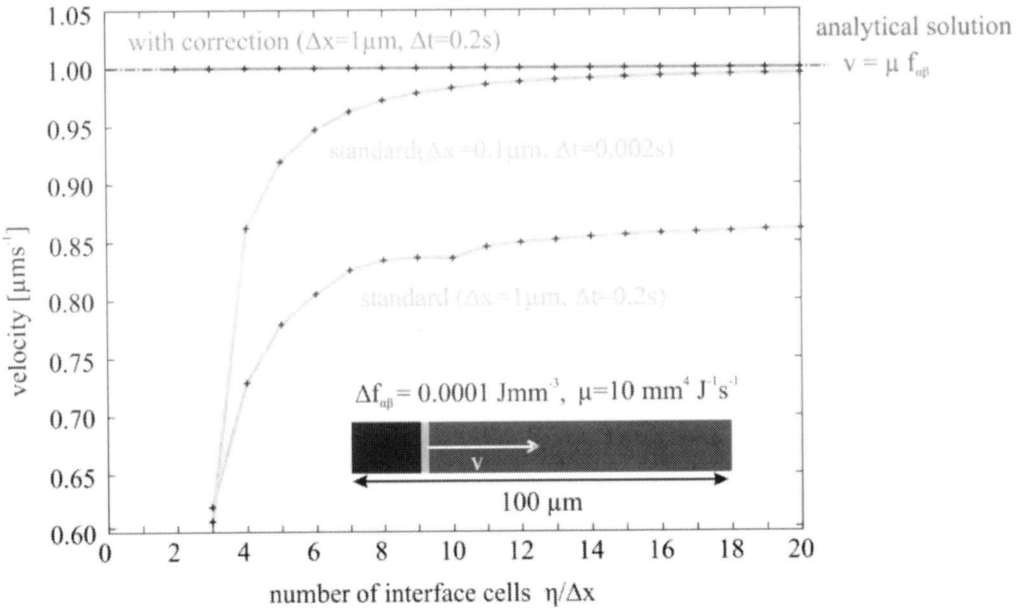

Figure 3. Benchmark simulations of a planar interface moving in one dimension. Constant values have been assigned to both the kinetic coefficient μ and the driving force $\Delta f_{\alpha\beta}$. The sharp-interface solution is given by $v = \mu\ \Delta f_{\alpha\beta}$. The accuracy of the standard formulation strongly depends on the interface width and grid spacing, while the corrected formulation precisely reproduces the sharp-interface solution, regardless of the choice of the numerical parameters.

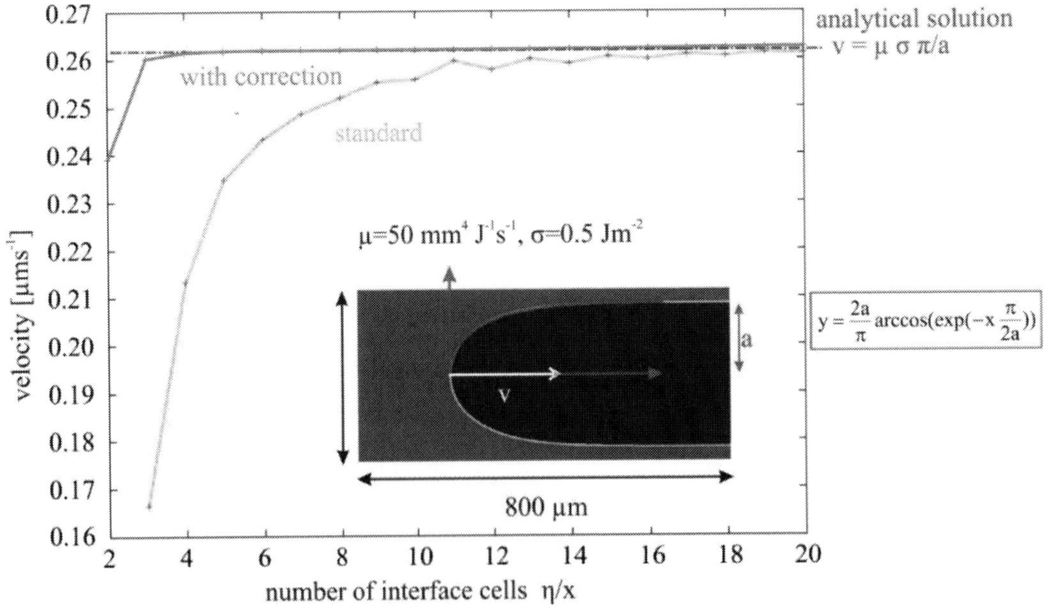

Figure 5. Benchmark simulations of a curvature driven interface motion in two dimensions. The sharp-interface solution for directional steady-state shrinkage is given by vx= μ σπ/a. The accuracy of the standard formulation strongly correlates with the number of interface cells. Using the corrected formulation, a discretization error is observed for an extremely small number of interface cells, only.

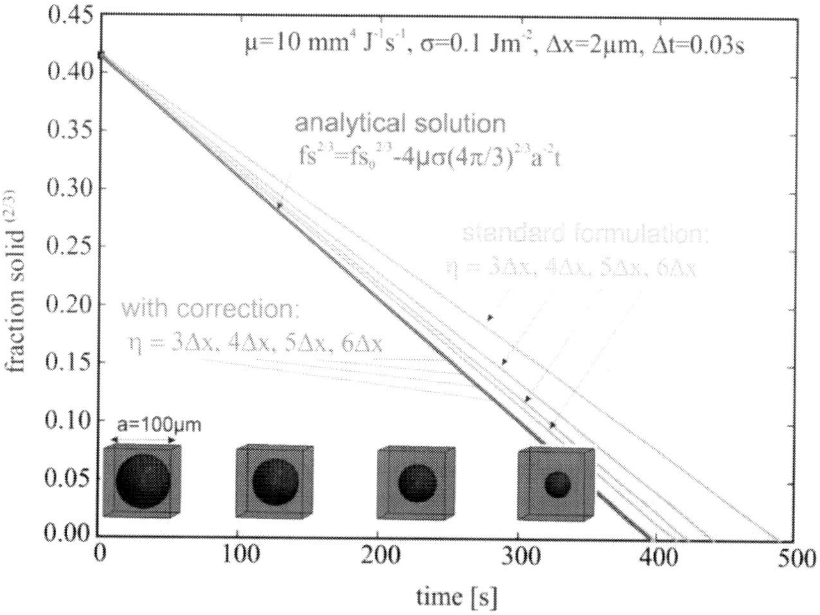

Figure 7. Benchmark simulations of a three-dimensional shrinking sphere. Like in the previous example, the standard FD-PF approach underestimates the kinetics of shrinkage for small numbers of interface cells. The new FPF approach, in contrast, well reproduces the sharp interface solution ($fs^{2/3} = fs_0^{2/3} - 4\mu\sigma\,(4\pi/3)^{2/3}\,a^{-2}t$) for all studied numerical resolutions ranging from three to six numerical cells in the interface.

Benchmark simulations proved the applicability in multi-dimensional space and demonstrated the significant gain in accuracy in comparison with the standard numerical solution. Using the new formulation, excellent agreement with the analytical sharp-interface solutions can be obtained for three or four interface cells only, which strongly improves numerical performance. As part of the current MICRESS® release, the new formulation already proved to be beneficial for various applications to multicomponent and multiphase systems with complex thermodynamics and anisotropies, and also to run numerically stable in combination with automatic time stepping.

Acknowledgement
This work has been supported by the German Research Foundation (DFG) in the Priority Programme "DFG-SPP 1296".

References

[1] Boettinger W J, Warren J A, Beckermann C and Karma A 2002 *Annu. Rev. Mater. Sci.* **32** 163
[2] Mecozzi M G, Eiken J, Apel M and Sietsma J 2011 *Comp. Mat. Sci.* **50** (6), 1846
[3] Eiken J, Böttger B and Steinbach I 2006 *Phys. Rev. E* **73** (6), 066122
[4] Eiken J 2010 *Int. J. Mater. Res.* 4, 503
[5] Beckermann C, Diepers H J, Steinbach I, Karma A and Tong X 1999 *J Comput Phys* **154** (2), 468-496
[6] MICRESS®, the microstructure evolution simulation software, *www.micress.de*

IOP Conf. Series: Materials Science and Engineering **33** (2012) 012106 doi:10.1088/1757-899X/33/1/012106

Growth of a free dendrite in pure substances under modulated flow conditions

H Neumann-Heyme, K Eckert, A Voigt and S Odenbach

Institute for Fluid Dynamics, Chair of Magnetofluiddynamics, Technische Universität Dresden, D-01062 Dresden, Germany

E-mail: Hieram.Neumann-Heyme@tu-dresden.de, Kerstin.Eckert@tu-dresden.de

Abstract. We study the free dendrite growth of a pure substance under the influence of a parallel flow, modulated in time using two different approaches, by means of two-dimensional phase-field simulations. In a previous work amplified side-branching was shown to occur for velocity pulses of appropriate duration (H. Neumann-Heyme et al., IOP Conference Series: Materials Science and Engineering 2012 [1]) which is extended for the case of modulated temperature. To efficiently detect the resonant pulse duration corresponding to the naturally preferred sidebranch frequency we make use of a technique in which the pulse duration is ramped during the growth of the dendrite. Geometrical features of the dendrites grown under the different modulation conditions are discussed.

1. Introduction

The application of temporally modulated rotating magnetic fields (RMFs) to tailor the flow in solidification processes has led to a grain refinement in combination with a reduction of the macrosegregations [2, 3]. Further improvements also require a better understanding of the response of the dendritic microstructure toward the flow oscillations associated with the RMF modulation. For the purely diffusive growth of free dendrites, and even to some extent for the presence of flow the operating state of the dendrite tip under *steady-state* conditions is well described by theoretical models [4, 5, 6, 7] (and references therein). However, despite the practical relevance, only a small number of works analyze dendrite growth under transient conditions [8, 9, 10, 11]. It has been demonstrated that by means of a temporally periodic modulation of physical quantities a highly regular side-arm branching of a dendrite can be achieved in resonance situations. This has been shown for a series of laser pulses [12], for an oscillating flow [13] or for both pressure and electric current modulation [14]. However, with exception of [13, 1] the impact of flow variations has not yet been addressed systematically. Therefore, the focus of this communication is on analyzing the response of a free dendrite to a modulated convection. To reduce complexity, we restrict ourselves to a pure substance, i.e. we consider a thermal dendrite. As an extension of our previous work [1] where a sequence of velocity impulses was used, we additionally include a parallel flow whose temperature is modulated in time. We show that for both situations a regime of resonant forcing exists in which a very regular sidebranch structure may be achieved. By means of a modulation scheme using a linear variation of pulse durations, an efficient strategy is furthermore proposed to identify the resonant frequencies. The geometrical growth features are discussed for different pulse durations.

Finally we note that in contrast to a global forcing as described in [14], convection-driven forcing acts locally at the flow-facing dendrite tip. Therefore the influence of disturbances in the tip state can be studied isolated from varying conditions at the sidebranches. This may promise a closer analogy to established theories, which assume local disturbances to constitute the major mechanism in sidebranch generation [4].

2. Phase-field model with modulated convection

To model the solidification of a pure substance we apply the phase-field model by Karma and Rappel [15], originally developed for the purely diffusive case. The evolution of the phase-field variable Ψ ranging from -1 in the liquid to 1 in the solid phase is described by

$$\begin{aligned}\tau(\boldsymbol{n})\partial_t\Psi &= (1-\Psi^2)\left[\Psi-\lambda u(1-\Psi^2)\right]+\nabla\cdot\left[W(\boldsymbol{n})^2\nabla\Psi\right]\\ &+\partial_x\left[|\nabla\Psi|^2W(\boldsymbol{n})\frac{\partial W(\boldsymbol{n})}{\partial(\partial_x\Psi)}\right]+\partial_y\left[|\nabla\Psi|^2W(\boldsymbol{n})\frac{\partial W(\boldsymbol{n})}{\partial(\partial_y\Psi)}\right]\end{aligned} \tag{1}$$

for a two-dimensional problem. Here $u \equiv (T-T_m)/(L/c_p)$ refers to the dimensionless temperature where T_m, L, c_p are the melting temperature, latent heat and the specific heat capacity, respectively. The anisotropic interface energy and kinetics are functions of the interface normal and are given as $W(\boldsymbol{n}) = W_0 a_s(\boldsymbol{n})$ and $\tau(\boldsymbol{n}) = \tau_0 a_s(\boldsymbol{n})^2$. The anisotropy function $a_s(\boldsymbol{n})$ for a fourfold crystal symmetry reads

$$a_s(\boldsymbol{n}) = 1-3\epsilon_4+4\epsilon_4\frac{(\partial_x\Psi)^4+(\partial_y\Psi)^4}{|\nabla\Psi|^4}, \tag{2}$$

with ϵ_4 representing the anisotropy strength of the surface energy. In the case of vanishing interface kinetics the phase-field parameters are related to physical quantities in the thin-interface limit [15], namely to the thermal diffusivity, D, and the capillary length, d_0, according to

$$\lambda = a_1W_0/d_0, \qquad \tau_0 = a_1a_2\frac{W_0^3}{d_0D}, \tag{3}$$

where $a_1 = 0.8839$ und $a_2 = 0.6267$ are numerical constants. To incorporate the melt flow, we follow the approach of Beckermann et al. [16] according to whom the conservation equations for mass and momentum, assuming constant material properties for the solid and liquid phase, are given by

$$\nabla\cdot(\Phi\boldsymbol{v}) = 0, \tag{4}$$

$$\partial_t(\Phi\boldsymbol{v})+\Phi\boldsymbol{v}\cdot\nabla\boldsymbol{v} = -\Phi\nabla p/\rho+\nu\nabla^2(\Phi\boldsymbol{v})-2\nu h\Phi(1-\Phi)^2\boldsymbol{v}/W_0^2, \tag{5}$$

where $\boldsymbol{v}$, p, ρ and ν stand for the flow velocity, pressure, density and kinematic viscosity of the melt. In these equations, the re-written phase-field variable $\Phi = (1-\Psi)/2$, which can be interpreted as a liquid fraction, implicitly describes the flow domain. The last term in (5) introduces a dissipative drag force acting over the diffuse interface region, which gives a diffuse representation of the usual boundary forces resulting in "no-slip" walls. For a choice of $h = 2.757$ the corresponding sharp interface solution is matched independently of the diffuse interface width [16]. Finally, for the energy equation the following holds:

$$\partial_t u+\Phi\boldsymbol{v}\cdot\nabla u = D\nabla^2u-\partial_t\Phi. \tag{6}$$

Initial and boundary conditions: A seed of size $10d_0$, with its crystal axes aligned with the coordinate axes, is placed in the middle of the square domain. It is exposed to an undercooled melt which enters the domain of size $L \times L$ ($L = 1200\,W_0$) as a parallel flow with a uniform velocity U. Two different modeling approaches are used, schematically shown in Fig. 1. In the first one, 1/2 domain is simulated considering the symmetry of the problem with respect to the horizontal center line. In the second approach a further reduction to 1/8 domain (Fig. 1) is undertaken. This is motivated by the fact that our primary interest is the sidbranching behaviour of the streamwise growing dendrite arm, in a region where the interaction with lateral arms becomes negligible. In the 1/8 domain the flow is subject to an open boundary condition on the diagonal dashed line in Fig. 1. Besides a reduction of the numerical effort, the flow conditions are closer to a free needle crystal because the artificial flow acceleration due to blocking by the perpendicular main branch is omitted. The simulation results for the 1/2 and 1/8 domains agree perfectly for an unperturbed smooth dendrite. For modulation-induced sidebranch growth, the deviation between the morphologies vanishes in the first half of the dendrite after the tip. In the remaining part the sidebranch length in the 1/8 domain is slightly smaller than in the 1/2 domain since the thermal boundary layer is less compressed in the 1/8 domain for the aforementioned reasons. Furthermore, all qualitative observations are consistent for both cases. At the flow inlet boundary either velocity or temperature are subject to a temporal modulation scheme which is explained in more detail later.

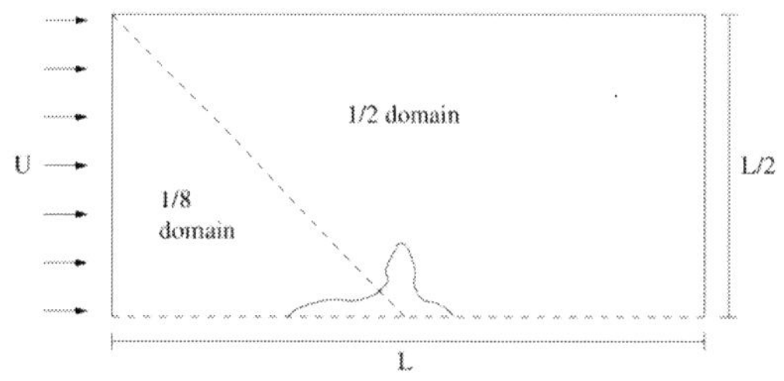

Figure 1. Model setup for the 1/2 and the 1/8 domain approach.

Modeling details: The entire problem was implemented using the FEM toolbox AMDiS, which enables the application of sophisticated parallelization and mesh adaptation techniques [17]. A single triangular mesh was employed for all equations, locally refined by bisection to meet the maximum spatial resolution required by the equations involved. For the phase-field a constant element size of $\Delta x/W_0 = 0.6$ is forced inside an interface region of fixed width. The hierarchical mesh coarsening normal to the interface, together with a maximum limit for the element size, guarantees an adequate resolution for both the flow and advected temperature equations.

Semi-implicit time stepping schemes were used for the phase-field and flow equations. Without the stability limit present in explicit codes it was possible to choose a time step size of $\Delta t = 0.25\tau_0$ in order to reach converged values for the operating state of the fastest-growing tip. Because of the large discrepancy between the different time scales it was found that in accordance with other work [5] a time step greater by a factor of ten than for the phase-field equations was sufficient to obtain accurate results for the flow problem. In addition, the flow problem was treated in a quasi-stationary manner, as in the case of constant external flow the timescale of geometry changes is much larger than the characteristic time of inertial flow features. Even transient flow states have shown not to significantly affect the flow boundary layer around the inflow facing branch, particularly in the most relevant near-tip region. This allows a faster convergence of the flow algorithm, which enables the choice of larger time steps. Also degenerate volume conservation behaviour of the diffuse boundary flow representation could be omitted for situations with rapid changes in flow conditions.

In the following all variables are given in dimensionless form using d_0, d_0^2/D and D/d_0 as scales for length, time and velocity, respectively. The simulations were performed with unified values for $\tau = W_0 = 1$ together with the following physical parameters: $\Delta = 0.55$, $\epsilon = 0.05$, $D = 2$ and a Prandtl number of $Pr = 23.1$. The code was validated in [1], where the agreement of the tip velocities with [5] for different flow directions was shown.

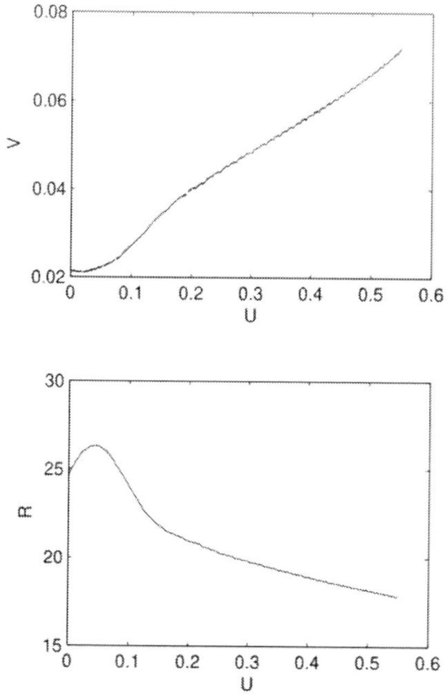

Figure 2. Tip velocity and radius for continuously increasing flow velocity.

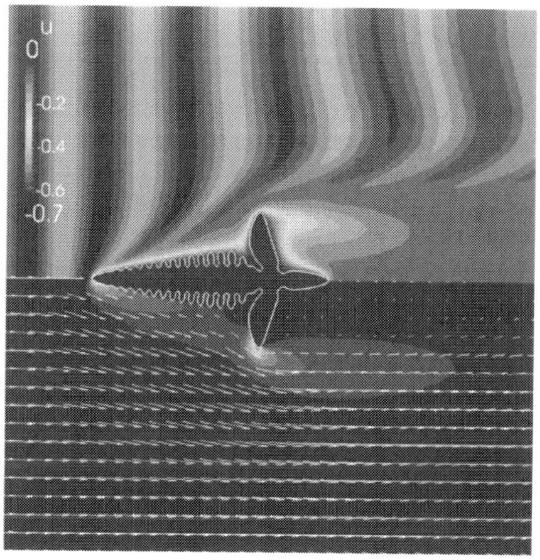

Figure 3. Dendrite growth under modulation of the inflow temperature with resonant pulse duration $t_p = t_{pr}$ at $t = 77400$ in the full computational domain. Upper half: temperature field, lower half: vectors and vorticity of the flow field.

Modulation schemes First, we apply a modulation of the inflow velocity in form of sinusoidal pulses with a fixed amplitude U, a period t_p and a mean velocity $\bar{U}$ for which we consider $\bar{U} = 0.3$ and 0.15. Since this method exerts only a moderate impact on the growth, here the maximum amplitude is applied, leading to variations between 0 and $2\bar{U}$. In the second method the temperature of the melt entering the model domain is modulated while the inflow velocity is kept constant. In the latter method one has to consider the time delay of the temperature pulses traveling towards the tip together with a correction of the pulse duration at the inflow according to the relative tip velocity $\bar{V}$, giving an effective pulse duration of $t_p/(1 - V/\bar{U})$. In case of a temperature modulation at $\bar{U} = 0.3$, an amplitude of 0.2Δ is sufficient to achieve a similar extent of sidebranch amplification compared to velocity modulation. Both methods, however, have some limitations in decreasing flow velocity to arbitrary small values. Whereas temperature modulation requires sufficiently large flow velocities to guarantee that the advection of heat dominates diffusion, for velocity modulation the maximum available amplitude is constrained by the mean flow velocity. In future investigations some of these shortcomings may be circumvented by involving a moving domain approach, allowing the distance between tip and flow inlet to be kept constant.

3. Results

3.1. Tip operating state depending on flow velocity

As a basis for further analysis, knowledge of the tip operating state depending on the mean flow velocity is essential. Therefore we have quasi-steadily increased the flow velocity according to $U = at$, where $a \approx 2\bar{U}V/L$, and examined the behavior of tip velocity and radius. In Fig. 2 we observe that after passing a transient initial state the velocity enters a linear regime, followed by a final small deviation from linearity, which results from boundary effects caused by the dendrite tip approaching the inflow region. Thus we can assume that inflow and tip velocity obey a sufficiently linear relationship throughout the range relevant to our study. Further, we note that in this range the tip radius behaves roughly according to $R \propto V^{-1/2}$. This can be used as an estimate for the general scaling of the geometrical features which are approximately proportional to R [5].

3.2. Determination of resonant forcing by sweeping over pulse duration

Besides showing the potential of temperature modulation to achieve a highly amplified and regular sidebranching, Fig. 3 serves to demonstrate characteristic features of our setup: enhanced side branching under modulated flow occurs exclusively in the flow-facing direction. Morphology evolution in the wake direction is inhibited due to a stretching of the thermal boundary layer.

With the information from Fig. 2 about the admissible velocity amplitude we next expose the nucleus in the center of the domain to a temporarily modulated flow of a constant average velocity, given by $U(t) = U_0 sin[2\pi t_1/t_{p1} log(t_{p1}t)]$. This corresponds to a pulse duration increasing linearly with time from zero to t_{p1} at $t = t_1$. This strategy provides an efficient way to determine the specific pulse duration at which a maximum amplified sidebranch growth occurs.

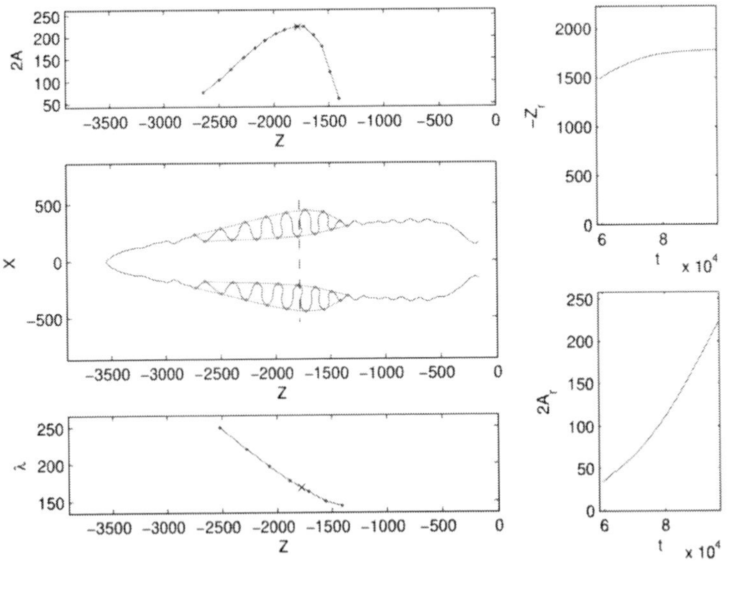

Figure 4. Left-hand images: sweep over pulse duration t_p from zero to 10^4 at $t = 10^5$ for a modulated velocity $\bar{U} = 0.15$. Right images: Position and magnitude of the maximum in the top left figure over time.

In the middle section of Fig. 4 the morphology at the final state of a dendrite grown under such ramped modulated flow conditions can be seen, exhibiting a maximum sidebranch length. For a quantitative evaluation the sidebranch *amplitude* is determined by subtracting the linearly

Table 1. Sidebranch spacings and optimum pulse lengths under different conditions.

	U=0.3 veloc. mod.	U=0.3 temp. mod.	U=0.15 veloc. mod.	U=0.15 temp. mod.
λ_r	154.6	157.8	168.4	208.1
$t_{pr}(\lambda_r)$	3382	3457	4733	5849
$t_{pr}(t_r)$	3672	3689	4998	6429

interpolated upper and lower envelopes of the branching boundary, resulting in the upper diagram. The Z position of the amplitude maximum is then obtained by second-order finite differences. The lower picture shows the *spacing* λ of the sidebranches along the Z coordinate measured in terms of the horizontal distance of the sidebranch minima.

Given the maximum Z position, obtained from the top diagram of Fig. 4, the resulting optimum pulse durations can be estimated in two different ways. In the first approach the evaluation of the spacing λ_r at the resonant Z position together with the mean tip velocity $\bar{V}$ gives the pulse duration at resonance as $t_{pr}(\lambda_r) = \lambda_r/\bar{V}$. The mean tip velocity can be estimated from Fig. 2. The second approach correlates the tip position data to the resonance position to obtain the respective point in time $Z_t(t_r) = Z_r$ and the related pulse duration $t_{pr}(t_r)$.

The results of the two methods are shown in Tab. 1 for temperature and velocity modulation for two different mean flow velocities. For $U = 0.3$ we find a good agreement of the values comparing velocity and temperature modulation results. The values for temperature modulation are, however, slightly higher, which may be attributed to the general damping behaviour inherent to this method. In further analysis a value of $t_{pr} = 3400$ is chosen, and verified to accurately match the resonance criteria given in the following section. This indicates the reliability of the value given by $t_{pr}(\lambda_r)$, whereas $t_{pr}(t_r)$ overestimates the optimum pulse duration by about 7%. In the case of $U = 0.15$ the values for temperature and velocity modulation deviate significantly. A plausible explanation may be that due to the reduced flow velocity, temperature modulation suffers from a substantial diffusion-induced decay of the pulse amplitude, affecting the position of the sidebranching maximum. Hence, velocity modulation appears to be more suitable to give accurate results down to lower flow velocities.

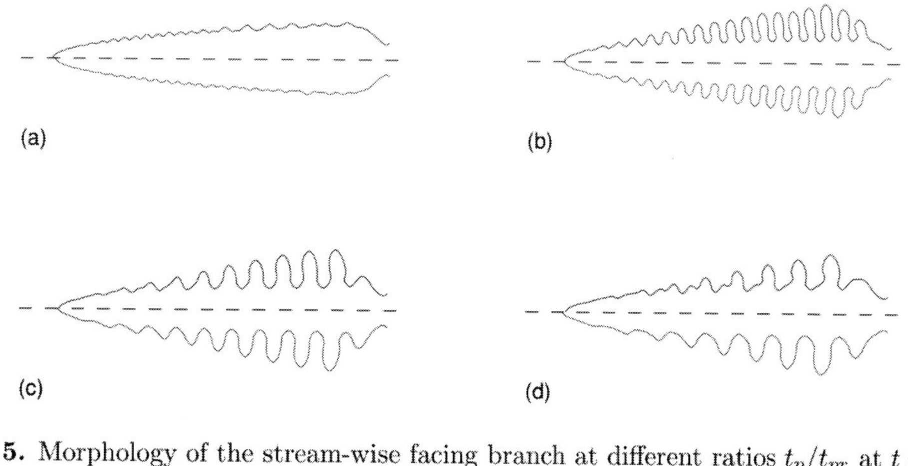

Figure 5. Morphology of the stream-wise facing branch at different ratios t_p/t_{pr} at $t = 70000$, $t_{pr} = 3400$: (a) 2/3, (b) 1, (c) 5/3, (d) 2. Upper half in each picture: velocity modulation, lower half: temperature modulation.

Once the optimum pulse duration t_{pr} has been obtained, the dendrite is exposed to different constant pulse durations $t_p = n\, t_{pr}$. The results are shown in Fig. 5 for $\bar{U} = 0.3$ where the upper half of each picture represent results from velocity modulation and the lower half those from temperature modulation, respectively. As can be observed from Fig. 4 the amplitude maximum is asymmetrical, i.e. is approached faster coming from smaller wavelengths (corresponding to smaller Z magnitudes) which leads to the choice of $t_p = 2/3t_{pr}$ as the smallest value, lying closer to the optimum at $t_p = t_{pr}$ than the preceding value of $t_p = 5/3t_{pr}$.

One interesting feature in Fig. 5 (a) is the instability of the sidebranch wavelength which occures at some distance away from the tip and shows a quick transition between wavelengths of factor 1 and factor 2 with respect to external forcing. In Fig. 5 (c) and (d) resulting from pulse durations $t_p > t_{pr}$ we see the increasing destabilizing effect on sidebranches whose spacing is determined by the external forcing. This leads to the temporary splitting of one sidebranch. Subsequently the additional sidebranch is superseded again by its dominant neighbor. Generally there is a close similarity between the impact on the morphology as resulting from velocity and temperature modulation. However, the velocity modulation exhibits a more pronounced response to higher frequencies in the development of the boundary, while temperature modulation shows a stronger tendency towards smoothing of morphological features. The latter may be attributed to the fundamentally limited response towards variation of thermal conditions related to the diffusive process.

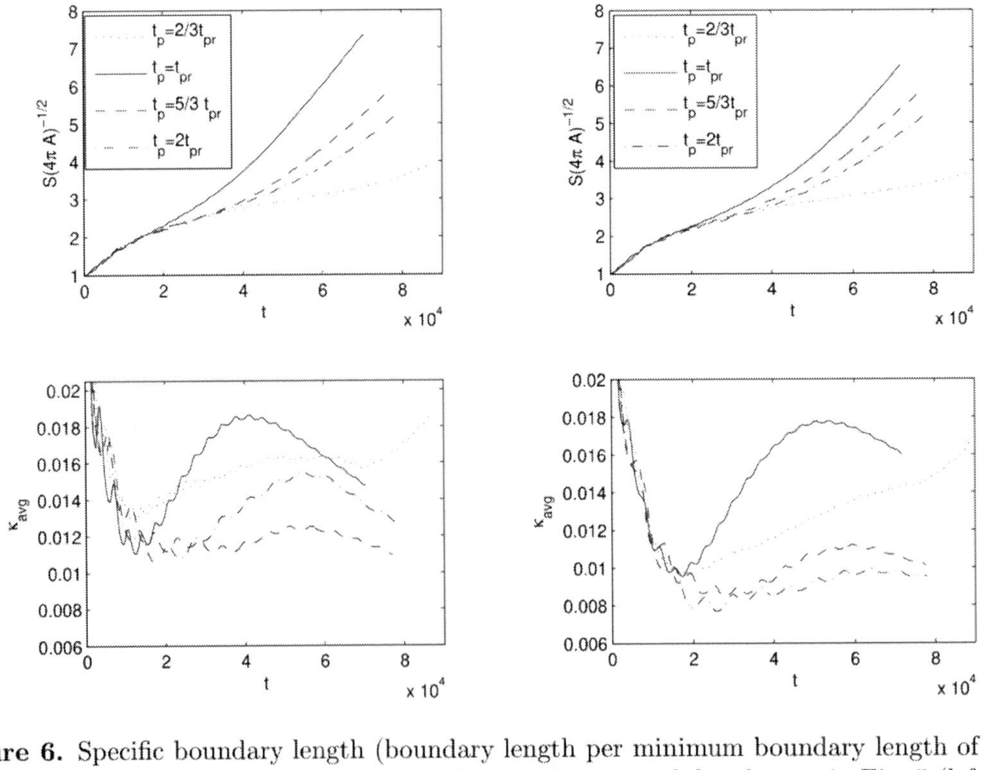

Figure 6. Specific boundary length (boundary length per minimum boundary length of area) and boundary-averaged curvature for the different velocity-modulated cases in Fig. 5 (left) and temperature-modulated cases (right).

In order to find a quantitative description for the differences in morphological development seen in Fig. 5 two global properties are evaluated, delivering characteristic indicators for resonant behavior. On the one hand, there is the specific boundary length, given by $S(4\pi A)^{-1/2}$, which defines the ratio of the actual boundary length S to the minimal boundary length at the given area A of the dendrite. This quantity can be seen as a measure for the development of geometrical complexity. Initially having a value of unity according to the circular seed shape the specific boundary length can in its convex section be described by an exponential law with respect to time. On the other hand we evaluate the magnitude of curvature averaged over the boundary κ_{avg}. Here, an initial uniform decay is followed by a distinct increase which indicates the onset of sidebranching. Additionally, a maximum appears, which can be interpreted as the point where initial sidebranches reach a steady state and no further contribute to global curvature generation. In both quantities we find clear support for the extremal properties in morphology development associated with the resonant pulse length. Hence, in the case of $t_p = t_{pr}$ a minimal time is needed to reach onset and steady state of sidebranching combined with a maximum in the slope of the specific boundary length and of the variation in averaged curvature.

To conclude, it was shown that a flow modulation with an optimum pulse duration t_{pr} may lead to resonant sidebranching, by means of which a highly periodic array of sidebranches with elongated amplitudes can be achieved. To adapt the investigation of resonant features toward different materials or growth conditions it is important to understand the influence of the key parameters undercooling, anisotropy and mean flow velocity on the operating state of the tip and the resulting natural wavelength. This needs further attention in the future.

Acknowledgments

Financial support by the Deutsche Forschungsgemeinschaft in the form of the Collaborative Research Center "Electromagnetic Flow Control in Metallurgy, Crystal Growth and Electrochemistry" (SFB 609) is gratefully acknowledged. We especially thank T. Witkowski and S. Aland for their support in using their software package AMDiS, and S. Eckert for numerous fruitful discussions. We further thank the Center for Information Services and High Performance Computing (ZIH) at the TU Dresden for computational and software support.

References

[1] Neumann-Heyme H, Eckert K and Odenbach S 2012 *IOP Conference Series: Materials Science and Engineering* **27** 012045
[2] Eckert S, Nikrityuk P, Raebiger D, Eckert K and Gerbeth G 2007 *Metallurgical and Materials Transactions B-Process Metallurgy and Materials Processing Science* **38** 977–988
[3] Willers B, Eckert S, Nikrityuk P, Raebiger D, Dong J, Eckert K and Gerbeth G 2008 *Metallurgical and Materials Transactions B-Process Metallurgy and Materials Processing Science* **39** 304–316
[4] Bouissou P, Perrin B and Tabeling P 1989 *Phys. Rev. A* **40** 509–510
[5] Tong X, Beckermann C, Karma A and Li Q 2001 *Phys. Rev. E* **63** 061601
[6] Jeong J H, Dantzig J and Goldenfeld N 2003 *Metallurgical and Materials Transactions A* **34**(3) 459–466 ISSN 1073-5623
[7] Ramirez J and Beckermann C 2005 *Acta Materialia* **53** 1721–1736
[8] Steinbach I, Diepers H J and Beckermann C 2005 *Journal of Crystal Growth* **275** 624 – 638
[9] Koss M, LaCombe J, Chait A, Pines V, Zlatkowski M, Glicksman M and Kar P 2005 *Journal of Crystal Growth* **279** 170 – 185
[10] Fell M and Bilgram J 2007 *Phys. Rev. E* **75**(6) 061603
[11] Dougherty A and Nunnally T 2007 *Journal of Crystal Growth* **300** 467 – 472 ISSN 0022-0248
[12] Qian X W and Cummins H Z 1990 *Phys. Rev. Lett.* **64** 3038–3041
[13] Bouissou P, Chiffaudel A, Perrin B and Tabeling P 1990 *EPL (Europhysics Letters)* **13** 89
[14] Börzsönyi T, Tóth-Katona T, Buka A and Gránásy L 1999 *Phys. Rev. Lett.* **83** 2853–2856
[15] Karma A and Rappel W J 1998 *Phys. Rev. E* **57** 4323–4349
[16] Beckermann C, Diepers H J, Steinbach I, Karma A and Tong X 1999 *Journal of Computational Physics* **154** 468 – 496
[17] Vey S and Voigt A 2007 *Comput. Vis. Sci.* **10**(1) 57–67

IOP Conf. Series: Materials Science and Engineering **33** (2012) 012107 doi:10.1088/1757-899X/33/1/012107

Phase-field modelling of microstructure formation during the solidification of continuously cast low carbon and HSLA steels

B Böttger[1], M Apel[1], B Santillana[2] and D G Eskin[3]

1) Access, Aachen, Germany
2) Tata Steel RD&T, POBox 10000, 1970CA, IJmuiden, The Netherlands
3) BCAST, Brunel University, Uxbridge UB8 3PH, U.K.

E-mail: b.boettger@access.rwth-aachen.de

Abstract. Cracking in continuous casting of steels has been one of the main problems for decades. Many of the cracks that occur during solidification are hot tears. To better understand the factors leading to this defect, microstructure formation is simulated for a low carbon (LCAK) and two high strength low alloyed (HSLA) steel grades during the initial stage of the process where the first solidified shell is formed inside the mould and where breakouts typically occur. 2D simulation is performed using the multiphase-field software MICRESS [1], which is coupled to the thermodynamic database TCFE6 [2] and the mobility database MOB2 [2], taking into account all elements which may have a relevant effect on the mechanical properties and structure formation during or subsequent to solidification. The use of a moving-frame boundary condition allows travelling through the entire solidification history starting from the slab surface, and tracking the morphology changes during growth of the shell. A heterogeneous nucleation model is included to permit the description of morphological transitions between the initial solidification and the subsequent columnar growth region. Furthermore, a macroscopic one-dimensional temperature solver is integrated to account for the transient and nonlinear temperature field during the initial stage of continuous casting. The external heat flux boundary conditions for this process were derived from thermal process data of the industrial slab caster. The simulation results for the three steel grades have been validated by thickness measurements of breakout shells and microstructure observation of the corresponding grades. Furthermore, the primary dendrite spacing has been measured across the whole thickness of the shell and compared with the simulated microstructures. Significant microstructure differences between the steel grades are discussed and correlated with their hot-cracking behavior.

1. Introduction

Despite of the outstanding practical relevance, there is still relatively little known about the exact mechanisms of hot cracking in continuous casting of steels. It is generally assumed that crack initiation occurs if a local strain in the semi-solid zone cannot be compensated by inflowing melt [3]. Consequently, the permeability and depth of the semi-solid zone, determined by the morphology of the dendrites and the specific relation of the fraction liquid with temperature, should be key issues that can be estimated from exact modeling of the solidification microstructure and used for the prediction of the hot cracking susceptibility.

Phase-field models have become very popular for the simulation of microstructure evolution during solidification. The software MICRESS® [1] is based on the phase-field concept for multiphase systems [4] which has been applied to binary alloys [5] and consequently extended to multicomponent systems by direct coupling to thermodynamic databases [6,7]. Since then the software has been developed further and also applied to steels [8-10]. Through the integration of a macroscopic one-dimensional temperature solver as a flexible external temperature boundary condition, transient and nonlinear thermal fields similar to those present in continuous casting can be addressed [10,11].

This study is motivated by specific problems with hot cracking in industrial continuous casting. Three grades, a low carbon low alloyed (LCAK) and two micro-alloyed (HSLA, LR-HSLA) steels (table **1**), showing different cracking behavior were selected. Casting statistics indicate a high risk for LCAK, an even slightly higher risk for the LR-HSLA grade, but substantially less hot-cracking problems with the HSLA steel in the initial casting stage.

The scope of this work is to find out whether microstructure simulation is capable to reveal and quantify the small differences in the solidification microstructure of these rather similar steel grades, and whether the results can help in understanding their hot-cracking behavior. With this aim, phase-field modeling of the solidification of the first few millimeters of the solid shell which are critical for breakouts was performed.

Table 1: Typical chemical composition of three steel grades

Steel grade	C (wt %)	Mn (wt %)	V (wt %)	Nb (wt %)	N (ppm)	
					aim	max
LCAK	0.045	0.22	-	-	-	50
LR-HSLA	0.045	0.8	0.04	0.013	80	100
HSLA	0.045	0.8	0.13	0.013	130	150

2. Multiphase-field model and MICRESS® software

The multiphase-field theory describes the evolution of multiple phase-field parameters $\phi_\alpha(\vec{x},t)$ in time and space. They reflect the spatial distribution of multiple grains with different orientation and/or of multiple phases with different thermodynamic properties. At the interfaces, the phase-field variables change continuously over an interface thickness η which can be defined as being large compared to the atomic interface thickness but small compared to the microstructure length scale. Their time evolution is calculated by a set of phase-field equations deduced by the minimization of the free energy functional [4,7]:

$$\dot{\phi}_\alpha = \sum_\beta M_{\alpha\beta}(\vec{n})\left(\sigma^*_{\alpha\beta}(\vec{n})\, K_{\alpha\beta} + \frac{\pi}{\eta}\sqrt{\phi_\alpha\phi_\beta}\;\Delta G_{\alpha\beta}(\vec{c},T)\right) \tag{1}$$

$$K_{\alpha\beta} = \phi_\beta \nabla^2\phi_\alpha - \phi_\alpha\nabla^2\phi_\beta + (\pi/\eta)^2(\phi_\alpha - \phi_\beta) \tag{2}$$

In equation (1), $M_{\alpha\beta}$ is the mobility of the interface as a function of the interface orientation, described by the normal vector $\vec{n}$. $\sigma^*_{\alpha\beta}$ is the anisotropic surface stiffness, and $K_{\alpha\beta}$ is related to the local curvature of the interface. The interface, on the one hand, is driven by the curvature contribution $\sigma^*_{\alpha\beta}$ $K_{\alpha\beta}$ and on the other hand by the thermodynamic driving force $\Delta G_{\alpha\beta}$. This driving force which is a function of temperature T and local composition $\vec{c}$ couples the phase-field equations to the multi-phase diffusion equations

$$\dot{\vec{c}} = \nabla \sum_{\alpha=1}^{N} \phi_\alpha \vec{D}_\alpha \nabla \vec{c}_\alpha \qquad \text{with } \vec{c} \text{ defined by} \qquad \vec{c} = \sum_{\alpha=1}^{N} \phi_\alpha \vec{c}_\alpha \tag{3}$$

and $\vec{D}_\alpha$ being the multicomponent diffusion coefficient matrix for phase α. $\vec{D}_\alpha$ is calculated online from databases for the given concentration and temperature.

The above equations are implemented in the software package MICRESS® [1] being used for the simulations throughout this paper. Direct coupling to thermodynamic and mobility databases is accomplished via the TQ-interface of Thermo-Calc Software [2]. The thermodynamic driving force ΔG and the solute partitioning are calculated separately using the quasi-equilibrium approach [7], and are introduced into the equation for the multiple phase-fields (equation (1)).

3. Boundary conditions and numerical parameters

In continuous casting, during the initial formation of a solid shell, the local temperature field is highly transient and non-linear. To obtain a realistic temperature boundary condition for the phase-field simulation, a one-dimensional temperature solver is integrated. According to the high diffusivity of heat λ compared to solutes, a lower spatial resolution is required. This allows the temperature field to range deeply into the melt, ideally up to the centre of the casting, without causing too much computational effort (figure 1). By this means, the temperature T can be explicitly solved in the 1D field, including release of latent heat in the simulation domain [11]:

$$\dot{T} = \frac{1}{\overline{C}_p}\left(\nabla \sum_{\alpha} f_\alpha \lambda_\alpha \nabla T + \sum_{\alpha} H_\alpha \dot{f}_\alpha\right) \tag{4}$$

Here, f_α denotes the average fraction of phase α, $\overline{C}_p$ the average heat capacity and H_α the enthalpy per phase in each temperature grid cell. The bottom boundary condition of the 1D temperature field is the slab surface temperature (figure 2), which was obtained from a simulation with the heat transfer model CON1D [12].

The solidification process and thus the latent heat release not always occur completely inside the simulation domain (darker rectangular zone in figure 1). To assure a correct calculation of latent heat outside this domain in the region of the one-dimensional temperature solver, an iterative homoenthalpic approach was used which assumes a uniform H(T) relation in the region where temperature is solved, and thus allows finding a macroscopic temperature solution which is consistent with the microscopic latent heat release [11].

Due to performance reasons, the simulation domain cannot span the whole solidified length. Instead, the solidification front was tracked by using a moving-frame boundary condition. Tracking criterion was a constant distance of the highest dendrite tip from the top of the simulation domain.

The boundary conditions for the concentration fields and the phase boundaries (phase-field parameter field) at the different sides of the simulation domain are of utmost importance. To obtain the most realistic selection behavior, isolated boundary conditions were use on the right and left side as well as at the bottom of the simulation domain. For the top, a fixed concentration condition was used to assure that the far-field concentration remains constant (and equal to the alloy specification).

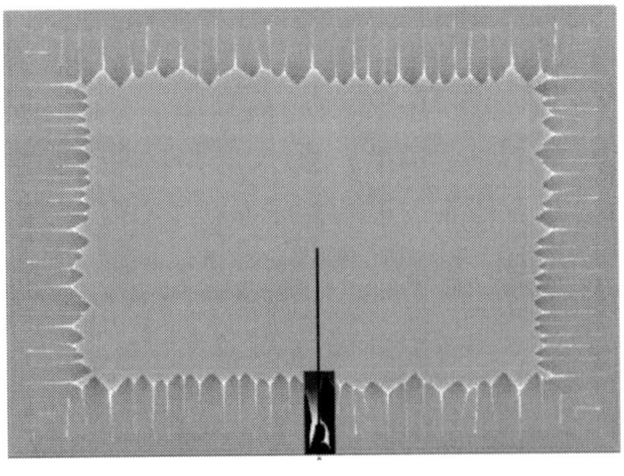

Figure 1: Schematic representation of the position of the simulation domain (dark rectangular region) and the external 1D temperature field (black line) in the cross-section of the slab.

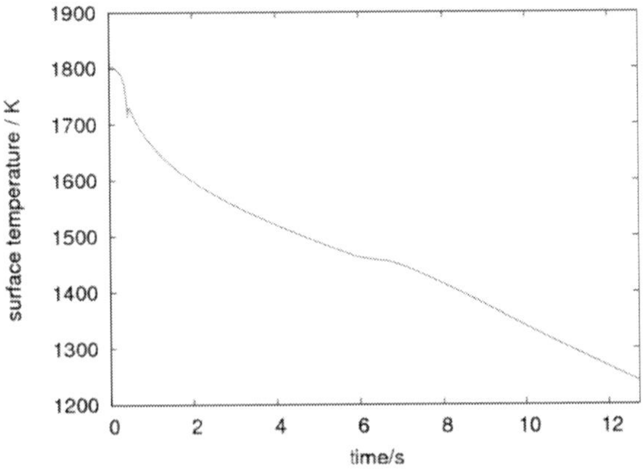

Figure 2: Slab surface temperature which was used as thermal boundary condition for the bottom of the 1D temperature field

The microscopic simulation domain was 2000 phase-field grid cells high and had a width of 1000 or 5000 cells, the grid size was $\Delta x=0.333\ \mu m$. The interface mobility of the liquid-bcc interface was calibrated by comparison with high resolution results ($\Delta x=0.05\ \mu m$) for a small test domain, a value of $0.065\ cm^4/(Js)$ was found to reproduce the correct diffusion limited interface kinetics. For the one-dimensional temperature field, 3000 temperature grid cells with a grid size of 15 µm were chosen.

Thermodynamic data were obtained directly by coupling to the thermodynamic database TCFE6, diffusion coefficients were taken from the mobility database MOB2 [2].

4. Nucleation model

As nucleation can affect grain selection in columnar growth, a heterogeneous model for nucleation from the melt was used. The approach is based on the assumption of a density distribution of heterogeneous nuclei of different diameter *d*. The critical nucleation undercooling is determined by

$$\Delta T_{Nuc} = \frac{4\sigma}{\Delta S\, d} \tag{5}$$

where ΔS is the entropy of fusion, and σ the solid–liquid interfacial energy [13]. This idea is extended to a fully spatially resolved model which further accounts for an inhomogeneous composition [14] and temperature [11] distribution. At the beginning of each simulation run, the potential nucleation sites are randomly distributed over the domain, according to the given seed density distribution, and attributed with the critical nucleation undercooling given by the nucleant size according to equation (5). If this critical undercooling is reached locally, a grain with random orientation is set.

The seed density distribution is defined by an arbitrary number of seed classes which have different radii and as such different critical nucleation undercoolings. In this work, a three parameter log-normal distribution function was used:

$$\frac{dN}{dr} = \frac{N}{sr\sqrt{2\pi}} exp\left(\frac{1}{2}\left[\frac{ln(r)-\mu}{s}\right]^2\right), \qquad \mu = ln(r_0) - \frac{s^2}{2} \tag{6}$$

with $r_0=0.05\ \mu m$, $s=0.14$ and $N=10^5$, the class width was 0.05 µm.

5. Simulation results and comparison to experimental findings

Figure 3 shows a time series of results for LCAK for the carbon concentration distribution. The solid-liquid interface and its development are clearly visible over the time range up to 12.75 seconds. The first seeds grow from the melt (a), but rapidly convert to a columnar structure (b-d). Later (e), the nucleation of some equiaxed grains ahead of the front is visible. The reason for this is the strong cooling at this time (figure 2) which also leads to an increase of the temperature gradient and to a narrowing of the semi-solid zone. The moving frame starts action after about 0.5 seconds (f). During the later stages, a coarsening of the dendrites due to slower cooling as well as a decrease of the temperature gradient is observed (figure 3g-j).

In figure 4, the grain structure obtained for a bigger domain size is shown as orientation distribution. It can be seen from the colors of the grains that a strong grain selection occurs, and that after selection only grains with orientations close to 0° with respect to the temperature gradient are left.

In figure 5 (left), a micrograph for LCAK shows a predominantly columnar structure, although some nucleation events seem to have occurred as new orientations also appear inside the slab. In figure 5 (right), this experimentally observed microstructure is compared to the result of a simulation with a simulation domain of 5000 x 2000 phase-field grid cells. A good qualitative agreement can be seen with respect to the general dendritic pattern, taking into account that the experimental micrograph is a cut through a 3D structure, and that after solidification other phase transformations have occurred which left their fingerprints.

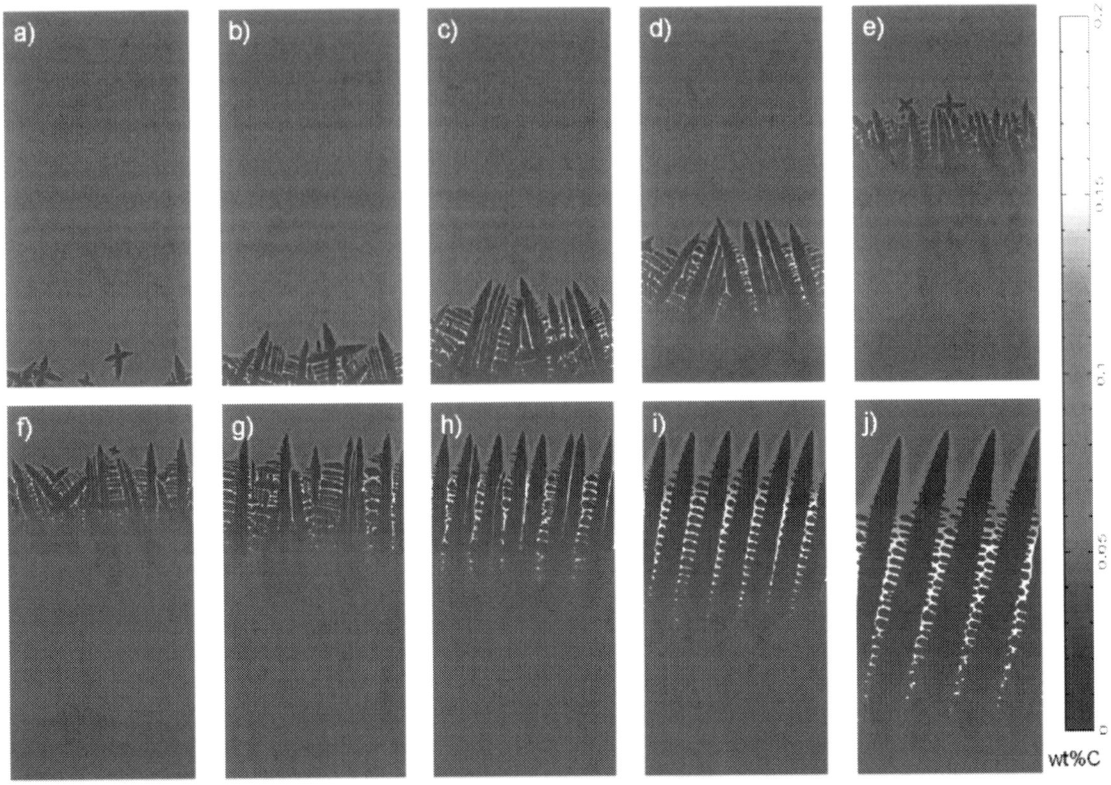

Figure 3: Time sequence for solidification of LCAK. The C distribution is shown after **a)** 0.18 s, **b)** 0,21 s, **c)** 0.28 s, **d)** 0,36 s, **e)** 0,46 s, **f)** 0.5 s, **g)** 1.0 s, **h)** 1.5 s, **i)** 3.0 s, **j)** 12.75 s.

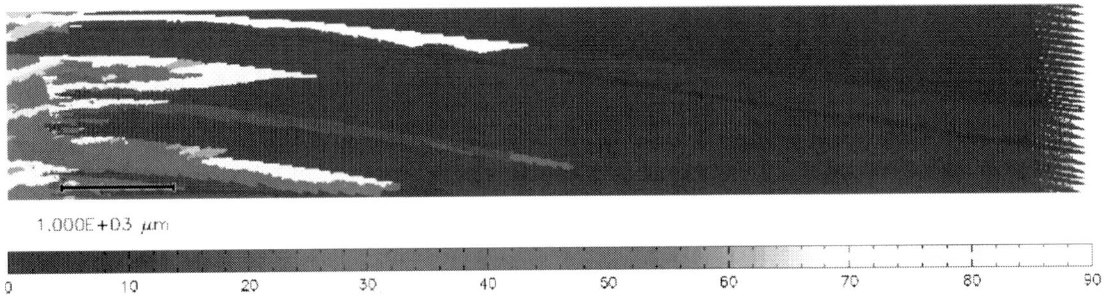

Figure 4: Orientation distribution (in degrees) for LCAK up to 8.9 s solidification time, large simulation domain of 5000 x 2000 phase-field grid cells.

A check of the growth velocity of the dendrite front was done by comparing the solidification length in the simulation to the results of the heat transfer model CON1D [12] which had been calibrated for the case of several breakout shells of LCAK [15]. Figure 6 shows the obtained shell thickness (up to the dendrite tips) in comparison to the front position predicted by the phase-field simulation. As can be seen, the fit is nearly perfect up to about 7 mm and does not exceed 5% at the mould exit. This demonstrates that the 1D approximation of the temperature field is a useful approach under these conditions, and that fluid flow does not significantly affect the temperature field at this stage.

IOP Conf. Series: Materials Science and Engineering 33 (2012) 012107 doi:10.1088/1757-899X/33/1/012107

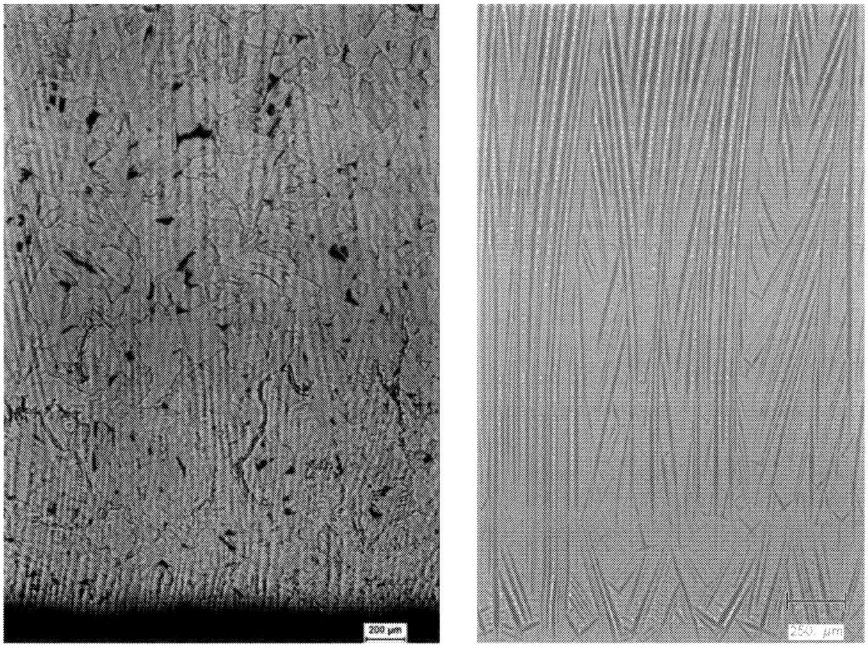

Figure 5: Comparison between a composed micrograph of the "chill zone" of LCAK (left) and simulation results (Mn distribution in grey scale, right).

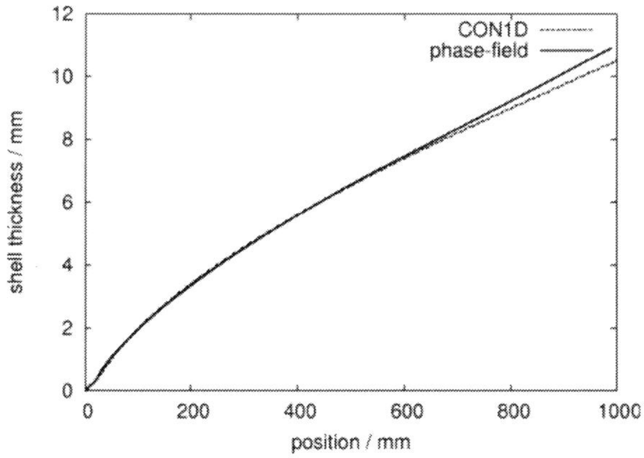

Figure 6: Shell thickness comparison for a LCAK slab between CON1D and phase-field simulation

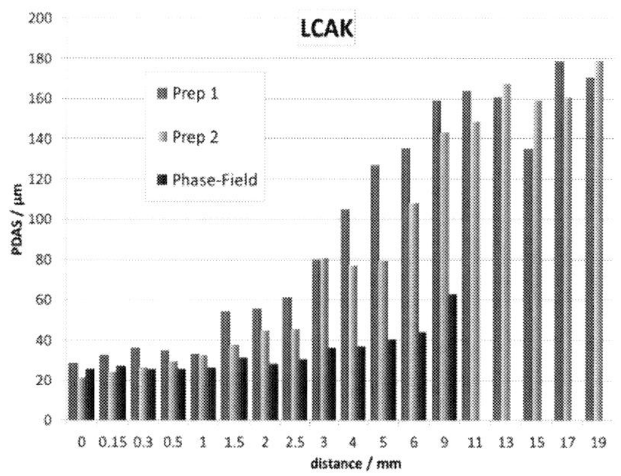

Figure 7: Comparison of the PDAS measured for two LCAK samples and simulation results, as function of the distance from the shell surface.

Finally, a comparison of the PDAS (Primary Dendrite Arm Spacing) between experiment and simulation is shown (figure **7**). The measured PDAS values for two samples (Prep 2 and Prep 3) are opposed to the results of the phase-field simulation. Although the PDAS values are in quite good agreement for about the first 1 mm, there is a clear discrepancy for higher positions (>2 mm) in the solidified shell. Even uncertainties like insufficient calibration or measurement errors cannot explain why there is a strong increase in the PDAS between about 2 and 6 mm in the experimental results, which is not found in the simulation. The strong bending of dendrites typically found in micrographs under these conditions suggests that a strong fluid flow is present in this region which could have altered the PDAS and which was not taken into account in the simulation.

6. Comparison between LCAK, LR-HSLA and HSLA

The microstructure differences between LR-HSLA and HSLA are by far smaller than those between HSLA and LCAK. This can be seen in the simulated solidification microstructures in figure 8. Shown is the Mn distribution for all three steel grades at the same stage of solidification.

In order to quantify the results with respect to possible effects on the permeability and the length of the mushy zone, the length of surface per area between the dendrite α and the liquid l is evaluated and displayed in Figure 9 in arbitrary units. Each individual value is obtained by integrating the phase-field gradient over one row in x direction which corresponds to an isothermal line:

$$S = \frac{1}{A}\int \sqrt{\phi_\alpha \phi_l}\, dA \tag{7}$$

A smoothened plot of S is shown in Figure 9. The curves obtained for the three alloy grades allow evaluating the thickness and the fineness of the mushy zone which both have a strong impact on the permeability for melt flow. For both quantities, a slight increase is visible for HSLA with respect to LR-HSLA. LCAK, in contrary, shows a clearly coarser and smaller semisolid region.

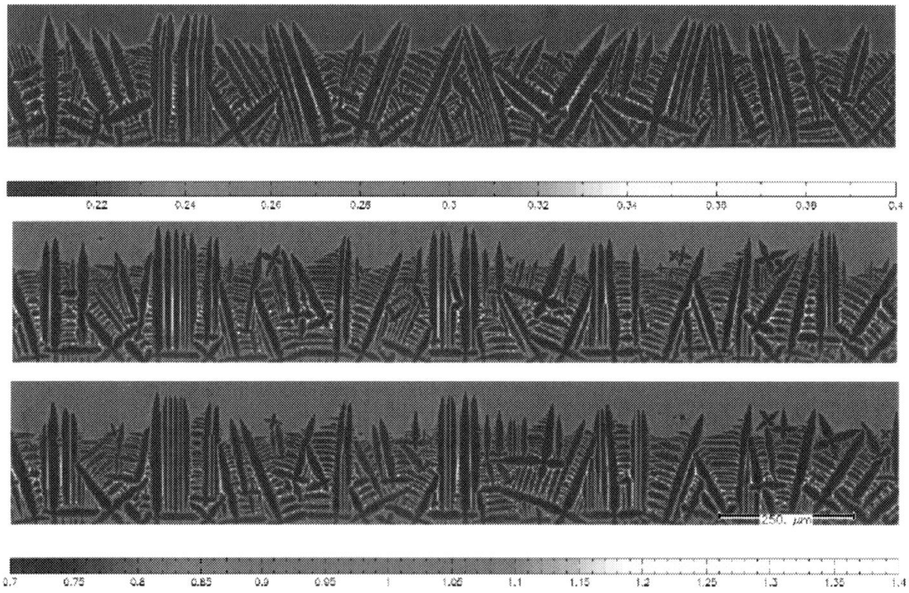

Figure 8: Microstructure of LCAK (top) after 0.32 s, HSLA (center) after 0.36 s and LR-HSLA (bottom) after 0.35 s, shown is the Mn distribution. Domain size 5000 x 2000 phase-field grid cells.

Finally, average fraction liquid–temperature curves from 6 different outputs between 1.0 and 1.25 s for LR-HSLA and HSLA are plotted and compared in figure 10. Most striking is the shift of the curve for LR-HSLA by about 2 K to higher temperatures, which is reasonable considering its lower composition in V and N. Furthermore, this shift is slightly stronger for lower temperatures, corresponding to a reduction of the width of the mushy zone in the critical fraction liquid range below 10%. This is consistent with the slightly shorter semisolid region found for LR-HSLA in figure 9, but would indicate an (if at all) small decrease of the hot-cracking risk compared to HSLA rather than an increase.

All in all, the observed microstructure differences do not conform with the observed hot-cracking characteristics of the steel grades. Even though the small observed increase of the outbreak risk of LR-HSLA relative to the LCAK grade can be explained by the higher relative surface and thickness of the mushy zone, the pronounced drop of the susceptibility for hot-cracking for HSLA is not at all consistent with the simulation results.

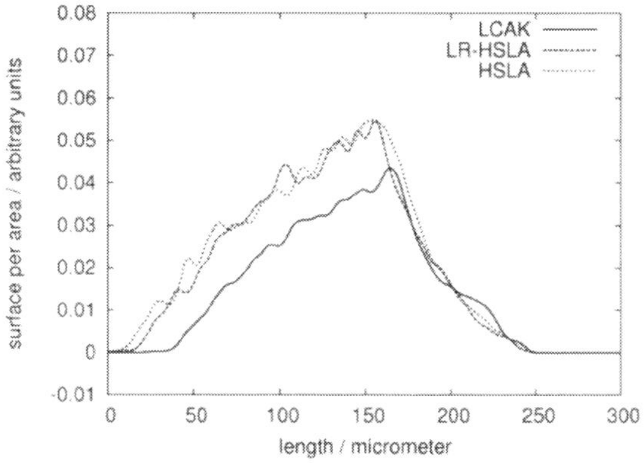

Figure 9: Comparison of the specific surface per area for LCAK, LR-HSLA and HSLA, averaged along the x-direction in figure 8.

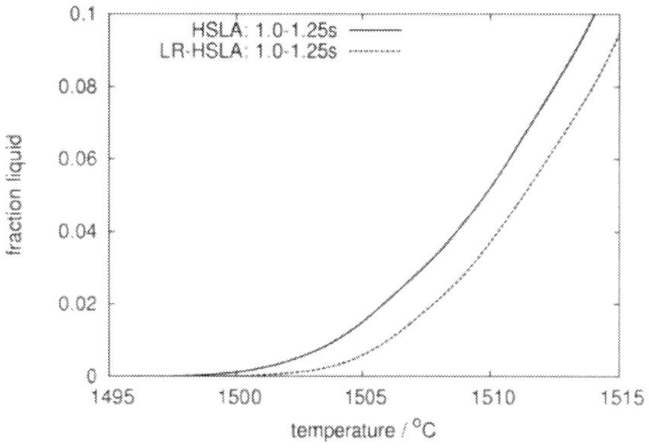

Figure 10: Comparison of fraction liquid-temperature curves for LR-HSLA and HSLA, aggregated for results between 1.0 and 1.25 s.

7. Conclusions and Outlook

It has been shown that the multiphase-field method coupled to thermodynamic databases and combined with a 1D temperature field as external thermal boundary condition is suitable for simulating microstructures of low-alloyed and HSLA steels, as well as for identifying the small differences in the microstructure evolution between steel grades of very similar composition: Not only the clearly different microstructure of LCAK, but also the slight variation between the two HSLA grades, which only differ in the V and N content, could be evaluated and quantified.

Nevertheless, this analysis is not conclusive for explaining the observed differences in the hot tearing behavior between the two HSLA grades. We assume that, apart from the purely solutal effects which have been investigated in this paper, micro-alloying such as V, Nb, Ti etc. can lead to the formation of precipitates at high temperatures, which also may be very important for the hot tearing susceptibility. The role of such precipitates in breakouts during continuous casting will be investigated in future work.

References

[1] http://www.micress.de
[2] Themo-Calc Software http://www.thermocalc.se
[3] Rappaz M, Drezet J-M and Gremaud M 1999 *Met. and Mat. Trans.* A **30** 449
[4] Steinbach I, Pezzolla F, Nestler B, Seeßelberg M, Prieler R, Schmitz G J and Rezende J J L 1996 *Physica D* **94** 135
[5] Tiaden J, Nestler B, Diepers H J and Steinbach I 1998 *Physica D* **115** 73
[6] Böttger B, Grafe U, Ma D and Fries S G 2000 *Mater Sci Technol* **16** 1425
[7] Eiken J, Böttger B and Steinbach I 2006 *Phys Rev E* **73** 066122
[8] Nakajima K, Apel M and Steinbach I 2006 *Acta Mater* **54** 3665
[9] Böttger B, Apel M, Eiken J, Schaffnit P and Steinbach I 2008 *Steel Research Int.* **79** No.8 608-616
[10] Böttger B, Stratemeier S, Subasic E, Göhler K, Steinbach I and Senk D 2010 *Adv. Eng. Mater.* **12**(4) 101
[11] Böttger B, Eiken J and Apel M 2009 *J. Comput. Phys.* **228** 6784
[12] Santillana B, Hibbeler L C, Thomas B G, Kamperman A A and van der Knoop W 2011 *ISIJ International* **48** No 10 1380
[13] Maxwell I and Hellawell A 1975 *Acta Metall.* **23** 229
[14] Böttger B, Eiken J and Steinbach I 2006 *Acta Mater* **54** 2704
[15] Santillana B, Thomas B G, Botman G and Dekker E, Conference contribution to the 7th ECCC, held in Düsseldorf, Germany on 27th June - 1st July 2011

IOP Conf. Series: Materials Science and Engineering **33** (2012) 012108 doi:10.1088/1757-899X/33/1/012108

Fractal characteristics of dendrite in aluminum alloys

K Ohsasa, T Katsumi, R Sugawara and Y Natsume
Akita University, Tegata Gakuenmachi, Akita, 010-8502, Japan

E-mail:sasa@gipc.akita-u.ac.jp

Abstract. The fractal dimensions of dendrites in Al-Si and Al-Cu binary alloys were measured under furnace cooling and casting experiments. The fractal dimension of the Al-Si alloy increased from 1.275 to 1.495 along with increase in Si content. The fractal dimension of the Al-Cu alloy increased from 1.139 to 1.486 along with increase in Cu content. The fractal dimension of the binary alloys also increased with increase in cooling rate during solidification. Phase-field simulations for the evolution of the dendrites in the binary aluminium alloys were carried out and a same tendency as the experimental results was obtained. The permeability of an Al-5mass%Si alloy was estimated from the measured fractal dimension of an experimentally observed dendrite structure. The estimated permeability agreed well with reported permeability of an Al-Si alloy.

1. Introduction

Control of the macro-segregation in castings and ingots is very important because the macro-segregation has influence on the mechanical properties of the final products. One of the major factors causing the macro-segregation is an interdendritic fluid flow during solidification. The fluidity of the solute rich residual liquid in solidifying dendrite arrays can be restricted by complex dendrite morphology. Conventionally, the dendrite morphology has been expressed by using dendrite arm spacing (SDAS).

Several attempts have been carried out for applying the fractal theory to describe the morphology of materials structure[1-4]. Yang *et al.*[4] calculated the fractal characteristic of dendrite and cellular structures of a Ni-based superalloy under various cooling conditions. Sanyal *et al.*[5] described the mushy zone of an alloy as a network of continuous fractal structure, and they calculated the permeability of the mushy zone based on the fractal dimensions of the dendrite network. Fractal dimension may be an effective parameter for representing the complexity of dendrite array of an alloy in comparison with the SDAS. First aim of this study is to examine the factors affecting the fractal dimension of dendrites based on the experiments and the phase-field simulations for Al-Si and Al-Cu binary alloys

The fluidity of the solute rich residual liquid in a solidifying dendrite array can be expressed as the permeability. Several measurements of the permeability have been carried out by experimental methods for some metallic[6-12] and borneol-paraffin organic alloy systems[13-15]. However, the experimental measurements of the permeability are time consumption works and high cost would be required to obtain the permeability data of many commercial multi-component alloys. The second aim of this study is to develop a new method to estimate the permeability from an experimentally observed dendrite structure by measuring the fractal dimensions of solidified dendrite structure.

2. Experiment

Furnace cooling and casting experiments were carried out for the Al-Si and Al-Cu binary alloys. In the furnace cooling experiment, a sample of 50g in mass was set in an alumina crucible and melted in an electric furnace, and then the furnace was switched off and the sample was cooled to room temperature in the furnace. To examine the effect of the solute content on dendrite morphology, Si content was changed from 1mass% to 9mass% in the Al-Si binary alloy and Cu content was changed from 1mass% to 20mass% in the Al-Cu binary alloy. Solidified samples were cut longitudinally at the center of the samples, then polished, and etched with an acid reagent. In the casting experiment, Al-5mass%Si and Al-4mass%Cu alloys were used. Samples of 1kg in mass were melted in the electric furnace, and poured into a steel mold with 80mm in inner diameter, 100mm in outer diameter and 120mm in height. The superheat of molten samples at the pouring is 110K. Microstructures at the center, middle and near the mold positions in the longitudinal cross section of the casting were metallographically examined. The fractal dimensions of dendrites in the samples were measured by the box counting method as will describe below.

3. Phase-field simulation

A phase-field model for an alloy with the thin interface limit condition proposed by Kim *et al.* was used (KKS model)[16]. The governing equations of the phase-field method for a dilute binary alloy are as follows:

$$\frac{\partial \phi}{\partial t} = M\left(\varepsilon^2 \nabla^2 \phi - f_\phi\right) \tag{1}$$

$$\frac{\partial c}{\partial t} = \nabla \cdot \left(\frac{D(\phi)}{f_{cc}} \nabla f_c\right) \tag{2}$$

where ϕ is phase-field, c is mole fraction, M and ε are phase-field parameters, $D(\phi)$ is the diffusivity of solute and the subscripts under f denote the first and second derivatives of corresponding variables.
The free energy density, $f(c, \phi)$, is defined as the sum of free energies of liquid and solid phases and imposed parabolic potential, $g(\phi)$. The solute composition in the interface region, c, is determined to be a fraction-weighted sum of liquid and solid compositions.

$$f(c,\phi) = h(\phi) f^S(c_S) + [1-h(\phi)] f^L(c_L) + Wg(\phi) \tag{3}$$

$$c = h(\phi) c_S + [1-h(\phi)] c_L \tag{4}$$

where $h(\phi) = \phi^2(3-2\phi)$, $g(\phi) = \phi(1-\phi)$, W is also the phase-field parameter and the subscripts of S and L denote solid and liquid phases, respectively.

The phase-field parameters of ε and W are related to the interfacial energy, σ, and interfacial thickness, 2λ, and are given as follows:

$$\sigma = \frac{\varepsilon\sqrt{2W}}{8}\pi \tag{5}$$

$$2\lambda = \frac{\varepsilon}{\sqrt{2W}}\pi \tag{6}$$

The phase-field mobility, M, derived by the thin interface limit condition depends on the kinetic coefficient and the solid and liquid compositions at the interface.

$$M = (M_k + M_d)^{-1}$$
$$M_k = \frac{\varepsilon^2}{\sigma}\frac{RT}{V_m}\frac{1-k_0}{m}\frac{1}{\mu_k} \tag{7}$$
$$M_d = \frac{\varepsilon^3}{D\sigma\sqrt{2W}} f_{cc}^S(c_S^e) f_{cc}^L(c_L^e)(c_L^e - c_S^e)^2 \int_0^1 \frac{h(\phi_0)[1-h(\phi_0)]}{[1-h(\phi_0)]f_{cc}^S(c_S^e) + h(\phi_0) f_{cc}^L(c_L^e)} \frac{d\phi_0}{\sqrt{\phi_0(1-\phi_0)}}$$

where R is gas constant, T is temperature, V_m is molar volume, k_0 is equilibrium partition coefficient, m is liquidus slope, μ_k is linear kinetic coefficient and the subscript of e denotes the equilibrium condition, respectively. In the vanishing kinetic coefficient condition ($\mu_k^{-1} = 0$), the mobility is rewritten as $M = M_d^{-1}$. .

For numerical calculation, Eqs. (1) and (2) were discretized into uniform square grids and solved by using the explicit finite difference method.

The anisotropy of the S/L interface is introduced to the phase-field parameter, ε, which is related to the interfacial energy. Four-fold symmetry at two-dimensional condition is expressed as follows:

$$\varepsilon(\theta) = \varepsilon[1 + \varepsilon_4 \cos(4\theta)] \tag{8}$$

where ε_4 is the magnitude of anisotropy, which have to be $\varepsilon_4 < 1/15$, and θ is the angle between the axis normal to the S/L interface and the x axis.

In the simulation, a calculation domain with rectangular shape containing 1000×1000 grids was used. The grid size is 0.01×10^{-6} m. A planar solid was put at the left side of the domain as an initial condition to simulate the evolution of a dendrite. Zero flux condition was set at the left and right sides of the domain and a periodic condition was set at the upper and lower sides of the domain.

4. Evaluation of dendrite morphology

4.1. Fractal dimension

In this study, "Fractal dimension" was used to evaluate the complexity of the dendrite morphology of alloys quantitatively. The fractal dimensions of observed and simulated dendrites were evaluated by using the box counting method[17]. The procedure of the box counting method is as follows; an area including dendrites is divided into square boxes with the size of r. Then, the number of boxes, N, in which the S/L interface is included, is counted. Then, the size of the box, r, is changed and the same procedure is repeated. If following relationship holds true between the number of the boxes, N, and the size of the box, r, the geometry of the dendrite shows fractal characteristic.

$$N(r) = r^D \tag{9}$$

The exponent, D, in the equation (9) is called "Fractal dimension" and its value can be obtained from the slope of the plot line of log r vs. log N. In the measurement of the Fractal dimension, 8 box sizes were used and total number of boxes was changed from 32 to 42658 during measurement.

4.2. Dimensionless perimeter of dendrite

In addition to the fractal dimension, a new index "Dimensionless perimeter" of a dendrite was introduced for evaluating the morphology of dendrites. "Dimensionless perimeter" was calculated as follows; first, the perimeter of a simulated dendrite *L1* is measured. Then the perimeter *L2* of a circle whose area is same as the perimeter of the dendrite is measured. The "Dimensionless perimeter" of the dendrite was defined as the ratio of *L1*/*L2*. The dimensionless perimeter increases when dendrite morphology becomes complex. Therefore, dimensionless perimeter is another index for showing the complexity of dendrite shape. Because the measurement of dimensionless perimeter is difficult for experimentally observed dendrites, dimensionless perimeter was measured only for simulated dendrites.

5. Results and discussion

5.1. Experimental results

5.1.1 Effect of solute content

Change in the dendrite morphology of the Al-Si binary alloy along with increase in Si content is shown in Fig.1. Left side photographs of the Figure 1 are unretouched structures. When Si content increases, primary α dendrites become fine and eutectic regions become large. Eutectic structures consist of α solid solution and eutectic Si with acicular shape. Direct measurement of the fractal dimension from the unretouched structures in the left side of the Figure 1 by using the box counting method is inaccurate because the boxes which include the interface between α solid solution and eutectic Si phase are also counted. Hence, the outlines of primary α dendrites were traced and eutectic regions were painted out. Furthermore, grey primary α dendrites were modified to a simple image with white and black dichotomous colors. The right side figures of the Figure 1 show retouched photographs. This modification makes it possible to measure the fractal dimensions of the primary α dendrites accurately. Numbers in the right side in the Figure 1 are measured fractal dimensions from the modified photographs. The correlation coefficient for determining the Fractal dimension from the slope of the plot line of log r vs. log N is -0.9999. The fractal dimension of the Al-Si alloy increases with increase in Si content. The morphology of primary dendrites becomes complex with increase in the fractal dimension.

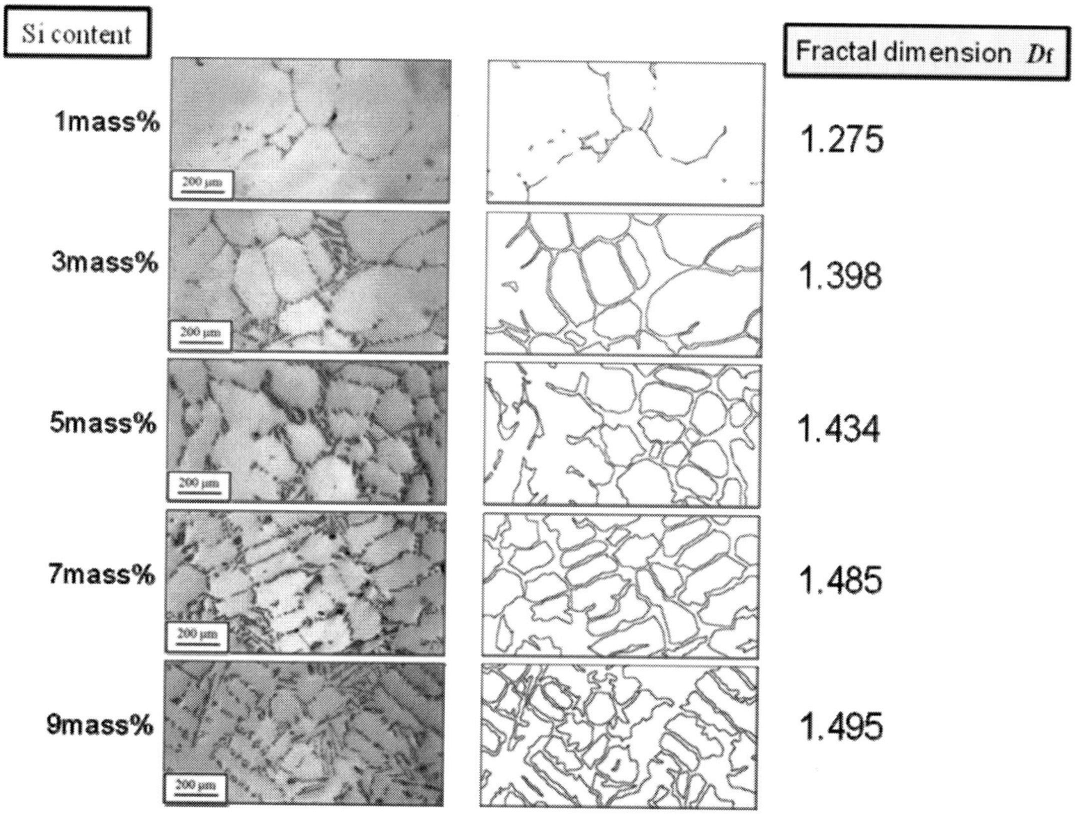

Figure 1. Change in the solidified structure of a furnace cooled Al-Si binary alloy along with the increase in Si content. Left side photographs are unretouched structures and right side images are retouched structures for the measurement of the fractal dimension. Numbers at the right side of the Figure show measured fractal dimensions from the retouched dendrite structures.

Figure 2 shows change in the fractal dimension of primary α dendrites with increase in Si content. The fractal dimension increase until 7 mass%Si content, and the fractal dimension becomes almost constant value over 7mass%Si content. Figure 3 shows change in the fractal dimension of the Al-Cu alloy with increase in Cu content. Same tendency as that of the Al-Si alloy is seen. The fractal dimension increase until 10 mass%Cu, and the fractal dimension becomes almost constant value at over 10mass%Si content. These results can be explained as follows;

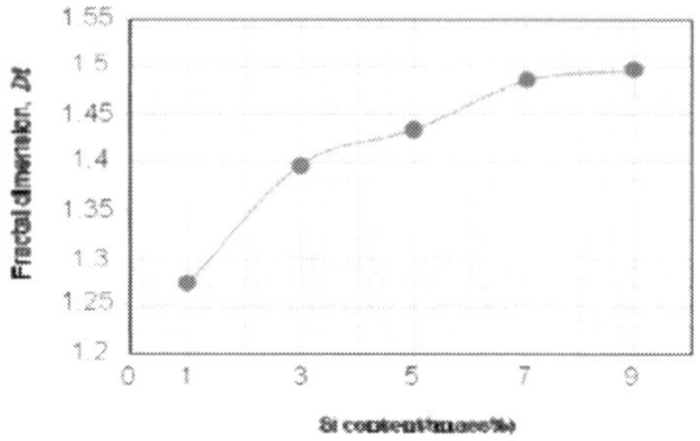

Figure 2. Change in the fractal dimension of dendrites in an Al-Si binary alloy along with the increase in Si content.

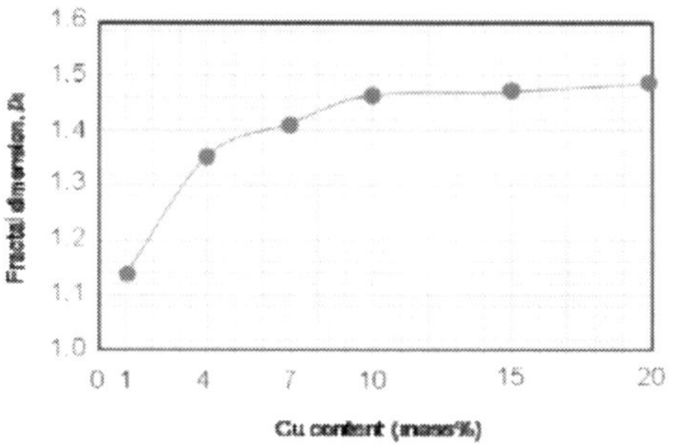

Figure 3. Change in the fractal dimension of dendrite in an Al-Cu binary alloy along with increase in Cu content.

The increase in solute content enhances the "constitutional super cooling" and the instability of a S/L interface increases. Complex S/L inter face produces large amount of the S/L interface energy which restrain the instability of the S/L interface. Finally a balance of both driving forces holds true and the increase in the fractal dimension of dendrites halts.

5.1.2. Effect of cooling rate

Figure 4 shows observed and retouched dendrite structures in an Al-5mass%Si alloy casting. Measured local cooling rate at a position near the mold wall was larger than that at the center of the casting, and fractal dimension of dendrites near the mold wall region is greater than that at the center of the casting.

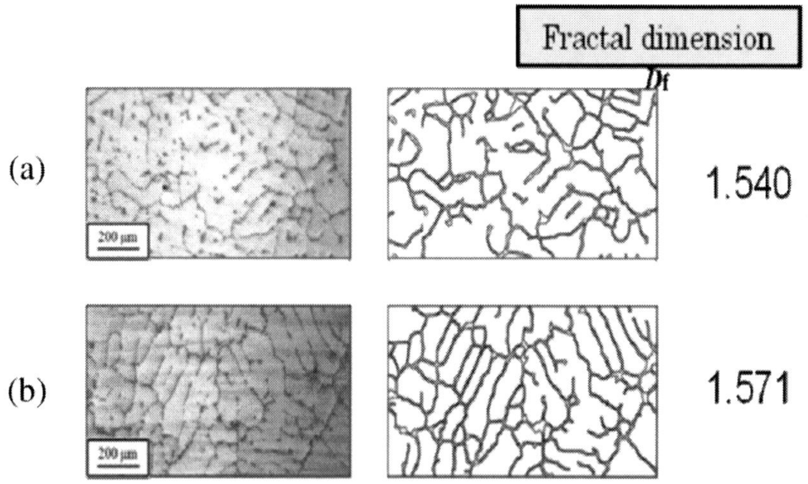

Figure 4. The dendrite morphology in an Al-5mass%Si alloy casting along with the dendrite outline for fractal analysis. Images (a) correspond to a position at the center of the casting while (b) correspond to a position near the mold wall.

Figure 5 shows the relationship between cooling rate and fractal dimension of dendrites measured from the furnace cooled and casting samples. The fractal dimension increases with increase in the cooling rate.

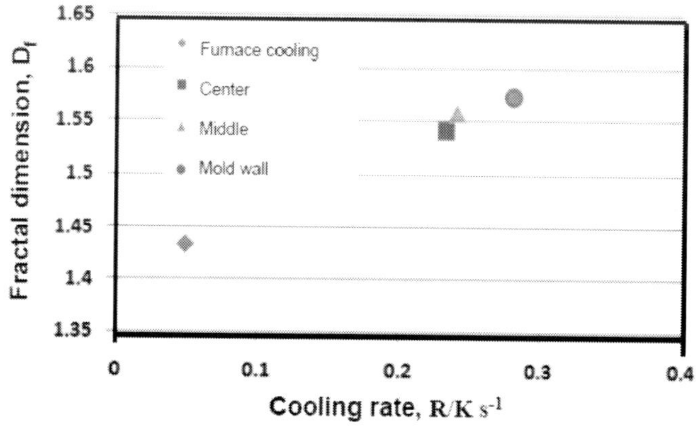

Figure 5. Change in the fractal dimension along with the increase in cooling rate in an Al-5mass%Si alloy. The fractal dimensions were measured from the furnace cooled sample and three positions in the casting sample.

5.2. Simulated dendrite morphology

Figure 6 shows simulated dendrites of Al-1mass%Si and Al-3mass%Si binary alloys by the phase-field method. It is shown that the dendrite becomes more complex when Si content increases from 1mass% to 3mass%. Comparison between the fractal dimensions of simulated and observed dendrites for Al-Si and Al-Cu binary alloys is shown in Fig. 7. The fractal dimension of simulated dendrite increases with increase in solute content and these results agree well with experimental results.

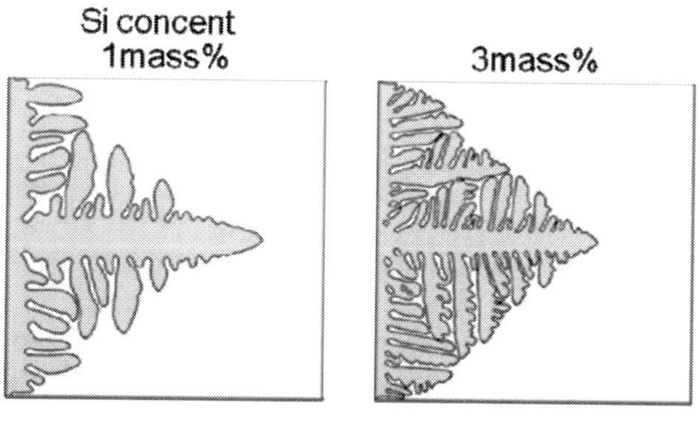

Figure 6. Simulated dendrites of Al-Si binary alloys by using the phase-field method.

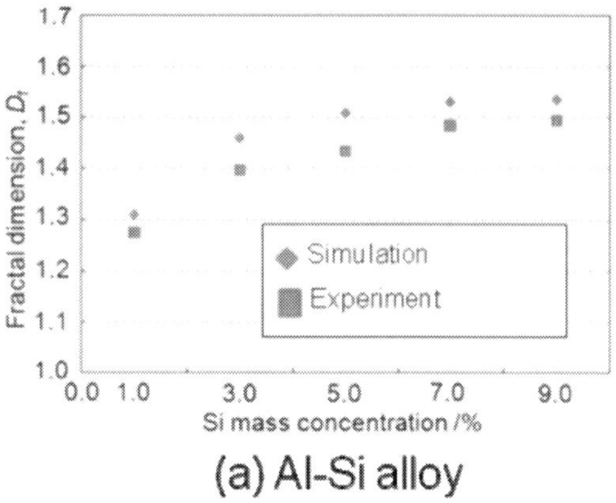

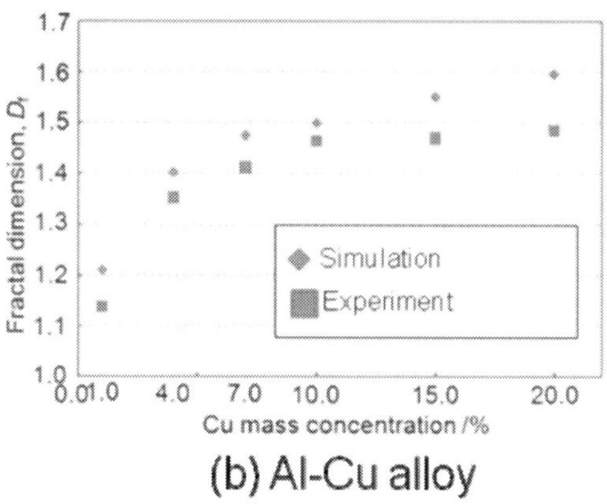

Figure 7. Comparison between the fractal dimensions of simulated and observed dendrites in Al-Si and Al-Cu binary alloys.

6. Permeability

Fluid flow through a porous medium is described by the Darcy's law, and mean flow velocity, v, is expressed as follows.

$$v = \frac{K}{\mu L} \Delta P \tag{10}$$

where K is permeability (m^2), μ is viscosity of liquid (Pa·s), L is the length of the porous medium (m), ΔP is the pressure drop (Pa). Until now, many studies have been carried out for investigating the interdendritic fluidity of a solidifying alloy. In those studies, the interdendritic flow was analyzed by using the Darcy's law, and the fluidity of a dendrite network is expressed as permeability.

Piwonka *et al.* expressed the permeability, K, of the mushy zone in a solidifying alloy as a following form by regarding the mushy zone as a medium having many flow channels[6)];.

$$K = \frac{g_L^2}{8\pi n \tau^3} \tag{11}$$

where g_L is volume fraction of liquid, n is number of flow channels per unit area (m^2), τ is tortuosity factor. The tortuosity factor was introduced to account for the fact that the flow channels are not straight and symmetrical.

If one assumes that the number of channels is equal to the number of regions between dendrite arms, and the spacing between these channels is equal to the dendrite arm spacing, λ_2, then $n=1/(\lambda_2)^2$ and

$$K = \frac{g_L^2 \lambda_2^2}{8\pi \tau^3} \tag{12}$$

It can be expected that the tortuosity factor, τ corresponds to the complexity of the dendrite morphology, i.e., "Fractal dimension". However, the fractal dimension can not be regarded as the tortuosity factor because the value of fractal dimension is relatively small with range from 1 to 2 in 2D figures in comparison with the value of the tortuosity factor. We showed that the dimensionless perimeter is an appropriate parameter to regard as the tortuousity factor in Fe base alloys[18)].

In this study, permeability of an Al-Si alloy was estimated based on an experimentally observed dendrite structure. Fractal dimension of experimentally observed dendrites can be measured as described above. However, the measurement of the dimensionless perimeter of an observed dendrite is difficult because the observed 2D dendrite is the cross section of a 3D structure. Hence, relationship among the fractal dimension, fraction of solid and the dimensionless perimeter was evaluated based on the results of the phase-field simulation;

$$L = 3.40 D_f^{3.22} f_S^{0.27} \qquad (13)$$

where L is dimensionless perimeter, D_f is fractal dimension and f_s is fraction of solid. The measured value of the fractal dimension of Al-5mass%Si alloy shown in the Fig.1 was converted into a value of dimensionless perimeter through the Eq. (13). A dendrite arm spacing, λ_2 of the Al-5mass%Si alloy shown in the Fig.1 was estimated. Fraction solid of the Al-5mass%Si alloy at the eutectic temperature was obtained from an Al-Si binary phase diagram. The values of the dendrite arm spacing, λ_2, fraction of liquid at the eutectic temperature, gL and dimensionless perimeter L as the tortuousity factor, τ, were substituted into the Eq.(12), and the value of the permeability of the Al-5mass%Si alloy at the eutectic temperature was calculated. Figure 8 shows the comparison between calculated and reported permeability of the Al-Si alloy. A mark "P" in the figure denotes the case of parallel flow to primary dendrites, and a mark "N" denotes the case of normal flow to the primary dendrites. The permeability without marks in the Fig.8 corresponds to the condition with equiaxed dendrite structure. The permeability reported by Apellian et al.[7)] was measured in Al-4mass%Si alloy with equiaxed dendrite morphology which is same morphology as present study. Hence, it is found that the value of the estimated permeability in the present study agrees well with the reported value for an Al-Si alloy[7)].

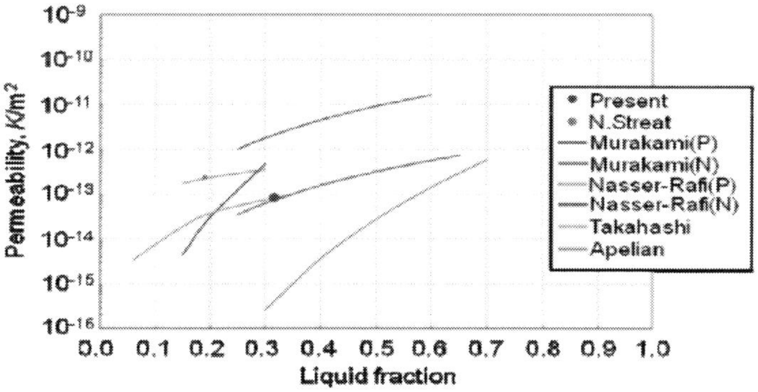

Figure 8. Comparison between the estimated value of the permeability in present study and the experimentally measured permeability.

Conclusions

The fractal dimensions of Al-Si and Al-Cu binary alloys were measured under casting and furnace cooling experiments. A phase-field simulation was carried out and the factors affecting dendrite morphology was examined. Obtained results are as follows;

1) Fractal dimension of an Al-Si alloy increased from 1.275 to 1.495 along with increase in Si content from 1 to 9mass%.
2) Fractal dimension of an Al-Cu alloy increased from 1.139 to 1.486 along with increase in Cu content from 1 to 20mass%.
3) Fractal dimension of Al base binary alloys also increased with increase in cooling rate
4) The phase-field simulation showed that the fractal dimension of dendrites in Al-Si and Al-Cu alloys increase with increase in solute content.

5) Measured value of the fractal dimension of the Al-5mass%Si alloy was converted into the value of dimensionless perimeter though an evaluated relationship by the phase-field simulation and the permeability of the Al-5mass%Si alloy was estimated by regarding the dimensionless perimeter as the tortuosity factor of interdendritic liquid channels. Obtained permeability agreed well with reported values of an Al-Si alloy.

References

[1] Kleiser T and Bocek M 1986 Z. Metallkd. **77** 582
[2] Hornbogen E 1987 Z. Metallkd. **78** 622
[3] Tanaka M J 1922 Mater. Sci. **27** 4712
[4] Yang Y, Xiong Y and Liu L 2001 Sci. Tech. Adv. Mat. **2** 101
[5] Sanyal D, Ramachandrarao P and Gupta O P 2006 Chem. Eng. Sci. **61** 307
[6] Piwonka T S and Flemings M C 1966 Trans. metall. Soc.A.I.M.E. **236** 1157
[7] Apelian D, Flemings M C and Mehrabian R 1974 Metall. Trans. **5** 2533
[8] STREAT N and WEINBERG F 1976 Metall. Trans. **7B** 417
[9] Takahashi T, Kudoh M and Yodoshi K 1979 Nippon-Kinzoku-Gakkaishi **43** 1086
[10] Takahashi T, Kudoh M and Nagai S 1982 Tetsu-to-Hagane **68** 623
[11] NASSER-RAFI R, DESHMUKH R, and POIRIER D R 1985 Metall. Trans. **16A** 2263
[12] POIRIER D R 1987 Metall. Trans. **18B** 245
[13] MURAKAMI K, SHIRAISHI A and OKAMOTO T 1983 Acta Metall. **31** 1417
[14] MURAKAMI K, SHIRAISHI A and OKAMOTO T 1984 Acta Metall. **32** 1423
[15] MURAKAMI K and OKAMOTO T 1984 Acta Metall. **32** 1741
[16] Kim S G, Kim W T and Suzuki T 1998 Phys. Rev E **57** 4323
[17] Matsushita N 2002 "Furakutaru no Buturi" Shokabou 36
[18] Ishida H, Natsume Y and Ohsasa K 2009 ISIJ International **49** 37

IOP Conf. Series: Materials Science and Engineering **33** (2012) 012109 doi:10.1088/1757-899X/33/1/012109

Phase-field modelling of rapid solidification in alloy systems: Spontaneous grain refinement effects

A M Mullis
Institute for Materials Research, University of Leeds, Leeds, LS2-9JT, UK.

E-mail: A.M.Mullis@leeds.ac.uk

Abstract. Phase-field modelling of rapid alloy solidification, in which the rejection of latent heat from the growing solid cannot be ignored, has lagged significantly behind the modelling of conventional casting practises which can be approximated as isothermal. This is in large part due to the fact that if realistic materials properties are adopted the ratio of the thermal to solute diffusivity (the Lewis number) is typically 10^3 - 10^4, leading to severe multi-scale problems. However, use of state-of-the-art numerical techniques such as local mesh adaptivity, implicit time-stepping and a non-linear multi-grid solver allow these difficulties to be overcome. Here we describe how the application of this model, formulated in the thin-interface limit, can help to explain the long-standing phenomenon of spontaneous grain refinement in deeply undercooled melts. We find that at intermediate undercoolings the operating point parameter, σ^*, may collapse to zero, resulting in the growth of non-dendritic morphologies such as doublons and 'dendritic seaweed'. Further increases in undercooling then lead to the re-establishment of stable dendritic growth. We postulate that remelting of such seaweed structures gives rise to the low undercooling instance of grain refinement observed in alloys.

1. Introduction

Dendritic solidification is a subject of enduring interest within the scientific community, both because dendrites are a prime example of spontaneous pattern formation and due to their pervasive influence on the engineering properties of metals. One long-standing problem with regard to the dendritic solidification of metals has been that of spontaneous grain refinement in undercooled pure melts, first reported to occur in Ni by Walker [1] in 1959. At a well defined undercooling, $\Delta T^* = 140$ - 150 K, Walker observed an abrupt transition from a coarse columnar grain structure to a fine equiaxed structure, with a reduction in grain size of at least one order of magnitude. Similar behaviour was found in Co, with a value for ΔT^* of ≈ 180 K. This effect has subsequently been identified in other pure metals [2, 3, 4] and in a range of alloy systems [5, 6, 7, 8, 9, 10, 11, 12], in which a more complex evolutionary sequence is often observed as the undercooling is increased. At low undercooling an initially columnar growth pattern is observed which gives way to an equiaxed grain structure as the undercooling is increased beyond a critical value, ΔT_1^*. At yet higher undercooling a second region of columnar growth is observed which, in most systems, is replaced by a second region of equiaxed growth, the critical undercooling for this second grain refinement transition being ΔT_2^*.

In many of the systems in which spontaneous grain refinement is observed simultaneous measurement of the dendrite growth velocity has also been undertaken. As first demonstrated by Willnecker *et al.* [2] in levitation melted Ni, such measurements reveal an apparent correlation between the velocity-undercooling relationship and grain refinement. Below ΔT^*, growth velocity can be adequately represented by current dendrite growth models, with $V \propto \Delta T^\beta$, $\beta > 1$. Above ΔT^* the velocity-undercooling relationship is approximately linear. In alloy systems this transition is usually observed to occur coincident with the ΔT_2^* transition. In most cases the transition to a grain refined microstructure at the lower undercooling, ΔT_1^*, does not appear to have any obvious signature in the velocity undercooling curve. Moreover, differences between the as-solidified microstructures of the two grain refined regions indicate that there may be subtle differences in the underlying grain refinement mechanism. In the grain refined structures found at high undercooling the solute segregation pattern is spheroidal about a small dendrite fragment, whereas in grain refined materials formed at low undercooling some of the grains appear to contain small, equiaxed structures [13].

The origins of the effect are controversial. Early theories suggested a range of mechanisms which included nucleation ahead of the solidification front induced by the pressure pulse associated with solidification [1], recrystallisation, either during or immediately after solidification [14, 15], or the role of minor solute additions [16, 17, 18]. However, more recent theories regarding the origin of this phenomenon tend to invoke fragmentation of the primary crystal, either during [19, 20] or immediately after [21, 22] recalescence. We have argued previously that the occurrence of a discontinuity in the velocity-undercooling curve [2, 23] coincident with the onset of grain refinement would tend to suggest the former of these possibilities is more likely, as would an apparent change in the morphology of the growth front revealed by high-speed imaging [24]. Specifically, we have suggested that a tip-splitting instability at the dendrite tip leads to the growth of an unstable 'dendritic seaweed' structure, which subsequently remelts to give the observed grain-refinement. Phase-field modelling of the thermally controlled solidification of a pure material at high undercooling [25] appear to support this suggestion, as does the observation, in deeply undercooled ultra-high purity Cu, of a 'frozen in' seaweed morphology [26] above some critical undercooling. Moreover, above this critical undercooling a discontinuity in the velocity-undercooling curve was observed identical in character to that found in materials that display spontaneous grain refinement.

However, the dendritic fragmentation model of Schwarz *et al.* [21, 22] is perhaps still the most widely accepted model for spontaneous grain refinement. The model postulates that two characteristic timescales can be defined for dendritic growth from an undercooled melt. The first is the breakup time, τ_{bu}, which is the time required for fragmentation of the side branches due to remelting and Rayleigh instability and is a monotonic function of the dendrite trunk radius, with small radii giving short breakup times. If, as seems likely from observations of dendritic growth in transparent systems such as xenon [27], dendrites are self-similar when scaled by the tip radius, ρ, this is equivalent to τ_{bu} being a monotonic function of ρ. The second timescale is the plateau time, τ_{pl}, which is the time the melt remains at, or around, the melting temperature during recalescence and is determined by the macroscopic heat extraction rate. The theory postulates that grain refinement occurs when $\tau_{bu} < \tau_{pl}$, which corresponds to the tip radius being below some critical value, ρ^*, determined by the heat extraction rate. Consequently, if it is assumed that ρ varies with undercooling as predicted by marginal stability theory [28, 29] the model appears to offer a natural explanation for there being a single grain refinement transition in pure materials and two transitions in alloy systems (see Fig. 1).

The dependence of the dendrite tip radius upon undercooling as predicted by marginal stability theory for alloy systems, with its characteristic local minimum followed by a local maximum, has very much been a cornerstone of rapid solidification theory for the past 20 years. However, since the advent of microscopic solvability theory [30, 31] there can be considered to be no theoretical basis for marginal stability theory and the experimental evidence in support of the existence of either a local minimum, or a local maximum, in the tip radius is scant. Transparent analogue casting alloys, such as succinonitrile-acetone, in which direct measurement of the dendrite tip radius is possible [32], can only be undercooled by very small amounts so that the predicted undercooling range in which a local minimum might be observed is not accessible. In metallic systems only an indirect estimate of the tip radius is possible, generally by assuming that some characteristic microstructural length-scale, such as the grain size or dendrite trunk radius, where observable, scales as a constant multiple of the tip radius. However, although there is plentiful evidence of an initial decrease in microstructural length scale in the low undercooling region, it has proved almost impossible to make a continuous extension of such an analysis into the high undercooling regime where a local minimum might be inferred.

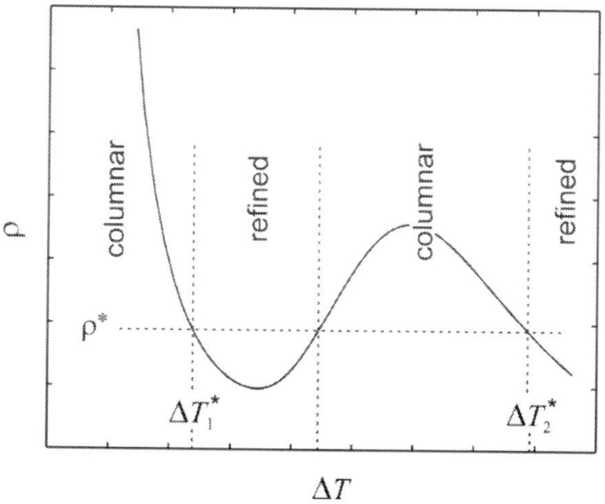

Figure 1. Schematic representation of the radius of a dendrite growing under coupled thermo-solutal control as predicted by marginal stability theory and how this related to the Schwarz model for grain refinement. It is assumed that for a particular macroscopic heat extraction rate there will be a critical radius, below which remelting of the dendrite will occur. If the form of the radius curve obtained from marginal stability is correct this would lead to two grain refinement transitions.

Recently though it has become possible to use quantitative phase-field modelling to simulate the growth of dendrites under coupled thermo-solutal control [33, 34], breaking the previous reliance on marginal stability models for estimating the tip radius in alloy systems where growth is sufficiently rapid that the isothermal approximation is no longer valid. Such coupled thermo-solutal simulations are extremely computationally challenging due to the severe multi-scale nature of the problem and it is only with the application of advanced numerical techniques such as adaptive meshing, implicit time-stepping and efficient multigrid solvers [35] that investigations into dendritic growth in undercooled alloy systems have become feasible over significant regions of the available parameters space, including at high Lewis number [36] (Lewis number, *Le*, is the ration of the thermal to solutal diffusivity, α/D). In fact, such studies appear to indicate that, although the occurrence of a local minimum in ρ as the undercooling is increased is indeed ubiquitous in the solidification of alloy systems, as predicted by marginal stability, a subsequent local maximum in ρ at yet higher undercooling is not [37]. In fact, in a systematic study of the calculated radius-undercooling behaviour of alloy systems as a function of alloy concentration, partitioning strength and Lewis number [38] were unable to identify any part of the studied parameter space in which a maximum in ρ could be observed as the undercooling was increased.

These calculations of the dendrite tip radius in undercooled alloy systems present a potentially serious problem for the accepted model of spontaneous grain refinement in alloy systems. If the radius does not display a local maximum as the undercooling is increased, then it is difficult to reconcile how a break-up time that scales monotonically with tip radius can predict two region of grain refined microstructure, nor indeed why the growth of grain refined structures above ΔT_1^* should give way to columnar growth as the undercooling is increased. However, these calculations also revealed that the tip selection parameter, σ^*, did show this pattern of local minimum followed by local maximum as the undercooling were increased. In fact, qualitatively, the similarities between the behaviour of σ^* as predicted by the phase-field model and the form of the curve shown in Fig. 1 are remarkable. In this paper we explore whether this behaviour observed in σ^* is related to spontaneous grain refinement.

2. Description of the Model

The model adopted here is based upon that of [33] in which, following non-dimensionalization against characteristic length and time scales, W_0 and τ_0, the evolution of the phase-field, ϕ, and the dimensionless concentration and temperature fields U and θ are given by

$$A^2(\psi)\left[\frac{1}{Le}+Mc_\infty\left[1+(1-k_E)U\right]\right]\frac{\partial\phi}{\partial t}=\nabla\cdot\left(A^2(\psi)\nabla\phi\right)+\phi(1-\phi^2)$$
$$-\lambda(1-\phi^2)^2(\theta+Mc_\infty U)-\frac{\partial}{\partial x}\left(A(\psi)A'(\psi)\frac{\partial\phi}{\partial y}\right)+\frac{\partial}{\partial y}\left(A(\psi)A'(\psi)\frac{\partial\phi}{\partial x}\right) \quad (1)$$

$$\left(\frac{1+k_E}{2}-\frac{1-k_E}{2}\phi\right)\frac{\partial U}{\partial t}=\nabla\cdot\left(D\frac{1-\phi}{2}\nabla U+\frac{1}{2\sqrt{2}}\left|1+(1-k_E)U\right|\frac{\partial\phi}{\partial t}\frac{\nabla\phi}{|\nabla\phi|}\right)+\frac{1}{2}\left(\left|1+(1-k_E)U\right|\frac{\partial\phi}{\partial t}\right) \quad (2)$$

$$\frac{\partial\theta}{\partial t}=\alpha\nabla^2\theta+\frac{1}{2}\frac{\partial\phi}{\partial t} \quad (3)$$

where, for 4-fold growth, $A(\psi) = 1 + \varepsilon.\cos(4\psi)$, d_0 is the chemical capillary length, k_E is the partition coefficient L and c_p are the latent and specific heats respectively and λ is a coupling parameter given by $\lambda = D/a_2 = a_1W_0/d_0$ with a_1 and a_2 taking the values $5\sqrt{2}/8$ and 0.6267 respectively [39]. U and θ are related to physical concentration, c, and temperature, T, via

$$U=\frac{1}{1-k_E}\left(\left(\frac{2c/c_\infty}{1+k_E-(1-k_E)\phi}\right)-1\right) \text{ and } \theta=\frac{\Delta T-mc_\infty}{L/c_p} \quad (4)$$

where m is the slope of the liquidus line, which has dimensionless form

$$M=\frac{|m|(1-k_E)}{L/c_p} \quad (5)$$

The governing equations are descritized using a finite difference approximation based upon a quadrilateral, non-uniform, locally-refined mesh with equal grid spacing in both directions. This allows the application of standard second order central difference stencils for the calculation of first and second differentials, while a compact 9-point scheme has been used for Laplacian terms, in order to reduce the mesh induced [40] anisotropy. To ensure sufficient mesh resolution around the interface region and to handle the extreme multi-scale nature of the problem at high Lewis number local mesh refinement (coarsening) is employed when the weighted sum of the gradients of ϕ, U and θ exceeds (falls below) some predefined value.

It has been shown elsewhere that if explicit temporal descretization schemes are used for this problem the maximum stable time-step is given by $\Delta t \leq Ch^2$, where $C = C(\lambda, Le, \Delta T)$, with C varying from ≈ 0.3 at $Le = 1$ to $C \leq 0.001$ at $Le = 500$ [41], leading to unfeasibly small time-steps at high Lewis number. Consequently, an implicit temporal descretization is employed here based on the second order Backward Difference Formula with variable time-step.

When using implicit time discretisation methods it is necessary to solve a very large, but sparse, system of non-linear algebraic equations at each time-step. Multigrid methods are among the fastest available solvers for such systems and in this work we apply the non-linear generalization known as FAS (full approximation scheme [42]). The local adaptivity is accommodated via the multilevel algorithm originally proposed by Brandt [43]. The interpolation operator is bilinear while injection is used for the restriction operator. For smoothing the error we use a fully-coupled nonlinear weighted Gauss-Seidel iteration where the number of pre- and post-smoothing operations required for optimal convergence is determined empirically [41]. Full details of the numerical scheme are given in [35, 44].

3. Results

Figure 2 shows the variation of (a) the dendrite tip velocity, V, and (b) the equivalent parabolic [34] tip radius, ρ, as a function of the dimensionless undercooling, Δ, for three values of the anisotropy parameter ε, namely 0.020, 0.015 and 0.010. Fig. 3 shows the effective radius selection parameter, σ*, calculated using the methodology described in [34]. Except for the anisotropy parameter all three sets of simulations were run under identical conditions with $Mc_\infty = 0.05$, $Le = 200$, $\lambda = 1$ and $k_E = 0.3$. All simulations are run on a domain of $\Omega = [-1600{:}1600]^2$ using a maximum of 12 levels of refinement, giving a minimum mesh size, h, of 0.78. This is equivalent, were a uniform mesh to have been used, of a mesh size which is $2^{12} \times 2^{12}$. For each parameter set simulations were run over the undercooling range Δ = 0.2-0.8. Below Δ = 0.2 growth is very slow leading to excessive computation times while above Δ = 0.8 the requirement that $W_0V/D << 1$ is not satisfied.

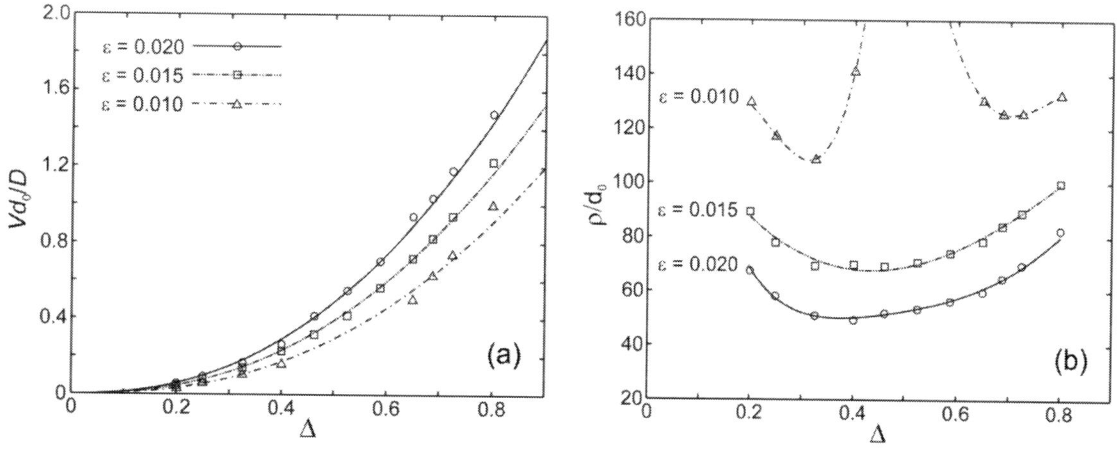

Figure 2. The dendrite growth velocity (a) and tip radius (b) as predicted by the phase-field model as a function of undercooling and anisotropy strength.

At $\varepsilon = 0.020$ the results are as previously reported, with the velocity increasing monotonically with undercooling and displaying, to a very good approximation, a power-law dependence with exponent ≈ 2.3. The radius displays a minimum at intermediate undercooling and subsequently increases at high undercooling. This results in an operating point parameter, σ*, which initially decreases with increasing undercooling before passing through a local minimum to increase with undercooling. At yet higher undercooling σ* passes through a local maximum so at the highest undercoolings studied σ* is decreasing rapidly with increasing undercooling. We have previously argued [38] that this behaviour can be rationalised if the competition between solutal and thermal control of dendritic growth is manifest not in the tip radius, as for instance is evident in marginal stability models of dendrite growth [29], but in the operating point parameter, σ*. We also note that in the limit Δ → 0, we find σ* ≈ 0.07, in good agreement with [34].

If we now consider reducing the anisotropy strength from $\varepsilon = 0.020$ to $\varepsilon = 0.015$ the effect on the radius and velocity is broadly in line with what we might expect. The radius follows the same trend as at the higher anisotropy level, but with a larger radius being observed at all undercoolings. This is to be expected as σ* is a monotonically decreasing function of ε. The velocity is correspondingly reduced, in line with the expectation that the Peclet number is only very weakly dependent upon ε. σ* is reduced at all undercoolings and, like the tip radius, displays the same general form as at the higher anisotropy level, although in this case we note that the difference between the minimum and maximum values is also significantly reduced.

Finally we consider a further reduction in the anisotropy strength to $\varepsilon = 0.010$, wherein a significant change in behaviour is observed. If we consider the behaviour of σ^* first we observe that, in line with expectation, the value of σ^* as $\Delta \to 0$ is reduced, and as with the curves for $\varepsilon = 0.020$ and 0.015, the value of σ^* initially decreases as Δ is increased. However, for undercooling between $\Delta = 0.4625$ and 0.5875 stable dendritic growth was not observed, with the solid nuclei used to seed solidification initially developing a preferred four-fold growth morphology but then experiencing a bifurcation at the tip which ultimately leads to a tip-splitting instability in the growing crystal. As a consequence of this tip splitting instability, in the undercooling range $\Delta = 0.4625$ - 0.5875 neither a value for ρ nor σ^* could be obtained. At undercoolings above $\Delta = 0.5875$ stable dendritic growth is once again established. From the values of σ^* either side of this undercooling range it appears that this growth instability is consistent with a collapse to zero in the value of σ^*. For undercoolings either side of this unstable range the measured tip radius appears to be abnormally large, which is also consistent with the hypothesised collapse of σ^*, with small σ^* giving rise to large values of ρ. In the unstable growth regime we have not attempted to determine a characteristic growth velocity as the instantaneous growth rate is subject to significant fluctuations as the morphology of the tip changes. However, once stable dendritic growth is re-established in the high undercooling regime, we note that the measured velocities are consistent with a power-law relationship with the same exponent as in the low undercooling regime.

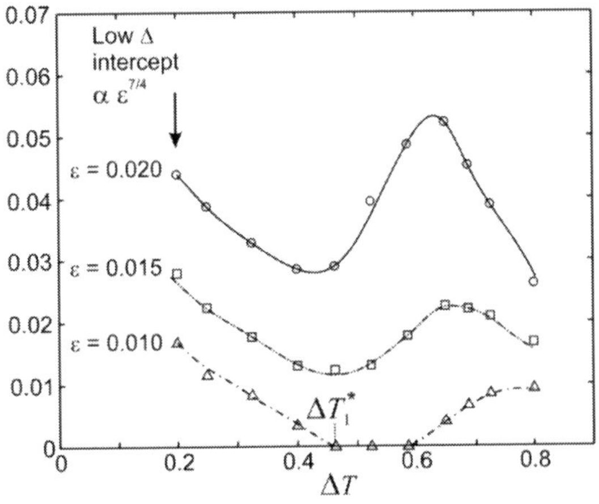

Figure 3. Calculated operating point parameter, σ^*, as predicted by the phase field model as a function of undercooling and anisotropy strength. At $\varepsilon = 0.01$ stable dendritic growth is not observed in the undercooling range $0.4625 \leq \Delta \leq 0.5875$. Instead a tip-splitting instability leads to the growth of 'dendritic-seaweed'. It is suggested that remelting of this seaweed structure gives rise to the grain-refined microstructures observed at low undercoolings.

4. Discussion

The breakdown of dendritic growth in phase-field simulations of solidification at high undercooling has been noted by a number of groups and has generally been attributed to a competition between capillary and kinetic anisotropies, with capillary effects dominating at low undercooling and kinetic effects dominating at high undercooling. In the case where these anisotropies are oppositely directed, doublon or dendritic seaweed morphologies, which are characteristic of growth at low anisotropy, may be observed when the competing effects are of similar magnitude. However, we do not believe that this is the case here. Firstly, the model has been constructed in the thin-interface limit, wherein the choice of parameters adopted here should eliminate all kinetics from the model and, secondly, the stable dendritic growth observed in the high undercooling regime has the *same orientation* as in the low undercooling regime (i.e. in both cases the dendrite arms are oriented towards the sides of the computational domain, which is the direction of the applied capillary anisotropy). This is inconsistent with an oppositely directed kinetic anisotropy which, if present, would result in growth directed towards the corners of the computational domain. By setting $\varepsilon = 0$ the model has also been tested at all undercoolings considered here to determine whether there are effects resulting from mesh induced anisotropy, and this appears not to be the case.

For the $\varepsilon = 0.010$ case given in Fig. 3, the maximum in σ* observed at higher anisotropies is quite poorly developed, possibly by virtue of being shifted to somewhat higher undercooling than for the higher anisotropy simulations. Therefore, in the particular case considered here a second region of dendritic seaweed growth was not observed at higher undercoolings within the undercooling range considered. This may be an artefact of the rather low Lewis number ($Le = 200$) used in these simulations. In most metals Le is typically of the order 10^4 while in [38] we report that increasing the Lewis results in a more pronounced maximum in σ* and a steeper decline at high undercooling. It is currently a matter for investigation as to whether or not such simulations do result in a second region of doublon formation, although the simulations become much more computationally intensive as the Lewis number is increased due to the multi-scale nature of the problem.

5. Summary and Conclusions

A phase-field model of model of coupled thermo-solutal growth, constructed in the quantitative thin-interface limit has been used to study the solidification of undercooled melts over a range of undercoolings and anisotropy strengths. At moderate to high anisotropy we observe dendritic growth at all undercoolings. The tip radius initially decreases with undercooling before passing through a local minimum beyond which it subsequently increases with undercooling. However, despite studying undercoolings up to 0.8 of the hypercooling limit we found no cases in which the calculated radius-undercooling curve also displayed a local maximum. This is in distinct contrast to marginal stability theories of dendritic growth in which the radius should decrease with undercooling at very high undercooling. Consequently, if the calculated radius-undercooling curves are typical of the behaviour in metals, it is unlikely a simple decrease in the tip-radius below some limit indicative of remelting could account for spontaneous grain refinement effects in undercooled alloys. Specifically, although the calculated minimum in the radius could give rise to the low-undercooling region of grain-refinement observed in many alloys (ΔT above ΔT_1^* in Fig. 1), it could not account for grain-refinement at high-undercooling (ΔT above ΔT_2^* in Fig. 1), as the radius would be increasing with undercooling.

We also found that at sufficiently low anisotropy a transition of the form dendritic -> doublon/dendritic-seaweed -> dendritic was observed as the undercooling was increased. This can be understood by considering the behaviour of the radius selection parameter, σ*, which we find to be a non-monotonic function of the undercooling. At moderate to high anisotropy, after initially decreasing with increasing undercooling, σ* displays a local minimum, beyond which it increases with further increases in undercooling until, at yet higher undercooling it displays a local maximum. Beyond this second stationary point σ* declines smoothly with undercooling. However, as the anisotropy is decreased σ* is shifted to lower values until a point is reached at which σ* collapses to zero for the intermediate undercoolings for which it had displayed a minimum. It is these regions that correspond to the observed instance of doublon or seaweed growth, rather than the growth of dendrites. We speculate that it is this transition in growth morphology that gives rise to the observed transition to a grain-refined microstructure in alloy melts at low undercooling. Noting the strong similarity between the form of the curves for σ* as a function of undercooling as calculated by the phase-field model and that for radius as a function of undercooling in marginal stability models of solidification we further speculate that grain refinement at high undercooling might be a consequence of a second region of doublon/dendritic seaweed growth resulting from a further drop in σ* at high undercooling. This has not yet been observed in the model but may exist at high Lewis number.

References

[1] Walker J L 1959, in *The Physical Chemistry of Process Metallurgy*, part 2, ed. G. R. St. Pierre, Interscience
[2] Wilnecker R, Herlach D M and Feuerbacher B 1989, *Phys. Rev. Lett.* **62** 2707
[3] Battersby S E, Cochrane R F and Mullis A M 1999, *J. Mater. Sci.* **34** 2049
[4] Dragnevski K, Cochrane R F and Mullis A M 2004, *Mater. Sci. Eng. A*, **375-377** 479
[5] Battersby S E, Cochrane R F and Mullis A M 2000, *J. Mater. Sci.* **35** 1365
[6] Zhang T, Liu F, Wang H F and Yang G C 2010, *Scripta Mater.* **63** 43
[7] Li J F, Zhou Y H and Yang G C 2000, *Mater. Sci. Eng. A* **277** 161
[8] Li J F, Jie W Q ,Yang G C and Zhou Y H 2002, *Acta Mater.* **50** 1797
[9] Zhao S, Li J, Lui L & Zhou Y H 2009, *J. Crystal Growth* **311** 1387
[10] Wang N, Gao J R, Wei B 1999, *Scripta Mater.* **41** 959
[11] Lu Y, Liu F, Yang G C, Wang H P and Zhou Y H 2007, *Mater. Lett.* **61** 987
[12] Lu S Y, Li J F and Zhou Y H 2007, *J. Crystal Growth* **309** 103
[13] Kattamis T Z and Flemings M C 1967, *Mod. Casting* **52** 97
[14] Southin R T and Weston G M 1973, *J. Aust. Inst. Met.* **18** 74
[15] Ovsiyenko D E, Maslov V V and Dneprenko V N 1979, *Russ. Met.* **6** 124
[16] Jones B L and Weston G M 1970, *J. Aust. Inst. Met.* **15** 167
[17] Jones B L and Weston G M 1970, *J. Aust. Inst. Met.* **15** 189
[18] Powell G L F and Hogan L M 1969, *Trans. AIME* **245** 407
[19] Mullis A M and Cochrane R F 1997, *J. Appl. Phys.* **82** 3783
[20] Mullis A M and Cochrane R F 1998, *J. Appl. Phys.* **84** 4905
[21] Schwarz M, Karma A, Eckler K and Herlach D M 1994, *Phys. Rev. Lett.* **73** 1380
[22] Karma A 1998, *Int. J. Non-equilibrium Processing* **11** 201
[23] Cochrane R F, Battersby S E and Mullis A M 2001, *Mater. Sci. Eng.* **A304-306** 262
[24] Matson D 1997, in: *Solidification 1998*, TMS, Warrendale, PA. 233-244, ISBN 0-87339-396-1
[25] Mullis A M and Cochrane R F 2001, *Acta Mater.* **49** 2205
[26] Dragnevski K, Cochrane R F and Mullis A M 2002, *Phys. Rev. Lett.* **89** 215502
[27] Bisang U & Bilgram J H 1996, *Phys. Rev. E* **54** 5309
[28] Lipton J, Glicksman M E and Kurz W 1987, *Metall. Trans. A* **18** 341
[29] Lipton J, Kurz W and Trivedi R 1987, *Acta Metall.* **35** 957
[30] Ben-Jacob E, Goldenfeld N D, Kotliar B G and Langer J S 1984, *Phys. Rev. Lett.* **53** 2110
[31] Langer J S 1986, *Phys. Rev. A* **33** 435
[32] Chopra M A, Glicksman M E and Singh N B 1988, *Metall. Trans. A* **19** 3087
[33] Ramirez J C, Beckermann C, Karma A and Diepers H-J 2004, *Phys. Rev. E* **69** 051607
[34] Ramirez J C and Beckermann C 2005, *Acta Mater.* **53** 1721
[35] Rosam J, Jimack P and Mullis A 2007, *J. Comp. Phys.* **225** 1271
[36] Rosam J, Jimack P K and Mullis A M 2009, *Phys. Rev. E* **79** 030601
[37] Rosam J, Jimack P K and Mullis A M 2008, *Acta. Mater.* **56** 4559
[38] Mullis A M 2011, *Phys. Rev. E* **83** 061601
[39] Karma A 2001, *Phys. Rev. Lett.* **87** 115701
[40] Mullis A M 2006, *Comp. Mater. Sci.* **36** 345
[41] Rosam J 2007, 'A fully implicit, fully adaptive multigrid method for multi-scale phase-field modelling', Ph.D. thesis, University of Leeds
[42] Trottenberg U, Oosterlee C and Schuller A 2001, Multigrid, Academic Press
[43] Brandt A 1977, *Math. Comp.* **31** 333
[44] Mullis A M 2010, *J. Cryst. Growth* **312** 1891

IOP Conf. Series: Materials Science and Engineering **33** (2012) 012110 doi:10.1088/1757-899X/33/1/012110

Phase field simulation of precipitation in a Mg-Al alloy using two techniques of approximation

G M Han [1], Z Q Han [1,4], A A Luo [2], A K Sachdev [2] and B C Liu [1,3]
[1]Key Laboratory for Advanced Materials Processing Technology (Ministry of Education), Department of Mechanical Engineering, Tsinghua University, Beijing 100084, China
[2]Chemical Sciences and Materials Systems Laboratory, General Motors Global Research and Development Center, Warren, MI 48090-9055, USA
[3]State Key Laboratory of Automotive Safety and Energy, Department of Automotive Engineering, Tsinghua University, Beijing 100084, China

E-mail: zqhan@tsinghua.edu.cn

Abstract. In this paper, the precipitation of β-$Mg_{17}Al_{12}$ in aging process of Mg-9%Al (mass fraction) alloy was studied using a diffuse-interface phase field model. In the model, the precipitate phase and the matrix are distinguished by a structural order parameter, and the interface region is assumed to be a mixture of the precipitate phase and the matrix. The mixture composition was calculated using a weighted average method. Two techniques of approximation for the chemical free energy of precipitate phase and matrix were employed, where the variation characteristics of the free energy with the mole fraction of solute were included. In the simulation, the temperature and diffusion mobility coefficient were assumed to be constant. The effects of the solution approximations on the morphology evolution and growth kinetics of the precipitate were discussed. The effects of the interface mobility coefficient on the interface growth kinetics were examined, and the effects of the interface anisotropy between the precipitate phase and the matrix on the morphology of the precipitate were also discussed. It is demonstrated that the solution approximation taking into account the interaction between atoms for a binary system provides a good description for the chemical free energy. The simulation results showed that the precipitate phase growing from a supersaturated solid solution has a plate-like shape, which is in agreement with experimental observations. The growth kinetics of the precipitate phase is significantly influenced by the interface mobility coefficient.

1. Introduction

Magnesium alloys are important lightweight structural materials and have received increasing attention from electronic, automobile and aerospace industries [1]. Mg-Al series alloys are commonly used because of their good castability and high specific strength. According to the Mg-Al binary phase diagram, eutectic β phase dissolves during solution treatment, and forms precipitates during the subsequent aging process. The β phase is a stoichiometric intermetallic compound with a composition of $Mg_{17}Al_{12}$ and an α-Mn-type cubic unit cell structure [2]. Since the β phase is an important

[4] Correspondent: Z Q Han, Ph.D, Associate professor, Department of Mechanical Engineering, Tsinghua University, Haidian District, Beijing 100084, China.

strengthening phase in Mg-Al based alloys, it is critical to model β phase precipitation in order to develop simulation tools for mechanical property prediction of magnesium alloys.

The phase field method is a powerful tool for microstructure simulation, especially for the precipitation process. This method describes microstructure evolution based on a set of conserved fields (e.g., composition field) and non-conserved fields (e.g., structural order parameter field). The microstructure evolution can be predicted by solving a series of governing equations for the fields closely related to energy relaxation. Some precipitation simulations using phase field methods have been reported in literatures. Li et al. [3] investigated the shape of a rhombohedral coherent $Ti_{11}Ni_{14}$ precipitate in a cubic matrix during aging using a time-dependent Ginzburg-Landau kinetic model. Zhu et al. [4] studied the coarsening kinetics of γ' precipitates in a binary Ni–Al alloy by using a 3D phase field model coupled with CALPHAD (CALculation of PHAse Diagram) approaches. Chen et al. [5] simulated the growth and dissolution of α precipitates in Ti–Al–V systems by coupling with CALPHAD thermodynamic data. In general, the microstructure evolution is closely related to energy evolution, so the approximation of free energy plays an important role in phase field modeling. Ginzburg-Landau polynomial model linking with first principle calculations and the solution approximation model using CALPHAD thermodynamic data are two distinctive methods for free energy approximations.

In a binary system, typical approximation models for free energy include ideal solution approximation, regular solution approximation, sub-regular solution approximation and sub-lattice approximation. The selection of free energy model for a phase should be based on the physical and chemical properties of the phase, such as order-disorder transition, crystallography, bonding type, etc. [6].

The aim of the present work is to use the phase field method to investigate the growth kinetics of β-$Mg_{17}Al_{12}$ precipitates during aging Mg-Al alloys. Two approximation techniques of free energy were used for the precipitate phase and the matrix and their effects on the morphology of the precipitate were examined. The effects of the interface mobility coefficient on interface growth kinetics, as well as the effect of interface anisotropy between the precipitate phase and matrix on precipitate morphology, are discussed.

2. Model description

2.1. Phase field model

In this paper, the thermodynamic description proposed by Kim *et al.* [7] was used to simulate the precipitation of β-$Mg_{17}Al_{12}$ during aging. The interface region was assumed to be a mixture of the precipitate and the matrix. The mixture composition was calculated using a weighted average method. For the Mg-Al system, despite the many variants of β precipitates with different orientations reported [2], only the dominant basal variant was chosen in the simulation. The elastic energy was not considered in this model. The precipitate phase and the matrix are distinguished by a structural order parameter η defined as 1 in the precipitate phase and 0 in the matrix.

The total free energy of the system $F(c,\eta,T)$ including chemical free energy and interfacial energy are defined as:

$$F(c,\eta,T) = \int_v \left[f(c,\eta,T) + \frac{\epsilon^2(\varphi)}{2}(\nabla\eta)^2\right] dv \qquad (1)$$

where F is the total free energy, f is the free energy density of the system, T is absolute temperature, $\epsilon(\varphi)$ is the gradient energy coefficient related to anisotropic interfacial energy, and φ is an angle used to describe the interfacial anisotropy.

The chemical free energy density of the system is defined as:

$$f(c,\eta,T) = h(\eta)f^\beta(c_\beta,T) + (1-h(\eta))f^\alpha(c_\alpha,T) + wg(\eta) \qquad (2)$$

where c_α and c_β are the molar fractions of Al atoms in the matrix and β-$Mg_{17}Al_{12}$ precipitate phase respectively. $f^\alpha(c_\alpha, T)$ and $f^\beta(c_\beta, T)$ are the chemical free energy density of the matrix and β-$Mg_{17}Al_{12}$ precipitate phase respectively. $g(\eta)$ is a double-well potential, and w is the height of the double-well potential. $h(\eta)$ is a monotonous function from 0 to 1. In the present work, $g(\eta)$ and $h(\eta)$ are chosen as:

$$h(\eta) = 3\eta^2 - 2\eta^3 \tag{3}$$
$$g(\eta) = \eta^2(1-\eta)^2 \tag{4}$$

In the phase field model, the evolution of conserved fields, such as composition c of the system, is governed by the Cahn-Hilliard equation [8]. The evolution of non-conserved fields, such as structural order parameter, η, is governed by the Allen-Cahn equation [8]:

$$\frac{\partial \eta}{\partial t} = -M(\varphi)\frac{\delta F}{\delta \eta} \tag{5}$$
$$\frac{\partial c}{\partial t} = \nabla \cdot \left(\frac{D(\eta,T)}{f_{cc}} \nabla(f_c)\right) \tag{6}$$

where f_c and f_{cc} are the first order and second order derivatives of f with respect to composition c, respectively. $D(\eta, T)$ is solute diffusivity, and $M(\varphi)$ is the interface mobility coefficient with anisotropy.

The compositions of α and β phase satisfy the following constraint conditions [7]:

$$c = (1 - h(\eta))c_\alpha + h(\eta)c_\beta \tag{7}$$
$$\frac{\partial f^\alpha(c_\alpha,T)}{\partial c_\alpha} = \frac{\partial f^\beta(c_\beta,T)}{\partial c_\beta} = \frac{\partial f(c,\eta,T)}{\partial c} \tag{8}$$

The first condition means that the composition of the system is calculated using a weighted average method with the weight $(1 - h(\eta))$ for c_α and $h(\eta)$ for c_β. The second constraint is a condition of equal chemical potential.

According to Kim *et al* [7], the phase field equations (5) and (6) can be simplified by using the expressions of f_c and f_{cc}, as follows:

$$\frac{\partial \eta}{\partial t} = M(\varphi)\left(\epsilon^2(\varphi)\nabla^2\eta - f_\eta\right) \tag{9}$$
$$\frac{\partial c}{\partial t} = \nabla \cdot (D(\eta,T)\nabla c) + \nabla \cdot \left(D(\eta,T)h'(\eta)(c_\alpha - c_\beta)\nabla\eta\right) \tag{10}$$

where,

$$f_\eta = \frac{\partial f(c,\eta,T)}{\partial \eta} = h'(\eta)\left(f^\beta(c_\beta,T) - f^\alpha(c_\alpha,T)\right) + h'(\eta)(c_\alpha - c_\beta)\frac{\partial f^\alpha(c_\alpha,T)}{\partial c_\alpha} + wg'(\eta) \tag{11}$$

$h'(\eta)$ and $g'(\eta)$ are the first order derivatives of $h(\eta)$ and $g(\eta)$ with respect to the structural order parameter η respectively.

2.2. *Phase field model parameters*

In the phase field equations (9), (10) and (11), there are four parameters to be determined, i.e. the interface mobility coefficient, $M(\varphi)$, the gradient energy coefficient, $\epsilon(\varphi)$, the height of the double-well potential, w, and the solute diffusivity, $D(\eta, T)$. The relationship between the gradient energy coefficient, $\epsilon(\varphi)$, the interface energy $\sigma(\varphi)$, and the interface width 2λ was established under one-dimension equilibrium condition [7]:

$$\epsilon(\varphi) = \sqrt{\frac{6\lambda\,\sigma(\varphi)}{2.2}} \tag{12}$$

For w, we can similarly obtain its expression

$$w(\varphi) = \frac{6.6\sigma(\varphi)}{\lambda} \tag{13}$$

The interface mobility coefficient $M(\varphi)$ is obtained using thin-interface limit analysis [7]:

$$M(\varphi) = \frac{\sigma(\varphi)\, D(\eta,T)\sqrt{2w(\varphi)}}{\epsilon^3(\varphi)\xi} \tag{14}$$

$$\xi = \frac{RT}{V_m}\left(c_\alpha^e - c_\beta^e\right)^2 \int_0^1 \frac{h(\eta)[1-h(\eta)]}{[1-h(\eta)]c_\alpha^e(1-c_\alpha^e)+h(\eta)c_\beta^e\left(1-c_\beta^e\right)}\frac{d\eta}{\eta(1-\eta)} \tag{15}$$

where R is the gas constant, V_m is the molar volume, c_α^e and c_β^e are the equilibrium compositions in the matrix and precipitate phase, respectively. A detailed description of the interface anisotropy can be found in reference [9]. For solute diffusivities in the matrix and precipitate, we assume that they are the same and only depend on temperature.

2.3. Approximation of chemical free energy

Two techniques for free energy approximation were used in the simulation. In the first technique called ideal solution approximation, the ideal solution approximation was adopted for both the matrix and the precipitate. In the second technique, a regular solution approximation was applied to the matrix while a sub-lattice approximation was employed for the precipitate phase, which we called mixing approximation.

In the ideal solution approximation, the free energy of a phase depends on the composition and temperature under atmospheric pressure [6]. For the two phases in the Mg-Al system, the general formulas are given as:

$$G_m^\alpha = c_\alpha {}^0G_{Al}^\alpha + (1-c_\alpha)\, {}^0G_{Mg}^\alpha + RT(c_\alpha ln c_\alpha + (1-c_\alpha)\, ln(1-c_\alpha)) \tag{16}$$

$$G_m^\beta = (1-c_\beta){}^0G_{Al}^\beta + c_\beta {}^0G_{Mg}^\beta + RT\left((1-c_\beta)ln(1-c_\beta) + c_\beta ln c_\beta\right) \tag{17}$$

where G_m^α and G_m^β are the molar free energy of the matrix and precipitate, respectively, ${}^0G_{Al}^\phi$ and ${}^0G_{Mg}^\phi$ are the molar free energy of pure element Al and Mg in phase ϕ, respectively, and their formulas can be found in reference [10].

Zhong et al. [11] gave an Al-Mg thermodynamic model based on the first principles calculation. For the matrix, a regular solution approximation is used, and the excess free energy is expressed in the Redlich–Kister polynomial [12]. For the precipitate phase, they chose the sub-lattice approximation with consideration of the three-sublattice model, i.e. $(Mg)_5(Al, Mg)_{12}(Al, Mg)_{12}$ as reference [13]. The molar free energies of the matrix and precipitate phase are as follows:

$$G_m^\alpha = c_\alpha {}^0G_{Al}^\alpha + (1-c_\alpha)\, {}^0G_{Mg}^\alpha + RT(c_\alpha ln c_\alpha + (1-c_\alpha)\, ln(1-c_\alpha)) + {}^{ex}G_m^\alpha \tag{18}$$

$${}^{ex}G_m^\alpha = c_\alpha(1-c_\alpha)\left({}^0L_{Al,Mg}^\alpha + {}^1L_{Al,Mg}^\alpha(2c_\alpha - 1) + {}^2L_{Al,Mg}^\alpha(2c_\alpha - 1)^2\right) \tag{19}$$

$$G_m^\beta = {}^{ref}G_m^\beta + {}^{mix}G_m^\beta + {}^{ex}G_m^\beta \tag{20}$$

where ${}^{ex}G_m^{\alpha}$ is the excess free energy, ${}^{j}L_{Al,Mg}^{\alpha}$ is the jth binary interaction parameter of the matrix [11], and ${}^{ref}G_m^{\beta}$, ${}^{mix}G_m^{\beta}$ and ${}^{ex}G_m^{\beta}$ are the reference free energy, the mixture free energy and the excess free energy of the precipitate phase, respectively [11].

3. Results and discussion

In this work, phase field equations (9), (10) and (11) and the two approximation techniques of free energy for the precipitate phase and the matrix were developed under the constraint condition equations (7) and (8) to simulate a single β precipitate on the basal plane at an aging temperature of 441K without considering the elastic energy at this time. At the beginning of the simulation, a round particle was placed as nuclei in the center of the simulation domain divided into 500×500 cells. The initial conditions were set as η=1.0 and c=0.4138 in the nuclei place while η=0.0 and c=0.0827 in other places. The interfacial energy for the semi-coherent interface parallel to the basal plane was assumed to be 60mJ/m^2. The interface thickness is $2\lambda=6\Delta x$, where Δx is the cell size. In the simulation, the composition of the precipitate (c=0.4138) was obtained from the equilibrium phase diagram of Mg-Al system, while the composition of the matrix was assumed to be the composition of Mg-9%Al, i.e., c=0.0827.

At a given temperature and atmospheric pressure, the chemical free energy only depends on the phase composition. The free energy versus molar fraction of Al for the two approximations is shown in Figure 1. It can be seen that the free energies of the precipitate phase are quite different in the two approximations. In Figure 1(a), the profile of the free energies of the precipitate phase and matrix are similar, and the free energies of the precipitate phase and matrix are just calculated results of the ideal solution approximation model, which are just for making a comparison. In Figure 1(b), the free energy curve of the matrix does correspond to Eq. 18 - 19. It looks like a straight line just because of the coordinate scale we chose to cove the free energy value of the precipitate. According to reference [2], β-$Mg_{17}Al_{12}$ precipitate is an intermetallic compound with a stoichiometric composition, and its free energy is defined only at c=0.4138, however, for the sake of numerical computation, the free energy of the precipitate in Figure 1(b) was treated as a parabolic approximation within a very narrow composition band. The mixing approximation technique is expected to provide a good description for the chemical free energy of precipitate phase.

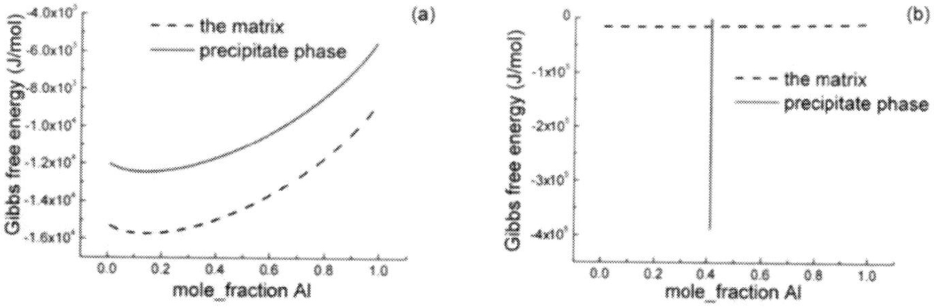

Figure 1. Two approximations for the chemical free energy of the matrix and precipitate, (a) the ideal solution and (b) the mixing approximation.

Figure 2 shows the profile of the order parameter for the two approximations. For the ideal solution approximation, the simulation results (Figure 2(a)) indicated that the precipitate does not grow significantly. For the mixing approximation, the results showed that the interface between the precipitate and the matrix is a quite thin band with a very steep profile of the order parameter. It's not an absolute sharp transition. Such a steep transition of the order parameter in Figure 2(b) is just because the free energy of the precipitate is defined only at c=0.4138. Figure 3 shows a comparison of the simulation results by using the mixing approximation with experimental results. It can be seen that the simulated morphology of the precipitate (Figure 3(a)) is in agreement with reported experimental observations [2] (Figure 3(b)).

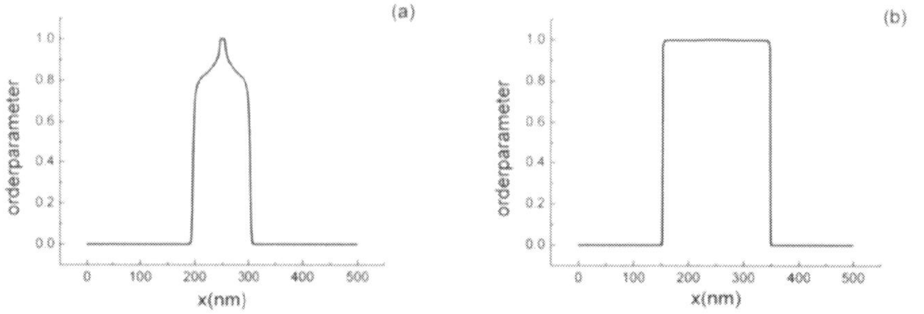

Figure 2. Order parameter profile along the length direction of the precipitate, (a) using the ideal solution approximation and (b) using the mixing approximation.

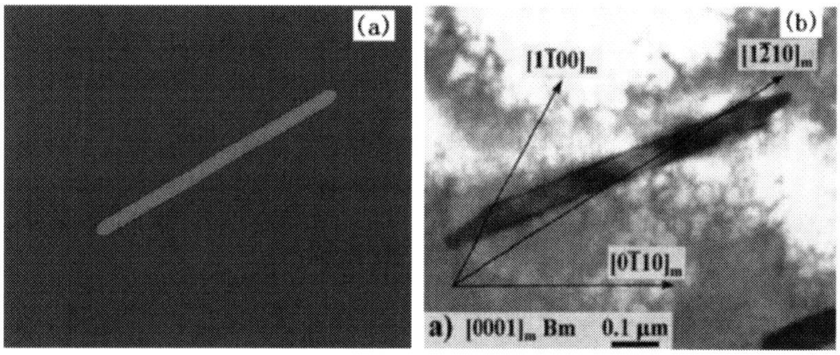

Figure 3. Simulation results of a single β-$Mg_{17}Al_{12}$ precipitate (a) and experimental observation [2] (b).

The morphology of the precipitate is closely related to interface anisotropy. Figure 4 shows the effect of interface anisotropy on growth kinetics, which clearly shows the precipitate to demonstrate isotropic growth without interface anisotropy. In the simulation of Figure 4(a), the interfacial energy for the semi-coherent interface was assumed to be $60mJ/m^2$, and a detailed approach for the interface anisotropy can be found in reference [9], while for Figure 4(b), the interfacial energy in all directions was assumed to be $60mJ/m^2$.

The effect of interface mobility coefficient on interface growth kinetics is shown in Figure 5. It can be seen that when the value of the interface mobility coefficient is very small, the precipitate almost doesn't grow, while it grows rapidly when the interface mobility coefficient is given a large value.

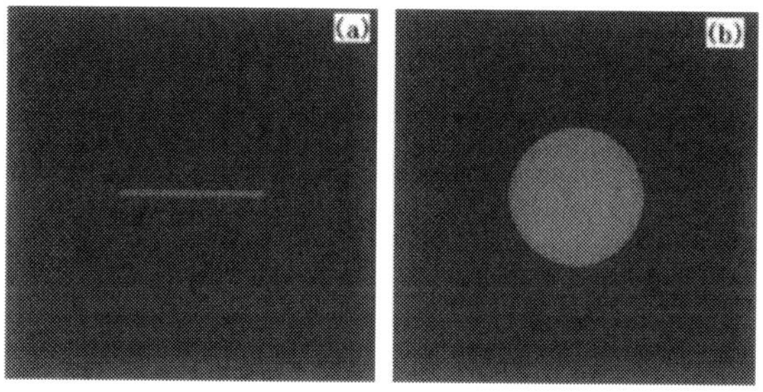

Figure 4. The effect of interface anisotropy on precipitate growth kinetics, (a) with interface anisotropy and (b) without interface anisotropy.

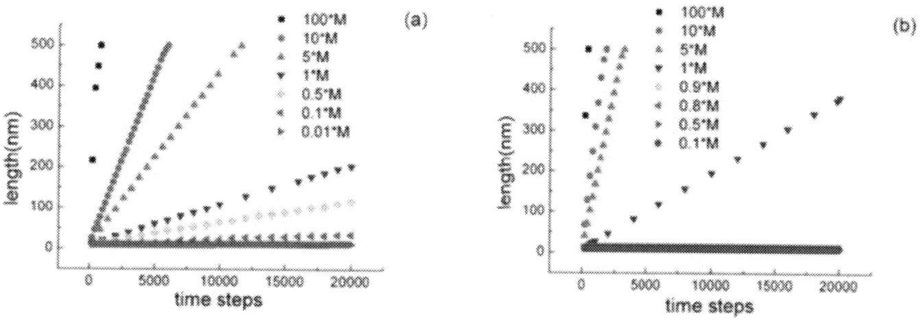

Figure 5. The effect of interface mobility coefficient on interface growth kinetics, (a) the ideal solution approximation and (b) the mixing approximation, time step 1.0e-4s.

4. Conclusions

The precipitation of β-$Mg_{17}Al_{12}$ during aging of Mg-9%Al (mass fraction) alloy was studied using a diffuse-interface phase field model. Two techniques for free energy approximation were used to simulate the precipitation of β-$Mg_{17}Al_{12}$. An ideal solution approximation was used for both the matrix and precipitate phase in the first technique, while in the second technique a regular solution approximation was used for the matrix while a sub-lattice approximation was employed for the precipitate phase, which we called the mixing approximation. The results demonstrated that the mixing approximation provides a good description for the free energy of the precipitate phase as a stoichiometric compound. Interface anisotropy has a major impact on the growth morphology of the precipitate phase. The precipitate phase showed isotropic growth characteristics without interface anisotropy. The effect of the interface mobility coefficient on the interface growth kinetics of the precipitate was also investigated. The results showed that the interface mobility coefficient affects the growth of the precipitate significantly.

Acknowledgements

This work is funded by the National Natural Science Foundation of China (Grant No. 51175291), the General Motors Global Research and Development Center, Tsinghua University Initiative Scientific Research Program, and the MOST (Ministry of Science and Technology) of China under the contract of No.2010DFA72760. The authors would also like to appreciate Dr. Ruijie Zhang for his valuable discussion and help in the research.

References

[1] Mordike B L and Ebert T 2001 *Mater. Sci. and Eng.* A **302** 37
[2] Celotto S 2000 *Acta Mater.* **48** 1775
[3] Li D Y and Chen L Q 1997 *Acta Mate.* **45** 2435
[4] Zhu J Z, Wang T, Ardell A J, Zhou S H, Liu Z K and Chen L Q 2004 *Acta Mater.* **52** 2837
[5] Chen Q, Ma N, Wu K S and Wang Y Z 2004 *Scripta Mater.* **50** 471
[6] Lukas H, Fries S G and Sundman B 2007 *Computational Thermodynamics: The Calphad Method* (London: Cambridge University Press)
[7] Kim S G, Kim W T and Suzuki T 1999 *Phys. Rev.* E **60** 7186
[8] Chen L Q 2002 *Annu. Rev. Mater. Res.* **32** 113
[9] Hu S Y 2004 *models of microstructure evolution in a system with elastic inhomogeneity and defects* (the Pennsylvania State University)
[10] Dinsdale A T 1991 *CALPHAD* **15** 317
[11] Zhong Y, Yang M, Liu Z K 2005 *CALPHAD* **29** 303
[12] Redlich O, Kister A T 1948 *Ind. Eng. Chem.* **40** 345
[13] Ansara I, Chart T G, Guillermet A F, Hayes F H, Kattner U R, Pettifor D G, Saunders N and Zeng K 1997 *CALPHAD* **21** 171

IOP Conf. Series: Materials Science and Engineering **33** (2012) 012111 doi:10.1088/1757-899X/33/1/012111

Phase-field modeling of the dendrite orientation transition in Al-Zn alloys

Jonathan Friedli, Paolo Di Napoli, Michel Rappaz and Jonathan A. Dantzig

Laboratoire de simulation des matériaux, Institut des matériaux, Ecole Polytechnique Fédérale de Lausanne, Station 12 - 1015 Lausanne, Switzerland

E-mail: `jonathan.friedli@epfl.ch`

Abstract. With a few exceptions, phase-field simulations of dendritic growth in cubic materials have been modeled using simple expressions for the interfacial energy anisotropy and with strong anisotropy. However, recent experimental results show that the Dendrite Orientation Transition (DOT) observed in Al-Zn alloys by Gonzales and Rappaz [Met. Mat. Trans. A37 (2006) 2797] occurs at weak anisotropy, and modeling these results requires at least two anisotropy parameters. In the present work, we solve the corresponding phase-field model on an adaptive grid, after measuring and compensating for the grid anisotropy. A systematic scan of equiaxed growth simulations was performed in the range of the anisotropy parameter space where the transition is expected. We find separate domains of existence of $\langle 100\rangle$ and $\langle 110\rangle$ dendrites, similar to those previously reported by Haxhimali *et al.* [Nat. Mat. 5 (2006) 660] for pure materials. In the so-called hyperbranched regime, lying between the $\langle 100\rangle$ and $\langle 110\rangle$ regions, we observe a competition between $\langle 100\rangle$ and $\langle 110\rangle$ growth directions, but no seaweed structures. Directional solidification simulations showed the stabilizing effect of the thermal gradient on the twofold splitting of $\langle 110\rangle$ dendrites, and the importance of the choice of anisotropy parameters. We also found a strong dependence between the orientation of the crystal axes with respect to the thermal gradient and the actual growth direction. Finally, 3-dimensional seaweed microstructures were modeled for the first time, demonstrating that this pattern is a result of not only the values of anisotropy parameters, but also a consequence of directional solidification.

1. Introduction

It is commonly taught that dendrites in cubic metals grow along $\langle 100\rangle$ directions, since these directions correspond to maxima of $\gamma_{s\ell}$. Although this is indeed the case in many systems, such as steel, Ni- base superalloys, and Cu-base alloys, aluminum alloys display a wider variety of dendrite orientations and morphologies [1–6]. In particular, it has been shown recently that Al-Zn alloys experienced a Dendrite Orientation Transition (DOT) from $\langle 100\rangle$ to $\langle 110\rangle$ when the zinc composition is increased [7, 8]. This was attributed to a modification of the weak anisotropy of the solid-liquid interfacial energy of aluminum by the strongly anisotropic zinc solute element [9]. In this work, we use phase-field simulations to further explore the role of interfacial energy anisotropy on the selection of dendrite orientation.

We express the interfacial energy $\gamma_{s\ell}$ as an expansion in cubic harmonics [10], given by

$$\gamma_{s\ell}(\vec{n}) = \gamma_{s\ell}^0 \left[1 + a_1\left(Q - \frac{3}{5}\right) + a_2\left(3Q + 66S - \frac{17}{7}\right) + \dots\right] \qquad (1)$$

IOP Conf. Series: Materials Science and Engineering **33** (2012) 012111 doi:10.1088/1757-899X/33/1/012111

where $\vec{n}$ is the vector normal to the interface, $\gamma_{s\ell}^0$ is the orientation-averaged part of the interfacial energy, and the cubic harmonics Q and S are defined in terms of the Cartesian components of $\vec{n}$ as

$$Q = n_x^4 + n_y^4 + n_z^4 \quad ; \quad S = n_x^2 n_y^2 n_z^2 \tag{2}$$

Note that if the series is truncated at the first term, only $\langle 100 \rangle$ dendrites can appear. We will show that the observed growth behavior in the Al- Zn system, including the transition between $\langle 100 \rangle$ and $\langle 110 \rangle$ growth, can be reproduced in simulations when the first two terms in the series are included.

1.1. Phase-field modeling

In principle, one could solve the so-called sharp interface problem, in which the interface between the solid and liquid phases evolves through diffusion of heat and solute in the respective phases, and the solid liquid interface satisfies the conditions of local equilibrium of a curved surface. In practice, however, the shape that evolves is too complex to be tracked accurately, especially in 3-D. We use instead the phase-field formalism for binary alloys introduced by Karma [11], and later extended by Echebarria *et al.* [12]. The model is strictly applicable to the case where the solid diffusivity D_s is negligible compared to the liquid diffusivity D_ℓ. We also adopt the "frozen temperature" approximation, in which the temperature field is assumed to be known *a priori*, which is applicable when the thermal diffusivities in both phases $\alpha_{s,\ell} \gg D_\ell$.

The phase-field $\psi \in [-1, 1]$ is defined such that $\psi = +1$ corresponds to the solid, $\psi = -1$ to the liquid, and intermediate values are associated with the solid-liquid interface. The concentration c is represented via $\tilde{U}$, the local dimensionless supersaturation with respect to the reference point c_ℓ^0, measured in units of the equilibrium concentration gap at that point,

$$\tilde{U} = \frac{1}{1-k_0}\left(\frac{c}{c_\ell^0} - 1\right) \tag{3}$$

In isothermal solidification, $c_\ell^0 = (T_M - T_0)/m_\ell$ is the equilibrium concentration of the liquid at a temperature T_0, T_M is the melting point of the pure substance, and m_ℓ and k_0 are the (constant) liquidus slope and partition coefficient, respectively. The evolution equations for ψ and $\tilde{U}$ are given by

$$\left[1 - (1-k_0)\tilde{\Theta}\right] \eta^2(\vec{n}) \frac{\partial \psi}{\partial \tau} = \nabla\left[\eta^2(\vec{n})\nabla\psi\right] + \nabla\left(|\nabla\psi|^2 \eta(\vec{n}) \frac{\partial \eta(\vec{n})}{\partial(\nabla\psi)}\right) + \psi - \psi^3 - \lambda(1-\psi^2)^2(\tilde{U} + \tilde{\Theta}) \tag{4}$$

$$\left(\frac{1+k_0}{2} - \frac{1-k_0}{2}\psi\right)\frac{\partial \tilde{U}}{\partial \tau} = \nabla\left(\tilde{D}_\ell \frac{1-\psi}{2}\nabla\tilde{U} + \vec{j}\right) + \frac{1}{2}\left[1 + (1-k_0)\tilde{U}\right]\frac{\partial \psi}{\partial \tau} \tag{5}$$

Eqs. (4) and (5) have been scaled in terms of the nominal interface width W_0 and relaxation time scale τ_0. The scaled diffusivity $\tilde{D}_\ell = D_\ell \tau_0 / W_0^2$. $\tilde{\Theta} = \tilde{x} - \tilde{v}_P \tau / \tilde{l}_T$ is used to couple the phase-field to the frozen temperature field in directional solidification, where the sample is pulled through a fixed temperature gradient G at velocity v_P. The dimensionless position along the growth axis $\tilde{x} = x/W_0$ is thus reduced by $v_P \tau_0 / W_0$ and scaled by the "thermal length" $\tilde{l}_T = l_T / W_0 = m_\ell (k_0 - 1) c_\ell^0 / G W_0$. $\eta(\vec{n})$ represents the anisotropic part of the interfacial energy (Eq. 1). Eq. (5) includes an "anti-trapping" current $\vec{j}$, introduced in order to cancel artificial trapping of solute due to the diffuse interface, given by

$$\vec{j} = \frac{1}{2\sqrt{2}}\left[1 + (1-k_0)\tilde{U}\right]\frac{\partial \psi}{\partial \tau}\frac{\nabla\psi}{|\nabla\psi|} \tag{6}$$

where $\nabla\psi/|\nabla\psi|$ expresses the normal vector to the interface.

The chemical capillary length $d_0 = \Gamma_{s\ell}/(|m_\ell|(1-k_0)c_\ell^0)$, where $\Gamma_{s\ell}$ is the Gibbs-Thomson coefficient, and the interface attachment coefficient β are related to the phase-field parameters via

$$d_0 = \alpha_1 W_0/\lambda \quad ; \quad \beta = \alpha_1 \frac{\tau}{\lambda W_0}\left[1 - \alpha_2 \frac{\lambda W_0^2}{\tau_0 D}\right] \tag{7}$$

where $\alpha_1 = 0.8839$ and $\alpha_2 = 0.6267$ are chosen to ensure that the phase-field model converges to the sharp interface model. See Echebarria *et al.* [12] for further details. Finally, the anisotropy of the interfacial energy is included in the model by defining the interface width $W = W_0\eta(\vec{n})$ and the relaxation time $\tau = \tau_0\eta^2(\vec{n})$. We note that in this model, one cannot guarantee that the kinetic coefficient vanishes, in contrast to the "two-sided" model, where $D_s = D_\ell$.

The length scales of the grid spacing ($\Delta x \approx W_0$ to resolve the interface) and the simulation domain ($L_{x,y,z} \gg D_\ell/v$ to avoid interaction between the solute field and the boundaries) are widely different. We therefore solve these equations on an adaptive finite element grid [13–18]. Local error estimators are used to refine or coarsen the mesh where needed, allowing the tracking of the interface as well as the resolution of gradients in the solute fields. Please see the original papers for the details about the adaptive techniques [13]. For the simulations of directional solidification, where initial transients can be very long, we solve the equations in a moving Lagrangian frame that translates at the pulling speed, and the time derivatives in Eqs. (4) and (5) are appropriately augmented with advective terms.

1.2. Parameter selection

Unlike the experiments reported in [7, 8] where the DOT is observed by changing the Zn composition, the simulations presented here focus on a single alloy of Al-10 wt pct Zn which presents a relatively short freezing range, hence a rapid development of the structure upon cooling. The anisotropy parameters a_1 and a_2 will be varied for this alloy while the other properties are given in Table 1. The diffusion coefficients were purposely lowered by a factor approximately 2 to allow larger time steps to be used in the explicit solution scheme.

Table 1. Material properties chosen for the phase-field simulations and values found in literature for the composition and temperature considered.

Property	Value in simulation	Literature	Units
c_0	0.1	-	wt fraction Zn
m_ℓ	-170	-170 [19]	K
k_0	0.45	0.45 [19]	-
T_M	660	660 [19]	°C
D_ℓ	1.2×10^{-9}	2.75×10^{-9} [20]	m^2/s
D_s	1.2×10^{-12}	2.7×10^{-12} [20]	m^2/s
$\Gamma_{s\ell}$	1.05×10^{-7}	1.05×10^{-7} [21, 22]	K m
ρL_f	1×10^9	1.07×10^9 [22]	J/m^3
ρc_p	3×10^6	2.93×10^6 [23]	J/m^3 K

Echebarria *et al.* [12] showed that mesh-converged results can be obtained with W_0/d_0 as large as 50. With the properties above and the assumption that $c_\ell^0 = c_\infty/k_0 = c_0/k_0$, we have $d_0 \approx 5 \times 10^{-9}$ m. The smallest expected tip radius is around 1×10^{-6} m, and using the rule that the tip radius should be at least ten times the interface width, we have $R_{\text{tip}}/W_0 = 10$, $W_0 = 1 \times 10^{-7}$ m, and thus $W_0/d_0 = 20$. We then obtain $\tilde{D}_\ell = 5.5$, which implies that $\tau_0 = 4.6 \times 10^{-5}$ s. All other physical values, such as the thermal length, the pulling velocity and the simulation box size are then non-dimensionalized with these values of W_0 and τ_0.

Since our study is focused on the role of anisotropy, it is essential that we measure and correct for grid anisotropy. To do this, we grew dendrites whose crystallographic axes were rotated by various angles with respect to the Cartesian grid. We then measured the difference in growth speed in the different directions, and adjusted the grid anisotropy in order to eliminate the differences. The amount of grid anisotropy depends on the grid spacing. In the calculations given in Section 2, we use $dx_{\min} = 0.75W_0$, and grid correction $a_1^{\text{grid}} = -0.019$. The time step in the explicit scheme was $0.05\tau_0$.

2. Results and Discussion

2.1. Equiaxed Growth

Our survey of the anisotropy parameter space for equiaxed growth was performed in one octant of a cubic domain, with all faces having zero-flux boundary conditions. The $\langle 100 \rangle$ crystallographic axes were aligned with the Cartesian directions of the grid and an initial one-eighth spherical seed of radius $5 - 10 \times W_0$ was set at one corner. The far-field supersaturation, defined as an input parameter, was either $\Omega = 0.4$ or $\Omega = 0.25$ in the results presented below. The initial concentration field was then defined as:

$$c = \begin{cases} k_o c_0 & r < R_0 \\ c_0\left(1 + \Omega(1-k_0)\left(\frac{R_0}{r} - 1\right)\right) & r \geq R_0 \end{cases} \tag{8}$$

The computational domain was 768^3 with $dx_{\min} = 0.75W_0$. In a uniform grid, this would represent 4.5×10^8 grid points, whereas in our simulations, the grid typically started with about 5000 nodes, and at the end of the simulations, there were $1 - 4 \times 10^6$ nodes. The calculations were run on both 2.93 and 3.30 GHz Intel Xeon CPUs. Run times to generate fully developed microstructures at $\Omega = 0.4$ (12 000 time steps) were less than one day.

Haxihmali *et al.* [9,24] examined the DOT in pure materials using phase-field methods. They performed line scans at several fixed values of $a_2 = (-0.005, -0.01, -0.02)$. At low values of a_1, $\langle 110 \rangle$ dendrites were found, while at high values of a_1, $\langle 100 \rangle$ dominated, and in the intermediate region, mixed structures which they termed "hyperbranched" were obtained. We surveyed the anisotropy parameter space using a similar approach, i.e., using line scans. Since all of the available experimental and MD data [25] show that these parameters are in the range $0 \leq a_1 \leq 0.12$ and $-0.01 \leq a_2 \leq 0$, the (a_1,a_2) couples were chosen as indicated by Fig. 1, along with a sampling of the observed structures.

We find similar results to those reported by Haxhimali *et al.*, with respect to the transitions between the three types of microstructures. Looking first at the line scan $a_2 = -0.01$, we see in Fig. 1 that $\langle 110 \rangle$ dendrites form when $a_1 = 0$. Notice that the tip has an elliptic rather than circular shape in a transverse section, reflecting the underlying symmetry of the $\langle 110 \rangle$ direction. As a_1 increases, the strength of the anisotropy along $\langle 100 \rangle$ increases, as a result, the $\langle 110 \rangle$ tips become more elliptic in (100) planes and are slowed down. If a_1 is further increased, the tips become sufficiently flat that they split along the $\langle 110 \rangle$ axis (see the case $a_1 = 0.04$, $a_2 = -0.01$ in Fig. 1). These tips continue to grow, and eventually produce hyperbranched structures. As a_1 increases along this line scan, the tips split earlier and thus grow larger (cf. $a_1 = 0.06$), and then as one approaches the boundary between the hyperbranched and $\langle 100 \rangle$ dendrites, the dendrites present four tips split along the $\langle 100 \rangle$ axis, until finally they unite into a single tip oriented along the $\langle 100 \rangle$.

We note that although the tip growth velocity has converged for the extreme cases with clear $\langle 100 \rangle$ and $\langle 110 \rangle$ dendrites, the speed of the double and quadruple tips are still in the initial transient at the end of the simulation time considered here. Thus, our simulations should be

IOP Conf. Series: Materials Science and Engineering **33** (2012) 012111 doi:10.1088/1757-899X/33/1/012111

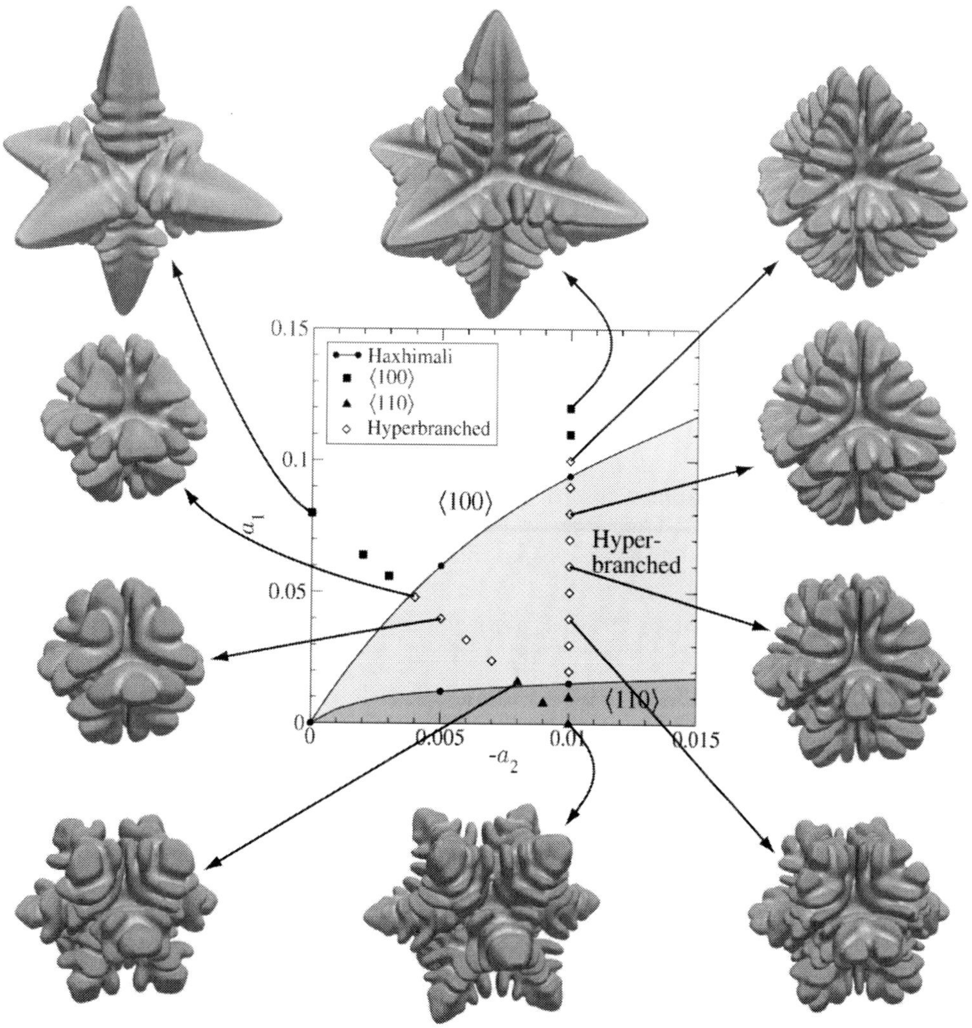

Figure 1. Equiaxed growth phase-field results for varying anisotropy parameters, with $\Omega = 0.4$. The scale and orientation of all dendrites is identical. The solid lines in the center graph represent the limits between the domains ⟨100⟩, hyperbranched and ⟨110⟩ structures reported by Haxhimali *et al.* [9]. The present calculations show very similar results for the locations of the boundaries.

performed for longer times to verify that those structures persist further on and are stable along the found directions at steady state.

Although the locations of the boundaries between the different morphologies are remarkably similar to those reported by Haxhimali *et al.* for thermal dendrites, the details of the dendritic structures in the hyperbranched region are somewhat different. Haxhimali *et al.* observed less branching in the off-axis directions. We attribute this difference to the fact that the simulations presented here correspond to solutal dendrites growing at large supersaturation ($\Omega = 0.4$), whereas Haxhimali *et al.* considered a pure material at very low undercooling, $\Delta = 0.05$. Figure 2 compares equiaxed dendrites computed for the parameter set $a_1 = 0.08$, $a_2 = -0.01$, using two different supersaturation values, $\Omega = 0.4$ and $\Omega = 0.25$. Although the envelopes of the two dendrites are similar, the amount of "activity" in the ⟨110⟩ directions is significantly less at lower supersaturation. We note that the computations become increasingly expensive as the supersaturation decreases, because the dendrites grow much more slowly, and thus the simulations must be run for a longer time, as noted in the figure caption.

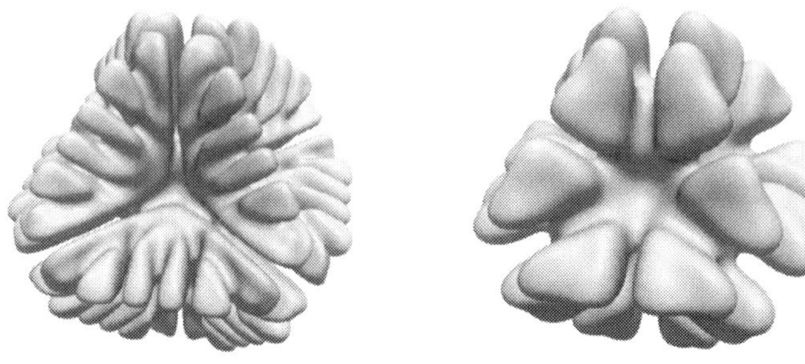

(a) $\Omega = 0.4, t = 14\,000$ (b) $\Omega = 0.25, t = 45\,000$

Figure 2. Comparison of computed equiaxed dendrites with $a_1 = 0.08$, $a_2 = -0.01$, at two different supersaturations Ω. The times were chosen from the two simulations such that the dendrite tips were at approximately the same distance from the center.

We would like to be able to present a mechanistic explanation for dendrite orientation selection based on these results. Haxhimali *et al.* [9] suggested that the dendrites would grow along the direction of minimum interfacial stiffness, which corresponds to the line given by $a_1 = -20a_2/3$. Another possible criterion, suggested by Eq. (1), is that there should be transition between ⟨100⟩ and ⟨110⟩ dendrites on opposite sides of the line $a_1 = -3a_2$, because along this line the coefficient of Q in Eq. (1) is zero. Thus, below this line, the coefficient of Q is negative, which favors ⟨110⟩ dendrites, and above it the coefficient is positive, favoring ⟨100⟩ growth. Both criteria are shown in Fig. 3, superimposed on the tableau of the phase-field results. One can see that although the proposed criteria are suggestive of the selection, clearly they are not predictive.

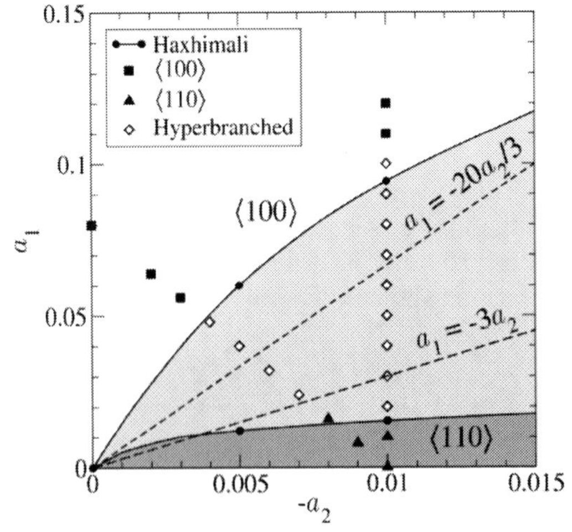

Figure 3. Dendrite growth direction selection criteria based on minimum of the interfacial stiffness (upper dashed line) and maximum of the interfacial energy, Eq. 1 (lower dashed line), superimposed on the results of the phase-field calculations for $\Omega = 0.4$, from Fig. 1.

2.2. *Columnar Growth*

Directional solidification (DS) simulations were performed in order to obtain a better comparison with the experimental results [7, 26] , in particular to see if seaweed-type morphologies can be modeled at intermediate zinc compositions (*i.e.*, for anisotropy parameters producing hyperbranched equiaxed dendrites). The parameters used for these simulations are listed in Table 2. These simulations were performed in a rectilinear domain, extended in the pulling direction, which is aligned with the x−axis. The $x-z$ faces were periodic, and the $x-y$ faces had symmetry boundary conditions. For cases where we expected to see $\langle 100\rangle$ dendrites, the computational domain in the $y-z$ plane was $192\times 96\ dx_{min}$, whereas for cases where we expected to see $\langle 110\rangle$, the computational domain in the $y-z$ plane was $168\times 48\ dx_{min}$. These dimensions permit a crystal of the corresponding crystalline orientation to grow periodically without being overly constrained by the boundaries.

Table 2. Dimensionless parameters used in the DS simulations, along with their dimensional equivalents. All parameters that are not specifically mentioned here are the same as the values used in the equiaxed growth simulations. $\tilde{v}_F$ is the velocity of the reference frame.

Dimensionless parameter	Value	Dimensional equivalent	Units
$\tilde{l}_T$	1×10^4	$G = 20\,777$	K/m
$\tilde{v}_P$	0.25	$v_P = 543$	μm/s
$\tilde{v}_F$	0.25	$v_F = 543$	μm/s
Ω_b	0.45	-	-

In DS, the tip speed is controlled by the pulling velocity. However, as the simulation domain follows the dendrite tip during its growth, we can define a supersaturation Ω_b of the bottom boundary of the domain (or equivalently its temperature). If $\Omega_b = 1$, and providing the box is long enough along the x-direction, the whole mushy zone will be simulated. In order to strike a balance between having enough of the dendrite in the domain to be interesting, and higher computational cost, we chose $\Omega_b = 0.45$ for the results presented below. We again started with a small spherical seed, this time placed at $(x, y, z) = (0, y_{max}/2, 0)$, i.e., along one edge of the base of the domain. Most of the simulations were run in a fixed reference frame ($\tilde{v}_F = 0$) for an initial period to allow the seed to develop, then the simulation was restarted from that point with $\tilde{v}_F = \tilde{v}_P$.

If the preferred crystallographic growth direction, determined from the simulations of equiaxed growth, was aligned exactly with the pulling direction, the expected dendritic growth patterns appeared. It is more interesting to consider cases where the crystallographic axes are rotated by an angle θ about the z−axis, such that neither the $\langle 100\rangle$ nor the $\langle 110\rangle$ direction is parallel to a coordinate axis. In this way, we can observe the interaction of the preferred growth direction with the constraint imposed by the thermal gradient.

Figures 4 and 5 present results for three cases, with parameter values chosen from values along the diagonal scan in Fig. 1. Each figure shows two fully developed dendrites and two outlined boxes, side-by-side. One of the dendrites is colored by the surface concentration to enhance contrast, whereas the other is simply a reflecting surface. The simulations were performed in just half of one of those boxes, with the results repeated across the periodic face, and reflected across the back face for clarity. Note that this is consistent with the applied periodic and symmetry conditions on the respective faces.

The first case corresponds to the parameter set $a_1 = 0.064$, $a_2 = -0.002$, $\theta = 5°$. From the results from our study of equiaxed growth, we would expect to find $\langle 100\rangle$ dendrites. Figure

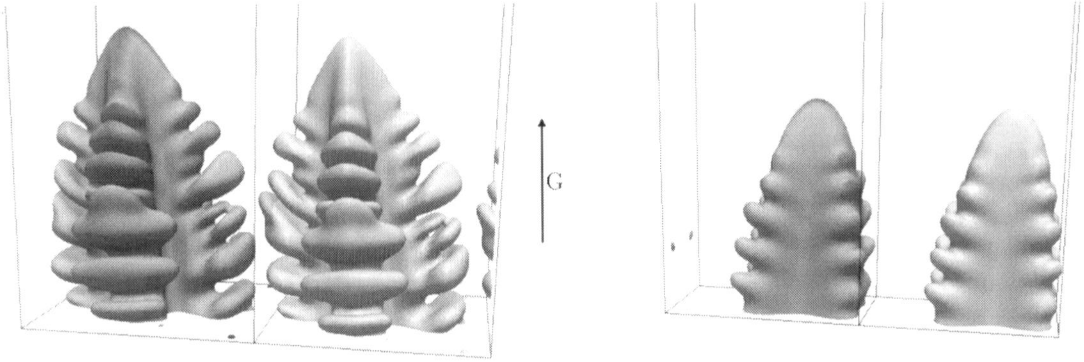

(a) $a_1 = 0.064,\ a_2 = -0.002,\ \theta = 5$ (b) $a_1 = 0.016,\ a_2 = -0.008,\ \theta = 40$

Figure 4. Computed DS dendrites for parameter sets that produce (a) $\langle 100\rangle$, or (b) $\langle 110\rangle$ dendrites in equiaxed growth. Note the difference in the size of the computational domain for the two cases.

4(a) shows that the dendrite is indeed of $\langle 100\rangle$ type, but its growth direction is slightly biased towards the direction of the gradient, *i.e.* between $\langle 100\rangle$ and $\vec{G}$. Note that the combination of the growth non-parallel to the pulling direction and the imposed moving frame causes the dendrite to translate along the $y-$axis, where it re-enters the domain upon reaching the periodic boundary. Figure 4(b) shows the results of simulation with $a_1 = 0.016$, $a_2 = -0.008$, $\theta = 40°$. This parameter set produced $\langle 110\rangle$ dendrites in equiaxed growth (see Fig. 1). As for the preceding case, this simulation produces the expected orientation, a $\langle 110\rangle$-type dendrite, growing at an intermediate angle between the $\langle 110\rangle$ and the pulling direction, as we saw previously for the $\langle 100\rangle$ dendrite in Fig. 4(a).

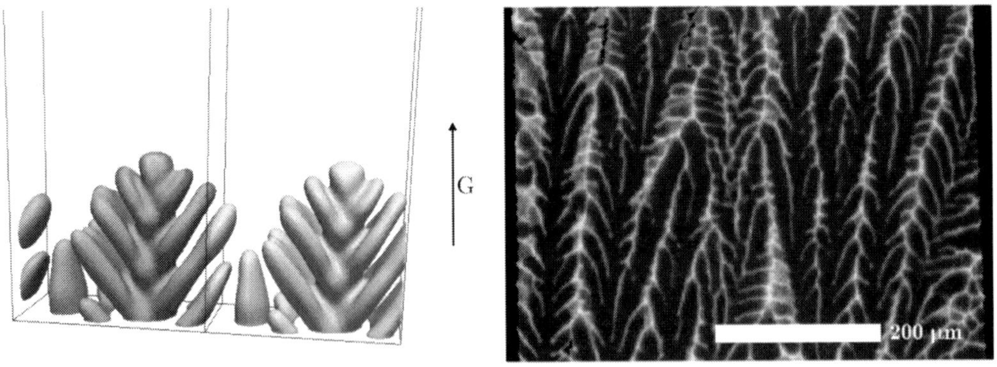

(a) $a_1 = 0.04,\ a_2 = -0.005,\ \theta = 40$ (b) X-ray tomography (100) section

Figure 5. (a) Computed DS dendrites for a parameter set that produces a hyperbranched structure in equiaxed growth. (b) X-ray micro-tomographic section (parallel to a (100) plane) of a directionally solidified Al-50 wt pct Zn alloy.

The most interesting case, shown in Fig. 5(a), corresponds to the parameter set $a_1 = 0.040$, $a_2 = -0.005$, $\theta = 40°$, which lies in the middle of the hyperbranched region in equiaxed growth. The resulting growth pattern produces a primary trunk that tries to follow $\langle 110\rangle$, but continually splits and produces a branch that returns toward the pulling direction. This structure looks remarkably similar to the microstructure obtained from X-ray tomographic measurements on a directionally solidified Al-50 wt pct Zn alloy, shown in Fig. 5(b) [26].

3. Conclusion

We have presented phase-field calculations for both equiaxed and directional solidification to investigate the role of interfacial energy anisotropy in the selection of dendrite orientation. Working in the range of parameters consistent with experimental measurements, we find distinct regions with ⟨100⟩ and ⟨110⟩ primary dendrites, and an intermediate region where both characters are observed, resulting in hyperbranched dendrites. We show that the hyperbranched structures result from tip splitting of the ⟨100⟩ or ⟨110⟩ primary directions due to the influence of the competing ⟨110⟩ or ⟨100⟩ character, respectively. In directional solidification, the constraints of the imposed thermal gradient and growth direction have a strong effect, especially in the hyperbranched region where neither character is dominant. In particular, in the region of anisotropy parameter space where hyperbranched dendrites are found in equiaxed growth, under DS conditions we observe alternating sidebranches, consistent with experimentally observed microstructures in Al-Zn alloys.

Acknowledgments

The authors gratefully acknowledge the support of the Swiss National Fund (Grant Nos. 200020-113260 and 200020-121598), and the European Community's Seventh Framework Program under grant agreement No. 229108 (MIntWeld), and the US National Aeronautics and Space Administration (NASA) under grant NNX10AJ73G.

References

[1] Henry S, Jarry P and Rappaz M 1998 *Metallurgical and Materials Transactions A: Physical Metallurgy and Materials Science* **29** 2807–2817
[2] Henry S, Minghetti T and Rappaz M 1998 *Acta Materialia* **46** 6431–6443
[3] Sémoroz A, Durandet Y and Rappaz M 2001 *Acta Materialia* **49** 529–541
[4] Salgado-Ordorica M, Desbiolles J L and Rappaz M 2011 *Acta Materialia* **59** 5074–5084
[5] Salgado-Ordorica M, Burdet P, Cantoni M and Rappaz M 2011 *Acta Materialia* **59** 5085–5091
[6] Niederberger C, Michler J and Jacot A 2006 *Physical Review E - Statistical, Nonlinear, and Soft Matter Physics* **74** 021604
[7] Gonzales F and Rappaz M 2006 *Metallurgical and Materials Transactions A: Physical Metallurgy and Materials Science* **37** 2797–2806
[8] Friedli J 2011 *Interfacial energy anisotropy and growth morphologies in aluminum-zinc alloys* Ph.D. thesis EPFL
[9] Haxhimali T, Karma A, Gonzales F and Rappaz M 2006 *Nature Materials* **5** 660–664
[10] Fehlner W R and Vosko S H 1976 *Canadian Journal of Physics* **54** 2159–2169
[11] Karma A 2001 *Physical Review Letters* **87** 115701
[12] Echebarria B, Folch R, Karma A and Plapp M 2004 *Physical Review E - Statistical, Nonlinear, and Soft Matter Physics* **70** 061604
[13] Jeong J H, Goldenfeld N and Dantzig J 2001 *Physical Review E - Statistical, Nonlinear, and Soft Matter Physics* **64** 416021–4160214
[14] Jeong J H, Dantzig J and Goldenfeld N 2003 *Metallurgical and Materials Transactions A: Physical Metallurgy and Materials Science* **34 A** 459–466
[15] Burman E, Jacot A and Picasso M 2004 *Journal of Computational Physics* **195** 153–174
[16] Provatas N, Greenwood M, Athreya B, Goldenfeld N and Dantzig J 2005 *International Journal of Modern Physics B* **19** 4525–4565
[17] Athreya B, Dantzig J, Liu S and Trivedi R 2006 *Philosophical Magazine* **86** 3739–3756
[18] Narski J and Picasso M 2007 *Computer Methods in Applied Mechanics and Engineering* **196** 3562–3576
[19] Mey S a and Effenberg G 1986 *Zeitschrift fuer Metallkunde/Materials Research and Advanced Techniques* **77** 449–453
[20] Du Y, Chang Y, Huang B, Gong W, Jin Z, Xu H, Yuan Z, Liu Y, He Y and Xie F Y 2003 *Materials Science and Engineering A* **363** 140–151
[21] Morris J, Mendelev M and Srolovitz D 2007 *Journal of Non-Crystalline Solids* **353** 3565–3569

[22] Valencia J and Quested P 2010 *Metals Process Simulation* (*ASM Handbook* vol 22B) (Materials Park, Ohio: ASM International) chap Thermophysical Properties, pp 18–32
[23] Smithells C and Brandes E 1983 *Smithells Metals Reference Book* (London: Butterworths) chap The physical properties of pure metals sixth ed
[24] Haxhimali T 2006 *Phase-field simulation study of dendritic crystal growth morphologies for cubic and hexagonal symmetries* Ph.D. thesis Northeastern University
[25] Asta M, Beckermann C, Karma A, Kurz W, Napolitano R, Plapp M, Purdy G, Rappaz M and Trivedi R 2009 *Acta Materialia* **57** 941–971
[26] Friedli J, Fife J, Di Napoli P and Rappaz M 2012 *Proceedings of MCWASP 2012* ed Ludwig A

Modeling and simulation of dendrite growth in solidification of Al-Si-Mg ternary alloys

SHI Yufeng, Zhang Yan, XU Qingyan[1]**, LIU Baicheng**
Key laboratory for Advanced Materials Processing Technology, Ministry of Education, Department of Mechanical Engineering, Tsinghua University, Beijing, 100084, China
CUI Hongbao, MI Guofa
School of Materials Science and Engineering, Henan Polytechnic University, Jiaozuo, 454150, China

Abstract. Based on a previously cellular automaton (CA) model for Al-Si binary alloys, a modified cellular automaton (MCA) model is developed to numerically simulate the dendrite growth and microsegregation of Al-Si-Mg ternary alloys. By direct coupling PanEngine, the MCA model can obtain thermodynamic data for dendrite growth of ternary alloys in real time, such as equilibrium liquidus temperature, partition coefficient and liquidus slope of two solute elements. In the model, the normal velocity of the interface cell can not be derived only by solving the solute conservation equation, and must be simultaneously coupled with dimensionless solute supersaturation equation for each alloying element and local equilibrium temperature equation at the S/L interface. All the equations of the MCA model are solved by finite difference method on a uniform mesh. By considering local curvature and constitutional undercooling, solute diffusion and solute redistribution at the S/L interface, the MCA model can simulate the columnar dendrite growth and the solidification path of the equiaxed dendrite growth.
Key words. MCA, ternary alloys, dendritic growth, Microsegregation

1. Introduction

It is a hot research topic to investigate the solidification path, the concentration profile along dendrite arms and the formation of microsegregation in material science and technology. Recently, the cellular automaton (CA) approach has been a powerful and efficient approach for the simulation of dendrite morphologies and microsegregation during the solidification of alloys, and has also been used as a means of assessing theories of dendrite growth, but it is still limited to binary dilute alloys[1-12]. There is one way to extend the binary CA model to multicomponent systems by coupling the CA model with the CALPHAD softwares, such as Thermo-Calc and Pandat. A CA-PanEngine model[13,14] for ternary alloys was developed to simulate dendrite growth and microsegregation patterns in Al-15%Cu-1%Mg and Al-4.5%Cu-0.5%Mg-1%Si alloys under different conditions of cooling rate. However, in order to calculate the growth velocity of the interface, a hypothetical interface kinetic coefficient was prescribed in the model. Michelic et al. [15] have developed a two-dimensional VFT model based on the approach of CA for multicomponent systems, which was used to simulate equiaxed dendrite growth and microsegregation of columnar dendrites in Fe–Si–Mn–P–S alloy. However, in the model, the approach to calculate the interfacial equilibrium composition was based on a functional

[1] Correspondent: XU Qingyan, Tel: 86-10-62795482, Fax: 86-10-62770190, E-mail: scjxqy@tsinghua.edu.cn.

extrapolation towards the discontinuity at the interface, which was solved by the complicated non-linear Gauss–Newton regression.

In the present paper, a modified CA model is developed to simulate two-dimensional microstructure evolution in tertiary alloys. The model was directly coupled with PanEngine to calculate the thermodynamic properties. A new and efficient method is used to calculate the equilibrium compositions of the two solute elements at the SL interface, based on a solute balance coupled with supersaturation equations associated with the two solute elements. Consequently, the interfacial growth velocity can be easily obtained. Then the model was applied to investigate the microstructure, microsegregation and the variation of the solidification path in Al-Si-Mg alloys under different thermal conditions of cooling rate.

2. Model description

2.1. Basic assumption

In the present model, the following assumptions have been given: (1) Only primary dendrites are simulated, and the simulation ends when the compositions of the two solute elements are down to the point just before where binary eutectic phase appears; (2) Off-diagonal diffusion is several orders of magnitude lower than the self diffusion of the solute elements. Therefore, cross diffusion is neglected. Meanwhile, the solute diffusion in solid is also neglected, because the solid diffusion coefficient is also several orders of magnitude lower than the liquid diffusion coefficient, and its influence on the growth velocity of the interface is very small.

2.2. Solute diffusion

During the solidification of primary phase in a ternary alloy, two solute elements are defined as B and C, respectively, and the solute fields of B and C in liquid and solid is calculated by

$$\frac{\partial C_{\mathrm{L},i}}{\partial t} = \nabla\bullet(D_{\mathrm{L},i}\nabla C_{\mathrm{L},i}) + C_{\mathrm{L},i}(1-k_i)\frac{\partial f_{\mathrm{S}}}{\partial t} \tag{1}$$

$$C_{\mathrm{S},i} = k_i C_{\mathrm{L},i} \tag{2}$$

where $C_{\mathrm{L},i}$ and $C_{\mathrm{S},i}$ are the composition of the element i (B or C) in the liquid and solid, respectively; $D_{\mathrm{L},i}$ denotes the liquid solute diffusion coefficient of the element i; k_i is the partition coefficient for the element i, which is obtained by the thermodynamic and phase equilibrium calculation package-PanEngine, developed by CompuTherm LLC; f_{S} denotes the solid fraction. The third term on the right-hand side of equation (1) represents the solute rejection at the solidification interface due to the increments of the solid fraction of the primary dendrite phase. In the current model $D_{\mathrm{L},i}$ is assumed to be temperature dependent.

2.3. Interface growth kinetics

In the present model, a two-dimensional rectangular domain is meshed into M×N squared cells, which can be marked as follows: liquid ($f_{\mathrm{S}}=0$), solid ($f_{\mathrm{S}}=1$) or interface ($0<f_{\mathrm{S}}<1$).

For the interface cells, local thermodynamic equilibrium can be obtained by

$$T = T_{\mathrm{L}}(C_{\mathrm{L,B}}, C_{\mathrm{L,C}}) - \Delta T_{\mathrm{C}} - \Delta T_{\mathrm{R}} \tag{3}$$

where $T_{\mathrm{L}}(C_{\mathrm{L,B}},C_{\mathrm{L,C}})$ is the liquidus temperature of the primary phase at the liquid composition $C_{\mathrm{L,B}}$ and $C_{\mathrm{L,C}}$, ΔT_{R} and ΔT_{C} are the curvature and solutal undercooling, respectively.

2.3.1. Curvature undercooling

In two-dimensional condition, considering the Gibbs-Thomson effect, the curvature undercooling of ternary alloys is given by[4]

$$\Delta T_{\mathrm{R}} = \Gamma\left[1-15\varepsilon\cos\left(4(\psi-\psi_0)\right)\right]\cdot K \tag{4}$$

where Γ is the Gibbs–Thomson coefficient, ε is the degree of anisotropy of the interfacial energy, ψ is the inclined angle of the normal to the interface with respect to the coordinate axis, ψ_0 is the angle between the preferred growth direction and the coordinate axis, and K is the curvature of the interface. The average curvature for an interface cell is affected by its neighboring cells, and is calculated with the following expression[3]

$$K = [1 - 2(f_S + \sum_{j=1}^{N} f_S^j)/(N+1)]/L \tag{5}$$

where L is the cell size, f_S^j and N are the solid fraction and the number of the neighboring cells, respectively. In the present model, N equals 8.

2.3.2. Solutal undercooling

Due to the difference between the solid and liquid solubility of the two solute elements at the interface, the solutal undercooling of the ternary alloys is influenced by more than one alloying element, and can be expressed by

$$\Delta T_C = m_{L,B}(C_{L,B} - C^*_{L,B}) + m_{L,C}(C_{L,C} - C^*_{L,C}) \tag{6}$$

$$m_{L,B} = \frac{\partial T_L(C_{L,B}, C_{L,C})}{\partial C_{L,B}}, \quad m_{L,C} = \frac{\partial T_L(C_{L,B}, C_{L,C})}{\partial C_{L,C}} \tag{7}$$

where $m_{L,B}$ and $m_{L,C}$ are the slope of the liquidus surface along the solute element B and C, respectively, $C^*_{L,B}$ and $C^*_{L,C}$ are the liquid equilibrium composition of B and C, respectively. During the solidification of ternary alloys, $m_{L,B}$ and $m_{L,C}$ are influenced by the temperature and the liquid composition at the interface, and can not be considered as constant. $C^*_{L,B}$ and $C^*_{L,C}$ can not be obtained only by solving the equation (6).

2.3.3. Solute balance at the S/L interface

Neglecting the solute redistribution on the solid side, the solute balance must be satisfied at the interface as follows[3]

$$v_n C^*_{L,i}(1-k_i) = -D_{L,i} \nabla C_{L,i} \cdot \mathbf{n} \tag{8}$$

where v_n is the normal velocity, and $\mathbf{n}$ is the unit normal vector, which is expressed as[4]

$$\mathbf{n} = -\nabla f_S / \|\nabla f_S\| = n_x \vec{i} + n_y \vec{j} \tag{9}$$

$$\begin{cases} n_x = \left(-\partial f_S / \partial x[(\partial f_S / \partial x)^2 + (\partial f_S / \partial y)^2]^{-1/2}\right) = \cos\psi \\ n_y = \left(-\partial f_S / \partial y[(\partial f_S / \partial x)^2 + (\partial f_S / \partial y)^2]^{-1/2}\right) = \sin\psi \end{cases} \tag{10}$$

By coupling the equations (8), (9) and (10), the normal velocity in two-dimensional Cartesian coordinates is given by

$$v_n C^*_{L,i}(1-k_i) = -D_{L,i}\left(\frac{\partial C_{L,i}}{\partial x} n_x + \frac{\partial C_{L,i}}{\partial y} n_y\right) \tag{11}$$

2.3.4. Solute supersaturation at the interface

Neglecting the cross diffusion of the two alloy elements, the growth of the multicomponent dendrite tip is also governed by the following equations for each solute element[16]

$$\Delta_i = \frac{C^*_{L,i} - C_{0,i}}{C^*_{L,i}(1-k_i)} = Iv(Pe_i) \tag{12}$$

where Δ_i, $C_{0,i}$ and Pe_i are the supersaturation, nominal composition and Péclet number of the solute element i (B or C), respectively.

During the solidification of primary phase of the ternary alloys, the dendritic growth is controlled by solutal effects in the low Péclet number regime (Pe_i<1). Therefore, the Ivantsov function can be approximated as[17]

$$Iv(Pe_i) = -Pe_i \ln(Pe_i)\text{-}0.5772Pe_i \tag{13}$$

2.3.5. Increment of solid fraction

After calculating v_n, the solid fraction increment of the interface in one time step Δt is calculated using the following equation[4]

$$\Delta f_S = (v_n \Delta t) / L \tag{14}$$

$$L = \Delta s / [\max(|\cos\psi|, |\sin\psi|)] \tag{15}$$

$$\text{then } f_S^{n+1} = f_S^n + \Delta f_S \tag{16}$$

where Δs is the cell size, $\Delta s=\Delta x=\Delta y$, Δf_S is the increment of solid fraction, L is the length of the line along the normal, f_S^n and f_S^{n+1} are the solid fraction at current and next time, respectively.

3. Application to Al–Mg–Si alloys

Table 1. Thermophysical properties of Al-Si-Mg alloys[18]

Parameter and symbol	Unit	Value
Thermal diffusion coefficient, α_T	m^2/s	2.6×10^{-5}
Liquid diffusion coefficient of Si, $D_{L,Si}$	m^2/s	$0.11\times10^{-4}\exp(-2856/T)$
Liquid diffusion coefficient of Mg, $D_{L,Mg}$	m^2/s	$0.99\times10^{-4}\exp(-8610/T)$
Gibbs-Thomson's coefficient, Γ	K·m	1.7×10^{-7}
Anisotropic coefficient, ε	-	0.04
Specific heat, c_p	J/(Kg·K)	1070
Latent heat, L_h	KJ/Kg	385

3.1. Columnar dendrite growth of Al-7%Si-0.4%Mg alloy

In order to investigate the influence of different pulling rates on the primary dendrite arm spacing of the columnar dendrites, the MCA model was used to simulate the directional solidification of columnar dendrite of Al-7%Si-0.4%Mg alloy. $m_{L,Si}$ and $m_{L,Mg}$ of Al-7%Si-0.4%Mg alloy are -6.85 and -3.70 K, respectively. The calculation domain was 460×340μm^2 and the cell size was 1um. At the beginning, 45 seeds with different preferred growth directions were set at the bottom boundary, and controlled by a constant temperature gradient G from the bottom to the top, but a changeable pulling rate V.

As Figure 1 shown, at lower pulling rate, such as 100μm/s, only a few columnar dendrites grew ahead and the others were blocked. With the increasing V, the primary dendrite arm spacing λ_1 decreased, because more columnar dendrites grew up.

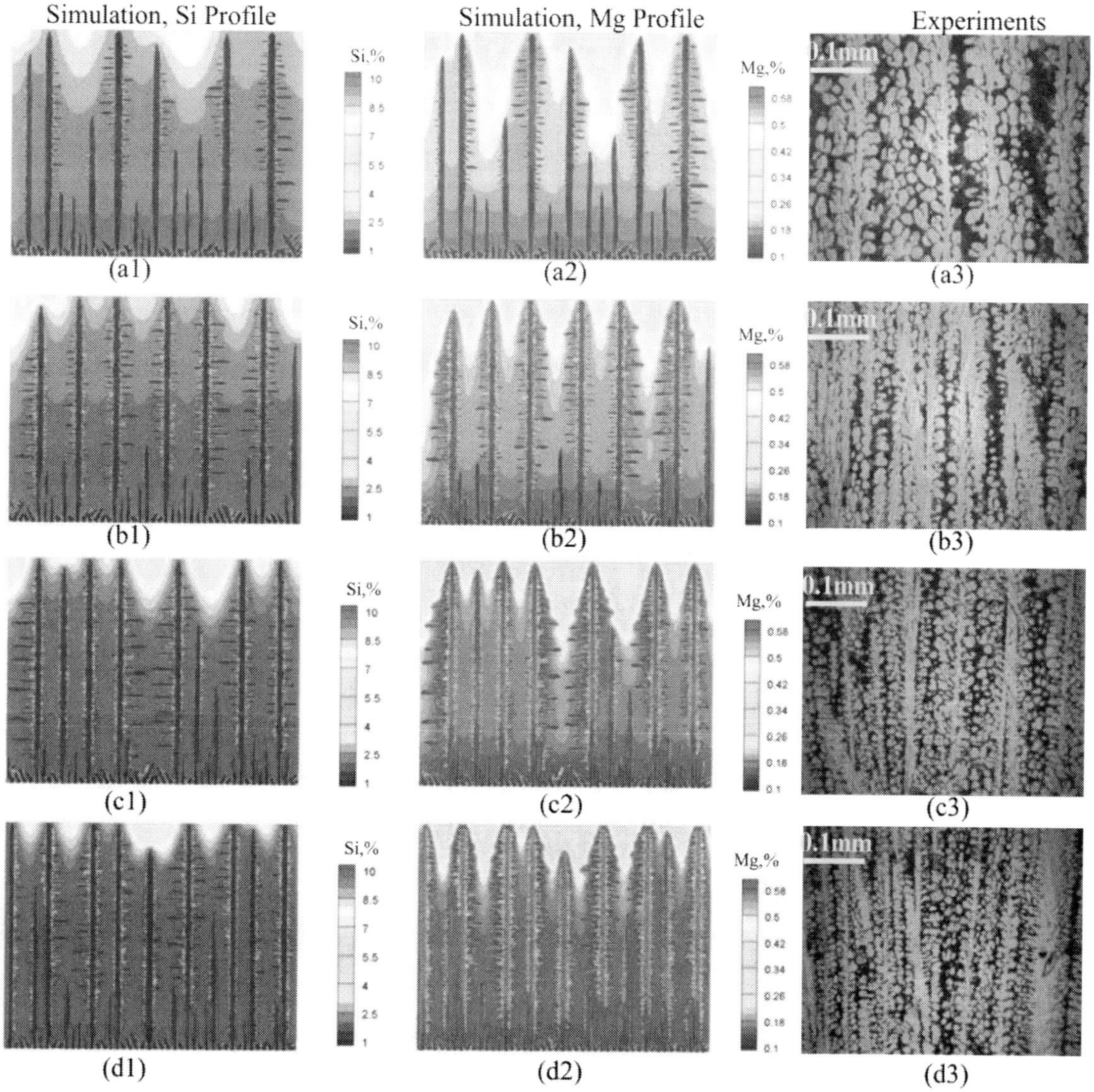

Figure 1. Comparisons of simulated and experimental results of two dimensional columnar dendrite growth of Al-7%Si-0.4%Mg with a constant temperature gradient 16.8K/mm under different pulling rates of 100μm/s (a1-a3), 200μm/s (b1-b3), 300μm/s (c1-c3) and 400μm/s (d1-d3)

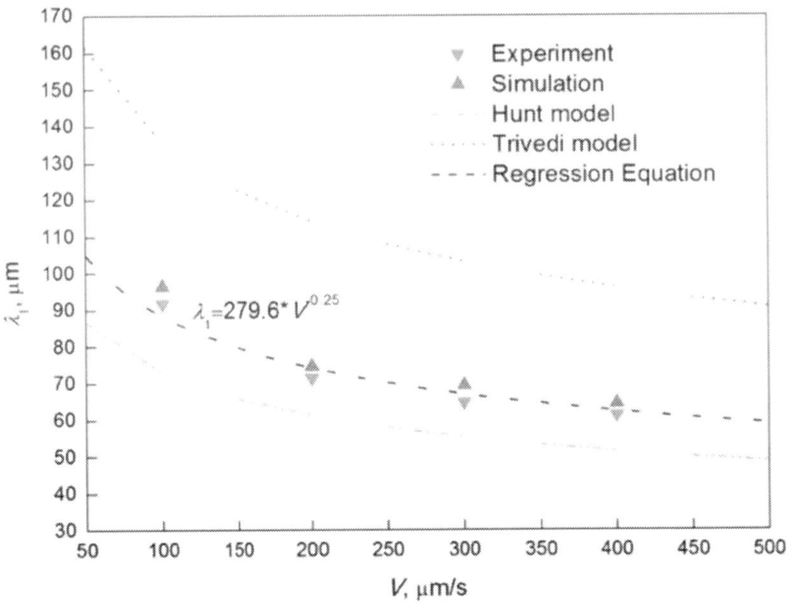

Figure 2. Comparison between simulated and experimental primary dendrite arm spacing and the nonlinear regression equation of λ_1-V relation

As Figure 2 shown, simulated results of λ_1 are compared well with the experimental results, and they are in the range of Hunt model and Trivedi model. The nonlinear regression equation was $\lambda_1=279.6\times V^{0.25}$. From both the simulated and experimental results, it could be seen that the columnar dendrites of Al-7%Si-0.4%Mg alloy transited from sparse to dense with the increasing pulling velocity, therefore, the MCA model was reliable to simulate the dendrite morphology evolution of the Al-Si-Mg alloys.

3.2. Simulation of the dendrite growth of Al-7%Si-3%Mg alloy

The MCA model was also used to predict evolution of primary dendrite growth in an Al-7%Si-3%Mg alloy. The growth of eutectic phase is not considered in the model. Therefore, when the temperature reached eutectic point, the calculation stopped. In order to investigate the effect of different cooling rates on the growth kinetics of the primary dendrites, 4 solid seeds with random preferred growth orientations were nucleated randomly in a 500μm×500μm calculation domain meshed by 1μm cell size. The calculations began from the liquidus temperature, cooled down by different constant cooling rates and stopped when a monovariant line of Al-Si-Mg alloys was reached. The thermodynamic data were obtained from the PanEngine software and table 1.

Figure 3 illustrates the simulated dendrite morphologies and the composition field of Mg at different cooling rates of 1, 10, 100K/s during solidification. In the case of low cooling rate (see Figure 3(a), 1K/s), it can be noted that the Mg composition is uniform in the liquid due to enough time for the Mg solute diffusion among the dendrites. With an increase in cooling rate, larger variations of Mg composition occur in the liquid, due to short time for the diffusion of rejected solute from the dendrite tip. Therefore, the solute buildup at the tip leads to an interface destabilization, which can cause the denser side branches as shown in Figure 3(b) and (c).

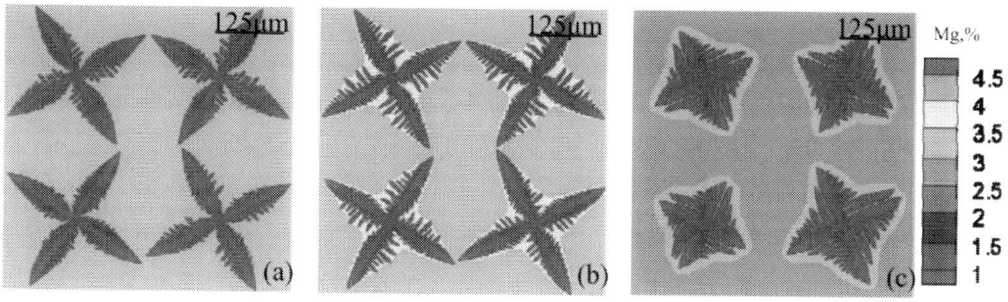

Figure 3. Simulated dendrite morphology and solute field of Mg for an Al-7wt.%Si-3wt.%Mg alloy (total solid fraction is 22%) under different cooling rate
(a) 1K/s, t=71.87s (b) 10 K/s, t=7.44s (c) 100 K/s, t=1.06s

The higher the cooling rate, the larger the ratio of Si/Mg in the liquid, which leads to the difference of the calculated solidification paths among the three cases as shown in Figure 4. These three regions are separated by monovariant lines which meet at a ternary eutectic point (830.6 K). At low cooling rate, it can be noted that the ratio of Si/Mg is smaller and the calculations ended when the solidification paths hit the Al-Mg_2Si-liquid monovariant line. When the cooling rate increases, the phase Si could precipitate. It can be concluded that the difference in the cooling rates could lead to different second phases.

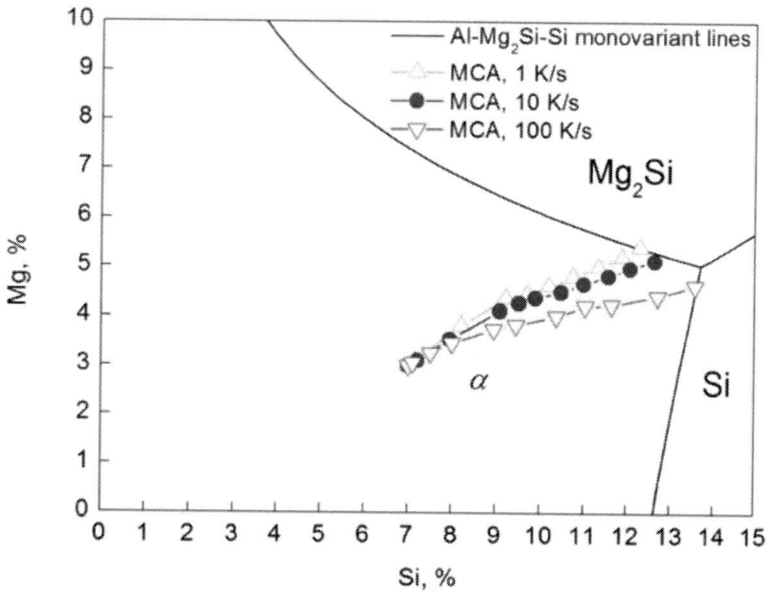

Figure 4. Simulated solidification paths of the Al-7Si-3Mg alloy under different cooling rates

4. Conclusion

Coupling with PanEngine, an MCA model has been developed to simulate the evolution of dendrite morphologies and microsegregation in a ternary alloy system. The model was used to simulate the columnar dendrite growth of Al-7%Si-0.4%Mg alloy, and the simulated primary dendrite arm spacing compared well with the experimental results. Therefore, the MCA model was reliable. Meanwhile, equiaxed multi-dendritic growth of Al-Si-Mg alloys with different cooling rate was also simulated by the MCA Model, and the simulated results can describe the primary and secondary dendrite arms, predict the microsegregation profiles quantitatively, and demonstrate that increase of cooling rate can change the precipitation of second phases for some critical compositions.

Acknowledgments
This research is funded by the National Basic Research Program of China (No. 2005CB724105 and 2011CB706801), the National Natural Science Foundation of China (No. 10477010 and 51171089), the National High Technology Research and Development Program of China (No. 2007AA04Z141) and the important national science & technology specific projects (No. 2009ZX04006-041-04 and 2011ZX04014-052).

5. References

[1] Rappaz M and Gandin C A 1993 *Acta Metall.* **41** 345
[2] Gandin C A and Rappaz M 1994 *Acta Metall.* **42** 2233
[3] Nastac L 1999 *Acta Mater.* **47** 4253
[4] Beltran S L and Stefanescu D M 2004 *Metall. Mater. Trans.* A **35** 2471
[5] Wang W, Lee P D and McLean M 2003 *Acta Mater.* **51** 2971
[6] Liu Y, Xu Q Y and Liu B C 2006 *Tsinghua Science and Technology* **11** 495
[7] Liu B C, Xiong S M and Xu Q Y 2007 *Metall. Mater. Trans. B* **38** 525
[8] Li B, Xu Q Y, Pan D, Liu B C, Xiong Y C, Zhou Y J and Hong R Z 2008 *Acta Metall Sin* **44** 243
[9] Yu J, Xu Q Y, Cui K and Liu B C 2007 *Acta Metall Sin* **43** 731
[10] Pan S Y and Zhu M F 2010 *Acta Mater.* **58** 340
[11] Shi Y F, Xu Q Y, Gong M and Liu B C 2011 *Acta Metall. Sin.* **47(5)** 620
[12] Liu B C, Xu Q Y, Jing T, Shen H F and Han Z Q 2011 *JOM* **63(4)** 19
[13] Zhu M F, Cao W, Chen S L, Hong C P and Chang Y A 2007 *J. Phase Equilib. Diffus.* **28** 130
[14] Dai T, Zhu M F, Chen S L, Cao W S and Hong C P 2008 *Acta Metall. Sin.* **44** 1175
[15] Michelic S C, Thuswaldner J M and Bernhard C 2010 *Acta Mater.* **58** 2738
[16] Rappaz M and Boettinger W J 1999 *Acta Mater.* **47** 3205
[17] Trivedi R and Kurz W 1994 *Int Mater Rev* **39** 49
[18] Du Q and Jacot A 2005 *Acta Mater.* **53** 3479

IOP Conf. Series: Materials Science and Engineering **33** (2012) 012113 doi:10.1088/1757-899X/33/1/012113

Molecular dynamics calculation of thermodynamic properties of iron solidification

J Liu[1] and **H B Dong**[1, 2]
[1] Department of Engineering, University of Leicester, Leicester, LE1 7RH, UK

E-mail: h.dong@le.ac.uk

Abstract. The aim of this study is to identify the best available inter-atomic potentials for molecular dynamics (MD) calculation of solidification of iron and then to use the best potential to calculate thermodynamic properties such as equilibrium melting temperature, enthalpy, heat capacity and solid-liquid interfacial free energy. Our study reveals that embedded atom method (EAM) potential developed by Ackland et al. [2004 *J. Phys.: Condens Matter.* **16** S2629] appears to be the most accurate model for MD simulation of iron solidification. Simulations with the above EAM model predict the equilibrium melting temperature of iron is 1790K, the solid-liquid interfacial energy 214 mJ/m2. The difference with the experimental data is 1.2%, and 4.9% respectively.

1. Introduction

Thermodynamic properties of solid-liquid interface affect solidification structure and properties of the solidified components [1]. How to obtain accurate thermodynamic properties of metal solidification is crucial for both industry application and fundamental research. During the last decades, various experimental technologies [2,3] such as electromagnetic, electrostatic and acoustic levitation are employed to measure the thermodynamic data with high accuracy. However, measuring the thermodynamic properties is expensive and some data such as solid-liquid interfacial free energy are difficult to measure [4]. As a consequence, experimentally measured thermodynamic data at solid-liquid interface is rather limited. To overcome the experimental difficulties, computer simulations based on atomistic models such as molecular dynamics (MD) simulations have been developed and becoming a powerful tool for gaining fundamental knowledge of materials processes and properties over the past two decades [1,5].

Atomistic simulation starts with establishing a potential model which describes atomic interactions in the material of interest [5]. The accuracy of the simulation results is determined by the potential models. Due to the industrial importance of iron and steel, in recent years, several semi-empirical potential models for iron [6-9] have been developed and used to calculate solid-liquid interface properties including solid-liquid interfacial free energy [10,11]. However, the calculated properties vary with the potentials selected for the calculation. For example, the calculated equilibrium melting temperatures using Finnis-Sinclair potential[6] and EAM potential developed by Mendelev et al. [9] are 2400K, and 1772K, respectively. The difference is significant compared to the experimental value 1811K [11]. The main purpose of the present work is to first identify an appropriate inter-atomic potential for MD simulations of iron solidification and then use the potential to calculate the thermodynamic properties of iron solidification.

[2] To whom any correspondence should be addressed.

2. EAM potential model

The potential model is the basis of molecular dynamic simulation. Currently one of the most widely used potential formats for metallic systems is given by the EAM model proposed by Daw and Baskes [12] on the basis of the density functional theory (DFT). In this study, we will use an EAM potential recently developed by Ackland *et al* [13] to calculate the thermodynamic properties of iron solidification. In this model, the total energy of iron system is expressed as

$$E_{tot} = \sum_i F_i(\rho_i) + \frac{1}{2}\sum_{i\neq j}\phi_{ij}(r_{ij}), \tag{1}$$

where $F_i(\rho_i)$ is the embedding energy which includes many-body effects. $F_i(\rho_i)$ is taken as [13]

$$F_i(\rho_i) = -\sqrt{\rho_i} + a_2\rho_i^2 + a_4\rho_i^4 \tag{2}$$

ρ_i is the electronic density at atom i due to the surrounding atoms:

$$\rho_i = \sum_{j\neq i} f_j(r_{ij}) \tag{3}$$

where $f_j(r_{ij})$ is the spherically averaged electron density at atom i due to atom j at a distance r_{ij}. In this study, $f_j(r_{ij})$ was calculated by [13]

$$f(r_{ij}) = \sum_K A_k(R_K - r_{ij})^3 H(R_K - r_{ij}) \tag{4}$$

where *H(x)* is the Heaviside step function.

The second term of the EAM, $\phi_{ij}(r_{ij})$, describes a two-body interaction potential was chosen as[13]

$$\phi(r_{ij}) = \begin{cases} \sum_K a_k(r_{ij} - r_k)^{n_k} H(r_k - r_{ij})H(r_{ij} - 2.05) & r_{ij} > 2.05 \\ \exp(B_0 + B_1 r_{ij} + B_2 r_{ij}^2 + B_3 r_{ij}^3) & 1.0 < r_{ij} \le 2.05 \\ H(1.0 - r_{ij})\dfrac{Q_i Q_j}{r_{ij}}\xi(\dfrac{r_{ij}}{r_s})/r_{ij} & r_{ij} \le 1.0 \end{cases} \tag{5}$$

where Q_i and Q_j are the nuclear charges; and where $r_s = 0.4683766/(Q_i^{2/3} + Q_j^{2/3})$;
$\xi(x) = 0.1818e^{-3.2x} + 0.5099e^{-0.9423x} + 0.2802e^{-0.4029x} + 0.02817e^{-0.2016x}$.

In the above equations, the B_i coefficients are determined by continuity of the potential and its derivative. A_k and a_k are fitted by using *ab initio* calculation and experimental data [13].

The quality of this EAM potential in describing the atomic interactions of Fe was checked by simulating radial distribution functions (RDF) and density of the system. The NPT (constant number, pressure, and temperature ensemble) algorithm in combination with periodic boundary conditions was applied in the process of simulations. The pressure is set to zero. The time step is 5fs.

The RDF is defined in equation (6)

$$g(r) = \frac{V\langle n_i(r, r+\Delta r)\rangle}{4\pi r^2 \Delta r N} \tag{6}$$

Here $\langle n_i(r, r+\Delta r)\rangle$ is the average number of atoms surrounding the ith atom in a spherical shell between r and $r+\Delta r$. N is the total number of atoms involved in the system and V is the volume of the system. The calculated RDFs of liquid iron based on different potentials at 1820 K are shown in figure 1 together with experimental data [9]. As indicated in the figure, the RDF predicted by current EAM model fits the experimental data best.

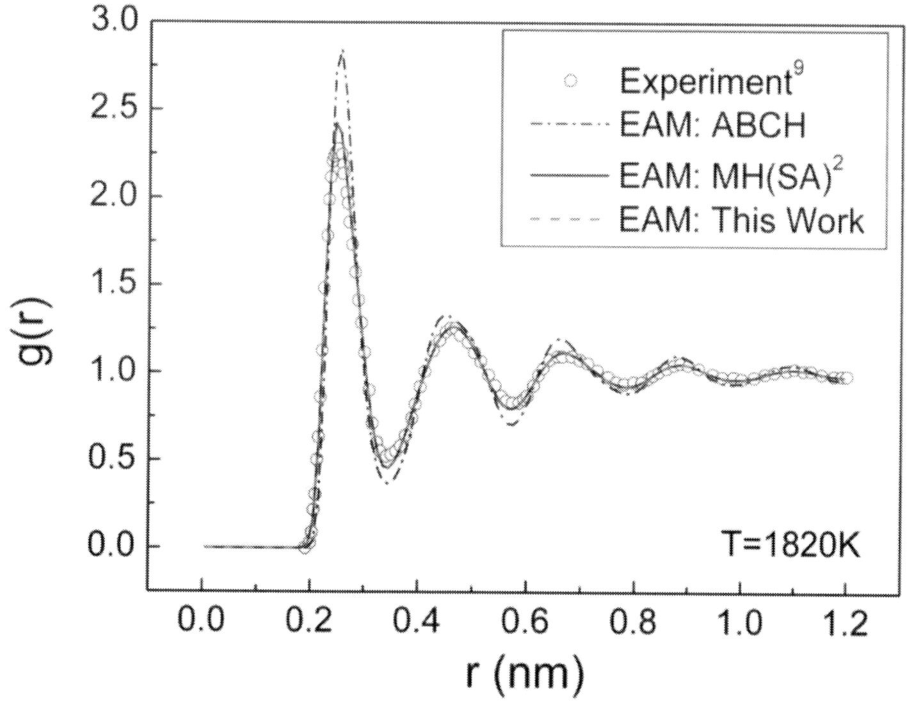

Figure 1. Simulated and experimental RDF of liquid Fe at a temperature of 1820K

Figure 2 shows the density data obtained from simulations for pure solid and liquid BCC Fe at different temperatures. By fitting the simulated data, the temperature dependent densities of solid and liquid phases can be approximated by the second order polynomial functions. The density of the bulk solid phase at a temperature T is

$$\rho_s(T) = 7987.9 - 0.099T - 1.1563\times10^{-4}T^2 \text{ kg/m}^3, \quad (7)$$

and the density of the bulk liquid phase is

$$\rho_l(T) = 7901.4 - 0.3815T - 4.625\times10^{-5}T^2 \text{ kg/m}^3. \quad (8)$$

Experimental data from reference 11, 14, 15 and 16 were also superimposed to figure 2. The difference between the simulated and experimental values of density is less than 5%, indicating the current EAM potential used in our simulation work is a good choice.

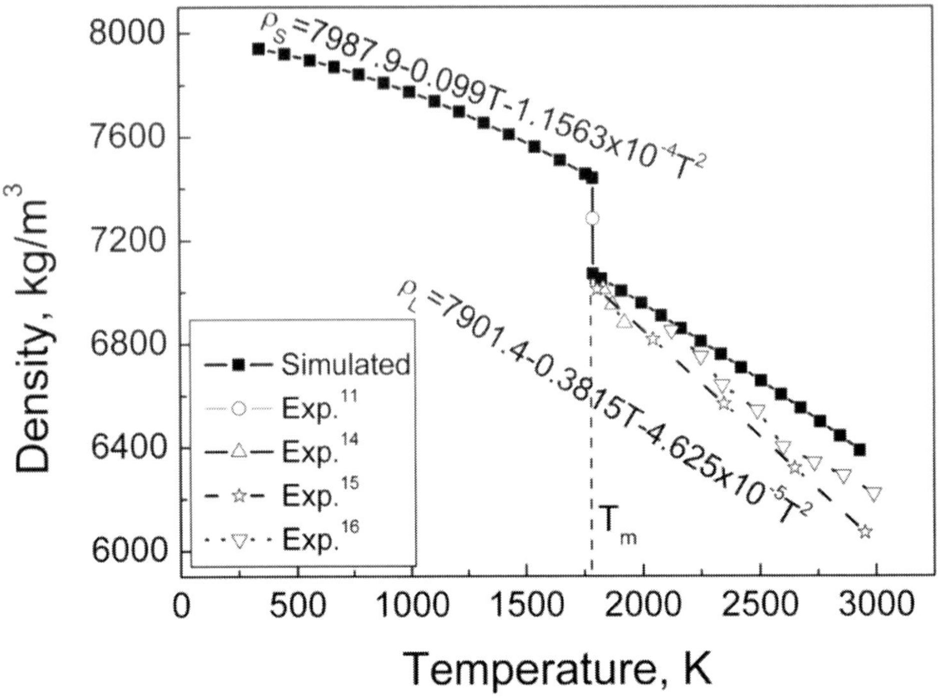

Figure 2. Density of solid and liquid Fe versus temperature

3. Results and discussions

3.1. Calculation of equilibrium melting temperature

The equilibrium melting temperature is a crucial reference point for studying material processes and properties. The accuracy of calculated T_m strongly affects the accuracy of results of other thermodynamic and kinetic parameters. During calculation of the T_m, an orthorhombic box subject to the periodical boundary conditions was filled with 43200 atoms. Half of the atoms were fixed and the others were maintained at high temperature of 3000K employing a Nose-Hoover thermostat [17]. The simulation system will contain two solid-liquid interfaces. Keeping the solid atoms fixed, the liquid atoms were subsequently quenched to an estimated melting temperature and equilibrated over 100 ps. After that, the whole system was allowed to relax up to 500 ps with the temperature maintained at the estimated value of T_m. The equilibrium melting temperature was then determined from the evolution of potential energy of the solid-liquid coexistence system, and the details will be described in next paragraph.

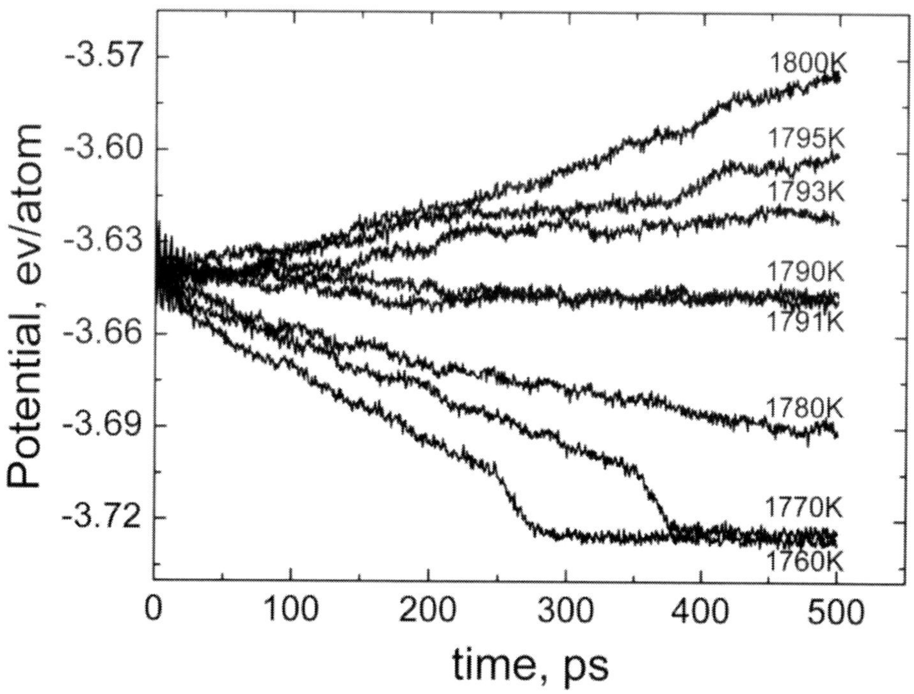

Figure 3. Estimating melting temperature from the potential energy of the solid-liquid coexistence system at different temperature

Table 1. Calculated T_m and Latent heat by using different potential models together with experimental data

Potential	T_m, K	$(T_{Exp}-T_m)/T_{Exp}$, %	L, kJ/mol	$(L_{Exp}-L)/L_{Exp}$, %
Experiment [11]	1811	0	13.8	0
EAM [13]: This work	1790±1	1.2	16.3	-18.1
FS [6]	2400±10 [10]	-32.5	21.2	-53.6
EAM [7]: ABCH	2328.7±4 [11]	-28.6	21.0	-52.2
EAM [9]: MH(SA)^2	1772±2 [11]	2.2	15.6	-13.0
Pair [8]	2311.8±3 [11]	-27.7	25.0	-81.2

Figure 3 shows that the potential energies increase with time at temperature above 1790K, indicating that the system is melting; while the potential energies decrease with time at temperature below 1790K, which means the system is freezing. At 1790K, the potential energy of the system remains unchanged which means the solid and liquid atoms coexist in the system. The equilibrium melting temperature was estimated to be about 1790±1 K, which agrees well with the experimental value of 1811K [11]. The difference is only 1.2%. Comparing with previous potential models [6-9], the EAM potential model used in present study appears to be the most accurate one. Table 1 summarized the calculated T_m by using different potential models together with the experimental data. It shall be noted that for temperatures at 1770K and 1760K, there are plateaus in the potential energy curves near the end of simulation, which means all liquid has solidified fully.

3.2. enthalpy, specific heat and latent heat

The enthalpies of solid and liquid phases were obtained from homogeneous solid and liquid phases respectively. The single phase system containing 16,000 atoms at different temperatures was equilibrated over 500 ps. Then the temperature dependence of enthalpies of the solid and liquid phases were calculated by

$$H(T) = E + PV, \tag{9}$$

where $H(T)$ is the enthalpy as a function of temperature T, E the internal energy; P is the pressure, and V is the volume of the system.

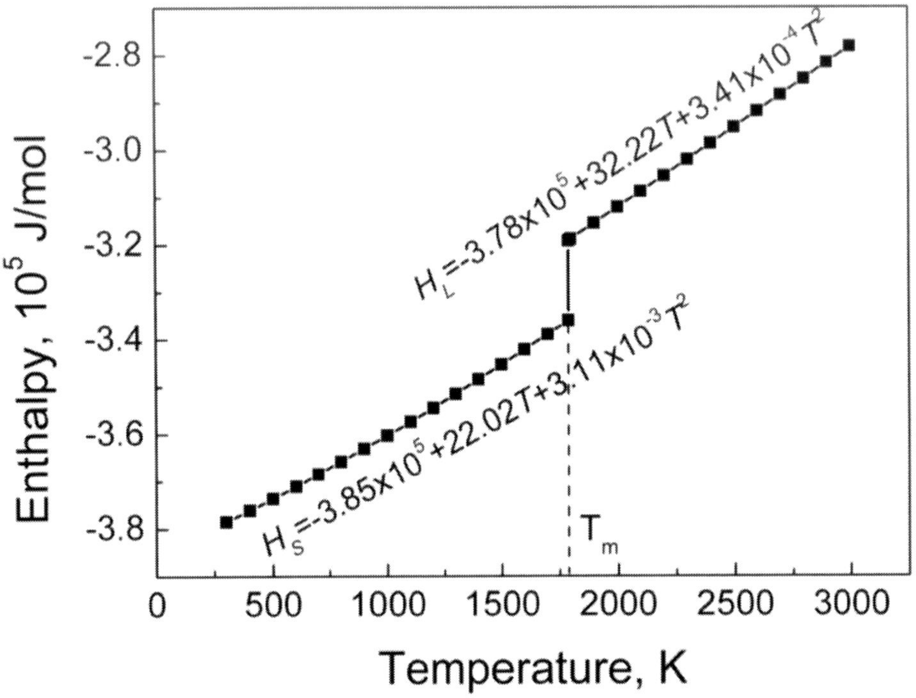

Figure 4. Calculated enthalpy of liquid and solid Fe versus temperature

Calculated enthalpies for solid and liquid Fe are shown in figure 4. It can be seen that the enthalpies increase with temperature. The second order polynomial fit of the data for the solid phase gives

$$H_S(T) = -3.85\times10^5 + 22.02T + 3.11\times10^{-3}T^2 \text{ J/mol}, \tag{10}$$

and for the liquid phase gives

$$H_L(T) = -3.78\times10^5 + 32.22T + 3.41\times10^{-4}T^2 \text{ J/mol}. \tag{11}$$

The latent heat was obtained from the enthalpy difference between solid and liquid phases at the melting temperature, that is,

$$L_V = H_S(T_m) - H_L(T_m) \tag{12}$$

The calculated latent heat is 16.3 kJ/mol. For a comparison, the calculated data using different potentials and the experimentally measured value are summarized in Table 1.

The isobaric specific heat can be obtained by the differential of temperature dependent enthalpy,

$$C_p = \frac{dH(T)}{dT} \tag{13}$$

The calculated isobaric specific heat for the solid phase is

$$C_{pS} = 22.02 + 6.22\times10^{-3}T \text{ J/mol/K}, \tag{14}$$

and the specific heat for the liquid phase is

$$C_{pL} = 32.22 + 3.41\times10^{-4}T \text{ J/mol/K}. \tag{15}$$

Hixson, Winkler and Hodgdon reported the specific heat of liquid Fe for $2125 \le T \le 3950$K to be 45.7J/mol/K at a pressure of 2kbar [16]. In this temperature range, our results slightly increase from 32.9J/mol/K to 33.6J/mol/K. The experimental value from Beutl, Pottlacher, and Jager's work is about 46.2J/mol/K which is measured under a pressure from atmospheric pressure up to 3800 bar [15]. Trevorton and Margrave reported a value of 42.7J/mol/K [18]. In contrast, our result is a little smaller than the reported value. The simulation provides a much wider temperature range for the enthalpy and specific heat. This can be used in the theoretical analysis of solidification and metastable liquid structure.

3.3. Solid-liquid interfacial free energy

The solid-liquid interfacial energy was calculated from Gibbs-Thomson coefficient in relation between critical nucleus sizes and the inverse of the critical undercooling temperature by treating spherical crystal [10,19]. The relation is defined by

$$\Delta T = \Gamma K, \quad K = \frac{1}{r_1} + \frac{1}{r_2}, \tag{16}$$

where Γ is the Gibbs-Thomson coefficient; K is the curvature, and r_1 and r_2 are the principal radii of curvature. In the case of the curvature of a sphere, the Gibbs-Thomson equation (16) can be rewritten as

$$\Delta T = 2\Gamma\frac{1}{r^*} = \frac{2\gamma_{SL}T_m}{L_V}\frac{1}{r^*}, \tag{17}$$

where r^* is the critical radius of a spherical nucleus. Therefore, the solid-liquid interfacial free energy can be calculated by

$$\gamma_{SL} = \frac{\Gamma L_V}{T_m} \tag{18}$$

In order to obtain the critical nucleus size r^*, a series of spherical crystal clusters were generated from a large BCC block of Fe, using various spherical cut-off radii r ranging from 1nm to 5nm, and embedded into a liquid matrix. Then the system was quenched to a desired temperature and relaxed for 500ps to investigate the structure evolution of the embedded solid sphere. According to classical nucleation theory (CNT), if the radius of the solid sphere r is smaller than the critical radius r^* at a certain undercooling, the solid sphere dissolves in the liquid; if $r > r^*$, the solid sphere grows; if $r = r^*$, the solid sphere and the liquid can coexist [19].

Figure 5 shows the size dependence of structure evolution of a spherical BCC crystal Fe embedded in the undercooled liquid at T=1600K. The total number of atoms in the system is 250000. It is found that when the radius of the solid sphere is 1.89nm, the spherical crystal grows in the undercooled liquid until the whole system becomes solid. However, when the radius is 1.75nm, the sphere melts and dissolves into the liquid surroundings. So the critical radius of the spherical nucleus can approximately be taken as 1.82nm at 1600K. The error in estimating the critical radius is within ± 0.07nm.

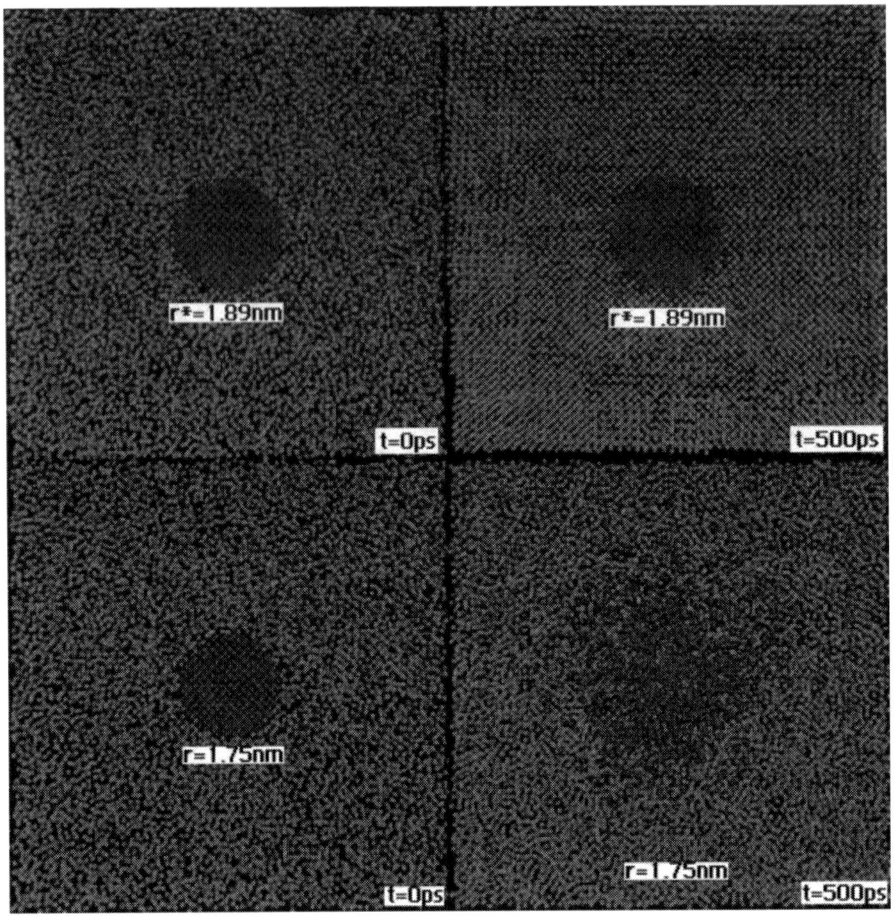

Figure 5. Snapshots of the structural evolution of the constructed spherical BCC crystalline with radius of 1.89nm and 1.75nm at t=0 and 500ps at 1600K.

The critical radii r^* as a function of temperature are presented in figure 6. It clearly indicates that the critical radius has an inverse relationship with the undercooling of the system. A linear fitting of the calculated data between ΔT and $1/r^*$ produces $\Delta T = 3542/r^*$ as indicated in figure 6(b). The Gibbs-Thomson coefficient for spherical Fe nucleus defined in equation (16) was found to be about 1.8×10^{-7} mK, which agrees with the experimental value 1.9×10^{-7} mK of iron measured by Turnbull and previous results calculated for the Finnis-Sinclair (FS) potential [10]. Different sizes of simulation systems containing 54000, 128000 and 250000 atoms were used to test the size effect on our final results. The three systems yield similar results as shown in figure 6(a). The system size has no obvious effect on the critical radius, which has also been confirmed in previous research work [10,19].

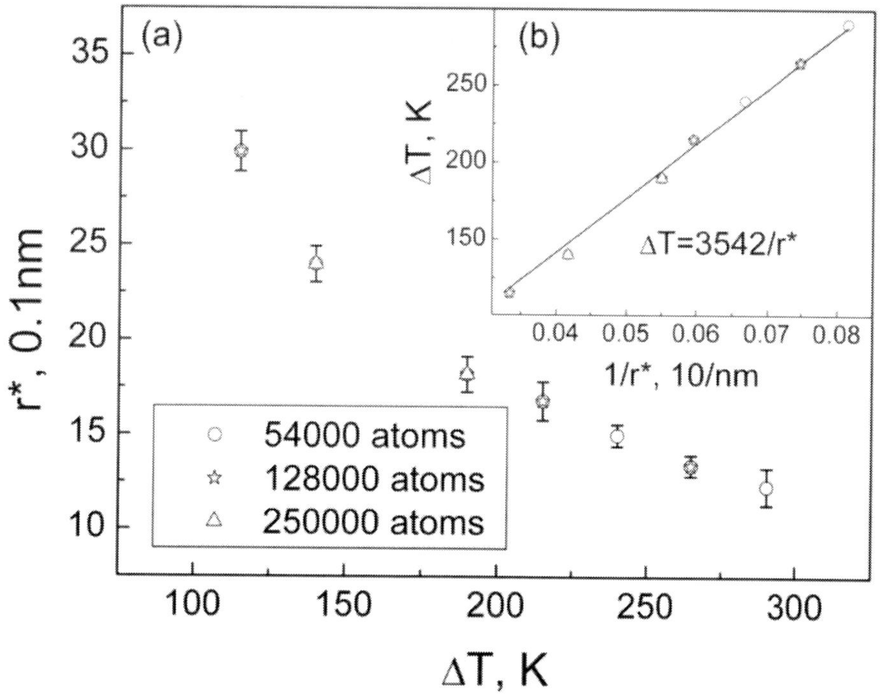

Figure 6. Temperature dependence of the estimated critical radius r^* for spherical crystal nucleus in the undercooled liquid

The solid-liquid interfacial energy at melting temperature was calculated from equation (18). Our simulations give a value of 214mJ/m^2, which is about 4.9% higher than the experimental value 204mJ/m^2. The FS potential used in reference 10 generates a value of 170mJ/m^2 based on the same simulation method. It is about 16.7% lower than the experimental one [10], which can contribute to a relatively high overestimation (about 32.5%) of the melting temperature (2400K vs 1811K). A summary of spherically averaged values of solid-liquid interfacial energy at T_m calculated by different potentials and methods is presented in table 2. The ABCH potential predicts the closest value to the experimental data by using capillary fluctuation method. However, considering the value was obtained at the melting temperature 2328.7K, which is 28.6% higher than experimental one, we think the value calculated by using current potential should be more reasonable.

Table 2. Solid-liquid interfacial energy of pure Fe at T_m estimated by using different potentials and different methods as well as experimental value. CFM means Capillary Fluctuation Method.

Potential	Method	γ_{SL}, mJ/m^2	$(\gamma_{SL\,Exp} - \gamma_{SL})/\gamma_{SL\,Exp}$, %
Experiment [11]	CNT	204	0
EAM [13]: This work	CNT	214	-4.9
FS [6]	CNT	170 [10]	16.7
EAM [7]: ABCH	CFM	206±10 [11]	-1.0
EAM [9]: MH(SA)^2	CFM	175±11 [11]	14.2
Pair [8]	CFM	221±14 [11]	-8.3

IOP Conf. Series: Materials Science and Engineering **33** (2012) 012113 doi:10.1088/1757-899X/33/1/012113

4. Conclusions

Molecular dynamics simulations (MD) based on the embedded atom method (EAM) potential were used to calculate the density, enthalpy, specific heat and solid-liquid interfacial free energy. The calculated results show reasonable agreement with the experimental data. It indicates that the present EAM potential model developed by Ackland *et al.* appears to be the most accurate model for MD simulation of iron solidification. Simulations with the above EAM model predict the equilibrium melting temperature of iron is 1790K, the solid-liquid interfacial energy 214 mJ/m2. The difference with the experimental data is 1.2%, and 4.9% respectively.

Acknowledgement

The research leading to these results has received funding from the European Community's Seventh Framework Programme under grant agreement No: 229108

References

[1] Lv Y J, Chen M 2011 *Int. J. Mol. Sci.* **12** 278
[2] Brillo I, Egry I 2003 *Int. J. Thermophys.* **24** 1155
[3] Okada J T, Ishikawa T, Watanabe Y, Paradis P F, Watanabe Y, Kimura K 2010 *Phys. Rev.* B.**81** 140201
[4] Hoyt J J, Sadigh B, Asta M, and Foiles S M 1999 *Acta Mater.* **47** 3181
[5] Mishin Y, Asta M, Li J 2010 *Acta Mater.* **58** 1117
[6] Finnis M W, Sinclair J E 1984 *Philos. Mag.* A **50** 45
[7] Ackland G J, Bacon D J, Calder A F, Harry T 1997 *Philos. Mag.* A **75** 713
[8] Carlsson A E 1990 *Solid State Phys.* **43** 1
[9] Mendelev M I, Han S, Srolovitz D J, Ackland G J, Sun D Y, Asta M 2003 *Philos. Mag.* **83** 3977
[10] Watanabe Y, Shibuta Y, Suzuki T 2010 *ISIJ Inter.* **50** 1158
[11] Sun D Y, Asta M, Hoyt J J 2004 *Phys. Rev.* B **69** 174103
[12] Daw M S, Baskes M I 1983 *Phys. Rev. Lett.* **50** 1285
[13] Ackland G J, Mendelev M I, Srolovitz D J, Han S, Barashev A V 2004 *J Phys: Condens Matter.* **16** S2629
[14] Yokoyama I, Arai T 2001 *J. Non-Cryst. Solids* **293-295** 806
[15] Beutl M, Pottlacher G, Jager H 1994 *Int. J. Thermophys.* **15** 1323
[16] Hixson R S, Winkler M A, and Hodgdon M L1990 *Phys. Rev.* B. **42** 6485
[17] Hoover W G 1985 *Phys. Rev.* A **31** 1695
[18] Treverton J A, Margrave J L 1971 *J. Chem. Thermodyn.* **3** 473
[19] Bai X M, Li M 2006 *J Chem Phys* **124** 124707

IOP Conf. Series: Materials Science and Engineering **33** (2012) 012114 doi:10.1088/1757-899X/33/1/012114

Application of particle method to the casting process simulation

N Hirata[1], **Y M Zulaida**[1] **and K Anzai**[1]
[1] Department of Metallurgy, Graduate School of Engineering, Tohoku University
6-6-11-1009, AobaAramakiaza, Aoba-ku, Sendai-shi, Miygai-ken, 980-8579, JAPAN

Abstract. Casting processes involve many significant phenomena such as fluid flow, solidification, and deformation, and it is known that casting defects are strongly influenced by the phenomena. However the phenomena complexly interacts each other and it is difficult to observe them directly because the temperature of the melt and other apparatus components are quite high, and they are generally opaque; therefore, a computer simulation is expected to serve a lot of benefits to consider what happens in the processes. Recently, a particle method, which is one of fully Lagrangian methods, has attracted considerable attention. The particle methods based on Lagrangian methods involving no calculation lattice have been developed rapidly because of their applicability to multi-physics problems. In this study, we combined the fluid flow, heat transfer and solidification simulation programs, and tried to simulate various casting processes such as continuous casting, centrifugal casting and ingot making. As a result of continuous casting simulation, the powder flow could be calculated as well as the melt flow, and the subsequent shape of interface between the melt and the powder was calculated. In the centrifugal casting simulation, the mold was smoothly modeled along the shape of the real mold, and the fluid flow and the rotating mold are simulated directly. As a result, the flow of the melt dragged by the rotating mold was calculated well. The eccentric rotation and the influence of Coriolis force were also reproduced directly and naturally. For ingot making simulation, a shrinkage formation behavior was calculated and the shape of the shrinkage agreed well with the experimental result.

1. Introduction

Nowadays, computer simulation has acquired an important role in predicting casting defects. However, the casting process involves various complicated phenomena which interacts each other. Therefore it is quite difficult to predict the casting defects. Recently, particle methods based on Lagrangian methods involving no calculation lattice have been developed rapidly because of their applicability to multi-physics problems[1]. This characteristics of the fully Lagrangian method has attracted considerable attention as a powerful tool to simulate complicated phenomena occurring in the casting processes.

The SPH (smoothed particle hydrodynamics) method and the MPS (moving particle semi-implicit) method are the most commonly used particle methods[2]. Discrete objects (particles) are used as the calculation elements in these methods. Because particles can be placed and moved freely in space, particle methods are fundamentally free from spatial constraint. This feature allows the particle methods to simulate the heat and mass transfer phenomena that are observed during the solidification process more easily and directly than other methods that use the calculation lattice. In addition, an algorithm using the MPS method is simple and similar to the conventional FDM (finite difference method); therefore, multi-physical simulations based on the MPS method have been developed rapidly in various fields[3–8]. However, few studies have been conducted in the field of casting. Cleary et al. reported on the application of the SPH method to the mold filling problem. In their study, SPH simulations that combined the effects of flow, heat transfer and solidification were used and showed good agreement with the experimental results[9]. Recently, a heat transfer and solidification simulation program that was based on the MPS method and incorporated interfacial heat resistance was reported by Hirata et al. They demonstrated that the accuracy of their program was as high as a commercial solidification simulation software based on FVM (finite volume method)[10]. Hirata et al. also reported a flow simulation program based on the MPS method[11], and the coupling simulation of flow and solidification simulation programs[12].

In this study, the fluid flow and solidification simulation programs based on the MPS method were applied to various casting processes. Powder entrapment problem, fluid flow in centrifugal casting and solidification shrinkage formation behavior in the ingot making were demonstrated to show the applicability of particle method on the casting simulation.

2. Numerical method

2.1. The MPS method

2.1.1. Weight Function In the MPS method, governing equations consist of particle interaction equations. All interactions between particles are limited to a finite distance. The strength of interaction can be described using a weight function of distance between two particles. The gradient or other operators can be described using the weight function. Whereas numerous types of weight function have been proposed, equation (1) was used in this study. Equation (1) is based on the kernel function of SPH method, and Ataie-Ashtiani reported that equation (1) gives better stability for fluid flow analysis[13].

$$w(r, c_k r_0) = \begin{cases} \dfrac{40}{7\pi c_k^{\,2}}\left(1-6\left(\dfrac{r}{c_k r_0}\right)^2+6\left(\dfrac{r}{c_k r_0}\right)^3\right) & (0 \le r < 0.5c_k r_0) \\ \dfrac{10}{7\pi c_k^{\,2}}\left(2-2\dfrac{r}{c_k r_0}\right)^3 & (0.5c_k r_0 \le r < c_k r_0) \\ 0 & (c_k r_0 \le r) \end{cases} \tag{1}$$

Here, r is the distance between the neighboring particles. r_0, which denotes the specific size of the particle, has the same significance as the lattice size in FDM. c_k is the kernel size coefficient and usually varies between 2 and 4. Particle number density n_i of the particle i is used in the MPS method to calculate the interaction between particle i and the surrounding particles. n_i is the sum of the weight function of the particles surrounding the particle i, and is expressed as follows.

$$n_i = \sum_{i \ne j} w\left(\left|\boldsymbol{r}_j - \boldsymbol{r}_i\right|, c_k r_0\right) \tag{2}$$

Here, $\boldsymbol{r}_i$ and $\boldsymbol{r}_j$ are the position vectors of particles i and j, respectively. n_i is equal to n_0 in the case that a particle has no surface particle around it. Assuming that the fluid is incompressible and

IOP Conf. Series: Materials Science and Engineering **33** (2012) 012114 doi:10.1088/1757-899X/33/1/012114

considering that the particle number density n_i is directly related to the fluid density, we can use n_0 for the mass conservation condition in the incompressible flow analysis using the MPS method[1].

2.1.2. Particle interaction models Particle interaction models are used to describe differential operator in the particle method [3]. If ϕ_i and ϕ_j are arbitrary scalars at positions $\boldsymbol{r}_i$ and $\boldsymbol{r}_j$, then the particle interaction models for the differential operators can be expressed as follows.

$$\nabla\phi_i = \frac{d}{n_i}\sum_{i\neq j}\frac{(\phi_i-\phi_j)}{|\boldsymbol{r}_j-\boldsymbol{r}_i|}\cdot\frac{(\boldsymbol{r}_j-\boldsymbol{r}_i)}{|\boldsymbol{r}_j-\boldsymbol{r}_i|}w(|\boldsymbol{r}_j-\boldsymbol{r}_i|, c_k r_0) \tag{3}$$

$$\nabla^2\phi_i = \frac{2d}{n_i}\sum_{i\neq j}\frac{(\phi_i-\phi_j)}{|\boldsymbol{r}_j-\boldsymbol{r}_i|^2}w(|\boldsymbol{r}_j-\boldsymbol{r}_i|, c_k r_0) \tag{4}$$

Equation (3) is a gradient model and equation (4) is a Laplacian model. Suffixes i and j represent the assigned numbers of particles. d is the number of space dimensions.

2.2. Fluid flow Analysis

The governing equations for incompressible flow consist of the continuity equation and Navier-Stokes equation as follows.

$$\frac{D\rho}{Dt} = 0 \tag{5}$$

$$\frac{D\boldsymbol{u}}{Dt} = -\frac{1}{\rho}\nabla p + \nu\nabla^2\boldsymbol{u} + \boldsymbol{f} \tag{6}$$

Here, ρ indicates density (kg/m^3); $\boldsymbol{u}$, velocity vector (m/s); p, pressure (Pa); ν, kinematic viscosity (m^2/s); and $\boldsymbol{f}$, the body force vector including gravity.

Flow calculation by the MPS method is based on the predictor-corrector method similar to that in the case of the FDM or other conventional methods. The major differences between the MPS method and the FDM are the formulation of the Poisson equation of pressure and the calculation of particle position. In the MPS method, the tentative particle position is calculated using tentative velocity in the prediction phase; therefore, the correction in position is calculated along with the correction in velocity in the correction phase. The Poisson equation for pressure in the MPS method is described using the tentative particle number density n^* and n_0 as follows[1]. n^* is the particle number density calculated using the tentative particle positions.

$$\nabla^2 p^{k+1} = -\frac{\rho}{\Delta t^2}\frac{n^*-n_0}{n_0} \tag{7}$$

2.3. Heat Transfer Analysis

Heat transfer equations are described as follows.

$$\frac{DH}{Dt} = \lambda\nabla^2 T \tag{8}$$

$$\Delta Q = \frac{\Delta t}{R}(T_1 - T_2) \tag{9}$$

Here, H represents the enthalpy per unit volume (J/m^3); t, time (s); λ, thermal conductivity (W/m·K); and T, temperature (K). ΔQ (J/m^2) is the thermal flux per unit area in time Δt (s), R is heat resistance (m^{2}K/W), and T_1(K) and T_2(K) are the surface temperatures of materials 1 and 2 at the interface between the materials.

Heat transfer equations (equation (8) and (9)) were transformed using the Laplacian model (equation (4)), and the enthalpy at the next time step k+1 was obtained from the temperature distribution at the time step k as follows. Superscripts k and k+1 signify time steps in the following equation.

$$H_i^{k+1} = H_i^k + \Delta t\left(\frac{2d}{n_i}\sum_{i\neq j}\frac{1}{\dfrac{1}{2\lambda_i}+\dfrac{1}{2\lambda_j}+\dfrac{R}{\left|\boldsymbol{r}_j^k-\boldsymbol{r}_i^k\right|}}\frac{T_j^k-T_i^k}{\left|\boldsymbol{r}_j^k-\boldsymbol{r}_i^k\right|^2}w\left(\left|\boldsymbol{r}_j^k-\boldsymbol{r}_i^k\right|,c_k r_0\right)\right) \tag{10}$$

The enthalpy method was used to calculate the effect of solidification.

2.4. *Temperature dependence of density*

Temperature-dependent r_0 variation was introduced to describe shrinkage using the following equation.

$$r_{0,i} = \left(\frac{M_i}{\rho_i}\right)^{1/d} \tag{11}$$

Here, $r_{0,i}$, M_i and ρ_i signify specific size (m), mass (kg) and density (kg/m^3) of the particle i, respectively. Assuming that the masses of each particle are constant, we can calculate the solidification shrinkage using the temperature-dependent density as described in equation (11). Particles having various values of r_0 can be handled simultaneously in the MPS method using the modified weight function. The modified weight function w_i at particle i is described as follows.

$$w_i = \frac{r_{0,j}^d w\left(\left|\boldsymbol{r}_j-\boldsymbol{r}_i\right|,c_k r_{0,i}\right)+r_{0,i}^d w\left(\left|\boldsymbol{r}_j-\boldsymbol{r}_i\right|,c_k r_{0,j}\right)}{2r_{0,i}^d} \tag{12}$$

3. Application

3.1. *Powder entrapment in continuous casting*

3.1.1 Calculation Model Firstly, mold powder entrapment problem was simulated. Mold powder entrapment is one of the fatal defects in continuous casting, which occurs by the vortex or complex flow at the melt-powder interface. In particle method, plural kinds of fluid particles are easily handled only by defining the properties for each particle. In this case, two kinds of fluid were defined; A, heavy fluid (representing the melt); and B, light fluid (representing the powder). The values of density were lower than the actual values of melt and powder for the calculation stability. The calculation was performed using half model of the mold in the continuous casting as shown in the figure 1. A forced convection region was defined to represent the melt inflow from the nozzle, and a baffle representing the nozzle was set above the convection region. If a particle comes into the convection region, the velocity of the particle was fixed at 1.5 m/s to a 30 degree downward direction.

3.1.2. Results Figure 2 shows the example of the result of the mold powder entrapment calculation. The result was visualized using ray-tracing software POV-Ray 3.6. In the visualization, the melt was set to be transparent and only the powder was visible. As a result of 2-phase flow calculation, the meniscus curve of melt-powder interface could be observed (figure 2 (a)). In figure 2 (b), complex behavior of the powder was calculated well; the flow was disturbed by the baffle, the powder was entrapped by the vortex, and the surfacing of the powder by the buoyancy force was taken into account. Because the behavior of the powder was described by the movement of particles (i.e. calculation elements themselves), the entrapped powder particles never disappear by numerical diffusion (figure 2 (c)) and could be traced well. The simulation program could also reproduce the influence of flow condition on the entrapment qualitatively. The amount of entrapped powder varied depending on the viscosity of fluid or density difference between melt and powder; higher viscosity of fluids and the larger difference of density reduce the amount of entrapment.

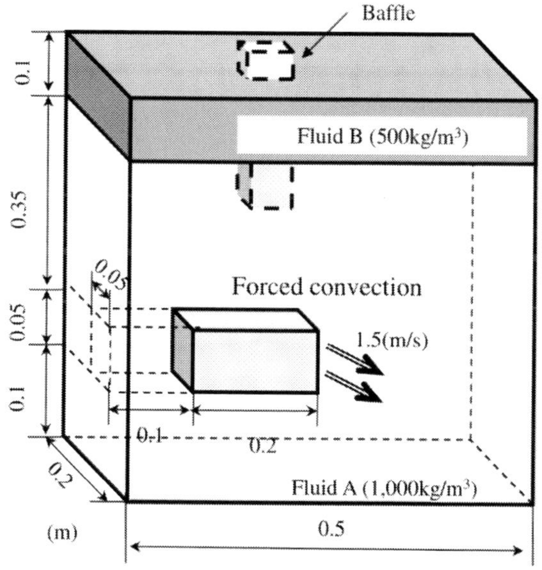

Figure 1. Calculation model : half model of the continuous casting mold with a baffle.

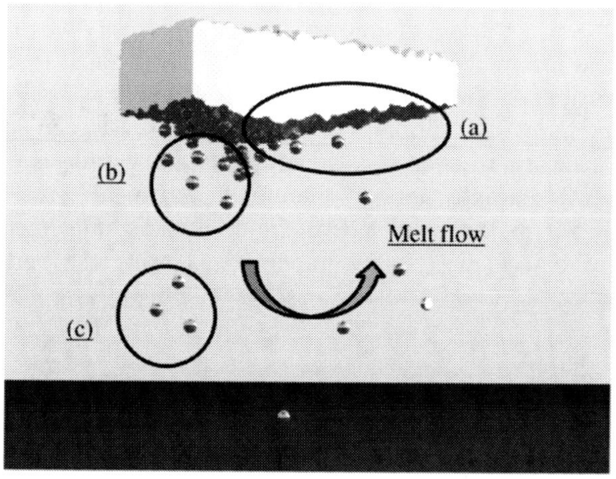

Figure 2. Calculated result of mold powder entrapment problem. Only powder particles are indicated using ray-tracing software POV-Ray 3.6. (a) 2-phase flow reproducing the shape of the powder-melt interface. (b) Influence of the vortex and the buoyancy on the entrapment of the powder. (c) Tracking the entrapped powder.

3.2. Centrifugal casting

3.2.1. Calculation Model Centrifugal casting is the process using a mold with high rotational speed. The melt flow is induced by the rotating mold, therefore the process is generally calculated using the special boundary condition at the mold surface. However, in the conventional method using calculation lattice, it is difficult to introduce the eccentric rotation or other complex movements of components. In this section, the mold was smoothly modeled along the shape of the real mold, and the fluid flow and the rotating mold are simulated directly.

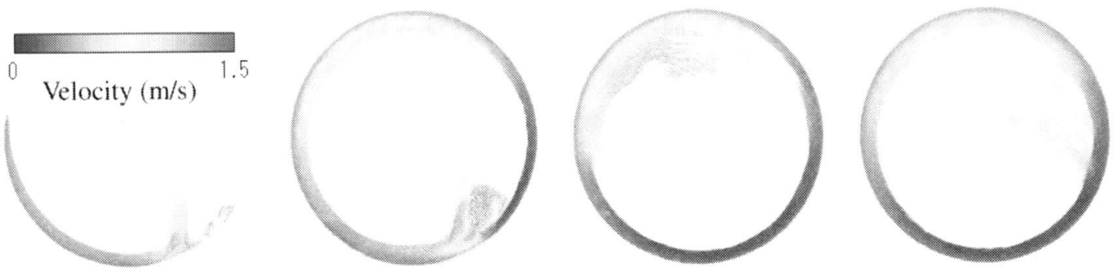

(a) 0.1 s (b) 0.3 s (c) 0.5 s (d) 2.0 s

Figure 3. Fluid flow of centrifugal casting : 250rpm.

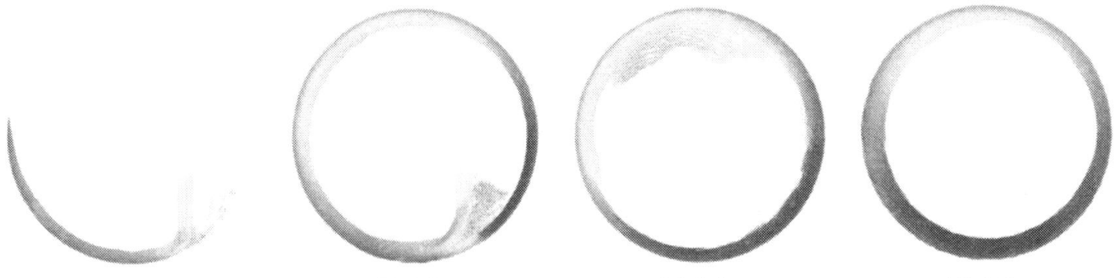

(a) 0.1 s (b) 0.3 s (c) 0.5 s (d) 2.0 s

Figure 4. Fluid flow of centrifugal casting : 250rpm, with eccentric rotation.

3.2.2. Results Figure 3 and 4 show the results of the 2-dimensional calculation. Inner diameter of the mold was 100 mm, rotation speed was 250 rpm. Viscous fluid (the kinematic viscosity was 1.0 x 10^{-5} m^2/s) flowed into the cavity by 1.0 m/s, and inflow duration was 0.3 s. The centrifugal force is weak under this condition, therefore an abrasion of the melt occurred. The abrasion could be observed even after 2 s (figure 3 (d)) under the normal condition. In contrast, no abrasion could be observed after 2 s in the figure 4 (d), in which eccentric rotation is considered; rotation was decentered by 2 mm. In this case, radius of rotation was partially longer, therefore centrifugal force and subsequent rotation speed increased. As shown in the results, Coriolis force was implemented naturally, and the influence of movements of objects was easily implemented using the particle method.

3.3. Shrinkage formation

3.3.1. Calculation Model Finally, a shrinkage formation behavior was simulated. Inner and outer shrinkage are sometimes separated at the upper part of the casting, and the boundary is called bridging. The solidification behavior of the upper part of casting is complicated; at which heat transfer, solidification and shrinkage flow occur simultaneously. In this section, the simulation programs for solidification and flow were combined considering temperature dependency of density. The combined program was applied to cylindrical shaped casting, and the calculated result was compared with the experimental result. Pure Sn was used for melt, sand for a mold. Air particles were used to consider the influence of air-cooling.

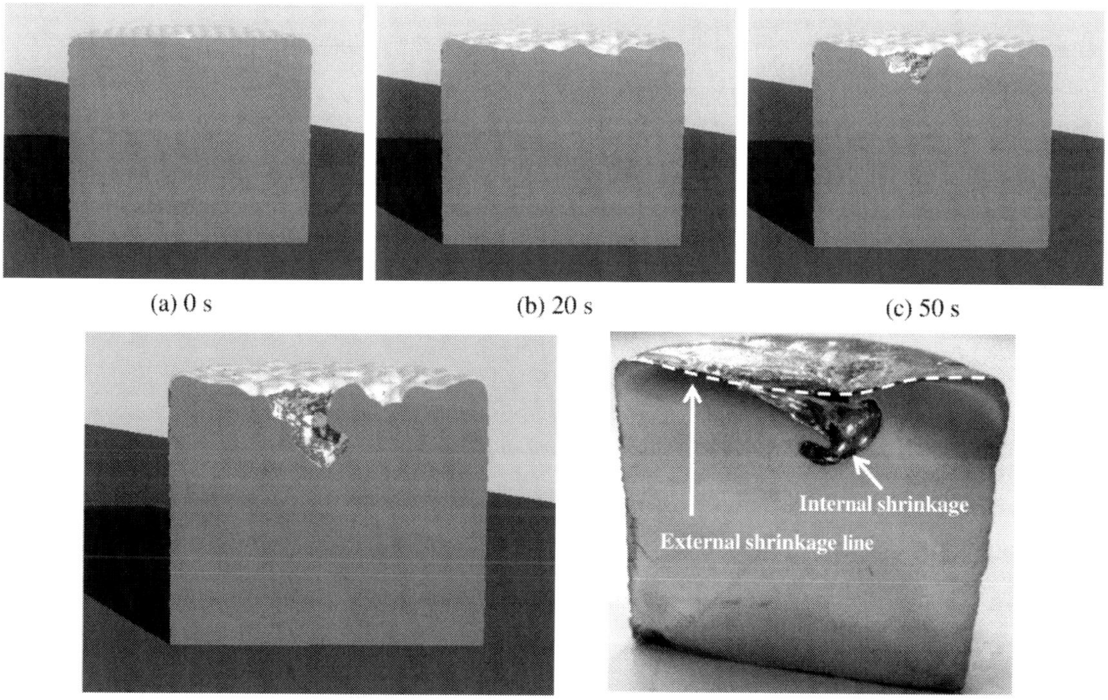

(a) 0 s (b) 20 s (c) 50 s

(d) 85 s (e) Experimental result.

Figure 5. Calculated shrinkage formation behavior and the experimental result.

3.3.2. Results Figure 5 (a) to (d) shows the bridging formation behavior that separated inner and outer shrinkage of the casting. The results were visualized using ray-tracing software POV-Ray 3.6. Figure 5(e) is the experimental result. The figures show the cross section of the casting. In figure 5 (a) to (d), the shrinkage formation behavior was described realistically, and the bridging at upper part of the casting was observed; the influence of air-cooling was necessary to reproduce the bridging formation. The particle method could take into account the surface drop by the solidification shrinkage, and the influence of subsequent shape variation on the heat transfer pathway between the casting and the air, easily and directly.

4. Conclusion

The flow and solidification simulation program based on the MPS method was applied to the casting processes, and the applicability of the particle method on the casting simulation were discussed. Firstly, the mold powder entrapment problem in continuous casting was calculated. As a result, complex behavior of the melt and the powder around the meniscus was calculated and the tracking ability of trapped powder was demonstrated. The results show the mass conservation ability of the particle method and the ease of the handling the multi-phase flow. Next, the flow in the centrifugal casting was simulated. The melt flow dragged by the rotating mold was described directly, and influence of eccentric rotation was demonstrated. The particle method could calculate the movement of the mold directly and Coriolis force was implemented naturally. Finally, the shrinkage formation was calculated using the combined simulation program for solidification and flow. As a result, the shrinkage formation behavior in the pure Sn casting was simulated realistically, and the influence of air-cooling on the bridging formation was reproduced well. In conclusion, the particle method has high capability to simulate complex phenomena involving multi-phase and multi-physics with the large movement of the objects, and is expected to simulate complicated casting processes.

5. References

[1] Koshizuka S 2005 *Ryushihou* (Tokyo: Maruzen)
[2] Koshizuka S 2008 *Ryushihou Simulation* (Tokyo: Baifukan)
[3] Koshizuka S, Tamako H and Oka Y 1995 *Computational Fluid Dynamics J.* **4** 29-46.
[4] Koshizuka S and Oka Y 1996 *Nucl. Sci. Eng.* **123** 421-434
[5] Koshizuka S, Nobe A and Oka Y 1998 *Int. J. Numer. Methods Fluids* **26** 751-769
[6] Koshizuka S and Oka Y 1996 *Nucl. Sci. Eng.* **123** 421-434
[7] Belytchko T, Krongauz Y, Organ D, Fleming M and Krysl P 1996 *Comput. Methods appl.Mech. Eng.* **139** 3-47
[8] Shao S and Lo E Y M 2003 *Adv. Water Resour.* **26** 787-800
[9] Cleary P W and Ha J 2006 *Applied Mathematical Modelling* **30** 1406-1427
[10] Hirata N and Anzai K 2008 *Journal of Japan Foundry Society* **80** 81-87
[11] Hirata N and Anzai K 2011 *Journal of Japan Foundry Society* **83** 259-267
[12] Hirata N and Anzai K 2011 *Materials Transactions* **52** 1931-1938
[13] Ataie-Ashtiani B and Farhadi L 2006 *Fluid Dynamics Research* **38** 241-256

Exploration of the double-diffusive convection during dendritic solidification with a combined volume-averaging and cellular-automaton model

A Kharicha, M Stefan-Kharicha, M Wu, A Ludwig
Chair for Simulation and Modelling of Metallurgical Processes, Dept. of Metallurgy, Univ. of Leoben, A-8700 Leoben, Austria

E-mail: abdellah.kharicha@notes.unileoben.ac.at

Abstract. A two-dimensional model is built and used to study thermo-solutal (double diffusive) convection generated during the solidification of a binary mixture in a rectangular enclosure cooled from bottom and side walls. In order to catch the smallest solute plumes scale the solidification model simulates directly the envelope of the columnar dendrites with a cellular automaton model. The mushy interior of the dendrite is modelled with a volume averaging method. The model predicts the occurrence of several flow regimes during solidification, such as turbulent, stratified and meandering flows. As a result or origin of the two dimensional meandering flow pattern, the concentration field is found to be organised in horizontal layers of uniform concentration. Those layers are separated by very thin boundary layers so that the concentration varies vertically in staircases.

1. Introduction

The study of double-diffusive natural convection has many environmental and industrial applications including oceanography, petrochemical applications processes, oil and gas extraction, and solidification processes. In a number of experimental and numerical studies, it has been shown that the interaction between thermal and compositional buoyancy, phenomena such as oscillatory flow, chaotic and turbulent flows can be generated. Turner [1] described double-diffusive convection as a phenomenon that happens when there are gradients of two or more properties with different molecular diffusivities. The double-diffusive instability represents an efficient phenomenon that releases potential energy stored in one of the density components, even when the density distribution is statically stable. There are mainly two types of double-diffusion, which are known as the plumes (or fingers) regime and the diffusive convection regime.

The plumes instability [2] appears as a close-packed array of up-and downward flowing convection cells which exchange large quantity of heat laterally, but only small amount of solute. This results in an advective transport of solute and, to a lesser extent, heat, in the vertical direction. This form of instability is, however, of limited interest, since diffusive oscillations are sometimes replaced by the relatively steady patterns, characterized by a layered distribution of temperature and (or) concentration [3]. Well-mixed layers separated by sharp diffusive interfaces are responsible for stair-like appearance of the temperature profiles in the central Arctic thermocline. Such profiles are referred to as thermohaline staircases. There exist numerous studies on formation of staircase-like structures. Its mathematical modelling, including fluid dynamics, and in oceanography is supported by many laboratory experiments with a large amount of experimental data [4-7]. Many aspects of the origin of such structures, in spite of long history of its studies, are still unclear and are a subject for speculations.

During solidification, natural convection arises from the temperature and concentration gradients. Concentration gradients are caused by the preferential incorporation or rejection of solute element at the solid-liquid interface. Due to the fact that direct observation of solidification and convection is not possible in metallic alloys, many researchers in the field of metallurgy have used transparent analogue systems [8-10], such as ammonium chloride salt. Using such transparent analogues it has been possible to visualize the effects of the convections on the solidification pattern and the resulting macrosegregation. Using a shadowgraph method during the solidification of an ammonium chloride-water solution [8, 9] has described a variety of double-diffusive phenomena including plumes and layering. More recently PIV measurements of the flow velocity revealed the existence of a "meandering flow regime" [10]. The liquid was found to flow along a snaky path from the top of the cavity to the bottom near the mushy zone. The velocity was mainly horizontal in the bulk, the flow was downward only at the vicinity of the lateral mushy zones. A staircase layering of the concentration field could be at the possible origin of this snaky flow regime. Unfortunately the concentration field could not be directly observed.

Double diffusive phenomena considerably influence the local solidification rate and in some cases a part of the mushy region can even be re-melted [8-10]. Remelting is believed to be the most probable origin of the chimney (freckle). Worster [11] studied the linear instability of directional solidification. Assuming a simple equation for permeability, he demonstrated the existence of two modes of instability. At large wavelengths convection cells are able to penetrate deep into the mushy zone, while at shorter wavelengths the convection cells remain confined in the liquid region. The long wavelengths were assumed to be at the origin of chimneys. Modelling of the evolution of convection cells and chimneys demands an analysis that captures all nonlinearities in the governing equations.

Simulations of the double diffusive convection during solidification were performed for various alloy composition [12-14]. However the formation of plumes or layers could rarely be captured. The main reasons lie on the limited grid size, but also on the fact that variations of concentrations are generated at discrete positions separated by a typical distances which could be related to the primary arm spacing. In fact, solidification (or melting) introduces an additional parameter compared to studies related to oceanography where most of the analysis focuses simply on the mixing or formation of solute layers. During columnar solidification, if the density of the rejected solute is lower, each dendrite tip launches convective plumes into the upper liquid bulk region.

The present paper presents a numerical model which aims to study the generation of the "double diffusive turbulence" and the occurrence of layering as observed experimentally by Kharicha [10] in a thin enclosure (80 mm high, 100 mm wide, 10 mm thick). Since very small scale of the dendrite envelope needs to be captured, a cellular automaton model is employed. Evolution of the solid fraction in dendrites, the flow and the species transport are modelled with a volume averaging approach [15].

2. Numerical Model

The model describes conservation of mass, momentum, energy and species for a binary mixture, fluid-solid system. The shallow layer 2D model is assumed and the equations are written in terms of volume-averaged quantities. We assume that the total buoyancy is the sum of thermal and solutal buoyancies. The dendritic solidification model is divided in two parts. The first part aims to predict the large scale morphology of the solidification. It consists in using a tuned cellular automaton to model the dendrites envelope. The inner volume of the dendrite is not directly simulated but modelled as a mushy region. The model is based on the following assumptions: (1) no diffusion occurs in the solid phases, and the solute reaches local equilibrium concentration at the solid/liquid interface; (2) equiaxed crystals are neglected, no nucleation occurs in the bulk liquid, only the growth of the seeds placed at the cooled walls is considered.

2.1. Shallow layer model and volume-averaging equations

The volume shrinkage/expansion is ignored, and an incompressible Boussinesq approximation is employed to consider the thermo-solutal buoyancy. We assume a constant thermal α and solutal β expansion coefficient around the reference temperature T_0 and concentration C_0. The 3D cell (geometry) is simplified as a 2D shallow layer. The mixture velocity $\vec{u}$ and mixture temperature T are assumed to be parabolic within the thickness (d) of the cell. The 2D equations governing conservation of mass, momentum, energy and species concentration are written as follows:

$$\left.\begin{aligned} &\frac{\partial(\rho_l \vec{u})}{\partial t} + \nabla\cdot(\rho_l \vec{u}\vec{u}) = \nabla(\mu_l \nabla\vec{u}) - \nabla P + \rho_l f_l \vec{g}\left[\alpha(T - T_0) + \beta(C_l - C_0)\right] - \frac{\mu_l}{\Pi}\vec{u} \\ &\nabla\cdot\vec{u} = 0 \end{aligned}\right\} \quad (1)$$

$$\frac{\partial(\rho_l f_l C_l)}{\partial t} + \nabla\cdot(\rho_l C_l \vec{u}) = \nabla\cdot(f_l \rho_l D_{eff} \nabla C_l) - \rho_s C_s \frac{\partial f_s}{\partial t} \quad (2)$$

$$\rho_s \frac{\partial f_s}{\partial t} = M_{CA} + M_{la} \quad (3)$$

$$\bar{\rho}\, C_p \frac{\partial T}{\partial t} + \bar{\rho}\, C_p \nabla\cdot(T\vec{u}) = \nabla(D_T \nabla T) + \rho_s L \frac{\partial f_s}{\partial t} - 2Q_{LG} \quad (4)$$

where $P, f_l, f_s, C_l, \rho_l, \rho_s, \mu_l$ are the static pressure, the liquid and solid fractions, the concentration of the solute (NH_4Cl), the liquid and solid densities, and the liquid viscosity. $\Pi, D_{eff}, D_T, \bar{\rho}, C_p$, L are the local permeability, the effective solute diffusivity, the mixture heat conductivity, the mixture density, the mixture heat capacity, and the heat of fusion. M_{CA} is the mass transfer due to the growth of the dendrite envelop, M_{la} the mass transfer due to the growth of secondary arms, and Q_{LG} is the heat lost through the front and back glass.

The glasses temperature evolves according to the exchanged heat with the bulk liquid Q_{LG} and with the ambient air Q_{Air}:

$$\rho_G C_{p\_G} \frac{\partial T_G}{\partial t} = \nabla(D_{T\_G} \nabla T_G) + Q_{LG} - Q_{Air} \quad (5)$$

In the above equation ρ_G, C_{p_G}, D_{T_G} are the glass density, heat capacity, and heat conductivity. Q_{Air} is the heat lost (or gained) through contact with air. The heat exchange terms takes the form:

$$Q_{LG} = \frac{D_T}{2d^2}(T - T_G) \quad (6)$$

The exchange with air is modelled with a constant heat transfer coefficient h:

$$Q_{Air} = -h\left(T_G - T_{Ambient}\right) \quad (7)$$

The temperature of the ambient air is assumed to be constant and equal to 20°C.

2.2. Cellular Automaton model for the dendrites

The envelope of the dendrite is assumed to solidify a fixed quantity F_{si} of solid. This quantity can represent the volume of solid contained in some primary and secondary tip length. Here F_{si} is set to be equal to 3%, this means that when a cell has been crossed by the envelop surface, 97% of the cell is still liquid. The algorithm of envelope growth consists in stating that each cell has three possible states: liquid ($f_s = 0$), interface ($0 < f_s / F_{Si} < 1$), or mush ($f_s / F_{Si} > 1$).

At the interface the equilibrium liquid concentration C_L^i is defined as:

$$C_L^i = C_0 - \frac{1}{m_L}\left[T_L - \Gamma_i \bar{k} - T(t)\right] \quad (8)$$

Where T_L is the equilibrium liquidus temperature of the alloy, m_L the liquidus slope in the phase diagram, $\bar{k}$ the mean curvature of the mush/liquid interface, Γ_i is a modified Gibbs-Thomson coefficient and C_0 is the initial concentration in the liquid. A nucleus that forms at a particular location will grow based on the growth velocity of the mush/liquid interface given by equation (9):

$$M_{CA} = V_n \left(\rho_l C_L^i - \rho_s C_S\right) F_{Si} = \begin{cases} M_{LGK} & (\bar{k} > 0.95/\Delta x) \\ D_{eff} \rho_l F_{Si} \dfrac{C_l - C_L^i}{\Delta x} & (\bar{k} \le 0.95/\Delta x) \end{cases} \quad (9)$$

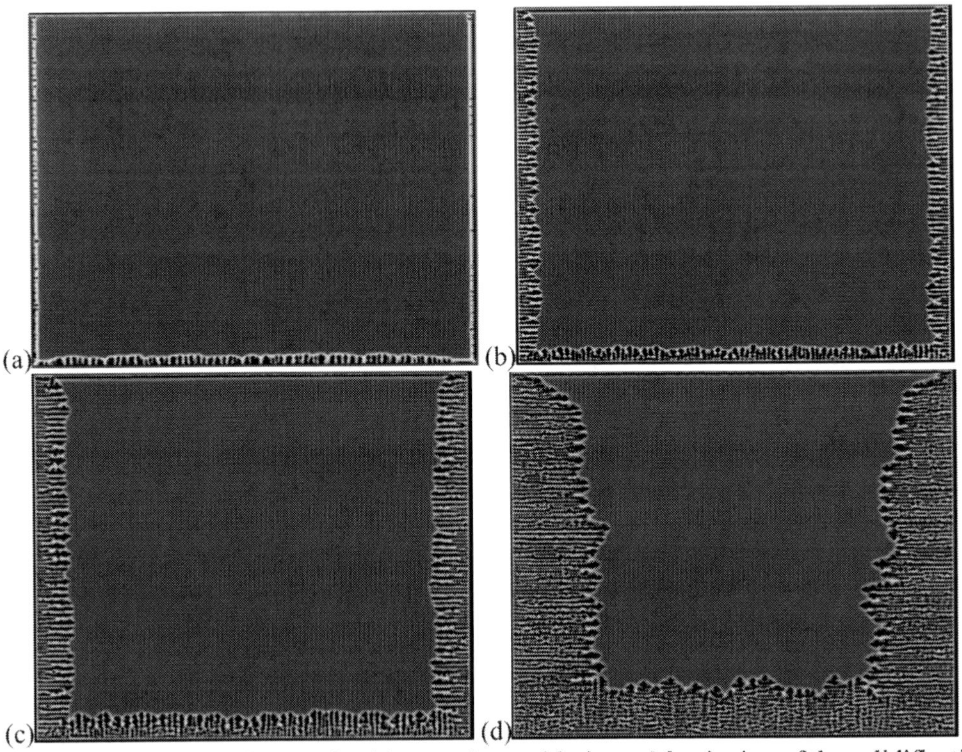

Figure 1. Evolution of the columnar dendrite envelope with time. a) beginning of the solidification (t~5 min); b) early stage of solidification (t~9 min); c) mid-stage of solidification (t~14 min); d) late stage (t ~30 min).

where V_n is the normal velocity of the interface, M_{LGK} is the species transfer, Δx the grid size, C_S the solute concentration in the solid of the interface, D_{eff} is the effective liquid solute diffusivity, $\vec{n}$ the vector normal to the interface. Since grid size Δx is too large to resolve the dendrite tip, the mass transfer rate is artificially adapted so that V_n equals the Lipton-Glicksman-Kurz (LGK) solution for columnar tip growth. The cell in which the dendrite tip is located is identified by its high level of curvature almost equal to the maximum possible curvature for the present grid size ($\bar{k}_{max} \approx 1/\Delta x$). Outside the tip region, the growth is assumed to be controlled through diffusion. However since the cell size is coarse, the effective diffusivity of the solute must be increased by relatively large factor. In the present study the possibility of growth anisotropy due to crystallographic orientation is not expressively included, some anisotropy is naturally introduced by the grid orientation. The columnar dendrites modelled by the present CA model must achieve a defined tip velocity and a main primary arm spacing. The tip velocity is automatically fulfilled by the use of the LGK model. The primary arm spacing will depend on the proper choice of modelling parameters, for the present case is better to choose a combination which is not too far from the original physical parameters. We simply assume that the present dendrite growths as if the scale was N (N=100) times larger, so the curvature and the effective diffusivity equation (9) are simply multiplied by the same factor. The summary about physical and effective properties used in the present investigation, are presented in Table 1.

2.3. Mass transfer in the inner dendritic mushy region

For the process to be studied, the mushy region never solidifies completely. If we can imagine that the mesh topology can be represented as parallel or crossed cylindrical sticks, we can use the multiphase volume-averaging approaches developed by Wu et al. [15]. Originally this model was developed for periodic alignment of primary arm columnar trunks. Since the primary arm spacing λ_1 is already modeled with the cellular automaton model, the model of Wu et al.[15] is used in the present analysis with the secondary arms spacing λ_2. The size of the sticks must represent an average of the secondary arms diameters. For the volume elements that has been passed by the envelop front, a diffusion-controlled growth model around cylindrical trunks is used. The growth velocity in the radial direction of such a cylindrical trunk is:

$$V_a = \frac{D_{eff}}{R_a}\frac{C_L^i - C_l}{C_L^i - C_s}\ln^{-1}(\frac{R_f}{R_a}) \quad 10)$$

where R_a is the average radius of the dendritic trunks, $R_f = \lambda_2/\sqrt{\pi}$ is the maximum packing radius of a secondary dendrite arm trunk. For the sake of simplicity, the trunks are assumed to be parallel and staggered, so that the surface area of the secondary arm trunks per volume is $S_A = \left(4/\sqrt{3}\right)\pi R_a/\lambda_2^2$.

The net volume-averaged mass-transfer rate for elements within the mushy region is:

$$M_{la} = V_a\left(\frac{4}{\sqrt{3}}\pi\frac{R_a}{\lambda_2^2}\right)\rho_s f_l \quad (11)$$

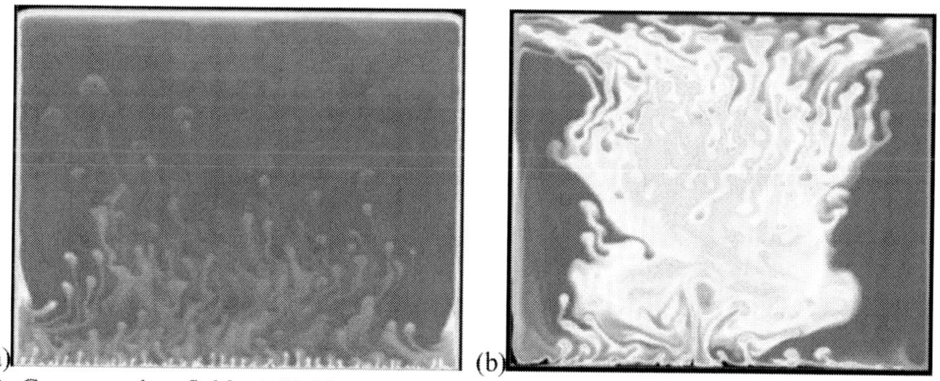

Figure 2. Concentration field. a) [0.29 wt.%, 0.3 wt.%] t~2 min (not shown in Figure 1); b) [0.26 wt.%, 0.3 wt.%] t~5 min (see Figure 1a).

2.4. Total Permeability

Permeability is typically modelled as varying with the solid fraction from zero at *fs*=0 to infinity at *fs*=1. In a shallow layer the permeability of the system includes the permeability of the shallow cell itself, $d^2/12$ [16]. The canonical choice for the mush permeability is the Kozeny-Carmen relationship. However by considering the morphology of the mush it is probably more appropriate to use the alternative form that represents a flow along or perpendicular to parallel cylinders in hexagonal configuration [17]:

$$\Pi_s = -\frac{4R_a^2}{32f_s}(\ln\frac{1}{f_s} - 1.497 + 2f_s - \frac{f_s^2}{2} - 0.739f_s^4 + \frac{2.534f_s^5}{1+1.2758f_s}) \quad (12)$$

The total effective permeability takes the form:

$$\Pi = (\frac{12}{d^2} + \frac{1}{\Pi_s})^{-1} \quad (13)$$

2.5. Numerical method

IOP Conf. Series: Materials Science and Engineering **33** (2012) 012115 doi:10.1088/1757-899X/33/1/012115

The model was implemented within the commercial CFD software FLUENT. The time step used in the calculation depends on the maximum velocity of the liquid and of the maximum growth velocity of the dendrite envelop:

$$\Delta t = \frac{1}{5}\min(\frac{\Delta x}{U_{\max}}, \frac{\Delta x}{V_{Max}}, 10^{-2}) \qquad (14)$$

In practice, it is the flow which limited most often the size of the time step. Typical calculation time for is in the order of 50-70 hours on an PC computer.

Table 1: Boundary conditions and effective parameters used for simulations

Property/Symbol	Value	Property/Symbol	Value
C_0	0.3 kg/kg	ρ_l	1078 kg/m^3
T_0	40°C	ρ_s	1500 kg/m^3
cooling rate	1.2 C/min	mixture $\overline{\rho}$	$\rho_l f_l + \rho_s f_s$
T_f	7 °C	mixture thermal conductivity, D_T	$0.54 f_l + 2.2 f_s$ W/m K
β_T	0.00021	glass thermal conductivity, D_{T_G}	0.96 W/m K
Δx	0.2 mm	L	3.5e5 W/kg
secondary arm spacing λ_2	0.1 mm	μ_l	0.0014
eff. solute diffusive coeff. D_{eff}	100*4.9e-9 m^2/s	C_p	$1000 f_l + 3500 f_s$
eff. Gibbs Thomson coeff. Γ_i	5.05e-5 K m	C_{p_G}	500 W/kg/K
h	100 W/K m^2	C_S	1 kg/kg
β_c	-0.3	ρ_G	2500 kg/m^3
F_{Si}	0.03	liquidus slope, m_L	4.8 K/mass fraction

3. Results and discussion

The numerical results are presented in Figures 1- 4. Figure 5 presents selected experimental flow patterns obtained with a PIV technique. Four pictures representing the evolution in time of the dendrites envelopes shows a columnar-like type of growth (Figure 1). Before the beginning of solidification the temperature distribution shows the presence of temperature boundary layers at the cooling walls, which drives the flow in the direction of natural convection. The solidification is found to generate 4 different patterns of thermo-solutal convection, a regular plume regime, a chaotic regime, and two layered regimes.

(a)

(b)

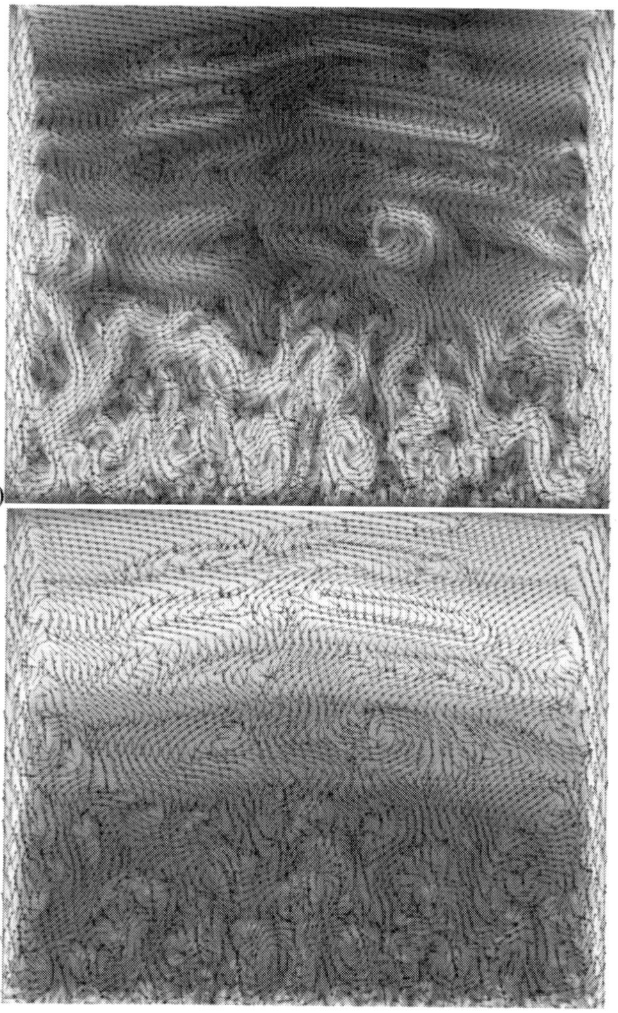

Figure 3. Velocity vectors during stratification of the concentration field t~9-15 minute. a) velocity magnitude and vectors [10^{-4} - 0.02 m/s]; b) velocity vectors and concentration field 0.2 wt% (blue) - 0.27 wt% (red).

At the start of solidification, the rejected solute disturbs the established thermo-buoyant convective flow by launching light solutal plumes (Figure 2a). The plume state displays two basic regimes. The initial one (Figure 2a) is dominated by the birth and rise of small plumes in vicinity of the bottom horizontal wall. During this early process the main flow path and bulk concentration stay almost unchanged. The duration is of the order of minutes.

The chaotic regime (Figure 2b and Figure 5a) is characterized by larger plumes generated within the liquid bulk and by an irregular formation of plumes near the horizontal solidification front. Due to the fast thermal diffusivity compared to the solutal one, the plumes generated in the liquid bulk consist of hot fluid with lower solute concentration. In opposite to previous regime large concentration gradient exists in the liquid bulk. During this regime no clear flow pattern can be observed, liquid flows successively in all directions at relatively high velocity magnitude. It can be assumed that this regime is the fruit of higher or balanced competition between thermal and solutal buoyancy. This unsteady behaviour changes the dynamic of plumes formation at the solidification front, typical distance between plumes seems to be no more related to the primary arm spacing as in early stage

(Figure 2a).The plumes are in fact emitted from an unstable and irregular boundary layer. The duration of this second regime is of the order of 10 minutes. In its course the bulk plume regime becomes increasingly disordered till it fades away giving birth to the stratified regimes.

(a)

(b)

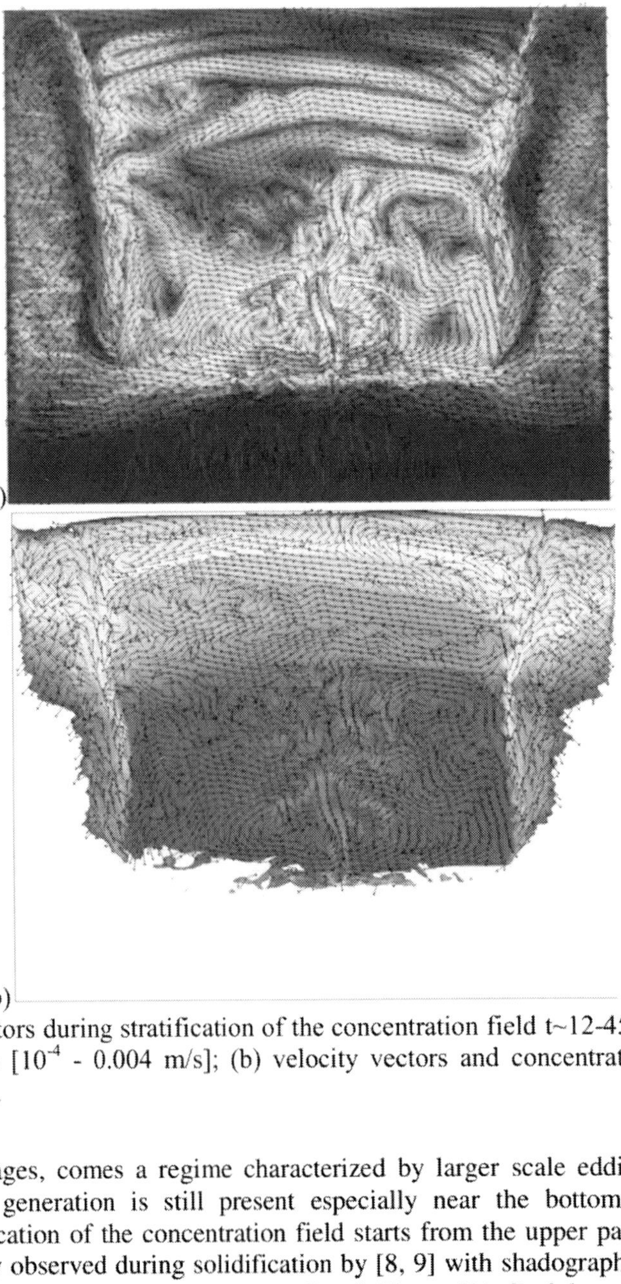

Figure 4. Velocity vectors during stratification of the concentration field t~12-45 minutes. a) velocity magnitude and vectors [10^{-4} - 0.004 m/s]; (b) velocity vectors and concentration field 0.211 wt% (blue) - 0.22 wt% (red).

After the plume stages, comes a regime characterized by larger scale eddies (Figure 3 and 4). However the plumes generation is still present especially near the bottom solidification front. Progressively a stratification of the concentration field starts from the upper part of the cell. One or two layers were clearly observed during solidification by [8, 9] with shadography technique. Here up to 5 layers of quasi-uniform concentration were predicted (Figure 3b). Such layers are characterised by horizontal convective cells. During PIV experiment similar thin convection cells were observed (Figure 5b), unfortunately the concentration field could not be simultaneously measured. However, by superimposing the predicted velocity and concentration field (Figure 3b), we can assume that each of the horizontal cells is associated with a staircase of concentration. With time such convection cells layers merge into a single meandering flow path (Figure 4). Experimentally this path fills the liquid region from left to right with typically 4 but up to 8 layers. In present simulations only 4 layers were formed (Figure 4a and b). It can be seen a good agreement between numerical and PIV results (Figure 5c).

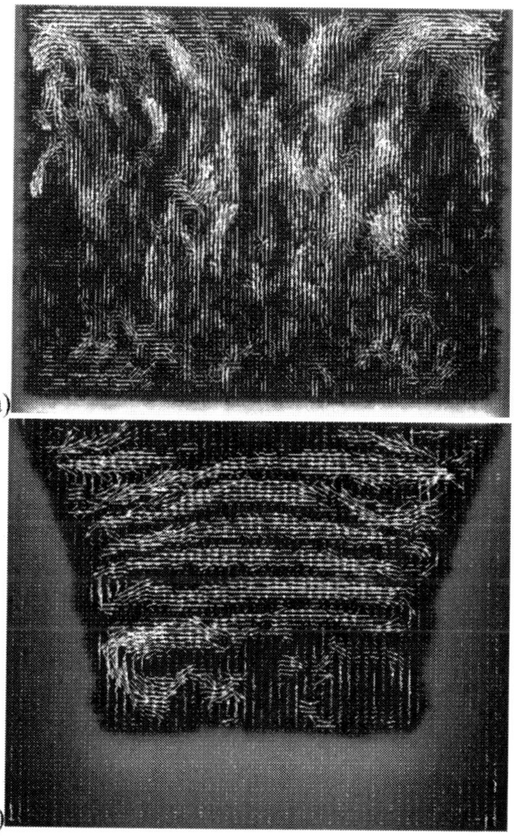

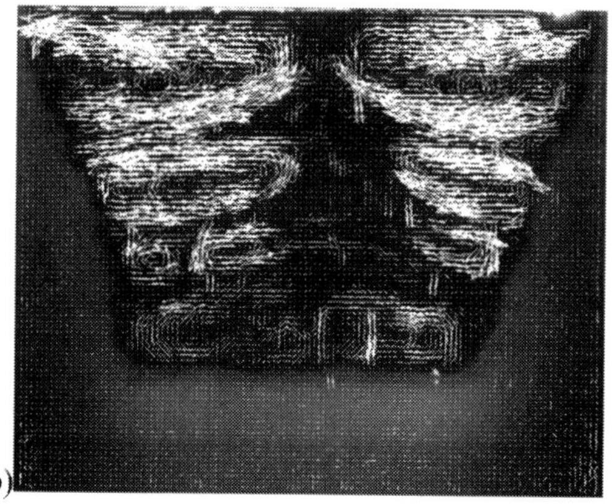

Figure 5. Experimental PIV measurements of liquid velocity during a) thermo-solutal turbulent regime (t~5 minutes), b) the layered convection regime (t~20 minutes)and c) the meandering flow regime(t>25 minutes).

The predicted solidification front fits qualitatively with the observed one. The solidification starts from all the walls with a planar front (Figure 1a). The growth morphology changes quickly into cellular, and finally bifurcates into a columnar dendritic type. The fact that solidification was slower in the earlier stages at the vertical walls is attributed to the strong downwards jet flow which thinner the heat boundary layers. Later the reorganisation of the flow, especially during the chaotic regime, redistributes the solute which then allows a faster solidification rate at the vertical walls. Solidification looses its symmetry during the stratified regimes (Figure 1d) which is not the case in the experiments (Figure 5)..Since new nucleation during the process is not allowed during the growth it was not possible to simulate the change in direction of the growing dendrites with the constitutional temperature gradient. In the future this quality will be implemented within the model.

The importance of solutal buoyancy can be stated as follows: in the present system the ratio of solutal to thermal buoyancy effects ($\beta_C \Delta C / \beta_T \Delta T$) is of order of unity, whereas the same ratio can be an order of magnitude higher for many metal-alloy systems. It is clear that a stronger solutaly driven convection can be expected in actual metal-alloy systems than in the NH_4Cl analogue. Moreover, solutal convection introduces kinetic energy at various wave lengths, which leads to the double diffusive chaos or turbulence. During solidification the smallest wave lengths are probably more related to the primary than to the secondary arm spacing. The present model predicted primary arm spacing larger but still in the same order as the observed ones (~1-3 mm). For industrial cast system, it is not soon that the primary arm spacing will be simulated as it is with the present approach. This means that correct numerical predictions needs either a grid size of λ_1 order, or the use of good closure laws. Unfortunately no universal closure laws exist which relates the thermo-solutal field to the turbulence parameters. This closure problem is very well known by the oceanographic community which tries to predict and explain how layering in salt concentration can be generated at the scale of the ocean. Nevertheless, we believe that the present model can be used for scales where a single cellular automaton dendrite can represent several (*N* times) dendrites. Instead of introducing waves of λ_1 length as reality, the model will introduce waves of $N\lambda_1$ length. The number *N* can probably be as large as 10 or 100. In the field of numerical simulation using volume averaging techniques, grid independency is often required as a guaranty of quality. However, it is clear that in the present model the proper choice of F_{Si}, N, D_{eff}, and Γ_i is grid size dependent. Similarly than in the field of turbulence modeling, unless resolutions of the smallest scales are achieved, grid independent results are neither realizable nor desirable. What is desirable is the ability to catch new physical mechanisms which can hardly be predicted with classical volume averaged techniques. With the use of additional proper closure laws, the model might give in the future good predictions even for large scale solidification processes.

4. Conclusion

A numerical model is developed to study the double diffusive convection during solidification of a binary alloy. The model uses a cellular automaton approach to simulate the dendrite envelope. A volume average model is applied to tackle the solidification inside the dendrite envelope. It was shown that various flow phenomena are induced by the liquid density change ahead of the solidification front. Even if the shape of the solidification front is simple, the flow patterns are rather complex. It is predicted that during the solidification, solutal buoyancy has the potential to overcome thermal buoyancy at certain locations, giving rise to turbulent solute-driven vortices. In the later stage strong stratification of the concentration field occurs. Hydrodynamics, associated with these variations of solute in staircases, are characterised by thin convective layers or by a meandering flow. The liquid flow follows an almost horizontal snaky path from the top to the bottom of the cavity. During those layered paths the coupling of the fluid flow with the solid dissolution produces irregular patterns at the solid-liquid interface. Such complex phenomena are expected to be prominent in actual metal–alloy systems, so more effort must be made to explore double diffusive convection during solidification.

Acknowledgements
The research was funded by the Austrian Science Fund (FWF): P22614-N22.

References

[1] Turner J S 1973 *Buoyancy Effects in Fluids* (Cambridge, England: Cambridge University Press)
[2] Stern M E 1960 *Tellus* **12** 172-175
[3] Prikasky I J Ph.D. Naval Postgraduate School Monterey, California 2007
[4] Fernando H J S 1987 *J. Fluid Mech.* **182** 525–541
[5] Fernando H J S 1989 *J of Physical Oceanography* **19** 1707–1715
[6] Linden P F 1979 *Geophysical and Astrophysical Fluid Dynamics* **13** 3–23
[7] Turner J S 1968 *J. Fluid. Mech.* **33** 183–200
[8] Beckermann C and Viskanta R 1988 *Int. J. Heat Mass. Transfer* **31** 2077-2089
[9] Beckermann C and Viskanta R 1989 *Chemical Eng. Communications* **85** 135-156
[10] Kharicha A, Stefan-Kharicha M, Ludwig A and Wu M 2012 MCWASP XII to be print
[11] Worster M G 1992 *J. Fluid Mech.* **237** 649-669
[12] Beckermann C, Fan C and Mihailovic J 1991 *Int. Video J. of Eng. Res.* **1** 71-82
[13] Li J, Wu M, Hao J and Ludwig A 2012 *Comp. Mater. Sci.* **55** 407-418
[14] Li J, Wu M, Hao J, Kharicha A and Ludwig A 2012 *Comp. Mater. Sci.* **55** 419-429
[15] Wu M and Ludwig A 2007 *Metall Mater Trans A* **38** 1465
[16] Bear J 1972 *Dynamics of fluids in porous media* (New York, USA: American Elsevier)
[17] Drummond J E and Tahir M I 1984 *Int. J. Multiph. Flow* **10** 515–540